2018 IEEE International Electron Devices Meeting (IEDM 2018)

San Francisco, California, USA
1-5 December 2018

Pages 460-938

IEEE Catalog Number: CFP18IED-POD
ISBN: 978-1-7281-1988-5

2018 IEEE International Electron Devices Meeting (IEDM 2018)

San Francisco, California, USA
1-5 December 2018

Pages 460-938

IEEE Catalog Number: CFP18IED-POD
ISBN: 978-1-7281-1988-5

**Copyright © 2018 by the Institute of Electrical and Electronics Engineers, Inc.
All Rights Reserved**

Copyright and Reprint Permissions: Abstracting is permitted with credit to the source. Libraries are permitted to photocopy beyond the limit of U.S. copyright law for private use of patrons those articles in this volume that carry a code at the bottom of the first page, provided the per-copy fee indicated in the code is paid through Copyright Clearance Center, 222 Rosewood Drive, Danvers, MA 01923.

For other copying, reprint or republication permission, write to IEEE Copyrights Manager, IEEE Service Center, 445 Hoes Lane, Piscataway, NJ 08854. All rights reserved.

*** *This is a print representation of what appears in the IEEE Digital Library. Some format issues inherent in the e-media version may also appear in this print version.*

IEEE Catalog Number: CFP18IED-POD
ISBN (Print-On-Demand):   978-1-7281-1988-5
ISBN (Online):   978-1-7281-1987-8
ISSN: 0163-1918

Additional Copies of This Publication Are Available From:

Curran Associates, Inc
57 Morehouse Lane
Red Hook, NY 12571 USA
Phone: (845) 758-0400
Fax: (845) 758-2633
E-mail: curran@proceedings.com
Web: www.proceedings.com

TABLE OF CONTENTS

4TH INDUSTRIAL REVOLUTION AND BOUNDRY: CHALLENGES AND OPPORTUNITIES 1
ES Jung

VENTURING ELECTRONICS INTO UNKNOWN GROUNDS 11
G. Fettweis ; K. Leo ; B. Voit ; U. Schneider ; L. Scheuvens

FUTURE COMPUTING HARDWARE FOR AI 21
J. Welser ; J. W. Pitera ; C. Goldberg

SCALING TRENDS IN NAND FLASH 27
K. Parat ; A. Goda

ANALYSIS AND REALIZATION OF TLC OR EVEN QLC OPERATION WITH A HIGH
PERFORMANCE MULTI-TIMES VERIFY SCHEME IN 3D NAND FLASH MEMORY 31
C.C. Lu ; C. C. Cheng ; H.P. Chiu ; W.L. Lin ; T.W. Chen ; S.H. Ku ; Wen-Jer Tsai ; T.C. Lu ; K.C. Chen ; Tahui Wang ; Chih-Yuan Lu

IMPLEMENTING SPIKE-TIMING-DEPENDENT PLASTICITY AND UNSUPERVISED
LEARNING IN A MAINSTREAM NOR FLASH MEMORY ARRAY 35
G. Malavena ; A. S. Spinelli ; C. Monzio Compagnoni

A NOVEL VOLTAGE-ACCUMULATION VECTOR-MATRIX MULTIPLICATION
ARCHITECTURE USING RESISTOR-SHUNTED FLOATING GATE FLASH MEMORY DEVICE
FOR LOW-POWER AND HIGH-DENSITY NEURAL NETWORK APPLICATIONS 39
Yu-Yu Lin ; Feng-Min Lee ; Ming-Hsiu Lee ; Wei-Chen Chen ; Hsiang-Lan Lung ; Keh-Chung Wang ; Chih-Yuan Lu

VERTICAL FERROELECTRIC HFO$_2$ FET BASED ON 3-D NAND ARCHITECTURE:
TOWARDS DENSE LOW-POWER MEMORY 43
K. Florent ; M. Pesic ; A. Subirats ; K. Banerjee ; S. Lavizzari ; A. Arreghini ; L. Di Piazza ; G. Potoms ; F. Sebaai ; S. R. C. McMitchell ; M. Popovici ; G. Groeseneken ; J. Van Houdt

HYBRID 1T E-DRAM AND E-NVM REALIZED IN ONE 10 NM NODE FERRO FINFET DEVICE
WITH CHARGE TRAPPING AND DOMAIN SWITCHING EFFECTS 47
Qing Luo ; Tiancheng Gong ; Yan Cheng ; Qingzhu Zhang ; Haoran Yu ; Jie Yu ; Haili Ma ; Xiaoxin Xu ; Kailiang Huang ; Xi Zhu ; Danian Dona ; Jiahao Yin ; Peng Yuan ; Lu Tai ; Jianfeng Gao ; Junfeng Li ; Huaxiang Yin ; Shibing Long ; Qi Liu ; Hangbing Lv ; Ming Liu

HIGH-PERFORMANCE (EOT<0.4NM, JG~10^{-7} A/CM^2) ALD-DEPOSITED RU\SRTIO$_3$ STACK
FOR NEXT GENERATIONS DRAM PILLAR CAPACITOR 51
M. Popovici ; A. Belmonte ; H. Oh ; G. Potoms ; J. Meersschaut ; O. Richard ; H. Hody ; S. Van Elshocht ; R. Delhougne ; L. Goux ; G. Sankar Kar

EXPLOITING HYBRID PRECISION FOR TRAINING AND INFERENCE: A 2T-1FEFET BASED
ANALOG SYNAPTIC WEIGHT CELL 55
Xiaoyu Sun ; Panni Wang ; Kai Ni ; Suman Datta ; Shimeng Yu

ANALOG COMPUTING FOR DEEP LEARNING: ALGORITHMS, MATERIALS &
ARCHITECTURES 59
W. Haensch

HARDWARE ACCELERATION OF SIMULATED ANNEALING OF SPIN GLASS BY RRAM
CROSSBAR ARRAY 63
Jong Hoon Shin ; Yeon Joo Jeong ; Mohammed A. Zidan ; Qiwen Wang ; Wei D. Lu

DEMONSTRATION OF GENERATIVE ADVERSARIAL NETWORK BY INTRINSIC RANDOM
NOISES OF ANALOG RRAM DEVICES 67
Yudeng Lin ; Huaqiang Wu ; Bin Gao ; Peng Yao ; Wei Wu ; Qingtian Zhang ; Xiang Zhang ; Xinyi Li ; Fuhai Li ; Jiwu Lu ; Gezi Li ; Shimeng Yu ; He Qian

ERROR-RESILIENT ANALOG IMAGE STORAGE AND COMPRESSION WITH ANALOG-
VALUED RRAM ARRAYS: AN ADAPTIVE JOINT SOURCE-CHANNEL CODING APPROACH 71
Xin Zheng ; Ryan Zarcone ; Dylan Paiton ; Joon Sohn ; Weier Wan ; Bruno Olshausen ; H. -S. Philip Wong

DEMONSTRATION OF 50-MV DIGITAL INTEGRATED CIRCUITS WITH
MICROELECTROMECHANICAL RELAYS 75
Z. A. Ye ; S. Almeida ; M. Rusch ; A. Perlas ; W. Zhang ; U. Sikder ; J. Jeon ; V. Stojanovic ; T.-J. K. Liu

HIGHLY SENSITIVE SPINTRONIC STRAIN-GAUGE SENSOR BASED ON MAGNETIC
TUNNEL JUNCTION AND ITS APPLICATION TO MEMS MICROPHONE 79
Yoshihiko Fuji ; Yoshihiro Higashi ; Shiori Kaji ; Kei Masunishi ; Akiko Yuzawa ; Tomohiko Nagata ; Kazuaki Okamoto ; Shotaro Baba ; Tomio Ono ; Michiko Hara

INTERMIXING OF MOTIONAL CURRENTS IN SUSPENDED CNT-FET BASED RESONATORS 83

Lalit Kumar ; Laura Vera Jenni ; Miroslav Haluska ; Cosmin Roman ; Christofer Hierold

GLOWING GRAPHENE NANOELECTROMECHANICAL RESONATORS AT ULTRA-HIGH TEMPERATURE UP TO 2650K 87

Fan Ye ; Jaesung Lee ; Philip X.-L. Feng

MONOLITHIC INTEGRATION OF MICRON-SCALE PIEZOELECTRIC MATERIALS WITH CMOS FOR BIOMEDICAL APPLICATIONS 91

C. Shi ; T. Costa ; J. Elloian ; K. L. Shepard

A NANO-MECHANICAL RESONATOR WITH 10NM HAFNIUM-ZIRCONIUM OXIDE FERROELECTRIC TRANSDUCER 95

M. Ghatge ; G. Walters ; T. Nishida ; R. Tabrizian

COMPREHENSIVE OPTICAL LOSSES INVESTIGATION OF VLSI SILICON OPTOMECHANICAL RING RESONATOR SENSORS 99

L. Schwab ; P.E. Allain ; L. Banniard ; A. Fafin ; M. Gely ; O. Lemonnier ; P. Grosse ; M. Hermouet ; S. Hentz ; I. Favero ; B. Legrand ; G. Jourdan

INTERCONNECT DESIGN AND TECHNOLOGY OPTIMIZATION FOR CONVENTIONAL AND EMERGING NANOSCALE DEVICES: A PHYSICAL DESIGN PERSPECTIVE 103

D. Prasad ; A. Naeemi

MECHANISMS OF ELECTROMIGRATION DAMAGE IN CU INTERCONNECTS 107

C.-K. Hu ; L. Gignac ; G. Lian ; C. Cabral ; K. Motoyama ; H. Shobha ; J. Demarest ; Y. Ostrovski ; C. M. Breslin ; M. Ali ; J. Benedict ; P. S. McLaughlin ; J. Ni ; X. H. Liu

INTERCONNECT METALS BEYOND COPPER: RELIABILITY CHALLENGES AND OPPORTUNITIES 111

K. Croes ; Ch. Adelmann ; C.J. Wilson ; H. Zahedmanesh ; O. Varela Pedreira ; C. Wu ; A. Lesniewska ; H. Oprins ; S. Beyne ; I. Ciofi ; D. Kocaay ; M. Stucchi ; Zs. Tokei

MICROSTRUCTURE EVOLUTION AND EFFECT ON RESISTIVITY FOR CU NANOINTERCONNECTS AND BEYOND 115

Szu-Tung Hu ; Linjun Cao ; Laura Spinella ; Paul S. Ho

INTEGRATING GRAPHENE INTO FUTURE GENERATIONS OF INTERCONNECT WIRES 118

Ling Li ; H.-S. Philip Wong

INTERCONNECT TREND FOR SINGLE DIGIT NODES 122

Mehul Naik

DEVICE CHALLENGES FOR NEAR TERM SUPERCONDUCTING QUANTUM PROCESSORS: FREQUENCY COLLISIONS 126

Markus Brink ; Jerry M. Chow ; Jared Hertzberg ; Easwar Magesan ; Sami Rosenblatt

SCALABLE QUANTUM COMPUTING WITH ION-IMPLANTED DOPANT ATOMS IN SILICON 129

A. Morello ; G. Tosi ; F.A. Mohiyaddin ; V. Schmitt ; V. Mourik ; T. Botzem ; A. Laucht ; J.J. Pla ; S. Tenberg ; R. Savytskyy ; M. Madzik ; F. Hudson ; A.S. Dzurak ; K.M. Itoh ; A.M. Jakob ; B.C. Johnson ; J.C. McCallum ; D.N. Jamieson

QUBIT DEVICE INTEGRATION USING ADVANCED SEMICONDUCTOR MANUFACTURING PROCESS TECHNOLOGY 133

R. Pillarisetty ; N. Thomas ; H.C. George ; K. Singh ; J. Roberts ; L. Lampert ; P. Amin ; T.F. Watson ; G. Zheng ; J. Torres ; M. Metz ; R. Kotlyar ; P. Keys ; J.M. Boter ; J.P. Dehollain ; G. Droulers ; G. Eenink ; R. Li ; L. Massa ; D. Sabbagh ; N. Samkharadze ; C. Volk ; B. P. Wuetz ; A.-M. Zwerver ; M. Veldhorst ; G. Scappucci ; L.M.K. Vandersypen ; J.S. Clarke

SILICON ISOTOPE TECHNOLOGY FOR QUANTUM COMPUTING 137

Satoru Miyamoto ; Kohei M. Itoh

TOWARDS SCALABLE SILICON QUANTUM COMPUTING 141

M. Vinet ; L. Hutin ; B. Bertrand ; S. Barraud ; J.-M. Hartmann ; Y.-J. Kim ; V. Mazzocchi ; A. Amisse ; H. Bohuslavskyi ; L. Bourdet ; A. Crippa ; X. Jehl ; R. Maurand ; Y.-M. Niquet ; M. Sanquer ; B. Venitucci ; B. Jadot ; E. Chanrion ; P.-A. Mortemousque ; C. Spence ; M. Urdampilleta ; S. De Franceschi ; T. Meunier

MAJORANA QUBITS 145

Leo Kouwenhoven

FIRST DEMONSTRATION OF 3D STACKED FINFETS AT A 45NM FIN PITCH AND 110NM GATE PITCH TECHNOLOGY ON 300MM WAFERS 149

A. Vandooren ; J. Franco ; Z. Wu ; B. Parvais ; W. Li ; L. Witters ; A. Walke ; L. Peng ; V. Deshpande ; N. Rassoul ; G. Hellings ; G. Jamieson ; F. Inoue ; K. Devriendt ; L. Teugels ; N. Heylen ; E. Vecchio ; T. Zheng ; E. Rosseel ; W. Vanherle ; A. Hikavyy ; G. Mannaert ; B. T. Chan ; R. Ritzenthaler ; J. Mitard ; L. Ragnarsson ; N. Waldron ; V. De Heyn ; S. Demuynck ; J. Boemmels ; D. Mocuta ; J. Ryckaert ; N. Collaert

BREAKTHROUGHS IN 3D SEQUENTIAL TECHNOLOGY 153

L. Brunet ; C. Fenouillet-Beranger ; P. Batude ; S. Beaurepaire ; F. Ponthenier ; N. Rambal ; V. Mazzocchi ; J-B. Pin ; P. Acosta-Alba ; S. Kerdiles ; P. Besson ; H. Fontaine ; T. Lardin ; F. Fournel ; V. Larrey ; F. Mazen ; V. Balan ; C. Morales ; C. Guerin ; V. Jousseaume ; X. Federspiel ; D. Ney ; X. Garros ; A. Roman ; D. Scevola ; P. Perreau ; F. Kouemeni-Tchouake ; L. Arnaud ; C. Scibetta ; S. Chevalliez ; F. Aussenac ; J. Aubin ; S. Reboh ; F. Andrieu ; S. Maitrejean ; M. Vinet

HYBRID BONDING FOR 3D STACKED IMAGE SENSORS: IMPACT OF PITCH SHRINKAGE ON INTERCONNECT ROBUSTNESS 157

J. Jourdon ; S. Lhostis ; S. Moreau ; J. Chossat ; M. Arnoux ; C. Sart ; Y. Henrion ; P. Lamontagne ; L. Arnaud ; N. Bresson ; V. Balan ; C. Euvrard ; Y. Exbrayat ; D. Scevola ; E. Deloffre ; S. Mermoz ; A. Martin ; H. Bilgen ; F. Andre ; C. Charles ; D. Bouchu ; A. Farcy ; S. Guillaumet ; A. Jouve ; H. Fremont ; S. Cheramy

EMBEDDED SELECT IN TRENCH MEMORY (ESTM), BEST IN CLASS 40NM FLOATING GATE BASED CELL: A PROCESS INTEGRATION CHALLENGE 161

S. Niel ; F. La Rosa ; A. Regnier ; M. Mantelli ; F. Trenteseaux ; G. Ghezzi ; A. Marzaki ; Q. Hubert ; J. Delalleau ; T. Cabout ; F. Maugain ; E. Lepape ; L. Baron ; A. Champenois ; D. Galpin ; N. Cherault ; S. Audran ; L. Parmigiani ; P. Gouraud ; B. Duclaux ; Y. Escarabajal ; F. Baudin ; E. Beche ; B. Saidi ; V. Arnal

HIGHLY RELIABLE FERROELECTRIC $HF_{0.5}ZR_{0.5}O_2$ FILM WITH AL NANOCLUSTERS EMBEDDED BY NOVEL SUB-MONOLAYER DOPING TECHNIQUE 165

T. Yamaguchi ; T. Zhang ; K. Omori ; Y. Shimada ; Y. Kunimune ; T. Ide ; M. Inoue ; M. Matsuura

INTERFACE DIPOLE MODULATION IN HFO2/SIO2 MOS STACK STRUCTURES 169

Noriyuki Miyata ; Jun Nara ; Takahiro Yamasaki ; Kyoko Sumita ; Ryousuke Sano ; Hiroshi Nohira

GE-BASED NON-VOLATILE LOGIC-MEMORY HYBRID DEVICES FOR NAND MEMORY APPLICATION 173

Na Wei ; Bing Chen ; Zejie Zheng ; Zhimei Cai ; Rui Zhang ; Ran Cheng ; Shiuh-Wuu Lee ; Yi Zhao

0.63 $M\Omega CM^2$ / 1170 V 4H-SIC SUPER JUNCTION V-GROOVE TRENCH MOSFET 177

T. Masuda ; Y. Saito ; T. Kumazawa ; T. Hatayama ; S. Harada

FIRST DEMONSTRATION OF DYNAMIC CHARACTERISTICS FOR SIC SUPERJUNCTION MOSFET REALIZED USING MULTI-EPITAXIAL GROWTH METHOD 181

S. Harada ; Y. Kobayashi ; S. Kyogoku ; T. Morimoto ; T. Tanaka ; M. Takei ; H. Okumura

CHANNEL ENGINEERING OF 4H-SIC MOSFETS USING SULPHUR AS A DEEP LEVEL DONOR 185

M. Noguchi ; T. Iwamatsu ; H. Amishiro ; H. Watanabe ; K. Kita ; N. Miura

DEMONSTRATION OF 1200V SCALED IGBTS DRIVEN BY 5V GATE VOLTAGE WITH SUPERIORLY LOW SWITCHING LOSS 189

T. Saraya ; K. Itou ; T. Takakura ; M. Fukui ; S. Suzuki ; K. Takeuchi ; M. Tsukuda ; Y. Numasawa ; K. Satoh ; T. Matsudai ; W. Saito ; K. Kakushima ; T. Hoshii ; K. Furukawa ; M. Watanabe ; N. Shigyo ; K. Tsutsui ; H. Iwai ; A. Ogura ; S. Nishizawa ; I. Omura ; H. Ohashi ; T. Hiramoto

2.44 KV GA_2O_3 VERTICAL TRENCH SCHOTTKY BARRIER DIODES WITH VERY LOW REVERSE LEAKAGE CURRENT 193

Wenshen Li ; Zongyang Hu ; Kazuki Nomoto ; Riena Jinno ; Zexuan Zhang ; Thieu Quang Tu ; Kohei Sasaki ; Akito Kuramata ; Debdeep Jena ; Huili Grace Xing

MULTIDOMAIN DYNAMICS OF FERROELECTRIC POLARIZATION AND ITS COHERENCY-BREAKING IN NEGATIVE CAPACITANCE FIELD-EFFECT TRANSISTORS 197

Hiroyuki Ota ; Tsutomu Ikegami ; Koichi Fukuda ; Junichi Hattori ; Hidehiro Asai ; Kazuhiko Endo ; Shinji Migita ; Akira Toriumi

MODELING OF MULTI-DOMAIN SWITCHING IN FERROELECTRIC MATERIALS: APPLICATION TO NEGATIVE CAPACITANCE FETS 201

A. Dasgupta ; P. Rastogi ; D. Saha ; A. Gaidhane ; A. Agarwal ; Y. S. Chauhan

ON THE MICROSCOPIC ORIGIN OF NEGATIVE CAPACITANCE IN FERROELECTRIC MATERIALS: A TOY MODEL 205

Asif Islam Khan

EFFECT OF POLYCRYSTALLINITY AND PRESENCE OF DIELECTRIC PHASES ON NC-FINFET VARIABILITY 209

Yen-Kai Lin ; Ming-Yen Kao ; Harshit Agarwal ; Yu-Hung Liao ; Pragya Kushwaha ; Korok Chatterjee ; Juan Pablo Duarte ; Huan-Lin Chang ; Sayeef Salahuddin ; Chenming Hu

A SIMULATION BASED STUDY OF NC-FETS DESIGN: OFF-STATE VERSUS ON-STATE PERSPECTIVE 213

T. Rollo ; H. Wang ; G. Han ; D. Esseni

1.5µM DUAL CONVERSION GAIN, BACKSIDE ILLUMINATED IMAGE SENSOR USING STACKED PIXEL LEVEL CONNECTIONS WITH 13KE-FULL-WELL CAPACITANCE AND 0.8E-NOISE 217

Vincent C. Venezia ; Alan Chih-Wei Hsiung ; Kelvin Ai ; Xiang Zhao ; Zhiqiang Lin ; Duli Mao ; Armin Yazdani ; Eric A. G. Webster ; Lindsay A. Grant

A 0.68E-RMS RANDOM-NOISE 121DB DYNAMIC-RANGE SUB-PIXEL ARCHITECTURE CMOS IMAGE SENSOR WITH LED FLICKER MITIGATION 221

S. Iida ; Y. Sakano ; T. Asatsuma ; M. Takami ; I. Yoshiba ; N. Ohba ; H. Mizuno ; T. Oka ; K. Yamaguchi ; A. Suzuki ; K. Suzuki ; M. Yamada ; M. Takizawa ; Y. Tateshita ; K. Ohno

A 24.3ME- FULL WELL CAPACITY CMOS IMAGE SENSOR WITH LATERAL OVERFLOW INTEGRATION TRENCH CAPACITOR FOR HIGH PRECISION NEAR INFRARED ABSORPTION IMAGING 225

M. Murata ; R. Kuroda ; Y. Fujihara ; Y. Aoyagi ; H. Shibata ; T. Shibaguchi ; Y. Kamata ; N. Miura ; N. Kuriyama ; S. Sugawa

A HDR 98DB 3.2µM CHARGE DOMAIN GLOBAL SHUTTER CMOS IMAGE SENSOR 229

A. Tournier ; F. Roy ; Y. Cazaux ; F. Lalanne ; P. Malinge ; M. Mcdonald ; G. Monnot ; N. Roux

HIGH PERFORMANCE 2.5UM GLOBAL SHUTTER PIXEL WITH NEW DESIGNED LIGHT-PIPE STRUCTURE 233

Toshifumi Yokoyama ; Masafumi Tsutsui ; Yoshiaki Nishi ; Ikuo Mizuno ; Veinger Dmitry ; Assaf Lahav

BACK-ILLUMINATED 2.74 µM-PIXEL-PITCH GLOBAL SHUTTER CMOS IMAGE SENSOR WITH CHARGE-DOMAIN MEMORY ACHIEVING 10K E-SATURATION SIGNAL 237

Y. Kumagai ; R. Yoshita ; N. Osawa ; H. Ikeda ; K. Yamashita ; T. Abe ; S. Kudo ; J. Yamane ; T. Idekoba ; S. Noudo ; Y. Ono ; S. Kunitake ; M. Sato ; N. Sato ; T. Enomoto ; K. Nakazawa ; H. Mori ; Y. Tateshita ; K. Ohno

CONFORMAL, WAFER-SCALE AND CONTROLLED NANOSCALE DOPING OF SEMICONDUCTORS VIA THE ICVD PROCESS 241

Jae Hwan Kim ; Hong Keun Park ; Kwan Yong Pak ; Alexander Yoon ; Yun Sang Kim ; Sung Gap Im ; Wan Sik Hwang ; Byung Jin Cho

LOW TEMPERATURE SPUTTERED GRAPHENIC CARBON ENABLES HIGHLY RELIABLE CONTACTS TO SILICON 245

M. Stelzer ; M. Jung ; U. Wurstbauer ; A.W. Holleitner ; F. Kreupl

LOCATION-CONTROLLED-GRAIN TECHNIQUE FOR MONOLITHIC 3D BEOL FINFET CIRCUITS 249

Chih-Chao Yang ; Tung-Ying Hsieh ; Po-Tsang Huang ; Kuan-Neng Chen ; Wan-Chi Wu ; Shih-Wei Chen ; Chia-He Chang ; Chang-Hong Shen ; Jia-Min Shieh ; Chenming Hu ; Meng-Chyi Wu ; Wen-Kuan Yeh

NOVEL MATERIALS AND PROCESSES IN REPLACEMENT METAL GATE FOR ADVANCED CMOS TECHNOLOGY 253

Ruqiang Bao ; Steven Hung ; Miaomiao Wang ; Kisup Chung ; Soumendra Barman ; Siddarth A Krishnan ; Yixiong Yang ; Wei Tang ; Luping Li ; Yongjing Lin ; Michael S Chan ; Zhebo Chen ; Xin Miao ; Marinus Hopstaken ; Richard A Conti ; Hemanth Jagannathan ; Michael P Chudzik ; Dalea McHerron ; Bala S Haran ; Sanjay Natarajan

WHY GEO$_2$ GROWTH ON GE IS SUPPRESSED AND GEO$_2$/GE STACK IS MUCH IMPROVED IN HIGH PRESSURE O$_2$ OXIDATION? 257

Xu Wang ; Akira Toriumi

EUV LITHOGRAPHY AT THRESHOLD OF HIGH-VOLUME MANUFACTURING 261

Anthony Yen ; Hans Meiling ; Jos Benschop

HALF PITCH 14 NM DIRECT PATTERING WITH NANOIMPRINT LITHOGRAPHY 265

T. Nakasugi ; T. Kono ; K. Fukuhara ; M. Hatano ; H. Tokue ; M. Komori ; H. Tsuda ; T. Komukai ; K. Takahata ; H. Kato ; K. Kobayashi ; A. Mitra ; S. Kobayashi ; S. Inoue ; T. Higashiki ; T. Motokawa ; M. Saito ; S. Kanamitsu ; M. Itoh ; T. Imamura ; K. Matasunaga ; K. Hashimoto ; Y. Kim ; J. Cho ; W. Jung

ALL CMOS INTEGRATED 3D-EXTENDED METAL GATE ISFETS FOR PH AND MULTI-ION (NA$^+$, K$^+$, CA^{2+}) SENSING 269

J.-R. Zhang ; M. Rupakula ; F. Bellando ; E. Garcia Cordero ; J. Longo ; F. Wildhaber ; G. Herment ; H. Guérin ; A.M. Ionescu

HIGH RESOLUTION ION DETECTOR (HRID) BY 16NM FINFET CMOS TECHNOLOGY 273

Peng-Chun Liou ; Tsung-Han Lee ; Chien-Ping Wang ; Yu-Lun Chueh ; Yue-Der Chih ; Jonathan Chang ; Chrong Jung Lin ; Ya-Chin King

HIGHLY PERFORMANT INTEGRATED PH-SENSOR USING THE GATE PROTECTION DIODE IN THE BEOL OF INDUSTRIAL FDSOI 277

G. T. Ayele ; S. Monfray ; S. Ecoffey ; F. Boeuf ; J-P. Cloarec ; D. Drouin ; A. Souifi

VERY LARGE SCALE INTEGRATION OPTOMECHANICS: A CURE FOR LONELINESS OF NEMS RESONATORS? 281

Maxime Hermouet ; Marc Sansa ; Martial Defoort ; Louise Banniard ; Sergio Dominauez-Medina ; Shawn Fostner ; Ujwol Palanchoke ; Alexandre Fafin ; Marc Gely ; Louis Hutin ; Christophe Plantier ; Emmanuel Rolland ; Claude Tabone ; Giulia Usai ; Thomas Ernst ; Patrick Villard ; Gérard Billiot ; Paul Mattei ; Guillaume Nonglaton ; Caroline Fontelaye ; Charlie Barrois ; Olivier Castany ; Eduardo Gil Santos ; Pierre E. Allain ; Emeline Vernhes ; Pascale Boulanger ; Ariel Brenac ; Christophe Masselon ; Ivan Favero ; Thomas Alava ; Guillaume Jourdan ; Sébastien Hentz

SIC-FET-TYPE NOX SENSOR FOR HIGH-TEMPERATURE EXHAUST GAS..284
Y. Sasago ; H. Nakamura ; T. Odaka ; A. Isobe ; S. Komatsu ; Y. Nakamura ; T. Yamawaki ; C. Yorita ; N. Ushifusa ; K. Yoshikawa ; K. Ono ; Y. Anzai ; S. Machida ; M. Kinoshita ; K. Fujisaki ; T. Usagawa ; K. Okishiro ; Y. Sugiyama

A SI FET-TYPE GAS SENSOR WITH PULSE-DRIVEN LOCALIZED MICRO-HEATER FOR LOW POWER CONSUMPTION 288
Yoonki Hong ; Seongbin Hong ; Dongkyu Jang ; Yujeong Jeong ; Meile Wu ; Gyuweon Jung ; Jong-Ho Bae ; Jun Shik Kim ; Ki Soo Chang ; Chan Bae Jeong ; Cheol Seong Hwang ; Byung-Gook Park ; Jong-Ho Lee

ECRAM AS SCALABLE SYNAPTIC CELL FOR HIGH-SPEED, LOW-POWER NEUROMORPHIC COMPUTING 292
Jianshi Tang ; Douglas Bishop ; Seyoung Kim ; Matt Copel ; Tayfun Gokmen ; Teodor Todorov ; SangHoon Shin ; Ko-Tao Lee ; Paul Solomon ; Kevin Chan ; Wilfried Haensch ; John Rozen

SOC LOGIC COMPATIBLE MULTI-BIT FEMFET WEIGHT CELL FOR NEUROMORPHIC APPLICATIONS 296
K. Ni ; J. A. Smith ; B. Grisafe ; T. Rakshit ; B. Obradovic ; J. A. Kittl ; M. Rodder ; S. Datta

EXPERIMENTAL DEMONSTRATION OF FERROELECTRIC SPIKING NEURONS FOR UNSUPERVISED CLUSTERING 300
Zheng Wang ; Brian Crafton ; Jorge Gomez ; Ruijuan Xu ; Aileen Luo ; Zoran Krivokapic ; Lane Martin ; Suman Datta ; Arijit Raychowdhury ; Asif Islam Khan

NEAR HYSTERESIS-FREE NEGATIVE CAPACITANCE INGAAS TUNNEL FETS WITH ENHANCED DIGITAL AND ANALOG FIGURES OF MERIT BELOW VDD=400MV 304
Ali Saeidi ; Anne S. Verhulst ; Igor Stolichnov ; Alireza Alian ; Hiroshi Iwai ; Nadine Collaert ; Adrian M. Ionescu

AN EXPERIMENTAL STUDY OF HETEROSTRUCTURE TUNNEL FET NANOWIRE ARRAYS: DIGITAL AND ANALOG FIGURES OF MERIT FROM 300K TO 10K 308
T. Rosca ; A. Saeidi ; E. Memisevic ; L-E. Wernersson ; A.M. Ionescu

HIGH THERMAL TOLERANCE OF 25-NM C-AXIS ALIGNED CRYSTALLINE IN-GA-ZN OXIDE FET 312
Hitoshi Kunitake ; Kazuaki Ohshima ; Kazuki Tsuda ; Noriko Matsumoto ; Hiromi Sawai ; Yuichi Yanagisawa ; Shiori Saga ; Ryo Arasawa ; Takako Seki ; Ryo Tokumaru ; Tomoaki Atsumi ; Kiyoshi Kato ; Shunpei Yamazaki

INTEL 22NM FINFET (22FFL) PROCESS TECHNOLOGY FOR RF AND MM WAVE APPLICATIONS AND CIRCUIT DESIGN OPTIMIZATION FOR FINFET TECHNOLOGY 316
H.-J. Lee ; S. Rami ; S. Ravikumar ; V. Neeli ; K. Phoa ; B. Sell ; Y. Zhang

GAN HEMTS FOR 5G BASE STATION APPLICATIONS 320
Shigeru Nakajima

100-340GHZ SYSTEMS: TRANSISTORS AND APPLICATIONS 324
M.J.W. Rodwell ; Y. Fang ; J. Rode ; J. Wu ; B. Markman ; S. T. Šuran Brunelli ; J. Klamkin ; M Urteaga

CONSIDERATIONS ON DESIGN OF HIGHLY-INTEGRATED MILLIMETER-WAVE TRANSCEIVERS IN SIGE HBT 328
V. Issakov ; S. Trotta

BAW FILTERS FOR 5G BANDS 332
R. Aigner ; G. Fattinger ; M. Schaefer ; K. Karnati ; R. Rothemund ; F. Dumont

TUNABLE FILTER TECHNOLOGIES FOR 5G COMMUNICATIONS 336
Dimitrios Peroulis

ULTRA-LOW POWER 3D NC-FINFET-BASED MONOLITHIC 3D+ -IC WITH COMPUTING-IN-MEMORY FOR INTELLIGENT IOT DEVICES 340
Fu-Kuo Hsueh ; Wei-Hao Chen ; Kai-Shin Li ; Chang-Hong Shen ; Jia-Min Shieh ; Chun Ying Lee ; Bo-Yuan Chen ; Hsiu-Chih Chen ; Chih-Chao Yang ; Wen-Hsien Huang ; Kun-Ming Chen ; Guo-Wei Huang ; Peng Chen ; Yung-Ning Tu ; Srivatsa Srinivasa ; Vijaykrishnan Narayanan ; Meng-Fan Chang ; Wen-Kuan Yeh

FIRST DEMONSTRATION OF GE FERROELECTRIC NANOWIRE FET AS SYNAPTIC DEVICE FOR ONLINE LEARNING IN NEURAL NETWORK WITH HIGH NUMBER OF CONDUCTANCE STATE AND GMAX/GMIN 344
Wonil Chung ; Mengwei Si ; Peide D. Ye

STT-MRAM DESIGN TECHNOLOGY CO-OPTIMIZATION FOR HARDWARE NEURAL NETWORKS 348
Nuo Xu ; Yang Lu ; Weiyi Qi ; Zhengping Jiang ; Xiaochen Peng ; Fan Chen ; Jing Wang ; Woosung Choi ; Shimeng Yu ; Dae Sin Kim

A 68 PARALLEL ROW ACCESS NEUROMORPHIC CORE WITH 22K MULTI-LEVEL SYNAPSES BASED ON LOGIC-COMPATIBLE EMBEDDED FLASH MEMORY TECHNOLOGY 352
M. Kim ; J. Kim ; C. Park ; L. Everson ; H. Kim ; S. Song ; S. Lee ; C. H. Kim

INTERCHANGEABLE HEBBIAN AND ANTI-HEBBIAN STDP APPLIED TO SUPERVISED LEARNING IN SPIKING NEURAL NETWORK 356
Che-Chia Chang ; Pin-Chun Chen ; Boris Hudec ; Po-Tsun Liu ; Tuo-Hung Hou

STOCHASTIC INFERENCE AND LEARNING ENABLED BY MAGNETIC TUNNEL JUNCTIONS 360

Abhronil Sengupta ; Gopalakrishnan Srinivasan ; Deboleena Roy ; Kaushik Roy

IN-MEMORY COMPUTING PRIMITIVE FOR SENSOR DATA FUSION IN 28 NM HKMG FEFET TECHNOLOGY 364

K. Ni ; B. Grisafe ; W. Chakraborty ; A. K. Saha ; S. Dutta ; M. Jerry ; J. A. Smith ; S. Gupta ; S. Datta

EXPERIMENTALLY VALIDATED, PREDICTIVE MONTE CARLO MODELING OF FERROELECTRIC DYNAMICS AND VARIABILITY 368

C. Alessandri ; P. Pandey ; A. C. Seabaugh

SCALABILITY STUDY ON FCRROCLCCTRIC-HFO2 TUNNEL JUNCTION MEMORY BASED ON NON-EQUILIBRIUM GREEN FUNCTION METHOD WITH SELF-CONSISTENT POTENTIAL 372

Fei Mo ; Yusaku Tagawa ; Takuya Saraya ; Toshiro Hiramoto ; Masaharu Kobayashi

ROLE OF OXYGEN VACANCIES IN ELECTRIC FIELD CYCLING BEHAVIORS OF FERROELECTRIC HAFNIUM OXIDE 376

C. Liu ; F. Liu ; Q. Luo ; P. Huang ; X. X. Xu ; H. B. Lv ; Y. D. Zhao ; X.Y. Liu ; J. F. Kang

FIRST-PRINCIPLES PERSPECTIVE ON POLING MECHANISMS AND FERROELECTRIC/ANTIFERROELECTRIC BEHAVIOR OF $HF_{1-X}ZR_XO_2$ FOR FEFET APPLICATIONS 380

Sergiu Clima ; S.R.C. McMitchell ; K. Florent ; L. Nyns ; M. Popovici ; N. Ronchi ; L. Di Piazza ; J. Van Houdt ; G. Pourtois

CHARACTERIZATION METHODOLOGY AND PHYSICAL COMPACT MODELING OF IN-WAFER GLOBAL AND LOCAL VARIABILITY 384

Krishna Pradeep ; Thierry Poiroux ; Patrick Scheer ; André Juge ; Gilles Gouget ; Gérard Ghibaudo

TOO NOISY AT THE BOTTOM? —RANDOM TELEGRAPH NOISE (RTN) IN ADVANCED LOGIC DEVICES AND CIRCUITS 388

Runsheng Wang ; Shaofeng Guo ; Zhe Zhang ; Qingxue Wang ; Dehuang Wu ; Joddy Wang ; Ru Huang

COMPREHENSIVE STUDY ON THE "ANOMALOUS" COMPLEX RTN IN ADVANCED MULTI-FIN BULK FINFET TECHNOLOGY 392

Jiayang Zhang ; Zhe Zhang ; Rusheng Wang ; Zixuan Sun ; Zuodong Zhang ; Shaofeng Guo ; Ru Huang

AN UNIQUE METHODOLOGY TO ESTIMATE THE THERMAL TIME CONSTANT AND DYNAMIC SELF HEATING IMPACT FOR ACCURATE RELIABILITY EVALUATION IN ADVANCED FINFET TECHNOLOGIES 396

S. Mukhopadhyay ; A. Kundu ; Y.W. Lee ; H. D. Hsieh ; D.S. Huang ; J.J. Horng ; T.H. Chen ; J.H. Lee ; Y.S. Tsai ; C.K. Lin ; Ryan Lu ; Jun He

7NM FINFET PLASMA CHARGE RECORDING DEVICE 400

Yi-Pei Tsai ; Jiaw-Ren Shih ; Ya-Chin King ; Chrong Jung Lin

DEVELOPMENT OF X-RAY PHOTOELECTRON SPECTROSCOPY UNDER BIAS AND ITS APPLICATION TO DETERMINE BAND-ENERGIES AND DIPOLES IN THE HKMG STACK 404

Pushpendra Kumar ; Charles Leroux ; Florian Domengie ; Eugenie Martinez ; Virginie Loup ; Denis Guiheux ; Yves Morand ; Jean-Michel Pedini ; Claude Tabone ; Frederic Gaillard ; Gerard Ghibaudo

IN-SITU INVESTIGATION OF THE IMPACT OF EXTERNALLY APPLIED VERTICAL STRESS ON III-V BIPOLAR TRANSISTOR 408

Y. Liu ; G. Hiblot ; M. Gonzalez ; K. Vanstreels ; D. Velenis ; M. Badaroglu ; G. Van der Plas ; I. De Wolf

MRAM AS EMBEDDED NON-VOLATILE MEMORY SOLUTION FOR 22FFL FINFET TECHNOLOGY 412

O. Golonzka ; J. -G. Alzate ; U. Arslan ; M. Bohr ; P. Bai ; J. Brockman ; B. Buford ; C. Connor ; N. Das ; B. Doyle ; T. Ghani ; F. Hamzaoglu ; P. Heil ; P. Hentges ; R. Jahan ; D. Kencke ; B. Lin ; M. Lu ; M. Mainuddin ; M. Meterelliyoz ; P. Nguyen ; D. Nikonov ; K. O'brien ; J.O Donnell ; K. Oguz ; D. Ouellette ; J. Park ; J. Pellegren ; C. Puls ; P. Quintero ; T. Rahman ; A. Romang ; M. Sekhar ; A. Selarka ; M. Seth ; A. J. Smith ; A. K. Smith ; L. Wei ; C. Wiegand ; Z. Zhang ; K. Fischer

DEMONSTRATION OF HIGHLY MANUFACTURABLE STT-MRAM EMBEDDED IN 28NM LOGIC 416

Y. J. Song ; J. H. Lee ; S. H. Han ; H. C. Shin ; K. H. Lee ; K. Suh ; D. E. Jeong ; G. H. Koh ; S. C. Oh ; J. H. Park ; S. O. Park ; B. J. Bae ; O. I. Kwon ; K. H. Hwang ; B.Y. Seo ; Y.K. Lee ; S. H. Hwang ; D. S. Lee ; Y. Ji ; K.C. Park ; G. T. Jeong ; H. S. Hong ; K. P. Lee ; H. K. Kang ; E. S. Jung

ENABLEMENT OF STT-MRAM AS LAST LEVEL CACHE FOR THE HIGH PERFORMANCE COMPUTING DOMAIN AT THE 5NM NODE 420

S. Sakhare ; M. Perumkunnil ; T. Huynh Bao ; S. Rao ; W. Kim ; D. Crotti ; F. Yasin ; S. Couet ; J. Swerts ; S. Kundu ; D. Yakimets ; R. Baert ; HR. Oh ; A. Spessot ; A. Mocuta ; G. Sankar Kar ; A. Furnemont

TRULY INNOVATIVE 28NM FDSOI TECHNOLOGY FOR AUTOMOTIVE MICRO-CONTROLLER APPLICATIONS EMBEDDING 16MB PHASE CHANGE MEMORY 424

F. Arnaud ; P. Zuliani ; J.P. Reynard ; A. Gandolfo ; F. Disegni ; P. Mattavelli ; E. Gomiero ; G. Samanni ; C. Jahan ; R. Berthelon ; O. Weber ; E. Richard ; V. Barral ; A. Villaret ; S. Kohler ; J.C. Grenier ; R. Ranica ; C. Gallon ; A. Souhaite ; D. Ristoiu ; L. Favennec ; V. Caubet ; S. Delmedico ; N. Cherault ; R. Beneyton ; S. Chouteau ; P.O. Sassoulas ; A. Vernhet ; Y. Le Friec ; F. Domengie ; L. Scotti ; D. Pacelli ; J.L. Ogier ; F. Boucard ; S. Lagrasta ; D. Benoit ; L. Clement ; P. Boivin ; P. Ferreira ; R. Annunziata ; P. Cappelletti

A COST-EFFICIENT 28NM SPLIT-GATE EFLASH MEMORY FEATURING A HKMG HYBRID BIT CELL AND HV DEVICE 428

R. Richter ; M. Trentzsch ; S. Dünkel ; J. Müller ; P. Moll ; B. Bayha ; K. Mothes ; A. Henke ; M. Mazur ; J. Paul ; P. Krottenthaler ; J. Poth ; S. Jansen ; R. Hüselitz ; H. Kim ; A. Zaka ; T. Herrmann ; E.M. Bazizi ; S. Beyer ; P. Ghazavi ; H. Om'mani ; S. Lemke ; Y. Tkachev ; F. Zhou ; J. Kim ; X. Liu ; V. Tiwari ; N. Do

A BI-STABLE 1- /2-TRANSISTOR SRAM IN 14 NM FINFET TECHNOLOGY FOR HIGH DENSITY / HIGH PERFORMANCE EMBEDDED APPLICATIONS 432

Yuniarto Widjaja ; James Wilson ; Tu Nguyen ; Jin-Woo Han ; Christopher Norwood ; Dinesh Maheshwari ; Stefan Lai ; Pieter Vorenkamp ; Zvi Or-Bach ; Yoshio Nishi

SIC DEVICES FOR MAINSTREAM ADOPTION 436

Peter Friedrichs

THE CURRENT STATUS AND FUTURE PROSPECTS OF SIC HIGH VOLTAGE TECHNOLOGY 440

A. Mihaila ; L. Knoll ; E. Bianda ; M. Bellini ; S. Wirths ; G. Alfieri ; L. Kranz ; F. Canales ; M. Rahimo

PROGRESS IN HIGH AND ULTRAHIGH VOLTAGE SILICON CARBIDE DEVICE TECHNOLOGY 444

Y. Yonezawa ; K. Nakayama ; R. Kosugi ; S. Harada ; K. Koseki ; K. Sakamoto ; T. Kimoto ; H. Okumura

EFFECTS OF BASAL PLANE DISLOCATIONS ON SIC POWER DEVICE RELIABILITY 448

R. E. Stahlbush ; K. N. A. Mahakik ; A. J. Lelis ; R. Green

GAN DEVICES FOR AUTOMOTIVE APPLICATION AND THEIR CHALLENGES IN ADOPTION 452

Tetsu Kachi

BARRIERS TO THE ADOPTION OF WIDE-BANDGAP SEMICONDUCTORS FOR POWER ELECTRONICS 456

I.C. Kizilyalli ; E.P. Carlson ; D.W. Cunningham

GAN POWER COMMERCIALIZATION WITH HIGHEST QUALITY-HIGHEST RELIABILITY 650V HEMTS-REQUIREMENTS, SUCCESSES AND CHALLENGES 460

P. Parikh ; Y. Wu ; L. Shen ; R. Barr ; S. Chowdhury ; J. Gritters ; S. Yea ; P. Smith ; L. McCarthy ; R. Birkhahn ; M. Moore ; J. McKay ; H. Clement ; U. Mishra ; R. Lal ; P. Zuk ; T. Hosoda ; K. Shono ; K. Imanishi ; Y. Asai

40× RETENTION IMPROVEMENT BY ELIMINATING RESISTANCE RELAXATION WITH HIGH TEMPERATURE FORMING IN 28 NM RRAM CHIP 464

Xiaoxin Xu ; Lu Tai ; Tiancheng Gong ; Jiahao Yin ; Peng Huang ; Jie Yu ; Da Nian Dong ; Qing Luo ; Jing Liu ; Zhaoan Yu ; Xi Zhu ; Xiu Long Wu ; Qi Liu ; Hangbing Lv ; Ming Liu

CHARACTERIZING ENDURANCE DEGRADATION OF INCREMENTAL SWITCHING IN ANALOG RRAM FOR NEUROMORPHIC SYSTEMS 468

Meiran Zhao ; Huaqiang Wu ; Bin Gao ; Xiaoyu Sun ; Yuyi Liu ; Peng Yao ; Yue Xi ; Xinyi Li ; Qingtian Zhang ; Kanwen Wang ; Shimeng Yu ; He Qian

IN-DEPTH CHARACTERIZATION OF RESISTIVE MEMORY-BASED TERNARY CONTENT ADDRESSABLE MEMORIES 472

D. R. B. Ly ; B. Giraud ; J-P Noel ; A. Grossi ; N. Castellani ; G. Sassine ; J-F Nodin ; G. Molas ; C. Fenouillet-Beranger ; G. Indiveri ; E. Nowak ; E. Vianello

MIXED-SIGNAL NEUROMORPHIC INFERENCE ACCELERATORS: RECENT RESULTS AND FUTURE PROSPECTS 476

M. Bavandpour ; M.R. Mahmoodi ; H. Nili ; F. Merrikh Bayat ; M. Prezioso ; A. Vincent ; D.B. Strukov ; K.K. Likharev

TEMPORAL SEQUENCE LEARNING WITH A HISTORY-SENSITIVE PROBABILISTIC LEARNING RULE INTRINSIC TO OXYGEN VACANCY-BASED RRAM 480

J. Doevenspeck ; R. Degraeve ; A. Fantini ; P. Debacker ; D. Verkest ; R. Lauwereins ; W. Dehaene

IN-MEMORY AND ERROR-IMMUNE DIFFERENTIAL RRAM IMPLEMENTATION OF BINARIZED DEEP NEURAL NETWORKS 484

M. Bocquet ; T. Hirtzlin ; J.-O. Klein ; E. Nowak ; E. Vianello ; J.-M. Portal ; D. Querlioz

A NEW HARDWARE IMPLEMENTATION APPROACH OF BNNS BASED ON NONLINEAR 2T2R SYNAPTIC CELL 488

Z. Zhou ; P. Huang ; Y. C. Xiang ; W. S. Shen ; Y. D. Zhao ; Y. L. Feng ; B. Gao ; H. Q. Wu ; H. Qian ; L. F. Liu ; X. Zhang ; X. Y. Liu ; J. F. Kang

GE CMOS GATE STACK AND CONTACT DEVELOPMENT FOR VERTICALLY STACKED LATERAL NANOWIRE FETS 492

M.J.H. van Dal ; G. Vellianitis ; G. Doornbos ; B. Duriez ; M.C. Holland ; T. Vasen ; A. Afzalian ; E. Chen ; S.K. Su ; T.K. Chen ; T.M. Shen ; Z.Q. Wu ; C.H. Diaz

ADVANTAGE OF NW STRUCTURE IN PRESERVATION OF SRB-INDUCED STRAIN AND INVESTIGATION OF OFF-STATE LEAKAGE IN STRAINED STACKED GE NW PFET 496

H. Arimura ; G. Eneman ; E. Capogreco ; L. Witters ; A. De Keersgieter ; P. Favia ; C. Porret ; A. Hikavyy ; R. Loo ; H. Bender ; L.-Å. Ragnarsson ; J. Mitard ; N. Collaert ; D. Mocuta ; N. Horiguchi

TUNABILITY OF PARASITIC CHANNEL IN GATE-ALL-AROUND STACKED NANOSHEETS 500

S. Barraud ; B. Previtali ; V. Lapras ; C. Vizioz ; J.-M. Hartmann ; S. Martinie ; J. Lacord ; M. Cassé ; L. Dourthe ; V. Loup ; G. Romano ; N. Rambal ; Z. Chalupa ; N. Bernier ; G. Audoit ; A. Jannaud ; V. Delaye ; V. Balan ; O. Rozeau ; T. Ernst ; M. Vinet

VOLTAGE TRANSFER CHARACTERISTIC MATCHING BY DIFFERENT NANOSHEET LAYER NUMBERS OF VERTICALLY STACKED JUNCTIONLESS CMOS INVERTER FOR SOP/3D-ICS APPLICATIONS 504

P.-J. Sung ; C.-Y. Chang ; L.-Y. Chen ; K.-H. Kao ; C.-J. Su ; T.-H. Liao ; C.-C. Fang ; C.-J. Wang ; T.-C. Hong ; C.-Y. Jao ; H.-S. Hsu ; S.-X. Luo ; Y.-S. Wang ; H.-F. Huang ; J.-H. Li ; Y.-C. Huang ; F.-K. Hsueh ; C.-T. Wu ; Y.-M. Huang ; F.-J. Hou ; G.-L. Luo ; Y.-C. Huang ; Y.-L. Shen ; W. C.-Y. Ma ; K.-P. Huang ; K.-L. Lin ; S. Samukawa ; Y. Li ; G.-W Huang ; Y.-J. Lee ; J.-Y. Li ; W.-F. Wu ; J.-M. Shieh ; T.-S. Chao ; W.-K. Yeh ; Y.-H. Wang

VERTICALLY STACKED GATE-ALL-AROUND SI NANOWIRE CMOS TRANSISTORS WITH REDUCED VERTICAL NANOWIRES SEPARATION, NEW WORK FUNCTION METAL GATE SOLUTIONS, AND DC/AC PERFORMANCE OPTIMIZATION 508

R. Ritzenthaler ; H. Mertens ; V. Pena ; G. Santoro ; A. Chasin ; K. Kenis ; K. Devriendt ; G. Mannaert ; H. Dekkers ; A. Dangol ; Y. Lin ; S. Sun ; Z. Chen ; M. Kim ; J. Machillot ; J. Mitard ; N. Yoshida ; N. Kim ; D. Mocuta ; N. Horiguchi

2D MATERIALS: ROADMAP TO CMOS INTEGRATION 512

C. Huyghebaert ; T. Schram ; Q. Smets ; T. Kumar Agarwal ; D. Verreck ; S. Brems ; A. Phommahaxay ; D. Chiappe ; S. El Kazzi ; C. Lockhart de la Rosa ; G. Arutchelvan ; D. Cott ; J. Ludwig ; A. Gaur ; S. Sutar ; A. Leonhardt ; D. Marinov ; D. Lin ; M. Caymax ; I. Asselberghs ; G. Pourtois ; I.P. Radu

FIRST DEMONSTRATION OF WSE2 BASED CMOS-SRAM 516

Chin-Sheng Pang ; Niharika Thakuria ; Sumeet Kumar Gupta ; Zhihong Chen

STEEP SLOPE P-TYPE 2D WSE$_2$ FIELD-EFFECT TRANSISTORS WITH VAN DER WAALS CONTACT AND NEGATIVE CAPACITANCE 520

Jingli Wang ; Xuyun Guo ; Zhihao Yu ; Zichao Ma ; Yanghui Liu ; Masun Chan ; Ye Zhu ; Xinran Wang ; Yang Chai

TOWARD HIGH-MOBILITY AND LOW-POWER 2D MOS$_2$ FIELD-EFFECT TRANSISTORS 524

Zhihao Yu ; Ying Zhu ; Weisheng Li ; Yi Shi ; Gang Zhang ; Yang Chai ; Xinran Wang

3D MONOLITHIC STACKED 1T1R CELLS USING MONOLAYER MOS$_2$ FET AND HBN RRAM FABRICATED AT LOW (150°C) TEMPERATURE 528

Ching-Hua Wang ; Connor McClellan ; Yuanyuan Shi ; Xin Zheng ; Victoria Chen ; Mario Lanza ; Eric Pop ; H.-S. Philip Wong

ATOMRISTORS: MEMORY EFFECT IN ATOMICALLY-THIN SHEETS AND RECORD RF SWITCHES 532

Ruijing Ge ; Xiaohan Wu ; Myungsoo Kim ; Po-An Chen ; Jianping Shi ; Junho Choi ; Xiaoqin Li ; Yanfeng Zhang ; Meng-Hsueh Chiang ; Jack C. Lee ; Deji Akinwande

AN ULTRA-FAST MULTI-LEVEL MOTE$_2$-BASED RRAM 536

F. Zhang ; H. Zhang ; P.R. Shrestha ; Y. Zhu ; K. Maize ; S. Krylyuk ; A. Shakouri ; J.P. Campbell ; K.P. Cheung ; L.A. Bendersky ; A. V. Davydov ; J. Appenzeller

FIRST CRYOGENIC ELECTRO-OPTIC SWITCH ON SILICON WITH HIGH BANDWIDTH AND LOW POWER TUNABILITY 540

F. Eltes ; J. Barreto ; D. Caimi ; S. Karg ; A. A. Gentile ; A. Hart ; P. Stark ; N. Meier ; M. G. Thompson ; J. Fompeyrine ; S. Abel

HIGH SPEED (F$_{3-DB}$ ABOVE 10 GHZ) PHOTO DETECTION AT TWO-MICRON-WAVELENGTH REALIZED BY GESN/GE MULTIPLE-QUANTUM-WELL PHOTODIODE ON A 300 MM SI SUBSTRATE 544

Shengqiang Xu ; Wei Wang ; Yi-Chiau Huang ; Yuan Dong ; Saeid Masudy-Panah ; Hong Wang ; Xiao Gong ; Yee-Chia Yeo

QUADRATIC ELECTRO-OPTICAL SILICON-ORGANIC HYBRID RF MODULATOR IN A PHOTONIC INTEGRATED CIRCUIT TECHNOLOGY 548

P. Steglich ; C. Mai ; A. Peczek ; F. Korndörfer ; C. Villringer ; B. Dietzel ; A. Mai

SILICON PHOTONICS: A SCALING TECHNOLOGY FOR COMMUNICATIONS AND INTERCONNECTS 552

P. Dong ; K. W. Kim ; A. Melikyan ; Y. Baeyens

INAS/GAAS QUANTUM DOT LASERS MONOLITHICALLY INTEGRATED ON GROUP IV PLATFORM 556
Keshuang Li ; Mingchu Tang ; Mengya Liao ; Jiang Wu ; Siming Chen ; Alwyn Seeds ; Huiyun Liu

HETEROGENEOUSLY INTEGRATED LIGTHT SOURCES ON BULK-SILICON PLATFORM 560
Dongjae Shin ; Jungho Cha ; Yonghwack Shin ; Kyoungho Ha ; Chang Bum Lee ; Changgyun Shin ; Dongshik Shim ; Byoung Lyong Choi ; Hyeongsun Hong ; Kyupil Lee ; Ho-Kyu Kang

INTERFACIAL THERMAL CONDUCTIVITY OF 2D LAYERED MATERIALS: AN ATOMISTIC APPROACH 564
Kamyar Parto ; Arnab Pal ; Xuejun Xie ; Wei Cao ; Kaustav Banerjee

COMPUTATIONAL DESIGN OF SILICON CONTACTS ON 2D TRANSITION-METAL DICHALCOGENIDES: THE ROLES OF CRYSTALLINE ORIENTATION, DOPING LEVEL, PASSIVATION AND INTERFACIAL LAYER 568
Xiaolei Ma ; Zhiqiang Fan ; Jixuan Wu ; Xiangwei Jiang ; Jiezhi Chen

FIRST PRINCIPLES STUDY OF MEMORY SELECTORS USING HETEROJUNCTIONS OF 2D LAYERED MATERIALS 572
Linsen Li ; Blanka Magyari-Köpe ; Ching-Hua Wang ; Sanchit Deshmukh ; Zizhen Jiang ; Haitong Li ; Yi Yang ; Huanglong Li ; He Tian ; E. Pop ; Tian-Ling Ren ; H.-S. Philip Wong

CAN KINETIC INDUCTANCE IN LOW-DIMENSIONAL MATERIALS ENABLE A NEW GENERATION OF RF-ELECTRONICS? 576
Kunjesh Agashiwala ; Arnab Pal ; Wei Cao ; Junkai Jiang ; Kaustav Banerjee

A SURFACE POTENTIAL- AND PHYSICS- BASED COMPACT MODEL FOR 2D POLYCRYSTALLINE-MOS$_2$ FET WITH RESISTIVE SWITCHING BEHAVIOR IN NEUROMORPHIC COMPUTING 580
Lingfei Wang ; Lin Wang ; Kah-Wee Ang ; Aaron Voon-Yew Thean ; Gengchiau Liang

DESIGN AND OPTIMIZATION OF ß-GA$_2$O$_3$ ON (H-BN LAYERED) SAPPHIRE FOR HIGH EFFICIENCY POWER TRANSISTORS: A DEVICE-CIRCUIT-PACKAGE PERSPECTIVE 584
Bikram K. Mahajan ; Yen-Pu Chen ; Woojin Ahn ; Nicolò Zagni ; Muhammad Ashraful Alam

DECONVOLUTING CHARGE TRAPPING AND NUCLEATION INTERPLAY IN FEFETS: KINETICS AND RELIABILITY 588
Milan Pesic ; Andrea Padovani ; Stefan Slcsazeck ; Thomas Mikolajick ; Luca Larcher

IMPACT OF SELF-HEATING ON RELIABILITY PREDICTIONS IN STT-MRAM 592
S. Van Beek ; B. J. O'Sullivan ; P. J. Roussel ; R. Degraeve ; E. Bury ; J. Swerts ; S. Couet ; L. Souriau ; S. Kundu ; S. Rao ; W. Kim ; F. Yasin ; D. Crotti ; D. Linten ; G. Kar

INVESTIGATING THE STATISTICAL-PHYSICAL NATURE OF MGO DIELECTRIC BREAKDOWN IN STT-MRAM AT DIFFERENT OPERATING CONDITIONS 596
J.H. Lim ; N. Raghavan ; A. Padovani ; J.H. Kwon ; K. Yamane ; H. Yang ; V.B. Naik ; L. Larcher ; K.H. Lee ; K.L. Pey

TRAP REDUCTION AND PERFORMANCES IMPROVEMENTS STUDY AFTER HIGH PRESSURE ANNEAL PROCESS ON SINGLE CRYSTAL CHANNEL 3D NAND DEVICES 600
A. Subirats ; A. Arreghini ; R. Delhougne ; E. Rosseel ; A. Hikavyy ; L. Breuil ; S. Vadakupudhu Palayam ; G. Van den bosch ; D. Linten ; A. Furnémont

22-NM FD-SOI EMBEDDED MRAM TECHNOLOGY FOR LOW-POWER AUTOMOTIVE-GRADE-L MCU APPLICATIONS 604
K. Lee ; R. Chao ; K. Yamane ; V. B. Naik ; H. Yang ; J. Kwon ; N. L. Chung ; S. H. Jang ; B. Behin-Aein ; J.H. Lim ; B. Liu ; E. H. Toh ; K. W. Gan ; D. Zeng ; N. Thiyagarajah ; L. C. Goh ; T. Ling ; J. W. Ting ; J. Hwang ; L. Zhang ; R. Low ; R. Krishnan ; L. Zhang ; S. L Tan ; Y. S. You ; C. S. Seet ; H. Cong ; J. Wong ; S. T. Woo ; E. Quek ; S. Y. Siah

14NS WRITE SPEED 128MB DENSITY EMBEDDED STT-MRAM WITH ENDURANCE>10^{10} AND 10YRS RETENTION@85°C USING NOVEL LOW DAMAGE MTJ INTEGRATION PROCESS 608
H. Sato ; H. Honjo ; T. Watanabe ; M. Niwa ; H. Koike ; S. Miura ; T. Saito ; H. Inoue ; T. Nasuno ; T. Tanigawa ; Y. Noguchi ; T. Yoshiduka ; M. Yasuhira ; S. Ikeda ; S.- Y. Kang ; T. Kubo ; K. Yamashita ; Y. Yagi ; R. Tamura ; T. Endoh

STT-MRAM DEVICES WITH LOW DAMPING AND MOMENT OPTIMIZED FOR LLC APPLICATIONS AT OX NODES 612
Luc Thomas ; Guenole Jan ; Santiago Serrano-Guisan ; Huanlong Liu ; Jian Zhu ; Yuan-Jen Lee ; Son Le ; Jodi Iwata-Harms ; Ru-Ying Tong ; Sahil Patel ; Vignesh Sundar ; Dongna Shen ; Yi Yang ; Renren He ; Jesmin Haq ; Zhongjian Teng ; Vinh Lam ; Paul Liu ; Yu-Jen Wang ; Tom Zhong ; Hideaki Fukuzawa ; PoKang Wang

MICROWAVE NEURAL PROCESSING AND BROADCASTING WITH SPINTRONIC NANO-OSCILLATORS 616
P. Talatchian ; M. Romera ; S. Tsunegi ; F. Abreu Araujo ; V. Cros ; P. Bortolotti ; J. Trastoy ; K. Yakushiji ; A. Fukushima ; H. Kubota ; S. Yuasa ; M. Ernoult ; D. Vodenicarevic ; T. Hirtzlin ; N. Locatelli ; D. Querlioz ; J. Grollier

HIGH ENDURANCE PHASE CHANGE MEMORY CHIP IMPLEMENTED BASED ON CARBON-DOPED GE$_2$SB$_2$TE$_5$ IN 40 NM NODE FOR EMBEDDED APPLICATION 620
Z. T. Song ; D. L. Cai ; X. Li ; L. Wang ; Y. F. Chen ; H. P. Chen ; Q. Wang ; Y. P. Zhan ; M. H. Ji

A 40NM LOW-POWER LOGIC COMPATIBLE PHASE CHANGE MEMORY TECHNOLOGY 624
J.Y. Wu ; Y.S. Chen ; W. S. Khwa ; S. M. Yu ; T. Y. Wang ; J.C. Tseng ; Y.D. Chih ; Carlos H. Diaz

8-BIT PRECISION IN-MEMORY MULTIPLICATION WITH PROJECTED PHASE-CHANGE MEMORY 628
I. Giannopoulos ; A. Sebastian ; M. Le Gallo ; V.P. Jonnalagadda ; M. Sousa ; M.N. Boon ; E. Eleftheriou

SYSTEM PERFORMANCE: FROM ENTERPRISE TO AI 632
A. Kumar ; L. Chang ; G.E. Tellez ; L.A. Clevenger ; J.L. Burns

DESIGN-TECHNOLOGY CO-OPTIMIZATION OF STANDARD CELL LIBRARIES ON INTEL 10NM PROCESS 636
Xinning Wang ; Ranjith Kumar ; Somashekar Bangalore Prakash ; Peng Zheng ; Tai-Hsuan Wu ; Quan Shi ; Marni Nabors ; Srinivasa Chaitanya Gadigatla ; Simeon Realov ; Chin-Hsuan Chen ; Ying Zhang ; Kaizad Mistry ; Andrew Yeoh ; Ian Post ; Chris Auth ; Atul Madhavan

AN ACCURATE FINFET'S VMIN ESTIMATION METHOD FOR EXTREME LOW OPERATION VOLTAGE DESIGN 640
H. W. Choi ; S. K. Kim ; H. Jung ; D. R. Chang ; S. Park ; Y. Yasuda-Masuoka ; J.S. Yoon

TACKLING FUNDAMENTAL CHALLENGES OF CARRIER TRANSPORT AND DEVICE VARIABILITY IN ADVANCED SINFINFETS FOR 7NM NODE AND BEYOND 644
Ming-Huei Lin ; Vincent S. Chang ; Jen-Hsiang Lu ; Shu-Hui Wang ; Shyh-Horng Yang

EXTENDABLE AND MANUFACTURABLE VOLUME-LESS MULTI-VT SOLUTION FOR 7NM TECHNOLOGY NODE AND BEYOND 648
Ruqiang Bao ; Huimei Zhou ; Miaomiao Wang ; Dechao Guo ; Bala S Haran ; Vijay Narayanan ; Rama Divakaruni

CHANNEL GEOMETRY IMPACT AND NARROW SHEET EFFECT OF STACKED NANOSHEET 652
Chun Wing Yeung ; Jingyun Zhang ; Robin Chao ; Ohseong Kwon ; Reinaldo Vega ; Gen Tsutsui ; Xin Miao ; Chen Zhang ; Chang-Woo Sohn ; Bum Ki Moon ; Ali Razavieh ; Julien Frougier ; Andrew Greene ; Rohit Galatage ; Juntao Li ; Miaomiao Wang ; Nicolas Loubet ; Robert Robison ; Veeraraghavan Basker ; Tenko Yamashita ; Dechao Guo

3NM GAA TECHNOLOGY FEATURING MULTI-BRIDGE-CHANNEL FET FOR LOW POWER AND HIGH PERFORMANCE APPLICATIONS 656
Geumjong Bae ; D.-I. Bae ; M. Kang ; S.M. Hwang ; S.S. Kim ; B. Seo ; T.Y. Kwon ; T.J. Lee ; C. Moon ; Y.M. Choi ; K. Oikawa ; S. Masuoka ; K.Y. Chun ; S.H. Park ; H.J. Shin ; J.C. Kim ; K.K. Bhuwalka ; D.H. Kim ; W.J. Kim ; J. Yoo ; H.Y. Jeon ; M.S. Yang ; S.-J. Chung ; D. Kim ; B.H. Ham ; K.J. Park ; W.D. Kim ; S.H. Park ; G. Song ; Y.H. Kim ; M.S. Kang ; K.H. Hwang ; C.-H. Park ; J.-H. Lee ; D.-W. Kim ; S-M. Jung ; H.K. Kang

A CMOS PROXIMITY CAPACITANCE IMAGE SENSOR WITH 16μM PIXEL PITCH, 0.1AF DETECTION ACCURACY AND 60 FRAMES PER SECOND 660
M. Yamamoto ; R. Kuroda ; M. Suzuki ; T. Goto ; H. Hamori ; S. Murakami ; T. Yasuda ; S. Sugawa

3D EXPANDABLE MICROWIRE ELECTRODE ARRAYS MADE OF PROGRAMMABLE SHAPE MEMORY MATERIALS 664
Ruoyu Zhao ; Xin Liu ; Yichen Lu ; Chi Ren ; Armaghan Mehrsa ; Takaki Komiyama ; Duygu Kuzum

BIO-INSPIRED 3D NEURAL ELECTRODES FOR THE PERIPHERAL NERVES STIMULATION USING SHAPE MEMORY POLYMERS 668
Yingchao Zhang ; Ning Zheng ; Yinji Ma ; Tao Xie ; Xue Feng

FABRICATION AND CHARACTERIZATION OF 3D MULTI-ELECTRODE ARRAY ON FLEXIBLE SUBSTRATE FOR IN VIVO EMG RECORDING FROM EXPIRATORY MUSCLE OF SONGBIRD 672
Muneeb Zia ; Bryce Chung ; Samuel J. Sober ; Muhannad S. Bakir

BIOELECTRONICS AT THE SINGLE MOLECULE LEVEL 676
O. Tolga Gul ; Kaitlin M. Pugliese ; Yongki Choi ; Arith J. Rajapakse ; Calvin J. Lau ; Narendra Kumar ; Kristin N. Gabriel ; Denys Marushchak ; Tivoli J. Olsen ; Deng Pan ; Gregory A. Weiss ; Philip G. Collins

SI NANOWIRE BIOSENSORS USING A FINFET FABRICATION PROCESS FOR REAL TIME MONITORING CELLULAR ION ACTITIVIES 679
Qingzhu Zhang ; Hailing Tu ; Huaxiang Yin ; Feng Wei ; Hongbin Zhao ; Chunling Xue ; Qianhui Wei ; Zhaohao Zhang ; Xiao Zhang ; Shaoming Zhang ; Qin Han ; Yudong Li ; Robert Chunhua Zhao ; Jiang Yan ; Junfeng Li ; Wenwu Wang

A FLEXIBLE, HETEROGENEOUSLY INTEGRATED WIRELESS POWERED SYSTEM FOR BIO-IMPLANTABLE APPLICATIONS USING FAN-OUT WAFER-LEVEL PACKAGING 683
G. Ezhilarasu ; A. Hanna ; R. Irwin ; A. Alam ; S. S. Iyer

PARALLEL-PLANE BREAKDOWN FIELDS OF 2.8-3.5 MV/CM IN GAN-ON-GAN P-N JUNCTION DIODES WITH DOUBLE-SIDE-DEPLETED SHALLOW BEVEL TERMINATION 687
T. Maeda ; T. Narita ; H. Ueda ; M. Kanechika ; T. Uesugi ; T. Kachi ; T. Kimoto ; M. Horita ; J. Suda

DEMONSTRATION OF AVALANCHE CAPABILITY IN POLARIZATION-DOPED VERTICAL GAN PN DIODES: STUDY OF WALKOUT DUE TO RESIDUAL CARBON CONCENTRATION 691
C. De Santi ; E. Fabris ; K. Nomoto ; Z. Hu ; W. Li ; X. Gao ; D. Jena ; H. G. Xing ; G. Meneghesso ; M. Meneghini ; E. Zanoni

SUPPRESSED HOLE-INDUCED DEGRADATION IN E-MODE GAN MIS-FETS WITH CRYSTALLINE GAO$_X$N$_{1-X}$ CHANNEL 695
Mengyuan Hua ; Xiangbin Cai ; Song Yang ; Zhaofu Zhang ; Zheyang Zheng ; Jin Wei ; Ning Wang ; Kevin J. Chen

RECENT ADVANCEMENT OF GAN HEMT WITH INALGAN BARRIER LAYER AND FUTURE PROSPECTS OF A1N-BASED ELECTRON DEVICES 699
J. Kotani ; A. Yamada ; T. Ohki ; Y. Minoura ; S. Ozaki ; N. Okamoto ; K. Makiyama ; N. Nakamura

POWER GAN HEMT DEGRADATION: FROM TIME-DEPENDENT BREAKDOWN TO HOT-ELECTRON EFFECTS 703
M. Meneghini ; A. Barbato ; M. Borga ; C. De Santi ; M. Barbato ; S. Stoffels ; M. Zhao ; N. Posthuma ; S. Decoutere ; O. Haeberlen ; T. Detzel ; G. Meneghesso ; E. Zanoni

NEW INSIGHTS INTO THE PHYSICAL ORIGIN OF NEGATIVE CAPACITANCE AND HYSTERESIS IN NCFETS 707
Huimin Wang ; Mengxuan Yang ; Qianqian Huang ; Kunkun Zhu ; Yang Zhao ; Zhongxin Liang ; Cheng Chen ; Zhixuan Wang ; Yuan Zhong ; Xing Zhang ; Ru Huang

A CRITICAL EXAMINATION OF 'QUASI-STATIC NEGATIVE CAPACITANCE' (QSNC) THEORY 711
Z. Liu ; M. A. Bhuiyan ; T. P. Ma

DIRECT RELATIONSHIP BETWEEN SUB-60 MV/DEC SUBTHRESHOLD SWING AND INTERNAL POTENTIAL INSTABILITY IN MOSFET EXTERNALLY CONNECTED TO FERROELECTRIC CAPACITOR 715
Xiuyan Li ; Akira Toriumi

ASSESSMENT OF STEEP-SUBTHRESHOLD SWING BEHAVIORS IN FERROELECTRIC-GATE FIELD-EFFECT TRANSISTORS CAUSED BY POSITIVE FEEDBACK OF POLARIZATION REVERSAL 719
Shinji Migita ; Hiroyuki Ota ; Akira Toriumi

EXPERIMENTAL STUDY ON THE ROLE OF POLARIZATION SWITCHING IN SUBTHRESHOLD CHARACTERISTICS OF HFO$_2$-BASED FERROELECTRIC AND ANTI-FERROELECTRIC FET 723
Chengji Jin ; Kyungmin Jang ; Takuya Saraya ; Toshiro Hiramoto ; Masaharu Kobayashi

DEMONSTRATION OF HIGH-SPEED HYSTERESIS-FREE NEGATIVE CAPACITANCE IN FERROELECTRIC HF$_{0.5}$ZR$_{0.5}$O$_2$ 727
M. Hoffmann ; B. Max ; T. Mittmann ; U. Schroeder ; S. Slesazeck ; T. Mikolajick

NEGATIVE-CAPACITANCE FINFET INVERTER, RING OSCILLATOR, SRAM CELL, AND FT 731
Kai-Shin Li ; Yun-Jie Wei ; Yi-Ju Chen ; Wen-Cheng Chiu ; Hsiu-Chih Chen ; Min-Hung Lee ; Yu-Fan Chiu ; Fu-Kuo Hsueh ; Bo-Wei Wu ; Pin-Guang Chen ; Tung-Yan Lai ; Chun-Chi Chen ; Jia-Min Shieh ; Wen-Kuan Yeh ; Sayeef Salahuddin ; Chenming Hu

EXTREMELY STEEP SWITCH OF NEGATIVE-CAPACITANCE NANOSHEET GAA-FETS AND FINFETS 735
M. H. Lee ; K.-T. Chen ; C.-Y. Liao ; S.-S. Gu ; G.-Y. Siang ; Y.-C. Chou ; H.-Y. Chen ; J. Le ; R.-C. Hong ; Z.-Y. Wang ; S.-Y. Chen ; P.-G. Chen ; M. Tang ; Y.-D. Lin ; H.-Y. Lee ; K.-S. Li ; C. W. Liu

OPTOCOUPLING IN CMOS 739
V. Agarwal ; S. Dutta ; A. J. Annema ; R. J. E. Hueting ; J. Schmitz ; M.J. Lee ; E. Charbon ; B. Nauta

HIGH VOLTAGE GENERATION USING DEEP TRENCH ISOLATED PHOTODIODES IN A BACK SIDE ILLUMINATED PROCESS 743
F. Kaklin ; J. M. Raynor ; R. K. Henderson

THROUGH-SILICON-TRENCH IN BACK-SIDE-ILLUMINATED CMOS IMAGE SENSORS FOR THE IMPROVEMENT OF GATE OXIDE LONG TERM PERFORMANCE 747
A. Vici ; F. Russo ; N. Lovisi ; L. Latessa ; A. Marchioni ; A. Casella ; F. Irrera

HIGH-PERFORMANCE GERMANIUM-AN-SILICON LOCK-IN PIXELS FOR INDIRECT TIME-OF-FLIGHT APPLICATIONS 751
N. Na ; S.-L. Cheng ; H.-D. Liu ; M.-J. Yang ; C.-Y. Chen ; H.-W. Chen ; Y.-T. Chou ; C.-T. Lin ; W.-H. Liu ; C.-F. Liang ; C.-L. Chen ; S.-W. Chu ; B.-J. Chen ; Y.-F. Lyu ; S.-L. Chen

CMOS-INTEGRATED SINGLE-PHOTON-COUNTING X-RAY DETECTOR USING AN AMORPHOUS-SELENIUM PHOTOCONDUCTOR WITH 11×11-µM^2 PIXELS 755
A. Camlica ; A. El-Falou ; R. Mohammadi ; P. M. Levine ; K. S. Karim

TRANSPORT MODELS BASED ON NEGF AND EMPIRICAL PSEUDOPOTENTIALS: A COMPUTATIONALLY VIABLE METHOD FOR SELF-CONSISTENT SIMULATION OF NANOSCALE DEVICES 759

Marco G. Pala ; Oves Badami ; David Esseni

FIRST PRINCIPLES SIMULATION OF ENERGY EFFICIENT SWITCHING BY SOURCE DENSITY OF STATES ENGINEERING 763

Fei Liu ; Chenguang Qiu ; Zhiyong Zhang ; Lian-Mao Peng ; Jian Wang ; Zhenhua Wu ; Hong Guo

UNIVERSAL SWING FACTOR APPROACH FOR PERFORMANCE ANALYSIS OF LOGIC NODES 767

M. Ali Pourghaderi ; Anh-Tuan Pham ; Seungkyu Kim ; Hyein Chung ; Zhengping Jiang ; Hesameddin Ilatikhameneh ; Hong-hyun Park ; Seonghoon Jin ; Jongchol Kim ; Won-Young Chung ; Uihui Kwon ; Woosung Choi ; Dae Sin Kim ; Shigenobu Maeda

MULTI-DOMAIN PROCESS MODELING FOR ADVANCED LOGIC AND MEMORY DEVICES: FROM EQUIMPMENTS TO MATERIALS 771

Inkook Jang ; Hyoungsoo Ko ; Alexander Schmidt ; Sae-Jin Kim ; Moonhyun Cha ; Hyoshin Ahn ; Honglae Park ; Dae Sin Kim ; Ho-Kyu Kang

ENTIRE BIAS SPACE STATISTICAL RELIABILITY SIMULATION BY 3D-KMC METHOD AND ITS APPLICATION TO THE RELIABILITY ASSESSMENT OF NANOSHEET FETS BASED CIRCUITS 775

Wangyong Chen ; Yun Li ; Linlin Cai ; Pengying Chang ; Gang Du ; Xiaoyan Liu

A PHYSICS-BASED THERMAL MODEL OF NANOSHEET MOSFETS FOR DEVICE-CIRCUIT CO-DESIGN 779

Linlin Cai ; Wangyong Chen ; Pengying Chang ; Gang Du ; Xing Zhang ; Jinfeng Kang ; Xiaoyan Liu

UNDERSTANDING THE INTRINSIC RELIABILITY BEHAVIOR OF N -/P-SI AND P-GE NANOWIRE FETS UTILIZING DEGRADATION MAPS 783

Adrian Chasin ; Erik Bury ; Jacopo Franco ; Ben Kaczer ; Michiel Vandemaele ; Hiroaki Arimura ; Elena Capogreco ; Liesbeth Witters ; Romain Ritzenthaler ; Hans Mertens ; Naoto Horiguchi ; Dimitri Linten

BTI RELIABILITY IMPROVEMENT STRATEGIES IN LOW THERMAL BUDGET GATE STACKS FOR 3D SEQUENTIAL INTEGRATION 787

J. Franco ; Z. Wu ; G. Rzepa ; A. Vandooren ; H. Arimura ; L. -Å Ragnarsson ; G. Hellings ; S. Brus ; D. Cott ; V. De Heyn ; G. Groeseneken ; N. Horiguchi ; J. Ryckaert ; N. Collaert ; D. Linten ; T. Grasser ; B. Kaczer

CHARACTERIZATION AND UNDERSTANDING OF SLOW TRAPS IN GEO$_X$-BASED N-GE MOS INTERFACES 791

M. Ke ; P. Cheng ; K. Kato ; M. Takenaka ; S. Takagi

SOFT ERROR TRENDS IN ADVANCED SILICON TECHNOLOGY NODES 795

B. Bhuva

CMOS-COMPATIBLE DOPED-MULTILAYER-GRAPHENE INTERCONNECTS FOR NEXT-GENERATION VLSI 799

Junkai Jiang ; Jae Hwan Chu ; Kaustav Banerjee

TIME DEPENDENT EARLY BREAKDOWN OF AIGAN/GAN EPI STACKS AND SHIFT IN SOA BOUNDARY OF HEMTS UNDER FAST CYCLIC TRANSIENT STRESS 803

Bhawani Shankar ; Ankit Soni ; Sayak Dutta Gupta ; Swati Shikha ; Sandeep Singh ; Srinivasan Raghavan ; Mayank Shrivastava

TOWARD HIGH PERFORMANCE SIGE CHANNEL CMOS: DESIGN OF HIGH ELECTRON MOBILITY IN SIGE NFINFETS OUTPERFORMING SI 807

C. H. Lee ; R. G. Southwick ; S. Mochizuki ; J. Li ; X. Miao ; M. Wang ; R. Bao ; I. Ok ; T. Ando ; P. Hashemi ; D. Guo ; V. Narayanan ; N. Loubet ; H. Jagannathan

ADVANCED ARSENIC DOPED EPITAXIAL GROWTH FOR SOURCE DRAIN EXTENSION FORMATION IN SCALED FINFET DEVICES 811

S. Mochizuki ; B. Colombeau ; L. Yu ; A. Dube ; S. Choi ; M. Stolfi ; Z. Bi ; F. Chang ; R. A. Conti ; P. Liu ; K. R. Winstel ; H. Jagannathan ; H.-J. Gossmann ; N. Loubet ; D. F. Canaperi ; D. Guo ; S. Sharma ; S. Chu ; J. Boland ; Q. Jin ; Z. Li ; S. Lin ; M. Cogorno ; M. Chudzik ; S. Natarajan ; D. C. McHerron ; B. Haran

EXTERNAL RESISTANCE REDUCTION BY NANOSECOND LASER ANNEAL IN SI/SIGE CMOS TECHNOLOGY 815

Oleg Gluschenkov ; Heng Wu ; Kevin Brew ; Chengyu Niu ; Lan Yu ; Yasir Sulehria ; Samuel Choi ; Curtis Durfee ; James Demarest ; Adra Carr ; Shaoyin Chen ; Jim Willis ; Thirumal Thanigaivelan ; Fee-li Lie ; Walter Kleemeier ; Dechao Guo

PARASITIC RESISTANCE REDUCTION STRATEGIES FOR ADVANCED CMOS FINFETS BEYOND 7NM 819

H. Wu ; O. Gluschenkov ; G. Tsutsui ; C. Niu ; K. Brew ; C. Durfee ; C. Prindle ; V. Kamineni ; S. Mochizuki ; C. Lavoie ; E. Nowak ; Z. Liu ; J. Yang ; S. Choi ; J. Demarest ; L. Yu ; A. Carr ; W. Wang ; J. Strane ; S. Tsai ; Y. Liang ; H. Amanapu ; I. Saraf ; K. Ryan ; F. Lie ; W. Kleemeier ; K. Choi ; N. Cave ; T. Yamashita ; A. Knorr ; D. Gupta ; B. Haran ; D. Guo ; H. Bu ; M. Khare

SUB-10[-9] Ω -CM[2] SPECIFIC CONTACT RESISTIVITY ON P-TYPE GE AND GESN: IN-SITU GA DOPING WITH GA ION IMPLANTATION AT 300°C, 25°C, AND –100°C 823

Ying Wu ; Lye-Hing Chua ; Wei Wang ; Kaizhen Han ; Wei Zou ; Todd Henry ; Xiao Gong

SELECTIVE FIN TRIMMING AFTER DUMMY GATE REMOVAL AS THE LOCAL FIN WIDTH SCALING APPROACH FOR N5 AND BEYOND 827

Toshihiko Miyashita ; Shiyu Sun ; Sushant Mittal ; Myung Sun Kim ; Ashish Pal ; Angada Sachid ; Kalpana Pathak ; Matt Cogorno ; Nam Sung Kim

FIRST EXPERIMENTAL DEMONSTRATION OF A SCALABLE LINEAR MAJORITY GATE BASED ON SPIN WAVES 831

Florin Ciubotaru ; Giacomo Talmelli ; Thibaut Devolder ; Odysseas Zografos ; Marc Heyns ; Christoph Adelmann ; Iuliana P. Radu

SPINTRONIC DEVICES FOR LOW ENERGY DISSIPATION 835

Kang L. Wang ; Hao Wu ; Seyed Armin Razavi ; Qiming Shao

ROOM TEMPERATURE HIGHLY EFFICIENT TOPOLOGICAL INSULATOR/MO/COFEB SPIN-ORBIT TORQUE MEMORY WITH PERPENDICULAR MAGNETIC ANISOTROPY 839

Qiming Shao ; Hao Wu ; Quanjun Pan ; Peng Zhang ; Lei Pan ; Kin Wong ; Xiaoyu Che ; Kang L. Wang

SCALED SPINTRONIC LOGIC DEVICE BASED ON DOMAIN WALL MOTION IN MAGNETICALLY INTERCONNECTED TUNNEL JUNCTIONS 843

E. Raymenants ; D. Wan ; S. Couet ; O. Zografos ; V.D. Nguyen ; A. Vaysset ; L. Souriau ; A. Thiam ; M. Manfrini ; S. Brus ; M. Heyns ; D. Mocuta ; D.E. Nikonov ; S. Manipatruni ; I. A. Young ; T. Devolder ; I. P. Radu

BINARY AND TERNARY TRUE RANDOM NUMBER GENERATORS BASED ON SPIN ORBIT TORQUE 847

Huiming Chen ; Shuai Zhang ; Nuo Xu ; Min Song ; Xin Li ; Ruofan Li ; Yi Zeng ; Jeongmin Hong ; Long You

HIGH-PERFORMANCE, COST-EFFECTIVE 2Z NM TWO-DECK CROSS-POINT MEMORY INTEGRATED BY SELF-ALIGN SCHEME FOR 128 GB SCM 851

Taehoon Kim ; Hyejung Choi ; Myoungsub Kim ; Jaeyun Yi ; Donghoon Kim ; Sunglae Cho ; Hyunmin Lee ; Changyoun Hwang ; Eung-Rim Hwang ; Jeongho Song ; Sujin Chae ; Yunseok Chun ; Jin-Kook Kim

A HIGHLY EFFICIENT AND SCALABLE MODEL FOR CROSSBAR ARRAYS WITH NONLINEAR SELECTORS 855

An Chen

ULTRA-HIGH ENDURANCE AND LOW IOFF SELECTOR BASED ON ASSEGE CHALCOGENIDES FOR WIDE MEMORY WINDOW 3D STACKABLE CROSSPOINT MEMORY 859

H. Y. Cheng ; W. C. Chien ; I. T. Kuo ; C. W. Yeh ; L. Gignac ; W. Kim ; E. K. Lai ; Y. F. Lin ; R. L. Bruce ; C. Lavoie ; C.W. Cheng ; A. Ray ; F. M. Lee ; F. Carta ; C. H. Yang ; M. H. Lee ; H. Y. Ho ; M. BrightSky ; H. L. Lung

OPTIMIZED READING WINDOW FOR CROSSBAR ARRAYS THANKS TO GE-SE-SB-N-BASED OTS SELECTORS 863

A. Verdy ; M. Bernard ; J. Garrione ; G. Bourgeois ; M. C. Cyrille ; E. Nolot ; N. Castellani ; P. Noé ; C. Socquet-Clerc ; T. Magis ; G. Sassine ; G. Molas ; G. Navarro ; E. Nowak

FORMING-FREE MOTT-OXIDE THRESHOLD SELECTOR NANODEVICE SHOWING S-TYPE NDR WITH HIGH ENDURANCE (> 10[12] CYCLES), EXCELLENT V_TH STABILITY (5%), FAST (< 10 NS) SWITCHING, AND PROMISING SCALING PROPERTIES 867

T. Hennen ; D. Bedau ; J. A. J. Rupp ; C. Funck ; S. Menzel ; M. Grobis ; R. Waser ; D. J. Wouters

FULLY MULTI-FUNCTIONAL GAN-BASED MICRO-LEDS FOR 2500 PPI MICRO-DISPLAYS, TEMPERATURE SENSING, LIGHT ENERGY HARVESTING, AND LIGHT DETECTION 871

Zhaojun Liu ; Ke Zhang ; Yibo Liu ; Siwa Yan ; Hoi Sing Kwok ; Jamal Deen ; Xiaowei Sun

ENVIRONMENTALLY FRIENDLY QUANTUM DOTS FOR DISPLAY APPLICATIONS 875

E. Jang

SOLUTION PROCESSED HIGH PERFORMANCE SHORT CHANNEL ORGANIC THIN-FILM TRANSISTORS WITH EXCELLENT UNIFORMITY AND ULTRA-LOW CONTACT RESISTANCE FOR LOGIC AND DISPLAY 879

L. Feng ; Y. Huang ; J. Fan ; J. Zhao ; S. Pandya ; S. Chen ; W. Tang ; S. Ogier ; X. Guo

RECORD STATIC AND DYNAMIC PERFORMANCE OF FLEXIBLE ORGANIC THIN-FILM TRANSISTORS 883

James W. Borchert ; Ute Zschieschang ; Florian Letzkus ; Michele Giorgio ; Mario Caironi ; Joachim N. Burghartz ; Sabine Ludwigs ; Hagen Klauk

HYBRID STRUCTURE OF SILICON NANOCRYSTALS AND 2D WSE_2 FOR BROADBAND OPTOELECTRONIC SYNAPTIC DEVICES 887

Zhenyi Ni ; Yue Wang ; Lixiang Liu ; Shuangyi Zhao ; Yang Xu ; Xiaodong Pi ; Deren Yang

HIGH PERFORMANCE 2D PEROVSKITE/GRAPHENE OPTICAL SYNAPSES AS ARTIFICIAL EYES 891

He Tian ; Xuefeng Wang ; Fan Wu ; Yi Yang ; Tian-Ling Ren

FIRST TRANSISTOR DEMONSTRATION OF THERMAL ATOMIC LAYER ETCHING: INGAAS FINFETS WITH SUB-5 NM FIN-WIDTH FEATURING IN SITU ALE-ALD 895

Wenjie Lu ; Younghee Lee ; Jessica Murdzek ; Jonas Gertsch ; Alon Vardi ; Lisa Kong ; Steven M. George ; Jesús A. del Alamo

INGAAS-ON-INSULATOR FINFETS WITH REDUCED OFF-CURRENT AND RECORD PERFORMANCE 899

C. Convertino ; C. Zota ; S. Sant ; F. Eltes ; M. Sousa ; D. Caimi ; A. Schenk ; L. Czornomaz

BALANCED DRIVE CURRENTS IN 10–20 NM DIAMETER NANOWIRE ALL-III-V CMOS ON SI 903

Adam Jönsson ; Johannes Svensson ; Lars-Erik Wemersson

HIGH PERFORMANCE QUANTUM WELL INGAAS-ON-SI MOSFETS WITH SUB-20 NM GATE LENGTH FOR RF APPLICATIONS 907

C. B. Zota ; C. Convertino ; Y. Baumgartner ; M. Sousa ; D. Caimi ; L. Czornomaz

HIGH PERFORMANCE INGAAS GATE-ALL-AROUND NANOSHEET FET ON SI USING TEMPLATE ASSISTED SELECTIVE EPITAXY 911

S. Lee ; C. -W. Cheng ; X. Sun ; C. D'Emic ; H. Miyazoe ; M. M. Frank ; M. Lofaro ; J. Bruley ; P. Hashemi ; J. A. Ott ; T. Ando ; W. Spratt ; G. M. Cohen ; C. Lavoie ; R. Bruce ; J. Patel ; H. Schmid ; L. Czornomaz ; V. Narayanan ; R.T. Mo ; E. Leobandung

SCALING ACOUSTIC FILTERS TOWARDS 5G 915

Yansong Yang ; Ruochen Lu ; Songbin Gong

PHYSICS OF HOLE TRAPPING PROCESS IN HIGH-K GATE STACKS: A DIRECT SIMULATION FORMALISM FOR THE WHOLE INTERFACE SYSTEM COMBINING DENSITY-FUNCTIONAL THEORY AND MARCUS THEORY 919

Yue-Yang Liu ; Xiangwei Jiang

PARASITIC SURFACE REACTIONS IN HIGH-ASPECT RATIO VIA FILLING USING ALD: A STOCHASTIC KINETIC MODEL 923

T. Muneshwar ; G. Shoute ; D. Barlage ; K. Cadien

PHYSICS-BASED MODELING OF VOLATILE RESISTIVE SWITCHING MEMORY (RRAM) FOR CROSSPOINT SELECTOR AND NEUROMORPHIC COMPUTING 927

W. Wang ; A. Bricalli ; M. Laudato ; E. Ambrosi ; E. Covi ; D. Ielmini

ANALYTIC MODEL FOR STATISTICAL STATE INSTABILITY AND RETENTION BEHAVIORS OF FILAMENTARY ANALOG RRAM ARRAY AND ITS APPLICATIONS IN DESIGN OF NEURAL NETWORK 931

P. Huang ; Y. C. Xiang ; Y. D. Zhao ; C. Liu ; B. Gao ; H. Q. Wu ; H. Qian ; X. Y. Liu ; J. F. Kang

EVIDENCE OF MAGNETOSTRICTIVE EFFECTS ON STT-MRAM PERFORMANCE BY ATOMISTIC AND SPIN MODELING 935

K. Sankaran ; J. Swerts ; R. Carpenter ; S. Couet ; K. Garello ; R. F. L. Evans ; S. Rao ; W. Kim ; S. Kundu ; D. Crotti ; G. S. Kar ; G. Pourtois

Author Index

GaN Power Commercialization with Highest Quality-Highest Reliability 650V HEMTs-Requirements, Successes and Challenges

P. Parikh[a], Y. Wu[a], L. Shen[a], R. Barr[a], S. Chowdhury[a], J. Gritters[a], S. Yea[a], P. Smith[a], L. McCarthy[a], R. Birkhahn[a], M. Moore[a], J. McKay[a], H. Clement[a], U. Mishra[a], R. Lal[a], P. Zuk[a], T. Hosoda[b], K. Shono[b], K. Imanishi[b], Y. Asai[b].
pparikh@transphormusa.com, 805-456-1307.
[a]Transphorm Inc., 75 Castilian Drive, Goleta, CA, USA, [b]Transphorm Japan, Shin-Yokohama, Yokohama, Japan.

Abstract—Gallium Nitride (GaN) is now a popular choice for power conversion. High voltage (HV) GaN HEMTs (GaN FETs) in the range of 650-900 volts are emerging as the next standard for power conversion. This paper highlights key successes in efficient and compact converters/inverters ranging from high performance gaming/crypto-mining power supplies, titanium class server power, servo drives, PV inverters, and automotive OBCs, dc-dc converters, pole charges. The reasons for market success including unmatched quality & reliability, high volume GaN on Si manufacturing, robust performance in applications as well as challenges to achieve the full potential of GaN FETs are presented.

I. INTRODUCTION

With its proven ability to reduce size (form factor) and save energy (high efficiency) 650V GaN FETs have now been adopted in the mass market. GaN provides cost-competitive, easy-to-embed solutions that reduce energy loss by >50 percent, shrink system sizes by >40 percent, to simplify converter/ inverter design and manufacturing, also contributing to system cost reduction. An un-compromised need for power market is quality and reliability. We have achieved this through appropriate choice of device, package and outgoing quality standards. Next, overall product performance in application including uniformity and repeatability across millions of parts is a must. We tackle this via vertically integrated manufacturing and Si-like production discipline. Third, strong application support and deep understanding of GaN key circuit topologies is required. Last but not least, clear system or circuit level value proposition/cost benefit is essential for adoption.

II. HIGHEST ROBUSTNESS NORMALLY OFF GAN

A. Normally off product configuration

We have designed, qualified and mass produced a 2-chip normally off 650V GaN platform, integrating a low voltage Si FET input/drive stage with a high voltage GaN output stage (Fig. 1). Among the approaches to make a normally-off high voltage GaN switching device, this approach [1] offers strongest gate robustness over the p-GaN alternatives [2] with low safety margins (Table 1) and much lower complexity than the multi-chip direct-drive based designs [3]. It is noteworthy

that every Si-MOSFET consists of a normally-off input portion (gate control) with a normally-on output portion (high voltage drift region), that happen to be integrated in one device [4]. We have integrated two separate die in one package in a die-on-die configuration with minimum inductances to achieve best of both worlds- highest gate/input strength of the proven Si LV-MOS (with its ideal $Si-SiO_2$) dielectric gate and the high-performance high-voltage GaN HEMT to deliver low loss, high voltage switching and reliability. The result is a normally-off power device package with hard to beat combination of reliability, robustness, design margin and performance.

B. Package Choices- Thermal management is key

Along with the device, a robust and easy to use package is key for a power product. The basic concept that heat from any semiconductor die is removed via the package through the system heat sink is many times overlooked. Whether it is surface mount products to benefit from GaN's high frequency capability (reaching 100s of KHz or MHz scale) or the more classic TO packages to get kilo-watt class power that avail of GaN's high efficiency/low loss switching capability; a solid thermal interface without undue system complexity is a must. Industry work-horses such as the TO220 & TO247, their surface mount equivalents as the D^2Pak & D^3Pak or DIP style and top side cooled modules offer robust package environments (Fig. 2). These have been coupled with simple but powerful high frequency/high speed switching design philosophies, to result in GaN parts with stable operation at multi kW at high-speed/multi-100 KHz to MHz and performing better (Fig.3).

III. QUALITY AND RELIABILITY

Suppliers are adopting common standards for reliability needs. While being part of this industry effort, our philosophy is to establish comprehensive reliability testing of products that achieve industry firsts in qualifying 650V GaN products for the marketplace. Instead of academic arguments of JEDEC standards applicability to GaN, the question customers ask is if GaN products are not passing JEDEC tests then what is wrong with those GaN products? Our approach is that JEDEC qualification for GaN (existing standards) is necessary but not sufficient and must be backed up by other comprehensive tests suited for GaN products (Fig. 4)

A. Qualification and Intrinsic Reliability

In addition to achieving successful JEDEC and AEC Q101 qualification of the 650V GaN platform [5], intrinsic lifetest along with associated failure modes and acceleration factors was also reported with both voltage and temperature based acceleration factors for the critical high voltage reverse bias failure mode [6], shown in Fig 5. Use plots indicate wear out of the device at 480V does not begin before 1 million hours, at any temperature within the ratings.

B. Quality levels/FIT rates

A key consideration is that parts may fail prior to intrinsic lifetime due to defects not screened out in the manufacturing/ test process including infant mortality or random failures during the useful life. Practically, potential fails in the first 10^5 hours for a given mission profile need understanding. The robustness of our 650V GaN FETs to well over 1000 Volts enables testing to failure using voltage acceleration with a 2 dimensional reliability matrix against voltage and temperature. Both FIT (Failure in Time) for constant failure rate (also equivalent to MTBF) and PPM (Parts Per Million) testing was done per the detailed methodology based on JEDEC std. 74A Annex G [7], reported elsewhere [8]. The results (Fig 6) predict FIT of 1.3, along with an annual failure rate of 0.001% (~ 10ppm) at 520V, 100C. This preliminary but comprehensive study gives a strong proof point of quality levels and field performance predictability for GaN devices.

C. Application centric (Switching) reliability considerations

We have also subjected our devices to switching stresses including comprehensive HTOL testing under actual power conversion operation at 175C, 300 KHz for 3000 hours (Fig 7) and accelerated switching at 150 KHz, Room Temp-125C at more than 1000 volts. No significant change is observed during as well as before/after HTOL testing. Initial accelerated switching tests indicate switching lifetime >10^8 hours (Fig 8).

IV. Manufacturing scale and metrics

It is a requirement that for any technology to exhibit the highest levels of quality and reliability, high levels of wafer-fab and product yields with adequate process capabilities and control must be established. Transphorm's commercial JEDEC and AEC qualified GaN HEMT power products are manufactured in an automotive grade 6-inch Si CMOS wafer fab. We have previously reported Silicon like manufacturing, yields (Fig 9) and process capabilities for our HV-GaN manufacturing [9]. Further, complete SPC tracking and control charts for hundreds of manufacturing/test parameters is routine.

V. Application Performance and System Value

The ultimate proof of the accepted reliability and robustness standards comes from use of the high voltage GaN products from various suppliers by a wide customer base. This has already started and many leading companies are now in production with high voltage GaN devices. As examples, Corsair has introduced best in class 1.6kW gaming power supplies with unparalleled performance at par cost based on the Totem Pole architecture enabled by 650V/50mohm GaN in

TO247 packages (Fig. 10), Bel Power Systems has introduced these 50mohm TO247 products in their 3kW TET3000 Titanium class Server Power Supplies [10] and Yaskawa Electric has introduced TO220 based products in an innovative integrated servo amplifier-motor drive system [11]. With strong circuit benefits (Fig 11) and ability to reduce size, weight and improve miles per gallon, GaN FETs are being designed in EV & HEV applications such as on & off board chargers and auxiliary power conversion in the 3-10kW range.

VI. Challenges- System Considerations

Challenges for wide adoption lie in the areas of system integration, robust supply, standards and cost improvements.

In systems, GaN offers high value within the AC to DC bridgeless totem-pole PFC, which unlike the well-established analog based classic boost PFC use digital programming. The Totem-pole PFC though using similar control techniques requires ground up development with additional needs such as soft start control and polarity definition with firmware design knowledge. Focused reference design such as the 3.3 kW PFC reference (Fig 12) assist customers get over this digital control hump. Having GaN solutions with sufficient drive margin (gate robustness), preferably with off the shelf standard drivers also aids system design experience.

As another challenge, customers designing with GaN must be assured performance in actual ac operation during power conversion as claimed on the datasheet and not suffer from any dynamic effects. In addition to designing for and 100% monitoring for such dynamic effects control, we ensure that datasheet ratings and maximums include such effects. In other words, what you see should be what you get. The good news for the GaN industry is now that this standard is widely agreed and various manufacturers have embraced it.

A robust supply chain with multiple sources enables user confidence. Transphorm has partnered with Fujitsu to establish 6-inch GaN wafer manufacturing in an automotive certified Si-wafer foundry in Aizu-Japan (now beginning to offer 3rd generation products) while others like GaN Systems, Navitas and EPC avail external foundries in Taiwan. Like Transphorm, companies like Infineon, Panasonic control their manufacturing lines, a distinct advantage in the first 10-15 years of any new semiconductor ramp. Transphorm is also committed to expand partnerships, including establishing credible second sources for its products. Finally, GaN device costs are on course for continuous reduction to approach and in some cases already beat Si Super-junction, especially at the system level.

VII. Summary and Acknowledgements

As GaN FETs launch into an exciting growth phase with multiple suppliers, continuous focus on reliability, supply assurance, value/cost will enable user confidence that will let GaN challenge the existing slots and be a forerunner for new growing power conversion applications like automotive.

The authors sincerely appreciate the contributions of the broader Transphorm team, our vendors, manufacturing partners notably Aizu-Fujitsu foundry, customers and financial partners in making this happen.

REFERENCES

[1] P. Parikh, et. al., "Commercialization of 600V GaN HEMTs," 2014 SSDM, Tsukuba, Japan (2014).

[2] D. Marcon, et. al., "Direct comparison of GaN-based e-mode architectures (recessed MISHEMT and p-GaN HEMTs) processed on 200mm GaN-on-Si with Au-free technology", Proc. SPIE 9363, GaN Materials and Devices (2015).

[3] P. Brohlin, et. al, "Direct-drive configuration for GaN devices", http://www.ti.com/lit/wp/slpy008/slpy008.pdf, 2016.

[4] Lazlo, B. "Design and Application Guide for High Speed MOSFET Gate Drive Circuits", (2006).

[5] K. Smith and R. Barr, Reliability Lifecycle of GaN Power Devices, 2017 and continuous updates at " http://www.transphormusa.com/wpcontent/uploads/2016/02/reliability-lifcycle-at-transphorm.pdf

[6] K. Smith etal, Microelectronics Reliability 58(2016)197-203

[7] JESD74A Early Life Failure Calculation Procedure for Semiconductor Components

[8] R. Barr et al, "High Voltage GaN Switch Reliability", WIPDA, 2018.

[9] S. Chowdhury, et. al., "650V Highly Reliable GaN HEMTs on Si Substrates over Multiple Generations: Matching Silicon CMOS Manufacturing Metrics and Process Control", IEEE CSICS, Austin, Texas, (2016).

[10] S. Taranovich, "PFC Totem Pole Architecture and GaN combine for high power and efficiency", Electronic Design News (EDN), June 14 (2017).

[11] Yaskawa Electric Company, "Yaskawa to Launch the World's First GaN Power Semiconductor Equipped Servo Motor with Built-in Amplifier, A New Addition to the AC Servo Drives Σ (Sigma)-7 Series", May 26, 2017.

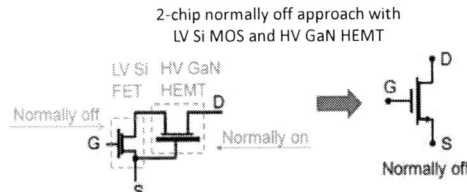

Fig. 1. Robust normally off GaN achieved by integration of LV Si MOSFET and HV GaN

Device	Threshold Voltage (V)	Gate Plateau Voltage(V)	V_{GS} Rating (V)	Gate Safety Margin (V)
Transphorm GaN, Gen-3, 2-chip normally off	4.0	6	20	20 − 6 = 14
1-chip normally off, A	1.7	3	7	7 − 3 = 4
1-chip normally off, B	1.2	2.2	4.5	4.5 − 2.2 = 2.3
1-chip normally off, C	1.4	2.7	6	6 − 2.7 = 3.3

Table 1. Common approaches for normally-off: 2-chip integrated solution provides +4V threshold, high design margin and gate robustness

Fig. 2 Array of packaged products both lead based, surface mount that are amenable to meeting stringent qualification standards: TO-247, TO-220, D3PAK, D2PAK (Auto Q101 standards) and PQFN (JEDEC/Industrial standards.)

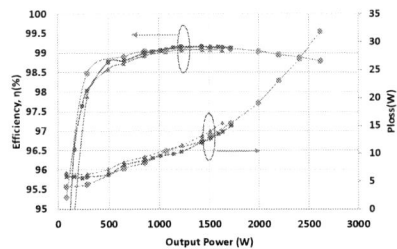

A well designed 70 mohm GaN 2-chip normally off in a robust TO220 package outperforms single chip E-mode on the market- higher power with smaller die (enabled by thermally robust package and GaN die with excellent dynamic switching characteristics

Fig. 3.Superior performance from a robust package GaN 650V product- 99% efficiency, 2kW, 50 KHz, 200V input

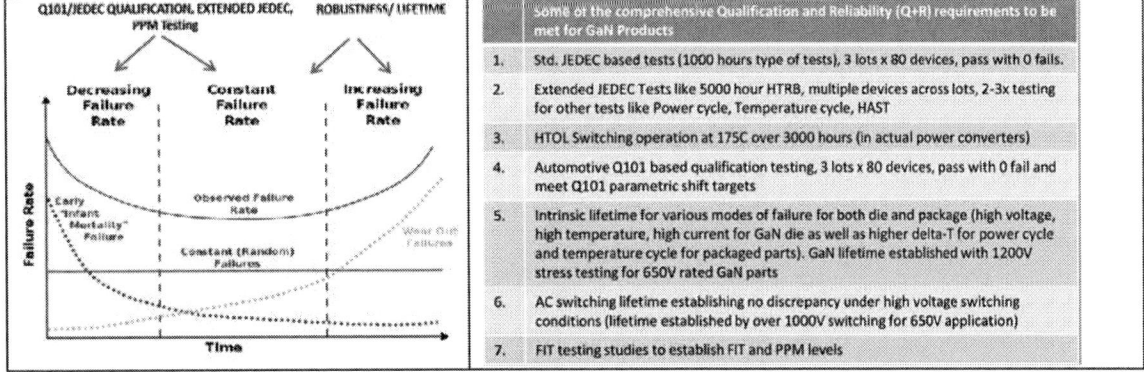

Fig. 4. Complete suite of quality and reliability for 650V GaN in volume production

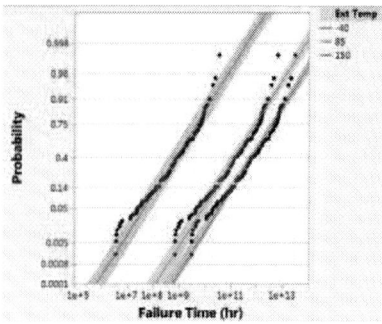

Fig. 5. Use plot from intrinsic life testing based on 1000V+ HVOS accelerated testing: GaN device wear-out does not begin before 1Million hours, MTTF much higher.

Average PPM		Voltage		
		400	480	520
Temp	25	16.8	78.6	169.5
	50	8.6	40.3	86.8
	100	3.0	13.8	29.8
	150	1.3	6.1	13.2
Average MTBF		Voltage		
		400	480	520
Temp	25	5E+08	1E+08	5E+07
	50	1E+09	2E+08	1E+08
	100	3E+09	6E+08	3E+08
	150	7E+09	1E+09	7E+08
Average Annual Failure Rate		Voltage		
		400	480	520
Temp	25	0.001680%	0.007860%	0.016949%
	50	0.000860%	0.004030%	0.008680%
	100	0.000300%	0.001380%	0.002980%
	150	0.000130%	0.000610%	0.001319%

Fig. 6. Average PPM per year, MTBF and Failure rate based on a pessimistic view of test data that includes infant mortality (typically screened out). This conservative view still predicts 480V/100C (use condition assumption) sub 20 PPM, 6E8 years MTBF and 0.0014% average annual failure rate

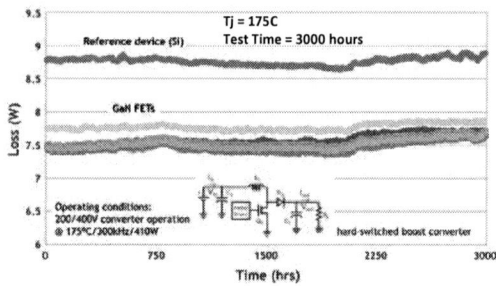

Fig. 7. High temperature operating life under hard switch actual boost converter with accelerated (temperature) conditions- no performance degradation as shown and also minimal change in before-after key datasheet parameters

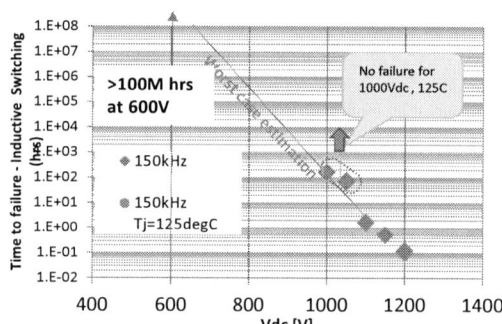

Fig.8. Preliminary AC Switching Accelerated high voltage testing at 1000V+ and 150 KHz: Indicative lifetimes of 1 Billion hours. Key conclusion- No adverse lifetime related phenomena during AC switching of Transphorm GaN FETs.

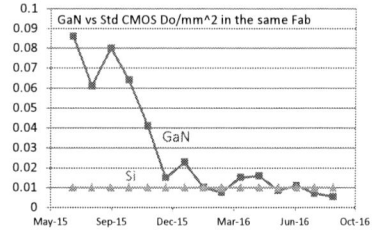

Fie. 9. Yields (represented by Do/defect density) for GaN have been on par with Si CMOS run side-by-side in our 6-inch Aizu Japan Wafer Foundry

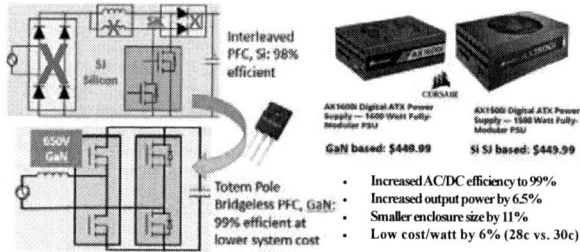

Fig. 10. Key application value, end user product: GaN provides unique benefits in Totem Pole PFCs=>99%+ efficiency, cost effective system by reducing part count. Higher performance power supply product at lower cost per watt are in market

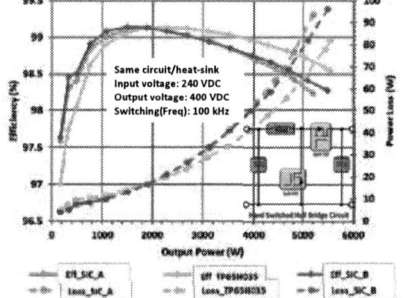

Fig. 11. GaN excels in both hard-switch (shown here) and resonant topologies. E.g. shows TP65H035WS (35 mohm GaN) outperforming 35mohm (A) and 65mohm (B) SiC MOSFETs

Fig. 12. Easy to adopt reference designs that illustrate breakthrough GaN performance/features like the 3.3kW /50-150 KHz Totem Pole PFC above are essential in increasing market traction

978-1-7281-1988-5/18 $31.00 © 2018 IEEE

40× Retention Improvement by Eliminating Resistance Relaxation with High Temperature Forming in 28 nm RRAM Chip

Xiaoxin Xu[†1,2], Lu Tai[†1,3], Tiancheng Gong[1,2], Jiahao Yin[1,2], Peng Huang[4], Jie Yu[1,3], Da Nian Dong[1,2], Qing Luo[1,2], Jing Liu[1,2], Zhaoan Yu[1,2], Xi Zhu[1,2], Xiu Long Wu[3], Qi Liu[1,2], Hangbing Lv[1,2*], Ming Liu[1,2*]

[1]Key Laboratory of Microelectronics Devices and Integrated Technology, Institute of Microelectronics, Chinese Academy of Sciences, Beijing, China;[2]University of the Chinese Academy of Sciences, Beijing, China. [3]School of Electronics and Information Engineering& Anhui University, Hefei, Anhui, China. [4]Institute of Microelectronics, Peking University, Beijing, China. Email: lvhangbing@ime.ac.cn, liuming@ime.ac.cn

Abstract— In this work, we proposed a high temperature forming scheme for 28 nm 1Mb RRAM test chip. Compared with room temperature forming scheme, the average forming voltage performed at 125 °C could be greatly reduced from 2.5 V to 1.7 V. Resistance relaxation resulted from the recombination of Vo and O^{2-} that generally occurred after programming was effectively eliminated as the residual O^{2-} in the filament was highly decreased. Benefit from this, retention improvement of more than 40× times was successfully achieved.

INTRODUCTION

Resistive random access memory (RRAM) is considered as one of most promising candidates to replace e-Flash for embedded memory application owing to its high speed, good endurance and excellent compatibility with CMOS process [1]. However, the initial forming step with the forming voltage higher than the subsequent switching voltage generally occurs. That will not only complicate and burden the periphery circuit [2], but also degrade the device performance. Up to now, there have been many reports on addressing how to eliminate the forming voltage, including material doping [3], interface engineering [4], thickness reduction [5], et al. However, most of them were proved to be at the cost of process complexity and compromise of other performance.

In this work, we adopted a high temperature forming scheme to tackle the forming issue. The average forming voltage was successfully lowered from 2.5 V to 1.7V, which could greatly facilitate the periphery circuit design. The resistance relaxation was effective eliminated due to the decrease of the residual O^{2-} inside the filament, thus greatly improve the retention characteristics. Moreover, the distribution of resistance was also much tightened thanks to avoiding of the kink-off effect of the access transistor under high forming voltage.

RESULTS AND DISCUSSION

A. *High Forming Voltage Issue in Array Operation*

Fig. 1a shows the schematic of peripheral circuit for the RRAM array. The operation voltage of core device at 28nm node is around 1V. As the forming voltage of the selected cell higher than 2.5 V, the unselected access transistors in the same active bitline (BL) will generate high leakage current (>10 nA)

due to the reverse bias of $V_{drain-sub}$ (Fig. 1b). The leakage current flowing through the selected BL would converge to the decoder transistor. The V_{drive} is the sum of voltage drop on decoder transistor and the V_{Form} of the selected cell, and it can be as high as 6 V if the V_{Form} reaches 2.5 V in the case of 1 k cells in bitline. That is much higher than the operation voltage of I/O transistor of 28 nm node (2.5 V). In order to guarantee the decoder transistor working in a safe voltage range, the forming voltage of the RRAM cell should be less than 2 V (Fig. 2). Fig. 3 shows the kink off effect of the transistor that generally occur when the V_{DS} is suffered from high voltage. Considering the forming voltage has certain variation, the compliance current will become uncontrollable, resulting in the wide distribution of LRS after forming. Scaling the switching layer thickness is the most effective way to reduce the forming voltage (Fig. 4a). However, the experimental results show that endurance degradation is also accompanied with the decrease of the switching layer thickness, as shown in Fig. 4b. [3-5]

B. *High Temperature Forming Scheme*

Before using the high temperature forming scheme, the temperature dependence of the leakage current and breakdown voltage of the transistor should be evaluated. As shown in Fig. 5, there is only slight increase of the leakage current when the temperature increase from RT to 125 °C and the breakdown voltage of the core device nearly has no change as the atmosphere temperature increased from 27 °C to 150 °C.

Fig. 6a and 6b shows the I-V curves of forming process under RT and 100 °C. Obvious decrease and more tight distribution of the forming voltage were observed under high temperature. Fig. 7 shows the statistical distribution of forming voltage of 3 kb mini-array under RT, 100 °C and 125 °C. The average V_{Form} is 2.5 V at room temperature and 1.7 V at 125 °C. More tight distribution of Ron after high temperature forming was observed, benefited from avoiding the kink-off effect of the access transistor and more precise control of the compliance current (Fig. 8a). The distribution of subsequent SET/RESET voltage is shown in Fig. 8b. The improved uniformity of V_{RESET} is related to the well-controlled filament by high temperature forming. Besides, the Ron and Roff distribution between the device to device are also much improved, as shown in the statistics data of 100 programming cycles in 1kb cells (Fig. 9). Based on the cell forming voltage and leakage current of the

978-1-7281-1988-5/18 $31.00 © 2018 IEEE

access transistor at different temperature conditions, the required V_{drive} of the periphery circuit could be plotted against the block size (**Fig. 10**). Compared with the RT-forming condition, the block size could be improved by 100 times under high temperature forming at 125 °C.

C. *Relaxzation Behavior of LRS*

Fig. 11a shows the data loss evolution of LRS baked at 150 °C. Obvious tail bits were generated after initial 10 min baking. The resistances after 2 days and 10 days show no further change. More clearly trend could be seen in the failure rate plot as the baking time (**Fig. 11b**). An abrupt failure about 8% is observed at the first stage. The insert of **Fig. 11b** shows the normalized failure rate by removing the failure cells at the initial 10 min. Only 0.5% failure was observed even after 10 days baking. This behavior indicates the resistance relaxation in the short period after programming plays an important role on retention failure. We thus carried out the LRS relaxation measurement in 1 kb array as the testing flow chart shown in **Fig. 12a**. The LRS after programming was checked after short period of 1 ms, 100 ms and 1 s. Tail bits could be clearly observed even after 1 ms relaxation.

In order to elucidate the cause of the LRS relaxation, the distributions of Vo and O^{2-} under RT and high temperature forming were simulated based on the simulation platform developed in our previous work [6]. In the case of RT forming, large amount of O^{2-} observed inside the conductive filament, as well as its surrounding (**Fig. 13a and 13b**). As the O^{2-} is in free-state, it is easy to recombine with the Vo in the filament, causing the degradation of LRS. In the case of high temperature forming, few quantity of residual O^{2-} was observed in the switching layer. The reason may lie in the fact that the O^{2-} could be easily driven into the oxygen reservoir under the high temperature and electric filed. As a result, the LRS relaxation could be greatly improved. **Fig. 14a** shows the 10 min relaxation of LRS programmed at 125 °C, no tail bit was observed, indicating the LRS relaxation was successfully suppressed by high temperature forming. The lateral LRS retention was also evaluated. After the specific forming at RT, 100 °C and 125 °C, all the cells undergo 5 cycles at room temperature. As shown in **Fig. 15**, after 250 h baking at 150 °C, the cells with RT forming shows 5% failure bits and the cells with 125 °C forming shows no failure bits, indicating great LRS retention improvement by high temperature forming.

D. *Relaxzation Behavior of HRS*

Fig. 16a shows the HRS relaxation trend for the cells formed at RT. Resistance decay towards lower value was observed. After 100 ms relaxation, about 12% bits fell below the reference criteria. **Fig. 16b and 16c** show the relaxation behavior for the cells performed by high temperature forming. After 100 ms relaxation, very few tail bits were observed, but still above the reference criteria, indicating that the high temperature forming is also helpful to improve the HRS relaxation. The relaxation of HRS could be understood by the Vo diffusion into the tunnel gap, resulting in gap length narrowing and HRS decay. For the device formed at RT, the high forming voltage tends to create more Vo in the filament,

making the residual Vo remain in the tunnel gap. Since the electron conduction of HRS is related with variable-range hopping (VRH). The conductivity at a low electrical field could be described by Mott equation (1) [7] and the density of Vo in the gap ($N(E_F)$) could be calculated by the equation (3) and (4), which is inversely proportional with the slope of lnI and 1/KT. As shown in **Fig. 17**, the fitting result indicates that there are more traps in the cells formed by RT forming. The further HRS retention measurement (150 °C baking) on the three groups of HRS with the corresponding cells formed at different temperature shows that the failure rate for the cells formed under RT temperature was up to 16% after 250 h, while only 0.4% failure bit was observed for the cells treated by 125 °C forming. A 40× times HRS retention improvement was achieved.

CONCLUSION

In this work, a high temperature forming scheme was proposed to mitigate the forming issue in chip design. The avarage forming voltage performed at 125 °C was sucessfully reduced from 2.5 V to 1.7 V, with negilible influence on the leakage current and breakdown voltage of the access transistor. Both of HRS and LRS relaxation were effectively supressed by high temperature forming treatment, thus greatly improve the data retention and bit yield. More than 40× retention improvement was demonstrated.

ACNOWLEDGMENT

This work was supported in part by the MOST of China under Grants 2016YFA0203800, 2016YFA0201803, 2018YFB0407502, and in part by the National Natural Science Foundation of China under Grants 61522408, 61334007, 61521064, and in part by Beijing Municipal Science & Technology Commission Program (Z161100000216153) and by Huawei Data Center Technology Laboratory. [†]These authors contribute to this work equally.

REFERENCES

[1] H. B. Lv, X. X. Xu, P. Yuan, D. N. Dong, T. C. Gong, J. Liu, Z. A. Yu, P. Huang, K. Zhang, C. X. Huo, C. B. Chen, Y. L. Xie, Q. Luo, S. B. Long, Q. Liu, J. F. Kang, D. Yang, S. Yin, S. F. Chiu and M. Liu, "BEOL based RRAM with one extra-mask for low cost, highly reliable embedded application in 28 nm node and beyond," Proc. IEEE International Electron Devices Meeting IEEE, 2017:2.4.1-2.4.4.

[2] H. T. Li, P. Huang, B. Gao, X. Y. Liu, J. F. Kang and H.-S. P. Wong. "Device and circuit interaction analysis of stochastic behaviors in cross-point RRAM arrays," IEEE Transactions on Electron Devices, 2017, 64(12):4928-4936.

[3] K. Wonjoo, H. Alexander, R. Christian, M. Stephan, J. W. Dirk, H-E Susanne, B. Dan, W. Rainer, and R. Vikas. "Forming-free metal-oxide ReRAM by oxygen ion implantation process," Electron Devices Meeting IEEE, 2017:4.4.1-4.4.4.

[4] S. Z. Rahaman, S. Maikap, T. C. Tien,H.Y.Lee, W. S. Chen, F.T. Chen, M. J. Kao and M. J. Tsai, "Excellent resistive memory characteristics and switching mechanism using a Ti nanolayer at the Cu/TaOx interface," Nanoscale Research Letters, 2012, 7(1):345.

[5] W. Kim, D. J. Wouters, S. Menzel , S. Menzel, C. Rodenbücher, R, Waser and V. Rana. "Lowering forming voltage and forming-free behavior of Ta2O5 Vikas Rana ReRAM devices," Solid-State Device Research Conference. IEEE, 2016:164-167.

[6] J. F. Kang, B. Gao, P. Huang, H. T. Li, Y.D. Zhao, Z. Chen, C. Liu, L.F. Liu, and X.Y. Liu, "Oxide-based RRAM: requirements and challenges of modeling and simulation," IEEE International Electron Devices Meeting. IEEE, 2015:5.4.1-5.4.4.

[7] S. Choi, Y. Yang and W. Lu, "Random telegraph noise and resistance switching analysis of oxide-based resistive memory," Nanoscale, 2014, 4(1) 400-404.

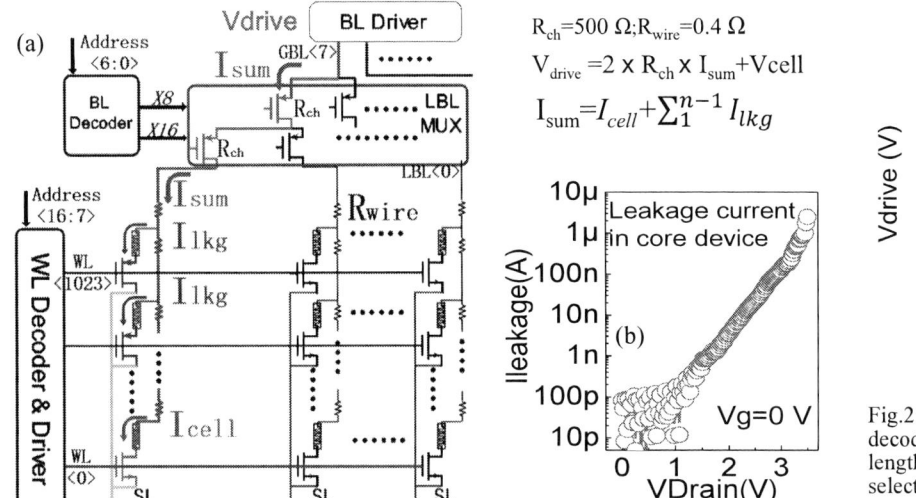

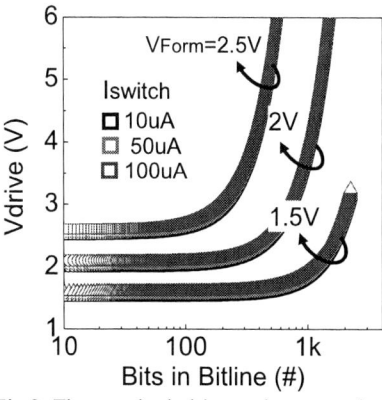

$R_{ch}=500\ \Omega; R_{wire}=0.4\ \Omega$

$V_{drive}=2\times R_{ch}\times I_{sum}+Vcell$

$I_{sum}=I_{cell}+\sum_{1}^{n-1}I_{lkg}$

Fig.2 The required drive voltage on the decoder transistor with different bitline length. The larger forming voltage of the selected cell, the higher V_{drive} required.

Fig.1 (a) The schematic of peripheral circuit for the RRAM array. The access transistors on the selected bitline generate leakage current due to the reverse bias between drain and substrate. The V_{drive} is the sum of voltage drop on decoder transistor and the V_{Form} of the selected cell. (b) The leakage current of the core device at 28 nm node. The leakage is higher than 10 nA if the Forming voltage of the selected memory cell is larger than 2.5 V.

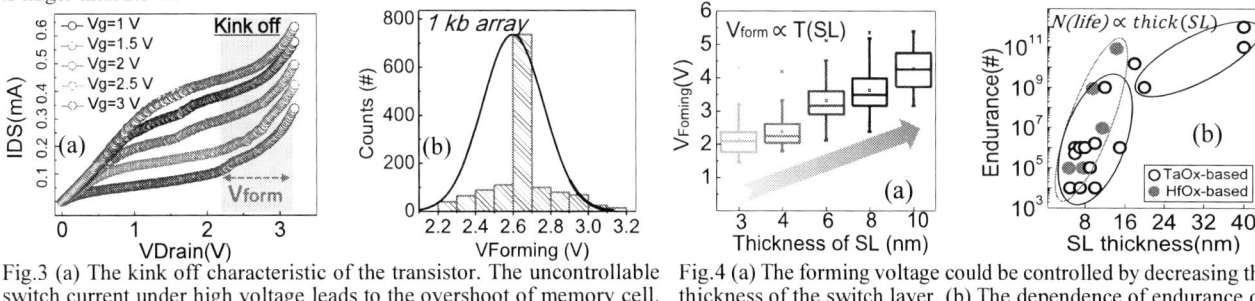

Fig.3 (a) The kink off characteristic of the transistor. The uncontrollable switch current under high voltage leads to the overshoot of memory cell. (b) The distribution of the $V_{forming}$ in the memory array.

Fig.4 (a) The forming voltage could be controlled by decreasing the thickness of the switch layer. (b) The dependence of endurance on the switching layer thickness.

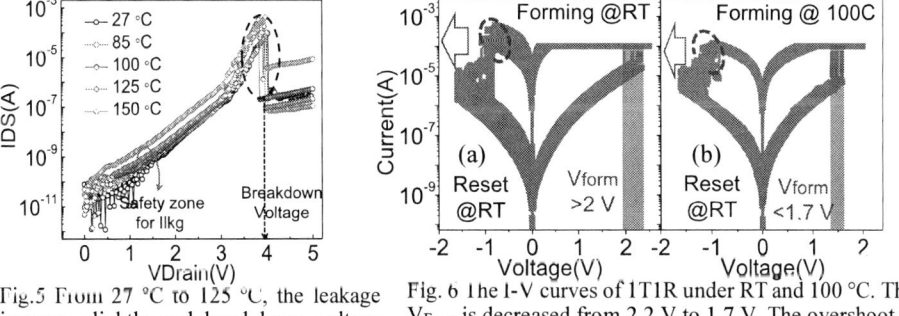

Fig.5 From 27 °C to 125 °C, the leakage increase slightly and breakdown voltage keeps almost constant.

Fig. 6 The I-V curves of 1T1R under RT and 100 °C. The V_{Form} is decreased from 2.2 V to 1.7 V. The overshoot is mitigated by the high temperature forming.

Fig.7 The distribution of V_{Form} in the memory array under 27 °C, 100 °C and 125 °C.

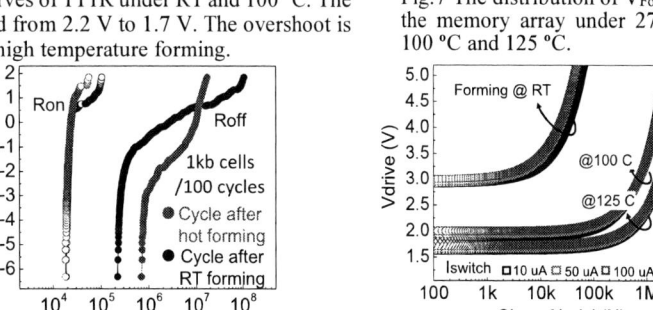

Fig.8 (a) The distribution of Ron after forming under different temperature. (b) The distribution of subsequent SET/RESET voltage.

Fig.9 The uniformity of the Ron/Roff is improved between the device to device.

Fig.10 The required V_{drive} based on the parameters of the access transistor and memory array at different temperature conditions.

978-1-7281-1988-5/18 $31.00 © 2018 IEEE

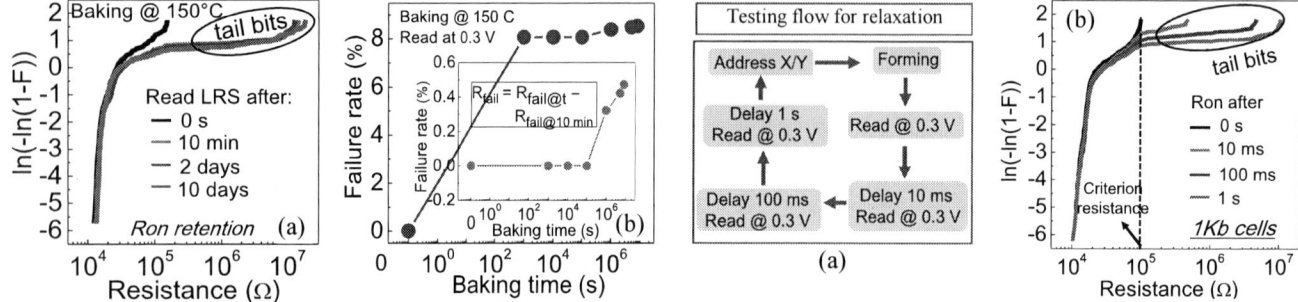

Fig. 11(a) The data loss evolution of the LRS. The obvious tail bits are generated just 10 min later, and then no more bits further fail. (c) The failure rate as the baking time. The inset is the normalized failure rate by removing the failed cells at the initial 10 min

Fig. 12 (a) The testing flow chart for the LRS relaxation measurement in 1 kb array. (b) The resistance distribution for LRS during the relaxation. The LRS degrade seriously after 1 s relaxation.

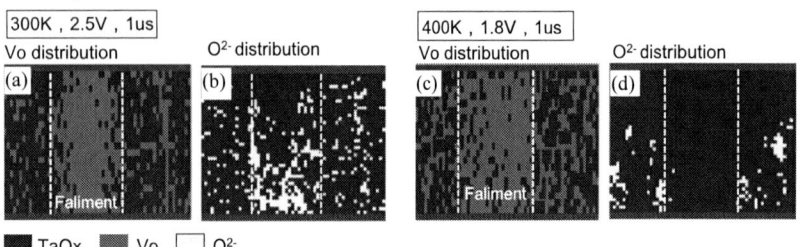

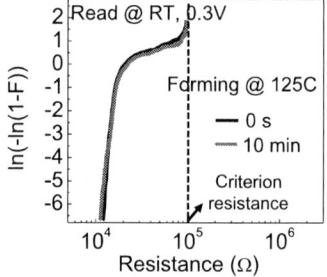

Fig. 13 The distribution of Vo and O^{2-} in the filament and surrounding for RT forming ((a) and (b)) and high temperature forming ((c) and (d)). Large amount of O^{2-} observed inside the conductive filament and its surrounding in (b). Few quantity of O^{2-} was observed in the filament and switching layer in (d).

Fig. 14. The 10 min relaxation of the LRS forming at 125 °C, no tail bit is observed.

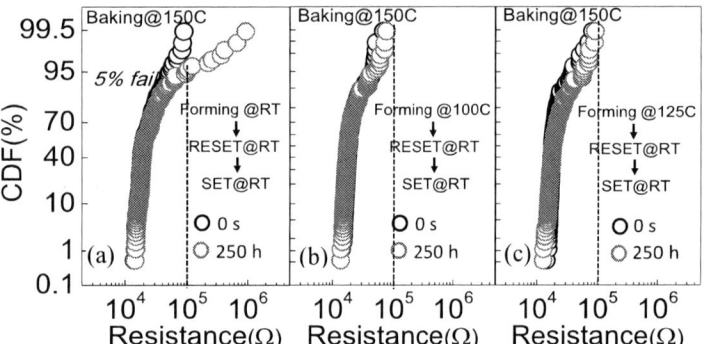

Fig. 15 The LRS retention of the cells under different forming condition: (a) room temperature, (b) 100 °C and (c)125 °C. All the cells undergo 5 cycles at room temperature after the specific forming process. The failure rate reduced from 5% to 0 by high temperature forming.

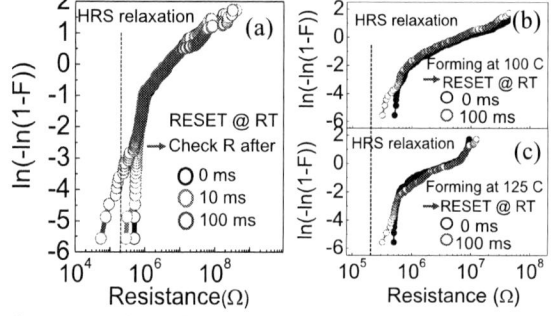

Fig. 16 (a) The resistance distribution for HRS during the relaxation. Obvious tail bits are generated for 100 ms seconds later. (b)(c) For the case of high temperature forming, there are weaker bits generated after 100 ms relaxation.

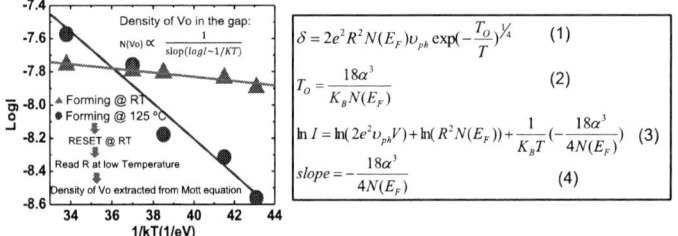

$$\delta = 2e^2 R^2 N(E_F) \upsilon_{ph} \exp(-\frac{T_O}{T})^{1/4} \qquad (1)$$

$$T_O = \frac{18\alpha^3}{K_B N(E_F)} \qquad (2)$$

$$\ln I = \ln(2e^2\upsilon_{ph}V) + \ln(R^2 N(E_F)) + \frac{1}{K_B T}(-\frac{18\alpha^3}{4N(E_F)}) \quad (3)$$

$$slope = -\frac{18\alpha^3}{4N(E_F)} \qquad (4)$$

Fig. 17 The density of Vo ($N(E_F)$) in the gap is extracted for HRS after different forming condition. In HRS, the conductivity under a low electrical field is described as equation (1). Following the Mott's approach, density of Vo in the gap could be calculated by the (3). The fitting result shows there are more traps in the cells with RT forming.

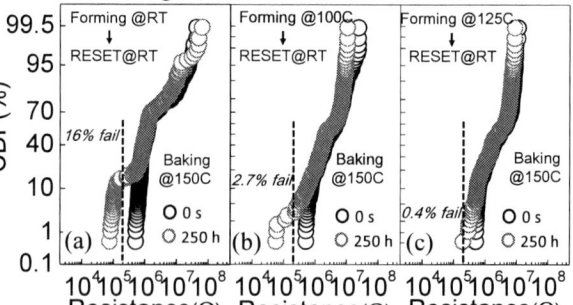

Fig. 18 The HRS retention with different forming condition: (a) room temperature, (b) 100 °C and (c) 125 °C. The failure rate decreased from 16% to 0.4% by high temperature forming.

978-1-7281-1988-5/18 $31.00 © 2018 IEEE

Characterizing Endurance Degradation of Incremental Switching in Analog RRAM for Neuromorphic Systems

Meiran Zhao[1], Huaqiang Wu[1]*, Bin Gao[1], Xiaoyu Sun[2], Yuyi Liu[1], Peng Yao[1], Yue Xi[1], Xinyi Li[1], Qingtian Zhang[1], Kanwen Wang[3], Shimeng Yu[2], and He Qian[1]

[1]Institute of Microelectronics, Tsinghua University, Beijing 100084, China;
[2]Georgia Institute of Technology, Atlanta, GA 30332, USA;   [3]Huawei Technologies CO., LTD.
Email: wuhq@tsinghua.edu.cn; gaob1@tsinghua.edu.cn;

Abstract—Resistive random access memory (RRAM) is attractive for neuromorphic computing systems as synaptic weights. In the neural network training, incremental switching occurs between the analog conductance states, thus the analog RRAM devices have unique endurance degradation behaviors compared to the convention digital memory application. In this work, a fast measurement platform is developed to characterize the endurance of incremental switching in analog RRAM. It is found that under weak weight update pulses, the incremental switching cycles of RRAM can be increased for more than 5 orders of magnitude compared with full window switching under strong programming pulses. The 10^{11}-cycle endurance of analog RRAM is proved to be sufficient for training neural networks online for various datasets (from MNIST to ImageNet). However, the nonlinearity and dynamic range of analog RRAM degrade during cycling, which may influence the learning accuracy of the neural network when it re-trains with new datasets.

I. INTRODUCTION

To train a RRAM based neural network online, the conductance of RRAM, which represents synaptic weight (Fig. 1), is required to be tunable for a large number of updates, e.g. 10^5 for MNIST, and 10^7 for ImageNet (Table I) [1-4]. However, it is well known that RRAM for data storage application has a limited endurance, typically ranged between 10^5 to 10^7 [5]. It seems that such endurance cannot support online training of a neural network. On the other hand, it should be noted that the weight update of analog RRAM for neural network training is quite different from the write operation for memory application. For memory write operation, a large resistance window is typically achieved by strong programming pulses. Endurance failure occurs when the resistance window collapses (Fig. 2). Whereas, for neural network training, the conductance of analog RRAM only changes incrementally by a weak programming pulse (Fig. 3), which is called incremental switching or analog switching [6]. In this case, the damage of each pulse on the material properties is much less significant than the memory write, thus the lifetime is expected to be much longer. To our best knowledge, no prior work has studied the endurance of analog switching. It is not clear how to characterize the unique behaviors of endurance degradation on analog switching, and how much difference between the endurance of digital memory switching and analog switching.

There are two reasons why characterizing endurance of analog RRAM is challenging: 1) It is difficult to control the analog switching within a designated resistance window during repeated cycling, and it lacks of an effective evaluation criterion for endurance degradation; 2) Due to the large variability of RRAM, endurance test on single device is insufficient, but statistical measurement requires a long time. In this work, we developed a testing platform to characterize the endurance behaviors of analog switching. We found that the endurance degradation behaviors on analog switching are different from the memory switching. The impact of endurance degradation on neuromorphic system performance is also discussed.

II. ENDURANCE CHARACTERIZATION METHOD

We fabricated 1T1R array where the RRAM cells are integrated on top of a transistor's drain contact. The transistor offers accurate control of various resistance states by tuning gate voltage during SET/RESET operations, which makes the characterization on analog switching possible. We also developed a fast and effective test platform for testing the 1T1R cell with Tektronix 4200 system. The test scheme of analog switching endurance is shown in Fig. 4. The main difference to the traditional endurance test scheme lies in setting the upper and lower boundaries of each state with variable voltage pulses.

The analog RRAM cell under test is TiN/ETML/HfO$_x$/TiN stack. HfO$_x$ is switching layer, and electro-thermal-modulation-layer (ETML) contributes to excellent linear analog switching behaviors [7]. The devices are demonstrated good uniformity of analog switching from cycle to cycle and device to device (Fig. 5), which is important for studying endurance behaviors.

III. ENDURANCE OF ANALOG SWITCHING

Firstly, we investigate how switching window influences endurance cycle. Three typical windows with different resistance levels show different endurance cycles (Fig. 6). The full window (high resistance state, HRS~1MΩ, low resistance state, LRS~20kΩ) shows the worst endurance. Most of the cells fail after 10^5 cycling. The high-R window case (HRS~1MΩ, LRS~100kΩ) also shows poor endurance, typically 10^6 cycles. The low-R window (HRS~100kΩ, LRS~20kΩ) shows much better endurance, greater than 10^8 cycles. When fixing low resistance state (LRS) around 20kΩ and changing high resistance state (HRS) from 1MΩ to 100kΩ, endurance shows improvement from 10^5 to 10^9 (Fig. 7). The endurance cycle increases significantly as the window reduces. However, when fixing HRS around 1MΩ or 200kΩ, and changing LRS from 20kΩ to 100kΩ, the endurance cycle changes less than 10 times (Fig. 8 & 9). Based on the statistical measurement for different switching windows (Fig. 10), it is concluded that higher resistance state influences endurance more significantly. This is attributed to the change of current mechanisms correlated with different morphology of oxygen vacancy (Vo) based conductive filament (CF) (Fig. 11). To get better endurance, the analog RRAM should be controlled within the relatively lower resistance range during training process. When the device fails, the resistance may be stuck randomly at LRS, HRS, or intermediate state (Fig. 12).

The above test is still similar as memory switching. To mimic the online training process in a neuromorphic system, the switching window is controlled within a very small range using weak programming pulse (update pulse), and the incremental conductance change after each pulse is recorded. With this test method, the analog switching does not fail even when $>10^{11}$ update pulses are applied (Fig. 13). Here, we use update number instead of cycle number to distinguish this type of endurance test from conventional memory endurance test. It should be noticed that we stop at 10^{11} only because of the measurement time limit. 10^{11} is not the limit of update number.

Although the device does not fail, the analog switching performance degrades as update number increases. Different from memory switching, analog switching is usually evaluated with some new metrics, such as asymmetry/nonlinearity in weight update, dynamic range, variability, etc [8, 9]. We measure the analog switching in a full dynamic range after certain update numbers (Fig. 14), and we see that the dynamic range of analog switching decreases after 10^6 update pulses (Fig. 15). The nonlinearity of analog switching also becomes worse after 10^7 update pulses. The conductance change cannot keep a constant value throughout the dynamic range for both SET and RESET process (Fig. 16). In this case, although the neural network can still be trained, the learning accuracy will sacrifice.

Physical mechanism of the endurance degradation in analog switching is speculated (Fig. 17). Initially, multiple weak CFs are formed, contributing to the good linear analog switching [10]. When switching in low-R window, CF gap does not form, which is confirmed by the Ohmic current at resistance levels lower than $500k\Omega$ (Fig. 11). In this case, electric field distributes uniformly in the switching layer. When switching in high-R/full window, CF gap forms, and large electric field concentrates at the gap region [10]. The large electric field damages the RRAM material properties seriously, resulting in poor endurance. After cycling, the multiple weak CFs morphology gradually transforms to a strong CF morphology, thus the nonlinearity and dynamic range degrade.

IV. IMPACT ON NEURAL NETWORK

A two-layer fully connected neural network for the MNIST dataset learning is used to investigate the impact of endurance degradation of analog switching (Fig. 18). The nonlinearity, dynamic range, endurance degradation, as well as variations, are all taken into consideration in the device behavioral model based on the measured results. When only considering the nonlinearity degradation, the learning accuracy degrades gradually by $>6\%$ at 10^9 update number (Fig. 19a). And when only considering the nonlinearity degradation, the learning accuracy remains unchanged until 10^7 update number, and then

begins to degrade rapidly by $>15\%$ at 10^9 update number (Fig. 19b). When considering both nonlinearity and dynamic range degradation, the accuracy starts to decrease at 10^6 update number, but the degradation degree is similar (Fig. 19c). If we map the synaptic weight to different resistance window, by considering the endurance failure, the accuracy degradation behaviors are different (Fig. 20). It is found that the low-R window (10uS~50uS) is the best condition for online training thanks to its highest accuracy and longest endurance lifetime. Meanwhile, even though some of the cells in the network fail, the neural network can still keep a reasonable accuracy with different weight patterns (Fig. 21). This is due to the redundancy and error tolerance of neural network.

Here we define a new parameter, accumulated ΔG, to characterize the endurance of analog switching. Accumulated ΔG means the sum of the absolute value of conductance change under each update pulse. Assuming the dynamic range of RRAM is 5uS~50uS, the accumulated ΔG for a complete training of the MNIST dataset distributes around 250mS (Fig. 22). Although variation exists, most of the RRAM cells in the network show similar accumulated ΔG with this algorithm. Then we calculated the measured accumulated ΔG before obvious endurance degradation, which is around 6kS. With comparison of device measurements and network simulations (Fig. 23), it is proved that the analog RRAM can meet the requirement of online training. If we need to re-train the network for different datasets, the RRAM based network can re-train for $>10^4$ times. Even for a more complex dataset such as ImageNet shown in Table I, we can conclude that such device can also support re-training for many times.

V. CONCLUSION

For the first time, we investigated the endurance behaviors of incremental switching in analog RRAM for neuromorphic computing. Key achievements include: 1) We developed a test platform to characterize the endurance of analog switching; 2) We demonstrated the endurance of analog RRAM is sufficient for online training; 3) We found the nonlinearity and dynamic range of analog switching degrade with cycling, and the endurance cycle is dependent on the switching window. This work provides valuable guidelines for evaluation and optimization of RRAM based neuromorphic systems.

ACKNOWLEDGMENT

This work is supported in part by the MOST of China (2016YFA0201801), Beijing Innovation Center for Future Chips (ICFC), Beijing Municipal Science and Technology Project (D161100001716002, Z181100003218001), and NSFC (61674087, 61674089, 61674092, 61076115).

REFERENCES

[1] A. Conneau *et al.*, *EACL* 2017, 1107.[2] Y. Li, *arXiv* 2017, 1701.07274.[3] K. Simonyan *et al.*, *Computer Science*, 2014. [4] D.-A. Clevert *et al.*, *Computer Science*, 2015.[5] G. Sassine *et al.*, *IRPS* 2018, P-MY.2-1. [6] P. Y. Chen *et al.*, *IRPS* 2018, 5C.4-1. [7] W. Wu *et al.*, *VLSI* 2018,103. [8] H. Wu *et al.*, *IEDM* 2017, 11.5.1. [9] H. Hwang *et al.*, *EDL* 2016. [10] M. Zhao *et al.*, *IEDM* 2017, 39.4.1.

978-1-7281-1988-5/18 $31.00 © 2018 IEEE

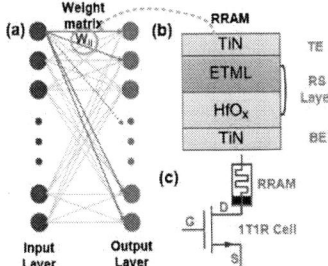

Fig. 1. Schematic diagram of (a) neural network; (b)RRAM; (c)1T1R cell. The conductance of RRAM defines weight in the network.

Table. I. Typical requirement of weight update numbers for online training of represenative datasets.

Task	DataSet	Update Number
Image Identification	MNIST	10^5
	CIFAR	10^5
	ImageNet	10^6-10^7
Natural language processing	-	10^7
Reinforcement learning	-	$>10^8$

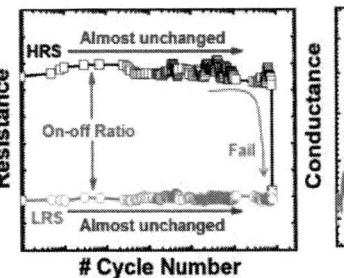

Fig. 2. Typical endurance failure behavior of RRAM for memory application. Resistance window closes after many cycles.

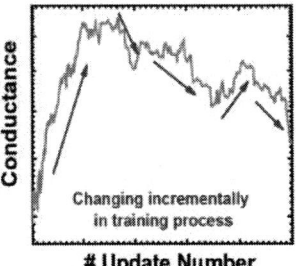

Fig. 3. Typical conductance change trace of analog RRAM during weight update process in a neural network.

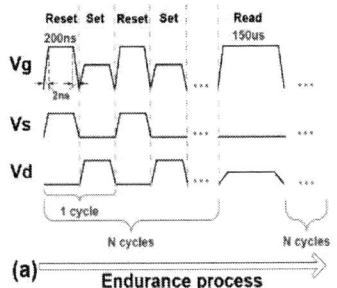

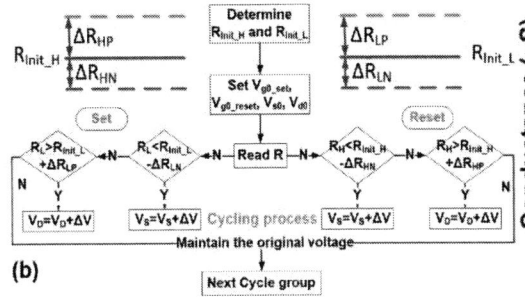

Fig. 4. Illustration of the analog switching endurance test process. (a) The waveform scheme and pulse condition of fast continuous cycling process; (b) The flow chart of determining the resistance window with verification opreation.

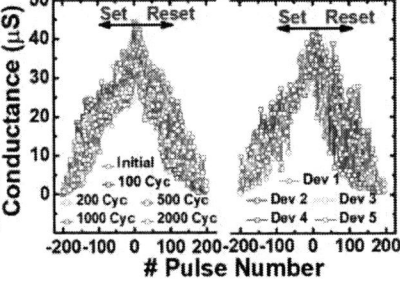

Fig. 5. The uniformity of analog switching behaviors of C2C and D2D. Pulse condition: width=50ns, Vset=1.4V, Vreset=-1.45V.

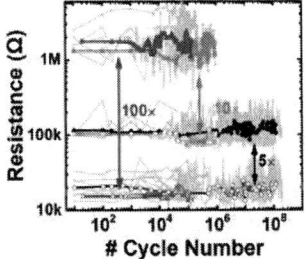

Fig. 6. Endurance failure bebavior of three windows. Grey lines are raw data. Colored lines are mean value.

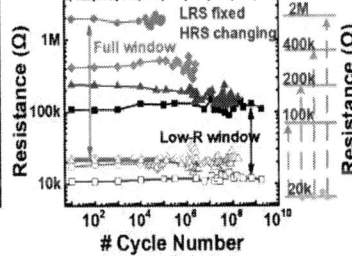

Fig. 7. Endurance failure behaviors with fixed LRS and various HRS (100k to 1MΩ). Measurement stops when the device fails.

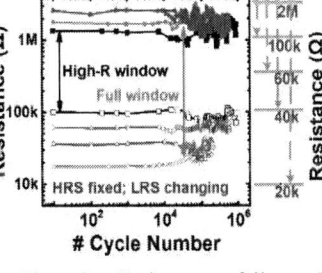

Fig. 8. Endurance failure behaviors with fixed HRS (1MΩ) and various LRS (20, 40, 60, 80 and 100kΩ).

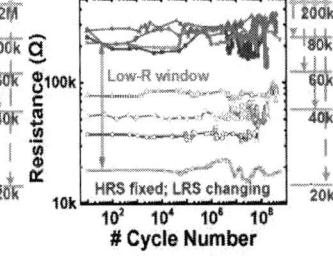

Fig. 9. Endurance failure behaviors with fixed HRS (200kΩ) and various LRS (20, 40, 60, and 80kΩ).

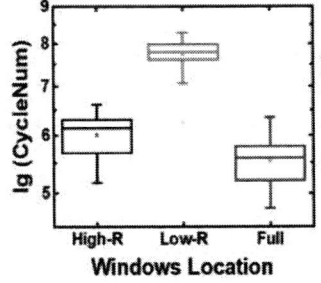

Fig. 10. Distribution of endurance lifetime for different switching windows.

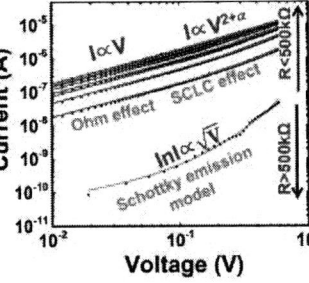

Fig. 11. I-V fitting for different resistance levels. Ohmic current is obscrved when R<500kΩ.

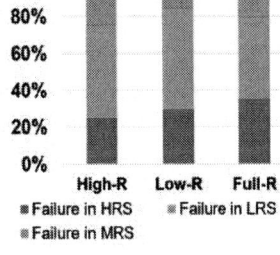

Fig. 12. Statistics of the stuck resistance levels when endurance failure occurs.

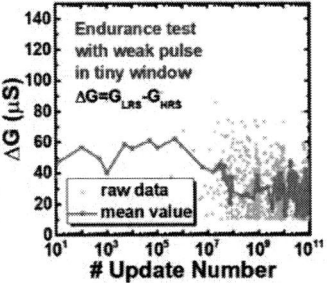

Fig. 13. Measured incremental conductance change within a small window during weak pulse cycling.

978-1-7281-1988-5/18 $31.00 © 2018 IEEE

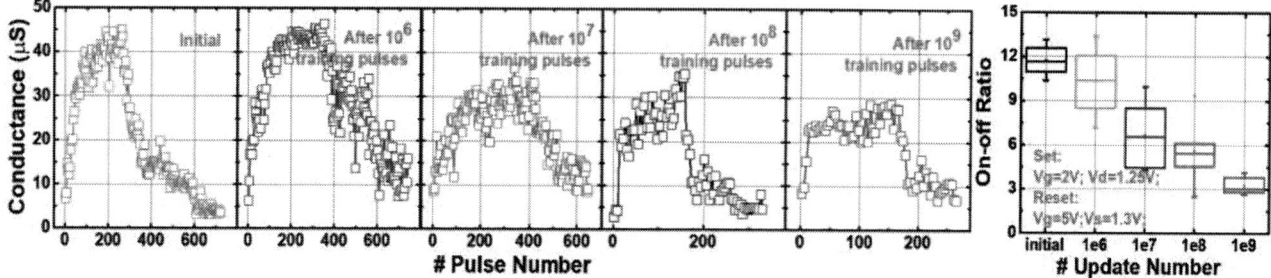

Fig. 14. Analog switching of full dynamic range with identical weight update pulses. The same RRAM device is measured after different numbers of update pulses, including initial, 10^6, 10^7, 10^8, and 10^9. Dynamic on/off ratio and nonlinearity of analog switching degrade as pulse number increasing.

Fig. 15. Distribution of dynamic on/off ratio of analog switching after different numbers of update pulse.

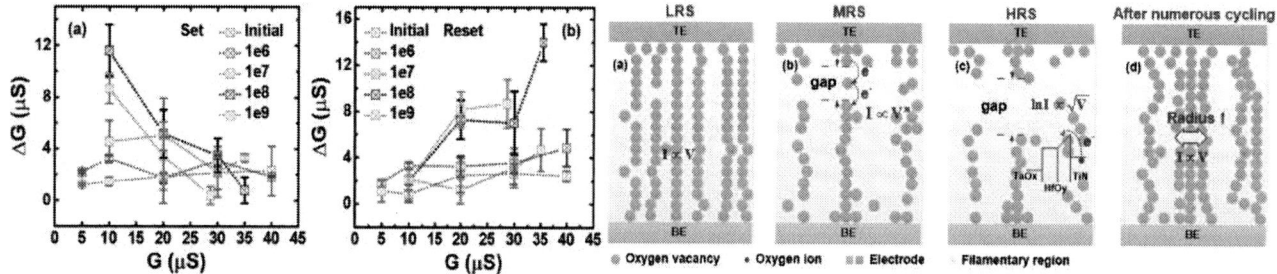

Fig. 16. Conductance change with one update pulse as a function of conductance level during (a) SET process and (b) RESET process. Closer to 0 means linear update, larger vaule means nonlinear update.

Fig. 17. Schematic of physical mechanism of endurance degradation for analog switching with different resistance windows.

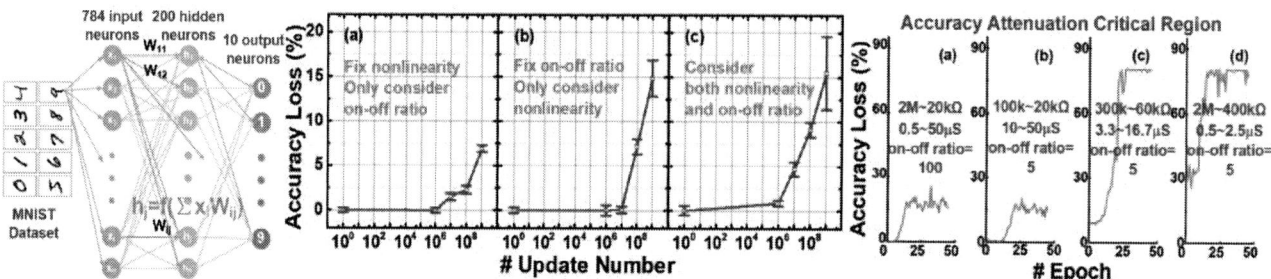

Fig. 18. Schematic of a two-layer fully-connected neural network for online learning and classifying MNIST dataset.

Fig. 19. Learning accuracy loss as a function of update number when only considering (a) on/off ratio; (b) nonlinearity and (c) both nonlinearity and on/off ratio.

Fig. 20. Learning accuracy loss with the degradation of device endurance. Different resistance ranges for weight mapping are compared.

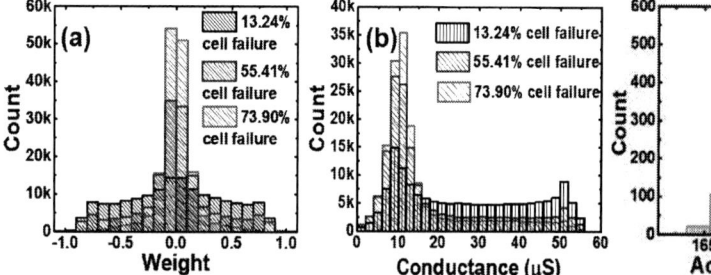
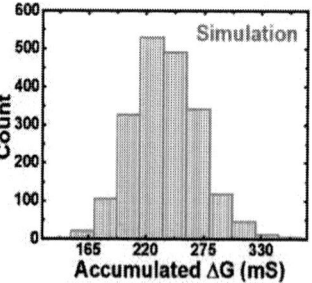
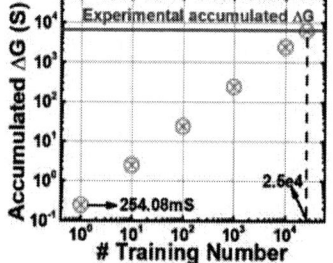

Fig. 21. Distribution of (a) algorithm weight and (b) RRAM conductance in the neural network after online training with different cell failure rate. Here, cell failure is only represented as resistance stuck at LRS, but the nonlinearity and ratio do not change. There is no obvious accuracy loss.

Fig. 22. Distribution of different RRAM's accumulated ΔG in the neural network after online training.

Fig. 23. Comparion of measured accumulated ΔG and simulated accumulated ΔG with different numbers of re-training of new datasets.

978-1-7281-1988-5/18 $31.00 © 2018 IEEE   471

In-depth Characterization of Resistive Memory-Based Ternary Content Addressable Memories

D. R. B. Ly[1], B. Giraud[1], J-P Noel[1], A. Grossi[1], N. Castellani[1], G. Sassine[1], J-F Nodin[1], G. Molas[1], C. Fenouillet-Beranger[1], G. Indiveri[2], E. Nowak[1] and E. Vianello[1]

[1]Univ. Grenoble Alpes, CEA, LETI, 38000 Grenoble, France, email: denys.ly@cea.fr ; elisa.vianello@cea.fr

[2] Institute of Neuroinformatics, University of Zurich and ETH Zurich

Abstract—Resistive Memory (RRAM)-based Ternary Content Addressable Memories (TCAMs) were developed to reduce cell area, search energy and standby power consumption beyond what can be achieved by SRAM-based TCAMs. In previous works, RRAM-based TCAMs have already been fabricated, but the impact of RRAM reliability on TCAM performance has never been proven until now. In this work, we fabricated and extensively tested a RRAM-based TCAM circuit. We show that a trade-off exists between search latency and reliability in terms of match/mismatch detection and search/read endurance, and that a RRAM-based TCAM is an ideal building block in multi-core neuromorphic architectures. These ones would not be affected by long latency time and limited write endurance, and could greatly benefit from their high-density and zero standby power consumption.

I. INTRODUCTION

Ternary Content Addressable Memories (TCAMs) provide a way of searching large data set using masks that indicate ranges. Therefore, they are very attractive for complex routing and big data applications, where an exact match is not often necessary [1]. As opposed to classic memory systems, where a memory cell stored information is retrieved by its physical address, TCAM circuits allow to search a stored information by its content. Fig. 1 depicts the search principle in a TCAM word of 3 bits. The Match Line (ML) is first precharged at a voltage VDD_ML. During the search phase, the ML is left floating and starts discharging. If the stored data in the TCAM word matches with the input searched data (Fig. 1 (a)), the ML slowly discharges through leakage currents $I_{ML,m}$. If at least one bit of the stored data mismatches the input searched data (Fig. 1 (b)), the ML quickly discharges through the mismatching bit with a high discharge current $I_{ML,mis}$.

Conventional SRAM-based TCAM circuits are usually implemented with 16 transistors (16T) [2]. This limits storage capacity of TCAM circuits to tens of Mbs [2-3] in standard memory structures, and takes up valuable silicon real-estate in neuromorphic computing spiking neural network chips [4-5]. In order to increase storage density, RRAM-based TCAM cells have been proposed [6-10]. However, one drawback of RRAM-based with respect to SRAM-based TCAMs is the relatively small ON/OFF current ratio of the memory elements ($\sim 10^5$ for MOSFET compared to 10-100 for RRAMs). Therefore, for long TCAM words, the sum of leakage currents $I_{ML,m}$ in case of a match can become comparable to a mismatching current $I_{ML,mis}$. Another challenge in designing

TCAM circuits with RRAMs is the limited endurance of RRAMs. While SRAMs can sustain an endurance up to 10^{16} cycles, it is difficult to reach endurance higher than 10^6 write cycles for RRAMs. However, endurance in SRAM refers to both write and search/read operations while in RRAM-based TCAMs, write and search/read operations are well distinguished, and must be characterized separately [11].

In this paper, an extensive characterization of search/read and write operations in a RRAM-based TCAM circuit is presented. To date, only the impact of the search/read voltage on the search latency time has been presented (Table 1). The expected impact of RRAM has been evaluated only by simulations [11]. Strong RRAM programming conditions associated with low search/read voltage allow to improve TCAM reliability (Time Ratio>5 and search/read endurance> 10^6 cycles) at the expense of lower performance (longer latency time and lower write endurance). Therefore, we propose to use the RRAM-based TCAM for routing in neuromorphic circuit where long match times (from few tens to hundreds of µs) are required to be compatible with spike length [4-5]. Moreover, this application features long idle times, frequent search and few write operations, thus taking full advantage from the zero standby power consumption.

II. FABRICATED RRAM-BASED TCAM CIRCUIT

Fig. 2 (a) presents the schematic of the fabricated RRAM-based TCAM circuit, composed of a Search word register (SL), a bit cell matrix and a read circuit (Sense Amplifier, SA). The TCAM bit cell is composed of two HfO_2-based RRAMs and two transistors in a 2T2R structure (Fig. 2 (b)). Fig. 3 (a) and (b) show a photo of the fabricated circuit and a SEM cross section of the integrated RRAMs, respectively. RRAMs are integrated in the Back End Of Line of a 130nm CMOS process, on top of the fourth metal layer.

Forming, Set and Reset operations are performed as in single 1T1R RRAM cell by applying the required top electrode voltage on the ML, the bottom electrode voltage on the BL while activating the gate voltage with SLT (resp. SLF). SLF (resp. SLT) signal must be at 0V in order to activate each 1T1R structure independently. During a search operation (Fig. 4), ML voltage is sensed with the SA. In order to evaluate the discharge time of the ML, the applied voltage is compared with a reference voltage V_{TRIP}. The Time Ratio (TR), defined as the ratio between the discharge time when all the word bits are matching ($t_{discharge,m}$) and when only 1 bit of the word is mismatching ($t_{discharge,mis1b}$), has to be maximized to guarantee a sufficient search margin. Different

978-1-7281-1988-5/18 $31.00 © 2018 IEEE

capacitances can be added on the ML signal with the signal CAP_CALIB to slow down the discharge of the ML and facilitate its measurement. A capacitance of 315pF is used in the following. Fig. 5 (a) and (b) show the impact of the ML capacitance on the discharge time $t_{discharge}$, in case of match (green) and mismatch (red). The discharge time increases for higher capacitance values. However, the TR is independent of the ML capacitance (Fig. 5 (c)).

III. TIME RATIO DEPENDENCIES

The ideal TCAM should have a short search latency time (short discharge time) while maximizing the Time Ratio (TR). Fig. 6 (a) presents the pristine, Low Resistance State (LRS) and High Resistance State (HRS) cumulative distributions directly measured on the TCAM cells. HRS distributions are obtained using the soft and strong programming conditions in Fig. 6 (b). The pristine resistance distribution can be used if the TCAM is programmed only one time. The Memory Window (MW) is defined as the ratio between the HRS and LRS values at 2.5σ of the distributions. Fig. 7 shows the impact of the search/read voltage ΔV, applied between the ML and the BL across the bit cells during the search phase, on the discharge time for soft and strong HRS. Higher ΔV enables lower $t_{discharge}$ and thus better performance (lower latency). However, TR is independent of ΔV (Fig. 8). Fig. 9 shows the impact of the MW on the TR. Stronger programming conditions (larger MWs) increase the TR. As expected, TR decreases with the TCAM word length WDL (Fig. 10). To ensure a reliable search taking into account spatial and transient variability, TR cannot go below a sensing limit of 2. Thus, a TCAM row of 128 bit cannot be programmed with soft HRS and it is necessary to use strong HRS. Fig. 11 shows discharge time as a function of the number of mismatching bits, n. Since the discharge time decreases with n, it can be used as an estimation of the Hamming distance between searched and stored data. This block is useful for hardware implementation of brain-inspired hyperdimensional computing systems [12].

IV. SEARCH/READ RELIABILITY

During a search operation, a positive voltage ΔV is applied on the RRAM top electrode in the same polarity as a Set operation (Fig. 12 top). Therefore, unwanted switching from HRS to LRS can occur as shown in Fig. 12 bottom. Constant Voltage Stress (CVS) measurements have been performed on a 4kbit 1T1R RRAM array to extract the Set switching time $t_{switching}$ for different ΔV. Fig. 13 presents the measured $t_{switching}$ as a function of ΔV (black lines). The different curves correspond to different percentages of switched cells. For comparison, we reported the discharge time $t_{discharge,m}$ as a function of the search voltage ΔV in case of a match (green lines), for different word lengths. To avoid unwanted switching during search operations, the search voltage must be diminished so that the discharge time is lower than $t_{switching}$ (green region in Fig. 13). As shown in Fig. 14 (a) top, a reduction of ΔV from 0.6V to 0.4V increases from 90k to 450k the number of reliable searches (defined as the

number of search operations before a RRAM device switches from the HRS to the LRS, Fig. 14 (b)). Another way to improve search reliability is to adopt strong HRS. For $\Delta V=0.6V$, 400k cycles are performed (Fig. 14 (a) bottom). These results are obtained in the worst case scenario, since the 315pF line capacitance artificially increases the search time. By using a capacitance of 90pF (Fig. 14 (a), bottom), we have an endurance higher than 10^6 cycles.

V. PROGRAMMING RELIABILITY

Fig. 15 shows a write endurance characterization measured on a 4kbit 1T1R RRAM array, for soft HRS. This case allows to improve endurance performance [11]. After 10^4 Set/Reset cycles, some cells remain stuck in HRS with a probability $p_{HRS\ stuck}$. At 10^6 cycles, breakdown failures (cells stuck in LRS) occur with a probability $p_{breakdown}$. For a TCAM circuit, HRS stuck failures increase the discharge time and therefore they have no impact on matches. However, this can lead to a mismatch failure with a probability depending on the word length (Fig. 16 (a)). On the other hand, breakdown failures decreases the discharge time. This leads to a match failure whatever the impacted matching TCAM cell, *i.e.* with a probability independent of the word length (Fig. 16 (a)). Fig. 16 (b) shows the probability of failures for a match (red) and a mismatch (blue) as a function of the number of write operations. An endurance in programming of 10^6 cycles can be reached, as the probability of mismatch failures is negligible ($\sim 10^{-38}$).

VI. DISCUSSION AND CONCLUSION

In this work, we experimentally show the impact of RRAM reliability on a TCAM circuit. To improve the search margin (Time Ratio) and search/read endurance, strong RRAM programming conditions (high HRS), low search voltage and limited word length have to be adopted. This comes at the expense of lower performance in terms of longer search latencies and lower write endurance (Table 2). The performance reduction can be a critical limiting factor in standard TCAM-based applications. However, multi-core neuromorphic computing architectures would not be affected by these problems and could greatly benefit from their high density. For example, the processing cores of the NeuRAM3 DYNAP-SEL neuromorphic chip recently proposed in [5] comprise multiple TCAM cells per neuron, to implement memory-optimized source-address routing schemes (see [4] for details). These TCAM cells are typically small in size (e.g. 22 bit in the DYNAP-SEL chip) and are programmed only at network configuration time. Assuming future neuromorphic computing architectures of this type will have thousands of cores, the non-volatility feature of the proposed TCAM circuits will provide an additional crucial benefit, as it will require the user to upload all the configuration bits only the first time the network is configured, and will be able to skip this potentially time-consuming process every time the chip is reset or power-cycled. Finally, long match times are required to be compatible with spike length.

ACKNOWLEDGMENTS: This work has been partially supported by the European H2020 NeuRAM 687299 project.

I. Introduction

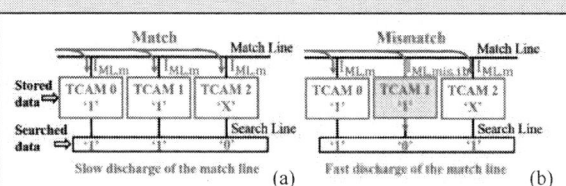

Fig. 1: Example of search/read operation. The Match Line (ML) is initially precharged, then it is left floating. If the stored data matches with the searched data, the ML slowly discharges (a). If the stored data mismatches with the searched data, the ML quickly discharges (b).

	[6] - 2T2R	[7] - 2.5T1R	[8] - 5T2R	[9] - 4T2R	[10] - 3T1R	This Work - 2T2R
TCAM circuit	8×2048×64 bit 90 nm CMOS	64×256 bit 90 nm CMOS	128×64 bit 90 nm CMOS	128×32 bit 180 nm CMOS	2×64×64 bit 90 nm CMOS	3×128 bit 130 nm CMOS
Search Latency	1.9 ns @ 0.75 V	1 ns @ 0.45 V	1.9 ns @ 0.75 V	1.2 ns @ 1.4 V	0.96 ns @ 0.48 V	90 ns @ 0.6 V
Measured results	Impact of *Search Voltage* on Search Latency					• Search Latency • Match/mismatch search margin • Search/Read endurance (>10⁶) • Programming endurance (> 10⁶)

Table 1: Silicon verified RRAM-based TCAM circuits presented in the literature. Few electrical characterization results have been presented and the impact of RRAM reliability on the TCAM performance has not been studied before.

II. Fabricated RRAM-based TCAM circuit

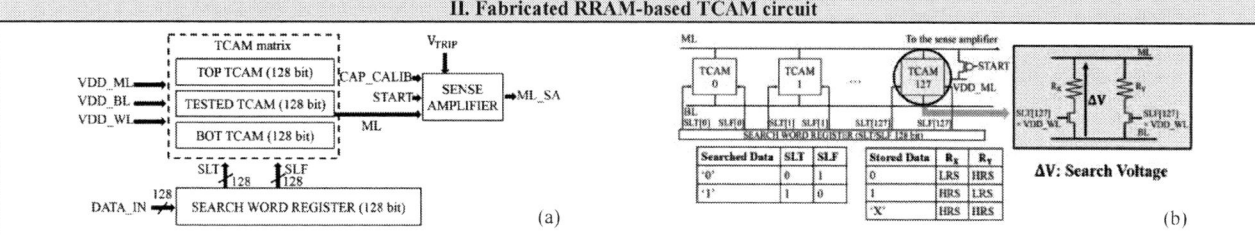

Fig.2: (a) RRAM-based TCAM circuit schematic. The Search word registers (SLs) take as input the searched data DATA_IN and send it to the TCAM matrix via the signals SLT=DATA_IN and SLF. The TCAM matrix comprises 3 rows of 128 bits. (b) RRAM-based TCAM row and bit cell schematics, and states definition.

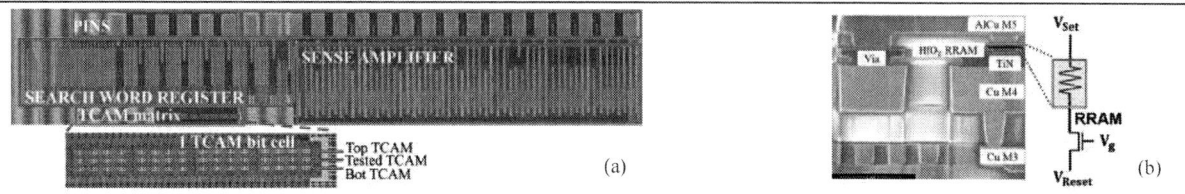

Fig. 3: (a) Die photo. (b) SEM cross section of the integrated TiN/HfO$_2$/Ti/TiN RRAM. Both HfO$_2$ and Ti layers feature a 10 nm thickness.

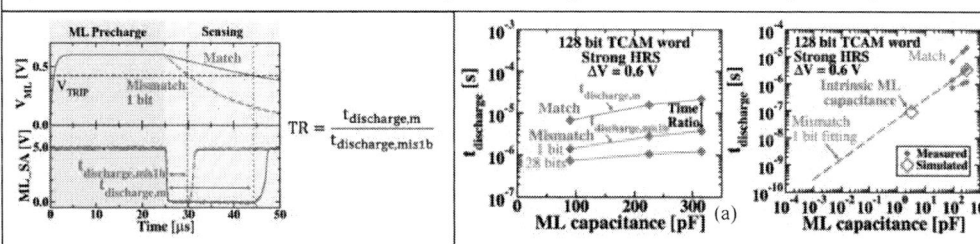

Fig. 4: Waveforms of the search operation for match (green) and mismatch (red) cases. The Match Line (ML) is initially precharged. During sensing, the ML is left floating and its voltage V_{ML} decreases. The output signal of the Sense Amplifier (SA), ML_SA, stays at 0V as long as V_{ML} remains higher than V_{TRIP}.

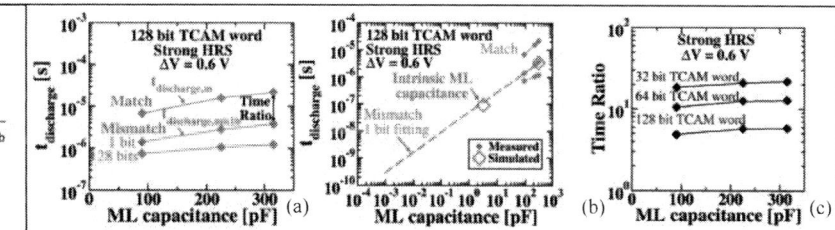

Fig. 5: (a) Discharge time as a function of the Match Line (ML) capacitance for the match (green) and mismatch (red) cases for a 128 bit TCAM word and (b) comparison with simulated results. (c) Time Ratio (TR) as a function of the ML capacitance for different word lengths. TR is independent of the ML capacitance. Strong programming conditions defined in Fig. 6 are used for (a), (b) and (c).

III. Time Ratio Dependencies

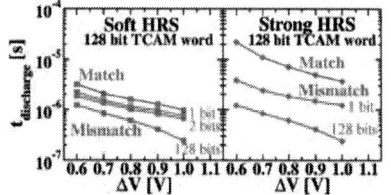

	LRS	Soft HRS	Strong HRS
ML	2.0 V	GND	GND
BL	GND	2.5 V	2.5 V
VDD_WL	1.3 V	2.5 V	3.0 V
R_{median} (Ω)	2.5 k	198 k	1.27 M
$R_{\pm 2.5\sigma}$ (Ω)	3.0 k	66 k	594 k

Fig. 6: (a) Pristine, LRS and HRS cumulative distributions and (b) associated programming conditions. Memory Window (MW) is defined as the ratio between the HRS and LRS values at ± 2.5 σ of the distributions.

Fig. 7: Discharge time as a function of the search voltage (ΔV in Fig. 2 (b)) for soft (a) and strong (h) programming conditions.

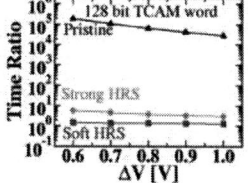

Fig. 8: Time Ratio as a function of the search voltage.

978-1-7281-1988-5/18 $31.00 © 2018 IEEE

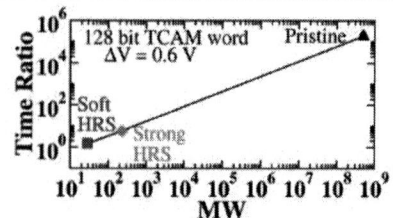

Fig. 9: Time Ratio as a function of the Memory Window. A large time ratio guarantees the correct detection of matched and mismatched words.

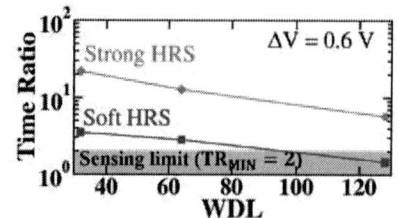

Fig. 10: Time Ratio as a function of the TCAM Word Length (WDL).

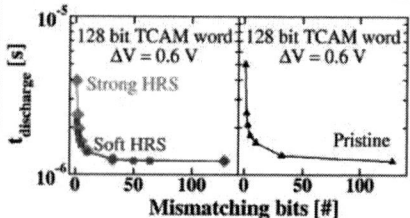

Fig. 11: Discharge time as a function of the number of mismatching bits for a 128 bit TCAM word.

IV. Search/Read Reliability

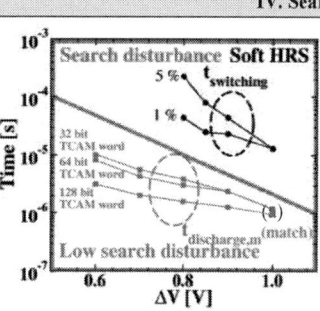

Fig. 12: During a search operation, ΔV is applied on the two 1T1R structures (top). This can cause an unwanted switching from HRS to LRS (bottom).

Fig. 13: Set switching time as a function of the applied voltage (black lines) extracted from CVS measurements on a 4kbit 1T1R array. The discharge time in a match case is reported for comparison (green lines). The discharge time has to be lower than the switching time.

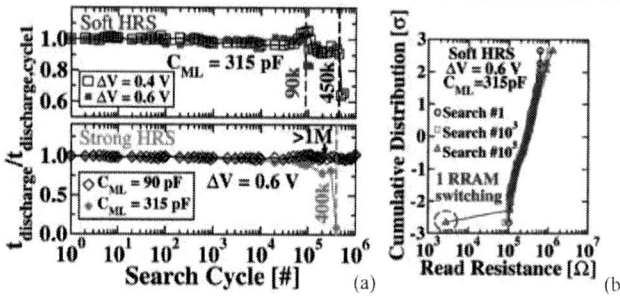

Fig. 14: (a) (Top plot) Read/search endurance with soft HRS and a match line capacitance $C_{ML} = 315$ pF, for $\Delta V = 0.6$ V and $\Delta V = 0.4$ V. (a) (Bottom plot) Search/read endurance with strong HRS and $\Delta V = 0.6$ V, for $C_{ML} = 315$ pF and $C_{ML} = 90$ pF. (b) HRS cumulative distribution after 1, 1k and 100k search operations, for the soft HRS and $\Delta V = 0.6$V. Measurements are performed on a 128 bit TCAM word in a match case configuration (all the cells in HRS).

V. Programming Reliability

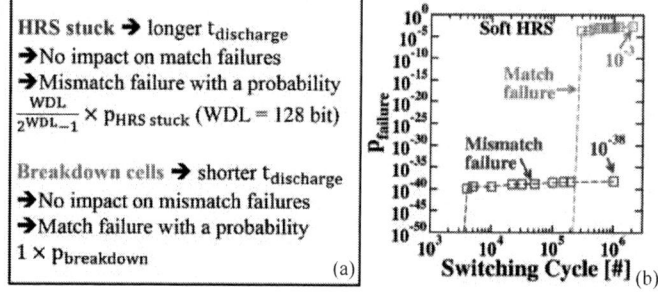

Fig. 15: Endurance characterization measured on a 4 kbit 1T1R array, for soft HRS. HRS stuck and breakdown failures are observed. HRS stuck failures occur much earlier than breakdown failures.

Fig. 16: (a) Impact of HRS stuck and breakdown failures on RRAM-based TCAM circuits. (b) Probability of mismatch and match failures due to the HRS stuck and breakdown failures respectively, for soft HRS.

VI. Conclusion

	$\Delta V \uparrow$	$R_{HRS} \uparrow$	WDL $\uparrow$
Search Latency (Discharge Time)	↓ ☺	↑ ☹	↓ ☺
Search Margin (Time Ratio)	↓ ☹	↑ ☺	↓ ☹
Comments	Search/read endurance ↓ ☹	➢ Search/read power ↓ ☺ ➢ Search/read endurance ↑ ☺ ➢ Programming endurance ↓ ☹	➢ Memory capacity ↑ ☺

Table 2: Summary of the study. There is a trade-off between Search Latency (Discharge Time) and search margin reliability (Time Ratio).

REFERENCES

[1] R. Karam et al., *Proc. IEEE*, 2015, vol. 103, no. 8, pp 1311-1330
[2] I. Hayashi et al., *A-SSCC*, 2012, pp. 65-68
[3] L. Nii et al., *ISSCC*, 2014, pp. 240-241
[4] Moradi et al., *TBCAS*, 2018, vol. 12 , no. 1, pp.106-122
[5] Qiao et al., *Proc. IEEE BioCAS*, 2016, pp. 552-555
[6] J. Li et al., *JSSC*, 2014, vol. 49, no. 4, pp 896-907
[7] C-C Lin et al., *ISSCC*, 2016, pp. 136-137
[8] M-F. Chang et al., *ISCAS*, 2016, pp. 1142-1145
[9] M-F. Chang et al., *JSSC*, 2016, vol. 51, no. 11, pp. 2786-2798
[10] M-F. Chang et al., *JSSC*, 2017, vol. 52, no. 6, pp. 1664-1679
[11] A. Grossi et al., *TVLSI*, 2018
[12] H. Li et al., *VLSI-TSA*, 2017, pp. 1-2

Mixed-Signal Neuromorphic Inference Accelerators: Recent Results and Future Prospects

M. Bavandpour[*], M.R. Mahmoodi[*], H. Nili, F. Merrikh
Bayat, M. Prezioso, A. Vincent, and D.B. Strukov[#]

UC Santa Barbara, Santa Barbara, CA 93106-9560, U.S.A.
[*] equal contributions, [#] strukov@ece.ucsb.edu

K.K. Likharev

Stony Brook University, Stony Brook, NY 11794-3800, U.S.A.
Konstantin.Likharev@stonybrook.edu

Abstract- Recent advances in dense, continuous-state nonvolatile memories have enabled extremely fast, compact, and energy-efficient analog and mixed-signal circuits. Such circuits are perfectly suited, in particular, for hardware implementations of the inference operation in advanced neuromorphic networks, which requires massive amounts of dot-product operations with low-to-medium precision. In this paper, we first review typical implementations of such mixed-signal circuits. We then describe some recent experimental demonstrations of prototype mixed-signal neuromorphic networks by our team, in particular, a mixed-signal inference accelerator with unprecedented speed and energy efficiency. The paper is concluded by outlining some urgently needed work, in particular the development of high-performance general-purpose inference accelerators, and discussing our preliminary results in this direction.

I. INTRODUCTION

The rapidly growing range of applications of machine learning for image classification, speech recognition, and natural language processing have led to an urgent need in specialized neuromorphic hardware. Of that, there is much more demand for fast, low-precision inference accelerators than for higher-precision systems for network training [1].

Though the vast majority of demonstrated accelerators from industry [1-3] and academia [4, 5] are digital, the most natural approaches, however, are based on analog and mixed-signal circuits [6-13]. Though the core principles of analog computing had been developed almost four decades ago [14, 15], its efficient implementations were enabled only recently by the appearance of novel continuous-state, nonvolatile memory devices [16] - the most crucial elements of analog circuits.

II. MIXED-SIGNAL CIRCUITS USING EMERGING MEMORIES

Fig. 1 shows typical mixed-signal circuits for the implementation of the vector-by-matrix multiplication (VMM), i.e. the most important operation in inference accelerators and other neuromorphic tasks, while Fig. 2 provides their qualitative comparison. Specifically, due to their superior integration density, VMMs based on passive crossbars with resistive nonvolatile devices (Fig. 1.I) [6], including metal-oxide memristors, conductive-bridge and phase-change memories, might be the most promising in the long term. Passive integration is, however, significantly more challenging, since in this case the distribution in the memristors' effective switching thresholds should be narrow enough to avoid the disturbance of already tuned devices at

their half-selection (Fig. 3a,b). Additional gate lines in active crossbars with 1T1R cells (inset of Fig. 1.I) solve the half-select problem [8, 12] and allow for either higher device variations at synaptic weight tuning (Fig. 3c), or higher precision of the finite weights, or both. (The cell's selector functionality, the main advantage of the 1T1R approach for digital memories, is less important for neuromorphic inference applications, since writes are typically very infrequent.)

Though the integration density of the floating-gate (FG) circuits (Figs. 1.II and 1.III) is comparable with that of systems using 1T1R cells, the fabrication technology available for the latter approach is more scalable. The main relative advantage of the former approach is the FG cell's amplification, that relaxes the requirement for gain of sensing circuitry, and enables very compact peripheral circuits.

Note also that each of options I-III may also operate with time encoding, which allows for better computing precision, for the price of certain speed reduction.

Finally, the lack of continuously tunable devices in the switch capacitor approach (Fig. 1.IV) typically enables only 'near-memory' computing (instead of 'in-memory' computing possible with other options), and leads to inferior density and other metrics.

III. EXPERIMENTAL DEMONSTRATIONS

Because of still immature device fabrication technology, memristor-based inference circuit demos have been limited in complexity, and/or not fully integrated [6, 8, 12]. Fig. 4 shows a recent result from our collaboration – a small-scale, one-hidden-layer perceptron classifier implemented entirely in integrated hardware. This specific network used two passive 20×20 crossbar arrays with on $Pt/Al_2O_3/TiO_{2-x}/Pt$ memristors (Fig. 4a), board-integrated with discrete CMOS components [6]. The network was successfully trained (both in-situ and ex-situ) to perform classification of 4×4 pixel images (Fig. 4c). The successful demonstration was facilitated by improvements in memristor fabrication technology lowering device-to-device variations, and thus enabling accurate individual state tuning (Fig. 4d, e).

The situation is much better for mixed-signal circuits based on floating-gate crosspoint devices, due to the availability of advanced industrial-grade flash-memory technologies. Our team has recently designed, fabricated, and tested a prototype mixed-signal, 28×28-binary-input, 10-ouput, 3-layer neuromorphic network (Fig. 5a) based on embedded nonvolatile FG cell arrays, redesigned from a commercial 180-nm NOR flash memory [13]. Each array performs a very fast and energy-efficient analog VMM operation. All functional components of the prototype circuit, including 2 synaptic crossbar arrays with 101,780 floating-gate synaptic cells, 74 analog neurons, and peripheral circuitry for weight adjustment and I/O operations, have a total area below $1~mm^2$. Its testing on the MNIST benchmark set has shown a classification fidelity of 94.65%, close to the 96.2% obtained in simulation (Fig. 5b). Most importantly, the classification of one pattern takes time less than 1 µs (Fig. 5c) and energy below 20 nJ – both numbers at least $10^3\times$ better than at a digital implementation of the same task, with similar fidelity, fabricated using a much more advanced process [3].

Moreover, there are still many reserves for improving the performance and energy efficiency of such circuits. For example, Fig. 6 shows preliminary results for a much larger network-specific inference accelerator with more advanced circuitry. This chip was designed and fabricated in a 55-nm process, adapted for analog computing applications [9].

IV. FUTURE WORK AND SUMMARY

For the memristor-based approach, the most important goal is the development of foundry-grade, highly uniform fabrication technology, which would allow for monolithic integration of much larger, denser crossbars with CMOS circuits. Hopefully, this work would piggyback on the recent industrial efforts toward digital resistive memories.

For the FG-based approach, the preliminary experimental results for the chip-to-chip statistics, long-term drift, and temperature sensitivity of the 55-nm [9] and 180-nm [13] prototypes showed no evident showstoppers towards much more complex deep neuromorphic networks. This is why the major focus of future work in this direction may be on the system-level design. In this context, while ASICs have important application niches, general-purpose inference accelerators [2, 3, 18] may be more useful at the moment, in part due to the continuing evolution of neuromorphic algorithms and architectures.

Fig. 7a shows one such architecture, currently being developed by our group. The core of this design is four $M\times N$ rectangular blocks of $K\times K$ VMM crossbar arrays, with front-end digital-to-analog converters (DAC) and back-end sensing circuitry. The array outputs can be connected, via programmable analog buses, to implement larger-size VMMs. Other components of the accelerator include an instruction memory, a controller for decoding instructions and orchestrating the data flow, a small memory buffer for keeping frequently-used data close to the processing unit, and the main memory based on embedded DRAM for storing input, output, and intermediate data.

The performance of the proposed processor was simulated for three representative neural network architectures [19-21] (Fig. 7b). The results show that the mixed-signal VMM blocks take the largest fraction of the chip area, while communications, i.e. sending data across the VMM blocks, often dominates its energy consumption. This fact highlights the importance of in-memory computing using very dense memories for storing weights, as well as of an efficient design of peripheral VMM circuits and configurable busses, which allows fine-grain mapping of network models.

Our preliminary estimates show (Fig. 7c) that general-purpose mixed-signal inference accelerators may retain at least the same large advantage, in speed and energy efficiency, over their digital counterparts, that has been demonstrated in our first, network-specific experiments. The experimental verification of these estimates, as well as the refinement of cons and pros of various approaches to this key task of neuromorphic computing are very important goals for the nearest work.

ACKNOWLEDGMENT

This work was supported by DARPA's UPSIDE program under contract HR0011-13-C-0051UPSIDE via BAE Systems and NSF grant CCF-1528502. Authors are grateful to B. Chakrabarti, X. Guo, I. Kataeva, and M. Klachko for useful discussion and technical support.

REFERENCES

[1] NVIDIA Corp. Investor day presentation (2017).
[2] ARM ML processor (https://www.arm.com/products/processors/machine-learning); Intel Mobileye (https://www.mobileye.com/en-us/); Google Edge TPU (https://cloud.google.com/edge-tpu/).
[3] P. A. Merolla et al. A million spiking-neuron integrated circuit with a scalable communication network & interface. Science 345 668 (2014).
[4] Y. H. Chen et al. Eyeriss: An energy-efficient reconfigurable accelerator for deep convolutional neural networks. ISSCC 262 (2017).
[5] B. Moons et al. Envision: A 0.26-to-10 TOps/W subword-parallel dynamic-voltage-accuracy-frequency-scalable convolutional neural network processor in 28nm FDSOI. ISSCC 246 (2017).
[6] F. Merrikh Bayat et al. Implementation of multilayer perceptron network with highly uniform passive memristive crossbar circuits. Nature Comm. 9 2331 (2018).
[7] M. J. Marinella et al.. Multiscale co-design analysis of energy, latency, area, and accuracy of a ReRAM analog neural training accelerator. JETCAS 8 86 (2018).
[8] C. Li et al. Analogue signal and image processing with large memristor crossbars. Nature Electron. 1 52 (2018).
[9] X. Guo et al. Temperature-insensitive analog vector-by-matrix multiplier based on 55 nm NOR flash memory cells. CICC 1 (2017).
[10] M. Bavandpour et al. Energy-efficient time-domain vector-by-matrix multiplier for neurocomputing and beyond. arXiv:1711.10673 (2017).
[11] E. H. Lee et al. A 2.5 GHz 7.7 TOps/W switched-capacitor matrix multiplier with co-designed local memory in 40nm. ISSCC 418 (2016).
[12] G. W. Burr et al. Experimental demonstration and tolerancing of a large-scale neural network using phase-change memory as the synaptic weight element. TED 62 3498 (2015).
[13] X. Guo et al. Fast, energy-efficient, robust, and reproducible mixed signal neuromorphic classifier based on embedded NOR flash memory technology. IEDM 6.5.1 (2017).
[14] C. Mead, Analog VLSI and Neural Systems (1989).
[15] J. Hasler et al. Finding a roadmap to achieve large neuromorphic hardware systems. Front. Neurosci. 7 118 (2013).
[16] H. S. P. Wong et al. Memory leads the way to better computing. Nature Nanotechnol. 10 191 (2015).
[17] M. R. Mahmoodi et al. Breaking POp/J barrier with analog multiplier circuits based on nonvolatile memories. ISLPED 124 (2018).
[18] A Shafiee et al. ISAAC: A convolutional neural network accelerator with in-situ analog arithmetic in crossbars. Computer Architecture News 44 14 (2016).
[19] C. Szegedy et al. Going deeper with convolutions. CVPR 1 (2015).
[20] K. He et al. Deep residual learning for image recognition. CVPR 770 (2016).
[21] Y. Wu et al. Google's neural machine translation system: Bridging the gap between human and machine translation. arXiv:1609.08144 (2016).

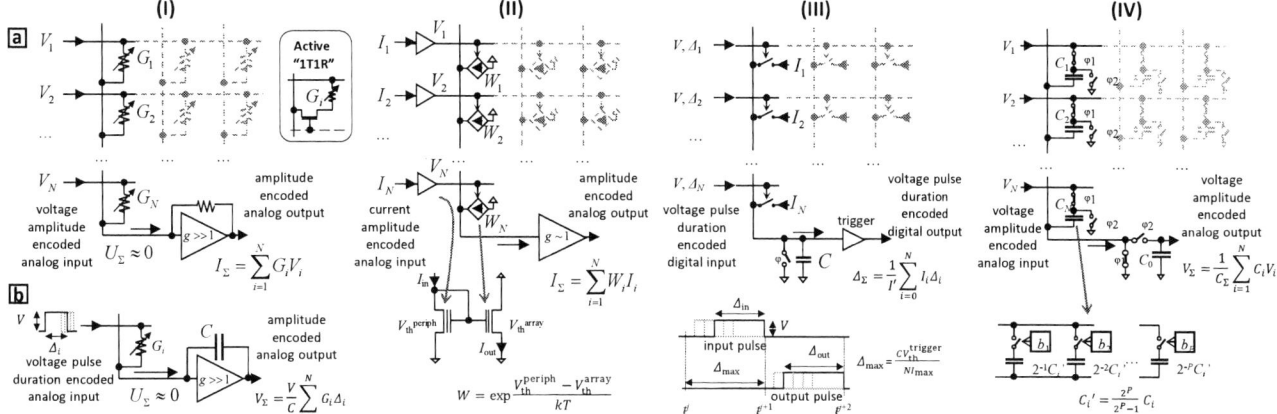

Fig. 1. Major types of mixed-signal VMM circuits. In (I), the matrix elements ('synaptic weights') are represented by continuous states of adjustable nonvolatile resistive devices (e.g., memristors), while the input signals are encoded with either (a) amplitudes, or (b) durations of voltage pulses. The top right inset shows an active ('1T1R') cell, which may be also used in circuits (a, b). In (II, III), each weight is stored in subthreshold-mode floating-gate (FG) cells, implemented as either (II) a current mirror pair formed by peripheral and array FG transistors, or (III) a voltage-gated current source. In (III), both inputs and outputs are encoded by the duration of pulses, generated within the corresponding time frame t, as shown at the bottom of panel III. In the switch capacitor approach (IV), P-bit weights are typically stored in binary-weighted fixed-value crosspoint capacitors, and the computation is performed by controlling the capacitor charge/discharge, using the switches φ_1 and φ_2.

Fig	Xpoint		Input/output	Density	Precision	Speed	Energy Efficiency	Maturity	Ref
Ia	0T	R	amp/amp	++	+	+++	++	-	[6]
Ib		R	time/amp	+++	++	++	+++	-	[7]*
Ia	1T	R	amp/amp	+	++	++	+	+	[8]
Ib		R	time/amp	++	+++	+	++	+	[7]*
II	FG		amp/amp	+	++	++	+	++	[9]
III	FG		time/time	++	+++	+	++	++	[10]
IV	C		amp/amp	-	+	++	++	+++	[11]

Fig. 2. Approximate comparison of features of the VMM approaches outlined in Fig. 1: '+++' - the best, '-' – the worst. The score for precision is based on a combination of the input, weight, and computing accuracies. The scores for density, speed, and energy efficiency (EE) reflect contributions from both the arrays and the peripheral circuits. Besides the maturity, all scores are for the expected level of each technology after it has been matured, rather then for its current state-of-the-art. *Ref. [7] describes, in particular, the additional circuitry for the conversion to the time-encoded output signals.

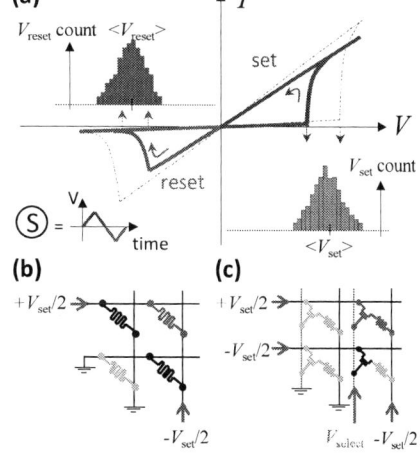

Fig. 3. Switching threshold variations in memristors: (a) A typical hysteretic dc *I-V* curve, for a symmetric voltage sweep (lower bottom inset). The inset histograms show typical variations in the switching voltages at which the effective conductance changes by more than a certain amount. (b, c) Four-device crossbar fragments with (b) passive '0T1R' and (c) active '1T1R' cells. The applied voltages show a specific example of the "half-biasing" technique for increasing the conductance of the selected device (shown in red). Solid black lines show the half-selected memristors, while the gray color is used to show unselected devices, with no applied voltage.

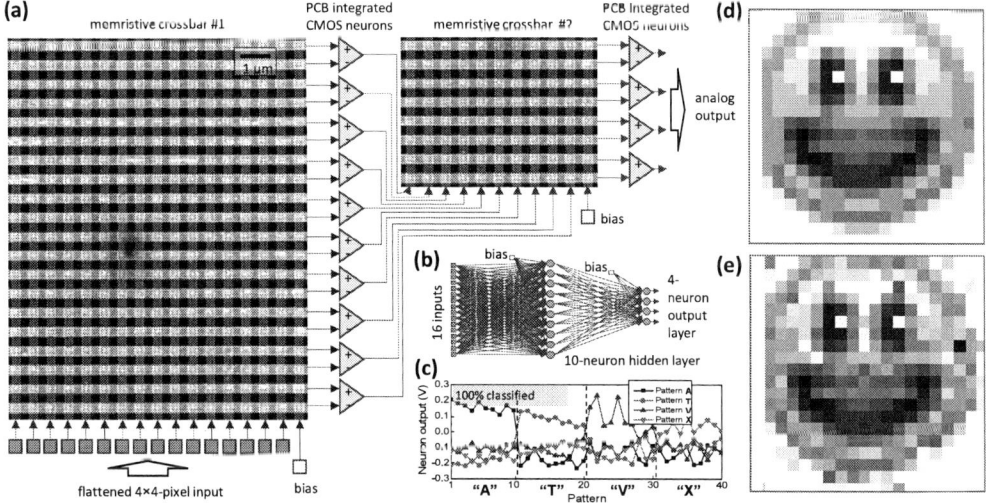

Fig. 4. MLP classifier demo based on passively integrated metal-oxide memristors [6]: (a) A perceptron diagram showing (as SEM images) the crossbar portions used in the experiment; (b) The implemented network's graph; (c) An example of measured output voltages for the ex-situ-trained network, tested on a set of 4 stylized 4×4-pixel letters; (d, e) An example of memristor tuning, showing (d) the desired 'smiley face' pattern, quantized to 10 gray levels, and (e) the actual resistance values measured after tuning all devices in a 20×20 memristive crossbar with the nominal 5% accuracy, using an automated tuning algorithm. The white / black pixels correspond to effective resistances 96.6 / 7.0 kΩ, measured at 0.2 V.

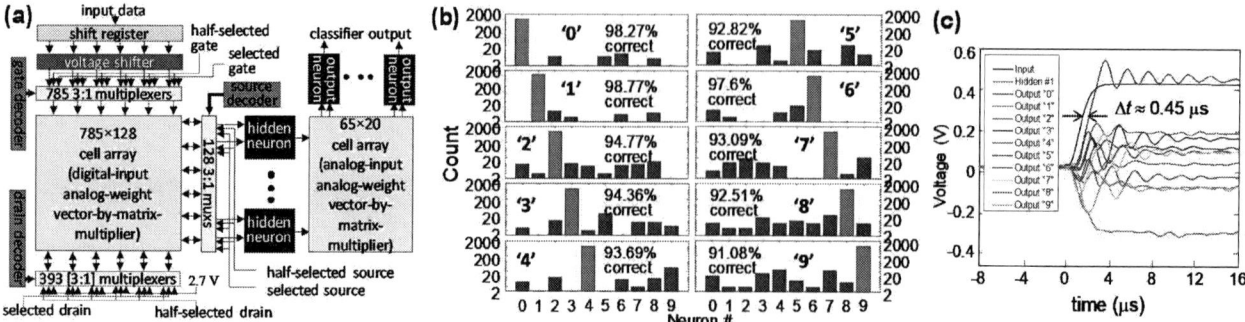

Fig. 5. MLP classifier demo in the 180 nm ESF1 process with 100+K FG cells [13]: (c) High-level architecture, designed for classification of MNIST benchmark images. (Weight tuning circuitry for the 2nd array is not shown for clarity). (b) Histograms of the experimentally measured largest output voltages from the ex-situ trained network for 10,000 MNIST test set patterns, showing that the correct outputs (red bars) always dominate. (c) Typical signal dynamics after an abrupt turn-on of the voltage shifter's power supply, measured simultaneously at the network input, at the output of a sample hidden-layer neuron, and at all network's outputs.

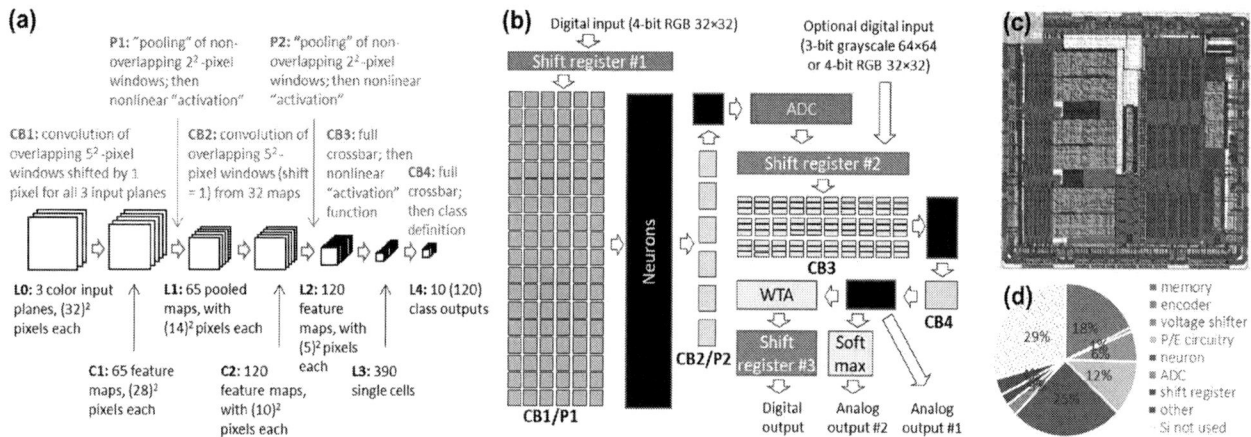

Fig. 6. Network-specific mixed-signal image classifier accelerator: (a) The architecture of the implemented deep convolutional neural network, (b) its block diagram, (c) the 55-nm CMOS chip layout, with ~ 13M embedded ESF3 NOR flash cells, and (d) the corresponding area breakdown. Some circuitry (e.g., for testing and cell tuning) is not shown for clarity. An advanced design [17] has enabled a reduction of the neuron circuits to ~6% of the chip area, while the FG arrays take ~30% of it.

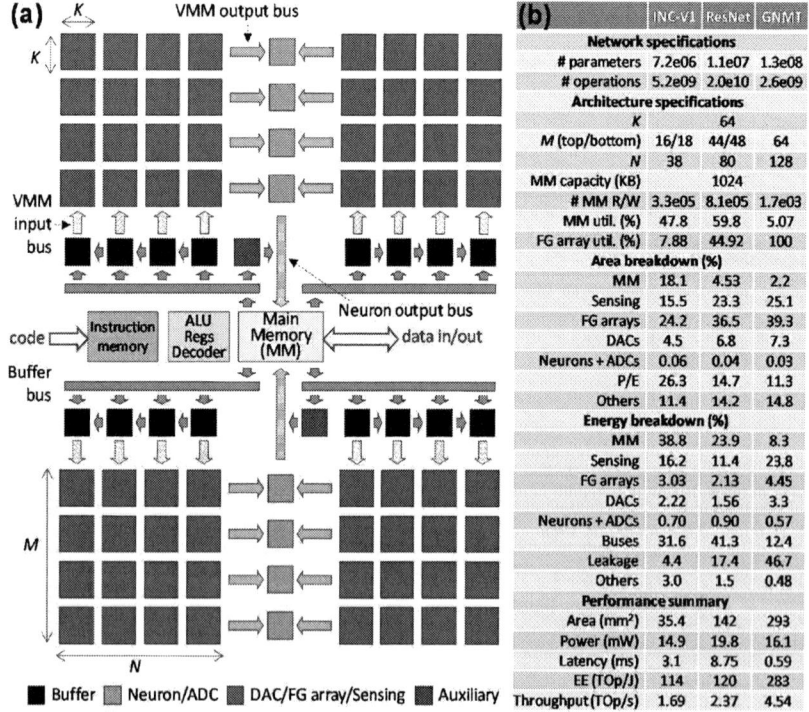

(b)

	INC-V1	ResNet	GNMT
Network specifications			
# parameters	7.2e06	1.1e07	1.3e08
# operations	5.2e09	2.0e10	2.6e09
Architecture specifications			
K		64	
M (top/bottom)	16/18	44/48	64
N	38	80	128
MM capacity (KB)		1024	
# MM R/W	3.3e05	8.1e05	1.7e03
MM util. (%)	47.8	59.8	5.07
FG array util. (%)	7.88	44.92	100
Area breakdown (%)			
MM	18.1	4.53	2.2
Sensing	15.5	23.3	25.1
FG arrays	24.2	36.5	39.3
DACs	4.5	6.8	7.3
Neurons + ADCs	0.06	0.04	0.03
P/E	26.3	14.7	11.3
Others	11.4	14.2	14.8
Energy breakdown (%)			
MM	38.8	23.9	8.3
Sensing	16.2	11.4	23.8
FG arrays	3.03	2.13	4.45
DACs	2.22	1.56	3.3
Neurons + ADCs	0.70	0.90	0.57
Buses	31.6	41.3	12.4
Leakage	4.4	17.4	46.7
Others	3.0	1.5	0.48
Performance summary			
Area (mm²)	35.4	142	293
Power (mW)	14.9	19.8	16.1
Latency (ms)	3.1	8.75	0.59
EE (TOp/J)	114	120	283
Throughput (TOp/s)	1.69	2.37	4.54

(c)

	Chip area (mm²)	Latency (ms)	EE (TOp/J)
Digital CMOS	321	607	0.35
Mixed-signal FG	293	0.59	283

Fig. 7. aCortex, a multi-purpose mixed-signal inference accelerator: (a) the block-level architecture, and (b) the simulated performance metrics for three representative neuromorphic tasks - image classification and natural language translation, based on two deep convolutional neural networks (INC-V1 [19] and ResNet [20]) and a deep recurrent network GNMT [21]. Though the results are reported for smallest hardware resources required for each network, the use of configurable buses enables achieving, for smaller networks (INC-V1 and ResNet), essentially the same performance by their mapping on the largest architecture (used for GNMT). (c) Preliminary comparison of key metrics of the mixed-signal and digital versions of the aCortex for the GNMT network. In the 4-bit digital version, synaptic weights are stored on-chip in an SRAM (~92% of the chip area), while the mixed-signal processing arrays are replaced with arrays of 64×64 4-bit digital VMM units (~6% of the area). The dot product in each digital VMM is implemented with array-based multipliers and parallel tree adders. All performance and area numbers are for a 55 nm CMOS process. All latencies are per one classification / translation task. Note that both designs were optimized for the energy efficiency, while being suboptimal in terms of latency / throughput. (Faster operation may be achieved by performing more operations in parallel, with some energy efficiency degradation.)

Temporal sequence learning with a history-sensitive probabilistic learning rule intrinsic to oxygen vacancy-based RRAM

J. Doevenspeck[1,2], R. Degraeve[1], A. Fantini[1], P. Debacker[1], D. Verkest[1], R. Lauwereins[1,2] and W. Dehaene[1,2]

[1]imec, Leuven, Belgium, email: jonas.doevenspeck@imec.be, [2]KU Leuven ESAT, Leuven, Belgium

Abstract

Widely spread and low value resistance distributions inhibit the use of filamentary resistive RAM (RRAM) at low currents for deep learning training and inference. An entirely different approach which employs RRAM as *active computational elements* is proposed. For this means, the history-sensitive probabilistic reset in Tantalum-Oxide (TaOx)-based RRAM is characterized and explained. This intrinsic RRAM effect is used as a local learning rule in a novel temporal sequence learning algorithm.

Introduction

Neuromorphic computing opens interesting opportunities for novel memory technologies [1]. Specifically, for 'deep learning'-type of neural networks, several publications suggest that RRAM can serve as a weight storage element, while a crossbar array can be used to perform a fast and low power vector-matrix multiplication by coding the input as a voltage or time vector and collecting the output as a current [2]. In practice, it is, however, very challenging to program RRAM to multiple R-levels while keeping sufficiently low operation current. Particularly, filamentary RRAM devices have a widely spread R-distribution inhibiting their use as weight storage element at low current [3].

In this paper, we propose an entirely different approach. We demonstrate that filamentary RRAM devices (and specifically oxygen vacancy-based RRAM or OxRRAM) can be used in machine learning applications by enabling them as *active computational elements* instead of merely storage elements. In our approach, the OxRRAM device is read in a binary fashion with respect to a fixed R-threshold. The learning capabilities reside in the *probability* of a successful RESET operation at low reset voltage and its large sensitivity to the program history. Using only 1 SET and 1 RESET pulse width and amplitude without any read-verify loops, we design a *non-deep learning algorithm* tailored for optimally using the statistical properties of OxRRAM devices. This approach targets continuous on-chip learning of context-sensitive time-dependent correlations in a data sequence.

TaOx – based RRAM for statistical learning

In this section, we characterize our OxRRAM device as a probabilistic local learning rule. Previous work [4,5], shows a constant RESET probability, but we show how multiple SET pulses drastically lower the RESET probability at low voltage, creating a history-sensitive probabilistic learning rule.

A. Devices and measurements

Data are collected from a 1Mb TaOx-based RRAM array fabricated with 65nm CMOS. The memory pillar was integrated between metal layer 3 and 4 (Fig. 1) and consists of a 3nm-thick TaOx-based layer and 5nm-thick Ta capping between the TiN bottom and top electrode (BE and TE resp.).

The measurement procedure is presented in Fig. 2 and measurement conditions in Table I. First, the devices are formed at the same current compliance I_C used during SET pulses. Second, all devices are subject to a hard RESET pulse to bring them to the high resistive state (HRS). Third, consecutive SET pulses are applied, s in total. Finally, consecutive soft RESET pulses are applied and the resistance is read after each pulse.

B. Soft RESET behavior

The cumulative distribution functions (CDFs) for consecutive RESETs after one SET are shown in Fig. 3. Starting from LRS, the CDF gradually moves to HRS with increasing number of RESET pulses. Note that the R-distributions are largely overlapping. While the distributions shift gradually, individual devices show a stochastic behavior with unpredictable resistance changes (up as well as down) as a function of the number of RESET pulses (Fig. 4).

C. Interpretation with hourglass model

The stochastic nature of the RESET process is interpreted in the context of the hourglass model (Fig. 5) [6]. First, the 1-SET pulse LRS and 1-hard RESET pulse HRS distributions (Fig. 6) are modeled. In the hourglass model, the conduction through the filament is calculated using a saddle surface potential describing the geometrically narrowest part of the filament (i.e. the 'constriction') [7]. This saddle surface has two parameters: ω_x and $\omega_y = \omega_{y,0}/n_c$ (n_c = number of constriction vacancies), inversely proportional to the constriction 'length' and 'width' respectively (Fig. 5). As elaborated in [6], two sources cause filament resistance variability: (i) the number of constriction vacancies n_c, and (ii) constriction shape variations.

(i) The number of oxygen vacancies is dynamically modeled by a Monte Carlo simulation of the SET/RESET transient. (ii) The shape variations are modeled by introducing a bivariate Gaussian distribution of the saddle surface parameters ω_x and $\omega_{y,0}$ (Fig. 7). Note that LRS and HRS in Fig. 6 are described by the same (ω_x, $\omega_{y,0}$) distribution, evidencing that both states have the same physical nature but merely differ in constriction dimensions.

Between consecutive soft RESET pulses in Fig. 3, the filament shape changes, causing the resistance to fluctuate. However, due to the low voltage of the RESET pulse, ω_x and $\omega_{y,0}$ are not independently sampled from the bivariate Gaussian distribution. Rather, the change in ω_x and $\omega_{y,0}$ is described by a kinetic process identical to the post-programming relaxation as reported in [8]. In [8], large post-programming filament shape variations are described through linearly auto-correlated changes of $\log(\omega_x)$ and $\log(\omega_{y,0})$ combined with small uncorrelated post-programming variations. In the present paper, we demonstrate that the exact same kinetic process with the same parameter settings also explains the stochastic low-voltage RESET behavior.

We proceed as follows (Fig. 8): First, we extract the number of constriction particles n_c from the median resistance (Fig. 9). Second, we calculate the resistance after resampling of ω_x and $\omega_{y,0}$ values using the same formula's as in [8]. As a result, the magnitude and frequency of the measured soft-reset fluctuations are very well captured as shown in Fig. 10.

We conclude that the (low voltage) RESET should be interpreted as a probabilistic event. Our analysis proves that low-voltage RESET behavior is intrinsically stochastic and therefore uncorrelated with device-to-device processing variations (experimentally confirmed, but not shown here).

D. RESET probability after multiple SETs

In Fig. 11, the measured cumulative probability for a successful RESET above a fixed R-threshold R_{th} is plotted as a function of the number of RESET pulses starting from an increasing number of SET pulses. The more SETs are applied, the more resilient the filament becomes to soft RESET. A sigmoid was fitted and the derivative gives the conditional reset probability. The SET-history sensitivity can be tuned by selecting V_{RES} (Fig. 12) and I_C (Fig. 13). Although the median LRS resistance after 1 and 50 SET pulses is only 14% different, the conditional RESET probability is much more affected: up to 83% at 150µA (Fig 14). In the following section, this property is exploited as local probabilistic learning rule.

Sequence learning algorithm using probabilistic OxRRAM

In this section, we briefly describe an algorithm that exploits the history-sensitive probabilistic SET/RESET behavior of OxRRAM. First, the concept is presented, then the hardware mapping to a memory array is shown and finally the learning algorithm is presented.

A. Conceptual description

A continuous stream of data is decomposed in a number of base vectors (=features). In the simplest case these are quantized states of the signal, but a set of semantically meaningful features is preferred. The feature sequence is stored in a sequence memory that aims at learning temporal correlations between the features. The approach is based upon but modified from [9]. Each feature is coded as an ensemble of N neurons. A subset of active neurons in the ensemble forms a context-sensitive representation of the feature. The active neurons at sequence time t are physically connected to the active neurons at time t+Δ. After sufficient training, the time-dependent correlations in the data are transformed into a network of connected neurons. The learning algorithm consists on the one hand of making and strengthening correct connections between active neurons, and on the other hand weakening or removing unlikely or erroneous connections.

B. Mapping to memory array

Each neuron at time t corresponds to an input line in the memory array, while each neuron at t+Δ corresponds to a column (Fig. 15). The ensembles are grouped in the periphery of the array. Active neurons at t can be connected through an OxRRAM filament by applying a SET pulse. Existing unwanted connections are weakened using a soft RESET pulse, resulting in a probability to increase the filament resistance above the read R_{th} (Fig 16). Connections that are often confirmed become intrinsically resilient to soft RESET.

C. Learning

The learning algorithm works as follows. A read operation (binary read at V_{read} with respect to a R_{th}) consists of powering the input lines corresponding to the active neurons at time t. This identifies which active neurons at t+Δ are already connected to the representation of the data at time t. This 'prediction' is overlaid with the actual data at t+Δ (Fig 17). Correctly predicted active neurons are selected for a SET pulse. To improve the learning performance, the selected neurons at t+Δ can on their turn be applied at the input lines, resulting in a prediction for t+2Δ. This step can be repeated in order to down-select those neurons at t+Δ that can predict future states as far as possible. Once the best representing neurons for the data at t+Δ have been chosen, the OxRRAM devices connecting the representation at t with that at t+Δ receive a SET pulse. The unselected connections receive a soft RESET pulse to weaken or remove them.

Note that our approach is an example of complete in-memory computing. The data representation and their time-dependent correlations are stored as positions in the memory array, the learning is achieved through the OxRRAM probabilistic SET/RESET features, and the training algorithm as described above can be implemented in the periphery of the array.

D. Verification

After training, the 2D-network topology, stored in the memory array, is used as a model to generate a new data sequence. To obtain this sequence, starting from any state, one or more predicted features are taken as new input to generate the next time step in the sequence. The goodness of the learned model is evaluated by comparing the generated sequence with the original training sequence.

Examples

Two illustrative examples are presented.

(i) Fig. 18: A simple sinus function is quantized and learned. A context-sensitive representation is needed to represent identical values in the rising and falling part of the sinus function. When 25% of the input values are replaced by random noise, the generated data after sufficient training has less noise than the input, because only repeated (functional) dependencies are trained. This example shows how one can build a noise filter for arbitrary functions. Also, it shows the benefit of using a history-sensitive learning rule. The learning without history fails to filter out the noise.

(ii) Fig. 19: A music score containing monophonic tunes in classical style is coded as note positions and tone intervals. Tone intervals rather than absolute pitches are used – this is an example of semantically meaningful coding. After training the sequence of these features, new music can be generated. The evaluation is largely subjective. If considered 'good', the newly composed music can be added to the model as training material. In this way the model is adapted to generate a user-specific output in accordance to the user's preferred style.

Summary and conclusions

We characterized and explained the probabilistic low voltage RESET behavior of filamentary OxRRAM. We demonstrated the increased resilience against RESET after multiple SETs. Using this property as a local learning rule, we propose and demonstrate an in-memory algorithm for continuous learning of correlations in time sequences.

Acknowledgments

This research or part of this research is conducted within the imec IIAP entitled "Machine Learning".

References

[1] G. Burr et. al, Advances in Physics: X, vol. 2 no.1 pp 89 (2016)
[2] S. Yu, et. al, IEEE Proc., vol. 106 no. 2 pp 260 (2018)
[3] A. Fantini et. al., IMW (2013)
[4] M. Suri, et. al, IEEE TED, vol. 60 no. 7 pp 2402 (2013)
[5] S. Yu, et. al, frontiers in Neuroscience, vol. 7 pp. 186 (2013)
[6] R. Degraeve, et. al, IPFA Proc., pp. 254 (2014)
[7] M. Büttiker, et. al, Phys. Rev. B, vol. 41 no. 11 pp 7906 (1990)
[8] R. Degraeve, et. al, IRPS Proc., (2016)
[9] Y. Cui, et. al, Neural Computation, vol. 28, no. 11, pp2474 (2016)

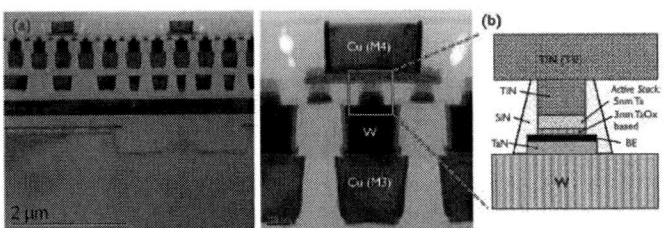

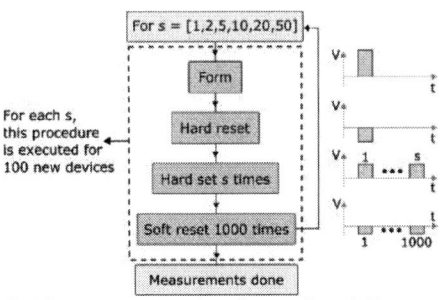

Fig. 1. (a) TEM cross section of part of the 1Mbit TaOx-based OxRRAM array, showing the 60nm active memory element and the electrically inactive dummies at 200nm pitch. (b) Composition of the TaOx-based memory stack.

Fig. 2. Measurement procedure for one I_C and $V_{RESET,SOFT}$. For each combination of different number of preceding SET pulses s, I_C and $V_{RESET,SOFT}$, 100 new devices were formed and measured. After each soft reset, the resistance is measured.

Table I: Measurement parameters for procedure in

Parameters	Values
I_C	50/150 μA
V_{SET}	1.5 V
$V_{RESET, hard}$	1.5 V
$V_{RESET, soft}$	-0.7 – -1.0 V
V_{form}	3.3 V
$V_{WL, RESET}$	3.3 V
V_{read}	0.1 V
T_{SET}	100 ns
$T_{RESET, hard}$	100 ns
$T_{RESET, soft}$	100 ns
T_{form}	100 ns

Fig 2.

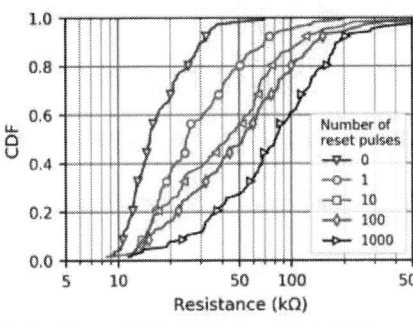

Fig. 3. Resistance CDFs gradually shift to higher resistances after soft RESET pulses. The LRS was obtained with one hard SET pulse at I_C = 50μA. CDFs are shown for logarithmically increasing number of soft RESET pulses at -0.8V.

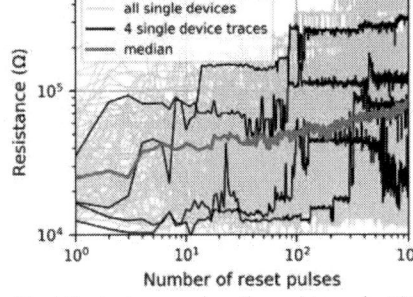

Fig. 4 Device traces and median resistance for 1000 soft RESET pulses at -0.8V after one hard SET pulse at I_C = 50μA. While individual devices follow stochastic RESET paths, the median increases gradually.

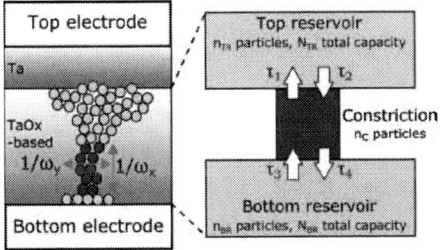

Fig. 5. Hourglass model [6]: Oxygen vacancies are depicted as single particles. The filament is abstracted into three reservoirs: a top reservoir with n_{TR} particles and max. N_{TR} particles, a conducting constriction with n_C particles, a bottom reservoir with n_{BR} particles and max. N_{BR} particles. Particle transition time constants are determined by an activation energy E_a

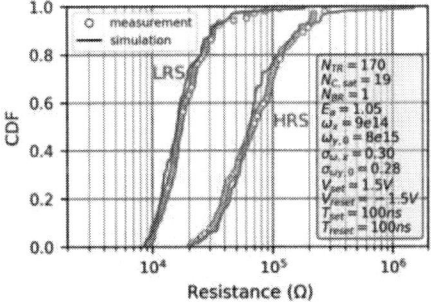

Fig. 6. Experimental data for the LRS and HRS at I_C = 50uA, V_{SET} = 1.5V, V_{RES} = -1.5V. Simulated distributions are obtained with the hourglass model and agree well with the measurements. With a single parameter set, both the LRS and HRS can be simulated.

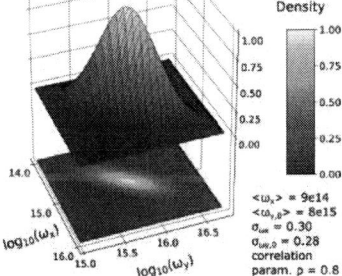

Fig. 7. Bivariate Gaussian distribution describing the variability of the saddle surface potential determining the filament resistance. Parameters are obtained by fitting LRS and HRS with the Hourglass model (Fig 6).

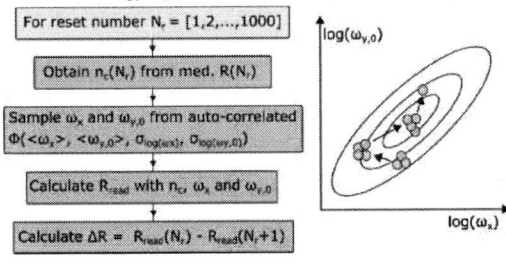

Fig. 8. Simulation methodology for one device. The resistance jumps are simulated by extracting n_C from the median resistance and resampling ω_x and $\omega_{y,0}$ to account for post-programming relaxation after each RESET pulse [7].

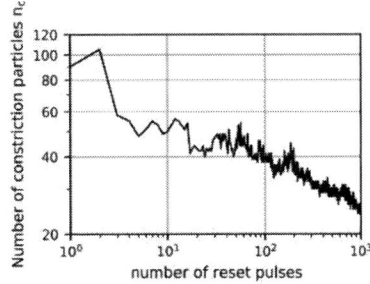

Fig. 9. After each RESET pulse (-0.8V), the number of constriction particles n_C is obtained from the median resistance for a filament with average shape parameters $<\omega_x>$ and $<\omega_{y,0}>$.

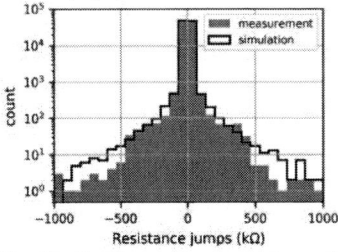

Fig. 10. Resistance jumps are shown for 1000 RESET pulses and 100 devices. The simulated resistance jumps agree well with the measurements, indicating that post-programming relaxations account for the stochastic reset behavior.

978-1-7281-1988-5/18 $31.00 © 2018 IEEE

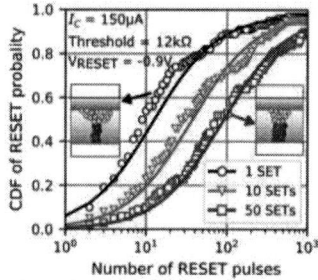
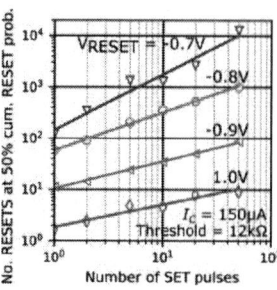
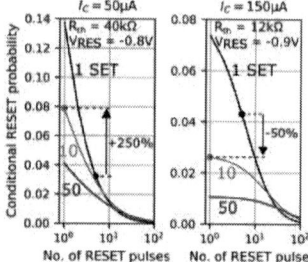
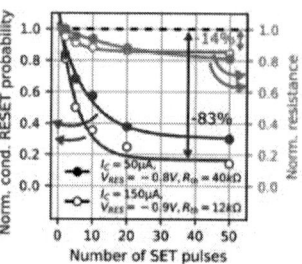

Fig. 11. RESET prob. after increasing no. of RESET pulses for different SET histories. An increasing number of preceding SET pulses lowers the cumulative RESET probability by widening the OxRRAM filament.

Fig 12. No. of RESET pulses required to bring 50% of devices above the threshold for various RESET voltages.

Fig. 13. Conditional RESET prob. for multiple SET histories. As an example: the difference between RESET prob. after 10 SETs + 1 RESET and after 1 SET + 5 RESETs is indicated. The ratio can be strongly modified.

Fig. 14. Normalized conditional reset prob. after 1 RESET and resistance after different number of preceding SET pulses. While after 50 SET pulses at I_C = 150µA, the resistance decreases only with 14%, the reset prob. decreases with 83%.

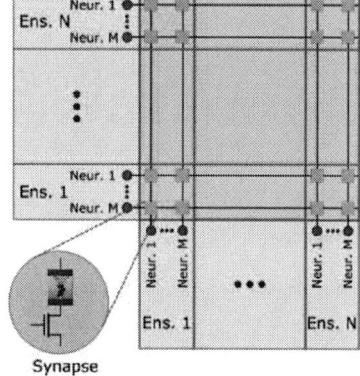

Fig. 15. Hardware mapping of neurons and synapses to wordlines, bitlines and TaOx-based RRAM devices on a crossbar array. Neurons are grouped in ensembles in the periphery.

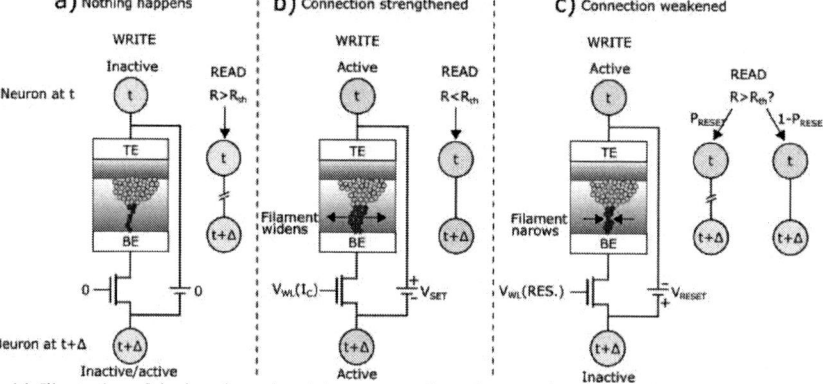

Fig. 16. Illustration of the learning rule. **a)** A neuron at time t is not active: nothing happens. **b)** A neuron at time t is active and another neuron at time t+Δ is also active: A SET pulse is applied which result in the creation (if synapse was above threshold) or strengthening (if the synapse was below the threshold) of the connection. **c)** A neuron at time t is active and another neuron at time t+Δ is not active: a soft RESET pulse is applied which removes the connection with a certain probability P$_{RESET}$.

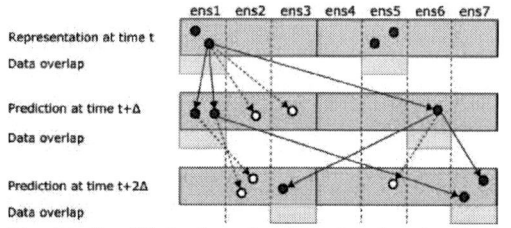

Fig. 17. Simplified schematic of the learning algorithm. Neurons which correctly predict the data in a time window t+Δ, t+2Δ, ... are connected with neurons active at the current timestep t. Only connections starting from one neuron are shown for simplicity.

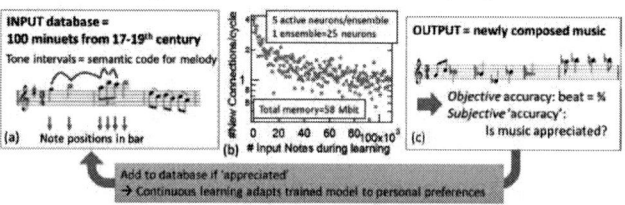

Fig. 19: Example of complex data training. (a) As input a continuous stream of randomly selected music scores from a database of 100 scores in beat ¾ is presented. The scores are coded as a sequence of intervals and note positions. (b) The algorithm converges and after 1000 input examples were presented, (c) new music scores are generated. Objective accuracy (beat=3/4) scores 100%. Subjective quality of the music can be evaluated by a human user. Appreciated music can be added to the database in order to modify the trained model in accordance to the user's preference.

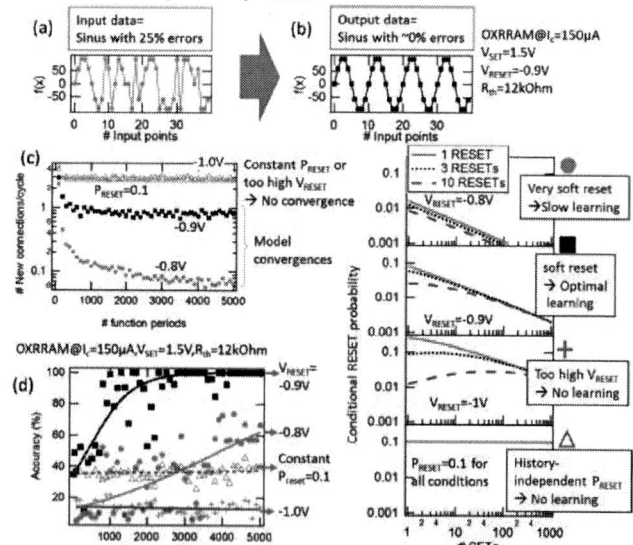

Fig. 18: Illustrative example demonstrating the function of the OXRRAM local learning rule. (a) As input a simple quantized sinus is presented, with 25% of the input points random errors. (b) Since only functional relations are repeatedly confirmed, the output is noise-free for optimal V$_{RES}$. (c) The convergence of the algorithm is witnessed by a decreasing number of newly generated connections corresponding to (d) an increasing output accuracy. If a SET-history independent constant P$_{RESET}$ = 0.1 or a too aggressive RESET is used, no algorithm convergence nor noise filtering is obtained.

978-1-7281-1988-5/18 $31.00 © 2018 IEEE

In-Memory and Error-Immune Differential RRAM Implementation of Binarized Deep Neural Networks

M. Bocquet[1*], T. Hirztlin[2*], J.-O. Klein[2], E. Nowak[3], E. Vianello[3], J.-M. Portal[1] and D. Querlioz[2]

[1]Aix Marseille Univ, Université de Toulon, CNRS, IM2NP, Marseille, France
[2]C2N, Univ Paris-Sud, CNRS, Orsay, France, email: damien.querlioz@u-psud.fr
[3]CEA, LETI, Grenoble, France   [*]These authors contributed equally to the work

Abstract—RRAM-based in-Memory Computing is an exciting road for implementing highly energy efficient neural networks. This vision is however challenged by RRAM variability, as the efficient implementation of in-memory computing does not allow error correction. In this work, we fabricated and tested a differential HfO_2-based memory structure and its associated sense circuitry, which are ideal for in-memory computing. For the first time, we show that our approach achieves the same reliability benefits as error correction, but without any CMOS overhead. We show, also for the first time, that it can naturally implement Binarized Deep Neural Networks, a very recent development of Artificial Intelligence, with extreme energy efficiency, and that the system is fully satisfactory for image recognition applications. Finally, we evidence how the extra reliability provided by the differential memory allows programming the devices in low voltage conditions, where they feature high endurance of billions of cycles.

I. INTRODUCTION

Deep neural networks are currently the most widely investigated architecture in Artificial Intelligence (AI) systems, with incredible achievements in image recognition, automatic translation, Go or Poker games. Unfortunately, when operated on central or graphics processing units (CPUs or GPUs), they consume considerable energy, in particular due to the intensive data exchanges between processors and memory [1,2]. Neural networks using in-memory computing (iMC) with RRAM are widely proposed as a solution to the Von-Neumann bottleneck [1]. However, RRAMs are prone to variability [3], and using Error Correcting Codes (ECC) as in more standard memories would ruin the benefits of iMC. ECCs indeed require large decoding circuits [4], which would need to be replicated multiple times in the case of iMC. This last point is the key challenge that we have to face for reliable neural networks on large RRAM memory arrays. In this paper, an experimental RRAM array with differential memory bit-cell (2T2R) based on HfO_2 devices, including all peripheral and a differential sensing scheme is fully characterized. This differential approach completely solves the key reliability challenge of large neural network implemented on RRAM using iMC concept. Due to its differential structure, our memory has intrinsically reduced errors. For the first time, we show that it improves reliability similarly to ECC with the same bit-cell count, whereas it considerably reduces CMOS overhead in the sensing scheme resulting in a clear gain in sensing speed. Additionally, we show that this structure allows the natural implementation of one of the most modern concepts of deep learning: Binarized Neural

Networks (BNN) [5,6]. Such neural networks, can achieve state of the art AI performance, with very reduced memory requirements. Additionally, these networks use RRAMs as purely binary memories. We also show that the level of reliability achieved with our differential approach is fully appropriate as BNNs have an intrinsic tolerance to errors. We finally evidence that the robustness brought by our approach allows us to program RRAM devices at low voltage, where the devices feature very high endurance.

II. DIFFERENTIAL MEMORY STRUCTURE: AN IDEAL ARCHITECTURE FOR IN-MEMORY COMPUTING

For this work, we fabricated memory arrays with a differential memory structure in a HfO_2-based OxRAM process, integrated in the BEOL of a 130 nm CMOS logic process [7], on top of the fourth metal layer (Cu) (Fig. 1). The OxRAM devices correspond to $TiN/HfO_2/Ti/TiN$ stacks. The thickness of both HfO_2 and Ti layers is 10 nm. Each bit is stored in a 2T2R structure in a complementary fashion: the two devices are programmed to complementary states (LRS/HRS or HRS/LRS) (Fig. 2). Each column features a differential precharge sense amplifier (PCSA) [8] (Fig. 2 and 3), which operates by comparing the resistance of the two memory devices. We fabricated and tested several structures with 2k devices, associated sense amplifiers and row and column decoders on chip. Fig. 4 first shows statistics of the forming process of the devices: all of them are formed, and the two devices do not influence each other. Fig. 5 shows the programming distribution in a low 55 µA Sct compliance current (I_c) situation, prone to a high 1.2 % bit error rate in a 1T1R memory. Fig. 6 shows the response of all devices programmed in the same condition in a kbit 2T2R array as measured by our differential sense: only 0.2 % bit error is seen. Fig. 7 validates the functionality of the differential sense circuit with comprehensive testing. In previous works, 2T2R RRAM differential memories have already been fabricated, but their benefits on reliability have never been proven until now [9], [10], therefore we characterized our arrays extensively. Fig. 8 presents the mean number of bit errors on kbits array. We see that this number depends extensively on the programming conditions. Measurements with hundred millions of cycles on a single device, where the state of the devices is measured at each cycle (Fig. 9), evidence that LRS and HRS become less differentiated when the device ages, and that the 2T2R structure has much lower error rate than 1T1R in this situation. Overall, Fig. 10(a) shows that 2T2R always decreases the bit error rate with regards to 1T1R in diverse regimes. This Figure associates

978-1-7281-1988-5/18 $31.00 © 2018 IEEE

full array (device-to-device) measurements taken in a low compliance current regime, to address high error rates, and cycle-to-cycle experimental results in higher compliance current, to address lower error rates. The black curve is a theoretical result, assuming the PCSA has an ideal behavior, and therefore shows the minimum error rate achievable by the 2T2R approach. It is insightful to compare the benefits of 2T2R with the approach of ECC used in non-iMC contexts. Fig. 10(b) and Fig. 10(c) show the reliability benefits of various Single Error Correction (SEC) and Single Error Correction Double Error Detection (SECDED) codes. Interestingly, a code with the same memory redundancy as our approach ("SECDED(8,4)") leads to similar improvement in error rate. However, ECC decoding brings considerably more CMOS overhead than our approach: it needs logic circuitry to detect if an error occurred, and complex circuitry to detect the position of the error and correct it, requiring hundreds to thousands of logic gates [4]. This cost is unacceptable in iMC, as ECC decoders would need to be replicated for each memory array in the system, which can be hundreds. By contrast, our approach only uses a sense circuit that has no added complexity with regards to 1T1R solutions. Our work therefore extends the state of the art of RRAM iMC, where previous works do not propose a differential approach and are not compatible with technologies with errors [11,12]. In our approach, it is also possible to extend the sense amplifier to perform part of the logic, and thus to limit the CMOS overhead even further. For example, the circuit in Fig. 11 reads an RRAM cell, and at the same time performs a XNOR operation.

III. Use of The Differential Array Structure for Implementing Deep Neural Networks

Binarized Neural Networks (BNN) are ideally suited for exploiting our memory structure. They are conventional artificial neural networks, but weights and neuron activations are binary values instead of real numbers (Fig. 12). These systems require no multipliers, as they are replaced by XNOR logic operations, while additions are replaced by Popcount gates. BNNs can perform state of the art AI, with very reduced memory requirements [5,6]. This makes BNN ideal candidates for iMC. Fig. 13 shows how our memory structure can naturally implement iMC inference on BNN, associating a collection of kbit differential 2T2R memory arrays, all devices programmed in a binary fashion, with lightweight digital CMOS circuitry. XNOR operations can be performed directly in the PCSAs, or in separate logic gates. Popcount operations are based on 7-bits digital CMOS counters. Unlike most previous designs of neural networks with RRAM [13,14], this design is entirely digital, avoiding the need of high area operational amplifier or analog-to-digital converters. It does not require any multiplier, which allows extreme energy efficiency. We have designed the whole system based on synthesizable Verilog descriptions with Cadence IC design tools, and simulated it using the measured results on the memory arrays to model the memory blocks, and appropriate Value Change Dumps (VCD) inputs. This analysis was done with the design kit of a commercial 28 nm CMOS technology, to evaluate its potential on current technology. Fig. 14, which includes all CMOS overhead, highlights the amazing power efficiency of our design: it requires only nanoJoules to recognize one handwritten digit, while GPUs or CPUs-based AI requires micro to milliJoules.

IV. Robustness to Device Variability, Possibility to Use the Devices in High Endurance Regimes

We now investigate the impact of RRAM variability on an iMC BNN. We simulate our system for two tasks: handwritten character recognition task (MNIST) (Fig. 15ab), and a much more complicated photograph recognition task (CIFAR10), with a more complex deep neural network (Fig. 15cd). Without errors, our system can recognize 98.4% of the handwritten digits, and 87% of the photographs. Fig. 16 shows the impact of RRAM bit errors on the performance of the two tasks. Although errors change weight values between +1 and -1, up to $\sim 2 \times 10^{-3}$ bit error rate can be tolerated with negligible impact on the performance of the neural network in both tasks. This is in contrast with most digital computing tasks, where the errors are catastrophic. The low demands of BNN in terms of errors, as well as the reliability brought by the 2T2R memory structure means that we can actually use RRAM devices in weak programming regimes where they individually are prone to errors. Fig. 17 shows that devices programmed in such weak conditions (reset voltage of 1.5V), with a 2T2R array bit error rate of 2×10^{-3}, can show outstanding endurance of twenty billion cycles, that has very little impact on BNN performance.

V. Conclusion

In this work, we showed experimentally that the 2T2R differential memory is a simple way to decrease the effect of RRAM variability, allowing comparable gains than SECDED error correction with a similar memory overhead, but without the associated area, time and energy overhead. The differential memory is also an ideal building block for in-memory BNN. We also showed that the relaxed requirements of BNNs in terms of errors, as well as the reliability benefit of the differential memory allows using RRAM devices in a low voltage regime, which implies extended endurance up to billions of cycles. These results highlight that although in-memory computing cannot rely on ECC, if a differential memory architecture is chosen, this does not have to translate into stringent requirements on device variability.

Acknowledgment

This work is supported by ERC grant NANOINFER (715872).

References

[1] S. Yu, *Proc. IEEE*, vol. 106, n° 2, p. 260-285, 2018.
[2] Editorial, *Nature*, vol. 554, n° 7691, p. 145, 2018.
[3] D. Ielmini *et al.*, *Nat. Electron.*, vol. 1, n° 6, p. 333-343, 2018.
[4] S. Gregori *et al.*, *Proc. IEEE*, vol. 91, n° 4, p. 602-616, 2003.
[5] I. Hubara *et al.*, in *Proc. NIPS*, 2016, p. 4107–4115.
[6] M. Rastegari *et al.*, in *Proc. ECCV*, 2016, p. 525-542.
[7] A. Grossi *et al.*, in *IEDM Tech. Dig.*, 2016, p. 4.7.1-4.7.4.
[8] W. Zhao *et al.*, *IEEE TCAS I.*, vol. 61, n° 2, p. 443-454, 2014.
[9] Y.-H. Shih *et al.*, *Proc SSDM*, p. 137, 2017.
[10] W.-T. Hsieh *et al.*, *Proc SSDM*, p. 171, 2017.
[11] W. H. Chen *et al.*, in *IEDM Tech. Dig.*, 2017, p. 28.2.1-28.2.4.
[12] W. H. Chen *et al.*, in *Proc. ISSCC*, 2018, p. 494-496.
[13] S. Yu *et al.*, in *IEDM Tech. Dig.*, 2016, p. 16.2.1-16.2.4.
[14] S. Ambrogio *et al.*, *Nature*, vol. 558, n° 7708, p. 60-67, 2018.

978-1-7281-1988-5/18 $31.00 © 2018 IEEE

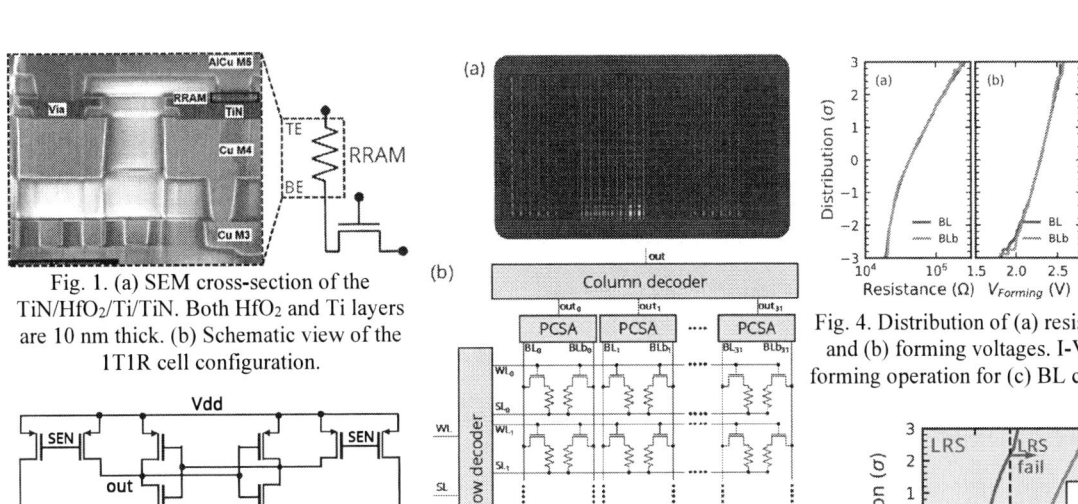

Fig. 1. (a) SEM cross-section of the TiN/HfO₂/Ti/TiN. Both HfO₂ and Ti layers are 10 nm thick. (b) Schematic view of the 1T1R cell configuration.

Fig. 2. Schematic of 2T2R precharge sense amplifier (PCSA).

Fig. 3. (a) Photography and (b) schematic of the 2T2R array.

Fig. 4. Distribution of (a) resistance after forming and (b) forming voltages. I-V characteristics of forming operation for (c) BL cell and (d) BLb cell.

Fig. 5. Example of error rate extraction in 1T1R mode base on LRS/HRS distribution. Inset: bit error rate extracted. $V_{appReset}$=2.5V, t_{Pulse}=1µs, Ic=55µA.

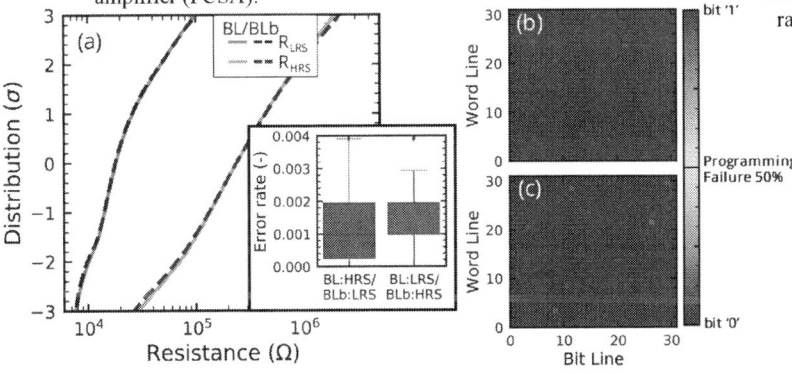

Fig. 6. (a) Distribution of resistance for bit '1' (BL:HRS/BLb:LRS) and bit '0' (BL:LRS/BLb:HRS) in the case of a checkerboard type of programming the memory array. (b-c) Failure rate on 100 programming according differential sense for two checkerboards configuration. $V_{appReset}$=2.5V, t_{Pulse}=1µs, Ic=55µA.

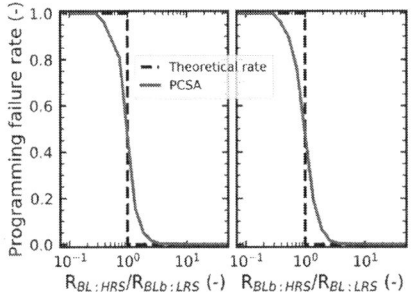

Fig. 7. Rate of programming failure indicated by the PCAS circuit as function of R_{HRS}/R_{LRS} ratio obtained by a high-resolution resistance measurement @ Vread = 0.1V.

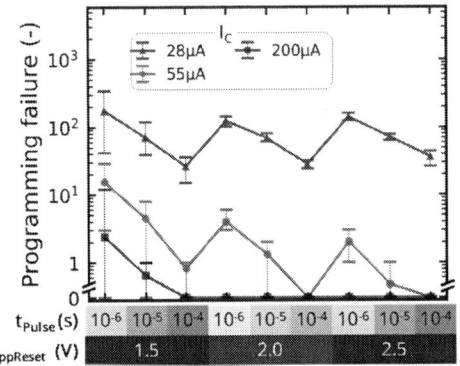

Fig. 8. Programming failure for different programming conditions for 2T2R configuration (PCSA) on a kbit 2T2R array.

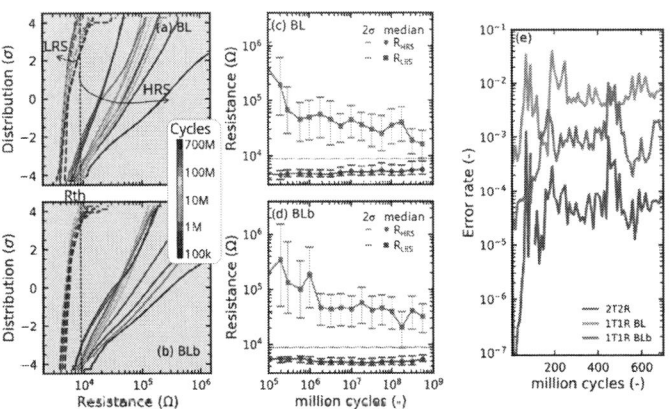

Fig. 9. a-b) The distribution of the resistance values, (c-d) the average value and (e) average error rate over 10 million cycles according to 2T2R configuration as function of number of cycles. $V_{appReset}$=2.5V, t_{Pulse}=1µs, Ic=200µA.

978-1-7281-1988-5/18 $31.00 © 2018 IEEE   486

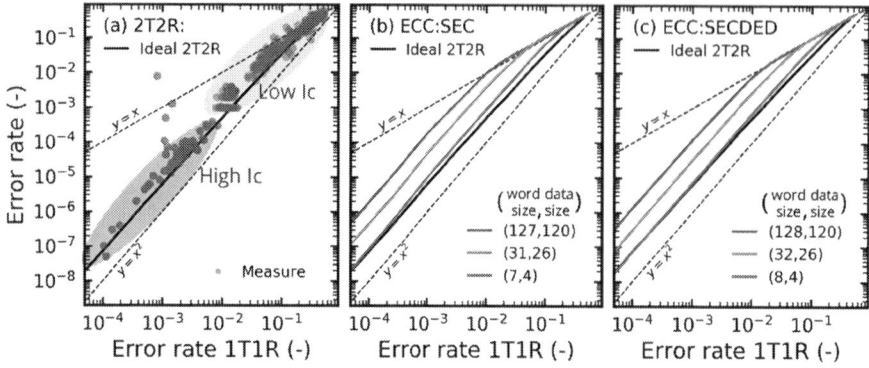

Fig. 10 (a) Experimental bit error rate of the 2T2R array as a function of the bit error rate on the individual (1T1R) RRAM devices. Bit error rate obtained with (b) SEC and (c) SECDED ECC as a function of the error rate on the individual devices.

Fig. 11. Adaptation of the PCSA circuit to perform a XNOR operation with the A input at the same time as READ operation.

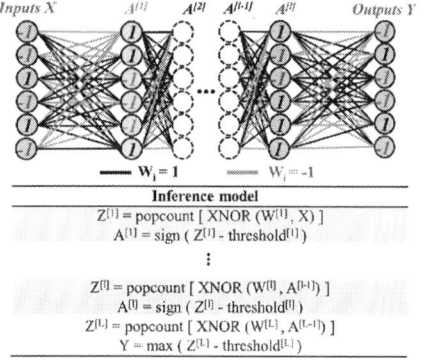

Inference model

$$Z^{[1]} = \text{popcount} [\text{XNOR} (W^{[1]}, X)]$$
$$A^{[1]} = \text{sign} (Z^{[1]} - \text{threshold}^{[1]})$$
$$\vdots$$
$$Z^{[l]} = \text{popcount} [\text{XNOR} (W^{[l]}, A^{[l-1]})]$$
$$A^{[l]} = \text{sign} (Z^{[l]} - \text{threshold}^{[l]})$$
$$Z^{[L]} = \text{popcount} [\text{XNOR} (W^{[L]}, A^{[L-1]})]$$
$$Y = \max (Z^{[L]} - \text{threshold}^{[L]})$$

Fig. 12. Basic principle of a BNN. Synaptic weights W and Neural Activations A are binary values.

RRAM In-memory BNN (this work)	25nJ
RRAM In-memory 8-bit fixed point	80nJ
Analog Phase Change Memory* [14]	~56nJ
GPU (Tesla V100)	~µJ
CPU (Xeon E5)	~mJ

Fig. 14. Comparison of the energy to recognize one handwritten digit, including all CMOS overhead. RRAM results are computed for a commercial 28 nm technology. *Taking into account inference-only, and scaled to the size of our neural network.

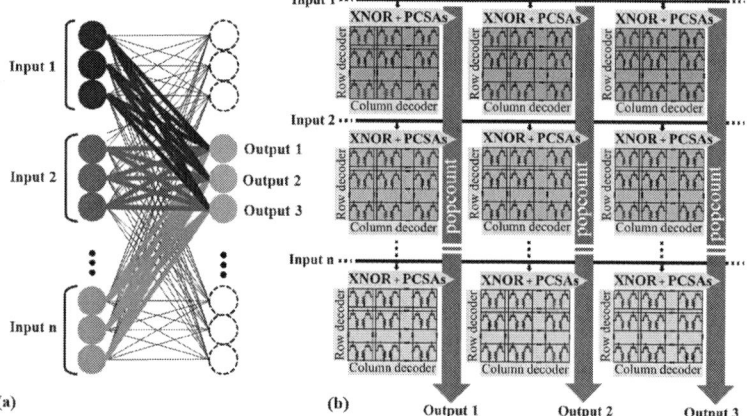

Fig. 13. Simplified architecture of an iMC BNN associating kbit 2T2R RRAM arrays with lightweight CMOS logic. The colors indicate the correspondence between formal neural network and hardware resources.

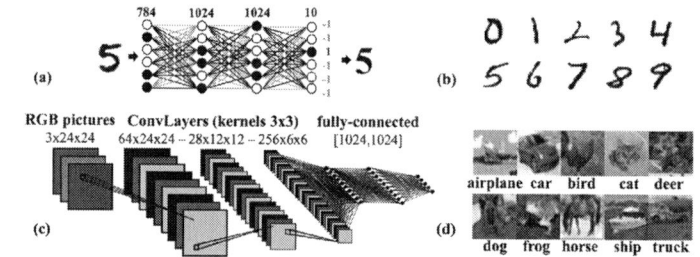

Fig. 15. Neural networks used for (a) digits recognition (MNIST) (c) photograph recognition (CIFAR10) – examples of digits (b) and photographs (d) to recognize.

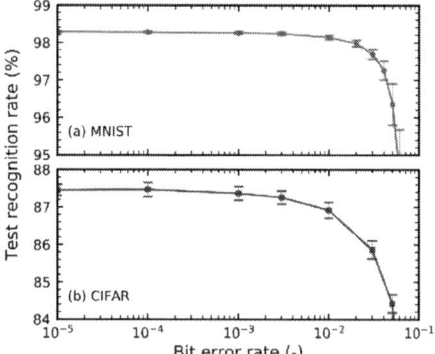

Fig 16. Dependence of the recognition rate of our BNN with the error rate of the memory arrays, for the MNIST and CIFAR10 tasks.

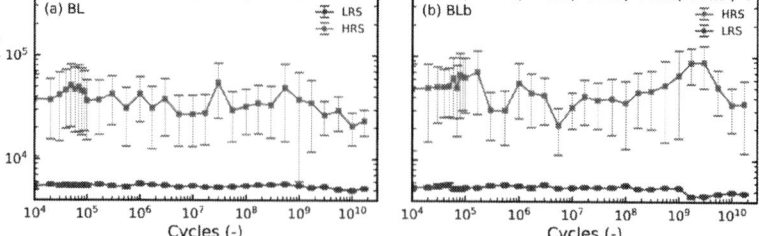

Fig. 17. Endurance measurement on two devices programmed at low voltage (V_{RESET}=1.5V), programming time of 1 µs and compliance current 200 µA. In this regime, the whole array 2T2R bit error rate is 2×10^{-3}.

978-1-7281-1988-5/18 $31.00 © 2018 IEEE

A new hardware implementation approach of BNNs based on nonlinear 2T2R synaptic cell

Z. Zhou[1], P. Huang[1*], Y. C. Xiang[1], W. S. Shen[1], Y. D. Zhao[1], Y. L. Feng[1], B. Gao[2], H. Q. Wu[2], H. Qian[2], L. F. Liu[1], X. Zhang[1], X. Y. Liu[1], and J. F. Kang[1#]

[1]Institute of Microelectronics, Peking University, Beijing 100871, China;
[2]Institute of Microelectronics, Tsinghua University, Beijing 100084, China;
E-mail: [*]phwang@pku.edu.cn; [#]kangjf@pku.edu.cn

Abstract —For the first time, we propose a new hardware implementation approach which can utilize the non-linear synaptic cells to build a Binarized-Neural-Networks (BNNs) for online training. A 2T2R-based synaptic cell is designed and demonstrated by the fabricated RRAM array to achieve the basic functions of synapse in BNNs: binary weight ($sign(W)$) reading and analog weight updating ($W+\Delta W$). The performance of BNNs based on 2T2R synaptic cells is evaluated by MNIST, and the recognition accuracy of 97.4% can be achieved. A novel refresh operation is proposed to enhance the network performance.

I. Introduction

In recent years, the capability of Artificial Intelligence (AI) is further enhanced by rapid developments in Deep Neural Networks (DNNs) [1]. Substantial research efforts are invested in accelerating DNNs by optimizing algorithm, device performance and hardware architecture [2]. The resistive random access memory (RRAM) has emerged as one of the most promising candidates of synapses due to its superior features, low programming consumption and excellent scaling ability [3-4]. It is charming to modulate the conductance (weight) of synaptic device directly by identical pulses without reading operation, which needs a relatively simple peripheral circuit. However, owing to the non-linearity of RRAM in potentiation and depression (conductance vs. pulse), this scheme has to meet a great challenge to reproduce weight accumulation ($W+\Delta W$), which is a critical function in algorithm. To solve this problem, previous researches mainly focus on the improvement of linearity of the synapse through optimizing device structure and operation scheme [5-7]. Fortunately, BNNs algorithm [8], one type of modified DNN, was proposed recently to reduce the memory access by binarizing the weight in calculation while maintaining high accuracy, which may open up new solutions to achieve the online training.

In this work, for the first time, we propose and demonstrate a new hardware implementation approach which offers the feasibility of the non-linear device applied as the synapse in BNNs for online training. A 2T2R-based synaptic cell is designed and demonstrated to exhibit the required synaptic behaviors, which enables both analog weight updating by identical pulses and binary ($sign(W)$) reading as BNNs required. For further acceleration, the XNOR logic ($sign(W)\cdot sign(Input)$) is also designed and demonstrated based on the 2T2R-synaptic cell. After that, we built a five-layer BNN to evaluate the requirements of the 2T2R-based synaptic cells and the impacts on the system performance of BNNs. Furthermore, a novel refresh operation is proposed to improve the operation robustness of the synaptic cells in long time training or device fast saturation scenarios.

II. Operation Principle

Fig. 1a shows the schematic of BNNs algorithm. The training process of BNNs consists of three stages: forward propagation, backward propagation and weight update. The binary form of weight ($sign (W)$) is used in the forward and backward propagation stages and the analog form of weight (W) is used in the updating stage. It is worth noted that, the analog or multi-level weight gradient (ΔW, calculated in backward propagation) is still needed in BNNs. Therefore, to implement BNNs, the synaptic cell should have the following two critical functions as shown in **Fig. 1b**: a) analog ΔW accumulation and b) binary reading.

To implement the functions, we propose a novel scheme with two non-linear devices (**Fig. 2**), in which analog ΔW accumulation can be realized by accumulating pulses in devices and binary reading can be realized by comparing the conductance of the devices. In the scheme (**Fig. 2a**), the weight (W) is represented by the conductance difference between the devices ($G_S=G_P-G_D$, where G_P&G_D is the conductance of device P&D). W gradient (ΔW) is quantified into pulses N_P ($N_P \propto \Delta W$, when $\Delta W>0$) and N_D ($N_D \propto |-\Delta W|$, when $\Delta W<0$), which is used to modulate G_P and G_D. The comparator is used to judge the sign of G_S in binary reading stage. The working principle is described as follows: W can be considered as $\Sigma (\Delta W)$ in training process, which is converted to $\Sigma (N_P)$ & $\Sigma (N_D)$ and stored in the devices in the form of conductance ($G_S=f(\Sigma (N_P)) - f(\Sigma (N_P))$), where $f(x)$ is the non-linear function of devices (conductance vs. pulses)). Due to the non-linearity, G_S cannot reflect the value of W. However, when the non-linearity function monotonous increasing, the sign of G_S is equal to the sign of W identically, which fulfills the requirement of BNNs. Two typical examples are shown in **Fig. 2c**, where $sign(G_S)$ keeps its sign when W back to its initial state during training process. Therefore, ΔW accumulation is realized when G_P&G_D increase with N_P&N_D, and binary reading is realized by detect $sign(G_S)$ in circuits.

To implement the scheme, a novel 2T2R synaptic cell and the corresponding operation rules are proposed in **Fig. 3**, and both binary reading and analog ΔW accumulation can be realized. The synaptic cell weight is denoted by G ($G=G_A-G_B$, G_i is the conductance of device i). For binary reading, both

978-1-7281-1988-5/18 $31.00 © 2018 IEEE

transistors maintain ON state with 0.1V/-0.1V reading voltage in the bit-lines (BL_A/BL_B). The sign of conductance is detected by the comparator, which could reuse along the source-line. For analog ΔW accumulation (LTP/LTD operation), the device A/B is selected by the word-line (WL) through controlling the gate voltage of transistors, and SET pulses are applied to the corresponding BL. Thus, G could be modulated by accumulating pulses in device A/B. Based on the synaptic structure, we fabricate the 1Kb 1T1R array (**Fig. 4**) in 130 nm technology node followed by package, which is measured through the customized testing board. Non-linearly analog characteristics can be measured in the device among the array, in which the conductance increases/decreases with the accumulation of identical pulses in SET/RESET process (**Fig. 5**). The analog behavior could be reproduced by our physics-based model in [9]. To demonstrate synaptic functions, a measured method is designed (**Fig. 6**), in which the device A and B are modulated alternately by pulses with increasing number. The amount of pulses follows arithmetic progression, and the difference is set as 5. The weight sign conversion (the green points) occurs with increasing pulses, as measured in the array (**Fig. 7**). It indicates that, by using the proposed synaptic cell, the non-linear RRAM realizes ΔW accumulation and binary reading, which enable it applied to BNNs for online training. Moreover, we tested various kinds of NVMs [10-13] (**Fig. 8**), and the results indicate that the synaptic cell is applicable to NVM devices with different non-linearity. Besides, in BNNs, both the activations and weights are binarized (+1/-1), which enables that the multiplication (floating point) between weights and their inputs in DNNs could be replaced by 1-bit XNOR logic. Therefore, we propose a read mode to further speed up the BNNs by realizing XNOR logic in the synaptic cell directly. In the read mode, the inputs (+1&-1) are delivered into the synaptic cell with two kinds of voltages as shown in **Fig. 9**. The results is detected by judging the current direction. The experimental results is consistent with the prediction.

III. DEVICE & SYSTEM CODESIGN

To investigate the impacts of the synaptic cell on system, we design a 5-layer fully connected network based on BNNs algorithm to handle MNIST learning and recognition tasks. Sub-algorithms, including weight initialization [14], batch normalization [15], and Adam algorithm [16] are considered in the network with key parameters shown in **Fig. 10**. To implement BNNs in hardware, the first step is to quantify ΔW, which is represented by pulse number. **Fig. 11** shows the impact of ΔW quantization on recognition accuracy. The network has high accuracy (97.4%, red-line, which is close to the ideal value, 97.6%, black-line), when quantization interval (Q_{int}) is 1×10^{-3} (weight is normalized to -1~1). As well as, the accuracy is still above 92% when $Q_{int}=1\times10^{-2}$. Then, the impacts of device performances including non-linearity, pulse to pulse variation and device to device variation are discussed as follow. Here we divide the non-linearity into 5 levels (#1-#5), and the relationship (conductance vs. pulses) is shown in the inserted figure of **Fig. 12**. The device non-linearity has little impact on the accuracy, mainly due to the fact that the non-linearity does not affect the judgment of *sign(W)*. The pulse to pulse variation is normalized to the domain of conductance change ($G_{max}-G_{min}$). As observed in **Fig. 13**, the accuracy decreases from 96% (10% variation) to 92.5% (50% variation). The device to device variation has little impact on the accuracy (**Fig. 14**). The reason is that limited number of synapse cells convert those sign frequently (**Fig. 15**), and the absolute conductance value of most synaptic cells increases with increasing epochs. Thus, the reduced precision of weights sign conversion caused by high device to device variation is negligible in the network. Furthermore, we checked the total pulse number applied to devices under different quantized intervals (**Fig. 16**). The total pulse number increases during the training, and the raising rate is directly related to Q_{int}. It is worth noted that the conductance of device will saturate with massive pulses, which is attributed to the limited conductance range of the device. If both devices in synaptic cell tend to saturation (**Fig. 17**), due to the fact that conductance change is lower than device variation, sign conversion error occurs more frequently, which damages the accuracy seriously. To address this issue, a refresh operation is introduced to refresh the device. The **Fig. 18** shows the device response to the refresh operation measured in array. The conductance of device decreases with multiple RESET pulses. To verify the function of refresh operation, an extreme case is introduced (**Fig. 19**), in which a small Q_{int} (3×10^{-3}) is applied and a device (saturation after 100 pulses, 10% variation) is used. The performance can be enhanced by the refresh operation. Accordingly, the refresh operation could solve the device saturation problem, such as long time training, small Q_{int}, and limited conductance range etc. For convenient visualization, the features of the synaptic cell is summarized and listed in **Fig. 20**.

IV. CONCLUSION

For the first time, we proved the feasibility of non-linear RRAM as synapses in BNNs for online training, and proposed a novel implementation approach of BNNs based on the 2T2R synaptic cell. For MNIST learning and recognition tasks, the network reached 97.4% accuracy. Refresh operation is proposed to improve network performances.

ACKNOWLEDGMENT

This work was supported by National Natural Science Foundation of China No.61421005, No.61604005 and No. 61334007.

REFERENCES

[1] W. J. Dally, et al., VLSI 2018, pp. 3-6.
[2] B. Fleischer, et al., VLSI 2018, pp. 35-36.
[3] W. H. Chen, et al., IEDM 2017, PP. 657-660.
[4] H. Lv, et al., IEDM 2017, pp. 36-39.
[5] P. Y. Chen, et al., IEDM 2017, 135-138.
[6] W. Wu, et al., VLSI 2018, 103-104.
[7] Y. Li, el al., VLSI 2018, 25-26.
[8] M. Courbariaux, et al., arXiv preprint, 2016, arXiv: 1602.02830.
[9] P. Huang, el al., TED 2017, 64 (2), 614-621.
[10] I-T. Wang, et al., IEDM 2014, pp. 665-668.
[11] S. Park, et al., IEDM 2013, pp. 625-628.
[12] S. H. Jo, et al., Nano Letters 2010, pp. 1297-1301.
[13] O. Bichler, et al., TED 2012, pp. 2206-2214.
[14] X. Glorot, et al., AISTATS 2010, pp. 249-256.
[15] S. Ioffe, et al., arXiv preprint, 2015, arXiv: 1502.03167.
[16] D. P. Kingma, arXiv preprint, 2014, arXiv: 1412.6980.

(a) BNNs Algorithm

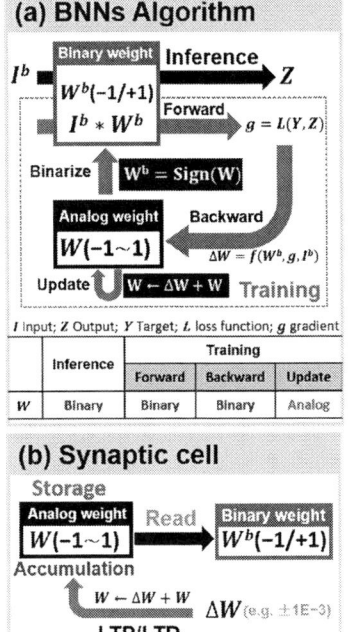

I Input; Z Output; Y Target; L loss function; g gradient

	Inference	Training		
		Forward	Backward	Update
W	Binary	Binary	Binary	Analog

(b) Synaptic cell

Fig. 1. (a) Schematic of BNNs algorithm. Analog weight is used in update stage binary weight is in the others stages. **(b)** Requirements of synapse: analog ΔW accumulation and binary reading.

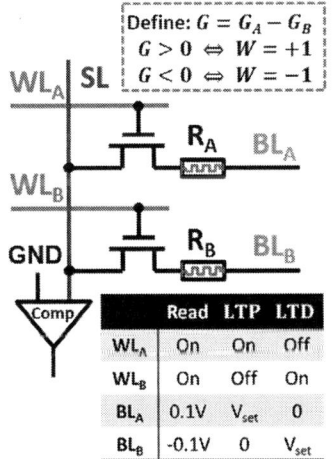

Fig. 3. Structure of 2T2R synaptic cell and corresponding operation rules. ΔW is accumulated and stored in R_A&R_B by LTP&LTD. Read realizes binarization.

Functional verification method

$$W_0 \xrightarrow{-\Delta W} W_1 \xrightarrow{2\Delta W} W_2 \xrightarrow{-3\Delta W} W_3 \cdots$$

Prediction:
$$Sign(W_{2k-1}) = -1;\ Sign(W_{2k}) = 1 \quad (k=1,2,\ldots n)$$

Fig. 6. The measured method to verify the functions of the synaptic cell. Apply increasing number of pulse to devices, and detect the sign(G_S).

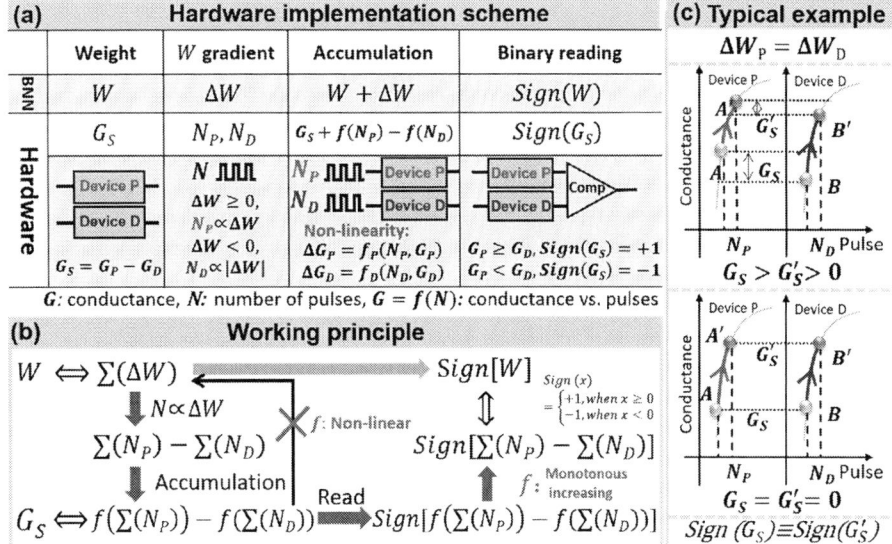

Fig. 2. The implementation scheme. **(a)** The relationship of parameters between the algorithm and the hardware scheme. ΔW is converted into two parameters N_P&N_D, owing that the pulses is non-negative. ΔW is accumulated by applying N_P&N_D to devices and is stored in the form of conductance (ΔG_P&ΔG_D). Binary reading is realized by comparing the conductance of the devices (G_P&G_D) **(b)** The working principle. If the $f(N)$ monotonous increasing with increased pulses, the sign of G_S equals to the sign of weight identically. Therefore, we could precisely reflect the $sign(W)$, the only parameter needed in BNNs algorithm, through detecting the $sign(G_S)$ in hardware. **(c)** Typical examples of ΔW accumulation process ($W+\Delta W_P-\Delta W_D$). When $\Delta W_P=\Delta W_D$, G_S maintains its sign.

Fig. 4. 1Kb (32×32) 1T1R RRAM (TiN/TaOx/HfO2/TiN) array is fabricated and used to demonstrate proposed hardware implementation of BNN algorithm

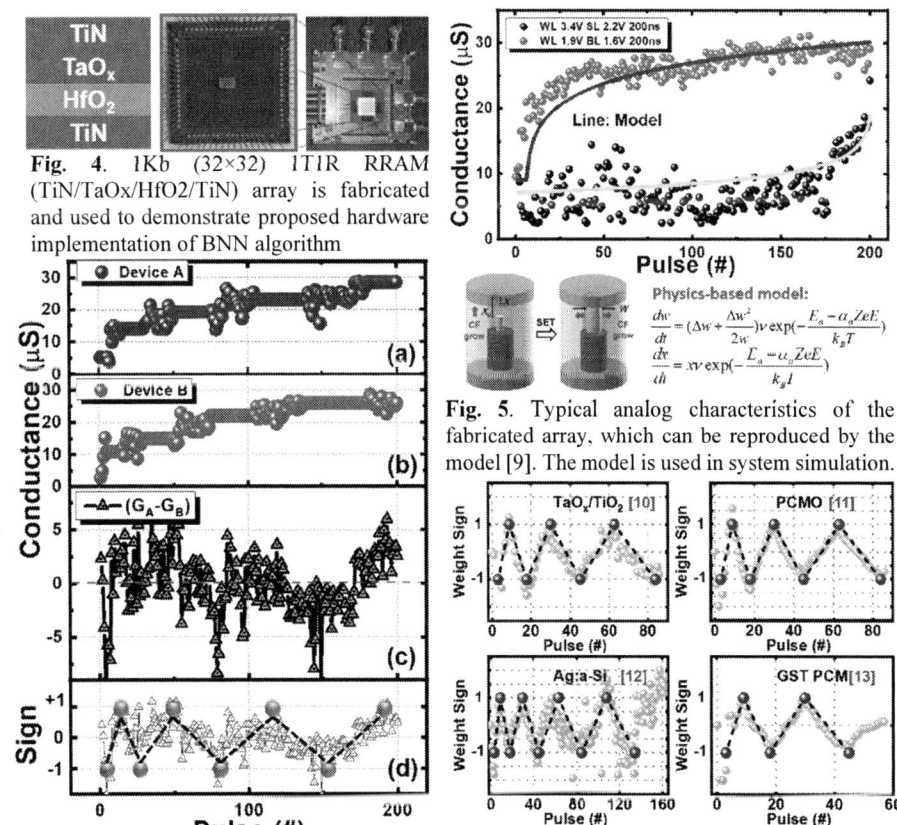

Fig. 5. Typical analog characteristics of the fabricated array, which can be reproduced by the model [9]. The model is used in system simulation.

Fig. 7. Experimentally demonstrate of the functions of the synaptic cell. The sign is converted by accumulating pulses in devices.

Fig. 8. NVMs with different non-linearity are tested based on our scheme. The results indicate that the synaptic cell is applicable to NVMs.

978-1-7281-1988-5/18 $31.00 © 2018 IEEE

Read	WL$_{A,B}$	BL$_A$	BL$_B$
Input = +1	on	+0.1V	−0.1V
Input = −1	on	−0.1V	+0.1V

XNOR logic	Weight=+1 $G_A>G_B$	Weight=−1 $G_A<G_B$
Input = +1	$I_{SL}>0$ (+1)	$I_{SL}<0$ (−1)
Input = −1	$I_{SL}<0$ (−1)	$I_{SL}>0$ (+1)
Testing result	$G_A>G_B$	$G_A<G_B$
Input = +1	1.22 µA (+1)	-1.55 µA (-1)
Input = −1	-1.35 µA (-1)	1.33 µA (+1)

Fig. 9. Read mode and measured results of the XNOR logic in synaptic cell.

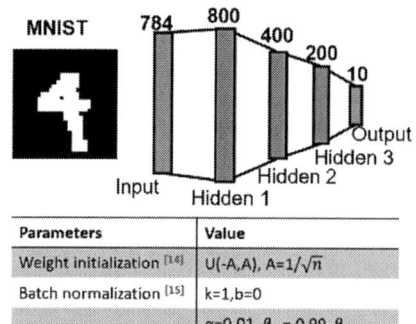

Parameters	Value
Weight initialization [14]	U(-A,A), A=1/$\sqrt{n}$
Batch normalization [15]	k=1,b=0
Adam algorithm [16]	α=0.01, β_1 = 0.99, β_2 =0.999 and ε =1e-8

Fig. 10. Structure of BNNs for simulation and key parameters used in the network.

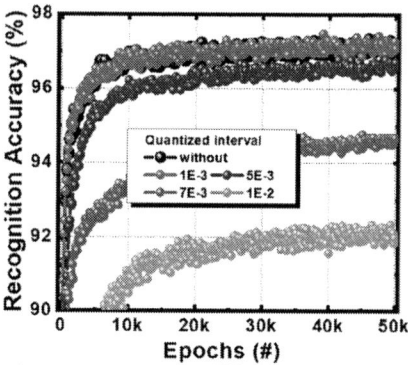

Fig. 11. Quantize the $\varDelta W$ into pulses. Higher accuracy achieved with smaller interval.

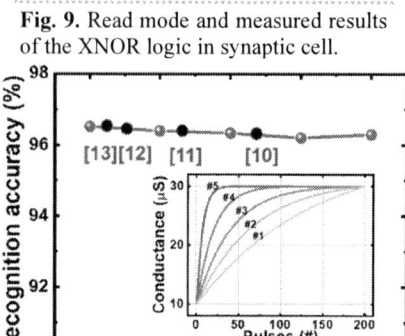

Fig. 12. The impact of the non-linearity on the recognition accuracy. Devices with different non-linearity levels have similar accuracy.

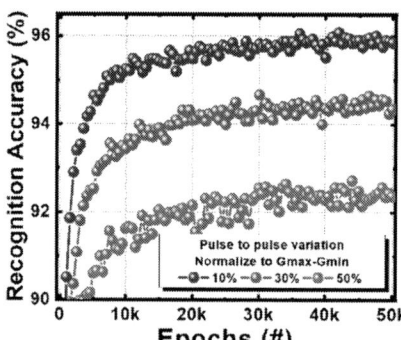

Fig. 13. The impact of pulse to pulse variation on the recognition accuracy. Accuracy decreased with increasing variation.

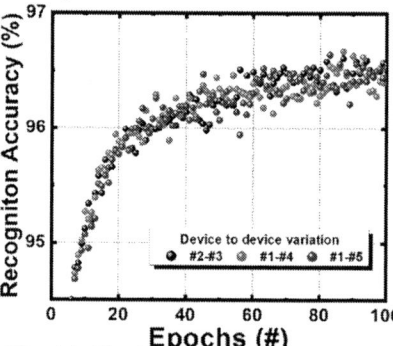

Fig. 14. The impact of device to device variation on the recognition accuracy. The network could tolerate the variation.

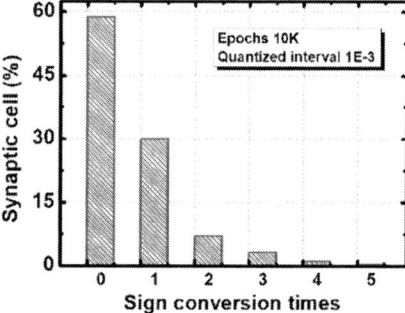

Fig. 15. Statistics of the number of sign conversion times of synaptic cells over the training process. limited number of synaptic cell convert its sign frequently.

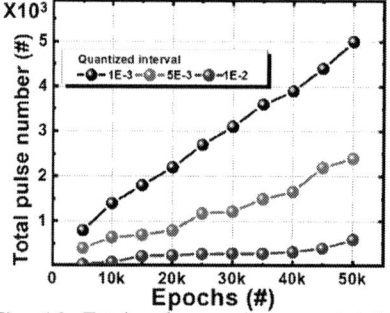

Fig. 16. Total pulse numbers vs. training epochs. The pulses applied on the devices increases as the epoch increase, which leads to the saturation of device conductance.

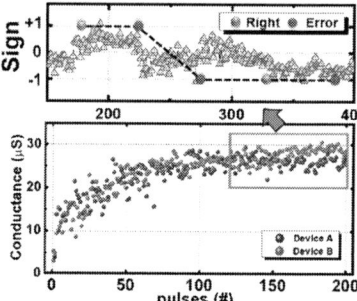

Fig. 17. Sign conversion error occurs in rear of analog behavior. The reason is that the conductance change lower than device variation.

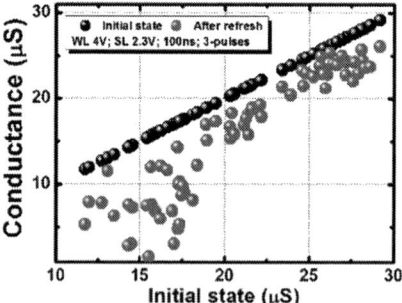

Fig. 18. Measured results of conductance state before and after refresh operation. The conductance of saturated device can be refreshed by RESET.

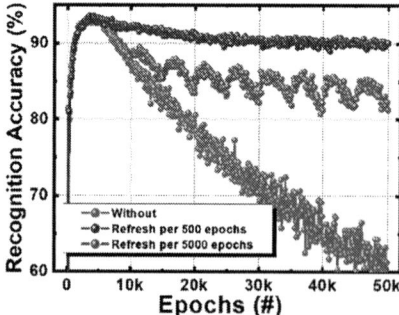

Fig. 19. Accuracy before and after refresh operation. The refresh operation could enhance the network performance in extreme cases: long time training process or device fast saturation.

Features	This work
Training function	Online training
Analog W storage	2T2R
Device linearity	Monotonic increasing
Supervised updating	Not required
Weights accumulation	In synaptic cell
Accuracy (MNIST)	97.4%
Pulse-pulse variation	Tolerate to > 50%
Generalization ability	Compatible with DNNs

Fig. 20. The summary of this work. The online training can be realized in non-linear 2T2R based synaptic cell, which do not need read state during update stage.

978-1-7281-1988-5/18 $31.00 © 2018 IEEE

Ge CMOS gate stack and contact development for Vertically Stacked Lateral Nanowire FETs

M.J.H. van Dal[1], G. Vellianitis[1], G. Doornbos[1], B. Duriez[1], M.C. Holland[1], T. Vasen[1], A. Afzalian[1],
E. Chen[2], S.K. Su[2], T.K. Chen[2], T.M. Shen[2], Z.Q. Wu[2], C.H. Diaz[2]

TSMC Corporate Research, [1]Kapeldreef 75, Leuven B-3001 Belgium, [2]TSMC, Hsinchu 308-44 Taiwan
email: mark_van_dal@tsmc.com

Abstract— We present (i) a novel, thermally stable Atomic Layer Deposition (ALD) high-k dielectric stack that, for the first time, has the potential to meet all gate stack requirements for both n- and p-channel Ge FETs, (ii) record low contact resistivity for n-Ge/metal contacts using an implant-free contact scheme with successful implementation into a single nanowire (NW) Ge nFET baseline, (iii) single NW Ge pFETs with short-channel effect (SCE) immunity down to 24 nm physical gate length, of which electrical data show excellent agreement with calibrated models and (iv) demonstration of Ge-channel vertically stacked lateral NW FETs using a 300 mm VLSI compatible platform.

I. INTRODUCTION

Ge is a promising material to replace the Si channel owing to (i) high electron and hole mobility, and (ii) high density-of-states avoiding source starvation, which limits performance of short channel III-V MOSFETs [1]. While Ge pFETs have shown great potential [2], building high-quality Ge nFETs has proven more challenging [3], mainly due to the nature of defects in the Ge lattice and at its interfaces. Point defects in Ge create acceptor states, countering n-type doping [4], Ge/metal contacts show Fermi-level pinning close to the valence band (E_v) resulting in high contact resistivity for n-Ge/metal contacts [5], and charge neutrality levels at Ge/insulator interfaces are close to E_v hampering nMOS inversion [6]. Therefore, in this work we have focused our efforts to alleviate these n-Ge related problems. In [2] we introduced a gate stack with low interface trap density (D_{it}) resulting in p-Ge FinFETs with record performance at 0.5 V supply voltage V_{dd}. In this work we used the same dielectric to validate our in-house multi subband Boltzmann equation solver on well-behaved single NW Ge pFETs and investigate on-performance of vertically stacked lateral Ge NW FETs.

II. DEVICE FABRICATION

Single and vertically stacked NW Ge pFETs with different NW diameter and cross-sectional shape were fabricated using the low D_{it} gate stack from [2] in a replacement-gate (RPG) scheme. 300 mm Ge or $Si_{0.25}Ge_{0.75}$ virtual substrates were used as starting material, on which multiple epitaxial $Si_{0.25}Ge_{0.75}$/Ge layers were grown. Thickness and number of the individual Ge and SiGe layers were tuned to obtain the targeted NW number, shape and spacing. SiGe/Ge multilayer fins were patterned by self-aligned double patterning and etched at a fin pitch of 45 nm. In some cases, pFETs have boron-implanted extensions, and for both n- and pFETs epitaxial, *in situ* doped source and drains were grown. After RPG, the sacrificial SiGe layers were selectively removed using the chemistry from [7] forming suspended Ge nanostructures after which gate dielectric and TiN metal were deposited. Finally W deposition and chemical mechanical polishing (CMP) were used to fill the gate openings and planarize. Back-end-of-line was processed up to the first metal level M1.

III. RESULTS

A. Demonstration of thermally stable, hysteresis-free and low D_{it} CMOS Ge gate stack

Gate stacks for future technology nodes need to simultaneously fulfill a number of requirements: low D_{it}, sub-1 nm equivalent oxide thickness (EOT), sub-2 nm physical thickness, low gate leakage, and good reliability. Here we introduce a new gate dielectric stack comprising a novel ALD interlayer (IL) and a novel ALD high-k dielectric having the potential of satisfying all requirements for both n-Ge and p-Ge FETs. Fig. 1 shows multiple frequency *C-V*s and hysteresis loops for both n-Ge and p-Ge MOS capacitors after optimizing the post metal anneal conditions including a high-resolution TEM micrograph of the gate stack. Low D_{it} and hysteresis are obtained for both n- and pMOS capacitors and the electrical properties are preserved up to 550°C (Fig. 2). Fig. 3 shows split *C-V* extracted mobility at V_{ds} of +/−50mV for planar Ge n- and pFETs having a L_g of 1 μm. Fig. 4 shows the electron and hole mobility vs. EOT benchmark. The novel dielectric shows competitive values at scaled EOT for both electron and hole mobility [2, 8-17]. For the first time, we demonstrate a thermally stable hysteresis-free ALD gate stack that passivates both n-Ge and p-Ge substrates leading to high e-mobility and h-mobility at scaled EOT (485 and 124 cm²/Vs at $5 \cdot 10^{12}$ cm⁻² sheet carrier density).

B. Record nGe/metal contact resistivity by implant-free process

To achieve low contact resistivity between n-Ge and metal contacts, several approaches have been explored: (i) increasing dopant activation at the Ge/metal interface to promote tunneling through the Schottky barrier [3], (ii) inserting a thin conductive dielectric (*e.g.* ZnO) between Ge and metal to reduce metal-induced gap states and unpin the Fermi level [18]. Here we have implemented a multi-layer *in situ* heavily doped epitaxy. Dopants are electrically activated in a nanosecond laser anneal process using a XeCl Excimer (308 nm) laser. After activation, a Ti/TiN bi-layer was used as contact metal. We applied the multi-ring circular transmission line method (CTLM) (Fig. 5a) from [20] to access ρ_c below 10^{-8} Ω·cm². Fig. 5b shows CTLM resistance as a function of CTLM spacing. After optimizing epitaxy and annealing conditions, we achieved a contact resistivity as low as $1.6 \cdot 10^{-9}$ Ω·cm² (Fig. 6). Our data represent the lowest ρ_c on n-Ge to date (Fig. 7) [3, 18, 19, 21-25]. We integrated the novel implant-free contact process in a Ge single-NW GAA baseline using the RPG gate stack optimized in previous work for Ge pFETs [2]. We omitted extension implants to reduce SCE. Fig. 8 shows transfer characteristics of a well-behaved 1NW Ge nFET having L_g of 70 nm. To the best of our knowledge, this is the first demonstration of a VLSI compatible Ge NW nFET without large S/D structures.

C. Single Ge NW gate-all-around (GAA) pFETs with 6 nm diameter for model calibration.

For a target technology with L_g shorter than 15 nm, NWFETs require a NW diameter below 10 nm to maintain good SCE control

978-1-7281-1988-5/18 $31.00 © 2018 IEEE

[26]. At these dimensions non-stationary effects and the impact of structural confinement become too significant for semi-classical TCAD to remain predictive. Development of more advanced models is required [27], which, especially for non-Si CMOS, is hampered by the lack of reliable data for model validation. To obtain the required clean electrical data on 6 nm diameter Ge NWFETs without parasitic channel we used SiGe virtual substrates to eliminate the parasitic bottom Ge fin [7], reduced the NW diameter to 6 nm by tuning the SiGe/Ge thickness and fin width, and omitted the extension implantation to reduce SCEs. For this experiment we integrated the gate stack from [2] for which the planar and FinFET device behavior is well established. Fig. 9 shows a cross-section TEM micrograph of a 1NW Ge pFET after full processing. As part of this experiment, we also investigate the impact of inserting 4 Si monolayers by Chemical Vapor Deposition (CVD) prior to high-k deposition, often used to passivate Ge for low D_{it} gate stacks for pFETs [7,14,15,28], on the device characteristics. Fig. 10a shows the short channel transfer curves and TEM cross-section of single NW Ge pFET with (Type A) or without (Type B) Si passivation IL after full processing. Comparing the TEM micrographs of device A and B, we observe no significant difference in NW shape. At V_{ds} = -0.5 V, device B $I_{d,off}$ drops comfortably below 0.1 nA/fin, while $I_{d,min}$ for device A is limited by gate leakage due to the high V_t and not by bulk or direct source/drain tunneling. Strikingly, except for a large V_t-shift of 580 mV, device performance is very similar. The V_t shift is in quantitative agreement with the prediction of an interfacial dipole at the Ge/Si interface by *ab initio* simulations [29], resulting in ~0.5 eV band discontinuity. Fig. 10b shows the near ideal output characteristics (Device A). Both device types are immune to SCEs as apparent from the (lack of) L_g-dependence of V_t and S (Fig. 11). It is important to note that these values are averaged over a large population of devices (~100 per gate length) measured over the entire 300 mm wafer (*e.g.* 28 nm L_g devices of type B have 100% functional yield) with a $\sigma(V_t)$ < 5 mV. We used the Device B data to validate our multi subband Boltzmann equation solver that uses modified $k.p$ parameters fitted to atomistic models. [26]. We demonstrate excellent agreement on IV characteristics without fitting any parameters (Fig. 12). These experimentally calibrated models were recently used to benchmark to Si equivalent NW devices and a 2x improvement in intrinsic on-performance of Ge over Si was found at L_g = 12 nm [27].

D. Vertically Stacked NW Ge pFETs

Analogously to fin height for a FinFET technology, vertically stacking wires is an important technology knob to increase the current density per footprint for a NW technology. Fig. 13 shows a TEM micrograph of a 6-fin, 3 vertically stacked NW pFET having a NW cross-section of about 14 nm² built on a Ge virtual substrate. At L_g of 34 nm, high transconductance g_m of 40 and 196 µS/fin and subthreshold swing S of 76 and 114 mV/dec at V_{ds}=−0.05 and −0.5V was measured (not shown), however the best I_{on}/I_{off} trade-off is obtained for longer (L_g = 70 nm) devices, with a record performance I_{on} of 40 µA/fin at I_{off} of 7 nA/fin and V_{dd} of 0.5 V (Fig. 14, including data from [2,7,28,30-32]). This is due to a significant loss of SCE control below 100 nm (Fig. 15). To elucidate this behavior, we deployed semi-classical 3D TCAD, with parameters calibrated to our Ge FinFET data [2]. Carefully replicating the TEM in the full 3D model (Fig. 16), TCAD reproduces transfer curves of the 70 nm device well. Using the TCAD to split the contributions by the individual NWs and by the underlying parasitic fin to the total current, we can conclude that (i) despite the n-doping, the parasitic fin significantly contributes to the on-current of the device (~25%) and

(ii) the loss in SCE control is solely caused by the poor gate control of the parasitic Ge fin and is not related to the NWs.

Altering epitaxial SiGe and Ge thickness prior to fin etching and, in particular, applying thinner layers results in high aspect-ratio NWs, so-called nano-sheets (NS). We fabricated tall vertically stacked nanostructure devices, with 5 stacked Ge NSs. As for the 1NW device, we also demonstrate the impact of Si IL gate passivation process on the 5-NS device characteristics. Fig. 17 shows transfer curves of 70 nm devices with or without Si passivation and their corresponding TEM cross-sections. Comparing the transfer curves and TEM images, two key findings are observed. Firstly, the suspended Ge structures reshape into nearly perfectly rounded Ge NWs with a diameter of 7–8 nm. This is due to an enhanced surface mobility of Ge atoms during Si CVD leading to reflow of the Ge surface [7]. Secondly, as observed for the 1NW pFET, except for the large V_t difference, performance of the two devices is very similar.

IV. CONCLUSIONS

In this work, we tackled key challenges for Ge nFET (gate stack and nGe/metal contacts), demonstrated single NW Ge pFET without SCEs down to L_g of 24 nm matching well with advanced physical transport model and addressed the impact of the NW number and shape on the performance of vertically stacked Ge NWFETs integrated on a 300 mm platform, paving the way for possible implementation of Ge CMOS in future technology nodes.

ACKNOWLEDGMENT

IMEC, LAM Research, ASM and Screen are acknowledged for processing capabilities and technical support. SiGe virtual substrates are provided by Siltronic AG, Germany. The authors would like to thank Dr. Y. C. Sun for management support.

REFERENCES

[1] A. Toriumi and T. Nishimura, *Jap. J. Appl. Phys.* 57, 010101 (2018).
[2] B. Duriez *et al.*, *IEDM Tech. Dig.*, 522 (2013).
[3] M.J.H. van Dal *et al.*, *IEEE Trans. Electron Devices* 62(11), 3567 (2015).
[4] J. Kim *et al.*, *Appl. Phys. Lett.* 101, 112107 (2012).
[5] T. Nishimura *et al.*, *Appl. Phys. Lett.* 91, 123123 (2007).
[6] A. Dimoulas *et al.*, *Appl. Phys. Lett.* 89, 252110 (2006).
[7] L. Witters *et al.*, *IEEE Trans. Electron Devices* 64(11), 4587 (2017).
[8] C.H. Lee *et al.*, *VLSI Symp. Tech. Dig.*, 28 (2013).
[9] R. Zhang *et al.*, *IEEE Trans. Electron Devices* 60(3), 927 (2013).
[10] R. Zhang *et al.*, *IEDM Tech. Dig.*, 642 (2011).
[11] X. Gong *et al.*, *IEDM Tech. Dig.*, 231 (2014).
[12] H. Arimura *et al.*, *IEDM Tech. Dig.*, 588 (2015).
[13] R. Xie *et al.*, *IEDM Tech. Dig.*, 393 (2008).
[14] R. Pillarisetty *et al.*, *IEDM Tech. Dig.*, 150 (2010).
[15] J. Mitard *et al.*, *VLSI Symp. Tech. Dig.*, 82 (2008).
[16] Y. Shin *et al.*, *VLSI Symp. Tech. Dig.*, 102 (2014).
[17] S.H. Yi *et al.*, *IEEE Electron Device Lett.* 38(5), 544 (2017).
[18] P. Manik *et al.*, *Appl. Phys. Lett.* 101, 182105 (2012).
[19] H. Yu *et al.*, *IEDM Tech. Dig.*, 604 (2016).
[20] H. Yu *et al.*, *IEEE Electron Device Lett.* 36(6), 600 (2015).
[21] A. Firrincieli *et al.*, *Appl. Phys. Lett.* 99, 242104, (2011).
[22] Z. Li *et al.*, *IEEE Electron Device Lett.* 34(9) 1097 (2013).
[23] M. Shayesteh *et al.*, *IEEE Trans. Electron Devices* 60(7), 2178 (2013).
[24] K. Gallacher *et al.*, *Appl. Phys. Lett.* 100(2), 022113 (2012).
[25] S. Gupta *et al.*, *J. App. Phys.* 113, 234505 (2013).
[26] S. Bangsaruntip *et al.*, *IEEE Electron Device Lett.* 31(9), 903 (2010).
[27] S. K. Su *et al.*, *accepted for 23rd SISPAD* (2018).
[28] J. Mitard *et al.*, *VLSI Symp. Tech. Dig.*, 138 (2014).
[29] G. Pourtois *et al.*, *Appl. Phys. Lett.* 91, 23506 (2007).
[30] E. Capogreco *et al.*, *VLSI Symp. Tech. Dig.* (2018).
[31] H. Mertens *et al.*, *VLSI Symp. Tech. Dig.*, 58 (2016).
[32] P. Hashemi *et al.*, *VLSI Symp. Tech. Dig.*, 120 (2017).

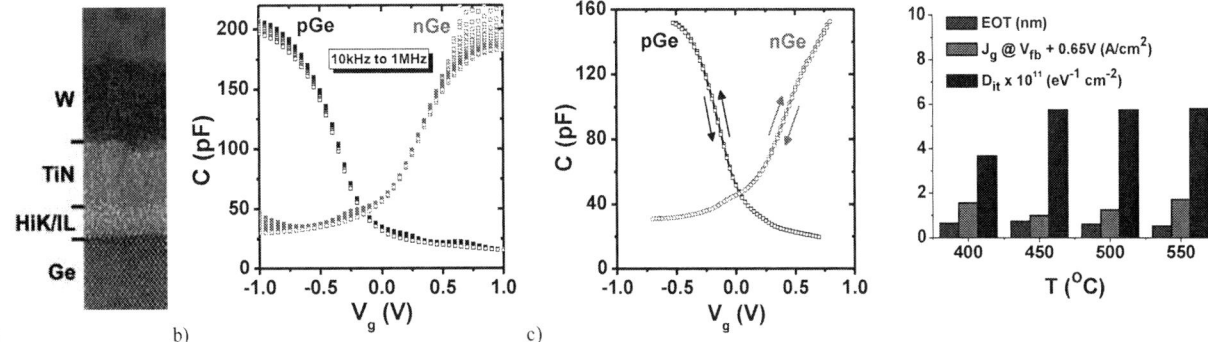

a) b) c)

Fig. 1. a) High Resolution TEM cross-section of novel gate dielectric stack; b) Multi-frequency (10kHz – 1MHz) C-V of Ge n- and pMOS capacitors. For both n and pMOS, D_{it}, determined by conductance method, are < 5×10^{-11} eV^{-1}cm^{-2}; c) C-V Hysteresis loop showing ΔV_t < 10 mV for both n- and pMOS capacitors.

Fig. 2. Temperature dependence of EOT, gate leakage J_g, and D_{it} (5 min anneals) nMOS capacitors. Similar results are found for pMOS capacitors.

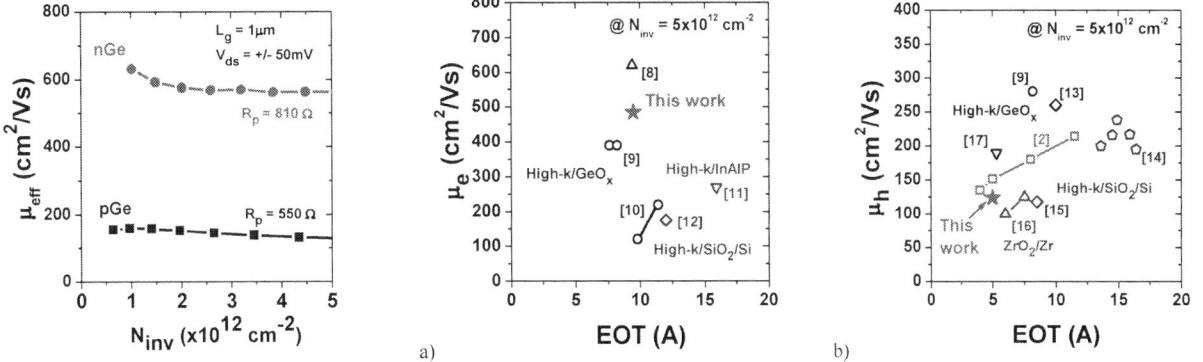

a) b)

Fig. 3. Effective mobility extracted by split-C-V method at V_{ds}= +/-50mV for L_g=1μm Ge planar nMOSFETs and pMOSFETs.

Fig. 4. a) Electron and b) hole mobility $vs.$ EOT benchmark on (100) Ge surfaces at inversion carrier density N_{inv}=5·10^{12} cm^{-2} including data from [2, 8-17]. Red stars correspond to novel common n/p Ge gate stack.

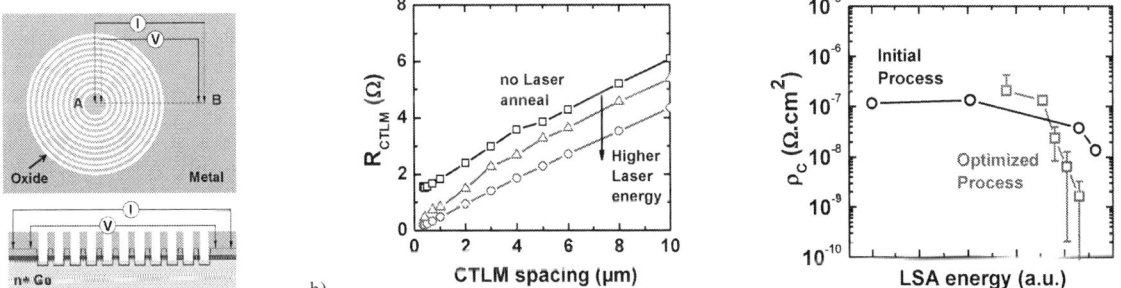

a) b)

Fig. 5. a) Schematic of multi-ring CTLM [20] and b) R $vs.$ CTLM spacing for multi-layer epi / metal contacts without activation anneal (black) and after different laser anneal conditions (red).

Fig. 6. Contact resistivity $vs.$ laser anneal energies for multi-layer epi / metal contacts.

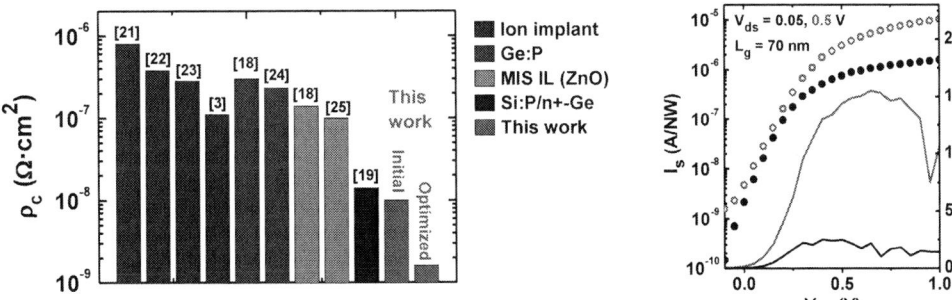

a) b)

Fig. 7. Contact resistivity benchmark including state-of-the-art nGe/Metal from [3, 18, 19, 21-25]. Several techniques have been investigated such as conventional ion implantation, in-situ doped Ge, Metal-Insulator-Semiconductor (MIS) contacts.

Fig. 8. a) I_s-V_g in A/NW and g_m in μS/NW at V_{ds} = 0.05 and 0.5V 1NW Ge nFET at L_g= 70nm with novel implant-free contact scheme; b) TEM cross-section of a single Ge 9nm NW nFET.

978-1-7281-1988-5/18 $31.00 © 2018 IEEE

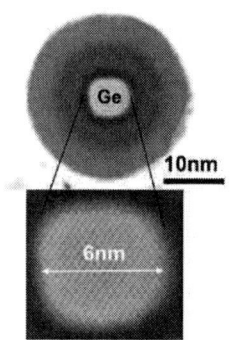

Fig. 9. High resolution cross-section TEM of 1NW Ge pFET after full processing.

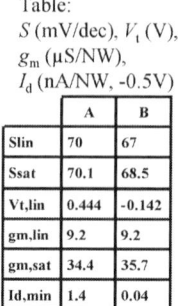

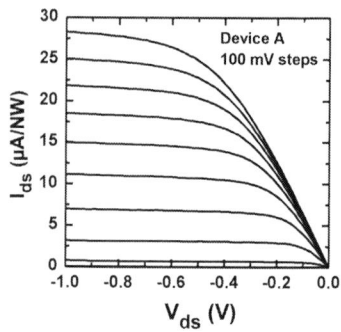

Table: S (mV/dec), V_t (V), g_m (µS/NW), I_d (nA/NW, -0.5V)

	A	B
Slin	70	67
Ssat	70.1	68.5
Vt,lin	0.444	-0.142
gm,lin	9.2	9.2
gm,sat	34.4	35.7
Id,min	1.4	0.04

Fig. 10. a) I-V_g in A/NW (symbols I_s, solid lines I_d) at V_{ds} = -0.05 and -0.5V at L_g = 28nm and TEM cross-section of single 6nm NW Ge pFET with (Device A) or without (Device B) Si interlayer passivation; b) I_d-V_d in µA/NW of L_g = 28nm (Device A).

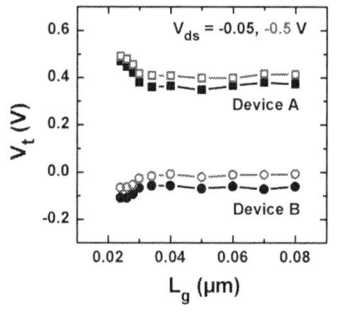

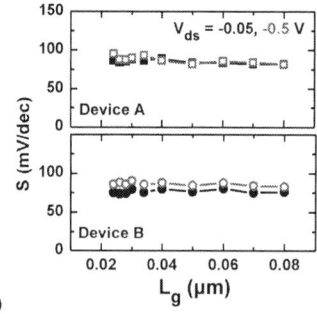

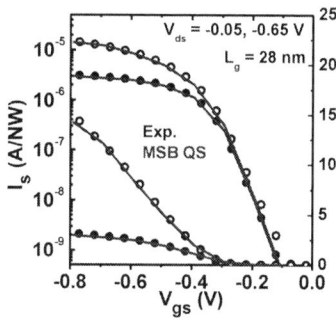

Fig. 11. a) Average V_t vs. L_g and b) average S vs. L_g at V_{ds} = -0.05 and -0.5V for 1NW Ge device with (device A) or without (Device B) Si cap passivation.

Fig. 12. I_s-V_g in A/NW at V_{ds} = -0.05 and -0.65V of 1NW pFET at L_g = 28nm (Device B) comparing experimental and simulated data (symbols) using our Multi Subband Quantum Solver (MSB QS) (line).

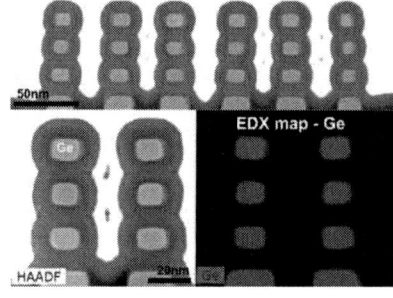

Fig. 13. TEM High Angle Annular Dark Field (HAADF) cross-section and Electron Diffraction Map of 6-fin 3NW Ge FET with ~14 nm² on Ge virtual substrates.

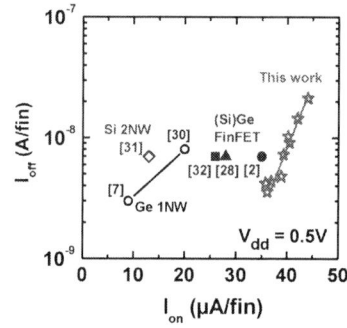

Fig. 14. I_{on}/I_{off} pFET benchmark at V_{dd} = 0.5V including best Ge HGAA 3NW pFETs and state-of-the-art (Si)Ge FinFET [2,28,32] and NWFET [7,30,31].

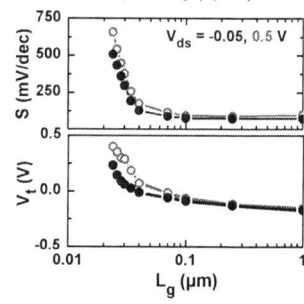

Fig. 15. Average S and V_t roll-off curves at V_{ds} = -0.05 and -0.5V of 6-fin 3NW HGAA pFET.

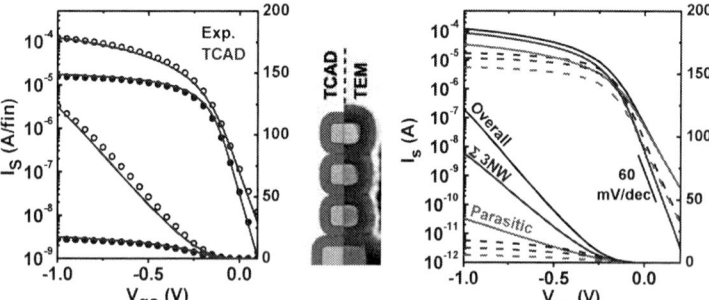

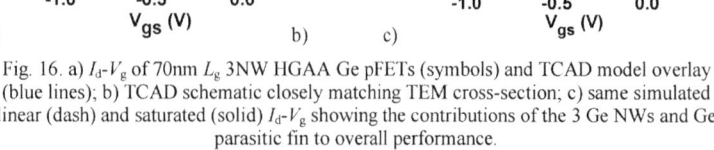

Fig. 16. a) I_d-V_g of 70nm L_g 3NW HGAA Ge pFETs (symbols) and TCAD model overlay (blue lines); b) TCAD schematic closely matching TEM cross-section; c) same simulated linear (dash) and saturated (solid) I_d-V_g showing the contributions of the 3 Ge NWs and Ge parasitic fin to overall performance.

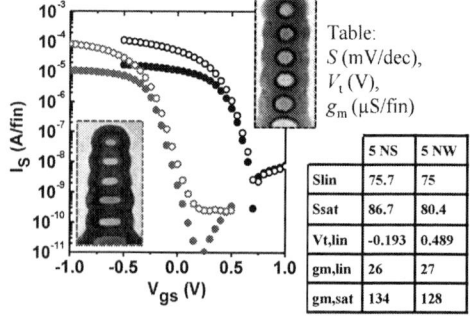

Table: S (mV/dec), V_t (V), g_m (µS/fin)

	5 NS	5 NW
Slin	75.7	75
Ssat	86.7	80.4
Vt,lin	-0.193	0.489
gm,lin	26	27
gm,sat	134	128

Fig. 17. Transfer curves and TEM of 5 NW device (L_g = 70nm) at V_{ds} = -0.05 and -0.5V with (black) or without Si IL (red) as part of the gate stack. Si process leads to rounding of the Ge NW and a large V_t shift.

978-1-7281-1988-5/18 $31.00 © 2018 IEEE

Advantage of NW structure in preservation of SRB-induced strain and investigation of off-state leakage in strained stacked Ge NW pFET

H. Arimura, G. Eneman, E. Capogreco, L. Witters, A. De Keersgieter, P. Favia, C. Porret, A. Hikavyy, R. Loo, H. Bender, L.-Å. Ragnarsson, J. Mitard, N. Collaert, D. Mocuta, N. Horiguchi

imec, Leuven, Belgium, email: Hiroaki.Arimura@imec.be

Abstract—Nanowires (NW) and nanosheets (NS) are promising channel structure for future technology nodes as they can offer better electrostatics than FinFETs. In this paper, we show another advantage of strained Ge NW pFET over strained Ge FinFET, which lies in the preservation of Strain-Relaxed-Buffer (SRB)-induced strain through fin cut and S/D recess. This benefit comes from the presence of the sacrificial SiGe layers. Lowering the Ge concentration in the SiGe sacrificial layer is a way to further suppress the strain loss. Furthermore, a comparison of Ge NW pFETs integrated on Ge SRB and SiGe SRB reveals that SiGe SRB provides a huge advantage not only in the strain engineering but also in I_{OFF} control. These are key enablers in maximizing the performance while minimizing the I_{OFF} of strained Ge NW pFETs.

INTRODUCTION

High-mobility Ge channel is a valid option as performance booster for future technology nodes. It has been shown that to outperform strained Si(Ge) pFETs, strain engineering for Ge pFET is also indispensable, with the use of SiGe SRB as one of the most efficient stressor [1]. Next to that, the demand for continuous Lg scaling requires changing the device architecture from FinFET to NW or NS for better electrostatic control. Recently we have demonstrated strained Ge NW pFETs integrated on SiGe SRB showing a comparable performance to the best reported (Si)Ge pFinFETs [2, 3]. For SiGe p-channel FinFETs, it has been shown that substrate-induced strain is partially lost after fin cut especially near the fin edge, causing the layout-dependent effect [4]. Minimizing the strain loss over the integration flow guarantees high level strain in the final devices. We show that the Ge NW structure can maintain higher strain after the fin cut and S/D recess as compared to Ge FinFETs, thanks to the presence of the sacrificial SiGe layers.

As a disadvantage, Ge devices suffer from high Gate-Induced Drain Leakage (GIDL) current due to the small bandgap. The GIDL is attributed to band-to-band tunneling (BTBT) and trap-assisted tunneling (TAT). To tackle this issue, a further understanding of the source of the leakage is required. Unlike a Ge NW integration using Germanium-On-Insulator (GOI) substrates, a NW integration starting from SRB wafers leaves a parasitic device on the substrate under the NWs [5], which needs to be kept inactive during the NW device operation. In this paper, we compare Ge NW pFETs integrated on Ge SRB and SiGe SRB with respect to I_{OFF}. While using the Ge SRB is an option in view of Ge CMOS integration, an advantage of SiGe SRB over Ge SRB is shown to lie in suppressing the I_{OFF}. Finally, a further I_{OFF} reduction by focusing on the TAT component will be discussed.

DEVICE INTEGRATION

The key fabrication steps of the stacked Ge NW pFET on 300 mm Si (100) wafers are summarized in Fig. 1 [2, 5]. After well and groundplane (GP) doping in the Ge or $Si_{0.3}Ge_{0.7}$ SRB, an epitaxial SiGe/Ge/SiGe/Ge/SiGe superlattice structure was grown, followed by fin patterning. The Ge content of the sacrificial SiGe is 70%, unless mentioned otherwise. Embedded Ge S/D was grown after spacer deposition and S/D recess. In the Replacement Metal Gate (RMG) module, after dummy oxide removal, the sacrificial SiGe is then etched selectively to the Ge NW, followed by our standard gate stack process using Si-cap [6]. Fig. 2 shows the median SS,sat of single [2] and double Ge NW pFETs integrated on a SiGe SRB. Thanks to the smaller diameter of the stacked wires (~9.5 nm) as compared to the single NW (~13 nm), the electrostatic control is improved for sub-40 nm Lg. STEM confirmed that similar Ge NWs have been formed on Ge and SiGe SRBs, except that some bottom wires on Ge SRB were not fully released (Fig. 3). The wire perimeter was used for the data normalization.

BENEFIT OF NW STRUCTURE IN STRAIN PRESERVATION

Fig. 4 shows the median Gm,lin of 85-nm-Lg stacked Ge NW pFETs on Ge and SiGe SRB with the same embedded Ge S/D. The peak Gm,lin is significantly higher for the SiGe SRB, which can be explained by the increased hole mobility thanks to the compressive strain from SiGe SRB. When the peak Gm is compared among 75-nm-Lg devices with different device layout, a strong layout dependence is found only for the Ge NW pFETs integrated on a SiGe SRB (Fig. 5(a)). The major difference among the two groups of device layout is the presence of dummy gates at 220 nm pitch for D and E as illustrated in Fig. 5(d), which is known to enhance the efficiency of the S/D stressor [7]. A positive Vt shift (Fig. 5(b)) observed for the layout D and E are qualitatively in agreement with the expected band structure change by the increased compressive strain [8, 9]. Our previous study has shown that while the strain in Ge NW is partially lost after S/D recess, it is reintroduced by the growth of the embedded Ge S/D [2]. This new result helps our understanding about the strain evolution, i.e. an appropriate S/D stressor is also necessary together with the SiGe SRB to maximize the strain in the final Ge NW devices. Minimizing the strain loss over the integration flow is another way to guarantee a high strain level, which requires a better understanding of the strain relaxation in a Ge NW structure. As seen on strained SiGe FinFETs [4], compressively strained Ge grown on $Si_{0.3}Ge_{0.7}$ SRB also shows elastic strain relaxation towards fin edge after fin cut according to the Nano-Beam Diffraction (NBD) measurement (Fig. 6). Lattice mismatch of +1.3% relative to the $Si_{0.3}Ge_{0.7}$ SRB suggests full relaxation of the Ge lattice. To investigate the impact of the superlattice structure for Ge NW integration, Scntaurus-Process [10] TCAD simulations were performed after fin cut and S/D recess. Fig. 7 shows the simulated longitudinal stress after fin cut of the superlattice fin, and it is compared to a reference fin with the same total fin height. While the stress is significantly reduced towards the top edge of the reference fin, Ge wires show significantly higher stress after fin cut. In case of the Ge NW, the strain in the Ge lattice is maintained by the SiGe sacrificial layer. at the interface. For various fin lengths (220-660 nm), it is confirmed that the stress relaxation at the fin center is reduced to about half in the Ge NWs as compared to the reference fins (Fig. 8). In addition, in case of the Ge fin, a significant stress difference is seen between the top and bottom part of the fin, however, Ge NW shows reduced difference. Fig. 9 shows the longitudinal stress after the S/D recess, which causes an additional Lg-dependent stress relaxation. For the assumed 40-nm-Lg device in the center of the 220-nm long fin, the remaining stress for the reference fin is only ~1/3 of the initial stress, while the Ge NW can maintain ~2x higher stress after the S/D recess (Fig. 10). The reduced layout sensitivity and better strain preservation are additional advantages of the NW over FinFETs next to better electrostatics. Note that a similar benefit can be expected for a strained NS as well as a strained SiGe NW/NS on Si substrate with Si or low-Ge-content SiGe sacrificial layers.

Furthermore, the merit of having a SiGe sacrificial layer can be further strengthened by lowering the Ge concentration. The $Si_{0.5}Ge_{0.5}$ sacrificial layer is shown by simulation to be more effective to preserve the longitudinal stress after S/D recess as compared to the $Si_{0.3}Ge_{0.7}$ one (Fig. 11). This is visualized in Fig. 12, by comparing the deformation of the structure, where the sacrificial SiGe layer helps to keep the initial structure. The advantage of using a $Si_{0.5}Ge_{0.5}$ sacrificial layer is experimentally confirmed by NBD on post-S/D-recess NW structures (Lg~40 nm) (Fig. 13). In this comparison, $Si_{0.5}Ge_{0.5}$ sacrificial layers were used for the single Ge NW, whereas $Si_{0.3}Ge_{0.7}$ sacrificial layers were used for the double Ge NW. Despite the shorter fin length (440 nm vs 660 nm), higher strain is maintained with $Si_{0.5}Ge_{0.5}$ sacrificial layers. An additional benefit of the $Si_{0.5}Ge_{0.5}$ sacrificial layer over the $Si_{0.3}Ge_{0.7}$ one is the easier removal due to the larger difference in Si content of the SiGe and Ge NW.

I_{OFF} CONTROL BY THE CHOICE OF SRB

Fig. 14 compares I_D(/fin)-V_G characteristics of stacked Ge NW pFETs on Ge SRB and SiGe SRB. To see the response of the parasitic channel in the substrate, a positive back bias (V_B) was applied. At a Lg of 85 nm and V_D of -50 mV (Fig. 14(f)), the I_D-V_G in accumulation behaves similarly, except the current value is higher for Ge SRB. With a Ge SRB, the Vg-dependent off current, which is the feature of GIDL, is clearly modulated by V_B as seen on planar devices [11], while the subthreshold and inversion region is insensitive to V_B, indicating that the majority of carrier transport is in the NW isolated from the substrate. Thus, the GIDL through the parasitic channel in the Ge SRB is seen as the I_{OFF} source at low V_D. With SiGe SRB, the I_{OFF} is lower and almost insensitive to V_B, suggesting that the leakage current of the parasitic channel in the SiGe SRB is sufficiently low, thanks to the wider bandgap of the SiGe reducing GIDL. When Lg decreases, I_{OFF} increases due to the increased contribution from the parasitic channel in the Ge SRB, showing V_B-dependent I_{OFF} with poor subthreshold slope. This can be explained by the inferior electrostatics in the parasitic channel as compared to the Ge NW, resulting in a Vt roll-off with degraded SS at short Lg. In contrast, a constant I_{OFF} can be seen for Ge NW on SiGe SRB.

At high V_D, all characteristics become less sensitive to V_B, since the GIDL in Ge NWs becomes more dominant. While the I_{OFF} is higher and increasing with Lg scaling for the devices on the Ge SRB, the I_{OFF} on SiGe SRB is ~0.3 nA/fin, thus meeting the high performance (HP) I_{OFF} target (~3 nA/fin), and remains constant down to 35-nm Lg. To decouple the contribution of the Ge NWs and the parasitic channel in the SRB, each component was simulated by replacing the Ge NW or SRB with SiO_2 (Fig. 15). Note that the strain effect is not taken into account in this case. The current density at off state (Vg=0 V) shows that a significant off current flows in the Ge SRB at Lg of 25 nm. The simulated I_D-V_G characteristics show that the Vt of the parasitic channel in SiGe SRB is sufficiently separated from that of the Ge NW (Fig. 16). However, the I_D-V_G of the parasitic channel in Ge SRB appears close to the Ge NW with poor SS, resulting in the increase in I_{OFF}, similar to what is seen in Fig. 14. The large difference between the parasitic channel in the Ge and SiGe SRB in terms of Vg (Fig. 16) is mainly coming from ΔE_V (~0.2 eV) and ΔEg between the Ge and SiGe, a higher phosphorous activation in SiGe SRB as compared to Ge SRB, and worse electrostatics of the Ge mainly due to the smaller band gap [12]. The advantages of SiGe SRB over Ge are summarized in Fig. 14 (right). This TCAD together with the electrical characterization suggests that (1) the parasitic channel in the Ge SRB is the cause of the higher I_{OFF}, and (2) for the Ge NW on SiGe SRB, defects near the Ge NW / drain interface causing TAT GIDL need to be eliminated for further I_{OFF} reduction.

Here, we discuss two possible ways to passivate such defects. One is to perform an efficient passivation anneal, such as high-pressure anneal (HPA) as reported previously [6]. The I_{OFF} and its temperature dependence are reduced after a HPA (Fig. 17), indicating the defect passivation. Another way is to minimize process-induced damages in/near the S/D. Fig. 18 compares the I_D-V_G of three types of embedded B-doped Ge S/D, i.e. (a) using GeH_4 precursor ([B] = 3×10^{19} cm^{-3} on blanket), (b) with additional B implantation and anneal, and (c) using a low temperature Ge_2H_6 in cyclic deposition-etch (CDE) mode ([B] = 2×10^{20} cm^{-3} on blanket), without B implantation [13]. While the additional implantation increases the I_{ON}, the I_{OFF} is also increased by 1 order of magnitude, indicating that defects are generated by ion implantation damage [14]. However, when the highly in-situ B-doped Ge S/D is used without relying on an additional ion implantation, I_{OFF} is again reduced to the original level without compromising I_{ON}. An extension-less Ge NW pFET device is expected to further reduce I_{OFF}.

CONCLUSION

We have found that a higher level of SRB-induced strain in Ge NWs can be maintained after fin cut and S/D recess as compared to a Ge FinFET thanks to the SiGe sacrificial layers keeping the lattice of Ge NWs together. Furthermore, Ge NW integration on SiGe SRB is not only advantageous in strain engineering but also for off-state leakage control at short Lg, as compared to using Ge SRB. Optimization of the Ge content in the SiGe SRB and the sacrificial layers, and the use of implantation-free in-situ-doped S/D are expected to further improve the on-state performance and off-state leakage of strained Ge NW pFETs.

978-1-7281-1988-5/18 $31.00 © 2018 IEEE

REFERENCES

[1] G. Eneman et al., in Proc. IEDM 2012, p.131. [2] E. Capogreco et al., in Proc. VLSI 2018, p.193. [3] J. Mitard et al., in Proc. VLSI 2018, p.83. [4] G. Tsutsui et al., in Proc. VLSI 2017, p.122. [5] L. Witters et al., TED **64** (11) p.4587 (2017). [6] H. Arimura et al., in Proc. VLSI 2017, p.196. [7] M. Garcia Bardon et al., in Proc. VLSI 2013, p.114. [8] Y. S. Choi et al., J. Appl. Phys. **103**, 064510 (2008). [9] G. Eneman et al., in Proc. VLSI 2013, p.92. [10] TCAD Sentaurus Suite, Version M-2016.12 (Synopsys). [11] Y.-K. Lin et al., TED **64** (10) p.3986 (2017). [12] G. Eneman et al., in Proc. IEDM 2013, p.320. [13] C. Porret et al., ECS Trans. (2018) accepted. [14] V. A. Tiwari et al., TED **61** (5) p.1270 (2014).

ACKNOWLEDGEMENT: The authors thank the imec core CMOS program members, the European commission, local authorities and the imec pilot line for their support. Acknowledgement also goes to E. Chiu from HPSP for the HPA demo support, ASM for the Intrepid®, and Siltronic for providing the SiGe SRB wafers.

- Ge SRB or $Si_{0.3}Ge_{0.7}$ SRB
- Well + GroundPlane implantations
- **SiGe / Ge / SiGe / Ge / SiGe epi**
- SADP fin patterning
- Low temperature STI and recess
- Dummy gate patterning
- Extension implantations
- Spacer + **S/D recess**
- **Embedded S/D epitaxy**
- ILD0 and dummy gate removal
- Dummy oxide removal
- Sacrificial layer etch
- Si cap + oxidation + HK + WF metal
- Metal gate fill and CMP
- LI1 + LI2 + V0 + BEOL
- H_2 sintering

Fig. 1 Integration flow of Ge NW pFETs on a 300 mm Si wafer.

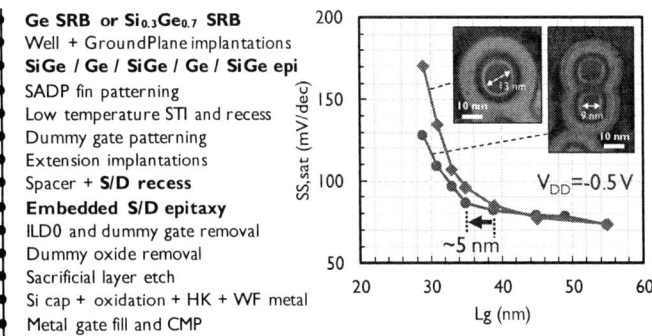

Fig. 2 Median SS,sat vs Lg of sGe single NW [2] and double NW pFETs. Electrostatic control is improved for the double NW thanks to the smaller wires.

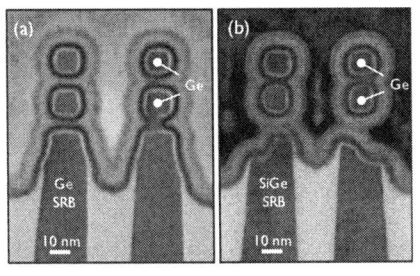

Fig. 3 TEM images of double Ge NW pFET on (a) Ge SRB and (b) $Si_{0.3}Ge_{0.7}$ SRB at the end of the process. Data was normalized by wire perimeter.

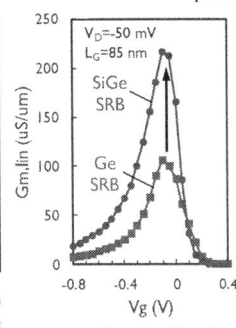

Fig. 4 Median Gm,lin of Ge NW pFETs on Ge SRB and SiGe SRB (Lg=85 nm).

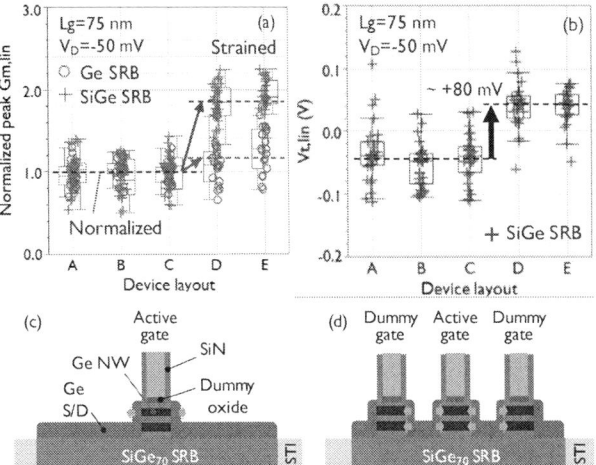

Fig. 5 (a) Normalized peak Gm,lin and (b) Vt,lin of 75-nm-Lg double Ge NW pFETs. Gm was normalized by the mean of layout A, B, C. ~60% Gm increase is seen for the Ge NW on SiGe SRB, indicating the increased mobility by higher strain in layout D, E. The layout A, B, C have single gate on a long fin (c), while the layout D, E have multiple gates (three for D, five for E) at 220 nm pitch (d). Slight Gm increase for layout D and E of Ge SRB case could be coming from lower R_{EXT}. Positive Vt shift for D and E also indicates the increased compressive strain.

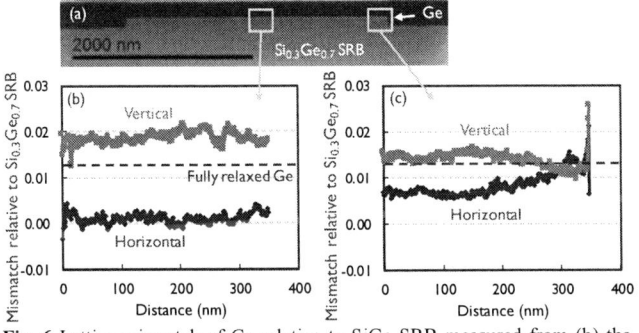

Fig. 6 Lattice mismatch of Ge relative to SiGe SRB measured from (b) the center and (c) the edge of strained Ge fin by using NBD.

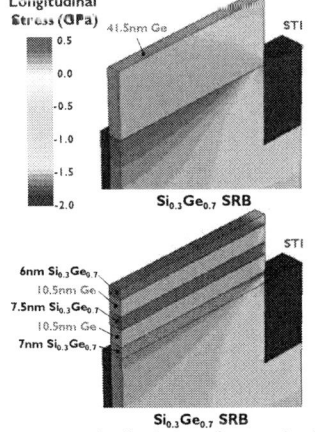

Fig. 7 Longitudinal stress for a strained Ge fin and superlattice Ge/SiGe structure after fin patterning. Fin width = 17 nm. Fin length = 220 nm.

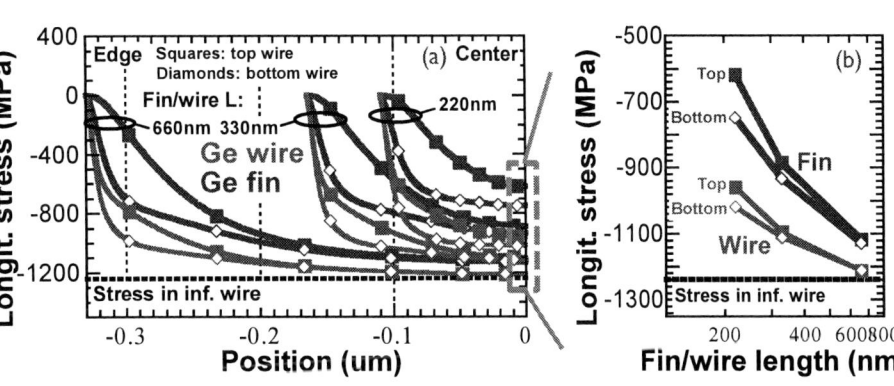

Fig. 8 (a) Longitudinal stress in strained Ge fin and strained double Ge NW after fin cut for 660 nm, 330 nm and 220 nm fins, corresponding to Fig. 7. (b) Longitudinal stress at the center of top and bottom fin/wire as a function of fin length.

978-1-7281-1988-5/18 $31.00 © 2018 IEEE

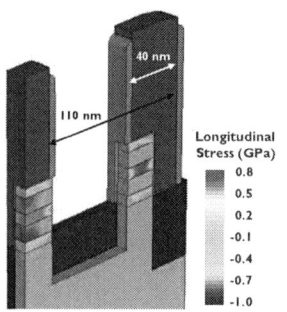

Fig. 9 Longitudinal stress contour for strained Ge NW after S/D recess. Fin width = 17 nm. Fin length = 220 nm.

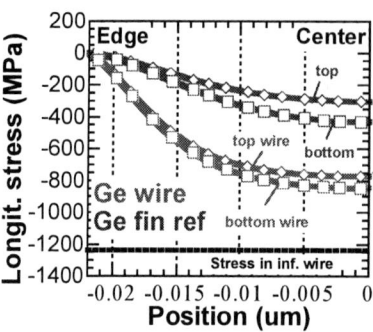

Fig. 10 Longitudinal stress for strained Ge fin and Ge NW structure after S/D recess. Fin width = 17 nm, fin length = 220 nm, and Lg = 40 nm.

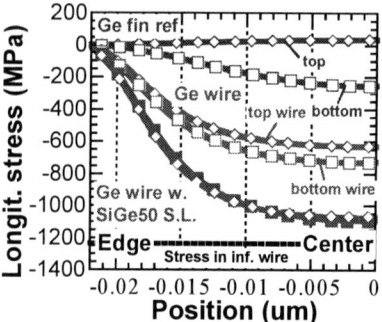

Fig. 11 Longitudinal stress in the post S/D recess sGe fin, sGe NW with 50% and 70% SiGe sacrificial layers on 50-nm-long fins. Lg=40 nm.

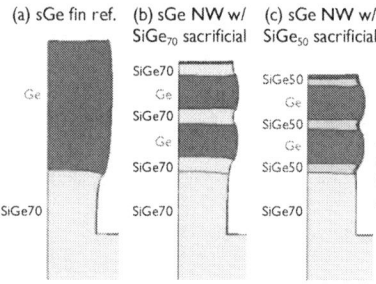

Fig. 12 Simulated deformation (enhanced 30×) of sGe fin and sGe NW with different Ge concentration in SiGe sacrificial layers after S/D recess.

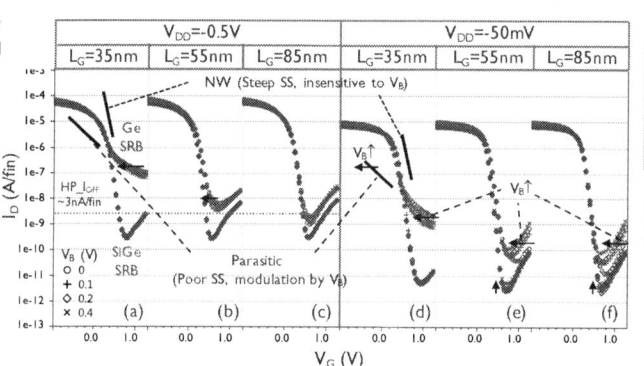

Fig. 14 I_D(/fin)-Vg of 35-85-nm-Lg Ge NW pFETs integrated on Ge SRB and SiGe SRB, measured with various V_D and V_B (back bias, applied to the substrate as shown in the schematic). The absence of I_{OFF} modulation by V_B indicates that there is no clear I_{OFF} contribution from the parasitic channel in SiGe SRB, whereas significant contribution from the parasitic channel in the Ge SRB is present especially at the short Lg (35 nm). Benefits of using SiGe SRB over Ge SRB are listed (right).

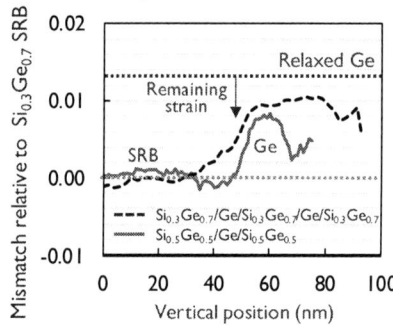

Fig. 13 NBD measurements of longitudinal lattice mismatch relative to $Si_{0.3}Ge_{0.7}$ SRB of single Ge NW with 50% SiGe sacrificial layer (Lfin=440 nm, Lg=40 nm) and double Ge NW with 70% SiGe sacrificial layer (Lfin=660 nm, Lg=44 nm) after S/D recess.

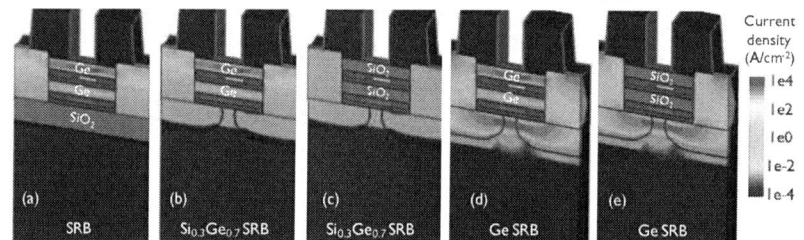

Fig. 15 S-process TCAD simulation showing off-state (Vg=0 V) current density of device structures, (a) Ge NW only, Ge NW + parasitic channel in (b) SiGe SRB and (d) Ge SRB, and parasitic channel in (c) SiGe SRB and (e) Ge SRB only. Lg=25 nm and width and height of wires are 10 nm. Same doping and activation level in Ge and SiGe SRB is assumed.

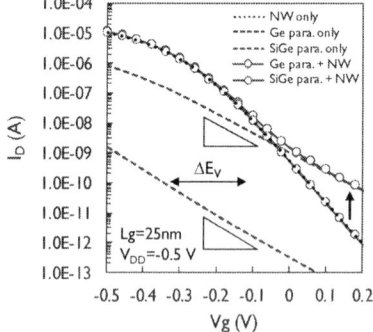

Fig. 16 Simulated I_D-V_G for devices corresponding to devices in Figs. 15. I_{OFF} is increased due to the parasitic channel in Ge SRB, while it can be suppressed by using SiGe SRB.

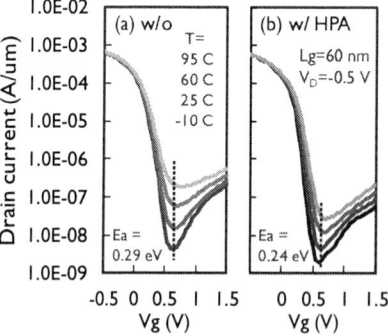

Fig. 17 Temperature dependence of I_D-V_G characteristics of strained Ge single NW pFETs [6] (a) without and (b) with H_2 HPA at 450°C for 20 min.

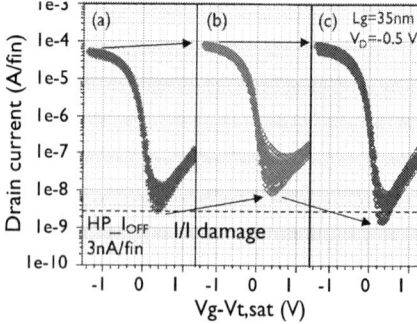

Fig. 18 I_D-V_G characteristics of strained Ge double NW pFETs with three different in-situ B-doped Ge S/D processes (a) using GeH4, (b) using GeH4 with additional implantation and anneal, (c) using Ge2H6 in cyclic deposition-etch mode.

978-1-7281-1988-5/18 $31.00 © 2018 IEEE   499

Tunability of Parasitic Channel in Gate-All-Around Stacked Nanosheets

S. Barraud, B. Previtali, V. Lapras, C. Vizioz, J.-M. Hartmann, S. Martinie, J. Lacord, M. Cassé, L. Dourthe, V. Loup, G. Romano*, N. Rambal, Z. Chalupa, N. Bernier, G. Audoit, A. Jannaud**, V. Delaye, V. Balan, O. Rozeau, T. Ernst, M. Vinet

CEA-LETI, Minatec campus, and Univ. Grenoble Alpes, 38054 Grenoble, France. (*) STMicroelectronics, 38926 Crolles, France. (**) SERMA Technologie, Minatec campus, 38054 Grenoble, France. E-mail: sylvain.barraud@cea.fr

Abstract— For the first time, a comprehensive study going from the integration of 3D stacked nanosheets Gate-All-Around (GAA) MOSFET devices to SPICE modeling is proposed. Devices have been successfully fabricated on SOI substrates using a replacement high-κ metal gate process and self-aligned-contacts. Back-biasing is herein efficiently used to highlight a drastic improvement of electrostatics in the upper GAA Si channels. Advanced electrical characterization of these devices enabled us to calibrate a new version of physical compact model (LETI-NSP) in order to assess the performance of ring oscillators for different configurations of GAA FETs integrating up to 8 vertically stacked Si channels.

I. INTRODUCTION

After being proposed and developed a little more than ten years ago [1-5], stacked NanoSheets (NS) Gate-All-Around (GAA) MOSFETs are becoming an industrial reality [6]. Hence, after FinFET, GAA multi-channels are a new class of advanced CMOS delivering higher performances, consuming less and pushing further scaling limits. Research conducted over the last three years has allowed to make major advances [7-14] to now place this novel technology in the roadmap of leading chipmakers [6].

In this paper, GAA stacked-NS field-effect-transistors are fabricated with a replacement metal gate (RMG) process and self-aligned contacts (SAC). Variable sheet width (W) up to 50 nm offers a superior capability in driving a load capacitance. The impact of size (W, gate length) and substrate orientation ([110] and [110]) on electrostatics and the overall performance of stacked wires *n*-FETs is studied. The ability to back-bias the Triple-Gate (TG) lower Si channel is used to promote electron transport in the GAA upper Si channel. Finally, SPICE simulations based on LETI-NSP compact model calibrated on the experimental results are used to assess the performance of ring oscillators of several geometries of stacked NS.

II. DEVICE FABRICATION

The transistors are processed on 300 mm SOI substrates. The process flow used for the fabrication of Si stacked-wires *n*-FETs is shown in **Fig. 1**. An epitaxial growth of ($Si_{0.7}Ge_{0.3}/Si$) multilayers is performed in order to have two stacked Si channels. The thickness of the Si channel and of the $Si_{0.7}Ge_{0.3}$ sacrificial layers is 9 nm. The patterning of ($Si_{0.7}Ge_{0.3}/Si$) multilayers used to define fins is based on a sidewall image transfer (SIT) technique. As a result, a fin pitch of 35 nm is achieved. After fin patterning, a SiO_2/Poly-Si dummy gate is defined with nitride spacers prior to the anisotropic etching of the ($Si_{0.7}Ge_{0.3}/Si$) multilayers in the sources/drains regions. Only a

few nm of Si are kept before the heavily in-situ phosphorus-doped Si selective epitaxial growth [15]. The Si nanosheets are released up to 50 nm width during the RMG fabrication module thanks to selective etching of $Si_{0.7}Ge_{0.3}$. This selective etch is followed by the conformal deposition of a 2 nm thick HfO_2 dielectric, and a metal gate made of TiN (2.2 nm) and Tungsten (W). Finally, self-aligned-contacts made of $Ni_{0.9}Pt_{0.1}$ (5 nm RF-PVD), Ti (5 nm RF-PVD) and a W liner (5 nm) are deposited. A standard Copper (Cu) back-end-of-line process is used up to the M1 metal level. The width and gate length mentioned in this work are calibrated on SEM and Transmission Electron Microscopy (TEM) imaging. Cross-sectional TEM images of stacked nanosheets along the source-to-drain direction are shown in **Fig. 2**. The excellent crystallinity of the upper GAA Si channel is evidenced. The conformal double-layered metal gate (TiN and W) has been characterized by Energy Dispersive X-ray Spectroscopy (EDS). High-Angle Annular Dark-Field scanning TEM of the same device and deformation maps acquired by Precession Electron Diffraction (PED) [16] in the (ε_{xx}), (ε_{zz}) and (ε_{xz}) directions are shown in **Fig. 3**. The high spatial resolution of PED highlights a highly tensile strained-Si area located between the W contacts and the TiN/W metal gate. This results in a slight tensile stress in the upper GAA Si channel which can be used to boost *n*-FET performance.

III. DEVICE PERFORMANCE WITHOUT BACK-BIASING

Split-CV measurements were carried out for the extraction of equivalent oxide thickness (EOT). Different channel widths were considered. The measurements obtained from a multi-fingers gate and an array (#120) of stacked-NS, show a consistent linear dependence of gate-to channel-capacitance on NS width. An EOT of 0.9 nm is extracted from **Fig. 4** measurements. Basic electrostatics properties of stacked-NS devices are discussed for different sheet widths and lengths. The Drain-Induced-Barrier-Lowering (DIBL) *vs* L_G for stacked-NS *n*-FETs with W = 10 nm is shown in **Fig. 5**. A DIBL of 80 mV/V is achieved at L_G = 20 nm. As the width increases, the DIBL becomes stronger and tends towards the TG limit (W→∞) of the lower Si channel. As expected, the control of Short-Channel-Effects (SCE) is more difficult when the channel length is small and the sheet width large, as shown in **Fig. 6**. The linear subthreshold slope (SS_{LIN}) plotted in **Fig. 7** as a function of L_G for W = 10 nm shows once more that immunity to SCE is decreased below L_G = 50 nm. This results in a SS_{LIN} of 80 mV/dec for L_G = 25 nm and a width dependence in line with DIBL measurements. The maximum of transconductance ($G_M=\partial I_{DS}/\partial V_{GS}$ with V_{DS} = 50 mV) versus L_G

978-1-7281-1988-5/18 $31.00 © 2018 IEEE

is shown in **Fig. 8**. In the linear regime, $G_{M,MAX}$ is inversely proportional to L_G and shows little dependence on W, in line with the low-field electron mobility behavior shown in **Fig. 9**. In the saturation regime (V_{GT}-V_{TH}=0.65V and V_{DS} = 0.9 V), the drain current $I_{SAT,VT}$ (normalized by the wire perimeter) is significantly enhanced for narrow W (**Fig. 10**). This can be explained by a lower on-state resistance (R_{ON}), as shown in the inset of **Fig. 9**. In addition, it should be noted that the $I_{SAT,VT}$ / $I_{LIN,VT}$ ratio is perfectly conform to the theoretical law $(V_{GT} \times V_{DS} - V_{DS}^2/2)/(V_{GT}^2)$ (**Fig. 11**). The impact of substrate orientation shown in **Fig. 12a** highlights a divergence of $I_{SAT,VT}$ for short L_G between [100] and [110] *n*-FET nanosheets structures, which is explained by a change of R_{ON} resistance (**Fig. 12b**).

IV. DEVICE PERFORMANCE WITH BACK-BIASING

In this section, back-biasing is used to tune the threshold voltage of the lower TG Si channel and focus on the electrostatics properties and transport in the upper GAA Si channel (**Fig. 13**). The effect of substrate biasing on $I_{DS}(V_{GS})$ is shown in **Fig. 14** for W = 10 and 50 nm. Due to the thick Buried-OXide (145nm), a reverse body bias (V_{BB}) up to -30 V is applied. In agreement with wide planar devices (Fully-Depleted-SOI), the best body-factor ($\Delta V_T/\Delta V_{BACK}$) is obtained in devices with a large sheet width. The $V_{T,SAT}$ change with V_{BB} is more efficient for high coupling, as shown in **Fig. 15**. The threshold voltage of W = 50 nm devices becomes very close to that of W = 10 nm devices for high V_{BB}. This is not the case for V_{BB} = 0 V. This enables us to differentiate electrostatic properties of lower and upper Si channels. In **Fig. 16**, the minimum of the saturation subthreshold slope (SS_{SAT}), extracted as $\ln(10) \times dV_{GS}/d\ln(I_{DS})$ highlights the impact of parasitic channel (TG) at V_{BB} = 0 V. However, under high back-biasing, SS_{SAT} is significantly reduced and converges toward that of the upper GAA channel, reaching a SS_{SAT} value of 66 mV/dec only for W = 50 nm. In **Fig. 17**, the DIBL *vs* L_G of *n*-FET stacked-NS also evidences a strong reduction of SCE for W = 50nm at V_{BB} = -30 V. Effectiveness of reverse back-biasing channel modulation results in a drastic improvement of electrostatics properties in GAA structures, as demonstrated in **Fig. 18**. This allows to consider an additional lever arm to offer more power/performance flexibility in 3D stacked channels.

V. COMPACT MODELING AND SPICE SIMULATION

In this last part, a revised version of LETI-NSP compact model [17] was implemented in order to consider the lower TG and the upper GAA Si channels. This model is able to handle nanosheets of arbitrary shapes in lateral or vertical GAA MOSFETs. It yields an accurate description of quantum confinement effects, carrier mobility and parasitic effects [17]. The efficiency and robustness of this new implementation was assessed by comparing gate capacitance (C_{GG}) and drain current computations with experimental results for large ranges of applied biases and device geometries. The agreement between the model and C_{GG} measurements for NS width ranging from 10 nm to 30 nm (**Fig. 19**) is excellent. Other comparisons of $I_{DS}(V_{GS})$ and $I_{DS}(V_{DS})$ curves, transconductance (G_M) and conductance ($G_D=\partial I_{DS}/\partial V_{DS}$) are provided in **Figs. 20** to **22** for W = 10 nm and L_G = 50 nm. Once again, this model agrees well with measurements for longer L_G (210 nm), as shown in **Fig. 23**. It highlights the accuracy of our physical compact model to predict GAA stacked-NS MOSFET operation. High Performance devices (with lower V_{TH}) were then simulated using the LETI-NSP model. SPICE simulations were performed to assess the performance of ring-oscillators (RO) for GAA structures S1 and S2 described schematically in **Fig. 24**. The layout footprint (LF), the stack thickness as well as the fin-to-fin space (S={LF-N×W_{Fin}}/{N-1}) were assumed constant for both devices. Gate-to-source/drain parasitic capacitances are extracted at V_{GS} = 0 V. Values are given in **Fig. 24**. The C_{gd0} *vs* V_{GS} curves obtained through small signal simulations were obtained by 3D TCAD simulation [18] for each type of structure. Simulations of RO (**Fig. 25**) yield higher performances for the S1 structure, including a full GAA configuration. As shown in **Fig. 26**, the replacement of the lower TG Si channel by a GAA Si wire leads to a delay reduction (×0.44 at V_{DD}=0.9V) thanks to a higher effective current. The Energy-Delay-Product (EDP) *vs* V_{DD} shown in **Fig. 27** is also significantly reduced for the S1 structure due to lower delay and dynamic leakage current. Finally, we show in **Figs. 28 to Fig. 30** that a higher number of stacked Si channels does not yield a better EDP. This is due to a saturation of delay per stage and an increase of dynamic and static leakage current as the number of stacked Si channels is increased.

VI. CONCLUSION

In the near future, 3D stacked nanosheets GAA MOSFET devices will likely succeed to FinFET technology to offer higher performance with a greater design flexibility. In this work, such devices are first successfully fabricated using a replacement high-κ metal gate process and SAC. Then, lower Si channel is modulated by back-biasing to offer a significant improvement of electrostatics properties, suggesting more flexibility for lower power consumption or much higher performance. Finally, SPICE simulations of RO based on a revised version of LETI-NSP model calibrated on experimental results show an optimal energy efficiency for GAA structures integrating up to 3 stacked Si channels.

ACKNOWLEDGMENT

This work was partly funded by the French Public Authorities through the NANO 2017 program and EQUIPEX FDSOI11. It is also partially funded by the SUPERAID7 (grant N° 688101) project.

REFERENCES

[1] S.-Y. Lee *et al.*, VLSI, 10.1109/VLSIT.2004.1345478, 2004, [2] T. Ernst *et al.*, IEDM, 10.1109/IEDM.2006.346955, 2006, [3] L.K. Bera *et al.*, IEDM, 10.1109/IEDM.2006.346841, 2006, [4] C. Dupré *et al.*, IEDM, 10.1109/IEDM.2008.4796805, 2008, [5] E. Bernard *et al.*, VLSI, pp. 16-17, 2008, [6] https://news.samsung.com/global/, [7] H. Mertens *et al.*, VLSI, 10.1109/VLSIT.2016.7573416, 2016, [8] S. Barraud *et al.*, IEDM, 10.1109/IEDM.2016.7838441, 2016, [9] N. Loubet *et al.*, VLSI, 10.1109/VLSIT.2017, 2017, [10] Y.M. Lee *et al.*, IEDM, pp. 29.3.1-29.3.4, 2017, [11] N. Yoshida *et al.*, IEDM, pp. 22.2.1-22.2.4, 2017, [12] J. Zhang *et al.*, IEDM, pp. 22.1.1-22.1.4, 2017, [13] S. Barraud *et al.*, IEDM, pp. 29.2.1-29.2.4, 2017, [14] M. Garcia Bardon *et al.*, VLSI, 2018, [15] J.M. Hartmann *et al.*, Semicond. Science and Technology 32, p. 104003, 2017, [16] D. Cooper *et al.*, Nano Lett.,vol. 15, p. 5289, 2015, [17] O. Rozeau *et al.*, IEDM, pp. 7.5.1-7.5.4, 2016, [18] J. Lacord *et al.*, IEEE TED, 63, pp. 781-786, 2016.

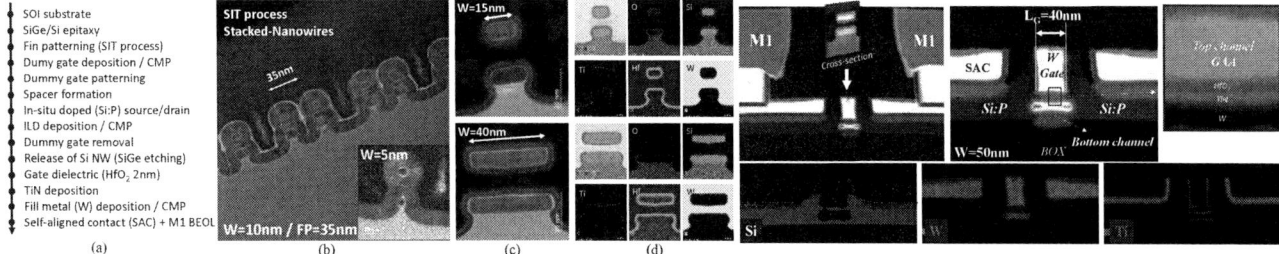

- SOI substrate
- SiGe/Si epitaxy
- Fin patterning (SIT process)
- Dumy gate deposition / CMP
- Dummy gate patterning
- Spacer formation
- In-situ doped (Si:P) source/drain
- ILD deposition / CMP
- Dummy gate removal
- Release of Si NW (SiGe etching)
- Gate dielectric (HfO₂ 2nm)
- TiN deposition
- Fill metal (W) deposition / CMP
- Self-aligned contact (SAC) + M1 BEOL

Fig.1: (a) Process flow for the fabrication of stacked nanosheets *n*-FETs including SIT process, Si:P in-situ doped sources and drains, RMG module and self-aligned contacts. (b) TEM images of (SiGe/Si) fins obtained by SIT which have a fin pitch of 35nm and a W=10nm width. Ultra-scaled NWs are demonstrated with a diameter of 5nm in the inset of fig.1b. (c,d) TEM images of NS (W=15 and 40nm). A wide range of W is considered. Conformal deposition of high-κ and metal layers are also confirmed by EDS analysis.

Fig.2: Cross-sectionnal TEM images of stacked nanosheets along the source to drain direction. In this example, the NW width is 50nm and the gate length is 40nm. The crystallinity of the GAA channel is evidenced. Conformal deposition of TiN/W metal gate between two stacked Si channels and NiPt/Ti/W self-aligned contacts are obvious on the EDS elemental maps.

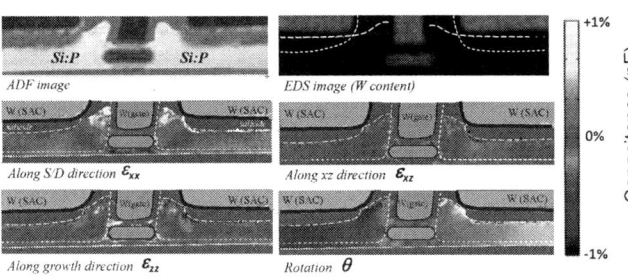

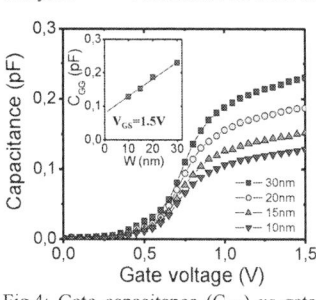

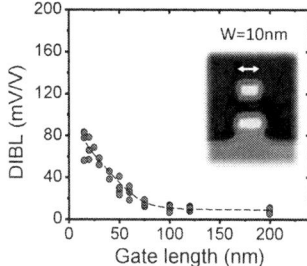

Fig.3: HAADF STEM image of stacked-NS *n*-FET and deformation maps acquired by PED in the (ε_{xx}), (ε_{zz}) and (ε_{xz}) directions. A spatial resolution of about 1.5nm is achieved. Strain is measured after the M1 level. A strong tensile stress is generated between the contacts and the metal gate.

Fig.4: Gate capacitance (C_{GG}) *vs* gate voltage (V_{GS}) for stacked-NS *n*-FETs with a width varying from 10nm up to 30nm. Here, L_G=500nm.

Fig.5: Median of DIBL *vs* gate length (L_G) for stacked-NS *n*-FETs with W=10nm. A DIBL of 80mV/V is achieved for a L_G=20nm gate length.

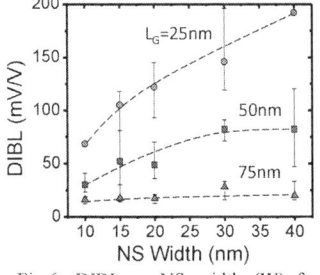

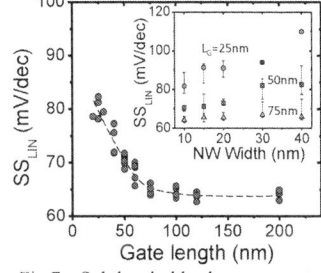

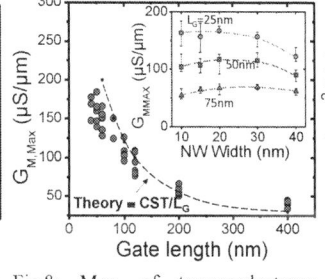

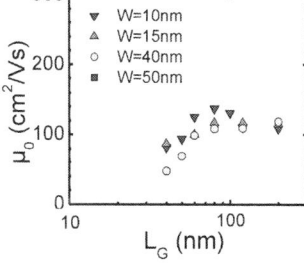

Fig.6: DIBL *vs* NS width (W) for stacked-NS *n*-FETs with different gate lengths: L_G=25, 50 and 75nm.

Fig.7: Subthreshold slope *vs* gate length (L_G) for stacked-NS *n*-FETs with W=10nm. Inset shows the width dependence for L_G=25, 50 and 75nm.

Fig.8: Max. of transconductance ($G_{M,MAX}$) *vs* gate length (L_G) for W=10nm. Inset shows the width dependence for L_G=25, 50 and 75nm.

Fig.9: Low-field electron mobility extracted with the Y-function method for different sheet widths W (from 10nm to 50nm).

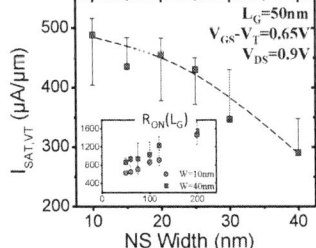

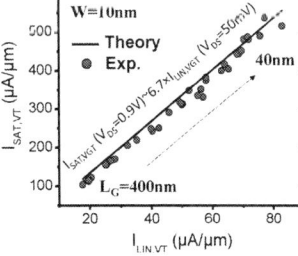

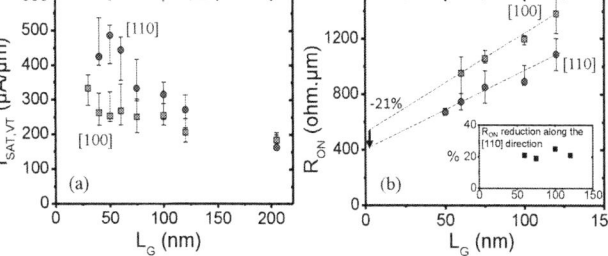

Fig.10: Saturation current ($I_{SAT,VT}$) *vs* NS width (W) for L_G=50nm. The saturation current is measured at V_{GS}-V_T=0.65V and V_{DS}=0.9V. The inset shows the $R_{ON}(L_G)$ for W=10nm and W=40nm.

Fig.11: (a) Saturation current (V_{DS}=0.9V) *vs* linear current (V_{DS}=0.05V) of stacked-NS *n*-FETs measured on a large range of L_G (from 400nm down to 40nm). Good agreement with theoretical results.

Fig.12: (a) Saturation current (V_{DS}=0.9V) *vs* gate length (L_G) for W=10nm. Measurements are performed for [100]- and [110]-stacked-NS *n*-FETs. (b) On-resistance (R_{ON}) *vs* gate length (L_G) along the [110] and [100] transport directions. R_{ON} is extracted at V_{GS}-V_T=0.65V and V_{DS}=0.05V. A lower access resistance is obtained for [110]-stacked-NWs *n*-FETs (R_{ON}~400Ω.µm).

978-1-7281-1988-5/18 $31.00 © 2018 IEEE   502

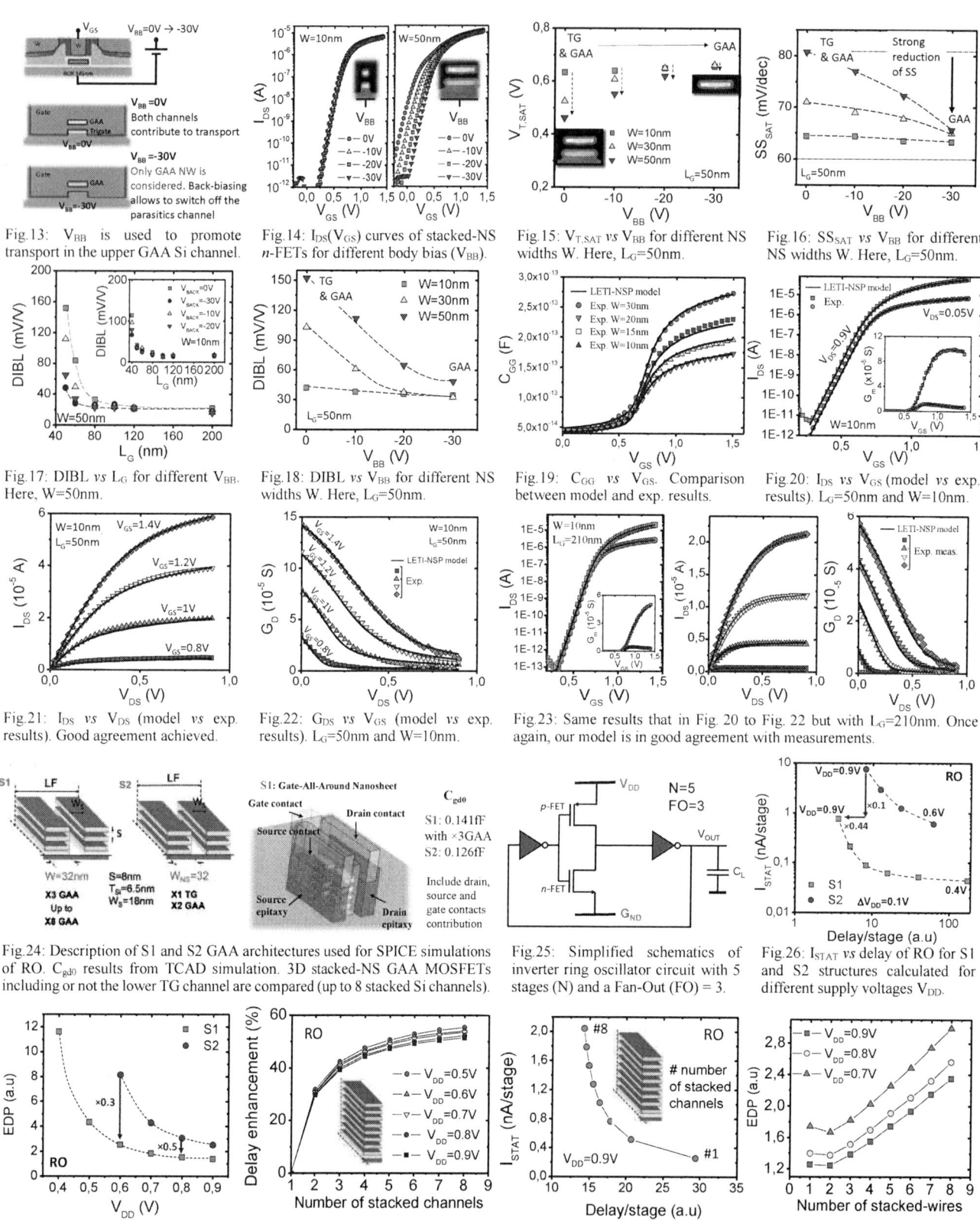

Fig.13: V_{BB} is used to promote transport in the upper GAA Si channel.

Fig.14: $I_{DS}(V_{GS})$ curves of stacked-NS n-FETs for different body bias (V_{BB}).

Fig.15: $V_{T,SAT}$ *vs* V_{BB} for different NS widths W. Here, L_G=50nm.

Fig.16: SS_{SAT} *vs* V_{BB} for different NS widths W. Here, L_G=50nm.

Fig.17: DIBL *vs* L_G for different V_{BB}. Here, W=50nm.

Fig.18: DIBL *vs* V_{BB} for different NS widths W. Here, L_G=50nm.

Fig.19: C_{GG} *vs* V_{GS}. Comparison between model and exp. results.

Fig.20: I_{DS} *vs* V_{GS} (model *vs* exp. results). L_G=50nm and W=10nm.

Fig.21: I_{DS} *vs* V_{DS} (model *vs* exp. results). Good agreement achieved.

Fig.22: G_{DS} *vs* V_{GS} (model *vs* exp. results). L_G=50nm and W=10nm.

Fig.23: Same results that in Fig. 20 to Fig. 22 but with L_G=210nm. Once again, our model is in good agreement with measurements.

Fig.24: Description of S1 and S2 GAA architectures used for SPICE simulations of RO. C_{gd0} results from TCAD simulation. 3D stacked-NS GAA MOSFETs including or not the lower TG channel are compared (up to 8 stacked Si channels).

Fig.25: Simplified schematics of inverter ring oscillator circuit with 5 stages (N) and a Fan-Out (FO) = 3.

Fig.26: I_{STAT} *vs* delay of RO for S1 and S2 structures calculated for different supply voltages V_{DD}.

Fig.27: EDP *vs* V_{DD} of RO for S1 and S2 GAA structures.

Fig.28: Delay *vs* number of stacked Si channels (for S1 GAA structure).

Fig.29: I_{STAT} *vs* delay of S1 structure calculated up to 8 stacked Si channels.

Fig.30: EDP *vs* number of stacked channels (for S1 GAA structure).

Voltage Transfer Characteristic Matching by Different Nanosheet Layer Numbers of Vertically Stacked Junctionless CMOS Inverter for SoP/3D-ICs applications

P.-J. Sung[1,2], C.-Y. Chang[2], L.-Y. Chen[3], K.-H. Kao[3], C.-J. Su[1], T.-H. Liao[4], C.-C. Fang[4], C.-J. Wang[1], T.-C. Hong[2], C.-Y. Jao[4], H.-S. Hsu[4], S.-X. Luo[3], Y.-S. Wang[3], H.-F. Huang[3], J.-H. Li[3], Y.-C. Huang[2], F.-K.Hsueh[1,2], C.-T.Wu[1], Y.-M.Huang[1],F.-J. Hou[1], G.-L. Luo[1],Y.-C. Huang[1],Y.-L. Shen[1],W. C.-Y. Ma[4], K.-P. Huang[5], K.-L.Lin[1], S. Samukawa[6], Y. Li[7], G.-W Huang[1], Y.-J. Lee[1,*], J. -Y. Li[1,8], W.-F. Wu[1], J.-M.Shieh[1], T.-S. Chao[2], W. -K. Yeh[1], Y.-H. Wang[3,9]

[1]National Nano Device Laboratories, Hsinchu, Taiwan; [2]Dept. of Electrophysics, National Chiao Tung University, Hsinchu, Taiwan; [3]Dept. of Electrical Engineering, National Cheng Kung University, Tainan, Taiwan; [4]Dept. of Electrical Engineering, National Sun Yat-sen University, Kaohsiung, Taiwan; [5]Mechanical and Systems Research Laboratories, Industrial Technology Research Institute, Hsinchu, Taiwan; [6]Institute of Fluid Science, Tohoku University, Sendai, Japan; [7]Dept. of Electrical and Computer Engineering, National Chiao Tung University, Hsinchu, Taiwan; [8]Dept. of Electrical Engineering and Graduate Institute of Electronics Engineering, National Taiwan University, Taipei, Taiwan; [9]National Applied Research Laboratories, Taipei, Taiwan; Tel: +886,-3-5726100-7793, Fax: +886-3-5722715, *Email: yjlee@narlabs.org.tw

Abstract

For the first time, CMOS inverters with different numbers of vertically stacked junctionless (JL) nanosheets (NSs) are demonstrated. All fabrication steps were below 600 °C, and 8-nm thick poly-Si NSs with smooth surface roughness were formed by a dry etching process. Compared to single channel devices, stacked n/p-channel FETs exhibit higher on-current with low leakage current. Furthermore, a common-gate process was performed for the fabrication of CMOS inverters. By adjusting the NS layer numbers for n/pFETs, respectively, the voltage transfer characteristics (VTCs) of the CMOS inverter can be matched much better to reduce the noise margin due to on-current matching without area penalty. This work experimentally demonstrates a new configuration of CMOS inverters on stacked NSs, which is promising for System-on-Panel (SoP) and 3D-ICs applications.

Introduction

For the applications of SoP and 3D-ICs, high-performance transistors are fabricated on a glass substrate or buried oxide at low temperatures [1-2]. Although a fin structure can offer better gate controllability, fin pitch and width restrict the FinFET scaling. To further improve the device performance, Gate-all-around (GAA) nanowire or NS FETs are promising candidates to outperform FinFETs to continue the device scaling for the N5 logic technology or beyond [3-5]. A vertically stacked NS structure can enhance the device performance by increasing the effective device width compared GAA and FinFETs in a given footprint [4]. Moreover, compared to enhance-mode FETs, JL FETs are attractive due to the possess simplicity and immunity of mobility degradation of carriers scattering at the channel/oxide interface [6].

In this work, we demonstrated a novel design for CMOS inverters with the stacked NS devices with different channel numbers to achieve a better noise margin. The NS devices were fabricated at low temperature (≤ 600 °C). Without area

penalty, a CMOS inverter of gain ~18 V/V is achieved by stacking different NS layer numbers for nFETs and pFETs.

Device Fabrication

Figs. 1 and 2 illustrate the process flow and schematic of vertically stacked NSs poly-Si CMOS fabrication. (a) First, a 45-nm amorphous-Si layer was deposited on SiO_2 followed by ion implantation and thin-down steps. After the film thinned down to 8 nm by dry etching steps with 30 nm PECVD oxide for interlayers isolation, top layer was prepared by repeating Fig. 2(a). Since surface roughness and grain sizes are very critical for the electrical performance of poly-Si FETs, the etching-back step was carried out to form NSs below 600 °C. After the hard mask deposition, vertically stacked NSs were defined by lithography and dry etching processes. After the interlayer oxide removed by dilute HF, gate stacks were defined with high k/metal gate, as shown in Fig. 2(e), followed by oxide passivation and metallization (Fig. 2(f) and 2(g)). While the gate always controls the two stacked channels at the same time, the S/D contacts are defined such that the stacked devices can be operated together or only the top one is contributing the drain current (bottom channel floating, see Fig. 2(h) and 2(i)). Fig. 2 (j) shows the cross-sectional SEM image for a contact hole. Contact area of the top layer would influence the performance of stacked NSs FETs, and isotropic wet etching process was used after the dry etching step to increase the contact area.

Fig. 3 shows the cross-sectional TEM image the stacked NSs. The sheet thickness is 8 nm and 30 /40 nm for top/bottom sheets width. Narrow top sheet width is due to dry etching process.

Results and Discussion

Dopant Diffusion and Surface Roughness:

After solid-phase crystallization (SPC) and activation at 600 °C, the SIMS profiles show no significant diffusion of phosphorus (^{31}P) and boron (Fig. 4) for one or two annealing cycles (one cycle : 24-hr SPC and 4-hr dopant activation). In Fig. 5, the AFM images of (a) 45-nm and (b) 10-nm poly-Si films by SPC process, and (c) 10-nm poly-Si by dry etching thin-down process from 45-nm poly-Si are illustrated. RMS roughness reveals that the thinned-down poly-Si in Fig. 5(c) is smoother and better for device fabrication. Fig. 5(d) shows the XRD of different poly-Si films. Both XRD and the diffraction pattern (inset in Fig. 3) confirm the recrystallization of the poly-Si film.

Comparison between Top and Stacked Channels:

Figs. 6 and 7 show the transfer curves of JL poly-Si NS n/pFETs, respectively, with two channels (2-ch) and one channel (1-ch). Higher on-current is achieved for the stacked NS JL FETs without a trade-off in leakage current and DIBL. In addition, the inset in Fig. 6 is the champion device exhibiting a subthreshold swing of 68 mV/dec and an I_{on}/I_{off} ratio of 10^7. Output characteristics in Figs. 8 and 9 present 70% and 50% current enhancement for nFET and pFETs, respectively, due to the stacked NSs of two layers. Figs. 10 and 11 display the relation between I_{on} and I_{off} for n/pFETs respectively, with a single top channel and stacked NSs with 2 channels. The latter shows an average I_{on} improvement about 50% without I_{off} degradation. In addition, high-energy ^{31}P implantation could increase I_{on} significantly (up to about 160% as compared to low-energy ^{31}P implantation) due to higher conductivity in the channel (Fig. 10).

Fig. 12 presents the VTCs of a stacked JL poly-Si NSs CMOS inverter at various drain-to-source voltages from 0.8 to 1.2 V. Higher ^{31}P implanted energy for nFETs improves the noise margin of the inverter with an increased gain from 5.5 to 11 V/V, as shown in the inset.

Different NS layer numbers in n/pFETs for CMOS Inverter:

To simplify the process flow, single gate electrode for a CMOS inverter is needed, but the asymmetric behavior of $|V_{TH}|$ between the n/pFET would degrade the voltage transfer characteristics (VTCs) under operation. To overcome this issue, the effect of different NS layer numbers for the n/pFET on the inverter performance are investigated. Fig. 13(a) shows the top-view SEM image of the vertically stacked poly-Si CMOS and the zoom-in images (b) a one-channel pFET and (c) a two-channel stacked nFET. Fig. 14 shows the transfer and output curves of the one-channel pFET and two-channel nFET in a CMOS inverter. Fig. 15 presents the VTCs of the inverter, at V_D from 0.6 to 1.2 V. with a maximum gain up to 18 V/V (inset). In Table 1, the performance of the stacked devices in this work is benchmarked with the prior works. The CMOS inverter's VTC is improved due to on-current matching without area penalty.

Numerical Simulation for Stacked Structure

Fig. 16 shows the SEM images of non-optimized NS structure. The top-sheet may collapse after the interlayer oxide removal when the separation between two contact pads

is too long and the gap between two sheets is too narrow. Therefore, it is needed to further investigate the impact of the distance between NSs on the device performance. Self-consistent numerical simulations solving Poission and current equations have been carried out to investigate the electrical characteristics of devices with different S/D contact configurations (Fig. 17). The employed models include Fermi statistic, doping-dependent mobility and bandgap narrowing, SRH and nonlocal BTBT [7].

Fig. 18 shows more pronounced enhancement of on-currents for the device (A) when increasing the distance between two channels. It is attributed to the weakened the electrostatic coupling of the stacked channels as shown in Fig. 19. Simulations hint that stacked devices should not leave floating channel, which degrades others' electrical performance by electrostatic coupling. And coupling is stronger when channels are closer.

Conclusion

For the first time, a CMOS inverter is demonstrated on vertically stacked JL nanosheets with different sheet numbers for n/p FETs. Even though the recrystallized poly Si layer is thinned down to 8 nm, the grain size and surface roughness of poly-Si can still be retained. Compared to the device with a single channel, the stacked n/pFETs show higher on-current without leakage current degradation. This is due to a larger effective width of the vertically stacked nanosheets. While single metal gate electrode for a CMOS inverter could simplify the process, the asymmetry of $|V_{TH}|$ between n/pFETs can degrade the VTC. In addition, from simulation results, stacked devices should not leave floating channel due to the coupling effect. By adjusting the nanosheet layer numbers for n/pFETs, the ideal VTCs of CMOS inverter can be achieved to much reduce the noise margin without area penalty, which is beneficial for SoP/3D ICs applications.

ACKNOWLEDGMENT

This work was performed by the National Nano Device Laboratories facilities and supported by the Ministry of Science and Technology under grant numbers 107-2636-E-006-004, 105-2628-E-492-002-MY3, 105-2221-E-492-029-MY2, 106-2221-E-492-034, 107-2628-E-492-001-MY3 and-107-2633-E-009-003. The authors are grateful for the support by National Center for High-performance Computing (NCHC), Taiwan and Hitachi High-Technologies Corp. Japan.

REFERENCES

[1] C.-C. Yang et al., IEDM Tech. Dig., 2014, pp.410-413. [2] K. Ota et al., Proc. Symp. VLSI Tech., 2015, pp. 214-215. [3] H. Mertens et al., IEDM Tech. Dig., 2017, pp.828-831. [4] N. Loubet et al., Proc. Symp. VLSI Tech., 2017, pp. 230-231. [5] S. Barraud1 et al., IEDM Tech. Dig., 2016, pp.464-467.[6] J.-P. Colinge et al., Nature Nanotechnology, 2010, pp.225-229. [7] Sentaurus TCAD, Synopsys, 2017. [8] L.-C. Chen et al., IEEE Electron Device Lett., 2017, pp. 1256-1258.[9] C. C.-C. Chung et al., IEEE Trans. Electron Device, 2018, pp.756-762. [10] P.-Y. Kuo et al., Proc. Symp. VLSI Tech., 2018, pp. 21-22. [11] M.-S. Yeh et al., IEDM Tech. Dig., 2014, pp.618-621.

1. Bottom-layer a-Si deposition (45 nm)
2. Solid phase crystallization at 600 °C
3. CMOS lithography & Implantation :
4. Bottom-channel JL-pFETs :
   ■ BF₂, 5×14 cm⁻², 30 KeV.
5. Bottom-channel JL-nFETs:
   ■ ³¹P, 5×14 cm⁻², 16 KeV/20 KeV.
6. Dopant activation (600 °C /4 hrs)
7. Poly-Si thinning down to 8 nm by dry etching.
8. TEOS deposition for isolation
9. Top-layer a-Si deposition (45 nm)
10. Repeating steps 2 to 7 for top-layer
11. Define active region by lithography
12. HF dip to release nanosheets
13. Gate stacks formation:
    ■ Al₂O₃ deposition by ALD.
    ■ TiN deposition
14. Gate patterning
15. Passivation
16. Contact holes formation
17. Metallization for CMOS

Fig. 1. Process flow of vertically stacked poly-Si JL NSs CMOS fabrication. All process temperature are below 600 °C

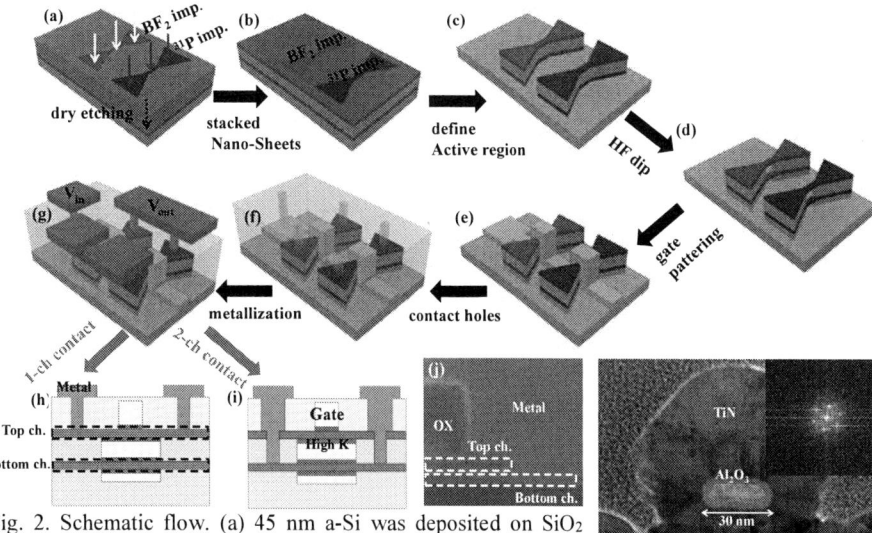

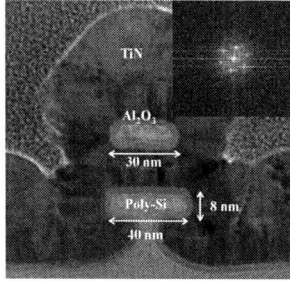

Fig. 2. Schematic flow. (a) 45 nm a-Si was deposited on SiO₂ followed by implantation and thin-down steps. (b) After film thinned down to 8 nm by dry etching process, top- and bottom-channels were prepared repeatedly. (c) & (d) Vertically stacked NSs were released by interlayer oxide removal. (e) Gate stack formation with high k/metal gate deposition. (f) & (g) Oxide passivation and metallization. (h) & (i) illustrate the S/D contacts are deliberately connected to the top channel only (bottom channel floating) and two channels, respectively. (j) SEM cross-section image for a contact hole.

Fig. 3. TEM image of vertically stacked nanosheets. The sheet thickness is 8 nm and 30 /40 nm for top/bottom sheets width. The inset is the diffraction pattern.

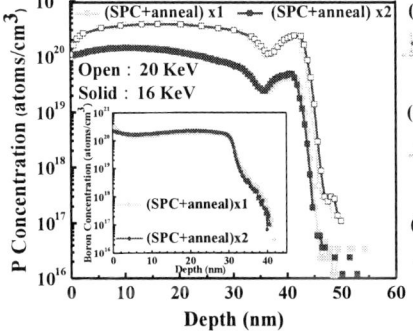

Fig.4. After SPC and activation anneal at 600 °C, SIMS profiles show no significant diffusion of ³¹P and boron (inset) between one and two thermal cycles.

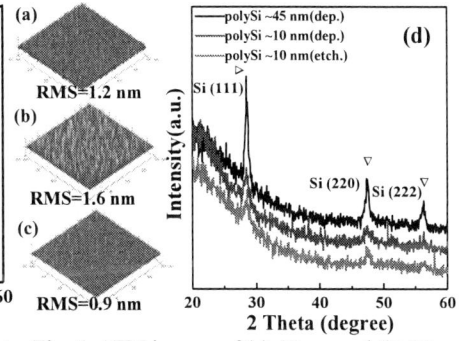

Fig. 5. AFM images of (a) 45 nm and (b) 10 nm poly-Si after SPC process, (c) poly-Si thinned down by dry etching process from 45 to 10 nm. (d) XRD of poly-Si 45 nm, 10 nm, and etched poly-Si.

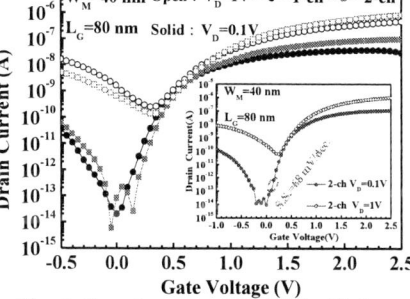

Fig. 6. Transfer curves of nFETs. Higher on currents could be achieved with stacked (2 channels) NSs. W_M denotes bottom-channel width. The inset is the champion device, which shows 68 mV/dec. S.S. and 7 orders of I_{on}/I_{off}.

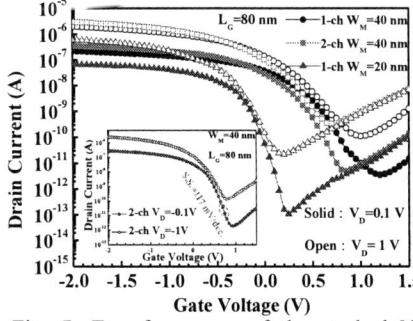

Fig. 7. Transfer curves of the stacked NSs pFETs with one and two channels. S.S. and on/off ratio are improved significantly as W_M shrinks down to 20 nm. The inset is the champion device, which shows 117 mV/dec. with 5 orders of I_{on}/I_{off}.

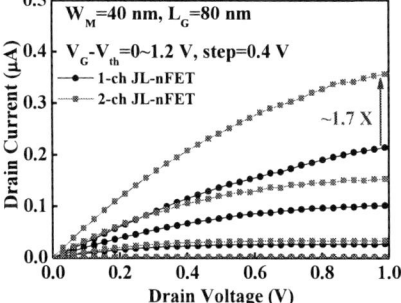

Fig. 8. Comparisons of I_D-V_D curves of nFETs with one and two channels. 70 % on-current improvement is achieved due to the stacked 2 channels.

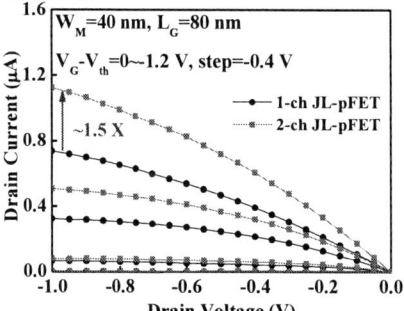

Fig. 9. Comparisons of I_D-V_D curves of stacked NS pFETs with one and two channels. 50 % on current improvement is achieved due to the stacked 2 channels.

978-1-7281-1988-5/18 $31.00 © 2018 IEEE

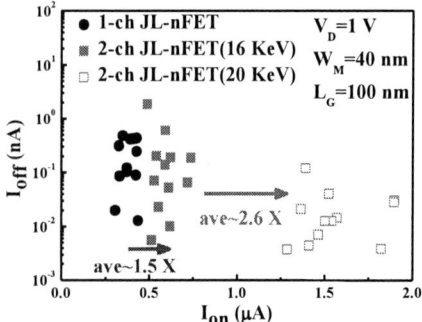

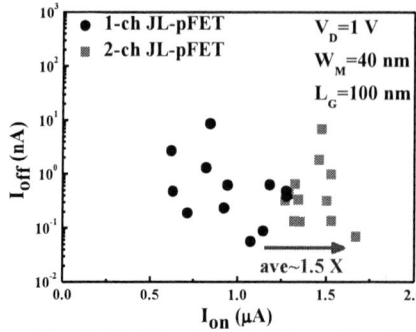

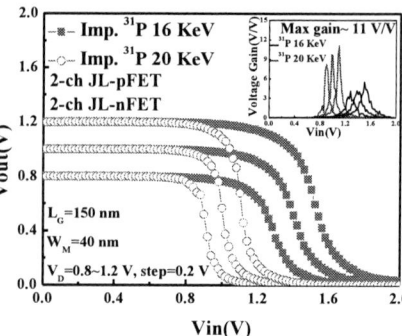

Fig. 10. Relation between I_{on} and I_{off} for nFETs with single top-channel and stacked NSs (2-ch), and the later shows the average I_{on} improvement about 50 % without I_{off} degradation. Higher energy ^{31}P implantation increases I_{on} significantly (up to 160 %) due to higher dopant concentration in the channel.

Fig. 11. Relation between I_{on} and I_{off} for pFETs with single top-channel and stacked NSs, and the later shows the average I_{on} improvement about 50% without I_{off} degradation.

Fig. 12. V_{OUT} versus V_{IN} of the stacked JL poly-Si NS CMOS inverter, at V_D from 0.8 to 1.2 V. Higher implanted energy for nFETs improves the CMOS inverter performance with increased gain, as shown in the inset.

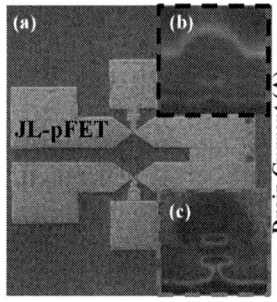

Fig. 13. (a) Top-view SEM image of the vertically stacked poly-Si CMOS. Zoom-in images of (b) a one-channel pFET and (c) a stacked NSs nFET.

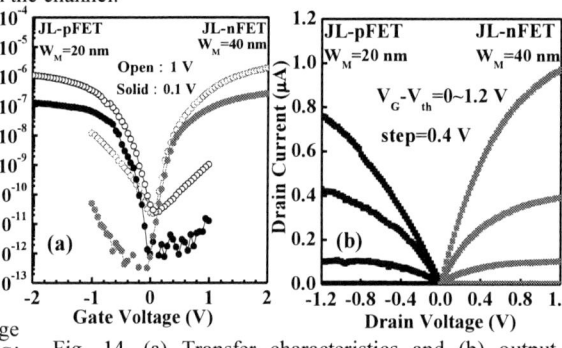

Fig. 14. (a) Transfer characteristics and (b) output curves of the one-channel p-and two-channel nFETs in a CMOS inverter.

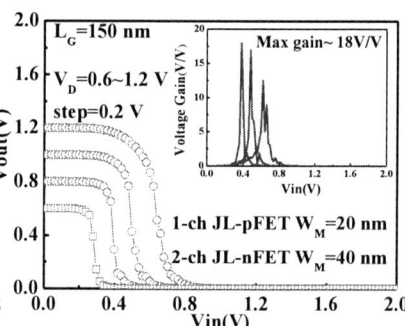

Fig. 15 V_{OUT} versus V_{IN} of the stacked JL poly-Si NS CMOS inverter, at various V_D from 0.8 V to 1.2 V with max. gain up to 18 V/V in the inset.

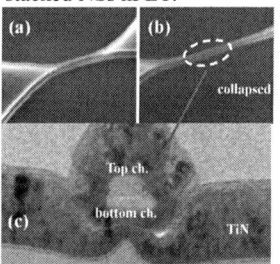

Fig. 16. The top-channel may collapse after the oxide removal for a long pads distance and a narrow gap between NSs.

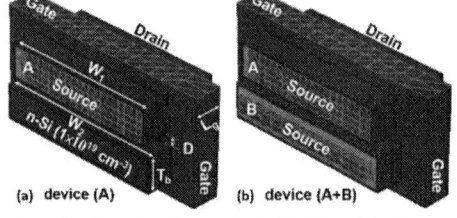

Fig. 17. The simulated NS JLFETs with (a) top channel (device A) and (b) stacked channels (device A+B) for current flow. All channels have the same body thickness T_b (10 nm) and uniform doping concentration (1×10^{19} cm^{-3}). The grid regions represent the contact terminals. The gate workfunction of 4 eV and EOT of 1 nm are defined in all simulations. (W_1 = 55 nm, W_2 = 60 nm, L_g = 15 nm.)

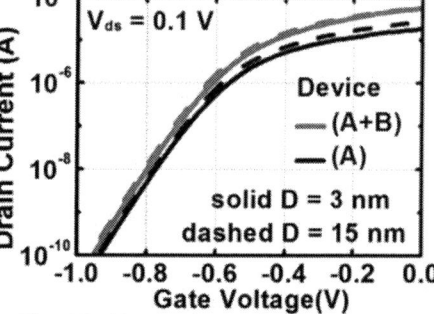

Fig. 18. Simulated transfer characteristics of devices (A) and (A+B) with different distance D.

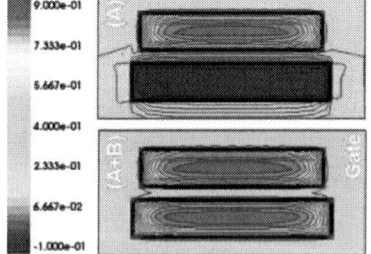

Fig. 19. Simulated cross sectional electrostatic potential (V) contour of devices (A) and (A+B) obtained at Vds = 0.1 V, Vgs = − 0.6 V and with distance D = 3 nm and Lg = 15 nm.

Table I. Performance comparison of stacked NS JLFFTs in this work with other publications.

Ref.		This Work	[8]	[9]	[10]	[11]	[1]
Structure		stacked NS JL	stacked NS	stacked JL NW	a-IWO NS JL	Trench JL	Tri-gate
Fin Height		8 nm	17 nm	12 nm	4 nm	2.4 nm	14 nm
CMOS		Y	N	N	N	N	Y
W/L(μm/μm)		0.03/0.08	0.084/1	0.04/0.35	80/40	0.07/0.5	0.02/0.03
S.S.	(N-type)	68	168	150	150	99	96
	(P-type)	117	N	N	N	N	127
I_{on}/I_{off}	(N-type)	>10^7	>10^7	>10^5	>10^8	>10^7	>10^5
	(P-type)	>10^5	N	N	N	N	>10^5

978-1-7281-1988-5/18 $31.00 © 2018 IEEE   507

Vertically Stacked Gate-All-Around Si Nanowire CMOS Transistors with Reduced Vertical Nanowires Separation, New Work Function Metal Gate Solutions, and DC/AC Performance Optimization

R. Ritzenthaler*, H. Mertens, V. Pena[1], G. Santoro[1], A. Chasin, K. Kenis, K. Devriendt, G. Mannaert, H. Dekkers,
A. Dangol, Y. Lin[2], S. Sun[2], Z.Chen[2], M. Kim[2], J. Machillot[1], J. Mitard, N. Yoshida[2], N. Kim[2], D. Mocuta, N. Horiguchi
IMEC, Kapeldreef 75, 3001 Leuven, Belgium email: *romain.ritzenthaler@imec.be
[1] Applied Materials, Leuven, Belgium, [2] Applied Materials, 3050 Bowers Avenue, Santa Clara, CA 95053, USA.

Abstract—We report on vertically stacked gate-all-around (GAA) Si nanowire (NW) MOSFETs, integrated in a CMOS dual Work Function Metal Replacement Metal Gate (RMG) flow. The integration of a lower temperature STI module and a SiN liner, designed to mitigate the oxidation-induced NW size loss and improve the width/height aspect ratio and NW controllability, is validated electrically. Additionally, Si GAA devices with reduced vertical nanowire spacing are demonstrated. The challenges in terms of Work Function Metal thickness scaling are highlighted, and a thinner nMetal process with low V_{TH} capability and no J_G/PBTI lifetime penalty is proposed. Electrically, these process innovations lead to a large improvement of I_{ON}/I_{OFF} performance and short channel margin. Finally, a ring oscillator circuit demonstration is shown, with a improvement of gate delay from 24ps down to 10ps at matched V_{DD} demonstrated.

I. INTRODUCTION

Due to their optimal electrostatic control of the channel, GAA MOSFETs [1,2] are considered as promising candidates to extend the gate length and gate pitch scaling beyond the FinFET limits [3]. Lateral nanowires offer the advantage of a process flow relatively comparable to FinFETs, and vertical stacking allows to maximize the drive current for a given footprint on the wafer.

Compared to our previous works [4,5,6], we report process improvements and electrical validation for shallow trench isolation (STI), reduced vertical nanowires spacing, integration of a thinner nMetal process with low V_{TH} capability, and DC/AC performance improvement.

II. DEVICE FABRICATION

A. Process flow

The process flow for the fabricated devices is illustrated in **Fig. 1**, with the new modules with respect to our previous report [6] highlighted. The vertically stacked GAA Si NW-FETs are fabricated on standard 300-mm Si wafers. After ground plane formation and Si/SiGe layers epitaxial growth, the Si/SiGe fin superlattice is patterned. The STI fill module is then performed, followed by dummy gate patterning. In-situ doped S/D are made up of highly-doped Si:P for NMOS (**Fig. 2**) and SiGe:B+Si:B liner (used as a protection barrier for S/D during NW release) for PMOS (**Fig.3**), with the flow available in both NMOS or PMOS Epi S/D 1st integration schemes. After dummy gate removal, the SiGe layers are selectively removed (performed by Applied Materials Selectra™ etch) in order to release the nanowires. A Dual Work Function RMG scheme is then performed with Applied Materials Endura®, with the rest of the process flow completed by a conventional Li1/Li2/V0/BEOL scheme.

B. Minimization of SiGe/Si fin Oxidation

The SiGe oxidation and Ge diffusion during the STI densification lead to a significant rounding of the Si NW corners, and SiGe/Si fin oxidation brings difficulties to control the NW shape (**Fig. 4.a**). By reducing the STI thermal budget and introducing a SiN STI liner, a reduced NW size loss and better shape controllability are demonstrated (**Fig. 4.a**). Thanks to reduced STI oxidation, larger wires (featuring a width of 15nm and height of 11nm) with improved width/height aspect ratio can also be demonstrated (**Fig. 4.b and 4.c**). Electrically, the comparison of long channel transfer characteristics with and without SiN STI liner shows no subthreshold slope degradation, indicating that no interface trap density increase is observed (**Fig. 5.a**). Good subthreshold characteristics are also maintained for short channels (**Fig. 5.b**), validating the integration of the SiN liner.

C. Reduction of vertical nanowires separation

In order to maximize the available active volume per fin height and to minimize the parasitic capacitance, a reduction of the vertical nanowires separation below 10nm is desirable [7]. This can be achieved by thinning down the SiGe sacrificial layer in the Si/SiGe superlattice. However, it may come at the price of increased difficulties to conformally deposit and remove the gate stack materials (gate oxide and work function metals) in the dual work function RMG flow. Additionally, SiGe selective removal may need additional optimization for SiGe thickness change. In **Fig. 6.a**, it is shown that a vertical nanowire separation thinning from 12nm down to 8nm has been successfully achieved, with correct SiGe selective removal and no modification of NW dimensions. However, it is also observed by energy dispersive X-ray spectroscopy (EDS) that the gap between the nanowires could be entirely filled by pMetal if it is not sufficiently removed before nMetal deposition in a PMOS-first integration scheme (**Fig. 6.b**). Assuming the pMetal can be effectively removed, the space between the nanowires still leaves little room for nMetal fill. In this context of reduced available space to fit in the gate stack, a reduction of the Work

978-1-7281-1988-5/18 $31.00 © 2018 IEEE

Function Metal thickness without threshold voltage penalty should therefore be considered. A new nMetal process is proposed, and demonstrates low V_{TH} capability (**Fig.7.a**) for a physically thinner layer and no gate leakage (**Fig.7.b**) or PBTI lifetime penalty (**Fig.7.c**). It is verified that low V_{TH} capability integrity is maintained down to short channels (**Fig.8**).

III. ELECTRICAL CHARACTERISTICS

Transfer characteristics for short channels (L_G=26nm) Si NW-FET are shown in **Fig.9**. Good electrostatic control is obtained for both NMOS and PMOS devices. Besides, the short channel margin, despite the NW dimensions increase, is improved w. r. t. Best Known Method reference (**Fig. 10**). This indicates that the device's effective gate length L_{EFF} is not reduced with the introduction of the improved in-situ doped S/D module. I_{ON}/I_{OFF} PMOS performance comparison (with current normalization made by footprint on wafer) between previous best known method and newly reported devices is shown in **Fig. 11**. A large I_{ON} improvement at fixed I_{OFF} is demonstrated with the improved process flow reported in this work. Such I_{ON}/I_{OFF} improvement is related to a drive current (I_{ON}) improvement, as demonstrated by the $I_{ON}(V_{TH,SAT})$ plot shown in **Fig. 12**. On the other hand, the off-state leakage degradation is very limited (**Fig. 14**), as seen in the $I_{OFF}(V_{TH,SAT})$ plot (**Fig. 13**). When normalizing the drive current per effective channel width, PMOS I_{ON}/I_{OFF} performance is also improved (**Fig. 14**). It demonstrates that the reported drive current improvement is not solely due to the increase of effective channel width in the stacked NW structure, but is linked as well to an improvement of short channel mobility and/or external resistance. A similar exercise is done for NMOS devices, and a large improvement of I_{ON}/I_{OFF} performance is also demonstrated (normalization is made per footprint, **Fig. 15**). However, I_{ON}/I_{OFF} improvement disappears when the current is normalized by the effective channel width (**Fig. 16**), demonstrating that for NMOS devices the drive current improvements are exclusively related to the increase of NW dimensions.

In order to further investigate the root cause for the drive current improvement in the PMOS case, external resistance is extracted with the $R_{ON}(1/V_{GT})$ method (**Fig. 17** [8]). It is shown that external resistance is not improved between previously reported results and this work, and therefore cannot explain the drive current improvements. Additionally, short channel electrical characteristics are compared against the Virtual Source Model (**Fig. 18**, [9]). It is shown that effective hole injection velocity is improved with regard to previously reported results [5], while electrons exhibit identical values. The physical underlying mechanisms might be linked to a modification of injection velocity related to the NW shape and dimensions due to better subband occupation and less surface roughness scattering [6], and/or higher strain with the optimized in-situ doped SiGe:B S/D (PMOS).

Finally, owing to the DC performance improvements, a strong ring oscillator gate delay (41 stages, F.O. 1) improvement is reported (**Fig. 19**). At matched V_{DD}, a gate delay improvement from 24ps down to 10ps is demonstrated.

IV. CONCLUSIONS

In this work, we report on process innovations to improve the performance of vertically stacked CMOS Si nanowire MOSFETs. By reducing the STI thermal budget and integration of a SiN liner, the fin oxidation during STI densification is mitigated. This leads to a large improvement of nanowire size and shape controllability, without degradation of electrical performance due to the liner integration. Si GAA devices with reduced vertical nanowire spacing (12 to 8 nm) were also successfully fabricated. Challenges related to the gate stack conformal deposition and removal due to reduced vertical NW separation are highlighted, and a new nMetal is proposed to circumvent those issues. Those process optimizations lead to large improvement of I_{ON}/I_{OFF} performance and short channel margin for both NMOS and PMOS devices, regardless of the chosen normalization method. A ring oscillator gate delay improvement from 24ps down to 10ps at matched V_{DD} is finally demonstrated.

ACKNOWLEDGMENT

The imec sub-10nm program members, the European commission and local authorities, the imec pilot line, and amsimec test labs are acknowledged for their support.

REFERENCES

[1] C. Dupre et al., "15nm-diameter 3D Stacked Nanowires with Independent Gates Operation: PhiFET", IEDM Tech. Dig. 2008, pp.1-4 (2008).

[2] S.G. Hur *et al.*, "A Practical Si Nanowire Technology with Nanowire-on-Insulator structure for beyond 10nm Logic Technologies", IEDM Tech. Dig. 2013, pp.649-652 (2013).

[3] K. J. Kuhn, "Considerations for ultimate CMOS scaling," IEEE Trans. Electron Devices, vol. 59, no. 7, pp. 1813-1828, Jul. 2012.

[4] H. Mertens et al., "Gate-All-Around MOSFETs based on Vertically Stacked Horizontal Si Nanowires in a Replacement Metal Gate Process on Bulk Si Substrates", 2016 Symposium on VLSI Technology Digest of Technical Papers, pp. 158-159 (2016).

[5] H. Mertens et al., "Vertically Stacked Gate-All-Around Si Nanowire CMOS Transistors with Dual Work Function Metal Gates", IEDM Tech. Dig. 2016, pp.524-7 (2016).

[6] H. Mertens et al., "Vertically Stacked Gate-All-Around Si Nanowire Transistors: Key Process Optimizations and Ring Oscillator Demonstration", IEDM Tech. Dig. 2017, pp.828-831 (2017).

[7] M. Garcia Bardon et al., "Extreme Scaling enabled b y 5 Tracks Cells: Holistic design-device co-optimization for FinFETs and Lateral Nanowires", IEDM Tech. Dig. 2016, pp. 687-690 (2016).

[8] A. Paul et al., "Comprehensive study of effective current variability and MOSFET parameter correlations in 14nm multi-Fin SOI FINFETs", IEDM Tech. Dig. 2013, pp.361-4 (2013).

[9] A. Khakifirooz *et al.*, "A Simple Semiempirical Short-Channel MOSFET Current–Voltage Model Continuous Across All Regions of Operation and Employing Only Physical Parameters", IEEE Trans. Electron Dev., vol. 56, no. 8, pp. 1674-1680 (2009).

- Bulk Si wafer
- Ground plane formation
- SiGe/Si epitaxy **(NEW)**
- SADP fin patterning **(partially**
- STI fill + fin reveal ⟶ **presented**
- Dummy gate patterning **in [6])**
- Spacer formation +
  embedded S/D epitaxy **(optimized)**
- ILD0 fill + CMP
- Dummy gate removal
- Nanowire release
- Dual WFM integration **(NEW)**
- Metal gate fill + CMP
- LI1 + LI2 +V0 + BEOL

Fig. 1: Process flow for the fabrication of CMOS GAA Si NW-FETs, with the new modules highlighted.

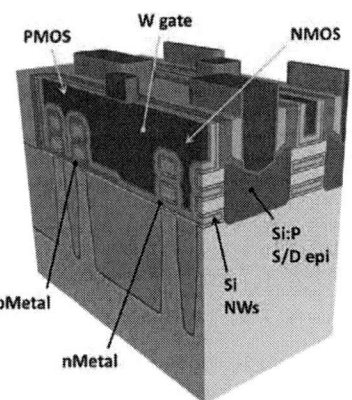

Fig. 2: Schematic representation (up to Li1 local interconnects etch) of the vertically stacked CMOS Si NW-FETs reported in this work (Coventor® image). In-situ doped HD Si:P Source/Drain (NMOS) and dual work function metal gates are illustrated.

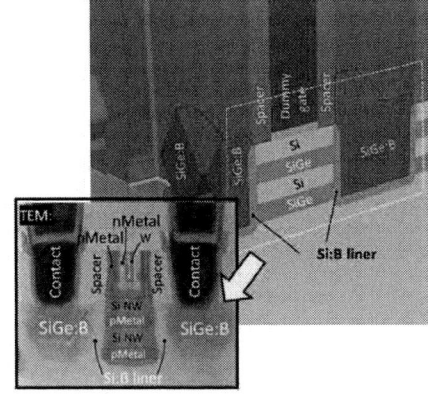

Fig. 3: Schematic representation of the process flow for PMOS devices (Coventor® image) after the embedded S/D epitaxy step. Si:B liner and in-situ doped Si:Ge:B Source/Drain are illustrated. Inset shows a TEM cross-section of corresponding devices at end of process.

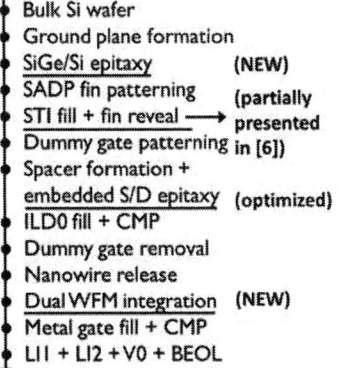

1. Old BKM

2. Reduced STI thermal budget

3. SiN STI liner

Fig .4.(a)

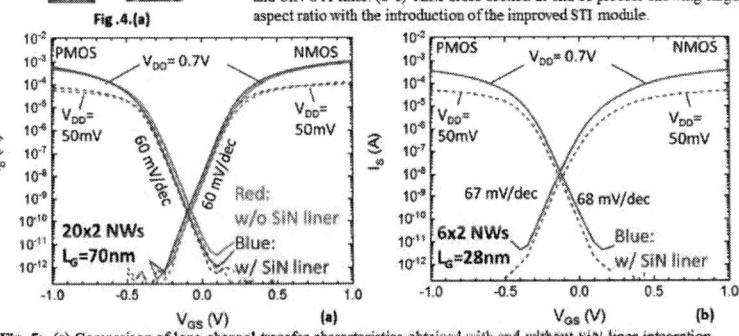

Fig. 4: (a) STEM images of fins after STI fill (left) and TEM cross-section at end of process (right), revealing reduced NW size loss and better shape controllability with the integration of lower temperature STI module and SiN STI liner. (b-c) TEM cross-section at end of process showing larger wires and improved width/height aspect ratio with the introduction of the improved STI module.

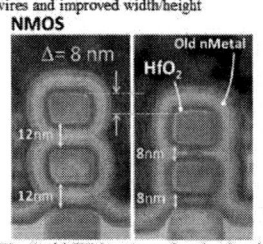

Fig. 6: (a) TEM cross-section showing the stacked NW total height reduction obtained by the thinning of SiGe sacrificial layers. (b) energy dispersive X-ray spectroscopy (EDS) mapping showing that the NW gap is filled with TiN before nMetal deposition.

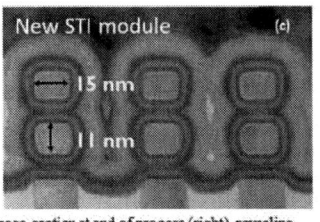

Fig. 5: (a) Comparison of long channel transfer characteristics obtained with and without SiN liner integration, exhibiting no subthreshold slope degradation (=interface trap density increase) with SiN liner integration. (b) short channel transfer characteristics for devices integrated with SiN liner, showing excellent short channel behavior.

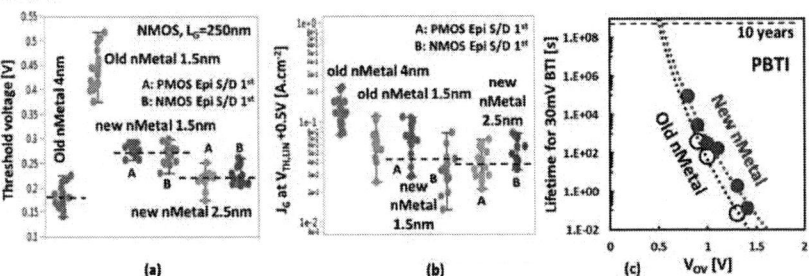

Fig. 7: (a) Linear threshold voltage, (b) gate leakage, and (c) PBTI lifetime for NMOS Si NWFETs with the old nMetal process and with the new nMetal reported in this work. The workfunction metal thickness reduction (4nm down to 1.5nm) with the old nMetal leads to a severe threshold voltage increase, while the new nMetal integration demonstrates low threshold voltage for a physical thickness of 1.5nm for no gate leakage/PBTI lifetime penalty.

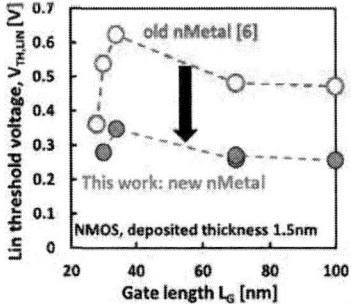

Fig. 8: Linear threshold voltage for the previously reported NMOS Si NWFETs (open symbols, [6]) and for the new nMetal process reported in this work (closed symbols). Low V_{TH} is maintained at short L_G with the implementation of the new nMetal process (physical thickness: 1.5nm).

978-1-7281-1988-5/18 $31.00 © 2018 IEEE

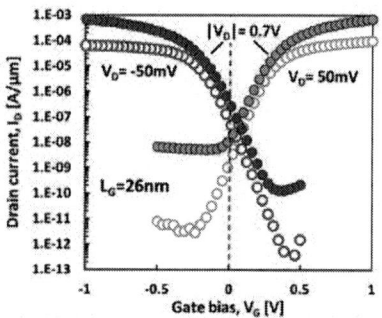

Fig. 9: Short-channel $I_D(V_{GS})$ for NW-FETs with HD Si:P (NMOS) and SiGe:B (PMOS) in-situ doped S/D structures, and improved STI module.

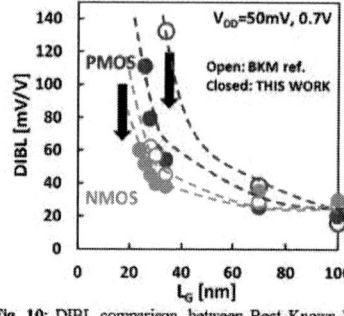

Fig. 10: DIBL comparison between Best Known Method reference (open symbols) and improved GAA Si NW-FETs. (closed symbols). Drive current improvements shown in following figures are obtained for improved short channel margin.

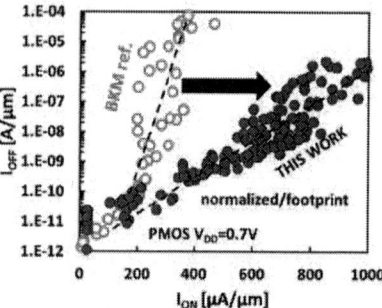

Fig. 11: PMOS I_{ON}/I_{OFF} comparison between Best Known Method reference (open symbols) and improved process flow reported in this work (closed symbols). A large performance improvement is reported when normalizing current per footprint on wafer.

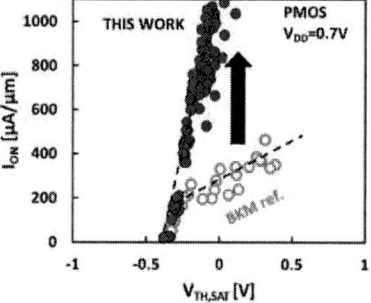

Fig. 12: PMOS $I_{ON}(V_{TH,SAT})$ comparison between Best Known Method reference (open symbols) and improved process flow reported in this work (closed symbols). Performance improvement is related to drive current improvement (normalization per footprint on wafer).

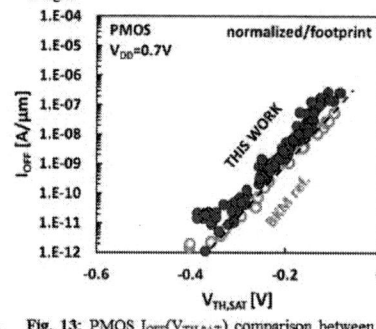

Fig. 13: PMOS $I_{OFF}(V_{TH,SAT})$ comparison between Best Known Method reference (open symbols) and improved process flow reported in this work (closed symbols). Off-state leakage degradation with NW size increase is very limited (normalization per footprint on wafer).

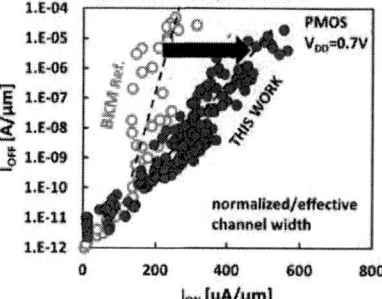

Fig. 14: PMOS I_{ON}/I_{OFF} comparison between Best Known Method reference (open symbols) and improved process flow reported in this work (closed symbols). When normalized per effective channel width. PMOS drive current improvement is also reported.

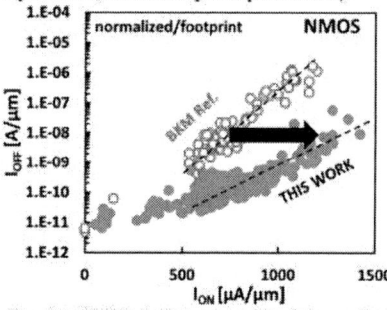

Fig. 15: NMOS I_{ON}/I_{OFF} comparison between Best Known Method reference (open symbols) and improved process flow reported in this work (closed symbols). A large performance improvement is reported when normalizing current per footprint on wafer.

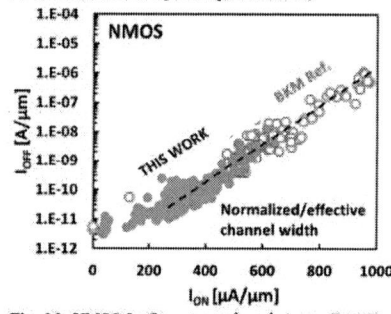

Fig. 16: NMOS I_{ON}/I_{OFF} comparison between Best Known Method reference (open symbols) and improved process flow reported in this work (closed symbols). When normalized per effective channel width, no NMOS drive current improvement is reported.

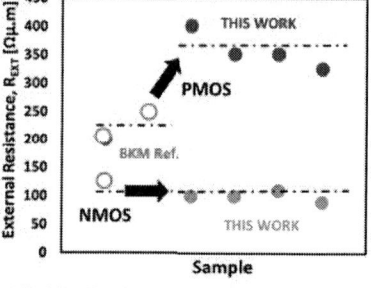

Fig. 17: NMOS and PMOS External resistance R_{EXT} (extracted with the $R_{ON}(1/(V_G-V_{TH})$ method [8]) comparison between Best Known Method reference (open symbols) and improved process flow reported in this work (closed symbols). Drive current improvements are not related to S/D R_{EXT} reduction.

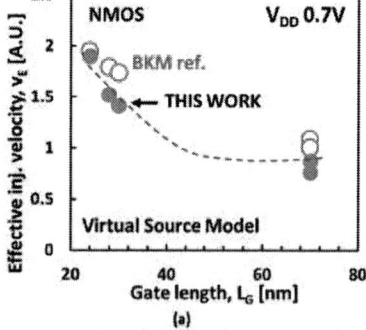

Fig. 18: Effective injection velocity extracted with the Virtual Source Model [9] for (a) NMOS and (b) PMOS devices. Drive current improvement for PMOS devices are linked to an improvement of short channel injection velocity that might be related to larger channel dimensions and/or shape, or strain.

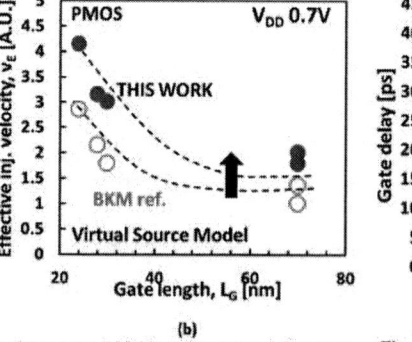

Fig. 19: Ring Oscillator gate delay comparison between Best Known Method reference and improved CMOS process flow. A gate delay decrease from 24 down to 10ps at matched V_{DD} is reported with the improved CMOS process flow.

978-1-7281-1988-5/18 $31.00 © 2018 IEEE   511

2D materials: roadmap to CMOS integration

C. Huyghebaert[1](chuygeb@imec.be), T.Schram[1], Q. Smets[1], T.Kumar Agarwal[1], D.Verreck[1], S.Brems[1],
A.Phommahaxay[1], D.Chiappe[2], S. El Kazzi[1], C.Lockhart de la Rosa[1,3], G. Arutchelvan[1,3],D.Cott[1], J.Ludwig[1,3],
A.Gaur[1,3], S.Sutar[1], A. Leonhardt[1,3], D. Marinov, D. Lin[1], M.Caymax[1], I.Asselberghs[1], G.Pourtois[1] and I.P. Radu[1]

[1]imec, Leuven, Belgium, [2]currently@ASM Europe, Helsinki, Finland, [3]KULeuven, Belgium

Abstract—**To keep Moore's law alive, 2D materials are considered as a replacement for Si in advanced nodes due to their atomic thickness, which offers superior performance at nm dimensions. In addition, 2D materials are natural candidates for monolithic integration which opens the door for density scaling along the 3rd dimension at reasonable cost. This paper highlights the obstacles and paths to a scaled 2D CMOS solution. The baseline requirements to challenge the advanced Si nodes are defined both with a physical compact model and TCAD analysis, which allows us to identify the most promising 2D material and device design. For different key challenges, possible integrated solutions are benchmarked and discussed. Finally we report on the learning from our first lab to fab vehicle designed to bridge the lab and IMEC's 300mm pilot line.**

Si CMOS scaling is facing 2 fundamental hurdles [1]. 1.) The decreasing charge mobilities in ultra-thin channels like fins, nano-sheets or -wires introduced to maximize electrostatic control at scaled dimensions. 2.) A slow V_{dd} downscaling dictated by the 60mV/dec thermal limit. These limitations can be addressed by replacing the channel materials and device design. Charge mobility values up to few 1000 $cm^2/V.s$ have been predicted for monolayer 2D semiconductors and their inherent self-passivated nature promises better immunity to short channel effects.

I. PERFORMANCE BENCHAMRK AND CIRCUIT SIMULATION

Over the past 10 years several semiconducting 2D materials have been proposed for transistor applications, such as transition metal dichalcogenides (TMDC) [2], [3], [4], [5] and black phosphorus [6]. Different 2D materials excel with their specific properties, such as carrier mobility, ease of growth, EHS (Environment, Health and Safety), stability, existence of a native oxide, etc. However, as of today none of them combines all of these properties and stands out as an obvious candidate. Therefore, in search of a potent contender for iN3 and beyond, the potential of integrating various 2D materials under different CMOS device designs has been investigated and the performance numbers are benchmarked against the incumbent technologies. A recent study [7] focused on benchmarking different 2D FET architectures against the advanced Si FinFET platform (Fig1) with different TMDCs using a physical compact model. The study suggested that 2D materials, with WS_2 as the best candidate, could meet N3 performance requirements with an improved energy consumption compared to Si-FinFET (Fig2). A second study [8] which combines a calibrated MoS_2 FET device TCAD

model with a three-level Cu-TaN/Ru interconnect scheme showed that to achieve similar performance with a 2D FET circuitry, the dominant R_{drive} in 2D FETs needs to be balanced through a reduction of the wire dimensions in the interconnect compared to Si FinFET. A reduction of R_{drive} through the addition of a second gate results in reduced delay but at the cost of a slightly increased power (Fig3). Both studies come to similar conclusions: only double gated 2D FET architectures can compete against the Si FinFET, given that the carrier mobility can be increased to several 100's $cm^2/V.s$ and the device parasitic resistance can be lowered to ~100 $\Omega.\mu m$ (Fig3-4).

II. MATERIAL QUALITY ON TARGET WAFER

For 2D materials to be compatible with conventional semiconductor standards, upscaling to wafer level and maintaining single-crystal flake-like quality is one of the key challenges. Typically, two different routes are pursued.

The first and preferred approach is a direct growth approach. Thermal budget notwithstanding, direct growth of TMDC's on amorphous dielectrics has been demonstrated for which electrical mobility values are almost on par with micro-exfoliated flakes [9], [10]. Unfortunately, the high temperature and the harsh growth ambient also affects the substrate oxide quality and reliability. Cointegration on existing devices therefore implies limitations towards temperature and template choice. These limitations have an impact on the achievable material quality and high TMDC defectivity is reported if the temperature is kept below 450°C [11], [12].

Alternatively, 2D materials can be grown without any temperature restriction and on the desired template to optimize the crystal properties. Several methods are reported to achieve large area high quality growth at wafer level [13], [14]. Subsequently the 2D layer will need a transfer to a target wafer for device processing and integration.

During the transfer step, the 2D layer will interface with the manipulating polymers and liquid media. Even today most of the proposed transfer methods are typically lab based relying on craftsmanship rather than on automated and controlled manipulation. This results in a variable success rate as residues lead to doping effects, and cracks and wrinkles form as unwanted by-product of the mechanical forces applied during lamination. The electrical properties of 2D materials may also be significantly impacted by the transfer process because of all the interfacial interactions. Local strain variations in 2D materials can also result in local bandgap variations [15]. The control of the electrostatic potential landscape at the interface between the 2D layer and the dielectric, is critical to reducing variability in 2D FETs over the wafer as can be seen in Fig5. Leveraging technology developed for 3D wafer level integration on 2D material transfer is one of the opportunities to be researched in the coming years. The combination of controlled force and manipulation together with a controlled ambient during stacking allows for an automated transfer. The method could also build hetero-stacks with 2D materials of choice with pristine and ultra-flat interfaces at wafer level. Fig7 shows a schematic of a 300mm WS_2 transfer carried out in 3D wafer stacking infrastructure.

III. EOT SCALING

Depositing ALD oxides on the 2D surface is a true challenge due to the lack of dangling bonds [16]. Alternatively, interfacial Si has been introduced as a seed layer for ALD oxides. This sub-nm Si layer deposited by either PVD or MBE can be oxidized in ambient air and serves as a nucleation layer. In Fig8, the top-gate sub-threshold swing (SS) numbers are plotted against the current on-off ratio for two groups of MOSFETs. Reducing the Si layer thickness from 0.58 nm to 0.41 nm leads to noticeable improvement in SS. However, discrepancies between the nominal EOT and measured CET suggest the 2D surface is susceptible to process induced contamination and residues. Fig6 is the benchmarking plot of the top-gated 2D MOSFETs. Notice the imec data is on large area CVD materials while most of the literature data focus on flakes. Another EOT study with back-gated devices examine the scaling trends of MIMCAPs and MoS_2 MOSCAPs built on the same HfO_2-TiN-SiO_2-Si substrates, as shown in Fig9. Three different types of capacitors show similar scaling trends in measured CET but have different offsets. MoS_2 capacitors show the highest CET at the same HfO_2 thickness. This illustrates possible transfer-related residue and adsorption at the bottom MX_2 surface. The cleanliness of the 2D surfaces is critical to EOT scaling and needs to be addressed during gate stack engineering of the 2D MOS devices.

IV. PARASITIC RESISTANCE ENGINEERING

The main engineerable contributors in the device parasitic resistance of 2D FETs are the contact resistance (R_C) and the access resistance. Several models have been proposed to explain the carrier injection from the top metal to the 2D semiconducting material [17]. From a calibrated TCAD model we could establish the balance between the injection of carriers under the contact and lateral injection from the metal contact directly to the channel (Fig10). For films below 5 layers lateral injection is dominant and thus the transfer length is limited to about 5nm [18], [19]. Two approaches have been followed to improve R_C on 2D semiconductors: contact interface engineering [20], [21], [22] and doping of the 2D material. For doping, the increase of carriers of the 2D film close to the metal contact border is the most successful approach specially for thin films where lateral injection is the most relevant mechanism [18]. Theoretically, the lowest R_C that can be achieved by improving both the interface and doping, is around 80 $\Omega.\mu m$ [4]. The most common doping techniques are electrostatic [23], substitutional [24], [25] and surface doping [26], [27]. Electrostatic doping requires the presence of multiple gates, and scaled embedded structures that complicate

the processing of the devices and increase parasitic capacitances and current leakages. For the substitutional doping, the insertion of species replacing atoms in the 2D film damage the 2D film reducing the carrier mobility [25]. Only when the substitutional doping is achieved by occupying deficit defects in the 2D layer substitutional doping can be achieved without degrading the mobility [24]. However, this method is limited by the non-intentional defects in the 2D film. In surface doping, the dopant species are absorbed at the surface of the film and due to dipole interaction and (or) partial electron exchange [28], [29] with the adsorbed molecule 30], 27].

V. INTEGRATION

Finally we report on the first test vehicle that was set up in our 300m pilot line which bridges learnings from lab to fab. The device architecture is a double-gated WS_2 FET, as shown in the SEM/TEM in Fig11. The integration, including the low temperature PEALD WS_2 deposition directly grown on the dielectric, was entirely performed on full-scale 300 mm wafers in the imec pilot line. The side contacts are fabricated in a classical damascene module, which can be used in a configuration where metal interconnect lines are used for gating the 2D materials [31]. Although this device still has a long channel length >200nm and large EOT, it helps us in debugging the addressed challenges. The back gate overlaps the entire channel including the contact regions, eliminating the need to dope the 2D materials chemically in the extension region and reducing the access resistance. The PEALD WS_2 channel shows ambipolar conduction and current modulation by both top and back gates as seen in Fig12(a). The integration scheme demonstrates robustness as devices with transferred material (CVD MoS_2, sulfurized MoS_2 and graphene) have been produced with an unchanged integration scheme (Fig12f(b)). CVD MoS_2 channels with Ti side contacts currently show the best performance, and further improvements are expected with reduced channel length and EOT.

VI. CONCLUSION

Calibrated TCAD and compact models demonstrate that 2D materials are a competitive alternative for ultimately scaled Si technology nodes if the mobility, the device parasitic resistance and the variability can be optimized. Breakthroughs are essential in the growth of uniform synthetic crystalline material, in the interface control between 2D and dielectric environment, and in the reduction of parasitic resistances in the contact towards the material. A complete 300mm platform is established at imec, to create the ecosystem to allow to enforce integration solutions and demonstrate the viability of 2D technology in CMOS processing environments.

ACKNOWLEDGMENT

The authors gratefully acknowledge the logic Imec Industrial Affiliation Program and funding from the EC's Graphene Flagship initiative.

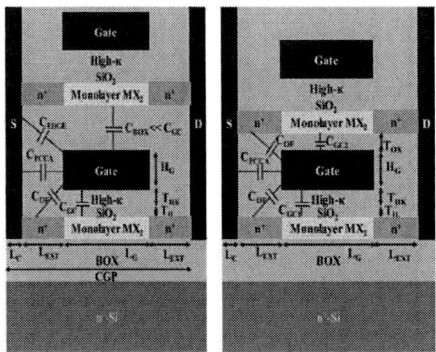

Fig1: Designs of benchmarked monolithic 3D integrated MX2 FETs, with single gate using thick SiO_2 between two stacked transistors (SG) with double gate using 2nm $SiO2$ as back gate oxide with double gate with high-K on both sides (DG).

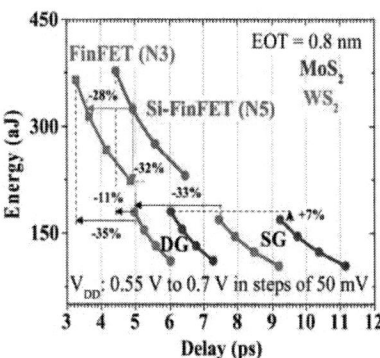

Fig2: Energy delay comparison of Single Gate (SG), Double Gate (DG) MX_2 FETs and Si FinFET for fixed wire load with N5 technology assumptions, computed using effective capacitance and ON current.

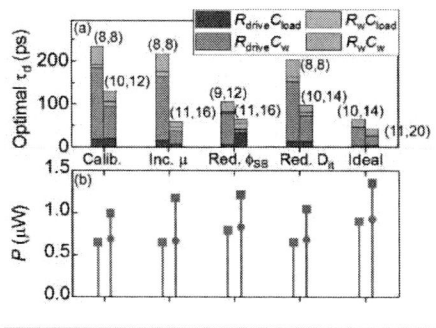

Fig3: (a) Optimal delay and contributions of the terms of the circuit and the device for SG and DG in various scenarios, with the associated device parameter values listed in Table (b).

Parameter	L_{ul}	μ	D_{it}	ϕ_{SB}	R_{drive}	C_{load}
Unit	nm	cm²/Vs	cm⁻²eV⁻¹	eV	kΩ	fF
Calib.	0	14.5	6x10¹²	0.72	173 | 61.3	0.148 | 0.496
Spacer	3	14.5	6x10¹²	0.72	2.13x10⁴ | 115	0.105 | 0.457
Inc. μ	0	100	6x10¹²	0.72	156 | 20.3	0.148 | 0.496
Red. D_{it}	0	14.5	5x10¹¹	0.72	144.7 | 42.5	0.149 | 0.486
Red. ϕ_{SB}	0	14.5	6x10¹²	0.2	61.2 | 21.8	0.148 | 0.585
Ideal	0	100	5x10¹¹	0.2	30.9 | 11.5	0.147 | 0.585
FinFET ref	/	/	/	/	12.8	0.500

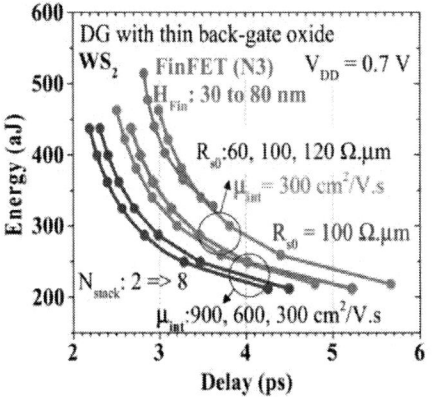

Fig4: Impact of mobility and device parasitic resistance on performance of stacked WS_2 based double-gate FETs with thin BG oxide. These calculations were performed with fixed wire load.

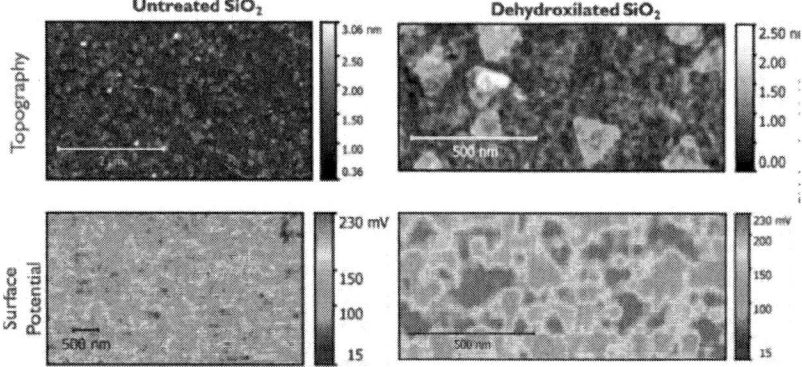

Fig5: Surface topography (top, measured by Atomic Force Microscopy) and surface potential (bottom, measured by Kelvin Probe Force Microscopy) of synthetic MoS_2, grown by MOCVD on sapphire substrates and transferred to SiO_2. On the right, the surface of the target wafer was prepared through thermal treatments before transfer. The synthetic MoS_2 was peeled from sapphire wafer with the help of a PMMA/Thermal Release Tape (TRT) stack and after a dip in ultrasonic DI-water bath. The lamination of the MoS_2/PMMA/TRT was done in a glovebox in order to reduce the H_2O

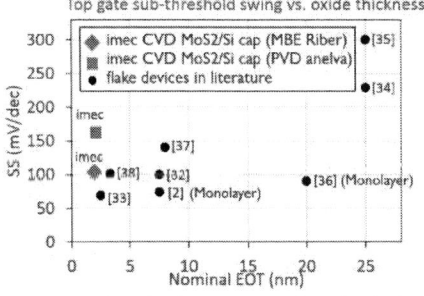

Fig6: Benchmark of our top-gated MoS_2 devices with the best results achieved on flakes in literature.

References [1] M. T. Bohr and I. A. Young, IEEE Micro, vol. 37, no. 6, pp. 20–29, Nov. 2017. [2] B. Radisavljevic et al, Nat. Nanotechnol., vol. 6, p. 147, Jan. 2011. [3] D. Ovchinnikov et al., ACS Nano, vol. 8, no. 8, pp. 8174–8181, Aug. 2014. [4] S. Manzeli et al. Nat. Rev. Mater., vol. 2, p. 17033, Jun. 2017. [5] M. J. Mleczko et al., Sci. Adv., p. 22, 2017. [6] S. J. Choi et al., Nano Lett., vol. 16, no. 7, pp. 3969–3975, Jul. 2016. [7] T. Agarwal et al.,Conference proceedings of IEDM2017, pp. 5.7.1-5.7.4. [8] D. Verreck et al., IEEE, no. IITC 2018 Conference Proceedings, p. 3. [9] K. Kang et al., Nature, vol. 520, no. 7549, pp. 656–660, Apr. 2015. [10] X. Liu et al., Nanotechnology, vol. 28, no. 16, p. 164004, 2017. [11] B. Groven et al., J. Vac. Sci. Technol. A, vol. 36, no. 1, p. 01A105, Nov. 2017. [12] B. Groven et al., Chem. Mater., vol. 29, no. 7, pp. 2927–2938, Apr. 2017. [13] D. Dumcenco et al., ACS Nano, vol. 9, no. 4, pp. 4611–4620, Apr. 2015. [14] M.-Y. Li et al., Science, vol. 349, no. 6247, pp. 524–528, Jul. 2015. [15] B. G. Shin et al., Adv. Mater., vol. 28, no. 42, pp. 9378–9384, Nov. 2016. [16] H. Zhang et al., J. Chem. Phys., vol. 146, no. 5, p. 052810, Nov. 2016. [17] A. Allain et al., Nat. Mater., vol. 14, p. 1195, Nov. 2015. [18] G. Arutchelvan et al., Nanoscale, vol. 9, no. 30, pp. 10869–10879, 2017. [19] A. Prakash et al. Sci. Rep., vol. 7, p. 12596, Oct. 2017. [20] C. D. English et al., Nano Lett., vol. 16, no. 6, pp. 3824–3830, Jun. 2016. [21] R. Kappera et al., Nat. Mater., vol. 13, p. 1128, Aug. 2014. [22] H.-J. Chuang et al., Nano Lett., vol. 16, no. 3, pp. 1896–1902, Mar. 2016. [23] G. V Resta et al., Sci. Rep., vol. 6, p. 29448, Jul. 2016. [24] L. Yang et al., Nano Lett., vol. 14, no. 11, pp. 6275–6280, Nov. 2014. [25] R. Murray et al, ECS J. Solid State Sci. Technol., vol. 5, no. 11, pp. Q3050–Q3053, 2016. [26] D. Kiriya et al., J. Am. Chem. Soc., vol. 136, no. 22, pp. 7853–7856, Jun. 2014. [27] C. J. Lockhart de la Rosa et al., Nanoscale, vol. 9, no. 1, pp. 258–265, 2017. [28] S. Najmaei et al., Nano Lett., vol. 14, no. 3, pp. 1354–1361, Mar. 2014. [29] R. Phillipson et al., Nanoscale, vol. 8, no. 48, pp. 20017–20026, 2016. [30] C. J. Lockhart de la Rosa et al., Appl. Phys. Lett., vol. 109, no. 25, p. 253112, Dec. 2016. [31] T. Schram et al., ESSDERC 2017 [32] Nano Lett. 14. 5905-5911 (2014) [33] Adv. Mater. 26, 6255-6261 (2014) [34] J. Mater. Chem. C 2, 8023-8028 (2014) [35] Nano Lett. 12, 3695 3700 (2012) [36] Nanoscale 5, 548-551 (2013) [37] IEEE Electron Dev. Lett. 33, 546-548 (2012) [38] Scientific Reports 5, No. 11921 (2015)

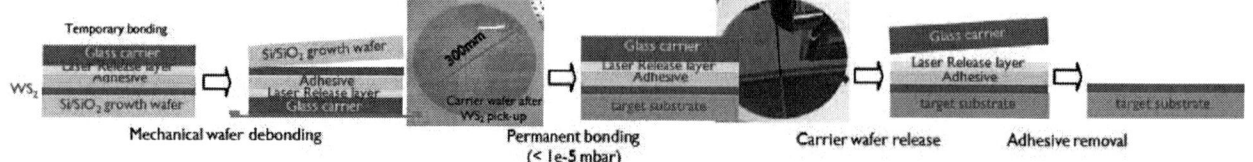

Temporary bonding

Mechanical wafer debonding   Permanent bonding   Carrier wafer release   Adhesive removal
(< 1e-5 mbar)

Fig7: Schematic process sequence of 300mm level 2D-material transfer by leveraging the 3D debonding and bonding technology. The 300mm WS_2 material is grown on SiO_2/Si wafer. A temporary adhesive/ laser release layer stack is spun on the glass wafer, which is bonded to the WS_2 layer under vacuum conditions. This stack allows to peel of the WS_2 from the SiO_2 growth substrate. This stack is then bonded to the target wafer in a permanent bonding chamber again under vacuum conditions. Finally the temporary stack is released by a laser in order to reduce the sheering forces to zero. Having access to 300mm grown material allows to use established wafer handling methods and increase the level of control in transfer engineering.

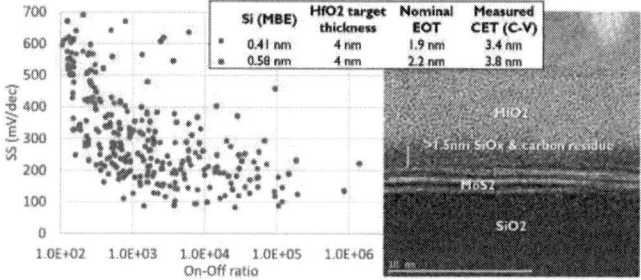

Fig8: The SS performance statistics of the top-gated long-channel MoS_2 MOSFETs illustrate the impact of EOT (Si cap thickness). The lowest SS (~100mV/dec) corresponds to a D_{it} level of ~5×10^{12}/cm^2-eV. However, capacitor C-V measurements give higher CET estimate.

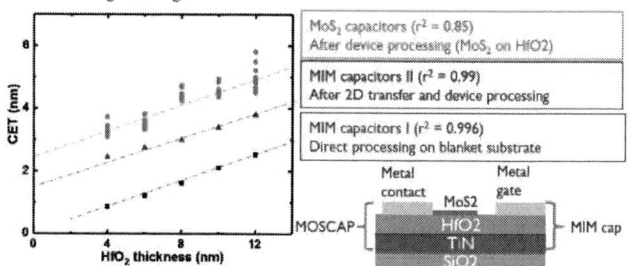

Fig9: The scaling trends of two types of MIM (Metal-insulator-metal) capacitors and the MoS_2 capacitors are compared based on the same substrate (HfO_2/TiN/SiO_2/Si). 2D transfer and subsequent processing appears to introduce interfacial layers between the top metal and the HfO_2 layer, as indicated by the 1.74nm additional CET between the two types of MIM caps. The additional CET difference between the MoS_2 and the MIM capacitors can be attributed to the transfer-related polymer residues, oxidation and water adsorption at the MoS_2 bottom surface.

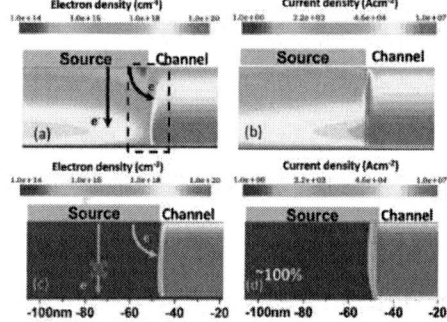

Fig10: Simulation by semi-classical model to identify the main mechanisms and trajectories for carrier injection at MoS_2 contacts. (a) Overview of the electron distribution in the structure for multilayer (>2) MoS_2, (b) The contribution of injection at the contact edge and injection under the contact increase with lateral and perpendicular fields. (c) The depleted MoS_2 in the contact area for monolayer and bilayer MoS_2 structures (d) The carriers are predominantly injected at the edge of the contact metal.

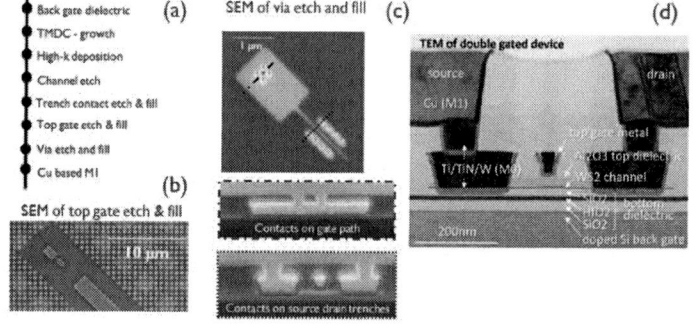

Fig11: (a) Overview of 300mm integration flow for top-gated WS_s FET. (b) SEM image of top gated device after gate etch and metal fill; (c) SEM image of top gated device after via processing. Cross-sections show the contacts to the gate path (top) and source drain trench (bottom). (d) TEM of double gated PEALD WS_2 FET integrated with Damascene side contacts

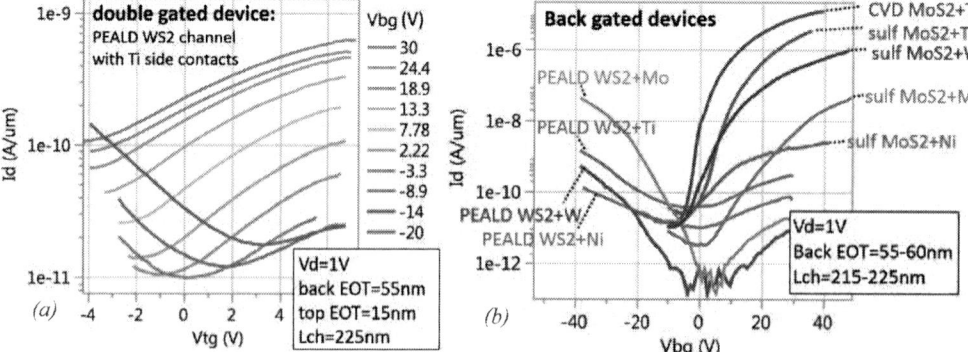

Fig12: (a) Double gated FETs with PEALD WS_2 channel and Ti side contacts show ambipolar conduction and current modulation by both top and back gates.

(b) Back gated FETs with PEALD WS_2 channel are ambipolar while MoS_2 channels have n-type conduction. Different side contact metals (deposited by PVD) are evaluated, with Ti showing the best performance for MoS_2. The EOT=60nm (PEALD WS_2 and sulfurized MoS_2) or 55nm (CVD MoS_2).

978-1-7281-1988-5/18 $31.00 © 2018 IEEE   515

First Demonstration of WSe$_2$ Based CMOS-SRAM

Chin-Sheng Pang[1,2], Niharika Thakuria[1], Sumeet Kumar Gupta[1], and Zhihong Chen[1,2*]

[1]School of Electrical and Computer Engineering and [2]Birck Nanotechnology Center, Purdue University,
West Lafayette, IN; *email: zhchen@purdue.edu

Abstract—In this work, we demonstrate a CMOS static random-access-memory (SRAM) using WSe$_2$ as a channel material for the first time, providing comprehensive DC analyses for transition metal dichalcogenide (TMD) material-based memory applications. A tri-gate design is adopted for the n-type MOSFET, while an air-stable, oxygen plasma induced doping scheme is introduced to implement the p-type MOSFET. DC measurements of SRAM cells demonstrate a unique dynamic tunability enabled by modulating the n-FET doping level through electrostatically gating the extended source/drain regions. Furthermore, with various read/write assist techniques, SRAM operation at low V_{DD} of 0.8V is achieved. Our low power demonstration and its 2D ultra-thin material nature suggest promising applications of WSe$_2$ for flexible electronics and Internet of Things (IoT).

INTRODUCTION

Two-dimensional (2D) semiconducting transition metal dichalcogenides (TMDs) have attracted enormous attention for the coming era of flexible electronics and Internet of Things (IoT), by virtue of their ultra-thin body, mechanical flexibility, and excellent transport properties [1-2]. Before 2D TMDs were re-discovered for electronic applications in the recent years, organic thin films, poly or amorphous Si, metal oxide thin films, and carbon nanotubes (CNTs) have been extensively explored for thin film transistor technologies. However, except CNTs, other channel materials' low charge mobility, large body thickness limiting the device scaling, or lack of p-type organic and metal-oxide thin film transistors to realize CMOS, leads to low device/circuit performance, low packing densities or excessive power consumption [3-4]. In the context of CNTs, obtaining pure semiconducting tubes with similar electronic properties remains challenging [5]. On the other hand, flexible 2D nanoelectronics using TMDs as the channel materials is believed to offer solutions to the above-mentioned challenges, providing a unique and compelling technology road map.

WSe$_2$ is an air-stable TMD with a sizeable bandgap. The Fermi levels of most contact metals are pinned close to the mid-gap, allowing access to both electron and hole carriers. Consequently, most prior arts adopting Schottky barrier FET structures with source/drain metals in direct contact with the channel always show ambipolar device characteristics, which unavoidably leads to higher static power consumption [6-7]. Attempts to implement MOSFET type of designs on WSe$_2$ in order to achieve unipolar characteristics have been made. However, most cases focus on improving single device performance [8] and few have demonstrated CMOS inverters [9-10] for basic logic operation. Here, we demonstrate, for the first time, a CMOS static random-access-memory (SRAM) using 6 WSe$_2$ MOSFETs, with the p-FETs being chemically doped and the n-FETs being electrostatically doped. We demonstrate a dynamic tunability in noise margin by

electrostatically controlling the doping level in the extended source/drain of the n-FETs, which is critical for resolving the read/write (R/W) conflict in SRAM operation, given the mismatch in threshold voltages (V_{TH}) and global variations in the strength of FETs. In addition, by exploiting R/W assist techniques, our design yields enhanced SRAM stability, enabling operation at low $V_{DD} = 0.8V$. This is the first CMOS SRAM demonstrated on any TMDs up to date, which is one step forward in the 2D materials based flexible electronics field.

CMOS IMPLEMENTATION

To realize true CMOS logic with low standby power, MOSFET device designs are adopted for WSe$_2$ FET fabrication. A novel tri-gate structure is implemented for n-FETs to obtain the desired n/i/n doping profile. First, pairs of Ti (2nm) /Au (8nm) source/drain (S/D) contacts were patterned, followed by micromechanical exfoliation of thin WSe$_2$ (~6nm) flakes, as shown in Fig. 1(a). A gate stack of ~7nm ALD HfO$_2$ dielectric with 1.5nm Al seeding layer and 30nm e-beam evaporated Ni electrode was then fabricated through a lift-off process to form two side gates. Finally, another ~9nm ALD HfO$_2$ with a Ni gate was patterned in the middle region of the channel to form the tri-gate structure, as shown in Fig. 1(b). The side gates were connected and positively biased ($V_{TGN} > 0V$) to provide the desired n/i/n doping profile. The colored scanning electron microscope (SEM) image is shown in Fig. 1(c). Transfer characteristics curves with distinct V_{TGN} are presented in Fig. 1(d), showing n-type unipolar characteristics with tunable on-state currents, more than 6 orders of magnitude on/off ratios, a near ideal subthreshold swing (S.S.) of 69mV/dec, and negligible hysteresis of 2mV.

For the p-FET fabrication, S/D contacts were defined on top of the WSe$_2$ channel, followed by a lift off process to form a gate stack of ALD HfO$_2$ /Ni in the middle with the extended S/D regions exposed, as shown in Fig. 2(a). We then performed an O$_2$-plasma treatment to form a WO$_{3-x}$ layer on top of the exposed WSe$_2$ portion, which served as a strong p-type dopant due to its high electron affinity [11-12]. Fig. 2(b) shows V_{BG}-dependent carrier injection behaviors in a pristine device. In contrast, after the treatment with V_{BG} being floating, the device displayed p-type unipolar characteristics with improved on-state, S.S. of 90mV/dec and small hysteresis of 30mV.

SRAM DESIGN AND DC MEASUREMENTS

A pull-up (PU) p-FET, a pull-down (PD) n-FET, and an access n-FET were integrated on the same substrate with proper internal connections to form a half-SRAM cell. The evaluation of read/write/hold noise margins (NM) of SRAM is based on our measurements of two distinct half-cells in the next sub-section, while the assist techniques consider symmetric SRAM.

A. DC Characteristics of SRAM Cell

We fabricated two sets of half-cells noted as "Left" and "Right". Fig. 3(a) shows a design approach in which V_{DD} and

978-1-7281-1988-5/18 $31.00 © 2018 IEEE

V_{TGN} share the same voltage source (required due to the limited number of probes in our measurement system). Voltage transfer characteristics (VTCs) for distinct SRAM operations are shown in Fig. 3(b) along with applied bias voltages. To obtain read/hold static noise margins (SNM), we fit the largest square in the "eye" of the read/hold butterfly curve. For the write margin (WM), the smallest fitted square is obtained for curves at V_{BL}=1.5V and 0V. Those extracted values are shown in Fig. 3(c). NMs for these cells indicate a write-favored SRAM behavior with large WM (~0.435V) achieved at the cost of read SNM.

B. Read Assist Using Word Line (WL) Under-drive

To explore the possibilities of enhancing read stability, we analyze WL under-drive as a read assist technique considering symmetric SRAM that involves lowering the WL voltage [13]. This technique weakens the access n-FET compared to the PD n-FET. The measured VTCs and NM values are shown in Fig. 4(a) and (b) with enhanced read stability up to 0.39V at V_{WL} = 0.8V without sacrificing WM.

C. Dynamically Tunable SRAM via V_{TGN} Modulation

The unique capability of dynamically modulating the n-FET strength in the demonstrated WSe₂ devices offers appealing design options for SRAMs. Fig. 5(a) shows the schematic of a design, in which we employ a common voltage source for V_{DD} and V_{BL} for hold operation to allow the freedom of tuning V_{TGN} (while keeping the number of probes the same in our experimental setup). Fig. 5(b) shows the optical image for a half-cell. The VTCs and hold SNM of the SRAM can be dynamically tuned via V_{TGN} (Fig. 5(c)). To characterize the read and write operation on the same half-cell, we physically disconnected V_{DD} and V_{BL} in the original design and connected V_{DD} to V_{WL}, as illustrated in Fig. 6(a). VTCs with distinct n-FET strengths at V_{BL}=1.5V (Fig. 6(b)) and 0V (Fig. 6(c)) were used to characterize the read/write (R/W) stabilities (considering symmetric half-cells), with varying V_{TGN} applied on both PD n-FET and access n-FET (Fig. 6(d)). No square could be defined for the black curve of the WM indicating a failed write operation at V_{TGN} = 1V. However, by increasing V_{TGN}, write functionality is established. An explicit conflict with regard to the R/W stability is shown in Fig. 6(e) as a strong n-FET (large V_{TGN}) leads to a robust write but a vulnerable read, while a weak n-FET (small V_{TGN}) results in an opposite outcome. We find V_{TGN} = 1.25V yields balanced R/W stabilities with read SNM = 0.19V and WM=0.21V.

D. V_{TGN} Optimziation with Negative V_{BL} Write Assist

In order to further enhance the write margin and the overall stability of the cell, we employ the negative V_{BL} technique [13]. This write assist enhances the strength of the access n-FET with respect to the PU p-FET. VTCs presented in Fig. 7(a) show obvious differences under distinct V_{BL} values. The extracted NMs are plotted in Fig. 7(b) with an 87-242% of improvement of write stability using the negative V_{BL} technique. With increased WM, V_{TGN} can be lowered to increase read stability. With V_{BL}= -0.2V and V_{TGN} = 1.1V, we demonstrate read SNM = 0.38V and WM = 0.39V.

E. V_{DD} Scaling for SRAM Cell

Operating SRAMs at scaled V_{DD} is essential for low power applications [14-15]. However, low V_{DD} operation leads to an increase in failures [16]. Fig. 8(a) and (b) show VTCs and NMs at V_{DD} ranging from 1.8 to 1.2V. A limited SNM for read prevents V_{DD} from scaling further (Fig. 8(c)).

To enhance SRAM stability for low voltage operation, we adopt a ground (GND) lowering technique as read assist [13] and the previously described negative V_{BL} technique as write assist. The GND lowering utilizes an enhancement in the strength of the PD n-FET over the access n-FET to improve the read stability. As shown in the read butterfly curves in Fig. 9 at various V_{DD}, an improved NM for read is achieved. Employing this in conjunction with the negative V_{BL} technique for write, stable SRAM operation for scaled V_{DD} down to 0.8V is achieved with extracted NMs in Fig. 10. While this paper focuses on experimentally analyzing the DC behaviors of WSe₂ SRAMs, the transient analysis (including the effect of R/W assist on array operation) will be a focus for our future study.

SIMULATION ANALYSIS AND BENCHMARKING

SRAM transient analysis is performed through simulations of read/write time and power based on a Verilog A 2D FET model [17]. We calibrate the model using device dimensions and parameters from the experiments. Further, we match the device ON/OFF ratios to experimentally measured values and obtain similar read currents as those from experiments. Our results show read time ranging between 0.35ns -7ns (Fig. 11) with V_{DD}= 1.5V-2.5V for the 32x32/64x64/128x128 bit arrays. Write time of ~10.8-16.8ns is obtained as well (Fig. 11). Read power of 0.72-9.92µW and write power of 0.39-1.89µW are estimated for V_{DD} =1.5-2.5V. Finally, we compare our WSe₂ SRAM at V_{DD} = 2V with available TFT based SRAM demonstrations in Table 1. Our SRAM shows substantially lower read/write time and power compared to other technologies as those normally employ devices with long channels, large overlap capacitance and low mobility channel materials [18].

CONCLUSION

We have demonstrated WSe₂ based SRAM for the first time with comprehensive DC analyses. Combining tunable n-FETs with R/W assist techniques, stable SRAM operation at low V_{DD} of 0.8V is achieved. Superior speed and power compared to other technologies is predicted, suggesting a promising application in flexible electronics and IoT.

ACKNOWLEDGEMENT: This work was supported in part by the Semiconductor Research Corporation (SRC) as the NEW LIMITS Center and NIST through award number 70NANB17H041.

REFERENCES: [1] W. Zhu et al., *Flex. Print. Electron.*, **2**, 043001 (2017); [2] L. Cai et al., *Nano Lett.*, **17**, 3854 (2017); [3] K. Myny, *Nat. Electronics*, **1**, 30 (2018); [4] M. L. Geier et al., *Nat. Nanotech.*, **10**, 944 (2015); [5] D.-M. Sun, *Small*, **9**, 1888 (2013); [6] M. Tosun et al., *ACS Nano*, **8**, 4948 (2014); [7] J. Pu et al., *Adv. Mater.*, **28**, 4111 (2016); [8] H. Fang et al., *Nano Lett.*, **12**, 3788 (2012); [9] C.-S. Pang et al., *IEEE DRC* (2018); [10] G. V. Resta et al., *IEEE DRC* (2018); [11] B. Liu et al., *ACS Nano*, **10**, 5153 (2016); [12] P. R. Pudasaini et al., *Nano Research*, **11**, 722 (2018); [13] B. Zimmer et al., *IEEE TCAS II*, **59**, 853 (2012); [14] H. N. Patel et al., *International Symp. on Quality Electronic Design* (2016); [15] E. Morifuji et al., *IEEE TED*, **53**, 1427 (2006); [16] C.-H. Lin et al., *VLSI-TSA.* (2006); [17] S. V. Suryavanshi et al., S2DS Transistor Model. doi: 10.4231/D3ZC7RV9X; [18] D. Roose., *IEEE JSSC*, **52**, 3095 (2017); [19] K. Fukuda et al., *Adv. Funct. Mater.*, **21**, 4019 (2011); [20] J. A. Avila-Nino et al., *Organic Electronics*, **31**, 77 (2016).

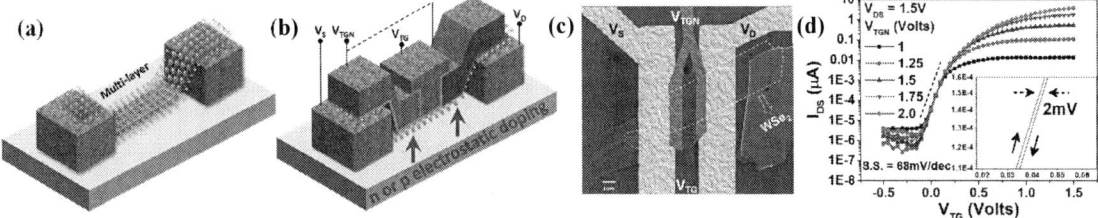

Fig. 1. (a) to (b) Schematics of fabrication processes of a novel tri-gate structure for n-FET and (c) an SEM image. (d) Transfer characteristics of an n-FET operation with negligible hysteresis of 2mV shown in the inset.

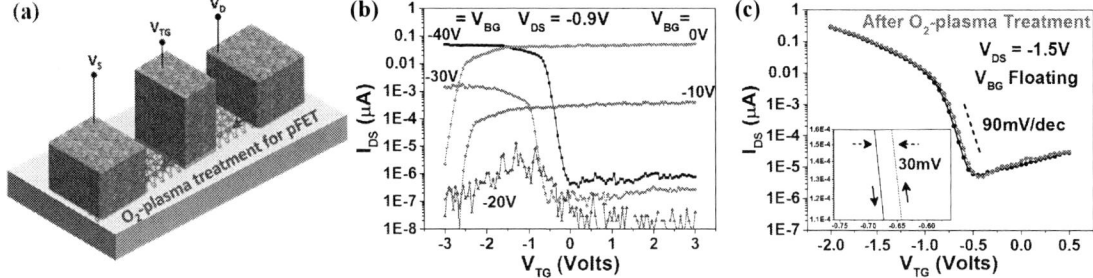

Fig. 2. (a) Illustration of p-FET with exposed extended S/D regions for O_2-plasma treatment. Transfer characteristics of (b) a pristine device with V_{BG}-dependent current injection and (c) a p-FET after O_2-plasma treatment with V_{BG} being floating.

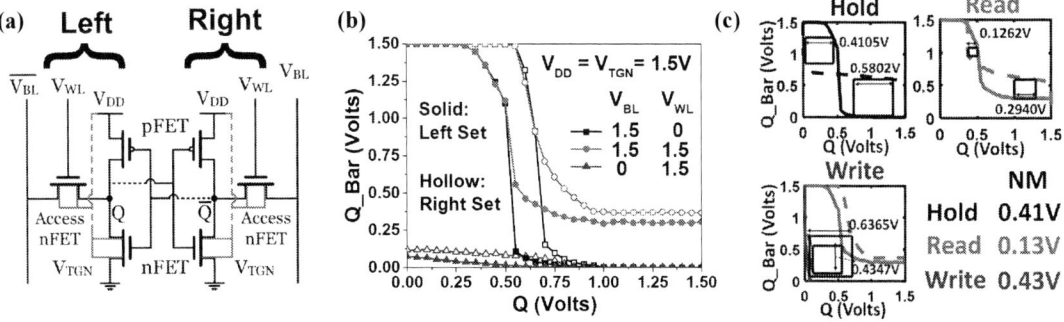

Fig. 3. (a) SRAM cell design wit V_{DD} connected to V_{TGN}. (b) VTCs of two different device sets. (c) NM extraction of hold and read from butterfly curves, and from write curves. The individual NM values indicate a write-favored SRAM behavior.

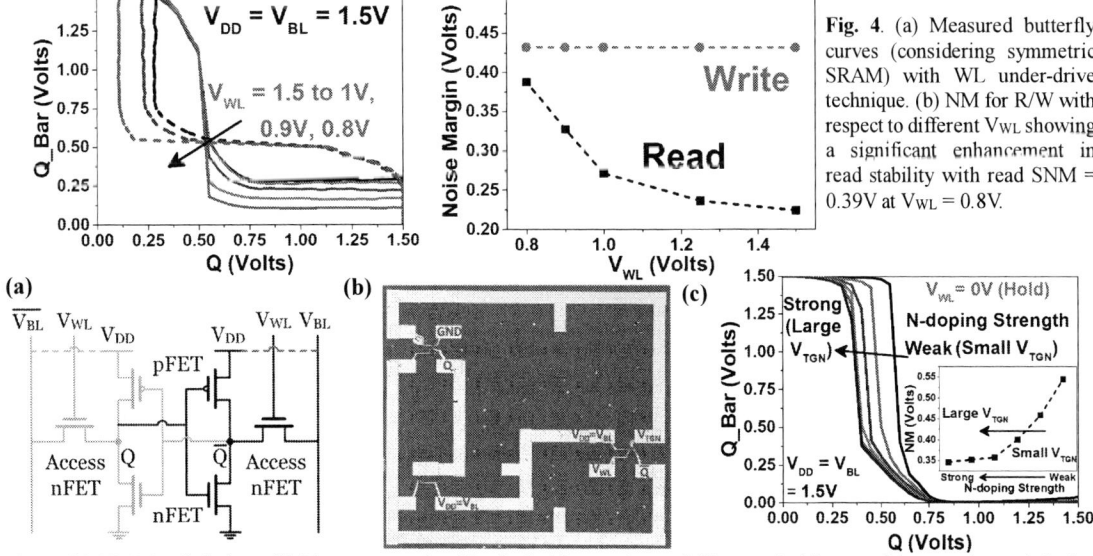

Fig. 4. (a) Measured butterfly curves (considering symmetric SRAM) with WL under-drive technique. (b) NM for R/W with respect to different V_{WL} showing a significant enhancement in read stability with read SNM = 0.39V at V_{WL} = 0.8V.

Fig. 5. (a) SRAM cell design with V_{DD} connected to V_{BL} for hold operation and (b) an optical image of the experimental device set. (c) VTCs and inset shows dynamically tunable NMs extracted from butterfly curves, resulted from different doping strength of the n-FET by tuning V_{TGN}.

978-1-7281-1988-5/18 $31.00 © 2018 IEEE   518

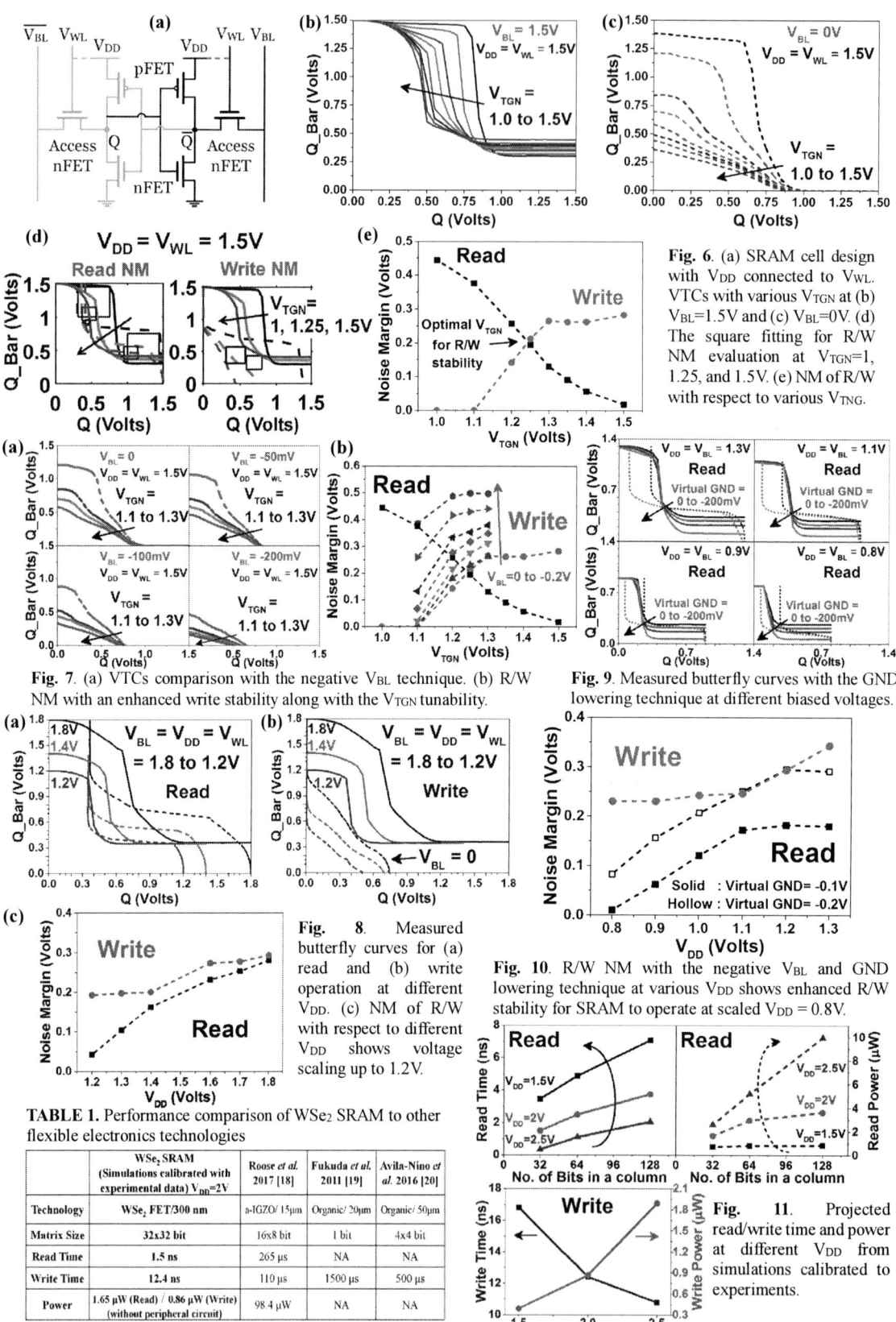

Fig. 6. (a) SRAM cell design with V_{DD} connected to V_{WL}. VTCs with various V_{TGN} at (b) V_{BL}=1.5V and (c) V_{BL}=0V. (d) The square fitting for R/W NM evaluation at V_{TGN}=1, 1.25, and 1.5V. (e) NM of R/W with respect to various V_{TNG}.

Fig. 7. (a) VTCs comparison with the negative V_{BL} technique. (b) R/W NM with an enhanced write stability along with the V_{TGN} tunability.

Fig. 9. Measured butterfly curves with the GND lowering technique at different biased voltages.

Fig. 8. Measured butterfly curves for (a) read and (b) write operation at different V_{DD}. (c) NM of R/W with respect to different V_{DD} shows voltage scaling up to 1.2V.

Fig. 10. R/W NM with the negative V_{BL} and GND lowering technique at various V_{DD} shows enhanced R/W stability for SRAM to operate at scaled V_{DD} = 0.8V.

Fig. 11. Projected read/write time and power at different V_{DD} from simulations calibrated to experiments.

TABLE 1. Performance comparison of WSe₂ SRAM to other flexible electronics technologies

	WSe₂ SRAM (Simulations calibrated with experimental data) V_{DD}=2V	Roose *et al.* 2017 [18]	Fukuda *et al.* 2011 [19]	Avila-Nino *et al.* 2016 [20]
Technology	WSe₂ FET/300 nm	a-IGZO/ 15µm	Organic/ 20µm	Organic/ 50µm
Matrix Size	32x32 bit	16x8 bit	1 bit	4x4 bit
Read Time	1.5 ns	265 µs	NA	NA
Write Time	12.4 ns	110 µs	1500 µs	500 µs
Power	1.65 µW (Read) / 0.86 µW (Write) (without peripheral circuit)	98.4 µW	NA	NA

Steep Slope p-type 2D WSe$_2$ Field-Effect Transistors with Van Der Waals Contact and Negative Capacitance

Jingli Wang[1], Xuyun Guo[1], Zhihao Yu[2], Zichao Ma[3], Yanghui Liu[1], Masun Chan[3], Ye Zhu[1], Xinran Wang[2] and Yang Chai[1*]

[1]Department of Applied Physics, The Hong Kong Polytechnic University, Hong Kong, China, *email: ychai@polyu.edu.hk
[2]National Laboratory of Solid State Microstructures, School of Electronic and Engineering, and Collaborate Innovation Center of Advanced Microstructures, Nanjing University, Nanjing 210093, China
[3]Department of Electronic and Computer Engineering, The Hong Kong University of Science and Technology, Hong Kong, China

Abstract - Steep-slope p-type 2D WSe$_2$ back-gated field-effect transistors (FETs) are realized by using van der Waals Pt-WSe$_2$ contact and HfZrO$_2$/Al$_2$O$_3$ as the dielectric layer. The van der Waals Pt-WSe$_2$ contact is free from disorder and Fermi level pinning and decreases the subthreshold slope. The WSe$_2$ NCFET with van der Waals contact shows low subthreshold slope for both forward and reverse gate voltage sweep (the minimum SS$_{forward}$ = 18.2 mV/dec and SS$_{reverse}$ = 44.1 mV/dec) with a hysteresis as small as 20 mV at subthreshold region.

I. INTRODUCTION

With the downward scaling of the transistors, efficient gate control and ultra-thin body is required to reduce power consumption and subthreshold slope (SS) [1]. Recently, CMOS compatible HZO was used in negative capacitance (NC) FETs to achieve efficient gate control and SS less than 60 mV/dec [2-5]. Two dimensional (2D) transition metal dichalcogenides (TMDs) are semiconductors with ultrathin body and considered as promising candidates for future low power electronics [5-8]. Tungsten diselenide (WSe$_2$) has balanced conduction and valence band edges, providing an appropriate channel material for realizing complementary FETs [5-6]. However, pristine multilayer WSe$_2$ FETs usually exhibits large Schottky barrier (SB) and ambipolar behavior [5, 6, 9]. The disorder at the contact region results in Fermi level pinning effect and influences the band alignment of the metal/semiconductor [7, 9]. The SB reduces the on-state current, and significantly affects the efficient carrier injection, which gives rise to a large SS at the subthreshold region [7-9]. Thus, it is important to reduce the SB to maintain high thermionic current and improve the effective SS and ON current at the same applied gate voltage.

In this work, we fabricated van der Waals (vdWs) metal/semiconductor contact by transferring metal electrode on top of WSe$_2$, which enables an interface without chemical bonding and disorder [7]. The electrical contact without Fermi level pinning reduces the Schottky barrier and the subthreshold slope. With HZO/Al$_2$O$_3$ gate stack, sub thermionic SS for both forward and reverse gate voltage sweep is obtained with minimum SS$_{forward}$ = 18.2 mV/dec and SS$_{reverse}$ = 44.1 mV/dec. The drain current can be modulated by 5×10^4 within 220 mV, making it promising for low power device applications.

II. FABRICATION OF DEVICES

Fig. 1 is the schematic illustration of the fabrication process flow. We prepared 30-nm-thick Pt electrode on a Si substrate using standard photolithography and electron-beam lithography. The deposition rate is set to be 0.1 Å/s. Dry transfer method was applied to fabricate the van der Waals contact to 2D materials. A PVA film was used to mechanically pick up the electrodes from the Si substrate. The temperature was set to be 55 °C for 3 min to avoid strong interaction between the electrode and the Si substrate. The PVA film was attached to a PDMS stamp together with the electrode. The PDMS stamp was then attached to a glass. WSe$_2$ flakes were mechanically exfoliated onto the substrate using scotch tape. After we aligned the electrode and the WSe$_2$ flakes using a 2D materials transfer platform, the PVA thin film with electrodes was physically contacted to the WSe$_2$ flake. After the heat treatment, the PVA film was washed away with deionization water and isopropanol.

Atom layer deposition (ALD) was used to deposit HZO film on p^{++} silicon at 180 °C. We used tetrakis (dimethylamido) hafnium as Hf source, tetrakis (dimethylamido) zirconium as Zr source and H$_2$O as oxidant. The HfO$_2$/ZrO$_2$ ratio was controlled to be 1:1. For capacitance matching and gate leakage reduction, 4-nm-thick Al$_2$O$_3$ was deposited by using trimethylaluminium as Al source and H$_2$O as oxidant. A rapid thermal annealing in nitrogen environment was performed to crystallize the HZO and enhance the ferroelectricity.

III. VAN DER WAALS CONTACT

Fig. 2 shows the transfer characteristics of a typical WSe$_2$ FET on 15 nm ZrO$_2$ with the V$_{ds}$ from - 0.1 V to - 1.0 V at room temperature. Obviously, the device shows an ambipolar transport behavior. Its p-branch current is almost 3 orders of magnitude larger than that of the n-branch. Fig. 3 is the corresponding output characteristics of the WSe$_2$ device. The I$_d$-V$_{ds}$ exhibits nonlinear behavior at the low V$_{ds}$ region, indicating a relative high Schottky barrier Φ_p for holes [8, 9].

For the WSe$_2$ devices with transferred vdWs Pt electrode, its transfer curve is given in Fig. 4. The ambipolar effect is greatly suppressed. The p-branch is enhanced while the n-branch vanishes, indicating a higher barrier Φ_n for electron and reduced Schottky barrier Φ_p for hole [7, 8]. Fig. 5 shows the corresponding output characteristics of the device with vdWs contact. The linear behavior of the output curve indicates small barrier between the Pt electrode and the valence band of WSe$_2$.

978-1-7281-1988-5/18 $31.00 © 2018 IEEE

With reduced hole Schottky barrier, the SS is also greatly decreased. Fig. 6 shows the point SS versus I_d of the devices. The average SS of the FETs with vdWs Pt is 134 mV/dec; while the device with the evaporated Pt electrode exhibits a SS of 226 mV/dec. The SS in the low current region is similar; while in the high current region, the devices with vdWs contact show much smaller SS than that with the evaporated electrode. The device with vdWs contact exhibits minimum forward $SS_{forward}$ = 96 mV/dec and reverse $SS_{reverse}$ = 71 mV/dec, as shown in Fig. 7. The point SS analysis is given in Fig. 8. Fig. 9 is the transfer characteristic of a WSe_2 FET with transferred Au electrode. The device shows significant ambipolar behaviors. The nonlinear behavior of the output characteristics at the low V_{ds} region indicating that the Fermi level of Au electrode lies in the middle of the WSe_2 band gap, and Schottky barrier exists for both electron and hole (Fig. 10). Compared with vdWs Au, vdWs Pt is more suitable for a p-type contact.

Fig. 11 illustrates the simplified band diagrams of the FETs with evaporated and transferred Pt electrode. Fig. 12 is the corresponding transfer curve. In the thermionic region ($|V_{gs}|$ < $|V_{fb}|$), the band position changes equivalent to V_{gs} (irrespective of depletion and interface trap capacitance) [8, 10]. The slope is given by SS = $k_bT/q \cdot \ln 10$ = 60 mV/dec, which is the thermionic limit. When $|V_{fb}| < |V_{gs}| < |V_{th}|$, the tunneling barrier decreases the current and the Slope is given by SS = $\eta k_b T/q \cdot \ln 10$ (η is the ideal factor related to barrier width, η > 1), which is larger than in the thermionic region. With smaller SB, the V_{fb} is close to V_{th} and the tunneling region is reduced, resulting in a small SS within all the sub threshold region (Fig. 12), which is the key factor in low power electronics. This phenomenon is also observed in Fig. 6. The SS in both cases are small in low current region. With high current, the devices with vdWs electrode maintain a smaller SS, while the FETs with evaporated electrodes start to increase.

IV. NEGATIVE CAPACITANCE GATE

Then we use negative capacitance gate to further reduce the SS. Fig. 13 is the high angle annular dark field scanning transmission electron microscopy (HAADF-STEM) image and the energy dispersive spectrum (EDS) of the NCFET gate stack, confirming the element distribution. High-resolution cross-sectional TEM image clearly reveals the structure of the $HZO/AlO_x/WSe_2$ stack with clean interfaces. After the annealing process at 450 °C, the HZO layer is highly crystalline, as shown in Fig. 14 (a). Benefiting from the vdWs contact, the transferred metal/WSe_2 junctions feature an atomically sharp and clean interface within a few angstrom (Fig. 14 (b)). For the deposited metal/TMD junction, there are considerable defects, strain and disorder at the interface [7]. These interfacial defects cause metal-induced gap states and interfaces dipoles and influence the Schottky barrier [7, 9].

Fig. 15 is the polarization versus electric field hysteresis loop of the HZO film. After the annealing at 450 °C, the transition from dielectric to ferroelectric characteristics is clearly observed (measured at 1 kHZ). Fig. 16 is the transfer characteristics of the WSe_2 NCFET with van der Waals contact. The device exhibits minimum forward $SS_{forward}$ = 44 mV/dec and reverse $SS_{reverse}$ = 18 mV/dec. The $SS_{forward}$ and $SS_{reverse}$ is

given in Fig. 17. The drain current can be modulated by 5×10^4 within 220 mV, which is among the smallest in published NCFETs, especially for p-type FET. Fig. 18 is the transfer characteristics with various V_{ds} from - 0.1 V to - 0.5 V. The DIBL is observed, the shift is calculated to be 200 mV/V. The DIBL effect conclusively confirm the realization of negative capacitance effect in this gate stack [3, 5]. Fig. 19 is the output characteristics of the NCFET. The output characteristics of the NCFET shows good saturation at V_{ds} = 1 V.

Fig. 20 is the on state current versus on/off ratio at $V_{dd} = V_{ds}$ = V_{gs}(on)-V_{gs}(off) of the WSe_2 devices with negative capacitance gate. At V_{dd} = 0.5 V, the WSe_2 devices exhibit the on/off ratio over 10^5, which is much better than that of the device with ZrO_2 ones (Fig. 21). Compared with previous works, the performance of our NCFET is outstanding, especially for p-type NCFETs (Table I). The device can achieve 5×10^4 modulation within 220 mV, among the best devices in published p-type NCFETs [2-5].

V. CONCLUSION

In summary, we fabricated WSe_2 NCFET devices with HZO/Al_2O_3 gate stack. Van der Waals Pt contact is used to reduce the Fermi level pinning at the metal/WSe_2 interface. This fermi level pinning free contact turns an ambipolar FET into a unipolar one and reduce the SS. With HZO/Al_2O_3 gate stake SS less than 60 mV/dec is obtained (minimum 18.2 mV/dec) and the hysteresis in the sub threshold region is as small as 20 mV. Moreover the device can be modulated by 5×10^4 within 220 mV, which makes it a promising candidate in ultra-low power electronics.

ACKNOWLEDGMENT

This work is supported by Research Grant Council of Hong Kong (PolyU 152016/17E) and National Natural Science Foundation of China 61734003.

REFERENCES

[1] H. Ilatikhameneh, et al., "Scaling Theory of Electrically Doped 2D Transistors," IEEE Electron Device Lett., vol. 36, pp. 726-728, 2015.

[2] S. Dasgupta et al., "Sub-kT/q switching in strong inversion in $PbZr_{0.52}Ti_{0.48}O_3$ gated negative capacitance FETs,". IEEE J. Exploratory Solid-State Comput. Devices Circuits, vol. 1, pp. 43-48,2015

[3] Z. H. Yu et al., "Negative Capacitance 2D MoS_2 Transistors with Sub-60mV/dec, Subthreshold Swing over 6 Orders, 250 uA/um current density, and Nearly-Hysteresis-Free," in IEDM Tech. Dig., pp. 577–580, 2017

[4] J. Zhou et al., "Ferroelectric $HfZrO_x$ Ge and GeSn PMOSFETs with Sub-60 mV/decade subthreshold swing, negligible hysteresis, and improved I_{ds}," in IEDM Tech. Dig., pp. 310–313, 2016

[5] M. W. Si et al., "Steep-Slope WSe_2 Negative Capacitance Field-Effect Transistor", Nano Lett., vol 18, pp 3682-3687, 2018

[6] A. Allain and A. Kis, "Electron and Hole Mobilities in Single-Layer WSe_2," Acs Nano, vol. 8, pp. 7180-7185, 2014.

[7] Y. Liu et al., "Approaching the Schottky–Mott limit in van der," in Nature, vol 557, pp 696-700, 2018

[8] A. Prakash et al., "Understanding contact gating in Schottky barrier transistors from 2D channels," Sci. Rep., vol. 7, pp. 12596, 2017

[9] M. Chhowalla et al., "Two-dimensional semiconductors for transistors," Nat. Rev. Mater., vol. 1, p. 16052, 2016.

[10] M. Houssa et al. 2D Materials for Nanoelectronics, Boca Raton: CRC Press, p. 211, 2016

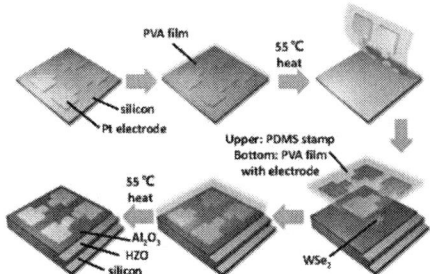

Fig. 1. Schematic process flow for transferring metal electrode onto the top of WSe₂ flakes. The WSe₂ and metal forms the van der Waals contact.

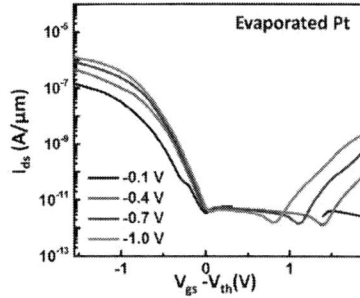

Fig. 2. Transfer characteristics of the WSe₂ FET evaporated Pt electrode. The device shows ambipolar behavior. Channel length is 5 μm. Channel thickness is 10 nm.

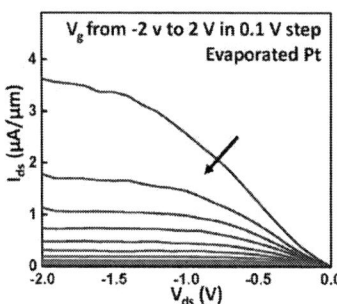

Fig. 3. Output characteristics of the WSe₂ FET evaporated Pt electrode. The output curve is nonlinear at small bias, indicating Schottky contact.

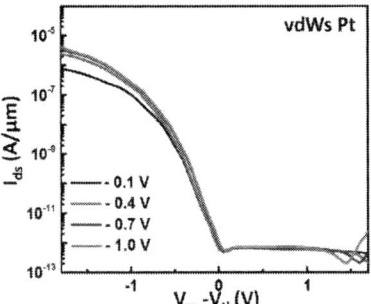

Fig. 4. Transfer characteristics of the WSe₂ FET with transferred vdWs Pt electrode. The device shows unipolar p-type behavior. Channel length is 5 μm.

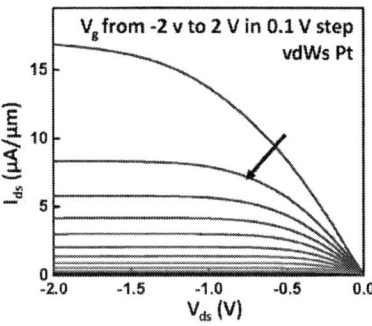

Fig. 5. Output characteristics of the WSe₂ FET with van der Waals Pt electrode. The output curve is linear at small bias.

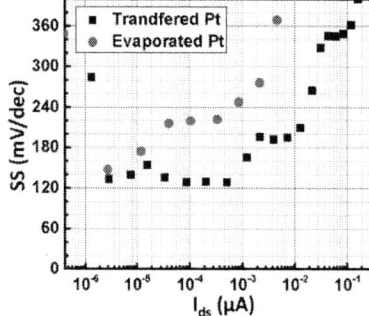

Fig. 6. Point SS of the WSe₂ FET with evaporated and vdWs Pt electrode. The SS are similar in low current region and the vdWs one is smaller with higher current.

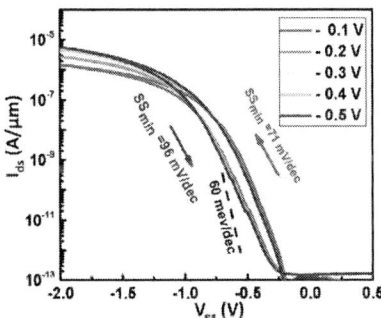

Fig. 7. Transfer characteristics of the WSe₂ FET transferred vdWs Pt electrode with minimum SS.

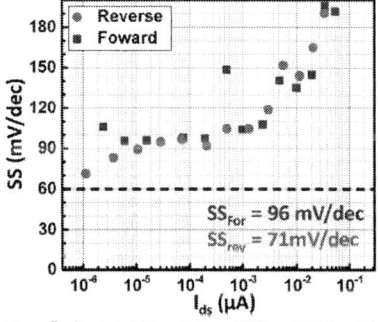

Fig. 8. Point SS of the WSe₂ FET with vdWs Pt electrode.

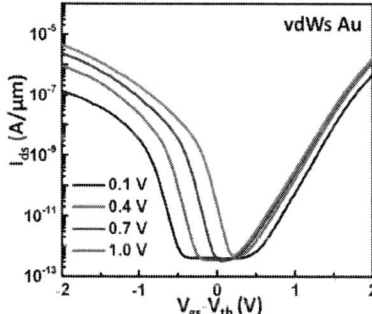

Fig. 9. Transfer characteristics of the WSe₂ FET with vdWs Au electrode. Channel length is 5 μm.

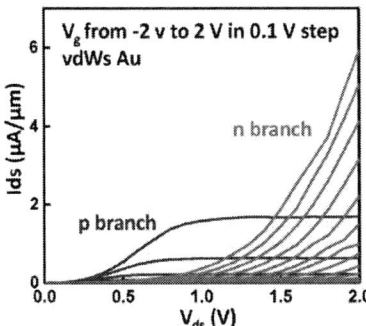

Fig. 10. Output characteristics of the WSe₂ FET with vdWs Au electrode.

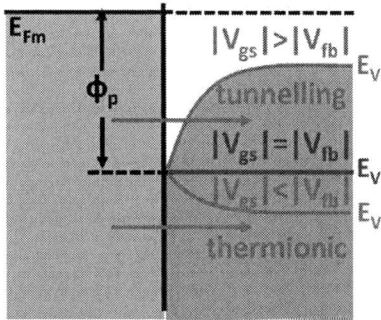

Fig. 11. Simplified band diagram of the FET with evaporated and transferred vdWs Pt electrode.

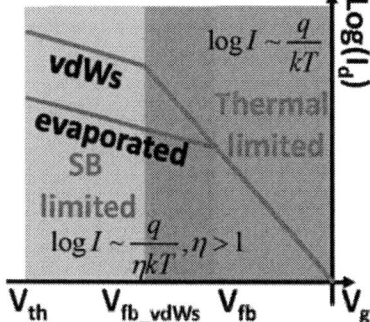

Fig. 12. Corresponding transfer curve of the FET for evaporated and vdWs Pt electrode with different band alignment.

978-1-7281-1988-5/18 $31.00 © 2018 IEEE

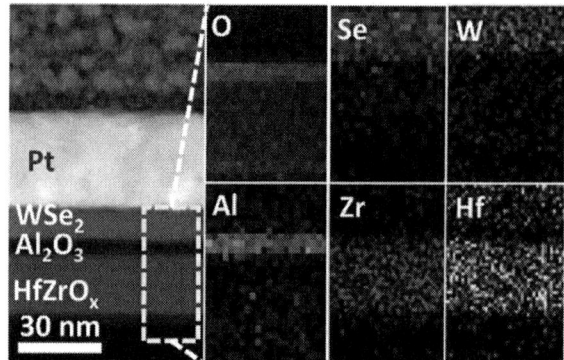

Fig. 13. (a) HAADF-STEM image of the WSe2 NCFET and (b) EDS mapping of the element distribution within the marked region.

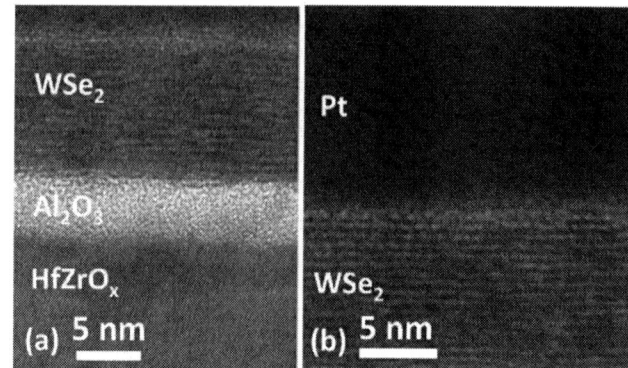

Fig. 14. (a) High-resolution TEM of the gate stack of a WSe2 NCFET. The HZO is highly crystallized. (b) Interface of the Pt/WSe2 contact region.

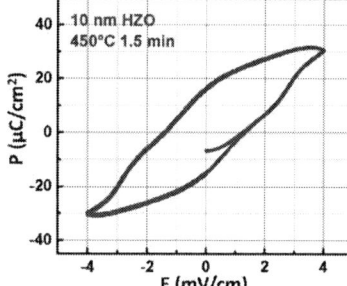

Fig. 15. P-E lop of HZO stack. Describe sth.

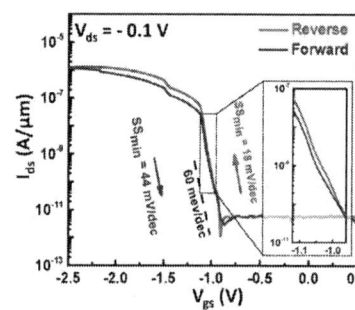

Fig. 16. Transfer characteristics of the WSe2 NCFET with vdWs Pt electrode.

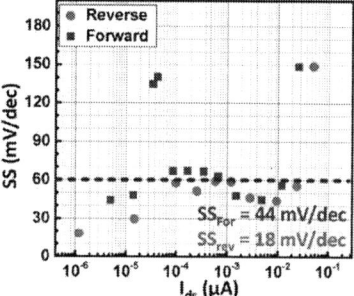

Fig. 17. Point SS of the WSe2 NCFET with transferred vdWs Pt electrode.

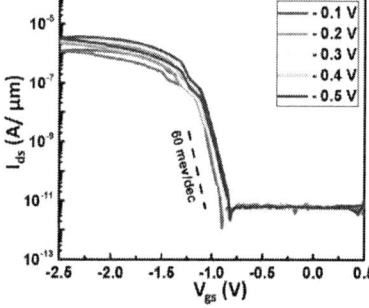

Fig. 18. transfer characteristics of the vdWs contact WSe2 NCFET with various V_{ds} from -0.1 V to -0.5 V. DIBL is observed with 200mV/V.

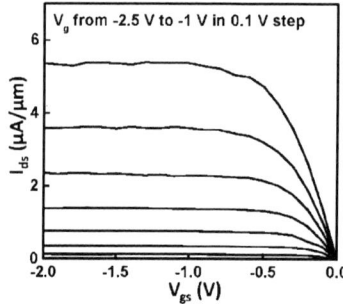

Fig. 19. Output characteristics of the vdWs contact WSe2 NCFET. The output curve shows linear behavior at small bias.

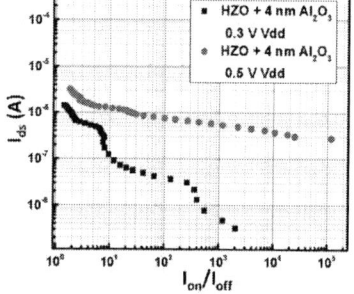

Fig. 20. On state current versus on/off ratio at $V_{dd} = V_{ds} = V_{gs}$ (on)-V_{gs} (off) of the WSe2 NCFET.

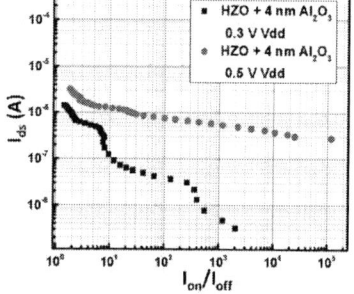

Fig. 21. On state current versus on/off ratio at $V_{dd} = V_{ds} = V_{gs}$ (on)-V_{gs} (off) of the WSe2 FET on ZrO2.

Table I comparison of this work with published NCFET devices

Device type	Minimum SS	Hysterysis	Drive magnitude
HZO MoS2 n-NCFET [3]	23 mV/dec	24 mV	232 mV for 10^5
PZT, Si n-NCFET [2]	32 mV/dec (forward), 13 mV/dec (reverse)	10 V	2.15 V for 10^5
HZO.Ge&GeSn p-NCFET [4]	47 mV/dec (foward), 43 mV/dec (reverse)	40 mV	~ 0.3 V for 10^3
HZO with MIG, WSe2 p-NCFET [5]	41.2 mV.dec (forward) 14.4 mV/dec (reverse)	120 mV	360 mV for 10^4
This work	44.1 mV/dec (forward), 18.2 mV/dec (reverse)	20 mV	220 mV for $5*10^4$

978-1-7281-1988-5/18 $31.00 © 2018 IEEE

Toward High-mobility and Low-power 2D MoS$_2$ Field-effect Transistors

Zhihao Yu[1][†], Ying Zhu[1][†], Weisheng Li[1], Yi Shi[1], Gang Zhang[2], Yang Chai[3] and Xinran Wang[1]*

[1]National Laboratory of Solid State Microstructures, School of Electronic Science and Engineering, and Collaborate Innovation Center of Advanced Microstructures, Nanjing University, Nanjing 210093, China, *email: xrwang@nju.edu.cn

[2]Institute of High Performance Computing, 1 Fusionopolis Way, 138632, Singapore

[3]Department of Applied Physics, The Hong Kong Polytechnic University, Hong Kong, China

[†]These authors contribute equally to this work

Abstract—2D semiconductors are promising candidates for future electronic device applications due to their immunity to short-channel effects (SCE), but many issues regarding mobility, contact, interface and power consumption still remain (Fig. 1). We develop a low-field model to calculate the mobility of monolayer MoS$_2$ FETs. Guided by the model, high carrier mobility of 150 cm^2/Vs and saturation current over 450 µA/µm are realized in long-channel monolayer MoS$_2$ FETs, through a series of interface optimization by high-κ dielectric and thiol chemical treatment. For low-power applications, we demonstrate hysteresis-free MoS$_2$ negative capacitance FETs (NCFETs) using ferroelectric HfZrO$_x$(HZO) as gate dielectric, achieving sub-60mV/dec subthreshold slope (SS) over 6 orders of I$_D$, minimum SS of 24 mV/dec and 10^7 on/off ratio under V_{dd}=0.5V. We further study the high frequency performance and show that sub-60mV/dec is maintained at least to 10 kHz without signs of degradation. Finally, by performing different gate sweeps we conclude that the steep slope is indeed due to NC effects rather than ferroelectric switching of HZO.

I. INTRODUCTION

As the scaling of Si CMOS down to 10 nm node, challenges to reduce device static power consumption is demanding. 2D semiconductors represented by transition-metal dichalcogenides are promising alternative channel materials for their large bandgap, atomic channel thickness, environmental stability and possible large-area synthesis. In spite of these advantages, currently the performance of 2D FETs is still far behind theoretical limit and does not meet the requirement of the roadmap. For example, the reported mobility is much lower than theoretical limit [1], and saturation current density is only 830 µA/µm even for short-channel devices [2]. On the other hand, ITRS 2023 requires ultralow V_{dd}=0.62V for high-performance applications, so strategies to overcome the Boltzmann limit of 60 mV/dec are actively pursued. Here, taking MoS$_2$ as a prototypical example, we present a systematic study on 2D FETs, with particular attention to interface properties, mobility and negative capacitance effects.

II. INTERFACE ENGINEERING AND MOBILITY BOOSTING IN MoS$_2$

In this section we focus on monolayer MoS$_2$, which represents the 2D limit of channel thickness. It came to our attention that most monolayer MoS$_2$ FETs reported so far have relatively low mobility ~1-100 cm^2/Vs [3]. To improve mobility, we first identify the major scattering sources in the channel of a typical backgated FET (Fig. 1) and calculate the scattering rate using Boltzmann transport equation and the relaxation time approximation [4]. Fig. 4 shows the calculated intrinsic and surface optical (SO) phonon-limited mobility as a function of temperature for several common dielectrics. μ_{SO} is lower on high-κ substrate like HfO$_2$ due to lower phonon energy. Our calculations show that phonon scattering is not important except near RT, but even at RT, mobility below 100 cm^2/Vs is not limited by phonons. Rather, the mobility in most experiments is limited by Coulomb impurity (CI) scattering, which refer to charges residing at the oxide interface or in the channel. μ_{CI} depends on impurity density n_{CI}, oxide dielectric constant, carrier density n and temperature T (Figs. 5 and 6). Since screening effects can reduce Coulomb scattering, low n_{CI}, high κ and n are desirable.

Guided by these calculations, we experimentally use thiol chemical treatment (which can repair the sulfur vacancies in MoS$_2$ and reduce n_{CI}) and high κ substrate to reduce the CI scattering and improve the mobility of monolayer MoS$_2$ [1,5]. Fig. 7 shows that these interface engineering approaches can significantly boost the four-probe mobility (removing contact effects) compared to bare SiO$_2$ substrate, reaching 150 cm^2/Vs at RT. Figs. 2 and 3 show the transfer and output characteristics of a typical 1.8 µm-long device on Al$_2$O$_3$ high κ dielectric, with two-probe mobility of 41 cm^2/Vs and saturation current over 450 µA/µm. We note that these MoS$_2$ FETs still have large contact resistance R_c over 1 kΩ·µm even at on-state, which is an order larger than typical value in Si CMOS (100 Ω·µm). Achieving good electrical contacts for MoS$_2$ FETs is a prerequisite to pursue high current at ultimate scaling.

Furthermore, we need to account for the localized trap states (n_{it}) manifested in the commonly observed metal-insulator-transition (MIT) behavior [5] (Fig. 2) and capacitance measurements [6]. The trap states effectively reduce mobility, and for simplicity are modeled by a Gaussian distribution below the bandgap [7]. By using n_{CI} and n_{it} as fitting parameters, we are able to reproduce the experimental $\mu(T,n)$ relationship nicely for a wide range of MoS$_2$ FETs fabricated in our lab (Fig. 7). The improved mobility after thiol treatment and high κ interface engineering is indeed captured by our model as a consequence of reduced n_{CI} and n_{it} and dielectric screening [1]. Our model can also provide insights on the main factors leading

978-1-7281-1988-5/18 $31.00 © 2018 IEEE

to the low mobility. In addition, we apply our model to analyze the literature mobility data for monolayer MoS$_2$ and reveal a universal trend even though the devices come from several different groups (Fig. 8) [4]. We conclude that interface engineering can effectively reduce n_{CI} and n_{it} compared to SiO$_2$, and the best interface so far is achieved by BN encapsulation. Currently we still need to find a scalable process to achieve high interface quality.

III. LOW-POWER MoS$_2$ NCFETs

The ultrathin body immune to SCE makes 2D semiconductors ideal for low-power logic applications. We recently start to work on NCFETs using few-layer MoS$_2$ as channel and ferroelectric HZO/Al$_2$O$_3$ as gate dielectric (see Fig. 9 for schematic and cross-section TEM image). The HZO is deposited by alternating HfO$_2$ and ZrO$_2$ cycles in ALD followed by RTA [8]. The key to realize negative capacitance is the ferroelectric property of the HZO. To this end, we perform PFM and P-E measurements on 22nm thick HZO, which show stable ferroelectric and piezoelectric properties with single-point switching voltage about ± 5V and maximum remanent polarization of ~26 μC/cm^2. The polarization direction after RTA was uniform, but it can be controllably programed by the tip voltage of PFM (Fig. 10). We further performed pulsed charging and discharging measurement (800 kHz) on FE and regular gate stack with Au top electrodes. Fig. 11 shows that while the regular HfO$_2$ capacitors always show monotonic RC charging behavior, the HZO/Al$_2$O$_3$ stack with the same thickness shows non-monotonic behavior suggesting that $C_{FE} = \delta_Q/\delta_{V_{FE}} < 0$ for a brief period, consistent with NC transient response.

Our MoS$_2$ NCFETs show steep switching characteristics at RT. Fig. 12 compares the RT transfer curves of NCFETs with different HZO thickness (t_{FE}) and a regular FET with Al$_2$O$_3$ dielectric. Both NCFETs show sub-60mV/dec over 5 orders of I_D (Fig. 13), small V_{th} (-0.25 V and -0.19 V) and nearly zero hysteresis. The minimum SS is 24 mV/dec (41 mV/dec) for device on 22 nm (11 nm) HZO, and remains below Boltzmann limit even at 100 K (Fig. 13). Fig. 15 plots the transfer curves of 15 NCFETs fabricated in different batches showing excellent reproducibility, and all devices show sub-60 mV/dec SS over 3 orders of I_D. Next, we benchmark the performance of our MoS$_2$ NCFETs against current literatures using on/off ratio as a function of $\Delta V_{GS} = V_{GS} - V_{on}$ (Fig. 14, V_{on} is defined as the start of subthreshold region log-scale transfer curve).[8-14] Under V_{DD}= 0.5 V, our MoS$_2$ NCFET can realize on/off ratio over 10^7, which is over 40 times larger than reported 2D NCFETs [10-12]. It is also 70 times larger than 10 nm Si Fin-FET and 3.8 times larger than 14 nm Si NC Fin-FET [8,9].

High frequency operation is also critical for the application of NCFETs. Fig. 17-19 shows the transfer characteristics using different sweeping speed and pulse measurements (Fig. 16) down to 1μs pulse width. Under pulse mode, the detection limit of our setup increases due to short integration time. Even so, the

transfer curves above the detection limit all overlap with DC sweep up to 1MHz, without any sign of degradation. At 10 kHz, our NCFET shows hysteresis-free transfer with on/off ratio of 10^5 and minimum SS of 54mV/dec. These pulse measurements suggest that our MoS$_2$ NCFETs can potentially operate at higher frequency than 1 MHz. However, we are currently limited by the low I_{off} to probe the frequency limit.

Although NCFETs show steep slope in DC and high-frequency measurements, the switching mechanism is still controversial between NC and FE switching. To gain more insight into the mechanism, we vary the V_{GS} sweeping range. We obverse two different switching behaviors as V_{GS} range increases. For small V_{GS} range below ± 4V, no hysteresis is observed consistent with the data in Fig. 12 (black and red, Fig. 17). Beyond ± 4V however, a clockwise hysteresis appears presumably due to carriers filling the deep trap of the gate oxide. The direction of hysteresis is opposite to the counter-clockwise direction for FE switching, which suggest that FE switching did not occur within the V_{GS} range and that the total capacitance is positive after capacitance matching. To further confirm the NC mechanism, we apply a large V_{GS} pulse (± 8V) to pre-charge the dielectric before sweeping V_{GS} in a confined region. As shown in Fig. 21, the V_{th} shift is consistent with the trap filling picture. What's more, the device shows sub-60mV/dec with negligible hysteresis under either positive or negative V_{GS} which cannot induce FE switching, indicating that FE switching is not responsible for the observed steep slope in our NCFETs.

IV. CONCLUSION

We show that interface engineering is important for improving the mobility and current density of 2D FETs. By combining HZO in the gate stack, our MoS$_2$ FETs show hysteresis-free steep switching characteristics up to 10 kHz, with sub-60mV/dec over 6 orders of I_D, due to NC effect rather than FE switching.

Acknowledgment

We thank Dr. Wei Cao for fruitful discussions. This work is supported by NSFC 61734003, 61521001, National Key Basic Research Program of China 2013CBA01604, 2015CB921600 and Research Grant Council of Hong Kong PolyU 152016/17E.

REFERENCES

[1] Z. Yu et al, Adv. Mater., 28, 547, 2016. [2] Y. Liu et al, Nano Lett., 16, 6337, 2016. [3] S. Li et al, Chem. Soc. Rev., 45, 118, 2016. [4] Z. Yu, et al., Adv. Func. Mater., 27, 1604093, 2017.[5] Z. Yu et al, Nature Commun., 5, 5290, 2014. [6]X. Chen et al, Nature Comm. 6, 6088, 2016. [7] Z. Yu et al, IEDM 2017, 23.6. [8] Z. Krivokapic et al, IEDM 2017, 15.1. [9] C. Auth et al. IEDM 2017, 29.1. [10] M. Si et al., Nature Nanotech., 18, 3682, 2017. [11] X. Wang et al., npj 2D Mater. & Applications. 1, 38, 2017. [12] M. Si et al., Nano Lett., 18, 3682, 2018. [13] J. Zhou et al., IEDM 2016, 12.2 [14] M. H. Lee IEDM 2017, 23.3

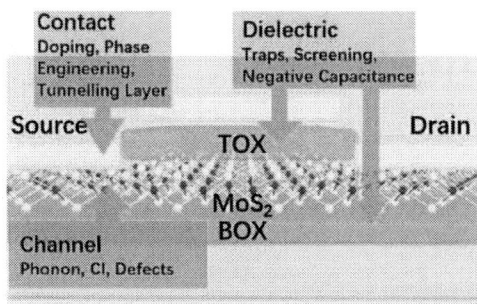
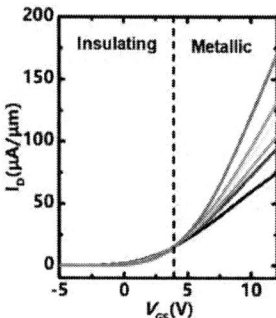
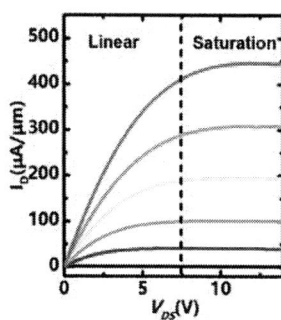

Figure 1. The 3D schematic of dual-gate MoS2 FET. The dominated factors of device performance listed on the top of the figure is divided into three parts of contact, channel and dielectric.

Figure 2. A typical I_D-V_{GS} curves of MoS2 FET on 30nm Al_2O_3 as gate dielectric, which is measured under 20K~300K with a crossover at V_{GS}=4.2V, indicating the insulating and metallic transport region.

Figure 3. A typical I_D-V_{DS} curves of device in Fig. 2 at 300K. The largest current density is over 450μA/μm at saturation region.

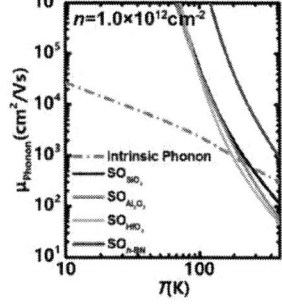
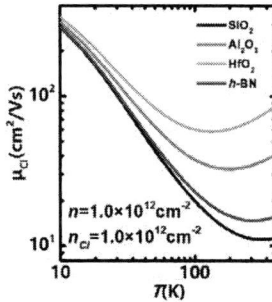
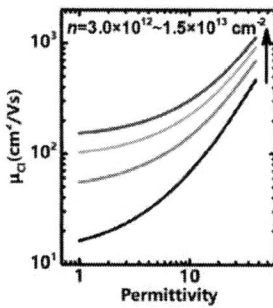
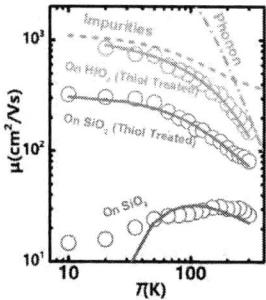

Figure 4. The magenta dash dot line is intrinsic phonon-limited mobility as a function of temperature. The solid line is SO-limited mobility as a function of temperature on SiO_2 (black), Al_2O_3 (red), HfO_2 (green) and h-BN (bule) at n=1.0×10^{12}cm^{-2}.

Figure 5. CI-limited mobility as a function of temperature on SiO_2 (black), Al_2O_3 (red), HfO_2 (green) and h-BN (bule) at n=n_{CI}=1.0×10^{12}cm^{-2}.

Figure 6. CI-limited mobility as a function of permittivity at n=3.0×10^{12} ~1.5×10^{13}cm^{-2}, n_{CI}=1.0×10^{12}cm^{-2}.

Figure 7. Mobility as a function of temperature for bare SiO_2, Thoil-treated SiO_2, and Thoil-treated HfO_2 substrate, together with the best theoretical fittings (solid lines,), the calculated CI-limited mobility (orange dashed lines), and the calculated phonon-limited mobility (orange dashed short lines)

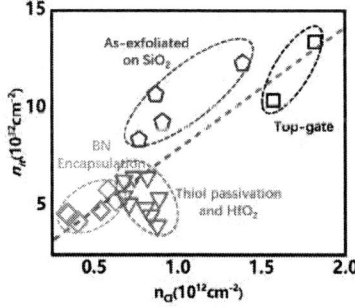
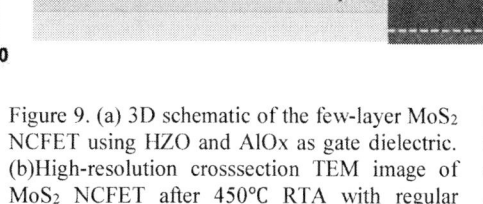
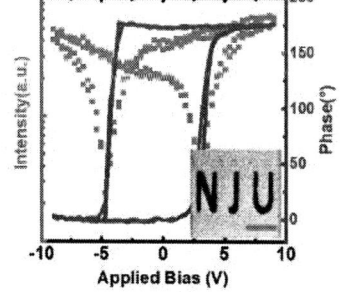

Figure 8. n_{it} and n_{CI} distribution extracted through fitting devices data in literatures with linear fitting. Each color region represents similar fabrication processes.

Figure 9. (a) 3D schematic of the few-layer MoS2 NCFET using HZO and AlOx as gate dielectric. (b)High-resolution crosssection TEM image of MoS2 NCFET after 450°C RTA with regular crystallographic orientation with regular crystallographic orientation.

Figure 10. Single-point PFM characterization with 180° phase transition of HZO showing ferroelectric behavior of the HZO layer after 450°C RTA. The insert is phase difference pattern of NJU using large scale PFM litho. Scale bar is 3 μm.

978-1-7281-1988-5/18 $31.00 © 2018 IEEE

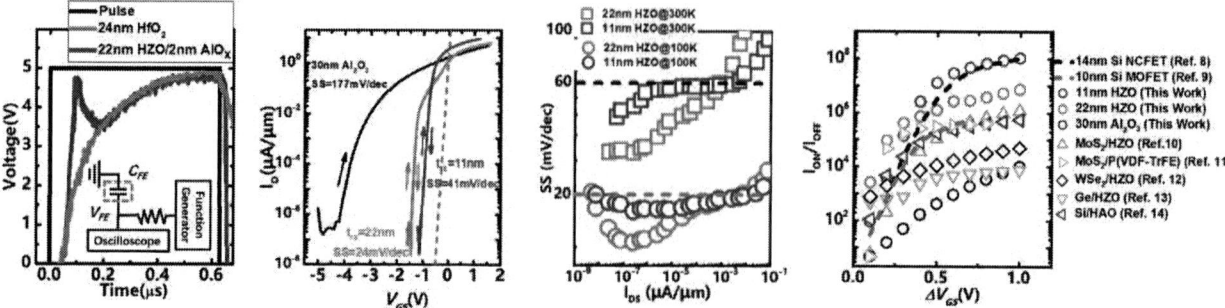

Figure 11. Real-time transient response across the FE capacitor and regular capacitor at a pulse voltage of ±5V, 800kHz. The insert is schematic of pulse measurements

Figure 12. Double sweep I_D-V_{GS} curves of a MoS$_2$ NCFET on 22nm HZO/2nm AlO$_x$ (red), 11nm HZO/2 nm AlO$_x$(blue) and control device(black) at V_{DS} = 0.1V. The NCFETs shows the steep average SS 24 and 41 mV/decade for 22nm and 11nm HZO respectively. The control device shows 177mV/decade, over 7(4) times larger than the NCFETs.

Figure 13. Point SS of the MoS$_2$ NCFETs vs. I_D in Fig. 13 at RT and 100K. The sub-60 mV/dec region spans 5 orders of magnitude in I_{DS}.

Figure 14. on/off ratio as a function of $\Delta V_{GS} = V_{GS} - V_{on}$ extracted from literatures compared with ours. Here, V_{on} is defined as the start point of drain current in log-scale transfer curve. In our work, we achieve highest on/off ratio within 1V gate drive voltage, which is comparable with the state-of-the-art processes.

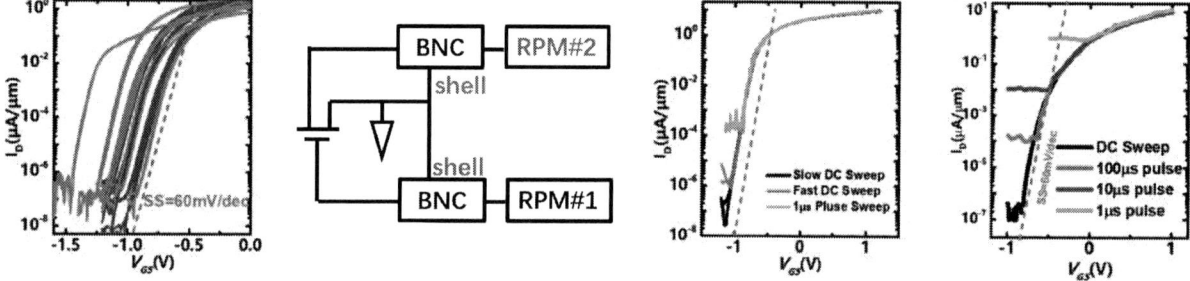

Figure 15. I_{DS}-V_G curves for 15 devices of t_{FE}=11nm and 22nm NCFETs.

Figure 16. High speed synchronization pulse measurement setup using Keithley 4200SCS and HR-RPM. We can achieve 10 ns pulse output using this setup. I_D is collected when V_{GS} and V_{DS} are driven by synchronization pulse voltage.

Figure 17. Double sweep I_{DS}-V_G curves for t_{FE}=11nm NCFET under different sweep speed of slow DC (black), fast DC (red) with different integration time and ultra-fast pulsed (green) sweep.

Figure 18. I_{DS}-V_G curves for t_{FE}=22nm NCFET under DC sweep and different pulse speed. All pulse measurements show excellent agreement with DC down to the detection limit. The reason for increased off-state current is due to the detection limit of the instrument under high-speed pulse.

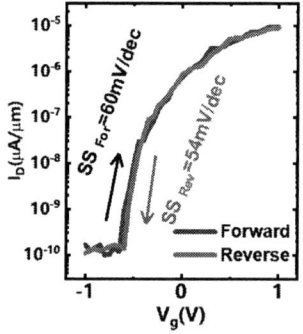 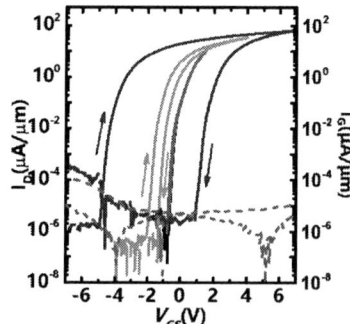 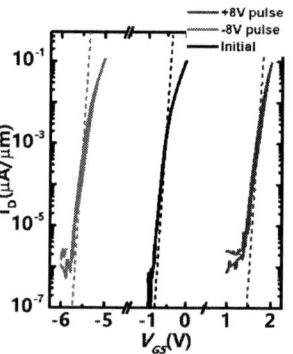

Figure 19. Double sweep I_{DS}-V_G curves under 10KHz for t_{FE}=22nm NCFET in Fig. 20 (black and red line: forward and reverse sweep respectively). A reasonable on/off ratio of 105 and smaller SS of 60mV/dec and 54 mV/dec are realized even under high speed sweep.

Figure 20. Double sweep I_D-V_{GS} curves with different gate range of ±1V (black), ±2V (red), ±4V (green) and ±7V (blue) with different hysteresis behaviors. The grey dash show the gate leakage current of ±7V V_{GS}.

Figure 21. Double sweep I_D-V_{GS} curves with pre-charged oxide in Fig. 20. The dash lines label the SS=60mV/dec operation.

978-1-7281-1988-5/18 $31.00 © 2018 IEEE

3D Monolithic Stacked 1T1R cells using Monolayer MoS$_2$ FET and hBN RRAM Fabricated at Low (150°C) Temperature

Ching-Hua Wang[1*], Connor McClellan[1], Yuanyuan Shi[3], Xin Zheng[1], Victoria Chen[1],
Mario Lanza[3], Eric Pop[1,2] and H. -S. Philip Wong[1#]

[1]Department of Electrical Engineering, [2]Materials Science & Engineering, Stanford University, Stanford, CA, 94305, USA
[3]Institute of Functional Nano & Soft Materials, Soochow University, China. *chwang9@stanford.edu, #hspwong@stanford.edu

Abstract— We demonstrate 3D monolithically integrated two-level stacked 1-transistor/1-resistor (1T1R) memory cells, using monolayer MoS$_2$ transistors and few-layer hBN RRAMs, fabricated at temperatures below 150 °C. The stacking process is scalable to an arbitrarily large number of layers and on any substrate material without foreseeable physical limitations. The 1T1R cells can be switched with programming current < 130 µA and voltage < 1 V, close to typical CMOS logic voltages. These cells are promising for in-memory and neuromorphic computing because (1) the hBN RRAM has gradual set and reset switching due to multiple weak-filaments formed along local defects and (2) the MoS$_2$ transistor has low off-current due to the large band gap of monolayer MoS$_2$ ($E_g >$ 2 eV). We also show that the linearity of RRAM resistance change is well-controlled by the gate voltage of the transistor.

I. INTRODUCTION

Today's data-driven applications, such as big data analytics, neural networks, and machine learning, require huge memory and computing resources. 3D monolithic integration offers opportunities to immerse logic with memory on the same chip to increase bandwidth and reduce latency of memory access [1]. Further energy efficiency can be gained through the use of in-memory [2] and neuromorphic computing [3] that capitalize on the analog properties of resistive memories and fine-grained logic-memory integration. For example, arrays of analog programmable resistive memory can perform massively parallel vector-matrix multiplication [4]. The amount of analog weights (resistive memory) required is large, often in the 10's of GByte range [5]. 3D monolithic integration can increase the memory density in a scalable fashion (scaled up in the number of 3D layers) and reduce parasitic interconnect resistances [6].

Resistive Random-Access Memory (RRAM) is a promising analog memory element for in-memory and neuromorphic computing as the resistance can be controlled using electrical signals. Previous works have shown intriguing resistive switching in hexagonal boron nitride (hBN) with Ti Top Electrode (TE) due to boron vacancies and Ti metal ion diffusion through local defects into the hBN stacks [7, 8]. Here we show that hBN with Atomic Layer Deposition (ALD) passivation layers displays linear changes in resistance in response to an identical pulse train, a desired behavior for analog weight changes in neural networks. However, to integrate this hBN RRAM into a 3D monolithic memory array, a stackable selector element must be developed to reduce the sneak current path. In this work, we demonstrate two-level stacks of 1T1R cells fabricated at low (150°C) temperature

using a 2D monolayer semiconductor MoS$_2$ transistor as an energy-efficient selector.

II. FABRICATION PROCESS

Figure 1(a) shows the cross-section schematic of our two-level stacked 1T1R cells. To avoid topography issues for this initial experiment, we offset the 2nd level 1T1R 1 mm horizontally from the 1st level 1T1R. Topography issues can be solved in the future by Chemical-Mechanical Planarization (CMP). The local back-gate dielectric for 1st and 2nd level are 15 nm and 20 nm respectively using ALD Al$_2$O$_3$ deposited at 150 °C (our highest temperature processing step). The monolayer MoS$_2$ is grown on a 90 nm SiO$_2$/Si substrate using a solid-source CVD process [9] and then transferred on top of the local back-gate dielectric, followed by Au source/drain deposition [10]. Few layer hBN is grown on Cu foil and then transferred after forming the source/drain contacts of the 1st level MoS$_2$ transistor. The fabrication process is the same for the subsequent 1T1R level, meaning no new process steps are needed to add additional 1T1R levels onto the 3D chip.

Figure 1(b,c,d) are the TEM cross-section images of 1st level MoS$_2$ transistor (in the channel area outside the hBN region), 2nd level MoS$_2$ transistor (with hBN in the channel area) and 2nd level hBN RRAM with the labeled film stack respectively corresponding to the Fig. 1(a). Figure 1(e) is the TEM zoomed-in image of hBN RRAM with 10 − 13 layers of hBN, which is true for all hBN RRAM in this work. . In this image, we can see the hBN layered structure has several defective paths that resulted from growth, transfer and ALD thermal process. Figure 1(f, g) are the topview SEM images of the hBN RRAM and MoS$_2$ transistor in this study, respectively. The size of hBN RRAM is between $0.5 - 1.5$ µm^2 and transistor channel length is 2 µm in this work.

To transfer the MoS$_2$ and hBN, we use a partial dry transfer process adapted from [11] and [12] for MoS$_2$ and hBN respectively, as illustrated in Fig. 2. As reducing moisture introduced during the 2D material transfer process is critical for improving transistor yield and eliminating hysteresis, we place the samples on a 110 °C hot plate for 5 minutes before releasing the 2D material onto the target chip. As we use CVD grown 2D materials, our 3D monolithic process can be extended to wafe-scale dimensions.

III. 1T1R STRUCTURE FOR ARRAY

Figure 3(a) shows the structure of a two-level stacked 2×2 1T1R array with local back-gate MoS$_2$ transistor and hBN RRAM. In this array architecture, as shown in Fig. 3(b), the Word Lines (WL) are perpendicular to the Bit Lines (BL, labeled V_1, V_2). Each Source Line (SL) outputs a current summation (I_1, I_2) which is equal to the matrix-vector multiplication of weights (RRAM conductance) and input voltages (V_1, V_2) of the cells on the same column. Since the local back-gate serves as a buried WL in an array design, this 1T1R has similar area-efficiency as traditional top-gated planar 1T1R (Fig. 3(c)). Compared to the simple cross bar array architecture [4], our 1T1R array not only reduces the sneak current path, but can also control the linearity of the resistance change by optimizing the V_{GS} (Fig. 15 – 19).

IV. hBN RRAM (1R) GRADUAL SWITCHING

Figure 4 shows 50 consecutive DC switching cycles of the hBN RRAM (1R) with ALD Al_2O_3 capping. All measurements in this work were performed in vacuum ($<10^{-4}$ torr). This hBN RRAM was initially in the Low Resistance State (LRS) without requiring a forming process and switched with voltage less than 0.8 V. Figure 5(a) shows a few representative set curves to illustrate the transition from abrupt to gradual setting. This phenomenon suggests the resistive switching is initially dominated by one (or a few) filament but evolves to form multiple weak-filaments over cycling. We observed that the ALD process may introduce more defective paths and help form multiple weak-filaments in the hBN RRAM. Zhao et al. [13], assessed the performance metrics for the cases of one strong filament, multiple weak-filaments, and non-filamentary RRAM for neuromorphic computing (Fig. 5(b)). Multiple weak-filamentary RRAM is a good compromise, appropriately balancing retention, speed and analog behavior. The retention test (Fig. 6) shows no obvious resistance change for different resistance states after 5000 seconds at 125 °C.

The gradual set/reset process enables linear resistance change. In Fig. 7 and 8, we measured the resistance change with the pulse number applied for set and reset operation using an identical pulse train. The pulse widths are 1 μs for set and 10 μs for reset. Lower set/reset voltage gives better weight update linearity but has reduced resistance window.

V. CVD MONOLAYER MoS₂ TRANSISTOR

Figure 9 shows the I_D-V_{GS} data of the local back-gate MoS_2 transistor at several points of the device fabrication sequence. Although the MoS_2 transistor current decreased after the hBN transfer, the current is mostly restored after ALD Al_2O_3 capping, possibly due to doping from reducing moisture or by sub-stochiometric AlO_x [14]. Figure 10 is the I_D-V_{GS} curve after the final ALD capping (the last step of our process), demonstrating off-current $\approx$ 1 pA/μm with V_{DS} = 1 V (after properly shifting the threshold voltage). This low off-current is due the large bandgap of MoS_2 ($E_G > 2V$) and can be lowered further with improved processing [15]. The I_D-V_{DS} data (Fig. 11) shows little hysteresis, indicating a clean MoS_2 transfer.

VI. 1ST & 2ND LEVEL 1T1R ELECTRICAL RESULTS

Figure 12 shows 50 consecutive DC switching cycles of our two-level stacked 1T1R cells. Both 1st and 2nd level show similar gradual set and reset switching with operation voltage less than 1.5 V. The probability distributions of resistive states are summarized in Fig. 13. Figure. 14 shows that the required reset current in 1T1R cell can be reduced by lowering the V_{GS} for the set operation. This cell also shows DC switching cycles with programming current < 130 μA and voltage < 1V. This low voltage operation is essential for CMOS logic compatibility and low energy consumption because the CV^2 energy for charging the wires of the memory array can be substantial [16].

In Fig. 15 and 16, we demonstrate that the resistance change linearity is well-controlled by V_{GS} (using the 2nd level 1T1R cell measured in Fig. 12). Lowering the V_{GS} results in more linear resistance change for both set and reset processes but with a smaller resistance window. To compare 1T1R with the 1R hBN RRAM, we normalized the curves in Fig. 7, 8, 15 and 16 by the maximum resistance window observed in each dataset of the respective figures (e.g. 1 ≈ 17.5 kΩ and 0 ≈ 2.5 kΩ for Fig. 7). For the reset operation shown in Fig. 17(a), the 1T1R configuration can either increase resistance change linearity with the same resistance window or have similar linearity with a larger resistance window compared to 1R. In Fig. 17(b) we show the same observation for set operation.

In Figure 18, we used a series of short pulses to perform a linear resistance change for set/reset. We observe that the gradual change in resistance and magnitude of the resistance window are controlled by V_{GS} during reset. The resistance values return to the same LRS and HRS for multiple cycles, displaying low cycling hysteresis. The normalized resistance changes with reset V_{GS} = 1 V, 0.75 V, and 0.5 V are compared in Fig. 19, showing that the resistance change linearity during reset is improved with lower V_{GS}.

VII. CONCLUSIONS

We have demonstrated sequential 3D monolithically integrated two-level stacked 1T1R cells using fabrication temperature < 150 °C, scalable to many levels by repeating the same fabrication steps. The hBN RRAM shows forming-free, < 1V gradual set and reset that is preferable for linear weight updating in neuromorphic computing using low-power CMOS. We show the linearity of resistance change for both set and reset can be improved in 1T1R configuration by controlling the gate bias. The array demonstration so far is limited by yield, which can be improved by a cleaner transfer process and uniform large area growth of materials in an industrial environment.

Acknowledgements

This work is supported in part by NSF EFRI E3AL (1542883), NSF EFRI 2-DARE, AFOSR MURI (FA9550-16-1-0031), ASCENT (one of six centers in JUMP, a Semiconductor Research Corporation (SRC) program sponsored by DARPA), and member companies of the Stanford SystemX Alliance and the Stanford Non-Volatile Memory Technology Research Initiative (NMTRI). M. L. acknowledges support from the Young 1000 Talent program of China and the NSFC (61502326, 41550110223, 11661131002). We also acknowledge Professor Deji Akinwande and Ruijing Ge for help with the MoS₂ transfer process.

Reference

[1] M. M. S. Aly, *et al.*, *Computer* **48**, 12, pp. 24 (2015) [2] D. Ielmini, *et al.*, *Nat. Electron.* **1**, 6, pp. 333 (2018) [3] S. Yu, *et al.*, *IEEE Trans. on Elec. Dev.* **58**, 8, pp. 2729 (2011) [4] S. Yu, *et al.*, *IEDM* (2015) [5] W. Hwang, *et al.*, *ISCAS* (2018) [6] P. A. Merolla, *et al.*, *Science* **345**, 6197, pp. 668 (2014) [7] F. M. Puglisi, *et al.*, *IEDM* (2016) [8] Y. Shi, *et al.*, *IEDM* (2017) [9] K. K. Smithe *et al.*, *2D Materials* **4**, 1, pp. 011009 (2016) [10] C. D. English, *et al.*, *Nano Lett.* **16**, 6, pp. 3824 (2016) [11] R. Ge *et al.*, *Nano Lett.* **18**, 1, pp. 434 (2017) [12] S. Sinha *et al.*, *2D Materials* **3**, 3, pp. 035010 (2016) [13] M. Zhao *et al.*, *IEDM* (2017) [14] C. J. McClellan *et al.*, *DRC* (2017) [15] C. U. Kshirsagar, *et al.*, *ACS Nano*, **10**, 9, pp. 8457 (2016) [16] H.-S. P. Wong, *DRC* (2018).

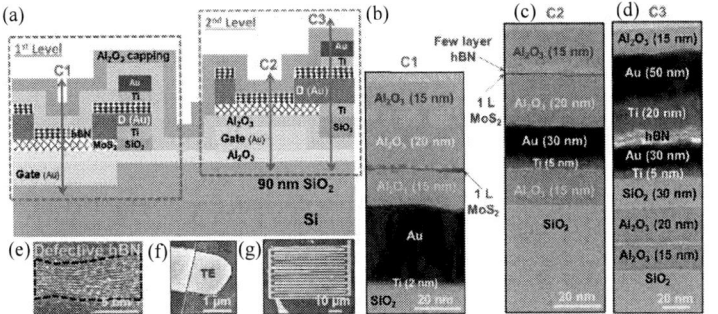

Fig. 1. (a) The cross-section schematic of two-level stacked 1T1R cells (b) The TEM images of the 1st and (c) 2nd level MoS₂ transistors, and (d) the 2nd level hBN RRAM. (e) The TEM zoomed-in image of hBN layered structure (10-13 layers) with multiple defective paths. (f, g) The SEM images of RRAM and MoS₂ transistor, respectively.

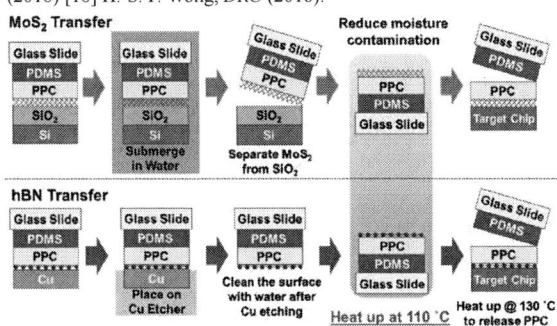

Fig. 2. The partial transfer process of hBN and MoS₂ are adapted from [11] and [12] respectively. The critical step introduced for our stacked 1T1R cells is heating up the 2D material before transferring onto the target chip to avoid moisture contamination, which degrades the gate dielectric of local back-gate MoS₂ transistor and causes transistor hysteresis.

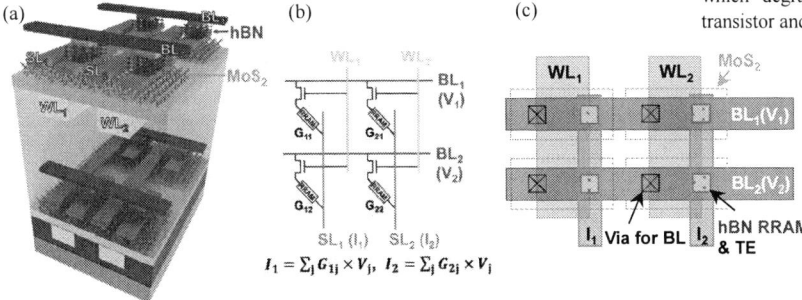

Fig. 3. (a) The schematic of the two-level 1T1R cells with local back-gate MoS₂ transistors. (b) An array architecture for matrix-vector multiplication with 1T1R cells to reduce sneak current. Each column outputs a current sum (I_1, I_2) of the row input voltage (V_1, V_2) weighted by the device conductance at each cross-point. (c) Layout schematic of a 2×2 back-gated 1T1R array with density similar to top-gated planar 1T1R array.

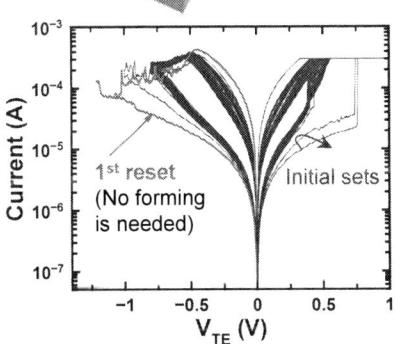

Fig. 4. 50 DC switching cycles of the 1R cell. The cell was initially in LRS after process completion.

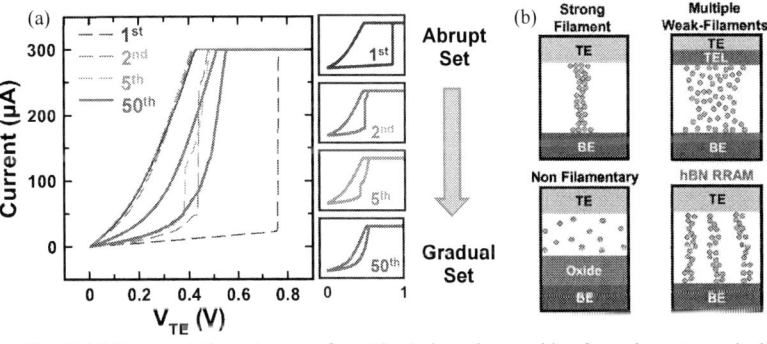

Fig. 5. (a) Representative set curves from Fig. 4 show the transition from abrupt to gradual set. (b) The schematic of different filament forming RRAM [13]. Filaments in hBN are more independent than multiple weak-filaments in RRAM as they are defined by defects in the crystalline structure.

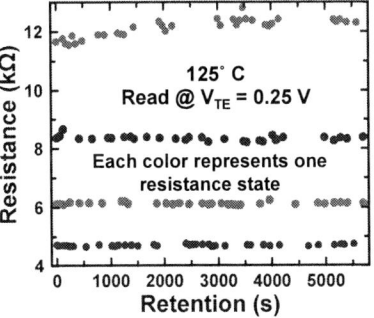

Fig. 6. The retention data of hBN RRAM shows no obvious resistance change, demonstrating reasonable retention for analog memory.

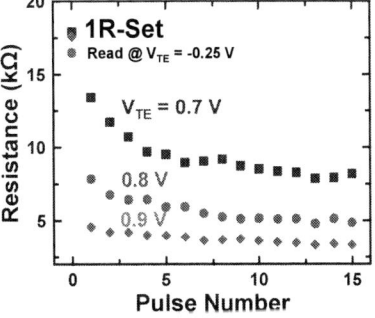

Fig. 7. Resistance change with pulse number for setting in 1R. (pulse width = 1 μs, fall/rise time = 0.1 μs)

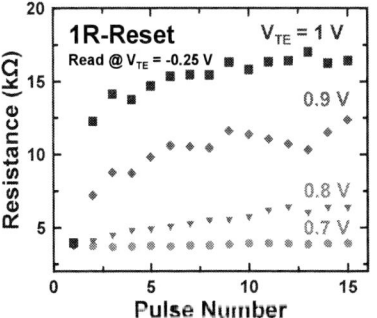

Fig. 8. Resistance change with pulse number for resetting in 1R. (pulse width = 10 μs, fall/rise time = 0.1 μs)

978-1-7281-1988-5/18 $31.00 © 2018 IEEE

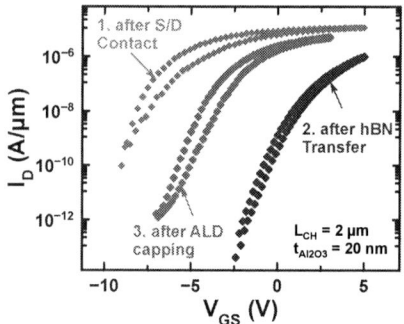

Fig. 9. I_D-V_{GS} curves of the MoS_2 transistor after S/D contact deposition, hBN transfer and ALD Al_2O_3 capping. The ALD process shifts V_t to negative and improves current density.

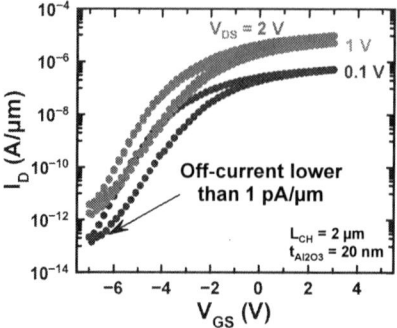

Fig. 10. I_D-V_{GS} curves after the final ALD Al_2O_3 step with various V_{DS}. Off-current is lower than 1 pA/μm (after properly shifting the threshold voltage).

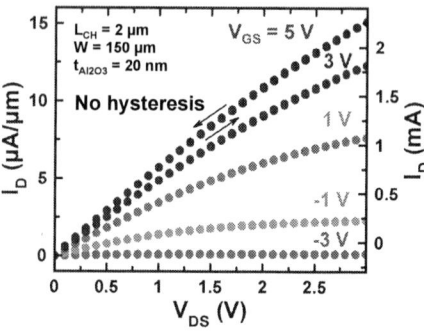

Fig. 11. I_D-V_{DS} curves of the device in Fig. 10 after the final ALD step with various V_{GS} show little hysteresis, indicating a clean MoS_2 transfer process.

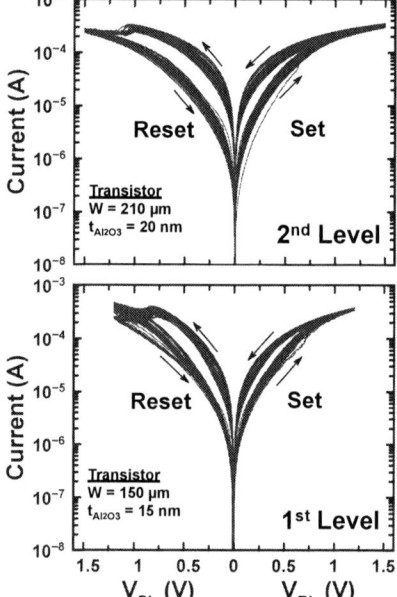

Fig. 12. 50 DC switching cycles of the 1st and 2nd level 1T1R. Both devices show similar bi-directional gradual switching. The 1T1R cells can be switched less than 1.5 V for set and reset. (2nd level: Set @ V_{GS} = 2 V; Reset @ V_{GS} = 3 V; 1st level: Set @ V_{GS} = 0.5 V; Reset: V_{GS} = 3 V)

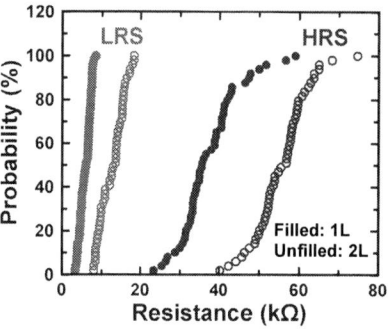

Fig. 13. Probability distribution of LRS and HRS from Fig. 12. Similar resistance level for 1st and 2nd level 1T1R cells. (Read @ V_{GS} = 3 V and V_{SL} = 0.25 V)

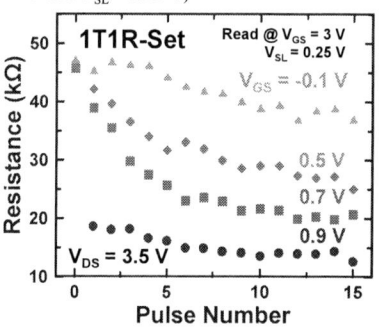

Fig. 15. Resistance change with pulse number for set in 1T1R. Lower V_{GS} improves linearity. (Same pulse conditions as in Fig. 7)

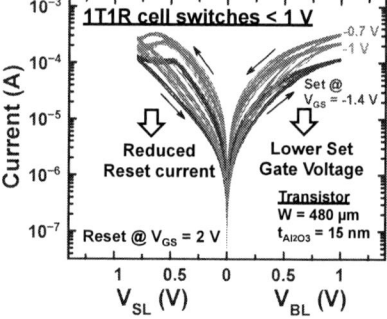

Fig. 14. The reset current can be lowered by reducing the compliance current during set which also results in a lower on/off ratio. This 1T1R cell can be switched with less than 1 V.

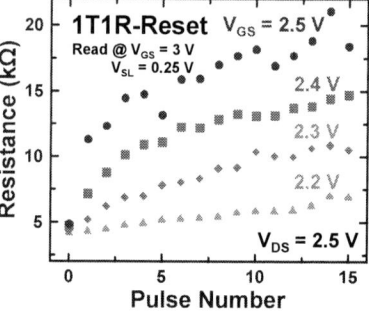

Fig. 16. Resistance change with pulse number for reset in 1T1R. Lower V_{GS} improves linearity. (Same pulse conditions as in Fig. 8)

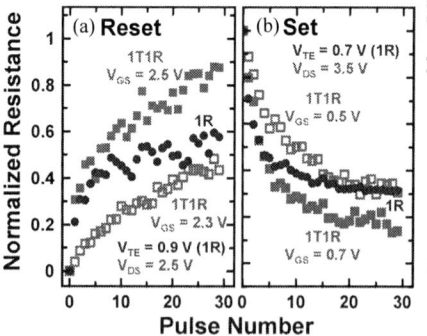

Fig. 17. Summary of normalized data from Fig. 7, 8, 15 and 16, showing 1T1R can achieve either better linearity or higher resistance window for both set and reset.

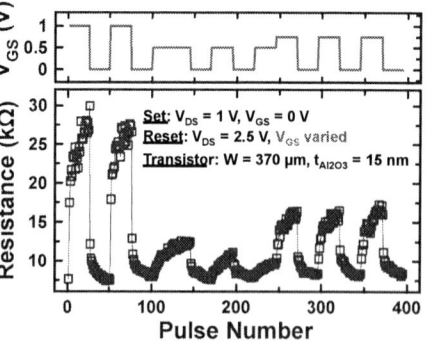

Fig. 18. Resistance change with series of identical set/reset pulses (V_{GS} shown in the top figure). Varied V_{GS} during resetting. (Pulse width = 60 ns for set/reset, fall/rise time = 2 ns; Read @ V_{GS} = 3 V and V_{SL} = 0.25 V)

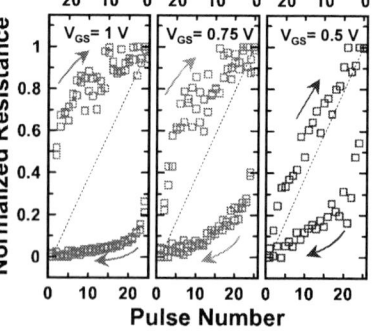

Fig. 19. Resistance change with pulse number with varied V_{GS} during reset. The resistance change linearity is improved with lower V_{GS}. Dotted lines represent linear resistance change (ideal case).

978-1-7281-1988-5/18 $31.00 © 2018 IEEE

Atomristors: Memory Effect in Atomically-thin Sheets and Record RF Switches

Ruijing Ge*[1], Xiaohan Wu*[1], Myungsoo Kim*[1], Po-An Chen[1,2], Jianping Shi[3], Junho Choi[4], Xiaoqin Li[4], Yanfeng Zhang[3], Meng-Hsueh Chiang[2], Jack C. Lee[1] and Deji Akinwande[1]

[1]Microelectronics Research Center, University of Texas at Austin, Austin, TX 78758. *equal contribution. deji@ece.utexas.edu
[2]Institute of Microelectronics, Department of Electrical Engineering, National Cheng Kung University, Tainan 701, Taiwan
[3]Department of Materials Science and Engineering, College of Engineering, Peking University, Beijing 100871, China
[4]Department of Physics, University of Texas at Austin, Austin, TX 78712

Abstract—Non-volatile resistive switching (NVRS) has been recently observed with synthesized monolayer molybdenum disulfide (MoS_2) as the active layer and termed atomristors [1]. In this paper, we demonstrate the fastest switching speed (<15 ns) among all crystalline two-dimensional (2D) related NVRS devices to the best of our knowledge. For the first time, ab-initio simulation results of atomristors elucidate the mechanism revealing favorable substitution of specific metal ions into sulfur vacancies during switching. This insight combined with area-scaling experimental studies indicate a local conductive-bridge-like nature. The proposed mechanism is further supported by sulfur annealing recovery phenomenon. Moreover, *exfoliated* MoS_2 monolayer is demonstrated to have memory effect for the first time, expanding the materials beyond synthesized films. State-of-the-art non-volatile RF switches based on MoS_2 atomristors were prepared, featuring 0.25 dB insertion loss, 29 dB isolation (both at 67 GHz), and 70 THz cutoff frequency, a record performance compared to emerging RF switches. Our pioneering work suggests that memory effect maybe present in dozens or 100s of 2D monolayers similar to MoS_2 paving the path for new scientific studies for understanding the rich physics, and engineering research towards diverse device applications.

I. INTRODUCTION

2D materials are promising candidates to overcome vertical scaling obstacle in non-volatile resistive switching (NVRS) based on conventional metal-insulator-metal (MIM) structure [2,3]. The resistance can be switched between high resistance state (HRS) and low resistance state (LRS) via electrical bias, and the state retained absent any power supply [4]. Recently, researchers have reported NVRS in various 2D materials [3, 5-8]. Among those 2D systems, atomristor stands out due to the atomic thinness of the active layer, low switching voltage, forming-free characteristic and large on/off current ratio [1]. In this work, MoS_2 atomristor MIM devices are fabricated using CVD-grown or high-quality exfoliated MoS_2 with various process realization techniques. Ab-initio simulation results reveal the underlying mechanism based on the interaction of metal (Au) ions and sulfur vacancies. We found that as the sulfur vacancies occupied by metal ions increase, the monolayer becomes conductive owing to a finite and increasing density of states at the Fermi level, which helps to explain the switching from HRS to LRS. As an evidence of the participation of sulfur defects, device recovery phenomenon has been discovered after annealing in sulfur ambient. These atomristors are also suitable for flexible memory and RF switches, with the latter featuring record performance compared to emerging RF switch devices. For instance, we demonstrate a record 15-ns pulse operation among crystalline 2D related NVRS devices. The NVRS behavior in 2D monolayers enables future applications in ultra-scaled memory fabrics, and RF switches for communication systems.

II. DEVICE STRUCTURE AND FABRICATION

The process flows of (i) crossbar, (ii) litho-free and transfer-free, and (iii) exfoliated atomristors are shown in **Fig. 1** (a,b,c). (i) Crossbar: CVD-synthesized monolayer MoS_2 was grown on SiO_2/Si substrates. For device fabrication, Au (60 nm) bottom electrodes (BE) were patterned and deposited by electron beam lithography (EBL) and e-beam evaporation, respectively. The monolayer MoS_2 was transferred onto the BE. Then top electrodes (TE) were prepared using the same fabrication process as BE. (ii) Litho-free and transfer-free: monolayer MoS_2 was directly grown on Au foils [9]. TE was then deposited through shadow mask without any patterning. (iii) Exfoliation: monolayer MoS_2 was mechanically exfoliated from bulk crystal onto the deposited Au film on SiO_2/Si substrate. TE was patterned using EBL. Importantly, the litho and transfer free device represents a near ideal clean device and serves to confirm the memory effect is an intrinsic property.

To identify the layer thickness and quality of exfoliated flake, Raman spectrum, photoluminescence (PL) spectrum and PL mapping (**Fig. 2**) measurements were performed. Scanning tunneling microscopy (STM) was utilized to check the atomic structure of as-grown MoS_2 on Au foil (**Fig. 1e**).

III. NON-VOLATILE SWITCHING BEHAVIOR

A. Nov-Volatile Resistive Switching of Monolayer MoS_2

Fig. 1 (g,h,i) shows representative I-V curves of (i) crossbar (ii) litho-free and transfer-free and (iii) exfoliated monolayer MoS_2 atomristors. Taking the I-V curve of crossbar device as an example, the voltage sweeps from 0 to 1V and back to 0V, then opposite bias is applied from 0 to -1V. At about 1V, the current abruptly increases, indicating a transition (SET process) from HRS to LRS. While at about -1V, the device switches from LRS to HRS (RESET process). *To rule out the possible influence from polymer residue due to patterning and transfer, litho-free and transfer-free clean devices were prepared based on CVD MoS_2 grown directly on Au foil*, and used in atomristors as the active layer. All the three devices show NVRS behavior, indicating that memory effect is an intrinsic property of monolayer MoS_2 independent of the fabrication process and

978-1-7281-1988-5/18 $31.00 © 2018 IEEE

material source. It's worthwhile to mention that the variation of the switching characteristics could be due to several sources of variability including materials, processes, and dimensional fluctuations, which definitely requires further research.

Considering the small area capability, crossbar devices were mainly used in the following experiments. Stable switching characteristics have been achieved with good retention (above 1 month) and DC cycling (above 120 cycles) (**Fig. 3**). Besides DC operation, 15 ns pulse SET operation is demonstrated as shown in **Fig. 4**, which is the fastest switching speed in crystalline 2D related NVRS. A high on/off current ratio of 10^8 is achievable for small areas. A comparison of the atomristor with other representative 2D NVRS is presented in **Table I**. Besides fast switching speed, the atomristor has the advantages including thinnest active layer, relatively low switching voltage and forming-free characteristic.

B. Simulation and Mechanism

For the first time, ab-initio simulation is implemented to help elucidate the switching mechanism (in MoS_2) using the Atomistix ToolKit (ATK)®. It is found that Au^{1+} tends to occupy and then reduce at the sulfur vacancies with captured electrons in MoS_2 (**Fig. 5**), while neutral Au, Au^{3+} and Au^{1-} are not favorable for substitution. Furthermore, monolayer materials were investigated with sulfur replaced by gold atoms at different percentages (**Fig. 6**), which are consistent with the final states of gold ion chemisorption at sulfur vacancies in MoS_2. The increasing density of states around Fermi level as the Au percentage rises leads to the electronic transition from HRS to LRS. At high replacement percentage, a number of states around the Fermi level allude to conductive characteristic at LRS, which theoretically validates our initial hypothesis [1].

The proposed mechanism is supported by current sweep, area dependence and sulfur annealing recovery. In current sweep SET operation (**Fig. 7**), each sudden drop in voltage is related to a resistance decreasing process (SET). Four separate SET may be caused by the discrete increasing number of Au atoms replacement. This multi-level characteristics might enable applications in multi-bit storage and reconfigurable amplifiers and attenuators. According to the area dependence (**Fig. 8**), the relatively flat LRS profile is consistent with the theory of a single or a few Au conductive bridges. Besides, as the device area increases, defined by the overlapped region between TE and BE, more defects exist in the monolayer MoS_2 between two electrodes. Thus, it is easier for Au ions to occupy sulfur vacancies, leading to a lower SET voltage. The annealing recovery with the supply of sulfur (**Fig. 9**) further proves the essential role of sulfur defects, especially sulfur vacancies. The failed devices cannot be RESET and thus retained at LRS even after applied negative bias as shown in Fig. 9(a). Then, the sample was annealed in a sulfur-rich atmosphere (sublimation of sulfur powder) at 150 ℃ for one hour. After annealing, failed devices can be recovered and switched as shown in Fig. 9(b). For the failed devices without RESET capability, a possible reason is that there are too many Au atoms occupied at sulfur vacancies, and thus makes it difficult to drive sufficient amount of Au atom back to electrodes. The annealing in the sulfur atmosphere can be used to replace some Au atoms with sulfur and restore a balance between sulfur vacancies and Au atoms.

Undoubtedly, further studies are warranted to investigate all plausible mechanisms via advanced theory and experiments.

IV. APPLICATIONS

Applications including RF switches and flexible memory can take unique advantage of the low LRS resistance and mechanical flexibility, respectively. Small-signal performance of the RF switch was characterized by calibrated network analyzer measurements aided by an equivalent circuit model. The switch figure of merit (FOM) cutoff frequency ($f_c=1/2\pi R_{ON}C_{OFF}$) was determined using the de-embedded ON-state resistance (R_{ON}) and OFF-state capacitance (C_{OFF}) from the circuit model. **Fig. 10** shows the intrinsic experimental RF characteristics of monolayer MoS_2 switch with promising results of ~0.25 dB insertion loss in the ON-state and isolation <29 dB in the OFF-state up to 67 GHz. Bilayer MoS_2 RF switch shows similar but higher insertion loss in the ON state (**Fig. 11**). The extracted cutoff frequencies for monolayer and bilayer MoS_2 RF switches are 70 THz and 8.4 THz, respectively. The mono- and bi- layer samples have different areas. Notably, the monolayer MoS_2 RF switch achieved a record cutoff frequency value compared to emerging solid-state, MEMS, and phase change (PC) material-based switches (**Table II**), with the added benefit of smaller feature size and frequency scalability without compromising insertion loss. These new results are also superior to our initial work with almost 7x higher FOM [10].

The flexible atomristors were also investigated on polyimide substrate revealing that both HRS and LRS afford retention after 1000 bending cycles with reproducible switching curves (**Fig. 12**). These results reveal the promising potential of atomristors in high performance RF switches, flexible memory devices and other novel applications yet to be developed.

V. CONCLUSION

In summary, atomristors based on CVD and exfoliated atomically-thin MoS_2 were investigated, featuring low switching voltage down to ~0.6 V, forming-free characteristic, and high on/off ratio up to 10^8. Record 15 ns pulse operation has been demonstrated among crystalline 2D NVRS memories. Ab-initio simulations reveal that the NVRS can be attributed to the interactions between metal ions and sulfur vacancies. In addition, small-area MoS_2 RF switches offer record 70 THz figure of merit. This work indicates the new potential of 2D monolayers for NVRS memory and RF switches.

ACKNOWLEDGMENT

The authors acknowledge the group of Prof. Nanshu Lu of The University of Texas, Austin (UT-Austin) for bending cycle facility. We appreciate fruitful discussions with Ying-Chen Chen of UT-Austin, and Jesse Tice and Xing Lan of Northrop Grumman. This work is funded in part by a PECASE award, NSF ECCS-1809017 and MRSEC DMR-1720595 grants.

REFERENCES

[1] R. Ge et al., *Nano Lett.*, vol. 18, pp. 434-441, 2018.
[2] G. R. Bhimanapati et al., *ACS Nano*, vol. 9, no. 12, pp. 11509-11539, 2015.
[3] C. Tan et al., *Chem. Soc. Rev.*, vol. 44, pp. 2615-2628, 2015.
[4] H. −S. P. Wong et al., *Proceedings of the IEEE*, vol. 100, May 2012.
[5] A. A. Bessonov et al., *Nat. Mater*, vol. 14, pp. 199−204, 2014.
[6] D. Son et al., *Adv. Mater.*, vol. 28, pp. 9326−9332, 2016.
[7] F. M. Puglisi et al., 2016 *IEDM.*, pp 34.8.1-34.8.4
[8] C. Hao et al., *Adv. Funct. Mater.*, vol. 26, pp. 2016−2024, 2016.
[9] J. Shi et al., *ACS Nano*, vol. 8, pp. 10196-10204, 2014.
[10] M. Kim et al., *Nat. Commun.*, vol. 9, pp. 2524, 2018.

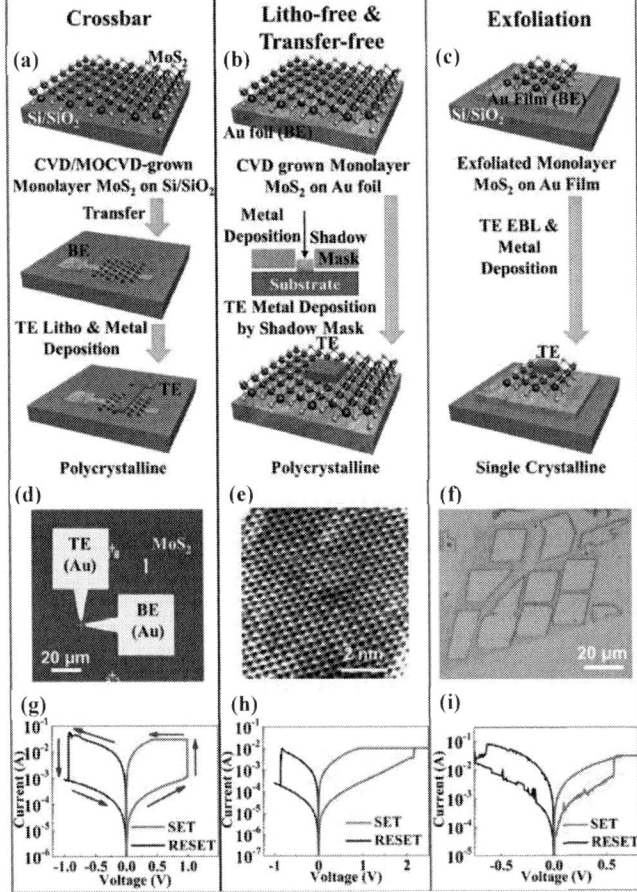

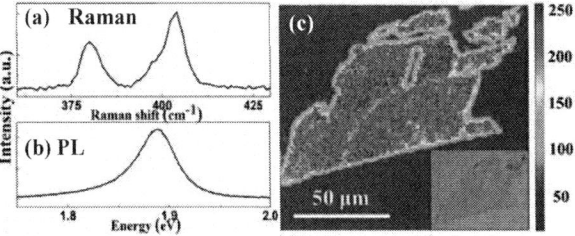

Fig. 2. (a) Raman spectrum, (b) photoluminescence (PL) spectrum and (c) PL intensity mapping of exfoliated monolayer MoS₂ flake. Inset: Optical image of the same flake. These indicate a uniform, high-quality exfoliated monolayer MoS₂ flake.

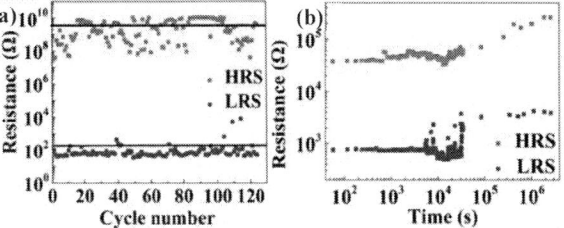

Fig. 3. (a) Endurance and (b) retention time of selected MoS₂ atomristors. 120 manual DC switching cycles and more than 4 weeks retention can be achieved.

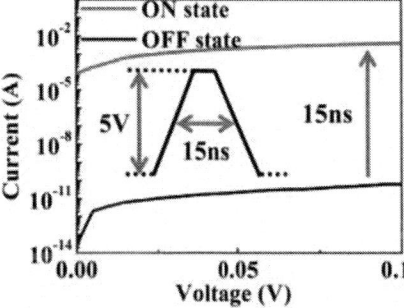

Fig. 4. Record 15-ns pulse SET demonstration among crystalline 2D related NVRS devices. The I-V characteristics before and after pulse driver clearly show the switching from HRS to LRS.

Fig. 1. (a, b, c) Process flow, (d, e, f) microscopy characteristics and (g, h, i) typical switching I-V curves of crossbar, litho-free and transfer-free, and exfoliation samples. (d) and (f) are the device optical images, while (e) is the STM image of as-grown monolayer MoS₂ on Au foil. The non-volatile resistance switching (NVRS) behavior in all three samples indicates a universal intrinsic property in various device structures and MoS₂ material sources. The switching in litho-free and transfer-free devices rules out the possible influence from process residues.

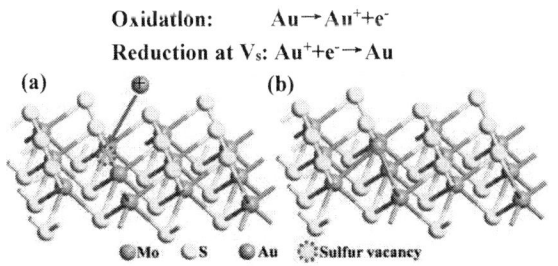

Fig. 5. The ab-initio simulation results of (a) the initial state and (b) the final state of optimization process for Au⁺ ion and sulfur vacancies (Vₛ) in monolayer MoS₂. The result shows that the Au⁺ tends to move to the Vₛ, possibly resulting in the conductive bridge formation and SET process, while same simulations based on neutral Au atom, Au³⁺, Au¹⁻ ions are unfavorable for occupation of Vₛ.

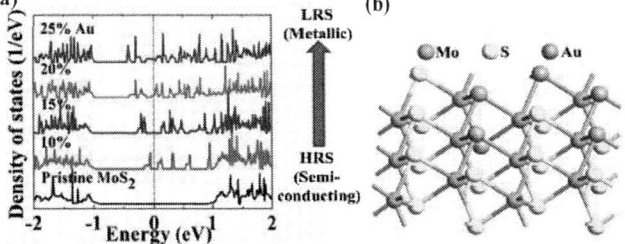

Fig. 6. (a) Density of states for different Au replacement percentage of sulfur (dash line: Fermi level). The results indicate Au replacement can lead to switching to LRS. (b) An illustration for density of states calculation, where 25% sulfur atoms are replaced by gold atoms.

978-1-7281-1988-5/18 $31.00 © 2018 IEEE

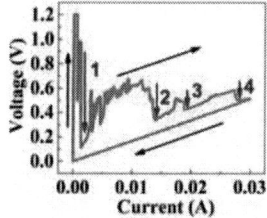
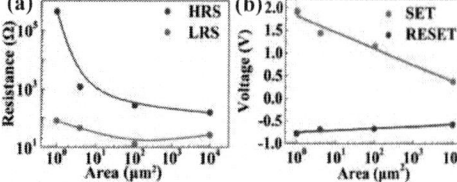
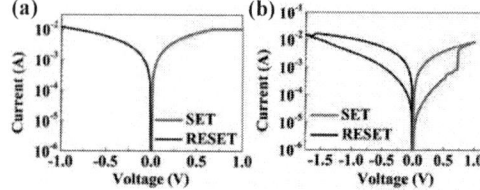

Fig. 7. Current sweep SET operation. Arrows show the current sweep direction. Four voltage drops indicate four separate resistance decreasing (SET) processes, corresponding to the existence of multiple defects.

Fig. 8. Area dependence of (a) HRS & LRS resistance and (b) SET & RESET voltages. HRS and SET voltage have decreasing trend as the area increases. While LRS and RESET voltage has relatively weak variation. The area dependence reveals a mechanism based on 1D conductive bridge.

Fig. 9. Typical switching I-V curve of (a) failed device retained at LRS (cannot be RESET) and (b) recovered device after annealing with supply of sulfur vapor at 150°C for 1 hour. The novel recovery phenomenon indicates that sulfur defects, especially sulfur vacancies, play an essential role in switching.

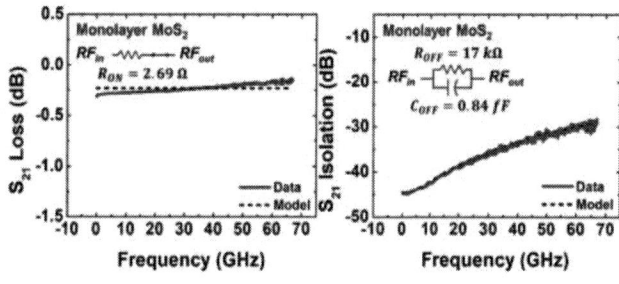
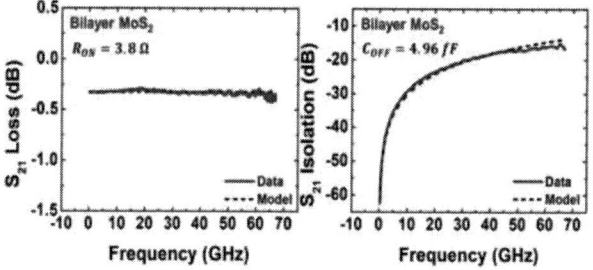

Fig. 10. Measured and extracted S-parameter data in both the ON-state (insertion loss) and OFF-state (isolation) of an RF switch based on 0.15×0.2 μm² monolayer MoS₂ atomristor. The extracted R$_{ON}$, C$_{OFF}$, and f$_c$ values are 2.69 Ω, 0.84 fF, 70 THz respectively.

Fig. 11. S-parameter data of an RF switch based on 0.5×0.5 μm² bilayer MoS₂ atomristor. R$_{ON}$ is 3.8 Ω and C$_{OFF}$ is 4.96 fF. The extracted f$_c$ is 8.4 THz.

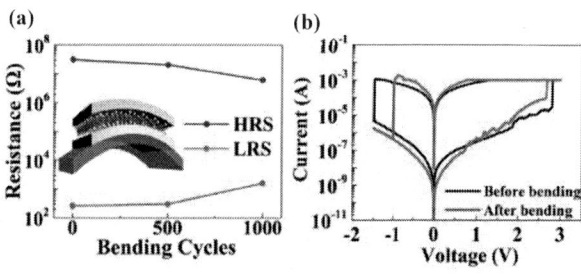

Fig. 12. (a) Stable resistance of both HRS and LRS and (b) typical switching I-V curve before and after 1000 cycles at 1% strain of MoS₂ crossbar device. Inset: schematic of flexible device. These results show a promising application for flexible memory devices based on MoS₂ atomristors.

Table 1. Comparison of this work with other representative 2D NVRS publications.

Reference	Active Layer Materials	Active Layer Thickness	Device Structure	Switching Voltage	Switching Time
This Work	MoS₂	~0.65 nm	Vertical	Down to 0.6 V	15 ns
Sangwan et al., Nature, 2018	MoS₂	~0.65 nm	Planar	> 20 V	1 ms
Wang et al., Nat. Electron., 2018	MoS₂₋ₓOₓ	~ 40 nm	Vertical	~ 1V	100 ns
Pan et al., Adv. Funct. Mater., 2017	hBN	~1.8-2.5 nm	Vertical	Down to 0.4 V	Not reported
Zhao et al., Adv. Mater., 2017	BNOₓ	0.9-2.3 nm	Vertical	0.6 V-1.7 V	1 ms
Cheng et al., Nano Lett., 2016	1T MoS₂	550 nm	Vertical	~ 0.1 V	Not reported
Hao et al., Adv. Funct. Mater., 2016	Degraded BP	~ 10 nm	Vertical	1 - 2 V	Not reported

Table 2. Comparison of this work with other representative RF switch papers.

Reference	Device Technology	Non-volatility	Control Voltage	Cutoff frequency	Operating Environment	Switching time	Dimension (single device W x L)
This Work	MoS₂ Switch	Yes	~0.5 – 1.5V	70 THz	Ambient condition	15 ns	0.15 um x 0.2 um
Leon et al., IMWS-AMP, 2017	GeTe Phase-change Switch	Yes	3.5 V	17 THz	Heater needed	< 0.5 us	0.8 um x 0.5 um
Madan et al., IEDM, 2015	VO₂ Phase-change Switch	No	~1.5 V	26.5 THz	Ambient condition	25 ns	1 um x 0.1 um
Pi et al., Nat. Commun., 2015	Memristive switch	Yes	3V	35.2 THz	Ambient condition	Not reported	0.11 um x 0.035 um
Stefanini et al., J. Microelectromech Syst., 2011	MEMS Switch	No	~65 V	3.8 THz	Hermetic packaging	2.2 us	300 um x 24 um

An Ultra-fast Multi-level MoTe$_2$-based RRAM

F. Zhang[1], H. Zhang[2,3], P.R. Shrestha[2,3], Y. Zhu[1], K. Maize[1], S. Krylyuk[2,3], A. Shakouri[1], J.P. Campbell[3],
K.P. Cheung[3], L.A. Bendersky[3], A.V. Davydov[3] and J. Appenzeller[1*]

[1]Purdue University, West Lafayette, IN, USA, *email: appenzeller@purdue.edu,
[2]Theiss Research, Inc., La Jolla, CA, USA, [3]National Institute of Standards and Technology, Gaithersburg, MD, USA,

Abstract — We report multi-level MoTe$_2$-based resistive random-access memory (RRAM) devices with switching speeds of **less than 5 ns** due to an electric-field induced 2H to 2H$_d$ phase transition. Different from conventional RRAM devices based on ionic migration, the MoTe$_2$-based RRAMs offer intrinsically better reliability and control. In comparison to phase change memory (PCM)-based devices that operate based on a change between an amorphous and a crystalline structure, our MoTe$_2$-based RRAM devices allow faster switching due to a transition between *two* crystalline states. Moreover, utilization of atomically thin 2D materials allows for aggressive scaling and high-performance flexible electronics applications. Multi-level stable states and synaptic devices were realized in this work, and operation of the devices in their low-resistive, high-resistive and intrinsic states was quantitatively described by a novel model.

I. INTRODUCTION

RRAM has promises of being an emerging technology due to its potential scalability, high operation speed, high endurance and ease of process flow. However, reliable and repeatable operation is a potential challenge in future applications since switching involves the uncontrollable motion of individual atoms. In this work, we present a new switching mechanism for MoTe$_2$-based RRAM. An electric field induces the phase transition from the stable semiconducting 2H phase to a more conductive 2H$_d$ phase, which provides a potential path towards better stability. The newly formed 2H$_d$ is structurally close to the 2H phase, which holds the promise for faster switching if compared with the significant migration of ions in conventional RRAM [1] or the amorphous-to-crystalline transition in PCM [2] (see Fig.1). Initial pulse measurements show impressive switching speed of less than 5 ns. Moreover, multi-level states can be programmed into the devices by applying proper set/reset voltages, which allows to gradually changing the device resistance with multiple pulses, creating a "synaptic device".

II. SWITCHING IN MoTe$_2$ RRAM DEVICES

Fig. 1 illustrates the advantages of this new type MoTe$_2$-based RRAM as compared to conventional RRAM and PCM due to switching by an electric-field induced phase transition. Fig. 2(a) shows schematically a vertical MoTe$_2$ device. First, a bottom electrode Ti/Au (10 nm/25 nm) was deposited onto a 90 nm silicon dioxide (SiO$_2$) layer covering a highly doped silicon wafer. Next, MoTe$_2$ (2D Semiconductor) layers were exfoliated onto this electrode using standard scotch tape techniques, followed by thermal evaporation of 55 nm SiO$_2$ insulating layer. The device fabrication was finished by the deposition of a Ti/Ni (35 nm/50 nm) top electrode. Different

from previously reported CVD grown 2D material based RRAM devices [3,4] whose operation is mediated by uncontrollable defects/grain boundaries in the device structure, our active material is a single-crystalline layer, where the observed RRAM behavior is due to the intrinsic properties of MoTe$_2$. Fig. 2(b) shows an AFM image of a MoTe$_2$ vertical device. The active region is about 0.1 μm^2. Fig. 3(a) displays I-V curves of a pristine device, the device forming process and successive cycling through its high resistive state (HRS) and low resistive state (LRS). Stable and reproducible bipolar RRAM behavior was observed. Fig. 3(b) shows the I-V curves for MoTe$_2$ devices with different layer thicknesses, and Fig. 3(c) summarizes how the forming and set voltages scale with the MoTe$_2$ layer thickness.

Thermoreflection microscope images were acquired to map the location of the filament on the device after forming. Surface temperature maps with 50 mK temperature resolution and submicron (diffraction limited) spatial resolution can be acquired within a few minutes [5]. Fig. 4(a) and (b) show self-heating hotspots on the nickel electrode surface superimposed with optical images for two representative devices, indicating the position of the filaments. The occurrence of a single hotspot for each device with characteristic hotspot full-width-at-half-maximum of ~ 200 nm is consistent with joule self-heating from a source as small as a MoTe$_2$ filament. In six out of eight devices imaged the hotspot was located at the edge or corner of the active region as can be seen in Fig. 4(a) and (b). We speculate that this preferential occurrence is a result of stronger electric fields at "sharp" topological features due to patterning during the fabrication that enhances the filament formation. Fig. 4(c) shows the calibrated temperature change map for the hotspot in device (b). The detected temperature change on the filament portion is ~ 15 K.

In order to understand the filament formation mechanism in MoTe$_2$, scanning transmission electron microscopy (STEM) of cross-sectional samples was utilized. As Fig. 5 shows, in the LRS, a distorted 2H$_d$ phase was identified in the regions extending vertically throughout the MoTe$_2$ layer. The 2H$_d$ phase was identified as a distorted modification of the 2H structure – a transient state with atoms displaced to the sites of a lower symmetry, but still within atomic arrangements of the 2H structure. A detailed analysis of the structure can be found in ref. [6]. Fig. 6 shows an energy dispersive spectrometry (EDS) scan along the filament region of a device in its LRS. Almost no Ti and Au signals were detected within the filament, which – in particular when also considering the unavoidable ion-milling contamination during FIB sample preparation – implies that the switching mechanism is *not* related to the migration of metal ions. To further confirm this point, graphene was used to replace the metal top and bottom electrodes. Fig. 7 shows the "typical" bipolar RRAM behavior

978-1-7281-1988-5/18 $31.00 © 2018 IEEE

observed here in a Graphene-MoTe$_2$-Graphene device. Note that this is the first demonstration of an entirely 2D materials-based RRAM. Based on previous studies [7], graphene is a good diffusion barrier for metal ions. This 2D RRAM excludes completely the possibility of migration of metal ions as a source for the resistive switching observed by us. Based on these results, an electric-field induced phase transition from 2H to a more conductive 2H$_d$ state is believed to be responsible for the RRAM behavior in vertical MoTe$_2$ devices.

III. A PHYSICAL MODEL

To fully understand the vertical transport through the pristine, LRS and HRS states in MoTe$_2$ devices and to explore the properties of the new 2H$_d$ phase, a physical model was constructed. The barrier height Φ_{2H} and Φ_{2Hd} shown in figure 8(a) were extracted by utilizing the numerical model from ref. [8]. In this model, two different transport mechanisms are considered: thermal diffusion at low voltages and Fowler-Nordheim (FN) tunneling at higher voltages, as illustrated for a pristine data set in Fig. 8(b). In our model the 2H phase has a larger barrier height than the 2H$_d$ phase as evident when their current values in the low voltage range are compared. An excellent fit can be obtained for the pristine I-V characteristics (Fig. 8(b)), as well as for the LRS and HRS (Fig. 8(c)) by employing the band diagrams and parameters shown in Fig. 8(a). In the pristine state, the MoTe$_2$ is in its 2H phase with a large barrier height of $\Phi_{2H} = 0.38$ eV. On the other hand, in the LRS, a filament of 2H$_d$ was created through the setting process with an extracted barrier height of $\Phi_{2Hd} \approx 0.07$ eV. The HRS is characterized by formation of the 2H/2H$_d$ heterojunction due to rupture of the 2H$_d$ filament during the reset process. The thickness of newly formed 2H segment in the filament can be estimated to be ~ 1.8nm. Thus, the simulation results are consistent with the notion that a new semiconducting 2H$_d$ state with a smaller barrier height is formed during the set process that is responsible for the higher conductivity of the LRS compared with the HRS.

IV. PERFORMANCE STUDY AND PULSE MEASUREMENTS

A. Performance study

Fig. 9 illustrates the pulse switching behavior in MoTe$_2$ based RRAM devices. The pulse width is 80 µs. By applying set/reset pulses, the device can switch between the LRS and HRS. Fig. 9(c) shows the read out current per cycle in the LRS and HRS. Fig. 9(d) is a retention measurement. All performance studies indicate stable and reproducible RRAM behavior.

B. Pulse measurements

To test the switching speed of our MoTe$_2$ RRAM devices, the experimental measurement setup shown in the inset of Fig. 10(b) was utilized. The current through the device was measured using a 50 Ω termination at the oscilloscope [9]. Note that no current compliance was used in this setup. The switching was controlled by varying the pulse width or pulse amplitude. A current and voltage versus time (t) plot of one such SET operation is shown in Fig. 10(a). The full width at half max (FWHM) of the applied voltage pulse is 5 ns. Fig.

10(b) shows the same data as in Fig. 10(a) but plotted as the current versus voltage. This clear change in resistance during the 5 ns voltage pulse is further evidence that the switching speed in MoTe$_2$ based RRAM is less than 5 ns.

The devices were reset (switched OFF) using negative pulses. Multiple pulses were required for a gradual reset process. Fig. 10(c) shows 10 pulses used to reset the device. Current versus voltage plots of cycle 1 (1st pulse), cycle 5 (5th pulse) and cycle 10 (10th pulse) are shown in Fig. 10(d). The gradual resistance change is a desirable feature for neuromorphic computing. Fig. 11 shows the characteristics of a device that was programmed by a series of positive pulses (1.1 V, 80 µs) followed by a series of negative voltage pulses (-1.2 V, 80 µs). The resistance of the MoTe$_2$ device gradually decreased and increased, which is similar to the potentiation and depression of biological synapses.

By carefully tuning the set/reset voltages, multi-level states can be programmed into the devices. Fig. 12(a) shows stable resistive states after various short (80 µs) and long (560 µs) voltage pulses that were read at a 0.2 V level. Long pulses result in a more substantial change (training) of the resistive state of the system. Fig. 12(b) shows the switch on/off behavior in each state. All the pulse measurements hint at an additional application space of this class of vertical TMD devices in the realm of neuromorphic computing.

ACKNOWLEDGMENT

This work was supported in part by the Semiconductor Research Corporation (SRC) as the NEWLIMITS Center and NIST through award number 70NANB17H041. H.Z. acknowledges support from the U.S. Department of Commerce, NIST under the financial assistance awards 70NANB15H025 and 70NANB17H249. S. K. acknowledges support from the U.S. Department of Commerce, NIST under the financial assistance award 70NANB18H155.

REFERENCES

[1] H. S. P. Wong, *et al.*, "Metal-Oxide RRAM," *Proceedings of the IEEE*, vol. 100, pp. 1951-1970, 2012.

[2] H. S. P. Wong, *et al.*, "Phase Change Memory," *Proc. IEEE*, vol. 98, pp. 2201-2227, 2010.

[3] V. K. Sangwan, *et al.*, "Gate-tunable memristive phenomena mediated by grain boundaries in single-layer MoS$_2$," *Nat. Nanotechnol.*, vol. 10, pp. 403-406, 05//print 2015.

[4] F. M. Puglisi, *et al.*, "2D h-BN based RRAM devices," in *2016 IEEE International Electron Devices Meeting (IEDM)*, 2016, pp. 34.8.1-34.8.4.

[5] K. Maize, *et al.*, "Transient Thermal Imaging Using Thermoreflectance," in *2008 Twenty-fourth Annual IEEE Semiconductor Thermal Measurement and Management Symposium*, 2008, pp. 55-58.

[6] F. Zhang, *et al.*, "Electric field induced semiconductor-to-metal phase transition in vertical MoTe2 and Mo$_{1-x}$W$_x$Te$_2$ devices," *arXiv preprint arXiv:1709.03835*, 2017.

[7] R. Mehta, *et al.*, "Transfer-free multi-layer graphene as a diffusion barrier," *Nanoscale*, vol. 9, pp. 1827-1833, 2017.

[8] Y. Zhu, *et al.*, "Vertical charge transport through transition metal dichalcogenides–a quantitative analysis," *Nanoscale*, vol. 9, pp. 19108-19113, 2017.

[9] P. R. Shrestha, *et al.*, "Compliance-free pulse forming of filamentary RRAM," *ECS Transactions*, vol. 75, pp. 81-92, 2016.

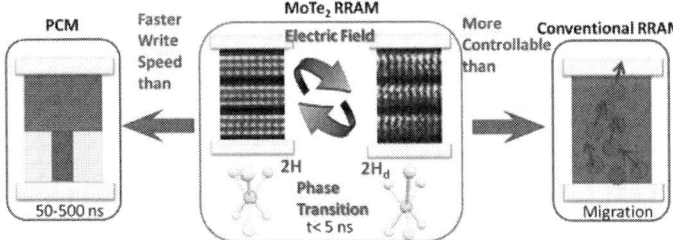

Fig. 1. Highlight features of the MoTe₂ based RRAM.

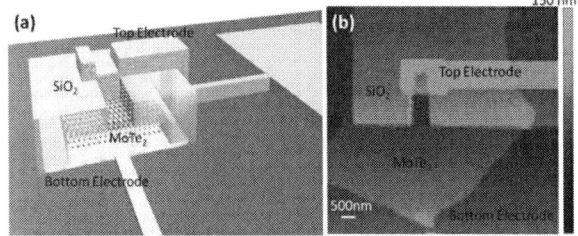

Fig. 2. (a) Schematic diagram of a vertical metal/MoTe₂/metal device. (b) AFM image of a vertical MoTe₂ device.

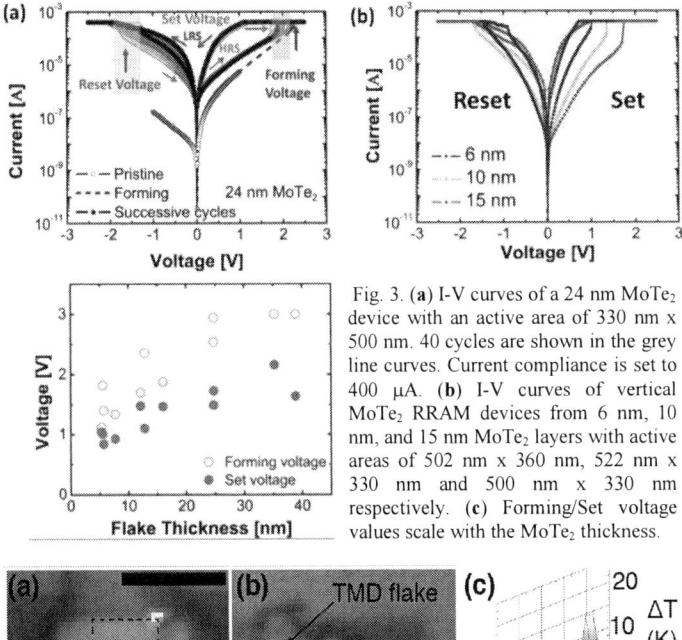

Fig. 3. (a) I-V curves of a 24 nm MoTe₂ device with an active area of 330 nm x 500 nm. 40 cycles are shown in the grey line curves. Current compliance is set to 400 μA. (b) I-V curves of vertical MoTe₂ RRAM devices from 6 nm, 10 nm, and 15 nm MoTe₂ layers with active areas of 502 nm x 360 nm, 522 nm x 330 nm and 500 nm x 330 nm respectively. (c) Forming/Set voltage values scale with the MoTe₂ thickness.

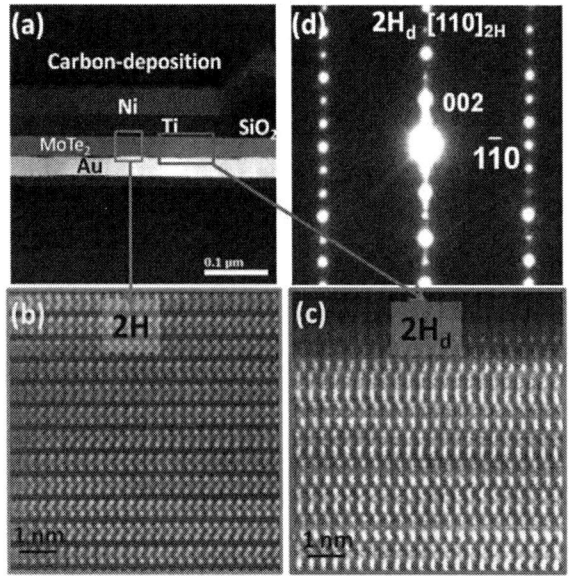

Fig.5. (a) HAADF-STEM image showing the cross-section of a MoTe₂ device. Higher magnification HAADF image from the region defined by the blue/red box in (a) showing (b) 2H and (c) a distorted structure (2H$_d$) taken along the [110]$_{2H}$ zone-axis. (d) Corresponding nano-beam diffraction pattern taken from the distorted 2H$_d$ area.

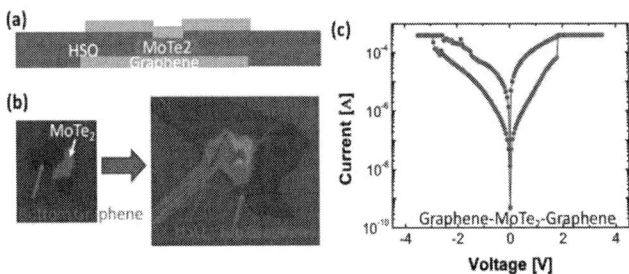

Fig. 4.(a-b) Thermoreflectance images showing the location of the filament. The scale bar is 5 μm. (c) The calibrated temperature change map for the filament in the device (b).

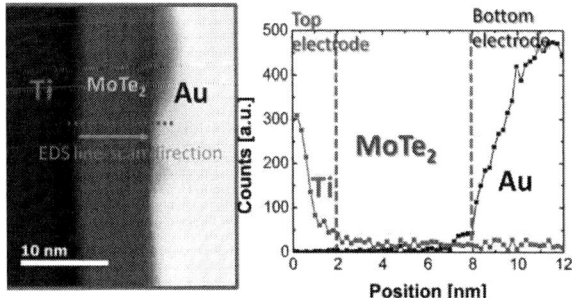

Fig. 6. EDS line-scan analysis of the filament region of a device in its LRS. Little Ti/Au signals were detected in the filament region due to ion-milling contamination during sample preparation.

Fig. 7. (a) Schematic and (b) optical images of a Graphene-MoTe₂-Graphene device. (c) I-V curve of the RRAM device in (b).

978-1-7281-1988-5/18 $31.00 © 2018 IEEE   538

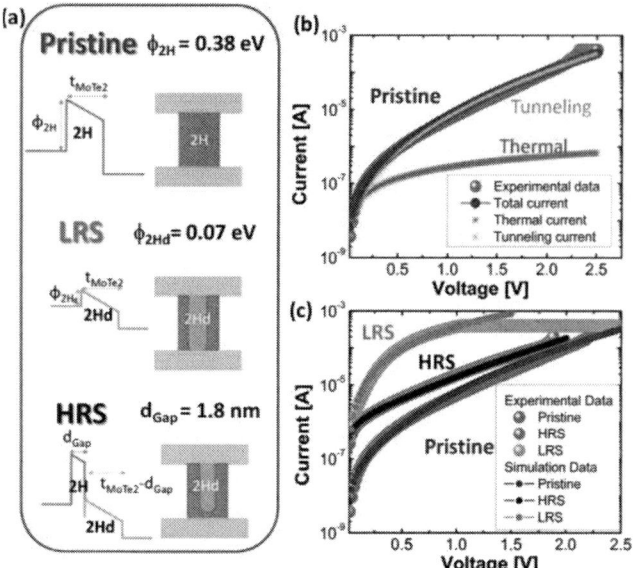

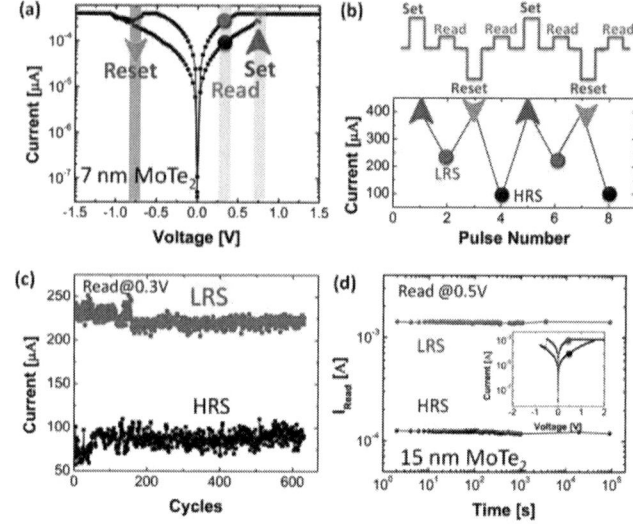

Fig. 8. (**a**) Schematic band diagrams of the RRAM in its various states.

(**b**) Experimental and simulation data of a pristine device. (**c**) Experimental and simulation data for the LRS, HRS and pristine state.

Fig. 9. (**a**) DC characteristic of a 7 nm thick MoTe$_2$ layer RRAM device. (**b**) shows the various current levels after pulse switching. (**c**) Current versus cycle in the LRS and HRS at a read voltage of 0.3 V. Current compliance is set to 400 μA. (**d**) Retention of the HRS and LRS for a 15 nm thick MoTe$_2$ RRAM device. Current compliance is set to 1.2 mA.

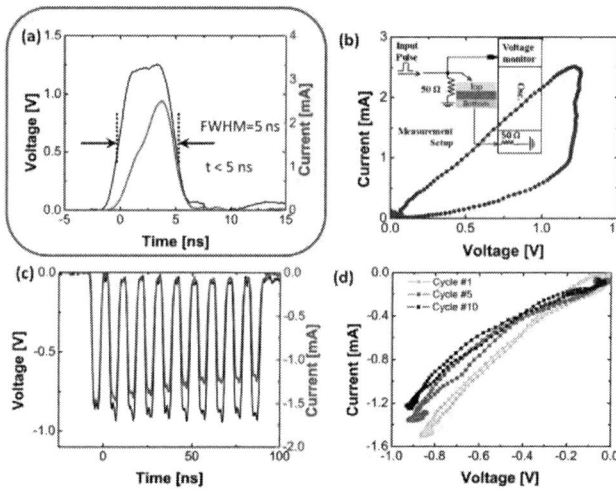

Fig. 10. (**a**) I/V vs. time plot showing switching of the device within a 5 ns voltage pulse. (**b**) I-V plot of the data shown in (**a**) to show the change in resistance during the applied pulse. The inset figure displays the experimental setup for the pulse measurement. (**c**) I/V vs. t plot of 10 reset pulses. (**d**) I-V plot of data plotted in (**c**) of pulses 1, 5 and 10.

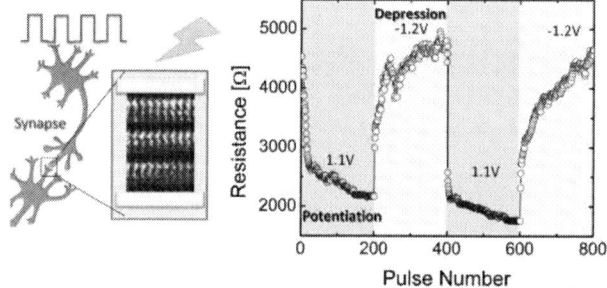

Fig. 11. Synaptic device from an MoTe$_2$ RRAM.

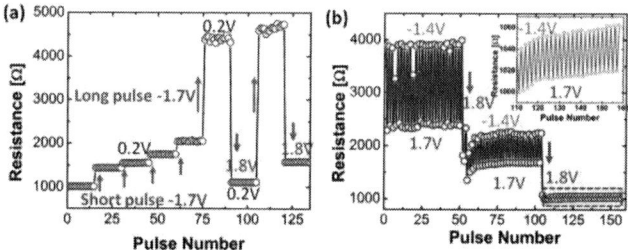

Fig. 12. (**a**) Multiple stable states of a device after various set and reset voltage pulses had been applied. Read out occurs at a voltage of 0.2 V. Every state is characterized by 15 subsequent read outs. A short pulse (80 μs) and longer pulse (560 μs) of a reset voltage of -1.7 V result in different changes of the resistance. (**b**) Multi-level characteristics of a vertical MoTe$_2$ device. Each level exhibits stable switch on/off at set/reset voltages of 1.7 V and -1.4 V respectively. A third state can be "dialed in" through yet another 1.8 V pulse. The inset figure is the zoom-in of the red dashed part.

978-1-7281-1988-5/18 $31.00 © 2018 IEEE

First cryogenic electro-optic switch on silicon with high bandwidth and low power tunability

F. Eltes[1]*, J. Barreto[2], D. Caimi[1], S. Karg[1], A. A. Gentile[2], A. Hart[2], P. Stark[1], N. Meier[1], M. G. Thompson[2], J. Fompeyrine[1], S. Abel[1]

[1]IBM Research – Zurich, Säumerstrasse 4, 8803 Rüschlikon, Switzerland, *email: fee@zurich.ibm.com
[2]Quantum Engineering Technology Labs, University of Bristol, Bristol, UK

Abstract— We demonstrate the first electro-optic switch operating at cryogenic temperatures of 4 K with a high electro-optic bandwidth of >18 GHz. Our novel technology exploits the Pockels effect in barium titanate thin films co-integrated with silicon photonics and offers low losses, pure phase modulation, and sub-pW electro-optic tuning.

I. INTRODUCTION

To achieve scalable quantum computing, or supercomputers based on rapid single-flux quantum technology, integrated electrical [1, 2] and optical [3, 4] circuits operating at cryogenic temperatures are required. A common challenge among all technologies is the signal interface between the cryogenic processors and the outside world at room temperature (**Fig. 1**) [2]. Currently, the signal transmission relies on electrical wires connecting multiple stages at different temperatures. Here, electrical connections have three major limitations. First, cooling the chip to temperatures at 4 K and below becomes demanding due to the high thermal conductivity of metals and the limited cooling power available. Second, since the bandwidth of single electrical connections is limited, parallel wires would be needed to increase the data throughput, increasing further the total thermal conductivity. Third, increasing the carrier frequency in electrical connections results in larger radio-frequency losses. Such losses are dissipated as heat, which requires a larger fraction of the totally available cooling power.

Fiber-optic connections to transmit signals could solve these issues by providing low power, high-bandwidth data links and low heat-leakage connections at the same time. While some of the building blocks of integrated optical circuits such as light-sources and detectors perform better at cryogenic temperatures, a viable solution for modulators and switches – both key components in optical links – have not yet been found.

Previous demonstrations of integrated electro-optic (EO) switches operated at cryogenic temperatures rely either on thermo-optic phase shifters [5], or on the plasma-dispersion effect, the most common effect in silicon photonic modulators [6]. Both of these approaches suffer from intrinsic limitations: The thermo-optic switches exploit Joule heating to change the refractive index, which requires significant cooling power and which has a low bandwidth of less than MHz [7]. The plasma-dispersion switches rely on very high doping levels in order to compensate for freezing out of charge carriers. Because of the high doping levels only very small micro-disk resonators can be used in this technology. Tunable low-loss wavelength filters or broadband switches are not feasible due to large insertion losses. In addition, the carrier freeze-out, and therefore a high series resistance, limits the micro-disk resonators to a low bandwidth of <5 GHz at 4.8 K.

The EO Pockels effect does not suffer from the intrinsic limitations of the thermo-optic and plasma-dispersion effects at cryogenic temperatures. However, the Pockels effect requires materials with specific properties, something which has only recently become available in integrated devices compatible with Si photonic platforms and large-scale substrates [8]. Using integrated components based on the Pockels effect it is possible to achieve efficient modulation, low loss, and high bandwidth at cryogenic temperatures. Here, we demonstrate Si photonics-compatible high-speed EO switches based on single-crystal barium titanate ($BaTiO_3$, BTO) thin-films operating at 4.3 K. We show that losses from BTO are negligible and that static power consumption is minimal. The EO bandwidth exceeds 18 GHz – the highest value achieved to date for an integrated device at cryogenic temperature. Our results prove that integrated BTO-devices are highly suitable for applications in cryogenic photonics (**Fig. 2**).

II. DEVICE STRUCTURE AND FABRICATION

Our electro-optic switches rely on the integration of a thin-film Pockels material in a strip-loaded waveguide geometry (**Fig. 2b**). We use BTO as it is known to have a large Pockels coefficient [9], can be grown on large size Si substrates, and can be integrated in Si photonic platforms [8] (**Fig. 3**). To exclude any electro-optic effects from charge-carriers in a Si waveguide, we employ SiN as the waveguide material. We used an 80-nm-thick BTO layer and a 150-nm-thick SiN layer, structured into a 1.1 μm wide waveguide. This geometry results in 19% mode overlap with BTO for the first order TE mode at 1550 nm (**Fig 2c**). We conservatively designed electrodes separated by 9 μm to ensure the absence of any loss from absorption in the metal. By designing a smaller electrode spacing and improving the electro-optic mode overlap e.g. by thicker BTO layers and different strip-waveguide dimensions, the devices can be optimized for maximal performance. However, the current implementation allows us to evaluate all relevant aspects of BTO as a technology platform for cryogenic EO switches.

978-1-7281-1988-5/18 $31.00 © 2018 IEEE

We deposited the single-crystal BTO epitaxially on $SrTiO_3$-buffered 200 mm SOI substrates using a previously reported process [9]. X-ray diffraction shows that we indeed have a high-quality epitaxial BTO layer (**Fig. 4**). Subsequently, the BTO layer was transferred onto a SiO_2 terminated host wafer via direct wafer bonding, with thin alumina layers serving as bonding interface [10]. The excellent bonding strength yields a transfer ratio close to 100%. After removal of the donor wafer, we deposited 150 nm SiN which was patterned by dry-etching to form the waveguides. We used two-levels of metallization and a 1.5-μm-thick SiO_2 cladding to add electrodes to the devices (**Fig. 5**). A high-resolution electron micrograph shows the final cross-section of the BTO/SiN waveguides (**Fig. 5f**).

III. DEVICE PERFORMANCE

We characterized the electrical and electro-optic properties of our devices in a cryogenic probe station at a temperature of 4.3 K. We measured the electrical and passive optical properties as well as the electro-optic functionality of the devices. By showing broad-band switches, low-power tunable filters, and high-speed modulators, we demonstrate components necessary for a cryogenic photonics platform.

The ferroelectric nature of the BTO film (**Fig. 6**) can be clearly seen in the electrical current measured between two co-planar electrodes (**Fig. 7**). The current through our devices at low temperature is extremely low, <500 fA/mm in the operating range. This very small current results in an extremely low static power consumption, which is critical for cryogenic operation. This low leakage current allows us to probe the current generated by ferroelectric domain switching (**Fig. 7**). Additionally, the electric field dependence of the capacitance also shows the typical hysteretic behavior of ferroelectric materials (**Fig. 7**) [11].

We extracted propagations losses at 1550 nm of 5.6±0.3 dB/cm in the waveguides using cut-back measurements performed at room temperature (**Fig. 8**). In resonant devices we do not observe any increase in losses at 4 K. The propagation losses are only limited by the waveguide fabrication process, as evidenced by identical losses in SiN reference waveguides fabricated with the same process but without any BTO layer (5.4±0.2 dB/cm). Waveguides with propagation losses <1 dB/cm are available on advanced photonics platforms [12]. Unlike EO switches based on doped Si which suffer high absorption losses, the BTO-photonic technology has no such intrinsic limitation [13].

To demonstrate switching functionality, we employ asymmetric MZMs (**Fig. 9**) with 500 μm long phase shifters. Using 2x2 multimode interference splitters we divide the power from the MZM into two outputs (**Fig. 10**). By applying an electric field to one arm of the MZM we can switch the optical power between the two outputs. In the full range required to switch between the two outputs, the DC power consumption stays below 10 pW (**Fig. 11**), a record-low number.

Besides low power optical switches, tunable wavelength filters are another important component for quantum optics,

which can be implemented with a tunable resonator (**Fig. 12**). We use a racetrack resonator with a radius of 50 μm and 75 μm long straight phase shifters. We can select the filter wavelength by shifting the resonance position via the electric field bias (**Fig. 13**). When tracking the resonance position as a function of applied electric field, we see the same hysteretic behavior characteristic of a ferroelectric material as visible in the capacitance measurement (**Fig. 14**). This hysteresis also enables novel device types and applications in the form non-volatile optical switches [14].

Our technology is capable of high-speed EO modulation at cryogenic temperatures as shown by the S_{21} response of a racetrack modulator. The low Q-factor (Q ~ 1800) of the device used for the experiments (**Fig. 15**) ensures a high intrinsic bandwidth and allows to probe the high-speed EO response of the BTO/SiN phase shifters. We do not reach the 3-dB cut-off of our device within the range of our measurement equipment (18 GHz) (**Fig. 16**). The high bandwidth in our switches makes the BTO/SiN technology the fastest optical switching technology, to date, for integrated EO devices operating at 4 K (**Fig. 17**).

IV. CONCLUSION

We have shown the first demonstration of cryogenic operation of integrated EO switches exploiting the Pockels effect. The performance of our devices – low power, low loss, high bandwidth – shows that BTO-based switches can meet all the necessary requirements for optical interconnects in cryogenic systems.

ACKNOWLEDGMENT

This work has received funding from the European Commission under grant agreements no. H2020-ICT-2015-25-688579 (PHRESCO), from the Swiss State Secretariat for Education, Research and Innovation under contract no. 15.0285 and 16.0001, from the Swiss National Foundation project no 200021_159565 PADOMO, from EPSRC grants EP/L024020/1 and EP/K033085/1, the Quantum Technology Capital grant: QuPIC (EP/N015126/1), and ERC grant 2014-STG 640079.

REFERENCES

[1] J. M. Gambetta *et al.*, *npj Quantum Inf.*, vol. 3, no. 1, p. 2, 2017.
[2] D. S. Holmes *et al.*, *IEEE Trans. Appl. Supercond.*, vol. 23, no. 3, p. 1701610, 2013.
[3] J. W. Silverstone *et al.*, *IEEE J. Sel. Top. Quantum Electron.*, vol. 22, no. 6, p. 6700113, 2016.
[4] J. L. O'Brien *et al.*, *Nat. Photonics*, vol. 3, no. 12, pp. 687–695, 2009.
[5] A. W. Elshaari *et al.*, *IEEE Photonics J.*, vol. 8, no. 3, 2016.
[6] M. Gehl *et al.*, *Optica*, vol. 4, no. 3, p. 374, 2017.
[7] N. C. Harris *et al.*, vol. 22, no. 9, pp. 83–85, 2014.
[8] F. Eltes *et al.*, in *2017 IEEE International Electron Devices Meeting (IEDM)*, 2017, p. 24.5.1-24.5.4.
[9] S. Abel *et al.*, *Nat. Commun.*, vol. 4, p. 1671, 2013.
[10] N. Daix *et al.*, *APL Mater.*, vol. 2, no. 8, p. 086104, 2014.
[11] D. Bolten *et al.*, *Ferroelectrics*, vol. 221, no. 1, pp. 251–257, 1999.
[12] P. De Dobbelaere *et al.*, in *2017 IEEE International Electron Devices Meeting (IEDM)*, 2017, p. 34.1.1-34.1.4.
[13] F. Eltes *et al.*, *ACS Photonics*, vol. 3, no. 9, pp. 1698–1703, 2016.
[14] S. Abel *et al.*, in *2017 IEEE International Conference on Rebooting Computing (ICRC)*, 2017.

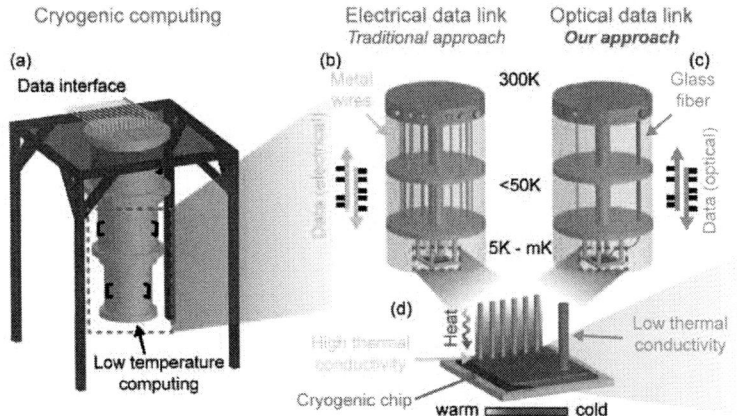

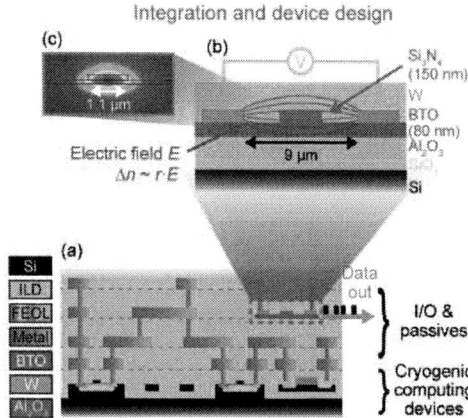

Fig. 1. (a) Schematics of a cryostat typically used for operating cryogenic computing systems. The computing chip is isolated from room temperature via multiple thermal stages with (b) electrical and (c) optical data connections. (d) The high thermal conductivty of multiple electrical connections locally increases the chip temperature and requires more cooling power than thermally insulating, high-bandwidth optical fiber links.

Fig. 2. (a) Integration concept of BTO optical switches in a CMOS compatible process flow as demonstrated in [8]. (b) Cross section of the SiN/BTO optical switch as used in this work. (c) Simulated power distribution of the optical mode in the waveguide center.

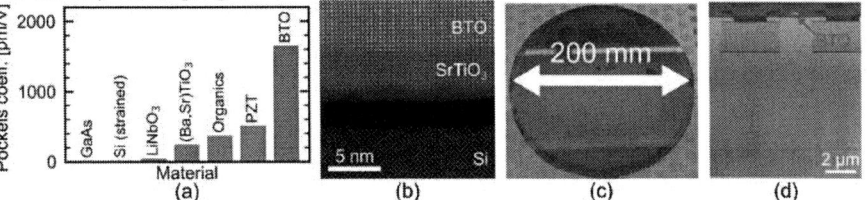

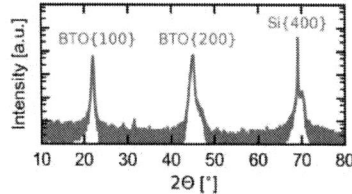

Fig. 3. (a) Comparison of Pockels coefficients in various materials. (b) High-resolution STEM image showing epitaxial BTO film on Si. (c) Demonstration of a wafer bonding transfer of BTO film using 200 mm wafers. (d) Integration of Si-BTO modulators in BEOL of an advanced silicon photonics platform. [8]

Fig. 4. X-ray diffraction spectrum showing good crystal quality, and epitaxial relationship of the BTO on Si used in this work.

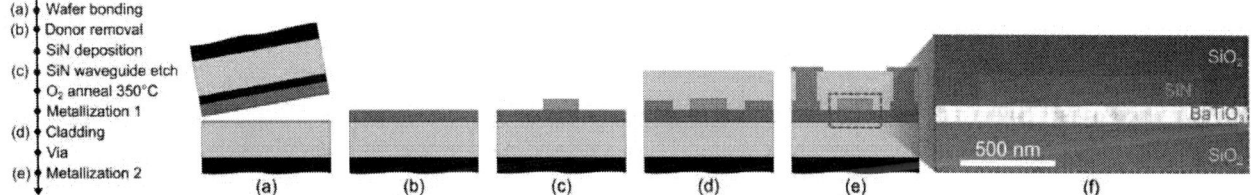

Fig. 5. (a-e) Process flow for fabrication BTO/SiN devices based on single-crystal BTO layers transferred by wafer bonding. (f) STEM image showing the cross-section of a finished BTO/SiN waveguide.

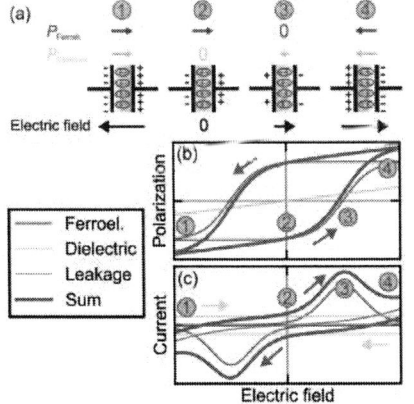

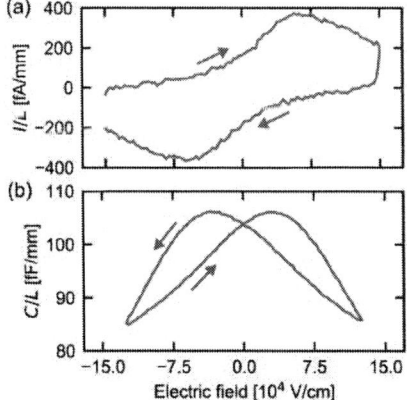

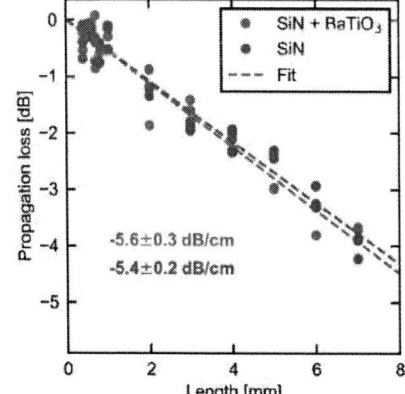

Fig. 6. Change in polarization and resulting electrical current in BTO capacitors. (a) Polarization states in capacitor at different field strengths. The calculated (b) polarization and (c) current show contributions from ferroelectric domain switching, dielectric charging, and leakage.

Fig. 7. Normalized (a) current and (b) capacitance across BTO measured at 4.3 K between two electrodes separated by 2 μm. Ferroelectric domain switching and capacitive charging result in the hysteretic behavior (see Fig. 6). The arrows indicate the sweep direction.

Fig. 8. Propagation losses in the BTO/SiN waveguides, and a reference chip with SiN waveguides fabricated without the bonded BTO layer, both measured at room temperature. The fitted dashed lines correspond to the propagation losses.

978-1-7281-1988-5/18 $31.00 © 2018 IEEE

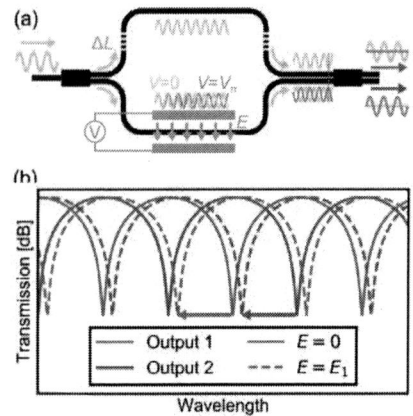

Fig. 9. (a) Concept and operating principle of an asymmetric MZI switch. (b) Schematics of the transmission spectrum through both MZI outputs for two different electric field strengths.

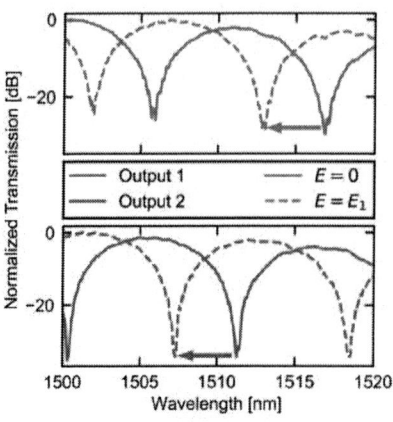

Fig. 10. Transmission spectrum of asymmetric Mach-Zehnder interferometer at 4.3 K showing the power transmitted to the two output ports at different wavelengths.

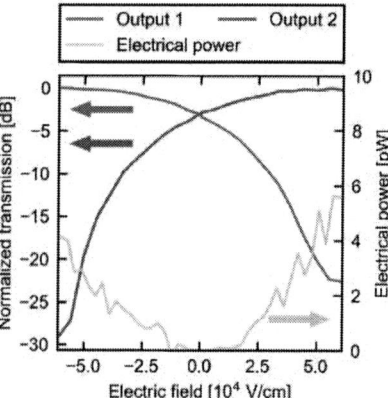

Fig. 11. Electro-optic switching between the two output ports of a Mach-Zehnder interferometer with 500-μm-long phase shifters at 4.3 K. The static power consumption stays below 10 pW.

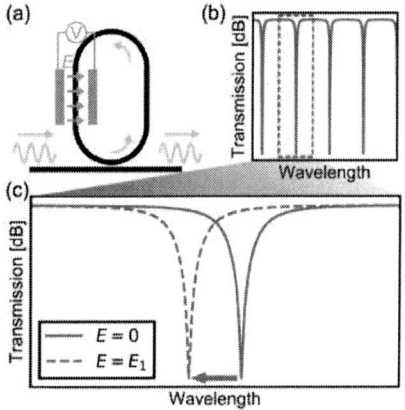

Fig. 12. (a) Schematic of a racetrack resonator with straight phase shifter to induce a resonance shift via applying an electric field (green arrows). (b) Schematics of the transmission spectrum of a resonator and (c) induced resonance shift from applying an electric field in the phase shifter.

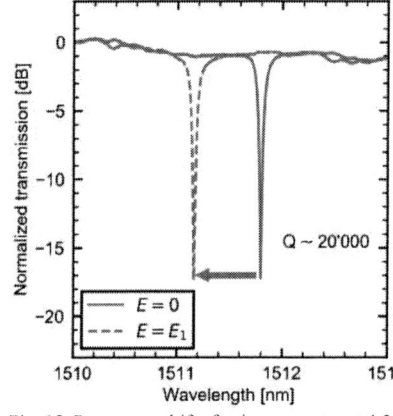

Fig. 13. Resonance shift of a ring resonator at 4.3 K when applying an electric field. This shift allows the device to be used as a wavelength selective filter for various applications, at extremely low power (<5 pW).

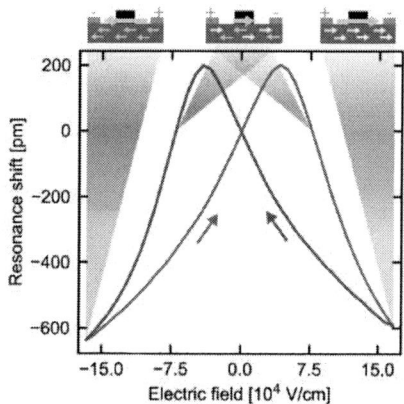

Fig. 14. Hysteretic behavior of the resonance shift in a BTO/SiN device at 4.3 K. The arrows indicate the sweep direction. The hysteresis is due to the ferroelectric domain (yellow arrows) switching in the BTO at different electric fields, as illustrated schematically on top.

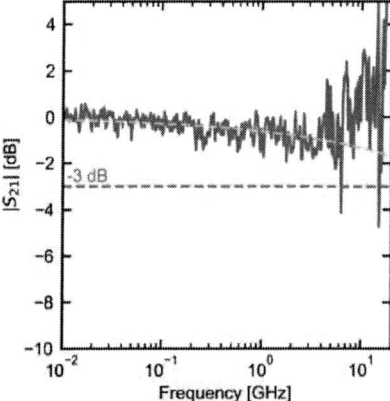

Fig. 15. Transmission spectrum of low Q resonator used for high-speed bandwidth measurements.

Fig. 16. Frequency response of a BTO/SiN ring modulator at 4.3 K showing a bandwidth >18 GHz. The dashed yellow line is provided as a guide to the eye.

	This work	**Gehl et al.** [6]	**Elshaari et al.** [5]
Modulation mechanism	**Pockels effect**	Carrier depletion	Thermo-optic
$P_{\pi, DC}$ [W]	**10^{-11}**	10^{-9}	0.5
4 K EO bandwidth [GHz]	**>18**	4.5	<1
Pure phase modulators	**YES**	NO	YES
Low loss switches	**YES**	NO	YES

Fig. 17. Comparison of results reported in this work with previously reported technologies for cryogenic integrated electro-optic switches.

High Speed ($f_{3\text{-}dB}$ above 10 GHz) Photo Detection at Two-micron-wavelength Realized by GeSn/Ge Multiple-quantum-well Photodiode on a 300 mm Si Substrate

Shengqiang Xu,[1] Wei Wang,[1] Yi-Chiau Huang,[2] Yuan Dong,[1] Saeid Masudy-Panah,[1] Hong Wang,[3]
Xiao Gong,[1,*] and Yee-Chia Yeo.[1,*]

[1]Department of Electrical and Computer Engineering, National University of Singapore, 117576 Singapore.
[2]Applied Materials Inc., Sunnyvale, California 94085, United States.
[3]School of Electrical and Electronic Engineering, Nanyang Technological University, 639798 Singapore.
*Email: elegong@nus.edu.sg and eleyeoyc@nus.edu.sg

ABSTRACT—High speed photo detection at two-micron-wavelength has been achieved with a GeSn/Ge multiple-quantum-well (MQW) photodiode (PD), demonstrating a 3-dB bandwidth ($f_{3\text{-}dB}$) above 10 GHz for the first time. The device layer stack was grown on a standard 300 mm (001) Si substrate using RPCVD, showing potential for large-scale integration. Radio frequency (RF) characterization was performed using 2-μm RF optical measurement setup. To our knowledge, this is also the first PDs on Si with direct RF measurement to quantitatively confirm the high speed functionality at 2 μm.

I. INTRODUCTION

Recent progress in low-loss hollow-core photonic band-gap fibers and thulium doped fiber amplifiers with high gain and low noise makes the two-micron-wavelength range a new and promising spectral window for telecommunications [1,2]. This potential window could enhance the telecommunication capacity to meet the requirement of steeply increasing volume of data transmission. A high speed photo detector is the key component of an optical receiver for telecommunication. Current optical receivers covering this range are mainly based on III-V materials. Integrating photonics components on Si platform could explore the full potential of well-established Si processing technology. Although heterogeneous integration of III-V on Si through wafer-bonding or direct growth has been investigated [3-5], there has been no report on the high speed III-V on Si photo detector operating at 2-μm range. Monolithic solution based on Group-IV materials could offer potential cost advantage and ease of integration with mainstream CMOS technology. Jason Ackert *et al.* contributes, to date, the only reported high speed photo detector at 2-μm range by introducing divacancies into Si through low-dose inert ion implantation [6]. However, avalanche mode operation of this detector suffers from high reverse bias of 15 to 27 V and low response caused by weak interaction between sub-bandgap light and defects requires device length to be several hundreds of micrometers.

In this work, we demonstrate the first high speed GeSn/Ge MQW PD with $f_{3\text{-}dB}$ above 10 GHz at two-micron-wavelength with low operation reverse bias and compact device dimension. The active MQW layer stack was grown on a 300 mm (001)-oriented Si substrate using a high quality 1 μm thick Ge buffer, showing great potential for large-scale integration. Ultra-low leakage of 44 mA/cm² (at a bias of -1 V) and extended photo detection beyond 2 μm were achieved, enhancing the detection capabilities of Ge-based or Si-based photonics. In addition, MQW structure employed in this work holds the potential for light emission and modulation, enabling integrated optical interconnections by GeSn MQW on Si substrate for two-micron-wavelength applications.

II. DEVICE DESIGN AND FABRICATION

Fig. 1 shows the cross-sectional schematic of the GeSn/Ge MQW PD with key highlights. The microscope image and the cross-sectional TEM of the as-grown sample are shown in Fig. 2(a) and (b), respectively. Fig. 2(c) shows the zoomed-in view XTEM in the MQW well region, showing clear interface between GeSn and Ge. AFM scan in Fig. 3 reveals a smooth surface with RMS roughness of 0.636 nm for a scan area of 10 × 10 μm². The HRXRD (004) rocking curve [Fig. 4(a)] of the GeSn/Ge MQW sample shows two main peaks and a series of satellite peaks, corresponding to Si substrate, Ge buffer, and $Ge_{0.92}Sn_{0.08}$/Ge MQW with excellent crystalline quality and periodic uniformity. (115) RSM mapping [Fig. 4(b)] shows that the GeSn/Ge MQW is fully strained to Ge buffer, as revealed by the same in-plane reciprocal lattice vector (Q_x). h_c of $Ge_{1\text{-}x}Sn_x$ on Ge decreases with increase of x. Fig. 5 plots the h_c as a function of x for bulk and GeSn/Ge MQW (25 nm well/35 nm barrier) on Ge/Si virtual substrate, predicted by P-B model [7]. Implementation of a GeSn/Ge MQW structure could increase pseudomorphic GeSn thickness. Thickness of GeSn/Ge MQW in this work is less than predicted h_c for MQW, as highlighted in Fig. 5, ensuring high quality of epitaxial MQW to achieve low leakage. The well thickness and number of QWs are carefully chosen to consider weak quantum confinement effect and carrier trapping effect for long wavelength detection and high speed operation. Fig. 6 shows calculated band diagram for GeSn/Ge QW using model-solid theory [8]. The parameters and equations used in the band structure calculation are summarized in Table 1 and Fig. 7 [9-13]. Type-I band alignment is achieved. Ground state energy is calculated to be 0.0113 and 0.0022 eV for Γ-valley electron and heavy hole as related to band edge of strained GeSn by solving 1D time-independent Schrödinger equation, indicating a cutoff wavelength (λ_c) of 2.05 μm.

Fig. 8 shows key process steps for fabricating the GeSn/Ge MQW PD. 3D schematic of the PD is shown in Fig. 9 with a double-mesa structure. Fig. 10(a) shows the cross-sectional SEM image of the GeSn/Ge MQW after first mesa formation. The second mesa was implemented and formed using F-based RIE to expose the Si substrate. This second mesa etch is critical for realization of high speed PDs as it allows the electrode pads to be formed on the intrinsic Si substrate rather than on the heavily-doped p⁺-Ge to reduce the parasitic capacitance (C_p). Tilted-view SEM of the PD after second mesa formation is

978-1-7281-1988-5/18 $31.00 © 2018 IEEE

shown in Fig. 10(b). SiO_2 was then deposited as passivation and isolation layer by PECVD. After that, the contact was patterned and opened by dry etching followed by wet etching using DHF. Finally, aluminum electrodes were formed by sputtering and Cl-based ICP etch. The electrodes were designed in standard GSG configuration. Fig. 10(c) shows the top-view SEM image of one finished PD with a diameter (D) of 20 μm. Fig. 11(a) shows the tilted-view SEM image of the GeSn/Ge MQW PD with FIB cut line indicated. TEM images of the PD with two mesa sidewalls, zoomed-in view at the MQW sidewall, and interface between GeSn and Ge with clear lattice fringes are shown in Fig. 11(b), (c), and (d), respectively. Chemical composition of the MQW was verified using 1D energy dispersive X-ray (EDX) scan in Fig. 12 with the EDX scan line indicated in Fig. 11(c).

III. RESULTS AND DISCUSSION

A. I_{dark} of the GeSn/Ge MQW PD

Fig. 13(a) shows I_{dark}-V_{bias} of the GeSn/Ge MQW PD with D = 20 μm at various T from 270 to 330 K. Arrhenius plot in Fig. 13(b) shows insight into the leakage mechanism. Linear fitting using weighted least-square method yields a straight line with the gradient corresponding to activation energy (E_a). The extracted E_a is replotted as a function of reverse bias voltage (V_{re}) as shown in Fig. 13(c). E_a decreases with increasing V_{re} as increasing V_{re} enlarges band-bending of intrinsic region which leads to enhanced trap-assisted (TAT) and band-to-band tunneling (BTBT) [14]. Fig. 14 shows dark current density J_{dark}-$1/D$ characteristics with a low bulk current density (J_{bulk}) and surface current density (J_{surf}) of 42 mA/cm^2 and 5.7 μA/cm, respectively. Fig. 15 benchmarks J_{dark} for reported GeSn-on-Si p-i-n PDs at V_{bias} = -1 V [15-30]. A J_{dark} among the lowest and comparable to conventional Ge-on-Si PDs was achieved for a relative high Sn composition of 8%. It should also be noted that higher leakage is usually expected for $Ge_{1-x}Sn_x$ with higher x.

B. DC photo response from 1550 nm to 2000 nm

Steady-state photo response characteristics of the GeSn/Ge MQW PD was measured from 1550 to 2000 nm. Fig. 16(a) presents the dark and illuminated I-V_{bias} characteristics of the fabricated PD with D = 40 μm at λ of 1550 nm. The arrow indicates the direction of increasing incident light power (P_{in}). Photocurrent (I_{ph}) increases linearly with P_{in} and no current saturation is observed. Flat photo response is also achieved even at zero bias, indicating good collection efficiency of photo-generated carriers. Photo response beyond traditional communication bands was also investigated [Fig. 16(b)]. Three DFB laser diodes with emission λ of 1742, 1877, and 2000 nm were used for the optical measurement. P_{in} is fixed at 3.6 dBm. I_{ph} is reduced at longer λ. Fig. 17 shows the optical responsivity spectrum at V_{re} of 1 V from 1550 to 2000 nm. Photo response is observed for all these wavelengths, covering not only conventional telecommunication bands (O to U-band), but also the 2-μm window. R_{op} of 0.214, 0.047, 0.029, and 0.015 A/W were achieved for λ of 1550, 1742, 1877, and 2000 nm, respectively. R_{op} could be further improved by introducing photon-trapping microstructures [31].

C. RF photoresponse at two-micron-wavelength

Fig. 18 illustrates optical setup for 2-μm RF measurement. The light source was a fiber-coupled FP laser diode (Thorlabs FPL2000S) at λ of 2000 nm. A polarization controller was implemented before the connection of 10 GHz LiNbO$_3$ EO MZI modulator (Photline MX2000-LN-10). The modulator was driven by DC and RF signal generated from a DC source meter and a VNA (Agilent N5244A), respectively. The DC signal provided a stable V_{bias} for intensity modulation at a certain operation point. The modulated light was guided to the surface window of the PD through a SM fiber. The coplanar GSG electrical in-/out-put were connected to the second port of the VNA through a coaxial cable. A built-in bias tee was utilized to provide DC supply to PD. The measurements were calibrated to a commercial 2 μm InGaAs detector (ET5000) with a bandwidth larger than 12.5 GHz to de-embed the effect from modulator, cables, and other system components. Fig. 19 shows the small signal frequency response of the GeSn/Ge MQW PD with D = 20 μm. $f_{3\text{-}dB}$ increases with increasing V_{re}, which could be due to the relatively high background doping of the epitaxial GeSn. At V_{re} of 7 V, photo detection beyond 10 GHz is achieved. Because of the frequency limitation of the modulator used, it was not possible to measure the frequency response at higher speeds. It should be noted that this is the first high speed photo detection directly characterized at λ = 2 μm. Future optimization in GeSn material growth to reduce the background doping will further enhance the device performance.

IV. CONCLUSION

For the first time, we demonstrated high speed photo detection beyond conventional telecommunication bands and reaching at two-micron-wavelength realized by GeSn/Ge MQW PD monolithically integrated on Si substrate. Low leakage of 44 mA/cm^2 is obtained at V_{bias} = -1 V, which is among the lowest reported values for GeSn-on-Si PDs and is comparable to that of Ge-on-Si PDs. The PD shows extended sensitivity to λ greater than 2000 nm and we expect significant improvement by introducing photon-trapping microstructures. Beyond 10 GHz functionality is experimentally demonstrated directly at two-micron-wavelength. This work offers a very promising option for high speed photo detection at 2 μm, paving way for telecommunication at this new spectral window.

ACKNOWLEDGEMENT

This work at NUS was supported by MOE (R-263-000-B50-112) and NUS Trailblazer (R-263-000-B43-733). The authors acknowledge Ms. Xin Guo at Silicon Technologies, Centre of Excellence (Si-COE) in NTU for RF measurement.

REFERENCES

[1] H. Zhang et al., ECOC, 2014, p. 5.20. [2] Z. Li et al., OE 21, 9289 (2013). [3] R. Wang et al., OE 24, 8480 (2016). [4] S. Chen et al., Nat. Photonics 10, 307 (2016). [5] Y. Wan et al., Optica 4, 940 (2017). [6] Jason J. Ackert et al., Nat. Photonics 9, 393 (2015). [7] R. People and J. Bean, APL 47, 322 (1985). [8] C. G. Van de Walle, PRB 39, 1871 (1989). [9] S.-W. Chang et al., JQE 43, 249 (2007). [10] Y.-H. Li et al., PRB 73, 245206 (2006). [11] T. Brudevoll et al., PRB 48, 8629 (1993). [12] H. Lin et al., APL 100, 102109 (2012). [13] I. A. Fischer et al., OE 23, 25048 (2015). [14] M. Gonzalez et al., ME 125, 33 (2014). [15] S. Su et al., OE 19, 6400 (2011). [16] M. Oehme et al., APL 101, 141110 (2012). [17] H. Tseng et al., APL 103, 231907 (2013). [18] M. Oehme et al., OE 22, 839 (2014). [19] M. Oehme et al., OL 39, 4711 (2014). [20] Y.-H. Peng et al., APL 105, 231109 (2014). [21] Y. Dong et al., OE 23, 18611 (2015). [22] T. Pham et al., OE 24, 4519 (2016). [23] Y.-H. Huang et al., OL 42, 1652 (2017). [24] M. Morea et al., APL 110, 091109 (2017). [25] Y. Dong et al., OE 25, 15818 (2017). [26] J. Mathews et al., APL 95, 133506 (2009). [27] J. Mathews et al., ECS Transactions 33, 765 (2010). [28] J. Werner et al., APL 98, 061108 (2011). [29] W. Wang et al., JAP 119, 155704 (2016). [30] H. Cong et al., PJ 8, 1 (2016). [31] Y. Gao et al., Nat. Photonics 11, 301 (2017).

❑ Highest f_{3-dB} directly measured at $\lambda = 2$ μm for next generation telecommunication

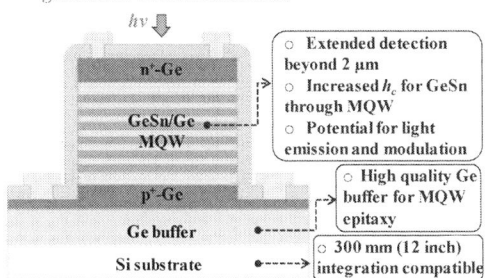

❑ **Extended detection beyond 2 μm**
❑ **Increased** h_c **for GeSn through MQW**
❑ **Potential for light emission and modulation**

❑ **High quality Ge buffer for MQW epitaxy**

❑ **300 mm (12 inch) integration compatible**

Fig. 1. Cross-sectional schematic and key highlights of the high speed GeSn/Ge MQW PD. The MQW epitaxy was realized on 300 mm Si substrate. High f_{3-dB} of beyond 10 GHz was achieved at 2 μm, directly using a 2-μm RF optical system.

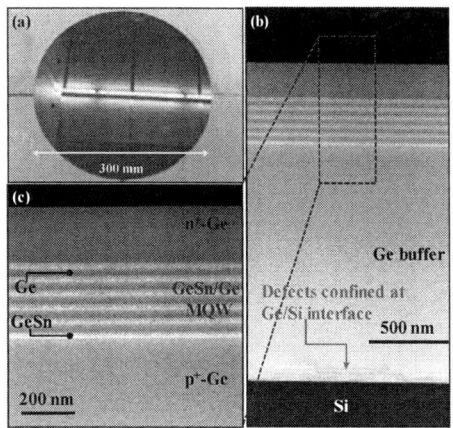

Fig. 2. (a) Microscope image of the 300 mm as-grown MQW on Ge/Si virtual substrate. (b) XTEM depicts high quality of the epitaxial layer stack. (c) Zoom-in view of the XTEM clearly shows the QWs with sharp interface between GeSn and Ge. We have confirmed good uniformity achieved across the 300 mm wafer, showing great potential for large scale integration.

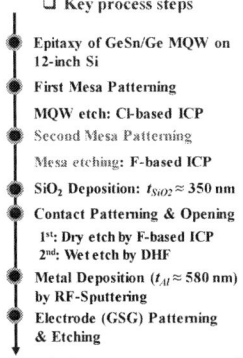

Fig. 3. AFM of sample surface after epitaxy of GeSn/Ge MQW. RMS roughness is 0.636 nm with a scan area of 10×10 μm².

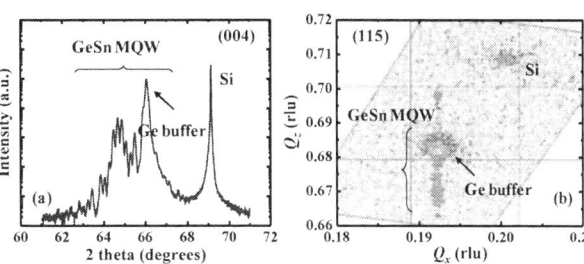

Fig. 4. (a) HRXRD (004) rocking curve of the GeSn/Ge MQW sample. (b) The (115) reciprocal space map (RSM) of the GeSn/Ge MQW sample. The GeSn quantum wells are fully strained to the Ge buffer.

Table 1. Parameters of Ge and α-Sn for electronic band structure calculation.

	Ge	α-Sn
a_{bulk} (Å)	5.6573[a]	6.4892[a]
$E_{v,ave}$ (eV)	0[a]	0.69[a]
Δ_0 (eV)	0.29[a]	0.8[a]
$E_{g,\Gamma}$ (eV)	0.7985[a]	-0.413[a]
$E_{g,L}$ (eV)	0.664[a]	0.092[a]
C_{11} (GPa)	128.53[a]	69[a]
C_{12} (GPa)	48.26[a]	29.3[a]
a_v (eV)	1.24[a]	-4.7[a]
b_v (eV)	-2.9[a]	-2.7[a]
a_c^Γ (eV)	-8.24[a]	-6[b]
a_c^L (eV)	-1.54[a]	-2.14[c]

[a][9], [b][10], [c][11].

$$\varepsilon_\parallel = \frac{a - a_{bulk}}{a_{bulk}} \quad (1) \qquad \varepsilon_\perp = -2\frac{C_{12}}{C_{11}}\varepsilon_\parallel \quad (2)$$

$$\Delta E_{HH} = a_v(2\varepsilon_\parallel + \varepsilon_\perp) + b_v(\varepsilon_\parallel - \varepsilon_\perp) \quad (3)$$

$$\Delta E_{LH} = a_v(2\varepsilon_\parallel + \varepsilon_\perp) - \frac{1}{2}b_v(\varepsilon_\parallel - \varepsilon_\perp) - \frac{1}{2}\Delta_0$$
$$+ \frac{1}{2}\sqrt{\Delta_0^2 - 2\Delta_0 b_v(\varepsilon_\parallel - \varepsilon_\perp) + 9b_v^2(\varepsilon_\parallel - \varepsilon_\perp)^2} \quad (4)$$

$$\Delta E_c^\eta = a_c^\eta(2\varepsilon_\parallel + \varepsilon_\perp) \quad (5)$$

Fig. 7. Equations used in electronic band structure calculation. $\varepsilon_\parallel$: in-plane strain; $\varepsilon_\perp$: normal strain; Δ_0: spin-orbit split-off energy; a_v and a_c: hydrostatic deformation potential for valence and conduction band; b_v: shear deformation potential. η refers to Γ or L conduction valley. Linear interpolation is taken for all parameters except bandgap with bowing parameters of 2.42 [12] and 0.99 [13] eV for direct and indirect bandgap.

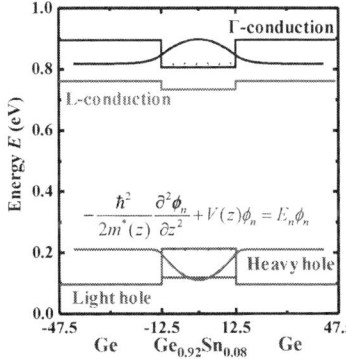

Fig. 6. Calculated band diagram of the GeSn QW. Ground state energy was solved for Γ-valley electron and heavy hole.

Fig. 5. Theoretical critical thickness for bulk GeSn and GeSn/Ge MQW grown on Ge by P-B model.

❑ **Key process steps**

● Epitaxy of GeSn/Ge MQW on 12-inch Si
● **First Mesa Patterning**
  MQW etch: Cl-based ICP
● **Second Mesa Patterning**
  Mesa etching: F-based ICP
● SiO₂ Deposition: $t_{SiO2} \approx 350$ nm
● **Contact Patterning & Opening**
  1ˢᵗ: Dry etch by F-based ICP
  2ⁿᵈ: Wet etch by DHF
● **Metal Deposition** ($t_{Al} \approx 580$ nm) by RF-Sputtering
● Electrode (GSG) Patterning & Etching

Fig. 8. Key process steps for fabricating the GeSn/Ge MQW high speed PD. Double-mesa structure was implemented.

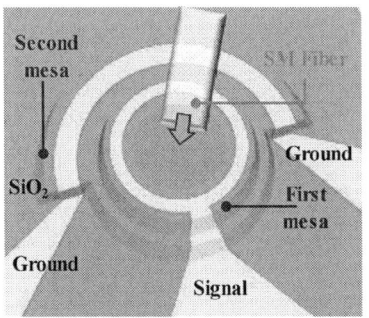

Fig. 9. Three-dimensional (3D) schematic of the GeSn/Ge MQW high speed PD. PECVD SiO₂ is deposited as the passivation and isolation layer for GSG electrodes. Double mesa structure is used for reducing parasitic capacitance (C_p).

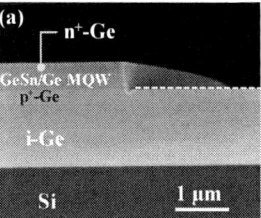
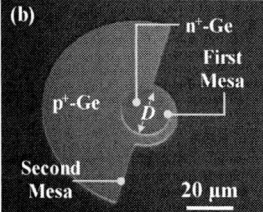
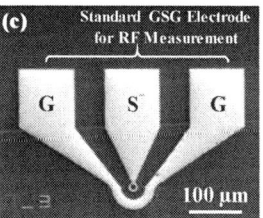

Fig. 10. (a) Cross-sectional SEM image of the GeSn/Ge MQW after the first mesa etch. (b) Tilted-view SEM image of the GeSn/Ge MQW photodiode after double-mesa formation. (c) Top-view of one completed photodiode with standard GSG electrode.

978-1-7281-1988-5/18 $31.00 © 2018 IEEE   546

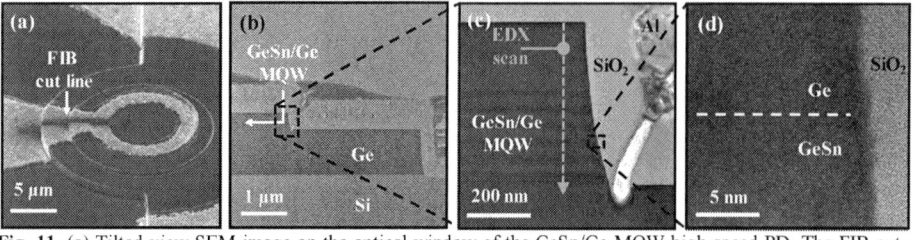
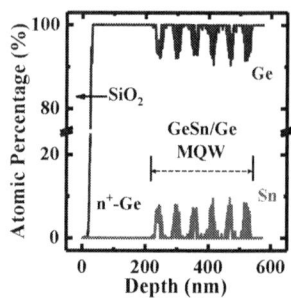

Fig. 11. (a) Tilted-view SEM image on the optical window of the GeSn/Ge MQW high speed PD. The FIB cut line across two sidewalls is indicated. (b) XTEM image of the PD showing two mesa sidewalls. (c) The zoom-in view of the XTEM image clearly shows GeSn quantum wells and smooth etching sidewall. EDX scan was performed from the SiO$_2$ to the MQW, as indicated in orange line. (d) High resolution TEM (HRTEM) image shows the interface between GeSn and Ge at the sidewall. Clear lattice fringes can be observed.

Fig. 12. 1D EDX scan of the GeSn/Ge MQW high speed PD at the MQW region.

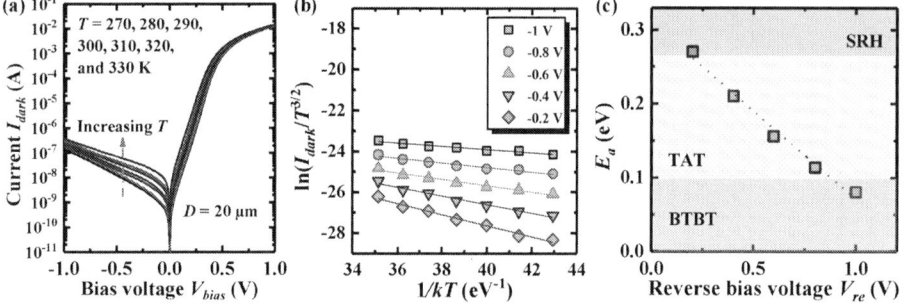
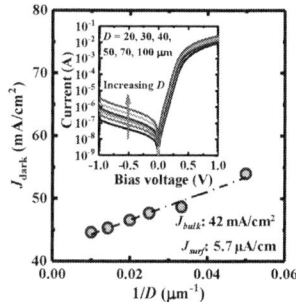

Fig. 13. (a) Temperature-dependent dark I-V characteristics of the GeSn/Ge MQW high speed PD. T ranges from 270 to 330 K with an increasing step of 10 K. (b) Plot of $\ln(I_{dark}/T^{3/2})$ as a function of $1/kT$ for the photodiode at various reverse bias voltages V_{re}. (c) Extracted activation energy from the linear fitting in (b) $vs.\,V_{re}$.

Fig. 14. J_{dark}-$1/D$ characteristics of the PDs at V_{bias} = -1 V. Inset shows the I-V curves of PDs with various D.

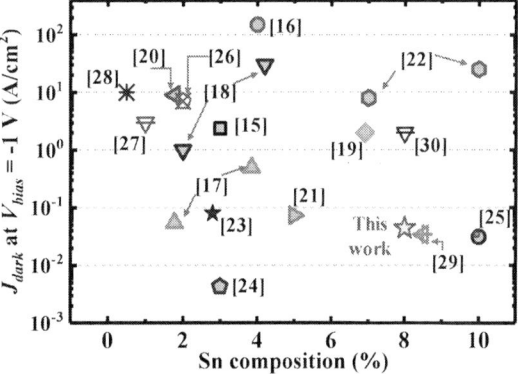
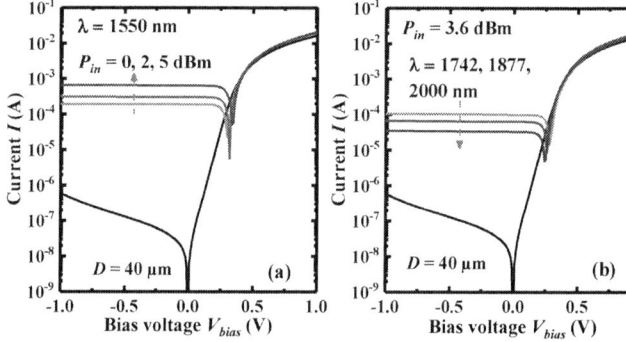

Fig. 15. Benchmarking of dark current density J_{dark} of GeSn-on-Si p-i-n photodiodes at V_{bias} = -1 V. A J_{dark} among the lowest reported values was achieved for a relative high Sn concentration of 8%.

Fig. 16. (a) I-V_{bias} characteristics of the GeSn/Ge MQW photodiode at illumination wavelength of 1550 nm with incident power P_{in} ranging from 0 to 5 dBm or decibel-milliwatts. (b) I-V_{bias} characteristics of the photodiode at illumination wavelength of 1742, 1877, and 2000 nm with a fixed P_{in} of 3.6 dBm.

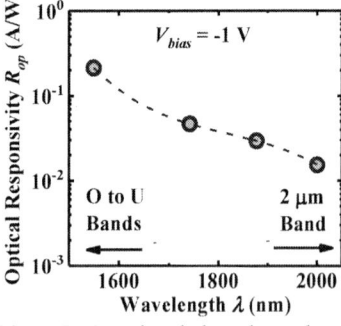
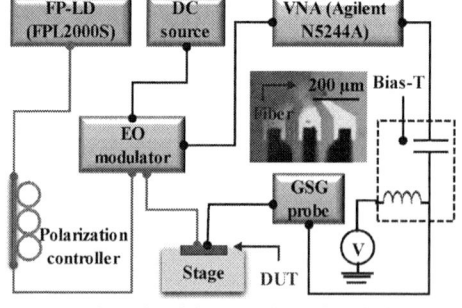
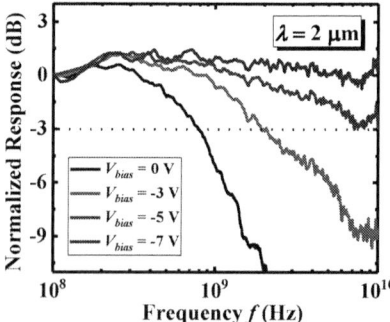

Fig. 17. Wavelength-dependent photo responsivity of the GeSn MQW photodiode ranging from 1550 to 2000 nm at V_{re} of 1 V. Obvious cut-off detection beyond 2 μm is observed, consistent with the theoretical calculation.

Fig. 18. Schematic of the experimental optical setup for RF measurement of the GeSn/Ge MQW PD at two-micron-wavelength. The red and black lines indicate the optical and electrical connections, respectively. Inset shows top-view microscope image of the PD device under testing (DUT).

Fig. 19. Normalized optical response versus frequency of the 20-μm-diameter GeSn/Ge MQW photodiode at two-micron-wavelength. The bias voltage (V_{bias}) ranges from 0 to -7 V. At reverse bias larger than 5 V, the photodiode shows a 3-dB bandwidth beyond 10 GHz.

978-1-7281-1988-5/18 $31.00 © 2018 IEEE

Quadratic electro-optical silicon-organic hybrid RF modulator in a photonic integrated circuit technology

P. Steglich[1,2], C. Mai[1], A. Peczek[1], F. Korndörfer[1], C. Villringer[2], B. Dietzel[2], and A. Mai[1,2]

[1]IHP, Frankfurt(Oder), Germany, email: steglich@ihp-microelectronics.com, mai@ihp-microelectronics.com
[2]Technical University of Applied Sciences Wildau, Wildau, Germany

Abstract— For the first time, an integrated electro-optical RF modulator based on the quadratic electro-optical effect with CMOS compatible sub-volt driver voltages is presented. As unique feature, this modulator provides an amplitude tuning of the modulated carrier wave. The silicon-based modulator was fabricated using process steps of an established photonic integrated circuit technology and covered by a nonlinear optical polymer in a post-process. We demonstrate a device tunability of up to 350 pm/V, surpassing state-of-the-art silicon modulators with an order of magnitude. Moreover, the ring resonator is designed to have an ultra-low per-bit energy consumption of 87 aJ/bit demonstrating the potential for high-performance photonic devices with low energy consumption.

I. INTRODUCTION

A major issue for the implementation of silicon photonic modulators is that silicon as a material lacks efficient electro-optical (EO) effects [1]. Various new materials have been explored to overcome this bottleneck. A highly successful approach has been to integrate organic materials, which exhibit off-resonant EO properties and allow the manipulation of amplitude and phase independently. The linear electro-optical effect (LEOE) in polymers provides large bandwidth, high-speed operation and low energy consumption, which has been exploited in silicon-organic hybrid (SOH) modulators [2]. However, there has been concern about the practical use of the LEOE. The LEOE requires high-voltage poling of the polymer due to the necessary non-centrosymmetric molecular orientation. In turn, this leads to thermal and long-term stability issues due to orientational relaxation dynamics [3]. In contrast, the quadratic electro-optical effect (QEOE) in polymers is present in any molecular orientation avoiding poling procedures and providing long-term stability. On the other side the challenge for the use of QEOE is the need of high electric field strengths and therefore this approach was long time considered to be impracticable for low voltage applications. However, slot waveguides have been demonstrated to enhance electric field strengths at moderate voltages (**Fig. 1**) making the QEOE feasible for on-chip operation [4]. In this work, we present for the first time RF modulation at CMOS compatible driver voltages for a QEOE based modulator. We also demonstrate linear amplitude tuning of the modulated carrier wave by adding an offset voltage. The new modulator therefore does not require poling and is intrinsically thermally stable. Further, we demonstrate the integration of polymers with QEOE using a PIC-compatible process flow.

II. DESIGN AND FABRICATION

A partially slotted ring resonator (PSRR) [4-7] design is employed in this work. A schematic of the PSRR is depicted in **Fig. 2a**. The PSRR is connected through tungsten vias to metal electrodes, as shown in **Fig. 2b**. A slot waveguide is used to exploit the QEOE in an EO polymer because it provides a large overlap between the optical (**Fig. 2c**) and the electrical field (**Fig. 2d**). However, a strip waveguide is used at the bending part of the ring to reduce optical losses. Here, we use a strip waveguide with a waveguide width of $w = 500\ nm$ and a slot waveguide with a slot width of $s = 150\ nm$ and a length of $L_{slot} = 12\ \mu m$. The chip-layout with the complete ring resonator and grating coupler is shown in **Fig. 3**.

The device is fabricated in a PIC technology as a modular part of a complex EPIC-SiGe-BiCMOS platform [8]. The process flow was interrupted for simplicity after structuring the first metal layer. To access the silicon slot waveguide the SiO_2 cladding was removed locally above the PSRR using a wet etch process with resist mask (**Fig. 4**). Scanning-electron microscopic images of the fabricated device are shown in **Fig. 5**. The etched cavity is highlighted in blue in **Fig. 5a** and a magnification is shown in **Fig. 5b**. **Fig. 5c** shows a further magnification of the strip-to-slot mode-converter and **Fig. 5d** a cross-sectional view of the slot waveguide. SEM images of the released strip-to-slot mode-converter is shown in **Fig. 6**. After chip separation and a cleaning procedure with acetone and 2-propanol, the EO polymer is directly spun onto each chip at 80 rps. We employed a side-chain system from Sigma Aldrich (570427 Aldrich) whereby the azo benzene dye Disperse Red 1 (DR1) is grafted at a 25% mass concentration as side-chain to a poly(methyl methacrylate) (PMMA) main-chain polymer backbone. The chemical structure is shown in **Fig. 7**. The polymer was solved in 1.1.2.2-tetrachloethane and filtered using a PTFE-membrane filter. To remove the solvent after deposition we have dried the samples at 80°C in a vacuum oven.

III. ELECTRO-OPTICAL PERFORMANCE AND DISCUSSION

The optical transmission of the ring resonator is measured in a direct-detection fiber-grating-coupler set-up. The observed transmission spectrum of the ring resonator is shown in **Fig. 8**. Applying a DC voltage to the ground-signal-ground (GSG) electrodes leads to a red shift of the resonance peaks due to the QEOE (**Fig. 9**). The extinction ratio varies less than 0.6% and the full width at half maximum less than 2.2% indicating that only the phase is altered due to the off-resonant QEOE making it suitable for higher modulation schemes. This is a clear advantage compared to the plasma dispersion effect (PDE) in silicon, where both the phase and amplitude are influenced by carrier injection.

The observed resonance wavelength shifts as function of the applied voltage are plotted in **Fig. 10**. This graph clearly reflects the quadratic nature of the DC Kerr effect and, hence, figure of merits (FOM) like the voltage-length product or device tunability are not directly applicable since they presume a linear response. Therefore, we deduce such FOMs from a certain operation point and within a small voltage range in order to ensure an approximately linear response. We deduced a voltage-length product of 1 Vcm and a device tunability of 350 pm/V in the voltage range between 4.2 V and 4.8 V. This large device tunability allows the use of a bias voltage to fine-tune the operation point, which avoids power consuming heaters. A further advantageous aspect of the PSRR is the ultra-low capacitance C_{slot}, which drastically decreases the per-bit energy consumption ($W_{bit} = C_{slot}U_{pp}^2/4$). We have estimated an extremely low per-bit energy consumption of nominally 87 aJ/bit assuming a peak-to-peak voltage of $U_{pp} = 1\,V$. This value is about 25 times smaller compared with that of state-of-the-art silicon modulators based on the PDE [9].

After modulating the optical carrier signal with an electrical input, an optical output signal is generated and delivered to a photodiode that converts the modulated optical signal to an electric output signal ($U_{out} = U_{opt}\sin(\omega t)$). We can distinguish between two cases: the electrical input signal i) has no offset voltage ($U_{in} = U_m\sin(\omega t)$) and ii) has an offset voltage ($U_{in} = U_{DC} + U_m\sin(\omega t)$), as illustrated in **Fig. 11**. The latter case allows the development of an adjustable amplitude modulator without the need of additional photonic devices representing a unique feature of the QEOE. Consequently, the output amplitude U_{opt} can be tuned by changing either the modulation amplitude U_m or the offset voltage U_{DC} of the input signal. **Fig. 12** shows exemplary a modulated signal at different offset voltages.

For the first time, we present a small signal frequency experiment. Measured EO scattering parameter S$_{21}$ as a function of the frequency are plotted in **Fig. 13**. From this figure we inferred a 3dB-bandwidth of 1.34 GHz. This demonstrates the feasibility of the QEOE for RF photonics at low voltages. The RF performance is currently limited by parasitics coming from etch induced damage of the thin silicon striploads inducing a high sheet resistance as well as from impedance mismatch between the off-chip driving system and the device. Thus, an optimization of the local etching procedure and an optimized electrode layout may improve the 3dB-bandwidth. The device performance can be further improved by using cladding materials with stronger QEOE [10]. Beside the small signal frequency experiments, high-speed data transmission of pseudo random bit signals (PRBS) are currently performed and will be presented at the conference. **Table 1** gives a summary of relevant device metrics. The reported device performance demonstrates, for the first time, the capability of RF signal processing using a QEOE SOH modulator.

IV. CONCLUSION

We have presented an efficient, integrated EO modulator for RF photonics fabricated with a PIC-technology. An ultra-low per-bit energy consumption of 87 aJ/bit is estimated and a voltage-length product of 1 Vcm as well as a device tunability of 350 pm/V are deduced from DC experiments. We demonstrated, for the first time, the capability of RF modulation with a 3dB-bandwidth of 1.34 GHz and an amplitude tuning of the carrier wave. Since no poling procedure is required and due to its inherently long-term stability, this approach eases the integration of organic materials with strong EO effect in an EPIC-technology. The presented integration scheme as well as the PSRR concept give rise to ultra-low power photonic building blocks; e.g. tunable filter, switches and RF modulators.

ACKNOWLEDGMENT

This project has received funding from the European Regional Development Fund under grant agreement no. 10.13039/501100008530. The authors gratefully acknowledge the support of IHP cleanroom stuff.

REFERENCES

[1] J. Leuthold et al., "Silicon-organic hybrid electro-optical devices." IEEE Journal of Selected Topics in Quantum Electronics, vol. 19, no. 6, pp. 114-126, Nov. 2013.

[2] W. Heni et al., "Silicon–Organic and Plasmonic–Organic Hybrid Photonics", ACS Photonics, vol. 4, no. 7, pp. 1576-1590, June 2017.

[3] F. Michelotti et al., "Study of the orientational relaxation dynamics in a nonlinear optical copolymer by means of a pole and probe technique." Journal of applied physics, vol. 80, no. 3, pp. 1773-1778, Mar. 1996.

[4] P. Steglich et al., "Quadratic electro-optic effect in silicon-organic hybrid slot-waveguides," Optics Letters, vol. 43, no. 15, pp. 3598-3601, Aug. 2018.

[5] P. Steglich et al., "Novel Ring Resonator Combining Strong Field Confinement With High Optical Quality Factor," in IEEE Photonics Technology Letters, vol. 27, no. 20, pp. 2197-2200, Oct. 2015.

[6] P. Steglich et al., "Partially slotted silicon ring resonator covered with electro-optical polymer," Proc. SPIE 9891, Silicon Photonics and Photonic Integrated Circuits V, 98910R (13 May 2016)

[7] P. Steglich *et al.*, "Hybrid-Waveguide Ring Resonator for Biochemical Sensing," in *IEEE Sensors Journal*, vol. 17, no. 15, pp. 4781-4790, Aug. 2017.

[8] D. Knoll et al., "High-performance photonic BiCMOS process for the fabrication of high-bandwidth electronic-photonic integrated circuits," 2015 IEEE International Electron Devices Meeting (IEDM), Washington, DC, 2015, pp. 15.6.1-15.6.4.

[9] Raphaël Dubé-Demers et al., "Ultrafast pulse-amplitude modulation with a femtojoule silicon photonic modulator," Optica, vol. 3, no. 6, pp. 622-627, 2016.

[10] A. Narayanan and M. Thakur, "Quadratic electro-optic effect in the nonconjugated conductive polymer iodine-doped poly(β-pinene) measured at longer wavelengths including 1.55µm", Solid State Communications, vol. 150, no. 7, pp. 375-378, 2010.

[11] D. Patel et al., "Design, analysis, and transmission system performance of a 41 GHz silicon photonic modulator," Opt. Express, vol. 24, pp. 14263-14287, 2015.

[12] S. Koeber et al., "Femtojoule electro-optic modulation using a silicon–organic hybrid device." Light-Sci. Appl., vol. 4, p. e255, 2015.

[13] F. Eltes et al., "A novel 25 Gbps electro-optic Pockels modulator integrated on an advanced Si photonic platform," 2017 IEEE International Electron Devices Meeting (IEDM), San Francisco, CA, 2017, pp. 24.5.1-24.5.4.

[14] K. Alexander et al., "Broadband electro-optic modulation using low-loss PZT-on-silicon nitride integrated waveguides," 2017 Conference on Lasers and Electro-Optics (CLEO), San Jose, CA, 2017, pp. 1-2.

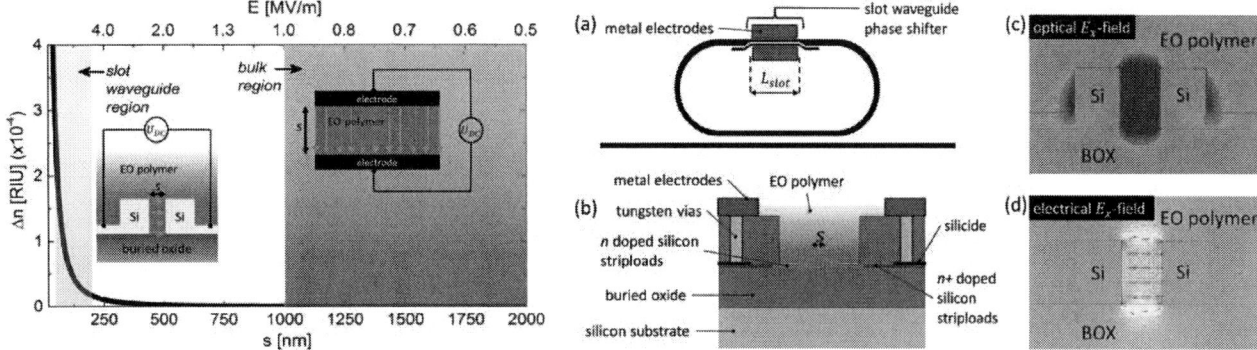

Fig. 1. Refractive index change induced by the QEOE as function of the electrode distance s and the electric field E [4].

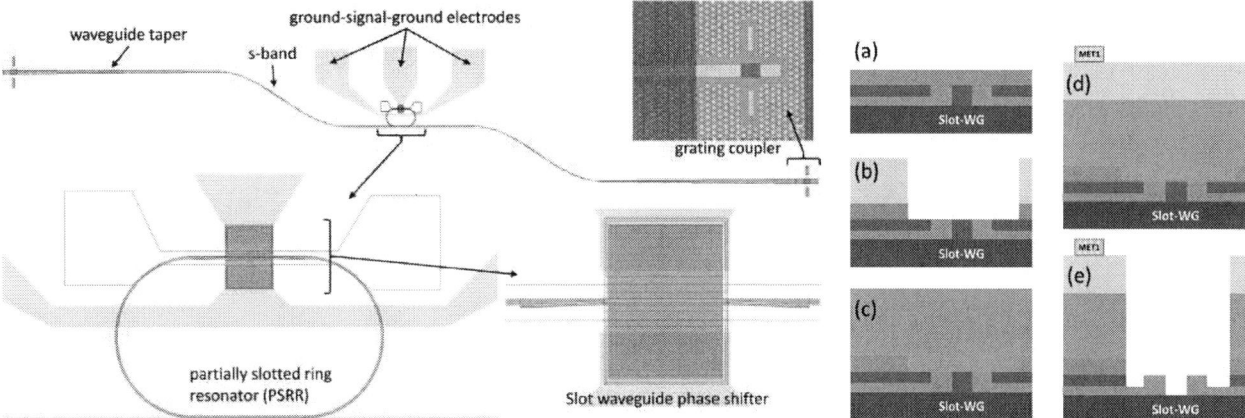

Fig. 2. (a) Schematic of the PSRR. (b) Cross section of the slot waveguide. (c) Optical E_x-field of the quasi TE-mode. (d) Electrical E_x-field (red arrows) and electric potential (false colors) inside the slot waveguide.

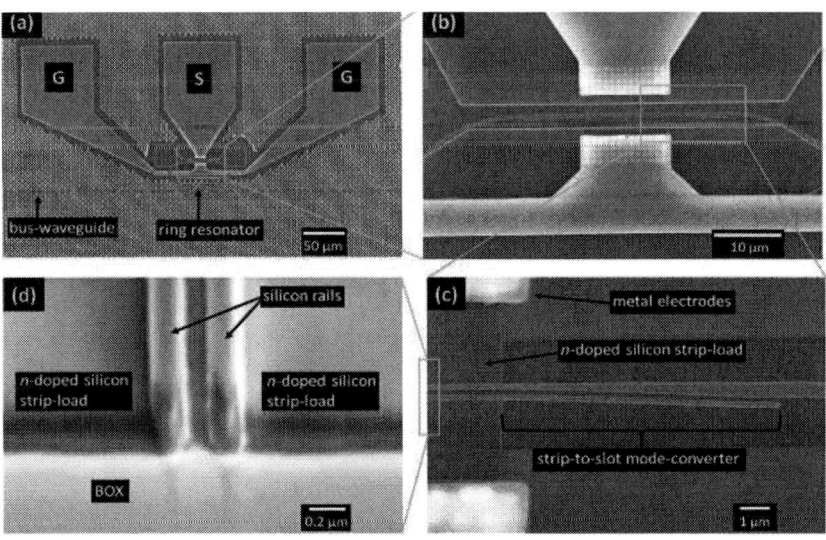

Fig. 3. Chip-layout: Grating coupler are designed to couple the light from a vertical fiber or from a beam of light to the silicon waveguides in the plane of the chip. A waveguide taper confines the light in a strip-waveguide and a s-band is separating the input grating coupler and the output grating coupler by a distance of 250 μm.

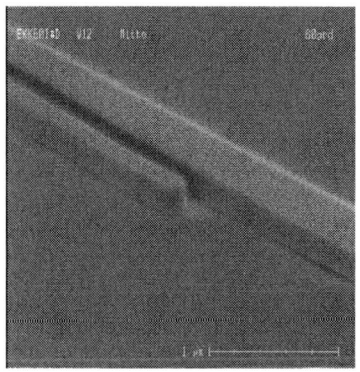

Fig. 4. a) Structured slot waveguide. b) Removing nitride above slot waveguide by wet etching with oxide hard mask. c) Deposition of oxide and chemical mechanical planarization (CMP) process. d) Deposition of inter-layer dielectric oxide and realization of first metal layer. e) Removing oxide cladding above slot waveguide over resist mask.

Fig. 5. (a) SEM image of the GSG aluminium electrode from the top view with a trench (highlighted in blue) to functionalize the slot waveguide with an EO polymer. The ring resonator and bus waveguide are illustrated as dashed red line. (b) Part of the ring resonator with strip-to-slot mode-converter and strip-loaded slot waveguide. (c) Magnification of the strip-to-slot mode-converter. (d) Strip-loaded silicon slot waveguide [6].

Fig. 6. SEM image of the fabricated strip-to-slot mode-converter for 150 nm wide slot.

978-1-7281-1988-5/18 $31.00 © 2018 IEEE   550

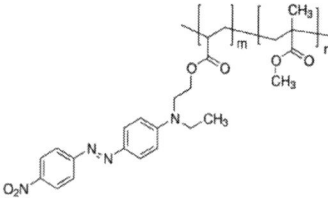

Fig. 7. Chemical structure of the employed EO polymer Poly[(methyl methacrylate)-co-(Disperse Red 1 methacrylate)] from Sigma Aldrich.

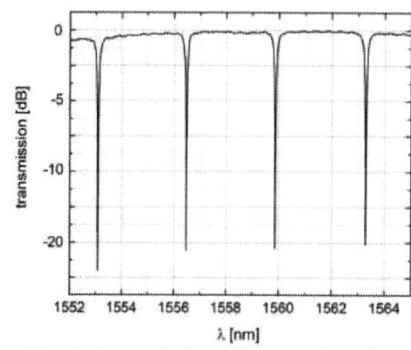

Fig. 8. Transmission spectrum of the ring resonator. The free spectral range is about 3.4 nm and the full width at half maximum is 150 pm.

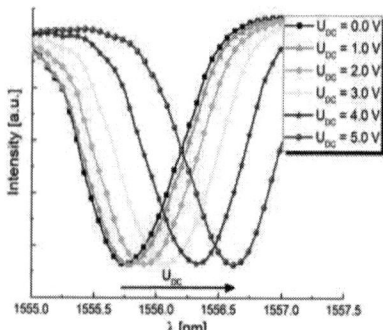

Fig. 9. Transmission spectra of the ring resonator at different applied voltages.

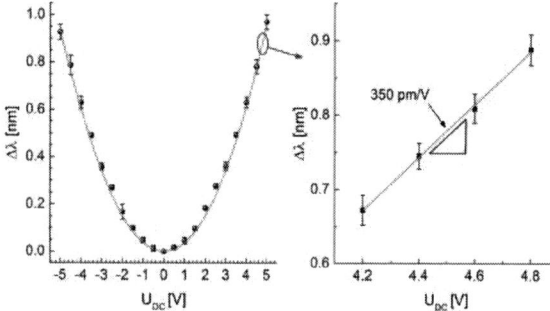

Fig. 10. Resonance wavelength shift of the hybrid-waveguide ring resonator as a function of the applied DC voltage (left) as demonstrated in Ref. [4] and a small voltage range to demonstrate an approximately linear resonance shift (right).

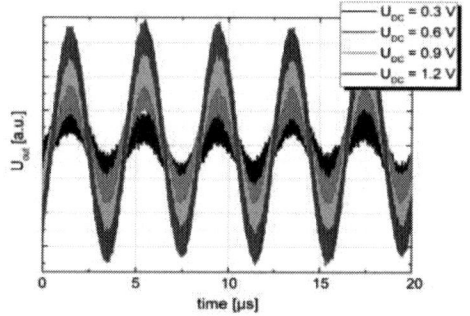

Fig. 12. Modulated carrier signal at different DC offset voltages. The amplitude increases with larger DC voltages, while the modulation amplitude of the electrical input signal is fixed to 50 mV.

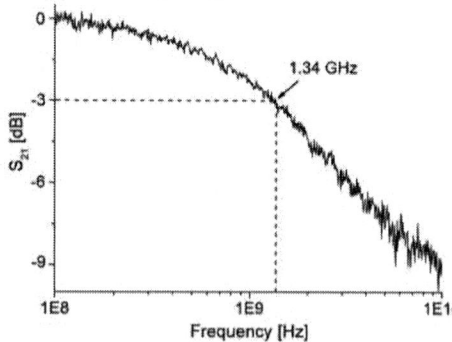

Fig. 13. Small signal frequency response of the ring resonator at 1556.4 nm with an offset voltage of 5 V and 3 dBm modulation amplitude.

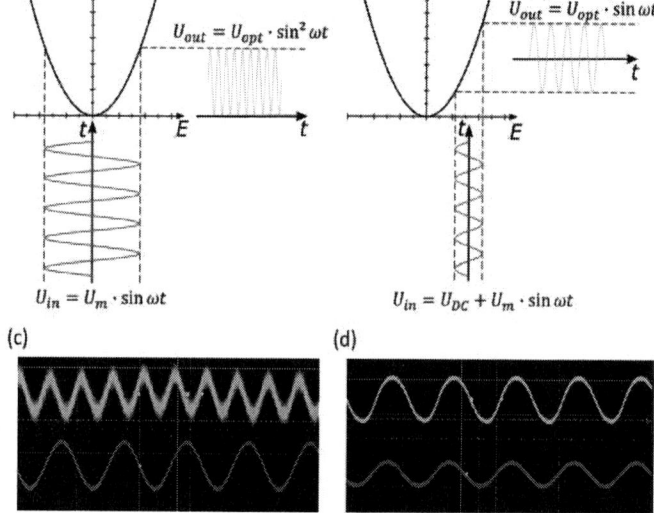

Fig. 11. Modulation schemes: (a) Without offset voltage (b) Applying additionally a DC voltage (offset voltage), leads to a linearization. (c) and (d) show the observed oscilloscope traces without and with offset voltage, respectively.

	This Work	**[11]**	**[12]**	**[13]**	**[14]**
	SOH (QEOE)	Doped Si (PDE)	SOH (LEOE)	BTO (LEOE)	PZT (LEOE)
Poling voltage	0	0	40	2	40
Device tunability [pm/V]	350	20	-	-	17
Long-term stability [h]	Very high	Very high	low	Very high	-
Thermal stability	high	Very high	low	Very high	high
Voltage-length product [Vcm]	1	2.8	0.1	0.3	1

QEOE = quadratic electro-optic effect LEOE = linear electro-optic effect PDE = plasma dispersion effect
SOH = silicon-organic hybrid BTO = barium titanate PZT = lead zirconate titanate

Table 1. Comparison of relevant figure of merit for state-of-the-art integrated modulators taken from Refs. [11-14].

Silicon Photonics: a Scaling Technology for Communications and Interconnects

P. Dong[1], K. W. Kim[1], A. Melikyan[1], and Y. Baeyens[2]
[1]Nokia Bell Labs, Holmdel, NJ, USA, email: po.dong@nokia-bell-labs.com
[2]Nokia Bell Labs, Murray Hill, NJ, USA

Abstract—Silicon photonics exploits CMOS foundry processes to fabricate passive and active photonic circuits on silicon substrates. This technology offers superior scalability in terms of integration level and energy efficiency, two key metrics to obtain sustainable capacity growths in future telecom, datacom, and chip-scale interconnects. We illustrate the advantages of compactness and low power consumption by describing two novel silicon photonic based devices, namely, microring resonators and directly reflectivity modulated lasers.

I. INTRODUCTION

We are living in the era of information. Various network traffic growth studies reveal compound annual growth rates (CAGRs) between 25% and 80% in many different regions and applications and for at least the past two decades. These growth rates were supported largely by the continued exponential complexity increases in information technologies for the generation and processing of data, e.g., data centers, supercomputers, microprocessors, core routers, and memory/storage devices. Thanks to Moore's Law, these silicon based technologies offer performance or capacity growth with CAGRs in the range between ~40% and ~90%. However, the primary technology to interconnect these systems, optical fiber communications, consistently showed a CAGR of only 20% over the past three decades. These scaling disparities, as observed in [1], are mainly attributed to the lack of a Moore's scaling Law in analog opto-electronic device technology.

Very recently, silicon photonics has emerged as a disruptive opto-electronic device technology with a scaling capability on integration and energy efficiency. Built on silicon-on-insulator (SOI) substrates, the device cross section commonly used in silicon photonics foundries is presented in Fig. 1. Apart from the lasers and optical amplifiers, monolithic CMOS processes allow passive optical components such as splitters and couplers, high-speed modulators based on silicon *pn* junctions, and high-speed photo detectors (PDs) based on epitaxial grown germanium films. Heterogeneous integration of III-V light sources is feasible as well. In the next sections, we argue that this technology provides a superior scaling capability for future communication applications.

II. SCALABILITY OF PHOTONICS INTEGRATION

The capability of integrating many optical elements on a single millimeter-sized or centimeter-sized chip provides the required I/O bandwidth densities to the processors and switch ASICs. Conventional optical devices are bulky, based on different materials and processes, and therefore hard to integrate. The large index difference of silicon waveguides permits a tight bending radius less than 5 μm, resulting in extremely compact optical components and hence flexible circuit design, which are crucial for very large-scale photonic integration. Wafer-scale CMOS fabrication further enables silicon photonics to be a viable platform for very large-scale photonic integrated circuits (PICs) with high yield and low cost, by exploiting existing and mature fabrication facilities and processes. Furthermore, monolithic integration of silicon PICs with CMOS drivers leads to more complex optical functions being practical, while at the same time lowering power consumption and packaging cost.

The exponential growth of the complexity of silicon photonic chips over the past 15 years is highlighted in Fig. 2, depicting the published number of components monolithically integrated on a single chip. Despite the recent emergence of this technology [2-7], photonic circuits monolithically integrating over 100 different elements have already been demonstrated. Two of our early demonstrations are presented in Fig. 3 and Fig. 4, which show the chip photographs for optical coherent transmitters and receivers enabling dual-polarization optical coherent communications [4, 5]. Further exponential data capacity growth will require strong parallelism in optical circuits, either by using many wavelength channels or by exploiting spatial diversity in multi-core fibers [1]. The high yield allowed by CMOS processes will be crucial in meeting this demand by providing highly complex photonic circuits with thousands of parallel channels.

III. SCALABILITY OF ENERGY EFFICIENCY

Future exascale computing systems exploiting strong parallelism face a critical challenge to move the data among numerous computer resources. The role of the interconnect is central to chip and system performance. Figure 5 illustrates the growing energy costs of data communication with transmission distance. The data for this plot was extracted in [8]. For conventional electronic interconnect technology, the energy per bit for off-chip communication exceeds 1 pJ and becomes 100 pJ to communicate between systems. This high power consumption is the main constraint on the ultimate performance of future computer systems. Revolutionary

978-1-7281-1988-5/18 $31.00 © 2018 IEEE

technologies are required and optical interconnect is the most promising. Optical interconnects promise scalable capacity, transparency, low latency, and low energy per bit. Because of thermal reasons, energy efficiency is the most critical parameter to implement optical technology into chip-scale and board-level communications. The energy efficiency for photonic links, as shown in Fig. 5, is expected to be less than 1 pJ/bit for board-to-board and 0.1 pJ/bit for intra-chip communications [9-10]. However, for commercially available transceiver modules, the energy consumption is currently more than 10 pJ/bit (Fig. 6), which implies that an order-of-magnitude improvement is required for the applications of photonics in off-chip and board-level interconnects.

To break the barrier of 1 pJ/bit, silicon photonics offers two important technical advantages, compared with other photonic technologies such as VCSELs or III-V photonic circuits. First, the use of silicon substrates enables a close 2.5D or 3D integration of silicon photonics with processors and switch ASICs, due to comparable thermal expansion and mechanical stability from the substrates. An alternative researched by some teams is the fully monolithic integration of digital ASICs and silicon photonics. Additionally, silicon photonics enables the fabrication of innovative devices with unprecedented energy efficiency as illustrated in the next few paragraphs with two examples.

The high index contrast of silicon waveguides drastically reduces the size of many optical devices and thanks to the lower device capacitance also improves the energy efficiency. One example is silicon microring resonator. Microrings are tiny optical notch filters that can be used both as add-drop filters and as modulators with energy efficiency on the order of fJ/bit [11-13]. Wavelength-division multiplexing (WDM) transmitters can be easily constructed by cascading microring modulator arrays with a multi-wavelength light source, as shown in Fig. 7. This greatly simplifies the WDM architecture as microrings simultaneously provide multiple functionalities including wavelength de-multiplexing, modulation, and wavelength multiplexing. WDM receivers can be also built if the rings are used as add-drop filters (Fig. 7). Typically, thermally tunable rings are employed with a very low tuning power by carefully engineering the thermal structure around the microrings.

However, the resonant wavelength of a microring is highly sensitive to fabrication variations and ambient temperature changes. A low-power and robust solution to actively tune and lock the ring resonances to the desired wavelengths is required. The procedure of wavelength locking involves monitoring the rings' response and adapting the heater power based on the information of the monitoring signal. In [12] and [13], we demonstrated that a single monitoring signal can simultaneously lock a photonic system comprising of many rings, as illustrated in Fig. 8. At the transmitter side, we demonstrated that a simultaneous locking of multiple ring modulators is feasible by detecting the RF components of the wavelength-multiplexed optical signal in the through port [12]. We also realized a closed-loop control of an eight-ring WDM receiver chip by continuously monitoring and minimizing the DC component of the optical signal at the through port [13]. The use of a single monitoring

signal can reduce the power consumption, simplify the control circuitry, and increase the system scalability.

Another low-power device we recently demonstrated is the directly reflectivity modulated laser (DRML) [14-15]. While conventional directly modulated lasers (DMLs) are compact and low-power devices, they suffer from narrow wavelength tunability, high chirp, and allowing only intensity modulation. Additionally, the lasers relaxation oscillation frequency strongly limits the modulation speed. In [14], we proposed and experimentally demonstrated a novel DRML, where the laser output is modulated through an actively tunable Michelson interferometer (MI) based mirror. The demonstrated DRML allows for a high-speed and low-chirp modulation, a wide wavelength tunability, and a possibility of both intensity and phase modulations. In addition, the device still retains the key benefit of traditional DMLs by requiring a low RF drive power about 100 fJ/bit. In [15], we further demonstrate that it is feasible to simultaneously modulate the reflectivities of two mirrors in a single laser (see Fig. 9), allowing for generation of two independent 20 Gb/s on-off-keying (OOK) signals using a hybrid silicon/III-V platform (Fig. 10). The measured eye diagrams and bit error ratios (BERs) of the signals are presented in Fig. 11. The reported DRML are critical for energy-efficient transmitters in telecommunications and optical interconnects.

REFERENCES

[1] P. J. Winzer and D. T. Neilson, "From scaling disparities to integrated parallelism: a decathlon for a decade," J. Lightwave Technol. 35, 1099-1115 (2017).

[2] M. J. R. Heck et al, "Hybrid silicon photonic integrated circuit technology," IEEE J. Sel. Topics Quantum Electron. 19, pp. 6100117 (2013).

[3] E. Agrell et al., "Roadmap of optical communications," J. Opt., vol. 18, 2016, Art. no. 063002.

[4] P. Dong et al., "224-Gb/s PDM-16-QAM modulator and receiver based on silicon photonic integrated circuits," in Proc. 2013 Optical Fiber Communication Conference, paper PDP5C.6.

[5] P. Dong et al., "Monolithic polarization diversity coherent receiver based on 120-degree optical hybrids on silicon," Opt. Express 22, 2119-2125 (2014).

[6] P. Dong et al. "Reconfigurable 100 Gb/s silicon photonic network-on-chip," J. Opt. Commun. Netw. 7, A37-A43 (2015).

[7] L. Chen et al., "Monolithic silicon chip with 10 modulator channels at 25 Gbps and 100-GHz spacing," Opt. Express 19, B946-B951 (2011).

[8] ASCAC Subcommittee 2014, "Top ten exascale research challenges", US Department Of Energy Report.

[9] Y. S. J. Ben, "The role of photonics in future computing and data centers," IEICE Trans. Commun. E 97-B 1272–80 (2014).

[10] A. V. Krishnamoorthy et al., "Energy-efficient photonics in future high-connectivity computing systems," J. Lightw. Technol. 33, pp. 889–900, 2015.

[11] Q. Xu et al., "Cascaded silicon micro-ring modulators for WDM optical interconnection," Opt. Express 14(20), 9431–9435 (2006).

[12] P. Dong et al, "Simultaneous wavelength locking of microring modulator array with a single monitoring signal," Opt. Express 25, 16040–16046 (2017).

[13] R. Gatdula et al., "Simultaneous four-channel thermal adaptation of polarization insensitive silicon photonics WDM receiver," Opt. Express 25, 27119-27126 (2017).

[14] P. Dong, A. Maho, R. Brenot, Y.-K. Chen, and A. Melikyan, "Directly reflectivity modulated laser," J. Lightwave Technol. 36, 1255-1261 (2018).

[15] P. Dong, K. Kim, and A. Melikyan, "Generating two optical signals from a single directly reflectivity modulated laser," ECOC 2018.

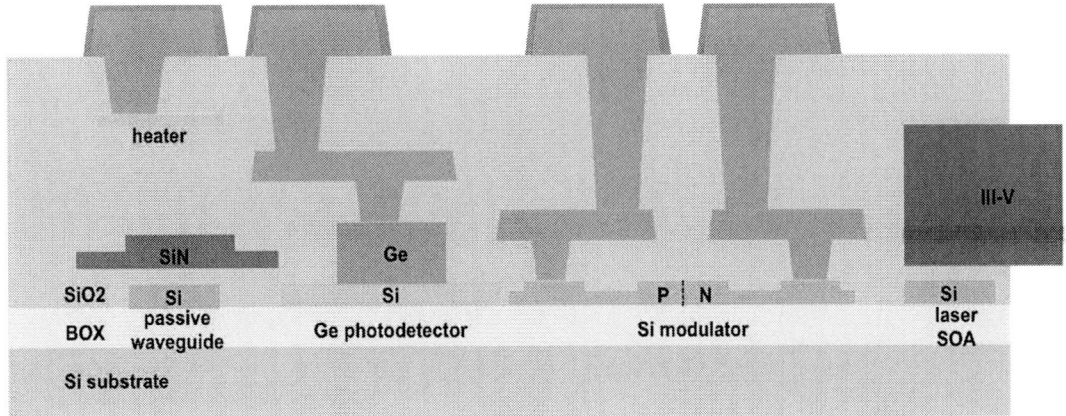

Fig. 1. Device cross section for a typical silicon photonics foundry.

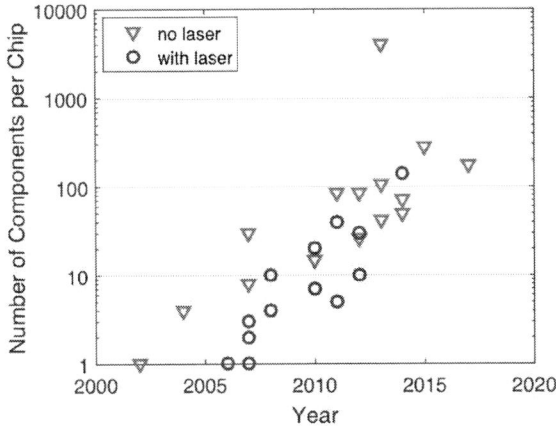

Fig. 2. Development of chip complexity measured as number of components per chip for silicon photonics. Data are obtained from [2-7].

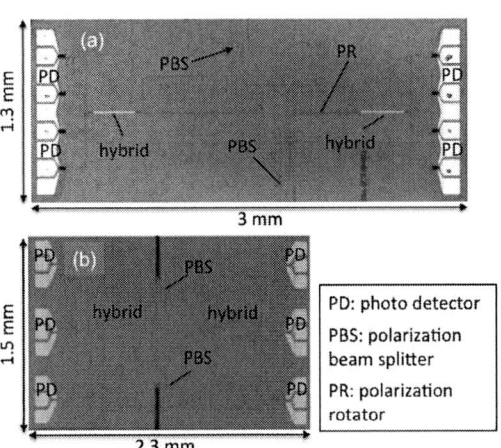

Fig. 3. Optical microscope pictures of two monolithic silicon photonic coherent receiver chips.

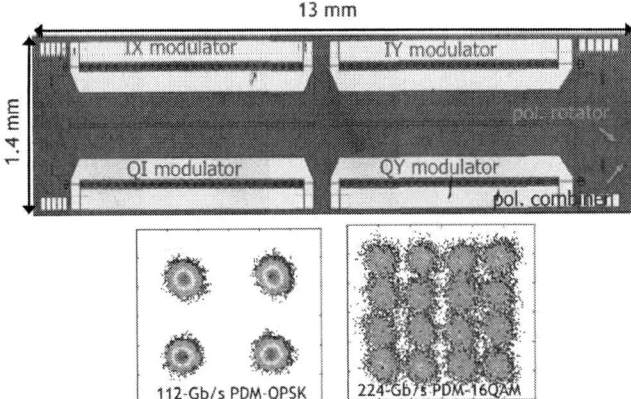

Fig. 4. Optical microscope picture of a monolithic dual-polarization in-phase/quadrature modulator to generate coherent signals [4].

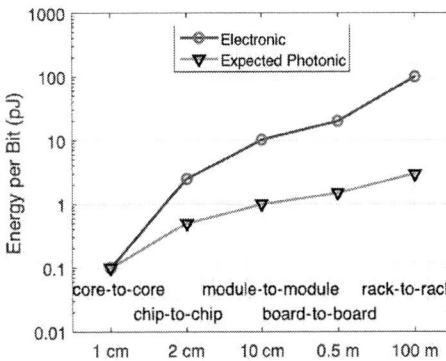

Fig. 5. Energy per bit for electronic interconnects and expected photonic interconnects.

978-1-7281-1988-5/18 $31.00 © 2018 IEEE

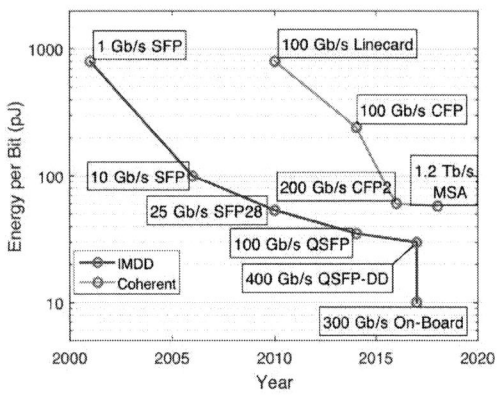

Fig. 6. Energy per bit for commercially available optical transceivers for both intensity-modulation direct detection (IMDD) and coherent transmission.

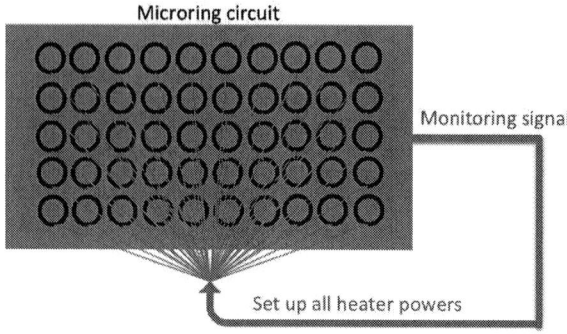

Fig. 8. Closed-loop control of microring circuits with a monitoring signal.

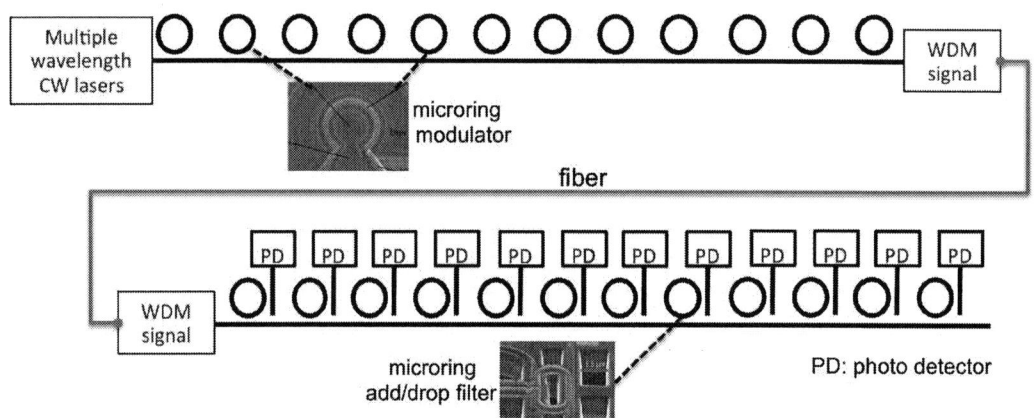

Fig. 7. WDM transmitter and receiver based on mircoring modulators and add-drop filters.

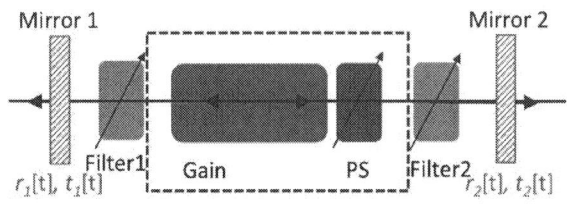

Fig. 9. Directly reflectivity modulated laser with two outputs.

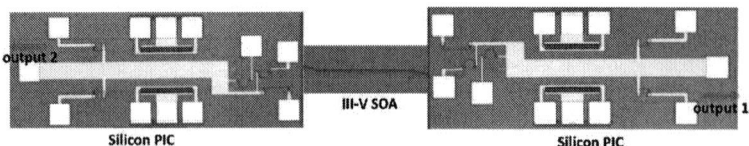

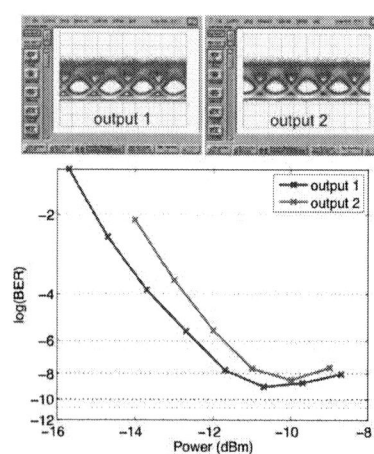

Fig. 11. Eye diagrams of the dual-output DRML and bit error ratios (BERs) as a function of receiving power.

Fig. 10. A dual-output DRML is realized in a hybrid silicon/III-V platform. PIC: photonic integrated circuit; SOA: semiconductor optical amplifier.

978-1-7281-1988-5/18 $31.00 © 2018 IEEE

InAs/GaAs Quantum Dot Lasers Monolithically Integrated on Group IV Platform

Keshuang Li, Mingchu Tang, Mengya Liao, Jiang Wu, Siming Chen, Alwyn Seeds,and Huiyun Liu
Department of Electronic and Electrical Engineering, University College London, London, email: Huiyun.liu@ucl.ac.uk

Abstract—III-V quantum dot lasers monolithically integrated on silicon platform attracts intensive interests due to its advantages on providing a promising solution for reliable and efficient light source to integrated on photonics and electronics circuits. Compared to wafer bonding technique, monolithic integration its more attractive for large scale, low cost and streamline fabrication. In this paper, we give a brief review on our recent progress of III-V quantum dot lasers monolithically integrated on 4° offcut and exact (001) Si substrates for the silicon photonic integration.

I. INTRODUCTION

Monolithic integration of III-V reliable and efficient electrically pumped continues wave (CW) semiconductor laser on silicon platform is an ultimate solution for producing complex Si based optoelectronics integrated circuits. Directly epitaxial growth for III-V semiconductor material on silicon substrate gives a promising solution to provide a low cost, large scale and high yield on-chip light source. However, large material dissimilarities between group IV and group III-V materials cause many significant challenges for monolithic integration. For example, large material mismatch, thermal expansion ecoefficiency difference and polar on non-polar growth which cause various types of defects including threading dislocations (TDs), microcracks and antiphase boundaries (APBs). These may generate non-radiative recombination centres which will dramatically decrease the performance of fabricated devices. To overcome these challenges, many approaches including dislocation filter layers (DFLs) and thermal annealing (TA) have been considered in our results in order to improve device performance. Considering the active region, quantum dot structure attracts extensive research interests in past several decades for the monolithic integration of III-V on Si. It has been widely used as active layers in semiconductor lasers due to the advantages of low threshold current density (J_{th}), insensitive to TDs, and high temperature stability. In this paper, we will give a review of our recent progress on III-V quantum dot lasers monolithically integrated on Si platform.

II. QUNATUM DOT LASERS MONOLITHICALLY INTEGRATED ON OFFUCT SILICON SUBSTRATE

In this work, InAs/GaAs quantum dot lasers structure was directly grown on 4° offcut to [011] Si substrate using solid-source molecular beam epitaxy (MBE) system to prevent the formation of APBs. In our previous work, we have reported the first electrically pumped InAs/GaAs quantum dot laser directly grown on Si substrate by using a optimized GaAs nucleation layer on Si and a InGaAs/GaAs strained layer superlattice (SLS) as DFLs [1]. After that, we consistently improve the devices performance in our following works. Eventually, due to the high-quality GaAs film we realized on offcut Si substrate by applying combined strategies as AlAs nucleation layer, optimized InGaAs/GaAs DFLs and in-situ thermal annealing. A high performance 1310 nm InAs/GaAs quantum dot laser monolithically grown on offcut Si substrate with record low threshold current density 62.5 A/cm^2 at room temperature (RT) has been demonstrated [2]. The maximum output power of single facet is about 105 mW and the maximum operation temperature is up to 75°C under CW mode. We will give a more detailed description in the following two sessions.

A. Epitaxial structire

The epitaxial structures of quantum dot lasers were deposited by Veeco solid-source MBE GEN 930. After the preparation of offcut Si substrate, a thin film of AlAs nucleation layer was deposited on Si substrate by migration enhanced epitaxy (MEE) technique at relative low temperature about 350°C. As AlAs nucleation layer suppressed the three-dimensional growth, a good interface was formed for the following growth. After this, GaAs buffer was formed by three-step growth technique in different growth temperature. First, 30 nm GaAs was deposited at the same temperature as AlAs; then, 170 nm GaAs film was formed afterwards at higher temperature 450°C; at the end, 800 nm high temperature GaAs was deposited at 590°C. After depositing the GaAs buffer, 10 nm In$_{0.18}$Ga$_{0.82}$As/10 nm GaAs SLS was used as DFLs to further reduce the threading dislocation density (TDD). Four repeats of DFLs were performed on GaAs buffer and each DFL included five periods of SLS and separated by 300 nm GaAs. In-situ thermal annealing was performed in each DFL after the formation of five periods of SLS. This will help to increase the mobility of TDs and then lead to the annihilation before the subsequently growth. Fig. 1 (a) (b) give a clear explanation on the TDD according to the different position of epitaxial structure. Most of the TDs were stopped at first 200 nm of the buffer layer. However, TDD is still at the level of 10^9 cm^{-2} at position 1. After four repeats of DFLs, the TDD was clearly reduced and down to the level of 10^{5}cm^{-2} after the last 300nm GaAs space layer. We successfully demonstrated a high-quality buffer for the following laser structure growth. Subsequently, a typical InAs/GaAs DWELL laser structure will be deposited on well optimized low TDD GaAs virtual substrate on Si.

B. Device performance

To date, we achieved low threshold current density and high optical output power quantum dot lasers on offcut Si substrate.

978-1-7281-1988-5/18 $31.00 © 2018 IEEE

Fig. 2 (insert) gives an AFM image of typical uncapped quantum dots layer grown under same condition with the laser devices. Good uniformity with $3 \times 10^{10} cm^{-2}$ quantum dot density is obtained. Fig. 2 also provides a photoluminescence (PL) image measured of the uncapped quantum dot layer which emitting at ~1300 nm with full-width at half maximum (FWHM) ~ 29 meV. Broad area lasers were fabricated based on the laser structure we developed. Eventually, we demonstrated a high reliability, high performance electrically pumped continuous wave 1.3 μm QD laser on 4° offcut Si (001) substrate with J_{th} of 62.5 A/cm^2 at room temperature and maximum operating temperature 75°C at CW mode. Fig. 3 shows the light-current-voltage (LIV) characteristic measurement for our InAs/GaAs quantum dot laser monolithically grown on Si. High optical output power of 105mW was observed under injection current density of 650 A/cm^2.

III. QUANTUM DOTS LASER MONOLITHICALLY INTEGRATED ON EXACT (001) SILICON SUBSTRATE

Based on our previous discussion, we have successfully demonstrated the III-V quantum dot lasers grown on offcut Si substrate to prevent the formation of APBs. However, although the offcut substrate can annihilate the APBs caused by polar III-V materials grown on non-polar group IV materials, it is not fully compatible to the industrial standard fabrication. Therefore, most recently, we have demonstrated the 1.3μm electrically pumped CW quantum dot lasers on exact Si (001) substrate which is industrial compatible. A J_{th} down to 425 A/cm^2 at room temperature with maximum operation temperature 36 °C under CW mode was achieved [3]. Detailed Epitaxial structure and devices performance will be discussed in following sessions.

A. Epitaxial structure

The schematic diagram of our epitaxial structure on exact (001) Si substrate is shown in Fig. 4. 400 nm GaAs buffer was grown in steps by MOCVD on an 300 mm diameter industrial standard Si (001) substrate without offcut. Firstly, a 40 nm thin film GaAs nucleation layer was deposited at low growth temperature (400-500°C). Then, a 360 nm GaAs buffer was grown at higher temperature (600-700°C) [4]. Eventually, as the AFM image shown in Fig. 5 (a), an APB free GaAs surface and small root mean square (RMS) surface roughness of ~0.86 nm were achieved. Afterwards, sample was moved to MBE chamber for the following laser structure growth. 600 nm GaAs buffer was deposited firstly and following by five repeats of DFLs consist of five periods of optimized InGaAs/GaAs SLS separated by 300 nm GaAs. Then, a InAs/GaAs quantum dot laser structure was deposited on it started with 1.4 μm n-type AlGaAs cladding layer and ended with 300 nm p-type GaAs contact layer. The active region consists of five repeats of undoped InAs/GaAs DWELL structure separated by 50 nm GaAs space layers.

B. Device performance

These quantum dot laser samples grown on exact (001) Si substrate were fabricated into broad area lasers and took various relative measurements. Fig. 5 (b) shows the room temperature PL comparison for these two samples. They present comparable emission intensity at ~1285 nm with ~32 meV FWHM. The intensity of GaAs based quantum dots sample is about 1.3 times greater than Si (001) based sample. The surface morphology investigation was done by AFM for an uncapped InAs quantum dots layer grown on same condition with laser structure Fig. 6 presents the 1×1 μm^2 AFM image of uncapped InAs QDs grown on Si (001) substrate which shows good uniformity and the dot density are ~$3.5 \times 10^{10} cm^{-2}$. The LIV characteristic comparison is presented in Fig. 7, the J_{th} is 210 A/cm^2 and 425 A/cm^2 for GaAs based, and Si based lasers respectively. The slope efficiencies are calculated as 0.12 W/A and 12.7% for GaAs based devices and 0.068 W/A and 7.2% for the Si based devices. From Fig. 8, the maximum output power of single facet is 43 mW achieved under 1332 A/cm^2 injection current density at CW mode and only small rollover presented. For pulsed mode, the maximum output power is 134 mW under 2kA/cm^2 injection current density. Fig. 9 presents the emission spectrum under different injection current density under CW mode. When the injection current density increases to 425A/cm^2, typical lasing behavior presented. Fig. 10-11 show the light-current (LI) characteristic under pulsed and CW mode respectively. Under pulsed mode, the characteristic temperature T_0 is ~32K between 16°C to 102°C. And for CW mode, lasing operation was observed when heatsink heated up to 36°C.

IV. CONCLUSION

We reviewed the recent results about III-V quantum dots lasers monolithically integrated on offcut Si and exact (001) Si substrate in this paper. They all present promising potential for high efficient and reliable on-chip light source for Si based optoelectronics integrated circuits.

ACKNOWLEDGMENT

This work was supported by UK Engineering and Physical Sciences Research Council (EPSRC) [grant numbers EP/J012904/1, EP/P000886/1, EP/P006973/1] and EPSRC National Epitaxy Facility; S.C. thanks for the Royal Academy of Engineering for funding his Research Fellowship.

REFERENCES

[1] T. Wang, H. Liu, A. Lee, F. Pozzi, and A. Seeds, "1.3- μm InAs/GaAs quantum-dot lasers monolithically grown on Si substrates," Opt. Express, vol. 19, pp. 11381-11386, 2011.

[2] S. Chen, W. Li, J. Wu, Q. Jiang, M. Tang, S. Shutts, S. Elliott, A. Sobiesierski, A. Seeds, I. Ross, P. Smowton, and H. Liu, "Electrically pumped continuous-wave III–V quantum dot lasers on silicon," Nat. Photonics, vol. 10, pp. 307-311, 2016.

[3] S. Chen, M. Liao, M. Tang, J. Wu, M. Martin, T. Baron, A. Seeds, and H. Liu, "Electrically pumped continuous-wave 1.3μm InAs/GaAs quantum dot lasers monolitically grown on on-axis Si (001) substrates," Opt. Express, vol. 5, pp. 4632-4639, 2017..

[4] R. Alcotte, M. Martin, J. Moeyaert, R. Cipro, S. David, F. Bassani, F. Ducroquet, Y. Bogumilowicz, E. Sanchez, Z. Ye, X. Bao, J. Pin, and T. Baron, "Epitaxial growth of antiphase boundary free GaAs layer on 300 mm Si (001) substrate by metalorganic chemical vapour deposition with high mobility," APL Mater. 4(4), 046101 (2016).

978-1-7281-1988-5/18 $31.00 © 2018 IEEE

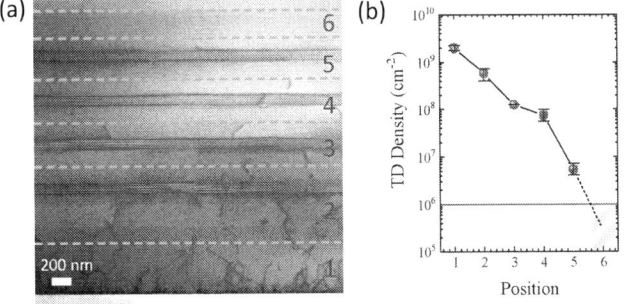
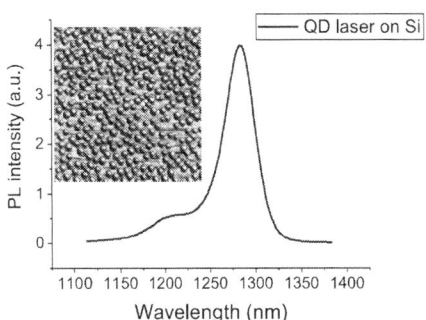

Fig. 1. (a) Bright-field scanning TEM image of DFLs; (b) Dislocation density measured at different positions in (a).

Fig. 2. PL spectrum for QD grown on Si emitting at ~1300nm; Inset: representative AFM image of uncapped QD grown on Si

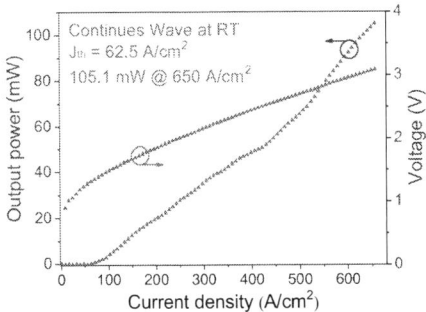

Fig. 3. LIV characteristics for a 50 μm × 3200 μm InAs/GaAs QD laser grown on Si substrate under CW operation at 18 °C.

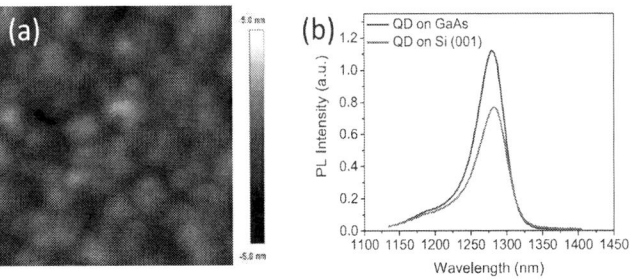

Fig. 5. (a) 5 × 5 μm² AFM image of a 400 nm GaAs buffer direct grown on exact Si (001) substrate; (b) Room temperature PL comparison of InAs/GaAs QD laser structure grown on exact Si (001) and GaAs substrate.

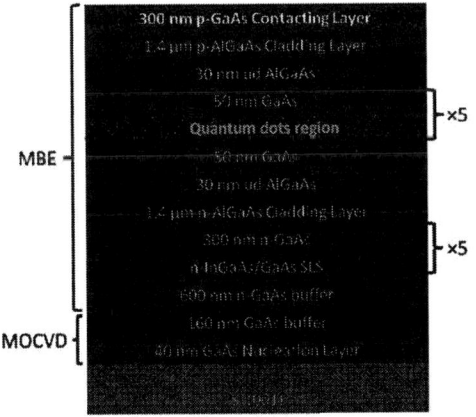

Fig. 4. The schematic diagram of the QD laser structure grown on the on-axis Si (001) substrate.

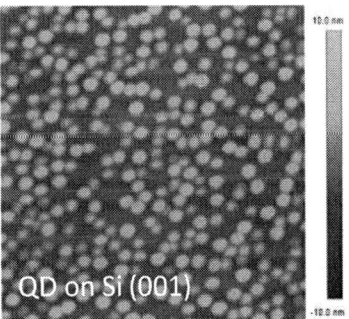

Fig. 6. 1 × 1 μm² AFM image of uncapped InAs QDs grown on exact Si (001) substrate.

978-1-7281-1988-5/18 $31.00 © 2018 IEEE

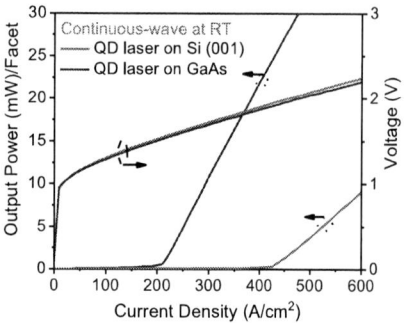

Fig. 7. LIV characteristic comparison of a InAs/GaAs QD laser grown on Si (001) and native GaAs substrate at room temperature under CW operation.

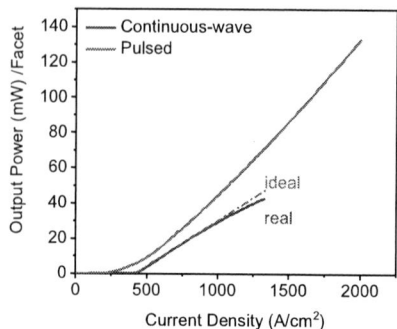

Fig. 8. LI comparison of InAs/GaAs QD laser grown on Si (001) substrate under CW and pulsed conditions at room temperature.

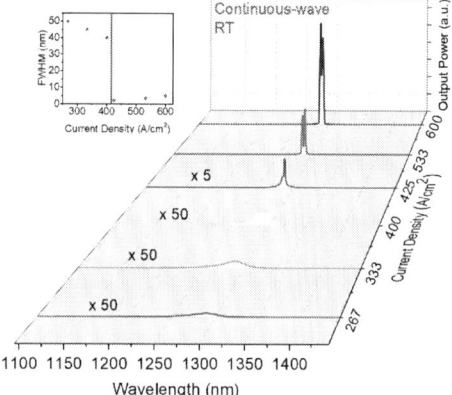

Fig. 9. Emission spectra for InAs/GaAs QD laser on Si (001) substrate at various injection current densities at room temperature. The inset shows the FWHM change as a function of injection current density.

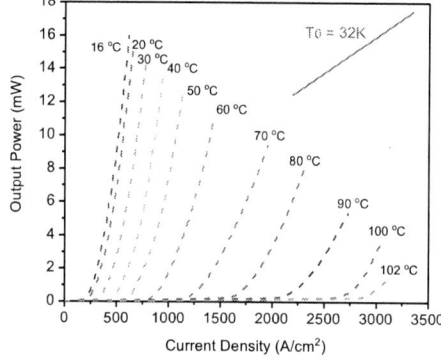

Fig. 10. Single facet LI curve for a InAs/GaAs QD laser grown on Si (001) substrate at different heat sink temperatures under pulsed condition. The inset shows the natural logarithm of current density against temperature in the ranges of 16 - 102°C.

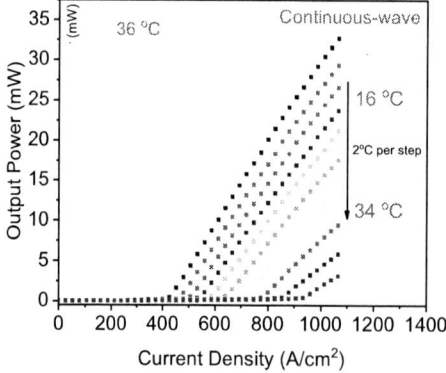

Fig. 11. Single facet LI curve for the same InAs/GaAs QD laser on Si (001) as a function of temperature under CW operation. The inset shows the LI curve for this QD laser at a heat sink temperature of 36°C.

Heterogeneously integrated ligtht sources on bulk-silicon platform

Dongjae Shin[1], Jungho Cha[1], Yonghwack Shin[1], Kyoungho Ha[1], Chang Bum Lee[2], Changgyun Shin[2], Dongshik Shim[2], Byoung Lyong Choi[2], Hyeongsun Hong[1], Kyupil Lee[1], and Ho-Kyu Kang[1]

[1]Advanced Technology Development Team, Semiconductor R&D center, Samsung Electronics, Hwaseong, Korea,
[2]Imaging Device Lab., Samsung Advanced Institute of Technology, Samsung Electronics, Suwon, Korea,
email: dongjae.shin@samsung.com

Abstract—Heterogeneously integrated single-λ and λ-tunable light sources on bulk-silicon platform are presented. Thanks to the thermal advantage of the bulk-silicon platform, the single-λ source showed WPE of 8% up to 70°C, feasibility of 25Gb/s direct modulation, and 70°C MTTF of ~46000h. The λ-tunable source showed 42.2nm tuning range. This result completes the optical device library suite for the bulk-silicon platform used in most semiconductor products.

I. INTRODUCTION

As applications requiring photonic functions or high-speed interconnects have been emerging, silicon photonics has been drawing great attention from CMOS industry because it can leverage the matured CMOS manufacturing infrastructures to deliver next-generation photonic integrated circuits(PICs) and electro-photonic integrated circuits(EPICs) [1-2]. Since the light source for the PICs is still not viable from the traditional CMOS process, its solution has been either hybrid assembly of discrete III/V laser diodes(LDs) or heterogeneous integration(HI) of III/V gain materials to the CMOS process [3-4]. The hybrid assembly approach resorts to the existing process, but the packaging and assembly cost may increase. On the other hand, the HI approach is expected to promise low-cost high-volume manufacturing in the long run in spite of initial III/V-related process investment.

So far, previous studies and developments on the HI LDs have been based on the silicon-on-insulator(SOI) platform because its buried oxide(BOX) layer provides optical isolation preventing optical leakage to the substrate. However, the same BOX layer also causes unwanted thermal isolation hindering the heat dissipation from the laser active region to the substrate. Therefore, the SOI platform is not optimal for temperature-sensitive HI LDs in the thermal perspective.

In this paper, we present single-λ and λ-tunable HI light sources on the bulk-silicon(BS) platform for the first time to authors' knowledge. In addition to the cost advantage of the BS wafers by a factor of ~10, we present decent performance and reliability achieved from the BS-based light sources, taking advantage of better heat dissipation of the BS platform. Following the BS-based silicon-photonics devices previously reported by authors [2, 4], this result completes the optical device library suite for the PIC on the BS platform.

II. DEVICE STRUCTURE

A. Cross-sectional structure

The cross-sectional structure of the HI light sources on the BS platform is shown in Fig. 1. It includes the III/V epitaxial layer directly bonded to the pre-patterned silicon wafer, the current confinement with the proton-implanted III/V mesa structure, and the optical confinement with the III/V mesa on top of the silicon rib waveguide [1-3, 6-8]. The unique feature of the BS platform is the local BOX structure below the rib waveguide with the lateral offset. The silicon layer on top of the local BOX was formed by our proprietary solid phase epitaxy(SPE) process [2, 4]. The cross-sectional scanning electron microscopy(SEM) image in Fig. 1(b) confirms the local BOX structure instead of the global BOX layer in the SOI platform.

B. Planar structure

The planar structure of the single-λ and λ-tunable HI LDs are illustrated in Fig. 2(a) and 2(b). The active length of the single-λ LD was 210um and the mode converter connecting the active region and the passive waveguide was 10um. The grating-based cavity was asymmetric to enhance the front-side optical output. The grating couplers for on-wafer tests were designed not only for its coupling loss to optical fiber probes, but also for reducing its reflection back to the single-λ LD. The λ-tunable LD had 2 active sections for optical amplification and 2 ring-resonator-based mirrors for cavity formation. The 2 ring mirrors were designed to have slightly different radius for a selective resonance of a single longitudinal mode through the Vernier effect [8]. The lasing wavelength was tuned by the embedded heater thermally shifting the lasing comb of the ring mirror.

III. MEASUREMENTS AND ANALYSIS

A. Optical power and WPE of single-λ HI LD

Fig. 3(a) and 3(b) show measured front-side optical output powers and corresponding WPEs as a function of current at the chuck temperatures of 25, 55, and 70°C. At an optimal operating current of ~50mA, the optical power reached 7mW at 70°C, which translated to WPE of 8%. The small ripples in the light powers were due to wavelength-dependent coupler loss variation. A proprietary single-λ LD model was constructed in a semi-heuristic approach as illustrated in Fig.

3(c) [5]. The model outputs showed a decent agreement with the measurements.

B. Optical spectrum of single-λ HI LD

Fig. 4(a) shows the spectra of the single-λ LD over increasing current at 25°C. The peak wavelengths and side-mode suppression ratio(SMSR) in Fig. 4(b) were extracted from the spectra. The single-λ operation was maintained over increasing current range from 20 to 180mA with SMSR >45dB.

Fig. 5(a) shows the feasibility of wavelength division multiplexing(WDM) from an array of 11 single-λ LDs with increasing grating periods. The channel spacing of 5.69nm was achieved with 1nm spacing of the grating period as shown in Fig. 5(b).

C. Thermal analysis of single-λ HI LD

The thermal advantage of the BS platform was verified in thermal impedance measurements. Fig. 6(a) shows the lasing wavelength shift over increasing LD power, providing the dλ/dP. Fig. 6(b) shows the lasing wavelength shift over increasing LD temperature, providing the dλ/dT. The thermal impedance was found to be 2.03 (K/W)*cm, which is ~40% lower than the previous report on the similar SOI-based LD structures [9].

The impact of the reduced thermal impedance on the LD performance was calculated in the single-λ LD model. Fig. 6(c) shows the calculations for the four platforms which are of interest in the thermal perspective. As thermal impedance decreases, the maximum output power(Pmax) at 70°C rapidly increases, but the maximum WPE(WPEmax) at 70°C slowly increases and saturate at ~2(K/W)*cm. In this modeling analysis, the BS platform is expected to provide 52%(1.8dB) higher Pmax and 13% higher WPEmax at 70°C compared to the SOI counterpart. If the silicon substrate is thinned down to 100um, the thermal impedance is simulated to be ~1.14(K/W)*cm and the corresponding 70°C Pmax enhancement is expected to be 143%(3.8dB). This indicates that the BS platform is well suited for applications where a single high-power laser feeds many devices.

D. Direct modulation of single-λ HI LD

Direct modulation is desired for low-cost implementation of optical transmitters. Fig. 7 shows the eye diagrams of the directly modulated single-λ LD. The eye opening was verified up to 25Gb/s, but the eye quality was degraded beyond 16Gb/s as shown in Fig. 7(a) and 7(b). Fig. 7(c) shows that the eye quality was improved at lower temperature, implying that lower thermal impedance of the BS platform also contributed to the 25Gb/s modulation feasibility [6].

E. Reliability of single-λ HI LD

There has been reliability concern on the HI LD mainly because the directly bonded III/V and silicon materials have different thermal expansion coefficients, and 70°C mean time to failure(MTTF) of ~40000h has been reported on the SOI platform [7]. Fig. 8 summarizes the high-temperature operating lifetime(HTOL) tests on the HI LD on the BS platform over 1900h at 70°C, and the 70°C MTTF was found to be ~46000h.

F. λ-tunable HI LD

The λ-tunable LD plays an important role in WDM system and optical sensor applications [8]. The 2 ring resonators in the λ-tunable LD were 311 and 320um long with a designed free spectral range(FSR) of 1.3nm. Fig. 9 shows the tuning performance of the λ-tunable LD on the BS platform. The tuning range of 42.2nm with SMSR >30dB was achieved.

IV. Summary

Light sources heterogeneously integrated on the silicon platform are of great interest for volume-manufacturing of next-generation PICs or EPICs. Contrary to the popular HI light sources on the SOI platform, we adopted the BS platform to develop the single-λ and λ-tunable light sources. The single-λ LD presented 8% WPE and 7mW front-side output power at 70°C with SMSR >45dB over wide operating conditions. The WDM feasibility from the array of 11 LDs was also demonstrated. The thermal advantage of the BS platform was verified in the thermal impedance measurement. The modeling analyses show that the BS platform is expected to outperform the SOI platform by 1.8dB in Pmax and 13% in WPEmax at 70°C. 70°C MTTF of ~46000h was also achieved in the HTOL test. The λ-tunable LD presented the wide tuning range of 42.2nm. This BS-based single-λ and λ-tunable light sources complete the device library suit for the BS photonic integration platform. Since the BS platform is dominant in the CMOS industry, the BS platform offers better opportunities towards the vision of EPICs embedding the PICs in CMOS ICs with minimal changes to the CMOS industry.

Acknowledgment

The authors would like to thank Taek Kim for helpful discussions over applications and industry impact.

References

[1] J. E. Bowers, "Heterogeneous photonic integration on silicon," *Proc. of IPC2016*, WC2.1 (2016).

[2] D. Shin, K. Cho, H. Ji, B. Lee, S. Kim, J. Bok, S. Choi, Y. Shin, J. Kim, S. Lee, K. Cho, B. Kuh, J. Shin, J. Lim, J. Kim, H. Choi, K. Ha, Y. Park, and C. Chung, "Integration of silicon photonics into DRAM process," *Proc. of OFC2013*, OTu2C (2013).

[3] D. Liang, and J. E. Bowers, "Recent progress in lasers on silicon," Nature photonics, **4**, 511-517 (2010).

[4] D. Shin, J. Cha, S. Kim, Y. Shin, K. Cho, K. Ha, G. Jeong, H. Hong, K. Lee, and H. Kang, "O-band DFB laser heterogeneously integrated on bulk-silicon platform," *Opt. Express* vol. 26 no. 11, pp. 14768-14774, May 2018.

[5] L. A. Coldren, S. W. Corzine, M. L. Mashanovitch, *Diode lasers and photonic integrated circuits*, 2nd ed. (Wiley-Interscience, 2012)

[6] Y. L. Cao, X. N. Hu, Y. B. Cheng, H. Wang, and Q. J. Wang, "Optimization of hybrid silicon lasers for high-speed direct modulation," *IEEE Photon. J.* vol. 7, no. 2, Apr. 2015.

[7] S. Srinivasan, N. Julian, J. Peters, D. Liang, and J. E. Bowers, "Reliability of hybrid silicon distributed feedback lasers," *IEEE J. sel. Topics Quantum Electron.* vol. 19, no. 4, Jul. 2013.

[8] J. C. Hulme, J. K. Doylend, and J. E. Bowers, "Widely tunable Vernier ring laser on hybrid silicon," *Opt. Express* vol. 21 no. 17, pp. 19718-19722, Aug. 2013.

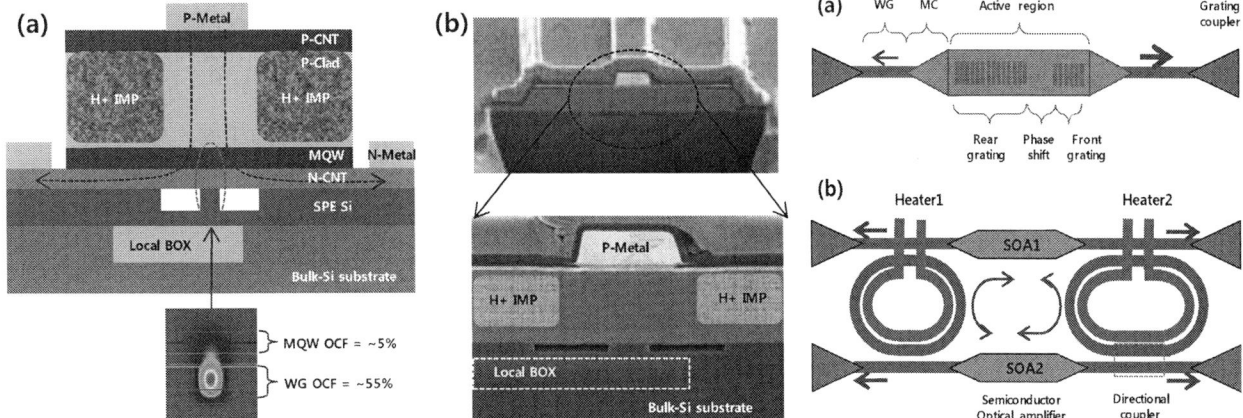

Fig. 1. (a) Conceptual illustration of HI LD on BS platform. The inset is a simulated optical mode. (b) SEM image of a fabricated device.

Fig. 2. Conceptual illustration of planar structures of single-λ HI LD(a) and λ-tunable HI

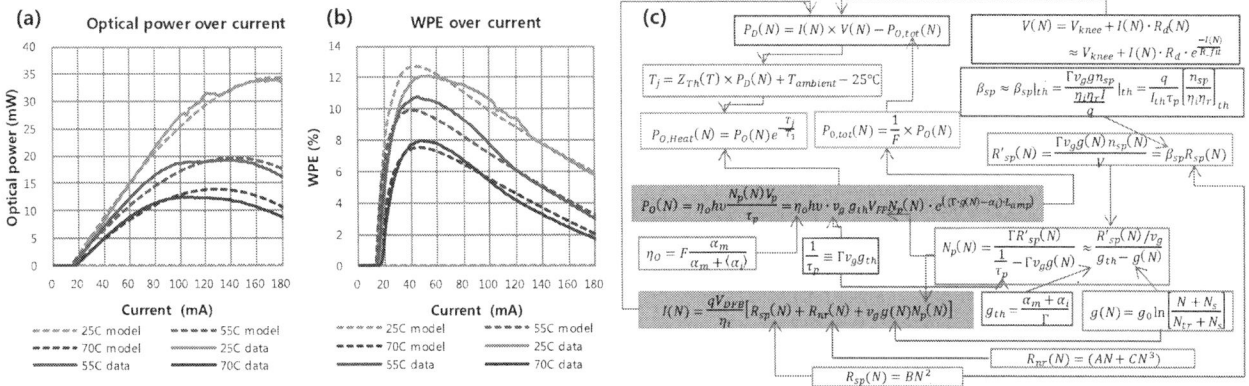

Fig. 3. Optical power(a) and WPE(b) over driving current at 25, 55, 70°C. Dotted lines are from the single-λ LD model. (c) Schematic illustration of the single-λ model. All acronyms and notations follow those of ref. 5.

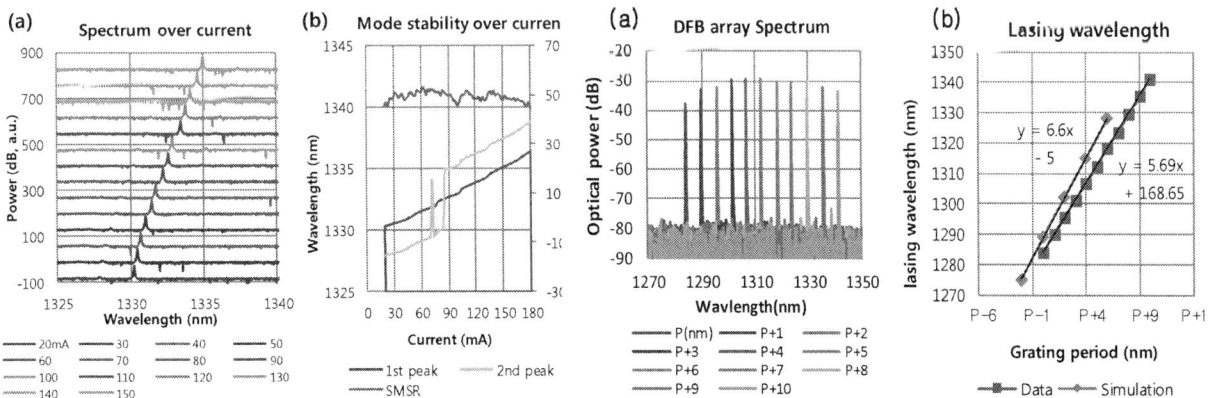

Fig. 4. (a) Optical spectra measured over increasing current at 25°C. (b) Peak wavelength and SMSR extracted from the spectra.

Fig. 5. (a) Superimposed optical spectra from 11 single-λ LDs with increasing grating period. (b) Lasing wavelength from the data and simulation.

978-1-7281-1988-5/18 $31.00 © 2018 IEEE   562

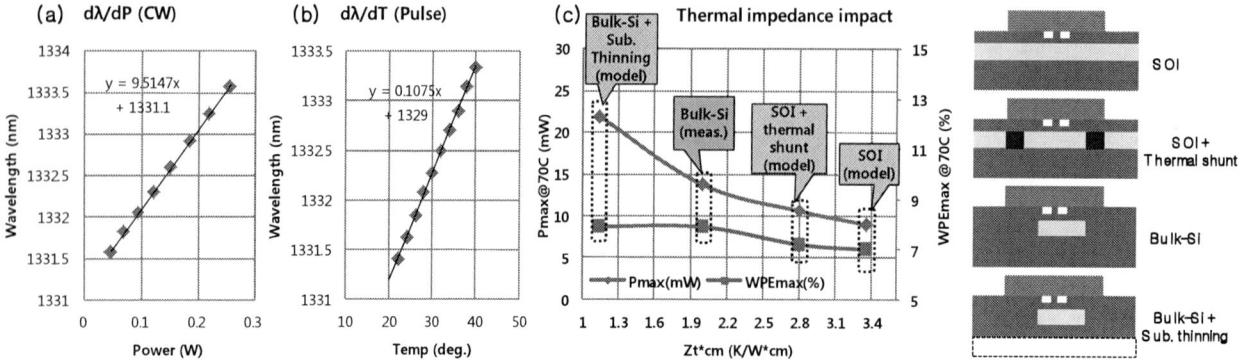

Fig. 6. Thermal impedance measurements and analyses. (a) Lasing wavelength shift over increasing power in continuous operation. (b) Lasing wavelength shift over increasing temperature in pulsed(duty 0.1%) operation. (c) Maximum output power and maximum WPE at 70°C calculated using the single-λ LD model for 4 platforms. The calculation is calibrated with the measured data of the bulk-Si platform.

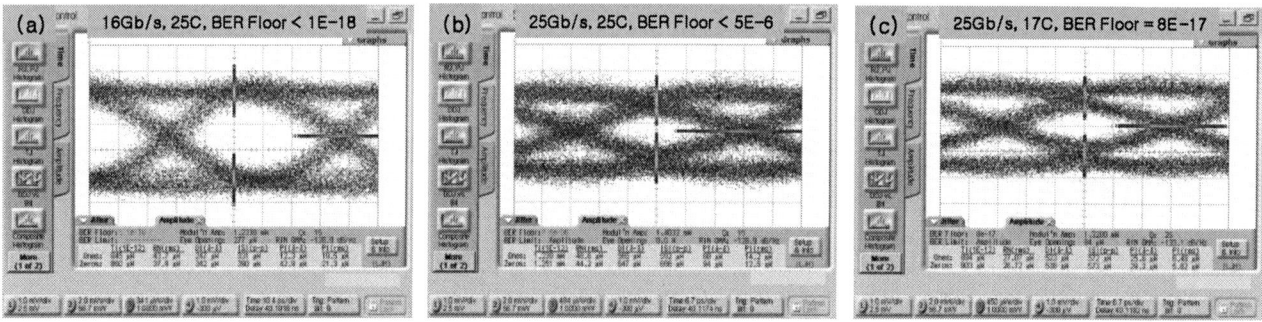

Fig. 7. Direct modulation of the single-λ HI LD. (a) 16Gb/s eye diagram measured at 25°C. (b) 25Gb/s eye diagram measured at 25°C. (c) 25Gb/s eye diagram measured at 17°C.

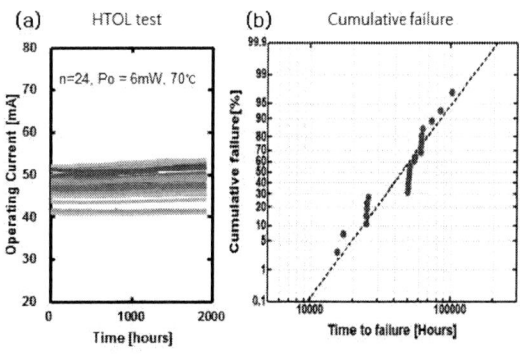

Fig. 8. (a) 70°C HTOL test result over 1900h with 24 single-λ HI LDs. The output power was set to be 6mW. (b) Cumulative failure plot for lifetime estimation.

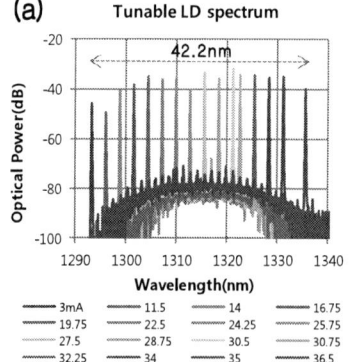

Fig. 9. (a) Superimposed optical spectra over increasing heater current from a single λ-tunable HI LD. (b) Lasing wavelength and SMSR extracted from the λ-tunable LD spectra.

978-1-7281-1988-5/18 $31.00 © 2018 IEEE   563

Interfacial Thermal Conductivity of 2D Layered Materials: An Atomistic Approach

Kamyar Parto, Arnab Pal, Xuejun Xie, Wei Cao and Kaustav Banerjee*

[1]Department of Electrical and Computer Engineering, University of California, Santa Barbara, CA; *Email: kaustav@ece.ucsb.edu

Abstract—This paper presents the first comprehensive modeling and analysis of thermal transport across both lateral and vertical interfaces to two-dimensional (2D) layered materials. Using an ab-initio Atomistic Green's Function (AGF) approach that accurately accounts for the interface geometry including the van der Waals gap, as well as interatomic force constants and interface phonon scatterings, we provide estimation of crucial interfacial thermal properties including thermal conductivity that are invaluable for assessing the performance, scaling, and reliability limits of all emerging 2D based nano-devices, interconnects, circuits and non-planar (monolithic 3D) integration schemes.

I. INTRODUCTION

2D layered materials such as graphene, MoS_2, h-BN, etc., and their heterostructures offer wide range of properties [1] and unique opportunities in nanoelectronics including novel FETs [2-4], interconnects [5], flexible electronics [6] and sensors [7]. Thermal transport in these materials plays a crucial role in determining the performance, scalability and reliability of the nano-devices and circuits that incorporate them. A summary of the key thermal challenges in 2D applications are presented in **Fig.1**. While most 2D materials, due to strong in-plane covalent bonds, offer high in-plane thermal conductivity w.r.t bulk electronic materials, transport in the out-of-plane direction and through lateral or vertical interfaces to other 2D or 3D materials is relatively less efficient due to the presence of a van der Waals gap (vdW) or disordered atomic structure. Hence, in all practical 2D based nano-structures and devices, such out-of-plane thermal properties, including the Interfacial Thermal Conductivity (ITC) is critical to determine their scaling, performance and reliability limits.

Even though several studies have modeled in-plane thermal properties of 2D materials, very limited information is available on the interfacial thermal transport. In this work, for the first time, we conduct a comprehensive and comparative study of the thermal properties of all possible interfaces of 2D materials. Furthermore, we introduce and assess important criteria affecting the ITC, and present a computational framework to find their values. These results offer 2D device engineers a new perspective on thermal transport and provides guidelines for the appropriate choice of materials and interface configurations that maximize the device performance.

Fundamental phonon transport models that can account for arrangement of atoms and their force constants at the interface includes the Boltzmann Transport Equation (BTE), Molecular Dynamics (MD) and Atomistic Green's Function (AGF) [8]. BTE's implementation for different interfaces requires modeling of individual scattering processes which makes it an impractical approach. MD method employs time domain analysis and does not explicitly provide information about transmittivity of individual phonon frequencies, a crucial factor in the design of Field-Effect Transistors (FETs), where

the accumulation of high-energy optical phonons near the drain interface poses reliability issues **(Fig.1a.(iii))**. In comparison, AGF working in the frequency domain provides transmission probabilities of all phonon modes across the interface, efficiently capturing the physics of the interface atomic arrangement and their associated scattering events, thereby making it our preferred choice. In this study, AGF is applied to study various interfaces of 2D materials, from lateral 2D-2D in-plane heterostructures to 2D-3D edge interfaces as well as monolayer 2D vertical interfaces to metals and dielectrics, relevant to their applications.

II. METHODOLOGY

At atomistic level, heat transport across an interface is mainly a two-phonon elastic scattering process, where an incoming phonon with a given wave vector and frequency couples to an out-going phonon in the other medium with the same frequency but not necessarily the same wave vector **(Fig.2a, b)**. The probability of this coupling depends upon the atomic arrangement and binding strength of the interface. Therefore, the first step in evaluation of ITC is a qualitative comparison between Phonon Density of States (PDOS) of each material **(Fig.2d)**. A high PDOS overlap indicates presence of significant mutual frequency phonon pairs, signaling high thermal conductance, while a high mismatch signifies high thermal resistance. It should be noted that PDOS of two materials can change at their interface due to strain and interface atomic disorder, but our results show that even for covalently bonded lateral interfaces, this shift is negligible, thereby validating this approach. Next, by considering the physics of the interface, transmission probability of each mutual phonon pair is calculated through the AGF method. Optimized geometry of the interface is determined within Density Functional Theory (DFT) framework **(Fig.2e)** and interatomic force constants are extracted to construct the dynamical matrix of the system **(Fig.2f)**. Effect of interface geometry and its binding strength are inherent in this matrix. It is then used as an input to the AGF framework to obtain the transmission spectrum **(Fig.2g)**, which determines the contribution of each phonon mode to ITC, and to calculate PDOS **(Fig.2h)**. Finally, ITC is calculated by summing over contributions of all phonon modes in Landauer like formalism as in **Fig.2c**.

Eight different interfaces to monolayer MoS_2, the most prevalent 2D semiconductor with attractive physical properties for FETs and optical devices [9], are selected as representatives of various possible 2D interfaces. FETs fabricated with 2D materials can have lateral, edge **(Figs.3-6)** and vertical interfaces **(Figs.8-11)** with dielectrics (such as h-BN) and metals (Graphene, 1T MoS_2, Ti, Au). Also, light emitting diodes can be fabricated with the MoS_2-WSe_2 p-n junction, a seamless interface (no structural atomic disorder). All interfaces under study were realized within their lowest

978-1-7281-1988-5/18 $31.00 © 2018 IEEE

formation energy configurations. Supercell sizes were adequately chosen to minimize the strain due to the two lattice mismatched materials. DFT calculations were performed in "QuantumATK" [10] using Local Density Approximation (LDA) [11] as exchange correlation functional and Double Zeta Polarized (DZP) basis set with 150 Rydberg energy mesh cut-offs and maximum force tolerance of 0.01 eV/Å. The Brillouin zone is sampled by $8\times8\times8$ and $8\times1\times8$ k-points for vertical and lateral interfaces respectively. LDA performs well, w.r.t high-level functionals with vdW corrections, and closely reproduces interlayer distance and force constants of vdW interfaces and is widely used in phonon calculations [12].

AGF method requires both materials to be semi-infinite in the heat transport direction. Hence, vertical ITC of monolayer or few-layer 2D materials cannot be obtained directly due to their inherent atomically-thin body. Also, ITCs obtained by assuming an interface to a bulk like 2D material (see **Fig.7a.(i)**) are not applicable to monolayer 2D materials since the presence of out-of-plane (k_c) acoustic phonons in both mediums results in overestimation **(Fig.7a.(ii))** and it also excludes the physical effects of the substrate (suppressed ZA phonons in the 2D monolayer [13]). In **Fig.7b** we have proposed a method that applies AGF to monolayers while accounting for the substrate effects.

III. DISCUSSION

For each interface, PDOS of the isolated materials and phonon transmission spectrum is presented, these two plots along with the phonon band structures provide information about the contribution of each phonon mode to ITC. Transmission of same frequency phonons from multiple branches can yield a total transmission probability (T) greater than one. **Table.1** reports the final ITC values. Only the ITC of MoS₂-graphene lateral interface has been studied previously with MD [14], for which our method provides a perfect match.

Effect of interface geometry: Although graphene is the most thermally conductive material, its interface with MoS₂ suffers from low ITC compared to 1T-2H MoS₂ and MoS₂-WSe₂ interfaces **(Fig.12a)**. Also, the high ITC of MoS₂-WSe₂ interface, even with its high mismatch in PDOS **(Fig.5b)**, reflects the importance of interface atomic disorder in phonon transport. Therefore, seamless interfaces **(Fig.5a)** are most efficient in thermal transport.

Effect of vdW gaps: Vertical interfaces with vdW gap **(Figs.8,9,11)**, due to low interfacial atomic force constants, showed suppressed phonon transmission at all frequencies w.r.t covalently bonded lateral interfaces. This results in about a ~5x degradation in ITC for vertical interfaces. The presence of this vdW gap also degrades the ITC of MoS₂-Au interface w.r.t MoS₂-Ti **(see Table.1)**, both vertical metal interfaces, because of the presence of covalent bonds in the latter which increases the interlayer coupling.

Optical phonon transmission: MoS₂'s vertical and edge interfaces to metals (Au, Ti) show negligible transmission of optical phonons. **(Figs.6,9,10,11)** Low PDOS bandwidth of these metals results in no available mutual frequency optical phonon pairs at the interface **(Figs.6b,8b)**. In addition to this, the low ITC of vertical interfaces, because of the presence of vdW gap, makes them most susceptible to failure especially in FETs with drain hotspot. However, lateral interfaces to 1T MoS₂ and graphene show high optical phonon transmissions

(Figs.3d,4c), alleviating the problem of phonon accumulation, and making them more conducive to heat transmission. This is more notable in graphene interface, where coupling of MoS₂ low velocity optical phonons to graphene's high velocity acoustic phonons **(Fig.3d)** can further increase the rate of the heat flow away from the interface.

Vertical 2D interfaces to MoS₂ with h-BN substrate: A common trait across all vertical configurations with h-BN substrate is the transmission of low energy acoustic phonons between 0-7 meV **(Figs.8c,9c,11c)** which corresponds to out-of-plane $\Gamma - A$ phonon branches of bulk h-BN. These phonons can directly transmit through the thin 2D monolayer due to their long wavelength and high velocity, making them the main heat carriers in Metal-MoS₂-h-BN configurations.

Strong optical phonon couplings in h-BN-MoS₂-h-BN: Out-of-plane transmission of optical phonons in MoS₂ encapsulated with h-BN configuration **(Fig.8c)** is considerable due to the strong coupling of transverse optical modes of individual layers, resulting in a large increase of ITC as a function of temperature **(Fig.12b)**.This gives rise to critical temperatures at which conductance through the encapsulated MoS₂ exceeds that of the h-BN-MoS₂-metal interface, thereby drastically changing the thermal dissipation pathways in the device. **(Fig.12b)**

IV. SUMMARY

Thermal transport across all possible interfaces of 2D vdW materials is comprehensively studied using ab-initio atomistic Green's functions. We identify important factors that affect ITC which provides guidelines for 2D device engineers to maximize their device performance. For instance, we found that lateral "seamless interfaces" can achieve maximum ITC. Foundational analysis provided in this work can be further used to determine scaling, performance, and reliability limits of 2D devices.

ACKNOWLEDGMENT

This work was supported by the ARO (grant W911NF1810366).

REFERENCES

[1] P. Ajayan, P. Kim, and K. Banerjee, "Two-dimensional van der Waals materials," *Physics Today*, vol. 69, no. 9, pp. 38-44, 2016.

[2] W. Liu, J. Kang, D. Sarkar, Y. Khatami, D. Jena and K. Banerjee, "Role of metal contacts in designing high-performance monolayer n-type WSe₂ field-effect-transistors," *Nano Letters*, vol. 13, no. 5, pp. 1983-1990, 2013.

[3] W. Cao, J. Kang, D. Sarkar, W. Liu and K. Banerjee, "2D semiconductor FETs-projections and design for sub-10 nm VLSI," *IEEE Transactions on Electron Devices*, vol. 62, no. 11, pp. 3459-3469, 2015.

[4] D. Sarkar, X. Xie, W. Liu, W. Cao, J. Kang, Y. Gong, S. Kraemer, P. Ajayan and K. Banerjee, "A subthermionic tunnel field-effect transistor with an atomically thin channel," *Nature*, vol. 526, pp. 91-95, 2015.

[5] J. Jiang, J. Kang, W. Cao, X. Xie, H. Zhang, J. Chu, W. Liu and K. Banerjee," Intercalation Doped multilayer-graphene-nanoribbons for next-generation Interconnects," *Nano Letters*, vol. 17, no. 3, pp.1482-1488, 2017.

[6] D. Akinwande, N. Petrone and J. Hone, "Two-dimensional flexible nanoelectronics," *Nature Communications*, vol. 5, no. 5678, 2014.

[7] D. Sarkar, W. Liu, X. Xie, A. Anselmo, S. Mitragotri and K. Banerjee, "MoS₂ field-effect transistor for next-generation label-free biosensors," *ACS Nano*, vol. 8, no. 4, pp. 3992-4003, 2014

[8] W. Zhang, T. S. Fisher, and N. Mingo, "The atomistic green's function method: an efficient simulation approach for nanoscale phonon transport," *Numerical Heat Transfer Part B-Fundamentals*, vol. 51, no. 4, pp. 333-349, 2007.

[9] B. Radisavljevic, A. Radenovic, J. Brivio, V. Giacometti, and A. Kis, "Single-layer MoS₂ transistors," *Nature Nanotechnology*, vol. 6, no. 3, pp. 147–150, 2011.

[10] Atomistix Toolkit version 2018.06, Synopsys QuantumWise.

[11] J.P. Perdew and A. Zunger, "Self-interaction correction to density-functional approximations for many-electron systems," *Physical Rev. B*, vol. 23, no. 10, pp. 5048–5079, 1981.

[12] Z. Yan, L. Chen, M. Yoon and S. Kumar, "The role of interfacial electronic properties on phonon in two-dimensional MoS₂ on metal substrates," *ACS Appl. Mater. Interfaces*, vol. 8, no. 48, pp. 33299-33306, 2016.

[13] J. H. Seol, I. Jo, A. L. Moore, L. Lindsay, Z. H. Aitken, M. T. Pettes, X. Li, Z. Yao, R. Huang, D. Broido, N. Mingo, R. S. Ruoff and L. Shi, "Two-dimensional phonon transport in supported graphene," *Science*, vol. 328, no. 5975, pp. 213-216, 2010.

[14] X. Liu, J. Gao, G. Zhang and Y. Zhang, "MoS₂-graphene in-plane contact for high interfacial thermal conduction," *Nano Research*, vol. 10, no. 9, pp. 2944-2953, 2017.

[15] E. Yalon *et al.*, "Energy dissipation in monolayer MoS₂ electronics," *Nano Letters*, vol. 17, pp. 3429-3433, 2017.

Fig. 1. Applications of 2D Materials and Interface Thermal Challenges. (a) 2D FETs: (i) Schematic of a 2D FET. (ii) Schematic of a double gate FET with lateral metal contacts. (iii) Typical heat map of a FET with red color marking high temperature regions. High electric field at the drain increases the energy of electrons leading to generation of optical phonons. Accumulation of these low velocity optical phonons in the channel creates drain hot spots. MoS₂ channel breaks down when drain hot spot exceeds MoS₂'s oxidation threshold at around 380°C. [15] **(b) Flexible 2D Electronics:** Poor thermal stability of flexible substrates poses critical reliability issues and limits the device performance. **(c) Monolithic 3D Integration:** Thermal transport across vertical interfaces becomes even more crucial for vertically stacked 3D integrated designs as the power density increases further. **(d) Interfaces to 2D materials.** (i) MoS₂-Metal (2D-3D) edge interface, (ii) MoS₂-2D lateral interface, (iii). Dielectric-MoS₂-Dielectric vertical interface, (iv) Metal-MoS₂-Dielectric interface. **(e) Anisotropic thermal conductivity in 2D materials.** 2D materials have high in-plane thermal conductivity w.r.t common semiconductors, whereas their out-of-plane thermal conductivity is low.

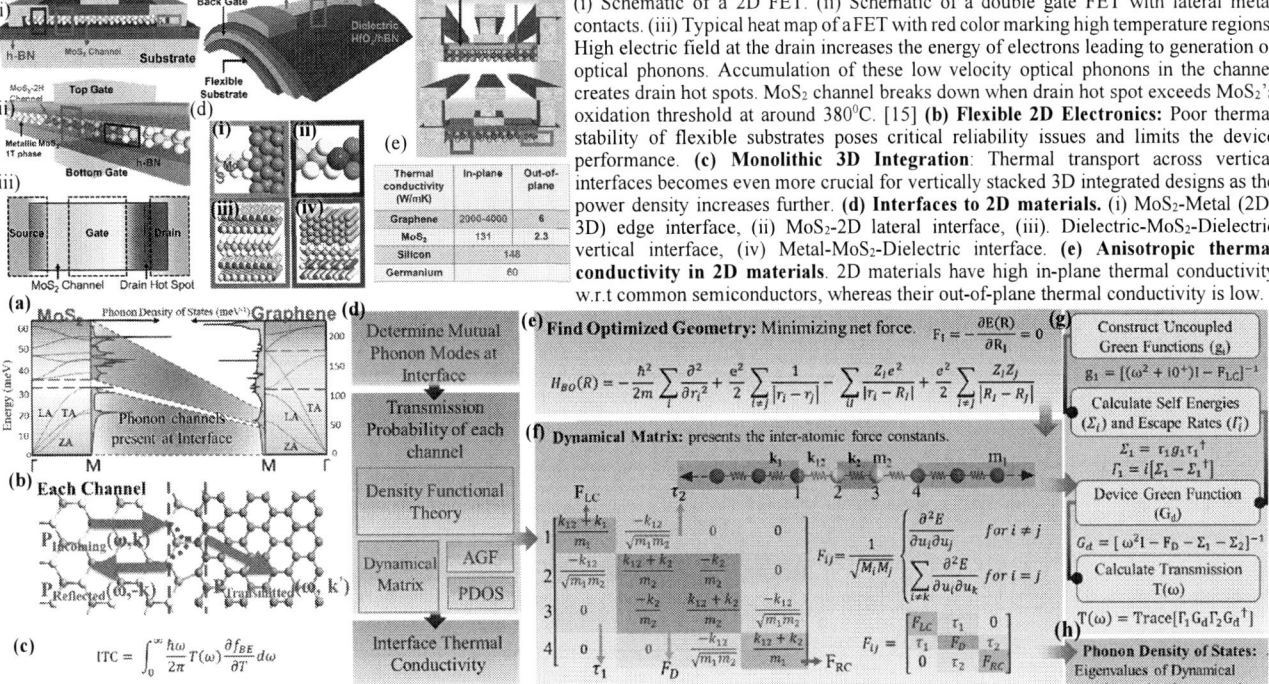

Fig.2. Mesoscopic Heat Transport. (a) Overlap of the mutual frequency phonon pairs of two different materials (MoS₂/Graphene) at their interface. Phonons with the same frequency couple together and conduct energy. **(b)** A two-phonon scattering event at the interface. Transmission probability of this event is determined via Atomistic Green's Function (AGF) method **(c)** Interfacial Thermal Conductivity (ITC) is calculated by summing over contributions of all phonon modes. $T(\omega)$ is the transmission spectrum and f_{BE} is Bose-Einstein distribution function. **(d)** Framework for calculation of ITC and transmission spectrum. **(e)** Interface optimized geometry is obtained using DFT. $E(R)$ is the eigenenergy of Born-Oppenheimer Hamiltonian H_{BO}. R_I and r_i refer to position of ions and electrons respectively and Z_I is the charge of the ion. First term in H_{BO} represents kinetic energy of the electrons, the second term is the electron-electron interaction, the third term denotes the electron-ion interaction and the fourth term is the ion-ion interaction. First derivative of $E(R)$ w.r.t ion position is the force (F_I) exerted on the ion. **(f)** Dynamical matrix is the second-order partial derivatives of $E(R)$ w.r.t displacement of ions scaled by the atomic masses. Entries are proportional to spring constant (k) between two atoms. Dynamical matrix of a 1D atomic chain is presented considering only the nearest neighbor interaction. This matrix is the input of **(g)** AGF method and **(h)** Phonon DOS (PDOS) calculations.

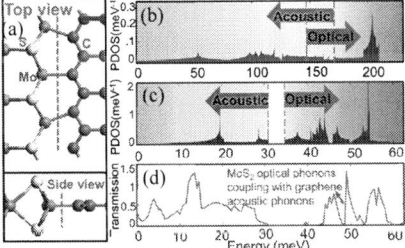

Fig.3. MoS₂-Graphene Lateral Interface. (a) DFT optimized geometry **(b)** Graphene PDOS **(c)** MoS₂ PDOS **(d)** Transmission Spectrum: Graphene's high PDOS energy range provides mutual phonon modes for all MoS₂ phonons. Heat is mainly carried by acoustic phonons due to their high occupation probability at room temperature. Average acoustic phonon transmission rate is 0.2 (Three acoustic phonon branches exist at low acoustic frequencies; hence maximum transmission is 3 **(see Fig.2a)**). Transmission rate for MoS₂ optical phonons is of the same order. This is notable since MoS₂'s low velocity optical phonons couple to graphene's high velocity acoustic phonons, signaling the promising thermal properties of graphene-MoS₂ interface.

Fig.4. 1T MoS₂ Metallic Phase – 2H MoS₂ Semiconducting Phase, Lateral Homogeneous Interface. (a) DFT optimized geometry showing the relaxation of 1T phase to 1T' at the interface. **(b)** 1T MoS₂ and 2H MoS₂ PDOS. Negative phonon modes found in 1T phase (not shown) show its inherent stress and the metastable nature. **(c)** Transmission Spectrum: misalignment of phonon gaps between their PDOS plots suppresses the phonon transmission in the 16-36 meV range. Transmission probabilities for phonons below 15 meV (acoustic phonons) and above 35 meV (optical phonons) are high. For low energy (high occupation) acoustic phonons, transmission is higher than that of graphene; hence ITC of this interface can exceed graphene. Notably for low energy optical phonons (~35-42 meV), complete transmission (T=2) of both low energy optical branches of 2H MoS₂ (TO₁, LO₁) is observed. Also, non-zero transmission for frequencies above 50 meV (where no mutual frequency phonon pairs are available) is justified since due to relaxation of 1T phase to 1T' at the interface, actual PDOS plots would shift to higher energies.

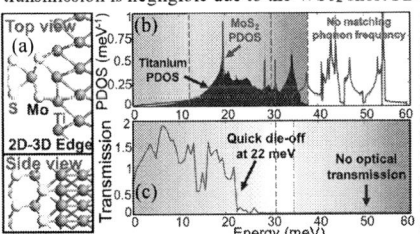

Fig.5. MoS₂-WSe₂ Lateral Heterogeneous Interface. (a) DFT optimized geometry **(b)** MoS₂ and WSe₂ PDOS **(c)** Transmission Spectrum: Acoustic phonons below 17 meV have high transmission (T). For frequencies between 3-13 meV, T=3 indicates that all 3 acoustic branches fully transmit, signaling the high heat carrying capacity of the interface. This result shows seamless interfaces (interfaces with no structural atomic disorder) are most efficient in interface thermal transport. At optical frequencies, transmission is negligible due to the WSe₂ short PDOS energy range and misalignment of phonon gaps

Fig.6. MoS₂-Titanium Edge (100-plane) Interface (a) DFT optimized geometry **(b)** MoS₂ and Ti (alpha) PDOS **(c)** Transmission Spectrum: Phonon transmission at MoS₂ optical frequencies is suppressed by low bandwidth of titanium's PDOS (0-38 meV) and the positioning of MoS₂ phonon gap (31-35 meV). While transmission is high for low energy acoustic phonons its quick die-off at frequencies above 23 meV can lead to accumulation of high energy phonons at the interface that cause hot spots and reliability issues.

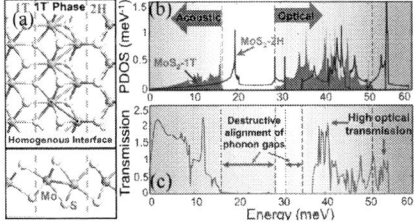

978-1-7281-1988-5/18 $31.00 © 2018 IEEE

Fig.7. Modeling interfacial thermal conductance for vertical interfaces of monolayer 2D materials. (a) AGF limitation and the invalid approach:(i) Vertical interface between bulk MoS₂ and semi-infinite Titanium. Vertical ITC of this configuration is not the same w.r.t interface with monolayer MoS₂ due to the absence of out-of-plane phonon modes in the latter. (ii) Phonon band structure of bulk MoS₂ highlighting the missing phonon branches (Γ-A) in monolayer 2D materials. (iii) Monolayer and bulk MoS₂ PDOS. Bulk MoS₂ acoustic phonon PDOS is almost twice that of the monolayer. Also, out-of-plane phonon modes are normal to the vertical interface and will have high transmission rates. These

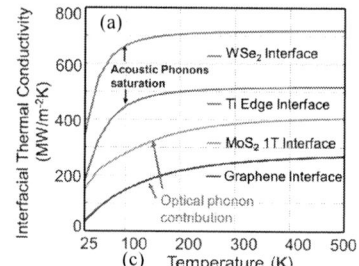
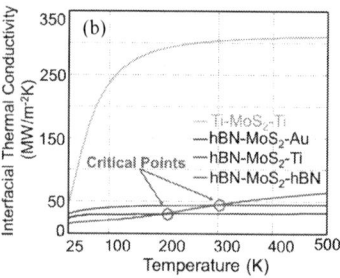

factors result in different values for vertical ITCs of monolayer 2D w.r.t bulk 2D materials. **(b) Proposed method for calculating vertical ITC of monolayer 2D materials.** (i) In self-heating modeling, ITC of each interface is necessary to determine the dissipated heat in each direction. (ii) By sandwiching a 2D monolayer between two bulk materials, total thermal conductivity (G_{tot}) through the 2D layer is calculated. In this setup, total thermal resistivity (G_{tot}^{-1}) is the sum of individual interface thermal resistances (G_{int1}^{-1}, G_{int2}^{-1}). (iii) To find individual ITC's from G_{tot} for the heterogenous configuration, we approximate the ratio of ($G_{int1}^{-1}/G_{int2}^{-1}$) from the individual thermal conductivities obtained when MoS₂ is sandwiched between each of these materials in two separate homogeneous configurations. In this way, using AGF framework, we calculate vertical ITC's of monolayer 2D materials in presence of substrate.

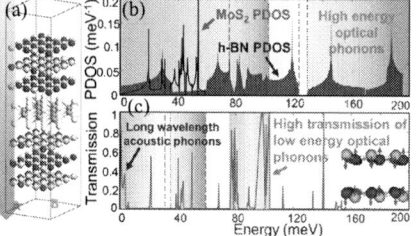

Fig.8. h-BN-MoS₂-h-BN Interface. (a) DFT optimized geometry **(b)** Hexagonal-BN and MoS₂ PDOS **(c)** Transmission Spectrum: Interface is weakly bonded by van der Waals forces that result in low inter-atomic force constants at the interface. The atomically thin MoS₂ acts as a scattering barrier between the two bulk like h-BN dielectrics. In this configuration, a phonon needs to first transmit from top h-BN to MoS₂ and then transmit from MoS₂ to bottom h-BN (3 phonon scatterings). As a result, transmission spectrum becomes heavily suppressed and selective for all frequency ranges. However, for acoustic phonons below 7 meV high transmission is found. These are the long wavelength (high velocity) acoustic phonons that couple directly from top to bottom h-BN. Transmission is negligible for high energy optical phonons (130-195 meV) but for h-BN low energy optical branches (80-100 meV) it becomes considerable. These optical branches (TO phonons) correspond to the out-of-plane vibrations of h-BN atoms in unit cell (**see Fig.8c**). High transmission is also observed for low energy MoS₂ optical frequencies (~40-50 meV). Contribution of optical phonons to ITC in this configuration is significant and cannot be overlooked.

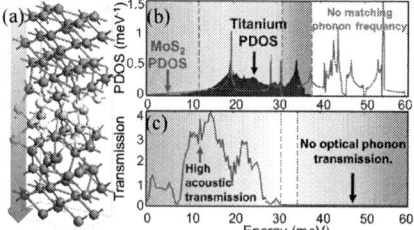

Fig.10. Ti-MoS₂-Ti Interface (a) DFT optimized geometry **(b)** Ti (alpha) and MoS₂ **(c)** Transmission Spectrum: Titanium forms covalent bonds with both interfaces of MoS₂. High interatomic force constants drive MoS₂ to a hybrid phase. Acoustic phonons between 10-20 meV have high transmission. Result of this simulation is needed to extract the thermal conductance of MoS₂-Ti and MoS₂-h-BN interfaces from configuration in **Fig.9**.

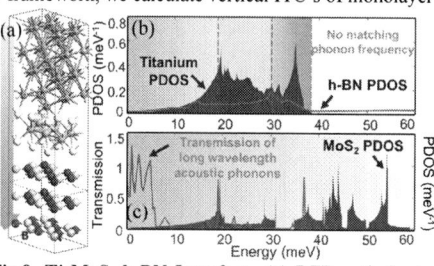

Fig.9. Ti-MoS₂-h-BN Interface. (a) DFT optimized geometry **(b)** Ti (alpha) and h-BN PDOS **(c)** Transmission Spectrum: Titanium forms covalent bonds with MoS₂, resulting in high interatomic force constants at the MoS₂-Ti interface. Atomically thin MoS₂ can be treated as scattering region between titanium and h-BN. Titanium has a limited PDOS energy range (up to 38 meV) w.r.t h-BN (up to 195 meV) hence, there is no transmission for frequencies above 38 meV. Weakly bonded vdW gap at MoS₂-h-BN interface suppresses most of the phonon couplings for energies below 38 meV. Acoustic phonons below 7 meV have high transmissions. These phonons can penetrate through the van der Waals gap owing to their high velocity and long wavelength.

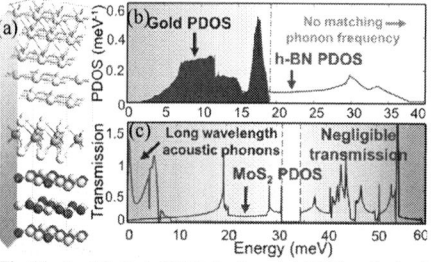

Fig.11. Au-MoS₂-h-BN Interface (a) DFT optimized geometry **(b)** Gold and h-BN PDOS. **(c)** Transmission Spectrum: Gold does not form covalent bonds with MoS₂ and van der Waals gaps are still present at the interface. Also, Gold PDOS range is very small (up to 20 meV) compared to h-BN (up to 200 meV). These should lead to a lower ITC w.r.t Titanium interface. Long wavelength acoustic phonons below 7 meV are responsible for heat conduction across the interface. Transmission of these high-velocity (long wavelength) phonons through the van der Waals gap is a repeating characteristic of vertical interfaces of 2D h-BN.

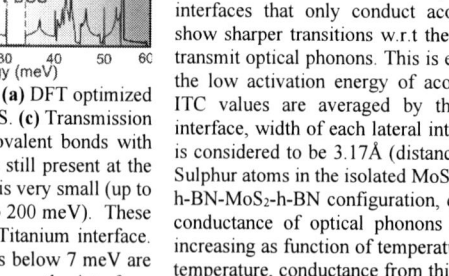

Fig.12. Interfacial Thermal Conductivity vs Temperature. (a) Lateral and Edge Interfaces. **(b)** Vertical Interfaces. ITC is calculated for different temperatures using the formula in **Fig.2c**. As temperature increases more phonon modes become occupied and thermal conduction increases. ITC starts to saturate, when all available phonon modes have been occupied. Therefore, interfaces that only conduct acoustic phonons show sharper transitions w.r.t the interfaces that transmit optical phonons. This is expected due to the low activation energy of acoustic phonons. ITC values are averaged by the area of the interface, width of each lateral interface to MoS₂ is considered to be 3.17Å (distance between two Sulphur atoms in the isolated MoS₂ unit cell). For h-BN-MoS₂-h-BN configuration, due to the high conductance of optical phonons ITC keeps on increasing as function of temperature. Near room temperature, conductance from this configuration surpasses conductance through the h-BN-MoS₂-Metal configurations, an important hint for thermal management of 2D devices that incorporate h-BN either as dielectric or substrate.

Table.1. Room temperature(T=300K) ITC values.

Room temperature	Lateral			Edge	Vertical Interfaces			
Interface	MoS₂-Graphene	MoS₂ 1T-2H	MoS₂-WSe₂	MoS₂-Ti Edge	hBN-MoS₂-hBN	Ti-MoS₂-Ti	Ti-MoS₂-hBN	hBN-MoS₂-Au
ITC (MW/m⁻²K)	244.9	384.9	714	511.4	44.87	300.5	44.15	30.57

MoS₂-Metal ITC	342.7	77.04
MoS₂-hBN ITC	50.68	~50

The individual ITC's of Ti-MoS₂-h-BN and Au-MoS₂-h-BN are extracted using the proposed method in **Fig.7b**.

Computational Design of Silicon Contacts on 2D Transition-Metal Dichalcogenides: The Roles of Crystalline Orientation, Doping Level, Passivation and Interfacial Layer

Xiaolei Ma[1,2], Zhiqiang Fan[2], Jixuan Wu[1,2], Xiangwei Jiang[2,+], Jiezhi Chen[1,*]

[1]School of Information Science and Engineering, Shandong University, Qingdao, P. R. China, *email: chen.jiezhi@sdu.edu.cn
[2]Institute of Semiconductors, Chinese Academy of Sciences, Beijing, P. R. China, +email: xwjiang@semi.ac.cn

Abstract—Systematic numerical simulations based on density functional theory (DFT) and non-equilibrium Green's function (NEGF) formalism have been carried out for comprehensive understanding of the physical properties of silicon (Si) contacts on monolayer transition metal dichalcogenides (TMDs). The effects of different contact crystalline orientations including Si (001), (110) and (111), Si doping levels, possible surface passivation such as H- and F-, as well as interfacial layer (IL) engineering using BN and Graphene are thoroughly discussed. On the one hand, it is found that the contact properties of different crystalline orientations follow similar trend, and the doping modulation of the Schottky barrier height (SBH) remains inappreciable in a practical range of doping level. On the other hand, H- and F- passivation are found to be effective ways to diverge the intrinsic contact into *n*- and *p*- type contacts, respectively. In addition, it is found that surprisingly good *p*-type contact with vanishing *p*-SBH could be formed by using monolayer BN as the IL.

1. Introduction

As traditional Si channel metal-oxide-semiconductor (MOS) device approaches its scaling limit, alternative materials are needed in the post-Si era. In past decade, low-dimensional materials such as carbon nanotube, graphene, transition-metal dichalcogenides (TMDs) emerged with promising electronic properties for possible replacements of silicon (Si). Single-layer MoS_2 based field-effect transistors have been fabricated and intensively investigated [1]. Ultimately scaled gate length down to 1 nm has been demonstrated utilizing atomically thin MoS_2 [1] channel and carbon nanotube gate. New operating principles have been proposed to overcome the power scaling limit, and recently been combined with 2D semiconductors, such as MoS_2 based tunnel FETs, negative capacitance FETs, as well as the graphene and carbon-nanotube based Dirac source FETs. Despite the excellent intrinsic electronic properties of the above ultimate-scaled 2D semiconductors and fruitful physics regarding the beyond-CMOS working principles, electrical contact problem remains unavoidable and most likely the common bottleneck of all 2D semiconductor devices [2]. Indeed, it is the contact determines carrier injection and connect the 2D transistor to 3D interconnect in an IC chip volume. Since such, tremendous efforts have been made to understand the contact physics so that engineering the carrier injection to 2D transistor. There were numerous previous studies [2-6] focusing on traditional metal contacts, suggesting appreciable metals for certain 2D semiconductors (Table.1), distinguishing van der Walls contacts from "strong coupled" ones, as well as proposing interfacial layer (IL) to avoid fermi level pinning (FLP) [7, 8]. Nevertheless, in a sense

of Si compatible fabrication, it would be extremely interesting and useful to understand and engineer the Si-based contacts to the above 2D semiconductors. Unfortunately, little is known so far. In a previous attempt, Tang et al carried out first-principle simulation on the Si (111) contact to MoS_2, concluding that the Schottky barrier height (SBH) to MoS_2 channel can be modulated effectively by doping Si source and drain [9]. However, the effect of broken bonds in bare Si surface have not been considered, which might change the conclusion since the un-passivated dangling bonds could strongly influence the band structure as well as the location of the fermi level.

In this paper, we report a comprehensive study on the Si contacts to monolayer TMDs, taking into account the previously unexplored effects of the crystalline orientations, Si surface passivation and interfacial layers. By fully *ab initio* simulations based on DFT, different conclusion is drawn regarding the doping modulation on Si-SBHs, and new design options through surface passivation and interfacial layer are provided.

2. Modeling and simulation approach

Si (001), (110), (111) contacts on monolayer MoS_2 in a ultra-scaled FET sketched in Fig. 1(b) are targeted in this study. Numerical simulations are carried out by Atomistix Toolkit (ATK2017), which is based on DFT in combination with NEGF formalism. The flow chart of the device simulation is shown in Fig. 2, where the quantum transport property and transfer characteristics of a FET can be eventually output. It should be noted that one can tell the a priori contact property (SBH etc.) by calculating the band structure of the composite system as shown in Fig. 3. The Si-TMDs contact structures in Fig. 3 are constructed with minimum strain, so that the band gap of TMDs will not be affected by lattice mismatch. The geometry optimization is performed using generalized gradient approximation (GGA) functional and Perdew Burke Ernzerhof variant (PBE) pseudopotential, until total energy is converged to less 10^{-5} eV, and the maximum force on each atom is less than 0.01 eV/Å.

To avoid the underestimation of LDA on semiconductor band gaps, GGA PBE with DFT-1/2 is chosen in this work for the electronic structure calculations and transport simulation of a junction device. The calculated band gaps E_g=1.8 eV of monolayer MoS_2 and E_g=1.6 eV of monolayer WSe_2 are both consistent with experimental results [10, 11]. The SBH of Si (111)-MoS_2 contact system extracted from band structure projection is found to be 0.62 eV as a *n*-type contact, which is consistent with previous theoretical simulation [9]. Hence, GGA PBE DFT-1/2 can give accurate band gaps for TMDs and right SBH for Si-MoS$_2$ contact and is used throughout this simulation.

3. Simulation results and discussion
A. SBH of Si-TMDs contact system

Fig. 4 shows the band structures of Si-MoS$_2$ contact system with different Si orientations: (001), (110) and (111). n-type SBH is defined as the potential difference between the fermi energy of the contact system and the conductance band minimum (CBM) of MoS$_2$, and p-type SBH is defined as the potential difference between the fermi energy of the contact system and the valence band maximum (VBM) of MoS$_2$. For example, in Si(110)-MoS$_2$ interface, the system fermi level is close to the CBM of MoS$_2$, indicating that Si-MoS$_2$ performs as n-typed contact with 0.4 eV SBH. Similarly, SBH of Si(001)- and Si(111)-MoS$_2$ contact are estimated as 0.68 eV and 0.62 eV, respectively. Considering that the electronic properties of MoS$_2$ is easily distorted in the bare Si contact due to the dangling bonds, we use hydrogen (H) and fluorine (F) to passivate Si surface. Our calculating results show that, the passivation can effectively adjust MoS$_2$ fermi level to be closer to MoS$_2$ CBM by H-passivation or closer to MoS$_2$ VBM by F-Passivation. Calculated SBH are summarized in Figs. 4(d), (e) and (f). Obviously, SBH will be much lowered by using H-passivation, and $\Phi_{SB,N}$ in Si-MoS$_2$ contact system will drop to 0.16~0.21 eV. More importantly, for the first time, it is found that F-passivation on Si surface changes the original n-type Si-MoS$_2$ contact to p-type contact and SBH is extremely low. Specially, SBH is almost zero in Si(001)-MoS$_2$ contact. To confirm our findings, PDOS are calculated and Φ_{SB} is then estimated by comparing the energy difference between E_F and E_c (E_v) of MoS$_2$ (Fig. 5) and similar results are obtained. Si-MoS$_2$ contact is n-type contact if there is no surface treatment. Then, E_F will be shifted towards CBM by using H-passivation and E_F will be shifted to VBM by using F-passivation. Because F-passivation can change the original n-type Si-MoS$_2$ contact to p-type contact, this finding will provide a new way to design Si-TMDs compatible devices with vanished p-type Schottky barriers.

Next, Si doping effects on SBH in Si-MoS$_2$ contact are investigated. Fig. 6 summarizes the SBH change of MoS$_2$-Si contact with different doping concentrations. It is found that doping effects is weak until the doping concentration is as high as 1E22 e/cm^3. In the case of no passivation, heavy doping can lower the SBH to ~0.2eV in Si(110)-MoS$_2$ and Si(111)-MoS$_2$. However, in H-passivated Si-MoS$_2$, p-type heavy doping will increase the SBH on the contrary, especially in Si(111)-MoS$_2$ contact system. As to F-passivated Si-MoS$_2$, though the doping effect is almost ignorable, p-type heavy doping could obtain vanished SBH in Si(110)-MoS$_2$ and Si(111)-MoS$_2$.

To understand whether our finding is a special case or if it could be applied to other TMD materials, we studied other six TMDS and same results are obtained. H-passivation will enhance the n-type doping level (E_F is closer to CBM) and F-passivation will change the n-type contact to the p-type contact. Furthermore, it is found that (110) orientated Si is helpful to achieve the lowest SBH of Si-MX_2 contact systems for most

TMDs, except Si-MoSe$_2$/Si-WSe$_2$/Si-MoS$_2$ with H-passivation and Si-MoS$_2$ with F-passivation (Fig. 7). Then, the band gap E_g and the effective mass (m_e, m_h) as well as SBH of monolayer MX_2 are summarized in Table 2, and calculated results of the Schottky currents are shown in Fig.8. It is observed that the Schottky current is higher in the contact systems of Si-MS_2 and Si-MSe_2 (M=Mo, W), which can be explained by their lower SBH (~0.1-0.2 eV) and relatively heavy m_e (~0.5-0.6 m_0). Finally, I-V curves in the constructed monolayer MoS$_2$ FET with different Si surface treatments are calculated (Fig. 9). Obviously, the channel current (I_D) can be largely improved with H-passivation. Again, it is confirmed that F-passivation treatment can change n-type Si-MoS$_2$ contact to p-type contact.

B. Effects of interfacial layer

Inserting an interfacial layer in 2D material contact with metal has been demonstrated as an effective approach to lower the SBH. For example, Farmanbar et al. proposed a new way to achieve low work function metal contacts to TMDs [7-8], wherein a monolayer h-BN is inserted between metal surface and MoS$_2$ [7]. To check if this scheme can be applied in our Si-MoS$_2$ contact system, we theoretically calculated the SBH of Si-MoS$_2$ with inserted h-BN monolayer between the Si surface and MoS$_2$. It is found that different Si orientations had no effect on our results. As shown in Fig. 10(b), E_F is close to the original E_v of Si-MoS$_2$ [compared to Fig. 5(f)] and the system is a p-type contact. In Fig. 10(c), we can find that the SBH of Si(110)-BN-MoS$_2$ contact is 0 eV [compared to Fig. 4(b)], which indicates that inserting a BN layer between the Si and MoS$_2$ can form p-type vanishing Schottky barrier. Then, we inserted monolayer graphene as the interfacial layer for comparison and different results are obtained. As shown in Fig 10(e) and (f), graphene interfacial layer has no effect on the Schottky barrier. As we know, in the contact system of metals and TMDs, lower p-type SBH can be obtained if the metal has high work function (WF). Similarly, vanishing Schottky barrier in Si-BN-MoS$_2$ can be understood because h-BN on Si will increase Si WF of ~2.1eV, while Si WF does not change in Si-Graphene-MoS$_2$.

4. Conclusions

Si contacts on 2D TMDs are systematically investigated in this work, with main focus on Schottky barrier engineering by including the impacts of crystalline orientations, doping levels, surface passivation and interfacial layers. It is found that, for most TMDs, (110)-orientated Si could be useful to obtain low SBH in Si-MX_2 contact systems, and the doping effects on SBH is weak until the doping level is extremely high. More importantly, it is found that n- and p- type contacts can be realized by H- and F- passivation, respectively. Furthermore, with monolayer BN as the interfacial layer, surprisingly good p-type contact with vanishing p-SBH can be formed.

Acknowledgment: This work is supported by MOST of China (2016YFA0201802, 2018YFA0306101), and the National Natural Science Foundation of China (11574304, 11774338).

Reference: [1] B. Radisavljevic, et al., Nature nanotech. vol. 6, p.147, 2011; [2] J. Kang, et al., PRX, vol. 4, 031005, 2014; [3] I. Popov, et al., PRL, vol.108, 156802, 2012; [4] Z.-Q. Fan et al., PRB, vol.96, 165402, 2017; [5] Zhong H, et al., Sci. Rep., vol. 6, 21786, 2016; [6] Shi X, et al. Small, vol. 14, 1704526, 2018; [7] Farmanbar M, et al., PRB, vol. 91, 161304, 2015. [8] Su J, et al., PCCP, vol. 18, p. 16882, 2016; [9] Y.-T. Tang et al., IEDM, p. 14.3, 2016; [10] K. F. Mak, PRL, vol. 105, 136805, 2010; [11] D. Vob, PRB, vol. 60, 14311, 1999.

MoS₂ contact	Reported work							This work
Metallic	Ti	Ag	In	Ni	Au	Pt	Si(111)	Si(001),(110),(111)
Experient	0.3-0.35[2]	-	-	0.66[6]	-	-	-	
DFT calculation	0.35[2][5] N-type	0.21[5] N-type	0.47[5] N-type	0.633[5] N-type	0.62[2] N-type	0.520[5] P-type	0.26 N-type 0.20 P-type (w/o surface treatment)[9]	surface treatment: H/F passivation & interfacial layer

Table. 1. A summary of previous work on Schottky barrier height (SBH) of MoS₂-Metalic contacts, including the simulation data and experiment data.

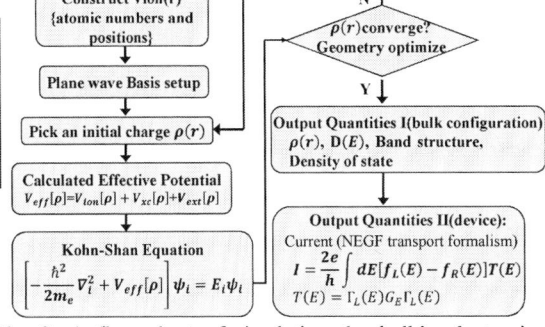

Fig. 2. A flow chart of simulating the bulk's electronic structures and the 4nm channel length MoS₂ FET simulation.

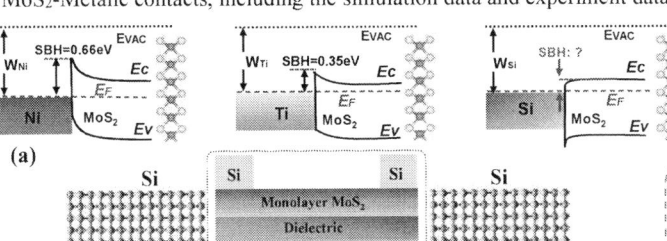

Fig.1. (a) Band structures of MoS₂-Ni and MoS₂-Ti are shown as examples; (b) A schematic of 4nm channel length monolayer MoS₂ FET with Si-MoS₂ contacts.

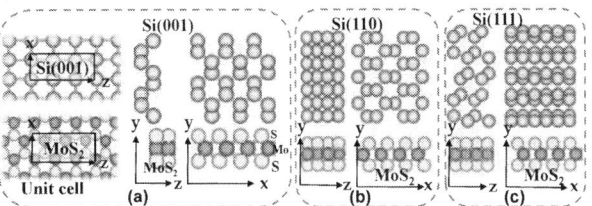

Fig. 3. Optimized geometries of Si-MoS₂ contacts, (a) Si(001)-MoS₂. (b) Si(110)-MoS₂ and (c) Si(111)-MoS₂ in different views.

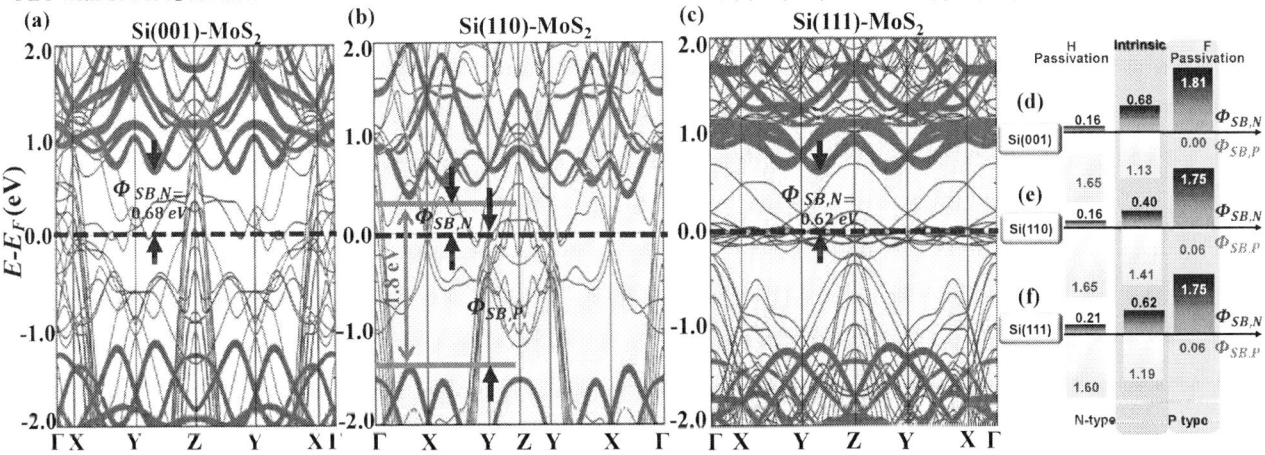

Fig. 4. Band structures of (a) Si(001)-MoS₂, (b) Si(110)-MoS₂, (c) Si(111)-MoS₂ contact. And Schottky barrier change of (d) Si(001)-MoS₂, (e) Si(110)-MoS₂, (f) Si(111)-MoS₂ contact with H-passivation, no passivation, and F-passivation treatment, respectively.

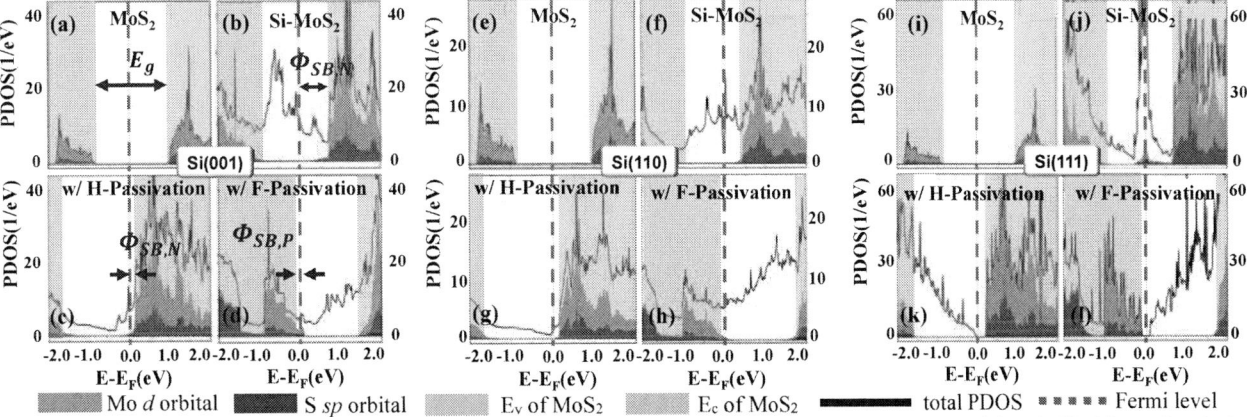

Fig. 5. Partial density of states (PDOS: DOS on specified atoms and orbits, for example, d orbital on Mo) of (b)-(d) Si(001)-MoS₂, (f)-(h) Si(110)-MoS₂, (j)-(l) Si(111)-MoS₂ contacts with H-passivation, intrinsic, F-passivation, respectively. (a)(e)(i) are PDOS of MoS₂.

978-1-7281-1988-5/18 $31.00 © 2018 IEEE

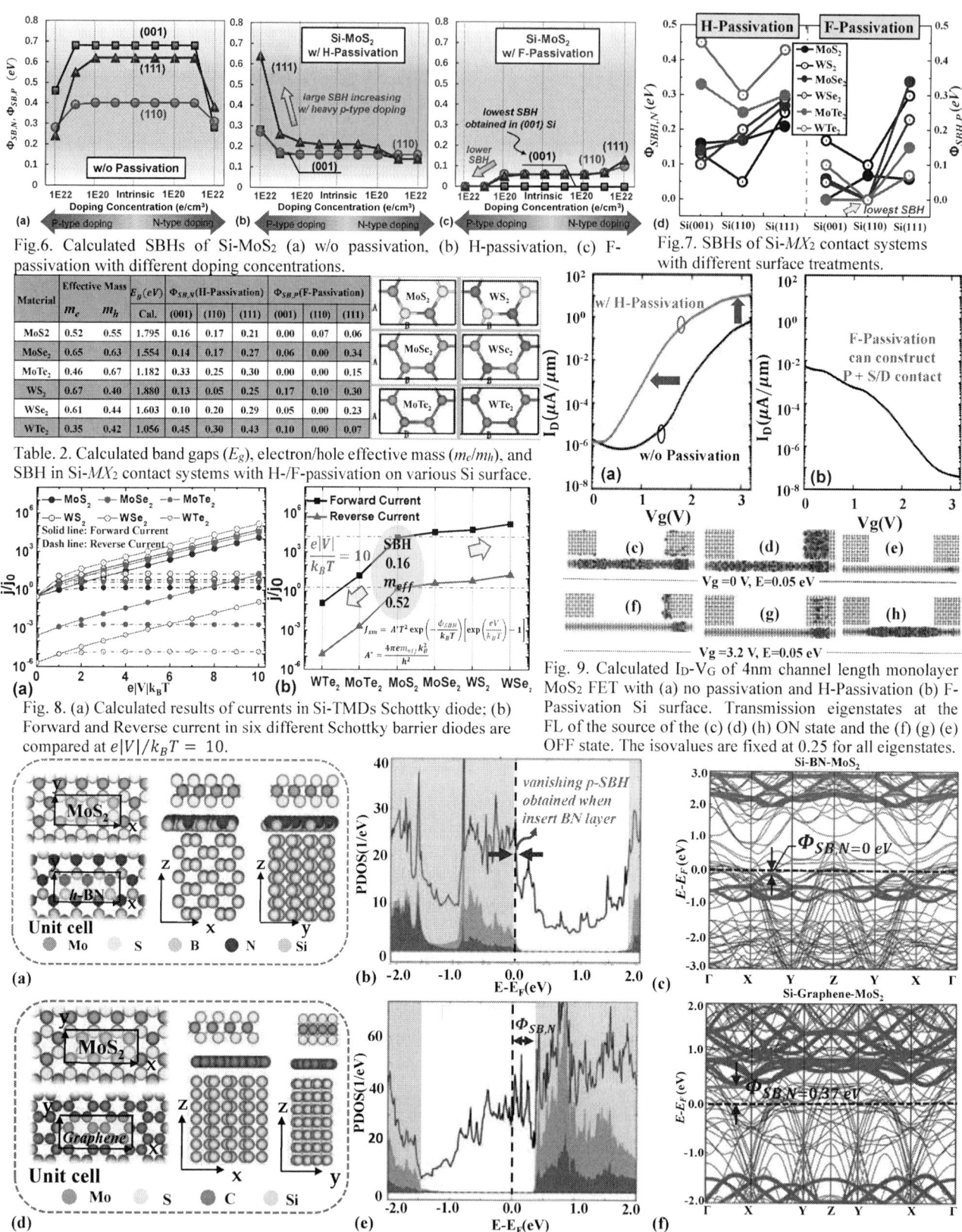

Fig.6. Calculated SBHs of Si-MoS$_2$ (a) w/o passivation, (b) H-passivation, (c) F-passivation with different doping concentrations.

Fig.7. SBHs of Si-MX_2 contact systems with different surface treatments.

Material	Effective Mass		E_g(eV)	$\Phi_{SB,N}$(H-Passivation)			$\Phi_{SB,P}$(F-Passivation)		
	m_e	m_h	Cal.	(001)	(110)	(111)	(001)	(110)	(111)
MoS2	0.52	0.55	1.795	0.16	0.17	0.21	0.00	0.07	0.06
MoSe$_2$	0.65	0.63	1.554	0.14	0.17	0.27	0.06	0.00	0.34
MoTe$_2$	0.46	0.67	1.182	0.33	0.25	0.30	0.00	0.00	0.15
WS$_2$	0.67	0.40	1.880	0.13	0.05	0.25	0.17	0.10	0.30
WSe$_2$	0.61	0.44	1.603	0.10	0.20	0.29	0.05	0.00	0.23
WTe$_2$	0.35	0.42	1.056	0.45	0.30	0.43	0.10	0.00	0.07

Table. 2. Calculated band gaps (E_g), electron/hole effective mass (m_e/m_h), and SBH in Si-MX_2 contact systems with H-/F-passivation on various Si surface.

Fig. 8. (a) Calculated results of currents in Si-TMDs Schottky diode; (b) Forward and Reverse current in six different Schottky barrier diodes are compared at $e|V|/k_BT = 10$.

Fig. 9. Calculated I$_D$-V$_G$ of 4nm channel length monolayer MoS$_2$ FET with (a) no passivation and H-Passivation (b) F-Passivation Si surface. Transmission eigenstates at the FL of the source of the (c) (d) (h) ON state and the (f) (g) (e) OFF state. The isovalues are fixed at 0.25 for all eigenstates.

Fig. 10. (a)(d) Different views of the optimized structures to the Si-BN-MoS$_2$ interface and Si-Graphene-MoS$_2$ interface, (b) PDOS: DOS on Mo element and d orbital, (c) Band structure of Si-BN-MoS$_2$ interface and (e)(f) Si-Graphene-MoS$_2$ interface. The red color measures a projection of the band structure on MoS$_2$ orbitals.

First Principles Study of Memory Selectors using Heterojunctions of 2D Layered Materials

Linsen Li[1,3], Blanka Magyari-Köpe[3], Ching-Hua Wang[3], Sanchit Deshmukh[3], Zizhen Jiang[3], Haitong Li[3], Yi Yang[1], Huanglong Li[2], He Tian[1], E. Pop[3], Tian-Ling Ren[1,*], H.-S. Philip Wong[3,*]

[1]Institute of Microelectronics & Beijing National Research Center for Information Science and Technology (BNRist), Tsinghua University, Beijing 100084, China, *Email: RenTL@tsinghua.edu.cn
[2]Department of Precision Instrument, Tsinghua University, Beijing 100084, China
[3]Department of Electrical Engineering, Stanford University, CA 94305, USA, *E-mail: hspwong@stanford.edu

Abstract— **Two-dimensional (2D) tunnel heterojunctions with an H-shaped energy barrier could serve as ultrathin memory selectors with good symmetry, non-linearity, and high endurance. Atomically thin 2D layered materials can potentially deliver high on-state tunneling current density. We explore the design space for H-shaped memory selectors using heterojunctions of 2D layered materials, using physical modeling and first principles density functional theory (DFT) quantum transport simulations. The difference between simulations and the few existing experiments is also discussed. A selector must be designed to suit the resistive memory (1R) characteristics. We evaluate the H-shaped selector in the one-selector-one-resistor (1S1R) configuration and provide design guidelines for the heterojunction (metal/nL hBN/nL 2D material/nL hBN/metal) design to match with the 1R characteristics.**

I. INTRODUCTION

Two-terminal selector devices are required to suppress the unwanted leakage paths of cross-point resistive memory arrays by adding non-linearity to the memory cell [1-3]. Unlike many conventional selectors, a tunnel selector based on quantum tunneling (an electronic process) is expected to have longer endurance than other selectors that are based on the physical motion of atoms or ions [4]. Moreover, the use of atomically thin 2D materials should enable large tunneling current density. The 2D heterojunction has an atomically sharp interface which can lead to high non-linearity with a proper design [5,6]. Some simple symmetric selector structures for 1S1R memory arrays are shown in Fig. 1, but they cannot reach sufficient non-linearity.

Here, we study the H-shaped barrier selector with the salient feature that there are insulator layers between the semiconductor in the middle and the metal electrodes on either side, as shown in Fig. 2(a). The corresponding band diagram is shown in Fig. 2(b), resembling the letter 'H'. This design reduces the leakage current with the high insulator barriers but maintains high on-state current with strong tunneling at high bias [7,8].

Our simulated selector with the structure Au/1L hBN/1L WS$_2$/1L hBN/Au shown in Fig. 2(c) can achieve a non-linearity over 500 with an on-state current density of 171 MA/cm^2 at supply voltages of 2 V. (In this paper, "nL" signifies n layers of the 2D material.) The nonlinearity of 1S1R is over 100, which can meet the requirement for a Tbit-class 3D RRAM [9]. We

explore the design space of the selector with the H-shaped barriers shown in Fig. 2(d) for a variety of thickness and material combinations for the insulator, semiconductor and metal layers.

II. SIMULATION APPROACH

The flowchart of the simulation is shown in Fig. 3. The atomic structure for the 2D heterojunction was relaxed with the Vienna Ab initio Simulation Package (VASP) [10] and Atomistix Toolkit version 2017 (ATK2017). For open systems, the density matrix is computed using non-equilibrium Green's functions (NEGF) [11]. The transmission coefficients are calculated using ATK2017 within the framework of DFT [12]. The Generalized-Gradient Approximation (GGA-PBE) exchange-correlation function is applied.

NEGF simulations are compared to experimental data for tunneling through hBN in Fig. 4(a). We simulated the M/I/M (Au/nL hBN/Au) structure using n = 1 to 4 layers. The trend of the tunneling current under different biases and the magnitude difference among different hBN thicknesses are consistent with the experimental data from reference [13]. The absolute value of the current can be matched by accounting for the van der Waals (vdW) gap correction factor: $\lambda = \exp(-k\Delta d)$. The tunneling current will decrease exponentially as the vdW gap increases, as shown in Fig. 4(b) and the coefficient k can be extracted. In Fig. 4(a), the correction factor: $\lambda = 0.03$ corresponds to $\Delta d = 0.14$ nm for this structure. The Δd is the total vdW gap difference, which can be explained by interfaces being atomically flat in simulations, but having some finite surface roughness (~nm) in the experiments.

III. RESULTS AND DISCUSSION

A. Architecture: Symmetric H-Shaped 2D Selector

The structure of the selector should be symmetric for the 1S1R [14]. Two symmetric structures under 1 V bias are compared as an example to illustrate the necessity for the insulator of the H-shaped barrier. The transmission spectrum T(E) for these two structures are shown in Fig. 5(a) and Fig. 5(b) respectively. The imaginary red line always indicates the Fermi function difference between the right and the left electrode, which determines the integration window for T(E) to obtain the system current. With the introduction of the insulator layer, the T(E) is 10^3 times smaller within the integration window (Fig. 5b). Such T(E) will have a small contribution to the current in low bias but a relatively large contribution in high bias, which helps achieve high non-linearity for the selector. Fig. 5(c)

978-1-7281-1988-5/18 $31.00 © 2018 IEEE

shows the transmission eigenstates plot for two models. Eigenstates with a higher transmission eigenvalue will also have a larger amplitude, which reflects the higher transmission probability of an incoming state.

The J−V curves for various thicknesses of hBN in the H-shaped tunnel barrier are shown in Fig. 6(a). The drive current density requirement for the selector is larger than 1 MA/cm^2. Therefore, single-layer hBN will be a preferred design for reaching high non-linearity while achieving high current density. It is possible to observe the negative differential resistance region since the symmetric insulator barrier layer will introduce the resonant tunneling effect [15]. Fig. 6(b) is the iso-surface plot for transmission eigenstates of two circled simulation data in Fig. 6(a). The iso-value we choose corresponds to the amplitude of the wavefunction, and the phase is used to color the iso-surface. The resonant tunneling effect can be observed in the atomistic structure where the wavefunction tunnels through the 2L hBN across a 0.5 V bias shown in Fig. 6(b). There are some resonant regions in the integration window of T(E) in Fig. 6(c) with 0.5 V bias, but not in Fig. 6(d) with 1 V bias.

B. Principle: Physical Model with NEGF Simulation

Considering the resonant tunneling effect, the total current density J can be considered as the sum of the resonant tunnel related current J_1 and the normal tunnel current density J_2, where J_1 can be calculated using the model in reference [15]. The normal tunnel current J_2 can be modeled using the transmission spectrum shown in Fig. 7(a). When $V < \Delta E$, the tunnel current J_2 increases exponentially with the bias V. We study the structure shown in Fig. 2(a) as an example. After the sum of the J_1 and J_2, the H-shaped selector J−V characteristics is shown with the red line in Fig. 7(b) along with the little circles from NEGF simulation. It can be inferred that the H-shaped selector physical tunnel model is consistent with the NEGF simulation.

The transmission barrier, ΔE, shown in Fig. 7(a), has a positive correlation with the band gap of the 2D semiconductor layer (here estimated by simulations). If the band gap is smaller, the selector will have a lower transmission barrier. Fig. 7(c) shows J−V characteristics for various 2D materials in our discussion. The inset in Fig.7(c) shows the calculated density of states (DOS) of WTe$_2$, WS$_2$, and MoS$_2$, marked with the triangle, star, square respectively. It can also be observed that materials with heavier elements can have stronger electron localization leading to the stronger nonlinearity of the J−V curve. The thicker the 2D semiconductor layer, the lower the current level as shown in Fig. 7(d). The metal should also be carefully chosen so that the transmission spectrum and the final characteristics of the J−V curve have good symmetry. Examples are shown in Fig. 8 that compare the same H-shaped barrier 1L hBN/1L WS$_2$/1L hBN with different metal electrodes like TaN in Fig. 8(c) and Au in Fig. 8(d). It can be inferred from Fig. 8(b) that Au is a suitable metal for such a structure to attain good symmetry, while the data from Fig. 8(a) show that TaN has broken the symmetry with improper work function.

C. Application: 2D Selector Designed for 1S1R

We can simplify the selector model as a piecewise linear model described by a maximum supply voltage V_{1S} and a turn-on voltage V_{TURN} in the logarithmic scale shown in Fig. 9(a). From Fig. 7(a), we notice that the transmission barrier ΔE separates the J_2 into two regions. It is better to place $V_{TURN} < \Delta E$ and V_{1S} around ΔE to fully utilize the exponentially increasing part of the selector I-V to reach high nonlinearity. Once we have the I-V curve of the RRAM, we have the maximum 1S1R voltage $V_{MAX} = V_{1R} + V_{1S}$. If ΔE is larger, the V_{1S} will also need to be larger so the selector can have high non-linearity, which is less influenced by the RRAM. A single-layer 2D semiconductor with the larger band gap can enable the larger ΔE with the proper choice of the metal electrodes.

Note that the non-linearity of the 1S selector is ***different*** from the 1S1R combination. For example, a non-linearity requirement for the 1S1R (defined as NL = $R(V_{MAX}/2) / R(V_{MAX}) >$ 100) [9], requires a 1S selector non-linearity of $J(V_{1S}) / J(V_{TURN})$ > 200. This is because the 1R and 1S are in series. There is also a current requirement for read: $I(V_{read})_{LRS}/I(V_{read})_{HRS} \geq 5$ for $0.8V_{MAX} \geq V_{READ} \geq 0.65V_{MAX} > V_{TURN}$. The V_{TURN} should be > $0.5V_{MAX}$ so that the 1S1R has better non-linearity. A smaller V_{1R}, proper V_{TURN}, and larger ΔE for V_{1S} are preferred for the 1S1R. We take the SET branch of the RRAM I-V curve from the reference [16] as an example of 1S1R design. The selector itself prefers higher values of on-state current density (J_{MAX}), supply voltage (V_{1S}), and non-linearity (NL). The selector, shown in Fig. 9(a), meets all requirements for 1S1R shown in Fig. 9(b). In Fig. 9(c), we compare the relative merits (normalized to compare with the selector of Fig. 9(a)) of the materials studied (hBN, WS$_2$, WTe$_2$, MoS$_2$, Au, TaN) with respect to the three important parameters (J_{MAX}, V_{1S}, NL) for the selector.

IV. CONCLUSION

This work provides foundational research on H-shape tunnel barriers for the design of memory selectors using 2D material heterojunctions. Using a single layer (1L) of hBN as the insulator layer can achieve both high nonlinearity and high J_{MAX}. A single layer semiconductor with larger band gap as the middle layer can have higher V_{1S} and J_{MAX} for selector design. A semiconductor with heavier elements like Te, W can enhance NL for stronger electron locality. Au is one of the suitable contact metals for the 1L hBN/1L WS$_2$/1L hBN barrier for the specific RRAM example we used as the illustration. The insulator, semiconductor, and metal layers must be further tuned for the specific RRAM characteristics to meet 1S1R performance requirements. Our design can achieve enough nonlinearity to meet the requirements for a Tbit-class 3D RRAM.

V. ACKNOWLEDGMENT

This work was supported by Stanford UGVR Program, National Key R&D Program (2016YFA0200400), National Natural Science Foundation (61574083, 61434001, 61704096), Natural Science Foundation of Beijing Municipality (4164087) and National Basic Research Program (2015CB352101) of China. The authors are also thankful for the support of the Research Fund from Beijing Innovation Center for Future Chip, the Independent Research Program of Tsinghua University (2014Z01006) and Shenzhen Science, Students Research Training Program of Tsinghua University (1822T0062) and Technology Program (JCYJ20150831192224146). Stanford authors acknowledge the support of the Non-volatile Memory Technology Research Initiative (NMTRI) at Stanford University.

978-1-7281-1988-5/18 $31.00 © 2018 IEEE

REFERENCES

[1] Leqi Zhang, et al., IEDM, 6.8.1-6.8.4, (2014).
[2] Sung Hyun Jo, et al., IEDM, 6.7.1-6.7.4, (2014).
[3] Euijun Cha, et al., IEDM, 10.5.1-10.5.4, (2013).
[4] Z. Tan, et al., Appl. Phys. Lett., 93.24: 242109, (2008).
[5] Jeong H, et al., Nano Letters, 16(3):1858, (2016).
[6] Wang J, et al., Nano Letters, 17(8): 5156, (2017).
[7] H. Lue, et al., IEEE T-DMR, 10.2: 222-232, (2010).
[8] S. Verma, et al., IEDL, 29(3), 252-254, (2008).
[9] Zizhen Jiang, et al., VLSI, 107-108, (2018).
[10] J. P. Perdew, et al., Phys. Rev. Lett., 77: 3865, (1996).
[11] J. Taylor, et al., Phys. Rev. B, 63, 121104, (2001).
[12] Atomistix Toolkit 2017.12, Synopsys QuantumWise (www.quantumwise.com).
[13] Britnell L, et al., Nano Letters, 12(3): 1707-1710, (2012).
[14] Jiun-Jia Huang, et al., IEDM, 31.7.1-31.7.4, (2011).
[15] Schulman J. N, et al., IEDL, 17(5), 220-222, (1996).
[16] B. Govoreanu, et al., ICICDT, 1-4, (2015).

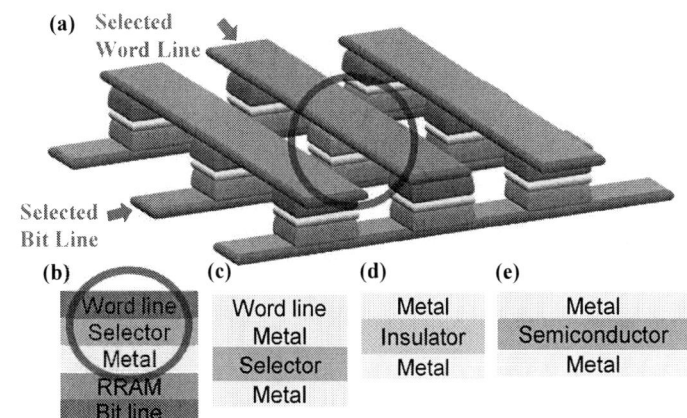

Fig. 1 **Selector application and structure:** **(a)** The schematic of a 1S1R crossbar array. **(b)** The schematic of the selected cell circled in (a). **(c)** The schematic of the selector circled in (b). **(d)** The M/I/M structure. **(e)** The M/S/M Structure.

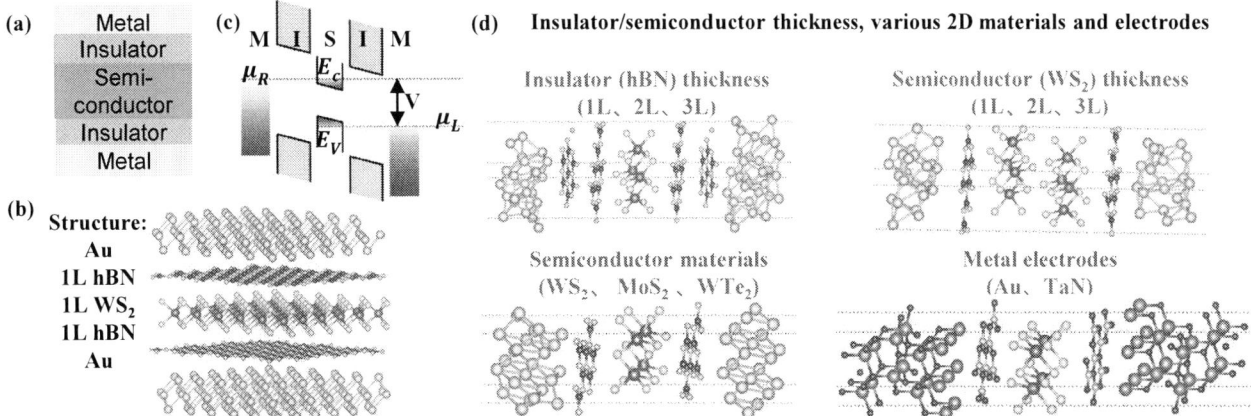

Fig. 2 **H-shaped tunnel barrier:** **(a)** The M/I/S/I/M structure (H-shaped) for the selector. **(b)** The schematic of the atomistic structure of (a). **(c)** Band diagram of the H-shaped tunnel barrier. The green region with a large bandgap is the insulator. The middle region with a small bandgap is the semiconductor. **(d)** Materials studied: the insulator, semiconductor, metal layer of the H-shaped selector.

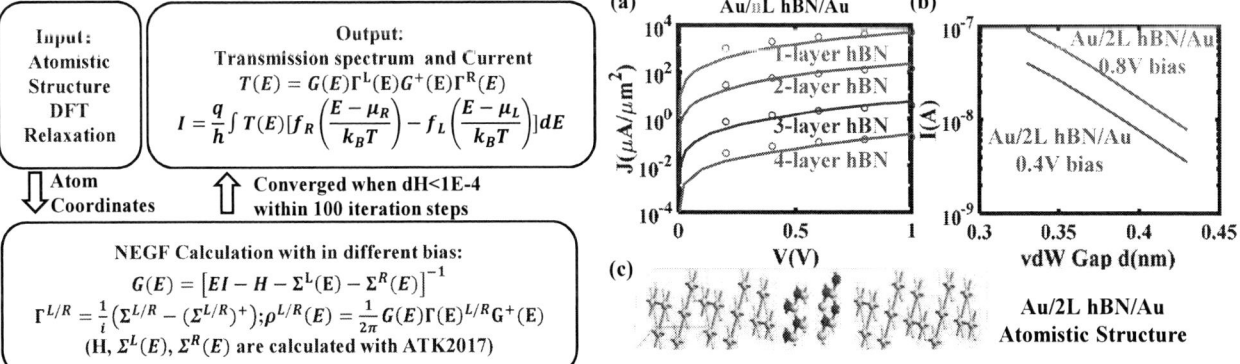

Fig. 3 **The flow chart of the simulation:** First, the atomistic structure is built and relaxed with DFT calculation using VASP and ATK2017. For open systems, the density matrix is calculated using NEGF. The transmission coefficient is calculated by ATK2017 within the framework of DFT. We can analyze the transmission spectrum and get current under specific bias simulation.

Fig. 4 **The NEGF simulation compared with experiment:** **(a)** The red, magenta, blue, green lines show the experimental results [13] of Au/nL hBN/Au, n=1,2,3,4 respectively. The scattered points show the corrected NEGF simulation results correspond to the same color line. **(b)** The influence of the vdW gap on system current. The blue line and red line shows the current of the Au/2L hBN/Au under the bias of 0.4V and 0.8V respectively. **(c)** Atomistic structure for Au/2L hBN/Au.

978-1-7281-1988-5/18 $31.00 © 2018 IEEE

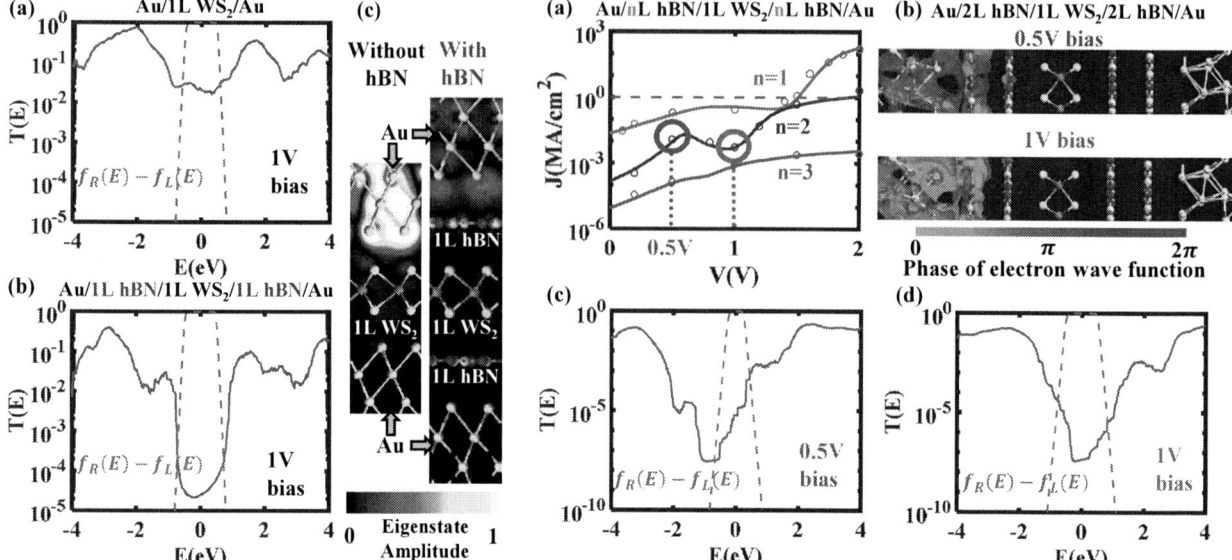

Fig. 5 **Structure Comparison: (a), (b)** The transmission spectrum without and with hBN layer respectively. **(c)** The transmission eigenstates plot for (a) and (b).

Fig. 6 **Insulator thickness: (a)** The J-V curve with different thicknesses of the hBN. **(b), (c), (d)** The iso-surface plot for transmission eigenstate and the transmission spectrum of two circled simulation data in (a) respectively.

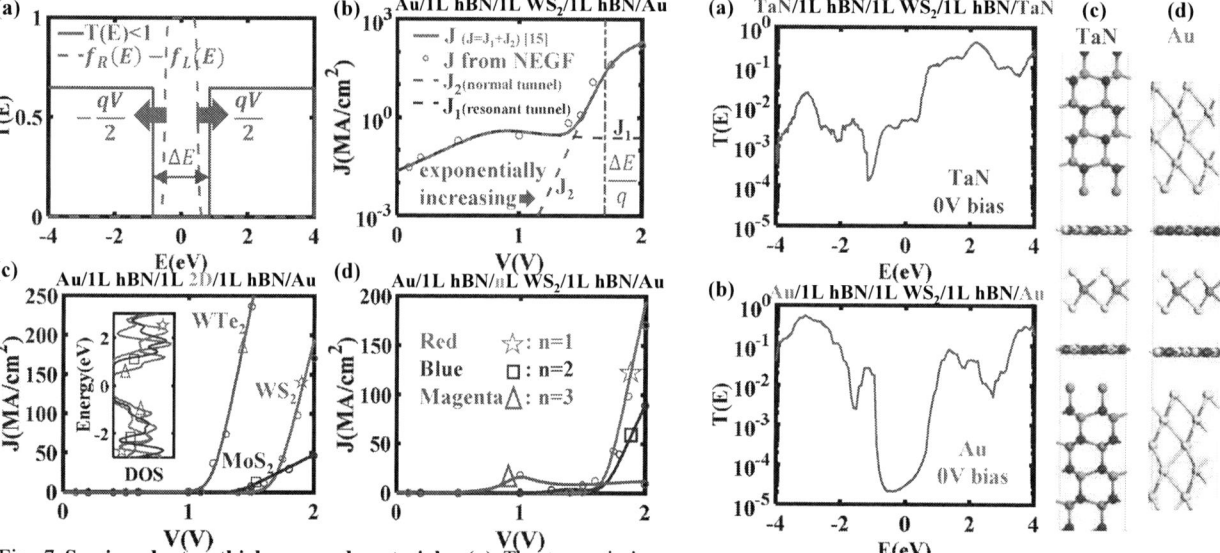

Fig. 7 **Semiconductor thickness and materials: (a)** The transmission spectrum of the model in linear scale. **(b)** H-shaped selector model compared with NEGF simulation, using Au/1L hBN/1L WS₂/1L hBN/Au as an example. **(c), (d)** J-V curve with various 2D materials and thickness. Little circles are NEGF simulation. Lines are model fitting the simulation.

Fig. 8 **Metal materials: (a), (b)** The transmission spectrum for Metal/1L hBN/1L WS₂/1L hBN/Metal, using metal TaN and Au respectively. **(c), (d)** Two atomistic structures for simulation respectively.

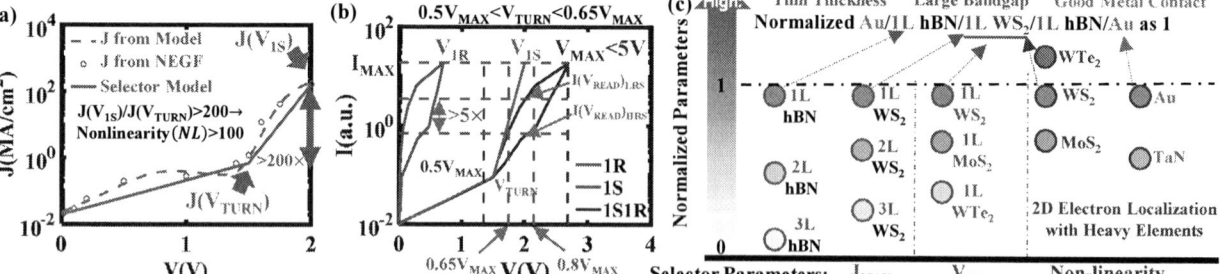

Fig. 9 **2D selector designed for 1S1R: (a)** Model for 2D tunnel selector. **(b)**1S1R Design Verification. The IV curves of the RRAM from the reference [16] (green), Selector from (a) (red), and 1S1R (blue). Dash lines show $0.65V_{MAX} < V_{READ} < 0.8V_{MAX}$,which meets the requirement: $I(V_{READ})_{LRS}/I(V_{READ})_{HRS} > 5$. **(c)** Relative merits of the H-shaped selectors designed with various material and thickness choices. The higher J_{MAX}, V_{1S}, and non-linearity are preferred.

978-1-7281-1988-5/18 $31.00 © 2018 IEEE

Can Kinetic Inductance in Low-Dimensional Materials Enable a New Generation of RF-Electronics?

Kunjesh Agashiwala, Arnab Pal, Wei Cao, Junkai Jiang and Kaustav Banerjee*

ECE Department, University of California, Santa Barbara, CA 93106; *Email: kaustav@ece.ucsb.edu

Abstract- Kinetic Inductance has been recently exploited at room temperature to create materials with inductance densities that exceed the traditional Faraday limit [1],[2]. In this work, for the first time, we develop a rigorous theoretical framework to uncover the physics and origin of kinetic inductance to identify/engineer the materials for addressing inductance-density and performance requirements in next generation passive devices for RF-ICs. Employing three different approaches (Drude model, Fermi sphere and the Boltzmann Transport Equation), we provide guidelines for optimally exploiting kinetic inductance in natural as well as engineered low-dimensional materials to simultaneously achieve maximum inductance-density and performance required for next-generation RF/wireless applications.

I. INTRODUCTION

The growth of the Internet of Things (IoT) coupled with the upcoming 5G/6G/THz technologies promises unprecedented connectivity between people and devices by 2020, with faster speed, larger bandwidths and higher data rates. This necessitates the realization of wireless devices (particularly passives) that demand scalability, flexibility and ease of integration despite the high-frequency issues that haunt the current 4G/LTE technology (**Fig. 1a**). On-chip inductors, a significant passive component in almost all wireless circuits (**Fig. 1b**), can occupy up to 50% of the total chip area. This is because the size of the inductors, determined by their inductance density, is limited by Faraday's law of electromagnetic induction which requires large surface area to achieve desirable values of inductance. The recent developments aimed at addressing the issue of large inductor area either uses magnetic materials [3], or stacked inductor topology [4], and are either not CMOS compatible or significantly increase the Back-End-of-Line (BEOL) process complexity. This has significantly limited the scaling of inductors, unlike transistors and interconnects (**Fig. 1c**) and has become the ultimate bottleneck to RF-IC miniaturization, resulting in a critical need for addressing this issue to support the development of future discrete IoT devices. Besides inductor scaling, increased frequency of operation has led to significant high frequency parasitic effects such as the skin effect (**Fig. 1d**), proximity effect, and eddy current losses to the substrate, which cause undesirable increase in the resistance of the conductors [5], thereby degrading the Quality factor (Q) (or performance) of the inductors (**Fig. 1e**).

A significant breakthrough in addressing the inductance scaling and high-frequency issues has been recently achieved [1], where kinetic inductance (L_K) in Br_2-intercalated multi-layer graphene (MLG) was exploited to obtain higher inductance densities and better quality-factors (up to 12) in on-chip spiral inductors at room temperature. **Fig. 1e** also shows the simulated improvement in inductance and Q values of a spiral inductor in the presence of L_K. L_K manifests itself in the inertia associated with the kinetic energy of the electrons, which results in a finite response time under the presence of an oscillating electric field. **Fig. 2** illustrates the basic concepts of magnetic inductance (L_M) and L_K. Even though the phenomenon of L_K has been known to occur in superconducting materials at cryogenic temperatures (due to the large inertia of Cooper-pairs) [6], there have been very limited attempts to fully understand its origin and the physics of L_K in conventional conducting materials. This is primarily because L_K is negligible in conventional conductors such as Cu. However, the recent demonstration that materials can indeed be engineered to evoke significant L_K, which can be uniquely exploited for overcoming fundamental problems in passive devices [1] calls for rigorous re-examination of L_K, particularly in low-dimensional materials. This work, for the first time, attempts to provide such a comprehensive theoretical understanding of L_K by modeling and analyzing it from various physical viewpoints and finally illustrating the underlying connections between them (**Fig. 3**). Various low-dimensional (1D and 2D) materials that are known to exhibit significant L_K have been explored as possible inductor materials (and compared with the existing technology) for achieving highest possible inductance densities to address the scaling/performance challenges of next generation on-chip inductors and RFIDs, even under increasing frequency to support future 5G/6G and THz applications. Our theoretical framework also provides a foundational platform for developing compact models incorporating L_K for exploring and enabling a wider variety of passives for future generations of RF-ICs. Moreover, *intercalation* of 2D layered materials, introduced in [1] for enhancing L_K has been examined in more details in terms of the type of intercalants and the extent of intercalation, along with dimensional scaling and the band-structure of the resulting materials. Thus, this work constitutes one of the earliest attempts toward a systematic development of engineered nanomaterials for overcoming some of the fundamental challenges in the design of passives in emerging IoT devices.

II. METHODOLOGY

Fig. 4 describes a flowchart for evaluating the L_K of any material using the Boltzmann Transport Equation (BTE) approach. In this model, we have neglected the effect of any external electrostatic potential ($U(x,t) = 0$) and also assumed a constant Fermi level ($E_F(x,t) = E_F$), free of any spatial or temporal dependence. Using the BTE, L_K of various materials such as graphene nanoribbons (GNR) (1D), carbon nanotubes (CNT)(1D), monolayer (1L) graphene (2D), MLG (2D), monolayer (1L) MoS_2/WSe_2 (2D) and copper (3D) have been calculated. The band-structure of 1D-GNR is obtained from [7], whereas a conventional linear band relation is assumed for the CNT. For 1L/ML Graphene, the band-structure is obtained from [8], while for MoS_2/WSe_2 (2D) and copper (3D), 2D/3D parabolic bands with an effective mass m^* are respectively employed.

The band-structure of intercalated materials are obtained from ab-initio Density Functional Theory (DFT) simulations. To consider the van-der-Waals (vdW) interaction between adjacent layers in vdW materials, a DFT-D2 approach is used where a semi-empirical dispersion potential described by a pair-wise force is added to the conventional Kohn-Sham DFT energy. These calculations are performed using Generalized Gradient Approximation (GGA) as the exchange correlation potential and Perdew-Burke-Ernzerh (PBE) as the basis set with a density mesh cut-off of 200 Rydberg and a maximum force constant of 0.05 eV/Å for geometry optimizations. The temperature of the system is set to be 300 K. Once the band-structure and the position of the Fermi level are known, the L_K of intercalated structures can be calculated using the

978-1-7281-1988-5/18 $31.00 © 2018 IEEE

methodology described in **Fig. 4**. FeCl$_3$ [9] and Br$_2$ [1] are the two intercalation agents chosen for analysis in this report as they have been found to be relatively stable in air without getting oxidized or desorbed from the host material.

Note that under the approximation applied in **Fig. 3d**, the BTE approach of deriving L_K leads to the Drude model, which expresses L_K as the ratio of the scattering time (τ) to the DC conductivity (σ_{DC}). This DC conductivity, being proportional to τ under the Relaxation Time Approximation (RTA), therefore renders L_K independent of τ in both the approaches. However, in general, under the absence of RTA, σ_{DC} might be a non-linear function of τ, making L_K dependent on τ.

III. RESULTS AND DISCUSSION

Inductance Density (L_M+L_K) for Various Materials: As we shrink the dimension of a material (3D to 2D to 1D), we reduce the number of conducting channels (or charge carriers). A lesser number of carriers would require more velocity to maintain the same current as compared to the case with larger number of carriers. Since a larger velocity implies larger inertia, L_K of 1D materials is higher than that of 2D followed by 3D materials as shown in **Fig. 5**, where CNT exhibits the highest inductance density among all the materials. However, it is difficult to realize CNT based inductors due to fabrication challenges [5]. GNR shows the second-best L_K (and hence inductance density). However, the extremely small width of GNR significantly increases the series resistance of the fabricated inductor, resulting in small Q. Hence, the optimal candidate for next generation of inductors is Graphene (or MLG), which also has been realized experimentally [1].

Effect of Number of Layers on L_K: Increase in the number of layers of a 2D layered material enhances the interlayer interaction, which causes a reduction of the carrier velocity, and hence, the carrier inertia. This results in an overall decrease in L_K. This dependence is shown in **Fig. 6a** for the case of MLG, which shows a reduction in L_K from 2 nH/µm (1 layer) to ~0.15 nH/µm (10 layers).

Effect of Temperature on L_K: Increase in the temperature of the system leads to an increase in the mobile charge carrier concentration (Fermi-Dirac statistics). This results in an increase of the total charge carriers in the system, which causes a decrease in L_K, as observed in **Fig. 6b**, where the L_K of various 1D/2D/3D materials has been plotted.

Effect of Intercalation Doping on L_K: The inter-layer interactions in multi-layered materials limit the carrier velocity, i.e., inertia, of the carriers as compared to their monolayer counterparts. Intercalation doping has been suggested in [1] to enhance the L_K by eliminating this inter-layer coupling by increasing the gap between adjacent layers (**Fig. 7**). Physical vapor transport (diffusion) is a favorable method of intercalating layered materials as compared to a chemical solvent, which possess the threat of introducing defects or disrupting the surface, thereby affecting the band-structure and reducing L_K.

Effect of Stage Number of Intercalation on L_K: The extent of intercalation in a 2D system is quantified by the stage number of intercalation, which is defined by the number of 2D layers sandwiched between two consecutive layers of intercalant atoms/molecules. The extent of this intercalation, and hence, the stage number is a function of the intercalant diffusion time (**Fig. 8a**). Intercalation not only restores the monolayer pristine band-structures from the bulk configuration, leading to an increase in L_K, but also transfers charge to the system modulating its Fermi level. **Fig. 8b** shows the variation in L_K as a function of the stage number for two different intercalants. The presence of linear band-structures results in the highest L_K for stage-1 intercalated compounds. DFT simulations in

Fig. 8c and **Fig. 8d** confirms the restored linear and parabolic band-structures for stage-1 (monolayer graphene) and stage-2 (bi-layer graphene) FeCl$_3$ intercalated MLG respectively.

Effect of Sample Dimension(width/thickness) on L_K: An increase in the sample dimension implies an increase in the number of conducting channels (or charge carriers), which decreases L_K as shown in **Fig. 9**. Intercalated compounds provide ~10^3-10^4 improvement in inductance density (by improving L_K) at dimensions, which are much smaller than the threshold limit for Cu scaling, making them optimal candidates for next generation inductor materials.

Effect of Fermi Level on L_K: **Fig. 10a** shows the variation in L_K as a function of position of the Fermi level of the system. The decrease in L_K with an increasing value of $|E_F|$ is due to an increase in the number of charge carriers.

Even though the BTE model provides a significant understanding of L_K, it contains certain approximations which need to be addressed. As explained in **Fig. 4**, the estimation of L_K requires the knowledge of the band-structure of a material in the entire Brillouin zone, which can be challenging for new low-dimensional materials. The dependence of L_K on the geometry of the system also needs to be understood. Additionally, since intercalation is an extremely non-uniform process, the position of the Fermi level varies throughout the sample [1], which is not captured in this model. A thorough first principle understanding of the intercalant diffusion and charge transfer process is therefore necessary for accurately estimating L_K in intercalated compounds.

IV. SUMMARY

This work provides the first comprehensive understanding of kinetic inductance by uncovering its physics and developing an analytical model for the same. Various factors of design (**Fig. 10b**) have been studied and the understanding has been extended to provide materials and device engineers guidelines for optimally exploiting kinetic inductance to achieve the highest inductance densities for next generation of wireless/IoT applications. Among the various materials analyzed, intercalated MLG has been identified as the best inductor material to provide the maximum benefit of L_K for addressing the 5G/6G and THz technology requirements.

Acknowledgement: This work was supported by the ARO (grant W911NF1810366) and the UC-MRPI (grant MRP-17-454999).

REFERENCES

[1]. J. Kang, Y. Matsumoto, X. Li, J. Jiang, X. Xie, K. Kawamoto, M. Kenmoku, J. H. Chu, W. Liu, J. Mao, K. Ueno, and K. Banerjee, "On-chip intercalated-graphene inductors for next-generation radio frequency electronics," *Nature Electronics*, vol. 1, no. 1, pp. 46-51, 2018.

[2]. E. Siegel, 'The last barrier to ultra-miniaturized electronics is broken, thanks to a new type of inductor', 2018. [Online].Available: https://www.forbes.com/sites/startswithabang/2018/03/08/breakthrough-in-miniaturized-inductors-to-revolutionize-electronics/#56e59a69779e . [Accessed: 8- Mar- 2018].

[3]. D. S. Gardner, G. Schrom, F. Paillet, B. Jameison, T. Karnik, and S. Borkar, "Review of on-chip inductor structures with magnetic films," *IEEE Transactions on Magnetics*, vol. 45, no. 10, pp. 4760-4766, 2009.

[4]. J. Chen, and J. J. Liou, "On-chip spiral inductors for RF applications: An overview," *Semiconductor Tech. and Science*, vol. 4, no. 3, pp. 149-167, 2004.

[5]. H. Li, and K. Banerjee, "High-frequency analysis of carbon nanotube interconnects and implications for on-chip inductor design," *IEEE Trans. on Electron Devices*, vol. 56, no. 9, pp. 2202-2214, 2009.

[6]. R. Meservey, and P. M. Tedrow, "Measurement of the kinetic inductance of superconducting linear structures," *Journal of Applied Physics*, vol. 40, no. 5, pp. 2028-2034, 1969.

[7]. K. Wakabayashi, K. Sasaki, T. Nakanishi, and T. Enoki, "Electronic states of graphene nanoribbons and analytical solutions," *Science and Technology of Advanced Materials*, vol. 11, no. 5, pp. 054504, 2010.

[8]. F. Guinea, A. H. Castro Neto, and N. M. R. Peres, "Electronic states and landau levels in graphene stacks," *Physical Review B*, vol. 73, no. 24, pp. 245426, 2006.

[9]. W. Liu, J. Kang, and K. Banerjee, "Characterization of FeCl$_3$ intercalation doped CVD few-layer graphene," *IEEE Electron Device Letters*, vol. 37, no. 9, pp. 1246-1249, 2016.

[10]. H. Greenhouse, "Design of planar rectangular microelectronic inductors," *IEEE Trans. on Parts, Hybrids, and Packaging*, vol. 16, no. 2, pp. 100-104, 1996.

978-1-7281-1988-5/18 $31.00 © 2018 IEEE

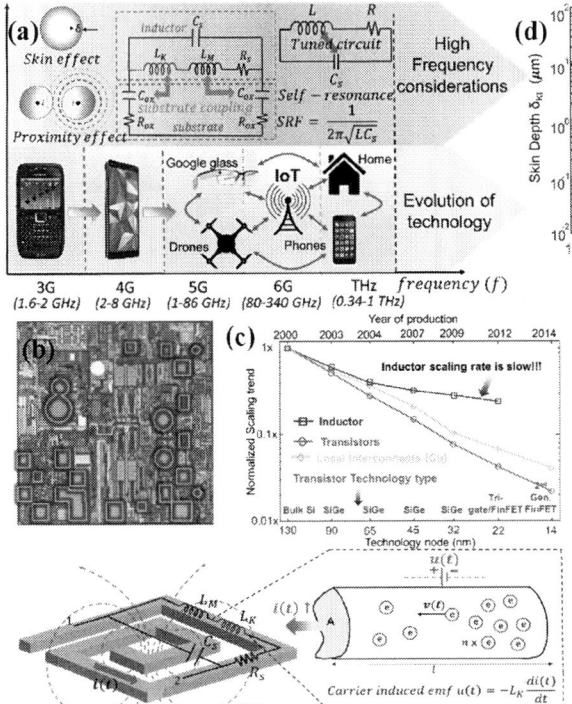

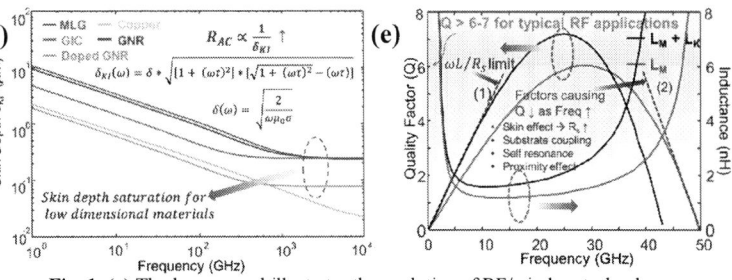

Fig. 1: **(a)** The lower panel illustrates the evolution of RF/wireless technology across various generations: from 3G to 4G (current smart phones), and to the upcoming 5G, and future 6G and THz frequencies to support the emerging paradigms of the IoT and Internet of Everything (IoE). The top panel enlists all the effects that we typically encounter in a conductor as we attempt to operate in the higher frequency range (> 20 GHz) **(b)** Photo of a Qualcomm RF-IC used in iPhone-7Plus with more than 20 on-chip inductors (orange colored). **(c)** Normalized scaling trend of the area of inductors, area of transistors (= gate pitch x metal pitch) and width of interconnects (Metal 1) w.r.t 130 nm technology node. **(d)** Variation in skin depth (δ_{kl}) as a function of frequency for copper and low-dimensional materials (multi-layer graphene (MLG), graphite intercalated compounds (GIC), graphene nanoribbons (GNR) and doped-GNR). The reduction in skin depth at high frequencies increases the resistance (R_{AC}) which degrades the quality-factor (Q) of an inductor as shown in (e). The lesser prominent effects such as proximity effect, substrate coupling, and self-resonance also contribute in lowering the quality-factor at those high frequencies. **(e)** Simulation of the quality factor as a function of frequency for an inductor including kinetic inductance (L_M+L_K) and excluding kinetic inductance (L_M). Higher inductance density and better quality-factor in the high frequency range of consideration can be obtained by exploiting L_K (excluding any layout optimizations/magnetic cores) [1]. This simulation was performed with L_M = 1 nH, R_S = 20 Ω, C_S = 10 fF and an experimentally verified [1] value of L_K = 0.4 nH (40% of L_M).

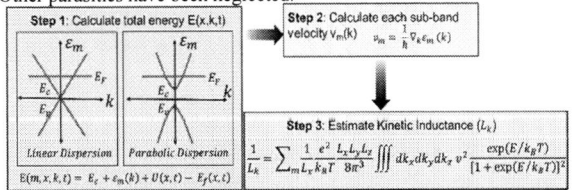

Fig. 2: **Difference between magnetic and kinetic inductance:** A spiral inductor has its magnetic inductance (L_M) arising due to flux linkages within the conductor itself (L_{Self}) and due to flux linkages with other conductors (L_{mutual}). These time varying flux linkages result in an electromagnetic emf $e(t)$. Kinetic inductance (L_k) is an additional inductance, intrinsically present, which arises due to the inertia of the charge carriers to move with a constant velocity resulting in a finite reaction time in response to changing current $i(t)$. This emf is the carrier induced emf $u(t)$. Equivalent circuit of a spiral inductor (without substrate) showing L_M, L_K, C_S (inter-turn coupling capacitance) and R_S (series resistance) is embedded within the spiral inductor schematic. Other parasitics have been neglected.

Fig. 4: **Flowchart describing the evaluation of L_k using the BTE approach:** The first step involves the calculation of the Energy (E) of the system with the knowledge of the energy-band dispersion relation $\varepsilon_m(k)$ (m denotes the sub-band index), external potential ($U(x,t)$), position of the Fermi level ($E_F(x,t)$), and bottom of the conduction band (E_C). Step 2 involves calculating the group velocity of each sub-band. Using step 1 and step 2, L_k of the material can be estimated in step 3 by integrating the expression over the first brillouin zone summed over all sub-bands. Note that the above expression is for 3D materials, equivalent expressions for 1D/2D materials can be derived by reducing the order of integration.

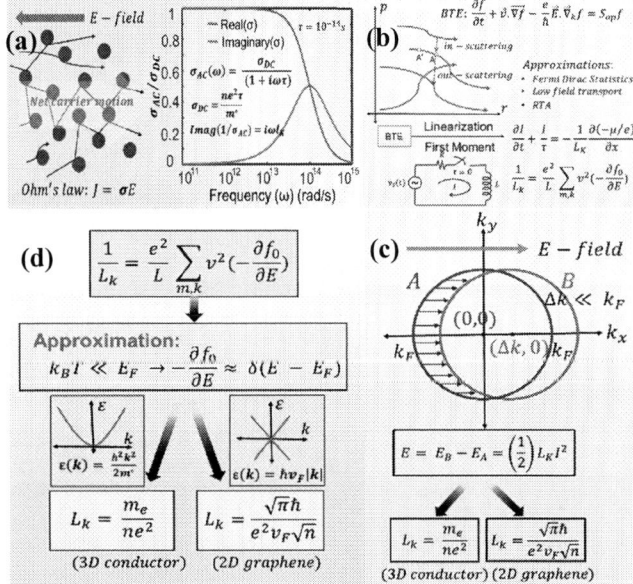

Fig. 3 Understanding kinetic inductance (L_k) via different approaches (a) Drude model: L_k naturally originates from the Drude approach by assuming a non-zero relaxation (collision) time (τ) which results in a phase lag in voltage with respect to the applied a.c. field. This results in a complex a.c. conductivity, whose imaginary part contains information regarding L_K. The plot on the right shows the real and imaginary parts of the normalized a.c. conductivity (σ_{AC}/σ_{DC}) as a function of frequency typical metal (such as Cu) with $\tau = 10^{-14}$ s. **(b) BTE approach:** Assuming a low field transport and Relaxation Time Approximation (RTA), the BTE is simplified to an expression representing conservation of energy in a forced RL circuit. The equivalent inductance (L_K) extracted through this procedure depends on the specific band-structure and the environment of the material, making it a fundamental approach for analyzing L_K **(c) Fermi Sphere approach:** The Fermi sphere represents the system (2D graphene or 3D generic conductor) at rest. An externally applied electric field imparts a net velocity to the carriers, as shown by the displaced red sphere. Kinetic inductance is calculated by evaluating the excess kinetic energy between the spheres (A & B) and equating it to $L_k I^2/2$. The unique dispersion relations for 2D graphene and a 3D conductor lead to different analytical expressions of L_K. **(d) Connection of BTE with Drude model:** BTE approach boils down to the exact same analytical expression (for both 3D conductor and graphene) as the Drude model under the approximation $k_B T << E_F$. This can be fundamentally related to the fact that Drude model assumes Maxwell-Boltzmann (MB) statistics, while BTE assumes Fermi-Dirac (FD) statistics and they behave identically either at high energies or low temperatures.

978-1-7281-1988-5/18 $31.00 © 2018 IEEE

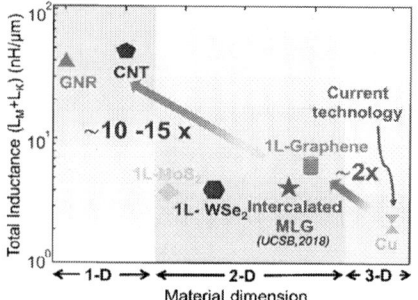

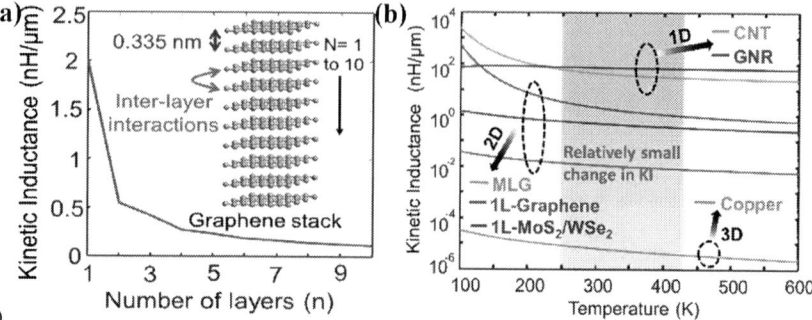

Fig. 5: Calculation of total inductance density ($L_M + L_K$) for various 1D/2D/3D materials in a microstrip configuration. The results have been normalized to per unit length values for comparison with 1D materials. Magnetic inductance is estimated using the Greenhouse Algorithm [10] and kinetic inductance is calculated using the BTE approach. Experimentally reported inductance density for intercalated MLG is also shown [1]. There was no external potential assumed ($U(x,t)$ =0 in Step 1 of **Fig. 4**) and the Fermi level is at its intrinsic position.

Fig. 6: **(a)** Kinetic inductance as a function of the number of layers for undoped MLG using the methodology developed in **Fig. 4**. The analytical expression of the band-structure for MLG is obtained from [8]. **(b)** Kinetic inductance as a function of temperature for various 1D/2D/3D materials was calculated using the procedure shown in **Fig. 4**. L_K almost remains constant in the practical temperature range varying from 250-450 K. For both Fig. 6a and 6b, it was assumed that there is no external potential and the Fermi level is at its intrinsic position.

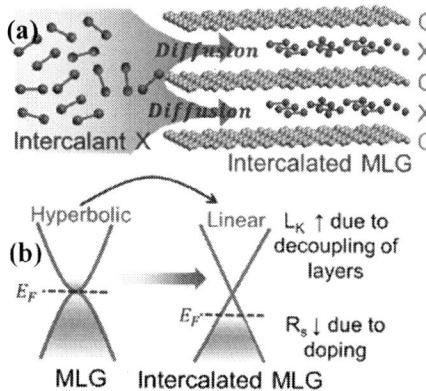

Fig. 7: **(a) Phenomenon of Intercalation doping**: The process of intercalation involves the reversible inclusion of an atom/molecule (Intercalant X) in a layered material (e.g., MLG, multilayer-TMDs). This expands the vdW gap between two adjacent layers, which de-couples the layers. **(b) Effect of intercalation on Band-structure:** The transfer of charge from the host intercalant to the layered material results in a shift of the Fermi level (doping effect), which reduces R_S and thus enhances the Q factor of the inductor. The reduction in inter-layer interaction increases the relaxation time (τ) of the carriers which increases L_K.

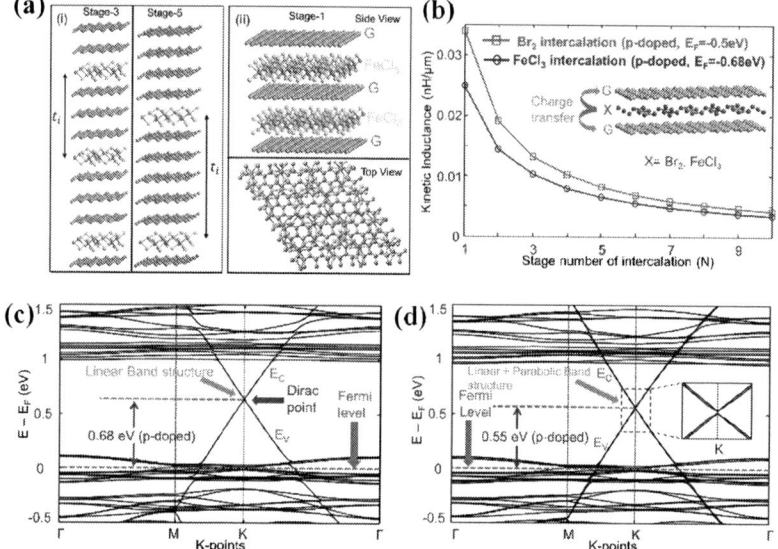

Fig. 8: **(a)** Schematic showing different stages of intercalation (i) Stage-3 and Stage-5 intercalated MLG (GIC). Stage number = number of graphene layers between two consecutive intercalant (FeCl$_3$) layers. t_i represents the thickness of a unit cell (ii) Side view and top view of Stage-1 intercalated MLG. **(b)** Kinetic inductance as a function of stage number of intercalation for various intercalants for a 0.5 μm thick MLG. DFT calculations are performed to obtain the position of the Fermi level for various intercalants. This information is utilized to obtain the L_K using the methodology outlined in **Fig. 4**. The green arrow shows the phenomenon of charge transfer from the intercalant layer to the graphene layer. DFT is employed to calculate the band-structures of **(c)** Stage-1 and **(d)** Stage-2 intercalated MLG.

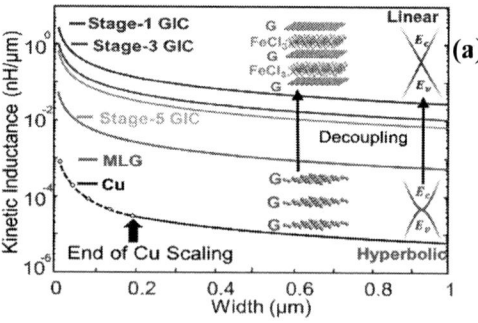

Fig. 9: The variation of kinetic inductance as a function of the sample dimension (width) is studied for various stages of intercalation doped MLG (2D) and compared with undoped MLG (2D) and copper (3D).

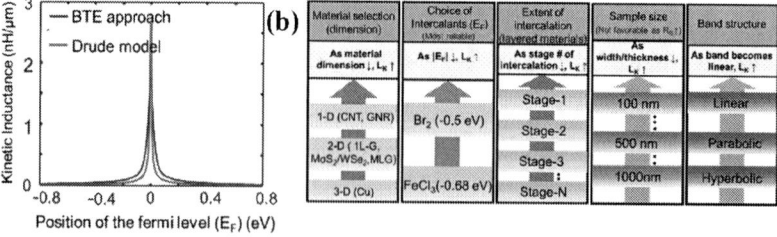

Fig. 10: **(a)** Effect of the variation of the E_F on the kinetic inductance of monolayer graphene of width w =1 μm in a microstrip configuration using the BTE and Drude approaches, and the almost similar curves easily validate the approximation shown in **Fig. 3d**. **(b)** Illustration of the various factors affecting kinetic inductance, serving as a guideline for optimally exploiting L_K to achieve highest inductance density (without using any VLSI incompatible magnetic materials or structures) for next generation wireless applications.

978-1-7281-1988-5/18 $31.00 © 2018 IEEE

A Surface Potential- and Physics- Based Compact Model for 2D Polycrystalline-MoS₂ FET with Resistive Switching Behavior in Neuromorphic Computing

Lingfei Wang, Lin Wang, Kah-Wee Ang, Aaron Voon-Yew Thean, Gengchiau Liang*

National University of Singapore, Singapore. Email: lwang@u.nus.edu; elelg@nus.edu.sg

Abstract—For the first time, a surface potential- and physics-based compact model for two dimensional (2D) polycrystalline- molybdenum disulfide (MoS₂) field effect transistors (FETs) with resistive switching (RS) behavior is developed and verified by experimental data. This model is incorporated with the theories of thermal activation transport, grain boundary (GB) barrier and space charge limited current (SCLC). Based on the GB induced disorders, the grain size, low temperature and high electrical field dependent characteristics are studied. The predicted transfer and output characteristics have excellent quantitative agreement with experimental results. Furthermore, considering the hopping process induced defect- (i.e., sulfur vacancy) redistribution, the GB (e.g., intersecting or bisecting GB) dependent resistive switching behavior is physically investigated. Finally, this model is implemented to simulate the synaptic activity such as short-term/long-term plasticity, which indicates the possibility of using 2D-FETs for neuromorphic computing applications.

I. INTRODUCTION

Two-dimensional (2D) polycrystalline molybdenum disulfide (Poly-MoS₂) field effect transistors (FETs) have presented analog resistive switching (RS) behavior for potential neuromorphic computing application [1].The accurate physics-based compact models are highly needed to evaluate circuit or RS performance. Especially, the extensive grain boundary (GB) effects, lacked in the previous compact models, are expected to dominate the transport behavior in Poly-MoS₂. Effects such as GB energy barrier (i.e., charging effects [2]) and space charge limited current (SCLC) are important to RS in these devices, and play critical roles in resistance state transition during the migration of defects (e.g., sulfur vacancy) in GBs.

In this work, we develop a physics-based compact model to capture disorders induced transport within Poly-MoS₂ FETs. We have applied this model to various grain size and channel length FETs under different temperatures. Besides, the incorporation of defect hopping process is investigated for different resistive switching behaviors with bisecting and intersecting GBs. Furthermore, we predict the influence of the gate voltage, disorder parameters and statistical hopping on RS properties and postsynaptic current (PSC) simulation. It is found that tunable hopping process and gate voltage are able to modulate the short-term/long-term plasticity, which indicates the potential applications for neuromorphic computing.

II. DEVICE MODELING

A. Surface Potential Calcualtion

For 2D polycrystalline materials, the trap states are mainly induced by the defects in the grain boundary. Based on effective medium approximation (EMA), the exponential band tail with grain length (L_g) dependent trap density ($N_t \sim 1/L_g$) is utilized as shown in Fig. 1. Due to Fermi Dirac distribution, the analytical free-(N_F) and localized-(N_L) carrier density are obtainable in Eqs. (1, 2). Using capacitance divider schematic, the surface potential (φ_s) is described by the charge convergence formula, which is a transcendental function [Eq. (3)]. To gain analytical solutions, an imprecise variable φ_{si} is worked out by neglecting the exponential term in N_L and expanding the N_F by 2nd order Taylor series. Consequently, the Schroder Series is employed to reduce the error for the analytical φ_s in Eq. (4) [3].

B. Mobility and Current Model

Analogy to polycrystalline organic FETs, the GB energy barrier (E_b) can be induced by the depletion between GBs and Grains in MoS₂. Based on the GBs or other structural disorders induced exponential band tail, E_b can be calculated by Eq. (5) combining the 2D material junction barrier [4] and $N_{F, L}$ dependent incomplete depletion width [5]. However, below φ_{sc} at $N_F=N_L$, E_b will decrease sharply, which may lead to abrupt increase of current in the sub-threshold region. To improve such behavior, E_b is kept constant below φ_{sc} in Eqs. (6, 7). Thus, the GB based mobility with the thermal activation transport is provided in Eq. (8). In other cases (e.g., high V_{ds} or short channel), the ultra-high electrical field may lead to the obvious SCLC. Such current ($I \sim V_{ds}^m$) is characterized by transition from ohmic liner region ($m=1$) to power law region ($m>1$). Due to previous studies, the coefficient $\sim(1+V_{ds}/V_c)^{T_0/T}$ [6] provides a simple and physics-based insight into such behavior. Thus, the SCLC based mobility as a function of V_{ds} is given in Eq. (9). For small V_{ds}, the coefficient tends to be 1, and μ_{SCLC} will be dominated by the thermal activated transport. Using Eqs. (8, 9), the surface potential based current is calculated by Eq. (11) [3].

C. Resitive Swiching Model

The GB defect model- and SCLC- based RS behaviors were previously demonstrated in ZnO based varistor and RRAM, respectively. The GB defect model [7, 8] is interpreted by motion of the vacancies towards the GB interface, which finally modulates E_b. SCLC based RS is also attributed to the modulation of traps in the device. To incorporate such effects, N_t is regarded as the state variable, and the simple vacancy hopping process is described by Eq. (12) [9]. It is dependent on the thermal activation energy ($E_{1,2}$) and applied voltage (V_{ds} or V_{gs}). v_0 is a constant rate of density change and $\gamma_{1,2}$ are the hopping distance related fitting parameters. During small time intervals, the dynamic N_t is easily described by considering a constant dN_t/dt in Eq. (13). Moreover, the boundary condition should be satisfied during the Set/Reset process. By using the window function in Eq. (14) [10], the dN_t/dt will tend to be zero at the stable high resistance state (HSR) or low resistance state.

III. MODEL VALIDATION

The schematic and optical image of our fabricated single layer polycrystalline MoS₂ FET are illustrated in Figs. 1(a) and 1(b), correspondingly. Due to the presence of GBs, the GB barrier will be formed by the depletion effects in Fig. 1(c). In high electrical field region, the traps may result in the SCLC as shown in Fig. 1(d). Both mechanisms are dominated by N_t, which can be regarded as a state variable to simulate the RS behaviors. Based on Eq. (3), the analytical and numerical results of φ_s are compared in Fig. 2 (a), with error less than 0.8%. Different from φ_b ($\sim E_b/q$) in Eq. (5), the modified φ_b in Eq. (7) saturates in the lower V_{gs} region, cf. Fig. 2(b). As a result, the analytical output and transfer current characteristics show good agreement with the experimental data ($\sim$1-4 mm grains) in Fig. 2(c) and Fig. 2(d), respectively. Furthermore, the current characteristics of the device with 600 nm (S2) grains are also demonstrated in Fig. 3(a, b), which match well with our model. By decreasing the length of grains to 200 and 20 nm [11], the increase of N_t finally results in the mobility degradation in Fig. 3(c). The transfer currents under different temperatures indicate the thermal activated transport (i.e, $\ln(I_{ds})\sim T^{-1}$) in Fig. 3(d). However, if the channel is further decreased to the single-grain size (S3: 500 nm or S4: 160 nm), SCLC based transport mechanism tends to emerge [12]. The transfer and non-linear (before saturation) output current characteristics (for S3 and S4) are shown in Fig. 4(a-d). More importantly, the transition from linear to power law relation is observed in Fig. 4(d) under different temperatures. All the current curves will intersect at a critical voltage V_c ($\sim$ 3 V). Due to such transport behaviors, key disorder parameters such as trap density N_t, disorder strength T_0, maximal depletion width W_m and critical voltage V_c can be extracted by the temperature dependent current characteristics in Fig. 5. The parameter values are listed in Table 1.

IV. NEUROMORPHIC COMPUTING

A. Resitive Swiching Behaviour

The RS behavior is based on the hopping process [Eq. 12] and the window function [Eq. (14)]. As SCLC based model is sensitive to both N_t and T_0, we utilized the GB barrier based model to neglect the complicated correlation between N_t and T_0. Firstly, the important disorder parameters are extracted from the transfer current without RS in Fig. 6(a), where the analytical results show good agreement with the experiments (S5: single intersecting GB). Based on these parameters, Set/Reset process can be observed in a continuous cycle from 20 to -20 V in Fig. 6(b). Compared with the set process, the switching rate of reset process is lower, which is reflected by a higher activation energy ($\sim$ 1.53 eV). By altering the plot scale in Fig. 6(c) and its inset, the set process can be shown more clearly, which indicates the potential SCLC effect [1]. Then, to incorporate the gate tunable behaviors, the activation energy was modulated by V_{gs} to fit the current at HRS more accurately [Fig. 6(d)]. The good agreement of our model and experimental results suggests that the set voltage can be tunable by V_{gs}. For another device (S6: single bisecting GB), the RS process is reverse that the reset occurs in positive V_{ds} region. Based on Eq. (12), by changing the sign of V_{ds}, the RS behaviors (from -20 to 60 V) are simulated at forming and electroformed states in Fig.7 (a) and Fig. 7(b), respectively. The forming process shows lower switching rate than the formed one. Besides the GB barrier based simulations, we investigate the alternative SCLC effect on RS as shown in Fig. 8(a). It is suggested that both T_0 and N_t play the important roles in the RS process, which may require

a deeper insight into the correlation (of N_t and T_0). In addition, all the RS parameters are shown in Fig. 8(b).

B. Postsynaptic Current Simulation

Prior to PSC simulation, the RS behavior is optimized by reducing the activation energy (E_2) and enhancing the electrical field strength (γ_2) for the reset process in Fig. 9(a). For RS$_1$, the switching behavior is optimal by generating a balanced loop of N_t in Fig. 9(b), where the boundary is determined by the window function. Then, based on RS$_1$, the additional random variable β was implemented into dN_t/dt in Eq. (12). The RS current and state variable N_t in continuous 100 cycles are shown in Fig. 9(c) and Fig. 9(d), respectively. Such statistical effects may originate from the random hopping process, temperature or electrical crosstalk, and largely affect the Set/Reset voltage.

Due to RS$_2$, the asymmetry switching behavior results in long term depression (LTD) as shown in Fig. 10(a). The potentiation was established in first 20 pulses. However, it cannot recover in the following 140 negative pulses. If the simulation is based on the RS$_1$, the depression term will be shorter which indicates a potential transition from LTD to STD. As the gate tunable RS was observed, we utilize the V_{gs} dependent E_1 to simulate PSC in Fig. 10(c). Since the set voltage is dependent on V_{gs}, the potentiation duration is tunable under same pulses. At lower V_{gs}, the potentiation process will be shorter but with lower amplitude of PSC. Finally, based on the statistical effects, we simulate the PSC in 5 cycles [Fig. 10(d)], which is indicative of a large amplitude variation of PSC.

V. CONCLUSION

A surface potential- and physics- based compact model for polycrystalline MoS$_2$ FETs is successfully developed and extensively verified by experimental data under different grain sizes and temperatures. By incorporating defect hopping process, this model can predict both gate-tunable and grain boundary-based resistive switching behaviors with statistical effects. Therefore, it serves as a practical and reliable tool to evaluate the potential application in neuromorphic computing.

ACKNOWLEDGEMENT

This work was supported in part by the Ministry of Education of Singapore (Grant No. MOE2017-T2-1-114) and A*STAR Science and Engineering Research Council (Grant No. 152-70-00013).

REFERENCES

[1] V. K. Sangwan, *et al.* "Gate-tunable memristive phenomena mediated by grain boundaries in single-layer MoS$_2$." *Nat. nanotechnology*, 10(5): 403, 2015.
[2] C. Yan, *et al.*, "Charging effect at grain boundaries of MoS$_2$." *Nanotechnol.* 29(19): 195704, 2018.
[3] L. Wang, *et al.* "A unified surface potential based physical compact model for both unipolar and ambipolar 2D-FET", *IEDM*: 31-4, 2017.
[4] H. Ilatikhameneh, *et al.* "Dramatic Impact of Dimensionality on the Electrostatics of PN Junctions." *IEEE Trans. Nanotechnol.*, 17(2):293-298, 2018.
[5] H. He, *et al.* "On the grain boundary barrier height and threshold voltage of undoped polycrystalline silicon thin-film transistors." *IWJT. IEEE*, 2012.
[6] A. Sussman, "Space-Charge-Limited Currents in Copper Phthalocyanine Thin Films". *Journal of Applied Physics*, 38(7): 2738-2748, 1967.
[7] T. K. Gupta, *et al.* "A grain-boundary defect model for instability/stability of a ZnO varistor." *Journal of materials science*, 20(10): 3487-3500, 1985.
[8] R. Tararam, *et al.* "Resistive-switching behavior in polycrystalline CaCu$_3$Ti$_4$O$_{12}$ nanorods." *ACS applied materials & interfaces*, 3(2): 500-504, 2011.
[9] P. Chen, Yu S. "Compact modeling of RRAM devices and its applications in 1T1R and 1S1R array design". *IEEE Trans. Electron Devices.* 62(12):4022, 2015.
[10] J. Zha, *et al.*, "A novel window function for memristor model with application in programming analog circuits." *TCSII*, 63(5):423-427, 2016.
[11] J. Zhang, *et al.* "Scalable growth of high-quality polycrystalline MoS$_2$ monolayers on SiO$_2$ with tunable grain sizes." *ACS. Nano.*, 8(6): 6024-6030, 2014.
[12] S. Ghatak, *et al.*, "Observation of trap-assisted space charge limited conductivity in short channel MoS$_2$ transistor". *Appl. Phys. Lett.*, 103(12): 122103, 2013.

$$N_F(\varphi_S) = D_0 k_B T \exp\left(-\frac{E_c - q\varphi_S}{k_B T}\right) \quad \textbf{free} \qquad D_0 = \frac{m^*}{\pi\hbar}\sum_n H(E - E_n) \tag{1}$$

$$N_L(\varphi_S) = A_t k_B T_0 \exp\left(-\frac{E_c - q\varphi_S}{k_B T_0}\right) \quad \textbf{localized} \qquad A_t = \frac{N_t}{k_B T_0}\frac{\pi T}{\sin(\pi T/T_0)T_0} \tag{2}$$

Surface Potential Model   **Taylor expansion**

$$Q_{net}(N_F, N_L) = C_{ox}(V_g - V_0 - V_x - \varphi_S) \Longrightarrow \varphi_{Si} = Sign(-E_c + x_{gd})(x_s k_B T_0 + E_c) \tag{3}$$

$\Longrightarrow$ **Schroder series:** $\quad d\varphi = \frac{-f}{f^{(1)}}\left(1 + \frac{f^{(2)}}{2f^{(1)}}\frac{f}{f^{(1)}}\right) \qquad \varphi_S = \varphi_{Si} + d\varphi \tag{4}$

$$f = -Q_{net}(\varphi_{Si}) + C_{xg}(V_g - V_0 - \varphi_{Si}) \; ; \; x_{gd} = \frac{V_g - V_0 - V_x}{k_B T_0} \; ; \; c_1 = \frac{qD_0 T}{C_{ox}T_0} \; ; \; c_2 = \frac{A_t q}{C_{ox}}$$

$$k_B T T_0^2 x_s = (\sqrt{T_0^2(E_c - x_{gd})^2 - (2k_B T^2\ln(c_1) + 2k_B^2 T^2\ln(c_3) + (-E_c + k_B T + x_{gd})^2)}$$
$$-E_c^2 T_0 + E_c k_B T T_0 + 2E_c T_0 x_{gd} - k_B T T_0 x_{gd} - T_0 x_{gd}^2) \; ; \; c_3 = \ln((x_{gd} - E_c)^2 + 10^{-5})$$

Mobility Model: GB barrier effect and SCLC effect

$$E_b(\varphi_s) = \frac{q^2\ln(4)N_L}{\pi\varepsilon_s\varepsilon_0}\frac{N_L}{N_L + N_F} \quad (5) \qquad \varphi_{sc} = \frac{1}{q}\left(E_c + \ln\left(\frac{D_0 T}{A_t T_0}\right)\frac{k_B T T_0}{T - T_0}\right) \tag{6}$$

$$E_{bm}(\varphi_s) = \frac{Tanh((\varphi_s - \varphi_{sc})10^4) + 1}{2}E_b(\varphi_s) \qquad \frac{Tanh((\varphi_{sc} - \varphi_s)10) + 1}{2}E_b(\varphi_{sc}) \tag{7}$$

$$\mu_{GB} = \mu_0 e^{-\frac{E_{bm}}{k_B T}} \quad (8) \qquad \mu_{SCLC} = \mu_0 e^{-\frac{E_a}{k_B T}}(1 + |V_{ds}|/V_c)^{\frac{T_0}{T}} \tag{9}$$

Current characteristic Model

$$g(\varphi_S) = -\int q N_e(\varphi_S)\left(1 + \frac{dQ_{net}(\varphi_S)}{C_{ox}d\varphi_S}\right)d\varphi_S \tag{10}$$

$$I_{ds} = \frac{W}{L}\left(\mu(\varphi_{sd})\cdot g(\varphi_{sd}) - \mu(\varphi_{ss})\cdot g(\varphi_{ss})\right) \tag{11}$$

R-S Model (state variable N_t)

$$\frac{dN_t}{dt} = -v_0\left(e^{\frac{-E_1 \pm q\gamma_1 V_{ds}}{k_B T}} - e^{\frac{-E_2 - (\pm q\gamma_2)V_{ds}}{k_B T}}\right)\cdot Win \tag{12}$$

V_{ds}	'+' region	'-' region
'+' in Eq. (12)	set	reset
'-' in Eq. (12)	reset	set

$$N_t(t_0 + \Delta t) = N_t(t \to t^-) + \frac{dN_t(t)}{dt}\Delta t \tag{13}$$

Window function:

$$f(x) = 1 - \left(\frac{(x - Step(V_{ds}))^2 + 3}{4}\right)^2 \tag{14}$$

$Win = f\left(|\frac{N_t(t) - N_{t\_min}}{N_{t\_max} - N_{t\_min}}|\right)$ ⟸ boundary condition

T_0: Disorder Parameter for band tail;
N_t: Defects or GB induced trap states
$f^{(x)}$: xth order derivative; $N_e = N_F + N_L$;
Q_{net}: Net charge in the channel.
$E_{b(m)}$: (Modified) GB energy barrier;
φ_{sc}: Critical φ_s for maximal E_b;
W_m: Maximal depletion width in grains;
E_a: Thermal activation energy;
V_c: Critical voltage of SCLC behavior;
$\mu_{SCLC(GB)}$: SCLC (GB) based mobility;
V_0: Offset voltage; T: Temperature;
C_{ox}: Gate oxide capacitance;
$\varphi_{sd(s)}$: Surface potential at drain (source);
Δt: Small time intervals for simulation;
$E_{1,2}$: Generation or recombination energy;
$\gamma_{1,2}$: Local electrical field strength;
Step, Sign can be expressed by *Tanh*;
$q\varphi_s$ and E_c refer to the intrinsic Fermi level;
All integrals and equations are analytical.

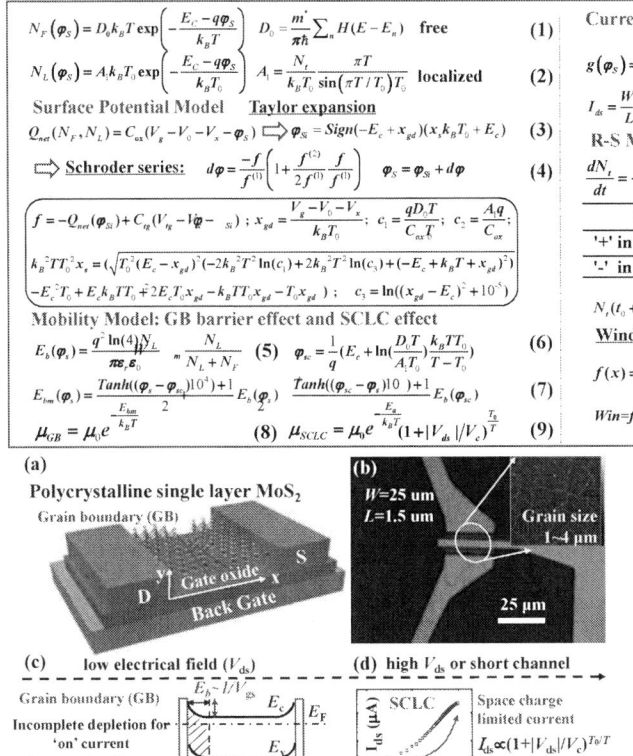

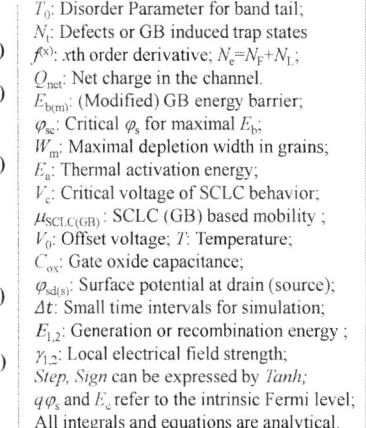

Fig. 1. (a) Schematic for polycrystalline MoS2-FET with GBs. (b) Optical image of our fabricated MoS2 FET with 20 nm HfO2 as the top-gate dielectric. Schematic of GB barrier based transport (c) and SCLC based transport (d). The resistance variation based on GB barrier and SCLC are previously introduced in ZnO based varistor and RRAM, respectively. Since both transport mechanisms are based on the exponential trap distribution, N_t is regarded as a state variable due to defect hopping process.

Fig. 3. Agreement between our model and experiments (from [11]) for output (a) and transfer characteristics (b). Transfer characteristics under different L_g (c) and T (d) simulated at V_{ds}=2V. When L_g=200(20) nm in (c), N_t=3.4(14) ×10^{15} m^{-2}, T_0=920(1700) K and V_0=45(40) V. In (d), $I_{ds} \sim T^{-1}$ is inserted, and μ_0 is modified as 8×(300/T)$^{1.3}$ cm^{-2}/Vs.

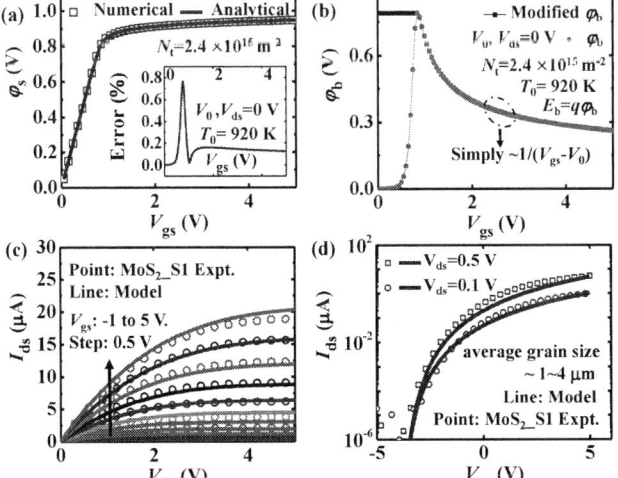

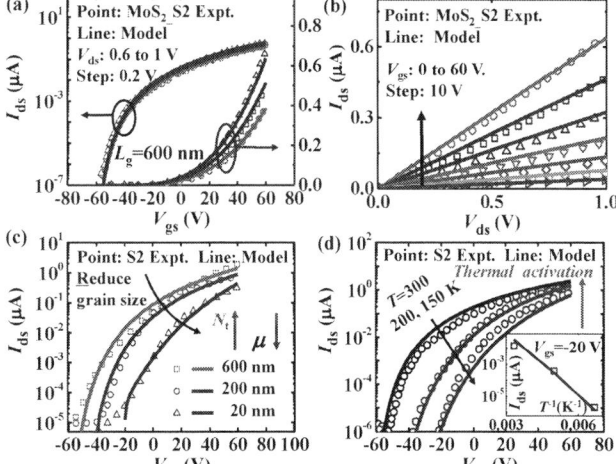

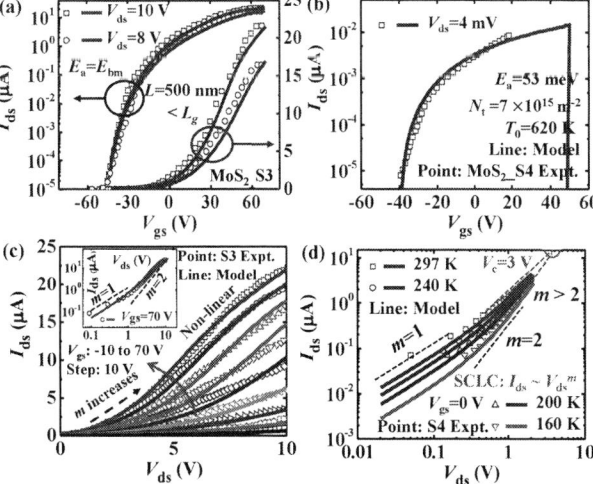

Fig. 2. (a) Agreement between analytical and numerical results for surface potential. Error is inserted. (b) Comparison of φ_b and modified φ_b varying with V_{gs}. Agreement between our model and experimental data (S1) for output (c) and transfer (d) characteristics.

Fig. 4. Agreement between our model and experimental data for transfer characteristics of S3 (a) and S4 (b) and output characteristics of S3 (c) and S4 (d). For saturation effect in (c), $I_{ds}=(I_{sclc}^{-5}+I_s(\mu)^{-5})^{-1/5}$, where $\mu=\mu_0 e^{(-E_a/k_B T)}$ and E_a=0.12 eV. $I_{ds} \sim V_{ds}$ in log-log scale is inserted in (d). S3 (S4) experimental data are from [11] ([12]).

Table 1. Parameter values for different MoS₂ samples.

Channel & Mobility	$W/L(\mu m)$	$\mu_0(cm^2/Vs)$	$T_0(K)$	$N_t(10^{15}m^2)$	$V_c(V)$	$V_0(V)$	$W_m(nm)$
S1: 1~4 μm grains (μ_{GB})	25/1.5	2.4	920	2.4	---	-4	702
S2: 600 nm grains (μ_{GB})	5/20	8	920	2	---	-60.1	600
S3: single grain(μ_{SCLC})	0.5/0.5	6	920	1.98	9	-54.1	500
S4: short channel (μ_{SCLC})	~1.9/0.16	0.418	620	7	3	-40	---
S5: intersecting GB(μ_{GB})	~ 8/7	0.28	1820	1.6(t=0)	---	-75	1200
S6: bisecting GB(μ_{GB})	~ 7/7	0.032~ 0.07	3020	1.2(t=0)	---	-45	1200

PS: E_g=1.8 eV; E_c=$E_g/2$; ε_{r_MoS2}=3.51; D_0=10^{18} m^{-2}eV^{-1}; C_{ox} = 6.6 ×10^{-3} (S1), 0.8×10^{-4} (S2-S6) F/m^2

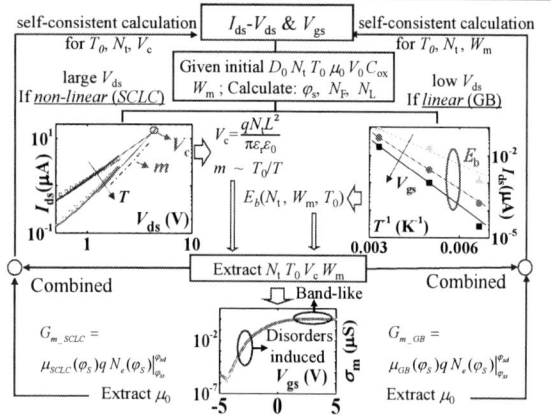

Fig. 5. Temperature dependent transfer and output current based extraction flow of key disorder parameters (i.e., N_t, T_0, V_c and W_m).

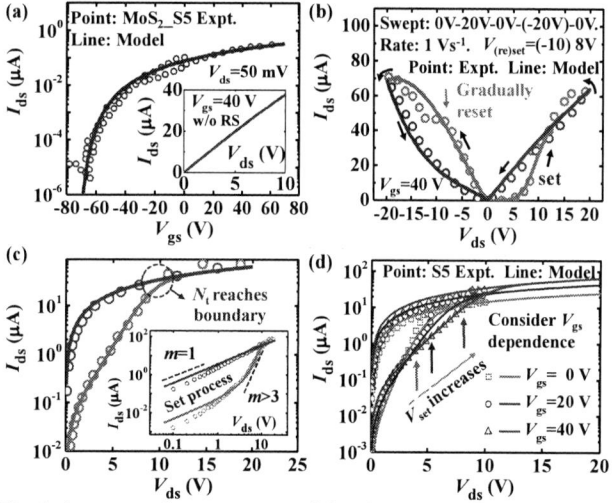

Fig. 6. Agreement between our model and experimental data (from [1]) of transfer current (a), RS current (b), RS current in log-scale (c) and RS current under different V_{gs} (d). The output current without RS is shown in the inset of (a). Besides, the log-log scale of RS current is shown in the inset of (c).

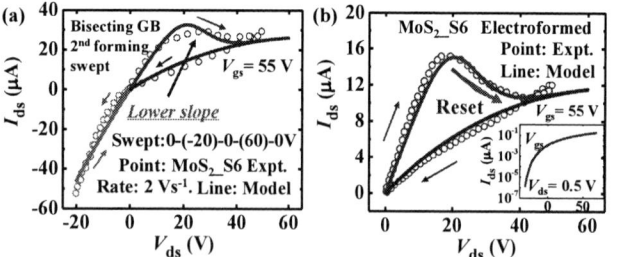

Fig. 7. Agreement between our model and experimental data (from [1]) of RS current in 2nd forming cycle (a) and electroformed cycles (b). At forming (formed) state, μ_0=0.070 (0.032) cm^2/Vs. The simulation result of transfer current is shown in the inset of (b).

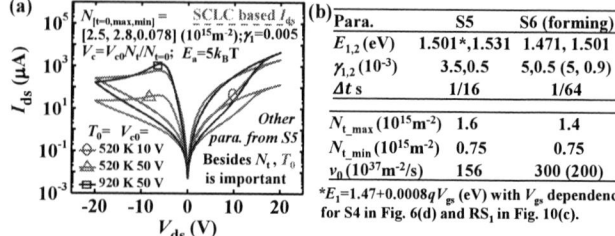

Fig. 8. (a) Predication of SCLC based RS current under different T_0 and V_{c0}. (b) Parameters used for simulating the RS behaviors of S5 and S6. The gate tunable behavior is achieved by considering V_{gs} dependent E_1, which dominates the set voltage.

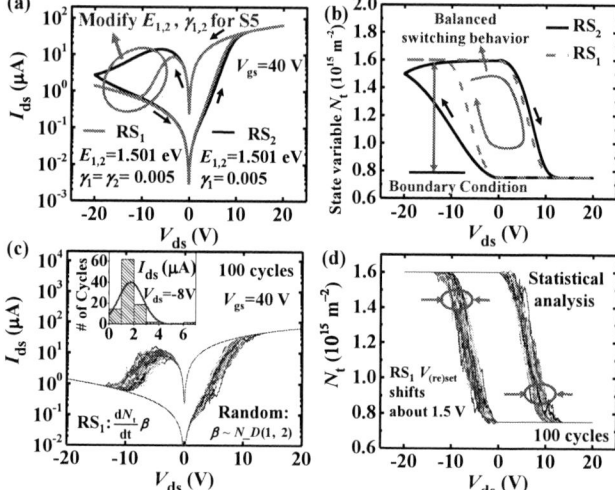

Fig. 9. Simulation results of RS current (a) and state variable (b) for RS₁ and RS₂ with different sets of $E_{1,2}$ and $\gamma_{1,2}$. Simulation results of statistical hopping effects on RS current (c) and state variable (d). β obeys the normal distribution with mean value (~ 1) and standard deviation (~ 2). The statistical distribution of current is inserted in (c).

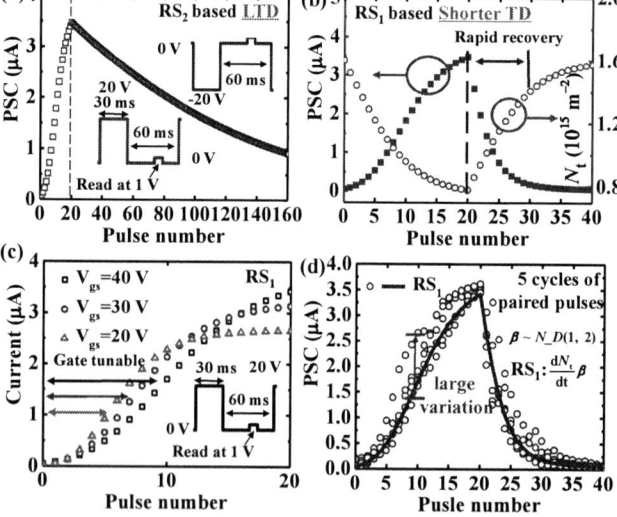

Fig. 10. Simulation results of postsynaptic current (PSC) based on RS₂ (a) and RS₁ (b) under 20 potentiation pulses and 160 (20 for RS₁) depression pulses. The corresponding state variable of RS₁ is shown in (b). (c) Simulation results of the potentiation performances under different gate voltages for RS₁. (d) Simulation results of the PSC with statistical effects.

Design and Optimization of β-Ga₂O₃ on (h-BN layered) Sapphire for High Efficiency Power Transistors: A Device-Circuit-Package Perspective

Bikram K. Mahajan[+], Yen-Pu Chen, Woojin Ahn, Nicolò Zagni and Muhammad Ashraful Alam[*]

Bikram K. Mahajan, Yen-Pu Chen, and Woojin Ahn contributed equally to this study.

Department of ECE, Purdue University, West Lafayette, IN 47907 USA, Email: [+]bmahaja@purdue.edu, [*]alam@purdue.edu

Abstract – Despite exceeding the Baliga's Figure of Merit (BFOM) by 400% and Huang's Chip Area Manufacturing FOM (HCAFOM) by 330% [1], the performance of existing β-Ga₂O₃ FETs is inferior to that of GaN, primarily due to extreme self-heating. Self-heating effect (SHE) has emerged as an important concern for device performance, output power density, run-time variability and reliability for modern logic transistors. The effects are even more severe for high-power transistor where the channel material may be a poor thermal conductor, e.g. β-Ga₂O₃. Very high internal electric fields, extreme temperature and mechanical stresses associated with these transistors drive electrochemical reactions [2], influence atomic processes [3], and accelerate multiple non-equilibrium effects [4]. A device-circuit-package, multi-physics, multi-scale simulation is needed to capture these effects self-consistently, but such a model has not yet been developed. In this paper, we (i) **develop** the first self-consistent device (TCAD), circuit (HSPICE), and package (COMSOL) model considering SHE which predicts FET performance on variety of substrates accurately; (ii) use the model to **propose** a novel hexagonal-Boron Nitride (h-BN) based β-Ga₂O₃ FET with 30% (cf. Sapphire substrate) and 80% (cf. SiO₂ substrate) reduction in thermal resistance (R_{th}); (iii) **demonstrate** the performance of boost converter (with parameters extracted from our TCAD model) with h-BN based β-Ga₂O₃ FET, which outperforms the existing β-Ga₂O₃ FETs, achieving an efficiency within 10-15% of highest performing enhancement mode (E-mode) GaN FET; (iv) **propose** h-BN based FinFET which exceeds the I_ON of the existing β-Ga₂O₃ FET by more than 500%; and (v) **develop** a Faraday-cage type novel packaging strategy for effective heat dissipation and efficient system performance in β-Ga₂O₃ FETs.

I. INTRODUCTION AND MOTIVATION

Background: β-Ga₂O₃ power transistors are projected to be less expensive (the substrate is cheaper) but achieve comparable/higher performance than GaN and SiC devices (Fig. 1). The impressive experimental demonstrations include: a kV-class Schottky barrier diode [5], lateral FETs [6] and D-mode FETs with high breakdown [7]. It also has potential for low power loss during high frequency switching in GHz regime [1]. Unfortunately, β-Ga₂O₃ have a poorer electron mobility (μ), and much lower thermal conductivity (κ) compared to SiC and GaN. These intrinsic material properties cannot be changed, but device-circuit-package modeling can identify the bottlenecks and novel device design strategies can alleviate them.

Shortcomings of existing models: Device-circuit-package models provide a critical interface between performance of the device and the ultimate system performance, however, it has been a challenge to accurately model β-Ga₂O₃ FETs. Existing TCAD models [8] calculate $I_D - V_G$ characteristics with effective parameters, but the failure to include SHE makes the model ineffective in predicting FET response on substrates with various κ. Another model does account for SHE and predicts channel thickness dependent R_{th}, however, it has not

been calibrated extensively with experimental data [1]. Most importantly, neither types of models predict device-circuit-package implications of the design.

Key Contributions: In this paper, we develop a self-consistent TCAD model to predict the I-V characteristics for β-Ga₂O₃ FETs with different substrates. We extract thermal time constants ($\tau = R_{th} C_{th}$) from the TCAD model and calibrate the HSPICE model with experimental data. We compare the boost converter performance of β-Ga₂O₃ FET using HSPICE simulation. We propose a novel design on h-BN layered Sapphire substrate to relieve the thermal choke-point. We demonstrate that the proposed device outperforms traditional transistors fabricated on SiO₂ and Sapphire. Finally, an innovative packaging strategy is proposed that allows for efficient heat dissipation in power modules.

II. TCAD AND HSPICE MODEL

A. TCAD model: The TCAD model must account for the specific device physics of wide band gap (WBG) β-Ga₂O₃. For the calibration against experimental data, we considered β-Ga₂O₃ FET fabricated on two substrates: Sapphire and SiO₂ [9]. The substrate-specific simulations ensure that the thermal models are calibrated, decoupled from the transport models for electrons/holes. Fig. 2 shows the device structure (taken from [9]) for TCAD simulation. Once the model is calibrated with experimental data, a wide variety of transistors can be simulated using the TCAD model. The calibration of temperature-dependent mobility is particularly important. Note that the Ga-O bond is strongly ionic which gives rise to strong Frölich interaction and low optical phonon energies [10]. This specific mobility model does not exist in classical device simulators, but the corresponding power-law can be approximated by using an empirical exponent to reproduce temperature-dependent mobility within 300-500K [3]. For our simulations, μ of 30 cm²/V-s with exponent of 0.5 and 20 cm²/V-s with exponent of 0.8 were used for Sapphire and SiO₂ substrates, respectively. We assume an idealized device, where the effect of trapping/detrapping is not significant. The doping concentration has been kept constant at 2.7 x 10¹⁸/cm³ [9]. The SHE depends on the thermal environment. κ of various layers shown in Fig. 2(b) are summarized in Fig. 3. For this model calculation, we have thermally grounded all the contacts of the device (source, drain, gate, and substrate).

The comparison between TCAD and HSPICE model predictions and the experimental data is shown in Fig. 4. The on-currents (Fig. 4a and 4b) as well as the threshold characteristics (Fig. 4c) are precisely reproduced. The temperature rise (ΔT) predicted from the TCAD model matches the experimental data reasonably well (Fig. 5). Since this calibration device is a D-mode FET, significant negative

978-1-7281-1988-5/18 $31.00 © 2018 IEEE

voltage (~-28 V) must be applied to turn the transistor off. D-mode FETs are undesirable because they require complex gate drive circuit for fail-safe operation [8]. Fortunately, the threshold voltage can be tuned by varying β-Ga$_2$O$_3$ thickness, as shown in Fig. 6. A sub-20nm film may be necessary to create a traditional E-mode transistor.

B. HSPICE FEOL Compact Model: A HSPICE model is needed to determine the circuit performance of a power transistor. First, the TCAD simulation results discussed above is used to calculate R_{th}, using $R_{th} = \Delta T/P$, where ΔT is the lattice temperature difference and P is the power. The average R_{th} obtained for the FET on Sapphire and SiO$_2$ substrates were 705.25 and 2447.2 K/W, respectively. Similarly, the thermal capacitance, $C_{th} \sim C_v \rho V$, (where C_v is the specific heat, ρ is the density, V is the heated volume) was calculated [11]. These values, along with temperature-dependent μ and geometrical parameters, were used to refine and recalibrate MIT Virtual Source GaNFET-HV HSPICE model [12]. The validity of the self-consistent electro-thermal HSPICE model is shown in Fig. 4, where it anticipates the experimental data very well.

III. BOOST CONVERTER

The calibrated HSPICE model can now be used to analyze the performance of a typical power electronics circuit (such as a boost converter, see Fig. 7) based on different β-Ga$_2$O$_3$ FETs. Since V_{th} varies with channel thickness, a feasible range of V_{th} for the converter operation must be determined. From Fig. 8, we find that for $V_{th} = -4\ to\ -10$ (corresponding to thickness of 10-40nm) gives the highest efficiency ($\eta = P_{out}/P_{in}$) for FET on Sapphire. For SiO$_2$, however, the efficiency is essentially independent of thickness, due to extremely high SHE associated with the substrate. Fig. 9 shows the efficiency for a wide range of duty cycles. For high conversion ratio (V_{out}/V_{in}) as well as high efficiency, $D = 40\%$ has been chosen for the subsequent simulations and the corresponding converter transient response is shown in Fig. 10. These simulations confirm that SHE limits the β-Ga$_2$O$_3$ FET efficiency, hence new thermal shunts and/or innovative packing strategies e.g. thermal Faraday cage are needed for improved performance.

IV. THERMAL SHUNTS: A H-BN BASED FET SOLUTION

h-BN is an emerging WBG material ($E_g \sim 6eV$) with sp^2 hybridized Boron and Nitrogen atoms arranged in graphite-like layers. It has very high κ (bulk: 400W/m.K [13]) and has been previously used as dielectric layers [14] in electronics. h-BN growth on a Sapphire substrate has already been achieved [15] and can be an excellent choice for β-Ga$_2$O$_3$ FETs. Fig. 11 shows the schematic of the proposed h-BN based FET. Figs. 12 and 13 show that even though the drain current of a h-BN based FET is comparable to that of Sapphire based FET, but ΔT (and correponding R_{th}) decreases considerably for the new transistor. Fig. 14 shows that the efficiency of D-mode ($V_{th} = -4V$) h-BN based β-Ga$_2$O$_3$ FETs is within 10-15% of highest performing E-mode GaN boost converters [16]. With additional R_{th}-engineering, such as top edge thermal shunting [1], the efficiency will match that of the GaN devices.

If E-mode devices are required, h-BN based FETs must use a surround-gate configuration, see Fig. 15. The $I_D - V_G$ and transconductance curves are shown in Fig. 16 and the subthreshold slope has been determined to be 85 mV/dec. At least 500% enhancement in I$_{ON}$ has been observed for h-BN based FET (Fig. 17) compared to the existing FET (inset) [17].

V. THERMAL FARADAY CAGE: A PACKAGING SOLUTION

To determine the overall system efficiency of power transistors with β-Ga$_2$O$_3$, one must account for SHE in packaging modules to get system level reliability. The ambient temperature (T_a, for the TCAD model) must be derived from the COMSOL FEM analysis of a packaged device, see Figs. 18 and 19. In particular, we compare classical side-by-side inductor-transistor packaging with a novel thermal "Faraday-cage" packaging (Fig. 19) to reduce inductor-IC mutual heating. In the classical configuration, Faraday caging does not reduce the temperature significantly. However, Faraday caging reduces inductor temperature, T_{ind} significantly in the reduced-footprint, stacked inductor-transistor configuration. The temperature rise is similar to that in GaN commercial packaging modules [18]. Reduced T_{ind} lowers the inductor loss so that the efficiency can be further enhanced [19]. Fig. 20 summarizes the implication of various packaging configuration on T_{ind} and T_{IC} for various configurations. We realize that the combination of two innovative ideas proposed in this paper (e.g. h-BN thermal shunt and thermal Faraday caging) leads to maximum reduction of SHE. As mentioned previously, the corresponding decrease in R_{th} (Fig. 22, including the heat dissipation in the inductor, PCB board, and encapsulants) and τ will significantly increase the efficiency (Fig. 23) and operating speed of β-Ga$_2$O$_3$ FETs.

VI. CONCLUSIONS

Despite the potential in terms of BFOM and HCAFOM, and despite the promising experimental demonstration of variety of power devices (e.g. Schottky diode, FETs), we conclude that the low μ and poor κ will ultimately limit the efficiency and frequency response of β-Ga$_2$O$_3$ FETs far below GaN or SiC FETs. An in-depth device-circuit-package model quantify the magnitude of self-heating and performance loss for a wide variety of configurations. Here, we suggest two innovative concepts based on h-BN dielectric layer and thermal Faraday-cage packaging to reduce SHE and increase performance. Together with the significantly lower substrate cost, the new device-circuit-package structure is likely to deliver the performance improvement originally promised by β-Ga$_2$O$_3$ FET.

VII. REFERENCES

[1] Jessen *et al.*, *DRC*, 2017.[2] Gao *et al.*, *IEEE TED*,2014.[3] Tsao *et al.*, *Adv. Electron. Mater*,2018.[4] Christensen *et al.*, *Ph.D. Thesis, GaTech.*, 2009.[5] Konishi *et al.*, *APL*, 2017.[6] Wong *et al.*,*IEEE EDL*, 2016.[7] Huang *et al.*, *IEEE TED.*,2004.[8] Wong *et al.*, *ISPSD*, 2018.[9] Zhou *et al.*, *ACS Omega*, 2017.[10] Ma *et al.*, *APL*, 2016.[11] Lienhard *et al.*, *A Heat Transfer Texbook*, 2016.[12] Radhakrishna *et al.*, *IEDM*, 2013.[13] Zhang *et al.*, *J. Mater Chem. C*, 2017.[14] Jeong *et al.*, *ACS Nano*, 2015.[15] Yang *et al.*, *J. Cryst. Growth*, 2018.[16] Das *et al.*,*IEEE EDL*, 2011.[17] Chabak *et al.*, *APL*, 2016.[18] GaN Systems, Application Note: GN005, 2014. [19] Ho *et al.*, *IEEE TPE*, 2012.

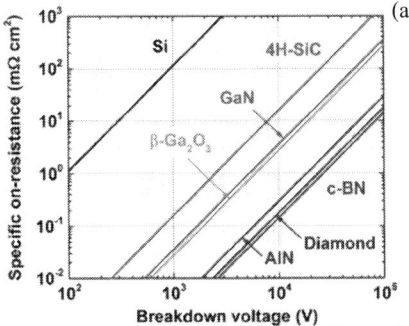

(a)

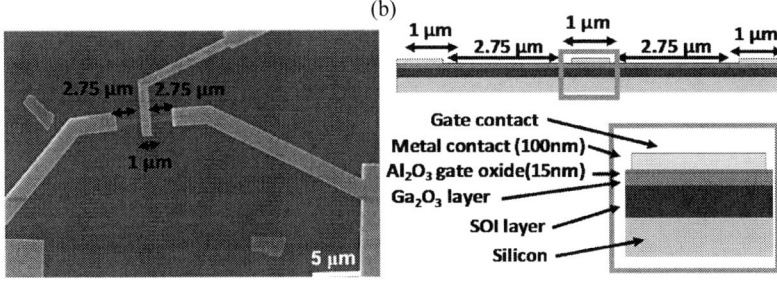

(b)

Fig. 1: Theoretical unipolar performance limits of R_{on} as a function of V_{BR} for β-Ga$_2$O$_3$ and other major semiconductor power devices, taken from [3].

Fig. 2: (a) Top view scanning electron microscope image of the device with dimensions, taken from [9]. Exact dimensions are critical for achieving a good match with TCAD simulation. (b) Device structure simulated in Synopsis. For the case where β-Ga$_2$O$_3$ layer is on Sapphire, SiO$_2$ layer and Silicon are replaced with Sapphire.

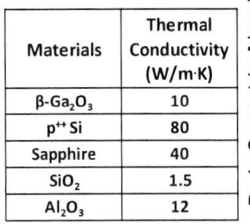

Materials	Thermal Conductivity (W/m·K)
β-Ga$_2$O$_3$	10
p^{++} Si	80
Sapphire	40
SiO$_2$	1.5
Al$_2$O$_3$	12

Fig. 3: Thermal conductivity values for various substrates used in the simulations.

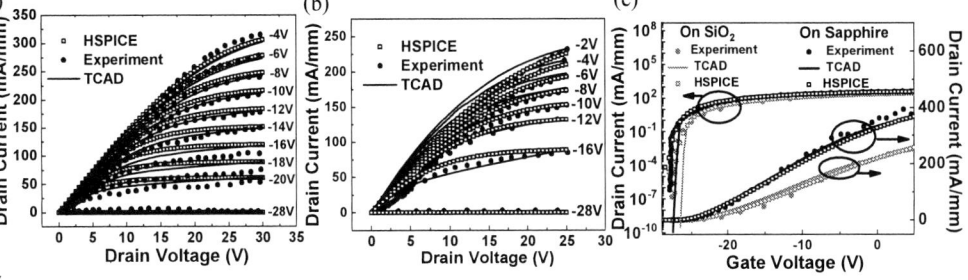

Fig. 4: TCAD and HSPICE model calibration with experimental $I_D - V_D$ data with β-Ga$_2$O$_3$ FET on (a) Sapphire and (b) SiO$_2$ substrates. (c) Model calibration with experimental $I_D - V_G$ data with β-Ga$_2$O$_3$ FET on Sapphire and SiO$_2$ substrate in both linear and log scales.

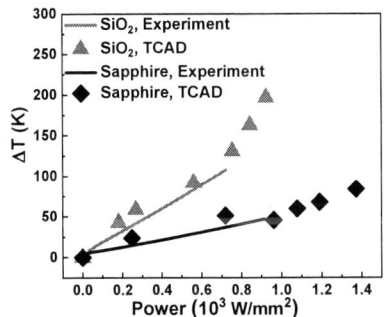

Fig. 5: The increase of temperature with power (V_{DS} x I_{DS}) shows good match with experimental data. The comparison validates the self-consistent electro-thermal TCAD model.

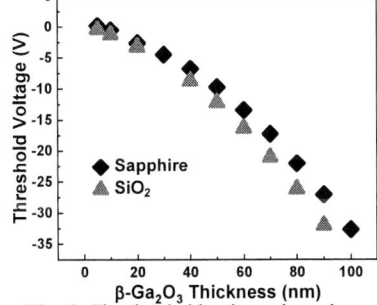

Fig. 6: The threshold voltage dependence on channel thickness. The standard D-mode FETs must be pinched-off to be turned off, therefore thinner epitaxial layer is required to fully deplete the channel at 0V.

Parameters	Value	Parameters	Value
Mobility	30/20 cm^2/V-s	SD Distance	6.5 μm
Vt	-4	Gate Length	1 μm

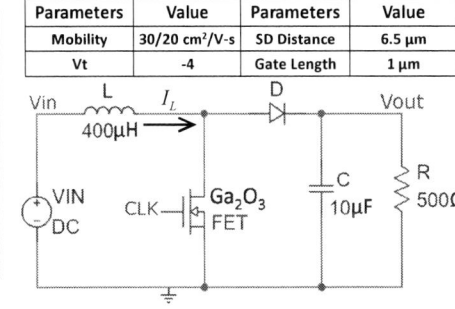

Fig. 7: Boost converter is an important and widely used power electronics circuit. In this figure a boost converter circuit based on β-Ga$_2$O$_3$ FET is presented. The parameters are extracted from TCAD and taken from the actual device [9]. [CLK: clock; I$_L$: inductor current; D: diode]

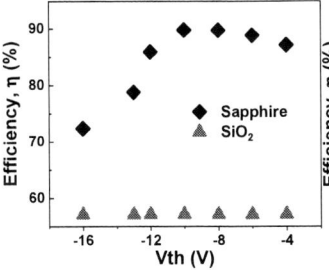

Fig. 8: Threshold voltage dependence of the efficiency of the boost converter. From this figure we conclude that threshold voltage of -10 to -4V, (corresponding to 10-40nm thickness of β-Ga$_2$O$_3$ as channel) will be ideal for a boost converter.

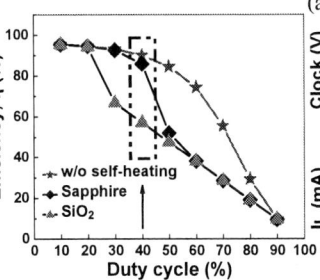

Fig. 9: Duty Cycle dependence of the efficiency η of the β-Ga$_2$O$_3$ based boost converter. A trade-off exists between the efficiency and the conversion ratio and hence D=40% has been chosen as the duty cycle for all simulations, to get the best performance.

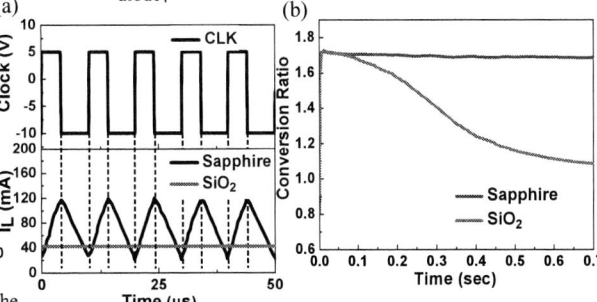

Fig. 10: (a) Boost converter transient behavior (Duty Cycle of 40%). The driving voltage (CLK) and inductor current (I$_L$) are shown for FETs on Sapphire and SiO$_2$. I$_L$ is suppressed due to severe SHE. (b) Transient response of conversion ratio (V_{out}/V_{in}) of β-Ga$_2$O$_3$ FET for Sapphire and SiO$_2$ substrate. We can see that SHE severely limits the output voltage. [CLK: clock; I$_L$: inductor current]

978-1-7281-1988-5/18 $31.00 © 2018 IEEE

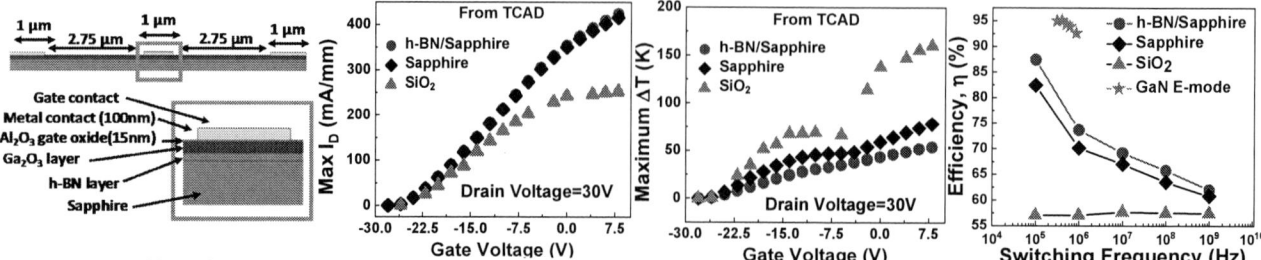

Fig. 11: Proposed h-BN based FET. First the h-BN layer is grown on Sapphire, and then the β-Ga₂O₃ is exfoliated and placed on top, followed by deposition of the contacts and the Al₂O₃ layer.

Fig. 12: Maximum drain current as a function of gate voltage (V_{DS}=30V) for β-Ga₂O₃ FET on h-BN/Sapphire, Sapphire and SiO₂. The current for h-BN/Sapphire and Sapphire are comparable because the mobility is presumed similar.

Fig. 13: Maximum ΔT rise for h-BN/Sapphire is almost half as that of Sapphire, and almost 1/6 of that of SiO₂ (V_{DS} =30V)

Fig. 14: Efficiency comparison of boost converters based on different FETs. The efficiency of h-BN is within 10-15% of E-mode GaN FET. GaN Data obtained from [16]

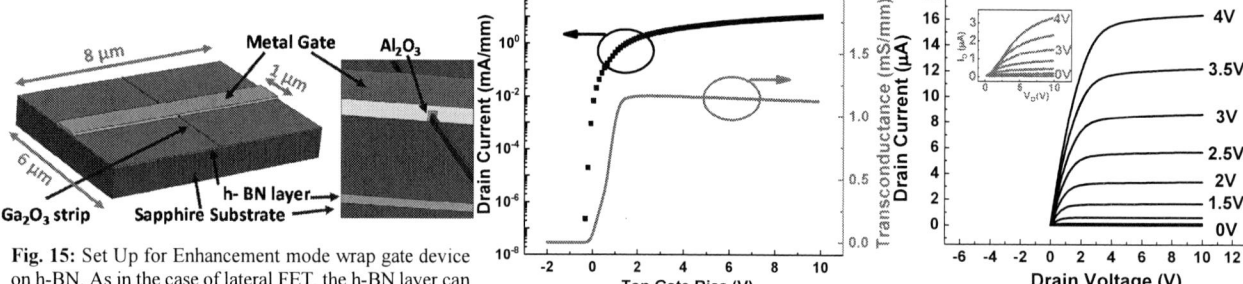

Fig. 15: Set Up for Enhancement mode wrap gate device on h-BN. As in the case of lateral FET, the h-BN layer can be deposited on Sapphire, followed by deposition of β-Ga₂O₃ layer, which can be etched into fins using a hard mask as in [17]. The contacts and the oxide layer can then be deposited.

Fig. 16: $I_D - V_G$ of the β-Ga₂O₃ FET on h-BN coated Sapphire substrate, as obtained from TCAD simulation. The subthreshold slope is approximately 85 mV/dec.

Fig. 17: $I_D - V_D$ data with β-Ga₂O₃ FET on h-BN coated Sapphire substrate, obtained from TCAD, which is about 500% better than the experimental data from [17], given in inset.

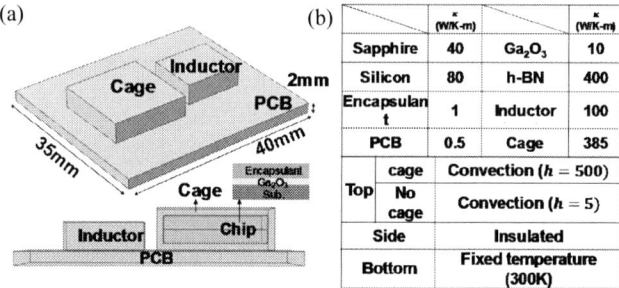

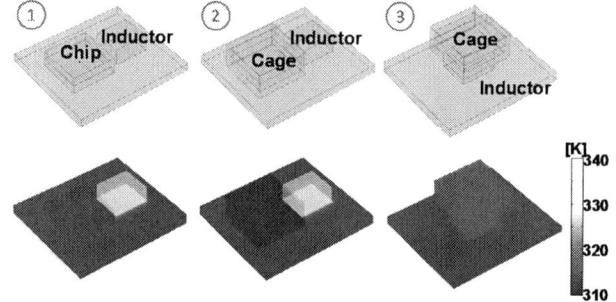

Fig. 18: (a) System level simulation schematic of β-Ga₂O₃ FET on various substrates. 'Cage' refers a "thermal Faraday Cage" which allows heat flux shielding. **(b)** The table summarizes the material properties, and boundary conditions for FEM simulations.

Fig. 19: Temperature distribution comparison among different packaging configuration. Inductor temperature of vertical stack shows a considerable reduction in mutual heating effect through the use of a thermal Faraday cage.

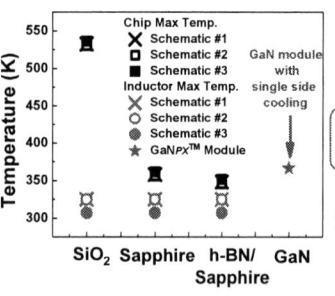

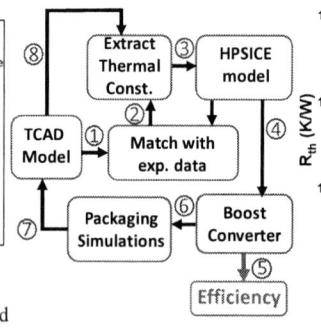

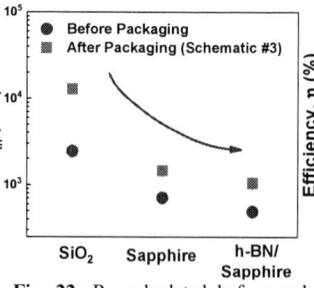

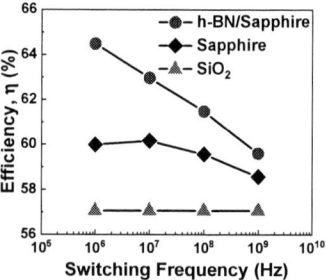

Fig. 20: Chip (β-Ga₂O₃ FET) and inductor temperature comparison among various packaging strategies, (Schematic #s) device structure. GaN data taken from [18]

Fig. 21: Schematic showing the flow chart of analysis done in this paper.

Fig. 22: R_{th} calculated before and after packaging of the β-Ga₂O₃ FET using the strategy proposed in schematic #3. This strategy provides the highest output while minimizing the IC area.

Fig. 23: Efficiency of β-Ga₂O₃ FET on various substrates after packaging. The efficiency is reduced due to the thermal resistance associated with lower κ materials used for packaging.

978-1-7281-1988-5/18 $31.00 © 2018 IEEE

Deconvoluting charge trapping and nucleation interplay in FeFETs: Kinetics and Reliability

Milan Pesic[1], Andrea Padovani[1], Stefan Slesazeck[2], Thomas Mikolajick[2,3], and Luca Larcher[4]

[1]MDLSoft Inc, Great America Parkway, Santa Clara, CA, USA, email: milan.pesic@mdlsoft.com
[2]NaMLab gGmbH, Dresden, Germany, [3]IHM, TU Dresden, Germany, [4]DISMI, University of Modena and Reggio Emilia, Italy

Abstract—Discovery of ferroelectric (FE) behavior in HfO_2 removed the compatibility roadblocks between the state-of-the-art CMOS and FE memories. Even though FE FETs (FeFETs) are scaled into 22 nm nodes and beyond, the limits of the technology as well as the physical mechanisms and reliability are still under research. In this paper we successfully developed a multiscale modeling platform to understand the interplay between the FE switching and charge trapping. Starting from the nucleation theory and rigorous charge transport modeling we present for the first time a self-consistent modeling framework we used for investigation of reliability and variability in FeFETs.

I. DEVICE AND MATERIAL CHARACTERIZATION

The Ferroelectric FETs (FeFETs) represent the core of both low-power non-volatile memory (NVM) solutions and the low-power negative capacitance (NC) logic. Despite the ferroelectricity (FE) in HfO_2 has triggered diverse research and enabled demonstration on 22nm technology nodes [1], the comprehensive understanding of both materials and devices is still under debate. Particularly, the interplay between charge trapping and ferroelectric switching, and its impact on the switching kinetics as well as on the reliability and variability of the device is not clear. Thus, the quantitative understanding of the charge trapping and nucleation interplay is crucial. In detail, the key questions to be answered are: (1) How does the interplay between ferroelectric switching and the charge trapping in HfO_2 influence the performance and scalability of FeFETs; (2) what are the boundary conditions for a reliable operation; (3) how can the material properties be tailored to achieve the maximum reliability for such devices?

Polarization reversal due to the application of the gate voltage sets the FeFET into high and low V_{TH} state (Fig.1). The characterization of FeFET (28nm bulk FET under test) considered for NVM applications shows that during endurance testing the closure of the memory window (MW=V_{TH+}-V_{TH-}) starts at the point when the charge trapping begins to dominate the FE switching process (Fig.2). To investigate the complexity of the switching process and find optimal operation conditions, we recorded the program/erase (PRG/ERS) matrices, Fig. 3. Within these maps, three regions can be identified: (1) a region of negative MW where charge trapping (CT) dominates; (2) a region of negligible MW (CT compensates the FE switching); (3) and a region of positive MW where the FE switching dominates. Fig. 4 shows the evolution of these regions with field cycling, exhibiting a drift of the switching region towards pulses with lower amplitude, stabilizing at amplitudes of

around -4 and -5 V and short pulses (1 µs). After 10^5 PRG/ERS cycles the CT completely compensates the MW of the FeFET resulting in negative or zero MW. Moreover, the evolution of density of interfacial states reported in [3] indicates that bipolar stress generated by PRG/ERS and the subsequent charging and discharging (carriers tunnel back and forth) degrade the interface properties. Consequently, the increased defect density and deteriorating interface properties may impact the switching kinetics of the FeFET.

II. MULTI-SCALE MODELING PLATFORM

To capture the aforementioned FE switching - CT interplay and reproduce the FE-device kinetics as well as to understand the physical mechanisms behind, we extended a multiscale modeling platform (Fig. 5). This modeling platform is comprised of two main parts, consistently connected, which address the charge transport and stress-induced material modifications, respectively. These are crucial to model device aging and reliability phenomena [4]. The capability to handle individual contributions of atomic species and defects (i.e. interstitial ions and vacancies) is included, along with their diffusion and generation processes. In parallel, the FE properties of the film responsible for switching processes and kinetics are also accounted for. The electric potential within the FE-device is calculated by solving the Poisson equation consistently with time dependent Ginzburg-Landau formalism enriched with the domain interaction and depolarization term. Parameters used in the study are listed in [5,6].

III. CHARGE TRAPPING AND NUCLEATION INTERPLAY

A. Charge trapping

Multiphonon charge transport mechanisms dominated by trap-assisted-tunneling are used to investigate the charge transport within the device whose geometry and doping is depicted in Fig. 6a. We investigated the FE-device behavior by simulating the I_D-V_G evolution during a PRG/ERS cycle. First a device under test was preconditioned and set in the low V_{TH} state to maintain the reference state. As shown in Fig.6b, charge trapping results in a clockwise I_D-V_G curve hysteresis (CW) and negative MW, whereas the FE switching exhibits a counter CW (CCW) hysteresis. The typical clockwise (CW) hysteresis of I_D-V_G curves upon the charge trapping is successfully reproduced by simulating a single pulse (V_{TH}-extracted on rising/falling edges). To investigate the defect distribution within the stack, we consistently model capacitance-voltage (C-V) and I_G-V_G curves of a multistructure- devices (large total gate area) together with parametrized charge trapping experiment (see

Fig.7a-c) in turn extracting a defect map responsible for the CW hysteresis (Fig.7d).

A high-operation field results in a high field drop across the low-k interface yielding a much higher generation of defect states within the interfacial layer (IL), see Fig.8a-c. To investigate the impact of the PRG/ERS on the bulk defects in the HfO$_2$ layer, we monitored and simulated (using a TAT model) the leakage current [7], which allows extracting the defect density evolution with stress shown in Fig.8b-c. This increase remains limited as the current transport is dominated by conduction band conduction. Evolving concentration is omitted deliberately due to the combined effect of defect concentration and switching state of the device (details in section B).

B. *Interplay between charge trapping and nucleation*

The impact of interfacial/bulk defect states on FeFET variability is assessed through Monte Carlo simulations, allowing to investigate the impact of the scaling and defects of the polycrystalline stack on the V$_{TH}$ distribution (Fig.9a). The impact of the defects and trapped charge on the potential within the gate stack is shown in Fig.9b.

To assess the impact of the FE-HfO$_2$ material changes on FE-device performances (including variability and reliability), we show in Fig. 10 the simulation of the evolution of the FE domain orientation over time upon the application of the external excitation field. These domain maps represent a top view of the gate of the FE-transistor with source and drain on the left and right side, respectively. The 2D domain map shows a clear domain reorientation from the initial random condition to a common oriented state, which proves that the model captures nucleation occurring upon application of the electric field. The model accounts self-consistently for the potential changes due to the domain nucleation, as shown in the corresponding potential maps in Fig.11a. Multiple PRG/ERS operations followed by I$_d$-V$_g$ readout results in distributions of V$_{TH}$+, V$_{TH}$- and MW (Fig.11b), which is nicely captured by the simulations. Additionally, impact of the different grains (FE switching properties) along the width of device on single device performance was investigated (considering the average grain size of 20-30 nm, 30x80nm FeFET has in average 3 grain along the channel; see [8]). It can be seen that the worst domain (lowest MW when alone) determines the MW of the device (without trapping effects included). Still averaging effect is observable when portion of device is non-FE or pinned; in turn impacting the device to device variation of the FeFET.

In the next step we switched on previously calibrated charge trapping into the model. The band diagrams extracted from the grain 1 and grain 2 (Fig.11a) are shown in Fig.13a. Different field over the device changes the dynamics of the charge carrier injection (Fig.13b-c) which in turn impacts the nucleation of domains along the width of the channel (Fig. 14). Same impact has a band bending and injection level change due to the variation of the program pulse amplitude. Using the developed model, we show that gate leakage is determined by the conduction band transport (similar to FE tunnel junction), thus

the polarized device is characterized with a higher leakage compared to the non-switched device (13b). To gain insight into the kinetics of the trapping, we set the device in different PRG states and simulated a charge trapping (Fig.13c-d) and gate transients (V$_G$, I$_G$) upon application of a single pulse. The comparison of charge trapping kinetics simulated considering both conventional non-switching HfO$_2$ FET and programmed FeFET shows that the increased band-bending (induced by FE switching) impedes the charge trapping (for this scanning voltage and extracted defect distribution) within the stack, decreasing the trapped- ΔV$_{TH}$, but increases the gate leakage.

To gain insight into the kinetics of the device and competing processes of charge trapping and FE switching we simulated the FE-gate stack response upon application of the positive gate voltage pulse. Compared to the reference device (without defects and charge trapping), the device with CT effects included (Fig. 15a-b) shows that trapping pins the field at the interface, inducing a stronger change in the field across the FE:HfO$_2$, which forces an earlier onset of FE switching (mono-domain) and ability to reach previously inactive "*slow*" domains (Fig. 14) in multi-domain case. However, with cycling this increased field and repeatedly charge tunneling back and forth through the IF layer results in a complete destruction of the interface and defect generation within the bulk. Finally, this degradation results in increased leakage current. The defect generation and consecutive trapping pins the domains and closes the MW of FeFET (Fig. 15c). The model of the competing processes and degradation is shown in Fig. 16.

IV. CONCLUSION

Within this paper we investigated the interplay between the ferroelectric switching and charge trapping in FeFET devices. We purposely extended a multiscale device modeling platform enabling the self-consistent assessment of this entanglement based on the nucleation FE theory and multiphonon mediated trap assisted transport. We show that the defect generation and subsequent charge trapping strongly impacts the switching kinetics within the FeFETs, thus influencing the distribution of the V$_{TH}$ and directly impacting the MW. With the help of simulations we point out that the degradation of the dielectric and consequent charge trapping results in: 1) a slow-down of switching kinetics and 2) a reduction of the memory window of the FeFET. In summary we demonstrated that the proposed comprehensive model can capture the central physics of the FeFET and as such paves the path towards deeper understanding of the device and further improvement of its already excellent memory properties.

ACKNOWLEDGMENT

The GLOBALFOUNDRIES Dresden Module One LLC & Co. KG is gratefully acknowledged for providing the hardware.

REFERENCES

[1] Trentzsch et. al. IEDM2017; [2] Pesic et al. Adv.Func.Mat.2016; [3] Pesic et al IEEE-TDMR2018; [4] www.mdlsoft.com; [5] Saha et al IEDM2017; [6] Hoffman et al. Adv.Func.Mat.2016; [7] Sereni et al. IEEE-TED2015. [8] Mulaosmanovic et al. IEDM 2015. [9] E. Yurchuk PhD Thesis 2014.

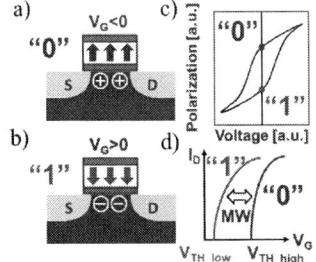

Figure 1. Schematic representation of the FeFET gate stack comprised of $Si/SiO_2/FE:HfO_2/TiN$ upon application of a) negative and b) positive pulse on the gate. Resulting c) ± polarization state yields d) high ("0") and low ("1") V_{TH} of FeFET

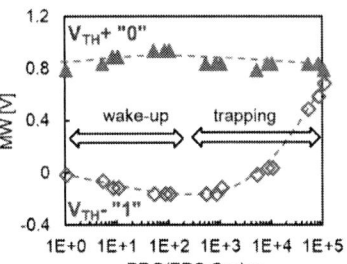

Figure 2. Endurance characteristics of FeFET. The difference between $V_{TH}+$ and $V_{TH}-$ defines the MW of the device. At higher cycle count the MW closes because charge trapping becomes dominant compared to FE switching.

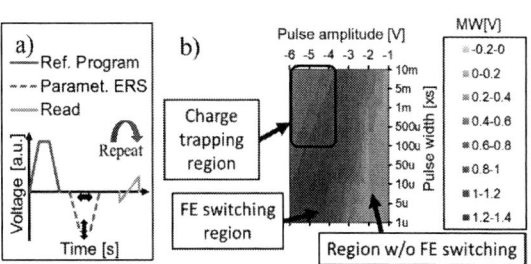

Figure 3. a) Reference pulse and parametrized erase pulse. b) Color intensity graph of erase matrix of woken up device with SiO_2 interface buffer layer. Three sub-regions can be seen: 1) charge trapping region 2) ferroelectric switching region 3) region without ferroelectric switching. The optimum operating conditions can be deduced from such graphs.

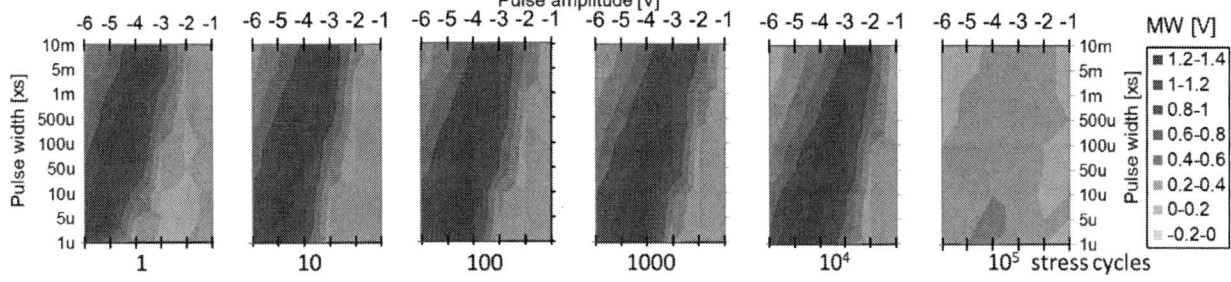

Figure 4. Evolution of the MW as a function of pulse width and pulse amplitude (erase matrix) with cycling. The different colors as shown on the right-hand side denote the memory window in volts.

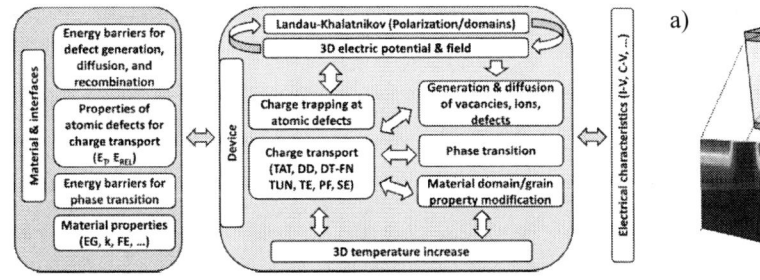

Figure 5. Flow chart of the multi-scale simulation framework accounting for the physical phenomena occurring in electron devices. The FE nucleation is selfconsistently coupled with charge trapping in discrete defects together with a Poisson solver.

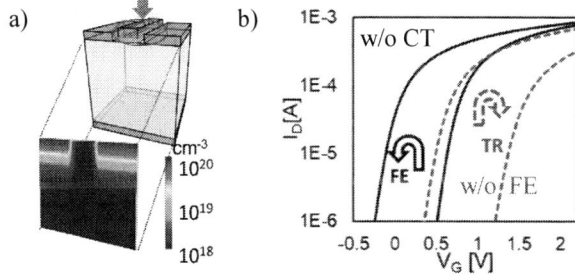

Figure 6. a) Device structure geometry of the simulated bulk FET with the 9nm thick HfO_2 and 1nm thick interfacial SiO_2 buffer layer. b) Simulated semilogarithmic transfer characteristics. The opposing impact of the ferroelectric switching and charge trapping within the FeFET is illustrated here showing CW hysteresis behavior of the charge trapping (when going from negative to positive voltages and back) and CCW behavior of the FE switching

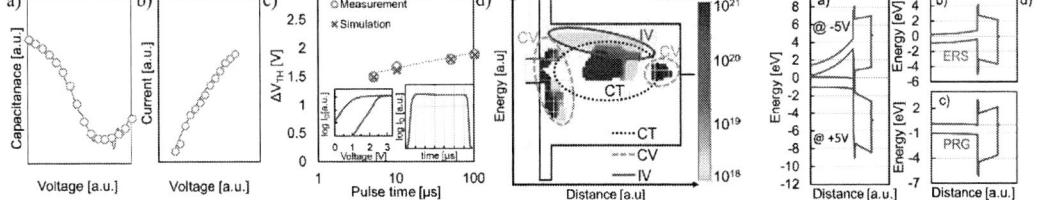

Figure 7. Measurements and corresponding simulations of a) CV, b) I_G-V_G [9] and c) ΔV_{TH} obtained from the single pulse measurements (see for inset I_D-V_G and $I_D(t)$) of a large gate area FeFET. d) Obtained defect map with sensitivity regions within band gap.

Figure 8. Band diagrams a) @ ±5V stress; b) ERS and c) PRG state @ 0V and d) Measured [9] and simulated (solid) leakage evolution during PRG/ERS stressing of the device.

978-1-7281-1988-5/18 $31.00 © 2018 IEEE   590

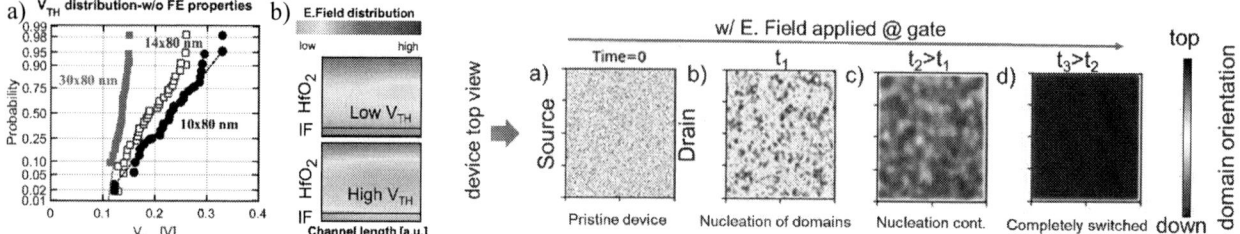

Figure 9. a) Simulated variability and impact of scaling on V_{TH} distribution of the control device with 9nm HfO$_2$ without FE properties. b) Local field distribution between two devices with high and low V_{TH}, respectively.

Figure 10. Simulated domain dynamic in 2D illustrating the nucleation processes within the device (top view) while setting it into the erase state. a) initial randomly distributed polarization (without applied electric field); b)-c) initial coalescence of the domains and nucleation; d) whole device area set into erased-set (negative polarization state).

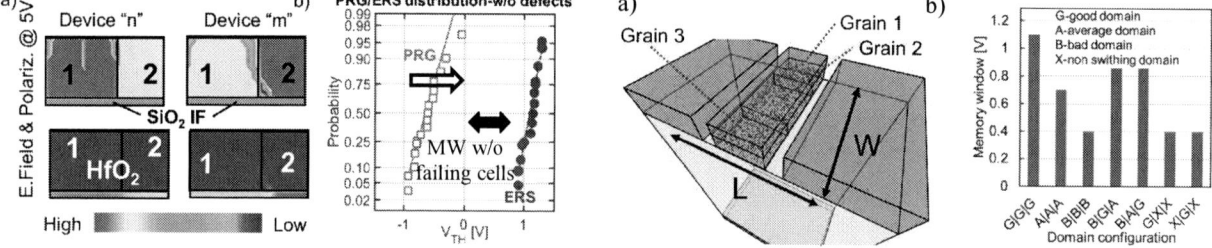

Figure 11. a) Local potential changes within the gate stack SiO$_2$(IF)/HfO$_2$ due to the nucleation of the domain. b) Statistical simulation of the V_{TH} distribution after PRG and ERS of the 30x80 nm FeFET device. Randomization of k-value; Landau parameters. 1 and 2 denote grain 1 and 2. Arrow indicates the shift with inclusion of trapping and degradation.

Figure 12. Impact of the FE properties variation along the channel. a) Geometry of the multigrain FeFET with different switching characteristics along the channel (with indicated 3 grains along the device W (30x80nm FeFET)) and c) MW dependence of the available domain configuration. The worst domain determines the V_{TH} causing the averaging effect.

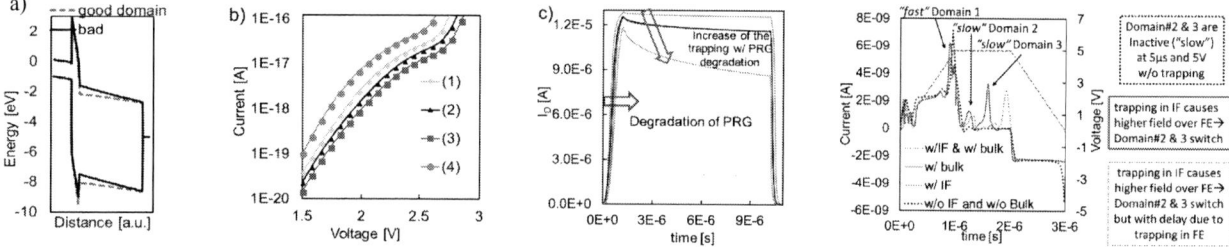

Figure 13. Simulated a) Band diagrams extracted from the grain 1 and grain 2 (Fig.12) and b) leakage current in (1) completely (2) partially switched 3) non-switched and 4) non-switched degraded FeFET. PRG state of the FeFET is considered. Different field over the device changes the dynamics of the injection of electrons (and nucleation of domains) along the width of the channel; c) I$_D$-transient of the degrading FeFET (equivalent to (1)-(4)).

Figure 14. Simulated switching kinetics of the 3-domain based FeFET depending on its interaction with interface and bulk defects.

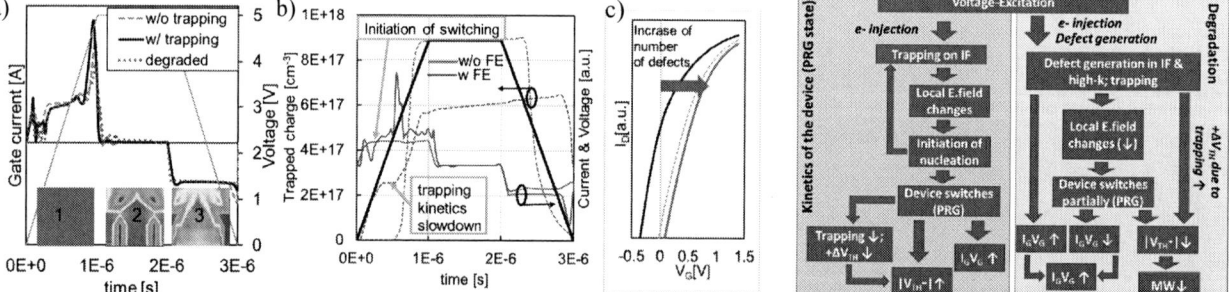

Figure 15. Kinetics of the switching. I$_G$(t) simulated w/ and w/o presence of the a) CT b) FE switching upon application of the positive V_G. IF trapping (in fresh FeFET) and change the field over the FE:HfO$_2$ induces switching (1) earlier compared to device w/o trapping (2). Increase of the defect concentration and trapping alters the electric field over the device and pins the domain out (3) closing the MW of the FeFET. c) PRG V_{TH} degradation with time.

Figure 16. Model of the (left) switching kinetics and trapping interplay and (right) degradation processes within the FeFET.

Impact of self-heating on reliability predictions in STT-MRAM

S. Van Beek[1], B. J. O'Sullivan, P. J. Roussel, R. Degraeve, E. Bury, J. Swerts, S. Couet, L. Souriau, S. Kundu, S. Rao, W. Kim, F. Yasin, D. Crotti, D. Linten and G. Kar

imec, Leuven, Belgium, [1]also at KU Leuven, Dept. ESAT, Leuven, Belgium. email: simon.vanbeek@imec.be

Abstract—At breakdown conditions, large current flows in STT-MRAM devices. We experimentally show that this large current causes significant self-heating of 200-300°C, which impacts the reliability extrapolation to operating conditions. By measuring and analyzing breakdown at various temperatures and on different MgO thickness, we successfully incorporate self-heating into the breakdown model. We find that the 10 year lifetime is underestimated by a factor 10^3 at 63-percentile, to even 10^7 when applying percentile scaling to 1 ppm.

I. INTRODUCTION

Spin-Transfer-Torque Magnetic Random Access Memory (STT-MRAM) is a key emerging non-volatile memory technology. The magnetic tunnel junction (MTJ) consists of 1 nm thick MgO dielectric tunnel barrier sandwiched in between 2 ferromagnetic layers. Information is stored in the magnetic orientation of the free layer (FL) and reference layer (RL). A high current density of more than 1 MA/cm² is pushed through the MgO when switching the FL. Besides causing significant self-heating [1], this high current can cause breakdown of the MgO barrier, limiting the endurance. Currently, predictions of the time-to-breakdown rely on acceleration models used in CMOS gate dielectrics, but do not correctly take into account the impact of self-heating [1].

In the present paper, we extend the breakdown model by incorporating the effect of self-heating. The self-heating effect is decoupled from the voltage acceleration by using thicker MgO (1.7 nm) with negligible self-heating. As such, it is possible to correct the apparent temperature acceleration seen in the industrially relevant thin MgO (1.0 nm) and estimate the temperature at breakdown conditions at 200-300°C. The corrected 63% lifetime extrapolation is more optimistic and can differ by a factor of 1000. Moreover, this factor increases after percentile scaling.

II. EXPERIMENTAL

The perpendicular MTJ-stacks with dual free layer (FL) MgO interfaces (Fig. 1) have been deposited on 300 mm wafers. The stack is patterned by either a reactive ion etch (RIE) or an ion beam etch (IBE) into circular pillars with electrical diameters (Ø) from 45 nm to 140 nm. The thickness of the MgO tunnel barrier is either 1.0 nm (industrially relevant thickness) and 1.7 nm (used for self-heating analysis), resulting in a resistance area product (RA) of 10 and 1000 Ωµm², respectively. The 1.7 nm stack has been fabricated into 1024x1024, Mbit, arrays (65 nm underlying CMOS technology). Each cell in the Mbit array can be measured using 4-point sensing (Fig. 2), with a pulsed constant voltage stress (CVS) or pulsed ramped voltage stress (RVS). The pulse width depends on the number of repeats of the clock cycle (Fig. 3). The resistance of the cells is read after each stress cycle. Also the 0T1MTJ devices are measured with 4-point sensing (Fig. 1). Breakdown is triggered as an abrupt transition of the MTJ resistance.

III. THERMAL SIMULATIONS

A simplified 2.5D thermal simulation of the stack allows assessing the time constants of heating and estimating the steady-state temperature. Fig. 4(a) illustrates the impact of the distance of the thermal boundaries on the maximum temperature in MgO for Ø 50 and 140 nm. When the thermal boundary, which is fixed at 25°C, is placed at 10 nm from the MTJ pillar, steady-state temperature is reached after 10 ns. When the full dimensions of the 0T1MTJ structure are considered, i.e. width of the isolation and total length of the TE and BE, there is additional heating with a slower pace, similar as in [2]. The additional heating is more significant in larger devices (20% for Ø 140 nm vs. 5% for Ø 50 nm). For simplification the thermal conductivity of the metallic multilayers used in the stack is given by $\sigma_{th,pillar}$. The value of the thermal conductivity of the metallic multilayers layers has an important impact on the total thermal resistance (Fig. 4(b)) and is by default chosen to be $\sigma_{th,pillar}$ = 10 W/(K.m).

In Fig. 5, the self-heating at steady-state is plotted as a function of stress voltage for 3 different areas and 2 MgO thickness (1.0 and 1.7 nm). There is more self-heating in the thin MgO. In addition, for the same voltage (dashed line), a large temperature difference $\Delta T_{V=cst}$ is observed between the different areas. For 1.0 nm MgO, this $\Delta T_{V=cst}$ results in breakdown improvement "beyond area scaling", i.e. small areas have higher V_{BD} than expected from area scaling of large devices (see Fig. 6). Here we make use of Poisson area scaling, see formula in inset of Fig. 6. The ratio A_x/A_y is calculated with the electrical resistance, assuming RA constant, which results in a ratio of parallel resistances R_y/R_x. Note that in RVS the resulting breakdown voltage (V_{BD}) can be converted into an equivalent breakdown time [3]. Also in this case, self-heating will significantly impact the lifetime extrapolation and needs to be incorporated into the breakdown model. However, since the thermal simulations use only rough estimates of the thermal conductivities of the numerous materials in the stack, we will first attempt to assess the impact of self-heating by direct experimental means. This can be accomplished by investigating breakdown behavior in 1.7 nm thick MgO.

IV. EXPERIMENTAL OBSERVATION OF AREA SCALING IN 1.7 NM MGO

As is demonstrated below, 1.7 nm MgO is expected to have negligible self-heating. In Fig. 7(a), the power (=current x stress voltage) right before breakdown is significantly lower for 1.7 nm MgO compared to 1.0 nm. As such, self-heating at breakdown is considerably reduced (Fig. 7(b)). Investigating the area scaling in these thick devices, results in the experimental observation of perimeter scaling for the RIE samples and area scaling for the IBE samples, suggesting edge breakdown in the former and uniformly distributed breakdown in the latter case. In addition, the breakdown parameters are extracted at different external temperatures (25, 50, 75 and 100°C). We make use of a maximum likelihood method that *simultaneously* fits the breakdown data t_i for a Weibull slope β, 63% breakdown time η_{ref} at reference voltage V_{ref} and the voltage power law extrapolation exponent n, as presented in [4]:

$$W_i = \ln[-\ln(1 - F_i)] = \beta \left[\ln t_i - \ln \eta_{ref} - n \cdot \ln\left(\frac{V_i}{V_{ref}}\right)\right] \quad (1)$$

The low Weibull slope β, observed for RIE confirms the hypothesis of edge breakdown by process-induced damage (Fig. 8(a)). The Weibull slopes have only a small temperature dependence, see Fig. 9(a). On the other hand, $\ln(\eta_{ref})$ and power-law exponent n show an Arrhenius behavior with activation energy α_T and ϕ_T, respectively (Fig. 9(b,c)). In Fig. 9(d) the breakdown voltage distribution for 4 different temperatures is plotted, illustrating the process improvement using IBE.

V. INCORPORATING SELF-HEATING IN THE BD-MODEL

When measuring the temperature acceleration of breakdown, the external temperature is assumed to be the device temperature (T_{BD}=T_{ext}):

$$W = \beta \left[\ln t - \ln \eta_{ref} - n \cdot \ln\left(\frac{V}{V_{ref}}\right) + \alpha_T \left(\frac{1}{kT_{BD}} - \frac{1}{kT_{ref}}\right)\right] \quad (2)$$

Thermal simulations, however, indicate that this assumption is not valid for 1.0 nm MgO, with self-heating temperatures expected above 200°C at accelerated breakdown conditions (Fig. 7(b)). Therefore, the extracted temperature acceleration α_T has to be corrected with the unknown self-heating contribution T_{SH} as follows: (i) the temperature acceleration α_T and temperature dependence of the power law $n(T)$ are measured for 1.7 nm MgO. (ii) Based on an empirical model, supported by large datasets measured for SiO$_2$ gate dielectrics, the voltage dependence of α_T is extrapolated to V_{BD}-conditions of 1.0 nm MgO, i.e. $\alpha'_T(V)$. (iii) The thermal resistance is fitted, such that the underestimated temperature acceleration matches with $\alpha'_T(V)$.

(i) The activation energy of $n(T)$ is similar in SiO$_2$ and MgO (Fig. 10(a)). The temperature dependence of $n(T)$ results in a logarithmic voltage dependence of the temperature acceleration α_T (Fig. 10(b)). This is only observed at voltages below 4 V, because in this range direct tunneling (DT) dominates over Fowler-Nordheim tunneling (FN) (Fig. 10(c)). The increase in temperature acceleration is not related to oxide thickness (Fig. 10(d)) [5]. (ii) Extrapolating this voltage trend from the 1.7 nm MgO to 1.0 nm MgO results in an $\alpha'_T \approx 1.7$ eV at 1 V (Fig. 11(a)). This extrapolation assumes no self-heating. The difference between the experimentally extracted α_T is corrected by adding the self-heating-induced temperature increase T_{SH} to T_{ext} (see inset Fig. 11(a)). (iii) For the 1.0 nm MgO dataset it results in a T_{SH} of 250°C at breakdown. Since $T_{SH} = P \cdot R_{th}$, we are able to find the effective thermal resistance R_{th} (Fig. 11(b)). The experimentally extracted temperatures at breakdown are shown in Fig. 12(a) and are between 200-300°C. The thermal resistance corresponds well with the simulation (Fig. 12(b)). Ultimately, incorporating the self-heating into the breakdown model results in an elevated temperature at breakdown condition and a reduced temperature at operating condition. The temperature differences will affect the power law exponent and the temperature acceleration. In Fig. 13, both effects are taken into account (red curves) and result in up to a factor of 1000 difference in 10 year lifetime. When extrapolating from the 63-percentile to 1 ppm this difference increases up to 10^7.

To further confirm our observation, we demonstrate the impact of the self-heating at high voltage and very short breakdown time. At these conditions, increased self-heating reduces the absolute value of n. In contrast, higher temperatures will accelerate breakdown, again increasing $|n|$. We experimentally demonstrate these dependencies with an "ultrafast" RVS technique based on the rise time of a voltage pulse and monitoring the voltage ramp with 2 oscilloscope channels, as depicted in Fig. 14. Breakdown distributions are acquired for rise times ranging from 20 ns to 200 ms and corresponding ramp-rates (RRs) between 10 and 10^8 V/s (Fig. 15). Plotting RR as a function of 63% V_{BD} allows extracting the exponent n [3]. In a typical RVS study the range of RR is only 3 orders of magnitude and a constant n fits the data well. In our experiments, RRs range over 7 decades so that fitting with a constant n is not possible for 1.0 nm MgO because of the excessive self-heating (Fig. 16(a)). On the other hand, in 1.7 nm MgO a constant n is observed, consistent with our earlier hypothesis of negligible self-heating (Fig. 16(b)). Moreover, since there is no significant self-heating for 1.7 nm MgO, raising the *external* temperature clearly lowers $|n|$, whereas for 1.0 nm MgO the instantaneous n is determined by (i) the temperature acceleration α_T and (ii) the temperature dependence of n given by ϕ_T.

VI. CONCLUSION

We successfully incorporated self-heating into the breakdown model used for ultra-thin MgO in STT-MRAM structures. The extracted temperature is estimated at 200-300°C at typical breakdown conditions. An empirical model, based on breakdown data as a function of temperature and oxide thickness, combines (i) the temperature dependence of the power law extrapolation exponent n as well as (ii) the temperature acceleration at elevated temperature due to self-heating. The model allows extracting the thermal resistance. Furthermore, the corrected lifetime extrapolation can deviate by a factor of 1000 at the 63-percentile and a factor of 10^7 at 1 ppm. Incorporating self-heating is important to correctly predict lifetime, but also to study variability in arrays, certainly when going to lower RA products (3-5 $\Omega\mu m^2$) with increased self-heating.

978-1-7281-1988-5/18 $31.00 © 2018 IEEE

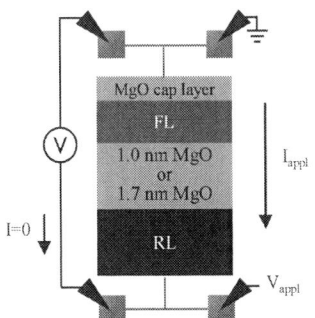

Fig. 1: Simplified schematic of a 0T1MTJ 4-point sensing structure with a simplified MTJ-stack. The thickness of the MgO barrier is either 1.0 (industrial relevant) or 1.7 nm (used for self-heating analysis).

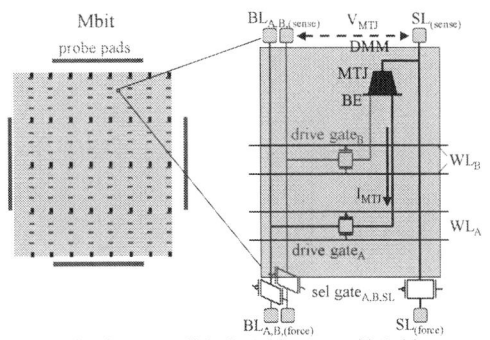

Fig. 2: Schematic of 1 MTJ cell in the Mbit array, with 2 drive transmission gate transistors and 3 selector transmission gate transistors. V_{MTJ} is measured using 4-point sensing via the digital multimeter (DMM). The applied current I_{MTJ} flows from the source line (SL) through the device and the bottom electrode (BE), through drive transistors A and to the bit line (BL_A).

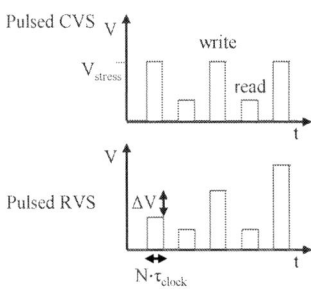

Fig. 3: Breakdown measurements in Mbit array.

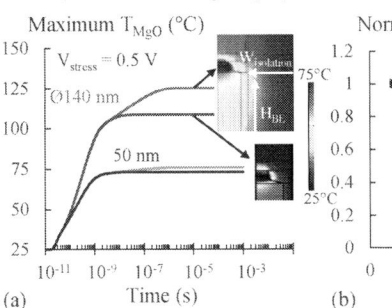

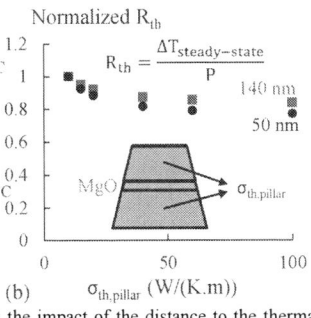

Fig. 4: (a) 2.5D thermal simulation shows the impact of the distance to the thermal boundary for Ø 50 and 140 nm device. If the thermal boundary is close to the device (10 nm from BE and TE, bottom inset), $T_{steady-state}$ is reached after 10ns. Including the actual isolation width and length of the BE and TE (top inset), results in additional heating (20% for Ø 140 nm vs. 5% for Ø 50 nm). (b) Normalized R_{th} as a function of the unknown thermal conductivity $\sigma_{th,pillar}$ of the metallic multilayers next to MgO.

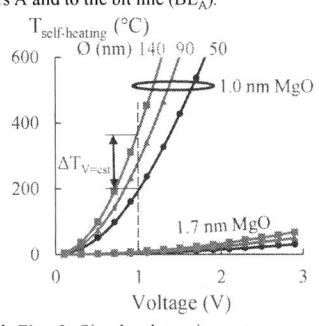

Fig. 5: Simulated maximum temperature increase in 1.0 and 1.7 nm MgO as a function of voltage (Ø 50, 90 and 140 nm). At the same voltage there is a significant temperature difference $\Delta T_{V=cst}$ for 1.0 nm, but not for 1.7 nm MgO.

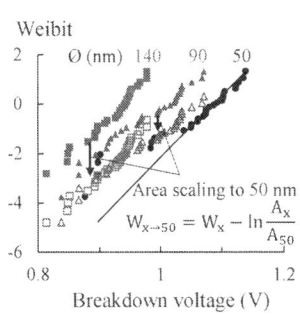

Fig. 6: Illustration of "beyond area scaling" in 1.0 nm MgO 0T1MTJ devices (Ø 50, 90, 140 nm). Open symbols correspond to BD-distributions rescaled to the 50 nm size and do not overlap.

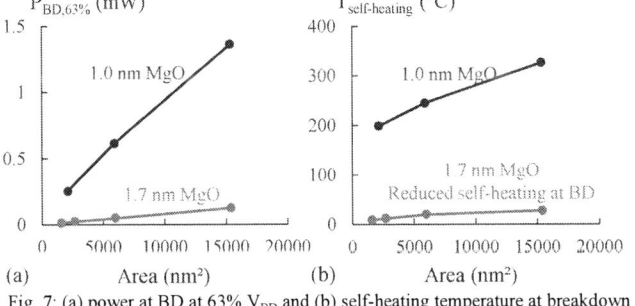

Fig. 7: (a) power at BD at 63% V_{BD} and (b) self-heating temperature at breakdown condition in 1.0 nm and 1.7 nm MgO. These results indicate the reduced effect of self-heating in thicker MgO. The 1 nm data comes from the same datasets used in Fig. 6, and the 1.7 nm dataset from the IBE samples (Fig. 9).

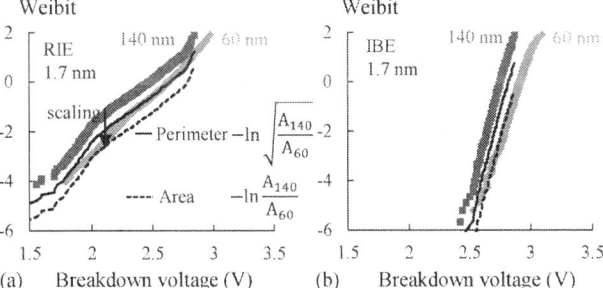

Fig. 8: BD-distribution for Mbit arrays with 1.7 nm thick MgO with Ø 60 nm (pink diamonds) and 140 nm (red, squares). Ø 140 nm is rescaled to 60 nm (perimeter scaling = solid line and area scaling = dashed line). (a) For RIE, the bulk of the distribution fits better with perimeter scaling, indicating preferential breakdown around the perimeter, (b) for IBE area scaling is observed.

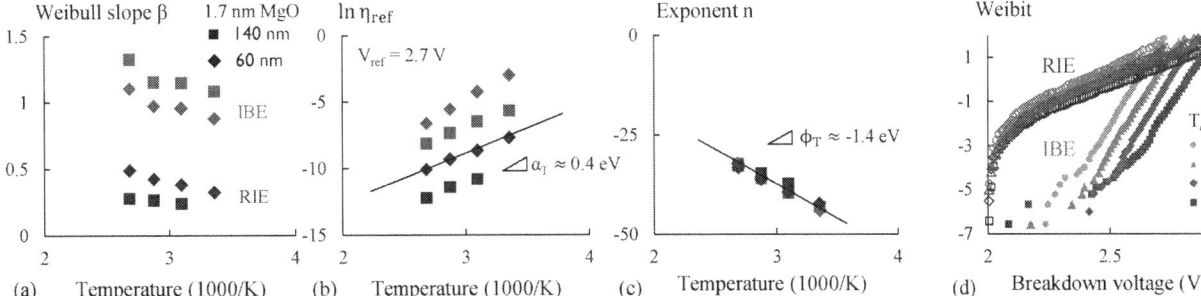

Fig. 9: Temperature acceleration in Mbit arrays with 1.7 nm thick MgO for Ø 60 nm (diamonds) and Ø 140 nm (squares) and for a RIE process (purple) and IBE process (blue). (a) Weibull slope β has only a limited temperature dependence. The low β seen with RIE is an indication of edge damage. (b) ln η_{ref} shows an Arrhenius trend with activation energy $\alpha_T \approx 0.4$ eV. (c) Also the power law exponent n shows Arrhenius behavior with a slope $\phi_T \approx -1.4$ eV. (d) Breakdown voltage distributions for the four external temperatures (25, 50, 75 and 100°C).

978-1-7281-1988-5/18 $31.00 © 2018 IEEE

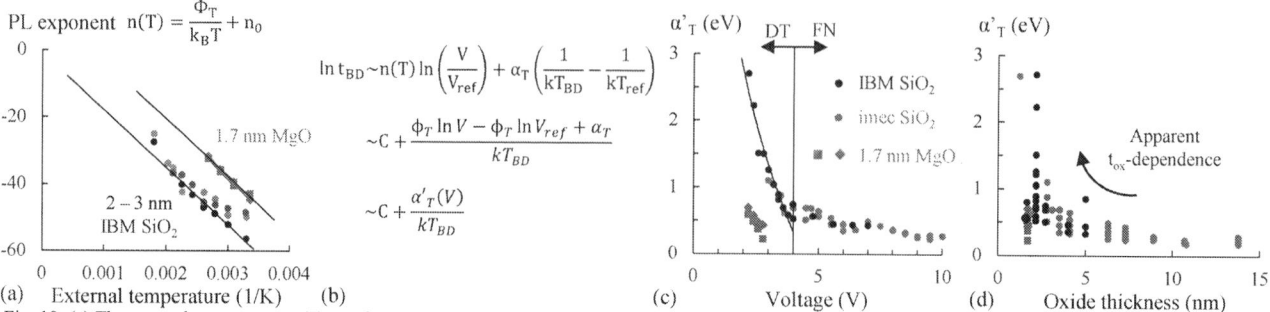

Fig. 10: (a) The power-law exponent $n(T)$ as a function of temperature can be approximated with an Arrhenius law. (b) Formulas that take into account the Arrhenius trend of $n(T)$ result in a voltage dependent temperature acceleration α'_T. (c,d) Each point is the extracted slope of an Arrhenius fit of $\ln(t_{BD})$ data from SiO_2 and MgO data for different oxide thickness and stress voltage as a function of voltage (c) and oxide thickness (d). At low voltage the direct tunneling (DT) takes over from Fowler-Nordheim (FN) tunneling. In this range the $\ln(V)$-dependence of α'_T is observed. This effect is not related to oxide thickness. SiO_2 data from [5], [6], [7].

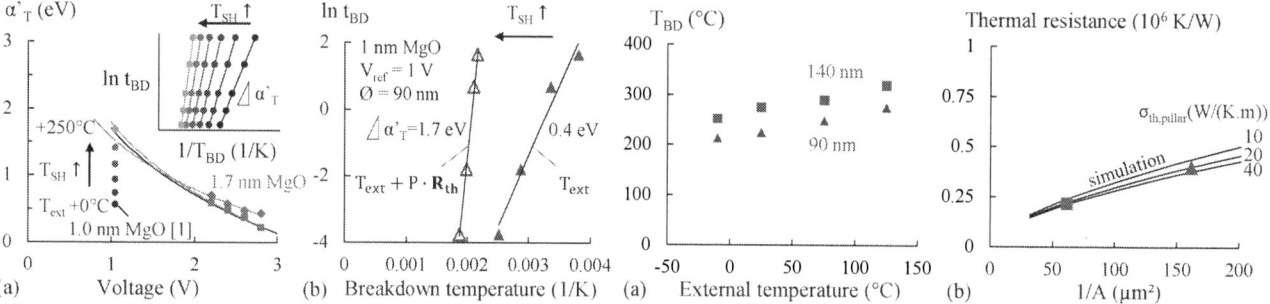

Fig. 11: (a) Breakdown temperature acceleration slope α'_T for MgO data. The solid line is a fit assuming $\alpha'_T(V) \sim \ln V$ from the breakdown voltage range of the 1.7 nm MgO to the one of the 1.0 nm MgO. This extrapolation assumes no self-heating. Inset: adding self-heating temperature T_{SH} tot the data increases the Arrhenius slope α'_T. $T_{SH} \approx 250°C$ results in an activation energy of $\approx 1.7eV$. (b) Methodology where R_{th} is fitted for a specific breakdown dataset to result in an $\alpha'_T \approx 1.7eV$.

Fig. 12: (a) Estimated temperature at breakdown condition as a function of external temperature for RVS measurements on 1.0 nm MgO and 0T1MTJ devices (Ø 90 and 140 nm). (b) The corresponding fitted thermal resistance as a function of $\frac{1}{Area}$, correlates well with our simulation.

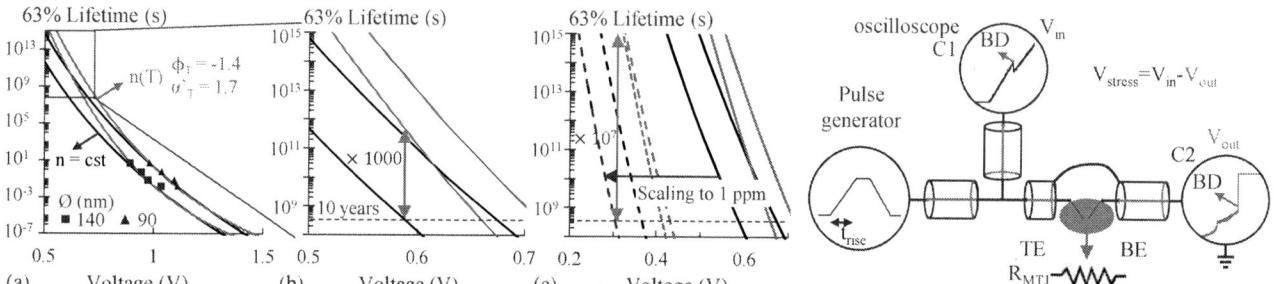

Fig. 13: (a) 63% lifetime extrapolation of 1.0 nm MgO devices without self-heating and a constant n (black), and with self-heating and (i) a temperature dependent $n(T)$, determined by ϕ_T, and (ii) temperature acceleration, determined by α'_T (red lines). (b) Zoom-in to 10 year lifetime region, for the Ø 140 nm data the lifetime is underestimated by a factor 1000. (c) This factor increases to 10^7 when performing percentile scaling to 1 ppm.

Fig. 14: Experimental setup for ultrafast RVS. The ramp is generated by the rising edge of a pulse. The rise time t_{rise} is changed between (20 ns – 200 ms). An oscilloscope monitors the incoming and outgoing rising edge, to determine the stress voltage and detect the breakdown.

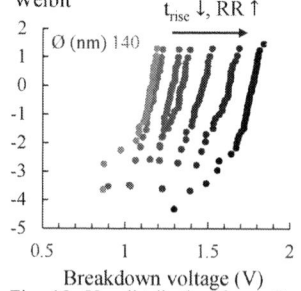

Fig. 15: V_{BD}-distribution for a Ø 140 nm device measured with ultrafast RVS for different rise times (t_{rise} = 200 ms, 32 ms, 1.6 ms, 80 µs, 4 µs, 200 ns, 20 ns).

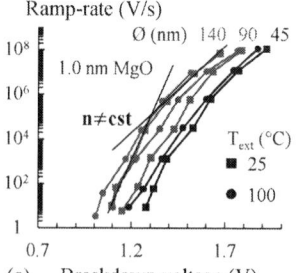

Fig. 16: V_{BD} measured with ultrafast RVS on 0T1MTJ structures for (a) 1.0 nm MgO and (b) 1.7 nm MgO. (a) Ramp-rate as a function of 63%-V_{BD} for Ø 45, 90 and 140 nm devices at 25°C (squares) and 100°C (circles). The traces cannot be fitted with a constant PL exponent n, because of the elevated temperatures at higher V_{BD}. (b) In 1.7 nm MgO there is negligible self-heating, resulting in a constant n, which decreases when increasing T_{ext}.

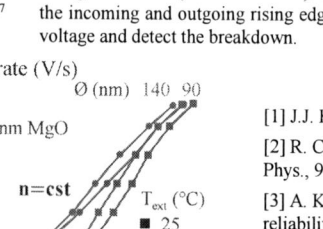

REFERENCES

[1] J.J. Kan et al., IEDM, 2016

[2] R. C. Sousa et al., J. Appl. Phys., 99, p. 08N904, 2006

[3] A. Kerber, Microelectronics reliability, 47, p. 513, 2007

[4] S. Van Beek, ESSDERC, 2018 (accepted)

[5] E. Wu et al., Microelectronics reliability, 45, p. 1809, 2005

[6] E. Wu et al., J. Appl. Phys., 144, p. 014103, 2013

[7] B. Kaczer et al., IEEE TED, 47, p. 1514, 2000

978-1-7281-1988-5/18 $31.00 © 2018 IEEE

Investigating the Statistical-Physical Nature of MgO Dielectric Breakdown in STT-MRAM at Different Operating Conditions

J.H. Lim[1, 2, ♣], N. Raghavan[1], A. Padovani[3], J.H. Kwon[2], K. Yamane[2], H. Yang[2], V.B. Naik[2], L. Larcher[4], K.H. Lee[2] and K.L. Pey[1]

[1]Singapore University of Technology and Design (SUTD), Singapore. [2]GLOBALFOUNDRIES Singapore Pte. Ltd., Woodlands, Singapore.
[3]MDLSoft Inc., Santa Clara, CA, USA. [4]Università di Modena e Reggio Emilia, Italy. ♣E-mail: jiahao_lim@mymail.sutd.edu.sg

Abstract

Ultra-thin dielectric breakdown (BD) has been studied in-depth for SiO_2 and HfO_2 in CMOS devices in the past. In general, the degradation physics and model governing BD in these materials are assumed to hold true for MgO. This study provides evidences that this assumption may not be true by investigating in detail the statistical nature of BD in MgO dielectrics for wide range of operating conditions, relevant to its application as spin transfer torque magnetic random access memory (STT-MRAM). Our analysis shows that - MgO BD is polarity dependent; lifetime is lower for bipolar (AC) stress; defect generation is clustered in space and time; self-heating dominates for low frequencies; temperature within the percolation path exhibits fast transients (thermal runaway); Weibull model does not apply to BD statistics and defect generation (F^+) is charge fluence driven (and field assisted) with power law model being most suited for lifetime extrapolation.

I. Introduction, Objective and Novelty

The choice of the right statistical model to describe the distribution of dielectric BD in MgO layer is crucial for robust STT-MRAM chip design and reliability qualification. The memory element of an STT-MRAM cell is a magnetic tunnel junction (MTJ), consisting of two ferromagnetic layers and a tunnel barrier sandwiched between them. Polycrystalline MgO is commonly used as the tunnel barrier [1, 2]. MTJ operates as a binary storage device through the relative spin orientation of the reference layer (RL) and free layer (FL). A large current density, (4-7 MA/cm^2), is needed for current induced magnetization switching [3]. To reduce power consumption, a low resistance-area product (RA) is considered, which requires tunnel barrier to be very thin (~1nm) [4]. However, such thin oxides are susceptible to thermal stress, process variations, extrinsic failure mechanisms and interface/bulk trap sites, whose energetics are not fully understood yet. These factors complicate the BD process in MgO. The objective of this study is to examine the statistical distributions and physics of failure of ~1nm MgO for wide range of operating conditions using pulsed TDDB tests over large sample size. The novelty of our approach lies in the multiphysics link that supports our statistical inferences using compact defect generation models and simulations.

II. Test Structure and Electrical Characterization

MTJ stack was deposited by magnetron sputtering on 300mm Si wafers. Ferromagnetic CoFeB-based layers were used for FL and RL. Magnetization of FL can be switched by current. For RL, it is pinned with synthetic anti-ferromagnet (SAF) layers and remains fixed within a range of switching current for FL. The MTJ devices are cylindrical with nominal diameter of 70nm. The RA of the MTJ stack was 10Ω-µm². STT-MRAM bit-cell arrays were subject to pulsed voltage TDDB tests with a fixed pulse period (ranging from 40ns to 5µs) to characterize MgO dielectric endurance. Positive polarity direction is defined as electron flowing from FL to RL. MTJ devices are set to the anti-parallel (AP) state during TDDB testing with positive voltages (Fig. 1).

III. Reliability Test Results – Data Analysis and Discussions

Fig. 2 shows the typical trend of resistance degradation observed in the tested STT-MRAM devices. MTJ resistance (sensed here in the parallel (P) state (R_P) after different number of pulses) abruptly drops by almost an order of magnitude when MgO BD occurs. Note that in all the results plotted, we define t_{BD} to be the cumulative time for which MgO was subjected to non-zero voltage pulses up to BD.

Wafer Level TDDB Analysis - Fig. 3 plots the wafer-scale TDDB data for MTJ stack. While the distribution is mono-modal without extrinsic tails at low percentile (indicating good process quality), contrary to our expectation, the distribution does not obey the Weibull model well and shows concavity at high percentile region. While these observations may normally be attributed to process variations [5], we checked TDDB failure distributions at die-level to get better insights.

Die Level TDDB Analysis - Fig. 4 shows the wafer plot of the TDDB distributions in all the dies tested. It is interesting to note that the concavity in the data is still clearly evident in the dies irrespective of their spatial location across the wafer. This leads us to infer that the shape of the distribution must be related more to the intrinsic defect generation kinetics of MgO, not to the process variations.

Non-Poisson Defect Clustering - Fig. 5 plots the typical t_{BD} data sets obtained from a single MTJ array (within same die at center of the wafer) for different combinations of pulse width and bipolar stress voltage (V_{stress}). Due to concavity observed at high percentile, the quality of data fit using a Weibull distribution was poor. When the defect clustering model (Eqn. 1 in Table I) proposed by Wu et al. [5] is used for fitting the same set of data (Fig. 5), the Akaike Information Criterion (AIC) value improved significantly for cluster fit (in spite of the penalty incurred by using an additional parameter, α_C), which justifies its validity. The value of α_C (cluster factor) ranged from $0.55 - 1$, implying strong spatial correlation of defect generation. In contrast to SiO_2 and HfO_2 where the first BD event is generally well described by the Weibull [6, 7], the clustering model has become a necessity for MgO TDDB studies, as also used by Van Beek et al. previously [8]. The need for clustering model could be due to spatially correlated defect generation within MgO due to local thermal hotspots, grain boundaries that are thermodynamic vacancy sinks or lowering of activation energy (E_A) for defect nucleation near pre-existing defects. Physical failure analysis is needed to ascertain the cause for this non-Poisson behavior.

Polarity Dependence of Breakdown – Fig. 6 shows the Weibull plot of t_{BD} for endurance cycling done with (a) bipolar and (b) unipolar sequences of current stressing at different V_{stress}. The bipolar stress scheme was found to yield shorter t_{BD} than unipolar stress for same voltage magnitude (contrary to trends in HfO_2 for high frequencies [9]). This could be attributed to alternate trapping and detrapping of charges at each successive pulse during bipolar stressing in addition to field-induced thermochemical bond breaking of Mg-O, causing larger charge modulation during the delay between pulses, thereby accelerating BD of MgO barrier. The changing polarity also creates alternate tensile and compressive strain at MgO interface, as nanoscale MgO film behaves as a piezoelectric for non-zero ξ-field [10], causing E_A for Mg-O bond breakage to reduce. The non-overlapping TDDB distributions for different polarities suggest that thermochemical ξ-model alone cannot explain degradation physics in MgO.

Under the framework of the clustering model, the α_C value is very low (implies high degree of clustering) for bipolar stress mode and is independent of V_{stress} (Fig. 7). This is again due to charge trapping, interfacial dipoles and charge modulation effects that locally enhance defect generation kinetics (tunneling current density is sufficiently high even at normal operating conditions for STT-MRAM). On the contrary, the values of α_C are very high for unipolar stress (Fig. 7), suggesting that the Weibull distribution holds true (as evident from Fig. 6(b)), as interfacial dipoles no longer intensify the field in the oxide locally (therefore, defect generation is more random in space and time). Reduction in α_C from very high values with higher voltage for unipolar stress (Fig. 7) is not important here as any values of $\alpha_C > 3$ can be considered as "almost Weibull". When Weibull slope (β) values were plotted with V_{stress} (Fig. 8), the trend of β evolution for unipolar and bipolar case was very different. Increase in β for bipolar case suggests more time confined defect generation due to positive feedback between current density induced charge trapping, interfacial mechanical strain and activation barrier for Mg-O bond breakage. The value of β is insensitive to stress for unipolar case due to negligible dipole effects, which is the general trend we observe in SiO_2 / HfO_2 [11].

Pulse-Width Dependence of Breakdown – Fig. 9 presents the Weibull plot of t_{BD} for three pulse widths ranging from 40ns to 5μs for $V_{stress} = 0.90 - 0.98V$ in the bipolar mode. Clearly, the distributions of the data get more concave for lower pulse widths, implying that Weibull model fails for higher frequency of pulsing. This trend may be explained by the self-heating effects that take about a microsecond to saturate in these devices (Fig. 10). For very short pulse widths, self-heating plays a negligible role, causing defect generation to be more confined to the interfaces as determined by the field and charge modulation effects. For longer pulse widths exceeding 1μs, self-heating effect kicks in and "heats up" the full device resulting in "more uniform and spatially random" defect generation leading to a Weibull trend. Temperature rise due to self-heating is estimated to range between 180-220K.

Fig. 11(a) plots the extracted mean-time-to-failure (MTTF) for different pulse-widths and bipolar V_{stress}. At each voltage, the increasing insensitivity of MTTF to longer pulse widths clearly supports the role of self-heating. Also, when β and α_C values are plotted with pulse width in Figs. 11(b) and (c) for all V_{stress}, both parameters increase in value for longer pulse widths. This may be attributed to the area-wide domination of thermal stress induced by self-heating, causing t_{BD} values to be constrained tightly, irrespective of the initial defect configuration and setting. In other words, the roles of local field and current are overshadowed by self-heating effects at large pulse width conditions.

Temperature Dependence of MgO Breakdown – Fig. 12 shows the Weibull plot of t_{BD} for four temperatures, $T = \{25^0C, 55^0C, 85^0C, 125^0C\}$ at two bipolar pulse V_{stress} of 0.88V and 0.94V, keeping pulse period fixed at 200 ns. Temperature plays a critical role due to thermally enhanced bond breakage probability and inelastic trap-assisted tunneling (TAT) current. The Arrhenius plot in Fig. 13 reflects this and from the data plotted as a function of ξ-field (Fig. 14), we extract the field-free activation energy (E_A) in Fig. 15 to range between 2.2-2.4 eV and the dipole moment $p_0 \sim 4.5e\text{Å}$ (close to the theoretical one of $\sim 5.23e\text{Å}$, computed using Eqn. (2) [12], [13]). This corresponds to a BD field of $\xi_{BD} \sim 13$-14 MV/cm (making use of Eqn. (3)). These values are in good agreement with other studies on MgO dielectrics as well [14, 15]. Comparing extreme temperatures of $T = 25^0C$ and 125^0C in Fig. 16, we note that β value increases with T at all V_{stress}. This could be attributed to enhanced TAT current induced charge trapping and associated barrier reduction for new defect nucleation at high temperatures, which increases trap density time exponent in percolation model. The increase of β with temperature was not observed for SiO_2 and HfO_2 [11].

Multiphysics Simulation of MgO Breakdown – To explain the results observed from electrical tests and investigate the intrinsic cause for non-Weibullian behavior in MTJ, we perform electro-thermal simulations of defect generation and leakage current evolution towards percolation BD using Ginestra™ [16], a multi-scale device physics simulation platform that self-consistently describes all the main physical mechanisms occurring in dielectric layers subjected to electrical stress. Of the three common oxygen vacancies (V_0) defect states for MgO → F^0, F^+ and F^{2+}, only F^+ centers seem to be the suitable one that yields t_{BD} values similar to our test observations with a reasonable relaxation energy of $E_{REL} = 0.9eV$ [17] and trap depth of $E_T = 3.2$-4.2eV. Fig. 17 shows one instance of the defect evolution pattern for a constant voltage stress ($V_g = 0.9V$) applied to a 20×20 nm^2 device area from initial time all the way to TDDB and subsequent progressive BD, considering CoFeB work function of 4.8eV, MgO thermal conductivity of 45 Wcm^{-1}K^{-1} and MgO bandgap of 7.8eV. It is interesting to note that while the defects were initially spatially uncorrelated, they tend to become more clustered from 2.3μs onwards. This clustering is observed here due to internal local

field enhancement around the F^+ centers [17] and also possibly due to reduction in activation energy for Mg-O bond breakage at neighboring sites of existing vacancy due to charge trapping. Take note that the simulations here are performed without considering self-heating effects, so as to observe the sole contribution of the defect generation process to the electro-thermal effects during TDDB stress. The evidence of spatio-temporal correlation in defect generation process towards later stages of degradation lends support to the concavity in TDDB distributions observed for short pulse endurance measurements in Fig. 9(a) for 40 ns pulse (where self-heating effects are minimal as it takes some μs for temperature saturation during self-heating [18, 19]). The temperature of hotspot for multiple BD simulation instances is plotted in Fig. 18. An increase of temperature by around $\Delta T_{BD} \sim 110$-230K at the percolation region is observed and the temperature evolution is abrupt, suggesting that thermal runaway here is due to spatio-temporal defect clustering. The temperature of the hotspot is overshadowed by significant self-heating observed for longer pulse TDDB tests, which explains the return to a "more Weibullian" behavior in Figs. 9(b, c). The increase of α_C from $T = 25^0C$ to 125^0C for TDDB further proves this claim (Fig. 19).

TDDB Extrapolation Model and Degradation Physics – To determine the suitable extrapolation model for MgO BD in the absence of TDDB data spanning 6-8 orders of magnitude, we use the temperature dependence trend of the field acceleration factor (γ) (for the ξ and $1/\xi$ models) or the voltage exponent, n (power-law model). From the MTTF versus ξ-field plot in Fig. 14, the trend of γ versus T is plotted in Fig. 20(a), along with the theoretical expectation (based on thermochemical ξ-model (Eqn. (4))), considering $p_0 \sim 5.23\text{Å}$. In order to fit the theoretical trend, the extracted data will need to be shifted laterally leftwards by $\Delta T = -200K$, which is unphysical as self-heating should only require a rightward shift of the data, if the model were true. This suggests that the pure ξ-model fails to describe the physics of MgO BD, as also evident from the polarity dependent t_{BD} in Fig. 6. A similar analysis using γ for $1/\xi$ model (Eqn. 5) in Fig. 20(b) shows that the extracted γ does not align well with the theory of anode hole injection (AHI). This could be because AHI postulates that tunneling carriers have high energy to cause impact ionization and creation of new electron-hole pairs at the anode, which is very unlikely for the low voltage operation of STT-MRAM (0.5– 0.9V). Lastly, we explore the power law model (Eqn. 6) which has a characteristic drop of n with T [20]. As shown in Fig. 20(c), we did observe a consistent reduction of n with T with n ranging from 42-55 (similar to values reported for HfO_2 [21]). The power law model may be the most suited one for STT-MRAM as it is based on anode hydrogen release (AHR) model [22] that requires several electrons to collectively assist in defect generation at Si-H interfaces in MOSFETs. In MTJ, given the high current density of operation ($\sim$4-7 MA/cm^2) with low energy electrons, it appears logical for a collective impact of several charge carriers (similar to AHR) to assist in the bond breaking of defective Mg-O region at the interface (by reducing effective E_A). Moreover, as stated by Wu et al. [23], the power-law model serves as a good "collective" model that captures the combined role of field and current.

IV. Conclusions

The collective role of field, current and temperature as well as different operating conditions (polarity and pulse frequency) on the endurance of STT-MRAM devices were examined in this study. In contrast to SiO_2 and HfO_2, the MgO dielectric in the MTJ stack experiences significant self-heating and clustered defect generation, with charge carriers playing a more dominant role in the degradation process (due to high current density and low voltage of operation). Bipolar pulse conditions resulted in a worse TDDB lifetime for MgO, in striking contrast to improved lifetime in HfO_2. This unique behavior is attributed to the strain developed at the piezo-natured MgO interface due to dipoles. The active traps in MgO were found to reside 3-4 eV deep inside the bandgap. The power law model is postulated to be the best representative for MgO BD lifetime extrapolation. Further use of physical analysis techniques, coupled with in-depth multiphysics, multi-scale simulations are needed to unveil the true defect chemistry in MgO.

978-1-7281-1988-5/18 $31.00 © 2018 IEEE   597

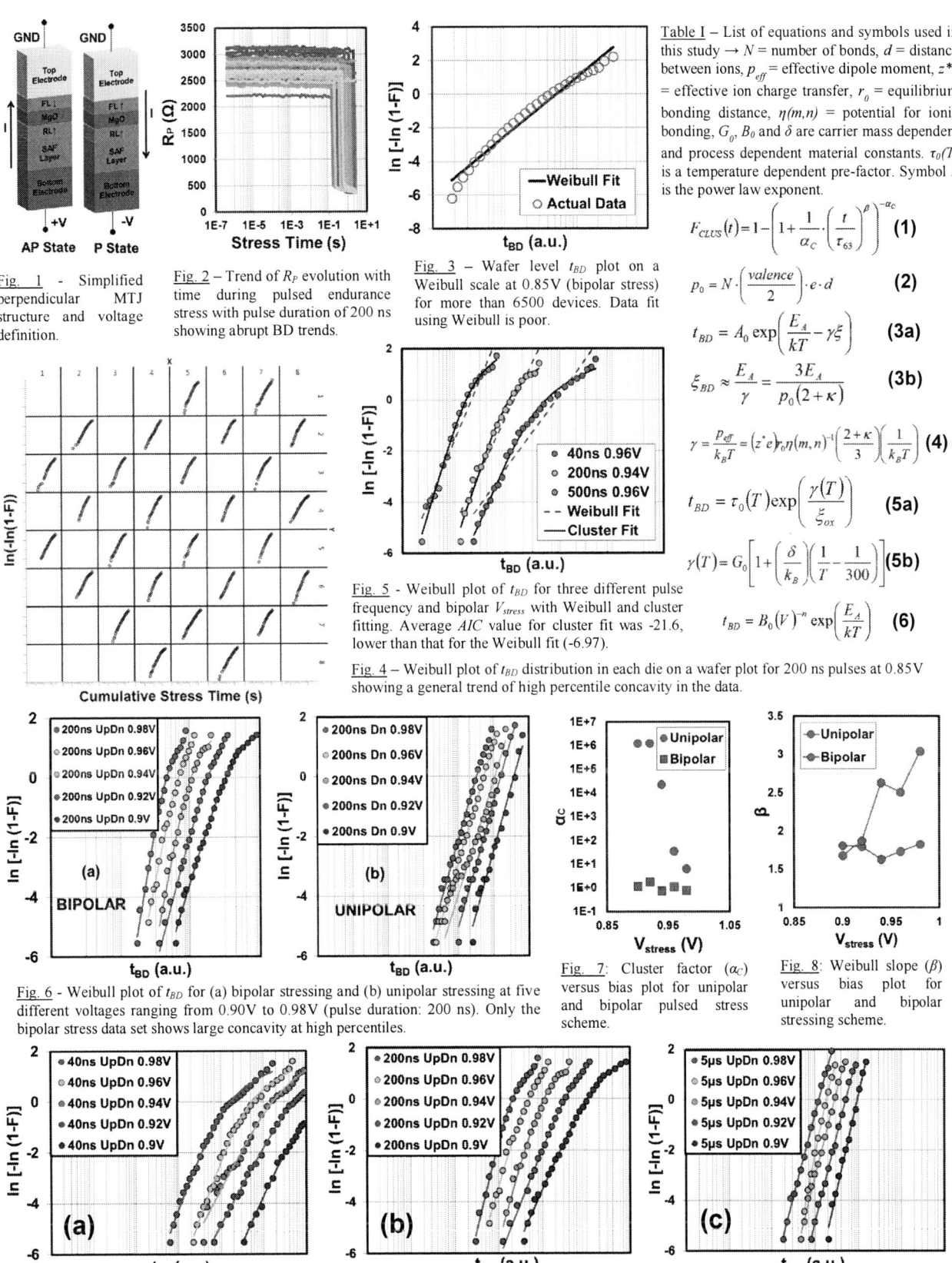

Fig. 1 - Simplified perpendicular MTJ structure and voltage definition.

Fig. 2 – Trend of R_P evolution with time during pulsed endurance stress with pulse duration of 200 ns showing abrupt BD trends.

Fig. 3 – Wafer level t_{BD} plot on a Weibull scale at 0.85V (bipolar stress) for more than 6500 devices. Data fit using Weibull is poor.

Table I – List of equations and symbols used in this study → N = number of bonds, d = distance between ions, p_{eff} = effective dipole moment, $z*e$ = effective ion charge transfer, r_0 = equilibrium bonding distance, $\eta(m,n)$ = potential for ionic bonding, G_0, B_0 and δ are carrier mass dependent and process dependent material constants. $\tau_0(T)$ is a temperature dependent pre-factor. Symbol n is the power law exponent.

$$F_{CLUS}(t) = 1 - \left(1 + \frac{1}{\alpha_C} \cdot \left(\frac{t}{\tau_{63}}\right)^\beta\right)^{-\alpha_C} \quad (1)$$

$$p_0 = N \cdot \left(\frac{valence}{2}\right) \cdot e \cdot d \quad (2)$$

$$t_{BD} = A_0 \exp\left(\frac{E_A}{kT} - \gamma\xi\right) \quad (3a)$$

$$\xi_{BD} \approx \frac{E_A}{\gamma} = \frac{3E_A}{p_0(2+\kappa)} \quad (3b)$$

$$\gamma = \frac{p_{eff}}{k_BT} = (z^*e)r_0\eta(m,n)^{-1}\left(\frac{2+\kappa}{3}\right)\left(\frac{1}{k_BT}\right) \quad (4)$$

$$t_{BD} = \tau_0(T)\exp\left(\frac{\gamma(T)}{\xi_{ox}}\right) \quad (5a)$$

$$\gamma(T) = G_0\left[1 + \left(\frac{\delta}{k_B}\right)\left(\frac{1}{T} - \frac{1}{300}\right)\right] \quad (5b)$$

$$t_{BD} = B_0(V)^{-n}\exp\left(\frac{E_A}{kT}\right) \quad (6)$$

Fig. 5 - Weibull plot of t_{BD} for three different pulse frequency and bipolar V_{stress} with Weibull and cluster fitting. Average AIC value for cluster fit was -21.6, lower than that for the Weibull fit (-6.97).

Fig. 4 – Weibull plot of t_{BD} distribution in each die on a wafer plot for 200 ns pulses at 0.85V showing a general trend of high percentile concavity in the data.

Fig. 6 - Weibull plot of t_{BD} for (a) bipolar stressing and (b) unipolar stressing at five different voltages ranging from 0.90V to 0.98V (pulse duration: 200 ns). Only the bipolar stress data set shows large concavity at high percentiles.

Fig. 7: Cluster factor (α_C) versus bias plot for unipolar and bipolar pulsed stress scheme.

Fig. 8: Weibull slope (β) versus bias plot for unipolar and bipolar stressing scheme.

Fig. 9: Weibull plot of MTJ MgO BD for pulse periods of (a) 40 ns, (b) 200 ns and (c) 5 μs, showing more Weibull-like trend lines for longer pulses due to more uniform and saturated self-heating. Transient self-heating is dominant at shorter pulses (40 ns and 200 ns).

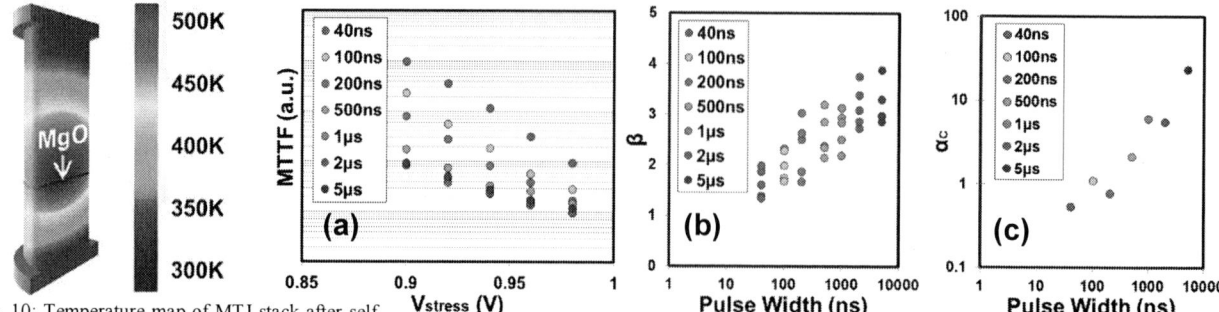

Fig. 10: Temperature map of MTJ stack after self-heating saturation for bias of 0.90V. The thermal enhancement (ΔT_{SH-MAX} = 180-220K) is almost uniform across entire 70 nm MgO device.

Fig. 11: Evolution of (a) MTTF, (b) Weibull slope (β) and (c) cluster factor (α_C) for the various pulse width conditions of MTJ BD ranging from 40 ns to 5µs.

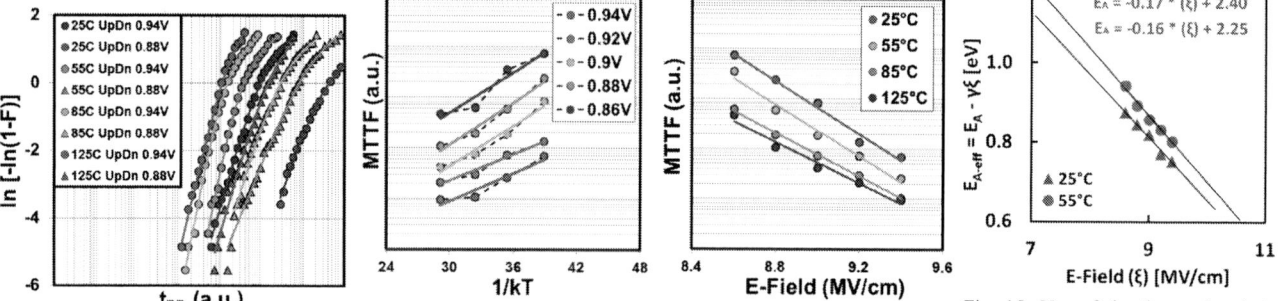

Fig. 12: Weibull plot of t_{BD} for 70 nm MTJ device at different temperatures ranging from 25-125°C, keeping pulse period fixed at 200 ns.

Fig. 13: Arrhenius plot for five different symmetric (bipolar) voltage conditions for T = 25-125°C.

Fig. 14: MTTF versus field plot extracted from bipolar pulsed TDDB tests in Fig. 13.

Fig. 15: Use of the thermochemical model framework to extrapolate and estimate the zero-field activation energy (E_A).

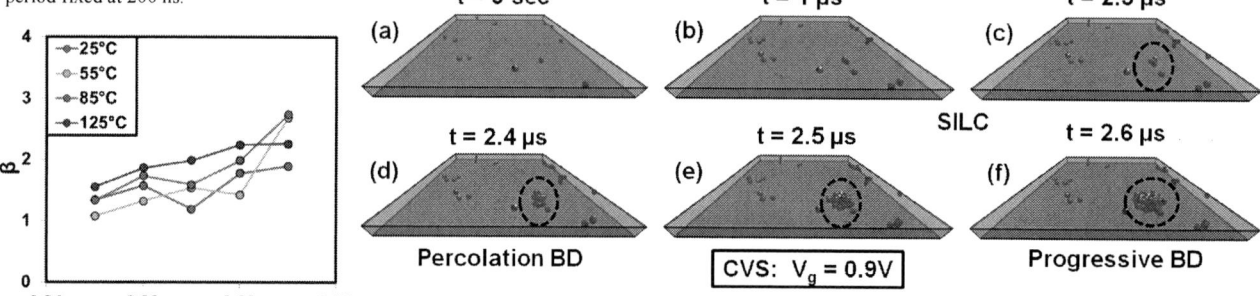

Fig. 17: Simulation of the evolution of defects through self-consistent calculation of stress induced leakage current (SILC) (not shown here for brevity) in 1 nm MgO at different instants of time, resulting in a percolation BD event, as shown by (d).

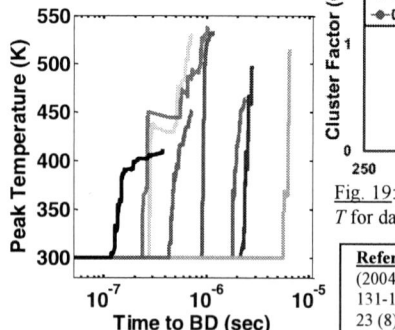

Fig. 16: Trend of β over voltage stress at 25-125°C. Comparing the two extreme temperatures, β is consistently higher at T = 125°C.

Fig. 18: Evolution of peak temperature with time in MgO on multiple devices during defect generation, TDDB and post-BD. Temperature rise is abrupt during the percolation transient stage.

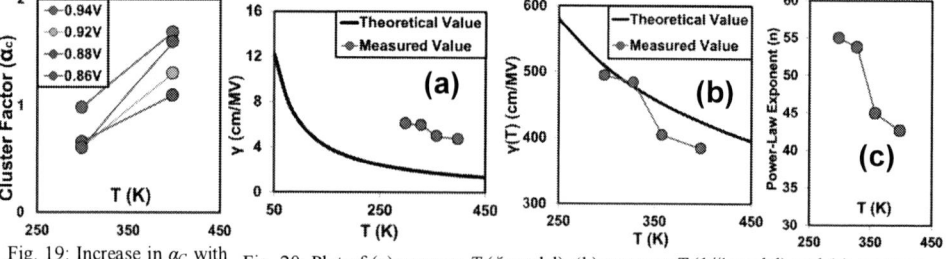

Fig. 19: Increase in α_C with T for data set in Fig. 12.

Fig. 20: Plot of (a) γ versus T (ξ-model), (b) γ versus T (1/ξ-model) and (c) n versus T (power-law model). Power law model is best obeyed by the MgO MTJ data set.

References: [1] J. Bean et al., Nature Sci. Rep. 7, 45594 (2017); [2] Parkin, S. S. et al, Nature Materials. 3, pp. 862–867 (2004); [3] D. H. Lee et al., APL, 92 (13), 233502 (2008); [4] K. H. Lee et al., IEEE Transactions on Magnetics, Vol. 47, pp. 131-136 (2011); [5] E.Y. Wu et al., APL, 152907, (2013); [6] Y.H. Kim et al., EDL, 24(1), (2003); [7] E.Y. Wu et al., EDL, 23 (8), (2002); [8] S. Van Beek et al., IRPS, MY.4, (2015); [9] M. Rafik et al., IRPS, 4A.3, (2018); [10] V.B. Naik et al., APL, 052403, (2014); [11] E.Y. Wu et al., TED, 56(7), (2009); [12] J. McPherson et al., APL, 82 (13), 2121 (2003); [13] A. Padovani et al., JAP 121 (15), 155101 (2017); [14] Al-Moatasem El-Sayed et al., PRB, in press; [15] C. Yoshida et al., IRPS 2009; [16] www.mdlsoft.com; [17] N.A. Richter et al., PhD Thesis, TU Berlin, Chapter 6, (2013); [18] K. Hosotani et al., JJAP, 04DD15, (2010); [19] C.M. Choi et al., Semi.Sci. Tech., Vol. 31, 075004, (2016); [20] E.Y. Wu et al., IEDM, pp.653-656, (2012); [21] E.Y. Wu et al., TED, 49(12), pp.2244-2253, (2002); [22] A. Haggag et al., Micro. Rel., Vol. 45, pp.1855-1860, (2005); [23] E.Y Wu et al., IEDM, pp. 529-532, (2017).

978-1-7281-1988-5/18 $31.00 © 2018 IEEE

Trap Reduction and Performances Improvements Study after High Pressure Anneal Process on Single Crystal Channel 3D NAND Devices

A. Subirats, A. Arreghini, R. Delhougne, E. Rosseel, A. Hikavyy, L. Breuil, S. Vadakupudhu Palayam, G. Van den bosch, D. Linten and A. Furnémont

IMEC, Kapeldreef 75, B3001 Leuven, Belgium. Email: alexandre.subirats@imec.be

Abstract – **We study the impact of HPAP on SCC 3D NAND devices. We show that the process can reduce trap density but is leaving trap impact on devices V_T unaffected. It is also shown, both by simulations and measurements, that further scaling could lead to the increase of single trap impact. Finally, we measure that despite largely improving devices electrical parameter, HPAP has no effect on memory performances (Program/Erase) or could slightly degrade it (Retention).**

Introduction

In order to increase device density, the FLASH non-volatile memory industry took the direction of moving from planar to 3D NAND [1]-[2]. The current technology is featuring Poly-Silicon as the material for the device channel [3]. Although easy and cost effective to integrate, the material comes with multiple limitations. One of them is the presence of grains and grain boundaries. These areas are intrinsically limiting the current conduction in the devices [4] and are a showstopper to additional vertical gate stacking. The defects present in the grain boundaries can limit even further the devices performances and cause devices instability [5]-[6]. Engineering the Poly-Si Channel (PSC) to improve the devices performances can be a challenging task. Another possible solution could be to use Single Crystal Channel (SCC) instead of Poly-Si, thus removing the main source of defects inside the channel and therefore improving device performance. An Epitaxial-Silicon (Epi-Si) growth in a Full Channel architecture [7] and more recent results using a replacement channel scheme in a Macaroni architecture [8] were shown to clearly improve devices electrical performances by providing the SCC morphology. Furthermore, High Pressure Anneal Process (HPAP) performed after full wafer processing has shown to improve PSC Macaroni devices performances [9]-[10].

The combination of those two processes (Epi-Si growth and HPAP) is studied here through Precise IV measurements [7] and the effect of those processes on devices performances and trapping characteristics is explained both from experimental and simulation points of view.

PSC and SCC electrical performances before & after HPAP

The effects of three different processes on device performance and trapping behavior are studied in this paper: (1) a PSC reference process, using a Macaroni architecture, with fabrication process details that can be found in [11], (2) a Full Channel architecture, using an Epi-Si process, with fabrication details that can be found in [12], (3) an Epi-Si Macaroni process, using a replacement channel approach [8], with main steps pictured in Fig. 1. All samples are featuring a 3-gates architecture shown in Fig. 2. The O/N/O thickness layer consists of ~6/5/4nm stack (blocking oxide/trapping layer/tunnel oxide), the gate length is ~50nm while the Inter Gate Spacing is ~30nm. In the Macaroni

architecture, channel thickness is ~10nm. A HPAP final step can be added to the 3-different integration scheme. The anneal is performed in an H2 ambient at 20 atmospheres and 450°C for 30 minutes after full wafer processing.

Id-Vg transients for the three integration schemes, before & after HPAP, are shown in Fig. 4. Normal Dist. of devices electrical parameters (V_T, I_{ON} and STS) are extracted in Fig. 5. One can first note the better performances of the SCC devices over the PSC devices. The HPAP is improving even further these performances. The benefits of the HPAP was already proven for PSC Macaroni in [9]-[10]. These significant improvements were due mainly to a curing of traps present in the grain boundaries. These results confirm the efficiency of the process for devices with SCC morphology.

Precise IV measurements to evaluate trap impact on device performances

To further investigate the benefits of the HPAP on SCC devices, both in Full Channel and Macaroni architecture, Precise IV measurements are performed [7]. The measurement consists in a typical Id-Vg but using small ΔVg steps (here $\Delta Vg_{step}=500\mu V$). An example of the IV transient obtained by the technique is shown in Fig. 6. It allows to observe trap activity directly on the IV measurement (see Inset of Fig. 6) and correct for the trapping when extracting the Gm parameter [13]. The Gm distributions extracted for the SCC Macaroni are shown in Fig. 7. One can see a global reduction of the Gm at higher temperature which is a confirmation of the SCC morphology [5]. Scatter plot of Gm (extracted at 25°C) vs. Gm Activation Energy (Ea) is shown in Fig. 8. Due to imperfect Epi-Si process in the Full Channel integration scheme, some devices are showing positive activation energy [7]. However, after HPAP, those devices are then showing reduced Ea. Fig. 9 is showing a benchmark of Gm & Ea median value (note that for the Full Channel case, the devices are separated between the ones having Ea>0 and Ea<0 as their channel morphology differ). One can see that in the case of Ea>0 (PSC morphology) a strong Ea reduction and Gm improvement is observed with HPAP. This is due to the curing of the defects present in the grain boundaries [10]. On the other hand, the effect of HPAP on SCC devices (Ea<0) is showing limited effect for the Full Channel and only marginal Gm improvement for the Macaroni. Single trap impact on device V_T (ΔV_{T-ST}) is extracted from Precise IV measurements and shown as Complementary CDF (CCDF) in Fig 10. First, one can see that the ΔV_{T-ST} is very similar between the Macaroni and the Full Channel structures (~15mV & ~16.5mV) in SCC morphology. This result is interesting as one could have expected that the traps present in the Macaroni devices could have a stronger impact due to smaller channel thickness. The second important point is the HPAP that is not reducing the impact of the traps as it was typically observed previously in PSC [10]. The absence of effect is likely explained by the conduction inside the channel which is already uniform before HPAP (due to the devices SCC morphology) in contrast to the presence of percolation paths in the PSC. In such morphology, the trap curing in the channel (grain boundaries traps) can then change percolation paths and reduce $\Delta V_{T\text{-}ST}$. Trap density is extracted for Full Channel and Macaroni devices in Fig. 11. It is confirmed that the HPAP is clearly reducing the number of traps present in the SCC devices. It is also worth noting that the Macaroni architecture is showing larger number of traps than the Full Channel one (either before or after HPAP). The higher defect density is expected due to the presence of a second interface between the channel and the inner oxide. Trap spectroscopy is also extracted in Fig. 12 to confirm the absence of traps in the channel. The SCC Macaroni is not showing another peak at positive $V_{G,Switching}$-V_T (Band #2 of defect) which typically corresponds to channel trap response [14] as it is seen for the PSC devices. Furthermore, the HPAP is clearly reducing the trap density close to $V_{G,Switching}$-$V_T \sim 0$ (Band #1 of defects) corresponding to interface and tunnel oxide traps. Finally, yield measurements performed at various Hole Critical Dimension (CD) are confirming the good stability Macaroni replacement channel process as it can be seen in Fig. 13. Therefore, $\Delta V_{T\text{-}ST}$ is evaluated on 6 different dimensions (from a CD of 70nm up to 150nm) and $\Delta V_{T\text{-}ST}$ CCDF are shown in Fig. 14. First, it can be seen that, for all dimensions, exponential distribution well describes the measurements. Then, even with SCC morphology, trap impact is increasing as the hole CD is decreasing. This can be better seen in Fig. 15 where average trap impact (η) is plotted vs. Hole CD.

3D TCAD simulations of $\Delta V_{T\text{-}ST}$

To explain the $\Delta V_{T\text{-}ST}$ results obtained in the previous section, 3D TCAD simulations have been performed [15]. In the simulations, 3D structure has been recreated and $\Delta V_{T\text{-}ST}$ is studied. Example of the channel potential variation caused by a single charge is shown in Fig. 16. Local potential drop caused by the filled charge is clearly visible. For meaningful comparison, statistical simulations where trap position is randomized are required. Therefore, a Monte Carlo process described in Fig. 17 is used. The simulations are always comparing IV characteristic of a fresh device (*i.e.* without trap) and a device with a single trap. The corresponding $\Delta V_{T\text{-}ST}$ caused by the trap at a specific position can then be extracted. For the Macaroni architecture, the simulations will be done also for different Hole CD. The ONO (6/5/4 nm) and channel thickness (T_{Ch}=10nm) remain unchanged, only the thickness of the Inner Oxide filler is modified. CCDF of the $\Delta V_{T\text{-}ST}$ for Full Channel and Macaroni devices are shown in Fig. 18. For the Macaroni case, inner oxide traps have smaller impact than the ones in the tunnel oxide. This is understandable as conduction in the channel is occurring closer to the Tunnel Oxide/Channel interface. Furthermore, the impact of the Tunnel Oxide traps is similar between Macaroni and Full Channel architectures confirming the experimental results of Fig. 10. As device

size is scaled, $\Delta V_{T\text{-}ST}$ is increasing for both Tunnel Oxide trap and Inner Oxide trap as it can be seen in Fig. 19. The average impact is also extracted as a function of Hole CD in Fig. 21 and thus confirming qualitatively the experimental results showing the increase of trap impact at smaller dimensions. The origin of this increase is coming from trap "scattering area" that is becoming relatively bigger as devices dimensions decrease and might have to be considered for further potential scaling of devices dimensions.

SCC devices memory performances before & after HPAP
We showed in the previous part that HPAP was efficient in reducing trap density and boosts electrical performances of SCC devices (in both Macaroni and Full Channel architecture). However, it is also interesting to evaluate the influence of such process on standard memory operation. ISPP and ISPE from program state are shown in Fig. 21 for both PSC & SCC Macaroni. Except for initial V_T, the Program is not affected by the change of channel morphology. The HPAP is only curing initial defects in the stack and the Program curves between SCC before and after HPAP are quickly merging. On the other hand, the Erase is slightly worse in the case of SCC devices. It is worth noting that PSC & SCC gate stacks are the same and not particularly optimized for Erase performances. Fig. 22 is showing correlation between the ΔV_T after a 20V Program pulse and initial V_{T0} for SCC Macaroni (before & after HPAP). A strong correlation is seen before HPAP (and not seen after) confirming the presence of Pre-Charging [16] in our initial Macaroni devices (that is cured when using the HPAP). Finally, in Fig. 23, retention at 85°C is measured after the devices were programmed at a target of 4V. One can note that before HPAP, minor retention loss is measured for all the different architectures (SCC Full Channel, SCC & PSC Macaroni). However, a small degradation of the retention is occurring at the latest time on the devices that received the HPAP. The process might need some optimization in Macaroni as it was not seen in PSC Full Channel [16].

Conclusion

We report the improvement for 3D NAND devices by using SCC, both in Macaroni or Full Channel. The better electrical performances of SCC are pushed even further using HPAP. It is shown that the anneal has only an effect on trap density and not on their impact due to an already uniform channel conduction. In SCC Macaroni devices, we observe strong increase in $\Delta V_{T\text{-}ST}$ with devices scaling which is reproduced by 3D TCAD simulations. Finally, we show that HPAP does not alter P/E performances but is slightly detrimental for Retention in Macaroni devices and still needs some fine optimization.

Acknowledgement

HPSP is acknowledged for the HPAP. GTS (Global TCAD Solutions) is acknowledged for 3D TCAD software support.

References

[1] H. Tanaka, et al., *VLSI* 2007. [2] H. Aochi, *IMW* 2009. [3] J. Min-Kyu, et al., *VLSI* 2012. [4] W.-J. Tsai, et al., *IEDM* 2016. [5] R. Degraeve, et al., *IEDM* 2015. [6] M. K. Jeong et at., *VLSI* 2012. [7] A. Subirats, et al., *IEDM* 2017. [8] R. Delhougne, et al., *VLSI* 2018. [9] A. Arreghini, et al., *IMW* 2017. [10] A. Subirats, et al., *IRPS* 2017. [11] G. Congedo, et al., *IMW* 2014. [12] E. Capogreco, et al., *TED* 2017. [13] M. Toledano-Luque, et al., *IEDM* 2012. [14] M. Toledano-Luque, et al., *IEDM* 2013. [15] www.globaltcad.com. [16] L. Breuil et al., *IMW* 2016.

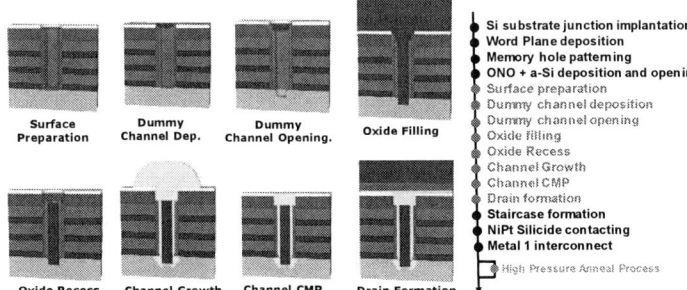

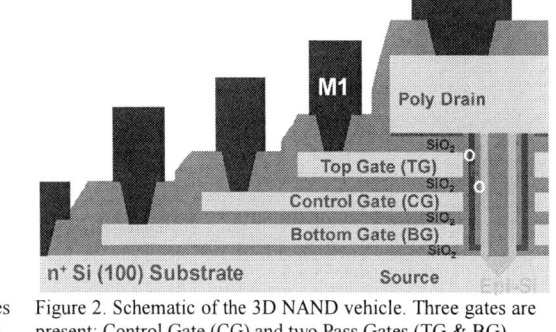

Figure 1. Epi-Si replacement channel process to make SCC 3D NAND Macaroni devices [8]. The integration can then be followed by a last HPAP step after full wafer processing

Figure 2. Schematic of the 3D NAND vehicle. Three gates are present: Control Gate (CG) and two Pass Gates (TG & BG)

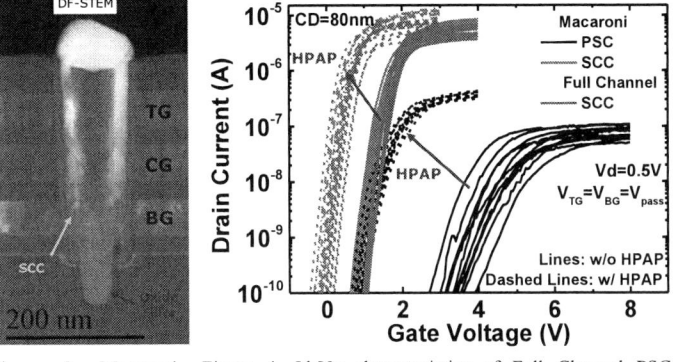

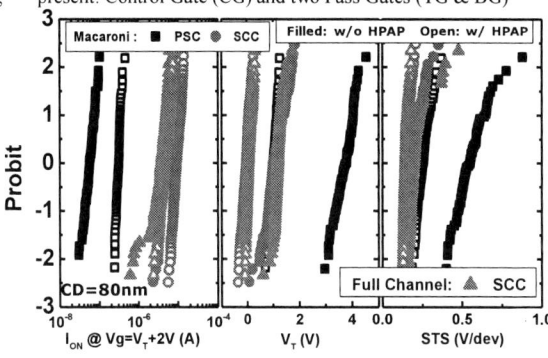

Figure 3. Macaroni SCC device DF-TEM. No channel grains are visible

Figure 4. Id-Vg characteristics of Full Channel PSC devices, SCC and PSC Macaroni devices (with & without HPAP). HPAP improves clearly improve devices performances

Figure 5. Probit of electrical parameters (I_{ON}, V_T, STS) measured on Macaroni PSC, SCC Full Channel and Macaroni devices (with & without HPAP). Devices electrical parameters are all improved with HPAP. Full Channel and Macaroni have similar performances

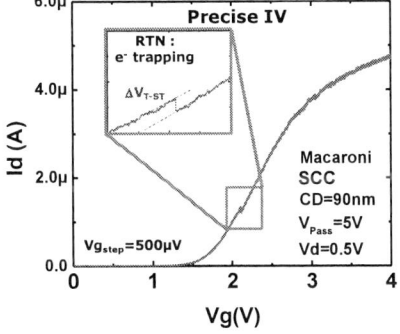

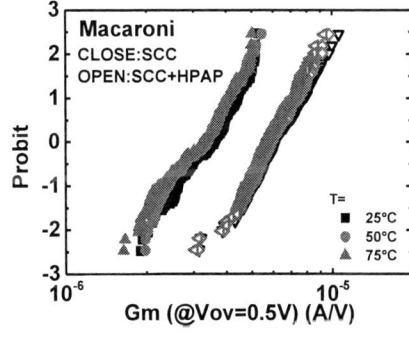

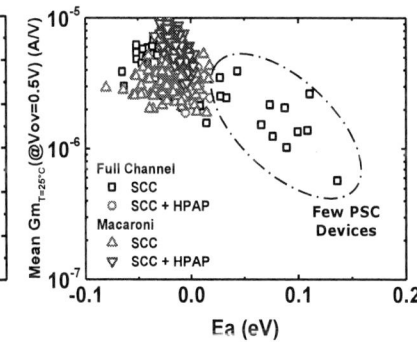

Figure 6. Example of Precise IV measurement performed on SCC Macaroni device (V_{pass} is applied to TG and BG during meas)

Figure 7. Gm extracted from Precise IV meas. at different T for SCC Macaroni. Gm decreases at higher T (confirming SCC morphology behavior)

Figure 8. Gm vs. Ea for SCC devices with & without HPAP. Ea>0 for SCC Full Channel shows in fact few devices with PSC morphology

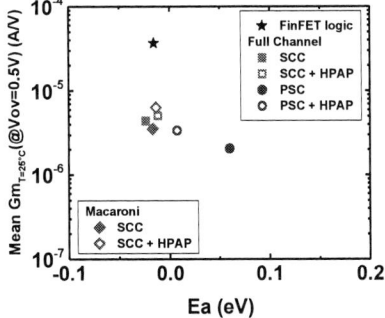

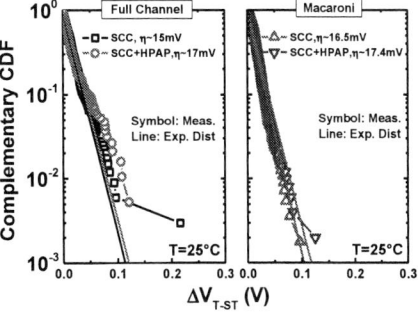

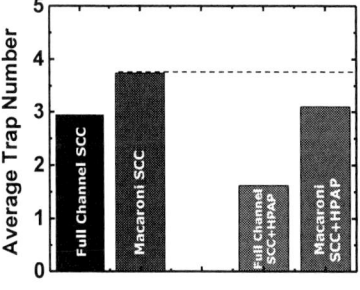

Figure 9. Benchmark of Gm (@Vov=0.5V) vs Ea for different channel morphology. HPAP does not change Ea for single grain channel morphology

Figure 10. ΔV_{T-ST} on SCC Macaroni (left) and Full Channel (right). Similar ΔV_{T-ST} is obtained in Full Channel & Macaroni. HPAP has no effect on ΔV_{T-ST}

Figure 11. Average trap number is reduced after HPAP. Macaroni devices have more traps than Full Channel.

978-1-7281-1988-5/18 $31.00 © 2018 IEEE

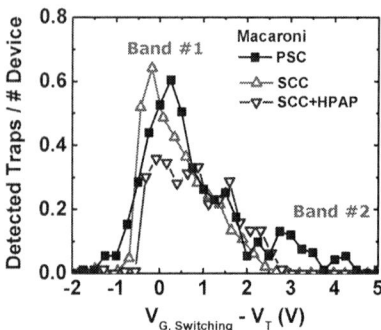

Figure 12. Trap density for PSC and SCC Macaroni. PSC are showing similar trap in the Band #1 of defects but higher trap density in Band #2 (corresponding to channel traps)

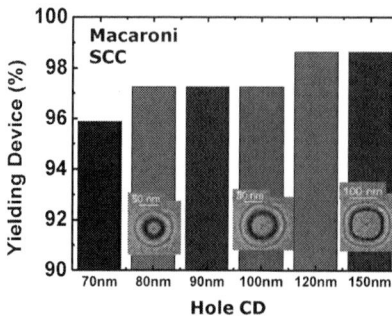

Figure 13. Percentage of functional devices w.r.t Hole CD in SCC Macaroni devices & CD-SEM for 3 dimensions after channel removal. More than 95% functional devices at all CD are measured.

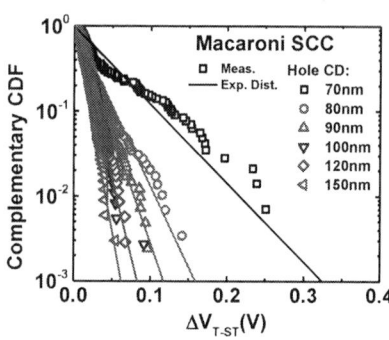

Figure 14. (Symbols) ΔV_{T-ST} from Precise IV measurements on SCC Macaroni devices at different Hole CD. (Line) Exp. Dist. Exp. Dist. is well fitting the measurements at all dimensions

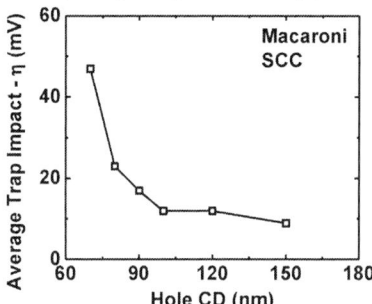

Figure 15. Average trap impact (η) plotted as a function of SCC Macaroni Hole CD. The impact of the trap is clearly decreasing at larger device dimension

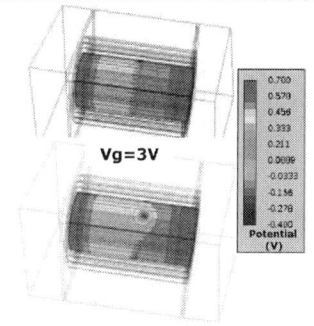

Figure 16. 3D TCAD simulations of channel potential (for Vg=3V) with & without a filled trap in the tunnel oxide. The single charge creates local potential drop in the channel

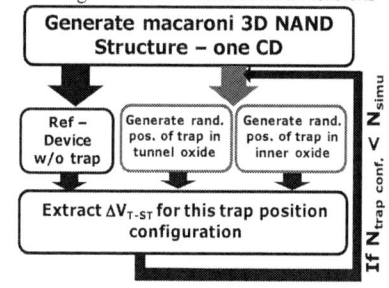

Figure 17. Monte-Carlo procedure used for the 3D TCAD simulations to evaluate ΔV_{T-ST} depending of its position in the oxide. 500 simulations are used for each device dimension

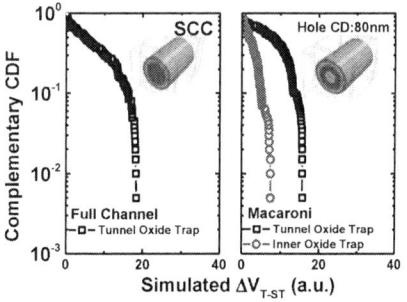

Figure 18. TuOx ΔV_{T-ST} in Full Channel (Left) and Macaroni (Right) are similar. In Macaroni, TuOx traps have a larger impact than InOx traps due to the proximity of the carrier conduction

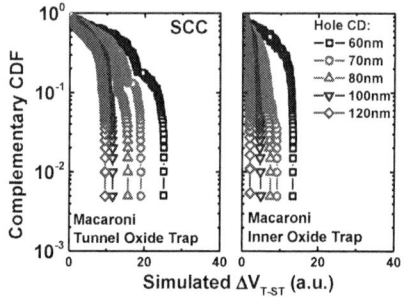

Figure 19. In both Tunnel Oxide (Left) and Inner Oxide (Right) the impact of single trap is increasing as the hole CD of the device decreases (Channel thickness remain constant, T_{Chan}=10nm)

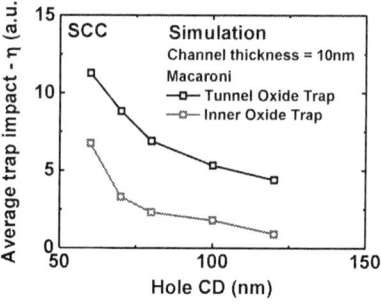

Figure 20. The average trap impact (η) is decreasing as the devices dimensions increase. The relative impact of a trap is getting smaller as the dimension of the device is increased

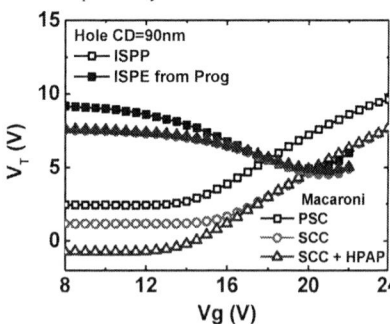

Figure 21. ISPP & ISPE transient for PSC & SCC Macaroni (before & after HPAP). Program and Erase performances are similar between PSC & SCC and HPAP has no effect on P/E on SCC.

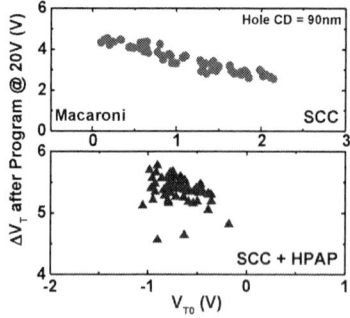

Figure 22. Correlation between ΔV_T after 20V Program Pulse & Initial V_{T0}. Correlation due to Pre-Charging is observed in SCC Macaroni. The correlation disappears after trap curing by HPAP

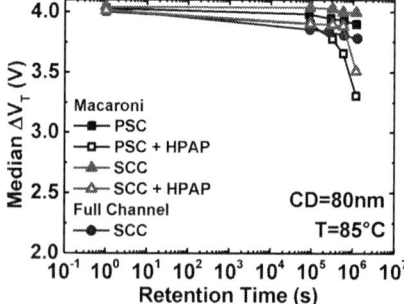

Figure 23. Retention measurement at 85°C. Good retention is achieved for both PSC and SCC morphologies whereas the HPAP is showing a degradation of the retention at the latest time

978-1-7281-1988-5/18 $31.00 © 2018 IEEE   603

22-nm FD-SOI Embedded MRAM Technology for Low-Power Automotive-Grade-1 MCU Applications

K. Lee, *Senior Member*, R. Chao, K. Yamane, V. B. Naik, H. Yang, J. Kwon, N. L. Chung, S. H. Jang, B. Behin-Aein, J. H. Lim, S. K, B. Liu, E. H. Toh, K. W. Gan, D. Zeng, N. Thiyagarajah, L. C. Goh, T. Ling, J. W. Ting, J. Hwang, L. Zhang, R. Low, R. Krishnan, L. Zhang, S. L Tan, Y. S. You, C. S. Seet, H. Cong, J. Wong, S. T. Woo, E. Quek, S. Y. Siah
GLOBALFOUNDRIES, Singapore, email: kangho.lee@globalfoundries.com

Abstract—We demonstrate 22-nm FD-SOI 40Mb embedded MRAM (eMRAM) macros for automotive-grade-1 (Auto-G1) MCU applications, highlighting sub-ppm t0 bit error rate and zero failure after 1M endurance cycles across Auto-G1 operating temperature range (-40~150 °C). Read disturbance characterization with external field also reveals that 40Mb eMRAM macro is capable of active-mode magnetic immunity > 500 Oe at 150 °C. In addition, based on 22-nm eMRAM macro data, we review the effects of magnetic tunnel junction (MTJ) size on reliability and examine scalability of eMRAM technology beyond 22 nm.

I. INTRODUCTION

With the advent of a new era of computing driven by artificial intelligence, there have been growing interests for low-power embedded non-volatile memory (eNVM) technologies. In particular, as always-connected and autonomous vehicles are getting closer to reality, demands for high-density and energy-efficient eNVM solutions are expected to grow tremendously in automotive applications.

To enable an emerging eNVM technology for automotive applications, it is crucial to verify that eNVM is capable of meeting sub-ppm bit error rate (BER) and endurance / data retention requirements across the automotive-grade operating temperature range. In this paper, we present the world-first demonstration of the functionality and reliability of a 22-nm FD-SOI eMRAM macro across the Auto-G1 temperature range (-40~150 °C). Based on 40Mb macro data, we also examine scalability of magnetic tunnel junction (MTJ) technology beyond 22 nm.

II. EMBEDDED MRAM MACRO

Figure 1 shows the TEM cross section and die photo of a 40Mb eMRAM macro. MTJ was integrated between 1x and 2x metal layers, using standard 400 °C BEOL processes. The functionality and data retention performance of the macro have been demonstrated previously [1]. The chips size is ~4 mm^2, including internal bias generator, timing control and ECC.

III. DEVICE/ARRAY CHARACTERISTICS

Figure 2 shows the bit-cell resistance distributions of state "0" and "1" from one 128Kb sub-array. No resistance outliers were observed. The magnetoresistance ratio (MR), defined by $(R_{ap} - R_p)/R_p$, decreases over increasing bias and temperature (Fig. 3). R_{ap} (state "1") and R_p (state "0") are the MTJ resistances in the anti-parallel (AP) and parallel (P) states, respectively. At 150 °C, MR at 100 mV decreased by 28% relative to MR at 25 °C, degrading the read margin. The improved MTJ stack enhanced MR by > 10%, improving MR/σ(R) at 150 °C above 25.

Figure 4 shows normalized switching voltage (V_c) vs. median bit error rate (BER) after 5x reflows from various MTJ stack/etch splits. To improve write margins while meeting the solder reflow compatibility, we have defined spin-transfer-torque (STT) efficiency as energy barrier (E_B) @ 260 °C / V_c,@ 25 °C and optimized MTJ stack/etch processes. The data points in the shaded area met the median post-reflow BER < 1 ppm and also the macro write margin target simultaneously.

IV. MACRO READ/WRITE PERFORMANCE

Figure 5 shows the write shmoo data at -40 and 150 °C. The pass/fail criterion is raw BER out of 40Mb < 1 ppm without ECC. The write pulse width is 200 ns. The internal bias system controls write bias conditions to track the temperature sensitivity of MTJ devices, resulting in comparable write margins across the Auto-G1 temperature range. V_c increases at lower temperature due to less thermal assistance during the STT switching process [2]. Still, zero failure out of 40Mb is achieved at -40 °C (Fig. 6).

Figure 7 shows the read shmoo data at 25, 125 and 150 °C. The same pass/fail criterion as in the write shmoo plots was applied. Due to the MR reduction at high temperature (Fig. 3), the read margin window shrinks over increasing temperature. Still, sub-ppm BER at 150 °C was achieved down to read access time of 19 ns (target = 22 ns) at the target read bias (tick 26) without ECC.

V. RELIABILITY

A. Data Retention

Figure 8 shows BER after 5x solder reflows for three different MTJ processes. Process C showed median post-reflow BER of 2.4E-8 (1 bit failure out of 40Mb). Process B was chosen to balance device performance between E_B and V_c. To estimate stand-by data retention at 125 °C, we baked the wafer at 280-300 °C for 1hr and measured post-bake BER. The effective E_B of eMRAM macros were extracted by the Arrhenius equation [3]. The estimated E_B at 125 °C is 86.3 k_BT, which significantly surpasses the Auto-G1 requirement (Fig. 9).

B. Endurance

Cycling endurance was tested across -40 ~ 150 °C using the same wafer (Process B). Bipolar pulses (program/erase in one cycle) were applied to 1Mb sub-arrays. The operating write bias conditions were set to 10% overdrive from the minimum bias setting to meet t0 BER < 1 ppm. No endurance failure was observed after 1M cycles across -40 ~ 150 °C (Fig. 10). Figure 11 shows the MR distribution of a 128Kb sub-array before and after 1M cycles at 150 °C. The MR distribution remains unchanged, assuring the measured cycling endurance performance.

C. Read Disturbance

In eMRAM, read current is restricted to be significantly less than write current to prevent read disturbance. Although high-E_B MTJ stack generally suppresses read disturbance, read-disturbance-free read operations at 150 °C has not been validated to date. In this macro, the read bias is applied to the bitline, hence read current can potentially cause only AP→P ("1"-to-"0") disturbance.

To characterize the read disturbance rate (RDR), 40Mb arrays were reset to state "1" and read failures were monitored while ramping up read bias and external magnetic field. The effective dwell time per each field step was ~1 sec. Figure 12 shows the measured RDR at 150 °C for two read bias conditions. No read failure was observed up to 400 Oe. Also, RDR was not modulated within the read bias conditions tested.

D. Stand-by Magnetic Immunity

Stand-by magnetic immunity was tested by counting read failures out of 40Mb after exposing a macro to external perpendicular field at 150 °C for 1 hr (Fig. 13). For the BER criterion of 0.01 ppm, the stand-by magnetic immunity of the eMRAM macro is 400 Oe-1hr at 150 °C. The domain wall propagation E_B model can be used to fit the BER trend over external field [4]. Using this model, effective magnetic immunity can be estimated for various field profiles.

VI. SCALABILITY OF MTJ DEVICE

A. MTJ Scaling Effects on Read Margin and Reliability

As electrical MTJ size (eCD) decreases, $\sigma(R_p)$ increases while MR remains nearly constant. Hence the figure of merit for eMRAM read margins, $MR/\sigma(R_p)$, decreases at smaller eCD. Figure 14 shows $MR/\sigma(R_p)$ at 150 °C for different MTJ stack/etch processes. The etch process improvement reduced sidewall damages, increasing eCD for the same design CD and CD dependence of $MR/\sigma(R_p)$. Assuming that $MR/\sigma(R_p) \sim 20$ is required at temperature for a given sense amplifier, the high-MR stack combined with the optimized etch process allows ~20% MTJ eCD scaling from the typical eCD

Figure 15 shows the eCD dependence of effective E_B at 260 °C. Considering global/local process variations, the E_B target for solder reflow compatibility was set to 40.6 k_BT. Process B has room for additional 10% eCD scaling with the best STT efficiency. With Process C, 17% eCD scaling is possible at the expense of increased write power consumption.

Lastly, the eCD dependence of intrinsic MgO TDDB was examined for two MTJ etch processes (Fig. 16). Bipolar pulses with effective voltages across MTJ (V_{mtj}) of 0.9-1.0V were applied at 25 °C. The optimized MTJ etch process with less sidewall damages improved $t_{63}@0.9V$ by ~3×. Reducing eCD improves intrinsic TDDB margin due to the areal dependence of TDDB and less Joule heating [5].

B. Scalability of eMRAM Bit-cell beyond 22 nm

Figure 17 shows normalized V_{mtj} as a function of MTJ eCD for the current 22-nm eMRAM bit-cell (bit-cell A) and 15% smaller bit-cell (bit-cell B) based on Spice simulation data (P→AP switching, SS NMOS corner at -40 °C). The operating voltage (V_{op}) requirement, indicated by the dotted line, is calculated based on 22-nm device data from Process B. Scaling MTJ eCD increases the write margin ($V_{mtj} - V_{op}$). To maintain the write margin for bit-cell B, MTJ eCD needs to be scaled down by 12%. As discussed in the previous section, MTJ scaling is limited by read margin at the high temperature corner and solder reflow compatibility. Process B with minor tuning is still capable of 15% bit-cell size scaling (bit-cell B). Further MTJ process optimizations for STT efficiency are expected to improve yield and macro power consumption.

Based on GLOBALFOUNDRIES 14-nm FinFET (14LPP-FF) design rules, we also examined 14-nm FinFET eMRAM bit-cells. The MTJ pitch at 14LPP-FF is 25% smaller than 22 nm FD-SOI due to poly pitch scaling. With SRAM design rules applied, the cell size for 14LPP-FF eMRAM bit-cell with double fins can be scaled down up to 50% compared to bit-cell B. Spice simulation results indicate that ~25% eCD scaling is needed for the minimum-size 14LPP-FF eMRAM bit-cell (data not shown). Process C with further process optimizations can be applied to achieve aggressive bit-cell size scaling at 14LPP-FF.

VII. CONCLUSION

22-nm FD-SOI eMRAM technology is capable of Auto-G1 MCU applications. We have demonstrated sub-ppm BER, data retention, endurance and magnetic immunity across the Auto-G1 operating temperature range and confirmed no fundamental technology barriers to serve Auto-G1 MCU applications. In addition, the current MTJ technology used in 22 nm is scalable and can drive eMRAM bit-cell size reduction beyond 22 nm. Simulation results indicate that 14-nm FinFET allows disruptive bit-cell size reduction up to 50% with further MTJ scaling and STT efficiency improvement.

ACKNOWLEDGMENT

The authors gratefully acknowledge the contributions of Everspin Technologies.

REFERENCES

[1] K. Lee, et. al, *Symp. VLSI Tech.*, T17-2 (2018)
[2] Z Li, et al., *Phys. Rev. B*, vol. 69, 134416 (2004).
[3] W. F. Brown, *Phys. Rev.*, vol. 130, pp. 1677-1686 (1963)
[4] L. Thomas et al., *IEDM Tech. Dig.*, pp 26.4.1-4 (2015)
[5] J. J. Kan, et al., *IEEE Trans. Elec. Dev.*, vol. 64, pp. 3639-3646 (2017)

Fig. 1. (a) TEM cross section and (b) die photo of 22-nm FD-SOI eMRAM macro.

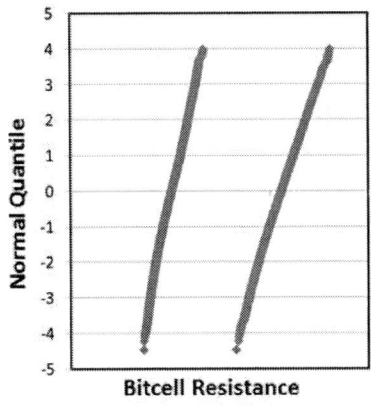

Fig. 2. Bit-cell resistance distributions of state "0" and "1" from actual 128Kb cell array

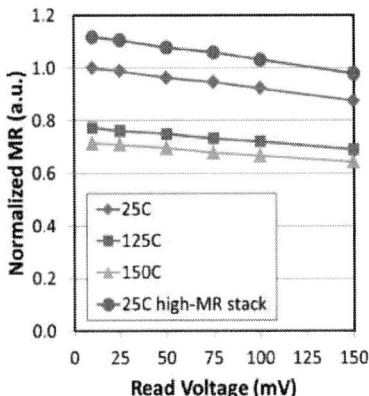

Fig. 3. Bias and temperature dependence of MR at 25, 125, and 150 °C.

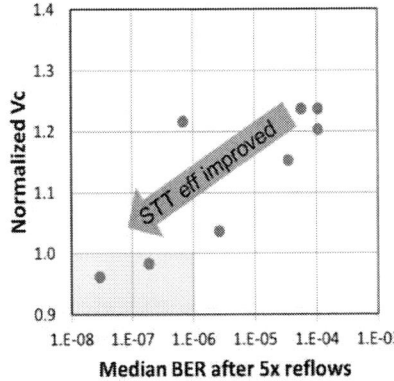

Fig. 4. Normalized switching voltage (V_c) vs. median BER after 5x reflows. The data points in the shaded region met post-reflow BER < 1ppm and Vc target simultaneously.

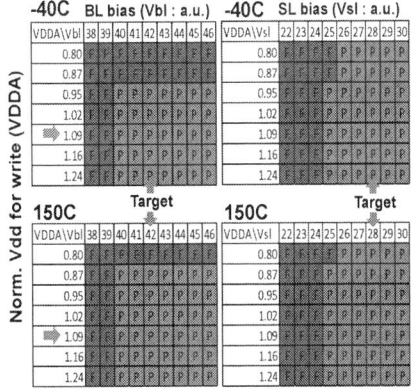

Fig. 5. Write shmoo data at -40 and 150 °C. The P/F criteria is BER < 1E-6. The internal write bias controlled by the bias system.

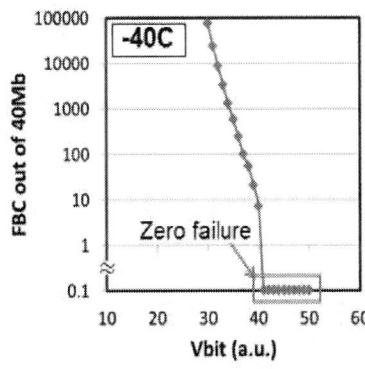

Fig. 6. Failure bit count (FBC) out of 40Mb at -40 °C as a function of write bias (Vbit). Zero failure observed above a certain Vbit threshold.

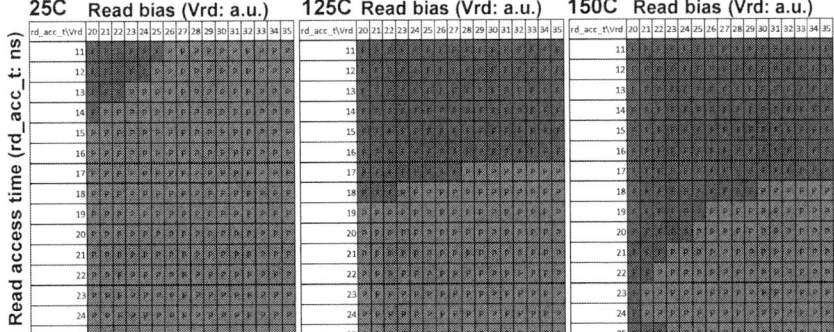

Fig. 7. Read shmoo data at 25, 125 and 150 °C. The P/F criteria is BER < 1E-6. For the typical read bias tick of 26, read access time down to 19 ns demonstrated at 150 °C.

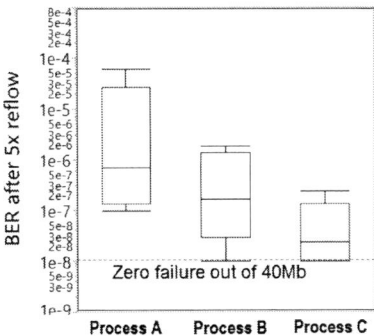

Fig. 8. Post-reflow BER for three different MTJ processes.

978-1-7281-1988-5/18 $31.00 © 2018 IEEE

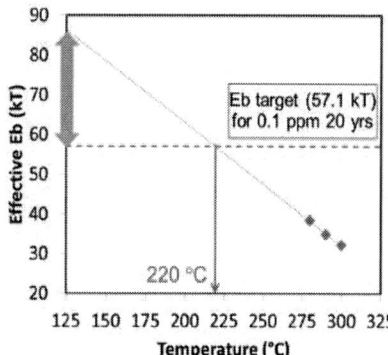

Fig. 9. Sufficient stand-by data retention margin at 125 °C. The maximum temperature for 20 year data retention and 0.1 ppm BER criteria is 220 °C.

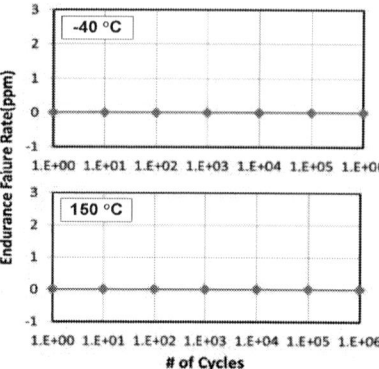

Fig. 10. Cycling endurance data at -40 and 150 °C. Zero failure demonstrated after 1M cycles with bipolar pulses.

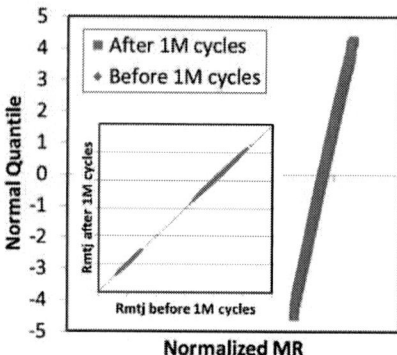

Fig. 11. MR distributions of 128Kb sub-array before and after 1M cycles at 150 °C. The inset shows the corresponding R_p and R_{ap} distributions.

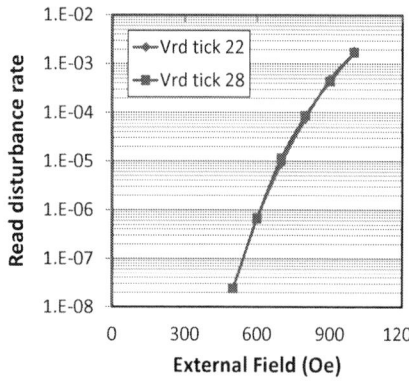

Fig. 12. Read disturbance rate as a function of external magnetic field at 150 °C for two read bias settings (typical 26).

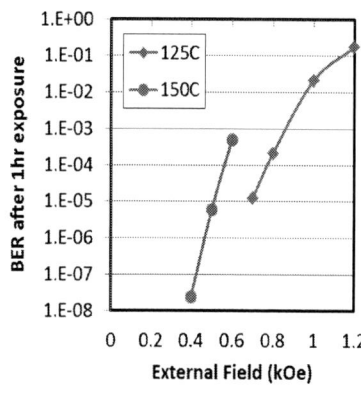

Fig. 13. Stand-by magnetic immunity at 150 °C for 1-hr exposure. Zero failure out of 40Mb below 400 Oe.

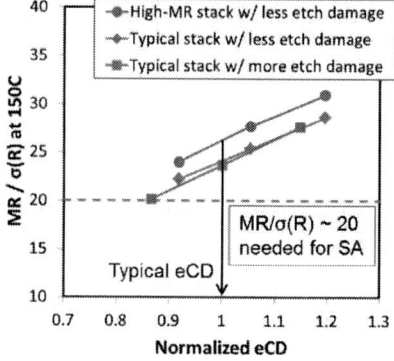

Fig. 14 MR@100mV / $\sigma(R_{low})$ as a function of MTJ eCD at 25, 125, and 150 °C.

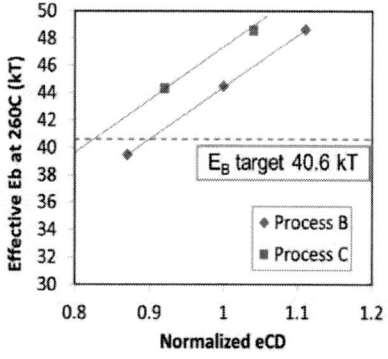

Fig. 15. Effective E_B at 260 °C as a function of MTJ eCD. Process B and C are capable of 10% and 17% MTJ eCD scaling.

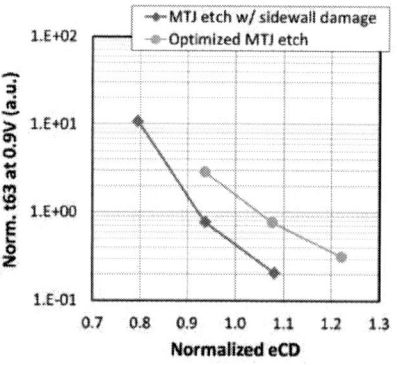

Fig. 16. t_{63} at V_{mtj} = 0.9V as a function of MTJ eCD. Bipolar stress was applied. The optimized MTJ etch process improved t_{63}@0.9V by 3×.

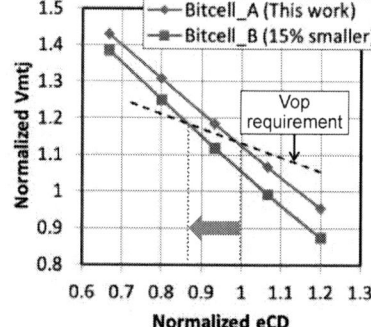

Fig. 17. Write margin comparison for two 22-nm eMRAM bit-cells. Bit-cell_B is 15% smaller than Bit-cell_A. The operation voltage (V_{op}) requirement was extracted from MTJ device/array data.

14ns write speed 128Mb density Embedded STT-MRAM with endurance>10^{10} and 10yrs retention @85°C using novel low damage MTJ integration process

H. Sato[1-4], H. Honjo[1], T. Watanabe[1], M. Niwa[1], H. Koike[1], S. Miura[1], T. Saito[5], H. Inoue[1], T. Nasuno[1], T. Tanigawa[1], Y. Noguchi[1], T. Yoshiduka[1], M. Yasuhira[1], S. Ikeda[1-4], S.- Y. Kang[6], T. Kubo[6], K. Yamashita[6], Y. Yagi[6], R. Tamura[7], and T. Endoh[1-5]

[1]CIES, Tohoku University Japan, [2]CSRN, Tohoku Univ., [3]RIEC, Tohoku Univ., [4]Center for Science and Innovation in Spintronics (Core Research Cluster), Tohoku Univ., [5]Graduate School of Engineering, Tohoku Univ., [6]Tokyo Electron Ltd., [7]Advantest corp.

Abstract—Novel damage control integration process technology has been developed through development of new low-damage MgO deposition process, low-damage RIE process, and low temperature SiN-cap process. Application of the developed damage control integration process technology to MTJ fabrication enabled us to demonstrate an improvement of TMR ratio, thermal stability factor, and switching efficiency. Moreover, it is shown that the endurance of the fabricated MTJs is over 10^{10}, although thermal stability factor drastically increased. Finally, with the developed 37-nm p-MTJ technology and the damage control integration process technology, 128Mb density embedded STT-MRAM was fabricated. By using our 128Mb density STT-MRAM, 14ns write speed at V_{dd} of 1.2V was successfully demonstrated. This result will contribute to low power MCU/IoT chip solution and so on.

I. INTRODUCTION

STT-MRAM with perpendicular MTJs (p-MTJs) has been intensively developed [1-4] and are now about to enter risk mass production [5-7]. For achieving high-density, high-speed, and high reliable embedded STT-MRAM for SRAM-like application and eFlash-like one, higher performance MTJ with RIE process should be developed as well as CMOS/MTJ-hybrid circuit technology. Especially, trade-off among high thermal stability, tough endurance, and high-speed R/W operation should be improved. Development of high performance MTJ stack with exotic material set is one of ways. However, damage-suppression of MTJ process under RIE patterning process is more effective STT-MRAM development schemes, because one can keep continuity of current mass production technology in which CoFeB/MgO interfacial-perpendicular-magnetic-anisotropy type p-MTJ system is used. Moreover, many research results of STT-MRAM patterned with IBE have been reported, as damage of IBE process is lower than that of RIE, although RIE is needed for STT-MRAM scaling.

In this study, we developed low damage process technologies constructed by MTJ stack process, RIE process, and SiN cap process and demonstrated high performance MTJ patterned by RIE process. Moreover, with applying the MTJ and process technologies, we investigated the MTJ with endurance > 10^{10} and 10yrs retention @ 85°C and 128Mb density embedded STT-MRAM with 14ns write speed.

II. EXPERIMENTAL PROCEDURE

Figure1 shows process flow for fabrication of the MTJ with on-Via MTJ structure on CMOS. The CoFeB/MgO-based p-MTJs where the perpendicular-anisotropy originates from interfacial anisotropy at CoFeB/MgO interface [8] were fabricated on Via using 300 mm process tools. The CoFeB/MgO-based p-MTJs with double CoFeB/MgO interface structure for the free layer and synthetic ferrimagnetic reference layer have capability to withstand 400°C annealing for 1 h [9-12].

We have successfully fabricated 37 nm-MTJ with RIE patterning, its dot array for evaluation of magnetic properties, and 128Mb density embedded STT-MRAM chip with 40 nm CMOS technology as shown in Fig. 2.

III. IMPROVEMENT OF MTJ PROPERTIES WITH DEVELOPED LOW-DAMAGE PROCESSES

A. MgO deposition process

First, we developed RF-MgO deposition process in PVD system. With focusing on negative oxygen ion with high energy and composition ratio between Mg and O [13], we developed new low-damage MgO deposition process. Fig. 3 shows normalized coercivity H_C of the free layer with the developed MgO process compared with conventional MgO deposition process. By utilizing the developed MgO deposition process, H_C can be increased by a factor of ~1.7. This increase in H_C leads to an enhancement of thermal stability factor, determining retention time. The increase in H_C should be attributed to low damage feature of the developed MgO process. By the results, an impact of the developed MgO deposition process on MTJ properties is clarified.

B. Reactive ion etching process

We show how the RIE process affects the properties of the MTJs. As shown in Fig. 4, we focus on MTJ patterning damage induced by hydrogen and nitrogen. In previous-RIE1, hydrogen and nitrogen are contained in etching gas. It was reported that hydrogen gives rise to damage on MgO [14]. For nitrogen, it can diffuse into CoFeB from side wall of the MTJ as Fe and B can form stable nitride, by which the magnetic anisotropy could degrade as the presence of impurity, nitrogen in the present case, deteriorates interfacial anisotropy at CoFeB/MgO interface. Next, in pervious-RIE2, we reduce content of hydrogen and eliminate nitrogen in etching gas [15].

978-1-7281-1988-5/18 $31.00 © 2018 IEEE

From these results, we developed novel low-damage RIE process without hydrogen and nitrogen for fabrication of MTJ and 128Mb density embedded STT-MRAM. The low-damage RIE process uses dynamic process using gas pulsing that is developed under Tohoku University and Tokyo Electron collaboration.

Figure 5 shows cross-sectional TEM image of the MTJs using previous-RIE1, previous-RIE2, and the developed RIE. The developed RIE maintained the similar tapered shape with less bird's beak at MgO interface and less degradation at metal hardmask, although MTJ diameter is scaled down from 63 nm/55 nm to 37 nm. On the other hand, we have observed striking difference of H_C as shown in Fig. 6. By developing RIE technologies, H_C, which is related to thermal stability, can be increased by a factor of ~6, indicating the low damage feature of the developed RIE.

C. SiN capsulation process

As we showed in previous RIE section, nitrogen deteriorates the properties of the MTJs. Because we used SiN capsulation after the MTJ etching, SiN capsulation process could be also a damage source for the MTJs. In order to suppress the N incorporation into CoFeB during SiN deposition, we developed low temperature deposition for SiN-cap as shown in Fig. 7.

Figure 8 shows the HAADF-STEM image and EELS mapping for nitrogen diffusion in the MTJ with the developed SiN-cap process in comparison with conventional one. We did not see notable change for the shape of the MTJs from the comparison in HAADF-STEM images. On the other hand, in case of previous SiN-cap process, nitrogen diffusion is detected at CoFeB/MgO junction around side wall facing SiN-cap by EELS result. However, in case of developed low temperature SiN-cap process, the degree of N diffusion is successfully reduced to undatable level by EELS.

Figure 9 shows an impact of H_C improvement of the free layer in the MTJs with the developed SiN-cap process in both cases of previous-RIE2 with hydrogen and without nitrogen and the developed low-damage RIE without hydrogen and nitrogen. H_C related to thermal stability factor is improved by a factor of 1.2-1.3 due to suppression of nitrogen diffusion at SiN-cap deposition by the developed low temperature SiN-cap process for both RIE processes without nitrogen. Moreover, H_C of MTJ with the developed RIE is larger than one with previous-RIE2, which is attributed to the reduction of hydrogen at patterning process. From those results, with low damage integration process technology from RIE process to capping process, degradation of MTJ properties is minimized by suppression of nitrogen and hydrogen attack.

IV. HIGH PERFORMANCE MTJ WITH THE DAMAGE CONTROL INTEGRATION PROCESS TECHNOLOGY

As shown so far, we have developed damage control integration process technology using the developed MgO deposition process, low-damage RIE process without hydrogen and nitrogen, and low temperature SiN-cap process.

We show how the development of the damage control integration process affect the properties of single-bit MTJ. First, we show junction CD dependence of H_C and TMR ratio

in Fig. 10. By utilizing developed integration process-A, TMR ratio and H_C increased by a factor of 1.7 and 5, respectively. On the other hand, in case of developed integration process-B, only TMR ratio effectively increased and H_C improvement was small. Fig. 11 shows normalized thermal stability (Δ) and switching efficiency (Δ/I_{C0}) at room temperature. With the developed integration process-A, thermal stability drastically increased over criteria for 10yrs retention time at 85°C with 1FIT error in 128Mb STT-MRAM. Switching efficiency also drastically increased by a factor of 9 compared with conventional integration process. Figure 12 shows endurance test results. All the MTJs pass the endurance test with the number of write cycles up to 10^{10}. In particular, although the developed integration process-A improves Δ by a factor of 9, endurance of its MTJ exceeds at least 10^{10} as large as conventional integration process.

As can be seen, comparing the developed integration process A and B, to achieve high-level damage control for MTJ performance, integration process technology is important besides each low-damage unit process such as PVD, RIE, and CVD.

V. 128MBIT DENSITY STT-MRAM WITH 37 NM P-MTJ

By using the developed processes, we have fabricated 128Mb density STT-MRAM. Figure 13 shows measured shmoo plot of subarray in 128Mb density STT-MRAM. 7ns write speed is achieved at V_{dd} of 1.8 V. In addition, at V_{dd} of 1.3V/1.2V, 10ns/14ns write speed was achieved, respectively.

VI. CONCLUSION

With the developed damage control integration process technology, TMR ratio, thermal stability factor, and switching efficiency increased by a factor of 1.7, 9, and 9, respectively. Moreover, the endurance of the fabricated MTJs is over 10^{10}, although thermal stability factor drastically increased. Finally, with using developed damage control integration process technology, for the first time, 128Mb density embedded STT-MRAM with 7ns/10ns/14ns write speed at V_{dd} of 1.8V/1.3V/1.2V was successfully demonstrated, respectively.

ACKNOWLEDGMENT

This work was supported by STT-MRAM R&D program under Industry-Academic collaboration of CIES consortium, JST-ACCEL, and JST-OPERA.

REFERENCES

[1] T. Endoh *et al.*, IEEE Journal on Emerging and Selected Topics in Circuits and Systems, Vol. 6, 109 (2016).
[2] Y. J. Song *et al.*, IEDM2016, p. 27.2.1 (2016).
[3] D. Shum *et al.*, 2017 Symp. on VLSI Technology, T15-4 (2017).
[4] H. Sato *et al.*, IMW 2018, P. 135.
[5] Y.-C. Shih *et al.*, 2018 Symp. on VLSI Circuits C8-1 (2018).
[6] Y. K. Lee *et al.*, 2018 Symp. on VLSI Technology, T17-1 (2018).
[7] K. Lee *et al.*, 2018 Symp. on VLSI Technology, T17-2 (2018).
[8] S. Ikeda *et al.*, Nat. Mater. **9**, 721 (2010).
[9] H. Honjo *et al.*, Symp. VLSI Technol. Dig. Tech. Papers, T160 (2015).
[10] H. Sato *et al.*, Appl. Phys. Lett. **105**, 062403 (2014).
[11] H. Sato *et al.*, Appl. Phys. Lett. **101**, 022414 (2012).
[12] H. Sato *et al.*, IEEE Trans. Magn. **49**, 4437 (2013).
[13] K. Ono *et al.*, Jpn. J. Appl. Phys. **50**, 023001 (2011).
[14] J. Jeong and T. Endoh, Jpn. J. Appl. Phys. **54**, 04DM07 (2015).
[15] S.-Y. Kang, T. Kubo, and T. Endoh, in abstract of 62nd Magnetism and Magnetic Materials, BE-04.

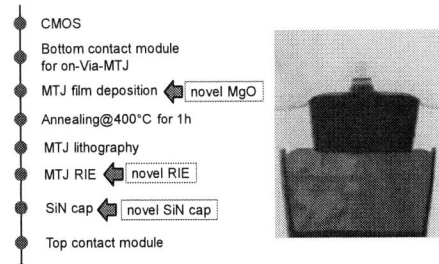

Fig. 1 Developed total low-damage integration process constructed by novel MgO PVD process, RIE process, and SiN cap-process for on-Via type 37nm p-MTJ fabrication.

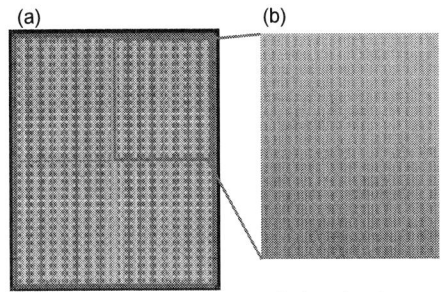

Fig. 2 (a) Chip floor plan for 128Mb density STT-MRAM. (b) photograph of its subarray.

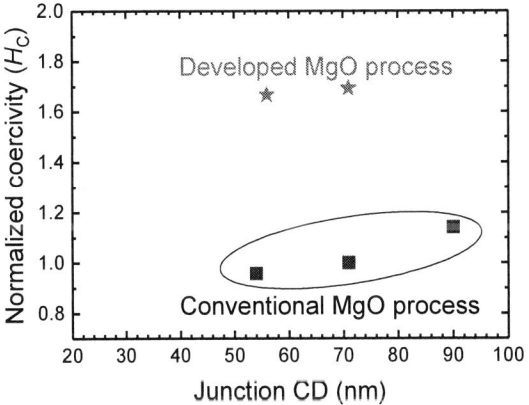

Fig. 3 An impact of the developed low-damage MgO deposition process on MTJ properties.

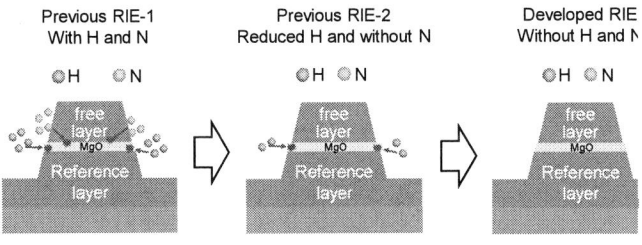

Fig. 4 Novel low damage RIE process without hydrogen and nitrogen for fabrication of MTJ and 128Mb STT-MRAM.

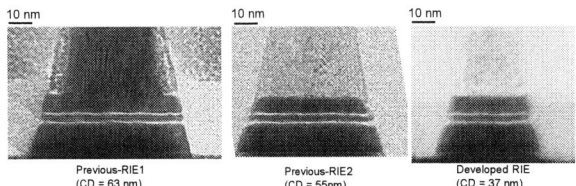

Fig. 5 Cross-sectional TEM image of patterned MTJ using (a) previous-RIE1, (b) previous-RIE2, and (c) developed low damage RIE. By the developed RIE, bird's beak at MgO interface is reduced and taper angle is kept while MTJ size is scaled down to 37 nm.

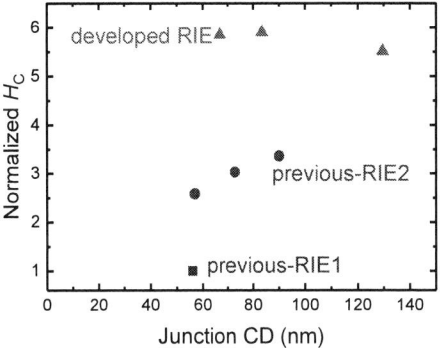

Fig. 6 An effect of the developed low-damage RIE process on the MTJ properties.

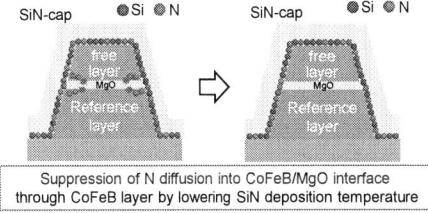

Fig. 7 Concept of the developed low-damage SiN-cap process compared with conventional SiN-cap process. By low temperature SiN-cap process, additional nitrogen diffusion can be suppressed after patterning with the developed RIE process without hydrogen and nitrogen.

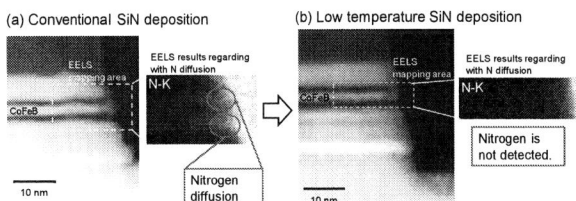

Fig. 8 HAADF STEM image and EELS mapping result with K-edge for nitrogen for (a) conventional SiN-cap and (b) low temperature SiN-cap deposition.

978-1-7281-1988-5/18 $31.00 © 2018 IEEE   610

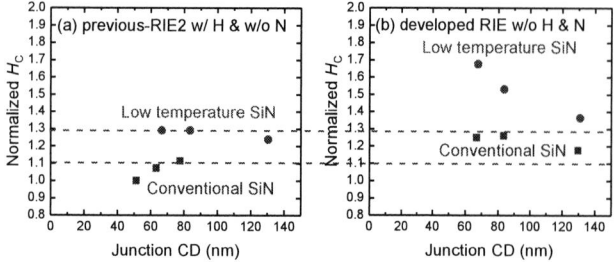

Fig. 9 An effect of the developed low-damage SiN-cap process on the MTJ properties with (a) previous-RIE2, (b) developed low damage RIE. Coercivity (H_C) related to thermal stability factor is drastically improved due to suppression of nitrogen diffusion in addition to hydrogen diffusion.

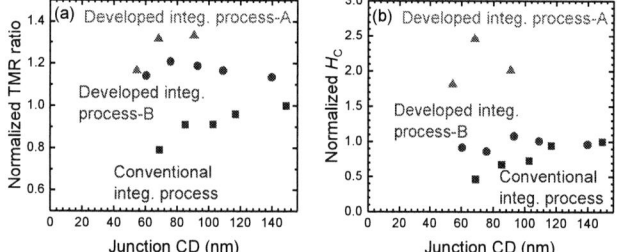

Fig. 10 An impact of damage control integration process on (a) TMR ratio and (b) coercivity (H_C). By utilizing developed integration process-A, TMR ratio and H_C increased by a factor of 1.7 and 5, respectively.

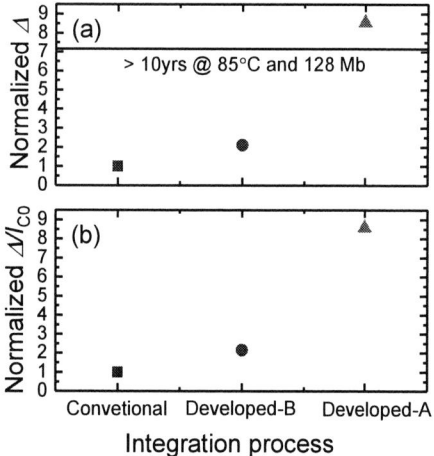

Fig. 11 An impact of damage control integration process on (a) thermal stability factor (Δ) and (b) switching efficiency (Δ/I_{C0}). With the developed integration process-A, thermal stability drastically increased over criteria for 10yrs retention time at 85°C for 1FIT error in 128Mb STT-MRAM. Switching efficiency also drastically increased by a factor of 9 compared with conventional integration process.

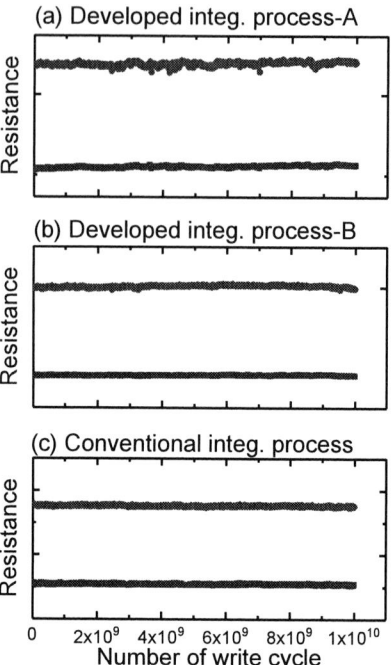

Fig. 12 Endurance results of the MTJ with (a) developed integration process-A, (b) developed integration process-B, and (c) conventional integration process. Although the developed integration process-A improves the thermal stability factor by a factor of 9, endurance of its MTJ exceeds at least 10^{10} as the same as conventional integration process.

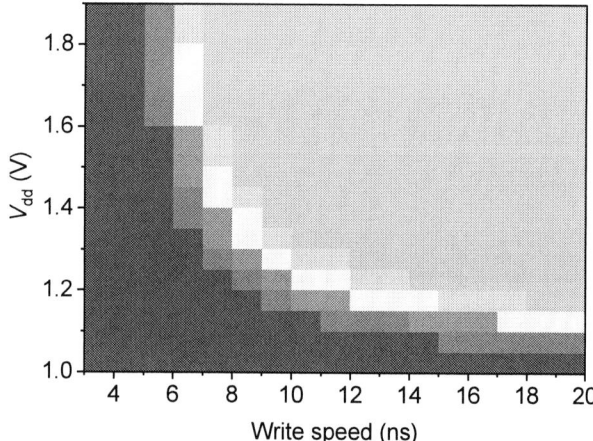

Fig. 13 Shmoo plot of subarray in 128Mb density STT-MRAM. By using the fabricated 128Mb density STT-MRAM with damage control integration process technology, the developed STT-MRAM cell achieved 7 ns write speed at V_{dd} of 1.8 V. Our STT-MRAM cell also achieved 10ns/14ns write speed at V_{dd} of 1.3V/1.2V, respectively.

STT-MRAM devices with low damping and moment optimized for LLC applications at 0x nodes

Luc Thomas, Guenole Jan, Santiago Serrano-Guisan, Huanlong Liu, Jian Zhu, Yuan-Jen Lee, Son Le, Jodi Iwata-Harms, Ru-Ying Tong, Sahil Patel, Vignesh Sundar, Dongna Shen, Yi Yang, Renren He, Jesmin Haq, Zhongjian Teng, Vinh Lam, Paul Liu, Yu-Jen Wang, Tom Zhong, Hideaki Fukuzawa, and PoKang Wang
TDK- Headway Technologies, Inc., 463 S. Milpitas Boulevard, Milpitas CA 95035, USA.
email: luc.thomas@headway.com

Abstract— Last-Level-Cache applications at 0X technology nodes require devices switching reliably in less than 10ns at currents smaller than 50uA, while preserving data retention up to 85°C. In this paper, we show that both low Gilbert damping and low magnetic moment are the primary factors for efficient writing at nanosecond time scales. We report comprehensive device-level measurements of damping using both conventional free layer designs and an optimized free layer that combines low damping and low moment and meets LLC requirements.

I. INTRODUCTION

As major foundries have announced production schedules of Perpendicular Spin-Transfer-Torque Magnetic Random Access Memories (pSTT-MRAMs) for embedded memory applications at 2X lithography nodes [1-3], there is growing interest in using this technology for SRAM replacement at 0X nodes, in particular for Last Level Cache (LLC) applications [4]. Indeed, the 1T-1MTJ STT-MRAM bit cell is more compact and scalable than the 6T SRAM cell. However, LLC applications pose a different set of challenges from those faced by first-generation STT-MRAM devices, which were mostly focused on Flash-replacement applications. While data retention at high temperature was the primary requirement [1-3], at the expense of relatively high write current (100's of μA) and moderate write speed (10's of ns), LLC applications require low write current (<50uA) and power, and write speed of 10ns or faster. In this paper, we show that this target can only be met by optimizing the STT-MRAM stack for low current writing. We combine STT switching data as a function of current pulse length with comprehensive device-level damping measurements using Spin-Torque Ferromagnetic Resonance (ST-FMR) to show that conventional free layer designs using a metallic cap or a metallic insertion are unable to meet LLC requirements. We propose a new ODM-FL structure having both low damping and low moment, allowing for writing down to sub-ppm error rates using 10ns current pulses below 50 μA for 30nm devices [4].

II. DESIGN PARAMETERS FOR OPTIMIZED WRITING

STT efficiency is widely used as a figure of merit of device performance. It is defined as ratio of the thermal stability factor Δ and the write current I_W (written here using the expression valid in the high-speed precessional regime [5] relevant to LLC applications):

$$\frac{\Delta}{I_W} = \left(\frac{\alpha}{P} \frac{4e}{\hbar} \left[1 + \frac{\tau_D}{2t_P} \ln\left(\frac{\pi^2 \Delta}{4P_{BER}} \right) \right] \right)^{-1} \quad (1)$$

where t_P is the write pulse length and P_{BER} the write error rate, α is the Gilbert damping, P the spin polarization factor, e the electron charge and $\hbar$ the reduced Planck constant. The characteristic timescale τ_D is given by $\tau_D = 1/(\gamma \alpha H_K)$, with γ the gyromagnetic ratio and H_K the anisotropy field. While according to (1), I_W should be proportional to Δ, experiments often show no such correlation, mostly because Δ is not an intrinsic property, but rather exhibits a complex dependence on the magnetization switching mechanism, the device size and even the testing method [6]. We use the LLG Micromagnetics Simulator™ package to illustrate this point. Fig. 1a&b show the magnetization and the energy during STT-driven switching for two 30nm devices having exactly the same properties except for the exchange stiffness A. Since switching is mediated by domain wall (DW) propagation, Δ during reversal depends on the DW energy (which is proportional to $A^{1/2}$) and thus, is different for the two devices (Fig. 1b). By contrast, the switching current is almost the same (Fig. 2). In order to correlate STT switching current with intrinsic device parameters, Equation (1) is rewritten in terms of conservation of spin angular momentum:

$$I_W = \frac{M_S V e}{P g \mu_B} \left[\frac{2}{\tau_D} + \frac{1}{t_P} \ln\left(\frac{\pi^2 \Delta}{4P_{BER}} \right) \right] \quad (2)$$

where V is the FL volume, μ_B the Bohr magneton and g the Landé factor. Equations (1) and (2) are equivalent for uniform switching, in which case $\Delta = H_K M_S V/2$, but the spin-angular momentum conservation described by (2) still applies for non-uniform switching, as illustrated by Fig. 1 & 2. Equation (2) has important consequences for the design of STT-MRAM devices suitable for LLC applications, as shown in Fig.3. First, the write current is proportional to the FL moment, $M_S V$, irrespective of pulse length or write error rate. Second, the write current "floor" in the long pulse limit is proportional to $1/\tau_D$, i.e. to α and H_K. Third, the current increase in $1/t_P$ at

978-1-7281-1988-5/18 $31.00 © 2018 IEEE

short pulse lengths depends primarily on the FL magnetic moment, rather than anisotropy or damping. Thus, both low magnetic moment and low damping are necessary for low switching current at deep error rate regime using nanosecond long write pulses.

III. EXPERIMENTAL RESULTS

STT-MRAM stacks studied in this paper include a MgO tunnel barrier and a synthetic antiferromagnet (SAF) pinned layer (PL) [7]. We have investigated three types of FL designs (Fig. 4): i/ CoFeB-based FL with metallic cap ii/ CoFeB-based FL with metallic insertion and MgO cap iii/ optimized low damping and low moment FL (ODM-FL). All devices are submitted to 400°C annealing for 3½ hours after integration. Fig. 5 shows the cross-sectional transmission electron microscope (TEM) image of a typical 30nm ODM-FL device. Devices diameters quoted below are electrical diameter derived from resistance measurements. Switching current measurements are obtained from the switching probability as a function of write pulse length and voltage. BER curves measured on a 30nm ODM-FL device are shown in Fig. 6, showing sub-ppm errors rates well below 300 mV for 10ns pulses. The resistance vs. magnetic field hysteresis loop measured on a different but similar device is shown in Fig. 7. Δ is derived from the switching field distributions using the macrospin model (Fig.8). Δ=59 is obtained at zero field by averaging data for both read current polarities, to cancel the effect of STT, and for both P to AP and AP to P switching. Note that the macrospin fit might underestimate Δ if reversal is mediated by a DW [6]. However, the discrepancy is expected to be small for 30nm devices.

ST-FMR is used to quantify H_K and α. This technique takes advantage of the rectification effect resulting from the resistance oscillations of a device submitted to microwave excitations, which gives rise to a dc voltage across the device [8-9]. This voltage is measured as a function of magnetic field and microwave excitation frequency, and exhibits clear resonance peaks when FMR conditions are satisfied (Fig. 9). FL resonance peaks are fitted with the sum of symmetric and antisymmetric Lorenztian functions [9], allowing us to determine the FMR resonance frequency f_R and linewidth Δf as a function of magnetic field (Fig. 10). H_K is given by $f_R = (\gamma/2\pi)(H+H_K)$, where H_K=5kOe for this device (Fig. 10b). Note that H_K values quoted below are measured at magnetic fields lower than the SAF PL switching field. α is defined as $\Delta f = \alpha f_R + \Delta f_0$, where Δf_0 is the inhomogeneous broadening resulting from non-uniformities in the device. While Δf_0 is often assumed to be constant, Fig. 10c clearly indicates a more complex behavior involving additional broadening not proportional to f_R. In most of the devices we have tested, peak broadening is associated with changes in the magnetic configuration (e.g. above 25 GHz in Fig. 10c, which corresponds to reversal of the SAF PL top layer at 4.5kOe). In order to enable the automated measurement of α, regardless of this complex non-linear dependence on f_R, we compute the cumulative distribution function (CDF) of the ratio ($\Delta f/f_R$)

over the entire experimental range (Fig. 11). Without additional broadening contributions, the CDF should be a step function centered on α. To minimize the effect of broadening while mitigating the influence of noise and fitting errors, we define α at CDF=10%. Note that this is an upper bound, since it still includes the smallest broadening contributions, if any. However, we find this method to be much more robust than fitting Δf vs. f_R, which in many cases leads to unrealistically small or even negative values of α. We have carried out automated ST-FMR measurements and data analysis as a function of FL structure, device diameter and temperature, allowing us to determine H_K and α distributions (Fig. 12). Fig. 13 shows data as a function of device CD and temperature obtained using a slightly different ODM-FL structure. Interestingly, α is weakly dependent on device diameter, whereas H_K increases for smaller CDs, as already reported [7]. The temperature dependence of H_K, α and Δ is shown in Fig. 14 for 30nm devices of the same wafer. Δ=49 is obtained at 85°C, high enough to ensure fewer than 1ppm retention errors after 1 hour, even after accounting for distributions of Δ [8].

Fig.15 shows the comparison of ST-FMR properties of 30nm devices from various wafers using the ODM-FL (red) and conventional metal cap (blue) and metal insertion (cyan) FLs. Different data points display wafers having different design parameters within each class of FLs. Metal insertion and cap FLs show higher damping than ODM-FL wafers, for which α~0.005. Such low damping allows for highly efficient STT switching. Fig. 16 shows resistance vs. dc voltage for 30nm devices using metal cap (a) and ODM-FL (a). Not only is the switching voltage reduced significantly for the ODM-FL, but a tunnel magnetoresistance ratio larger than 180% is also achieved. As discussed above, low damping must be combined with low moment for low current switching at short pulses. Fig. 17 shows the 50% P to AP switching current as a function of pulse length for the ODM-FL, compared with metal cap and insertion FLs. Data are in good agreement with (2). In the case of metal cap, high damping gives rise to large current I_{w0} for longer pulses, but relatively small increase at shorter pulse because of the small magnetic moment. By contrast, the metal insertion wafer shows lower I_{w0} because of lower damping, but larger increase due to higher moment. Only the ODM-FL combines lowest damping with low moment, leading to write current well below 50 μA for pulses of 10ns or shorter. We believe the concepts underlying this ODM-FL design can be developed further, opening the way to even lower power operation of STT-MRAM for highly efficient computing.

REFERENCES

[1] M. C. Shih *et al.*, Dig. Tech. Pap., Symp. VLSI Technology, **2016**, p. 114
[2] Y. K. Lee *et al.*, Dig. Tech. Pap., Symp. VLSI Technology, **2018**, p. 195
[3] K. Lee *et al.*, Dig. Tech. Pap., Symp. VLSI Technology, **2018**, p. 183
[4] G. Jan *et al.*, Dig. Tech. Pap., Symp. VLSI Technology, **2018**, p. 65
[5] J. He, J. Z. Sun and S. Zhang, J. Appl. Phys. 101, 09A501 (2007)
[6] L. Thomas *et al.*, Tech. Dig. - Int. Electron Devices Meet. **2015**, p 672
[7] L. Thomas *et al.*, J. Appl. Phys. **115**, 172615 (2014).
[8] L. Thomas *et al.*, Tech. Dig. - Int. Electron Devices Meet. **2017**, 38.4
[9] C. Wang *et al.*, Phys. R. B **79**, 224416 (2009)

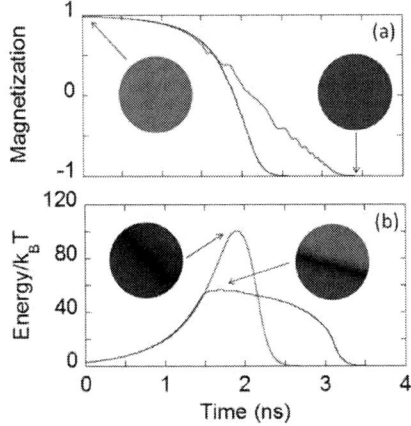

Fig.1. Micromagnetic simulations of STT-driven switching of 30nm devices having the same properties except for the exchange stiffness. Color maps show magnetization distributions in the up (red) and down (blue) states, as well as in the demagnetized domain wall states.

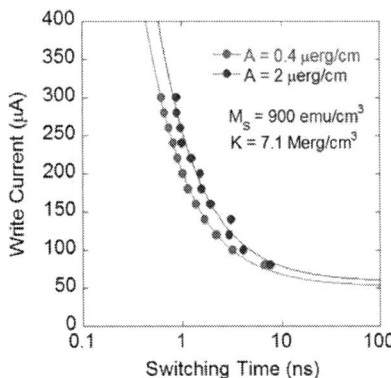

Fig. 2. Write current I_W vs. switching time t_{SW} for the two devices of Fig.1. Note that simulations are performed at constant current, and t_{SW} is defined such that $M(t_{SW})<-0.8$. Simulations results are well fitted to (2): $I_W=I_{W0}+I_1/t_{SW}$.

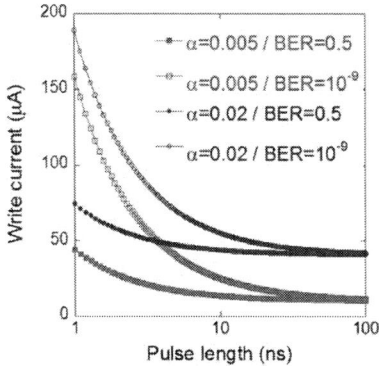

Fig. 3. Write current vs. pulse length calculated using (2) for the values of damping α and write error rate indicated on the figure. Note that while the write current for long pulse lengths depends on α, the $1/t_p$ current increase for short pulses is independent of α.

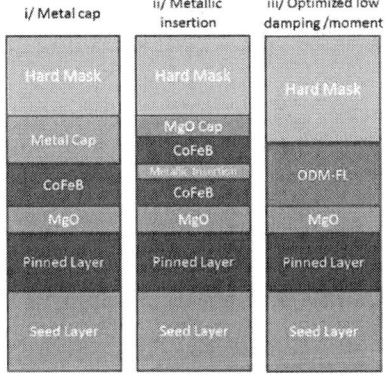

Fig. 4. Schematics of the three STT-MRAM stacks designs used in this study.

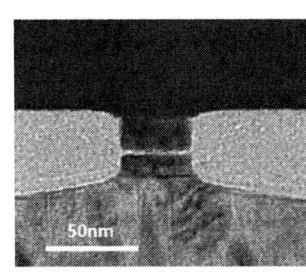

Fig. 5. Cross-sectional TEM image of 30nm ODM-FL device.

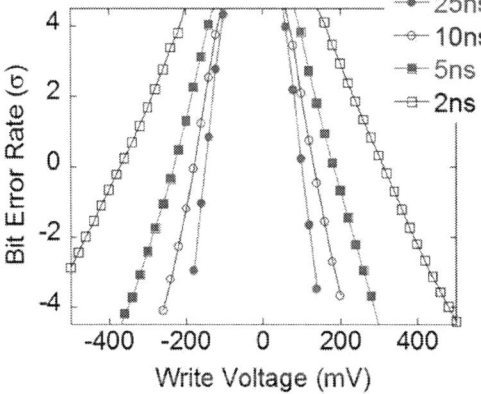

Fig. 6. Bit error rate in quantile scale for a 30nm ODM-FL device measured at 25°C at various pulse lengths.

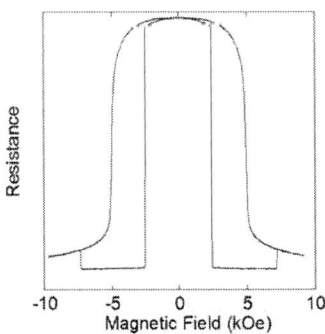

Fig. 7. Resistance vs. magnetic field hysteresis loops measured for a 30nm ODM-FL device.

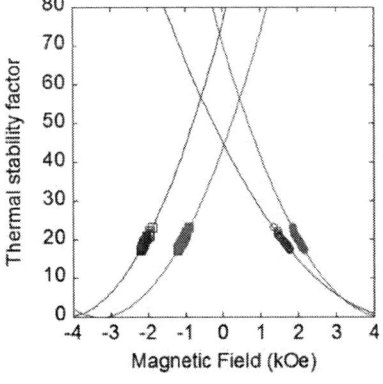

Fig. 8. Thermal stability factor Δ derived from the switching field distribution for small positive (blue) and negative (red) read currents. Solid lines are fits to the macrospin model $\Delta(H)=\Delta(1-H/H_K)^2$.

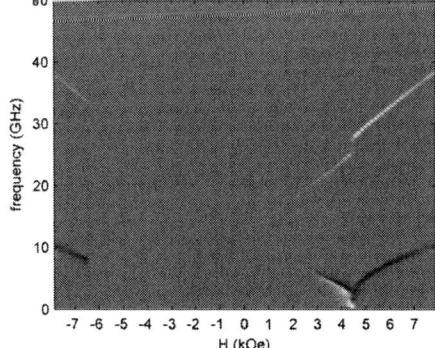

Fig. 9. Contour plot of the ST-FMR signal for the same device shown in fig 7&8 as a function of the microwave excitation frequency and applied magnetic field H. FL and PL resonance branches correspond to white and black contrasts, respectively.

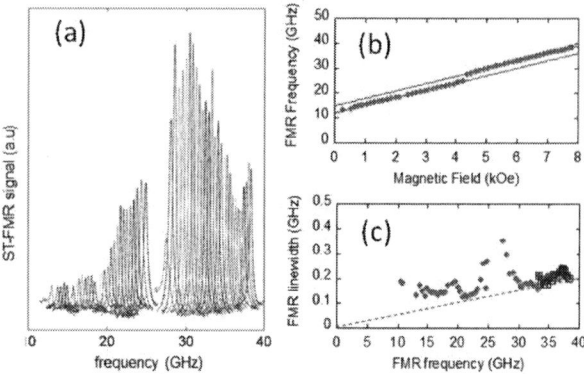
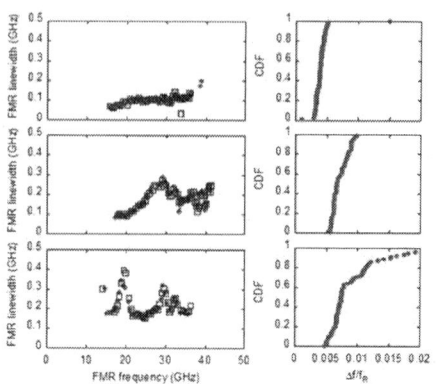
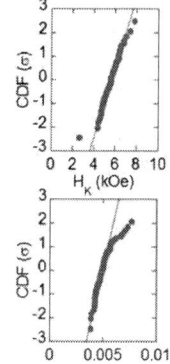

Fig. 10. (a) ST-FMR data of Fig. 7 are fitted to a Lorenzian line shape (solid lines) to extract the resonance frequency (b) and linewidth (half width at half maximum) (c). Red and blue symbols in (c) show data measured at positive and negative fields, respectively. Dashed line shows the linear relationship expected without additional broadening.

Fig. 11. FMR linewidth Δf vs. resonance frequency f_R for three devices with different linewidth broadening. The same features are observed for both positive (red) and negative (blue) fields. α is defined as the 10% value of the CDF of $\Delta f/f_R$ (right panels).

Fig. 12. Distributions of α and H_K for 75 nominally identical devices across a wafer.

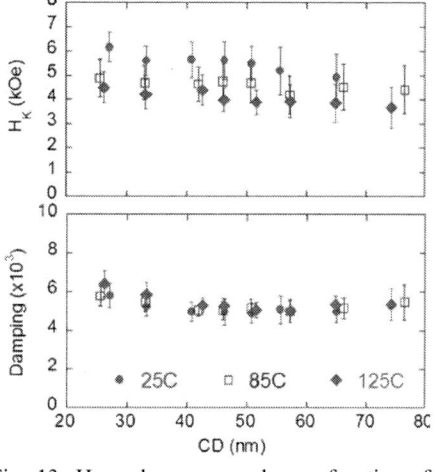
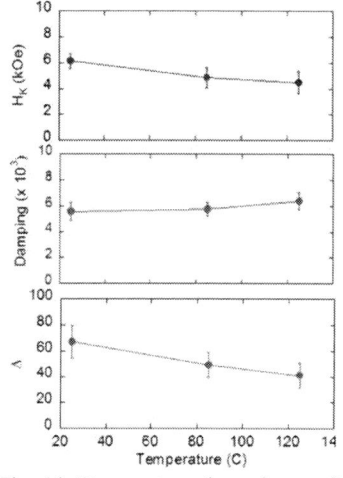
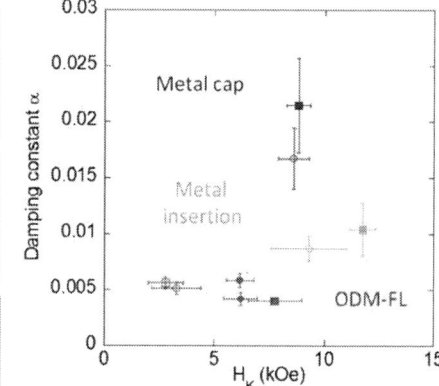

Fig. 13. H_K and α measured as a function of device diameter at indicated temperatures for an ODM-FL. 20 to 80 devices have measured for each condition. Symbols and error bars show the median and standard deviation.

Fig. 14. Temperature dependence of anisotropy field HK, damping and thermal stability factor of 30nm devices for the ODM-FL of Fig. 13.

Fig. 15. H_K and damping of 30nm devices measured for different classes of FL structures. Data points and error bars show median and standard deviation values for each wafer. Different data points represent different design parameters within each class of FLs.

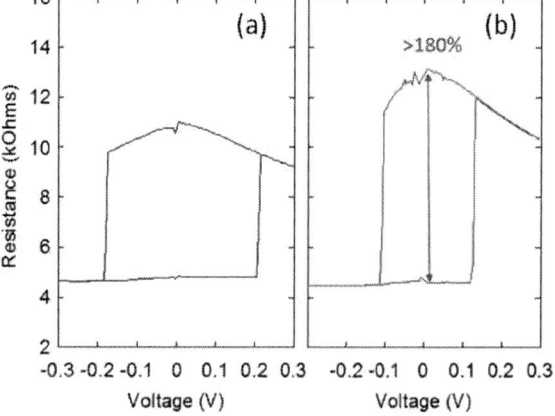
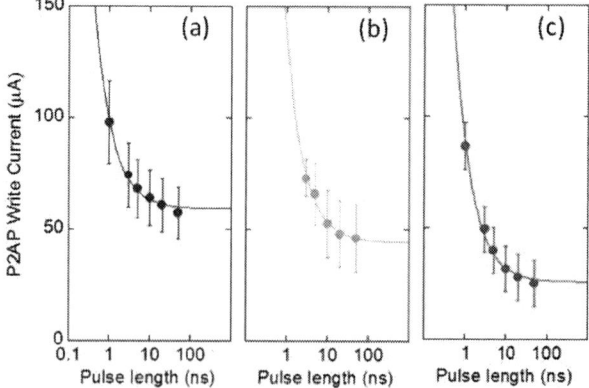

Fig. 16. Resistance vs. dc voltage hysteresis loops for 30nm devices using (a) metal cap and (b) ODM-FL. TMR larger than 180% is observed for the ODM-FL device. Both stacks have a resistance-area product below 4 Ohm.μm^2.

Fig. 17. Current for P to AP switching with 50% probability measured on 30nm devices for different FL designs (a) Metal cap (b) Metal insertion (c) ODM-FL. Data points and errors bars show median and standard deviations of tens to hundreds of devices.

978-1-7281-1988-5/18 $31.00 © 2018 IEEE

Microwave Neural Processing and Broadcasting with Spintronic Nano-Oscillators

P. Talatchian[1], M. Romera[1], S. Tsunegi[2], F. Abreu Araujo[1,3], V. Cros[1], P. Bortolotti[1], J. Trastoy[1], K. Yakushiji[2], A. Fukushima[2], H. Kubota[2], S. Yuasa[2], M. Ernoult[1,4], D. Vodenicarevic[4], T. Hirtzlin[4], N. Locatelli[4], D. Querlioz[4], J. Grollier[1]

[1]Unité Mixte de Physique, CNRS, Thales, Univ. Paris-Sud, Université Paris-Saclay, France, email: julie.grollier@cnrs-thales.fr
[2]National Institute of Advanced Industrial Science and Technology (AIST), Spintronics Research Center, Japan
[3]Institute of Condensed Matter and Nanosciences, UCLouvain, Belgium
[4] Centre de Nanosciences et de Nanotechnologies, CNRS, Univ. Paris-Sud, Université Paris-Saclay, France

Abstract— Can we build small neuromorphic chips capable of training deep networks with billions of parameters? This challenge requires hardware neurons and synapses with nanometric dimensions, which can be individually tuned, and densely connected. While nanosynaptic devices have been pursued actively in recent years, much less has been done on nanoscale artificial neurons. In this paper, we show that spintronic nano-oscillators are promising to implement analog hardware neurons that can be densely interconnected through electromagnetic signals. We show how spintronic oscillators maps the requirements of artificial neurons. We then show experimentally how an ensemble of four coupled oscillators can learn to classify all twelve American vowels, realizing the most complicated tasks performed by nanoscale neurons.

I. SPINTRONIC NANO-OSCILLATORS

Spintronic nano-oscillators are magnetic tunnel junctions, whose CMOS-compatible technology is essentially identical to magnetic memory cells that can be fabricated by billions on a single chip [1]. They are cylinder shaped, with a diameter that can be reduced below 10 nm. The central part is a tunnel barrier (usually MgO) separated by two ferromagnetic layers (often based on cobalt, iron and boron alloys). As illustrated in Fig. 1, when a direct current is sent through the cylinder, it gets spin-polarized by crossing the first magnetic layer, tunnels to the other one, to which it transfers its excess of angular momentum by exerting a torque on the magnetization [2]. This spin-torque can then induce sustained magnetization precessions which are converted through magnetoresistance to voltage oscillations that can reach up to a few millivolts across the junction (Fig. 2) [3]. The power consumption of spin-torque nano-oscillators, which decreases with their diameter, is around one microwatt for a diameter of ten nanometers. The frequency of the oscillations varies from hundreds of megahertz to tens of gigahertz. While the power consumption per se is not weak weak compared to slower CMOS-based neurons, the energy can therefore be very small, below a hundred attojoules per oscillation [4]. This means that hundreds of millions of oscillators could be assembled on a chip and used for computing.

II. SPIN-TORQUE NANO-OSCILLATORS CAN COMPUTE A NON-LINEARITY AND BROADCAST THE RESULT

A. Spin-torque nano-oscillators are non-linear nano-radios

The key feature of spin-torque nano-oscillators that makes them interesting as neurons is their non-linear response. As can be observed in Fig. 3, when the dc current through the oscillator is varied, the angle with which magnetization precesses varies widely, resulting in a non-linear dependence of the oscillation voltage amplitude. This means that, just like formal neurons in deep networks, spin-torque nano-oscillators can non-linearly transform their input (here the injected current). Moreover, the output of the computation is a microwave voltage or magnetic field that can be emitted. In other words, spin-torque-nano-oscillators can non-linearly process information like neurons, then broadcast the result to other neurons.

B. Using the non-linear response of spin-torque nano-oscillators for computing

The non-linear response of spin-torque nano-oscillators combined with their stability and long life time enables neuromorphic computing. This has been experimentally demonstrated recently by our team and collaborators [3]. We have used time-multiplexing to emulate a full neural network of 400 neurons with a single spin-torque nano-oscillator (Fig. 4). The temporal connections between neurons come from the finite relaxation time of oscillator due to magnetization relaxation. Using the framework of reservoir computing [5], we have demonstrated that the multiplexed nano-oscillator could recognize spoken digits from 0 to 9 with a precision up to 99.6 %, which is as good as much larger neurons and software simulations. These results show that spin-torque nano-oscillators can be readily used as neurons.

III. SPIN-TORQUE NANO-OSCILLATORS CAN LEARN TO CLASSIFY THE MICROWAVE SIGNALS THEY RECEIVE

A. Spin-torque nano-oscillators response is highly sensitive to incoming microwave signals

Spin-torque nano-oscillators feature an outstanding tunability [6]. For instance, their frequency can be varied by more than 50 % by changing the injected direct current or the applied magnetic field. Due to this property, their dynamics can easily be influenced by weak external microwave currents or magnetic fields. In particular, spin-torque nano-oscillators have a high ability to phase lock to external microwave signals (Fig. 5) and to mutually synchronize (Fig. 6). From a neural network point of view, this means that spintronic oscillators as neurons in layer $k+1$ have the ability to adapt their response to the microwave outputs broadcasted by neurons in layer k. In other words, microwave inter-neuron communication is possible in spin-torque nano-oscillator networks, opening a path to ultra-fast processing. How such a network would function and compute is an open question, but many possibilities exist. Fig. 7 gives an insight on how this could work: if neuron i in layer k and neuron j in layer $k+1$ are synchronized, they share the same frequency, which means that the synapse w_{ij} that connects them is strong. On the other hand, if neuron i and j have very different frequencies, the synaptic weight w_{ij} is weak.

B. Learning to classify microwave inputs

We have recently given a proof of concept of microwave signal classification through synchronization [4]. The experimental network is composed of four spin-torque nano-oscillators interconnected by millimeter-long wires (Fig. 8). The microwave emission from each oscillator propagates through this electrical loop and in turn modifies the dynamics of all the other oscillators. The oscillators are all-to-all coupled though this mechanism, which emulates synaptic connections. The frequency of the individual oscillators can be controlled by injecting different direct currents in the junctions. The oscillator microwave emissions are detected with a spectrum analyzer. As illustrated in Fig. 9, microwave inputs are injected in an antenna (a strip line) located just above the oscillators. Through the antenna, the microwave inputs generate a microwave magnetic field that modifies the dynamics of the oscillators. Depending on the relative frequency of oscillators and inputs, the frequency of the oscillator is pulled by the ensemble of input signals, or it can synchronize to one of the inputs. These emerging synchronization configurations can be used to classify inputs, if a given class of inputs always gives rise to the same synchronization configuration (for instance, the same set of synchronized oscillators). This behavior can be obtained by training the neural network through a gradual modification of the direct currents injected in the oscillators, which results in frequency tuning.

C. Vowel recognition

We have recently shown that this neural network of four spin-torque nano-oscillators could be trained to classify seven American vowels (https://youtu.be/IHYnh0oJgOA). The frequencies characteristics of the vowels are accelerated and combined to generate two input signals in the microwave domain. The experimental recognition rate after training is 89% on the test data (84% after cross validation). This performance is comparable and even slightly better than that of a multilayer perceptron trained on the same task with a similar number of parameters. Indeed, in our scheme the coupled oscillators cooperate to decide which vowel is recognized. This is not the case in perceptrons where intra-layer neurons are not connected. This result demonstrates that spin-torque nano-oscillators dynamical properties can be finely tuned to learn. It also shows that their coupling and synchronization properties can be harnessed to classify.

To go further, we show here that our scheme can be extended to classify all twelve American vowels. For this we use a larger number of the twenty experimentally observed synchronization states and combine them to recognize a vowel. The currents injected in the oscillators, their frequency and the recognition rate during training are plotted in Fig. 10. The final map of synchronization states and corresponding vowels is shown in Fig. 11. We reach a recognition rate of 68.4 % on train and test datasets. This recognition rate can be largely increased in the future by increasing the number of oscillators in the system. Indeed, the number of synchronization regions that can be used and combined for recognition scales as N^2 where N is the number of oscillators [7].

IV. CONCLUSION AND PERSPECTIVES

We have shown that spin-torque nano-oscillators can non-linearly transform inputs and broadcast the result of their computation. We have shown that they can also learn to classify the microwave signals they receive. Two main challenges remain in order to build large scale deep neural networks with these devices: to fabricate synapses that can tune the inter-neuron microwave communication, and to densely interconnect neurons through these synapses. In future work, we will investigate how the multifunctionality of spintronic building blocks [8] and the possibility to assemble them in 3D [9] can enable solutions to these challenges.

ACKNOWLEDGMENT

This work was supported by the European Research Council ERC under Grant bioSPINspired 682955, the French National Research Agency (ANR) under Grant MEMOS ANR-14-CE26-0021 and by a public grant overseen by the ANR as part of the "Investissements d'Avenir" program (Labex NanoSaclay, reference: ANR-10-LABX-0035).

REFERENCES

[1] S. W. Chung et al., *2016 IEEE International Electron Devices Meeting (IEDM)*, 2016, p. 27.1.1-27.1.4.
[2] N. Locatelli, V. Cros, et J. Grollier, *Nat. Mater.*, vol. 13, n° 1, p. 11-20, janv. 2014.
[3] J. Torrejon et al., *Nature*, vol. 547, n° 7664, p. 428-431, juill. 2017.
[4] M. Romera et al., *ArXiv171102704 Cond-Mat Q-Bio*, nov. 2017.
[5] L. Appeltant et al., *Nat. Commun.*, vol. 2, p. 468, sept. 2011.
[6] A. Slavin et V. Tiberkevich, *IEEE Trans. Magn.*, vol. 45, n° 4, p. 1875-1918, avr. 2009.
[7] D. Vodenicarevic, N. Locatelli, F. A. Araujo, J. Grollier, et D. Querlioz, *Sci. Rep.*, vol. 7, p. 44772, mars 2017.
[8] J. Grollier, D. Querlioz, et M. D. Stiles, *Proc. IEEE*, vol. 104, n° 10, p. 2024-2039, oct. 2016.
[9] A. Fernández-Pacheco, R. Streubel, O. Fruchart, R. Hertel, P. Fischer, et R. P. Cowburn, *Nat. Commun.*, vol. 8, p. 15756, juin 2017.

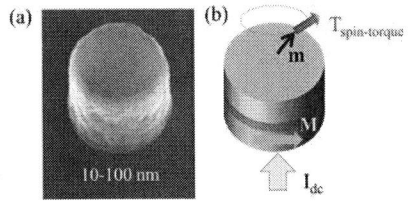

Fig. 1. (a) SEM picture of a 400 nm diameter resist dot before etching the magnetic tunnel junction stack. (b) Illustration of magnetization dynamics under spin-torque in a magnetic tunnel junction.

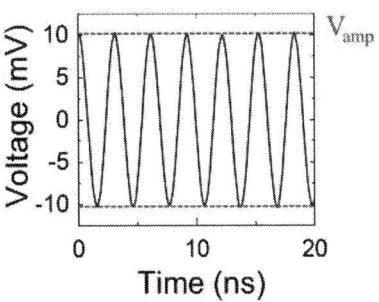

Fig. 2. Spin-torque induced voltage oscillations in a magnetic tunnel junction. The amplitude of the oscillations is indicated in red. The central stack is based on a CoFeB 2.4 nm pinned layer, MgO 1nm and FeB 6 nm free layer. The ground state is a magnetic vortex.

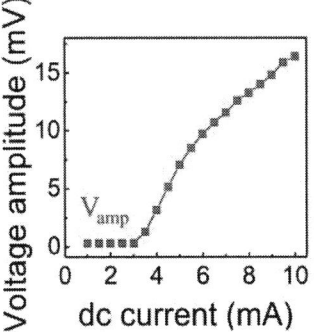

Fig. 3. Voltage amplitude as a function of the injected direct current for the junction of Fig.2.

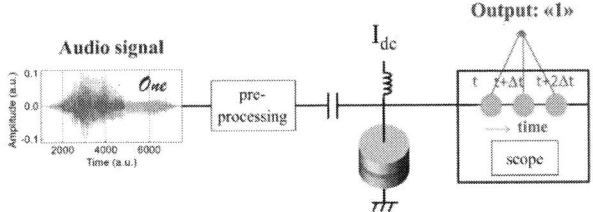

Fig. 4. Illustration for spoken digit recognition with a single time-multiplexed spin-torque nano-oscillator through reservoir computing.

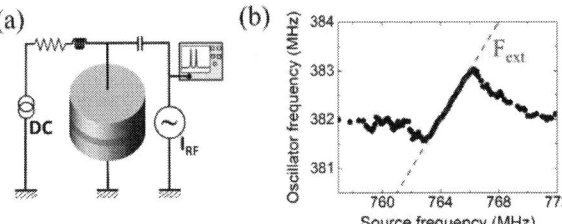

Fig. 5. (a) Schematic of the magneto-transport set-up for phase-locking (b) Oscillator frequency versus source frequency. When the oscillator is phase-locked, its frequency becomes equal to the source frequency.

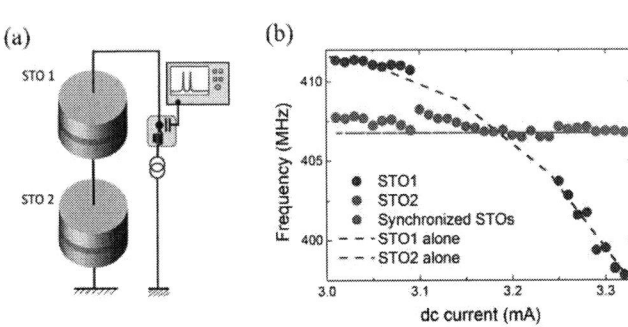

Fig. 6. a) Schematic of the magneto-transport set-up for the electrical mutual synchronization of two oscillators (b) Oscillators frequency as a function of dc current. In the locking-range, the two oscillator frequencies become identical.

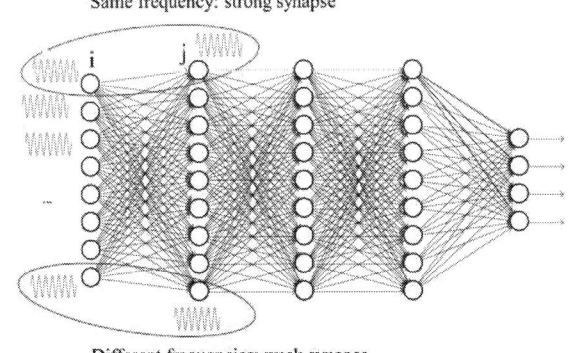

Fig. 7 Illustration of the working principle of a microwave neural network.

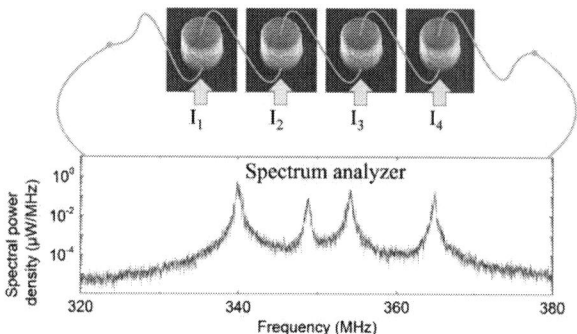

Fig. 8. Schematic of the neural network composed of four interconnected spin-torque nano-oscillators. Independent direct currents can be applied to the different oscillators, and their microwave emissions are detected with a spectrum analyzer.

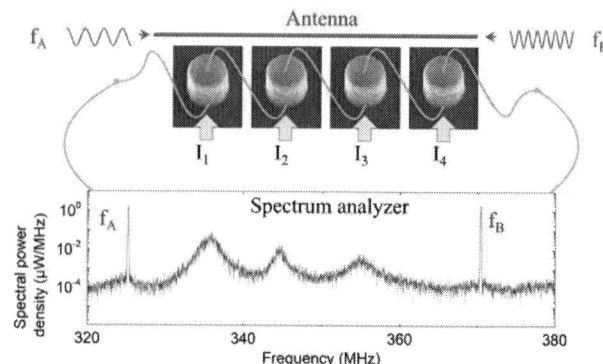

Fig. 9. Schematic of the spin-torque oscillator neural networks with microwave inputs are applied through a strip line above the oscillators. In that case, oscillator 4 is synchronized to frequency f_B.

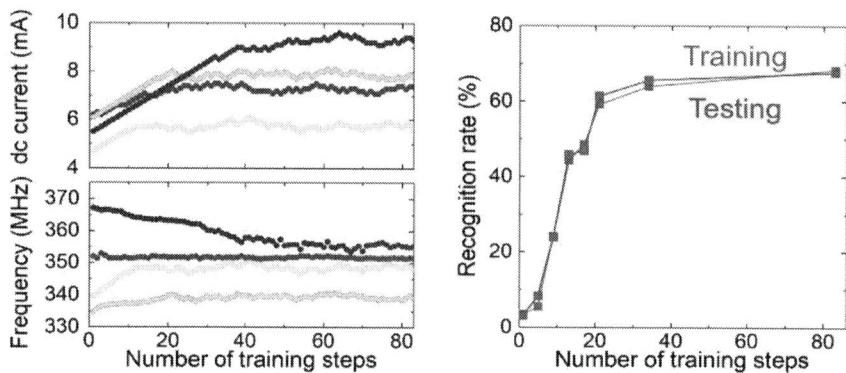

Fig. 10. Frequencies, dc currents through the oscillators and recognition rates on the train and test datasets as a function of training step.

(a) Before learning   (b) After learning

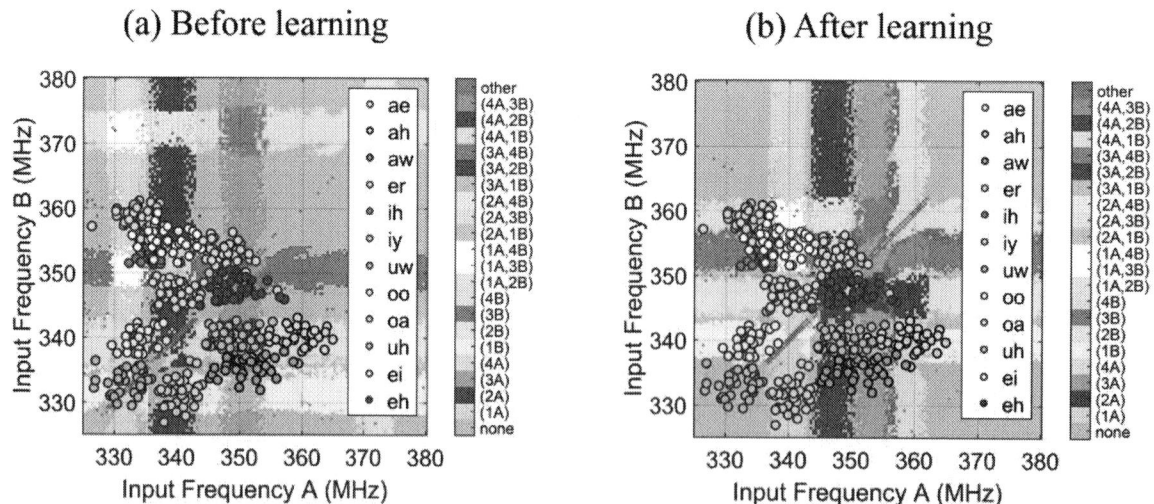

Fig. 11 Maps of synchronization regions before and after learning. The label (XA,YB) indicates that oscillator X is synchronized to source A and oscillator Y is synchronized to source B. Vowels are indicated as circles with the color of the synchronization regions in which they should be classified.

High Endurance Phase Change Memory Chip Implemented based on Carbon-doped $Ge_2Sb_2Te_5$ in 40 nm Node for Embedded Application

Z. T. Song[1], D. L. Cai[1], X. Li[1], L. Wang[2], Y. F. Chen[1], H. P. Chen[1], Q. Wang[1], Y. P. Zhan[2], M. H. Ji[2]

[1]State Key Laboratory of Functional Materials for Informatics, Shanghai Institute of Micro-system and Information Technology, Chinese Academy of Sciences, Shanghai, China, email: ztsong@mail.sim.ac.cn; caidl@mail.sim.ac.cn
[2]Semiconductor Manufacturing International Corporation, Shanghai, China

Abstract— In this work, we present the results of a highly reliable phase change memory (PCM) based on Carbon-doped $Ge_2Sb_2Te_5$ material in 40 nm node. The large Reset/Set resistance ratio of more than 2 orders of magnitude is achieved. The chip exhibits excellent data retention, endurance characteristics, and the sensing window is even larger after 260 ℃ soldering test. It is estimated that the PCM could retain data for 10 years at 128 ℃. In a 128 Mb test chip over 10^8 cycles is achieved. PCM is suitable for applications requiring high thermal stability and cycling endurance.

I. INTRODUCTION

Phase change Memory (PCM) is widely investigated as one of the most promising candidate for non-volatile memory [1]. High speed, high density, low voltage, and compatibility with standard CMOS technology performance are the PCM superiority [2]-[5]. But the poor retention property and failure of solder flow test restrict its application in embedded systems. It is possible to improve the PCM performance by store material [6]. In contrast to the $Ge_2Sb_2Te_5$ (GST) material, the faster, lower-power phase transition and high thermal stability by using Carbon-doped $Ge_2Sb_2Te_5$ (CGST) material is proposed. In this work, we realized Single-Level-Cell (SLC) 128Mb PCM chips using the CGST in 40 nm node and carried out a comprehensive reliability study. The results show that SLC PCM of the CGST are suitable for applications of embedded systems that require high performance at high thermal circumstances [7-8]. And over 10^8 cycles endurance is achieved in the chip. The 128 Mb capacity can meet the capacity requirements of most embedded applications. **Table 1** compares embedded nonvolatile memories technology [9]. Moreover, the preparation of 40nm 128 Mb PCM has laid a solid foundation for the PCM integration in the 28nm high-k/metal gate process and beyond.

II. MATERIAL AND BEC CHARACTERIZATION

Fig. 1(a) shows the temperature-dependent sheet resistance curves of CGST and GST films. As shown in this figure, with the increase of temperature, a slow and gradual drop of sheet resistance Rs is observed. These films present a semiconductor-like behavior in the low temperature range. After an initial monotonic decrease, until the crystallization temperature (Tc), the resistance values produce a sudden precipitous drop, corresponding to Tcs, happen at 150.6 ℃, and 300.6 ℃ for GST, and CGST, respectively. This value is much higher than that of conventional GST (~150 ℃). The high Tc can inhibit the spontaneous crystallization and can make the amorphous state more stable, which is proved in **Fig. 1(b)**. **Fig. 1(b)** presents the data retention characteristics for the CGST and GST samples. The curves of Rs versus time for all samples at different ambient temperatures are studied, and the failure time is defined as the time required for the Rs falling to half of its initial value at a specific temperature. As shown in the **Fig.1(b)**, low failure time is obtained with GST samples. By extrapolation from the failure time, on the basis of the Arrhenius law, the fail temperatures after 10 years for CGST and GST films are found to be 167 and 80 ℃, respectively. The result shows that the data retention of CGST is much better than that of GST. It is obviously that the data retention (thermal stability) can be improved by doping C atoms into GST material. Thus, the CGST based PCM device can store information much longer and safer than the GST-based one.

Fig. 2(a) shows the cross-section of memory cell integrated based on the 40 nm node. The PCM unit is built between CT and Metal 1 by adding only three extra masks. The blade bottom electrode contact (BEC) area ~ 4×63 nm² and the typical metal-oxide semiconductor field effect transistor (MOSFET) acted as the selector. A lance-shape CGST film was placed above on BEC. The top and the bottom electrode made by metal tungsten (W) and the TiN film as adhesive layer. Current–voltage characteristics of PCM cell using CGST material with threshold voltage 1.08 V is shown in **Fig. 2(b)**. The specific PCM device structure is for the 128 Mb test chip which layout is shown as **Fig. 2(c)**. The die photo is shown as **Fig. 2(d)**. **Fig. 3** shows the dependence of resistance on the width of the blade TiN BEC. As the width of BEC decreasing, the BEC resistance increases. The BEC resistance has good linear characteristics at a same width. The BEC resistance thermal reliability could affect that of the PCM device. The resistance distribution of such TiN BECs is well concentrated $2.25 \sim 4$ KΩ. **Fig. 4** shows the Voltage - Current (V - I) curves of BEC in the case of with different ambient temperature. From the figure, the BEC resistance is not variational at different ambient temperatures. The results show that the TiN BEC has good linear resistance performance under the thermal tress.

978-1-7281-1988-5/18 $31.00 © 2018 IEEE

A write endurance test is performed by repeating the alternate writing of operation until write operation fails for the TiN BEC. The pulses for operation is 0.5 mA/50 ns. As shown in **Fig. 5**, different width BEC are operated and the results indicate that the endurances are over 10^{11} cycles. It has been confirmed that the endurance of TiN BEC is excellent for applying in PCM.

III. Test Chip Characterization

Fig. 6 shows the current-resistance curves with different electrical pulse widths. From **Fig. 6**, we can see that CGST-based PCM cells have a high resistance larger than 10^6 Ω and a resistance ratio of about 1 orders of magnitude. The cells can be reversibly switched by using 20 ns width pulses. This means that, with better thermal stability, PCM cells using CGST alloy are more suitable for embedded PCM application than GST-based ones.

The Set resistance is found closely dependent on the Set pulse width, as shown in **Fig. 7**, which is programmed by pulses with a constant height of 0.4 mA. As the pulse width increases, the Set tends to decrease obviously. But the Reset is not closely dependent on the Reset pulse height, as shown in **Fig. 8**, which is programmed by pulse with a constant width 50ns. As the pulse height increases, the Reset increase is not obvious. **Fig. 9** shows the Reset and Set resistance distribution of 32Kb cells in the 128 Mb array by applying 0.64 mA/50 ns pulse for Reset and 0.4 mA/500 ns pulse for Set, respectively. The Set resistance is between 31 and 100 kOhm. And the Reset resistance is between 1000 and 10000 kOhm. The minimal Reset/Set resistance ratio is about 1 orders of magnitude and the maximal ratio is over 2 orders of magnitude. So large ratio is enough sensed by circuit. **Fig. 10** shows the chip yield map of the 128Mb chip applying operation parameters from **Fig. 9**. Form the figure, 93.3% chip yield can be achieved and only on the central point and the edge of wafer the fail chips occur.

The ambient temperature can affect the PCM performance. **Fig. 11** shows the resistance distribution at the ambient temperature from 25~100 °C. Form **Fig. 11**, we can see that the Reset and Set resistance decrease with the ambient temperature increasing. The memory window also gets smaller, but the Reset/set ratio is still over 1 order of magnitude at 100 °C. The results show that the PCM based CGST alloy can be used in some high temperature circumstance. **Fig. 12** shows the Reset and Set distribution before and after the soldering test at 260 °C. After 5 minutes baking, there still exists obvious detectable window. The Set state drifts down and Reset state still keeps at the same level, which means the sensing window is even larger after soldering test. The results show that the PCM based CGST alloy can be used to need preprogram in chip before package.

PCM based on CGST has good thermal stability. The resistance variation is measured for the thermal reliability test after cycling stress. The purpose is to understand the data retention performance after cycling stress. **Fig. 13(a)** shows the resistance read out results of pre 1 cycle, 1k cycles, 10k cycles and 100k cycles. **Fig. 13(b)** shows the result after

150 °C baking 1 hour. **Fig. 13(c)** shows the result after rewriting into data and then read out. After baking, the Reset resistance value decrease and the Set resistance value increase, so the memory windows get small. For pre 100k cycles device, the Reset resistance distribution becomes narrower and shows the overlap with the Set distribution. The memory window maintains well about 1 order of magnitude for pre 10K cycles device. This indicates that cells with high cycling stress shows poor thermal stability. Rewriting into data for the baked device, the memory window is almost back to before baking as shown in **Fig. 13(c)**. The results show that the operation of baking after 100k cycles can affect the storage state, but it can't affect the storage performance. The fail velocity of the Reset state cell at the same temperature baking depends on the Reset pulse height. The fail velocity decreases with the Reset pulse height at the same width (50 ns) increasing as shown in **Fig. 14**. **Fig. 15** shows the life time estimation base on temperature stress method using three temperatures 130 °C, 140 °C, and 150 °C. Although the high cycling number degrades the data retention, the estimated temperatures is 128 °C for 10-year data retention.

Next, endurance characteristics of these device cells are also investigated, as shown in **Fig. 16**. the results indicate that the Reset/Set endurance is over 10^8 cycles. The typical endurance parameter for FLASH is about 10^6 cycles and some endurance for PCM is about 10^6 [10], 10^7 [11] cycles. It has been confirmed that the endurance of PCM based on CGST is very superior to FLASH technology.

IV. Conclusions

We introduce the CGST phase change materials to enhance the thermal reliability, data retention and endurance. Our example in this work, the CGST-based PCM device of 128 Mb PCM test chip in 40 nm node demonstrates the properties including: over 10^{11} cycle endurance of the TiN blade BEC, Reset/Set ratio over 2 orders of magnitude, 10-year 128 °C data retention, and over 10^8 cycles. It has great potential to be extendable to embedded applications.

Acknowledgment

This work was supported by the National Key Research and Development Program of China (2017YFA0206101, 2017YFB0701703), "Strategic Priority Research Program" of the Chinese Academy of Sciences (XDA09020402), National Integrate Circuit Research Program of China (2009ZX02023-003), National Natural Science Foundation of China (61376006, 61401444, 61504157, 61622408).

References

[1] H.-S. P. Wong et al., Nat. Nanotech., 10, pp. 191–194, 2015.
[2] S. Lai et al., IEDM Tech. Dig. pp. 803-806, 2001.
[3] F. Rao et al., Science. 358, pp. 1423-1427, 2017.
[4] A. V. Kolobov, et al., Nat. Mater. 3(10), pp. 703-708. 2004.
[5] D.H. Im et al., IEDM, Tech. Dig. pp. 1-4, 2008.
[6] W. S. Khwa, et al., IEDM, Tech. Dig. pp. 29.8.1 - 29.8.4, 2014.
[7] W. S. Khwa, et al., ISSCC, pp. 134 - 135, 2016.
[8] W.C. Chien, et al., IEDM, Tech. Dig. pp. 552 - 555, 2016.
[9] G. D. Sandre, et al., IEEE J. Solid-State Circuits. 46(1), 52-62, 2011
[10] D. Kau et al., IEDM Tech. Dig. pp. 617–620,2009.

	This work	eFLASH	EEPROM
Cell Size(F^2)	**50**	16~22	90~110
P/E Voltage (V)	**3.3**	~10	~10
Set Time	**500 ns**	20~100 ms	1~2 ms
Reset Time	**50 ns**	1~20 μs	1~2 ms
Data Rentention	**10 yrs@128 ℃**	10 yrs@85~125 ℃	10 yrs@105~150 ℃
Endurance	**10^8**	10^4~10^5	10^5

Table 1. Comparison Embedded Nonvolatile Memories Technology.
[11] S. Kang et al., IEEE J, Solid-state Circuits, 42, pp.210-218, 2007.

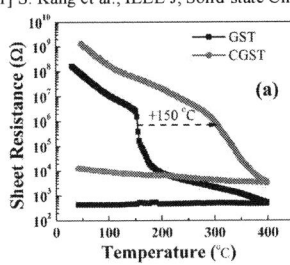

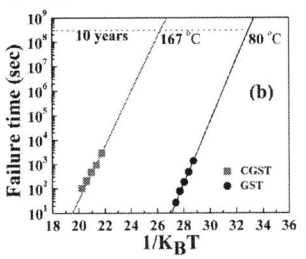

Fig.1(a). Measured sheet resistance as a function of annealing temperature for CGST and GST. (b) The Arrhenius extrapolation at 10 years of data retention for CGST films and GST films.

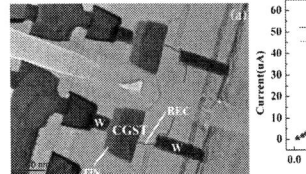
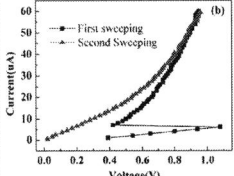

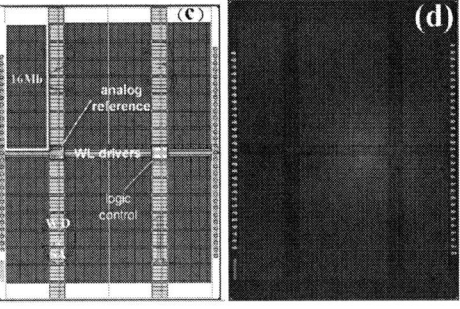

Fig.2. (a) Cross sectional TEM view of CGST-based cell in array; (b) Current–voltage characteristics of PCM cell using CGST material with threshold voltage 1.08 V; (c)The layout of 128 Mb PCM chip; (d) The die photo of 128 Mb PCM chip.

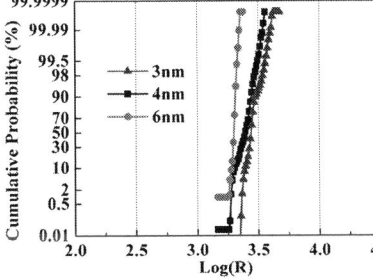

Fig. 3. TiN blade BEC resistance distribution with different width from 4k cells measured results for each kind size. The resistance distribution of such TiN BECs is well concentrated 2.25 ~ 4 KΩ.

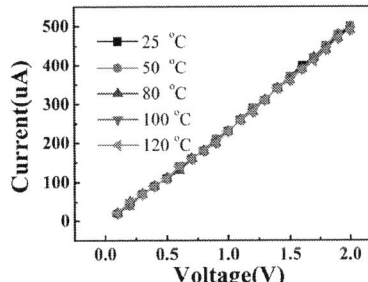

Fig. 4. Measured Voltage-current (V-I) curves of the TiN blade BEC at 25 ~ 120 °C. TiN BEC has good linear resistance performance under the thermal tress.

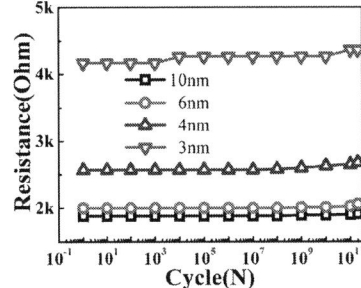

Fig. 5. Measured endurance of the TiN blade BEC with different width. The resistance varies little over 10^{11} cycles.

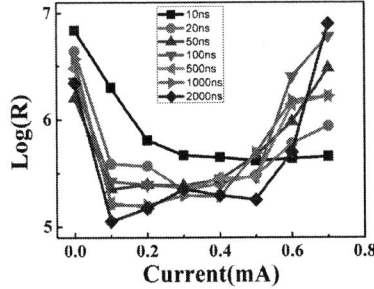

Fig. 6. Measured current-resistance curves with different electrical pulse widths. The cells can be reversibly switched by using 20 ns width pulses.

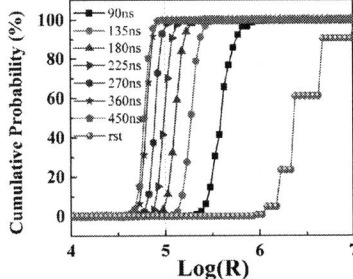

Fig. 7. Set resistance closely depends on the Set pulse width from 16 Kb cells. As the pulse width increases, the Set tends to decrease obviously.

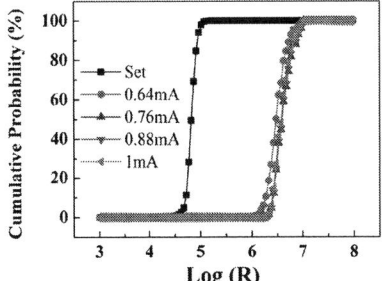

Fig. 8. Reset resistance is not closely dependent on the Reset pulse height from 16 Kb cells. As the pulse height increases, the Reset increase is not obvious.

978-1-7281-1988-5/18 $31.00 © 2018 IEEE

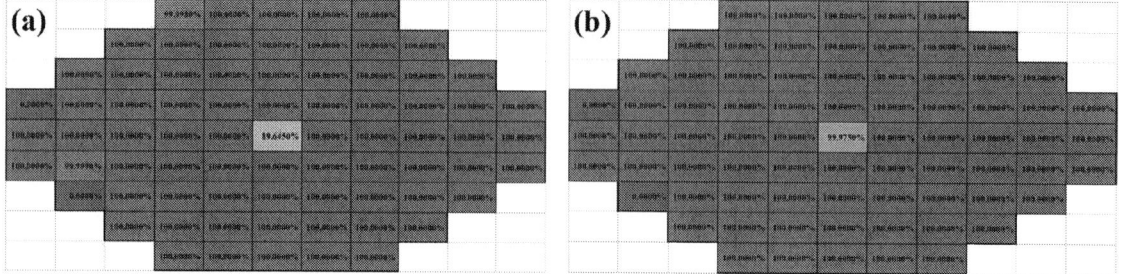

Fig. 10. Chip yield map of the 128Mb chip applying operation parameters from Fig. 9. 93.3% chip yield can be achieved and only on the central point and the edge of wafer the fail chips occur. (a) Reset operation; (b) Set operation.

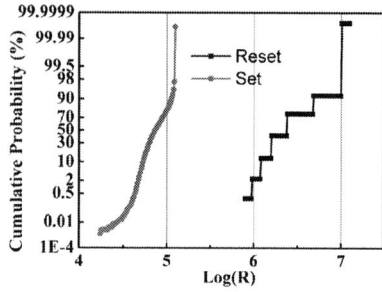

Fig. 9. Reset and Set resistance distribution of 128Kb cells in the 128 Mb array by applying 0.64 mA/50 ns pulse for Reset and 0.4 mA/200 ns pulse for set, respectively.

Fig. 11. Ambient temperature affects the PCM performance. Resistance distribution (4 Kb cells) at the ambient temperature from 25 ~ 100 °C.

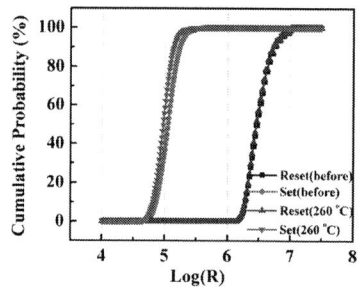

Fig. 12. Reset and Set distribution (16 Kb cells) before and after the soldering test at 260 °C. After 5 minutes baking, there still exists obvious detectable window.

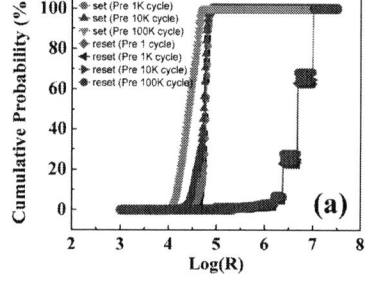
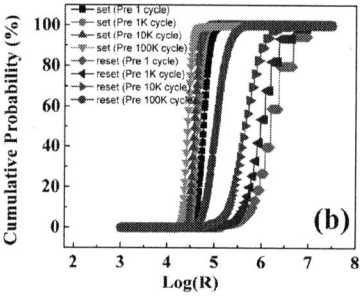
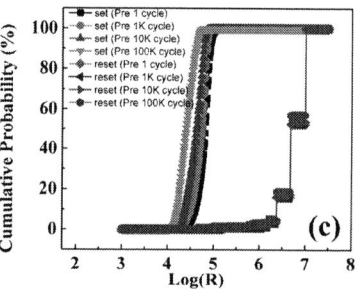

Fig. 13. Resistance variation is measured form 32 Kb cells for the thermal reliability test after cycling stress. (a) before baking, (b) after baking, and (c) rewriting into data and then read out.

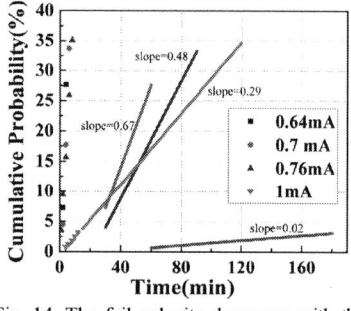

Fig. 14. The fail velocity decreases with the Reset pulse height increasing at the same pulse width.

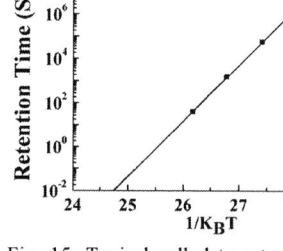

Fig. 15. Typical cell data retention of the CGST-based PCM 128 Mb chip. 10-year 128 °C data retention is achieved.

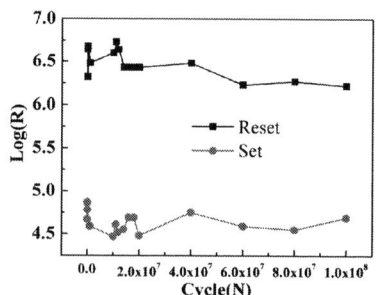

Fig. 16. Typical cell cyclability of the CGST-based PCM 128 Mb chip. The endurance is over 10^8 cycles.

978-1-7281-1988-5/18 $31.00 © 2018 IEEE

A 40nm Low-Power Logic Compatible Phase Change Memory Technology

J.Y. Wu, Y.S. Chen, W.S. Khwa, S.M. Yu, T.Y. Wang, J.C. Tseng, Y.D. Chih and Carlos H. Diaz

Taiwan Semiconductor Manufacturing Company Ltd., Corporate Research, 168, Park Ave. 2,

Hsinchu Science Park, Taiwan

Tel: 886-3-5636688, email: jywuz@tsmc.com

Abstract

An embedded phase change memory technology in 40nm low-power logic platform is demonstrated with minimal added process complexity - two non-critical additional masks over standard logic. Specially designed hard mask and etching process was used to achieve 50% shrinkage of the memory cell bottom electrode dimension with same lithography tooling as the 40nm logic platform. Bottom electrode CD shrinkage along with optimization of the electrode materials in terms of electrical and thermal conductivity enabled significant (~4x) write current reduction attaining competitive levels of ~300 µA at 40nm BE CD. Embedded PCM cells reported in this work demonstrated over 100x memory window - (RESET/SET resistance switching ratio), over 200k cycling endurance with extrapolated 10 year retention at 120 ℃. In this work not only large switching resistance ratios but also highly-controllable resistance values that are almost independent of the PCM starting resistance state are presented along with the corresponding programing pulse requirements. The switching resistance ratio and resistance value controllability are key features for neural network and compute-in-memory applications. In this work, their benefits on design margins for energy efficient high-density binary neural network for inference applications aiming accuracy levels of well over 90% is asserted over an MNIST dataset.

Introduction

Phase Change Memory (PCM) has been reported as a good candidate for non-volatile memory applications [1-5]. Regardless of the application, low power operation, high density, and low process complexity – cost are key requirements to be met; for embedded memory applications compatibility with underlying logic process is also a upmost importance. Tackling the power challenge at fundamental memory cell level relates to minimizing the current required to promote phase change and the overall voltage drop across the memory element and its access device. Minimizing operating current entails PCM material optimization along with the memory cell structure electro-thermal design. Primary cell structures can be categorized as confined and non-confined [6-7]. Confined cells have the potential of smallest footprint at expense of challenging phase change material bottom-up gap-fill inside small holes or requiring complex patterning schemes [7-9]. In this work we focus on non-confined structures for 1T1R embedded memory arrays where transistor as access device enables write/reads at lower operating power supply as compared to ovonic threshold switches (OTS) as selector devices [10-11]. We report on operating current reduction while preserving solid memory switching windows as demonstrated on 1Mb test arrays by cost-effective bottom electrode dimension scaling and electro-thermal design, memory cell PCM material and etch optimization (un-doped to doped GST); results show good scalability potential and compare very well with published work from other groups [7,12-13]. This work also reports on our findings pertaining memory cell programing schemes aimed attain dependable and well controllable resistance switching independent of the initial state of the memory cells. Finally, models based on this work's experimental results are utilized to study the design space and guidelines the PCM technology potential benefits for energy-efficient binary neural network applications. These results complement well other reports on PCM as a key technology candidate for artificial intelligence (AI) applications [14-15].

Device Structure and Thermal Simulation

Figure 1 shows the TEM image of mushroom structure PCM device fabricated between M4 and M5 of 40nm CMOS logic technology. Figure 2 shows the BE process flow for shrinking 50% BE CD from original logic design rule. Dual metal-1 and metal-2 used in BE process flow are capable of filling in 40nm BE CD. CMP process is performed to decide final BE height. Doped GST is then deposited onto small scale BE, followed by a top electrode (TE) layer. TE/GST are patterned and connected to M5 through top Via. In order to optimize etching condition for TE and GST individually, two step etching process is compared in Fig. 3. Composition change induced by etch process is observed on TE#1 (high Cl_2 flow) and TE#2 (low Cl_2 flow) is the key to achieve uniform GST composition. Optimized halogen gases for GST etching is also important to minimize etch damage as shown in Fig. 4 [16-17]. In Fig. 5, dual metal thicknesses ratio (BE#1 with Met-1/Met-2=1/8 and BE#2 with Met-1/Met-2=8/1) are designed for engineering BE structure through TCAD thermal simulation [18]. For BE#2, more heat can be generated and stay in the GST/BE interface for reducing write current requirement. In addition, melting volume for BE#2 is smaller as comparing to BE#1. In Fig. 6 and Fig. 7, electrical resistivity of Met-2 and thermal conductivity of Met-1 are studied on two different BE structures. From simulation result, BE#2 can combine both advantages for dramatically write current reduction as comparing to BE#1.

Device Performance

In Fig. 8, BE#2 can demonstrate 300 µA write current for full amorphous state that is ~4x reduction comparing to

978-1-7281-1988-5/18 $31.00 © 2018 IEEE

BE#1. In Fig. 9, RESET current reduction with BE area scaling is studied and experimental result follows the trend of $I \sim A^{0.8}$, better than literature's results ($I \sim A^{0.65}$) [19]. Fig. 10 shows RT and 150℃ resistance drift behavior and estimated RESET R drift coefficient is 0.1. Designed 1T1R PCM 1Mb test-chip on 40nm low-power logic platform is shown in Fig. 11. In Fig. 12, good HRS and LRS switching result is demonstrated with over 100x switching ratio.

Figure 13 shows the resistance distribution of HRS and LRS with various drain voltages (Vd). There is a great improvement of the chip operated to higher resistance by larger Vd due to larger amorphous volume formed while LRS resistance slightly increases owing to some residual amorphous regions existed after SET. In Fig. 14, SET speed is characterized with programming the cell back to the HRS before applying each SET pulse. The 100ns SET speed is defined by reaching 100x resistance switching window on doped GST. Through waveform modulation by increasing falling tail pulse width, symmetry of RESET/SET conductance as a function of write voltage can be demonstrated in Fig. 15. PCM program characteristic features advantage for precisely resistance control and it is critical for improving inference accuracy in computing-in-memory architecture. Fig. 16 demonstrated good cycling endurance up to 200K with applying 200ns pulse width without read verification. Fig. 17 is retention result for doped GST material. Calculated Ea is ~2.9 eV for predicting 10 year data retention at 120 ℃. Through write-and-verification scheme, standard deviation of LRS and HRS resistance can be improved to < 3.5%.

COMPUTING IN MEMORY

To improve the data utilization efficiency in CMOS-based deep neural network (DNN) accelerators, parallelized computation across multiple processing-elements (PE) are highly preferred [20-21]. Computing-in-memory (CIM) is an alternative a possible approach that integrates the computation into the memory array. In CIM, the matrix multiplication could be performed efficiently by activating multiple wordlines (WL) simultaneously and comparing the accumulated bitline current (I_{BL}) or voltage (V_{BL}) [22-23]. From the characterization result on 40nm PCM test-chip, the impact of cell resistance standard deviation (std) and resistance ratio (R_{ratio}) on inference accuracy are simulated using a LeNet-5 binary neural network (BNN) with MNIST dataset [24-25]. We focused on the last two fully-connected (FC) layers, because they are technically more challenging due to more WL activations.

Fig. 18 shows the product-sum result distributions of the last two FC layers, with number of positive one and number of negative one on the x-axis and y-axis, respectively. A {x=18, y=13} point would indicate this MNIST test image yields a +5 product-sum result. The average product-sum result from FC-1 and FC-2 are +4.37 and +8.2, respectively. This implies that FC-1 would face more technical challenge due to its narrower margin. Fig. 19 shows the impact analysis of std and R_{ratio} on inference accuracy. To achieve a >90% accuracy on a two-layer FC (64x64x10), a minimal R_{ratio} of 10 is required with 3% and 4% std on FC-1 and FC-2, respectively. Inference accuracy is very sensitive to std.

Increasing the std from 3% to 4%, with a R_{ratio} of 10, on FC-1 increases the error rate by >2.9X. If R_{ratio} could be increased to 143, then the std can be relaxed to 5%. Larger R_{ratio} is highly favorable and could relax the std requirement. Fig. 20 shows a case analysis of PCM versus two other emerging memory candidates. Even for BNN, in which both the weights and neuron activations are binarized to +1/-1, it is still critical to have a sufficiently large R_{ratio} to compensate for the intrinsic memory device variation. In this regard, PCM benefits tremendously from its ability to fine-tune cell resistance and its high R_{ratio}.

CONCLUSION

A 40nm CMOS-compatible PCM technology is demonstrated with cost-effective BE reduction method and damage-free TE/GST patterning processes. 300 µA write current is achieved with 4x reduction through optimizing dual metal thicknesses ratio. Over 100x resistance ratio, good resistance controllability, reliable cycling and good high temperature data retention are also demonstrated. Proposed PCM technology is a promising candidate of DNN hardware accelerator for handwritten MNIST accuracy over 90% with binary neural network.

Acknowledgements

The authors would like to thank the support of Cheng-Chung Chien and Julian R. Lee in process modules as well as Jeff Wu and Pike Chang for TCAD simulation.

References

[1] G.Servalli, "A 45nm Generation Phase Change Memory Technology", IEDM Tech. Dig., p113, 2009.
[2] F. Pellizzer1, et. al., "A 90nm Phase Change Memory Technology for Stand-Alone Non-Volatile Memory Applications", VLSI Tech. Symp., 2006.
[3] Daewon Ha, et. al. "Recent Advances in High Density Phase Change Memory (PRAM)", IEDM Tech. Dig., 2007.
[4] Scott W. Fong., et. al., "Phase-Change Memory—Towards a Storage-Class Memory", IEEE TED, Vol 64, No.11, pp.4374-4385, 2017.
[5] H.-S. Philip Wong, et. al., "Phase Change Memory", IEDM Tech. Dig., p2201, 2010.
[6] Y.J. Song, et. al., "Highly Reliable 256Mb PRAM with Advanced Ring Contact Technology and Novel Encapsulating Technology", VLSI Tech. Symp., 2006.
[7] M. J. Kang, et. al., "PRAM cell technology and characterization in 20nm node size", IEDM Tech. Dig., p39, 2011.
[8] M. Breitwisch, et. al.,"Novel Lithography-Independent Pore Phase Change Memory", Symp. VLSI Tech., p100, 2007.
[9] J.I. Lee, et. al., "Highly Scalable Phase Change Memory with CVD GeSbTe for Sub 50nm Generation ", Symp. VLSI Tech., p102, 2007.
[10] DerChang Kau, et. al., "A stackable cross point phase change memory", IEDM Tech. Dig., p617, 2009.
[11] H. Y. Cheng, et. al., "An Ultra High Endurance and Thermally Stable Selector based on TeAsGeSiSe Chalcogenides Compatible with BEOL IC Integration for Cross-Point PCM", IEDM Tech. Dig., p28, 2016.
[12] Y. H. Ha, et. al., "An Edge Contact Type Cell for Phase Change RAM Featuring Very Low Power Consumption", Symp. VLSI Tech., p175, 2003.
[13] W.C. Chien, et. al., "Reliability Study of a 128Mb Phase Change Memory Chip Implemented with Doped Ga-Sb-Ge with Extraordinary Thermal Stability", IEDM Tech. Dig., p662, 2016.
[14] S. Yu, et. al., "Neuro-inspired computing with emerging nonvolatile memorys," in Proceedings of the IEEE, vol. 106, no. 2, pp. 260-285, 2018.
[15] G. W. Burr, et. al., "Experimental Demonstration and Tolerancing of a Large-Scale Neural Network (165000 Synapses) Using Phase-Change Memory as the Synaptic Weight Element", IEEE TED, vol. 62, no. 11, pp. 3498-3507, 2015.
[16] E. A. Josepht, et. al., "Patterning of N:Ge2Sb2Te5 Films and the Characterization of Etch Induced Modification for Non-Volatile Phase Change Memory Applications", VLSI-TSA, 2008.
[17] Se-Koo Kang, et. al., "Etch Damage of Ge2Sb2Te5 for Different Halogen Gases", Jpn. J. Appl. Phys., 086501-1, 2011
[18] J.Y. Wu, et. al., "A Low Power Phase Change Memory Using Thermally Confined TaN/TiN Bottom Electrode", IEDM Tech. Dig., p43, 2011.
[19] D. Ielmini, et. al., "Phase change materials in non-volatile storage", Materialstoday, vol. 14, pp. 600-607, 2011.
[20] Y.H. Chen, et. al., "Eyeriss: an energy-efficient reconfigurable accelerator for deep convolutional neural networks," IEEE ISSCC, p262, 2016.
[21] N. P. Jouppi, et. al., "In-datacenter performance analysis of a Tensor Processing Unit," ACM/IEEE ISCA, p1, 2017.
[22] A. Biswas, et. al., "Conv-RAM: An energy-efficient SRAM with embedded convolution computation for low-power CNN-based machine learning applications," IEEE ISSCC, pp. 488-490, 2018.
[23] S. K. Gonugondla, et. al., "A 42pJ/decision 3.12TOPS/W robust in-memory machine learning classifier with on-chip training," IEEE ISSCC, pp. 490-492, 2018.
[24] W. S. Khwa, et. al., "A 65nm 4Kb algorithm-dependent computing-in-memory SRAM unit-macro with 2.3ns and 55.8TOPS/W fully parallel product-sum operation for binary DNN edge processors," IEEE ISSCC, pp. 496-498, 2018.
[25] R. Liu, et. al., "Parallelizing SRAM Arrays with Customized Bit-cell for Binary Neural Networks", Design Automation Conference (DAC), pp. 21.1-21.6, 2018.

978-1-7281-1988-5/18 $31.00 © 2018 IEEE

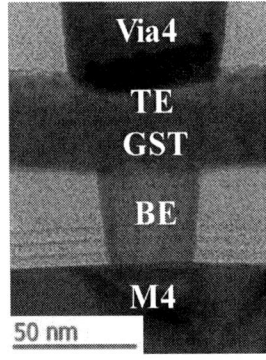

Fig. 1 TEM image for intergraded PCM on 40nm low-power logic platform. PCM is fabricated between M4 and M5.

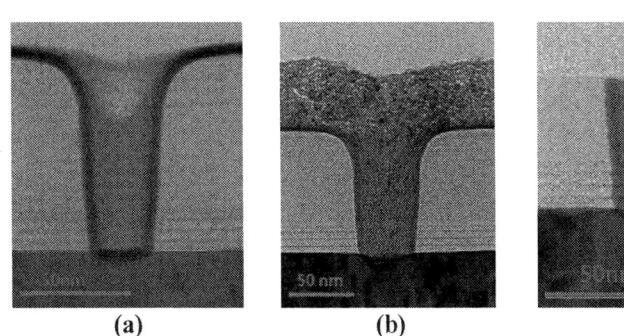

Fig. 2 Post BE etch TEM images with shrinking 50% BE CD from logic design rule after (a) metal-1 deposition, (b) metal-2 deposition and (c) CMP.

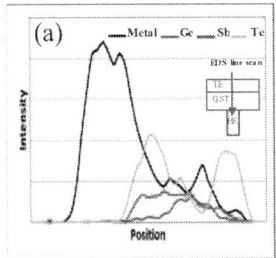 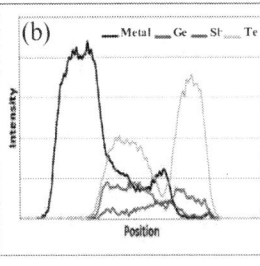 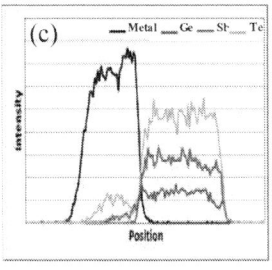

Fig. 3 TE-Metal/Ge/Sb/Te composition uniformity is studied by EDS line scan (vertical) with different TE and GST etching conditions that includes (a) TE#1/GST#1, (b) TE#1/GST#2 and (c) TE#2/GST#2. TE metal etch is as important as GST etch since both etch conditions can affect Ge/Sb/Te composition uniformity.

Fig. 4 Ge/Sb/Te composition uniformity in GST is studied by EDS line scan (horizontal) with optimized TE/GST etching conditions.

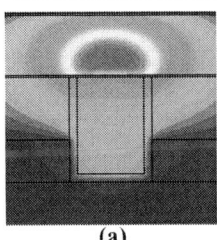

Fig. 5 TCAD thermal simulation for BE with engineering dual metal thicknesses ratio. Met-1 and Met-2 are outer and inner layers, respectively. (a) BE#1 with Met-1/Met-2=1/8 and (b) BE#2 with Met-1/Met-2=8/1.

Fig. 6 BE engineering by changing electrical resistivity of Met-2 (inner layer). Higher electrical resistivity Met-2 can reduce RESET current.

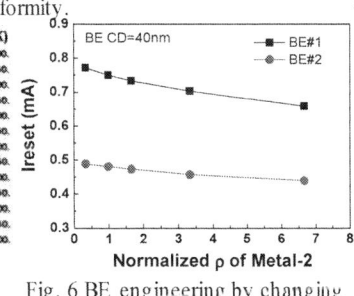

Fig. 7 BE engineering by changing thermal conductivity of Met-1 (outer layer). Low thermal conductivity Met-1 can reduce RESET current.

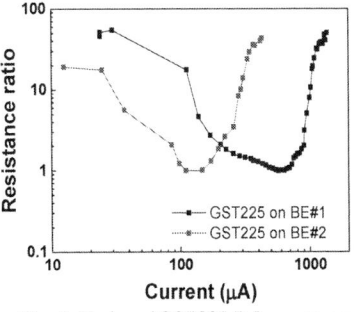

Fig. 8 Undoped GST225 R-I curves are measured on different BE structures. RESET current as low as 300uA is demonstrated on BE#2.

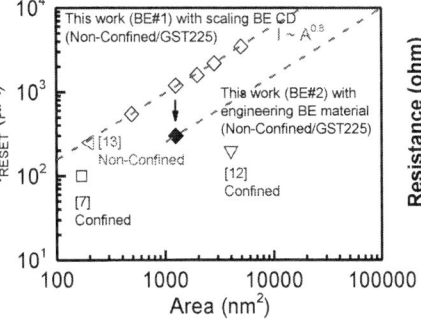

Fig. 9 RESET current as a function of BE area is plotted from experiment results. Similar to thermal simulation result, write current can be further reduced on BE#2.

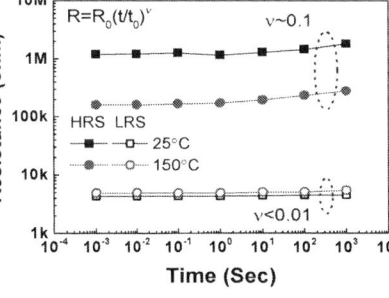

Fig. 10 Data retention result at RT and 150 °C. LRS is stable with time (v<0.01) while HRS starts to drift after 1 sec with v~0.1. v is defined as drift coefficient.

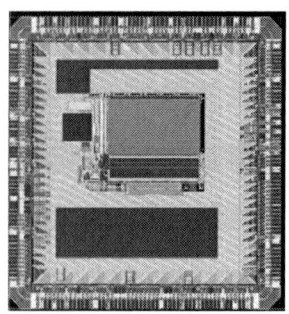

Fig.11 The 1Mb 1T1R PCM test-chip on 40nm low-power logic platform.

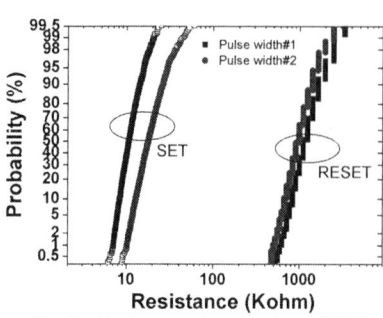

Fig.12 Resistance distribution of HRS and LRS are compared with different write speeds. Longer PW#2 can achieve larger memory window than PW#1.

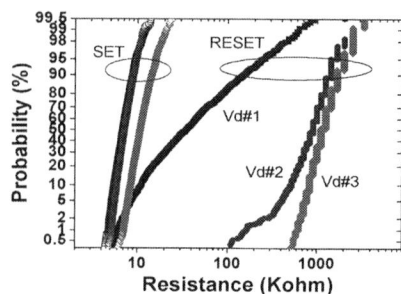

Fig.13 Resistance distribution of HRS and LRS with various drain voltages (Vd). A great improvement of HRS by a larger Vd while LRS resistance only slightly increases.

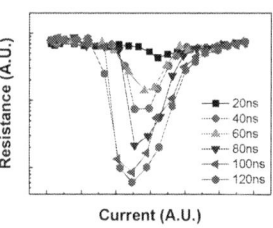

Fig.14 R-I curve is measured for RESET and SET operation. Estimated SET speed are 100ns for reaching 100x R switching ratio.

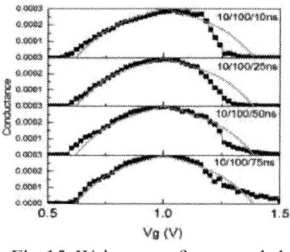

Fig.15 Write waveform modulation can achieve symmetry of RESET/SET conductance as a function of write voltage.

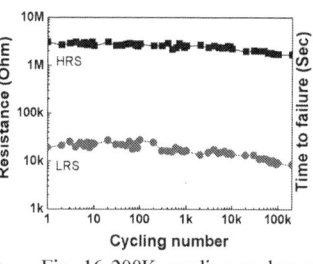

Fig. 16 200K cycling endurance of 1T1R PCM is demonstrated. RESET and SET are operated with 200ns pulse width.

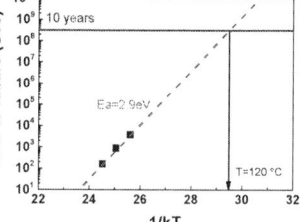

Fig. 17 Data retention result for extracting Ea on 1T1R PCM device for predicting 10 year lifetime at 120 ℃.

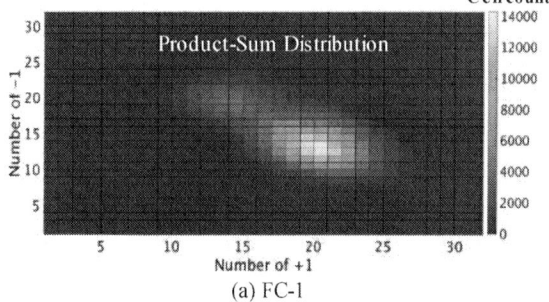

(a) FC-1

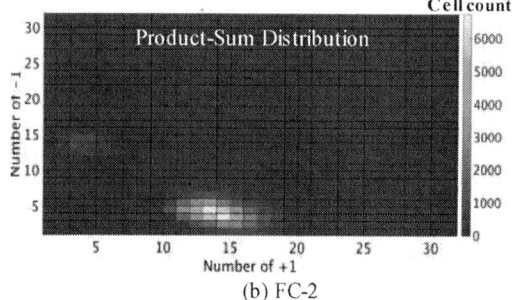

(b) FC-2

Fig. 18 Simulated product-sum distributions from MNIST dataset show the last two FC layers will have different requirement. The average product-sum result for (a) FC-1 and (b) FC-2 are +4.37 and +8.2, respectively. FC-1 layer faces more stringent requirement due to its narrower margin from having a smaller difference between the number of +1 and -1.

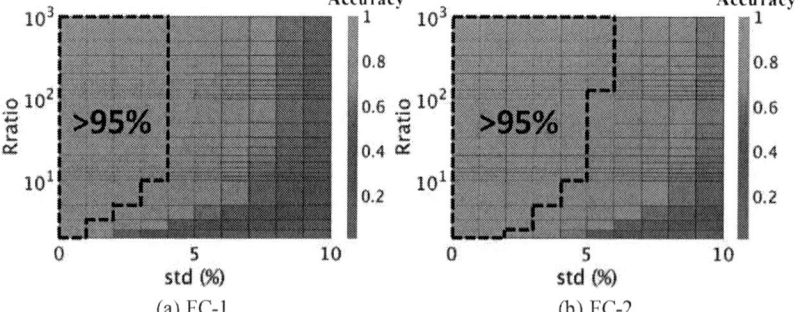

(a) FC-1

(b) FC-2

Fig. 19 Simulation analysis shows both standard deviation (std) and resistance ratio (R$_{ratio}$) are critical for MNIST accuracy. In particular, for a FC size of 64x64x10, a minimum R$_{ratio}$ of 10 is suggested with 3% and 4% std on FC-1 and FC2, respectively. If R$_{ratio}$ is increased to 143, the std could be relaxed to 5%.

PCM: Rratio = 100, std = 3.5%
Memory 1: Rratio = 10, std = 6%
Memory 2: Rratio = 2, std = 5%

	Accuracy %
PCM (FC-1)	95.303
Memory 1 (FC-1)	51.780
Memory 2 (FC-1)	0.013
PCM (FC-2)	99.122
Memory 1 (FC-2)	77.494
Memory 2 (FC-2)	22.274

Fig. 20 MNIST accuracy simulation analysis shows that even for binary neural network, a large R$_{ratio}$ is still critical to compensate the intrinsic memory device variation.

8-bit Precision In-Memory Multiplication with Projected Phase-Change Memory

I. Giannopoulos*, A. Sebastian*, M. Le Gallo, V.P. Jonnalagadda, M. Sousa, M.N. Boon, and E. Eleftheriou

IBM Research – Zurich, 8803 Rüschlikon, Switzerland, email: nno@zurich.ibm.com, ase@zurich.ibm.com

Abstract— In-memory computing is an emerging non-von Neumann approach in which certain computational tasks such as matrix-vector multiplication are performed using resistive memory devices organized in a crossbar array. However, the conductance variations associated with the memory devices limit the precision of this computation. Here, we demonstrate that the so-called projected phase-change memory (Proj-PCM) devices can achieve 8-bit precision while performing scalar multiplication. The devices were fabricated and characterized using electrical measurements and STEM investigation. They are found to be remarkably immune to conductance variations arising from structural relaxation, $1/f$ noise and temperature variations. Moreover, it is possible to compensate for the temperature-dependent conductance variations in a crossbar array using a simple model. Finally, we experimentally demonstrate a neural network-based image classification task involving 30 such Proj-PCM devices.

I. INTRODUCTION

In-memory computing is an emerging computing paradigm that has the potential to increase the performance and area/energy efficiency of several artificial intelligence related computational tasks [1-4]. However, the limited precision of in-memory computing remains a key challenge. Much of the research effort is focused on system-level or architectural solutions to address this problem [5,6]. Here, we propose a device-level solution to address this challenge based on the concept of projected phase-change memory (Proj-PCM) [7,8]. In a Proj-PCM device, there is a non-insulating projection segment in parallel to the phase-change segment. By exploiting the highly non-linear IV characteristics of phase-change materials, we can ensure that during the write process, the projection segment has minimal impact on the operation of the device. However, during read, conductance values of programmed states are mostly determined by the projection segment that appears parallel to the amorphous phase-change segment. Hereby, we demonstrate the efficacy of these devices with respect to in-memory multiplications.

II. DEVICE FABRICATION AND CHARACTERIZATION

We fabricated Proj-PCM devices based on a lateral device geometry (Fig. 1). GeTe serves as the phase-change layer, while the projection layer consists of a metal nitride. By applying appropriate programming pulses that cause a melt-quench process on the as-fabricated crystalline GeTe, it is possible to modulate the amorphous region's size. During the read process, one measures the resistance of the projection layer, which is in parallel to the amorphous region. The resulting programing curve is shown in Fig. 2. To validate our assumption, that we have a single contiguous amorphous

region separated by crystalline GeTe, we performed extensive STEM studies on these devices (Fig. 3). These studies also indicate that the amorphous region is not perfectly centered, which is indicative of additional thermo-electrical effects that have been shown to play a significant role in these nanoscale devices [9]. Scalar multiplication using resistive memory devices rely on Ohm's law. Hence it is of interest to study the field dependence of electrical transport in Proj-PCM devices. It can be seen that compared with conventional PCM devices, Proj-PCM devices show a much weaker field dependence (Fig. 4). The precision associated with scalar multiplication and subsequently matrix-vector multiplication is strongly determined by the conductance variations associated with these devices. For example, conventional PCM devices exhibit a temporal evolution of conductance (drift) attributed to the structural relaxation of the amorphous phase [10]. Proj-PCM devices show a 50-fold reduction in drift (Fig. 5). Besides drift, there are conductance fluctuations arising from the $1/f$ noise which is also found to be substantially lower in our devices (Fig. 6). The variation in conductance arising from temperature fluctuations is another key challenge given the highly thermally activated nature of electrical transport in phase-change materials. What is even more detrimental is that the activation energy tends to vary for different programmed states and as can be seen later, this poses key challenges for developing effective temperature compensation schemes. The Proj-PCM devices on the other hand show a substantially weaker dependence on temperature variations (Fig. 7).

III. IN-MEMORY MULTIPLICATION

A. Scalar Multiplication

To perform the scalar multiplication operation, $\beta = \alpha \cdot \xi$, the variable ξ is mapped proportionally to a read voltage and α into a conductance state of the Proj-PCM device. Due to Ohm's law, one can obtain an approximate result $\hat{\beta}$ of β from the resulting read current. We performed 20,000 scalar multiply operations on 12 conductance states of a Proj-PCM device (Fig 8a). Due to the analog nature of the programming curve, it is possible to program the device to a desired conductance state with high precision using iterative programing. The achieved precision of the scalar multiply operation is comparable to 8-bit fixed point arithmetic at room temperature (Fig. 8b,c). This remarkable result is attributed to the significantly low conductance variations associated with the programmed states. The average dissipated power in a Proj-PCM device for scalar multiplication is 60 nW and the average energy consumption, assuming a 100 ns read time provided by an integrated readout circuit, is 6 fJ (Fig. 9). The latter is 33x lower than an 8-bit digital multiplication in 45 nm (0.2 pJ) [11]. A comparative study was done using non-

978-1-7281-1988-5/18 $31.00 © 2018 IEEE

projected PCM devices and one could observe substantially reduced precision (Fig. 10). This is due to the significantly higher drift, $1/f$ noise and non-Ohmic transport behavior.

B. Temperature compensation method

Remarkably, the 8-bit precision can be retained at elevated temperatures aided by a simple compensation scheme. The projection material's resistivity exhibits a well-defined length-independent temperature dependence. A compensation scheme was devised by multiplying the read current with a single-variable equation that describes the temperature dependence of the projection material, that is $f(T) = 1 + \alpha_p(T - T_0)$. The same value of α_p (-3.0×10^{-3} K^{-1}) was used independent of the device conductance state. We repeated the scalar multiplication experiment while varying the ambient temperature as a sinusoidal profile between 25 and 55°C. Fig. 11a shows the resulting error and its elimination by the compensation scheme in scalar multiplication, recovering the 8-bit fixed point arithmetic comparable precision (Fig. 11b). A slight shift to higher $\hat{\beta}$ at high temperatures arises from the fact that the compensation scheme assumes zero contribution of the amorphous PCM in the total temperature behavior. This could be tackled with a more complex compensation scheme.

C. Matrix-vector multiplication

By invoking the Kirchhoff's current summation rule in addition to the Ohm's law, one can multiply a matrix by a vector. If resistive memory devices are organized in a crossbar configuration, the matrix-vector multiplication $A \cdot x = b$ can be performed by mapping the elements of A to conductance values and the elements of x to read voltages applied to the rows of the crossbar (Fig. 12). Subsequently, the elements of b are computed from the column currents. The temperature compensation scheme proposed earlier can be applied to the column current. First, we simulated a 256×256 crossbar and tested the temperature compensation method in both PCM and Proj-PCM devices (Fig. 14a). The temperature dependence of the phase-change material was captured by a model based on the experimental results of Fig. 7. The activation energy for electrical transport was assumed to be normally distributed around the mean value $\overline{E}_\alpha = 0.2$ eV with a standard deviation of 15 meV. For the Proj-PCM case, the projection material was modelled as a parallel resistor with a single temperature coefficient (Fig. 13). Because of the parallel current path with weaker temperature dependence, the Proj-PCM crossbar outperforms the PCM one. More importantly, it is impossible to compensate for ambient temperature variations at a crossbar array level with PCM due to the significant variations in the activation energy values (Fig. 14b). Proj-PCM devices, on the other hand, with their state/device-independent temperature dependence of electrical transport are much more amenable to such a compensation scheme. In addition to the simulation studies, we experimentally emulated 2000 matrix-vector multiplications employing 12 Proj-PCM devices arranged in a 4×3 virtual crossbar configuration. An equivalent experiment was repeated under temperature variations spanning from 25 to 55°C. The column currents were translated to $\hat{b}_i$ and plotted against the exact result b_i for constant and varying temperature (Fig. 15 & 16, respectively). The precision loss in the latter case was recovered by the compensation scheme.

IV. PATTERN CLASSIFICATION

A single-layer neural network was experimentally emulated using 30 physical Proj-PCM devices arranged in a 10×3 crossbar (Fig. 17b). The network was trained to classify 3×3 pixel images to 3 classes [12]. It consists of 9 input nodes in which the pixel values are fed in, mapped as read voltages, while a 10th input neuron serves as bias. Being a single layer network, the input and output neurons are directly connected, with the conductance of each of the 30 fully connected devices corresponding to the synaptic weight. The dot-product of the weights and inputs is calculated at each output neuron, which in our case is the column read current. Fig. 17a is a schematic representation of this network next to the training set, which consists of 3 images, the numbers 4, 1 and 0. The network was trained offline using the back-propagation algorithm. The resulting weights were mapped to conductance values and the ranges adjusted to match the dynamic range of the Proj-PCM devices (Fig. 17c). The classification accuracy was obtained on 2 test sets of 27 images under 2 scenarios of noise at the input neurons: 1) analogue noise introduced as a Gaussian distribution of pixel colors between 0 and 1 in the grayscale, and 2) digital noise that had one pixel of each original image flipped (Fig 18a). For the Gaussian noise a standard deviation value of 0.2 was chosen for the experiment (Fig. 17d). In both scenarios the classification accuracy was 100% at room and elevated temperature, even without the need to apply the temperature compensation method (Fig 18b & c).

V. CONCLUSIONS

We have conclusively shown 8-bit precise and low-power (60 nW) in-memory multiplication using Proj-PCM devices. We demonstrated scalar and matrix-vector multiplication with 8-bit precision, along with a method that corrects for the temperature variations and recovers the constant temperature precision. We also successfully implemented a single-layer neural network with 30 hardware Proj-PCM devices capable of errorless pattern classification at elevated temperatures. The 8-bit precision requires highly accurate conductance tuning as well as low-offset/low-noise analog circuitry, lest one of those factors become the actual limit on effective precision. Future work will aim at decreasing the absolute device conductance and enlarging the conductance window for easier integration in large neuromorphic crossbars, which should be achievable through material engineering and device scaling.

We acknowledge partial financial support from ERC grant 682675.

REFERENCES

[1] M. Hu et al., Adv. Mater. 30.9, 1705914, 2018. [2] D. Ielmini et al., Nat. Electron. 1.6, 333, 2018. [3] G.W. Burr et al., Adv. in Phys.: X 2.1, 89-124, 2017. [4] A. Sebastian et al., Nat. Commun. 8.1, 1115, 2017. [5] M. Le Gallo et al., Nat. Electron. 1.4, 246-253, 2018. [6] I. Boybat et al., Nat. Commun. 9.1, 2514, 2018. [7] S. Kim et al., Proc. IEDM, 30.7.1-30.7.4, 2013. [8] W. Koelmans et al., Nat. Commun. 6, 8181, 2016. [9] J. L. M. Oosthoek et al., Rev. Sci. Instrum. 86, 033702, 2015. [10] M. Le Gallo et al., Adv. Electron. Mater., 1700627, 2018. [11] M. Horowitz, Proc. ISSCC, 10-14, 2014. [12] M. Prezioso et al., Nature 521, 61-64, 2015.

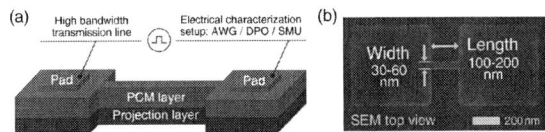

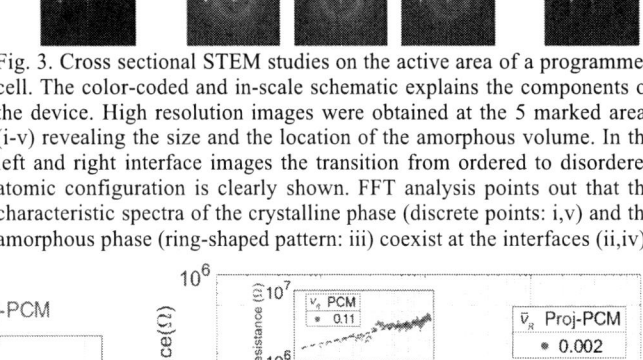

Fig. 1. (a) Schematic 3D view of a lateral Proj-PCM device. The bi-layer stack is encapsulated by SiO2 and 2 pads connect it to a characterization setup. (b) SEM image of a device during fabrication. Different active-area dimensions within the noted width/length range were fabricated and characterized.

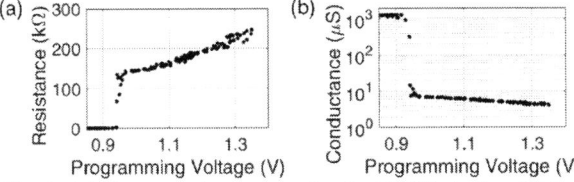

Fig. 2. (a) Programming curve of a Proj-PCM device. Higher voltage amplitudes increase the amorphous volume. (b) Any desired conductance state within the dynamic range can be achieved using iterative programing.

Fig. 3. Cross sectional STEM studies on the active area of a programmed cell. The color-coded and in-scale schematic explains the components of the device. High resolution images were obtained at the 5 marked areas (i-v) revealing the size and the location of the amorphous volume. In the left and right interface images the transition from ordered to disordered atomic configuration is clearly shown. FFT analysis points out that the characteristic spectra of the crystalline phase (discrete points: i,v) and the amorphous phase (ring-shaped pattern: iii) coexist at the interfaces (ii,iv).

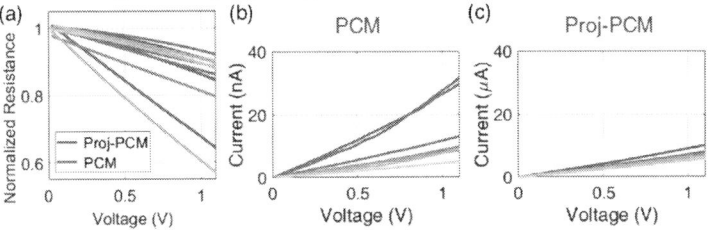

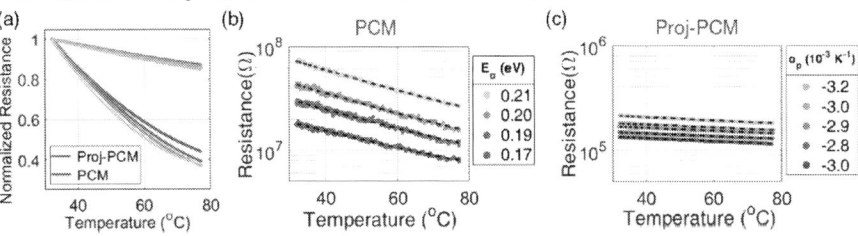

Fig. 4. (a) Normalized resistance versus voltage for different programmed states in PCM and Proj-PCM, showing the strong field dependence of PCM, and its weakening in Proj-PCM. Proj-PCM has consistently an Ohmic behavior over a wider range in the low-field regime, i.e. during read. (b) Current-voltage characteristics of PCM. (c) Current-voltage characteristics of Proj-PCM.

Fig. 5. Resistance drift of 6 resistance states in Proj-PCM. The drift coefficient v_R is determined by a power-law fit and shows a 50-fold reduction in Proj-PCM compared to PCM (inset).

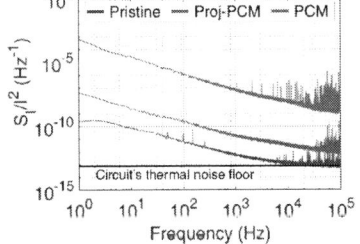

Fig. 6. Normalized spectral density of the read-current noise in the crystalline phase and in a programmed state for PCM and Proj-PCM, where it is 10^4 times lower.

Fig. 7. (a) Normalized resistance versus temperature for different annealed states in PCM and Proj-PCM. (b) The exponential temperature dependence of resistance in PCM is fitted by an Arrhenius equation $R(T) = R^* \exp{(E_\alpha/k_B T)}$. Activation energy E_α is determined by the slope. (c) The temperature dependence of Proj-PCM can be described by a simple linear approximation $R(T) = \rho[1 + \alpha_p(T - T_0)]$. Temperature coefficient of resistance α_p is extracted via linear fit.

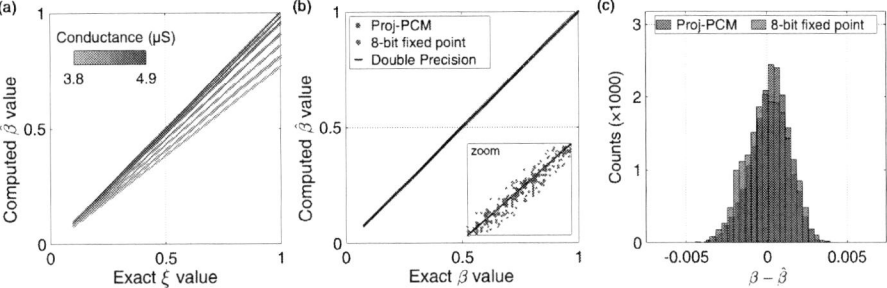

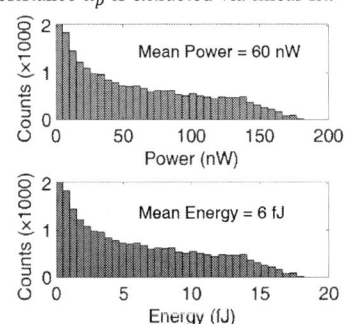

Fig. 8. (a) Scalar multiplication result $\hat{\beta}$ obtained by the read current using Ohm's law against ξ which is mapped in read voltage values, for Proj-PCM conductance states between G_{min} (3.8 μS) and G_{max} (4.9 μS). (b) Scalar multiplication result $\hat{\beta}$ computed using the Proj-PCM against both 8-bit fixed point arithmetic and the exact result β. (c) Error distribution of the 20000 scalar multiplication results for the Proj-PCM compared with the 8-bit fixed point arithmetic.

Fig. 9. Distribution of power per scalar multiply operation. Energy is calculated assuming 100 ns read time provided by an integrated readout circuit.

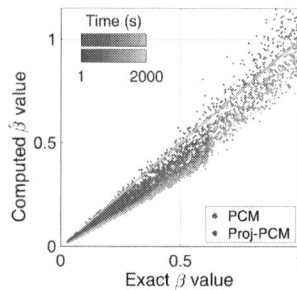

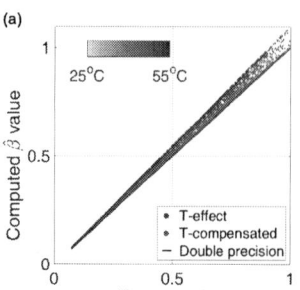

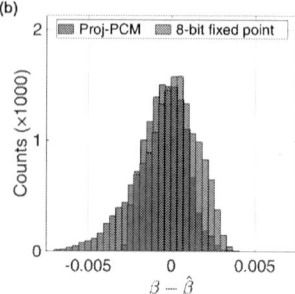

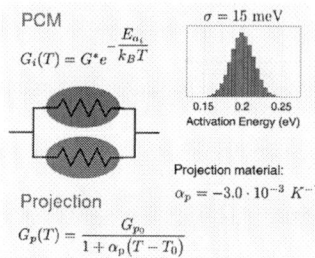

Fig. 10. Scalar multiplication result $\hat{\beta}$ in PCM and Proj-PCM against exact result β. Drift and $1/f$ noise in PCM cause errors.

Fig. 11. (a) The effect of temperature in the experiment of Fig 8b. The temperature compensation model is used to correct read current and recover precision loss. (b) Error distribution for the temperature compensated Proj-PCM compared with 8-bit fixed point arithmetic.

Fig. 13. Resistance network model of the amorphous PCM in parallel with the projection layer. Based on experimental data, E_α was normally distributed, while α_p was the same.

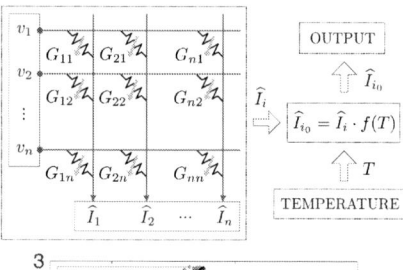

Fig. 12. Temperature compensation procedure in a crossbar that comprises Proj-PCM devices. Column current is corrected by the compensation equation: $f(T) = 1 + \alpha_p(T - T_0)$, in which temperature is the only required input.

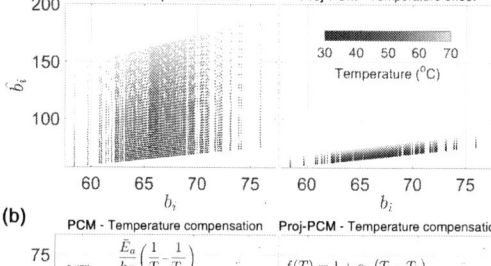

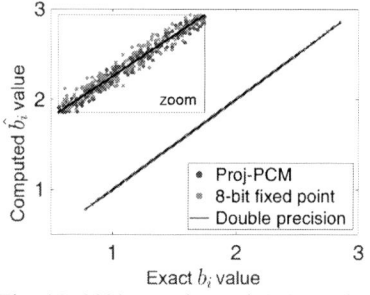

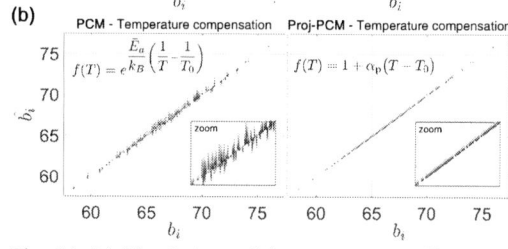

Fig. 15. 2000 experimental 4×3 matrix-vector multiplication results $\hat{b}_i$ computed with Proj-PCM against the exact result b_i and the 8-bit fixed point arithmetic.

Fig. 16. 2000 experimental 4×3 matrix-vector multiplications at various elevated temperatures. Precision loss is recovered by the compensation scheme.

Fig. 14. (a) Simulations of the temperature effect on a matrix-vector multiplication in a 256x256 crossbar. (b) The inapplicability of a crossbar array level compensation scheme in PCM due to significant variations in the activation energy values, compared to Proj-PCM. For each case we used the corresponding compensation equation $f(T)$, as described in Fig. 12.

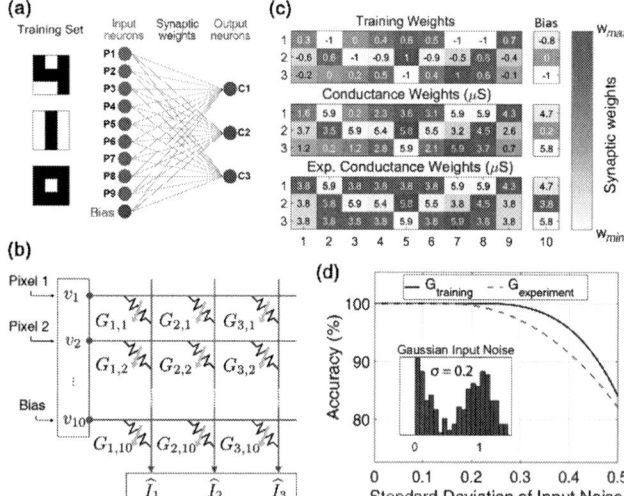

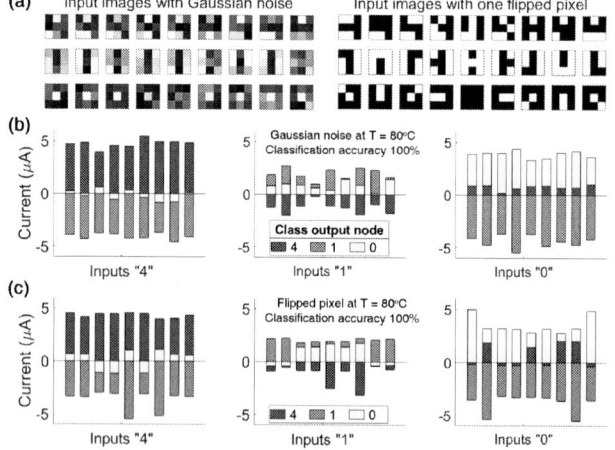

Fig. 17. (a & b) A set of 30 Proj-PCM cells make up a 10×3 crossbar that serves as a neural network that is trained on a set of 3 images. (c) Training weights are mapped to conductance values and ranged to match the dynamic range of the devices. (d) The latter affects the classification accuracy in the Gaussian noise scenario.

Fig. 18. Classification of 27 images in 2 scenarios of noise at the input neurons. (a) Analogue noise was introduced as a Gaussian distribution of pixel colors between 0 and 1 in the grayscale ($\sigma = 0.2$) whereas the set for digital noise had one pixel of each original image flipped. (b & c) In both scenarios the classification accuracy was 100% at room and elevated temperatures, without the need to apply the temperature compensation scheme.

978-1-7281-1988-5/18 $31.00 © 2018 IEEE

System Performance: From Enterprise to AI

A. Kumar, L. Chang, G.E. Tellez, L.A. Clevenger, and J.L. Burns

IBM Research, Yorktown Heights, NY, USA email: {arvkumar,lelandc,tellez,lacleven,jlburns}@us.ibm.com

Abstract— System performance has shown many decades of continuous improvement. After first reviewing historical trends and the current outlook, in this paper we discuss the challenges and opportunities in future computing systems due to the disruptive confluence of stalled scaling and emerging AI workloads. Heterogeneous integration is highlighted as a key means to future systems performance growth.

I. INTRODUCTION

Continued improvement in system performance to satisfy an increasingly diverse set of applications requires progress across many fronts. For example, some applications have strict latency requirements (e.g., fraud detection, autonomous driving) and benefit directly from gains in single-thread performance. Other workloads may be highly parallelizable (e.g., HPC, AI) and therefore benefit most from gains in throughput, which depend on a combination of thread performance, the number of cores, and the interconnectivity between them. Energy efficiency is universally beneficial across applications – important not only for the mobile user concerned about battery life, but also for the datacenter user able to enjoy more compute capability at lower cost.

System-level performance gains in the forms of single-thread performance, throughput, and power efficiency have historically been fueled by performance, density, and power improvements from device and wiring scaling. Today this picture is undergoing dramatic change primarily due to two disruptive forces: (1) the stalling of scaling, shifting dependency to other performance drivers, and (2) emerging AI workloads, which are imposing new demands and fundamentally reshaping computing architectures. This paper examines the changing role of technology in system performance in the face of these challenges.

II. HISTORICAL SCALING TRENDS

Until the early 2000s, Dennard scaling coupled with Moore's Law enabled a predictably steady increase in CMOS device performance and density at lower power [1]. The rise in passive leakage currents, which limited reduction of the supply voltage, abruptly ended ideal Dennard scaling, but materials innovations allowed scaling to continue at a slower pace [2]. As gate length scaling eventually stalled, 3D FinFETs enabled continued density scaling [3], although performance gains became increasingly limited by worsening parasitics.

From a system performance perspective, the dramatic gains from ideal Dennard scaling led to a robust increase in single thread performance, essentially the operating frequency. Further single thread performance gain was possible at the architecture level: the abundance of transistors unconstrained by energy considerations allowed more complex cores capable of techniques such as branch speculation and out-of-order execution. As the ability to scale frequency diminished, increasing the number of cores on a chip became a much more promising path to system performance improvement, making power efficiency a key concern. Compensated by parallelism, operating frequency decreased to remain within power constraints. These constraints tended to rein in techniques beneficial to single thread performance while favoring ones beneficial to throughput such as multithreading [4].

Increasing the number of cores also led to a memory hierarchy with extensive on-chip caches to avoid the high latency and energy cost of accessing external memory [5]. The high area cost of large last level caches was mitigated by the introduction of embedded DRAM in the 45 nm node, providing a 3x density boost compared to 6T-SRAM along with substantial power and SER immunity benefits [6].

III. CURRENT OUTLOOK

The current state of affairs is summarized by the well-known plot in Fig. 1 [7]. It shows clearly the continuation of density scaling, saturation in single thread performance, the weak decline in frequency, power per chip constrained by the socket, and increasing number of cores supporting increased throughput. Today, increasing density is of paramount importance, even though direct performance gains from scaling are becoming more marginal. This Section discusses the current technology trends to support continued system-level performance improvement due to the device and wiring levels.

As FinFET scaling ends, stacked Gate All Around (GAA) nanosheet is a promising device structure for continued density scaling to the 5 nm node and beyond [8]. A three-sheet single stack configuration results in 30% more active width per footprint than FinFET. Notably the constraints imposed by width quantization in FinFETs, such as assist circuitry to support Vmin reduction [9], are lifted. GAA devices may provide better leakage control, which would be advantageous for low Vt devices, benefitting system performance for both high frequency and low voltage applications.

The BEOL wiring is becoming a critical bottleneck to system performance improvement because RC parasitic delay is not scaling. Resistance is increasing disproportionately node-to-node, accounting for a higher fraction of total delay, and the capacitance is not decreasing [10]. This drives a need for additional levels of metal not just for signal routing, but also power distribution. Fig. 2(a) shows the impact of increased parasitics, plotting frequency degrade due to increases in line resistance, via resistance, and capacitance [11]. At the 3 nm node with 24 nm wiring pitch, exponential increases in Cu resistivity will lead to a catastrophic RC limit (Fig. 2(b)) requiring innovations such as Co [12] and ULK dielectric [13].

978-1-7281-1988-5/18 $31.00 © 2018 IEEE

The delayed introduction of EUV has led to complex multi-patterning optical solutions and made density scaling very challenging. Cell area scaling through Design Technology Co-Optimization (DTCO) [14, 15] has been extensively used to supplement lithographic scaling. Self-Aligned Double Patterning (SADP) for the 1x wiring levels has imposed severe restrictions on allowed widths and spacings in the 7 nm node, leading to a significant tradeoff between performance and density. Fig. 3 shows an example of how wide wires in restricted positions can be used to benefit power distribution, as needed for a high-performance design. Although wire delay is benefitted, this comes at a high cost to routability (number of signal routing tracks in a fixed space).

IV. EMERGING WORKLOADS

Traditional enterprise computing, a great beneficiary of the scaling trends above, is being augmented by rapidly changing emerging workloads largely driven by AI. With explosive growth in the collection of data across all domains, machine learning and deep learning algorithms are becoming key components of system performance metrics. In response, specialized accelerators such as FPGAs, GPUs, and ASICs have gained broad use on these tasks as they can significantly outperform CPUs. In contrast with many traditional workloads that depend strongly on low latency and single-thread performance, many AI workloads are parallelizable and can thus leverage energy efficiency gains to improve throughput. Moderate voltage scaling can therefore be an important technique to improve AI system performance [16].

For deep learning, key acceleration drivers have been efficient compute engines supported by high bandwidth connectivity to memory [17] and the targeted use of reduced precision computation [18], which leverages the quadratic scaling of compute energy and area with the number of bits. Fig. 4 shows the benefit of reduced precision for deep learning. Since memory access patterns of deep learning workloads are well characterized, specialized accelerators with dataflow architectures can achieve very high compute utilization through the judicious use of on-chip memory scratchpads to optimize data flow and maximize data reuse [17].

From a system performance perspective, it is critical to assess practical workloads to achieve balanced parallelism across accelerator cores within algorithm, on-chip communication bandwidth, and memory capacity constraints. Analysis frameworks such as described in [19] are essential to optimize system hardware to maximize performance for real neural networks. Future system performance gains can only be realized if input data can be efficiently fed to a large number of parallel compute engines, highlighting the importance of compute utilization as a metric for system performance [19]. While continued technology density scaling can provide large, on-chip memories to increase data reuse, off-chip communication bandwidth must also be increased commensurately for both compute engine to compute engine and compute engine to memory interfaces. Otherwise, sustained performance can be substantially lower than peak performance [20]. Fig. 5 illustrates this point by using the simulation tool in [19] to compare the effect on system performance of improving compute performance vs. memory

bandwidth on a multi-chip network of AI accelerators performing training of the Resnet-50 network, which is typical of large-scale deep learning datacenter workloads. The x-axis is the relative increase in compute performance or memory bandwidth. Compute performance can be increased either by increasing the number of cores at fixed performance or by increasing the performance for each core keeping the number of cores fixed. For this memory-bound problem, increasing the compute performance at fixed bandwidth is of little benefit, while increasing external bandwidth provides some benefit. The greatest benefit is achieved for a balanced increase in compute performance and bandwidth.

V. HETEROGENEOUS INTEGRATION

Emerging workloads are strongly driving systems to become much more architecturally diverse -- composed of a mix of heterogeneous components including CPUs, GPUs, specialized accelerators and memories -- that will require high interconnectivity to maintain system performance growth. Heterogeneous Integration (HI), which seeks to achieve enhanced performance and functionality through advanced packaging techniques [21,22], is thus becoming a key enabler for performance in future systems. The motivation for HI goes beyond its ability to provide high bandwidth solutions for AI. For instance, the diminishing node-to-node returns from scaling are leading to the breakup of large chips into smaller chiplets, giving a significant yield benefit [23]. An additional benefit is the ability to optimize selection of the technology node for each chiplet based on cost and functionality [24].

High-performance GPUs and AI accelerators are leveraging 2.5D interposers in recent product generations to access High Bandwidth Memory [25], responding to the growing bandwidth needs of AI workloads. Compared to 2D or 2.5D, 3D die-to-die stacking allows much denser direct connectivity, enabling in turn the possibility of much higher bandwidth (Fig. 6) with the potential to alleviate the memory bottleneck discussed in the previous section. While 3D stacking of low power components such as memory dies is in production today, 3D stacking to closely integrate compute and memory connected by very high-bandwidth interfaces will likely require the use of advanced cooling techniques. For example, dual-side heat removal using a convective Si interposer (Fig. 7) is capable of effectively cooling a three-tier stack including CPU, GPU, and memory [26]. Targeted local cooling within the chip stack can be achieved by two-phase cooling [27,28], in which a coolant flows in embedded microchannels between stacked high-power active layers (Fig. 8). Finally, HI could facilitate introduction of promising analog-based accelerators [29,30].

VI. SUMMARY AND CONCLUSION

Computing systems are becoming far more heterogeneous in response to a greater variety of workloads and the reduced scaling-driven performance. Technology will continue to play a pivotal role in growing system performance in this new era. However, its focus will shift away from scaling and device performance towards providing the high interconnectivity needed to support increasingly heterogeneous systems. Heterogeneous integration will therefore be a key enabler both to counter the diminishing returns of scaling and to meet the bandwidth demands of data-intensive AI workloads.

ACKNOWLEDGMENT

The authors gratefully acknowledge the contributions of R. Divakaruni, W. Haensch, N. Lanzillo, and S. Venkatarami.

REFERENCES

[1] R. Dennard et al., *IEEE Solid-State Circuits Soc. Newsletter*, vol. 12, p. 35, 2007.

[2] W. Haensch et al., *IBM J. Research and Development*, vol. 50, pp. 339-361, 2006.

[3] K.-I. Seo et al., *IEEE Symposium on VLSI Technology Dig. Tech. Papers*, 2014.

[4] S. Chaudhry et al., *IEEE Micro*, vol. 25, pp. 32-45, 2005.

[5] M. Horowitz, *IEEE Solid-State Circuits Conf. Dig. Tech. Papers*, 2014.

[6] S. Iyer, www.gsaglobal.org/events/2012/0426/docs/ keepingmooreslawalivekeynote-webpdf_000.pdf

[7] K. Rupp, www.karlrupp.net/2018/02/42-years-of-microprocessor-trend-data

[8] N. Loubet et al., *IEEE Symposium on VLSI Technology Dig. Tech. Papers*, 2017.

[9] T. Song et al., *IEEE Solid-State Circuits Conf. Dig. Tech. Papers*, 2016.

[10] L.-C. Lu, *Proc. 2017 ACM on International Symposium on Physical Design*, 2017.

[11] L. Clevenger, *IEEE International Interconnect Technology Conf./Advanced Metallization Conf.*, 2016.

[12] J. Kelly et al., *IEEE International Interconnect Technology Conf./Advanced Metallization Conf.*, 2016.

[13] A. Grill, *J. Vacuum Science and Technology*, vol. 34, p. 020801, 2016.

[14] L. Liebmann et al., *IEEE Symposium on VLSI Technology Dig. Tech. Papers*, 2016.

[15] C. Auth et al., *IEEE Electron Devices Meeting Dig. Tech. Papers*, 2017.

[16] L. Chang et al., *Proc. of the IEEE*, vol. 98, pp. 215-236, 2010.

[17] B. Fleischer et al., *IEEE Symposium on VLSI Technology Dig. Tech. Papers*, 2018.

[18] S. Gupta et al., *International Conf. on Machine Learning*, 2015.

[19] S. Venkataramani et al., *IEEE 26th International Conference on Parallel Architectures and Compilation Techniques (PACT)*, 2017.

[20] S. Williams et al., *Comm. of the ACM*, vol. 52, pp. 65-76, 2009.

[21] S. Iyer, *IEEE Trans. on Components, Packaging and Manufacturing Technology*, vol. 6, pp. 973-982, 2016.

[22] J. Lau, *J. Electronic Packaging*, vol. 138, p. 030802, 2016.

[23] L.Su et al., *IEEE Electron Devices Meeting Dig. Tech. Papers*, 2017.

[24] T. Haruta et al., *IEEE Solid-State Circuits Conf. Dig. Tech. Papers*, 2017.

[25] J. Macri, *IEEE Hot Chips 27 Symposium*, 2015.

[26] T. Brunschwiler et al., *IEEE 17th Intersociety Conf. on Thermal and Thermomechanical Phenomena in Electronic Systems*, 2018.

[27] A. Bar-Cohen et al., *CS ManTech Conf.*, 2013.

[28] M. Schultz et al., *IEEE 16th Intersociety Conf. on Thermal and Thermomechanical Phenomena in Electronic Systems*, 2017.

[29] S. Ambrogio et al., *Nature*, vol. 558, p. 60, 2018.

[30] S. Nandakumar et al., *IEEE International Symposium on Circuits and Systems*, 2018.

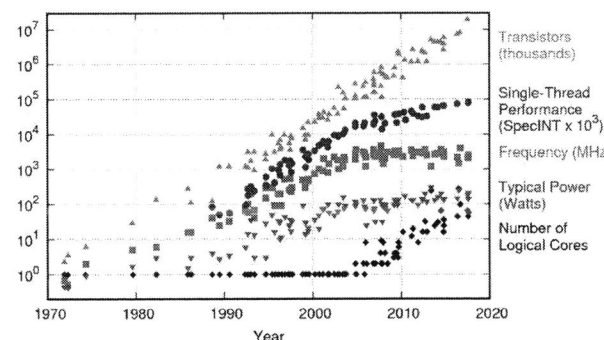

Fig. 1. Microprocessor trend data over a 42 year period, showing sustained density scaling, saturation in single thread performance, the weak decline in frequency, constrained power per chip, and increasing number of cores From [7].

Fig. 3. Tradeoff of routability with scaled wire delay.

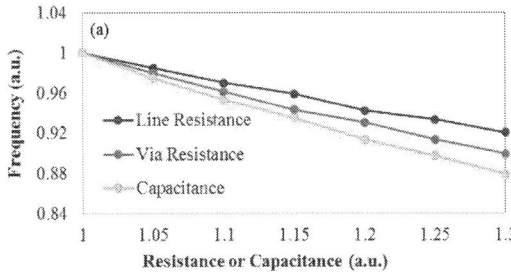

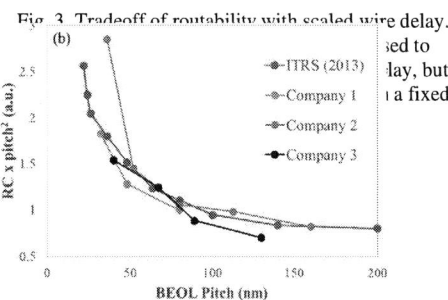

Fig. 2. (a) Frequency degrade due to increase in line resistance, via resistance, or capacitance. (b) Resistance increase with decreasing pitch from various sources, indicating catastrophic increase around 24 nm wiring pitch.

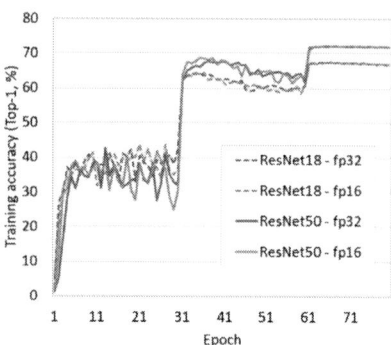

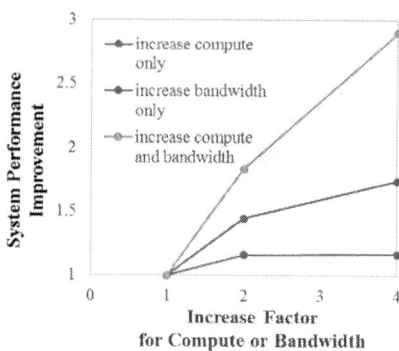

Fig. 4. Effect of reduced precision on training accuracy of a deep learning problem. Study indicates no loss of accuracy for computation using special-format 16-bit floating point (fp16) in place of full 32-bit floating point (fp32) computation. From [17].

Fig. 5. System performance improvement as either compute or bandwidth alone is increased, or the two are increased commensurately.

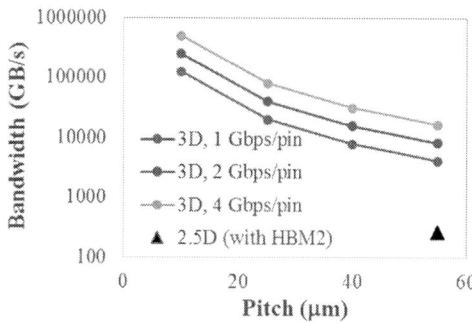

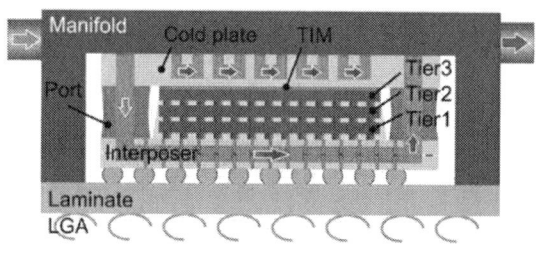

Fig. 6. Bandwidth possible in a 100 mm² area using 3D stacking, as a function of pitch, for various pin speeds. Compared to current 2.5D, 3D offers the possibility for much higher bandwidths.

Fig. 7. Cross-sectional view of a module with dual-side cooling containing a convective interposer. A three-tier stack is sandwiched between the interposer and a cold plate. From [26].

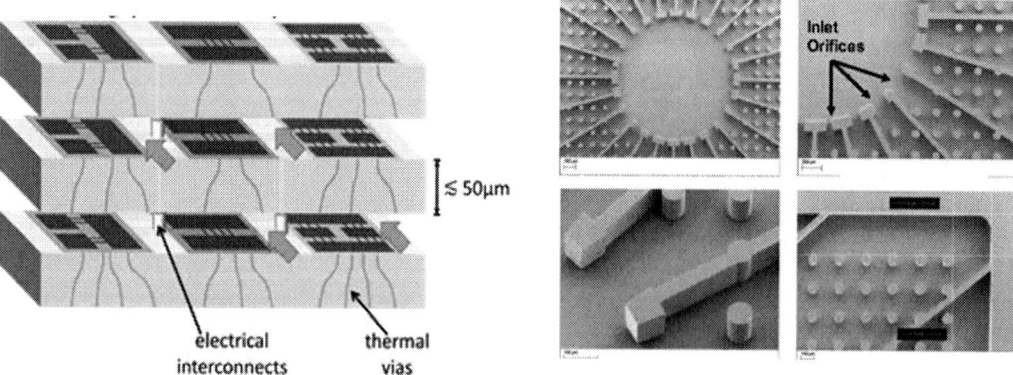

Fig. 8. Left shows coolant flow between high-performance active layers in 3D stacked configuration, from [27]. Right shows SEM image of orifices and radial microchannels between stacked chips, from [28].

Design-Technology Co-Optimization of Standard Cell Libraries on Intel 10nm Process

Xinning Wang, Ranjith Kumar, Somashekar Bangalore Prakash, Peng Zheng, Tai-hsuan Wu,
Quan Shi, Marni Nabors, Srinivasa Chaitanya Gadigatla, Simeon Realov, Chin-hsuan Chen,
Ying Zhang, Kaizad Mistry, Andrew Yeoh, Ian Post, Chris Auth, Atul Madhavan
Intel Corporation, Hillsboro, OR, USA, email: xinning.wang@intel.com

Abstract: This paper highlights the co-optimization of process technology, std. cell library offerings and block-level TFM on Intel 10nm node to enable unprecedented scaling opportunity for products ranging from high performance client/server to low power mobile/IoT segments. The 10nm short height library enables 2.7x transistor density scaling going from 14nm counterpart. The taller height libraries are optimized to meet performance and reliability requirements of Intel's leading edge client/server products. PPA trade-offs are analyzed both at std. cell level and block level on an industry standard Core IP design.

I. INTRODUCTION

Despite the challenges of lithography scaling, transistor density scaling has accelerated significantly on Intel 14nm [1] and 10nm [2] nodes as a result of aggressive pitch scaling, novel technology features, innovative std. cell architecture designs and extensive co-optimization of design rules, std. cell layout and block level tool-flow-methodology (TFM). Driven by time-to-market, cell-based SoC design methodology leveraging automated synthesis and place & route EDA tools is the leading implementation choice of product designs on advanced technology nodes. Std. cell libraries and block-level TFM play a crucial role in translating technology scaling benefit to product goodness. Std. cell height is a critical modulator for realizing area vs. performance trade-offs. To support a wide range of product performance-power-area (PPA) landing zones in different market segments (Fig. 1), std. cell libraries are designed at different heights based on Intel 10nm technology. These different cell layout architectures are co-optimized along with corresponding process design rules (Fig. 2). Block level implementation choices such as power grid design, routing track pattern selection and flow recipes are fine-tuned to realize PPA benefit meeting technology/product scaling expectations.

II. STD CELL LIBRARY ARCHITECTURE DESIGN

Fig. 2 presents Intel's 10nm std. cell library offerings including three different cells heights (CH) to cater to the diverse product needs across the various market segments. The figure also provides details about the corresponding Metal0 (M0) and Metal2 (M2) templates. The different library CHs are decided based on per leg fin count, power delivery, and track count requirements for efficient intra/inter cell signal routing. 272CH, 340CH and 408CH library architectures are setup to efficiently support 2-fins/leg, 3-fins/leg and 4-fins/leg, respectively, for both PMOS and NMOS. Furthermore, all CHs also offer an option for fitting an additional NMOS fin which is useful for implementing certain logic functions.

M0 and M2 templates for each library are defined based on the target application. In the 272CH library which is targeted for highest density, power (VCC/VSS) M0 and M2 rails are located to be shared between two adjacent standard cell rows. This choice enables 5 M0 signal tracks within the CH for efficient intra cell routing. However, the taller 340CH and 408CH libraries have higher performance demand compared to 272CH. Therefore 340CH & 408CH M0 templates are carefully architected for supporting higher performance power delivery without compromising intra cell routing efficiency. As compared to 272CH, the 340CH and 408CH libraries have 2x and 3x higher power M0 track density, respectively. The novel contact over active gate (COAG) feature in Intel's 10nm technology [2] effectively enables all the signal M0 tracks for making gate connections as shown in Fig. 3. The corresponding M2 templates associated with each of the library CHs are carefully chosen for enabling efficient inter cell routing. A minimum of 5 signal M2 tracks are provided in each library architecture. The taller 408CH supports two different M2 template options allowing for design optimization trade-off between higher routing track density vs. wider wires for higher performance.

The tightest metal pitch in Intel's 10nm technology is on M1 layer [2]. Fig. 4 compares Intel's 14nm and 10nm in terms of M1 track count and placement within a gate pitch span. The additional tracks provided by the tighter pitch M1 on 10nm node contributes to the overall cell and block area shrink resulting in hyper scaling going from 14nm technology. The misalignment between the contacted gate pitch and M1 pitch creates the need for innovative library design solution for realizing best overall block density. In order to freely place cells without any abutment constraints, all libraries include two M1 variations of every cell layout. Fig. 5 illustrates the concept of two M1 cell layout versions, wherein M1 center is offset with respect to the cell origin between the two versions. Furthermore, Intel's 10nm technology also enables single dummy gate isolation between active devices (Fig. 6), a critical feature for enabling enhanced cell area reduction and overall block level cell utilization boost.

Together with pitch scaling as indicated by [2], efficient M0/M2 architecture choices and increased M1 track density, Intel 10nm 272CH achieves 2.7x higher transistor density compared to the shortest CH in Intel 14nm technology [1]. The aggressive cell area shrink together with hyper pitch scaling on 10nm node also demands higher EM current density commit compared to traditional scaling. Use of Cobalt in layers below M1 significantly improves std. cell EM capability while at the same time improving cell area and interconnect capacitance. Furthermore, the increased metal track density allows for

978-1-7281-1988-5/18 $31.00 © 2018 IEEE

multiple via placement in high drive cells to improve power IR drop, signal path resistance and EM driving capability.

III. STD CELL LEVEL PPA

This section illustrates std. cell level PPA trade-offs among the above mentioned 3 libraries (272CH, 340CH, and 408CH). All data presented here is based on inverter cells sampled from corresponding libraries benchmarked under self-loaded fan-out of 3 condition.

Fig. 7 and Fig. 8 demonstrate performance and power trade-offs across cell drive range between the three CHs under iso-transistor-leg condition. With increased CH, larger transistor sizes can be accommodated per leg inside the std. cell. Taller 408CH library allows for achieving higher performance at the cost of higher power and area. Shorter 272CH high-density library is geared towards lower power with reduced performance. 340CH in the middle is more suited for mid-range performance and power domains. The wide span of PPA design trade-offs between these CHs enables a diverse portfolio spanning from high-performance client/server products to low-power mobile/IoT products.

Fig. 9 and Fig. 10 show intrinsic performance and power differences between all the 3 CHs measured under iso-transistor-fin condition. Taller 408CH library accommodates transistor fins in a higher performance configuration in the mid to high cell drive range. The taller CH also allows for lowest intrinsic cell-level interconnect resistance on power and signal sides to achieve best in class performance. Shorter 272CH aims at achieving lowest capacitance and area in the low cell drive range. In newer technology nodes like 10nm, lower drive cells from the different CHs are seeing variations in layout impact from process design rules leading to non-monotonic power/performance trends.

IV. STD CELL, BLOCK AND TFM CO-OPTIMIZATION

In order to realize the complete benefit of increased M1 track density, a novel block level power hook-up flow is enabled for Intel 10nm libraries. As shown in Fig. 11, when cells with internal power stubs (a) are used as is, they block more signal routing tracks as shown in (b). Alternatively, Intel 10nm library cells are designed without power M1 stubs and the power hook-up is completed at block level for maximizing track density as shown in (c). Furthermore, the two M1 layout versions supported for all library cells enables no cell abutment restrictions thus achieving high block density. The block-level power hook-up feature in combination with multiple M1 version layout placement is now supported seamlessly inside APR flows after tool/flow co-optimization with EDA tool vendors. Specifically, one variant of each std. cell is used during block level optimization to speedup design runtime, as the timing delta between M1 layout variants are well controlled. The flow starts off by legalizing placement of all cells to their optimal location. A custom algorithm is then kicked-in to swap each cell to its corresponding most appropriate layout variant based on placement location. One of the block-level PPA studies demonstrates that this methodology is able to achieve an additional 5% floorplan area improvement in already high density designs.

The Intel 10nm std. cell, block and TFM co-optimization effort also proposes a novel predefined global fabric grid (Fig. 12) to address design rule complexities on pin share layers when std. cells abut at block level. All the potential wire segment ending locations are predefined in this fabric. Both std. cell pins and router then honor this gridded system to establish a design rule free and more efficient routing solution. With this gridded routing system, the design rule search repair runtime is greatly reduced by 1.6x, while achieving better design rule convergence (Fig. 13).

V. BLOCK LEVEL PPA

An industry standard Core IP design is adopted for block level PPA analysis for the purpose of benchmarking the different 10nm std. cell library offerings. Various custom recipes including hybrid-clock scheme, concurrent clock and data optimization, and layer promotion for critical nets are implemented to push PPA metrics until saturated. Floorplan and routing track patterns are wisely chosen and optimized to reveal true critical paths. Std. cell content together with cell drive progression are customized to yield best PPA results.

Fig. 14 demonstrates block level performance and power trade-offs for the 3 library CHs. 408CH library can achieve highest performance at the cost of higher power. 272CH library, on the other hand, has best power profile at mid-low performance regime. 340CH library offers a good compromise between the two extremes. Although performance delta between 340CH and 408CH is insignificant, 340CH is able to achieve 18% lower area for this specific Core IP block.

VI. CONCLUSION

This paper presents innovative std. cell libraries designed based on industry-leading Intel 10nm process technology. With extensive design-process co-optimization, std. cell libraries are developed in multiple heights and with different architecture considerations to meet a wide range of PPA design trade-off requirements across the product spectrum. Our densest 10nm std. cell library delivers an unprecedented 2.7x transistor density scaling over its corresponding 14nm counterpart by taking advantage of contact-over-active-gate and single dummy gate technology features, and creative cell layout techniques. The block-level PPA trade-offs between the different library offerings are also demonstrated on an industry standard CORE IP design by co-optimizing physical design choices in light of the process design rules, and applying innovative APR recipe customization.

ACKNOWLEDGMENT

The authors gratefully acknowledge Dan Murray, Moonsoo Kang, Digvijay Rajurkar, Ameya Abhyankar, production library and AD library teams.

REFERENCES

[1] S. Natarajan, et al., IEDM Technical Digest, p. 71-73 (2014)
[2] C. Auth, et al. IEDM Technical Digest, p.673-676 (2017)

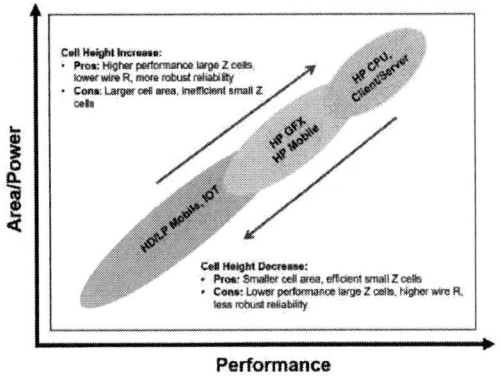

Fig. 1. Multiple library heights for diverse PPA trade-offs.

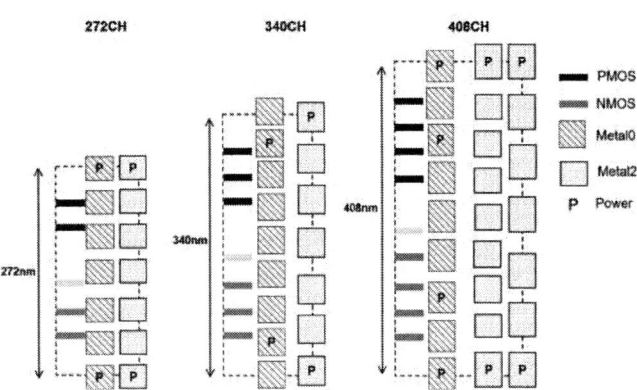

Fig. 2. Intel 10nm Library Architecture offerings.

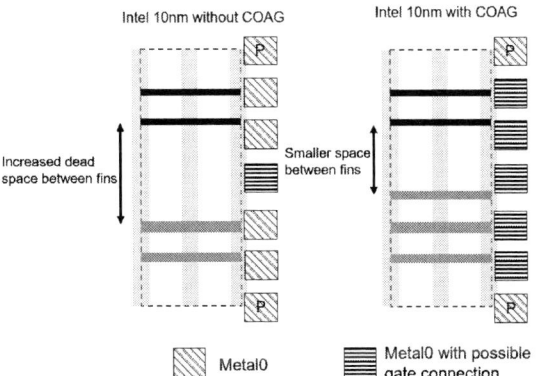

Fig. 3. Scaling benefits of COAG.

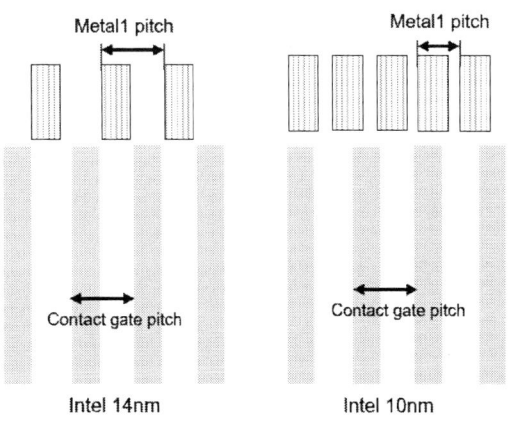

Fig. 4. Increased Metal1 track density in Intel 10nm

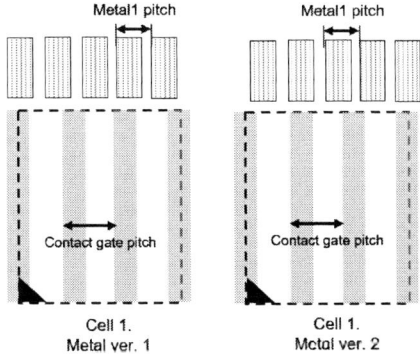

Fig. 5. Two cell variants due to Metal1 misalignment to gate

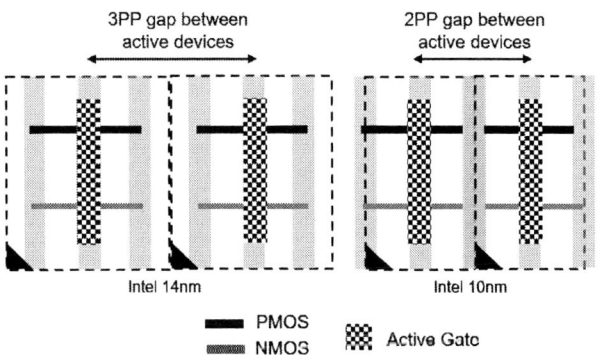

Fig. 6. Scaling benefit of single dummy gate

978-1-7281-1988-5/18 $31.00 © 2018 IEEE 638

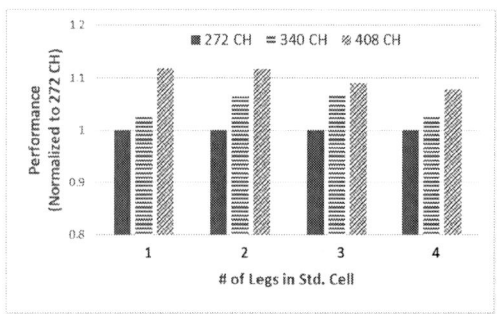

Fig. 7. Iso-leg performance benchmarking across
3 library heights vs. cell drive.

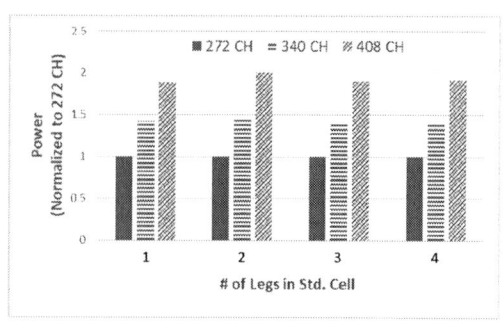

Fig. 8. Iso-leg power benchmarking across
3 library heights vs. cell drive.

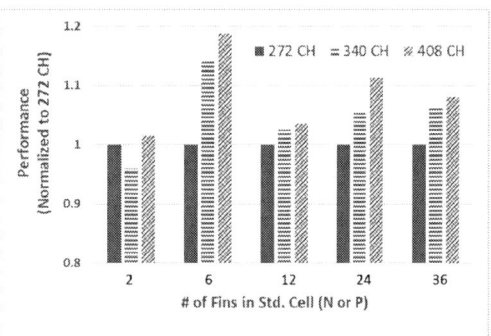

Fig. 9. Iso-fin performance benchmarking across
3 library heights vs. cell drive.

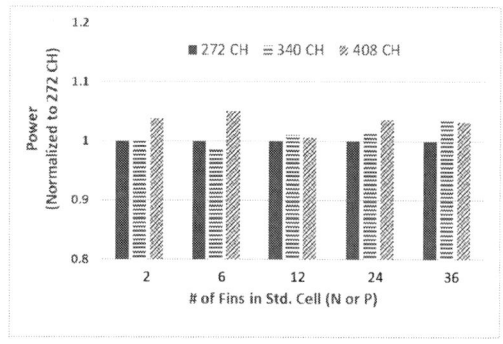

Fig. 10. Iso-fin power benchmarking across
3 library heights vs. cell drive.

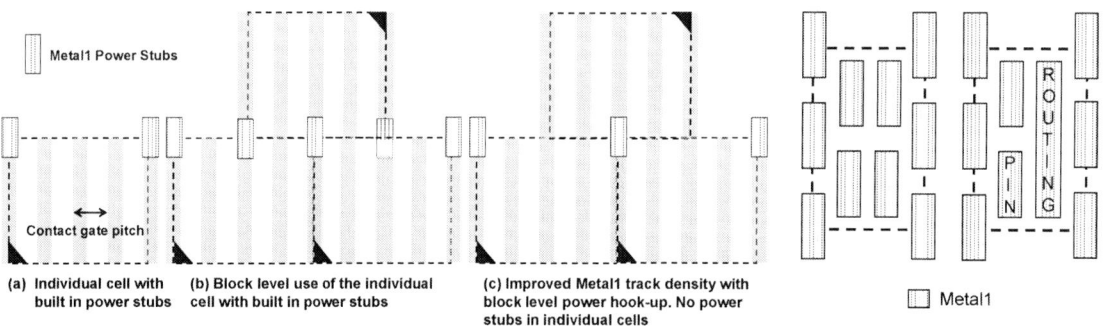

Fig. 11. Block level power hook-up scheme for improved density.

Fig. 12. Fabric gridded M1 routing system.

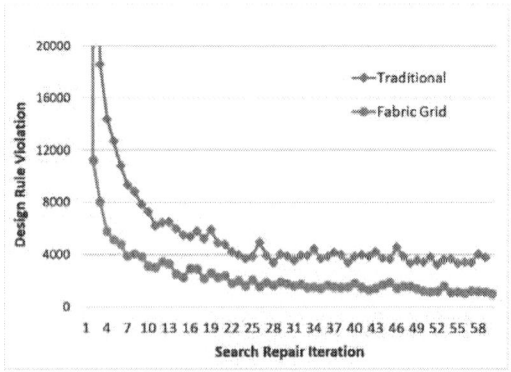

Fig. 13. Benefit of Metal1 fabric grid

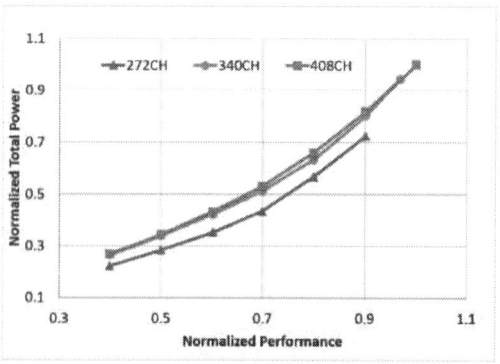

Fig. 14. Block level power-performance trade-offs

978-1-7281-1988-5/18 $31.00 © 2018 IEEE   639

An Accurate FinFET's Vmin Estimation Method for Extreme Low Operation Voltage Design

H. W. Choi, S. K. Kim, H. Jung, D. R. Chang, S. Park and Y. Yasuda-Masuoka, and J.S. Yoon

Foundry Business, Samsung Electronics Co. Ltd., Yongin 17113, South Korea, email: hwooks.choi@samsung.com

Abstract— In this paper, the minimum operating voltage (Vmin) estimation methodology for advanced FinFET technology is newly proposed with a manufacturability consideration. The experiment depicts that key factors to determine Vmin are the sum of threshold voltages of n-FET/p-FET, beta-ratio (n/p-FET strength ratio), and random variation. The new equation successfully captures the key electrical features, which is verified by both Monte-Carlo simulation and advanced 11nm/8nm FinFET experimental data. Based on the new model, the paper also provides the guideline for threshold voltage and local mismatch strategy for future advanced FinFET Vmin improvement.

I. INTRODUCTION

A new era of high performance computing (HPC) with big data, driven by the proliferation of the internet of things (IoT) and deep learning, demands both high performance and the utmost energy efficiency. The substantial increasing energy consumption has been considered as the most crucial constraint that hinders further steps in HPC system development. The most promising solution for reducing power consumption is scaling down the supply voltage (Vdd) [1]. The Vdd reduction, however, could severely cause functional fail on CMOS logic circuits without a careful technology assessment, as well as a circuit consideration. To design the technology for the extreme low voltage operation, the estimation of the minimum operating voltage (Vmin) is intensely essential. There have been many investigations and modeling for planar-FET [2-3]. However, for Advanced FinFET, although Si demonstration was shown [4], there is no Vmin estimation model yet.

In this paper, key transistor parameters impacting on Vmin were discussed using advanced sub-10nm CMOS process including SPICE and STD-Cell libraries. On the basis of the analysis, we propose the accurate expression for Vmin estimation as a figure of merit for CMOS technology, which is useful for transistor design guide suitable for low power application in early stage of technology development.

II. VMIN IMPACT OF DEVICE PARAMETERS

Fig. 1 shows the logic gate delay and switching energy versus the supply voltage for advanced FinFET technology. Based on CMOS Inverter (INV), even though the energy optimum supply voltage (Vopt) shows a sweet spot in lower Vdd region, the actual technology Vmin is 200mV higher due to the logic circuit functional fail. The Vmin deviation from

Vopt is an obstacle to overcome, which has strong connection with process parameters, such as global/local variability, and n-/p-FET balance.

The relative Vmin sensitivity with respect to key process parameters for INV and D Flip-Flop (DFF) is shown in Fig. 2. The basic functionality of INV is strongly influenced by drivability of transistors and a balance between n-/p-FET's strength, whereas the most critical factor for the dependency on Vmin of DFF is local Vth variation (Vtmm). The FF can be considered as a synchronous system which consists of cross-coupled inverters (latch), master and slave latches, and data propagation and clock paths. Because of the FF feature, the random variability component causes a functional failure in FFs. Overall, Vtmm plays a key role in the entire system. On the other hand, some performance boosting knobs induce a tradeoff to Vtmm due to short cannel effect control margin etc. (Fig.3). As a result, Vmin estimation methodology to define the technology scheme to give a guideline becomes critical.

III. MODELING AND EXPERIMENTAL RESULT

A. Vmin modeling

In early stage of design and technology developments, estimating accurate Vmin can be a powerful guide for path-finding. Vmin estimating equation has already been reported by Fuketa et al [2]. However, there are some limitations on applying this to nano-scale devices and capturing entire technology, because of negligent in global variability and Vth absolute value. In this paper, we propose the effective formula for estimating Vmin based on process parameters as follows:

$$ V_{min} = \frac{Vt_{sum} + \sigma_{glob}}{\sqrt{2}} + \frac{2\sigma_{mm}}{3H}\sqrt{\frac{\pi \ln(N/2)}{2}} + \frac{C + |B|}{H}, \quad (1) $$

where,

$$ \sigma_{glob} = \sqrt{\left(\sigma Vtglob_n^2 + \sigma Vtglob_p^2\right)}, \sigma_{mm} = \sqrt{\sigma Vtmm_n^2 + \sigma Vtmm_p^2}, $$

$$ Vt_{sum} = Vth_n + Vth_p, $$

$$ H = 1 + \lambda, \lambda = \Delta Vt / \Delta Vds, $$

$$ B = \frac{n}{4}\phi_t \ln\left(\beta_p / \beta_n\right), C = 20 \cdot n\phi_t \ln[1 - (2/n)], $$

$$ n = SS / 60, SS = \Delta Vgs / \Delta Id, $$

$\sigma Vtglob$ 3σ tolerance for global variation,

$\sigma Vtmm$ 3σ tolerance for local variation,

ϕ_t thermal voltage .

978-1-7281-1988-5/18 $31.00 © 2018 IEEE   640

β corresponds to the strength of transistor and the *I-V* characteristics of a transistor in sub-threshold regions is given by [2]

$$I_d = \beta \cdot e^{\frac{Vgs + \lambda \cdot Vds}{n\phi t}} \left(1 - e^{\frac{Vds}{\phi t}} \right), \quad (2)$$

$$\beta = I_0 \frac{W}{L} \cdot e^{-\frac{VT}{n\phi t}}, \quad (3)$$

where I_0 is the mobility-dependent parameter. This strength parameter can be derived using the critical threshold voltage current (*IVT*) in constant current extraction method and leakage source current I_{soff} as follows

$$\beta = \frac{IVT \cdot W/L - I_{soff}}{e^{(\lambda Vdd)/(n\phi t)} \left(e^{Vtsat/(n\phi t)} - 1 \right)}. \quad (4)$$

To capture the dependency of Vth and global variability, the proposed expression has been improved. As shown in Fig. 4, the proposed equation shows a good agreement with Mont-Carlo simulation for 3 key electrical parameters, compared to the previous model [2]. In particular, n-/p-FET strength dependence improvement contributes a key role, as shown in Fig. 4(d). The previous methodology [2] focuses only on n-/p-FET strength ratio in Fig.5, rather than the absolute n-/p-FET strength value, which cannot fully describe the Vmin characteristic with Vth lowering of the superior n-FET or p-FET (Fig.4 (a/d)). As shown in Fig. 6, Vmin has a strong correlation with the sum of Vth for n-/p-FETs, as well as with n-/p-FET drive strength. These two characteristics have been implemented in the proposed equation.

Another key concern is the manufacturability for extreme low Vdd circuit with the global variation, which is also incorporated in the new equation. The dependence of Vmin on global variation is shown in Fig 7. Vmin is directly proportional to the root-sum-squared global tolerance.

B. Experimental data and Vmin design guideline

Monte-Carlo simulation has been performed on INV RO with different number of stages from 11-stages to 101-stages in advanced sub-10nm CMOS process SPICE and STD-Cell libraries. Fig. 8 compares simulation result and Vmin by the new model for different RO stages. The calculated Vmin of INV RO for (a) various n-/p-FET balances and (b) global variation tolerances agrees very well with the Monte-Carlo simulation. For Si experimental data verification, the INV RO with 101 stages were compared using 11nm [5] and 8nm [6] logic technologies. Vmin was extracted from the measured RO frequency using a linear regression analysis as shown in Fig. 9. The measured Vmins for various n-/p-FET balances is shown in Fig. 10. The Vmin contour plot indicates that sum of threshold voltages of n-/p-FET is dominant over the skew of n-/p-FET's drive strength. Fig. 11 shows the comparison of Vmin between Si measurement and calculation using proposed and previous models. Vmin calculated using the new proposed expression is well matched with Si

experimental results, which proves that the proposed equation is widely applicable to various technologies and n-/p-FET balances.

Lastly, Fig.12 provides Tr. design guideline for extreme low Vdd operation to set Vth position and AVT value with a target Vmin reduction (ΔVmin). As shown in Fig.12, Vth -20mV is equivalent to AVT -0.1 to obtain the same Vmin reduction. This indicates that AVT reduction is crucial for the future technology to improve the energy optimum supply voltage (Vopt), which is determined by the balance of leakage energy and dynamic energy. In this reason, we also investigated AVT improvement with FinFET process using 3D TCAD simulator as shown in Fig.13. At the fixed performance, an impurity profile and structure optimization can bring down local Vth mismatch value for both n-/p-FET successfully, which can contribute to 40~60mV Vmin reduction for logic circuits. Local mismatch driven Tr design is one of key considerations for the future HPC era.

IV. CONCLUSION

We have presented the investigation on the minimum operating voltage (Vmin) versus key process parameters using advanced sub-10nm CMOS process, showing that the significant contributors to Vmin are the sum of Vth for n-/p-FETs and random variability in respect of INV and FF, respectively. Based on the result, new Vmin estimation model has been proposed, which accurately predicts Vmin of logic CMOS circuit. The new equation has been verified through Monte-Carlo simulation and measurements for various n-/p-FET balances and different technology nodes.

REFERENCES

[1] V. De. S. Vangal, and R. Krishnamurthy, "Near-Threshold Voltage (NTV) Computing," IEEE Design and Test Nature, vol. 34, pp. 24–30, 2017

[2] H. Fuketa, S. Iida, T. Yasufuku, M. Takamiya, M. Nomura, H. Shinohara, and T. Sakurai, "A closed-form Expression for Estimating Minimum Operating Voltage (VDDmin) of CMOS Logic Gates," in *Proc. Design Automation Conference (DAC)*, pp. 984-989, 2011

[3] T. Yasufuku, S. Iida, H. Fuketa, K. Hirairi, M. Nomura, M. Takamiya, and T. Sakurai, "Investigation of Determinant Factors of Minimum Operation Voltage of Logic Gates in 65-nm CMOS," in *Proc. International Symposium on Low Power Electronics and Design (IDLPED)*, pp. 22-26, 2011

[4] F. Crupi, M. Alioto, J. Franco, P. Magnone, M. Togo, N. Horiguchi, and G. Groeseneken, "Understanding the Basic Advantages of Bulk FinFETs for Sub- and Near-Threshold Logic Circuits From Device Measurements," *IEEE Transactions on Circuits and Systems, II: Express Briefs*, Vol. 59, No. 7, pp. 439-442, 2012

[5] H.J. Kim, B.H. Choi, Y.H. Lee, J.H. Ahn, Y.S. Bang, Y.D. Lim, J.H. Do, J.H. Jung, T.J. Song, Y. Yasuda-Masuoka, K.C. Park, S.D. Kwon, and J.S. Yoon, "Highly Manufacturable Low Power and High Performance s11LPP Platform Technology for Mobile and GPU Applications," *2018 Symposium on VLSI Technology*, pp. 213 - 214, June, 2018

[6] H.S. Rhee, I.R. Kim, J.H. Jeong, N.K. Son, H.B. Hong, S.I. Cho, Y.M. Park, D.W. Kim, Y.K. Choi, J.H. Ahn, S.G. Kang, K.H. Yeo, J.T. Kim, E.C. Lee, J.M. Youn, and J.S. Yoon, "8LPP Logic Platform Technology for Cost-Effective High Volume Manufacturing," 2018 Symposium on VLSI Technology, pp. 217 - 218, June, 2018

978-1-7281-1988-5/18 $31.00 © 2018 IEEE

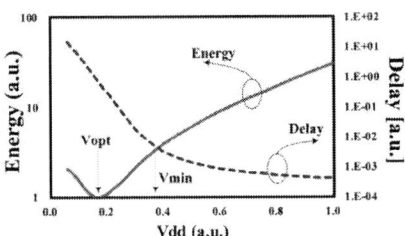

Fig. 1. Switching Energy and delay versus supply voltage Vdd for advanced FinFET CMOS logic circuit (Inverter, FO4).

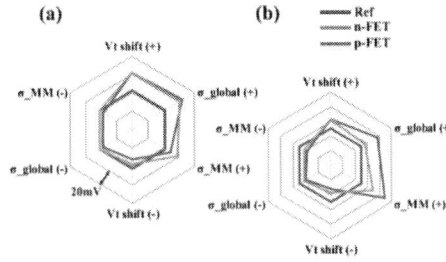

Fig. 2. Vmin sensitivity radar chart of key electrical parameters (ΔVth =25mV): (a) Inverter and (b) D-FlipFlop.

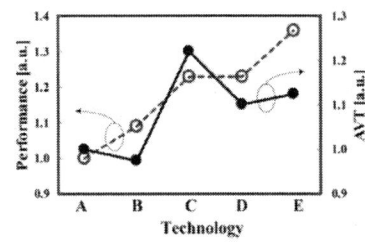

Fig. 3. Tr. AC performance and Vth mismatch (AVT) dependence with various processes conditions.

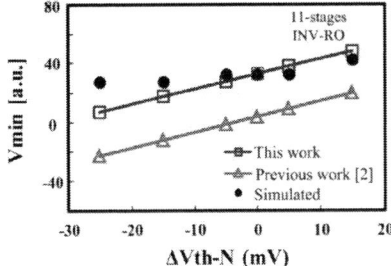

(a) n-FET: Vmin dependence on Vth shift (ΔVth-N).

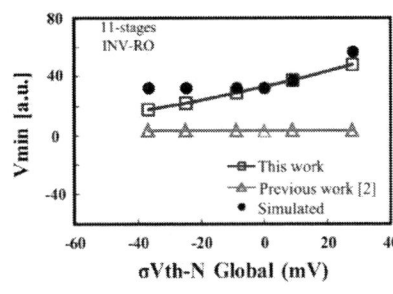

(b) n-FET: Vmin dependence on global variation (σVth-N Global).

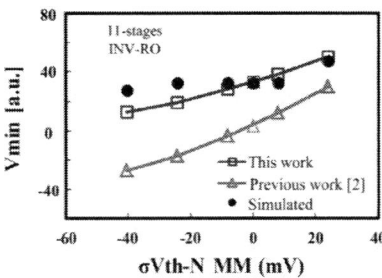

(c) n-FET: Vmin dependence on local variation (σVth-N MM).

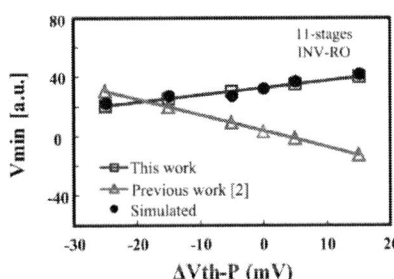

(d) p-FET: Vmin dependence on Vth shift (ΔVth-N).

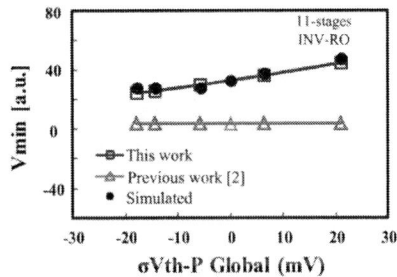

(e) p-FET: Vmin dependence on global variation (σVth-P Global).

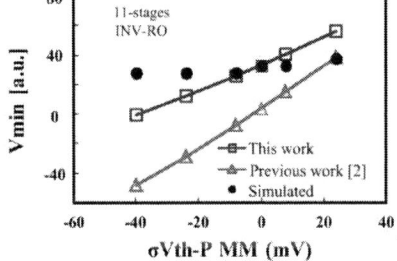

(f) p-FET: Vmin dependence on local variation (σVth-P MM).

Fig. 4. Vmin dependence on each n-/p-FET's key electrical parameters, compared to SPICE simulation, previous expression [2] and newly proposed model.

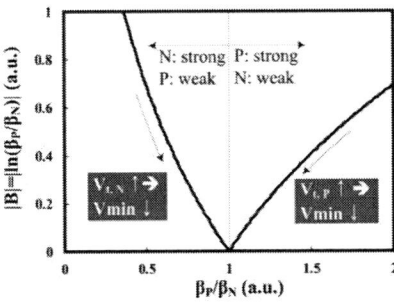

Fig. 5. |B| versus n-/p-FETs strength

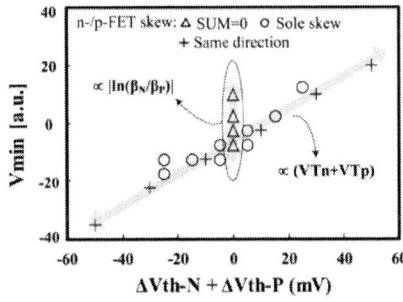

Fig. 6. Vmin dependence on summation of n/p-FET Vth shift (ΔVth-n + ΔVth-p)

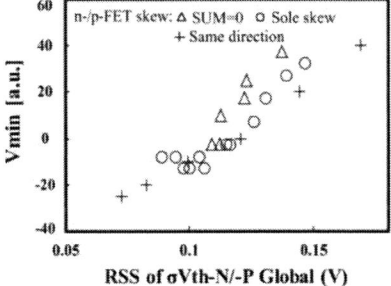

Fig. 7. Vmin dependence on RSS of n-/p-FET Vth global variation (σVth-n/-p global)

978-1-7281-1988-5/18 $31.00 © 2018 IEEE

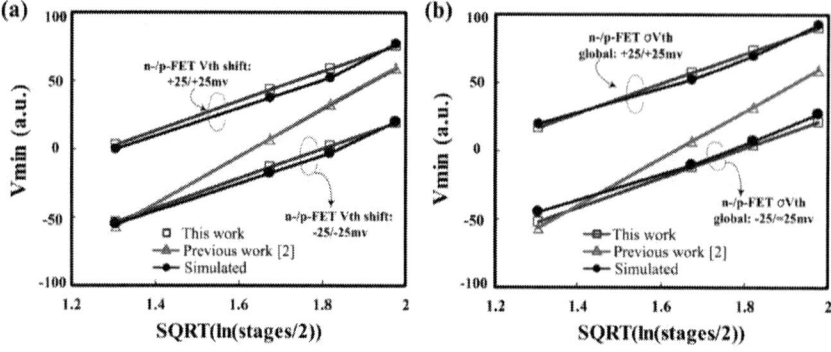

Fig. 8. Vmin of Inverter R.O. for various (a) Vth shifts and (b) global variation, compared to SPICE simulation, previous expression [2] and newly proposed model.

Fig. 9. Vmin extraction from experimental measurement data.

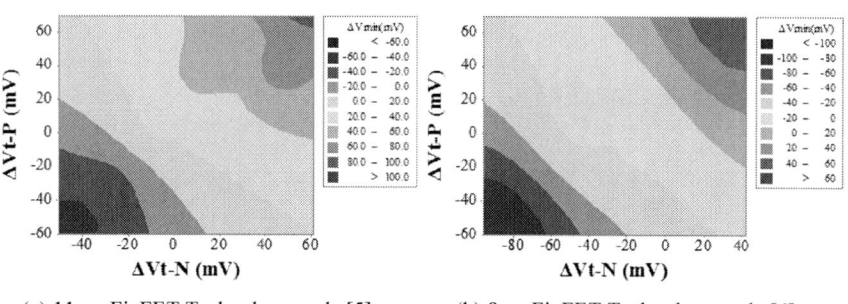

(a) 11nm FinFET Technology node [5]

(b) 8nm FinFET Technology node [6]

Fig. 10. Experimental Vmin data with various Vth position in different advanced 11/8nm FinFET Technology nodes [5-6].

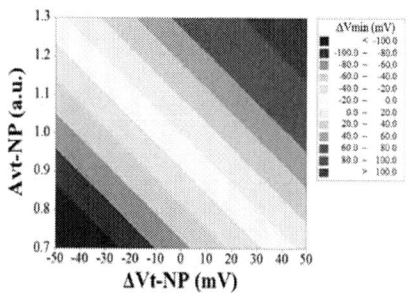

Fig. 12. Vmin dependence on Vth and random variation (proposed equation)

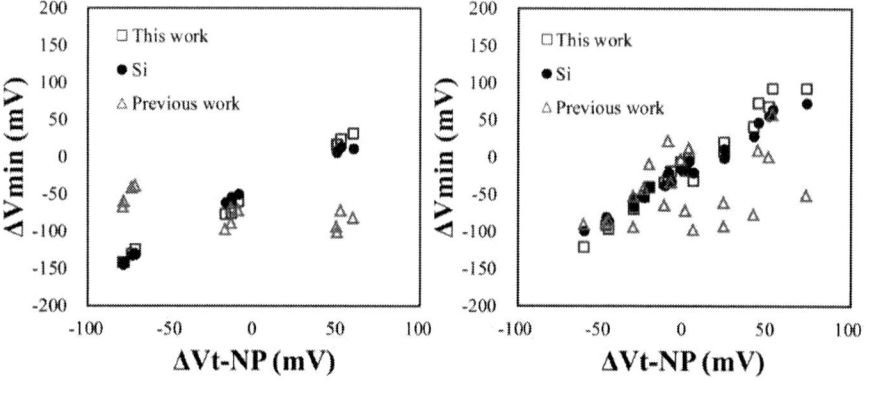

(a) 11nm Technology node [5]

(b) 8nm Technology node [6]

Fig. 11. Vmin estimation comparison between Si experimental data and newly

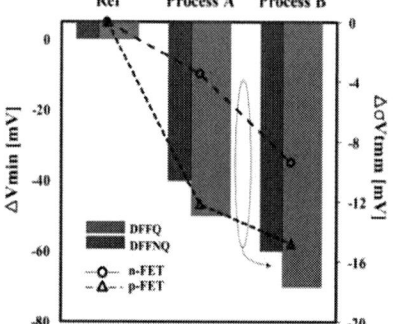

Fig. 13. Vmin and local mismatch value with various Fin process improvement.

Tackling Fundamental Challenges of Carrier Transport and Device Variability in Advanced Si nFinFETs for 7nm Node and Beyond

Ming-Huei Lin, Vincent S. Chang, Jen-Hsiang Lu, Shu-Hui Wang, and Shyh-Horng Yang

Technology Development, Taiwan Semiconductor Manufacturing Company, Hsinchu, Taiwan, R.O.C.

Tel: +886-3-5636688 ext.712-2897 E-mail: mhlinzr@tsmc.com

Abstract—We demonstrated that the fundamental scaling challenges of carrier transport and device variability can be tackled by S/D epitaxy and HK/MG RPG optimizations on the leading-edge 7nm Si nFinFETs, paving the way for continuous scaling. Mitigations of S/D long-range Coulomb interactions and gate-corner work-function roll-up enhance I_{DSAT} by 18% and 9% respectively at constant gate overdrive, translating to a 13% speed-power enhancement in the ring oscillator. These techniques show larger I_{DSAT} enhancements than that of I_{DLIN}. By using an improved characterization method, their unique transport characteristics are clarified.

I. INTRODUCTION

As the dimension of the source/drain (S/D) and the gate is scaled down to nanoscale regime, two fundamental challenges that impact drive current (I_{DSAT}) emerge (**Fig. 1**), namely the S/D long-range Coulomb interactions (LCI) [1], [2] and the effect of gate-corner work-function roll-up (GWR).

The heavily-doped and highly-scaled S/D regions no longer behave as ideal electrodes, but introduce additional scatterings that degrade the transport within the channel. Mechanisms such as LCI, neutral defects [3], and drain scattering [4] have been verified to explain the degradations of apparent mobility (μ_{app}) or virtual source velocity (υ_{X0}) in nanoscale FETs. An experimental work [5] further points out that LCI prevails over neutral defects for sub-25nm devices. The LCI phenomenon can be observed in Si nFinFETs with non-optimized S/D (**Fig. 2 (a)**). The υ_{X0} reaches the peak at L_{eff} ~30 nm, and then drops within the sub-20nm regime. Our experimental observation coincides with the pioneering Monte Carlo simulations [1], [2].

The GWR effect, which was previously referred as "weak corner turn on" in [6], adds challenges to gate length (L_g) scaling with high-κ/metal-gate (HK/MG) RPG process. The GWR raises the potential of the top-of-barrier (ToB) and increases the threshold voltage (V_T) and the critical length for order of kT drop from ToB. **Fig. 2 (b)** shows the effect of GWR on V_T in non-optimized RPG. The ratio of the mid-gap eWF region (L_{GWR}) to the effective channel length (L_{eff}) increases with scaled L_g. GWR is pronounced in the sub-30nm regime, since L_{GWR} overlaps with the underneath ToB. The raised V_T and increased critical length are deleterious to I_{DSAT}.

Although these limiting factors for L_g scaling have been identified, the solutions to overcome them have yet been examined systematically.

In this study, we employed Si nFinFETs to validate the techniques for improving carrier transport and device

variability without compromising parasitic resistance (R_P), short channel effects (SCE), and effective capacitance (C_{eff}). Their signatures and influences on transport parameters are investigated through virtual source model (VSM) [7], in terms of thermal velocity (υ_{TX}), mean free path (λ), critical length (L_C), and the parameter α [8] for the additional scatterings associated with S/D. Note that various α values were obtained for different device types [8]: ~1 for III-V FETs, 0.8 for ETSOI nFETs, and 0.35 for Si nFinFETs.

II. EXPRIMENTAL

Devices under test are 7nm Si nFinFETs [9] with embedded S/D epitaxy (EPI), multi-V_T HK/MG RPG, and stressors. The design of experiments is summarized in **Table I**. For LCI, we compared three EPI splits to study the impact of doping profile. For GWR, an optimized RPG was tailored with band-edge gate-corner eWF and shorter L_{GWR}.

To study the transport characteristics, we use split C-V method to determine the μ_{app} and υ_{X0}, in which the effects of parasitic components and drain-induced barrier lowering (DIBL) are carefully calibrated. Then we use an improved VSM-based characterization method, in which the carrier degeneracy and the additional scatterings associated with S/D are taken into account. The key procedures and formulas are outlined in **Fig. 3**. This generalized method extracts υ_{TX}, λ, L_C, and α without assuming υ_{TX}, presuming the temperature dependences of transport parameters [10], or neglecting the drain bias dependence of λ [11].

III. RESULTS AND DISCUSSION

A. S/D Long-range Coulomb interaction (LCI)

Fig. 4 shows the υ_{X0}-L_{eff} of different EPI designs. For the splits with lower doping concentration (N_D) near the junction, the υ_{X0} still increases monotonically with scaled L_{eff}. This indicates the LCI penalty can be suppressed by proper doping design, leading to the υ_{X0} enhancement by 11% (split B vs. A). It is intriguing that, by optimizing N_D near contact, the υ_{X0} of split C further increases by 6% w.r.t. split B. Despite the near-contact region being away from channel, it still affects the scattering. The μ_{app}-L_{eff} dependence (**Fig. 5**) of the optimized EPI decreases slower than the non-optimized. The α enhancement from 0.27 (non-optimized) to 0.34 (optimized) points out that α is also a function of S/D doping. In **Fig. 6**, the optimized EPI process recovers the abrupt λ reduction at L_{eff} ~30 nm, and corresponds to the less decay in λ down to L_{eff} ~13 nm. This signals the reduced LCI in optimized EPI. The λ degradation behavior is also a warning against the risk of "naive υ_{X0} projection", which extrapolates

978-1-7281-1988-5/18 $31.00 © 2018 IEEE

υ_{X0}–L_{eff} without understandings in λ. And the correlation between α and λ suggests that the α values can be used for process control monitoring. **Fig. 7** shows the υ_{TX} is affected by LCI, implying the effect of S/D plasmons "drag" down the velocity of injected carriers near ToB.

We next examine other perspectives. **Fig. 8** shows that it is possible to reduce channel resistance (R_{CH}) without compromising R_P. Note that, for nanoscale FET the R_{CH} is strongly associated with υ_{TX} and α, because the contribution of "ballistic" mobility (μ_B) outweighs that of the diffusive mobility (μ_{eff}). **Fig. 9** shows that the optimized EPI process does not compromise DIBL and C_{eff}. **Fig. 10** further shows that the local variability of optimized EPI process is improved, particularly for short-channel devices. The improvement in V_T mismatch can be attributed to the reduction in dopant fluctuation. The improvement in current factor mismatch is likely owing to the reduced LCI variation near ToB.

B. Gate-edge Work-function Roll-up (GWR)

Optimized RPG solves the V_T roll-up problem (**Fig. 11**). Regarding the GWR effect on I_{DSAT}, the I_{DSAT}–I_{OFF} performance degradation was attributed to the increase in the access resistance (R_{ACC}) induced by lower inversion [12]. But, in our case, we found that the impact of L_C is more pronounced than R_{ACC}. As shown in **Fig. 12**, the difference in R_P is small due to sufficient gate overlaps, and the difference in intrinsic V_{GS} is below 5 mV (i.e., <1% I_{DSAT}). However, **Fig. 13** shows the υ_{X0} gain by 9% in the shortest L_g device. Moreover, the comparable μ_{app} in **Fig. 14** shows the signature of GWR reduction: the υ_{X0} of short-channel devices is boosted without μ_{app} enhancement.

We compare the extracted transport parameters to understand its mechanism under high-field transport. Comparable α values and υ_{TX} values for both splits are found in **Fig. 15**. The fact that comparable α values are found indicates that the GWR reduction does not affect additional scatterings associated with S/D. The unchanged υ_{TX} values suggest that the υ_{X0} gain is the result of ballisticity (B_{SAT}) enhancement, which corresponds to the L_C shortening by ~13%. The υ_{X0} gain due to GWR reduction can be explained by the change in the source-end potential profile.

The GWR reduction also improves variability (**Fig. 16**). The merits of V_T and current factor mismatch are likely due to the reduction in the ToB variation incurred by L_{GWR} variation. **Fig. 17** shows that the well-engineered GWR reduction does not degrade DIBL or C_{eff}. In **Fig. 18**, the comparable gate leakage and capacitive effective thickness indicate that the gate oxide integrity is maintained.

C. Correlation between Velocity and Mobility

To illustrate the features of LCI and GWR reductions presented in this work, the relative change in velocity ($\Delta\upsilon_{X0}$) vs. the relative change in mobility ($\Delta\mu_{app}$) of transport boosters are presented in **Fig. 19**. For strain engineering techniques, the slope ranging from ~0.8 to 1 for both NFETs and PFETs can be found in [13], and the slope of 0.85 is further confirmed in Si pFinFETs [14]. The sensitivity of gate dielectrics [15] for short channel devices is ~0.43. They both show $\Delta\upsilon_{X0} \leq \Delta\mu_{app}$.

Interestingly, the LCI reduction features $\Delta\upsilon_{X0} \sim$ 2X $\Delta\mu_{app}$. This can be explained by the potential profiles at varied V_{DS}. In linear regime, the ToB is near the center of channel. As V_{DS} increases, ToB moves toward the source where LCI is pronounced. Therefore, the LCI reduction yields more improvement in υ_{X0}.

For the GWR reduction, the decoupling of high-field velocity and low-field mobility emerges in short channel devices, because the modulation on potential profile increases the B_{SAT} and υ_{X0} by shortening L_C, whereas it may not affect the υ_{TX}. Similar results can be found in [16] related to source-end potential engineering.

Fig. 20 shows that the $\Delta I_{DSAT} > \Delta I_{DLIN}$ features of LCI and GWR reductions correspond to $\Delta\upsilon_{X0} > \Delta\mu_{app}$, indicating that they are efficient I_{DSAT} boosters. In **Fig. 21**, the reduced LCI and GWR enhance the speed-power performance of ring oscillator by 13% with improved variability, evidencing the successful integration of EPI and RPG co-optimization.

IV. CONCLUSION

LCI and GWR reductions, featuring improved carrier transport and device variability, are crucial for continuous device scaling to 7nm node and beyond, and are demonstrated on advanced Si nFinFETs by EPI and RPG optimizations in this work. Their influences on transport parameters are comprehensively investigated in the framework of VSM. For LCI reduction, the υ_{TX} and α enhancements lead to 18% I_{DSAT} net gain by the enlarged λ. The signature of $\Delta I_{DSAT} \sim$ 2X ΔI_{DLIN} can be explained by the position of ToB at varied V_{DS}. For GWR reduction, V_T reduction and 9% I_{DSAT} net gain are achieved by reducing L_C without shrinking L_{eff} to avoid λ and SCE degradations. With EPI and RPG co-optimization, the ring oscillator shows a 13% speed-power enhancement with improved variation.

ACKNOWLEDGMENT

The authors are grateful to TD integration and ATMD for technical support. M.-H. Lin also thanks Ming-Hung Han, Ta-Chun Lin, Yung-Chih Wang, Chung-Pin Huang, and Chuan-Li Chen for fruitful discussions.

REFERENCES

[1] M. Fischetti and S. Laux, *J. Appl. Phys.* vol. 89, no. 2, p. 1205, 2001.
[2] M. Fischetti *et al.*, *IEEE TED*, vol. 54, no. 9, p. 2116, 2007.
[3] A. Cros *et al.*, *IEDM Tech. Dig.*, p. 663, 2006.
[4] K. Natori *et al.*, *J. Appl. Phys.* vol. 118, no. 23, p. 234502, 2015.
[5] S.-H. Hsieh *et al.*, *Proc. Silicon Nanoelectronics Workshop*, p. 32, 2016.
[6] W.-S. Ho *et al.*, US Patent 9583362B2, Feb. 28, 2017.
[7] M. Lundstrom and D. Antoniadis, *IEEE TED*, vol. 61, no. 2, p. 225, 2014.
[8] D. Antoniadis, *IEEE TED*, vol. 63, no. 7, p. 2650, 2016.
[9] S.-Y. Wu *et al.*, *IEDM Tech. Dig.*, p. 2.6.1, 2016.
[10] M.-J. Chen *et al.*, *IEDM Tech. Dig.*, p. 39, 2004.
[11] A. Majumdar *et al.*, *IEDM Tech. Dig.*, p. 8.3.1, 2012.
[12] L.-Å. Ragnarsson *et al.*, *Symp. VLSI Tech.*, p. 1, 2016.
[13] A. Khakifirooz and D. Antoniadis, *IEDM Tech. Dig.*, p. 667, 2006.
[14] T.-C. Lin *et al.*, *Proc. VLSI-TSA*, 2016.
[15] K. Tatsumura *et al.*, *IEDM Tech. Dig.*, p. 465. 2009.
[16] M.-H. Han *et al.*, *Proc. VLSI-TSA*, 2015.

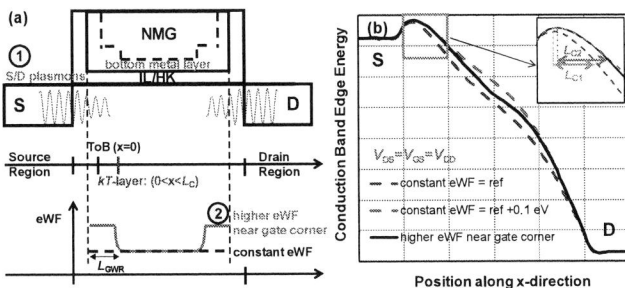
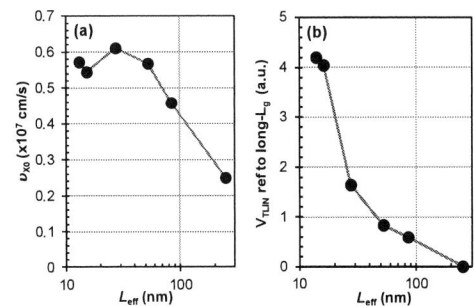

Fig. 1 (a) Schematic diagram of LCI and GWR for *n*FET. **(b)** Illustrative example of the conduction band edge energy of constant (without GWR) and non-uniform (with GWR) eWF profiles along longitudinal direction.

Fig. 2 Phenomena of **(a)** LCI and **(b)** GWR in non-optimized Si *n*FinFETs. In this work, the υ_{X0} and μ_{app} of different splits are compared at same gate overdrive ($N_S \sim 8 \times 10^{12}$ cm^{-2}).

TABLE I DESIGN OF EXPERIMENTS

S/D Epitaxy DOE for LCI Study

Doping	A Non-opti.	B	C Optimized
near junc.	ref.	lower	lower
near contact	ref.	comparable	optimized

RPG DOE for GWR Study

	Non-opti.	Optimized
gate-corner eWF	ref.	band-edge
region with high eWF	ref.	shorter
effective channel length	ref.	comparable

Fig. 3 Outline of the new extraction method. Eqn. (1) considers the carrier degeneracy and the effect of additional S/D scatterings on apparent mobility at low V_{DS}. Eqn. (3) can be obtained by the apparent mobility within kT-layer, the Einstein relation, and $\upsilon_{X0}^{-1} = \upsilon_{TX}^{-1} + (D/L_C)^{-1}$, where D is the effective diffusivity within the kT-layer.

Extraction Flow

1. I-V & C-V data of DUT w/ varied L_{eff}
→ determine μ_{app} & υ_{X0}
2. plot: (a) μ_{app}^{-1} vs. L_{eff}^{-1}
→ deduce αK_B from the slope by (4)
(b) L_{eff} / μ_{app} vs. υ_{X0}^{-1}
→ calculate L_C / L_{eff} from the slope by (5)
3. iteratively obtain α & υ_{TX} by self-consistent method [comparing β']

μ_{app}: apparent mobility
μ_{eff}: diffusive mobility
μ_B: "ballistic" mobility
υ_{X0}: virtual source velocity
υ_{TX}: thermal velocity
λ_{LIN}: mean free path in linear regime
λ_{SAT}: mean free path in saturation regime
L_C: critical length for order of kT drop form ToB
L_{eff}: effective channel length
α: parameter for S/D scatterings (low V_{DS})
β: parameter for S/D scatterings (high V_{DS}) in kT-layer
f: ratio of the Fermi-Dirac integrals of -1/2 and 0 order
ϕ_T: thermal voltage (kT/q)
s, i: slope and intercept of L_{eff} / μ_{app} vs. υ_{X0}^{-1} plot
$K_B \equiv \upsilon_{TX} f / 2\phi_T$

$$\frac{1}{\lambda_{LIN}} = \frac{K_B}{\mu_{app}}\left(1 - \frac{\mu_{app}}{\alpha K_B L_{eff}}\right) \quad (1)$$

$$\frac{1}{\lambda_{SAT}} = \frac{1}{2L_C}\left(\frac{\upsilon_{TX}}{\upsilon_{X0}} - 1\right) \quad (2)$$

$$f / \lambda_{SAT} = 1/(\beta L_{eff}) + 1/(\lambda_{LIN}) \quad (3)$$

$$1/\mu_{app} = 1/\mu_{eff} + 1/\mu_B \quad (4)$$

By (1), (2), and (3):

$$\frac{L_{eff}}{\mu_{app}} = \frac{\phi_T L_{eff}}{L_C}\frac{1}{\upsilon_{X0}} + \frac{\phi_T}{\upsilon_{TX}}f^{-1}\left[2\left(\frac{1}{\alpha} - \frac{1}{\beta'}\right)\right] - \frac{L_{eff}}{L_C} \quad (5a)$$

$$\text{where } \frac{1}{\beta'} \equiv \frac{1}{\beta} + (1-f)\frac{L_{eff}}{\lambda_{SAT}} \quad (5b)$$

$$\frac{1}{\alpha} - \frac{1}{\beta} = \frac{\frac{s^2}{2i\phi_T\upsilon_{X0}} + \frac{fs}{2\phi_T} + \frac{(\alpha K_B)L_{eff}}{\alpha\mu_{app}} + \frac{(1-f)s^2}{2i\phi_T\upsilon_{X0}}(\frac{\upsilon_{TX}}{\upsilon_{X0}} - 1)}{1 + \frac{s}{i\upsilon_{X0}}} \quad (6)$$

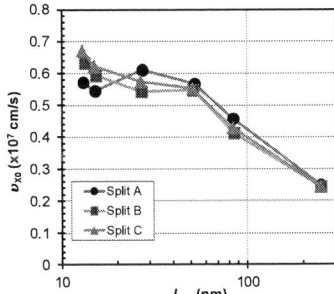

Fig. 4 For optimized EPI process, υ_{X0} still increases monotonically with L_{eff} scaling.

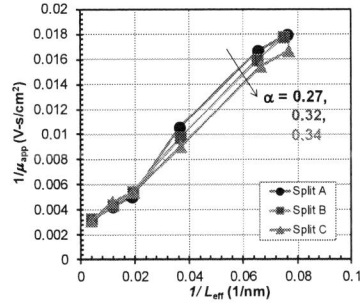

Fig. 5 The μ_{app} difference occurs in sub-30nm regime. The splits with higher α correspond to less μ_{app} degradation in short channel devices.

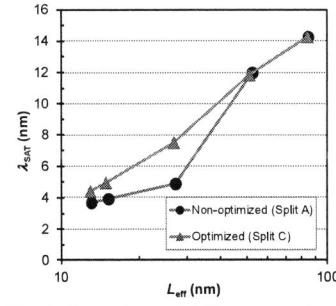

Fig. 6 Non-optimized EPI process shows abrupt λ_{SAT} decay, while the optimized EPI process recovers such penalty.

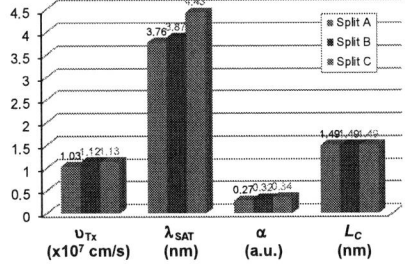

Fig. 7 Extracted transport parameters show that the optimized EPI process improves υ_{TX}, λ_{SAT}, and α, while L_C is unchanged due to comparable SCE (Fig. 9) and same RPG process.

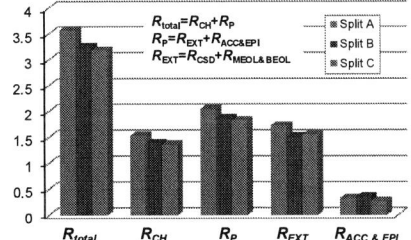

Fig. 8 Partition of resistance components. Here R_{total} is total resistance; R_{CH} is channel resistance; R_P is the sum of parasitic resistance components; R_{EXT} is the external resistance including contact resistance (R_{CSD}) and interconnect ($R_{MEOL\&BEOL}$); $R_{ACC\&EPI}$ is the combination of access region and S/D epitaxy.

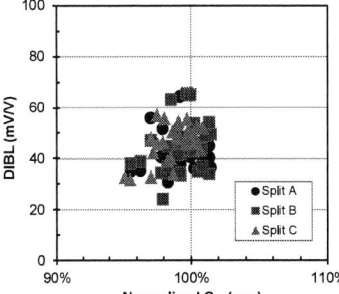

Fig. 9 Comparable DIBL and C_{eff} of short channel devices indicate that the υ_{X0} gain is not due to compromising SCE.

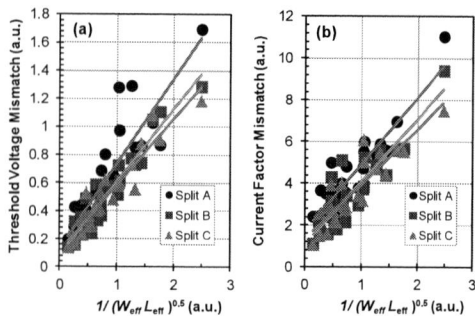

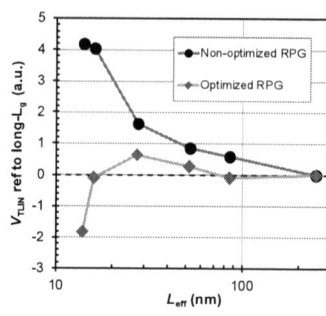

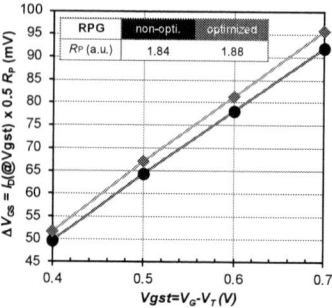

Fig. 10 The optimized EPI process improves the local variability of V_T and current factor.

Fig. 11 More V_T reduction in short channel devices suggests the reductions in gate-corner eWF and L_{GWR}.

Fig. 12 Assessment of R_P difference on the IR drop (ΔV_{GS}). The difference in ΔV_{GS} is small even at higher gate overdrive.

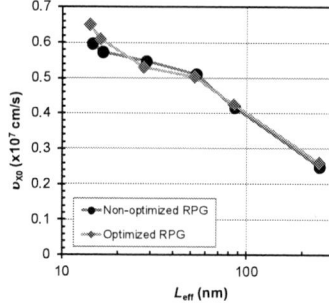

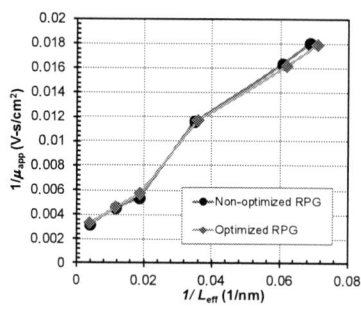

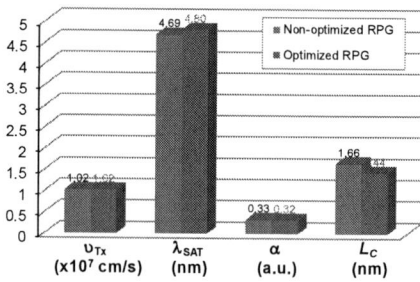

Fig. 13 The υ_{X0} enhancements occur in sub-20nm regime, in which more than +6% enhancements are shown. The shortest L_g device shows +9% gain.

Fig. 14 Minor μ_{app} change in short channel devices is observed, suggesting the decoupling of high-field velocity and low-field mobility.

Fig. 15 Extracted transport parameters show that the optimized RPG process enhances the ballisticity in terms of L_C and λ_{SAT}, while the α and υ_{TX} are comparable.

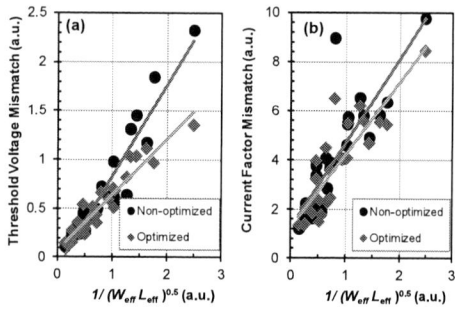

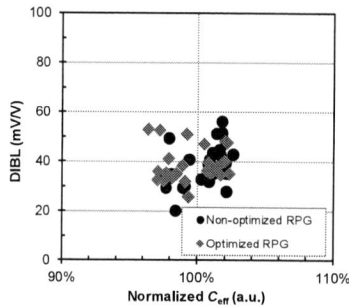

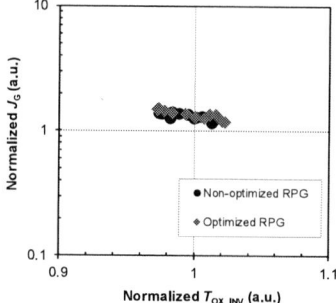

Fig. 16 The optimized RPG process improves the local variability of V_T and current factor.

Fig. 17 Comparable DIBL and C_{eff} indicate that the GWR reduction by well-engineered RPG process does not affect SCE.

Fig. 18 Gate leakage current vs. capacitive effective thickness. The RPG optimization does not alter T_{OX} nor gate oxide integrity.

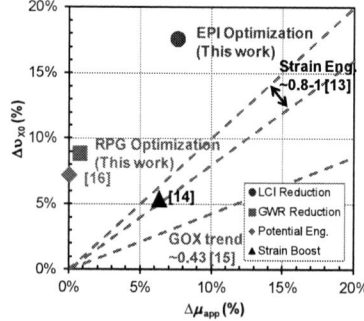

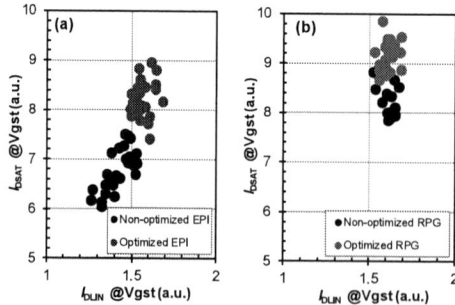

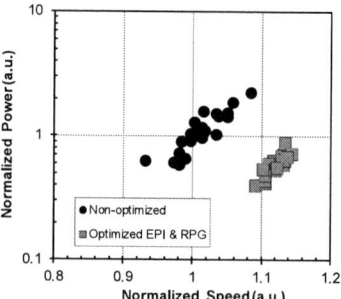

Fig. 19 Correlation between $\Delta \upsilon_{X0}$ and $\Delta \mu_{app}$. LCI and GWR reductions show $\Delta \upsilon_{X0} > \Delta \mu_{app}$, which distinguishes them from the others.

Fig. 20 Correlation between I_{DSAT} and I_{DLIN} at constant gate overdrive. (a) For LCI reduction by EPI optimization, ΔI_{DSAT} (18%) is ~2X ΔI_{DLIN} (10%). (b) For GWR reduction by RPG optimization, I_{DSAT} is improved by 9% with comparable I_{DLIN}.

Fig. 21 Demonstration of ring oscillator speed-power performance enhancement by 13 % with improved variation.

978-1-7281-1988-5/18 $31.00 © 2018 IEEE   647

Extendable and Manufacturable Volume-less Multi-Vt Solution for 7nm Technology Node and Beyond

Ruqiang Bao, Huimei Zhou, Miaomiao Wang, Dechao Guo, Bala S Haran, Vijay Narayanan, Rama Divakaruni

IBM Semiconductor Technology Research, Albany, NY, email: rbao@us.ibm.com

Abstract—We demonstrated more than 3 pairs of threshold voltage (Vt) devices by volume-less multiple Vt (multi-Vt) scheme plus dual work function metals (WFM) without performance and reliability degradation on 20nm gate length FinFET CMOS devices. Vt shifts over 200 mV were achieved for both nFET and pFET. The volume-less nature of this multi-Vt scheme relieves replacement metal gate (RMG) challenges and opens the path to offer multi-Vt solution for future highly scaled technologies.

I. INTRODUCTION

Multi-Vt enabled by RMG processes providing both performance improvement and Vt tunability in thin body transistors is essential for advanced CMOS technologies [1-5]. Several methods like gate metal treatments, gate metal thickness, and gate dipole can be applied to modulate the Vt in RMG devices. However, gate metal treatments like implant and plasma treatment have loading effects and are thus difficult to control for product manufacturing. Gate metal thickness and gate dipole are viable technology options due to conformal and uniform gate metal and dipole films enabled by atomic layer deposition (ALD).

The demand to scale the transistors for higher density, improved performance, reduced power consumption, and lower cost per transistor drives the contact poly pitch (CPP) scaling, as shown in Fig.1 and Fig.2, which consequently requires gate length shrinking to release more space for contact resistance reduction. Continuous scaling in RMG raises the multi-Vt offering challenges for 7nm technology and beyond due to limited space for metal gate patterning. Furthermore, self-aligned contact (SAC) becomes a critical element for device yield enhancement at future technology node. Therefore, a simplified RMG stack integration scheme is required to ensure good gate recess control and uniform SAC encapsulation. Multi-Vt options enabled by different gate metal thicknesses (metal multi-Vt) will face scalability challenges at aggressively scaled pitch. In this work, we present a volume-less multi-Vt solution to define all Vt flavors with different dipole layer thicknesses. The oxide dipole layer interacts with the SiOx based interfacial layer (IL) to produce Vt shifts concomitant with their group electronegativity difference [6]. The proposed scheme is demonstrated to be compatible with dual WFM process and can be applicable to highly scaled devices and novel device architectures due to its volume -less nature. To integrating multiple dipole thicknesses on the same chip is very challenging since the dipole thickness is very thin and the channel could suffer from patterning damage. In this paper, we

provide the solution to the above issue by demonstrating the proposed volume-less multi-Vt scheme on Si/SiGe FinFET CMOS, proving the independence of proposed multi-Vt solution on substrate material.

II. EXPERIMENTS

Si nFET, Si pFET, and SiGe pFET devices were fabricated using a standard high-κ RMG FinFET process. All performance and dipole patterning are evaluated on the short channel FinFETs with gate length (L_g) of 20 nm. The ring oscillator (RO) is composed of FinFETs with the same geometry for device level study. The dipole multi-Vt module is conducted with ALD La_2O_3 followed by drive-in anneal before metal multi-Vt module (Fig. 3). To study the extendibility of this dipole multi-Vt, we also compared against a single La_2O_3 thickness experiment, where only one thickness dipole is used and the dipole is only exposed to dipole T1 patterning. Current focus is given to Si/SiGe CMOS because SiGe is new channel material for performance and reliability improvement [7] and RMG processes for multi-Vt applicable to Si/SiGe CMOS are transparent to Si CMOS. Moreover, metal multi-Vt, performed in RMG metal multi-Vt module, is employed for device comparison. Table 1 summarizes the detail information for each Vt definition. FEOL reliability is also examined.

III. RESULTS AND DISCUSSION

A. Volumelessness for better integration

Dipole multi-Vt offers the volume-less multi-Vt solution. Fundamental concept of metal multi-Vt is to employ pWFM as a barrier underneath nWFM and thereby to modulate the oxygen vacancy concentration in HfO_2 to provide different Vts [8,9]. In this method, nWFM is essential and needs to be on top of pWFM. Low pVt devices require thicker pWFM to push nWFM away to reduce the impact of nWFM on the oxygen vacancy concentration in HfO_2 and thus becomes the most challenging devices for SAC enablement due to thickest WFM stack among all devices. Fig. 4 shows the TEM cross section after WFM chamfering, as an example. The WFM pinch-off in low pVt device has different WFM recess depth from low nVt device which has thinnest WFM stack. Therefore, additional WFM chamfering processes and passes are required to have good integration compatibility and thus increase transistor cost and process control difficulty. However, in dipole multi-Vt, sub-nm dipole thicknesses are employed (Fig.5). As a result, there is no metal stack difference between all same polarity devices (i.e. all nFETs or all pFETs), shown in Fig. 6 as an example. This scheme also solves the gate length dependent issue which is faced in metal multi-Vt scheme.

978-1-7281-1988-5/18 $31.00 © 2018 IEEE

B. Substrate independent

Fig. 7-9 show the I_d-V_g curves for Si nFET, SiGe pFET and Si pFET. As shown, La_2O_3 dipole shifts nFET to lower Vt but shifts both Si pFET and SiGe pFET to higher Vt. Therefore, this dipole can be used for both Si and SiGe channel, i.e. substrate independent. Because nVt shifts toward lower Vt, it reduces the need of thicker nWFM to provide low nVt device when no dipole is used, because thinner nWFM is preferred for device performance improvement discussed in next section. On the other hand, no nWFM is required to be on top of pWFM to achieve high pVt device if the dipole is used, which makes the integration simpler and SAC enablement easier. The near midgap or midgap nWFM for nFET and no scavenging metal for pFET improve the gate leakage for all nVt devices and high pVt devices and thus provide better breakdown voltage (VBD) than metal multi-Vt scheme [9]. Improved T_{inv} is also observed for dipole devices (Fig. 10).

C. Mutli-Vt by multiple dipole thickness

Fig. 11-12 show the Vt shift for nFET and pFET with different La_2O_3 thickness for 20nm L_g CMOS devices. As shown, there is more than 200mV Vt shift from base Vt devices by using different La_2O_3 thickness. It also demonstrates that La_2O_3 T2b, which is thicker than La_2O_3 T2a, can give another Vt for nFET and pFET on the same chip. Vt is relatively fat along the gate length (Fig. 13). Fig.14 shows the I_{eff} versus I_{off} for different Vt devices defined by dipole multi-Vt for nFET. Solid symbol is for multi-dipole thickness experiment, but open symbol is for single dipole thickness experiment. As shown, multi-dipole thickness experiment shows the same performance as single dipole thickness experiment. Similar result is observed for SiGe pFET (Fig. 15). Fig. 16 shows that dipole multi-Vt can be combined with metal multi-Vt to provide more Vts. Broad range RO Vt devices can be provided by using dual WFM plus different La_2O_3 thicknesses as shown in Fig. 17, with the performance following the trendline as single dipole thickness.

D. Gate resistance reduction

As shown in Fig. 18, no gate resistance difference between the same polarity devices is observed for both nFET and pFET, as expected. Because thinner nWFM can be combined with La_2O_3 to achieve the lowest nVt and offer lower gate resistance by providing more room for low resistivity material in gate since currently available nWFM is a high resistive material, dipole multi-Vt opens the opportunity for gate resistance reduction. As shown in Fig.19, thinner nWFM can offer lower gate resistance for Si nFET, which becomes more important for small gate length. Also, as discussed above, there is no longer need for either nWFM on top of pWFM layer or multiple pWFM layers deposited by multiple times underneath nWFM to define pVts. The reduction of metal layer interfaces and metal layer oxidation favors gate resistance reduction. When nWFM is completely skipped on pFET device, it gives significant gate resistance reduction for pFET as shown in Fig. 20. Therefore, dipole multi-Vt offers the opportunity to improve RO performance by gate resistance reduction [10].

E. Reliability improvement

PBTI and NBTI induced ΔVt for devices with no dipole and three different dipole thicknesses are plotted in Fig. 21 (a) and 21(b) respectively (T2b>T2a>T1). VBD reliability for the same set of devices is shown in Fig. 22. It is worth noting that thicker dipole thickness provides better immunity against PBTI and VBD in Si nFET and NBTI in SiGe pFET. Impact of dipole thickness on NBTI reliability in Si pFET is illustrated in Fig. 23, showing less NBTI degradation with incorporation of La_2O_3 in the gate stack for the high Vt devices.

IV. EXTENDIBILITY

Dipole multi-Vt enables RMG scalability and offers the future direction for multi-Vt definition in advanced technologies as described below:

A. Extendible for future technologies

In our scheme, total thickness for La_2O_3 patterning is thinner than both nWFM and pWFM thickness, independent of whether pWFM first or nWFM first is used for dual WFM definition. Therefore, it is not dipole patterning but dual WFM patterning that becomes the most challenging step in RMG for dipole multi-Vt scheme. Because of no volume and easy integration, dipole multi-Vt can be extended to future highly scaled technologies and new architecture device technologies.

B. Extendible for more Vts offering

Inserting patterning steps in dipole multi-Vt module can easily enable multiple dipole thicknesses on the same chip to define more Vts for both nFET and pFET simultaneously without interrupting the gate metal structure. As demonstrated in this paper, the dipole multi-Vt can also be combined with metal multi-Vt to have more flexibility to offer more Vts.

C. Extendible for more dipole materials

Our scheme opens the compatibility to have another dipole material. Therefore, this scheme can have more than 2 materials to provide multi-Vt for future CMOS technologies.

V. CONCLUSIONS

We have demonstrated that more than two La_2O_3 thicknesses can be integrated together on the same chip to define at least 3 pairs of Vts for Si/SiGe CMOS without device performance and reliability degradation. Our scheme can also be used to integrate more than two dipole materials to provide volume-less multi-Vt solution for future CMOS technologies.

ACKNOWLEDGMENT

This work was performed by the Research and Development Alliance Teams at various IBM Research and Development Facilities.

REFERENCES

[1] C-H. Lin et al., *IEDM*, 2014, p74. [2] H.-J. Cho et al., *VLSI*, 2016, p12. [3] Shien-Yang Wu et al., *IEDM*, 2016, p43. [4] C. Auth et al., *IEDM*, 2017, p673. [5] S. Narasimha et al., *IEDM*, 2017, p689. [6] H. Jsgannathan et al., *ECS Trans.* 16, 2008, p19. [7] D. Guo et al., *VLSI*, 2016, p14. [8] M. Togo et al., *VLSI*, 2018, p81. [9] R. Bao et al., *VLSI*, 2018, p115. [10] R. Bao et al., *IEDM*, 2015, p.884.

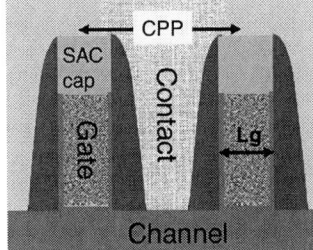

Fig. 1. Typical modern transistor structure.

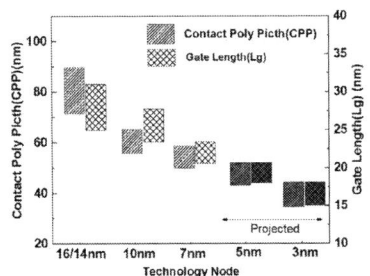

Fig. 2. CPP and gate length scaling trend.

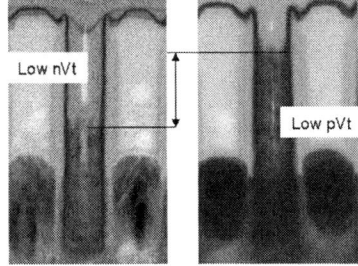

Fig.3 FinFET multi-Vt process flow with dipole multi-Vt module and metal multi-Vt module.

		nFET (Si)	pFET (SiGe)	Comments
Single dipole thickness	Dual WFM	RVt	SLVt	No dipole device
	La₂O₃ T1+ Dual WFM	SLVt	RVt	La₂O₃ thickness 1
Multi-dipole thickness	Dual WFM	RVt	SLVt	No dipole device
	La₂O₃ T1+ Dual WFM	LVt	LVt	La₂O₃ thickness 1
	La₂O₃ T2a + Dual WFM	SLVt(1)	RVt(1)	La₂O₃ thickness 2a
	La₂O₃ T2b+ Dual WFM	SLVt(2)	RVt(2)	La₂O₃ thickness 2b
Single dipole thickness + Metal Multi-Vt	La₂O₃ T1 + Metal Multi-Vt	Dipole Multi-Vt	Metal Multi-Vt + Dipole Multi-Vt	Metal Multi-Vt defined by metal thickness

Table 1. Experimental conditions to define Vts to evaluate device performance (thickness1 < thickness 2a < thickness 2b; SLVt < LVt < RVt).

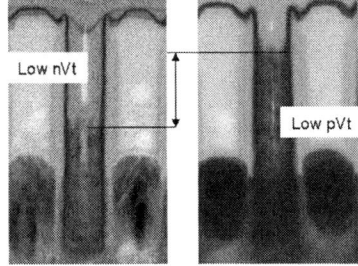

Fig. 4. Metal gate chamfering challenges in metal multi-Vt because of different metal thickness for each device.

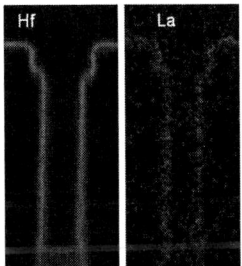

Fig. 5. EELS mapping to show the Hf and La profile in dipole device after patterning.

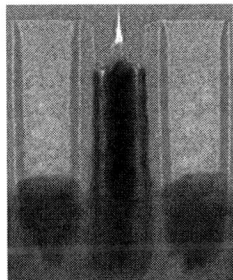

Fig. 6. TEM image for both dipole and no dipole devices which have simple integration.

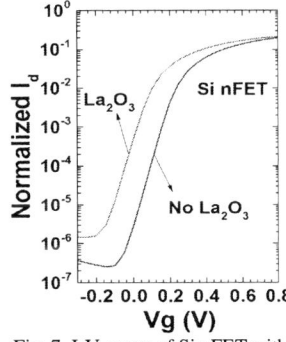

Fig. 7. I-V curve of Si nFET with or without La₂O₃ which shifts Vt to lower side.

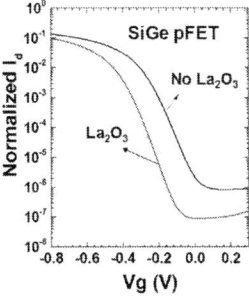

Fig. 8. I-V curve of SiGe pFET with or without La₂O₃ which shifts Vt to higher side.

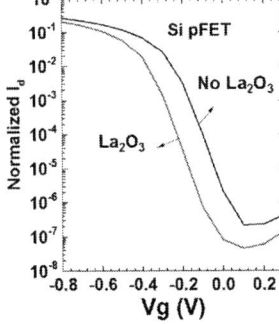

Fig. 9. I-V curve of Si pFET with or without La₂O₃ which shifts Vt to higher side.

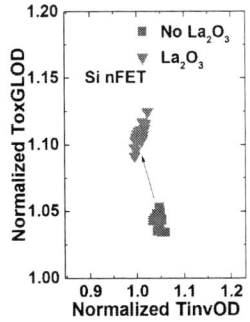

Fig. 10. Normalized Tinv and ToxGL at over drive (OD). Improved gate leakage and capacitance is observed.

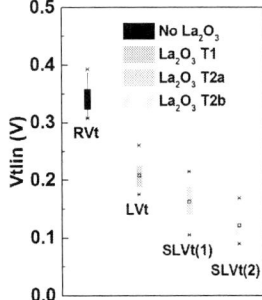

Fig. 11. Si nFET Vt for each device with or without La₂O₃.

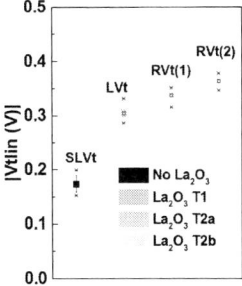

Fig. 12. SiGe pFET Vt for each device with or without La₂O₃.

978-1-7281-1988-5/18 $31.00 © 2018 IEEE

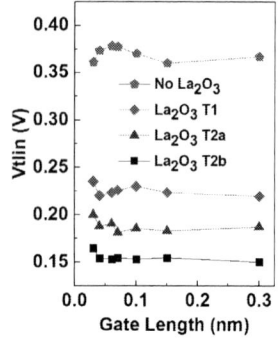

Fig. 13. Si nFET Vt as a function of gate length.

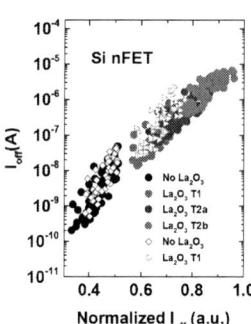

Fig. 14. Si nFET I_{eff} vs I_{off}. (Open - single dipole thickness, solid - multi-dipole thickness).

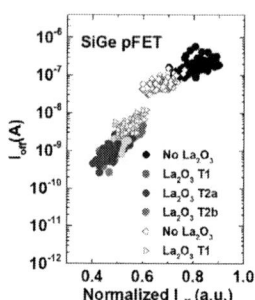

Fig. 15. SiGe pFET I_{eff} vs I_{off}. (Open - single dipole thickness, solid - multi-dipole thickness).

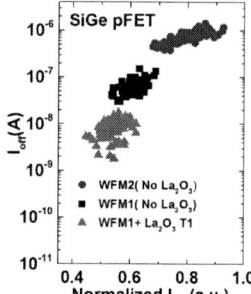

Fig. 16. SiGe pFET I_{eff} vs I_{off} for the case by combining metal multi-Vt and dipole multi-Vt.

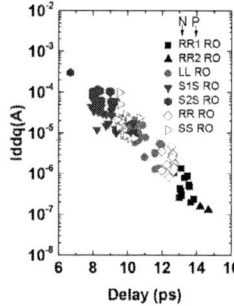

Fig. 17. Delay versus Iddq for different RO Vt favors. (Open - single dipole thickness, solid – multi-dipole thickness)

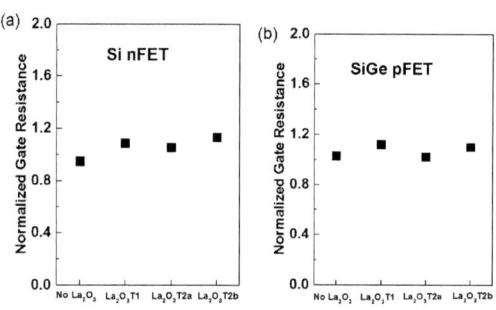

Fig. 18. nFET and pFET gate resistance comparison for different La₂O₃ thickness devices.

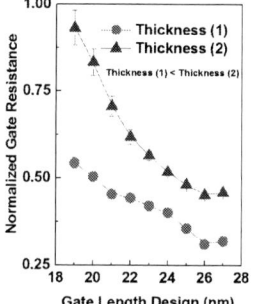

Fig. 19. Effect of nWFM thickness on nFET FinFET gate resistance.

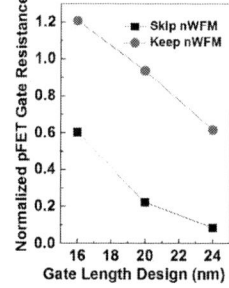

Fig. 20. Comparison of pFET gate resistance with nWFM or skip nWFM.

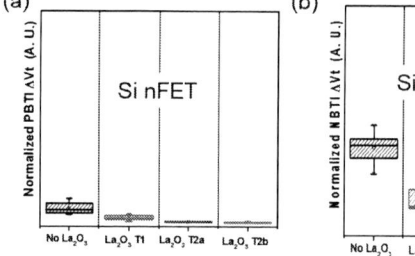

Fig. 21. Si nFET positive bias temperature instability (PBTI) and SiGe pFET negative bias temperature instability (NBTI) comparison for different dipole thickness devices.

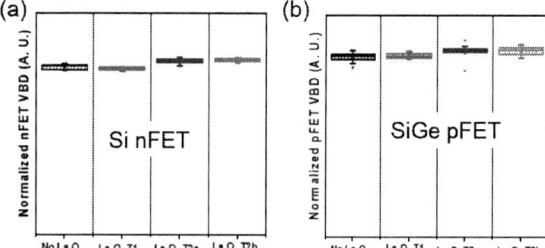

Fig. 22. Effect of La₂O₃ on breakdown voltage (VBD) for both Si nFET and SiGe pFET.

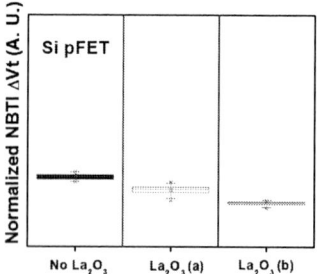

Fig. 23. Effect of La₂O₃ thickness on Si pFET NBTI (a<b).

978-1-7281-1988-5/18 $31.00 © 2018 IEEE

Channel Geometry Impact and Narrow Sheet Effect of Stacked Nanosheet

Chun Wing Yeung[1], Jingyun Zhang[1], Robin Chao[1], Ohseong Kwon[2], Reinaldo Vega[1], Gen Tsutsui[1], Xin Miao[1], Chen Zhang[1], Chang-woo Sohn[2], Bum Ki Moon[3], Ali Razavieh[3], Julien Frougier[3], Andrew Greene[1], Rohit Galatage[3], Juntao Li[1], Miaomiao Wang[1], Nicolas Loubet[1], Robert Robison[1], Veeraraghavan Basker[1], Tenko Yamashita[1], Dechao Guo[1]

[1]IBM, [2] SAMSUNG ELECTRONICS, [3] GLOBALFOUNDRIES, 257 Fuller Rd, Albany, NY 12203
Email: cyeung@us.ibm.com

Abstract

The characteristics of Stacked Nanosheet are investigated, focusing on channel geometry. For the first time, "*narrow sheet effect*" on carrier transport is observed. By comparing measured electron (μ_e) and hole (μ_h) mobilities, and the n-type/p-type opposite transconductance (gm) trends versus sheet width ($Wsheet$), we show that the mobility dependency on $Wsheet$, is attributed to reduced (100) plane conduction contribution as $Wsheet$ shrinks.

Introduction

Horizontally Stacked Gate-All-Around Nanosheet (Nanosheet) is a promising candidate as FinFET replacement, because it offers excellent short channel control and increased effective channel width ($Weff$) per foot print [1-3]. We have demonstrated Nanosheet at extremely scaled dimensions (44/48CPP) with aggressive sheet-to-sheet spacing [2], validated the possibility of modulating Nanosheet threshold voltage (Vt) using different work function metals by replacement metal gate (RMG) process, and proposed multi-Vt solutions for Nanosheet technology [3].

However, several fundamental questions remain unaddressed for Nanosheet to become a viable technology:

(1) Continuously variable $Wsheet$ has been proposed [2]. This offers power-performance co-optimization advantages over FinFET, which is constrained by discretized $Hfin$ defined $Weff$. Therefore, the characterization of $Wsheet$ impact is critical for device-circuit interaction.

(2) Due to the uniqueness of the integration process, where Si channel is formed by epitaxy and release of sacrificial SiGe layers [2], the measured mobilities need to be examined and compared with the entitlements.

(3) Tsi reduction is needed along with Lg scaling to maintain proper electrostatics. Quantization effects for $Tsi < 5nm$ becomes severe in ultrathin body SOI [4] and FinFET [11,12]. It is therefore instrumental to study the impacts of thin Tsi to Nanosheet.

In this paper, for the first time, above questions will be addressed by experimental data. The measured Nanosheet μ_e and μ_h and their dependency on $Wsheet$ are reported, and "*narrow sheet effect*" is first observed. Next, since hole transport on Nanosheet will be dominated by (100) plane, which has lower μ_h than (110) plane, it is prudent to have comprehensive study of Tsi impact on μ_h and Vt of PFET,

and the data will be discussed in the second part. (μ_e impact is more complicated [7] and beyond the scope of this paper).

Experimental Setup

Nanosheet devices are fabricated on (100) bulk Si wafers with <110> transport direction. The main fabrication process steps are described in [2]. A channel length of 100 nm is chosen for IV and CV measurements for mobility extraction, to highlight long channel behavior while also assuring no deformation or bending of suspended sheets [2].

Fig. 1(a) shows the schematic top-down view of Nanosheet. $Wsheet$ is defined as the width of the sheet perpendicular to the S-D direction. Fig. 1 (b) is the perpendicular sheet cross-section schematic, showing three suspended Si channels, with Tsi as the Si channel thickness. The sheet-to-sheet spacing, or the suspension spacing, is defined as $Tsus$. It is to note that after RMG process, the Si channel will be surrounded by high-k and WFM, thus filling the suspension space. Fig. 1(c) shows the perpendicular sheet TEM. Note that Tsi is uniform across the sheet width.

Results and Discussions

1) Mobilities and Narrow Sheet Effect

Fig. 2 shows the measured μ_e and μ_h with nominal Tsi, $Wsheet$ = 20 nm and Lg = 100 nm. The electron-to-hole mobility ratio is close to 2:1, which is higher than FinFET, and expected to have $Wsheet$ dependency. This is attributed to Nanosheet channel geometry having more (100) than (110) plane conduction, which has higher μ_e than μ_h .

μ_e is lower than pure (100) plane mobility and μ_h is higher than pure (100) plane reported in [10], and can be explained by Nanosheet conduction has contributions from both the (100) surface and the (110) sidewall (or other plane orientations if the corner is rounded).

Figs. 3 and 4 show the IV and gm of NFET and PFET, respectively, for $Wsheet$ = 20 nm and 40 nm. NFET peak gm decreases with $Wsheet$, while PFET shows the opposite trend. This can be understood by reduction in (100) plane contribution for narrow $Wsheet$.

The observed "*narrow sheet effect*" is unique in Nanosheet due to channel geometry, but is expected to diminish as $Wsheet$ increases. This effect needed to be modelled, and may impose new device-circuit interactions and circuit design considerations. However, as mentioned above, continuously variable $Wsheet$ offering provides flexibility and creates unique opportunities for current matching, circuit design, and power-performance co-optimization.

978-1-7281-1988-5/18 $31.00 © 2018 IEEE

2) T_{si} impact on hole mobility

To study the impact of Tsi on μ_h, different Si channel thicknesses are grown by a tightly controlled and optimized epitaxial process. The final Tsi thickness is based on TEM measurements. The schematic is shown in Fig. 5. $Tsus$ is kept constant across all Tsi splits, to avoid Vt variations [3] and impact to inner-spacer formation. The small differences in total stack height (due to change in epitaxial Si thickness) are well absorbed by robust process margin downstream. The design of this experiment is highly immune to extraneous variables, and as shown in Fig. 1(c), Tsi is very uniform along $Wsheet$ (unlike FinFETs which have a tapered fin profile), making this design of experiment ideal for fundamental study of Tsi and $Wsheet$ impact.

The source/drain extension region thickness and external resistance ($Rext$) under the spacer/inner-spacer is also determined by Tsi, thus affecting mobility readout without proper $Rext$ correction. In addition, for FinFET, mobility extraction can be measured on long channel devices; for Nanosheet, Lg is conservatively set to 100 nm, to avoid mechanical sheet bending. As a result, $Rext$ is a non-insignificant portion of total resistance, and could therefore affect the extraction of mobility.

2a) External resistance correction method

To minimize the impact of $Rext$, on state linear resistance (Vds = 50 mV) is plotted vs. $1/Vod$ (where Vod is the overdrive voltage) and the y-intercept is the gate-bias-independent external resistance. This intercept is subtracted from the curve to isolate gate-bias-dependent resistance, and can be referred to as channel resistance or Rch. It is assumed here that any gate-bias-dependent $Rext$ is a small fraction of the remainder, owing to the large Lg = 100 nm. Mobility extraction is calculated with and without $Rext$ correction. Fig. 7 shows the overdrive current vs. high field mobility without $Rext$ correction, which shows scattering of data points. Fig. 8 shows the same after $Rext$ correction. A highly correlated trend and clear separation between splits are observed. This indicates $Rext$ correction is needed and a better methodology to study and compare intrinsic mobility.

2b) Results

Fig. 6 shows the measured transfer characteristics of p-type Nanosheet devices with different Tsi. Inset shows the linear IV curves. Fig. 9 shows delta Vt vs Tsi is in good agreement with [11,12]. As Tsi reduces, Vt increases due to quantum confinement causing ΔE in the valence band and degradation of hole mobility [4-6]. Vt increases more rapidly for Tsi below certain thickness, due to increase in confinement effects. Fig. 10 shows near ideal sub-threshold swing, indicating excellent gate dielectric interface quality which does not corrupt the observed change in Vt with Tsi.

Normalized Ron, Rch, and $Rext$ are plotted vs Tsi in Figs. 11-13. Ron increases as Tsi decreases, due to an increase in both Rch and $Rext$. Rch increases due to mobility degradation. $Rext$ starts to increase for Tsi below certain thickness, because the extension thickness is dictated by

grown Si epitaxy. As Tsi shrinks, current crowding increases in the extension region, thus increasing $Rext$.

Fig. 14 shows the measured room temperature CV with different Tsi. Fig. 15 shows the first derivative of CV (dCg/dV) vs Vg. The increase of derivative of Cg is expected with thinner Tsi [6,8]. Fig. 16 plots the μ_h vs inversion carrier charge density and Fig. 17 plots the peak and high field μ_h for different Tsi. The degradation of μ_h is attributed to increase phonon scattering with thinner Tsi [4,9].

Fig. 18 shows PFET Vod at $Ninv$ = 1E13/cm^2 for different Tsi. The reduction of Vod is attributed to better electrostatics with thinner Tsi. Fig. 19 shows delta $Tinv$ vs. $Vod@Ninv$ = 1E13/cm^2. The smaller Vod required to achieve the same inversion charge with thinner Tsi correlates with extracted $Tinv$. Fig. 20 shows the simulated hole density at the same Vod for different Tsi, illustrating better elestrostatics with reducing Tsi. For short channel devices, co-optimization of mobility and SCEs by Tsi with will be critical.

Another factor affecting μ_h is $Wsheet$. Fig. 21 plots the percentage contribution of (100) plane to total $Weff$ as a function of Tsi, assuming a rectangular sheet. The (100) plane contribution increases as Tsi reduces, which is an additional factor for μ_h degradation unique to Nanosheet, besides increase in phonon scattering. As a result, the percent-wise degradation in μ_h vs. Tsi for $Wsheet$ = 40 nm will be smaller than $Wsheet$ = 20 nm, due to the different slope shown in Fig. 21. This is purely due to geometric difference. Phonon scattering with thinner Tsi should have no $Wsheet$ dependency if $Wsheet$ is wide enough.

Fig. 22 (a) (b) plots the measured PFET gm,max at various Tsi for $Wsheet$ = 20 nm and 40 nm normalized to $Wsheet$ = 20 nm, Tsi = 9 nm. The change in gm,max with Tsi is larger for $Wsheet$ = 20 nm than $Wsheet$ = 40 nm, supporting the assertion that both phonon scattering and geometry effects play a role in the change in hole mobility with thinner Tsi.

Conclusion:

Measured mobilities of Stacked Gate-All-Around Nanosheet at Lg = 100 nm and $Wsheet$ = 20 nm are reported with ~ 2:1 electron-to-hole mobility ratio, and shown to be $Wsheet$ dependenent because of *"narrow sheet effect"*. This presents unique opportunities when combined with continuosuly variable $Wsheet$ offering for current matching, circuit design, and power-performance co-optimization.

Acknowledgment

This work was performed by the Research Alliance teams at various IBM Research and Development Facilities.

References

[1] S.D. Kim et al., *S3S*, p.1-3, 2015. [2] N. Loubet et al., *Symp. VLSI Tech.*, T230, 2017. [3] J. Zhang et al., *IEDM.*, p22-1, 2017 [4] K. Uchida eta al, IEDM., p29-4, 2001. [5] K. Uchida et al., *IEDM,* p47-50, 2002. [6] G. Tsutsui et al., IEDM, p836-838, 2005 [7] S. Takagi et al., *IEDM*, p219-222, 1997 [8] Z. Ren et al, *EDL* p609-611, 2002. [9] D. Esseni et al., IEDM, p671-674, 2000 [10] M. Yang et al., *EDL*, p339-341 [11] X. He et al, IEDM, p20-2, 2017 [12] JB Chang et al., *VLSI Tech.*, p12-13, 2011.

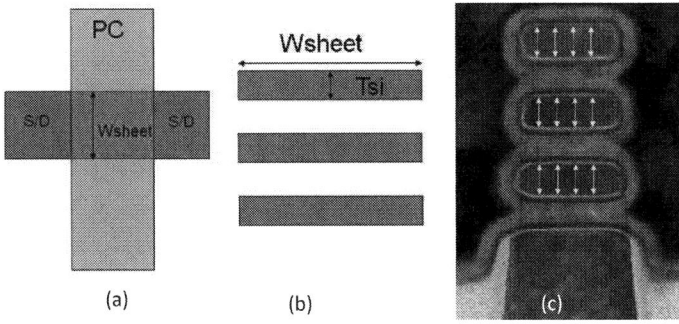

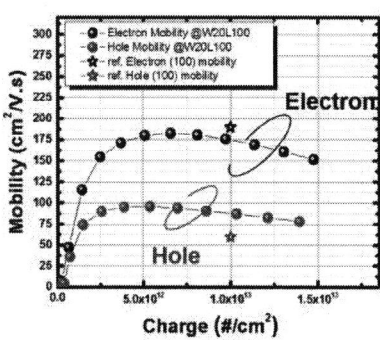

(a) (b) (c)

Fig. 1 **(a)** Schematic of topdown view of Nanosheet. (b) Schematic of across sheet corss-section with three stacked nanosheets. The channel thickness is defined as *Tsi*, and the width of the sheet is named *Wsheet*. (c) TEM cross-section of nanosheet. The sheet width is uniform across the *Wsheet* direction.

Fig. 2 Measured electron and hole mobilities versus inversion carrier density with nominal *Tsi* thickness.

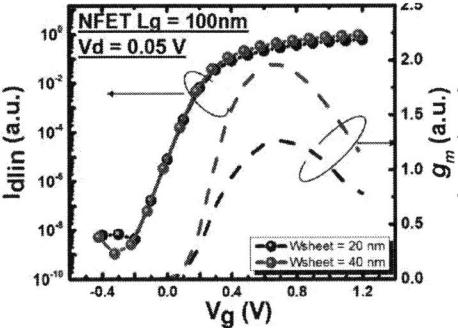

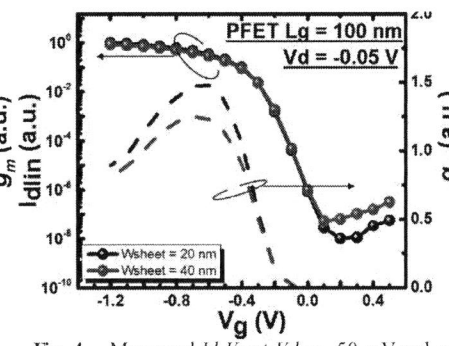

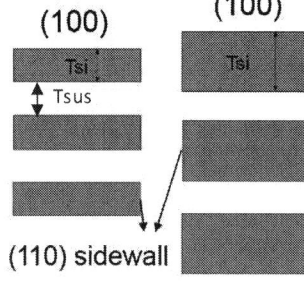

Fig. 3 Measured *Id-Vg* at *Vds* = -50 mV and *gm* of NFET with *Wsheet* 20 nm and 40 nm. Note: *gm* is higher for *Wsheet* = 40 nm than for 20 nm, attributed to more (100) plane conduction.

Fig. 4 Measured *Id-Vg* at *Vds* = -50 mV and *gm* of PFET with *Wsheet* = 20 nm and 40 nm. Note: *gm* is lower for *Wsheet* = 40 nm than for 20 nm, attributed to more (100) plane conduction.

Fig. 5 Schematic cross-sectional view for two Nanosheet stacks with different *Tsi*.

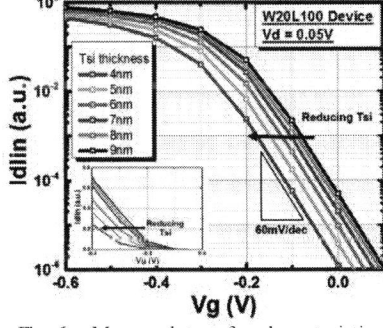

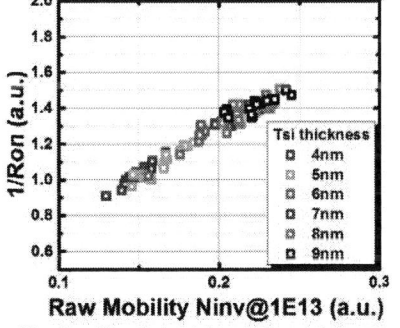

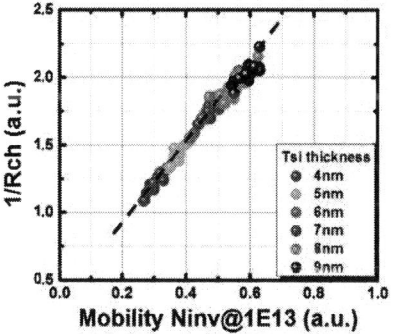

Fig. 6 Measured transfer characteristics of p-type Nanosheet devices with varying *Tsi* from 4 nm to 9 nm. Inset shows linear *Id-Vg*.

Fig. 7 Correlation chart of overdrive current vs. high field mobility (*Ninv*@1E13) without *Rext* correction.

Fig. 8 Correlation chart of overdrive current vs. high field mobility (*Ninv*@1E13) after *Rext* correction.

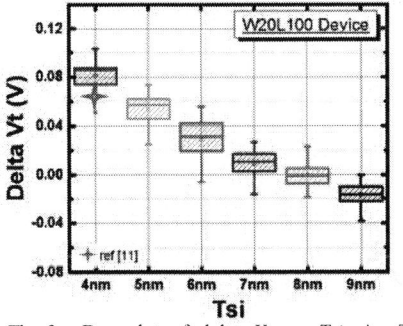

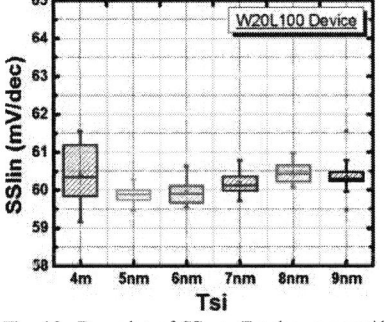

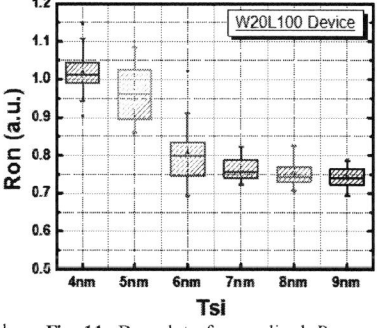

Fig. 9 Box plot of delta *Vt* vs. *Tsi*. As *Tsi* reduces, *Vt* increases rapidly for *Tsi* < 7 nm.

Fig. 10 Box plot of *SS* vs. *Tsi* shows near ideal sub-threshold swing indicating excellent gate dielectric interface quality.

Fig. 11 Box plot of normalized *Ron* vs *Tsi*.

978-1-7281-1988-5/18 $31.00 © 2018 IEEE   654

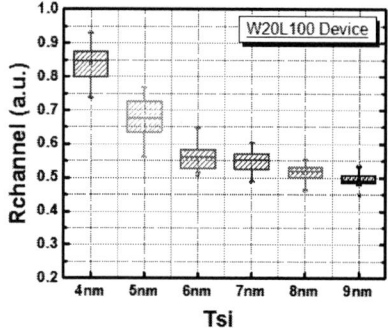

Fig. 12 Box plot of normalized *Rch* vs. *Tsi*.

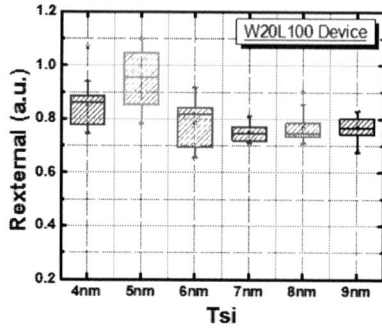

Fig. 13 Box plot of normalized *Rext* vs. *Tsi*.

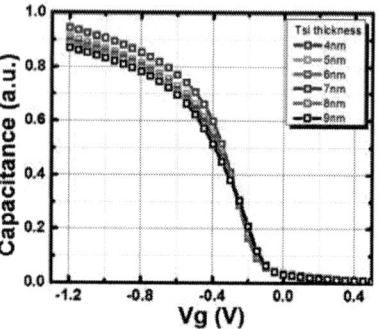

Fig. 14 Measured room temperature *C-V* for different *Tsi*. *Lg* = 100 nm, *Wsheet* = 20 nm.

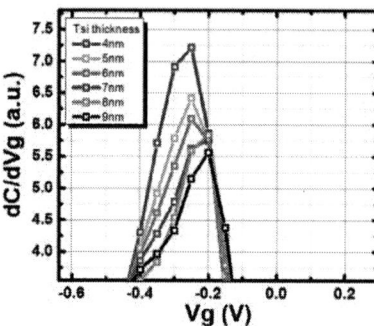

Fig. 15 *dC/dVg-Vg*. The peak value increases with *Tsi*.

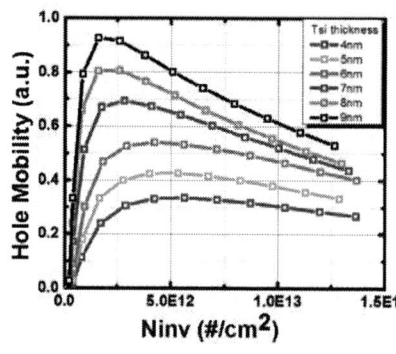

Fig. 16 Measured room temperature hole mobility (normalized) vs. inversion carrier charge density.

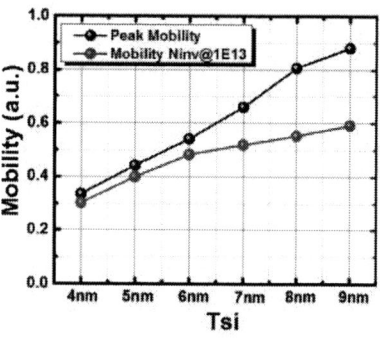

Fig. 17 Extracted peak hole mobility and hole mobility for *Ninv*@1E13/cm².

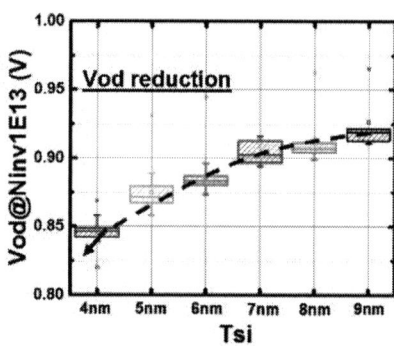

Fig. 18 Measured *Vod* @*Ninv* = 1E13/cm².

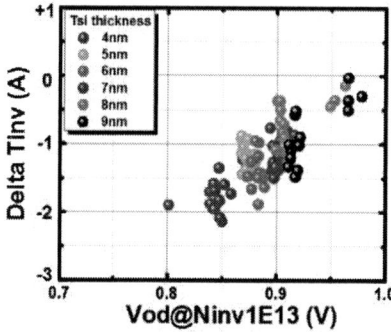

Fig. 19 Measured pFET delta *Tinv* at *Vod*@*Ninv* = 1E13/cm².

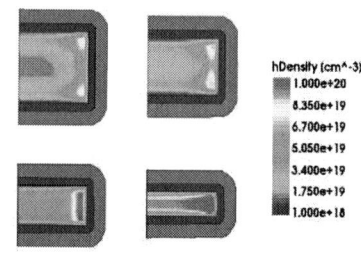

Fig. 20 Simulated hole density at *Vod* = 0.7 V for *Tsi* from 9 nm to 3 nm.

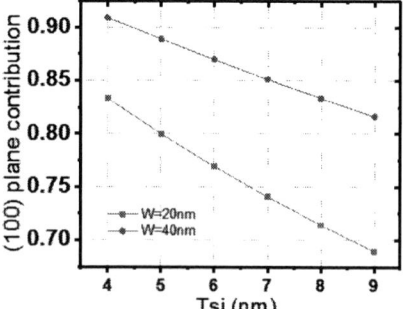

Fig. 21 Calculated (100) plane contribution to total *Weff* as a function of *Tsi*.

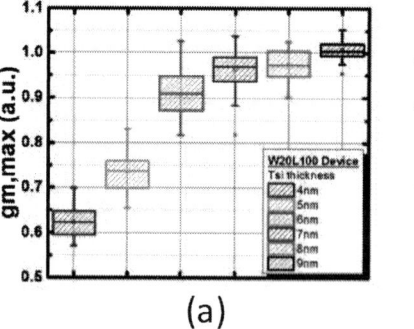

Fig. 22 Measured PFET *gm,max* vs. *Tsi* for (a) *Wsheet* = 20 nm, (b) *Wsheet* = 40 nm. Less *gm,max* degradation with thinner *Tsi* is observed for *Wsheet* = 40 nm.

3nm GAA Technology featuring Multi-Bridge-Channel FET for Low Power and High Performance Applications

Geumjong Bae, D.-I. Bae, M.Kang, S.M.Hwang, S.S.Kim, B.Seo, T.Y.Kwon, T.J.Lee, C.Moon, Y.M.Choi, K.Oikawa, S.Masuoka, K.Y.Chun, S.H.Park, H.J.Shin, J.C.Kim, K.K.Bhuwalka, D.H.Kim, W.J. Kim, J.Yoo, H.Y.Jeon, M.S.Yang, S.-J.Chung, D.Kim, B.H.Ham, K.J.Park, W.D.Kim, S.H.Park, G.Song, Y.H.Kim, M.S.Kang, K.H.Hwang, C.-H.Park, J.-H.Lee, D.-W. Kim, S-M.Jung, H.K.Kang

Samsung R&D Center, Samsung Electronics, Hwasung-City, Gyeonggi-Do, Republic of Korea,
email: goldenbell.bae@samsung.com

Abstract— As the most feasible solution beyond FinFET technology, a gate-all-around Multi-Bridge-Channel MOSFET (MBCFET) technology is successfully demonstrated including a fully working high density SRAM. MBCFETs are fabricated using 90% or more of FinFET processes with only a few revised masks, allowing easy migration from FinFET process. Not only on-target but also multiple Vt is achieved in challengingly limited vertical spacing between channels. Also, reliability of MBCFETs is shown to be comparable to that of FinFETs. Three representative superior characteristics of MBCFET compared to FinFET have been demonstrated — better gate control with 65 mV/dec sub-threshold swing (SS) at short gate length, higher DC performance with a larger effective channel width (Weff) at reference footprint, and design flexibility with variable nanosheet (NS) widths. The optimization of the standard cell design by using variable NS width is evaluated. The usefulness of MBCFET as a multi-purpose performance provider is proven by the modulation of effective capacitance (Ceff), effective resistance (Reff) and frequency by Weff control. Finally, mass production feasibility with MBCFET is proven through a fully working high density SRAM circuit.

I. INTRODUCTION

FinFET technology which was introduced to overcome limitations of planer technology is facing critical scaling challenges. As fabrication becomes increasingly complex, performance gain with continued scaling poses many difficulties and roadblocks. For the extension of FinFET technology, Fin width scaling for better gate controllability, Fin pitch scaling for capacitance reduction, and Fin height increase for DC current increase have been crucial factors [1-3]. Currently they are all facing the lack of process margin in areas of Fin etching, cutting, leaning and so on. Meanwhile, effective channel width (Weff) increase is continuously demanded for the performance improvement, although the cell height is reduced in the scaled technology nodes (Fig. 1). This has been achieved mainly by Fin height increase. However, the resulting performance improvement is less than expected because of parasitic capacitance increase. In addition, having a discrete number of Fins restricts the design optimization of standard cells and SRAM cells in terms of the balance between effective capacitance (Ceff) and effective resistance (Reff). Multi-Bridge-Channel MOSFET (MBCFET), having a vertically stacked NS and a Gate-All-Around (GAA) structure, has been proposed as a promising candidate for replacing FinFETs in the future nodes [4-7]. MBCFETs can have superior DC performance and better short channel control compared to FinFET due to the uniform channel thickness, the larger Weff and the GAA structure. The wide range of variable nanosheet (NS) widths, as opposed to the discrete number of Fins for FinFET, can also give the additional benefit in the design flexibility. Not only the performance gains but also the easy migration from the previous node should be considered for the mass production. The reusability of process and circuit design is also considered in this paper.

II. DEVICE DESIGN AND FABRICATION

The structural difference between FinFET and MBCFET is compared in Fig. 2. Contrary to vertically placed 3-dimensional channels of FinFETs, GAA channels are horizontally placed in MBCFETs. As a limitation of vertically placed channel, all Fins have a fixed height, and only a discrete number of Fins can be placed in a standard cell. On the contrary, various NS widths of MBCFET can be fabricated in the same wafer using direct patterning (Fig. 3). In addition, thin uniform thickness channel formed by epitaxial growth results in less variation than channel formed by etching.

Considering easy migration of process from previous technology, most FinFET processes are reused. On top of the FinFET baseline process, less than 10% of the processes are modified for MBC structure formation. The fabrication steps are as follows. Several pairs of Si channel and sacrificial layers are sequentially grown by epitaxial growth. Conventional Fin formation processes are used for channel definition and STI process. After Fin formation, similar processes are used for dummy poly gate (PC) formation. During the conventional source-drain process, inner-spacer is formed for parasitic capacitance reduction and internal gate length definition at the source-drain module [8]. Sacrificial layers located between the channels are selectively removed at the Replacement Metal Gate (RMG) module for channel formation [9]. For the evaluation of sacrificial layer removal, gate conductance curves are compared with two different removal conditions (Fig.4). Contrary to humped curve induced by residue of sacrificial layer, the optimized process has clean conductance curve which means the full removal of sacrificial layer. Conventional IL/high-k and Work Function Metal (WFM) processes are used to meet the Vt target of the CMOS and multi Vt [10]. The following MOL and BEOL processes are with the same as the conventional FinFET ones.

III. ELECTRICAL CHARACTERISTICS

The excellent gate controllability with very short gate length for both N- and PMOS devices is demonstrated by the MBCFET GAA structure as shown in Fig. 5. N- and PMOS sub-threshold swing (SS) of 65 and 67 mV/dec are achieved, respectively. In addition, improved N- and PMOS DC performance relative to FinFET is also demonstrated with MBCFET due to larger Weff and higher carrier mobility as well as better short channel control of the GAA structure. Fig. 6 shows a typical N- and PMOS Idsat-Idoff correlation, demonstrating 31% and 26% Idsat improvements, respectively, relative to FinFET with the same CPP and standard cell height. The reason for smaller DC improvement of PMOS compared to that of NMOS comes from the mobility difference due to the change in channel orientation from (110) to (100) [11].

For the vertically stacked NS structure, minimum vertical space between NS to NS is the key parameter to minimize the parasitic capacitance [12]. By full elimination of sacrificial layers and WFM and deposition of different types of WFM in a limited space, CMOS Vt targeting has been demonstrated (Fig. 7). On top of CMOS Vt process, three different flavors of Vt for N- and PMOS are achieved with modulation of WFM engineering (Fig. 8). High selectivity and a stable etch rate during the etch process for the various combination of WFM are inevitable processes to make different flavors of Vt.

Considering easy migration, modified standard cell from FinFET is used for evaluation. Contrary to restricted cell design of FinFET induced by the discrete number of Fins and fixed Fin pitch, flexible cell design with MBCFET is possible utilizing variable NS widths [13]. Considering the purpose of each device, wider or narrower NS widths can be used within the active area. Modification of NS size and position within the active region is also conceivable method for optimized cell design.

Modulations of DC performance and frequency using various NS widths are demonstrated on a single wafer (Figs.9, 10). Due to Weff increase without short channel effect degradation, DC performance can be improved by NS width increase. Frequency can be also improved by NS width increase because of the larger reduction of AC Reff than the increase in Ceff with increased Weff as shown in Fig. 11. Due to the relatively increased parasitic resistance at wider NS

width, however, speed gain continuously decreases as NS width increases further. Using larger modulation of the frequency induced by freedom of circuit design, multi-purpose performance provider such as low power device or high performance device can be easily supported.

As shown in Figs. 12, comparable time dependent dielectric breakdown (TDDB) to those of FinFET is achieved with a reliable MBC formation. In addition, self-heating of MBCFET is investigated with 3D TCAD simulation by using the spherical harmonics expansion Boltzmann transport equation (SHEBTE) model. Channel peak temperature is analyzed at various test frequencies, while applying periodic pulses. Similar to SOI devices, the peak temperature of MBCFET converges to that of FinFET as the pulse frequency increases, resulting in no significant difference at a typical circuit speed (Fig. 13).

IV. MBCFET SRAM

To check the feasibility of mass production with MBCFET, high density SRAM macro is fabricated, in which all the 6T SRAM and peripheral circuits are composed of MBCFETs. The butterfly curves of the HC SRAM cell are measured at different voltages ranging from 0.4 to 1.1V. The Static Noise Margin (SNM) of 0.18V and 0.13V is achieved for 0.7V and 0.5V operation, respectively, which is comparable to that of FinFET SRAM with a similar footprint (Fig. 14). The Shmoo plot of high density HC SRAM macro illustrates full read and write capability with large V_{DD} margin (Fig. 15).

V. CONCLUSIONS

Highly manufacturable 3nm MBCFET technology is presented. Three representative superior characteristics compared to FinFET are demonstrated with hardware—near ideal SS of 65 mV/dec and 31% higher on current with a larger Weff at reference footprint and design flexibility with variable NS widths. Easy process and cell design migration from FinFETs are demonstrated using optimization based on the Fin process and mask set. Key reliability characteristics of MBCFET are also confirmed with reliable GAA processes. Finally, mass production feasibility with MBCFET at the 3 nm technology node is proven through a fully working high density SRAM circuit.

REFERENCES

[1] H.-J. Cho et al., Symposium on VLSI, 2016
[2] Daewon Ha et al., Symposium on VLSI, 2017
[3] W.C.Jeong et al., Symposium on VLSI, 2018
[4] S-Y Lee et al., Trans on Nanotech, 2003
[5] S-Y Lee et al., Symposium on VLSI, 2005
[6] M.Kim et al., Symposium on VLSI, 2006
[7] N. Loubet et al., Symposium on VLSI, 2017
[8] H. Mertens et al., IEDM, 2017
[9] S.Barraud et al., IEDM, 2016
[10] H. Mertens et al., IEDM, 2016
[11] Guangyu Sun et al., Journal of Appllied Physics, 2007
[12] S-D Kim et al., SOI-3D-Subthreshold Microelectrics, 2015
[13] S.Barraud et al., IEDM, 2017

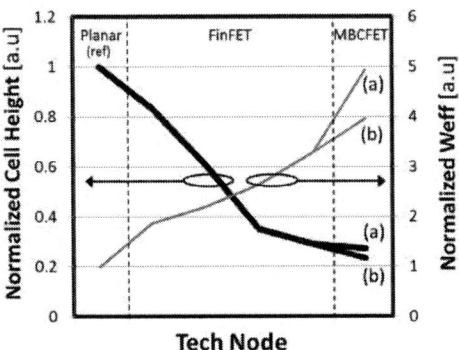

Fig. 1. Cell height and effective width trend with technology node scaling. (a) MBCFET with same design rule as FinFET (b) MBCFET with optimized standard cell design by variable NS width.

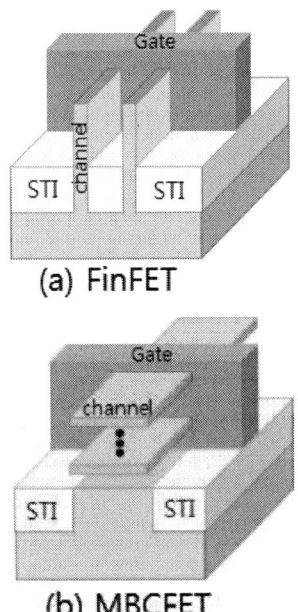

Fig. 2. Schematic comparison between (a) FinFET with vertical channel and (b) MBCFET with horizontal channel. Multiple numbers of NSs are vertically stacked.

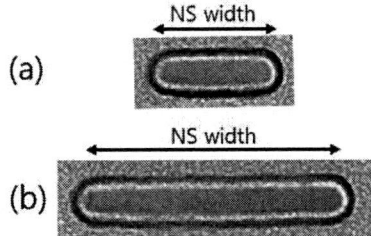

Fig. 3. Variable NS width formed by direct patterning. (a) Narrow width NS for high density SRAM or low power devices, (b) Wide width NS for high performance devices.

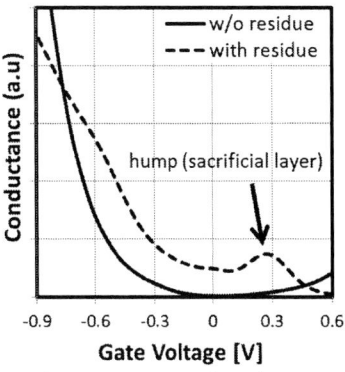

Fig. 4. Gate conductance comparison between two different sacrificial layer removal conditions. Clear curve without hump is proving the full removal of sacrificial layers.

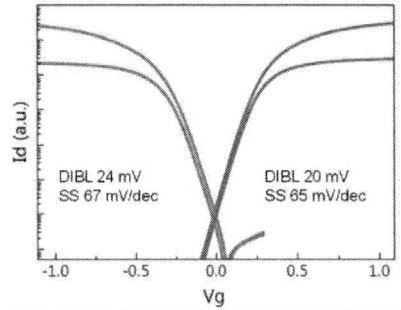

Fig. 5. Superior short channel effect with Gate-All-Around structure; N- and PMOS subthreshold swing of 65mV/dec and 67mV/dec, respectively.

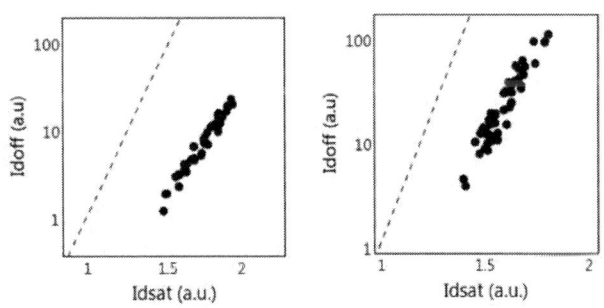

Fig. 6. MBCFET showing Idsat improvement of NMOS 31% and PMOS 26% compared to FinFET with the same CPP.

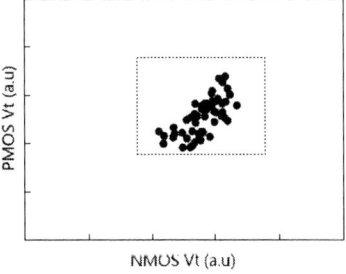

Fig. 7. Both NFET and PFET meet Vt targets with conventional WFM processing.

978-1-7281-1988-5/18 $31.00 © 2018 IEEE   658

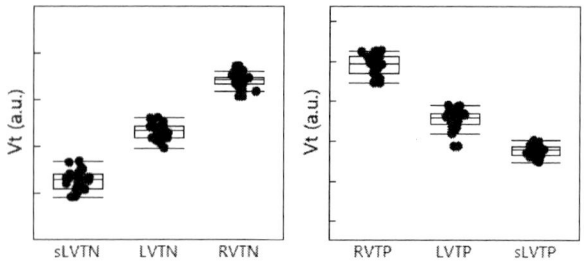

Fig. 8. Three different flavors of Vt are demonstrated with different WFMs.

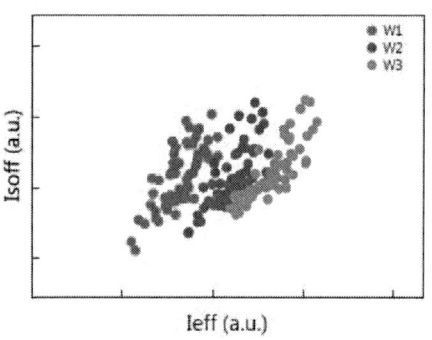

Fig. 9. DC performance modulation as a function of NS widths.

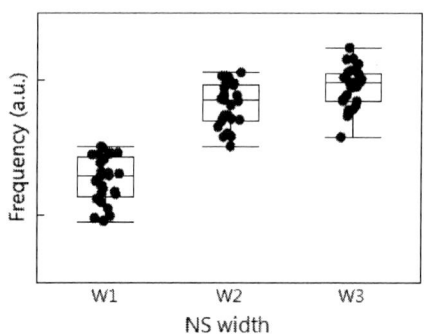

Fig. 10. Frequency modulation as a function of NS widths.

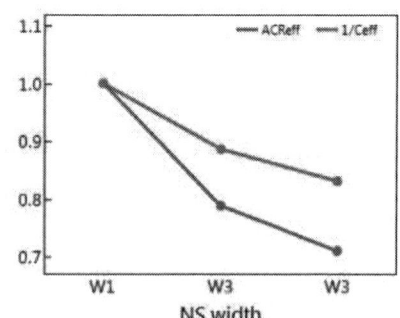

Fig. 11. ACReff and Ceff modulated by NW widths.

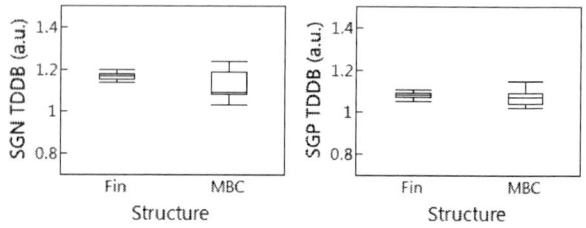

Fig. 12. Comparable TDDB characteristics of MBCFET NMOS and PMOS to those of FinFET.

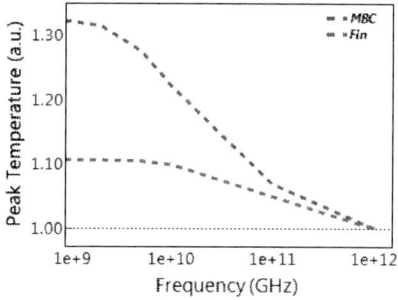

Fig. 13. Comparison of peak temperature between FinFET and MBCFET by 3D TCAD simulation.

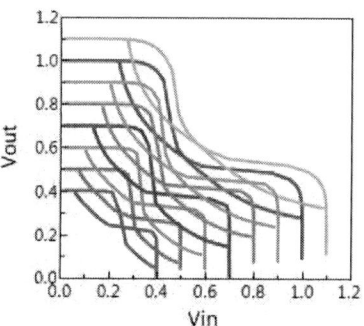

Fig. 14. SNMR characteristics of MBCFET SRAM cell. All the 6T SRAM and peripheral circuits are composed of MBCFETs

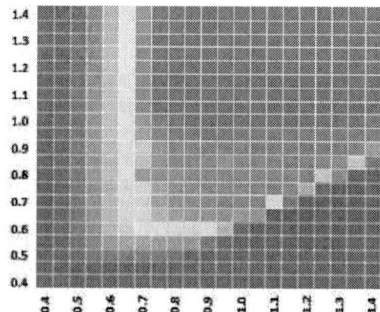

Fig. 15. Shmoo plot of high density SRAM macro having MBCFETs for SRAM cells and periphery devices.

978-1-7281-1988-5/18 $31.00 © 2018 IEEE   659

A CMOS Proximity Capacitance Image Sensor with 16μm Pixel Pitch, 0.1aF Detection Accuracy and 60 Frames Per Second

M. Yamamoto[1], R. Kuroda[1], M. Suzuki[1], T. Goto[2], H. Hamori[3], S. Murakami[3], T. Yasuda[3] and S. Sugawa[1,2]

[1]Graduate School of Engineering, Tohoku Univ., Sendai, Miyagi, Japan, email: masahiro.yamamoto.q6@dc.tohoku.ac.jp
[2]New Industry Creation Hatchery Center, Tohoku Univ., Sendai, Miyagi, Japan, [3]OHT Inc., Fukuyama, Hiroshima, Japan

Abstract— A 16μm pixel pitch 60 frames per second CMOS proximity capacitance image sensor fabricated by a 0.18μm CMOS process technology is presented. By the introduction of noise cancelling operation, both fixed pattern noise and kTC noise are significantly reduced, resulting in the 0.1aF (10^{-19}F) detection accuracy. Proximity capacitance imaging results using the developed sensor are also demonstrated.

I. INTRODUCTION

Capacitance sensors are used in various application fields such as consumer appliances, industrial equipment, inspection equipment, biomedical fields and so on [1-3]. Among various types of applications, proximity capacitance sensors are utilized for inspection of flat panel displays (FPDs), electrical capacitance tomography (ECT) and sensing of chemical/ biological reactions [3-6]. In those usage, a high spatial resolution in μm order, high detection accuracy in aF order and high measurement speed for real-time measurement are highly desired for sensors. Although some high sensitivity capacitance sensors exhibit aF order detection accuracies [4-5], spatial resolution and measurement speed are limited low. In order to simultaneously achieve above mentioned requirements, a high speed CMOS area sensor architecture with high sensitivity small pixels is expected to be useful. A proximity capacitance sensor using a 6.6Mpixel CMOS image sensor has been proposed for printed circuit board inspection usage [6]. In this sensor, although fixed pattern noise (FPN) was removed by an offset cancelling circuit operation [7], temporal random noise was relatively high due to the kTC noise arising in pixels.

This work presents a newly developed CMOS proximity capacitance image sensor with capacitance to voltage conversion pixels and an advanced noise cancelling operation to reduce kTC noise and FPN. A prototype sensor chip with 16μm pitch $256^H \times 256^V$ pixels was designed and fabricated by a 0.18μm CMOS process technology and its measured performances are described. Proximity capacitance imaging for 40μm pitch line/space pattern as well as real-time imaging of aqueous solution were experimented.

II. DEVELOPED CMOS CAPACITANCE IMAGE SENSOR

A. Basic structure and operation principle

Fig.1(a-c) shows the schematic illustrations of the proximity capacitance measurement method used in this work. Inside the sensor chip formed a floating detection electrode per pixel having a detection capacitance (C_C) between the constant voltage node (in this case, ground). One of the unique features of the developed sensor is to set a reference electrode in the measurement target side which induces a coupling capacitance

(C_S) between the reference and the detection electrodes to be measured. When measuring a conductive solid such as metal wiring of FPD shown in Fig.1(a), the conductive solid itself operates as the reference electrode and the C_S to each pixel is measured as illustrated in Fig.1(c). When measuring targets in liquid medium such as biological cells, the reference electrode is inserted to the liquid and C_S due to the targets is measured.

Figs.2(a) and 2(b) show a simplified schematic of the capacitance measurement circuit and a pulse timing diagram for operation, respectively. The circuit consists of C_C and C_S, a detection electrode, a reference electrode with voltage pulse supply of which amplitude is V_{IN}, a reset switch connected to the detection electrode, a source follower (SF) amplifier circuit and two parallel signal outputs for V_{OUTN} and V_{OUTS}, respectively. Here, V_{OUTN} and V_{OUTS} are the reference and signal voltage levels, respectively. At first the reference electrode is set to the first voltage level and the reset switch is turned ON to reset the detection electrode. When turning off the reset switch, kTC noise remains at the detection electrode node. Then V_{OUTN} is readout by the SF circuit. Here the V_{OUTN} contain the kTC noise and the fixed pattern noise (FPN) due to the V_{th} variation of SF transistors. Next, the reference electrode voltage is set to the second voltage level with the amplitude of V_{IN}. At this time, the voltage level of the detection electrode changes due to the series connection capacitances of C_C and C_S. Then, V_{OUTS} is readout by the same SF circuit. By taking the difference of V_{OUTN} and V_{OUTS}, the signal voltage V_{OUT} expressed by the following equation is obtained.

$$V_{OUT} = V_{OUTN} - V_{OUTS} = \frac{C'_S}{C_C + C_S} \cdot V_{IN} \cdot G_{SF} \quad (1)$$

Here, G_{SF} is the gain of SF. The kTC noise and FPN are cancelled out in V_{OUT}. By this advanced noise cancelling operation synchronized with the reference electrode voltage pulse, a high capacitance detection accuracy can be obtained.

B. CMOS proximity capacitance image sensor design

The structure and operation principle explained above were implemented into the pixels of developed CMOS proximity capacitance image sensor. Fig.3(a-b) shows the top and the cross sectional diagrams of metal wiring arrangement for the detection electrode and the guard ring in-between the pixels, and Fig.3(c) shows the pixel layout diagram, respectively. In order to improve the capacitance detection accuracy, the layout was carefully designed to minimize the value of C_C. Also, the guard ring was placed in-between the detection electrodes of pixels to improve the spatial resolution. The passivation layer above the detection electrode consists of 800nm thick SiO_2 and

978-1-7281-1988-5/18 $31.00 © 2018 IEEE

200nm thick SiN films on the top. A pixel consists of a detection electrode, C_C, a reset gate (R), a SF driver (SF), and pixel select switch (X). Here the SF is inside the isolated p-well in the n-well for the following two reasons; to obtain the unity gain of SF (G_{SF}=1) by connecting the body and the source electrodes, and to form bipolar protection diodes connected to the detection electrode. An excess amplitude V_{IN} is limited by the introduction of these protection diodes. In this work the pixel pitch was set to 16μm in order to place the detection electrode with sufficiently large area ($12μm^H \times 12μm^V$) which determines the value of C_S. However, the pixel size can be miniaturized as the pixel layout has enough margin.

Fig.4 shows the circuit block diagram of the developed prototype CMOS proximity capacitance image sensor. In order to achieve a high frame rate with scalable architecture, the readout circuitry similar to a CMOS active pixel image sensor was employed. The sensor circuit consists of an array of $256^H \times 256^V$ pixels, vertical and horizontal scanning circuits, column parallel current sources for the SF circuits, column parallel sample/hold circuits for V_{OUTN} and V_{OUTS}, and output buffers. The developed sensor chip has a flexible design scalability, the number of pixels can be increased in accordance to the application. The prototype sensor chip was fabricated by a 0.18μm node 1-poly-Si 5-metal layer CMOS process technology. Fig.5 shows the micrograph of the fabricated CMOS proximity capacitance image sensor. The die size is $4.8mm^H \times 4.8mm^V$. The power supply voltage is 3.3V.

III. MEASUREMENT RESULTS

Fig.6 shows the developed setup to measure the sensor performances. It consists of a head board with the fabricated sensor chip mounted face up, an analog front-end (AFE) circuit board with voltage regulators and a 14bit differential ADC directly connected to V_{OUTN} and V_{OUTS}, a FPGA board to supply operation pulses to the sensor chip, and a PC. A manipulator with tungsten probe was set as a reference electrode.

Figs.7 and 8 show the background signal images without a target, FPN and temporal random noise without and with the introduced noise cancelling operation, respectively. Without the noise cancelling operation, the input referred FPN and temporal random noise mainly due to the SF V_{th} variation and the kTC noise remained at the detection electrode were 13.5mV$_{rms}$ and 3.00mV$_{rms}$, respectively. By the introduction of the noise cancelling operation, the FPN and temporal noise were reduced to 56.1μV$_{rms}$ and 321μV$_{rms}$, respectively. Significant reductions in both FPN and temporal random noise were achieved by the introduced noise cancelling operation. In order to measure the detection accuracy of the developed sensor chip, the temporal random noise was measured as a function of signal averaging number as shown in Fig.9. The input referred temporal random noise in root mean square decreases as the averaging number increases and it reaches to 40.6μV$_{rms}$ at the averaging number of 1000. In order to characterize the measurement range of the developed sensor chip, the capacitance was measured when changing the distance between the probe and the sensor surface or, using saline solution as measurement target as shown in Fig.10. Here, when using the conductive saline solution, the physical distance between the reference electrode and the detection electrode is assumed to be the same as the passivation film thickness. Fig.11 shows the input referred signal voltage as a function of V_{IN} for different conditions of measurement targets. The V_{IN} was varied from 0.1 to 10.0 or 20.0V by 0.1V step. From the results, the measurement range was from 0.1aF to 10fF. Also the result indicates that a higher detection accuracy is to be obtained by increasing the V_{IN} value.

Next, the spatial resolution was examined by measuring the capacitance of line/space metal wire patterns shown in Fig.12 having 20μm width for both line and space. Here each wire is controllable to apply or not to apply the reference electrode pulse. Fig.13 shows the measured capacitance image and its cross section when the reference electrode pulse was supplied to only one of the line and others were grounded. Fig.14 shows the same set of results when the center line was grounded while the reference electrode pulse was supplied to the peripheral 6 lines on the both sides, respectively. This condition imitates that a driving pulse wire of FPD is electrically shorted to ground. Device simulations were also conducted to calculate the C_S between pixel detection electrodes and measurement target. The measured results agree well with the simulation results, validating the operation of the developed CMOS proximity capacitance image sensor and its high spatial resolution.

Last but not least, captured images of a drop of saline solution on the sensor surface are shown in Fig.15. The process of the saline solution drying out and remained salt on the sensor surface were clearly visualized with good temporal and spatial resolutions. In this experiment, the reference electrode pulse was not supplied to the saline solution. The reason why the clear images were obtained is most likely due to the capacitive coupling of the guard ring, measurement target and detection electrode but the detailed mechanism is under investigation.

The performance summary of the developed sensor chip is shown in table I.

IV. CONCLUSION

A 16μm pixel pitch 60 frames per second CMOS proximity capacitance image sensor fabricated by a 0.18μm CMOS process technology was demonstrated. By the introduction of the noise cancelling operation synchronized with the reference electrode voltage pulse, 0.1aF detection accuracy with very high spatial resolution was confirmed. Furthermore, a feasibility study for measurement of proximity target without applying the reference electrode pulse showed a promising result. The developed sensor is expected to be utilized in FPD inspection and biomedical fields.

REFERENCES

[1] Y. Ye et al., "Capacitive Proximity Sensor Array With a Simple High Sensitivity Capacitance Measuring Circuit for Human–Computer Interaction," IEEE Sensors J. vol.18, pp.5906-5914, 2018.

[2] B. Prashan et al., "Real-Time Measurements of Cell Proliferation Using a Lab-on-CMOS Capacitance Sensor Array," IEEE Trans. Biomed. Circ. S., vol.12, pp.510-520, 2018.

[3] M. Koerdel et al., "Contactless Inspection of Flat-Panel Displays and Detector Panels by Capacitive Coupling," IEEE T-ED, vol.58, pp.3453-3462, 2011.

[4] C. Elbuken et al., "Detection of microdroplet size and speed using capacitive sensors," Sens. Actuators A, Phys., vol.171, pp. 55–62, 2011.

[5] K. Mohammad et al., "Integrated 0.35 μm CMOS Capacitance Sensor with atto-Farad Sensitivity for Single Cell Analysis," 2016 IEEE BioCAS, pp.22-25, 2016.

[6] D. Scheffer et al., "A 6.6Mpixel pixel CMOS image sensor for electrostatic PCB inspection," IEEE Workshop on CCD and AIS," pp.145-148, 2001.

[7] B. Dierickx et al., "Offset free offset correction for active pixel sensors," IEEE Workshop on CCD and AIS," pp.R13-1-4, 1997.

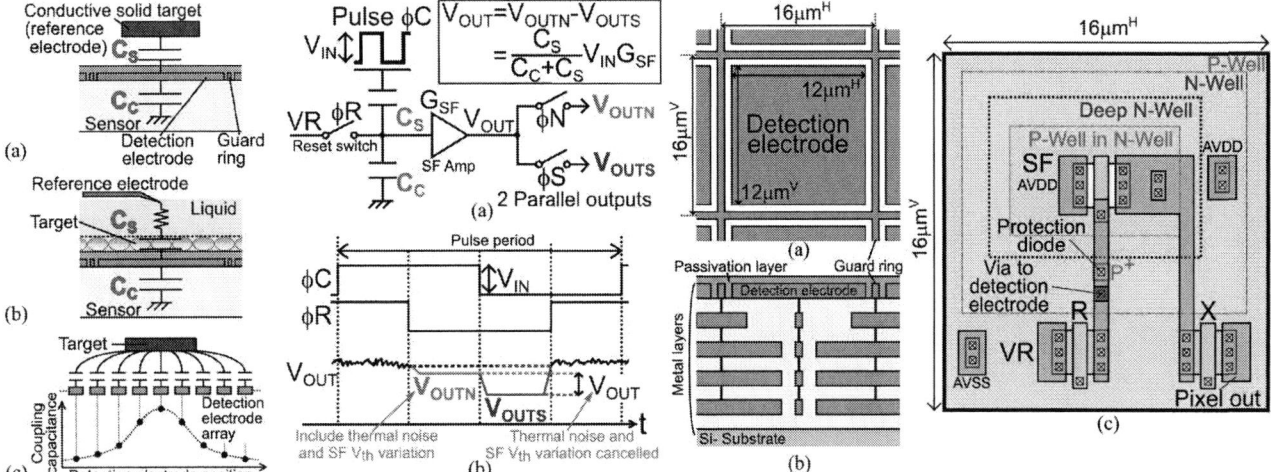

Fig. 1. Illustration of capacitance sensing method when measuring (a) a conductive solid target and (b) targets in liquid medium, and (c) shows the schematic illustrations of the proximity capacitance measurement method used in this work.

Fig. 2. Principle of sensing capacitance with noise canceling capability: (a) A simplified circuit schematic of the capacitance measurement circuit and (b) the pulse timing diagram.

Fig. 3. (a) The top and (b) the cross sectional diagrams and (c) the layout diagrams of the pixel up to the 1st metal layer.

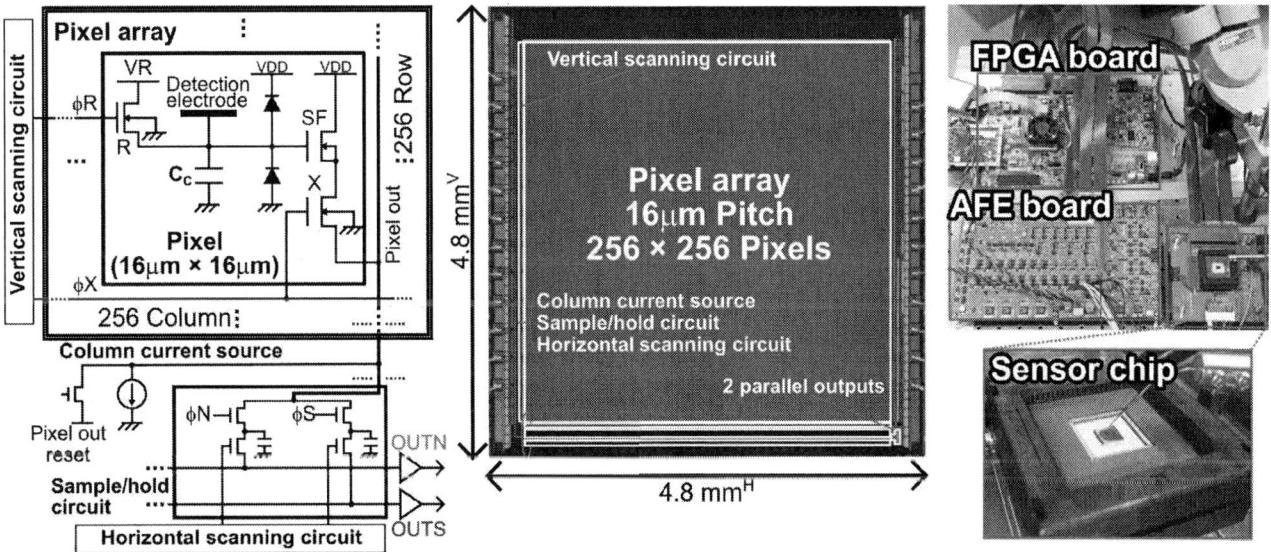

Fig. 4. The circuit block diagram of the developed prototype CMOS proximity capacitance image sensor.

Fig. 5. The micrograph of the fabricated CMOS proximity capacitance image sensor.

Fig. 6. The developed setup to measure the performance of fabricated sensor chip.

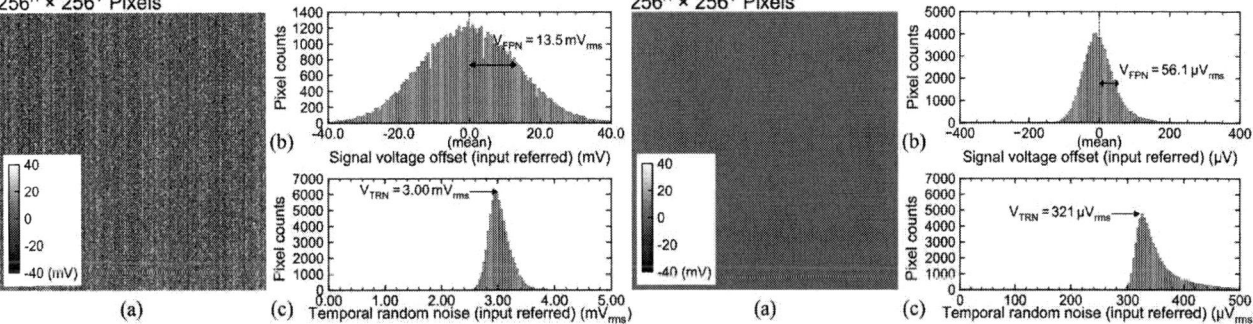

Fig. 7. (a) The background signal images without the noise canceling operation, distributions of (b) input referred signal voltage offset to show FPN, and (c) input refereed temporal random noise, respectively.

Fig. 8. (a) The background signal images with the noise canceling operation, distributions of (b) input referred signal voltage offset to show FPN, and (c) input referred temporal random noise, respectively.

978-1-7281-1988-5/18 $31.00 © 2018 IEEE

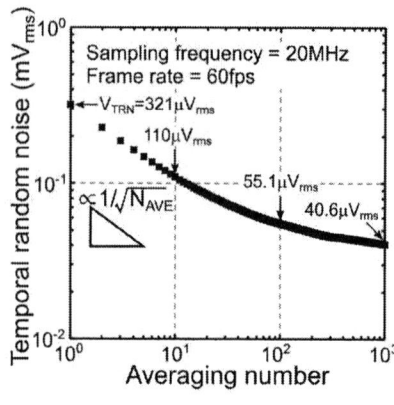

Fig. 9. The input referred temporal random noise in root mean square as a function of the averaging number.

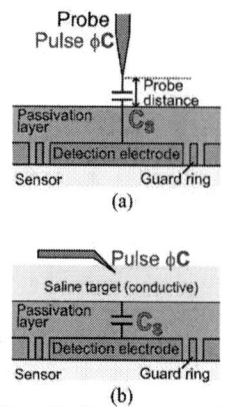

Fig. 10. Illustration of the measurement method to characterize the measurement range of developed sensor using (a) a metal probe and (b) saline solution.

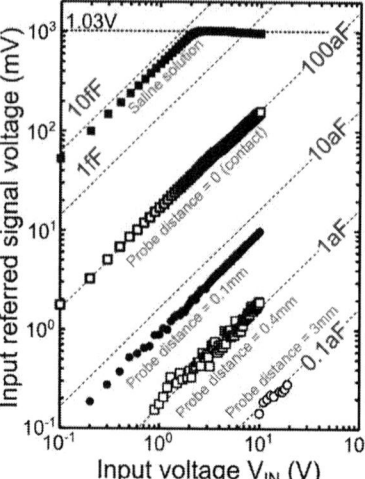

Fig. 11. The input referred signal voltage as a function of input voltage for different conditions of measurement targets.

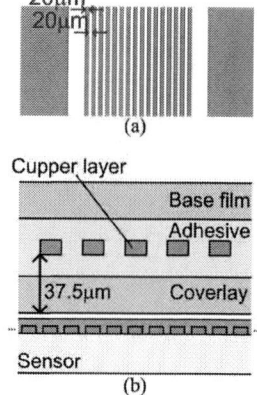

Fig. 12. The 40μm pitch line/space pattern to examine the spatial resolution of the developed sensor, (a) top view and (b) cross section view.

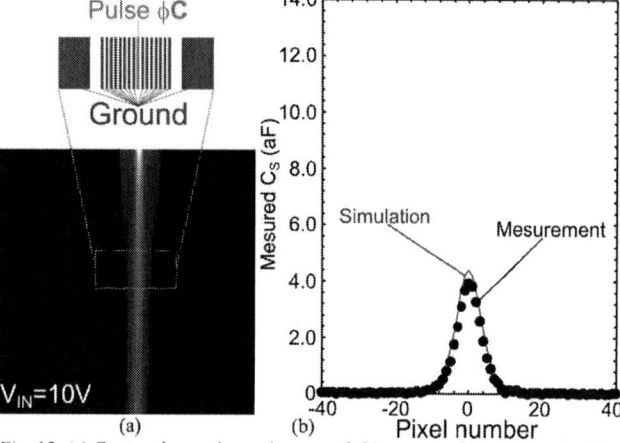

Fig. 13. (a) Captured capacitance image and (b) cross section compared with simulation result when the reference electrode pulse was supplied to only one of the line and others were grounded.

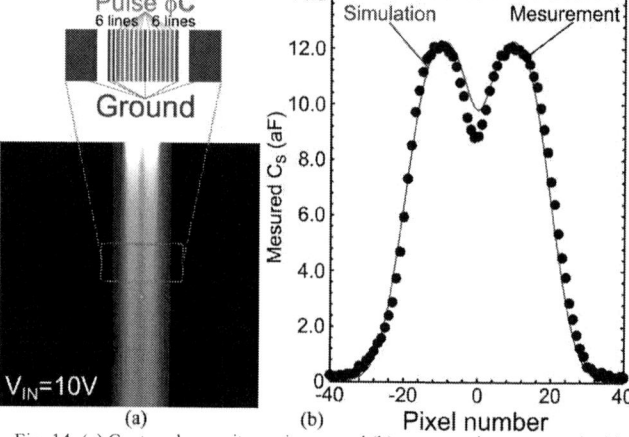

Fig. 14. (a) Captured capacitance image and (b) cross section compared with simulation result when the center line was grounded and the reference electrode pulse was supplied to the peripheral 6 lines on the both sides.

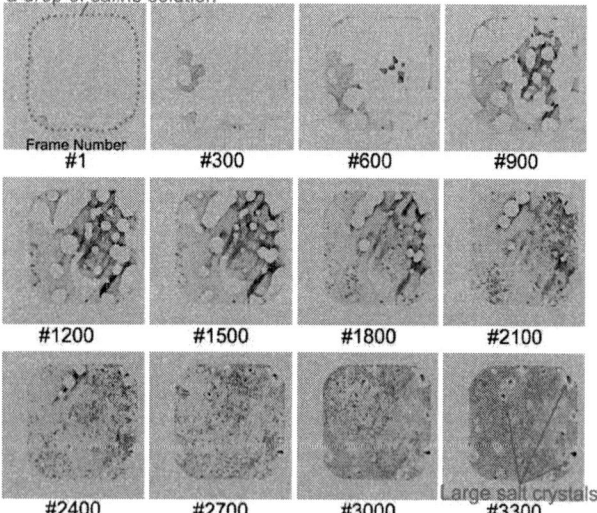

Fig. 15. Captured proximity capacitance images of a drop of saline solution on the sensor surface drying out as time advances captured by the developed sensor with 16um pitch $256^H \times 256^V$ pixels.

Table. 1. Performance summary of the developed chip.

Process	0.18μm 1P5M CMOS
Die size	$4800\mu m^H \times 4800\mu m^V$
Pixel area	$4096\mu m^H \times 4096\mu m^V$
# of pixels	$256^H \times 256^V$
Pixel size	$16\mu m^H \times 16\mu m^V$
Detection electrode size	$12\mu m^H \times 12\mu m^V$
Sampling frequency	20MHz
Frame rate	60fps
Saturation signal	1.03V (input referred)
Temporal random noise (input referred)	321μVrms (without averaging) 55.1μVrms (100 frames average) 40.6μVrms (1000 frames average)
Detection accuracy	0.1aF

978-1-7281-1988-5/18 $31.00 © 2018 IEEE

3D Expandable Microwire Electrode Arrays Made of Programmable Shape Memory Materials

Ruoyu Zhao[1*], Xin Liu[1*], Yichen Lu[1], Chi Ren[2], Armaghan Mehrsa[1], Takaki Komiyama[2] and Duygu Kuzum[1]

[1] Department of Electrical and Computer Engineering, University of California San Diego, La Jolla, CA, USA,
[2] Neurobiology Section and Department of Neurosciences, University of California San Diego, La Jolla, CA, USA.
Email: dkuzum@ucsd.edu, *These authors contributed equally

Abstract—Nitinol, a biocompatible material with shape memory effect and superelasticity, has been used in various biomedical applications. Here we demonstrate a 3D expandable nitinol microwire electrode array that can be programmed to the desired shape to conform to the brain vasculature, minimizing the vessel damage during implantation. We developed a fabrication process for precisely setting the shape of nitinol microwires and assembling them to form electrode arrays. We tested our nitinol microwire array in *in vivo* animal experiments and successfully demonstrated that our array can detect single spikes as well as local field potentials with minimum tissue and vessel damage.

I. INTRODUCTION

Implantable microelectrode arrays have been widely employed for recording various types of neural activities in basic neuroscience research and clinical studies. However, chronic stability of these implantable arrays has been significantly impeded by reactive tissue response as a result of implantation damage. Part of the tissue response arises from the vasculature damage in the brain. Insertion of electrodes leads to disruption of the blood brain barrier and hemorrhages from disrupted brain blood vessels. These hemorrhages are specifically detrimental for chronic long-term recordings. Bleeding from the blood vessels around the electrodes are known to cause extensive neuronal loss [1]. As a result, implantable array loses its ability to reliably record neural activity over time, as observed numerous times with widely adopted Utah arrays used in brain computer interface studies [2]. Therefore, it is extremely important to address vascular damage to improve the chronic reliability of implantable microelectrodes. Novel neurotechnologies, which penetrate into the neural tissue without puncturing blood vessels, are particularly needed. In order to minimize the blood vessel damage, microelectrode arrays conforming to the structure of the brain vasculature can be built (**Fig. 1**). 3D vasculature of the brain can be imaged in great detail using 2-photon microscopy [3]. A replica of major vasculature can be constructed using microfabrication or 3D printing techniques. Expandable 3D microelectrodes conforming to the surrounding vasculature, without damaging them during insertion can be developed. However, that requires electrode materials that are programmable to specific shapes and deformable to fit in small volumes for implantation.

In this work, we investigate programmable microwire electrode arrays made of a shape memory alloy, nitinol, which is a Nickel-Titanium alloy exhibiting shape memory effect and

superelasticity (**Fig. 2**). At high temperature, superelasticity allows nitinol to undergo large strain, while still able to return to the original shape. At low temperature, nitinol can be deformed to different shapes arbitrarily. When heated (i.e. body temperature), shape memory effect allows it to return to its original shape [4]. Furthermore, nitinol is biocompatible and MRI compatible [5]. Therefore, it has already been widely used in many different biomedical devices ranging from orthopaedic wires and screws, bone staples of electrodes, stents, surgical instruments, and cardiac implants [6, 7].

Here, we demonstrate nitinol microwire arrays programmable into desired shape via current application. We fabricated nitinol microwire electrodes with a diameter of ~30 μm. We performed systematic studies to understand the effect of current programing to shape memory effect and super-elasticity. The electrochemical characteristics of the microwires were characterized using impedance spectroscopy and cyclic voltammetry. In *in vivo* experiments with mice, we demonstrated successful recording of local field potentials and single neuron spiking activity from cortical layer IV neurons. Tissue damage analyses were used to examine the brain damage induced by 3D expandable microelectrode implantation.

II. NITINOL MICROWIRE ELECTRODE FABRICATION

The 16-electrode microwire bundles were prepared using cold worked nitinol wires with 23μm diameter (NiTi#1-CW, Fort Wayne Co). In order to change the cold worked wire into super elastic condition, current annealing was done by applying a 90 mA current through the wire for 20 s. After annealing, the wires became straight and possessed superelasticity. A PDMS mold with designated grooves that are 150 μm wide and 150μm deep was prepared, as shown in **Fig. 3**. The nitinol wires placed in the grooves were heated up for shape setting by applying current through joule effect. In order to find the optimal current amplitude and duration for shape-setting, we investigated the effect of current amplitude and duration on the bending angles. As shown in **Fig. 4**, the bending angle increases towards the target angle with higher current amplitude and duration. However, if the amplitude and the duration are too large, the wires overheat and lose superelasticity. Therefore, we set the current amplitude and duration to 155 mA and 10 s to achieve reliable shape-setting. As shown in **Fig. 5**, the bending angles have small variance across different wires. After shape-setting, 4.5 μm thick Parylene-C was coated on the wires as the insulation layer. The integrity of Parylene-C insulation layer is inspected by scanning electron microscopy (SEM) and electrochemical characterization. **Fig. 6a** shows a SEM picture

978-1-7281-1988-5/18 $31.00 © 2018 IEEE

of the nitinol bundle. **Fig. 6b** shows the diameter of the microwire before and after coating. It can be seen that the Parylene-C layer has the desired thickness of 4.5 μm. A 1 cm long Hamilton stainless steel needle with 210 μm inner diameter was used to bundle the nitinol wires so that they were constrained within a small space for implantation to the brain. The Hamilton needle does not penetrate the tissue; it only provides mechanical support for the bundle to implant microwires to the brain with minimal damage. Parylene-C coating was removed from the wire tips. A picture of the microwire array is shown in **Fig. 7**. A 3D printed microelectrode holder was designed to provide mechanical support for the microwire array and to fix the PCB board with epoxy. The nitinol/solution interface was characterized by electrochemical impedance spectroscopy (EIS) and cyclic voltammetry (CV). The EIS results in **Fig. 8a & 8b** show that the nitinol electrode exhibits impedance values in a reasonable range for electrophysiological recordings. The CV result in **Fig. 8d** shows no redox reactions at the electrode/electrolyte interface. As shown in **Fig. 8c**, the mean impedance is ~1.03 MΩ measured at 1 kHz. In order to test the 3D expansion of the array, we 3D printed a brain vasculature model using real data from NIH 3D print exchange database and constructed a brain phantom by immersing it into the agarose. As shown in **Fig. 9**, the nitinol array successfully penetrates and avoids the vessels.

III. IN VIVO ANIMAL EXPERIMENT

We validated the nitinol microwire bundle in *in vivo* animal experiments with mice during anesthesia and wakefulness. During the surgery, a wild-type mouse was anesthetized with 1%–2% isoflurane and a circular piece of scalp was removed to expose the skull. After cleaning the soft tissue on top of the bone, a head-bar was implanted with cyanoacrylate glue and cemented with dental acrylic. A craniotomy (~1mm in diameter) was made over the primary visual cortex. The ground/reference screws were implanted on the cerebellum. **Fig. 10** shows the experimental setup. The nitinol bundle was connected to a custom PCB, which was held by a custom-made holder attached to a micromanipulator (MP-285, Sutter Instrument). The electrodes were inserted to the visual cortex with an angle of 45° to the horizontal plane. After the experiment, we perfused the animal and sliced the brain to examine the tissue damage. Compared to the intact contralateral visual cortex without insertion, only small dents could be observed on brain surface and minimal damage caused by electrodes could be detected at recording site. **(Fig. 11)**.

IV. NEURAL DATA ANALYSIS

Fig. 12a shows a typical raw electrophysiological data recorded by one of the microelectrodes in layer IV (400 um deep) of the mouse visual cortex during anesthesia. To investigate the signal in frequency domain, we compute the spectrogram using wavelet transform **(Fig. 12b)**. It can be seen that, during anesthesia, the local field potential (LFP) exhibits transient oscillations in different frequency bands lasting between 2 – 10 seconds. As shown in **Fig. 12c & Fig. 12d**, we observed theta oscillations that have a central frequency around 7 Hz and have peak-to-peak amplitude of ~100 μV. Also, as

shown in **Fig. 12e & Fig. 12f**, there are 12 Hz alpha band oscillations that emerge randomly with similar amplitude as the theta oscillation. Finally, we observed activities that resemble the burst suppression waveforms that are commonly observed in LFP recordings during anesthesia (**Fig. 12g & Fig. 12h**) [8]. The waveform consists of a large biphasic waveform, coupled with oscillations between 5 to 15 Hz. Also, right before and after the bursting activity, the recorded electrical signals are flat, which is distinct from other time segments. These results confirm that our nitinol microwire electrode array can successfully record the LFPs of various dynamics with very low noise and high fidelity.

Besides the anesthesia, we also investigate the neural activities in awake state. **Fig. 13a** shows electrical recordings from a representative channel. Different from anesthesia, the electrical signals during awake state have larger amplitude. The spectrogram in **Fig. 13b** shows that compared to anesthesia, the power in almost all frequency bands increases. Also, there are no obvious oscillations in single frequency bands. Finally, during the awake state, we recorded high frequency multiunit activities and single spikes, which reflect the firing of far-away and nearby neurons respectively [9]. To see this, we filtered the data at 500 Hz using an 8th order Butterworth high-pass filter (**Fig. 13c**). Then we applied an amplitude threshold and a time window of 2 ms to extract the spikes. To assign different spikes to the neurons, we perform k-means clustering on the spike data. **Fig. 13d** shows an example of the clustering result from one of the channels. Different colors label the spikes that come from different neurons. These results show that the nitinol microwire can detect the single spikes and multiunit activities with high fidelity. Low noise observed in both anesthetized and awake recordings allows probing rich dynamics exhibited by single neurons and neuronal populations.

V. CONCLUSION

In this work, we developed a shape-programmable nitinol microwire array that conforms to the brain vasculature to minimize the damage to the blood vessels during implantation. We demonstrate that our nitinol microwire bundle can reliably record both local field potential and spiking activities in *in vivo* experiments. The developed nitinol microwire bundle provides new opportunities for vessel-damage free neural interfaces in chronic animal research to achieve stable long-term electrical recordings with minimum implantation damage.

ACKNOWLEDGMENTS

The authors acknowledge Office of Naval Research (N000141612531) and National Science Foundation (ECCS-1752241, ECCS-1734940) for funding.

REFERENCES

[1] W. M. Grill, et al., *Annu. Rev. Biomed. Eng.*, vol. 11, p. 1-24, 2009.
[2] V. S. Polikov, et al., *J Neurosci Methods,* vol. 148, p. 1-18, 2005.
[3] M. Thunemann, et al., *Nat Commun*, vol. 9, p. 2035, 2018.
[4] Y. Guo, et al., *CIRP ANN-MANUF TECHN*, vol. 62, p. 83-86, 2013.
[5] T. Duerig, et al., *Mater. Sci. Eng., A*, vol. 273, p. 149-160, 1999.
[6] M. Geetha, et al., *Prog. Mater Sci.*, vol. 54, p. 397-425, 2009.
[7] A. Bose, et al., *Stroke*, vol. 38, p. 1531-1537, 2007.
[8] K. K. Sellers, et al., *J Neurophysiol.*, vol. 110, p. 2739-2751, 2013.
[9] G. Buzsaki, et al., *Nat Rev Neurosci*, vol. 13, p. 407-20, 2012.

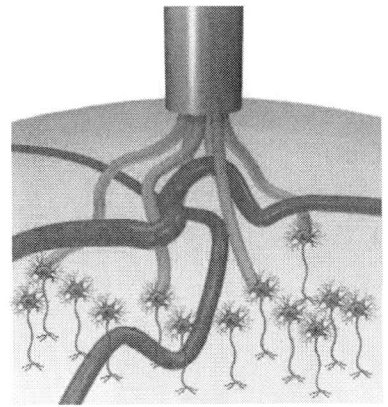

Fig. 1. A schematic showing the programmable nitinol wires penetrating to the brain, avoiding the blood vessels.

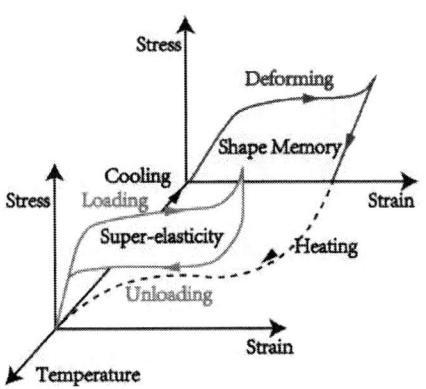

Fig. 2. A schematic showing the super-elasticity effect and the shape memory effect of the nitinol alloy. Adapted from [4].

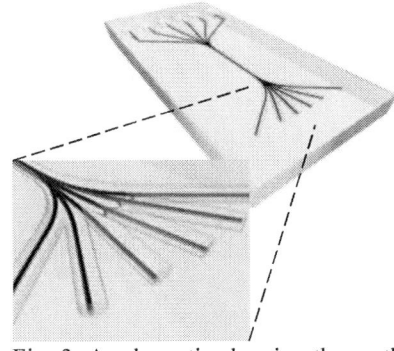

Fig. 3. A schematic showing the method of wire shape-setting using PDMS mold. The inset figure shows the nitinol wires embedded in the grooves.

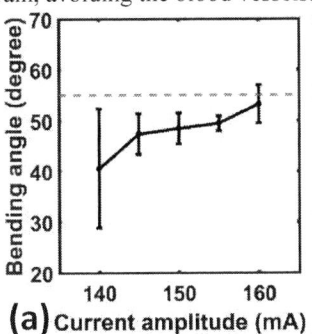

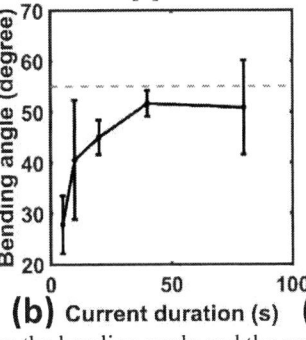

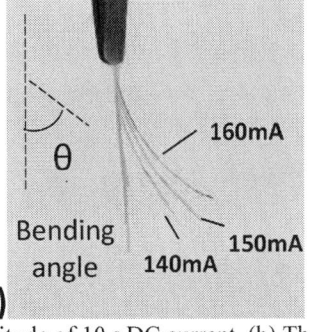

Fig. 4. (a) The relationship between the bending angle and the amplitude of 10 s DC current. (b) The relationship between the bending angle and the duration of the 140 mA DC current. Red line shows target angle. (c) The representative shape of the wires after shape-setting using different currents.

Fig. 5. The bending angles of 30 wires after shape setting with 155 mA current for 10 s.

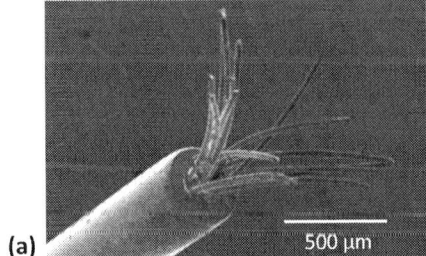

Fig. 6. (a)Scanning electron microscopy image shows Niti microwires expanded from the the tip of the Hamilton needle. (b) The wire diameters before and after coating are shown in the right diagram.

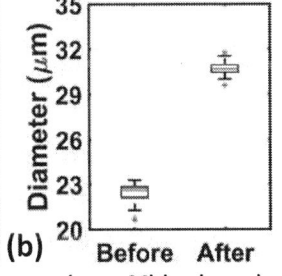

Fig. 7. (a) A photo of the device that shows Niti microwire bundle, the microwire holder, and the custom PCB. (b) A zoom-in picture showing the tip of the electrode bundle.

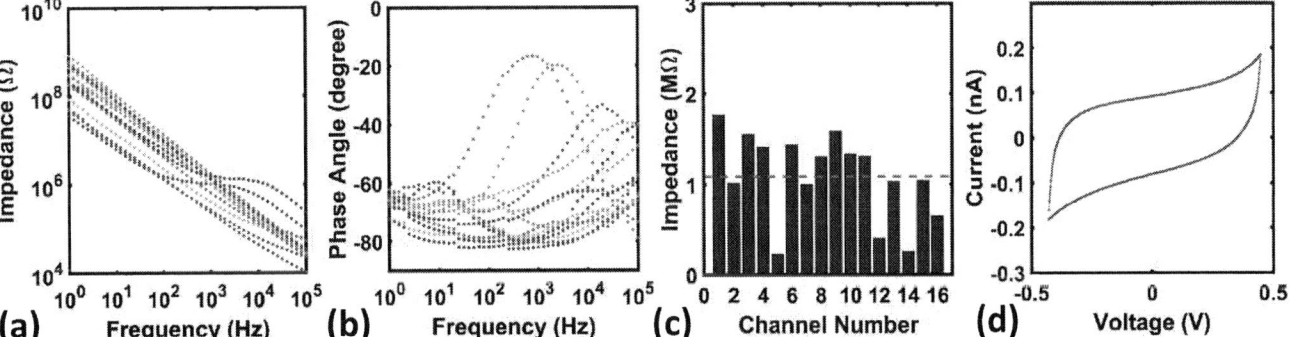

Fig. 8. (a) (b) The EIS results of all the 16 channels. (c) The impedances of all the channels measured at 1 kHz. The red line indicates the average impedance. (d) The CV curve of one representative channel.

978-1-7281-1988-5/18 $31.00 © 2018 IEEE

Fig. 9. A picture of the 3D printed brain vasculature phantom with nitinol wire bundle inserted. The wire conforms with the curvature of blood vessels.

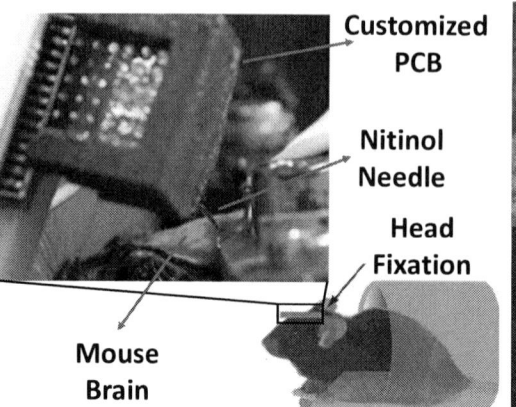

Fig. 10. A picture of the experimental setup, showing the board, the nitinol bundle, and the brain. The lower right cartoon shows the awake and head-fixed animal during the experiment.

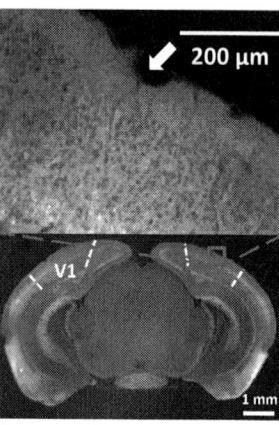

Fig. 11. Brain slice imaging shows the minimal damage caused by the nitinol microwires. The arrow shows penetrating position and angle.

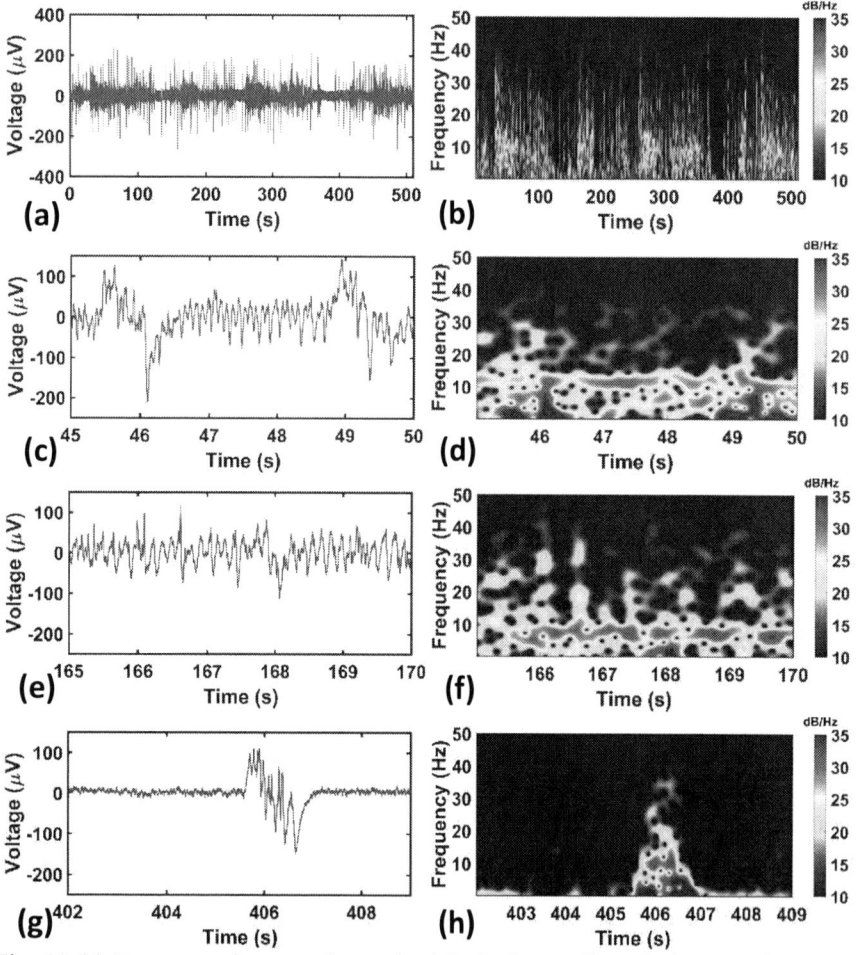

Fig. 12 (a) Representative raw electrophysiological recordings during anesthesia. (b) Spectrogram for the signals shown in (a). (c) Example time series of alpha oscillations and (d) its spectrogram. (e) Example time series of theta oscillations and (f) its spectrogram. (g) Example burst/suppression wave and (h) its spectrogram.

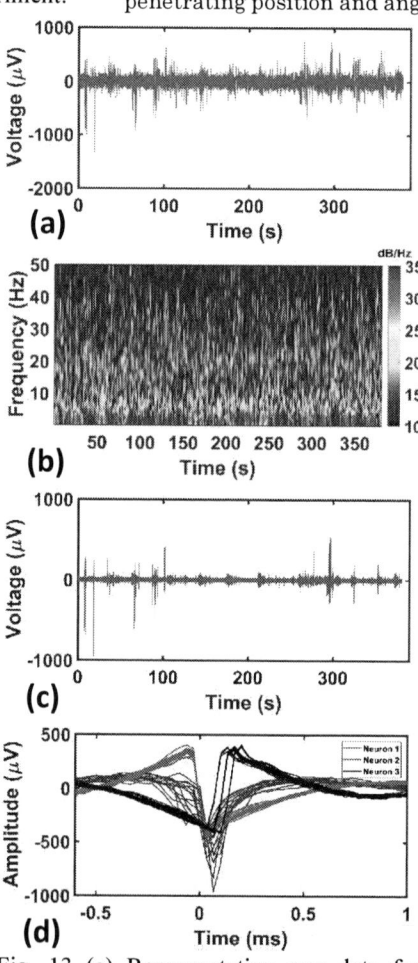

Fig. 13 (a) Representative raw data from awake recordings. (b) Spectrogram for data shown in (a). (c) High-pass filtered data showing single spike and multi-unit activities. (d) spike sorting results using the data recorded by a typical channel.

978-1-7281-1988-5/18 $31.00 © 2018 IEEE   667

Bio-inspired 3D neural electrodes for the peripheral nerves stimulation using shape memory polymers

Yingchao Zhang [1,2], Ning Zheng [3], Yinji Ma [1,2], Tao Xie [3] and Xue Feng [1,2,*]

[1] AML, Department of engineering Mechanics, Tsinghua University, Beijing 10084, China, email:fengxue@tsinghua.edu.cn
[2] Center for Flexible Electronics Technology, Tsinghua University, Beijing 10084, China
[3] State Key Laboratory of Chemical Engineering, College of Chemical and Biological Engineering, Zhejiang University, Hangzhou 310027, China.

Abstract—Peripheral nerves stimulation has been widely used in clinical practices, such as the vagus nerve stimulation (VNS) for heart failure, and motor nerve stimulation for controlling the prosthetics. However, the nerve injuries induced by the large mechanical and geometrical mismatch and complex surgical implantation process have restricted the further applications. Here, inspired by twining plants such as morning glories, we developed a 3D neural electrode that integrates the nano-gold film on flexible shape memory polymer (SMP) substrate from 2D planar state. Upon the response to 50°C normal saline, the flattened neural electrodes can self-climb to the 3D peripheral nerves with the aid of the shape memory effect. Two *in vivo* animal experiments are used to demonstrate the clinical practicality, i.e., VNS for the control of the heart rate (HR) and sciatic nerve stimulation for the control of the leg's movements. This technology offers a paradigm that fabricating the 3D bioelectronics in 2D planar state to match the 3D biological tissues by utilizing smart materials, and shows great potentials in clinical practices.

I. INTRODUCTION

Introduction of information into the peripheral nerves system (PNS) has been widely used in clinical practices to treat neurological disorders and motor function, such the VNS for the partial epilepsy and motor nerve stimulation for the controlling of muscles, which is enabled by the neural interfaces between the abiotic neural electrodes and biotic nerves [1]. Several forms of the neural electrodes have been developed with the aim to decrease the neural damage and increase the mechanical stability, which can be classified into the following two categories [2], namely, (1) intraneural electrodes such as invasive Utah electrodes and Michigan electrodes; and (2), extraneural electrodes such as Cuff electrodes, helical electrodes and flat interface nerve electrodes (FINE). The invasive intraneural electrodes can cause irreversible damage to the nerves thus are rarely used in the clinical practice. The extraneural electrodes may lead to less damage, although promising, these existing extraneural electrodes have at least one of the two drawbacks: pre-defined stiff structure that can hardly conformal contact with the nerves; planar flexible structure that complicated surgical implantation is required. Both of the above two drawbacks of the existing extraneural electrodes may lead to mechanical or electrical damages to the nerves, physiology performances including inflammation, axonal degradation and defective nerve conduction. Here, inspired by the growth process of the twining plants such as morning glories, who cling a supporter by circumnutational movements to achieve vertical growth to capture more sunshine and other resources (Fig. 1), we proposed a novel 3D neural electrode that integrates the nano-gold film (thickness: 200nm) on flexible SMP substrate (thickness: 100μm). The chosen materials for the substrates are smart materials capable of temporarily adopting programmed shapes and recovering their permanent shape upon the external stimulation [3]. With the aid of the shape memory effect, the initial planar electrodes that fabricated from 2D state can self-climb to the nerves and form conformal contact, once the 3D neural electrodes are triggered by 50°C normal saline. The *in vivo* experiments on a rabbit, i.e., VNS for the control of the heart rate (HR) and sciatic nerve stimulation for the control of the leg's movements demonstrate the potential practicality in clinical.

II. FABRICATION PROCESS

Fabrication of the 3D neural electrodes started with the synthesis of the SMPs. Here, we chose the SMPs network with thermally distinct elasticity and plasticity, which are capable of permanent shape reconfigurability via thermos-plastic. The synthesis process have been detailed described in our previous work [3]. Briefly, the SMPs were synthesized by a radical initiated reaction between two precursors, polycaprolactone (PCL, M_w=10000g/mol, Sigma-Aldrich, 10g) and 2-isocyanatoethylacrylate (ICEA, M_w=141.13g/mol, TCI, 0.34g), under the catalysis of dibutyltin dilaurate (DBTDL, M_w–631.57g/mol, TCI, 0.5wt %). The mixture was poured into a glass mold (thickness: 100μm) and irradiated under ultraviolet light for 5min (photoinitiator, BPO, M_w=242.23g/mol, TCI, 1wt%) to obtain the SMPs film. Then the initial planar SMP substrate (Fig. 2a) was heated above 50°C (the transition temperature, Tg) and twined on a glass rod with radius of r and pitch angle β, with the two-ends fixed by PI tape (Fig.2b). The permanent shape of the SMP substrate was reshaped to helical shape via thermo-plastic (130°C for 1h) (Fig. 2c). Afterwards, the helical SMP substrate was flattened at 50°C and cooled down to room temperature to obtain the flattened SMP substrate (Fig. 2d). Then, 100nm of Au was deposited at 0.1nm/s using electron beam evaporation (Fig. 2e), before which a shadow mash was attached on the surface of the flattened SMP substrate to define the metal patterns (Figs. 3a,b). Finally, the 3D neural electrode can be introduced into the body in 2D flattened state, and climb to the nerve driven by 50°C normal saline (Fig. 2f). Figs. 2c and 2d

978-1-7281-1988-5/18 $31.00 © 2018 IEEE

gives the 3D neural electrode in 2D flattened state and 3D helical state before and after recovery, respectively. Fig. 3 schematically illustrates the physical molecular principle and the corresponding network transformations during the fabrication process, where Figs. 3a-d give the principle of the thermos-plastic (i.e., corresponding to Figs. 1a-c), while Figs. 3e-g show the principle of the elastic recovery (i.e., corresponding to Figs. 1d-f). The recovery process of the 3D neural electrode from the flattened state to helical state was illustrated in Fig. 4, where the temperature of the water is 50°C. We have also performed the analog experiment (Fig. 5), where the flattened neural electrode was twined on a glass rod (imitate the nerve) driven by the 50°C water.

III. CHARACTERIZATION AND STRUCTURE DESIGN

General material characterizations of the synthesized SMPs were conducted using a Dynamic Mechanical Analyses (DMA) machine (Q800, TA instruments) to give the glass transition temperature (Tg,) the plasticity temperature (Tp) and the shape memory cycles. The DMA curve (Fig. 7a) shows that the SMPs possess a stable rubbery plateau above their corresponding glass transition temperature (Tg =50°C). Fig. 7b shows that complete stress relaxation at 140°C takes about 50 min. Therefore, the plasticity temperature can be chosen as 140°C. The consecutive shape memory cycles (Fig. 7c) show that the shape fixity and shape recovery ratio are both above 97% within each cycle, i.e., the synthetic SMPs possess excellent elastic recovery. These advantages shows that the SMPs are appropriate for the fabrication of the 3D neural electrodes and clinical practices, where several recovery cycles are required (Figs. 2c-f). Electrochemical characterizations were also conducted to evaluate the fabricated 3D neural electrode (CS350, Corrtest Instrument, China). The impedance spectroscopy, cyclic voltammogram are shown in Fig. 8. The impedance magnitude of the 3D neural electrode at 1kHz is about $50\Omega/cm^2$. The charge of delivery capacity (CDC) obtained from the CV curve is about 0.90mC/cm², which is defined as $CDC = \frac{1}{v}\int_{V_c}^{V_a}|i|\,dV$, where v is the scan rate. V_a and V_c are the anodic and cathodic potential limits (i.e., -0.6V to +0.8V) respectively. i is the measured current density (mA/cm²), V is the electrode potential versus Ag/AgCl. Both of the impedance spectroscopy and the CDC are comparable with the literatures. To decrease the damage to the nerves and increase the mechanical stability of the 3D neural electrodes, lower bending stiffness EI and maximum strain in the metal layer ε_{Au}^{max} are preferred. The mechanics model is shown in the inset of Fig. 9, from which we can obtain the EI and the ε_{Au}^{max}(the bending radius is chosen as 1mm). Results show that the EI increase with increasing the thickness of the SMPs (h_{SMP}) and the Au film (h_{Au}), while the ε_{Au}^{max} increase with the increasing h_{SMP} but decrease with the increasing h_{Au} (Fig. 7). The yield limitation of Au ε_{Au}^{yield} is 0.3%. Therefore, comprehensive consideration of the above aspects and the achievability of the fabrication process, the thickness of the SMPs and the Au film is chosen as 100μm and 200nm, respectively.

IV. IN VIVO ANIMAL EXPERIMENTS

We performed two *in vivo* animal experiments on New Zealand rabbits to demonstrate the clinical practicality. All the animal experiments were performed in Beijing Medical Services Biotechnology (Beijing, China), and were approved by the Ethics Committee of Beijing Medical Services Biotechnology (MDSW-2018-010C). Firstly, the 3D neural electrodes were twined around the right vagus nerve of anesthetized rabbit for the controlling of heart rate (Fig. 10a). The twining process is similar to the process shown in Fig. 6f-h. The electrocardiography (ECG) recorded before electrical stimulation (Fig. 8b) shows that the normal HR of the anesthetized rabbit is about 240bpm. During the electrical stimulation (current amplitude 0.3mA, wave width 100μs, frequency 10Hz), the HR was decreased to 180bpm (Fig.8b). Next, the 3D neural electrodes were twined around the left sciatic nerve of another anesthetized rabbit, and the electrocorticography (ECoG) was also synchronous recorded (Fig. 11a-c). The sciatic never stimulation can be used for the controlling of the leg's movements. During the stimulation (current amplitude 0.5mA, wave width 100μs, frequency 10Hz), the movements of the left leg were recorded (Figs. 11d-f). The resting and evoked potentials of ECoGs are shown in Fig. 11g and Fig. 11i, respectively. The root mean square (RMS) spectra (Fig. 10) shows that, after the electrical stimulation was introduce into the sciatic nerve, the electrical activity of the rabbit brain was raised at the frequency bands of 1-20Hz.

V. CONCLUSIONS

In summary, we introduced a processing diagram of 3D neural electrodes by utilizing shape memory effect, which are compatible with the traditional planar processing. The developed 3D neural electrodes can minimize the geometrical mismatch with the 3D peripheral nerves and greatly facilitate the surgical procedure. The potential clinical practicality is demonstrated by the two *in vivo* animal experiments. This work opens up new prospects for the peripheral nerves stimulation via the 3D neural electrodes.

ACKNOWLEDGMENT

We gratefully acknowledge the support from the National Basic Research Program of China (Grant No. 2015CB351900) and National Natural Science Foundation of China (Grant Nos. 11625207, 11320101001, 11222220).

REFERENCES

[1] K. Famm, B. litt, K. J. Tracey, E. S. Boyden, and M. Slaoui, "Drug discovery: A jump-start for electroceuticals," Nature, vol. 496, pp. 159-161, 2013.

[2] D. K. Leventhal and D. M. Durand, "Chronic measurement of the stimulation selectivity of the flat interface nerve electrode," IEEE Trans Biomed Eng, vol. 51, no. 9, pp. 1649-58, Sep 2004.

[3] Q. Zhao, W. Zou, Y. Luo, and T. Xie, "Shape memory polymer network with thermally distinct elasticity and plasticity," Science advances, vol. 2, no. 1, pp. e1501297, 2016.

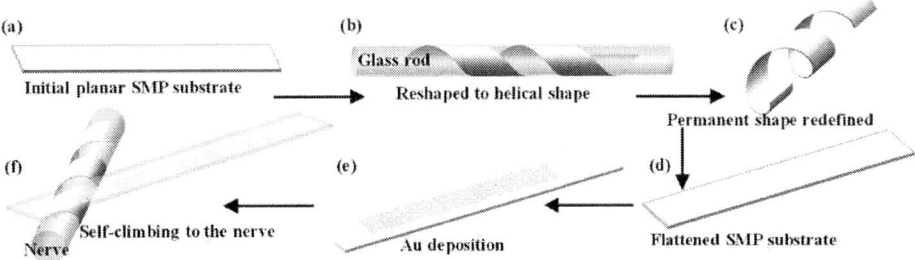

Fig. 1. Photograph image of twining plants.

Fig. 2. Schematic diagrams of the fabrication process; (a-c) the permanent shape redefined process via thermo-plastic; (d) the helical SMP substrate is flattened and temporarily fixed; (e) Au deposition on the flattened SMP substrate; and (f) diagram of the self-climbing process via elastic recovery.

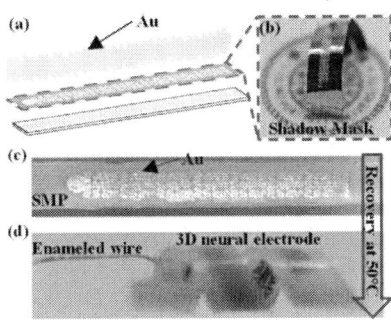

Fig. 3. Schematic diagrams of Au deposition via shadow mask (a, b); and the fabricated 3D neural electrode before (c) and after recovery (d).

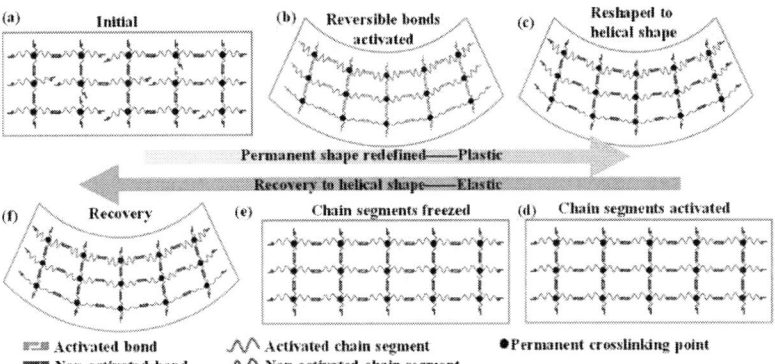

Fig. 4. Schematic illustrations of the physical molecular principle. (a-c) the permanent shape redefined process via thermo-plastic; and (d-f) elastic recovery process.

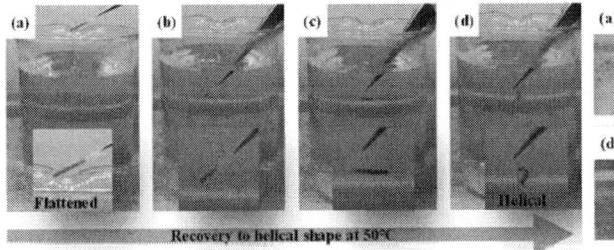

Fig. 5. The recovery process of the 3D neural electrode from the flattened state (a) to the helical state (d).

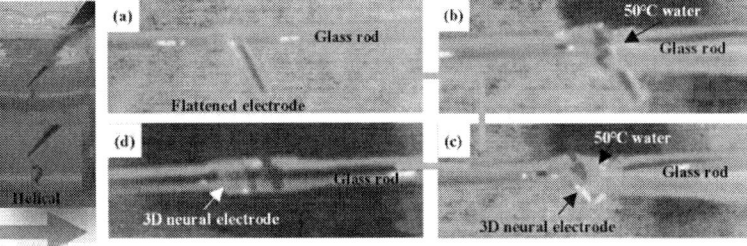

Fig. 6. Video screenshots of the process that the 3D neural electrode is twined on a glass rod upon the driven of 50°C water.

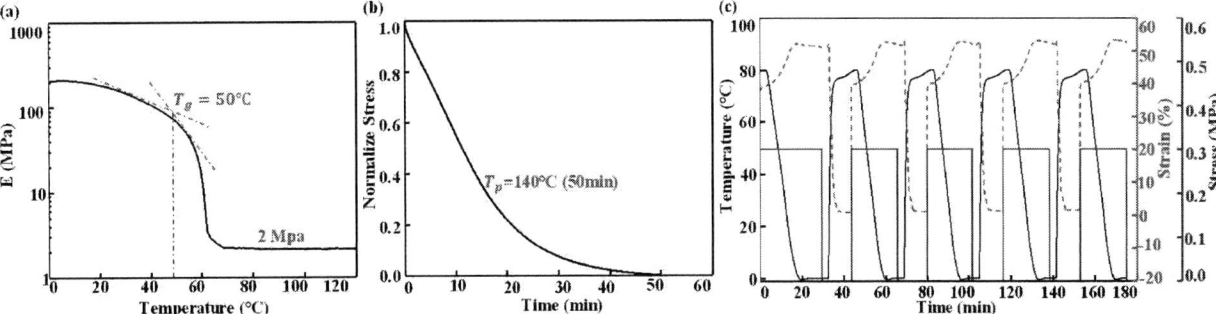

Fig. 7. Characterizations of the synthesized SMPs. (a) DMA curve; (b) stress relaxation curve; and (c) elastic shape memory cycles.

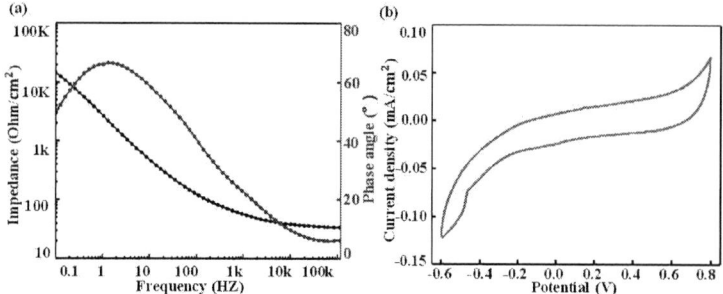

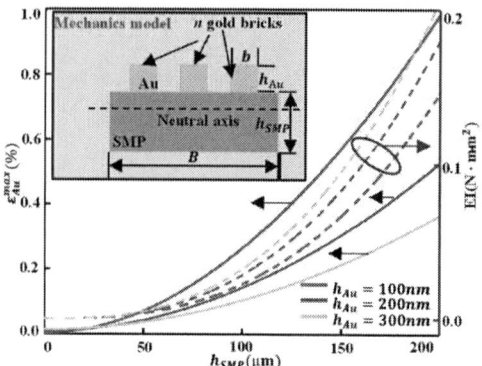

Fig. 8. Electrochemical characterizations. (a) The impedance and phase angel spectroscopy; and (b) the cyclic voltammogram (CV) curve, sweep rate 50mV/s. Phosphate buffered saline (PBS, PH:7.2-7.4).

Fig. 9. The maximum strain in the gold film and the bending stiffness versus the SMP thickness at different gold film thickness, where the inset gives the mechanics model.

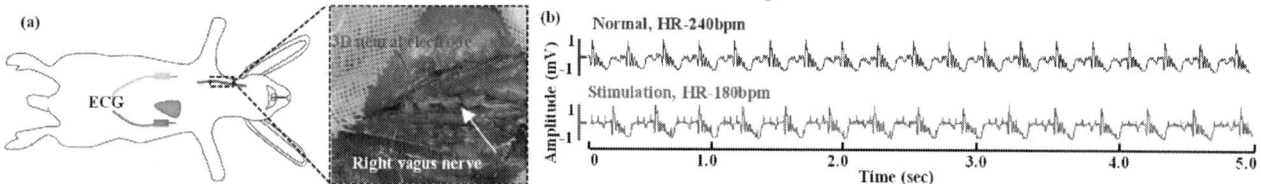

Fig. 10. The *in vivo* animal experiments of the VNS. (a) The schematic diagram of the VNS and recording of ECG (left), and the images of the implanted 3D neural electrodes (right); and (b) the ECGs of the anaesthetized rabbit in normal state (above) and during the stimulation (below).

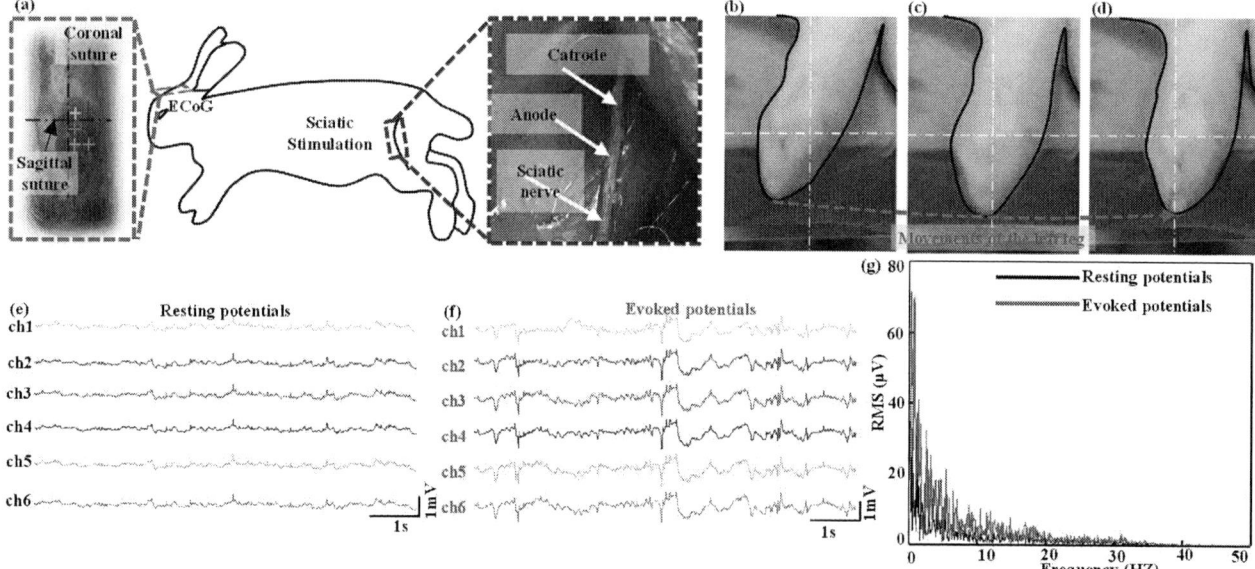

Fig. 11. The *in vivo* animal experiments of the sciatic stimulation and the recording of ECoGs. (a) The locations of the six stainless-steel screw electrodes for the ECoGs recording (right), and the implanted 3D neural electrodes (left); (b-d) the video screenshots of the movements of the left leg during the stimulation; (e) the resting potentials; (f) the evoked potentials during the stimulation; and (g) the root mean square of resting and evoked potentials versus the frequency.

978-1-7281-1988-5/18 $31.00 © 2018 IEEE   671

Fabrication and Characterization of 3D Multi-Electrode Array on Flexible Substrate for *In Vivo* EMG Recording from Expiratory Muscle of Songbird

Muneeb Zia[1], Bryce Chung[2], Samuel J. Sober[2], and Muhannad S. Bakir[1]

[1]Georgia Institute of Technology, Atlanta, GA, USA, email: muneeb.zia@gatech.edu

[2]Emory University, Atlanta, GA, USA

Abstract— This work presents fabrication and characterization of flexible three-dimensional (3D) multi-electrode arrays (MEAs) capable of high signal-to-noise (SNR) electromyogram (EMG) recordings from the expiratory muscle of a songbird. The fabrication utilizes a photoresist reflow process to obtain 3D structures to serve as the electrodes. A polyimide base with a PDMS top insulation was utilized to ensure flexibility and biocompatibility of the fabricated 3D MEA devices. SNR measurements from the fabricated 3D electrode show up to a 7x improvement as compared to the 2D MEAs.

I. Introduction

Recent advances in data analysis methods in neuroscience have provided new insights on how a nervous system controls complex behaviors such as vocal learning and song production in songbirds [1-2]. Recent evidence [3,4] has pointed to the importance of precise timing of individual motor units for controlling behavior and showed that EMG activity can be used to understand how nervous systems produce behaviors.

Understanding how nervous systems produce behaviors requires recording devices and algorithms that can identify individual motor events, called muscle potentials. Among the challenges involved with obtaining high fidelity recordings suitable for neural analyses are: biological compliance of recording devices [5] and the signal to noise ratio. In addition, characterizing single motor unit activity requires a stable, reliable EMG recording for a duration long enough to produce sufficient data for advanced computational analyses [6,7].

Polyimide, PDMS and parylene-C have been widely used for the fabrication of high-density mutli-electrode arrays [3, 8-13]. To increase the signal fidelity, three-dimensional neural and muscular recording devices have also been explored [9-13]. However, these involve complex processing methodologies increasing the fabrication complexity, cost and time.

To address these challenges, this work presents fabrication and characterization of a flexible 3D MEA utilizing a simple photoresist reflow process [14] to obtain the 3D electrodes. Polyimide is the base substrate for better metal adhesion and PDMS is the top insulation layer as it is more affordable, easier to etch and can be diluted to obtain thin top insulation layer. The height of the 3D electrodes can easily be modulated by changing the film thickness of the spin coated photoresist. *In*

vivo EMG measurements from an anesthetized songbird are also presented. The fabricated 3D MEAs provide up to 7x SNR improvement over the 2D [15] array allowing detection of small units which can otherwise get lost in noise.

II. Fabrication of 3D Multi-Electrode Arrays

The fabrication process of the 3D MEAs is outlined in Fig. 2. Polyimide (PI-2611 from HD Microsystems) is spin coated @ 450 rpm on a carrier wafer and subsequently cured to get a thick polyimide film. Photoresist is then spin coated and patterned followed by metallization and a lift-off process. A Ti/Au layer of 30nm / 200nm is deposited using an evaporation process. Thick photoresist (AZ-40XT) is then spin coated and patterned using photolithography which is then reflowed to form the hemispherical structures as shown in Fig. 2 (e). Double-reflow process described in [14] can also be utilized here to obtain multi-height 3D electrodes in the same fabrication flow. An electroplating seed layer consisting of 50 nm of Ti and 300 nm of Cu is subsequently sputtered. Photoresist is then spray coated and the electroplating mold is formed. Nickel electroplating (10 μm thick) is then performed followed by the removal of the underlying photoresist and the seed layer to give the free-standing 3D MEAs. Electroless gold plating is then performed to passivate the electrode surface and prevent oxidation. To obtain the top insulation layer, a thin coating of PDMS (Sylgard 184, 1:10 ratio) diluted with toluene (0.9% weight ratio) is obtained and then cured. A reactive-ion etching (RIE) process is then used to etch the PDMS to expose the 3D electrodes. SF_6 and O_2 were used as the etching gases with a flow rate of 90 and 6 sccm respectively while the RF power was 300 watts. The etch rate obtained for the PDMS was ~ 170 nm/ min. The final 3D electrodes obtained are shown in Fig. 4.

III. EMG Measurements and SNR Comparison

EMG and air pressure data were collected as outlined in Fig. 1(c). Rhythmic muscle activity generates air pressure during breathing. Analog signals are detected by the flexible MEA and an air pressure sensor. EMG activity from the flexible MEA is amplified and digitized by an Intan RHD2216 bipolar amplifier chip. Air pressure data was also simultaneously collected using a pressure sensor connected to a tube inserted into the air sac of the anesthetized songbird. The Intan RHD 2000 evaluation board records digital signals for both EMG and air pressure data for analysis. Spike sorting is used to distinguish individual motor units (Fig. 1) from background noise and mutual information is then used to determine the correlation between neural activity and behavior. Both 3D and 2D [15] arrays were used to record EMG activity from the expiratory muscles of anesthetized songbirds. All procedures were approved by the Emory University Institutional Animal Care and Use Committee. EMG recordings for the flexible MEA devices were collected using 16 contacts arranged in a 4x4 matrix. Example EMG units recorded on one of the 16 contacts were chosen based on their physiological properties including the type of spiking exhibited during breathing cycles (i.e., spiking at a constant, moderate firing rate versus only a few spikes at a higher firing rate) and the relative amplitude in comparison to other EMG units that were simultaneously recorded.

The data collection sequence of experiments for comparison of the SNR was conducted as follows: A total of four recordings of at least 30 minutes each were carried out alternating between 2D and 3D MEAs. After each recording and before the placement of the next MEA, a drop of saline was poured on the exposed expiratory muscle. Care was taken to ensure that the electrode placement was roughly over the same area of the expiratory muscle of the bird. Fig. 5 shows the EMG measurements along with the pressure data from the air sac of the bird. As seen from the figure, the 3D MEAs pick up more activity from the muscle as compared to the 2D MEA and give a more consistent recording across trials. On the other hand, the 2D array's signal deteriorates from one trial to the next; this can be attributed to poor contact with the muscle due to protein or other unwanted material build up as the EMG recording duration increases; this is also exacerbated by the fact that the electrodes are inset below the surface of the top PDMS insulation layer. SNR calculations were performed by taking the ratio of the root-mean-square (RMS) amplitude of the waveforms during periods of activity with large or small units (signal) over periods of noise. Fig. 6 shows the SNR for the 2D and 3D MEAs over the course of the four different trials that were carried out. An average SNR was calculated in 3-minute intervals with 4 measurements during each interval used to determine the standard error of the mean (shaded region in Fig. 6).

Table I summarize the SNR measurements for the 2D and 3D electrodes over the different trials carried out for small and large amplitude unit activity; the 3D MEAs provide significant improvement in SNR for both small and large units with a 3.5x SNR improvement 5 minutes after the array placement and more than 7x SNR improvement 25 minutes into the recording for the larger unit. Improvement of SNR within a trial over time can be explained by better contact of the electrodes with the muscle as the saline dries out.

IV. CONCLUSION

A flexible 3D MEA for high SNR in vivo EMG recordings is presented. The process flow allows for easy height modulation of 3D electrodes by changing the film thickness of the photoresist. An Intan RHD2000 evaluation board and an RHD2216 amplifier chip is used to record expiratory muscle EMG activity and air pressure data. The 3D MEAs yielded

higher SNR measurements over a longer duration of time as compared a 2D array. This is particularly important for detecting and analyzing smaller units which are otherwise lost in noise. Using the 3D arrays, an SNR of up to 7x was achieved and some of the improvement may have been due to better electrical isolation as excess liquid dried around the recording site. With better signal fidelity, individual units can be identified more reliably and for longer periods of time, which will allow more advanced analysis techniques that can be used to understand how nervous systems control behavior.

ACKNOWLEDGMENT

The authors would like to acknowledge HD Microsystems for providing the polyimide used for the fabrication of the 3D MEAs described in this paper. The authors would also like to acknowledge Mingu Kim for helpful discussion for the process development.

REFERENCES

[1] Mackavicius EL, Fee MS. Building a state space for song learning. Current Opinion in Neurobiology, vol. 49, pp. 59-68 , April 2018.

[2] Srivastava KH, Elemans CPH, Sober SJ. Multifunctional and context-dependent control of vocal acoustics by individual muscles. J Neurosci, vol. 35, iss. 42, pp. 14183-14194, October 2015.

[3] Srivastava KH, et al. Motor control by precisely timed spike patterns. PNAS, vol. 114, iss. 5, pp. 1171-1176, January 2017.

[4] Tang C, Chehayeb D, Srivastava K, Nemenman I, Sober SJ. Millisecond-scale motor encoding in a cortical vocal area. PLoS Biology, vol. 12, iss. 12, December 2014.

[5] Ahuva Weltman, James Yoo and Ellis Meng, "Flexible, Penetrating Brain Probes Enabled by Advances in Polymer Fabrication", Micromachines 2016, 7(10), 180

[6] Kraskov A, Stogbauer H, Grassberger P. Estimating mutual information. Phys Rev E Stat Nonlin Soft Matter Phys, vol. 69, iss. 6, pt. 2, 066138, June 2004.

[7] Nemenman I, Bialek W, de Ruyter van Steveninck. Entropy and information in neural spike trains: Progress on the sampling problem. Physical Review E, vol. 69, iss. 5, 056111, May 2004.

[8] Kim O, et al. "Novel neural interface electrode array for the peripheral nerve," in Rehabilitation Robotics (ICORR), 2017 International Conference on, IEEE, pp. 106-1072, July 2017.

[9] Lin K, Wang X, Zhang X, Wang B, Huang J, Huang F. An FPC based flexible dry electrode with stacked double-micro-domes array for wearable biopotential recording system. Microsystem Technologies, vol. 23, iss. 5, pp.1443-1451, May 2017.

[10] Guvanasen GS, et al. A stretchable microneedle electrode array for stimulating and measuring intramuscular electromyograhpic activity. Neural Systems and Rehabilitation Engineering, Transactions on, IEEE, no. 9, pp. 1440-1452, September 2017.

[11] Kim JM, Im C, Lee WR. Plateau-shaped flexible polymer microelectrode array for neural recording. Polymers, vol. 9, iss. 12, p. 690, December 2017.

[12] Metallo C, White RD, Trimmer BA. Flexible parylene-based microelectrode arrays for high resolution EMG recordings in freely moving small animals. J Neurosci Methods, vol. 195, iss. 2, pp. 176-184, February 2011.

[13] Nandra MS, Lavrov IA, Edgerton VR, Tai YC. "A parylene-based microelectrode array implant for spinal cord stimulation in rats," in Micro Electro Mechanical Systems (MEMS), 2011 IEEE 24th International Conference on, IEEE, pp. 1007-1010, January 2011.

[14] C. Zhang, H. S. Yang, and M. S. Bakir, "A double-lithography and double-reflow process and application to multi-pitch multi-height mechanical flexible interconnects," J. Micromech. Microeng., vol. 27, no. 2, p. 025014, 2017.

[15] M. Zia, B. Chung, S.J. Sober and M.S. Bakir, "In Vivo EMG recording from breathing muscle of Songbird using hybrid polyimide-PDMS flexible multi-electrode arrays", to be published.

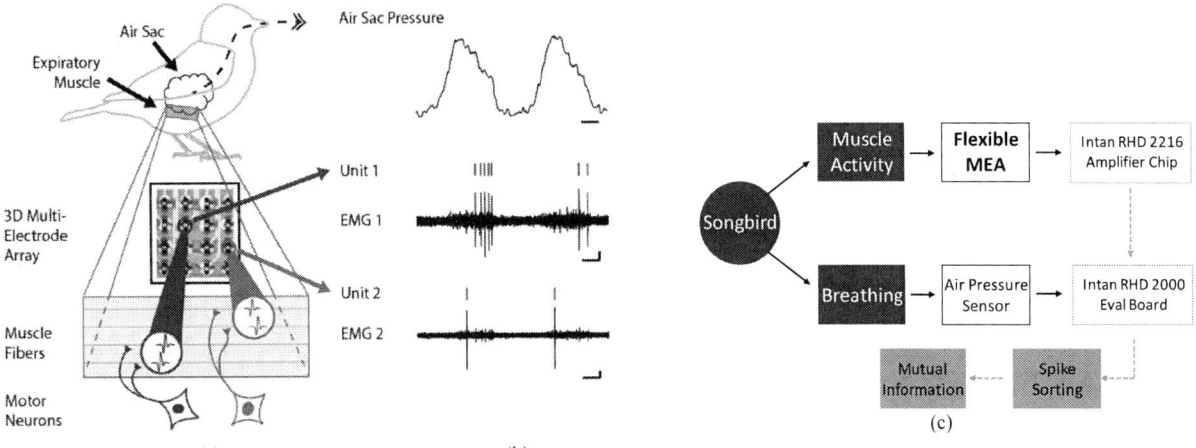

(a)   (b)

Fig. 1. Schematic of experimental set up and EMG activity recorded during breathing. (a) The exhaling phase of breathing in songbirds is controlled by expiratory muscles that contract around an air sac. Motor neurons excite individual muscle fibers that cause the expiratory muscle to contract. Multi-electrode arrays are used to record electromyography (EMG) activity, (b) Increases in air pressure occur when the expiratory muscles contract. Spike sorting algorithms are used to detect individual spikes (Unit 1, Unit 2) from recorded muscle activity (EMG 1, EMG 2). Tick marks above physiological traces indicate spike times. Time scale: 100 ms. Vertical scale: 30 µV, (c) Data collection flow chart; the EMG data from the flexible MEA is amplified by the using the intan RHD 2216 chip which is then fed into the evaluation board along with the air pressure data.

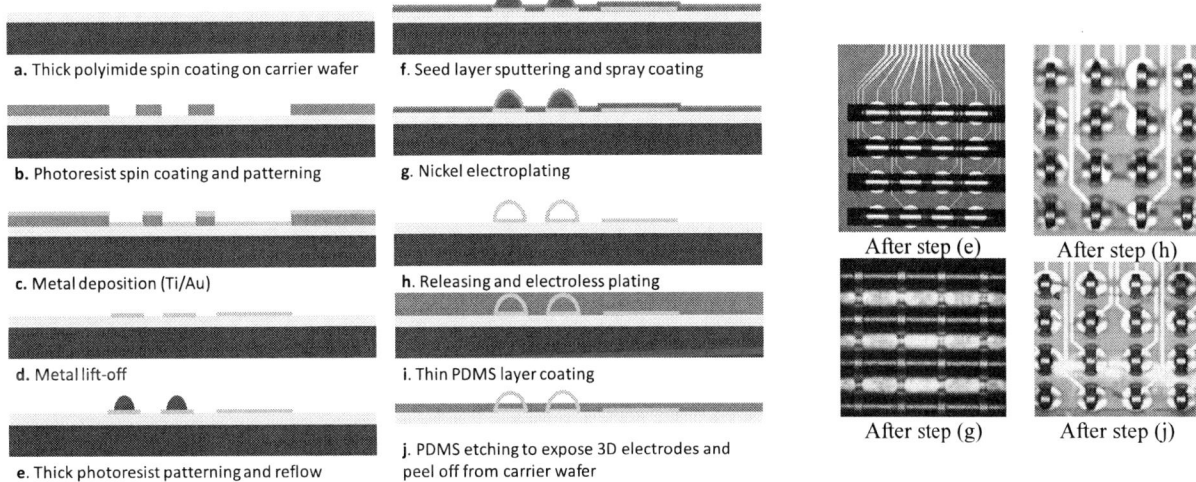

Fig. 2. Fabrication process flow for the flexible 3D MEAs

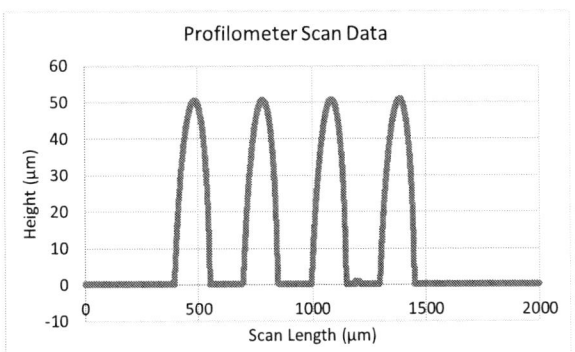

Fig. 3. Profilometer scan data after reflow of photoresist to form the hemispherical domes (step (e) in the process flow). A 50 µm dome height was obtained for the 3D MEAs and can be modulated by changing the photoresist film thickness.

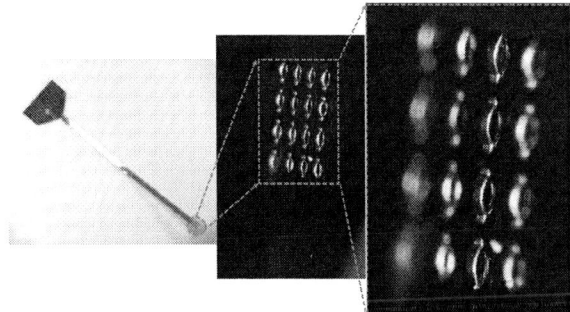

Fig. 4. 3D optical images of the fabricated 3D MEAs

978-1-7281-1988-5/18 $31.00 © 2018 IEEE   674

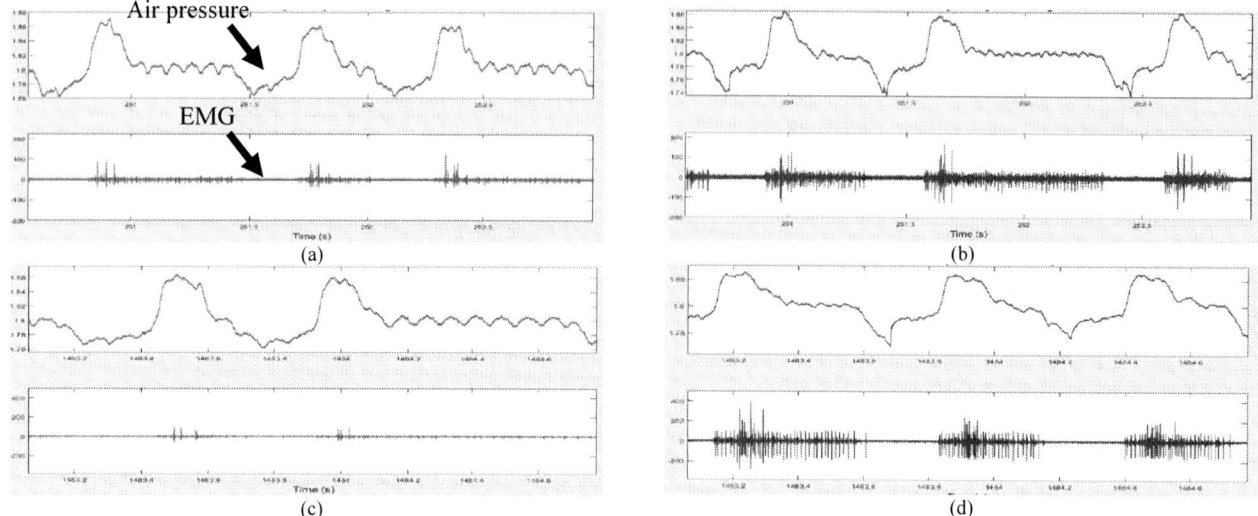

Fig. 5. Example air pressure and electromyograph recordings. (a,b) EMG recording after 5 minutes of array placement on the expiratory muscle of the songbird with (a) flexible 2D MEA, and (b) flexible 3D MEA. (c,d) EMG recording after 25 minutes of array placement on the expiratory muscle of songbird with (c) flexible 2D MEA, and (d) flexible 3D MEA.

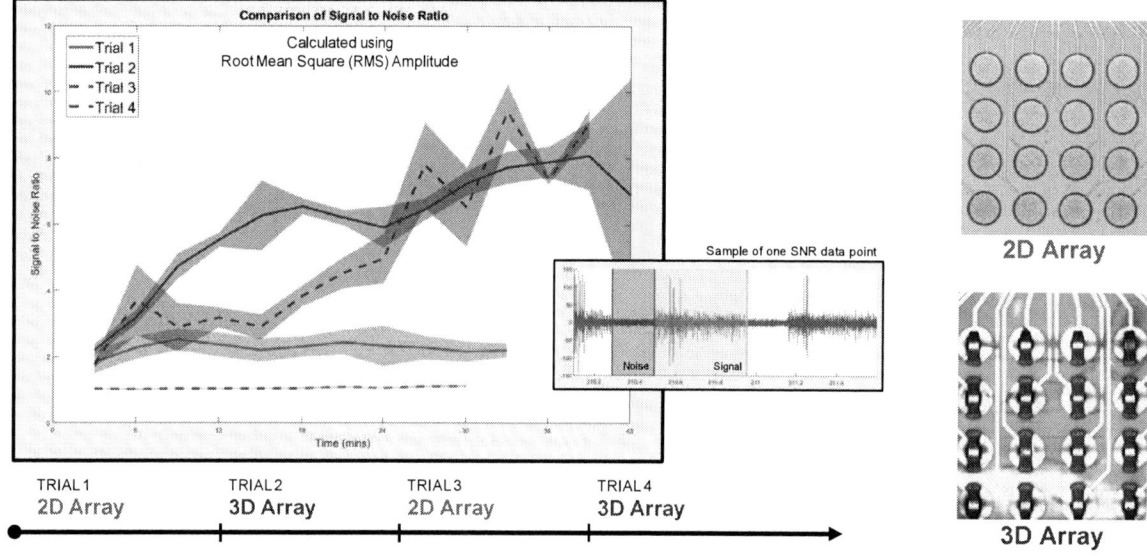

Fig. 6. Comparison of signal-to-noise ratio (SNR) using multi-electrode arrays with either 2D (red) or 3D (blue) electrode sites. Recordings for each trial were collected over at least 30-minute periods and alternated between 2D and 3D MEAs to control for non-stationary factors of the in vivo preparation. An average (solid or dashed line) SNR was calculated every 3 minutes with 4 measurements during each minute to determine a standard error of the mean (shaded regions).

Trial #	Contact Type	Time Since Array placement (mins)	Large Unit SNR	Small Unit SNR
1	2D	5	3.76	1.74
		25	3.58	1.79
2	3D	5	4.68	2.77
		25	8.99	5.70
3	2D	5	1.02	N/A
		25	1.18	N/A
4	3D	5	3.59	2.12
		25	8.45	6.82

Table I. SNR comparison of EMG signal recorded using the 2D and 3D MEAs. N/A is listed where signal was not discernable from noise.

Bioelectronics at the Single Molecule Level

O. Tolga Gul[1,2], Kaitlin M. Pugliese[3], Yongki Choi[2], Arith J. Rajapakse[2], Calvin J. Lau[2], Narendra Kumar[2], Kristin N. Gabriel[4], Denys Marushchak[2], Tivoli J. Olsen[3], Deng Pan[2], Gregory A. Weiss[3,4], and Philip G. Collins[2*]

[1]Department of Physics, Ankara Hacı Bayram Veli University, Polatlı 06900, Turkey

Departments of [2]Physics and Astronomy, [3]Chemistry, and [4]Molecular Biology and Biochemistry[3], University of California at Irvine, Irvine, California 92697, U.S.A. *Corresponding author, email: collinsp@uci.edu

Abstract—Bioelectronic devices built with single molecules of a protein, enzyme, or aptamer represent a new class of hybrid electronics. When biofunctionalization of nanoscale conductors is reduced to one molecule, that molecule's dynamic activity can be transduced into a large amplitude, high bandwidth electronic output. Using DNA polymerase I as an example, we show that single-molecule bioelectronics reveal biochemical activity with bond-by-bond resolution.

I. INTRODUCTION

The motivation and vision for *molecular electronics* is usually stated in terms of the miniaturization and scaling limits of conventional semiconductor technologies [1]. Ongoing research in molecular electronics promises the eventual production of molecule-based memories, switches, and circuits. However, such conventional elements would certainly shortchange the broader opportunities of fabricating electronic devices at molecular scales. Individual molecules can perform sophisticated functions, and the far-reaching, transformative opportunity for molecular electronics is to bridge solid state electronics with the world of chemical activity and biological complexity. Commercial electronic devices do not yet incorporate single biomolecules as functional components, but molecule-based architectures make such hybrids possible.

Beyond switching and memory storage, these hybrid devices could create new application spaces throughout the life sciences and health care. Consider, for example, arrays of devices directly reporting the dynamic activity of proteins as they respond to changing biological conditions. Beyond mere biosensing, single-molecule devices can capture the dynamic signals generated as target molecules arrive, activate, or bind, and quantitatively reveal the timing, thermodynamics, and rate distributions of complex chemical events like protein-protein recognition, enzymatic activation, or pharmaceutical inhibition.

The past decade has seen tremendous progress towards realizing these goals, especially by moving away from the traditional molecular-electronics architecture of nanometer-scale tunneling gaps. One-dimensional (1-D) conductors like silicon nanowires (SiNWs) and single-walled carbon nanotubes (SWNTs) have provided a new architecture for electrically communicating the state of a biomolecule to distant connective electrodes [2-5]. Fig. 1 depicts a generic scheme of one enzyme attached to a SWNT. The enzyme's motions and catalytic activity generate conductance fluctuations by electrostatically perturbing and gating a conventional source-drain current flowing in the underlying SWNT [5, 6].

This single-molecule architecture has many attractive features. Firstly, large currents and high bandwidths are achieved by electrostatically perturbing a high-conductivity channel. The scheme is generalizable to virtually any biomolecule, since the electrostatic mechanism eliminates the problem of transport within the biomolecule itself. The single-molecule architecture of Fig. 1 is stable at room temperature and compatible with being submerged in electrolytes like physiological buffers. The intrinsic amplification associated with one scattering site in an otherwise pristine, 1-D channel makes transduction insensitive to the exact placement of the active biomolecule along the channel; consequently, molecular-level precision is not required for successful fabrication. And lastly, numerous biochemical linkage schemes exist, including a noncovalent linkage highlighted in Fig. 1. Our recent, single-molecule biophysics research using this technique has been summarized in two review articles [7, 8].

II. ILLUSTRATIVE SINGLE-MOLECULE SIGNALS

The Klenow Fragment (KF) of DNA polymerase I is an excellent example enzyme for illustrating single-molecule bioelectronics and its possible applications. In biological cells, DNA polymerases like KF bind to single-stranded DNA (ssDNA) and move down the strand base-by-base while incorporating complementary nucleotides. For each nascent base pair, KF briefly closes upon the ssDNA to catalytically bind a new nucleotide before reopening and translocating to the next unpaired base [9].

Each catalytic cycle involves allosteric conformational changes that extend through the entire KF molecule. Thus, the enzyme's machinery for detecting complementarity at the active site results in concerted motions of amino acid sidechains on the enzyme's outer surface. When KF is attached to a sensitive conductor like a SWNT, charged sidechains can gate the SWNT and induce conductance fluctuations.

Figs. 2 and 3 show example signals $\Delta I(t)$ acquired from KF in a buffer solution containing 1 nM poly(dA)$_{42}$ ssDNA. In the presence of complementary triphosphate nucleotides (i.e. 10 μM dTTP), current excursions $\Delta I(t)$ of 5-10 nA occur with the creation of each new base pair (Fig. 2) [10]. Catalytic activity ceases when the complementary dTTP is absent or replaced by a noncomplementary nucleotide, and so do the $\Delta I(t)$ excursions (Fig. 3). Consequently, $\Delta I(t)$ excursions can be confidently identified as base incorporations. Analysis of these signals is useful for the enumeration of bases [8] or for the statistical analysis of KF's kinetics [10, 11]. The signals generated by heteropolymeric ssDNA containing all four nucleotides

978-1-7281-1988-5/18 $31.00 © 2018 IEEE

increases in complexity (Fig. 4), opening the door to DNA sequencing based on solid-state electronics. More broadly, single-molecule bioelectronics provides a feasible route to low-cost, massively parallel arrays having a wide range of healthcare applications, from DNA sequencing to drug discovery and disease detection and monitoring.

III. OPPORTUNITIES AND CHALLENGES

Many DNA polymerases make fewer than one misincorporation per million base pairs [12]. The mechanisms responsible for this high fidelity remain poorly understood. However, our lack of understanding is no barrier to incorporating nature's elegant solution: DNA polymerases are the foundation of all existing technologies for DNA sequencing even though human engineering remains decades away from designing or building comparable nanoscale machinery.

The principle of using components with complex biological function to build useful new technologies is the primary opportunity for single-molecule bioelectronics. Enzymes, antibodies, proteins, and other molecules have evolved to perform functions that include long- and short-term memory, linear and nonlinear response to stimuli, logic and decision making, and construction and demolition. New technologies can exploit molecular biology's dense layers of functionality by transducing these functions into accessible physical signals. As one example, single-molecule light microscopy is already flourishing as a new tool for life science research despite the physical limitations of diffraction and single-photon shot noise. Single-molecule bioelectronics constitutes a complementary toolkit based on solid-state electronics, and this electronic transduction will inherit the scaling and noise characteristics of modern semiconductor processing.

Precise control of SWNT band structure and fabrication of SWNT devices has not been an outstanding hurdle to prototyping these bioelectronic applications. In fact, modest arrays of 100-1000 devices have been fabricated and tested. In practice, insensitivity to SWNT variability arises because semi-metallic and semiconducting SWNTs both exhibit enough gate sensitivity to transduce protein activity [5]. Fig. 5 compares the gate sensitivity of example devices fabricated from semi-metallic and semiconducting SWNTs to illustrate the similar transconductance obtained after each was labeled with KF. Protein attachment creates a local perturbation on SWNTs which, similar to covalent defects, produces enhanced sensitivity that is self-aligned with the attachment itself.

The real challenge to exploiting this self-alignment is the proper orientation of the protein on the SWNT transducer. In typical physiological buffers, screening limits the SWNT's sensitivity to charged amino acid sidechains that are immediately adjacent to the protein's attachment site. Using mutagenesis, we can create single-cysteine enzyme variants that attach to the maleimide linkers in the same orientation on each device. However, the signals generated by these enzymes are quite sensitive to selection of the attachment site.

All of the $\Delta I(t)$ data in Figs. 2-4 was generated by a particular KF variant (L790C), in which the leucine at position 790 was converted to a cysteine to direct the protein's orientation and attachment to SWNT devices. Fig. 6 shows comparison data generated by three alternate cysteine positions. An arginine-to-cysteine conversion at position 794 (R794C) gave KF a very similar orientation, and it generated similar $\Delta I(t)$ signals. However, a wild-type cysteine at position 907 (907C) placed KF in a different orientation that generated no signal in the underlying SWNT. A cysteine introduced at the N-terminus produced devices that transduced random, uncoordinated KF motions unrelated to ssDNA processing. These examples illustrate that KF cannot be treated as a monolithic signal-generating element. Successful bioelectronic designs will require either success with a shotgun approach trying different orientations one-by-one or else precise knowledge of the enzyme's structure and structure-function relationships. Detailed information is available for a widely studied enzyme like KF but not for every molecule that might conceivably be studied with this bioelectronic technique.

ACKNOWLEDGMENT

We acknowledge support for this work from NIH 1R01GM106957 and 5R01HG009188.

REFERENCES

[1] M. Ratner, "A brief history of molecular electronics," *Nature Nanotechnology*, vol. 8, no. 6, pp. 378-381, 2013.

[2] F. Patolsky, B. P. Timko, G. H. Yu, Y. Fang, A. B. Greytak, G. F. Zheng, and C. M. Lieber, "Detection, stimulation, and inhibition of neuronal signals with high-density nanowire transistor arrays," *Science*, vol. 313, no. 5790, pp. 1100-1104, Aug, 2006.

[3] F. Patolsky, G. F. Zheng, and C. M. Lieber, "Nanowire-based biosensors," *Analytical Chemistry*, vol. 78, no. 13, pp. 4260-4269, Jul, 2006.

[4] S. Sorgenfrei, C.-y. Chiu, R. L. Gonzalez, Y.-J. Yu, P. Kim, C. Nuckolls, and K. L. Shepard, "Label-free single-molecule detection of DNA-hybridization kinetics with a carbon nanotube field-effect transistor," *Nature Nanotechnology*, vol. 6, no. 2, pp. 126-132, 2011.

[5] Y. Choi, I. S. Moody, P. C. Sims, S. R. Hunt, B. L. Corso, G. A. Weiss, and P. G. Collins, "Single-Molecule Lysozyme Dynamics Monitored by an Electronic Circuit," *Science*, vol. 335, pp. 319-324 2012.

[6] Y. Choi, T. J. Olsen, P. C. Sims, I. S. Moody, B. L. Corso, M. N. Dang, G. A. Weiss, and P. G. Collins, "Dissecting Single-Molecule Signal Transduction in Carbon Nanotube Circuits with Protein Engineering," *Nano Letters*, vol. 13, pp. 625-631, 2013.

[7] Y. Choi, G. A. Weiss, and P. G. Collins, "Single Molecule Bioelectronics," *Comprehensive Bioelectronics*, S. Carrera and K. Iniewski, eds., Cambridge: Cambridge University Press, 2015.

[8] O. Gül, K. Pugliese, Y. Choi, P. Sims, D. Pan, A. Rajapakse, G. Weiss, and P. Collins, "Single Molecule Bioelectronics and Their Application to Amplification-Free Measurement of DNA Lengths," *Biosensors*, vol. 6, no. 3, pp. 29, 2016.

[9] C. M. Joyce, "Techniques used to study the DNA polymerase reaction pathway," *Biochimica Et Biophysica Acta-Proteins and Proteomics*, vol. 1804, no. 5, pp. 1032-1040, May, 2010.

[10] T. J. Olsen, Y. Choi, P. C. Sims, O. T. Gul, B. L. Corso, C. Dong, W. A. Brown, P. G. Collins, and G. A. Weiss, "Electronic Measurements of Single-Molecule Processing by DNA polymerase I (Klenow fragment)," *Journal of the American Chemical Society*, vol. 135, no. 21, pp. 7855-7860, 2013.

[11] K. M. Pugliese, O. T. Gul, Y. Choi, T. J. Olsen, P. C. Sims, P. G. Collins, and G. A. Weiss, "Processive Incorporation of Deoxynucleoside Triphosphate Analogs by Single-Molecule DNA Polymerase I (Klenow Fragment) Nanocircuits," *Journal of the American Chemical Society*, vol. 137, pp. 9587-94, 2015.

[12] P. McInerney, P. Adams, and M. Z. Hadi, "Error Rate Comparison during Polymerase Chain Reaction by DNA Polymerase," *Molecular Biology International*, vol. 2014, pp. 8, 2014.

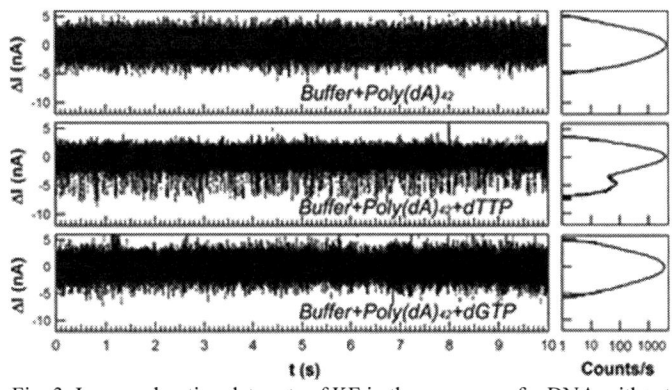

Fig. 1. Schematic device architecture for a single, active biomolecule transducing electrical signals in a 1-D conductor. In the 1-D limit, source and drain electrodes can be fully passivated and far from the molecular attachment. Inset highlights one noncovalent scheme [5] for biofunctionalization of a SWNT using pyrene-maleimide linkers to a cysteine residue on the enzyme. The protein depicted here is DNA polymerase I from the bacillus phage phi29.

Fig. 3. Longer-duration data sets of KF in the presence of ssDNA without nucleotides (top), with complementary nucleotides (middle), and with noncomplementary nucleotides (bottom). Adapted from Olsen *et. al.* [10].

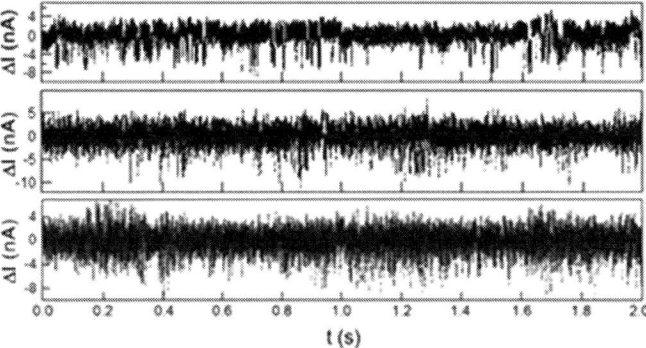

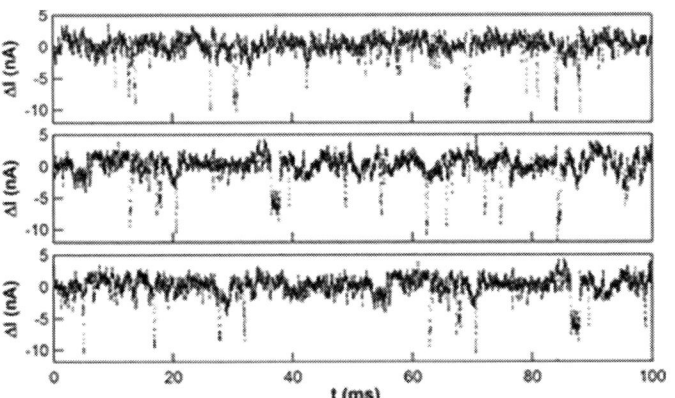

Fig. 4. Direct comparison of single-molecule KF signals generated by processing ssDNA templates with increasing complexity: (dA)$_{42}$ (top), (AC)$_{21}$ (middle), and (GTCR$_{10}$A)$_4$.

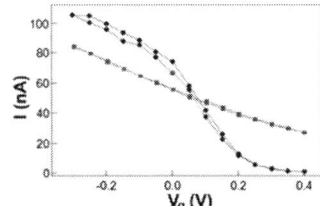

Fig. 5. Electrolyte-gated response of two protein-labelled SWNT devices, one having a semiconducting SWNT (black) and the other one having a semimetallic SWNT (red).

Fig. 2. Example snapshots of current fluctuations generated by KF while processing single-stranded, homopolymeric DNA template (poly dA$_{42}$ in the presence of dTTP). Adapted from Olsen *et. al.* [10].

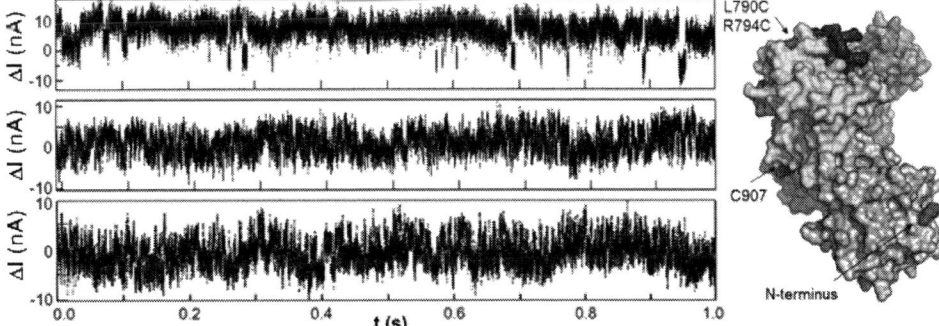

Fig. 6. Example signals generated by three different KF devices processing poly dA$_{42}$. Each panel represents a different orientation of the enzyme on a SWNT device, where the orientation is designed into the protein using mutagenesis. R794C (top panel), C907 (middle), N-terminus (bottom).

Si Nanowire Biosensors Using a FinFET Fabrication Process for Real Time Monitoring Cellular Ion Actitivies

Qingzhu Zhang[1, 2], Hailing Tu[1], Huaxiang Yin[2, 5], Feng Wei[1], Hongbin Zhao[1], Chunling Xue[3], Qianhui Wei[1], Zhaohao Zhang[2], Xiao Zhang[1], Shaoming Zhang[1], Qin Han[3], Yudong Li[2], Robert Chunhua Zhao[3], Jiang Yan[4] Junfeng Li[2] and Wenwu Wang[2]

[1]State Key Laboratory of Advanced Materials for Smart Sensing, General Research Institute for Nonferrous Metals, Beijing, China; (email: tuhl@grinm.com; weifeng@grinm.com). [2]Key Laboratory of Microelectronics Devices & Integrated Technology, Institute of Microelectronics, CAS, Beijing, China; (email: yinhuaxiang@ime.ac.cn). [3] Institute of Basic Medical Sciences Chinese Academy of Medical Sciences, School of Basic Medicine Peking Union Medical College, Beijing, China; (email: zhaochunhua@ibms.pumc.edu.cn). [4]College of Electronic and Information Engineering, North China University of Technology, Beijing, China; (email: yanjiang@ncut.edu.cn). [5]University of Chinese Academy of Sciences, Beijing, China.

Abstract—In this paper, a biocompatible biosensor based on horizontal Si nanowire (NW) array field-effect transistor (FET) has been fabricated by the feasible spacer image transfer (SIT) process. The Si NW FET as biosensor is proposed for the real-time cellular Ca^{2+} monitoring for mesenchymal stem cells (MSCs), which presents fast-responded and high-sensitive characteristics. Compared with the conventional sensing techniques, the Si NW biosensor exhibits non-invasive, biocompatible and reliable advantages. This will help us to further understand the mechanism of cellular ion activities and provides a promising method for the cell-level diagnose and therapy.

INTRODUCTION

With the developing of cell-based therapy techniques for a variety of diseases, efforts towards all-electrical device for electrophysiology have been made to expand high-precision intracellular recording methods for electrogenic cells. These versatile methods, such as patch clamp, substrate-integrated microelectrode arrays (MEAs) and multi-transistor arrays (MTAs) have been successfully applied to provide activity recording, electrical stimulation, and chemical stimulation [1–2]. Nowadays, complementary metal-oxide-semiconductor (CMOS) circuit based methodologies have revealed many promising advantages in recording the reaction under the cell-level activity. Especially, Si NW FETs with the merits of high surface-to-volume (S/V) ratio, excellent biocompatibility and being a CMOS-compatible fabrication in large scale, have been of particular interests for their significantly increased recorded signal strength [3]. However, most of the Si NW FETs sensors are fabricated with the direct-write electron beam process with large landing pads or Si NWs are fabricated with vertical arrays, which bring difficulty and low efficiency in mass production or get invasive to cells. Although it is recognized that Si NW in nanoscale are suitable to fabricate high-sensitivity sensors for non-invasive recording cell activities, up to date, there are few reports on this field.

In this work, we developed horizontal Si NWs array with high efficiency by spacer image transfer (SIT) process and fast-responded, high-sensitive Si NW sensors to determine the cellular activities for mesenchymal stem cells (MSCs). Traditionally, optical methods for interrogation combining Ca^{2+} indicators and potentiometric dyes are used as intracellular recording. Si NW FETs, as the non-invasive method, are successfully used as cell sensors with high-sensitive characteristics on intracellular activities. Comparing with the typical optical methods, this method shows the remarkably consistent results, which further identify the mechanism of the intracellular activities for MSCs by monitoring the extrusion state of Ca^{2+}.

DEVICE FABRICATION & CHARACTERIZATION

The Si NWs array sensors were fabricated on 200 mm silicon on insulator (SOI) wafers with a 55 nm thick boron-doped p-type Si (8-12 $\Omega\cdot$cm). The process flow based on the conventional main stream FinFET fabrication is shown in **Fig. 1** [4-5]. Firstly, the thickness of the top silicon were reduced to 30 nm through sacrificial oxidation followed by SiO_2 removal using DHF (**Fig.1 (b)** and **(c)**). Then a sequential deposition of multi-layer SiO_2/amorphous Si (α-Si)/Si_3N_4 were performed (**Fig. 1 (d)**). A conventional lithography process (i-line stepper) followed by conventional dry etch processes for the Si_3N_4 and α-Si films were carried out to form rectangular masks arrays, (**Fig. 1(e)**. Next, the top Si_3N_4 hard masks (HMs) was removed by hot H_3PO_4 solution (**Fig. 1 (f)**). A Si_3N_4 film with 30 nm thickness was deposited and then the corresponding reactive ion etch (RIE) of Si_3N_4 was performed to form spacers (**Fig. 1 (g)** and **(h)**). Afterwards, the inner core α-Si material between two Si_3N_4 spacers was removed by tetramethylammonium hydroxide (TMAH) and then nanometer-size Si_3N_4 spacers HMs arrays were left on the top of SiO_2 film (**Fig.1 (i)**). Following by the dry etch processes of oxide and ultra-thin top Si, the Si NWs arrays were formed after the removal of the top HMs using DHF (**Fig.1 (j)** and **(k)**). Compared with the direct-write electron beam process, the SIT process indicates higher efficiency without any landing pads. It is beneficial to provide an excellent solution for reducing the cost of NWs fabrication and achieving smaller volume of integrated sensors. Next, the electrodes were patterned by negative resist process and deposited 100 nm thick Ti/Au layers by sputtering followed by lift-off processes (**Fig.1 (l)**). The electrodes were designed to 2 mm in length to isolate the test probes from drop of cells. To

978-1-7281-1988-5/18 $31.00 © 2018 IEEE

ensure a signal recording with higher sensitivity in liquid environment, a conformal 10 nm thick biocompatible high-k HfO_2 layer was grown by atomic layer deposition (ALD) process (**Fig.1 (m)**). The non-invasive recording of the MSCs is carried out when a good adherence is achieved between the cells and the surface of the Si NW sensors (**Fig.1 (n)**).

The cross-sectional and top views were observed using Hitachi scanning electron microscope (SEM), and transmission electron microscope (TEM) to characterize the morphological profiles of the formed structures. The electrical characterization was performed using a Keithley 4200 semiconductor parameter analyzer after the adherent of MSCs to the Si NWs sensors.

RESULTS AND DISCUSSITION

The images of the fabricated Si NWs sensors by the SIT process are shown in **Fig. 2**. **Fig.2 (a)** and **(b)** are the top views of Si NW arrays by optical microscope and corresponding SEM. Highly uniform Si NWs arrays with 30 nm in thickness are formed without any landing pads, which are suitable to integrate more Si NW sensors or achieve smaller size of the devices. **Fig.2 (c)** shows cross sectional TEM image of the Si NW sensors. A very conformal and uniform HfO_2 layer is observed, which is contributed to a good isolation between electrode and the solution environment. The Si NW is about 30 nm in height and 25 nm in width, and provides a very high S/V ratio. **Fig.2 (d)** and **(e)** are top views of sensors consist with Si NW arrays. The electrodes are extended 2 mm in length to avoid a large leakage through the solution and the Si NWs sensors during the cells measurement. The gate lengths (L_Gs) of the Si NW FETs are fabricated in 30 μm, 20 μm and 10 μm respectively. Si NWs are connected together with one Au electrodes pair. **Fig.2 (f)** indicates that the Au electrodes are well defined and have perfect controlled boundaries.

The transfer and output curves of 30 μm L_G p-type SOI Si NWs sensors are plotted in **Fig. 3** and **Fig. 4** respectively. The results show the Si NWs could be used as promising FETs sensors with a good modulation by bias gate voltage. **Fig. 5** illustrates the typical curves of the current as a function of applied voltage for 30 μm L_G p-type Si NWs sensors and the resistances of sensors are extracted. The non-linear characteristic may be caused by the direct contact between Au and Si NWs with light doping, which needs a further process optimization in the future. The statistical resistance values for the Si NW sensors with different L_Gs are shown in **Fig. 6**. The values of the resistance owe a good uniformity and drops with the decreasing of L_Gs. **Fig. 7** shows the optical image of the self-made Si NW sensors. **Fig. 8** demonstrates the reliability results of our integrated Si NW sensors after introducing deionized water for 6000 s. It is demonstrated that the integrated Si NW sensors are fully reliable, ensuring long term stable operation.

Fig. 9 (a) presents the recording results of the MSCs monitoring by Si NW FETs sensor and the multiple additions of Ca^{2+} stimulator is performed during the test. 1μL MSC solution with a 10^{15} MSCs per litre is dropped on the active area. It is found that the MSCs emerge adherence to the active area surface of the fabricated Si NW sensors (inserted figure in **Fig.9 (a)**). A long time for monitoring the MSCs activities is recorded

using Keithley 4200 with 0.1 s intervals. As shown in **Fig. 9 (a)**, we observed sharply oscillations of the recording current by addition of Ca^{2+} stimulator after about 100 s. The great oscillations indicate the drastic ion activities for the tested MSCs. In general, this phenomenon can be recorded about two or three times for the tested MSCs after several addition of Ca^{2+} stimulator and there are no response for continuous addition of Ca^{2+} stimulator. **Fig. 9 (b)** and **(c)** show the amplified current signals during the great oscillations period. We believe that these observed oscillations with varying frequency and amplitude could be used to investigate specific activities of the MSCs. **Fig. 9 (d)** shows the result of blank test sample for comparison, where the Ca^{2+} stimulator was added into the dropped nutrient solution without MSCs. It is found that there are no any oscillation periods appearing (see **Fig. 9 (d)**), which further demonstrates that the new generated oscillations are caused by the activities of the MSCs.

The two-photon fluorescence microscope was used to monitor intracellular calcium changes in real time after Ca^{2+} stimulator treated with MSCs. After addition of Ca^{2+} stimulator, intracellular Ca^{2+} significantly increases so that calcium-associated protein expression will be increased [6]. **Fig. 10 (a)** shows fluorescence images of MSCs developed by Ca^{2+} stimulator for 0 s, 70 s, 100 s, 150 s and 200 s, respectively. **Fig. 10 (b)** shows the intensity of the fluorescence as a function of the time after addition of Ca^{2+} stimulator. After stimulation about 100 s, it has shown much better and clearly imaging quality of the fluorescence and abrupt higher increasing of fluorescence intensity caused by larger intracellular Ca^{2+} concentration.

These results recorded by Si NW sensors are in good agreement with the results recorded by the conventional optical method, as shown in **Fig. 9**. However, it is worth noting that the recorded results from Si NW sensors provide higher precision with electrical signals. It is believed that these Si NWs sensors can monitoring various activities of the electrogenic cells, which are helpful for further investigation on cell-based monitoring and corresponding mechanism.

SUMMARY AND CONCLUSION

In summary, Si NW array FETs biosensors are fabricated through the conventional main stream FinFET process. The SIT processes used for the formation of Si NW provide the capacities of higher efficiency and integrated level to fabricate the Si NW FET biosensor, which have been successfully applied to the real-time monitoring cellular Ca^{2+} activities of the MSCs with high sensitivies, indicating a promising candidate for low-cost and ultra-sensitive cell-based monitoring and therapy technique.

REFERENCES

[1] Bertholet, Ambre M., et al., in *Cell metabolism* Vol.25, pp. 811-822. [2] A. Casanova1, M-C. Blatche1, F. Mathieu et al., in *IEDM Tech. Digest*, 2017, pp.641-644. [3] Chen K I, Li B R, Chen Y T. in *Nano Today*, 2011, Vol. 6, pp. 131-154. [4] S. Natarajan, M. Agostinelli, S. Akbar et al. , in *IEDM Tech. Digest*, 2014, pp.71-73. [5] Q.zhang, H. Tu, H. Yin et al., in *Microelectronic Engineering*, 2018, vol. 198, pp. 48-54. [6] S. Wang, X. Qu, R. C. Zhao et al. , in *Journal of hematology & oncology*, 2012, vol.5 pp. 1-9.

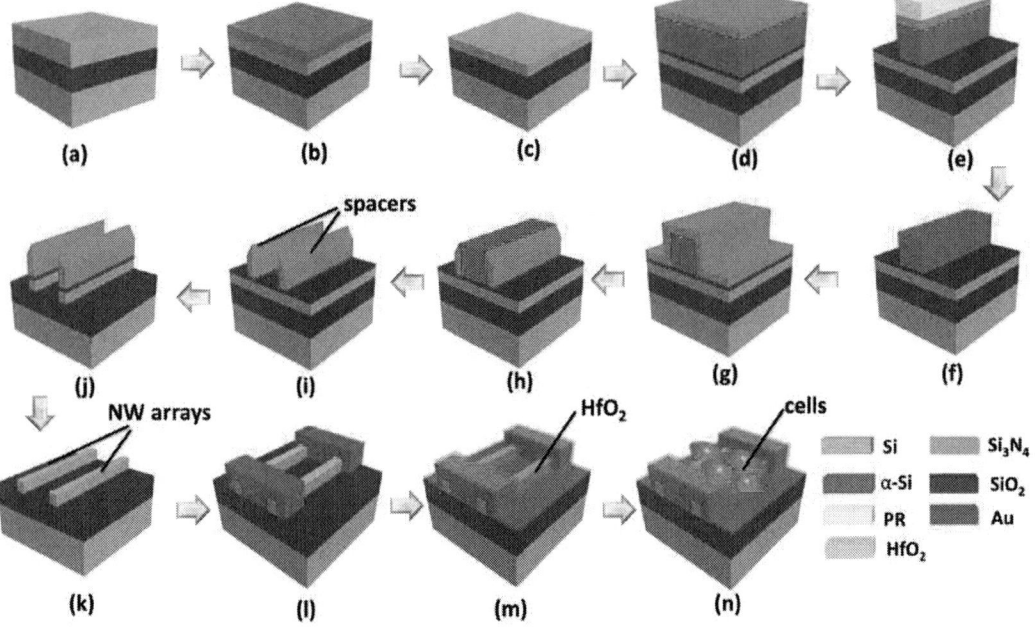

Fig. 1. Process flow of the Si NWs arrays formation. **(a)** 200 mm SOI wafers with p-type (100) crystal face, **(b)** 30 nm sacrificial oxidation of top silicon, **(c)** removal of the top SiO_2, **(d)** 25 nm Oxide/100 nm α-Si/ 25 nm Si_3N_4 deposition, **(e)** lithography and Si_3N_4/α-Si RIE, **(f)** Si_3N_4 hard mask removal, **(g)** Si_3N_4 film deposition, **(h)** Si_3N_4 spacers hard mask formation by RIE, **(i)** removal of α-Si dummy gate, **(j)** oxide and top Si RIE, **(k)** Si_3N_4 spacers hard masks removal **(l)** Au contact by lift-off process, **(m)** 10 nm thick HfO_2 dielectric deposition by ALD process, **(n)** introduce cells adherence to the HfO_2 dielectric surface.

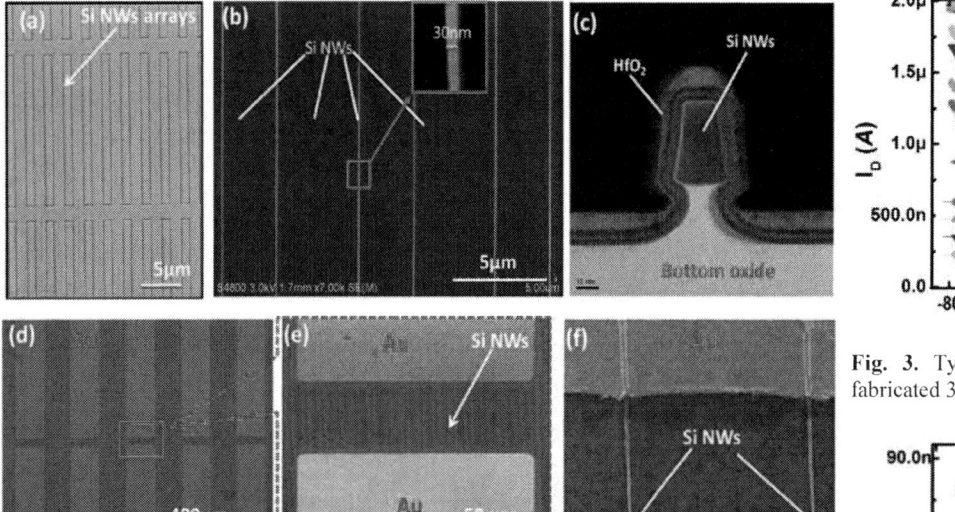

Fig. 2. Fabrication of Si NWs sensor: **(a)** top view of Si NW arrays by optical microscope, the length of NWs is about 50 μm **(b)** corresponding SEM image of Si NW arrays, the width of NWs is about 30 nm, **(c)** cross sectional TEM image of Si NW sensors, conformal and uniform HfO_2 layer are observed, which is attributed to a good isolation between electrode and the solution of cells. (d) & (e) top views of Si NW arrays sensors, the electrodes are extended 2 mm in length to avoid a large leakage through the solution and the Si NWs sensors are with 30 μm in L_G and Si NWs connected to Au electrode, (f) the Au pads are well defined and have well controlled boundaries.

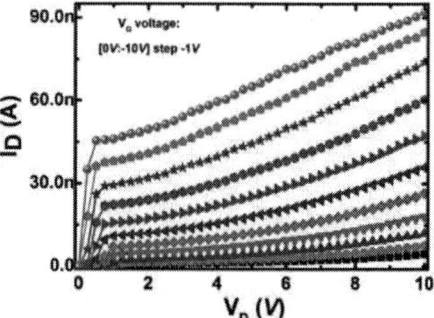

Fig. 3. Typical transfer characteristics of the fabricated 30 μm L_G Si NW sensors.

Fig. 4. Typical output characteristics of the fabricated 30 μm L_G Si NW sensors.

978-1-7281-1988-5/18 $31.00 © 2018 IEEE   681

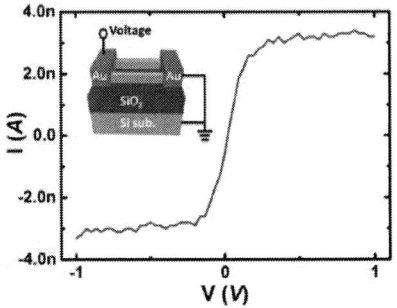

Fig. 5. Typical curves of the current as a function of applied voltage for Si NWs sensors.

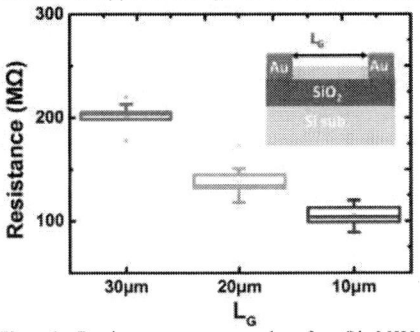

Fig. 6. Resistance test results for Si NWs sensors with different L_{GS}.

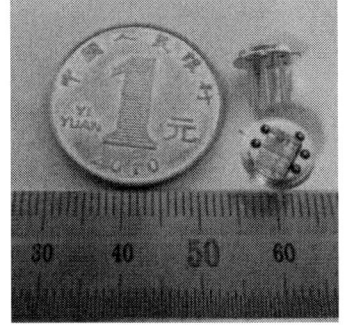

Fig.7. Photo of the self-made Si NW sensors.

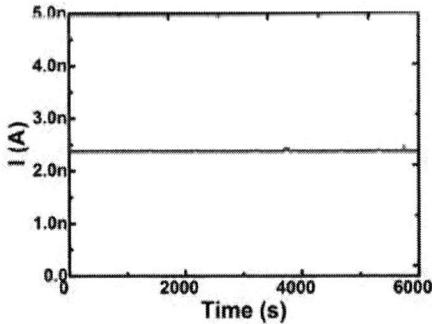

Fig.8. Stability of the fabricated Si NWs sensors while perfusing deionized water.

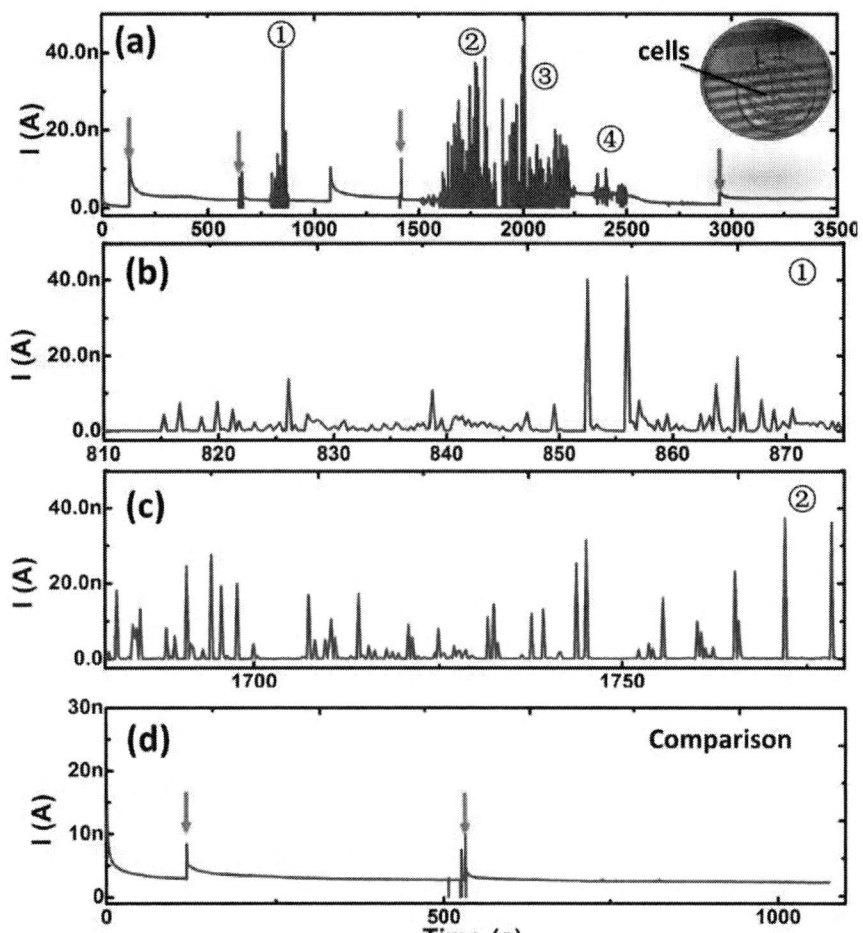

Fig. 9. (a) Long time recording of cells activities by the Si NWs sensors, the red arrows represent the addition of additions of Ca^{2+} stimulator. **(b)** and **(c)**, the amplified during the great oscillations period, **(d)** additions of Ca^{2+} stimulator into the nutrient solution for comparison.

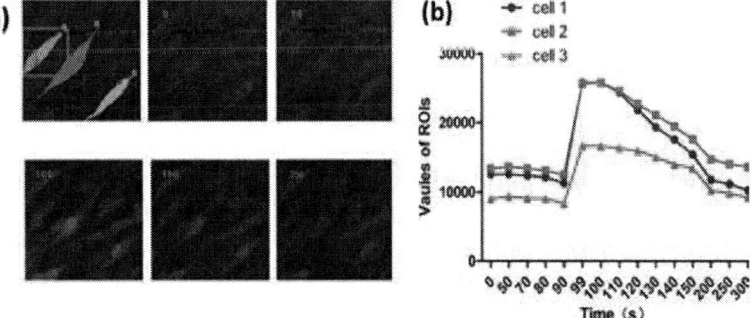

Fig. 10. (a) Fluorescence images of MSCs developed by Ca^{2+} stimulator, **(b)** the intensity of the fluorescence as a function of the time after addition of Ca^{2+} stimulator.

A Flexible, Heterogeneously Integrated Wireless Powered System for Bio-Implantable Applications using Fan-Out Wafer-Level Packaging

G. Ezhilarasu[1], A. Hanna[1], R. Irwin[1], A. Alam[1], and S. S. Iyer[1]

[1]Center for Heterogeneous Integration and Performance Scaling, Department of Electrical Engineering, University of California, Los Angeles, CA90095, email: goutham93@g.ucla.edu

Abstract— Fan-Out Wafer-Level Packaging (FOWLP) is used to fabricate a near field wireless implantable system on an ultra-flexible (~5mm bending radius) & biocompatible elastomeric substrate. A µLED is powered wirelessly with an efficiency > 15% @ 1cm transmit distance. The implantable system is only ~535µm thick with a diameter <2cm.

I. INTRODUCTION

In recent years, there has been a surge of interest in the development of flexible wireless medical implants for healthcare monitoring and stimulation therapy. Conventional implants are bulky and made of rigid materials that often cause patient discomfort and infection with prolonged use [1]. Mostly battery operated without a wireless recharging capability, they also have a limited lifetime hence requiring the patient to go through periodic surgical procedures for replacements [2]. Flexible implants, however, can be made with organic materials that closely match the mechanical flexibility of human tissue and can thus conform to any surface inside the body causing minimal discomfort to the patient [3]. They can also have sub-millimeter thicknesses allowing easy access to constrained spaces in the body without causing infections. By making such implants wireless, it is possible to increase their lifetime significantly and hence avoid the necessity of periodic surgery [2].

Wireless medical implants are often highly complex and heterogeneous systems consisting of several active chips and passives. The complexity of these systems has historically been a bottle neck for their high density integration on flexible platforms which typically use a flex-rigid PCB or a conventional Flexible Hybrid Electronics (FHE) approach [4]. Flex-rigid PCBs use mounted packaged chips and are only semi-flexible while conventional FHE relies on integration of chips using printed interconnects which have coarse interconnect pitches in hundreds of µm range [4]. Hence both techniques severely limit the system integration density. As an alternative approach, we adapted FOWLP techniques to FHE to develop a novel platform called FlexTrate™ [5]. In this platform, we use standard Si BEOL processing techniques to interconnect heterogeneous dies embedded in PolyDiMethylSiloxane (PDMS) at very fine interconnect pitches of < 40µm. This technology thus opens new doors to building high performance implantable systems that have small form factors and high flexibility. PDMS is also typically used as an encapsulation material in medical implants since it has a low Young's modulus that is comparable to human tissue [6].

In this paper, we demonstrate the versatility of this flexible packaging scheme by heterogeneously integrating upto five different commercially available components (see Fig. 1(b)) along with a resonant magnetic coupling coil at 40µm interconnect pitch to show a fully functioning wireless powered implantable system (shown schematically in Fig. 1(a). A circuit diagram of the complete system is given in Fig. 1(c). The system provides a regulated DC voltage, using an active Low Drop Out (LDO) regulator, to a green (AlGaInP, λ=570nm) µLED. This technology has potential applications as a subdermal implant for Optogenetic stimulation therapy to treat neurodegenerative diseases [7]. A detailed experimental analysis of the power transfer link efficiency and system performance is conducted in different environments and bending conditions. To the best of our knowledge, this is the first demonstration of a fully functional ultra-flexible & thin wireless powered subdermal implant based on FOWLP with an interconnect pitch < 40µm and total package thickness of only 535 µm.

II. DESIGN AND ANALYSIS OF WIRELESS POWER TRANSMISSION LINK

A. Electromagnetic design and charactarization

We designed a two Coil resonant magnetic Wireless Power Transmission (WPT) link operating at a frequency of 13.56 MHz for an optimal coupling distance of 1cm using analytic techniques specified in literature, [8]. Results of the optimized design are shown in Fig. 2 (a). For resonance at 13.56 MHz, the external coil requires a 50pF series capacitance while the implant coil requires a 150pF parallel capacitance. The optimized coils were then fabricated on FlexTrate™. The corrugations in the metallization which are needed for mechanical reliability (as discussed in [5]) do not affect the electromagnetic behavior of the coils. The fabricated coils were characterized using a Vector Network Analyzer (VNA) and compared with HFSS simulations as shown in Fig. 2 (b & c). A good agreement between experimental results and simulation is found for the designed coils. To compare the performance of the WPT link in different environments, we measured the Power Transfer Efficiency (PTE) as a function of frequency & coupling distance with a VNA in air and a human body phantom using Phosphate Buffered Saline (PBS) in a glass beaker. The results of the measurement in both environments is shown in Fig. 3 (a, b, & c). Although degradation in PTE is observed at 1cm coupling distance from ~30% in air to ~15% in PBS, it is still within acceptable limits. We also studied the effect of bending on the performance of the link by measuring the PTE as a function of the implant coil bending radius in both air and PBS. The results of the measurement as shown in Fig. 3 (d) demonstrates a PTE >12% & 4% even at an exaggerated bending of 5mm radius in air & PBS respectively.

978-1-7281-1988-5/18 $31.00 © 2018 IEEE

B. Cyclic Bending Reliability

The mechanical reliability of the corrugated implant coil was studied by measuring its inductance and quality factor after cyclic bending at 10 mm and 5 mm bending radius for a total of 1000 cycles each as described in [5]. The results of this measurement are shown in Fig. 4 (a). The inductance has < 1% and quality factor < 5% change after the cyclic bending test. The relatively large decrease in quality factor at the initial onset of bending is suspected to be because of elongation of the metallization perpendicular to the axis of bending causing an increase in wire resistance.

III. OPTIMAL DIE PLACEMENT

A total of eight dies (three 100pF Caps, two 1μF Caps, Schottky Diode, LDO & μLED) are needed for the full implant system. To address the optimal placement of the heterogeneous dies ensuring flexibility without compromising form factor, we performed FEA to evaluate minimum optimal die pitch for various die sizes and thicknesses. The simulation assumed Quarter symmetry with a 15mm x 4mm x 0.5mm sample of Polydimethylsiloxane (PDMS) with embedded Si dies bent around a 5mm radius surface until full conformity is reached. The mechanical properties of PDMS and Si were taken from literature [9], [10].

Strain was measured from the edge of the outer die to the edge of the center die along the line of symmetry on the top surface of the PDMS keeping die thickness fixed. The result of this simulation is plotted in Fig. 4 (b). We find that the larger the size of the die, the higher is the peak strain at the die edge for a given inter-dielet spacing. Hence to reduce the peak strain in PDMS, it is necessary to space the dies further apart. We also performed a simulation of peak strain in the die (Si) as a function of inter-dielet spacing for variable die thickness keeping die size fixed (see Fig. 4 (c)). We find that for die thickness in the 100-400μm range, there is a negligible dependence of peak strain on inter-dielet spacing when it exceeds approx. half the die side. Based on these results, as a simple rule of thumb, we assumed the minimum inter-dielet spacing to be equal to the dimension of the largest die side.

IV. HETEROGENEOUS INTEGRATION FOR WIRELESS POWERED SYSTEM

We achieved successful integration of eight heterogeneous dies at 40μm interconnect pitch along with the optimized implant coil on FlexTrate[TM] using the process flow shown in Fig. 5. Optical images of the fabricated system are given in Fig. 6 (a & b). The process features two electroplated Copper metallization layers connected through a Via: Layer 1 at 40μm pitch, thickness of 7μm that interconnects the dies & Layer 2 with the implant coil of 20μm thickness. The two layers are prevented from shorting at crossings by a 10μm thick SU-8 bridge as shown in Fig. 6 (a). The entire implantable system is only 2 cm in diameter with a total thickness of ~535μm. We would like to point out that the Deep Trench (DT) Cap dies (~400μm thick) required a 500μm thick PDMS for encapsulation. If die thinning techniques are employed, an even thinner package is possible which could be in the sub 100μm range. To the best of our knowledge, this is first time

Si DT capacitors were integrated with other components on a flexible substrate using a completely solder-free process.

A. Experimental evaluation of system performance

To evaluate the performance of the WPT system integrated on FlexTrate[TM], we performed a comparative analysis with the same system on a breadboard. The output voltage (V_o of LDO) is measured with a multimeter as a function of coupling distance at 13.56 MHz in both air and the human body phantom. Results of the measurement are shown in Fig. 7 (a & b). There is < 5% drop in output voltage in the FlexTrate[TM] system compared to the breadboard system at 1cm coupling distance in both air & saline confirming the successful integration. Fig 7(c) shows a series of images of the integrated system successfully operating at different bending conditions in air. The μLED has very good luminescence even at a 5mm bending radius of the implant.

B. Comparison with similar systems in literature

A comparison of the system features and performance metrics between this work and similar studies on WPT in literature is reported in Table 1. Our technique achieves the finest interconnect pitch using a completely solder-free integration. Also, it is to be noted that a comparison based on the full system size is not an accurate representation of the integration density achieved by our system due to the relatively large area occupied by the implant coil for electromagnetic reasons. In terms of number of dies per unit area, our approach certainly has an upper edge.

V. CONCLUSION

We have successfully demonstrated a fully functional medical implant powered wirelessly using resonant magnetic coupling with upto eight heterogeneous dies integrated (solder-free) at 40μm interconnect pitch on FlexTrate[TM]. The overall implant system has a maximum thickness of only ~535μm and remains operational even at a significant bending of 5mm radius. Such an ultra-flexible and thin wireless system can find immense application for subdermal implants where conforming to curvilinear surfaces and tight spots is a necessity. The standalone system with the single green μLED can find potential application in Wireless Optogenetics.

ACKNOWLEDGMENT

The authors gratefully acknowledge the contributions of UCLA CNSI and Nanolab cleanroom staff. This work was supported in part by AFRL/NBMC, DARPA, SRC, UCLA CHIPS Consortium & the UC system.

REFERENCES

[1] L. Luan et al., *Science Adv.*, Vol.3, p. e1601966, 2017. [2] J. E. Ferguson et al., *Expert Rev Med Devices*, July 2011; 8(4): 427-433. [3] Kenry et al., *Microsystems & Nanoengineering* (2016) 2, 16043. [4] R. H. Reuss et al., *Proc. IEEE*, Vol 103, No. 4, pp.491-496, Apr. 2015. [5] A. Hanna et al., *Proc. of Electronic Components and Technology Conference* (*ECTC* 2018). [6] M. A. McClain et al., *Biomed Microdevices* (2011) 13:361-373. [7] J. D. Ordaz et al., *Neural Regen Res*, Aug 2017; 12(8): 1197-1209. [8] U. M. Jow, *Conf Proc IEEE Eng Med Biol Soc.* 2009; 2009:6387-6390. [9] A. P. Roman, (2004) *Scholar Archive*, 3088. [10] M. A. Hopcroft, *journal of MEMS*, Vol.19, No.2, Apr 2010. [11] Shin *et al.*, *Neuron* 93, 509-521, Feb 8, 2017. [12] J. Kim et al., *Adv. Funct. Mater*, 2015, 25, 4761. [13] K. Okabe et al., *Sensors* 2015, 15(12), 31821-31832.

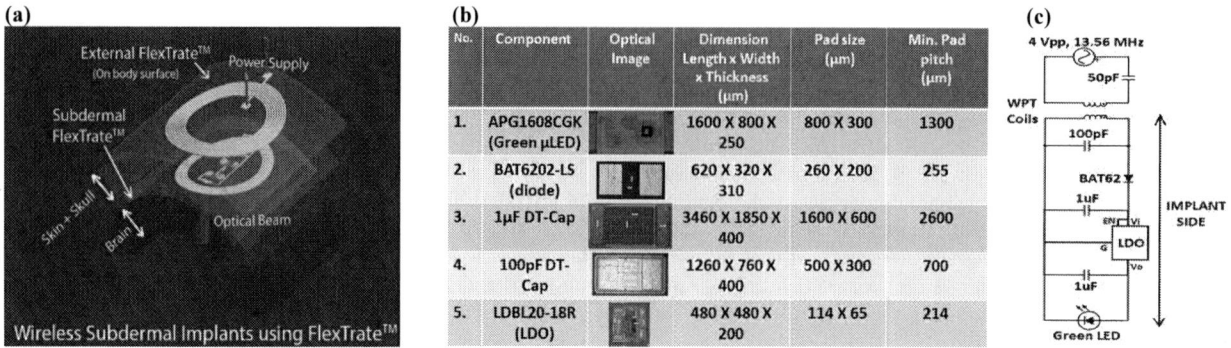

Fig. 1. (a) Schematic of system concept, (b) Table of components used in fabrication and (c) Circuit diagram of System

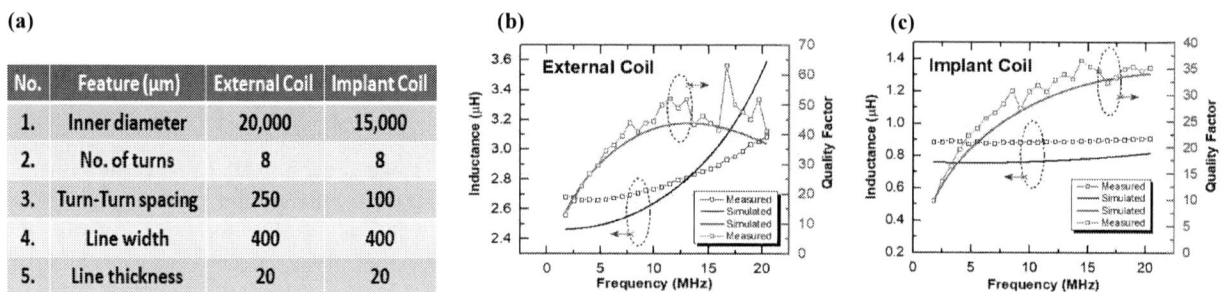

Fig. 2. (a) Optimized design parameters for WPT coils. Inductance (in μH) & Quality Factor vs. Frequency, both simulated and measured, for (b) External coil and (c) Implant coil

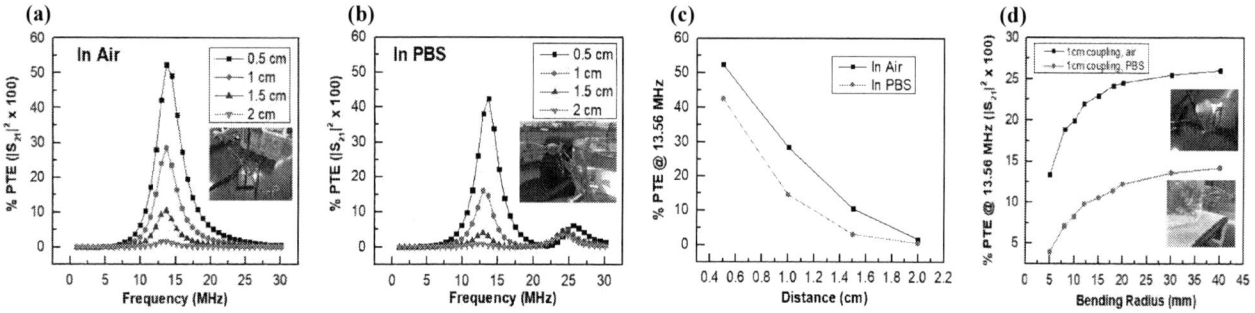

Fig. 3. PTE (Power Transfer Efficiency in %) of resonant WPT link vs. Frequency for different coupling distances in (a) air & (b) PBS. The second resonance observed at ~25MHz in (b) is suspected to be due to parasitic capacitances in the high K medium (ε_r~70-100 for PBS in HF band [5]). Comparison of PTE at 13.56 MHz as a function of (c) distance in air & PBS and (d) implant coil bending in air & PBS at a fixed coupling distance of 1cm. Insets in each diagram show experimental setup for measurement; for the coupling experiments in PBS, the external coil is placed on the outer surface of the glass beaker.

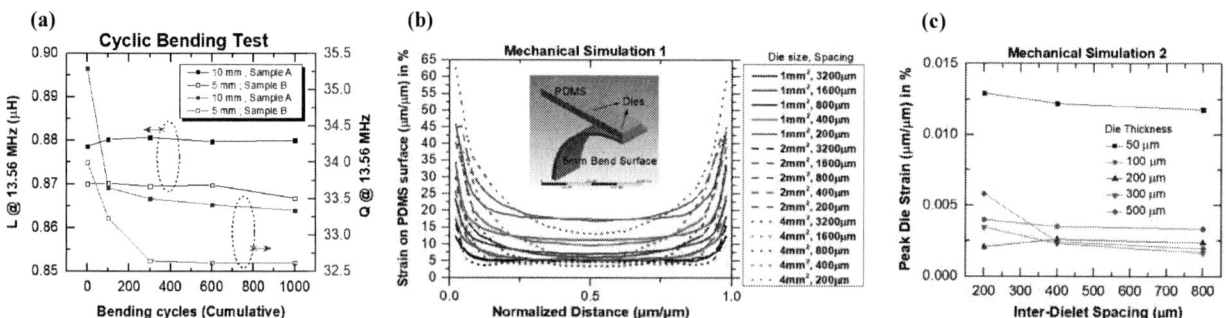

Fig. 4. (a) Inductance (in μH) and Quality factor of two Implant Coils, Sample A&B, plotted as a function of cumulative number of bending cycles for 10mm and 5mm bending radii respectively for upto 1000 bending cycles. Results of mechanical bending simulations for (b) variable die sizes & inter-dielet spacing with a fixed die thickness of 200μm; Strain on PDMS (0.5mm thick) surface vs. Normalized distance along symmetry line and (c) variable die thickness keeping die size fixed at 1mm by 1mm; peak strain on Si die vs. inter-dielet spacing. Inset in (b) shows simulation setup with Quarter Symmetry.

978-1-7281-1988-5/18 $31.00 © 2018 IEEE

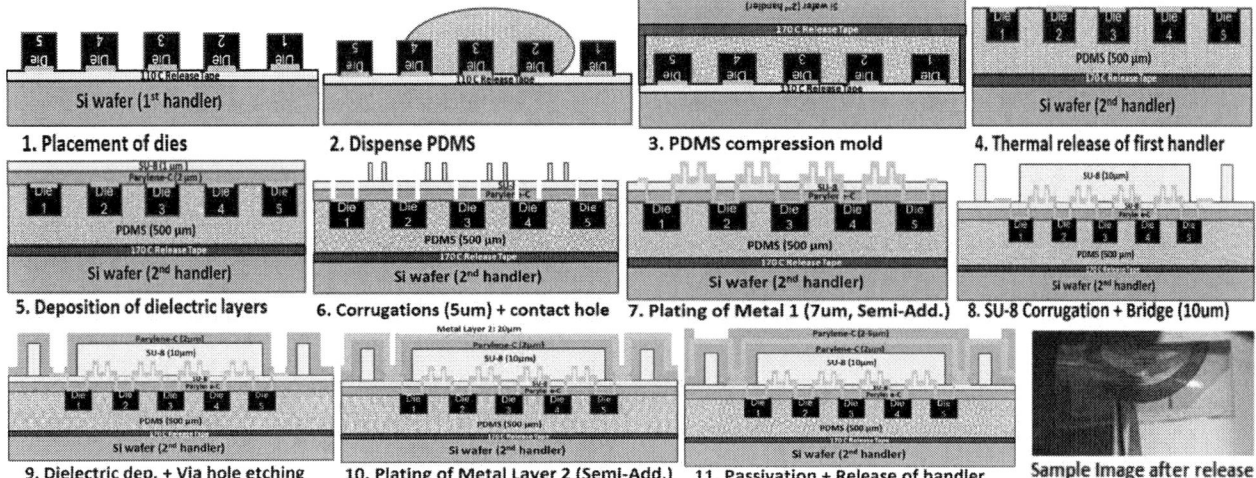

Fig. 5. Process flow for fabrication of FlexTrate™

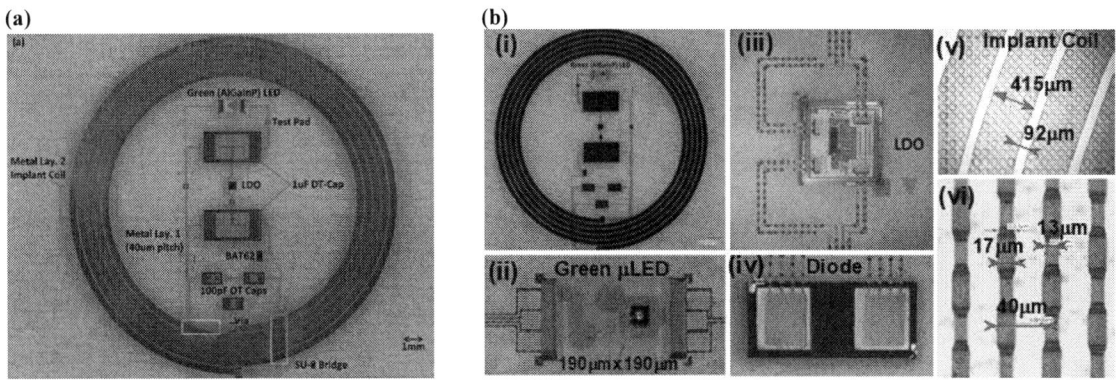

Fig. 6. (a) Optical image (Top-side) of the fabricated WPT system. **(b)** Optical image **(i)** shows the bottom-side of the same sample looking through the PDMS; zoomed in images of Green μLED (bottom-side), LDO & Schottky Diode (Top-side) are given in **(ii)**, **(iii)** & **(iv)** respectively. Optical images of corrugated metal layer 2 & 1 are shown in **(b)**, **(v)** & **(vi)** respectively.

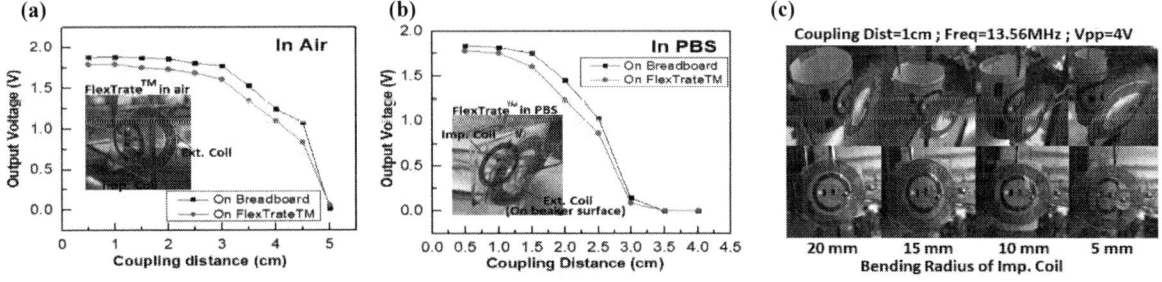

Fig. 7. Comparison of output voltage (V₀ of LDO, in Volts) vs. coupling distance for the breadboard test system and actual integrated system on FlexTrate™ in **(a)** air and **(b)** PBS. **(c)** shows a series of images for the FlexTrate™ system operating at different bending conditions in air for a coupling distance of 1cm (center-center). Insets in **(a)** & **(b)** are images of the integrated system operating in air & PBS respectively at 1cm coupling.

Ref.	Type of WPT	Freq. of WPT (MHz)	Size of Transmit antenna	Operating distance	PTE @ operating dist. (%) in air	PTE @ operating dist. (%) in Body medium	Minimum bending radius of coil (mm)	No. of components integrated	Solder-free integration?	Size of integrated system	Inter-connect pitch (µm)	Max. thickness of system (µm)	Application
[11]	NRIC	13.56	Rect., 30cm x 30cm	up to 30cm	NR	NR	9	4	N	Circular, dia. ~ 9.3mm	>140	1300	Wireless Optogen.
[12]	NIC	14-15	Rect., 5cm x 3cm	Up to 2cm	NR	NR	5	2	N	Circular, dia. ~ 4-7mm	>400	~130-150 (thinned dies)	Wearable
[13]	FFEM	825	NR	Up to 10cm	NR	Max. of 0.497% @ 5cm dist.	NR	1	Y, but ACP used	Rect., 2.7cm by 0.5cm	N/A	~410	Neural Implant
This Work	NRIC	13.56	Circular, ~ 3cm dia.	Upto 2cm	~30% @ 1cm dist.	~15% @ 1cm dist.	5	5	Y	Circular, dia. ~ 20mm	~40	~535 (un-thinned dies)	Wireless Optogen.

Table 1. Comparison of this work with others in literature. Abbreviations: NRIC: Near-field Resonant Inductive Coupling; NIC: Near-field Inductive Coupling; N/A: Not Applicable; FFEM: Far Field Electro-Magnetic; NR: Not Reported by author; ACP: Anisotropic Conductive Paste

978-1-7281-1988-5/18 $31.00 © 2018 IEEE

Parallel-Plane Breakdown Fields of 2.8-3.5 MV/cm in GaN-on-GaN p-n Junction Diodes with Double-Side-Depleted Shallow Bevel Termination

T. Maeda[1*], T. Narita[2], H. Ueda[2], M. Kanechika[2], T. Uesugi[2], T. Kachi[3], T. Kimoto[1], M. Horita[1], and J. Suda[1, 3]

[1]Kyoto University, Kyoto, Japan, [2]Toyota Central R&D Labs., Inc., Aichi, Japan, [3]Nagoya University, Aichi, Japan

*Email: maeda@semicon.kuee.kyoto-u.ac.jp

Abstract - We report homoepitaxial GaN p-n junction diodes with novel beveled-mesa structures. The n-layers and p-layers, the doping concentrations of which are comparable, were prepared. We found that electric field crowding does not occur in the structure using TCAD simulation. The fabricated devices showed the breakdown voltages of 180-480 V, small leakage currents, and excellent avalanche capabilities. The breakdown voltages increased at elevated temperature. At the breakdown, nearly uniform luminescence in the entire p-n junctions was observed in all the devices. These results are strong evidences that the uniform avalanche breakdowns occurred in the devices. We carefully characterized the depletion layer width at the breakdown, and the parallel-plane breakdown electric fields of 2.8–3.5 MV/cm were obtained, which are among the best of the reported non-punch-through GaN vertical devices.

I. INTRODUCTION

Owing to its high critical electric field ($E_{cr} \sim 3$ MV/cm), GaN has attracted much attention as a material for the next-generation power devices. There have been many reports on GaN vertical power devices [1–23]. To suppress a premature breakdown due to electric field crowding at a device edge, these devices have edge termination structures: such as field plates [3, 9], ion implantation isolations [7, 8], and ion implanted bilayer edge terminations [10, 11]. These edge termination structures realized high breakdown voltages (V_b) and excellent figure-of-merits. However, electric field crowding still remained and the parallel-plane breakdown electric fields (E_b) seemed not to reach E_{cr} of GaN in the reported devices.

In this study, we propose GaN p-n junction diodes (PNDs) with small-angle beveled-mesa structures. We performed TCAD simulation to investigate the electric field distributions in beveled-mesa edges with various doping profiles, and found that electric field crowding does not occur when the donor and acceptor concentrations in n- and p-layers are comparable. We fabricated the GaN PNDs with four different non-punch-through epitaxial structures. The reverse current-voltage (I-V) characteristics of the fabricated devices were investigated. The parallel-plane E_b of 2.8, 2.9, 3.2, and 3.5 MV/cm were obtained for the devices with V_b of 480, 385, 250, and 180 V, respectively.

II. DEVICE DESIGN AND FABRICATION

There are two configurations of beveled-mesa structure; the positive bevel and the negative bevel. In the positive bevel, which is defined as one where more material is removed from the edge when progressing from the heavily doped side to the lightly doped side of the p-n junction, electric field crowding does not occur in theory, since the depletion layer at the mesa surface extends wider than that at the bulk [24].

In the previous studies on GaN vertical power devices, a lightly Mg-doped p-GaN was not employed, since the growth of such p-GaN was difficult. Recently, the growth technique of a lightly doped p-GaN has been developed [25]. The control of Mg doping of $\sim 10^{17}$ cm^{-3} is now possible.

In this study, we modify the concept of the positive bevel to improve the breakdown voltages. We performed the two-dimensional numerical simulations of the electric field distributions for the beveled-mesa structures with various epitaxial structures using Sentaurus TCAD [26]. We found that lower acceptor concentrations in upper-side p-layers than donor concentrations in lower-side n-layers are not necessary for eliminating electric field crowding in the case of a shallow bevel angle. For example, $E_{pp}/E_{max} > 99\%$ can be realized when $N_a < 2N_d$ and $\theta < 15°$ for a non-punch-through condition. This double-side-depleted shallow bevel can be fabricated by the current p-GaN growth technique, and this edge termination is expected to realize GaN power devices without electric field crowding.

Fig. 1 shows the device structure of the GaN PND with the double-side-depleted shallow bevel termination. Four samples with various n- and p-layers (PN1–4) were prepared. The p$^+$-layers, p-layers, and n-layers were grown by metal-organic vapor phase epitaxy (MOVPE) on GaN bulk substrates grown by hydride vapor phase epitaxy (HVPE). **Fig. 2** shows the depth profiles of the Mg and Si concentration measured by secondary ion mass spectrometry (SIMS) in the epilayer in PN1 representatively. Mg in the p-layer and Si in the n-layer are uniformly distributed along the depth direction. **Fig. 3** shows the capacitance-voltage (C-V) characteristics of PN1 representatively. Excellent linearity was observed in $1/C^2$-V plots, indicating that net acceptors and net donors are also uniformly distributed along the depth direction. For all the devices, reduced net donor concentrations $N_a N_d/(N_a+N_d)$ are very close to [Mg][Si]/([Mg]+[Si]). The thicknesses, [Si], [Mg], and $N_a N_d/(N_a+N_d)$ in PN1–4 are listed in **Table I**.

978-1-7281-1988-5/18 $31.00 © 2018 IEEE

Since we employed the growth conditions which reduce residual carbon, the compensations by carbon and/or other impurities [27] are negligible. The beveled-mesa structures were formed by Cl_2-based inductive coupled plasma-reactive ion etching (ICP-RIE) with a photoresist mask [28, 29]. Smooth mesa surfaces without any structural defects were observed by scanning electron microscopy (SEM). The mesa angles and mesa heights were 12° and 3.6 μm, respectively.

III. EXPERIMENTAL RESULTS

Fig. 4 shows the reverse *I-V* characteristics of PN1–4. Small leakage currents were observed in all the devices, which may be caused by variable-range-hopping through dislocations [12, 30]. The V_b of approximately 480, 385, 250, and 180 V were obtained for PN1, PN2, PN3, and PN4, respectively. The variations of the V_b among the devices in the same wafer were very small. All the devices showed excellent avalanche capabilities; a catastrophic breakdown did not occur and the same characteristics can be reproduced many times. **Fig. 5** shows the electric field distributions of PN1–4 at breakdown. In the calculations the dielectric constant ($\varepsilon_s = 10.4\varepsilon_0$) [31] was used. The parallel-plane E_b of about 2.8, 2.9, 3.2, and 3.5 MV/cm were obtained for PN1, PN2, PN3, and PN4, respectively. It should be noted that these values are very reliable, since we confirmed that the net doping concentrations were uniformly distributed along the depth directions by both SIMS and *C-V* measurements.

Fig. 6 shows the electric field distribution calculated by TCAD simulation for the beveled-mesa edge of PN1 under the reverse bias voltage of 480 V. The depletion layer extends to both the p-layer and the n-layer, since the Mg concentration in the p-layer is comparable to the Si concentration in the n-layer. The electric field at the mesa surface is reduced due to the small mesa angle [21]. The electric field profile at the p-n junction interface is shown in **Fig. 7**. The maximum electric field is almost equal to the electric field at the parallel-plane region. In other words, electric field crowding does not occur.

Fig. 8 shows the reverse *I-V* characteristics in PN1–4 at 223, 298, and 373 K. **Fig. 9** shows the temperature dependence of the breakdown voltages in PN1–4. The breakdown voltages almost linearly increased with rising temperature. For example, in PN4, V_b increased from 182 V at 298 K to 203 V at 573 K. Parallel plane E_b reached 3.74 MV/cm at 573 K. These are signatures of an avalanche breakdown; the impact ionization coefficients decreased at elevated temperature due to the increase in the phonon scattering rate.

The strong luminescence can be observed at the breakdown, since many electron-hole pairs are generated by impact ionizations in the depletion layer and large current flows. When an avalanche breakdown occurs at the electric-field-crowding point, localized luminescence is observed there [32]. **Fig. 10** shows the pictures of PN4 at (a) 0 V, (b) −180 V, −100 μA, and (c) −180 V, −1 mA. For taking these pictures, ring electrodes (the junction diameter of 450 μm) were used. At the breakdown, nearly uniform white luminescence in the entire p-n junction was observed in PN1–4. This is the strong evidence that the uniform avalanche breakdowns occurred, i.e., successful suppression of electric field crowding.

Fig. 11 shows the benchmark plots of V_b, E_b vs. doping concentration for previously reported GaN power devices and this work. The E_b in this work are among the best of the reported non-punch-through GaN devices for relatively high doping concentrations (low breakdown voltages).

IV. CONCLUSIONS

In this study, we designed and fabricated the GaN PNDs with the double-side-depleted shallow bevel termination, in which electric field crowding does not occur. The breakdown voltages of 480, 385, 250, and 180 V were obtained. The devices showed excellent avalanche capabilities, the increases in the breakdown voltages with elevating temperature, and nearly uniform luminescence at the breakdown. These are strong evidence that the uniform avalanche breakdown occurred in the devices. Based on careful characterization of the doping concentrations in the epitaxial layers in the devices, the parallel-plane breakdown fields of 2.8–3.5 MV/cm were obtained, which are among the best of the reported non-punch-through GaN devices. This edge termination structure is considered to be useful for relatively low V_b (< 1 kV-class) devices as well as the devices for fundamental studies of the breakdown characteristics of GaN.

ACKNOWLEDGMENT

This work was supported by the Council for Science, Technology and Innovation (CSTI), the Cross-ministerial Strategic Innovation Promotion Program (SIP), and Next-generation power electronics (funding agency: NEDO)

REFERENCES

[1] Z. Z. Bandic *et al., Appl. Phys. Lett.* **74**, 9 (1999).
[2] J. W. Johnson *et al., IEEE Trans. Electron Devices* **46**, 1 (2002).
[3] Y. Saitoh *et al., Appl. Phys. Express* **3**, 081001 (2010).
[4] W. Li *et al., IEEE Trans. Electron Devices* **64**, 4 (2017).
[5] Y. Zhang *et al., IEEE Electron Device Lett.* **38**, 8 (2017).
[6] K. Nomoto *et al.,Phys. Status Solidi A* **208**, 7 (2011).
[7] I. C. Kizilyalli *et al., IEEE Electron Device Lett.* **35**, 2 (2014).
[8] I. C. Kizilyalli *et al., IEEE Electron Device Lett.* **36**, 10 (2015).
[9] K. Nomoto *et al., IEEE Electron Device Lett.* **37**, 2 (2016).
[10] J. R. Dickerson *et al., IEEE Trans. Electron Devices* **63**, 1 (2016).
[11] J. Wang *et al.*, in *IEDM Technical Digest*, 2017. P. 9.6.1.
[12] S. Usami *et al., Appl. Phys. Lett.* **112**, 182106 (2018).
[13] M. Kanechika *et al., Jpn. J. Appl. Phys.* **46**, 21 (2007).
[14] M. Kodama *et al., Appl. Phys. Express* **1**, 021104 (2008).
[15] T. Oka *et al., Appl. Phys. Express* **7**, 021002 (2014).
[16] T. Oka *et al., Appl. Phys. Express* **8**, 054101 (2015).
[17] S. Chowdhury *et al., IEEE Electron Device Lett.* **33**, 1 (2012).
[18] H. Nie *et al., IEEE Electron Device Lett.* **35**, 9 (2014).
[19] D. Shibata *et al.*, in *IEDM Tech. Digest*, 2016. P.10.1.1.
[20] M. Sun *et al., IEEE Electron Device Lett.* **38**, 4 (2017).
[21] Y. Zhang *et al.*, in *IEDM Tech. Digest*, 2017. P. 9.2.1.
[22] C. Gupta *et al., IEEE Electron Device Lett.* **38**, 3 (2017).
[23] D. Ji *et al.*, in *IEDM Tech. Digest*, 2017, p. 9.4.1.
[24] B. J. Baliga, *Fundamentals of Power Semiconductor Devices*, (Springer, 2008), pp. 91-166.
[25] T. Narita *et al., J. Appl. Phys.* **123**, 161405 (2018).
[26] *Sentaurus J-2014.09*, Synopsis Inc.
[27] N. Sawada *et al., Appl. Phys. Express* **11**, 041001 (2018).
[28] F. Yan *et al., Mater. Sci. Forum* **389**, 1305 (2002).
[29] H. Niwa *et al., IEEE Trans. Electron Devices* **62**, 10 (2015).
[30] Y. Zhang *et al.*, in *IEDM Tech. Digest*, 2015, p.35.1.1.
[31] A. S. Barker, Jr. and M. Ilegems, *Phys. Rev. B* **7(2)**, 743 (1973).
[32] T. Hiyoshi *et al., IEEE Trans. Electron Devices* **55**, 8 (2008).

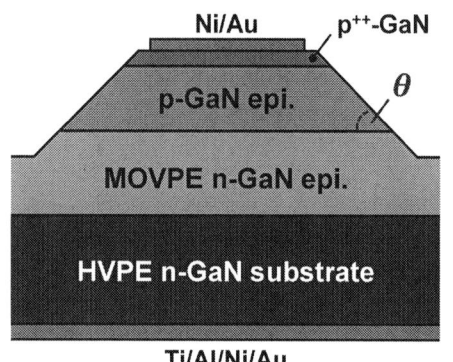

Fig. 1. The schematic cross section of the GaN PND with the double-side-depleted shallow bevel termination. The smooth mesa surfaces without any structural defects were obtained. The angles were θ~12°. The thicknesses and doping concentrations of n- and p-layers are summarized in **Table I**.

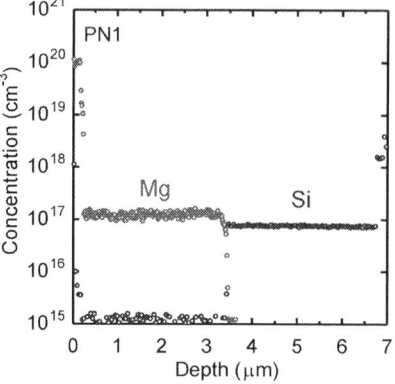

Fig. 2. The depth profiles of the Mg and Si concentration in the epitaxial layer in PN1 obtained by SIMS. Mg in the p-layer and Si in the n-layer are uniformly distributed along the depth direction.

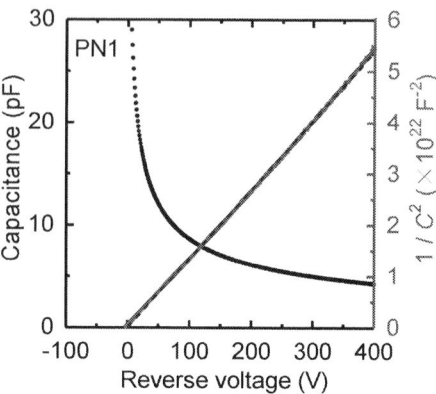

Fig. 3. The *C-V* characteristics in PN1 shown as the *C-V* plot (black dots) and the $1/C^2$-*V* plot (red dots). The clear linearity was observed in the $1/C^2$-*V* plot, indicating that the space charges are uniformly distributed along the depth direction.

Table. I. The thicknesses of the p-layers (d_p), the Mg concentrations in the p-layers ([Mg]), the thicknesses of the n-layers (d_n), and the Si concentrations in the n-layers ([Si]) in the GaN PNDs (PN1–4). [Mg][Si]/([Mg]+[Si]) and $N_a N_d/(N_a+N_d)$ are almost same for all the devices.

	PN1	PN2	PN3	PN4
d_p (µm)	3.3	2.5	2.0	1.5
[Mg] (cm⁻³)	1.2×10^{17}	1.4×10^{17}	2.7×10^{17}	3.9×10^{17}
d_n (µm)	3.3	2.5	2.0	1.5
[Si] (cm⁻³)	7.6×10^{16}	1.2×10^{17}	2.1×10^{17}	4.1×10^{17}
$\frac{[Mg][Si]}{[Mg]+[Si]}$ (cm⁻³)	4.7×10^{16}	6.5×10^{16}	1.2×10^{17}	2.0×10^{17}
$\frac{N_a N_d}{N_a+N_d}$ (cm⁻³)	4.7×10^{16}	6.4×10^{16}	1.2×10^{17}	1.9×10^{17}

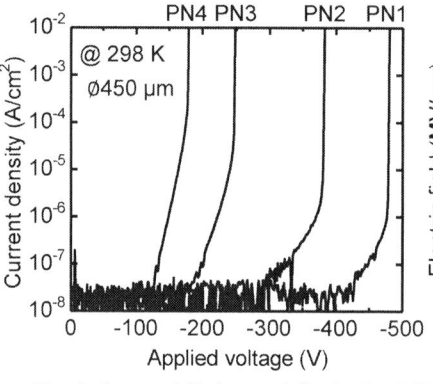

Fig. 4. Reverse *I-V* characteristics in the GaN PNDs (PN1–4). The excellent avalanche capabilities were observed in all the devices. The higher breakdown voltages were observed at lower doing concentrations.

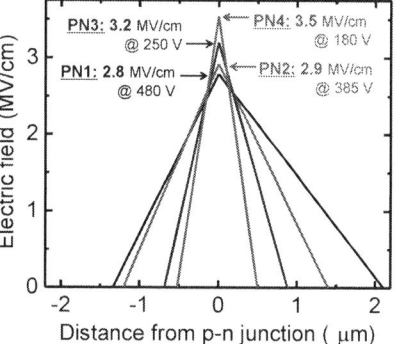

Fig. 5. Electric field distributions in the depletion layers in the GaN PNDs (PN1–4). The dielectric constant of GaN ($\varepsilon_s = 10.4\varepsilon_0$) was used in the calculations.

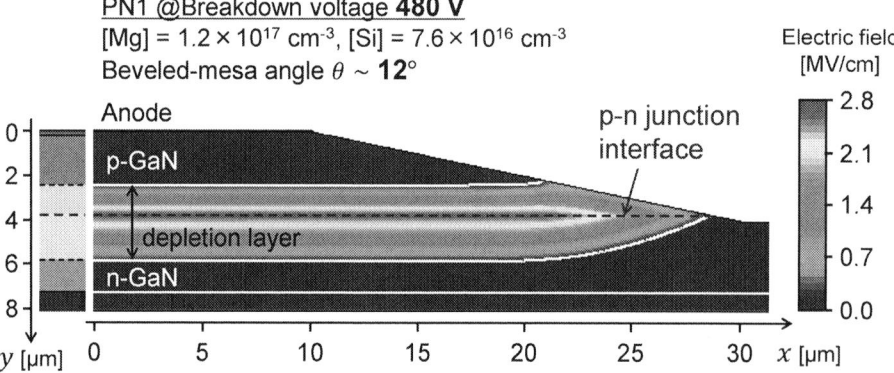

Fig. 6. The 2-D electric field distribution in the beveled-mesa edge in PN1 under the reverse bias voltage of 480 V calculated by TCAD simulation. The depletion layer extends to both the p-layer and the n-layer, since the Mg concentration in the p-layer is almost equal to the Si concentration in the n-layer. Then, the electric field at the surface is reduced and the uniform electric field distribution at the p-n junction interface (as shown in **Fig. 7**) can be obtained.

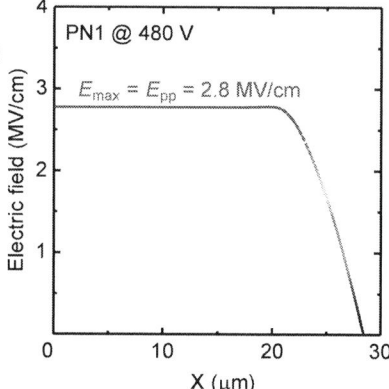

Fig. 7. The electric field profile at the p-n junction interface (along the blue broken line in **Fig. 6**). The maximum electric field (E_{max}) is almost equal to the electric field at the parallel-plane region (E_{pp}).

978-1-7281-1988-5/18 $31.00 © 2018 IEEE   689

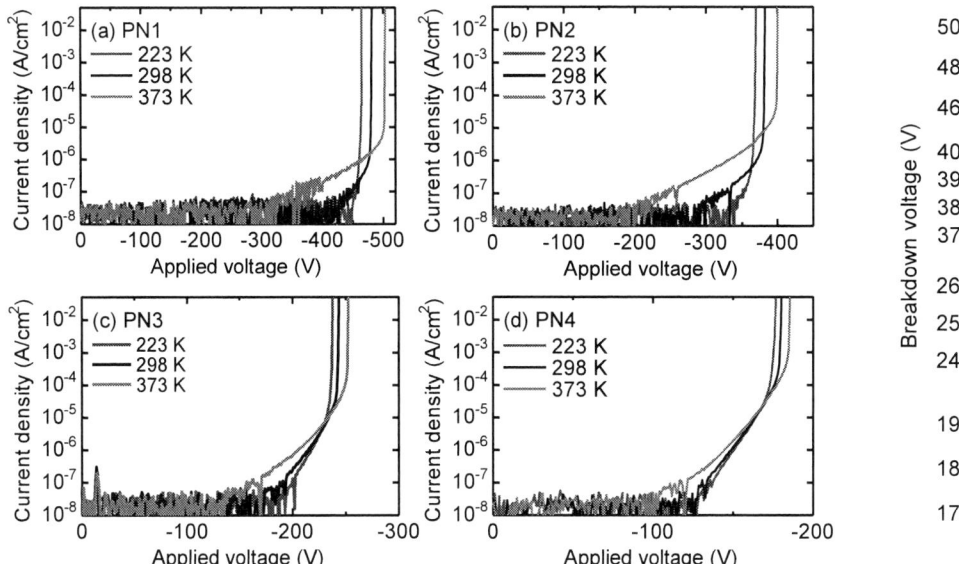

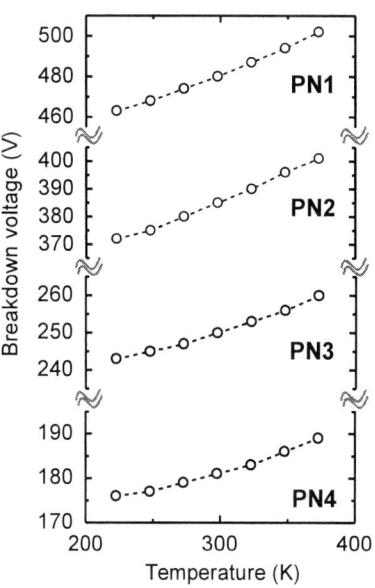

Fig. 8. Reverse *I-V* characteristics in the GaN PNDs (PN1–4) at 223, 298, and 373 K. The breakdown voltages in all the devices increased with elevating temperature. This temperature dependence is a signature of the avalanche breakdown.

Fig. 9. Temperature dependence of the breakdown voltages in the GaN PNDs (PN1–4).

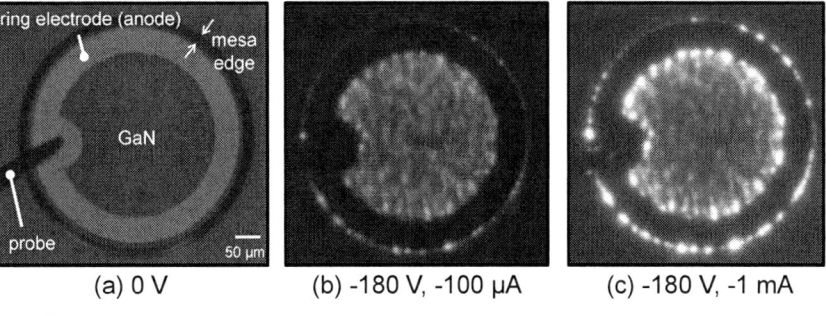

(a) 0 V   (b) -180 V, -100 μA   (c) -180 V, -1 mA

Fig. 10. Pictures of PN4 at (a) 0 V, (b) −180 V, −100 μA, and (c) −180 V, −1 mA taken by a cooled-CMOS camera. To observe luminescence, ring electrodes were used only this measurement. At the breakdown voltage, nearly uniform electroluminescence in the p-n junction was observed. This breakdown luminescence was also observed in PN1–3. This is the strong evidence that the uniform avalanche breakdowns occurred.

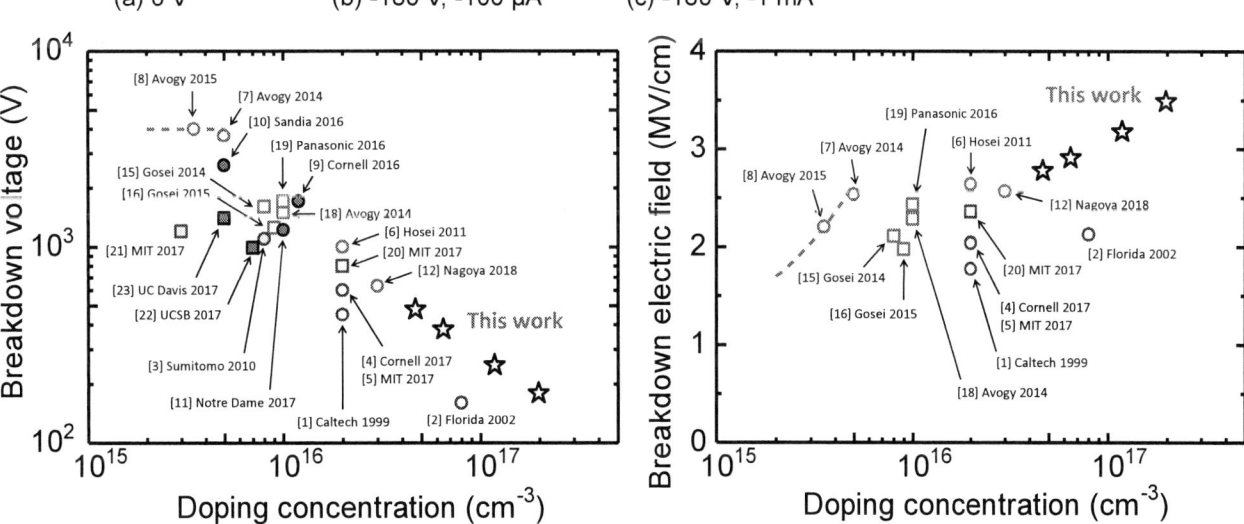

Fig. 11. Benchmark plots of breakdown voltage, breakdown field vs. doping concentration for reported GaN power devices (SBDs and JBSs: ○, PNDs: ○, Fin FETs: □, MOSFETs, CAVETs, and OG-FETs: □) and this work: ☆. The open marks and the filled marks mean non-punch-through and punch-through, respectively. The breakdown fields in this work are among the best of the reported values, and these are the uniform breakdown. For lower doping concentration than the middle of 10^{16} cm^{-3}, it seems to be difficult to determine the accurate *net* doping concentration due to the compensation of carbon (e.g. see [8]).

Demonstration of avalanche capability in polarization-doped vertical GaN pn diodes: study of walkout due to residual carbon concentration

C. De Santi[1], E. Fabris[1], K. Nomoto[2], Z. Hu[2], W. Li[2], X. Gao[3],
D. Jena[2], H. G. Xing[2], G. Meneghesso[1], M. Meneghini[1], and E. Zanoni
[1]University of Padova, Padova, Italy, email: carlo.desanti@dei.unipd.it, matteo.meneghini@dei.unipd.it
[2]Cornell University, Ithaca, NY, USA, [3]IQE, Somerset, NJ, USA

Abstract—**For the first time, we demonstrate and investigate the avalanche capability in vertical GaN-on-GaN pn diodes with polarization doping. Specifically: (i) we prove that the analyzed devices have avalanche capability, and we describe the dependence of breakdown voltage and leakage on temperature and monochromatic illumination; (ii) we demonstrate the presence of avalanche walkout, i.e. a recoverable increase in breakdown voltage induced by stress in avalanche conditions; (iii) we describe the time-dependence of avalanche walkout as a function of temperature, and demonstrate that walkout is caused by charge trapping due to residual carbon; (iv) we calculate the related activation energy and propose a model able to explain the experimental data. The reported results are of the utmost importance for the improvement in performance of high-voltage avalanche-capable GaN diodes.**

I. INTRODUCTION

Demonstrating vertical GaN-on-GaN diodes with avalanche capability is of fundamental importance for the development of high-voltage devices based on GaN. However, only little is known on the behavior of GaN vertical devices under avalanche conditions. We present **the first comprehensive investigation on this topic**, by demonstrating the avalanche capability of vertical GaN-on-GaN pn diodes with polarization doping and investigating the related physical mechanisms.

II. AVALANCHE BREAKDOWN

The devices under test are vertical pn GaN diodes grown by metal-organic chemical vapor deposition (MOCVD) [1] with a polarization-doped p-side [2] to overcome the low ionization of Mg:GaN (see Fig. 1 for the structure). Such devices show a very low leakage current up to V_{BD}>1300 V, when a sudden current increase takes place. The breakdown voltage has a **positive temperature coefficient** (Fig. 2), supporting the hypothesis that the increase in current is related to avalanche breakdown [3], [4]. The temperature coefficient of 0.5 V/°C (Fig. 3) is compatible with previous reports [3], [4], and consistent with the higher phonon scattering with increasing temperature.

In principle, an abrupt increase in vertical leakage could also be explained by a field-assisted conduction through dislocations, known to generate the leakage in bulk GaN [5]. In this second hypothesis, the positive temperature coefficient would originate from a thermally-assisted electron capture inside the dislocations, limiting the current flow at higher temperature. **In order to confirm the avalanche capability of the devices**, we carried out reverse breakdown measurements under external illumination (Fig. 4). As can be noticed, light is able to reach the active region, since an increase in current in the pre-avalanche region is visible, even when the photon energy is below the energy gap and no band-to-band (365 nm) absorption takes place. This phenomenon may be caused by defect-related absorption or by the Franz-Keldysh effect. Since the absorption coefficient changes with the electric field (e.g. for the curve under 405 nm illumination) and since the spectral dependence is compatible with the high-field absorption tails reported in [6], we conclude that **Franz-Keldysh effect is causing the photon absorption and increased leakage** in this case. By analyzing the high-field region, we notice that the breakdown voltage and the equivalent resistance in the high-current region are not influenced by the external illumination, excluding a possible role of electron trapping and **giving an optical confirmation to the assumption that an avalanche process is present.**

III. BREAKDOWN WALKOUT

The stability of the breakdown voltage is important for high reverse voltage operation. As can be seen in Fig. 5, when the device is operated for a long time in reverse bias mode **a time-dependent mechanism, responsible for the increase in the breakdown voltage of several (30) volts**, takes place. This behavior is **not related to a permanent degradation** of the device, since after some rest time the device slowly returns to its previous current and breakdown voltage levels (Fig. 6 and 7). This process, called **breakdown walkout**, was already demonstrated to be possible and permanent in devices based on the Si [7] and GaAs [8] material system, but recent reports on GaAs pseudomorphic high electron mobility transistors show that it can also be caused by charge trapping [9], in this case at interface states. **Up to now no report on avalanche breakdown walkout is present in the literature for vertical GaN devices.**

Taking into account the significant power dissipated by those devices and by the neighboring ones in a power supply and/or converter, the effect of the operating temperature cannot be neglected. By repeating the reverse-bias stress experiment at different ambient temperatures (Fig. 8), it was possible to demonstrate that **the breakdown walkout process becomes**

978-1-7281-1988-5/18 $31.00 © 2018 IEEE

stronger and faster at higher temperature, a behavior compatible with the aforementioned charge trapping process. **The activation energy of the walkout process was found to be 0.52 eV**. Deep levels with a similar signature (Fig. 9) were already reported in the past [10]–[13] and related to the presence of carbon at the nitrogen site (C_N) [14]. Typically, C_N defects have higher activation energy (0.8-0.9 eV); in our case, the high electric field (close to the breakdown limit of GaN, >3 MV/cm) may significantly lower this activation energy, due to Poole-Frenkel effect. In the devices under test no intentional carbon doping is present, but a residual concentration ($2\text{-}6 \times 10^{16}$ cm^{-3} in the whole structure) was detected by secondary ion mass spectroscopy (SIMS), likely originating from the metal-organic precursors. It is worth noting that the carbon level in these device epitaxial layers can be lowered to be below the detection limit similar to the optimized epitaxial layers employed in our previously reported GaN p-n diodes [4].

IV. ROLE OF RESIDUAL CARBON

In order to identify if a deep level related to the residual carbon is really present in the devices, we developed a new setup for capacitance deep level transient spectroscopy (C-DLTS). In common systems, the maximum voltage that can be applied to the device under test is limited to ±10 V and the maximum filling and measure times are in the order of hundreds of microsecond. The high-voltage nature of the reported diodes requires higher voltages in order to effectively probe their performance in a region affected by high-voltage operation, therefore we increased the maximum voltage to ±40 V. Beyond this limit the leakage is too high and the capacitance measure is not reliable. Additionally, given the depth of carbon-related energy states in the GaN gap, the capture and emission time constants can be relatively long. In most cases this issue is addressed by high-temperature measurements, which are not possible in this case due to the limited reliability of such pre-maturity research samples. For this reason, in our new setup we were able to achieve no maximum limits for the filling and measure phase, which we chose to be 10 s and 1200 s, respectively.

This long measure time is needed to accurately record the capacitance transient of a minority carrier trap at 300 K (Fig. 10). To make sure that the detected transient is not related to surface states or parasitic effects, the figure reports the results of tests carried out at different voltages, i.e. by analyzing different active volumes. The amplitude of the capacitance transient, proportional to the total number of deep levels inside the probed region, increases when a more negative measure voltage is applied. This confirms that **the deep levels generating** the transient are located near the junction and that their position is not limited to a specific layer or interface but **is distributed in the whole volume, compatibly with the hypothesis on the residual point-defect (carbon) concentration**. The results of temperature-dependent measurements are reported in Fig. 11. **By comparing the extrapolated Arrhenius plot with previously published papers (Fig. 12), an excellent agreement with other carbon-related deep level signatures can be noticed.**

Based on the experimental findings, the breakdown walkout process can be modeled as follows. When a strong reverse bias is applied, **hole emission** (i.e. electron capture) from residual carbon takes place, as demonstrated by C-DLTS. The increased (flowing) electron-to-(trapped) electron scattering causes **a reduction in the electron mean free path**, increasing the required applied voltage to start the avalanche multiplication.

V. CONCLUSIONS

In summary, we unambiguously demonstrated, by means of electrical and optical measurements, that high-voltage GaN-based devices with polarization doping exhibit avalanche breakdown. The breakdown voltage is time-dependent, affected by a recoverable walkout mechanism. It is caused by hole emission from residual carbon, as confirmed by temperature-dependent and DLTS measurements. All these phenomena were never reported for GaN before, and their understanding is necessary in order to design reliable and well-performing devices able to exploit the good characteristics of the GaN material system for power and high-voltage operation.

ACKNOWLEDGMENT

This research activity was partly funded by project "Novel vertical GaN-devices for next generation power conversion", NoveGaN (University of Padova), through the STARS CoG Grants call. The devices analyzed were grown and fabricated under the ARPAe SWITCHES project.

REFERENCES

[1] K. Nomoto et al., "GaN-on-GaN p-n power diodes with 3.48 kV and 0.95 mΩ-cm²: A record high figure-of-merit of 12.8 GW/cm²," in *2015 IEEE International Electron Devices Meeting*, 2015, p. 9.7.1-9.7.4.

[2] H. G. Xing et al., "Unique opportunity to harness polarization in GaN to override the conventional power electronics figure-of-merits," in *2015 73rd Annual Device Research Conference (DRC)*, 2015, pp. 51–52.

[3] I. C. Kizilyalli et al., "High voltage vertical GaN p-n diodes with avalanche capability," *IEEE Trans. Elec. Dev.*, vol. 60, no. 10, pp. 3067–3070, 2013.

[4] Z. Hu et al., "Near unity ideality factor and Shockley-Read-Hall lifetime in GaN-on-GaN p-n diodes with avalanche breakdown," *Appl. Phys. Lett.*, vol. 107, no. 24, p. 243501, Dec. 2015.

[5] S. Usami et al., "Correlation between dislocations and leakage current of p-n diodes on a free-standing GaN substrate," *Appl. Phys. Lett.*, vol. 112, no. 18, p. 182106, Apr. 2018.

[6] G. Franssen et al., "Photocurrent spectroscopy as a tool for determining piezoelectric fields in $In_xGa_{1-x}N$/GaN multiple quantum well light emitting diodes," *Phys. Rev. B*, vol. 69, no. 4, p. 045310, Jan. 2004.

[7] Wei Lian Guo et al., "Walkout in p-n junctions including charge trapping saturation," *IEEE Trans. Electron Devices*, vol. 34, no. 8, p. 1788, 1987.

[8] P. C. Chao et al., "Breakdown walkout in AlAs/GaAs HEMTs," *IEEE Trans. Electron Devices*, vol. 39, no. 3, pp. 738–740, Mar. 1992.

[9] W.-B. Tang et al., "Reversible Off-State Breakdown Walkout in Passivated AlGaAs/InGaAs PHEMTs," *IEEE Trans. Device Mater. Reliab.*, vol. 6, no. 1, pp. 42–45, Mar. 2006.

[10] J. Osaka et al., "Deep levels in n-type AlGaN grown by hydride vapor-phase epitaxy on sapphire characterized by deep-level transient spectroscopy," *Appl. Phys. Lett.*, vol. 87, no. 22, pp. 1–3, 2005.

[11] M. Meneghini et al., "Time-and field-dependent trapping in GaN-based enhancement-mode transistors with p-gate," *IEEE Electron Device Lett.*, vol. 33, no. 3, pp. 375–377, 2012.

[12] M. Caesar et al., "Generation of traps in AlGaN/GaN HEMTs during RF- and DC-stress test," *IEEE Int. Reliab. Phys. Symp.*, pp. 1–5, 2012.

[13] K. Tanaka et al., "Effects of Deep Trapping States at High Temperatures on Transient Performance of AlGaN/GaN Heterostructure Field-Effect Transistors," *Jpn. J. Appl. Phys.*, vol. 52, p. 04CF07, 2013.

[14] U. Honda et al., "Deep levels in n-GaN doped with carbon studied by deep level and minority carrier transient spectroscopies," *Jpn. J. Appl. Phys.*, vol. 51, p. 04DF04, 2012.

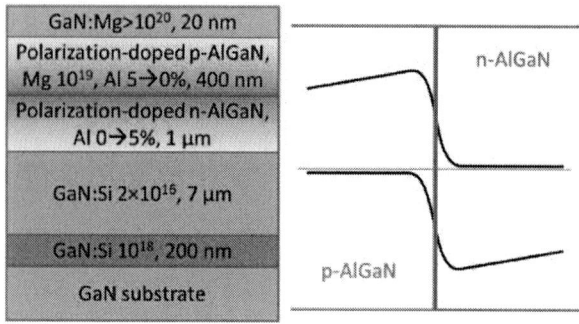

Figure 1. Structure of the devices under test and sketched band diagram.

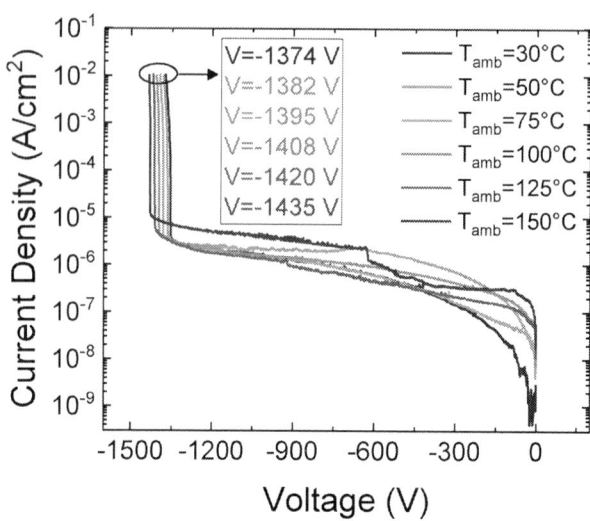

Figure 2. Reverse-bias breakdown measurements at various temperatures. The analyzed devices show avalanche capability, and a positive temperature coefficient is observed

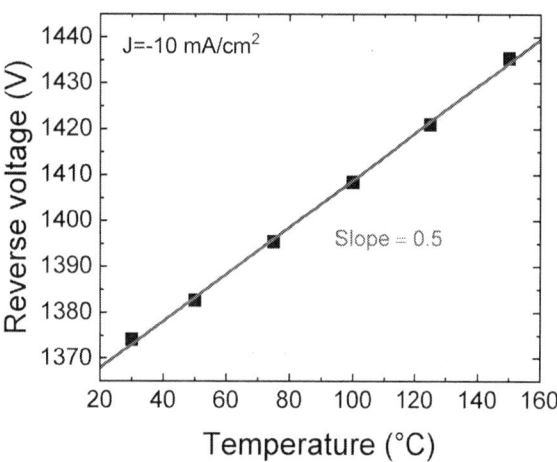

Figure 3. Positive temperature coefficient of the avalanche breakdown: dependence of V_{BD} on temperature.

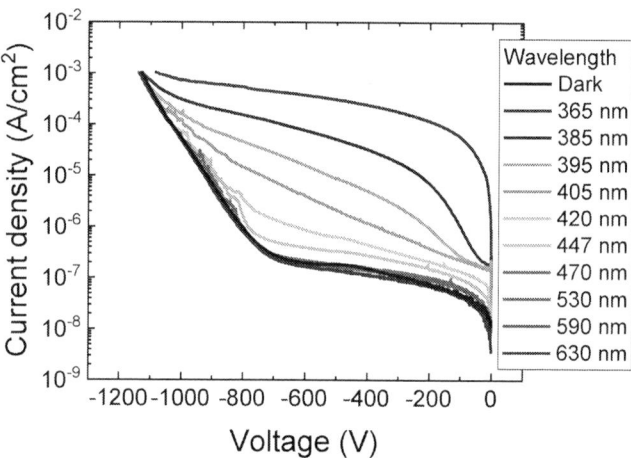

Figure 4. Dependence of reverse current on monochromatic illumination wavelength.

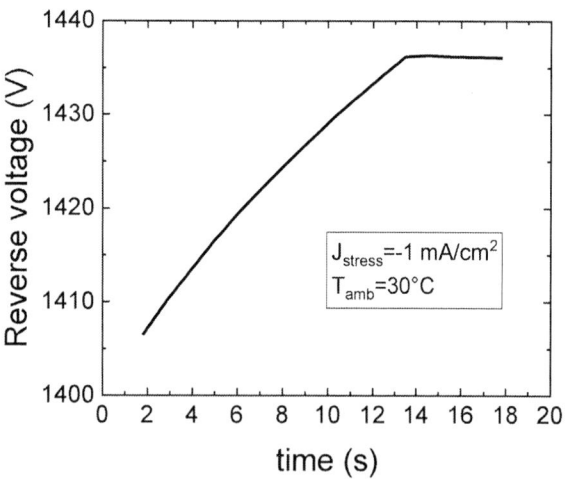

Figure 5. Time-dependent variation of the breakdown voltage during reverse-bias stress at -1 mA/cm², 50 °C (i.e. during stress in avalanche regime). A breakdown walkout mechanism is observed, leading to a significant increase in V_{BD}

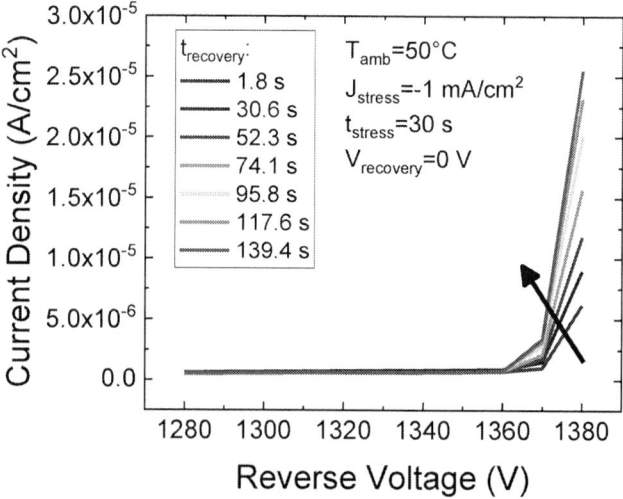

Figure 6. Recovery of the breakdown voltage after reverse-bias stress at -1 mA/cm², 50 °C. The observed breakdown walkout is fully recoverable

978-1-7281-1988-5/18 $31.00 © 2018 IEEE   693

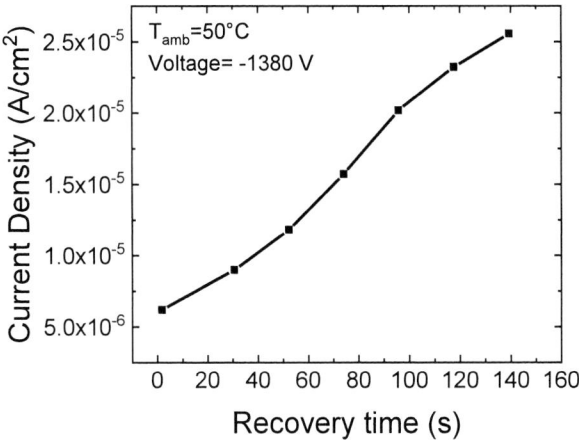

Figure 7. Recovery of the reverse current after reverse-bias stress at -1 mA/cm², 50 °C. Breakdown walkout is fully recoverable, with slow de-trapping kinetics

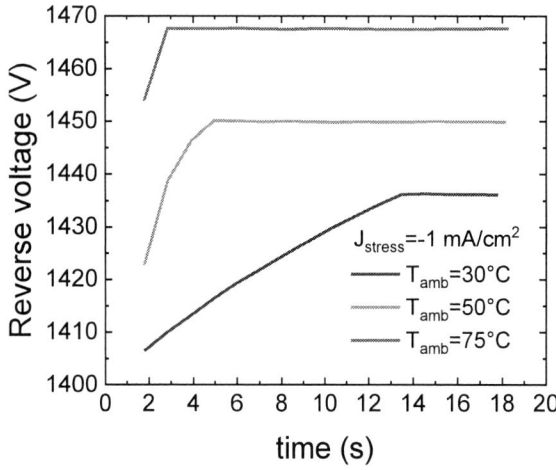

Figure 8. Dependence on temperature of the breakdown walkout process. Increasing temperature leads to a stronger and faster breakdown walkout

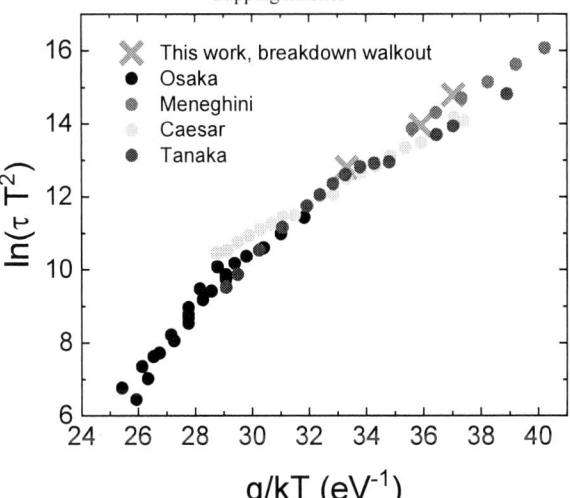

Figure 9. Arrhenius plot of the breakdown walkout process and comparison with deep level signatures from the literature. A correspondence is found with deep-levels related to carbon at nitrogen site

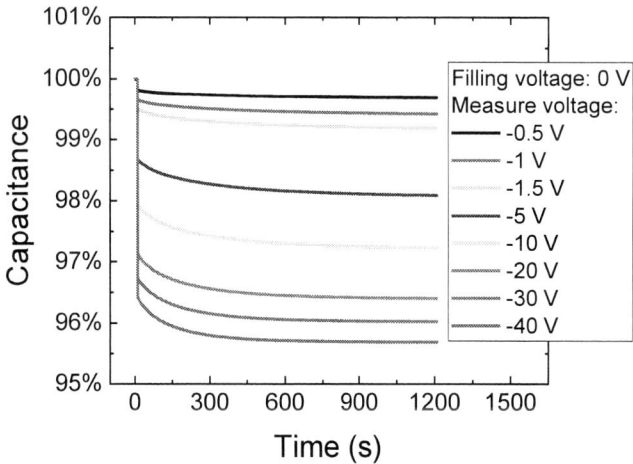

Figure 10. C-DLTS tests at different measure voltage. The results indicate the presence of the deep level in the whole analyzed volume

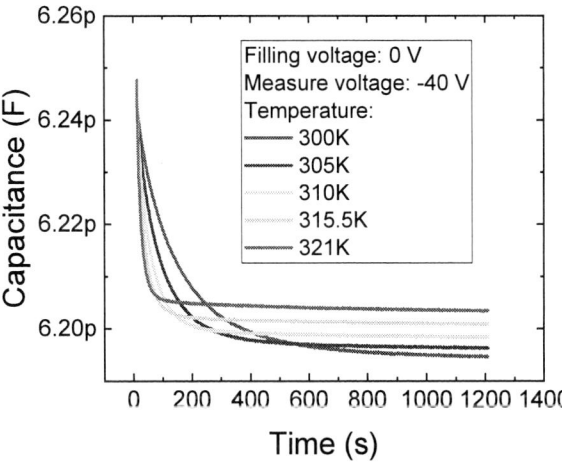

Figure 11. C-DLTS tests at various temperatures. Slow de-trapping processes are dound to take place after exposure to negative bias

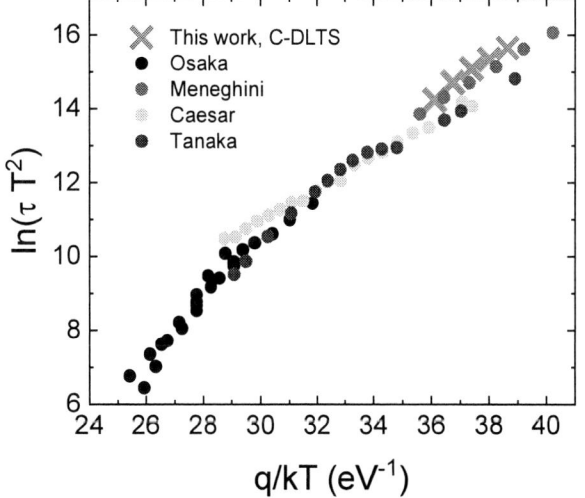

Figure 12. Activation energy of the detected deep level and comparison with deep level signatures from the literature.

Suppressed Hole-Induced Degradation in E-mode GaN MIS-FETs with Crystalline GaO$_x$N$_{1-x}$ Channel

Mengyuan Hua[1], Xiangbin Cai[2], Song Yang[1], Zhaofu Zhang[1], Zheyang Zheng[1], Jin Wei[1], Ning Wang[2], and Kevin J. Chen[1]

[1]Department of ECE, [2]Department of Physics, The Hong Kong University of Science and Technology, Hong Kong, China
email: mhua@connect.ust.hk

Abstract—Under reverse-bias stress with a high drain voltage, hole-induced gate dielectric degradation in the E-mode GaN MIS-FETs could lead to non-recoverable V_{TH} shifts and devastating time-dependent breakdown. Such a degradation can be effectively suppressed by converting the GaN channel into a crystalline GaO$_x$N$_{1-x}$ channel in the gated region. The valence band offset between GaO$_x$N$_{1-x}$ and the surrounding GaN creates a hole-blocking ring around the gate dielectric, preventing holes from flowing to the gate dielectric and therefore mitigating the hole-induced degradation.

I. Introduction

With GaN power devices being commercialized progressively, reliability and stability issues are becoming the focus of many on-going investigations [1]–[3]. Bias temperature instability (BTI), time-dependent dielectric breakdown (TDDB) and reverse-bias stress instability (e.g. HTRB) associated with the input-node (i.e. the gate) are some of the most critical issues. For GaN devices with a metal-insulator-semiconductor (MIS) gate, high gate dielectric reliability can be achieved with a suitable dielectric film, such as SiN$_x$ prepared by low-pressure chemical vapor deposition (LPCVD) at around 800 °C [4], [5]. The type-II alignment between SiN$_x$ and GaN yields a large conduction band offset $\Delta E_C \sim 2.3$ eV, resulting in small gate leakage and large forward gate breakdown voltage. In addition, the high-temperature LPCVD process could produce a SiN$_x$ film with low defect density, leading to long TDDB lifetime. For E-mode MIS-FETs with recessed gate structure, interface protection technique has been developed to prevent the etched-GaN surface from degradation at high temperature [5], [6]. Consequently, small BTI has been obtained in the E-mode GaN MIS-FETs simultaneously with long TDDB lifetime.

HTRB reliability remains a serious challenge to E-mode GaN MIS-FET [7]. Under high drain bias stress, impact ionization could occur in the high electric field region near the gate edge [8], resulting in the electron-holes generation and the subsequent hole-drift toward the gate [9]. Under high gate-to-drain reverse bias (V_{GD}), holes tend to drift through the gate dielectric, leading to a generation of new defects in the gate dielectric during a long-term stress. Without effective suppression of the hole-induced degradation, non-recoverable positive V_{TH} shifts and devastating time-dependent breakdown could happen [10]. A promising solution to this problem is to introduce a hole-blocking layer around the channel underneath the gate dielectric, and prevent the holes from flowing to the gate side.

In this work, a hole blocking scheme was proposed and demonstrated by implementing a crystalline GaO$_x$N$_{1-x}$ channel layer with a wider bandgap (~4.1 eV) under the gate of the E-mode GaN MIS-FETs (Fig. 1). With stronger oxidation of the exposed GaN surface (than that previously developed for the formation of a thin interfacial layer [6]) and subsequent high-temperature *in-situ* annealing, a crystalline GaO$_x$N$_{1-x}$ channel of ~5.6 nm-thickness has been formed with some of N atoms replaced by O atoms. The GaO$_x$N$_{1-x}$ channel has high thermal stability and a valence band offset with GaN. The effective hole blocking around the gate dielectric and subsequent suppression of the hole-induced gate dielectric degradation is then obtained.

II. Crystalline GaO$_x$N$_{1-x}$ Channel

The E-mode MIS-FETs with GaO$_x$N$_{1-x}$ (GaON) channel and LPCVD-SiN$_x$ gate dielectric were fabricated on a GaN-on-Si wafer with device dimensions of $L_{GS}/L_G/L_{GD} = 2/1.5/15$ μm. The fabrication commenced with a passivation layer deposition, followed by gate window opening. After gate recess etching, the exposed-GaN surface was oxidized in an ICP chamber with an O$_2$ flow of 40 sccm and plate/coil power of 10/10 W for 10 minutes. Afterward, the sample was *in-situ* annealed in the LPCVD chamber at 780 °C in NH$_3$ ambience to form the GaON channel prior to SiN$_x$ deposition. The remaining process steps are the same as those in ref. [6].

A. Formation of Crystalline GaO$_x$N$_{1-x}$

The aberration-corrected scanning transmission electron microscopy (ACSTEM) enables an atomic-scale visualization of the arrangements of Ga, N and O atoms in GaON. Fig. 2 shows the high-angle annular dark field and annular bright field images of the LPCVD-SiN$_x$/GaON and LPCVD-SiN$_x$/GaN stacks with an optimal spatial resolution of 0.83 Å. After the high-temperature LPCVD process, the GaN surface without surface protection layer was severely degraded with Ga and N vacancy near the surface due to the inferior thermal stability (Fig. 2 (d)), while the GaON maintains a sharp interface with SiN$_x$ and a clear crystalline structure (Fig. 2 (a)). The arrangement of Ga atoms in GaON is identical with that in the GaN (Fig. 2 (b) and (e)), while the position of the small atoms (N or O) changes every other layer along the c-axis direction (Fig. 2 (c) and (f)).

EDS (energy dispersive X-ray spectroscopy) elemental mapping analysis was carried out to roughly obtain the element distribution in the LPCVD-SiN$_x$/GaON stack (Fig. 3 (a)). A clear and sharp interface without Si or Ga inter-diffusion was observed, while some O atoms diffused into GaN for a few nanometers to form the GaON crystalline. The thickness of the

978-1-7281-1988-5/18 $31.00 © 2018 IEEE

GaON layer was measured to be ~ 5.6 nm with SIMS (secondary-ion mass spectrometry) (Fig. 3 (b)).

B. Band Alignment with SiN$_x$ and Atomic Structure

The valence band offset (ΔE_V) of GaON/SiN$_x$ and GaN/SiN$_x$ was measured with XPS (X-ray photoelectron spectroscopy) analysis (Fig. 4) [11]. The energy difference between VBM (valence band maximum) and Ga 2p core level in GaON is 0.3 eV smaller than that in GaN (Fig. 4 (a) (b)), indicating that part of Ga states change from Ga^{3+} to Ga^{2+} by forming bond with O. Following the alignment between the core levels and VBM, the ΔE_V was estimated from the relative energy difference between Si 2s and Ga 2p core levels at the interface (Fig. 4 (e) (f)), yielding a value of −1.1 eV in GaON/SiN$_x$ and −0.5 eV in GaN/SiN$_x$, respectively. Thus, a ΔE_V of 0.6 eV exists at the interface between GaON and GaN to block the hole flow from GaN at the drain-side gate edge.

The conduction band offset (ΔE_C) between GaON and SiN$_x$ was estimated with the FN tunneling fitting (Fig. 5). With high electric field, the FN tunneling dominates the electrons transportation through SiN$_x$ without temperature dependence (Fig. 5 (a)) [4]. The barrier height (i.e. ΔE_C) can be extracted from the slope of FN plots (Fig. 5 (b)), which is ~ 2.2 eV between SiN$_x$ and GaON. Therefore, the band gap of GaON is ~4.1 eV, about 0.7 eV larger than that of GaN (3.4 eV).

According to a first-principles calculation, GaON with N replaced by O every other layer along the c-axis direction (Type I) has lower formation energy than that with O distributed in every layer (Type II) (Fig. 6). The random distribution of O in Type I has small effects on the formation energy. Therefore, the Type I with N randomly replaced by O every other layer is easier to be formed than Type II. The most possible structure of GaON with the lowest formation energy is shown in Fig. 7, of which the Ga-O bonds at the surface could provide a higher thermal stability than GaN. After replacing N atoms, the O atoms could move away from the original positions of N to cause the distorted atomic arrangement as shown in Fig. 2 (c).

III. E-MODE MIS-FETS WITH GAO$_X$N$_{1-x}$ CHANNEL

A. Static Performance and V_{TH} Stability at (Semi-)on State

Compared with the high performance E-mode MIS-FET with GaN channel and interfacial protection layer [5], the MIS-FET with GaON channel also delivers desirable performance, including a positive V_{TH} (+1.3 V @ I_D = 1 µA/mm), small V_{TH} hysteresis, small subthreshold swing SS (~ 90 mV/dec) (Fig. 8 (a)) and small R_{ON} (~12 Ω·mm) (Fig. 8 (b)). Both devices have a high off-state breakdown voltage of 650 V, limited by the vertical substrate leakage (Fig. 8 (c)).

Long-term stress tests were conducted to evaluate the V_{TH} stability. The positive BTI (PBTI) stress tests were conducted with a constant V_{GS} of 10 V and V_{DS} of 0 V at various temperatures (Fig. 9 (a)). The small V_{TH} shifts reflect the low traps density at and near both the SiN$_x$/GaN and SiN$_x$/GaON interfaces. Under a constant source-to-drain current stress at semi-on state with strong hot-electron generation [9] (Fig. 9

(b)), the small ΔV_{TH} indicates that both GaON and GaN channels can sustain a large number of hot electrons.

B. Suppressed Hole-induced Degradation

To investigate the hole-induced gate dielectric degradation, transfer characteristics were monitored during reverse-bias stress with different $V_{GS,}$ while maintaining V_{DS} = 200 V for each test at 25 °C and 200 °C (Fig. 10). For both MIS-FETs, negligible R_{ON} degradation was observed, while the strong dependence of ΔV_{TH} on the negative gate bias was observed in the MIS-FETs with GaN channel. According to the hole-induced degradation model in [10], holes can be generated by impact ionization under high reverse-bias stress that forms a high electric field at the drain-side gate edge. The generated holes would then drift to the source and gate following the electric potential distributions. With a more negative gate bias (V_{GS}< −5 V), the flow of holes through the gate dielectric to the gate electrode is enhanced, leading to stronger defect generation in the gate dielectric and subsequent electron trapping and V_{TH} shift. For the MIS-FETs with GaON channel, such a hole-induced degradation was greatly suppressed because of the presence of the hole barrier between the GaON channel and the surrounding GaN region, resulting in smaller ΔV_{TH}. It is noted that ΔV_{TH} of MIS-FETs with GaON channel is negative and slightly enhanced with more negative V_{GS}. This could stem from the de-trapping of electrons that are originally trapped in the low-density deep levels (e.g. Si^{3+}-dangling bond) in LPCVD-SiN$_x$ [12].

To verify the suppressed hole-induced degradation, reverse-bias stress was conducted with a higher V_{DG} of 320 V and V_{GS} of −20 V (i.e. a harsh condition for accelerated stress test) until a hard breakdown reached (Fig. 11). The continuous transport of holes through the gate dielectric could induce time-dependent breakdown. Both the MIS-FETs show a TDDB behavior. The t_{BD} at a failure rate of 63% of MIS-FET with GaON channel is 5 times longer than that of MIS-FET with GaN channel (Fig. 12), which validates the suppression of hole-induced degradation by GaON channel.

C. Mechanism of Suppressed Hole induced Degradation

A positive valence band offset indicates that there is a hole barrier between the GaON channel and the adjacent GaN (Fig. 13 (a)) where impact ionization takes place. Thus, holes are prevented from going through the gate dielectric at high revers-bias stress, (Fig. 13 (b)). The hole-induced leakage current to the gate side was measured with the method described in ref. [10], which is ~4 times smaller in MIS-FETs with GaON channel than that with GaN channel (Fig. 14).

IV. CONCLUSION

E-mode GaN MIS-FETs with suppressed hole-induced degradation and enhanced HTRB reliability have been demonstrated with a GaO$_x$N$_{1-x}$ (E_g ~ 4.1 eV) channel replacing the conventional GaN (E_g ~ 3.4 eV) channel. The new channel presents an effective barrier to the holes generated by impact ionization in the adjacent GaN region, and block the holes from transporting through and degrading the gate dielectric. As a result, enhanced HTRB reliability is obtained.

978-1-7281-1988-5/18 $31.00 © 2018 IEEE

Acknowledgement: This work was supported by Hong Kong Innovation and Technology Fund under grant ITS/412/17FP.

References: [1] K. J. Chen, *et al.*, *IEEE TED.*, **64**, 779 (2017). [2] C. Ostermaier, *et al.*, *Microelectron. Reliab.*, **82**, 62 (2018). [3] G. Meneghesso, *et al.*, *Semicond. Sci. Technol.*, **31**, 9 (2016). [4] M. Hua, *et al.*, *IEEE TED*, **62**, 3215, (2015). [5] M. Hua, *et al.*, *IEDM* '16, p. 10.4.1. [6] M. Hua, *et al.*, *ISPSD* '17, p. 10A-1. [7] S. Warnock, *et al.*, *IRPS* '17, p. 4B-3.1. [8] M. Meneghini, *et al.*, *IRPS* '12, p. 2C-2.1. [9] M. Hua, *et al.*, *IEEE TED*, early access (2018). [10] M. Hua, *et al.*, *IEDM* '17, p. 33.2.1. [11] H. Sun, *et al.*, *Appl. Phys. Lett.*, **111**, 16 (2017). [12] W. L. Warren, *et al.*, *J. Electrochem. Soc.*, **139**, 881 (1992).

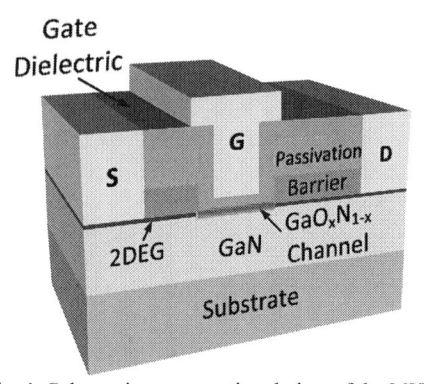

Fig. 1. Schematic cross-sectional view of the MIS-FETs with recessed gate structure and GaO_xN_{1-x} (simplified as GaON) channel.

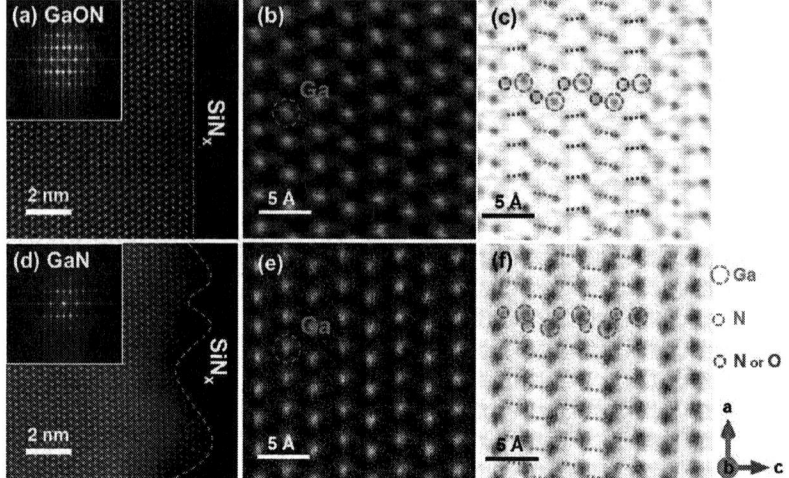

Fig. 2. Cross-sectional ACSTEM images of the (a) (b) (c) LPCVD-SiN$_x$/GaO$_x$N$_{1-x}$ and (d) (e) (f) LPCVD-SiN$_x$/GaN stacks. (Inset: Diffraction pattern of (a) GaO$_x$N$_{1-x}$ and (d) GaN). (a) (b) (d) (e) are high-angle annular dark field images that are sensitive to heavy elements and (c) (f) are annular bright field images that are sensitive to light elements.

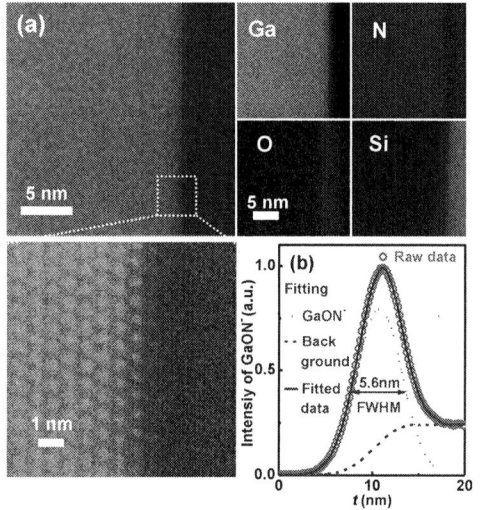

Fig. 3. (a) Cross-sectional energy dispersive X-ray spectroscopy (EDS) mapping analysis of the LPCVD-SiN$_x$/GaON stack. (b) Depth profile of the GaON$^-$ intensity measured with secondary-ion mass spectrometry (SIMS) on LPCVD-SiN$_x$/GaON stack.

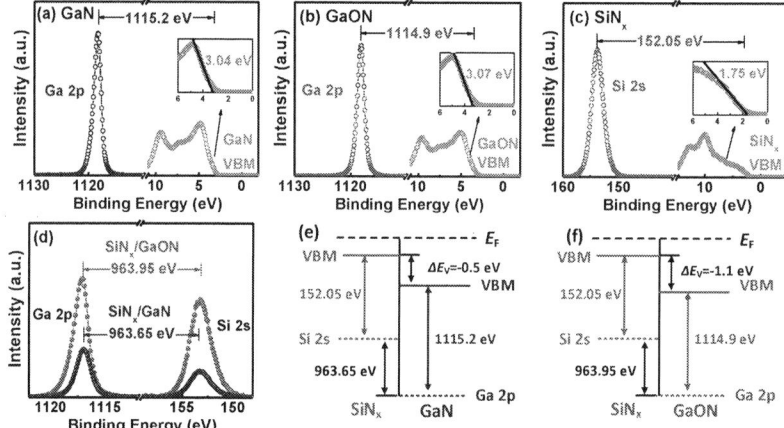

Fig. 4. X-ray photoelectron spectroscopy (XPS) core-level and valence band spectra of (a) GaN (b) GaON and (c) LPCVD-SiN$_x$ bulk. (d) XPS core-level spectra at the interface of LPCVD-SiN$_x$/GaN and LPCVD-SiN$_x$/GaON. Valence band offset between (d) GaN and (e) GaON and LPCVD-SiN$_x$.

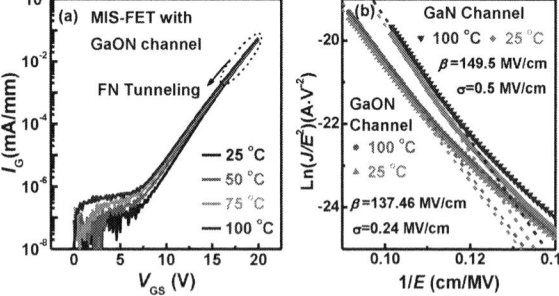

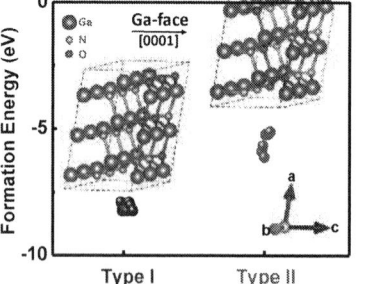

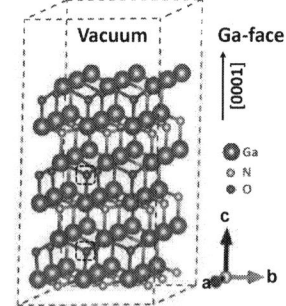

Fig. 5. (a) I_G–V_G characteristics of the LPCVD-SiN$_x$ MIS-FET with GaON channel. (b) FN plots at 25 °C to 100 °C of the MIS-FET with GaON and GaN channel (10 devices have been measured to obtain the standard deviation (σ)).

Fig. 6. The formation energy with GaN referred as 0 eV of two types GaON: with N randomly replaced by O at (I) every other layer and (II) every layer.

Fig. 7. Simulated structure of GaON. The position of small atoms changes with N replaced by O every other layer.

978-1-7281-1988-5/18 $31.00 © 2018 IEEE

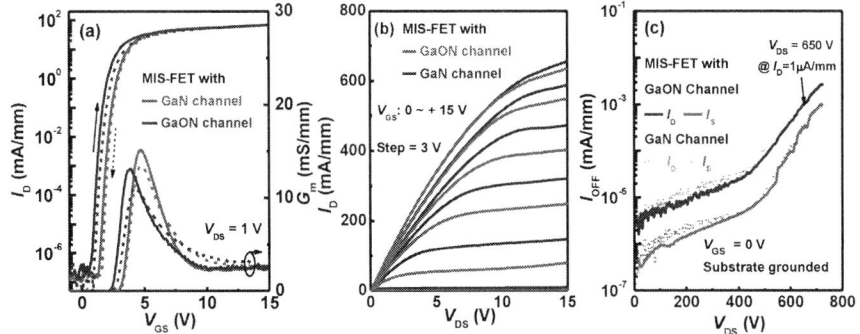

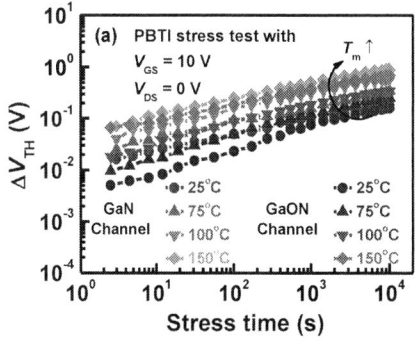

Fig. 8. (a) Transfer, (b) output and (c) OFF-state leakage characteristics of E-mode MIS-FETs with GaON channel and GaN channel.

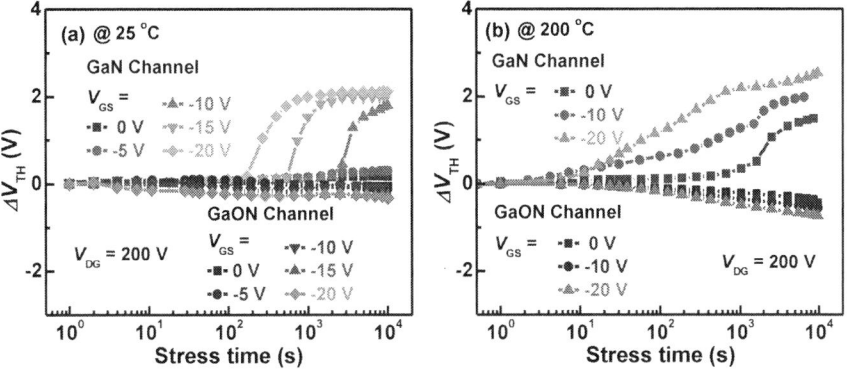

Fig. 10. Threshold voltage shifts of LPCVD-SiN$_x$ MIS-FETs with GaN channel and GaON channel during reverse-bias stress with a constant V_{DG} of 200 V and various V_{GS} at (a) 25 °C and (b) 200 °C, respectively.

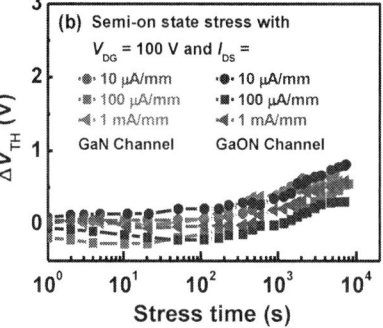

Fig. 9. Threshold voltage shifts of MIS-FETs with GaN and GaON channel during (a) PBTI stress with V_{GS} = 10 V and V_{DS} =10 V at various temperature and (b) semi-on state stress with I_{DS} of 10 μA/mm, 100 μA/mm and 1 mA/mm, respectively.

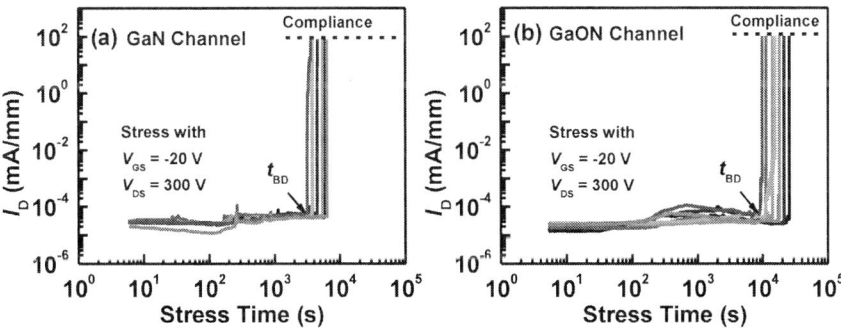

Fig. 11. Time-dependent off-state breakdown of MIS-FETs with (a) GaN channel and (b) GaON channel biased with V_{GS} = −20 V and V_{DS} = 300 V.

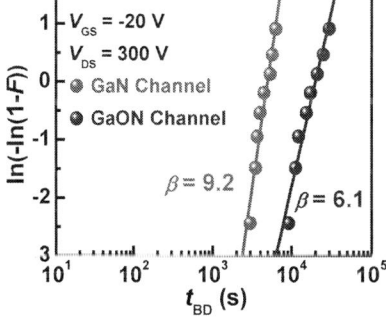

Fig. 12. Time-to-breakdown distributions in Weibull plot of MIS-FETs with GaN channel and GaON channel.

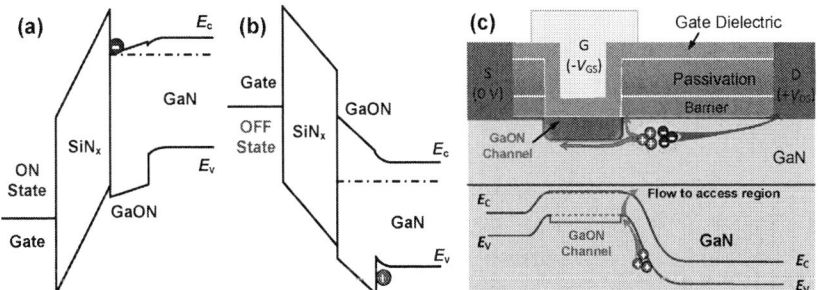

Fig. 13. Schematic band-diagram in the gate region of MIS-FET with GaON channel at (a) on-state and (b) off-state. (c) Hybrid cross-section and schematic band-diagram along the channel. Holes are prevented away from the gate dielectric by the GaON layer.

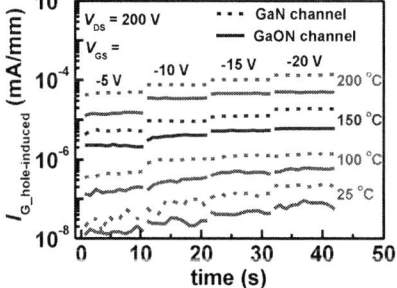

Fig. 14. Reduced hole-induced gate leakage current in MIS-FET with GaON channel measured with V_{DG} = 200 V and various V_{DG}.

978-1-7281-1988-5/18 $31.00 © 2018 IEEE

Recent advancement of GaN HEMT with InAlGaN barrier layer and future prospects of AlN-based electron devices

J. Kotani[1,2], A. Yamada[1,2], T. Ohki[1,2], Y. Minoura[1,2], S. Ozaki[1,2], N. Okamoto[1,2], K. Makiyama[1,2], and N. Nakamura[1,2]

[1] Fujitsu Ltd., [2]Fujitsu Laboratories Ltd., 10-1 Morinosato-Wakamiya, Atsugi-shi, Kanagawa, Japan
email: kotani.junji-01@jp.fujitsu.com

Abstract—The high-power operation of InAlGaN/GaN high-electron-mobility transistor (HEMT) amplifiers in the wide-frequency range from the S-band to the W-band has been achieved. A re-grown n$^+$-GaN contact layer and an InGaN back-barrier layer was employed for the W-band GaN HEMT amplifiers. For the S-band, 2-dimensional electron gases (2DEG) mobility was improved using atomically flat AlGaN spacer layers. This technology allows us to reduce the 2DEG densities maintaining the low access resistance, which contributes to the lower electric-field concentration at the edge of gate electrodes i.e. enables high voltage operation. Furthermore, a diamond heat spreader was introduced to decrease the thermal resistance and we successfully confirmed the further improvement in the output power density. Finally, AlN-based next generation devices are proposed and the lower thermal resistance are expected compared to the conventional GaN/SiC structures.

I. INTRODUCTION

Since the first proposal of InAlN/GaN high-electron-mobility transistor (HEMT) structures, numerous studies have been devoted to the high-frequency devices [1,2]. Contrary to the superior high-frequency performances, critical problems have been left behind, such as large gate leakage current, off-state drain leakage current, and low breakdown voltage. These issues must be solved before the commercial use of InAlN/GaN HEMTs is possible. We believe the low growth temperature of InAlN is one of the causes of these issues. Therefore, we have recently employed InAlGaN barrier layer as the InAlGaN can be grown at higher temperature compared to the InAlN [3].

In this paper, we demonstrate the high-power operation of InAlGaN/GaN HEMTs in the wide-frequency range from the S-band to the W-band. Finally, AlN-based electron devices are proposed.

II. HIGH-POWER GAN HEMT WITH INALGAN BARRIER LAYER

A. High-power W-band InAlGaN/GaN HEMT amplifiers

We developed 80-nm-gate-length InAlGaN/GaN HEMT employing a re-grown n$^+$-GaN contact layer and InGaN back-barrier layer. Fig. 1 compares the I_d-V_{ds} characteristics of the AlGaN/GaN HEMT and InAlGaN/GaN HEMT. A maximum drain current of 1.7 A/mm was confirmed thanks to the low on-resistance. Furthermore, off-state drain leakage current was well suppressed by the existence of the InGaN back-barrier. Fig. 2 and Fig. 3 show the measured power characteristics and the benchmark for high-frequency/high-power MMIC amplifiers. The maximum output power density of our InAlGaN/GaN HEMT MMIC amplifier reached to 4.5 W/mm, which corresponds to more than double of the conventional AlGaN/GaN HEMT MMIC amplifiers.

B. Electron mobility enhancement for InAlGaN/GaN HEMT structures

Low-access resistance is indispensable for high current operation as demonstrated in the W-band GaN HEMT amplifiers; however, increasing the 2-dimensional electron gas (2DEG) density is accompanied by a strong electric-field concentration, as shown in Fig. 4. This degrades off-state breakdown voltage and hampers increasing the output power of GaN HEMTs. Therefore, it should be noted that the reduction of the access resistance must be achieved by increasing the electron mobility not by increasing electron density.

To increase the electron mobility of 2DEG, surface/interface-related electron scattering must be minimized. Fig. 5 shows atomic-force microscope (AFM) images of the surfaces of the MOVPE-grown GaN channel, AlGaN spacer, and InAlN. It was found that the surface pits appeared only when the InAlN layer was grown on the AlGaN spacer layer, clearly indicating that the AlGaN spacer layer is responsible for the formation of the surface pits [4]. Based on these results, we focused on the growth of the AlGaN spacer layer to realize high-electron-mobility 2DEG. Fig. 6 shows the surface pit ratio depending on ammonia (NH$_3$) supply rate during the growth of AlGaN spacer layer. It was revealed that a high NH$_3$ supply rate was essential for smooth surfaces. We achieved high electron mobility of 1550 cm^2/Vsec at the high 2DEG density of 1.94×10^{13} cm^{-2}, as shown in Fig. 7 [4]. Further electron mobility enhancement is expected by a reduction of the 2DEG density as the optical-phonon scattering rate is decreased [5].

C. High-power S-band InAlGaN/GaN HEMT amplifiers with diamond heat spreader

For S-band GaN HEMT amplifiers, we must pay more attention on thermal issues because the increased heat generation hinders stable operation of GaN HEMTs and seriously degrades long-term reliability. In addition to the regrown n^+-GaN layer and the InGaN back-barrier layer used for the W-band GaN HEMT amplifiers, we employed a single-crystal diamond substrate as a heat spreader as schematically shown in Fig. 8(a). To maximize the heat dissipation effect, the SiC substrate was thinned to 50 μm and chemical-mechanical polishing (CMP) was applied to the back-side surface prior to the bonding of diamond wafer [6, 7]. Fig. 8(c) shows I_d-V_{gs} characteristics of the diamond-bonded device, where no degradation of the DC characteristics was observed for wafer-bonding process. Furthermore, it was confirmed that the output power density was increased significantly with the diamond heat spreader, which proved the effect of heat dissipation with the diamond (not shown here).

III. ALN-BASED ELECTRON DEVICES

GaN substrates have become widely available in the past decade, and GaN on GaN structures are expected to push performance limits as the crystal quality is drastically improved. In contrast to the positive outlook, there are concerns about GaN material limitations. As a possible candidate for next generation devices, we investigated the AlN- or high Al composition AlGaN-based electron devices.

A. 2DEG channel with asymmetric electron density

To solve the trade-off between maximum drain current and breakdown voltage, an asymmetric 2DEG channel is desired, i.e., increasing the 2DEG density for the source-side contributes to a high drain current and decreasing the 2DEG density for the drain-side improves off-state breakdown voltages. Applying a strain-field into the barrier layer by using a passivation layer is an effective approach to control the 2DEG density in the vicinity of the gate electrodes [8]. This local electron density modulation is appropriate to reduce electric-field concentration at the gate edge maintaining the low access resistance. As possible candidates for strained-passivation layer we investigated AlO_x and MgO, as their small electron affinities are also effective to form an electron barrier even for nitride-based wide-bandgap materials. Fig. 9 shows the post-annealing temperature dependence of the measured stress in AlO_x and MgO films. It was revealed that the stress can be controlled up to +2GPa. To clarify the effect of the stress on the 2DEG density quantitatively and find the effect of the barrier layer thickness, the strain simulation was carried out by varying the thickness of the $Al_xGa_{1-x}N$ barrier layer. The Al composition of the thin barrier layer was increased to keep the low access resistance at the source-side. Fig. 10 shows the simulated electron density at the drain-side for the AlGaN/GaN HEMT depending on the barrier thickness. It was revealed that the 2DEG density was well controlled especially for the thinner barrier layer, probably because strong strain penetrated into the AlGaN/GaN interface as shown in Fig. 11. These results indicate that the thin AlGaN barriers with high Al compositions are suitable for high-power devices with an asymmetric 2DEG density.

B. Thermal analysis of AlN-based electron devices

As we experimentally demonstrated S-band GaN HEMT amplifiers, thermal management can be a crucial issue for increasing the output power density of devices. Thus, the diamond-bonded heat spreader will be a promising solution for the thermal issues. To realize AlN/diamond bonding, the thinning and planarization of the AlN substrate must be achieved to maximize the heat spreading effect. Then, we carried out mechanical polish and CMP for the back-side surface of AlN substrates. It was confirmed that a 64 μm-thick AlN substrate was obtained with relatively low Ra of 1 nm with CMP, as shown in Fig. 12. Further planarization technology is under development for AlN/Diamond wafer bonding. To estimate the potential of AlN-based electron devices from the viewpoint of thermal issues, we compared the thermal resistance of the AlN/AlN structures with the conventional GaN/SiC, as shown in Fig. 13. Recently, high thermal conductivity of 341 W/mK has been reported for AlN substrate [9]; therefore, it was expected that the thermal resistance of AlN/AlN structures would be 26% lower than that of the conventional GaN on SiC structures. Our results indicate that high thermal conductivity of AlN will be a large advantage for future high-power devices.

IV. CONCLUSION

We successfully demonstrated high-power operation of InAlGaN/GaN HEMTs by employing a re-grown n^+-GaN layer, an InGaN back-barrier and high-electron-mobility 2DEG channel with AlGaN spacer layer. In addition, a diamond heat spreader further improved the output power density. Finally, future technologies for AlN-based electron devices were proposed and it was revealed that high thermal conductivity of AlN will be a large advantage for the future high-frequency/high-power devices.

ACKNOWLEDGMENT

This work was partially supported by Innovative Science and Technology Initiative for Security, ATLA, Japan.

REFERENCES

[1] D. S. Lee, X. Gao, S. Guo, D. Kopp, P. Fay, and T. Palacios, "300-GHz InAlN/GaN HEMTs with InGaN back barrier," IEEE Electron Device Lett., vol. 32, pp. 1525 1527, Nov. 2011.

[2] Y. Yue et al., "Ultrascaled InAlN/GaN high electron mobility transistors with cutoff frequency of 400 GHz," Jpn. J. Appl. Phys., vol. 52, no. 8S, p. 08JN14, 2013.

[3] K. Makiyama et al., "Collapse-free high power InAlGaN/GaN-HEMT with 3 W/mm at 96 GHz," in IEDM, Tech. Digest, pp.9.1.1, 2015

[4] A. Yamada, T. Ishiguro, J. Kotani, and N. Nakamura, "Electron mobility enhancement in metalorganic-vapor-phase-epitaxy-grown InAlN high-electron-mobility transistors by control of surface morphology of spacer layer," Jpn. J. Appl. Phys., vol. 57, no. 1S, p. 01AD01, 2018.

[5] M. Gurusinghe, S. Davidsson, and T. Andersson, "Two-dimensional electron mobility limitation mechanisms in $Al_xGa_{1-x}N$/GaN heterostructures," Phys. Rev. B, vol. 72, no. 4, p. 045316, Jul. 2005.

[6] Y. Minoura et al., "Surface activated bonding of SiC/Diamond for thermal management of GaN devices," IEEE Semiconductor Interface Specialists Conf., #5.26, 2017.

[7] N. Okamoto et al., "Thermal Analysis of GaN-HEMT/SiC on Diamond by Surface Activated Bonding," CS MANTECH Conf., #5.1, 2018.

[8] K. Osipov, I. Ostermay, M. Bodduluri, F. Brunner, G. Trankle, and J. Wurfl, "Local 2DEG Density Control in Heterostructures of Piezoelectric Materials and Its Application in GaN HEMT Fabrication Technology," IEEE Trans. Electron Devices, vol. 65, no. 8, pp. 1–9, 2018.

[9] R. Rounds et al., "Thermal conductivity of single-crystalline AlN," Appl. Phys. Express, vol. 11, no. 7, p. 071001, 2018.

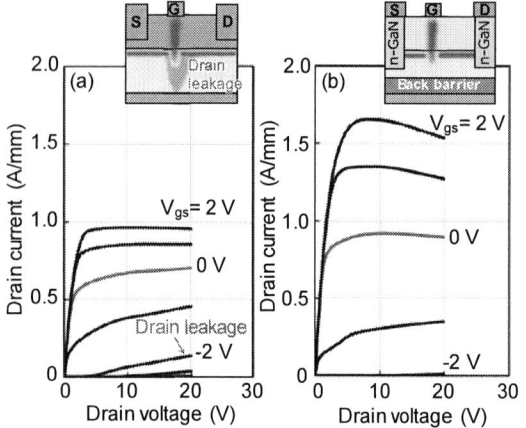
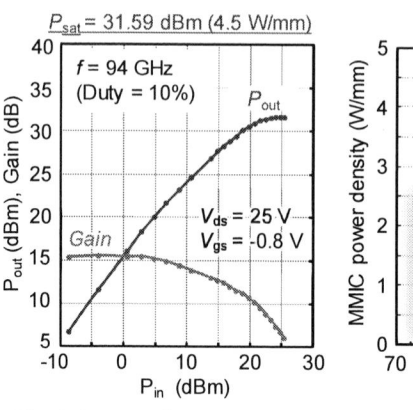
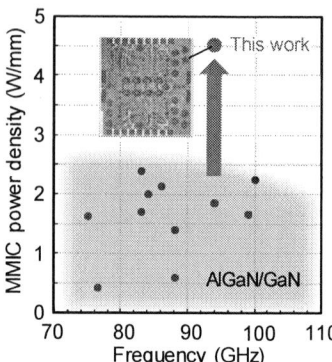

Fig. 1. I_d-V_{ds} characteristics of (a) AlGaN/GaN and (b) InAlGaN/GaN HEMTs. The InAlGaN/GaN HEMT exhibits high maximum drain current with a low drain-leakage current.

Fig. 2. Measured P_{in}-P_{out} characteristics of InAlGaN/GaN HEMT MMIC at 94 GHz under pulsed operation.

Fig. 3. MMIC power density vs. operating frequency.

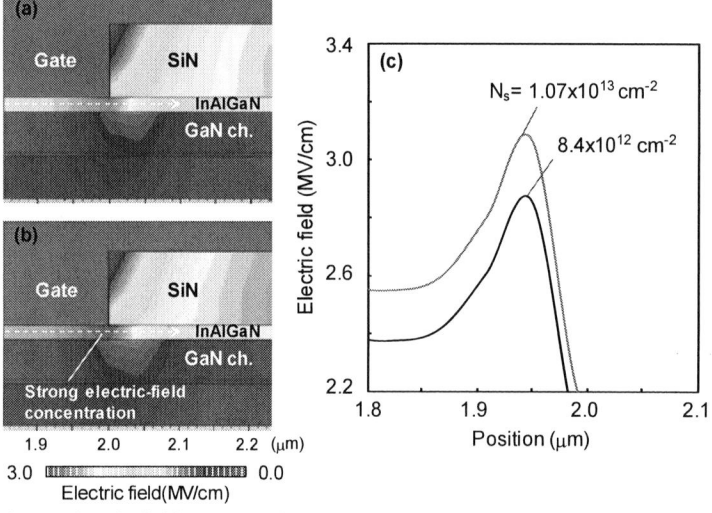
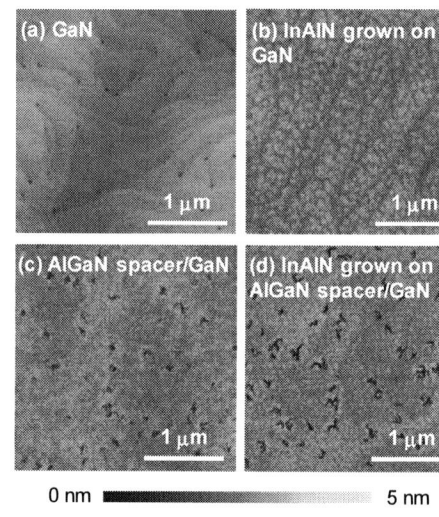

Fig. 4. Electric-field concentration at the drain-side of the gate electrodes for different 2DEG densities of (a) 8.4×10^{12} cm^{-2} and (b) 1.07×10^{13} cm^{-2}. (c) Electric-field strength in the InAlGaN barrier layer along the arrows, indicating the reduced electric-field concentration with the lower electron density.

Fig. 5. AFM images of surfaces of (a) GaN, (b) InAlN/GaN, (c) AlGaN spacer/GaN and (d) InAlN/AlGaN spacer/GaN. Surface pits appeared only when AlGaN spacer layer was inserted, clearly indicating AlGaN spacer is responsible pit formation.

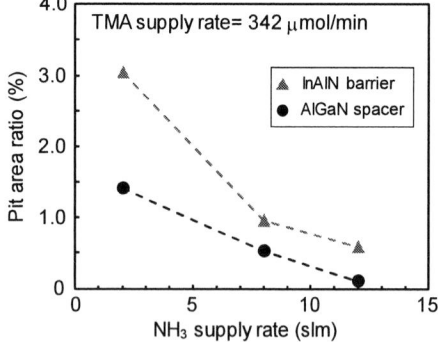
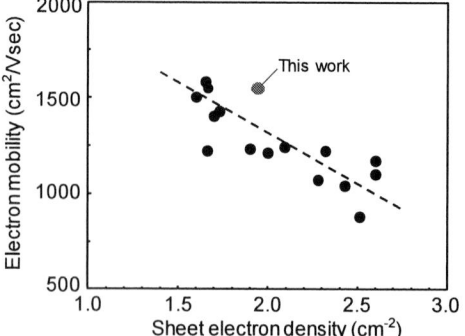

Fig. 6. Pit area ratios of AlGaN spacers and InAlN barriers depending on NH$_3$ supply rate during the growth of AlGaN spacer layer. Increasing the NH$_3$ supply rate is effective in improving surface morphologies of the AlGaN spacer layer.

Fig. 7. Electron mobility vs. sheet electron density of MOVPE-grown InAlN/GaN HEMT structures. Our InAlN/GaN HEMT structure shows superior electron mobility as compared to other previously reported InAlN/GaN HEMT structures.

978-1-7281-1988-5/18 $31.00 © 2018 IEEE

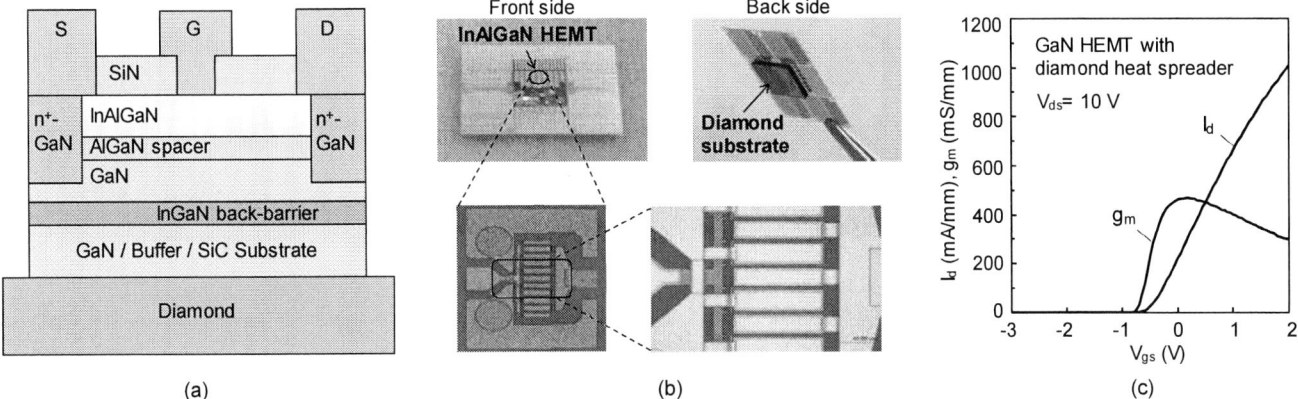

Fig. 8. (a) Schematic illustration of high-power S-band InAlGaN/GaN HEMT with a diamond heat spreader. An n+-GaN regrown contact layer was formed to reduce the contact resistance and InGaN back-barrier layer were employed to suppress drain-leakage current. (b) Photographs of InAlGaN/GaN HEMT bonded to the diamond heat spreader. (c) Measured I_d-V_{gs} curve of the InAlGaN/GaN HEMT with diamond heat spreader. No degradation was observed in the wafer-bonding process.

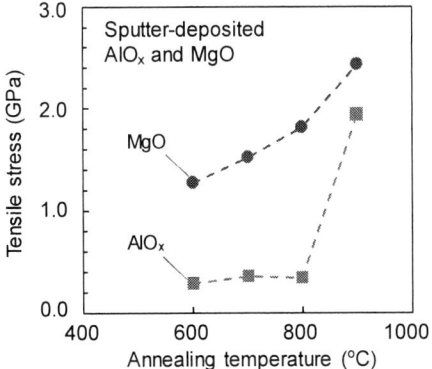

Fig. 9. Measured internal stress of sputter-deposited AlO_x and MgO films depending on the post-annealing temperature.

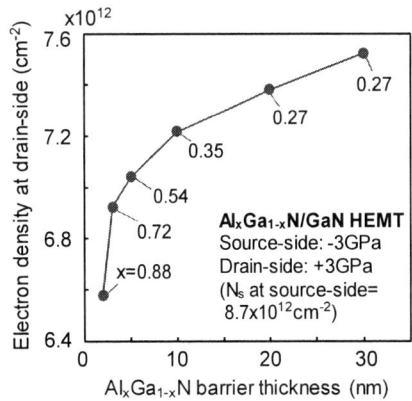

Fig. 10. Stress-induced electron density modulation for the $Al_xGa_{1-x}N$/GaN HEMT structures depending on the thickness of the barrier layer. The Al composition of the thin barrier was increased to keep electron density at the source-side at 8.7×10^{12} cm^{-2}.

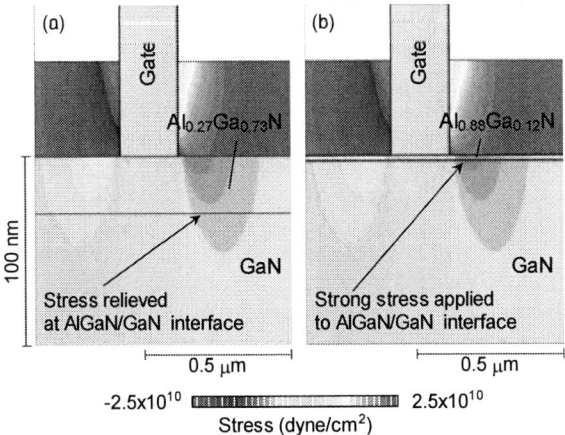

Fig. 11. Simulated strain distribution for AlGaN/GaN HEMTs with (a) 3-nm- and (b) 30-nm-thick AlGaN barrier layers. Stress in the passivation layer is -3GPa and +3GPa for the source-side and drain-side, respectively.

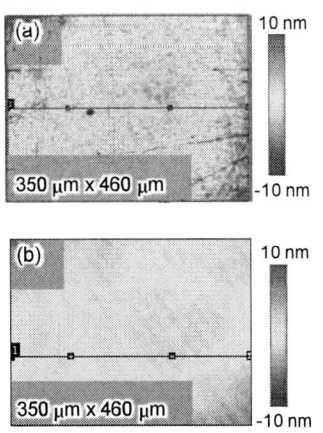

Fig. 12. Surface morphologies of the back-side surface of the AlN substrate: (a) before and (b) after CMP.

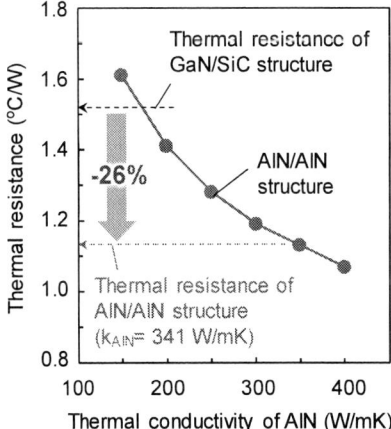

Fig. 13. Simulated thermal resistance of AlN devices on AlN substrates and the conventional GaN on SiC substrates.

978-1-7281-1988-5/18 $31.00 © 2018 IEEE

Power GaN HEMT degradation: from time-dependent breakdown to hot-electron effects

M. Meneghini[1], A. Barbato[1], M. Borga[1], C. De Santi[1], M. Barbato[1], S. Stoffels[2], M. Zhao[2], N. Posthuma[2], S. Decoutere[2], O. Haeberlen[3], T. Detzel[3], G. Meneghesso[1], E. Zanoni[1]

[1]Univ. of Padova, Dept. of Information Engineering, via Gradenigo 6/B 35131 Padova (Italy), email: matteo.meneghini@unipd.it
[2]imec, Kapeldreef 75, 3001 Heverlee (Belgium) [3]Infineon, Siemensstraße 2, 9500 Villach, Austria

Abstract—This paper describes our most advanced results in the field of GaN-HEMT degradation, with focus on power devices. We discuss three main aspects: *(i)* the first part of the paper analyzes the dependence of breakdown voltage on substrate and buffer properties, by reporting the results obtained on wafers with different substrate resistivities and superlattice thickness. *(ii)* the second part of the paper demonstrates the existence of time-dependent breakdown of GaN buffer submitted to high vertical stress, and describes the related process. *(iii)* in the third part of the paper, we focus on the role of hot-electrons in limiting the dynamic performance of the devices. We demonstrate that the exposure to hard switching transitions may lead to an increase in dynamic R_{on}. This effect is ascribed to the presence of hot electrons, which was verified by means of electroluminescence measurements.

INTRODUCTION

Over the past years, GaN has emerged as an excellent material for the fabrication of power devices [1]: the low on-resistance (R_{on}), the low $R_{on}*Q_g$ and $R_{on}*Q_{oss}$ products (Q_g is the gate charge, Q_{oss} is the output charge), the absence of reverse recovery charge and the high breakdown voltage of these devices allow to design converters with high switching frequency and efficiency [2]. For a successful diffusion of GaN transistors, the mechanisms that limit the performance and the reliability of these devices must be clearly understood. Over the past years, several papers have been published in the field of GaN reliability, focusing mostly on: the problem of dynamic-R_{on} [3], the reliability of the gate stack [4], the degradation mechanisms related to off-state and dielectrics [5]. The aim of this paper is to focus on three specific aspects that are not widely described in the literature: *(i)* the impact of substrate and buffer properties on the characteristics of the devices; *(ii)* the time-dependent breakdown of the semi-insulating GaN buffer, that can lead to a premature failure of the devices; *(iii)* the degradation of the dynamic properties of the transistors under hard switching conditions. These aspects have been widely explored within the ECSEL project PowerBase, and are of fundamental importance for the development of reliable and high performance GaN devices.

IMPACT OF SUBSTRATE PROPERTIES ON DEVICE PERFORMANCE

GaN power devices are typically grown on a 150 mm or 200 mm silicon substrate, in order to minimize fabrication costs and ensure compatibility with silicon manufacturing lines. The properties of the substrate can significantly impact on the performance and reliability of the devices [6-8]. To demonstrate this, we refer to two different experiments: first, we describe the impact of the resistivity of the substrate on the breakdown voltage/leakage characteristics; second, we investigate by simulations the vertical leakage of p⁻-Si/AlN heterostructures. Figure 1 reports the 2-terminal vertical I-V curves measured between drain and substrate on devices having *a)* different substrate resistivity (between 0.01 Ωcm and 6 Ωcm, p-type) and *b)* different thickness of the superlattice (25x or 50x). The following considerations can be made: *(i)* by increasing the thickness of the superlattice, an obvious increase in breakdown voltage is observed. *(ii)* the resistivity of the silicon substrate significantly impacts on the shape of the vertical I-V plot and on the breakdown voltage. When a low resistivity substrate is used, the vertical I-V plot has a single slope (in semi-log scale), and breakdown voltages (10^{-4} A/mm) are around 300 V (for 25x superlattice) and 700 V (for 50 x superlattice). The use of a resistive silicon substrate substantially modifies the vertical I-V plot, by inducing the appearance of a plateau. The length of such plateau significantly depends on the resistivity of the substrate, and the overall result (for highly-resistive substrates) is an increase in breakdown voltage. The origin of the plateau was investigated by specific experiments and numerical simulations: Figure 2 reports the vertical I-V plot measured on a simple p⁻-Si/AlN heterojunction grown on a highly resistive (p⁻) substrate and the related simulation. Also on this very simplified structure, the plateau is present. The results of the simulations are consistent with a recent study on the topic [6], and are summarized as follows: when the Si/AlN heterojunction is submitted to vertical bias, an inversion layer is formed at the Si-side of the p-Si/AlN interface. Conduction mainly occurs due to tunneling from the inversion layer towards the AlN, as shown in the inset of Figure 2 (V<30 V). For higher voltages (30 V<V<90 V), the p-type substrate starts depleting, and vertical leakage is limited by the availability of electrons generated in the p-type substrate. For high voltages (V>90 V), generation/impact ionization in the silicon substrate yields to an increase in the density of electrons in the silicon, and current starts increasing again. Figure 3 shows the simulated conduction band diagrams of the Si/AlN heterojunction at different voltages, i.e. 0 V (no applied bias), 20 V (before plateau), 40 V (start of plateau), 80 V (end of plateau) and 100 V (generation/impact-ionization regime). As can be noticed, the slope of the conduction band in the AlN (and thus the width of the tunneling barrier) does not significantly change during the plateau regime, while in the same voltage range a substantial depletion of the silicon substrate occurs.

978-1-7281-1988-5/18 $31.00 © 2018 IEEE

If on one hand the use of a resistive silicon substrate can increase the breakdown voltage, on the other hand it can enhance the trapping processes. Figure 4 reports the pulsed I-V plots measured up to V_{DS}=300 V on devices having substrate with low and high resistivity (these devices are from a 200 V buffer). As can be noticed, the transistors grown on a low-resistivity (0.01 Ωcm) substrate do not show a significant dynamic-R_{on} increase. On the other hand, identical devices grown on a resistive substrate (6 Ωcm) show a substantial increase in dynamic-R_{on} when submitted to high off-state bias. This effect is ascribed to a buffer trapping process that takes place when the substrate is depleted. In the plateau regime, the depleted substrate acts as a capacitor. When the devices are switched from off- to on-state conditions, the presence of this capacitor leads to a (temporary) positive backgating effect on the GaN buffer [7]. This brings to a significant charge trapping, since the C-GaN/uid-GaN pn diode becomes positively biased, and electrons are injected in the C-doped GaN buffer [9]. Based on the results described above, we can conclude that the use of a resistive substrate can positively impact on the breakdown voltage. However, the depletion of the silicon substrate can have a detrimental effect on dynamic performance, leading to V_{th}/R_{on} instabilities. For this reason, a trade-off between the two effects must be considered, when optimizing the properties of the substrate for GaN power devices.

TIME-DEPENDENT BREAKDOWN OF GAN

GaN is a wide bandgap semiconductor, with a high breakdown field of 3.3 MV/cm. When dealing with reliability, it is often important to study long-term reliability and the existence of time-dependent degradation processes. In several cases, long-term reliability is limited by the time-dependent failure of dielectrics, used for the gate-stack [10] or as a passivation layer [5]. Here we demonstrate that semi-insulating GaN itself can show a time-dependent breakdown mechanism. Figure 5 reports the results of a 2-terminal constant voltage stress experiment carried out at V_{DS}=800 V on a semi-insulating GaN buffer grown on a silicon substrate (buffer is designed for 200 V operation). As can be noticed, vertical leakage current is constant in the initial phase of stress, and then increases/becomes noisy. A catastrophic failure is reached for long times for all analyzed devices. In this experiment, no dielectric is involved, since GaN buffers are submitted to vertical stress. This means that the semi-insulating GaN itself is behaving as a dielectric, and showing a time-dependent failure. Figure 6 shows that time-to-failure is Weibull-distributed; EL measurements carried out during vertical stress indicate a significant light emission around the drain pad (current flows under the pad that shades breakdown luminescence) and failure takes place at a random spot within the contact area. The breakdown is ascribed to a defect generation process within the semi-insulating buffer, activated by the high electric field. Failure could be easier in proximity of dislocations, where the presence of a higher local leakage can accelerate the process of defect generation. A possible explanation for the time-dependent breakdown is the fact that GaN has polar bonding, as common dielectrics (SiN, SiO$_2$) [11]: when GaN is exposed to a high field, a significant distortion of the lattice takes place,

and the Gibbs potential of the lattice changes, leading to less stability. In this view, the time-dependent breakdown of GaN would be a stress-relaxation mechanism, that leads to the generation of defects. It is worth noticing that this is not a limiting mechanism for current GaN technologies: in a recent paper [12] we demonstrated a lifetime of 20 years with 1 % failure rate at 560 V for a 200 V technology. Possible methods for improving TTF are reducing the density of defects in the lattice and optimizing the vertical E-field profile.

DEGRADATION UNDER HARD SWITCHING

To date, the dynamic performance of GaN transistors is almost ideal, and devices without the Dynamic-R_{on} issue have been already demonstrated [3]. An example is reported in Figure 7 which shows the pulsed I-V curves of a state-of-the-art transistor measured after exposure to different quiescent bias points in the off-state. As can be noticed, no significant charge trapping is observed up to V_{DS}=600 V, indicating a good stability. In several cases, during normal operation GaN transistors can undergo hard switching transitions (consider for example the case of a boost power converter). It is therefore important to develop tools for analyzing the variation in on-resistance induced by hard switching transitions. To this aim, we developed a novel setup capable of investigating the hard-switching performance of the devices, by switching on a resistive load with an external capacitance (C_{ext}) in parallel to the transistor. The points of novelty of this setup are the following: *(i)* thanks to the use of a clamp circuit, it is possible to monitor both the current/voltage waveforms during the switching events and the dynamic on-resistance right-after switching; *(ii)* the circuit is mounted on a microscope, and allows on-wafer characterization under hard switching conditions. This is a great advantage, since it is not necessary to reach the packaging step to evaluate the switching performance. Figure 8 (a) shows a schematic representation of the circuit used to induce hard switching transitions, while Figure 8 (b) reports the variation of dynamic-R_{on} induced by hard switching at different voltages, for three different values of parasitic capacitance C_{ext}. As can be noticed, exposure to hard switching conditions leads to an increase in dynamic-R_{on}, that is not observed in soft-switching conditions (see Figure 7). It is worth noticing that – due to the parasitic and external capacitors being larger by two orders of magnitude compared to the device capacitance – the setup used here induces hard switching transitions much longer (few 100 ns) than real power converters (few ns), so the overall dynamic-R_{on} increase is overestimated with respect to realistic conditions. This allows for highly accelerated ageing investigations and fast feedback loops for process and device optimization. The origin of such dynamic-R_{on} increase can be understood by looking at Figure 9 that reports the voltage and current waveforms during a hard switching event induced by the setup in Figure 8. As can be noticed, for a substantial time interval there is the simultaneous presence of high drain current and high drain voltage, leading to maximum (instantaneous) power dissipations above 100 W/mm. A different view of the hard switching transition is given by the IV switching locus diagram in Figure 10 showing the points of the IV plot crossed by the devices during the turn-on event in Figure 9. When the device is turned on the current starts to rise while the voltage stays around 600 V. Only after the current has discharged the capacitor, the voltage starts decreasing and the device reaches the linear region. Hard-switching conditions can increase self-heating and (more importantly) lead to hot-electron trapping effects. The role of hot-electrons was verified by means of electroluminescence measurements, carried out under hard-switching conditions at different voltages. The setup was optimized in order to capture the weak luminescence signal emitted during the (very short) hard switching events. Figure 12 shows the dependence of the EL intensity on the switching voltage; as can be noticed, under the extreme hard switching conditions being used here, hot-electron luminescence starts being observed around 300 V and increases monotonically above this voltage. The increase in on-resistance reported in Figure 8 (b) can be ascribed to the injection of hot electrons towards trap states located either in the buffer or in the AlGaN barrier. Such effect can be minimized by reducing the electric field within the structure via suitable field plates.

CONCLUSIONS

GaN power devices have already demonstrated a great robustness and reliability. With this paper we have summarized our latest results on the impact of substrate properties on device reliability, on the existence of time-dependent buffer breakdown, and on the degradation of the dynamic properties of the devices under hard switching transitions. The results reported within this paper indicate that through proper optimization of device properties it is possible to fabricate transistors with excellent dynamic performance and reliability. Further challenges in GaN reliability will be on novel technologies, such as 1200 V GaN lateral devices and GaN-based vertical transistors.

REFERENCES

[1] H Amano et al., "The 2018 GaN power electronics roadmap", J. Phys. D: Appl. Phys. 51 163001 (2018)

[2] K. J. Chen et al., "GaN-on-Si Power Technology: Devices and Applications," in IEEE Transactions on Electron Devices, vol. 64, no. 3, pp. 779-795, March 2017

[3] P. Moens et al., "Negative dynamic Ron in AlGaN/GaN power devices", Proc. Int. Symp. Power Semiconductor Devices ISPSD 2018, pp. 97-100

[4] S. Stoffels et al., "Failure mode for p-GaN gates under forward gate stress with varying Mg concentration," 2017 IEEE International Reliability Physics Symposium (IRPS), Monterey, CA, 2017, pp. 4B-4.1-4B-4.9

[5] I. Rossetto et al., "Field-Related Failure of GaN-on-Si HEMTs: Dependence on Device Geometry and Passivation," in IEEE Transactions on Electron Devices, vol. 64, no. 1, pp. 73-77, Jan. 2017

[6] L. Sayadi et al., "The Role of Silicon Substrate on the Leakage Current Through GaN-on-Si Epitaxial Layers", in IEEE Transactions on Electron Devices, vol 65, no. 1, pp- 51-58, 2018

[7] M. Borga et al., "Impact of substrate resistivity on the Vertical Leakage, Breakdown, and Trapping in GaN-on-Si E-mode HEMTs", in IEEE Transactions on Electron Devices, vol 65, no. 7, pp. 2765-2770 (2018)

[8] X. Li et al., "Investigation on Carrier Transport Through AlN Nucleation Layer From Differently Doped Si(111) Substrates", IEEE Transactions on Electron Devices 65 (5) 1721 (2018)

[9] S. Stoffels et al., "The physical mechanism of dispersion caused by AlGaN/GaN buffers on Si and optimization for low dispersion," 2015 IEEE International Electron Devices Meeting (IEDM), Washington, DC, 2015, pp. 35.4.1-35.4.4

[10] S. Warnock et al., "Time-Dependent Dielectric Breakdown in High-Voltage GaN MIS-HEMTs: The Role of Temperature", in IEEE Transactions on Electron Devices, vol. 64, no. 8, pp. 3132-3138, 2017

[11] J. W. McPherson, "Brief History of JEDEC Qualification Standards for Silicon Technology and Their Applicability(?) to WBG Semiconductors", Proc. IEEE-IRPS 2018, pp. 3B.1-1- 3B.1-8

[12] M. Borga et al., "Evidence of Time-Dependent Vertical Breakdown in GaN-on-Si HEMTs," in IEEE Transactions on Electron Devices, vol. 64, no. 9, pp. 3616-3621, Sept. 2017

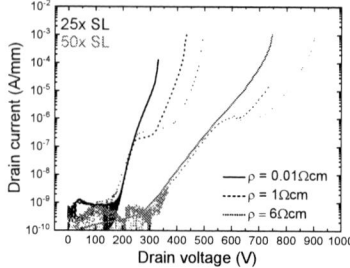

Figure 1: 2-terminal (drain-to-substrate) I-V curves measured on devices grown on substrates with different resistivities, and on buffers with different superlattice thicknesses

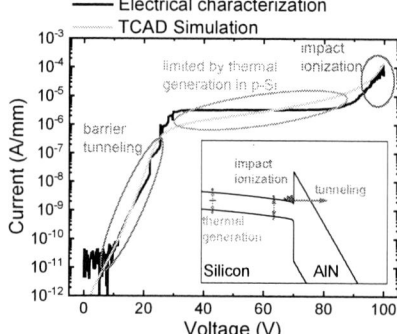

Figure 2: vertical I-V plot (black) of a silicon/AlN heterojunction and related simulation (green). Inset: schematic representation of the dominant tunneling path responsible for leakage conduction

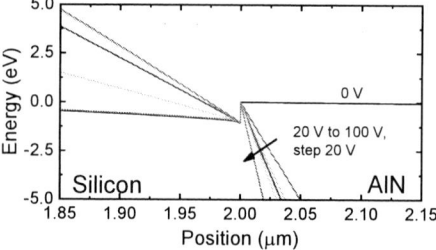

Figure 3: simulated conduction band diagrams of the Silicon/AlN heterojunction at different voltages (20 V, 40 V, 80 V, 100 V). All band diagrams have been shifted to match the energy of the inversion layer at the Si/AlN interface

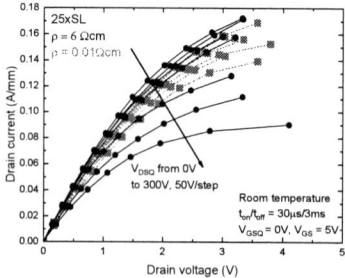

Figure 4: pulsed I-V measurements on samples with different substrate resistivities up to V_{DS}=300 V. Only samples with resistive substrate show Dynamic-R_{on}

978-1-7281-1988-5/18 $31.00 © 2018 IEEE

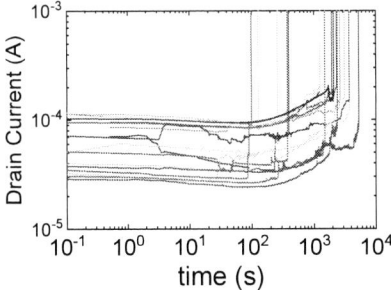

Figure 5: results of a 2-terminal (drain-to-substrate) stress experiment carried out at V_{DB}=800 V on a semi-insulating GaN buffer grown on a silicon substrate. The structure consists of Si substrate, 200 nm AlN, 1000 nm graded buffer, 2000 nm C-doped AlGaN, 300 nm uid GaN

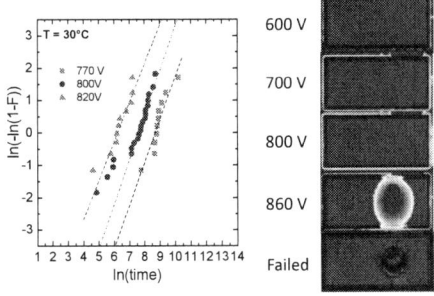

Figure 6: (left) Weibull plot of TTF for three different stress voltages (2-terminal stress on semi-insulating GaN buffers). (right) EL images collected during a vertical step-stress experiment until failure

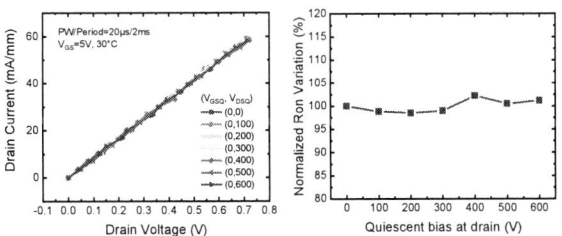

Figure 7: (left) pulsed I V curves measured starting from several quiescent bias points in the off-state, up to V_{DS}=600 V. (right) dependence of dynamic-R_{on} on the quiescent bias point. No significant variation in dynamic-R_{on} is observed even after 600 V off-state bias

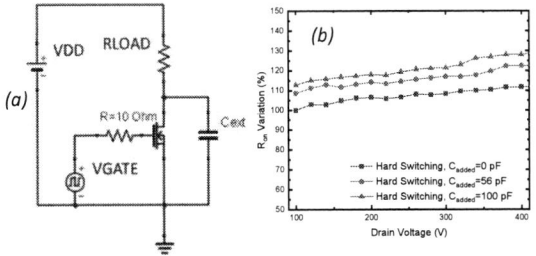

Figure 8: (a) schematic representation of the circuit used to study hard switching transitions (clamp circuit not shown for brevity). (b) dependence of dynamic-R_{on} on switching voltage for different values of capacitance C_{ext}. It is worth noticing that due to the physical implementation of the board, capacitance C_{added} adds to the parasitic capacitance $C_{parasitic}$ (~ 15 pF) of the circuit ($C_{ext}=C_{added}+C_{parasitic}$). $C_{parasitic}$ is already two orders of magnitude higher than the C_{oss} of a typical small test device with few mm gate length. C_{added} is used to make hard-switching even more stressful.

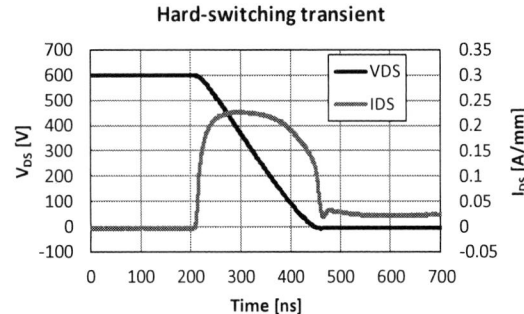

Figure 9: drain voltage and drain current waveforms measured during a hard switching transition induced by the setup in Figure 8 (a) (turn-on transition, C_{added}=100 pF)

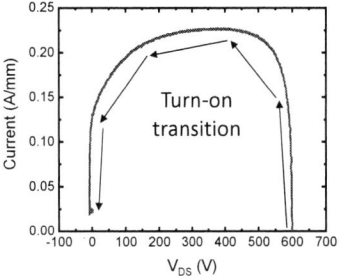

Figure 10: I-V switching locus diagram corresponding to the data in Figure 9 (turn-on transition, C_{added}=100 pF)

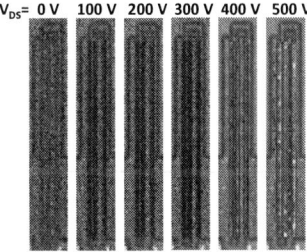

Figure 11: electroluminescence measurements carried out under hard-switching conditions at different voltages (setup in Figure 8 (a))

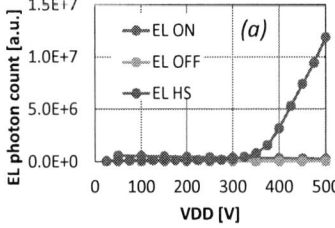

Figure 12: Dependence of EL signal on the switching voltage for different device conditions (ON: permanent on, OFF: permanent off, HS: hard switching). A significant increase in the EL signal is observed above 350 V, indicating the presence of hot-electrons during hard switching

Acknowledgments: This project has received funding from the Electronic Component Systems for European Leadership Joint Undertaking under grant agreement No 662133. This Joint Undertaking receives support from the European Union's Horizon 2020 research and innovation programme and Austria, Belgium, Germany, Italy, Netherlands, Norway, Slovakia, Spain, United Kingdom. This article reflects only the authors' view and the JU is not responsible for any use that may be made of the information it contains

New Insights into the Physical Origin of Negative Capacitance and Hysteresis in NCFETs

Huimin Wang[1], Mengxuan Yang[1], Qianqian Huang[1,2*], Kunkun Zhu[1], Yang Zhao[1], Zhongxin Liang[1],
Cheng Chen[1], Zhixuan Wang[1], Yuan Zhong[1], Xing Zhang[1], Ru Huang[1,2*]

[1]Institute of Microelectronics, Peking University, Beijing 100871, China (*Email: hqq@pku.edu.cn; ruhuang@pku.edu.cn)
[2]National Key Laboratory of Science and Technology on Micro/Nano Fabrication, Beijing 100871, China

Abstract—In this paper, direct experimental observation of negative capacitance (NC) in a standalone ferroelectric (FE) capacitor is reported for the first time, which proves that the physical origin of NC is the domain switching dynamics rather than the stabilized switching. Based on this origin, the "dynamic polarization (DP) matching", different from the traditional capacitance matching, is rigorously derived and verified to be the prerequisite for sub-60mV/dec subthreshold swing (*SS*) in NCFET. The proposed DP matching can accurately describe and predict the features of NCFET based on our developed device model, showing that the *SS* and hysteresis are highly sensitive to the input sweeping voltage and FE switching dynamics, as well as other device parameters. Moreover, an intrinsic conflict is found between hysteresis and *SS* optimization. This work provides new understanding for the NCFET mechanism.

I. INTRODUCTION

Negative capacitance (NC) FETs have triggered immense interest due to their capability of sub-60 mV/dec subthreshold swing (*SS*) which has been observed in numerous experimental studies [1-3]. However, the theoretical explanation of NC in the ferroelectric (FE) is still controversial. It has been presented very recently that the NC effect could be modeled as a transient phenomenon caused by the delay of FE polarization switching [4-6], rather than the prevailing theory of stabilized *S*-shaped polarization-electric field (*P-E*) curve [7]. Therefore, direct experimental evidence is in urgent need to confirm the actual cause of NC effect in the FE. Moreover, rigorous mechanism analysis and design optimization for SS and hysteresis in NCFETs, based on reliably physical-based FE switching model instead of *S*-curve model, are still absent and indistinct, since the previous FE switching models for NC explanation were oversimplified [4] and were not integrated with transistor [5-6].

In this work, we first directly measure the NC effect in the standalone FE capacitor, and propose that the NC is generated from FE polarization switching and can contribute to sub-60 *SS* in NCFET only when the proposed dynamic polarization (DP) matching condition is satisfied. A physical NCFET model is established based on the calibrated FE switching model. The physical origin of *SS* and hysteresis in NCFETs and their correlation with FE switching properties are thoroughly studied. Based on this, *SS* and hysteresis of NCFET optimization are reconsidered.

II. EXPERIMENTS AND NC EFFECT MEASUREMENTS

A. Experimental characterization of NCFET gate stack

$Hf_{0.5}Zr_{0.5}O_2$ (HZO) is explored as the FE film for realizing NCFETs in this work. A standalone FE capacitor with 5 nm HZO is fabricated on Si substrate (Fig.1a). The annealing process was performed for the crystallization of HZO, and the GI-XRD results in Fig. 1b further confirm the polycrystalline nature. The measured typical hysteretic loops of polarization-voltage (P-V_{FE}) show the ferroelectricity of HZO film after annealing (Fig. 1c). From negative polarization state to positive polarization state, the switching of FE occurs through domain nucleation and growth [8], and the dP/dV_{FE} is always positive.

However, the FE capacitance (C_{FE}) is not isolated in the NCFET gate stack, in which the C_{FE} is in series with positive capacitance (C_P) (Fig. 2a). Only with the negative differential FE voltage ($dV_{FE}<0$), can NCFET exhibit the steeper *SS* than MOSFET (Fig. 2b). A FE-Dielectric(DE) gate stack of HZO-HfO_2 is further fabricated to investigate the NC origin (Fig. 2c).

B. Direct NC measurement in standalone FE capacitor

The prevailing theory of *S*-curve points out that the NC can only be stabilized and extracted in a series system [7, 10]. To figure out the real NC, we first designed the V_{FE} measurement system for series FE-DE capacitance stack (Fig. 3). Under the bipolar triangular voltage pulse of V_G with period time of T, the V_{FE} in series stack shows "anomalous" decrease (or increase) when V_G is increasing (or decreasing), leading to negative dV_{FE}/dV_G (Fig. 4). By further programing the V_{FE}-t waveforms into the standalone FE (Fig. 5), the measured results demonstrate that the NC phenomenon (negative dP/dV_{FE}) can even exist with a single layer FE film, which indicates that the polarization is still switching with increasing charges (shown in P-t) when the capacitor voltage decreases (shown in V_{FE}-t) (Fig. 6). This is the first direct experimental proof of NC effect in standalone FE capacitor, which is considered unable to be measured due to thermodynamic stability according to the prevailing theory [7]. These experimental results further confirm that the cause of NC is originated from the domain switching dynamics of FE rather than the stabilized switching.

Moreover, the observed negative dV_{FE}/dV_G phenomenon in the gate stack is strongly related to the sweeping rate of gate voltage. Only within a small range of sweeping rate, the NC phenomenon in P-V_{FE} can be observed. In this experimental work of HZO, the NC effect is the most distinct when the T of input voltage pulse is 45ms (Fig. 7). It indicates that the generation of NC is also associated with the voltage sweeping.

978-1-7281-1988-5/18 $31.00 © 2018 IEEE

III. PHYSICAL ORIGIN OF SUB-60 *SS* IN NCFET

A. NCFET modeling based on FE domain switching

According to the measured results (Fig. 8), the nature of FE polarization is strongly dependent on both the V_{FE} and the time (t), which can be described by Kolmogorov-Avrami-Ishibashi (KAI) equation [9] based on multi-domain switching process. In this work, a physical-based NCFET model is established by self-consistently solving the calibrated KAI equation (Fig. 9) with baseline MOSFET model through charge balance equation (Fig.10). The simulated $P(V_{FE}, t)$ in NCFET is shown in Fig.11.

B. "DP matching" as the prerequisite for sub-60 SS

With charge balance equation considered, the expression of differential voltage amplification (A_V) in NCFET can be precisely derived. Instead of capacitance matching [7], which takes FE capacitance as a stabilized negative value, and never confirmed by experiments, "dynamic polarization matching" is derived based on our direct measurement and found to be the prerequisite for NC generation ($A_V > 1$) (Fig. 12), the physical meaning of which is that the time-induced increment of polarization charge (($\partial P/\partial t$)·dt) should exceed the charge increment of undelaying C_P in response to total input gate voltage ($C_P \cdot dV_G$).

For FE polarization switching, the $\partial P/\partial t$ drops rapidly with t increasing and also strongly depends on V_{FE} (Fig. 13). When sweeping V_G with different rates (dV_G/dt) or starting voltages (V_{sta}), the V_{FE} in NCFET varies non-linearly with time t (Fig. 14), leading to the diverse $\partial P/\partial t$ curves (Fig. 15). The extracted $\partial P/\partial t$ curves are further compared with $C_P \cdot dV_G/dt$ in NCFET to identify the DP matching situation. As shown in Fig. 16, the SS of NCFET can be smaller than MOSFET only when the DP matching condition is satisfied, and the more time-induced polarization charges, the steeper SS. If the sweeping rate is too slow for the quasi-static situation or too fast that the polarization charge cannot respond, the SS of NCFETs will both degrade (Fig. 16). Besides sweeping rate of input voltage, the sweeping range will also impact $\partial P/\partial t$ curves and thus the SS in NCFET. For forward sweeping, V_{sta} increasing will induce the steeper SS due to the larger $\partial P/\partial t$ (Fig. 17). For reverse sweeping, the SS is steeper than that for forward sweeping, which can be also attributed to the larger $\partial P/\partial t$ in the subthreshold region (Fig. 18).

IV. PHYSICAL ORIGIN OF HYSTERESIS IN NCFET

The hysteresis of NCFET is usually characterized by the difference between threshold voltages for forward sweeping (V_{th_F}) and reverse sweeping (V_{th_R}).

A. Threshold voltage for forward and reverse sweeping

From Fig. 16-18, it is shown that the V_{th_F} and V_{th_R} of NCFET are very different from the V_{th} of MOSFET (V_{th_MOS}). According to charge balance equation, the voltage difference between NCFET and MOSFET at the same drain current (I_{DS}) equals to the voltage across the FE film of NCFET (i.e., V_{FE}). Fig. 14 has shown that the V_{FE} in NCFET is strongly related to the sweeping rate and range of V_G, thereby the V_{th} of NCFET in Fig.16-18 differs correspondingly. For the threshold voltage difference between NCFET and MOSFET (ΔV_{th_F} for forward sweeping and ΔV_{th_R} for reverse sweeping), the value can be extracted through the intersection of the V_{FE}-t curve and the polarization contour line in $P(V_{FE}, t)$ when $P=Q_{MOS\_th}$ (Q_{MOS_th} is the total charge of MOSFET at threshold-state) (Fig. 19).

Based on the V_{th_MOS} and extracted ΔV_{th_F} (or ΔV_{th_R}), the V_{th_F} (or V_{th_R}) of NCFET can be obtained (Fig. 20).

B. Hysteresis in NCFET

Based on the above analysis, the hysteresis in NCFET (V_{th_F}-V_{th_R}) can be characterized by the difference between ΔV_{th_F} and ΔV_{th_R}, which also equals the V_{FE} difference between forward and reverse sweeping (V_{FE_F}-V_{FE_R}) when $P=Q_{MOS\_th}$ as shown in Fig. 20. For FE polarization forward and reverse switching in NCFET, positive V_{FE_F} and negative V_{FE_R} are indispensable, respectively. Therefore, the hysteresis of NCFET always exists theoretically and shows counter-clockwise behavior. Beside, $|V_{FE\_F}|$ and $|V_{FE\_R}|$ would be different with the different $|V_{sta}|$ of V_G (Fig. 21).

C. Intrinsic trade-off between hysteresis and SS

Both hysteresis and SS of NCFET are time-dependent. With the slower V_G sweeping rate or the faster FE switching response, the hysteresis of NCFET can be significantly reduced but cannot be entirely eliminated (Fig. 22a&c). However, sub-60 SS can only be obtained within a range of input sweeping rate or FE switching response time (Fig. 22b&d). It indicates that even if the FE could switch its polarization with ultrafast response, the SS of NCFET might not improve as long as the dynamic polarization is not matched.

Besides transient-related parameters, the sweeping voltage range and FE-material-related parameters will also impact the NCFET characteristics (Fig. 23). Based on the above analysis, smaller V_{FE} and larger $\partial P/\partial t$ are required for smaller hysteresis and steeper SS respectively. However, since the $\partial P/\partial t$ decreases with V_{FE} reducing, the SS and hysteresis of NCFET cannot be optimized simultaneously, indicating an intrinsic optimization conflict (Fig. 24). The experimental results of reported HZO-based NCFETs are benchmarked (Fig. 25), showing that ultra-steep SS with hysteresis-free behavior is still challenging for logic applications of NCFETs.

V. CONCLUSION

NC effect in a standalone FE capacitor is measured for the first time without the need of series system to stabilize, directly proving the polarization domain switching to be the physical origin of NC. It is found that the NC could be generated for steep SS in NCFET only when a new proposed dynamic polarization matching is achieved rather than capacitance matching, and the hysteresis behavior in NCFET cannot be simply eliminated by stabilizing PE loop of FE due to FE inherent dynamic behavior. Based on physical NCFET model coupled with calibrated FE switching model instead of S-curve model, an intrinsic conflict between SS and hysteresis optimization is also demonstrated, which is the most pressing challenge for logic applications of NCFETs.

ACKNOWLEDGMENT

This work was partly supported by NSFC (61421005 and 61604006). The authors thank Yue Peng and Prof. Genquan Han for HZO deposition, and Ji Ma and Prof. Jing Ma for the support of ferroelectric test system.

REFERENCES

[1] Z. Krivokapic *et al., IEDM*, p.357, 2017; [2] Q. H. Luc *et al., VLSI*, p.47, 2018, [3] J. Zhou *et al., IEDM*, p.373, 2017; [4] J. V. Houdt *et al., IEEE Electron Device Lett.*, vol. 39, p.877, 2018; [5] B. Obradovic *et al., VLSI*, p.51, 2018; [6] Y. J. Kim *et al., Nano Lett.*, vol. 17, p. 7796, 2017; [7] S. Salahuddin *et al., Nano lett.*,vol.8, p.405, 2008; [8] Y. Ishibashi *et al., JPSJ*, vol.31, p.506, 1971; [9] H. Orihara *et al., JPSJ*, vl.63, p.1031, 1994; [10] A.I. Khan *et al., Nature mat.*, p.14, 2015.

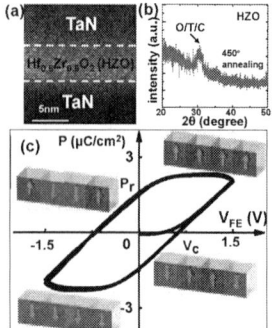

Fig. 1 (a) Cross-sectional HR-TEM and (b) GI-XRD of ferroelectric-HZO (FE-HZO) after annealing. (c) Measured P-V loop of HZO-based FE capacitor.

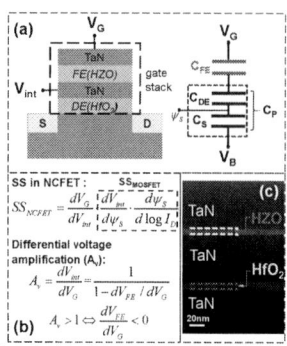

Fig. 2 (a) Schematic structure and equivalent circuit of NCFET; (b) voltage amplification in NCFET; (c) TEM image of FE-DE stack.

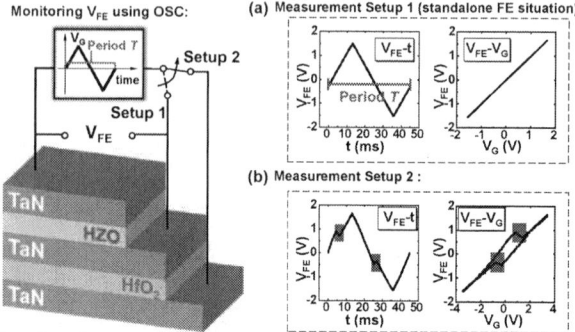

Fig. 3 Schematic experimental setup for the measurement of V_{FE} in FE-DE stack. *Setup 1* is for FE; *setup 2* is for FE-DE gate stack.

Fig. 4 Measured V_{FE} versus t and V_{FE} versus V_G curves under a triangular voltage pulse of V_G for (a) *setup 1* and (b) *setup 2* in Fig. 3.

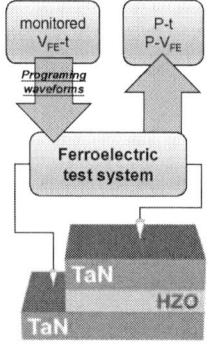

Fig. 5 Measurement flow diagram for P-V loop of standalone FE by programming the input voltages of V_{FE} in Fig. 3.

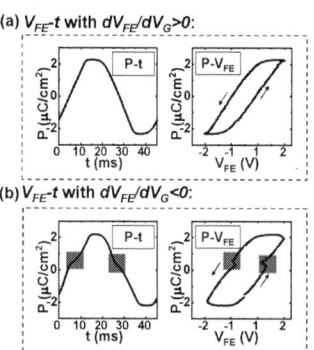

Fig. 6 Measured P-t and P-V_{FE} curves of standalone FE with different monitored V_{FE}-t.

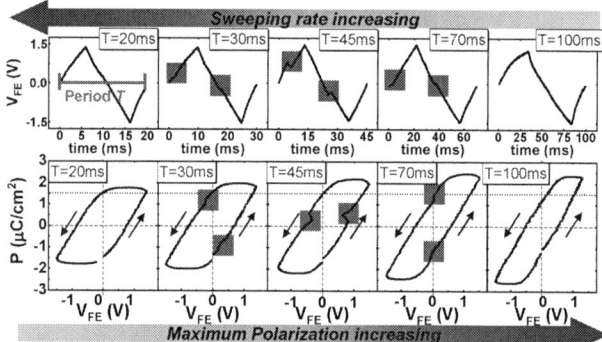

Fig. 7 Monitored V_{FE}-t waveforms and the corresponding measured P-V loop with various period time (T) of triangular voltage pulse V_G, showing the distinct NC phenomenon at T=45ms and NC vanishes when sweeping faster or slower.

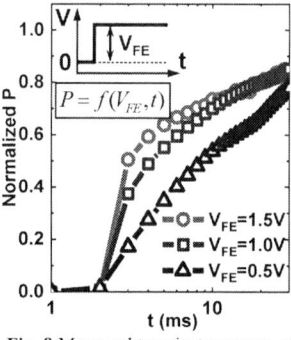

Fig. 8 Measured transient response of polarization to a voltage pulse in a standalone FE-HZO.

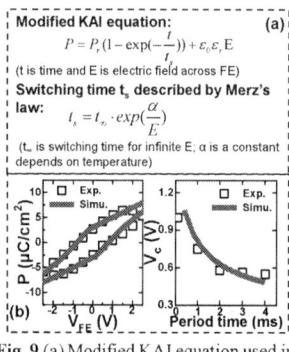

Fig. 9 (a) Modified KAI equation used in this work; (b) modeled P-V curves and V_C-T curves verified by experiments.

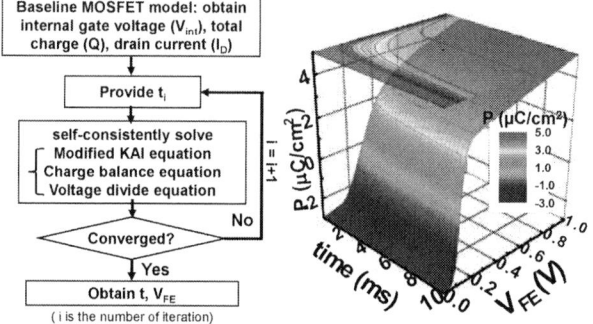

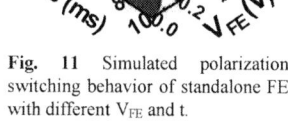

Fig. 10 Simulation flow of NCFET based on KAI equation integrated with baseline MOSFET model.

Fig. 11 Simulated polarization switching behavior of standalone FE with different V_{FE} and t.

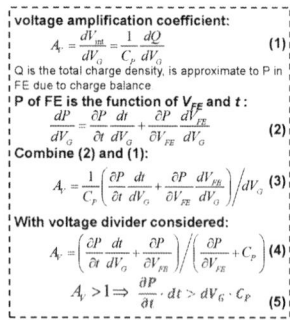

Fig. 12 Derivation of polarization matching condition for sub-60 SS of NCFET.

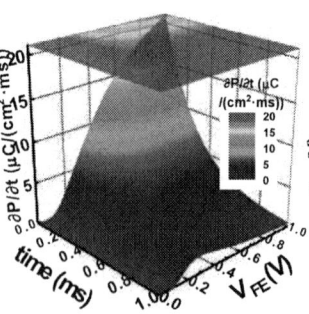

Fig. 13 Simulated $\partial P/\partial t$ of standalone FE under different V_{FE} and t.

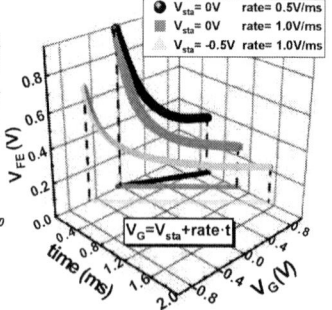

Fig. 14 V_{FE} versus t in NCFET at various V_G with different rates and V_{sta}.

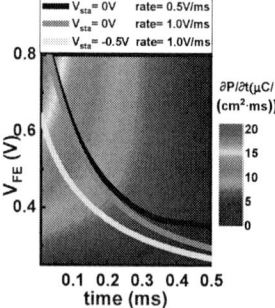

Fig. 15 $\partial P/\partial t$ of FE in NCFET under V_{FE}-t curves from Fig. 14.

978-1-7281-1988-5/18 $31.00 © 2018 IEEE

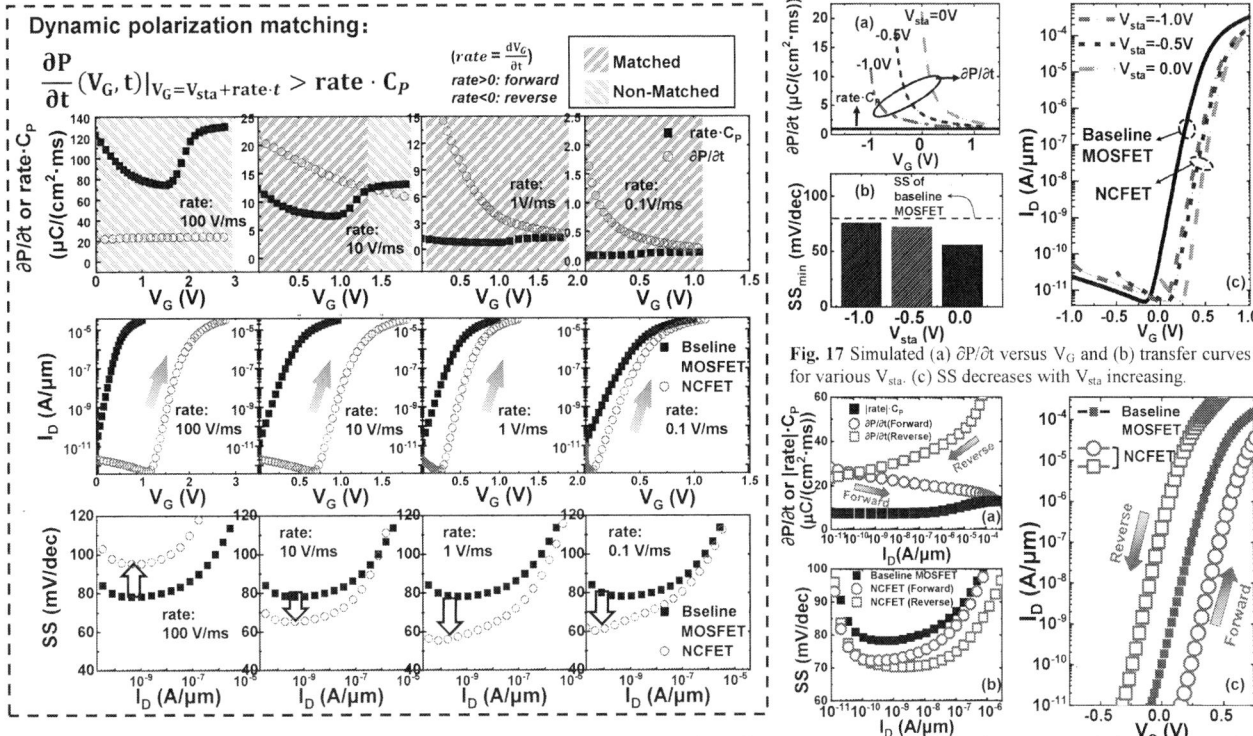

Dynamic polarization matching:

$$\frac{\partial P}{\partial t}(V_G, t)\big|_{V_G = V_{sta} + rate \cdot t} > rate \cdot C_P$$

$(rate = \frac{dV_G}{dt})$
$rate > 0$: forward
$rate < 0$: reverse

Matched / Non-Matched

Fig. 16 Extracted different polarization matching situations and the corresponding simulated transfer curves and SS characteristics with various sweeping rates of V_G in NCFET ($V_{sta} = 0$V).

Fig. 17 Simulated (a) $\partial P/\partial t$ versus V_G and (b) transfer curves for various V_{sta}. (c) SS decreases with V_{sta} increasing.

Fig. 18 (a) DP matching, (b) SS characteristics and (c) transfer curves for forward and reverse sweeping in NCFET.

Fig. 19 Extraction of V_{th_F} in NCFET by crossing point of V_{FE}-t curve and P=Q$_{MOS\_th}$ contour line.

Fig. 20 The relationship between the hysteresis of NCFET and the V_{th} of MOSFET.

Fig. 21 Extraction of $|V_{FE\_F}|$ for forward sweeping (V_{sta}=-0.5V) and $|V_{FE\_R}|$ for reverse sweeping (V_{sta}=1.5V).

Fig. 22 ΔV_{th_F} and SS characteristics of NCFET with different (a) (b) sweeping rate of V_G and (c) (d) switching time of FE.

Fig. 23 SS and hysteresis characteristics of NCFET with various (a) V_{sta}, (b) P_r, (c) V_C and (d) t_{FE}.

Fig. 24 Extracted ΔV_{th_F} in NCFET with the change of V_{sta}, P_r, V_C and t_{FE}. SS and hysteresis show opposite optimization direction.

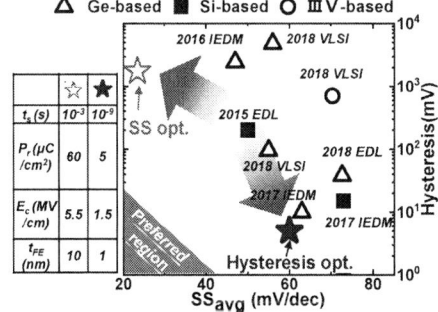

Fig. 25 Experimental performances benchmark of reported HZO-based NCFET. (star) simulated results for SS and hysteresis optimization with Fe parameters in the table.

	☆	★
t_s(s)	10^{-3}	10^{-9}
$P_r(\mu C /cm^2)$	60	5
E_c(MV /cm)	5.5	1.5
t_{FE} (nm)	10	1

978-1-7281-1988-5/18 $31.00 © 2018 IEEE

A Critical Examination of 'Quasi-Static Negative Capacitance' (QSNC) theory

Z. Liu, M. A. Bhuiyan, and T. P. Ma

Department of Electrical Engineering, Yale University, New Haven, CT 06511, USA

Email: zhan.liu@yale.edu, t.ma@yale.edu

Abstract— Recent years have seen a boom in the interest of using the alleged ferroelectric 'negative capacitance' to realize low subthreshold swing of MOSFETs. According to the "quasi-static negative capacitance' (QSNC) theory, a ferroelectric (FE) capacitor has an intrinsic but unstable 'negative capacitance' (NC) region. By adding a matching linear dielectric (DE) capacitor in series, it is possible to 'stabilize' the NC region. In this work, we examined the validity of the QSNC theory, and performed several key experiments to verify that experimental results are consistent with our theoretical assessment. Unfortunately, our overall results do not support the QSNC theory.

I. Introduction

The 'negative capacitance' (NC) FETs have recently drawn much attention for promises to providing room-temperature sub-60mV/decade subthreshold swing (SS). The original idea of 'quasi-static negative capacitance' (QSNC) was introduced by Salahuddin and Datta in 2008 [1]. They claimed that the Landau-Devonshire (LD) theory [2] would lead to an intrinsic NC region for a ferroelectric capacitor. By connecting a linear dielectric (DE) capacitor in series with a ferroelectric (FE) capacitor, it was claimed that this NC region could be effectively stabilized.

II. Derivation of the QSNC theory

According to the LD theory, the Gibbs free energy density of a ferroelectric material is:

$$U = \alpha P^2 + \beta P^4 + \gamma P^6 - E_{ext} P \tag{1}$$

Where P is the polarization, and α, β, and γ are expansion coefficients. This is coupled with the Landau-Khalatnikov (LK) equation:

$$\rho \, dP/dt + dU/dP = 0 \tag{2}$$

where ρ is the damping constant. Under the quasi-static condition, $dP/dt = 0$. Solving Eqs. (1) and (2), we get:

$$E_{ext} = 2\alpha P + 4\beta P^3 + 6\gamma P^5 \tag{3}$$

The above is the relationship between the external electric field and the ferroelectric polarization in the QSNC theory as derived in [1]. As shown in Fig. 1, this may be represented as an 'S' shaped curve. The NC region between $-E_c$ and E_c corresponds to the local maxima of the Gibbs free energy under different external field. The theory argues that a normal capacitor can be connected in series to stabilize the NC region. The theoretical derivation in [1] further assumed (erroneously, as to be shown later) that

$P_{FE} = Q_{FE} = Q_{DE} = Q$, which leads to the following two equations.

$$V_{FE} = (2\alpha Q + 4\beta Q^3 + 6\gamma Q^5) t_{FE} \tag{4a}$$

$$V_{APP} - V_{FE} = Q/C_{DE} \tag{4b}$$

As shown in Fig. 2, the linear capacitor C_{DE} in series is supposed to act as a load line. If C_{DE} is small enough to avoid intersecting with the 'positive capacitance' (PC) region, the intersection will locate in the NC region, and the NC region is supposed to be 'stabilized' by the linear capacitor. This is the so-called 'capacitance matching' in the QSNC theory.

The validity of the QSNC theory relies upon the following two assumptions: (1) There is an intrinsic NC region for a ferroelectric capacitor, and (2) This NC region can be stabilized by adding a series 'matching' linear capacitor, provided that $P_{FE} = Q_{FE} = Q_{DE}$.

Concerning the first assumption, the NC region corresponds to the set of local maximum states of the Gibbs free energy under different external electric fields. In the Landau theory, only the local minimum states of the Gibbs free energy are physically possible. Under the equilibrium condition, besides Eq. (2), $d^2U/dP^2 > 0$ is also necessary [3, 4]. Since it is obvious that $d^2U/dP^2 < 0$ in the NC region, it is apparent that the alleged NC region is not physical even in theory. Concerning the second assumption, as illustrated in Fig. 3, Q_{FE} is the free charge on the electrodes, while P_{FE} is the ferroelectric polarization inside the dielectrics. Therefore, in general $P_{FE} \neq Q_{FE}$ when $E_{ext} \neq 0$. This inequality is well known in the ferroelectric-gated FET (FeFET) community [5]. Thus, the Q-V curve shown in Fig. 2, which is a copy of the P-V curve shown in Fig.1 by using the erroneous assumption of $P_{FE} = Q_{FE} = Q_{DE}$, is not valid, and therefore the associated load-line analysis is incorrect.

III. Experimental verification of the theory

The following four key predictions of the QSNC theory can be verified experimentally, by measuring a ferroelectric capacitor (C_{FE}) in series with a linear dielectric capacitor (C_{DE}): (1) the negative capacitance of the ferroelectric capacitor should make the overall capacitance of the aforementioned series combination larger than the DE capacitor alone, (2) $dV_{DE}/dV_{APP} > 1$ should be observed; i.e., when ramping the applied voltage, the incremental voltage across the linear capacitor should exceed the total incremental applied voltage (this is the origin of the alleged sharp subthreshold slope for a ferroelectric-gated FET,

978-1-7281-1988-5/18 $31.00 © 2018 IEEE

where the linear capacitor is replaced by the semiconductor capacitance), (3) no hysteresis should be observed when a matching capacitor is selected, even when the amplitude of the applied voltage causes V_{FE} to exceed the coercive voltage of the ferroelectric capacitor, and (4) the $V_{DE} - V_{APP}$ relationship should not be influenced by the amplitude of the applied voltage, $V_{amp,APP}$.

To experimentally verify the above, we connected an $Hf_{0.5}Zr_{0.5}O_2$ (HZO)-based ferroelectric capacitor in series with a commercial-grade precision capacitance box that can be adjusted between 1pf and 1000pf. It should be noted that it is very important for this experiment that these two capacitors be separated by electrodes, although numerous papers published in the past erroneously used a composite gate stack consisting of a FE-DE bi-layer heterostructure to represent the series capacitor circuit mentioned above without any electrode separating the two layers, as shown in the left pane of Fig.4 Many of these publications actually observed an enhanced capacitance compared to the linear DE alone without the ferroelectric layer, and attributed to the NC effect [6, 7]. However, it has been shown that such composite dielectric stack involving a ferroelectric layer is likely to give rise to strong polarization coupling effects that significantly enhance the overall capacitance relative to that of the DE layer alone, without invoking the NC effect at all [8-13].

The HZO-based ferroelectric capacitor used in this work is shown schematically in Fig. 5. The thickness of the HZO layer is 20nm, with 30nm of TiN as electrodes. The radius of the capacitor is 100um. Fig. 6 (a) shows the hysteresis P-V curves measured using an AixACCT TF analyzer 2000. The applied waveform is triangular wave at 1kHz. Two Saturated loops and one unsaturated loop under various applied voltage amplitudes ($V_{amp,APP}$) are shown. Fig. 6 (b) shows the C-V curves of this capacitor measured at 10KHz. The total parasitic capacitance from wire, oscilloscope, and signal generator is very small, so it is reasonable to be ignored. To simulate $V_{DE} - V_{APP}$ curves based on the QSNC theory (using Eqs. (4a) and (4b)), we extracted the Landau expansion coefficients based on the measured hysteresis. As shown in Fig. 7, the corresponding 'S' curve well fits the saturated hysteresis loop.

The easiest experimental verification is to evaluate the total capacitance of the C_{DE} & C_{FE} in series. The C-V measurements were taken with an HP 4284A system at room temperature in a frequency range of 1KHz – 1MHz. As shown in Figs. 8 (a) – (c), the measured total capacitance is always smaller than C_{DE}. Although not shown here to save space, the measured total capacitance is smaller than C_{DE} over the entire range of the precision capacitance box. In fact, the total capacitance always matches the formula $C_{DE} = C_{DE} C_{FE} / (C_{DE} + C_{FE})$ as predicted by the classical circuit theory, with no trace of the NC. Fig. 7 (d) shows that C_{DE} is larger than C_{FE-DE} in the entire frequency range of 1KHz to 1MHz. These results are contrary to the prediction (1) listed in the first paragraph of this Section.

Figs. 9 (a) – (d) show the simulation and experimental results for a set of different C_{DE}'s. Besides QSNC based simulation, we replaced Eq. (4a) with our measured hysteresis loops and simulated the $V_{DE} - V_{APP}$ curves using Eq. (4b) as the control. One can see that dV_{DE}/dV_{APP} never exceeds 1 in all these cases (as well as all other cases measured between $C_{DE}=1pF$ and $C_{DE}=1000$ pF, not shown here). It is worth noting that the control curves match nearly perfectly with the experimental curves, including the hysteresis behavior, without any trace of the 'S' curve predicted by the QSNC theory. For C_{DE} as small as 100pF, the hysteresis behavior still exists while the QSNC theory predicts no hysteresis. These results are all contrary to the QSNC predictions (2) and (3) listed above.

Fig. 10 shows the $V_{DE} - V_{APP}$ curves for various $V_{amp,APP}$'s. Fig. 10 (a) shows that the $V_{DE} - V_{APP}$ hysteresis window shrinks when $V_{amp,APP}$ is reduced from 14V to 8V while the QSNC theory predicts no change. In Fig. 10 (b), we simulated the $V_{DE} - V_{APP}$ curves, and the results fit nearly perfectly the experimental curves. Fig. 10 (c) shows that for C_{DE} as small as 100pF, the hysteresis behavior is still observed for as long as V_{FE} exceeds the coercive voltage of the ferroelectric capacitance. These results are in direct contrary to the QSNC predictions (3) and (4) listed above.

IV. CONCLUSION

Theoretically, this paper has shown that the QSNC theory misapplies the Landau-Devonshire theory by introducing a non-existent 'negative capacitance' region, and, by using an incorrect assumption ($P_{FE} = Q_{FE}$) in the derivation, the QSNC theory draws the wrong conclusion that adding a matching linear capacitor could 'stabilize' the 'negative capacitance' region. Experimentally, this paper has demonstrated that the measured results are in direct contrast to four key predictions of the QSNC theory.

ACKNOWLEDGMENT

The authors gratefully acknowledge Prof. Peide Ye and Mengwei Si at Purdue University for providing the ferroelectric capacitors in this work, as well as Nanbo Gong at IBM and Peng Wu at Purdue University for inspiring discussions. This work has been partially supported by the National Science Foundation (NSF) under Award No: 1609162.

REFERENCE

[1] Salahuddin, S., and S. Datta. *Nano letters* 8.2 (2008): 405-410.
[2] Devonshire, A. F. *Advances in physics* 3.10 (1954): 85-130.
[3] Landau, L. D., and E. M. Lifshitz. "Statistical Physics, Part 1." (1980).
[4] Fridkin, V., and S. Ducharme. "Ferroelectricity at the Nanoscale Basics and Applications." 1st ed., Springer Berlin, 2014.
[5] Ma, T. P., and J. Han. *IEEE EDL* 23.7 (2002): 386-388.
[6] Islam Khan, A., et al. *Applied Physics Letters* 99.11 (2011): 113501.
[7] Appleby, Daniel JR, et al. *Nano letters* 14.7 (2014): 3864-3868.
[8] Sun, F-C., et al. *Journal of materials science* 51.1 (2016): 499-505.
[9] Ho Tsang, C., et al. *Journal of the Physical Society of Japan* 73.11 (2004): 3158 3165.
[10] Zhou, Y. *Solid State Communications* 150.29-30 (2010): 1382-1385.
[11] Dawber, M., et al. *Physical review letters* 95.17 (2005): 177601.
[12] Salev, P., et al. *Physical Review B* 93.4 (2016): 041423.
[13] Kittl, J. A., et al. *Applied Physics Letters* 113.4 (2018): 042904.

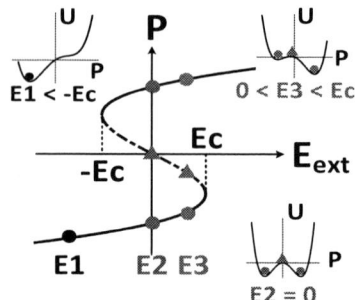

Fig. 1. $P - E_{ext}$ curve from QSNC theory. The insets are $U - P$ curves under different external electric fields. Circles and triangles correspond to local minimum and maximum, respectively, in $U - P$ curves.

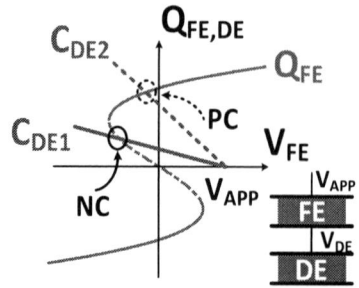

Fig. 2. Eqs. (4a) and (4b) plotted in the same figure, where $C_{DE1} < C_{DE2}$. C_{DE1} stabilizes the 'negative capacitance' region, while C_{DE2} stabilizes the 'positive capacitance' region.

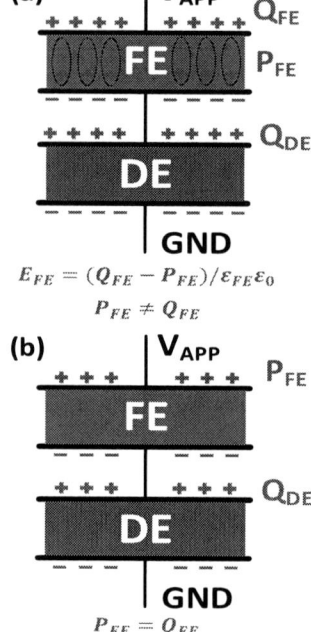

$$E_{FE} = (Q_{FE} - P_{FE})/\varepsilon_{FE}\varepsilon_0$$
$$P_{FE} \neq Q_{FE}$$

$$P_{FE} = Q_{FE}$$

Fig. 3. (a) Illustration of P_{FE}, Q_{FE} and Q_{DE}, where Q_{FE} and Q_{DE} are the free charge (density) on the electrodes, while P_{FE} is the polarization charge effectively located at the interface between electrode and ferroelectric layer. (b) Incorrect $P_{FE} - Q_{FE}$ relationship in QSNC theory, where the polarization charge is equal to the free charge on the electrode.

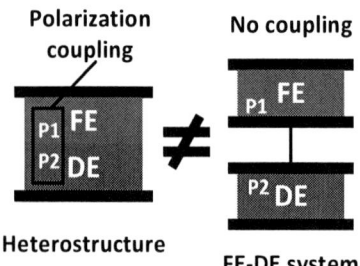

Fig. 4. Illustration of the difference between FE-DE bi-layer heterostructure and the series combination of the FE and the DE capacitors. The presence of electrodes between FE and DE layers screens polarization coupling.

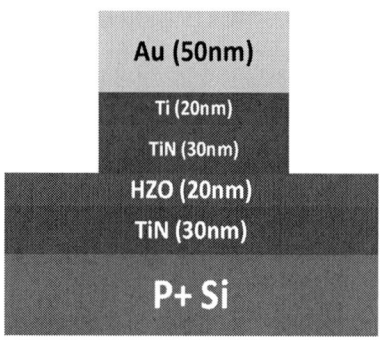

Fig. 5. Schematic sketch of the HZO ferroelectric capacitor used in this work. The radius of the capacitor is 100um.

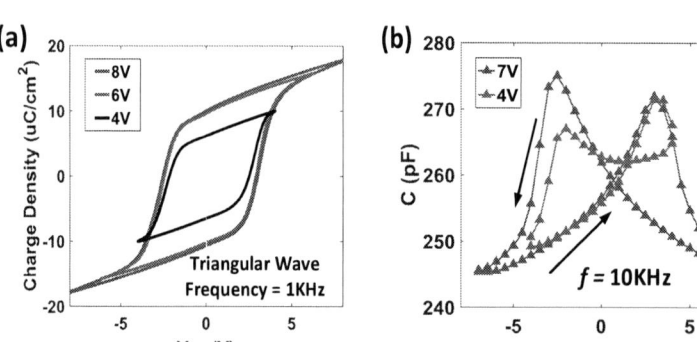

Fig. 6. (a) $Q - V$ Hysteresis loops of the ferroelectric capacitor used in this work measured with three different $V_{amp,FE}'s$. Large $V_{amp,FE}$ (6V and 8V) results in saturated loops, while small $V_{amp,FE}$ (4V) results in unsaturated loop (minor loop). (b) $C - V$ curves of the ferroelectric capacitor with $V_{amp,FE} = 7V$ and 4V. Black arrows indicate forward and backward sweeps.

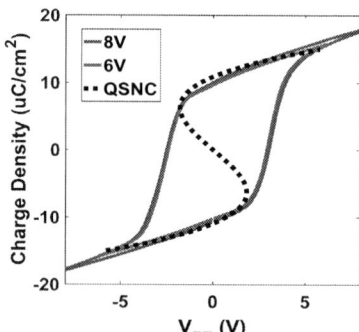

Fig. 7. The hypothetical 'S' curve extracted from the measured hysteresis loop.

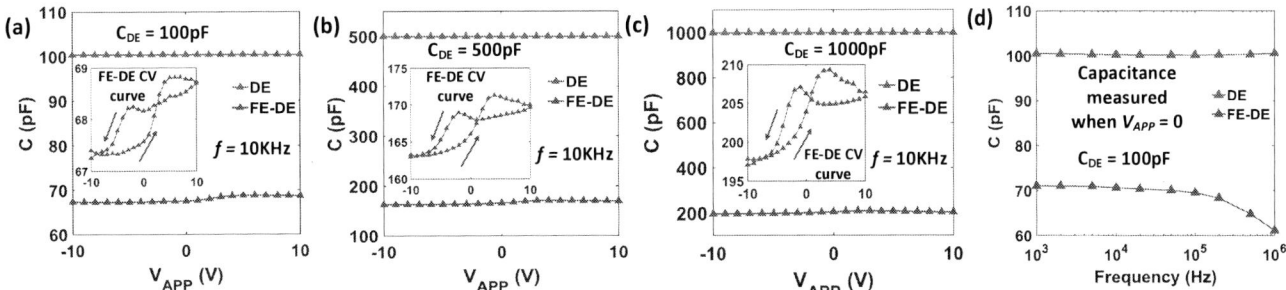

Fig. 8. Comparing $C - V$ curves of FE-DE series combination with that of DE alone, with $C_{DE} = $ (a) 100pF, (b) 500pF, and (c) 1000pF, measured at 10KHz. Note that $C_{DE} > C_{FE-DE}$ is observed for each case, and also all other C_{DE}'s ranging from 1pF to 1000pF (not shown here). The insets show the magnified $C_{FE-DE} - V_{APP}$ curves. (d) C_{FE-DE} and C_{DE} at $V_{APP} = 0$ measured at varies frequencies ranging from 1KHz to 1MHz. C_{DE} is larger than C_{FE-DE} in the entire frequency range.

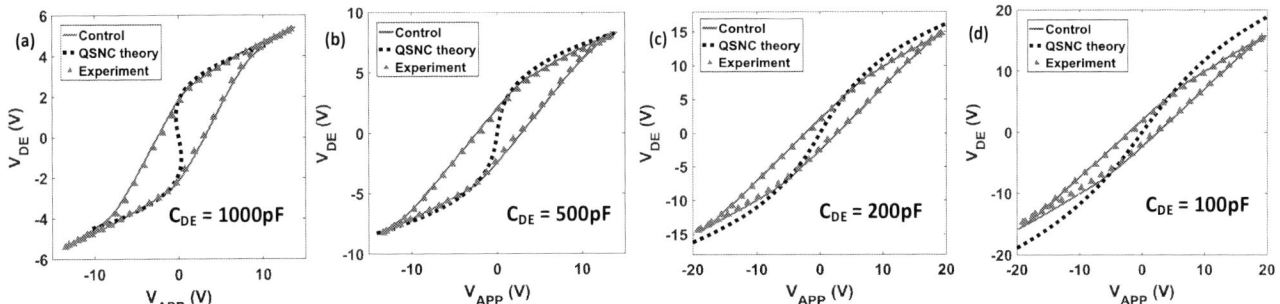

Fig. 9. $V_{DE} - V_{APP}$ curves corresponding to $C_{DE} = $ (a) 1000pF, (b) 500pF, (c) 200pF, and (d) 100pF. The red triangles are experiment results, black dashed curves are the simulation results based on the QSNC theory, and blue curves are the simulation results using the measured hysteresis loops that serve as the control curves. The QSNC theory predicts $dV_{DE}/dV_{APP} > 1$, which is not observed experimentally. The QSNC theory predicts no hysteresis while the experimental results all show hysteresis.

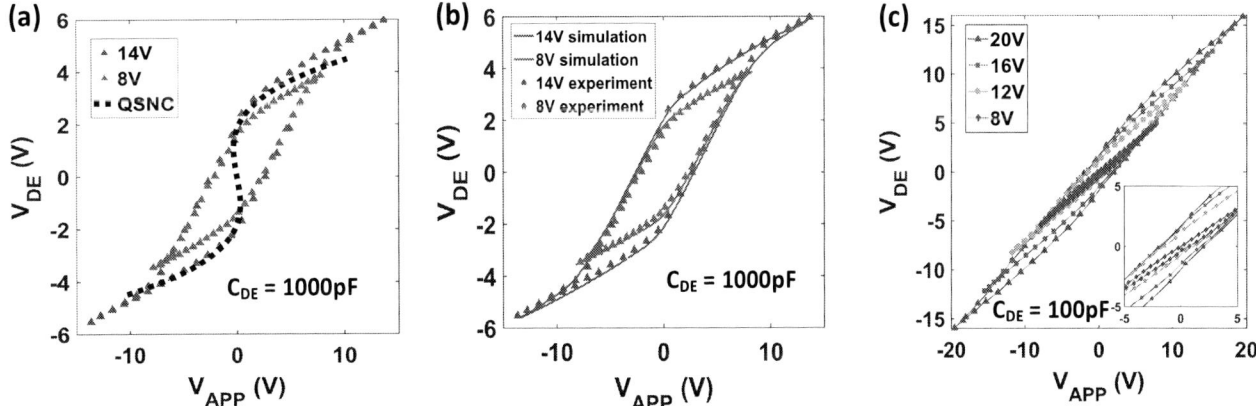

Fig. 10. (a) & (b) both show $V_{DE} - V_{APP}$ curves for two different $V_{amp,APP}$'s with $C_{DE} = 1000$pF, where the experimental curves all show dependence on $V_{amp,APP}$, which contradicts the prediction from the QSNC theory; in (b), the simulation curves fit very well with the experimental curves; in (c) $C_{DE} = 100$pF, and for $V_{amp,APP} = 20$V, 16V and 12 V, significant hysteresis is revealed in each curve, whereas for $V_{amp,APP} = 8$V, it shows almost no hysteresis, due to the fact that the voltage across the FE film is only slightly above the coercive voltage. The inset shows an amplified view of the 8V data near $V_{APP} = 0$, which still reveals some finite hysteresis, while $dV_{DE}/dV_{APP} < 1$ near $V_{APP} = 0$.

Direct relationship between sub-60 mV/dec subthreshold swing and internal potential instability in MOSFET externally connected to ferroelectric capacitor

Xiuyan Li and Akira Toriumi

Department of Materials Engineering, The University of Tokyo

Hongo, Tokyo 113-8656, Japan

Phone : +81-3-5841-1907, E-mail: xiuyan@adam.t.u-tokyo.ac.jp

Abstract— Steep subthreshold swing (SS) in ferroelectric (FE) FETs have been intensively discussed these years in terms of the negative capacitance(NC) effect, but still under debate. This paper demonstrates the direct correlation between sub-60 mV/dec SS and internal potential (V_{int}) enhancement in MOSFET externally connected to FE capacitor in DC mode through systematic experiments. It is shown that V_{int} enhancement only occurs in a limited voltage window, and that hysteresis-free steep SS is achievable by tuning the paraelectric capacitance. The present results support that the steep SS values so far reported are tightly related to FE domain switching around the coercive field rather than the ideal NC effect.

I. INTRODUCTION

Negative capacitance (NC) effect, using ferroelectric (FE) film as the gate dielectric for enhancing internal potential in FET, has been intensively discussed recently, because sub-60 mV/dec subthreshold swing(SS) might be achieved [1-6]. AC analysis has been reported to provide some evidences to NC effects [7,8]. But, steep SS effects are discussed in DC mode [2-6]. The objective of this paper is to experimentally demonstrate the direct relationship between steep SS and internal potential enhancement in DC mode, and then to shed light on the origin of steep SS results reported in literatures.

II. CHALLENGES IN THE MEASUREMENT OF INTERNAL POTENTIAL BETWEEN CAPACITORS IN SERIES

A. Charge conservation at the floating node

To accurately measure the internal potential, V_{int}, between two capacitors connected in series, the total charges at the floating node should be maintained. This requires no leakage through the capacitors. But, DC impedance is infinite in case of no leakage in capacitor, and it is impossible to estimate V_{int} experimentally. In the actual system, the finite leakage of capacitors enables V_{int} measurement within a time constant, τ. **Fig. 1(a)** shows the equivalent circuit for paraelectric (PE) capacitors (CAP) connect in series. τ is calculated in **Fig. 1(b)**. Only in case with $t \ll \tau$ (t is the waiting time after the voltage setting in measurement), the equivalent circuit can be regarded as capacitances in series. The system impedance as a function of t using typical parameters of FE CAP and MOSFET employed in our measurement was analyzed (**Fig. 2**), which indicates $t < 1$s is required in the measurement.

B. Input impedance requirement in DC measurement

The total amount of charges at the floating node is seriously affected not only by the leakage in the sample but also by the measurement system. **Fig. 3** shows the time to phase angle of 85, 87 and 89° of the system impedance in Fig.

1 as a function of the input impedance of the measurement system. To see the capacitance effect in DC mode, the input impedance, Z, should be $> 10^{12}$ Ω, conventional voltmeters with $Z \leq 10^{10}$ Ω will not provide a correct value of V_{int}. **Fig. 4** shows measured results of PE/PE system in two cases with high ($\sim 10^{16}$ Ω) and conventional (10^{10} Ω) Z voltmeters. Using the former one, an accurate measurement of V_{int} was assured down to 1pF. Therefore, it was used in the present experiment.

III. INTERNAL POTENTIAL MEASUREMENT AT FE/PE CAP INTERFACE

PZT films with Pt electrodes were used as the FE CAP. Its capacitance was about 150-200 pF, and leakage was about 5-20 pA at 1 V. **Fig. 5** shows the reason and equivalent circuit to measure V_{int} in FE/PE system. Leakage of PE CAP was negligibly small. The waiting time, t of 0.5 s, was set in each measurement. Initial charging/discharging in FE CAP should be specially cared. Otherwise, the initial characteristics were subject to unknown charges stored inside.

A. Correlation between FE properties and internal potential enhancement

P-V curves of three FE CAPs with different switching slope are shown in **Fig. 6(a)**. Fig 6(b) and (c) demonstrate V_{int}-V and $\delta V_{int}/\delta V$-V characteristics in corresponding FE/PE system. V_{int} jumps are seen in V sweeping in each case, and they correspond to $\delta V_{int}/\delta V > 1$. Namely, the internal potential enhancements are clearly observed in the DC mode. Interestingly, the maximum enhancement factor, α, and its voltage window closely correlate with the switching slope of polarization in FE CAP. With a sharper switching, α is bigger while the voltage window is narrower, and vice versa. But, the integrated area of $\delta V_{int}/\delta V > 1$ is universal, possibly because the saturated polarization is comparable in all cases (**Fig. 6(d)**). Note that an α~30 is achieved in case with the sharpest switching of FE CAP, while FE CAP showing a slow switching is actually not so disadvantageous, as it gives a wider window of V_{int} enhancement though α is not so big.

B. PE capacitance effect on internal potential enhancement

PE capacitance is varied from 33 nF to 20 pF to understand how it affects V_{int} in FE/PE system. The bias position showing α, defined as V_{amp} in Fig. 6(b), is changed with PE capacitance (**Fig. 7(a)**). An interesting point is that we can tune the position of V_{amp}, so its hysteresis between forward and backward V sweeping disappears with this effect. Moreover, α increases with the increase in PE capacitance (**Fig. 7(b)**). Concerning that a smaller PE capacitance gives a smaller τ, the charge decay is more serious within a given time in this case and V_{int} enhancement may be deteriorated by quick discharge from the floating node.

978-1-7281-1988-5/18 $31.00 © 2018 IEEE

C. Origin of internal potential enhancement using FE

To understand FE effect on V_{int} at FE/PE interface, V_{FE} change during V sweeping is discussed. V_{FE} is gotten as $V_{FE}=V-V_{int}$. A negative slope of V_{FE}, followed by small oscillations, is observed symmetrically in V-V_{FE} characteristic (**Fig. 8**). Note that V_{FE} showing negative slope corresponds to V at V_{amp} and it is consistent with the coercive voltage, V_c, of FE CAP. This indicates that V_{int} enhancement occurs during the polarization switching. Moreover, the negative slope and the hysteresis of V_{FE} are independent on PE capacitance (**Fig. 9**). Thus, it is inferred that the internal potential enhancement in the present experiment is not from the ideal S character-like trajectory of FE film but possibly comes from the local polarization (domain) switching, because the former mechanism should give a different negative slope and smaller hysteresis of V_{FE} with a smaller PE capacitance [9]. It is understandable how the domain switching enhances V_{int} from schematics in **Fig. 10**. At $V_{FE}<V_c$, ΔV_{int} is determined by capacitive charging/discharging, so $\Delta V_{int}<\Delta V$. While on the domain flipping, a huge polarization charges appear with the opposite direction at FE/PE interface, resulting in a big jump of V_{int}, hence $\Delta V_{int}>\Delta V$. Thus, the potential across FE CAP drops and a negative slope of V_{FE} appears. Since the present FE CAP exhibited a rather slow polarization switching, which may cause the following small but successive V_{FE} shift.

IV. SI/SIO₂-MOSFET WITH AN EXTERNAL FERROELECTRIC CAPACITOR CONNECTED IN SERIES

We next discuss whether V_{int} enhancement directly relates to the steep SS in MOSFET. FE CAP was connected through a cable to poly-Si gate/SiO₂ n-MOSFET. V_{int} on MOSFET was also measured using the high-Z voltmeter. The equivalent circuit is shown in **Fig. 11**. Typically, $t = 0.5$ s was set as well.

A. Direct correlation between internal potential enhancement and sub-60 mV/dec SS achievement

Fig. 12 and **13** show V_{int}-V and $\delta V_{int}/\delta V$-V relationship in FE/MOSFET system. By choosing a suitable FE CAP and MOSFET, $\alpha \sim 1.5$ with a voltage window about 0.5V was obtained near the threshold voltage (V_{th}) in the MOSFET in both forward and backward sweeping. I_{DS}-V_{GS} characteristics $w/$ and w/o FE CAP are shown in **Fig. 14**. In case $w/$ FE CAP, SS looks sharper than w/o case in a given region and on-current has no degradation even though the total gate capacitance is actually reduced. **Fig. 15** plots V_{GS} dependence of SS. In case $w/$ FE CAP, SS is significantly improved down to 45 mV/dec during both forward and backward sweeping. Note that the improvement factor of SS is consistent with amplification factor of V_{int}. These results clearly demonstrate that FE film definitely improves SS value below the Boltzmann limit by enhancing the gate potential in MOSFET.

B. Leakage effect on sub-60 mV/dec SS achievement

As indicated in part II and III, the leakage of the system should be seriously taken into account for considering the steep SS. Two kinds of experiments were carried out to understand this. First, t in measurements was changed from 0.1s to 2s. **Fig. 16** shows that SS is sharply improved with t

decrease, as expected from Fig. 1(b). With $t \geq 1s$, SS almost recovers to 60 mV/dec. Second, τ of the system was varied by changing channel area (**Fig. 17(a)**) or SiO₂ thickness (**Fig. 17(b)**). Larger channel area and thinner SiO₂ in MOSFET, both resulting in a larger capacitance, led to a lower SS (**Fig. 18**). Since all measured MOSFETs show a smaller leakage than FE CAP, similar to that in FE/PE system in part III, a smaller MOS capacitance leads to a smaller τ and requires the faster measurement as in Fig. 16. It should be noted that the gate leakage exponentially increases in ultra-thin gate oxides.

V. DISCUSSION

On the origin of steep SS: Based on our results, hysteresis-free steep SS is achievable via carefully tuning the PE capacitance. It is apparently similar to the expectation in "S-like" ideal NC effect. But, actually the internal potential enhancement behaves quite differently in a limit voltage window as schematically shown in **Fig. 19**. Nevertheless, we cannot conclude it is impossible to achieve ideal NC effects, but it is inferred that most of experimental results so far reported in SS improvement may be associated with the internal potential enhancement presented in this work, derived from the local domain switching in *FE* films [2-6].

On the internal electrode effect: The present FE/SiO₂ was electrically connected through each electrode. MOSFETs with no electrode between FE and PE should be something different. In the former case, each polarization switching interacts with others through the internal electrode, the effect is finally averaged. In addition, leakages in FE and/or PE film shape the charge conservation at the floating node for the worse (**Fig. 20 (a)**). In the latter case w/o internal electrode, however, each domain flipping and leakage path behaves locally and does not affect others in principle [10] (**Fig. 20 (b)**). In scaled FETs, it might be possible to utilize "NC" effect only at high frequency operation.

VI. CONCLUSION

This work has demonstrated the steep SS by connecting ferroelectric capacitor to SiO₂ MOSFETs in conjunction with the internal potential instability. By comparing two experimental results, it can be concluded that ferroelectric polarization flipping induces the steep SS. Most of experimental results in DC measurements so far reported in literatures can be explained by this effect. This fact can be negative capacitance effect but is not from the initially proposed NC one. In high frequency mode measurement of small MOSFETs, it is still under investigation whether the ideal NC effects is possible or not in an intrinsic sense.

ACKNOWLEDGMENT

This work was supported by JST-CREST (JPMJCR14F2). We thanks T. Nishimura and T. Yajima for discussion and suggestions.

REFERENCES

[1] S. Salahuddin *et al.*, Nano Lett., **8**, 405 (2008) [2] J. Jo *et al.*, EDL, **37**, 245 (2016). [3] Z. Krivokapic *et al.*, IEDM 2017. [4] G. Pahwa *et al.*, TED **65**, 867 (2018).[5]. A. Rusu *et al.*, Nanotech., **27**, 115201 (2016). [6] B. Obradovic *et al.*, VLSI 2018. [7] A. I. Khan *et al.*, Nature materials, **14**, 182 (2015). [8] P. Sharma *et al.*, EDL, **39**, 272 (2018) [9] A. K. Saha *et al*, JAP. **123**, 105102 (2018). [10] A. I. Khan *et al.*, TED **63**, 4416 (2016).

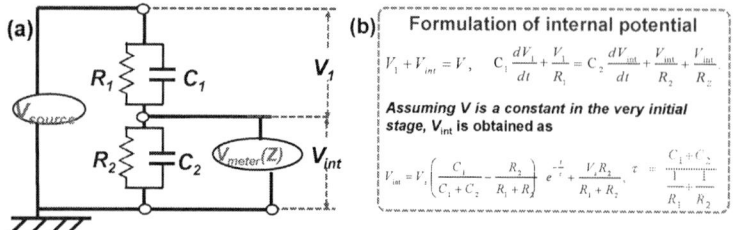

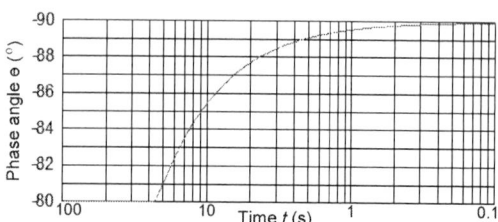

Fig. 1 **(a)** Equivalent circuits in the internal potential measurement in PE/PE CAPs connected in series. **(b)** Time dependent internal potential in the circuit in Fig. 1(a). τ was obtained and used for discussion in part III and IV.

Fig.2 Calculated phase angle of the system impedance in Fig. 1 as a function of time, t, with $R_1=R_2=10^{11}\Omega$ and $C_1=C_2=200$pF. $t<1$ sec is required to regard as the capacitance equivalent system.

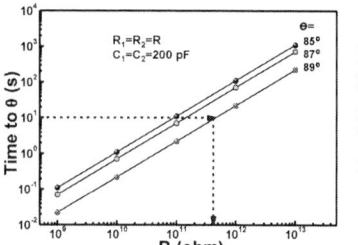

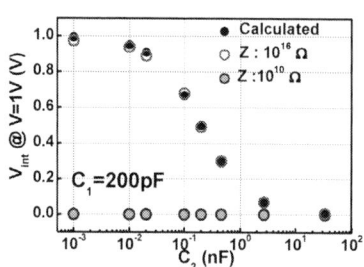

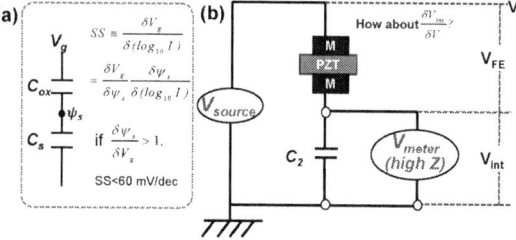

Fig.3 Time to phase angle of 85, 87 and 89° of the system impedance as function of input impedance in measurement system. A voltmeter, $Z>10^{12}\Omega$, should be used in DC mode.

Fig.4 V_{int} as a function of C_2 at $V=1$V measured using high and normal impedance DC voltmeters. Only the former one enables to estimate the internal potential.

Fig. 5 **(a)** Why we should care about internal potential in FE/PE system and what we should pay attention to. **(b)** Equivalent circuit for measuring the internal potential in *FE/PE* system.

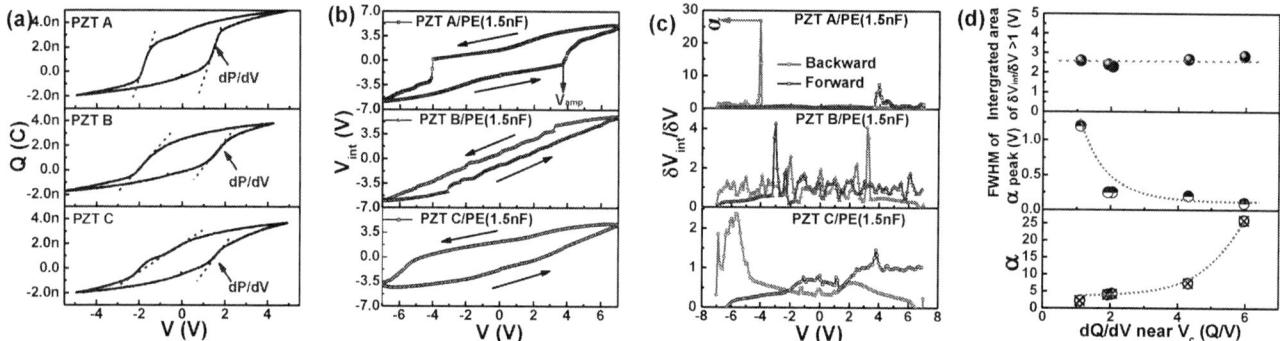

Fig. 6 **(a)** Q-V curves in FE CAP, **(b)** V_{int} -V and **(c)** $\delta V_{int}/\delta V$-V characteristic in FE/PE system, for three kinds of PZT samples. V_{int} jumps are observed at several points along with $\delta V_{int}/\delta V>1$. **(d)** Maximum amplification factor, α, and FWHM of α peak as well as the integrated area of $\delta V_{int}/\delta V>1$ as a function of dQ/dV. dQ/dV and α are defined in Fig. 6(a) and (c). α is for the sharpest peak while the integrated area includes all peaks of $\delta V_{int}/\delta V>1$. α and FWHM strongly depend on dQ/dV, while the integrated area seems to be universal.

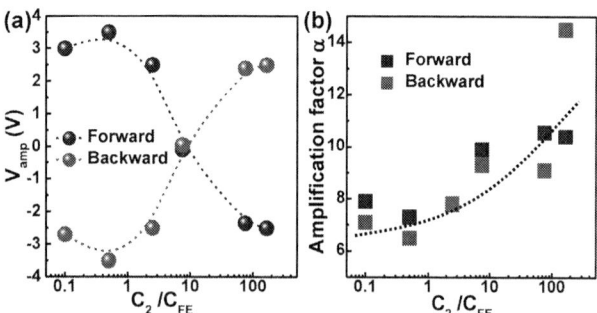

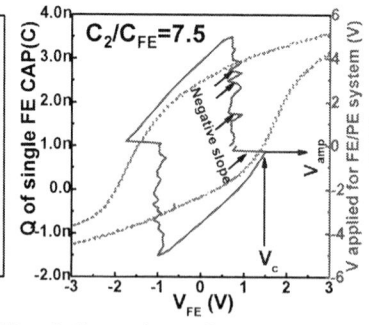

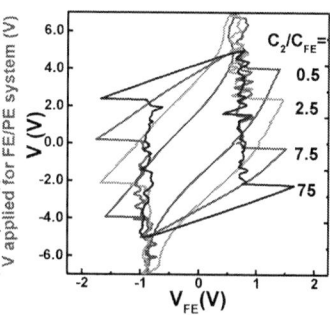

Fig. 7 Evolution of **(a)** V_{amp} and **(b)** α with PE capacitance, C_2, in FE/PE system. V_{amp} is defined in Fig. 6(b). By adjusting C_2, we can tune V_{amp} position with no hysteresis between forward and backward V sweeping. C_2 dependence of α may be due to the time constant difference of the system affected by C_2.

Fig. 8 Comparison of V-V_{FE} in FE/PE system and Q-V in a single FE CAP. A negative slope of V_{FE} is observed near V_c of FE, followed by small but successive jumps along with V_{int} enhancement.

Fig. 9 V-V_{FE} in FE/PE system with different PE capacitance, C_2. The negative slope of V_{FE} and hysteresis does not change irrespective of different C_2.

978-1-7281-1988-5/18 $31.00 © 2018 IEEE   717

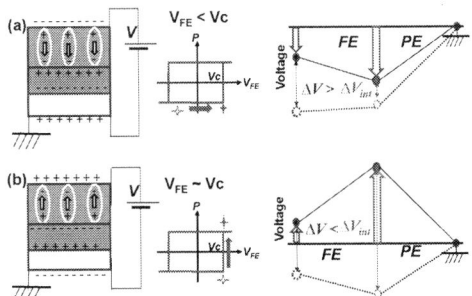

Fig. 10 Schematics for understanding how internal potential on PE is changed **(a)** before and **(b)** during domain switching in FE film.

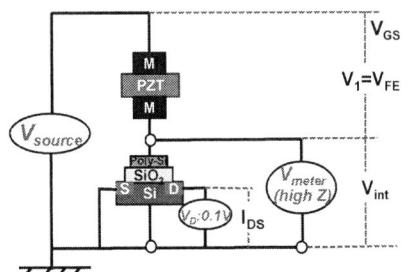

Fig. 11 Equivalent circuit for measuring the internal potential and I_{DS}-V_{GS} in *FE/MOSFET* system. Note that V_{int}-V_{GS} and I_{DS}-V_{GS} were measured separately for the same devices.

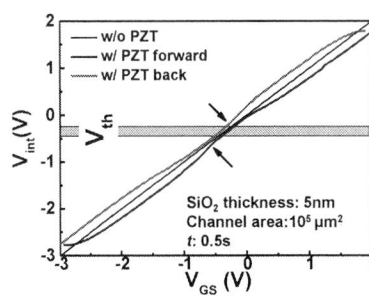

Fig. 12 V_{int}-V_{GS} characteristics in FE/nMOSFET system. Near threshold voltage of nMOSFET, V_{int} changes drastically with small hysteresis.

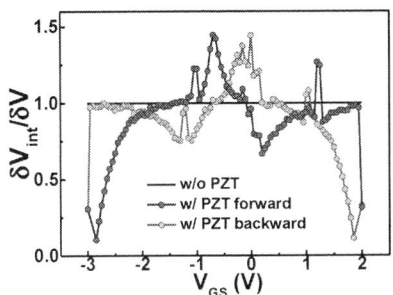

Fig. 13 $\delta V_{int}/\delta V$-V in FE/MOSFET system. $\delta V_{int}/\delta V$ of 1.5 with a width of 0.5V is observed, which overlaps between forward and backward sweeping.

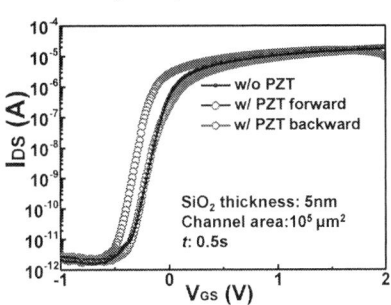

Fig. 14 I_{DS}-V_{GS} characteristic of MOSFET *w/* and *w/o* external FE CAP. SS looks shaper *w/* FE CAP, and on-current has no big change.

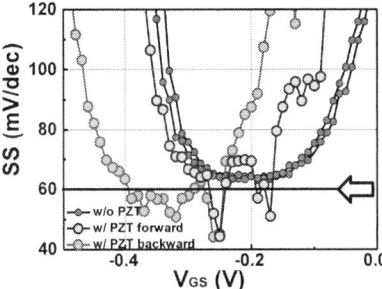

Fig. 15 V_{GS} dependence of SS in MOSFET *w/* and *w/o* FE CAP. SS is improved down to 45 mV/dec in case *w/* FE CAP.

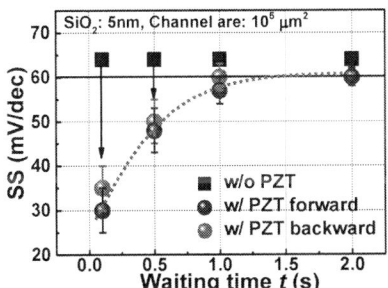

Fig. 16 Waiting time (t) dependence of SS in FE/MOSFET system. SS is improved with decrease of t drastically as expected in Fig. 1(b).

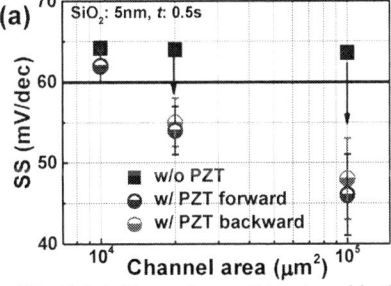

Fig. 17(a) Channel area **(b)** gate oxide thickness dependences of SS in FE/MOSFET system. SS is smaller in case with larger channel area or thinner gate oxide, both cases give a larger PE capacitance.

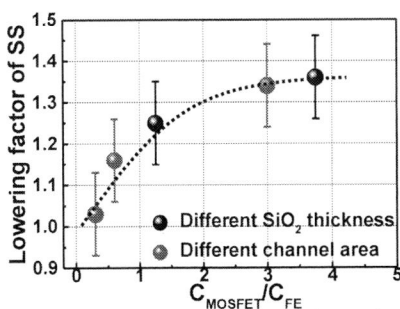

Fig. 18 Improvement factor of SS in FE/MOSFET system universally depends on the capacitance of MOSFET. Smaller capacitance gives a smaller time constant, hence a leakage effect may degrade SS in a given waiting time.

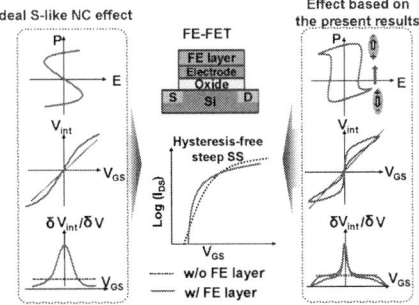

Fig. 19 Comparison of FE effect on SS in MOSFETs between ideal "S-like" NC expectation and the present results. Both may enable to achieve apparently the same SS, but mechanism behind is actually different in the P-E trajectory.

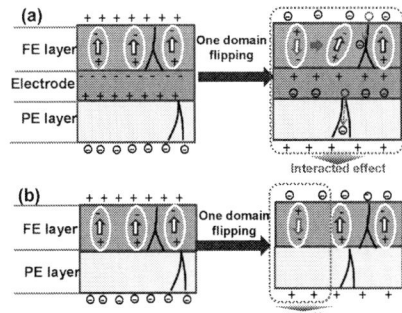

Fig. 20 FE effects on PE layer *w/* and *w/o* internal electrode. *w/* internal electrode, each domain switching and leakage path are interrelated to others, while the effect is local in case of *w/o* internal electrode.

978-1-7281-1988-5/18 $31.00 © 2018 IEEE   718

Assessment of Steep-Subthreshold Swing Behaviors in Ferroelectric-Gate Field-Effect Transistors Caused by Positive Feedback of Polarization Reversal

Shinji Migita[1], Hiroyuki Ota[1], and Akira Toriumi[2]

[1]National Institute of Advanced Industrial Science and Technology (AIST), Ibaraki, Japan, email: s-migita@aist.go.jp
[2]The University of Tokyo, Tokyo, Japan

Abstract—Steep-subthreshold swing (*SS*) behaviors in ferroelectric-gate field-effect transistors (Fe-FETs) are investigated using the metal-ferroelectric-metal-insulator-semiconductor (MFMIS) gates stack structures with different area ratios between MIS and MFM capacitors. It is analyzed that the capacitance matching between them by adjusting the area ratio is significant to efficiently utilize the polarization reversal behavior in the ferroelectric layer. In this work we explain the steep-SS behavior from viewpoint of positive feedback of polarization reversal. Furthermore it is discussed why steep-*SS* is observable in recent Fe-FETs.

I. Introduction

Steep-subthreshold swing (*SS*) technologies are strongly desired in logic transistors on LSI, in order to reduce the power consumption while maintaining the performance through the realization of low-operation voltage. Among many challenges of developments such as I-MOS, NEMS-switch, and tunnel field-effect transistor (tunnel-FETs), negative capacitance-FET (NC-FET) [1] is attracting attention now. It is theoretically induced from the L-K equation of ferroelectric formula, that an S-shape curve in the polarization–voltage character is available. The NC-FET consists of series capacitors of ferroelectric and paraelectric on the gate of FET. It is categorized as the ferroelectric-gate FET (Fe-FET). It is argued that in an optimized coupling of ferroelectric and paraelectric, the NC-FET realizes steep-SS and hysteresis-free.

Since the proposal of NC-FET, there are many experimental reports Fe-FETs that observed steep-SS, although they are operated at large gate voltage and hysteresis is obvious. They are concluded as the appearance of NC-effect. Hysteresis-free behavior is also reported. In addition, a fast response of the ferroelectric capacitor-resistor circuit is analyzed to be a trace of NC-effect [2]. Although these results are promising, there are counterarguments that NC-like behaviors are explainable by other effects [3-6].

In this work, we examined the behavior of Fe-FETs prepared in the style of MFMIS gate stack, integrated with Hf-Zr-O ferroelectric and SiO₂ paraelectric capacitors. In this MFMIS structure, we changed the area ratio of MIS and MFM capacitors. The change of the area ratio enables to utilize the saturated loop of the ferroelectric hysteresis [7]. In addition, we found that it contributes to the change of the apparent k-value of the ferroelectric capacitor. It is found that there exists an optimal area ratio in order to maximize the polarization

effect of ferroelectric. Through the experiment and analysis, we explain the steep-SS behaviors in Fe-FETs simply as a result of polarization reversal. This understanding also explain why steep-*SS* is observable in recent Fe-FETs.

II. Experimental Procedure

Schematics of MFMIS-type Fe-FETs fabricated in this work are shown in **Fig. 1**. The devices were fabricated in a manner of gate-last process, shown in **Fig. 2**. Thermal SiO_2 was prepared on p-type Si substrate and the thickness is 3.8 nm. MIS capacitor was formed by sputter deposition of TaN film on SiO_2. 10-nm-thick $Hf_{0.5}Zr_{0.5}O_2$ film prepared by sputtering was used as the ferroelectric. Crystallization was performed at 700 °C after the gate patterning etching. The important part of this device is that the area ratio between MIS and MFM capacitors are changed. It was realized by depositing the top TaN film into the contact area (size is 5 μm x 5 μm, typically 14 contact areas) for the electrode. The transistor sizes are between 10 μm and 100 μm in length and between 20 μm and 100 μm in width. In this work, the area ratio is defined as A (=S_{MIS}/S_{MFM}), and it could be changed between 1 and 29. In addition to the MFMIS Fe-FETs, MIS-FETs with 3.8-nm-thick SiO_2 and TaN electrode are fabricated in order to compare their electrical properties.

Cross-sectional TEM images of MFM capacitor and MIS capacitors are shown in **Fig. 3**. Interfaces are flat and the thicknesses are precisely the same as they were designed. Crystallization morphology is observable in the $Hf_{0.5}Zr_{0.5}O_2$ layer. Electrical properties of 3.8-nm-thick SiO_2 MIS capacitor measured in the MIS-FET are shown in **Fig. 4**. The C-V characteristic (10 kHz) shows proper curve with no-hysteresis. The accumulation capacitance corresponds with the SiO_2 thickness. I-V characteristic shows that the breakdown of SiO_2 occurs at 7.8 MV/cm. From the capacitance value and breakdown field, it is calculated that the maximum limit of the charge for SiO_2 layer is 2.7 μC/cm². Electrical properties of 10-nm-thick $Hf_{0.5}Zr_{0.5}O_2$ MFM capacitor are shown in **Fig. 5**. Polarization behaviors change with the sweep voltage. The coercive field of $Hf_{0.5}Zr_{0.5}O_2$ is 1 MV/cm. Thus, with the chance of sweep voltages (between -0.8 V and 0.8 V for case (1), -1.2 V and 1.2 V for case (2), and -3 V and 3 V for case (3)), the P-V curve (1 kHz) changes from a minor loop to a saturated loop. The maximum polarization also changes with the swing voltage. In case of the 3-V-swing, the maximum polarization reaches 30 μC/cm². I-V characteristic shows the breakdown at 4.1 MV/cm.

978-1-7281-1988-5/18 $31.00 © 2018 IEEE

III. DESIGN CONCEPT OF A-PARAMETER IN MFMIS

In the stack of MFM and MIS capacitors, there is a mismatch of charge density. The MFM capacitor can store as much as 30 $\mu C/cm^2$, while the MIS capacitor is limited to 2.7 $\mu C/cm^2$. Therefore, the MFMIS stacks must be designed with the consideration of SiO_2-breakdwon. A-parameter (S_{MIS}/S_{MFM}) contributes to smoothen this discrepancy. **Figure 6** shows P-V curves and load-lines of MIS for three cases of A-parameters. Because the MFM capacitor is designed to be small, the polarization values evaluated with the size of MIS capacitor become small. As a result, the criterion of Q_{MAX} for SiO_2 is satisfied in all cases. At the same time, the A-parameter also changes the apparent k-value of ferroelectric.

With the change of A-parameter, the ratio of ferroelectric charge (Q_{Ferro}) and paraelectric charge (Q_{Para}) changes (**Fig. 7**). The sharpness of polarization reversal (dQ/dV_F) is also governed by the A-parameter (**Fig. 8**). In case of the $Hf_{0.5}Zr_{0.5}O_2$ and SiO_2 couple, the maximum effect is obtainable at A=7 (**Fig. 9**).

IV. RESULTS AND DISCUSSION

Based on the experimental results of MFMIS-capacitors (10 kHz) and FETs (pulse IV method), we show the appearance of steep-*SS*. Then the origin of steep-SS is considered based on the abrupt polarization reversal nature of ferroelectric. Finally, we discuss the best ferroelectric material for steep-*SS*.

A. MFMIS-capacitors and FETs

C-V characteristics of MFMIS-capacitors are compared with the SiO_2 MIS-capacitor (**Fig. 10**). With the stack of MFM capacitor with small sizes, the capacitances decrease drastically. CET values are calculated and shown in **Fig. 11**. Owing to the uncertainty of wet etching process for contact-hole opening, CETs are shifted from the calculated values. In any case, the CET values are increased form 3.8 nm (SiO_2) to 4.9 nm (A=3), 9.4 nm (A=7), and 16.5 nm (A=14), respectively. Intuitively, these larger CET values than SiO_2 degrade the *SS*-factors.

I_D-V_G characteristic of 3.8-nm-SiO_2 MISFET is shown in **Fig. 12** together with the *SS* parameter. There is no hysteresis in the I_D trace and the minimum *SS* is calculated to be 70 mV/decade. It is a reasonable *SS*-value for 3.8-nm-thick SiO_2.

I_D-V_G characteristics of MFMIS Fe-FETs with different A parameters are shown in **Fig. 13**. In case of A=3, the hysteresis show clock-wise loop, indicating the existence of large number of charge traps in the $Hf_{0.5}Zr_{0.5}O_2$ film. Because the ratio of polarization charge is small in case of A=3 (Fig. 7), charge trap effect appeared as a major contribution. Improvement of the film quality is a next issue. In contrast in cases of A=7 and 14, counter-clockwise appeared clearly, indicating the ferroelectric behavior. The memory window is larger for the case of A=14, and it is comparable to the hysteresis window in P-V (Fig. 5).

SS-factors of MFMIS Fe-FETs against I_D are shown in **Fig. 14**. In all cases, steep-*SS* less than 60 appeared at some points. More importantly, *SS*-values are smaller than those expected from the CET. Using the CET values in Fig. 11, the *SS*-factors are calculated to be 72.9 (A=3), 84.8 (A=7), and 103.5

mV/decade (A=14), respectively. These values are shown by the dashed line. Smaller *SS*-values than calculation indicate the occurrence of steep-SS in all cases. Among them the steep-*SS* is prominent in the case of A=7, where the effect of ferroelectric polarization is the largest (Figs. 7-9).

B. Modeling of Steep-SS in Fe-FETs

We consider that this steep-*SS* is originated from the normal behavior of ferroelectric polarization reversal. It is discussed in **Fig. 15**. Because charge density at the subthreshold region of MIS capacitor is small, it corresponds to the polarization reversal region of the ferroelectric. In this region, assume a small change of gate voltage (ΔV_G). ΔV_G is divided into ΔV_{MFM} and ΔV_{MIS} in proportional to the inverse of respective dielectric constants (k_{Ferro} and k_{SiO2}). In addition to the change of charge ΔQ_{Para} caused by ΔV_G, ferroelectric polarization individually occurs caused by ΔV_{MFM} and induces the charge ΔQ_{Ferro}. The increment of total charge against the voltage change ($\Delta Q/\Delta V_G$) corresponds to the inverse of *SS*. In this way, owing to the additional charge caused by the ferroelectric polarization, the steep-*SS* behaviors appear in Fe-FETs, which is not expected in conventional MIS-FETs.

Now the question is, why steep-*SS* was not observed in previous Fe-FETs using PZT and SrBiTaO? There are two reasons. One is that the thicknesses of SiO_2 layers in those works were designed to be thick in order to prevent diffusion of metals. Another is that the dielectric constants of ferroelectrics are very large (**Table 1**). From viewpoint of Fe-FET design, small polarization and small k-value are promising. Discovery of HfO_2 ferroelectric thus brought the successful achievements of steep-*SS*.

V. CONCLUSIONS

Steep-*SS* behaviors become prominent when the area ratio of MIS and MFM capacitors in MFMIS Fe-FETs is optimally designed. This behavior is analyzed on the basis of ferroelectric polarization charge that emerges individually and additionally contributes to the change of the surface potential of MIS structure. This is the origin of steep-SS in Fe-FETs. In order to attain a superior steep-*SS*, ferroelectric material with small k-value and small polarization should be selected, and the SiO_2 layer should be thin.

ACKNOWLEDGMENT

This work was supported by JST CREST Grant Number JPMJCR14F2, Japan.

REFERENCES

[1] S. Salahuddin and S. Datta, Nano Lett. **8**, 405 (2008).

[2] A. I. Khan, K. Chatterjee, B. Wang, S. Drapcho, L. You, C. Serrao, S. R. Bakaul, R. Ramesh, and S. Salahuddin, Nat. Mater. **14**, 182 (2015).

[3] Sou-Chi Chang, Uygar E. Avci, Dmitri E. Nikonov, Sasikanth Manipatruni, and Ian A. Young , Phys. Rev. Applied **9**, 014010 (2018).

[4] B. Obradovic, T. Rakshit, R. Hatcher, J. A. Kittl, M. S. Rodder, Symp. on VLSI Tech. Dig. 2018, p.51.

[5] Atanu K. Saha, Suman Datta, and Sumeet K. Gupta, J. Appl. Phys. **123**, 105102 (2018).

[6] J. A. Kittl, B. Obradovic, D. Reddy, T. Rakshit, R. M. Hatcher, and M. S. Rodder, Appl. Phys. Lett. 113, 042904 (2018).

[7] Eisuke Tokumitsu, Gen Fujii and Hiroshi Ishiwara, Appl. Phys. Lett. **75**, 575 (1999).

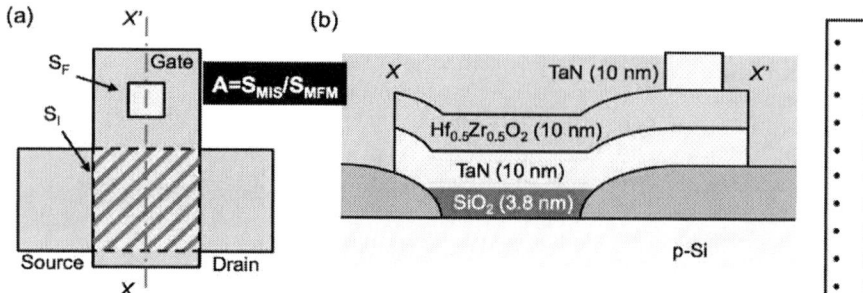

Fig. 1. Schematics of (a) top-view and cross-sectional view of MFMIS-FET. The area ratio between MIS capacitor and MFM capacitor is defined as A.

- p-Si, 1E16 /cm^3, Isolation
- S/D formation (Phos. I/I, RTA)
- SiO$_2$ (3.8 nm), thermal oxide
- TaN (10 nm), sputtering
- Hf$_{0.5}$Zr$_{0.5}$O$_2$ (10 nm), sputtering
- Gate patterning (RIE)
- Crystallization (700 °C, 1 min)
- ILD (TEOS), Contact opening
- Metal pads formation
- FGA (450 °C, 30 min)

Fig. 2. Process flow of MFMIS-FeFET.

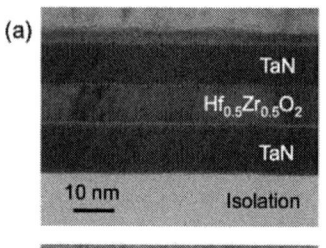

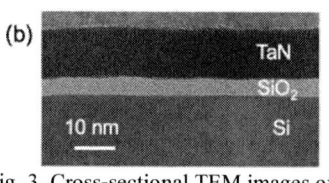

Fig. 3. Cross-sectional TEM images of (a) MFM and (b) MIS structures.

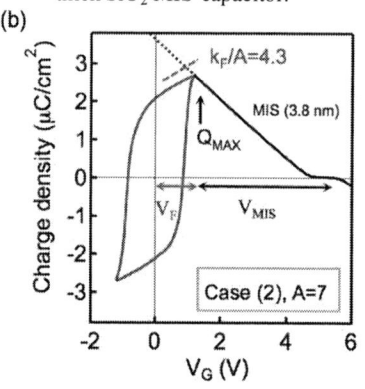

Fig. 4. C-V and I-V characteristics of 3.8-nm-thick SiO$_2$ MIS-capacitor.

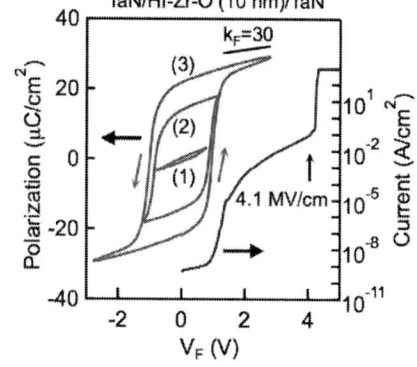

Fig. 5. P-V and I-V characteristics of 10-nm-thick Hf-Zr-O MFM-capacitor.

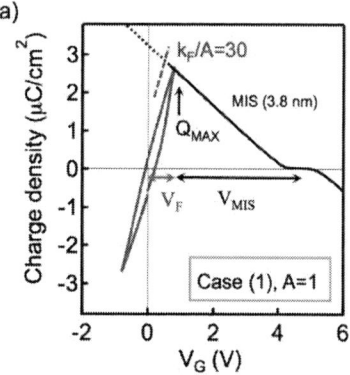

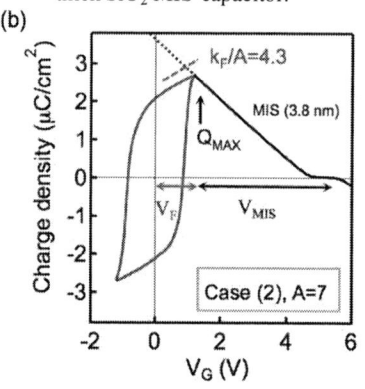

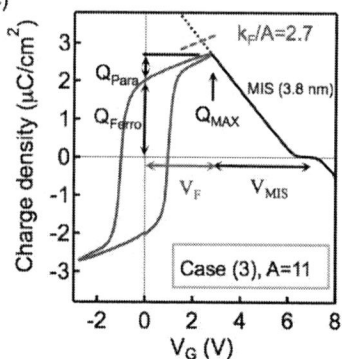

Fig. 6. Relationships between P-V curves of MFM (Fig. 5) and load lines of MIS with different A parameters. In all cases, the maximum charge is limited by the breakdown of SiO$_2$, Q$_{MAX}$ < 2.7 μC/cm^2 (Fig. 4).

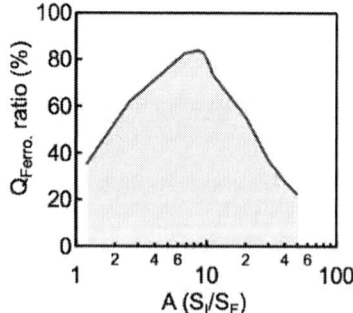

Fig. 7. The ratio of ferroelectric component with A parameter.

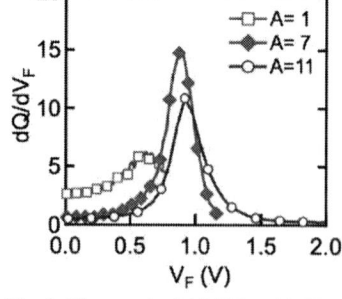

Fig. 8. The trend of dQ/dV$_F$ with V$_F$ at typical A parameter conditions.

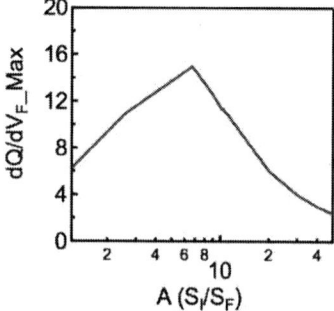

Fig. 9. The maximum dQ/dV$_F$ at respective A parameters.

978-1-7281-1988-5/18 $31.00 © 2018 IEEE

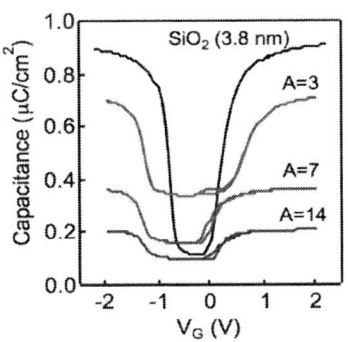

Fig. 10. C-V characteristics measured in transistors SiO2-FETand MFMIS structures with different A parameter.

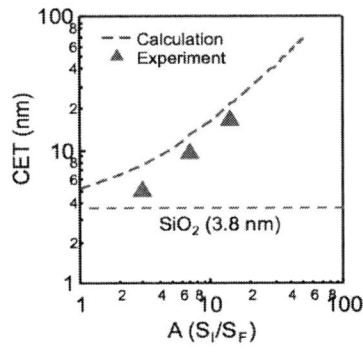

Fig. 11. CET values of MFMIS structures in Fig. 10 and calculated trend of CET with A parameter.

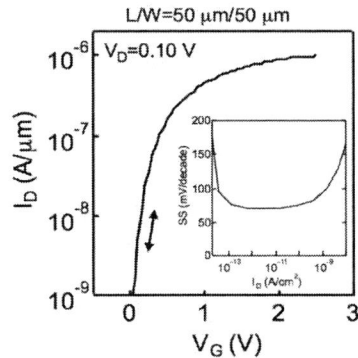

Fig. 12. I_D-V_G character of 3.8-nm-thick MISFET. Inset show SS-factor with drain current.

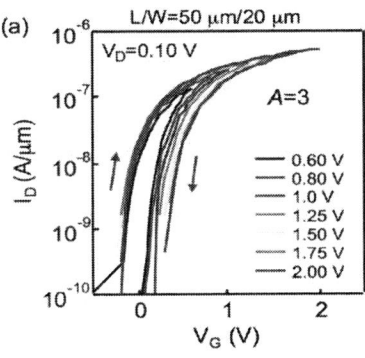
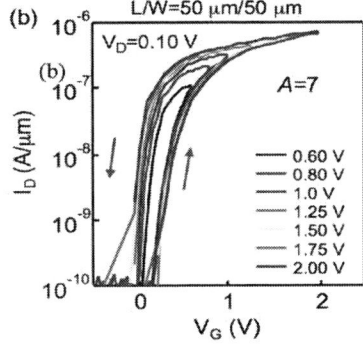
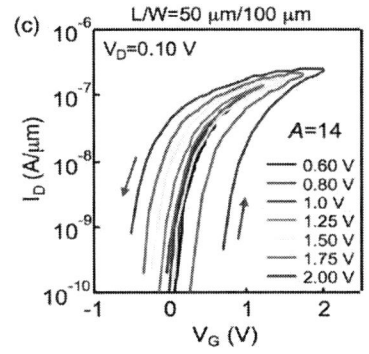

Fig. 13. I_D-V_G characteristics of MFMIS-FeFETs with different A parameters.

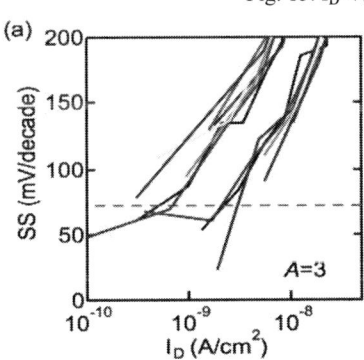
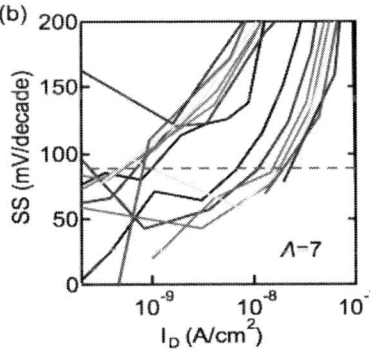
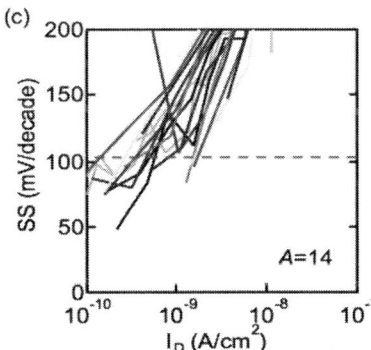

Fig. 14. SS-I_D characteristics of MFMIS-FeFETs with different A parameters, calculated using Fig. 13.

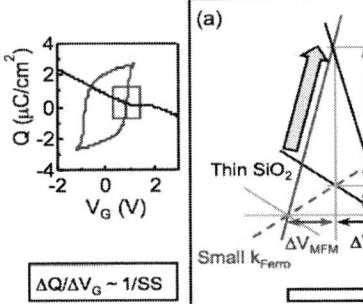
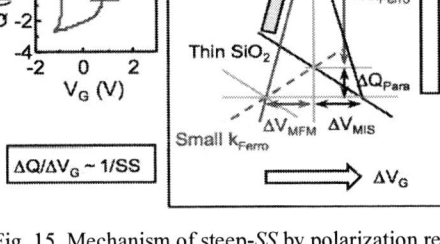
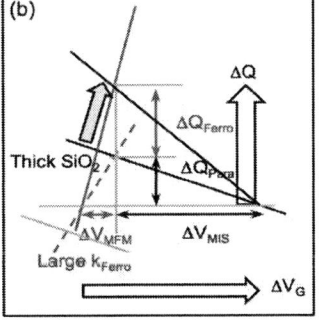

Table. 1. Dielectric constant and remnant polarization values of typical ferroelectrics.

	k	Pr (µC/cm²)
BaTiO	2000	8-25
PZT	300-600	30-50
SrBiTaO	100-200	14
HfO₂	30-40	10-20
PVDF	11	4-8

Fig. 15. Mechanism of steep-SS by polarization reversal, combinations of (a) thin SiO2 and small k_Ferro and (b) thick SiO2 and large k_Ferro.

978-1-7281-1988-5/18 $31.00 © 2018 IEEE

Experimental Study on the Role of Polarization Switching in Subthreshold Characteristics of HfO$_2$-based Ferroelectric and Anti-ferroelectric FET

Chengji Jin, Kyungmin Jang*, Takuya Saraya, Toshiro Hiramoto, and Masaharu Kobayashi
Institute of Industrial Science, The University of Tokyo, Tokyo, Japan, email: cjjin@nano.iis.u-tokyo.ac.jp
*Currently in Toshiba Memory, Japan

Abstract—We have experimentally studied and revealed the direct relationship between polarization switching and steep subthreshold slope (SS) characteristics of HfO$_2$-based ferroelectric FET (FeFET) and Anti-FeFET (A-FeFET) by systematically designing and fabricating devices, and monitoring I_g with high resolution, for the first time. In the circumstances that charge injection prevents polarization switching from occurring in subthreshold region of FeFET, we have obtained two major findings: (1) Sub-60 SS as low as 23.5 mV/dec is observed by adjusting V_g bias sequence, which is attributed to charge injection assisted by polarization switching. (2) Anti-ferroelectric facilitates to align polarization switching in subthreshold region and SS can be improved in A-FeFET as a consequence, which is directly observed by monitoring I_g.

I. INTRODUCTION

Ferroelectric field-effect transistor (FeFET) with sub-60 subthreshold slope (SS) caused by negative capacitance (NC) effect has been proposed to break through the physical limit of 60mV/dec, so called, "Boltzmann Tyranny" in conventional metal-oxide-semiconductor FET (MOSFET) for ultra-low power applications [1, 2]. Recent discovery of ferroelectricity in CMOS-compatible HfO$_2$ makes it one of the most promising steep-slope transistors [3]. Recently, FeFET with sub-60 SS has been experimentally demonstrated by many research groups [4-7]. However, the role of polarization switching in subthreshold characteristic of FeFET is still not clear and only a few studies experimentally report the relationship between polarization switching and steep SS [7, 8]. Normally, it is not straight forward to directly observe how ferroelectric characteristics and polarization switching affect subthreshold characteristic of FeFET by experiment. In this work, we systematically design/fabricate FeFET and anti-FeFET (A-FeFET), and then characterize the FETs by measuring gate current (I_g) with high resolution in order to directly capture the polarization switching and investigate its role in subthreshold characteristics.

II. DEVICE FABRICATION

FeFET/A-FeFET with metal-FE/AFE-metal-insulator-semiconductor (MFMIS/MAFMIS) structure is fabricated in this work (Fig.1). We choose this structure since it provides us a suitable platform to explore device physics from the following perspectives: (1) MFM capacitor and base MISFET can be independently designed and integrated in the same device. In particular, the area ratio (AR) of MFM capacitor to base MISFET can be varied to achieve different capacitance matching conditions [9]. (2) MFM capacitor and base MISFET can be characterized independently on the same device. In this way, we can get one-to-one relationship between FE property

and FeFET performance. Also, whether SS improvement occurs or not can be judged by comparing performance of FeFET and its corresponding base MISFET on the same device.

Fig. 1 and 2 show key process steps and structure for fabricated devices. Fabrication starts from p-type Si substrate. Initially, base MISFET is fabricated using gate last process up to gate stack formation. After that, 4.5 nm SiO$_2$ is thermally grown and TiN/HfZrO$_2$(HZO)/TiN (30nm/10nm/30nm) stack with different Zr % is deposited. Depending on Zr %, the structure becomes MFMIS or MAFMIS. After patterning, MF(AF)M capacitor with different pad size is formed on base MISFET. Then, source/drain (S/D) interconnects and contact pads are formed. Finally, rapid thermal annealing (RTA) is carried out in N$_2$ ambient at 500 °C for 30 s to crystallize HZO. Device parameters in this work are summarized in Fig. 3.

III. RESULTS AND DISCUSSIONS

A. The Strategies to Align and Directly Observe Polarization Switching in Subthreshold Region

We start from characterizing FeFETs. Fig. 4 shows Q_{fe}-V_{fe} and capacitances of MFM capacitors integrated in FeFET A, B, and C. By modulating Zr % of HZO, remanent polarization (P_r) and dielectric constant of FE (ε_{fe}) can vary in a wide region, which is important for realizing different capacitance and charge matching conditions. Fig. 5 shows I_d-V_g and I_g-V_g for base MISFET. Fig. 6, 7, and 8 plot I_g-V_g, I_d-V_g, and SS-I_d for FeFET A, B, and C, respectively, with the same AR. FeFET C showed ferroelectric-like counter-clockwise hysteresis in I_d. FeFET A showed clockwise hysteresis, because, as Zr % increases, MFM capacitance and charge increase and more voltage is applied in MISFET gate, which induces more gate current and charge injection into internal gate [10,11]. In fact, FeFET B has small hysteresis because ferroelectric hysteresis was compensated by charge injection. In order to probe polarization switching, I_g was monitored with high resolution. Polarization switching was clearly observed in I_g as two peaks in FeFET A and B. I_d also showed peaks correspondingly. However, polarization switching did not occur in subthreshold region of FeFET A, B, C, and SS was larger than base MISFET just as paraelectric capacitor is added in series to MISFET gate insulator and total capacitance becomes smaller. These results can be explained by operation point analysis in Fig. 9. While operation point is supposed to be in the middle of polarization switching without charge injection, however, if charge injection happens, the operation points shift away from polarization switching region by charge offset.

In order to align and directly observe polarization switching in subthreshold region and study their relationship, we propose

978-1-7281-1988-5/18 $31.00 © 2018 IEEE

two strategies for physical understanding. First, we adjust V_g bias sequence in accordance with monitored I_g peak (section B). Second, we use A-FeFET to align polarization switching in subthreshold region even with charge injection (section C).

B. Sub-60mV/dec Steep SS by Polarization Switching and Charge Injection in FeFET

By adjusting V_g bias sequence in accordance with monitored I_g peak, we can control polarization switching to occur in subthreshold region. Fig. 10 plots I_g-V_g and I_d-V_g of FeFET D with two different V_g bias sequences (Seq.1: +3→−3→+3 and Seq.2: +2→−2→+2). Base MISFET of FeFET D had relatively high gate-drain leakage current. For Seq.1, clockwise hysteresis in I_d-V_g and two current peaks caused by polarization switching in both I_g-V_g and I_d-V_d were observed, which is similar to the phenomenon described in previous section. For Seq. 2, we switched back V_g sweep just in the middle of polarization switching. As a result, Sub-60 SS was achieved for forward sweep (Fig. 10 (b) inset). It should be noted that no steep SS was observed in forward sweep for Seq. 1, since V_g sweep was switched back at −3 V where polarization switching (current peak) already finished (Fig. 10 (a)). Therefore, polarization switching plays an important role in steep SS behavior. Another important factor to achieve steep SS is charge injection assisted by polarization switching due to the fact that steep SS did not happen for FeFET whose base MISFET has very low leakage current (not shown), even though V_g sweep was switched back in the middle of polarization switching. To further confirm the role of charge injection in steep SS phenomenon. Fig. 11 (a) and (b) plot SS_{min} as function of V_g step (measurement time) and V_d for FeFET D with the use of Seq. 2. Smaller step (longer measurement time) and higher V_d lead to lower SS. With longer measurement time or higher V_d, larger amount of positive charge can be injected into internal gate after V_g sweep is switched back at −2 V, leading to lower SS. Fig. 12 summarizes the physical mechanism of steep SS at operation points (1), (2), and (3) as indicated in Fig. 10 (b). The key mechanism is that, when V_g sweep is switched back in the middle of polarization switching, the polarization switching has inertial force to assist charge injection likely between internal gate and drain, which enhances internal node voltage and rapidly turns on the channel.

C. SS Improvement by Polarization Switching in Subthreshold Region in A-FeFET

Since polarization switching occurs two times in non-zero charge region in AFE capacitor, it is possible to utilize polarization switching even with charge injection in A-FeFET. Fig. 13 plots $Q_{a\text{-}fe}$-$V_{a\text{-}fe}$ and capacitances of MAFM capacitors integrated in A-FeFET with 80, 90 and 100% Zr. Saturated polarization depends on Zr %. Fig. 14, 15 and 16 plot I_g-V_g and I_d-V_g for A-FeFET E, F, and G, respectively, with different AR at Zr 90%. For forward sweep, FeFET E and G were able to align the 1st and 2nd polarization switching in subthreshold region, respectively, while subthreshold region of FeFET F was between 1st and 2nd polarization switching. For reverse sweep, FeFET E, F and G align the same polarization switching in subthreshold region. Key observations are (a) SS is improved

by aligning polarization switching in subthreshold region compared to the case without alignment such as FeFET A-C and A-FeFET F in forward sweep, and (b) I_g peak appears to be split when polarization switching is aligned in subthreshold region. This is because MISFET has small depletion layer capacitance in transition from either accumulation or inversion region, large voltage is induced on MISFET, and thus polarization switching appears to be suppressed. As AR decreases, larger voltage is induced on MAFM capacitor and charge injection is less. Therefore, I_d hysteresis changes from clockwise to counter-clockwise. These results can be explained by operation point analysis as shown in Fig. 17 which show the schematics of possible operation points of A-FeFET for different amount of charge injection and offset. While operation point is not in the middle of polarization switching without charge injection, however, if charge injection happens, the operation points shift into polarization switching region by charge offset for A-FeFET unlike FeFET. Fig. 18 shows SS of A-FeFETs against base MISFET with different AR for 80, 90 and 100% Zr. These all nine cases can be understood based on the following facts by operation point analysis: (i) Higher Zr % has lower saturation polarization charge (Fig. 13) and causes less charge injection at the same AR. (ii) P-V curve in operation point diagram will be squeezed in vertical axis for lower AR. (iii) Charge injection and charge offset are smaller for smaller AR. Note that we do not observe NC effect in A-FeFET. NC effect will happen with faster sweep and larger polarization switching delay in transient [8,12,13]. Even in transient NC effect, however, the key to realize steep SS is to flow polarization switching current through depletion capacitance and cause depolarizing effect in subthreshold region. Therefore, this work revealed physical evidence for the SS improvement in quasi-static case through I_g monitoring.

IV. Conclusions

FeFET/A-FeFET with MFMIS/MAFMIS structure were fabricated and characterized by monitoring I_g to directly capture polarization switching. SS=23.5 mV/dec was achieved by switching back V_g sweep, which is attributed to the internal voltage amplification due to charge injection assisted by polarization switching. Polarization switching was aligned in subthreshold region for A-FeFET even with charge injection. SS was improved as a direct consequence of aligning polarization switching in subthreshold region.

Acknowledgment

This work is partly supported by JST PRESTO.

References

[1] S. Salahuddin and S. Datta, Nano Lett., vol. 8, no. 2, pp. 405–410, 2008.
[2] A. I. Khan, et al., in IEDM Tech. Dig., 2011, pp. 255-258.
[3] J. Müller et al., Nano Lett., vol. 12, no. 8, pp. 4318–4323, 2012.
[4] M. Lee et al., in IEDM Tech. Dig., 2017, pp. 565-568.
[5] J. Zhou et al., in IEDM Tech. Dig., 2016, pp. 310-313.
[6] M. Si et al., in IEDM Tech. Dig., 2017, pp. 573-576.
[7] P. Sharma et al., in Proc. Symp. VLSI Technol., 2017, pp. 154–155.
[8] B. Obradovic et al., in Proc. Symp. VLSI Technol., 2018, pp. 51–52.
[9] S. Migita et al., IEEE Silicon Nano Workshop 2018, pp. 11-12.
[10] E. Yurchuk et al., TED, vol. 63, no. 9, pp. 3501–3507, 2016.
[11] K. Ni et al., TED, vol. 65, no. 6, pp. 2461-2469, 2018.
[12] M. Kobayashi et al., in IEDM Tech. Dig., 2016, pp. 314-317.
[13] C. Jin, T. Hiramoto, and M. Kobayashi, in SSDM, 2018, pp. 199-200.

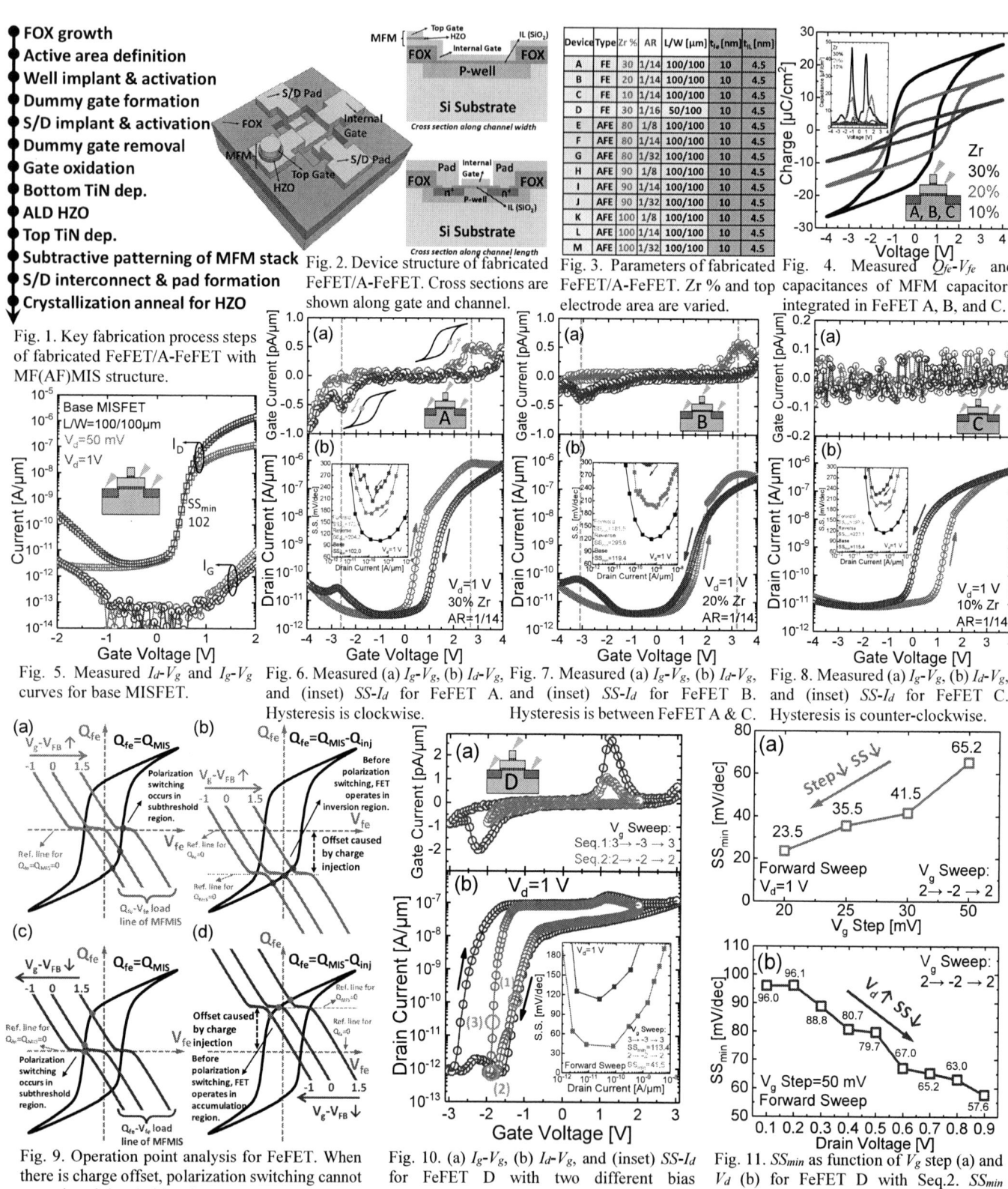

- FOX growth
- Active area definition
- Well implant & activation
- Dummy gate formation
- S/D implant & activation
- Dummy gate removal
- Gate oxidation
- Bottom TiN dep.
- ALD HZO
- Top TiN dep.
- Subtractive patterning of MFM stack
- S/D interconnect & pad formation
- Crystallization anneal for HZO

Fig. 1. Key fabrication process steps of fabricated FeFET/A-FeFET with MF(AF)MIS structure.

Fig. 2. Device structure of fabricated FeFET/A-FeFET. Cross sections are shown along gate and channel.

Device	Type	Zr %	AR	L/W [μm]	t_{fe} [nm]	t_{il} [nm]
A	FE	30	1/14	100/100	10	4.5
B	FE	20	1/14	100/100	10	4.5
C	FE	10	1/14	100/100	10	4.5
D	FE	30	1/16	50/100	10	4.5
E	AFE	80	1/8	100/100	10	4.5
F	AFE	80	1/14	100/100	10	4.5
G	AFE	80	1/32	100/100	10	4.5
H	AFE	90	1/8	100/100	10	4.5
I	AFE	90	1/14	100/100	10	4.5
J	AFE	90	1/32	100/100	10	4.5
K	AFE	100	1/8	100/100	10	4.5
L	AFE	100	1/14	100/100	10	4.5
M	AFE	100	1/32	100/100	10	4.5

Fig. 3. Parameters of fabricated FeFET/A-FeFET. Zr % and top electrode area are varied.

Fig. 4. Measured Q_{fe}-V_{fe} and capacitances of MFM capacitors integrated in FeFET A, B, and C.

Fig. 5. Measured I_d-V_g and I_g-V_g curves for base MISFET.

Fig. 6. Measured (a) I_g-V_g, (b) I_d-V_g, and (inset) SS-I_d for FeFET A. Hysteresis is clockwise.

Fig. 7. Measured (a) I_g-V_g, (b) I_d-V_g, and (inset) SS-I_d for FeFET B. Hysteresis is between FeFET A & C.

Fig. 8. Measured (a) I_g-V_g, (b) I_d-V_g, and (inset) SS-I_d for FeFET C. Hysteresis is counter-clockwise.

Fig. 9. Operation point analysis for FeFET. When there is charge offset, polarization switching cannot occur in subthreshold region.

Fig. 10. (a) I_g-V_g, (b) I_d-V_g, and (inset) SS-I_d for FeFET D with two different bias sequences. Sub 60 SS was observed in Seq. 2.

Fig. 11. SS_{min} as function of V_g step (a) and V_d (b) for FeFET D with Seq.2. SS_{min} becomes lower at smaller step or higher V_d,

978-1-7281-1988-5/18 $31.00 © 2018 IEEE

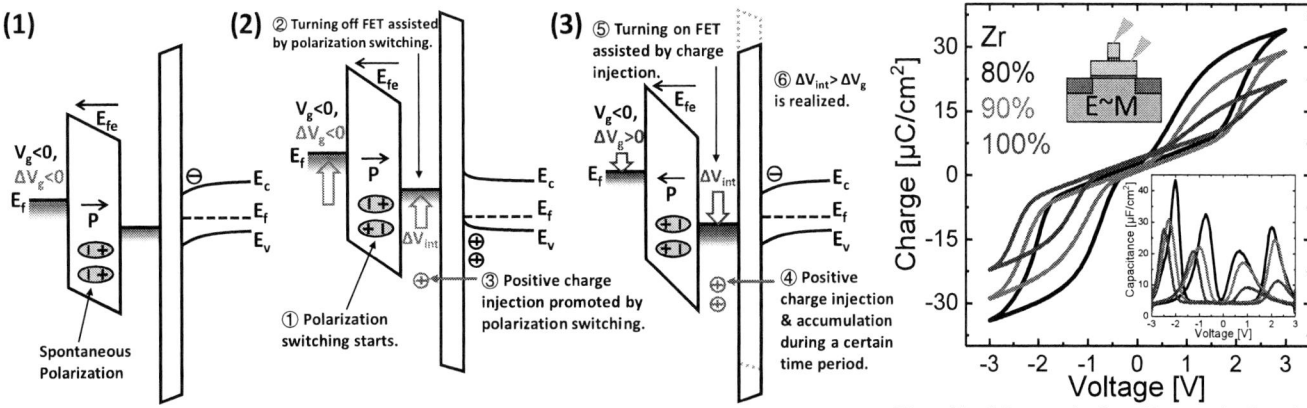

Fig. 12. Schematic band diagrams of MFMIS structure to illustrate physical mechanism of steep SS at operation points (1), (2), and (3) in Seq. 2, as indicated in Fig. 10 (b). polarization switching has an inertial force and promote charge injection likely between internal gate and drain.

Fig. 13. Measured $Q_{a\text{-}fe}$-$V_{a\text{-}fe}$ and (inset) capacitance for MAFM capacitors integrated in A-FeFETs with 80, 90 and 100% Zr.

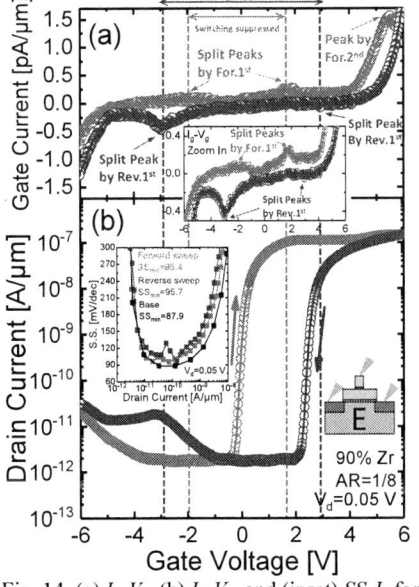

Fig. 14. (a) I_g-V_g, (b) I_d-V_g, and (inset) SS-I_d for A-FeFET E (90% Zr, AR=1/8). Polarization switching is aligned to sub-V_{th} region in both sweep directions where SS is as good as base.

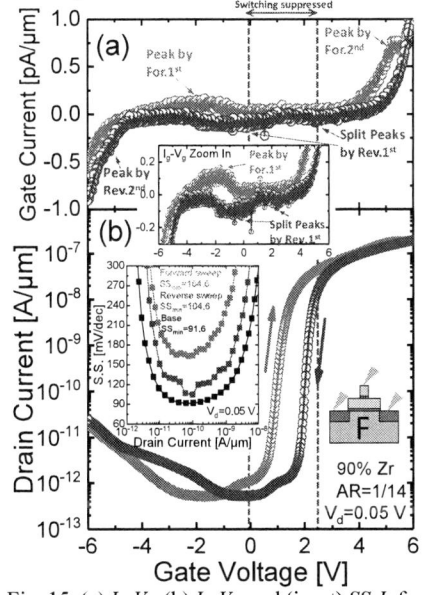

Fig. 15. (a) I_g-V_g, (b) I_d-V_g, and (inset) SS-I_d for A-FeFET F (90% Zr, AR=1/14). Polarization switching is aligned to sub-V_{th} region only in backward sweep where SS is improved.

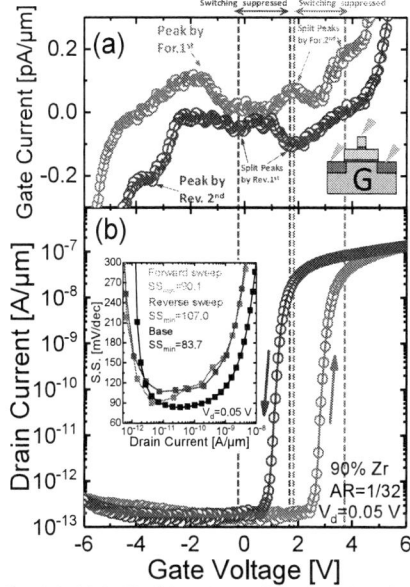

Fig. 16. (a) I_g-V_g, (b) I_d-V_g, and (inset) SS-I_d for A-FeFET G (90% Zr, AR=1/32). Polarization switching is aligned to sub-V_{th} region in both sweep directions where SS is as good as base.

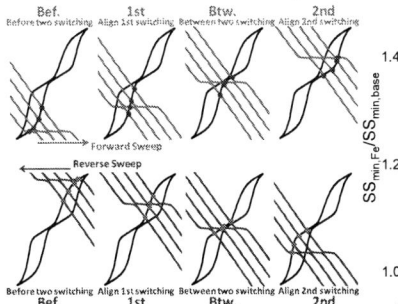

Fig. 17. Possible operation points of A-FeFET for different amount of charge injection and offset due to Zr % and AR.

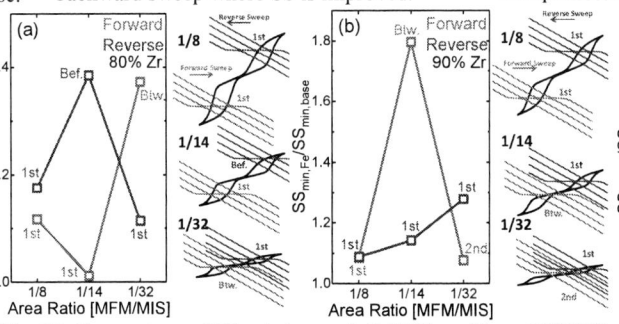

Fig. 18. Comparison of SS_{min} between A-FeFETs and base MISFET with different AR for (a) 80, (b) 90 and (c) 100% Zr. Their corresponding operation point diagrams are also shown. The operation points are in lower half-loop for forward sweep and upper half-loop for reverse sweep.

Demonstration of High-speed Hysteresis-free Negative Capacitance in Ferroelectric $Hf_{0.5}Zr_{0.5}O_2$

M. Hoffmann[1], B. Max[1,2], T. Mittmann[1], U. Schroeder[1], S. Slesazeck[1], and T. Mikolajick[1,2]

[1]NaMLab gGmbH, Noethnitzer Str. 64, 01187 Dresden, Germany, email: michael.hoffmann@namlab.com
[2]Chair of Nanoelectronic Materials, TU Dresden, Dresden, Germany

Abstract—We report the experimental observation of hysteresis-free negative capacitance (NC) in thin ferroelectric $Hf_{0.5}Zr_{0.5}O_2$ (HZO) films through high-speed pulsed charge-voltage measurements. Hysteretic switching is suppressed by the addition of thin Al_2O_3 layers on top of the HZO to prevent the screening of the polarization. We observe an S-shaped polarization-electric field dependence without hysteresis in agreement with Landau theory, which enables direct extraction of NC modeling parameters for ferroelectric HZO. Hysteresis-free NC is demonstrated down to 100 ns pulse widths limited only by our measurement setup. These results give critical insights into the physics of ferroelectric NC and practical NC device design using ferroelectric HZO.

I. Introduction

To overcome the fundamental limits of power dissipation in nanoscale transistors due to the minimum subthreshold swing $S > 60$ mV/dec at room temperature, several new device concepts have been proposed, e.g. the tunnel-FET and the negative capacitance (NC) FET [1]. The latter is based on the idea that ferroelectric materials exhibit an S-shaped polarization-electric field (P-E) dependence, which implies a region of NC (i.e. $C_F < 0$) without hysteresis according to Landau theory (see the red line in Fig. 1). However, in typical ferroelectric devices, large hysteresis and only positive capacitance $C_F > 0$ is observed as shown in the blue line in Fig. 1. To obtain the S-shaped P-E curve, it is necessary to add a dielectric (or semiconducting) layer in contact to the ferroelectric, to prevent the immediate screening of its spontaneous polarization P_S [2]. For hysteresis-free operation, the capacitance C_D of this dielectric layer must be matched to the ferroelectric NC region ($C_D < |C_F|$), see Fig. 1. The most promising ferroelectrics for NCFET applications are HfO_2 and ZrO_2 based materials due to their full CMOS process compatibility and high scalability [3]. Especially the equal mixture of $Hf_{0.5}Zr_{0.5}O_2$ (HZO) shows good ferroelectric properties even for thinner films and lower thermal budget integration. While NCFETs with $S < 60$ mV/dec and without hysteresis have been demonstrated using DC measurements [4], all reports of faster or even pulsed measurements have shown considerable hysteresis so far. Therefore, an unambiguous demonstration of hysteresis-free NC in HZO even during fast pulsed operation (which corresponds to the actual operating condition in a digital circuit) is urgently needed. To experimentally show this, here we fabricated capacitors using ferroelectric HZO with and without dielectric Al_2O_3 layers on top. We also varied the thickness of both HZO and Al_2O_3 layers to investigate the scalability of the

NC effect as well as capacitance mismatch ($C_D > |C_F|$) and charge trapping, which can give rise to undesirable hysteresis. By using a pulsed charge-voltage (Q-V) measurement approach [5], we can distinguish between NC and hysteretic switching and examine the speed limits of hysteresis-free NC in HZO. Furthermore, from the pulsed Q-V data we can reconstruct the S-shaped P-E curve and thus directly extract the Landau parameters for HZO, which are crucial for accurate NCFET device modeling.

II. Experimental

Metal-ferroelectric-metal (MFM) and metal-ferroelectric-insulator-metal (MFIM) capacitors were fabricated on Si substrates. TiN bottom electrodes of 12 nm thickness were reactively sputtered in a BESTEC physical vapor deposition tool at room temperature. Subsequently, 7.7 nm and 11.3 nm thin HZO films were grown by atomic layer deposition (ALD) in an Oxford Instruments OpAL ALD tool at 260 °C using the precursors TEMA-Hf, and TEMA-Zr with water as an oxidant. For MFIM samples, ALD of 0.5 nm to 4 nm Al_2O_3 was carried out directly after HZO deposition without breaking vacuum using TMA and water as precursors also at 260 °C. The film thicknesses were adjusted by varying the number of ALD cycles. TiN top electrodes were deposited in the same way as the bottom electrodes. The HZO layers were then crystallized by 600 °C annealing for 20 s in N_2 atmosphere. Capacitor pads were defined by evaporating 10 nm Ti and 30 nm Pt through a shadow mask. These Pt dots (~7000 μm^2) served as a hard mask during the wet etch (NH_4OH, H_2O_2, and H_2O solution) of the TiN top electrode. Similarly, $TiN/Al_2O_3/TiN$ reference capacitors were fabricated to extract the relative permittivity $\varepsilon_r = 8$ of the Al_2O_3. X-ray reflectometry and grazing-incidence X-ray diffraction (GIXRD) measurements were carried out on a Bruker D8 Discover (Cu-Kα radiation, $\lambda = 0.154$ nm) for structural analysis of the samples. Transmission electron microscopy (TEM) analysis was carried out on a Zeiss Libra 200 TEM. Electrical measurements were performed on a Cascade Microtech Probe Station with a Keithley 4200 SCS with a 4225-PMU and remote amplifier, an HP 8110A pulse generator and a Tektronix TDS7154B digital oscilloscope. Standard polarization-electric field hysteresis was measured by applying triangular voltage signals with 10 kHz frequency. Capacitance-voltage measurements were carried out using a small-signal amplitude of 50 mV and a frequency of 10 kHz. For pulsed charge-voltage measurements, capacitors were connected to the pulse generator while measuring the current and voltage via the oscilloscope with a 50 Ω and 1 MΩ input impedance, respectively.

978-1-7281-1988-5/18 $31.00 © 2018 IEEE

III. RESULTS AND DISCUSSION

To confirm ferroelectricity in our HZO films we first used GIXRD to determine their crystalline structure. As shown in Fig. 2, both 7.7 nm and 11.3 nm thin HZO films mainly consist of the ferroelectric orthorhombic phase. The electrical measurements of the fabricated MFM capacitors in Fig. 3 show excellent ferroelectric properties with coercive fields of ~1 MV/cm and a high remanent polarization P_r of up to 27 $\mu C/cm^2$ after 10^5 electric field cycles. Furthermore, symmetric butterfly-shaped capacitance-field characteristics (Fig. 4) and only small P_r changes with electric field cycling are observed (Fig. 5). To investigate the NC behavior of these HZO layers, we fabricated and characterized MFIM structures with dielectric Al_2O_3 layers of 0.5 nm to 4 nm thickness. Fig. 6 shows a TEM cross-section of an MFIM capacitor with 7.7 nm HZO and 4 nm Al_2O_3 to confirm the thickness of the individual layers. In the higher resolution TEM cross-section in Fig. 7, the polycrystalline nature of the HZO layer in contrast to the amorphous structure of Al_2O_3 can be seen. For all combinations of layer thicknesses (7.7 nm/11.3 nm HZO and 0.5-4 nm Al_2O_3) we carried out small-signal capacitance measurements to look for a capacitance enhancement $C > C_D$, which would indirectly prove NC in the HZO layer. The inverse capacitances C^{-1} are plotted as a function of the Al_2O_3 thickness in Fig. 8. As can be seen, no small-signal capacitance enhancement is observed for the MFIM layers. In fact, the HZO and Al_2O_3 layers behave exactly as two positive capacitors in series where $C^{-1} = C_D^{-1} + C_F^{-1}$. This means that no stabilized NC is observed in these structures, which seems to be related to fixed charges at the HZO/Al_2O_3 interface that screen the remanent polarization when no voltage is applied [5]. Therefore, pulsed Q-V measurements were necessary to access the ferroelectric NC region during switching. Using a pulse generator and oscilloscope, the applied voltage and current flowing were measured (see setup in Fig. 9), from which the charge on the capacitor was calculated. Fig. 10 shows the results of such a pulsed measurement on a 11.3 nm HZO/4 nm Al_2O_3 capacitor: In (a) the applied voltage waveform V_1 is shown with a pulse width of ~500 ns and increasing amplitude. In (b), the measured current I is shown during charging and discharging of the capacitor. From the integration of I, the charge is obtained, and maximum, released and residual charges are defined in Fig. 10(c) as Q_{max}, Q_D and Q_{res}, respectively. These charges are plotted against the maximum applied voltage $V_{1,max}$ in Fig. 10(d). Note that the released charge Q_D for higher voltages is larger than the charge expected for a 4 nm Al_2O_3 layer only, which means that the capacitance of the HZO layer C_F must be negative in this region [5]. The field in the HZO layer is given by $E_F = (V_{1,max}-Q_D/C_D-RI)/t_F$, where R is the resistance and t_F the ferroelectric thickness. If we plot E_F as a function of $Q_D \approx P$, we obtain the HZO P-E_F curve shown in Fig. 11. A region of NC is observed which starts for positive applied pulses V_1 around the coercive field of ~ 1 MV/cm. For negative applied pulses, only a linear dielectric response is observed, which shows that the HZO is initially in the negative P_r state due to compensating charges at the HZO/Al_2O_3 interface. Using Landau theory ($E_F = 2\alpha P_S + 4\beta P_S^3$ and $Q_D = \varepsilon_0\varepsilon_b E_F + P_S \approx P$),

we obtain an excellent agreement to the experimental data (α = -8.8·10^8 m/F, β = 1.3·10^{10} $m^5/(C^2F)$ and ε_b = 25). Note that the $P_r \approx 18$ $\mu C/cm^2$ is in good agreement with the values in Fig. 5 for the pristine sample. To confirm that this NC effect is hysteresis-free, we applied 500 ns ascending and descending pulses to the MFIM samples with 4 nm Al_2O_3 which is shown in Fig. 12. Fig. 13 shows the extracted P-E_F curve for the 7.7 nm HZO sample, which confirms the absence of hysteretic switching. Again, Landau theory can nicely fit the measured results (α = -1.1·10^9 m/F, β = 2.5·10^{10} $m^5/(C^2F)$ and ε_b = 25). To investigate hysteresis as a function of Al_2O_3 thickness, we repeated the experiment for the other MFIM samples. As exemplarily shown in Fig. 14, for 1 nm Al_2O_3 and 7.7 nm HZO, a hysteresis in the Q_D-$V_{1,max}$ characteristics emerges which is related to the capacitance mismatch $C_D > |C_F|$ of both layers. The maximum Q_D-hysteresis ΔQ_D is plotted as a function of the Al_2O_3 thickness in Fig. 15. For Al_2O_3 layers thinner than ~1.8 nm, $C_D > |C_F|$ which results in a large hysteresis. For 2-3 nm Al_2O_3 thickness (where $C_D < |C_F|$) we observe a transition region which might be related to some mismatched domains and/or charge trapping. Only for the 4 nm Al_2O_3 sample, negligible hysteresis is observed since all domains are matched and charge trapping is completely inhibited. To investigate the speed of NC in HZO, we shortened the pulse widths down 100 ns, where we still observed NC effects without hysteresis as shown in Fig. 16. For future work, investigations on the relation between fundamental switching kinetics [6] and NC observed in HfO_2 based ferroelectrics will be of interest. Lastly, Fig. 17 compares all hysteresis-free NC reports of MFIM devices from literature. While prior works used thicker perovskite ferroelectrics, this work demonstrates the superior scalability of HfO_2 based materials with the additional advantage of full CMOS process compatibility.

IV. CONCLUSIONS

Hysteresis-free NC in HZO/Al_2O_3 stacks has been demonstrated down to 100 ns short pulsed voltage operation. S-shaped P-E curves for direct fitting of Landau coefficients were extracted. For thinner Al_2O_3 films, increased hysteresis was observed. These results highlight pathways towards fast and hysteresis-free NC devices for digital applications.

ACKNOWLEDGMENT

This work has received funding from the Electronic Component Systems for European Leadership (ECSEL) Joint Undertaking under grant agreement No 692519. ECSEL receives support from the EU's Horizon 2020 research and innovation programme and Belgium, Germany, France, Netherlands, Poland, United Kingdom. Part of this work was supported by the EFRE fund of the European Commission and by the Free State of Saxony. We gratefully acknowledge U. Mühle of Fraunhofer IKTS for TEM analysis.

REFERENCES

[1] S. Salahuddin and S. Datta, *Nano Lett.*, vol. 8, pp. 405–410, 2008.
[2] P. Zubko et al., *Nature*, vol. 534, pp. 524–528, 2016.
[3] J. Müller et al., *Nano Lett.*, vol. 12, pp. 4318–4323, 2012.
[4] M. Si et al., *Nat. Nanotechnol.*, vol. 13, pp. 24–28, 2018.
[5] Y. J. Kim et al., *Nano Lett.*, vol. 16, pp. 4375–4381, 2016.
[6] H. Mulaosmanovic et al., *ACS Appl. Mater. Interfaces*, vol. 9, pp. 3792–3798, 2017.

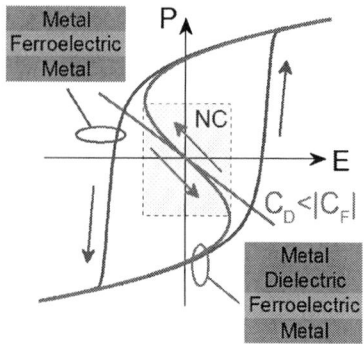

Fig. 1. Polarization-electric field (*P-E*) dependence of a ferroelectric based on Landau theory. Without dielectric, hysteresis and only positive capacitance C_F is expected while with dielectric, no hysteresis and negative capacitance (NC) $C_F < 0$ is achieved, if the dielectric capacitance C_D is matched in the NC region, i.e. $C_D < |C_F|$ (see green load line of the dielectric).

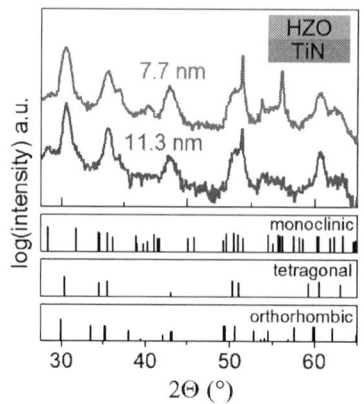

Fig. 2. Grazing-incidence X-ray diffraction patterns of ferroelectric HZO layers of different thickness compared to reference patterns for common phases in HfO_2. The HZO films are mostly orthorhombic.

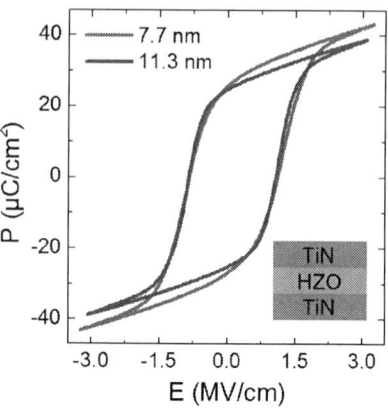

Fig. 3. Experimental *P-E* hysteresis curves measured at 10 kHz for TiN/HZO/TiN capacitors with different HZO thicknesses after wake-up cycling (10^5 cycles at 100 kHz).

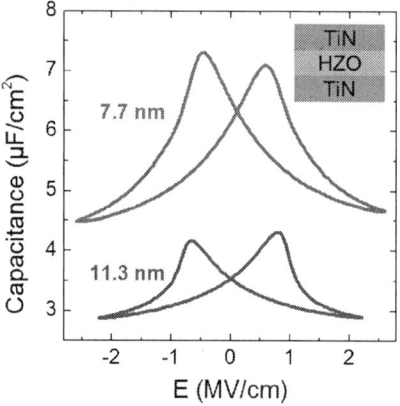

Fig. 4. Capacitance-electric field hysteresis for different HZO thicknesses measured at the frequency $f = 10$ kHz and 50 mV small-signal amplitude after wake-up cycling (10^5 cycles at 100 kHz).

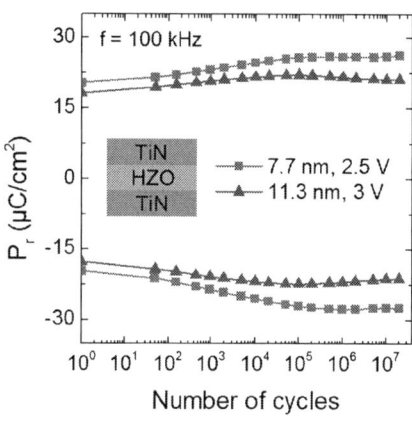

Fig. 5. Evolution of remanent polarization P_r with electric field cycling ($f = 100$ kHz) for different HZO thicknesses.

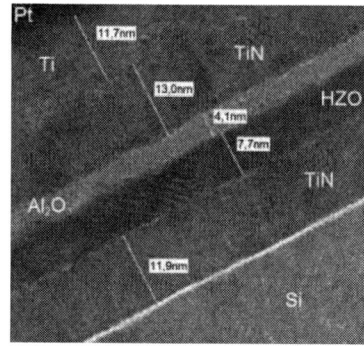

Fig. 6. Transmission electron microscope (TEM) cross-section of an Al_2O_3/HZO capacitor structure with about 4 nm Al_2O_3 and 7.7 nm HZO.

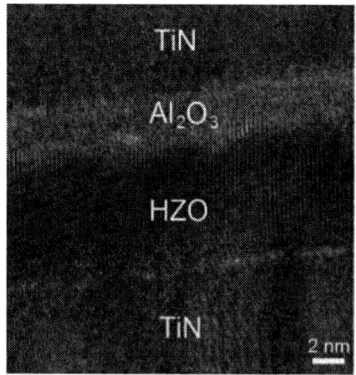

Fig. 7. High-resolution TEM of the 4 nm Al_2O_3 and 7.7 nm HZO capacitor. The polycrystalline structure of HZO and amorphous phase of the Al_2O_3 layer can be seen.

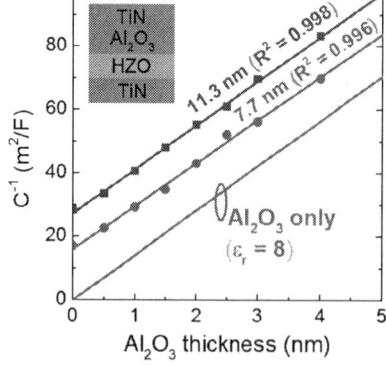

Fig. 8. Inverse small-signal capacitance C ($f = 10$ kHz, 50 mV amplitude) for different HZO and Al_2O_3 thicknesses. Solid blue and red lines show linear fits and the green line corresponds to the theoretical capacitance of Al_2O_3 without HZO.

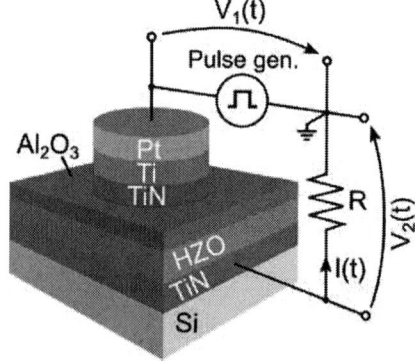

Fig. 9. Schematic sample structure and experimental setup for pulsed charge-voltage measurements. Short voltage pulses are applied to the top electrode while the current is measured through the 50 Ω input resistance of an oscilloscope. The pulse voltage is measured at the same time.

978-1-7281-1988-5/18 $31.00 © 2018 IEEE

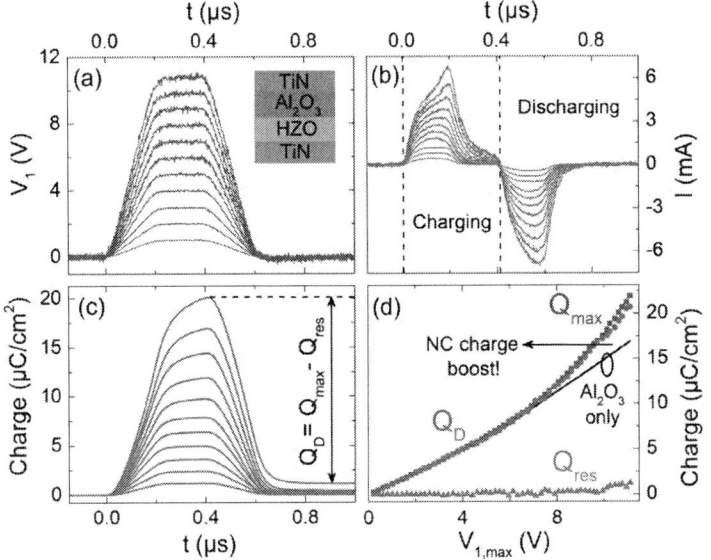

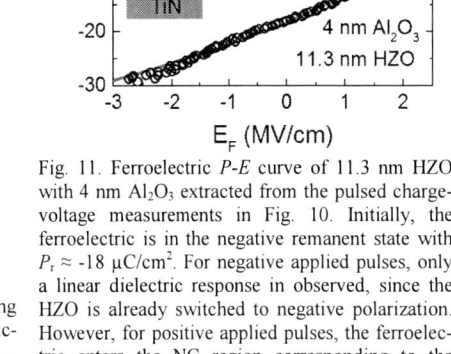

Fig. 10. (a) Applied voltage pulses with increasing amplitude. (b) Measured current during charging and discharging of the 11.3 nm HZO/4 nm Al_2O_3 capacitor. (c) Charge on the capacitor as a function of time t integrated from the current in (b). (d) Maximum, released and residual charges (Q_{max}, Q_D and Q_{res}) as a function of maximum applied voltage. Black line shows expected charge for 4 nm Al_2O_3 layer without HZO.

Fig. 11. Ferroelectric P-E curve of 11.3 nm HZO with 4 nm Al_2O_3 extracted from the pulsed charge-voltage measurements in Fig. 10. Initially, the ferroelectric is in the negative remanent state with $P_r \approx$ -18 $\mu C/cm^2$. For negative applied pulses, only a linear dielectric response in observed, since the HZO is already switched to negative polarization. However, for positive applied pulses, the ferroelectric enters the NC region corresponding to the negative slope of the S-shaped Landau P-E curve.

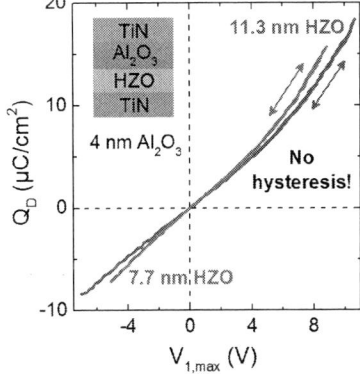

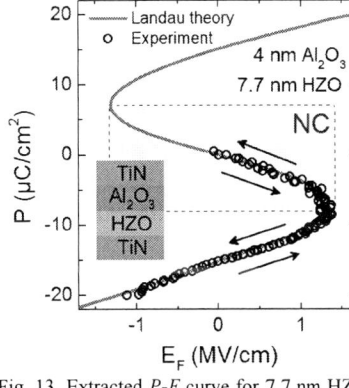

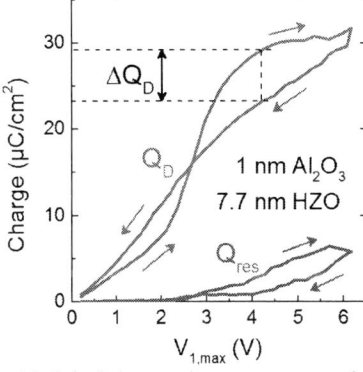

Fig. 12. Pulsed charge-voltage hysteresis measurement for capacitors with 4 nm Al_2O_3. No hysteresis is observed for ascending and descending voltage pulse trains with 500 ns pulse width.

Fig. 13. Extracted P-E curve for 7.7 nm HZO with 4 nm Al_2O_3 from Fig. 12. No hysteresis is observed in the S-shaped curve in accordance with fundamental Landau theory.

Fig. 14. Pulsed charge-voltage measurement for a capacitor with only 1 nm Al_2O_3. Hysteresis is observed due to capacitance mismatch. ΔQ_D is defined as the maximum Q_D hysteresis.

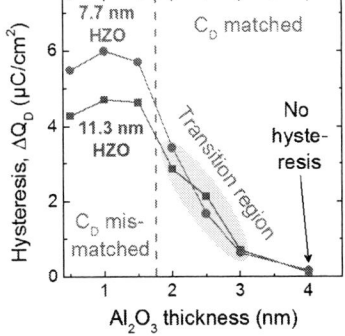

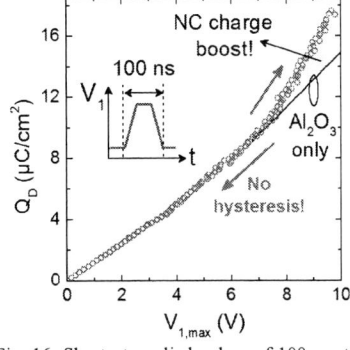

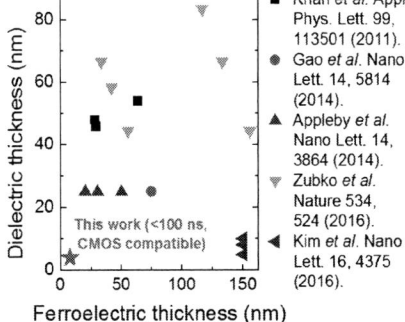

Fig. 15. Maximum hysteresis as a function of layer thicknesses. For Al_2O_3 layers thicker than 3 nm, no hysteresis is observed in the pulsed charge-voltage measurements with pulse widths of 500 ns. The transition region is due to partial domain mismatch and/or charge trapping effects.

Fig. 16. Shortest applied pulses of 100 ns still show hysteresis-free NC in the 11.3 nm HZO/4 nm Al_2O_3 sample. Applying even shorter pulses was not possible due to the RC-delay of the measurement setup ($RC \approx$ 30 ns).

Fig. 17. Comparison of hysteresis-free negative capacitance in ferroelectric/dielectric capacitors reported in literature. Besides this work, all other reports applied perovskite-based ferroelectrics which are not CMOS compatible. Additionally, HZO is much more scalable compared to perovskite ferroelectrics.

978-1-7281-1988-5/18 $31.00 © 2018 IEEE   730

Negative-Capacitance FinFET Inverter, Ring Oscillator, SRAM Cell, and Ft

Kai-Shin Li[1], Yun-Jie Wei[1], Yi-Ju Chen[1], Wen-Cheng Chiu[1], Hsiu-Chih Chen[1], Min-Hung Lee[2], Yu-Fan Chiu[1], Fu-Kuo Hsueh[1], Bo-Wei Wu[1], Pin-Guang Chen[1], Tung-Yan Lai[1], Chun-Chi Chen[1], Jia-Min Shieh[1], Wen-Kuan Yeh[1], Sayeef Salahuddin[4], Chenming Hu[3, 4]

[1] National Nano Device Laboratories, National Applied Research Laboratories, Hsinchu, Taiwan
[2] Institute of Elecro-Optical Science and Technology, National Taiwan Normal University, Taipei, Taiwan
[3] National Chiao Tung University, Hsinchu, Hsinchu, Taiwan
[4] Dept. of Electrical Eng. and Computer Science, University of California, Berkeley, USA;
Tel: +886-3-572-6100 ext. 7706, Fax: +886-3-572-6109, Email: ksli@narlabs.org.tw

Abstract

In this work, we use thermal-ALD to prepare ferroelectric $HfZrO_2$ (HZO) thin film with thickness from 3 to 7 nm for the NC-FinFET's gate stack. The subthreshold swing (SS) was as low as 5 mV/dec (SS_{min}) over 4 orders of I_D. Lower thermal budget process, CO_2 far-infrared laser activation and 400 °C Ni silicide are employed in the 2-level metal backend integration for maintaining the orthorhombic phase in HZO thin film and minimizing the hysteresis in IV. NC-FinFET inverter has 77% higher voltage gain compared to FinFET-inverter employing HfO_2 gate dielectric. NC-FinFET ring oscillator exhibit small speed and power advantages over FinFET oscillator. For the first time, NC-FET cut-off frequency (F_t) frequency is measured, 23.1 GHz or 23% higher than the control FET Ft. NC-FinFET SRAM was observed to exhibit large noise margin.

I. Introduction

Steep slope Negative Capacitance FET (NC-FET) [1-4] has been proposed for low power device applications. Inserting a ferroelectric film in the FinFET gate stack as a voltage amplifier was realized in our previous work [1]. HZO is often employed as the ferroelectric thin film material in NCFET.

The ferroelectric behavior in HZO thin film is sensitive to the Zr concentration, deposition condition, thermal treatment and strain that affect to the HZO phase composition. The phase transition of HZO, from monoclinic, orthorhombic to tetragonal, requires fine tuning of the above process condition [5]. In general, HZO is para-ferroelectric as deposited. After annealing at 450 °C~ 800 °C, ferroelectric property can be observed [6]. However, high annealing temperature -can cause tetragonal phase formation that leads to anti-ferroelectric behavior. In this study, we carefully kept the whole fabrication process of NC-FinFET circuit under low thermal budget to maintain ferroelectricity.

Especially, the high thermal process, 950 °C RTA activation and 650 °C Ti silicidation are replaced by localized CO_2 laser annealing activation and 400 °C Ni silicidation in fabricating the M2 NC-circuits. We fabricated NC-FinFET inverter, RF test structure, RO and SRAM cell to measure the voltage gain, power, gate delay, speed and cell noise margin.

II. NC-FinFET Fabrication and Material Analysis

NC-FinFETs with L_G = 60 nm and W_{fin} = 20 nm employed in this work are based on the NDL FinFET platform [8], which is described in Fig 2(1-8). The process flow is provided in Fig. 1. Fig. 3(a) shows the TEM image of 5nm HZO gate stack. Polycrystalline structure can be observed in HZO film annealed at 650 °C, 30sec as shown in Fig. 3(b). Fig. 4(a) shows the GI-XRD spectrums of HZO film on 50nm TiN layer before and after annealing at two different temperatures. Ferroelectric orthorhombic phase was detected with annealing temperatures above 650 °C. From X-ray photoelectron spectroscopy analysis (Fig. 4(b)), Hf and Zr atomic ratio in as deposited $HfZrO_2$ thin film is about 1:1. Fig. 5 shows the standard ferroelectric P-V hysteresis loop and corresponding C-V curve in thick 10 nm HZO MIM capacitor after RTA at 650 °C. Without the 10 nm TiN cap, HZO thin film on Si after annealing presents the para-ferroelectric behavior and partial monoclinic phase as shown in Fig. 6. In contrast, TiN provides HZO the necessary stress to form orthorhombic phase and ferroelectric behavior as shown in Fig. 6(a) and (b).

III. Results and Discussion

A. $HfZrO_2$ NC-FinFET

Fig. 7 shows that 7 nm HZO NC-FinFET, has ultra-low SS of 5mV/decade over 4 orders of magnitude of I_d through turn-off from 10^{-6} to 10^{-10} A/um. While this low SS is attractive, the large hysteresis in the turn-on-turn-off loop is troubling for circuit performance as is the large SS in the

978-1-7281-1988-5/18 $31.00 © 2018 IEEE

turn-on portion of the loop. The hysteresis can be reduced by thinning down the HZO film thickness as shown in Fig. 7. Fig. 8 shows that ΔV_{th} is reduced form 1.27 V to 0.09 V depending on HZO thickness.

Hysteresis can be reduced through other means, too. Besides using low thermal budge CO_2 laser S/D activation, the contact silicidation is another notable high thermal budget process. Particularly, 650 °C Ti silicide provides high enough thermal energy to cause phase transition in HZO thin film. Fig. 9 shows that the 400 °C Ni silicide process can maintain SS < 60 mV/dec and reduce the hysteresis. Through low thermal budget processing, the I_D-V_G curves of n-type and p-type 5 nm NC-FinFETs with small hysteresis could be fabricated. SS_{min} reaches 51 mV/dec for pNC-FinFET and 45 mV/dec for nNC-FinFET. 5 nm HfO_2 control FinFET, with 950 °C RTA and Ti silicide contact, are also presented in Fig. 10 for comparison. Fig. 11(a) shows the comparison of I_D-V_D between the 5nm HZO NC-FinFET and the 5nm HfO control FinFET. The 5nm HZO FinFET has larger Gm, 15.1mS, compared to the 5nm HfO FinFET, 11.7mS, and the 2.5nm HfO FinFET, 13.2mS, as shown in Fig.11 (b).

B. HZO NC-FinFET Circuit

Because the hysteresis depends on measurement conditions and its cause, e.g. charge trapping, is still not clear, it is impossible to predict the NCFET circuit behaviors solely based on the IV measurements. Fabricating circuits is essential. We fabricated several circuits using the low thermal budget 5nm HZO NC-FinFET and used the control 5nm HfO FinFET to directly study the NC-FinFET circuit performance. Fig. 12 plots the voltage transfer curves (VTC) of NC-FinFET and control FinFET inverters under different supply voltages. In Fig. 13, the maximum voltage gain in NC-FinFET inverter is about 43, which is much larger than the voltage gain, 24, of the control FinFET inverter because of the higher Gm and lower output conductance of NC-FinFET that are obvious in Fig. 11. The benefit of NC-FinFET is also observed in the RF test structure. The cut-off frequency (Ft), 23.1 GHz, of the NC-FinFET is 23% higher than the 18.8 GHz of the control FinFET as shown in Fig. 14(a). The cut-off frequencies are lower than expected for transistors with 60nm gate length employed here because the test structure has significant capacitance load. Fig. 15 shows the 31 stage ring oscillator (RO) output waveforms in NC-FinFET and control FinFET. The oscillation frequency is 113MHz for the NC-FinFET RO and 110MHz for the FinFET RO. The gate delay time is 143 ps for the NC-FinFET and 147 ps for the control FinFET. The measured power consumption is slightly lower in NC-FinFET RO than the control FinFET RO as shown in Fig. 16. Fig. 17 shows the butterfly curves of the fabricated 6T-SRAMs based on 5nm HZO NC-FinFET. Good static noise margin is achieved.

IV. Conclusions

The first HZO NC-FinFET with 5 mV/dec over 4 decades of drain current is reported in this work. Also reported for the first time are NC-FinFET F_t and SRAM. NC-FinFET circuits at M2 process are fabricated using low thermal budget techniques such as CO_2 laser annealing activation and 450 °C Ni silicidation. NC-FinFET has 77% higher voltage gain and 23% higher cut-off frequency. 31-stage ring oscillator employing NC-FinFET has similar speed and power consumption compared with control FinFET RO.

Acknowledgement

This work was performed by the National Nano Device Laboratories facilities and supported by the National Science Council, Taiwan and the "Center for the Intelligent Semiconductor Nano-system Technology Research" from The Featured Areas Research Center Program within the framework of the Higher Education Sprout Project by the Ministry of Education (MOE) in Taiwan.

References

[1] K.-S. Li et al., in *IEDM Tech. Dig.*, 2015, pp. 620-623.

[2] Daewoong Kwon, Yu-Hung Liao, Yen-Kai Lin, Juan Pablo Duarte, Korok Chatterjee, Ava J. Tan, Ajay K. Yadav, Chenming Hu, Zoran Krivokapic and Sayeef Salahuddin, in *VLSI-Tech*, T05-3, 2018.

[3] Zoran Krivokapic et al., in *IEDM Tech Dig.* 2017, 15.1.1

[4] Sayeef Salahuddin and Supriyo Datta, *Nano Letters* Vol. **8**, pp. 405-410, 2008.

[5] Patrick Polakowski and Johannes Mouller, *APPLIED PHYSICS LETTERS* **106**, 232905 (2015)

[6] M. H. Lee1, P.-G. Chen, C. Liu, K-Y. Chu, C.-C. Cheng, M.-J. Xi, S.-N. Liu, J.-W. Lee, S.-J. Huang, M.-H. Liao, M. Tang, K.-S. Li and M.-C. Chen, in IEDM Tech. Dig., 2016, 12.1

[7] Chenming Hu, Sayeef Salahuddin, Cheng-I Lin, Asif Khan, in *Device Research Conference (DRC)*, pp. 39-40, 2015.

[8] M.-C. Chen, C.-H. Lin, Y.-F. Hou, Y.-J. Chen, C.-Y. Lin, F.-K. Hsueh, H.-L. Liu, C.-T. Liu, B.-W. Wang, H.-C. Chen, C.-C. Chen, S.-H. Chen, C.-T. Wu, T.-Y. Lai, M.-Y. Lee, B.-W. Wu, C.-S. Wu, I. Yang, Y.-P. Hsieh, C.H. Ho, T. Wang, A.B. Sachid, C. Hu and F.-L. Yang, in VLSI Symp. Tech. Dig., 2013, pp. 218–219, 2013.

[9] Hiroyuki Ota, Tsutomu Ikegami, Junichi Hattori, Koichi Fukuda, Shinji Migita, and Akira Toriumi, in *IEDM Tech. Dig.*, 2016, 12.

- ● **Wafer clean & N,P Well Imp**
- ● **Fin Patterning & STI Process**
- ● **ALD HZO(7-3nm)/HfO$_2$ & PVD TaN/TiN**
- ● **Gate Patterning**
- ● **As & B Ion S/D Imp**
- ● Laser Annealing + RTA(650°C, 30sec)
- ● **ILD dep+ Contact**
- ● **650°C Ti or 400°C Ni Silicide**
- ● **M1 & M2 Process**
- ● **RF, Ring Oscillator, SRAM Test**

Fig. 1: Key processes of the ferroelectric NC-FinFET. Laser annealing is adopted for S/D activation. Simple NC-FinFET circuits were fabricated in this study.

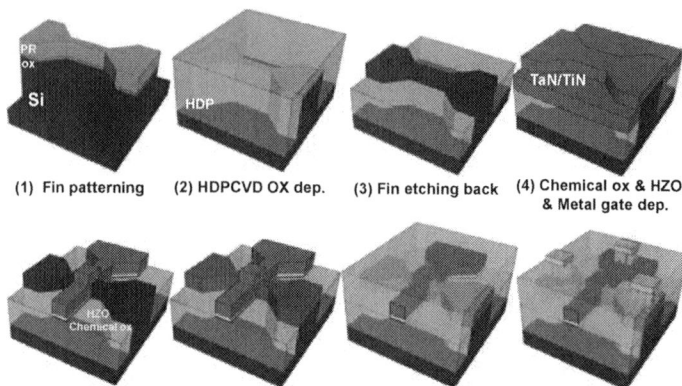

Fig. 2: Illustration of NC-FinFET device fabrication. (1)~(3) Fin formation (4) and (5) Ferroelectric gate formation and patterning (6) HZO crystallization and S/D activation by LSA/RTA (7) and (8) M1 process

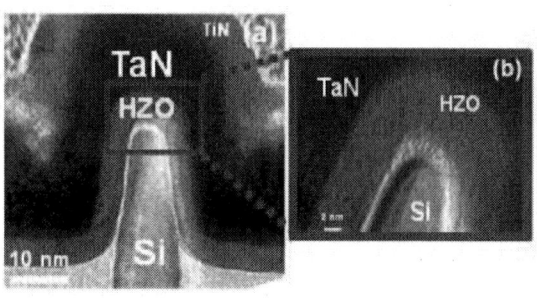

Fig. 3: (a) TEM Cross-section image of the NC-FinFET gate stack, **(b)** Zoom in image of crystalline HZO thin film.

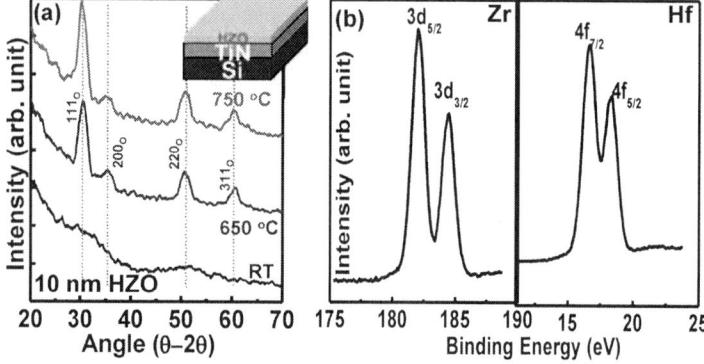

Fig. 4: (a) XRD results of the 10 nm HfZrO films after RT, 650 °C, 750 °C, 30 sec annealing. **(b)** Zr 3d and Hf 4f XPS spectra of the Hf$_{1-x}$Zr$_x$O$_2$ films. From the intensity analysis, x is about 0.5.

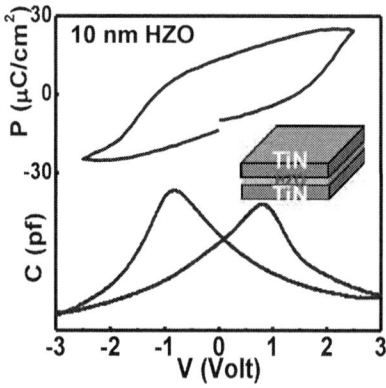

Fig. 5: C-V and PV curves measured on 10nm HZO MIM capacitor with 650 °C, 30sec annealing.

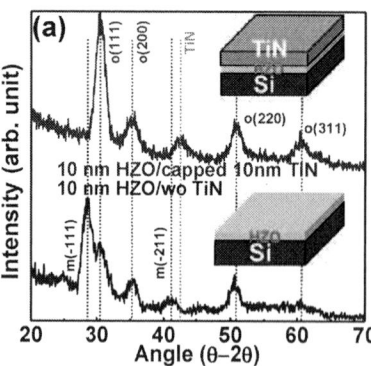

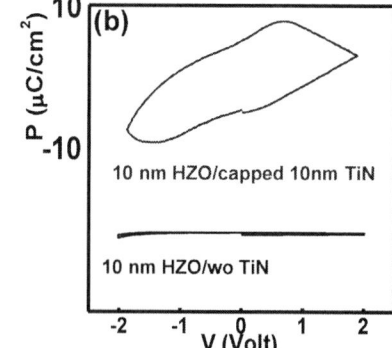

Fig. 6: (a) XRD of the 10 nm HfZrO film with and without 10 nm TiN cap during RTA 650 °C, 30 sec annealing. **(b)** PV hysteresis loops of 10 nm HZO MOSCAPs.

978-1-7281-1988-5/18 $31.00 © 2018 IEEE

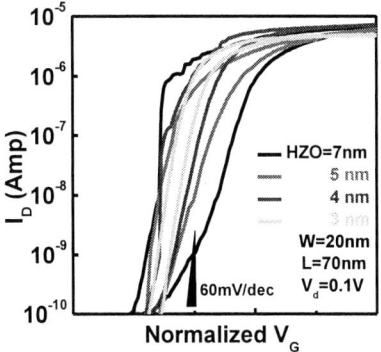

Fig. 7: NC-FinFET Id-Vg characteristics with 3,4,5,7nm HZO thickness.

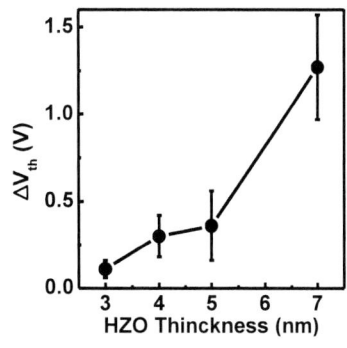

Fig. 8: S.S. and ΔV_{th} versus HZO thickness showing that Id-Vg hysteresis decreases with thinner HZO thickness.

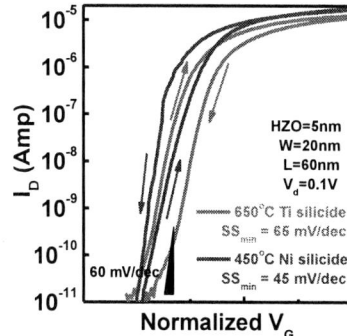

Fig. 9: I_D-V_G characteristics in NC-FinFET with 5 nm HZO gate oxide after Ti and Ni silicide process.

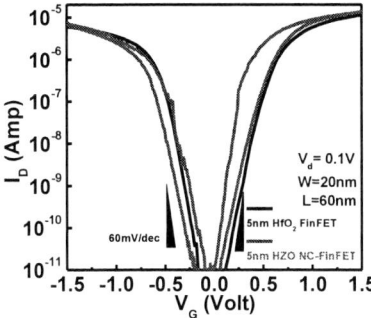

Fig. 10: I_D-V_G characteristics in 5 nm HZO NC-FinFET and 5nm HfO_2 FinFET.

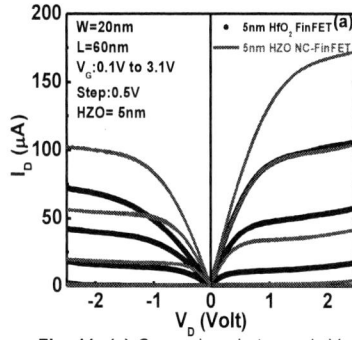

Fig. 11: (a) Comparison between I_D-V_D curves of NC-FinFET and control FinFET. **(b)** Gm of FinFETs with 5nm HZO, 2.5 nm HfO2 and 5 nm HfO2 gate oxide.

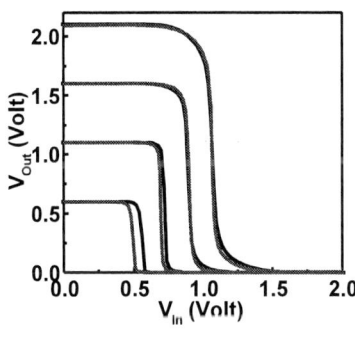

Fig. 12 Voltage Transfer Curve of FinFET and NC-FinFET inverters.

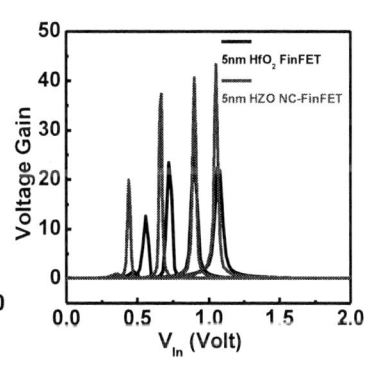

Fig. 13: Voltage gain of NC-CMOS and CMOS inverters.

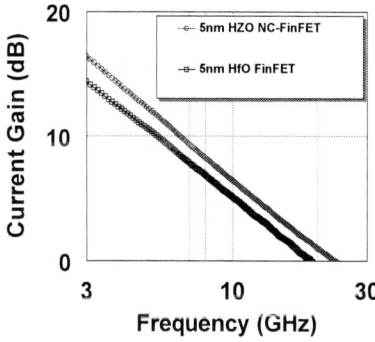

Fig. 14: RF measurements of NC-FinFET and control FinFET cut-off Frequencies.

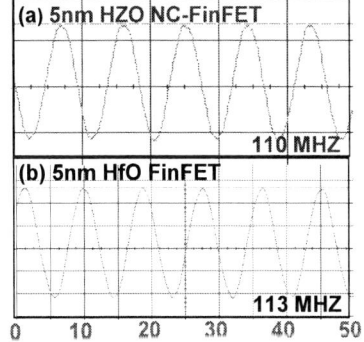

Fig. 15: Output waveforms of the **(a)** NC-FinFET and **(b)** control FinFET 31-stage ring oscillators.

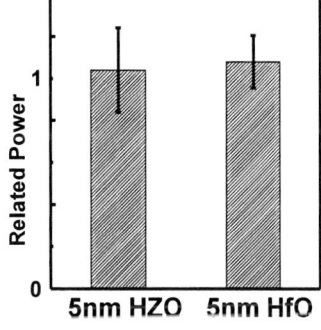

Fig. 16: Relative power consumption of NC-FinFET and control FinFET ring oscillators.

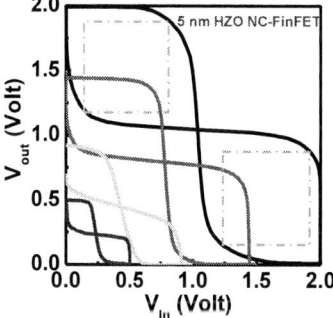

Fig. 17: Definition of the static noise margin (SNM) from the read voltage transfer characteristics (VCT) for NC-6T-SRAMs.

978-1-7281-1988-5/18 $31.00 © 2018 IEEE

Extremely Steep Switch of Negative-Capacitance Nanosheet GAA-FETs and FinFETs

M. H. Lee[1,*], K.-T. Chen[1], C.-Y. Liao[1], S.-S. Gu[1], G.-Y. Siang[1], Y.-C. Chou[1], H.-Y. Chen[1], J. Le[1], R.-C. Hong[1], Z.-Y. Wang[1], S.-Y. Chen[1], P.-G. Chen[2], M. Tang[3], Y.-D. Lin[4], H.-Y. Lee[4], K.-S. Li[2], and C. W. Liu[5]

[1] Institute of Electro-Optical Science and Technology, National Taiwan Normal University, Taipei, Taiwan
[2] National Nano Device Laboratories, Hsinchu, Taiwan
[3] Device Design Division, PTEK Technology Co., Ltd, Hsinchu, Taiwan
[4] Electronic and Optoelectronic System Research Laboratories, Industrial Technology Research Institute, Hsinchu, Taiwan
[5] Graduate Institute of Electronics Engineering, National Taiwan University, Taipei, Taiwan
*E-mail: mhlee@ntnu.edu.tw

Abstract—Extremely steep switch of negative-capacitance (NC) Nanosheet (NS) GAA-FETs and FinFETs are experimentally presented with SS_{avg}/SS_{min}=22/14mV/dec and SS_{avg}/SS_{min}=38/21mV/dec, respectively. The sub-60mV/dec current magnitude of sub-60mV/dec is >4 and ~5 decades for NC-NSGAA and NC-FinFET, respectively. Both NC-NSGAA and NC-FinFET exhibit extremely steep switch behavior due to FET scale down to nano-scale and comparable domain size of polycrystalline HZO. The dramatic current switch with steep slope is measured with only several dipole domains flipping over with gate voltage applied. The apparent Negative-DIBL and NDR (Negative Differential Resistance) are observed due to strong NC boost. The SS depends on W_{Fin}/L ratio, and $W_{Fin} < L$ is the solution to achieve sub-60mV/dec. The super-steep slope on current behavior still occurs after multiple DC sweep. The uniform size of each NS for stacked NC-NSGAA is an important issue to optimize the NC effect with SS=19mV/dec due to single T_{NS} for capacitance matching by modeling.

I. INTRODUCTION

Well gate control ability of FinFETs and Nanosheet (Fin-like) GAA(Gate-All-Around) FETs [1] is already the state-of-the-art technology for current node and candidate for future generation, respectively. Moreover, the negative-capacitance (NC) is one of possible solutions for future nodes path finding [2][3]. Recently, it tends to develop the integration of NC and multi-gate FETs (MuGFETs), including NC-FinFETs and NC-NWFETs[4-9], as well as boost optimization. Hf-based oxide for ferroelectric is investigated widely with polycrystalline [10][11]. The ultra-thin HfZrO$_2$ (HZO) with <10nm is demonstrated for ferroelectricity and steep switch on FETs performance, in which the domain size of grain is close to polycrystalline film thickness. The scaling FETs to nano-scale reaches the comparable domain size of polycrystalline HZO to characterize the ferroelectric ceramics with volume effect, domain effect and grain boundary effect according to their different length scales [12]. Therefore, the Nanosheet GAAFETs and FinFETs with comparable device dimension and domain size polycrystalline FE-HZO will be evaluated on current switch in this work, as well as on SS performance, reliability, layout dependence, and stacked Nanosheets for NC-

GAAFETs modeling. The feasible concept of coupling capacitances matching C_{FE} leads the practicability for NC-GAAFETs and NC- FinFETs.

II. DEVICES FABRICATION

For the Nanosheet GAAFETs process, standard 6-inch MOS-based line is employed and SOI wafer is used as the substrate. The schematic diagram of GAAFET with FE-HZO and fabrication process flow is shown in Fig. 1. The I-line stepper is used for lithography, and supplemented by trimming and oxidation process. Then, ALD (Atomic Layer Deposition) HZO is deposited with gate-first process and self-alignment for Source/Drain (S/D) definition. Finally, the HZO crystallization with ferroelectricity and S/D activation is performed by RTA (Rapid Thermal Annealing). After all process, W_{NS} (Nanosheet Width) < 100 nm and T_{NS} (Nanosheet Thickness) ~ 20 nm are observed by HR-TEM (Fig. 2). The physical thickness of FE-HZO is 5-5.5nm, and the lattice image is observed due to crystallization after annealing. This indicates the polycrystalline nature in HZO to form ferroelectric-phase. The FE-HZO and gate metal (TaN) are whole cladding the Nanosheet to form GAAFET. The EDS along the vertical direction of the middle Nanosheet shows the Si thickness 20-25nm, and TaN/HZO is deposited on both the top and the rear sides of Si (Fig. 3).

The FinFET on bulk-Si substrate with FE-HZO is fabricated by using FinFET platform in this study. The schematic diagram of NC-FinFET and the fabrication process flow are shown in Fig. 4. The gate-first process is employed with gate stack of interfacial layer and ALD HZO. The annealing process for the HZO crystallization with ferroelectricity and S/D activation is performed by RTA. Physical thickness 4.5-5.5 nm FE-HZO and 0.6-1.8nm IL are characterized by cross-sectional HR-TEM (Fig. 5). The thicker IL is formatted at the top as compared with the side wall. The lattice image of HZO is observed clearly to confirm the crystallization after annealing.

Thin films of HZO combined with HfO$_2$ and ZrO$_2$ by supercycles are grown using ALD. The P-V and extracted C-V (*dP/dV*) of MFM (Metal/Ferroelectric/Metal) with 5nm-thick FE-HZO show typical hysteresis loop from 0.1 V to 1.7V (Fig. 6), and there is no significant degradation after fatigue testing with 10^4 cycles. The same FE-HZO process is adopted for the proposed NC-GAAFETs and NC-FinFETs of this work.

III. RESULTS AND DISCUSSION

A. Nanosheet NC-GAAFETs

Fig. 7 shows transfer characteristics ($I_{DS}V_{GS}$) and SS vs. I_{DS} of NC-NSGAA (Nanosheet GAAFET) and HfO$_2$-NSGAA. The SS_{avg}=22mV/dec with more than 4 order current magnitudes and SS_{min}=14mV/dec are obtained for NC-NSGAA. Both forward and reverse sweeps show SS improvement as compared with HfO$_2$-NSGAA (Fig. 7(b)). The low gate leakage current (I_{GS}) ~ 10^{-13}A/NS (Nanosheet) is much lower than I_{DS}. The distribution of cumulative probability for sub-60mV/dec current order magnitudes shows the highest as 4 decades (Fig. 8). The NC is boosted for steep current switch with V_{GS} sweep range beyond ± 1.5V, which overcomes coercive voltage. The effect degrades until around ± 3V range due to gate leakage (Fig. 9). The extremely steep switch is existed after 10 times of DC sweep (Fig. 10). Note that V_T negative shift with sweep cycles is due to charge trapping by stress effect during sweep. Fig. 11 (a) shows $I_{DS}V_{GS}$ and SS vs. I_{DS} of Negative-DIBL (N-DIBL) for NC-NSGAA. The apparent N-DIBL is observed due to strong NC boost. The current level of steep slope is raised for higher V_{DS}, in which the SS=7 mV/dec keeps ~2 decades. The apparent NDR (Negative Differential Resistance) of NC-NSGAA is also observed (Fig. 11(b)). The output characteristics ($I_{DS}V_{DS}$) of NC-NSGAA presents the stable saturation current for reference (Fig. 11(b)). In comparison with previous studies of NC-GAAFETs (Table I), the Nanosheet GAA with NC in this work is proposed, as well as apparent N-DIBL and NDR. The NC-NSGAA shows extremely steep SS (SS_{avg}/SS_{min}=22/14mV/dec) and more than 4 decades of sub-60mV/dec.

B. NC-FinFETs

The $I_{DS}V_{GS}$ and SS vs. I_{DS} of NC-FinFET and BL(baseline)-FinFET are shown in Fig. 12, in which HfO$_2$ is used as BL-FinFET. The SS_{avg}=38mV/dec with ~ 5 order current magnitudes and SS_{min}=21mV/dec are obtained for NC-FinFET. Both NC-NSGAA and NC-FinFET exhibit extremely steep switch behavior due to FET scale down to nano-scale and comparable domain size of polycrystalline HZO, in which the grain size is 5-10nm (Fig. 2&5). Only several dipole domains flip over while gate voltage is applied; this makes dramatic current switch with steep slope. Besides, the lower SS on average and minimum for NSGAA is obtained due to well gate control ability with GAA. As high as 5 decades on current is obtained for sub-60mV/dec (Fig. 13). The extremely steep switch is existed after 100 times of DC sweep (Fig. 14). For the geometry effect, the sub-60mV/dec is achieved with W_{Fin}=20-35nm regardless of L, and long channel devices would benefit W_{Fin}=400nm (Fig. 15(a)&(b)). The results agree with ref. [7]. The SS strongly depends on W_{Fin}/L ratio based on total 193 measured SS, and W_{Fin} < L is the solution to achieve sub-60mV/dec (Fig. 15(c)). Note that the SS is around 60mV/dec for ratio=1 (W_{Fin}=L). The similar trend is obtained on DIBL for 114 measured data (Fig. 16). The negative value of DIBL (N-DIBL) is occurred for W_{Fin}/L < 1, which accompanies with sub-60mV/dec. The typical N-DIBL of NC-FinFET on $I_{DS}V_{GS}$ and SS vs. I_{DS} are shown in Fig. 17, which is similar with NC-NSGAA. The apparent N-DIBL is observed due to strong NC

boost and it is beneficial for the retardation of SCE (short channel effect). The steep switch region of current is raised for increasing V_{DS}, which is around 10 mV/dec for ~ 2 order magnitudes. The apparent NDR of NC-FinFET is also observed and operated with gate bias between subthreshold and weak inversion region (Fig. 17(b)). The output characteristics of NC-FinFET presents the stable saturation current for reference which is larger bias into strong inversion region (Fig. 17(b)). The NC-FinFET is also validated by pulse sweep measurement, and the steep switch is observed with pulse widths (Fig. 18). There is no significant difference in SS for DC and pulse sweep; this indicates the NC boost for steep SS independent on sweep modes. In comparison with previous studies of NC-FinFETs (Table II), the NC-FinFET shows SS_{avg}/SS_{min}=38/21mV/dec and ~ 5 decades of sub-60mV/dec in this work, as well as apparent N-DIBL and NDR.

C. Stacked NanoSheet NC-GAAFETs Modeling

Electron density at the middle of NS increases with scaling T_{NS} down due to quantum confinement (QC), where the Density-Gradient Quantization model is served for QC (Fig. 19). The capacitance increasing with thinner T_{NS} is due to higher electron density and limited depletion region in NS (Fig. 20). This indicates the reduced T_{FE} to match the capacitance of scaling T_{NS} down. For stacked NSs simulation by 7nm technology node (L_g=12nm)[1], the uniform size of each NS is an important issue to optimize the NC effect due to single T_{NS} for capacitance matching (Fig. 21).

IV. CONCLUSION

The NC-NSGAA and NC-FinFET are demonstrated with extremely steep switch due to comparable device dimension and domain size polycrystalline FE-HZO. The lower SS on average and minimum for NSGAA is obtained due to well gate control ability with GAA. The apparent N-DIBL and NDR are observed due to strong NC boost. The SS depends on W_{Fin}/L ratio, and W_{Fin} < L is the solution to achieve sub-60mV/dec. The uniform size of each NS is an important issue to optimize the NC effect. The proposed technology of NC-NSGAA (Fin-like) and NC-FinFET have the advantage of feasible concept to develop the super steep slope transistors.

ACKNOWLEDGMENTS

The authors are grateful for the funding support from the National Science Council (MOST 107-2218-E-003-004), process supported by National Nano Device Laboratories (NDL) & Nano Facility Center (NFC), computing supported by National Center for High-performance Computing (NCHC), Taiwan.

REFERENCES

[1] N. Loubet et al, in *Symp. on VLSI Technology and Circuits*, 2017, pp. T230-T231.
[2] S. Salahuddin and S. Datta, *NanoLetters*, vol. 8, no. 2, pp. 405-410, 2008.
[3] M. H. Lee et al, in *IEDM Tech. Dig.*, 2016, pp. 306-309.
[4] K. S. Li et al, in *IEDM Tech. Dig.*, 2015, pp. 620-623.
[5] Z. Krivokapic et al, in *IEDM Tech. Dig.*, 2017, pp. 357-360.
[6] W. Chung et al, in *IEDM Tech. Dig.*, 2017, pp. 365-368.
[7] H. Zhou et al, in *Symp. on VLSI Technology and Circuits*, 2018, pp. 53-54.
[8] W. Chung et al, in *Symp. on VLSI Technology and Circuits*, 2018, pp. 89-90.
[9] C.-J. Su et al, in *IEDM Tech. Dig.*, 2017, pp. 369-372.
[10] M. H. Lee et al, *IEEE J. of the Electron Device Society*, vol. 3, no. 4, pp. 377-381, 2015.
[11] M. H. Lee et al, *IEEE Electron Device Letter*, vol. 36, no. 4, pp. 294-296, 2015.
[12] Y. A. Genenko et al, *Materials Science and Engineering B*, vol. 192, pp.52-82, Feb. 2015.

978-1-7281-1988-5/18 $31.00 © 2018 IEEE

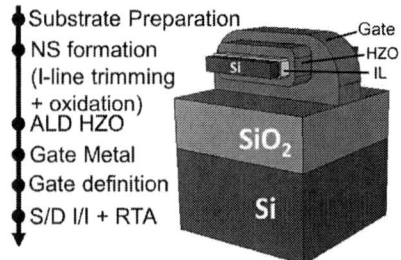

Fig. 1. Process flow and schematic diagram of Nanosheet NC-GAAFET with FE-HZO gate. The I-line stepper is used for lithography, and supplemented by trimming and oxidation process. The HZO crystallization with ferroelectricity and S/D activation is performed by RTA.

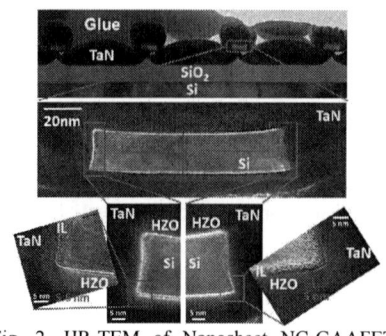

Fig. 2. HR-TEM of Nanosheet NC-GAAFET. After all processes, W_{NS} (Nanosheet Width) < 100 nm and T_{NS} (Nanosheet Thickness) ~ 20 nm. The physical thickness of FE-HZO is 5-5.5nm.

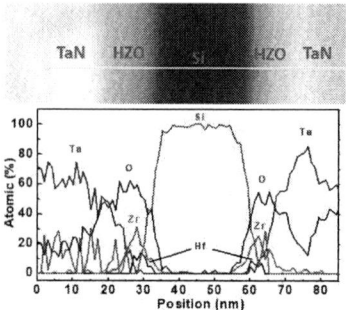

Fig. 3. EDS of Nanosheet NC-GAAFET along the vertical direction of the middle Si Nanosheet (Fig. 2). It shows the Si thickness 20-25nm, and TaN/HZO is deposited on both the top and the rear sides of Si.

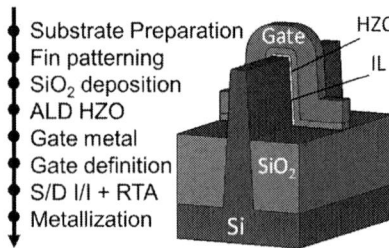

Fig. 4. Process flow and schematic diagram of NC-FinFET with FE-HZO gate. The gate-first process is employed with gate stack of interfacial layer and ALD HZO. The H_{Fin} (Fin Height) is ~30nm and W_{Fin} (Fin Width) is according to the design.

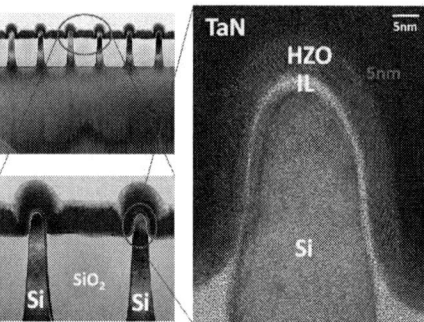

Fig. 5. Cross-sectional HR-TEM of NC-FinFET. HZO and IL are 4.5-5.5nm and 0.6-1.8 nm, respectively. The lattice image of HZO is observed due to crystallization after annealing.

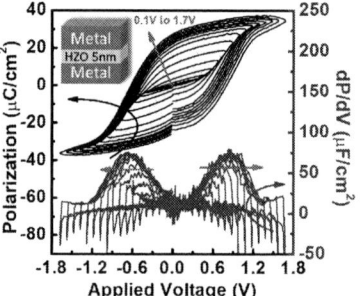

Fig. 6. P-V and extracted C-V (dP/dV) of MFM with HZO 5nm for applied voltages. The same FE-HZO process is adopted for the proposed NC-GAAFETs and NC-FinFETs of this work.

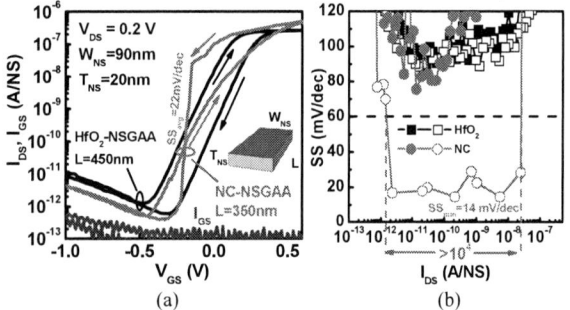

Fig. 7. (a) Transfer characteristics ($I_{DS}V_{GS}$) and (b) SS vs. I_{DS} of NC-NSGAA and HfO₂-NSGAA. The SS_{avg}=22mV/dec with more than 4 order current magnitudes and SS_{min}=14mV/dec are obtained for NC-NSGAA. The low gate leakage current (I_{GS}) ~ 10^{-13}A/NS (Nanosheet) is much lower than I_{DS}.

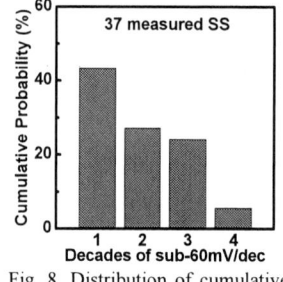

Fig. 8. Distribution of cumulative probability for sub-60mV/dec current order magnitudes. It shows the highest as 4 decades in 37 measured SS<60mV/dec of NC-NSGAA.

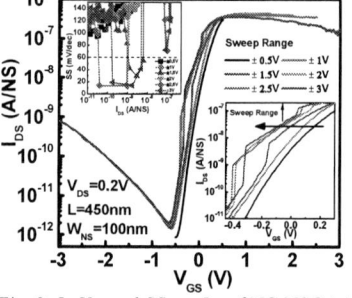

Fig. 9. $I_{DS}V_{GS}$ and SS vs. I_{DS} of NC-NSGAA with V_{GS} sweep ranges. The NC is boosted for steep current switch with V_{GS} sweep range beyond ±1.5V, which overcomes coercive voltage.

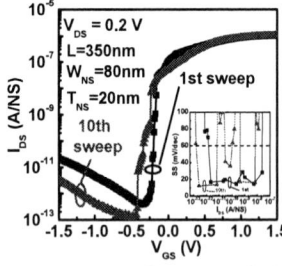

Fig. 10. $I_{DS}V_{GS}$ and SS vs. I_{DS} of NC-NSGAA with multiple DC sweep cycles. The extremely steep switch on current still occurs after sweep more than 10 cycles.

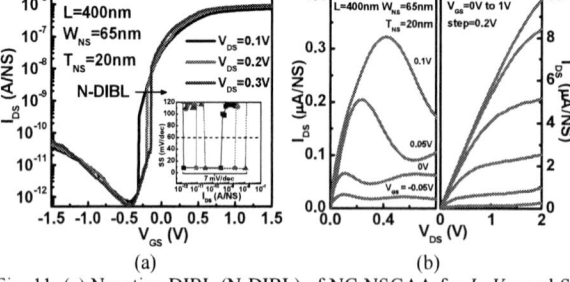

Fig. 11. (a) Negative-DIBL (N-DIBL) of NC-NSGAA for $I_{DS}V_{GS}$ and SS vs. I_{DS}. (b) Negative Differential Resistance (NDR) and output characteristics ($I_{DS}V_{DS}$) of NC-NSGAA. The apparent Negative-DIBL and NDR are observed due to strong NC boost. The current level of steep slope is raised for higher V_{DS}, in which the SS=7 mV/dec keeps ~2 decades.

Table I. Comparison with previous studies of NC-GAAFETs. In this work, the Nanosheet GAA with NC is proposed, as well as apparent N-DIBL and NDR. The NC-NSGAA shows extremely steep SS (SS_{avg}/SS_{min}=22/14mV/dec) and more than 4 decades of sub-60mV/dec.

NC-GAAFET (Gate Stack /Channel)	Nanosheet (HZO/IL/Si) [this work]	Nanowire (HZO/GeO$_x$ /Ge) [9]	Nanowire (Al$_2$O$_3$/HZO /Al$_2$O$_3$/GeO$_x$ /Ge) [8]
Type	N	N	P
t_{FE} (nm)	5	6	10
SS_{avg}/SS_{min} (mV/dec)	22/14	NA/54	NA/43
Decades (<60mV/dec)	>4	<1	4~5
N-DIBL	Yes	No	Yes
NDR	Yes	NA	Yes

978-1-7281-1988-5/18 $31.00 © 2018 IEEE

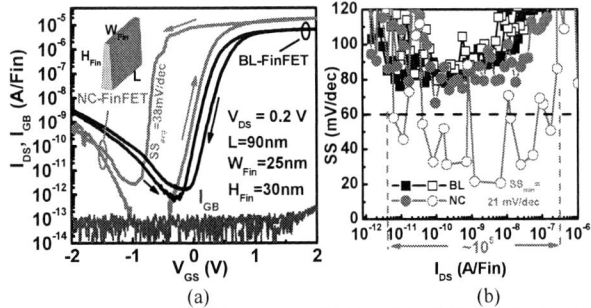

(a)

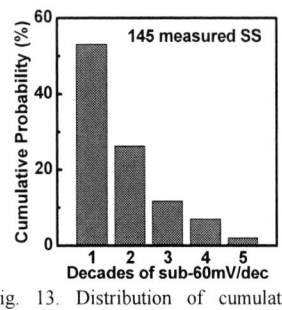

(b)

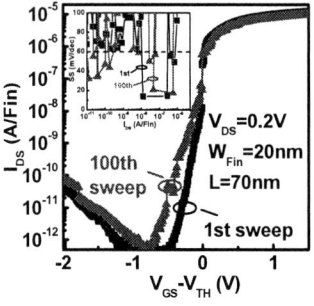

Fig. 12. (a) $I_{DS}V_{GS}$ and (b) SS vs. I_{DS} of NC-FinFET and BL(baseline)-FinFET (HfO$_2$-FinFET). The SS$_{avg}$=38mV/dec with ~ 5 order current magnitudes and SS$_{min}$=21mV/dec are obtained for NC-FinFET. Both forward and reverse sweeps show SS improvement as compared with BL-FinFET.

Fig. 13. Distribution of cumulative probability for sub-60mV/dec current order magnitudes. It shows the highest as 5 decades in 145 measured SS<60mV/dec of NC-FinFET.

Fig. 14. $I_{DS}V_{GS}$ and SS vs. I_{DS} of NC-FinFET with multiple DC sweep cycles. The extremely steep switch still occurs after sweep more than 100 cycles.

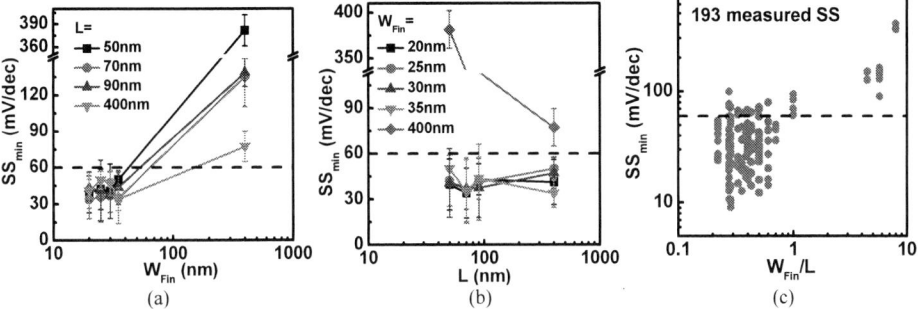

(a)   (b)   (c)

Fig. 15. Layout dependence of NC-FinFETs for (a) SS$_{min}$ vs. W$_{Fin}$, (b) SS$_{min}$ vs. L, (c) SS$_{min}$ vs. W$_{Fin}$/L. The sub-60mV/dec is achieved with W$_{Fin}$=20-35nm regardless of L, and long channel devices would benefit for W$_{Fin}$=400nm. The SS strongly depends on W$_{Fin}$/L ratio based on total 193 measured SS, and W$_{Fin}$ < L is the solution to achieve sub-60mV/dec. Note that the SS is around 60mV/dec for ratio=1 (W$_{Fin}$=L).

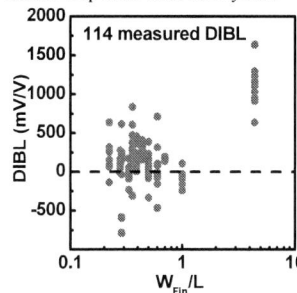

Fig. 16. DIBL vs. W$_{Fin}$/L of NC-FinFETs based on 114 measured data. The negative value of DIBL (N-DIBL) occurs for W$_{Fin}$/L < 1, which accompanies with sub-60mV/dec.

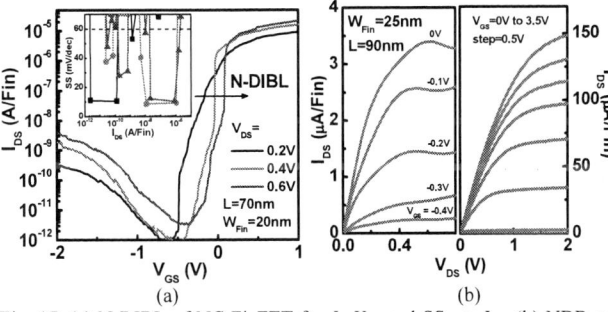

(a)   (b)

Fig. 17. (a) N-DIBL of NC-FinFET for $I_{DS}V_{GS}$ and SS vs. I_{DS}. (b) NDR and output characteristics ($I_{DS}V_{DS}$) of NC-FinFET. The apparent N-DIBL and NDR are observed due to strong NC boost, which is similar with NC-NSGAA. It is beneficial for the retardation of SCE (short channel effect).

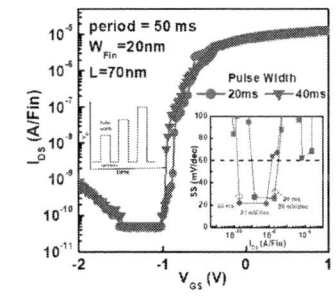

Fig. 18. Pulse sweep $I_{DS}V_{GS}$ and SS vs. I_{DS} of NC-FinFET. There is no significant difference in SS for DC and pulse sweep.

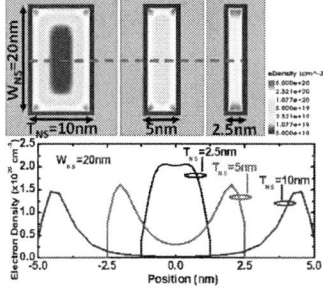

Fig. 19. Electron density distribution of NSs with T$_{NS}$=10nm, 5nm and 2.5nm. Due to quantum confinement effect, the electron density at the middle of NS increases with scaling T$_{NS}$ down.

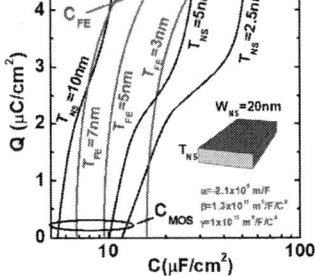

Fig. 20. Q vs. C of NSs with T$_{NS}$=10, 5, 2.5 nm, and T$_{FE}$=7, 5, 3 nm. The reduced T$_{FE}$ is necessary to match the capacitance of scaling T$_{NS}$ down. Note that the capacitance increasing with thinner T$_{NS}$ is due to higher electron density (Fig. 20) and limited depletion region in NS.

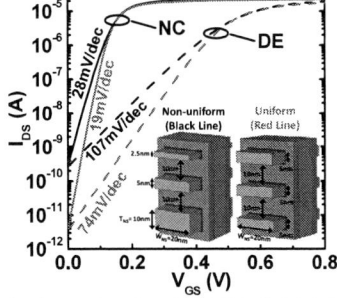

Fig. 21. $I_{DS}V_{GS}$ of stacked NS NC-GAA for uniform and non-uniform NSs. The optimized SS is 19mV/dec for uniform stacked NSs due to single T$_{NS}$ for capacitance matching. The inset shows the structure of stacked NS GAA with space 10nm.

Table II. Comparison with previous studies of NC-FinFETs. In this work, the NC-FinFET shows SS$_{avg}$/SS$_{min}$=38/21mV/dec and ~ 5 decades of sub-60mV/dec, as well as apparent N-DIBL and NDR. The lower SS on average and minimum for NSGAA (Table I) is obtained due to well gate control ability with GAA. The proposed technology of NC-NSGAA (Fin-like) and NC-FinFET have the advantages of feasible concept to develop the super steep slope transistors.

NC-FinFET (Gate Stack /Channel)	HZO/IL/Si [this work]	Si:HfO$_2$/IL/Si [5]		HZO/M/ HfO$_2$/Si* [4]	HZO /Al$_2$O$_3$ /HfO$_2$/Si [7]	Al$_2$O$_3$/HZO /Al$_2$O$_3$/GeO$_x$ /Ge [6]	
Type	N	N	P	N	N	N	P
t$_{FE}$ (nm)	5	3	3	5	4	2	10
SS$_{avg}$/SS$_{min}$ (mV/dec)	38/21	NA/60	NA/65	NA/58	54.5/NA	NA/43	NA/7
Decades (<60mV/dec)	~5	0	0	1	2	<1	1
N-DIBL	Yes	No	No	NA	Yes	No	Yes
NDR	Yes	NA	NA	NA	Yes	NA	Yes

* internal gate

978-1-7281-1988-5/18 $31.00 © 2018 IEEE   738

Optocoupling in CMOS

V. Agarwal[1], S. Dutta[2], A. J. Annema[1], R. J. E. Hueting[2], J. Schmitz[2], M.-J. Lee[3], E. Charbon[3] and B. Nauta[1]

[1]IC design group, University of Twente, Enschede, The Netherlands. Email: v.agarwal@utwente.nl
[2]MESA+ Institute for Nanotechnology, University of Twente, Enschede, The Netherlands.
[3] Advanced Quantum Architecture Lab (AQUA), EPFL, Neuchâtel, Switzerland.

Abstract— For on-chip data communication with galvanic isolation, a monolithically integrated optocoupler is strongly desired. For this purpose, silicon (Si) avalanche mode LEDs (AMLEDs) offer a great potential. However such AMLEDs have a relatively low internal quantum efficiency (IQE) and high power consumption. For the first time, in this work, data communication in a monolithically integrated optocoupler is experimentally demonstrated. The novelty of this work is the use of highly sensitive single-photon avalanche diodes (SPADs) for photo-detection to compensate for the low IQE of AMLEDs. We investigated our optocoupler realized in a standard 140 nm CMOS SOI technology, without post-processing, for various LED designs and points of operation. The power consumption of the AMLEDs is minimized through a novel AMLED design and employment of a low power LED driver circuit. The advantages of AMLEDs over forward biased Si LEDs are also demonstrated. For the best AMLED design, the achievable data rate is few Mbps and the energy consumption a few nJ/bit. The active area of the proposed systems is < 0.01 mm^2.

I. Introduction

Many smart power applications require data communication between different voltage domains with galvanic isolation. Currently this is achieved using inductive isolators, capacitive isolators or discrete optocouplers [1]. Such isolators have various disadvantages: they are large in size, source of extra complexity and costly. In addition, some are also sensitive to external interferences. Conversely, when monolithically integrated, optocouplers in standard CMOS technologies are largely free of such issues. Communication by means of light is also attractive because light is robust to external electrical or magnetic interferences.

So far, optocouplers have not been monolithically integrated in standard CMOS because of the poor overlap between the emission spectrum of a (traditional) forward biased Si LED and the responsivity of a Si photodetector (PD) (Fig. 1a). Therefore, in most discrete optocouplers, LEDs are designed using III-V compounds e.g. GaAs [2] and PDs are typically implemented using Si. This heterogeneous integration makes fabrication more complicated and thereby expensive.

In avalanche mode, Si AMLEDs have an emission spectrum in the visible wavelength region [3]; this spectrum has a significant overlap with the responsivity of Si PDs (Fig. 1b). The spectrum of AMLEDs is due to the radiative recombination of high-energy carriers during avalanche breakdown [3]. However, AMLEDs have a poor IQE ($\sim 10^{-5}$) [4]. Also, systems implemented with AMLEDs can be power hungry because of the avalanche operation under high voltage [4,5].

The significant development of SPADs in CMOS [6] opens new doors for monolithic integration of optocouplers. SPADs are highly sensitive PDs and can be used to compensate for the poor IQE of AMLEDs. SPADs also have a count rate of up to tens of Mbps, which is sufficient for many smart power applications. For the first time, we report the integration of Si AMLEDs-SPADs for optocoupling applications (Fig. 2).

It was shown that the breakdown voltage (V_{BR}) of AMLEDs can be reduced down to 6 V without degrading their IQE [7]. In this work, we have designed novel AMLED structures to reduce their V_{BR} for low power consumption and then we employed these AMLEDs for optocoupling focusing on power consumption, speed and bit-error-rate (BER). The LEDs and SPADs were designed in a 140 nm CMOS SOI technology; the LED driver and SPAD quenching-and-recharge circuits (QRCs) were implemented externally using off-the-shelf components. The proposed optocoupler (LEDs and SPADs) occupy much less area (~ 0.008 mm^2) compared to inductors (~ 1.9 mm^2 in [8]) and capacitors (~ 1.6 mm^2 in [9]) of inductive and capacitive isolators respectively. The LED driver and QRCs can be integrated on-chip at relatively low area requirements [4,6]. The results suggest a great potential of the AMLED-SPAD combination for monolithic optocoupling applications.

II. Experimental Devices

A. AMLEDs and SPAD designs

Fig. 3 shows schematic cross-sections of the two LEDs (A1,A2) and the SPAD (S1). A1 is a vertical n+-pwell diode. For optimized light emission with a minimal V_{BR}, A2 was implemented using closely spaced p+-n+ layers in an n-well. To reduce the dark count rate (DCR) of S1, the depletion region was isolated from the shallow trench isolation using a lowly doped guard ring [10]. The galvanic isolation was implemented using medium trench isolation (MTI); the MTI in this technology can provide an isolation of > 100 V [11]. Fig. 4 shows the micrograph of the devices A1-S1 and A2-S1. A "C"-shaped layout of A2 was designed to increase the coupling efficiency between A2 and S1.

Fig. 5 shows the reverse *I-V* characteristics of the devices. The high leakage current for A2 is most likely because of the presence of band-to-band tunneling. This was verified by the temperature dependence of the V_{BR}. For A2, V_{BR} was practically constant between $0 - 75$ °C: possibly because the negative temperature dependency of Zener breakdown counteracts the

978-1-7281-1988-5/18 $31.00 © 2018 IEEE

positive temperature dependency of avalanche breakdown [12]. Fig. 6 shows the AMLED light emission profiles and spectra.

The DCR of the SPADs were measured using a passive QRC. Fig. 7 shows the dependence of the DCR on the excess bias voltage V_{EX} (the excess voltage at which the SPAD is biased above breakdown) for different temperatures. The inset also shows the DCR measurement setup schematics. Fig. 8 shows the photon detection probability (PDP) of these SPADs, indicating their higher sensitivity in the visible wavelengths.

S1 was biased above breakdown and the avalanche firing rate (AFR) of S1 was measured when the corresponding LED was biased at the same DC current level in avalanche mode (AM) or in forward mode (FM), *i.e.* an FMLED. In dark conditions, AFR is equivalent to DCR. For the AM operation, the measured AFR was ~ 4x higher than that for the FM operation (Fig. 9). Despite that the IQE for an FMLED is about two orders of magnitude higher [13] than that of an AMLED, Fig. 9 demonstrates better spectral overlap for AMLED-SPAD than that for FMLED-SPAD. Fig. 10 shows the DC coupling between the A1-S1 and A2-S1 when S1 was operated as a "conventional" photodiode. A higher coupling efficiency (η_{CE}) is observed for A2-S1 than for A1-S1.

B. LED driver circuit schematics

Fig. 11 describes the equivalent circuit of our LED driver. It is a self-quenched driver circuit that limits the amount of charge-per-bit (Q_b) and energy-per-bit (E_b) through the LED. For more details about this driver circuit, please refer to [4] in which we present a fully integrated version.

C. Data modulation and BER definition

For SPADs, the data are encoded in their timing response [14]. Therefore, an appropriate modulation scheme for data communication using SPADs is a pulse position modulation (PPM) [14]. Fig. 12 shows the schematics of the LED driver control signals IN and RST for this modulation scheme. A PRBS data sequence with a length of 2^{10} bits was generated and the performance of the optical links was measured in response to this data sequence across all LEDs and operating conditions.

The received data is decoded from the timing response of the SPAD: if the SPAD is triggered only during the first half of the bit duration, data bit is demodulated as a "0", and if the SPAD is triggered only during the last half of the bit duration, data bit is demodulated as "1". For all other cases, data cannot be reliably decoded and those cases were counted as bit-error (Note: BER > 0.5 is possible as per this definition). Using the demodulated data, the BER of the link was estimated.

III. OPTICAL LINK PERFORMANCE

Fig. 13 shows an example of the measured transients of the control signals of the driver circuit, I_{AMLED} and output of QRC (V_{OUT}) at the specified operating conditions.

Figs. 14 and 15 show the estimated BER of the system for different operating conditions when the LEDs were operated in FM and AM respectively. Both in AM and FM, A2-S1 show a lower BER than A1-S1 because of the higher η_{CE} (see Fig. 10). At higher f_S, T_{ON} reduces (Fig. 11) and therefore Q_b decreases [4]; a lower Q_b reduces the amount of photons per bit (P_b) and thereby increases the BER.

In AM, a much lower BER is obtained compared to FM operation. At higher f_S, A2 shows a lower BER also because of its lower resistance (~ 25 Ω) than A1 (~ 850 Ω): for a fixed T_{ON}, a low resistance results in a high Q_b and consequently a high P_b, therefore a low BER [4]. Fig. 16 shows the measured BER of the optical link as the E_b through the LEDs was varied for several f_S in FM and AM. In AM, A2 requires ~ 2x lower E_b than A1 for a similar BER which demonstrates the advantages of our low V_{BR} AMLED design. The bare BER performance measured here can be improved using (e.g. Hamming) error-correction techniques, requiring a limited number of (low voltage) logic gates.

These results demonstrate that SPADs combined with AMLEDs are attractive for monolithically integrated optocouplers. The performance of the system has been measured up to 1 Mbps. The AMLEDs are expected to reach a high modulation speed; 10 Mbps using AMLEDs in this technology was shown in [4]. The designed SPADs in this technology require a deadtime of about 100 ns for a low afterpulsing probability [15]. This would limit the speed of SPADs up to ~ 10 Mbps. Therefore, the maximum expected single channel data rate of the proposed system is ~ 10 Mbps.

IV. CONCLUSIONS

In this work, for the first time, we have successfully demonstrated data communication using a monolithically integrated optical link in a standard CMOS technology without any post processing. The optical link was implemented using Si AMLEDs and SPADs. The advantages of AMLEDs over forward biased Si LEDs were shown. The coupling efficiency of LEDs with Si PDs was improved by a novel AMLED design. Poor quantum efficiency of AMLEDs was compensated by the use of highly sensitive SPADs. The energy consumption of the system was minimized by the design of novel AMLEDs and low power driver circuits. The results show strong potential of AMLEDs-SPADs combination for monolithic optical links.

ACKNOWLEDGMENT

We acknowledge Dr. D. Dochev, M. Swanenberg, Prof. dr. P.G. Steeneken and Henk de Vries for support and discussions, NXP semiconductors for silicon donation and NWO-TTW for funding (project 12835).

REFERENCES

[1]. K. Gingerich *et al.*, Appl. Report SLLA198, *Texas Instruments*, 2006.
[2]. IL4208 Optocoupler, Product datasheet, *Vishay Semiconductors*.
[3]. R. Newman, *Phys. Rev.*, **100**(2), 1955.
[4]. V. Agarwal *et al.*, *Opt. Express*, **25**(15), 2017.
[5]. N. Lodha *et al.*, *Proc. IISW*, 2013.
[6]. A. Rochas *et al.*, *Rev. Sci. Instrum.*, **74**(7), 2003.
[7]. S. Dutta *et al.*, *IEEE EDL*, **38**(7), 2017.
[8]. P. Lombardo *et al.*, *IEEE (ISSCC)*, pp. 300-301, 2016.
[9]. Y. Moghe *et al.*, *2012 IEEE SOI Conference*, pp. 1-2, 2012.
[10]. C. Veerappan *et al.*, *IEEE J. Sel. Top Q. Elec.*, **20**(6), 2016.
[11]. P. Wessels *et al.*, *Solid-state electronics*, **51**(2), 2007.
[12]. S.M. Sze and K. K. Ng, "Physics of semiconductor devices", Wiley, 2007.
[13]. V. Puliyankot *et al.*, *IEEE Trans. Elec. Dev.*, **59**(1), 2012.
[14]. C. Favi *et al.*, *Proc. 45th ACM/IEEE Des. Aut. Conf.*, pp. 343-344, 2008.
[15]. M.-J. Lee *et al.*, *Opt. Express*, **23**(10), 2015.

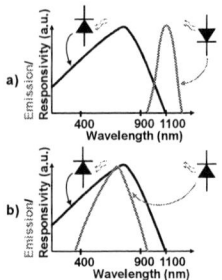

Fig. 1. Schematic illustration of a relatively (a) poor overlap between the emission spectrum of a forward biased Si LED (FMLED) and the responsivity of a Si PD. (b) good overlap between the emission spectrum of a Si avalanche mode LED (AMLED) and a Si PD. Sketches are derived using [4,7,12].

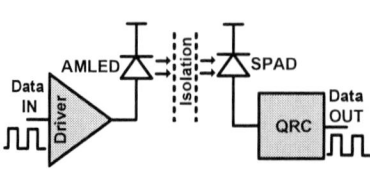

Fig. 2. Proposed optical link comprising of AMLEDs and SPADs. The LED driver circuit drives the LED in response to the incoming data to transmit photons across the isolation barrier. Some of these transmitted photons are detected across the barrier by the SPADs with quenching-and-recharge circuits (QRCs). The data is demodulated from the output of QRCs.

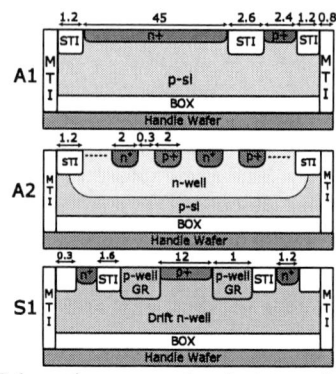

Fig. 3. Schematic cross-section (not to scale) of LEDs (A1, A2) and SPAD (S1); all dimensions in µm. Medium Trench Isolation (MTI) layers provide galvanic isolation between LEDs/SPADs.

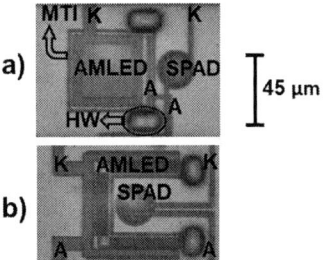

Fig. 4. Micrograph of (a) A1-S1 (b) A2-S1. "K" and "A" indicate cathode and anode respectively. Closest distance between the active areas of the LED and the SPAD in both cases is 10 µm. The "C" shaped layout of A2 was chosen to increase coupling efficiency (η_{CE}) between LED and PD (Fig. 10). A circular SPAD is used to reduce edge breakdown. "HW" is handle wafer contact, acting as heat sink.

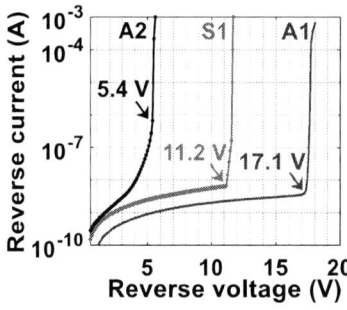

Fig. 5. Reverse I-V characteristics of the devices measured using an Agilent B2901A with 1 s integration time. V_{BR} (indicated) is defined as the voltage at which I_R starts to sharply increase. V_{BR} for A1 had a temperature coefficient of ~16 mV/K whereas V_{BR} for A2 showed negligible temperature dependency. The temperature coefficient of V_{BR} for S1 was measured to be ~ 8 mV/K.

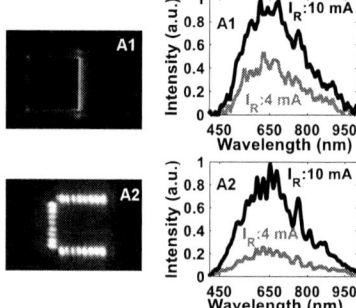

Fig. 6. (Left) Visible light emission profiles from A1 and A2 (on same scale as Fig. 4). AMLEDs were biased at a constant reverse DC current of 10 mA and the images were captured using a Nikon D3100 camera using 30 s integration time. For A1, light is emitted mostly from the edge closest to the p+ region due to high field and current crowding [4]. For A2, light is emitted from the regions between the n+ and p+ regions, indicating a lateral breakdown. (Right) Normalized emission spectra of the two AMLEDs for two I_R values.

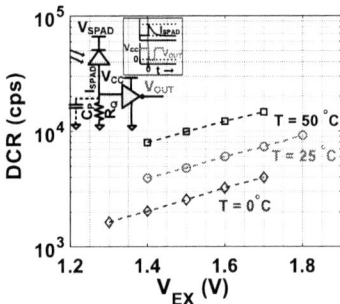

Fig. 7. Measured DCR for S1 (12 µm diameter) as a function of $V_{EX} = (V_{SPAD} - V_{BR})$ for different temperatures. At lower temperatures, DCR is reduced due to reduced thermal generation of dark carriers. Inset depicts the schematic of the measurement setup and illustrative transient waveforms. A passive QRC is used with a quenching resistance (R_Q) of 20 kΩ and V_{CC} = 2 V.

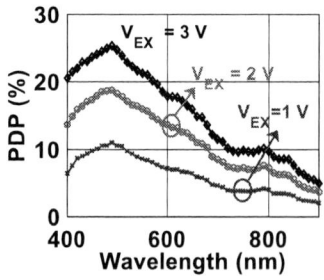

Fig. 8. Photon detection probability (PDP) of SPADs as a function of wavelength of incident photons for different SPAD excess bias [15]. A high sensitivity for photons in the visible spectrum can be observed.

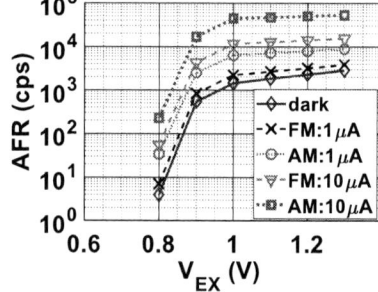

Fig. 9. Avalanche firing rate (AFR) as a function of V_{EX} when S1 was biased above breakdown and illuminated with the integrated Si LED (A1). When A1 was biased using a reverse current in avalanche mode (AM), a higher AFR was obtained compared to the case when A1 was biased using a same value of forward mode (FM) DC current. In dark conditions, AFR is equivalent to DCR.

978-1-7281-1988-5/18 $31.00 © 2018 IEEE

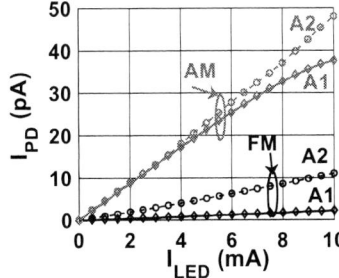

Fig. 10. DC coupling: S1 was operated as a "conventional" PD and its short circuit current (I_{PD}, which is reverse current at $V_R = 0$) was measured when the corresponding LEDs (A1/A2) were biased in AM and FM using DC currents (I_{LED}). A2-S1 shows a higher η_{CE} (= I_{PD}/I_{LED}) than A1-S1 (Fig. 4).

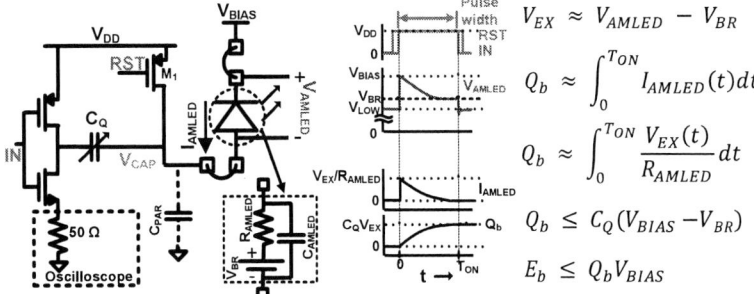

$$V_{EX} \approx V_{AMLED} - V_{BR}$$
$$Q_b \approx \int_0^{T_{ON}} I_{AMLED}(t)dt$$
$$Q_b \approx \int_0^{T_{ON}} \frac{V_{EX}(t)}{R_{AMLED}}dt$$
$$Q_b \leq C_Q(V_{BIAS} - V_{BR})$$
$$E_b \leq Q_b V_{BIAS}$$

Fig. 11. AMLED driver circuit with illustrative transients [4]. Initially, RST is set high, thus turning off the switch M1. Then IN is set high and initially voltage across an AMLED ($V_{AMLED} = V_{BIAS} - V_{CAP}$) increases to ~ V_{BIAS} (assuming quenching capacitance $C_Q \gg$ parasitic capacitance C_{PAR}). As $V_{AMLED} > V_{BR}$, avalanche is triggered and the avalanche current (I_{AMLED}) charges C_Q, thereby increasing V_{CAP} and reducing V_{AMLED}. As V_{AMLED} approaches V_{BR}, the avalanche is quenched. The total charge per bit through the AMLED (Q_b) is limited to $C_Q(V_{BIAS}-V_{BR})$. The energy-per-bit (E_b) is then limited to $Q_b V_{BIAS}$. I_{AMLED} is directly measured on a Keysight DSO-X 3024A oscilloscope. In our setup, Q_b and therefore E_b could be tuned by tuning C_Q and V_{BIAS} (with $V_{BIAS} - V_{DD} < V_{BR}$). A simplified model for an AMLED is shown in the right hand side dashed area [4].

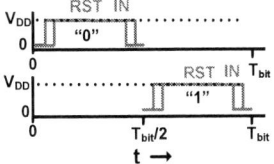

Fig. 12. Illustrative transient waveforms of the control signals IN and RST for bit "0" and for bit "1" when data were modulated using a two-level PPM. $T_{bit} = 1/f_s$ is the bit duration time where f_s is the data rate. For bit "0", photons are transmitted in the first half of T_{bit}, whereas for bit "1", photons are transmitted in second half.

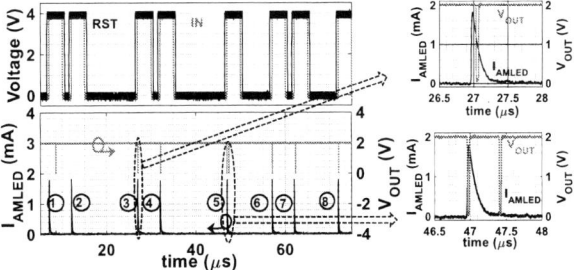

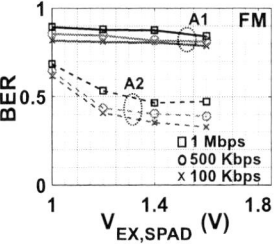

Fig. 13. For A1-S1, V_{BIAS} = 20 V, V_{DD} = 4 V, C_Q = 100 pF, f_s = 100 kbps, $V_{EX,SPAD}$ = 1 V: (Left) Example of measured transients of the control signals IN and RST (top); measured I_{AMLED} and SPAD QRC output (V_{OUT}) in response to those control signals (bottom). For marked bit 1, SPAD is triggered after some delay whereas SPAD is not triggered for bit 2 causing a bit-error. For bits 3,4,6,7 and 8, SPAD is triggered when AMLED is triggered. For bit 5, SPAD is triggered multiple times. (Right : a zoomed response of I_{AMLED} and V_{OUT} (scales enhanced for clarity)).

Fig. 14. Measured BER vs excess bias of SPAD ($V_{EX,SPAD}$) for different f_s with LEDs (A1,A2) operated in FM (driver circuit of Fig. 11 with cathode and anode of the LEDs interchanged). Operating conditions of the driver circuit: V_{BIAS} = 4 V, V_{DD} = 4 V and C_Q = 220 pF. A2-S1 shows a lower BER because of a higher η_{CE} (Figs. 4 and 10). BER vs E_b is shown in Fig. 16.

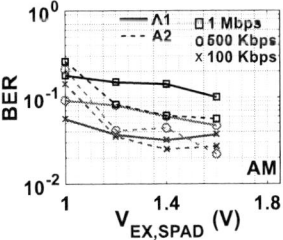

Fig. 15. Measured BER vs $V_{EX,SPAD}$ for different f_s when the LEDs are operated in AM. Note that y-axis is on a different scale than in Fig. 14. Operating conditions of the driver circuit (Fig. 11): V_{BIAS} = 20 V for A1 and 9 V for A2, V_{DD} = 4 V and C_Q = 220 pF. A much lower BER is obtained for AMLED-SPAD combination than for an FMLED-SPAD combination (Fig. 14). BER vs E_b is shown in Fig. 16.

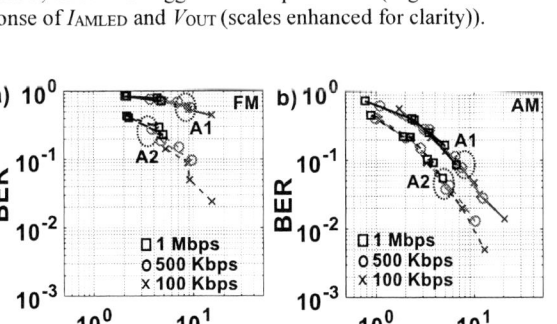

Fig. 16. Measured BER of the optical links as a function of E_b for (a) FM (b) AM LED operation; E_b was varied by varying C_Q and V_{BIAS} for several f_s. For both LEDs (A1,A2), BER reduces for higher E_b because of higher P_b. At similar E_b, A2 shows a lower BER due to a higher Q_b, P_b and η_{CE}. Similar BER were measured across all data rates for similar E_b. $V_{EX,SPAD}$ = 1.2 V for all measurements.

Fig. 17. Application example: IEEE logo (8-bit resolution) transmitted through our optical link A1-S1 (in AM) at 1 Mbps. From the received bits, image was reconstructed without any error-correction. Bits which could not be resolved from SPAD output were estimated as 0 or 1 randomly using uniform distribution.

978-1-7281-1988-5/18 $31.00 © 2018 IEEE

High Voltage Generation Using Deep Trench Isolated Photodiodes in a Back Side Illuminated Process

F. Kaklin[1,2], J. M. Raynor[2], R. K. Henderson[1]

[1]School of Engineering, Institute for Integrated Micro and Nano Systems, The University of Edinburgh, email: f.kaklin@sms.ed.ac.uk

[2]STMicroelectronics Imaging Division, Edinburgh, UK

Abstract—We demonstrate passive high voltage generation using photodiodes biased in the photovoltaic region of operation. The photodiodes are integrated in a 90nm back side illuminated (BSI) deep trench isolation (DTI) capable imaging process technology. Four equal area, DTI separated arrays of photodiodes are implemented on a single die and connected using on-chip transmission gates (TG). The TGs control interconnects between the four arrays, connecting them in series or in parallel. A series configuration successfully generates an open-circuit voltage of 1.98V at 1klux. The full array generates 423nW/mm^2 at 1klux of white LED illumination in series mode and 425nW/mm^2 in parallel mode. Peak conversion efficiency is estimated at 16.1%, at 5.7klux white LED illumination.

I. INTRODUCTION

Constant improvements in complementary metal-oxide semiconductor (CMOS) technology have allowed for designing devices with very low power consumption levels. Sensors operating with power consumption below 1μW have been reported [1]. This power level is comparable to that which can be generated using a silicon photovoltaic (PV) cell with an area of a few mm, opening up the possibility for mm^3 scale power supply independent devices. Since the underlying physical device used for both CMOS and photovoltaics is the same, the two could be fully integrated on the same die.

This high level of integration is challenging in part because of the relatively low voltage which is generated by on-chip photovoltaics, approximately 0.5V when in open-circuit and varying logarithmically with illumination. While specially designed circuits can operate with power supplies even lower than 0.5V [1], a higher voltage is often desirable. The PV output has been boosted using integrated charge-pumps with low start-up voltage in [2], up to 1.08V ad achieving peak 70% efficiency but also dropping significantly when the input power is low. At low illumination levels, for example in indoor conditions, an integrated PV and charge pump combination may operate far from its efficiency peak, wasting energy when it is already scarce. Passively boosting the output voltage instead may extend the operating range of a PV powered device into lower illumination levels.

Attempts at passive PV voltage boosting in standard CMOS processes by connecting multiple cells in series have highlighted the problem caused by the fact that all PVs share a common substrate [3], illustrated in Fig. 1. The deep diodes remain photoactive and load previous cells in the stack, drastically reducing the efficiency of such a structure, resulting in an expensive use of silicon area if the voltage is to be increased. These deep diodes are the most efficient junctions in a standard CMOS process, but generate a negative voltage with respect to a grounded bulk and require a voltage inverting charge pump step [2]. Complete physical and electrical isolation between consecutive cells has been achieved in [4] using air gaps to isolate areas of silicon from each other, but the 50μm wide trenches result in a significant overall silicon area loss which directly translates to a lower power generation per unit of chip area.

In this paper a deep trench isolation (DTI) processing step has been used to achieve isolation, leading to better area utilization, and power generation with higher efficiency at voltages beyond a single PV open circuit voltage (V_{OC}).

II. BSI INTEGRATED PHOTODIODES WITH DTI

The DTI capability of the chosen process technology can be used to electrically isolate distinct p regions of the substrate using an insulator. This eliminates the negative voltage generation problem from [2], and saves considerable area versus both [3] and [4]

DTI is performed by etching narrow trenches into the epitaxially grown layer of silicon, after which they are filled with an insulator. The entire wafer and is later flipped and background until individual p doped silicon regions are revealed before continuing with backside processing steps. The individual regions are doped to create p-n junctions with two floating terminals and no parasitic junction connections.

Because of BSI and the depth at which the depletion region is formed, these PVs have a higher quantum efficiency (QE) in the red and infrared region than PVs implemented in standard, front side illuminated (FSI) processes. Additionally, metal routing does not shade the silicon giving a near ideal optical fill factor.

Test chips with four different PV structures were manufactured. Their cross sections are shown in Fig. 2. PV a) is a pepi/n+ junction b) is a pwell/n+ junction, c) has two pn junctions: a pepi/nwell, and nwell/p+, d) is formed from custom doping layers available for pixel design and also has two junctions.. The simulated electrostatic potentials developed inside are shown in Fig. 3.

The extra junctions create more depletions regions where electron-hole-pair separation can occur, potentially leading to charge collection and hence power generation.

978-1-7281-1988-5/18 $31.00 © 2018 IEEE

III. Test Chip Design

All four structures were laid out to a 22.4μm pitch and then implemented in larger 8×32 arrays. A single PV cell therefore consists of 256 individual diodes connected in parallel. Four of these cells were implemented per chip and connected together using transmission gates (TG). A block diagram of the chip is shown in Fig. 4. The TGs control the internal routing and allow for reconfiguring the cells into a 4-in-series or 4-in-parallel mode with the circuit in Fig. 5. The voltage generated at each cell is routed to an analog output pad. The photovoltaic area of the chip is 0.51mm^2. Chips were wire bonded in 68CPGA packages with a glass lid. An external supply is used to bias the padring electrostatic discharge (ESD) protection diodes and control the transmission gates. A micrograph of the chip is in Fig. 6.

IV. Results

Figure 7 shows the open circuit voltages developed at each PV cell in the stack across varying illumination for the structure a) from Fig. 2. The voltage developed across a single cell varies logarithmically over the full illumination range, hence this variation is multiplied by the number of cells in the stack at its output V4. The plot shows that over 2.2V can be generated using 4 cells at 10klux from a 2700K LED light source. This is sufficient to operate thick oxide digital gates available in this process technology.

Figure 8 shows the IV and generated power curves of structure a) when in series and parallel mode. Measurements were taken at 1klux in an office setting with a white LED source. For a given photovoltaic area and illumination, the peak generated power is approximately constant, regardless if the PVs are connected in parallel or in series.

Figures 9 and 10 show the IV and power generation curves over increasing illumination levels for structure a). As expected, the short circuit current increases linearly and the peak power point shifts to higher voltages with higher light level. Results obtained for remaining structures are very similar and have been omitted from these figures for clarity. The short circuit currents, open circuit voltages and peak generated powers are instead summarized for all four structures in Table 1. PV structures a) and d) generate the highest open circuit voltage. At a higher light level of 5.7klux, these two structures also generate the most power − around 50nW more (3-4%) than structures c) or b). At this light level the parallel connection mode generates 0.5-1% more power for each test structure. The power loss in the series mode is likely because of the extra resistance of the transmission gates used for interconnects and the resistance of the PVs themselves. Power generation over increasing illumination is shown in Fig. 11. The differences are small and magnified in the inset graph.

The use of a second junction in structures c) and d) has not resulted in more power generation against single junction structures a) and b). The poorly performing structures b) and c) are more highly doped and the resulting higher recombination rates may reduce their power generation.

A. Estimating Efficiency

Many previous published results on integrated PVs have used a figure of merit (FoM) which mixed physical and photometric units: watts per lux per square mm [5]. This unit is inadequate for a fair comparison because the spectral sensitivity of silicon is unrelated to that of the human eye. An artificial, apparently white light source will likely have a spectral luminous efficacy of 300±50lm/W [6]. Knowing this, we estimate the irradiance at the sensor level using an AMPROBE LM-1000 light meter and dividing the reading in lux by the lumens per watt estimate, thereby obtaining a suitable FoM: photovoltaic conversion efficiency. Using this method we estimate a peak measured efficiency of 16.1±2.8% at 5.7klux. We also note that the conversion efficiency is not constant across illumination, and is lower at low light levels, as shown in Fig. 12.

V. Discussion

The series connected photovoltaics passively increase the output voltage, removing the need to implement voltage boosting circuits which have inherent inefficiencies that may limit the lower operation bound for a small autonomous device using integrated PVs. Powering these devices in low light environments remains a challenge primarily because of the low irradiance, but it is made even more difficult by the reduced efficiency in such conditions. Given that indoor lighting is in the range of 50-1000 lux, extra effort should be placed on maintaining high efficiency in this region.

VI. Conclusion

The chips presented here implement a power generation method suitable for a fully integrated energy harvesting mm^3 scale sensor. The DTI process step addresses the area inefficiencies present in previous passive voltage boosting attempts, while CMOS compatibility allows for dynamic PV array reconfiguration trading off voltage versus current with minimal peak power penalty. Differences in power generation across the tested PV structures have been minimal suggesting that any of them are suitable for implementation in an autonomous energy harvesting device.

References

[1] S. Hanson, Z. Foo, D. Blaauw and D. Sylvester, "A 0.5 V Sub-Microwatt CMOS Image Sensor With Pulse-Width Modulation Read-Out," in *IEEE J. Solid-State Circuits*, vol. 45, no. 4, pp. 759-767, April 2010.

[2] Z. Chen, M. Law, P. Mak and R. P. Martins, "A Single-Chip Solar Energy Harvesting IC Using Integrated Photodiodes for Biomedical Implant Applications," in *IEEE Trans. Biomed. Circuits Syst.*, vol. 11, no. 1, pp. 44-53, Feb. 2017.

[3] M. K. Law and A. Bermak, "High-Voltage Generation With Stacked Photodiodes in Standard CMOS Process," in *IEEE Electron Device Lett.*, vol. 31, no. 12, pp. 1425-1427, Dec. 2010.

[4] Y. Hung, Y. Cheng, M. Cai, C. Lu and H. Su, "High-Voltage 12.5-V Backside-Illuminated CMOS Photovoltaic Mini-Modules," in *IEEE J. Electron Devices Soc.*, vol. 6, pp. 135-138, 2018.

[5] S. Park, K. Lee, H. Song and E. Yoon, "Simultaneous Imaging and Energy Harvesting in CMOS Image Sensor Pixels," in *IEEE Electron Device Lett.*, vol. 39, no. 4, pp. 532-535, April 2018.

[6] T. W. Murphy, "Maximum spectral luminous efficacy of white light," *J. Appl. Phys.*, vol. 111, no. 10, 2012.

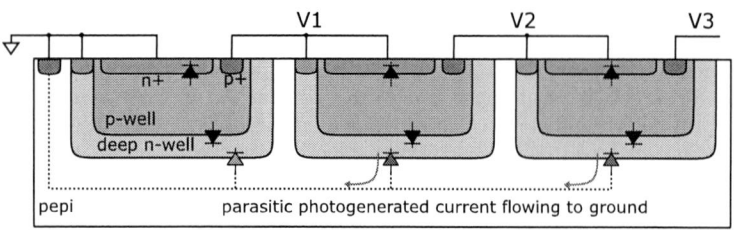

Fig. 1. Conventional diodes in a triple-well CMOS process connected in series. The deep nwell to pepi diodes indicate junctions which are considered parasitic.

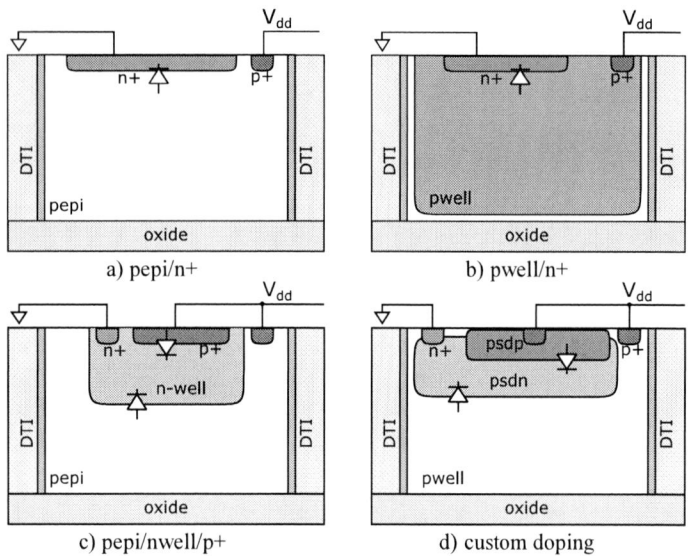

a) pepi/n+   b) pwell/n+

c) pepi/nwell/p+   d) custom doping

Fig. 2. Cross section diagrams of the DTI PVs. Light enters the structure through the bottom oxide. Each structure is implemented on a separate die.

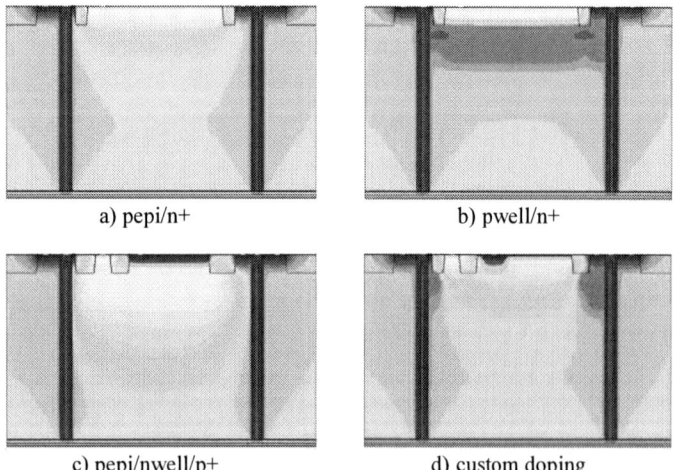

a) pepi/n+   b) pwell/n+

c) pepi/nwell/p+   d) custom doping

Fig. 3. TCAD simulation results of the electrostatic potentials generated inside the PVs. Labels a)-d) correspond to diagrams from Fig. 2. Dark blue indicates a low electrostatic potential; yellow indicates a high electrostatic potential.

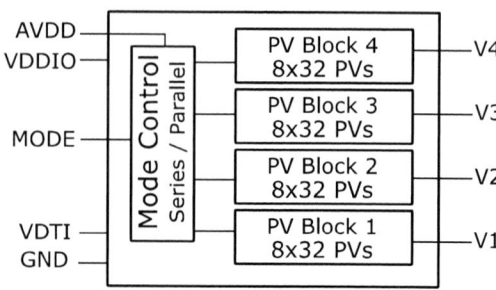

Fig. 4. Chip block diagram.

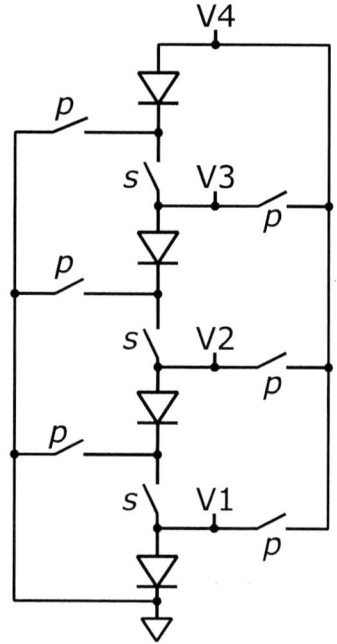

Fig. 5. Circuit diagram showing the switching interconnects between PVs. Switches implemented using transmission gates and controlled using the MODE signal. Switches marked *s* are closed to enable series mode; switches marked *p* are closed to enable parallel mode. Nodes V<1-4> are routed to pads.

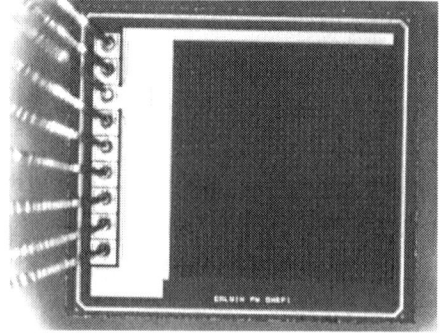

Fig. 6. Chip micrograph with structure a) PVs. All 4 chips were laid out and bonded identically. Die dimensions 1.18mm×1.18mm Array dimension 716μm×716μm

978-1-7281-1988-5/18 $31.00 © 2018 IEEE   745

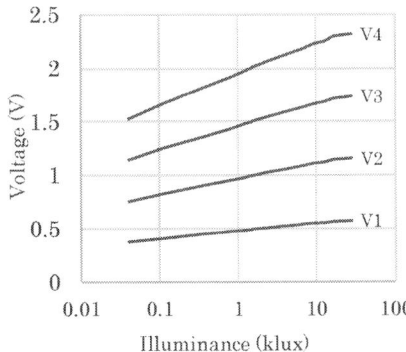

Fig. 7. Open circuit voltages generated at each cell of the stack under 2700K LED light. Structure a) from Fig.2 .

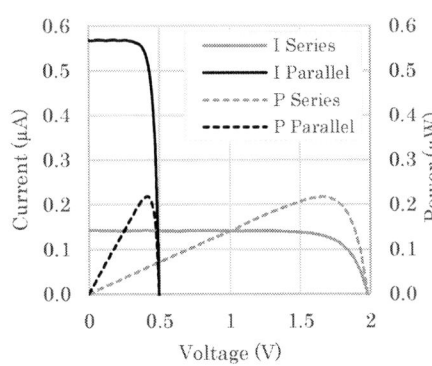

Fig. 8. Current-voltage and power-voltage curves for the structure in a) from Fig. 2. Taken at 1klux illumination from white LED source.

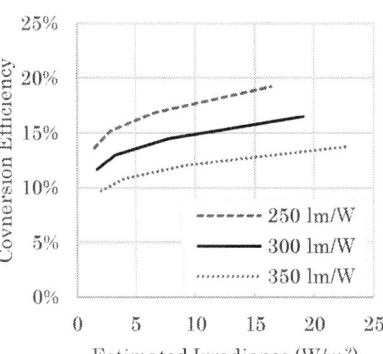

Fig. 12. Estimated photovoltaic conversion efficiency under white LED light using the spectral luminous efficacy of white light. True efficiency will be bounded by these curves.

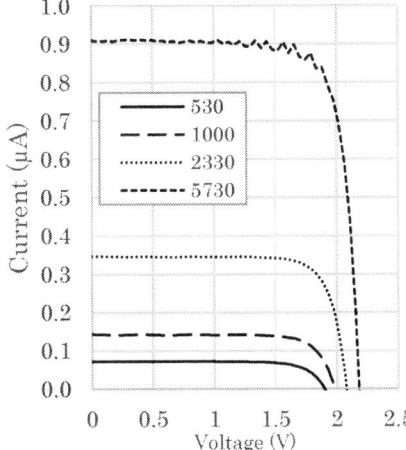

Fig. 9. IV curves for structure a) under increasing illumination in lux (see legend) of white LED light.

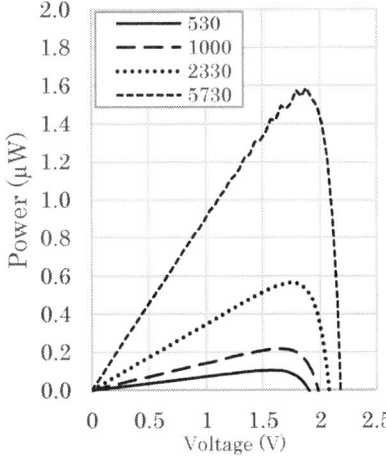

Fig. 10. Power generation in structure a) under increasing illumination in lux (see legend) of white LED light.

Illum. (lux)	530	1000	2330	5730
a) pepi/n+				
V_{OC} (V)	1.902	1.982	2.080	2.184
I_{SC} (nA)	72	143	347	909
P_{MAX} (µW)	0.103	0.217	0.566	1.59
b) pwell/n+				
V_{OC} (V)	1.863	1.939	2.039	2.139
I_{SC} (µA)	59	117	294	829
P_{MAX} (µW)	0.102	0.213	0.553	1.54
c) pepi/nwell/p+				
V_{OC} (V)	1.824	1.899	1.994	2.092
I_{SC} (µA)	72	142	344	922
P_{MAX} (µW)	0.100	0.209	0.542	1.54
d) custom doping				
V_{OC} (V)	1.891	1.975	2.079	2.192
I_{SC} (µA)	72	143	346	919
P_{MAX} (µW)	0.104	0.219	0.567	1.60

Table 1. Open circuit votlage (V_{OC}), short citcuit current (I_{SC}), and peak power generated (P_{MAX}) by each structure when operating in series.

Fig. 11. Peak power generated by the tested structures vs. illuminance. Inset is a magnification of the high end of the range.

Parameter	[2]	[3]	[4]	This work
Process	0.18µm CMOS	0.35µm CMOS	CMOS with trench etch	90nm BSI
PV Type	Pwell/n+ ‖ pwell/dnwell ‖ dnwell/psub	nwell/p+	BSI	DTI - type a) pepi/n+
Array Area (mm²)	1.54	0.0322	4	0.514
Optical Fill Factor (%)	-	-	6.25	98
PV Active Area (mm²)	1.3	0.0322	0.25	0.504
Illumination Source	Halogen	Halogen	Laser 980nm	White LED
Illumination (klux)	N/A	N/A	-	5.7 klux
Spectral Luminous Efficacy (lm/W)	-	-	0	300±50
Irradiance (µW/mm²)	1130	63.7 (555nm)	6000	19.1
Generated Power (µW)	322	0.0012	115*	1.58
Voltage Boost	Inverting charge pump	Ratio stack 64:8:1	Stack ×25	Stack ×4
V_{OC} (V) (boosted)	0.53 (1.08)	1.3	12.5	2.2
Array Efficiency (%)	12.1 (post charge pump)	0.06	0.5	16.1±2.8
PV Efficiency (%)	21.9	2.1	8.1	16.4±2.8

Table 2. Comparison against published results.

978-1-7281-1988-5/18 $31.00 © 2018 IEEE

Through-silicon-trench in back-side-illuminated CMOS image sensors for the improvement of gate oxide long term performance

A. Vici[1], F. Russo[2], N. Lovisi[2], L. Latessa[2], A. Marchioni[2], A. Casella[2], F. Irrera[1]

[1]DIET, Sapienza University of Rome, Italy – email: fernanda.irrera@uniroma1.it
[2]LFoundry, a SMIC Company, Avezzano (Aq), Italy.

Abstract—To improve the gate oxide long term performance of MOSFETs in back side illuminated CMOS image sensors the wafer back is patterned with suitable through-silicon-trenches. We demonstrate that the reliability improvement is due to the annealing of the gate oxide border traps thanks to passivating chemical species carried by trenches.

I. INTRODUCTION

A CMOS image sensor (CIS) is an array of light sensitive pixels. Each pixel consists of a photodiode (PD) and several control MOSFETs. In front side illumination (FSI) (Fig. 1a), the light reaches the PD active region through multiple front layers. However, coupling of light with the front layers has severe drawbacks, which cause reduction in the maximum available photons and then quantum efficiency degradation, especially in the blue and UV region. So, a few decades ago the back side illumination (BSI) was proposed [1], which was set aside for long time, while today is becoming popular with pixel size downscaling. During BSI manufacturing the wafer is flipped upside down so that the metallization and passivation layers are located beyond the PD with respect to impinging light (Fig. 1b). However, flipped manufacturing introduces additional degradation of the transistors reliability. A common approach in the semiconductor industry is to open in the wafer back a suitable pattern of deep-trenches that, passing through the passivation layers and the silicon substrate, land in the inter-level metal dielectrics (through silicon trench, TST). TST are supposed to favor the flow of passivating specie (H, ^{2}H) directly to the silicon active area. *In this work, we start from the experimental evidence that in BSI the gate oxide exhibit a noticeable degradation of the time-to-breakdown test (TDDB [2]), performed on wafer-level-reliability dedicated structures respect to FSI ones of the same node (180 nm gate length, 6.8 nm oxide thickness). A systematic characterization of traps in n-MOSFETs is performed here, which demonstrates that border traps [2] are mainly responsible for the marginal reliability in BSI.* In fact, although the microstructure of border and bulk traps is approximately the same, the border traps influence more heavily the device operation. Their proximity to the channel transitor increases the probability that channel carriers tunnel to them [3-12]. *We demonstrate that a quite effective annealing of the border traps is obtained when the TST is present, with a successful recovery of the long term performance of the gate oxide.*

II. DEVICES UNDER TEST (DUT) AND PRELIMINARY CHARACTERIZATION

A schematic cross-section of the periphery (the region all around the pixel array) of a typical BSI chip is sketched in Fig.2. A through Silicon VIA (TSV) is etched through the final passivation stack and the silicon substrate and then filled with aluminium in order to contact from the back the metal pads built in the front-side portion of the process flow. A preliminary comparison between the electrical behaviour of two n-MOSFETs realized in FSI and BSI technology respectively, the latter without using a TST approach, is shown in Fig.3. In particular, Fig.3a shows the steady-state high-frequency CV curves Fig.3b shows instead the steady state J_g-V_g curves. Dashed red line refers to FSI, open triangles to BSI. As one can see, both capacitance and trap-assisted-tunnel current curves are the same in the two configurations. This outlines that *the oxide bulk traps are the same in BSI and FSI.* On the contrary, the time dependent dielectric breakdown (TDDB), evaluated at T=125°C reveals that the BSI features show a dramatically degraded long term gate oxide performance respect to FSI. In Tab.I the normalized electrical parameters averaged over 20 wafers are listed for three sensor configurations: the FSI, which is our normalization reference (first row, reported in arbitrary units), the BSI (second row) and the engineered BSI-T (third row), which will be discussed in the following.

III. RESULTS AND DISCUSSION

The result obtained with the preliminary characterization indicates that the n-MOSFETs in BSI and FSI exhibit similar short term performances, but the BSI dramatically suffers from early breakdown. In order to understand the nature of the defects which penalize so heavily the BSI respect to the FSI sensor, we made a systematic investigation using several electrical techniques. This was fundamental to characterize, validate and possibly optimize the through-silicon-trench patterning approach, with recovered long term performance. In the following, we will show the results for two BSI samples, realized with and without the TST patterning approach respectively. Hereafter, the sample realized with the TST will be called BSI-T. In Fig.2 the open trench lies outside the square box and is depicted with white color. The role of

978-1-7281-1988-5/18 $31.00 © 2018 IEEE

this through-silicon-trench is thought to favor the transmission of the H and ^{2}H specie towards the silicon-oxide interface during the final annealing step, enhancing the trap recovery.

A. Charge pumping (CP)

Spectroscopic-CP (S-CP) measurements were performed on fresh n-MOSFETs in BSI and BSI-T, progressively increasing the gate bias from -3.5 V up to 0.5 V and superimposing a trapezoidal pulses train of 2.5 V amplitude, with rise and fall time in the range 8 ns-10 ms. This technique allows deriving information on the energy distribution of interface trap density [13]. The distributions obtained on the basic BSI and on the engineered BSI-T are sketched in Fig.4. As a result, we can say that the interface trap density is very similar in BSI and BSI-T, and its value lies between 10^{10}-10^{11} cm^{-3} around mid-gap and increases moving toward the band edge. Then, we performed frequency resolved charge pumping (f-CP), which is a conventional CP where the frequency of the train pulses is varied. The f-CP is characterized by a cut-off frequency f_0, beyond which only the fast interface traps can respond and below which also slower border traps can give their contribution. On the basis of previous S-CP results, we expect the same contribution from the interface traps. This is verified in Fig.5, where f_0 is around 3 KHz: above f_0, the curves are similar, but *below f_0 the basic BSI exhibits a noticeable increase corresponding to the progressively increasing contribution from border traps, which is absent in the engineered BSI-T.* Using literature models of f-CP [14], one can obtain the spatial distribution of the slow border traps from the low frequency behavior. For the basic BSI, it is depicted in Fig. 6, where the trap density increases exponentially moving inside the oxide. As an example, it comes out that the density of border traps 1nm far from the interface is slightly higher than 10^{11} cm^{-3}.

B. Flicker Noise

Flicker noise measurements were performed increasing the gate voltage of the n-MOSFETs from the threshold condition on. The normalized $S(I_d)/I_d^2$ spectrum is reported in Fig.7a and Fig.7b for the BSI and BSI-T, respectively. The 1/f trend is found (dashed lines). In Fig.7 is important to observe that the engineered BSI-T shows an overall noise lower than the basic BSI one (the same area of plot is comprised within the two dashed lines). In particular, in Fig. 8, the noise curves are drawn as function of the overdrive, from 100 mV up to 800 mV, and are compared for three different frequencies. As one can see, the difference between BSI and BSI-T is always evident around the threshold condition, when the channel charge is small, while at higher bias exactly the same noise is measured. Following the Hooge model of carrier mobility fluctuation [15], the I_d noise comes from the fluctuations of the carrier mobility through the variations of the scattering cross section entering the collision probability likely due to phonon number fluctuations. This gives a linear dependence of $S(I_d)/I_d^2$ on $1/I_d$. On the other hand, following the carrier number and correlated mobility fluctuations model [16] the noise results from the variations of the interfacial oxide charge after

dynamic trapping/detrapping of free carriers into slow oxide traps. To some extent, this gives a dependence of $S(I_d)/I_d^2$ on $(g_m/I_d)^2$. The measured data of $S(I_d)/I_d^2$ against I_d are reported with symbols in Fig.9 for BSI and BSI-T configurations. The solid line in the same figure represents the interpolation of the measured $(g_m/I_d)^2$ values. The dependence on the squared inverse is evident, *indicating a strong prevalence of a mechanism of trapping/detrapping in border traps.*

C. Random Telegraph Noise (RTN)

RTN measurements were performed on the n-MOSFETs measuring the I_d current in the bias condition V_{ds}=200 mV and V_{gs}-V_T=450 mV. The experiment lasted 5 seconds. An interval of 150 ms is shown in Fig. 10a for the basic BSI and in Fig.10b for the BSI-T. As one can see, in basic BSI there is a random switch between two current levels, namely $I_{d\text{-}high}$=4.625 μA and $I_{d\text{-}low}$=4.575 μA, with a gap ΔI_d= 50 nA, which corresponds to a carrier trapping of 3×10^{11} electrons/s, in agreement with the trap density at 1 nm far from the interface (Fig.6). On the contrary, in the BSI-T the current level remains at the value 3.725 mA and no current variation is appreciable on that scale (see Fig.10b). *This is a further evidence that in the engineered BSI-T border traps are recovered respect to basic BSI.* Getting deeper inside the RTN in BSI, the distributions of the currents is sketched in Fig.11. In the inset of the same figure, the time occurrence of the two current levels is plotted and the prevalence of $I_{d\text{-}high}$ is evident. In conclusion, we can affirm that $I_{d\text{-}high}$ is the current level with that bias and $I_{d\text{-}low}$ is an unstable level. The measurement was repeated with increasing the overdrive from 450 mV up to 600 mV, keeping V_{ds}=200 mV. The Lag plots reported in Fig.12 show two populated spots on the main diagonal, indicating the two states, while other points outside the diagonal correspond to transition between them. Increasing the overdrive, both the current levels grow keeping the gap unchanged, so that the metastability progressively seems to disappear, since ΔI_d= 50 nA becomes negligible in percentage respect to $I_{d\text{-}high}$ and $I_{d\text{-}low}$.

IV. CONCLUSIONS

In back-side illuminated CMOS image sensors border traps penalize the gate oxide long term performance respect to FSI. Due to their proximity to the interface, border traps impact noticeably the device operation thanks to the high probability that a channel carrier undergoes trapping and detrapping. We reported about the characterization of border traps for n-MOSFETs in BSI, using f-CP, 1/f noise and RTN technique. The role played by slow border traps in the BSI device operation was enlightened. The TST opening approach, supposed to promote the transport of annealing specie as H and ^{2}H toward the silicon/oxide interface, was demonstrated to favor a recovery of the long term performance. Samples processed with the TST patterning were measured with the same electrical techniques used for the basic (without TST) ones. As a result, contrary to what was observed in basic samples, they did not exhibit any appreciable contribution from border traps. Finally, as a confirm, in Tab.I it was added the excellent result on the TDDB test for the engineered BSI-T measured in the same conditions as BSI and FSI.

REFERENCES

[1] A. Lahav, A. Fenigstein, A. Strum in High Performance Silicon Imaging Cap.4, Editor: Daniel Durini, Woodhead Publishing, 2014, Pag 98-123.

[2] JEDEC Standard - JESD92 Procedure for Characterizing Time-Dependent Dielectric Breakdown of Ultra-Thin Gate Dielectrics

[3] C.E. Blat et al., J. Appl. Phys. 69 (1991) 1712–1720.

[4] D.M. Fleetwood et al., J. Appl. Phys. 73 (1993) 5058 5074.

[5] S. Ogawa et al, J. Appl. Phys. 77 (1995) 1137–1148.

[6] S. Ogawa et al. Phys. Rev. B 51 (1995) 4218–4230.

[7] L. Tsetseris et al., Appl. Phys. Lett. 86 (2005) 142103.

[8] A.E. Islam et al., IEEE Trans. Electron Devices 54 (2007) 2143–2154.

[9] T. Grasser et al., IEEE Trans. Electron Devices 58 (2011) 3652–3666.

[10] T. Grasser, Microelectron. Reliab. 52 (2012) 39–70.

[11] D.M. Fleetwood, Microelectronics Reliability 80 (2018) 266–277.

[12] D.M. Fleetwood, IEEE Trans. Nucl. Sci. 39 (1992) 269–271.

[13] Groeseneken, G. V., P. Heremans, and H. E. Maes, IEEE Transactions on Electron Devices 38.8 (1991): 1820-1831.

[14] Jakschik, Stefan, et al., IEEE Trans. on Electron Devices 51.12 (2004): 2252-2255.

[15] Hooge, F. N., Physica B+ C 83.1 (1976): 14-23.

[16] Ghibaudo G., T. Boutchacha. "Electrical noise and RTS fluctuations in advanced CMOS devices." Microelectron. Reliab. 42.4-5 (2002): 573-5

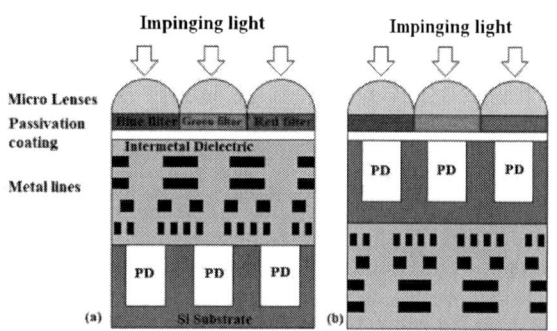

Fig.1. Sketch of (a) frontside and (b) backside illuminated sensor. In the BSI the wafer is flipped upside down.

Fig.2. Sketch of the DUT in BSI configuration with the dummy trench open by the side, to form the BSI-T.

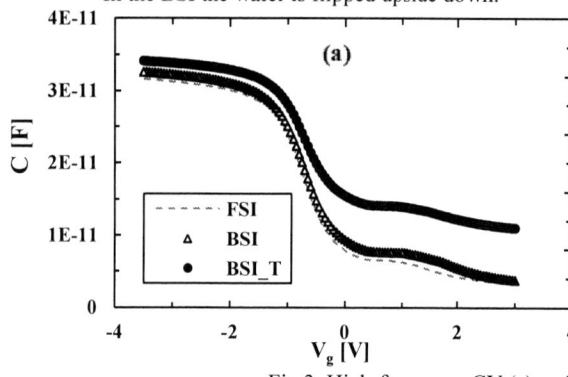

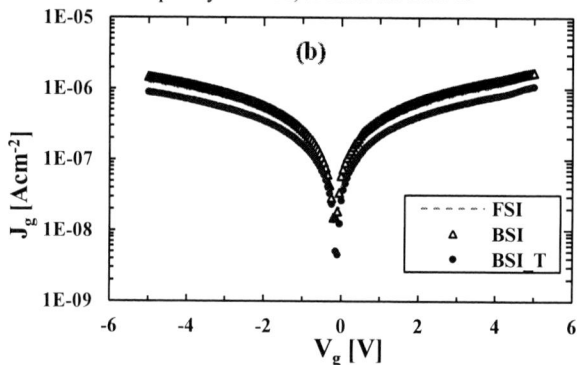

Fig.3. High-frequency CV (a) and I_g-V_g (b) curves for the three configurations.

	Norm. V_T (average over 20 wafers)	I_G leakage @0.45V (average over 20 wafers)	Norm. TDDB (average over 20 wafers)
FSI (A.U.)	1 (±2.1%)	1	1
BSI/FSI	1 (±2.6%)	1.12	10^{-4}
BSI-T/FSI	1 (±3.7%)	0.63	1

Tab. I. Short term and long term gate oxide electrical parameters, averaged over 20 wafers and normalized to the FSI values (reported in arbitrary units), for the three configurations.

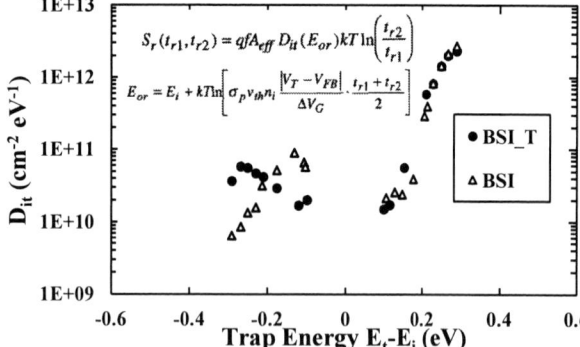
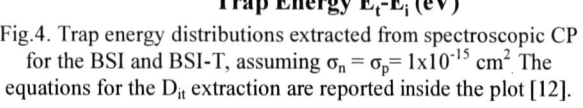

$$S_r(t_{r1}, t_{r2}) = qfA_{eff}\,D_{it}(E_{or})kT\ln\left(\frac{t_{r2}}{t_{r1}}\right)$$

$$E_{or} = E_i + kT\ln\left[\sigma_p v_{th} n_i \frac{|V_T - V_{FB}|}{\Delta V_G} \cdot \frac{t_{r1}+t_{r2}}{2}\right]$$

Fig.4. Trap energy distributions extracted from spectroscopic CP for the BSI and BSI-T, assuming $\sigma_n = \sigma_p = 1\times10^{-15}$ cm^2. The equations for the D_{it} extraction are reported inside the plot [12].

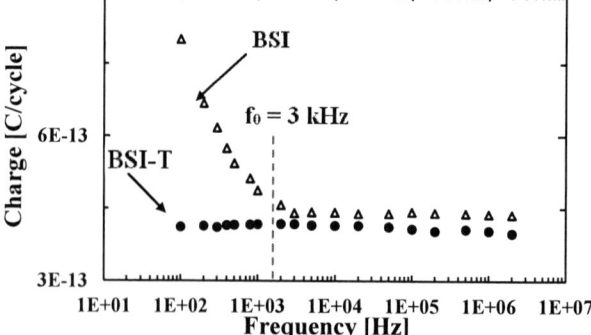

Fig.5. Charge Recombined per cycle vs. frequency extracted from frequency resolved CP, for the BSI and BSI-T. The increase of Q below f_0 in BSI is attributed to border traps [13].

978-1-7281-1988-5/18 $31.00 © 2018 IEEE

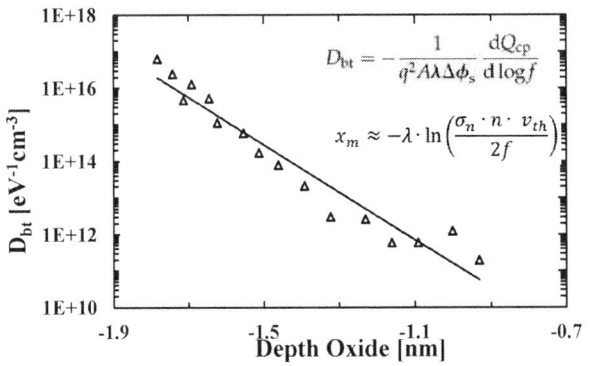

Fig.6. Border trap density in the oxide vs. the distance from the SiO_2/Si interface, assuming $\sigma_n = 1\times10^{-15}$ cm^2. The equations for the D_{bt} extraction are reported inside the plot [13].

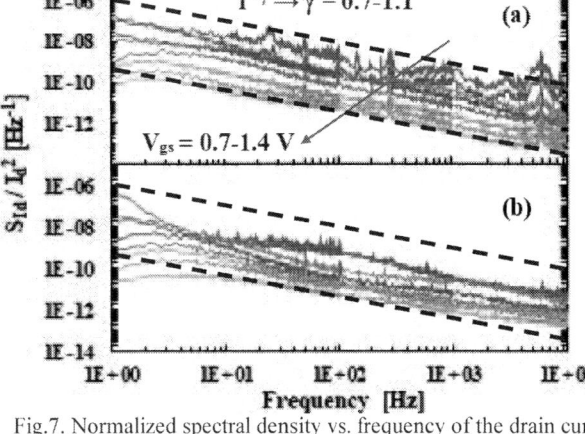

Fig.7. Normalized spectral density vs. frequency of the drain current noise measured varying the overdrive, for the BSI (a) and BSI-T (b) configurations.

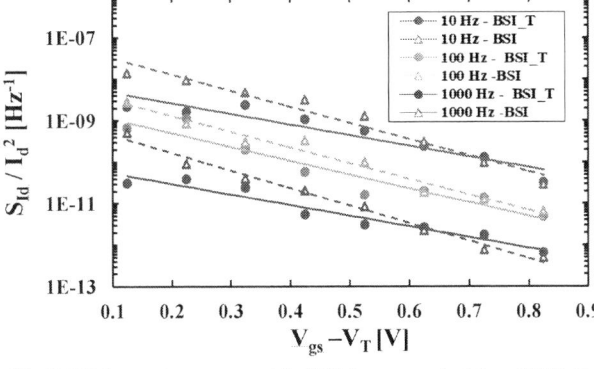

Fig.8. Flicker noise measured in BSI (open symbols) and BSI-T (closed symbols) as function of the overdrive at three different frequencies.

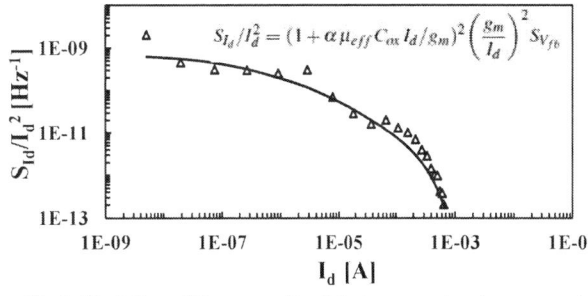

Fig.9. Variation of the normalized drain-current noise versus drain current for the BSI configuration. The quadratic dependence predicted in [15] is reported inside the plot.

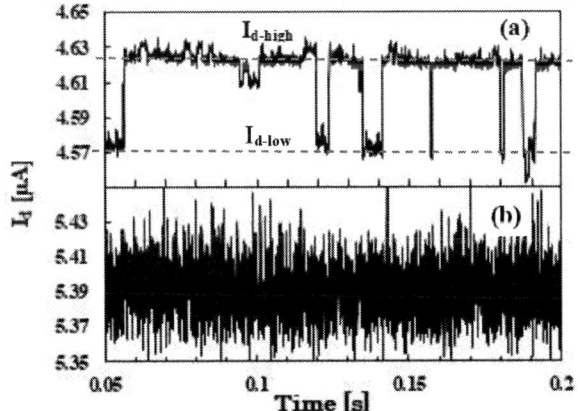

Fig.10. Zoom of the drain current RTN trace for the BSI (a) and BSI-T (b) measured at V_{ds}=200 mV with a 450 mV overdrive.

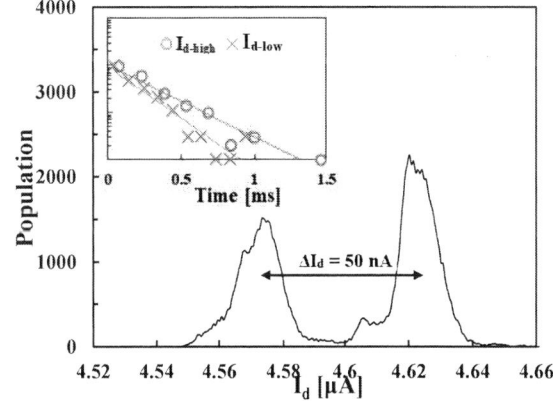

Fig.11. Current distributions derived from the RTN experiment. Inset: Time distributions of the drain current.

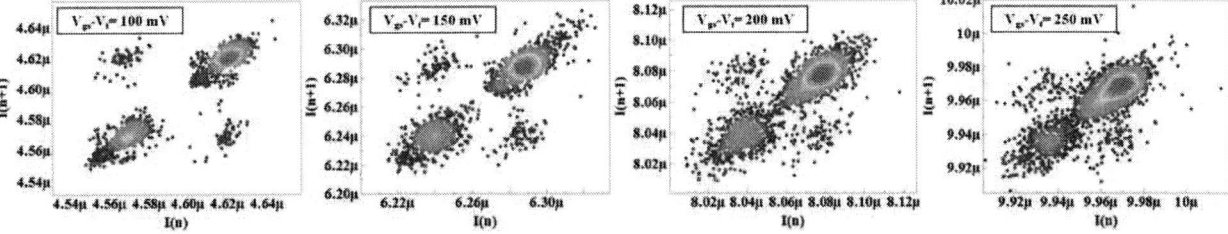

Fig.12. Time Lag Plots of the drain current obtained with the RTN experiment increasing the overdrive.

978-1-7281-1988-5/18 $31.00 © 2018 IEEE   750

High-Performance Germanium-on-Silicon Lock-in Pixels for Indirect Time-of-Flight Applications

N. Na*, S.-L. Cheng, H.-D. Liu, M.-J. Yang, C.-Y. Chen, H.-W. Chen, Y.-T. Chou, C.-T. Lin, W.-H. Liu, C.-F. Liang, C.-L. Chen, S.-W. Chu, B.-J. Chen, Y.-F. Lyu, and S.-L. Chen

Artilux Inc., Zhubei City, Hsinchu County, Taiwan ROC
* Email: rp@artiluxtech.com

Abstract—We investigate and demonstrate the first Ge-on-Si lock-in pixels for indirect time-of-flight measurements. Compared to conventional Si lock-in pixels, such novel Ge-on-Si lock-in pixels simultaneously maintain a high quantum efficiency and a high demodulation contrast at a higher operation frequency, which enable consistently superior depth accuracies for both indoor and outdoor scenarios. System performances are evaluated, and pixel quantum efficiencies are measured to be >85% and >46% at 940nm and 1550nm wavelengths, respectively, along with demodulation contrasts measured to be >0.81 at 300MHz. Our work may open up new routes to high-performance indirect time-of-flight sensors and imagers, as well as potential adoptions of eye-safe lasers (e.g. wavelengths > 1.4µm) for consumer electronics and photonics.

I. INTRODUCTION

1D depth sensors and 3D image sensors are vital to a variety of applications such as hand tracking, facial recognition, 3D model scanning, simultaneous localization and mapping for spatial navigations, surveillance, and light detection and ranging for autonomous vehicles, to name a few. There are at least four major techniques that are well studied in the literature, i.e., passive/active stereovision [1], structured light imaging [2], direct time-of-flight (TOF) sensing via single photon avalanche photodiodes and time-to-digital converters [3], and indirect TOF sensing via lock-in pixels [4]. In particular, the indirect TOF measurement features advantages such as direct acquisition of a depth information without additional computational algorithms, simple laser light projector without complex optical modules, and low power consumption with respect to pixel size/number scaling.

The principle of the indirect TOF measurement is essentially based on homodyne detection with radio-wave operation frequency f_o, in which the roles of a local oscillator/mixer/detector are replaced by a single lock-in pixel, to demodulate the transmitted laser light that is initially modulated at f_o, then reflected from an 3D object, and finally received by the lock-in pixel. Various Si lock-in pixels have been proposed and demonstrated, such as one-tap CCD [5], demodulated photogates [6], pinned photodiode with transfer gates [7], and current assisted photonic demodulator [8]. To achieve a high signal-to-noise ratio and hence a small depth error in an indirect TOF system, it is desired to increase the pixel quantum efficiency η_q and frequency bandwidth B_f. However, especially with the recent trend to shift the laser wavelength from 850nm to 940nm to take advantage of the

solar ambient spectrum dip at 940nm, η_q degrades further as Si features a weaker absorption at a longer wavelength. Limited by the efficiency-bandwidth-product, while it is possible to obtain a higher η_q by simply having a thicker Si absorption layer, the resultant smaller B_f may in return spoil the depth accuracy. Consequently, it has become very challenging to engineer a high-performance Si lock-in pixel.

In this paper, we present novel Ge-on-Si (GOS) lock-in pixels in a two-tap configuration as shown in Fig. 1, at 940nm (for indoor-to-outdoor applications) and 1550nm (for eye-safe applications) wavelengths. Due to the ~0.8eV direct bandgap of Ge, compared to Si, a stronger absorption manifests itself with cut-off wavelength up to 1.6µm wavelength [9]. Such a property boosts the pixel η_q and B_f, and reduces the depth error measured in an indirect TOF system. Moreover, our GOS lock-in pixels are developed via Taiwan Semiconductor Manufacture Company (TSMC) image sensor platform, demonstrating the technology can be incorporated into the manufacture of FSI/BSI CMOS image sensors.

II. EVALUATION OF SYSTEM PRFORMANCES

We first calculate and compare the depth errors of a Si pixel and a GOS pixel at 940nm. The parameters used are listed in Table 1, and their values are based on Ref. [10-11] and Ref. [12] for Si and GOS, respectively. In Fig. 2, we plot depth errors of the Si and GOS pixels as a function of depth, for both the indoor and outdoor cases, given the laser power equal to 2W. It is observed that, for the indoor case, the depth error of the GOS pixel is about the same as or better than that of the Si pixel; moreover, for the outdoor case, the depth error of the GOS pixel is significantly better than that of the Si pixel (for more details please see Ref. [13]). These results might be surprising at the first sight as the dark current I_d of the GOS pixel is set to be orders of magnitude larger than that of the Si pixel (the depth errors in Fig. 2 remain almost the same even if a lower Si pixel I_d is used). The reason lies on that, in an indirect TOF system, the dominant noise factor is in fact due to the indoor/outdoor ambient light and laser light for the given indoor/outdoor scenarios. This is in sharp contrast to the earlier efforts in replacing InGaAs with GOS to make SWIR sensors, in which the system noise is limited by pixel I_d because the application mainly emphasizes on weak light detection, but analogous to the recent efforts in replacing III-V compound semiconductors with GOS to make high-speed optical receivers [9], in which the system noise is limited by the input referred noise of a transimpedance amplifier instead of detector I_d.

978-1-7281-1988-5/18 $31.00 © 2018 IEEE

III. CHARACTERIZATION OF DEVICE PROPERTIES

The GOS lock-in pixels are fabricated in a BSI configuration and characterized on a wafer-level electrical-optical tester for various dc/ac parameters. In Fig. 3, we show the measured I_d at V_2 node from 4 different lots. One wafer is selected from each lot, in which Lot1 is prepared via a first design/process vehicle, and Lot2~Lot4 are prepared via a second design/process vehicle. The same pixel is measured over 37 dies with at least 3 repeats for each wafer. Except for a few outliers, it is observed that for each lot I_d features a tight distribution and is significantly improved by almost two orders of magnitudes from Lot1 to Lot2~Lot4, because of improved designs and Ge processes. Note that I_d in Lot2~Lot4 is consistent with the assumed I_d used in generating Fig. 2. In Fig. 4, the external η_q is measured and steadily improved from Lot2 to Lot4 up to ~85% (median value with ±5% measurement error), because of further design and Ge process optimizations. In Fig. 5, we prepare a scatter plot to correlate the measured I_d and η_q, in which the improvements from the 1st to the 2nd vehicle can be clearly seen. In Fig. 6 and 7, we plot I_d as a function of V_2 and ΔV, respectively. The apparent increase of I_d with V_2 is caused by the increased local electric field and so the corresponding carrier generation rate. Nevertheless, I_d changes little with ΔV as the switching region is distant from V_2 node. In Fig. 8 and 9, we plot η_q as a function of V_2 and ΔV, respectively. η_q slightly increases with both V_2 and ΔV because of a faster carrier transit time that reduces the chance of carrier recombination. Note that from Fig. 6 to 9, 2~3 dies with 60 repeats are measured for each wafer.

In Fig. 10, we show the measured average power P_a from 3 different lots. Two distinct device groups M and N are presented, in which group M features standard designs and group N features low-power designs. It can be seen that an effective ~5 times reduction is achieved in Lot2/Lot3. In Fig. 11, the dc demodulation contrast C_{dc} is measured from 3 different lots. It can be seen that C_{dc} in group M is slightly higher than that of group N in Lot2/Lot3, suggesting a trade-off between P_a and C_{dc}. This is further evidenced by the correlation shown in Fig. 12 scatter plot. Note that from Fig. 10 to 12, the low-power devices in Lot4 are designed to be less aggressive compared to Lot2/Lot3, and experimentally shown to feature a less clear P_a-to-C_{dc} trade-off. In Fig. 13 and 14, C_{dc} are plotted as a function of V_2 and ΔV, respectively, with only group M shown for simplicity. C_{dc} changes very little with V_2 as the switching region is distant from V_2 node. On the other hand, C_{dc} increases with ΔV in a linear/sublinear fashion between 0V~0.2V/0.2V~0.8V, and saturates beyond 0.8V, showing a good two-tap switching capability. Note that in Fig. 14 the slightly lower C_{dc} in Lot3 is analyzed and found to be induced by wafer-to-wafer process variation.

In Fig. 15 and 16, given 940nm/1550nm wavelengths and Lot1/Lot2 wafers, we present scatter plots between I_d and η_q for group M and N, respectively, over 37 dies with at least 3 repeats for each wafer. The spreads of experimental data mainly arise from including >2 pixel designs to investigate the wavelength sensitivity. It can be seen that the highest η_q at 1550nm is ~50%, suggesting a large tensile-strain enhanced Ge absorption [9]. From Fig. 17 to 20, scatter plots between P_a and C_{dc} are presented for group M and N, with the wafers of Lot1/Lot2 and the wavelengths of 940nm/1550nm, over 2~3 dies with at least 3 repeats for each wafer. The spreads of experimental data mainly arise from including >600 designs to cover a large design space. Similar distributions between the scatter plots at 940nm or 1550nm for the same pixel are observed, suggesting the dissimilar spatial carrier distributions (exponential at 940nm and nearly constant at 1550nm) caused by different absorption coefficients has a minimum impact. Moreover, C_{dc} is improved from the 1st to the 2nd vehicle, i.e., from Lot1 to Lot2, and again P_a is reduced from group M to N. Note that the long red tails in Fig. 19 and 20 suggest that there is a limit to the low-power design rendering switching capability degradation. In Fig. 21 and 22, scatter plots between P_a and ac demodulation contrast C_{ac} are presented for group M and N in Lot2, at 940nm and 1550nm, respectively. The trade-off between P_a and C_{ac} are observed, as in P_a and C_{dc}. Further evidence is provided in Fig. 23 using Lot4 data at 940nm, as C_{dc} and C_{ac} up to 0.97 and 0.83 show a positive correlation.

IV. SUMMARY

Novel GOS lock-in pixel is investigated at 940nm and 1550nm wavelengths and shown to be a strong contender against conventional Si lock-in pixel. The measured statistical data further demonstrate the technology yields good within-wafer and wafer-to-wafer uniformities that may be ready for mass production in the near future.

ACKNOWLEDGMENT

The authors gratefully acknowledge the support of device fabrication by TSMC, Hsinchu City, Taiwan ROC.

REFERENCES

[1] G. Bianco et al., "A Comparative Analysis between Active and Passive Techniques for Underwater 3D Reconstruction of Close-Range Objects," Sensors **13**, 11007-11031 (2013).

[2] J. Geng, "Structured-light 3D surface imaging: a tutorial," Advances in Optics and Photonics **3**, 128-160 (2011).

[3] M. Perenzoni et al., "Compact SPAD-Based Pixel Architectures for Time-Resolved Image Sensors," Sensors **16**(5), 1-12 (2016)

[4] S. Foix et al., "Lock in Time of Flight (ToF) Cameras: A Survey," IEEE Sensors J. **11**(9) 1917-1925 (2011).

[5] R. Lange et al., "Solid-State Time-of-Flight Range Camera," IEEE J. Quant. Electron. **37**(3), 390-397 (2001).

[6] C. S. Bamji et al., "A 0.13 μm CMOS System-on-Chip for a 512 × 424 Time-of-Flight Image Sensor With Multi-Frequency Photo-Demodulation up to 130 MHz and 2 GS/s ADC," IEEE J. Solid-State Circuits **50**(1), 303-318 (2015).

[7] S.-J. Kim et al., "A Three-Dimensional Time-of-Flight CMOS Image Sensor With Pinned-Photodiode Pixel Structure," IEEE Electron. Dev. Lett. **31**(11), 1272-1274 (2010).

[8] D. V. Nieuwenhove et al., "Photonic Demodulator With Sensitivity Control," IEEE Sensors J. **7**(3) 317-318 (2007).

[9] J. Michel et al., "High-performance Ge-on-Si photodetectors," Nature Photon. **4**, 527-536 (2010).

[10] Y. Kato et al., "320 × 240 Back-Illuminated 10-μm CAPD Pixels for High-Speed Modulation Time-of-Flight CMOS Image Sensor," IEEE J. Solid-State Circuits **53**(4), 1071-1078 (2018).

[11] S. Yokogawa et al., "IR sensitivity enhancement of CMOS Image Sensor with diffractive light trapping pixels," Sci. Rep. **7**, 3832 (2017).

[12] N. Na et al., "Proposal and demonstration of lock-in pixels for indirect time-of-flight measurements based on germanium-on-silicon technology," arXiv:1806.07972 [physics.ins-det], (2018).

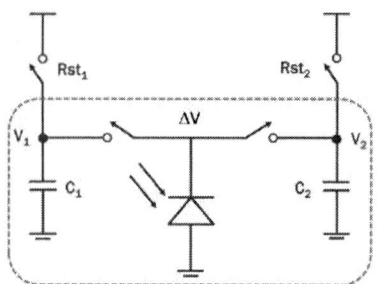

Fig. 1 The equivalent circuit of a two-tap lock-in pixel. Differential voltage ΔV is applied to control the two demodulation switches; voltages V_1/V_2 are applied to extract the photo-carriers.

Table 1 System parameters

Pixel Type	Si	Ge-on-Si
Laser Wavelength	940 nm	
Pixel Pitch	10 μm	
Camera Effective Focal Length	2 mm	
Camera Focal Number	1.1	
Absorption Efficiency	20 %	90 %
Operation Frequency	100 MHz	300 MHz
Demodulation Contrast at Operation Frequency	0.85	0.9
Effective Bandpass Filter Linewidth	60 nm	
Indoor & Outdoor Ambient Light Spectral Intensity	3 & 300 mW/m²/nm	

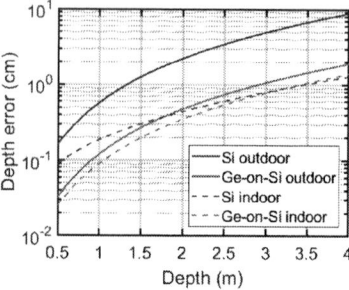

Fig. 2 Depth error plotted as a function of depth for Si and GOS lock-in pixels at indoor/outdoor ambient conditions.

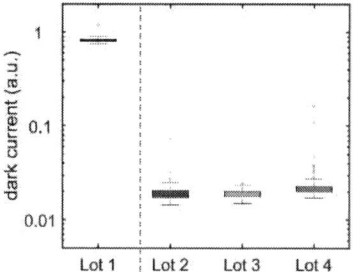

Fig. 3 I_d plotted with 4 different lots; Lot1 is from the 1st vehicle and Lot2~Lot4 are from the improved 2nd vehicle with I_d reduced by roughly 2 orders of magnitude.

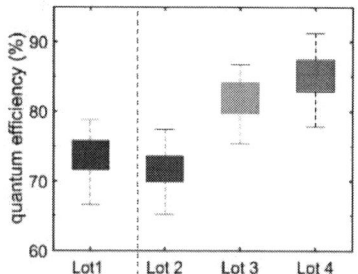

Fig. 4 η_q plotted with 4 different lots; Lot1 is from the 1st vehicle and Lot2~Lot4 are from the improved 2nd vehicle with η_q gradually improved by design/process optimizations.

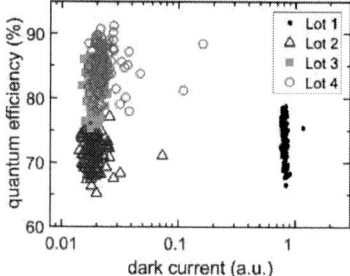

Fig. 5 η_q plotted as a function of I_d with 4 different lots.

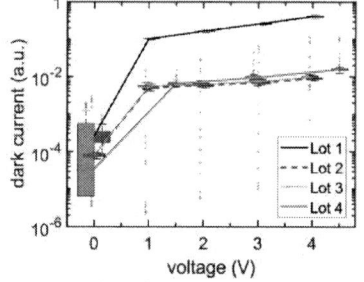

Fig. 6 I_d plotted as a function of V_2. Lines intercepting median values are drawn for visual comparison.

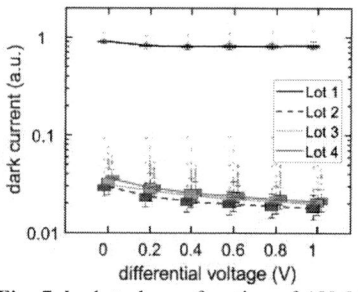

Fig. 7 I_d plotted as a function of ΔV. Lines intercepting median values are drawn for visual comparison.

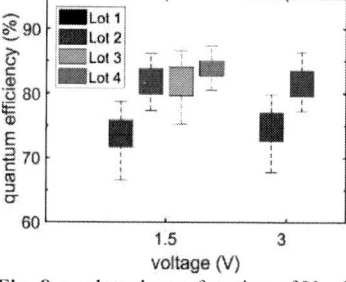

Fig. 8 η_q plotted as a function of V_2. Only data at 1.5V and 3V are shown.

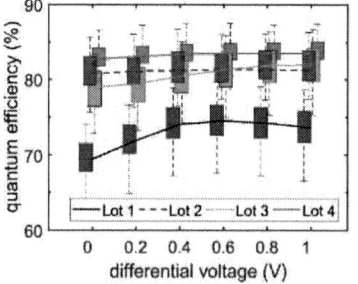

Fig. 9 η_q plotted as a function of ΔV. Lines intercepting median values are drawn for visual comparison.

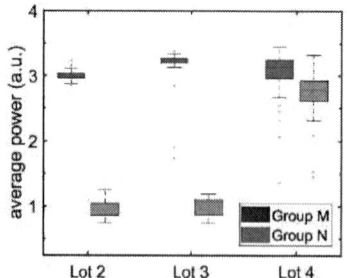

Fig. 10 P_a plotted with 3 different lots for group M/N standard/low-power designs. Note that the low-power designs in Lot4 are less aggressive.

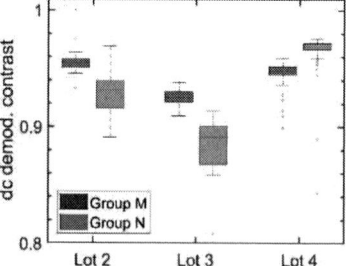

Fig. 11 C_{dc} plotted with 3 different lots for group M/N standard/low-power designs. Note that the low-power designs in Lot4 are less aggressive.

978-1-7281-1988-5/18 $31.00 © 2018 IEEE

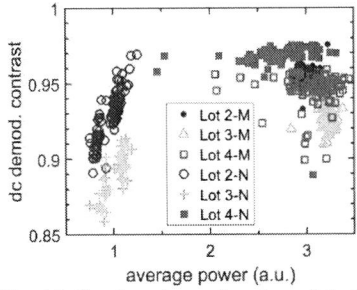
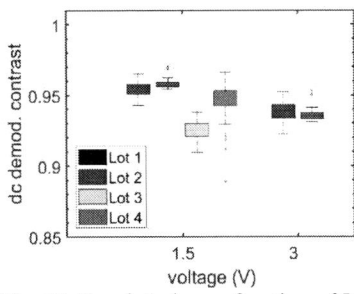
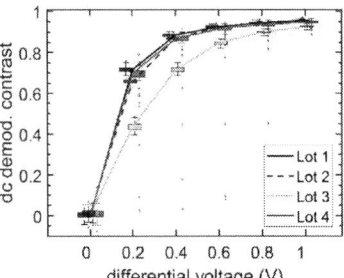

Fig. 12 C_{dc} plotted as a function of P_a. Note that the low-power designs in Lot4 are less aggressive.

Fig. 13 C_{dc} plotted as a function of V_2 for group M standard design. Only data at 1.5V and 3V are shown.

Fig. 14 C_{dc} plotted as a function of ΔV for group M standard design. Lines intercepting median values are drawn for visual comparison.

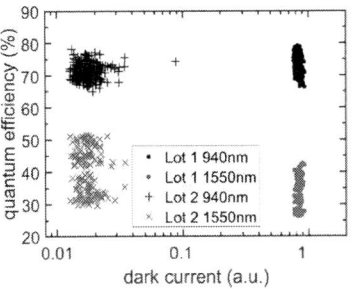
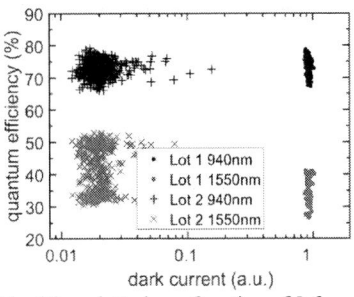
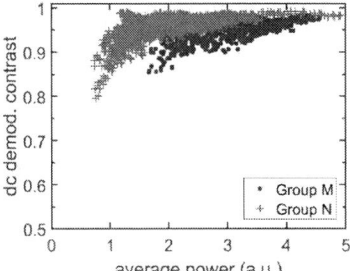

Fig. 15 η_q plotted as a function of I_d for group M standard design. Note that the 940nm data are selected from Lot1/Lot2 for visual comparison rather than the highest η_q.

Fig. 16 η_q plotted as a function of I_d for group N low-power design. Note that the 940nm data are selected from Lot1/Lot2 for comparison rather than the highest η_q.

Fig. 17 C_{dc} plotted as a function of P_a for Lot1 with >600 designs at 940nm to cover a large design space.

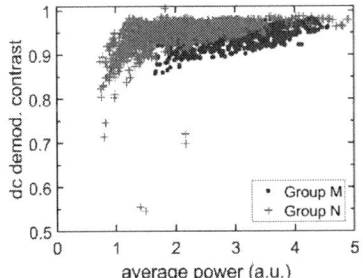
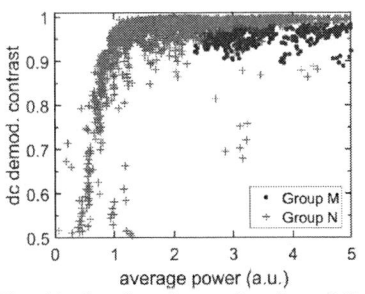
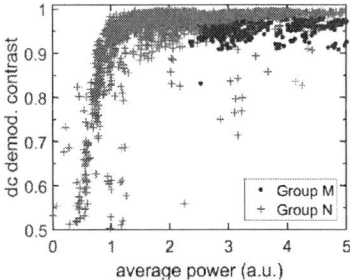

Fig. 18 C_{dc} plotted as a function of P_a for Lot1 with >600 designs at 1550nm to cover a large design space.

Fig. 19 C_{dc} plotted as a function of P_a for Lot2 with >600 designs at 940nm to cover a large design space.

Fig. 20 C_{dc} plotted as a function of P_a for Lot2 with >600 designs at 1550nm to cover a large design space.

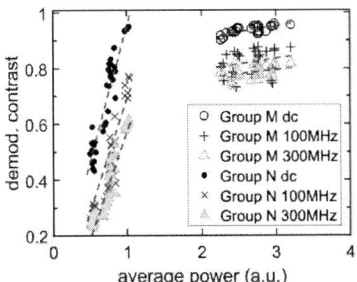
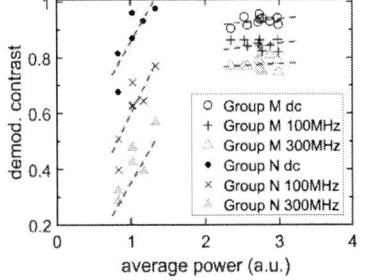
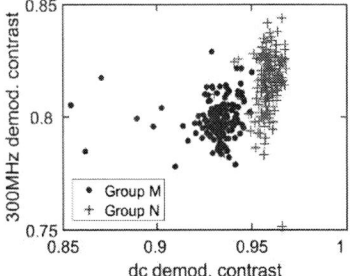

Fig. 21 C_{dc} and C_{ac} plotted as a function of P_a for Lot2 at 940nm. Note that compared to Lot4, the low-power designs in Lot2 are more aggressive.

Fig. 22 C_{dc} and C_{ac} plotted as a function of P_a for Lot2 at 1550nm. Note that compared to Lot4, the low-power designs in Lot2 are more aggressive.

Fig. 23 300MHz C_{ac} plotted as a function of C_{dc} for Lot4 at 940nm. Note that compared to Lot2, the low-power designs in Lot4 are less-aggressive.

978-1-7281-1988-5/18 $31.00 © 2018 IEEE

CMOS-Integrated Single-Photon-Counting X-Ray Detector using an Amorphous-Selenium Photoconductor with 11×11-μm² Pixels

A. Camlica[*], A. El-Falou[*], R. Mohammadi, P. M. Levine, and K. S. Karim

Department of Electrical and Computer Engineering, University of Waterloo, Waterloo, Ontario, Canada

Email: {acamlica, aelfalou}@uwaterloo.ca

Abstract— We report, for the first time, results from a single-photon-counting X-ray detector monolithically integrated with an amorphous semiconductor. Our prototype detector combines amorphous selenium (a-Se), a well-known X-ray photoconductive material suitable for large-area applications, with a 0.18-μm-CMOS readout integrated circuit containing two 26×196 photon counting pixel arrays. The detector features 11×11-μm² pixels to overcome a-Se count-rate limitations by unipolar charge sensing of the faster charge carriers (holes) via a unique pixel geometry that leverages the small pixel effect for the first time in an amorphous semiconductor. Measured results from a mono-energetic radioactive source are presented and demonstrate the untapped potential of using amorphous semiconductors for high-spatial-resolution photon-counting X-ray imaging applications.

I. INTRODUCTION

Large-area digital X-ray detectors are rapidly replacing X-ray film screen and computed radiography systems globally because of their better imaging performance and efficient work flows. However, X-ray detector technology still has considerable room for improvement in high-resolution and low-dose applications such as mammography and angiography, where lower radiation dose is essential for patient safety, and better contrast and spatial resolution can reduce medical diagnostic errors.

Single photon counting (SPC) X-ray detectors provide energy discrimination for improved image contrast and offer the advantages of lower noise and higher dynamic range compared to traditional integration-mode X-ray detectors [1,2]. However, current SPC X-ray detector technology is limited to small-area imaging applications due to scaling constraints on both the X-ray sensor material and the readout integrated circuit (IC). With the recent advances in column-buttable CMOS IC technology, IC scaling constraints are mitigated, making the X-ray sensor a limiting step [3].

Existing X-ray sensor materials demonstrated for SPC applications detect X-rays directly and make use of primarily crystalline (e.g., Si) or poly-crystalline (e.g., CZT, CdTe) materials. However, these materials are challenging to scale to large-area medical applications because of yield and cost issues associated with the growth and bonding technology needed to interface the sensor with the readout IC.

An alternate approach is to use a large-area-compatible direct-X-ray-detection sensor such as poly-crystalline HgI_2

[4]. However, HgI_2 is not yet commercially available. A commercially viable alternative is amorphous selenium (a-Se), already the predominant technology for large-area mammography X-ray detectors.

The technical challenges for photon counting with a-Se lie in overcoming (1) the slow carrier-transport properties of a-Se [5], which lead to count-rate limitations due to pile-up, and (2) low X-ray-to-charge conversion gain, which degrades SNR. In this paper, we report, for the first time, the design and preliminary characterization of an a CMOS-integrated SPC detector with an a-Se photoconductor. Our design features ultra-small 11×11-μm² pixels which enable us to overcome the count-rate limitations imposed by a-Se by leveraging the "small-pixel" geometry with amorphous semiconductors [6]. We also employ a unique pixel circuit design to achieve sufficient SNR for photon counting with a-Se.

II. SENSOR CHARACTERIZATION

We quantify the count-rate limitations of our proposed SPC detector and show that by leveraging the small-pixel effect, we could achieve adequate detector performance for mammography. To potentially achieve higher spatial resolution compared to existing detectors, we design our prototype using 11×11-μm² pixels. Based on the expected maximum incident flux of 10^8 photons/s/mm² for mammography [7], our pixels must support a maximum count rate C_{max} of ~12×10³ photons/s/pixel.

However, our time-of-flight (TOF) measurements of a 100-mm² a-Se sensor show that the long electron transport time would prevent us from meeting C_{max}. Based on our TOF results shown in Figs. 1(a) and (b), we compute hole and electron mobilities of 0.147 and 0.004 cm²/V·s, respectively. From these, we estimate hole drift time t_h and electron drift time t_e of 1.36 μs and 50 μs, respectively, for a 200-μm-thick a-Se sensor biased at a 10-V/μm electric field (typical for mammography). Targeting a maximum pile-up probability P_p of 20%, we calculate the per-pixel count rate of the detector $C_d = -\ln(1-P_p)/t_e \approx 1500$ photons/s/pixel, which does not meet the C_{max} requirement.

Consequently, we choose to operate our SPC detector in a unipolar charge sensing regime to observe the faster carriers (holes), thereby masking the extended rise time due to slower carriers (electrons). This small pixel effect (SPE) is noticeable when pixel geometry is made significantly smaller than the sensor thickness. Repeating the above C_d calculation, but with the hole drift time t_h, we estimate an achievable count rate of 55×10³ photons/s/pixel, exceeding C_{max}.

---

*These authors contributed equally to this work.

978-1-7281-1988-5/18 $31.00 © 2018 IEEE

III. DETECTOR DESIGN AND IMPLEMENTATION

The cross-section of our detector, shown in Fig. 2, consists of a 0.5-μm-thick continuous parylene stabilization layer deposited on individual top-metal Al pads (under an 8×10-μm^2 passivation opening) on the CMOS pixel array readout IC, a 70-μm-thick a-Se layer (deposited via thermal evaporation and shadow masking), and a 50-nm-thick Au top-contact to bias a-Se. As a proof-of-concept, we are employing an a-Se thickness about $3\times$ smaller than that used in a conventional mammography detector. The total applied bias voltage to a-Se is limited to 300 V (or ~4 V/um) to protect our CMOS IC from high-voltage breakdown since no encapsulation layer is utilized. Nevertheless, we expect sufficient photon capture to characterize our detector using a mono-energetic 60-keV radioactive source.

Our 4×3-mm^2 SPC detector IC, shown in Fig. 3, is designed in a 1.8-V-supply, 0.18-μm mixed-signal CMOS (non-CIS) process and integrates two 26×196 SPC-pixel arrays. Each 11×11-μm^2 pixel is connected to a unique off-array 5-bit counter, to register single-photon events.

We employ the pixel architecture shown in Fig. 4, with the timing diagram shown in Fig. 5. Generated charge from a captured photon is integrated on parasitic capacitance C_i, estimated at 20 fF, for a fixed period T_{int} before being reset (during a 0.3-μs pixel dead time). We threshold the input voltage signal against a reference voltage V_{th} corresponding to the desired photon energy using offset-corrected 1st-stage comparator PA1 followed by gain stage PA2. The resolved logic value from PA2 is then latched and used to increment or hold the counter value. In our design, pixel input-referred noise is dominated by reset noise $Q_{nr}= (kTC_i)^{0.5}/q= 57$ e$^-_{rms}$ (at T=300 K) and comparator input noise, leading to a total simulated noise Q_n of ~90 e$^-_{rms}$.

We use pulse-height spectroscopy (PHS) measurements to characterize the a-Se sensor layer. PHS results for a 70-μm-thick 0.7-mm^2 a-Se sensor biased at 300 V are shown in Fig. 6, and indicate an ionization energy $W_\pm$ of 78 eV for 60-keV photons. Based on these results, we estimate that a signal charge Q_t of 769 ehp is generated per captured 60-keV photon, leading to an estimated SNR for our SPC detector of $10\log_{10}[Q_t^2/(Q_t+Q_n^2)]$=18.2 dB. PHS measurements are also used to quantify the expected number of photon counts under the same conditions and parameters as for our integrated SPC detector. Semi-Gaussian output pulses, characterized with a 44-μs peaking time, are acquired from a multi-channel analyzer for 120 min. In Fig. 6, the spectrum is seen above the background noise component. The total number of counts, calculated by integrating the area under the spectrum, is 87 kcounts/mm^2. Considering the 121-μm^2 pixel area of our SPC detector, the expected number of counts for the same exposure time is estimated at 7.2 counts/pixel.

IV. EXPERIMENTAL RESULTS AND DISCUSSION

We demonstrate photon-counting operation of our SPC detector using the experimental setup shown in Fig. 7 and a mono-energetic 60-keV ^{241}Am source. We measure an average of 5.5 counts/pixel for 80 adjacent pixels in one row of the detector over a 120-min period, which is comparable to

the 7.2 counts/pixel estimated via PHS measurements. Fig. 8(a) displays the count results from these pixels with and without the source present. Hit probability versus estimated input charge for one pixel in the row is displayed in Fig. 8(b), showing an input-referred noise of 190 e$^-_{rms}$. This is greater than the simulated result and is caused by an anomalous source of interference on the particular chip under test. The detection histogram and variation in pixel threshold voltage are shown in Figs. 8(c) and (d), respectively.

Based on the results obtained, we hypothesize that the SPE is enabling unipolar charge sensing in our detector, mitigating the impact of slow electron drift times in a-Se. Although we observed a 44-μs electron drift time in a large-area 70-μm-thick a-Se sensor, SPE enables use of the shorter 9.7-μs integration time while maintaining a charge-collection efficiency that is close to unity in the integrated detector.

V. SUMMARY

We have demonstrated, for the first time, single-X-ray-photon counting using an a-Se photoconductor integrated with a CMOS readout IC. The presented results are also unique because, to the best of our knowledge, the 11×11-μm^2 pixel pitch is the smallest reported for X-ray photon counting and this work reports the first demonstration of SPE for amorphous semiconductors.

This research indicates that by leveraging a combination of synergistic device and circuit architectures, amorphous material shortcomings related to slow carrier transport, noise, and gain can be overcome. The counting operation demonstrated with an a-Se photoconductor in this research indicates that well-established large-area a-Se can meet the requirements of emerging medical imaging applications such as photon-counting mammography or angiography without resorting to new sensor materials or crystalline semiconductors that are challenging to scale up to larger areas.

ACKNOWLEDGEMENTS

We thank Dr. Denny Lee for his help with PHS measurements, Dr. Celal Con for help with packaging and processing, and Dr. Xu Di and Prof. Michael Mayer for IC wirebonding.

REFERENCES

[1] M. Aslund et al., "Physical characterization of a scanning photon counting digital mammography system based on Si-strip detectors," *Med. Phys.*, vol. 34, no. 6, pp. 1918-1925, 2007.

[2] E. Fredenberg et al., "Energy resolution of photon-counting silicon strip detector," *Nucl. Instrum. Meth. Phys. Res. A*, vol. 613, pp. 156-162, 2010.

[3] R. Ballabriga et al., "Medipix3: A 64k pixel detector readout chip working in single photon counting mode with improved spectrometric performance," *Nucl. Inst. Meth. A.*, vol. 633, pp.S15-S18, 2010.

[4] H.-S. Kim et al., "An asynchronous sampling-based 128×128 direct photon-counting X-ray image detector with multi-energy discrimination and high spatial resolution," *IEEE J. Solid-State Circuits*, vol. 48, no. 2, pp. 541–558, Feb. 2013.

[5] S. Kasap et al., "Amorphous and polycrystalline photoconductors for direct conversion flat panel X-ray image sensors", *Sensors*, vol. 11, no. 5, pp. 5112-5157, May 2011.

[6] H. H. Barrett et al., "Charge transport in arrays of semiconductor gamma-ray detector," *Phys. Rev. Lett.*, vol. 75, pp. 156–159, 1995.

[7] L. Abbene et al., "Performance of a digital CdTe X-ray spectrometer in low and high counting rate environment," *Nucl. Instr. Meth.*, vol. 621, pp. 447-452, 2010.

978-1-7281-1988-5/18 $31.00 © 2018 IEEE

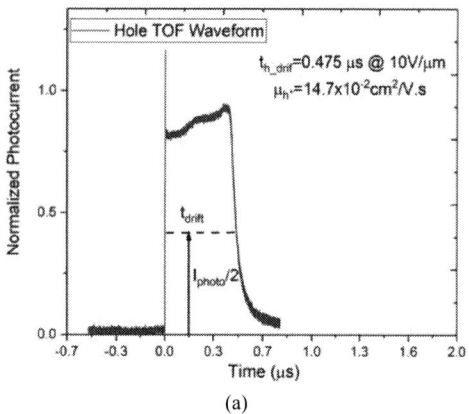 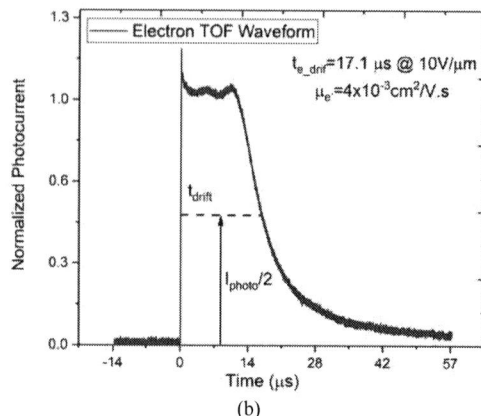

(a)   (b)

Fig. 1: Time-of-flight (TOF) transient photoconductivity measurements of an a-Se sensor showing the response of (a) holes and (b) electrons.

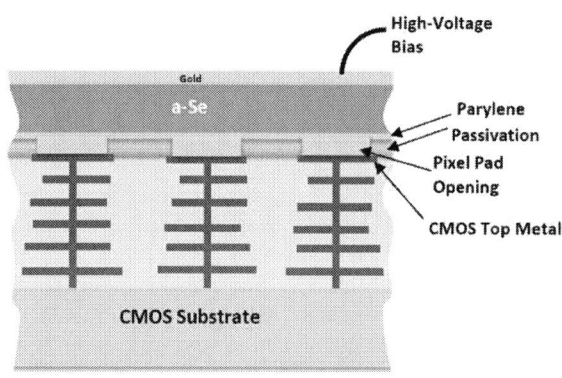

Fig. 2: Simplified cross-section of the CMOS-integrated a-Se SPC X-ray detector.

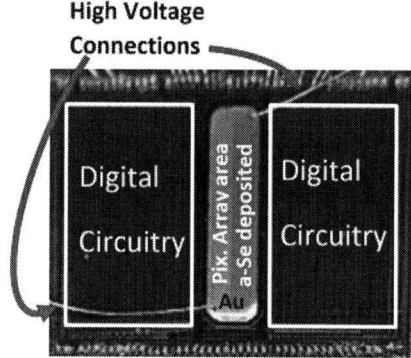

Fig. 3: Photograph of the 4×3-mm² SPC X-ray detector IC.

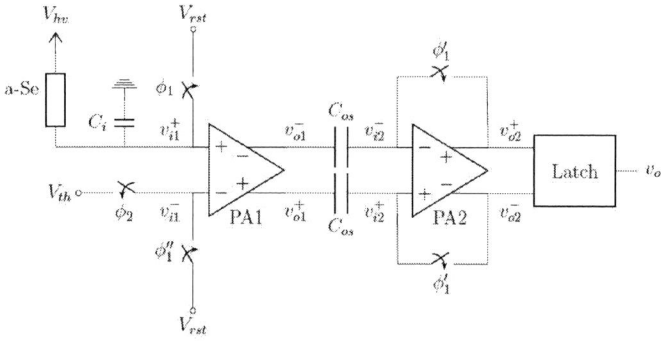

Fig. 4: CMOS pixel circuit schematic, consisting of two amplifier stages followed by a latch. Capacitor C_i represents the parasitic input capacitance of the pixel. Output-offset correction of amplifier PA1 is implemented with capacitors C_{os}.

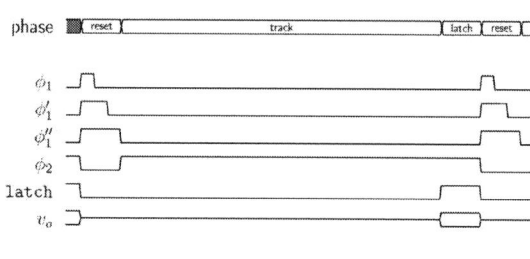

Fig. 5: Timing diagram showing the three-phase operation of the CMOS pixel circuit. Reset dead time is 0.3 µs and the integration (track) phase lasts for 9.7 µs.

978-1-7281-1988-5/18 $31.00 © 2018 IEEE   757

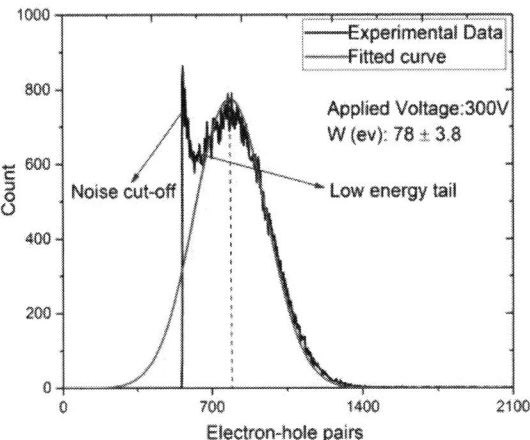

Fig. 6: Calibrated pulse-height spectrum from measurement of a 70-μm-thick a-Se sensor exposed to a mono-energetic 60-keV [241]Am source.

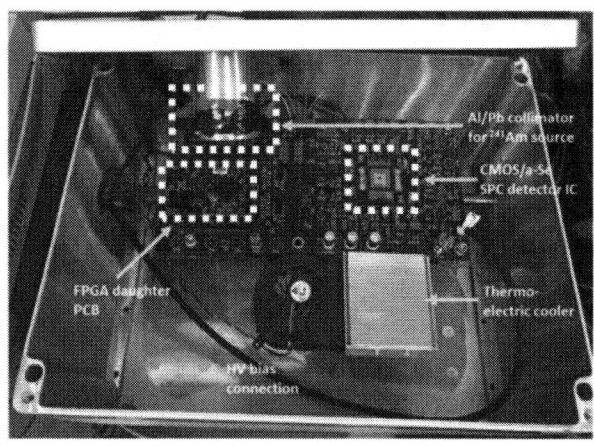

Fig. 7: Experimental setup of the integrated a-Se SPC detector. The Al/Pb collimation tube attached to the lid (containing the radiation source) is aligned with the SPC chip surface using a precision stepper. The source to detector distance is 7 mm.

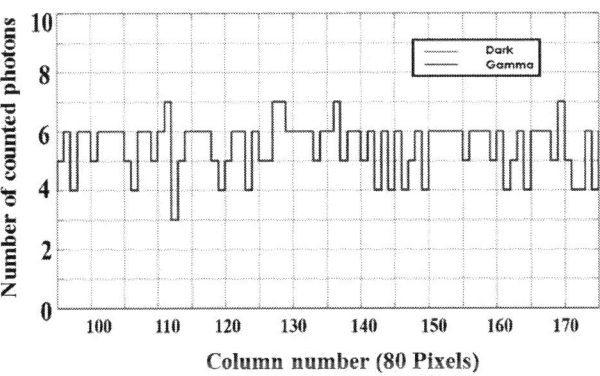

(a)   Counted photons per pixel with the radiation source present and absent (dark).

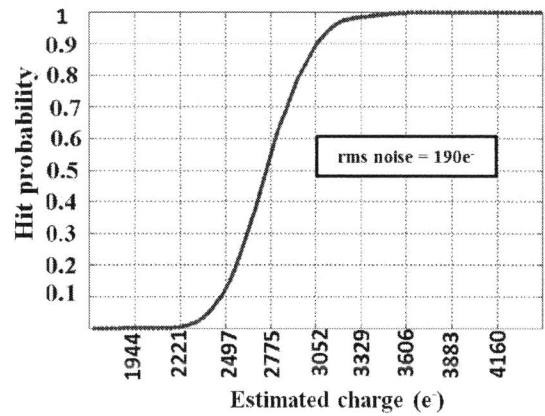

(b)   Hit probability versus input generated charge.

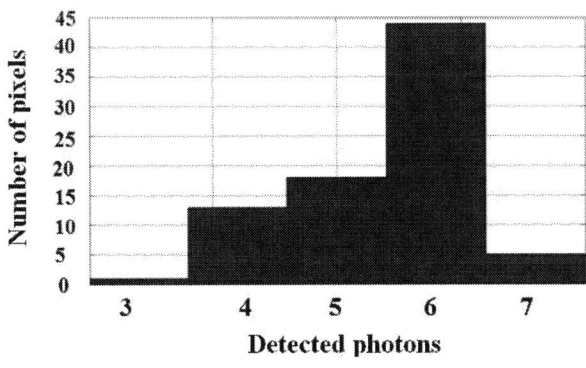

(c)   Histogram of detected photons.

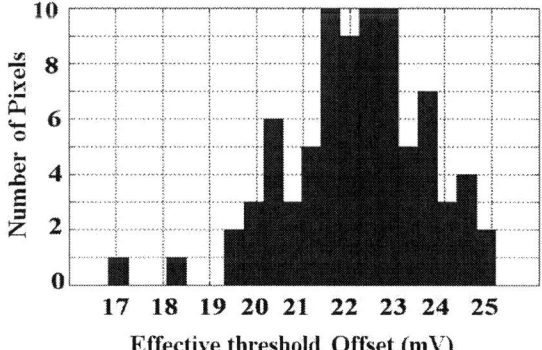

(d) Histogram counts for threshold offset.

Fig. 8: Measured counting results from 80 adjacent pixels in one row of the CMOS-integrated SPC detector when exposed to a mono-energetic 60-keV [241]Am source.

978-1-7281-1988-5/18 $31.00 © 2018 IEEE

Transport models based on NEGF and empirical pseudopotentials: a computationally viable method for self-consistent simulation of nanoscale devices.

Marco G. Pala[1], Oves Badami[2] and David Esseni[2]

[1] C2N, CNRS, Univ. Paris-Sud, Université Paris-Saclay, F-91405 Orsay, France; email: marco.pala@c2n.upsaclay.fr
[2] DPIA, University of Udine, Via delle Scienze 206, 33100 Udine, Italy, FAX:+39-0432-558251; email: david.esseni@uniud.it

I. Abstract

We present new theoretical developments and applications concerning Non-Equilibrium Green's Functions (NEGF) based transport modelling with an Empirical Pseudopotential (EP) Hamiltonian. We have extended the methodology to include arbitrary crystal orientations and strain conditions, and have reformulated quantum confinement and spatial discretization to improve the computational efficiency.

II. Introduction

Modern CMOS FETs resemble quantum constrictions connecting the source/drain carrier reservoirs [1], [2], and quantum effects have become prominent, ranging from simple subband splitting [3], to tunnelling phenomena [4], [5]. This is confirmed by the fact that silicon nanowires have been recently proposed as an industrial platform for quantum computing [6]. Quantum transport models including full band effects are theoretically and computationally challenging. We recently proposed an NEGF based simulation approach employing an EP Hamiltonian [7], [8], that leverages on and extends some previous contributions [9], [10], and it is an alternative method to tight-binding based models [11].

We present remarkable advancements in the development of the EP-NEGF simulation approach consisting in: a) transport for arbitrary crystal orientations; b) inclusion of strain; c) reformulation of the confining operator as a local operator and drastic improvement of computational efficiency. Our results qualify the EP-NEGF method as a physically sound, computationally viable method for full quantum simulation of nanoscale devices with technologically relevant dimensions.

III. EP based physical Model

A. Arbitrary crystal orientations and strain conditions

In our previous presentation of the EP-NEGF method we assumed that transport and confining directions are aligned with $\langle 100 \rangle$ directions of the underlying semiconductor [7], [8]. However, in electron devices simulations the Device Coordinate System (DCS) is frequently different from the Crystal Coordinate System (CCS) (see **Fig.1**). Because the confinement and transport directions are defined in the DCS, hereafter all equations are written in the DCS. The reciprocal lattice vectors in the DCS are given by $\mathbf{G}=\mathbf{R}_{CD}\mathbf{G}_c$, where $\mathbf{R}_{CD}$ is a 3×3 rotation matrix from CCS to DCS, and $\mathbf{G}_c$ are the well known lattice vectors in the CCS. We here use $\mathbf{k}$ for the wave-vectors in the reduced zone, and $\mathbf{K}=(\mathbf{k}+\mathbf{G})$ for the corresponding wave-vectors in extended zone (see **Eq.1**).

The $\mathbf{G}$ vectors in the DCS are important because they set the reduced zone of the bulk semiconductor to be used for transport calculations. For a 2D electron gas in the UTB FET

of **Fig.1**, for example, if we let $\mathbf{G}_{xx}$, $\mathbf{G}_{zz}$ denote the smallest $\mathbf{G}$ vectors aligned respectively with x and z, then the k_x range of the reduced zone is $-|\mathbf{G}_{xx}|/2 \leq k_x < |\mathbf{G}_{xx}|/2$, which ensures that, for any $\mathbf{K}_{yz}=[(k_y,k_z)+(G_y,G_z)]$, the corresponding $K_x=k_x+G_x$ components cover with no voids the entire extended K_x range [8]. Likewise the $|\mathbf{G}_{zz}|$ sets the k_z range in the confinement direction as $-|\mathbf{G}_{zz}|/2 \leq k_z < |\mathbf{G}_{zz}|/2$. The k_y range is finally established by the volume of the reduced zone, namely $4(2\pi/a_0)^3$ in the unstrained lattice. Some examples of reduced zones are shown in **Fig.1**.

The EP method is very suitable for the inclusion of strain [12]. If we denote with ε_c the 3×3 strain matrix in the CCS, the direct lattice, $\boldsymbol{a}$, reciprocal lattice vectors, $\boldsymbol{b}$, and unit cell volume, Ω, of the strained lattice are given by **Eq.2**, where $\boldsymbol{a}^0$ and Ω_0 are lattice vectors and unit cell volume of the unstrained lattice, and $\boldsymbol{I}_3$ is the 3×3 identity matrix.

The eigenvalue problem for a local EP Hamiltonian has the well known form shown in **Eqs.3, 4**, where $T(\mathbf{k}+\mathbf{G})=\hbar^2|\mathbf{k}+\mathbf{G}|^2/2m_0$ is the kinetic energy operator for a continuous real space [13], and $U_L(|\mathbf{G}|)$ are the form factors (for a diamond material). **Eq.3** also recalls the expression for the Bloch wave-function $\Phi_{n\mathbf{k}}(\boldsymbol{r})$ in terms of coefficients $B_{n\mathbf{k}}$. For an unstrained crystal the EP problem is solved by using only three non null $U_L(|\mathbf{G}|)$ components for $|\mathbf{G}|=\sqrt{3}$, $\sqrt{8}$, $\sqrt{11}$ (see table in **Fig.2**). In the strained lattice the $|\mathbf{G}|$ vectors take also different values and the form factors $U_L(Q)$ need to be interpolated. **Fig.2** shows the cubic spline interpolation used in this work. **Eq.4** also shows how strain affects the atomic basis vector $\boldsymbol{\tau}$. In the presence of shear strain additional adjustments of $\boldsymbol{\tau}$ have been suggested on the basis of *ab-initio* calculations [16]. However we verified that such corrections have a small effect and we thus neglected them in our model.

Fig.3 compares our calculations for the energy bandgap of biaxially strained silicon versus the Ge content of the virtual substrate: a good agreement is obtained with experiments [17], and previous calculations [18]. **Fig.4** reports equi-energy plots for the silicon valence band: a compressive stress along [110] direction results in a large reduction of the hole effective mass in the same direction, in close agreement with [20].

B. EP based calculation of the dielectric response

An advantage of the EP method is that it provides atomistic wave-functions, that for a bulk crystal are the Bloch wave-functions $\Phi_{n\mathbf{k}}(\boldsymbol{r})$ in **Eq.3**. The wave-functions allow one to calculate several important physical quantities, such as the phonon scattering with the rigid-ion approximation, and the dielectric response $\varepsilon(\mathbf{q},\omega)$ [14]. We here obtained $\varepsilon(\mathbf{q},\omega)$ by using **Eq.5** [14], where the matrix elements $|\langle \Phi_{n\mathbf{k}}|\Phi_{n'(\mathbf{k}+\mathbf{q})}\rangle|$

978-1-7281-1988-5/18 $31.00 © 2018 IEEE

were calculated from the eigenvectors $B_{n\mathbf{k}}$ of **Eq.3**.

Fig.5 reports the silicon dielectric function versus energy. The agreement with the experimental value for the static $\varepsilon_{si}=11.7\varepsilon_0$ is remarkably good considering that no adjustable parameters were used. Furthermore the peak of the imaginary part gives a good estimate of the average optical bandgap, which is in fact between 4 and 5 eV for silicon [14].

IV. NEGF based transport model

We here use a hybrid basis consisting of real space along x and plane waves $|\mathbf{K}_{yz}\rangle$ in the (y,z) plane, such that the Hamiltonian takes the block tridiagonal form shown in **Fig.6** [8]. The size of the blocks $\mathbf{H}_{l,l}$, $\mathbf{H}_{0,1}$ is crucial for both the memory and the CPU requirements of the recursive techniques based on the Dyson equation used to solve the NEGF transport model. In our original formulation each block was a_0 long due to the non local quantum confinement operator $\mathbf{H}_{cnf}$ and to the high order spatial discretization scheme [8]. We have modified both above aspects to improve numerical efficiency.

A. New local pseudopotential for quantum confinement

For the 2D electron gas in the UTB FET of **Fig.1** we introduce the local pseudopotential $V_{2D}(\boldsymbol{r})$ in **Eq.6**, where $V_{sc}(\boldsymbol{r})$ and $V_{ox}(\boldsymbol{r})$ are the EP of respectively the semiconductor and oxide region. Here $\Theta(z)$ is a unitary step function, such that $\Theta(z)=0$ for $|z|\leq T_{sct}/2$ and $\Theta(z)=1$ for $T_{sct}/2<|z|<L_z/2$, where T_{sct}, L_z are the semiconductor thickness and periodicity length along z. The representation of $\Theta(z)$ in $\mathbf{K}$ space is readily given by **Eq.7**. The $\mathbf{K}$ space form of $V_{2D}(\mathbf{K}-\mathbf{K}')$ in **Eq.8** can be derived from **Eq.6** by noting that the product $V_{cnf}(\boldsymbol{r})\Theta(z)$ in real space leads to the convolution between $V_{cnf}(\mathbf{K}-\mathbf{K}')$ and $\Theta(\mathbf{K}-\mathbf{K}')$, and then by using **Eq.4** to express $V_{sc}(\mathbf{K}-\mathbf{K}')$, $V_{cnf}(\mathbf{K}-\mathbf{K}')$. The oxide in **Eqs.6, 8** is in effect an artificial material having the same a_0 as the semiconductor, but whose EP parameters are chosen so that the energy bandgap is about 9 eV and the conduction and valence band edge are fairly independent of $\mathbf{k}$. The EP problem for a 2D gas reads as in **Eq.9**, where $\mathbf{k}_{xy}=(k_x,k_y)$ varies in the 2D reduced zone. **Fig.7** reports the conduction band minima of an ultra-thin silicon film and it shows that EP calculations with the new, local pseudopotential $V_{2D}(\boldsymbol{r})$ in **Eq.6** agree well with the results of the non local confinement operator [7], and with tight-binding.

B. Hamiltonian matrix, self-energies, NEGF algorithms

Because the 2D gas in an UTB FET is now described by the local pseudopotential V_{2D}, the number of x points in blocks $\mathbf{H}_{l,l}$, $\mathbf{H}_{0,1}$ of **Fig.6** is set by the real-space discretization and, in particular, for a $2p$ discretization order the number of x points is p [8]. In this work we decided to use a standard second order discretization (i.e. $2p=2$), such that $\mathbf{H}_{l,l}$, $\mathbf{H}_{0,1}$ can be reduced to a single x point (see **Fig.6**).

The drawback of a second order discretization is that the N_d points in an a_0 have to be increased to about 30 in order to have a correct description of the bands. The reduction in size of the blocks $\mathbf{H}_{l,l}$, $\mathbf{H}_{0,1}$ more than compensates the increase in the number of device sections, so that the overall

memory and CPU requirements are greatly improved by this new approach.

V. Self-consistent Device Simulations

We here report self-consistent simulations for ultrathin-body FETs and TFETs, with the device structure sketched in **Fig.1**. **Fig.8** shows I_{DS} versus V_{GS} curves for a $(001)/[110]$ p-FET for unstrained Si and for a compressive uniaxial stress in the [110] direction. The large I_{DS} values are due to the fact that neither scattering nor series resistance are included. As expected the stress increases I_{DS} in the on-state, which is ascribed to the reduction of the effective mass in **Fig.4**. Strained FETs, however, also have degraded sub-threshold swing, SS, and larger off current. This behavior is explained in **Fig.9**, reporting the subbands profile and the current spectral density $J_D(E)$. As it can be seen, the reduction of the effective mass implies an increase of source-to-drain tunnelling.

Then we analyzed strain in n-type InAs Tunnel FETs, where **Fig.10** shows the bandgap reduction due to uniaxial tensile stress for a $6a_0$ thick InAs film. **Fig.11** shows that the unstrained, defect free Tunnel FET has an SS well below 60mV/dec [5], and that strain improves I_{DS} at fixed I_{off}. However strain also degrades SS and leads to an undesirable ambipolarity of the I_{DS}-V_{GS} curve. **Fig.12** confirms the stress induced increase of $J_D(E)$ due to bandgap reduction.

VI. Conclusions

This paper has presented substantial new developments about the EP-NEGF methodology for quantum simulation of electron devices, comprising the inclusion of arbitrary crystal orientations and strain conditions. Several improvements in the numerical efficiency have been introduced, which made the simulation of technologically relevant devices computationally affordable. Our results qualify the EP-NEGF approach as a new method capable of a very promising balance between physical accuracy and numerical burden.

REFERENCES

[1] R. Kim *et al.*, *IEEE Trans. Electron Devices*, vol.62, no.3, pp.713, 2015.
[2] O. Badami *et al.*, *IEEE IEDM*, pp.13.2.1-4, 2017.
[3] D.Esseni *et al.*, Nanoscale MOS Transistors, *Cambr. Univ. Press*, 2011.
[4] A. C. Seabaugh *et al.*, *IEEE Proceedings*, vol.98,no.12,pp.2095,2010.
[5] D. Esseni *et al.*, *Semicond. Science and Techn.*, vol. 32, pp. 083005, 2017.
[6] S. De Franceschi *et al.*, *IEEE IEDM*, pp.339-342, 2016.
[7] M. Pala, O. Badami, D. Esseni, *IEEE IEDM*, pp.35.1.1-4, 2017.
[8] M.Pala and D.Esseni, *Phys. Rev. B*, vol.97, no.12, pp.125310, 2018.
[9] A.Garcia-Lekue *et al.*, Progress in Surf. Science,vol.90,pp.292,2015.
[10] J. Fang *et al.*, *Journal Appl. Phys.*, vol.19, pp.035701, 2016.
[11] G. Klimeck *et al.*, *IEEE Trans. Electr. Dev.*, vol.54, no.9, pp.2079, 2007.
[12] M. V. Fischetti *et al.*, *Journal Appl. Phys.*, vol.80, no.4, pp.2234, 1996.
[13] J. R. Chelikowsky *et al.*, *Phys. Rev. B*, vol.10, pp.5025, 1974.
[14] M.Cohen and J.R.Chelikowsky,*Electronic Structure and Optical Properties of Semiconductors* in Springer Series in Solid-State Sciences, 1988.
[15] M. C. Cohen *et al.*, *Phys. Rev. B*, vol.114, pp.789, 1966.
[16] Q. M. Ma *et al.*, *Phys. Rev. B*, vol.74, pp.1936, 1993.
[17] J. Munguia *et al.*, *Appl. Phys. Letters*, Vol.93, pp.102101, 2008.
[18] Stephan-Enzo Ungersböck, *PhD Dissertation*, 2007.
[19] S.M.Sze and Kwok K.Ng, Physics of Semiconductor Devices, *John Wiley & Sons*,2006.
[20] S. E. Thompson *et al.*, *IEEE IEDM*, pp.221-224, 2004.
[21] T. B. Boykin, *et al.*, Phys. Rev. B vol. 69, p. 115201-1 115201-10, 2004.
[22] F. Conzatti *et al.*, *IEEE Trans. Electr. Dev.*, vol.59, no.8, pp.2085, 2012.

Wavevectors, lattice vectors	$\mathbf{G}_c$, $\mathbf{G}=\mathbf{R}_{CD}\mathbf{G}_c$: Reciprocal lattice vectors in CCS and DCS $\qquad$ $\mathbf{K}=(\mathbf{k}+\mathbf{G})$ wave-vectors in DCS extended zone	(1)
Strained lattice	$\boldsymbol{a}=(\boldsymbol{I}_3+\boldsymbol{\varepsilon}_c)\boldsymbol{a}^0 \quad \Omega=\Omega_0(1+\varepsilon_{xx}+\varepsilon_{yy}+\varepsilon_{zz}) \quad \boldsymbol{b}_1=\frac{2\pi}{\Omega}(\boldsymbol{a}_2\times\boldsymbol{a}_3) \quad \boldsymbol{b}_2=\frac{2\pi}{\Omega}(\boldsymbol{a}_3\times\boldsymbol{a}_1) \quad \boldsymbol{b}_3=\frac{2\pi}{\Omega}(\boldsymbol{a}_1\times\boldsymbol{a}_2)$	(2)
Bulk crystal EP problem	$\sum_{\mathbf{G}'}\{T(\mathbf{k}+\mathbf{G})\delta_{\mathbf{G},\mathbf{G}'}+U_L(\mathbf{G}-\mathbf{G}')\}B_{\mathbf{k}}(\mathbf{G}') \qquad = \qquad E_b(\mathbf{k})\,B_{\mathbf{k}}(\mathbf{G}) \qquad \Phi_{nk}(\boldsymbol{r})=$ $\sum_{\mathbf{G}}B_{nk}(\mathbf{G})e^{i(\mathbf{k}+\mathbf{G})\cdot\boldsymbol{r}}$	(3)
Bulk crystal EP in $|\mathbf{K}\rangle$ space	$V_{3D}(\mathbf{K}-\mathbf{K}')=U_L(\mathbf{G}-\mathbf{G}')\delta_{\mathbf{k},\mathbf{k}'}, \qquad U_L(\mathbf{G}-\mathbf{G}')=U_S(|\mathbf{G}-\mathbf{G}'|)\cos[(\mathbf{G}-\mathbf{G}')\cdot\boldsymbol{\tau}], \qquad \boldsymbol{\tau}=\mathbf{R}_{CD}\boldsymbol{\tau}_c$, $\boldsymbol{\tau}_c=[\boldsymbol{I}_3+\boldsymbol{\varepsilon}_c]\boldsymbol{\tau}_c^0$	(4)
Dielectric response	$\varepsilon(\mathbf{q},\omega)=1-\lim_{\alpha\to0}\sum_{\mathbf{k},n,n'}\left\{|\langle\Phi_{nk}|\Phi_{n'(\mathbf{k}+\mathbf{q})}\rangle|^2\left[\frac{f_0(E_{n'}(\mathbf{k}+\mathbf{q}))-f_0(E_n(\mathbf{k}))}{E_{n'}(\mathbf{k}+\mathbf{q})-E_n(\mathbf{k}))-\hbar\omega-i\hbar\alpha}\right]\right\}$	(5)
2D gas: EP in real space	$V_{2D}(\boldsymbol{r})=V_{sc}(\boldsymbol{r})+[V_{ox}(\boldsymbol{r})-V_{sc}(\boldsymbol{r})]\Theta(z)=V_{sc}(\boldsymbol{r})+V_{cnf}(\boldsymbol{r})\Theta(z) \qquad V_{cnf}(\boldsymbol{r})=[V_{ox}(\boldsymbol{r})-V_{sc}(\boldsymbol{r})]$	(6)
Step function in $|\mathbf{K}\rangle$ space	$\Theta(\mathbf{K}-\mathbf{K}')=\Theta(K_z-K_z')\delta_{K_x,K_x'}\delta_{K_y,K_y'} \qquad \Theta(K_z-K_z')\approx-\left[\frac{\sin[(K_z-K_z')T_{sct}/2]}{2\pi(K_z-K_z')T_{sct}/2}\right]$	(7)
2D gas: EP in $|\mathbf{K}\rangle$ space	$V_{2D}(\mathbf{K}-\mathbf{K}')=U_{sc}(\mathbf{G}-\mathbf{G}')\delta_{\mathbf{k},\mathbf{k}'}+\sum_{G_z''}[U_{cnf}(\mathbf{G}_{xy}-\mathbf{G}_{xy}',G_z-G_z'-G_z'')\Theta(k_z-k_z'+G_z'')]\delta_{k_x,k_x'}\delta_{k_y,k_y'}$	(8)
2D gas EP problem	$\mathbf{H}_{\mathbf{k}_{xy}}(\mathbf{K},\mathbf{K}')=T(\mathbf{k}+\mathbf{G})\delta_{\mathbf{G},\mathbf{G}'}+V_{2D}(\mathbf{K}-\mathbf{K}')$	(9)

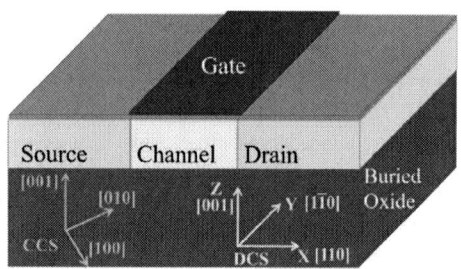

(Confinement)/[Transp.]	k_x $(2\pi/a_0)$	k_y $(2\pi/a_0)$	k_z $(2\pi/a_0)$
(001)/[100]	$[-1,1[$	$[-0.5,0.5[$	$[-1,1[$
(001)/[110]	$[-\sqrt{2},\sqrt{2}[$	$\left[-\frac{0.5}{\sqrt{2}},\frac{0.5}{\sqrt{2}}\right[$	$[-1,1[$

Fig. 1: Device (DCS) and Crystal Coordinate System (CCS) for an ultra-thin body (UTB) FET. The table reports examples of reduced $\mathbf{k}$ zones for different DCS. **(001)/[100]** corresponds to DCS≡CCS: $x=[100[$, $y=[010[$, $z=[001[$. **(001)/[110]** corresponds to $x=[110[$, $y=[1\bar{1}0[$, $z=[001[$.

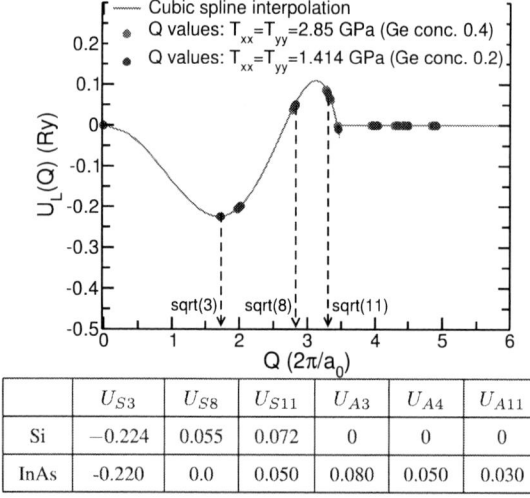

	U_{S3}	U_{S8}	U_{S11}	U_{A3}	U_{A4}	U_{A11}
Si	-0.224	0.055	0.072	0	0	0
InAs	-0.220	0.0	0.050	0.080	0.050	0.030

Fig. 2: Cubic spline interpolation curve used for the EP form factors $U_L(Q)$ employed in strained Si calculations, where we set $U_L(Q)=0$ for $Q=0$ and for $Q>\sqrt{12}$. Circles indicate examples of $Q=|\mathbf{G}|$ values other than $\sqrt{3}$, $\sqrt{8}$, $\sqrt{11}$ $[2\pi/a_0]$ that are necessary for strained Si calculations. The table reports the EP form factors (in Rydberg) used in all calculations for Si [13] and InAs [15]. For InAs both symmetric (U_{S3}, U_{S8}, U_{S11}) and antisymmetric form factors (U_{A3}, U_{A4}, U_{A11}) are reported. U_{S3} stands for $U_S(|\mathbf{G}|=\sqrt{3})$ and likewise for other similar symbols.

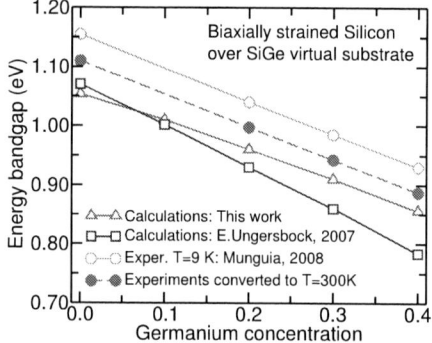

Fig. 3: Energy bandgap for biaxially strained silicon on a SiGe virtual substrate. Calculations of this work are in good agreement with experiments from [17], and with pseudopotential calculations from [18]. Experiments are converted from T=9K to T=300K by using $E_G(T)=E_G(T=0)-\alpha T^2/(T+\beta)$, with $\alpha=4.7\cdot10^{-4}eV/K$, $\beta=655$ K [19].

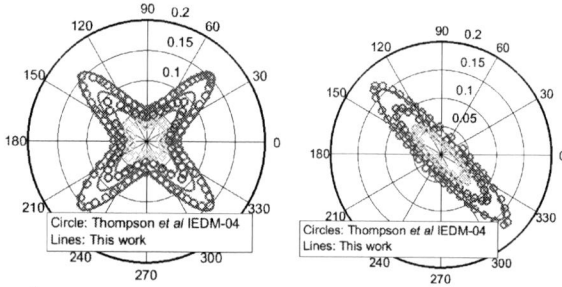

Fig. 4: Contour plots in the (k_x,k_y) plane corresponding to an energy 10, 25 and 50 meV below the top of the Si valence band and for $k_z=0$. Left graph; unstrained Si; Right graph: Si with compressive uniaxial stress of -1 GPa along [110] direction. A good agreement is found between our calculations (solid lines) and previous results (circles) from a $\mathbf{k}\cdot\mathbf{p}$ Hamiltonian in [20].

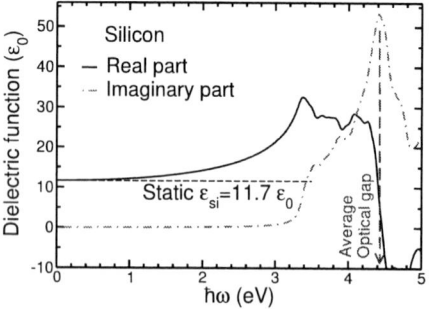

Fig. 5: Real and imaginary part of the dielectric function at $\mathbf{q}=0$ versus energy for silicon as calculated from **Eq.5**.

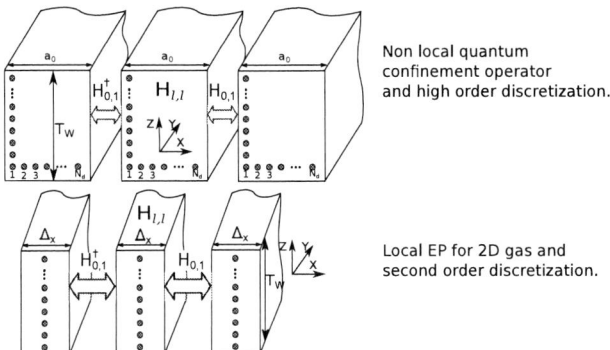

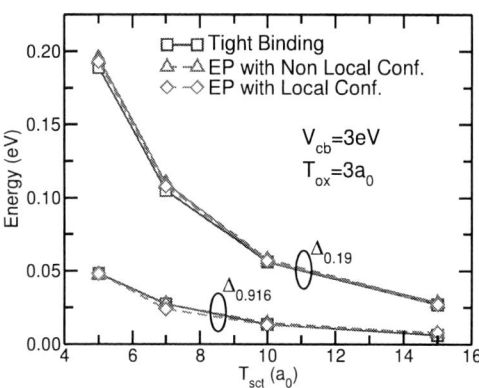

Fig. 6: Pictorial illustration for the block tridiagonal structure of the Hamiltoniana. Top: original formulation where each block includes N_d discretization points; Bottom: new formulation where each block includes a single discretization point.

Fig. 7: Lowest ($\Delta_{0.916}$) and second lowest conduction subband minimum ($\Delta_{0.19}$) located respectively at $(k_x,k_y)=(0,0)$ and $(k_x,k_y)=(0.85,0)$ (in units of $2\pi/a_0$) for ultra-thin silicon films versus T_{sct}. EP calculations are reported both for the local pseudopotential $V_{2D}(\boldsymbol{r})$ in **Eq.6**, and for the non local confinement operator [7]. EP calculations agree well also with tight-binding results with parameters from [21].

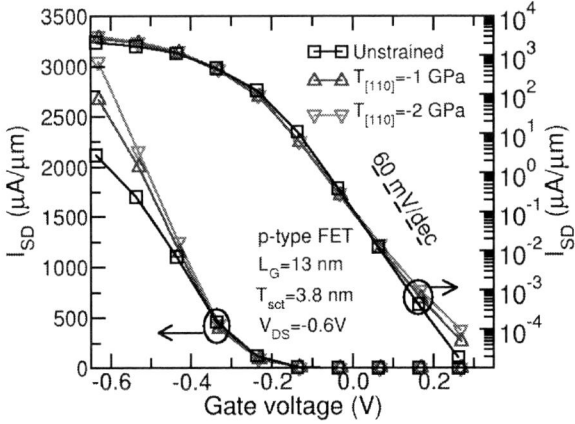

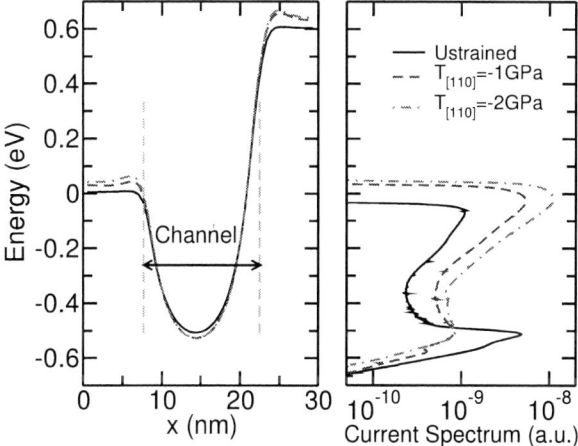

Fig. 8: Simulated I_{DS} versus V_{GS} characteristics at $V_{DS}=-0.6$ V for a p-type, Si FET with gate length $L_G \simeq 13$ nm and $T_{sct}=7a_0 \simeq 3.8$ nm. Quantization and transport directions are [001] and [110]. Results for unstrained Si and for compressive uniaxial stress in the [110] transport direction. Gate workfunction is 4.65 eV.

Fig. 9: Valence subband profile and corresponding current density plot for the p-type silicon FET of **Fig.8** and at $V_{GS}=+0.27$ V, $V_{DS}=-0.6$ V (OFF state). The source Fermi level E_{fS} is taken as the energy reference.

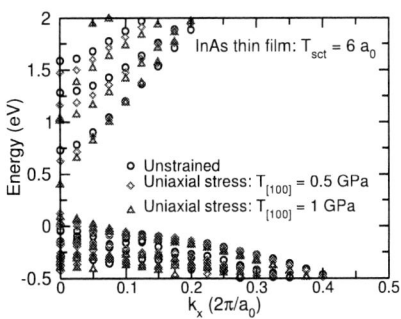

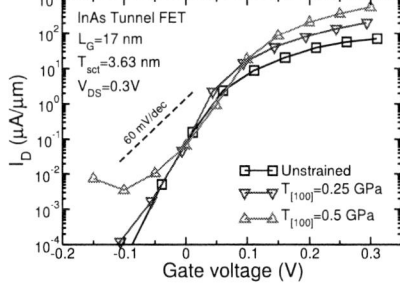

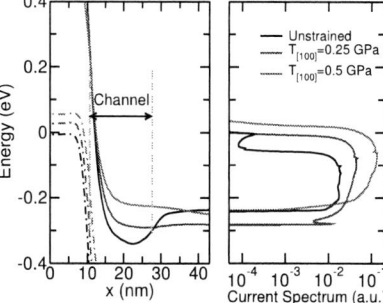

Fig. 10: Band-structure about the Γ point for an ultra-thin film of InAs with $T_{sct}=6\ a_0$. The uniaxial tensile stress along [100] tends to significantly reduce the energy gap.

Fig. 11: Simulated drain current versus gate voltage characteristics for an n-type, InAs Tunnel FET at $V_{DS}=0.3$ V with gate length $L_G \simeq 17$ nm and $T_{sct}=6a_0 \simeq 3.63$ nm. Gate workfunction is about 4.61 eV for the unstrained FET and it has been increased in strained FETs so as to have same $I_{off}=100$nA/μm at $V_{GS}=0$ V for all devices.

Fig. 12: Conduction and valence subband profile profile and corresponding current density plot for the InAs Tunnel FETs of **Fig.11** at $V_{GS}=V_{DS}=0.3$ V (ON state) and for different stress conditions. The source Fermi level E_{fS} is taken as the energy reference.

978-1-7281-1988-5/18 $31.00 © 2018 IEEE

First Principles Simulation of Energy efficient Switching by Source Density of States Engineering

Fei Liu[1,2], Chenguang Qiu[3], Zhiyong Zhang[3], Lian-Mao Peng[3], Jian Wang[2], Zhenhua Wu[4], and Hong Guo[2,5]

[1] Institute of Microelectronics, Peking University, Beijing 100871, China; [2] Department of Physics, The University of Hong Kong, Hong Kong, China; [3]Key Laboratory for the Physics and Chemistry of Nanodevices and Department of Electronics, Peking University, Beijing, China; [4] Key Laboratory of Microelectronics Devices and Integrated Technology, Institute of Microelectronics, Chinese Academy of Sciences, Beijing, China; [5]Department of Physics, McGill University, 3600 rue University, Montreal PQ, Canada H3A2T8; email:feiliu@pku.edu.cn, hong.guo@mcgill.ca

Abstract Achieving sub-60 mV/decade FET switching is critical for reducing power dissipation in integrated circuits. Here we propose and theoretically investigate steep slope switching made possible by a "cold source" that suppresses "hot" electrons at the thermal tail of the Fermi distribution. We show sub-60 mV/decade switching with: (i) using gapless/gapped graphene as injection source, (ii) introducing a band gap in the source of Si FET. The feasibility and design of the cold source are investigated by first principles on different metals, pocket doping and disorder.

INTRODUCTION

Power dissipation is a major issue facing modern transistor technology. Decreasing supply voltage is effective to reduce this issue. A steep switching that also maintains a reasonable on-off current ratio is very helpful [1,2]. Several steep slope devices exist which break the switching limit of conventional FETs, such as the tunneling FET (TFET) [3] and negative capacitance FET [4] etc. Very recently, we experimentally [5] and theoretically [6] demonstrated sub-60 mV/decade switching by injecting Dirac electrons from graphene. Such Dirac source FETs (DS-FET) with ballistic carbon nanotube (CNT) as the channel material have an average subthreshold swing (SS) of 40 mV/decade over four decades of current at room temperature, as well as promising on-state current [5]. The steep switching of DS-FETs is due to modulation of the source's density of states (DOS) via graphene – realizing a type of "cold source", together with the subsequent ballistic transport of injected carriers [5,6].

In this work, we investigate design considerations of cold source more generally. We compare physical mechanisms of conventional FET, TFET and the hereafter proposed cold source FET (CS-FET) which we show can simultaneously realize steep switching and high on-state current. By first principles atomic simulation we investigate CS-FETs with: (i) graphene as injection source and, (ii) a junction composed of a p-type Si (pSi), a thin metal (M) and a n-type Si (nSi) as injection source. The injected current density optimization of the pSi/M/nSi cold source is investigated by first-principles.

PRINCIPLE OF COLD SOURCE

Conventional FET is switched by modulating carriers over the channel barrier [Fig. 1(a)]. Electrons at the FET source follow the Fermi–Dirac distribution which has a tail at finite temperature, leading to the SS limit of $\ln(10)k_BT/q$ where k_B is the Boltzmann constant, T is the temperature and q the carrier charge. To compare, TFET switches by modulating quantum tunneling [Fig. 1(b)] rather than thermionic current, as a result its SS can break the 60 mV/decade physical limit.

For ballistic FETs, here we consider another possibility, namely by manipulating DOS of the source, as follows. The ballistic current is determined by Landauer-Büttiker formula:

$$I = \frac{2q}{h} \int dE\, T(E)D(E)[f(E - E_{FS}) - f(E - E_{FD})]$$

where $T(E)$ is the transmission probability, $D(E)$ the DOS of injected carriers; $f(E)$ the Fermi-Dirac distribution function, $E_{FS,D}$ the Fermi energy of the source (S) and drain (D). In n-type conventional FET, electrons are injected from metal where $D(E)$ is essentially independent of energy E. However, if injection is from some material having a strong energy dependent DOS (like graphene or having a band-gap near Fermi level), current can be switched-off faster because the thermal tail in $f(E)$ can be arranged to be "cut-off" by $D(E)$ in the above formula, thus breaking the switching limit of conventional FET. CS-FET implements this idea, Fig. 2. Compared with conventional FET, in CS-FET a cold source replaces the conventional source. One may think of at least three possible cold source structures [see Fig. 2(b)]: a Dirac material with linear dispersion [5,6]; a pSi/M/nSi junction; and a type-III heterojunction [7]. The latter two produce gapped cold source. Due to suppression of the thermal tail in the distribution [Fig. 2(c,d)], switching below 60 mV/decade becomes possible [Fig. 2(f)], as calculated by the Landauer-Büttiker formula with T(E)=1 (thermionic carriers) with various examples of energy dispersion [6] in Fig. 2(e).

CS-FET STRCTURE AND SIMULATION METHOD

Two kinds of CS-FETs are investigated: CNT FETs with graphene as the (gapless) injection source shown in Fig. 3(a), and Si FETs with a gapped pSi/M/nSi cold source [Fig. 7(a)]. The CNT CS-FETs with 3nm Y_2O_3 gate oxide have p-type graphene in the source. The source, drain and gate length are 20 nm. Electronic property of graphene-CNT heterostructure is calculated by density functional theory (DFT) with the VASP package. The calculated band gap of (10,0) CNT is 0.75 eV, and the energy difference between the Dirac point of the graphene and the CNT conduction band minimum (CBM) is 0.33 eV. For the pSi/M/nSi cold source, injection is determined by carrying out DFT within the nonequilibrium Green function (NEGF) formalism (NEGF-DFT) [8]. Finally, the CS-FET transistor simulation is done by self-consistently solving the Schrodinger and Poisson equations within NEGF, where a two-band Dirac model is used for graphene and CNT, four-band k•p model is used for Si and metal, and the band parameters were obtained by fitting to DFT results.

RESULTS AND DISCUSSION

CNT CS-FET. We first present the calculated transport property of CNT CS-FET in Fig. 3(a). The contact between graphene and CNT is the Schottky type [Fig. 3(b)]. P-type

978-1-7281-1988-5/18 $31.00 © 2018 IEEE

graphene is applied as the injection source and a p-n junction is formed in the graphene as shown in Fig. 4(a,b). Graphene p-n junction is realizable by voltage modulation or chemical doping. Pristine graphene is gapless, electrons can transport directly from the valence band on the p-type side to the conduction band on the n-type side. It was experimentally shown that a band gap up to 250 meV can be induced in metal-graphene contacts due to A-B sublattice symmetry breaking [9]. Hence, two types of Dirac source [Fig. 4(a, b)] are analyzed. Fig. 4(a) is for both p-type graphene and n-type graphene to be gapless. Fig.4(b) is for the situation of a band gap in the p-type graphene. Fig. 5 plots the calculated I_d-V_g of CNT CS-FETs. SS of 51 mV/decade is achieved by using p-type gapless graphene as the cold source. Theoretically, we derived [6] a SS formula of CS-FETs with the graphene cold source:

$$\frac{\partial V_G}{\partial \log_{10} I_{thermal}} = \frac{kT \ln 10}{q} \frac{1}{C_1} \left(1 - \frac{kT}{E_{Dirac} - \Phi_B}\right)$$

where C_1 describes the gate control; E_{Dirac} is the energy of the source's Dirac point; Φ_B the channel barrier. When $\Phi_B < E_{Dirac}$, SS as small as 40 mV/decade can be achieved from the above formula. Again, the reason is that due to the linear DOS of graphene, the number of injected carriers decreases concomitantly as the channel barrier is raised by the gate voltage [6]. Fig. 6(a) plots the band edge profiles and current density of CNT CS-FETs with gapless graphene at $V_g = 0.1$V. It is found that the top of the barrier is above the source Dirac point and thermionic current contributes about 95% of the total. As V_g is increased to 0.2V, the top of the barrier gets to below the source's Dirac point [Fig. 6(b)] and essentially 80% of total current is now thermionic. Fig. 6(c,d) plots band profiles and current density of CNT CS-FETs with *gapped* graphene have a gap E_g=100 meV: the SS is further reduced by the gap since the thermal tail is more effectively suppressed. The inset of Fig. 5 shows that SS decreases linearly with the size of the gap.

Injection by pSi/M/nSi cold source. We now investigate Si CS-FETs shown in Fig. 7(a): it is a conventional FET but with a pSi/M/nSi type cold source discussed above. Notably, electrons above the VBM of the pSi in the cold source is suppressed by its gap [Fig. 7(b)]. Such a cold source can be realized with different metals as shown in Fig. 7(c,d,e). Two dimensional semi-metals of graphene and 1T-Phase metal dichalcogenides are also applied to connect pSi with nSi, which avoid metal atom diffusion into Si. The interface distance between the metal and Si is relaxed by DFT. Transport property of pSi/M/nSi cold source is determined by NEGF-DFT [8]. The transport direction is assumed along the x-axis. Fig. 8(a) plots the local DOS along the transport direction. Electrons from pSi traverse through the thin metal to nSi and the transport channels are confined between the pSi VBM and the nSi CBM [Fig. 8(b)]. The transmission spectra of the cold source with different metals are compared in Fig. 9(a), while the calculated current densities of 8 nm Si cold source (along z-axis, Fin width of Intel 14 nm technology) by NEGF-DFT [8] are shown in Fig. 9(b). These current densities are larger than 1×10^3 μA/μm except using the 1T phase MoSe$_2$ as the thin metal. The injected

current densities can be further improved by using pocket doping [Fig. 9(c)]. Effects of Si vacancy and metal atom diffusion to Si are shown in Fig. 10(a,b). Transport calculations of four random configurations show that the current density is actually increased due to Si vacancy at the metal-nSi interface and fluctuates with different Au diffusion configurations.

Si CS-FET. The calculated device performance of Si CS-FETs shown in Fig. 7(a) is compared with conventional Si FETs and Si TFETs at V_d =0 .5 V, in Fig. 11(a). The Si CS-FETs have the lowest SS and largest on-state current. The I_{on} reaches 6.6×10^2 μA/μm with I_{off} fixed at 20 pA/μm. At V_g=0.06V, the current is mainly contributed by tunneling [Fig. 11(b)]. As V_g is increased to 0.16 V, the current is mainly contributed by thermionic electrons going over the channel barrier [Fig. 11(c)]. Finally, I_{on}/I_{off} ratio as a function of I_{on} is compared in Fig. 11(d). Si CS-FETs has the largest I_{on}/I_{off} ratio at the I_{on} between 1~8.3×10^2 μA/μm.

SUMMARY

Ballistic cold source FET is studied by quantum transport modeling and first principles materials simulation. Results reveal that sub-60 mV/decade switching can be achieved by replacing the source of conventional FETs with a cold source to filter out high energy electrons at the tail of the Fermi-Dirac distribution. CS-FETs using graphene and a pSi/M/nSi junction as the cold source are calculated. Very reasonable current density can be injected from the pSi/M/nSi cold source and further enhanced by pocket doping. Compared with conventional FETs and TFETs, CS-FETs using Si show very promising device performance at low supply voltage.

REFERENCES

[1] A. C. Seabaugh and Q. Zhang, "Low voltage tunnel transistors for beyond CMOS logic," Proc. IEEE, vol. 98, no. 12, pp. 2095-2110, Dec. 2010.

[2] A. M. Ionescu and H. Riel, "Tunnel field-effect transistors as energy efficient electronic switches," Nature, vol. 479, no. 7373, pp. 329-337, Nov. 2011.

[3] J. Appenzeller, Y. M. Lin, J. Knoch, and P. Avouris, "Band-to-band tunneling in carbon nanotube field-effect transistors," Phys. Rev. Lett., vol. 93, no. 19, p. 196805, Nov. 2004.

[4] S. Salahuddin and S. Datta, "Use of negative capacitance to provide voltage amplification for low power nanoscale devices," Nano Lett., vol. 8, no. 2, pp. 405-410, Feb. 2008.

[5] C. Qiu, F. Liu, L. Xu, B. Deng, M. Xiao, J. Si, L. Lin, Z. Zhang, J. Wang, H. Guo, H. Peng, and L.-M. Peng, "Dirac source field effect transistors as energy-efficient and high-performance electronic switches," Science, vol. 361, pp. 387-392, Jun. 2018.

[6] F. Liu, C. Qiu, Z. Zhang, L.-M. Peng, J. Wang, and H. Guo, "Dirac electrons at the source: breaking the 60-mV/decade switching limit," IEEE Trans. Electron Devices, vol. 65, no. 7, pp. 2736-2743, Jul. 2018.

[7] J. T. Smith, S. Das, and J. Appenzeller, "Broken-gap tunnel MOSFET: A constant-slope sub-60-mV/decade transistor," IEEE Electron Device Lett., vol. 32, no. 10, Oct. 2011.

[8] J. Taylor, J. Wang, and H. Guo, "Ab initio modeling of quantum transport properties of molecular electronic devices," Phys. Rev. B, 63, p. 245407, Jun. 2001. Software website: www.hzwtech.com

[9] S. Y. Kwon, C. V. Ciobanu, V. Petrova, V. B. Shenoy, J. Bareno, V. Gambin, I. Petrov, and S. Kodambaka, "Growth of semiconducting graphene on palladium," Nano Lett. vol. 9, no. 12, pp. 3985-3990, Sep. 2009.

[10] G. V. Hansson and S. A. Flodstrom, "Photoemission study of the bulk and surface electronic structure of single crystals of gold," Phys. Rev. B vol. 18, p. 1572, Aug. 1978.

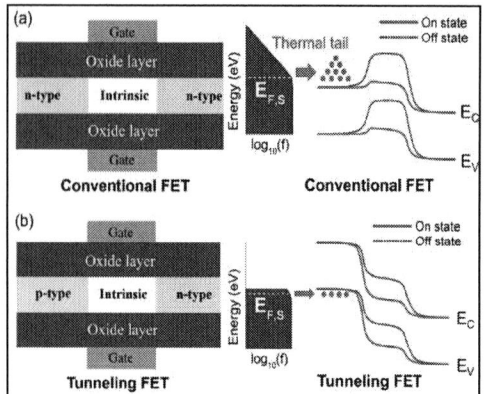

Fig. 1 Device structures and working mechanism of (a) conventional FET and (b) TFET. Conventional FET relies on thermionic current and subjected to the SS limit of ln(10)kT/q. This limit is broken in TFETs by quantum tunneling.

Fig. 2 Device structure and working mechanism of the cold source FET. (a) A cold source replaces the source of conventional FET. (b) Three cold source structures. Working mechanisms of FET with: (c) Dirac source (d) gapped cold source. (e) Various DOS vs energy. (f) SS of thermionic current versus gap for different DOS.

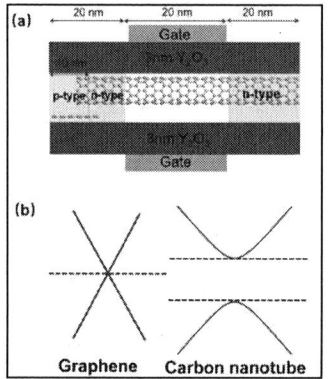

Fig.3. (a) Device structure of Dirac source FET using graphene and carbon nanotube (CNT). (b) Band alignment between graphene and CNT obtained by DFT.

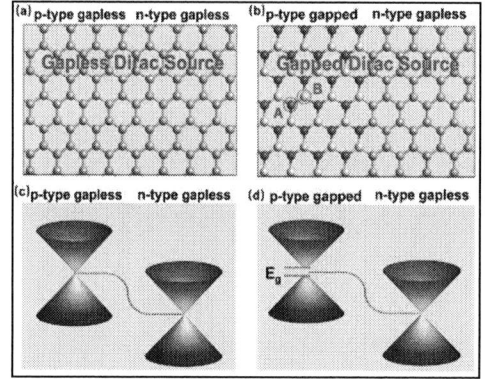

Fig.4. (a,b) atomistic structures and, (c, d) band alignments, of graphene p-n junctions as injection source. Graphene p-n junction can be realized by voltage modulation or chemical doping. By breaking A-B carbon atom symmetry, a small band gap is induced in graphene (b, d).

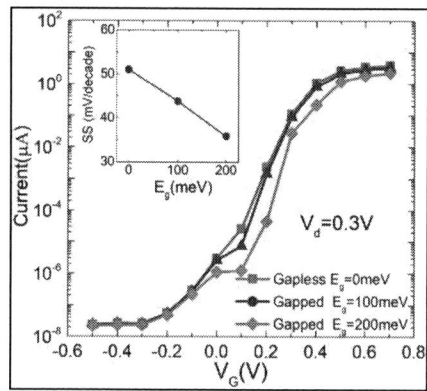

Fig. 5 Transfer characteristics of Dirac source FETs with gapless and gapped graphene. SS < 60 mV/decade is obtained which decreases further with the size of the gap.

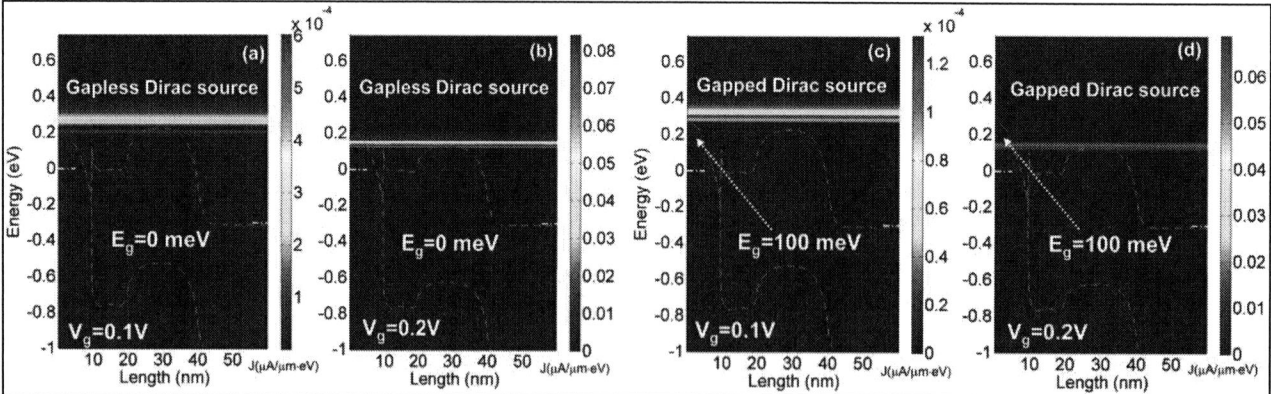

Fig.6 Local DOS of Dirac Source FETs with (a, b) gapless and (c, d) gapped graphene as the injection source. By increasing gate voltage, the top of channel barrier is switched below the energy of source's Dirac point or gapped graphene. Sub-60 mV/decade switching is achieved due to the linear DOS of graphene and the gapped graphene.

978-1-7281-1988-5/18 $31.00 © 2018 IEEE

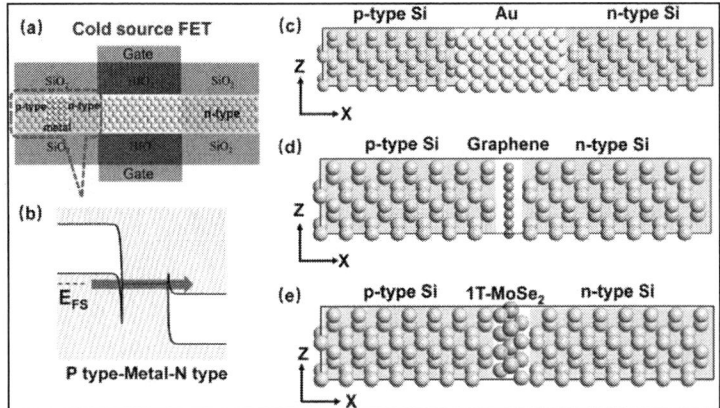

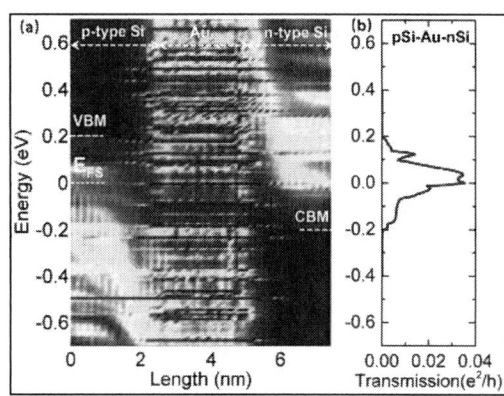

Fig. 7. (a) FET with a cold source made by pSi/M/nSi. (b) Band alignment in pSi/M/nSi. Such cold source can employ different metals: (c) Au, (d) graphene, (e) two-dimensional metallic material of 1-T phase MoSe2.

Fig.8. (a) Local DOS and, (b) transmission of pSi/Au/nSi cold source with 2.5 nm thickness pSi, 2.5 nm Au and 2.5nm nSi. Injected electrons are confined between VBM of pSi and CBM of nSi.

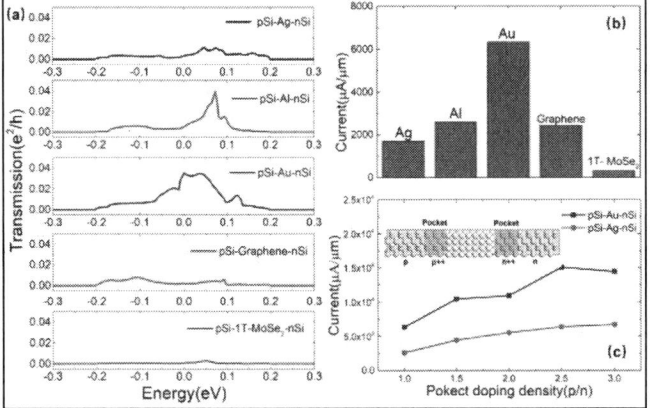

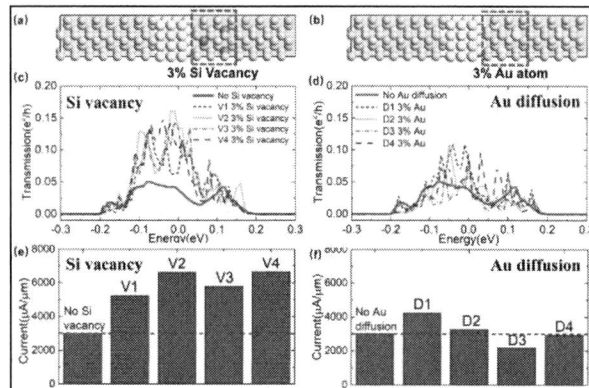

Fig. 9. Transport property of pSi/M/nSi cold sources. (a) Transmissions versus energy of pSi/M/nSi with different metals. (b) Projected injection current density of 8 nm-thick (Fin width of Intel 14 nm Fin-FET technology) pSi/M/nSi cold sources with different metals. (c) The injected current is improved by heavier pocket doping near the interfaces between Si and the metal.

Fig. 10. Disorder effects on transport property of pSi/Au/nSi cold source with 0.7 nm Au, 2.5 nm pSi and nSi. (a) Si vacancy and (b) Au atom diffusion appearing in 1nm nSi at the Au/nSi interface. (c, d) Transmission versus energy for four randomly generated samples (V1–V4, D1–D4). (e, f) The projected current densities of different disorder configurations.

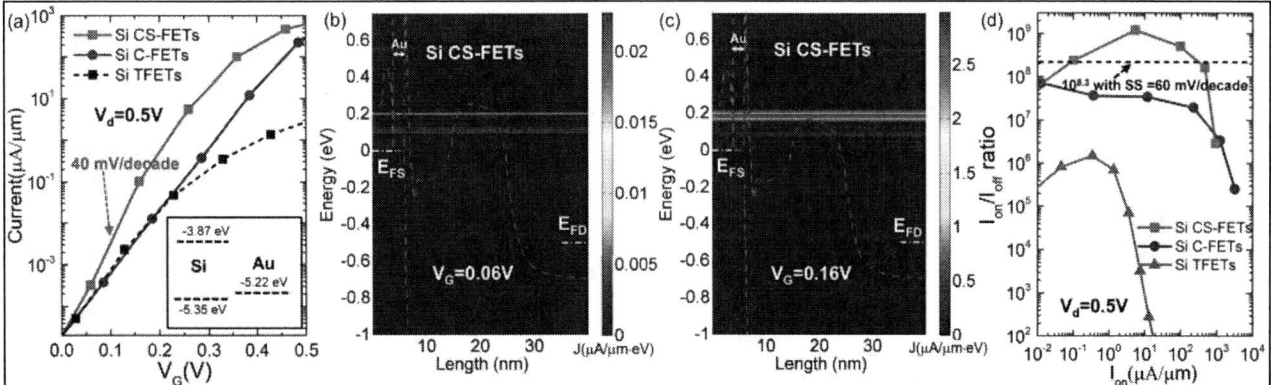

Fig. 11. Device performance of Si CS-FETs with pSi-Au-nSi cold source, 3nm thick Si, 10 nm gate length, 15 nm source and drain at V_d=0.5V. Band structure and work function of 3 nm Si film are calculated at the HSE hybrid functional level of DFT. The work function of gold film is -5.22 eV [10]. (a) I_d-V_g of Si CS-FETs, conventional FETs (Si C-FET) and TFETs. pSi-Au-nSi cold source has 4 nm pSi and 2 nm Au. (b, c) Band edge profiles of Si CS-FETs with pocket doping to boost injecting current. (d) I_{on}/I_{off} ratio as a function I_{on} for three different FETs. Si CS-FETs show the best switching property with I_{on} between 1 ~ 8.3×10^2 μA/μm.

978-1-7281-1988-5/18 $31.00 © 2018 IEEE   766

Universal Swing Factor Approach For Performance Analysis Of Logic Nodes

M. Ali Pourghaderi[1], Anh-Tuan Pham[2], Seungkyu Kim[1], Hyein Chung[1], Zhengping Jiang[2], Hesameddin Ilatikhameneh[2], Hong-hyun Park[2], Seonghoon Jin[2], Jongchol Kim[1], Won-Young Chung[1], Uihui Kwon[1], Woosung Choi[2], Dae Sin Kim[1] and Shigenobu Maeda[1]

[1]Semiconductor R&D center, Samsung Electronics, Hwasung-si Gyeonggi-do, South Korea, email: ali.p@samsung.com
[2]Device Lab, AHQ(DS) R&D, Samsung Semiconductor Inc., San Jose, CA, USA

Abstract—Deterministic Boltzmann-transport solver has been integrated in performance analysis of logic cells. Employing universal-swing-factor (USF) approach, our setup accurately entails quasi-ballistic transport effects. The injection current and carrier mean free path (MFP) have been extracted for various channel dimensions and interface qualities. The resulting database is used to study candidate architecture for logic nodes. In particular, performance of ring oscillator (RO) with tapered FinFET is presented. For a given junction profile and contact-poly-pitch (CPP), the optimum gate-length (Lg), spacer thickness and contact-CD (CCD) are evaluated. The feasible gain by lowering the spacer k-value and contact resistance is also reported.

I. INTRODUCTION

The relentless down scaling of logic cells has pushed the transistor dimensions to quasi-ballistic regime. While the simulation accuracy can be hardly overrated, the conventional drift-diffusion (DD) model is widely used for aggressively scaled devices. This is mainly due to the overwhelming computational burden of carrier transport solvers in handling the practical dimensions. As the remedy of this issue, variant modifications of DD model have been proposed [1, 2]. For rather simplistic structures, the up-graded DD can effectively reproduce the result of microscopic transport solvers. However, the model parametrization may not be transferable to practical complex structures. Previously, we have reported the potentials of USF method to handle arbitrary large cross section [3]. In this study, the model has been extended to cover whole IV curves. The updated form is integrated in design-technology co-optimization (DTCO) flow, Fig. 1 (a), which applies to practical devices. As it is shown in Figs. 1 (b)-(c), introduction of USF-unit corrects the ballisticity error, which may grow up to 20% depending on Lg and mobility.

To include low V_{DS} bias points, original USF form has been modified, as follows:

$$I_{DS} = \frac{I_{inj0} U_2(\Gamma) \times (1 - \exp(-qV_{DS}/kT))}{1 + \exp(-qV_{DS}/kT) + \frac{L_{Eff} + \Lambda}{MFP} U_1(\Gamma)} \quad (1),$$

where kT/q is thermal voltage and Γ is the ratio of effective channel and natural scaling length; L_{Eff}/λ. I_{Inj0} and MFP are long channel transmission properties representing effective injection and normalized mean free path. Λ is offset length accounting for modulation of L_{Eff} in on- and off-state. U_1 and U_2 are universal function calculated for each bias point. To extract MFP and I_{Inj0}, in-house Multi-Subband Boltzmann Transport Equation (MSBTE) solver has been employed [4].

The simulation sets have been arranged to cover relevant range of channel thicknesses (T_{Ch}) and rms heights of surface roughness (Δ). This database is employed to benchmarked DC performance of FinFET and stacked-nanowires (NW) in different CPPs. Moreover, the standard AC performance is reported for ring oscillator cell.

For all simulations in this study, I_{DS} is fixed as 0.1nA/μm at V_{GS}=0V and V_{DS}=0.7V. The junctions have Gaussian profile with abruptness and proximity of 5nm/Dec and 7nm, respectively. The peak of source/drain doping is 3×10^{20} cm^{-3}. The channel direction is <110>, while the dominant surface orientation is (110). The oxide structure is composed of 0.8nm SiO$_2$ and 1.5nm HfO$_2$ layers.

II. CALIBRATION

Advanced technology nodes inevitably deal with low dimensional systems. Consequently, modeling of such devices ought to consider the confinement effects meticulously. In particular, aggressive thinning of channel will enhance the surface phenomena and trigger drastic deviations from bulk parameters. To address these effects, detailed calibration of band structure and scattering parameters has been conducted for all presented dimensions. Figs. 2(a) and (b) show k.p band versus reference tight-binding simulations. In order to capture the mobility degradation of thin channels, surface-phonon and sound-velocity modulation have been taken into account. The subsequent scattering machinery can reproduce the measured electron- and hole-mobility data [5-6], as demonstrated in Figs. 2, (c)-(d). As it is shown in Fig. 3(a), thinner channels come with shorter λ and better electrostatics control, while mobility will be degraded. The dependency of mobility components to T_{Ch} has been shown in Fig. 3(b). As expected, the mobility drop of electrons and holes follows different patterns. Though, phonon-limited mobility (Ph-Mob) shows similar monotonic drop, surface roughness components behave differently for electrons and holes. Surface roughness mobility (SR-Mob) is proportional to power six of effective quantum well thickness; $\propto (QW)^6$ [7]. In case of surface-inversion, the total quantum well area, TQW, accounts for two parallel channels. Therefore, the effective quantum well thickness is proportional to TQW/2. In case of volume-inversion, quantum well is geometrically confined to device cross section, hence QW $\propto$ TQW $\propto$ T_{Ch}. By transition from surface- to volume-inversion, SR-Mob will softly switch from $(TQW/2)^6$ to $(TQW)^6$. This transition happens at different thicknesses for electrons and holes, Fig. 4, which accounts for the difference in SR-mob pattern.

III. USF APPLICATION FOR PERFORMANCE ANALYSIS

The first step in application of USF form is to find L_{Eff} and λ for target transistors. For a given junction and channel profile, the DD solver provides a SS factor over the desired range of Lg. Due to universality of SS for arbitrary device dimensions and geometries, the corresponding Γ factor can be deduced, Fig.5. The linear regression of Γ vs Lg yields the values of L_{Eff} and λ. Finally, DC current will be calculated using Eq. (1) for a given surface roughness value. The USF-corrected IV curves are identical to DD results in off-states and entail quasi-ballisticity corrections in on-state, as shown in Fig. 6. Typical DC performance for various channel cross sections has been shown in Fig. 7(a). The tradeoff between electrostatics boost and mobility degradation manifests itself in performance curves. For long channels, the electrostatics is well controlled and mobility value dominates DC performance. As the result, thick channels outperform the thinner one over all long Lg devices. For short channel devices, the performance is mainly swing-limited. Therefore, some portions of mobility can be sacrificed to recover the electrostatics. Effectively, this can be achieved by thinner channels or 2D confined cross sections, NWs. Since the mobility degradation is not drastic for pmos devices, the benefit of channel thinning is more pronounced for p-channels. The USF form can illustrate this tradeoff in a straightforward manner, Fig. 7(b). To avoid short channel effect, U_2 function has to be shifted in left direction of Γ axes. This can be achieved by scaling the natural length, i.e. stronger confinement, which in turn triggers MFP degradation. Therefore, confined cross sections will shift U_2 in desired direction, while they hamper the transmission rate. Since the on-current is product of U_2 and transmission rate (T-Fin/T-NW), a shift and scaled pattern will be emerged. Fig. 8 demonstrates the reduction of MFP by scaling of T_{Ch} or NW height. It is worth noting the resemblance of MFP profile and mobility response in Fig. 3(a). DC performance of stacked NW and corresponding FinFET device are benchmarked in Figs. 9 10. The Fin sidewall is assumed to be perfectly straight. Although such a Fin-shape is not practical, this structure can serve as the upper limit of FinFET electrostatics. The distinctive feature of FinFET and stacked NW is highlighted in Fig. 9. In short channels, FinFET performance is highly sensitive to Lg scaling, while NW result is dominated by surface roughness scattering. The dependency of DC performance to wire height, Fig.10, is consistent with the aforementioned shift and scale pattern. Results in Figs. 9-10 do not include contact resistance. For fair comparison, one should fix CPP and contact scheme. The typical nonlinear relation of contact resistance and CCD is shown in Fig. 11(a). Fixing the spacer thickness as 9nm, we have benchmarked FinFET and stacked nanowires in Fig. 11 (b). In the following, we consider three different CPP scales; CPP_A as nominal design, CPP_B=CPP_A-6nm and CPP_C=CPP_A-10nm. For CPP_A, there is enough room to place desired Lg and CCD as such that performance is not compromised. For such a relaxed node, FinFET will outperform NWs. As we move to tighter CPPs, a severe compromise between Lg and CCD will be

developed For FinFET devices. For CPP_C, Lg with acceptable swing squeezes CCD to very tight values, which results in high contact resistance. On the other hand, the design with desired CCD imposes very short Lg and inevitably performance becomes swing-limited. In such nodes, NW can effectively relieve the compromise. Due to superb electrostatics control, NWs can be scaled to very short Lg, opening enough room to place the contact. Any progress in contact and spacer scheme will improve the situation for FinFET. In the absence of such options, stacked NW might be an interesting candidate. In Fig. 12, the peak performance of stacked NWs is normalized to the corresponding FinFET results. Interestingly, NWs outperform FinFET at CPP_C for all relevant Δ. Finally, the performance of NOT gate in CPP_A is presented in Fig. 13. The transistor has tapered Fin shape. A representative MFP has been extracted by weighted average over the range of Fin thickness. The weight function in Fig. 8(b) and (c) is adopted consistently with current-density in DD simulations and thickness distribution along the Fin height. Since, the sidewall slope changes abruptly along the Fin height, the resulting weighting function is discontinuous. Following the steps in Fig. 1(a), DD simulation results are corrected in the USF-unit. Then, CVs and corrected-IVs are exported to the compact model unit. In parallel to these steps, parasitic RCs are extracted from standard layout and the results are passed to SPICE-unit. To find the optimum design, various combinations of CCD and spacer thickness have been simulated. For all considered devices, junction profile is fixed, independently of spacer profile. Fig. 13(a) shows the contour plot of RO frequency for default spacer k-value and contact resistivity profile, the black line in Fig. 11(a). For the given junction, the optimum design is laid along a constant L_{Eff}. If the spacer k-value is reduced by 20%, the peak frequency is increased by 4%, Fig. 13(b). If the contact resistance is reduced by 30%, the peak frequency will be increased by ~6%, Fig. 13(c). The superposition of spacer k-value and contact resistance reduction gives ~10% performance boost, as shown in Fig. 13(d).

IV. CONCLUSION

To incorporate quasi-ballistic effects in practical device simulations, a hybrid scheme has been devised. Universal-Swing-Factor unit is integrated into conventional DTCO flow; as such it can produce accurate IVs for short channel devices with complex cross sections. To study the interplay of electrostatics and mobility in scaled channels, the database of carrier mean-free-path has been extracted. This parameter set covers relevant channel dimensions and interface qualities. As an application, some basic CPP constraints on device performance are presented.

References: [1] A. Erlebach *et al., ESSDERC 2016*, Sept 2016, pp. 420–423. [2] O. Penzin *et al.,* TED vol. 64, NO.11, Nov. 2017 pp. 4599–4606. [3] A. Pourghaderi *et al.,* EDL vol. 39, Issue 2, Feb. 2018 pp. 168-171. [4] S. Jin *et al., SISPAD 2013.* pp. 348–351. [5] K. Uchida, *et al., IEDM Tech. Dig.,* 2002, p. 47. [6] K. Uchida, *et al., IEDM Tech. Dig.,* 2003, p. 805. [7] K. Uchida, *et al., APL.* vol. 82, Issue 17, pp 2916-2918. [8] C. Auth, *et al., IEDM Tech. Dig.,* 2017, p. 673.

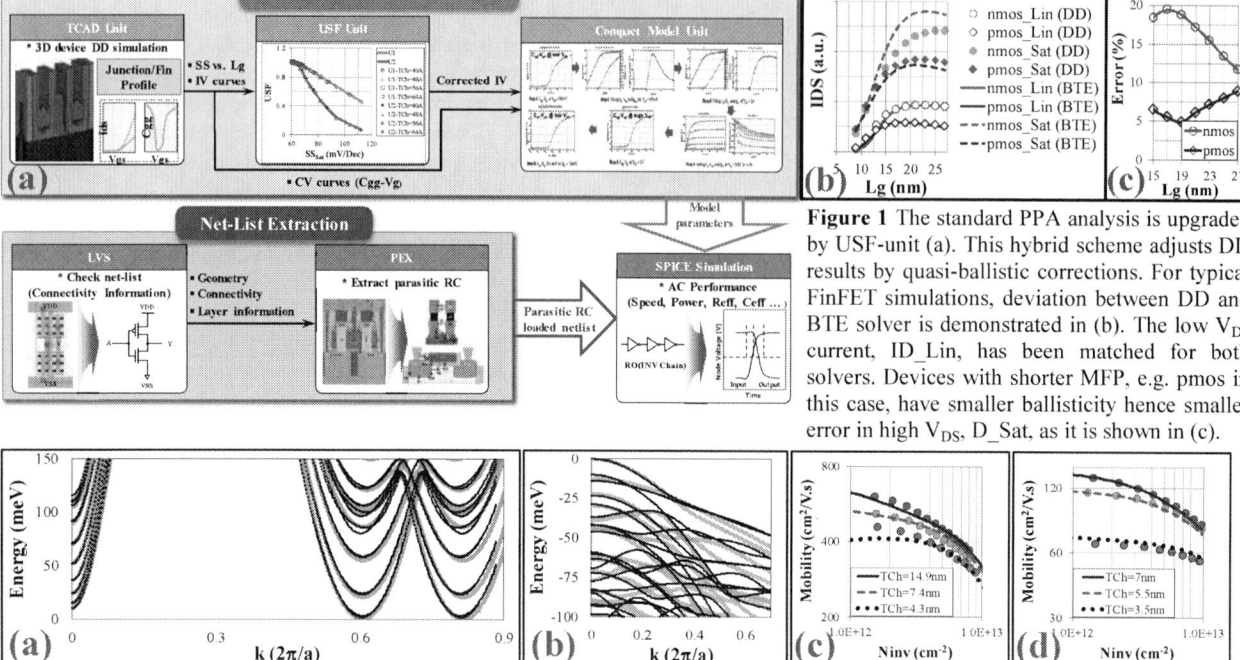

Figure 1 The standard PPA analysis is upgraded by USF-unit (a). This hybrid scheme adjusts DD results by quasi-ballistic corrections. For typical FinFET simulations, deviation between DD and BTE solver is demonstrated in (b). The low V_{DS} current, ID_Lin, has been matched for both solvers. Devices with shorter MFP, e.g. pmos in this case, have smaller ballisticity hence smaller error in high V_{DS}, D_Sat, as it is shown in (c).

Figure 2 k.p conduction and valance band calibration for 5x15nm wire are shown in (a) and (b), respectively. Dots refer to tight-binding results and lines are corresponding k.p bands. Mobility calibration is demonstrated in (c) and (d) for electron and hole in (100)/<110> surface/channel orientations. Lines refer measured data [5-6] and symbols are linearized MSBTE simulation results.

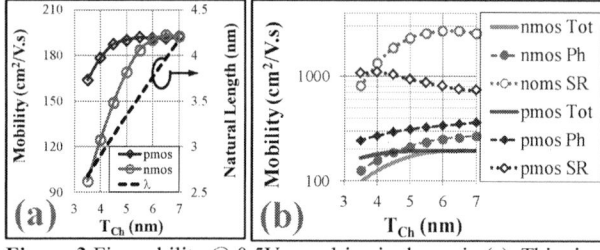

Figure 3 Fin mobility @ 0.5V overdrive is shown in (a). Thinning the channel will linearly reduce λ at the cost of mobility. Ph-and SR- mob are decomposed in (b). Phonon scattering is the dominant mechanism. For pmos devices, -1.2GPa uniaxial strain is applied.

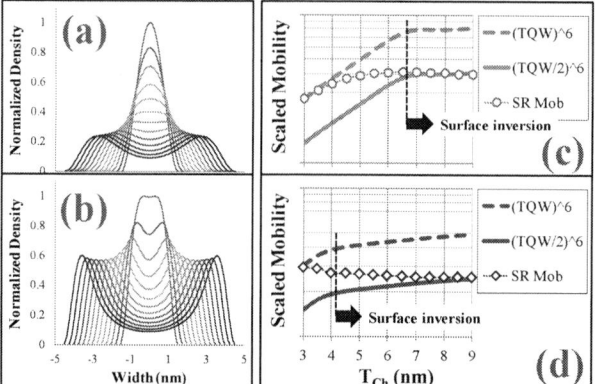

Figure 4 Inversion pattern of electron and hole are shown in (a) and (b), respectively. The overdrive is fixed as 0.5V and T_{Ch} is swept from 3 to 9nm with 0.5nm steps. SR-Mob and TQW profile for electron and hole is shown in (c) and (d), respectively.

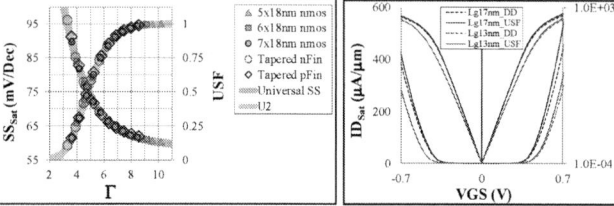

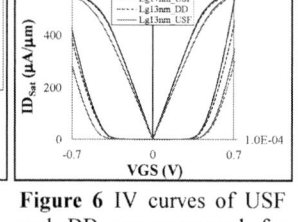

Figure 5 Employing universality of SS, the corresponding Γ for the given SS facctor is deduced. By linear regression of Γ vs. Lg, one can extract λ for target transistor. SS is obtained @ $I_{DS} = 0.1$nA/μm from DD results. Fixing λ value, one can evaluate USF functions for arbitrary Lg.

Figure 6 IV curves of USF and DD are compared for tapered Fin. As expected, IV curves are on top of each other in off-state. The deviations in on-state are due to quasi-ballistic corrections. For these results, Δ is fixed as 4Å. The corrected IVs, full-lines, will be used for CM parameter extraction.

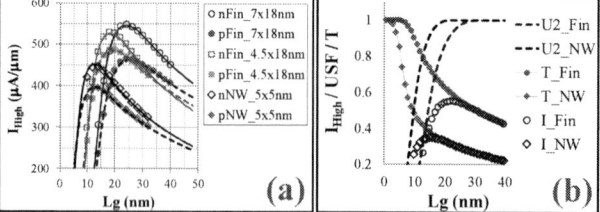

Figure 7 I_{High} for 7x18, 4.5x18 and 5x5 nm NWs are compared in (a). For these results, Δ is fixed as 4Å. Symbols are calculation by MSBTE and lines are USF results. Decomposition of USF components for 7x18 and 5x5nm nmos device is shown in (b). "High" corresponds to V_{GS}=0.7V & V_{DS}=0.35V bias points.

978-1-7281-1988-5/18 $31.00 © 2018 IEEE

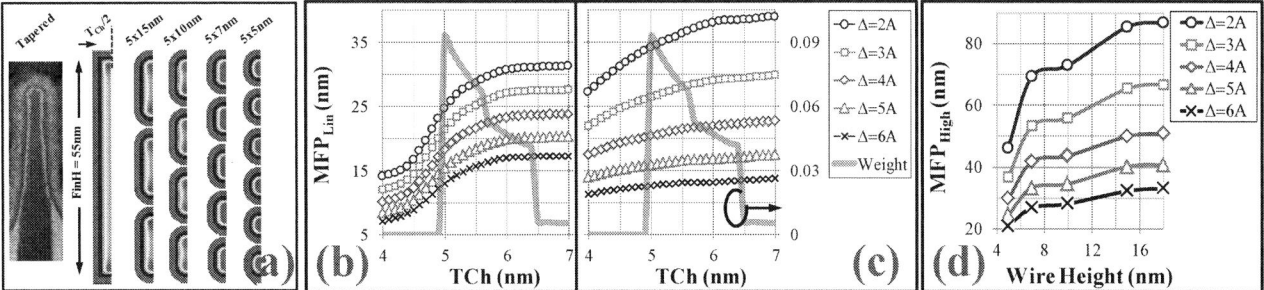

Figure 8 MFP for practical range of Δ and different channels in (a) are calculated. Tapered Fin figure is just schematic example from [8]. nFin and pFin MFP are shown in (b) and (c), respectively. The response to T_{Ch} scaling is similar to Fig. 3 (a). For tapered Fin, the weight function is used to extract the representative MFP. For pmos NW, the reduction of MFP due to wire height scaling is shown in (d). In these graphs "Lin" refers to $V_{GS}=0.7V$ & $V_{DS}=0.05V$.

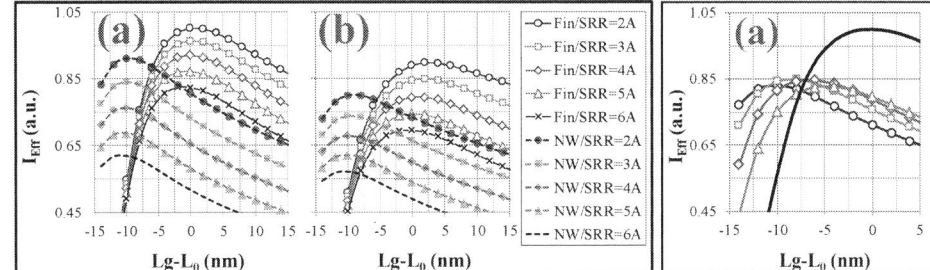

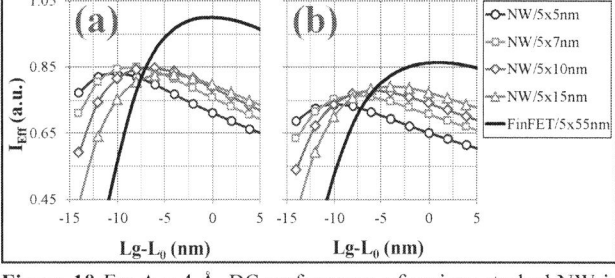

Figure 9 DC performances of 5x55nm FinFET and 5x5nm stacked NW are compared for different Δ. L_0 is a fix number. I_{Eff} is average of I_{High} and I_{Low}, where "Low" refers to $V_{GS}=0.35V$ & $V_{DS}=0.7V$. The results in (a) is for nmos devices and graph (b) represents pmos data. For pmos devices, -1GPa uniform uniaxial strain is applied.

Figure 10 For Δ = 4 Å, DC performance of various stacked NW is compared with 5x55nm FinFET. The results in (a) is for nmos devices and graph (b) represents pmos data. For pmos device, -1GPa uniform uniaxial strain is applied. These results exclude the impact of contact resistance.

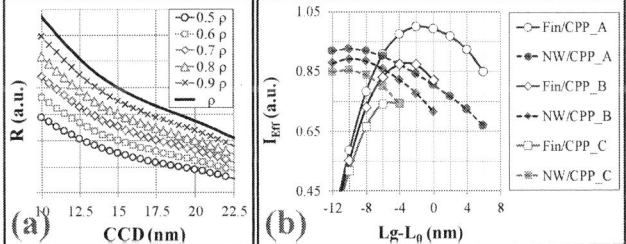

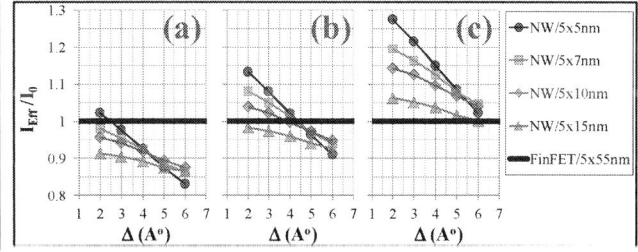

Figure 11 Nonlinear relation of contact resistance and CCD is shown in (a). DC performance of nmos 5x55nm FinFET and 5x5 nm stacked NW is compared in (b). The spacer thickness is 9nm and Δ is fixed as 4Å for all CPPs. It should be noticed that the peak performance of FinFET and NWs occurs in different Lg. The sensitivity to CPP scaling is larger for FinFET results.

Figure 12 The peak performance of nmos NWs are normalized to FinFET results. For CPP_A, CPP_B and CPP_C, the results are shown in (a)-(c), respectively. The slope of performance degradation by surface roughness is largest for 5x5nm cross sections. For all relevant Δ, NWs outperform FinFET @ CPP_C.

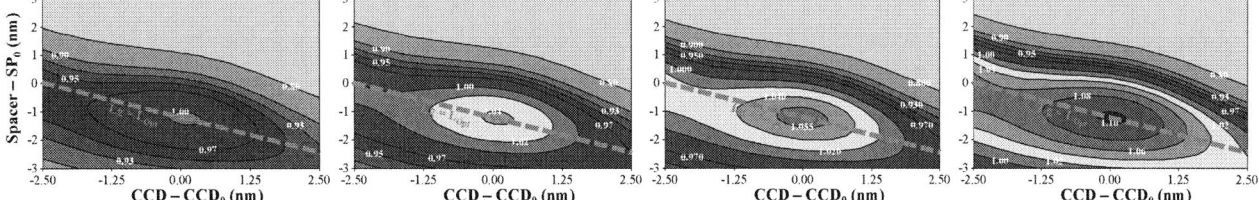

Figure 13 RO composed of 101 inverters with fan-out 3 for each stage has been considered @ CPP_A. The layout is prepared for various combinations of spacer thickness and CCD. -1GPa uniform uniaxial strain is applied to all pmos devices. Δ is fixed as 4Å. For each combination, the frequency value corresponding to delay per stage is measured. The contour (a) presents the results for nominal spacer material and contact technology, black line in Fig. 11(a). Results in (b) shows the performance boos due to 20% reduction in spacer k-value. In (c) contour, contact resistance is scaled by 0.7X, red line in Fig. 11(a). In (d) contour, both spacer k-value and contact resistance reduction is applied. Among all combinations, the best performance is laid around a constant L_{Eff}.

978-1-7281-1988-5/18 $31.00 © 2018 IEEE   770

Multi-domain process modeling for advanced logic and memory devices: from equimpments to materials

Inkook Jang[1], Hyoungsoo Ko[1], Alexander Schmidt[1], Sae-Jin Kim[1], Moonhyun Cha[1], Hyoshin Ahn[1], Honglae Park[1], Dae Sin Kim[1], and Ho-Kyu Kang[2]

[1]CAE Team, [2]Semiconductor R&D Center, Samsung Electronics Co. Ltd., Hwasung, Korea, email: inkook.jang@samsung.com

Abstract— For modern semiconductor devices, the level of details which we should investigate for predictive simulation is going extreme. Not only the atomistic simulation is required but equipment and transistor scale simulation is also needed to understand the formation of atomic scale feature. In this paper, practical applications of multi-domain simulations are introduced for advanced S/D process in logic, interface engineering in DRAM cell and cell stack ALD process of flash memory devices.

I. INTRODUCTION

For recent scaled devices, many ideal assumptions which have been used in conventional TCAD simulation might be significant detractor of simulation accuracy. One of the important origins of these is the effect of atomic scale feature [1]. For aggressively scaled devices the relative amount of atoms which belongs to the interface or surface region is not negligible anymore, and their properties are very different from the bulk atoms. Extended defects, crystallinity, surface roughness and discrete impurities can be other important atomic scale features to be treated explicitly. In order to consider these effects, the modeling should describe how those features are formed, what their structure look like, and what their physical and chemical properties could be. These works require several simulation methods with different scale and physics, which can include equipment simulation with chamber scale, process simulation with feature scale, and atomic scale simulation. In this paper, three applications of multi-domain process simulations will be introduced. They have been used to investigate real engineering issues of advanced logic, DRAM and flash devices.

II. SOURCE/DRAIN FORMATION IN 3D LOGIC DEVICES

The optimization of source and drain (S/D) epitaxy process is one of the most important steps of technology node development and a good example of multi-scale complexity: while epitaxial growth process is dominated by atomic-scale reactions of CVD precursors, it strongly depends on feature scale variation of local stress, chip level variation of layout and wafer level variation of temperature and gas flux. In addition, physical processes occurring at these different spatial scales have huge difference in time scales, making development of accurate and fast simulation even more complicated.

In order to overcome such problems, an integrated multi-scale and multi-physics approach has been adopted. Kinetic Monte Carlo (KMC) is one of most common atomistic simulation methods applicable for semiconductor processing [2] that treats the system evolution as a chain on atomic-level events and ignores small fluctuations. This allows drastic increase of simulation performance, but makes model reliant on empirical parameter values that can be extracted by *ab-initio* methods such as density functional theory (DFT).

KMC method used in this study considers simultaneous epitaxy (by explicit treatment of adsorption, desorption and reactions of precursors with the surface) with virtual lattice approximation [3] and also includes dopant and point defect diffusion and reaction in the bulk of material, which is done in an off-lattice like way. Examples of processes considered by KMC model are shown in Fig. 1. Since local atomic neighborhood is considered explicitly in the model, differences of reaction and deposition rate for (100), (110) and (111) surfaces lead to facet formation (Fig. 2).

Additionally, equipment level simulation is needed for extraction of key epitaxy process parameters: temperature and gas flow distributions over the wafer. Since the device scale is much smaller than the equipment, special atomistic particle tracing engine was developed to extract feature scale precursor attachment probability from the equipment simulation data. In Fig. 3, an example of combined topography and feature scale precursor scattering simulation is shown: depending on sidewall height both S/D shape and final volume as well as the amount of dopant precursor gas that reaches growing surface are changing. It leads not only to difference in final stress in the device, but also change in the S/D and contact resistances.

Once epitaxy simulation is complete, a Valence Force Field (VFF) method coupled with Finite Element Method (FEM) is applied to extract stress [4] (Fig. 4). Overall simulation flow is shown in Fig. 5. This integrated approach allows S/D shape and structure optimization by means of TCAD and eventually targeting the device performance depending on epitaxy recipe.

III. INTERFACE OPTIMIZATION OF DRAM CELL TRANSISTOR

The retention time (tRET) and the variable retention time (VRT) are key parameters of DRAM cell transistor performance. It has been reported that the VRT is strongly related with the passivation/de-passivation of interface trap (Nit) between Si and silicon oxide [5]. Interface quality depends on the gate oxide formation process. Assuming oxygen deficient Si/SiOx interface by Oxide-A process and oxygen rich condition by Oxide-B process, the differences of trap characteristics including spatial distribution calculated by DFT are presented in Fig. 6.

978-1-7281-1988-5/18 $31.00 © 2018 IEEE

Changes of trap density and level in various Si orientations by DFT are shown in Fig. 7 and 8. Fig. 9(a) shows that dissociated H-Si bond density which is closely correlated with interface trap depends on the oxygen flux during oxidation. As a result of different initial dangling bonds density depending on crystal orientation, suppressing sidewall trap density is more challengeable than bottom. TCAD model was built up for tRET with trap level, density, and Si-H passivation/de-passivation energy calculated from DFT simulation. Fig. 9(b) shows that the multi-scale TCAD simulation could be applicable to find optimal oxidation thickness within limited channel space.

The fundamental mechanism of gas treatment for controlling trap density is passivation of Si-dangling bonds by forming Si-X bonds. In contrast, de-passivation of the Si-H during operation is known as major source of VRT. Possible candidates for stable trap passivation including concentration effect were evaluated by DFT calculation, which is shown in Fig. 10. It could be mentioned that trap density decreased with increase gas at low concentration, but increased by over supply in all elements. For elements within same period in Fig. 10(a), sensitivity on concentration was increased on atomic number. Higher electron negativity and reactivity with Si-dangling bonds could results in higher sensitivity. Within same group in Fig. 10(b), a bigger element could easily diffuse into Si layer because of the higher stress level at interface, appeared as lower sensitivity in Nit. This suggests that fluorine gas can be good candidates for decreasing Nit as long as the concentration is well controlled.

IV. CELL STACK ALD PROCESS IN 3D FLASH DEVICE

Fluid analysis is playing a major role for process optimization of thin film deposition process since uniform gas distribution is a key factor for the process yield enhancement. However, fine 3D structures on the wafer surface causes drastic variations in correlation between gas distribution and process yield. Structures on the wafer modify effective surface area and surface reactions which in turn affect gas flow. Hence, classical fluid analysis has limitations for quantitative optimization of process conditions [6]. Challenges are to find out proper relations between modification of surface reaction and fine structures in wafer including consequent effects on the gas flow inside process chamber of equipment. These are the motivation of development of scale bridging simulation which relates feature scale geometry to equipment scale fluid simulation by means of iterative calculation (Fig. 11). The output of fluid simulation is fed into particle simulation as flux data. The latter calculates the amount of source consumption and byproduct generation and updates the boundary conditions of the former. Fig. 12 depicts that the scale bridging simulation is more corresponding to experimental measurement compared to conventional fluid only modeling.

Feature scale geometry responds to on-wafer fluxes in a non-linear way and alters chamber scale gas flow eventually. Gas molecules entering into the feature experience complex environment via many reactions and re-emissions in high aspect ratio geometry. Thus gas transport inside channel hole results in effectively high enough reactivity which can alter chamber scale gas flow even with very low rate reactions of ALD process. Geometry dependence of surface reaction addressed above requires feature scale simulation with sub-nanometer resolution. The simulation can be fast enough with proper abstraction of molecule-surface reaction with two-body interaction. Feature scale ALD simulation uses Monte Carlo (MC) particles representing gas fluxes entering channel hole by converting CFD result to fluxes, and launch MC particle from the top opening area of channel hole. Surface reactions are handled with a couple of half reactions with by-product removal to describe full ALD cycles, and surface saturation is well delineated in this simulation as shown in Fig. 13. Fig. 14 and 15 show simulation results corresponding to ALD process sub-steps of precursor feeding and inert gas purging, which are in good accordance with experimental results.

In order to ensure accuracy of surface reaction with abstracted two-body interaction, set of events at the surface has been captured by DFT calculations. Since an adsorption of a molecule on a large surface takes place along the multi-stage reaction path, it is hardly possible to define single reaction coefficient. However, if statistically large number of reaction trials can be assured like molecules inside channel hole in flash device, defining single reaction parameter as ensemble average is still effective. Multi-step reactions are decomposed into successive sequence of reaction steps, and detailed paths of each reaction steps are calculated by NEB method as in Fig. 16 and effective reaction coefficient is averaged from the partition functions. Non-local vdW functional is employed to calculate reaction energy (Fig. 17), and reaction coefficients are taken into account by calculating grand canonical ensemble average with partition function of physisorbed states.

V. CONCLUSIONS

Although multi-scale and multi-physics method is general approach nowadays, it is still challenging task to make it valid and effective in engineering simulation. In this work, it is shown that multi-domain simulations can be applicable to many complicate problems which cannot be dealt with convention TCAD simulation. It is expected that their impact on the semiconductor R&D will be increased as more sophisticated bridging technologies are developed.

REFERENCES

[1] K. -H. Lee, "Challenges and Responses for Virtual Silicon," SISPAD 2015.

[2] I. Martin-Bragado, R. Borges, J. P. Balbuena, M. Jaraiz, "Kinetic Monte Carlo simulation for semiconductor processing: A review," *Progress in Materials Science*, vol. 92, pp. 1-32, 2018.

[3] R. Chen, W. Choi, A. Schmidt, K.-H. Lee, and Y. Park. "A new kinetic lattice Monte Carlo modeling framework for the source-drain selective epitaxial growth process." SISPAD 2013.

[4] H.-H. Park, "Multiscale strain simulation for semiconductor devices base on the valence force field and the finite element methods." SISPAD 2015.

[5] G.A.M. Hurkx , H. C. de Graaff, W. J. Kloosterman, and M. P. G. Knuvers, "A New Analytical Diode Model Including Tunneling and Avalanche Breakdown," IEEE TED, vol. 9, pp. 2090-2098, Sep. 1992.

[6] M. Leskela, and M. Ritala, "Atomic layer deposition (ALD): from precursors to thin film structures," Thin Solid Films, vol. 409, pp. 138-146, April 2002.

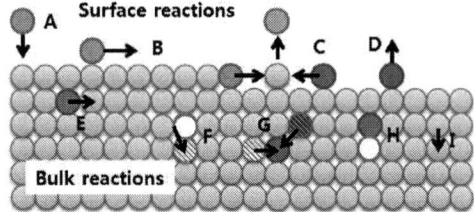

Fig. 1. Some of basic KMC processes. A, B, C, D: precursor adsorption, diffusion, chemisorption and desorption, respectively, E: interstitial diffusion, F: dopant-vacancy mobile complex formation, G: dopant cluster formation, H: defect recombination, I: direct dopant diffusion.

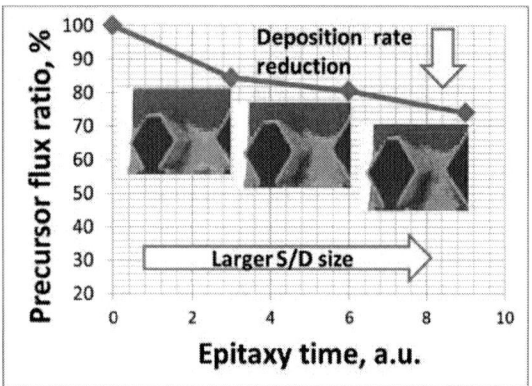

Fig. 3. As precursor deposition rate is reduced during the epitaxy process, the amount of dopant deposited at bottom side of S/D is changing.

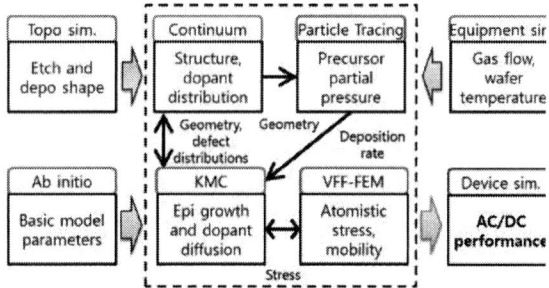

Fig. 5. The flow of multi-scale and multi-physics process simulation for logic devices.

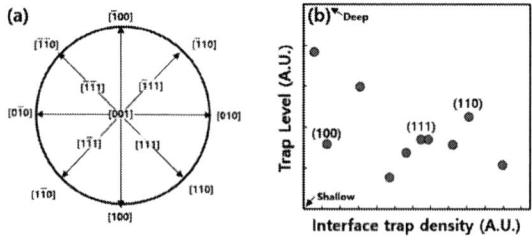

Fig. 7. (a) Simplified stereographic projection of (100) wafer which representing possible Si orientation perpendicular to z-direction, and (b) DFT calculation results of interface trap density and level for possible interface

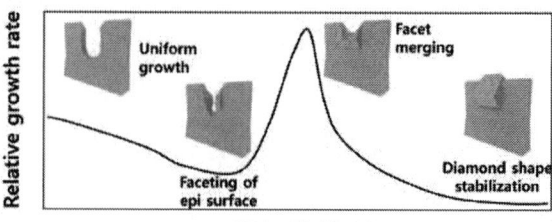

Fig. 2. Logic S/D epitaxy simulation with lattice KMC. Strongly non-uniform growth due to faceting is observed. Only Silicon is shown.

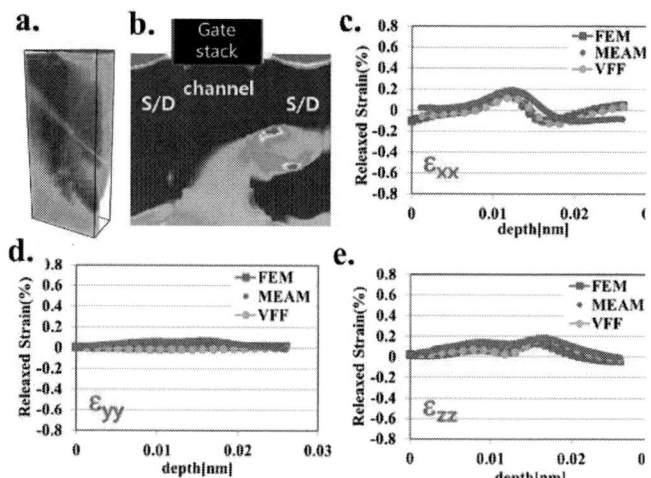

Fig. 4. (a) VFF+FEM stress simulation of perfect S/D region; (b) deatomization of stress with two opposite dislocations (DLs). Only the stress generated by DL cores is shown; (c), (d), (e) comparison of single 60 degree DL core stress: VFF, MD with modified embedded atom potential and FEM.

	Oxide-A	Oxide-B
Interface Stress	< 0.8 GPa	> 3 GPa
Trap Structure	Si(3)≡Si• (Pb0)	Si(2)O≡Si•
	High-Nit, Shallow Level	Low-Nit, Deep Level
Trap Properties		
	Si(3)≡Si• (Pb0)	Si(2)O≡Si• (Pb1)
Trap Structure		

Fig. 6. Atomic structure, spatial distribution of trap, and density of states inside band gap (as inset) at Si/SiOx interface. In case of Oxide-B, Si at interface neighbored with both Si and O resulted in trap near mid-gap, while Si at interface neighbored with only Si resulted in trap near band edge in Oxide-A. Trap density of Oxide-B was estimated lower than Oxide-A.

978-1-7281-1988-5/18 $31.00 © 2018 IEEE

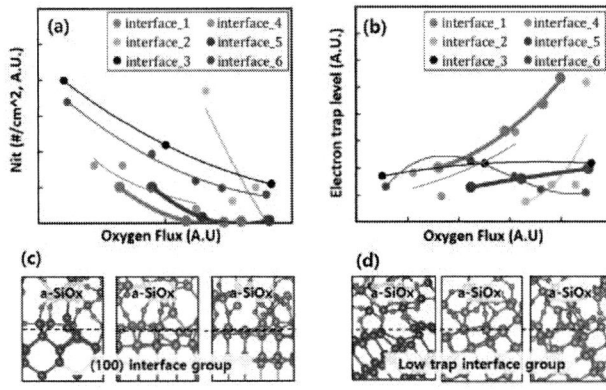

Fig. 8. Dependents of Si orientation on oxide flux vs. electron trap (a) density and (b) level from mid-gap. (c) and (d) are possible interface structures of Si (100) or low trap Si orientation wafer.

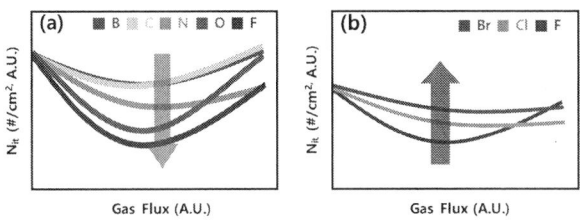

Fig. 10. Effects of gas flux on Nit within same (a) group or (b) period.

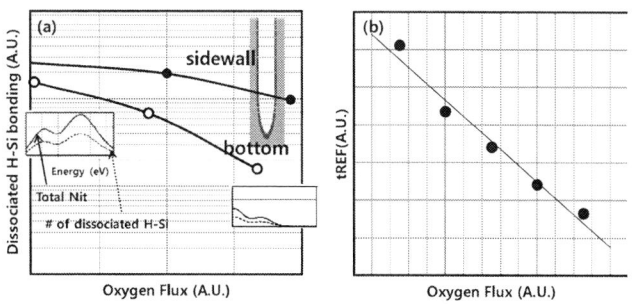

Fig. 9. Oxygen flux dependence of (a) number of dissociated H-Si bond at bottom and sidewall with changes of Nit as inset, and (b) tRET by DFT applied TCAD simulation.

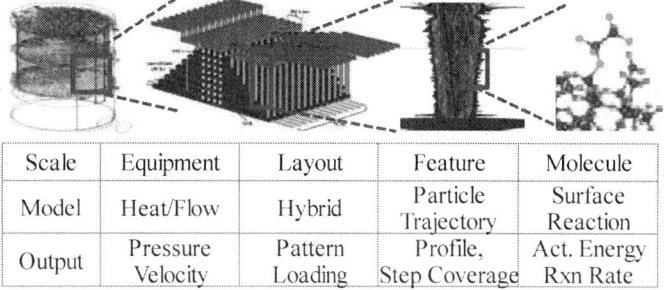

Scale	Equipment	Layout	Feature	Molecule
Model	Heat/Flow	Hybrid	Particle Trajectory	Surface Reaction
Output	Pressure Velocity	Pattern Loading	Profile, Step Coverage	Act. Energy Rxn Rate

Fig. 11. Conceptual description of scale bridging simulation including CFD-based equipment modeling, particle simulation in feature size, and atomistic surface reaction calculation

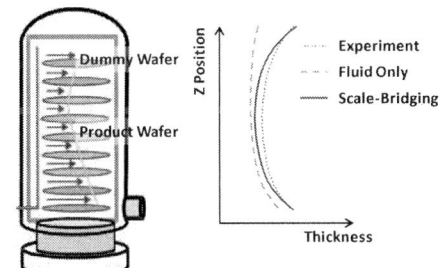

Fig. 12. Comparison between scale-bridging modeling and conventional fluid only simulation for ALD process.

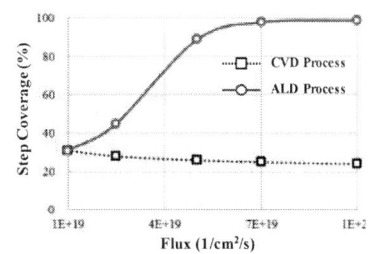

Fig. 13. ALD topography simulation showing step coverage enhancement according to the increased source flux in contrast to CVD

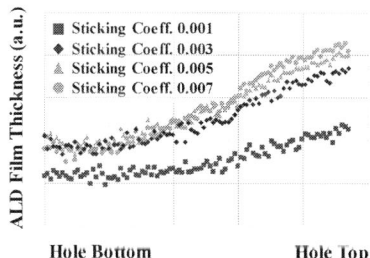

Fig. 14. Effect of precursor sticking coefficient on film deposition rate.

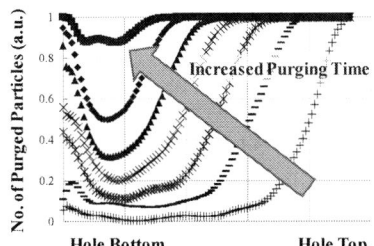

Fig. 15. Effect of purge time on the amount of purged molecules which are physisorbed on substrate.

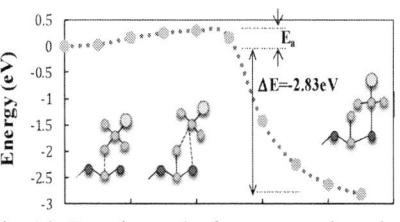

Fig. 16. Reaction path of precursor adsorption.

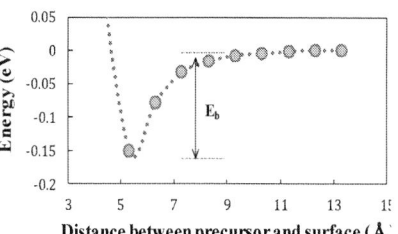

Fig. 17. Reaction energy of physisorption between precursor and surface.

978-1-7281-1988-5/18 $31.00 © 2018 IEEE

Entire Bias Space Statistical Reliability Simulation By 3D-KMC Method and Its Application to the Reliability Assessment of Nanosheet FETs based Circuits

Wangyong Chen, Yun Li, Linlin Cai, Pengying Chang, Gang Du* and Xiaoyan Liu*

Institute of Microelectronics, Peking University, Beijing 100871, China

*Email: xyliu@ime.pku.edu.cn, gangdu@pku.edu.cn

Abstract—The trap behaviors based 3D-Kinetic Monte Carlo (KMC) simulator is developed for statistical reliability assessment over the entire bias space. The main features include (i) physical insight into trap charging/discharging, coupling and generation/recombination behaviors for tracking trap-induced degradation of MOSFETs with multilayer gate dielectrics in the entire bias space. (ii) simulation of statistical reliability for the MOSFETs biased under arbitrary mixed stress conditions. (iii) assessment of reliability degradation in circuit operations with various Vg/Vd stress patterns and self-heating. The statistical reliability in nanosheet (NS) FETs and corresponding circuits are investigated. The impacts of the initial interface state and bulk trap density on the threshold voltage shift during the stress and relaxation phases are also analyzed.

I. INTRODUCTION

Time-dependent degradation and variability still remain a threat to the lifetime of device and circuit due to the prominent impacts of trap behaviors during operation, especially in deeply scaled technology node [1-2]. Generally, macroscopic metrics degradation biased by several typical gate voltage/drain voltage (Vg/Vd) stress combinations is measured to predict the device and circuit lifetime [3]. However, the CMOS circuits undergo a range of bias combinations with multiple reliability issues, such as Hot Carrier Injection (HCI), Bias Temperature Instability (BTI) and Off State Stress (OSS). These issues also interact with each other [4-5]. Therefore, the statistics of trap-induced degradation and time-dependent variability for reliability assessment is a challenge to design and optimize the CMOS ICs.

In this paper, we develop a 3D-Kinetic Monte Carlo (KMC) simulator based on trap behaviors to address the statistical degradation and variations over the entire bias space. The trap behaviors associated with carrier density, energy distribution, temperature and oxide electric field are self-consistently included in the simulation to provide physical insight into the reliability degradation in the MOSFETs with multilayer gate dielectrics under arbitrary Vg/Vd bias conditions. The simulator also allows to assess the reliability in circuit operations with various stress patterns and self-heating (SH). Statistical reliability of all bias regimes in nanosheet (NS) FETs is investigated from device to circuit level. The impacts of initial interface state density (N_{it}) and bulk trap density (N_{ot}) on stress and relaxation induced threshold voltage shift (ΔVth) and recovery are evaluated. Additionally, circuit reliability in NS FETs based inverter is systematically assessed by taking into account the operation frequency (f), duty cycle (d), and transition time (t_T) together with the self-heating effect.

II. 3D-KMC BASED RELIABILITY SIMULATION

The gate dielectric reliability can be attributed to the multi-trap behaviors, including charge trapping from channel/gate, charge detrapping to channel/gate, trap interaction between traps, trap generation and recombination, as sketched in the processes (1)-(5) of **Fig. 1(a)** [6-7]. **Fig. 1(b)** shows the energy band diagram of charge trapping in the nMOSFET with the interfacial layer (IL) and high k layer (HL) under stress considering the impact of carrier energy distribution. The two-state transition potentials between the empty trap and charged trap follow the non-radiative multi-phonon model, as illustrated in **Fig. 1(c)** [8]. For the process of Si-H bond dissociation shown in **Fig. 1(d)**, truncated harmonic oscillator model is applied to associate Si-H bond-breakage rate with carrier energy distribution, here two competing mechanisms (i.e. anti-bonded by a single particle and multi-vibration excitation via multi-particle) are included [9]. The corresponding equations to simulate fully-coupled trap behaviors are listed in **Table 1** [10]. The traps considered in the simulator include the interface state located at channel/IL as well as the bulk traps in IL, HL and IL/HL. **Fig. 2** shows the 3D-KMC based simulation framework for the given operation conditions (i.e. rise time (t_r), fall time (t_f), f, d, temperature $T(t)$, and bias sequence). The input of 3D electrostatic distributed quantities (i.e., electric field, potential, carrier density, and energy distribution) can be obtained from the BTE/MC device simulation [11-12]. Moreover, the macroscopic degradation and variability are obtained from charge statistics. The 3D KMC simulator is verified with the experimental data in our previous work [10]. In this work, the model parameters listed in **Table 2** are calibrated by HfO_2 based MOSFETs measurements under different Vg/Vd stress conditions [13-14], as shown in **Fig.3**. All of the following simulations are conducted using 200 samples.

III. ENTIRE BIAS SPACE STATISTICAL RELIABILITY SIMULATION FOR THE NS FETS

Fig. 4 shows the simulated three-layer horizontally stacked gate-all-around (GAA) NS FETs with SiO_2 and HfO_2 gate dielectrics, and also lists the corresponding structure parameters [15]. Comparison of the electrical performance between the simulation by TCAD and experimental results is presented in Ref. [16]. The initial pre-existing traps are randomly distributed in the gate dielectrics according to the initial defect density. The initial N_{it} and N_{ot} in the NS FETs are $2\times10^{10}cm^{-2}$ and $1\times10^{18}cm^{-}$

978-1-7281-1988-5/18 $31.00 © 2018 IEEE

[3], respectively. **Fig. 5** illustrates the newly-generated traps and charged traps after 100s stress during HCI, BTI and OSS modes. Trap number evolution in **Fig. 6** shows that the statistical average trap number in top-down (TD) gate dielectric is 5 times larger than that in left-right (LR) gate dielectric irrespective of HCI, BTI and OSS. Particularly, the generated trap number is maximum under HCI stress, followed by BTI and OSS modes. The trap activity (i.e., the repetition rate of trapping/detrapping and generation/recombination) along the gate dielectrics is shown in **Fig.7**. For BTI mode, due to the higher electric field near HL, traps located at the IL/HL interface are more active than that at the channel/IL interface. In contrast, for HCI mode, trap activity at the channel/IL interface is higher because of more interface state generation. It also reveals that the trap activity in the HL towards gate gradually decreases and then increases due to more charge emission from trap to gate. **Fig. 8** shows the cumulative distributions of ΔVth after 100s stress, and the separate contribution of bulk traps and interfacial traps to the total Vth shifts, respectively. It indicates the ΔVth variation induced by HCI is the largest. Interfacial and bulk defects are both involved in the stress-induced degradation under arbitrary Vg/Vd stress condition, but there is a difference between the trap number, location and relative contribution. Top view of statistical charge distributions from 200 samples in the gate dielectrics of NS FETs under HCI, BTI and OSS conditions is shown in **Fig. 9**. For HCI mode, interfacial charges are mainly close to the drain side and most bulk charges are generated near the source. These two contributions result in the uniform charge distribution in the dielectrics. For BTI mode, partial charges are distributed along the edge of the dielectric corner because of the strong corner effect. For OSS mode, the charges are mainly close to the drain side. **Fig. 10** and **Fig. 11** show the mapping of statistical average ΔVth together with the corresponding variance over the entire bias space stressed 100s at 398K, indicating the NS nFETs suffer from the most serious degradation ($\sim$130mV) and variability ($\sim$23mV) under Vg=Vd=1.5V condition. Besides, power-law exponent n during 100s stress is extracted according to ΔVth=$A \times t^n$ for NS nFETs in **Fig. 12**. The highest n is also at Vg=Vd=1.5V, close to the value of 0.49, but the n in pure BTI region remains below $\sim$0.20 during 100s stress at 398K. The simulated result is in accord with the measurement of high-k gate MOSFETs [17].

The impacts of the initial N_{it} at channel/IL interface and N_{ot} in the HL on ΔVth are shown in **Fig. 13**. The ΔVth induced by HCI, BTI and OSS exhibit severer degradation with the increased N_{it} and N_{ot}. Furthermore, the stress-induced ΔVth under HCI stress increases $\sim$23.1% with 5$\times$ N_{it} and $\sim$8.4% with 5$\times$ N_{ot}. And the slight recovery enhancement under BTI for the increased defect density case. Thus, reducing the initial N_{it} is a more effective way for reliability improvement. **Fig. 14** plots the ΔVth statistics of the width (ΔVth$_W$) and height (ΔVth$_H$) direction during the stress and relaxation phases, indicating the width direction plays a dominant role for Vth shifts since the average oxide electric field in the width side is larger. The temperature dependent ΔVth contribution under 50%-duty cycle in **Fig. 15** shows the ΔVth$_W$ is more sensitive to the

elevated temperature compared to ΔVth$_H$. As the temperature increases from 300K to 398K, the Vth shift and recovery can be up to $\sim$54mV and $\sim$13mV, respectively.

IV. RELIABILITY ASSESSMENT IN THE NS FETs BASED CIRCUITS

The simulator is capable of statistical reliability simulation for the MOSFETs stressed by arbitrary mixed bias patterns. An example of simulated ΔVth induced by mixed HCI, BTI, OSS and relaxation is plotted in **Fig. 16**. The CMOS circuits experience various Vg/Vd stress patterns, leading to mixed multi-reliability degradation. **Fig. 17(a)-(b)** shows the reliability issues occurring in the CMOS inverter and the schematic of the t_T definition. The quasi-static simulation with the adaptive time step is employed in the simulator as indicated in **Fig. 17(c)**. **Fig. 18** plots the ΔVth degradation induced by interfacial charges (ΔVth$_{it}$) and bulk charges (ΔVth$_{ot}$) in the NS FETs based inverter considering the effects of t_T and d. It is noted that the contribution of ΔVth$_{it}$ is relatively low owing to the small transition time, and ΔVth$_{ot}$ becomes the dominant factor. ΔVth$_{it}$ exhibits no dependence on d and slightly increases with t_T longer than 10^{-8}s (e.g. ΔVth$_{it}\sim$4mV for $t_T\sim$1μs). In contrast, ΔVth$_{ot}$ is strongly dependent on d and is hardly affected by t_T. **Fig. 19** shows the frequency-dependent ΔVth in the AC operations without and with SH considered. The self-heating is taken into account by integrating an analytical model developed in our previous [18] work into the simulator. It is found that ΔVth is independent of f when f >10Hz, and is weakly increased in the SH case due to the fact that SH is accompanied by HCI during the small-duty transition.

V. CONCLUSION

The trap behaviors based 3D-KMC simulator to assess the entire bias space statistical reliability is developed. The simulator enables to provide physical insight into trap behaviors induced degradation in MOSFETs with multilayer gate dielectrics and to investigate reliability degradation under arbitrary mixed stress conditions. Assessment of circuit reliability with various stress patterns and the consideration of self-heating can also be obtained. The simulations of reliability in the NS FETs reveal that the increased initial N_{it} leads to more stress-induced ΔVth compared to the N_{ot}. The ΔVth caused by the interfacial charges exacerbates with the long effective transition time t_T, whereas bulk charges contribution is duty-cycle dependent and dominant in the inverter operations.

ACKNOWLEDGMENT

This work is supported in part by 2016YFA0202101 , NSFC61674008 and 61421005.

REFERENCES

[1] C. Liu *et al.*, *IEDM*, 2014. [2] A. Chasin *et al.*, *IEDM*, 2017. [3] B. Kaczer *et al.*, *IRPS*, 2016. [4] F. Cacho *et al.*, *IRPS*, 2014. [5] M. Duan *et al.*, *IRPS* 2017. [6] Y. Li *et al.*, *IEDM*, 2016. [7] Y. Li *et al.*, *SISPAD*, 2015. [8] B. Kaczer *et al.*, *IRPS*, 2016. [9] T. Grasser *et al.*, *TED*, 2014. [10] Y. Li *et al.*, *TNANO*, 2017. [11] X. Y. Liu *et al.*, *ICSICT*, 2012. [12] S. Y. Di *et al.*, *JJAP*, 2017. [13] S. Zafar *et al.*, *VLSI*, 2006. [14] W. Zhang *et al.*, *MR*, 2016. [15] N. Loubet *et al.*, *VLSI*, 2017. [16] L. L. Cai *et al.*, *TED*, 2018. [17] S. Mukhopadhyay *et al.*, *TED*, 2017.[18] W. Y. Chen *et al.*, *TED*, 2018.

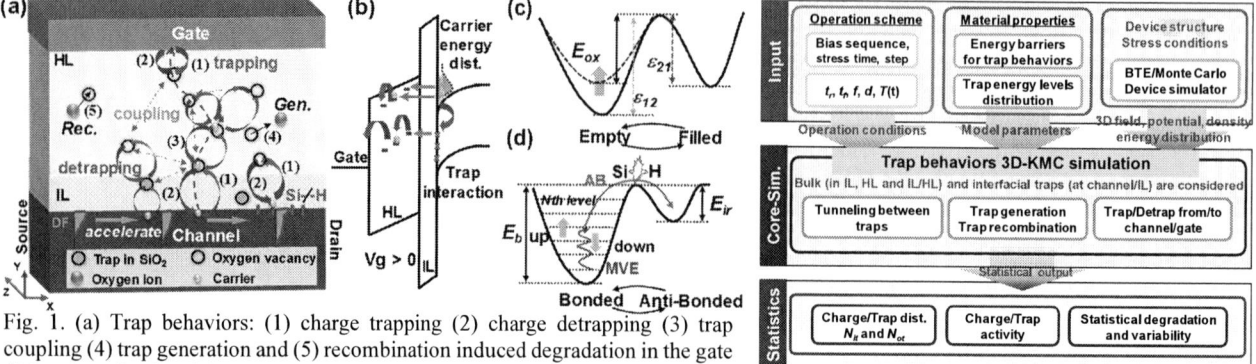

Fig. 1. (a) Trap behaviors: (1) charge trapping (2) charge detrapping (3) trap coupling (4) trap generation and (5) recombination induced degradation in the gate dielectrics simulated by 3D-KMC method. (b) Energy band diagram of a nMOSFET under stress. (c) Schematic of defect potentials for trapping charges. (d) Truncated harmonic oscillator model for Si-H bond-breakage.

Fig. 2. Framework of the trap behavior based 3D-KMC simulator for statistical reliability assessment.

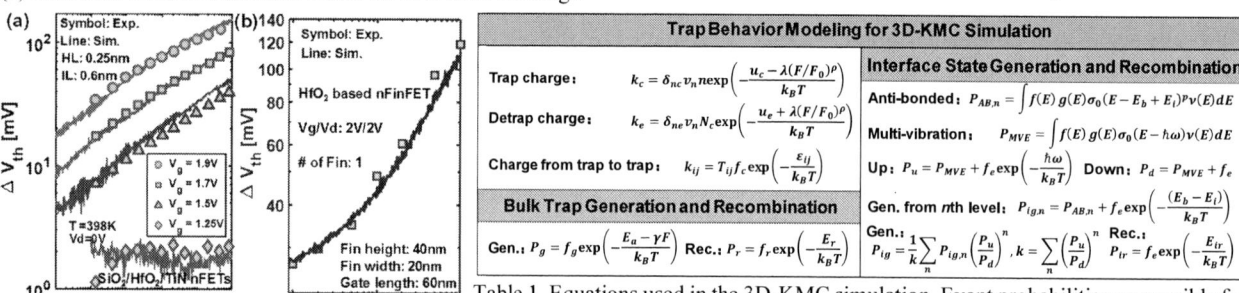

Fig. 3. Comparison of simulation results with experiments [13-14] under different Vg/Vd biases.

Table 1. Equations used in the 3D-KMC simulation. Event probabilities responsible for trap behaviors are driven by carrier density, energy distribution, temperature, oxide electric field. Traps in the interfacial layer (IL), high k layer(HL) and IL/HL interface are regarded as bulk traps. Interfacial traps are located at channel/IL interface.

Trap Behavior Modeling for 3D-KMC Simulation

Trap charge: $k_c = \delta_{nc} v_n n \exp\left(-\frac{u_c - \lambda(F/F_0)^\rho}{k_B T}\right)$

Detrap charge: $k_e = \delta_{ne} v_n N_c \exp\left(-\frac{u_e + \lambda(F/F_0)^\rho}{k_B T}\right)$

Charge from trap to trap: $k_{ij} = T_{ij} f_c \exp\left(-\frac{\varepsilon_{ij}}{k_B T}\right)$

Bulk Trap Generation and Recombination

Gen.: $P_g = f_g \exp\left(-\frac{E_a - \gamma F}{k_B T}\right)$ Rec.: $P_r = f_r \exp\left(-\frac{E_r}{k_B T}\right)$

Interface State Generation and Recombination

Anti-bonded: $P_{AB,n} = \int f(E) g(E) \sigma_0 (E - E_b + E_i)^p \nu(E) dE$

Multi-vibration: $P_{MVE} = \int f(E) g(E) \sigma_0 (E - \hbar\omega) \nu(E) dE$

Up: $P_u = P_{MVE} + f_e \exp\left(-\frac{\hbar\omega}{k_B T}\right)$ Down: $P_d = P_{MVE} + f_e$

Gen. from nth level: $P_{ig,n} = P_{AB,n} + f_e \exp\left(-\frac{(E_b - E_i)}{k_B T}\right)$

Gen.: $P_{ig} = \frac{1}{k}\sum_n P_{ig,n}\left(\frac{P_u}{P_d}\right)^n, k = \sum_n \left(\frac{P_u}{P_d}\right)^n$ Rec.: $P_{ir} = f_e \exp\left(-\frac{E_{ir}}{k_B T}\right)$

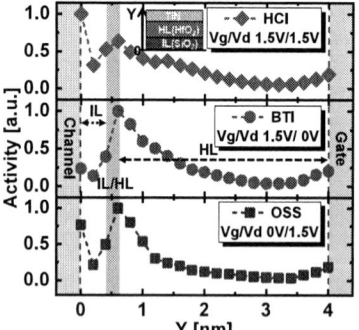

Fig. 4. Schematic of the three-layer stacked Nanosheet FET with IL/HL gate dielectrics and corresponding structure parameters used in this work.

Par.	Value [nm]
Lg	12
Hsh	5
Wsh	25
Space	10
Tox(HL)	3.7
Tox(IL)	0.4

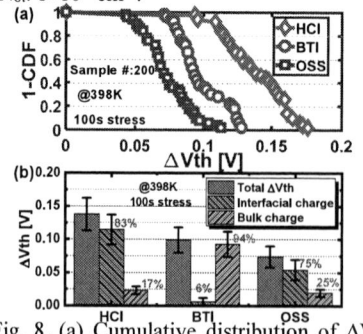

Fig. 5. Trap and charge distributions evolution after 100s stress under three bias conditions. Fresh case: traps are randomly distributed and the initial N_{it}: 2×10^{10} cm^{-2}, N_{ot}: 1×10^{18}cm^{-3}.

× Interfacial trap ● Charged bulk trap @398K
● Charged interfacial trap ○ Bulk trap 100s stress

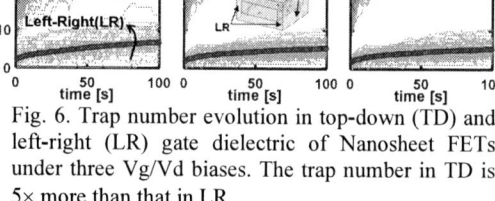

Fig. 6. Trap number evolution in top-down (TD) and left-right (LR) gate dielectric of Nanosheet FETs under three Vg/Vd biases. The trap number in TD is 5× more than that in LR.

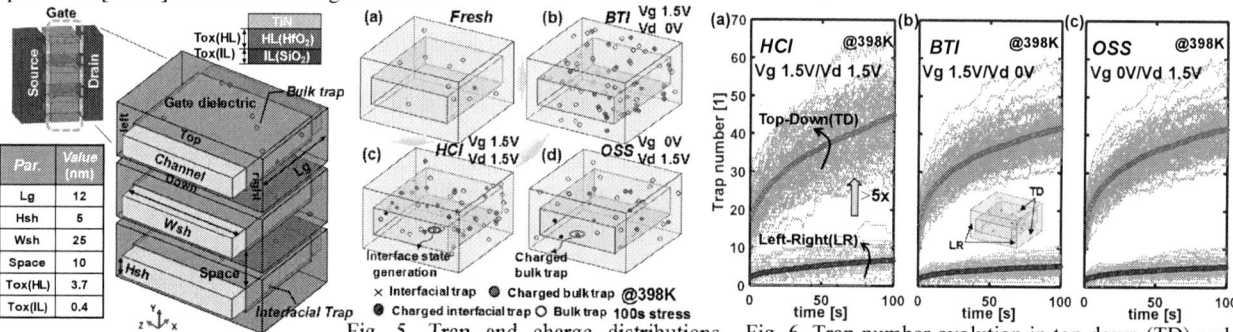

Fig 7. Trap activity along the thickness direction of the SiO$_2$/HfO$_2$ gate dielectrics.

Fig. 8. (a) Cumulative distribution of ΔVth and (b) charge contribution to ΔVth under BTI, HCI and OSS stresses. Error bar denotes standard error obtained from 200 samples.

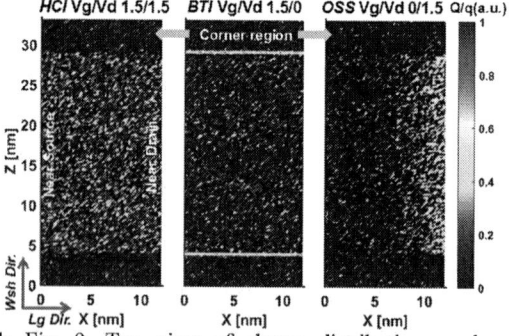

Fig. 9. Top view of charge distributions under different biases. Charges are uniform in HCI. For BTI and OSS, charges are mainly located along the dielectric edge and close to the drain, respectively.

978-1-7281-1988-5/18 $31.00 © 2018 IEEE

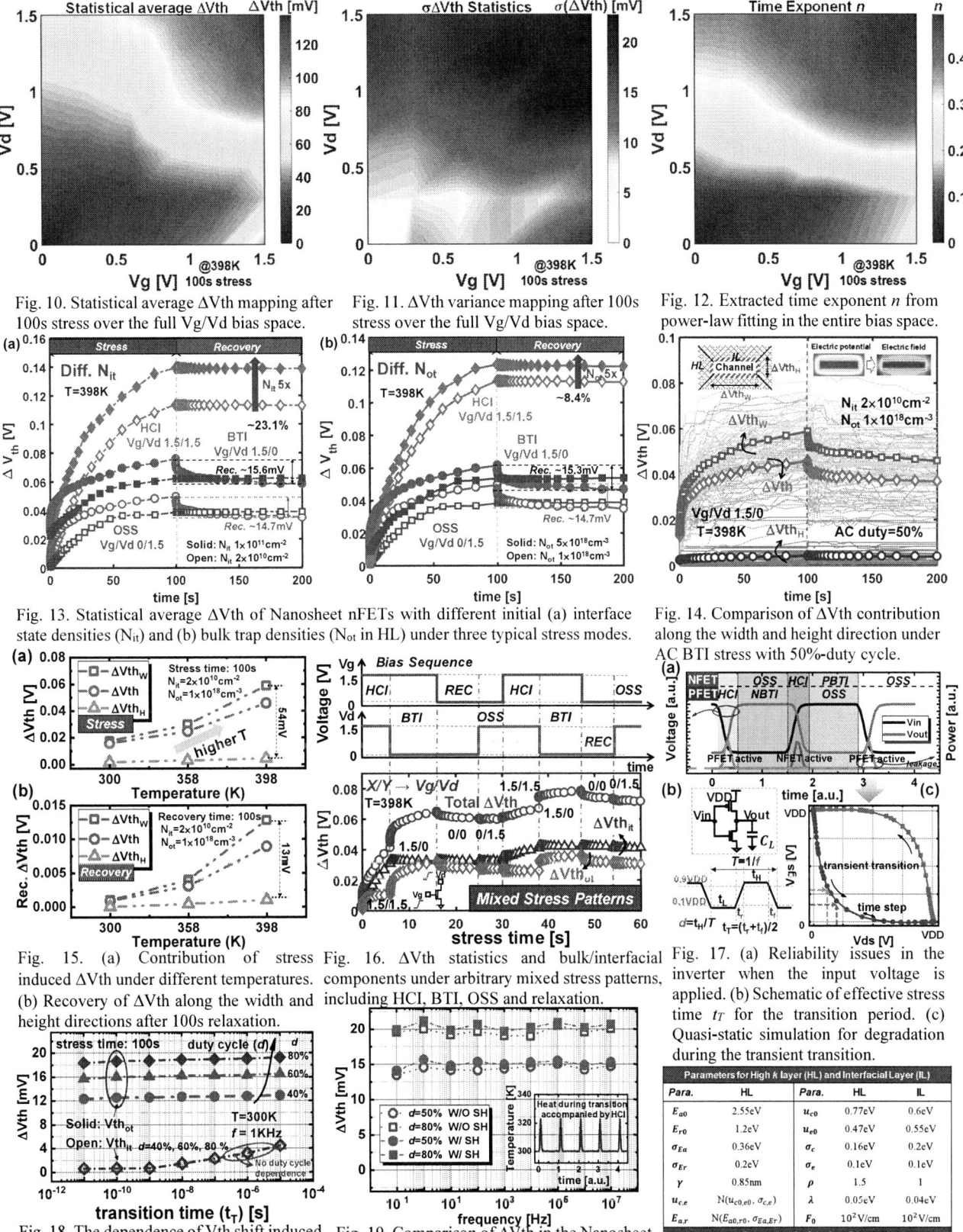

Fig. 10. Statistical average ΔVth mapping after 100s stress over the full Vg/Vd bias space.

Fig. 11. ΔVth variance mapping after 100s stress over the full Vg/Vd bias space.

Fig. 12. Extracted time exponent n from power-law fitting in the entire bias space.

Fig. 13. Statistical average ΔVth of Nanosheet nFETs with different initial (a) interface state densities (N_{it}) and (b) bulk trap densities (N_{ot} in HL) under three typical stress modes.

Fig. 14. Comparison of ΔVth contribution along the width and height direction under AC BTI stress with 50%-duty cycle.

Fig. 15. (a) Contribution of stress induced ΔVth under different temperatures. (b) Recovery of ΔVth along the width and height directions after 100s relaxation.

Fig. 16. ΔVth statistics and bulk/interfacial components under arbitrary mixed stress patterns, including HCI, BTI, OSS and relaxation.

Fig. 17. (a) Reliability issues in the inverter when the input voltage is applied. (b) Schematic of effective stress time t_T for the transition period. (c) Quasi-static simulation for degradation during the transient transition.

Fig. 18. The dependence of Vth shift induced by interface states and bulk charges on transition time for the inverter operation with different duty cycles, respectively.

Fig. 19. Comparison of ΔVth in the Nanosheet nFET based inverter with different AC frequency and duty cycle, respectively, with/without self-heating (SH).

Parameters for High k layer (HL) and Interfacial Layer (IL)			
Para.	HL	*Para.*	HL	IL
E_{a0}	2.55eV	u_{c0}	0.77eV	0.6eV
E_{r0}	1.2eV	u_{e0}	0.47eV	0.55eV
σ_{Ea}	0.36eV	σ_c	0.16eV	0.2eV
σ_{Er}	0.2eV	σ_e	0.1eV	0.1eV
γ	0.85nm	ρ	1.5	1.5
$u_{c,e}$	$N(u_{c0,e0}, \sigma_{c,e})$	λ	0.05eV	0.04eV
$E_{a,r}$	$N(E_{a0,r0}, \sigma_{Ea,Er})$	F_0	10^2V/cm	10^2V/cm
Parameters for Interface State			
$\hbar\omega$=0.075eV	p=11	E_b=1.5eV	f_e=0.1ps^{-1}

Table 2. Model parameters used in the 3D-KMC simulation.

978-1-7281-1988-5/18 $31.00 © 2018 IEEE   778

A Physics-based Thermal Model of Nanosheet MOSFETs for Device-Circuit Co-design

Linlin Cai, Wangyong Chen, Pengying Chang[+], Gang Du, Xing Zhang, Jinfeng Kang and Xiaoyan Liu[*]

Institute of Microelectronics, Peking University, Beijing 100871, China, Email: [+]pychang@pku.edu.cn; [*]xyliu@ime.pku.edu.cn

Abstract—A physics-based thermal model is developed to describe the self-heating effects (SHE) on nanosheet MOSFETs. Three stages of transient temperature response due to the anisotropic heat dissipation and asymmetrical temperature distribution are well understood by the thermal RC network model, providing the physical insight into frequency-dependent SHE in AC operation. The proposed model is further implemented into SPICE simulator for high-efficient thermal assessment in circuit level by the flexible BEOL. Layout design in inverter cell correlated with thermal behavior is investigated for static and transient operation downwards 3nm CMOS node. The SHE and thermal-aware reliability in inverter-based ring oscillator are predicted. The thermal model can be used as a device-circuit co-design tool to assess the thermal behavior accurately and efficiently.

I. INTRODUCTION

With technology node scaling down, the self-heating effect (SHE) in stacked gate-all-around (GAA) nanosheet (NS) MOSFETs becomes severer compared to FinFETs or nanowires due to the confined topological structure [1-3]. The rapidly increased lattice temperature if not be dissipated timely may lead to the degradation of performance and eventually induce reliability issues [4-5]. However, the experimental thermoreflectance technology [6] can only obtain the surface temperature of device without the specific local temperature in different regions. Furthermore, the fast frequency-dependent temperature response in AC operation is still difficult to capture by measurements. Thus, the modeling of SHE is necessary to bridge device to circuit design [7-8]. The thermal evaluation in circuits now still has two main problems: one is accuracy and the other is efficiency. In this work, we develop a physics-based thermal RC network model for stacked GAA NS MOSFETs to achieve high-efficient thermal simulation without loss of accuracy. The proposed model reveals that the transient temperature response undergoes three stages which are affected by thermal resistance (R_{th}) and thermal capacitance (C_{th}) in different regions respectively, corresponding to the finite-element modeling (FEM) simulation. This finding provides physical insight into frequency-dependent SHE in AC operation. By implanting the model into SPICE simulator, the SHE from device to circuit can be investigated with the flexible back end of line (BEOL). The model is verified and calibrated by FEM simulation both in device level and circuit level. The SHE in inverter cell and ring oscillator (RO) is investigated as an example for layout optimization and electromigration (EM) lifetime prediction.

II. THERMAL MODEL OF NS MOSFETs

Fig. 1 shows the self-heating in the stacked NS MOSFETs by FEM simulation considering anisotropic thermal conductivity with $67.5\mu W$ thermal power input [3], which is referred as a baseline in the following discussion. The channel regions of 3-layer nanosheets exhibit the asymmetrical temperature distribution, indicating the junction temperature (T_j) appears at the regions near the drain. The transient temperature response shown in **Fig. 2** by FEM simulation indicates that the thermal process can be divided into three stages according to the response time. Changing the parameters such as the thermal conductivity of source and drain (κ_{sd}) affects the transient SHE. **Fig. 3** shows the top view of time evolution of temperature distribution in NS MOSFETs corresponding to the three stages. It confirms that the heat generates from nanosheet ($\sim$10ps), then diffuses to drain, gate and source ($\sim$1ns), and finally to substrate ($\sim$0.1μs). Based on the heat flux dissipation paths, the equivalent thermal RC network model is proposed including the interface thermal resistance (IR) at contacts and vias, as illustrated in **Fig. 4**. The model can be packaged with four external ports as a sub-module for circuit level simulation. The related parameters for the baseline and the equations of R_{th} and C_{th} are listed in **Table I**. As shown in Fig. 2, the simulation result obtained from the thermal network model shows an excellent agreement with the FEM simulation, and provides higher accuracy compared with lump RC model.

The impacts of R_{th} on T_j: **Fig. 5** shows that in Stage I increased R_{th} of nanosheets ($R_{th}sh$) aggravates self-heating in the next Stage II and III, indicating the heat in Stage I is mainly dominated by $R_{th}sh$. Similarly, the heat in Stage II is mutually affected by $R_{th}S$, $R_{th}d$, $R_{th}g$ of source, drain and gate. In Stage III, the $R_{th}sub$ of substrate modulates the temperature and trends to steady. **Fig. 7** plots the impacts of multi-R_{th} in Stage II on T_j. The normalized R_{th} is the ratio relative to the baseline value. It is shown that $R_{th}d$ has a more significant effect on self-heating than $R_{th}s$ and $R_{th}g$.

The impacts of C_{th} on transient response: The similar effects of C_{th} in three stages can be found in **Fig. 6**. With fixed R_{th}, changing C_{th} in different regions dominates the speed of thermal response without skewing of steady temperature. The decreased $C_{th}sub$ in Stage III accelerates the transient response. Based on this phenomenon, the AC operation with different frequency according to three stages is shown in **Fig. 8**. **Fig. 8 (a)** illustrates that with reduced frequency, the temperature increases towards the thermal steady state in AC operation. From **Fig. 8 (b)-(c)**, the temperature in source (Ts), drain (Td), gate (Tg) and substrate (Tsub) cannot follow the fast frequency at 10G Hz, because the transient heat in Stage I is only affected by $R_{th}sh$ and $C_{th}sh$. However, when the frequency decreases to 1G Hz during Stage II, the Ts, Td and Tg respond to frequency except Tsub. Only at around 10M Hz in Stage III, all regions show the frequency-dependent thermal response.

III. MODELLING SHE FROM DEVICE TO CIRCUIT

Fig. 9 shows the flowchart of constructing thermal network model for self-heating evaluation from device to circuit level. The simulated thermal power can be obtained from the device characteristics. The thermal RC model is implemented into SPICE simulator as a packaged sub-module for thermal assessment in circuits. The key connections between device level and circuit level are determined by flexible BEOL such as interconnect materials, metal wires placement and patterning process. The equivalent thermal conductivity for BEOL is calculated by effective media theory model [9]. The reliability degradation can be further predicted based on the thermal behavior in circuits. **Fig. 10** shows the layout of NS based CMOS inverter. The nodes (Ng for input, Nd for output, Ns1 to GND and Ns2 to VDD) are marked in the figure. The related thermal RC network model is shown in **Fig. 11** using two packaged sub-modules for NFET and PFET. The networks of external interconnects are constructed with two metal layers (M1, M2) and top BEOL by Eq. 5-9 [10]. Using this model, the self-heating in inverter is evaluated in static operation with input in one device (NFET in Fig. 12). The modeling node temperature (T_{node}) of inverter by adjusting correction factors S_c in **Table II** shows a well agreement with the FEM simulation, as shown in **Fig. 12**. By using the thermal model (cpu time 1.05×10^2s), 4× improvement of efficiency can be obtained in simulating the SHE of inverter compared to the FEM method (cpu time 4.3×10^2s).

IV. RESULTS AND DISCUSSION

A. SHE in inverter

Layout optimization: **Fig. 13** shows the impacts of contacted poly pitch (CPP) on self-heating in inverter cell with scaled CMOS node. The anticipated scaling size of the standard cell is summarized in Table II [11]. The 2× increase of thermal power density causes about 20K (30K) temperature rise at 5nm (3nm) CMOS node. **Fig. 14** shows the influence of cell height associated with metal pitch (MP) and track number on SHE. The benchmark means the maximum track number in cell based on layout design rules for 7nm, 5nm and 3nm nodes [11]. Reducing the effective track number exacerbates the self-heating in inverter cell as the result of decreased metal capacity in BEOL. The T_j in inverter at 3nm node increases 40K compared to 7nm node. The SHE on the selections of via with advanced material technology [12-13] (eg. Ru, Ni, Co, W) for 7nm, 5nm and 3nm CMOS nodes are compared in **Fig. 15**. The Ru and W material show the advantages in suppressing thermal effects. Therefore, the optimal strategy to reduce self-heating in inverter cell is increasing the track number and using the Ru or W as the via material.

AC operation analysis: The frequency-dependent temperature response of the inverter is evaluated in **Fig. 16**. Due to the fact that the self-heating cannot respond sufficiently at high frequency, the T_j reduces with increased frequency. The different power duty cycles (ε) 1%, 10%, 50% are compared, indicating that the pulse with small duty cycle has low thermal

effects and transition frequency. The thermal crosstalk from the adjacent device can be observed at low frequency. From **Fig. 17**, increasing cell height for device isolation may be the better way to alleviate thermal crosstalk. **Fig. 18** indicates that the T_{node} marked in Fig. 10 responds only to the frequency ranging from 100K Hz to 1G Hz. The Nd has the highest temperature and the Ns2 in PFET because of the connection to VDD has the higher temperature than Ns1 in NFET.

B. SHE in RO and EM prediction

Fig. 19 shows the SHE in RO based on the inverters at 5nm CMOS nodes. The oscillation frequency of RO depends on the numbers of stage (# of stage) and # of fan-out, as illustrated in **Fig. 19(a)**. Corresponding T_j with different # of stage and fan-out are evaluated in **Fig. 19(b)**, respectively. It can be seen that although both the increased # of stage and fan-out decrease the frequency, the increased # of fan-out aggravates the self-heating while the increased # of stage alleviates the thermal effects. **Fig. 20** shows the T_{node} in 15 stage RO with advanced interconnect materials in M1. The output port (Nd) at the last stage suffers the worst self-heating and needs to pay more attention. Therefore, the following assessment is based on the worst case in M1. EM degradation of different materials [13] is predicted with or without SHE according to the Black's empirical equation [14], as shown in **Fig. 20 (b)**. The EM lifetime in Ru interconnects with SHE provides a 6.4× improvement compared to Cu. The mapping of EM lifetime in Cu M1 versus # of stage and fan-out shown in **Fig. 21** indicates EM degradation trends to stabilization with increased # of stage.

V. CONCLUSION

In this work, 1) a physics-based thermal RC network model is proposed to achieve accurate and efficient thermal simulation for GAA NS MOSFETs. 2) Three stages of transient temperature response affected by R_{th} and C_{th} in different regions is clarified to provide physical insight into frequency-dependent SHE in AC operation. 3) The proposed model is implemented into SPICE simulator as a sub-module to bridge device to circuit thermal assessment considering flexible BEOL. 4) The 4× improved efficiency is obtained by using the thermal model in simulating the SHE of inverter compared to the FEM. 5) Thermal-aware layout design and EM lifetime prediction in circuit level are evaluated for device-circuit co-optimization.

ACKNOWLEDGMENT

This work is supported by 2016YFA0202101 and NSFC 61674008, 61421005.

REFERENCES

[1] M. H. Liao et al., *IEEE TED*, vol. 64, pp. 646-648, Feb. 2017
[2] L. L. Cai et al., *IEEE VLSI-TSA*, pp. 125-126, 2018
[3] L. L. Cai et al., *IEEE T-ED*, vol. 65, pp. 2647-2654, June, 2018
[4] H. C. Sagong et al., *IEEE VLSI technology*, pp.121-122, 2018
[5] H. Jiang et al., *IEEE VLSI technology*, pp.136-137, 2017
[6] J-Y Park et al., *IEEE T-ED*, vol. 64, pp. 4393-4399, Nov. 2017
[7] W. Ahn et al., *IEEE IEDM*, pp. 330-333, 2017
[8] J-Y Yan et al., *IEEE VLSI technology*, pp. 113-114, 2018
[9] J. C. M. Garnett et al., *Philos. Trans. Roy. Soc*, pp. 385-420, Jan. 1904
[10] W. Ahn et al., *IEEE T-ED*, vol. 64, pp. 3555-3563, Sep. 2017
[11] A. Mocuta et al., *IEEE VLSI technology*, pp. 147-148, 2018
[12] C. Auth et al., *IEEE IEDM*, pp. 673-676, 2017
[13] O. V. Pedreira et al., *IEEE IRPS*, pp. 6B-2.1-2.7, 2017
[14] J. R. Black et al., *IEEE IRPS*, pp. 300-303, 1982

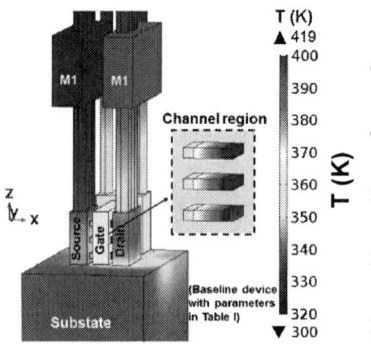

Fig.1 3D view of temperature distribution in nanosheet MOSFETs by FEM simulation. The channel region shows the asymmetrical temperature profile.

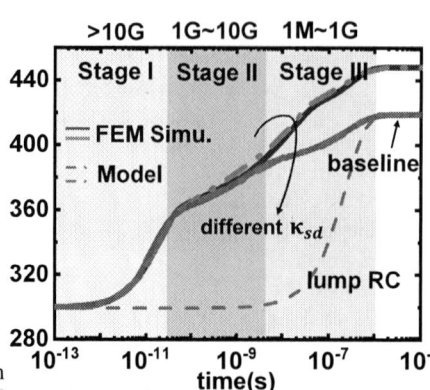

Fig. 2 FEM simulated transient temperature response in comparison to the proposed model in this work and the lump RC model.

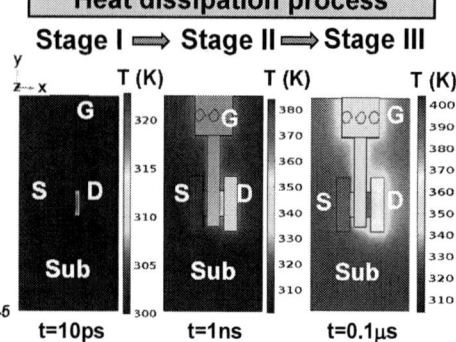

Fig. 3 Transient temperature response includes three physical processes: Stage I, II and III based on different heat dissipation paths.

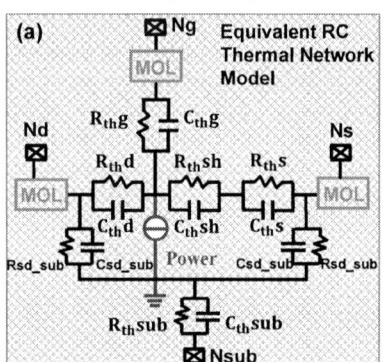

Fig. 4 (a) Equivalent thermal network model in the device level corresponding to the heat dissipation paths in Fig. 3. (b) RC network in middle of line (MOL) with interface thermal resistance (IR) (c) Packaged sub-module with four ports for circuit thermal evaluation.

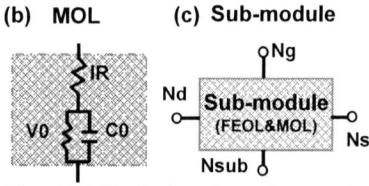

Fig. 5 The impacts of R_{th} in different regions on junction temperature (T_j) of nanosheet MOSFETs: Stage I is mainly dominated by $R_{th}sh$; Stage II is mutually affected by $R_{th}S$, $R_{th}d$ and $R_{th}g$; Stage III is influenced by $R_{th}sub$. (with fixed C_{th})

Fig. 6 The impacts of C_{th} in different regions on transient temperature response of nanosheet MOSFETs: Stage I can be modulated by $C_{th}sh$; Stage II depends on $C_{th}s$, $C_{th}d$ and $C_{th}g$; Stage III is influenced by $C_{th}sub$. (with fixed R_{th})

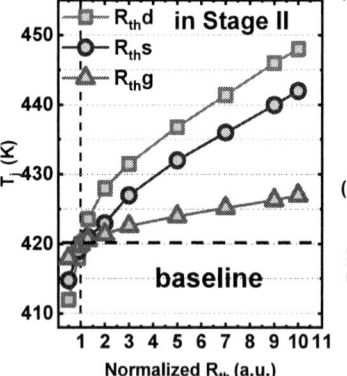

Fig. 7 Comparison of impacts of R_{th} in source, drain, gate region on self-heating, dominating the transient response of Stage II.

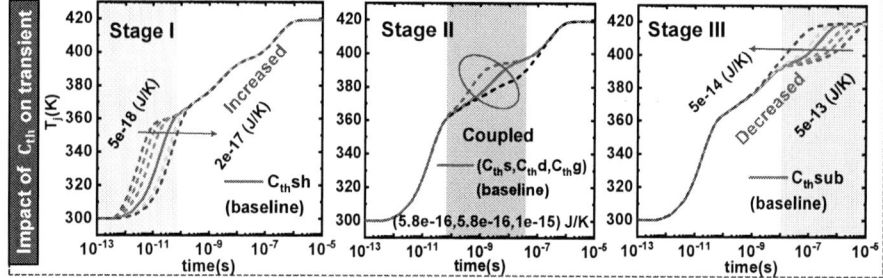

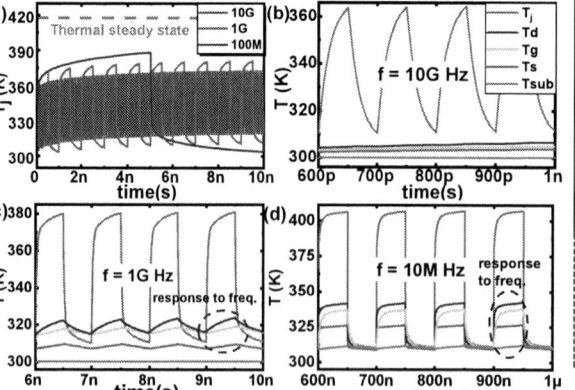

Fig. 8 (a) Self-heating in AC operation. Temperature response of different regions at (b) 10G Hz (c) 1G Hz and (d) 10M Hz, indicating the temperature in source, drain, gate and substrate (T_s,T_d,T_g,T_{sub}) cannot respond to high frequency (10G Hz).

Structure parameters		Thermal conductivity	
			κ_x, κ_y, κ_z
Lg	12nm	κ_{sh}	[18 18 2] W/K·m
Wsh	25nm	κ_{sd}	20 W/K·m
Hsh	5nm	κ_{sub}	140 W/K·m
Lsd	12nm	κ_{via}	174 W/K·m
Tox	0.8nm	κ_{ox}	0.4 W/K·m
Tsub	3μm	κ_{metal}	400 W/K·m

Dimension dependent thermal parameters calculation	
$R_{thi} = L_i / (\kappa_i \cdot S_i)$ (1)	R_{thi}: Thermal resistance
$C_{thi} = C_{pi} \cdot V_i$ (2)	C_{thi}: Thermal capacitance
	i: sh, s, d, g, sub…
$\kappa_i(z) = \kappa_0(T) \int_0^{\pi/2} \sin^3\theta\{1 - \exp(-l/2\lambda(T)\cos\theta)$	
$\times \cosh((l-2z)/(2\lambda(T)\cos\theta))\}d\theta$ (3)	
$Power(t) = I(t)V(t)$ (4)	
$[\kappa_x, \kappa_y, \kappa_z]$ denotes the anisotropic thermal conductivity	
L_i: the length along the heat conduction direction	
S_i: the cross-sectional area of heat flux	
V_i: the volume of the heat conduction	

Table I. Typical device parameters (baseline) and the equations for thermal RC network model.

978-1-7281-1988-5/18 $31.00 © 2018 IEEE   781

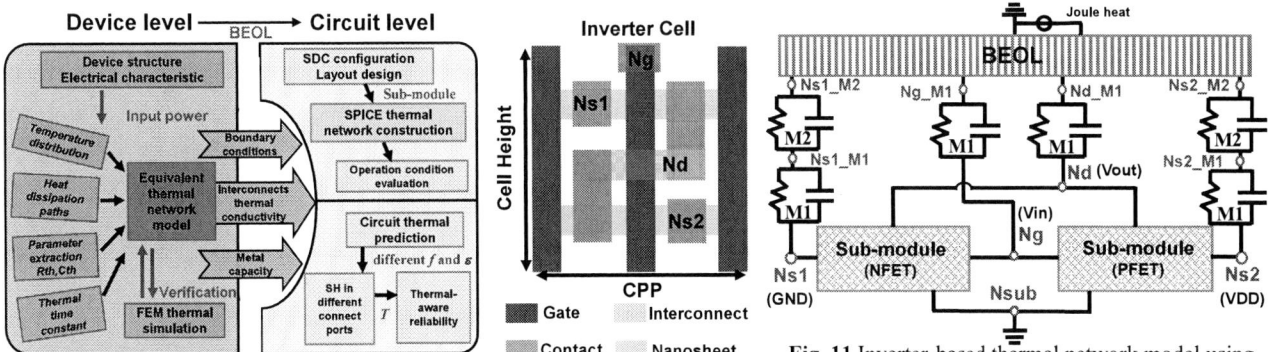

Fig. 9 Flowchart of thermal network construction for self-heating assessment from device to circuit level.

Fig. 10 Schematic of nanosheet-based CMOS inverter layout.

Fig. 11 Inverter-based thermal network model using packaged sub-modules with two metal layers (M1, M2) and top BEOL.

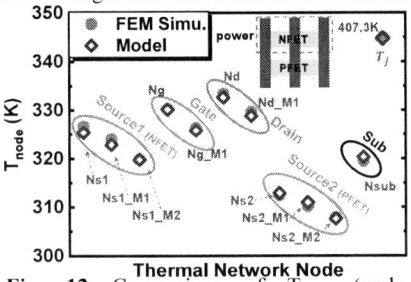

Fig. 12 Comparison of T_{node} (node temperature) between inverter-based thermal model and FEM simulation in static state with input in NFET.

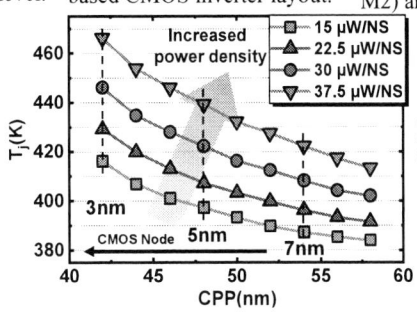

Fig. 13 Self-heating in the inverter cell with different CPP under increased power density.

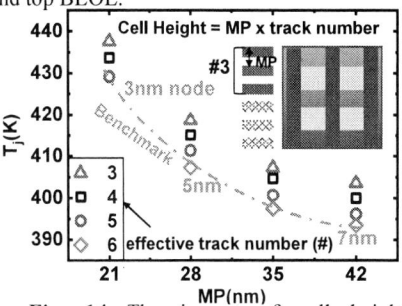

Fig. 14 The impact of cell height associated with MP and effective track number on self-heating in the inverter.

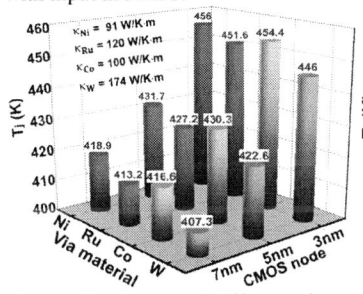

Fig. 15 The impact of different via materials on self-heating in the inverter at scaling technology node.

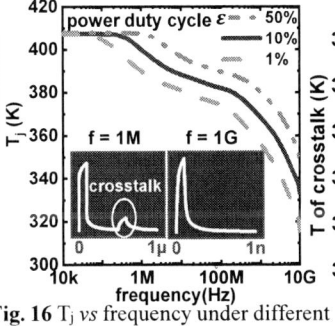

Fig. 16 T_j vs frequency under different power duty cycles(ε). Evident thermal crosstalk in low-frequency operation.

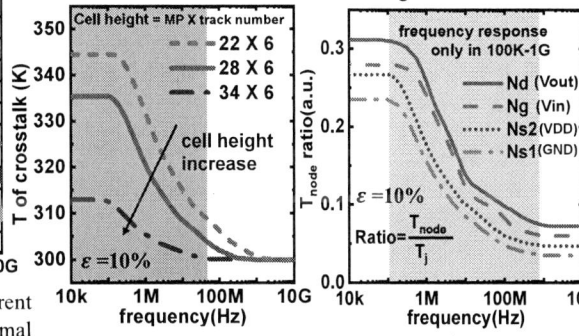

Fig. 17 Frequency dependent thermal crosstalk in the inverter cell with different cell heights.

Fig. 18 T_{node} in the inverter cell responds only to the frequency ranging from 100K Hz to 1G Hz.

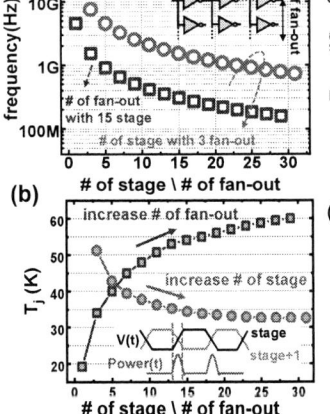

Fig. 19 (a) Relations between frequency and # of stage\fan-out in RO. (b) Model evaluation of T_j vs increased # of stage and fan-out.

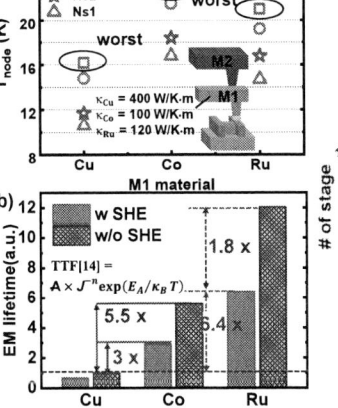

Fig. 20 (a) Predicted T_{node} in the RO with advanced interconnect materials. (b) EM lifetime prediction in M1 with the worst SHE.

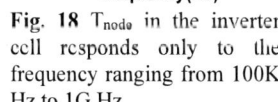

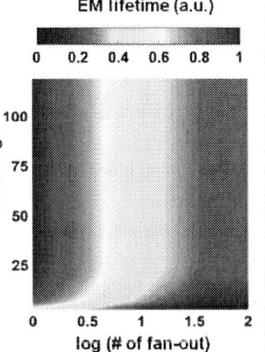

Fig. 21 Mapping of EM lifetime in Cu M1 with the worst SHE vs # of stage and fan-out.

Self-heating Modeling for Standard Cell (SDC)

$$k_{via} = k_m \phi_{via} + k_i(1 - \phi_{via}) \quad (5)$$

$$k_{metal} = k_i \frac{1 + \phi_{metal}[\beta_{xy}(1 - L_{11})]}{1 - \phi_{metal}} \cdot (\beta_{xy} \cdot L_{11}) \quad (6)$$

$$\phi_{metal} = \frac{S_c \cdot F_{metal}}{A_{SDC}} \quad (7)$$

S_c: Correction Factor
N_{via}: # of Via
$F_{via,metal}$: Footprint Area
A_{SDC}: SDC Area

$$\phi_{via} = \frac{N_{via} \cdot F_{via}}{A_{SDC}} \quad (8)$$

$$R_{via,metal} = \frac{H_{via,metal}}{k_{via,metal} \cdot A_{SDC}} \quad (9)$$

CMOS node	SDC Area		Scaling ratio
	CPP	Cell Height	
7nm	56nm	252nm	
5nm	48nm	168nm	42.86%
3nm	42nm	105nm	45.31%

Table II. Equations for thermal-equivalent BEOL and layout parameters of inverter cell used in this work.

978-1-7281-1988-5/18 $31.00 © 2018 IEEE

Understanding the intrinsic reliability behavior of *n-/p*-Si and *p*-Ge nanowire FETs utilizing degradation maps

Adrian Chasin*, Erik Bury, Jacopo Franco, Ben Kaczer, Michiel Vandemaele, Hiroaki Arimura, Elena Capogreco, Liesbeth Witters, Romain Ritzenthaler, Hans Mertens, Naoto Horiguchi, Dimitri Linten
imec, Leuven, Belgium, *adrian.chasin@imec.be

Abstract—We compare and model the main reliability limitations of stacked Gate-All-Around (GAA) *n-/p*-channel Silicon and strained *p*-channel Germanium Nanowire (NW) transistors. Stress measurements in the entire $\{V_G, V_D\}$ space allow to separate the different degradation modes and how they interact with each other. We show that these degradation modes are not universal, as they have a different relative weight depending on the considered technology, and that they can show different acceleration mechanisms. Moreover, we also discuss the impact of self-heating effects (SHE) by means of activation energy extraction in the entire $\{V_G, V_D\}$ map.

I. INTRODUCTION

Horizontal cylindrical GAA transistors can enable ultimate MOSFET scaling without the need of disruptive technology changes [1]. Recently, Silicon CMOS [2] and strained *p*-channel Germanium [3] stacked nanowires with diameters of 9nm were fabricated on bulk Si and SiGe Strain-Relaxed-Buffers (SRB) substrates, respectively (cross-sections in Fig.1). Si CMOS devices revealed excellent electrostatic control and effectively matched *n*- and *p*-Vth's via a nanowire compatible dual-work function metal integration scheme [2], while the *p*-channel Ge integration scheme resulted in device record performance [3].

Understanding the possible failure mechanisms of this novel architecture is essential to enable its commercial application. The novel device electrostatics, and the novel channel material in the Ge case can have a direct impact on the reliability and variability. A few studies have focused on the BTI/variability [4] or on hot-carrier degradation [5] aspects only, whereas recently we have studied the degradation mechanism of nMOS Si GAA devices within the entire $\{V_G, V_D\}$ bias space [6]. Here, we extend this methodology to *p*-Si and *p*-Ge NWs and compare to *n*-Si. We observe that previous findings for *n*-channel devices are not necessarily valid also for *p*-channel devices as other degradation mechanisms play a more relevant role in these devices. Furthermore, Ge based devices show a clearly different hot-carrier degradation acceleration as compared to Si.

This paper is structured as follows: first we characterize and compare the degradation maps of *n/p*-Si and *p*-Ge GAA devices. Second, the analysis of the degradation mechanisms is based on three independent metrics within the $\{V_G, V_D\}$ bias space: 1) the absolute $I_{D,Sat}$ degradations, 2) the amount of relaxation (i.e. degradation recovery) observed after each stress phase and 3) the time-to-failure (TTF) dependency on stress current. Finally, we extract the activation energy (E_A) of each process to obtain insights about SHE-related dependencies.

II. EXPERIMENTAL

Silicon *n*- and *p*- channel devices with gate lengths of 28nm and 100nm with 8 and 44 stacked parallel NW are used,

respectively. The Ge *p*-channel GAA devices have fixed channel length of 100nm and 44 stacked wires. The diameter of each wire is ~9nm (Fig.1). The gate stack of Si based devices consist of SiO_2 IL + HfO_2 with an EOT ~ 0.8nm. The same gate stack is fabricated on Ge GAA devices by employing a 6 monolayer Si-cap passivation of the Ge surface [3].

The degradation measurements in $\{V_G, V_D\}$ space comprise ~300 FETs, with 3 nominally identical devices stressed for each bias condition. The stress experiments are based on a modified eMSM scheme [7]. After each stress phase, first the typical relaxation is traced at fixed V_G for ~10s; afterwards, full I_D-V_G characteristics in linear, saturation, and reversed-source/drain (S/D) saturation regimes are recorded. All measurements are performed at T=125C, unless otherwise stated. The median values of the degradation parameters are reported. For the device performance metrics, we assume $|V_{DD}|$ equal to 0.9V and 0.5V for Si and Ge based devices, respectively.

III. DEGRADATION MECHANISMS

The initial characteristics of the devices are shown in Fig.2. *n*-Si ref. devices with L_G=28nm have the highest saturation current levels, followed by the *p*-Ge NWs despite the longer L_G=100nm and smaller V_{DD}. The high current level of such devices is mainly due to the superior hole mobility, revealed by its higher G_{m0}. Finally, Si GAA devices have initial subthreshold swing (SS_0) close to the ideal value of 80mV/dec (125^{0}C) which indicates excellent electrostatic control and low D_{it}. In subsection A, we characterize the Si based devices, focusing on the *p*-channel, followed by the *p*-Ge (subsection B).

A. Silicon NWs

Fig.3(a,b) depicts the $I_{D,Sat}$ and $I_{D,Sat,Rev}$ degradation maps of *p*-Si GAA after 530s of stress. A safe operating bias region is defined by adopting a failure criteria of 1% $I_{D,Sat}$ or $I_{D,Sat,Rev}$ degradation at end of stress (rescaled from a typical 10% at 10 years). For *p*-Si devices, this safe operating area—simply defined here as a rectangle with sides $V_D=V_G=V_{DD}$, with the maximum operating V_{DD} limited by the most degrading mechanism—is limited by NBTI degradation. Within the high V_D region, the degradation map of the I_{Sat} measured with reversed S/D terminals reveal even larger degradation. Interestingly, at {low V_G, high V_D} biases, I_{Sat} increases after stress. Fig.4(right) shows that this increase of I_{Sat} within the high V_D region is due to an opposite (positive) shift of V_{th}, which counter acts the normal V_{th} shift due to hole trapping. Consequently, at specific bias conditions, even with increasing V_G, it seems that device is not degraded as I_{Sat} does not shift. However, I_{off} and SS increase substantially, which indicates a degradation of the channel electrostatic control of the gate. Fig.5 maps the relaxation slope of *p*-Si GAA devices. Degradation induced by NBTI recovers much faster than the degradation induced by hot-carrier mode (higher V_D, for the same V_G), which indicates that the Multi-Vibrational Excitation (MVE) induces mainly "quasi-permanent" degradation. Off-state stress also reveal the same lack of recovery, which indicates that even though MVE and off-state stress are caused by trapping of opposite charges, the trap time constants are similar. Fig.6 shows the degradation resulting from NBTI and MVE is mostly uniform within the device channel, whereas for high V_D biases there are two distinctive regions: hole trapping close to drain at the channel pinch-off region induced by the Single-Vibrational Excitation (SVE) hot-carrier degradation mode, and the electron trapping also at the drain terminal at low V_G biases in off-state stress hot carrier mode. Note that the SVE mode is based on a high energetical incident hole with enough energy to excite the adsorbate resonance state, causing Si-H bond breaking [8].

It is known that NBTI aging also takes place during hot-carrier degradation [9] and therefore the two degradation mechanisms should be decoupled in order to independently characterize the hot-carrier degradation. Fig.7(a) reveals that at 125°C and V_G=1.3V, the residual NBTI is the most important degradation mechanism up to $V_D = 1$V, from where MVE dominates. These bias conditions are used to plot the "NBTI-free" TTF vs. I_{Stress} as shown in Fig.7(b). Even though Fig.6 shows slightly localized degradation by hole trapping at the drain side (SVE), all the degradation conditions align with a single power law dependency on I_{Stress}, which indicates that the MVE mechanism dominates the hot-carrier degradation in p-Si GAA. Contrary to pMOS, nMOS shows a clear dependency on $V_{D,stress}$ (Fig.8b) for $V_D \gg V_G$, where SVE degradation dominates.

Fig.9 maps E_A (extracted from measurements at 125°C, 75°C and 25°C assuming an Arrhenius process) for n-/p-Si GAA devices. pFET and nFET show a similar behavior for BTI and MVE. However, the E_A values are different within the high V_D region: nFETs have small T-deactivation due to SVE process, whereas pFETs still show small T-activation. Contrary to some previous studies, p-Si devices do not show a higher E_A at any bias conditions [10]. This is observed only at specific stress conditions (Fig.10), which indicates that the typically-adopted self-heating correction is not universal and should be done taking in account the specific stress biases.

Fig.11 reveals the ΔI_{Sat} at end of stress for both p-/n-Si GAA devices. At low V_D biases, NBTI is more important than PBTI, being the most limiting reliability aspect for Si CMOS devices. At high V_D biases, however, SVE in n-Si device becomes prominent and limits the CMOS reliability, whereas it is very small for p-Si GAA devices. SVE mode, therefore, dramatically increases the degradation of n-Si devices when compared to p-Si devices, in which its effect is negligible, as also shown in Fig.8. Most of the reliability studies present in literature indeed indicates that the limiting reliability modes for n- and p-MOS devices are HCI and NBTI, respectively [11]. However, the study of MVE mode is often neglected. Fig. 12(a) shows that contrary to the SVE process, MVE has a higher impact on pMOS devices (e.g. at the same stress current, TTF of pMOS is smaller). Moreover, the degradation acceleration (e.g. the power slope dependency of TTF on I_{Stress}) is the same for the two types of devices. This indicates that the number of vibrational states (N_{steps}) of the dissociating interface bond is the same in both devices [9] (the Si-H bond is the defect precursor in both cases).

B. Germanium NWs

Fig.13(a) depicts the $I_{S,Sat}$ degradation map of p-Ge GAA after 530s of stress. NBTI is significantly suppressed even at high $|V_{ov}|$ (Fig. 14(b)), and therefore hot-carrier degradation becomes the limiting reliability mechanism. Given the smaller band-gap of Ge, off-state stress is also more prominent, causing increase of I_{Sat} due to electron trapping at $|V_{D,Stress}|$ as low as 1V. The increase of I_{off} is also observed during hot-carrier degradation, as shown in Fig. 13(b), which indicates that I_{off} and SS increase are important figures-of-merit for Ge devices reliability and qualification. Fig12(b) compares the TTF of 100nm p-Si and p-Ge GAA devices. As discussed previously for p-Si devices, all bias conditions with sufficiently high V_D to avoid the NBTI influence can be fitted with a single power-law dependency of I_{stress}, independently from $V_{D,stress}$, which shows that SVE mode is also negligible in p-Ge NWs. An important contrast between these two devices is related to the MVE degradation acceleration: the slope of p-Ge GAA devices is about 2x larger than p-Si, which indicates that the number of interface bond harmonic oscillator steps (N_{steps}) of these two MOS technologies are different. N_{steps} is dependent on the bond breaking energy E_b and the energy of vibrational mode ($\hbar\omega$). Therefore, it may be related to the different interface states precursors: even though Ge GAA devices have a Si cap layer, SiGeOx suboxides may be present due to Ge segregation into the Si cap [12]. Different interface state precursors might exist in this system on the top of the Si-H bonds as present in Si devices. Finally, Fig.14(a) compares the ΔI_{Sat} of p-Si and p-Ge NWs at the end of stress. NBTI suppression in p-Ge is clear with MVE being the dominant degradation mechanism at $V_{D,Stress}$ as low as 0.4V, whereas for p-Si MVE is only dominant for $V_{D,Stress}$ > 1V. Moreover, even though off-state stress is present on both devices, p-Ge reveals to be more susceptive to $V_{D,stress}$ biases.

Fig.15 summarizes the main degradation components of each technology: n-Si is limited by SVE and MVE hot-carrier degradations, p-Si is limited by OSS and NBTI, and p-Ge is limited by OSS and MVE modes. p-Ge devices have overall better reliability, with OSS being the main reliability concern.

IV. CONCLUSIONS

We have mapped the degradation of n-, p- Si and p-Ge NWs in the entire bias space. By characterization of various degradation metrics, multiple active degradation mechanisms could be revealed. It was found that the same mechanisms are active in each of the studied technologies, albeit in different magnitudes. As such, it was found that SVE degradation is negligible in pMOS (both for Si and Ge channels) with respect to nMOS. The MVE degradation was found to be smallest for p-Ge, followed by p-Si and n-Si devices, but the MVE current acceleration factor was 2x higher for p-Ge NWs, indicating a different defect or bond-breaking path. Even though off-state degradation is present in both p-Si and p-Ge, it becomes a serious reliability concern in the latter as it is observed at short stress time at $|V_D|$ as low as 1V due to the reduced Ge bandgap.

REFERENCES

[1] J. P. Colinge and James C. Greer, Nanowire Transistors, Cambridge University Press, 2016; [2] H. Mertens *et al.*, in *Proc.* IEDM, p.524, 2016; [3] E. Capogreco *et al.*, in *Proc.* VLSI, p.193, 2018; [4] A. Chasin *et al.*, in *Proc.* IRPS, p. 5C-4.1, 2017; [5] A. Laurent *et al.*, in *Proc.* IRPS, p.6F 3-1, 2018; [6] A. Chasin *et al.*, in *Proc.* IEDM, p.159, 2017; [7] B. Kaczer *et al.*, in *Proc.* IRPS, p.20, 2008; [8] C. Guerin *et al.*, J. Appl. Phys., vol. 105, p. 114513, 2009; [9] C. Guerin *et al.*, in *Proc.* IRPS, p. 692, 2007; [10] M. Jin *et al.*, in *Proc.* IRPS, p.2A-2.1, 2016; [11] J. Franco *et al.*, in *Proc.* IRPS, p.6A.4.1-6, 2011; [12] K. Martens *et al.*, in *Proc.* ESSDERC, p.138, 2008

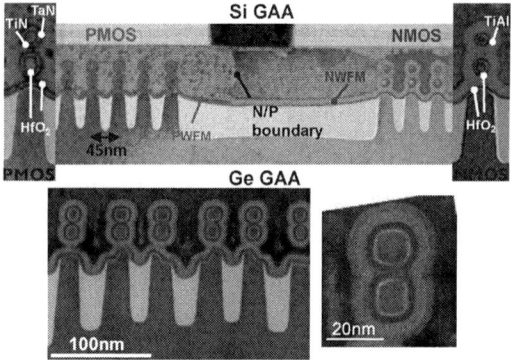

Fig. 1. Cross-sections of Si – CMOS (top) and Ge - pMOS (bottom) vertically staked Gate-All-Around devices. Both structures have similar diameters of ~9nm. Si and Ge devices are processed on bulk Si and $Si_{0.3}Ge_{0.7}$ SRB wafers, respectively.

Fig. 2. Initial characteristics of all 4 GAA devices (at 125C). $|V_{DD}|$ is equal to 0.9V and 0.5V for Si and p-Ge devices, respectively. p-Ge devices have high $SS_{0,Lin}$ when compared to Si devices, but much higher G_m and drive current.

Fig. 3. Current degradation ($\Delta I_S/I_{S0}$) measured at saturation (left) reversed-saturation (right) for p-Si GAA devices. Color scale is cropped at ±10% shift. The stars demarcate the biases where a failure criterion (±1% shift at end of stress) is reached for each of the main degradation modes. The red rectangles demarcate the safe operating area, limited by NBTI in p-Si devices.

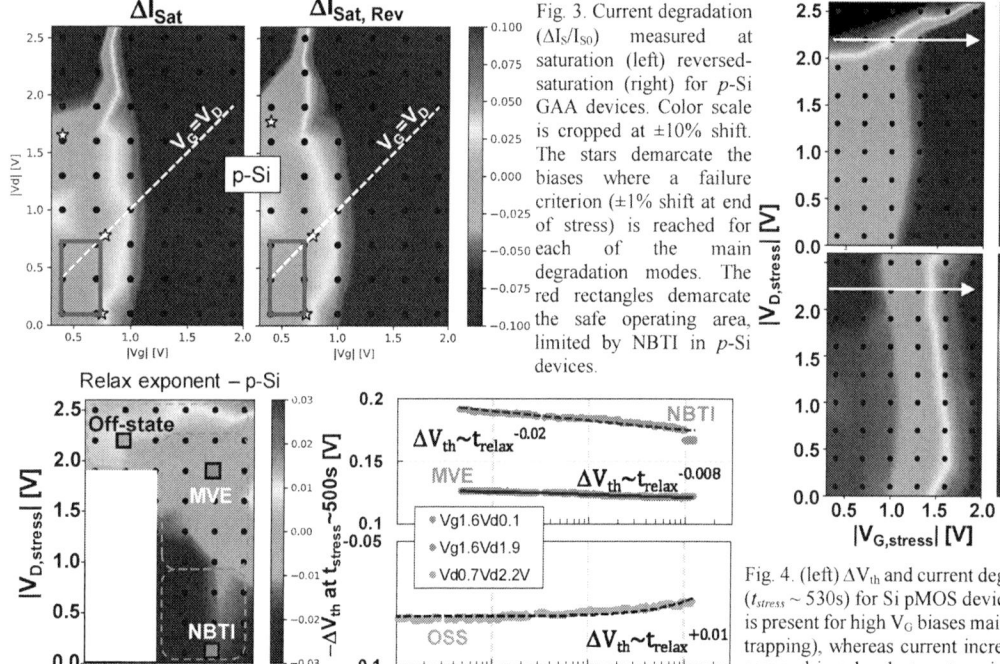

Fig. 4. (left) ΔV_{th} and current degradation ($\Delta I_S/I_{S0}$) at the end of stress ($t_{stress} \sim 530s$) for Si pMOS devices ($L_G=28nm$). Current degradation is present for high V_G biases mainly driven by negative V_{th} shift (hole trapping), whereas current increase is present at high V_D x low V_G corner driven by electron trapping which shifts V_{th} positively. (right) ΔV_{th} and ΔI_{Sat} as a function of t_{stress}, revealing counter-acting degradation effects from hot-carrier and off-state stresses, resulting in negligible I_{Sat} degradation for some specific biases.

Fig. 5. (left) Map of relaxation slope of p-Si GAA devices reveals stronger relaxation for BTI and a quasi-permanent degradation for both MVE and off-state stresses (OSS). Three specific relaxation behaviors from each of the degradation modes are depicted in the right graph.

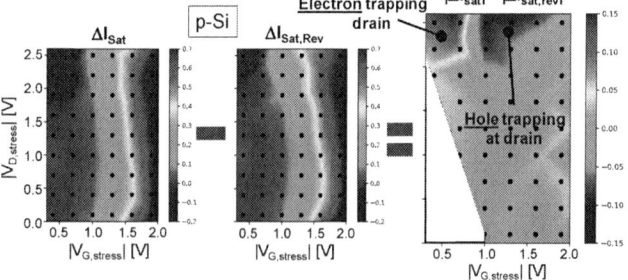

Fig. 6. Mapping of the difference between the degradation in saturation and reversed-saturation regimes at end of stress for p-Si GAA devices. Degradation occurs mainly uniformly along the channel under NBTI and MVE stresses, but it is strong localized at the drain side for SVE and off-state stress.

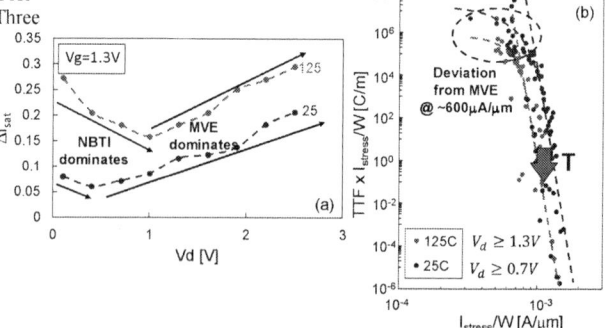

Fig. 7. (a) ΔI_{Sat} as a function of $V_{D,stress}$ for specific $V_{G,stress} = 1.3V$ at two different temperatures. NBTI is the most important degradation mode up to $V_D=0.4V$ and $V_D=1V$ for 25C and 125C, respectively. From this V_D levels, MVE starts to dominate. These bias conditions are then used in the TTF vs. I_{stress} plot (b), which shows that for p-Si GAA devices all stress conditions are driven by the MVE mode with the except of few points that deviate at low I_{stress} of 600 µA/µm.

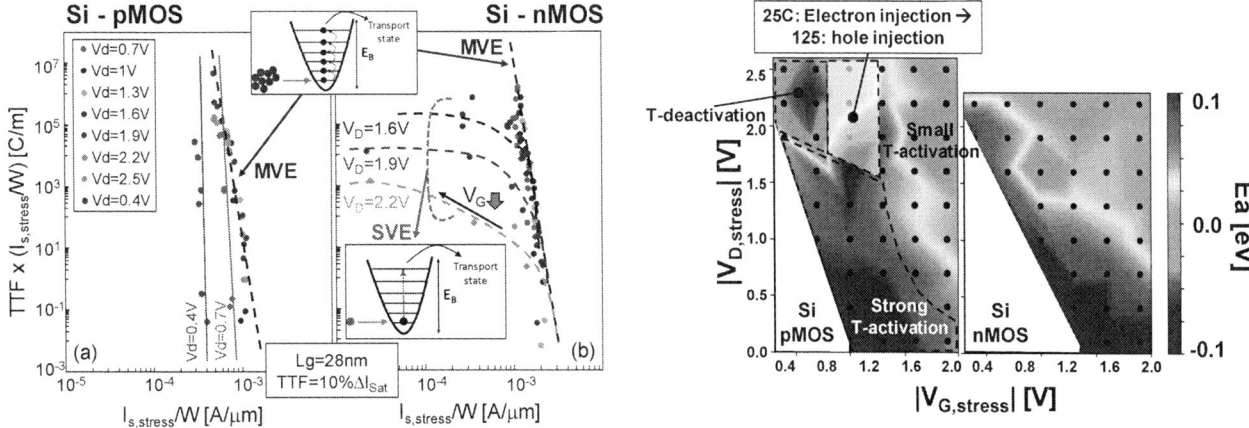

Fig. 8. TTF plots for p-Si (a) and n-Si (b) GAA devices. The clear deviation from the MVE of n-Si devices for high V_D X low V_G biases is not observed for p-Si.

Fig. 9. Normalized $\Delta I_{SAT,125C}$ and E_A maps for Si pMOS (left) and nMOS (right) GAA devices. pMOS device have higher E_A at some specific bias conditions.

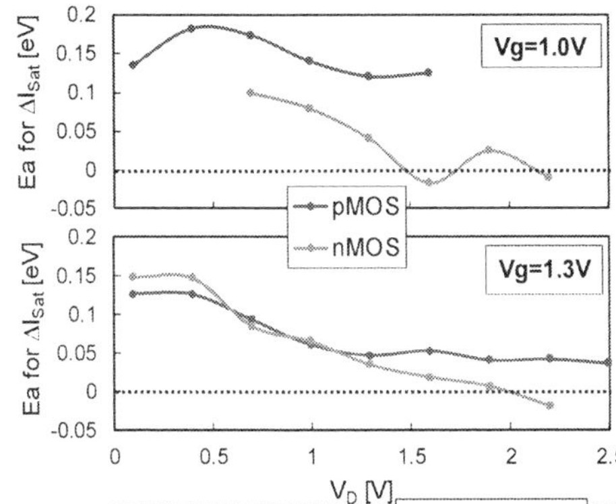

Fig. 10. Activation energy as a function of $V_{D,Stress}$ at $V_{G,Stress}=1V$ (top) and 1.3V (bottom) for both n-Si and p-Si GAA devices from Fig.9. E_A is the highest at pure BTI and decreases as a function of V_D for both devices. At low V_G biases, p-Si has higher E_A than n-Si devices.

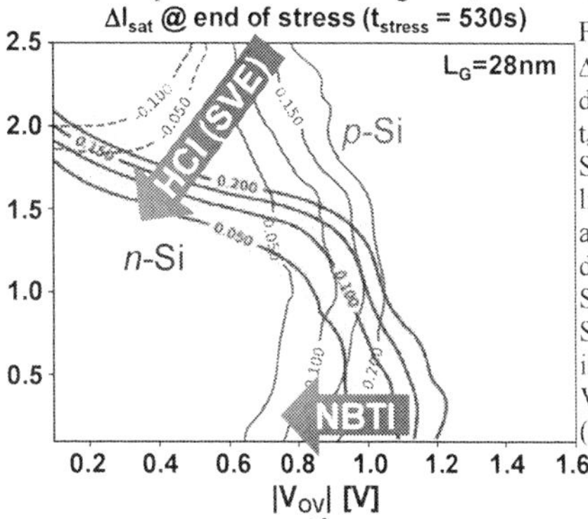

Fig. 11. Contour lines of ΔI_{Sat} for n/p-Si GAA devices at 125^0C after $t_{stress}=530s$. NBTI and SVE are the most limiting reliability aspects for p-Si and n-Si devices, respectively. SVE is negligible for p-Si. Off-state-stress (OSS) is measured only at high $V_{D,stress}$ for p-Si device (dashed contour lines).

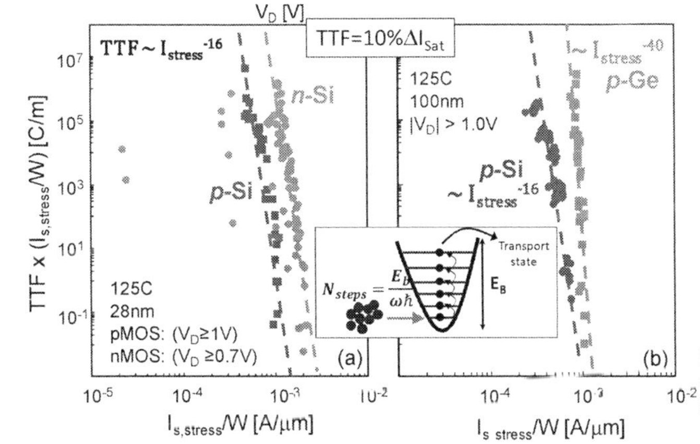

Fig. 12. TTF dependency on I_{Stress} is the same for both Si n and p-channel GAA devices (a), whereas Ge based device has a much higher slope (b). If only MVE is considered, p-Si devices have lower lifetime than n-Si counterpart, whereas p-Ge devices have higher lifetime at the same I_{Stress}.

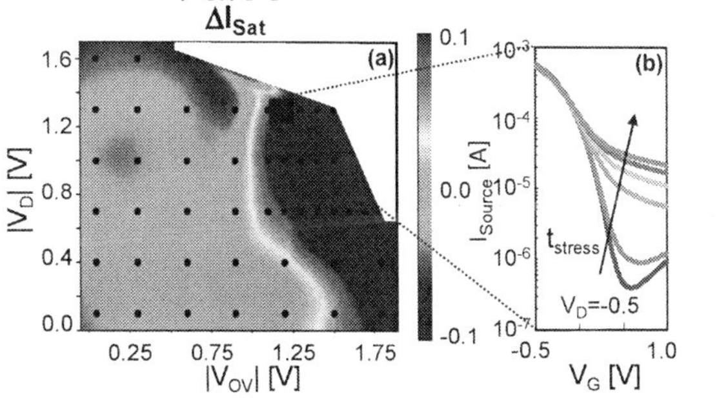

Fig. 13. ΔI_{Sat} map of p-Ge GAA devices have remarkable differences when compared to p-Si FETs (Fig. 3,4). Off-state stress is present at lower $|V_{D,stress}|$ biases and off-current increase is an important concern.

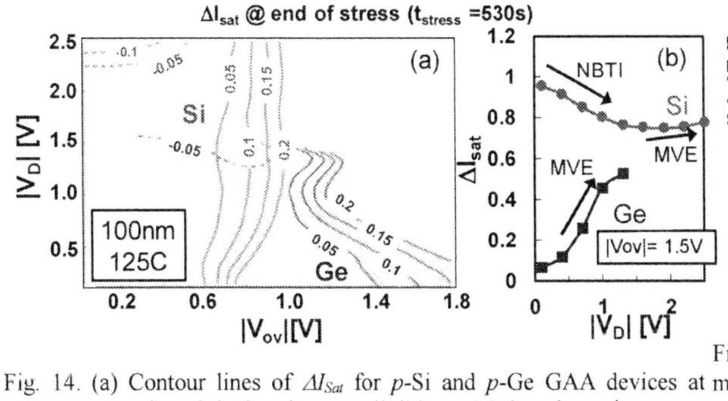

Fig. 14. (a) Contour lines of ΔI_{Sat} for p-Si and p-Ge GAA devices at $t_{stress}=530s$. Ge based devices have negligible NTBI, but shows important MVE and I_{off}-state degradation mechanisms.

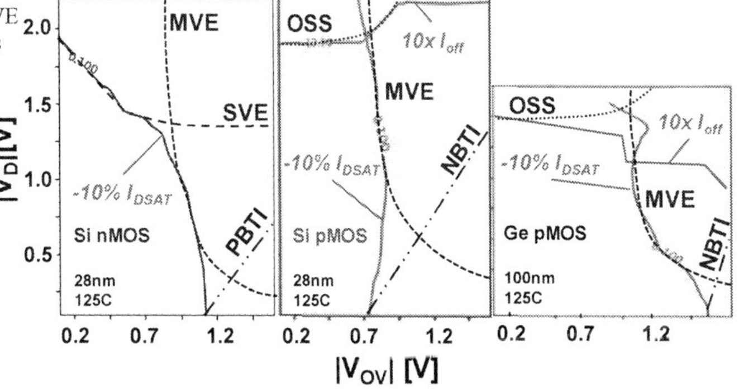

Fig. 15. Safe-Operating-Area of n-Si, p-Si and p-Ge GAA NWs with the degradation mechanisms schematically shown. SOA is defined by 10% ΔI_{Sat} decrease or 10x increase of $I_{Off,sat}$ measured at the end of stress (full lines). p-Ge devices have overall better reliability, with OSS being the main reliability concern.

978-1-7281-1988-5/18 $31.00 © 2018 IEEE

BTI Reliability Improvement Strategies in Low Thermal Budget Gate Stacks for 3D Sequential Integration

J. Franco[*], Z. Wu[1], G. Rzepa[2], A. Vandooren, H. Arimura, L.-Å Ragnarsson, G. Hellings, S. Brus, D. Cott, V. De Heyn, G. Groeseneken[1], N. Horiguchi, J. Ryckaert, N. Collaert, D. Linten, T. Grasser[2], B. Kaczer

imec, Leuven – Belgium, [1]also at KU Leuven – Belgium , [2]TU Wien – Austria [*]Jacopo.Franco@imec.be

Abstract—**Low thermal budget gate stacks will be required for novel integration schemes, such as 3D sequential stacking of CMOS tiers. We study the impact of a reduced thermal budget on BTI reliability, and we demonstrate two strategies to tolerate the inherently large high-k defect densities: i) replacing inversion mode devices with highly doped junction-less transistors, or ii) engineering dipoles at the interface between SiO₂ and HfO₂ to suppress the carrier-defect interaction. The latter approach is demonstrated for nMOS PBTI and, for the first time here, also for pMOS NBTI, as even this aging mechanism is controlled by high-k defects in ultra-thin EOT low thermal budget gate stacks.**

Introduction

3D sequential integration is a recently envisioned approach to increase CMOS functionality per die area by stacking transistors on top of each other (**Fig. 1**), or to co-integrate heterogeneous technologies on multiple tiers of the same wafer within a single fabrication flow [1,2]. Thermal budget management represents the most crucial challenge of this integration approach: thermal steps for the fabrication of the top tier should be limited to preserve the functionality of the bottom tier devices and interconnects (BEOL). Gate stack reliability is extremely sensitive to process temperature, particularly for high-k/metal gate (HKMG) technologies. In a gate-first integration flow, the high-k layer is exposed to the source/drain (S/D) activation anneal (~1000°C), which reduces the dielectric defect densities and minimizes Bias Temperature Instability (BTI) [3,4]. In more contemporary Replacement Gate (RMG) flows, a dedicated high-temperature 'reliability' anneal (~900°C) is typically performed after the deposition of the final gate stack to ensure sufficient stability [5,6]. Such high temperature steps are not suitable for top tier fabrication as they would degrade the BEOL of the bottom tier.

In this paper, we study first the impact of a limited thermal budget on the BTI reliability of the HKMG stack. By using our recently introduced physics-based BTI modeling framework ComPhy ("Compact Physical", [7]), we compare the oxide defect properties in an as-deposited low thermal budget SiO₂/HfO₂/TiN gate stack with the same gate stack exposed to a typical 'reliability' anneal, and with a commercial high-k-first 28nm HKMG technology, revealing larger oxide defect density at energy levels around the Si channel band edges. We then use the pathfinder predictive capabilities of ComPhy to define two possible strategies to guarantee sufficient BTI reliability despite the inherently large defect density in low thermal budget high-k dielectrics. The first strategy consists of replacing the standard inversion mode devices with junctionless devices, as the latter device type has been shown to offer improved reliability due to lower operating oxide fields [8]. We discuss how the reliability can be improved further by increasing the channel doping density in these devices, counter-intuitively to inversion mode transistors. The second strategy consists of engineering functional dipoles at the interface between SiO₂ and HfO₂ (by depositing a thin dipole-former interlayer with ALD) to 'shift up' or 'down' the energy levels of the high-k defects with respect to the Si conduction or valence band, for nMOS and pMOS reliability, respectively. We have demonstrated this latter strategy earlier to improve the PBTI reliability of Ge and InGaAs nMOSFETs [9,10] and of Si *n*-channel capacitors [11]. Here we demonstrate it on Si planar nMOSFETs, highlighting a positive correlation between channel electron mobility and PBTI reliability.

Furthermore, we show for the first time that dipole engineering can improve also pMOS reliability, as for low thermal budget thin-EOT gate stacks with <1nm SiO₂ interfacial layer (IL) NBTI is also limited by high-k traps. *All these findings are confirmed on Si hardware.*

The results presented here open up the reliable use of low thermal budget high-k gate stacks for 3D sequential integration, but also for other novel integration concepts as, e.g., the embedding of thin film transistors in the BEOL, or to conveniently re-arrange the integration flow of standard Si CMOS which is currently dictated by the highest temperature steps, e.g., the S/D contacts are fabricated only after the gate stack 'reliability' anneal instead of directly after S/D epitaxy.

Experimental

To study the impact of different thermal budget on the top and bottom tiers of a 3D sequential integration separately, we fabricated CMOS wafers without applying any thermal budget after gate stack deposition ("Top": as-deposited, **Fig. 2a**) or by performing a 2h long Post-Metal Anneal (PMA) at 525°C to mimic the additional thermal budget that bottom devices see during top tier fabrication ("Bottom", **Fig. 2b**). Fully stacked wafers were also fabricated (Fig. 1), with *n*- or *p*-channel SOI junction-less devices on top of a standard planar Si CMOS (**Fig. 2c**). For all devices the gate stack comprised a chemical SiO₂ IL (~0.6nm as-dep., 1nm w/ PMA), ~1.8nm HfO₂, and 5nm TiN.

Impact of low thermal budget on HKMG stack reliability

Top tier devices which did not receive any high temperature anneal after HKMG deposition show poor PBTI and NBTI reliability (**Fig. 3a,b**): extremely large threshold voltage shifts (ΔV_{th}: ~20× and ~10× larger than the acceptable target for nMOS and pMOS respectively) and detrimentally weak BTI voltage accelerations are observed, in contrast to the same gate stack exposed to a ~900°C-1s 'reliability' anneal, and to a commercial 28 nm HKMG technology. Interestingly, the same gate stack yields almost sufficient BTI reliability when used in bottom tier devices (**Fig. 3c,d**), suggesting that the additional thermal budget exposure might make the implementation of a dedicated reliability anneal unnecessary for these devices.

To understand the origin of the poor top tier reliability, we used ComPhy [7] to model the ΔV_{th} measured in nMOS and pMOS devices subjected to a complex stress/recovery waveform and to more conventional measure-stress-measure patterns [12], with various voltages and at different temperatures (**Fig. 4**). An excellent match of the modeled BTI kinetics to the experimental data is achieved by properly calibrating shallow and deep defect band properties for both the SiO₂ and HfO₂ dielectric; these defect properties are compared to the ones extracted in [7] from BTI measurements of a commercial 28 nm technology (**Table I**). The reduced thermal budget results in a ~2× larger defect density in HfO₂, and especially in a significantly reduced mean trap energy level and mean thermal barrier for capture for both the SiO₂ and HfO₂ traps, possibly due to a less stiff, more disordered amorphous oxide. Therefore, a larger amount of oxide defects can trap channel carriers at operating oxide electric field (E_{ox}=3MV/cm), as compared to a standard high thermal budget technology (**Fig. 5**).

We demonstrate two alternative strategies for sufficient low thermal budget reliability: i) the use of highly doped junctionless transistors, which operate at lower E_{ox} compared to inversion mode devices, or ii) the insertion of dipole-former interlayers between SiO₂ and HfO₂ to shift the defect levels in the latter w.r.t. the Si channel bandgap.

978-1-7281-1988-5/18 $31.00 © 2018 IEEE

Solution I: Junction-less devices optimized for minimum E_{ox}

Accumulation-mode junction-less devices offer a superior reliability as they operate close to the MOS flatband voltage [8]: on-state is achieved with a low E_{ox}, while an opposite E_{ox} is necessary to switch-off the current flow by depleting the thin channel of the device (**Fig. 6**) [note: the device V_{th} is therefore defined as the V_G required to fully deplete the channel [13], **Fig. 7**]. In a conventional inversion-mode MOSFET, the on-state E_{ox} is determined by the sum of the inversion charge (proportional to the gate overdrive voltage, $V_{ov}=V_G-V_{th}$) and the depletion charge: as such, a higher channel doping level results in a larger E_{ox} for a given V_{ov}, detrimental to reliability [14]. We argue that in a junction-less device instead, a higher doping can be beneficial for reliability: electrostatic simulations [15] show that for a given E_{ox} a larger V_{ov} (i.e., higher drive current) can be applied on a junction-less device for increasing doping densities (**Fig. 8**). Experimental data confirm that sufficient PBTI and NBTI reliability is achieved in junction-less devices with doping $>2\times10^{18}/\text{cm}^3$ (**Fig. 9**), despite the large defect density in the low thermal budget oxides. Note that the BTI trends are excellently reproduced in ComPhy (Fig. 9, lines), by using the same defect properties calibrated on low thermal budget inversion mode devices (Table I), and the specific $E_{ox}(V_{ov})$ relation implied by the junction-less electrostatics (cf. Fig. 8). We conclude that junction-less devices represent a convenient option for top tiers, as their low operating E_{ox} relieves the dielectric quality requirements. Moreover, these devices are inherently thermal-budget-friendly as they do not require S/D activation (note: channel doping is activated before bonding the top Si slab to the bottom tier passivation oxide [2]).

Solution II: dual interface dipole engineering for sufficient BTI reliability in inversion-mode CMOS

As discussed above, the main shortcoming of a low thermal budget HKMG stack is the low energy level of the oxide defects, which enhances charge trapping. Properly engineered interface dipoles can 'shift up' the shallow high-k defect band w.r.t. the Si conduction band (for improved PBTI), or 'shift down' the deep defect band w.r.t. the Si valence band (for improved NBTI), and therefore reduce the density of accessible defects to a level comparable to the high temperature HKMG stack of a commercial 28 nm technology (**Fig. 10**, cf. Fig. 5).

A.   ComPhy pathfinding study: nMOS and pMOS

ComPhy simulations show that a dipole at the SiO_2/HfO_2 interface is very effective to improve PBTI: a $\Delta V_{th}<50$mV is projected for 10-year operation at $V_{ov}=0.7$V at 125°C if the high-k shallow defect energy is increased by 0.4eV (**Fig. 11**). This improvement is virtually independent of the SiO_2 thickness, as PBTI is mostly controlled by high-k electron traps (SiO_2 electron traps are negligible, cf. Table I).

Interestingly, ComPhy simulations show that a similar approach can be effective also for improving NBTI, despite that in pMOS this mechanism is commonly ascribed to interfacial traps: in a low thermal budget HKMG stack, the deep defects in HfO_2 have a sufficiently high density to surpass the SiO_2 hole traps (**Fig. 12a**, dashed vs. dotted lines), and therefore a proper interface dipole shift can reduce charge trapping to a tolerable level. However, if a thick IL is used (**Fig. 12b**, 1nm vs. 0.6nm), the total density of hole traps in the larger SiO_2 volume becomes dominant, defeating the effectiveness of an interface dipole at the SiO_2/HfO_2 interface for improving the overall NBTI.

This pathfinding simulations suggest that by using a 0.6nm thin SiO_2 IL and by inducing a +0.4eV or a -0.4eV dipole shift at the SiO_2/HfO_2 interface, *PBTI- and NBTI-induced ΔV_{th} can be brought back to a level comparable to the (high thermal budget) commercial 28 nm ref. technology* (**Fig. 13**), guaranteeing >10-year reliable operation.

B.   Experiment: LaSiOx dipole for nMOS PBTI

We have recently shown [11] that a ~0.3nm thin $LaSiO_x$ layer ALD-deposited between SiO_2 and HfO_2 can induce a ~0.4eV V_{fb} shift in n-Si MOS capacitors. We reproduce this here on planar MOSFETs, demonstrating low nMOS V_{th} with a TiN high work function metal. Moreover, the insertion of the dipole forming interlayer is observed to be beneficial also for channel electron mobility (**Fig. 14**) due to reduced carrier-defect interaction. The PBTI ΔV_{th} measured on $LaSiO_x$-inserted nMOSFETs are significantly reduced (~8×) w.r.t. the reference low thermal budget "top tier" gate stack, in excellent quantitative agreement with the ComPhy prediction (**Fig. 15**). Note that when combined with the higher thermal budget of the bottom tier, exceptional PBTI reliability can be achieved (Fig. 15, squares); hence, this gate stack represents an interesting option also for bottom devices.

C.   Experiment: Al2O3 dipole for pMOS NBTI

To experimentally verify the effectiveness of interface dipoles for NBTI reliability, a negative dipole forming layer should be identified. It was previously reported [16], that Al_2O_3 is a strong negative dipole former on SiO_2. In order to achieve a sufficient interface dipole density for a ~0.4eV shift, a ~1nm thick Al_2O_3 layer is necessary (**Fig. 16**, quantitatively in-line with [16]). Additional trapping in such a thick layer might defeat the reliability improvement strategy. However, our previous BTI study of bulk Al_2O_3 oxide defects suggests a negligible trap density at energies close to the Si valence band (**Fig. 17**), making Al_2O_3 suitable for pMOS use (while its wide defect distribution around the Si conduction band makes it unsuitable for nMOSFETs [10]).

By depositing a 1nm Al_2O_3 layer on SiO_2 before HfO_2, a ~10× reduced NBTI ΔV_{th} is demonstrated (**Fig. 18**). In particular, a stronger NBTI voltage acceleration is observed when inserting Al_2O_3 (**Fig. 19**), quantitatively in line with the ComPhy prediction (solid lines) and comparable to best-in-class SiGe pMOSFETs [17], confirming an effective decoupling of the defect energy level w.r.t. the channel Fermi level. A ~0.2nm thin Al_2O_3, inducing ~0.24eV of dipole shift, is already sufficient to bring NBTI ΔV_{th} within specs, with a minimal EOT penalty (cf. Fig. 16a). We note the improvement achieved by inserting thicker Al_2O_3 layers is much larger than the E_{ox} reduction related to the thicker EOT (**Fig. 20**).

For proper pMOS V_{th} tuning, the effective work function shift related to the Al_2O_3 dipole might need to be compensated: this can be achieved by replacing the TiN metal with a more mid-gap work function metal such as TaN, maintaining the improved reliability (**Fig. 21**). Finally, we compare the reliability of the Al_2O_3-inserted gate stacks to the reference one at top- and bottom-tier thermal budgets (**Fig. 22**): in the as-deposited gate stack with a ~0.6nm native SiO_2 IL, a thicker Al_2O_3 layer results in larger eWF shift and maximum operating V_{ov}; in contrast, when applying a 525°C-2h PMA to mimic the thermal budget seen by the bottom devices, the NBTI reliability is not impacted by the Al_2O_3 dipole: this is due to i) the higher thermal budget, yielding a sufficient baseline reliability, and ii) a thicker SiO_2 IL (1nm vs 0.6nm, cf. Fig. 22b) reducing the contribution of high-k deep traps to the overall NBTI, in line with the ComPhy prediction (cf. Fig. 12b).

Conclusions

We have demonstrated two strategies to achieve sufficient gate stack BTI reliability at low thermal budget, relevant for novel integration schemes such as 3D sequential. The first approach consists of using highly doped junction-less devices in the top tier to reduce the operating oxide electric field, and therefore the density of charging oxide traps. The second approach utilizes ALD dipole-forming interlayers ($LaSiO_x$ for nMOS and Al_2O_3 for pMOS) to engineer the energy alignment of the high-k defect levels (i.e., the main contributors to both PBTI and NBTI ΔV_{th} in low thermal budget gate stacks) w.r.t. the Si channel band edges. Both strategies were identified by using the pathfinding predictive capabilities of our BTI simulation framework ComPhy, and were demonstrated experimentally on Si hardware.

Acknowledgements: research funded by imec's core partner program.

References

[1] P. Batude *et al.*, in *Proc.* VLSI Tech. 2015;
[2] A. Vandooren *et al.*, in *Proc.* VLSI Tech. 2018;
[3] E. Bury *et al.*, in *Proc.* IRPS 2013;
[4] H. Arimura *et al.*, in *Proc.* IRPS 2014;
[5] G. Rzepa *et al.*, in *Proc.* IRPS 2017;
[6] B. Linder *et al.*, in *Proc.* IRPS 2016;
[7] G. Rzepa *et al.*, in *Micr. Rel.* 85, pp. 49-65 , 2018;

[8] M. Toledano-Luque *et al.*, in EDL 35(12), 2014;
[9] H. Arimura *et al.*, in *Proc.* IEDM 2016;
[10] J. Franco *et al.*, in *Proc.* IEDM 2017;
[11] J. Franco *et al.*, in *Proc.* IRPS 2017;
[12] B. Kaczer *et al.*, in *Proc.* IRPS 2008;

[13] J.-P. Colinge *et al.*, Nature Nanotechnology 5, pp. 225–229, 2010;
[14] J. Franco *et al.*, in *Proc.* IRPS 2016;
[15] J. R. Hauser and K. Ahmed, in *Proc.* AIP Conference 449, 235 ,1998 (CVC version 5.0);
[16] Y. Kamimuta *et al.*, in *Proc.* IEDM 2007;
[17] J. Franco *et al.*, in *Proc.* IEDM 2013.

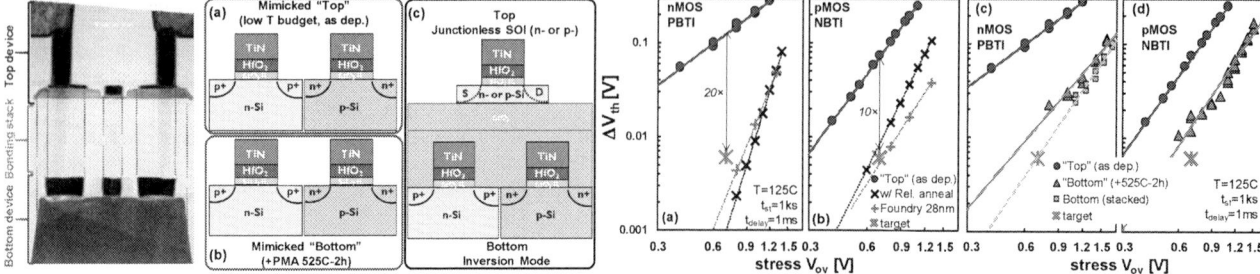

Fig. 1: TEM of a 3D structure showing stacked top and bottom tier devices with nanometric alignment [2].

Fig. 2: Used test structures: planar CMOS with **(a)** as-deposited gate stack (compatible with top tier, "top"), or **(b)** with a 525ºC-2h PMA mimicking the thermal budget seen by the bottom tier during top fabrication ("bottom"); **(c)** stacked structures with SOI junction-less devices on bulk CMOS [2].

Fig. 3: Low thermal budget HKMG stack ("top") shows poor **(a)** PBTI and, **(b)** NBTI compared to the same stack with a high temperature 'reliability' anneal, or compared to a 28nm commercial technology [5,7]. In contrast, thanks to the additional thermal budget of the top tier fabrication, the "bottom" stack almost meets the **(c)** PBTI and **(d)** NBTI targets (ΔV_{th}<6mV at V_{ov}=0.7V, T=125C, t_{st}=1ks, rescaled from 50mV at 10 years) without a dedicated reliability anneal.

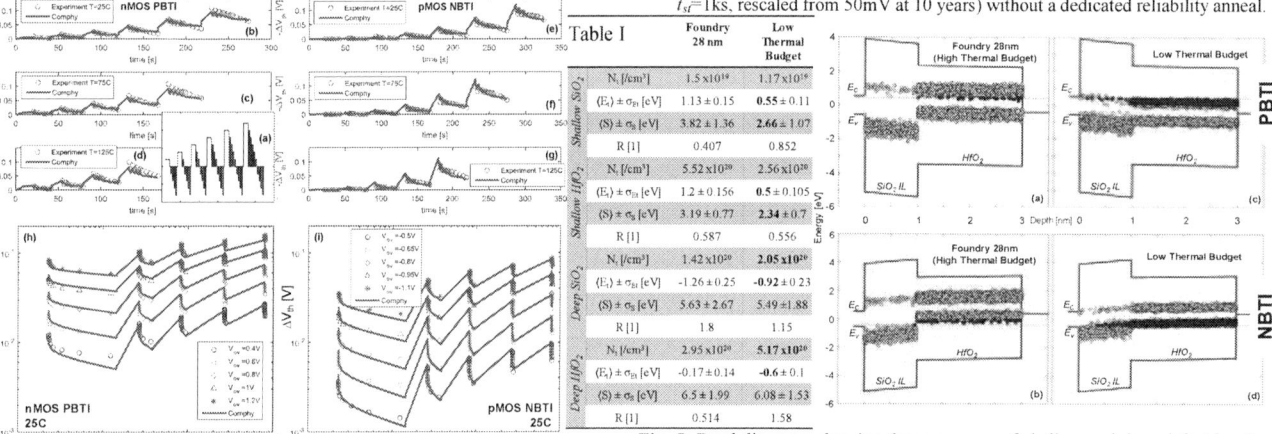

Table I

		Foundry 28 nm	Low Thermal Budget
Shallow SiO₂	N_t [/cm³]	1.5 x10¹⁹	1.17 x10¹⁹
	$\langle E_t \rangle \pm \sigma_{Et}$ [eV]	1.13 ± 0.15	0.55 ± 0.11
	$\langle S \rangle \pm \sigma_S$ [eV]	3.82 ± 1.36	2.66 ± 1.07
	R [1]	0.407	0.852
Shallow HfO₂	N_t [/cm³]	5.52 x10²⁰	2.56 x10²⁰
	$\langle E_t \rangle \pm \sigma_{Et}$ [eV]	1.2 ±0.156	0.5 ±0.105
	$\langle S \rangle \pm \sigma_S$ [eV]	3.19 ±0.77	2.34 ±0.7
	R [1]	0.587	0.556
Deep SiO₂	N_t [/cm³]	1.42 x10²⁰	2.05 x10²⁰
	$\langle E_t \rangle \pm \sigma_{Et}$ [eV]	-1.26 ±0.25	-0.92 ±0 23
	$\langle S \rangle \pm \sigma_S$ [eV]	5.63 ±2.67	5.49 ±1.88
	R [1]	1.8	1.15
Deep HfO₂	N_t [/cm³]	2.95 x10²⁰	5.17 x10²⁰
	$\langle E_t \rangle \pm \sigma_{Et}$ [eV]	-0.17 ±0.14	-0.6 ±0.1
	$\langle S \rangle \pm \sigma_S$ [eV]	6.5 ±1.99	6.08 ±1.53
	R [1]	0.514	1.58

Fig. 4: Defect models (see parameters in Table I) were calibrated in ComPhy [7] to excellently reproduce the nMOS PBTI (b,c,d,h) and (e,f,g,i) pMOS NBTI kinetics in a variety of stress/recovery patterns: **(a)** complex waveform comprising increasing stress voltages and decreasing discharge voltages, performed at **(b,e)** 25ºC, **(c,f)** 75ºC, **(d,g)** 125ºC; and **(h-i)** standard extended measure-stress-measure patterns [12] with various stress overdrive voltages (25ºC).

Fig. 5: Band diagrams showing the occupancy of shallow and deep defect bands (cf. Table I) in SiO₂ and HfO₂ at BTI stress condition ($E_{ox,IL}$=±3MV/cm), as calibrated in Comphy on **(a-b)** commercial 28nm nMOS and pMOS, **(c-d)** low thermal budget HKMG stack ("top", as deposited). All defect levels crossing the channel Fermi level due to the applied E_{ox} are depicted here as filled (as their charge state might change); note the larger density of charging defects in the low thermal budget stack, as compared to the commercial 28nm technology.

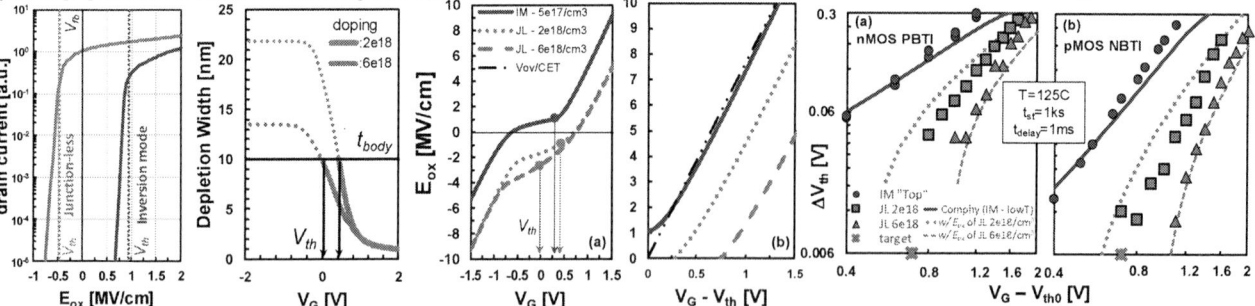

Fig. 6: Simulated I_D vs. E_{ox} for an inversion mode MOSFET and for a junction-less device (replotted from [8]). The latter operates close to V_{fb} with a low E_{ox}, while it requires a negative E_{ox} to switch off the current.

Fig. 7: Calculation [15] of the channel depletion width in a MOS stack for two different doping levels. In a junction-less device, the V_{th} is defined [13] as the V_G required to fully deplete the thin body (~10nm).

Fig. 8: Calculated E_{ox} for inversion- (N_d=5e17 /cm³) and accumulation-mode MOSFETs (N_d= 2-6e18/cm³) vs. **(a)** V_G, and vs. **(b)** $V_G - V_{th}$ (note: V_{th} defined at full depletion, cf. Fig. 7). At a given V_{ov}, junction-less show a reduced E_{ox} for increasing doping. Note the invalidity of the approx. E_{ox}~V_{ov}/CET for junction-less devices.

Fig. 9: **(a)** nMOS PBTI, and **(b)** pMOS NBTI ΔV_{th} measured in inversion mode and junction-less devices vs. V_{ov}. The same low thermal budget gate stack is used in all devices. The reduced ΔV_{th} in junction-less for increasing doping levels is quantitatively predicted by Comphy, accounting for the E_{ox} reduction in junctionless devices (cf. Fig.8b).

978-1-7281-1988-5/18 $31.00 © 2018 IEEE

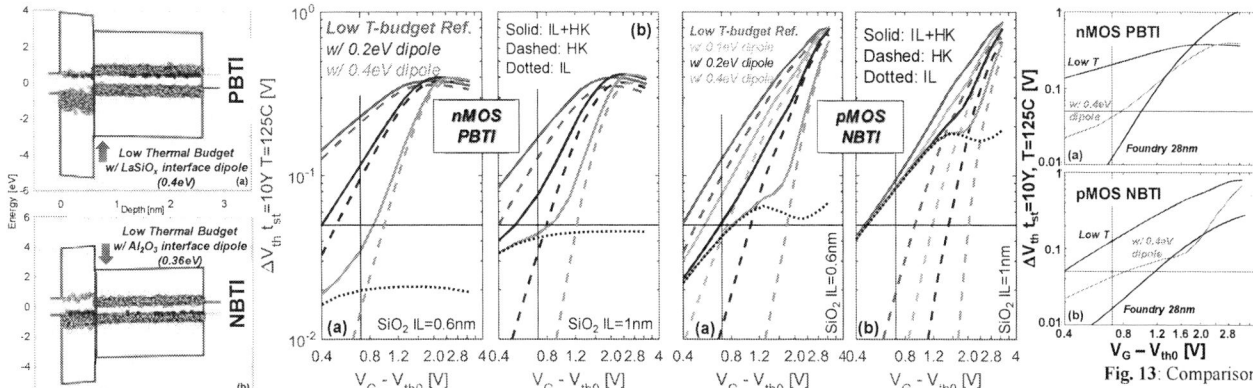

Fig. 10: Occupancy of shallow and deep defect bands in low thermal budget SiO₂ and HfO₂ at BTI stress bias ($E_{ox,IL}=\pm3MV/cm$), considering a ~0.4eV dipole **(a)** up-shift for improved PBTI, or **(b)** down-shift for improved NBTI.

Fig. 11: Comphy simulations of PBTI ΔV_{th} (10Y,125°C) for increasing V_{ov} (defect model for low thermal budget, cf. Table I). The impact of a dipole at the SiO₂/HfO₂ interface is emulated by shifting up the energy level of the high-k traps of 0.2 and 0.4eV. The dashed and dotted lines show the respective contributions of HfO₂ and SiO₂ traps. Two scenarios are considered **(a)** 0.6nm thin SiO₂ IL, **(b)** conventional 1nm SiO₂ IL. In both cases, a 0.4eV dipole is predicted to maintain PBTI ΔV_{th}(10Y,125°C) within a 50mV target at V_{ov}=0.7V.

Fig. 12: Same as Fig. 11, for NBTI. The dipole impact is emulated by shifting the energy level of the HfO₂ deep traps down of 0.1, 0.2 and 0.4eV, for **(a)** a 0.6nm SiO₂ IL, and **(b)** a 1nm SiO₂ IL. A dipole shift is effective in improving NBTI only in the former case, while in the latter the reliability is limited by hole traps in the 1nm SiO₂ IL. (Note: the kink in the contribution of SiO₂ traps is due to gate interaction on trap occupancy at high V_G).

Fig. 13: Comparison of **(a)** PBTI and **(b)** NBTI ΔV_{th} (10Y, 125°C) simulated for the low thermal budget ref. and the dipole-engineered stacks, compared to the projection for a commercial 28 nm technology.

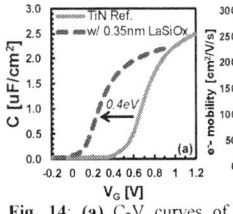

Fig. 14: **(a)** C-V curves of *n*-substrate MOS capacitors w/o and w/ LaSiO$_x$ interlayer. The insertion of LaSiO$_x$ between SiO₂ and HfO₂ improves the interface quality, resulting also in **(b)** enhanced electron mobility (both w/o and w/ 525°C-2h PMA).

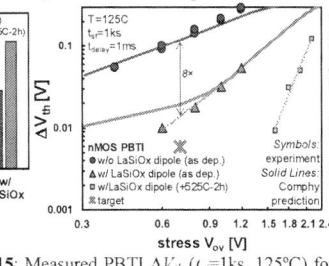

Fig. 15: Measured PBTI ΔV_{th} (t_{st}=1ks, 125°C) for increasing V_{ov}, on the 'as-dep.' low thermal budget gate stack, w/o and w/ the LaSiO$_x$ layer. A ~8× ΔV_{th} reduction is observed, quantitatively in line with the prediction based on dipole shift (solid lines). Further improvement is observed with a PMA.

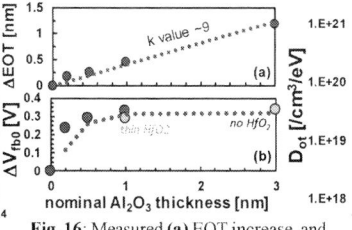

Fig. 16: Measured **(a)** EOT increase, and **(b)** V_{fb} tuning for increasing thicknesses of Al₂O₃ between SiO₂ and HfO₂. A ~0.24eV dipole shift is obtained with a minimum Al₂O₃ thickness with an EOT penalty of only ~1.8Å, while Al₂O₃ ≥1nm yields a max. shift of ~0.36eV.

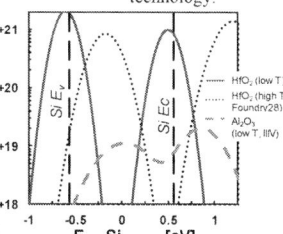

Fig. 17: Defect bands calibrated on HfO₂ (low and high thermal budget) compared to Al₂O₃ defect bands estimated in [10]; the latter shows negligible defect density around the Si valence band.

Fig. 20 (left): Estimated pMOS max V_{ov} for 10Y operation vs. the CET (~EOT+0.4Å) of each stack: the insertion of Al₂O₃ improves the reliability considerably more than expected due to the induced EOT increase, related to the beneficial dipole shift.

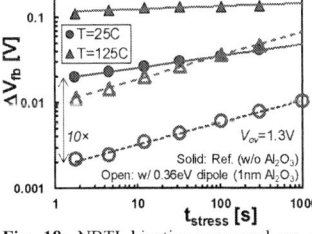

Fig. 18: NBTI kinetics measured on *p*-substrate capacitors at 25°C and 125°C (stress V_{ov}=1.3V), on the low thermal budget gate stack w/o and w/ Al₂O₃ interface dipole. Up to ~10× improvement is observed in the latter case.

Fig. 21 (right): Estimated max V_{ov} vs. eWF. The insertion of Al₂O₃ increases eWF due to the dipole shift: the eWF can be tuned back by replacing TiN with a lower work function metal (e.g. TaN), maintaining the reliability improvement related to the high-k defect energy shift.

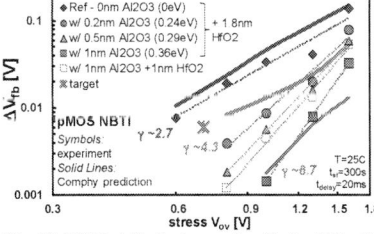

Fig. 19: NBTI shifts for increasing V_{ov} (t_{st}=300s, T=25°C) measured on the low thermal budget gate stack w/o and w/ Al₂O₃ interlayers of increasing thicknesses. The insertion of a ~0.2nm Al₂O₃ layer brings the reliability within specs. Further improvement is obtained for larger dipole shifts, quantitatively in line with the Comphy prediction (solid lines). Note the extremely favorable voltage acceleration factor γ~6.7 obtained with a ~0.36eV dipole shift, in line with best-in-class SiGe pMOS reliability [17].

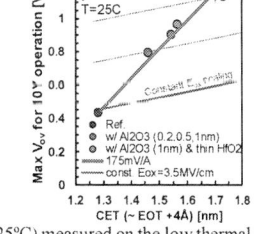

Fig. 22 (right): Max operating V_{ov} for the ref. gate stack and the various Al₂O₃-inserted stacks, both as-dep. ("top") or w/ 525°C-2h PMA ("bottom"), plotted vs. **(a)** the eWF and **(b)** the gate stack CET (note: the PMA induces a ~0.4nm SiO₂ regrowth). At low thermal budget, the insertion of Al₂O₃ increases eWF, CET and max V_{ov}; on the "bottom" gate stacks instead, Al₂O₃ does not yield additional reliability improvement despite similar eWF and CET modulations: this is related to i) the thicker SiO₂ IL reducing the impact of the dipole shift on NBTI (cf. Fig. 12b), and to ii) the inherently improved reliability at high thermal budget (cf. Fig. 3d).

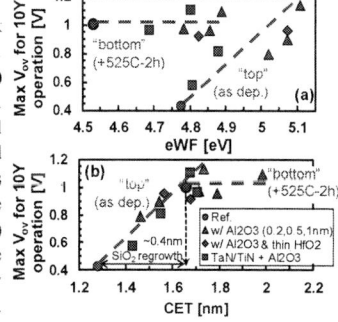

978-1-7281-1988-5/18 $31.00 © 2018 IEEE   790

Characterization and understanding of slow traps in GeO$_x$-based n-Ge MOS interfaces

M. Ke, P. Cheng, K. Kato, M. Takenaka, and S. Takagi

Department of Electrical Engineering and Information Systems, the University of Tokyo, Tokyo, Japan

Email: kiramn@mosfet.t.u-tokyo.ac.jp

Abstract— The properties of slow electron traps in n-Ge MOS interfaces over a wide range of electrical field across gate oxides (E_{ox}) are systematically investigated. It is found through careful examination of the C-V hysteresis that slow trapping under low E_{ox} conditions is attributed only to electron trapping into existing slow traps. Under large E_{ox} conditions, on the other hand, generation of slow electron traps and hole trapping are found to additionally affect the slow trapping characteristics. We propose a new measurement scheme to discriminate existing and generated slow electron traps and apply this method to the three different GeO$_x$-based MOS interfaces in order to clarify the nature of slow traps. It is revealed from this analysis that a pre-plasma oxidation process reduces existing slow electron traps and improves slow trapping in low E_{ox}. On the other hand, ultrathin Y$_2$O$_3$ insertion reduces generation of slow electron traps and improves slow trapping in high E_{ox}.

I. Introduction

One of the key technologies for realizing Ge CMOS is the formation of gate stacks with low defect densities. In order to reduce fast interface states, (HfO$_2$)/Al$_2$O$_3$/GeO$_x$/Ge interfaces realized by post plasma oxidation (Post-PO) are promising [1, 2]. However, a remaining critical issue is the existence of a large amount of slow traps [3-5], which can be an inherent problem for Ge gate stacks. It has been reported that Y-doped GeO$_x$ interfaces and Al$_2$O$_3$/GeO$_x$/Ge formed by pre plasma oxidation (pre-PO) can reduce slow trap density (N_{st}) [6-8]. However, reduction in N_{st} is not sufficient yet, particularly for electrons. Thus, understanding of physical origins of the slow electron traps and the carrier trapping properties is strongly required to establish a guideline for further reduction in N_{st} and a method of the oxide reliability prediction for Ge MOS interfaces. Here, an evaluation method of slow traps is an important issue for the proper understanding. One of the simple ways of evaluating N_{st} is to use hysteresis of C-V curves. Thus, full utilization of this evaluation method is effective in obtaining information of carrier trapping behaviors. However, such an investigation has not been performed yet.

In this study, we examine the physical meaning of the present hysteresis measurement for slow electron traps in n-Ge MOS interfaces. Then, it is found that the applied electric field during this measurement strongly affects the electron trapping properties and that hole trapping also occurs in the electric field higher than a critical one. We propose a new method to discriminate existing and generated electron traps and hole traps. By utilizing this technique, we examine the difference in the slow electron trap properties among three promising GeO$_x$-based MOS interfaces Finally, we also touch on the slow carrier trapping behaviors by using Ge n-MOSFETs.

II. Device Fabrication

Fig. 1 shows the process flows of fabricated MOS structures The first and third structures have 1.5-nm-thick Al$_2$O$_3$ only and 0.7-nm-thick Y$_2$O$_3$, 1.5-nm-thick Al$_2$O$_3$ by ALD at 300°C, respectively, followed by ECR post-PO. The second sample has pre-PO GeO$_x$, followed by 1.5-nm-thick Al$_2$O$_3$ ALD at 300°C. PDA was performed for 30 min at 400 °C in N$_2$ ambient, followed by 100-nm-thick Au gate electrodes [7-8]. Fig. 2 shows the C-V curves of these capacitors. We also fabricated Ge n-MOSFETs with Al$_2$O$_3$/GeO$_x$/Ge gate stacks by using the fabrication process described in ref. [9].

III. Slow trap properties under low V_g (E_{ox}) conditions

Recently, a simple and effective method to estimate N_{st} responsible for BTI reliability from MOS capacitors has been proposed [4, 5]. Fig. 3 schematically shows the procedure. Here, V_g is repeatedly scanned between the minimum voltage (V_{start}) and the maximum voltage (V_{stop}), with increasing V_{stop} (sequence I), as shown in Fig. 4. The amount of slow trap density responding to this scan (ΔN_{st}) can be estimated from

the amount of hysteresis (ΔV_{hys}) in the C-V measurement with forward and backward scans as a function of the maximum effective electric field across gate insulators (E_{ox}). Here, E_{ox} and ΔN_{st} can be given by $E_{ox} = |V_{stop} - V_{FB}|/CET$ and $q \Delta N_{st} = C_{ox} \Delta V_{hys}$ on the assumption that all traps locate very close to the MOS interfaces. Fig. 5 shows the experimental results. Here, values of V_{FB} in the forward and back scan, extracted from Fig. 5, are plotted as a function of E_{ox} in Fig. 6. It is confirmed that V_{FB} in the backward scan keeps increasing, while V_{FB} in the forward scan almost no change under the present low E_{ox}, meaning that only electron trapping and no hole trapping occur. It is verified, as a result, that the amount of electrons trapped in slow states increases with an increase in E_{ox}, because the difference of V_{FB} in the forward and backward scan corresponds to ΔV_{hys} and resulting ΔN_{st}. Next, the repeated scan with same V_{start} and V_{stop} (sequence II), shown in Fig. 7, is performed to confirm the quantitativeness of the measured ΔN_{st}. Fig. 8 and 9 show the C-V curves under this sequence II and the extracted V_{FB} values as a function of scan cycle number. No change in the C-V and V_{FB} means that ΔN_{st} are stable under a given condition and no generation of slow traps occurs during the present V_g scan with low E_{ox}.

Next, the influence of the C-V scan time is examined. Here, the hold time at the V_{min} and V_{max} points during C-V measurements is varied under constant V_{min} and V_{max} values. Fig. 10 show the C-V curves with changing the V_{start} and V_{stop} hold times. The C-V curves have no change with changing the V_{start} hold time, meaning that the occupancy of slow traps at V_{FB} in the forward scan is under the equilibrium condition. On the contrary, when the V_{stop} hold time increases, ΔV_{hys} becomes larger. Fig. 11 and 12 show the hold time dependence of V_{FB} in the forward and backward scan, and the estimated ΔN_{st} as a function of the V_{stop} hold time, respectively. Since V_{FB} only in the backward scan increases, the total amount of trapped electrons increases with the time. There is no saturation in ΔN_{st}, which can be represented by $\Delta N_{st} \propto t^{0.21}$. These results indicate that the time constant of electron trapping into slow traps is widely distributed and that traps with very long time constants exist. These characteristics of electron trapping can be qualitatively understood by trap distributions spread widely along both the energy and the depth directions, as shown in Fig.13. Here, only slow traps with the energy levels below E_F at V_{stop} and with the position having the time constant shorter than the C-V scan time can be filled with electrons during the forward and backward scan. Also, no saturation in ΔN_{st} means that we cannot detect total amounts of slow traps and that only traps locating in an energy range and a depth range contribute to ΔV_{hys}. As a result, the measured ΔN_{st} amounts to a part of total N_{st} as the effective one. On the other hand, as far as the measurement condition is fixed, the relative comparison in ΔN_{st} can still be meaningful.

Under the understanding of the physical meaning of the present evaluation method, the slow trap properties in the three types of the fabricated Ge MOS capacitors are compared. Fig. 14(a) shows the measured ΔN_{st}-E_{ox} relationship in the three types of the n-Ge MOS capacitors. It is confirmed for n-Ge MOS interfaces that insertion of Y$_2$O$_3$ can slightly decrease ΔN_{st} and that the pre-PO process leads to much lower ΔN_{st} than the post-PO process. Also, the comparison between p-Ge and n-Ge is shown in Fig. 14 (b). ΔN_{st} in p-Ge with both post-PO and pre-PO is much lower than that in n-Ge, meaning that the gate stack instability due to slow trap is much more serious in n-Ge. This is the reason why we focus on the slow traps in n-Ge MOS capacitors. In order to carefully examine the difference in ΔN_{st} of n-Ge between the pre- and post-PO process, ΔN_{st} is evaluated by C-V curves with changing the step time at each V_g step, shown in Fig. 15. It is confirmed that ΔN_{st} is higher in the post-PO capacitors, irrespective of the step time, indicating that the difference in ΔN_{st} between the pre- and post-PO process is attributed to the difference in the total slow trap density, not to the modulation of the time constant. Fig. 16 shows a schematic model to explain the increase in slow traps by the post-PO process. We can interpret that additional slow traps can be generated by the post-PO process, independent of the defects inherent to GeO_x, probably through any reaction and/or inter-diffusion between Al_2O_3 and GeO_x.

IV. Slow trap properties under high V_g (E_{ox}) conditions

As described, the hysteresis observed in the C-V scan with low V_{stop} (E_{ox}) is attributed to electron trapping into existing slow trap sites, which is the common interpretation of the hysteresis in Ge MOS interfaces. However, we have found that, when higher V_{stop} (E_{ox}) is applied, both hole trapping and generation of new electron slow traps happen. It should be noted here that the discrimination of existing and generated traps is important for identifying the physical origin of the slow traps. Fig. 17 shows the C-V curves under repeated scan (sequence I), where E_{ox} is increased up to values higher than a critical one ($E_{critical}$). It is observed that V_{FB} starts to shift toward negative V_g, which is totally different from the C-V curves under small E_{ox}. As seen in Fig. 18, when E_{ox} becomes sufficiently large (typically larger than 10 MV/cm), V_{FB} in the forward scan starts to decrease rapidly, while V_{FB} in the back scan increases gradually. Next, the repeated scan with constant V_{stop} and V_{start} (sequence II) is applied to capacitors under E_{ox} higher than $E_{critical}$ in Fig. 19. Fig. 20 summarizes the voltage shift of the C-V scan under low and high E_{ox} as a function of the scan cycle number. The negative shift of the forward scan increases with an increase in the cycle number, suggesting the increase in the amount of trapped holes, which has been reported in Ge p-MOSFETs after NBTI stress [10, 11].

The gate current (J_g) before and after the J_g measurement is shown in Fig. 21. J_g for the post-PO capacitors is higher than the pre-PO ones, suggesting the higher density of defects in the post-PO capacitors. Fig.22 shows J_g before and after the C-V scan. Although a small amount of stress-induced leakage current is observed, the significant degradation is not observed after applying E_{ox} higher than $E_{critical}$. In order to examine the de-trapping of holes, C-V curves after the C-V scan up to high E_{ox} are measured with changing the negative V_{start} values (Fig. 23). No change in the C-V curves means that trapped holes works as fixed charges and do not contribute to ΔV_{hys}. The present results can be explained by a model of Fig. 24. Under low E_{ox}, only electrons in Ge are trapped into slow traps during C-V measurements. Under high E_{ox}, on the other hand, hot holes created probably in the gate metal are injected into dielectrics and trapped into hole traps. These trapped holes do not come out and cause the negative V_{FB} shift as fixed positive charges. In order to determine $E_{critical}$ for hole trapping, the C-V scan using the sequence II with high E_{ox} is applied to the three types of capacitors. Fig. 25 shows the hole trap density after 1 and 2 cycles as a parameter of E_{ox}. The linear relationship between the hole trap density and the cycle number is clearly observed. Fig. 26 shows this slope as a function of E_{ox}. As a result, $E_{critical}$ is estimated to be 11.5 and 14.8 MV/cm for w/ and w/o Y_2O_3 n-Ge MOS interfaces, respectively. It is found that inserting the Y_2O_3 interfacial layer makes the interface more robust against hole trapping.

In addition to hole trapping, we have found that generation of slow electron traps occurs at E_{ox} higher than $E_{critical}$. In order to discriminate generated and existing electron slow traps, we propose a new measurement using sequence III, shown in Fig. 27. Here, after applying high V_{stop} once, ΔV_{hys} under the initial C-V scan condition with sufficiently low V_{stop} is re-measured. Since ΔV_{hys} does not change for the repeated measurements with low V_{stop} (Fig. 8), the increment ΔV_{hys} corresponds to ΔN_{st} for generated slow traps. Fig. 28 shows the change in V_{FB} by using sequence III. We can discriminate and determine $\Delta N_{st-total}$, $\Delta N_{st-generated}$, $\Delta N_{st-existing}$ (total, generated, existing electron slow traps density), and ΔN_h (hole traps density) under the proposed measurement as follows;

$$\Delta N_{st-total} = \Delta V_{hys2} \times C_{ox}/q, \quad \Delta N_{st-generated} = (\Delta V_{hys1} - \Delta V_{hys0}) \times C_{ox}/q,$$
$$\Delta N_h = \Delta V_{forward} \times C_{ox}/q, \quad \Delta N_{st-existing} = \Delta N_{st-total} - \Delta N_{st-generated}$$

Fig. 29 shows the V_{FB} values measured by sequence III. After the V_g scan up to high V_{stop}, V_{FB} in the forward scan decreases and V_{FB} in the backward scan slightly increases, resulting in the increase of ΔV_{hys} in comparison with the initial scan. This result means that trapped holes appear and electron slow traps are generated by applying higher V_{stop} (E_{ox}). By using this new measurement method, we compare the contributions of generated and existing electron slow traps and hole traps to total ΔN_{st} among the three types of n-Ge MOS capacitors. Fig. 30 shows $\Delta N_{st-total}$, $\Delta N_{st-existing}$, $\Delta N_{st-generated}$ and ΔN_h in the $Al_2O_3/GeO_x/Ge$ MOS interfaces with post-PO. It is found that $\Delta N_{st-existing}$ dominates $\Delta N_{st-total}$ in lower E_{ox}, while $\Delta N_{st-generated}$ is comparable to or higher than $\Delta N_{st-existing}$ in higher E_{ox}. Also, ΔN_h is comparable to $\Delta N_{st-existing}$ and $\Delta N_{st-generated}$ in high E_{ox}. These results indicate that the influence of slow trap generation and hole trapping must be taken into account in high E_{ox} as the physical origin of slow electron trapping in Ge MOS reliability. Fig. 31 shows the comparison of $\Delta N_{st-total}$, $\Delta N_{st-existing}$, $\Delta N_{st-generated}$ and ΔN_h among the post- and pre-PO $Al_2O_3/GeO_x/n$-Ge, and post-PO $Al_2O_3/Y_2O_3/GeO_x/n$-Ge gate stacks. It is found that $\Delta N_{st-existing}$ in high E_{ox} is almost the same, while pre-PO exhibits the lowest value in low E_{ox}, as shown in Fig. 4(a). Also, $\Delta N_{st-generated}$ is almost the same between pre- and post-PO $Al_2O_3/GeO_x/Ge$. These facts suggest that the slow electron trapping and generation in high E_{ox} are determined by a common nature of pure GeO_x. It is found, on the other hand, that $Al_2O_3/Y_2O_3/GeO_x/Ge$ shows much lower ΔN_h and $\Delta N_{st-generated}$, meaning that insertion of Y_2O_3 and doping of Y into GeO_x can be an effective way to improve the gate stack reliability in high E_{ox}. This result is consistent with the report on stabilization of GeO_2 network by Y incorporation [12]. As a result, we can interpret that lower $\Delta N_{st-total}$ of pre-PO in low E_{ox} and $Al_2O_3/Y_2O_3/GeO_x/Ge$ in high E_{ox} are attributed to the suppression of existing trap generation due to pre-PO and the increased robustness against electron slow trap generation and the hole traps, respectively. The present analyses also suggest that existing and generated electron slow trap and hole traps have the different physical origins.

V. Characteristics of slow traps in Ge n-MOSFETs

As described above, the slow traps measured by C-V hysteresis can be a part of existing slow traps. In order to estimate all of slow traps affecting device performance, the evaluation of time dependence of MOSFET currents is effective. Thus, the characteristics of channel currents of Ge n-MOSFETs with $Al_2O_3/GeO_x/Ge$ gate stacks are studied. Fig. 32 and 33 show I_s as a function of V_g and the time dependence, respectively. The decrease in I_s corresponds to the trapped electrons. The long-term I_s-t characteristics in Fig. 34 indicate that slow trapping has no saturation and is described by the $\log t$ dependence. We have also found that the trapping characteristics have the temperature dependence, as seen in Fig. 35. The activation energy of the slope in $\log t$ plot is estimated to be ~ 20 meV (Fig. 36). Thus, any temperature-dependent process could be incorporated into the slow trapping mechanism of channel electrons, as Fig. 37.

VI. Conclusion

The properties of slow traps in Ge MOS interfaces in small and large E_{ox} were systematically examined. It was found that only existing slow traps are responsible in low E_{ox}, while generation of slow traps and hole trapping additionally occur in high E_{ox}. Also, we proposed a new measurement scheme to discriminate existing and generated electron slow traps. The pre-PO and Y_2O_3 insertion have been found to reduce existing and generated slow electron traps, respectively, contributing to the reduction in total slow trap density.

ACKNOWLEDGEMENTS This work was partly supported by a Grant-in-Aid for Scientific Research (17H06148) from the Ministry of Education, Culture, Sports, Science, and Technology in Japan and JST-CREST, Grant Number JPMJCR1332, Japan

REFERENCES [1] R. Zhang et al., APL 98 (2011) 112902 [2] R. Zhang et al., TED 60 (2013) 927 [3] R. Zhang et al., IEDM (2011), 642 [4] J. Franco et al., IEDM (2013) 397 [5] G. Groesenken et al., IEDM (2014) 828 [6] C. Lu et al., JAP 116 (2014) 174103 [7] M. Ke et al., MEE 178 (2017), 132 [8] M. Ke et al., ESSDERC (2017) 296 [9] M. Ke et al., APL 109 (2016) 032101 [10] J. Ma et al., MEE 109 (2013) 43 [11] J. Ma et al., TED 61 (2014) 1307 [12] C. Lu et al., IEDM (2015) 370

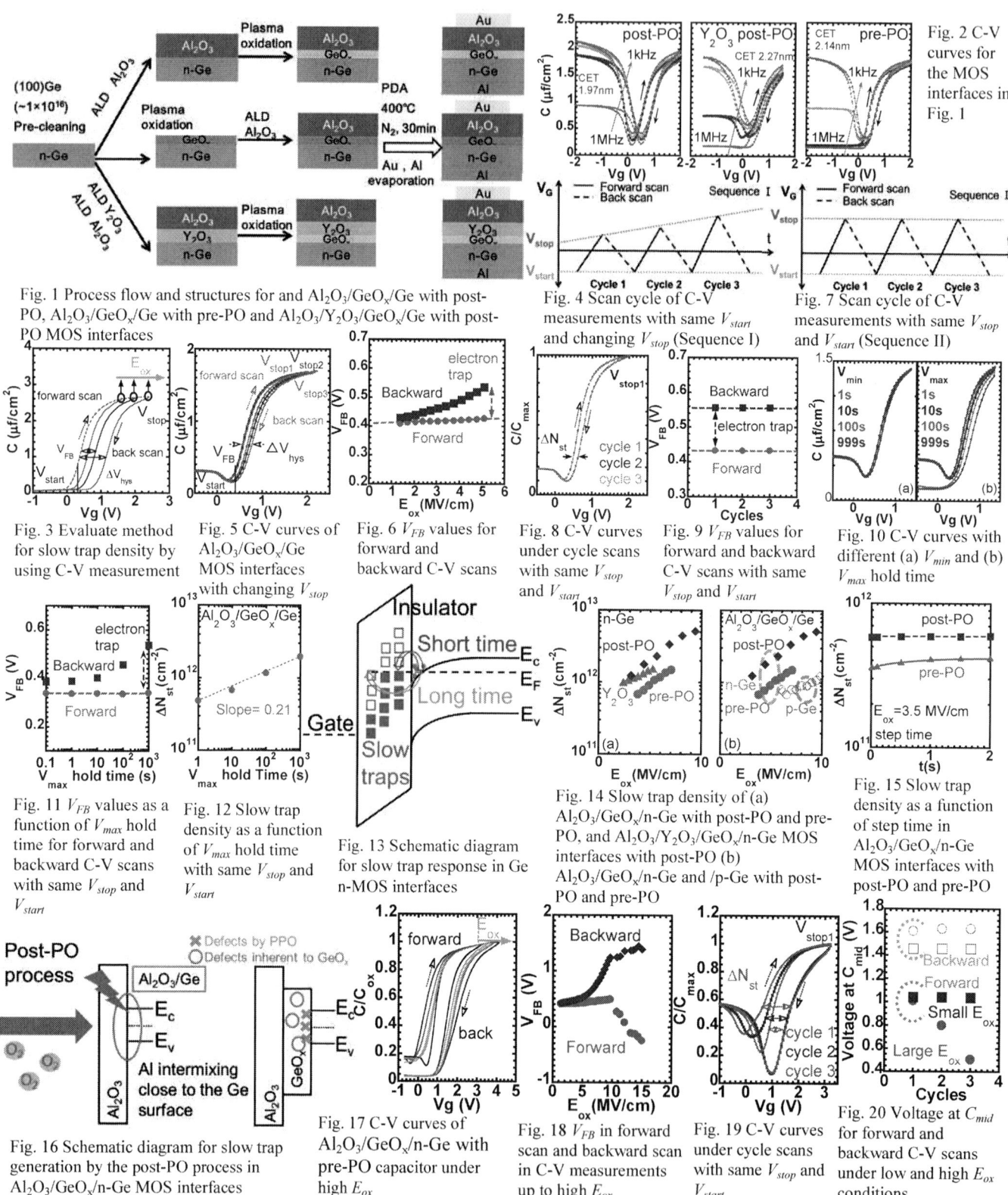

Fig. 1 Process flow and structures for and $Al_2O_3/GeO_x/Ge$ with post-PO, $Al_2O_3/GeO_x/Ge$ with pre-PO and $Al_2O_3/Y_2O_3/GeO_x/Ge$ with post-PO MOS interfaces

Fig. 2 C-V curves for the MOS interfaces in Fig. 1

Fig. 3 Evaluate method for slow trap density by using C-V measurement

Fig. 4 Scan cycle of C-V measurements with same V_{start} and changing V_{stop} (Sequence I)

Fig. 5 C-V curves of $Al_2O_3/GeO_x/Ge$ MOS interfaces with changing V_{stop}

Fig. 6 V_{FB} values for forward and backward C-V scans

Fig. 7 Scan cycle of C-V measurements with same V_{stop} and V_{start} (Sequence II)

Fig. 8 C-V curves under cycle scans with same V_{stop} and V_{start}

Fig. 9 V_{FB} values for forward and backward C-V scans with same V_{stop} and V_{start}

Fig. 10 C-V curves with different (a) V_{min} and (b) V_{max} hold time

Fig. 11 V_{FB} values as a function of V_{max} hold time for forward and backward C-V scans with same V_{stop} and V_{start}

Fig. 12 Slow trap density as a function of V_{max} hold time with same V_{stop} and V_{start}

Fig. 13 Schematic diagram for slow trap response in Ge n-MOS interfaces

Fig. 14 Slow trap density of (a) $Al_2O_3/GeO_x/n$-Ge with post-PO and pre-PO, and $Al_2O_3/Y_2O_3/GeO_x/n$-Ge MOS interfaces with post-PO (b) $Al_2O_3/GeO_x/n$-Ge and /p-Ge with post-PO and pre-PO

Fig. 15 Slow trap density as a function of step time in $Al_2O_3/GeO_x/n$-Ge MOS interfaces with post-PO and pre-PO

Fig. 16 Schematic diagram for slow trap generation by the post-PO process in $Al_2O_3/GeO_x/n$-Ge MOS interfaces

Fig. 17 C-V curves of $Al_2O_3/GeO_x/n$-Ge with pre-PO capacitor under high E_{ox}

Fig. 18 V_{FB} in forward scan and backward scan in C-V measurements up to high E_{ox}

Fig. 19 C-V curves under cycle scans with same V_{stop} and V_{start}

Fig. 20 Voltage at C_{mid} for forward and backward C-V scans under low and high E_{ox} conditions

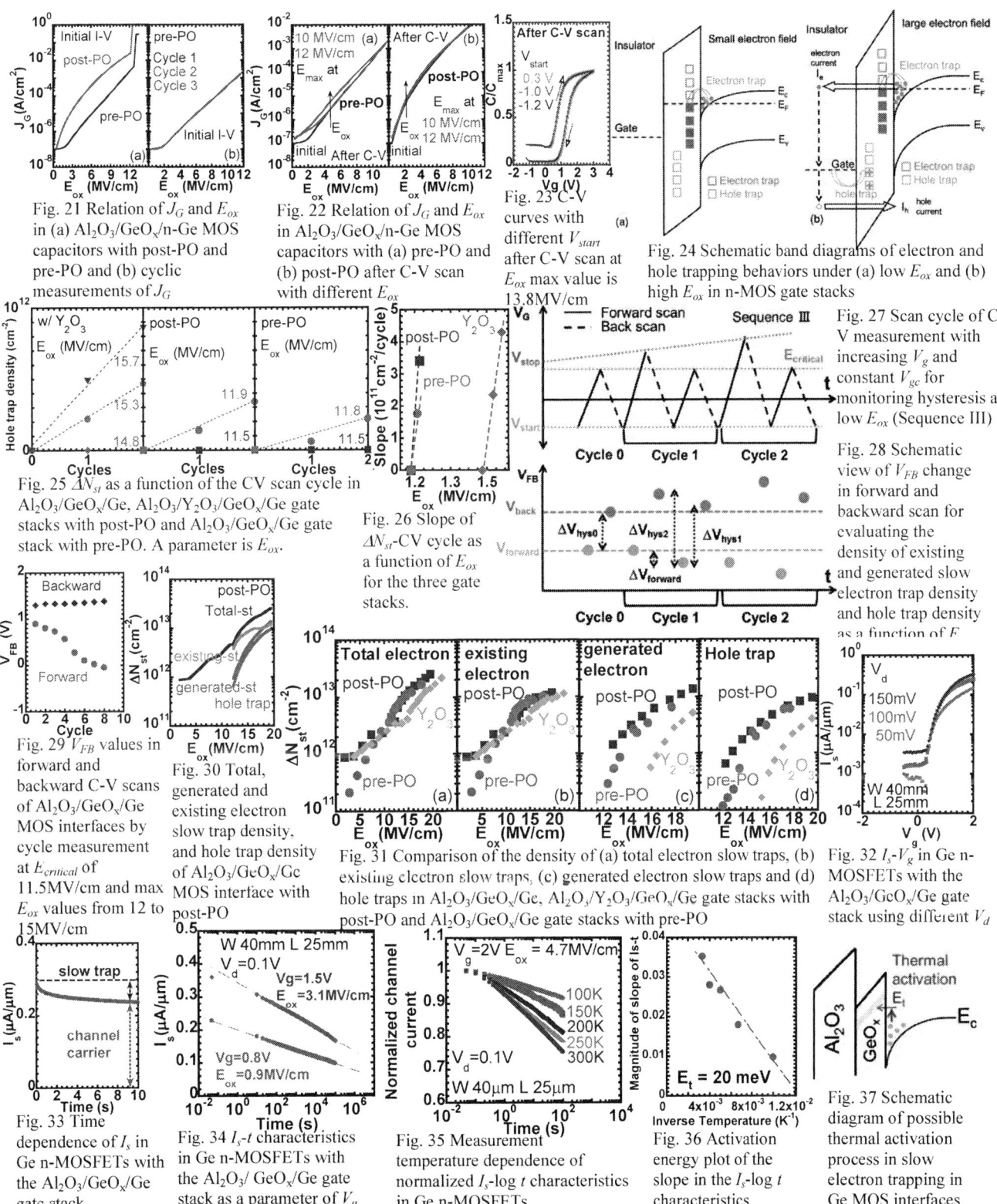

Fig. 21 Relation of J_G and E_{ox} in (a) Al_2O_3/GeO_x/n-Ge MOS capacitors with post-PO and pre-PO and (b) cyclic measurements of J_G

Fig. 22 Relation of J_G and E_{ox} in Al_2O_3/GeO_x/n-Ge MOS capacitors with (a) pre-PO and (b) post-PO after C-V scan with different E_{ox}

Fig. 23 C-V curves with different V_{start} after C-V scan at E_{ox} max value is 13.8MV/cm

Fig. 24 Schematic band diagrams of electron and hole trapping behaviors under (a) low E_{ox} and (b) high E_{ox} in n-MOS gate stacks

Fig. 25 ΔN_{st} as a function of the CV scan cycle in Al_2O_3/GeO_x/Ge, $Al_2O_3/Y_2O_3/GeO_x$/Ge gate stacks with post-PO and Al_2O_3/GeO_x/Ge gate stack with pre-PO. A parameter is E_{ox}.

Fig. 26 Slope of ΔN_{st}-CV cycle as a function of E_{ox} for the three gate stacks.

Fig. 27 Scan cycle of C-V measurement with increasing V_g and constant V_{gc} for monitoring hysteresis at low E_{ox} (Sequence III)

Fig. 28 Schematic view of V_{FB} change in forward and backward scan for evaluating the density of existing and generated slow electron trap density and hole trap density as a function of E

Fig. 29 V_{FB} values in forward and backward C-V scans of Al_2O_3/GeO_x/Ge MOS interfaces by cycle measurement at $E_{critical}$ of 11.5MV/cm and max E_{ox} values from 12 to 15MV/cm

Fig. 30 Total, generated and existing electron slow trap density, and hole trap density of Al_2O_3/GeO_x/Ge MOS interface with post-PO

Fig. 31 Comparison of the density of (a) total electron slow traps, (b) existing electron slow traps, (c) generated electron slow traps and (d) hole traps in Al_2O_3/GeO_x/Ge, $Al_2O_3/Y_2O_3/GeO_x$/Ge gate stacks with post-PO and Al_2O_3/GeO_x/Ge gate stacks with pre-PO

Fig. 32 I_s-V_g in Ge n-MOSFETs with the Al_2O_3/GeO_x/Ge gate stack using different V_d

Fig. 33 Time dependence of I_s in Ge n-MOSFETs with the Al_2O_3/GeO_x/Ge gate stack

Fig. 34 I_s-t characteristics in Ge n-MOSFETs with the Al_2O_3/GeO_x/Ge gate stack as a parameter of V_g

Fig. 35 Measurement temperature dependence of normalized I_s-log t characteristics in Ge n-MOSFETs

Fig. 36 Activation energy plot of the slope in the I_s-log t characteristics

Fig. 37 Schematic diagram of possible thermal activation process in slow electron trapping in Ge MOS interfaces

Soft Error Trends in Advanced Silicon Technology Nodes

B. Bhuva

Vanderbilt University

bharat.bhuva@vanderbilt.edu

Abstract—Soft errors for planar and FinFET nodes have shown different trends for various designer-controlled parameters. This paper examines effects of some of these parameters for the 20-nm Planar and the 16-nm FinFET technology nodes. Latchup vulnerability of FinFET node is also investigated through simulations.

Keywords- Terrestrial soft errors, neutrons, alpha particles, logic flip-flops, combinational logic, CMOS technology, tecnology scaling, soft errors.

I. INTRODUCTION

Soft errors (SE) have emerged as a major reliability threat for advanced technology nodes. In the case of older technologies, mitigation of SRAM soft errors was sufficient to bring overall soft error failure-in-time (FIT) rates in terrestrial environments to reasonable levels. However, for advanced technologies, the reduction of supply voltages and nodal capacitances leads to reduced critical charges (Q_{crit}) needed to corrupt the information stored at circuit nodes. On the other hand, with decreasing transistor sizes and increased doping densities, charge collection areas (usually referred to as sensitive area) per transistor and charge-collection efficiencies decrease. All these factors also depend on the physical structure of a transistor. With the transition from planar to FinFET structures at recent technology nodes, the focus has shifted to SE effects in FinFET technologies and how these effects differ from those at planar technologies. For planar technologies, SE related failures were expected to have the highest FIT rates compared to any other type of failures [1]. The 3-D structure of FinFET, as opposed to 2-D structure of planar transistors, significantly alters the charge generation and collection processes associated with an ion strike [2]. These effects have resulted in reduced SE vulnerabilities for memories and sequential logic circuits for FinFET nodes [3, 4].

Sequential circuits are usually binned into two categories; storage cells (such as latches, flip-flops and SRAM cells) and logic circuits. For the past several technology nodes, most mitigation efforts have focused on protecting storage cells (memories and latches) because of their much larger contribution to overall SE error rates compared to that from logic circuits. This was mainly because of the relatively low operating frequencies. Recent results at advanced technology nodes have shown that at the GHz range, logic SER may exceed memory soft-error rate (SER) [5-8]. Thus, for advanced technologies, the SER for storage cells (such as SRAM cells or flip-flops) and combinational logic must be addressed to meet the system-level reliability specifications. This paper presents effects of different parameters on SER for planar as well as FinFET technologies.

II. FACTORS AFFECTING SER

For any given technology, SER is largely determined by three primary factors. These are logic-gate delay, Single-Event Transient (SET) pulse width, and charge-sharing after a single event. Individual logic-gate delays are a function of nodal capacitances and transistor drive currents. For a minimum size Inverter, nodal capacitances are expected to reduce with scaling while transistor currents hover in the 100 μA range. Since most storage cells are made up of minimum size logic gates, these delays essentially dictate write times (or feedback loop delay) of each storage cell. SET pulses are the transient pulses that are generated when an ion deposits energy within a semiconductor region, creating electron-hole pairs. Due to the existing electric fields in the region, these carriers are separated and collected at a circuit node, resulting in a change in the voltage given by, to a first degree, $\Delta V = \Delta Q / C$, where ΔV is the change in voltage, ΔQ is the collected charge, and C is the nodal capacitance. Other transistors associated with the circuit node restore the nodal voltage back to its original value. The time taken to restore the original voltage is given by, to a first degree, $\Delta t = C \, \Delta V / I_{sat}$, where Δt is the SET pulse width, and I_{sat} is the saturation current of the restoring transistor. In general, for a storage cell, if the SET pulse width exceeds the feedback loop delay, a soft error will result. Thus, SER is a function of transistor currents, nodal capacitances, and collected charge. As nodal capacitances are reduced, feedback loop delay decreases along with SET pulse width. These two competing factors (feedback loop delay and SET pulse width) largely determine the overall SER for storage cells (SRAM cells and latches).

The last factor affecting SER is the charge sharing. Whenever an ionizing particles travels through a semiconductor region, the carriers created due to coulombic interactions diffuse throughout the semiconductor region. This means multiple circuit nodes may collect charges, resulting in multiple simultaneous SET pulses in the circuit. These pulses either defeat the mitigation schemes (such as DICE FF) or introduce multiple-cell upsets (MCU) (e.g. in SRAM arrays). Both of these effects are undesirable and increase the overall SER for the circuit. Extent of MCU also determines the effectiveness of Error correcting codes in SRAM ICs.

978-1-7281-1988-5/18 $31.00 © 2018 IEEE

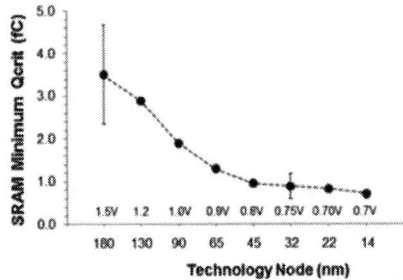

Fig. 1. Critical charge required to cause an upset for an SRAM cell as a function of technology node [4].

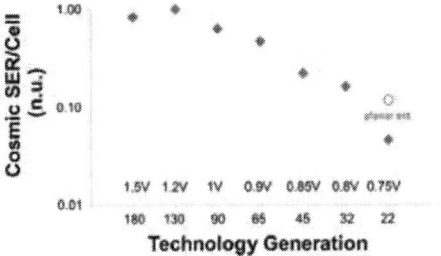

Fig. 2. SRAM SER as a function of technology node [5].

For planar and FinFET technologies, the impact on SER of these three parameters is different. Fig. 1 shows the minimum critical charge values for SRAM cells for recent technology nodes [4]. 3-D structure of FinFET results in almost the same nodal capacitances as the previous generation of planar transistor, resulting in similar critical charge for SRAM cells. Since FinFET transistor has smaller footprint, the sensitive area decreases as technologies switch from planar node to FinFET node, resulting in a decrease in SER, as shown in Fig. 2. Additionally, the planar transistor resides entirely within the substrate as seen in Fig. 3. However, FinFET has a narrow connection to the substrate. For planar transistor, this means, charges in the substrate can easily diffuse towards the drain region and get collected, increasing the total collected charge and an increase in the pulse width of the resulting SET pulse. For FinFET, narrow connection to the substrate means less charge diffuse through the narrow fin region for collection, resulting in lower collected charge and shorter SET pulse widths. Fig. 4 shows these effects though simulated results for charge collection under identical conditions for planar and FinFET nodes. Narrow connection to the substrate also reduces charge-sharing effects for FinFET nodes.

III. EFEFCTS OF SCALING ON SER

For the most part, fabrication process parameters affecting the SER are not user controlled. These factors are doping densities, fin height, fin width, gate insulator characteristics, parasitics, etc. [4]. Since these factors are determined with performance in mind and not SER, it is generally understood that designers will need to address SER through designer-controlled parameters, such as transistor threshold voltage and well selection along with circuit design. Most advanced nodes offer multiple threshold-voltage options for managing speed and power. These nodes also offer dual-well and triple-well

options for managing substrate noise for mixed-signal circuits.

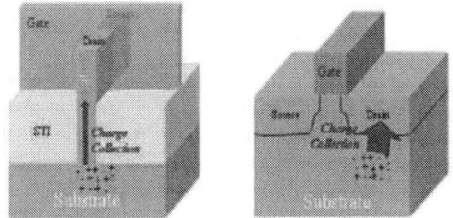

Fig. 3. Differences in charge collection mechanisms for Planar and FinFET nodes [5].

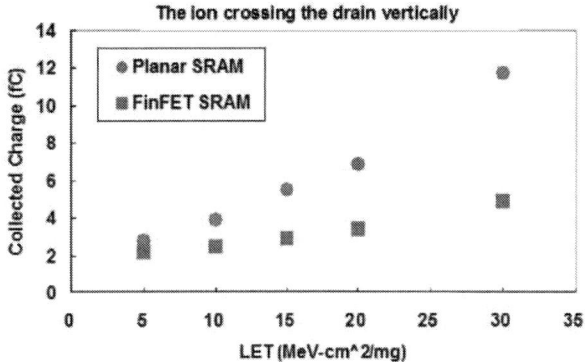

Fig. 4: Simulated collected charge for equivalent FinFET and planar node SRAM cells for normally-incident particles with different LET values [2].

Changes in threshold voltage affects SER through changes in feedback loop delay and SET pulse width. Increased transistor currents (assuming everything else remains the same) reduces feedback-loop delay and reduces SET pulse width. These two competing factors determine the overall trend for SER as a function of threshold voltage. Fig. 5 shows the SER trend for 20-nm planar node. For a D-FF, the highest threshold voltage results in the lowest SER. This is due to the fact that the increasing the V_T increases feedback loop delay much faster than the SET pulse width. On the other hand, for 16-nm FinFET node, exactly opposite trend is seen (as shown in Fig. 6). These trends exacerbate as supply voltage is reduced.

Simulation results were obtained using a commercial 20-nm PDK to show the percentage increase in SET pulse width and feedback-loop delay as V_T increases at nominal supply voltage ($V_{dd} = 850$ mV), as shown in Table I. A low-LET particle (with the LET value close to that of alpha particles, ~1 MeV-cm²/mg) was used to strike inverters with off-state NMOS to generate SET pulses for these simulations. The percentage changes were calculated with respect to the SET pulse width and feedback-loop delay of the ULVT design. For the 20-nm technology node, the increase in feedback-loop delay is higher than that for the SET pulse width as V_T increases from ULVT to HVT, resulting in lower upset probability for HVT and LVT designs. Therefore, for this 20-nm DFF design, increasing V_T can effectively reduce static power consumption without any penalty in SEU vulnerability. It must be kept in mind that different FF designs may yield different trends.

For 16-nm node, ULVT design shows the lowest SEU cross-section and the SVT design shows the highest SEU cross-section, which is completely opposite to what was observed for the 20-nm technology node. Alpha particle exposure results as a function of supply voltage are shown in Fig. 6 clearly support these simulations. As V_T increases from ULVT to SVT, the alpha particle-induced SEU cross-section increases by ~2x. For the 16-nm DFF designs, decrease in transistor current caused by increasing threshold voltage also results in an increase in SET pulse width and feedback-loop delay for the DFFs. A commercial 16-nm PDK was used to simulate the SET pulse width and feedback-loop delay for the DFF design at nominal supply voltage (V_{dd} = 800 mV).

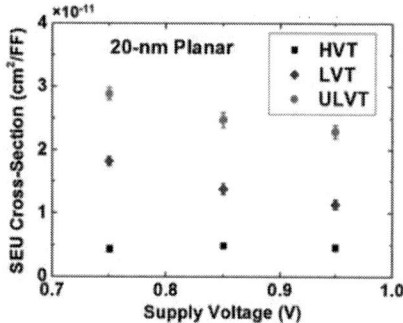

Fig. 5. SER as a function of VDD for different V_T options for 20-nm planar node [9]

TABLE I
PERCENTAGE CHANGE IN SET PULSE WIDTH AND FEEDBACK-LOOP DELAY
(RELATIVE TO ULVT DESIGN)

	HVT	LVT	ULVT
SET PW	39.7%	15.3%	0
FD	51.8%	18.2%	0

Similarly, the percentage changes were calculated with respect to the SET pulse width and feedback-loop delay of the ULVT design. Table II shows simulation results of the percentage change in SET pulse width and feedback-loop delay when threshold voltage increases from ULVT to SVT for the 16-nm bulk FinFET technology. Different SEU cross-

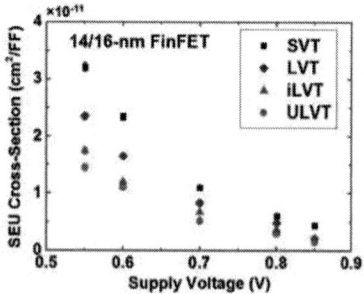

Fig. 6. SER as a function of VDD for different V_T options for 16-nm FinFET node [9]

section trends for 20-nm and the 16-nm D-FF designs are mainly due to the difference in the physical structure, transistor size, and spacing of transistors, which lead to

TABLE II
PERCENTAGE CHANGE IN SET PULSE WIDTH AND FEEDBACK-LOOP DELAY
(RELATIVE TO ULVT DESIGN)

	SVT	LVT	iLVT	ULVT
SET PW	116.5%	61.6%	37.0%	0
FD	78.9%	36.0%	15.6%	0

different rates of change of SET pulse width and feedback-loop delay.

With closely packed cells in an SRAM array, the important factor that determines the SRAM design is the size of Multi-Cell Upsets (MCU). With the underlying Silicon crystal structure remaining the same for all technology nodes, the spread of charge that causes MCU doesn't change significantly with scaling. As a result, the size of MCU cluster should not change with scaling, as seen in Fig. 7 [4]. With cell size decreasing with scaling and the size of MCU cluster remaining constant, the number of MCU in SRAM array increases with scaling. This will have significant impact on interleaving and ECC designs for SRAM ICs.

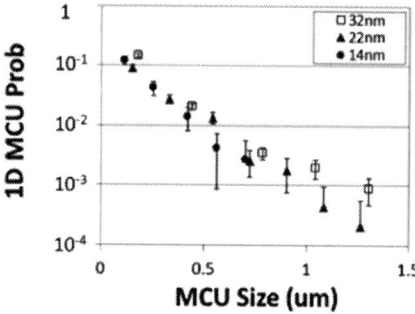

Fig. 7. SRAM MCU rate for different technology nodes [4].

While storage cell upsets are constant over frequency, combinational logic upsets are a linear function of operating frequency. Logic upsets occur when an ion deposits charge within a logic gate. The resultant SET pulse may propagate through the circuit and get latched into a storage cell. For older technologies, the operating frequencies were low

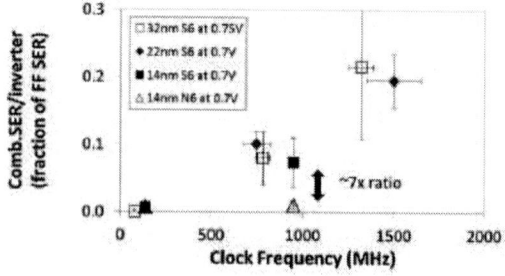

Fig. 8. Logic upsets vs technology node [4].

978-1-7281-1988-5/18 $31.00 © 2018 IEEE

enough that the storage cell upsets dominated overall SER. For ICs operating at GHz range of frequencies, the logic upsets are comparable to FF upsets in ASIC designs. Fig. 8 shows the ratio of logic SER per Inverter to D-FF SER for recent technology nodes [4]. The slope of the curve does not change with scaling significantly indicating that the logic SER remains a threat for all recent and future technology nodes.

IV. SYSTEM-LEVEL SER

Although the errors/bit is monotonically decreasing, soft errors are far from being insignificant when it comes to IC and system level FIT rate. The rate at which particles are incident for a given Si area is independent of underlying technology node. The crystalline Si lattice remains almost the same for all technology nodes. As a result, the probability of interactions between Si lattice and incident particles remains constant for all technology nodes. With scaling, the cell size decreases and critical charge needed to generate a given SET pulse width decreases. TCAD simulations show that the critical charge has decreased from 1.8 fC in 28-nm planar technology node to 0.9 fC in the 16-nm FinFET node for conventional D-FF design. Decreases in individual cell size reduces the probability that a given cell will be hit by an ion. Reduced critical charge increases sensitive region around an ion strike (cells further away from a hit location may still collect enough charge to cause an SET pulse).

On top of these factors, scaling increases the total number of cells per given area. All these factors eventually result in similar number of upsets at the IC level across technology nodes (assuming comparable Si area per IC). With the increased system-level complexities (and the increased number of ICs to achieve these complex functions), the system-level error rates continue to increase with each new generation. Fig. 9 shows the measured FF-, IC-, and system-level SER trend across recent technology nodes.

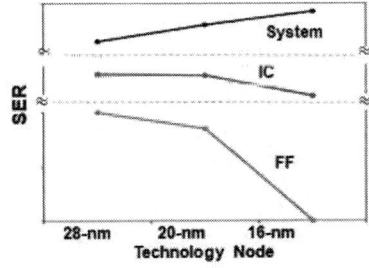

Fig. 9. SER at FF level decrease with technology, but remains constant at IC level and increases at the system level.

V. LATCHUP IN FINFET TECHNOLOGY

Recent work has demonstrated that the STI depth is shallower for FinFET nodes than that for planar nodes. This results in increased gain of parasitic bipolar transistors forming the latchup structure. Fig. 10 shows the current gain for npn and pnp transistors in FinFET and planar nodes [10]. This implies that I/O circuits in FinFET technologies that operate at ~1.8 V (or higher) may be vulnerable to latchup. With ionizing particles capable of depositing enough charge

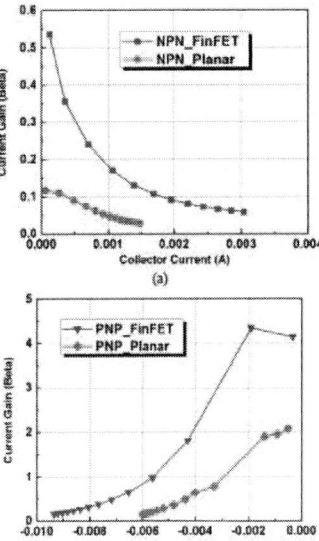

Fig. 10. Npn and pnp current gains for FinFET and planar nodes [10].

to initiate latchup, designers will need to take extra precautions for I/O circuits.

VI. SUMMARY

This paper examined effects of threshold voltage on soft error rate of conventional D-FF designs for FinFET and planar nodes, highlighting the differences in charge collection as well as circuit-level parameters. Logic soft error rates show similar trend across technologies as a function of frequency, making their contribution increase significantly in the GHz range of frequencies. Even though FF-level FIT rates are decreasing with scaling, IC-level FIT rates are not changing significantly with scaling. Simulation results show increasing gain of parasitic bipolar transistors may make FinFET technologies more susceptible to latchup than planar nodes.

REFERENCES

[1] SIA Roadmap 2014,
[2] Y-P. Fang and A. Oates, *IEEE Trans. Dev. and Mat. Reliab.*, Dec. 2011,
[3] P. Nsengiyumva, et al., *IEEE Trans. Nucl. Sci.*, vol. 63, no. 1, pp. 266-272, Feb. 2016.
[4] N. Seifert, et al., *IEEE Trans. Nucl. Sci.*, vol. 62, pp. 2570-2577, Dec. 2015.
[5] N. Seifert, et al., Nuclear Science, IEEE Transactions on, vol.59, no.6, pp.2666,2673, Dec. 2012.
[6] N. N. Mahatme, et al., *Proc. IEEE Int. Reliability Physics Symp.*, pp. 5.F.2.1-5.F.2.6 2014
[7] S. Jagannathan, et al., *IEEE Transactions on Nuclear Science*, Vol 59, pp 2796-2802, Dec 2012.
[8] I. Chatterjee, et al., *IEEE Trans. Nucl. Sci.*, vol. 61, no. 6, pp. 3512-3518, Dec. 2012.
[9] H. Zhang, et al., IEEE Trans on NS, pp. 457-463, Vol 64, January 2017.
[10] C.-T. Dai, et al., Proceedings of 2017 IEEE International Symposium on Reliability Physics, Monterey, CA.

CMOS-Compatible Doped-Multilayer-Graphene Interconnects for Next-Generation VLSI

Junkai Jiang, Jae Hwan Chu, and Kaustav Banerjee*

Department of Electrical and Computer Engineering, University of California, Santa Barbara, CA; *Email: kaustav@ece.ucsb.edu

Abstract—Cu interconnects suffer from steep rise in resistivity and severe reliability degradation for sub-20 nm line widths. Other metals, including Co and Ru, have been demonstrated with higher electromigration (EM) resistance, but exhibit lower electrical conductivity that degrades circuit performance. This work reports multilayer graphene (MLG) directly grown on SiO_2 substrate at 300 °C by a novel pressure-assisted solid-phase diffusion synthesis method, and, for the first time, demonstrates a CMOS-compatible intercalation doped graphene nanoribbon (DGNR) interconnect technology with smaller electrical resistivity than Cu, Co and Ru interconnects. The DGNR interconnect also exhibits < 4% conductivity degradation over 1000 hours at room temperature (RT) without any encapsulation or barrier layer, and negligible EM under 100 MA/cm^2 current stress test at > 100 °C.

I. INTRODUCTION

Cu interconnects suffer from significant size-effect at sub-20 nm width, because of strong carrier scatterings and low-conductive barriers, leading to degraded electrical conductivity/delay, as well as severe self-heating that reduces the EM lifetime and current-carrying capability [1]. Other metals with higher bulk melting points w.r.t Cu (Fig. 1(a)), including Co [2] and Ru [3],[4], have been explored as local interconnects for future technology nodes. While challenges in process integration of theses metals, including scaled liners/barriers and optimized chemical mechanical planarization (CMP), still exist, at 20-nm line width, the metals exhibit >80% higher electrical resistivity w.r.t. Cu [2],[3], which results in >60% circuit delay increase (Fig. 1(c)). Recently, intercalation-doped graphene nanoribbon (DGNR) was reported with comparable electrical conductivity down to 20-nm line width w.r.t Cu [5] and current carrying capacity of >200 MA/cm^2 [6], as well as multi-level capability of multilayer graphene (MLG) has been demonstrated with carbon nanotube (CNT) vias [7]. The high current carrying capability of DGNR allows more aggressive vertical scaling w.r.t Cu, Co, or Ru, leading to lower parasitics and significantly smaller switching energy (Fig. 1(d)). However, the MLG films were grown by chemical vapor deposition (at ~1000 °C) and transferred, which is not compatible with CMOS technology, both in terms of back-end-of-line (BEOL) thermal budget (<500 °C) and cost-effectiveness. In this work, we report CMOS-compatible DGNR interconnects with smaller electrical resistivity w.r.t Cu, Co and Ru interconnects (Fig. 1(b)), with excellent stability and EM resistance, obtained from MLG directly grown on SiO_2 substrate at 300 °C. Instead of CMP that introduces process complexity, potential voids in metal fill and wire thickness variations, subtractive etching process (Fig. 2) [4] is used to fabricate DGNR interconnect down to 20-nm width.

II. CMOS-COMPATIBLE GRAPHENE GROWTH

The reported graphene preparation/growth methods include: (1) mechanical exfoliation [8],[9], (2) chemical vapor deposition (CVD) method, which requires transfer process to desired dielectric substrate [5],[10], and (3) direct synthesis method on dielectric substrate [11]-[13]. The mechanical exfoliation offers best quality graphene at RT, but it is unable to control the graphene layer thickness and uniformity at wafer scale. The CVD method provides controlled thickness and uniformity at wafer scale, but the (wet) transfer method is not preferred for cost-effectiveness. The direct graphene growth on dielectric has been explored by chemical vapor deposition (CVD, @ >800 °C [13]), plasma-enhanced CVD (@ 650 °C [12]) and remote catalyzation (@ >850 °C [11]), but they all require high temperature that violates BEOL process thermal budget. The high-temperature process in traditional graphene growth is required for gas-phase precursor (e.g., CH_4 and C_2H_2) decomposition to supply carbon during growth. In this work, a novel CMOS-compatible pressure-assisted solid-phase diffusion synthesis method (@ 300 °C, Fig. 3(a)) that combines the advantages of a metal catalyst and direct growth on dielectric is proposed and characterized, and doped GNR interconnect that surpasses Cu, Ru and Co in terms of resistivity at 20-nm line width is demonstrated.

Growth Mechanism: Instead of cracking the gas-phase precursors in CVD methods, we exploit the high solubility of carbon (C) in nickel (Ni) [14] and the natural segregation of C to a Ni surface [15]. A solid-phase carbon source (in the form of graphite powder) is deposited on electron-beam-evaporation deposited (at RT) 100-nm thick polycrystalline Ni film on SiO_2 substrate, followed by C dissolution into the Ni and segregation to the Ni/SiO_2 interface, and graphene formation under 300 °C and uniformly applied 65 psi pressure condition (Fig. 3(b)) at the interface. The C diffusion (Fig. 3(c)) predominantly occurs through the grain boundaries (GBs, activation energy = 0.7 eV) rather than through bulk Ni (activation energy = 1.5 eV) at 300 °C [16], and the final graphene thickness of 20 nm in 30-min growth confirms that multilayer graphene is grown at the Ni/SiO_2 interface (Fig. 4). A commercial vacuum bonding machine is used to apply the pressure (65 psi) during the growth that assists the graphene growth by (1) assuring that the graphite powder is firmly attached to the Ni surface, and (2) assisting the decomposition of sp^2 bonds in the graphite powder and C dissolution into Ni. On the other hand, the applied pressure is small so that the C solubility in Ni stays unchanged as under atmospheric pressure [17], and no C or Ni diffusion into the SiO_2 dielectric was observed after the growth, according to the energy-dispersive X-ray spectroscopy (EDAX, Fig. 5), indicating no barrier for C or Ni is needed for the growth process. After the growth, excess C at the Ni top surface are removed by oxygen

978-1-7281-1988-5/18 $31.00 © 2018 IEEE

plasma, followed by etching away the Ni layer by $FeCl_3$ solution.

Temperature Dependency: As the temperature dependency of C diffusion coefficient and solubility in Ni follows the Arrhenius equation [17] (Fig. 4(a),(b)), the C flux to the Ni/SiO_2 interface and the graphene growth rate increase rapidly with the growth temperature (Fig. 4(c)). The graphene quality is determined by its growth rate, thus sensitive to the growth temperature. Insufficient C flux at low temperature (200 °C) results in discontinuous graphene islands that shows low G/D ratio in Raman spectrum (Fig. 6(b)). Excess C flux at high temperatures (400 °C and 500 °C) causes non-uniform graphene thickness (Fig. 6(d),(e)) and worse quality (Fig. 6(e)). Similarly, the C flux (Fig. 4(a)) and MLG quality is affected by the Ni thickness. The optimal growth condition is identified as 300 °C at 65 psi pressure for 30 min using a 100-nm thick Ni layer that gives multilayer graphene with G/D ratio of 3.41 (Fig. 6(c)). The experiments in the rest of the paper are on MLG grown at 300 °C that is BEOL compatible.

MLG Characterization: Fig. 7(a) shows the optical image of continuous MLG of > 5 mm^2 area, with uniform surface profile from the scanning electron microscope (SEM) image (Fig. 7(b)). The atomic force microscopy (AFM) profile illustrates a thickness of ~20 nm, and the sharp G peak (Fig. 7(d)) confirms highly ordered sp^2 hybridized hexagonal MLG sheets. MLG wires in 4-probe test structures with Au/Ni contacts (Fig. 8(a),(b)) are fabricated with the low-temperature grown MLG, and the measurements (Fig. 8(c)) indicate a resistivity of 193 $\mu\Omega$-cm, similar to that of the CVD MLG [5].

III. DGNR INTERCONNECT CHARACTERIZATION

The intercalation doped graphene nanoribbon (DGNR) interconnects down to 20 nm in width are fabricated (Fig. 9), as described in [5]. The MLG obtained by pressure-assisted low-temperature growth is patterned into GNRs by electron beam lithography and oxygen inductive coupled plasma (ICP) etching, followed by $FeCl_3$ intercalation doping in a high-pressure (1.5-2.0 atm pressure) reactor and Au/Ni contact formation (Fig. 10). The entire fabrication process is well within the BEOL process budget (Fig. 9).

Electrical Conductivity: The measured electrical resistivity of $FeCl_3$ DGNR from 4-probe measurements is within the range of 20-50 $\mu\Omega$-cm. The resistivity decreases as DGNR width decreases, because of higher intercalation doping efficiency at narrower widths [5]. The extracted contact resistance is < 20 Ω-μm, similar to that of CVD DGNR interconnects [5]. The measured DGNR resistivity is smaller than the resistivity reported for Cu, Co and Ru interconnects of the same thickness (20 nm) below 40-nm width. The doping level of the DGNR vary from -0.27 eV to -0.47 eV, smaller than the theoretical maximum $FeCl_3$ doping level (-0.6 eV) [5], indicating that even smaller resistivity is feasible by improving doping efficiency.

Stability and Reliability: The doping stability of DGNR are characterized by measuring its resistivity during a 6-week period (> 1000 hours, Fig. 11(a)) when the DGNR is stored in vacuum at RT with no dielectric passivation. The 4% resistivity increase in the first two weeks is the result of partial $FeCl_3$ dopant diffusion into vacuum, but the doping remained stable in the next 4 weeks. This diffusion can be eliminated by an interlayer dielectric (ILD) encapsulation of the DGNR.

DC current stress test (Fig. 11(b)) is performed on DGNR of 20-nm width and 2-μm length at stress current density of 100 MA/cm^2 and temperature of > 100 °C in vacuum. Resistance increase is less than 3% after 28 hours of stress, and no gradual resistance increase is observed, indicating absence of EM in DGNR, consistent with our previous EM characterization [6]. Due to DGNR's stress-free van der Waals interfaces, stress migration may not be an issue for DGNR.

IV. CONCLUSIONS

CMOS-compatible doped-graphene-nanoribbon (DGNR) interconnects fabricated by subtractive etching process, with smaller resistivity than that of Cu, Co and Ru interconnects, and negligible EM, are demonstrated for the first time for sub-40 nm line widths. The DGNR interconnect process is enabled by a novel pressure-assisted solid-phase low-temperature (300 °C) multilayer graphene growth. The DGNR promises ~4-folds smaller circuit delays than Co and Ru (Fig. 12), with plenty of room for further performance improvement by improving graphene quality. Moreover, the significantly higher melting point of DGNR compared to those of Co and Ru, can allow smaller aspect ratios, leading to smaller parasitics and significantly reduced switching energies.

ACKNOWLEDGMENT

This research was supported by the ARO (grant W911NF1810366) and the UC-MRPI (grant MRP-17-454999). Device fabrications were carried out using the facilities at the California NanoSystems Institute (CNSI) and the National Nanotechnology Infrastructure Network (NNIN) at UCSB.

REFERENCES

[1] K. Banerjee, et al., "Global (interconnect) warming," *IEEE Circuits and Devices Magazine*, vol. 17, no. 5, pp. 16-32, 2001.

[2] F. Griggio, et al., "Reliability of dual-damascene local interconnects featuring cobalt on 10 nm logic technology," *IEEE Int. Reliability Physics Symp. (IRPS)*, 2018, pp. 6E.3.1-6E.3.4.

[3] C.-K. Hu, et al., "Future on-chip interconnect metallization and electromigration," *IEEE Int. Reliability Physics Symp. (IRPS)*, 2018, pp. 4F.1.1-4F.1.4.

[4] D. Wan, et al., "Subtractive etch of ruthenium for sub-5nm interconnect," *IEEE Int. Intercon. Tech. Conf. (IITC)*, 2018, pp. 10-12.

[5] J. Jiang, et al., "Intercalation doped multilayer-graphene-nanoribbons for next-generation interconnects," *Nano Letters*, vol. 17, no. 3, pp. 1482-1488, 2017.

[6] J. Jiang, et al., "Characterization of self-heating and current-carrying capacity of intercalation doped graphene-nanoribbon interconnects," *IEEE Int. Reliability Physics Symp. (IRPS)*, 2017, pp. 6-B.1-6-B.6.

[7] J. Jiang, et al., "All-carbon interconnect scheme integrating graphene-wires and carbon-nanotube-vias," *IEEE Int. Electron Devices Meeting (IEDM)*, 2017, pp. 14.3.1-14.3.4.

[8] K. S. Novoselov, et al., "Electric field effect in atomically thin carbon films," *Science*, vol. 306, pp. 666-669, 2004.

[9] J. N. Coleman, "Liquid exfoliation of defect-free graphene," *Acc. Chem. Res.*, vol. 46, no. 1, pp. 14-22.

[10] W. Liu, et al., "Controllable and rapid synthesis of high-quality and large-area Bernal stacked bilayer graphene using chemical vapor deposition," *Chem. of Materials*, vol. 26, no. 2, pp. 907-915, 2014.

[11] P. Teng, et al., "Remote catalyzation for direct formation of graphene layers on oxides," *Nano Letters*, vol. 12, no. 3, pp. 1379-1384, 2012.

[12] D. Wei, et al., "Low temperature critical growth of high quality nitrogen doped graphene on dielectrics by plasma-enhanced chemical vapor deposition," *ACS Nano*, vol. 9, no. 1, pp. 164-171, 2015.

[13] M. Stelzer, et al., "Graphenic carbon-silicon contacts for reliability improvement of metal-silicon junctions," *IEEE Int. Electron Devices Meeting (IEDM)*, 2016, pp. 21.7.1-21.7.4.

[14] J. J. Lander, et al., "Solubility and diffusion coefficient of carbon in nickel: Reaction rates of nickel-carbon alloys with barium oxide," *Journal of Appl. Physics*, vol. 23, no. 12, 1952.

[15] J. C. Shelton, et al., "Equilibrium segregation of carbon to a nickel (111) surface: A surface phase transition," *Surf. Sci.* 43, 493-520, 1974.

[16] S. Hofmann, et al., "Low-temperature growth of carbon nanotubes by plasma-enhanced chemical vapor deposition," *Applied Physics Letters*, vol. 83, no. 135, pp. 135-137, 2003.

[17] M. Singleton, et al., "The C-Ni (carbon-nickel) system," *Bulletin of Alloy Phase Diagrams*, vol. 10, no. 2, pp. 121-126, 1989.

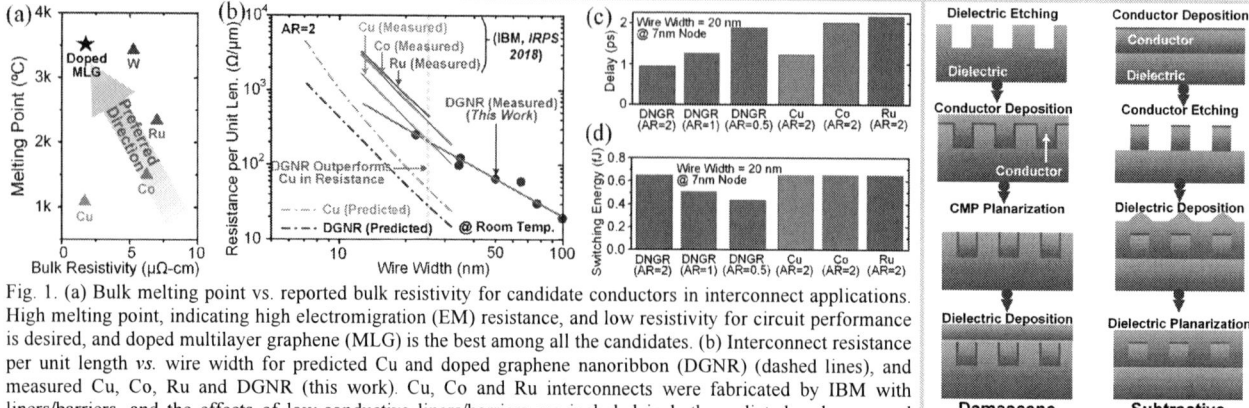

Fig. 1. (a) Bulk melting point vs. reported bulk resistivity for candidate conductors in interconnect applications. High melting point, indicating high electromigration (EM) resistance, and low resistivity for circuit performance is desired, and doped multilayer graphene (MLG) is the best among all the candidates. (b) Interconnect resistance per unit length vs. wire width for predicted Cu and doped graphene nanoribbon (DGNR) (dashed lines), and measured Cu, Co, Ru and DGNR (this work). Cu, Co and Ru interconnects were fabricated by IBM with liners/barriers, and the effects of low-conductive liners/barriers are included in both predicted and measured values for these metals. The wires are assumed to have aspect ratio (AR=thickness/width) of 2. The predicted DGNR assumes grain size of >1 μm, perfect edges and maximum FeCl₃ doping levels. The measured DGNR resistance outperforms Cu for beyond ~25 nm width. (c) and (d) are delay and switching energy for a unit-size inverter driving a fanout-of-4 load via 100×minimum-gate-pitch long local interconnect of 20-nm width from HSpice simulations at 7 nm node. Contact resistances to DGNR and via resistances are considered.

Fig. 2. Schematic of process steps for fabricating horizontal wires by Damascene process (left) and subtractive etching process (right).

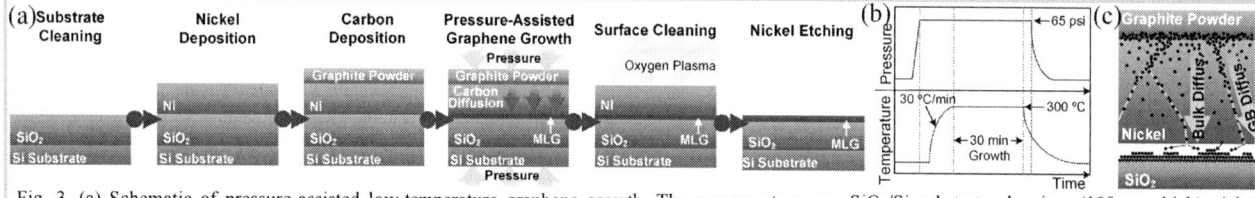

Fig. 3. (a) Schematic of pressure-assisted low-temperature graphene growth. The process steps are: SiO₂/Si substrate cleaning, (100-nm thick) nickel deposition by electron-beam evaporation, carbon (graphite powder) deposition, pressure-assisted graphene growth at nickel/SiO₂ interface, removal of carbon at top surface by oxygen plasma and nickel etching. (b) Pressure and temperature profiles for the low-temperature graphene growth. The growth occurs at 65 psi pressure and 300 °C for 30 min. (c) Schematic of the pressure-assisted graphene growth mechanism. The grain boundaries (GBs) are marked by grey lines in the nickel region, and carbon transports from nickel's top surface to nickel/SiO₂ interface by diffusion via bulk and GBs (marked by grey arrows), and acts as graphene growth precursor.

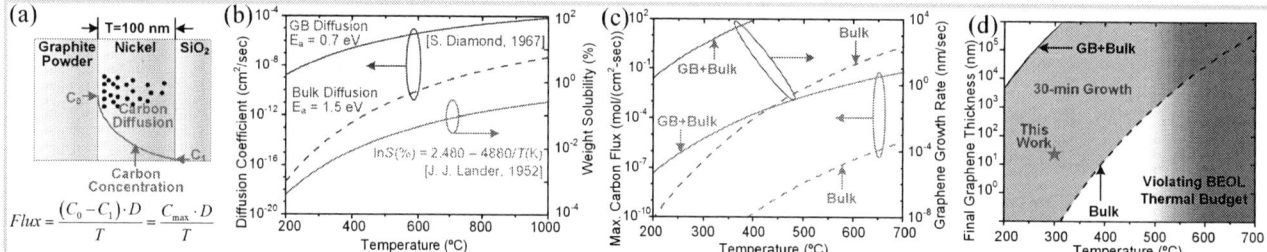

Fig. 4. (a) Schematic showing carbon from graphite powder diffusing through nickel because of concentration gradient. The maximum carbon flux can be calculated by concentration difference ($C_0 - C_1$), carbon diffusion coefficient (or diffusivity, D) and nickel thickness ($T = 100$ nm). (b) Diffusion coefficient for bulk and grain boundary (GB) diffusion modes, and weight solubility of carbon in nickel vs. temperature. (c) Maximum carbon flux (through bulk and bulk+GB) and corresponding graphene growth rate vs. temperature. (d) Final graphene thickness after 30 min growth vs. temperature, for bulk diffusion only and GB + bulk diffusion. Growth temperature > 500 °C violates BEOL process budget (marked by red region). The achieved graphene thickness (marked by red star) is ~20 nm for 30-min growth at 300 °C, indicating the existence of both bulk and GB diffusion of carbon during growth.

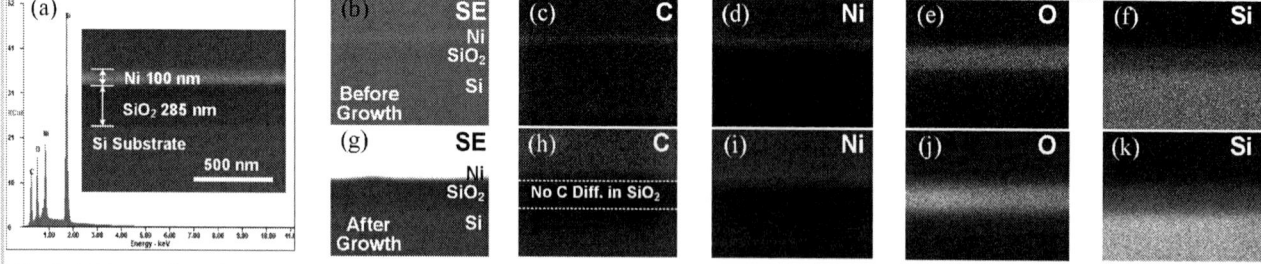

Fig. 5. (a) Energy-dispersive X-ray (EDAX) spectrum of cross section of Ni/SiO₂/Si stack (inset SEM image) before growth. The peaks (from left to right) correspond to carbon, oxygen, nickel and silicon, respectively. The existence of carbon is mainly from the deposited nickel layer. (b) and (g) are the SEM images by secondary electron (SE) mode on the cross section, before and after the growth. (c)-(f) and (h)-(k) are the EDAX element mapping images of carbon, nickel, oxygen and silicon before and after the growth, respectively. No carbon or nickel diffusion into SiO₂ is observed from (h) and (i) after the growth, which eliminates the need for a barrier layer. The extended carbon and nickel region (h-i) above the nickel layer is from the nickel clip, which is part of the sample holder to discharge accumulated static charge in SiO₂ from scanning electrons.

978-1-7281-1988-5/18 $31.00 © 2018 IEEE

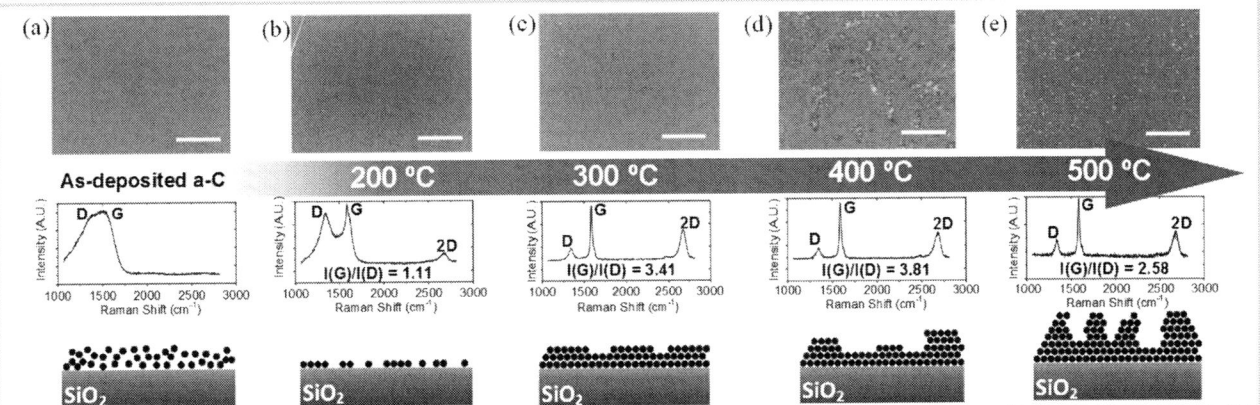

Fig. 6. Optical images, Raman spectrum and schematics of (a) as-deposited amorphous carbon (a-C) and graphene grown at (b) 200 °C, (c) 300 °C, (d) 400 °C and (e) 500 °C. The scale bars are 100 µm. (a) The amorphous carbon is deposited by electron-beam evaporation, and the corresponding Raman spectrum shows a combination of sp^2 and sp^3 bonding. (b) Lack of carbon at nickel/SiO$_2$ interface at 200 °C results in poor quality (I(G)/I(D) = 1.11). (c) 300 °C growth condition offers uniform and good quality (I(G)/I(D) = 3.41) multilayer graphene, compared with other growth conditions. (d) and (e) show non-uniform multilayer graphene, which is resulted from excess carbon at interface. (e) shows worse quality (I(G)/I(D) = 2.58) at 500 °C.

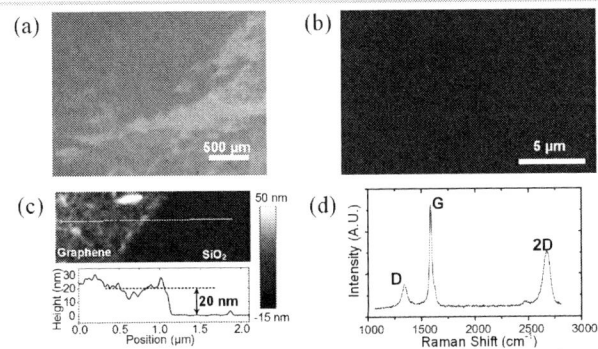

Fig. 7. (a) Optical image of low-temperature grown multilayer graphene (MLG) of area > 5 mm^2 on SiO$_2$. (b) SEM image of the MLG. (c) AFM profile of the MLG indicating 20-nm thickness. (d) Raman spectrum of the MLG showing sharp G peak, indicating good quality.

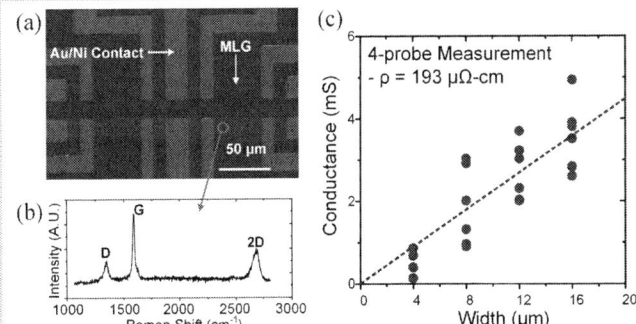

Fig. 8. (a) SEM image of undoped MLG wires in 4-probe test structure of 60 µm length and 4, 8, 12 and 16 µm width, with Au/Ni contacts. (b) Raman spectrum measured on the undoped MLG wire in (a). (c) Measured electrical conductance of the undoped MLG wires, showing a resistivity of 193 µΩ-cm.

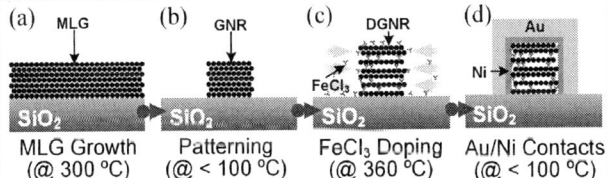

Fig. 9. Schematics showing (a) MLG on SiO$_2$ after the graphene growth at 300 °C, (b) GNR after oxygen plasma etching at < 100 °C, (c) FeCl$_3$ intercalation doping at 360 °C in Ar atmosphere and (d) Au/Ni (10 nm / 100 nm) contact formation by electron beam evaporation at < 100 °C. The corresponding process temperatures are well within the BEOL process thermal budget.

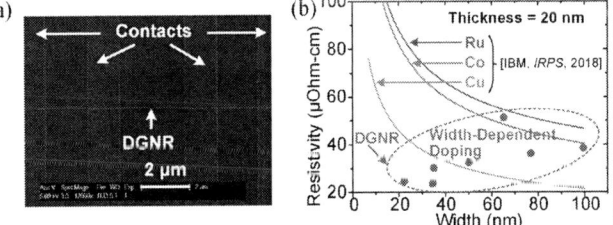

Fig. 10. (a) SEM image of DGNR of 20-nm width in 4-probe test structure. (b) Measured resistivity of the DGNR (doping level: -0.27 eV to -0.47 eV) vs. wire width from 4-probe measurements. Reported resistivities of Cu, Co and Ru wires (with liners/barriers) are plotted, assuming 20-nm wire thickness.

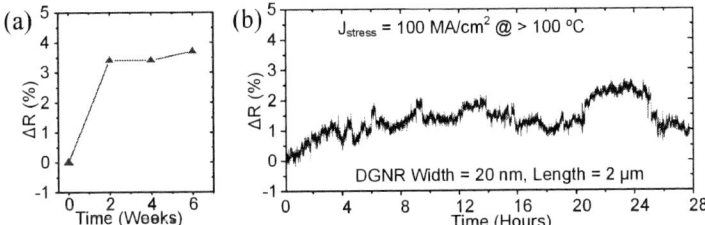

Fig. 11. (a) 20-nm wide DGNR (unpassivated) resistance stability test results during 6-weeks time, when DGNRs are placed at room temperature environment. The increase in resistance from FeCl$_3$ instability occurs mainly in the first 2 weeks and is less than 4%. (b) Resistance change during current stress test at 100 MA/cm^2 on the 20-nm wide DGNR at > 100 °C for 28 hours. No breakdown or gradual resistance increase is observed, indicating no EM.

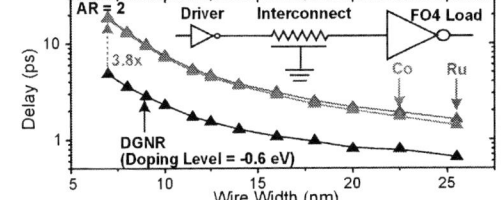

Fig. 12. Delay of a unit-sized inverter driving an FO4 load by local interconnect of 100× minimum gate pitch length vs. interconnect wire width for doped graphene nanoribbon (DGNR), Co and Ru (both with liners/barriers) from HSpice simulations. The DGNR is assumed to have the same quality (grain size, etc.) as in this work, with maximum (-0.6 eV) doping level.

978-1-7281-1988-5/18 $31.00 © 2018 IEEE

Time Dependent Early Breakdown of AlGaN/GaN Epi Stacks and Shift in SOA Boundary of HEMTs Under Fast Cyclic Transient Stress

Bhawani Shankar[1], Ankit Soni[1], Sayak Dutta Gupta[1], Swati Shikha[1], Sandeep Singh[1],
Srinivasan Raghavan[2] and Mayank Shrivastava[1]

[1]Department of Electronic Systems Engineering and [2]Center for Nanoscience and Engineering,
Indian Institute of Science, Bangalore, India, email: mayank@iisc.ac.in

Abstract—This experimental study reports first observations of (i) SOA boundary shift in GaN HEMTs and (ii) early time to fail of vertical AlGaN/GaN Epi stack under fast changing (sub-10ns risetime) cyclic transient stress conditions for a 600V qualified commercial grade HEMT stack. It is shown that a stack qualified for 10 years lifetime under DC stress, fails faster under cyclic transient stress. Integrated electrical and mechanical stress characterization routine involving Raman/PL mapping and CL spectroscopy reveals material limited unique failure physics under transient stress condition. Failure analysis using cross-sectional TEM investigations reveal signature of different degradation and failure mechanism under transient and DC stress conditions. A failure model is proposed for failure under cyclic transient stress.

I. INTRODUCTION

AlGaN/GaN HEMTs promise superior switching performance than their Si counterparts. However, their reliability, in particular, under switching conditions is less explored. In a typical power converter, during OFF-cycle the high blocking voltage at drain introduces significant lateral (drain to gate) and vertical (drain to substrate) stress. While the corresponding time dependent failure physics has been widely studied, its limited however to DC like stress [1]-[5]. As a result, failure physics under practical switching conditions [6] remain broadly unclear. To enhance widespread adoption of GaN HEMTs, it is crucial to understand the failure mechanisms and estimate the lifetimes under switching conditions [7]-[12]. This work aims to contribute to the existing understanding of OFF-state stress, however under fast & cyclic pulse transient conditions, emulating real converter circuit like stress scenario. This study highlights serious implications of OFF-state transient stress during switching cycles on device's Safe Operating Area (SOA), shift in SOA boundary with accumulative stress and very small time to fail under fast transient stress conditions. The underlying degradation physics is investigated, and a unique failure model is proposed.

II. DEVICES UNDER TEST

In this work two classes of HEMT devices were used: (i) commercially available e-mode GaN HEMT to study SOA and extract SOA boundary. (ii) In-house developed HEMT devices were used to study time dependent failure in GaN under vertical DC and fast transient cyclic pulse stress. Rise time (RT) of chain of voltage pulses was varied in sub-10ns range, which is typical in GaN based power converters. Stress measurements were done at different temperatures. Devices were also stressed at room temperature while giving intermediate heat treatment between stress cycles to study role of mechanical stress compensation. HEMTs were realized on a commercially available AlGaN/GaN layer stack on p-type silicon (111) substrate qualified for 600-V operation with 10-year DC lifetime (Fig. 1). Processed run had >90% device yield with variation less than ± 5%.

III. STRESS DEPENDENT SOA BOUNDARY SHIFT

Figure 2 depicts a typical SOA boundary of HEMT and its degradation under pulse stress [5]-[7]. Maximum device degradation can be seen under OFF-state stress condition (Fig. 2b), which justify OFF state stress conditions, as worst-case scenario, used in this work. Figure 3 shows a shift in SOA boundary when device was pre-stressed under OFF state before extracting SOA boundary. This is unusual as such a SOA shift doesn't exist in Si power devices and was never explored earlier for HEMTs. Such an unusual SOA boundary shift sets serious limitations to the use of GaN HEMTs and therefore requires detailed explorations as presented in subsequent sections.

IV. DEVICE CHARACTERIZATION

Experimental setup used in this work (Fig. 4a) tends to mimic OFF state stress scenario across HEMTs in a DC-DC converter (Fig. 4b) by applying cyclic fast transient pulse stress. High voltage pulse chain with ON time of 500ns and varying rise time was used to stress the drain contact with respect to substrate, keeping source and gate were left floating during the stress. It should be noted that OFF period between two consecutive pulses was in the order to 1s, to ensure reasonable time given for relaxation. Following measurements (Fig.4c) were carried out: (i) drain to substrate leakage current after each pulse, (ii) HEMT's DC I-V and C-V characterization, at regular intervals, to capture impact of applied stress on device parameters, (iii) μ-Raman mapping to record mechanical stress profile and (iv) UV-Photoluminescence (PL) mapping to monitor defect distribution in source-drain access with stress time, (v) Cathodoluminescence (CL) spectra at locations with high PL intensity of defect bands, to determine defect location in the epitaxial stack. Post failure SEM and TEM analysis was done routinely.

978-1-7281-1988-5/18 $31.00 © 2018 IEEE

Time to Vertical Breakdown under Pulse Stress: Time to fail (TTF) extracted using fast transient cyclic pulse stress shows exponential dependence on applied stress voltage (Fig. 5) however the TTF was found to be much smaller compared to the same under DC stress case. For instance, 600V GaN Epi-stack qualified for 10-year lifetime under DC stress, failed after ~100 pulses, each being 500ns wide. Interestingly, TTF was found to improve when pulse rise time was increased, which indicates transient time dependent failure to be related to rate of change in electrical stress. Unlike under DC stress case (Fig. 6b), under cyclic pulse stress the vertical leakage with stress time remain unchanged till the verge of failure (Fig. 6a). Rapid and more sever changes in device characteristics are observed under pulse stress compared to that under dc condition (Fig. 7). Drain to substrate vertical leakage hysteresis was found to increases with stress time under pulse stress (Fig. 8), which points to charge trapping in different layers present between drain and substrate. PL mapping of drain-source region (Fig. 9) reveals increased defect generation introducing deep level traps within energy range of E_C - 0.48 eV and E_C - 1.14 eV. Highest PL intensity is observed near the drain contact (Fig. 9c). CL spectra (Fig. 10) taken at same location verifies increase in Blue (BL) and Yellow luminescence (YL) intensities with stress and penetration depth of ~1.2 μm from top surface. Hence, defects in GaN buffer are located close to 1.2 μm below surface in the drain contact vicinity. Raman map confirms mechanical stress builds up at the drain contact edge after the device is stressed for 5000 cycles under pulse condition (Fig. 11). Since AlGaN is transparent to used visible (532 nm) laser, the stress signature received is from GaN buffer. This correlates well with penetration depth calculated from CL. Post failure cross-sectional SEM shows a catastrophic damage in GaN buffer, in vicinity of drain contact (Fig. 12). HR-TEM of defected region reveals fine cracks near GaN buffer and AlGaN transition interface (Fig. 12). However, under DC stress, damage is localized to device surface and no failure signature is observed in the bulk.

V. DISCUSSION AND NEW PHYSICAL INSIGHTS

Understanding Failure Under Cyclic Transient Stress: Under time dependent drain-to-substrate DC stress, the devices show a time-dependent failure (Fig. 1) with strong field dependence. Failure is believed to occur via percolation paths between drain and substrate formed by defect activation at high electric field [3]. Exponential dependence of TTF on pulse stress voltage, shows field dependent failure mechanism similar to DC stress condition. However, the defect percolation theory (borrowed from gate dielectric failure models) cannot explain significantly reduced TTF, its dependence on rise-time (Fig.5) and aggressive catastrophic failure (Fig. 12) under cyclic transient stress condition. These newly discovered trends and findings indicate an electrical shock-based fatigue phenomenon responsible for time dependent GaN Epi stack failure. Electrically developed fatigue was earlier reported for piezoelectric materials [13]. This can be explained as following: when, a voltage stress is applied across the stack vertical component of electric field (E_Z) introduces piezoelectric in-plain strain in AlGaN/GaN layers which is given by [14]: $\sigma_{XY} =$

$\left(e_{33} \frac{C_{13}}{C_{33}} - e_{31} \right) E_Z$. Here, σ_{xy} is in-plane strain, e_{33}, e_{31} are piezoelectric coefficients and, C_{13}, C_{33} elastic stiffness tensors. Piezoelectric strain generates defects in bulk region, in drain contact vicinity, where peak field lies and accumulates with increasing number of stress pulses (Fig. 11b). Furthermore, under pulse stress, the piezoelectric AlGaN/GaN stack undergoes cyclic loading via mechanical strain generation-relaxation as evident from the Raman map, captured before and after pulse stress (Fig. 11). Over time, cyclic stress causes changes in piezoelectric properties [13] and results in time dependent strain variation [15]; $\sigma_{XX}(t) = \sigma_0 \left(1 - e^{\frac{-t}{\tau}} \right)$. Such an electrical fatigue is accumulative in nature as evident from increased YL, BL bands in CL spectra (Fig. 10) and increased hysteresis (Fig. 8) over time. Electrical fatigue nucleates micro-cracks at GaN buffer/AlGaN transition interface which has highest residual stress (Fig. 12). High strain energy at crack tip under cyclic electrical loading propagates it towards device top and hits the surface to form pits and causes catastrophic failure (Fig. 12). Fatigue accumulation accelerates with rate of change of electric field (pulse rise time), resulting in drop in TTF for faster cyclic transients (Fig. 5).

Validation of Proposed Failure Model: Stress accumulation and defect generation in GaN epi layers under pulse condition occurs due to electrical cyclic loading via compressive piezoelectric stress. Heating in device introduces thermoelastic tensile stress [16] which partially relaxes the field induced piezoelectric stress, as confirmed using Raman mapping (Fig. 16). This resulted in TTF improvement when stress applied at higher temperature or heat treatment was given in between cyclic stress (Fig. 17).

VI. CONCLUSION

Shift in SOA boundary under pulse stress was found, which is uncommon in Si power devices. Vertical drain to substrate breakdown in GaN Epi-stack under fast and cyclic pulse stress was found to obey different degradation physics than DC stress. TTF extracted for GaN Epi-stack from fast, cyclic pulse stress measurements did not obey the lifetime predicted by DC stress. Time to fail, under pulse stress showed exponential dependence on stress voltage, which however increased with pulse rise time or temperature. Cyclic electrical loading invoked electric fatigue in buffer. Defect density increased in buffer with stress time due to accumulative nature of fatigue. Increased deep level defects deteriorated device performance. Under pulse stress, cracks nucleated at GaN-buffer/AlGaN-transition region, triggered device failure. Failure under DC stress occurred close to device surface and with no signature of damage in bulk.

References: [1] Warnock et. al. IEEE TED, pp. 3131, 2017, [2] M.Meneghini et. al. APL,pp. 33505(1), 2012, [3] M.Meneghini et. al. IEEE TED, pp. 2549, 2015, [4] M.Borga et. al. IEEE TED, pp. 3616, 2017, [5] M. Borga et.al. IEEE TED, pp. 1-6, 2018, [6] S.Bahl et. al. IRPS 2016, pp. 4A.3, [7] Ikoshi et. al. IRPS 2018, pp. 4E-1, [8] Huang et. al. ISPSD, 2014, pp 273-276, [9] A.Castellazzi et. al. IRPS 2018, pp. 4E-1, [10] B.Shankar et. al, IRPS 2017, pp.WB5-1, [11] B.Shankar et.al, IRPS 2018, 4E.3, [12] B.Shankar et. al, IRPS 2018, pp.4E.4, [13] Glaum et. al. J. Am. Ceram. Soc., pp. 665, 2014, [14] Sarua et. al, Semicond. Sci. Technol. 085004, pp. 1-8, 2010, [15] Chung et. al. Mater.Res.Exp.3, 105026, 2016, [16] J.P. Jones et.al, ITHERM 2014, pp. 959

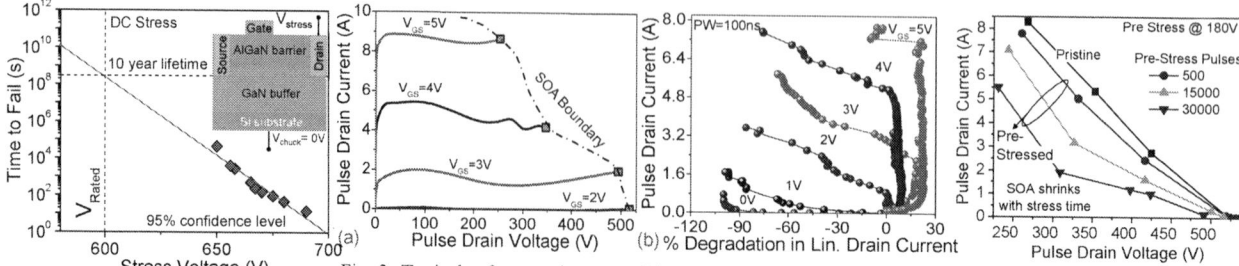

Fig. 1. Time-to-fail (TTF) for the commercially available 600V AlGaN/GaN stack (used in this work) demonstrating 10-year lifetime for 600V dc stress.

Fig. 2. Typical safe operating area (SOA) and resulting device degradation GaN HEMTs. (a) Fast pulse I-V characteristics depicting SOA boundary of a commercially available e-mode HEMT device. (b) Percentage change in linear drain current (on x-axis) measured after each voltage/stress pulse. Under OFF state stress condition, device shows highest degradation.

Fig. 3. SOA of commercial e-mode GaN HEMT, extracted at different stages of OFF-state stress cycles. Set of devices are pre-stressed for different no. of pulses before extracting SOA boundary. Shift in SOA boundary as a function of stress can be seen.

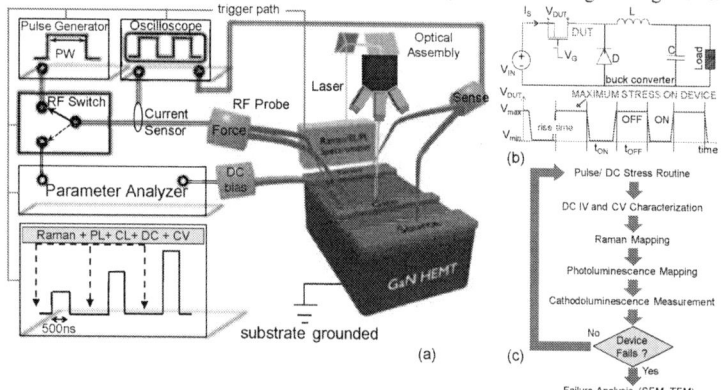

Fig. 4. (a) Schematic of experimental setup with integrated Raman, PL and sub-μs pulse generator used for on-the-fly electrical and material characterization of HEMT/Epi-Stack under dc and pulse stress conditions. Electrical stress routine (dc and pulse) was interrupted at regular intervals, to record, evolution of mechanical stress profile using visible Raman (λ=532nm) and defects density distribution using UV-PL (λ=325nm) across the active region. CL spectra was also captured in gate-drain region at different stress time. Inter-pulse dc IV and CV characterization was also done to record variation in device parameters with stress time. (b) Schematic representation of a buck power converter showing device voltage (V_{DUT}) waveform during switching. Maximum voltage stress appears across the device during the OFF-cycle of the converter. Hence, pulse stress degrades the device performance with each cycle.

Hence device's safe operating area shrinks over time under switching operation. (c) The flowchart represents the sequence of various types of device and material characterizations performed during pulse and dc stress routines.

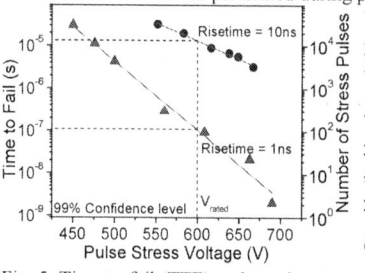

Fig. 5. Time to fail (TTF) under pulse stress conditions presented on left y-axis and the corresponding number of pulses required for device failure at different voltages, shown on right side y-axis. Here pulse width of 500ns and OFF state period between two consecutive pulses are 100μs. Figure depicts, TTF is significantly lowered under pulse operation compared to that under dc and exhibits dependence on the rise time (slew rate) of the stress pulse.

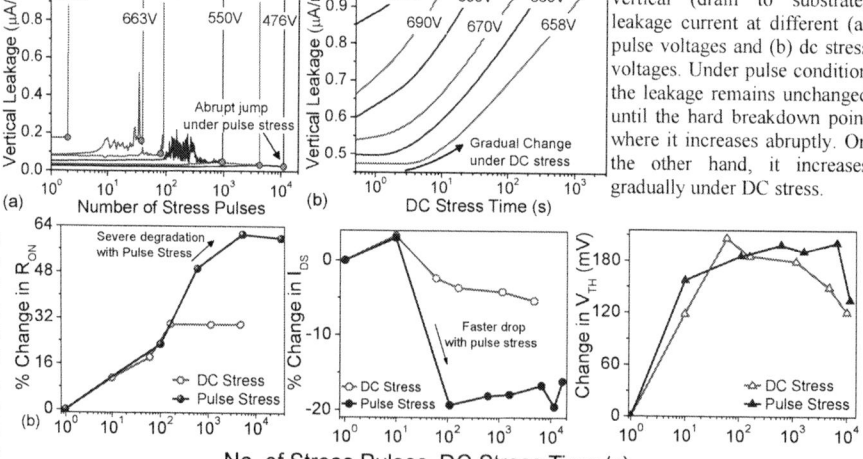

Fig. 6. (a) Time evolution of vertical (drain to substrate) leakage current at different (a) pulse voltages and (b) dc stress voltages. Under pulse condition the leakage remains unchanged until the hard breakdown point where it increases abruptly. On the other hand, it increases gradually under DC stress.

Fig. 7. Degradation in the device parameters recorded on-the-fly during dc and pulse stress measurements done at 185V. (a) ON-resistance (R_{ON}) increases and (b) ON-current (I_{ON}) decreases at a much faster rate under pulse stress condition when compared to DC stress. With stress time, the density of deep levels in GaN increases, as evident from increased Yellow luminescence and Blue luminescence in Fig. 10, which electrostatically influence 2DEG and increase R_{ON} and lowers I_{ON}. Threshold voltage instability gets accelerated under pulse stress. For instance, with just ten stress pulses applied on drain, the V_{TH} changed by 180mV as evident in (c).

Fig. 8. Hysteresis behavior recorded between drain and substrate current after different number of stress cycles (pulse stress voltage = 200V). Increase in loop area highlights charge /trap accumulation in buffer and transition regions with each stress pulse, attributed to increased density of defect in bulk GaN. This correspond to blue and yellow luminescence as visible from broadened FWHM of defect band (400-600nm) in CL spectra of GaN recorded close to drain edge. Carrier de-trapping is slow from deep levels, which results in accumulation of trapped charges in buffer layers with increasing stress cycles. This translates to increased hysteresis.

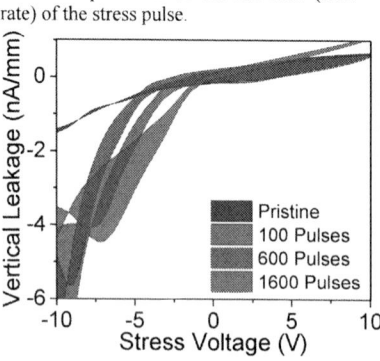

978-1-7281-1988-5/18 $31.00 © 2018 IEEE

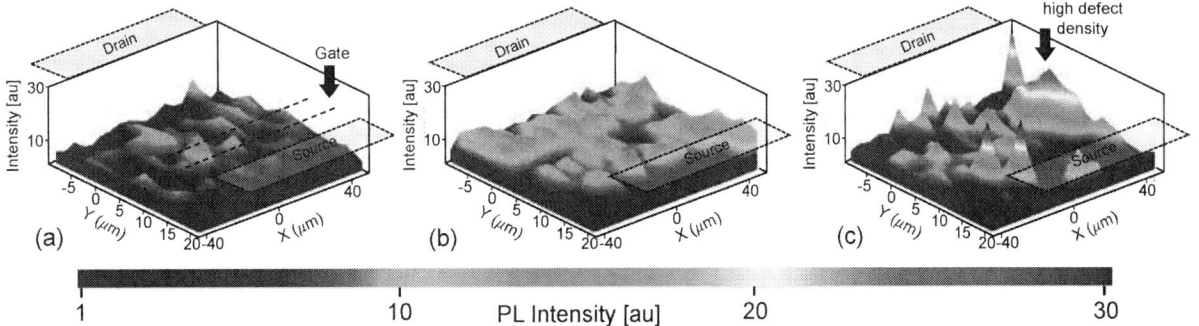

Fig. 9. PL intensity distribution for defect band (425-525nm) captured between source-drain region of the HEMT device (a) before stress, (b) after stressing with 5000 pulses and (c) at the verge of failure. Figure reveals that the density of mid-bandgap defects, increases with number of pulses (of 200V) with peak concentration close to drain contact edge.

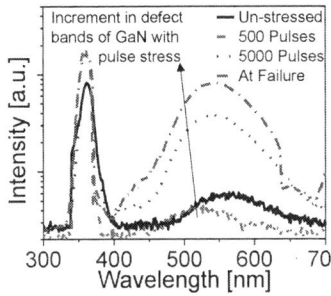

Fig. 10: CL spectra of GaN captured between gate - drain region. FWHM of the defect band (400nm-600nm) increases with number of stress pulses (@ 200V) applied at drain with substrate grounded. It reveals increased density of deep levels under pulse stress.

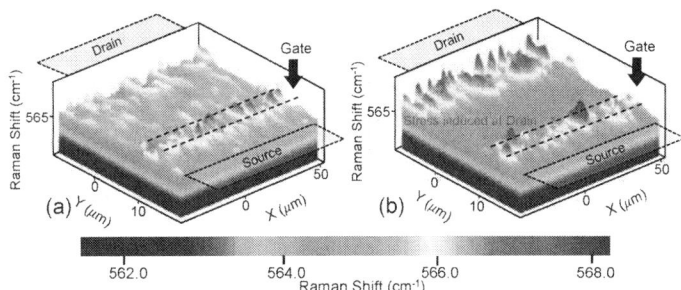

Fig. 11. (a) Raman map captured between source - drain region of pristine device, i.e. before applying stress. Figure reveals residual compressive stress in recessed region under gate. (b) Raman map of the device after application of 5000 stress pulses (@180V) between drain and substrate. Stress accumulation can be noticed near the drain contact edge, beside gate finger.

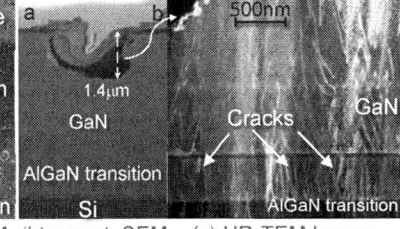

(a) Post Failure SEM (b) x-sect. SEM (c) HR-TEM Image

Fig. 12. (a) Top view SEM image depicting HEMT device failed under 600V pulse stress condition. Failure occurred with massive crack in the gate-drain region. (b) Cross-sectional SEM taken along line a-b reveals that the damage reached 1.4 micron deep into GaN buffer. (c) HR-TEM image of the region below damage area depicts cracking at the interface of GaN buffer/AlGaN-transition region.

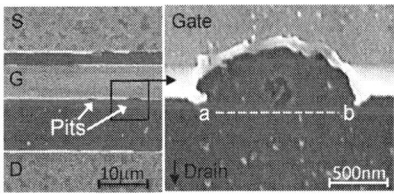
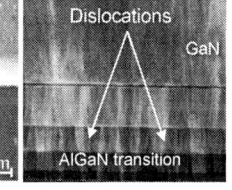

(a) Post Failure SEM (b) HR-TEM Image

Fig. 13. (a) Top view SEM images of HEMT device failed after 600V of DC stress depicts shallow damage confined to top surface. (b) Cross-sectional TEM along a-b line shows no damage in bulk along dashed line 'a-b', reveals the failure occurred in device's active region close to AlGaN/GaN interface at the top. No signature of bulk failure found. Similar localized failures observed in other dc stressed devices near top surface.

Transient Voltage with sub-10ns RT result in fast change in electric field
Piezoelectric GaN translates electric field shock into mechanical shock
Cyclic electrical shock loading of GaN induces electrical fatigue
Cumulative fatigue introduces cracking at GaN-buffer/AlGaN-transition interface
High energy crack tip propagates towards device surface
Deep defects created near channel trap carriers and deteriorate R_{ON} and V_{TH}
Cracks reach AlGaN barrier and surface
Catastrophic failure of AlGaN/GaN Stack

Fig. 15. The sequence of events which are responsible for Fast Transient Stress Assisted Time Dependent Breakdown and Shift in SOA Boundary of AlGaN/GaN Epi Stacks & HEMTs.

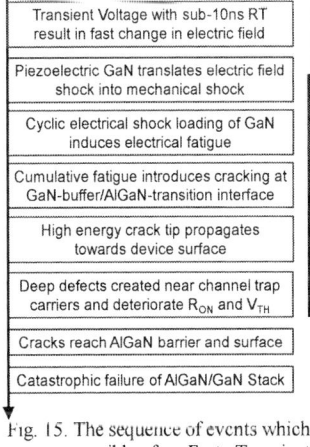

Fig. 14: TEM image of a pristine device taken under the drain depicting least dislocation density. Therefore, it can be said that failure under dc stress occurs via dislocations which are leakage path

Fig. 17. Variation in vertical time-to-fail under pulse stress, recorded at different temperatures treatment. The increase in temperature, improves the TTF.

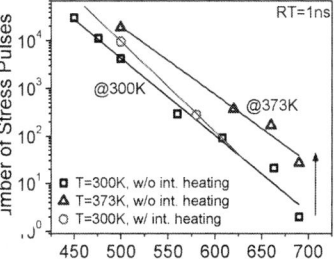

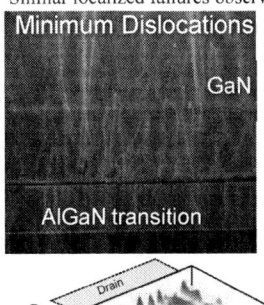

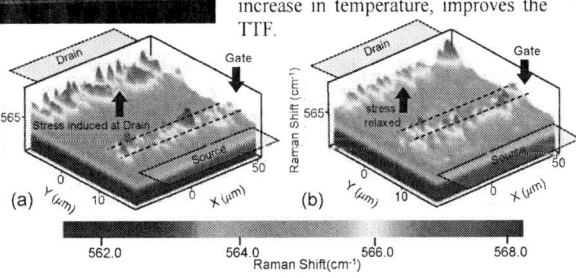

Fig. 16: (a) Post stress Raman map device stressed using 5000 pulses depicts stress accumulation around drain contact. (b) Relaxation in device stress, confirmed from Raman map captured, after the device was heated at 100°C for 10 mins post pulse stress

978-1-7281-1988-5/18 $31.00 © 2018 IEEE   806

Toward High Performance SiGe Channel CMOS:
Design of High Electron Mobility in SiGe nFinFETs Outperforming Si

C. H. Lee, R. G. Southwick III, S. Mochizuki, J. Li, X. Miao, M. Wang, R. Bao, I. Ok, *T. Ando,
*P. Hashemi, D. Guo, *V. Narayanan, N. Loubet, and H. Jagannathan

IBM Research and *IBM T. J. Watson Research Center
257 Fuller Road, Suite 3100, Albany, NY 12203, email: clee@us.ibm.com

Abstract—For the first time, high electron mobility in tensile-strained SiGe channel nFinFETs outperforming Si is reported to explore the feasibility of high performance SiGe CMOS. To examine the electron mobility behaviors in SiGe channel, a series of tensile-strained SiGe nFinFETs are fabricated on various strain relaxed buffer layers by taking into account the minimum threading dislocation density and strain engineering. For SiGe (Ge >20%) nFinFETs, we identify the existence of additional electron trapping site close to the conduction band edge in IL/HK, leading to the abnormal Vt shift, PBTI degradation, and low electron mobility. We also fabricated short-channel SiGe nFinFETs, which exhibit excellent cut-off behavior and electrostatics (SS ~65mV/dec and DIBL ~18mV at V_{DD}=0.7V). In addition, the dynamic performance of tensile-strained SiGe CMOS against Si CMOS is evaluated by TCAD simulation based on experimental data.

I. INTRODUCTION

Although Si FinFETs have firmly been established as today's main stream CMOS logic technology [1-3], the difficulty in further improving the conventional FinFET makes it prudent to search for alternative channel materials. The main concern with Si CMOS is predominantly associated with the Si pFET having significant NBTI, relatively lower hole mobility, and difficulty in lowering pFET Vt using band-edge effective work function metals. All of these issues are attributed to the intrinsic material properties of Si and its native oxide. From this point of view, dual channel Si/SiGe CMOS is a promising technology element to solve the issues mentioned above by replacing the Si with SiGe channel in pFETs. The benefit of increased hole mobility, lower NBTI, and Vt tunability in SiGe pFET over Si have been proved [4-6]. However, a conventional dual channel approach for integrating Si and SiGe in three-dimensional device structures (FinFETs and Nanosheets) poses co-integration challenges as Si and SiGe have different material properties such as etch chemistry and thermodynamic stability. In this sense, an all SiGe channel CMOS appears a reasonable option if SiGe nFET can provide high performance. Unfortunately, SiGe nFET outperforming Si has not been reported yet because of high interface trap states, low electron mobility, and lack of channel strain engineering (SiGe channel grown on Si gives a compressive strain, resulting in lower electron mobility than Si) [7,8]. The advantage and concern of channel material configuration for CMOS design are summarized in **Table I**.

In this work, we fabricated a series of tensile-strained SiGe nFinFETs on strain relaxed buffer (SRB) layers to explore the feasibility of high performance SiGe channel CMOS. The

electrical impacts of Ge content and tensile strain on the performance of SiGe nFinFET is thoroughly discussed. The short-channel (L_g ~40 nm) tensile-strained SiGe nFinFETs was fabricated to measure the immunity to the short-channel-effects. In addition, the dynamic performance evaluation of tensile-strained SiGe CMOS against Si CMOS based on the experimental data is performed by TCAD.

II. DEVICE FABRICATION AND CHARACTERIZATION

Relatively low Ge content SiGe channels were epitaxially grown on fully relaxed SiGe SRB layers to introduce tensile strain in the active SiGe channel. The tensile strain in the SiGe channel is confirmed by XRD analysis. The peak positions are extracted from XRD reciprocal space maps around Si 224 and SiGe 224 diffraction for active SiGe channels and SRB layers (**Fig. 1(a)**). SiGe nFinFETs on SRB virtual substrate was fabricated using a replacement metal gate (RMG) process to see the impact of Ge content and tensile strain on transistor electrical properties (**Fig. 1(b)**). A well-controlled gate stack (IL/HK/MG) for SiGe channel in RMG module was employed to reduce interface defects [9]. HRTEM and EELS elemental mapping (Si, Ge, Hf, Ti) of SiGe10% nFinFET fabricated on a SRB30% confirm the vertical profile of SiGe fin (110) sidewalls and gate stack (**Fig. 2**). Lattice deformation of SiGe10% nFinFET at M1 level was measured by using precession electron diffraction with 1 nm spatial resolution. The [001] out-of-plane lattice deformation of SiGe10% fin shows the compressive strain, indicating that there is a corresponding tensile strain along the carrier transport direction (**Fig. 3**). Note the bottom fin portion of SRB30% shows a fin liner/STI-induced residual strain.

I_D-V_G transfer curves of long channel SiGe nFinFETs fabricated on SRB virtual substrate are shown (**Fig. 4**), where Ge content in both active SiGe channel and SRB are varied to understand the role of Ge content and strain in the SiGe nFinFETs. The charge pumping method was used to measure the interface trap charge (N_{it}) at the SiGe/IL. Although N_{it} increases monotonically as Ge percentage in the SiGe channel increases, a minimum N_{it} of ~4e11 cm^{-2} up to SiGe20% is achieved, resulting in the decent subthreshold slope (SS) ~63 mV/dec in SiGe nFinFET (**Fig. 5**). Electron mobility as a function of inversion carrier density (N_{inv}) in SiGe nFinFETs is shown (**Fig. 6**). It suggests that at least 0.77% of tensile strain in the SiGe channel is required to outperform the electron mobility in Si nFinFET. Interestingly, the degradation of electron mobility is observed in higher Ge percentage in the SiGe channel despite the same level of tensile strain, indicating that extrinsic carrier scattering source might be generated.

978-1-7281-1988-5/18 $31.00 © 2018 IEEE

III. CARRIER TRANSRPORT IN CONDUCTION BAND

A. Strain response of electron mobility

It is well known that Si and Ge have effective mass anisotropy in the conduction band arising from the substantial dependence of the subband structures on surface orientation. The equi-energy valleys for a two-fold valley ($\Delta 2$), lighter conductive mass, and a four-fold valley ($\Delta 4$), heavier conductive mass, in SiGe (110) inversion layer with/without uniaxial tensile strain (corresponding to the sidewall surfaces of (110)/<110> SiGe FinFETs) is shown in **Fig. 7**. By applying uniaxial tensile strain along the carrier transport <110> direction, the electrons repopulate from $\Delta 4$ into $\Delta 2$, where the subband energy of $\Delta 2$ is lowered [10]. The tensile strain response of electron mobility in SiGe10% nFinFETs is investigated (**Fig. 8**). Low-field electron mobility shows higher sensitivity to tensile strain than high-field electron mobility because quantum confinement compensates for the tensile strain-induced subband splitting in the conduction band minima. The tensile strain benefit on electron mobility in higher Ge SiGe channels becomes less efficient (**Fig. 9** SiGe20% vs **Fig. 8** SiGe10%), which could be ascribed to an additional scattering source generated near the conduction band minima.

B. Additional electron trapping site

One of the most striking features in SiGe nFinFETs is that Vt constantly goes up as Ge percentage in the SiGe channel increases which cannot be simply explained by the conduction band offset (ΔEc) (< 0.1 eV) of SiGe relative to Si (**Fig. 10(a)**). Also, the Vt sensitivity to tensile strain in SiGe nFinFET is much weaker compared to Si nFinFET. To better understand the abnormal Vt behaviors in SiGe nFinFETs, ΔE_C and ΔVtlin are directly compared as a function Ge percentage in the SiGe channel (**Fig. 10(b)**), where ΔEc is extracted by fitting the quasistatic QM CV simulations to experimental full-CV curves in SiGe pFinFETs [9]. A wide discrepancy between ΔEc and ΔVtlin above SiGe20% suggests that gate stack related extrinsic factor causes the additional Vt shift in SiGe nFinFETs. In order to verify our hypothesis about additional Vt shift component, the DC Vt shift at a constant gate bias across SiGe compositions measured at 125C. The 10 yr DC projected end-of-life (EOL) PBTI Vt shift for SiGe nFinFETs is estimated based on the well-known power-law (**Fig. 11(a)**). It shows that PBTI performance starts degrading slightly up to SiGe20% due to the higher conduction band energy level of SiGe channel, which allows the inversion layer electron to interact with PBTI site, A, located near the conduction band edge in HfO_2. On the other hand, higher Ge percentage (Ge >20%) in SiGe nFinFET generates the additional electron trapping site in the IL/HK, B, whose energy level is lower than A, resulting in a significant degradation of PBTI performance (**Fig. 11(b)**). Moreover, it is considered likely to be negatively charged in nFET operation, leading to the positive Vt shift as well as electron scattering source.

IV. DESIGN OF SiGe CHANNEL CMOS

Short-channel SiGe nFinFETs at L_g ~40 nm were built on SRB virtual substrate to measure short-channel-effects. **Fig. 12** shows the linear and saturation I_D-V_G transfer

characteristics where the inset shows a dark-field TEM image of SiGe10% nFinFET fabricated on SRB30%. The devices exhibit excellent cut-off behavior and electrostatics (SS~65 mV/dec and DIBL~18mV at Vdd =0.7V for SiGe10% channel on SRB30%, SS ~66 mV/dec and DIBL ~20mV at Vdd=0.7V for SiGe20% channel on SRB40%). From SiGe channel CMOS perspective, both PBTI and NBTI in SiGe devices should be carefully considered since they show the opposite trend with respect to Ge percentage in the SiGe channel (**Fig. 13**). It is interesting to see that SiGe10% shows the lowest EOL total Vt shift (sum of PBTI and NBTI) and SiGe20% is a little better or similar to Si. It is worthy to note that NBTI performance is predominantly governed by IL thickness as hole trapping sites are located close to the valence band edge in the IL, which limits the Tinv scaling of Si CMOS. SiGe10% and SiGe20% channel provide ~2.5 times and ~9 times improvement of NBTI at a given Tinv, respectively, due to the higher valence band energy level compared to Si (**Fig. 14**). On the other hand, PBTI is not as sensitive to the IL thickness, indicating a SiGe channel would be beneficial in the scaled Tinv regime. We already reported that the Ge percentage in the SiGe channel pFET plays a critical role in determining D_{it}, Vt shift, and NBTI improvement, regardless of the channel strain, while hole mobility is improved by the compressive strain [11]. Note that tensile strain in SiGe pFET neither improves or degrades hole mobility. Thus, the tensile strained SiGe CMOS can enjoy the higher electron mobility without hole mobility penalty. The effective drive current (I_{eff}) [12] of tensile-strained SiGe10% CMOS and SiGe20% CMOS for benchmarking the dynamic performance is simulated by TCAD (**Fig. 15**). The key parameters for simulation are summarized in **Table II**. The Tinv value used for TCAD simulation is obtained based on NBTI/PBTI data. The tensile-strained SiGe10% and SiGe20% CMOS show about 10% improvement of I_{eff} and further improvement is expected with process optimization.

V. CONCLUSION

High electron mobility in tensile-strained SiGe nFinFETs outperforming Si is presented for the first time to explore the feasibility of high performance SiGe CMOS. Abnormal Vt shift, PBTI degradation, and low electron mobility in high Ge percentage (Ge >20%) SiGe nFinFET could be ascribed to the generation of additional electron trapping site close to the conduction band in IL/HK. The short-channel (L_g ~40 nm) SiGe nFinFETs on SRB virtual substrate exhibit excellent cut-off behavior and electrostatics (SS ~65mV/dec and DIBL ~18mV at V_{DD}=0.7V). In addition, the tensile-strained (0.77%) SiGe CMOS can provide ~10% improvement of effective drive current compared to conventional Si CMOS.

ACKNOWLEDGMENT

This work was performed by the research and development alliance teams at various IBM research and development facilities.

REFERENCES

[1] K. Seo, *et al.*, *VLSI Tech. Symp.*, 12 (2014). [2] D. Ha *et al.*, *VLSI Tech. Symp.*, T68 (2017). [3] C. Auth *et al.*, *IEDM*, p. 673 (2017). [4] D. Guo *et al.*, *VLSI Tech. Symp.*, 14 (2016). [5] G. Tsutsui *et al.*, *IEDM*, 456 (2016). [6] P. Hashemi *et al.*, *IEDM*, 824 (2017). [7] L. Hutin *et al.*, *VLSI Tech. Symp.*, 37 (2010). [8] J. Oh *et al.*, *VLSI Tech. Symp.*, 39 (2010). [9] C. H. Lee *et al.*, *IEDM*, 766 (2016). [10] T. Irisawa *et al.*, *TED*, vol. 56, 1651 (2009). [11] C. H. Lee *et al.*, *IEDM*, 820 (2017). [12] M. H. Na *et al.*, *IEDM*, 121 (2002).

978-1-7281-1988-5/18 $31.00 © 2018 IEEE

Table I. Channel material configuration for CMOS design

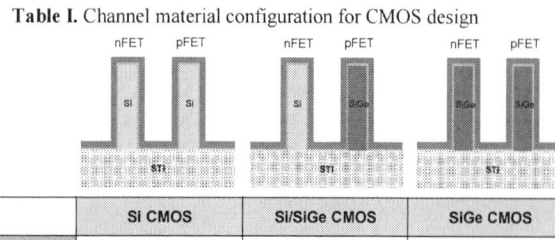

	Si CMOS	Si/SiGe CMOS	SiGe CMOS
Advantage	- Simple integration - Well-known channel material	- Superior NBTI - High hole mobility - pFET Vt tunability	- Simple integration (single channel) - NBTI/pFET Vt tunability
Concern	- Poor NBTI - Difficulty in lowering pVt - Low Hole mobility	- Process complexity/cost - Dual channel materials	- Poor Dit and SS - Low electron mobility - Not outperforming Si

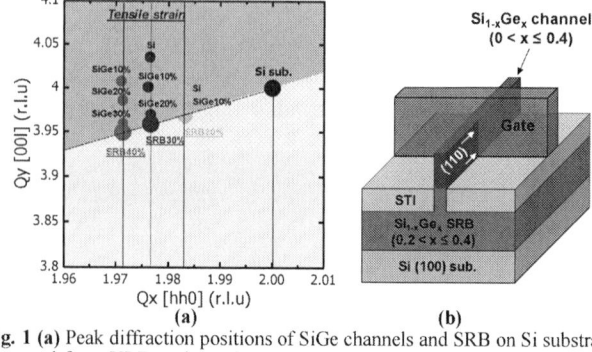

Fig. 1 (a) Peak diffraction positions of SiGe channels and SRB on Si substrate, extracted from XRD reciprocal space maps around Si 224 and SiGe 224. **(b)** Tensile-strained SiGe nFinFETs fabricated on various SRB virtual substrates.

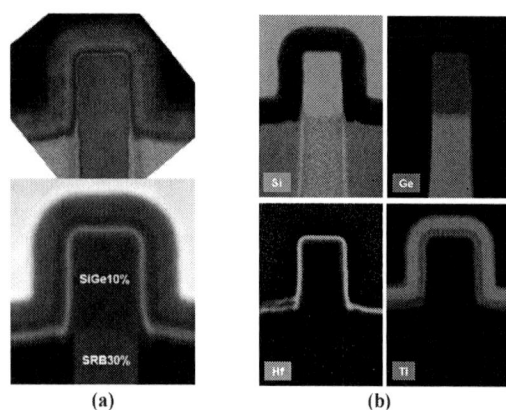

Fig. 2 (a) HRTEM analysis of tensile-strained SiGe10% nFinFETs fabricated on SRB30% at M1 level. **(b)** EELS elemental mapping (Si, Ge, Hf, Ti) of tensile-strained SiGe10% nFinFET fabricated on SRB30%.

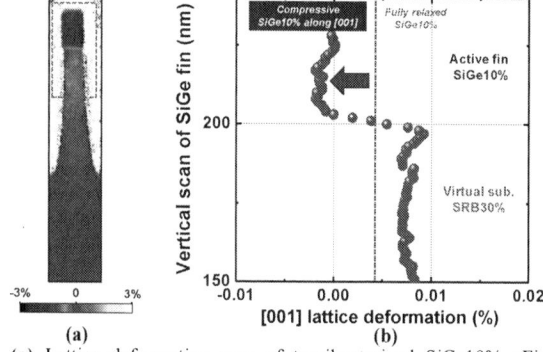

Fig. 3 (a) Lattice deformation map of tensile-strained SiGe10% nFinFET on SRB30% at M1 level measured by precession electron diffraction. **(b)** [001] out-of-plane lattice deformation profiles. Active SiGe10% fin shows the compressive strain, indicating that there is a corresponding tensile strain along the channel.

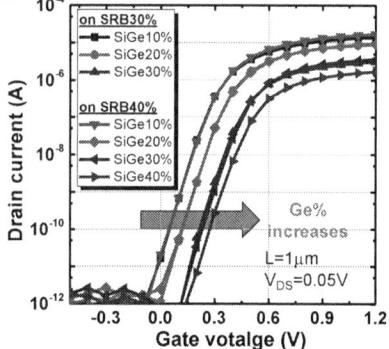

Fig. 4. I_D-V_G curves of long channel SiGe nFinFETs fabricated on SRB virtual substrates. Ge percentage in both active SiGe channel and SRB are varied to understand the role of Ge content and strain in the SiGe nFinFETs.

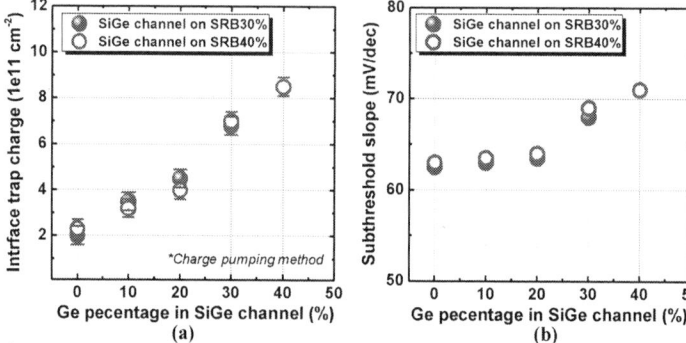

Fig. 5 (a) Interface trap charge (N_{it}) as a function of Ge percentage in the SiGe channel measured by charge pumping. **(b)** Subthreshold slope in SiGe nFinFETs on SRB30% and SRB40% virtual substrate. A subthreshold slope ~63 mV/dec in SiGe nFinFET (Ge <20%) is achieved, comparable to Si FinFET.

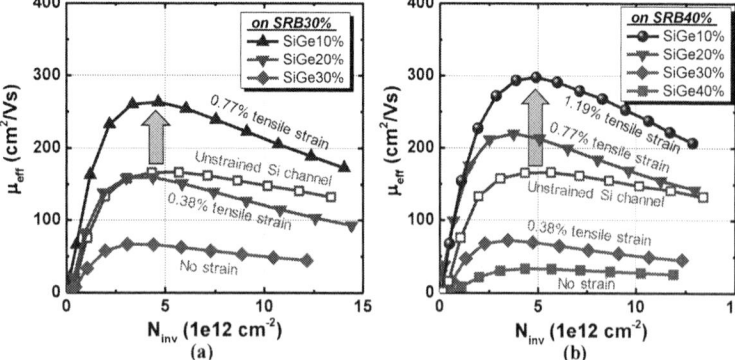

Fig. 6. Electron mobility as a function of inversion carrier density (N_{inv}) in **(a)** SiGe nFinFETs on SRB30% and **(b)** SiGe nFinFET on SRB40%.

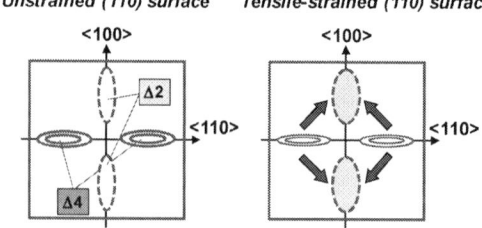

Fig. 7. Equi-energy valleys for a two-fold valley (Δ2), lighter conductive mass, and a four-fold valley (Δ4), heavier conductive mass, in SiGe (110) inversion layer with/without uniaxial tensile strain (corresponding to the sidewall surfaces of (110)/<110> SiGe FinFETs).

978-1-7281-1988-5/18 $31.00 © 2018 IEEE

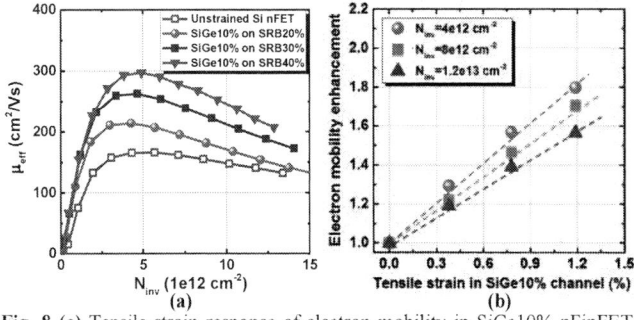
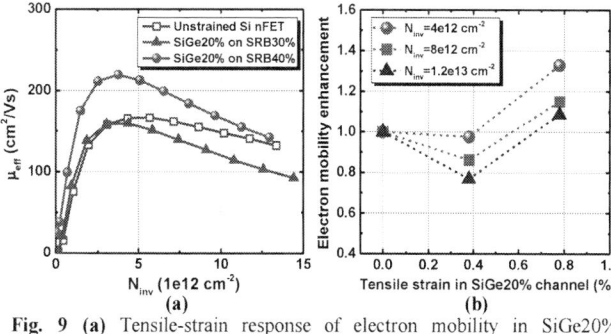

Fig. 8 (a) Tensile-strain response of electron mobility in SiGe10% nFinFET. **(b)** Electron mobility enhancement as a function of tensile strain in SiGe10% channel. Low-field electron mobility shows higher sensitivity to tensile strain than high-field electron mobility because quantum confinement compensates for the tensile strain-induced subband splitting in the conduction band.

Fig. 9 (a) Tensile-strain response of electron mobility in SiGe20% nFinFET. **(b)** Electron mobility enhancement as a function of tensile strain in SiGe20% channel. Tensile strain benefit on electron mobility in higher Ge percentage in the SiGe channel becomes less efficient.

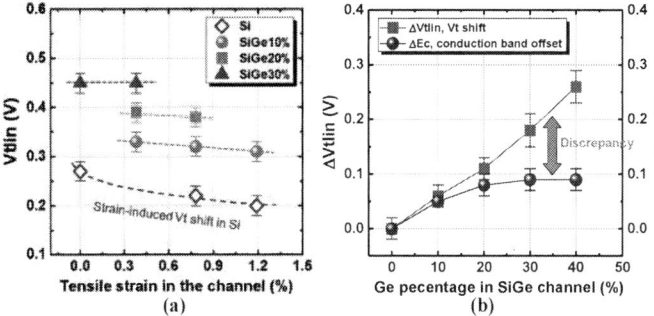
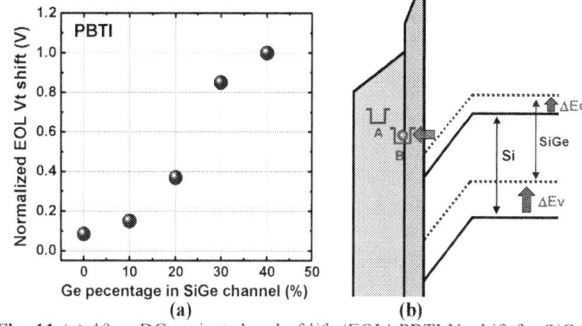

Fig. 10 (a) Vtlin in SiGe nFinFETs as function of tensile strain in the SiGe channel. **(b)** ΔVtlin and ΔEc [9] as a function of Ge percentage in the SiGe channel. A wide discrepancy between ΔEc and ΔVtlin above SiGe20% suggests that gate stack related extrinsic factor causes the additional Vt shift.

Fig. 11 (a) 10 yr DC projected end-of-life (EOL) PBTI Vt shift for SiGe nFinFETs. **(b)** Band diagram of higher Ge percentage (Ge >20%) SiGe nFinFETs under the device operation. The additional electron trapping site in the IL/HK, B, whose energy level is lower than initial PBTI site, A, is generated, resulting in a significant degradation of PBTI.

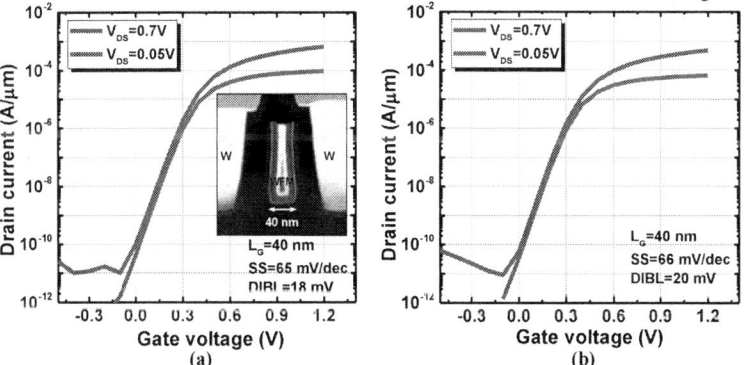
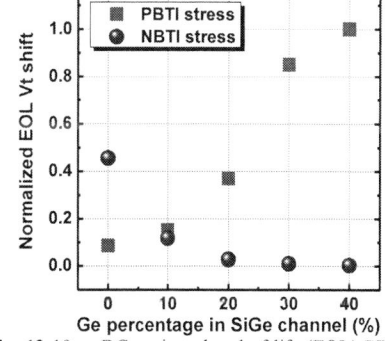

Fig. 12 (a) Typical I_D-V_G transfer curves of short-channel (L_G ~40 nm) tensile-strained SiGe10% nFinFET built on SRB30%, where the inset shows a dark-field TEM device image. **(b)** I_D-V_G transfer curves of short-channel (L_G ~40 nm) tensile-strained SiGe20% built on SRB40%.

Fig. 13 10 yr DC projected end-of-life (EOL) PBTI and NBTI Vt shift as a function Ge percentage in the SiGe channel. They show the opposite trend with respect to Ge percentage in the SiGe channel.

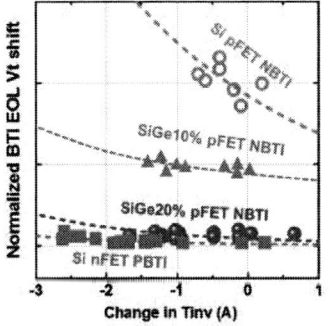
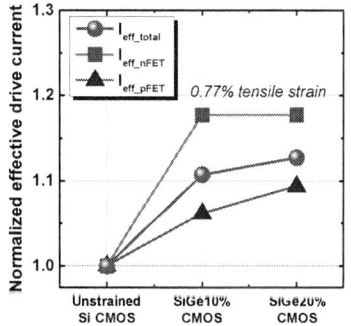

Fig. 14 Normalized NBTI/PBTI Vt shift as a function Tinv. SiGe10% and SiGe20% channel provide ~2.5 times and ~9 times improvement of NBTI at a given Tinv, respectively.

Fig. 15 Effective drive current (I_{eff}) of tensile-strained SiGe10% CMOS and SiGe20% CMOS for benchmarking the dynamic performance. 0.77% tensile-strained SiGe channel is assumed in the TCAD simulation.

Table II. Key parameters for TCAD simulation

	unstrained Si	t-strained SiGe10%	t-strained SiGe20%
Fin thickness	6 nm	6 nm	6nm
Fin height	45 nm	45 nm	45 nm
Channel length	15 nm	15 nm	15 nm
SRB	-	SRB30%	SRB40%
CPP	44 nm	44 nm	44 nm
T_{inv}	nFET: 12.5A pFET: 13A	nFET: 12 A pFET: 12.5 A	nFET: 11.5A pFET: 12A
D_{it}	Data in Fig. 5	Data in Fig. 5	Data in Fig. 5
Mobility	Data in Fig. 6	Data in Fig. 6	Data in Fig. 6

978-1-7281-1988-5/18 $31.00 © 2018 IEEE

Advanced Arsenic Doped Epitaxial Growth for Source Drain Extension Formation in Scaled FinFET Devices

S. Mochizuki[1], B. Colombeau[2], L. Yu[1], A. Dube[2], S. Choi[1], M. Stolfi [2], Z. Bi[1], F. Chang[2], R. A. Conti[1], P. Liu[2], K. R. Winstel[1], H. Jagannathan[1], H.-J. Gossmann[2], N. Loubet[1], D. F. Canaperi[1], D. Guo[1], S. Sharma [2], S. Chu[2], J. Boland[2], Q. Jin[2], Z. Li[2], S. Lin[2], M. Cogorno[2], M. Chudzik[2], S. Natarajan[2], D. C. McHerron[1] and B. Haran[1]

[1]IBM Research, 257 Fuller Road, Suite 3100, Albany, NY 12203, email: smochiz@us.ibm.com
[2]Applied Materials, 974 E Arques Avenue, Sunnyvale, CA 94085 USA, email: benjamin_colombeau@amat.com

Abstract— In this paper, we demonstrate a novel Source Drain Extension (SDE) approach to enable NMOS device scaling along with improved performance. For the first time, SDE formation with epitaxially grown As doped Si (Si:As) has been examined and compared to the current state-of-the-art SDE formation in FinFET at 10nm logic ground rules. It is found that a Si:As layer based SDE provides a clear improvement in the short channel effect and a significant device performance increase. It is also shown that a careful co-optimization of the Si:As layer and Source / Drain (S/D) lateral recess is required to achieve the optimum device gain. This paves the way for the ultimate nSDE formation for current and next generation CMOS devices.

I. INTRODUCTION

Formation and optimization of the Source Drain Extension (SDE) is one of the key factors for aggressively scaled devices. It is necessary to minimize the sheet resistance of the SDE (junction resistance) while balancing the short channel effect (SCE) with an abrupt junction. Having a low resistive doped material with low diffusivity underneath the spacer is the most effective way to achieve high performance devices.

SDE formation by an implantation technique and Plasma Doping have been reported [1-4]. However, extension doping for a 3D device structure, such as FinFET, is challenging due to the fin geometry. Therefore, SDE formed with selective doped epitaxial growth has been widely used for scaled FinFET devices due to its better sidewall coverage and higher active dopant concentration while eliminating crystalline damage introduction [5-6]. SDE is formed with dopant drive-in diffusion from the epitaxially grown S/D region. Phosphorus doped Si (Si:P) has been widely used as the epitaxial S/D material for nFET. However, degradation of SCE is a concern due to diffusion of phosphorus in Si during downstream processes. It is well known that carbon co-doping in Si:P (Si:CP) suppresses phosphorus diffusion, but the resistance of the film increases due to the presence of carbon.

In this work, SDE formation with epitaxially grown As doped Si (Si:As) has been examined. The SDE formed with Si:As (at different As concentration, thickness) has been compared with SDE formed with Si:P. In addition, the electrical nFinFET device characteristic validation has been performed with different S/D recess shape (lateral etch amount, depth) demonstrating that the shape highly impacts both junction abruptness and resistance.

II. EXPERIMENTAL

Fig. 1 shows the typical CMOS FinFET integration process and the SDE formation strategies evaluated in this study. Two major components of the flow are examined: 1) SDE formation with Si:As followed by Si:P S/D formation by in-situ doped selective epitaxial growth on Applied Materials RP Epi ™, 2) S/D recess shape modification. The S/D recess modification has been performed by S/D Fin recess with a RIE process followed by the Applied Materials Selectra ™ selective isotropic dry lateral etch process which etches the S/D region selective to the spacer material. All the short channel nFinFETs in this study have a gate length (L$_g$) of 22 nm.

Fig. 2 (a) and (b) show the hall carrier concentration and film resistivity of the P and As activated layers in Si:P and Si:As films as a function of additional milliseconds laser anneal temperature. With an increase in laser anneal temperature, the hall carrier concentration is increased. Active P and As concentration $> 6 \times 10^{20}$ atoms/cm^3 and film resistivity of Si:P and Si:As < 0.4 m$\Omega \cdot$cm are achieved by laser anneal (1250 °C). The depth profiles for both P and As in the Si:P and Si:As layers after spike rapid thermal annealing (sRTA) at 950 °C are shown in Fig. 3. While a marked P diffusion is observed, As diffusion is minimal due to its low diffusivity in Si. According to HRTEM analysis of epitaxial Si:As/Si:P layer with As and P doping concentration of 1.4×10^{21} atoms/cm^3 shown in Fig. 4, no visible defects in the Si:As layer and at the Si:As/Si substrate interface are observed. The uniform As distribution in the Si:As layer is confirmed by EELS map analysis. Fig. 5 shows cross section TEM images of the S/D region along the Fin post S/D Fin recess and lateral recess processes. The selective isotropic radical-based atomic-level etch process on Selectra™ resulted in a controllable recess of the active fin underneath the spacer. Fig. 6 shows the structural demonstration of a combination of the isotropic lateral recess and Si:As/Si:P selective epitaxial growth on a scaled FinFET device structure. Conformal Si:As layer growth at the SDE region underneath the spacer and Si:P layer growth at the S/D region is observed.

III. RESULTS AND DISCUSSION

A. SDE material study:

To better understand the role of SDE formed with Si:As in short channel device performance, electrical readouts in nFinFETs with 5 nm S/D lateral recess are compared between Si:P SDE and Si:As SDE. The Ron – DIBL characteristics are

978-1-7281-1988-5/18 $31.00 © 2018 IEEE

shown in Fig. 7. The Si:As SDE process delivers around 10% Ron reduction and DIBL benefit of 14 mV due to less dopant diffusion into the channel region. The Ioff – Ieff characteristics in Fig. 8 also confirm the large performance gains. Si:As SDE process gives 13% Ieff benefit at the same Ioff. The SS – DIBL characteristics shown in Fig. 9 indicate that the Si:As SDE process, by reduction of dopant diffusion, gives a significant channel quality improvement as seen in the smaller SS degradation by suppression of dopant diffusion.

B. S/D lateral recess effect:

The two conventional methods for optimization of the junction are: 1) thermal process optimization and 2) spacer thickness optimization. Thermal process optimization, e.g., sRTA temperature and ramp up/down rates, controls dopant diffusion from the S/D region into the channel region. However, changes in the Ron – DIBL trade-off with respect to the thermal process can be expected due to DIBL degradation caused by modulation of the junction abruptness at a higher thermal budget. Spacer thickness optimization controls the distance between the S/D and the channel region. However, increases in parasitic capacitance, such as gate to S/D and gate to contact capacitance, can be expected as the spacer becomes thinner. We adopted and further optimized the S/D lateral recess process in order to avoid the problems mentioned above.

The Ron vs DIBL of nFinFETs with various S/D lateral recess amounts with Si:As SDE are given in Fig. 10. A Ron – DIBL trade-off is observed when varying the S/D lateral recess amount. DIBL degrades with more lateral recess amount for Si:As SDE where 5% Ron degradation is observed with 8 nm lateral recess compared to 2 nm lateral recess due to dopant diffusion into the channel region. The Ioff – Ieff characteristics of the same set of nFinFETs in Fig. 11 also confirm the performance modulation with respect to the lateral recess amount. The Si:As SDE process gives 6% Ieff benefit compared to Si:P SDE in the case of no lateral recess and with the lateral recess process there is a further improvement of 8% (2 nm lateral recess) at the same Ioff. The performance improvement with respect to the lateral recess amount is shown in Fig. 12. The data indicates that a lateral recess amount of 2 nm is optimal for this specific device structure (e.g. spacer thickness, thermal budget in the downstream processes) where a ~15% Ieff is obtained over the baseline process (Si:P SDE with no lateral recess).

Short channel device characteristics improvements by changing SDE from Si:P to Si:As are summarized in Fig. 13. The improvements from Si:As SDE are more pronounced in the case of 5 nm lateral recess which has closer proximity structure compared to 0 nm lateral recess. These improvements prove the benefit of Si:As SDE in scaled/short channel device structures.

C. As dopant concentration study:

Two different As concentrations (7.0×10^{20} vs 1.2×10^{21} atoms/cm^3) in the SDE were compared. An Ron reduction of 7% without degrading DIBL with higher As doping concentration is observed as shown in Fig. 14. Fig. 15 compares DC performance where a 3% performance enhancement is

observed with higher As doping because of the higher active dopant at SDE (lower resistance underneath spacer) while keeping junction abruptness.

D. Si:As buffer layer thickness effect:

To investigate the thickness dependence of the Si:As layer on electrical properties, two different Si:As layer thicknesses (10 nm vs 20 nm) were compared. A thicker Si:As layer shows increased Ron by 8% with improved DIBL by 25 mV as shown in Fig. 16, indicating a Ron – DIBL trade-off. This result suggests that there is diffusion of P from the S/D to the channel region through the Si:As layer. Furthermore, the resistivity of the Si:As layer is slightly higher than that of Si:P layer (Fig. 2) and can be considered as the cause of the increase in Ron. The Ioff – Ieff characteristics for both Si:As layer thickness conditions are on the same trend as shown in Fig. 17, suggesting that the device performances are equivalent.

E. S/D recess depth effect:

In addition to S/D lateral recess, the S/D recess depth also affects the junction profile. Fig. 18 and 19 show the impact on Ron – DIBL and Ioff – Ieff characteristics when modulating the in-coming S/D recess depth. Deeper recess shows slightly lower Ron without degrading DIBL, and slightly better device performance. This performance benefit is due to the improved junction profile. In addition, it indicates that DIBL degradation due to the increase in recess depth is suppressed by the low diffusivity of As in the Si:As layer grown on the bottom of the recess.

F. Improved Vt roll-off behaviour:

Fig. 20 shows the ΔVt comparison between various SDE structures. ΔVt is defined as the Vt difference between the long channel device and the short channel device. Improved Vt roll-off characteristics are clearly observed for SDE formed with Si:As layers.

IV. CONCLUSIONS

We have investigated the influence of different approaches to form SDE on SCE and device performance. Junction optimization with SDE formed by Si:As layers successfully reduced Ron and improved DIBL compared to SDE formed by Si:P layers. In addition, we demonstrated that the S/D recess shape (lateral etch amount, depth), As concentration, and Si:As layer thickness play a critical role in modulating the SCE and device performance. The demonstrated elements and integration are essential and viable for scaled FinFET and Nanosheet technologies.

ACKNOWLEDGMENT

This work was performed by the joint research and development project teams at IBM/Applied Materials research and development facilities.

REFERENCES

[1] M. Togo et al., VLSI Tech. Symp. Dig., T196, 2013. [2] Y. Sasaki et al., IEDM, p542 (2013). [3] G. Zschätzsch et al., IEDM Tech. Dig., p841, 2011. [4] Y.Sasaki et al., VLSI Tech. Symp. Dig., T30, 2015. [5] T. Chiarella et al., ESSDERC, p131, 2016. [6] G. Tsutsui et al., IEDM Tech. Dig., p456, 2016.

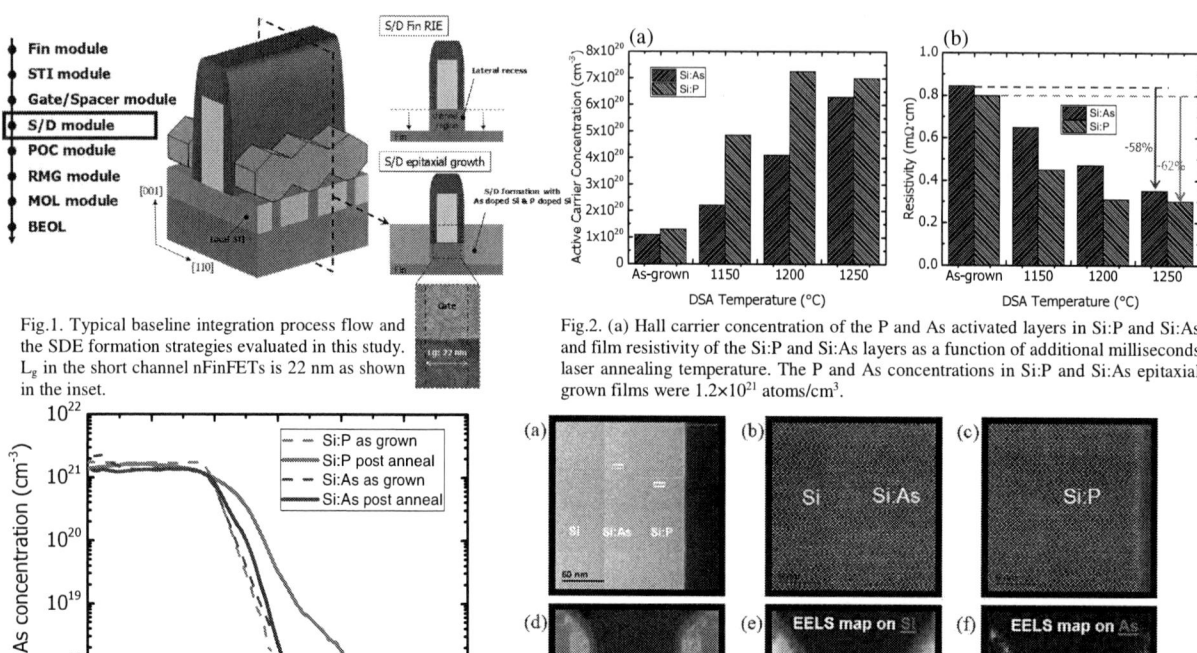

Fig.1. Typical baseline integration process flow and the SDE formation strategies evaluated in this study. L_g in the short channel nFinFETs is 22 nm as shown in the inset.

Fig.2. (a) Hall carrier concentration of the P and As activated layers in Si:P and Si:As and film resistivity of the Si:P and Si:As layers as a function of additional milliseconds laser annealing temperature. The P and As concentrations in Si:P and Si:As epitaxial grown films were 1.2×10^{21} atoms/cm³.

Fig.3. Depth profiles for both P and As in the Si:P and Si:As layers before and after sRTA at 950 °C.

Fig.4. HRTEM analysis of epitaxial Si:As layers on Si substrate. (a) – (c) Cross-sectional HRTEM images, and (d) – (f) EELS maps. There are no visible defects in the Si:As layers and Si:As/Si substrate interface. High crystallinity Si:As film with uniform As distribution is observed.

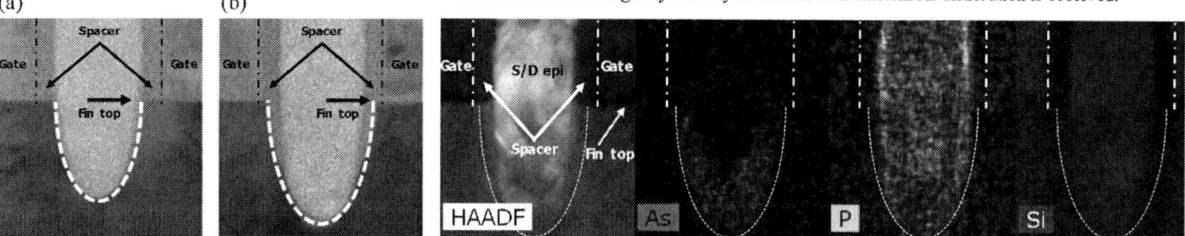

Fig.5. Cross section TEM images along the Fin showing S/D region. (a) post S/D Fin recess, (b) post conformal lateral etch.

Fig.6. Cross-sectional TEM and EDX elemental mapping of the S/D structure formed by a combination of lateral recess and Si:As/Si:P selective epitaxy on device structure.

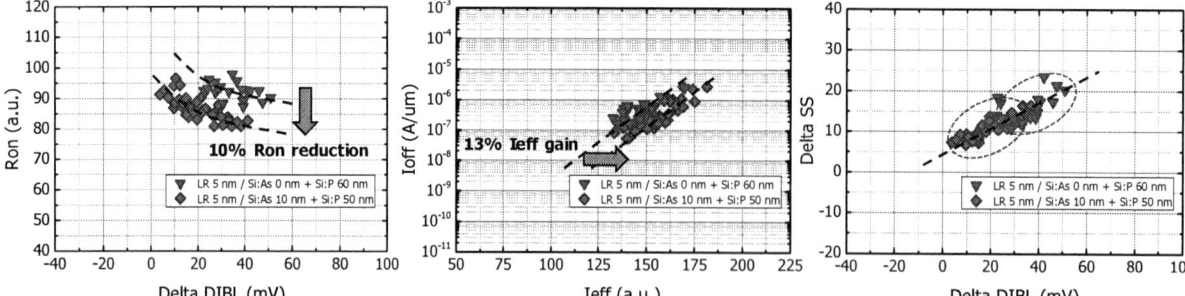

Fig.7. Ron – DIBL characteristics of short channel (L_g = 22 nm) nFinFETs with SDE formed with Si:P and Si:As. The S/D lateral recess (LR) amount is 5 nm.

Fig.8. Ioff – Ieff characteristics of the same sets of short channel nFinFETs in Fig. 7.

Fig.9.SS – DIBL characteristics of the same sets of short channel nFinFETs in Fig. 7.

978-1-7281-1988-5/18 $31.00 © 2018 IEEE

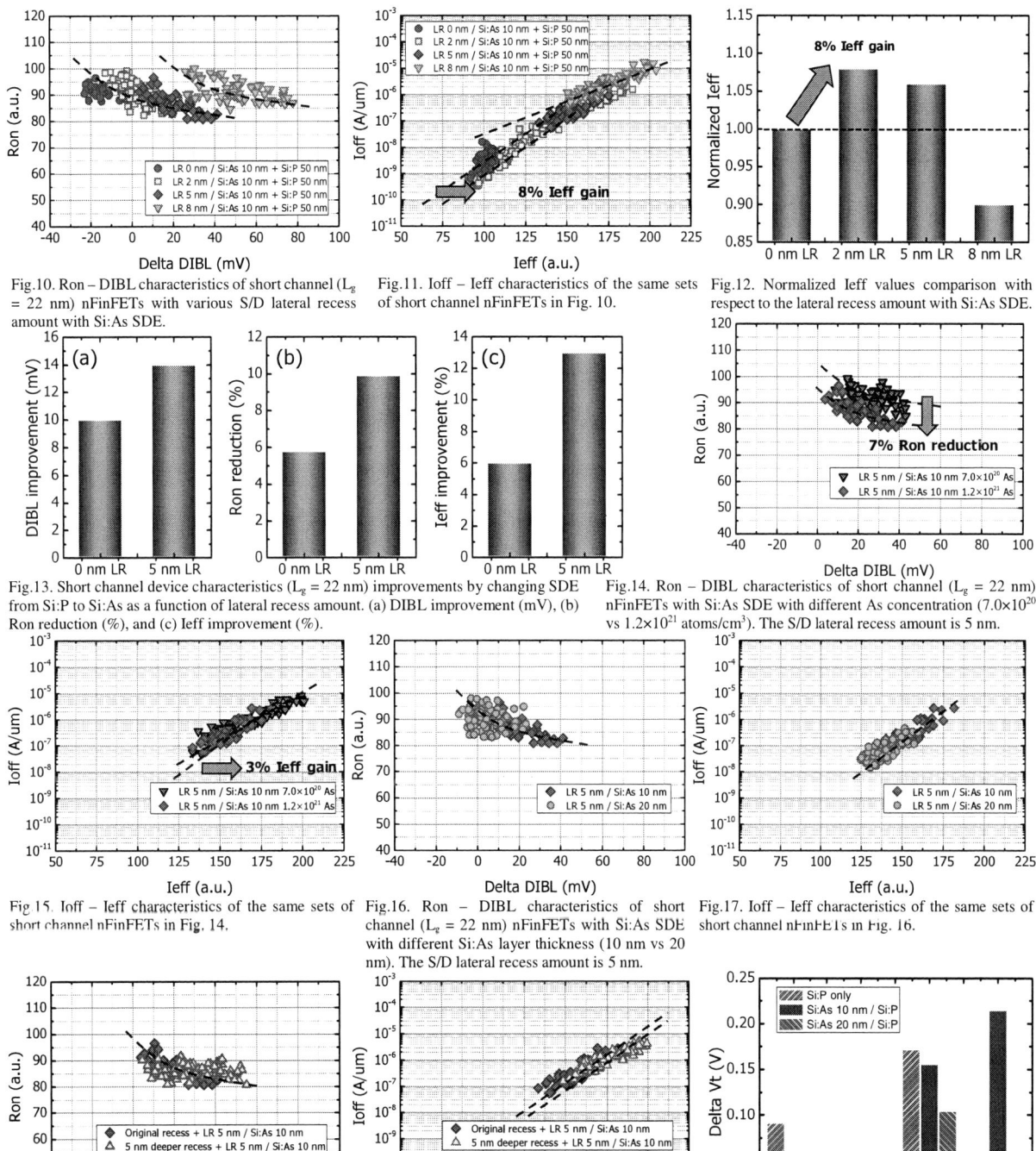

Fig.10. Ron – DIBL characteristics of short channel (L_g = 22 nm) nFinFETs with various S/D lateral recess amount with Si:As SDE.

Fig.11. Ioff – Ieff characteristics of the same sets of short channel nFinFETs in Fig. 10.

Fig.12. Normalized Ieff values comparison with respect to the lateral recess amount with Si:As SDE.

Fig.13. Short channel device characteristics (L_g = 22 nm) improvements by changing SDE from Si:P to Si:As as a function of lateral recess amount. (a) DIBL improvement (mV), (b) Ron reduction (%), and (c) Ieff improvement (%).

Fig.14. Ron – DIBL characteristics of short channel (L_g = 22 nm) nFinFETs with Si:As SDE with different As concentration (7.0×10^{20} vs 1.2×10^{21} atoms/cm³). The S/D lateral recess amount is 5 nm.

Fig.15. Ioff – Ieff characteristics of the same sets of short channel nFinFETs in Fig. 14.

Fig.16. Ron – DIBL characteristics of short channel (L_g = 22 nm) nFinFETs with Si:As SDE with different Si:As layer thickness (10 nm vs 20 nm). The S/D lateral recess amount is 5 nm.

Fig.17. Ioff – Ieff characteristics of the same sets of short channel nFinFETs in Fig. 16.

Fig.18. Ron – DIBL characteristics of short channel (L_g = 22 nm) nFinFETs with Si:As SDE with different S/D recess depth (original vs 5 nm deeper recess). The S/D lateral recess amount is 5 nm.

Fig.19. Ioff – Ieff characteristics of the same sets of short channel nFinFETs in Fig. 18.

Fig.20. ΔVt comparison between various SDE structures.

978-1-7281-1988-5/18 $31.00 © 2018 IEEE

External Resistance Reduction by Nanosecond Laser Anneal in Si/SiGe CMOS Technology

[1]Oleg Gluschenkov, [1]Heng Wu, [1]Kevin Brew, [2]Chengyu Niu, [1]Lan Yu, [1]Yasir Sulehria, [1]Samuel Choi, [2]Curtis Durfee, [1]James Demarest, [1]Adra Carr, [3]Shaoyin Chen, [3]Jim Willis, [3]Thirumal Thanigaivelan, [1]Fee-li Lie, [2]Walter Kleemeier, and [1]Dechao Guo

[1]IBM Research, 257 Fuller Road, Albany, NY 12203, USA, email: olegg@us.ibm.com
[2]GLOBALFOUNDRIES Inc., Albany, NY, USA, [3]ULTRATECH, a division of VEECO INSTRUMENTS Inc., San Jose, CA, USA

Abstract—We report on a significant pFET external resistance reduction (~40%) and corresponding 10% R_{ON} decrease by nanosecond laser annealing of S/D structures applicable to advanced technology nodes. Selective melting of pFET S/D elements is responsible for this improvement. Process window boundaries are defined by channel and junction melting at the upper end and by S/D SiGe melting at the lower end. Short channel characteristics are not degraded within the identified process window. Contacted gate pitch (CPP) and fin number dependence of the process window is assessed.

I. Introduction

MOSFET parasitic resistance degrades rapidly as transistor dimensions approach the scaling limit. Shrinking volume of conductive elements, reduced interfacial area between these elements, and increased effective channel width (W_{eff}) in 3D transistor architectures lead to a significant increase in the external resistance (R_{EXT}). Fig. 1 illustrates components of FinFET external parasitic resistances in cross gate and cross fin directions. The R_{EXT} can be partitioned into S/D epi resistance (R_{EPI}), contact resistance (R_C), and MOL metal stud resistance (R_{METAL}). Much attention has been recently given to R_C and R_{METAL} exploring various approaches such as dual silicide for R_C reduction [1, 2] and cobalt contact stud for R_{METAL} reduction [2, 3]. While these are vital, R_{EPI} and R_C reduction is equally critical.

Laser annealing at the contact level has also been explored for R_C improvement [4-10] via forming interfacial dopant-semiconductor supersaturated metastable alloys through solid or liquid phase epitaxial (SPE/LPE) re-growth. Such laser-induced SPE/LPE processes yielded a record low contact resistivity at or below 1×10^{-9} $\Omega \cdot cm^2$ for both n-type and p-type contacts. Nanosecond-scale (nSec) laser melt annealing is particular promising technique for achieving such low resistivity in metal-semiconductor contacts but its implementation is hindered by a small process window and various layout dependences [10]. Hierarchy of materials melting points and other catastrophic failures define the process window. Low melting point materials such as SiGe may allow for an improved process window [11] and exploring nSec laser melt annealing for pFETs with high/mid percent Ge in SiGe S/D structures is strategically important in the context of practical implementation for advanced CMOS technologies.

In this work, we systematically examine the effect of nSec laser melt annealing on external parasitic resistance in advanced pFinFETs, determine the hierarchy of melting thresholds and their effect onto R_{EXT} improvement and CMOS process window, and reveal key process window dependencies on CPP and number of fins. pFET R_{EXT} is significantly reduced (~40%) by selective S/D melting at the contact level positively affecting both R_{EPI} and R_C.

II. 1D Epitaxial Films and TLM Structures

Blanket one dimensional epitaxial films were used to elucidate basic electrical response of nSec laser melt annealing and dopant activation and redistribution in the molten phase. 40-80nm of mid/high percent Ge SiGe epitaxial films were grown on n-Si substrate and implanted with a p-type dopant. Top ~7-10nm were amorphized by the implantation process. The implanted film was laser annealed at different incident energy densities (ED) corresponding to different surface peak temperatures. The sheet resistance response of the annealed film is shown in Fig. 2. Three distinct regions correspond to re-crystallization of amorphous SiGe with a drop in R_S, no change in R_S upon raising surface temperature above amorphous SiGe re-crystallization threshold, further reduction in R_S upon melting crystalline SiGe underlayer with subsequent dopant redistribution and activation. This response is typical for this material system with a thin amorphized layer and qualitatively does not change with Ge content in SiGe and the type of p-type dopant as long as the dopant has a high solubility limit in SiGe and is present in abundance well above its solubility limit. In addition to R_S, the semiconductor-metal contact resistivity ρ_C has also been assessed for this basic material system using TLM structures. The SiGe epitaxial film is covered with a silicon oxide isolation (ILD) layer and contact trenches are etched in the ILD layer at different distances to each other forming the basis for the TLM measurement. The base epitaxial film is implanted through the contact trenches creating doped amorphous pockets as schematically shown in Fig. 3. The implanted film was laser annealed at different energy densities corresponding to different SiGe film temperatures. Fig. 4 shows evolution of amorphous pocket as it is subjected to progressively higher laser energy density. Fig. 4a shows initial amorphous pocket after implantation. Fig. 4b shows a partial re-growth at a low laser energy density. Fig. 4c shows a full re-growth at an intermediate laser energy density. Fig. 6 is a dark field STEM image of a formed contact corresponding to the laser energy density employed in Fig. 4c. Re-distribution of Ge can be seen in the re-crystallized pocket. Fig. 7 provides an elemental line scan of Ge distribution in re-crystallized a-SiGe pocket. Observed Ge segregation at the surface is an earmark of a-SiGe melting and LPE and is beneficial for reducing contact resistance. A further increase in the laser energy density leads to c-SiGe film melt. Fig. 5 shows the extracted contact resistivity including the resistance of contact metal stud. At a low energy density corresponding to partial re-growth, the ρ_C is high. At an intermediate energy corresponding to full re-growth and Ge segregation, the ρ_C is low. At a high energy corresponding to c-SiGe melting, the ρ_C increases again due to the contact dopant reduction in dopant redistribution process.

978-1-7281-1988-5/18 $31.00 © 2018 IEEE

III. Fin Laser Melt Threshold

FinFETs structures have drastically different laser energy density melting thresholds than those of blanket films and basic TLM structures. This is due to their different optical reflectance and thermal conductance. Accordingly, the melting thresholds need to be re-established. In our case, pFETs have 4 basic materials with progressively higher melting points as shown in Fig. 8. Amorphous SiGe contact pocket (A) has the lowest melting point followed by that of crystalline SiGe source/drain (B), then by that of channel SiGe (C), and the highest melting point occurs in Si subfin region (D). Gross melting of regions (C) and (D) has been found by STEM imaging after laser exposure. Fig. 9 shows bright and dark field STEM images for determining laser energy density for such gross melting. Figs. 9-1a/1b show an incoming fin structure with SiGe active fin and Si subfin region. Figs. 9-2a/2b show an onset of channel SiGe melting showing channel defects in the bright field view and Ge striations in the dark field view. Figs. 9-3a/3b show a complete channel melt with numerous channel defects and an onset of Si subfin melt with a Ge re-distribution from the channel into the subfin. Figs. 9-4a/4b show a complete melt of Si subfin with Ge re-distribution deep into the subfin region and numerous defects in it. The laser energy density and substrate pre-heat temperature corresponding to Figs. 9-2a/2b were taken as an upper limit for electrical hardware.

IV. pFinFET Electrical Response

After contact implantation, pFETs were exposed to different nSec laser energy density with the interval corresponding to ~50-60°C step in annealing temperature up to the channel melt condition at the higher end. Figs. 10 and 11 show pFET R_{ON} and R_{EXT} response, respectively. The reference cell is the SPE pocket re-growth induced by a millisecond-scale laser annealing. There are two distinct regions corresponding to a smaller improvement at low nSec laser energy densities and a large improvement at higher energy densities. The cell corresponding to the highest energy density is not shown due to an electrical short.

Figs. 12 and 13 show corresponding changes in short channel characteristics: ΔDIBL and ΔSsat, respectively. Only the high energy cell exhibits a change in these parameters suggesting that the dopants started to penetrate into the channel at this energy density. Fig. 14 shows corresponding pFET I_{DSAT}-V_G curves. The highest energy density cell results in the source-to-drain electrical short suggesting that the S/D dopants penetrated deep into the channel and shorted it. The cell with the degraded DIBL and Ssat slope is clearly seen in this chart and points at the onset of dopant penetration into the channel. The remaining cells show varying degree of improvements.

Fig. 15 shows I_{DSAT}-V_G curves for nFETs with Si channel and source/drain. No changes in nFETs behavior is seen even at the highest energy density consistent with the absence of Si fin melting. Absence of Si fin melting is due to the energy density limit set in section IV and the difference between melting points of SiGe and Si.

V. Melting Threshold Hierarchy

Fig. 16 pictorially summarizes the observed melting hierarchy and its impact in pFinFETs. Fig. 16-1 is the incoming pFETs structure. Fig. 16-2 shows melt and re-crystallization of amorphized pockets with re-distribution of dopants and Ge in the pockets. It is this process that is believed to correspond to a smaller improvement of R_{ON} and R_{EXT} at low nSec laser energy densities. Fig. 16-3 shows melt and re-crystallization of crystalline SiGe source/drain structures with dopant re-distribution and activation. It is this process that is believed to correspond to a large improvement of R_{ON} and R_{EXT} at intermediate energy densities. Fig. 16-4 shows melt and re-crystallization of junctions adjacent to the SiGe channel with the dopants moving into the channel near gate edges. The SiGe junction may have a higher Ge content than that of the channel but less than that of source/drain. It is this process that is believed to correspond to degrading short channel characteristics. Fig. 16-5 shows melt and re-crystallization of the SiGe channel with the dopants moving deep into the channel. It is this process that is believed to correspond to shorting pFET channel. Fig. 16-6 shows melt and re-crystallization of the Si subfin and channel with the dopants and channel Ge moving into the subfin region.

VI. Layout Dependencies

The onset of degrading short channel characteristics defines the upper end of laser energy density. Accordingly, ΔDIBL can be used to assess a catastrophic layout effect that may induce a severe degradation of short channel effect in some common layouts. Figs. 17 and 18 show dependencies of ΔDIBL on fin number and CPP, respectively. No CPP dependence has been observed within the studied range. Transistor with reduced number of fins are more susceptible to the onset of DIBL degrade suggesting that they are heated to a higher peak temperature, as shown in Fig. 17. This is likely caused by a low thermal conductance in a FinFET with reduced number of fins. Reducing laser energy density by one interval eliminates this DIBL degradation pointing to less than ~50°C difference in annealing temperature between these transistors.

VII. Conclusion

In this work, we systematically examine the effect of nSec laser melt annealing on external parasitic resistance and short channel characteristics in advanced pFinFETs. R_{EXT} is significantly reduced (~40%) by selective S/D melting at the contact level positively affecting both R_{EPI} and R_C. Process window boundaries are defined by channel and junction melting at the upper end and by S/D SiGe melting at the lower end. Short channel characteristics are not degraded within this process window. Contacted gate pitch (CPP) does not affect the process window within the studied range and the fin number may induce a shift in the process window by roughly ~50°C.

ACHNOWLEGEMENT

This work was performed by the Alliance Teams at various IBM Research and Development Facilities.

REFERENCES

[1] P. Adusumilli et al., *Proc. VLSI Technol. Symp.*, p.1, 2016
[2] C. Auth et al., *IEDM Tech. Dig.*, p.29.1.1, 2017
[3] V. Kamineni, *Proc. IEEE IITC/AMS*, p.105, 2016
[4] K. Goto et al., *IEDM Tech. Dig.*, p. 931, 1999
[5] H. Niimi et al., *IEEE Electron Device Lett. 37*, p.1371, 2016
[6] O. Gluschenkov, *IEDM Tech. Dig.*, p.17.2.1, 2016
[7] J-L. Everaert et al., *Proc. VLSI Technol. Symp.*, p.T214, 2017
[8] L. Date et al., *IEDM Tech. Dig.*, p.22.4.1, 2017
[9] H. Wu et al., *IEDM Tech. Dig.*, p.22.3.1, 2017
[10] Z. Liu et al., *Proc.VLSI Technol. Symp.*, p. T213, 2017
[11] O. Gluschenkov and H. Jagannathan, *ECS Transactions* 85(6), p. 11, 2018

978-1-7281-1988-5/18 $31.00 © 2018 IEEE

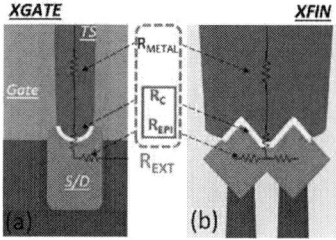 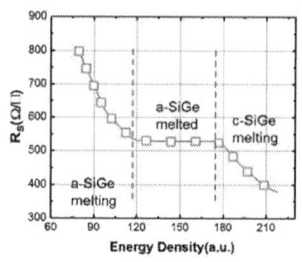 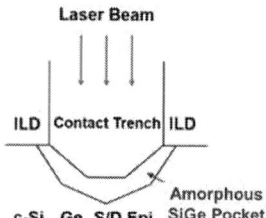

Fig. 1. External parasitic resistance (R_{EXT}) in FinFETs: (a) cross gate direction (b) cross fin direction. Focus of this work is on the epi (R_{EPI}) and contact (R_C) resistances.

Fig. 2. Sheet resistance of implanted and laser annealed SiGe epi layers. Increasing laser power density results in 3 distinct regions. Laser exposure duration is ~60 nsec.

Fig. 3. Schematic of TLM structure for contact resistance measurement. Implanted amorphous SiGe pocket is re-grown via laser-induced LPE.

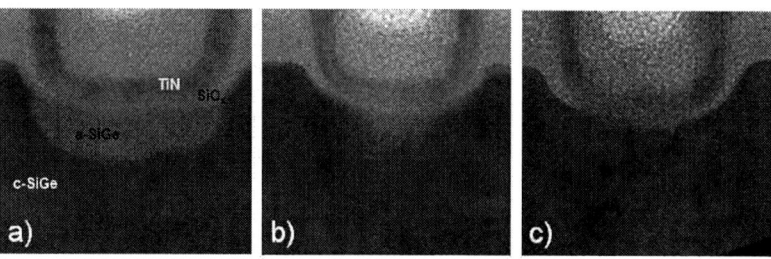 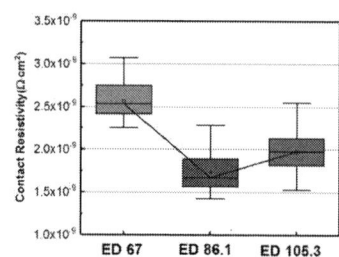

Fig. 4. TLM contact structures: a) after trench pocket amorphization; b) after laser exposure at an energy density (ED) of 67 - partially regrown; c) after ED = 86.1 exposure - fully regrown. ED refers to the incident energy density expressed in a.u. Laser exposure duration is ~60nsec. Substrate preheat did not result in a-SiGe recrystallization. TiN was used for STEM highlight without oxide removal.

Fig. 5. TLM contact resistance as the function of incident energy density at laser exposure duration of ~60nsec. Metal stud resistance is not subtracted.

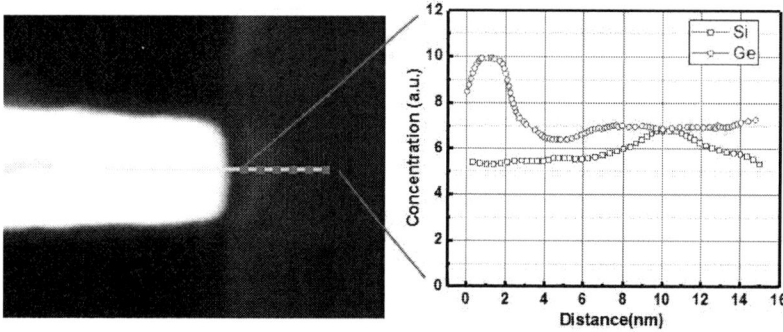 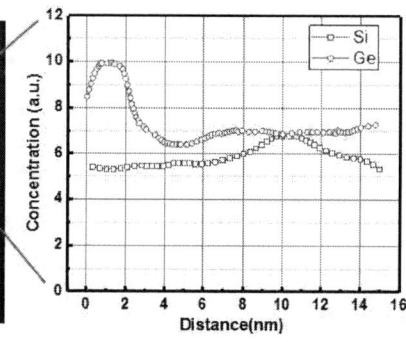 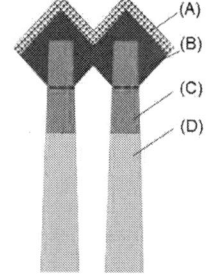

Fig. 6. Dark field STEM image of a formed contact using laser induced LPE at ED = 86.1.

Fig. 7. Elemental line scan across the re-crystallized a-SiGe pocket. Ge segregates near the surface.

Fig. 8. Schematic of pFinFET highlighting materials with different melting points: (A) amorphous SiGe pocket; (B) SiGe S/D epi; (C) SiGe active fin; (D) Si subfin.

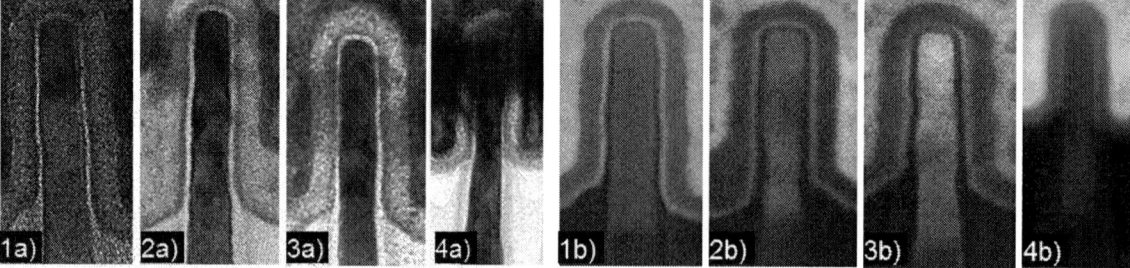

Fig. 9. Bright field ("a"-series) and dark field ("b"-series) STEM images of pFET fins: (1) prior to laser annealing; (2) after laser exposure at ED = 82.3 – signs of SiGe channel melting; (3) after exposure at ED = 105.3 – complete channel melting and signs of Si subfin melting; and (4) after exposure at ED = 95.7 and 100°C higher substrate base temperature – complete melting of fin and subfin regions and Si/SiGe mixing.

978-1-7281-1988-5/18 $31.00 © 2018 IEEE   817

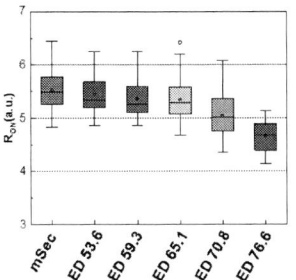

Fig. 10. pFET R_{ON} versus incident nSec laser energy density. Millisecond-scale (mSec) annealing serves as the reference.

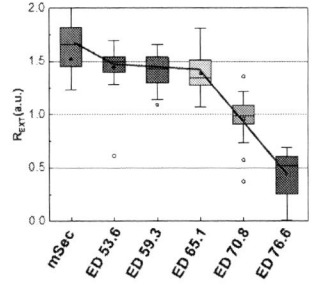

Fig. 11. pFET R_{EXT} versus incident nSec laser energy density. Millisecond-scale (mSec) annealing serves as the reference.

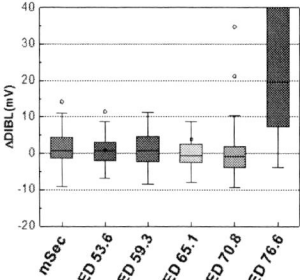

Fig. 12. pFET ΔDIBL versus incident nSec laser energy density. Millisecond-scale (mSec) annealing serves as the reference.

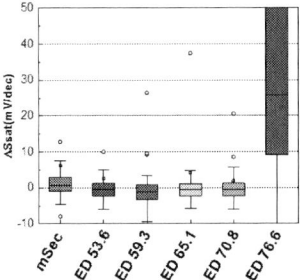

Fig. 13 pFET change in subthreshold slope (ΔSsat) versus incident nSec laser energy density. Millisecond-scale (mSec) annealing serves as the reference.

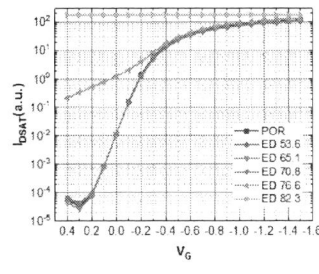

Fig. 14 pFET I_{DSAT}-V_G as the function of incident nSec laser energy density. Millisecond-scale (mSec) annealing serves as the "POR" cell.

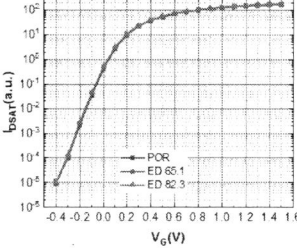

Fig. 15 nFET I_{DSAT}-V_G as the function of incident nSec laser energy density. Millisecond-scale (mSec) annealing serves as the "POR" cell.

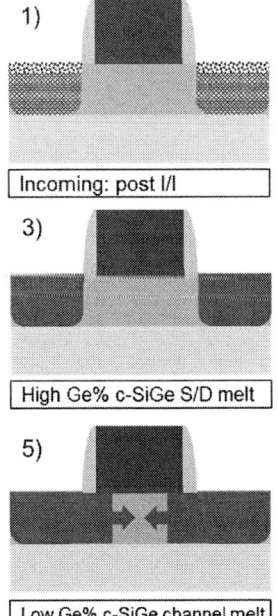

1) Incoming: post I/I

3) High Ge% c-SiGe S/D melt

5) Low Ge% c-SiGe channel melt

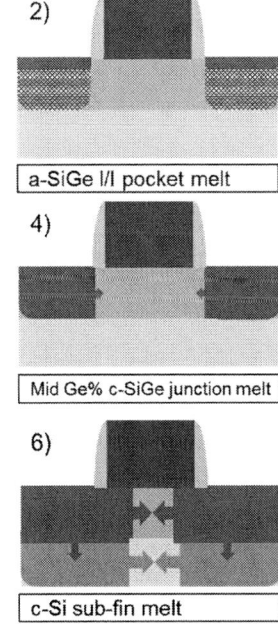

2) a-SiGe I/I pocket melt

4) Mid Ge% c-SiGe junction melt

6) c-Si sub-fin melt

Fig. 16. Schematic illustrating materials melting hierarchy: (1) incoming structure; (2) amorphous SiGe pocket melt; (3) crystalline high-percent-Ge SiGe source/drain melt; (4) crystalline mid-percent-Ge SiGe junction melt; (5) crystalline low-percent-Ge SiGe channel melt; (6) Si subfin melt.

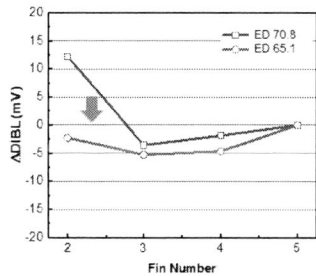

Fig. 17 pFET ΔDIBL dependence on the fin number. pFETs with reduced number of fins annealed at higher temperature. Lowering incident ED by one interval eliminates DIBL degrade.

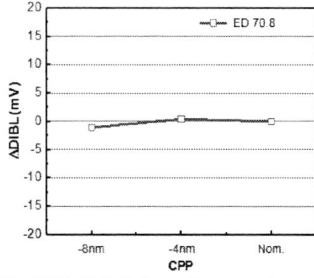

Fig. 18 pFET ΔDIBL dependence on the contacted gate pitch (CPP). No dependence is observed within the studied range suggesting that CPP dependence is weak.

Parasitic Resistance Reduction Strategies for Advanced CMOS FinFETs Beyond 7nm

H. Wu, O. Gluschenkov, G. Tsutsui, *C. Niu, K. Brew, *C. Durfee, *C. Prindle, *V. Kamineni, S. Mochizuki, C. Lavoie, *E. Nowak, Z. Liu,
J. Yang, S. Choi, J. Demarest, L. Yu, A. Carr, W. Wang, J. Strane, *S. Tsai, *Y. Liang, H. Amanapu, I. Saraf , *K. Ryan, F. Lie, *W. Kleemeier,
K. Choi, *N. Cave, T. Yamashita, *A. Knorr, D. Gupta, B. Haran, D. Guo, H. Bu, and M. Khare
IBM Semiconductor Technology Research, 257 Fuller Road, Albany, NY 12203, *GLOBALFOUNDRIES Inc.
Email: wuheng@us.ibm.com

I. Abstract

This work thoroughly investigates the external parasitic resistance in advanced FinFET technology. The optimization of the parasitic resistance is systematically examined in terms of 1) source/drain epi resistance, 2) contact resistance and 3) middle of line metal stud resistance. Various resistance reduction knobs have been experimentally explored in these three aspects and low contact resistivity of 1×10^{-9} and 7×10^{-10} $\Omega \cdot cm^2$ have been demonstrated on transistor level for NFET and PFET. By combining all the parasitic resistance reduction strategies, more than 70% and 60% reductions [1] in external parasitic resistance have been realized on NFET and PFET, respectively.

II. Introduction

With the transistor dimension approaching the physical limit of scalability, further boosting the device performance is very difficult. In the last decade, various approaches have been extensively explored in terms of channel property enhancement [1-3], junction engineering [4-5], external resistance reduction [7-14], parasitic capacitance minimization [15] and design process technology co-optimizations (DTCO) [16].

In particular, the external resistance (R_{EXT}) degrades rapidly not only because of the contact area shrinking linked to transistor scaling but also as a consequence of the 3D device structure where the contact widths are typically smaller than the channel widths. The R_{EXT} reduction is thus playing an increasing important role in advanced CMOS technology development.

The MOSFET's external resistance has been independently studied in terms of either source/drain (S/D) epitaxy growth [4, 8], contact/extension doping [7-9], contact silicide formation [10-11] or middle of line (MOL) metallization [13-14]. The optimization of each distinct module is very important but a unified co-optimization strategy for overall external resistance reduction is now critical and has rarely been reported.

In this work, we systematically examine the optimization of external parasitic resistance in advanced CMOS FinFETs. The contribution of each resistance component is first studied and it is found that the epi and contact resistances contribute most to the R_{EXT}. Each external resistance component is then studied and optimized. For the S/D epi, higher in-situ P doping for NFET and Si/Ge/B/C contents tuning for PFET effectively reduce the epi resistance by 60%/45%. For the contacts, solid phase epitaxy (SPE) process combined with high-P Si:P on NFET and high-Ge SiGe on PFET demonstrate very low contact resistivity (ρ_C) of 1×10^{-9} / 7×10^{-10} $\Omega \cdot cm^2$ on NFET / PFET. In addition, an enhanced contact cleaning is adapted and shows more effective area utilization without spacer erosion concerns. For the MOL metals, the interface between the TS and CA metals is crucial. By optimizing the etch/cleaning/metallization processes, more than 60% metal stud resistance reduction is obtained. Meanwhile, different metals of Co and W are also compared in terms of the TS/CA lateral line resistance where Co shows a 60% decrease in resistance. In the end, all the resistance reduction schemes have been integrated to deliver more than 70% and 60% external resistance reduction [1] for NFET and PFET.

III. External Parasitic Resistance Partition

For advanced CMOS FinFETs, the R_{EXT} is a considerable fraction of the transistor ON state resistance (R_{ON}) and becomes even higher at smaller dimensions. Fig. 1 shows the $I_{EFF}@I_{OFF}$ versus R_{EXT}/R_{ON} relationship for the FinFETs studied in this work. More external resistance dominance in R_{ON} shows worse performance, indicating that the R_{EXT} reduction is critical.

Figs. 2 a-b) illustrate the cartoons of a FinFET in cross gate and cross fin directions with each R_{EXT} element marked. The R_{EXT} can be electrically partitioned into S/D epi resistance (R_{EPI}), contact resistance (R_C) and MOL metal stud resistance (R_{METAL}). The R_{METAL} is composed of 1) TS vertical metal resistance (R_{TS}) 2) CA to TS metal interface resistance (R_{CA-TS}) and 3) CA vertical metal resistance (R_{CA}). Note that the R_{TS}, R_{CA-TS} and R_{CA} cannot be electrically separated in our layout design.

To fully understand the contribution of each element in R_{EXT}, Figs. 3 a-b) give the partition results of R_{EPI}, R_C and R_{METAL} in NFETs and PFETs studied, respectively. The epi and contact resistances represent more than 90% of the external resistance for both NFETs and PFETs with the non-optimized POR processes. The results with optimized processes to be discussed through this paper are also compared, showing great benefits. Fig. 3 c) gives the typical FinFET fabrication process flow. For external resistance reduction, the experiments have focused on the epi, RMG, contact and MOL modules.

Note that all the partitioned FET resistances are obtained on nominal devices reported in [1-2, 8-9] and are normalized into a consistent arbitrary unit, which is specifically noted as a.u.[device] through this paper. The contact resistivity in this work is extracted by 4-points kelvin measurements on nominal devices and calibrated with the contact area measured by TEM and the MOL metal resistance is subtracted.

IV. Epi Resistance Reduction

For epi resistance, the in-situ doping plays a critical role and R_{EPI} can be effectively reduced by introducing more dopants during the epitaxy. Fig. 4 shows the SIMS profile of P inside Si:P S/D of NFET for the POR epi [8] and optimized epi, confirming the 1.8 times higher in-situ doping in the optimized Si:P epi. This 80% increase in doping level reduces the device R_{EPI} by more than 60% as shown in Fig. 5.

In terms of the SiGe:B epi for PFET, R_{EPI} is not only controlled by in-situ doping but also affected by the Ge content in SiGe. Higher Ge% in SiGe has higher carrier mobility.

978-1-7281-1988-5/18 $31.00 © 2018 IEEE

However, due to the low solubility of B in Ge ($\sim 10^{19}$ cm^{-3} [8, 10-11]), higher Ge% also features a much lower active B doping. Thus, considering the tradeoff between carrier mobility and density, a co-optimization of Ge and B content is needed. Fig. 6 shows the epi sheet resistance with Ge%, indicating that optimal Ge content ranges from 40% to 70% for R_{EPI}. By combining the optimized SiGe:B epi with reduced C content, more than 45% R_{EPI} reduction is realized on PFETs (Fig. 7).

Beside the epi process, the downstream thermal budget must be considered, as it greatly affects the dopant activation [10], thus changes the R_{EPI}. Fig. 8 show the epi sheet resistance of Si:P epi with different downstream thermal processes. The low temperature and long-time furnace anneal deactivates the active dopants a lot and thus degrades the sheet resistance. By using the high temperature (>1200 °C) laser spike anneal (LSA) process, the epi sheet resistance is reduced by 35%.

V. Contact Resistance Reduction

As reported earlier [8-9], the SPE process can significantly lower the contact resistivity because of the formation of meta-stable alloys that are super saturated with dopants. Fig. 9 shows 60% NFET R_C reduction with the SPE process. For further ρ_C reduction, a high Si:P epi top layer process is developed where the in-situ doping reaches about 2.8 times of POR, which is confirmed by the P peak near the surface shown in Fig. 10. The electrical properties of the high Si:P epi top layer are also examined in Fig. 11, showing 10% lower epi sheet resistance with the same epi thickness as POR. Fig. 12 compares the ρ_C of NFETs with and without the high Si:P epi top layer. Thanks to the higher in-situ doping at the surface combined with SPE process, very low ρ_C of 1×10^{-9} $\Omega \cdot$cm^2 is obtained.

Reduction of ρ_C using the SPE process [9, 11] has been even more efficient on PFET where 80% ρ_C reduction is seen in Fig. 13. Due to the Fermi-level pinning for Ge near the valence band edge [10-11], higher Ge% can lower the Schottky barrier height of metal to SiGe, thus reducing ρ_C. However, lower B solubility in pure Ge can also degrade ρ_C. Fig. 14 shows the ρ_C relationship with Ge concentration in SiGe epi. Because of the trade-off between Schottky barrier height and active doping, ρ_C first decreases when Ge% increases to 70% and then degrades with Ge% approaching 100%. As a result, a high Ge SiGe ($\sim$75% Ge) epi top layer process is developed, which is confirmed by the Ge-enriched surface region in Fig. 15. With this 75% SiGe top layer process, a record low device level ρ_C of 7×10^{-10} $\Omega \cdot$cm^2 is obtained on PFETs (Fig. 16).

Besides lowering the ρ_C, the contact area can be optimized to further reduce the R_C [13]. To fully utilize the contact area, an enhanced selective cleaning process is introduced to more effectively remove the non-conductive residues between S/D epi diamonds. Electrically, 20% and 40% R_C reductions are obtained on NFET and PFET, as shown in Figs. 17-18. What's more, because of the high selectivity of this clean, the overlap capacitance (C_{OV}) and contact metal lateral line resistance in Figs. 19 a-b) show comparable results, confirming no spacer erosion or contact CD widening in cross gate direction.

VI. MOL Metal Resistance Reduction

Despite the relative small contribution to R_{EXT} in Fig. 3, the MOL metal resistance becomes more important as the other contributing elements are optimized and it also increases

rapidly at smaller trench size [13-14]. In the nominal devices studied here, the CA to TS metal interface resistance is found to be dominating in the overall R_{META}. To improve the metal interface quality, we first optimized the CA metal trench etch process to reduce the reactive ion etch (RIE) damages to the top surface of TS metal and followed with tuning the cleaning prior to CA metallization to remove RIE residues/damages.

Fig. 20 shows the device R_{METAL} results with POR CA RIE and optimized CA RIE processes where more than 30% R_{METAL} reduction is observed. Moreover, adjusting the cleaning before CA metallization gives another 10% R_{METAL} benefit. Besides CA-TS interface engineering, reducing the CA TiN liner thickness can also help reduce R_{METAL} by 1) leaving more space for the metal fill 2) reducing highly resistive interface layer thickness. As a result, in Fig. 21, R_{METAL} decreases by 40% when the CA TiN liner is thinned down by 30%.

In addition, the local interconnect wiring resistance at MOL level is important for circuit performance. At this level, replacing W with Co could greatly relieve the concerns of local wiring resistance due to the resistivity/process benefits of Co over W at small dimensions. Fig. 22 gives the metal lateral line resistances of Co and W with the same liner in CA and TS metal trenches. Significant resistance reduction is seen with Co because of better metal fill and lower metal resistivity from larger grain sizes. Overall, 60% line resistance reduction is seen on Co metal fill compared with W in CA and TS levels.

VII. Parasitic Resistance Reduction for Better Scalability

Ultimately, we integrate all the R_{EXT} reduction schemes. To evaluate the overall R_{EXT} reduction from this unified strategy, the R_{EXT} dependence on contact size are given in Figs. 23-24. More than 70% and 60% R_{EXT} reductions are realized for NFET and PFET at the nominal contact size by incorporating all the resistance reduction knobs. With contact CD scaled to 70% of the nominal size, while R_{EXT} expectedly increases, larger R_{EXT} benefits over R_{ON} are measured with this integrated approach at smaller dimensions.

VIII. Conclusion

In this work, we thoroughly investigate the external parasitic resistance in advanced FinFET technology. Higher in-situ P in NFET and tuning SiGe:B:C content in PFET effectively reduce the epi resistance by 60%/45%. High-P Si:P epi top layer and high-Ge SiGe epi top layer combined with SPE process deliver NFET ρ_C of 1×10^{-9} and PFET ρ_C of 7×10^{-10} $\Omega \cdot$cm^2. Optimization of CA to TS metal interface also reduces the whole MOL metal resistance by 60%. Each resistance reduction scheme is eventually integrated, delivering more than 70% and 60% R_{EXT} reduction on NFET and PFET.

IX. Reference

[1] D. Guo et al., *VLSI*, 2016. p.14 [2] G. Tsutsui et al., *IEDM*, 2016, p.456 [3] Y. Xuan et al., *EDL*, 2008, p.294 [4] F. Ducroquet et al., *IEDM*, 2014, p.437 [5] J. Adkisson et al., *ECS Transactions,* 2013, p.83 [6] P. Adusumilli et al., *VLSI*, 2016, p.1 [7] H. Yu et al., *VLSI*, 2016, p.66 [8] O. Gluschenkov, *IEDM*, 2016, p.448 [9] H. Wu et al., *IEDM*, 2017, p.22.3 [10] A. Carr et al., *Microelectronic Engineering*, 2017, p.173 [11] C. Niu et al., *ASMC*, 2016, p.320 [12] A. Razavieh et al., *DRC*. 2017, p.1 [13] V. Kamineni, *IITC*, 2016, p.105 [14] S. Fan, *IITC*, 2017, p.1 [15] K. Cheng, et al., *IEDM*, 2016, p.444 [16] J. Kye et al., *VLSI*, 2018, p.149

978-1-7281-1988-5/18 $31.00 © 2018 IEEE

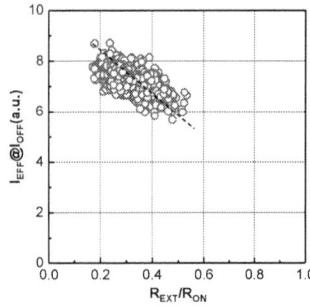

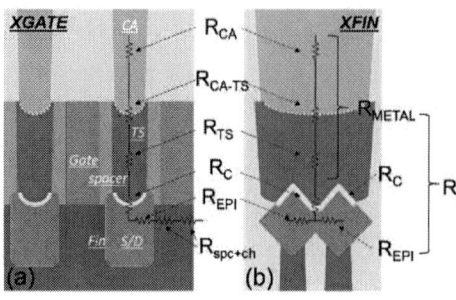

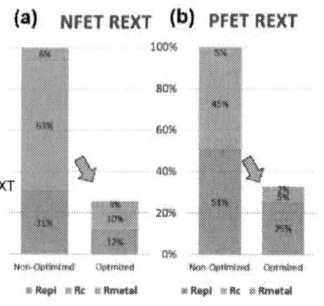

Fig. 1 $I_{EFF}@I_{OFF}$ v.s. R_{EXT}/R_{ON} relationship of FinFETs studied in this work. More dominance of R_{EXT} in R_{ON} has worse performance.

Fig. 2 parasitic resistances in FinFETs: a) cross gate direction b) cross fin direction. The R_{EXT} is composed of 1) epi resistance (R_{EPI}) 2) contact resistance (R_C) 3) MOL metal stud resistance $R_{METAL} = R_{CA}+R_{CA-TS}+R_{TS}$.

Fig. 3 Partition of R_{EXT} components in NFET(a) & PFET(b) with and without optimization. (c) Typical process flow and key modules for the external resistance optimization.

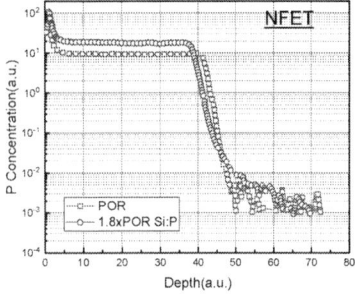

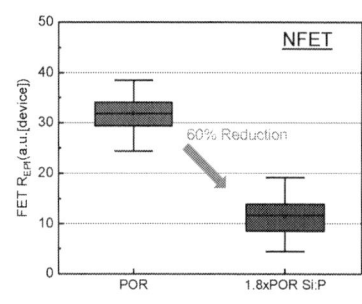

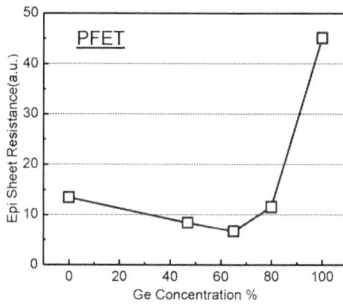

Fig. 4 SIMS profiles of P inside S/D epi for POR Si:P and optimized higher 1.8×POR Si:P.

Fig. 5 NFET R_{EPI} with the two epi processes in Fig. 4. 80% increase in doping density shows 60% R_{EPI} reduction.

Fig. 6 Epi sheet resistance v.s. Ge content in PFET's SiGe S/D epi. Ge% from 50%~70% is optimal.

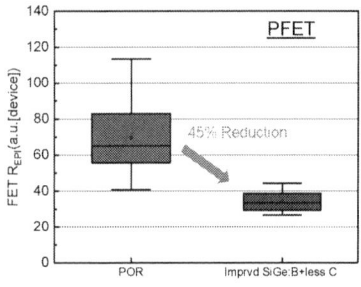

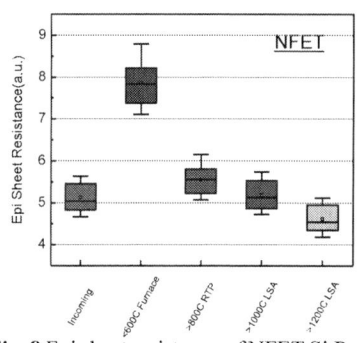

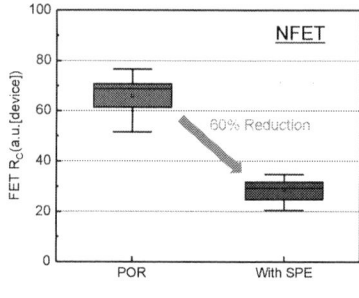

Fig. 7 PFET R_{EPI} for POR and optimized SiGe:B epi with reduced C content.

Fig. 8 Epi sheet resistance of NFET Si:P epi with different downstream thermal processes.

Fig. 9 NFET contact resistance with and without the SPE process.

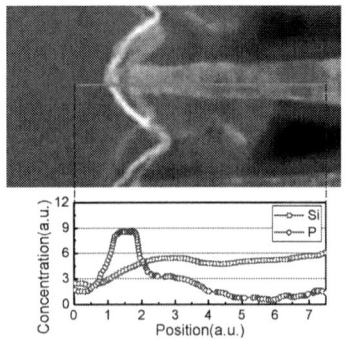

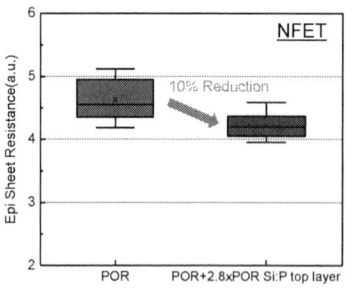

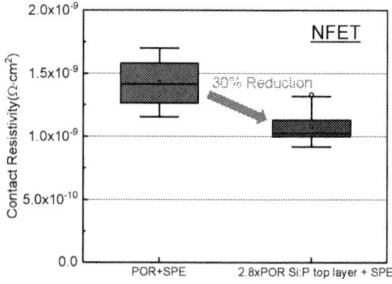

Fig. 10 NFET epi with 2.8×POR Si:P epi top layer and the elemental line scan of Si and P along the marked line.

Fig. 11 Epi sheet resistance of NFET epi with POR and 2.8×POR Si:P epi top layer described in Fig. 10.

Fig. 12 NFET Contact resistivity with and without the 2.8×POR Si:P epi top layer. Low ρ_c of 1×10^{-9} $\Omega\cdot cm^2$ is realized.

978-1-7281-1988-5/18 $31.00 © 2018 IEEE 821

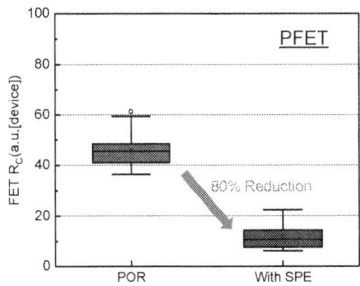

Fig. 13 PFET contact resistance with and without the SPE process.

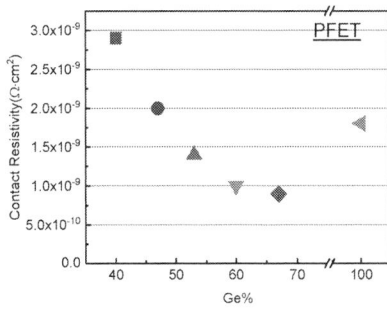

Fig. 14 Contact resistivity v.s. Ge content in SiGe epi. Optimal Ge% for ρc is around 70%.

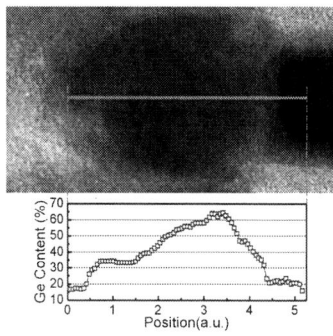

Fig. 15 PFET epi with SiGe75 top layer and the Ge content along the marked line. Slightly lower Ge% (~68%) seen is due to the projection effects.

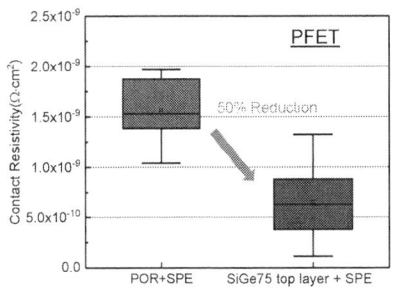

Fig. 16 PFET Contact resistivity with and without SiGe75 epi top layer. Record low ρc of 7×10^{-10} $\Omega \cdot cm^2$ is obtained.

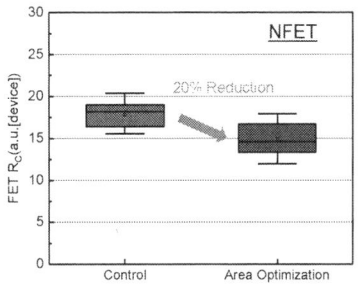

Fig. 17 NFET contact resistance with and without the enhanced cleaning for the contact area optimization.

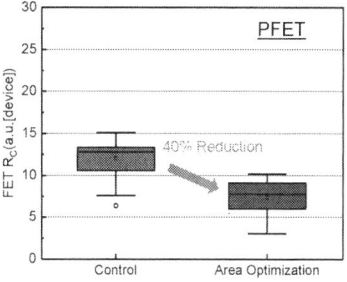

Fig. 18 PFET contact resistance with and without the enhanced cleaning for the contact area optimization.

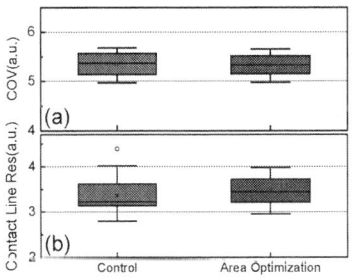

Fig. 19 S/D Overlap capacitance (a) and contact metal lateral line resistance (b) with and without the enhanced cleaning process. No spacer erosion and contact CD widening are confirmed.

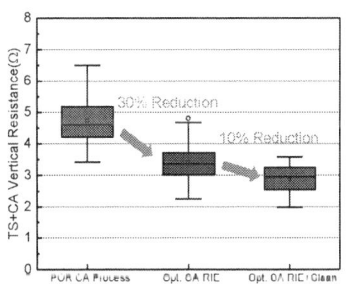

Fig. 20 FET MOL metal resistance with POR, optimized CA RIE and optimized CA metal clean to reduce the defects at the CA-TS metal interface.

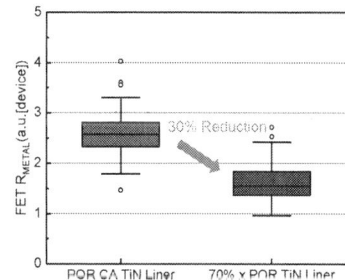

Fig. 21 FET MOL metal resistance with POR TiN liner and thinner 70% POR TiN liner in the CA metallization process.

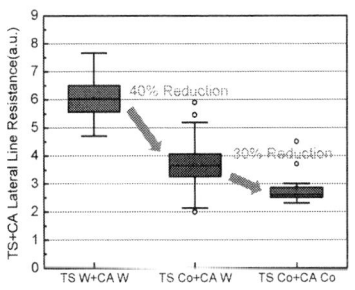

Fig. 22 TS+CA lateral line resistance with different metal combinations in TS and CA.

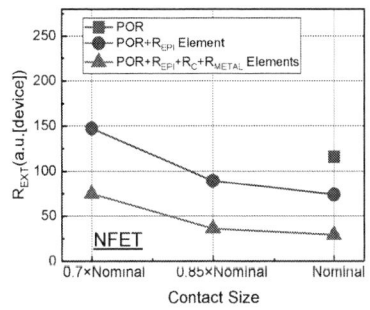

Fig. 23 Evolution of NFET external resistance at various contact sizes by inserting more resistance reduction knobs.

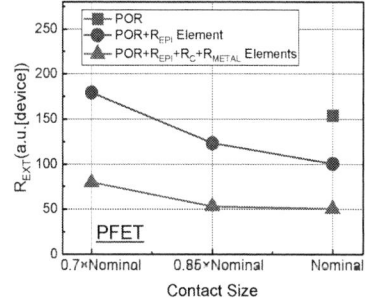

Fig. 24 Evolution of PFET external resistance at various contact sizes by inserting more resistance reduction knobs.

978-1-7281-1988-5/18 $31.00 © 2018 IEEE

Sub-10^{-9} Ω-cm^2 Specific Contact Resistivity on P-type Ge and GeSn: *In-situ* Ga Doping with Ga Ion Implantation at 300 °C, 25 °C, and -100 °C

Ying Wu,[1] Lye-Hing Chua,[2] Wei Wang,[1] Kaizhen Han,[1] Wei Zou,[2] Todd Henry,[2] and Xiao Gong[1, *]

[1] Department of Electrical and Computer Engineering, National University of Singapore (NUS), 117576 Singapore;

[2] Applied Materials-Varian Semiconductor Equipment, Gloucester, Massachusetts, 01930, USA.

*Phone: +65 6516-7871, E-mail: elegong@nus.edu.sg.

Abstract—for the first time, Ga ion implantation (Ga I/I) on *in-situ* Ga-doped Ge (Ge:Ga) and GeSn (GeSn:Ga) films at various temperatures (300 °C, 25 °C, and -100 °C) was investigated. It is found that cryogenic (-100 °C) and room temperature (RT, 25 °C) Ga I/I retains strain and the high quality of the GeSn layer after Ga activation while hot Ga I/I (300 °C) degrades the crystalline quality due to the implantation-induced defects. An ultra-low specific contact resistivity ρ_c of 8×10^{-10} Ω-cm^2 is achieved for Ti/p$^+$-GeSn contact by *in-situ* Ga doping followed by cryogenic or RT Ga I/I while ρ_c increases to 2.3×10^{-9} Ω-cm^2 using hot Ga I/I. An ultra-low ρ_c of 9×10^{-10} Ω-cm^2 is also demonstrated for *in-situ* Ga-doped Ge followed by RT Ga I/I. This is the first realization of sub-10^{-9} Ω-cm^2 ρ_c on non-laser-annealed p-type Ge. The sub-10^{-9} Ω-cm^2 ρ_c is thermally stable up to an annealing temperature of 500 °C.

I. INTRODUCTION

Due to the aggressive downscaling of dimensions of field-effect transistors (FETs), contact resistance (R_c) in the source/drain (S/D) regions becomes one of bottlenecks that limits the on-state current and switch speed of modern FETs [1, 2]. A specific contact resistivity ρ_c of less than 10^{-9} Ω-cm^2 is required for future technology nodes to alleviate the impact of R_c [3]. Ga is proposed as a potential p-type doping species for Ge-based materials such as SiGe, Ge, and GeSn, due to its high solid solubility (~5×10^{20} cm^{-3}) in Ge [4]. Moreover, the heavier atomic mass of Ga as compared to B is preferred for the formation of shallow junctions by minimizing the channeling effect. Superior contact properties have been demonstrated in both Ga-doped SiGe and Ge as compared with the B-doped counterparts [5, 6]. In most studies, Ga is used to increase the surface doping concentration in the epitaxially grown B-doped Ge or SiGe S/D regions. However, it has been found that the co-existence of Ga and B may degrade the ρ_c as compared with that doped by Ga only [6]. Strain relaxation induced by S/D implantation damages is another concern for p-channel FETs (PFETs) with S/D stressors. Novel implantation techniques capable of achieving high S/D doping concentration and maintaining strain are preferred. Therefore, the search for a pure Ga-doped S/D scheme without strain relaxation penalty for aggressively scaled SiGe or Ge PFETs with strained Ge or GeSn S/D regions (Fig. 1) is extremely important.

In this work, Ga ion implantation (Ga I/I, energy: 3 keV, dose: 5×10^{15} cm^{-2}) on the *in-situ* Ga-doped Ge and GeSn films is employed to boost surface Ga doping concentration, leading to an ultra-low ρ_c of 9×10^{-10} and 8×10^{-10} Ω-cm^2 for Ti/p$^+$-Ge and Ti/p$^+$-GeSn contacts, respectively. This is the first demonstration of sub-10^{-9} Ω-cm^2 ρ_c on non-laser-annealed p-Ge. In addition, we found, for the first time, cryogenic (-100 °C) and RT (25 °C) Ga I/I maintain the strain and high quality of grown GeSn layer after Ga activation while hot (300 °C) Ga I/I degrades the GeSn quality.

II. EPITAXIAL GROWTH OF IN-SITU GA-DOPED GE AND GESN FILMS

The *in-situ* Ga-doping Ge and GeSn films were grown by MBE. The total Ga doping concentration (N_T) is a function of the Ga cell temperature (T_{cell}) in the epi process [Fig. 2]. N_T increases from 8×10^{17} to 5.6×10^{20} cm^{-3} when T_{cell} increases from 540 to 820 °C. Fig. 3 shows the high-resolution XRD (HRXRD) curves of Ge:Ga and GeSn:Ga films grown with various T_{cell}. Ge:Ga and GeSn:Ga films with high-crystalline quality are maintained with an T_{cell} of 740 °C. The Sn composition determined from XRD results is 5%. Further increase in T_{cell} to 820 °C leads to a broadening of GeSn peak, indicating degradation of crystalline quality. Fig. 4 shows the SEM images of grown GeSn:Ga layers at various T_{cell}. Islands are observed on the surface of the sample grown with T_{cell} of 820 °C. The degradation of surface morphology is also confirmed by AFM (Fig. 5). The flat surfaces with a RMS roughness of 0.39 nm for Ge:Ga [Fig. 5(a)] and 0.46 nm GeSn:Ga [Fig. 5(b)] are maintained with T_{cell} of 740 °C. The RMS roughness increases to 8.77 nm when increasing T_{cell} to 820 °C for GeSn:Ga due to formed islands on the sample surface [Fig .5(d) and (e)].

Ga and Ge SIMS depth profiles of Ge:Ga, and Ga, Ge and Sn SIMS depth profiles of GeSn:Ga are shown in Fig. 6 and Fig. 7, respectively. For the rest part, the Ge:Ga and GeSn:Ga are grown at T_{cell} of 740 °C and the Sn composition of GeSn:Ga is 5%, unless further stated. Uniform Ga profiles are observed for both Ge:Ga and GeSn:Ga layers. The resistivity and carrier concentration are extracted using infrared spectroscopic ellipsometry based on Drude-like free carrier response in the dielectric function [7]. The measured resistivities are 0.5 and 0.8 mΩ-cm for Ge:Ga and GeSn:Ga, respectively (Fig. 8). The extracted active Ga concentration (Fig. 8) of Ge:Ga and GeSn:Ga films are 2.2×10^{20} and 1.6×10^{20} cm^{-3}, respectively.

III. GA ION IMPLANTATION AND ACTIVATION

Ga I/I with lower damage level is of great importance to reduce junction leakage. Implantation temperature is one of knobs to engineer the implantation damages. Kinetic Monte

978-1-7281-1988-5/18 $31.00 © 2018 IEEE

Carlo (KMC) simulation shows that decreasing the implantation temperature for Ge results in a thicker amorphous layer with a smoother amorphous/crystal (a/c) interface, and less unrepaired defects after the annealing (Fig. 9). Ge is used as the implantation species in KMC (energy: 3 keV, dose: 5×10^{15} cm^{-2}) here to mimic Ga implantation because the detailed defects information of Ga in Ge is still missing and similar implantation behavior was observed for Ga and Ge implantations into Ge substrate [8]. Fig. 10 shows the TEM images of Ga implanted GeSn:Ga at various temperatures. Single crystalline GeSn with a defect layer underneath the sample surface at the depth of 9 nm is observed after hot Ga I/I [Fig. 10(a) and (d)] while cryogenic Ga I/I leads to an amorphous GeSn layer with a smooth a/c interface [Fig. 10(c) and (f)]. TEM images of implanted GeSn after rapid thermal annealing (RTP) at 400 °C for 1 min are shown in Fig. 11(a) and (c), respectively. Unrepaired defects are observed for hot Ga I/I [Fig. 11(b)] after annealing while high-quality GeSn with clear lattice fingers for RT [Fig. 11(d)] and cryogenic Ga I/I (not shown here).

HRXRD curves of Ga-implanted GeSn after RTP annealing are shown in Fig. 12. The well-defined GeSn peak with clear thickness fringes indicates the fully strained GeSn with high quality after cryogenic and RT Ga I/I. The GeSn peak position shifts for different samples due to the variation of Sn compositions. Hot Ga I/I leads to the lowest GeSn peak intensity and vanished diffraction fringes, indicating the poor quality of the GeSn layer. Ga SIMS depth profiles of RT-implanted Ge:Ga [Fig. 13(a)] and GeSn:Ga [Fig. 13(b)] show negligible Ga in-diffusion after RTP annealing. Increasing the implantation temperature broadens the Ga profiles (Fig. 14). The depth at N_T(Ga) of 5×10^{20} cm^{-3} increases from 10 nm for cryogenic Ga I/I to 12.5 nm for hot Ga I/I.

IV. NANO-TLM TEST STRUCTURE AND CHARACTERIZATIONS

The ρ_c between metal and the Ga-doped Ge and GeSn is evaluated by the advanced nano-scale transmission line method (Nano-TLM). The 3D schematic of Nano-TLM structure is shown in Fig. 15(a). The zoom-in image of the mesa region of Nano-TLM is shown in Fig. 15(b), highlighting three key parameters of Nano-TLM, which are the mesa width W, the metal line width L_c, and the contact distance L_d. Key process steps for fabricating the Nano-TLM structure are listed in Fig. 16(a). The SEM image of fabricated Ti/Ge:Ga Nano-TLM is shown in Fig. 16(b). The well-defined 200 nm Ti lines are shown in Fig. 16(c). The HRTEM image of Ti/p$^+$-GeSn contact is shown in Fig. 17. Well defined Ti/p$^+$-GeSn contact with a flat Ti/GeSn interface is observed. The EDX element scan cross the contact interface (FF' in Fig. 17) is shown in Fig.18. A Ga peak induced by Ga I/I is observed near the GeSn surface.

V. RESULTS AND DISCUSSION

The method of ρ_c extraction using Nano-TLM structure is well established and documented in Ref. 9. Two measurement schemes, parallel and cross measurements, are used in Nano-TLM. The measured parallel terminal resistance R_p and cross terminal resistance R_x (symbols) from 4 sets of Ti/RT Ga I/I

GeSn followed by 400 °C annealing (denoted as 400-RT-GeSn) Nano-TLM with L_c of 200 nm and 150 nm are shown in Fig. 19(a), and (b), respectively. The best-fit curves of R_p and R_x are also shown in Fig. 19. The GeSn sheet resistance (R_{sh}) and ρ_c can then be obtained by the fitting of R_p and R_x. The R_{sh} of Ge:Ga and GeSn:Ga layers after RT Ga I/I followed by RTP annealing at 400 °C/1 min are shown in Fig. 20(a) and (b), respectively. The RT Ga I/I leads to 29% and 23% reduction of R_{sh} as compared with the in-situ Ga-doped Ge and GeSn, respectively. The influence of Ga I/I on ρ_c is shown in Fig. 21. RT Ga I/I decreases the ρ_c from 2.9×10^{-9} to 9×10^{-10} Ω-cm^2 for Ti/p$^+$-Ge contacts, and from 1.4×10^{-9} to 8×10^{-10} Ω-cm^2 for Ti/p$^+$-GeSn contacts, respectively.

ρ_c as a function of Ga implantation temperatures for Ti/p$^+$-Ge(GeSn) contacts is shown in Fig. 22. Increasing implantation temperature from -100 to 300 °C increases ρ_c from 8×10^{-10} to 2.3×10^{-9} Ω-cm^2 for Ti/p$^+$-GeSn contacts. For Ti/p$^+$-Ge contact, ρ_c increases from 9×10^{-10} to 4.8×10^{-9} Ω-cm^2 with raising the implantation temperature from 25 to 300 °C. ρ_c as a function of post-metal annealing temperature (T_{PMA}) for Ti contacts on RT-implanted GeSn:Ga (RT-GeSn:Ga) and cryogenic-implanted GeSn:Ga (Cryo-GeSn:Ga) is shown in Fig. 23. The annealing time is 30 s. For Ti/Cryo-GeSn:Ga contact, ρ_c decreases from 2×10^{-9} to 8×10^{-10} Ω-cm^2 with increasing T_{PMA} from 400 to 450 °C. The sub-10^{-9} Ω-cm^2 ρ_c is maintained when T_{PMA} is increased to 500 °C. For Ti/RT-GeSn:Ga contact, ρ_c is insensitive with T_{PMA} in range of 400 to 500 °C and stabilized at 8×10^{-10} Ω-cm^2.

Fig. 24 benchmarks the ρ_c of metal/p-Ge or GeSn contacts reported in the literature [10-17]. Sub-10^{-9} Ω-cm^2 ρ_c is achieved on both p-type Ge and GeSn by using in-situ Ga doping followed by Ga I/I. The ultra-low ρ_c of 8×10^{-10} Ω-cm^2 is comparable with the lowest result achieved using nano-second laser annealing [6].

VI. CONCLUSION

Ga ion implantation is a promising technique to boost the surface Ga doping concentration of the in-situ Ga-doped Ge and GeSn and reduces ρ_c. Cryogenic Ga I/I maintains the strain of grown GeSn:Ga and the high-crystalline quality, and leads to an ultra-low ρ_c of 8×10^{-10} Ω-cm^2 for Ti/p$^+$-GeSn contacts. Moreover, the sub-10^{-9} Ω-cm^2 ρ_c is thermally stable up to a 500 °C annealing. These key findings make in-situ Ga doping followed by cryogenic Ga I/I very attractive for advanced SiGe or Ge PFETs with strained Ge or GeSn S/D regions.

ACKNOWLEDGMENT

The authors acknowledge support from NUS Trailblazer Grant (R-263-000-B43-733), and Ministry of Education (MOE) Academic Research Fund (R-263-000-B50-112).

REFERENCES

[1] Y.-C. Yeo et al., IEDM 2015, pp. 28. [2] D. Yakimets et al., IEDM, 2017, pp. 501. [3] O. Gluschenkov et al., IEDM 2016, pp. 448. [4] F. A. Trumbore, BSTJ 39, pp. 205, 1960. [5] J. L. Everaert et al., VLSI 2017, pp. T214. [6] L.-L. Wang et al., IEDM 2017, pp. 549. [7] V. R. D'Costa et al., PRB 80, 125209, 2009. [8] G. Impellizzeri et al., APA, 103, pp. 323, 2011. [9] W. Liu et al., EDL 35, pp.178, 2014. [10] Y. L. Chao et al., TED 54, pp 2750, 2007. [11] H. Yu et al., TED 37, pp. 978, 2015. [12] L. Hutin et al., JEC 113, pp. H522, 2009. [13] P. Bhatt et al., EDL 35, pp. 69, 2014. [14] M. Hidenori et al., JJAP 53, 04EA05, 2014. [15] H. Miyoshi et al., VLSI 2014, pp. 978. [16] Y. Wu et al., VLSI 2017, pp. T218. [17] Y. Wu et al., VLSI 2018, pp. T77.

978-1-7281-1988-5/18 $31.00 © 2018 IEEE

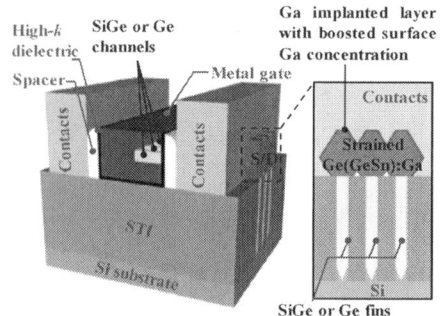

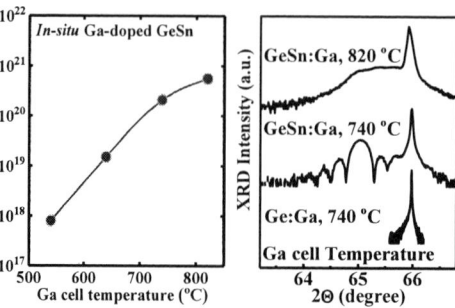

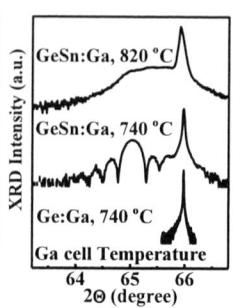

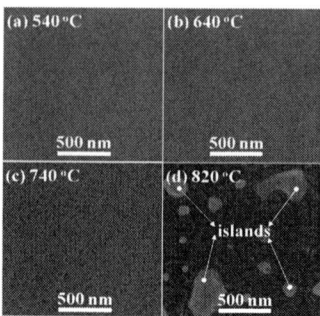

Fig. 1. Schematic of SiGe or Ge p-channel FinFETs with integration of Ga-doped Ge or GeSn source/drain (S/D) regions: *in-situ* Ga-doping followed by Ga ion implantation to boost Ga doping near surface in S/D regions.

Fig. 2. Total Ga doping concentration (N_T) of *in-situ* Ga-doped GeSn films grown at various Ga cell temperatures.

Fig. 3. High-resolution XRD curves of *in-situ* Ga-doped GeSn and Ge films grown at various Ga cell temperatures.

Fig. 4. SEM images of the GeSn:Ga films grown with T_{cell} of (a) 540 °C, (b) 640 °C, (c) 740 °C, and (d) 820 °C, respectively.

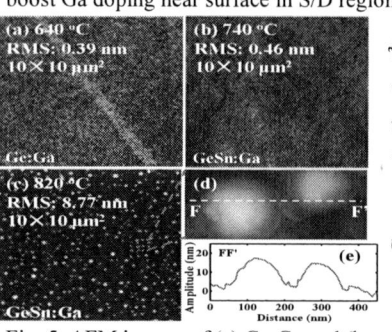

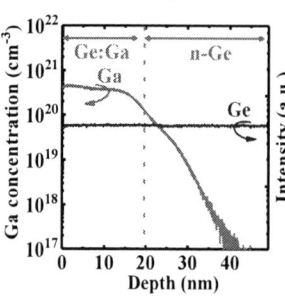

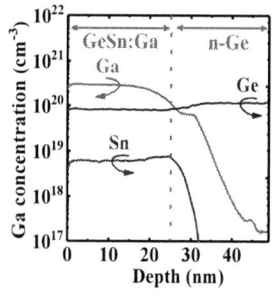

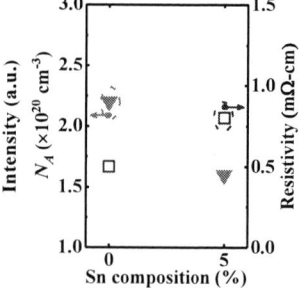

Fig. 5. AFM images of (a) Ge:Ga and (b-c) GeSn:Ga films grown with various T_{cell}. (d) and (e) show the islands formed on GeSn:Ga grown with T_{cell} of 820 °C.

Fig. 6. Ga and Ge SIMS depth profiles of Ge:Ga-on-Ge, showing a uniform Ga-doped Ge:Ga layer.

Fig. 7. Ga, Sn, and Ge SIMS depth profiles of GeSn:Ga-on-Ge, showing a N_T(Ga) of 3×10^{20} cm^{-3}.

Fig. 8. Extracted active Ga doping concentrations and resistivities of grown Ge:Ga and GeSn:Ga films.

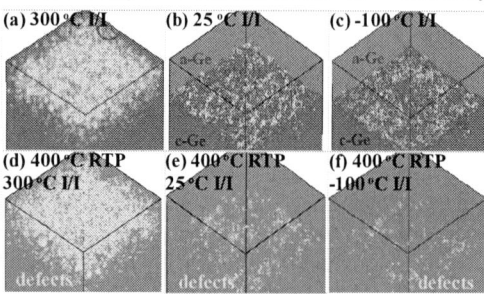

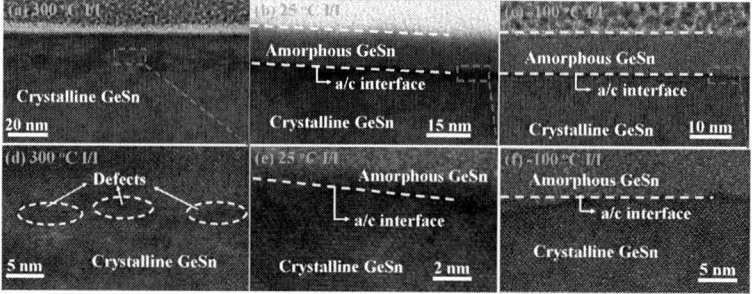

Fig. 9. Kinetic Monte Carlo simulation of Ge implantation at (a) 300 °C, (b) 25 °C, and (c) -100 °C. (d-f) The density of defect decreases with decreasing implantation temperatures after annealing at 400 °C.

Fig. 10. TEM images of GeSn:Ga after Ga I/I at (a) 300 °C, (b) 25 °C, and (c) -100 °C. HRTEM images of GeSn:Ga after Ga I/I at (d) 300 °C, (e) 25 °C, and (f) -100 °C. Crystalline GeSn is maintained after Ga I/I at 300 °C while an amorphous layer are observed for room-temperature and cryogenic Ga I/I.

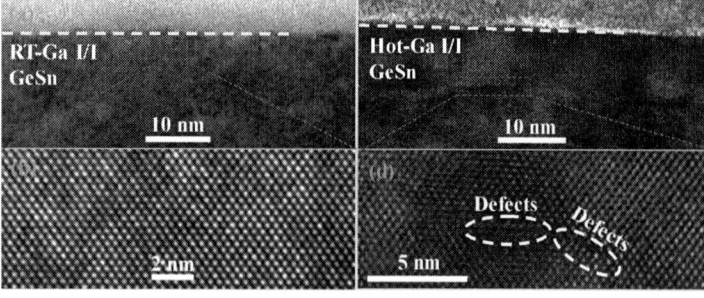

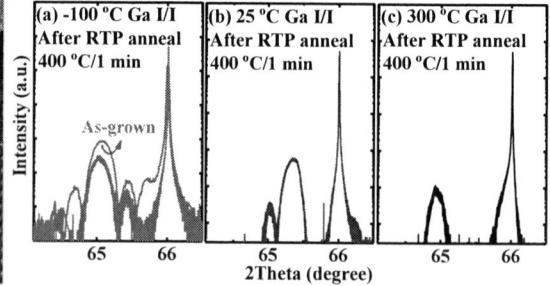

Fig. 11. TEM images of (a) RT-implanted and (c) hot-implanted GeSn:Ga after annealing at 400 °C for 1 min. HRTEM images of (b) RT-implanted and (d) hot-implanted GeSn:Ga after the 400 °C annealing for 1 min.

Fig. 12. HRXRD curves of GeSn after (a) cryogenic, (b) RT, and (c) hot Ga I/I followed by RTP anneal at 400 °C/1 min. Cryogenic and RT Ga I/I retains the strain and high quality of the grown GeSn.

978-1-7281-1988-5/18 $31.00 © 2018 IEEE

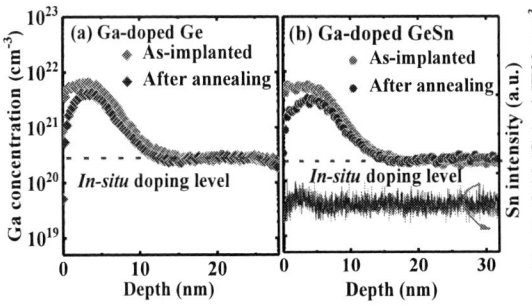
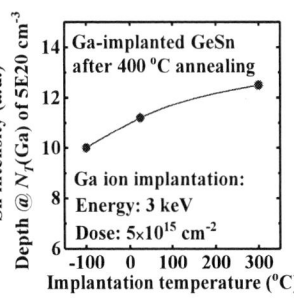
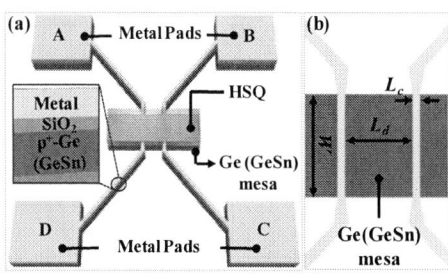

Fig. 13. Ga SIMS depth profiles of RT-Ga I/I (a) Ge and (b) GeSn before and after rapid thermal annealing (RTP) at 400 °C/1 min. The Sn composition before and after annealing is almost the same for RT-implanted GeSn.

Fig. 14. The depth at N_T(Ga) of 5×10^{20} cm^{-3} as a function of Ga implantation temperature.

Fig. 15. (a) The 3D schematic of the Nano-TLM structure. (b) The definition of L_c, L_d, and W in the mesa region of Nano-TLM structure.

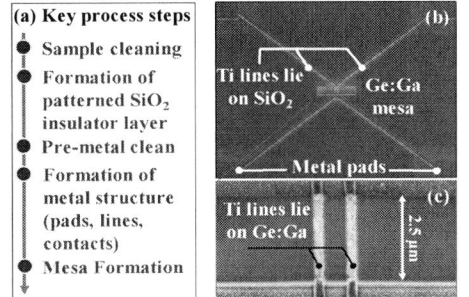
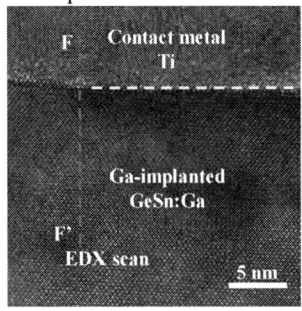
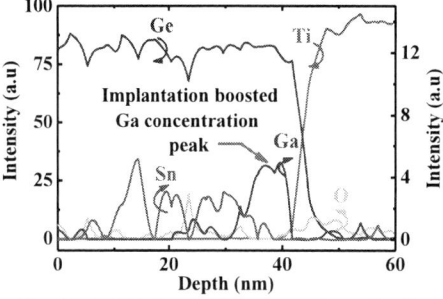

Fig. 16. (a) The key process steps of Nano-TLM fabrication, (b) SEM image of the fabricated Ti/Ge:Ga Nano-TLM structure, and (c) zoom-in image in the mesa region of Ti/Ge:Ga Nano-TLM.

Fig. 17. HRTEM images of the Ti contact on the Ga-implanted GeSn:Ga, indicating a flat and well-defined Ti/GeSn interface.

Fig. 18. EDX element line scan along the line FF' in Fig. 18. A clear Ga peak near the GeSn surface is observed induced by the Ga ion implantation.

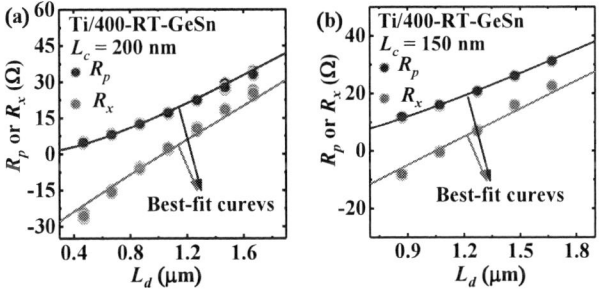
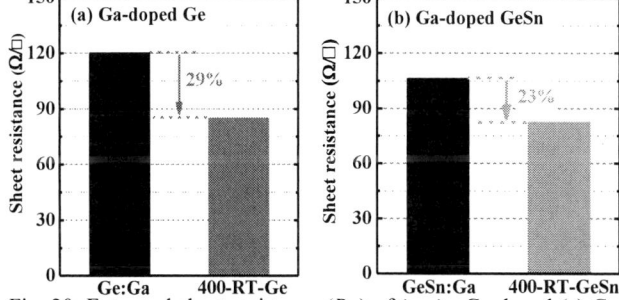

Fig. 19. Extracted R_p and R_x (symbols) at different L_d and best fitting curves of Ti/400-RT-GeSn Nano-TLM structures with L_c of (a) 200 nm and (b) 150 nm, respectively. R_p (R_x) are measured with the current flowing from ports A to B (D to C) and the voltage drop is measured between ports C and D (B to D).

Fig. 20. Extracted sheet resistance (R_{sh}) of in-situ Ga-doped (a) Ge and (b) GeSn with/without RT-Ga I/I followed by 400 °C RTP annealing, showing that RT-Ga I/I followed 400 °C RTP annealing leads to 29% and 23% reduction of R_{sh} as compared with the in-situ Ga-doped Ge (Ge:Ga) and GeSn (GeSn:Ga), respectively.

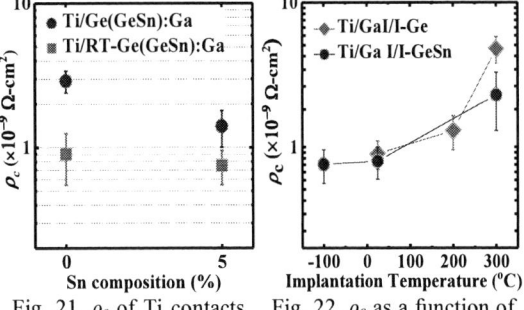
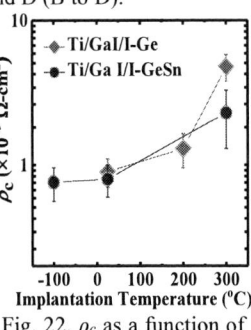
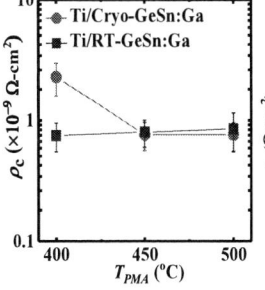
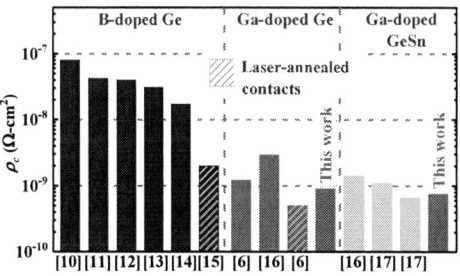

Fig. 21. ρ_c of Ti contacts on in-situ Ga-doped Ge and GeSn with/without RT Ga ion implantation.

Fig. 22. ρ_c as a function of implantation temperatures for Ti contacts to Ga implanted Ge:Ga and GeSn:Ga films.

Fig. 23. ρ_c as a function of post-metal annealing temperature (T_{PMA}) for Ti contacts to Ga implanted GeSn:Ga film.

Fig. 24. Benchmarking of reported ρ_c values on p-type Ge and GeSn. The in-situ Ga doping followed by Ga I/I leads to a sub-10^{-9} Ω-cm^2 ρ_c for metal contacts on p-Ge and GeSn. It is the lowest ρ_c achieved on non-laser-annealed p-Ge.

978-1-7281-1988-5/18 $31.00 © 2018 IEEE   826

Selective Fin Trimming after Dummy Gate Removal as the Local Fin Width Scaling Approach for N5 and Beyond

Toshihiko Miyashita, Shiyu Sun, Sushant Mittal, Myung Sun Kim, Ashish Pal,
Angada Sachid, Kalpana Pathak, Matt Cogorno, and Nam Sung Kim

Applied Materials Inc., 974 East Arques Avenue, Sunnyvale, California, USA, email: toshihiko_miyashita@amat.com

Abstract — Selective fin trimming after dummy gate removal is proposed as the local fin width scaling approach for further FinFET extension. In this approach, local fin trimming is selectively performed for the channel region after dummy gate removal in the replacement metal gate (RMG) module while preserving the original source and drain (S/D) fin width, in order to avoid the parasitic resistance increase that results from the global fin trimming approach. TCAD simulation shows clearly that the local fin trimming can improve gate electrostatics with narrower fins, while also providing the benefits of lower S/D resistance and PMOS high channel stress. Although, fin height reduction and parasitic capacitance increases are not preferable, overall gate delay is improved due to strong I_{on}-I_{off} boost. Selectra™ etch fin trimming results are also presented demonstrating good fin width controllability and smaller variations without any critical fin damages. This local fin trimming is promising for N5 and beyond to further extend FinFET technology.

I. INTRODUCTION

Density-driven scaling of key pitches is still the driving force for further CMOS scaling, where the fin pitch (FP), contacted poly pitch (CPP), and minimum metal pitch (MxP) are the most critical dimensions in current FinFET technology. Figure 1 shows the pitch scaling trend along with practical technology node metric, standard node [1]. It is shown that CPP scaling is slowing down, while logic standard cell size reduction relies more on MxP and/or track height scaling. In order to further scale CPP, gate length (L_g) scaling is essential because it is one of the key components of CPP expressed as;

$$CPP = L_g + 2 \times Spacer\_thickness + Contact\_CD \quad (1)$$

The challenge of L_g scaling here is degraded short-channel control, and new device architectures, such as horizontal gate-all-around (hGAA), have been under intense development [2] to enhance gate electrostatic control with surrounded-gate structures. However, hGAA requires complicated processes to achieve stacked nanowires/nanosheets for competitive effective channel width with FinFET [3]. Continuous FinFET extension is, therefore, still the candidate in CMOS scaling due to its highly matured technology.

On the other hand, fin profile is the most critical factor in FinFET devices, and it has evolved from shorter/wider/tapered to higher/narrower/vertical shape as shown in Fig. 2 [4-6]. Previous studies [7,8] explored global fin width scaling and demonstrated improved electrostatics. However, narrow fin at source and drain (S/D) region also brings higher parasitic resistance. In this paper, a local fin trimming approach is proposed after dummy gate removal in the replacement metal gate (RMG) module. It is shown by TCAD simulation that local fin

trimming can overcome the issues in global fin trimming, and that an optimal fin width and fin trimming window can be prescribed. We also demonstrated well controlled local fin trimming process with selective etching by Selectra™.

II. FIN WIDTH SCALING BY FIN TRIMMING

Figure 3 shows the definition of key physical fin profile parameters; fin height (H_{fin}), fin width (W_{fin}), FP, and fin taper angle, among which W_{fin} will be on the main focus in this work. To define narrower fin width at the end of full process, there are two approaches, namely self-aligned double patterning (SADP) fin spacer thickness scaling and fin trimming after initial fin width definition. As for the thinner SADP spacer approach, extremely small physical fin widths down to 1.6nm were successfully demonstrated in 14nm technology [8]; however, there remains the concern of mechanical strength and/or fin fragility which bring about fin bending and breaking during subsequent processes, especially for high-aspect-ratio fins. On the other hand, fin trimming could solve this issue, because it is basically performed after initial fin width definition, where the high-stress process steps of STI gap fill and densification have already been completed. Figure 4 illustrates the two different fin trimming approaches, (a) global and (b) local fin trimming. As for global fin trimming, both S/D and channel regions are simultaneously trimmed because it is performed after fin reveal. This leads to the significant increase of S/D resistance, R_{sd}, although gate electrostatics are improved with narrow channel fin. On the other hand, local fin trimming is performed after dummy gate removal in RMG module, and it enables selective channel fin trimming while maintaining the original fin width in the S/D region as shown in Fig. 5. Therefore, we propose this local fin trimming as the fin width scaling approach for next-generation FinFET.

III. RESULTS AND DISCUSSIONS

A. TCAD simulation

Local fin trimming impact on device performance was explored via TCAD. Calibration of the TCAD model was done in reference to fin width sensitivity data previously reported by X. He *et al.* [8]. Figure 6 shows the I_{on}-I_{off} performance as a function of fin width assuming global fin trimming where both S/D and channel fin widths are changed simultaneously. It is shown that DC performance has peak value for 6 - 7nm fin width, and it decreases with fin width reduction down to 2nm. This is due to both parasitic resistance increase and mobility degradation for narrower fins, and our TCAD simulation shows good agreement with the experimental data for both N and PMOS. Fin width impact on electrostatics is also shown in Fig. 7. As expected, drain induced barrier lowering (DIBL) decreases monotonically with decreasing fin width, also showing good agreement with experimental results. Figure 8 shows the global fin trimming impact on V_t, where the simulation, which accounts for quantum-confinement effects, produces a significant V_t increase and resultant I_{off} reduction with fin width reduction. From these results, it is concluded that our TCAD simulation is well calibrated and applicable to the local fin trimming simulation in the following discussions.

The local fin trimming approach was simulated comprehensively. Figure 9 shows the definition and the example of local fin trimming parameters used in this study, as well as corresponding electrical parameters. Fin width means final channel fin width after fin trimming, and fin trimming amount corresponds to total trimming where half of total trimming is etched for each fin side. It should be noted that the only the difference is S/D region fin width, if we take W_{fin}/trimming = 4/0 and 4/2nm cases as shown in Fig. 9. Figure 10 shows the simulated local fin trimming impact on I_{on}-I_{off} performance for different fin width cases. For both N and PMOS, I_{on}-I_{off} performance is improved with fin trimming, and a narrower fin shows more performance gain because absolute I_{on}-I_{off} performance is lower as shown in Fig. 6. In order to understand this DC performance behavior, possible local fin trimming effects are summarized in Fig. 11. Three major impacts can be considered, where R_{sd} reduction contributes to current increase while electrostatic degradation due to capacitance coupling and fin height reduction lead to current decrease as shown in the figure. This is completely different situation as compared with global fin trimming approach. Each factor shown in Fig. 11 was examined in the following simulations. Figure 12 shows the device resistance as a function of local fin trimming. It can be seen that R_{sd} decreases with fin trimming, while channel resistance, R_{ch}, keeps constant or rather increases due to fin height reduction. Furthermore, electrostatics are degraded with fin trimming as shown in the Fig. 13 plot of DIBL vs. fin trimming. Because S/D region fin width and resultant S/D epi volume become larger, capacitance coupling also becomes larger, leading to DIBL degradation. In Fig. 14, DC performance impacts are broken down into each component for NMOS with 4nm fin; the R_{sd} benefit is seen to be the dominant factor, and as a result, we can obtain an overall boost in DC performance. One extra benefit in PMOS is the channel stress increase as shown in Fig. 15, originated from relatively larger embedded S/D epi volume.

Gate-to-S/D parasitic capacitance degradation can be expected from DIBL degradation in Fig. 13. Figure 16 shows the simulated effective capacitance, C_{eff}, as a function of fin trimming. Due to increased S/D extension and epi volumes, C_{eff} increase with fin trimming is clearly observed, and it should affect device AC performance. Figure 17 shows the intrinsic gate delay as a function of fin trimming considering both DC performance improvement and parasitic capacitance increase. By implementing local fin trimming, intrinsic gate delay is improved for both N and PMOS, where optimal fin width and fin trimming are used. Figure 18 shows the overall (N and P) gate delay as a functions of fin width and fin trimming. From this figure, it is shown that the optimal fin trimming amount is around 2nm with fin width between 5 and 6nm. We can design

the device fin profile based on the guidelines with local fin trimming implemented in order to achieve improved performance.

B. Local fin trimming process demonstration

The local fin trimming concept was demonstrated experimentally using selective etching (Selectra™ etch) and an internal FinFET test vehicle. As shown in Fig. 4, local fin trimming is performed after dummy gate removal. Figure 19 shows the plan-view SEM of FinFET device in our internal test vehicle just after dummy gate removal. We can clearly observe gate grooves, where fin channels are also visible at the bottom of the grooves. Through these open windows, fin trimming was performed only for the channel region. Figure 20 shows Selectra™ etch actual fin trimming vs. target fin trimming amount. Well controlled selective fin trimming was demonstrated up to 5nm trimming. As shown in Fig. 21, TEM cross-sections after local fin trimming also demonstrated precise fin trimming without any fin damage. One of the concerns of this process is fin line-edge roughness (LER) degradation, however, Fig. 22 demonstrated no roughness degradation, or rather improved LER was observed, which enables fin width further scaling for next technology node.

IV. CONCLUSIONS

Local channel fin trimming after dummy gate removal is proposed for further fin width scaling. The calibrated TCAD simulation shows that local fin trimming produces a DC performance benefit with lower R_{sd} and PMOS improved stress. Even considering parasitic capacitance increase and slight short-channel effect degradation, overall gate delay shows the benefit in terms of strong I_{on} improvement. We have also demonstrated well controlled local fin trimming process using Selectra™ etch without any fin damage. This local fin trimming is promising for N5 and beyond to further extend FinFET technology.

REFERENCES

[1] See *e.g.*, https://www.semiwiki.com/forum/content/6895-standard-node-trend.html

[2] H. Mertens *et al.*, "Vertically Stacked Gate-All-Around Silicon Nanowire Transistors: Key Process Optimizations and Ring Oscillator Demonstration," 2017 IEDM Tech. Dig., pp. 828-831, 2017.

[3] S.-D. Kim *et al.*, "Performance Trade-offs in FinFET and Gate-All-Around Device Architectures for 7nm-node and Beyond," 2015 IEEE S3S, 2015.

[4] C. Auth *et al.*, "A 22nm High Performance and Low-Power CMOS Technology Featuring Fully-Depleted Tri-Gate Transistors, Self-Aligned Contacts and High Density MIM Capacitors," 2012 Symp. on VLSI Tech. Dig., pp. 131-132, 2012.

[5] C.-H. Jan *et al.*, "A 14 nm SoC Platform Technology Featuring 2nd Generation Tri-Gate Transistors, 70 nm Gate Pitch, 52 nm Metal Pitch, and 0.0499 um2 SRAM cells, Optimized for Low Power, High Performance and High Density SoC Products," 2015 Symp. on VLSI Tech. Dig., pp. T12-T13, 2015.

[6] C. Auth *et al.*, "A 10nm High Performance and Low-Power CMOS Technology Featuring 3rd Generation FinFET Transistors, Self-Aligned Quad Patterning, Contact over Active Gate and Cobalt Local Interconnects," 2017 IEDM Tech. Dig., pp. 673-676, 2017.

[7] Y.-S. Wu *et al.*, "Optimization of Fin Profile and Implant in Bulk FinFET Technology," VLSI-TSA 2016, T5-6, 2016.

[8] X. He *et al.*, "Impact of Aggressive Fin Width Scaling on FinFET Device Characteristics," 2017 IEDM Tech. Dig., pp. 493-496, 2017.

978-1-7281-1988-5/18 $31.00 © 2018 IEEE

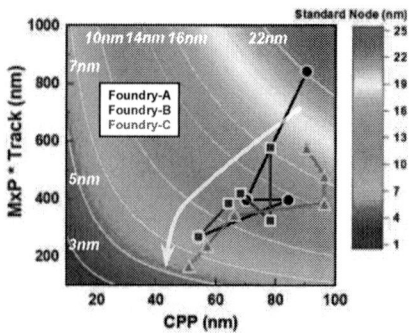

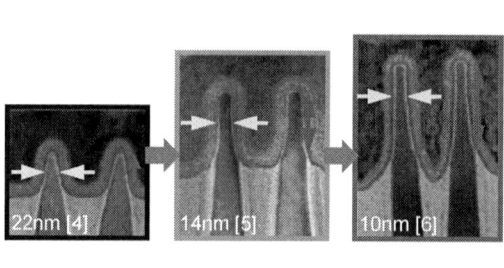

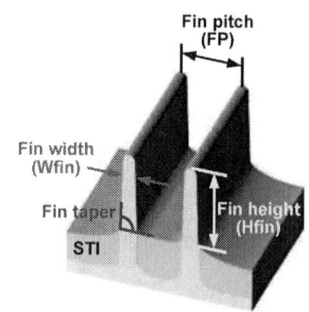

Fig.1 CPP and MxP*Track scaling trend along with practical technology node metric, standard node [1], for different foundry technologies.

Fig.2 Fin profile evolution from 22nm to 10nm node [4-6]. Higher, narrower, and vertical fin profile has been introduced in advanced technology for higher current drivability and better electrostatic control.

Fig.3 Schematic diagram of fin profile and the definition of key physical parameters in FinFET. Among these parameters, fin width is on the focused and discussed in this work.

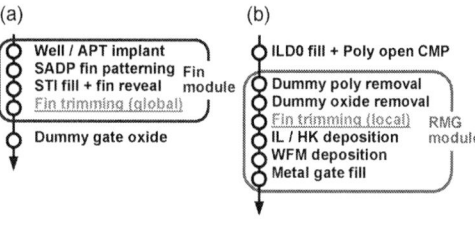

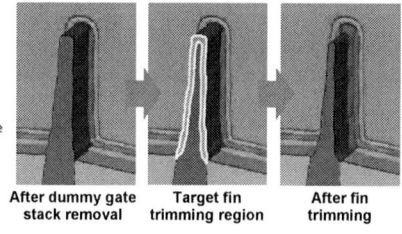

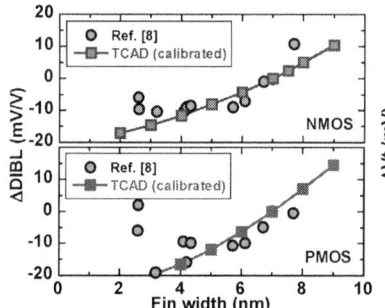

Fig.4 Process flow for two different fin trimming approaches; (a) global fin trimming after fin reveal and (b) local fin trimming after dummy gate removal. This paper focuses on fin trimming @RMG module.

Fig.5 Schematic concept of local fin trimming approach in RMG module. Fin channel is selectively trimmed to scale down channel fin width while maintaining the original fin width in the S/D region with embedded S/D epitaxy.

Fig.6 TCAD simulated fin width impact assuming global fin trimming on I_{on}-I_{off} performance comparing with experimental sensitivity reported in [8].

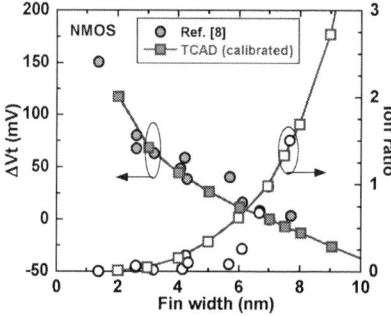

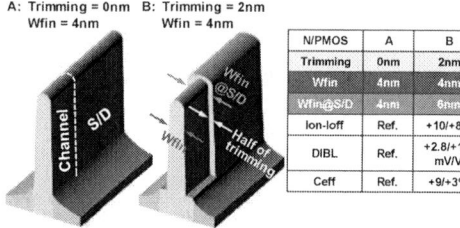

Fig.7 TCAD simulated fin width impact assuming global fin trimming on electrostatics (DIBL) comparing with experimental sensitivity reported in [8].

Fig.8 TCAD simulated fin width impact assuming global fin trimming on NMOS V_t and I_{off} comparing with experimental sensitivity reported in [8].

Fig.9 Definition and example of local fin trimming physical parameters in TCAD simulation, as well as key electrical simulated parameters. "A" and "B" correspond to W_{fin}/trimming = 4/0 and 4/2nm, respectively. The following simulation results follow this parameter definition.

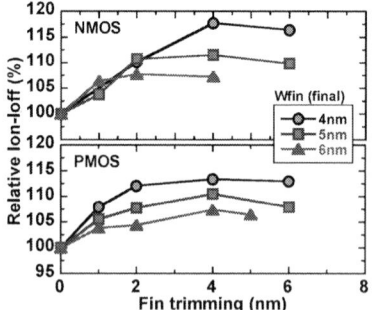

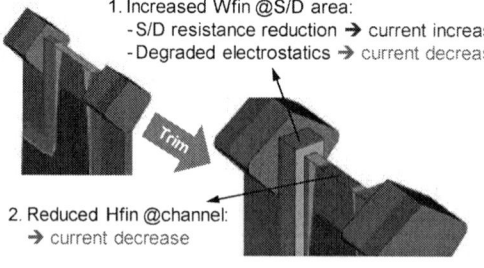

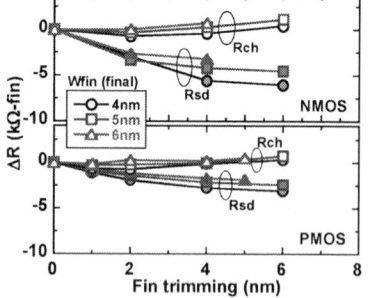

Fig.10 TCAD simulated local fin trimming impact on I_{on}-I_{off} performance for different final fin widths.

Fig.11 Local fin trimming impact on device DC performance. Parasitic resistance reduction contributes current drive increase, while electrostatics degradation and fin height reduction lead to current drive decrease.

Fig.12 TCAD simulated local fin trimming impact on device resistance. R_{sd} decreases with fin trimming, while R_{ch} keeps constant or rather increases due to fin height reduction.

978-1-7281-1988-5/18 $31.00 © 2018 IEEE

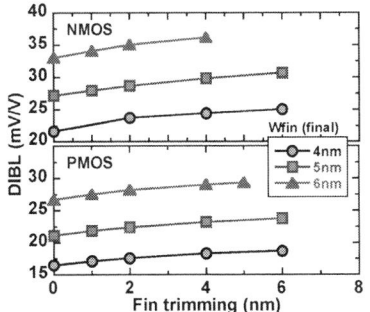

Fig.13 TCAD simulated local fin trimming impact on electrostatics (DIBL) for different final fin widths. DIBL degradation is observed with increasing fin trimming due to capacitance coupling with larger S/D region.

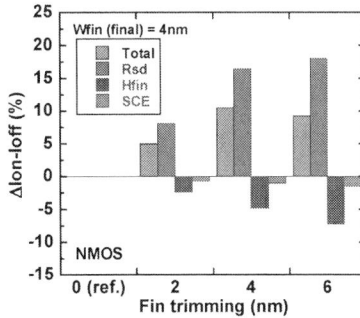

Fig.14 Breakdown of fin trimming impact on I_{on}-I_{off} performance for NMOS W_{fin} = 4nm case. R_{sd} benefit is the dominant factor for performance boost.

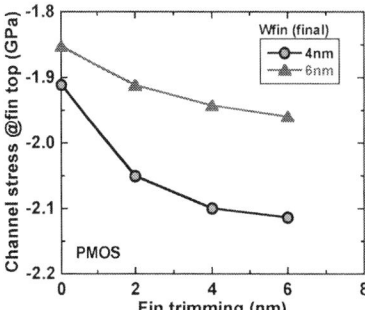

Fig.15 PMOS channel stress simulation results as a function of local fin trimming for W_{fin} = 4 and 6nm cases. Compressive channel stress becomes higher with increasing fin trimming due to larger effective S/D epi volume.

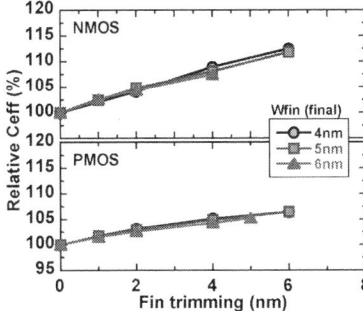

Fig.16 TCAD simulated C_{eff} as a function of local fin trimming. Due to increased S/D extension and epi volumes, C_{eff} increases with fin trimming. This penalty is still smaller than the benefit of I_{on}-I_{off} improvement.

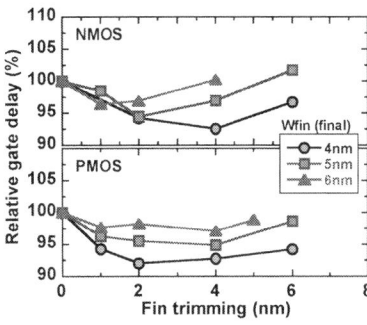

Fig.17 Intrinsic gate delay vs. local fin trimming for both N and PMOS. There exists optimal delay point with reflecting I_{on} vs. trimming (Fig.10), while C_{eff} changes monotonically against trimming (Fig.16).

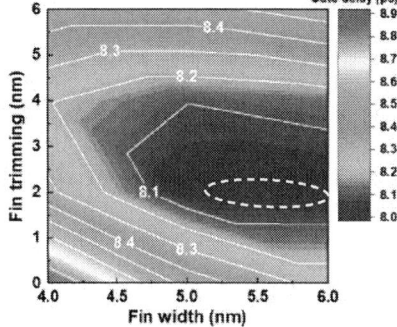

Fig.18 Overall gate delay (including both N and PMOS) as a functions of fin width and fin trimming. Optimal fin trimming amount is around 2nm with final fin width of 5 - 6nm.

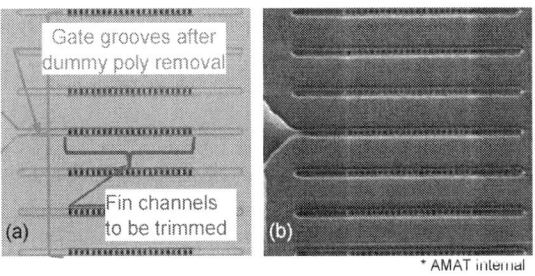

Fig.19 Plan-view (a) schematic layout and (b) corresponding SEM photograph of FinFET device after dummy gate removal. Fin trimming is performed through the open window of gate grooves before IL/HK/WF metal depositions.

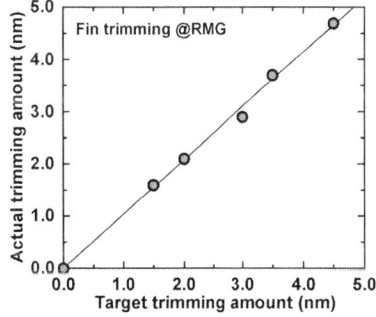

Fig.20 Actual fin trimming vs. target fin trimming amount demonstrating good process controllability. Well controlled selective fin trimming can be achieved by Selectra™ etch.

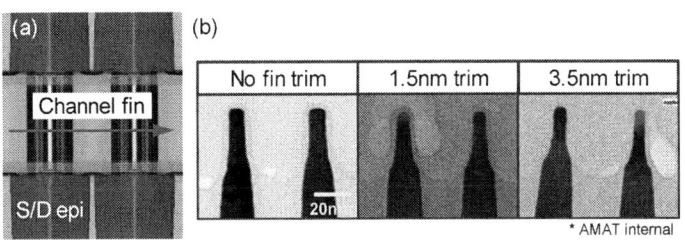

Fig.21 (a) Plan-view schematic layout of FinFET just after local fin trimming. (b) TEM cross-sections of channel fin region corresponding to red arrows in (a), for no fin trimming, 1.5nm trimming, and 3.5nm trimming cases.

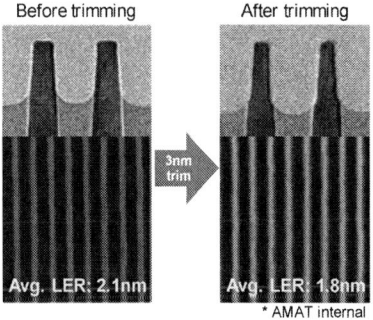

Fig.22 Fin LER before and after local fin trimming. No roughness degradation, or rather improved LER was observed without fin damage, which enables fin width further scaling.

978-1-7281-1988-5/18 $31.00 © 2018 IEEE   830

First experimental demonstration of a scalable linear majority gate based on spin waves

Florin Ciubotaru[1], Giacomo Talmelli[1,2], Thibaut Devolder[3], Odysseas Zografos[1], Marc Heyns[1,2],
Christoph Adelmann[1], and Iuliana P. Radu[1]

[1] imec, Leuven, Belgium, email: Florin.Ciubotaru@imec.be

[2] KU Leuven, Leuven, Belgium, [3] Centre de Nanosciences et de Nanotechnologies, Univ. Paris-Sud, Orsay, France

Abstract—We report on the first experimental demonstration of majority logic operation using spin waves in a scaled device with an in-line input and output layout. The device operation is based on the interference of spin waves generated and detected by inductive antennas in an all-electrical microwave circuit. We demonstrate the full truth table of a majority logic function with the ability to distinguish between strong and weak majority, as well as an inverted majority function by adjusting the operation frequency. Circuit performance projections predict low energy consumption of spin wave based compared to CMOS for large arithmetic circuits.

I. Introduction

Spintronic devices based on spin waves are promising alternatives to CMOS technology with high potential for power and area reduction per computing throughput [1,2]. The information can be encoded in either the amplitude or the phase of spin waves, while logic operation is based on their interference. Different spin-wave-based logic systems have been proposed, *e.g.* Mach Zender interferometers [3], magnonic transistors [4], or spin wave majority gates (SWMGs) [1]. The SWMG is the most promising concept as it possesses a higher expressive power than *e.g.* NAND or NOR gates and thus may reduce circuit complexity. The basic functionality of a such device has been proven at a mm scale using YIG films[5]. Device scalability to μm and nm dimensions has been predicted by micromagnetic simulations [6,7]. So far, the SWMG devices have been based on a "trident" shape. However, such a shape is difficult to scale due to increasing spin wave reflection at bends with μm and nm dimensions [6]. Moreover, such a shape is challenging to print at the nm scale using conventional lithography. In this work, we demonstrate a majority gate using spin wave interference in μm-sized ferromagnetic waveguides using a sequential "in-line" layout of input and output inductive antennas.

II. Fabrication and RF properties

The device consisted of a magnetic stripe that served as a waveguide for the spin waves. Three inductive antennas were used to excite spin waves and an addition antenna was used to detect the resulting wave after interference, *i.e.* after computation (see Fig. 1).

For the waveguide, a $Ta(3nm)/CoFeB(30nm)/Ta(3nm)$ film stack was sputtered onto 300 nm of SiO_2 on Si (100). The stack was then patterned into stripes with a width of 4

μm using ion-beam etching. The ferromagnetic CoFeB acted as the waveguide for the spin waves while the Ta layers served as seed and cap layers to prevent from oxidation of the magnetic film. The waveguide was finally covered by 40 nm of SiN_x for electric isolation (Fig. 2(a)). Subsequently, Ti/Au inductive antennas with a width of 500 nm were fabricated by electron-beam lithography and lift-off (Fig. 2(b)). The antennas were connected to microwave coplanar waveguides and contacted by RF picoprobes (Fig. 2 (c)). A vector network analyzer (VNA) was used for both the excitation and characterization of the spin waves.

Spin waves were generated in the waveguide by the Oersted field generated by the RF currents flowing in the U-shaped input antennas and detected by a single-wire output antenna. This design provided a weak electromagnetic parasitic coupling between adjacent input antennas (see Figs. 3(a) and (b)), as well as between the input and output antennas, as indicated both by experiments and electromagnetic simulations [8]. Electromagnetic simulations of the Oersted field distribution (Fig. 3(c)) show that the U-shaped antennas can efficiently excite spin waves with wavelengths down to 700 nm (Fig. 3(d)).

III. Spin Wave properties and interference

Spin wave transmission experiments were performed with the CoFeB waveguide magnetized by a magnetic field transverse to its long axis. In a first step, the propagation characteristics of spin waves emitted from each of the three input antennas towards the output antenna were determined. A schematic of the experimental configuration is shown in Fig. 4(a); Fig. 4(b) shows a typical transmitted signal from input I_1 towards output O, corresponding to a spin wave propagation distance of 4.8 μm. The minimum spin wave frequency is given by the ferromagnetic resonance, whereas the maximum frequency depends on maximum wavenumber that an antenna can excite. The full frequency-field dependence of the transmitted signal from input I_1 to output O is shown in Fig. 4(c). The device allowed for the generation and propagation of spin waves in a wide frequency range between 3 GHz and 22 GHz, depending on the external applied field. The dispersion relation calculated using parameters extracted from experiments (see Fig.4 (d)) demonstrates that the minimum spin wave frequency matches well the ferromagnetic resonance frequency, while the upper limit was set by the maximum wavevector ($k_{max} \sim 8.9$ rad/μm) that can be excited by the antenna (corresponding to a wavelength of $\lambda =$

978-1-7281-1988-5/18 $31.00 © 2018 IEEE

700 nm). The spin wave transmission from inputs I_2 and I_3 to the output O is shown in Fig. 5. The dephasing due to the different propagation distances could clearly be observed.

Subsequent experiments studied the interference of the spin waves generated simultaneously by multiple input antennas (see Fig. 6(a)). Microwave currents with the same frequency were applied to all three antenna inputs. The output signal was studied as a function of the input frequency, bias field, and relative phase difference between the input signals. For a given set of field-frequency parameters, an oscillatory signal was detected by varying the phase of the input signals corresponding to the constructive or destructive interference of the three generated spin waves. For example, Fig. 6(b) shows the detected signal for a phase rotation of up to 4π of the signal at input I_1, while I_2 and I_3 were kept in phase. The position of the maxima/minima could be tuned by varying the applied frequency, which changes the spin wave wavelength and thus modifies the interference pattern.

IV. LOGIC FUNCTIONS

Building logic functions based on spin wave interference requires the control of both the amplitude and the phase of the spin waves generated by each input. Signal matching at the output was obtained by phase shifters and attenuators in the microwave circuits of each input antenna (see Fig. 7(a)). Figure 7(b) shows that the amplitude and phase of the input signals could be synchronized over a 1 GHz bandwidth for a bias field of 40 mT.

Input 0 and 1 logic states of a SWMG were defined as the phase of the spin wave signals, *i.e.* as phases of 0 and π, respectively. The variation of the phase-sensitive S-parameter (here the imaginary part) measured by the VNA was used to define the logic output signal. Positive and negative variations correspond to output wave phases of π and 0, respectively (logic 1 and 0, respectively). Using the phase of one input as a control signal, for example setting the phase of I_3 to π (Fig. 8(a)), and changing the phase of I_1 and I_2 between 0 and π, a logic OR function could be demonstrated, as shown in Fig. 8(b). In addition, a logic AND gate could be demonstrated by setting the phase of the control input I_3 to 0.

By individually controlling the phase of each input, the truth table of the logic majority operation was demonstrated over a frequency bandwidth of ~300MHz, as shown in Fig. 9. Weak and strong majority states could be distinguished. The clear separation between the states suggests that adding additional inputs will allow to create an *n*-state logic. By tuning the applied frequency on the inputs, an oscillatory output signal was observed (Fig. 10(a)). This fact is explained by the variation of the global phase due to the dependence of spin wave wavelength on frequency, leading to a change of the interference pattern at the output for different frequencies. Thus, an inverted majority gate can be obtained in the same device by tuning the operation frequency (Fig. 10(b)).

Benchmarking

Spin wave logic concepts, and more specifically spin wave devices (SWDs) [1] have been benchmarked several times [9, 10]. All results show that using efficient voltage-driven spin

wave generation and detection, SWDs can outperform state-of-the-art CMOS technology in terms of energy consumption. To showcase this, based on previous benchmarking work [11], we adapt the energy calculations of the 10 designs described in Fig. 11. These benchmarks are combinational and represent a common subset of arithmetic designs used in digital integrated circuits. The energy consumption per operation of SWDs is compared in Fig. 12 to the 10nm CMOS technology node [12]. The energy consumption per operation of the SWD circuits is on average 7.6x times lower than for 10 nm CMOS. This benchmarking highlights the potential for ultralow-power logic built based on SWDs.

Weak & strong majority use cases

As shown in Fig. 9, the device allows for the distinction of strong and weak majority signals. This can be efficiently exploited by non-boolean or multilevel computational techniques. More specifically, it has been shown that strong/weak majority distinction can be applied to signal processing, such as pattern recognition [13]. Moreover, the above capability can be useful in applications where it is important to implement threshold functions (such as for neurons), where the thresholding sensitivity is more expressive than a binary component.

V. CONCLUSION

We have demonstrated a novel in-line spin wave majority gate concept that is both scalable and compatible with conventional CMOS patterning techniques. By individually controlling the phase and amplitude of signal at the three inputs, a full majority truth table was demonstrated. Due to the wave based nature of the operation, the output signal dependent on the applied frequency in an oscillatory way. This could be exploited to demonstrate an inverted majority function in the same device by adjusting the operation frequency. Circuit level benchmarking indicated the high potential of spin wave based devices for low-power electronics.

ACKNOWLEDGMENT

This work was performed as part of the imec IIAP program on Core CMOS and Beyond CMOS. Support from the H2020 project CHIRON (contract No. 692519) is gratefully acknowledged.

REFERENCES

[1] A. Khitun and K. Wang, J. Appl. Phys. 110, 034306 (2011)
[2] Radu et al., Proc. IEEE IEDM (2015)
[3] T. Schneider et al., APL 92, 022505 (2008)
[4] Chumak et al., Nature Comm. Vol. 5 No. 4700 (2014)
[5] T. Fisher et al., APL 110, 152401 (2017)
[6] S. Klinger et al., APL 105, 152410 (2014)
[7] O. Zogragfos et al., AIP Advances 7 (5), 056020 (2017)
[8] The simulations were performed using "ANSYS HFSS: High Frequency Electromagnetic Field Simulation" software
[9] O. Zografos et al., IEEE NANOARCH 2014, pp. 25-30.
[10] C. Pan et al., IEEE JxCDC 3 2017 101-110.
[11] O. Zografos et al., Chapter 7 - R. Topaloglu and P. Wong, Springer, 2018, ISBN 978-3-319-90384-2.
[12] J. Ryckaert et al., CICC 2014, pp. 1–8.
[13] S. Dutta et al., Scientific reports 7.1 (2017): 17866.

978-1-7281-1988-5/18 $31.00 © 2018 IEEE

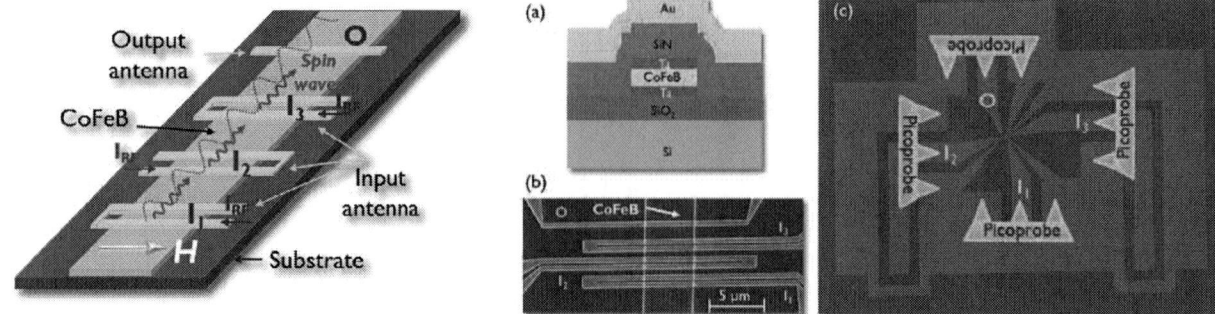

Fig.1. Schematic of a spin wave in-line majority gate

Fig. 2. (a) Schematic of device cross-section under an antenna. (b) SEM image of the active area of the device. The inputs are 3 U-shaped antennas and the output is a single wire antenna. (c) SEM image of the full microwave majority gate device with a sketch of the picoprobe connection to the transducers.

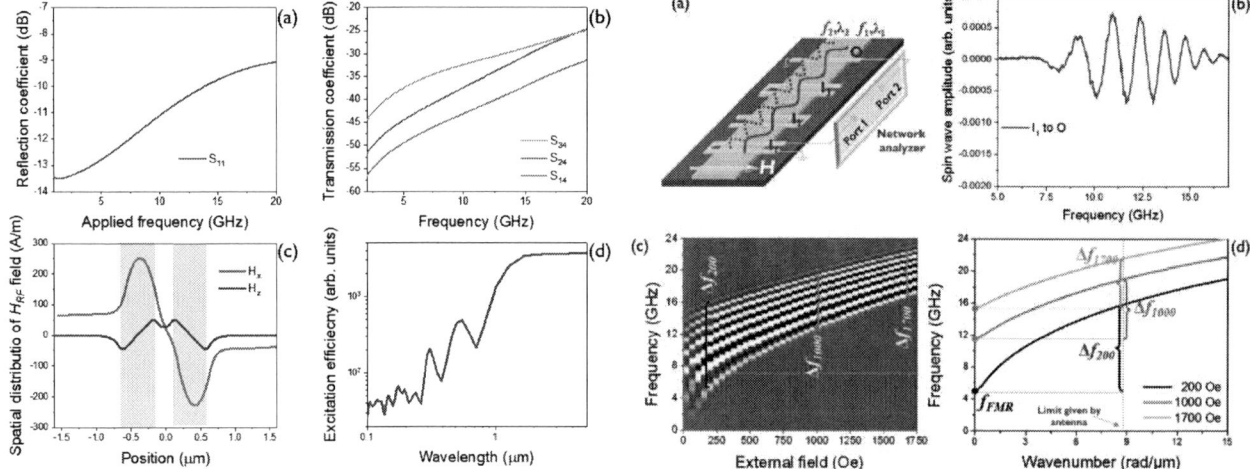

Fig. 3. Scattering parameters of the device simulated by HFSS: (a) reflection coefficient for U-shape antennas, and (b) direct parasitic coupling between every input and the output antenna. (c) Magnetic field components generated by a U-shaped antenna simulated by HFSS and (d) the resulting bandwidth of the device.

Fig. 4. (a) Schematic of the experiment for a single input and (b) the detected signal due to spin wave propagation at the output. (c) Frequency-field dependence of the spin wave transmission. Light blue color corresponds to zero spin-wave transmission, while the dark blue and the white band stands for propagating spin waves. (d) Spin wave dispersion relation calculated for three values of the magnetic field.

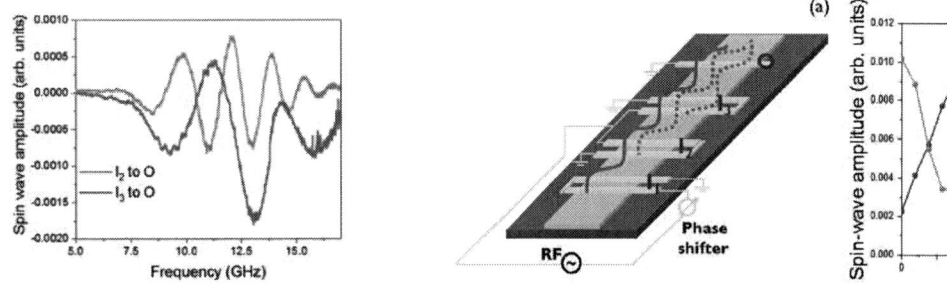

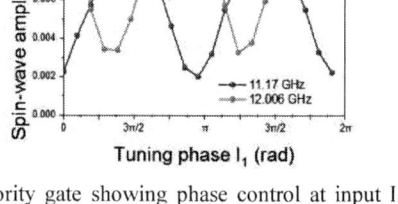

Fig. 5. Spin wave transmission from inputs I_2 and I_3 to the output O.

Fig.6. (a) Sketch of a spin wave in-line majority gate showing phase control at input I_1. (b) Output signal generated by the interference of the three spin waves. The change of the amplitude is determined by constructive and destructive interference. The bias field was set to 50 mT.

978-1-7281-1988-5/18 $31.00 © 2018 IEEE

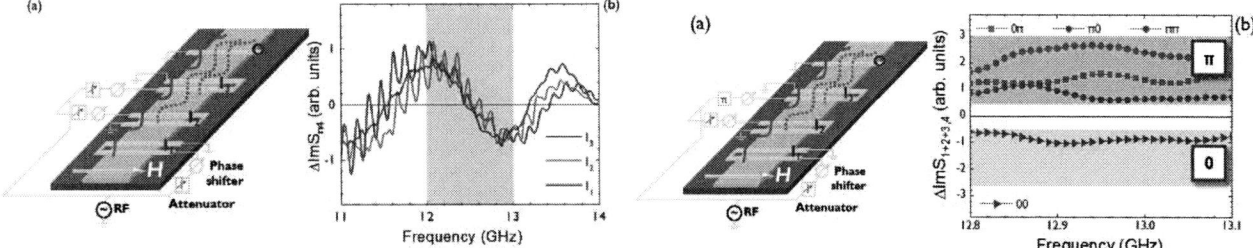

Fig. 7. (a) Schematic of a spin wave in-line majority gate with phase control at each transducer. (b) Output signal generated by each input showing the possibility to match phase and amplitude of the three input spin waves, as required to build logic gates.

Fig. 8. (a) Schematic of an OR gate where one input (I_3) is fixed at a phase of π as a control gate. (b) Experimental signal demonstrating the functionality as an OR gate,

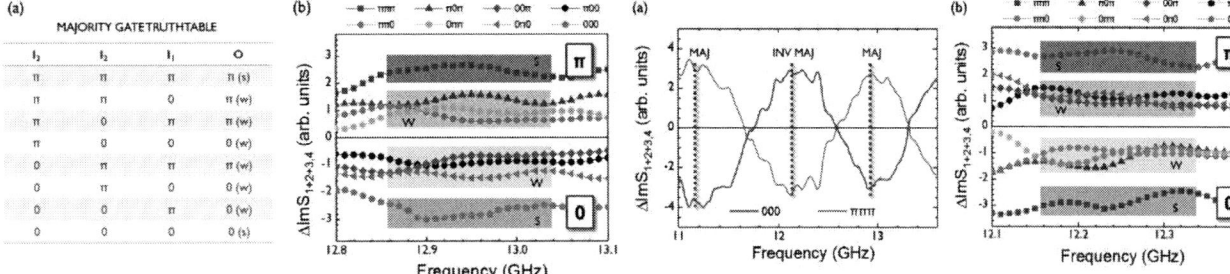

Fig. 9. (a) Majority Gate truth table, indicating cases of strong and weak majority (b) Spin wave transmission due to interference of the 3 waves at the output tranducer. The 8 cases can be observed as well as a clear separation between strong and weak majority gate in a 200 MHz frequency bandwidth.

Fig. 10. (a) Spin wave transmission due to interference of 3 waves at output in a broad frequency span for the two cases of strong majority. Due to the phase rotation it can be observed how a MAJ state transforms in a MIN (INV + MAJ) and then again in MAJ. (b) Experimental MIN truth table where strong and weak minority can be separated in a 200 MHz frequency range.

NAME	DESCRIPTION
CRC32	Cyclic redundancy check XOR tree
BKA264	2-operand 64-bit Brent-Kung Adder
GFMUL	Mastrovito multiplier for irreducible polynomial
CSA464	4-operand 64-bit Carry-Skip Adder
HCA464	4-operand 64-bit Han-Carlson Adder
WTM32	2-operand 32-bit Wallace tree Multiplier
MAC32	3-operand 32-bit (7,3) counter tree MAC
DTM32	2-operand 32-bit Dadda tree Multiplier
DIV32	2-operand 32-bit Divider
DTM64	2-operand 64-bit Dadda tree Multiplier

Fig. 11. Descriptions of benchmark designs used to compare spin wave logic and CMOS technologies.

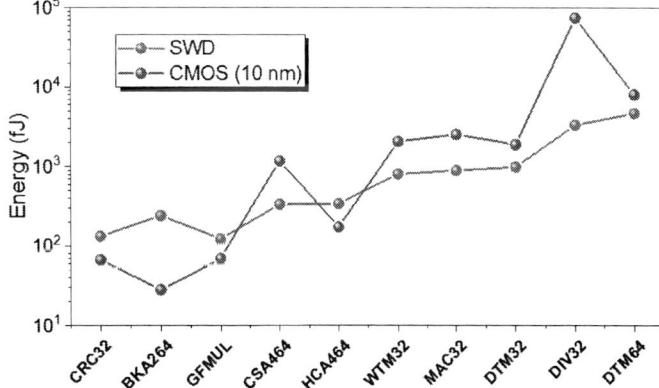

Fig. 12. Energy per operation of spin wave logic circuits and CMOS (10nm) as technology reference. Benchmarks are ordered in increasing circuit size. We observe that for sufficiently large circuits, wave logic operate at lower energy than state-of-the-art CMOS.

Spintronic devices for low energy dissipation

Kang L. Wang, Hao Wu, Seyed Armin Razavi, and Qiming Shao

Department of Electrical and Computer Engineering, and Department of Physics and Astronomy,
University of California, Los Angeles, California 90095, USA,
E-mail: wang@ee.ucla.edu

Abstract— Spintronic devices are considered as one of the best candidates for next-generation electronics to complement CMOS technology. First, we will briefly show the recent progresses on spintronic devices based on spin-transfer torque, spin-orbit torque and voltage-controlled magnetic anisotropy in reference to energy efficient applications. We will then discuss the recent progresses using antiferromagnets and topological insulators for applications in memory, logic and circuits. Finally, the new prospective of 2D materials for spintronic devices will be addressed.

I. INTRODUCTION

The next generation electronics to complement and to go beyond today's CMOS technology have been intensively studied. Among them, spintronic devices are considered to be one of the candidates to eliminate the standby power dissipation due to the advantage of non-volatility and to minimize the switching energy with an ultra-low energy consumption (< fJ), high speed (<10 ns), and almost unlimited endurance (>10^{16}). Recent advents on memory (magnetic random-access memory, MRAM) and logic devices have also drawn interests for applications in logic-in-memory circuits with non-volatile storage elements, which may overcome the latency of speed and energy dissipation in communication between logic and memory cells. Moreover, the parallel operations of logic-in-memory based on MRAM could dramatically increase the computer performance.

Recent progresses have demonstrated the ability to switch the magnetic moment of spintronic devices by current induced spin-orbit torque (SOT). Compared with today's spin-transfer torque (STT), the speed, endurance and energy consumption can be much reduced with SOT. Furthermore, the energy consumption could be further reduced by the use of voltage-controlled magnetic anisotropy (VCMA) or magnetoelectric random-access memory (MeRAM), where the magnetization is switched by the voltage without current induced Joule heating. New materials such as topological insulators and 2-dimensional (2D) materials have also been developed for spintronic devices, which could be used to further improve the performance and to offer additional different control mechanisms of magnetism for devices.

II. RESULTS AND DISCUSSION

A. Spin-transfer torque

To date, most spintronic devices are based on a magnetic tunnel junction (MTJ) built on a thin magnetic CoFeB fixed layer, a tunneling oxide (MgO) and a thin CoFeB free layer, (see Fig. 1) [1]. The first generation of the spintronic memory devices uses STT to switch the magnetic free layer in the structure. For switching, a vertical charge current is applied to the MTJ stack to transfer the angular momenta of electrons to the free layer from the magnetization of the fixed layer, and the current direction determines the final switched state (up or down), resulting in an on-off ratio or TMR, tunneling magneto-resistance ratio as $(R_{AP} - R_P)/R_P$. Today's STT-MRAM has achieved a switching energy on the order of ~100 fJ.

B. Spin-orbit torque

To further improve on STT, high spin-orbit coupling materials (such as heavy metals or ultimately topological insulators) are used for switching the magnetization of a free ferromagnetic layer. Upon applying an in-plane charge current through the high spin-orbit coupling layer, a spin-polarized current is created, and it exerts torques on the adjacent ferromagnet of the MTJ, switching it under proper conditions [2, 3] (see Fig. 2). This three terminal SOT-MRAM separates the reading and writing paths, and thus avoids passing a high writing current through the tunnel barrier, damaging the barrier as in the case of STT. Furthermore, SOT is more energy efficient.

C. Field-free SOT switching

Deterministic SOT switching usually requires an in-plane bias magnetic field to break the structural symmetry but this extra bias magnetic field is not desirable for practical applications. This issue was circumvented by using lateral structural asymmetry. The structural inversion symmetry can be broken with an out-of-plane effective field, enabling deterministic bias-field-free SOT switching (see Fig. 3 and Fig. 4) [4]. The bias field-free SOT switching has also been realized using an in-plane exchange bias at the antiferromagnet (AFM) (IrMn) /ferromagnet interface to replace the external bias field, as shown in Fig. 5 [5-7].

D. Antiferromagnetic SOT

Fig. 6 shows the SOT induced magnetization switching in the IrMn/CoFeB/MgO structure, where the SOT comes from the AFM IrMn, not from the conventional heavy metals (HMs). The spin Hall tangent and switching current density of IrMn are in the same orders of magnitude as of HMs. Also, the simultaneous presence of the in-plane exchange bias at the IrMn/CoFeB interface enables the field-free SOT switching.

E. Topological SOT from surface states

Spin-momentum locking in topological surface states provides a large charge-spin conversion efficiency. SOT from topological surface states for magnetization switching is a route to further improve the energy efficiency of SOT devices. Fig. 7 shows the giant SOT switching in a $(Bi_{0.5}Sb_{0.5})_2Te_3/(Cr_{0.08}Bi_{0.54}Sb_{0.38})_2Te_3$ bilayer heterostructure with a much reduced switching current density below 8.9×10^4 A/cm^2 at 1.9K [8] and in a Bi_2Se_3/GdFeCo structure at room temperature (300 K) with 2.0×10^6 A/cm^2, as shown in Fig. 8. This offers new opportunities for topological spintronics.

F. Voltage-controlled magnetic anisotropy

Another promising switching mechanism is based on VCMA, which is widely investigated because of its lower switching energy compared with STT or SOT [9-11]. In this method, a voltage is applied across the MTJ, which reduces the anisotropy energy barrier to allow for precessional switching of the free layer due to interface SOC. In VCMA, because of the charge accumulation or depletion at the ferromagnet/insulator interface, electron occupancy of different orbital levels is modified at the interface, resulting a change in magnetic anisotropy. At a reduced anisotropy, switching is realized with timing the voltage pulse (see Fig. 9). To improve the effectiveness of voltage control over magnetic anisotropy for scaling the voltage and energy as well as density, it is necessary to improve the VCMA coefficient > 1000 (fJV^{-1}m^{-1}), a 1000 challenge. VCMA switching can be more energy-efficient compared to current-driven switching, with the switching energy below the fJ/bit region, two orders of the magnitude smaller than STT or SOT (see Fig. 10) [12].

G. Spintronic programmable logic

A non-volatile spintronic programmable logic (SPL) based on a 3-teriminal MTJ is illustrated in Fig. 11. In this case, the SOT induced switching probability of MTJ can be reduced by the VCMA effect, or by applying a voltage across the MTJ [13]. Fig. 12 shows an example of the 2-input circuit design of SPL, which consists of a write circuit, a selection tree, and a current conveyer. Based on this logic circuit design, the SPL can realize any type of combinational and sequential logic functions by combining a flip-flop. Compared to SRAM and STT-MRAM-based logic circuit, the SPL based 2-input logic circuit can achieve 35% and 28% area reduction, as illustrated in Fig. 13.

H. 2D spintronics

Almost all the spintronic devices use MTJs as the storage element, where the on-off ratio is limited to ~200% over today's MTJs at room temperature. For large arrays, a large TMR of >1000% (another 1000 challenge) is required. Spintronic devices based on 2D materials have emerged with promising possibilities for future nonvolatile computing. Recently, layered 2D ferromagnetic materials, such as CrI$_3$ [14], Cr$_2$Ge$_2$Te$_6$ [15], and others have been demonstrated. Furthermore, the spin filtering effect from the 2D barrier (CrI$_3$) leads to a giant tunnel magnetoresistance ratio up to 20000% though at low temperature [16]. 2D spintronic materials can have AFM and topological properties, and may be used to forge new AFM and topological spintronics and

beyond. They may also offer solutions of the proposed VCMA 1000 challenge.

III. CONCLUSION

In conclusion, we present the recent research on spintronic devices with ultra-low power consumption. Several devices are discussed with current driven STT and SOT, the latter of which can further reduce energy consumption and endurance of spintronic devices. Furthermore, VCMA based spintronic devices show much higher energy efficiency than current-driven STT and SOT devices. We also show the programmable spin logic design based on SOT and VCMA. Two 1000 challenges are proposed for (i) improving the TMR of >1000% and (ii) the VCMA of >1000 (fJV^{-1}m^{-1}). 2D materials, AFMs and topological insulators offer additional potentials for further advancing spintronics and computing technology.

REFERENCES

[1] K. L. Wang, J. G. Alzate and P. K. Amiri, Journal of Physics D: Applied Physics **46** (7), 074003 (2013).
[2] L. Liu, C.-F. Pai, Y. Li, H. W. Tseng, D. C. Ralph and R. A. Buhrman, Science **336** (6081), 555-558 (2012).
[3] I. M. Miron, K. Garello, G. Gaudin, P.-J. Zermatten, M. V. Costache, S. Auffret, S. Bandiera, B. Rodmacq, A. Schuhl and P. Gambardella, Nature **476**, 189 (2011).
[4] G. Yu, P. Upadhyaya, Y. Fan, J. G. Alzate, W. Jiang, K. L. Wong, S. Takei, S. A. Bender, L.-T. Chang, Y. Jiang, M. Lang, J. Tang, Y. Wang, Y. Tserkovnyak, P. K. Amiri and K. L. Wang, Nature Nanotechnology **9**, 548 (2014).
[5] S. A. Razavi, D. Wu, G. Yu, Y.-C. Lau, K. L. Wong, W. Zhu, C. He, Z. Zhang, J. M. D. Coey, P. Stamenov, P. Khalili Amiri and K. L. Wang, Physical Review Applied **7** (2), 024023 (2017).
[6] S. Fukami, C. Zhang, S. DuttaGupta, A. Kurenkov and H. Ohno, Nature Materials **15**, 535 (2016).
[7] Y.-W. Oh, S.-h. Chris Baek, Y. M. Kim, H. Y. Lee, K.-D. Lee, C.-G. Yang, E.-S. Park, K.-S. Lee, K.-W. Kim, G. Go, J.-R. Jeong, B.-C. Min, H.-W. Lee, K.-J. Lee and B.-G. Park, Nature Nanotechnology **11**, 878 (2016).
[8] Y. Fan, P. Upadhyaya, X. Kou, M. Lang, S. Takei, Z. Wang, J. Tang, L. He, L.-T. Chang, M. Montazeri, G. Yu, W. Jiang, T. Nie, R. N. Schwartz, Y. Tserkovnyak and K. L. Wang, Nature Materials **13**, 699 (2014).
[9] P. K. Amiri and K. L. Wang, SPIN **02** (03), 1240002 (2012).
[10] W.-G. Wang, M. Li, S. Hageman and C. L. Chien, Nature Materials **11**, 64 (2011).
[11] T. Maruyama, Y. Shiota, T. Nozaki, K. Ohta, N. Toda, M. Mizuguchi, A. A. Tulapurkar, T. Shinjo, M. Shiraishi, S. Mizukami, Y. Ando and Y. Suzuki, Nature Nanotechnology **4**, 158 (2009).
[12] C. Grezes, F. Ebrahimi, J. G. Alzate, X. Cai, J. A. Katine, J. Langer, B. Ocker, P. Khalili Amiri and K. L. Wang, Applied Physics Letters **108** (1), 012403 (2016).
[13] H. Lee, F. Ebrahimi, P. K. Amiri and K. L. Wang, IEEE Magnetics Letters **7**, 1-5 (2016).
[14] B. Huang, G. Clark, E. Navarro-Moratalla, D. R. Klein, R. Cheng, K. L. Seyler, D. Zhong, E. Schmidgall, M. A. McGuire, D. H. Cobden, W. Yao, D. Xiao, P. Jarillo-Herrero and X. Xu, Nature **546**, 270 (2017).
[15] C. Gong, L. Li, Z. Li, H. Ji, A. Stern, Y. Xia, T. Cao, W. Bao, C. Wang, Y. Wang, Z. Q. Qiu, R. J. Cava, S. G. Louie, J. Xia and X. Zhang, Nature **546**, 265 (2017).
[16] T. Song, X. Cai, M. W.-Y. Tu, X. Zhang, B. Huang, N. P. Wilson, K. L. Seyler, L. Zhu, T. Taniguchi, K. Watanabe, M. A. McGuire, D. H. Cobden, D. Xiao, W. Yao and X. Xu, Science (2018).

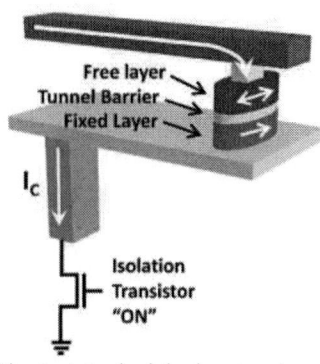

Fig. 1. A typical device structure for a 1 transistor – 1 MTJ memory cell, utilizing spin-transfer-torque (STT) for switching. A charge current, I_C, goes through the stack and switches the free layer magnetization [1].

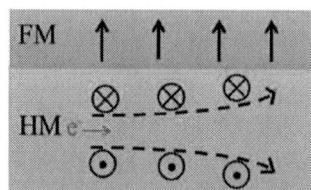

Fig. 2. Spin-orbit torque (SOT) for switching a perpendicular magnet. Upon applying a current through a heavy-metal (HM), electrons with different spins (up or down) are deflected in opposite directions, resulting in an accumulation of spin-polarized electrons at HM/ferromagnet (FM) interface. The spin accumulation exerts torque on the FM magnetization, and switches it under proper conditions.

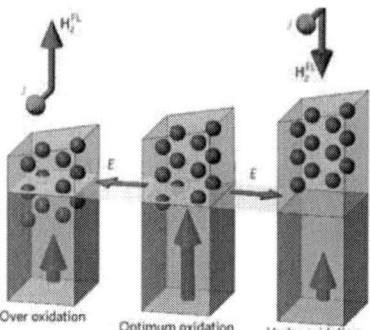

Fig. 3. Schematic illustrating a non-uniform oxygen distribution at the ferromagnet/oxide interface. The resulting non-uniform charge distribution induces an in-plane electric field (E) along the interface, which in turn contributes to the out-of-pane effective field H_z^{FL} [4].

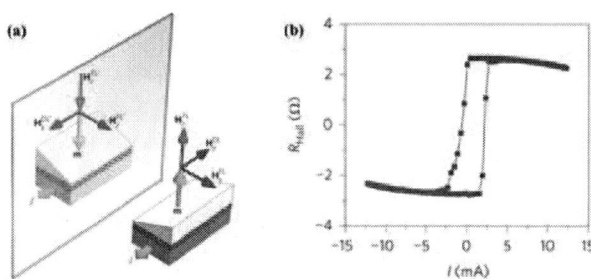

Fig. 4. Using structural asymmetry for bias-field-free SOT switching. (a) Structural asymmetry results in the creation of a current-induced out-of-plane effective magnetic field, H_z^{FL}, which determines the final switching direction. (b) Bias field-free SOT switching in an asymmetric structure, measured using anomalous Hall effect [4].

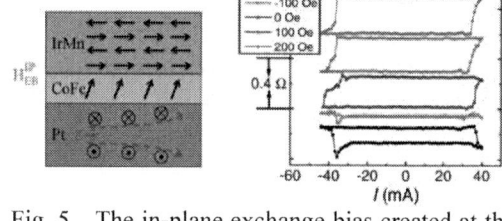

Fig. 5. The in-plane exchange bias created at the CoFe/IrMn interface replaces the external field and enables field-free SOT switching. The Pt layer provides SOTs [5].

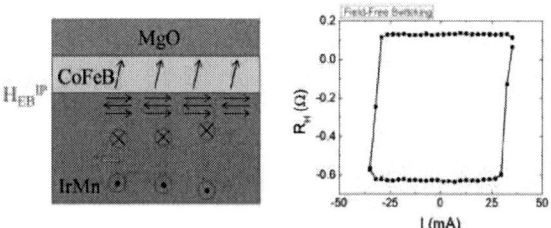

Fig. 6. A modified structure suitable for field-free SOT switching. In this system, a single antiferromagnet IrMn layer provides both the exchange bias and the SOT.

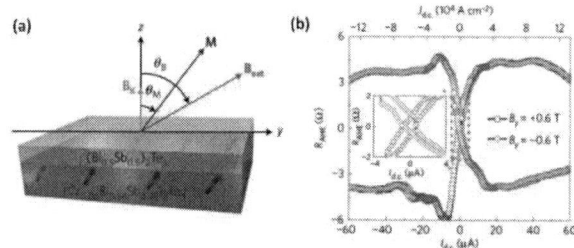

Fig. 7. SOT induced magnetization switching in $(Bi_{0.5}Sb_{0.5})_2Te_3/(Cr_{0.08}Bi_{0.54}Sb_{0.38})_2Te_3$ bilayer at 1.9 K in the presence of a constant y-axis in-plane external magnetic field with $\pm$ 0.6 T, respectively [8].

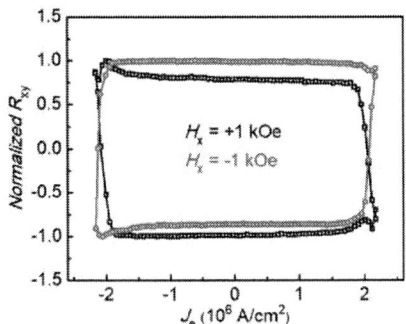

Fig. 8. Room temperature SOT induced magnetization switching in the Bi_2Se_3/GdFeCo structure with a ± 1 kOe in-plane magnetic field.

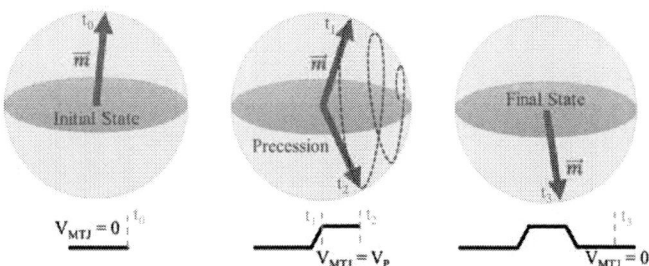

Fig. 9. Using voltage-controlled magnetic anisotropy (VCMA) for magnetization switching. Applying a voltage over the MTJ (V_P) reduces the anisotropy and starts the magnetization precession. Magnetization switching is achieved by timing the voltage pulse.

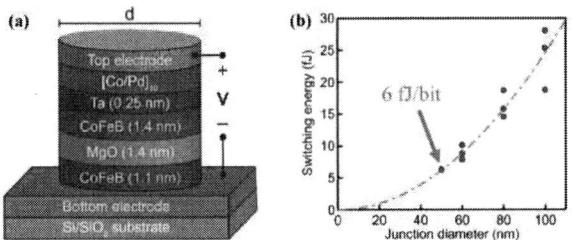

Fig. 10. Efficient magnetization switching using VCMA. (a) Schematics of a typical MTJ device, in which VCMA is used for switching. V is the applied voltage and d is the junction diameter. (b) VCMA switching energy with junction diameter. Switching energies are as low as 6 fJ/bit [12].

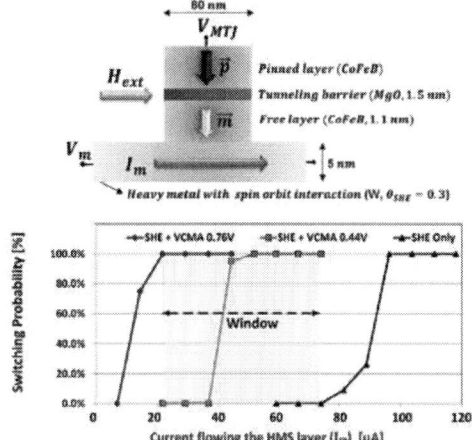

Fig. 11. Schematic and switching probability of the 3-terminal MTJ with combined SOT and VCMA [13].

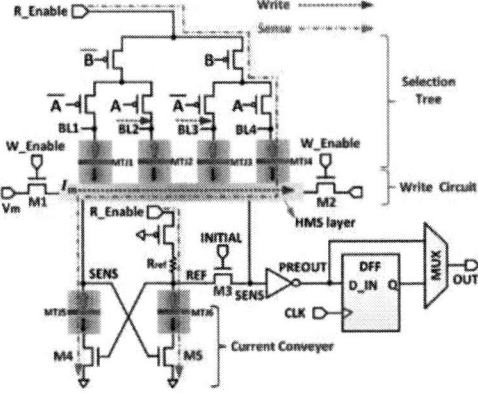

Fig. 12. Schematic of the proposed 2-input spintronic programmable logic (SPL) [13].

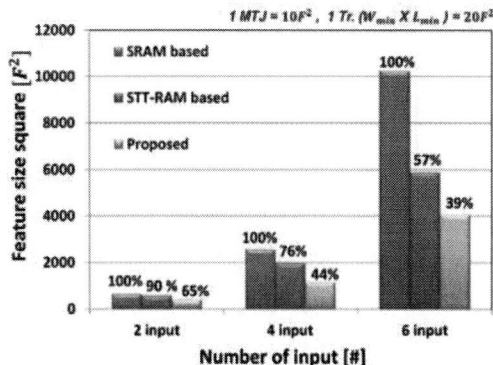

Fig. 13. Area comparison of different types of lookup tables (LUTs), which shows that SPL is the most area-efficient structure compared to SRAM and STT-RAM based LUTs [13].

978-1-7281-1988-5/18 $31.00 © 2018 IEEE

Room Temperature Highly Efficient Topological Insulator/Mo/CoFeB Spin-Orbit Torque Memory with Perpendicular Magnetic Anisotropy

Qiming Shao[*#], Hao Wu[*], Quanjun Pan[*], Peng Zhang, Lei Pan, Kin Wong, Xiaoyu Che and Kang L. Wang[#]

Department of Electrical and Computer Engineering, University of California, Los Angeles, CA 90095, USA

[*] These authors contributed to this work equally.

[#] Emails: sqm@ucla.edu and wang@ee.ucla.edu

Abstract—Spin-orbit torque (SOT)-MRAM is a promising candidate for future nonvolatile memory technology. Finding materials that have large SOT efficiency (ξ_{DL}) is critical for developing the SOT-MRAM. Topological insulators (TIs) have been shown to exhibit giant ξ_{DL} (>1) at room temperature. However, integration of high ξ_{DL} TIs with CoFeB with perpendicular magnetic anisotropy (PMA) at room temperature (RT) has not been achieved. In this work, we demonstrate a record-high ξ_{DL} (~2.66) in the $(BiSb)_2Te_3$ with PMA CoFeB and achieve magnetization switching with TI current density as low as $3\times10^9 A/m^2$ at RT. For the first time, we propose to insert a light metal spacer between TI and CoFeB to achieve resistance matching and thus reduce write energy. We show that without insertion, TI/CoFeB show in-plane magnetic anisotropy but TIs show high ξ_{DL}, consistent with previous reports. We then insert a Mo spacer to achieve PMA at RT. We accurately determine the ξ_{DL} using both second harmonic method and MOKE for the first time. We investigate the SOT-driven switching and discover a memristor-like behavior in the TI/Mo/CoFeB.

I. INTRODUCTION

Ever-present data require large and fast data processing capability. Traditional von Neumann architecture separates the computing and storage units, which cause significant delay and energy consumption over the data bus. Spin-transfer torque (STT)-MRAM and spin-orbit torque (SOT)-MRAM are promising for embedded memory technology thanks to their fast write and read (several ns) [1]. STT-MRAM single bit is a two-terminal device whereas SOT-MRAM single bit is three-terminal, where the write and read paths are separate. Since there is no charge current directly going through the magnetic tunnel junction (MTJ), the endurance and reliability of SOT-MRAM could be much higher. The switching current density (J_{sw}) of a SOT-MRAM device is given by $J_{sw} = 4ekT\Delta/(\hbar\xi_{DL}A_{MTJ})$, where Δ is the thermal stability factor and ξ_{DL} is the (anti-damping-like) SOT efficiency. Heavy metals (HMs), such as β-Ta [2], Pt [3-4], and β-W [5], are common channel materials used in SOT-MRAM for generating SOTs, whose ξ_{DL} range from 0.08 to 0.3 (see Table 1). Recently, topological insulators (TIs), like Bi_2Se_3 [6] and $(BiSb)_2Te_3$ [7], have emerged as potential channel materials thanks to their extraordinary large ξ_{DL} (>1). However, questions about the energy efficiency have been raised since TIs usually have a resistivity (ρ_{TI}) at the order of $10^3 - 10^4$ μΩ·cm and thus during switching, most of current flows in the CoFeB layer,

which is the most common MTJ ferromagnet and has a ρ_{CoFeB} ~ 170 μΩ·cm [1]. Also, perpendicular magnetic anisotropy (PMA) MTJ is essential for MRAM scaling [1]. But TIs with bulk PMA CoTb show a ξ_{DL} <1 [8]. Room temperature (RT) high ξ_{DL} TIs with PMA CoFeB has remain elusive.

In this work, we demonstrate a large ξ_{DL} ~ 2.66 in the TI/Mo/CoFeB/MgO structure with RT PMA using the second harmonic (2ω) method [9] and the magneto-optical Kerr effect (MOKE) [10]. Moreover, we show the SOT-driven magnetization switching with TI current density as low as 3×10^9 A/m². Interesting memristor-like behavior is observed.

II. SOT-MRAM UNIT CELL ENERGY MODELING

A. SOT-MRAM with different channel materials

Fig. 1 shows a schematic of a SOT-MRAM unit cell. For simplicity, we assume a square MTJ shape ($A_{MTJ} = R^2$), which is valid for the PMA case. Figs. 2 and 3 show the write energy (E_{write}) and write voltage (V_{write}) as a function of $\rho_{HM,TI}$, where we have two assumptions. First, we assume that switching time is 3 ns and $J_{sw} = 10^6$ MA/m² for R = 10 nm, Δ = 40, and ξ_{DL} = 0.1. Second, we assume that the SOT generation is dominated by the intrinsic mechanism [4], which means the $\xi_{DL} = \rho_{HM,TI}\cdot\sigma_{SH}$, where σ_{SH} is the intrinsic spin Hall conductivity. The parameters used for Figs. 2 and 3 are shown in Table 1. We observe that the E_{write} is lowest for Bi_2Se_3 thanks to its large ξ_{DL}. Meanwhile, the V_{write} is well below 0.3 V, which is feasible for practical applications. Fig. 4 shows that the E_{write} increases as the MTJ size scales down following $E_{write} \propto R^{-2}$.

B. Light metal spacer insertion for PMA and lower E_{write}

CoFeB is a widely used free layer due to its high TMR ratio in CoFeB/MgO/CoFeB MTJ. As we will show in the section IV, however, TI/CoFeB/MgO does not exhibit PMA. By inserting a light metal (LM) spacer between TI and CoFeB, interfacial PMA can be achieved. Note that the LM is for keeping the spin current and thus the large ξ_{DL} in TIs. Furthermore, Fig. 5 shows that by appropriately choosing the spacer thickness and ρ_{LM}, reduction of the E_{write} is achieved.

III. VALIDATION OF SOT MEASUREMENT METHODS

We employ the 2ω method [9] to detect the SOT effective field (B_{DL}) and thus determine the ξ_{DL}. Fig. 6 shows the measurement schematic and device image. The 2ω Hall resistance ($R_H^{2\omega}$) is given by [9]

978-1-7281-1988-5/18 $31.00 © 2018 IEEE

$$R_H^{2\omega} = R_P \frac{B_{FL}}{|B_{ext}|} \cos 2\varphi \sin \varphi + \left(\frac{R_A}{2} \frac{B_{DL}}{|B_{ext}|-B_K} + R_{th} \right) \sin \varphi, \quad (1)$$

where φ is the angle between the current and B_{ext}, R_A and R_P are anomalous and planar Hall resistance, B_{DL} and B_{FL} are anti-damping-like and field-like SOT effective field, R_{th} is the thermal contribution. To validate the 2ω method in our samples, we use the MOKE method [10], which is a direct measurement of the B_{DL}. There is no thermal contribution in MOKE measurements. The differential Kerr ($\Delta\theta_K$) is given by [10]

$$\Delta\theta_K = \theta_\parallel \frac{B_{FL}}{|B_{ext}|} \cos 2\varphi_P + \theta_\perp \frac{B_{DL}}{|B_{ext}|-B_K}, \quad (2)$$

where φ_P is the angle between the current and the polarization of the laser, $\theta_\perp$ and $\theta_\parallel$ are the first- and second-order MO coefficients that parameterize the strength of the coupling of the light to the out-of-plane (OOP) and the in-plane (IP) magnetization. Fig. 7 shows the measurement schematic and device image during the measurement. Figs. 8 and 9 show that the 2ω method and the MOKE method provide the same field dependence for the $R_H^{2\omega}$ and $\Delta\theta_K$. Note that this is happening only when $R_P \ll R_A$, the R_{th} is very small compared with the $R_A B_{DL}/(|B_{ext}|-B_K)$, and $\theta_\parallel \ll \theta_\perp$ [10].

IV. SOT DETERMINATION WITH IP AND OOP COFEB

We grow TI materials, Bi_2Se_3, Bi_2Te_3 and $(BiSb)_2Te_3$, using MBE and the growth method is shown in ref. [7]. Fig. 10 shows the typical RHEED image during growth, indicating the high quality of TIs.

A. TI/CoFeB

We deposit CoFeB thin films using magnetron sputtering at room temperature (RT). Unless specifically mentioned, all CoFeB thin films are capped with $MgO(2nm)/TaO_x(3nm)$ for protection. We do not observe PMA in all CoFeB thickness. This could be because there are atomic terraces on the surface of TI thin films due to the nature of van der Waals materials. Fig. 11a shows an OOP Hall resistance (R_H) hysteresis loop for a $Bi_2Se_3(6nm)/CoFeB(5nm)$ bilayer, where the anisotropy field (B_K) is estimated to around 1.3 T. We measure the $R_H^{2\omega}$ as a function of B_{ext} at different φ's (Fig. 11b), from which we can extract the B_{DL} for each φ using Eq. (1). Note that the R_{th} is significant for in-plane anisotropy systems. Fig. 11c shows that the φ-dependence of the B_{DL} satisfies a sine function, agreeing with the nature of the anti-damping-like SOT. Fig. 12 shows the current dependence of B_{DL} for different channel materials. We also summarize the results of ξ_{DL} in Table 1.

B. TI/Mo/CoFeB

Molybdenum (Mo) has very small ξ_{DL} and relatively long spin diffusion length. Thus, it does not affect the spin current generated from the HM layer [11]. Also, Mo/CoFeB/MgO has strong PMA. Motivated by the SOT-MRAM energy modeling and advantages of Mo, we insert 2 nm-thick Mo between $(BiSb)_2Te_3$ and CoFeB. Fig. 13 shows the OOP R_H hysteresis loops for different CoFeB thickness (t_{CoFeB}). The PMA exists in a narrower t_{CoFeB} window in $(BiSb)_2Te_3/Mo/CoFeB$, compared with the Ta/CoFeB and W/CoFeB cases.

Nevertheless, we achieve RT PMA in the $(BiSb)_2Te_3(6nm)/Mo(2nm)/CoFeB(1.02nm)$. Fig. 14 shows that the $(BiSb)_2Te_3(6nm)/Mo(2nm)/CoFeB(0.93nm)$ exhibits strong PMA with a sizeable coercive field when the sample is cooled down to 200 K. This suggests that the 0.93 nm-thick CoFeB on TI is superparamagnetic (multi-domain state) at RT. Fig. 15 show the SOT determination in the TI/Mo/CoFeB with PMA. Note that the R_{th} is negligible in PMA systems, which allows accurate determination of B_{DL}. The B_{DL} is linearly dependent on the current and the result of ξ_{DL} is summarized in Table 1.

V. SOT SWITCHING

A. RT SOT switching in TI/Mo/CoFeB

We realize current-driven magnetization switching in the $(BiSb)_2Te_3(6nm)/Mo(2nm)/CoFeB(1.02nm)$ Hall bar device as shown in Fig. 16. The channel width is 20 μm. The switching direction is opposite with opposite bias fields, agreeing with the nature of anti-damping-like SOT. We notice that the I_{sw} is around 4.5 mA and the J_{sw} in $(BiSb)_2Te_3$ is around $3\times10^9 A/m^2$. This low J_{sw} is consistent with large ξ_{DL}.

B. Memristor-like switching behavior

Interestingly, we observe memristor-like switching behavior in TI/Mo/CoFeB samples (Fig. 17). This could be due to the multi-domain state formation during the switching in this trilayer [12]. This effect could be potentially utilized for neuromorphic computing.

VI. CONCLUSION

Fig. 18 summarizes the research progress on channel materials towards an energy efficient SOT-MRAM with PMA. This work is among the first to demonstrate the combination of TIs and interfacial PMA CoFeB, which is industry-compatible. Moreover, we achieve a record-high ξ_{DL}.

Right before the submission of this work, we notice a very recent publication [13], in which sputtered Bi_xSe_{1-x}/Ta/CoFeB/Gd/CoFeB is used to obtain PMA.

ACKNOWLEDGMENT

The authors gratefully acknowledge the help of sample preparations from Koichi Murata and Guoqiang Yu, MOKE measurements from Bingqian Dai. This work is supported by SHINES under award #S000686 and MURI Grant Number W911NF-16-1-0472 and W911NF-15-1-10561.

REFERENCES

[1] K. L. Wang, et al., J. Phys. D Appl. Phys., 46, 7, 074003, 2013.
[2] L. Liu, et al., Science, 336, 6081, 555-8, 2012.
[3] I. M. Miron, et al., Nature, 476, 7359, 189-93, 2011.
[4] M. H. Nguyen, et al., Phys. Rev. Lett., 116, 12, 126601, 2016.
[5] C.-F. Pai, et al., Appl. Phys. Lett., 101, 12, 122404, 2012.
[6] A. R. Mellnik, et al., Nature, 511, 7510, 449-51, 2014.
[7] Y. Fan, et al., Nat. Mater., 13, 7, 699-704, 2014.
[8] J. Han, et al., Phys. Rev. Lett., 119, 7, 077702, 2017.
[9] Q. Shao, et al., Nano Lett., 16, 12, 7514-7520, 2016.
[10] M. Montazeri, et al., Nat. Commun., 6, 8958, 2015.
[11] D. Wu, et al., Appl. Phys. Lett., 108, 21, 212406, 2016.
[12] S. Fukami et al., Nat. Mater., 15, 5, 535-41, 2016.
[13] M. DC, et al., Nat. Mater., advance online publication, https://doi.org/10.1038/s41563-018-0136-z (2018)

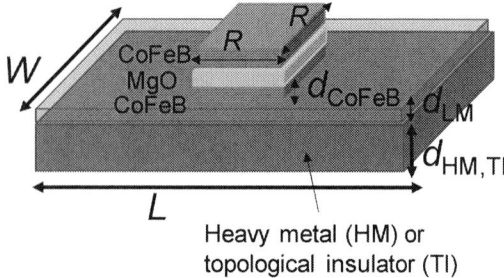

Fig. 1. Schematic of a single SOT-MRAM bit. For calculations in this work, we use $W = 2R$, $L = 3R$, ρ_{CoFeB} = 170 $\mu\Omega\cdot$cm, d_{CoFeB} = 1 nm, and $d_{HM,TI}$ = 6 nm.

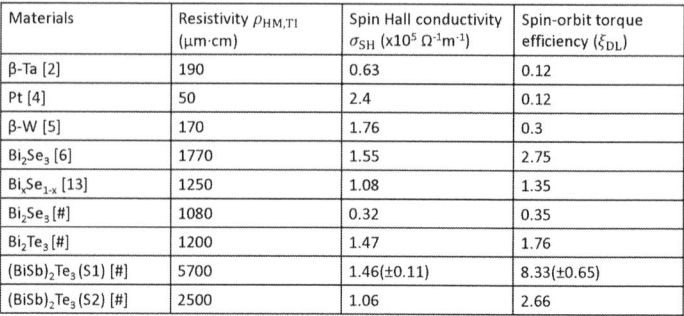

Materials	Resistivity $\rho_{HM,TI}$ (μm·cm)	Spin Hall conductivity σ_{SH} (x10⁵ $\Omega^{-1}m^{-1}$)	Spin-orbit torque efficiency (ξ_{DL})
β-Ta [2]	190	0.63	0.12
Pt [4]	50	2.4	0.12
β-W [5]	170	1.76	0.3
Bi₂Se₃ [6]	1770	1.55	2.75
BiₓSe₁₋ₓ [13]	1250	1.08	1.35
Bi₂Se₃ [#]	1080	0.32	0.35
Bi₂Te₃ [#]	1200	1.47	1.76
(BiSb)₂Te₃ (S1) [#]	5700	1.46(±0.11)	8.33(±0.65)
(BiSb)₂Te₃ (S2) [#]	2500	1.06	2.66

Table 1. Summary of RT resistivity, spin Hall conductivity and SOT efficiency for heavy metals and topological insulators. # This work.

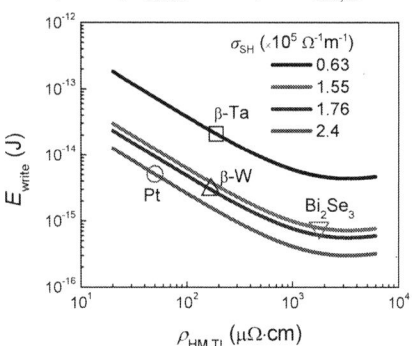

Fig. 2. Write energy as a function of channel material resistivity for different σ_{SH} (R = 10 nm)

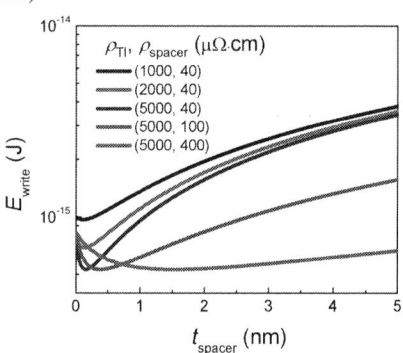

Fig. 3. Write voltage as a function of channel material resistivity for different σ_{SH} (R = 10 nm)

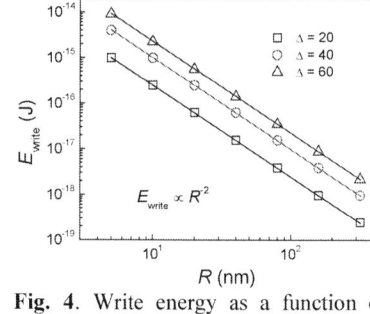

Fig. 4. Write energy as a function of MTJ radius for different Δ (ρ_{TI} =10³ $\mu\Omega\cdot$cm, σ_{SH} = 1.55 ×10⁵ $\Omega^{-1}\cdot m^{-1}$).

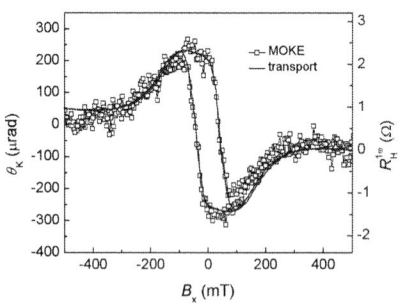

Fig. 5. Write energy as a function of spacer thickness for different TI and spacer material resistivities (R = 10 nm, σ_{SH} = 1.55 ×10⁵ $\Omega^{-1}\cdot m^{-1}$)

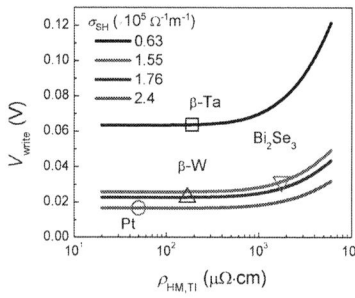

Fig. 6. 2ω Hall measurement setup and the optical image of the device.

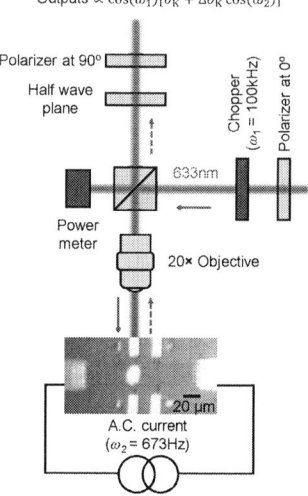

Fig. 7. MOKE setup and the optical image of the device with laser spot on it. Since the current and MOKE laser are modulated at different frequencies, the $\Delta\theta_K$ is free of any thermal contributions.

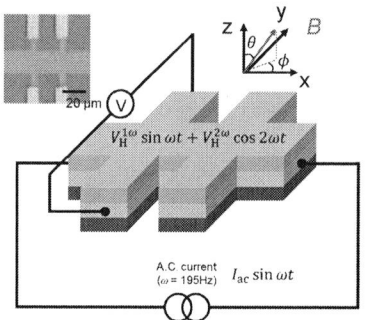

Fig. 8. Comparison between θ_K and first harmonic Hall resistance ($R_H^{1\omega}$) in the (BiSb)₂Te₃(6nm, S2)/Mo(2nm)/CoFeB(1.02nm)

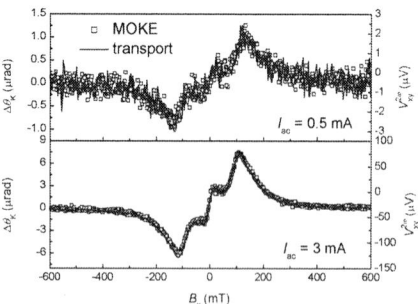

Fig. 9. Comparison between $\Delta\theta_K$ and second harmonic Hall voltage ($V_H^{2\omega}$) in the (BiSb)₂Te₃(6nm, S2)/Mo(2nm)/CoFeB(1.02nm)

978-1-7281-1988-5/18 $31.00 © 2018 IEEE

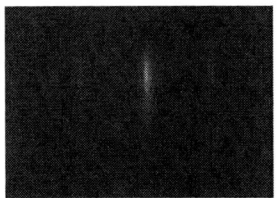

Fig. 10. RHEED for TI growth.

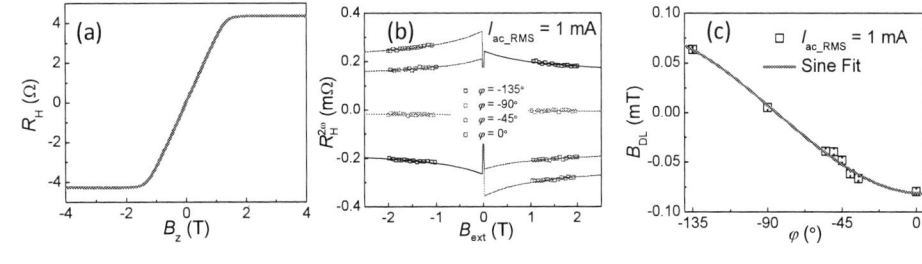

Fig. 11. SOT determination in the Bi$_2$Se$_3$(6nm)/CoFeB(5nm) sample with *in-plane magnetic anisotropy*. (a) OOP R_H hysteresis. (b) $R_H^{2\omega}$ as a function of external field at different φ's. (c) B_{DL} as a function of φ.

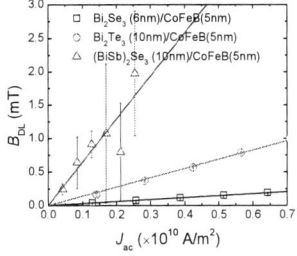

Fig. 12. SOT effective field as a function of TI current density for Bi$_2$Se$_3$, Bi$_2$Te$_3$ and (BiSb)$_2$Se$_3$ (S1).

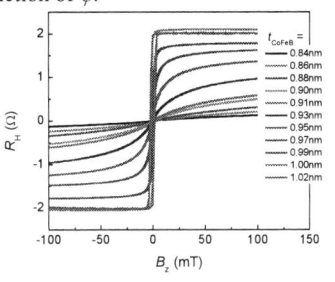

Fig. 13. OOP R_H hysteresis loops for (BiSb)$_2$Te$_3$ (6nm, S2)/Mo(2nm)/CoFeB(t_{CoFeB}).

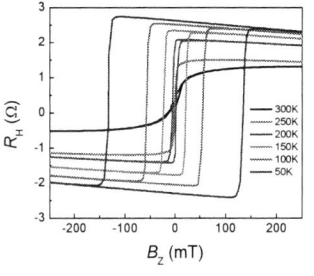

Fig. 14. OOP R_H hysteresis loops at different temperatures for (BiSb)$_2$Te$_3$ (6nm, S2)/Mo(2nm)/CoFeB(0.93nm).

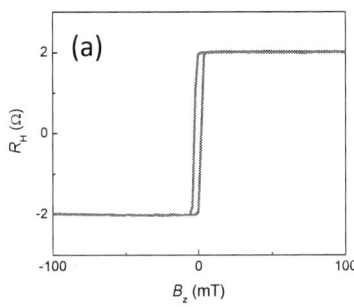
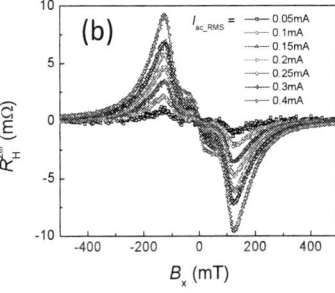
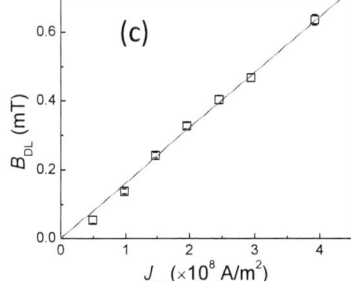

Fig. 15. SOT determination in the (BiSb)$_2$Te$_3$(6nm, S2)/Mo(2nm)/CoFeB(1.02nm) sample with *perpendicular magnetic anisotropy*. (a) OOP R_H hysteresis. (b) $R_H^{2\omega}$ as a function of external field at different currents. (c) B_{DL} as a function of TI current density.

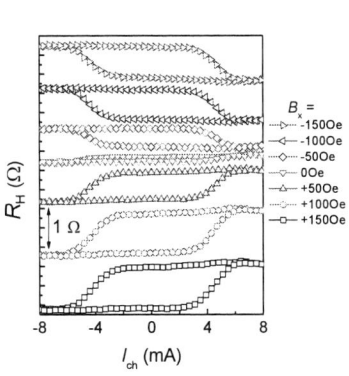

Fig. 16. RT SOT-driven magnetization switching in the (BiSb)$_2$Te$_3$(6nm, S2)/Mo(2nm)/CoFeB(1.02nm) sample

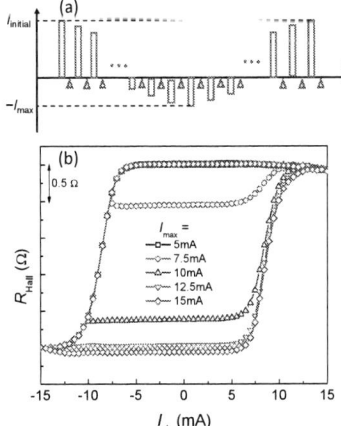

Fig. 17. (a) Red (write) pulse is 5 ms (1 s interval) and blue (read) pulse is 0.5 s (b) Memristor-like switching behavior in the (BiSb)$_2$Te$_3$(6nm,S2)/Mo(2nm)/CoFeB(0.93 nm) sample at 200 K

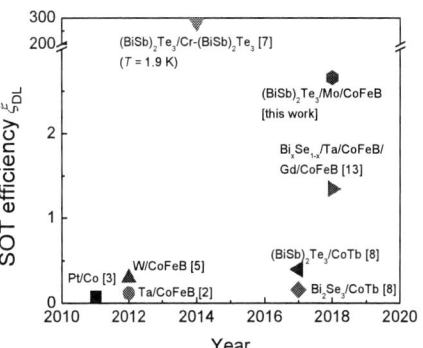

Fig. 18. Progress towards efficient SOT-MRAM with perpendicular magnetic anisotropy. All references are from RT unless specifically mentioned.

978-1-7281-1988-5/18 $31.00 © 2018 IEEE

Scaled spintronic logic device based on domain wall motion in magnetically interconnected tunnel junctions

E. Raymenants[1,2], D. Wan[1], S.Couet[1], O. Zografos[1],V.D. Nguyen[1], A.Vaysset[1], L. Souriau[1], A. Thiam[1], M. Manfrini[1], S. Brus[1], M. Heyns[1,2], D. Mocuta[1], D.E. Nikonov[3], S. Manipatruni[3], I. A. Young[3], T. Devolder[4] and I. P. Radu[1]

[1]imec, Leuven, Belgium, email: eline.raymenants@imec.be, [2]KU Leuven, 3001 Leuven, Belgium, [3]Intel Corporation, Hillsboro, OR, USA, [4]Centre de Nanosciences et de Nanotechnologies, CNRS, Univ. Paris-Sud, Univ. Paris-Saclay, C2N, 91405 Orsay cedex, France

Abstract— We present a scaled device based on magnetic domain wall (DW) transport for logic applications. The device consists of multiple magnetic tunnel junctions (MTJs) connected by the same magnetic free layer (FL). Magnetic domain walls are injected by spin-transfer torque (STT) at the input MTJs and are sensed by tunneling magnetoresistance (TMR) at the output MTJ after propagation through the FL. Logic functions can be built by merging several domain walls. By enabling real-time detection of long range DW transport, we demonstrate a spintronic component which can be used for either Boolean or non-Boolean logic.

I. INTRODUCTION

Conventional CMOS is becoming more difficult to scale and is potentially approaching fundamental scaling limits [1]. To circumvent these limitations, non-charge based logic technologies are increasingly explored. Spintronic concepts have the benefit of being intrinsically low-energy and non-volatile, therefore offering attractive properties for future logic devices [2]. One approach is to encode logic information in magnetic domain walls (DW). A DW-based three-terminal device, Fig. 1(a), was recently reported to perform buffer and inverter operations [3], and demonstrate shift register and full-adder circuits [4]. Furthermore, all basic logic functionalities were demonstrated in a magnetic network structure via gate controlled DW trajectory [5].

Here, we report on a scaled, non-volatile magnetic device where a 2nm thin CoFeB-based magnetic free layer can effectively transport information via DW motion between pillars with 150 nm pitch. Magnetic switching is initiated by STT below the first pillar to generate a DW and the motion of the DW is monitored and read in a series of adjacent pillars using the tunnel magneto-resistance (TMR). By generalizing this scheme to various numbers of pillars and geometries of the free layer, logic functions can be built. The Spin-Torque Majority Gate (STMG) in Fig. 1(b) is an example of such a logic device [6]. STMG could offer an advantage in area and energy compared to equivalent CMOS circuits.

II. DEVICE FABRICATION

Top-pinned perpendicular magnetic tunnel junctions (pMTJs) sharing a magnetic free layer (FL) were fabricated in imec's 300-mm CMOS fab with a BEOL-compatible flow (Fig. 2). Detailed information of the full integration flow is available in Ref [7]. Fig. 3(a) depicts the pMTJ stack consisting of a dual-MgO CoFeB-based FL and reference layer (RL) anti-ferromagnetically coupled to a Co/Pt-based hard layer (HL). This stack is the starting point of device fabrication. Of critical importance is the pillar patterning process, which utilizes ion-beam etching (IBE). During this step, the interpillar FL is exposed to ions which could deteriorate crystallinity of the MgO barrier and impact the interface-induced perpendicular magnetic anisotropy (PMA) in the FL. Previous results, showed strong DW pinning due to related process-induced damages [8,9].

We focus on devices consisting of three MTJs of 80 nm diameter separated by 70 nm each, arranged on a strip-shaped common FL of length 450 nm and width 150 nm. See Fig. 3b. A cross-sectional transmission electron micrograph (TEM) of the integrated device in Fig. 3(d) shows the FL spanning the entire device. Energy-dispersive X-ray spectroscopy (EDS) mapping for Mg in Fig. 3(c) confirms a continuous, damage-free dual-MgO FL in between the pillars. EDS mapping for Co indicates remaining RL materials which are necessary for preservation of PMA in the FL during pillar patterning. This is illustrated through wafer level magnetic characterization.

Fig. 4(a) shows magnetic hysteresis loops obtained by Magneto-Optical Kerr Effect (MOKE) microscopy after pillar patterning using different etch conditions (shorter vs longer). The FL is not yet patterned into the strip-shape and can be considered as continuous over the entire wafer. Over-etching (OE) results in a weak signal with high (> 60 mT) coercivity, indicating a fully etched interpillar spacing and magnetically isolated pillars. With a mild under-etch (UE1) condition, the FL is seen to be damaged with a partial in-plane signature. Only an even shorter etch (UE2) leads to an intact FL with acceptable PMA. Note that

978-1-7281-1988-5/18 $31.00 © 2018 IEEE

under-etch conditions, UE1 and UE2, appear structurally similar in Fig. 4(b), but result in very different magnetic response. About 3.5 nm of material (RL + HL) needs to remain above the MgO to protect it from IBE damage. After a subsequent post etch oxidation, only a small shorting path remains due to the presence of RL between the pillars. Vibrating sample magnetometry (VSM) hysteresis loops in Fig. 5(a) show the additional magnetic moment in the UE2 condition compared to the deposited dual-MgO FL. The extra magnetic moment corresponds to approximately 1 nm of magnetic CoFeB in the RL. In Fig. 5(b), µMOKE imaging in the UE2 condition demonstrates the ability to nucleate and expand a domain in the FL by short field pulses. This confirms that we have a magnetically active FL after pillar patterning (Fig. 2(a)) and that the remaining RL does not hinder domain wall propagation in the FL.

III. RESULTS AND DISCUSSION

A. Three-pillar device magnetic properties

Here, we demonstrate electrically that the interpillar spacing is also magnetically active after FL patterning (Fig. 2(b)) and the full integration. Fig. 6(a) presents the hysteresis curves, TMR versus field, for the three pillars sharing a strip-shaped FL. The parallel (P)- and antiparallel (AP)-state are denoted by ① and ③, respectively. The exact overlap of the switching field, H_C (coercivity), in the three pillars indicates that the FL is effectively one magnetic layer. Additionally, local AP regions can be nucleated by STT. This is the state marked by ② in Fig. 6(b), where the $P \to AP$ and $AP \to P$ reversals in P1 and P3 proceed as a function of voltage pulse amplitude. The attained resistance values after AP domain nucleation at P1 and P3 are lower than state ③, due to current spreading to the RL in between the pillars. The AP regions are nucleated close to P1 and P3 as no change in resistance is observed at P2. When starting the TMR versus field loop from state ② in Fig. 6(c), instead of from the full P-state ① in Fig. 6(a), we obtain a lower switching field. This is the depinning field H_D required to move DWs. The fact that $H_D < H_C$, proves successful DW motion through the shared FL from P1 (and/or P3) towards P2.

B. Time-resolved domain wall detection

By performing time-resolved measurements, DW-based nanosecond-scale switching in the shared FL can be captured. Separated MTJs locally probe magnetization in the FL, allowing us to estimate the DW velocity. We quantify the velocity at a low external field. After setting the field, an AP region is nucleated at P1 by STT (pulse width 1 ns). The domain expands along the strip-shaped FL driven by the field. A real-time trace shown in Fig. 7(a), captures the arrival of the domain at P2.

The DW velocity between P1 and P2 was measured ~12000 times to obtain a reliable value. From the histogram in Fig. 7 (b), the DW velocity is estimated to be 9 m/s on average. This velocity is in agreement with published data

on DW velocity in CoFeB-based systems [10]. It confirms that the overall integration scheme, including the critical pillar patterning step as well as remaining RL material, do not negatively impact the DW velocity of the magnetic conduit.

Additionally, in Fig. 8(a) we show real-time traces of domain expansion from P1 to both P2 and P3. An AP domain was nucleated at P1 and expands by the external field. DW propagation is sketched in Fig. 8(b). This experiment demonstrates the shift register operation '1-filled left shift' or the arithmetic function $2x + 1$, shown in Fig. 9. Note that each pillar is individually addressable for reading and writing, allowing for other shift operations (right and '0-filled'). Moreover, this device is the first demonstration of domain wall motion through an easily integrated and highly-scaled magnetic conduit shared by multiple MTJs at an aggressive pitch (150 nm). The conduit is perpendicularly magnetized, allowing further scalability and lower power consumption compared to in-plane systems. Reading and writing are performed by TMR and STT, respectively, identical to STT-MRAM technology. Other writing schemes, such as Spin-Orbit Torque (SOT) and Magneto-Electric (ME) polarization, could be implemented to improve the speed and energy efficiency of the devices. These first results show DW motion by external field. However, current-driven DW motion by STT and SOT can be applied as well. By careful stack engineering, the device can be optimized for higher DW velocities. DW velocities up to 750 m/s were reported in a synthetic antiferromagnetic structure driven by current [11] and field-driven DW motion with velocities up to 1700 m/s was achieved in ferrimagnetic materials [12]. These materials will be the excellent candidates for DW propagation based devices.

IV. CONCLUSION

We have demonstrated a fully integrated, scaled, and CMOS production compatible logic device based on magnetic domain wall transport. We present all necessary conditions for successful device fabrication and operation. Through quasi-static and time-resolved measurements, we show that domain walls can be controllably nucleated and propagated between MRAM-like pillars. The demonstration of this working magnetic device at the nanometer scale, enables a robust spintronic platform that will be used for functional silicon-integrated spintronic logic components and applications. This work paves the way towards the realization of spin-based beyond CMOS logic concepts.

ACKNOWLEDGMENT

This work was performed as part of the imec IIAP core CMOS and the Beyond CMOS program of Intel Corporation. The authors gratefully acknowledge V. De Heyn and the P-line for operational support, and MCA team for TEM images. E. Raymenants gratefully acknowledges FWO Flanders.

978-1-7281-1988-5/18 $31.00 © 2018 IEEE

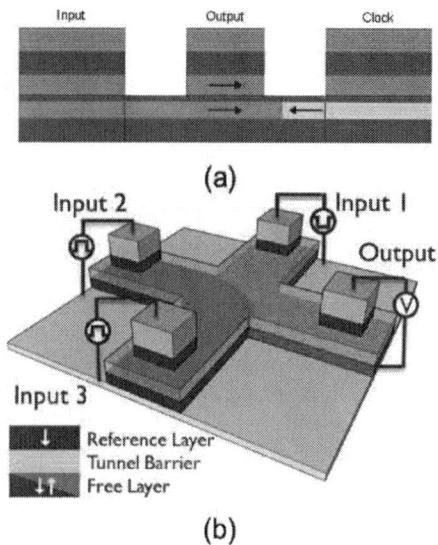

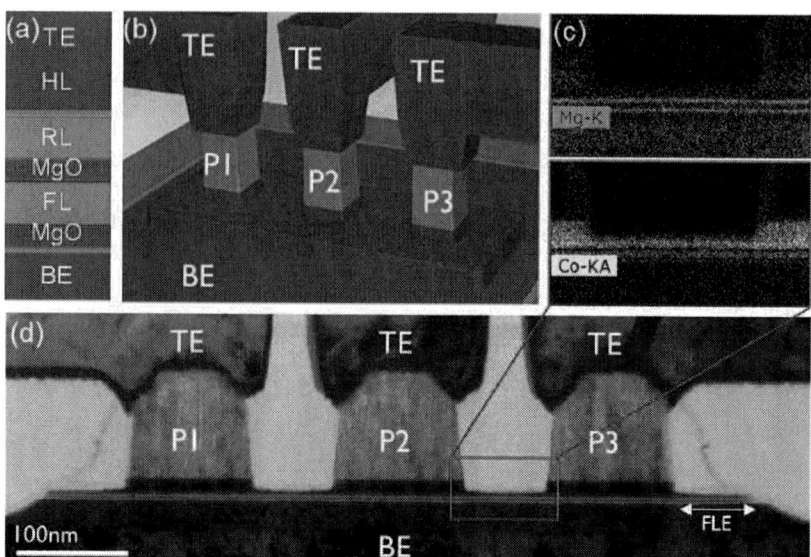

Fig. 1. Interconnected MTJ devices capable of Boolean operations.
(a) Three-terminal device
(b) Spin-Torque Majority Gate

Fig. 3. (a) Top-pinned pMTJ stack – CoFeB-based FL and RL anti-ferromagnetically coupled to Co/Pt based HL (b) 3D view of 3-pillar device (c) EDS mapping showing continuous FL in between pillars (d) TEM cross-section of 3-pillar device showing undamaged dual MgO FL interconnect.

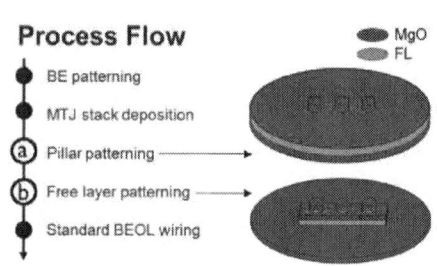

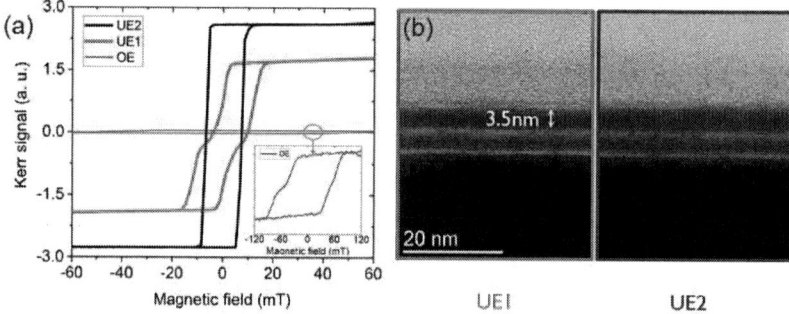

Fig. 2. Dedicated BEOL process flow for MTJs sharing the ferromagnetic FL. Critical step is pillar patterning, where the MgO layer should not be damaged to preserve PMA in the FL interconnect.

Fig. 4. (a) Magnetic hysteresis curve recorded on dense region of pillars after the critical pillar patterning step. Over-etch (OE) results in a fully patterned FL. A mild under-etch (UE1) condition damages the FL which has a partial in-plane character. A stronger under-etch (UE2) leads to an intact FL with good PMA. (b) UE1 and UE2 structurally look the same but have very different magnetic properties.

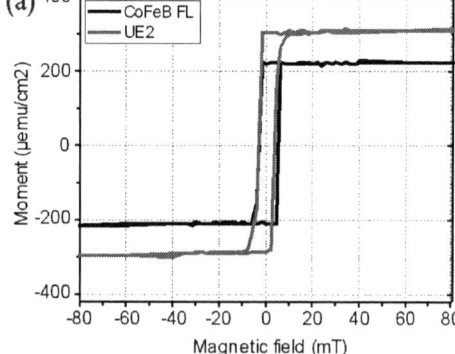

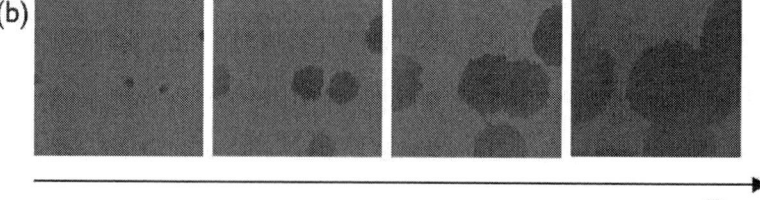

Fig. 5 (a) Hysteresis loops, moment versus field, as recorded by VSM. UE2 shows an additional moment coming for remaining RL, compared to a deposited CoFeB dual-MgO FL. (b) µMOKE image of UE2. A domain was nucleated and expanded by field in the FL. The remaining RL does not hinder the DW motion in the FL.

978-1-7281-1988-5/18 $31.00 © 2018 IEEE

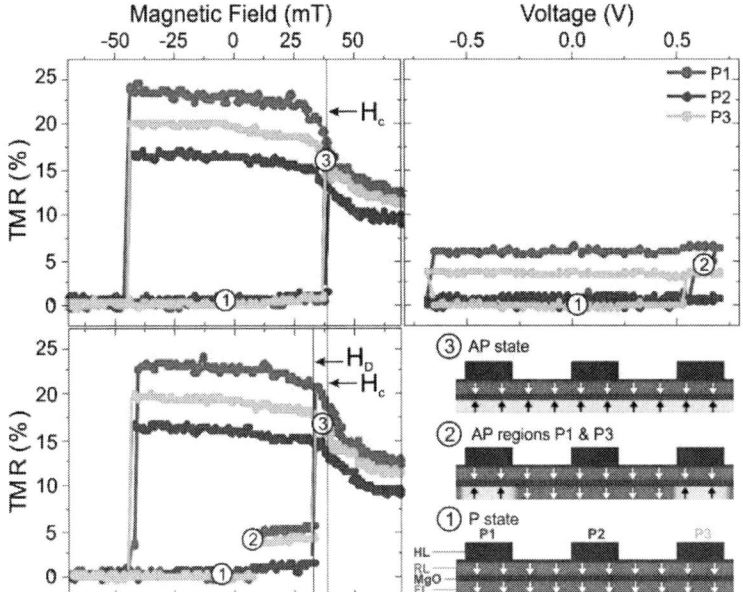

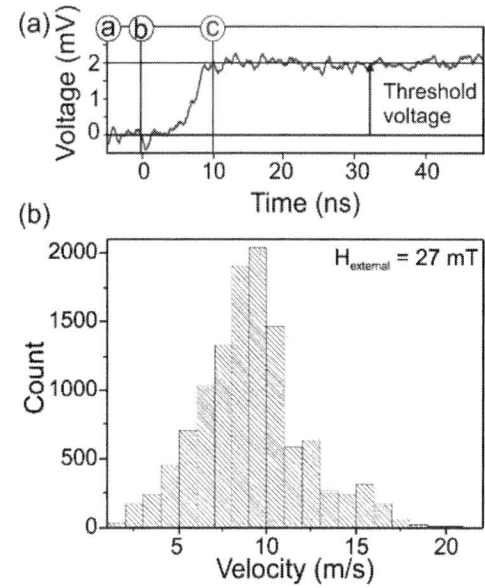

Fig. 6. (a) Hysteresis loop of FL of 3 pillars in the same device. P- and AP-state are denoted by ① and ③. All pillars have the same coercivity, meaning the FL is shared. (b) Local P-AP and AP-P switching at P1 and P3, ②. (c) Comparison of depinning field H_D and switching field H_C where $H_D < H_C$. (d) Sketch of magnetic states in FL and RL for figures (a), (b), (c).

Fig. 7. (a) Smoothed time trace recorded at P2 after an AP region was nucleated at P1 with a 1 ns pulse (at time zero). ⓐ pulse at P1 ⓑ DW arrival at P2 and ⓒ DW passed under P2. (b) Histogram of DW velocity at an external field of 27 mT, below the coercive field. Average DW velocity is 9 m/s between P1 and P2.

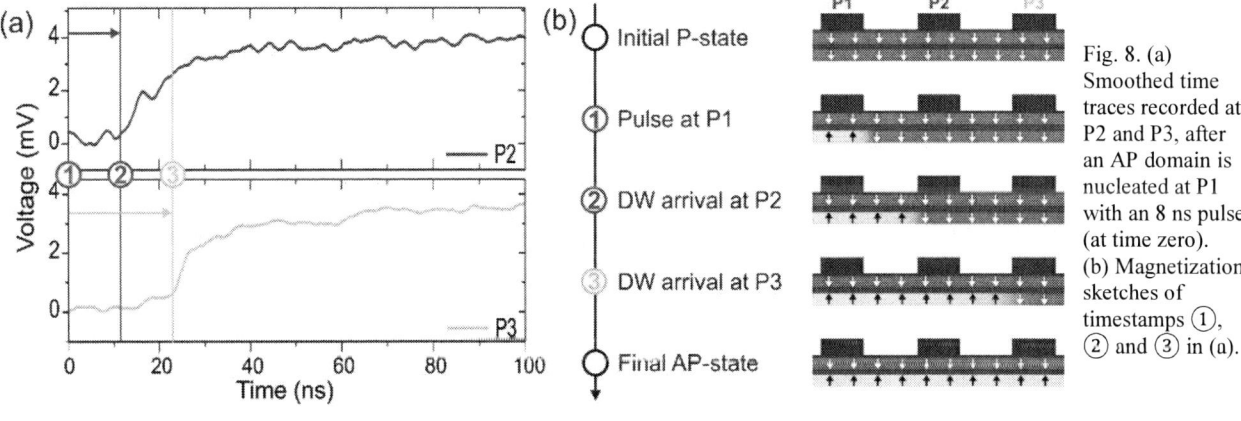

Fig. 8. (a) Smoothed time traces recorded at P2 and P3, after an AP domain is nucleated at P1 with an 8 ns pulse (at time zero). (b) Magnetization sketches of timestamps ①, ② and ③ in (a).

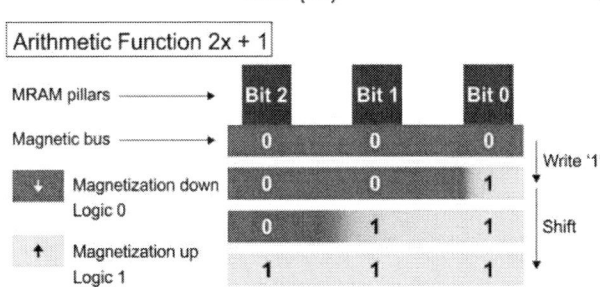

Fig. 9. Three-pillar device acting as a register. Magnetization 'down' and 'up' are logic '0' and '1' respectively. Each pillar is individually addressable for reading and writing. By writing a logic 1 to Bit 0 and domain expansion by field, we perform the arithmetic function 2x+1, as shown in the sketch.

REFERENCES

[1] Zhirnov, Victor V., et al. Proc. of the IEEE 91.11 (2003): 1934-1939.
[2] Nikonov D.E. et al., IEEE JXCDC 1, 3 (2015)
[3] Currivan-Incorvia, J. A., et al. Nature Communications 7 (2016): 10275.
[4] Currivan, J. A., et al. IEEE Magnetics Letters 3 (2012): 3000104-3000104.
[5] Murapaka, C., et al. Scientific Reports 6 (2016): 20130.
[6] Nikonov, D. E., et al. IEEE Electron Device Letters 32.8 (2011): 1128-1130.
[7] Wan, D., et al Japanese Journal of Applied Physics 57.4S (2018): 04FN01.
[8] Manfrini, M., et al. AIP Advances 8.5 (2018): 055921.
[9] Raymenants, E., et al. Journal of Physics D: Appl. Phys. 51.27 (2018): 275002.
[10] Burrowes, C., et al. Applied Physics Letters 103.18 (2013): 182401
[11] Yang S.H. et al., Nature Nanotechnology 10.3 (2015): 221.
[12] Kim, K-J., et al. Nature Materials 16.12 (2017): 1187

Binary and Ternary True Random Number Generators Based on Spin Orbit Torque

Huiming Chen[1,*], Shuai Zhang[1,*], Nuo Xu[2], Min Song[3], Xin Li[1], Ruofan Li[1], Yi Zeng[1], Jeongmin Hong[1], Long You[1§]

[1]School of Optical and Electronic Information, Huazhong University of Science and Technology, Wuhan 430074, China.
[2]Department of Electrical Engineering and Computer Sciences, University of California, Berkeley, CA 94720, USA.
[3]Faculty of Physics and Electronic Science, Hubei University, Wuhan 430062, China.
[*]authors contributed equally to this work; [§]Email: lyou@hust.edu.cn

Abstract—In this work, we have experimentally demonstrated the binary- and ternary- True Random Number Generators (B-TRNG and T-TRNG) based on the stochastic switching characteristics of the nano-scale Ta/CoFeB/MgO heterostructures with perpendicular magnetization anisotropy. For the first time, the random code generation utilizes the spin orbit torque (SOT) induced by current flowing in the heavy metal underneath the CoFeB layer. The 3-XOR post-processed random binary codes have passed the NIST SP800-22 test. Furthermore, the T-TRNG in the same ferromagnetic heterostructure with dual magnetic domains are also demonstrated, which provides a higher security level than its B-TRNG counterpart.

INTRODUCTION

Hardware True Random Number Generator (TRNG) is an important security primitive which can generate random codes used for cryptographically secure information storage and encrypted data transmission of the modern communication systems [1]. Traditional CMOS based TRNGs rely on physical noise such as thermal noise, burst noise and oscillatory jitters, which require extensive post-processing to ensure a high level of randomness, resulting in a severe power, latency and area overhead [2]. While Spin-Transfer-Torque Magnetic Random-Access Memory (STT-MRAM) devices have inherent switching stochasticity due to thermal fluctuation fields, can potentially contribute to TRNG implementations [3-5]. With additional advantages of low operating voltages and high device density, STT-MRAM has attracted considerable research interests [6-8]. However, as shown in Fig. 1, write success rate (WSR) based STT-MRAM TRNGs require picosecond resolution of pulse width as well as an accurate write voltage control to reach the 50% randomness, making it non-trivial for practical applications [3]. On the other hand, write time (WT) based STT-MRAM TRNGs rely on several operation cycles to code one random bit [4,5], imposes a non-reducible power consumption and latency. Furthermore, the electrical stress-induced MgO/CoFeB interface degradation [8,9] and relatively high device-to-device mismatch in scaled STT-MRAMs [9] hinders this emerging technology towards a reliable TRNG solution. Using the Spin Orbit Torque (SOT) generated in a Ta/CoFeB/MgO heterostructure with perpendicular magnetization anisotropy (PMA) to switch the ferromagnetic layer has been proposed for low-power logic and memory devices because of their superior power efficiency, fast switching speed [10] and improved endurance due to the separation of read and write paths in this structure [11]. As

shown in Fig. 2, when an in-plane current flows through the Ta layer along the x-direction, the spins will accumulate at the Ta/CoFeB interface due to the spin Hall effect (SHE). The switching is deterministic with the assistance of in-plane magnetic fields [12], exchange coupling [13] or breaking of geometrical symmetry [14], while under current-only operations on conventional SOT heterostructure, it results in a stochastic switching [15]. This stochastic behavior can be utilized as an ideal entropy generation source for a TRNG. In this work, for the first time, we proposed and experimentally demonstrated a TRNG based on SOT-induced stochastic switching in a PMA nano-ferromagnets (FM).

DEVICE CONCEPT AND FABRICATION

The principle of B-TRNG implemented by a FM layer with single domain structure is described in Fig. 3. First, a write current is applied to excite the FM to align its magnetization along the hard axis (in-plane directions). After the write current is removed, the magnetization orientation is driven to the easy axis (out-of-plane direction). Depending on the thermal fluctuations, the magnetization either goes 'upward' or 'downward', creating the random code. The switching speed of the PMA nanomagnet with single domain can be very fast, based on experimental results [16] and from our macro-spin simulations. As shown in Fig.4(a), one random switching operation requires no more than 3 ns. Furthermore, Fig. 4(b) and (c) indicate that the switching probability (P_{sw}) is insensitive to both current and temperature variation in a board range, proving its robustness under different operating conditions. As for the T-TRNG, the ternary state may originate from the domain wall (DW) propagation in dual-domain nano-FMs. The applied write current results in three random final states: (1) there are neither nucleation nor DW motion induced [Fig.5(d)]; (2) a DW is created but pinned at the middle of the FM layer during its motion [Fig.5(e)]; and (3) a DW is created and propagates to the end of the structure, causing a full magnetization reversal [Fig.5(f)].

A FM layer stack comprising Ta (10 nm)/CoFeB (1.2 nm)/MgO (1.6 nm)/Ta (5 nm) was first deposited on thermally oxidized Si substrate. As shown in Fig. 6, the DW propagation has been observed in the Ta/CoFeB/MgO thin film by magneto-optical *Kerr* (MOKE) microscopy experiments. The SOT induced effective field are shown in Fig. 7. Therefore, it is considered that the ternary state results from the DW propagation in a relatively-large area, a dual-domains FM device. The Ta/CoFeB/MgO thin film was further processed into the Hall bar structure by electron beam lithography (EBL)

978-1-7281-1988-5/18 $31.00 © 2018 IEEE

and argon-ion milling (AIM). The Hall bar contains the entire thin film stack and the region outside it was etched till the insulating SiO_2 substrate. A 10 nm thick hard mask (Ti) with sizes of 200×200 nm^2 and 500×500 nm^2 were grown at the center of the Hall bars by EBL and deposited by EB evaporation. AIM was then used to etch the stack outside the dot's region down to the bottom Ta layer. The smaller FM dot forms a single domain (for B-TRNG) while the lager one forms dual domains (for T-TRNG). All the fabrication was conducted at low temperature (from RT to less than 150°C). The optical and scanning electron microscope (SEM) images of B-TRNG and T-TRNG devices are shown in Fig. 8.

RESULTS AND DISCUSSION

A. Properties of the Ta/CoFeB/MgO Heterostructure

Fig. 9(a) shows the normalized R_{AHE}-H loops with different currents injected into the 200×200 nm^2 device. Clearly, the magnetic coercive field (H_c) decreases with the increasing current. The relationship between H_c and applied current is shown in Fig.9(b), indicating that current induced effective field (estimated as 2.5 Oe/(10^5A·cm^{-2})) favors the switching of Ta/CoFeB/MgO heterostructure. The SOT induced deterministic switching under external magnetic field is further shown in Fig. 9(c). On the other hand, the normalized R_{AHE}-H and R_{AHE}-I loops of the 500×500 nm^2 device are shown in Fig.10. The "rectangular" shape R_{AHE}-H loops suggest that both smaller- and larger-area devices have the dominant PMA with the easy axis along the out-of-plane (z) direction.

B. Performance of the Binary and Terbary TRNGs

The stochastic switching behavior of the B-TRNG device is shown in Fig. 11. The probability of upward and downward switching differs by 2.0% from measurement of 100 cycles. Fig. 12 shows the circuit schematics of the B-TRNG with detailed working principles elaborated in Fig. 13. A current pulse I_a [Fig. 13(a)] with fixed width is applied across the Hall bar along x axis, while the AHE voltage V_b [Fig. 13(b)] is measured along y axis. The write current has an amplitude of 0.5 mA (1.0×10^7 A/cm^2) and duration of 0.5s, and afterward a read current with an amplitude of 50 μA and duration of 1s is applied. V_b is determined by the magnetization orientation. When magnetization aligns upward, V_b is higher than the reference voltage (V_{ref}, the average of highest and lowest V_b), the output voltage (V_c) of the comparator goes to logic high level ('1'). In contrast, when magnetization aligns downward, V_b is lower than V_{ref} and V_c falls to logic low level ('0'). A sampling clock V_{clock} with a fixed duty cycle (Fig. 13(d)) and V_c are sent to an AND gate. The V_c in the period of the low I_a is sampled as the output of the B-TRNG. In this way, the output (a '0-1' code sequence) is distributed randomly due to the stochastic switching of the magnetization. A sequence of ~20k random bits generated by the B-TRNG with the raw data shown in Fig. 14(a) and the statistical results in Fig. 14(b), which indicates that the cases of upward and downward switching are different by 4.4%. After 3 times XOR post-processing (Fig. 15), the output codes can pass the NIST SP800-22 test [17] as shown in TAB I.

The stochastic switching behavior of the T-TRNG device is studied in Fig. 16. The P_{sw} of the three states varies with the current pulse amplitude. The operation window for T-TRNG

lies between 1.7 – 1.9mA. Therefore, 1.8 mA is chosen as the operation current level. Fig. 17 shows the circuit schematics of the T-TRNG with detailed illustrations in Fig. 18. Similar to B-TRNG, the write current pulse I_a is 1.8 mA/1s (Fig. 18(a)), while the read current is 50 μA/1s. Two comparators are used since the AHE voltage V'_b (Fig. 18(b)) is ternarily distributed. The resulted two reference voltages $V_{ref,h}$ and $V_{ref,l}$ are set as $3/4V'_{b,max}+1/4V'_{b,min}$ and $1/4V'_{b,max}+3/4V'_{b,min}$, as show in Fig. 17(c) and (d), respectively. The outputs of each comparators are sent to its associative AND gates. The sampling clock V_{clock} is used as the other inputs of the two AND gates, as shown in Fig.17. Two individual binary codes will be output from this setup as shown in (Fig. 18(e), and (f)). Then, the two sequences are sent to a decoder to generate ternary random sequence. A statistical result of 390k bits ternary sequence generated by T-TRNG is shown in Fig. 19, in which the maximum switching probability differences are ~1.88% from 390k measurements. The T-TRNG can provide a higher security than the B-TRNG with same sequence size. One can calculate the largest possible combination based on a 100-bit ternary sequence to be 5.15×10^{47}, which is 17 orders of magnitude more than that of a 100-bit binary sequence (with total combination numbers of 1.27×10^{30}).

C. Future Directions of SOT-TRNGs

TAB. II summarizes the work SOT-based TRNG's features, which are further compared against and compares to previously reported TRNGs [2-5]. Additionally, the current density for SOT devices can be reduced significantly through applying heavy metal with larger spin Hall angle (θ_{SH}). According to θ_{SH} = J_S/J_e, where J_e is the charge current density and [ℏ/(2e)]J_S is the spin current density arising from the SHE [12], required current decreases linearly with θ_{SH}. For example, by using β-W ($\theta_{SH, β-W} = 0.33 \pm 0.06$ [18], which is dozens of times that of compared to the Ta ($\theta_{SH,Ta} \sim 0.007$) [15]), operating current (power) can be lowered remarkably as demonstrated in this work. Besides, the low temperature fabrication of demonstrated TRNGs makes them promising for monolithic-3D integration into the standard CMOS back-end-of-line (BEOL) process flow. It is also feasible to integrate high-density SOT-TRNG arrays by accommodating the BEOL-embedded memory hierarchy, in which the logic peripheries (e.g. comparators and XOR gates, etc.) can be shared by a column of the TRNG bit array, to reduce the chip area.

CONCLUSIONS

The stochastic switching properties of the PMA MgO/CoFeB/Ta heterostructure have been demonstrated using current induced SOT. By tuning the device's size, either single or double domains can be formed within the CoFeB layer for Binary or Ternary TRNG applications, respectively. The random codes generated by B-TRNG with 3-stage XOR post-processing have passed the NIST SP800-22 test. Meanwhile, the code sequences produced by T-TRNG show high quality of randomness, guaranteeing a better security level. In comparison with other STT-MRAM based TRNG implementations, this work proposes a potential solution to achieve high device reliability and integration density, thanks to the SOT effect.

Acknowledgment: Authors acknowledges financial support from the National Natural Science Foundation of China (NSFC Grant No. 61674062 and No. 61821003), the Fundamental Research Funds for the Central Universities (HUST. 2018KFYXKJC019) and the Thousand Young Talents Program of China.

REFERENCES

[1] H. Jiang *et al.*, *Nat. Comm.*, 8, 882, 2017. [2] K. Yang *et al.*, *IEEE ISSCC Tech. Dig.*, pp. 280-282, 2014. [3] W. H Choi et al., *IEEE IEDM* pp. 12.5.1-12.5.4, 2014. [4] R. Carboni *et al.*, *IEEE EDL*, pp.951-954, 2018. [5] K. Yang *et al.*, *IEEE VLSI Symp. Circ.*, pp.171-172, 2018. [6] M. Manfrini *et al.*, *AIP Advances* 8.5, 055921, 2018. [7] DE. Nikonov *et al.*, *IEEE EDL* 32.8, pp.1128-1130, 2011. [8] R. Carboni *et al.*, IEEE IEDM Tech. Dig., pp.572-575, 2016. [9] N. Xu *et al.*, *IEEE VLSI Symp. Tech.*, pp.187-188, 2018. [10] G.Prenat et al., *IEEE Trans. Multi-Scale Computing Systems*, pp 49-60 ,2016. [11] G. Prenat et al., *Spintronics-based Computing.* Springer, Cham, pp.145-157, 2015. [12] L. Liu *et al.*, *Science*, pp. 555-558, 2012. [13] Y.-C. Lau *et al.*, *Nat. Nanotechnol.*, 11, pp. 758-762, 2016. [14] L. You *et al.*, *PNAS.* pp. 10310-10315, 2015. [15] G. Yu et al., *Nat. Nanotechnol.*, 9, pp. 548–554 ,2014. [16] M. Cubukcu et al., *IEEE Trans. Magnetics* 54.4 pp. 1-4, 2018. [17] A. Rukhin et al., NIST SP 800-22 test suites, 2010. [18] C. F. Pai et al., *Appl. Phys. Lett.*,101, 122404, 2012.

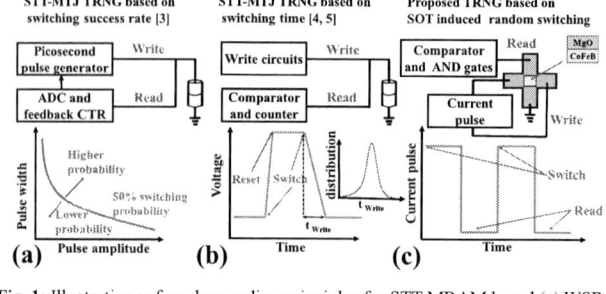

Fig. 1: Illustrations of random coding principles for STT-MRAM based (a) WSR-TRNG [3] and (b) WT-TRNG [4, 5]; and (c) proposed SOT-based TRNG.

Fig. 2: The schematic of the SOT induced switching of a PMA ferromagnet (FM). Switching current is applied in the underneath heavy metal layer.

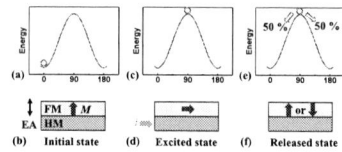

Fig. 3: Schematic of the principle of B-TRNG, with (a, b) as the initial state. In (c, d), a write current is applied to excite the FM to align its magnetization along the hard axis. When it is turned off (e, f), the magnetization is driven back to the easy axis by the PMA.

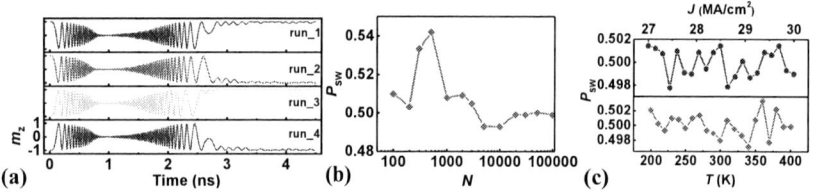

Fig. 4: Macro-spin simulation of random switching of a single FM domain, showing (a) four stochastic switching cases induced by the SOT; (b) the accumulative switching probability (P_{sw}) *vs.* number of switching cycles; and (c) P_{sw} as a function of temperature (T) and applied current density (J), showing only small fluctuations.

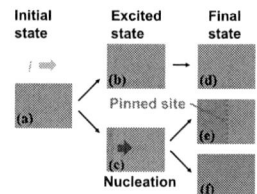

Fig. 5: Schematic of the principle of T-TRNG, with all possible stable states due to multi-domain effects and current induced DW propagation.

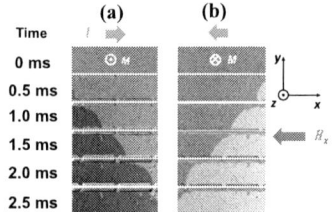

Fig.6: MOKE images of DW motion induced by SOT in a continuous Ta/CoFeB/MgO film.

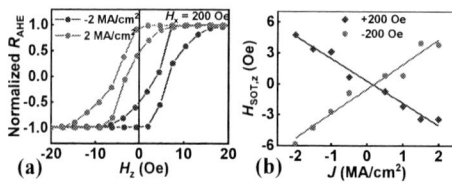

Fig. 7: Characterizations of the Ta/CoFeB/MgO thin film, showing (a) measured AHE loop shift induced by SOT effective filed, (b) the relationship between SOT induced effective field and current density, and (c) DMI effective field efficiency as a function of external field.

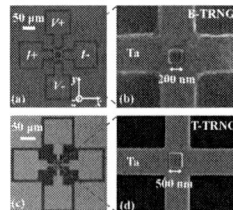

Fig. 8: Optical (a, c) and SEM (b, d) images of fabricated B-TRNG (a, b) and T-TRNG (c, d) devices.

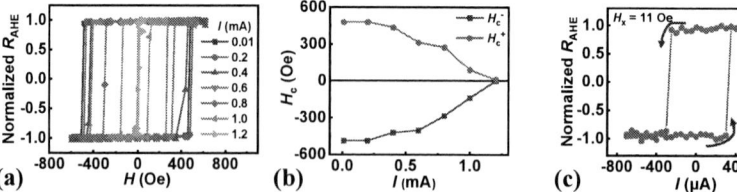

Fig. 9: Characterizations of the B-TRNG device, with (a) AHE loops under different currents, (b) H_c varies with currents, (c) SOT induced deterministic switching with the help of an external field.

978-1-7281-1988-5/18 $31.00 © 2018 IEEE

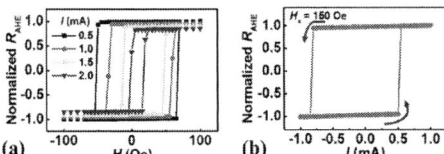

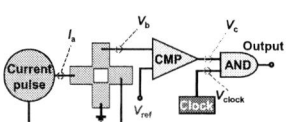

Fig. 10: Characterizations of the T-TRNG device, with (a) AHE loops under different switching currents and (b) SOT induced deterministic switching of T-TRNG with the help of an external field.

Fig. 11: SOT induced stochastic switching without external field, for 100 cycles.

Fig. 12: The schematic circuits of the B-TRNG.

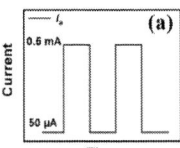 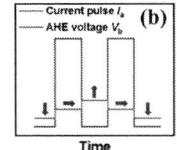 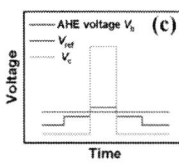 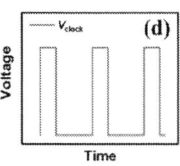 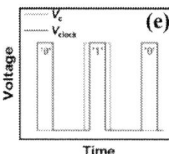

Fig. 13: Schematic waveforms at each stage of the B-TRNG circuit. (a) Input current pulse. (b) AHE voltage V_b obtained utilizing the read current of (a). Blue arrows represent magnetization. (c) V_b is transformed to logic voltage V_c by a comparator. (d) Sampling clock signal. (e) The random code bits are generated from the output of the AND gate.

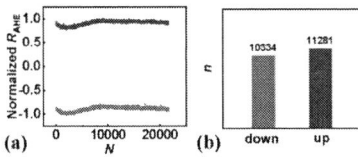

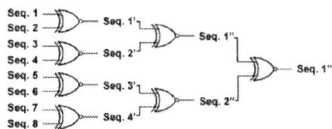

Fig. 14: The SOT induced stochastic switching of the B-TRNG without external field, with (a) the distribution of ~20k random switching results and (b) the switching probability.

Fig. 15: The schematic of the post-processing XOR operation for the B-TRNG.

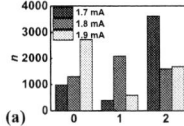 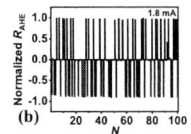

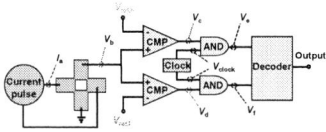

Fig. 16: The SOT induced stochastic switching of T-TRNG without external field, with (a) switching probability dependence on current pulse amplitude and (b) 100 random switching results of the T-TRNG.

Fig. 17: The schematic circuits of the T-TRNG.

	P-value	Pass rate	Success/failure
Frequency	0.709991	2/2	Success
Block Frequency	0.806341	2/2	Success
Cumulative Sums	0.876773	2/2	Success
Runs	0.733546	2/2	Success
Longest Run	0.443576	2/2	Success
Rank	0.333851	2/2	Success
FFT	0.036257	2/2	Success
Non Overlapping Template	-	290/296	Success
Overlapping Template	0.830234	2/2	Success
Serial	0.659356	4/4	Success
Linear Complexity	0.695952	2/2	Success

TAB I: The NIST SP800-22 test results after post-processing using XOR gates shown in Fig.14. (A 7k bits sequence was divided into two segments, P-value > 0.01, and proportion > 290/296)

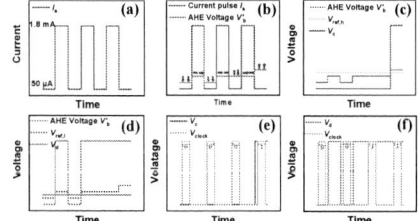

Fig. 18: Schematic waveforms at each stage of the T-TRNG circuit. Input current pulse. (b) AHE voltage V'_b obtained utilizing the read current of (a). Blue arrows represent magnetization. (c), (d) show the ternary V'_b is transformed to two logic voltage V_c and V_d by two comparators. (e), (f) The random bits are generated from the output of the AND gates.

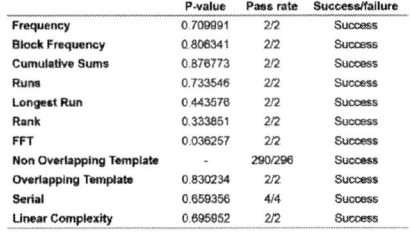

Fig. 19: The distribution of 390k random switching results from the T-TRNG. (a) Distribution of final states, with "0", "1" and "2" represent the low, medium and high resistance state, separately. (b) Distribution of switching events. "X Y" represents that 'X' switches to 'Y' state.

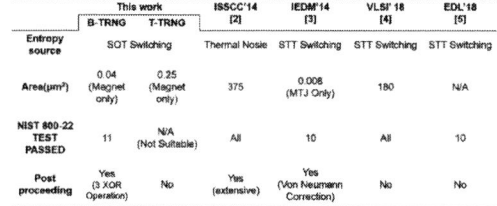

	This work		ISSCC'14 [2]	IEDM'14 [3]	VLSI'18 [4]	EDL'18 [5]
	B-TRNG	T-TRNG				
Entropy source	SOT Switching		Thermal Nosie	STT Switching	STT Switching	STT Switching
Area(μm²)	0.04 (Magnet only)	0.25 (Magnet only)	375	0.008 (MTJ Only)	180	N/A
NIST 800-22 TEST PASSED	11	N/A (Not Suitable)	All	10	All	10
Post proceeding	Yes (3 XOR Operation)	No	Yes (extensive)	Yes (Von Neumann Correction)	No	No

TAB II: Summary of the features of SOT-based TRNG in this work and STT MRAM based TRNGs reported in [2-5].

978-1-7281-1988-5/18 $31.00 © 2018 IEEE

High-performance, cost-effective 2z nm two-deck cross-point memory integrated by self-align scheme for 128 Gb SCM

Taehoon Kim, Hyejung Choi, Myoungsub Kim, Jaeyun Yi, Donghoon Kim, Sunglae Cho, Hyunmin Lee,
Changyoun Hwang, Eung-Rim Hwang, Jeongho Song, Sujin Chae, Yunseok Chun, Jin-Kook Kim
R&D Division, SK-Hynix, Icheon, Republic of Korea, email: taehoon12.kim@sk.com

Abstract— We demonstrate a high-performance and cost-effective cross-point memory (CPM) technology for two-deck 128 Gb storage class memory (SCM). The unit MAT size is 16 Mb consisting of a 2z nm 1S1M (one selector one memory) structure that is patterned by only two ArF-i steps per deck for a low cost per bit. The formidable task of self-align etch is enabled by the use of state-of-the-art etching and integration technology, which otherwise easily leads to hard fail or poor cell characteristics and reliabilities. New phase change materials (N-PCMs) are developed to have a large V_t window and a uniform V_t distribution for a sufficient read window margin (RWM) and a corresponding low raw bit error rate (RBER). New chalcogenide selectors (NCSs) are also developed to provide low V_t instability and very low leakage current. The new CPM is able to provide a sufficient RWM for 16 Mb MATs with very low latencies of write (set ≤ 300 ns) and read (≤ 100 ns). We also demonstrate its decent write disturbance and high reliabilities such as endurance and thermal retention.

I. Introduction

Ever-increasing demand for high-capacity and high-performance SCM for data centers and cloud computing systems has led to the pursuit of new types of SCM for both memory and storage applications [1]. The features desired of such memory are low latencies, byte-addressability, long endurance, and persistency at low cost. These requirements cannot be satisfied by the current NAND and DRAM technology [2]. In this paper, we introduce a high-performance, cost-effective two-deck CPM technology for 128 Gb SCM that is quite close to commercialization. We must emphasize that the data provided here does not represent the optimal single-cell performance but rather a typical 16 Mb arrays' performance for 128 Gb SCM. Therefore, this paper will provide the major issues and factors that must be considered at each development stage.

II. Results and Discussion

A. Structure and Integration

Fig. 1 shows a cross section of the two-deck cell array with a peri under cell (PUC) structure. A Cu multi-layer was used for high- speed data transmission between the cell array and the PUC. Fig. 2 shows the floor plan of the 128 Gb die. The unit MAT size is 16 Mb, consisting of an array of 2z nm pillar cells. Each pillar consists of a memory, a selector, and electrodes separating each material physically and optimizing its electrical properties, as shown in Fig. 3. To minimize the process cost per bit, it was necessary to perform integration with a self-align etch. Unfortunately, typical chalcogenide alloys are vulnerable to the conventional etching and cleaning processes. These technical difficulties have been directing PCM research towards the chemical vapor deposition (CVD) and atomic layer deposition (ALD) processes aiming damascene process for the last two decades. However, these processes are very slow and expensive, and the resulting film quality is poor both physically and electrically because of the porosity of the resulting structure, particularly at the sub-50 nm scale [3].

To overcome this, we used physical vapor deposition (PVD) for the cell stack deposition and developed new recipes for dry etching as well as cleaning for cross-point patterning. Integration schemes were also developed to minimize the process and integration damage. The self-align processes and integration schemes are shown in Fig. 3. The word lines (WLs) were patterned using a single ArF-i mask step after cell-material stack deposition. Both inter-layer dielectric (ILD) deposition and chemical mechanical polishing (CMP) were performed for bit line (BL) metal deposition followed by BL-patterning performed by another single ArF-i mask step.

B. Development of New PCM

Fig. 4(a) shows the IV behavior of a typical single cell, while Fig. 4(b) shows the V_t distributions of the array after the set and reset operations. A large read window margin (RWM) existed, even below -4.5σ, because of the N-PCM and NCS. Thus, a large RWM is a key enabler to satisfy the entire array's operations with a very low RBER and high production yield. To achieve this, the ΔV_t value between set and reset must be large enough, and the V_t distribution slope must be steep enough for the given array size, which can be expressed as

$$RWM = \Delta V_t - \sigma_{array} \times (\sigma_{Set} + \sigma_{Reset}), \quad (1)$$

where ΔV_t is the median V_t gap between reset and set, σ_{array} is the array size by normal quantile plot for the target BER, and σ_{Set} and σ_{Reset} are the V_t distribution slopes (standard deviations) of set and reset, respectively. A relatively easy way to increase ΔV_t is to increase the bandgap of the material by adding dopant. However, doing so will result in a longer t_{set} and a poor set distribution, as shown in Fig. 5(a). On the other hand, the N-PCM we developed exhibits a large ΔV_t without the disadvantage of t_{set}, as shown in Fig. 5(b). Furthermore, σ_{Set} and σ_{Reset} are also strong functions of the

978-1-7281-1988-5/18 $31.00 © 2018 IEEE

selector's V_t instability, because each cell's distribution can be expressed by the combination of PCM and NCS as shown in (2) and (3).

$$\sigma_{Set} = (\sigma_{Set\_PCM}{}^2 + \sigma_{Set\_NCS}{}^2)^{1/2} \tag{2}$$

$$\sigma_{Reset} = (\sigma_{Reset\_PCM}{}^2 + \sigma_{Set\_NCS}{}^2)^{1/2} \tag{3}$$

C. Development of New Selector

It is well known that, in CPM, the off-current of the deselected selectors (I_{off}) should be small enough to both avoid read/write errors and provide a decent V_t distribution with limited IR drop. Fig. 6 shows the normalized IV behaviors of the various selector materials after integration without PCM. The selector's vertical leakage should be small enough in the subthreshold region to avoid read/write failures. Compared to the conventional ovonic threshold switch (C-OTS), NCS exhibits a much lower I_{off} value (<1 nA between 0.7–0.8 V_t), which is sufficient for 16-Mb arrays and higher.

Another important factor is the selector's V_t instability, including drift and random telegraph noise (RTN). Drift is an intrinsic property of amorphous materials in which defects are annihilated spontaneously [4]. The RTN phenomenon that exists in amorphous materials is thought to be caused by either the current or the corresponding voltage fluctuations that occur when the carriers' hopping paths vary [5]. Fig. 7 depicts the ways of RWM consumption caused by these two mechanisms. Fig. 8 shows the drifts of various materials. There are huge differences among the materials. In particular, NCS exhibits very little drift, even at 55 °C, which guarantees several years with little RWM consumption. Fig. 9 shows the tradeoff between the drift and the RTN, which was defined using the standard deviation of 100 V_t reads. This result also indicates that a net improvement can be achieved by selecting a new material. However the tradeoff still exists. Therefore, these two parameters should be checked in determining the final RWM.

D. Read and Write Performances

Table 1 shows the overall performances of read and write, expressed via both the pulse widths and the latencies. The selector's quick turn-on and fast-charging characteristics make it possible to perform detections within 20–30 ns, which is a period short enough to guarantee a read latency shorter than 100 ns, including for both command and addressing. In the CPM structure, however, the set performance is more critical. Unlike the conventional line-type PCRAM in which crystal seeds surround the amorphous reset region, a self-align etched CPM has a confined structure that does not leave any crystal seeds behind after a full reset is performed, as shown in Fig. 10. Because of the nucleation step, particularly at the nano-scale, the set speed of the confined structure intrinsically becomes much slower than that of conventional PCRAM. Therefore, the set performance should be considered as carefully as the ΔV_t value. Fig. 11 shows that the set performance can easily be degraded by process damage, particularly in the low-probability region. In contrast, Fig. 12 shows that strategically integrated N-PCM can achieve a decent set distribution down to 300 ns without tail bits.

E. Disturbance and Reliability

The (thermal) write disturbance (WDT) is an unwanted reset to set transitioning of the adjacent (victim) cells during a reset write of the target (aggressor) cell. WDT is considered a major obstacle to PCRAM scaling. However, CPM's confined structure can assist in suppressing such disturbances. Decent integration for the uniform reset current distribution can also minimize the WDT by limiting the maximum reset current (I_{reset}). The benefit of such a solution is the elimination of the write-verify which means a huge advantage over traditional PCRAM on both write latency and power consumption. As explained, Fig. 13 shows no disturbed cell even at 1E5 cycle.

With respect to reliabilities, Fig. 14 shows the write (set/reset) cycle endurance, which is one of the major advantages of SCM over NAND flash memory. Our device maintained a sufficient RWM even after 1E7 cycle that is superior to NAND Flash at least more than three orders. The inset shows the variation of overall set/reset distribution including tail bits for each cycle. The RWM consumption by the write cycle is quite small. Finally, the thermal retention, which is determined by the crystallization of the PCM, was checked. N-PCM can maintain reset states for more than 10,000 hours at 85 °C, including the tail bits, as shown in Fig. 15. Although there is a tradeoff between the set speed and the retention, different activation energies for the crystallization at low and high temperatures can improve the retention margin without degrading the set performance.

F. Second-Deck Properties

The primary advantage provided by the two-deck structure is the doubling of the bit density while achieving the same number of net die per wafer. As a solution, a common BL structure was chosen to minimize the number of local/global BL-selection transistors. However, this resulted in a different polarity in the second deck, which caused an offset in the cell's characteristics. In addition, the differing thermal histories of the two decks can also cause an offset. This offset should be compensated for by modulating the integration between the two decks. Fig. 16 shows a comparison of the offsets in the set/reset V_t before and after the integration modulation, which clearly demonstrates that the offsets were successfully corrected for by the modulation.

III. CONCLUSION

In this letter, we demonstrated a cost-effective 128 Gb CPM technology that is nearly ready for commercialization. We developed a set of N-PCMs and NCSs for large RWM and V_t stability, which are both necessary factors for a low RBER and decent reliability margins. However, because of the intrinsic vulnerabilities of chalcogenide alloys, such characteristics can easily deteriorate or even be washed out after integration. To overcome this, robust materials, new etching, and cleaning recipes with novel integration schemes were developed.

In conclusion, functional 16 Mb MATs with sufficient RWM were successfully obtained. These MATs also exhibited great reliability in areas including write endurance, retention, and drift as well as little WDT.

REFERENCES

[1] S. Nazari, "Using storage class memory in next generation designs" *Flash memory summit keynote 10* (2017)

[2] G. W. Burr et. al, "Overview of candidate device technologies for storage-class memory," *IBM J. Res. Develop.*, vol. 52, no. 4, pp. 449-464, (2008)

[3] W. Kim et. al, "ALD-based Confined PCM with a Metallic Liner toward Unlimited Endurance," *IEDM Tech. Dig.*,pp. 4.2.1-4.2.4, (2016)

[4] D. Ielmini et. al, "Physical interpretation, modeling and impact on phase change memory (PCM) reliability of resistance drift due to chalcogenide structural relaxation," *IEDM Tech. Dig.*, pp. 939–942., (2007)

[5] D. Dong et. al, "The Impact of RTN Signal on Array Level Resistance Fluctuation of Resistive Random Access Memory" *IEEE Electron Device Lett.*, Vol. 39, (2018)

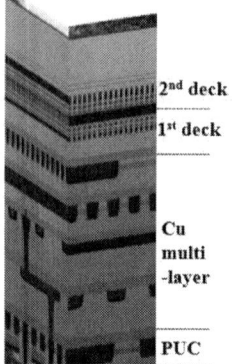

Fig. 1. TEM cross section of two-deck cell array with Cu multi-layer and PUC.

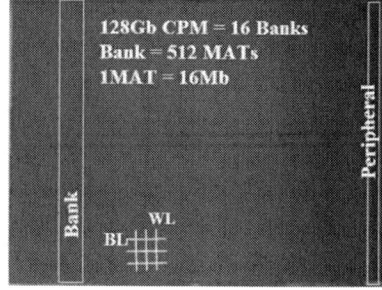

Fig. 2. Floor plan of 128 Gb die consisting of 16 banks.

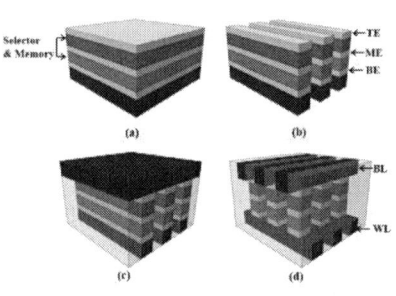

Fig. 3. Self-align process integration schemes: (a) cell stack material deposition, (b) after self-aligned WL patterning, (c) ILD deposition, CMP, and BL deposition, and (d) self-aligned BL patterning.

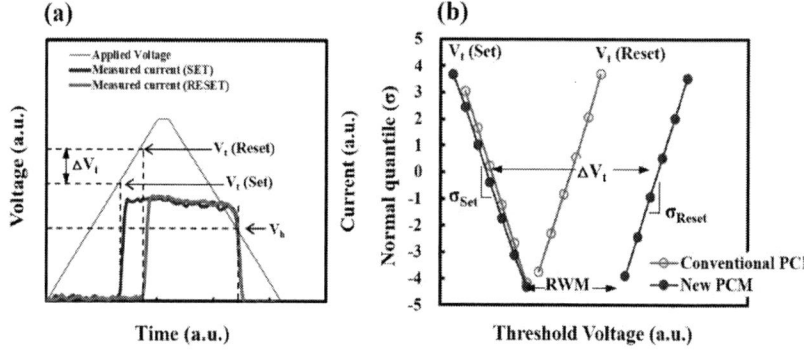

Fig. 4. (a) Voltage measurements by time for V_t and V_h (hold voltage) detection and (b) V_t distributions for set and reset. N-PCM exhibited much larger RWM than C-PCM.

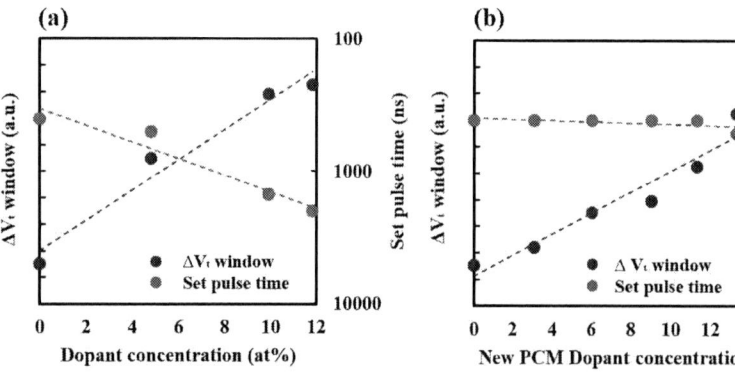

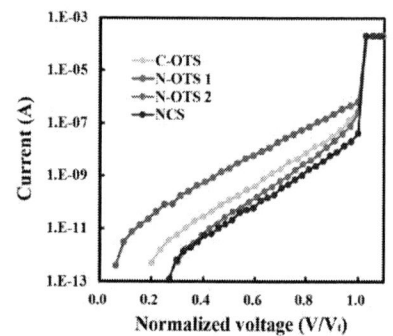

Fig. 5. Variations in ΔV_t by dopant concentration, which is proportional to bandgap: (a) conventional dopant results showing tradeoff between ΔV_t and t_{set} and (b) N-PCM dopant exhibiting increasing ΔV_t without t_{set} degradation.

Fig. 6. Normalized I-V behaviors of various selector materials after integration without PCM.

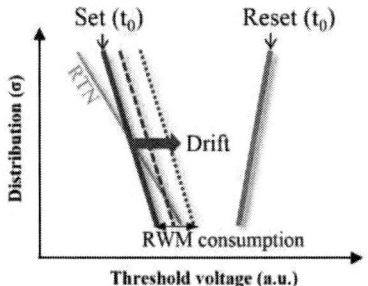

Fig. 7. Plot of set/reset V_t to explain RWM consumption by drift and RTN.

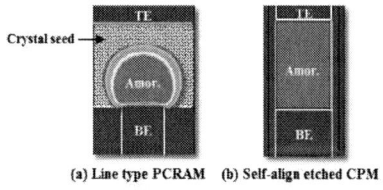

Fig. 8. Set V_t variation by time, representing drift from various selectors: C-OTS, N-OTS, and NCS.

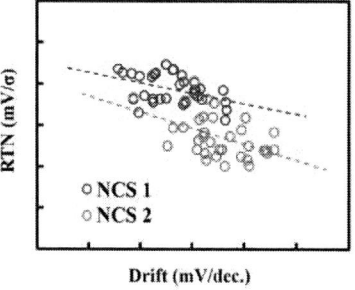

Fig. 9. Tradeoff between set drift and RTN at 55 °C, where RTN is defined as the standard deviation of 100 V_t reads.

Process technology	2z nm
Cell size	4F2 / two deck
Cell selector	NCS
Cell memory	N-PCM
Organization	8Gb* 16 Banks =128Gb
Read latency	≤100ns
Write latency	≤ 30ns (reset) ≤300ns (set)

Table 1. Basic die information, including structure and pulse widths (latencies).

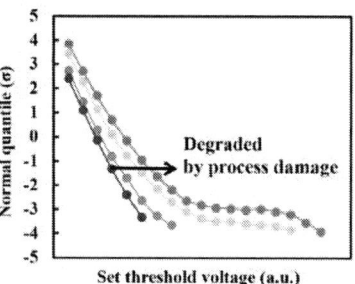

Fig. 10. Different environments for set operation (crystallization) in (a) line type and (b) confined structure. Note that there's no crystal seed in (b) after full amorphization, which requires additional nucleation step.

Fig. 11. Aggravating set tail by process damage from different etch recipe and integration.

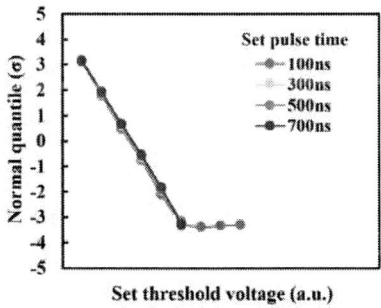

Fig. 12. Variations in set distribution by different set time. Note that tail starts to develop from 100 ns.

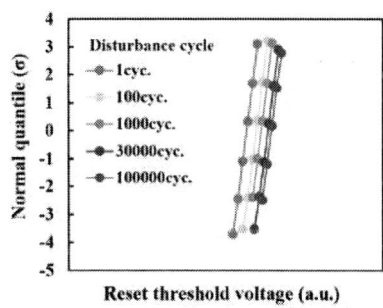

Fig. 13. Variations of victim cell's reset distribution by increasing aggressor's reset write cycle. For better visibility, X-axis offset is made intentionally.

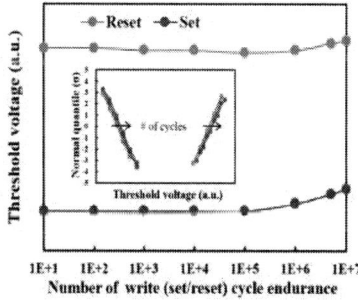

Fig. 14. Write cycle (set/reset) endurance of median cell (main) and entire array distribution, including tail bits (inset). Wide RWM is maintained even after 1E7 cycle.

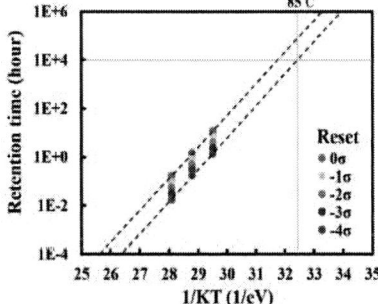

Fig. 15. Thermal retention of array. N-PCM can maintain reset state for more than 10,000 hours at 85 °C, even for the tail bits.

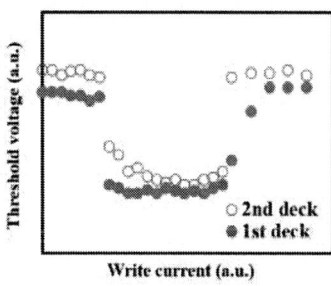

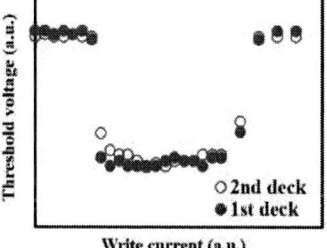

Fig. 16. Different V_t-I behaviors in first and second decks (a) before and (b) after integration modulation. Both decks exhibit similar behaviors after modulation.

978-1-7281-1988-5/18 $31.00 © 2018 IEEE   854

A Highly Efficient and Scalable Model for Crossbar Arrays with Nonlinear Selectors

An Chen

IBM Research Division, San Jose, CA 95120; Tel: (408) 927-1232; Email: chenan@us.ibm.com

Abstract

A scalable crossbar array model is proposed to solve arrays with various sizes and device characteristics. Computation is accelerated >1000× by array reduction, enabling array scaling analysis and extensive design space exploration. Computation errors can be reduced by extrapolation from calculations with different array reduction ratios. The impact of selectors on array performance is analyzed with this model, which can be applied to nonlinear, threshold switch, and rectifying selectors.

Introduction

Crossbar array (CBA) has a broad range of applications, including high-density storage, 3D memory, neuromorphics, *etc.* Fig. 1 illustrates a CBA of *m* word-lines (WLs) and *n* bit-lines (BLs). CBA of only linear resistors (1R) is not functional due to large number of sneak paths formed by unselected devices. The sneak paths can be suppressed by nonlinear or asymmetric selectors integrated with memory elements at every junction to form 1-selector-1-resistor (1S1R) arrays. Solving CBA with these nonlinear devices and line resistance (R_L) is important for the design of new memory and computing architectures and the assessment of memory and selector devices. A useful CBA model should incorporate all the array and device parameters as well as their inter-dependence (Fig. 2). A matrix-based CBA model describes array characteristics as matrices and solves the Kirchhoff equation at all junctions together for an accurate, complete array solution (Fig. 3) [1, 2]. Unfortunately, this accurate CBA solution is computationally intensive, which limits the practically solvable array sizes to several kbit. To solve larger arrays, it is necessary to adopt approximation. This paper presents a highly efficient and scalable CBA model applicable to various array designs and selector characteristics. Scaling of CBA up to Mbit sizes and the impact of selector and memory device parameters on CBA performance can be quantitatively analyzed with this model.

A Highly Efficient and Scalable Crossbar Array Model

The flowchart in Fig. 4 describes the proposed scalable CBA model. It is a common approach to lump up junctions and lines to solve a large CBA, which is also adopted in this model. Array reduction inevitably introduces errors. It was found that the deviation of solutions utilizing array reduction from the exact solutions without reduction correlates with the lumping size used in the array reduction (Fig. 5(a)). Therefore, a novel approach in the proposed model is to solve CBA with different lumping sizes and derive the final solution by extrapolation from the calculations with different lumping sizes. The derived final solution matches the exact solution very well (Fig. 5(b)). Fig. 6 illustrates how a large CBA is reduced. A unique approach in this model is to keep boundary lines separated from lumped groups, which serve as the basis for array expansion

back to the original size. Notice that in the worst scenario analysis where the selected device is the furthest away from voltage sources, the selected WL/BL are both boundary lines.

The CBA solution can be accelerated over 1000× by array reduction (Fig. 7(a)). A typical lumping ratio takes the square root of array length, e.g., 32 for a 1Mbit (1024×1024) array. Fig. 7(b) analyzes the error induced by array reduction by comparison with the exact solutions at smaller array sizes (e.g., 4kbit). The error is affected by many factors in CBA, e.g., R_L, bias schemes, selectors, *etc.* It can be kept well below 0.5% and in some cases below 0.01%. Essentially, the more functional the CBA (i.e., lower R_L and sneak leakage), the lower the errors.

To analyze CBA with selectors, some nonlinear selectors reported in [3-7] are simulated with a hyperbolic function and parameters estimated from measured I-V (Fig. 8). The black dashed line is a hypothetical selector used in array analysis.

Scaling Analysis of CBA with Nonlinear Selectors

The proposed CBA model enables array analysis up to Mbit sizes and extensive exploration of device and design space. The analysis of CBA scaling from 256bit to 1Mbit is presented in Fig. 9-11. Table 1 summarizes array and device parameters used in this paper. Fig. 9 compares the voltage delivered to a selected device furthest away from voltage sources and the maximum disturbance among unselected devices during writing, whose difference defines the ***writing voltage margin (WVM)***. For the worst scenario analysis, unselected devices are all in low-resistance-state (LRS), causing the highest sneak leakage. Ground bias scheme does not work even with selectors because the maximum disturbance occurs on unselected devices closest to the voltage sources and always exceeds the selected device voltage. Float bias and "1/2 bias" schemes are similar in terms of writing voltages and WVM. The "1/3 bias" has the highest WVM owing to the lowest disturbance. R_L has significant impact on CBA performance. For example, the maximum feasible CBA size (defined by WVM>0) may reach 1Mbit for R_L=1Ω (Fig. 9(e)); however, this size is reduced to 128kbit for R_L=10Ω. Notice that WVM>0 is a necessary but not sufficient condition for functional CBA, because of inevitable variation in memory writing voltage (V_{write}). The selected device voltage has to satisfy the highest V_{write} while the disturbance has to be below the lowest V_{write}. The writing power efficiency of different bias schemes is compared in Fig. 10. The float bias scheme has the highest efficiency due to the lowest leakage, while the ground bias scheme has the lowest efficiency. The "1/3 bias" scheme is less efficient than the "1/2 bias" scheme due to higher leakage allowed by the unselected junction bias of $\pm V_{dd}/3$.

Reading of the CBA uses a voltage-dividing scheme by a sensing resistor (R_S in Fig. 1). The ***sensing margin (SM)*** is the difference of V_{out}'s for the selected device in low- and high-

978-1-7281-1988-5/18 $31.00 © 2018 IEEE

resistance-state (LRS and HRS). As shown in Fig. 11, the ground bias scheme has decent SM but is more affected by R_L due to higher leakage. The float bias scheme is unsuitable for reading because all the unselected junctions contribute to sneak paths, reducing the influence of the selected device on V_{out}. Partial bias schemes improve SM. For memory and selector devices used in this simulation, 1Mbit CBA is feasible with appropriate bias schemes at reasonably low R_L. However, the increase of R_L will quickly decrease the feasible CBA size or require improved selector and/or memory characteristics.

Assessment of 1Mbit Arrays with Nonlinear Selectors

To assess the impact of selector parameters on CBA, a set of metrics are calculated for a 1Mbit CBA for a range of selector parameters, pre-factor J_0 and exponential coefficient α (Fig. 12). The increase of J_0 makes selectors more conductive and less effective in blocking sneak leakage; therefore, selected device voltage, WVM, and SM all decrease with increasing J_0. Higher α makes selector I-V steeper and is expected to improve CBA performance, which is true for SM (Fig. 12d). However, the selected device voltage and WVM first improve but then degrade with α. This behavior can be explained by the double effects of selectors: more conductive (steeper) selectors have less voltage-dividing effect with memories in a 1S1R junction (positive) but are less effective in suppressing sneak leakage (negative). In a large 1Mbit array, the negative effect of more conductive selectors (i.e., higher J_0 and α) dominates. In a smaller 1kbit array, the selected device voltage and WVM both improve with increasing J_0 and α, because the positive effect outweighs the negative effect in a small array (Fig. 13).

The voltage-dividing effect of selectors has important impact on CBA operation. Fig. 14 shows the percentage of a selected junction voltage taken by the selector, which decreases with increasing J_0 and α. This effect is more significant in larger arrays due to lower junction voltage that makes selector more resistive. For the device parameters used in this calculation, > 60% of junction voltage is taken by the selector. This percentage would decrease for memories with higher resistance or nonlinear characteristics. The increase of memory R_{LRS} does improve selected device voltage and WVM (Fig. 15(a)). As expected, higher memory on/off ratio improves SM (Fig. 15(b)). With increasing R_{LRS}, SM first improves (due to equivalently lower R_L and more favorable voltage-dividing effect for memory) but then degrades (because higher memory resistance equivalently makes selectors more conductive).

Higher array V_{dd} improves the selected device voltage, but increases the disturbance even faster and degrades the power efficiency (Fig. 16(a)). Therefore, array V_{dd} for writing should be chosen to be just high enough to write the selected device without any disturbance. Higher V_{dd} improves SM but also increases voltage on both selected and unselected devices (Fig. 16(b)). Therefore, array V_{dd} for reading should be chosen to provide sufficiently high SM without disturbing any devices.

Threshold Switch and Rectifying Diode Selectors

Threshold switch selectors have been studied extensively in recent years due to their high on/off ratio (Fig. 17) [8-16].

However, as switching devices, threshold selectors need to be balanced with memories in a 1S1R structure to avoid unstable (oscillatory) behaviors caused by voltage re-distribution after the switching of either selector or memory (Fig. 18). Because of the inevitable variation of voltage and resistance of both selectors and memories as well as unpredictable data patterns and voltage distribution in an array, finding and maintaining conditions for stable 1S1R behaviors are quite challenging. This has been largely ignored in device reports of threshold switch selectors. When the working conditions are satisfied, threshold switch selectors improve CBA performance more than nonlinear selectors, because their very high on/off ratio can suppress sneak paths effectively. The WVM and SM of CBAs with threshold selectors show less array size dependence (Fig. 19) than CBAs with nonlinear selectors (Fig. 9-11).

Rectifying diodes are another important category of selectors; however, they only work for unipolar switching memories. Rectifying diodes can be described by exponential I-V characteristics with J_0, n (non-ideality factor), and R_s (Fig. 20) [17-21]. Si p-n junction diodes have one of the best two-terminal selector characteristics, but require high-temperature processing. Fig. 21 assesses CBA performance with rectifying diode selectors. The "1/3 bias" scheme works better than the "1/2 bias" scheme with rectifying diodes, because unselected junctions are reversely biased in the "1/3 bias" scheme and can be blocked by the high reverse resistance of rectifying diodes.

Discussion and Summary

The proposed CBA model uses analytical selector device models, which can be extended to more complicated device behaviors, e.g., segmented models with varying parameters. This paper uses the worst-scenario analysis where uniform data patterns are assumed. The impact of device variation can be assessed by adopting the worst device parameters from their distributions, rather than exploring the entire statistical space of distribution. Linear extrapolation has been utilized in the proposed model to expand reduced arrays and to reduce errors induced by array reduction. More sophisticated extrapolation methods can be developed to further improve this model.

In summary, a highly efficient crossbar array solution is developed utilizing novel approaches to reduce array size and errors. Based on this method, array scaling and performance impact of selector and memory parameters are analyzed. The proposed method enables extensive device and design space exploration for various types of selectors and large array sizes.

References

[1] A. Chen, IEEE TED **60**, 1318 (2013); [2] A. Chen, IEDM, 746 (2013); [3] A. Kawahara, et al, ISSCC, 432 (2012); [4] W. Lee, et al, VLSI Tech., 37 (2012); [5] L. Zhang, et al, IEEE EDL **35**(2), 199 (2014); [6] B. Govoreanu, et al, IEEE EDL **35**(1), 63 (2014); [7] K.S. Li, et al, 385, ISCAS (2015); [8] H. Yang, et al, VLSI Tech., 130 (2015); [9] H. Yang, et al, IEDM, 836 (2017); [10] S.H. Jo, et al, IEDM, 160 (2014); [11] B. Govoreanu, et al, VLSI Tech., 92 (2017); [12] Q. Lin, et al, IEEE EDL **39**(4), 496 (2018); [13] H.Y. Cheng, et al, IEDM, 28 (2017); [14] W.G. Kim, et al, VLSI Tech., 138 (2014); [15] S. Kim, et al, VLSI Tech., 240 (2013); [16] M. Son, et al, IEEE EDL **32**(11), 1579 (2011); [17] Y. Sasago, et al, VLSI Tech., 24 (2009); [18] W.Y. Park, et al, Nanotech **21**, 195201 (2010); [19] A. Chasin, et al, IEEE EDL **35**(6), 642 (2014); [20] M.J. Lee, et al, IEDM, 771 (2007); [21] N. Huby, et al, Microelectronic Eng. **85**, 2442 (2008).

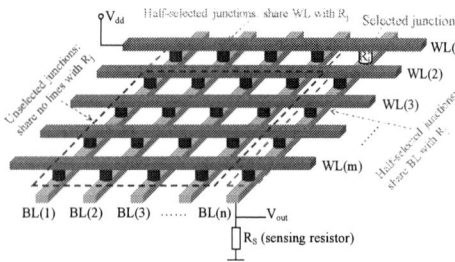

Fig. 1 Schematic of a crossbar array (CBA) with *m* word-lines (WLs) and *n* bit-lines (BLs).

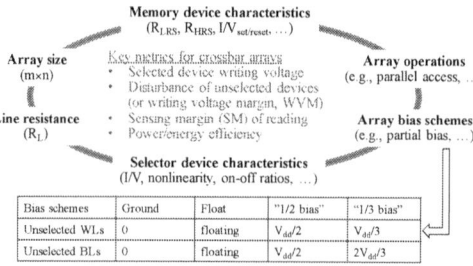

Fig. 2 Device/circuit parameters and key metrics for CBA analysis. **Array bias schemes** are defined by the bias of unselected WLs/BLs.

Bias schemes	Ground	Float	"1/2 bias"	"1/3 bias"
Unselected WLs	0	floating	$V_{dd}/2$	$V_{dd}/3$
Unselected BLs	0	floating	$V_{dd}/2$	$2V_{dd}/3$

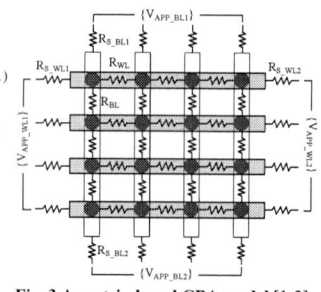

Fig. 3 A matrix-based CBA model [1-2].

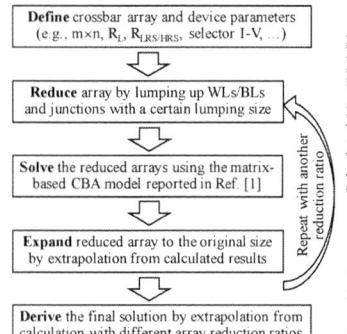

Fig. 4 Flowchart of the scalable CBA model.

- **Define** crossbar array and device parameters (e.g., $m \times n$, R_L, $R_{LRS/HRS}$, selector I-V, …)
- **Reduce** array by lumping up WLs/BLs and junctions with a certain lumping size
- **Solve** the reduced arrays using the matrix-based CBA model reported in Ref. [1]
- **Expand** reduced array to the original size by extrapolation from calculated results
- **Derive** the final solution by extrapolation from calculation with different array reduction ratios

Repeat with another reduction ratio

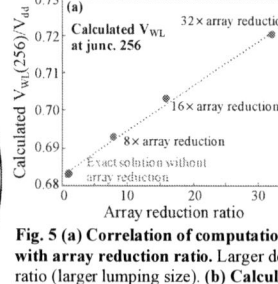

Fig. 5 (a) Correlation of computation error (deviation from exact solutions) with array reduction ratio. Larger deviation is observed with higher reduction ratio (larger lumping size). **(b) Calculated WL voltage *vs.* junction number for different array reduction ratios.** By extrapolating final results from calculations with different lumping sizes (e.g., 8 and 16 in the example of a 4kbit array), the final result matches well with the exact solution without array reduction.

Table 1. Parameters and values used in paper

	Parameters	Typical values used in the paper
Memory device	On/off ratio (R_{HRS}/R_{LRS})	10
	LRS resistance (R_{LRS})	50 kΩ
Nonlinear selector	I-V Model	$I(V) = I_0 \cdot \sinh(\alpha V)$
	Parameters I_0, α, R_s	$I_0 = 10^{-10}$A, $\alpha = 10$V^{-1}, $R_s = 100\Omega$
Threshold switch selector	On and off resistance	$R_{on} = 5$kΩ, $R_{off} = 1$GΩ
	Threshold and hold voltage	$V_{th} = 1.5$V, $V_h = 0.01$V
Rectifying diode selector	I-V model	$I(V) = I_0 \cdot [\exp(qV/nkT)-1]$
	Parameters I_0, n, R_s	$I_0 = 1.5 \times 10^{-14}$A, $n = 1.8$, $R_s = 1.5$kΩ
Array design	Array size	From 256bit to 1Mbit
	Line resistance (R_L)	1-10Ω (between junctions)
	Applied voltage (V_{dd})	Writing: 4V; reading: 2V

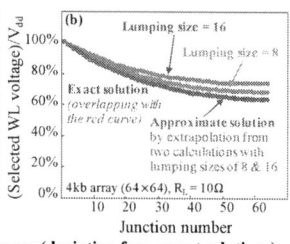

Fig. 7 (a) Computing acceleration by array reduction. Over 1000× speedup can be achieved with the proposed approach and simulation of arrays >1Mbit is feasible. **(b) Computation errors** induced by array reduction assessed on a 4kbit array that can be solved accurately without array reduction.

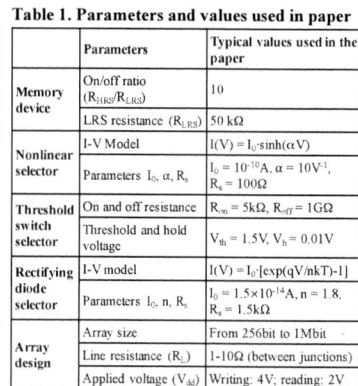

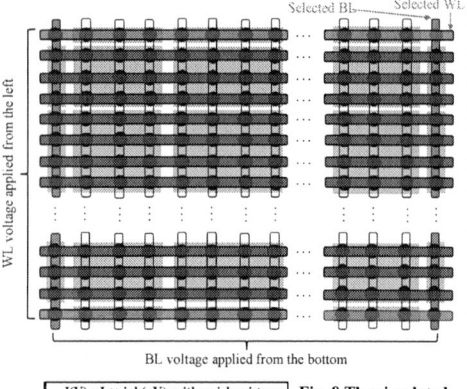

Fig. 6 Illustration of CBA reduction by lumping up certain number (lumping size) of junctions and lines. Junctions and lines in the yellow boxes are lumped together. Lines at boundaries are not lumped with any other groups. For the worst-scenario analysis, the selected device is the furthest away from voltage sources.

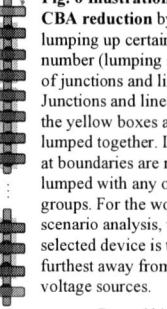

Fig. 8 The simulated I-V curves of some nonlinear selectors reported in [3-7], described by function $I(V) = I_0 \times \sinh(\alpha V)$ with serial resistance (R_s). A hypothetical selector model (black dashed line) is used to analyze scaling of CBA in Fig. 9-11.

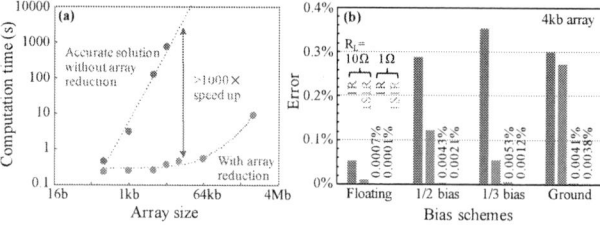

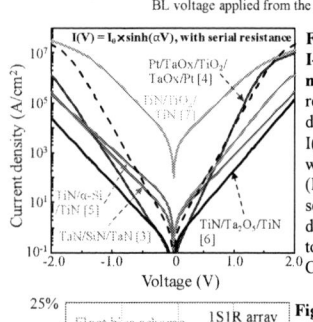

Fig. 10 Power efficiency of CBA with nonlinear selectors during writing. The efficiency is defined as the selected device power divided by the total array power.

Fig. 9 Writing performance of crossbar arrays with nonlinear selectors for four bias schemes: ground (a-b), float (c-d), "1/2 bias" (e-f), and "1/3 bias" (g-h). Red and blue curves are 1R and 1S1R arrays, respectively. Solid and dashed lines are the selected device voltage and the maximum disturbance of unselected devices, respectively. Black curves are the exact solutions without array reduction (up to 4kb). Unselected devices are in LRS. The selected device is the furthest away from voltage source and in LRS. Array V_{dd} is 4V.

Fig. 11 Reading performance of CBA with nonlinear selectors. Sensing margin (SM) is defined as the difference of readout voltage (V_{out} in Fig. 1) for the selected device in LRS and HRS. Unselected devices are all in LRS. Sensing resistance (R_S) is given by $(R_S)^2 = R_{LRS} \times R_{HRS}$. Array V_{dd} is 2V.

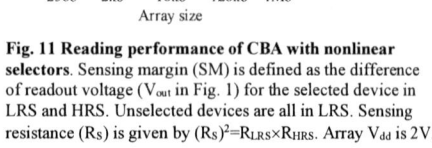

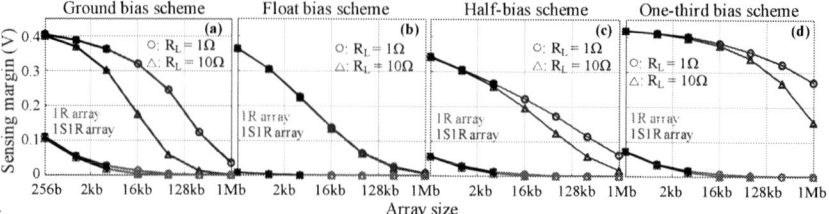

978-1-7281-1988-5/18 $31.00 © 2018 IEEE

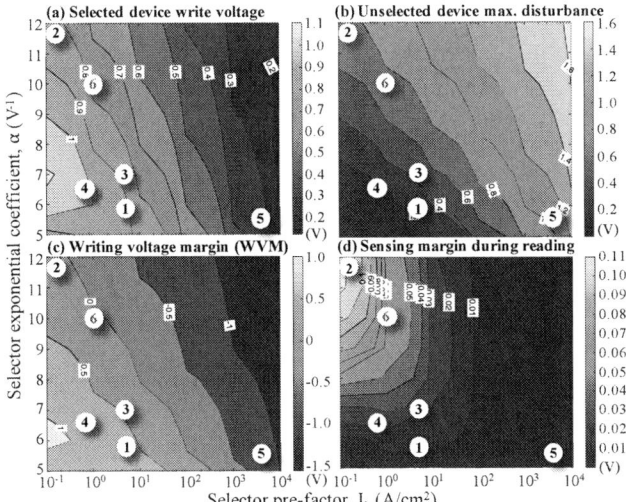

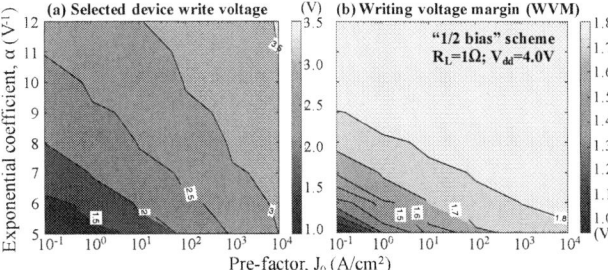

Fig. 13 Impact of nonlinear selector parameters on a 1kbit CBA: (a) selected device voltage; (b) writing voltage margin. Use "1/2 bias" scheme, R_L of 1Ω, and V_{dd} of 4V.

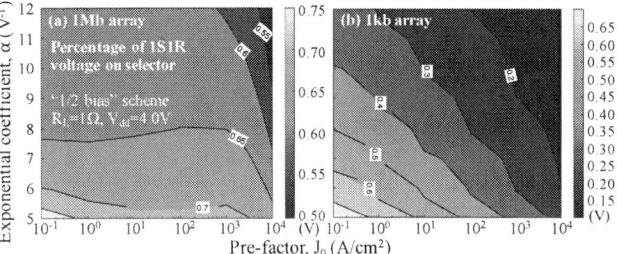

Fig. 12 Impact of nonlinear selector parameters (J_0, α) on a 1Mbit CBA measured by 4 metrics. Array V_{dd} is 4V for writing (a-c) and 2V for reading (d). The "1/2 bias" scheme is used and R_L is 1Ω. The six nonlinear selectors in Fig. 8 are marked in plots: (1) TaN/SiN/TaN; (2) Pt/TaO$_x$/TiO$_2$/TaO$_x$/Pt; (3) TiN/α-Si/TiN; (4) TiN/Ta$_2$O$_5$/TiN; (5) TiN/TiO$_2$/TiN; (6) the hypothetical nonlinear selector model.

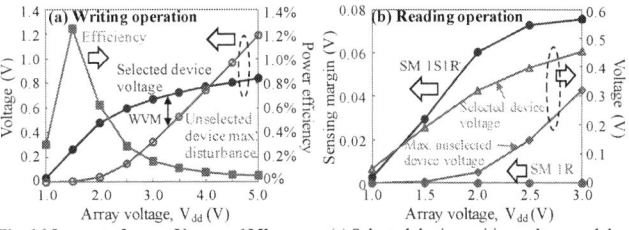

Fig. 14 Voltage-dividing effect of selectors in: (a) a 1Mb array; (b) a 1kb array.

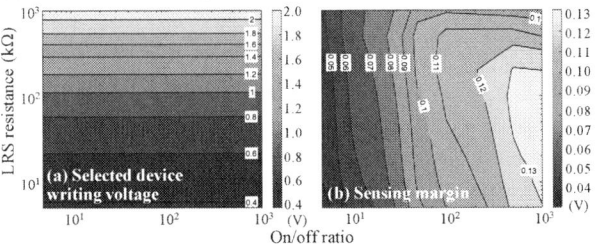

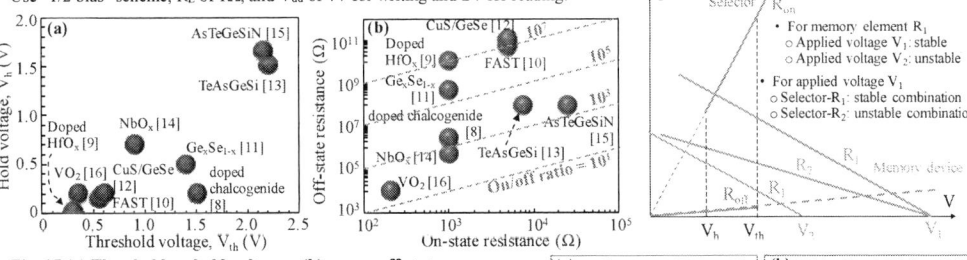

Fig. 15 Impact of memory parameters (R_{LRS}, on/off ratio) on a 1Mbit array: (a) the selected device writing voltage; (b) sensing margin (SM) during reading. Use "1/2 bias" scheme, R_L of 1Ω, and V_{dd} of 4V for writing and 2V for reading.

Fig. 16 Impact of array V_{dd} on a 1Mb array: (a) Selected device writing voltage and the maximum disturbance (left axis) and efficiency (right axis); (b) Sensing margin (left axis) and voltages of selected and unselected devices (right axis). "1/2 bias" scheme, R_L = 1Ω.

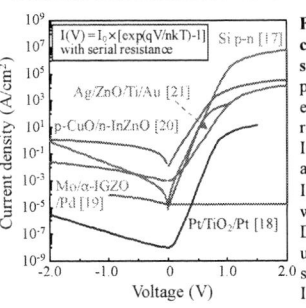

Fig. 17 (a) Threshold vs. hold voltages; (b) on- vs. off-resistance of some threshold switch selectors reported in [8-16].

Fig. 18 Voltage and resistance balance between threshold switch selectors and memory devices illustrated in a load-line analysis. V_1 and V_2 are two different voltages applied on a 1S1R structure, and R_1 and R_2 are two different memory resistance. Unbalanced voltage and resistance combinations between selectors and memories may lead to unstable (oscillatory) behaviors.

- For memory element R_1
  - Applied voltage V_1: stable
  - Applied voltage V_2: unstable
- For applied voltage V_1
  - Selector-R_1: stable combination
  - Selector-R_2: unstable combination

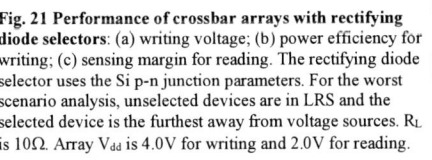

Fig. 20 Simulated I-V curves of rectifying diode selectors based on device parameters estimated from experimental I-V curves reported in [17-21]. Diode I-V can be described with an exponential function of I(V) = I$_0$×[exp(qV/nkT)-1] with serial resistance (R$_s$). Different parameters are used in +V and −V bias to simulate the asymmetric I-V characteristics.

I(V) = I$_0$×[exp(qV/nkT)-1] with serial resistance

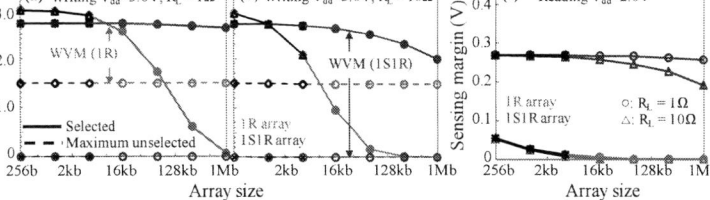

Fig. 19 Performance of crossbar arrays with threshold switch selectors: (a-b) writing, R_L=1Ω or 10Ω. (c) reading. The threshold switch selector parameters are V_{th}=1.5V, V_h=0.01V, R_{on}=5kΩ, and R_{off}=1GΩ. Red and blue curves are 1R and 1S1R arrays, respectively. Black curves are the exact solutions without array reduction (up to 4kbit). The "1/2 bias" scheme is used. Unselected devices are all in LRS.

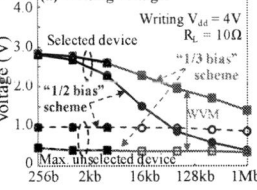

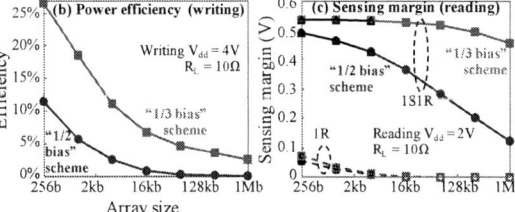

Fig. 21 Performance of crossbar arrays with rectifying diode selectors: (a) writing voltage; (b) power efficiency for writing; (c) sensing margin for reading. The rectifying diode selector uses the Si p-n junction parameters. For the worst scenario analysis, unselected devices are in LRS and the selected device is the furthest away from voltage sources. R_L is 10Ω. Array V_{dd} is 4.0V for writing and 2.0V for reading.

978-1-7281-1988-5/18 $31.00 © 2018 IEEE   858

Ultra-High Endurance and Low I_{OFF} Selector based on AsSeGe Chalcogenides for Wide Memory Window 3D Stackable Crosspoint Memory

H. Y. Cheng[1], W. C. Chien[1], I. T. Kuo[1], C. W. Yeh[1], L. Gignac[2], W. Kim[2], E. K. Lai[1], Y. F. Lin[1], R. L. Bruce[2], C. Lavoie[2], C. W. Cheng[2], A. Ray[2], F. M. Lee[1], F. Carta[2], C. H. Yang[1], M. H. Lee[1], H. Y. Ho[1], M. BrightSky[2] and H. L. Lung[1]

IBM/Macronix PCRAM Joint Project

[1]Macronix International Co., Ltd., Emerging Central Lab., 16 Li-Hsin Rd., Science Park, Hsinchu, Taiwan, ROC

[2]IBM T. J. Watson Research Center, P. O. Box 218, Yorktown Heights, NY 10598, USA

Tel: +1-914-945-2664, email: hymcheng@us.ibm.com; hycheng@mxic.com.tw

Abstract—New selector materials with very-low I_{OFF} and optimum V_{th} based on As-Se-Ge chalcogenides are studied. An optimized composition is proposed, which achieves a good trade-off between thermal stability and cycling endurance and it is successfully integrated with PCM in a 3D stackable pillar structure. SET/RESET operation are demonstrated with ~2V memory window. Selector is able to deliver 1mA ON current (7.9 MA/cm^2) and fast speed (10 ns). More than 1E12 read cycling endurance is achieved in 1S1R (OTS+PCM) device due to the excellent endurance of the selector.

I. INTRODUCTION

Storage class memory (SCM) technology requires NVM densely packed in "crosspoint" arrays to achieve 4F^2 footprint with selecting devices capable of delivering high current and power. When operating a crosspoint array by using "Half-V" select scheme [1], a set of voltages (V and zero (ground)) is applied at the edge of the array such that the desired operations (read and program) take place at the desired selected cells while all nonselected cells remain unperturbed due to the zero bias drop (inset in **Fig. 1(a)**). However, the cells that share same row or same column (half-selected) see half of the voltage drop across the selected array. In order to prevent program disturb for these half-selected cells, the threshold voltage of the selector in the crosspoint array is a key. **Figure 1(a)** shows schematic I-V characteristic of PCM (phase-change memory)+OTS (Ovonic Threshold Switching) device in SET and RESET state with sketched SET and RESET threshold voltage distributions. V (applied voltage) >V_{tRmax} (the maximum threshold voltage of the RESET state cell) is required to guarantee that the applied voltage can RESET the highest V_{tR} cell in the array. V/2<V_{tSmin} (the minimum threshold voltage of the SET state cell) is also required to assure that all of the half selected cells keep the same state. **Figure 1(b)** shows the calculated results with the above mentioned boundary conditions. The higher V_{tS} (~V_{th}, the threshold voltage of the selector), the higher memory window is guaranteed. For example, V_{tS} has to be higher than 3.4 V in order to have 1 V memory window. A large difference between V_{tR} and V_{tS} can allow cell-to-cell variation while retaining a good memory window. Therefore, choosing a relatively high threshold voltage of the selector is essential for crosspoint phase-change memory.

Te-based OTS selectors have been promising candidates [2-3] as access devices. **Figure 2** shows Te-based OTS [2] characteristics with different OTS thickness. By increasing OTS thickness to 80 nm, V_{th} can be increased to 3.1 V. However, such thick OTS layers have process difficulties to integrate into stackable pillar structure, therefore not are practical for high density cross point memory. GeSe-based materials have demonstrated good OTS characteristic and considered to be high V_{th} selectors especially with extra N doping [4-5]. However, insufficient cycling endurance means further improvements are needed for this material.

In this work, we replace main chalcogenide element of the selector (Te) with Se simplifying the system to a ternary compound (As-Se-Ge). With optimization of compositions (trade-off between cycling endurance and thermal stability), new selector shows promising characteristic with high V_{th} (3.5 V for 30 nm thick), ultra-low I_{OFF} (131 pA at 2 V) and high endurance of <1E10 demonstrated in the mushroom-type structure. It is then successfully integrated with PCM into pillar structure and demonstrated high memory window (~2 V) between RESET and SET state with low I_{OFF} (~10 nA at ½ V_{tR} (8V)) and superior read endurance (>1E12) in the PCM+OTS stacked memory.

II. OVONIC THRESHOLD SWITCHING MATERIAL DESIGN AND SELECTOR CHARACTERIZATION

Figure 3 shows the OTS materials studied in this paper in the As-Se-Ge (or Ge+Si)-Te quaternary phase diagram, as compared with Te-based OTS material [2]. Studied materials in this paper are located along As$_2$Se$_3$-Ge tieline. With increasing Ge content, thermal stability can be improved due to the increase of crosslink in the network. XRD (**Fig. 4**) confirms a stable amorphous phase of Material C from as-deposited to 450 °C indicating excellent thermal stability. To quickly verify new OTS materials characteristics, short turnaround time lift-off devices with 350 nm diameter W bottom electrode mushroom-type structure are used [2]. **Figure 5** shows the TEM image of the testing device with OTS Material C as an example. **Figure 6 and 7** show measured I-V characteristics of 30 nm Material C and

978-1-7281-1988-5/18 $31.00 © 2018 IEEE

Material D, respectively. Material C requires a forming voltage of 4.8 V; exhibits a high V_{th} of 3.5 V and extraordinary low leakage current (131 pA at 2V). V_{th} of the selectors can be successfully increased by changing OTS materials from Te-based to Se-based materials instead of directly increasing OTS thickness. This new OTS material presents an additional advantage of ultra-low I_{OFF} due to its high resistivity in amorphous phase (>1E8 ohm-cm). By increasing Ge content (Ge >30 at.%), I_{OFF} degrades significantly (I_{OFF} of Material D is 10 nA at 2V). **Figure 8** shows V_{th} and I_{OFF} characteristics of Material C with different thickness. V_{th} shows a good range at practical thickness for integration with PCM in 1S1R pillar configuration, providing a larger memory window in the stackable memory. With further decreasing Ge content (Material A and B), V_{th} and I_{OFF} show similar characteristics ($V_{th}/I_{OFF(at\ 2V)}$ ~ 3.1 V/170 pA and 2.9 V/500 pA, respectively, not show in this paper) to Material C, but thermal stability is much poorer (< 330 °C)(shows in **Fig. 3**). **Figure 9** compares the cycling endurance of studied materials based on AsSeGe system in mushroom-type structure. Material B with low Ge content exhibits superior cycling endurance (6.9E11), which is the best among studied materials. However, with further increasing Ge content (Material C), cycling endurance degrades. Material with higher than 30 at.% Ge (Material D), V_{th} becomes unstable during the cycling. Selector characteristic based on AsSeGe chalcogenides is sensitive to Ge content, higher Ge material guarantees thermal stability but unstable cycling endurance and high I_{OFF} while low Ge material exhibits extraordinary cycling endurance and low I_{OFF} but poor thermal stability. The trade-off has to be made when choosing the optimum composition. Material C, which shows good V_{th} and I_{OFF} and fair cycling endurance (<1E10), also exhibit excellent thermal stability (~ 450 °C) which meets BEOL process temperature requirements. **Figure 10** shows Raman spectra of amorphous AsSeGe materials studied in this paper. These show that the change in electrical characteristic is related to change in bonding characteristic of the material. Material D presents a high number of As-As, Ge-Ge and Se-Se homopolar bonds (defect bonds) [6-8], which maybe the root cause for unstable OTS characteristics.

III. OTS+PCM PILLAR DEVICE CHARACTERIZATION

We integrate a promising OTS (Material C) with doped GST PCM [9] in 1S1R pillar structure on top of the contact layer of standard CMOS logic devices, **Fig. 11**. New OTS material is successfully integrated with PCM in the stackable structure by carefully controlling the etching process. Using a buffer layer enables switching the etch chemistry between electrode and PCM/OTS layer [10]. TEM EDX images **(Fig. 12)** show uniform OTS and PCM compositions without element-segregation after complete integration.

Figure 13 shows I-V characteristics of 1S1R (OTS+PCM) device in RESET and SET state. The threshold voltage of RESET state (V_{tR}) is 7.4 V (50 ns RESET pulse was used) and the threshold voltage of SET state (V_{tS}) is 5.6 V. Almost 2V memory window (V_{read}) is achieved by using high V_{th} OTS

material. The reason for higher threshold voltage of OTS (~V_{tS}) in the pillar structure compared to the mushroom structure is due to extra buffer layer in series with OTS+PCM pillar structure and also the process temperature densification effect on the OTS material [2]. **Figure 14** shows distribution of the delta V_{th} (V_{tR} - V_{tS}) from a 1S1R array. Most of the cells show close to 2V memory window. **Figure 15** shows leakage current (I_{OFF}) at the bias condition for half-selected cell when the selected cell is in RESET and SET operation, respectively. Very low I_{OFF} (~10 nA) is demonstrated at high bias of 4V when the selected cell is at RESET operation (V_{tR}=8V). The low I_{OFF} selector of material C can prevent the aggregate leakage through all the unselected cells from dominating the overall system power budget. PCM+OTS device can be SET around 300 ns **(Fig. 16)**. **Figure 17** shows transient current of 1S1R device. The selector is able to turn ON/OFF with a short pulse of 10 ns and deliver a high current of 500 uA, which is sufficient for RESET operation. 1S1R device is still functional after 1E12 read endurance by using 1 mA ON current (7.9 MA/cm^2) to turn on the selector when the PCM is maintained in SET state and OFF current is read out at 4V **(Fig. 18)**. More than 10^5 ON/OFF ratio is achieved. Table I compares the OTS+PCM characteristics by using different selector material. The high read endurance and large ON/OFF ratio (I_{OFF} at ½ V_{tR}) of our 1S1R device is attributed to the extraordinary high endurance and low I_{OFF} of the new selector material based on As-Se-Ge chalcogenides.

IV. CONCLUSION

New Te-free OTS materials based on As-Se-Ge chalcogenides are studied. A selector with high V_{th} is suggested for OTS+PCM crosspoint memory for optimal performance in the "half-V" select scheme operation. New selector with optimized composition (Material C) provides good V_{th} (3.4 V at 30 nm thick), ultra-low I_{OFF} (131 pA at 2V) and high cycling endurance demonstrated in mushroom-type structure. It is then successfully integrated with PCM in pillar-type structure. RESET and SET operation are successfully demonstrated with large memory window (~2V), low I_{OFF} (both at ½ V_{tR} and V_{tS}) and ultra-high read cycling endurance (1E12). It is also exhibited 1mA ON current for the selector in 1S1R device.

REFERENCES

[1] G. Burr et al., J. Vac. Sci. Technol. B **32**, 040802 (2014).
[2] H.Y. Cheng et al., Tech. Dig.- Int. Electron Devices Meet. **2.2** (2017).
[3] M. J. Lee et al., Tech. Dig.- Int. Electron Devices Meet. **2.6** (2012).
[4] N. S. Avasarala et al., European Solid-State Device Research Conference (2017).
[5] A. Verdy et al., Tech. Dig.-Int. Memory Workshop (2017).
[6] L. O. Revuska et al., http://ela.kpi.ua/bitstream/ 123456789/20574/1/15.Revutska.%D1%81.60-63.pdf
[7] R. P. Wang et al., J. Appl. Phys. **106**, 043520 (2009).
[8] A. Prasad et al., Optics Express **16**, 2804 (2008).
[9] H. Y. Cheng et al., Tech. Dig.- Int. Electron Devices Meet. **30.6** (2013).
[10] C.W. Yeh et al., VLSI Tech. Dig. **19.3** (2018).
[11] G. Navarro et al., VLSI Tech. Dig. **7.3** (2017).

978-1-7281-1988-5/18 $31.00 © 2018 IEEE

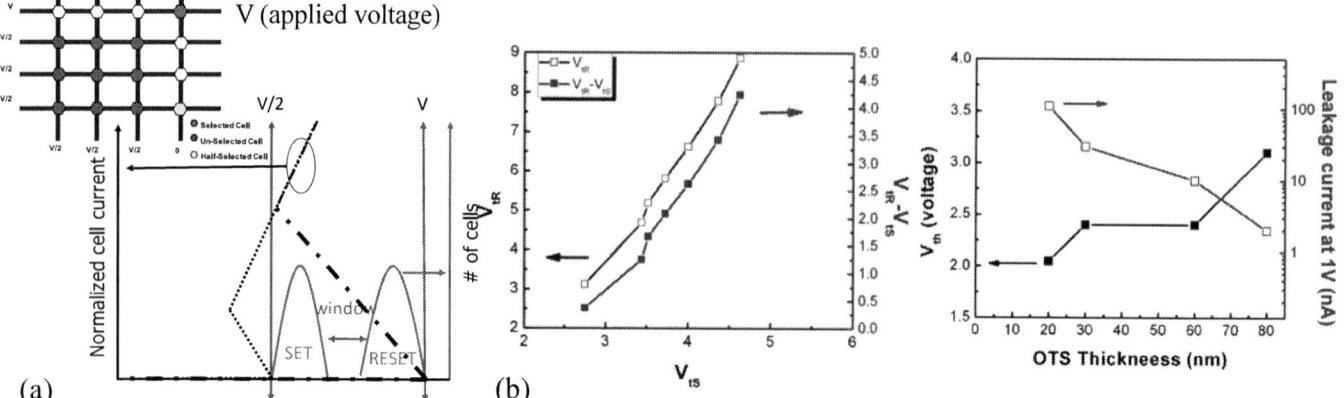

Fig. 1 (a) Schematic I-V Characteristics of a PCM+OTS cell in SET and RESET state with sketched SET and RESET threshold voltage distribution (# of cells is shown on the right ordinate). Voltage is normalized to the threshold voltage of the SET state, V_{tS}. Inset is the schematic view for "half-V" scheme for cross point array. (b) The calculation for V_{tS}, V_{tR} (the threshold voltage of SET, RESET state) and V_{tR}-V_{tS} (memory window) with satisfaction of two criteria: V (applied voltage) $>V_{tRmax}$ and V/2$<V_{tSmin}$. The higher V_{tS}, the higher memory window. Here SET and RESET distribution are assumed to be a Dirac δ-function.

Fig. 2 Te-based OTS selector device characteristics with different OTS thickness. V_{th} gradually increases with OTS thickness. However, even 80 nm thick OTS device only exhibits 3.1 V threshold voltage, which is still lower than minimum requirement (V_{th}>3.5 for 1V memory window (as shown in **Fig. 1(b)**)).

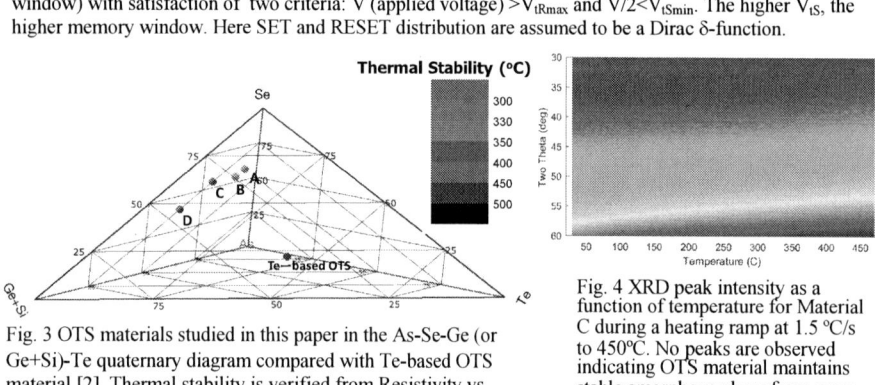

Fig. 3 OTS materials studied in this paper in the As-Se-Ge (or Ge+Si)-Te quaternary diagram compared with Te-based OTS material [2]. Thermal stability is verified from Resistivity vs. Temperature measurement [2]. With increasing Ge content, thermal stability is improved.

Fig. 4 XRD peak intensity as a function of temperature for Material C during a heating ramp at 1.5 °C/s to 450°C. No peaks are observed indicating OTS material maintains stable amorphous phase from room temperature up to 450 °C.

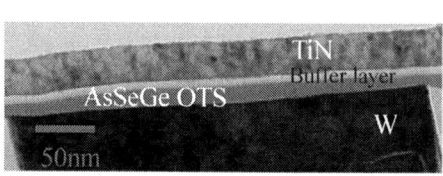

Fig. 5 Cross sectional TEM micrograph of OTS Material C sandwiched by buffer layer and W BE (bottom electrode) of 350 nm in diameter.

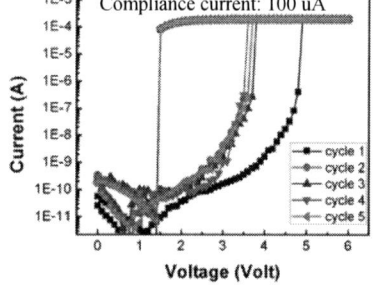

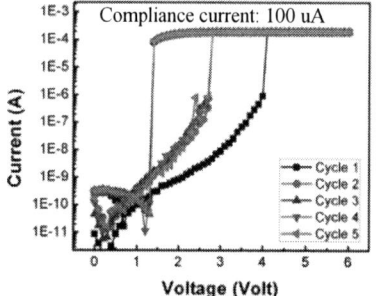

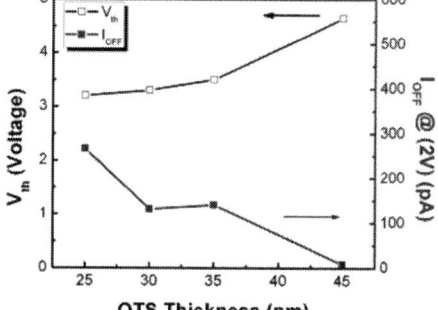

Fig. 8 OTS selector device characteristics of Material C with different thickness. AsSeGe OTS material exhibits relative high V_{th} where 45 nm device already shows V_{th} >4 V, compared to Te-based OTS material, which 80 nm thick OTS only reach V_{th} around 3V (in **Fig. 2**). High V_{th} would benefit memory window when OTS is integrated with PCM for cross-point operation.

Fig. 6 30 nm Material C OTS selector device characteristics with different cycles. Material C requires forming voltage of 4.8V and exhibits 3.5 V V_{th} and ultra low I_{OFF} (131 pA@2V)

Fig. 7 30 nm Material D OTS selector device characteristics with different cycles. By increasing Ge content in the As-Se-Ge material system, V_{th} slightly decrease, but I_{OFF} degrades to 10 nA @ 2V

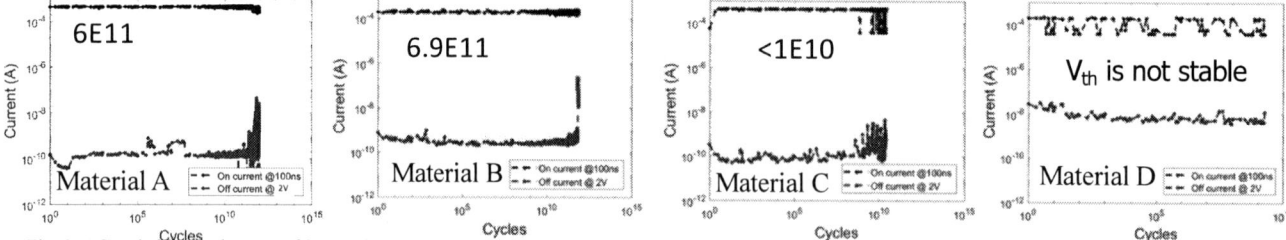

Fig. 9 AC switching endurance of 30 nm OTS Material A to D. Pulse with current (~200 uA) at100 ns was used to turn on the OTS. Low Ge-AsSeGe material (Material A) shows superior cycling endurance. Material B with slightly higher Ge than Material A exhibits the best cycling endurance (~6.9E11). However, with further increase of Ge, cycling endurance degrades (Material C). When Ge content is higher than 30 at. % (Material D), V_{th} becomes unstable during the cycling test. Ge content is very critical when designing the OTS materials based on AsSeGe chalcogenides along As_2Se_3-Ge tieline.

978-1-7281-1988-5/18 $31.00 © 2018 IEEE

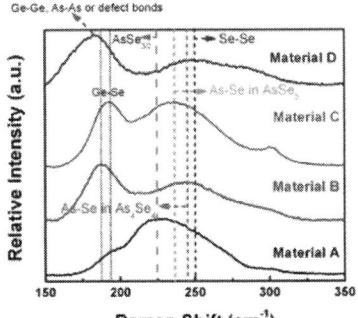

Fig. 10 Raman spectra of amorphous AsSeGe materials studied in this paper. The dominant peak in the Material A with low Ge content is at the band at 225 cm^{-1} which is assigned to the vibrations of the AsSe$_{3/2}$ pyramidal units [6-7]. With increasing Ge content, the dominant peaks are As-Se (245 cm^{-1}) and Ge-Se (190 cm^{-1}) where the band at 245 cm^{-1} and 190 cm^{-1} correspond to the motion of As-Se in As$_4$Se$_4$ structure and Ge-Se in GeSe$_4$ tetrahedrons, respectively [6, 8]. Material C has the structure similar to Material B with slightly band shift to 237 cm^{-1} which is attributed to the As-Se in AsSe$_3$ entities [6]. However, the dominate peak that associated with GeSe shifts to lower wavenumber as Ge keeps increasing (Material D); this band is attributed to defect bonds such as Ge-Ge, As-As [6]. In addition, the peak at high wavenumber also shifts to 250 cm^{-1}, which corresponds to the Se-Se vibrations in the Se-chains [7]. High number of defect bonds for high Ge version of AsSeGe OTS (Material D) is possible the reason to explain the unstable OTS characteristics during the cycling test.

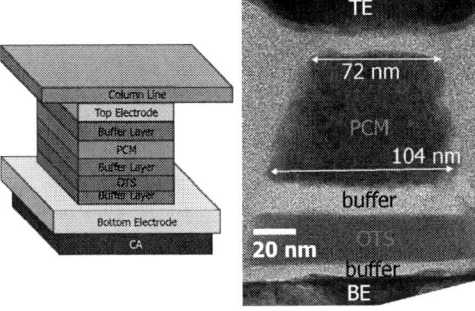

Fig. 11 Schematic 1S1R OTS-PCM pillar device structure. TEM image shows OTS Material C is successfully integrated with PCM in the pillar structure.

Fig. 12 Dark-field TEM micrograph of 1S1R OTS-PCM pillar device. TEM EDX image shows OTS Material C is successfully integrated with PCM in the pillar structure. Selector material C and GST are uniform without phase-segregation after complete integration.

Fig. 13 I-V characteristic for 1S1R device in RESET state and SET state. The threshold voltage of RESET state (V$_{tR}$) is 7.4 V (50ns RESET pulse was used). Threshold voltage of SET state (V$_{tS}$) is around 5.6 V. Almost 2V memory window (V$_{read}$) is achieved by using high V$_{th}$ OTS Material C. More than 100X resistance difference can be read out by V$_{read}$.

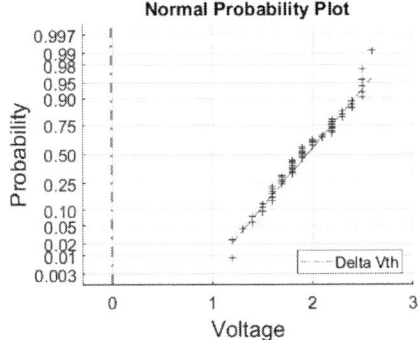

Fig. 14 Delta V$_{th}$ (V$_{tR}$-V$_{tS}$) distribution of 10x10 1S1R array. All the devices show more than 1V read window, most of them have close to 2V window.

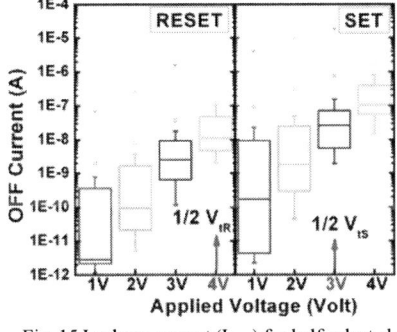

Fig. 15 Leakage current (I$_{OFF}$) for half-selected cells when the selected cell is at RESET (V$_{tR}$–8V) and SET (V$_{tS}$=6V) operation, respectively. I$_{OFF}$ increases with increasing applied voltage. I$_{OFF}$ of half-selected cell is around 10 nA (at 4V) and 26 nA (at 3V) when the selected cell is at RESET (8V) and SET operation (6V) operation, respectively.

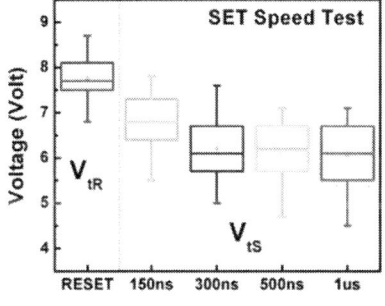

Fig. 16 SET speed of OTS+PCM device. 300 ns SET speed is demonstrated. Devices are programmed to RESET state at the pulse width of 100ns.

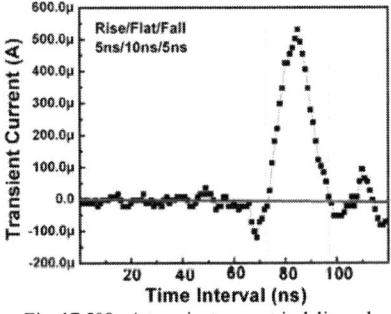

Fig. 17 500 uA transient current is delivered through 1S1R device with selector Material C and is able to turn on at 10 ns.

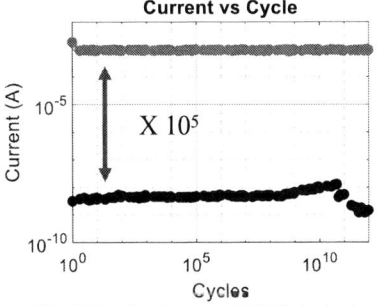

Fig. 18 Read endurance of 1S1R device is more than 1E12 cycles with turn on current of 1mA (7.9 MA/cm^2) indicating superior cycling endurance of the selector. OFF current is read at 4V (1/2 V$_{tR}$). ON/OFF ratio is more than 10^5.

Table I OTS+PCM Characteristic

	OTS+PCM [11]	OTS+PCM This work
Device	Mushroom	Pillar
Selector	Ge$_{58}$Se$_{42}$	GeAsSe
PCM	GeN/GST	Doped GST
ON/OFF ratio (OFF current at 1/2V$_{th}$)	150	10^5
Current density	1.5 MA/cm^2	7.9 MA/cm^2
Read endurance	1E9 (selector is OFF)	> 1E12 (selector is ON)

Optimized Reading Window for Crossbar Arrays Thanks to Ge-Se-Sb-N-based OTS Selectors

A. Verdy, M. Bernard, J. Garrione, G. Bourgeois, M. C. Cyrille, E. Nolot, N. Castellani,
P. Noé, C. Socquet-Clerc, T. Magis, G. Sassine, G. Molas, G. Navarro and E. Nowak

CEA, LETI, MINATEC Campus, 17 rue des Martyrs, 38054 Grenoble Cedex 9, France

Abstract—In this paper, we investigate the impact of Ovonic Threshold Switching (OTS) selector electrical parameters, such as the threshold and the holding current, on the reliability of the reading operation in 1S1R memory devices. Through physico-chemical analysis and electrical characterization of Se-rich Ge-Se-based OTS selectors, performed up to 400 °C, we demonstrate the possibility to reduce the fire voltage as well the leakage current thanks to N- and Sb-doping. Moreover, we describe the correlation that exists between the leakage current and the threshold current in OTS devices. We highlight the subsequent trade-off between the reading window and the array size in an OTS-based Memory Crossbar Array, evaluated up to an operating temperature of 150 °C. Finally, thanks to OTS engineering, we demonstrate how the reading window can be optimized for a target array size and application.

I. Introduction

Resistive Crossbar Memory arrays (RCM) receive nowadays an increasing interest for the design of new systems addressing revolutionary applications like Storage Class Memory (SCM) and neuromorphic computing. Back-End-of-Line (BEOL) access selector device featuring low leakage current becomes fundamental to achieve large memory arrays [1], that are affected by inherent sneak paths. Among the different threshold selector technologies, Ovonic Threshold Switching (OTS) represents a valuable solution [2] that demonstrated capability of high ON/OFF selectivity, good endurance and high switching speed.

The reliable access to the information stored in the memory (1R) through the selector (1S), in a 1S1R series configuration, is one of the main challenges of this integration, therefore different reading strategies have been proposed so far. A first one (ΔR), is based on the direct resistance reading of the memory after the switching of the selector [3]. Several studies which propose an OTS as a selector, on the contrary, take advantage for the reading operation of the shift of the threshold voltage (ΔV_{th}) which occurs between the two resistance states of the memory [4][5][6]. Moreover, a non-switching reading was proposed by probing the shift of the I-V characteristics of the 1S1R in the sub-threshold regime [7]. However, a detailed analysis of the OTS parameters that make one or the other strategy more suitable and reliable has not yet been reported. In this work, we show that holding current and threshold current are key parameters of an OTS selector to address the most suitable 1S1R reading strategy. Thanks to material engineering and detailed reliability analysis of Se-rich Ge-Se-based OTS, we achieve state-of-the-art selector performances, suitable for high 1S1R reading window. Finally, we demonstrate how the reading window can be optimized in RCM, addressing the specific array size of the target application.

II. OTS Parameters for 1S1R Reading Strategies

In **Fig. 1** we report an example of IV characteristic and we highlight the main electrical parameters measured for an OTS selector device proposed in this study. It can feature high ON/OFF selectivity ($>10^7$), and it requires a first initialization pulse (firing) characterized by a switching voltage (V_{fire}) higher than the one used for the following switching operations (V_{th}). After the firing, an expected increase of the leakage current can be observed.

The 1R device considered in our analysis is a Phase-Change Memory (PCM) [8]. If not differently specified by apex nomenclature (e.g. V_{th}^{PCM}) the following threshold parameters will always refer to the OTS device. In a ΔR strategy (**Fig. 2a**) the key parameter is the holding current I_h of the selector, that should be lower than the subthreshold current of the PCM in the RESET state (I_{sub}^{PCM}) to avoid data corruption. Moreover, in order to avoid the switching of half-selected cells during programming operations (in a V/2 operating scheme), the relation $V_{prog}/2 < V_{th}^{OTS} < V_{th}^{PCM}$ should be fulfilled (V_{prog} = programming voltage; V_{th}^{PCM} = PCM threshold voltage). As reported in **Table 1**, ΔR strategy is more indicated for OxRAM-based RCM targeting low power applications.

If on the contrary I_h is higher than I_{sub}^{PCM}, the switching of the OTS induces the degradation of the information stored in the PCM, and a ΔV_{th} strategy is more suitable (**Fig. 2b**). In this case, the OTS threshold current I_{th}^{OTS} becomes the key parameter (i.e. as well as the OTS threshold current density J_{th}^{OTS}), that should range within the values of I_{sub}^{PCM}. Indeed, this allows a reading window equal to ΔV_{th}, obtained by the shift of the threshold voltage of the 1S1R from a minimum value (i.e. when 1R is in the SET state) to the one achieved when the PCM is programmed in the RESET state. Then, ΔV_{th} likely corresponds to the voltage that drops on the RESET PCM at I_{th}^{OTS}. In **Fig. 3**, we report the ΔV_{th} evolution in the OTS+PCM device, as a function of J_{th}^{OTS}. As soon as J_{th}^{OTS} reaches the value of J_{th}^{PCM}, the reading window reaches its maximum value equivalent to V_{th}^{PCM}. Finally, if none of the previous conditions is met, the reading operation becomes not possible (**Fig. 2c**). RCM based on OTS+PCM, relying on a ΔV_{th} strategy, can be a relevant solution for SCM applications taking advantage of the high endurance capability of both OTS and PCM technologies (Table 1).

III. Se-rich Ge-Se materials engineering

OTS selectors based on Se-rich Ge-Se alloys are interesting because of their low leakage current; however, they suffer from a high threshold voltage and poor endurance performance [9]. We studied how to optimize $Ge_{30}Se_{70}$ alloys, in particular through Sb-doping (studied between 10% and 30%) in order to reduce V_{th}, and through N-doping (increased from x% up to w%) in order to improve the stability of the amorphous structure. The alloys were deposited by reactive co-sputtering from Sb and $Ge_{30}Se_{70}$ (GS) targets.

Raman spectroscopy performed on Sb-doped GS alloys (GS-Sb system) highlights the formation of Sb-Se and Sb-Sb bonds (**Fig. 4**). Adding N in GS doped with ~20% of Sb (GSS-N system) the formation of Ge-N bonds is favored wrt Ge-Se bonds. It leads to more Se atoms available for the formation of Sb-Se bonds revealed by the disappearing of the Sb-Sb modes, the apparition of Se-Se bonds at high N content (**Fig. 5**), and the probable formation of Sb-N bonds as shown in [10]. By a proper tuning of N concentration, an optimized composition (N_y) with almost no homopolar bonds can be finally achieved (GSSN).

978-1-7281-1988-5/18 $31.00 © 2018 IEEE

IV. ELECTRICAL CHARACTERIZATION OF GE-SE-SB-N-BASED OTS

Measurement of threshold parameters. In **Fig. 6**, we report the difference in terms of I_{th} (and V_{th}) obtained between DC and AC OTS I-V measurements. This is likely due to the time-dependent nature of the threshold mechanism [2] and it can affect, with an underestimation, the threshold parameters obtained by DC measurements. Thus, V_{fire}, V_{th} and I_{th} are measured in pulsed mode (AC). On the contrary, the OTS leakage current I_{leak} is measured at $V_{th}^{OTS}/2$ in DC mode.

ON-state analysis. In order to avoid the degradation of the cell, I_h and ON-state resistance (R_{ON}) are measured in pulsed mode. **Fig. 7** shows an I-V characteristics from which I_h can be easily extracted. The evolution of R_{ON} as a function of the injected current (I) is reported in **Fig. 8**. We found that R_{ON} and I are inversely proportional following the relation $R_{ON} \propto \beta_{ON}/I$ (β_{ON} is a material-dependent parameter).

Electrical parameters of GS-Sb and GSS-N OTS devices. OTS materials were integrated in single cell devices based on a plug of 350 nm of diameter. Tests were conducted on a population of about 50 devices for each alloy. In GS-Sb OTS selectors, the minimum V_{th} is obtained for 20% of Sb (GSS composition) (**Fig. 9**). An important increase of I_{leak} after the firing indicates a non-stable as-deposited material (**Fig. 10**). I_{th} shows a non-monotonous evolution increasing Sb-doping. N-doping permits to stabilize the GSS composition during the firing and to drastically reduce I_{leak} (**Fig. 11**). Indeed, we observe that for the optimized N content N_y (GSSN composition, without homopolar bonds) the dispersion of I_{leak} is highly reduced. On the contrary, the presence of Se-Se bonds at high N content looks detrimental for the functionality of the selector, inducing a high dispersion of I_{leak}. N-doping has also the effect to reduce I_{th} by about one decade (wrt GSS). However, even if V_{th} shows only a slight variation changing the N content, a considerable increase of V_{fire} appears when N is incorporated (**Fig. 12**).

I_{th} and V_{th} can also be tuned by scaling the thickness of the OTS layer. This is shown in particular for the optimized GSSN composition in **Fig. 13**. Nonetheless, thickness reduction should be properly limited to guarantee a low I_{leak}.

Infrared absorption spectroscopy on GSSN as-deposited material shows a high spread of the Ge-N absorption band, indicating that N atoms are randomly dispersed in the amorphous structure (**Fig. 14**). By a proper thermal treatment (400 °C 30min), a material reorganization occurs to form a homogeneous GeN$_x$ phase. Such annealing permits to reduce V_{fire} of GSSN to the same value found for GSS (**Fig. 15**), but without degrading I_{leak} that is maintained at the low value of 0.1 nA, close to the value obtained in as-fabricated samples. To be noted that I_{th} is increased by a factor ~4 between before and after annealing, that is beneficial for the reading window as described in section II (**Fig.16**).

The parameter β_{ON} ($\propto R_{ON}$) is decreased by Sb-doping while it is increased by N-doping. I_h remains almost stable at a value that does not allow the use of these OTS devices with a ΔR reading strategy (**Fig. 17**).

Fig. 18 summaries the trends observed for V_{fire} as a function of I_{leak}. Sb-doping permits to decrease V_{fire} but it is detrimental for I_{leak}. N-doping allows decreasing I_{leak} but with a consequent increase of V_{fire}. A proper annealing of GSSN composition finally enables to recover a low V_{fire}. The plot of I_{th} as a function of I_{leak} for all the compositions analyzed highlights a strong correlation (~ linear) between these two parameters (**Fig. 19**).

GSSN OTS electrical parameters at high operating temperatures. GSSN OTS electrical behavior was evaluated up to an operating temperature of 150 °C. While V_{fire} decreases with increasing of

temperature, V_{th} shows stability up to 150 °C (**Fig. 20**). A slight increase of I_h (that still does not allow a ΔR reading strategy) and a decrease of β_{ON} are measured (**Fig. 21**). Finally, I_{th} increases with temperature, suitable to keep a reliable reading window, whereas I_{leak} increases due to temperature activated Poole-Frenkel conduction (**Fig. 22**).

V. TRADE-OFF BETWEEN READING WINDOW AND MATRIX SIZE IN OTS-BASED RCM ARRAYS

The correlation between I_{leak} and I_{th} found in Fig. 19, and the one that naturally exists between the maximum matrix size and I_{leak} [11], added to the result of Fig. 3, permit to highlight a trade-off between the reading window ΔV_{th} and the RCM array size (**Fig. 23**). Moreover, the evolution of ΔV_{th} as a function of J_{th}^{OTS} at different temperatures (**Fig. 24**) shows that Poole-Frenkel conduction, that characterizes I_{sub}^{PCM}, induces a decrease of the reading window when temperature increases. From our measurements (Fig. 22), we simulated ΔV_{th} window and the maximum matrix size evolution with temperature for our optimized GSSN OTS device (**Fig. 25**). A leakage current below 1 nA is extrapolated at 70 °C with a ΔV_{th} ~ 0.2V, allowing a RCM array size of 1Mb.

Finally, we compare in terms of ΔV_{th} and maximum achievable matrix size, our GSSN device with other OTS material published in the literature (**Fig. 26**). GSSN material developed in this work features a very high threshold current allowing a ΔV_{th} reading window among the highest, still demonstrating a low value of leakage current (0.1 nA) to enable the largest matrix size among state-of-the-art OTS materials. GSSN device scaling will even improve such performances [12] (extrapolation is shown for a 50 nm OTS device in Fig. 26).

VI. CONCLUSIONS

We show that threshold and holding current are the key parameters in reading operations. Material engineering in Se-rich Ge-Se-based OTS leads to state-of-the-art performances of our selector devices, in particular featuring low leakage current below 0.1 nA and lower fire voltage. This result was achieved by limiting the presence of homopolar bonds, optimizing Sb- and N-doping. We highlight the trade-off between the reading window and the array size in an OTS+PCM-based RCM, evaluated up to an operating temperature of 150 °C. Finally, we demonstrate the capability of our optimized GSSN OTS technology to target best reading window and largest RCM array size wrt other OTS materials, making it suitable for SCM applications.

Acknowledgments: This work has been partially supported by the European PANACHE and WAKeMeUP projects.

REFERENCES

[1] G. W. Burr et al., *J. Vac. Sci. Technol. B*, vol. 32(4), pp. 040802, 2014.
[2] S.R. Ovshinsky, *Phys. Rev. Lett.*, vol. 21, pp. 1450-1453, 1968.
[3] S.H. Jo et al, *IEEE Trans. Electron Devices*, vol. 62, pp. 3477-3481, 2015.
[4] D.C. Kau et al., *in Proc. 2009 IEEE IEDM*, pp. 27.1.1-27.1.4, 2009.
[5] C. W. Yeh et al., *in Proc. 2018 VLSI*, pp. 205-206, 2018.
[6] M. Alayan et al., *in Proc. 2017 IEEE IEDM*, pp. 2.3.1-2.3.4, 2017.
[7] G. Navarro et al., *in Proc. 2017 VLSI*, T94-T95, 2017.
[8] O. Cueto et al., *in Proc. 2012 SISPAD*, 2012.
[9] S.D. Kim et al., *ECS Solid State Lett.* vol. 2(10), pp. Q75-Q77, 2013.
[10] A. Verdy et al., *2018 MRS Spring Meeting*, 2018.
[11] N. S. Avasarala et al., *in Proc. 2018 VLSI*, 209-210, 2018.
[12] H.-W. Ahn et al., *ECS Solid State Lett.* Vol. 2(9), pp. N31-N33, 2013.
[13] K. Gopalakrishnan et al., *in Proc. 2010 VLSI*, 205-209, 2010.
[14] H.Y. Cheng, *2017 IEEE IEDM*, pp. 2.2.1-2.2.4, 2017.
[15] Y. Koo et al., *IEEE Electron Device Lett.*, vol. 38 (5), pp. 568-571, 2017.
[16] H.-S. Choi, *Microelectron. Reliab.*, vol. 56, pp. 61-65, 2016.

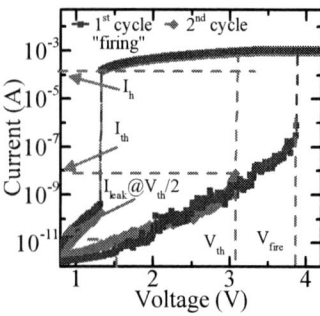

Fig. 1: I-V characteristic and main electrical parameters of an OTS device. V_{th}: threshold voltage. I_h & I_{th}: holding and threshold current. I_{leak}: leakage current measured at $V_{th}/2$. OTS requires a first initialization pulse (firing) characterized by a switching voltage (V_{fire}) higher than the one used for the following switching operations (V_{th}). An increase of the leakage current can also be observed.

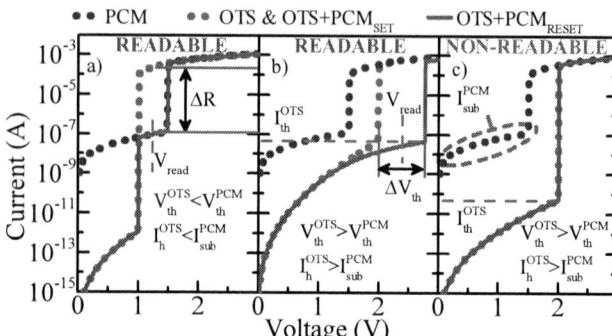

Fig. 2: Possible reading strategies in an OTS+PCM device (example of reading voltage V_{read} is reported when reading operation is allowed). a) ΔR reading is allowed if I_h is lower than the subthreshold current of the PCM in the RESET state (I_{sub}^{PCM}). b) ΔV_{th} reading is suitable when I_h is higher than I_{sub}^{PCM}. ΔV_{th} is equal to the voltage drop on the RESET PCM @ I_{th}^{OTS}. ΔV_{th} calculation has a low dependency on the switching voltage of 1S1R when the PCM is in the SET state, since this latter is $\sim V_{th}^{OTS}$. c) No reading is possible.

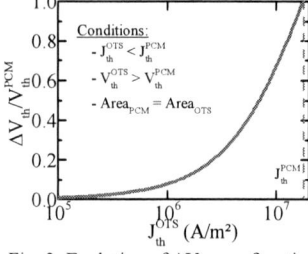

Fig. 3: Evolution of ΔV_{th} as a function of J_{th}^{OTS} in a ΔV_{th} reading strategy. When J_{th}^{OTS} reaches the value of J_{th}^{PCM}, the reading window reaches its maximum value $= V_{th}^{PCM}$.

Table 1: Summary of 1S1R reading strategies. The two reading strategies can target different applications and depends on the 1S and 1R technologies used. OxRAM, featuring low resistances for both RESET and SET states, coupled with MIEC selector technology can target low power applications (i.e. not requiring high endurance devices) using ΔR reading strategy. OTS combined with PCM, relying on a ΔV_{th} strategy, can be a relevant solution for SCM applications taking advantage of the high endurance capability of both OTS and PCM technologies.

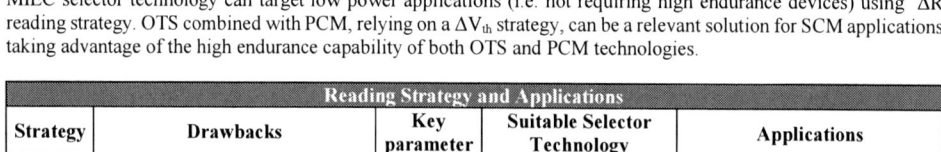

Reading Strategy and Applications				
Strategy	Drawbacks	Key parameter	Suitable Selector Technology	Applications
ΔR	Need of low V_{th}^{OTS} & low I_{leak} $V_{prog}/2 < V_{th}^{OTS} < V_{th}^{PCM}$	I_h	MIEC [13]	Low power applications (i.e. based on OxRAM)
ΔV_{th}	$V_{prog} \geq V_{th}^{OTS} + \Delta V_{th}$ (i.e. higher V_{prog})	I_{th}	OTS [14]	Storage Class Memory (i.e. based on PCM)

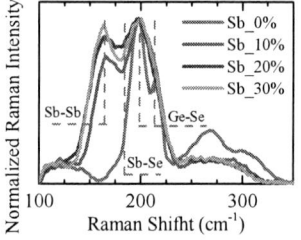

Fig. 4: Normalized Raman spectroscopy in GS-Sb alloys. Increasing Sb, Se in excess in GS creates Sb-Se bonds. Sb-Sb bonds are also formed.

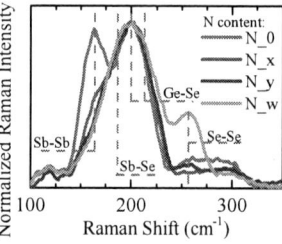

Fig. 5: Normalized Raman spectroscopy in GSS-N. Ge-N bonds reduce Sb-excess by liberating Se from Ge-Se. At high N content Se-Se bonds start to appear. No homopolar bonds are observed after N optimization in GSS-N_y (GSSN).

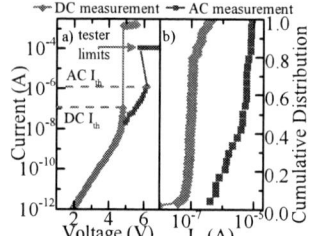

Fig. 6: a) Comparison of OTS I-V curves in DC & AC mode. DC stress induces lower threshold parameters. b) Distributions of I_{th}^{OTS} in AC & DC mode. DC measurements underestimate I_{th}^{OTS} by a factor ~10.

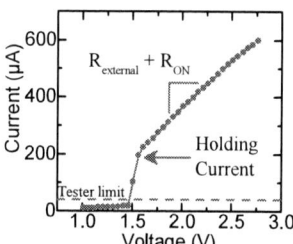

Fig. 7: ON-state I-V characteristic for an OTS device obtained by pulse mode. When current is lower than I_h, the OTS switches back to its OFF state. I_h is extracted from this measurement.

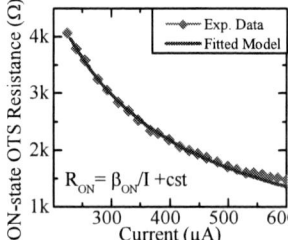

Fig. 8: From the I-V curves shown in Fig. 7, extracted R_{ON} and current I are found to be inversely proportional following the relation $R_{ON} \propto \beta_{ON}/I$.

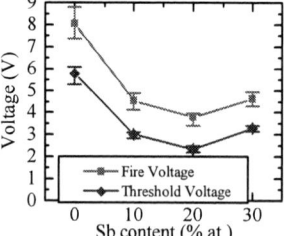

Fig. 9: Evolution of V_{fire} and V_{th} in GS-Sb OTS devices. Minimum V_{th} is achieved for Sb = ~20% (GSS composition).

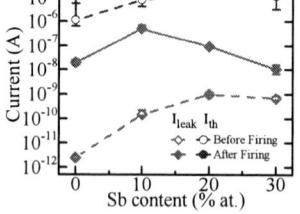

Fig. 10: Evolution of I_{th} and I_{leak}, before and after firing, in GS-Sb OTS. The increase of I_{leak} after firing indicates a non-stable as-deposited material. I_{th} can be tuned with Sb-doping in the range of 10^{-5}-10^{-4} A.

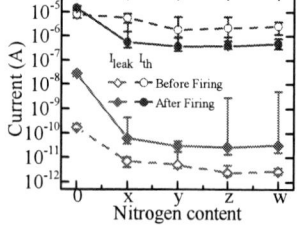

Fig. 11: Evolution of I_{th} and I_{leak}, before and after firing, in GSS-N selectors. N-doping drastically reduces I_{leak}. I_{leak} is more dispersed in highly N-doped compositions (z & w).

978-1-7281-1988-5/18 $31.00 © 2018 IEEE

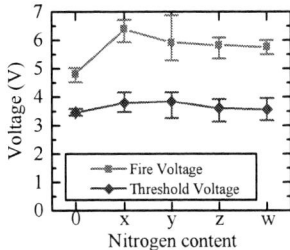
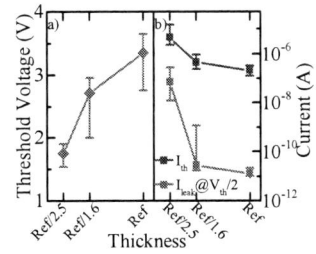
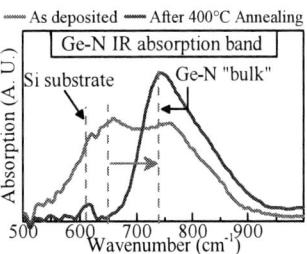
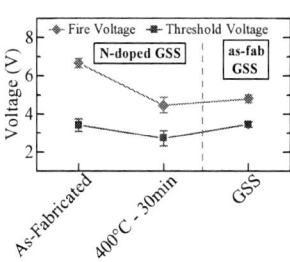

Fig. 12: Evolution of V_{fire} and V_{th} in GSS-N. N-doping induces an increase of the V_{fire}/V_{th} ratio by increasing the V_{fire}, whereas the V_{th} remains almost stable.

Fig. 13: Evolution of a) V_{th} and b) I_{leak} and I_{th} as a function of the GSSN OTS thickness. When the thickness is scaled down, I_{th} increases whereas V_{th} decreases allowing to tune the OTS parameters to optimize the 1S1R reading window. However, a too important reduction of the thickness degrades I_{leak}.

Fig. 14: IR absorption spectroscopy on GSSN showing the high spread of Ge-N absorption bands and the reorganization of Ge-N$_x$ phase after annealing at 400 °C.

Fig. 15: Reorganization of GeN$_x$ phase after annealing in GSSN permits to suppress the increase of V_{fire} when N is added to GSS.

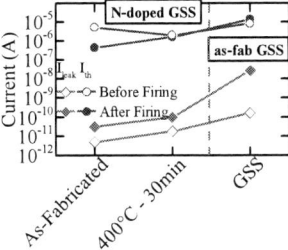
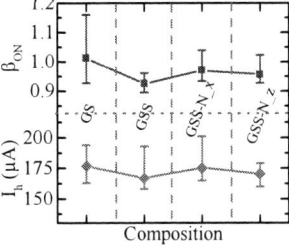
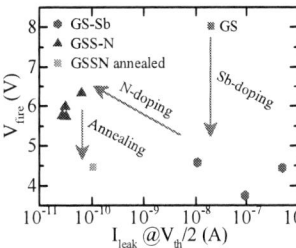
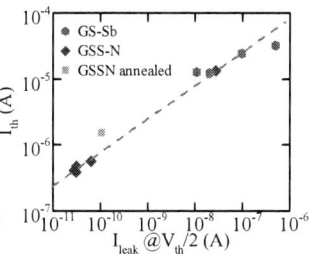

Fig. 16: After annealing at 400°C-30 min, I_{leak} is maintained at ~ 0.1 nA, whereas I_{th} increases by a factor ~ 4 wrt the as-fabricated devices (i.e. before annealing).

Fig. 17: Evolution of β_{ON} and I_h for different compositions. Only Sb doping permits to decrease β_{ON} while N-doping induces β_{ON} increase. I_h remains almost stable.

Fig. 18: V_{fire} vs I_{leak}, summarizing the material exploration. Sb permits to decrease V_{fire} but I_{leak} remains high. N permits to decrease I_{leak} but at the cost of V_{fire}, that is decreased with annealing.

Fig. 19: Correlation between I_{leak} and I_{th} in the Ge-Se-based studied materials. A reduction of I_{leak} implies the reduction of I_{th}.

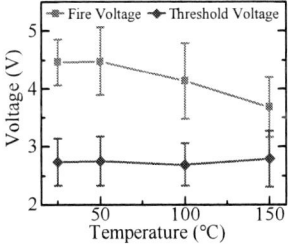
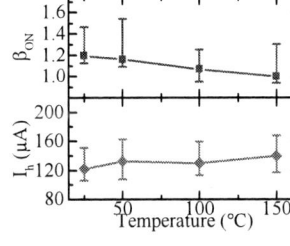
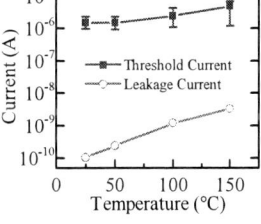
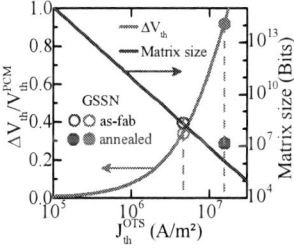

Fig. 20: Evolution of V_{fire} and V_{th} at operating temperatures up to 150 °C for GSSN devices (annealed at 400 °C-30 min). GSSN OTS exhibits reliable switching properties up to 150 °C.

Fig. 21: Evolution of β_{ON} and I_h of GSSN devices at high operating temperature up to 150 °C (after annealing at 400 °C-30 min). β_{ON} decreases with temperature whereas I_h slightly increases.

Fig. 22: Evolution of I_{th} and I_{leak} in GSSN, measured during tests performed for Fig. 20. I_{leak} increases due to Poole-Frenkel conduction. I_{th} increases rising the temperature, which is beneficial for reading operations.

Fig. 23: Trade-off between the reading windows ΔV_{th} and the maximum matrix size (for an OTS+PCM technology), reported as a function of J_{th}^{OTS} (correlation between I_{th} and I_{leak} found in Fig. 19 is considered). Annealing in GSSN leads to higher reading window still enabling large RCM array size.

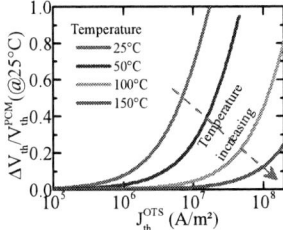
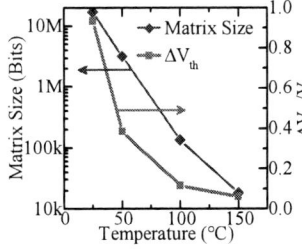
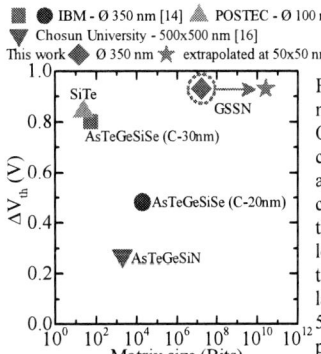

Fig. 24: Evolution of the normalized reading window ΔV_{th} simulated for an OTS+PCM device, for increasing operating temperatures, as a function of J_{th}^{OTS}. ΔV_{th} reading strategy is used.

Fig. 25: Evolution of the reading window ΔV_{th} and of the maximum matrix size for increasing operating temperatures, based on measurements performed on our optimized GSSN composition (i.e. after 400°C-30min annealing) reported in Fig. 22.

Fig. 26: Comparison of ΔV_{th} and maximum matrix size for different OTS materials found in the literature, considering an OTS+PCM architecture. It shows that GSSN composition, thanks to its high threshold current and very low leakage current can target at the same time a high reading window and a large matrix size. Scaling down to 50 nm can provide even better performances.

978-1-7281-1988-5/18 $31.00 © 2018 IEEE

Forming-free Mott-oxide threshold selector nanodevice showing s-type NDR with high endurance (> 10^{12} cycles), excellent V_{th} stability (< 5%), fast (< 10 ns) switching, and promising scaling properties

T. Hennen[1], D. Bedau[2], J. A. J. Rupp[1], C. Funck[1], S. Menzel[3], M. Grobis[2], R. Waser[1,3], and D. J. Wouters[1]

[1] IWE II, RWTH Aachen University, 52074 Aachen, Germany, email: t.hennen@iwe.rwth-aachen.de
[2] Western Digital San Jose Research Center, 5601 Great Oaks Pkwy, San Jose, CA 95119
[3] Peter Grünberg Institute, Forschungszentrum Jülich GmbH, 52428 Jülich, Germany

Abstract—In this work, thin film (down to 10 nm) $(V_{1-x}Cr_x)_2O_3$ Mott-oxide based nano-devices (electrode width down to 120 nm) are fabricated for the first time. The devices show volatile threshold switching and NDR caused by thermal feedback. Fast (< 10 ns) and very stable (< 5% variation) cycle to cycle threshold switching is obtained over 10^{12} cycles. Thickness and area dependence of the NDR curves are consistent with uniform volume switching and are explained with a thermal feedback model calibrated to the temperature dependent conductance of the $(V_{1-x}Cr_x)_2O_3$ films, enabling predictions for further scaled device geometries.

I. INTRODUCTION

Selector devices are crucial for the development of dense crossbar memories based on resistive switching devices. Of these, threshold switching devices are superior compared to non-linear diodes (e.g. based on tunneling) because of higher nonlinearity and larger read window [1].

The main candidate selector devices today are OTS switches [2]. However, these usually require a first-fire (forming) step, and show limited endurance and stability because of spontaneous crystallization and material segregation in these complex chalcogenide materials. Material optimization for OTS has resulted in a maximum demonstrated endurance of only 10^{10} cycles [3], but as selector devices also need to switch on every read operation, their endurance requirements are even more demanding than that of the memory device.

Volatile (as well as non-volatile) switching has been reported in Mott insulating materials, including chromium vanadium oxide $(V_{1-x}Cr_x)_2O_3$, where the switching was attributed to an electronically (avalanche process) induced Mott-Hubbard insulator to metal transition. This phenomenon was observed for single crystals at low temperature, and more recently for 880 nm and 100 nm polycrystalline films at room temperature [4].

In this work, we fabricated and characterized the smallest nano-devices of $(V_{1-x}Cr_x)_2O_3$ reported to date. We find attractive threshold switching properties, described for the first time by a mechanism of self-heating in this material.

II. DEVICE FABRICATION

Amorphous and polycrystalline $(V_{1-x}Cr_x)_2O_3$ thin films were deposited by RF magnetron sputtering from alloy targets (x=0.05, 0.15), with the substrate held at a constant temperature of 20 °C and 600 °C, respectively. The stoichiometry was optimized by tuning the oxygen flow rate, and was evaluated by transport measurements through the low temperature V_2O_3 metal-insulator phase transition (**Fig. 1**), which is known to depend very sensitively on the defect concentration in the sub 1% range [5]. For the 600 °C process, grazing incidence XRD confirms the expected corundum V_2O_3 crystallinity (**Fig. 2**). The deposition parameters have been described in more detail in a previous publication [6].

For our electrical characterization, we fabricated $(V_{1-x}Cr_x)_2O_3$-based nano-scale metal-insulator-metal (MIM) devices. These devices consist of full oxide films contacted from below by rectangular TiN vias and capped with 30 nm Pt top electrodes. The effective device area is defined by that of the TiN bottom electrode. A device cross-section for a 90 nm thick oxide can be seen in the SEM nanograph of **Fig. 3**. Different device widths of 120, 150, 250, 350, and 500 nm, and oxide thicknesses of 10, 30, and 90 nm were fabricated. We measured the expected inverse area scaling of the initial device resistance (**Fig. 4**), confirming oxide layer uniformity and intended device morphology. Our devices include integrated series resistors with values ranging from 0 to 100 kΩ.

III. ELECTRICAL CHARACTERISTICS

A. Temperature dependence of initial conductivity

To characterize the mechanism of conduction, we performed measurements of device current as a function of applied voltage and ambient temperature. For these measurements, the total power applied to the device was limited below the level of switching or self-heating effects. We observe a strong exponential temperature dependence of the resistance, of about one order of magnitude per 100 K. We obtained a reasonable fit of our conduction vs temperature data using a modified Poole-Frenkel-like equation (**Fig. 5**).

B. Negative Differential Resistance (NDR)

The devices with crystalline film show a volatile threshold switch following an s-type NDR under current control or voltage control combined with a suitable series resistor (1-20 kΩ) (**Fig. 6**). This fully reversible, continuous NDR effect appears without any forming step, for both polarities, and is very highly repeatable. We interpret this NDR effect as being due to self-heating, reminiscent of the effect identified in NbO_x [7, 8]. Under voltage control with small series resistance, the NDR manifests as abrupt threshold switching with a narrow hysteresis, due to the thermal runaway effect [9].

We measured a strong dependence of the point of NDR onset on the device dimensions, with lower area devices having

978-1-7281-1988-5/18 $31.00 © 2018 IEEE

a higher onset voltage, and thinner devices having a lower onset voltage (**Fig. 7**).

Interestingly, similar threshold switching characteristics were also obtained for amorphous films (**Fig. 8**), which we attribute to the same self-heating effect.

C. Non-volatile effects

Besides the forming-free, volatile threshold switching just described, we also observe a non-volatile (NV) process which can occur when the cell is subjected to higher current densities. The electrical result of this NV process is an irreversibly modified state with higher leakage current, but that continues to show s-type NDR curves with a lower NDR onset voltage (~1.7 V).

This NV process may consist of a sudden "forming" event under voltage control (**Fig. 9**), where the device area dependency of resistance thereafter shows strong evidence of filament formation (**Fig. 10**). Indeed, after forming we observe a similar resulting threshold switching characteristic for every device shape (**Fig. 11**). We have also seen that this NV process can be driven in a gradual way when using a current source or a larger series resistance. In this case, the effect appears to be equivalent to a uniform thinning of the oxide layer over the entire device area (**Fig. 12**).

Based on these observations, we hypothesize that the forming process results in a highly conductive pillar formation (**Fig. 13**), driven by heating and/or oxygen redistribution. This pillar could consist of an oxygen deficient or (re)crystallized oxide phase, or compressed metallic domain(s) having undergone a metal-insulator transition [10]. After forming, threshold switching proceeds at an effectively reduced volume in the pillar gap, but occurs in the same material and due to the same effect as in the forming-free operation.

D. Switching dynamics

From a measurement of sweep rate dependent *I-V* sweeps down to 100 ns duration, we confirm that the NDR effect persists on short time scales (**Fig. 14**), although the shape of the NDR curve is affected, and a hysteresis starts to open at high speeds. A standard voltage pulse response characterization found that threshold switching occurs in under 10 ns for formed devices (**Fig. 15**).

E. Switching endurance

Endurance measurements were performed by switching a formed device repeatedly with a 10 MHz square waveform (**Fig. 16**). Interleaved *I-V* loops lasting 1 ms were measured every 10^{10} cycles (**Fig. 17**) to evaluate drift of key selector parameters. We found less than 5% drift over 10^{12} cycles in the threshold voltage, leakage current, and NDR slope. Even over 10^{12} cycles, we saw little evidence of degradation (**Fig. 18**)

IV. THERMAL MODEL AND SCALING PROJECTION

To describe our experimental results, we developed a simple thermal feedback model, similar to [11, 12], calibrated to our fitted temperature dependent conduction equation. The model assumes a uniform Joule heating and temperature distribution inside the cell volume, and considers thermal transport through the cell boundaries, which are assumed to be held at room temperature. With the device temperature approximated through the concept of thermal resistance, a self-consistent equilibrium *V(I)* is obtained numerically (**Fig. 20**).

Informed by an analysis of the scaling behavior of the NDR onset power (**Fig 19**), we were able to closely reproduce the geometry dependence of our measured NDR curves using such a model assuming that lateral heat losses through the top electrode are dominant (**Fig. 21**). This model fully accounts for the observed width and thickness scaling behavior of the NDR onset. Using the parameters that fit our NDR measurements, we varied the device dimensions in our model to obtain the projected behavior for scaled devices (**Fig. 22**). We found a generally favorable scaling behavior, with the half-threshold leakage dropping to the 100 nA level while the threshold voltage stays in the range of 1-3 V for a 10×10×10 nm device.

Within this framework, considerations not only of the material properties, but of the thermal environment are critical for determining device behavior [13-15]. For our situation, the thermal dissipation appears to be limited by the lateral flow through the Pt top electrode. Adjusting the thermal isolation provides an engineering path for modifying the threshold voltage and leakage current (**Fig. 23**).

V. CONCLUSIONS

Our Pt / $(V_{1-x}Cr_x)_2O_3$ / TiN nanodevices show stable, fast volatile threshold switching which can be explained by a thermal feedback model. This model accounts for the geometry dependence of the measured NDR curves, and allows us to extrapolate to technologically relevant dimensions, where we find promising characteristics with respect to leakage and the threshold point. These devices do not require forming, switch over the entire volume, can be deposited at room temperature, and meet many key requirements for the selector application.

ACKNOWLEDGEMENTS

The authors acknowledge funding by the DFG (German Science Foundation) within the collaborative research centre SFB 917.

REFERENCES

[1] L. Zhang et al., IEEE 6th International Memory Workshop, pp. 1–4, 2014
[2] S. R. Ovshinsky, Physical Review Letters, vol. 21, no. 20, p. 6, 1968.
[3] H. Y. Cheng et al., IEDM 2017, pp. 2.2.1-2.2.4. 2017
[4] J. Tranchant et al., IEEE International Memory Workshop, pp. 1-4, 2018.
[5] S. Shivashankar et al., Phys. Rev. B, vol. 28, no. 10, pp. 5695–5701, 1983.
[6] J. Rupp et al., J. of App. Phys., vol. 123, no. 4, p. 044502, Jan. 2018.
[7] S. Slesazeck et al., RSC Adv. , vol. 5, no. 124, pp. 102318–102322, 2015.
[8] S. Kumar et al., Nature Communications, vol. 8, no. 1, Dec. 2017.
[9] C. Funck et al., Adv. Elec. Mat., vol. 2, no. 7, p. 1600169, Jul. 2016.
[10] V. Dubost et al., Nano Letters, vol. 13, no. 8, pp. 3648–3653, Aug. 2013.
[11] Z. Wang et al., App. Phys. Lett., vol. 112, no. 19, p. 193503, May 2018.
[12] G. A. Gibson et al., App. Phys. Lett., vol. 108, no. 2, p. 023505, Jan. 2016.
[13] C. Funck et al., J. Comp. Elec., vol. 16, no. 4, pp. 1175–1185, Dec. 2017.
[14] G. A. Gibson, Adv. Func. Mat., vol. 28, no. 22, p. 1704175, 2018.
[15] Z. Wang et al., App. Phys. Lett., vol. 112, no. 7, p. 073102, Feb. 2018.

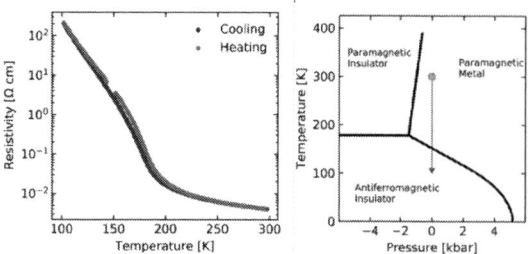

Fig. 1. Resistivity vs temperature for a 30 nm crystalline V_2O_3 film deposited at 600 °C (left). Single crystal V_2O_3 phase diagram showing the trajectory of the measurement (right).

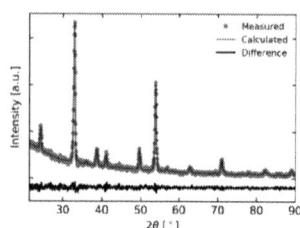

Fig. 2. Le Bail refinement of grazing-incidence XRD of a 100 nm crystalline V_2O_3 film (deposited at 600 °C).

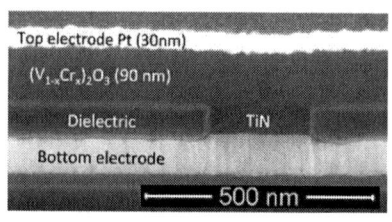

Fig. 3. Cross-sectional SEM of a $(VCr)_2O_3$ nanodevice, using a 90 nm thick oxide layer, 30 nm Pt top electrode, and 250 nm × 250 nm bottom point contact.

$$I/V = A \exp\left(\frac{-E_A}{k_B T}\right) \exp\left(c\sqrt{V}\right)$$

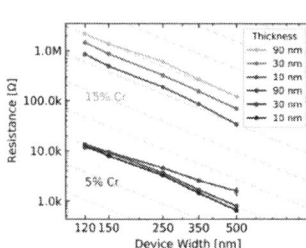

Fig. 4. Pristine device resistance vs device width, following the expected inverse area scaling (dotted lines).

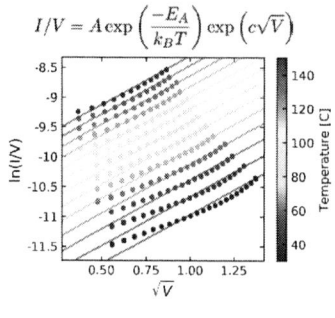

Fig. 5. Off-state conductivity of a 90×500×500 nm crystalline device vs voltage and temperature, fit using the empirical conduction equation given above.

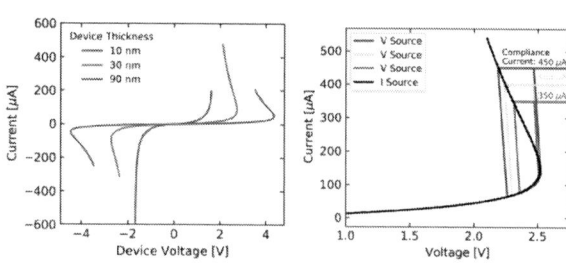

Fig. 6. Symmetric S-type NDR curves for crystalline films (left). Narrow hysteresis threshold switching under voltage sourcing with different compliance current levels (right).

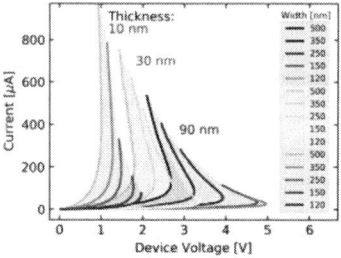

Fig. 7. Measured NDR characteristics for crystalline films. The dependence on device width and thickness is shown.

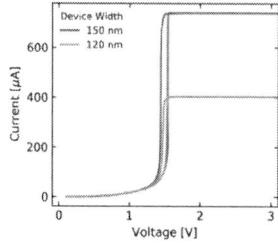

Fig. 8. Similar threshold switching behavior observed in 10 nm amorphous films, deposited at room temperature.

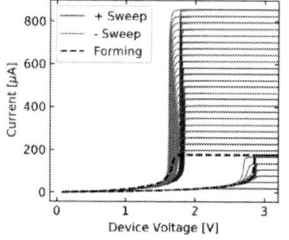

Fig. 9. Abrupt non-volatile change, or "forming". Current compliance level increased for each subsequent loop. Device thickness 30 nm, width 250 nm.

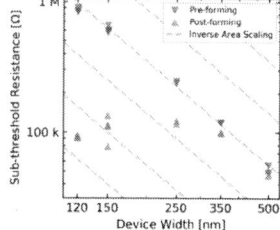

Fig. 10. Area independence after the sudden "forming" event, indicating a filamentary process. Device thickness 30 nm.

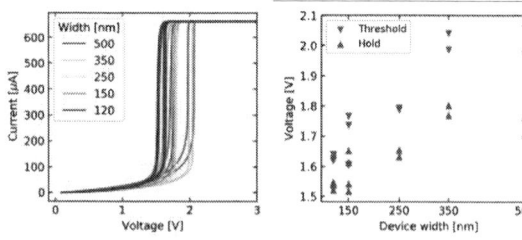

Fig. 11. Similar switching characteristics for different device sizes after forming (thickness 30 nm). The small dependence of threshold voltage on device width could be due to larger overshoots (different forming voltage needed).

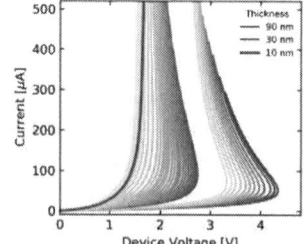

Fig. 12. Gradual forming case using a large series resistance. The effect of forming is qualitatively similar to reducing the oxide thickness.

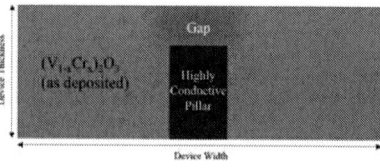

Fig. 13. Non-volatile conductive pillar formation model. Because of current limiting, a gap remains in the as-deposited oxide, which continues to show threshold switching due to thermal feedback in an effectively reduced switching volume.

978-1-7281-1988-5/18 $31.00 © 2018 IEEE

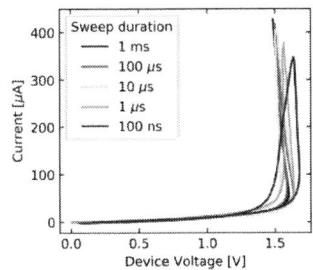

Fig. 14. Effect of sweep rate on NDR curves for a 30×250×250 nm formed device.

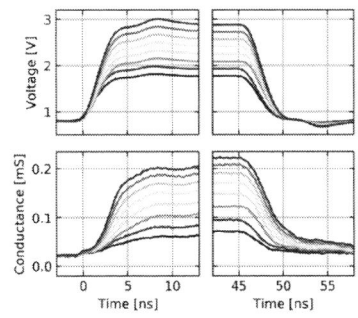

Fig. 15. Fast (< 10 ns) conductance switching in a 30×150×150 nm formed device. Colors correspond to different voltage pulse amplitudes.

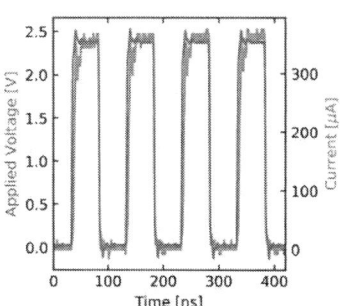

Fig. 16. 10 MHz cycling used for endurance measurement.

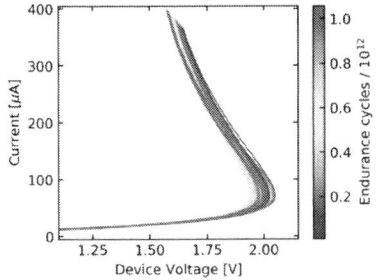

Fig. 17. Stability of NDR curves during 10^{12} cycle endurance. One NDR curve measured per 10^{10} cycles.

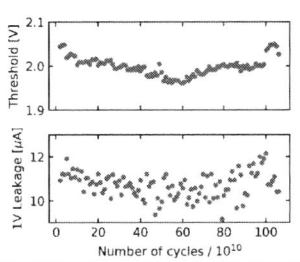

Fig. 18. Stability of NDR curves during 10^{12} cycle endurance measurement. The threshold voltage drifts by less than 5%.

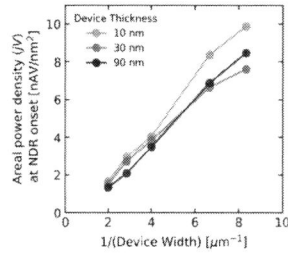

Fig. 19. Experimental indication that lateral heat loss through electrode is dominant. The areal power density at the NDR onset point roughly scales as 1/width, independent of film thickness.

Fig. 20. Details of the thermal model.

An empirical conduction equation

$$I(V,T) = A V \exp\left(\frac{-E_b}{k_B T}\right) \exp\left(c \sqrt{V}\right) \quad (1)$$

is used with parameters A = 4.17x10^{-2} S, E_A = 0.248 eV, and c = 1.65 V$^{-1/2}$, determined from a fit to a measurement (see Fig. 5). This equation is solved self-consistently with a self-heating approximation,

$$T(I, V) = T_0 + R_{th} IV, \quad (2)$$

where the thermal resistance, R_{th}, takes a value of 6×10^4 K/W for a 90×500×500 nm device.

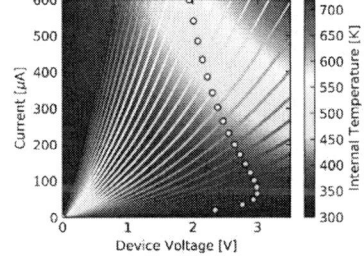

Graphical depiction of Eq. 1 (lines) and Eq. 2 (background). White circles indicate the common solution (line color equals background color), resulting in the observed NDR curve.

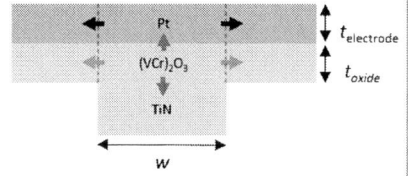

To calculate the geometry dependence of NDR curves (Fig. 21), we simply scaled R_{th} by $1/w$, as expected for the case of heat loss limited laterally through a top electrode of constant thickness.

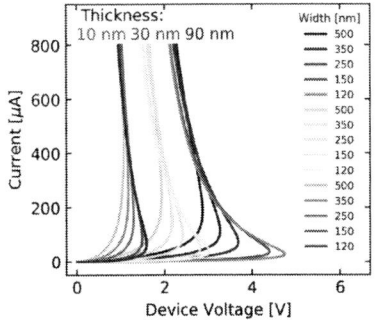

Fig. 21. Calculation of NDR curves reproduces the measured dependence on device geometry (compare with Fig. 7)

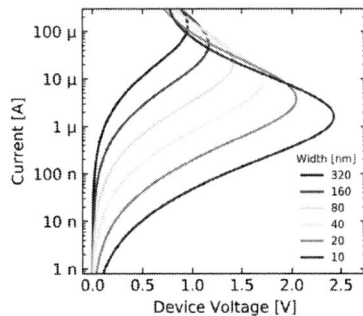

Fig. 22. Simulated effect of scaling a 10 nm thickness device to 10×10×10 nm. Leakage currents are drastically reduced, while the NDR onset voltage stays in a useful range.

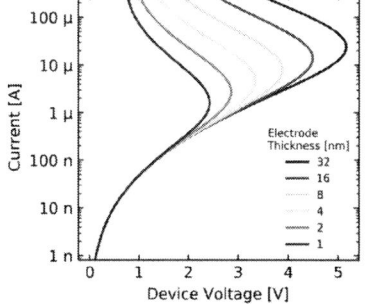

Fig. 23. Influence of the top electrode thickness on the NDR curve. Thinning the electrode is equivalent to reducing the heat losses.

Fully Multi-Functional GaN-based Micro-LEDs for 2500 PPI Micro-displays, Temperature Sensing, Light Energy Harvesting, and Light Detection

Zhaojun Liu[1*], Ke Zhang[1,2], Yibo Liu[1,2], Siwu Yan[1,2], Hoi Sing Kwok[1,2], Jamal Deen[1,3], and Xiaowei Sun[1]

[1]Southern University of Science and Technology, Shenzhen, China, email: (liuzj@sustc.edu.cn)
[2]Hong Kong University of Science and Technology, Hong Kong, China
[3] McMaster University, Hamilton, Ontario, Canada (jamal@mcmaster.ca)
*Tel: +86-755-88018506; Email: liuzj@sustc.edu.cn;

Abstract—GaN-based Micro-LEDs were developed for more than a decade and are considered as the next generation of display technology. Micro-LEDs have more versatile functions than displays. We report fully multi-functional Micro-LEDs for micro-displays, temperature sensing, light energy harvesting, and light detection. The 2500 pixel per inch (PPI) Micro-LED devices displayed animations and pictures in display mode driven by a Si complementary metal-oxide-semiconductor (CMOS) backplane. In temperature sensing mode, it showed a sufficient linearity with a resolution of 795 K/V. It also had a fill factor of 65.9% and efficiency of 15.5% in light energy harvesting mode and a sensitivity of 1240 and external quantum efficiency (EQE) of 33% in light detection mode. Results real that the proposed devices can be used as a self-sustainable micro-system for low cost and eco-friendly applications.

I. INTRODUCTION

GaN-based devices such as light emitting diodes (LEDs), Micro-LEDs, photodetectors and solar cells have been attracting more and more attentions for solid-stat lighting [1-2], displays [3-7], novel energy [8, 9], and light detection [10-12]. p-n junction, the core of these devices, can work in various modes according to the bias applied as illustrated in Fig.1. The superior properties of GaN such as wide bandgap, high carrier confinement, high quantum efficiency, and excellent reliability promise more potential of applications such as temperature sensing [13, 14]. We proposed a fully multi-functional GaN-based Micro-LEDs for micro-displays, temperature sensing, light energy harvesting, and light detection.

II. EXPERIMENT

A. System design

The GaN-based Micro-LED device had a resolution of 1200✕960 with a pixel pitch of 10 μm. It included three main parts: GaN Micro-LED arrays, Si CMOS active matrix (AM) backplane, and bump array for mass transferring bonding. The Micro-LED array and Si backplane were designed with the same pixel pitch correspondingly. A current control current source (CCCS) mode pixel circuits were designed to provide 8-bit grayscale to the Micro-LED pixels. The schematic

diagram of pixel circuit is shown in Fig. 2. A FPGA based periphery driving board was designed to program the Micro-LED device.

B. Fabrication of the GaN-based Micro-LED Arrays

The Micro-LEDs were fabricated on sapphire substrate with 10 periods of multiple quantum wells (MQWs). The mesa structure of the pixels was defined using photolithography and inductively coupled plasma (ICP) etching with SiO_2 as a hard mask. 10 nm of Ni/Au was evaporated onto the p-GaN area to serve as the current spreading layer (CSL). Then, a 200 nm Ti/Al/Ti/Au stack layer was evaporated onto the p-GaN and n-GaN as electrode. This Ti/Al/Ti/Au stack layer served as reflective mirror at the p-GaN light emitting region. A SiO_2 isolation layer was deposited by plasma enhanced chemical vapor deposition (PECVD) and patterned by reactive ion etching (RIE) and 350 nm Ni/Au was deposited on pad area of p-electrode and n-electrode for mass transferring bonding. The schematic diagram and scanning electron microscopic (SEM) image of the finished Micro-LED were shown in Fig. 3 and 4.

C. Fabrication of the Si CMOS AM backplane

The backplane was designed using mature Si CMOS technology. The output pad of each pixel was custom-designed with thick metal layer in circular shape. Post-CMOS process was done after the standard process in foundry. Photolithography steps were done and then a layer of Ni/Au was deposited as the bonding metal layer.

D. Fine-pitch and high-density mass-transferring

The solder bumps for mass transferring process were fabricated onto GaN Micro-LED wafer using photolithography and metal evaporation. A layer of Ni/Au (50nm/50nm) was deposited as the under ball metal (UBM) followed by thermal evaporation of indium (3μm). Then the sample was annealed in a reflow furnace with N_2 as protection gas. The indium formed solder bumps. Finally, the 420 μm-thick sapphire substrate was removed by laser lift-off (LLO) and the wafer was then diced into individual chips. The Micro-LED chips were then transfer bonded onto the Si backplanes with a flip-chip configuration. Another reflow step was applied to ensure stable connection and ohmic contact between the Micro-LEDs and the backplane.

III. RESULTS AND DISCUSSION

A. GaN-based Blue and Green Micro-LEDs

Fig.5 show the current-voltage (I-V) and current density-voltage (J-V) characteristics of GaN-based blue and green Micro-LEDs measured by Keysight B1500A semiconductor analyzer. Fig. 6 illustrates the photoluminescence (PL) and electroluminescence (EL) of the blue and green Micro-LEDs. The full width half maximum (FWHM) is 20 nm and 22 nm, respectively. From the transmission line measurement (TLM) results shown in Fig.7, the contact resistance between p-GaN, n-GaN and electrodes were measured as 2318 Ω and 2.61 Ω, respectively.

B. 2500 PPI Micro-LED micro-displays

Fig.8 shows the display results of the blue and green Micro-LED displays. Vivid animations and movies were displayed. The micro-displays can work under two modes: low brightness mode for direct-view and high brightness mode for projection. In low brightness mode, it has 30,000 nits for blue color and 50,000 nits for green color. The power consumption is 0.5W. In high brightness mode, it is ten times brighter for both blue and green chips, with a higher power consumption of 5W.

C. Temperature sensing

The relationship between temperature and voltage of a p-n junction was shown in (1) to (3).

$$I = I_s \left(e^{\frac{qV}{kT}} - 1 \right) \approx I_s * e^{\frac{qV}{kT}} \tag{1}$$

$$V = \frac{kT}{q} \left[\ln(I) - \ln(I_s) \right] \tag{2}$$

Assume $\ln(I_s)$ has weak temperature dependence. Then if I is fixed, V will change linearly with T. We will have a linear interpolation of

$$V = V_1 + \left(\frac{V_2 - V_1}{I_2 - I_1} \right)(I - I_1) \tag{3}$$

Fig.9 shows temperature dependent I-V curve of an 80µm $\times$ 80µm Micro-LEDs. The temperature varied from room temperature to 180℃. Figure 10 illustrates the relationship between temperature variation and voltage measured from the Micro-LED devices with a fixed bias current of 1µA. Consistent linearity can be read for varied device dimensions of Micro-LEDs. With a comparison of different device dimensions and the sensitivity of temperature sensing ($\Delta T/\Delta V$) in Fig. 13, it can be concluded that Micro-LEDs with device dimension of 80µm $\times$ 80µm have the better sensitivity ($\Delta T/\Delta V = 795K/V$).

D. Light energy harvesting

Fig.11 illustrates light energy harvesting characteristics of a green Micro-LED device with dimension of 50µm $\times$ 50µm and excited by blue light with various intensity. Under blue light with an intensity of 75 µW/cm^2, the I_{sc} and V_{oc} were measured as 1.79×10^{-10}A and 1.75V. The fill factor (FF) and external quantum efficiency (EQE) were calculated as 65.9% and 15.5%, respectively. Fig. 12 shows the current density-voltage and power-voltage curve of the Micro-LED device. Performance of Micro-LEDs with different dimensions are compared in Fig. 13, it can be concluded that Micro-LED with device dimension of 80µm $\times$ 80µm has the best performance of light energy harvesting.

E. Light detection

Fig.14 illustrates the relationship between photo-generated current and reverse bias voltage under various light illumination intensity. The sensitivity of the Micro-LED light detector was 1240. The light response curve is shown in Fig. 15. Fig. 16 shows the EQE of the Micro-LEDs as light detectors. Cut-off wavelength can also be read from the figure as 550nm.

F. Switching between function modes and driving method

As discussed above, the proposed fully multi-functional Micro-LED devices have various function modes, with properly bias conditions applied on the device. The Si CMOS backplane provides forward bias voltages and current for micro-display and lighting modes. It also serves as DC current source and read-out IC (ROIC) for temperature sensing mode. In light energy harvesting mode, the energy collected by the Micro-LED device will be stored in the capacitance of the backplane and then feed-back to the storage components on the periphery board. In light detection mode, the backplane will provide reverse bias voltage and work as ROIC again. Detailed parameters and function modes are listed in Table 1.

IV. CONCLUSION

We proposed a fully functional time-division multiplexing GaN-based Micro-LED devices. With properate bias conditions, the devices worked under various function modes such as display & lighting, temperature sensing, light energy harvesting, and light detection. The performances were measured superior than the latest references such as PPI, FF, efficiency, EQE, and temperature resolution, promising a great potential of highly-integrated device for consumable electronics.

ACKNOWLEDGMENT

The authors gratefully acknowledge the contributions of Prof. Kei May Lau from ECE department and NFF of HKUST.

REFERENCES

[1] B. L. Ahn, et al., Appl. Energy, vol. 113, pp. 1484–1489, 2014.
[2] T. Han, et al., Sci. Rep., vol. 7, pp. 1–7, 2017.
[3] K. Zhang, Z. Liu, et al., Dig. Tech. Pap. - SID Int. Symp., vol. 48, no. 1, pp. 358–361, 2017.
[4] M. Choi, et al., Adv. Funct. Mater., vol. 27, no. 11, 2017.
[5] K. Zhang, Z. Liu, et al., IEEE Trans. Electron Devices, vol. 63, no. 12, pp. 4832–4838, 2016.
[6] F. Olivier, et al., J. Lumin., pp. 1–5, 2016.
[7] H. S. El-Ghoroury, et al., SID Symp. Dig. Tech. Pap., vol. 46, no. 1, pp. 371–374, 2015.

[8] R. Dahal, et al., Appl. Phys. Lett., vol. 94, no. 6, pp. 2009–2011, 2009.
[9] D. H. Lien, et al., Nano Energy, vol. 11, pp. 104–109, 2015.
[10] X. Zhang, et al., J. Mater. Chem. C, vol. 5, no. 17, pp. 4319–4326, 2017.
[11] W. Y. Weng, et al., IEEE J. Sel. Top. Quantum Electron., vol. 17, no. 4, pp. 996–1001, 2011.
[12] F. González-Posada, et al., Nano Lett., vol. 12, no. 1, pp. 172–176, 2012.
[13] S. B. Khan, et al., Talanta, vol. 120, pp. 443–449, 2014.
[14] X. Ren, et al., Adv. Mater., pp. 4832–4838, 2016.

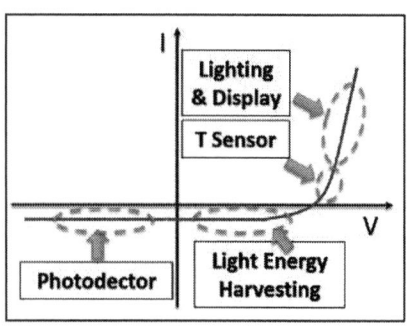

Fig. 1 Function modes of p-n junctions

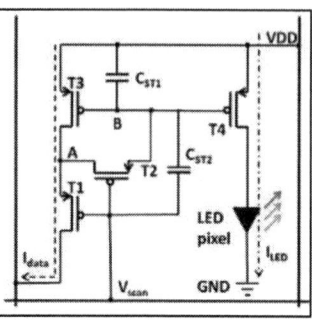

Fig. 2 Pixel circuit of the Si backplane for the Micro-LEDs

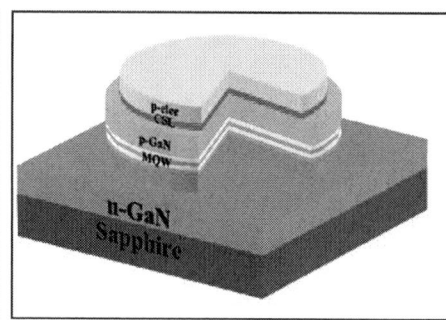

Fig. 3 Schematic diagram the Micro-LED pixel

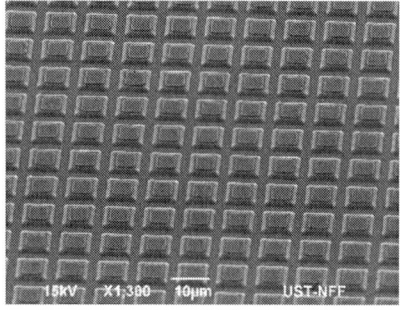

Fig. 4 SEM image the Micro-LED pixels

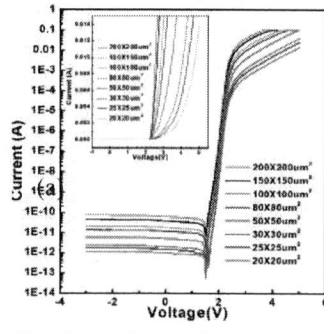

Fig. 5 (a) I-V and (b) J-V characteristics of Micro-LED pixels with various size

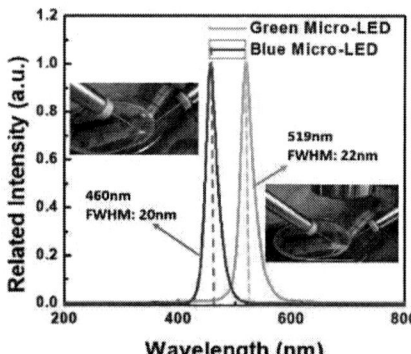

Fig. 6 PL of the Micro-LED pixels

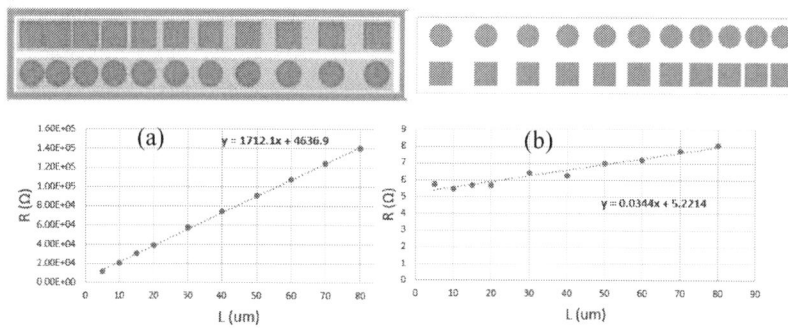

Fig. 7 TLM results of (a) p-GaN and (b) n-GaN of Micro-LEDs

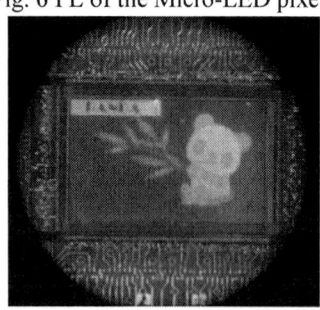

Fig. 8 Display results of (a) blue and (b) green Micro-LEDs micro-displays

Fig. 9 Temperature dependent I-V of Micro-LEDs

978-1-7281-1988-5/18 $31.00 © 2018 IEEE

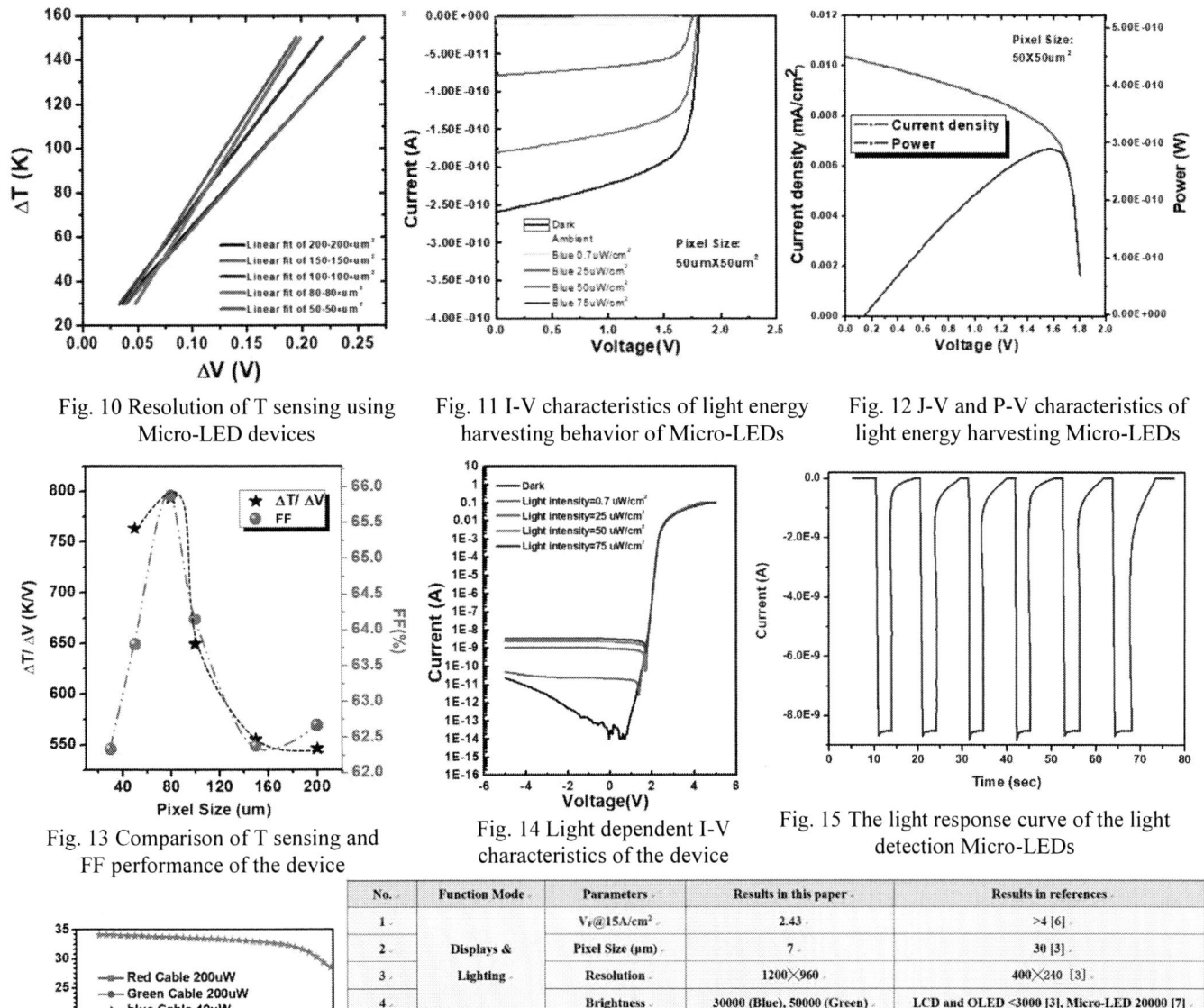

Fig. 10 Resolution of T sensing using Micro-LED devices

Fig. 11 I-V characteristics of light energy harvesting behavior of Micro-LEDs

Fig. 12 J-V and P-V characteristics of light energy harvesting Micro-LEDs

Fig. 13 Comparison of T sensing and FF performance of the device

Fig. 14 Light dependent I-V characteristics of the device

Fig. 15 The light response curve of the light detection Micro-LEDs

Fig. 16 EQE of the Micro-LEDs as light detectors. Cut-off wavelength can also be read from the figure as 550nm.

No.	Function Mode	Parameters	Results in this paper	Results in references
1	Displays & Lighting	V_F@15A/cm²	2.43	>4 [6]
2		Pixel Size (µm)	7	30 [3]
3		Resolution	1200×960	400×240 [3]
4		Brightness	30000 (Blue), 50000 (Green)	LCD and OLED <3000 [3], Micro-LED 20000 [7]
5	T Sensing	$\Delta T/\Delta V$ (K/V)	795	Not available
6	Light Energy Harvesting	J_{SC} (mA/cm²)	>3 for equivalent AM1.5	1.5 for AM 1.5 [8]; 0.4 for AM 1.5 [9]
7		V_{OC} (V)	1.75 @ 460nm illumination	2 [8]; 2.4 [9]
8		FF	65.9%	60% [8] [9]
9		Efficiency	15.5% @ 460nm	10% @450nm [8][9]
10	Light detection	Sensitivity	1240	971 [10]; 100 [11]
11		On/off ration	>1000; stable	101[10], 1000 [11]; unstable
12		Response time (s)	<5	>30 [10]; 40 [11]
13		EQE	33% for 50µW blue light, 66% for equivalent UV light	42.7 for equivalent UV light [11]
14		Responsivity (A/W)	0.118 for 50µW blue light, 0.236 for equivalent UV light	0.12 for equivalent UV light [11]
15		Cut-off Wavelength	~550nm (Green Micro-LED)	360nm [9]

Table. 1 A summary of the performance of the Micro-LEDs working as lighting & displays, T sensing, light energy harvesting, and light detection. Results are also compared with representative references.

Environmentally Friendly Quantum Dots for Display Applications

E. Jang

SAIT, Samsung Electronics, Suwon, Korea, email: ejjang12@samsung.com

Abstract— Ever since the physics of quantum dot (QD) was discovered, much research effort has been carried out for more than 30 years, and lots of applications adopting QDs have been proposed. Especially, wide color gamut displays using QDs as active light emitting materials have drawn much attention. And, the QD-based consumer displays such as LED TVs, tablets, and special monitors are now on the market. They provide best color gamut, reasonable power efficiency, and affordable price showing superior competitive edge to OLED technology. However, still there are issues and argues using Cadmium containing materials in practical consumer devices. In spite of the European RoHS Exemptions, we need to be aware the environmental risk of producing large quantity of Cd-containing materials and using them in the consumer electronics. And, this growing apprehension for environmental issues formed great limitation for QD's applications. Therefore, we have dedicated to develop more environmentally friendly InP based QDs that showed considerably high efficiency and saturated color spectrum compared to the Cd-containing materials. The structure of Cd-free QD was specially tailored for display applications and the synthetic process was optimized to produce reliable materials in commercial scales. In order to improve the efficiency and stability of the QDs in the devices operating under severe atmosphere, specific composite materials were designed and the fabrication process was optimized. From 2015, Samsung has released Cd-free QD adopted UHD TV for major product line-up which show the best color gamut among the current displays. Now we are trying to make additional breakthroughs in displays by using established QD material platform and broaden the technology to wider optoelectronic applications.

I. Introduction

After the quantum confinement effect was discovered in the early 1980's, many chemists tried to prepare ideal quantum dot materials and develop synthetic methods for uniform size, core/shells, and shape-controlled structures. Alongside the fundamental investigations, many potential applications were suggested such as laser, bio labeling, printable TFT, color converting LED, electrically driven LED, photovoltaics, as well as memory devices, and these are all patented by several leading technology groups. From 2010, industries started to produce prototypes and commercial products in display applications such as large sized TVs, tablets, laptops, and monitors. Fig.1 shows the recent industrial technology trends of QD display development and changes in exemptions in European RoHS. The first commercial product came out from Sony with QD Vision's

Color IQ technology, which was the glass tube optic using Cd-based QD. After that, TCL, Philips, BOE, Hisense and Konka applied the same technology for their TV and monitor products. Nanosys has been very active in commercializing QD-film through the joint development with film-fabricating companies such as 3M, LMS, Hitachi, and Nitto Denko. Starting from Amazon's Kindle fire tablet, QD color converting film, which is called as QDEF, was adopted in the ASUS's laptop and monitor, and TVs of Hisense, TCL, Vizio, AUO. In China, Najing technology started to provide QDs to film makers such as Sangbo and Poly-OE, and work with TCL and Hisense to produce TVs. Nanoco is one of the pioneer companies to have developed the Cd-free QDs from the beginning, and they have collaborated with Dow chemical and Merck for the mass-production.

We reported the fabrication of 46" prototype TV with white LED BLU using Cd-based QDs in 2010 [1], and completed the QD film technology using Cd-based QDs in 2012. However, we decided not to commercialize the QD TV products because we concluded not to use any environmentally harmful materials inside our products even though the minimal usage of Cd was still legitimate according to the RoHS exemptions. After the decision, we started the development of the QDs with no environmental problem to solve the health and safety issues over entire manufacturing processes. Aside from the environmental policy, it was the right time to start another challenge on the new materials because high levels of performances with Cd-QDs were already achieved. The RoHS exemption had been renewed several times to allow the usage of Cd-QDs in electronic devices for a while, but in 2018, the European Parliament finally decided not to extend the exemption after 2019.

II. Cd-free QD

Many researchers from academia and industries have been interested in the InP QD synthesis. In a quite early stage, InP QDs were prepared in strongly coordinating solvent and ligands, and as a result, the nanoparticle growth took more than days. Then, Peng [2] reported very effective reaction conditions using octadecene as a solvent, so that the crystal growth became very fast compared to the previous method. His continuous report [3], which was about the wavelength control of InP from 450nm~750nm and the possibility of making stable InP QDs in hydrophilic conditions, considerably alleviated the concerns about the practical applications using InP QDs. After that, subsequent reports disclosed the method to improve the quantum efficiency and FWHM as well as the understanding of the surface structures and growth mechanisms [4~6]. The progress of QDs was summarized in Fig. 2 in terms of QE according to the emission range. After the preparation of the first core/shell structure of Cd-QDs, which showed 50% QE, it took about 20 years to prepare QDs with 100% QEs, and it's still hard to find efficient blue light emitting one. For InP QDs, the progress was relatively very fast and it took less than 10 years to show higher than 85%. Now, our QDs showed higher than 95% of QE, and there are only small rooms for the improvement in each color. To obtain efficient InP QDs, we designed multi element core and gradient shell structure. ZnSe and ZnS composition gradient was controlled to reduce the lattice mismatch between InP and ZnS (Fig. 3). The oxidative interface was induced to remove defects on the interfaces of III-V core and II-VI shell. To enhance the crystallinity, we maintained relatively high temperature (about 340C) during the synthesis. Also, we tried to design the process to be as simple as possible to cut down the cost, and increase the production yield by controlling the uniformity of size, shape, and suppressing by-products.

We have optimized critical parameters in synthesis to make the process repeatable and reproducible in large scale. We have collaborated with Hansol chemical since 2013 and transferred our synthetic recipe after a verification of the reaction conditions with 20 L bench reactor in the lab. Currently, we prepared green and red QDs with FWHM less than 33nm in the lab, and these were reproduced as 35nm in the production scale (Fig. 4).

III. QLED TV

The current Samsung QLED TVs comprise InP QD color converting films and blue LEDs. After we prepared the green and red QDs, mixed them with photo-curable polymers, then made films with protective barrier against moisture and oxygen. Finally, the QD film was put in front of blue light guiding panel, then it became a white BLU which showed wider color gamut than conventional phosphor white LED-BLU (Fig. 5 & 6). There are several standards for color space, and we set the color coordination of green and red QDs to maximize the agreement to the DCI standard.

When we developed the TV set, we organized a cross functional team covering materials synthesis, interface design, product evaluation, and mass production. It was also necessary to handle the environmental issue, predict market size, estimate, cut down the cos, build the eco system, and make a strong patent position. Since we released InP based QD TVs in 2015, there have been progresses in brightness and color gamut of TVs. The most recent product showed average brightness of 800nit with 100% color volume. And, we are still working to reduce the FWHM of green QD to less than 30nm for targeting BT2020 as a next step.

IV. FUTURE DISPLAY APPICATIONS

There are other chances to use QD optical modules instead of the film such as emissive color filter using blue BLU or pixelated micro LEDs (Fig. 7). The advantage of using emissive color filter with backlights is the prefect viewing angle (Fig. 8), and, QD is the only light emitter for this application in terms of the size and stability. To make the QD color filter, we prepared QDs and mixed them with photoresist to make patterns through photo-lithography, or formulate ink composition to make patterns through jetting process. Compared to the conventional LCDs, the emissive color filter showed almost perfect viewing angle from 178 degree. Also, much attention was paid to the electrically driven QD-LEDs (Fig. 9). Unlike the color converting applications, a blue light emitting QD is required to be developed. For InP, the bandgap of the bulk is about 1.4eV, the diameter should be reduced to 2 nm for the blue emission. We prepared the InP QDs emitting at 460nm with relatively broad FWHM and 60% QE. Another promising candidate for blue emission is ZnSe QD, and now many people focus on this structure [7, 8]. In QD-LED device fabrication, it is important to control ligands of QDs for efficient charge transfer. Other charge transport layers should be optimized and stabilized with more robust structure. Fig. 10 shows the progress in QD-LEDs including both of Cd-QDs and Cd-free QDs. Currently, Cd based QD-LEDs show very high external QEs which are close to theoretical limit, and the performances of the Cd-free QD LED improves quite fast. However, QD-LED device fabrication process involves solution printing steps which have many latent issues of purity and reliability. As the soluble inkjet process develops, fundamental understandings of the interfaces are required in the nano/micro structures of layered LEDs.

V. SUMMARY

We have developed efficient Cd-free QD materials, and produced them in large scale. In display applications, InP based QDs have been used as color converters first, in the film and second in the color filter structure. In those applications, the formulation technology of QDs with polymers has been developed as well, and this can be expanded to patterning technologies such as photo lithography and inkjet printing. For the QD-LEDs, charge injection and recombination properties have been investigated to optimize the device performances. Beyond the display technology using QDs, the sensor devices using NIR light absorbing QD materials are one of our interests in the future.

REFERENCES

[1] E. Jang, S. Jun, H. Jang, J. Lim, B. Kim, and Y. Kim, "White-Light-Emitting Diodes with Quantum Dot Color Converters for Display Backlights", *Adv. Mater.*, 22, pp3076–3080, 2010

[2] D. Battaglia, and X. Peng, "Formation of High Quality InP and InAs Nanocrystals in a Noncoordinating Solvent" *Nano Letters*, 2 (9), pp 1027–1030, 2002

[3] R. Xie, D. Battaglia, and X. Peng, "Colloidal InP Nanocrystals as Efficient Emitters Covering Blue to Near-Infrared" *J. Am. Chem. Soc.*, 129 (50), pp 15432–15433, 2007

[4] L. Li and P. Reiss, "One-pot Synthesis of Highly Luminescent InP/ZnS Nanocrystals without Precursor Injection", *J. Am. Chem. Soc.*, 130 (35), pp 11588–11589, 2008

[5] J. Lim, W. Bae, D. Lee, M. Nam, J. Jung, C. Lee, K. Char, and S. Lee, "InP@ZnSeS, Core@Composition Gradient Shell Quantum Dots with Enhanced Stability", *Chem. Mater.*, 23 (20), pp 4459–4463, 2011

[6] P. Ramasamy, N. Kim, Y. Kang, O. Ramirez, and J. Lee, "Tunable, Bright, and Narrow-Band Luminescence from Colloidal Indium Phosphide Quantum Dots", *Chem. Mater.*, 29 (16), pp 6893–6899, 2017

[7] H. Shen, W. Cao, N. Shewmon, C. Yang, L. Li, and J. Xue, "High-Efficiency, Low Turn-on Voltage Blue-Violet Quantum-Dot-Based Light-Emitting Diodes" *Nano Lett.*, 15 (2), pp 1211–1216, 2015

[8] H. Shen, W. Cao, N. Shewmon, C. Yang, L. Li, and J. Xue, "High-Efficiency, Low Turn-on Voltage Blue-Violet Quantum-Dot-Based Light-Emitting Diodes" *Nano Lett.*, 15 (2), pp 1211–1216, 2015

978-1-7281-1988-5/18 $31.00 © 2018 IEEE

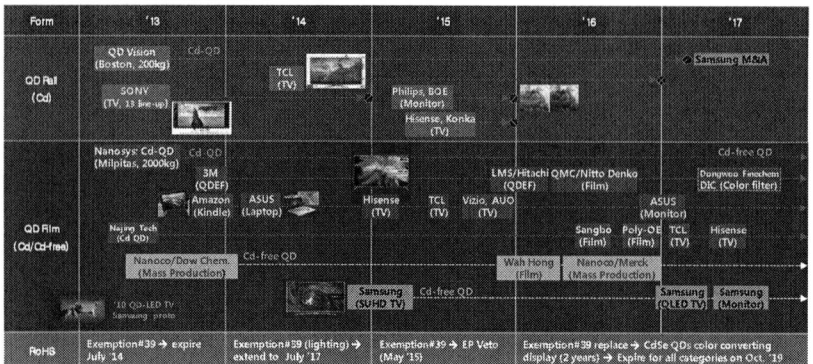

Fig. 1. Recent industrial developments of QD displays and exemptions in RoHS

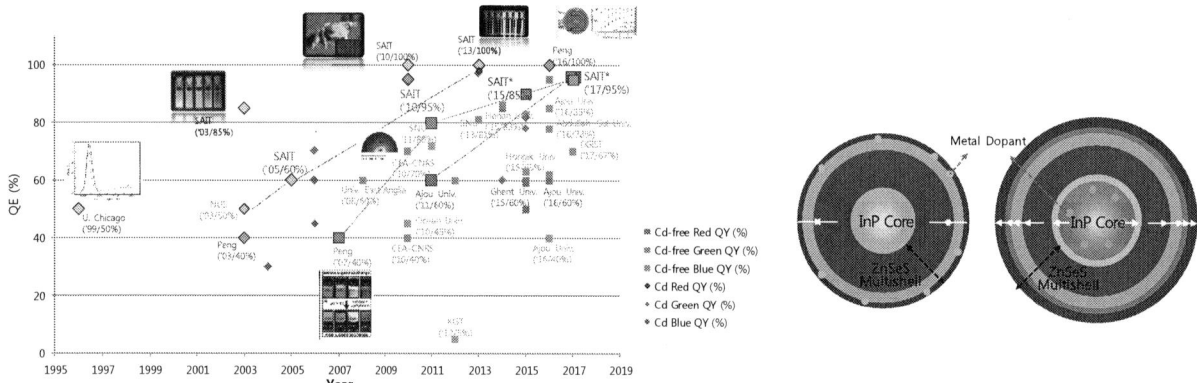

Fig. 2. Progress of QDs' quantum efficiency in the literatures

Fig. 3. Optimized structure of green (left) and red (right) light emitting InP/ZnSeS QD

Fig. 4 Photo luminescent spectra of green and red QDs prepared in lab scale reactor and mass production (left), 20 L bench scale reactor (middle), 100L pilot scale reactor (right)

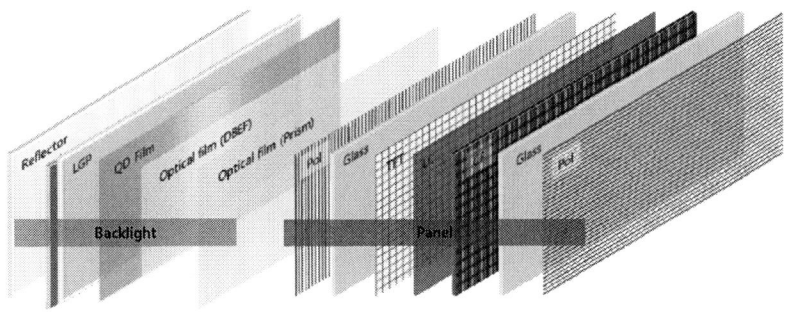

Fig. 5. Structure of QLED TV

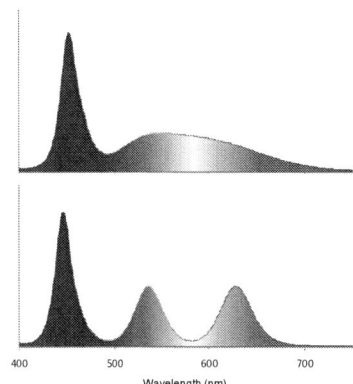

Fig. 6. The spectra of conventional phosphor BLU (up) and QLED BLU

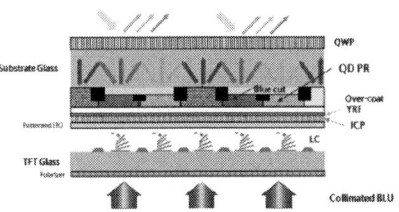

Fig. 7. Schematic drawing of QD color filter with conventional LCD backlight

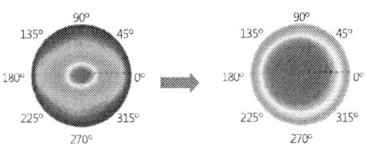

Fig. 8. Brightness at different viewing angle from the LCD display with conventional color filter (left) and QD color filter (right)

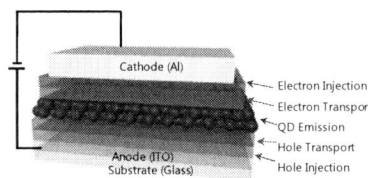

Fig. 9. Schematic drawing of current-driven QD-LED structure

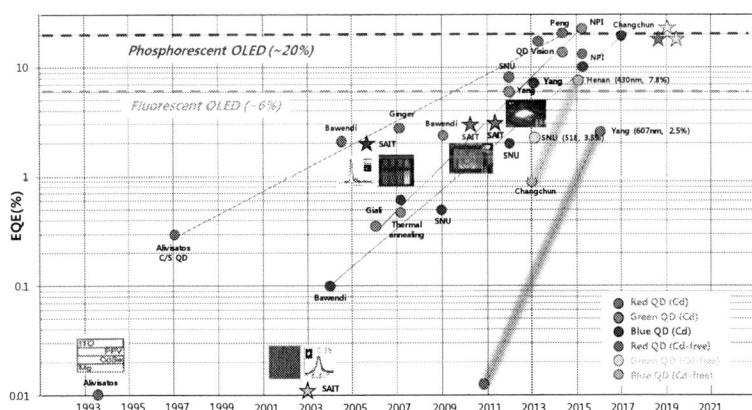

Fig. 10. The progress of QD-LEDs' external QE in the literature

Solution Processed High Performance Short Channel Organic Thin-Film Transistors with Excellent Uniformity and Ultra-low Contact Resistance for Logic and Display

L. Feng[1], Y. Huang[2], J. Fan[2], J. Zhao[2], S. Pandya[3], S. Chen[2], W. Tang[2], S. Ogier[1], and X. Guo[2]

[1]Wuhan Xinqu Chuangrou Optoelectronics Technology Co., Ltd, Wuhan, China, email: simon.ogier@neudrive.cn
[2]Department of Electronic Engineering, Shanghai Jiao Tong University, Shanghai, China, e-mail: x.guo@sjtu.edu.cn
[3]NeuDrive Limited, Macclesfield, Cheshire, UK

Abstract—**High performance organic thin-film transistors (OTFTs) are fabricated by using normal spin-coating processes with the highest processing temperature of 115 °C. The devices present negligible contact resistance and excellent uniformity over 4 inch area. An average mobility higher than 4 cm^2/V·s is achieved with 7 μm channel length devices. Ring oscillators and AMOLED displays based on the OTFTs are demonstrated. A dual gate structure is implemented to tune the threshold voltage of the OTFTs for improving the compensation and gray-level adjusting capability of an AMOLED in-pixel compensation circuit.**

I. INTRODUCTION

Organic thin-film transistors (OTFTs), made of solution processed organic semiconductor and polymer dielectric layers, have advantages of intrinsic mechanical flexibility, low temperature and low cost processing. Therefore, they are promising for developing low cost, truly flexible displays, sensors and integrated circuits to enable a wide range of emerging applications in the era of internet of things (IoT) [1]. In the past, significant efforts have been devoted to material design and processing for high mobility OTFTs However, most reported OTFTs of high mobility are characterized based on device architectures using oxide gate dielectric and top metal contacts with long channel length (several tens or more than one hundred μm) [2-4]. Some non-conventional processing techniques with narrow processing window and low efficiency have been used. Additionally, due to complex contact effects and carrier density-dependent mobility, a number of reports of high mobility OTFTs exhibit non-ideal field effect transistor behaviors, causing a substantial overestimation of the mobility [5]. In order to develop a competitive OTFT technology, high performance and good uniformity needs to be achieved at short channel lengths with manufacturing compatible structures.

In this work, OTFTs are fabricated using solution processed organic semiconductor and polymer dielectric layers in a manufacturable structure on glass and plastic substrates. With an optimized organic semiconductor solution and electrode surface treatment, high performance OTFTs of negligible contact resistance and excellent uniformity are obtained by using normal spin-coating processes over 4 inch area. An

average mobility higher than 4 cm^2/V·s is achieved with 7 μm channel length devices. Ring oscillators and AMOLED displays based on the OTFT are fabricated. With a dual gate structure, the threshold voltage of the OTFTs can be efficiently tuned to accommodate a wide range of circuit design.

II. DEVICE STRUCTURE AND PROCESSING

Fig. 1 illustrates the process flow of the fabricated top-gate bottom-contact (TGBC) OTFTs. After forming a thermally cross-linked polymer buffer layer on pre-cleaned glass or PEN substrate, about 50 nm thick gold (Au) source/drain (S/D) electrodes were deposited by sputtering and patterned using photolithography and wet etching. Thiol based SAMs were used to treat the electrode surface before depositing the organic semiconductor (OSC) layer. The OSC solution was spin-coated at 500 rpm for 10 s followed by 1000 rpm for 60 s and a further bake at 100°C for 60 s. The OSC solution is comprised a small molecule semiconductor and a high-k polymer semiconductor binder (Fig. 1(b)) [6]. Diluted CYTOP (CTL-809M) solution was spin-coated to deposit an about 300 nm thick organic gate insulator (OGI) layer (~6 nF/cm^2). A 50 nm gate metal electrode was then formed by thermal evaporation of Au, followed by photolithography and wet etching. The OSC and OGI layers were patterned by oxygen plasma etching using the gate metal as a hard-mask. A 850 nm thick SU8 passivation layer was deposited and patterned by UV photolithography to form via holes. 50 nm thick Au metal interconnect layer was sputtered and patterned using photolithography to create required electrical connections between the metal layers. All processes were carried out using standard micro-fabrication facilities. With the highest processing temperature of 115 °C for the SU8 passivation layer, the OTFT processes are compatible with plastic substrate of low temperature budget. The OTFT devices fabricated on a 25 μm thick PEN foil are shown in Fig. 2. All the electrical characterization were carried out in air ambient at room temperature.

III. RESULTS AND DISCUSSIONS

A. Electrical Properties and Uniformity

The measured typical transfer and output characteristics of the 7 μm channel length OTFTs are given in Fig. 3. Typical field effect transistor behaviors can be seen, including large ON/OFF ratio ($>10^8$) and well saturated output characteristics,. The linear I_D to V_{GS} dependence in the linear regime makes it acceptable to use the ideal MOSFET for mobility extraction. The extracted apparent mobility is above 4 cm^2/V·s for the 7 μm channel length OTFTs, which is compared to previously reported high mobility OTFTs with solution processed OSC layers in Fig. 4. Very few solution processed OTFTs of short channel length less than 10 μm have mobility exceeding 1 cm^2/V.s. For those reported OTFTs with mobility higher than 4 cm^2/V.s, they were all fabricated with long channel lengths (> 20 μm), and using inorganic gate dielectric layer based on a device structure not suitable for circuit integration. Special coating processes are also required, causing limited processing window and poor efficiency. This work achieves the highest mobility for solution processed OTFTs of short channel lengths (< 10 μm) using normal spin-coating processes. There is also much less dependence of the apparent mobility on the channel length compared to the normal OTFTs (Fig. 4). Excellent uniformity can also be observed from Fig. 5, which gives the measured I_D-V_{GS} curves for 36 devices in each channel length from 42 μm to 7 μm.

B. Contact Resistance

Based on the measured transfer curves (I_D-V_{GS}), the sum of the channel and contact resistances in the linear regime as a function of the channel length at different V_{GS} is plotted in Fig. 6 for contact resistance extraction based on the transmission line method (TLM). All the linearly extrapolated lines cross over the zero point, indicating ultra-low and negligible contact resistance with the devices. In Si FETs, excellent contact properties are realized with heavily doped source/drain regions for efficient charge tunneling from the metal to the semiconductor region and then charge transport from the source region to the channel. However, for OTFTs with simple electrode–semiconductor junctions, low resistive contacts are difficult to be formed, which limits the scaling of channel length for high performance OTFTs. In this work, the formulation of the small molecule and polymer semiconductor blend is optimized to achieve energy level match to the Au electrode and also good crystallization control through normal coating processes for low resistance contact regions. Further, choosing suitable SAMs for surface treatment of the Au source/drain electrodes before depositing the layer is also important, since it can affect the injection efficiency and the crystallization of the OSC layer on top. As shown in Fig. 7, devices using the 3-fluoro-4-methoxythiophenol (SAM_A) can achieve the highest mobility and the best uniformity.

C. Dual-gate for Threshold Voltage Modulation

With the OSC formulation, the fabricated OTFTs present depletion mode operation with a positive threshold voltage (V_{th}) (Fig.3). To modulate the V_{th} for accommodating wider range of circuit designs, a dual gate structure was implemented as illustrated in Fig. 9, with the other Au gate and polymer dielectric layer being formed beneath the channel. By applying different voltages at the back gate, the V_{th} can be modulated efficiently, while the whole shape of the transfer curve and the uniformity is not affected (Fig. 9).

D. Logic and Display Pixel Driver Circuits

5-stage ring oscillators with all p-type inverter configuration based on the single gate OTFTs were designed and fabricated (Fig. 10). The measurement result in Fig. 11 shows that the ring oscillator being operated with a frequency higher than 500 kHz, which is one of the best values for solution processed OTFTs. The operating frequency can be further improved with higher resolution photolithography patterning processes to reduce the channel length and gate to S/D overlaps.

Combination of the OTFT backplane and the OLED front plane would be an ideal technology choice for future development of low cost truly flexible displays. Based on a 2T-1C pixel circuit using the single-gate OTFT, a 64×64 OTFT driven AMOLED display panel was fabricated as shown in Fig. 13. In-pixel circuit compensation is generally needed for suppress the influence of spatial process variations and temporal variations during operation of the TFTs. Fig. 13 shows that, for a 6T-1C in-pixel compensation circuit, using the single gate OTFT of positive V_{th}, it cannot read out the V_{th} of the driving OTFT (T1) (upper of Fig. 13(b)), and thus the compensation function doesn't work properly (upper of Fig. 13(c)). With the dual-gate structure to tune the V_{th} to be negative, both the compensation and gray-level adjusting capability can be much improved (Fig. 13(b) and (c)).

IV. Conclusions

With an optimized OSC solution and proper electrode surface treatment, high performance OTFTs with negligible contact resistance and good uniformity are fabricated with normal coating processes in a manufacturable device structure. The achieved mobility above 4 cm^2/V·s is the highest reported result for solution processed OTFTs of short channel lengths (< 10 μm). With a dual-gate structure, the threshold voltage can be tuned to accommodate a wide range of circuit design. The developed OTFT technology is promising for developing low cost truly flexible displays and integrated circuits.

Acknowledgment

The authors gratefully acknowledge use of the fabrication facilities in Centre for Process Innovation (CPI), Sedgefield, UK, and funding support through National Key R&D Program of China (Grant No 2016YFB0401100).

References

[1] X. Guo, et al., IEEE Tran. Elec. Dev., **64**, 1906 (2017).
[2] B. Peng, et al., Adv. Func. Mat., **27**, 1700999 (2017).
[3] Z. Zhang, et al., Adv. Func. Mat., **27**,1703443 (2017).
[4] M. Uno, et al., Adv. Elec. Mat., **3**,1600410 (2017).
[5] H. H. Choi, et al., Nat. Mat., **17**, 2 (2018).
[6] S. Ogier, et al., Org. Elec., **54**, 40 (2018).

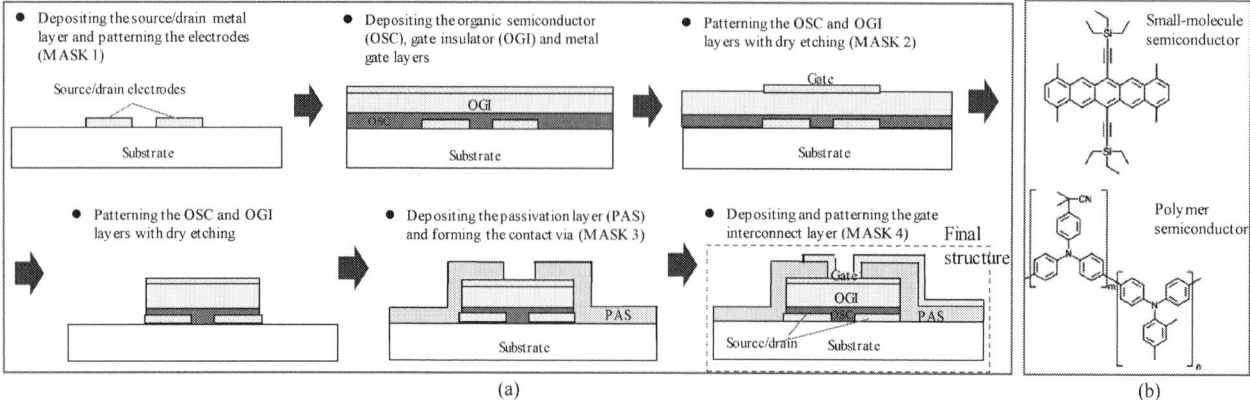

(a)   (b)

Fig. 1. (a) Illustration of the process flow for fabrication of the OTFTs and the final top-gate bottom-contact structure. Except the metal layers, all other layers are solution processed. The maximum processing temperature is 115 °C for the SU8 passivation layer. (b) Molecule structures of the small molecule organic semiconductor and the high-k polymer semiconductor binder used for the blend.

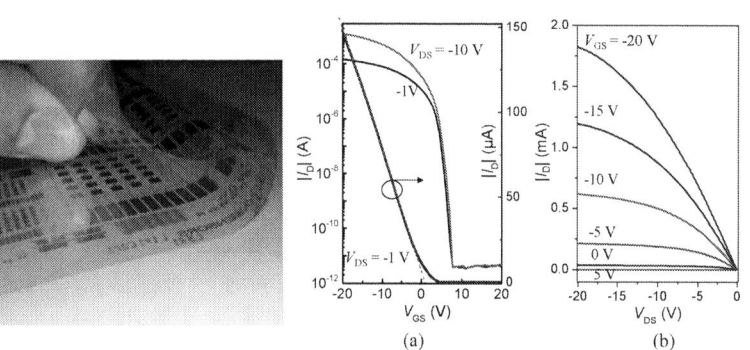

(a)   (b)

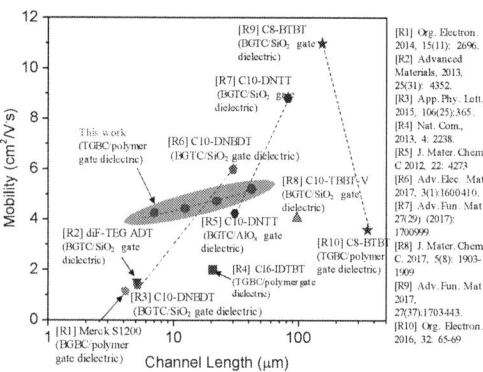

Fig. 2. Photo image of the fabricated flexible OTFT devices on a 20 μm thick PEN foil attributed to low processing temperature not exceeding

Fig. 3. The measured typical transfer (a) and output (b) characteristics of the OTFTs with a channel length of about 7 μm and a channel width of about 1100 μm, showing typical field effect transistor behaviors of large ON/OFF ratio ($>10^8$), linear I_D to V_{GS} dependence in the linear regime and well saturated output characteristics.

Fig. 4. Comparison of the extracted apparent mobility values of the fabricated OTFTs in this work with those of previously reported high mobility OTFTs using solution processed OSC layers. (BGTC: bottom-gate top-contact; TGBC: top-gate bottom-contact; BGBC: bottom-gate top-contact)

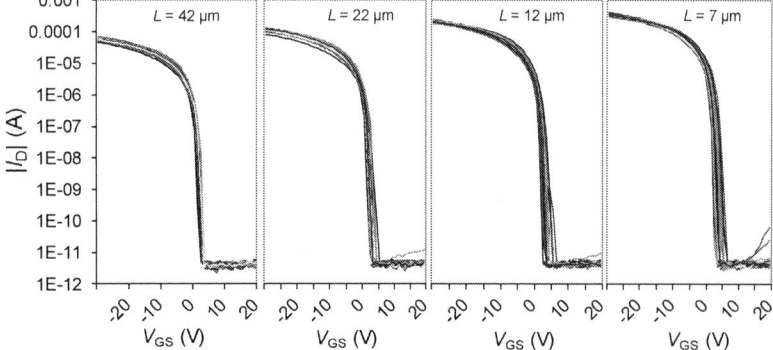

Fig. 5. The measured I_D-V_{GS} curves for 36 devices in each channel length from 42 μm to 7 μm at V_{DS} = -1 V, showing excellent uniformity. The channel width is about 1100 μm.

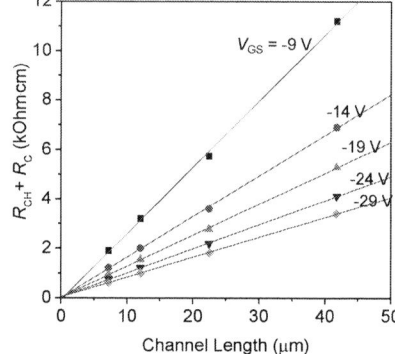

Fig. 6. The measured channel ($R_{CH} + R_C$) and contact resistance in linear regime for OTFTs of different channel lengths at different V_{GS} for contact resistance extraction based on the transmission line method (TLM).

978-1-7281-1988-5/18 $31.00 © 2018 IEEE   881

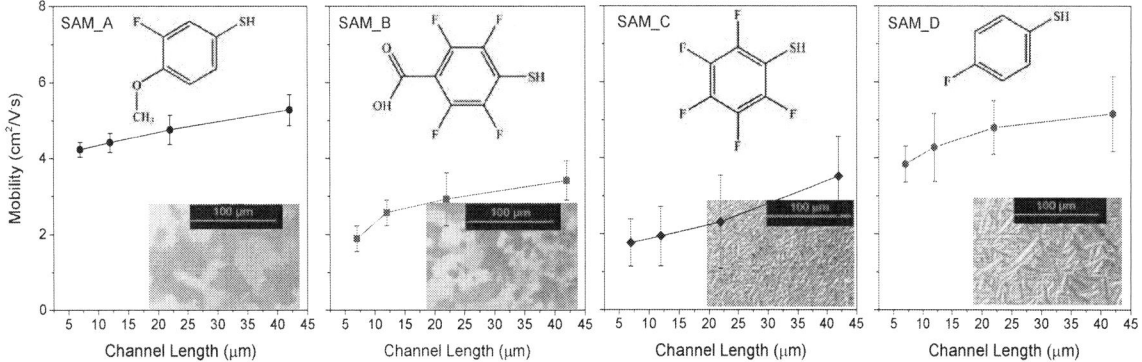

Fig.7 The extracted apparent mobility for OTFTs of different channel lengths fabricated using 4 different thiol based SAM materials to treat the source-drain electrodes. The OSC morphology images in the inset were taken under cross polarisers in reflectance. SAM_C and SAM_D tend to promote needle shape crystals in the channel area of the OTFT, which may contribute to the non-uniformity of the electrical results.

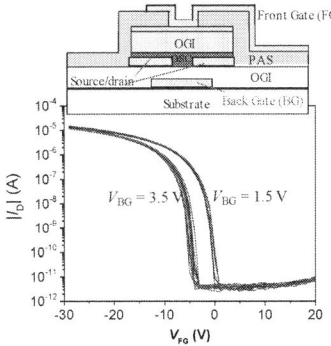

Fig. 9. Schematic of the structure of the fabricated dual-gate OTFT and the measured transfer curves for the dual-gate devices with 1.5V and 3.5V on the back gate (V_{BG}).

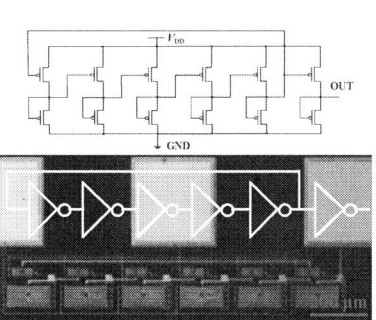

Fig. 10. Circuit schematic (upper) and photo image (bottom) of the fabricated 5-stage ring oscillator with all p-type inverter configuration based on the OTFTs.

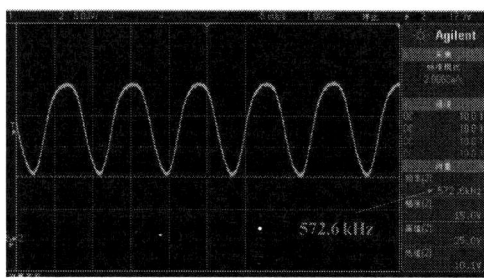

Fig. 11. Screen snapshot from the oscilloscope showing the fabricated 5-stage ring oscillator being operated with a frequency higher than 500 kHz, which is one of the best values for solution processed OTFTs.

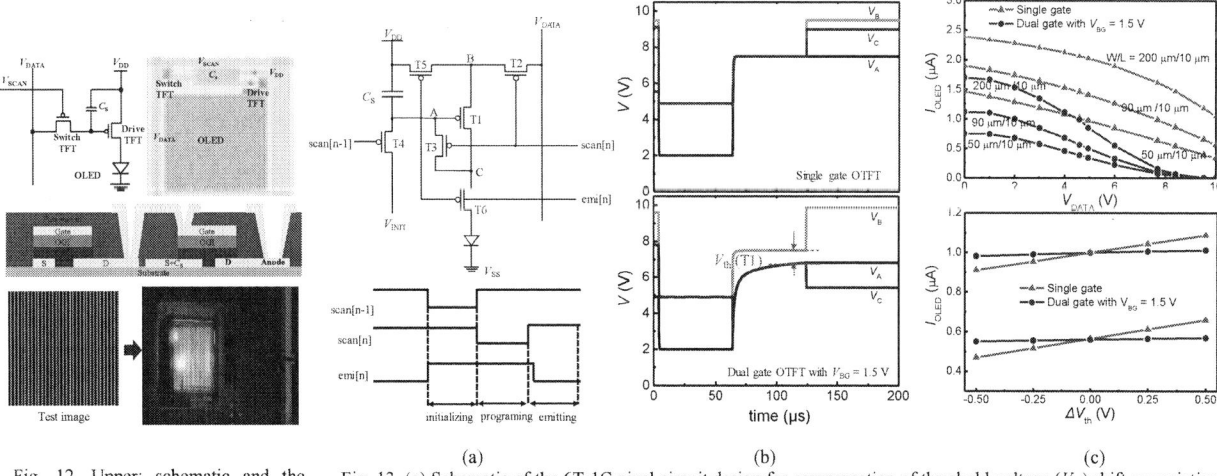

Fig. 12. Upper: schematic and the layout design of the OTFT 2T-1C pixel circuit for OLED displays (upper). Middle: the cross-sectional schematic of the integration structure. Bottom: photo image of the fabricated 64×64 OTFT driven AMOLED displays.

Fig. 13. (a) Schematic of the 6T-1C pixel circuit design for compensation of threshold voltage (V_{th}) shift or variation of the OTFT. (b) Simulation results show that the 6T-1C pixel circuit using the single gate OTFT with positive V_{th} cannot read out the V_{th} of the drive TFT (T1), while it working for the dual-gate device with a back gate bias (V_{BG}) of 1.5 V. (c) Simulation results show that the 6T-1C pixel circuit using the single gate OTFT cannot compensate V_{th} variations while it working for the dual-gate device with V_{BG} = 1.5 V (upper). It is also easier to modulate the final OLED current to lower current level with the dual-gate OTFT than that with the single gate one (bottom).

978-1-7281-1988-5/18 $31.00 © 2018 IEEE   882

Record Static and Dynamic Performance
of Flexible Organic Thin-Film Transistors

James W. Borchert[1,2], Ute Zschieschang[1], Florian Letzkus[3], Michele Giorgio[4,5],
Mario Caironi[4], Joachim N. Burghartz[3], Sabine Ludwigs[2] and Hagen Klauk[1]

[1]Max Planck Institute for Solid State Research, Stuttgart, Germany, email: J.Borchert@fkf.mpg.de
[2]Functional Polymers, Institute of Polymer Chemistry, Universität Stuttgart, Germany
[3]Institut für Mikroelektronik Stuttgart (IMS CHIPS), Germany
[4]Center for Nano Science and Technology @PoliMi, Instituto Italiano di Tecnologia (IIT), Milano, Italy
[5]Dipartimento di Elettronica, Informazione e Bioingegneria, Politecnico di Milano, Milano, Italy

Abstract—Organic thin-film transistors with record static and dynamic performance fabricated on flexible plastic substrates are presented. The TFTs operate with voltages of 3 V, have channel lengths as small as 0.6 µm and show on/off current ratios as large as 10^{10}, subthreshold slopes as steep as 59 mV/decade (at a temperature of 292 K), and signal delays measured in ring oscillators as low as 79 ns per stage. All these represent record figures for organic TFTs, enabled by small lateral device dimensions, a very small gate-dielectric thickness, and a small contact resistance of 29 Ωcm, which is also a record for organic field-effect transistors.

I. INTRODUCTION

Organic thin-film transistors (TFTs) can usually be fabricated at temperatures below 100 °C and thus on a variety of unconventional substrates, such as plastics, paper and textiles. However, organic TFTs have up to now been greatly limited by their high operating voltages (usually >10 V), poor static performance (on/off current ratio usually $<10^9$, subthreshold slope usually >80 mV/decade) and poor dynamic performance (signal delays >100 ns).

Here we demonstrate organic TFTs fabricated on low-cost, flexible, optically transparent polyethylene naphthalate (PEN) foil that have channel lengths as small as 0.6 µm, operate with gate-source and drain-source voltages of -3 V, and show on/off current ratios as large as 10^{10}, subthreshold slopes as steep as 59 mV/decade, signal propagation delays as low as 79 ns per stage in ring oscillators and 19 ns in individual inverters, and transit frequencies obtained from S-parameter measurements up to 6.7 MHz at gate-source and drain-source voltages of -3 V. These record performance characteristics are enabled mainly by a very small contact resistance of 29 Ωcm, which matches the contact resistance level of high-performance inorganic metal-oxide TFTs [1].

II. TFT FABRICATION

Organic TFTs can be fabricated either in the staggered or in the coplanar device architecture. Both architectures have advantages and disadvantages, but in terms of contact resistance, which is one of the critical factors limiting the dynamic performance of organic TFTs [2], the staggered architecture has typically provided better results [3,4].

However, recent device simulations by Zojer et al. [5] predict that coplanar organic TFTs will outperform otherwise identical staggered TFTs in terms of contact resistance (and thus also in terms of transit frequency) if the gate-dielectric thickness is sufficiently small, the energy barrier for charge injection at the metal-semiconductor interfaces is small, and the thin-film morphology of the organic semiconductor layer in the contact regions is similar to that in the channel region, so that efficient charge transfer between the contacts and the gate-induced carrier channel is facilitated.

To verify these simulation results, we have fabricated organic TFTs based on the small-molecule semiconductor dinaphtho-thienothiophene (DNTT) and its didecyl- and diphenyl-derivatives C_{10}-DNTT and DPh-DNTT [6] both in the bottom-gate, top-contact (inverted staggered) and the bottom-gate, bottom contact (inverted coplanar) architectures (see Fig. 1a), employing a gate dielectric with a thickness of 5.3 nm and a unit-area capacitance of 700 nF/cm². Stencil lithography based on high-resolution silicon stencil masks [7] was used to pattern all device layers (gate electrodes, source and drain contacts, organic semiconductor, interconnects) and define channel lengths as small as 0.6 µm and a total gate-to-contact overlap as small as 4 µm (see Fig. 1b).

For the bottom-contact (coplanar) TFTs, the key process step is the formation of a high-quality monolayer of pentafluorobenzenethiol (PFBT) on the surface of the Au source and drain contacts prior to the deposition of the organic semiconductor. Without this PFBT monolayer, the injection barrier would be too large and the semiconductor morphology on the contact surface would be different from that in the channel region, thus inhibiting charge transport across the interface [3]. Following the deposition of the source and drain contacts, the substrates were thus immersed into a PFBT solution, allowing a high-quality PFBT monolayer to form on the clean Au surfaces that induces a large interface dipole (minimizing the injection barrier) and reduces the surface energy of the Au contacts (enabling a favorable semiconductor thin-film morphology on the contact surface; see Fig. 1c,d). The highest process temperature is 90 °C, and all TFTs and circuits were fabricated on flexible polyethylene naphthalate (PEN) substrates. All measurements were performed in ambient air at room temperature (292 K).

III. STATIC TFT CHARACTERISTICS

Fig. 2 shows the measured transfer and output characteristics of a bottom-contact TFT with a channel length of 8 μm and a channel width of 200 μm. The transfer characteristics indicate an effective carrier mobility of 3.7 cm^2/Vs and an on/off current ratio of 10^{10}, which is the largest on/off current ratio reported to date for organic TFTs. The subthreshold slope is below 100 mV/decade over a range of four orders of magnitude in drain current and below 80 mV/decade over a range of three orders of magnitude in drain current, with a minimum of 59 mV/decade. This is the steepest subthreshold slope reported to date for an organic transistor measured at room temperature. For comparison, top-contact DPh-DNTT TFTs which we fabricated with the same dimensions have an effective carrier mobility of 3.8 cm^2/Vs, on/off ratios up to 10^9 and subthreshold slopes of 90 mV/decade.

Stencil lithography offers the possibility to fabricate small-channel-length organic TFTs on plastic substrates with high reproducibility. Fig. 3 shows an SEM image of a TFT with a channel length of 1.5 μm, a channel width of 7.5 μm and a total gate-to-contact overlap of 60 μm, as well as the measured transfer and output characteristics of 10 TFTs with these dimensions. These TFTs have a maximum width-normalized transconductance of 4.9 S/m, an average effective mobility of 5.2 ± 0.1 cm^2/Vs, on/off ratios above 10^8 and subthreshold slopes of 67 ± 5 mV/decade. These are the best values reported to date for organic TFTs on a plastic substrate with such a small channel length.

As mentioned above, the key requirement for a high transit frequency is a low contact resistance [1,2,4]. Using the transmission line method (TLM), we determined a contact resistance of 56 Ωcm for the top-contact TFTs, very similar to the record-low contact resistance reported by Yamamura et al. for this device architecture [4]. However, for the TFTs with the PFBT-treated bottom contacts, we obtained an even smaller contact resistance of 29 Ωcm (see Fig. 4), which is the smallest contact resistance reported to date for organic field-effect transistors; a smaller contact resistance has only been reported for organic electrochemical transistors [8]. This confirms both the beneficial effect of the PFBT contact treatment and the recent simulation results by Zojer et al. [5].

IV. DYNAMIC CHARACTERISTICS

For an 11-stage ring oscillator based on TFTs with a channel length of 1 μm and a total gate-to-contact overlap of 4 μm we have measured a signal propagation delay of 79 ns at a supply voltage of 4.4 V (see Fig. 5); this is the smallest signal propagation delay reported to date for an organic ring oscillator [9] and significantly smaller than the signal delay of the top-contact-TFT-based ring oscillators we fabricated for comparison, confirming the benefit of a small contact resistance for the dynamic TFT performance. Even for a supply voltage of 1.5 V, the signal delay is below 150 ns per stage.

By applying a square-wave signal to the input of an individual inverter based on TFTs with a channel length of 1 μm and a total gate-to-contact overlap of 4 μm and by fitting simple exponential functions to the transitions in the measured output signal, we have determined fall and rise time constants of 57 and 19 ns, respectively (see Fig. 6).

We have also performed S-parameter measurements [10] on individual TFTs with channel lengths of 0.6, 0.8 and 1 μm and with a total gate-to-contact overlap of 10 μm. The TFT with the smallest channel length (0.6 μm) shows an effective carrier mobility of 2 cm^2/Vs, which is the largest mobility reported to date for a submicron-channel-length organic transistor. For this TFT, the S-parameter measurements give a transit frequency of 6.7 MHz (see Fig. 7), which is the highest transit frequency reported to date for an organic transistor on a flexible substrate [11].

Normalized to the supply voltage of -3 V, it is also the highest voltage-normalized transit frequency reported to date for any organic transistor [4]. For organic transistors fabricated on glass substrates, the highest transit frequencies reported to date are 40 MHz at 8.6 V in a vertical permeable-base transistor [12], 27.7 MHz at 25 V in a bottom-gate, bottom-contact field-effect transistor with a channel length of 2 μm [13], and 20 MHz at 10 V in a bottom-gate, top-contact field-effect transistor with a channel length of 3 μm [4].

V. CONCLUSIONS

We have fabricated low-voltage (3 V) organic TFTs with channel lengths as small as 0.6 μm and carrier mobilities up to 6 cm^2/Vs in the bottom-gate, bottom-contact architecture on flexible plastic substrates. The TFTs show on/off ratios of 10^{10}, subthreshold slopes of 59 mV/decade, contact resistance of 29 Ωcm and signal delays of 79 ns in ring oscillators and 19 ns in inverters, all of which are records for organic field-effect transistors. S-parameter measurements indicate a transit frequency of 6.7 MHz, which is a record for flexible organic transistors.

REFERENCES

[1] N. Münzenrieder et al., *Appl. Phys. Lett.*, vol. 105, p. 263504, 2014.
[2] H. Klauk, *Adv. Electron. Mater.*, vol. 4, p. 1700474, 2018.
[3] D. J. Gundlach et al., *Nature Mater.*, vol. 7, p. 216, 2008.
[4] A. Yamamura et al., *Sci. Adv.*, vol. 4, p. eaao5758, 2018.
[5] K. Zojer et al., *Phys. Rev. Appl.*, vol. 4, p. 044002, 2015.
[6] U. Kraft et al., *Org. Electronics*, vol. 35, p. 33, 2016.
[7] T. Zaki et al., *IEEE J. Solid-State Circuits*, vol. 47, p. 292, 2012.
[8] D. Braga et al., *Appl. Phys. Lett.*, vol. 97, p. 193311, 2010.
[9] S. D. Ogier et al., *Org. Electronics*, vol. 54, p. 40, 2018.
[10] T. Zaki et al., *IEEE Electr. Dev. Lett.*, vol. 34, p. 520, 2013.
[11] S. G. Bucella et al., *IEEE. Trans. Electr. Dev.*, vol. 64, p. 1960, 2017.
[12] B. Kheradmand-Boroujeni et al., *Sci. Rep.*, vol. 8, p. 7643, 2018.
[13] M. Kitamura et al., *Jpn. J. Appl. Phys.*, vol. 50, p. 01BC01, 2011.

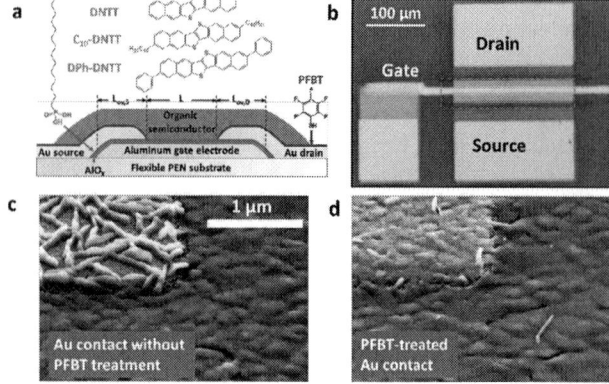

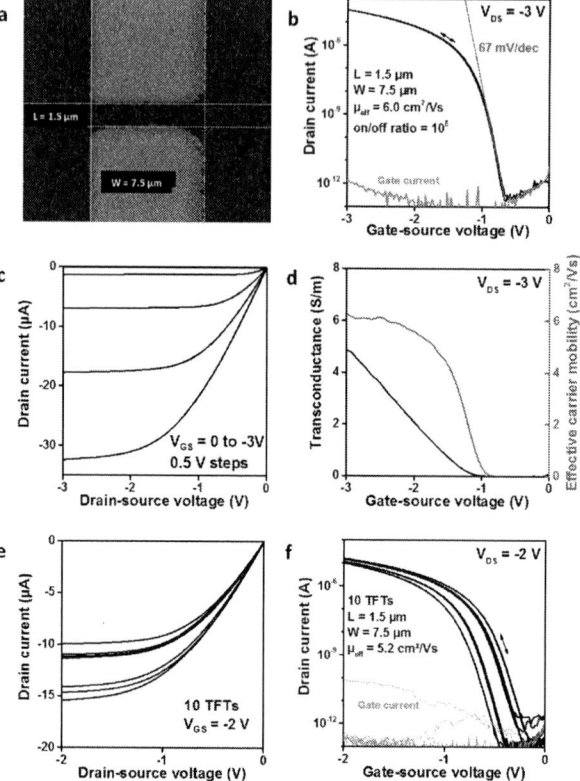

Figure 1. (a) Schematic cross-section of the organic TFTs fabricated in the bottom-gate, bottom-contact (inverted staggered) architecture on flexible, 125-μm-thick polyethylene naphthalate (PEN) foil.
(b) Photograph of a TFT with a channel length of 8 μm.
(c) SEM image of the edges of a Au contact without PFBT treatment after the deposition of the organic semiconductor layer. The organic semiconductor layer clearly shows poor film morphology on the surface of the untreated gold contact.
(d) SEM image of the edges of a PFBT-treated Au contact after the deposition of the organic semiconductor layer. Owing to the PFBT treatment and the associated decrease in the surface energy of the contact metal, the organic semiconductor layer shows significantly improved thin-film morphology on the contact surface and across the contact edges.

Figure 3. (a) SEM image of a TFT with L = 1.5 μm and W = 7.5 μm.
(b) Transfer and (c) output characteristics of a DPh-DNTT TFT with these dimensions.
(d) Width-normalized transconductance and effective mobility of the same TFT plotted as a function of the gate-source voltage.
(e) Output and (f) transfer characteristics of 10 such TFTs.

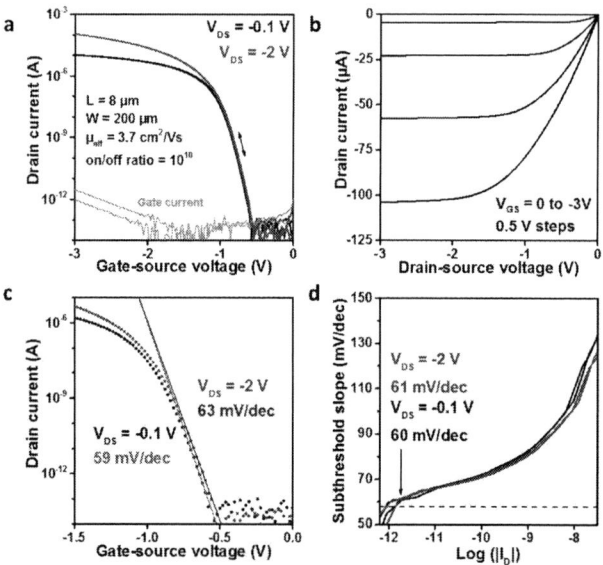

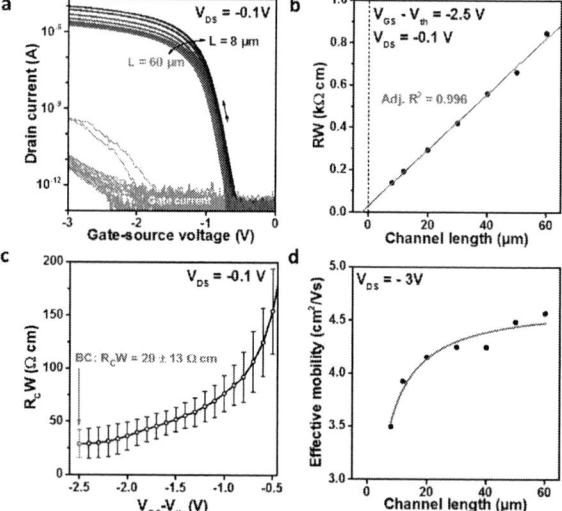

Figure 2. (a) Transfer characteristics of a DPh-DNTT TFT with a channel length of 8 μm and a total gate-to-contact overlap of 4 μm, showing an on/off current ratio of 10^{10} and a subthreshold slope of 59 mV/decade, both of which are records for organic TFTs.
(b) Output characteristics of the same TFT.
(c) Subthreshold slope evaluated by an exponential fit of the subthreshold region of the transfer curves.
(d) Pointwise calculation of the subthreshold slope as a function of the log of the drain current. The dashed line indicates the theoretical minimum of 58 mV/decade at the temperature at which the measurement was performed (292 K).

Figure 4. (a) Transfer characteristics of DPh-DNTT TFTs with channel lengths ranging from 8 to 60 μm employed for the TLM analysis.
(b) Total device resistance at a gate-overdrive voltage (V_{GS}-V_{th}) of -2.5 V.
(c) Width-normalized contact resistance as a function of V_{GS}-V_{th}. At V_{GS}-V_{th} = -2.5 V, the contact resistance is 29 Ωcm, which is the smallest contact resistance reported to date for organic TFTs.
(d) Effective carrier mobility plotted as a function of the channel length. The fit curve corresponds to Equation (1) in reference [6].

978-1-7281-1988-5/18 $31.00 © 2018 IEEE

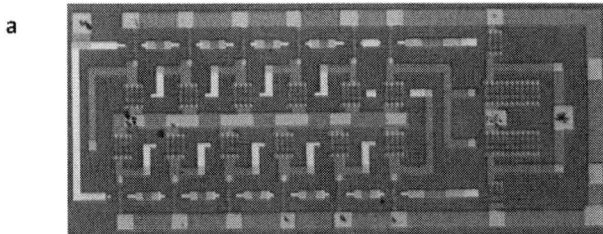

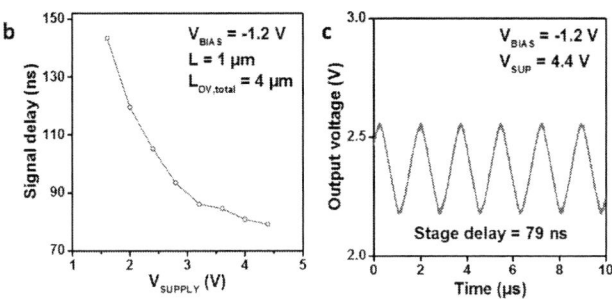

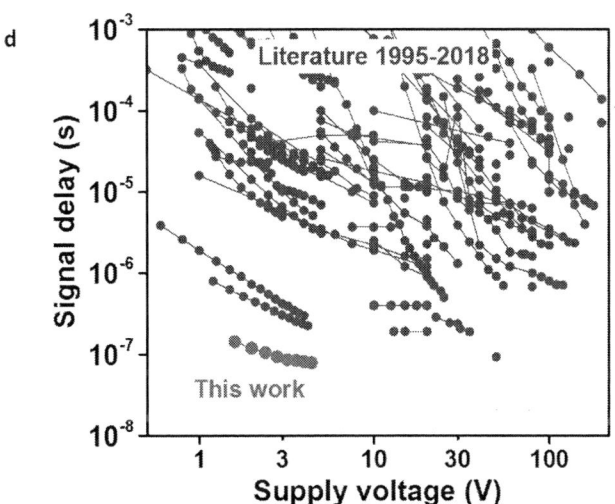

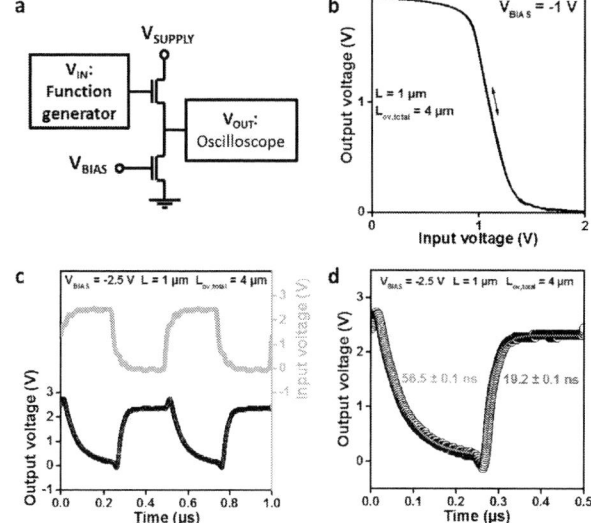

Figure 6. (a) Schematic diagram of the dynamic-measurement setup for determining the signal delay of an individual inverter.

(b) Static transfer characteristics of a biased-load inverter based on TFTs with L = 1 μm and $L_{ov,total}$ = 4 μm.

(c) In- and output waveforms recorded in a dynamic measurement performed on an inverter based on TFTs with these dimensions.

(d) Fit of the measured output waveform using simple exponential decay functions, showing rise and fall times of 19 ns and 57 ns.

Figure 5. (a) Photograph of an 11-stage ring oscillator based on TFTs with a channel length of 1 μm and a total gate-to-contact overlap of 4 μm fabricated on a flexible PEN substrate. The inverters utilize the biased-load circuit design.

(b) Signal propagation delay per stage of the ring oscillator measured at supply voltages ranging from 1.5 to 4.4 V.

(c) Output signal measured at a supply voltage of 4.4 V, showing a signal delay of 79 ns per stage, which is the shortest stage delay reported to date for an organic-TFT-based ring oscillator.

(d) Comparison with the signal delays of all organic-TFT-based ring oscillators reported to date.

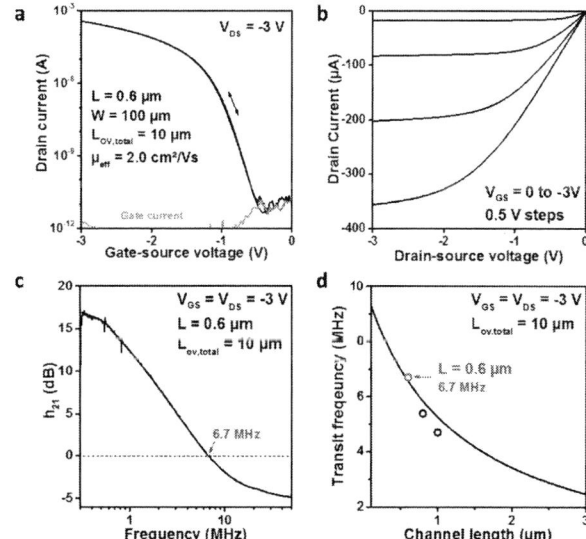

Figure 7. (a) Measured transfer characteristics of a TFT with L = 0.6 μm and $L_{ov,total}$ = 10 μm.

(b) Output characteristics of the same TFT.

(c) Result of an S-parameter measurement performed on the same TFT, indicating a transit frequency of 6.7 MHz, which is the highest transit frequency reported to date for flexible organic transistors.

(d) Transit frequencies obtained from S-parameter measurements performed on TFTs with channel lengths of 0.6, 0.8 and 1 μm. The curve corresponds to the calculated transit frequency using Equation (1) in reference [10], using the effective carrier mobility calculated for each channel length taking into account the contact resistance (45 Ωcm), the intrinsic channel mobility (4.4 cm²/Vs) and the threshold voltage (-1.2 V) of these TFTs.

978-1-7281-1988-5/18 $31.00 © 2018 IEEE

Hybrid Structure of Silicon Nanocrystals and 2D WSe$_2$ for Broadband Optoelectronic Synaptic Devices

Zhenyi Ni[1a], Yue Wang[1a], Lixiang Liu[2], Shuangyi Zhao[1], Yang Xu[2*], Xiaodong Pi[1*], and Deren Yang[1]

[1] State Key Laboratory of Silicon Materials and School of Materials Science and Engineering, Zhejiang University, Hangzhou, Zhejiang 310027, China

[2] College of Information Science and Electronic Engineering, Zhejiang University, Hangzhou, Zhejiang 310027, China

Emails: yangxu-isee@zju.edu.cn; xdpi@zju.edu.cn. [a] These authors have equal contributions

Abstract—As one of the most important technologies in the coming "More than Moore" era, neuromorphic computing critically depends on the development of synaptic devices. Here we take advantage of the synergy of the strong broadband optical absorption of boron (B)-doped silicon nanocrystals (Si NCs) and the efficient charge transport of two-dimensional (2D) WSe$_2$ to make synaptic devices based on the hybrid structure of Si NCs and 2D WSe$_2$. The Si-NC/WSe$_2$ synaptic devices can be optically stimulated in a broad spectral region from the ultraviolet (UV) to near-infrared (NIR), exhibiting important synaptic functionalities. The energy consumption of the Si-NC/WSe$_2$ synaptic devices may be as low as $\sim$ 75 fJ. This work has important implication for the development of synaptic devices by exploiting the abundant library of semiconductor NCs and 2D materials.

I. INTRODUCTION

Computers based on the conventional von Neumann architecture will soon struggle to meet the dramatically growing demand for highly energy-efficient and intelligent computing [1]. This demand currently leads to great efforts on the development of neuromorphic computing, which critically relies on high-performance synaptic devices mimicking the transmission of information between neurons in a biological neural system (Fig. 1).

Among all types of synaptic devices thin-film-transistor (TFT) synaptic devices have attracted significant attention given their excellent compatibility with the metal-oxide-semiconductor (CMOS) technology. It has been recently shown that two-dimensional (2D) materials such as graphene and MoS$_2$ may be used to fabricate TFT synaptic devices [2, 3], facilitating the realization of the ultralow energy consumption of the devices. Significantly, optical stimulation has been carried out for the TFT synaptic devices based on 2D materials. The optical stimulation has the great potential of broadening the bandwidth and mitigating the interconnect issues for synaptic devices [2-8]. However, the optical stimulation has been only performed with the ultraviolet (UV) and visible light up to now. In addition, the light absorption of 2D materials is routinely weak, rendering the requirement of the high-power optical stimulation. This actually gives rise to rather high energy consumption for the synaptic devices. It has been recently found that silicon nanocrystals (Si NCs) can efficiently absorb light from the UV to near-infrared (NIR) after they are heavily doped with boron (B) [9]. Such a finding encourages the

hybridization of Si NCs with 2D materials. The resulting hybrid structures feature the synergy of the efficient charge transport of 2D materials and the strong broadband optical absorption of Si NCs. It should be noted that the Si-NC-enabled optical stimulation in the NIR region may significantly facilitate the coupling of neuromorphic computing with optical communication.

In this work we use 2D WSe$_2$ to form a hybrid structure with B-doped Si NCs. The synaptic devices based on this hybrid structure can be optically stimulated in a broad spectral region from the UV to NIR. They exhibit important synaptic functionalities with rather low energy consumption.

II. DEVICE ARCHITECTURE

Fig. 2 shows the fabrication process of the Si-NC/WSe$_2$ synaptic devices. Cr/Au electrodes were first deposited to form the 10 µm × 120 µm (L × W) channels on a Si wafer with a 150 nm-thick SiO$_2$ layer. Mechanically exfoliated WSe$_2$ with a lateral length of $\sim$ 30 µm was then transferred onto each channel. B-doped Si NCs dispersed in ethanol was subsequently drop-cast on WSe$_2$ to form the hybrid structure of Si-NC/WSe$_2$. The device structure is schematically presented in Fig. 3. The atomic force microscopy image of WSe$_2$ in Fig. 4 shows that WSe$_2$ is $\sim$ 5 nm, indicating a $\sim$ 6-layer structure. The Si-NC film is $\sim$ 1.53 µm thick, as evidenced by the cross-section scanning electron microscopy image in Fig. 4. Si NCs used in this work were synthesized by non-thermal plasma [9], effectively absorbing light in a rather wide range from the UV to NIR ($\sim$ 350 – 2000 nm) (Fig. 5). The mean size of Si NCs is $\sim$ 6 nm, as shown in the inset of Fig. 5. Fig. 6 shows the red shift of the E_{2g}^1 peak at $\sim$ 248 cm^{-1} and the A_{1g} peak at 257 cm^{-1} for WSe$_2$ after the drop-casting of Si NCs [10]. Moreover, the transfer curve of the TFT shifts to the left after the incorporation of Si NCs onto WSe$_2$ (Fig 7). Both indicate that charge (hole) transfer from WSe$_2$ to Si NCs occurs when they from the hybrid structure. The resulting band alignment between Si NCs and WSe$_2$ is shown in the inset of Fig. 7.

III. RESULTS AND DISCUSSION

Fig. 8 shows the excitatory postsynaptic current (EPSC) of the Si-NC/WSe$_2$ synaptic device evoked by laser spikes in the spectral region from the UV to NIR. During the measurements the drain voltage and gate voltage are + 4 V and + 1V, respectively. The WSe$_2$ is depleted to minimize the energy consumption and increase the sensitivity to optical stimulation for the device. The values of the EPSC decay time (τ) for 375, 532 and 1342 nm laser spikes are ~ 18, 16 and 44 ms, respectively. They fall in the range for the typical decay time of a signal in a biological synapse. Since the bandgap of WSe_2 is ~ 1.31 eV in the current work, the EPSC of the device evoked by the 1342 nm laser spike exclusively originates from the B-doping-induced band-tail optical absorption of Si NCs and the subsequent transfer of the photogenerated holes from Si NCs to WSe_2. With the increase of the duration time of the spike, the EPSC initially increases and then exhibits a tendency of saturation (Fig. 9).

One of the most characteristic manifestations of the short-term plasticity (STP) of a synaptic device is its paired-pulse facilitation (PPF). Fig. 10 representatively shows the PPF with the interval time (Δt) of 40 ms between two successive laser spikes. The PPF index that stands for the ratio of the EPSC evoked by the second laser spiking to that evoked by the first one is a function of Δt (Fig. 11). The PPF index decays with the increase of Δt for all the laser spiking in the UV-to-NIR region. The decay of the PPF index is basically determined by the comparison of Δt with τ. Since τ varies with the laser wavelength, it is observed that the PPF index decays differently as the laser wavelength changes.

The long-term memory (LTP) of the Si-NC/WSe_2 synaptic device is evaluated by calculating the synaptic weight change (ΔS) (the EPSC evoked by the last spike compared with that evoked by the first one) for the stimulation with different spike numbers (Fig. 12) or frequencies (Fig. 13). It is seen that ΔS increases with the increase of the spike number or frequency, implying a STP to LTP transition of the Si-NC/WSe_2 synaptic device. The spike-timing-dependent plasticity (STDP) of the device is characterized by calculating the ΔS between the stimulations of the presynaptic and postsynaptic spikes with an interval time of $\Delta t_{pre-post}$. Fig. 14 shows the STDP behavior of the device when the laser spikes with the same wavelength are used for the presynaptic and postsynaptic stimulations. In this case, ΔS exhibits a symmetric distribution with respect to $\Delta t_{pre-post} = 0$, indicating an ideal symmetric Hebbian learning rule [11, 12]. In contrast, when the laser spikes with different wavelengths are used for the presynaptic and postsynaptic stimulations, ΔS shows an asymmetric distribution with respect to $\Delta t_{pre-post} = 0$ (Fig. 15). This is due to the different values of τ for laser spikes with difference wavelengths.

The energy consumption per synaptic event of optically stimulated synaptic devices can be calculated by using $dE = S \times P \times dt$ (S is the area of the device, P is the power density of the input light at the spike duration of t). Our calculation indicates that the Si-NC/WSe_2 synaptic device may work with rather low energy consumption. Among the three laser wavelengths considered in the current work the laser wavelength of 532 leads to the lowest energy consumption of the device because of the best balance between the optical absorption and photogenerated carrier diffusion in the Si-NC film. The P of the 532 nm laser is 1.25×10^{-3} $\mu W/cm^2$. For the shortest spike duration time of 20 ms in the current work the energy consumption per synaptic event of the Si-NC/WSe_2

synaptic device is ~ 75 fJ. This energy consumption can be further reduced to ~ 2.5 aJ by scaling down the dimension of the device to be similar to the size of a biological synapse, *i.e.*, ~ 100 nm. We have compared the current Si-NC/WSe_2 synaptic device with those reported in literature in terms of the energy consumption, as shown in Fig. 16. The wavelengths of light that has been employed to stimulate synaptic devices are additionally summarized in Table I. Clearly, the current Si-NC/WSe_2 synaptic device extends the optical stimulation into the NIR region for the first time. Therefore, the present results demonstrate that the Si-NC/WSe_2 synaptic device may prevail over previous optically stimulated synaptic devices given both the energy consumption and the spectral range of optical stimulation. The present hybridization strategy has important implication for exploiting the abundant library of 0D semiconductor NCs and 2D materials.

IV. CONCLUSIONS

We have fabricated optoelectronic synaptic devices by using the hybrid structure of B-doped Si NCs and 2D WSe_2. The optical absorption of B-doped Si NCs not only enables the synaptic devices to work with low-power optical stimulation, but also extends the optical stimulation of the synaptic devices into the NIR region. The current work signifies the great potential of the development of synaptic devices by using the hybrid structures of semiconductor NCs and 2D materials.

ACKNOWLEDGMENT

This work is supported by the National Key Research and Development Program of China (Grant No. 2017YFA0205700) and the NSFC (Grant Nos. 61774133 and 61674127).

REFERENCES

[1] M.M. Waldrop, "The chips are down for Moore's law," *Nature*, vol. 530, pp. 144-147, 2016.

[2] S. Qin, F. Wang, Y. Liu, Q. Wan, et al., "A light-stimulated synaptic device based on graphene hybrid phototransistor" *2D Mater.*, vol. 4, pp. 035022, 2017.

[3] R. A. John, F. Liu, et al., "Synergistic Gating of Electro-Iono-Photoactive 2D Chalcogenide Neuristors: Coexistence of Hebbian and Homeostatic Synaptic Metaplasticity" *Adv. Mater.*, vol. 30, pp. 1800220. 2018.

[4] S. Dai, X. Wu, D. Liu, Y. Chu, et al., "Light-Stimulated Synaptic Devices Utilizing Interfacial Effect of Organic Field-Effect Transistors," *ACS Appl. Mater. Inter.*, vol. 10, pp. 21472−21480, 2018.

[5] K. Pilarczyk, A. Podborska, M. Lis, M. Kawa, et al., "Synaptic Behavior in an Optoelectronic Device Based on Semiconductor-Nanotube Hybrid," *Adv. Electron. Mater.*, vol. 2, pp. 1500471, 2016.

[6] H. K. Li, T. P. Chen, P. Liu, S. G. Hu, et al., "A light-stimulated synaptic transistor with synaptic plasticity and memory functions based on $InGaZnO_x$–Al_2O_3 thin film structure," *J. Appl. Phys.*, vol. 119, pp. 244505, 2016.

[7] M. Lee, W. Lee, et al., "Brain-Inspired Photonic Neuromorphic Devices using Photodynamic Amorphous Oxide Semiconductors and their Persistent Photoconductivity" *Adv. Mater.*, vol. 29, pp. 1700951, 2017.

[8] X. Zhu, and W.D. Lu, "Optogenetics-Inspired Tunable Synaptic Functions in Memristors," *ACS Nano*, vol. 12, pp. 1242-1249, 2018.

[9] Z. Ni, L. Ma, S. Du, Y. Xu, et al., "Plasmonic Silicon Quantum Dots Enabled High-Sensitivity Ultrabroadband Photodetection of Graphene-Based Hybrid Phototransistors," *ACS Nano*, vol. 11, pp. 9854-9862, 2017.

[10] E. Corro, H. Terrones, A. Elias, C. Fantini, et al., "Excited Excitonic States in 1L, 2L, 3L, and Bulk WSe_2 Observed by Resonant Raman Spectroscopy," *ACS Nano*, vol. 8, pp. 9629−9635, 2014.

[11] D. Kuzum, S. Yu, and H.-S.P. Wong, "Synaptic electronics: materials, devices and applications," *Nanotechnology*, vol. 24, pp. 382001, 2013.

[12] Z. Xiao, and J. Huang, "Energy-Efficient Hybrid Perovskite Memristors and Synaptic Devices" *Adv. Electron. Mater.*, vol. 2, pp. 1600100, 2016.

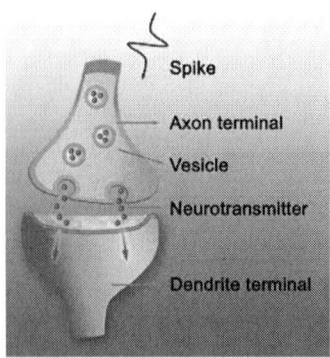

Fig. 1. Schematic diagram of a biological synapse.

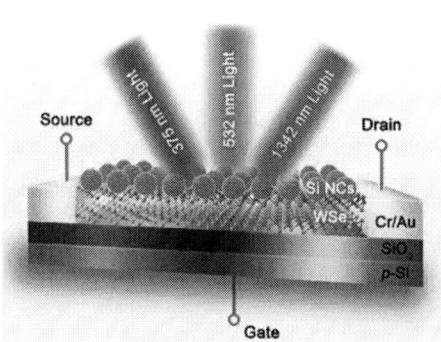

Fig. 2. Fabrication process of Si-NC/WSe₂ optoelectronic synaptic devices.

Fig. 3. Schematic of the Si-NC/WSe₂ synaptic device structure. During the measurements the drain voltage and gate voltage are +4 V and +1 V, respectively.

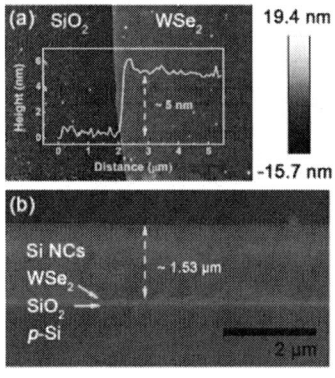

Fig. 4. (a) Atomic force microscopy image of 2D WSe₂ on a SiO₂/Si wafer with a line-scan profile. (b) Cross-section scanning electron microscopy image of the Si-NC film on WSe₂/SiO₂/Si.

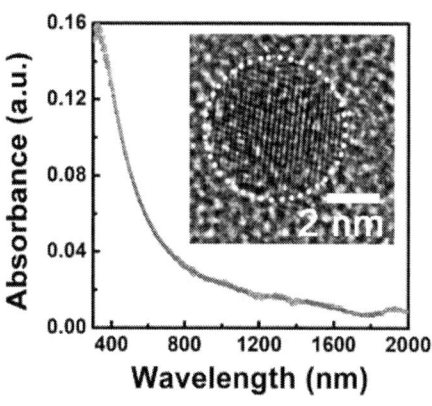

Fig. 5. Optical absorption spectrum of B-doped Si NCs. The inset shows the high-resolution transmission electron microscopy image of a ~ 6 nm B-doped Si NC.

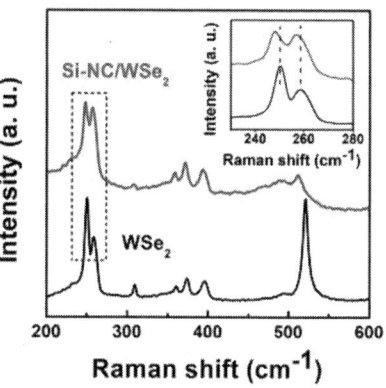

Fig. 6. Raman spectra of WSe₂ and Si-NC/WSe₂. The Raman shifts in the dashed rectangle are enlarged in the inset.

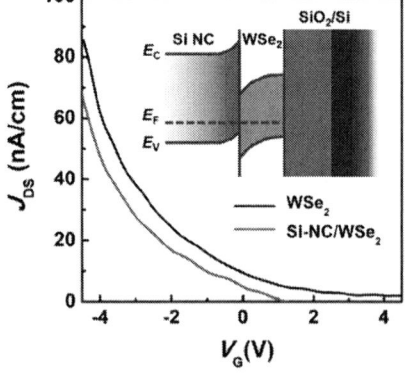

Fig. 7. Transfer curves of WSe₂ and Si-NC/WSe₂ TFTs. The inset shows the band alignment between Si NCs and WSe₂.

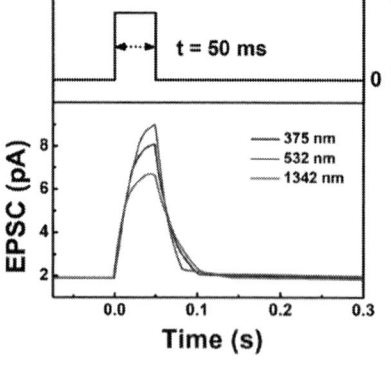

Fig. 8. EPSC of the Si-NC/WSe₂ synaptic device induced by laser spikes with different wavelengths.

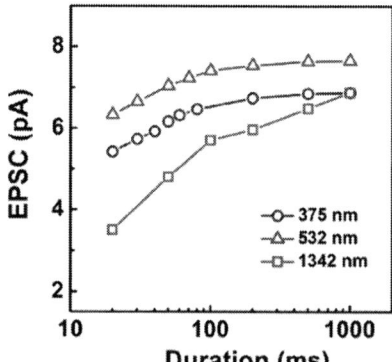

Fig. 9. Dependence of the EPSC on the duration time of the laser spikes with different wavelengths.

978-1-7281-1988-5/18 $31.00 © 2018 IEEE

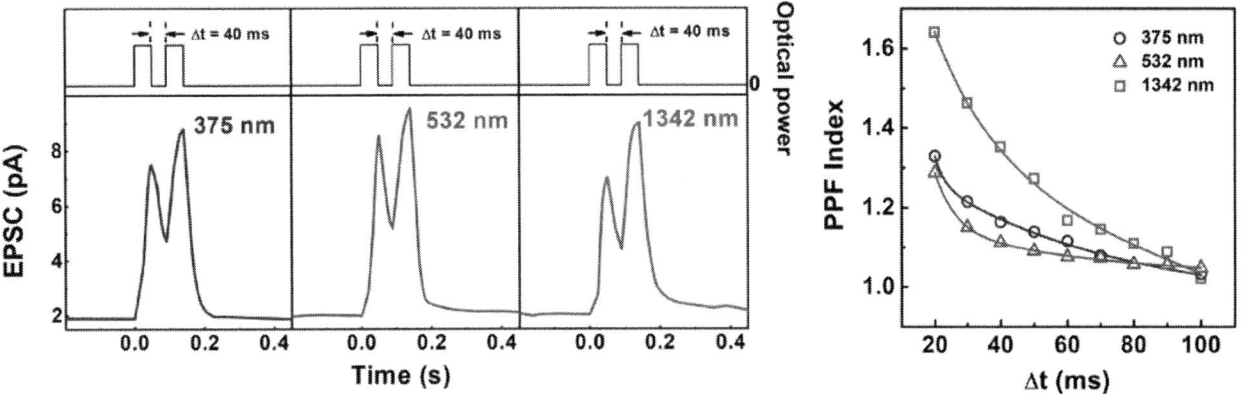

Fig. 10. EPSC of the Si-NC/WSe₂ synaptic device induced by two successive laser spikes with an interval time (Δt) of 40 ms.

Fig. 11. Dependence of the PPF index on Δt for the laser spikes with different wavelengths.

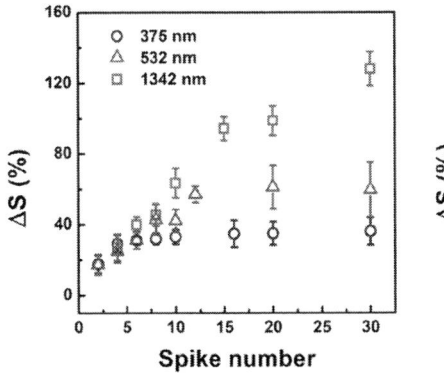

Fig. 12. Dependence of the synaptic weight change (ΔS) on the number of spikes for the Si-NC/WSe₂ synaptic device.

Fig. 13. Dependence of the synaptic weight change (ΔS) on the spike frequency for the Si-NC/WSe₂ synaptic device.

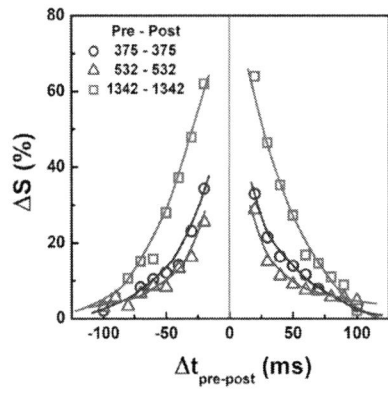

Fig. 14. Variation of the STDP-induced ΔS with respect to Δt$_{pre-post}$. The pre-synaptic and post-synaptic spikes are enabled by the lasers with the same wavelength.

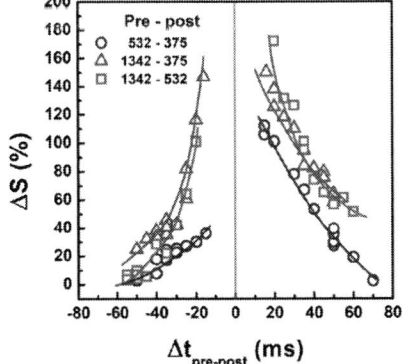
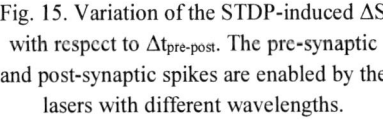

Fig. 15. Variation of the STDP-induced ΔS with respect to Δt$_{pre-post}$. The pre-synaptic and post-synaptic spikes are enabled by the lasers with different wavelengths.

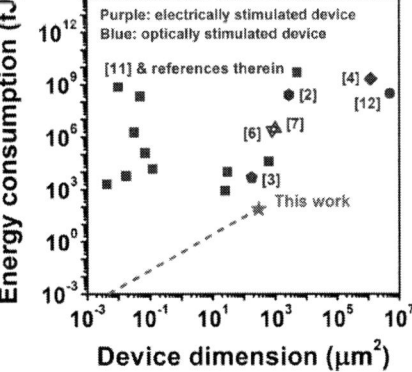

Fig. 16. Comparison of the current Si-NC/WSe₂ synaptic device (red star) with those reported in literature in terms of energy consumption

Table I. Wavelengths of stimulating light and available energy consumption for optically stimulated synaptic devices.

Structure	Spike wavelength (nm)	Energy consumption (J)	Ref.
Si-NC/WSe₂	375, 532 and 1342	75	This work
Carbon nanotubes/ graphene	405 and 532	2.50×10^8	[2]
MoS₂	445	4.80×10^3	[3]
C8-BTBT	360	2.16×10^9	[4]
CdS/carbon nanotubes	465	/	[5]
InGaZnOₓ-Al₂O₃	365	2.40×10^6	[6]
InGaZnOₓ/ InSrZnOₓ/ InSrOₓ/ InZnOₓ	380/460/ 530/630	3.00×10^6	[7]
CH₃NH₃PbI₃	500 and 635	/	[8]

978-1-7281-1988-5/18 $31.00 © 2018 IEEE

High Performance 2D Perovskite/Graphene Optical Synapses as Artificial Eyes

He Tian[1,*], Xuefeng Wang[1], Fan Wu[1], Yi Yang[1*], Tian-Ling Ren[1*]

[1]Institute of Microelectronics and Beijing National Research Center for Information Science and Technology (BNRist),
Tsinghua University, Beijing 100084, China
*Email: tianhe88@tsinghua.edu.cn, yiyang@tsinghua.edu.cn, RenTL@tsinghua.edu.cn

Abstract—Conventional von Neumann architectures feature large power consumptions due to memory wall. Partial distributed architecture using synapses and neurons can reduce the power. However, there is still data bus between image sensor and synapses/neurons, which indicates plenty room to further lower the power consumptions. Here, a novel concept of all distributed architecture using optical synapse has been proposed. An ultrasensitive artificial optical synapse based on a graphene/2D perovskite heterostructure shows very high photo-responsivity up to 730 A/W and high stability up to 74 days. Moreover, our optical synapses has unique reconfigurable light-evoked excitatory/inhibitory functions, which is the key to enable image recognition. The demonstration of an optical synapse array for direct pattern recognition shows an accuracy as high as 80%. Our results shed light on new types of neuromorphic vision applications, such as artificial eyes.

I. INTRODUCTION

Conventional von Neumann architectures feature large power consumptions due to large part of data movement between CPU and memory [1]. In order to overcome this drawback, neuromorphic computation with large scale synapses and neurons have been proposed, which can be regarded as a partial distributed architecture [2]. Here, inspired by retinal processes, we propose a new concept of an all distributed architecture (**Fig. 1**) by using optical synapses combined with neuron, which can further minimize data movement and reduce power consumption. Synapses play a key role between neurons as conveyers of information. Biological synapses in the retina have shown interesting light-evoked excitatory and inhibitory synaptic behaviors [3], which are key to enable image sensing and recognition. Basically, light-evoked excitatory (inhibitory) synapse means that the post-synaptic current (PSC) will increase (decrease) after the application of a light pulse, which is the basic component in retina. It was reported that the α ganglion cells [4] in dark-adapted mouse retinas under voltage-clamp conditions have shown light evoked excitatory cation current and inhibitory chloride current. Different kinds of electrical artificial synapses have been developed recently. However, most of the artificial synaptic devices operate under electrical input [5, 6]. An artificial synapse with light-evoked excitatory and inhibitory functions remains elusive.

Here, we show an optical synapse with a unique reconfigurable ability by using a layered 2D perovskite (Ruddlesden-Popper phase) material. Perovskites with mixed ionic and electronic transport represent a good analogy to the ionic transport in biological synapse. Moreover, large 2D perovskite crystals can be dry-transferred to form a sandwich structure with graphene. A graphene covering on a 2D perovskite layer provides good stability. The electron-hole pairs in 2D perovskite can be efficiently separated by the vertical graphene/2D perovskite/graphene structure, whereas the charge redistribution in 2D perovskite can be "read out" by the bottom graphene layer. Both short-term and long-term behaviors are investigated. Our device shows great potential to function as an "artificial eye" and might find application in fields such as robotic vision.

II. DEVICE FABRICATION

The device structure of the 2D perovskite-based optical synapse is shown in **Fig. 2** and fabricated as follows. The graphene on the bottom serves as a channel with ambipolar transport, which can be re-configured by switching the type of carrier collected. The layered 2D $(PEA)_2PbI_4$ single-crystal (PEA refers to phenethylammonium) perovskite was dry transferred onto the graphene in a N_2 glovebox (H_2O and O_2 less than 0.1 ppm). Finally, few-layer graphene is covered on the 2D perovskite to form a graphene/2D perovskite/graphene sandwich structure (**Fig. 2**). Conventional optical devices (i.e. photodetectors) are either based on in-plane or vertical structure. Our optical synaptic device uses a unique in-plane and vertical combined structure. This approach has two advantages. (1) In the vertical direction, electron-hole pairs can be effectively collected by an applied bias. Moreover, by tuning the polarity, the device can be reconfigured to collect electrons and holes in opposite directions. (2) In the lateral direction, the bottom graphene provides suitable amibpolar transport for detection. The left panel of **Fig. 3** shows the bio synapse in the retina region with the Ca^{2+} ions releasing under light illumination, which can pass information from one neuron to another. The 2D perovskite-based optical synapse makes a good analogy to the bio synapse in the retina (right panel of **Fig. 3**). In 2D perovskite, light absorption mobilizes I- ions (or iodide vacancies) which can be subsequently transported by the applied E-field. The transmission electron microscopy (TEM) cross-section and EDS line profile of sandwich structure is shown in **Fig. 4**.

III. PHOTO-DETECTION MEASUREMENT

In order to test the optical performance and understand the

978-1-7281-1988-5/18 $31.00 © 2018 IEEE

operation of the perovskite-based optical synapse, 520 nm laser with minimum 7.4 nW power is irradiated on the device. The device is measured in a vacuum chamber at a pressure of less than 10^{-4} Torr. In the dark, the transfer curve of the device is similar to a regular graphene FET. The photoresponsivity at -500 mV reaches ~730 A/W at 4 V gate voltage and -615 A/W at 20 V gate voltage (**Fig. 5**). The photoresponsivity in this device is higher than most previously reported solution-based 3D perovskite photodetectors. Photoresponsivity vs. power under -500 mV bias also shows a linear relation in log-scale in **Fig. 6**. The operation mechanism of bias polarization dependent behaviors are shown in the inset of **Fig. 6**. Negative bias can repel negative ions I⁻ moving toward bottom graphene surface. The accumulation of negative I⁻ ionic charges can be regarded as local gate for graphene FET, which can induce more holes in bottom graphene and make the transfer curve right shift. The stability of the 2D perovskite optical synapse is monitored with no obvious photo current change in 74 days (**Fig. 7**), which proves the good stability of the 2D perovskite due to the protection of graphene. The on/off light performance has also been measured (**Fig. 8**). At negative bias -500 mV condition, we have observed relative fast photo-response speed with 0.08 s rising time (**Fig. 9**), which is suitable for photo-sensing applications. Comparing with previous reported perovskite-based photodetector, the optical synapse shows very high photo-responsivity, long stability (**Fig. 10**). Moreover, the short-term and long-term decay of the photocurrent can enable the device operated as optical synapse (as discussed in the next section), which has rarely been investigated before.

IV. OPTICAL SYNAPSE OPERATION

Paired-pulse facilitation (PPF) effect is a very important short-term behavior in bio synapses which can be also mimicked by our optical synapse. As shown in **Fig. 10**, two optical pulses are applied to our device at V_g=10 V and V_{bias}=500 mV. It is expected to have higher peak of PSC after the second light pulse. As shown in **Fig. 11**, the second peak is higher than the first peak and the PPF index (The ratio of A_2/A_1) is 107.9% with 450 ms delay time. The PPF indexs with different delay time are also measured and summarized in **Fig. 12**. It shows a higher PPF index with a shorter delay time. Moreover, the PPF indexes drop faster at smaller time delay while drop slower at larger time delay, which indicates the present of two time constant. Spike-timing dependent plasticity (STDP) is also measured as long-term behavior, and we demonstrated a light and electrical coupled STDP behavior. As shown in **Fig. 13**, the light pulse is applied on the optical synapse as pre-synaptic input and electrical pulse is applied at drain as post-synaptic input. If the electrical signal is arrived before the light pulse 217 ms (**Fig. 14**), it shows LTP. While if the electrical signal is arrived after the light pulse 105 ms, it shows LTD. Different time delay between electrical signal and light pulse are summarized in **Fig. 15**. The time constant for pre- before post- synaptic inputs case is 37 ms while the pre- after post- synaptic inputs case is 96 ms. Time constants in the

tens of millisecond of our optical synapses can match well to the typical bio systems. Continuous potentiation and depression under light pulse is the key function to enable image recognition. **Fig. 16** and **Fig. 17** shows the continuous potentiation and depression under -500 mV and 500 mV respectively. The decay of the current shows two time constants. The short-term time constant can be related to electron or hole movements, while the long-term time constant can be related by the ionic movements (**Fig. 18**). A synaptic network with both excitatory and inhibitory synapses has been proposed (**Fig. 19**). **Fig. 20** shows the AFM image of two devices connecting in-parallel working as synaptic network. **Fig. 21** shows the overall current influenced by the synapse 1 and 2 under different V_g conditions with light input. The optical synapse in mouse retina can be tuned into excitatory or inhibitory by voltage clamp. Our optical synapse can also mimic tunable excitatory or inhibitory by gate voltage with very similar behaviors (**Fig. 22**). Moreover, the image recognition has been simulated based on a two-layer neural network to demonstrate the concept of artificial eyes with all distributed architecture (**Fig. 23**). **Fig. 24** shows the evolution accuracy vs. training epoch number. The accuracy can reach to ~80% based on such direct image recognition architecture.

V. CONCLUSION

In summary, for the first time, we have proposed the all distributed architecture based on optical synapses. An optical synapse based on layered 2D perovskite is demonstrated with unique configurability. Our reconfigurable optical synapses with 2D perovskite has very good analogy to bio optical synapse with light-evoked excitatory/inhibitory functions. It shows ultrahigh photo-responsivity up to 730 A/W based on unique photo-gating effect. The bio PPF, STDP behaviors, and synaptic network with both excitatory and inhibitory synapses have been demonstrate. Based on this optical synapse, the accuracy for direct image recognition is up to 80% by simulation. This artificial optical synapse provides a good path to realize neuromorphic vision and recognition in the future.

ACKNOWLEDGMENT

This work was supported by National Key R&D Program (2016YFA0200400), National Natural Science Foundation (61574083, 61434001), Beijing Natural Science Foundation (4184091) and National Basic Research Program (2015CB352101) of China. The authors are also thankful for the support of the Research Fund from Beijing Innovation Center for Future Chip, the Independent Research Program of Tsinghua University (2014Z01006) and Shenzhen Science and Technology Program (JCYJ20150831192224146).

REFERENCES

[1] P.A. Merolla et al., *Science*, vol. 345, pp. 668, 2014. [2] D. Lelmini et al., *Nature Electronics*, vol. 1, pp. 333, 2018. [3] T. Euler et al., *Nature*, vol. 418, pp. 845, 2002. [4] J.J. Pang et al., Journal of Neuroscience, vol 23, pp. 6063, 2003. [5] D Kuzum et al., *Nanotechnology*, vol. 24, pp. 382001, 2013. [6] H. Tian et al., *Nano Letters*, vol. 15, pp. 8013, 2015.

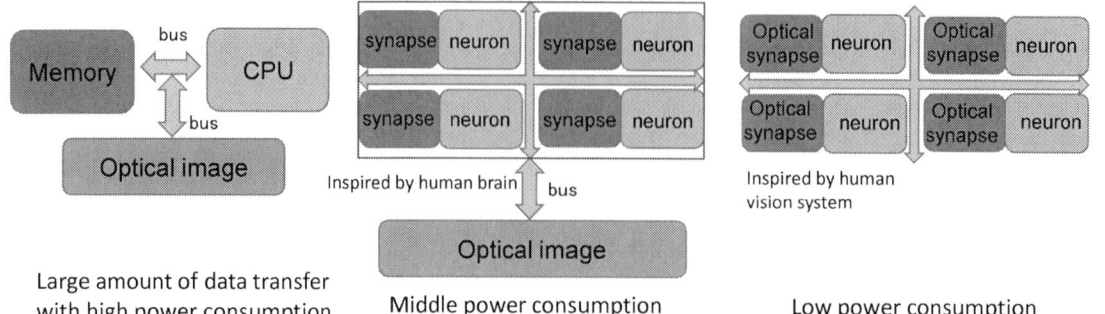

a Von Neumann Architecture **b** Partial distributed architecture **c** Our novel all distributed architecture

Large amount of data transfer with high power consumption   Middle power consumption   Low power consumption

Fig. 1 The schematic showing (a) Conventional Von Neumann architecture; (b) Synapse and neuron based partial distributed architecture; (c) Novel all distributed architecture enabled by replacing synapse into optical synapse. Most of the power consumes during the data movement at the data bus. Our novel all distributed architecture can operate without data bus. By combing optical synapse and neuron, direct image recognition can be realized with very low power.

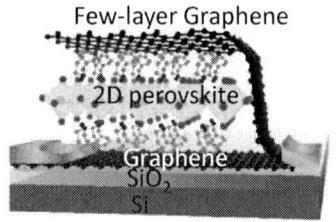

Fig. 2 Schematic of the optical synaptic device. The 520 nm laser pulses are applied as input synaptic signal. Output signals are measured at the drain side as post-synaptic terminal.

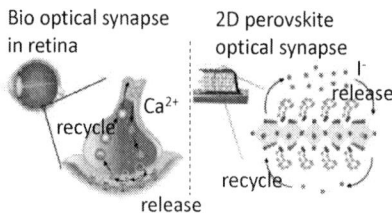

Fig. 3 The schematic showing the good analogy between bio optical synapse in retina region and 2D perovskite optical synapse.

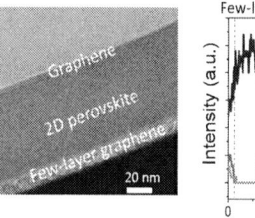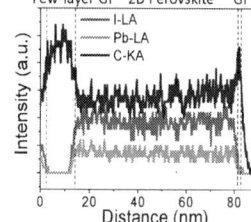

Fig. 4 The TEM cross-section of the graphene/2D perovskite/few-layer graphene heterostructure and the EDS line profile along the cross-section of the graphene/2D perovskite/few-layer graphene heterostructure.

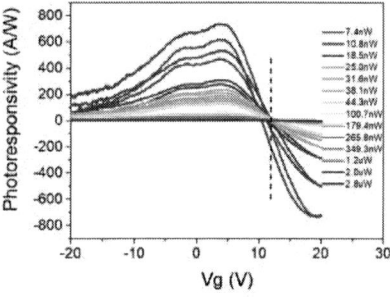

Fig. 5 The photoresponsivity vs. gate voltage showing high photoresponsivity up to 730 A/W has been achieved at -500 mV bias conditions.

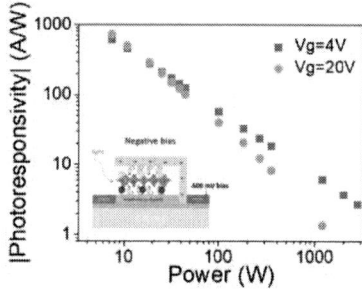

Fig. 6 The photoresponsivity vs. power showing the liner relation in log-scale at -500 mV bias conditions.

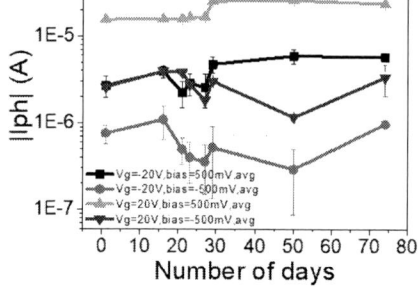

Fig. 7 The photocurrent vs. days showing great stability of the 2D perovskite synapses in 74 days.

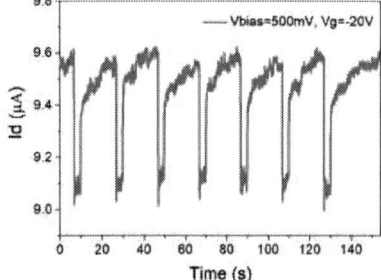

Fig. 8 The light on/off performance showing realatity good photo sensing speed.

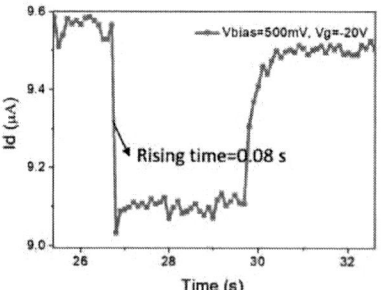

Fig. 9 The zoom-in image showing the fast rising speed of the photo-current.

Table I Performance comparision with other perovskite-based photodetectors. Our device has very outstanding photoresponsivity and stabiltiy.

Photosensitive material	R(A/W)	T$_r$(s)	Stability (hours)	Reference
Graphene/MAPbI$_3$	180	0.087	None	Lee et al., 2015.
Graphene/MAPbI$_3$	115	0.25	None	Dang et al., 2016.
rGO/MAPbI$_3$	0.0739	0.4	None	He et al., 2015.
MAPBI$_3$	1	20x10^{-6}	None	Li et al., 2015
MAPBI$_3$	203	None	500	Dong et al., 2015
MAPbI$_{3-x}$Cl$_x$	14.5	0.2x10^{-6}	100	Guo et al., 2015
Gr/2D perovskite/Gr	730	0.08	1776	This work

978-1-7281-1988-5/18 $31.00 © 2018 IEEE   893

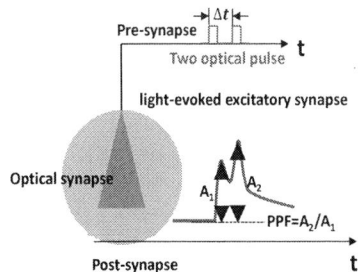

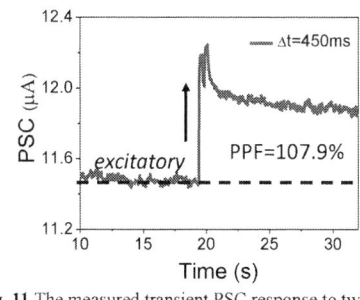

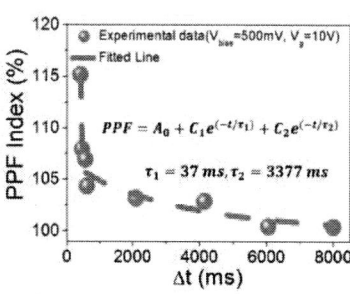

Fig. 10 The measurement setup for light-evoked excitatory synapse under 500 mV bias condition.

Fig. 11 The measured transient PSC response to two pulses with 450 ms delay time showing PPF effect.

Fig. 12 The summarized relation between PPF index and delay time at 500 mV bias and Vg=10 V condition.

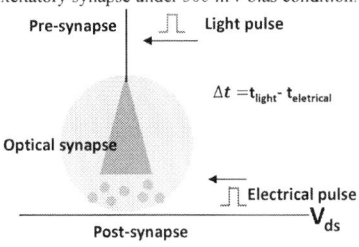

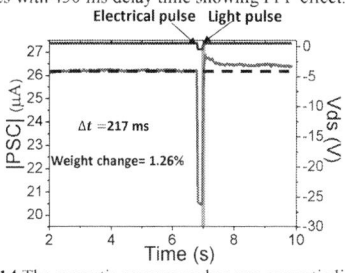

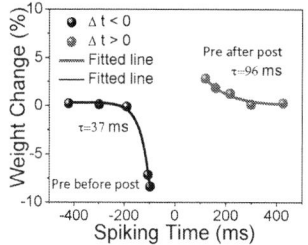

Fig. 13 The measurement setup by coupling light pulse and electrical pulse as pre- and post-synaptic input.

Fig. 14 The synaptic response when pre-synaptic light pulse is after the electrical pulse with 217 ms delay time showing LTP.

Fig. 15 The summarized weight change vs. spiking time showing similar STDP behavior to bio synapses.

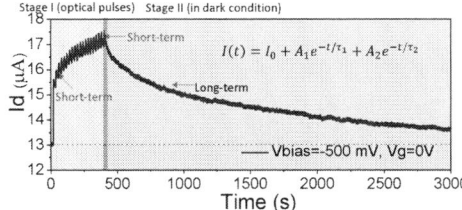

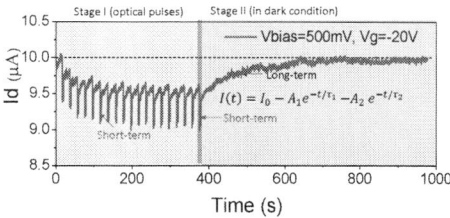

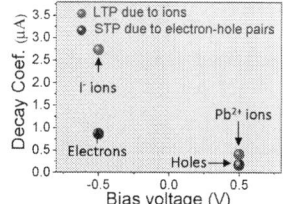

Fig. 16 Continuously light pulses applied to the optical synapses followed by keeping the device in dark condition at -500 mV bias condition. It shows continuously potentiation during the 20 pulses with instant short-term decay and slow long-term decay in dark condition.

Fig. 17 Continuously light pulses applied to the optical synapses followed by keeping the device in dark condition at 500 mV bias condition. It shows continuously depression during the 20 pulses with instant short-term decay and slow long-term decay in dark condition.

Fig. 18 The decay coefficent vs. bias voltage.

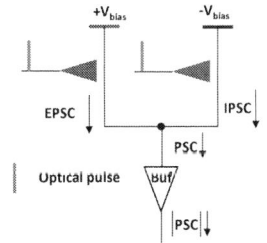

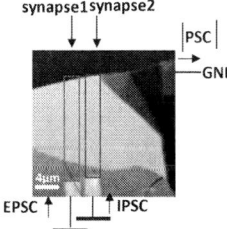

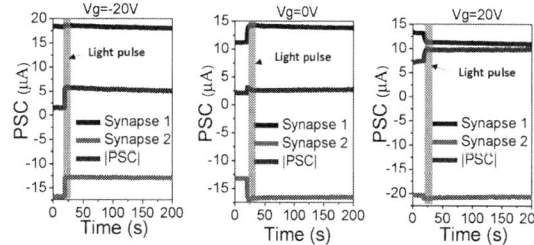

Fig. 19 The connection of showing both excitatory and inhibitory synapses are connected to the network.

Fig. 20 The AFM image of the synaptic network with related electrical connections.

Fig. 21 Synaptic network with both excitatory and inhibitory synapses. The output synaptic current after light pulse under Vg=-20 V, 0 V and 20 V conditions.

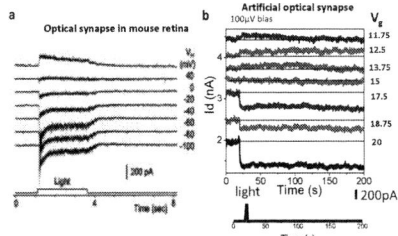

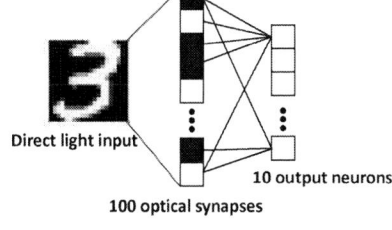

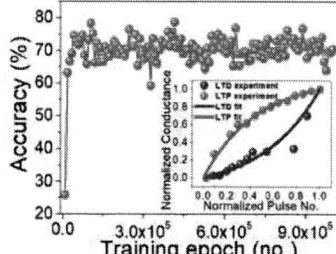

Fig. 22 (a) The bio synaptic behaviors measured in mouse retina [4] under ~3s light pulse by tuning voltage clamp V_H. (b) The mimicked synaptic behaviors of optical synapse with 520 nm green ~3 s light pulse by tuning V_g.

Fig. 23 A two layer neural network with a black-and-white input light signal for optical syaptic layer in the algorithm level to demo the concept of artificial eye with all distributed architecture.

Fig. 24 Accuracy vs. training epoch number based on our optical synapse for imaging recognition. The inset shows the training curves used for optical learning process.

978-1-7281-1988-5/18 $31.00 © 2018 IEEE

First Transistor Demonstration of Thermal Atomic Layer Etching: InGaAs FinFETs with sub-5 nm Fin-width Featuring *in situ* ALE-ALD

Wenjie Lu[1], Younghee Lee[2], Jessica Murdzek[2], Jonas Gertsch[2], Alon Vardi[1], Lisa Kong[1], Steven M. George[2], and Jesús A. del Alamo[1]

[1]Microsystems Technology Laboratories, MIT, Cambridge, MA, 02139, USA;
[2]Department of Chemistry and Biochemistry, University of Colorado, Boulder, CO, 80309, USA
E-mail: wenjie@mit.edu

Abstract—For the first time, thermal atomic layer etching (ALE) on InGaAs-based III-V heterostructures is demonstrated. Also, we report the first transistors fabricated by the thermal ALE technique in any semiconductor system. We further highlight one unique advantage of thermal ALE: its integration with atomic layer deposition (ALD) in a single vacuum chamber. Using *in situ* ALE-ALD, we have fabricated the most aggressively scaled self-aligned $In_{0.53}Ga_{0.47}As$ n-channel FinFETs to date, featuring sub-5 nm fin widths. The narrowest FinFET with $W_f = 2.5$ nm and $L_g = 60$ nm shows $g_m = 0.85$ mS/µm at $V_{ds} = 0.5$ V. Devices with $W_f = 18$ nm and $L_g = 60$ nm demonstrate $g_m = 1.9$ mS/µm at $V_{ds} = 0.5$ V. Subthreshold swings averaging $S_{lin} = 70$ mV/dec and $S_{sat} = 74$ mV/dec across the entire range of W_f, at minimum $L_g = 60$ nm have been obtained. These are all record results. The transistors demonstrated here show an average 60% g_m improvement over devices fabricated through conventional techniques. These results suggest a very high-quality MOS interface obtained by the *in situ* ALE-ALD process.

I. INTRODUCTION

As CMOS technology continues to scale down and device structures become more three-dimensional, manufacturing challenges compound. In recent years, 3D MOSFETs with sub-10 nm physical dimensions have been demonstrated in various material systems, such as Si, SiGe, and III-V's [1-5]. Further scaling progress demands fabrication technologies with Ångstrom-scale precision and fidelity. This is out of reach for mainstream plasma etching and wet digital etch techniques.

Atomic layer etching (ALE) is a novel technique that removes materials using sequential self-limiting processes [6-8]. There are two types of ALE. One uses energetic ions or neutrals, commonly assisted by plasma, and the etching is usually anisotropic. The other type is based on the chemical ligand-exchange, and it enables isotropic etching. This is usually referred to as "Thermal ALE", and its reaction sequence closely resembles that of an ALD process. Thermal ALE is still in its youth, and reports on thermal ALE are limited to etching of dielectrics, metals and some nitrides [9-11]. To our knowledge, there are no device demonstrations to date.

In this work, we report on the development of the first thermal ALE process for InGaAs-InAlAs heterostructures. InGaAs is a promising channel material for CMOS scaling and memory applications [12, 13]. The performance of advanced InGaAs FinFETs is still lacking, partly due to limitations in the

MOS stack quality [14]. Thermal ALE is a breakthrough technology that can address these problems. In this work, we demonstrate: (1) precise and highly controllable etching rate at Ångstrom/cycle-scale, (2) plasma-free conformal sidewall etching resulting in low damage and smooth surfaces, (3) material selectivity to enable fabrication of gate-all-around (GAA) structures, and, most importantly, (4) integration of ALE and ALD in an *in situ* process that completely prevents air exposure of the gate oxide-semiconductor interface. These unique attributes enable innovative transistor designs with remarkable ON and OFF-state performance.

We illustrate the device worthiness of our ALE technique by fabricating the most aggressively scaled InGaAs FinFETs to date with fin widths as narrow as 2.5 nm. Record device characteristics highlight the extraordinary device potential of the *in situ* thermal ALE-ALD process.

II. THERMAL ATOMIC LAYER ETCHING

Fig. 1 shows the schematic of the viscous flow ALD reactor in which both the thermal ALE and ALD processes are performed. **Fig. 2** shows the InGaAs/InAlAs heterostructure used to develop the thermal ALE process. It consists of 30 and 40 nm $In_{0.53}Ga_{0.47}As$ (with different doping) on an $In_{0.52}Al_{0.48}As$ buffer layer, on (100) InP substrate. **Fig. 3** shows the sequence of a complete cycle of thermal ALE of InGaAs. The first step is surface fluorination using HF-pyridine. The second step is a ligand-exchange process to remove the metal fluoride layer. For this we use dimethylaluminum chloride (DMAC) [15] at a partial pressure of 40 mTorr. The volatile metal etch products are then purged away, and the sequence is repeated in cycles. The entire process is performed at 300 °C. The reactor has a baseline vacuum of 5-10 mTorr and a working pressure of 1 Torr with N_2 flow. **Fig. 4** displays X-ray reflectivity scans of the substrate before and after 200 and 450 cycles of thermal ALE, showing an average etch rate for InGaAs of 0.21 Å/cycle for the first 200 cycles and 0.16 Å/cycle for the last 250 cycles.

Detailed ALE etch rate calibrations are obtained from fins and vertical nanowires (VNW) etched on the heterostructure of **Fig. 2** by RIE in a $BCl_3/SiCl_4/Ar$ ICP plasma [16]. After RIE, the samples are etched by thermal ALE for 250 cycles. We observe (**Fig. 5**) an average radial etch rate for $In_{0.53}Ga_{0.47}As$ and $In_{0.52}Al_{0.47}As$ of ~0.2 and ~0.6 Å/cycle, respectively, with smooth substrate and sidewall surfaces.

The 3:1 etching selectivity between InGaAs and InAlAs ALE can be exploited to create suspended InGaAs GAA fins or

978-1-7281-1988-5/18 $31.00 © 2018 IEEE

vertical nano-sheet MOS structures. The heterostructure is shown in **Fig. 7a**. **Fig. 6** shows TEM images of 50 nm tall InGaAs fins obtained by 250 cycles of ALE, with Al_2O_3 deposition *in situ* in the same reactor, finished by W metal ALD in a separate reactor. Fins with $W_f < 24$ nm are fully suspended, consistent with the ALE selectivity measured above. Fins as narrow as 3-4 nm are obtained. **Fig. 6** shows a remarkably sharp interface between InGaAs and Al_2O_3.

III. INGAAS FINFET FABRICATION

The starting heterostructure and cross-section schematics of the finished devices are illustrated in **Fig. 7**. The channel layer consists of 50 nm thick $In_{0.53}Ga_{0.47}As$ lattice matched to an InAlAs buffer on an InP substrate. A δ-doping layer is placed 5 nm below the channel ($N_d = 4 \cdot 10^{12}$ cm^{-2}). A 30 nm n$^+$-$In_{0.53}Ga_{0.47}As$ cap is placed above a 4 nm InP etch stopper.

Fig. 8 outlines the fabrication process. The process starts with Mo/W sputtering for the ohmic contacts. Then, CVD SiO_2 is deposited as contact spacer and hard mask to etch the gate foot. This is defined by e-beam lithography. After SiO_2 and Mo/W RIE and mesa lithography, the heavily doped InGaAs cap is recessed. This is performed in two steps. First, timed RIE is used to remove most of the InGaAs cap. This is followed by a 5 s citric acid:H_2O_2 wet etch to expose the InP etch stopper with minimal lateral etching. This results in a highly self-aligned geometry, with less than 5 nm extrinsic region to minimize series resistance (**Fig. 9a**).

The process follows with fin patterning by e-beam lithography using HSQ as the mask. 220 nm tall fins are etched in $BCl_3/SiCl_4/Ar$ ICP plasma [16]. Then, 4 cycles of alcohol-based digital etch, using methanolic H_2SO_4 and O_2 plasma, are carried out to shrink the fin width [17]. Following this, the samples are introduced into the ALE/ALD reactor. First, 162 cycles of thermal ALE are performed at 300 °C, as described in the previous section. After this, the substrate temperature is reduced to 250 °C, followed by ALD deposition of 3 nm of HfO_2 (EOT ≈ 0.8 nm). In a separate ALD reactor (due to limited precursor lines), 30 nm of W are deposited at 130 °C as the gate metal. **Fig. 9b** shows devices after gate stack formation. Fins with final $W_f < 10$ nm are suspended. The process continues with the gate head being defined by e-beam lithography and W patterning by SF_6/O_2 RIE. A backend process composed of inter-level dielectric (ILD) deposition, via etch and pad metallization completes the device fabrication.

Final W_f and L_g of the devices are measured by TEM and SEM, respectively. The final W_f ranges from 2.5 nm to 18 nm, L_g between 60 nm to 1 μm. **Fig. 10** shows the TEM cross-section of a finished device with $W_f = 2.5$ nm. The inset shows a close-up image of the upper portion of a fully suspended InGaAs channel. In the next section, unless indicated otherwise, all metrics are normalized by total conducting gate periphery.

IV. ELECTRICAL CHARACTERISTICS

Fig. 11 and **12** show electrical characteristics of the most scaled InGaAs FinFET with $W_f = 2.5$ nm and $L_g = 60$ nm (AR $= H_c/W_f = 20$). Classic MOSFET behavior is obtained, showing

excellent $S_{lin} = 62$ mV/dec, $S_{sat} = 68$ mV/dec, and DIBL = 40 mV/V. A maximum g_m of 0.85 mS/μm is obtained at $V_{DS} = 0.5$ V. This is the InGaAs FinFET with the thinnest fin width and highest aspect ratio ever demonstrated. **Fig. 13** shows electrical characteristics of a FinFET with $W_f = 6$ nm and $L_g = 60$ nm. It shows $S_{lin} = 61$ mV/dec, $S_{sat} = 72$ mV/dec, and DIBL = 50 mV/V. In the widest device ($W_f = 18$ nm, $L_g = 60$ nm), we demonstrate maximum $g_m = 1.9$ mS/μm at $V_{DS} = 0.5$ V (**Fig. 14**). In all these devices, the OFF-state current is limited by gate leakage.

Fig. 15 summarizes the scaling behavior of peak g_m ($V_{DS} = 0.5$ V) with W_f and L_g. Compared with InGaAs FinFETs fabricated without ALE (same heterostructure and EOT) [3], a consistent improvement of ~60% in peak g_m is obtained for $L_g = 60$ nm. **Fig. 16** summarizes S_{lin} and S_{sat} ($V_{DS} = 0.05$, 0.5 V, respectively) of $L_g = 60$ devices of different W_f, together with identical InGaAs FinFETs fabricated without ALE, with and without δ-doping under the channel [3]. Average $S_{lin} = 70$ mV/dec and $S_{sat} = 74$ mV/dec are obtained across the entire range of W_f at $L_g = 60$nm. **Fig. 17** shows the scaling of S_{lin} with L_g at $W_f = 9$-10 nm. A remarkable improvement in S in ALE-fabricated devices is obtained with values in the 60-80 mV/dec range and weak sensitivity to W_f. The extraordinary enhancement in electrostatic control confirms the very high interface quality obtained by *in situ* ALE-ALD.

Fig. 18 shows DIBL ($V_{DS} = 0.05$ and 0.5 V) vs. L_g for $W_f = 6$-7 nm and 18-19 nm. Excellent improvement in short-channel effects and electrostatics are demonstrated. **Fig. 19** shows the scaling behavior of R_{on} at fixed $V_{GS} = 0.6$ V, together with R_{on} of δ-doped and undoped FinFETs without ALE treatment ($L_g = 40 - 60$ nm). The very tight self-aligned process developed here yields a lower R_{on} that increases weakly as W_f decreases.

Fig. 20 benchmarks peak g_m, normalized by conducting gate periphery, and g_m/W_f, normalized by fin footprint, for InGaAs FinFETs from this work and [3], and from the literature ($V_{DD} = 0.5$ V), vs. W_f. For reference, g_m of Intel's Si FinFETs is also shown ($V_{DD} = 0.8$ V for 1st gen, and $V_{DD} = 0.7$ for the 2nd and 3rd gen). With the usual caveats when making comparisons of this kind, our InGaAs FinFETs match the performance of Intel's 14 nm node ($W_f = 7$ nm), in spite of the lower V_{DD} and longer L_g. At $W_f = 2.5$ nm, this work shows a record $g_m/W_f > 30$ mS/μm. Given that this is the first demonstration of III-V MOSFETs by ALE and the first demonstration of working III-V FinFETs at $W_f < 5$ nm, this work displays the great promise for both ALE technology and III-V devices.

V. CONCLUSIONS

We have developed thermal ALE for III-V heterostructures and demonstrated a FinFET fabrication process that incorporates thermal ALE in combination with *in situ* ALD to form the gate stack. To our knowledge, this is the first demonstration of thermal ALE in a transistor of any kind. We achieve the most aggressively scaled InGaAs FinFETs with $W_f = 2.5$ nm and a record AR ~ 20 among all existing FinFETs. *In situ* thermal ALE/ALD yields remarkable improvements of device performance and electrostatic control. Record g_m/W_f have been obtained in these devices.

Acknowledgement: This work was sponsored in part by DTRA (#HDTRA1-14-1-0057), SRC (#2016-LM-2655), and Lam Research. Devices fabrication was performed at the Microsystems Technology Laboratories and SEBL at MIT. The University of Colorado authors acknowledge support from Intel Corporation through SRC and additional support from NSF (CHE-1609554).

Reference: [1] C. Auth, *IEDM*, 2017. [2] P. Hashemi, *VLSI*, 2016. [3] A. Vardi, *IEDM*, 2017. [4] H. Hahn, IEDM, 2017. [5] W. Lu, *IEDM*, 2017. [6] S. M. George and Y. Lee, *ACS Nano*, 2016. [7] C. T. Carver, *JSS*, 2015 [8] K. J. Kanarik, *JVST. A*, 2015. [9] K. Ishikawa, *JJAP*, 2016. [10] Y. Lee and S. M. George, *Chem. Mater.*, 2017. [11] N. R. Johnson, *JVST. A*, 2016 [12] J. A. del Alamo, *Nature*, 2011. [13] E. Capogreco, *IEDM*, 2015. [14] J. A. del Alamo, *CSW*, 2018. [15] Y. Lee, *Chem. Mater.*, 2016. [16] X. Zhao and J. A. del Alamo., *EDL*, 2014. [17] W. Lu, *EDL*, 2017.

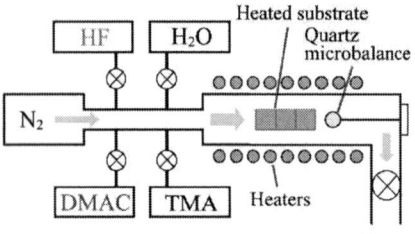

Fig. 1. Schematic of the hot wall viscous flow reactor used for ALE and ALD.

Fig. 2. Starting hetero-structure for the development of InGaAs/InAlAs thermal ALE process. The same structure is used for fin and VNW fabrication by ALE of Fig. 5.

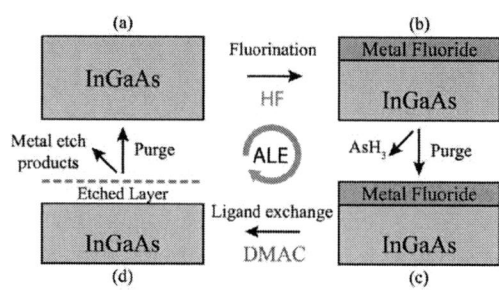

Fig. 3. Schematic presentation of the InGaAs thermal ALE process: (a)-(b) fluorination of InGaAs surface with HF. (c)-(d) ligand-exchange process by DMAC to remove the metal fluoride layer. The volatile etch products are then purged away.

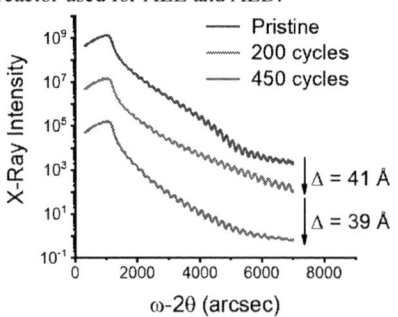

Fig. 4. X-ray reflectivity scan of InGaAs heterostructure after 200 and 450 cycles of DMAC/HF thermal ALE at 300°C. The average etch rate is 0.18 Å/cycle.

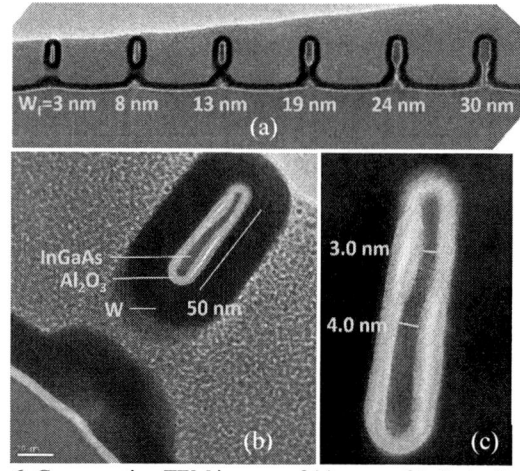

Fig. 6. Cross-section TEM images of (a) array of InGaAs fins (W_f = 3-30 nm) fabricated by *in situ* ALE-ALD process, (b) InGaAs suspended fin with minimum W_f of 3 nm, and (c) close-up image of the fin in (b).

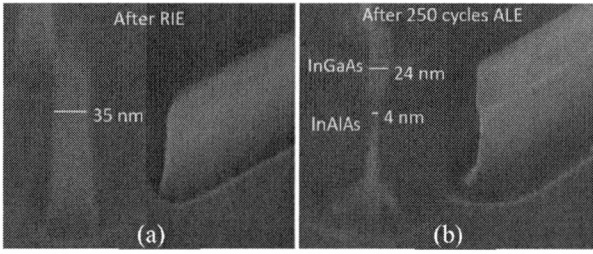

Fig. 5. InGaAs/InAlAs VNW and fin structures (a) after RIE (VNW has initial diameter of 35 nm), and (b) after 250 cycles of thermal ALE at 300 °C. The final diameter of InGaAs VNW is 24 nm (r = 0.2 Å/cycle), and that of InAlAs is 4 nm (r = 0.6 Å/cycle). The fin sidewall shows a smooth surface after the ALE process.

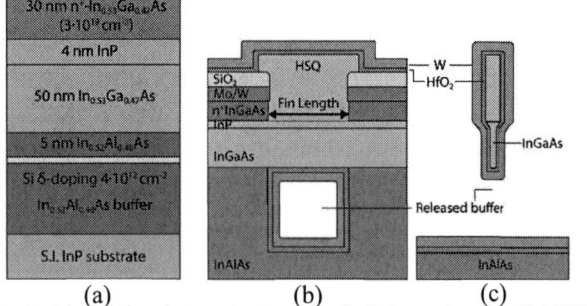

Fig. 7. (a) Starting heterostructure for InGaAs n-channel FinFET fabrication. Cross-section schematics of FinFETs: (b) along the fin length direction and (c) across the fin. For narrow fin widths, the InGaAs channel is fully suspended.

○ Sputtered Mo/W ohmic contact
○ CVD SiO₂ contact spacer/hard mask
● Gate EBL
○ Gate recess: SiO₂ & Mo RIE
● Mesa lithography, SiO₂ & Mo RIE
○ Gate recess I: timed RIE
○ Gate recess II: timed wet etch
● Fin EBL & Fin RIE
○ Alcohol-based digital etch
● Atomic layer etching
● In-situ ALD HfO₂ deposition
○ ALD W gate metal deposition
● Gate head photo and patterning
○ CVD SiO₂ ILD deposition
○ Via opening & Pad formation

Fig. 8. Process flow for InGaAs FinFET fabrication.

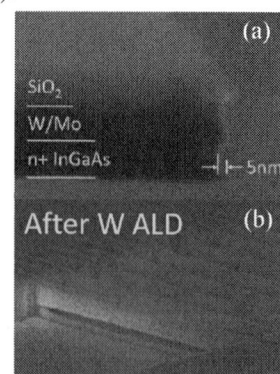

Fig. 9. SEM images of FinFETs after (a) gate recess, and (b) ALE-ALD gate process.

978-1-7281-1988-5/18 $31.00 © 2018 IEEE

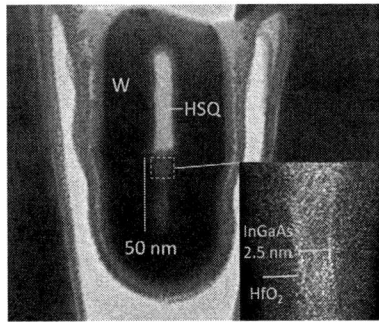

Fig. 10. Cross-section TEM image of finished device with $W_f = 2.5$ nm. Inset: close-up image of upper portion of fin.

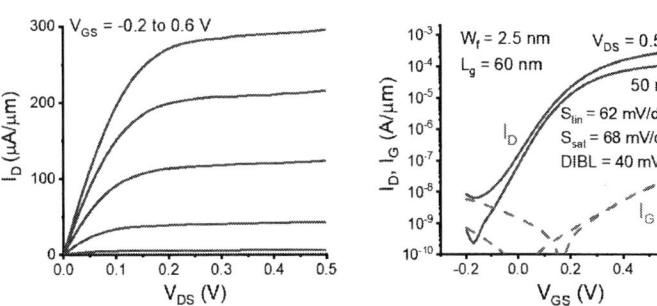

Fig. 11. (Left) output and (right) subthreshold characteristics of the most scaled InGaAs FinFETs with $W_f = 2.5$ nm and $L_g = 60$ nm.

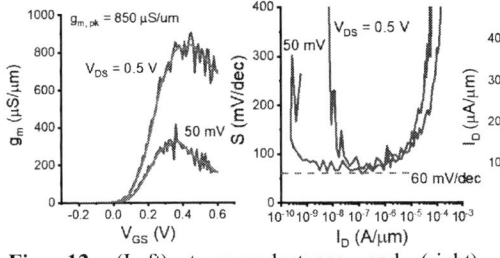

Fig. 12. (Left) transconductance and (right) subthreshold swing characteristics of InGaAs FinFETs with $W_f = 2.5$ nm and $L_g = 60$ nm. Maximum $g_m = 0.85$ mS/μm, and minimum $S_{lin} = 62$ mV/dec and $S_{sat} = 68$ mV/dec are obtained.

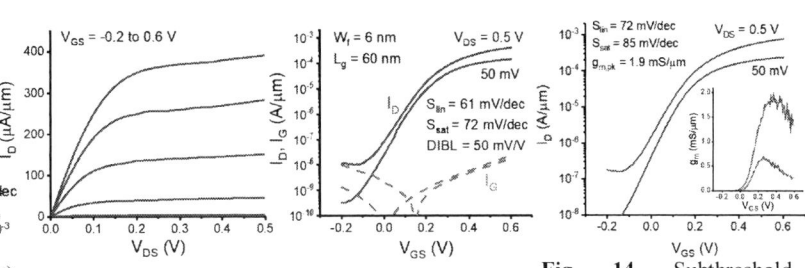

Fig. 13. (Left) output and (right) subthreshold characteristics of InGaAs FinFETs with $W_f = 6$ nm and $L_g = 60$ nm.

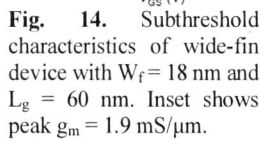

Fig. 14. Subthreshold characteristics of wide-fin device with $W_f = 18$ nm and $L_g = 60$ nm. Inset shows peak $g_m = 1.9$ mS/μm.

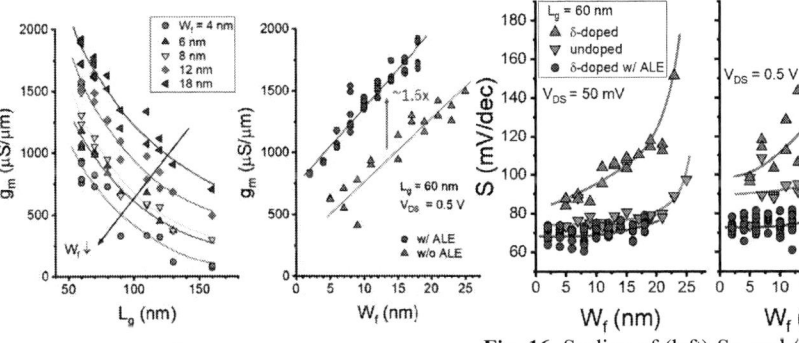

Fig. 15. (Left) scaling of g_m vs. L_g at various W_f, and (right) scaling of g_m vs. W_f at $L_g = 60$ nm of InGaAs FinFETs with and without thermal ALE [3] ($V_{DS} = 0.5$ V).

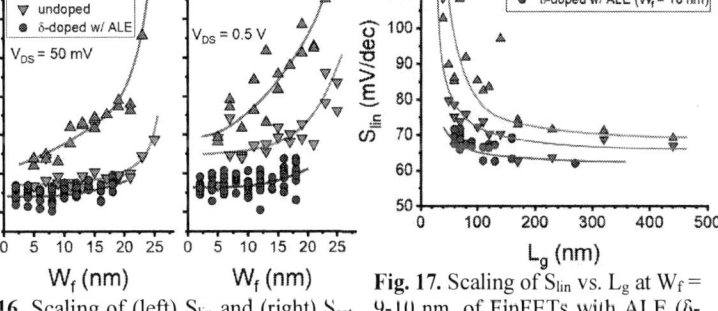

Fig. 16. Scaling of (left) S_{lin} and (right) S_{sat} vs. W_f at $L_g = 60$ and 100 nm, of FinFETs with ALE (δ-doped) and without ALE (δ-doped and undoped) [3].

Fig. 17. Scaling of S_{lin} vs. L_g at $W_f = 9$-10 nm, of FinFETs with ALE (δ-doped) and without ALE (δ-doped and undoped) [3].

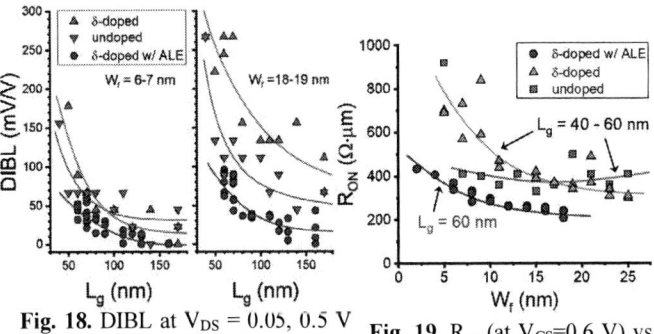

Fig. 18. DIBL at $V_{DS} = 0.05$, 0.5 V vs. L_g at (left) $W_f = 6$-7 nm, and (right) 18-19 nm, of FinFETs with ALE (δ-doped) and without ALE (δ-doped and undoped) [3].

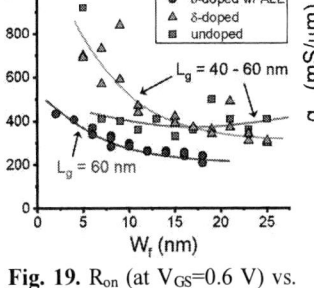

Fig. 19. R_{on} (at $V_{GS} = 0.6$ V) vs. W_f of FinFETs with ALE (δ-doped) and without ALE (δ-doped and undoped) [3].

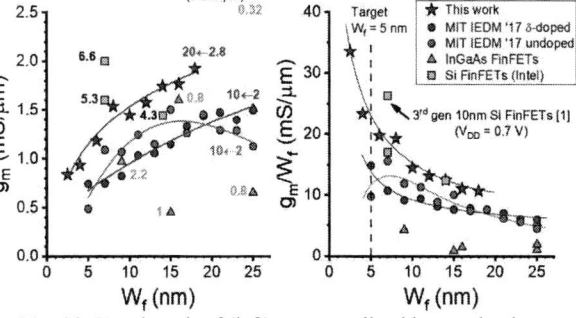

Fig. 20. Benchmark of (left) g_m normalized by conducting gate periphery, annotated by AR (H_c/W_f), and (right) g_m/W_f, as a function of W_f from this work and the InGaAs literature. State-of-the-art Si FinFETs are also included. $V_{DD} = 0.5$ V for InGaAs FinFETs and 0.7 V for the 2nd and 3rd generation Si FinFETs.

978-1-7281-1988-5/18 $31.00 © 2018 IEEE   898

InGaAs-on-Insulator FinFETs with Reduced Off-Current and Record Performance

C. Convertino[1], C. Zota[1], S. Sant[2], F. Eltes[1], M. Sousa[1], D. Caimi[1], A. Schenk[2] and L. Czornomaz[1]

[1]IBM Research Zurich, Switzerland, email: ino@zurich.ibm.com
[2]Integrated Systems Laboratory, ETH Zurich, Zurich, Switzerland

Abstract—In this work, we demonstrate InGaAs-on-Insulator FinFETs on silicon with optimized on/off trade-off showing record performance. This is achieved by using carefully designed source/drain spacers and doped extensions to mitigate the off-current, typically high in narrow band-gap materials, as part of a CMOS compatible replacement-metal-gate process flow. Using this technology, devices with L_G = 20 nm, spacers width of 10 nm and W_{fin} = 15 nm achieve record high on-current of 350 µA/µm (I_{OFF} = 100 nA/µm and V_{DD} = 0.5 V), for scaled III-V FETs on Si, enabled by an access resistance of 220 Ω.µm, SS_{sat} = 78 mV/decade and g_m = 1.5 mS/µm. We analyze the impact of spacers thickness, W_{fin} and L_G on device performance. 2D TCAD simulations provide further insights into device functionality and about the dominant off-state leakage mechanisms.

I. INTRODUCTION

High electron mobility compound semiconductors such as InGaAs are considered promising candidates to replace Si as the channel material in nFETs in advanced CMOS nodes [1-3]. InGaAs can be integrated on silicon in a 2D co-planar [4] or 3D monolithic integration scheme [5], e.g. by use of direct wafer bonding [6], which can provide high-quality scaled III-V layers on silicon substrates, as well as incorporate a buried-oxide-layer (BOX) to enhance electrostatic confinement.

At scaled dimensions, 3D channel geometries such as fins or gate-all-around (GAA) nanowires are necessary to match the node target performances in terms of both electrostatics and current density. However, the floating body of such structures may lead to accumulation of holes in the channel region, triggering a parasitic bipolar transistor effect (PBE). This effect amplifies band-to-band tunneling (BTBT) current already significant in narrow band-gap materials, which further increases off-current (I_{OFF}) for InGaAs FinFETs [7]. The introduction of source-drain spacers have been suggested by simulations to reduce the effect of the PBE and to decrease I_{OFF} [8]. Previous works reporting the use of source/drain spacers show an off-current reduction [9-10] but also an increase of access resistance: optimized doped extensions beneath spacers are indeed necessary to limit this effect. So far, there has been no balanced implementation of sidewall spacers and low access resistance in III-V FETs.

In the present work, we experimentally investigate the reduction of I_{OFF} using spacers and source/drain doped extensions by fabricating and comparing $In_{0.53}Ga_{0.47}As$ FinFETs with several different spacer designs. As a result, we obtain devices with improved off-state performance, yielding a new on-current record at scaled L_G for III-V-on-Si.

II. DEVICE FABRICATION

The InGaAs FinFETs are fabricated using a III-V-on insulator platform on silicon substrates with a replacement-metal-gate (RMG) process and raised-source-drain (RSD) modules, as schematized in **Fig. 1**. The fabrication starts with the integration of a 20-nm thick $In_{0.53}Ga_{0.47}As$ channel layer on Si by using direct wafer bonding. Fins are patterned by HSQ (Hydrogen Silsesquioxane) resist and dry etched down to the BOX. Dummy high-k and dummy gate are deposited, patterned by HSQ and etched with an optimized inductively-coupled plasma reactive ion etching process (ICP-RIE). SiN_x is then deposited with variable thickness and spacers are formed on the dummy gate sidewalls by dry etching. The extensions and the RSD epitaxy are formed in one growth step by metalorganic chemical vapor deposition (MOCVD). The position of the extension is varied and adjusted with nm precision by performing several digital etching (DE) cycles prior to the growth. DE is done by using diluted HCl and ozone, resulting in an undercut underneath the spacers. Afterwards, an encapsulating inter-layer dielectric (ILD0) is deposited and planarized by chemical-mechanical polishing (CMP). This step enables access to the top of the dummy gate that is removed by a selective dry etch process. A scaled bilayer of Al_2O_3/HfO_2 high-k dielectric is deposited on the InGaAs channel by plasma-enhanced atomic-layer deposition (PE-ALD) followed by *in-situ* TiN metal gate deposition. Subsequently, the gate is filled with W which is planarized by CMP. A second oxide layer (ILD0') is deposited and finally, contact vias on source, drain and gate are opened and filled with W.

A high resolution cross-section TEM of an L_G = 20 nm device is shown in **Fig. 2a** and close-ups on the channel/contacts and channel/oxide interfaces are shown in **Fig. 2b,c**. In this work, we compare three different spacer designs, as shown in **Fig. 3**: without spacers as well as with 4 and 10 nm SiN_x spacers. A high-resolution fin TEM cross-section is shown in **Fig. 4**, showing excellent crystal quality as well as allowing for accurate measurement of fin dimensions.

III. RESULTS AND DISCUSSION

First, we examine the influence of the S/D spacers on the off-current. **Fig. 5** shows I_{OFF}, defined as the minimum I_{DS}, versus gate length for FinFETs with W_{fin} = 25 nm, and with 0, 4 and 10 nm spacers. All data shown here are normalized to the gated periphery of the fins. At L_G = 100 nm, I_{OFF} is approximately one order of magnitude lower with 10 nm spacers compared to 4 nm spacers, and two orders of magnitude lower compared to no spacers. At L_G = 20 nm, this difference increases to three orders of magnitude.

978-1-7281-1988-5/18 $31.00 © 2018 IEEE

Fig. 6 shows access resistance (R_{access}) versus spacer thickness. $R_{access} = 220\ \Omega.\mu m$ is achieved for 10 nm spacers. Similar values of R_{access} are obtained for 4 nm ($210\ \Omega.\mu m$) and without spacers ($200\ \Omega.\mu m$), indicating that the doped extensions effectively mitigate increased R_{access} due to the ungated regions under the spacers, while still enhancing off-state performance. For devices with 10 nm spacers, the number of digital etches forming the RSD extensions was varied, resulting in 8 to 10 nm long extensions. R_{access} in turn varied by approximately $100\ \Omega.\mu m$ per nm of extension length below 10 nm, as shown in **Fig. 7**, which indicates the strong importance of a proper alignment of the doped extensions matching the width of the spacers. **Fig. 8** shows I_{ON}/I_{OFF} for the same devices. Here, I_{ON} is defined as I_{DS} at $V_{GS} = V_G(I_{OFF})+1$ V and $V_{DS} = 0.5$ V. 10 nm spacers improve I_{ON}/I_{OFF} by an order of magnitude compared to 4 nm spacers, indicating that I_{OFF} is reduced while maintaining I_{ON}.

To understand the influence of the spacers on the off-state performance and the origin of I_{OFF}, we perform 2D TCAD simulations of matching device structures using the same simulation set-up as in [8]. **Fig. 9** and **Fig. 10** show subthreshold characteristics of FinFET devices with $W_{fin} = 30$ nm and $L_G = 300$ and 100 nm respectively, all with 10 nm spacers. Experimental data (symbols) show a good match to simulated values (solid lines) in both on and off states for all gate lengths. Simulations indicate three sources of leakage currents in the off state [8]: (i) Trap-assisted tunneling (TAT) at the high-k/InGaAs interface on the drain side, (ii) source-to-drain tunneling (STDT), i.e. from the conduction band of the source to the drain, and (iii) BTBT, i.e. from the valence band of the channel to the drain. In addition, holes generated in the channel due to both BTBT and TAT lower the potential of the channel and increase the leakage through the forward-biased p-n junction on the source-side edge of the channel, which gives rise to the parasitic bipolar junction transistor effect (PBE) [11]. Since STDT is minimal in long-channel devices, due to the wide tunneling barrier, this effect of BTBT and TAT at gate-oxide/drain interface can be effectively suppressed by the 10 nm spacers utilized in this work. Note that traps may still be present at spacer/drain interface. Presence of a spacer eliminates formation of a triangular well at the interface, thereby inhibiting BTBT and TAT along the triangular well [8]. This eliminates major source of holes to the PBE and reduces off-state leakage. A similar effect is observed in the short channel devices as well. However, here the presence of STDT degrades the SS and slightly increases the off-state leakage.

Fig. 11 shows transfer characteristic of an $L_G = 20$ nm device, achieving $I_{ON} = 350\ \mu A/\mu m$ ($I_{OFF} = 100$ nA/μm and $V_{DD} = 0.5$ V), along with drain-induced barrier-lowering DIBL $= 30$ mV/V and subthreshold slopes in the linear and saturation regions, $SS_{lin} = 74$ and $SS_{sat} = 78$ mV/decade. The gate leakage (not shown) is lower than 1×10^{-10} A/μm. Peak transconductance reaches $g_m = 1.5$ mS/μm. **Fig. 12** shows output characteristics of the same device. Low output conductance is observed, as well as an on-resistance R_{ON} of

$300\ \Omega.\mu m$. $g_{m,peak}$ is shown versus L_G in **Fig. 13** for $W_{fin} = 25$ nm, exhibiting good scaling behavior and reaching 1.5 mS/μm at 20 nm, indicating strong resilience against short channel effects (SCEs). $g_{m,peak}$ is furthermore shown versus W_{fin} in **Fig. 14** for two gate lengths, 20 and 300 nm. At $L_G = 300$ nm, $g_{m,peak}$ remains approximately constant at 0.5 mS/μm, while at $L_G = 20$ nm, it scales with W_{fin}. Though increased surface scattering, reducing the electron mobility, is expected for scaled fins, the increase of $g_{m,peak}$ with W_{fin} can be explained by improved electrostatic control for narrow W_{fin}, which decreases g_d and improves extrinsic g_m. This is confirmed by **Fig. 15**, showing DIBL versus L_G for different W_{fin}, which strongly improves for narrow fins. **Fig. 16** and **17** show SS_{lin} and SS_{sat} versus L_G and W_{fin}, respectively. SS_{lin} reaches 65 mV/decade and SS_{sat} 78 mV/decade at $L_G = 20$ nm and $W_{fin} = 15$ nm. **Fig. 18** shows $I_{ON@IOFF,VDD}$, calculated at a fixed $I_{OFF} = 100$ nA/μm and $V_{DD} = 0.5$ V, versus L_G for the three different spacer thickness. Even at fixed I_{OFF}, the reduced minimum I_{OFF} through the use of wide spacers enables higher I_{ON} due to a steeper SS near the I_{OFF} target. **Fig. 19** shows a benchmark of I_{ON} (at $I_{OFF} = 100$ nA/μm and $V_{DD} = 0.5$ V) for various III-on-Si technologies [12-17]. Devices shown in this work achieve the highest reported $I_{ON} = 350\ \mu A/\mu m$ for III-V-on-Si FETs.

IV. CONCLUSION

We have demonstrated InGaAs FinFETs on silicon with optimized on/off trade-off showing record performance. This is achieved by using carefully designed source/drain spacers and doped extensions to mitigate the off-current. This enabled a reduction of the off-current by three orders of magnitude at scaled L_G, resulting in an improved SS_{sat} near the I_{OFF} target of 100 nA/μA while maintaining excellent R_{access}, leading to a record-high I_{ON} ($V_{DD} = 0.5$ V) of 350 μA/μm for III-V-on-Si FETs.

ACKNOWLEDGMENT

This work was funded by Horizon 2020 grant agreement numbers 688784 (INSIGHT) and 687931 (REMINDER). The authors gratefully acknowledge the support of the BRNC operations team as well as the MIND group.

REFERENCES

[1] J. A. Del Alamo, Nature, vol. 479, no. 7373. pp. 317–323, 2011
[2] C. B. Zota et al., IEDM, pp. 3.2.1-3.2.4, 2016.
[3] X. Sun et al., VLSI Techn. Symp., T3-4, 2017
[4] L. Czornomaz, et al., in VLSI Techn. Symp., T9-2, 2016
[5] V. Deshpande, et al., VLSI Tech. Dig.,T6-4 2017
[6] L. Czornomaz, et al., IEDM Tech Dig., p. 23.4.1, 2012
[7] J. Lin, et al., IEEE EDL, 35(12), p. 1203, 2014
[8] S. Sant, et al., IEEE TED, vol. 65, no. 6, pp. 2578-2584, 2018
[9] C. Y. Huang, et al. IEDM Tech Dig., p. 25.4.1, 2014
[10] V. Djara, et al, VLSI Techn. Symp., T176, 2015
[11] X. Zhao, et al., IEEE EDL, 39(4), p. 476, 2018
[12] H. Hahn, et al., IEDM Tech Dig., p. 17.5.1, 2017
[13] X. Zhou, et al., VLSI Techn. Symp., pp. 166-167, 2016
[14] C. Y. Huang, et al., VLSI Techn. Symp., 2015
[15] N. Waldron, at al., IEDM Tech Dig., p. 31.1.1, 2015
[16] V. Djara, et al., IEEE EDL., vol. 37, no. 2, pp. 169–172, 2016.
[17] C. Zota, et al., VLSI Techn. Symp., T15-5, 2018

InGaAs FinFET fabrication flow

- InGaAs-OI-Si substrate
- HSQ fins pattern and etch
- Dummy HK/gate deposition
- HSQ gate pattern and etch
- Spacers deposition and etch
- Digital etching for spacers undercut
- RSD n+InGaAs epitxy
- ILD0 deposition
- ILD0 CMP
- Dummy HK/gate etching
- HKMG/W deposition
- Metal CMP
- ILD0' deposition
- M1 contact patterning
- Ar/H2 anneal

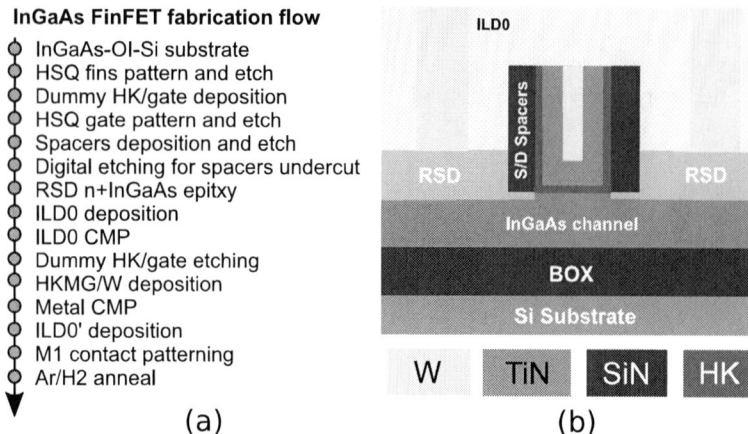

(a)

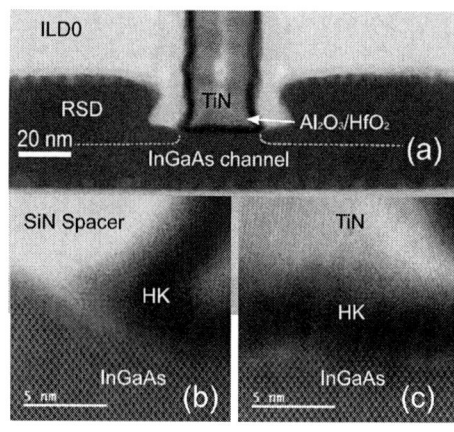

Fig. 1. (a) Process flow describing the self-aligned replacement-metal gate fabrication process for the InGaAs FinFETs presented in this work. (b) Schematic cross-section across the gate of the fabricated device. The InGaAs channel layer is 20 nm thick while the doped RSD InGaAs contacts are 25 nm thick. Doped extensions below the spacers are obtained by digital etching of the channel post-deposition of the spacers.

Fig. 2. (a) Cross-sectional STEM image of an InGaAs FinFET with L_G = 20 nm, showing SiN_x spacers and RSD contacts. (b) High resolution STEM close-up on the source side (c) and on the channel-HK interface.

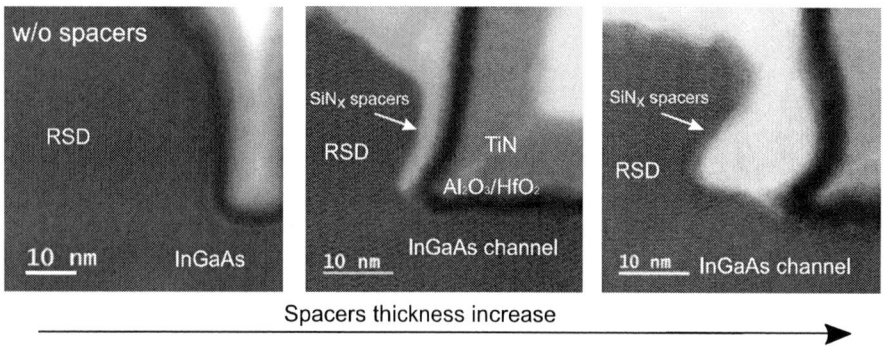

Fig. 3. Cross-sectional TEM images on the source-side gate region for the three spacer designs investigated in this work: (a) no spacers, (b) 4 nm spacers and (c) 10 nm spacers. The spacers are intended to reduce off-current leakage by reducing the parasitic bipolar effect through suppression of band-to-band tunneling as well as trap-assisted tunneling on the drain side.

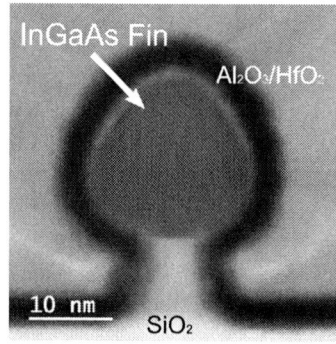

Fig. 4. STEM cross section of a fin with W_{FIN} = 20 nm. Excellent crystal quality is observed for the fin.

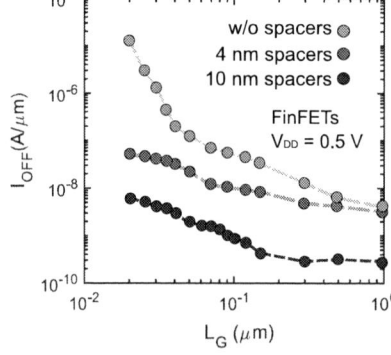

Fig. 5. I_{OFF} versus L_G. The I_{OFF} decreased by about 3 orders of magnitude at scaled L_G for devices with thicker spacers. The increase of I_{OFF} at smaller L_G is an indication of the presence of band-to-band tunneling and the resulting parasitic bipolar effect.

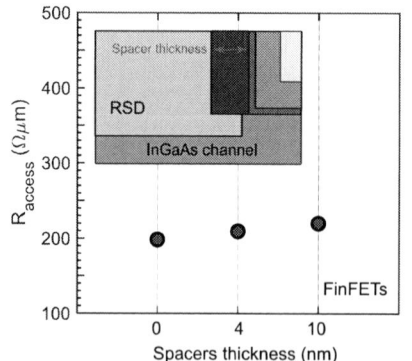

Fig. 6. Access resistance versus spacer thickness, as defined in the inset. The low R_{acc} value achieved for thick spacers devices indicates that the RSD spacer extensions effectively mitigate R_{acc} increase due to ungated regions under the spacers.

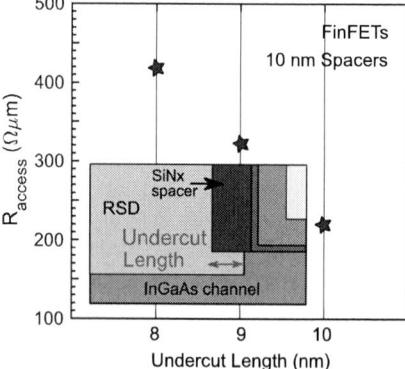

Fig. 7. Access resistance versus undercut length, as defined in the inset. The undercut length is set by the no. of digital etch cycles. This indicates the importance of carefully align the extensions region position to the sidewall spacers.

978-1-7281-1988-5/18 $31.00 © 2018 IEEE

901

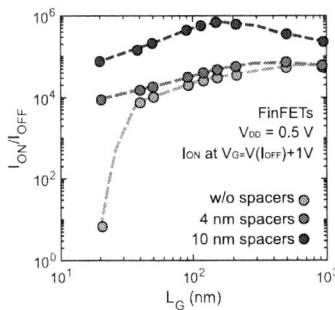

Fig. 8. I_{ON}/I_{OFF} versus L_G for FinFETs with and without spacers. The ratio is increased for shorter L_G. At long L_G, the higher I_{OFF} is balanced by the higher I_{ON} for the devices without spacers.

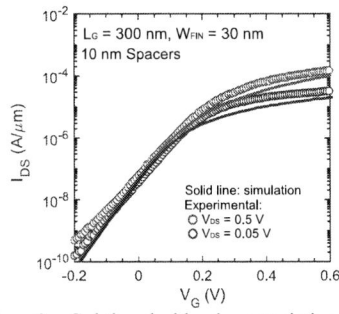

Fig. 9. Subthreshold characteristics of FinFET device with 10 nm spacers, W_{fin} = 30 nm and L_G = 300 nm. Solid traces show 2D TCAD simulations, that show excellent fit of experimental data (symbols).

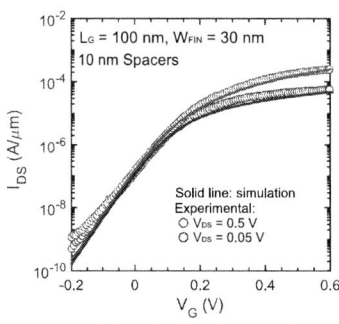

Fig. 10. Subthreshold characteristics of FinFET device with 10 nm spacers, W_{fin} = 30 nm and L_G = 100 nm. Solid traces show 2D TCAD simulations, that show excellent fit of experimental data (symbols).

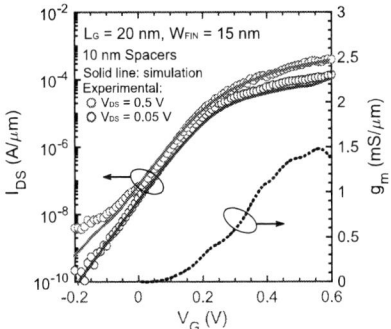

Fig. 11. I_D/V_G of FinFET device with 10 nm spacers, W_{fin} = 30 nm and L_G = 20 nm. This device exhibits record high I_{ON} of 350 $\mu A/\mu m$ (at I_{OFF} = 100 nA/μm, V_{DD} = 0.5 V).

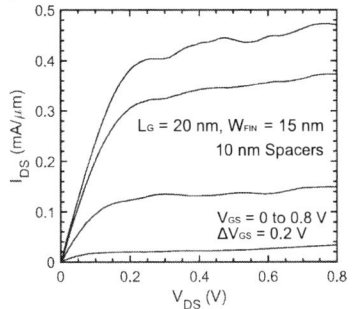

Fig. 12. Output characteristic of a L_G = 20 nm device with W_{FIN} = 15 nm. Low g_d is observed, along with R_{ON} = 300 $\Omega.\mu m$.

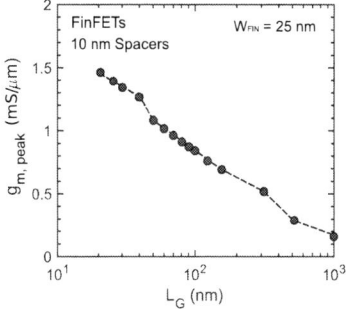

Fig. 13. Peak g_m versus L_G for FinFETs with 10 nm spacers, showing excellent scaling behavior down to L_G = 20 nm.

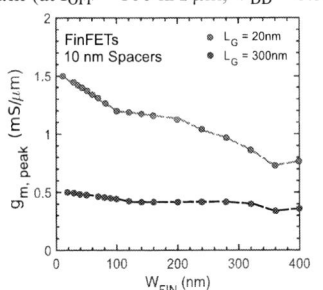

Fig. 14. Peak g_m versus W_{fin} for long and short channel devices. g_m increases in the latter due to reduction of g_d.

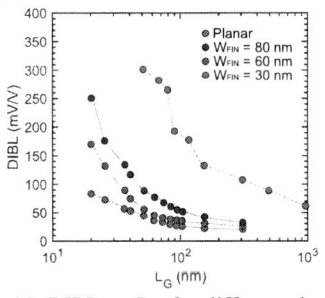

Fig. 15. DIBL vs L_G for different channel designs, showing strongly improved electrostatic control for narrow fins.

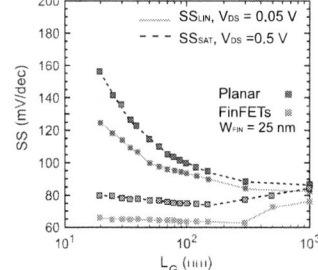

Fig. 16. Minimum SS in saturation and linear regime versus L_G for both planar and FinFET devices.

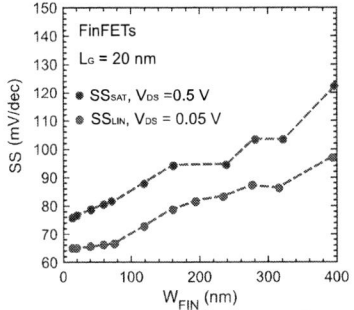

Fig. 17. Minimum SS in saturation and linear regime versus W_{fin} at L_G = 20 nm, showing the importance of fin width scaling. At minimum W_{fin}, SS_{sat} = 78 mV/decade is achieved.

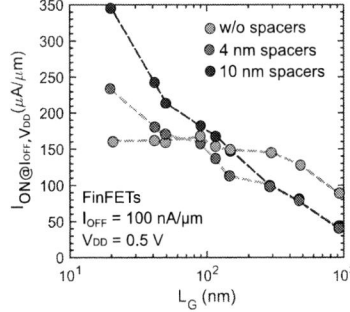

Fig. 18. On current versus gate length for III-V FinFETs devices with three different spacer designs. The presence of spacers generally improves I_{ON} at scaled L_G because of the lower SS close to the target I_{OFF} = 100

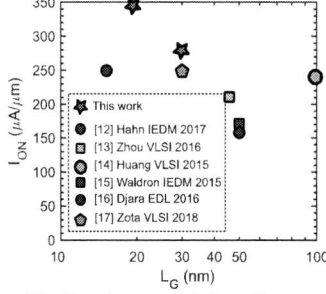

Fig. 19. Benchmark of I_{ON} (I_{OFF} = 100 nA/μm, V_{DD} = 0.5 V) versus L_G for different III-V-on-Si technologies. The value of 350 $\mu A/\mu m$ shown in this work represents the highest reported for this type of device.

978-1-7281-1988-5/18 $31.00 © 2018 IEEE

Balanced Drive Currents in 10-20 nm Diameter Nanowire All-III-V CMOS on Si

Adam Jönsson[1], Johannes Svensson[1], and Lars-Erik Wernersson[1]

[1]Department of Electrical and Information Technology, Lund University, Lund, Sweden, email: adam.jonsson@eit.lth.se

Abstract—We use a self-aligned, gate-last process providing n-type (InAs) and p-type (GaSb) MOSFET co-integration with a common gate-stack and demonstrate balanced drive current capability at about 100 µA/µm. By utilizing HSQ-spacers, control of gate-alignment allows to fabricate both n- and p-type devices based on the same type of vertical heterostructure InAs/GaSb nanowire with short gate-lengths down to 60 nm. Refined digital etch techniques, compatible with both sensitive antimonide structures and InAs, enable down to 16 nm diameter GaSb channel regions and 10 nm InAs channels. Balanced performance is showcased for both n- and p-type MOSFETs with I_{on} = 156 µA/µm, at I_{off} = 100 nA/µm, and 98 µA/µm, at $|V_{DS}|$ = 0.5, respectively.

I. INTRODUCTION

High mobility materials such as narrow band gap III-V compounds offer a possibility to increase the MOSFET performance for both logic and high-frequency devices. Material options such as InAs and GaSb present high bulk mobility for electrons and holes, respectively, which makes the combination attractive for CMOS implementation. GaSb based transistor performance is currently limited by the gate-stacks and the reactive nature of the antimony-compounds imposes challenges in both material growth and device fabrication. [1]

The continuation of the traditional down-scaling of MOSFETs for digital circuits has led to short channel effects due to deteriorated electrostatics [2]. 3D gate architectures are therefore proposed, and implemented, with gate-all-around (GAA) structures utilizing vertical nanowires as a strong candidate. Fundamentally, vertical nanowire MOSFETs presents a seamless way to decouple gate-length and contact geometry from the device footprint area. The small footprint also allows larger lattice mismatch without propagating defects, which simplifies integration of high mobility materials on top of Si substrates. [3]

In this work, we demonstrate a streamlined co-integration process for p- and n-type MOSFETs, with a common gate-stack, using a self-aligned, gate-last process. State-of-the-art vertical p-type GaSb MOSFET performance is demonstrated, with g_m = 230 µS/µm, co-integrated with a strong InAs n-type device showcasing good off-state with I_{on}=156 µA/µm at I_{off} = 100 nA/µm, all at $|V_{DS}|$ = 0.5 V (**Table 1**). The data includes 5x drive current improvement for the GaSb MOSFET combined with a 3x increase in g_m as well as a decreased SS_{min} as compared to previous results [4]. The improvement is attributed to adjustment in the aspect ratio (Diameter:L_g) for the n-type and p-type devices, from 2:5 and 2:4 to 2:30 and 2:6,

respectively, in order to achieve balanced drive currents. For the n-type device this has resulted in improved off-state characteristics reaching the I_{off} = 100 nA/µm limit and simultaneously the p-type current has been improved with a 5 times higher I_{on} of 98 µA/µm (**Table 2**).

II. DEVICE FABRICATION

The processed MOSFETs are based on vapor-liquid-solid (VLS) grown InAs-GaSb heterostructure nanowires overgrown with, a highly n-doped InAs shell. The implementation of an overgrown shell circumvents issues regarding etch selectivity and enables processing with hydrogen silsesquioxane (HSQ) allowing development of a self-aligned, gate-last process as the sensitive GaSb is protected by the InAs. Optimization of alcohol based digital etching, in conjunction with the gate last implementation, has enabled scaled diameters and selective digital etch of the channel region. Therefore, GaSb devices with diameters down to 16 nm have been achieved, which has proven crucial for improved performance.

Fig. 1 represents the critical fabrication steps for the co-integration process. The devices are based on 260 nm epitaxial InAs layer grown on p-type silicon (111) substrates. Subsequently, InAs-GaSb nanowires are grown by VLS from EBL defined 32 nm diameter Au discs. The top of the InAs segment and GaSb segment is doped by Sn and Zn, respectively. The nanowires are also overgrown with an InAs shell for improved etch selectivity (**Fig 1-a**).

After nanowire growth, an HSQ mask is applied whose thickness is controlled by the EBL exposure dose. The thickness control allows for varied gate-position along the nanowire, enabling p- and n-type devices to be fabricated from the same type of nanowires. The spacer is used as a template to align the top metal, which is applied by 200 nm sputtered W and 3 nm ALD TiN (**Fig. 1-b**). Prior to metal deposition, a citric acid dip is performed followed by HCL:IPA to remove the protruding InAs shell and restore the core-material. The applied metal is selectively removed from the planar surfaces by reactive ion etching leaving the finished top contact.

The HSQ mask, previously used for top contact alignment, is thinned by diluted HF 1:400 to form the bottom spacer and to expose the nanowire channel-region. The channel region is selectively digitally etched by 4 cycles of short ozone exposure followed by HCL:IPA 1:30 wet etch. The digital etch removes the InAs shell and further serves to trim the channel down to sub 25 nm diameters. A bilayer high-k is applied consisting of 6 cycles of Al_2O_3 and 36 cycles of HfO_2, corresponding to an

EOT of 0.85 nm (**Fig. 1-c**). The result after high-k deposition is shown in **Fig 2**, presenting before and after SEM images of a single nanowire p-type device.

Finally, 60 nm sputtered tungsten is used as gate metal and the top edge aligned vertically by a back etched S1813 resist as etch mask for an SF_6 dry etch. Afterwards an organic top spacer is defined followed by sputtering of the top contact consisting of Ni/Au (15/200 nm), see **Fig. 1-d**.

III. MEASUREMENTS

Fig. 3 and **Fig. 4** represent combined output and transfer characteristics for InAs (L_G = 150 nm, diameter = 10 nm) and GaSb (L_G = 60 nm, diameter 22 nm) channel devices, co-integrated on the same Si substrate. The data are showcasing well behaved characteristics with maximum g_m = 405 and 230 µS/µm, respectively, normalized to the total circumference, see **Table 1**. A high I_{on} = 156 µA/µm (at I_{off} = 100 nA/µm) is also achieved for the n-type device, representing a 14% improvement compared to previous vertical InAs MOSFETs [5]. Both the n- and p-type devices showcase good electrostatics with SS_{lin} = 72 and 175 mV/dec, attributed to the aggressive diameter scaling (**Fig. 5** and **Fig. 6**) and high-quality semiconductor/high-k interfaces. Also, the minimum subthreshold slope is maintained over a wide bias range for the n-type device (**Fig. 6**), demonstrated that the co-integration process does not introduce a drastic increase in D_{it} for InAs. Notice that the InAs transistor is fabricated from a 200 nm long InAs segment, which introduces significant constraint on contact formation contributing to a comparably high R_{on} = 1.4 kΩ·µm for the n-type MOSFET. For the p-type device, contributions to the contact resistance from rapid re-oxidation of GaSb and injection via a, not optimized, broken bandgap source serves to further limit the on-state performance. The limited off-state can be attributed to background doping in the GaSb channel (**Fig. 7**). [6]

To demonstrate the improved digital etch technique and technology scalability, a p-type device with diameter down to 16 nm, although with longer gate-length of 150 nm is shown in **Fig. 8**. Alcohol based digital etch techniques enable the aggressive diameter scaling. Notice the large difference between top contact diameter with respect to the channel region, which improves resistance originating from the drain contact. The device transfer characteristics is presented in **Fig. 9**, showing that a high transconductance of 87 µS/µm can be maintained also when the diameter is scaled. Notably, the performance of GaSb p-MOSFETs strongly depends on the gate length. Also the off-state performance is improved, quantified by SS_{sat} = 257 mV/dec (**Fig. 10**). The output characteristics for this device (**Fig. 11**) showcase an exponential behavior that indicates the presence of a potential barrier (**Fig. 12**) which can be resolved by further contact optimization.

To visualize the performance improvements as compared to previous GaSb, as well as InGaSb, devices a g_m versus SS_{sat} plot is presented in **Fig 12**. Here, importance of scaling the gate-length and diameter is clearly emphasized. With a balance between SS and g_m metrics also at scaled gate lengths, this work shows improved performance over state-of the-art GaSb MOSFETs, including InGaSb fin-FETs. [1]

IV. CMOS IMPLEMENTATIONS

Many alternative co-integration strategies have been proposed and implemented utilizing the same material combination, namely InAs and GaSb, see **Table 2**. One approach is to use nano-ribbons with a two-step transfer technique [7]. A variation of this technique with a one-step transfer was then showed [8]. A different approach was presented using a grown periodic InAs-GaSb planar structure where segments are selectively etched making separate lateral GAA InAs and GaSb devices [9]. The here used nanowire CMOS implementation based on vertical InAs-GaSb heterostructure nanowires, on top of Si, [10] has the potential to include heterostructure InAs-InGaAs segments [11] to reduce the off-state leakage and to further increase the transconductance. [4] In fact, we show a technology that can merge high transconductance n-type MOSFETs [11] with balanced CMOS implementation for high-speed logic and mixed applications.

From the benchmarking of various all-III-V CMOS implementations in **Table 2,** we note that the implementation presented in this work, represents the best set of combined metrics including g_m vs SS_{sat}. This work, shows that competitive device performance can be achieved within a co-integrated process, challenging other state-of-the-art III-V devices.

V. CONCLUSIONS

We present an all-III-V co-integration process scaled to aggressive gate-lengths (L_G = 60 nm) and diameters (D_{InAs} = 10 nm, D_{GaSb} = 16 nm). This has served to reach balanced drive-currents for the III-V CMOS at 156 and 98 µA/µm for the n- and p-type devices respectively (**Table 1**) as well as demonstration of competitive transistor performance.

ACKNOWLEDGMENT

This work was supported in part by the Swedish Research Council, in part by the Knut and Alice Wallenberg Foundation, in part by the Swedish Foundation for Strategic Research and in part by the European Union H2020 program INSIGHT (Grant Agreement No. 688784).

REFERENCES

[1] W. Lu *et al.*, *IEDM*, 2017, pp. 433–436.
[2] C. P. Auth *et al.*, *Device Lett.*, vol. 18, no. 2, pp. 74–76, Feb. 1997.
[3] Shadi A. Dayeh *et al.*, *Nano Lett.*, vol. 7(8), pp. 2486–249, 2007.
[4] A. Jonsson *et al.*, *IEEE Electron Device Lett.*, pp. 1–1, 2018.
[5] M. Berg *et al.*, *EDL*, vol. 37, no. 8, pp. 966–969, Aug. 2016.
[6] A. S. Babadi *et al.*, *APL*, vol. 110, no. 5, p. 53502, Jan. 2017.
[7] J. Nah *et al.*, *Nano Lett.*, vol. 12, no. 7, pp. 3592–3595, Jul. 2012.
[8] M. Yokoyama *et al.*, *VLSI*, 2014, pp. 1–2.
[9] K.-H. Goh *et al.*, *IEDM)* 2015, p. 15.4.1-15.4.4.
[10] J. Svensson *et al.*, *Nano Lett*, vol. 15, pp. 7898–7904, Dec. 2015.
[11] O.-P. Kilpi *et al.*, *IEDM*, 2017, p. 17.3.1-17.3.4.

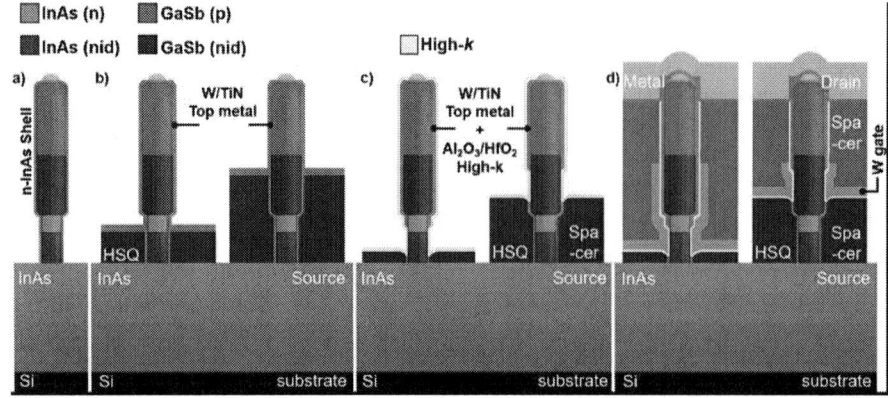

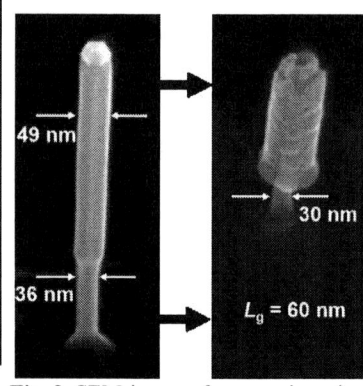

Fig. 1. Schematics of process flow, showing a) nanowire after VLS growth, b) top metal alignment, c) first spacer and high-k deposition, and d) final structure with contacts.

Fig. 2. SEM-image of a nanowire prior to processing and after first spacer and high-k deposition, see Fig. 1-c.

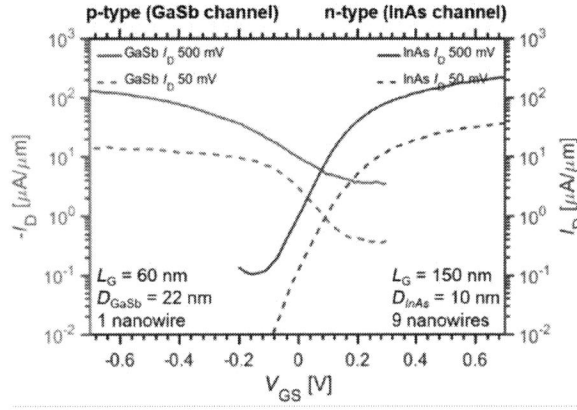

Fig. 3. Combined transfer characteristics for p-type single nanowire GaSb device (L_G = 60 nm, diameter 22 nm) and 9 nanowire n-type InAs device (L_G = 150, diameter 10 nm).

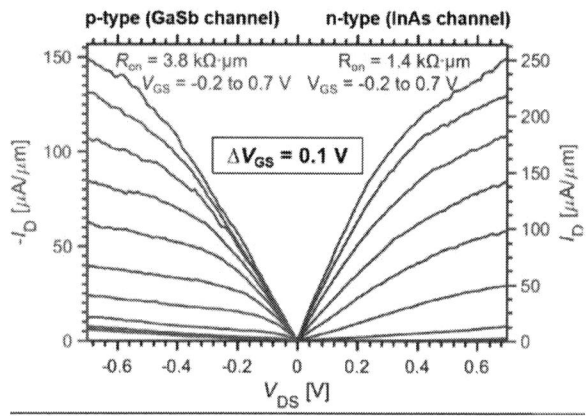

Fig. 4. Combined output characteristics for p-type single nanowire GaSb device (L_G = 60 nm, diameter 22 nm) and 9 nanowire n-type InAs device (L_G = 150, diameter 10 nm).

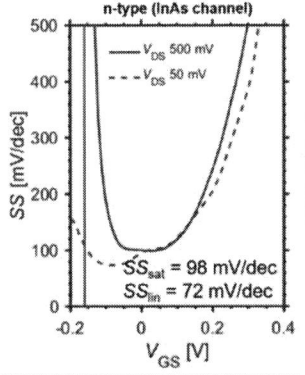

Fig. 5. SS_{sat} and SS_{lin} for the device consisting of 9 nanowires, with L_G = 150 nm and diameter 10 nm.

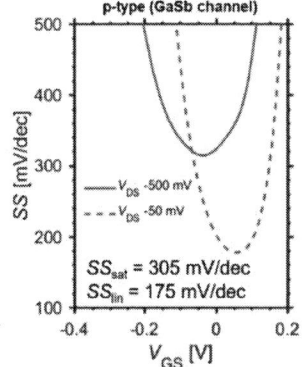

Fig. 6. SS_{sat} and SS_{lin} for the single nanowire device, with L_G = 60 nm and diameter 22 nm.

Fig. 7. Transfer characteristics and transconductance g_m for the single nanowire device with L_G = 60 nm and diameter 22 nm. A $g_{m,max}$ of 230 µS/µm is demonstrated.

	n-type	p-type
I_{on} [µA/µm]	156	98
g_m [µS/µm]	405	230
L_G [nm]	150	60
SS_{sat} [mV/dec]	98	305
SS_{lin} [mV/dec]	72	175

Table 1. Summary of DC-metrics for the all-III-V CMOS process. I_{on} defined at I_{off} = 100 nA/um for the n-type device and at V_{DS} = -0.5 V for the p-type device.

978-1-7281-1988-5/18 $31.00 © 2018 IEEE

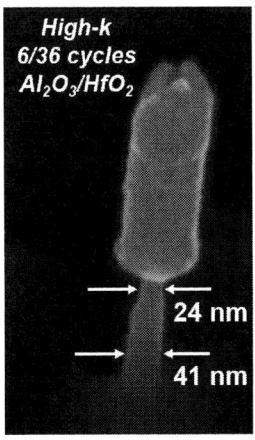

Fig. 8. SEM-image showcasing a diameter of 16 nm (+ 8 nm high-k) inside a 2-nanowire p-type device.

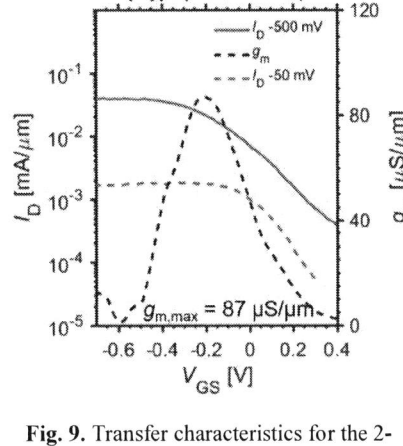

Fig. 9. Transfer characteristics for the 2-nanowire device, with $L_G = 150$ nm and diameter 16 nm.

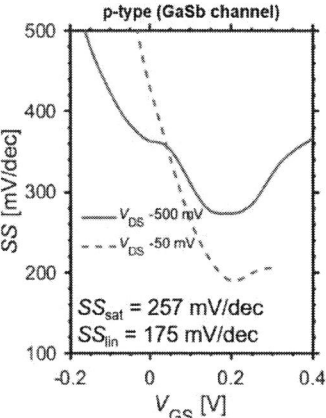

Fig. 10. SS_{sat} and SS_{lin} for the 2-nanowire device, with $L_G = 150$ nm and diameter 16 nm.

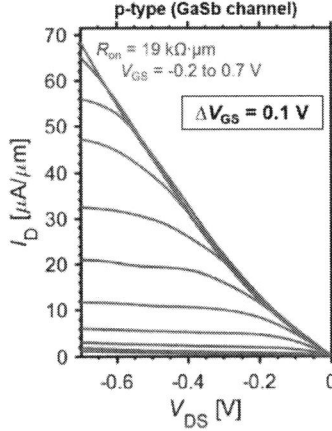

Fig. 11. Output characteristics for the 2-nanowire device, with $L_G = 150$ nm and diameter 16 nm.

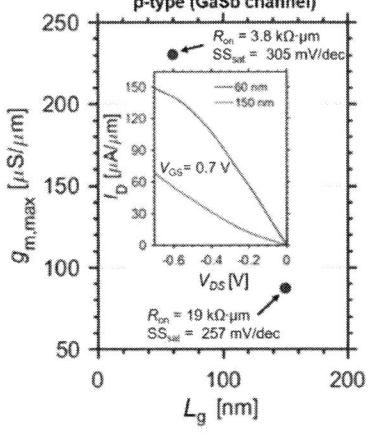

Fig. 12. The $g_{m,max}$ dependence of L_G for the single nanowire and the 2-nanowire p-type device. The inset highlights the difference at the on-state.

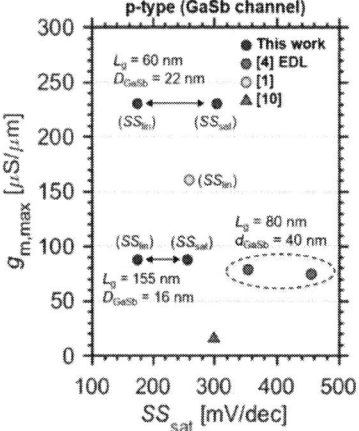

Fig. 13. Benchmarking with respect to GaSb and InGaSb p-type devices.

	n-type [This work]	n-type EDL [4]	n-type [9]	n-type [7]	n-type [8]	n-type [5]	p-type [This work]	p-type EDL [4]	p-type [9]	p-type [7]	p-type [8]	p-type InGaSb [1]
III-V CMOS							**III-V CMOS**					
I_{on} [μA/μm]	156				80	4	140	98	17	10	22	2.4	~100
g_m [μS/μm]	405	1200				640	230	74				160
L_G [nm] /Crit.Dim	150 /10	50 /20	500 /20	/13	/2.5	50 /28	60 /22	80 /40	500 /20	/7	/20	20 /10
SS_{sat} [mV/dec]	98	158	185	84		158	305	355		156		
SS_{lin} [mV/dec]	72	76					175	273				260

Table 2. Benchmarking table with devices from other III-V CMOS processes as well as key p- and n-type standalone processes. I_{on} for InAs devices defined at $I_{off} = 100$ nA/um limit and for GaSb/InGaSb p-type devices is defined at $V_{DS} = -0.5$ V. Blank spaces are due to incomplete data.

978-1-7281-1988-5/18 $31.00 © 2018 IEEE

High Performance Quantum Well InGaAs-On-Si MOSFETs With sub-20 nm Gate Length For RF Applications

C. B. Zota, C. Convertino, Y. Baumgartner, M. Sousa, D. Caimi and L. Czornomaz
*IBM Research GmbH Zürich Laboratory, Säumerstrasse 4, CH-8803 Rüschlikon, Switzerland
E-mail: zot@zurich.ibm.com

Abstract—We demonstrate RF-compatible quantum well InGaAs MOSFETs integrated on Si substrates, with L_G down to 14 nm and a Si CMOS compatible RMG fabrication flow. Devices exhibit simultaneously extrapolated f_t and f_{max} of 370 and 310 GHz, respectively, the highest reported combined f_t/f_{max} for III-V MOSFETs on Si. This is enabled by the scaled L_G, g_m of 1.75 mS/μm, 8 nm source and drain spacers and raised source and drain extensions maintaining low access resistance. The use of the $InP/In_{0.75}Ga_{0.25}As/InP$ quantum well offers three times higher electron mobility and a 60% increase of g_m, compared to reference devices.

I. INTRODUCTION

High-electron mobility III-V semiconductors, such as the $In_xGa_{1-x}As$ system, have been widely investigated as replacements for Si in CMOS technology, offering increased drive current at reduced V_{DD} [1][2]. They are also currently being used as the channel in state-of-the-art HEMTs for high-speed electronics, as well as considered for future mm-wave applications [3]. Integration of III-V layers on Si substrates presents challenges and opportunities in both cases. For the former, integration is necessary in order to utilize the infrastructure and technology established by Si CMOS [4]. For the latter, successful integration would enable e.g. combined RF and digital signal processing capabilities, reducing cost and enabling new functionalities, improving upon what is currently supplied by Si RF-CMOS technology [5]. Integration of HEMTs on Si is challenging due to in part having to transfer a complex heterostructure onto a Si substrate with minimal loss of electron mobility, in part due to that standard HEMT fabrication utilizes a flow which is typically not compatible with Si CMOS. In this work, we present InGaAs-on-Si quantum well MOSFETs for RF-applications, utilizing a Si CMOS compatible replacement metal gate (RMG) fabrication flow and achieving record RF performance for III-V-on-Si FETs. Using this highly scalable technology, we also demonstrate ultra-scaled RF-devices with L_G down to 14 nm, among the shortest ever reported for this type of device.

II. FABRICATION

Fig. 1 shows a cross-sectional schematic of the fabricated InGaAs-on-Si quantum well RF-MOSFET. Fabrication follows a Si CMOS compatible RMG process flow with self-aligned raised source and drain (RSD) epitaxially grown contacts [6]. Compared to our previous work, we introduce here an $InP/In_{0.75}Ga_{0.25}As/InP$ quantum well (QW) in the channel, an enhanced SiN_x spacer process, as well as RF-optimized device structure and layout. The fabrication proceeds with integration of the QW heterostructure on a Si wafer by direct wafer bonding, a technique compatible with large-area Si substrates as well as 3D sequential integration [7]. Reference devices without the InP top barrier layer, as well as FinFET RF-devices are also fabricated (the latter formed by a fin dry etch of the channel). Subsequently, the dummy gate is deposited and patterned, and the 8 nm SiN_x spacers are formed by atomic layer deposition (ALD) and reactive ion etching (RIE). RSD spacers extensions are then formed by several cycles of controlled oxidation and etching, which form a cavity under the spacers as well as remove the InP top barrier in the contact regions. Following, metal-organic chemical-vapor deposition (MOCVD) regrowth of 25 nm n+ InGaAs ($N_D \sim 1e19$ cm^{-3}) RSD contacts is performed, which fills the cavities under the spacers to form the RSD extensions. An interlayer dielectric layer (ILD) is then deposited and planarized by CMP, followed by stripping of the dummy gate. Subsequently, the Al_2O_3 & HfO_2 (EOT ~1 nm)/TiN high-k metal gate is deposited by ALD, Fig. 2(a) shows a cross-sectional STEM image of the device at this stage. Following, W is sputtered and patterned as the gate top-metal for reduced gate resistance. Next, a second ILD is deposited, and the M1 contacts are deposited and patterned. For devices with more than two gate fingers, a second metal level is deposited to connect the sources of the device. Fig. 2(b) shows a cross-sectional SEM image of a fabricated device with two gate fingers, as well as a top-view SEM image in Fig. 2(c), showing the layout of the device.

Fig. 3 shows a HRTEM image of the channel region, including an energy-dispersive X-ray spectroscopy (EDX) map of the QW heterostructure. From the EDX map, the dimensions of the QW are determined to be 2 nm InP/10 nm InGaAs/20 nm InP. The InGaAs layer shows high crystalline quality, as well as a near-perfect InP/InGaAs interfaces.

III. RESULTS

Devices are first characterized under DC conditions. Fig. 4 shows output characteristics of devices with L_G = 20, 60 and 120 nm, respectively. The presence of the 2 nm InP top barrier decreases resilience against short channel effects (SCE) (g_d of references devices without the top barrier is 50% lower), but L_G = 20 nm devices nevertheless show relatively healthy output behavior. The on-resistance, R_{ON}, for the three devices is 400, 470 and 525 Ωμm, respectively. Transfer characteristics for the L_G = 20 nm device is shown in Fig. 5. Peak transconductance, $g_{m,peak}$, reaches 1.25 mS/μm at V_{DS} = 0.5 V, as well as maximally 1.3 mS/μm at V_{DS} = 0.9 V. The g_m peaks at V_{GS} = 0

978-1-7281-1988-5/18 $31.00 © 2018 IEEE

V, which is optimal for RF-applications due to improved gate oxide reliability. Devices without the InP top barrier peak instead at 0.3 V, which indicates a reduced influence of the interface traps (i.e. Fermi level pinning) in the former. Fig. 6 further shows $g_{m,peak}$ versus L_G for devices with and without the 2 nm InP top barrier. In scaled devices, i.e. those operating in the quasi-ballistic regime, thus scaling with the transmission, rather than the mobility, $g_{m,peak}$ is ~60% higher with the top barrier [8]. For long-channel devices, i.e. those operating close to the drift-diffusion regime, $g_{m,peak}$ is approximately 3x higher using the top barrier, indicating a similar difference in electron mobility between the two types of devices. We calculate the mobility from the slope of R_{ON} versus L_G, as shown in Fig. 6, by approximating the oxide capacitance. 1500 and 500 cm^2/Vs is obtained for devices with and without the top barrier, respectively. This difference is assumed to be caused by reduced surface roughness and oxide defect scattering using the top barrier. From the y-axis intercept, the extrinsic resistance $R_{ext} \approx 400$ $\Omega\mu m$ is obtained. This parameter can be further analyzed considering the schematic of the total contributions to $R_S = R_{ext}/2 = R_C + R_A + R_{Sp}$ shown in the inset of Fig. 8. From TLM measurements (Fig. 8) we determine $R_C = 75$ $\Omega\mu m$ and $R_A = 25$ $\Omega\mu m$, and deduce the spacer resistance $R_{Sp} = 100$ $\Omega\mu m$. Device performance is thus limited by the R_{Sp}. Nevertheless, R_S is comparable to that of state-of-the-art HEMTs, which utilize modulation doping to obtain very low R_{Sp}, but incur a penalty in the vertical direction due to the presence of an InAlAs barrier in the contact regions [9].

RF-characterization was performed up to 45 GHz with a Keithley vector network analyzer. De-embedding using on-chip open and short structures was performed up to (but excluding) M1. Fig. 9 shows a gain plot of a two-finger, $L_G = 20$ nm device exhibiting cutoff frequency $f_t = 370$ GHz and maximum oscillation frequency $f_{max} = 310$ GHz at $V_{DS} = 0.9$ and $V_{GS} = 0$ V. Here, f_t and f_{max} are extrapolated at -20 dB/decade using a hybrid-π small signal model with excellent fit to the measured S-parameters [10]. This represents the highest combined f_t and f_{max} for a III-V MOSFET on Si. Fig. 10 and 11 show f_t and f_{max}, respectively, versus L_G. The inset of Fig. 10 shows a STEM image of the channel region in a $L_G = 14$ nm device, one of the shortest L_G III-V RF-MOSFETs or HEMT devices ever fabricated, enabled by the highly scalable process flow presented in this work. f_t peaks at 370 GHz for $L_G = 20$ nm, while f_{max} peaks at 360 GHz at $L_G = 35$ nm due to lower gate resistance, R_G. Fig. 12 shows g_m and g_d at $V_{DS} = 0.9$ V, obtained from the hybrid-π model at $f = 10$ GHz. g_m peaks at 1.75 mS/μm for $L_G = 30 - 35$ nm and is somewhat reduced for shorter L_G due to short channel effects. The g_m frequency dispersion is minimal, indicating only minor influence of oxide border traps compared to other reports for III-V MOSFETs. This is explained by an increased tunneling distance from the channel to the border trap due to the InP top barrier. Fig. 13 further shows the voltage gain $A_V = g_m/g_d$ for both FinFET and planar RF-devices. Planar devices exhibit $A_V - 5$ at optimal L_G, while FinFETs exhibit $A_V = 20$ to 30 at similar L_G. This shows a potential advantage of FinFETs for RF-applications. FinFETs fabricated here, however, show reduced $f_t/f_{max} = 150/150$ GHz,

due to increased parasitic capacitances coming from insufficient fin spacing scaling.

Fig. 14 shows C_{gs} and C_{gd} versus L_G. C_{gs} contains contributions from the parasitic capacitances $C_{gs,par}$ (primarily between the RSD and the gate, as well as the gate metal and the S/D W plugs), the oxide capacitance C_{OX} as well as the quantum capacitance C_Q, which becomes significant for very scaled EOT such as in this work. Thus, as can be seen, scaling L_G from 20 to 14 nm offers only a small reduction of C_{gs}, 0.7 to 0.65 fF/μm, due to C_{gs} being dominated by $C_{gs,par}$ and C_Q. Together with the reduction of g_m due to SCE, this explains why peak f_t is obtained at $L_G = 20$ rather than 14 nm. Fig. 15 shows f_t and g_m versus the distance d between the source/drain and the gate, from 200 to 1900 nm, at $L_G = 20$ nm. In this range, $C_{gs} + C_{gd}$ is reduced by 30% due to decoupling of the parasitic capacitance between the gate metal and the source/drain W plugs, while g_m is reduced from 1.45 to 1.05 mS/μm due to increased R_A. Peak f_t is obtained at $d = 400$ nm due to an optimal combination of $C_{gs} + C_{gd}$ and g_m.

Fig. 16 shows a benchmark of f_t and f_{max} for III-V-on-Si MOSFETs as well as state-of-the-art Si RF-CMOS [11]-[16]. Dashed traces show geometric means, $\sqrt{f_t \times f_{max}}$. The devices shown in this work represent the first demonstration of a Si CMOS compatible III-V technology clearly outperforming state-of-the-art Si RF-CMOS.

IV. CONCLUSIONS

We have demonstrated quantum well InGaAs RF-MOSFETs integrated on Si substrates using a Si CMOS-compatible self-aligned RMG process flow with L_G down to 14 nm. $L_G = 20$ nm devices exhibit extrapolated $f_t = 370$ GHz and $f_{max} = 310$ GHz, the highest reported for III-V MOSFETs on Si. This is the first demonstration of III-V-on-Si clearly outperforming Si RF-CMOS, showing that III-V's could make a significant impact in this field.

ACKNOWLEDGMENT

This work was funded by Horizon 2020 grant agreement numbers 688784 (INSIGHT) and 687931 (REMINDER). The authors gratefully acknowledge the support of the BRNC operations team as well as the MIND group.

REFERENCES

[1] H. Riel, et al., MRS Bulletin, vol. 39, no. 8, p. 668, 2014.
[2] X. Sun et al., in VLSI Techn. Symp., T3-4, 2017.
[3] X. Mei et al., IEEE Electron Device Lett., vol. 36, no. 4, p. 327, 2015.
[4] S.-H. Kim et al. IEDM Tech. Dig., p. 429, 2013.
[5] E.-Y. Jeong et al., in VLSI Techn. Symp., T11-2, 2017.
[6] C. Zota et al., in VLSI Techn. Symp., T15-5, 2018.
[7] L. Czornomaz et al., IEDM Tech. Dig., p. 23.4.1, 2012.
[8] R. Kim et al., IEEE Trans. Electron Devices, vol. 7, no. 6, p. 787, 2008.
[9] J. Wu et al., IEEE Electron Device Lett., vol. 39, no. 4, p. 472, 2018.
[10] I. Kwon et al., IEEE Trans. MTT., vol. 50, no. 6, p. 1503, 2002.
[11] J. Singh et al., in VLSI Tech. Dig., p. 31, 2017;
[12] B. Sell et al., IEDM Tech. Dig., p. 685, 2017.
[13] A. Leuther et al., Proc. EuMA., p. 130, 2017.
[14] S. Johansson et al., IEEE Electron Device Lett., vol. 35, no. 5, p.518, 2014.
[15] C. Zota et al., IEEE Trans. Electron Devices, vol. 61, no. 12, p.4078, 2014.
[16] D.-H. Kim et al., IEEE Electron Device Lett., vol. 34, no. 2, p. 196, 2013.

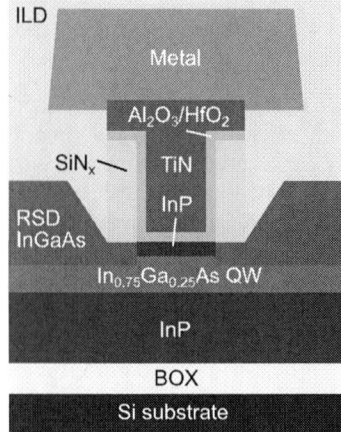

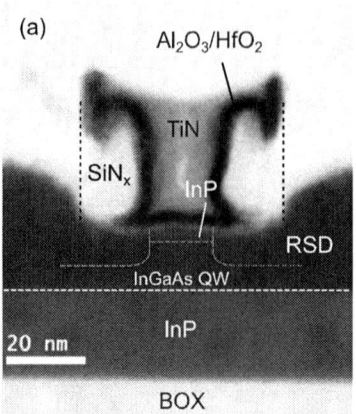

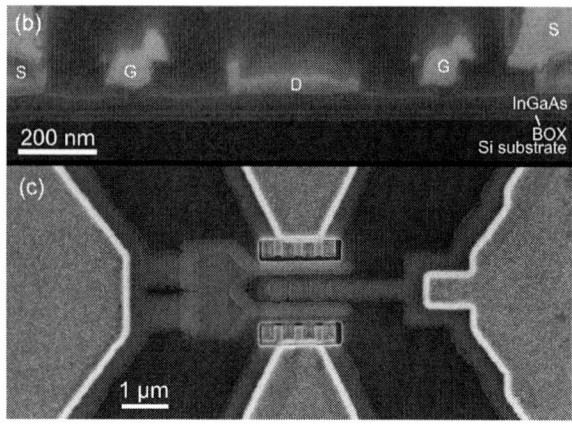

Fig. 1. Schematic cross-section of a fabricated device, with a HKMG, raised source and drain epi, SiNx spacers and a quantum well InGaAs channel.

Fig. 2. (a) Cross-sectional STEM pictograph of a fabricated device with L_G = 18 nm, before deposition of the M1 level. (b) Cross-sectional SEM pictograph of fabricated devices after M1. (c) Top-view SEM pictographs of a fabricated device with two gate fingers. Devices with 4 gate fingers were also fabricated, in which case an additional metal line connected the sources together.

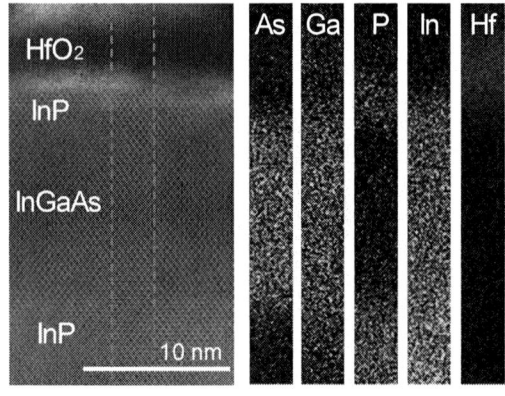

Fig. 3. HRTEM pictograph and EDX map of the channel region. 2 nm InP/10 nm $In_{0.75}Ga_{0.25}As$/20 nm InP is determined for the channel heterostructure. High crystal quality is observed for the InGaAs layer, as well as the InP/InGaAs interface, which gives rise to the strong increase of mobility as compared to devices without the InP top barrier.

Fig. 4. Output characteristics of fabricated devices with L_G = 20, 60 and 120 nm. The 20 nm device shows significant short-channel effects at high bias due to the presence of the 2 nm InP top barrier. The on-resistances are 400, 470 and 525 $\Omega\mu m$, respectively.

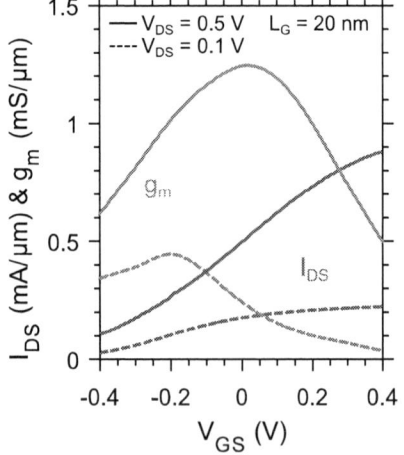

Fig. 5. Transfer characteristics of a device with L_G = 20 nm (same as in Fig. 4 and 11). Peak transconductance of 1.25 mS/μm is obtained at V_{DS} = 0.5 V.

Fig. 6. Average values of the on-resistance versus gate length, for devices both with and without the 2 nm InP top barrier. The extrinsic resistance R_{ext} is approximately equal, but the mobility is 1500 cm²/Vs and 500 cm²/Vs, with and without the InP, respectively.

Fig. 7. Peak transconductance for devices with and without the 2 nm InP top barrier. The increased mobility yields a 60% improvement at scaled L_G for devices with the QW, and a 3x improvement at long L_G, i.e. in the drift-diffusion regime, corresponding to the increase of mobility.

978-1-7281-1988-5/18 $31.00 © 2018 IEEE

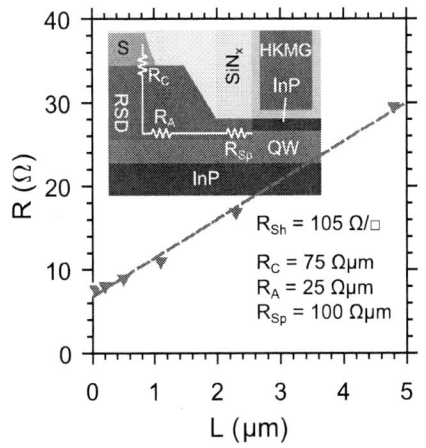

Fig. 8. TLM measurements of van der Pauw structures. Inset shows a schematic of the contact region with associated contributions to the total extrinsic resistance $R_{ext}/2 = R_C + R_A + R_{Sp}$.

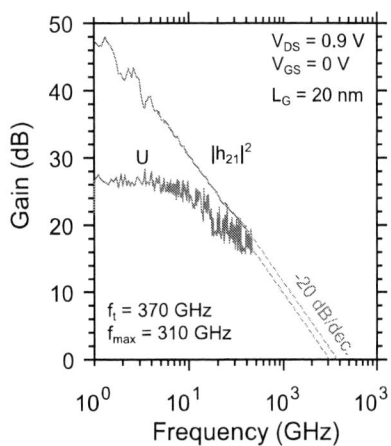

Fig. 9. Gain plot for the highest performing device. f_t and f_{max} are obtained from -20 dB/decade extrapolations confirmed by a small signal model with excellent fit to the measured S-parameters.

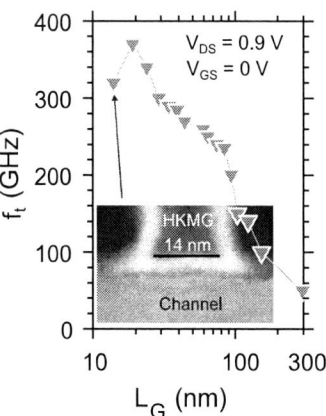

Fig. 10. f_t versus gate length. f_t peaks at 370 GHz for $L_G = 20$ nm. Inset shows a STEM image of the channel for the $L_G = 14$ nm device, which is among the shortest reported for an RF-MOSFET or HEMT.

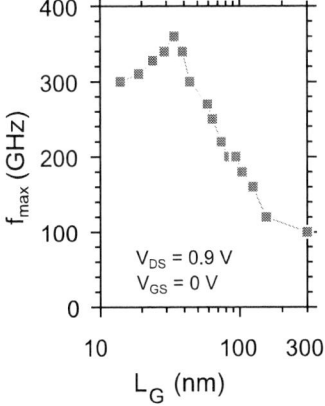

Fig. 11. f_{max} versus gate length. f_{max} peaks at 360 GHz for $L_G = 35$ nm. The reduction of f_{max} for shorter L_G is due to an increase of R_G, from ~30 Ω at maximum f_{max} to ~60 Ω at $L_G = 14$ nm.

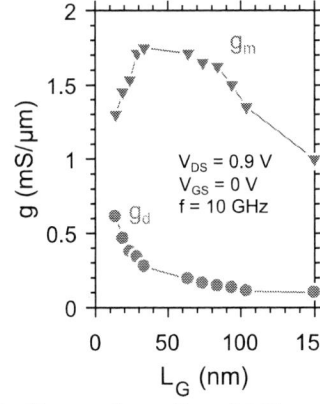

Fig. 12. g_m and g_d at $V_{DS} = 0.9$ V measured at 10 GHz versus L_G. For g_m, the frequency dispersion is minimal indicating only a minor impact from border traps. g_m peaks at 1.75 mS/µm but is reduced at shortest L_G due to short channel effects.

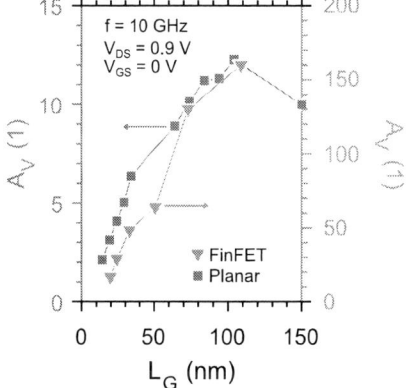

Fig. 13. Voltage gain $A_V = g_m/g_d$ for planar and FinFET RF-devices. At peak f_t, A_V is 4-5 for the planar devices, while being ~20 for the FinFETs due to strong reduction of g_d in the latter. However, FinFETs exhibit $f_t/f_{max} = 150/150$ GHz due to increased parasitics.

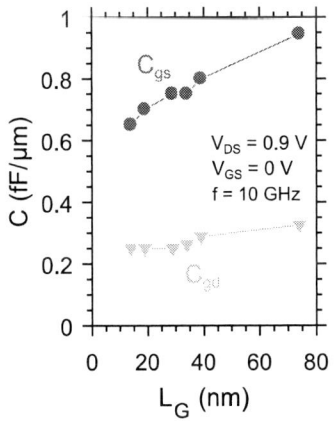

Fig. 14. C_{gs} and C_{gd} versus L_G. C_{gs} contains parasitic capacitances, as well as the oxide and quantum capacitances. Scaling of C_{gs} (through C_{ox}) with L_G is limited for this reason.

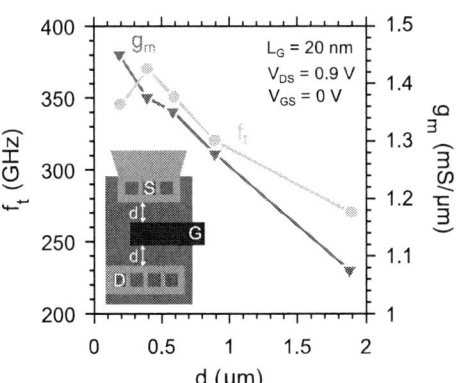

Fig. 15. f_t and g_m versus distance between source/drain and gate. In this range, g_m is reduced from 1.45 to 1.05 due to increased access resistance, while $C_{gs} + C_{gd}$ is reduced by 30%. The optimum between capacitance and g_m is found at $d = 400$ nm.

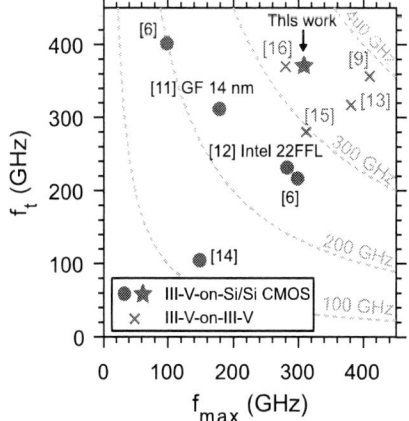

Fig. 16. Benchmark of III-V RF-MOSFETs as well as state-of-the-art Si RF-CMOS. The devices presented here for the first time clearly outperform Si RF-CMOS, as well as exhibit the highest combined f_t/f_{max} for a III-V-on-Si MOSFET.

978-1-7281-1988-5/18 $31.00 © 2018 IEEE

High Performance InGaAs Gate-All-Around Nanosheet FET on Si Using Template Assisted Selective Epitaxy

S. Lee[1], C. -W. Cheng[1], X. Sun[1], C. D'Emic[1], H. Miyazoe[1], M. M. Frank[1], M. Lofaro[1], J. Bruley[1], P. Hashemi[1], J. A. Ott[1], T. Ando[1], W. Spratt[1], G. M. Cohen[1], C. Lavoie[1], R. Bruce[1], J. Patel[1], H. Schmid[2], L. Czornomaz[2], V. Narayanan[1], R. T. Mo[1], and E. Leobandung[1]

[1] IBM T. J. Watson Research Center, 1101 Kitchawan Road, Yorktown Heights, NY 10598 USA,
Phone: 914-945-2518 Email: slee@us.ibm.com.
[2] IBM Research GmbH Zürich Laboratory, Säumerstrasse 4, CH-8803 Rüschlikon, Switzerland

Abstract—We report InGaAs gate-all-around nanosheet NFETs on Si substrate using template-assisted-selective-epitaxy (TASE) and a gate-last process with thermal budget advantages. Compared to our early report of the TASE process, in this paper we demonstrate that TASE can be scaled to a channel thickness of ~10 nm, which enables short gate devices without significant leakage. The defects and composition of the fabricated nanosheet FETs are also investigated. Enabled by this VLSI compatible process and a novel high-pressure deuterium annealing process, our 39 nm-L_g device shows a peak g_m of 1.37 mS/µm, a subthreshold slope in saturation of 72 mV/decade, and an I_{on} of 355 µA/µm at 0.5 V V_{gs}, the highest among reported sub-50 nm-L_g III-V FETs on Si.

I. INTRODUCTION

To meet the performance target beyond that of the 5 nm node, high mobility compound semiconductors, especially $In_{1-x}Ga_xAs$, are being studied as an alternative to strained Si for NFETs [1]-[9]. Despite well-matured material knowledge and many prototypes with good performance, the Si integration potential of the InGaAs channel is compromised by efforts to suppress non-uniformity and defects, and, by the required low thermal budget for this exotic compound material. With respect to overcoming these challenges, the TASE technique [10] provides advantages over other integration schemes: (1) less processing complexity and tighter channel pitch due to the absence of growth buffer required by other III-V-on-Si techniques, such as aspect-ratio-trapping (ART) [1], [9]; (2) compatibility with gate-all-around, stacked nanowires or nanosheets structure with the III-V layer isolated from Si; (3) flexible CMOS integration scheme because by using TASE, the low-thermal-budget III-V-based NFET process can be performed after the high-thermal-budget Si or SiGe PFET process. In this paper, we continue to report InGaAs nanosheet channel NFETs on Si substrate by TASE, highlighting a full-wafer uniform InGaAs nanosheet epitaxy as well as excellent device performance down to 39 nm L_g. To our knowledge, a record-high I_{on} is achieved for sub-50 nm-L_g III-V MOSFET devices directly grown on Si wafer. The crystal defects and material composition of fabricated FETs are also investigated.

II. TASE CHANNEL AND GATE-LAST PROCESS

The process flow of III-V nanosheet channel and raised source/drain(S/D) formation are described in Fig. 1. Devices are fabricated on (100) SOI substrate. The fabrication starts with thinning the Si on the BOX down to a target thickness. The template structures are aligned to the <110> direction and patterned into a Si nanosheet of ~20 nm thick and ~45 nm wide. An oxide shell is formed around the Si-nanosheet, and one tip of the shell is etched until the underlying Si is exposed. By selectively etching out the nanosheet from the opening of the shell, a hollow SiO_2 tunnel mold is formed with a Si seed at the end of the tunnel. MOCVD selectively grows III-V material from the exposed Si seed towards the opening of the shell, forming a single crystalline InGaAs nanosheet. After channel growth and shell removal, the dummy gate is formed. Following that, raised S/D consisting of heavily Si-doped $In_{0.53}Ga_{0.47}As$ is regrown on the InGaAs nanosheet. A thin liner is sequentially deposited and planarized. The dummy gate is pulled, and subsequently, the InGaAs channel is suspended from the BOX by a buffered HF dip. To achieve the target channel thickness of 10 nm, multiple cycles of digital etch are performed. An Al_2O_3/HfO_2 bilayer dielectric and TiN gate metal are deposited as a high-k-metal-gate by ALD. After filling W, the metal gate is then planarized. Gate and S/D contact holes are opened, followed by metal contacts (M1) formation. Prior to the electrical test, forming gas anneal (FGA) is performed to improve the metal-N+ InGaAs contact and the high-k/InGaAs interface. Fig. 2 (a) shows the cross-sectional schematic of the fabricated device. Fig. 2 (b) shows the TEM image of the cross-section of the InGaAs nanosheet for a 60 nm-L_g device. As shown, the channel is fully suspended from the BOX, resulting in a gate-all-around structure. The width and the thickness of the final nanosheet are 33 nm and 10 nm, respectively. Fig. 2 (c) shows the gate cross-section TEM for a 39 nm-L_g device. It should be noted that the digital etch expands the gate-length by 16 nm from the original 23 nm L_g in design. This could be resolved by adding a spacer prior to raised SD epitaxy.

III. CHANNEL EPI EVALUATION

Fig. 3 shows a top-down SEM image of InGaAs nanosheets grown on a whole wafer in ~45 nm width and ~20 nm thickness. The nanosheet is grown up to 1 µm in length and down to 250 nm in pitch. The growth of InGaAs can be initiated uniformly from the Si seed and be filled without any void in the tunnel mold across the wafer. Fig. 4 (a) shows HR-TEM along the

978-1-7281-1988-5/18 $31.00 © 2018 IEEE

nanosheet with a non-negligible density of [111] stacking faults. Based on the atomic-resolution STEM, these stacking faults are in the cubic phase with no evidence of antiphase boundaries. The stacking fault sequence shows a minimum of two layers before the next stacking sequence change. In Fig. 4 (b), further investigation by EDX line scan suggests that the $In_{1-x}Ga_xAs$ composition is Ga-rich (x~60%) near the Si seed, approximately even (x~50%) in the middle, and In-rich (x~40%) near the end of the nanosheet. This could be due to non-linear growth rate over time or transport of the precursor molecules in such a highly confined structure. Fig. 5 shows a HR-TEM image of the gate for the 39 nm-L_g device shown in Fig 2. (c). Twins and stacking faults, which are confirmed by FFT, are observed. The stacking faults are not aligned to [111] planes of the Si substrate, which indicates III-V is not strictly epitaxial to an [111] faceted Si seed. It is found that some of the stacking faults penetrate through the whole channel. The impact of these stacking faults on both on- and off-state should be moderate as evidenced by the devices with high g_m and low leakage described in section IV. However, it is still necessary to further investigate their effect on electron transport properties. Fig. 6 exhibits a Ga/In composition analysis of the 39 nm-L_g device. In Fig. 6 (a), a 2D EDX map of In (green) and Ga (red), the average Ga/In composition of the framed region is measured to be ~60/40. In Fig. 6 (b), an EDX line-scan also confirms that the channel composition is $In_{0.4}Ga_{0.6}As$.

IV. ELECTRICAL PERFORMANCE

Subthreshold and transfer characteristics of a 39 nm-L_g device after FGA are shown in Fig. 7 and Fig. 8. The device has SS of 72 mV/dec and 65 mV/dec at $V_{ds} = 0.5$ V and 50 mV, respectively, and a $DIBL$ of 17 mV/V at 1 µA/µm. Compared to the reported nanowire FET by TASE [10], this has significantly better off-state characteristics (i.e. I_{on}/I_{off} and SS_{sat} roll-off), which can be attributed to the thin nanosheet channel, Ga-rich channel composition and gate-all-around electrostatics. The device has 314 µA/µm of I_{on} at $I_{off} = 100$ nA/µm at $V_{ds} = 0.5$ V in Fig. 8. Fig.9 shows the output characteristics with R_{on} of 468 Ω-µm at $V_{ds} = 50$ mV and $V_{gs} = 0.8$ V. Fig. 10 shows $log(I_d)$-V_{gs} at various V_{ds} of the 39 nm L_g device. As seen, the minimum leakage current is exponentially proportional to the drain bias (inset), indicating that the minimum off-current could be limited by band-to-band-tunneling (or trap-assisted tunneling) under a high drain bias. With $V_{ds} = 0.5$ V, the SS_{sat} roll-off from SS_{lin} is negligible above the 100 nA/µm I_{off} limit, confirming that a low V_{ds} is preferred by the narrow band-gap channel. Fig. 11 and Fig. 12 show the R_{on} and SS_{sat} comparison with our 9 nm-wide InGaAs ART FinFETs reported in [1] at various L_g. In Fig. 11, the nanosheet device has a smaller R_{on} than the ART FinFETs in the same range of L_g, mainly due to the absence of spacers in the gate-last process. Fig. 12 shows a median SS_{sat} below 75 mV/dec for the nanosheet device, which is lower than that of the FinFET at similar L_g, indicating better electrostatics control in gate-all-around than in FinFET structures. As shown in Fig.13, SS_{lin} has an almost ideal median value of ~65 mV/dec, indicating excellent high-k/InGaAs interfaces free from thermal-induced degradation in our gate-

last process. $DIBL$ and V_T roll-off are also negligible at the L_g range of 39 nm in Fig. 14 and Fig. 15. Fig. 16 shows the median peak g_m increasing monotonically with a decreasing L_g. The highest peak g_m is ~1.1 mS/µm at L_g of 39 nm. To further enhance the device performance, we applied novel high-pressure deuterium (D_2) anneal [12-13] on the wafer. As shown in Fig. 17, the D_2 anneal reduces the minimum I_d by ~50% and improves the SS slope in the $I_{off} < 100$ nA/µm regime. This reduces the I_{off} level without SS_{sat} roll-off by an order from 100 nA/µm to 10 nA/µm under $V_{gs} = 0.5$V, which is favored by low power ICs. Our results suggest that the D_2 annealing is an effective way to suppress near-mid-gap interface states of InGaAs and therefore off-state leakage. Meanwhile, the I_d-V_{gs} and g_m are also improved by the D_2 annealing in Fig. 18 and Fig. 19. A higher 355 µA/µm I_{on} at $I_{off} = 100$ nA/µm and 1.37 mS/µm peak g_m at $V_{ds} = 0.5$ V is achieved. We can possibly attribute this to the suppression of border traps by D_2 [13] and contact resistance reduction from thermal annealing. Fig. 20 shows a benchmark of I_{on} at 100 nA/µm I_{off} and 0.5 V V_{gs}. Our devices demonstrate the highest I_{on} in the sub-50 nm-L_g regime among recently published CMOS-compatible InGaAs-based FETs integrated on Si substrates.

V. CONCLUSION

We have reported scaled (39 nm-L_g) InGaAs gate-all-around nanosheet FETs on Si using TASE and gate-last process. The integration is enabled by our wafer-level InGaAs epitaxy technique with promising uniformity and VLSI compatibility. The properties of defects and composition in the nanosheet channel are also investigated. Implementing a scaled channel has allowed us to achieve significantly improved off-state performance even at shorter L_g compared to our previous TASE work. A short-channel nanosheet device with 39 nm-L_g exhibits I_{on} of 355 µA/µm and SS_{sat} of 72 mV/dec. The results also suggest that the device performance can be further improved with more aggressive L_g scaling and at the same time, with optimizing the composition and thickness of the nanosheet channel.

ACKNOWLEDGMENT

We thank the staff of the IBM MRL for device fabrication. We also thank Dr. T. C. Chen and Dr. M. Khare for encouragement and management support. Poongsan/HPSP is acknowledged for performing the D_2 anneal.

REFERENCES

[1] X.Sun et al., in VLSI Techn. Symp., T3 - 4, 2017
[2] H. Hahn et al., in IEDM, pp. 17.5.1 - 17.5.4, 2017
[3] C.B.Zota etal., in IEDM, pp.3.2.1 - 3.2.4, 2016
[4] M. L. Huang et al., in VLSI Techn. Symp., pp.16-17, 2016
[5] L. Czornomaz et al., in VLSI Symp., 94-95, 2016
[6] V. Deshpande et al., in VLSI Techn Symp., T6-4, 2017
[7] O. Kilpi etal., in IEDM, pp. 17.3.1 - 17.3.4, 2017
[8] H. Schmid et al., in IEDM, pp. 3.6.1 - 3.6.4, 2016
[9] X. Zhou et al., in VLSI Symp., pp. 14-16, 2016.
[10] H. Schmid et al., Appl. Phys. Lett., vol. 106, 233101, 2015
[11] K. Shimamura et at., Apply Phys. Lett., vol 103, 022105, 2013
[12] M. M. Frank et al., IEEE SISC, 2017
[13] E. Cartier et al., WODIM, 2018

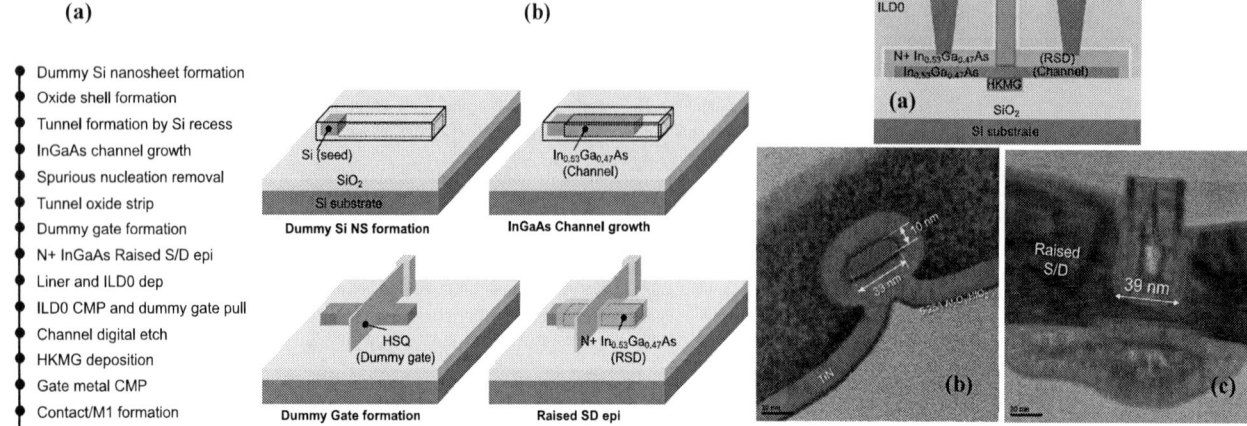

(a)

- Dummy Si nanosheet formation
- Oxide shell formation
- Tunnel formation by Si recess
- InGaAs channel growth
- Spurious nucleation removal
- Tunnel oxide strip
- Dummy gate formation
- N+ InGaAs Raised S/D epi
- Liner and ILD0 dep
- ILD0 CMP and dummy gate pull
- Channel digital etch
- HKMG deposition
- Gate metal CMP
- Contact/M1 formation
- H₂/N₂ anneal

Fig. 1. (a) Outlined process flow of InGaAs nanosheet N-FET. (b) schematics of the process flow for the nanosheet channel and raised S/D formation.

Fig. 2. (a) Cross-sectional schematic of the fabricated devices, (b) Cross-sectional TEM of the nanosheet channel (10 nm thickness x 33 nm width) for a 60 nm-L_g device, (c) Cross-sectional TEM of the TiN gate with L_g

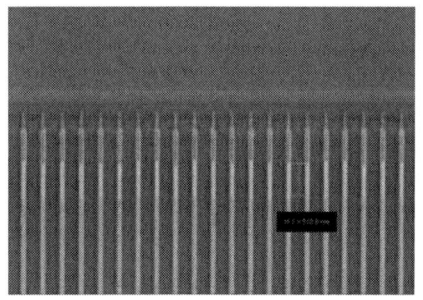

Fig. 3. Top-down SEM image after InGaAs nanosheet growth. The dimension of the tunnels is ~45 nm width x ~20 nm thickness x ~1 μm length with 250 nm pitch.

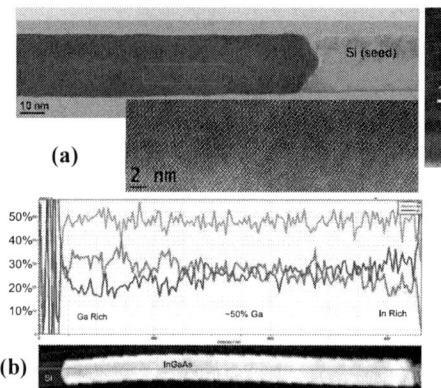

Fig. 4. (a) Cross-sectional HR-TEM images of a nanosheet channel growth, (b) EDX line-scan along the nanosheet for composition analysis.

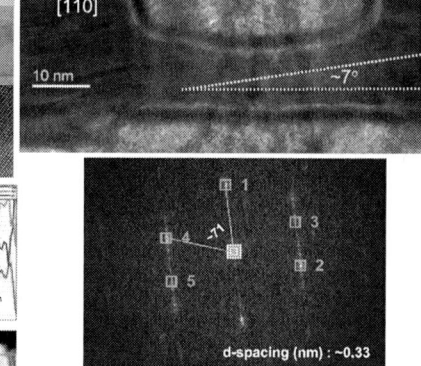

Fig. 5. HR-TEM image across the gate of a fabricated device with 39 nm-L_g and its FFT of the channel region.

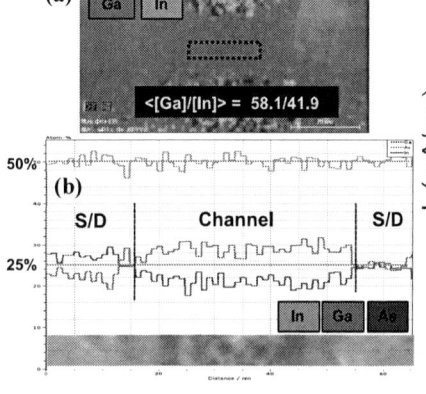

Fig. 6. EDX elemental analysis for a device with 39 nm-L_g (a) 2D mapping of Ga (red) and In (green). (b) line-scan of In, Ga, and As from the source to the drain

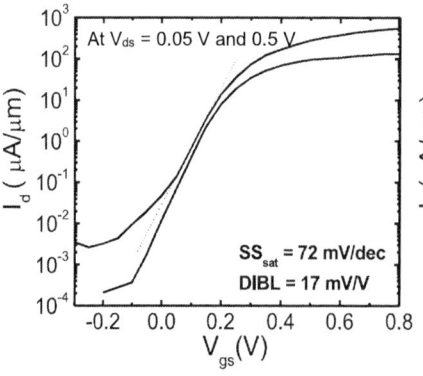

Fig. 7. Subthreshold characteristics, $\log(I_d)$-V_{gs}, of a 39 nm-L_g device. Its SS_{sat} and SS_{lin} are extracted to be 72 and 65 mV/dec, respectively.

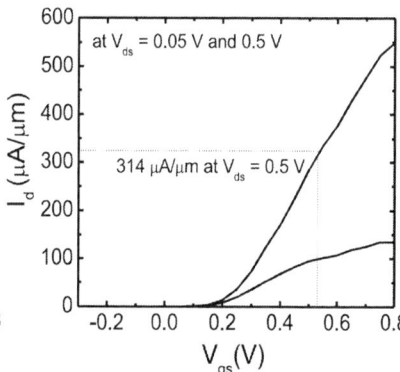

Fig. 8. Transfer characteristics, I_d-V_{gs}, of a 39 nm-L_g device. I_{on} at I_{off} = 100 nA/μm at V_{ds} = 0.5 V is measured to be 314

978-1-7281-1988-5/18 $31.00 © 2018 IEEE   913

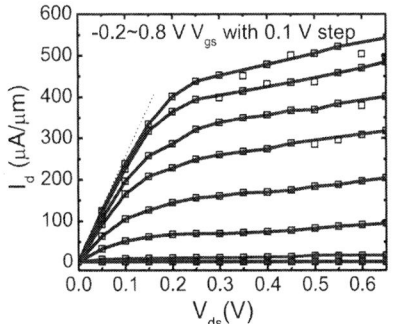

Fig. 9. Output characteristics (I_d-V_{ds}) of a 39 nm-L_g device. R_{on} is extracted to be 468 Ω-μm

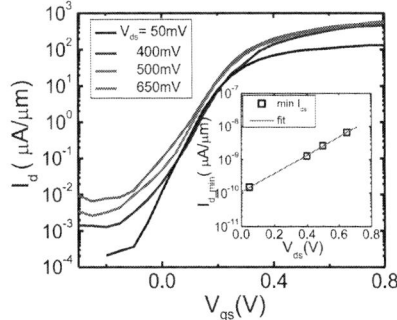

Fig. 10. log(I_d)-V_{gs} at various V_{ds} of a 39 nm-L_g device. (Inset) minimum I_d with respect to V_{ds}

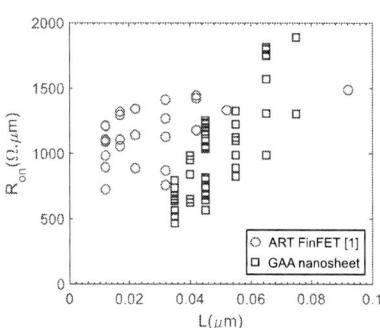

Fig. 11. R_{on} with respect to L_g, compared to the reported 9 nm-wide III-V FinFETs on ART [1]

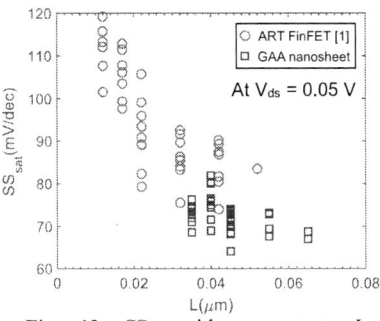

Fig. 12. SS_{sat} with respect to L_g, compared to the reported 9 nm-wide III-V FinFETs on ART [1]

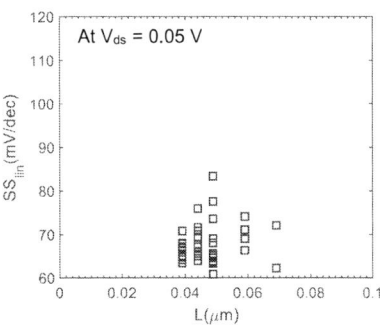

Fig. 13. SS_{lin} with respect to L_g.

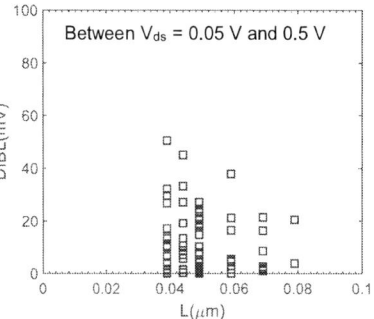

Fig. 14. DIBL with respect to L_g

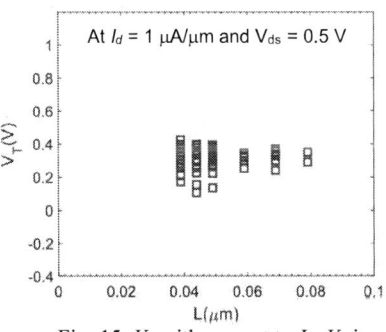

Fig. 15. V_T with respect to L_g. V_T is extracted at I_d = 1 μA/μm

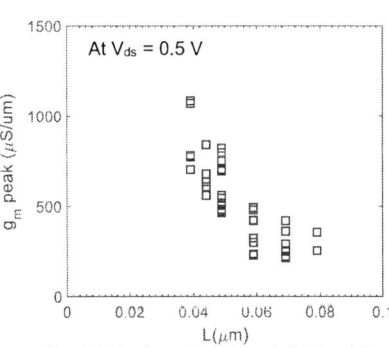

Fig. 16. Peak g_m (at V_{ds} = 0.5 V) with respect to L_g

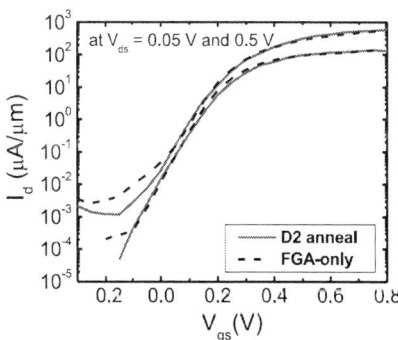

Fig. 17. Subthreshold characteristics (logI_d-V_{gs}) after D_2 anneal

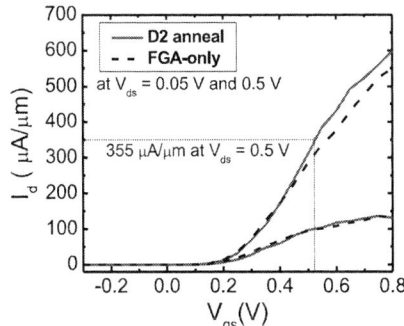

Fig. 18. Transfer characteristics of a 39 nm-L_g device after D_2 anneal

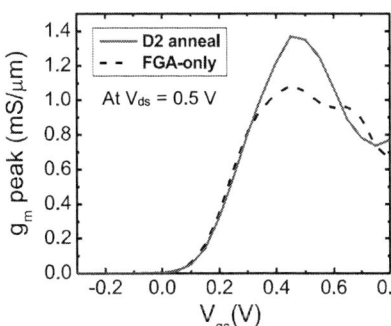

Fig. 19. Transconductance (g_m) of a 39 nm-L_g device after D_2 anneal

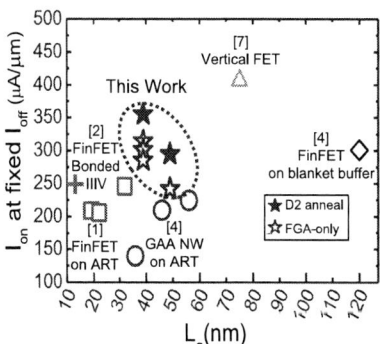

Fig. 20. I_{on} (at fixed I_{off}) vs. L_g Benchmark for CMOS-compatible InGaAs FETs integrated on a Si.

978-1-7281-1988-5/18 $31.00 © 2018 IEEE   914

Scaling Acoustic Filters Towards 5G

Yansong Yang, Ruochen Lu, and Songbin Gong

University of Illinois at Urbana Champaign, Urbana, IL, USA, email: yyang165@illinois.edu

Abstract— This paper presents a micro-electro-mechanical system (MEMS) filter at 10.8 GHz as the first step of scaling electromechanical filters towards fifth-generation (5G) frequencies beyond 6 GHz. The scaling of the center frequency to 10. 8 GHz is made possible by resorting to a higher order asymmetrical lamb wave mode (A3) in lithium niobate (LiNbO₃) MEMS resonators. The filter is then constructed using A3 resonator arrays in a ladder configuration. The fabricated resonator has demonstrated an electromechanical coupling (k_t^2) of 3.6% and a quality factor (Q) of 337. The Q is among the highest reported for piezoelectric MEMS resonators operating at this frequency range. The fabricated filter at 10.8 GHz has a 3 dB bandwidth of 70 MHz, a minimum insertion loss of 3.7 dB, an in-band ripple less than 0.1 dB, and a compact footprint of 0.7x0.5 mm².

I. INTRODUCTION

As the frequency bands below 6 GHz have already been fully allocated, the demand for more physical bandwidth to support increasing mobile data traffic has pushed 5G wireless systems towards higher frequencies. A mobile RF front-end beyond 10 GHz that can exceed the performance of its sub-6 GHz counterparts will be a key to the wireless connectivity promised by 5G. One key challenge in implementing front-ends beyond 10 GHz lies in the lack of high-performance miniature filters that can reject interference while allowing efficient access to the spectrum.

The commercial solutions for 4G front-end filters are surface-acoustic-wave (SAW) filters and film-bulk-acoustic-resonator (FBAR) filters. Their resonances are mostly limited to below 6 GHz [1, 2]. One of the promising candidates for enabling beyond-6 GHz acoustic or micro-electro-mechanical system (MEMS filters) is the recently-emerged higher-order asymmetric Lamb wave resonators based on lithium niobate (LiNbO₃) [3-5]. These resonators have been demonstrated with different orders of modes over a wide frequency range (1-30 GHz), showing great potential for enabling 5G front-end filters.

In this work, we aim to demonstrate beyond-10 GHz high-performance LiNbO₃ acoustic filters for the first time. To this end, the frequency scaling of this class of devices and its impact on electromechanical coupling (k_t^2) are first analyzed. Then, spurious modes response and energy confinement of asymmetric Lamb wave modes are studied. Based on these studies, a film thickness of 500 nm is chosen to scale the third-order asymmetric (A3) mode to 10.8 GHz. The fabricated resonator is measured with a k_t^2 of 3.6% and a Q of 337 at 10.8 GHz. The fabricated filter is measured with a 3 dB bandwidth of 70 MHz and an insertion loss (IL) of 3.7 dB at 10.8 GHz.

II. THIRD ORDER ASYMMETRIC LAMB WAVE RESONATORS

Asymmetric Lamb wave modes are a class of Lamb-wave modes characterized by their particular anti-symmetry about the median plane. The displacement mode shapes of the first and third order modes are shown in Fig. 1. The cross-section of the resonator can be treated as a two-dimensional cavity whose lateral and thickness dimensions set the resonance. The resonant frequency of an odd order mode, as discussed in previous work [5], can be determined by:

$$f_0^m = \frac{v_L}{2t} \sqrt{(\alpha m)^2 + (2\frac{t}{\lambda_L})^2} \tag{1}$$

where m is the mode order. v_L is the acoustic velocity in the longitudinal direction (+X axis in Fig. 1). t is the thickness of the cavity. α is the ratio between the velocities along the vertical and longitudinal directions. λ_L is the longitudinal wavelength.

Based on (1), it is apparent that the resonance of an A-mode device is predominantly set by the thickness and the mode order in the thickness direction. With a 500 nm thick Z-cut LiNbO₃ thin film and a mode order of 3, the asymmetric mode (A3) can be scaled beyond 10 GHz. The frequency scaling is also further confirmed by the Comsol-simulated phase velocity shown in Fig. 2.

The effect of frequency scaling through reducing the film thickness is also analyzed using Comsol for the A3 mode and shown in Fig. 2. A lower h/λ_L, i.e., a thinner film, also produces a higher k_t^2, therefore leading this work to focus on an h/λ_L below 0.1.

In addition to the resonance and coupling, spurious modes and Q are also considered. Spurious modes adjacent to the intended mode can introduce ripples in the passband of the comprising filter. In our asymmetric mode devices (Fig. 3) where acoustic energy is predominantly confined between electrodes, these spurious modes originate from insufficient confinement in the longitudinal direction. Therefore, Energy confinement of the A3 mode between interdigital transducers (IDTs) needs to be studied to ensure a spurious-free and high-Q response before their employment in a filter. As seen in Fig. 3, we approach the energy confinement by characterizing and comparing the wave numbers in the sections without (Medium 1) and with electrodes (Medium 2). The contrast between wave numbers would permit a quantitative understanding of the wave reflections at the interfaces between metalized and un-metalized regions, hence estimating the confinement between adjacent interfaces. To be consistent with the previous discussion of frequency scaling, the thickness (t) of the LiNbO₃ slab is set to be 500 nm and the covered metal is 100 nm thick aluminum. As seen in Fig. 4, the dispersion curves of the A1 and A3 waves in Medium 1 and 2 are calculated respectively using Comsol finite element analysis (FEA). For an A1 wave, the longitudinal wavenumber in Medium 2 ($\beta_{2,A1}$) is only several times of that in Medium 1 ($\beta_{1,A1}$), leading to insufficient confinement and spurious modes seen in Fig. 5. On the other hand, the difference between the longitudinal wavenumbers in these two mediums is significantly larger for the A3 mode [Fig. 4 (b)]. Therefore, the A3 wave has much better energy confinement between IDTs compared with the A1 wave, which would likely result in a spurious-free and high-Q response.

To further validate resonant characteristics and energy confinement, the resonator is simulated with 2D Comsol FEA. As shown in the cross-sectional mock-up (inset of Fig. 5), the device consists of a 3-electrode transducer on top of a mechanically suspended 500 nm thick Z-cut LiNbO$_3$ thin film. The design parameters are summarized in Table I. The simulated admittance response is shown in Fig. 5, including the displacement mode shapes of A1, A3, and the spurious modes adjacent to the A1 mode. Consistent with our theoretical analyses, the FEA results show that the A3 mode has a spurious-free response and less acoustic energy leakage to the LiNbO$_3$ sections covered by metal.

III. DESIGN OF ASYMMETRIC LAMB WAVE FILTER

To attain sharp roll-off and large fractional bandwidth, we use a simple ladder topology for the filter. The ladder filter consists of one series and two shunt resonator arrays. The resonators in each array are identical and should collectively have the same resonance. The array configuration is adopted to obtain sufficiently large static capacitances (C_0) to reduce the system impedance (Z_0) [Fig. 6(a)]. To achieve the maximum bandwidth allowed by the attained k_t^2, the resonant frequencies of series and shunt resonators arrays are designed with a frequency offset between them so that the parallel resonance of the shunt resonators closely aligns with the series resonance of the series resonators.

According to the dispersion curves in Fig. 4(b), the required resonant frequency offset can be attained by varying the distance between interdigitated electrodes (g) and hence the longitudinal wavelength. Thus, the electrodes gaps for the series and shunt resonators are chosen as 3 μm and 7 μm, respectively (Fig. 11). As seen in Fig. 6(b), the simulated responses of a single series resonator and a single shunt resonator show a frequency offset of 200 MHz.

IV. EXPERIMENTAL RESULTS AND DISCUSSION

To validate the analytical and modeling results, the designed resonators and filters were fabricated using a 500 nm transferred Z-cut LiNbO$_3$ thin film following the process in Fig. 7. The fabricated devices were characterized with a Keysight N5230A PNA-L network analyzer in dry air and at room temperature.

A fabricated standalone resonator is shown in Fig. 8. Similar to the simulation results in Fig. 5, the measurement

result shown in Fig. 10 exhibits two resonant frequencies at 4 GHz and 10.8 GHz, corresponding to the anticipated A1 and A3 modes respectively. The measurement is also compared to a multi-resonance MBVD model shown in Fig. 9 for extracting the key performance parameters listed in Table II. The MBVD modeled response is also shown in Fig. 10. As expected, the A3 mode shows a spurious-free response with a k_t^2 of 3.6 % and a Q of 337. The Q is among the highest reported for piezoelectric MEMS resonators operating at this frequency range.

The fabricated filter is shown in Fig. 11 with an overall footprint of 0.7×0.5 mm^2. The electrode gaps of the series and shunt resonators arrays are labeled in the zoomed-in SEM images [Fig. 11(b) and (c)]. The measured S21 and S11 responses, matched to a system impedance of 140 Ω, are shown in Fig. 12. Our first demonstration of scaling a LiNbO$_3$ filter beyond 10 GHz has shown an IL of 3.7 dB, an out-of-band rejection of -20 dB, an in-band ripple of below 0.1 dB, and a bandwidth of 70 MHz (Table III). The achieved bandwidth is lower than the intended frequency offset due to the large parasitic capacitance paralleled with the series resonators array, which can be mitigated by reducing the size of bus lines and lead lines. As seen in Fig. 13, the unwanted passband around 4 GHz induced by the A1 mode resonance can be suppressed by applying a shunt LC branch without compromising the performance of the intended passband.

V. CONCLUSION

In this work, the design of the third-order asymmetric Lamb wave mode resonator in LiNbO3 is first theoretically studied to compose acoustic filters at higher frequencies than 4G bands. The demonstrated filter at 10.8 GHz is the first step of scaling LiNbO$_3$-based acoustic devices towards the 5G frequency bands well beyond 10 GHz.

ACKNOWLEDGMENT

The authors would like to thank the DARPA NZERO program for funding support.

REFERENCES

[1] T. Tsutomu, *et al.*, "I.H.P. SAW Technology and its application to micro-acoustic components," *in 2017 IEEE International Ultrasonic Symposium (IUS)*, Sept. 2017.

[2] R. C. Ruby, R. Parker, D. Feld, "Method of Extracting Q Applied Across Different Resonator Technologies," IEEE Ultrasonics Symposium, 2008, pp. 1815-1818.

[3] Y. Yang, *et al.*, "5 GHz lithium niobate MEMS resonators with high FoM of 153," *2017 IEEE 30th Int. Conf. on MEMS*, Las Vegas, NV, 2017, pp. 942-945.

[4] Y. Yang, *et al.*, "1.7 GHz Y-cut Lithium niobate MEMS resonators with FoM of 336 and $f \cdot Q$ of 9.15×10^{12}," *in Microwave Symposium (IMS), 2018 IEEE MTT-S International*, June 2018, pp. 1-4.

[5] Y. Yang, *et al.*, "Towards Ka Band Acoustics: Lithium Niobate Asymmetrical mode Piezoelectric MEMS Resonators," *in 2018 IEEE International Frequency Control Symposium (IFCS)*, May 2018, pp. 1-4.

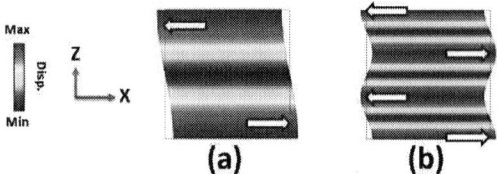

Figure 1. Displacement mode shapes of the (a) first-order (A1), and (b) third-order (A3) asymmetric modes. The arrows denote the displacement directions.

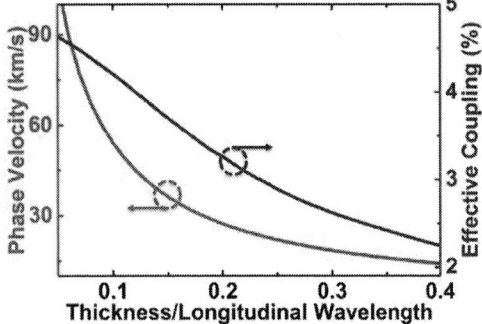

Figure 2. Simulated phase velocity and effective electro-mechanical coupling factor (k_{eff}^2) of the A3 mode *vs.* the ratio of film thickness to longitudinal wavelength (h/λ_L).

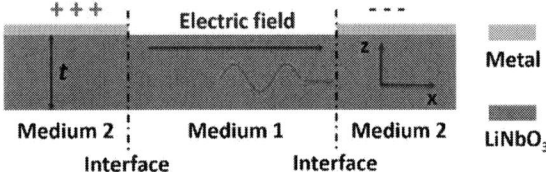

Figure 3. Simplified distribution of electrical field and model of 2-D acoustic wave propagating in A-mode LiNbO3 resonators.

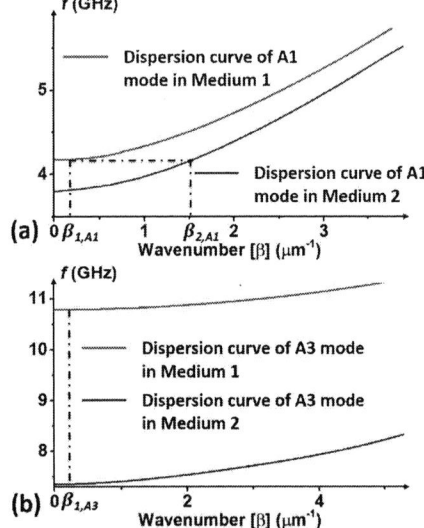

Figure 4. Dispersion curves of (a) A1, and (b) A3 modes in the two different mediums depicted in Fig. 3.

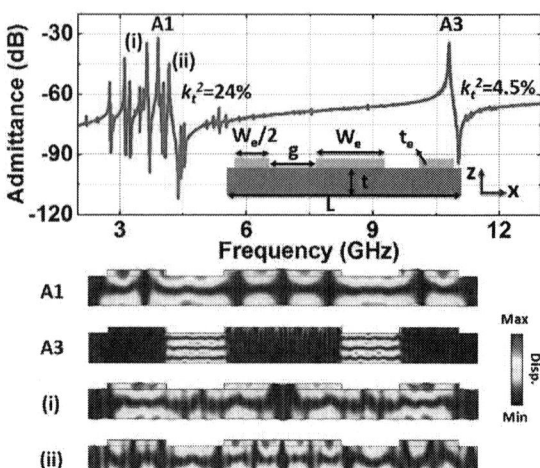

Figure 5. Simulated response of a Z-cut LiNbO₃ A-mode resonator. The simulated displacement mode shapes of A1, A3 and spurious modes (i & ii) are presented. The dimensions of the simulated device are listed in Table I.

Table I
Physical dimensions of the designed asymmetric resonator

Parameter	Description	Value
L	Resonator total length	100 μm
W	Resonator total width	20 μm
W_e	Electrode width	6 μm
g	Gap between electrodes	3 μm
t	Thickness of LiNbO₃	500 nm
t_e	Thickness of metal electrodes	100 nm
N	Number of electrodes	3

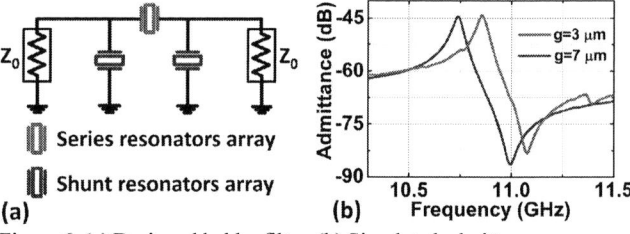

Figure 6. (a) Designed ladder filter. (b) Simulated admittance responses of the series and shunt resonators with different electrode gaps, showing a frequency offset of 200 MHz.

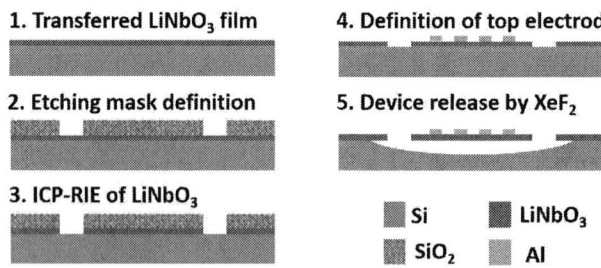

Figure 7. Fabrication process.

978-1-7281-1988-5/18 $31.00 © 2018 IEEE

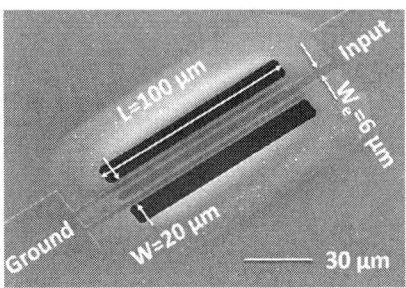

Figure 8. SEM image of the fabricated standalone resonator.

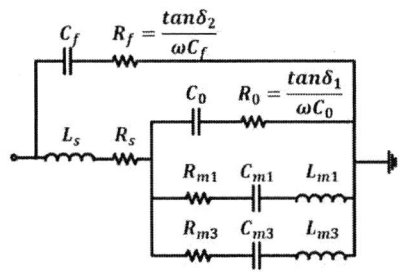

Figure 9. Multi-resonance MBVD model.

R_{m1}, C_{m1}, L_{m1}: Motional branch of A1 resonance
R_{m3}, C_{m3}, L_{m3}: Motional branch of A3 resonance
C_0: Static capacitance
R_0: Loss in LiNbO$_3$ film
L_s: Series inductance of the electrode fingers
R_s: Series resistance of the electrode fingers
C_f: Feedthrough capacitance
R_f: Loss in the Si substrate
$tan\delta_1$: Loss tangent of the LiNbO$_3$ film
$tan\delta_2$: Loss tangent of the Si substrate

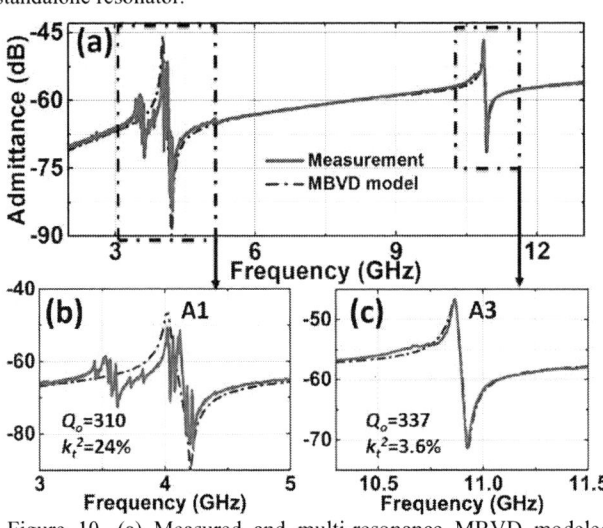

Figure 10. (a) Measured and multi-resonance MBVD modeled responses of the fabricated resonator. Zoomed-in views of the (b) A1, and (c) A3 mode resonances.

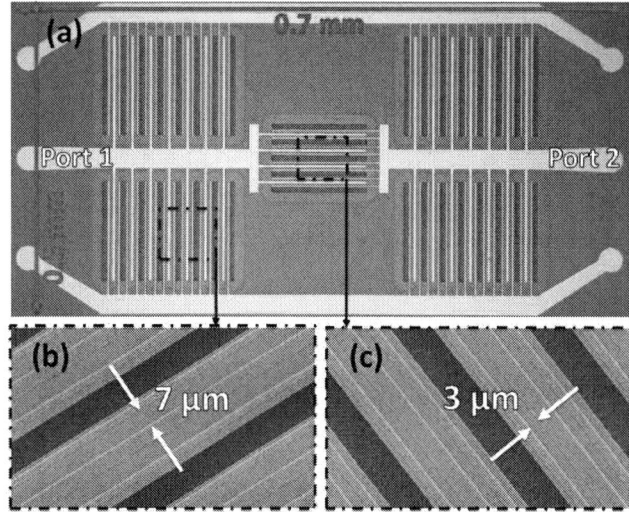

Figure 11. (a) Optical image of the fabricated ladder filter. SEM images of zoomed-in views of the (b) shunt resonators, and (c) series resonators.

Table II
Key measured and extracted parameters (from MBVD model) of the A3 mode resonance

Parameters	Value
f_0	10.8 GHz
Q_0	337
k_t^2	3.6%
FoM	12
R_{m3}	148 Ω
C_{m3}	0.21 fF
L_{m3}	1.04 μH
C_0	7.6 fF
R_s	40 Ω
C_f	12 fF
L_s	100 pF
$tan\delta_1$	0.05
$tan\delta_2$	0.005

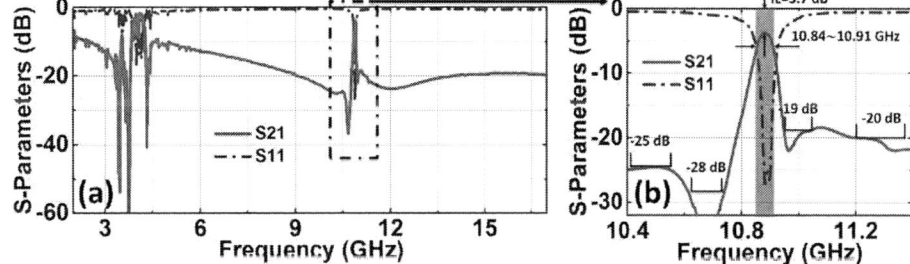

Figure 12. (a) Wide-frequency-range measured response of the fabricated ladder filter. (b) Zoomed-in view of the intended passband.

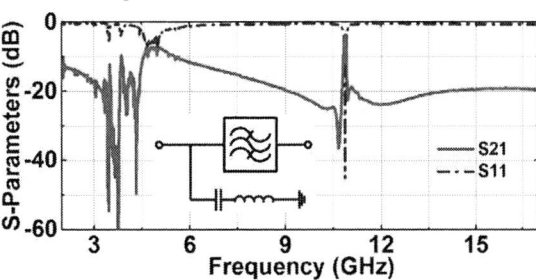

Figure 13. Wide-frequency-range response of the fabricated ladder filter with a shunt LC branch.

Table III
Key measured parameters of the fabricated filter

Parameters	Value
Center frequency	10.8 GHz
Z_0	140 Ω
Insertion loss	3.7 dB
3-dB bandwidth	70 MHz
OoB rejection	-20 dB
In-band ripple	<0.1 dB

978-1-7281-1988-5/18 $31.00 © 2018 IEEE

Physics of hole trapping process in high-k gate stacks: A direct simulation formalism for the whole interface system combining density-functional theory and Marcus theory

Yue-Yang Liu, Xiangwei Jiang[*]

Institute of Semiconductors, Chinese Academy of Sciences, Beijing, P. R. China, [*]email: xwjiang@semi.ac.cn

Abstract—Charge trapping defects in high-κ dielectrics and at their interfaces are known to be a challenging obstacle for the silicon based modern transistors. To facilitate the solution of such problems, a deeply physical understanding of the charge trapping process at atomistic scale is mandatory. As such, we propose, for the first time, a direct method to calculate the exact hole trapping rates explicitly in the high-κ gate stack consisting of silicon channel, SiO₂ interfacial layer and HfO₂. The physics of multiple-path (by trap locations) hole trapping processes is revealed by combining density-functional theory and Marcus theory. The roles of physical quantities including defect reorganization, coupling constant and Gibbs free energy are discussed. It is suggested that oxygen vacancies at high-κ interface with interfacial layer SiO₂ are dominant hole traps under NBTI stress. The developments and findings provide not only a deep physical insight into the hole trapping related reliability degradation mechanism, but also a new simulation framework.

Introduction

Replacing the conventional SiO₂/SiOₓNᵧ gate dielectric with high-κ (HK) materials such as HfO₂ has been widely accepted as a solution to reduce the gate leakage current caused by continuous size scaling [1-4]. However, while the reduction degree is satisfying, the high-κ materials on Si substrate suffers from much higher defect density than SiO₂, and thus leads to more severe charge trapping and threshold voltage shift as seen in the so-called bias temperature instability (BTI) phenomena. Due to the nature of quantum mechanics of such microscopic process, deep physical understanding of the charge trapping in Si/high-κ interface structures is in urgent demand to facilitate the solution of such problems and to improve the integration of high-κ materials in silicon CMOS technology.

Previous works have demonstrated that oxygen vacancies and oxygen interstitials are the most likely intrinsic defects in HfO₂ due to their low formation energies, and that oxygen vacancies in the HfO₂ layer are supposed to be the main charge traps in Si/HfO₂ systems because their energy levels are very close to the band edge of Si [5]. Although the above conclusion is illuminating, it is kind of rash because the trapping process of charge from Si to HfO₂ in reality is much more complicated. Firstly, the energy difference between the defect level and the Si band edge is not the only factor that determines the charge trapping rate, the coupling strength between the two states also plays an important role [6]. Coupling strength is determined by the distance and atomic structure between Si and the defect. It becomes especially important when two interfaces (Si/SiO₂ and SiO₂/HfO₂) exist as SiO₂ interfacial layer (IL) intentionally inserted. Secondly, the energy levels of oxygen vacancies in Si/SiO₂/HfO₂ interface structures variate significantly with

respect to their locations, i.e. at the SiO₂ layer, at the SiO₂/HfO₂ interface, or at the HfO₂ layer, instead of being a constant [7]. Thirdly, the defect levels in the dielectric layer depends greatly on the applied gate voltage, and so does the energy difference between the defect level and Si band edge. Last but not least, the existence of multi-interfaces makes the trapping process difficult to be described by phenomenological methods such as the WKB approximation, which is usually adopted in charge trapping studies in previous works [8, 9]. A more universal and accurate method is in urgent demand.

In this work, we propose a universal simulation method to capture the hole trapping process in Si/SiO₂/HfO₂ stack. Fully density-functional theory is applied to the whole stack system combining with Marcus theory for charge transfer. In this way, all the important factors of charge trapping process, including atomistic interfaces, defect levels and energy barriers, coupling strength, and oxide electric fields are taken into consideration.

Theory and Simulation of Charge Trapping

As schematically depicted in **Fig.1**, there are multiple paths of hole trapping process in a high-κ gate stack *p*-type MOSFET, namely the interfacial layer oxide hole traps, high-κ oxide hole traps and the (HK/IL) interface hole traps. To understand such a microscopic phenomenon, atomistic simulation of complex interfaces and charge trapping dynamic model are needed.

1.Marcus charge transfer theory

In 1950's, R. A. Marcus proposed an intuitive yet universal theory for charge transfer process in molecular systems [10]. In Marcus's physical picture, the sophisticated charge transfer process reduced into a single formula:

$$\tau^{-1} = \frac{2\pi}{\hbar}|V_C|^2 \sqrt{\frac{1}{4\pi k_B T}} \exp\left[-\frac{(\Delta G + \lambda)^2}{4\lambda k_B T}\right] \qquad (1)$$

where three key physical parameters determine the final charge transfer rate, namely the coupling constant V_C, the total Gibbs free energy change ΔG during the charge transfer reaction, and the reorganization energy λ. In the following, we will explain how those key physical quantities be obtained by fully density-functional theory as illustrated in the flowchart of **Fig.2**.

The schematic description of Marcus theory and the charge trapping process in a Si-dielectric defect system is shown in **Fig.3**. The horizontal axis represents different configurations of the coupled system, and the vertical axis is the energy of the configurations. The red line represents the energy profile when an electron/hole is at Si, while the blue line represents the case when the charge trapped into a defect of the gate dielectric. At the beginning of the hole trapping process, the hole lies on the valence band maxima of silicon channel (E_{VBM}), with a very delocalized wave function. Then the hole transfers to the defect by crossing an energy barrier that denoted by the green arrow in **Fig.3**. The amplitude of such a Landau-Zener transition as depicted in the anti-crossing Si VBM and defect state profile in **Fig.3** is determined by the coupling constant $|V_C|^2$. After that, the configuration will experience a structural relaxation due to the occupation of the defect level by a hole and the Gibbs free energy change of the system will be:

$$\Delta G = E_{trap} - E_{VBM} = E_{defect} - \lambda - E_{VBM} \qquad (2)$$

where E_{trap} represents the energy level of the charge trapping defect in a charge state, while E_{defect} represents its counterpart in neutral state. λ represents the total reorganization energy of the system, which reduces to the value of the charge trapping defect because of the vanishing reorganization for Si VBM.

Provided the above three key parameters, i.e. reorganization energy λ, Gibbs free energy change ΔG, and coupling constant V_C, the final charge trapping rate (from state to state) can be obtained through Marcus formula (**Eq.1**). In the following, we will provide the calculation framework of the three parameters based on fully density-functional theory.

2. Density-functional theory simulation

In our demonstration of the density-functional theory-based simulation, we constructed two sets of Si/SiO$_2$/HfO$_2$ interface structures, both with (001) orientations as shown in **Fig.4**. SiO$_2$ interfacial layers of 0.7 and 1.2 nm are considered, and various locations of the oxygen vacancies are calculated. In accordance with the most recent reports, the phases of the SiO$_2$ and HfO$_2$ are chosen as β-cristobalite and monoclinic, respectively **[8]**, and the side length of the unit cell is set as $10.86 \times 10.86 Å^2$. DFT calculations are carried out by the plane-wave package PWmat with GPU acceleration **[11]**. Hybrid HSE functional is used in all parts of the DFT calculation except the geometry optimization to ensure the accuracy of the band alignment and defect levels. HSE parameters are set separately with a mask function for different materials in the interface system. GGA-PBE functional is used for both interface structure relaxation and defects' local environment relaxation with a convergence criterion of 0.01eV/Å as the residual force.

The reorganization energy λ of the charge trapping defects is obtained by comparing the total energy of the system before and after a charge being trapped into the trapping center, which is depicted in **Fig.2** and **Fig.4**. The Gibbs free energy change ΔG can be obtained straightforwardly once the band alignment, defect levels and reorganization energy are known, by **Eq.2**.

The most important but difficult one is the coupling constant V_C. In the picture of Landau-Zener transition, we intentionally drive the two-level system ($E_{VBM} \sim E_{defect}$) into a crossing region by applying an electric field inside the oxide. As will be seen, an anti-crossing behavior occurs, and the coupling constant is exactly half of the anti-crossing energy gap.

Results and Discussion

1. Charge trapping defect reorganization

Fig.4 illustrate the reorganization process for three typical defect (oxygen vacancy V$_O$) locations. It is suggested that the hole trapping reorganization to V$_O$ center in SiO$_2$ remains weak, with a value of 0.31 eV. The reorganizations of hole trapping into V$_O$ centers in SiO$_2$/HfO$_2$ interface and HfO$_2$ bulk are essentially comparable, resulting to the value of reorganization

energy around 1 eV.

2. Band/defect level alignment and Gibbs free energy

Fig.5 and **Fig.6** depict the band/defect level alignment of the Si/SiO$_2$/HfO$_2$ interface system. By using the hybrid functional HSE in the self-consistent (SCF) DFT calculation, reasonable band gaps and band edge alignments are obtained. It is found that interface transition regions are not sharply recognized. The local atomic nature of interfaces must be considered in the physical simulation of the multi-interface gate stacks. **Fig.6** summarizes the defect level (eigen-level, E_{defect}) alignments of multiple-source V$_O$ centers. V$_O$ defects inside the SiO$_2$ IL are deeply bellow the Si VBM, so that not favorable for hole trapping. Those located at the HK/IL interface and inside HfO$_2$ are energetically more favorable, with the interface V$_O$ levels ranged in 0.64~1.40 eV bellow Si VBM which are most likely the hole trapping centers under NBTI stress, and HfO$_2$ V$_O$ levels ranged in 0.40~0.50 eV above VBM. From above results, Gibbs free energy change is ready to be obtained from **Eq.2**.

3. VBM-defect coupling

The coupling between defect state and silicon valence band state is the most difficult to simulate. This is done by DFT SCF calculations, intentionally applying a perturbation electric field to make the two levels cross. **Fig.7** and **Fig.8** depict the strong and weak coupling process. One can clearly see how the wave functions of the VBM state and defect state exchange during the trapping process. And the coupling constant V_C is extracted from the Landau-Zener anti-crossing energy gap.

In **Fig.9**, we distinguish different decay behaviors of the V_C of V$_O$ defects into three types. One is for V_C of SiO2 defects, another is for V_C of SiO$_2$/HfO$_2$ interface defects and HfO$_2$ border defects which are very close to the interface. Third is for HfO2 bulk defects. Although the first two cases both obey the exponential decay, it is surprised to find the decay length is largely different, 1.58 Å for SiO$_2$ and 1.92 Å for interface and HfO$_2$ border defects, indicating the WKB approximation is not valid for complex gate stack simulation.

4. Hole trapping rate

With all the decisive factors obtained, the hole trapping rate can be calculated by using **Eq.1**. In a FET, the hole trapping is always electric field-dependent. The oxide electric field F_{OX} will change the alignment of the defect level with respect to the Si VBM. We take this re-alignment into account, and the F_{OX} dependent hole trapping rate in the case of model-I and model-II for different defects are shown in **Fig.10**. It is confirmed the interface oxygen vacancy dominates the hole trapping during NBTI stress. Although the HfO$_2$ defects result in comparable state-to-state trapping rate when NBTI stress vanishes, the number of available holes in the channel becomes negligible. At last, the (110) oriented Si/SiO$_2$/HfO$_2$ gate stack is shown in **Fig.11**. And it is found the hole trapping in the (110) orientation is much less severe.

Conclusion and Outlook

Deep physical insights into the hole trapping process in high-κ gate stack, as well as a new simulation formalism are provided. Given the key physical parameters and fitting models in Table-I, this framework can be incorporated into reliability simulator.

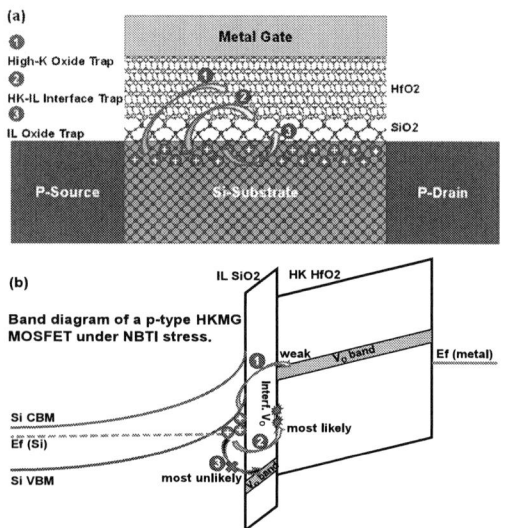

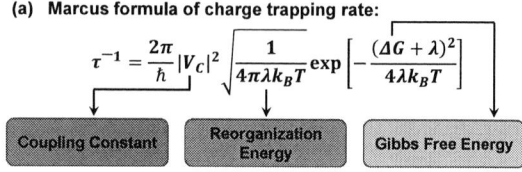

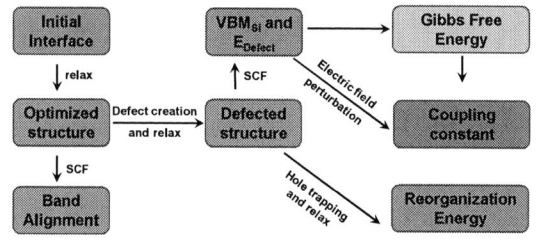

Fig. 1 Schematics of the multiple hole trapping sources: high-k traps, interface traps at high-k and interfacial layer, and SiO₂ interfacial layer traps. (a) is the typical device structure of a p-type FET, and (b) is the band diagram. From an energy level point of view, the interface defects are most likely the hole trapping centers.

Fig. 2 (a) The core equation of Marcus theory describing the charge trapping process, with its key parameters of coupling constant, reorganization energy and Gibbs free energy calculated through sophisticated density-functional theory. The calculation of these key parameters is summarized in a DFT flowchart (b).

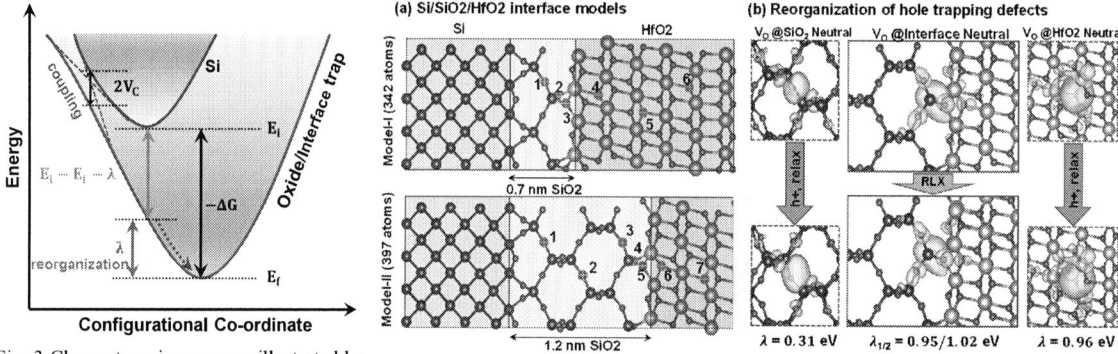

Fig. 4 (a) Two Si/SiO₂/HfO₂ gate stack models with 0.7 and 1.2 nm ILs, respectively. Six and seven oxygen vacancy positions are indicated. (b) Reorganization of the defect after charge trapping.

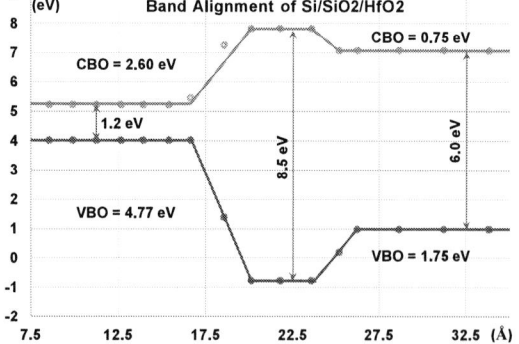

Fig. 3 Charge trapping process illustrated by an Energy Configuration diagram, explaining the key parameters of Marcus theory.

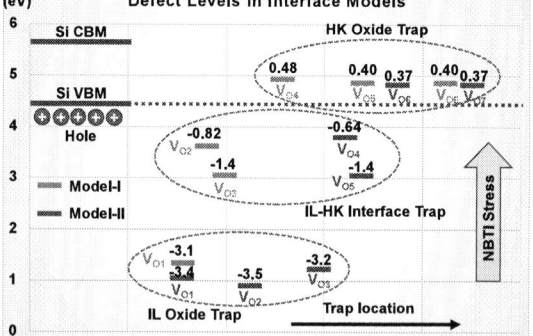

Fig. 5 The band alignment of Si/SiO₂/HfO₂ system calculated by density-functional theory with hybrid functional (HSE), which is extracted from a full DFT calculation of a 397-atom supercell.

Fig. 6 Various oxygen vacancy defect levels in interface model-I and model-II, with alignments to the silicon valence and conduction band edges. During NBTI stress, the defect levels go up, indicating the IL-HK interface defects are the most-likely hole trapping centers.

978-1-7281-1988-5/18 $31.00 © 2018 IEEE

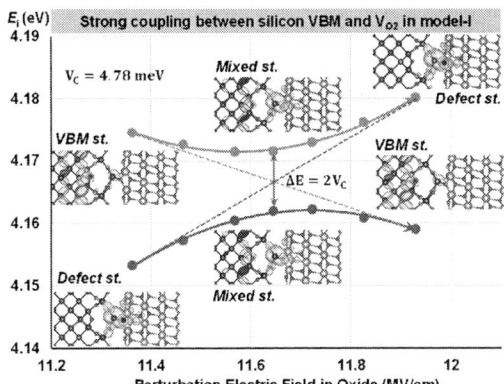

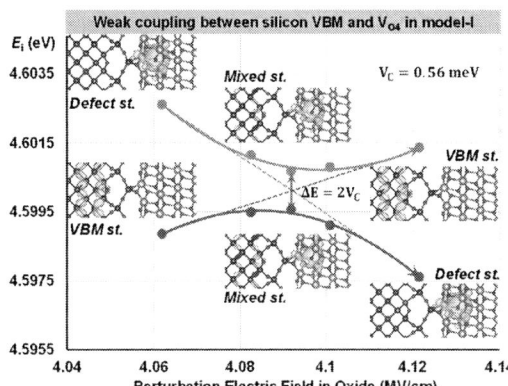

Fig. 7 Strong coupling between silicon VBM and HK/IL interface V_O state. As the two energy levels approaching each other when a perturbation electric field applied, the Landau-Zener anti-crossing occurs, resulting in an energy gap defining the coupling constant.

Fig. 8 Weak coupling between the silicon VBM state and the HfO2 oxygen vacancy state. One can clearly find the VBM-defect transition process in both branches. The wave functions exchange during the transition (trapping/de-trapping) process.

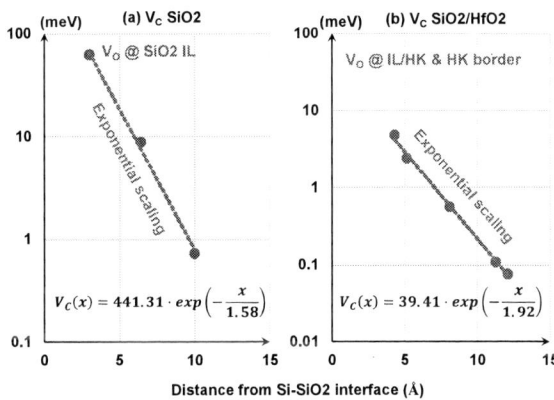

Fig. 9 The decay behavior of the VBM-defect coupling constant V_C. The coupling constants of V_O in SiO2 IL and SiO2/HfO2 interface both obey exponential decay, although the decay length turns to be very different. The coupling constants of VO in HfO2, however, do not obey the exponential law, probably due to the DFT accuracy.

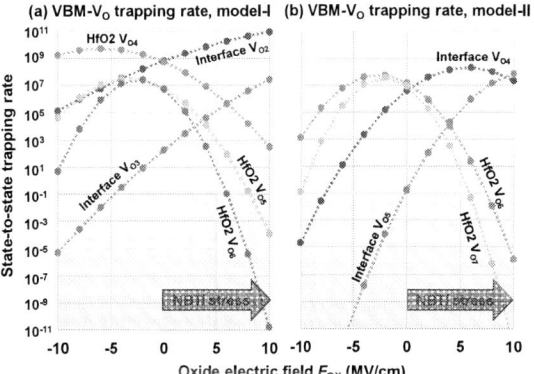

Fig. 10 The oxide field (F_{OX}) dependence of the hole trapping rate, calculated by Marcus formula with essential parameters extracted through DFT. During the NBTI stress, the HK/IL interface defect V_O dominates the hole trapping in both models, with its contribution increasing as NBTI stress goes higher.

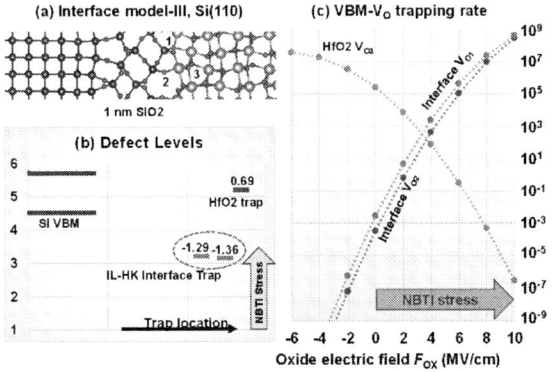

Fig. 11 Results for hole trapping defects in (110) oriented gate stack. (a) is the atomic model showing 3 typical oxygen vacancies: 2 at IL-HK interface, 1 in HfO2. (b) depicts the defect level alignment, and (c) is the field-dependent VBM-V_O hole trapping rate.

Table-I

Key parameters	@ SiO2	@ IL/HK or HK border	@HK
Coupling constant V_C	$441.31 \cdot exp\left(-\frac{x}{1.58}\right)$	$39.41 \cdot exp\left(-\frac{x}{1.92}\right)$	—
Reorganization energy λ	0.31 eV	0.95/1.02 eV	0.96 eV
Gibbs free energy ΔG	$E_{defect} + F_{OX} \cdot x - E_{VBM}$		

Acknowledgement

This work is supported by NSFC (grand nos. 11774338, 11574304), and 6J6011000. The authors would like to thank Dr. R. Wang, and Dr. J. Chen for helpful discussions.

References

[1] B. H. Lee, et al., IEDM, pp. 133-136, 1999. [2] E.P. Gusev, et al., IEDM, pp. 451-454, 2001. [3] F. Lazarevic, et al., IEEE Trans. Electron Devices 64, 5073-5080, 2017. [4] S. Zafar, et al., IEEE Trans. Device Mater. Rel. 5, 45-64, 2005. [5] K. Xiong, et al., Appl. Phys. Lett. 87, 183505, 2005. [6] K. Tarafder, et al., J. Am. Chem. Soc. 136, 5121-5131, 2014. [7] K. Takagi, et al., Jpn. J. Appl. Phys. 57, 066501, 2018. [8] T. Grasser, et al., Microelectron. Eng. 86, 1876-1882, 2009. [9] W.Goes, et al., Microelectron. Reliab. 87, 286-320, 2018. [10] R. A. Marcus, J. Chem. Phys. 24, 966, 1956. [11] W. Jia, et al., J. Comput. Phys. 251, 102-115, 2013.

Parasitic Surface Reactions in High-Aspect Ratio Via Filling using ALD: A Stochastic Kinetic Model

T. Muneshwar[1,3], G. Shoute[1,2], D. Barlage[2], and K. Cadien[3]

[1]Synthergy Inc., Edmonton, AB, CAN, email: muneshwar@synthergy.io
[2]ECE, University of Alberta, Edmonton, AB, CAN, [3]CME, University of Alberta, Edmonton, AB, CAN

Abstract—A scalable kinetic Monte-Carlo model (sKMC) of molecular transport for atomic layer deposition (ALD) for high aspect-ratio (AR) features is developed. Surface coverage is a critical parameter studied here in detail. The capabilities of the stochastic model provide insight into challenges in growing ALD films in high-AR via structures faced by the industry, including the effects of parasitic surface reactions resulting in poor coverage. Furthermore, we provide experimental results verifying the model's prediction by growing ALD SiN$_x$ on high-AR via structures. By compensating for the processing errors corroborated by the model, we experimentally improved sidewall coverage from 70% to 92%.

I. INTRODUCTION

Owing to the fundamental self-limiting characteristics of the relevant surface reactions, atomic layer deposition (ALD) is essential to sustaining the future of semiconductor device scaling and next generation memories. The continued advancement of three-dimensional (3D) architectures are limited by the number of vertically-stacked processing layers that can be achieved, requiring features such as superior coverage and uniformity afforded by ALD. As described in Fig. 1, ALD growth stems from the sequential exposure of precursor(s) and reactant(s) onto the substrate [1-2]. Since ALD process conditions are far from thermodynamic equilibrium, the overall surface saturation behavior is governed by the kinetics of elementary gas/solid interactions (e.g. physisorption, desorption, chemisorption, etc.). Attaining these optimum parameters poses a significant challenge when trying to balance the many kinetic dynamics.

While the solution can be obtained through trial and error, ascertaining these parameters has progressively become more difficult due to the increase of structural complexity. Yet, such models have been distinctly absent for high-AR structures with some recent exceptions [3-4]. For industry, stochastic models would be of great value especially with rising cost of test structures themselves, necessitating a more strategic approach to experimentation and in precursor selection, and more.

Coverage conformality, defined in Fig. 2, is an important and industry-relevant metric characterizing high-AR structures and is the focal point of the study. To understand the reported shortcomings of coverage [5], we develop a sKMC model for high-AR vias predicting the effects of growth conditions, such as non-ideal growth, that result in poor depositions. Using ALD silicon nitride (SiN$_x$), we experimentally support the findings of the simulations by improving side-wall coverage from 70% to 92%. We also further improve the grown SiN$_x$ with an OxPRE treatment to prohibit post-processed oxidation of the film.

II. STOCHASTIC KINETIC MODEL FOR ALD

A. Molecular Transport in 3D structures

The molecular transport within a via could be adequately described using ballistic transport mechanism if the via dimensions are significantly smaller than its mean free path in gas phase. Fig. 3 illustrates such transport, wherein the incident molecule may (i) directly imping and physisorb on the reactive site, (ii) undergo multiple reflections within the via until it finds a reactive site to physisorb, or (iii) exit the via following multiple reflections without undergoing physisorption. Since the initial position and trajectory of the incident molecule is random, a sKMC approach is ideal for simulating precursor-A and reactant-B transport within a via during ALD.

B. ALD Surface reactions

Partial ALD surface reaction during precursor exposure step could be represented by the equation:

$$A^{gas} + B^{che} \underset{k_A^{des}}{\overset{k_A^{in}}{\rightleftharpoons}} A^{phy} \xrightarrow{k_A^{che}} A^{che} \qquad (1)$$

where the incident gas-phase molecule A^{gas} would react at the B^{che} surface sites to form physisorbed A^{phy} units. These A^{phys} units could then either desorb back as A^{gas} or chemisorb to A^{che} units. In Eq. (1), k_A^{in} represents the impingement rate of A^{gas}, while k_A^{des}, and k_A^{che} are the kinetic rate constants for desorption and chemisorption of A^{phys} surface species.

Similarly, the ALD partial reaction during reactant exposure step could be represented as:

$$B^{gas} + A^{che} \underset{k_B^{des}}{\overset{k_B^{in}}{\rightleftharpoons}} B^{phy} \xrightarrow{k_B^{che}} AB^{ALD} + B^{che} \qquad (2)$$

where incident B^{gas} molecules interact with the surface A^{che} reaction sites, and k_B^{in} represents the impingement rate of B^{gas}, while k_B^{des} and k_B^{che} are the kinetic rate constants for desorption and chemisorption of B^{phys} surface species. In Eq. (2), AB^{ALD} is the unit of material deposited by ALD. These surface reactions are schematically illustrated in Fig. 4.

From Eq.(1) and (2) it could be seen that the overall kinetics of ALD process depend upon the kinetic factors k_A^{in}, k_A^{des}, k_A^{che}, k_B^{in}, k_B^{des}, and k_B^{che}. Moreover, the impingement rates (k_A^{in} and k_B^{in}) depend upon the respective partial pressures, whereas the rate of desorption (k_A^{des} and k_B^{des}) and chemisorption (k_A^{che} and k_B^{che}) vary with the substrate temperature (T_{sub}). For a uniform incident precursor/reactant flux, the kinetics of surface reactions (1) and (2) could be solved analytically [6]. However, stochastic models such as sKMC are most appropriate for simulating ALD in high-AR features.

978-1-7281-1988-5/18 $31.00 © 2018 IEEE

C. Reaction probablity profile along depth

Combining ballistic molecular transport and surface reaction kinetics, our calculations provided the probability of reaction on the sidewall as a function of depth from the surface, for a particular precursor/reactant exposure. Fig. 5 illustrates these reaction probabilities for an ALD half reaction in a via with a diameter of 300 units, and AR of 1, 3, and 10, calculated for model parameters provided in Table I.

Identical sKMC simulations for ALD on a flat surface (i.e. AR = 0) showed that an exposure of 0.06s was adequate to obtain a near complete (>99.5%) coverage with the chemisorbed species. As compared to this, the calculated reaction probability for 0.06s exposure at the bottom of the via was ~95%, ~56%, and ~10% for AR =1, 3, and 10 respectively.

On the other hand, for near uniform reaction along the sidewalls, vias with AR =1, 3, and 10 required an exposure of 0.10s, 0.40s, and 4.0s, respectively.

III. RESULTS

A. ALD growth in a via

An *ABAB...* pulsed deposition cycle consisting of sequential exposure of precursor A and reactant B with reaction probabilities as shown in Fig. 5, was essential for a particular 3D geometry. The probability of AB deposition along sidewall per deposition cycle was then obtained from the product of these reaction probabilities. Fig. 6 shows the calculated thickness profile of AB deposition in AR=5 via, for different combinations of exposures t_A and t_B. The full model parameters are listed in Table I, along with its simulation metrics.

Increasing t_B from 0.1s to 0.5s at t_A = 0.1s, the step coverage was observed to increase from 29% in (a) to 50% in (b) of Fig. 6, and further increase in t_B had negligible improvement in step coverage. However, an increase in t_A from 0.1s to 0.5s (at t_B = 0.5s) step coverage improved from 50% (b) to 93% (c) (Fig. 6).

B. CVD-like growth mode during ALD

Ideally, the ALD process is designed so that material growth results only from reactions governed by Eq. (1) and Eq. (2). However, it is not always possible to restrict every potential side reactions co-occurring and interfering with the self-limiting characteristics of ALD and resulting material properties [6-7].

One of the non-ALD mechanism, wherein the incident gas phase precursor molecule A^{gas} could react with the chemisorbed specie A^{che} in a CVD-like reaction could be represented as:

$$A^{gas} + A^{che} \underset{k_{A2}^{des}}{\overset{k_A^{in}}{\rightleftharpoons}} A_2^{phy} \xrightarrow{k_{A2}^{che}} A_2^{che} \qquad (3)$$

where, k_{A2}^{che} is the kinetic rate constant, and A_2^{che} is the reaction product of type -A-A containing two A-atoms [3]. These A_2^{che} surface units react with B^{gas} molecules introduced in the following reactant exposure step to give an excess growth which is not expected from a true-ALD process. Furthermore, since A_2^{che} formation is governed by the incident A^{gas} flux, density of A_2^{che} species is significantly higher near the via entrance. This phenomenon interferes with a conformal ALD growth, and consequently results in thicker sidewall deposits at the top as compared to the near bottom wall.

Fig. 7 shows the calculated AB deposition in an AR=5 via in presence of a non-ALD reaction in Eq. (2). The model parameters here are given in Table II. The simulation revealed that by increasing t_A from 0.1s to 0.3s, the step coverage was positively correlated beginning at 34% in (a) to 47% in (b) if Fig. 7, yet only marginally improves when t_A is increased to 0.5s. Further to that, the excess growth near the entrance hinders transport of precursor/reactant, adversely affecting deposition on the sidewall near bottom of via. Our calculated trends in sidewall step coverage is in agreement with reported studies, where increasing precursor exposure alone did not achieve conformal depositions [5].

A comparison between Fig. 6 and Fig. 7 highlights the detrimental effect of parasitic side surface reactions during ALD in non-planar. As the kinetic constant k_{A2}^{che} in Eq. (3) is temperature dependent, the extent of this undesired growth increases with T_{sub} as shown in Fig. 8.

C. Experiment Results

Simulations suggest that lower but appropriate T_{sub}, along with adequate dosage, are required for good coverage. Fig. 9 shows cross-sectional TEMs of SiN_x PEALD samples deposited on 3D trenches of AR ~5. SiN_x depositions were performed in ALD150LX Kurt Lesker, using tris-dimethylamino silane (3DMAS) and forming gas plasma. By choosing a low T_{sub} = 100 °C and adjusting t_{plasma} from 10s to 30s, in agreement with simulated trends, step coverage improved from 70% in (a) to 92% in (b). The impact of T_{sub} on the coverage SiN_x using 3DMAS affirms the importance of proper precursor selection.

Because a low T_{sub} was used, as revealed in Fig. 10(a), as-deposited SiN_x PEALD films at 100 °C were prone to ambient oxidation. To address this, we introduced an OXPRE surface treatment to the SiN_x film to prevent its degradation. Its successful implementation is shown in Fig. 10(b). Thus, the OxPRE treatment could potential serve as a practical solution to low T_{sub} grown films that are vulnerable to oxidization.

IV. CONCLUSION

We report a stochastic kinetic simulation (sKMC) for ALD growth in high-AR vias. The effects of process parameters and geometry of structure on conformality were discussed. We show that non-ALD reactions will adversely affect conformal depositions. In conclusion, our calculations show, that in order to achieve a step coverage of ~100% it is essential that parasitic side surface reactions must be restricted, either by operating at lower *T_{sub}* or adjusting dosages. These are consequently dependent on choosing appropriate precursors which a further developed model may potentially immensely benefit its users.

ACKNOWLEDGMENT

The authors thank D. Haussmann for providing 3D trench structures and helping with the imaging of our SiN_x samples.

REFERENCES

[1] S. Krishna, & D. Schepis, eds. *Handbook of thin film deposition*. William Andrew, 2018. pp. 359–377.
[2] C. Murray, *et al.*, *ACS Appl. Mat. Int.*, v.5, no.9, pp.3704–3715, 2013.
[3] Ylilammi, *et al.*, *Journ. of Appli. Phys.* v. 123, no. 20, pp.205301, 2018.
[4] M. Schwille, *et al.*, *Journ. of Vac. Sci. Tech. A*, v.35, no.6 ,01B118, 2017.
[5] T. Faraz *et al.*, *ACS Appl. Mater. Inter.*, v.9, no. 2, pp. 1858–1869, 2017.
[6] T. Muneshwar & K. Cadien, *Journ. Appl. Phys.*, (Review).
[7] T. Muneshwar, *et al.*, *J. Vac. Sci. Tech. A*, v.34, no.5, p. 050605, 2016.

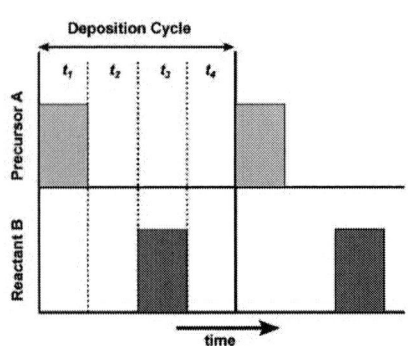

Fig. 1: Pulsing sequence of precursor A and reactant B in an ABAB... pulsed ALD.

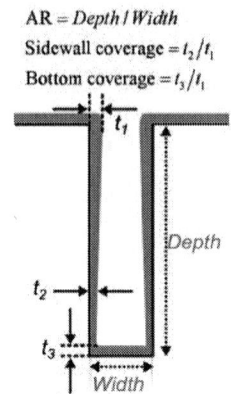

Fig. 2: Definition of coverage and aspect ratio (AR).

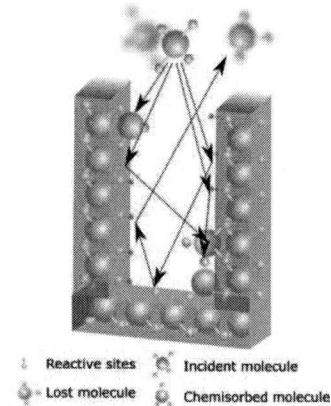

Fig. 3: Molecular transport in a simulated 3D via slice.

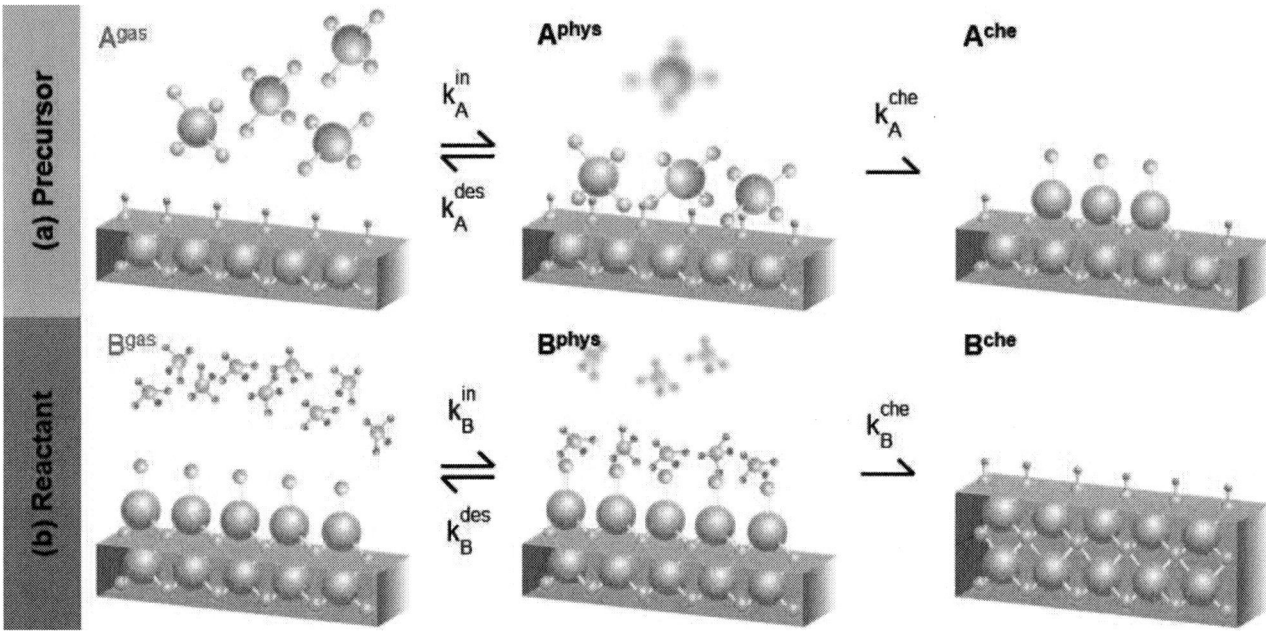

Fig. 4: Schematic of ALD surface reactions with respective kinetic coefficients. (a) Precursor exposure step where the precursor gas (A^{gas}) is released; (b) Reactant exposure step where the reactant gas (B^{gas}) is released.

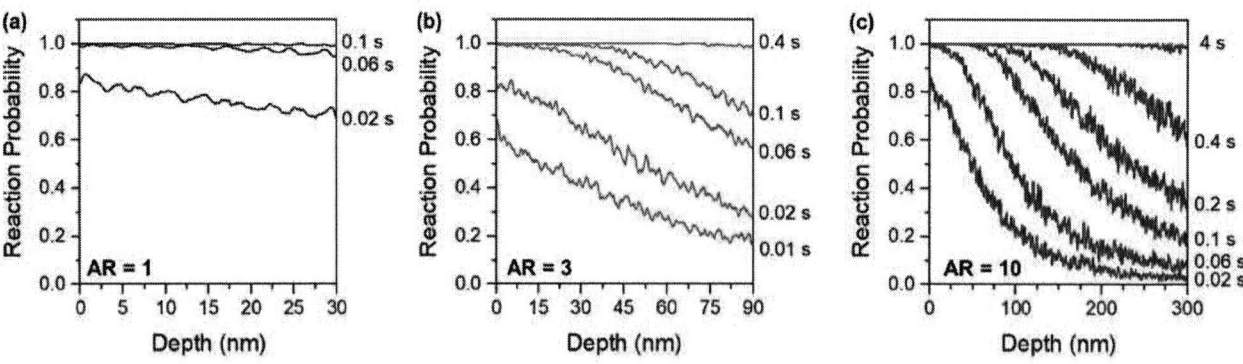

Fig. 5: sKMC-based reaction probabilities trends on sidewall against depth from surface for 3D feature with AR = (a) 1, (b) 3, and (c) 10 at increasing dosages. Model parameters were: $k^{phy} = 1\times10^4$ s^{-1}; $k^{des} = 1\times10^3$ s^{-1}; $k^{che} = 1\times10^2$ s^{-1}; and exposure lengths as shown with figure.

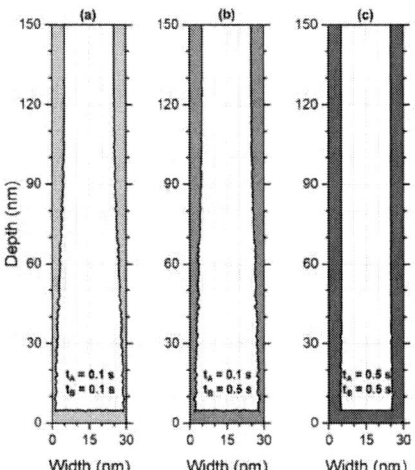

Fig. 6: Thickness profile in AR=5 feature for increasing precursor A and reactant B exposures determined by sKMC calculations. Simulation parameters in Table I.

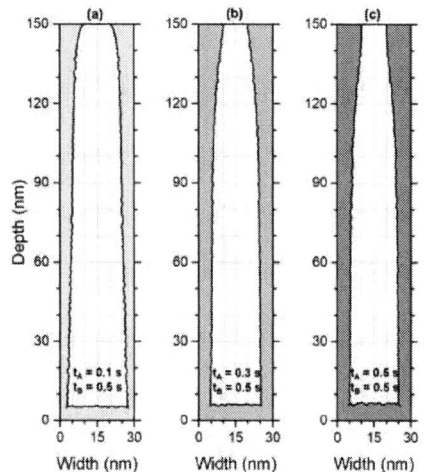

Fig. 7: Thickness profile in AR=5 feature for parasitic CVD reaction during precursor A exposures determined by sKMC calculations. Parameters given in Table II.

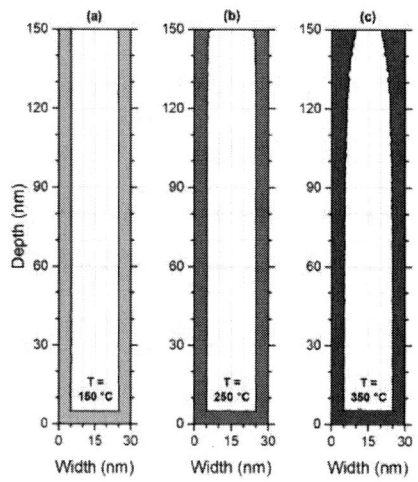

Fig. 8: Thickness profile in AR=5 feature for parasitic CVD reaction during precursor A exposures at different substrate temperatures determined by sKMC calculations. See Table III.

TABLE I
sKMC parameters for ideal ALD growth.

Precursor (A)		Reactant (B)	
Event	s^{-1}	Event	s^{-1}
k_A^{phys}	10^{+4}	k_B^{phys}	10^{+4}
k_A^{des}	10^{+3}	k_B^{des}	$5 \times 10^{+3}$
k_A^{che}	10^{+2}	k_B^{che}	10^{+3}

Coverage for AR = 5 for ideal ALD growth.

Simulation	(a)	(b)	(c)
Dosage	$t_A = 0.1$ s $t_B = 0.1$ s	$t_A = 0.1$ s $t_B = 0.5$ s	$t_A = 0.5$ s $t_B = 0.5$ s
Coverage			
t_2/t_1	29%	50%	93%
t_3/t_1	97%	97%	100%

TABLE II
sKMC parameters for non-ALD growth.

Precursor (A)		Reactant (B)	
Event	s^{-1}	Event	s^{-1}
k_A^{phys}	10^{+4}	k_B^{phys}	10^{+4}
k_A^{des}	$5 \times 10^{+3}$	k_B^{des}	$5 \times 10^{+3}$
k_A^{che}	10^{+2}	k_B^{che}	10^{+3}
k_A^{des2}	10^{+2}	k_B^{des2}	10^{+6}
k_A^{che2}	$2 \times 10^{+1}$	k_B^{che2}	2×10^{-5}

Coverage for AR = 5 for non-ALD growth.

Simulation	(a)	(b)	(c)
Dosage	$t_A = 0.1$ s $t_B = 0.5$ s	$t_A = 0.3$ s $t_B = 0.5$ s	$t_A = 0.5$ s $t_B = 0.5$ s
Coverage			
t_2/t_1	34%	47%	51%

TABLE III
sKMC parameters for subsrate temperature study.

Precursor (A)		Reactant (B)	
Event	s^{-1}	Event	s^{-1}
k_A^{phys}	10^{+4}	k_B^{phys}	10^{+4}
k_A^{des}	$5 \times 10^{+3}$	k_B^{des}	$5 \times 10^{+3}$
k_A^{che}	10^{+2}	k_B^{che}	10^{+3}
k_A^{des2}	10^{+2}	k_B^{des2}	10^{+6}
$k_A^{che2} = 4.5 \times 10^{-6} \left(-8.3/T_{SUB}\right)$		k_B^{che2}	2×10^{-5}

Coverage for AR = 5 with $t_A = 0.5$ s $= t_B$.

Simulation	(a)	(b)	(c)
T_{SUB}	150 °C	250 °C	350 °C
Coverage			
t_2/t_1	93%	70%	50%

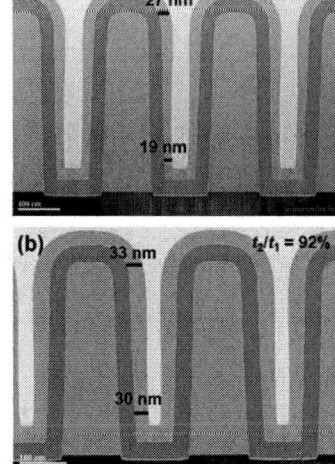

Fig. 9: PEALD SiN$_x$ grown at T$_{sub}$ = 100 °C with (a) t$_{plasma}$ = 10 s; (b) t$_{plasma}$ = 30 s.

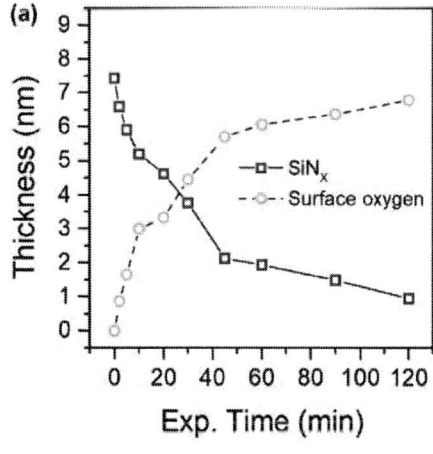

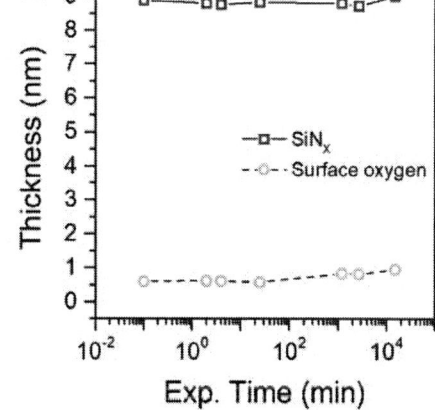

Fig. 10: Depth of surface oxidization in low temperature (T$_{sub}$ = 100 °C) PEALD grown SiN$_x$ over time with (a) no surface treatment; (b) after OxPRE treatment.

Physics-based modeling of volatile resistive switching memory (RRAM) for crosspoint selector and neuromorphic computing

W. Wang[1], A. Bricalli[1], M. Laudato[1], E. Ambrosi[1], E. Covi[1], and D. Ielmini[1*]

[1]Dipartimento di Elettronica, Informazione e Bioingegneria (DEIB), Politecnico di Milano and IU.NET, piazza L. da Vinci 32, 20133, Milano, Italy, *email: daniele.ielmini@polimi.it

Abstract—Volatile resistive switching memory (RRAM) is raising strong interest as potential selector device in crosspoint memory and short-term synapse in neuromorphic computing. To enable the design and simulation of memory and computing circuits with volatile RRAM, compact models are essential. To fill this gap, we present here a novel physics-based analytical model for volatile RRAM based on a detailed study of the switching process by molecular dynamics (MD) and finite-difference method (FDM). The analytical model captures all essential phenomena of volatile RRAM, e.g., threshold/holding voltages, on-off ratio, and size-dependent retention. The model is validated by extensive comparison with data from Ag/SiOx RRAM. To support the circuit-level capability of the model, we show simulations of crosspoint arrays and neuromorphic time-correlated learning.

I. INTRODUCTION

Resistive switching memory (RRAM) devices based on metallic filaments made of Cu and Ag generally offer high on-off ratio (> 10^7) [1, 2], steep switching slope (< 4 mV/dec), and high endurance [3]. Although they were originally proposed as non-volatile switching memories [4], it was soon evidenced that they tend to display *volatile* switching due to the spontaneous turn-off within a short retention time from few μs to few ms [5]. Thanks to its high nonlinearity (> 10^7), volatile RRAM has been proposed as select device in crosspoint arrays of memories or sensors [6], and as artificial synapse to mimic long/short-term memory phenomena in the brain [7]. However, a comprehensive understanding of the volatile switching mechanism is still missing [8]. Also, accurate compact models are needed to support all potential applications in memory and neuromorphic computing.

In this work, we introduce a new physics-based model for volatile RRAM. First, we study the switching process in volatile RRAM by molecular dynamics (MD) and finite-difference method (FDM) simulations of the spontaneous filament rupture. From this study, we derive a universal equation for size-dependent retention time. By combining the new retention model with a previous RRAM model for voltage-controlled filament growth/dissolution [9], we develop the first physics-based analytical model for volatile RRAM. The model accurately captures all device characteristics, such as DC and AC switching curves and retention. By including the spread of the activation energy, stochastic variations are also described. We finally show simulations of potential applications such as crosspoint selectors and neuromorphic synapses.

II. VOLATILE BEHAVIOR OF METALLIC FILAMENT

A. MD Simulations

Fig. 1a shows MD simulations of a nanoscale filament between two electrodes for increasing times at high temperature (800 K) using LAMMPS program [10, 11]. The simulation shows that the filament breaks up spontaneously without any external force after about 200 ps. This is because of the nanoscale "liquid-like pseudo-elasticity" [11], where metallic filaments change shape like a liquid

drop to minimize the total surface energy. Atoms in the bulk remain crystallized while surface atomic migration accounts for the shape evolution at the origin of filament disconnection.

B. Surface-tension Induced Filament Evolution

While highly illustrative of the disconnection process, the MD computational cost and time is not practical for simulating the switching and retention processes for typical temperature and times. FDM simulations were thus carried out by the equation:

$$J_s = -\left(\frac{D_s \gamma \delta^4}{kT}\right) \nabla_s \kappa, \quad (1)$$

where J_s is surface atomic flux along an arbitrary surface s, D_s is the surface diffusion coefficient, γ is the surface energy, δ is the interatomic distance of Ag atoms, k is the Boltzmann's constant, T is the temperature, and κ is the surface curvature [12]. FDM simulations at 300 K in Fig. 1b show that the filament evolution is divided in two phases: (1) The filament diameter ϕ decreases in the filament bottleneck (t < 0.2 ms), (2) the filament is disconnected and the gap length g increases (t > 0.2 ms). We thus define the retention time t_R as the time for the filament disconnection ($\phi = 0$ and $g = 0$). FDM simulations in Fig. 2 show that t_R strongly depends on the initial filament diameter ϕ_0 according to $t_R \sim \phi_0^4$.

C. Modeling of Filament Spontaneous Disconnection

Based on numerical simulations in Fig. 1b, the spontaneous filament disconnection at the origin of the volatile behavior can be modeled by a simple rate equation of the filament diameter ϕ, given by:

$$\left.\frac{d\phi}{dt}\right|_{sp} \propto -1 \Big/ \left.\frac{dt_R}{d\phi_0}\right|_{t=t_R, \phi=\phi_0}, \quad (2)$$

where $dt_R/d\phi_0$ can be obtained from Eq. (1), thus yielding $d\phi/dt|_{sp} \propto -\phi^{-3}$ at the origin of the $t_R \sim \phi_0^4$ dependence in Fig. 2.

III. COMPACT MODEL

By combining the voltage-controlled model of set/reset switching in RRAM [9] with the retention model of Eq. (2), we can derive the comprehensive analytical model of volatile RRAM in Tab. I. Here, Eq. (3) describes both voltage-driven set process [9] and spontaneous filament dissolution by $d\phi/dt|_{sp} \propto -\phi^{-3}$, the latter accounting for size-dependent retention. The increase of the filament gap g for t > t_R is described by Eq. (4), allowing to describe the post-dissolution evolution of the off-state current. The switching mechanism in the volatile RRAM is summarized in Fig. 3, showing the filament growth under an applied voltage, and the phenomena of *disconnection* (t < t_R) and *retraction* (t > t_R) arising spontaneously. In the model, the RRAM resistance was computed by the series model in Fig. 4a and b, which is analytically captured by Eqs. (5)-(8). Correspondingly, the effective temperature T during set/reset processes was computed by the Joule heating model in Fig. 4c, described by Eqs. (9)-(11). The model thus describes all essential phenomena of volatile RRAM, including switching and spontaneous disconnection.

978-1-7281-1988-5/18 $31.00 © 2018 IEEE

IV. Model Validation

A. DC and AC Characteristics

Experimental data were collected from RRAM devices with an Ag top electrode, a SiO_x dielectric layer with thickness $t_{ox} = 5$ nm, and a bottom electrode made of graphitic carbon [5, 13]. Fig. 5a shows the device structure, while Fig. 5b shows the measured and calculated I-V curves for compliance current $I_C = 35$ µA. The high on-off ratio ($> 10^7$) and steep switching slope are well captured by the model. As a positive threshold voltage V_{T+} is applied, the device switches to the on-state by increasing its current from 10^{-12} A to $\sim 10^{-5}$ A. The compliance current I_C was limited by an integrated select transistor in the one-transistor/one-resistor (1T1R) structure [13]. As the voltage is reduced below a characteristic holding voltage V_{H+}, the device spontaneously switches to the off-state by filament disconnection, marked by the steep drop of the current (Fig. 3). Thanks to the bidirectional behavior of the Ag/SiO_x device [13], switching to the on-state also occurs at negative voltage V_{T-}, with negative holding voltage V_{H-}. The bidirectional switching is well captured by the model with the parameters reported in Tab. II. The model could describe the I-V curves for any value of I_C from 0.1 µA to 80 µA. Fig. 6 shows the measured and calculated V_{T+}, V_{H+}, V_{T-} and V_{H-} as a function of I_C, supporting the accuracy of the model in a broad range of I_C.

Fig. 7 shows AC characteristics obtained by applying a triangular pulse of pulse width $t_p = 100$ µs, followed by a constant read voltage $V_{read} = 0.1$ V to monitor the time evolution of device resistance R (Fig. 7a). The current response (Fig. 7b) indicates a steep switching to the on-state at $I_C = 20$ µA, while the following read current shows a decay behavior with $t_R \sim 1.5$ ms. By this technique, we studied t_R as a function of the pulse width t_p in both experiments and simulations. Fig. 8 shows that t_R increases with t_p, thanks to the gradual increases of ϕ_0 with t_p and the consequently enhanced retention according to Fig. 2.

B. Cycle-to-cycle Variations

Statistical variations play a dominant role for volatile RRAM devices in both memory and computing applications. Cycle-to-cycle variability of V_T and V_H were computed by Monte Carlo simulations where we included a stochastic spread of the activation energy [14]. Fig. 9a shows the distributions of V_T and V_H from experiments and calculations, with the assumed spread of energy barriers (inset). The model can also predict the t_R distributions for various pulse width t_p, as shown in Fig. 9b.

The relatively long $t_R \sim 1$ ms of the volatile RRAM device is a concern for its application as selectors, due to half-selected cells remaining in the on-state after read. To capture this scenario, Fig. 10 shows a two-pulse experiment where the first pulse induces switching to the on-state, then a second pulse is used to probe the state of the cell after a certain delay time t_D. The RRAM device is still found in the on-state ($V_{T2} = 0$) for short time delay $t_D = 10$ µs (Fig. 10a), whereas the off-state is recovered for a longer delay $t_D = 10$ ms (Fig. 10b). Fig. 11 reports the threshold voltage V_{T2} observed in the second pulse as a function of t_D, indicating a recovery from on-state to off-state in about $t_R = 1$ ms under both positive and negative voltage. Our model reproduces the experimental results by 1,000 Monte Carlo simulations at each t_D. Fig. 12 shows the probability of finding the device in the off-state ($V_{T2} > 0$) as a function of delay time.

V. Applications as Select Device

Our model can be used to simulate the volatile RRAM within a one-selector/one-RRAM (1S1R) element of a crosspoint array

(Fig. 13a). Fig. 13b shows the calculated I-V curves for the memory element, obtained by our previous non-volatile RRAM model [15]. Fig. 13c shows the combined I-V curve of the 1S1R, clearly indicating memory set/reset transitions and selector on/off switching. By utilizing a V/2 reading scheme (Fig. 14a) with read voltage $V_R = 2.5$ V, the *static* window between a selected low-resistance state (A in Fig. 13d) and a half-selected state (B in Fig. 13d) is $\sim 10^7$, which translates in a feasible N×N array with $N \sim 10^7$ (Fig. 14b) [16]. However, *dynamic* simulations in Fig. 14c show that the read margin strongly decreases at increasing t_R, due to half-selected cells being recently accessed (C in Fig. 13d) and remaining in the on-state due to the long retention time.

VI. Applications in Neuromorphic computing

While limiting the applications in crosspoint memories, the retention time in the few-ms range well matches with the biological timescale for short-term plasticity, thus making volatile RRAM suitable for neuromorphic computing [17]. Fig. 15 shows the current response of a volatile RRAM to a train of spikes with frequency $f_{spike} = 2$ kHz using a series resistor compliance ($R_C = 10$ kΩ). The spiking stimulation results in paired-pulse facilitation (PPF), namely a gradual increase of the RRAM conductance due to filament growth during each spike. Reducing f_{spike} to 250 Hz in Fig. 16 causes the transition to paired-pulse depression (PPD) [7], as filament disconnection dominates over growth. The resulting spike-rate dependent plasticity (SRDP) is shown in Fig. 17a, highlighting the conditions for PPF and PPD.

The spiking frequency also affects the retention time, thus enabling tunable long-term memory (LTM, $t_R > 20$ ms) and short-term memory (STM, $t_R < 10$ ms) regimes (Fig. 17b). Unlike long-term plasticity, LTM is a transient memory which enables coupling of sensory information received at any time to the one received earlier, thus supporting speech recognition [18] and sequence learning [19]. The STM and LTM concepts are demonstrated in Fig. 18: high-frequency learning of the image '5' results in LTM (a), while low-frequency learning results in STM (b). A newly submitted pattern '0' can thus be coupled to the previous one '5' (Fig. 18c) which is essential for the time-correlated pattern learning in a spatio-temporal spiking neural network [19], while no sequence recognition is possible in the case of STM (Fig. 18d).

VII. Conclusion

In summary, we presented a physics-based analytical model for volatile RRAM, capturing DC and AC characteristics including switching and retention, and their stochastic variations. The model is validated against experimental data and deployed for the circuit-level simulation of crosspoint arrays and neuromorphic systems.

This work has received funding from the European Research Council (ERC) under the European Union's Horizon 2020 research and innovation programme (grant agreement No. 648635).

References

[1] Y. D. Zhao, *et al.*, IEEE IEDM, **2016**, 7.6. [2] Q. Luo, et al., IEEE IEDM, **2015**, 10.4. [3] X. Zhao, *et al.*, Adv. Mater., **30**, 1705193 (2018). [4] M. N. Kozicki, *et al.*, IEEE Trans. Nanotechnology, **3**, 331 (2005). [5] A. Bricalli, *et al.*, IEEE Trans. Electron Devices, **65**, 122 (2018). [6] M. Wang, *et al.*, Adv. Mater., 1802516 (2018). [7] Z. Wang, *et al.*, Nat. Mater., **16**, 101 (2017). [8] N. Shukla, *et al.*, IEEE IEDM, **2017**, 4.3. [9] D. Ielmini, IEEE Trans. Electron Devices, **58**, 4309 (2011). [10] S. Plimpton, J. Comput. Phys., **117**, 1 (1995). [11] J. Sun, *et al.*, Nat. Mater., **13**, 1007 (2014). [12] F. A. Nichols, *et al.*, J. Appl. Phys., **36**, 1826 (1965). [13] A. Bricalli, *et al.*, IEEE IEDM, **2016**, 4.3. [14] S. Ambrogio, *et al.*, IEEE Trans. Electron Devices, **61**, 2920 (2014). [15] S. Ambrogio, *et al.*, IEEE Trans. Electron Devices, **61**, 2378 (2014). [16] J.-J. Huang, *et al.*, IEEE Electron Device Lett., **32**, 1427 (2011). [17] T. Ohno, *et al.*, Nat. Mater., **10**, 591 (2011). [18] S. Hochreiter, *et al.*, Neural Comput., **9**, 1735 (1997). [19] W. Wang, *et al.*, Sci. Adv., Accepted (2018).

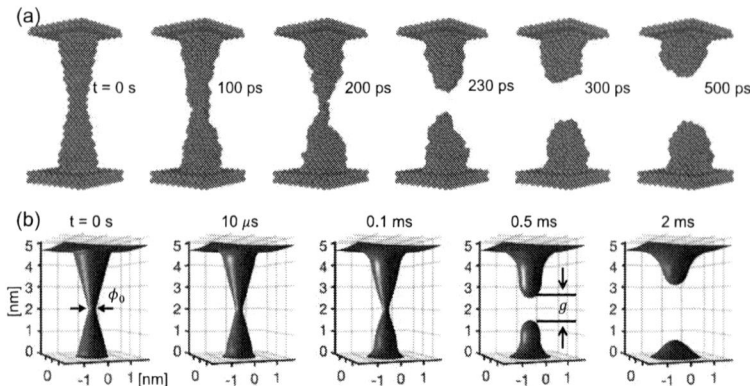

Fig. 1. (a) Accelerated (800K) molecular dynamic (MD) simulation of the relaxation process of a nanoscale filament between two electrodes. Spontaneous filament opening is observed due to surface diffusion. (b) Finite-difference method (FDM) simulation of the filament evolution.

(3) $\dfrac{d\phi}{dt} = A e^{\frac{E_{A0}-\alpha qV}{kT}} - \dfrac{C}{\phi^3} e^{\frac{E_{A1}}{kT}}$, $\quad \phi > \phi_a$	A, C: prefactors E_{A0}, E_{A1}: activation energy ϕ_a: single atom filament diameter
(4) $g = L(1 - e^{-t/\tau_{rt}})$, $\quad \phi < \phi_a$	τ_{rt}: stub retraction time constant

(5) $R = R_{top} + R_{bot} + R_{CF} \| R_{ox}$
(6) $R_{top} + R_{bot} \approx \rho_m \dfrac{4(L-g)}{\phi_0^2}$, $\quad R_{CF} \approx \rho_m \dfrac{4L}{\phi^2}$, $\quad R_{ox} \approx \rho_{ox} \dfrac{4g}{(\phi_0-\phi)^2}$
(7) $\rho_m = \rho_{m0} \dfrac{1 - p\, l_f}{1 + p\, \phi}$ $\quad$ (8) $\rho_{ox} = \dfrac{\rho_{ox0}}{1 + \gamma F}$

ρ_{m0}: bulk metal resistivity
ρ_{ox}: oxide resistivity

(9) $T = T_0 + \dfrac{J^2 \rho_m}{8\, k_{th}}(L^2 - g^2) + \dfrac{J^2 \rho_{eff}}{8\, k_{th,eff}} g^2$
(10) $k_{th,eff} = \phi^2/\phi_0^2 k_{th,m} + (1 - \phi^2/\phi_0^2) k_{th,ox}$
(11) $\rho_{eff} = \phi_0^2 (\rho_m/\phi^2 \| \rho_{ox}/(\phi_0^2 - \phi^2))$

T_0: ambient temperature
$k_{th,m}$: metal thermal conductivity
$k_{th,ox}$: oxide thermal conductivity

Table I. Equations used in the model

Param.	Value (Pos.)	Value (Neg.)	Param.	Value	Param.	Value
α	0.3	0.62	A	1×10^3 m/s	ρ_{ox}	2×10^4 Ω·m
E_{A0}	0.88 eV	0.58 eV	C	4×10^{-22} m^4/s	$k_{th,m}$	5×10^3 W/(m·K)
E_{A1}	0.59 eV	0.56 eV	ρ_{m0}	2 μΩ·m	$k_{th,ox}$	1 W/(m·K)

Table II. Relevant parameters for the calculation

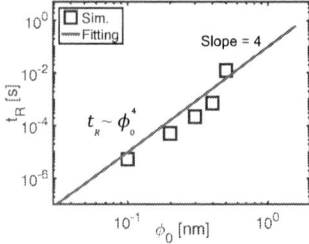

Fig. 2. The calculated retention time of a metallic filament, indicating a strong dependence on initial filament diameter ϕ_0.

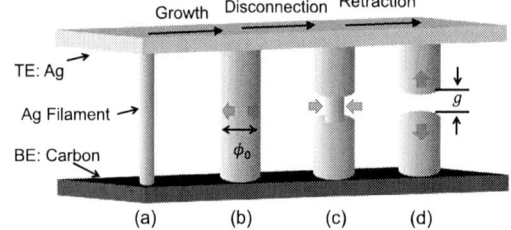

Fig. 3. Schematic view of the complete analytical model of volatile RRAM including: (a)-(b) growth of the filament, (b)-(c) disconnection, i.e., diameter decrease in the bottle neck, and (c)-(d) retraction, i.e., increase of gap length g.

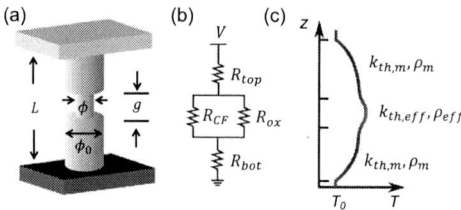

Fig. 4. Schematic views of morphological parameters of the filament (a), equivalent electrical circuit for resistance calculation (b), and parameters for Joule heating equations to obtain the filament temperature (c).

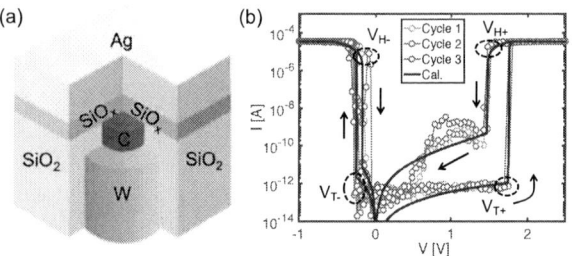

Fig. 5. (a) Device structure of the Ag-based volatile switching memory [5]. (b) Experimental and calculated DC I-V curves of the volatile RRAM device. Both filament disconnection and retraction are visible as abrupt on-state current decrease and gradual off-state current decrease, respectively.

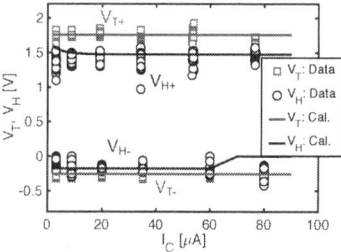

Fig. 6. Experimental and calculated threshold voltage (V_{T+} and V_{T-}) and holding voltage (V_{H+} and V_{H-}) as a function of compliance current I_C.

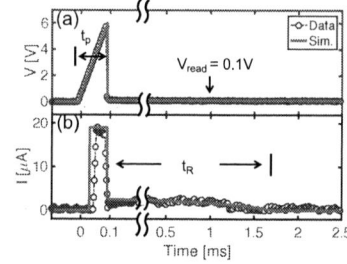

Fig. 7. Measured and calculated voltage (a) and current (b) indicating switching during the triangular pulse and retention/relaxation after t_R from the end of the pulse.

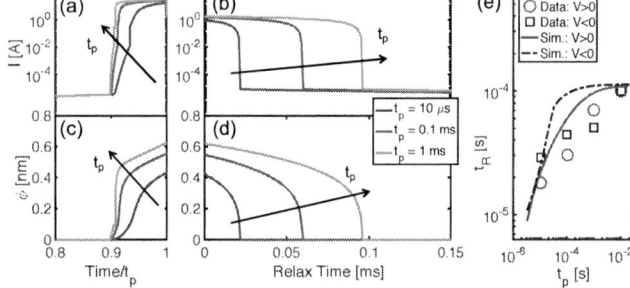

Fig. 8. Calculated current response (a, b) and filament diameter (c, d) for pulse experiments similar to Fig. 7, during either the triangular pulse (a, c) or the read phase (b, d) for various pulse width t_p. (e) Calculated t_R as a function of t_p for both positive and negative pulses, compared to experimental data from [13].

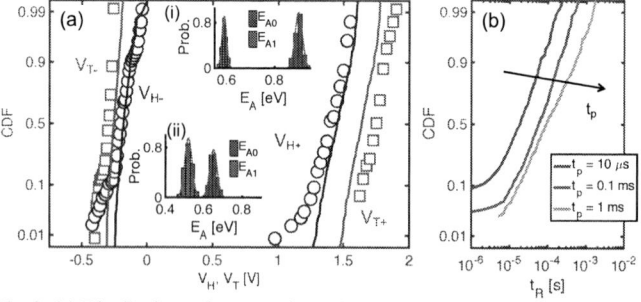

Fig. 9. (a) Distributions of measured (symbols) and calculated (lines) threshold voltage and holding voltage of the volatile RRAM, with insets showing the distributions of activation energies for positive bias (i) and negative bias (ii). (b) Distribution of the calculated retention time t_R for increasing pulse width t_p.

978-1-7281-1988-5/18 $31.00 © 2018 IEEE

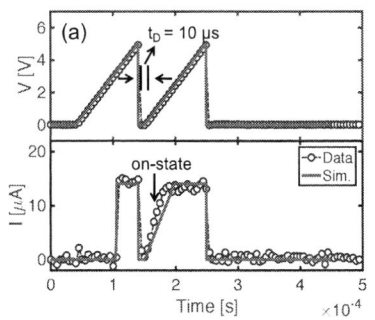
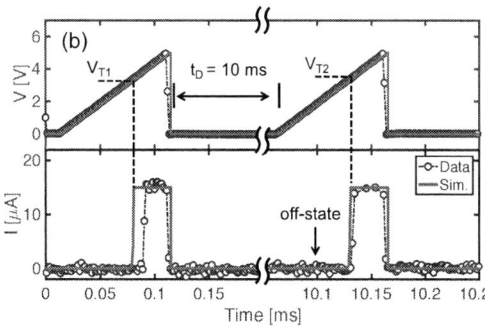
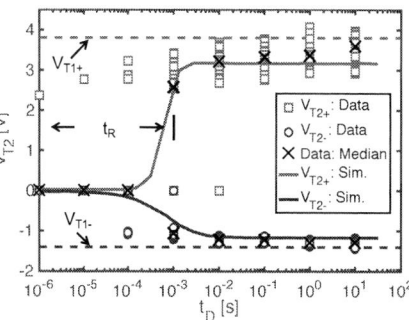

Fig. 10. Measured and calculated voltage and current for two-pulse experiments with delay time $t_D = 10$ µs (a) and 10 ms (b). The device is still found in the on-state for relatively short t_D, whereas the off-state is recovered for relatively long t_D, as revealed by the threshold voltage in the 2nd pulse V_{T2} being similar to the first one V_{T1}.

Fig. 11. Measured and calculated threshold voltage in the 2nd pulse for positive/negative pulses as a function of t_D indicating retention time t_R of about 1 ms.

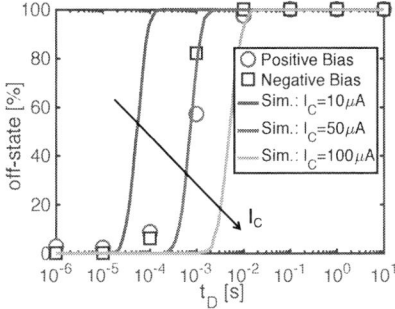
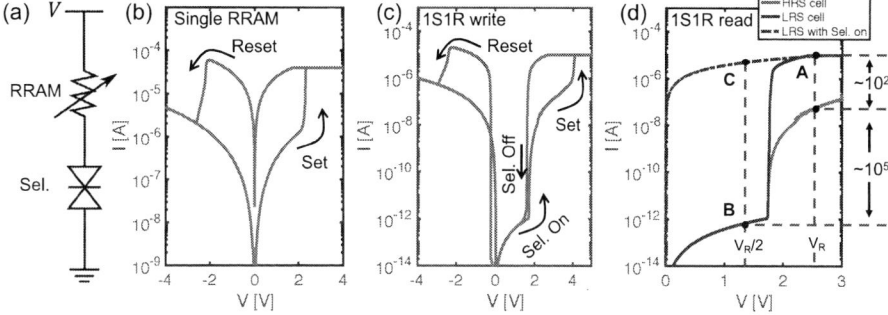

Fig. 12. Probability of finding the device in the off-state in correspondence of the 2nd pulse as a function of t_D in a two-pulse experiment as in Fig. 10 for increasing I_C.

Fig. 13. (a) 1S1R cell configuration. (b) Calculated I-V curve of single non-volatile RRAM with $I_C = 40$ µA by our previous RRAM analytical model [15]. (c) Calculated I-V curve of a 1S1R structure by combining the non-volatile RRAM analytical model and the developed volatile RRAM model. (d) Nonlinearities of the 1S1R cell to calculate read margins.

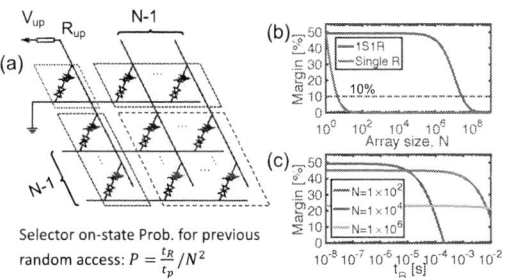
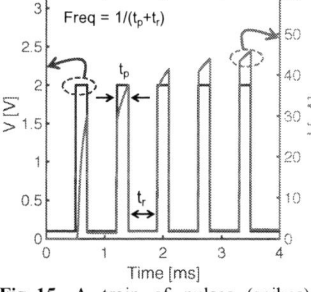
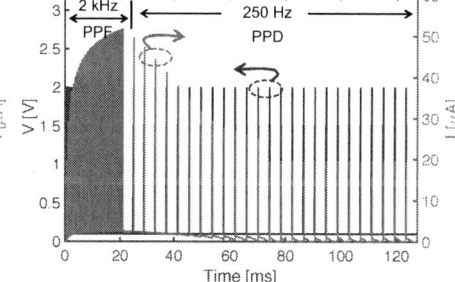

Fig. 14. (a) 1S1R crosspoint array with volatile RRAM selectors. (b) Static read margin as a function of array size. The size of the array is projected to be $N \sim 10^7$. (c) Dynamic read margin, where random access causes some selectors to remain in the on-state while read, thus strongly reducing the read margin for $t_p = 10$ ns.

Fig. 15. A train of pulses (spikes) induce gradual increase of current of the volatile RRAM. Relatively low pulse amplitude (2 V) is applied to limit the switching speed.

Fig. 16. A train of spikes causes paired-pulse facilitation (PPF) for relatively high frequency $f_{spike} = 2$ kHz, and paired-pulse depression (PPD) for relatively low frequency $f_{spike} = 250$ Hz. Similar experimental result has been reported in Ag-based volatile RRAM [7].

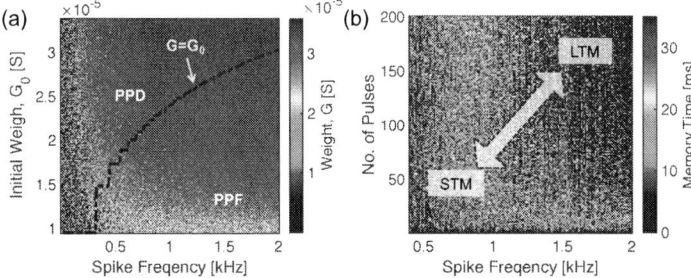
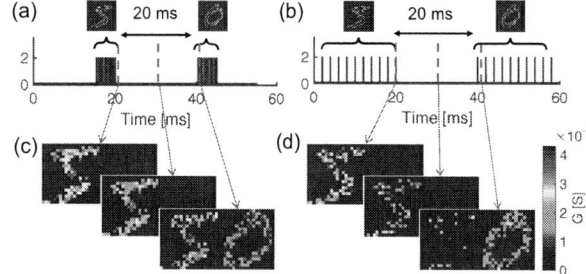

Fig. 17. (a) Color map of synaptic weight G as a function of spike frequency f_{spike} and initial weight G_0 after a train of 10 spikes, indicating the regimes for PPF and PPD, separated by the line of stable weight $G = G_0$. (b) Color map of retention time t_R as a function of f_{spike} and number of spikes, indicating regimes for long-term memory ($t_R > 20$ ms) and short-term memory ($t_R < 10$ ms).

Fig. 18. Simulation of visual sequence recognition in a synaptic array where '5' and '0' are sequentially submitted to recognize the number '50' (a, b). (c) Spiking at high f_{spike} leads to LTM, i.e. new pattern '0', can be linked to the previous one '5'. (d) Spiking at low f_{spike} leads to STM, which prevents the recognition of the full sequence due to lack of LTM.

978-1-7281-1988-5/18 $31.00 © 2018 IEEE 930

Analytic Model for Statistical State Instability and Retention Behaviors of Filamentary Analog RRAM Array and Its Applications in Design of Neural Network

P. Huang[1], Y. C. Xiang[1], Y. D. Zhao[1], C. Liu[1], B. Gao[2]*, H. Q. Wu[2], H. Qian[2], X. Y. Liu[1], J. F. Kang*[1]

[1] Institute of Microelectronics, Peking University, Beijing, China, email: kangjf@pku.edu.cn
[2]Institute of Microelectronics, Tsinghua University, Beijing, China, email: gaob1@mail.tsinghua.edu.cn

Abstract—For the first time, an analytic model is presented for the statistical state instability and retention behaviors of filamentary analog resistive random access memory (RRAM) array. In the model, the diffusion of oxygen vacancy (V_O), the Brownian-like hopping of V_O during the diffusion process and the recombination of V_O are considered. The statistical state instability and retention behaviors of different states under various temperatures are accurately described by the model, which is verified by the measured data of 1Kb filamentary analog RRAM (FA-RRAM) array. Furthermore, the analytic model is successfully implemented to evaluate and optimize the reliability of FA-RRAM based multi-layer neural network. Guided by the model, optimized synapse structures and refresh operation are proposed to significantly enhance the reliability of FA-RRAM based neural network.

I. INTRODUCTION

Neural networks have demonstrated success in cognitive tasks such as speech and image recognition [1, 2]. Today, software based neural network has been the norm, but significant benefits can be gained by adopting RRAM based neural network system due to its low energy consumption, high density, and CMOS compatibility [2-4]. Among various types of RRAM, filamentary analog RRAM (FA-RRAM) has emerged as one of the promising candidates for the synapse of future hardware neural network due to its high density for weight storage, fast processing speed for matrix multiplication, and good linearity for weight updating [5-7]. However, FA-RRAM suffers from serious state instability and retention degradation issues, which will degrade the performance of neural network [8]. Therefore, it is highly demanded to model the statistical state instability and retention behaviors aiming at evaluating and optimizing the reliability of FA-RRAM based neural network. However, there are still few works focusing on this issue.

In this work, an analytic model for statistical state instability and retention behaviors of FA-RRAM is presented for the first time by statistically modeling the variation of oxygen vacancy (V_O) concentration of filament region together with the percolation theory based conduction model. By considering the diffusion of V_O, the Brownian-like hopping of V_O during the diffusion process and the recombination between oxygen ion (O^{2-}) and V_O, the statistical state instability and retention behaviors of 1Kb FA-RRAM array are accurately described. The analytic model can be implemented to evaluate the reliability of FA-RRAM based neural network induced by state instability and retention degradation. Based on the presented model, optimized synapse structures and refresh operation are proposed to mitigate the performance degradation of FA-RRAM based neural network.

II. ANALYTIC MODEL

Fig. 1 shows the measured statistical instability and retention behaviors of 8 states in 1Kb TiN/thermal enhanced layer (TEL)/HfO$_X$(8nm)/TiN FA-RRAM array. Detailed information about the fabricated process and other basic electrical characteristics can be found in [6-8]. Irregular and abrupt current fluctuation (instability) can be observed in **Fig.**1(b). The distribution of read current (I_{read}) is spread with the increase of baking time, so overlap among neighboring resistance states emerges even though the change of mean conductance of 100 cells is slight. **Fig. 2** shows the physical mechanism accounting for the state instability and retention behaviors. The TEL and bottom electrode are connected through a conductive filament, which consists of a resistive switching (RS) region and a V_O rich (VR) region. The conductance of the device can be continuously modulated via tuning the V_O concentration ($C(V_O)$) in RS region. In a relatively short time, the hopping of individual V_O is similar to Brownian movement, which is random. The Brownian-like hoping of V_O at the critical site of current percolation path (CPP) will lead to irregular and abrupt fluctuation of conductance, namely state instability. In a relatively long time, the statistical hopping of a group of V_O is the diffusion, which is directional. The V_O diffusion along the radius direction and the recombination with O^{2-} released from the TEL lead to the decrease of $C(V_O)$ in RS/VR region and the conductance (Case I). While the diffusion of V_O from the VR region to the RS region will increase the conductance due to the fact that the resistance of cell mainly depends on the $C(V_O)$ in RS region (Case II). In summary, the diffusion and recombination of V_O will result in the decrease or increase of conductance, namely retention degradation. The hopping of V_O is investigated with 3D kinetic Monte Carlo simulator, which is an updated version of our previous 2D simulator [9]. The simulated results, as shown in **Fig. 3**, are consistent with the mechanism discussed above.

Fig. 4 schematically shows the 3D model for FA-RRAM. According to the percolation theory [10], the electrical characteristic of CPP is correlated with the concentration of defect. Therefore, we leverage the variation of $C(V_O)$ to represent the formation and rupture of CPP. Here the change of filament size due to the V_O diffusion is also depicted by the variation of $C(V_O)$ for the purpose of simplification. Hence, $C(V_O)$ in RS/VR region are the key parameters to characterize the state instability and retention behaviors. The initial $C(V_O)$ in RS/VR region can be obtained according to the operation schemes of Forming and last switching. By calculating the diffusion and recombination of V_O, the mean $C(V_O)$ as a function of time can be obtained. The statistical distribution of

978-1-7281-1988-5/18 $31.00 © 2018 IEEE

$C(V_O)$ can be deduced via modeling the Brownian-like hopping of V_O. According to the $C(V_O)$ in RS/VR region, conductance can be calculated by employing the percolation theory [10]. The equations and parameters used in the model are summarized in **Table I**.

III. MODEL VERIFICATION

The presented analytic model is verified by the measured statistical state instability and retention behaviors of 1Kb FA-RRAM array. The calculated state instability and retention behaviors of 100 RRAM cells are shown in **Fig. 5**, which is similar to the measured data as shown in **Fig. 1(b)**. **Fig. 6** shows the measured and calculated I_{read} distributions of 2μA level at different baking time. Both the calculated mean and standard deviation of I_{read} agree well with the measured data. The standard deviations of I_{read} of different states are calculated and compared with experimental data as shown in **Fig. 7**. Excellent agreement between the calculated and measured data indicates the validity of the developed analytic model to describe the statistical state instability behavior.

To further verify the model, the array is measured under higher temperature (175 °C) for a longer time (1.2×10^4 s). The calculated retention behavior agrees well with the measured results as shown in **Fig. 8**. It can be found that the mean I_{read} of high current states decrease with time, while the mean I_{read} of low current states show opposite trend. The evolution of I_{read} (2μA) distribution with time @175°C is calculated and compared with the measured results (**Fig. 9**). Although the variation of mean I_{read} is negligible at 10^4s, the state may overlap with neighboring states due to the spread of I_{read}. **Fig.10** shows the measured and calculated cumulative probability of I_{read} (3μA) at different baking time. Because the impacts of physical processes leading to the decrease and increase of I_{read} are matched, the median of the distribution keeps constant with time. The impact of temperature on the statistical state instability and retention behaviors is measured and calculated as shown in **Fig. 11**. It can be found that high temperature will accelerate the movement of V_O and aggravate the degradation and instability. The excellent agreements between the calculated and measured results confirm the validity of the developed analytic model.

IV. NEURAL NETWORK OPTIMIZATION

To display the potential of the presented analytic model for the evaluation and optimization of neural network, the impact of state instability and retention degradation on a $784 \times 100 \times 10$ fully connected neural network is investigated as shown in **Fig.12**. To represent the negative weight, a pair of FA-RRAM are used as one synapse cell to connect the neurons. Here, MNIST dataset is adopted to evaluate the performance of this small scale neural network. Notable degradation of recognition accuracy can be observed after a long time baking (**Fig. 13**). We found that the degradation of recognition accuracy is mainly induced by the state instability rather than the retention degradation. Because most of the weights in the neural network are 0, which correspond to low conductance, the mean conductance of synapses will increase with the baking time. Therefore, the energy consumption on the RRAM synapse increases with time as shown in **Fig. 14**. To

mitigate the recognition accuracy degradation caused by the state instability, two modified synapse structures are proposed (**Fig. 15**). The average conductance of two RRAM, which can decrease the relative variability, is used to store the positive or negative weight in structure A [11]. In structure B, the overlap of neighboring states is mitigated via decreasing the states of single RRAM. It can be found that the recognition accuracy can be improved by adopting the modified synapse at the cost of energy consumption (**Fig. 16**). Refresh operation is proposed to further enhance the reliability of FA-RRAM based neural network as shown in **Fig. 17**. The recognition accuracy of neural network constructed by synapse with structure B shows no degradation even after 10^7s, while noticeable decrease of recognition accuracy can be observed in the neural network constructed by synapse with structure A. The refresh principle is based on the prediction of the presented analytic model, which implies that training again or reloading the weight from the cloud is not necessary.

V. CONCLUSIONS

An analytic model for statistical state instability and retention behaviors of FA-RRAM array is presented for the first time by modeling the diffusion of V_O, the Brownian-like hopping of V_O during the diffusion process and the recombination of V_O. The calculated results show excellent agreements with the measured results of different states of 1Kb array under different temperatures, which demonstrates the validity of the presented model. Furthermore, we successfully implemented the analytic model to investigate and optimize the reliability of a multi-layer neural network. Guided by the model, optimized synapse structures and refresh operation are proposed to significantly enhance the reliability of FA-RRAM based neural network.

ACKNOWLEDGMENT

This work was supported in part by the NSFC (61421005 and 61604005).

REFERENCES

[1] P. A. Merolla, *et al.*, "A million spiking-neuron integrated circuit with a scalable communication network and interface," *Science*, vol.345, pp.668-673. 2014.

[2] S. B. Eryilmaz *et al.*, "Device and system level design considerations for analog non-volatile-memory based neuromorphic architectures," in *IEDM Tech. Dig.*, 2015, pp. 64–67.

[3] M. Prezioso *et al.*, "Training and operation of an integrated neuromorphic network based on metal-oxide memristors," *Nature*, vol. 521, no. 7550, pp. 61–64, 2015.

[4] D. Lee *et al.*, "Oxide based nanoscale analog synapse device for neural signal recognition system," in *IEDM Tech. Dig.*, 2015, pp. 91–94.

[5] L. Larcher *et al.*, "Multiscale modeling of neuromorphic computing: from materials to device operations," in *IEDM Tech. Dig.*, 2017, pp. 282–285.

[6] P. Yao, et al, "Face classification using electronic synapses," *Nature Communications*, 8, 15199, 2017.

[7] W. Wu, *et al.*, "A Methodology to Improve Linearity of Analog RRAM for Neuromorphic Computing," *Symp. VLSI Tech. Dig.*, 2018, pp.103-104.

[8] M. Zhao *et al.*, "Investigation of Statistical Retention of Filamentary Analog RRAM for Neuromophic Computing," in *IEDM Tech. Dig.*, 2017, pp. 872–875.

[9] J. F. Kang *et al.*, "Oxide-based RRAM: Requirements and Challenges of Modeling and Simulation," in *IEDM Tech. Dig.*, 2015, pp. 113–116.

[10] Mott NF. Conduction in non-crystalline materials. Oxford: Clarendon Press; 1987.

[11] S. Yu *et al.*, " Scaling-up Resistive Synaptic Arrays for Neuro-inspired Architecture: Challenges and Prospect," in *IEDM Tech. Dig.*, 2015, pp. 451–454.

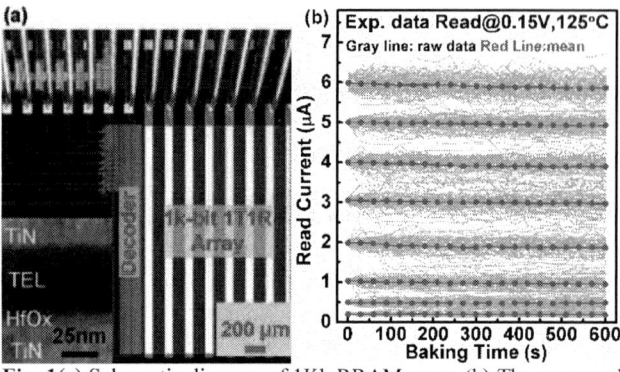

Fig. 1(a) Schematic diagram of 1Kb RRAM array. (b) The measured state instability and retention behaviors of 8 resistance states of 100 RRAM cells. Inset of **Fig.** 1(a) shows the device structure.

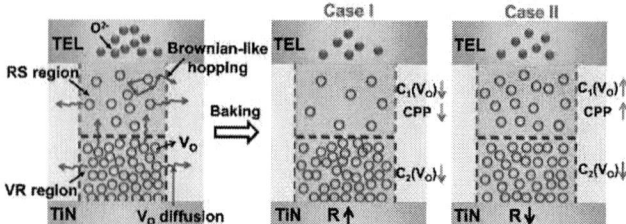

Fig. 2 Physical mechanism of state instability and retention degradation. Instability: the Brownian-like hopping of individual V_O will lead to the decrease/increase of V_O concentration ($C(V_O)$) and rupture/formation of current percolation path (CPP) in the resistive switching (RS) region. Retention behavior: resistance increases due to the V_O diffusion along the radius direction and recombination with O^{2-}, while the diffusion of V_O from the V_O rich (VR) region to the RS region leads to the decrease of resistance.

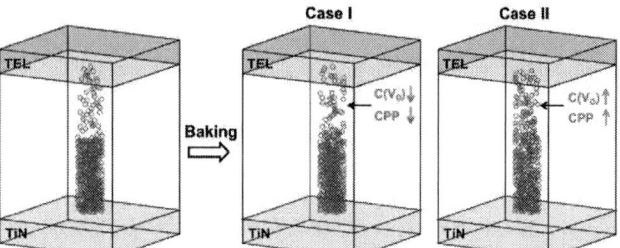

Fig.3 3D kinetic Monte Carlo Simulation of V_O movement. The V_O distributions and concentrations are different after the same baking time due to the randomness of Brownian-like hopping of V_O. The V_O outside the RS/VR region is not plotted for a better visualization.

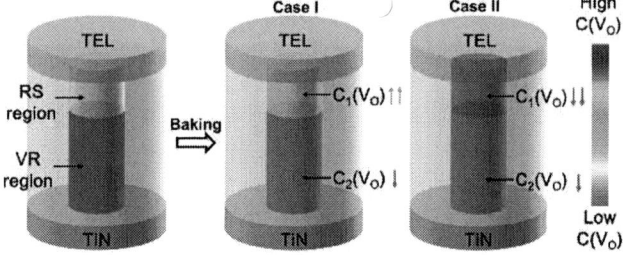

Fig. 4 3D model of RRAM state instability and retention behaviors. The formation and rupture of CPP is represented by the variation of $C(V_O)$ in the RS/VR regions. The evolution of mean $C(V_O)$ ($\mu(C(V_O))$) can describe the retentention behavior, the resistance instability is depicted by the standard deviation of $C(V_O)$ ($\sigma(C(V_O))$).

Table I Equations and Parameters used in this work

Equations		Parameter	Value	Unit
$\mu(C(V_O)) = C_0(V_O)\exp(-\frac{t}{\tau})^{1/\sqrt{2}}$	(1)	f	10^{13}	Hz
$\tau = \beta\exp(\frac{E_i}{k_BT})$	(2)	E_i	0.7	eV
$\sigma(C(V_O)) = \lambda_1(N_t C(V_O))^{0.5}$	(3)	ΔE	0.25	eV
$N_t = ft\exp(-\frac{E_i}{k_BT})$	(4)	E_0	0.55	eV
$\mu(\Delta N_{21}(V_O)) = \gamma(C_e - C_{10})(1 - \exp(-\frac{t}{\tau})^{1/\sqrt{2}})$	(5)	r_a	0.25	nm
$C_e = (C_{10}(V_O)V_1 + C_{20}(V_O)V_2) / (V_1 + V_2)$	(6)	G_0	10^6	S/m
$\sigma(\Delta N_{21}(V_O)) = \lambda_2(N_t(C_2(V_O) - C_1(V_O)))^{0.5}$	(7)	ε_{HfOx}	20	
$R = C(V_O)^{-1/3}$	(8)	α	1.33×10^9	m^{-1}
$G = G_0\exp(-\frac{E_P}{E_0})(R \le r_{th})$	(9)	β_1	1.8×10^9	s
$E_P = \frac{e^2}{4\pi\varepsilon_{HfOx}\varepsilon_0}(\frac{1}{r_a} - \frac{1}{R - r_a})$	(10)	β_2	1.5×10^9	s
		λ_1	1.2×10^8	$m^{-1.5}$
$G = G_0\exp(-2\alpha R)\exp(-\frac{\Delta E}{k_BT})$ $(R > r_{th})$	(11)	λ_2	1.2×10^{-17}	$m^{1.5}$
		γ	1.6×10^{-26}	m^3
		r_{th}	2.5	nm

Equations (1)-(4) describe the state instability and retention degradation caused by the diffusion of V_O along the radius direction and the recombination between V_O and O^{2-}, $C_0(V_O)$: initial $C(V_O)$, E_i: energy barrier of V_O hopping, t: baking time, f: vibration frequency of oxygen atom. Considering the recombination is mainly occurred in RS region, β is different for RS and VR region. Equations (4)-(7) are used to calculate the V_O exchange between RS and VR regions. $\Delta N_{21}(V_O)$: the number of V_O hopping from VR region to RS region, V_1/V_2: volume of RS/VR region. Conductance can be calculated based on percolation theory according to $C(V_O)$ by using (8)-(11). r_{th}: percolation threshold, R: average distance of V_O.

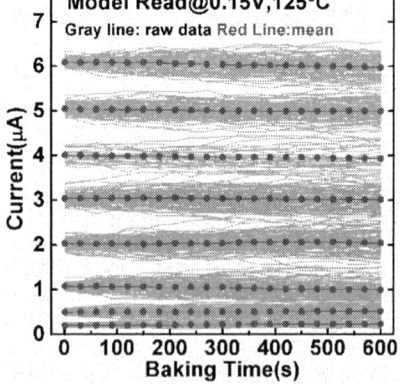

Fig. 5 The calculated state instability and retention behaviors (experimental data is shown in **Fig. 1**).

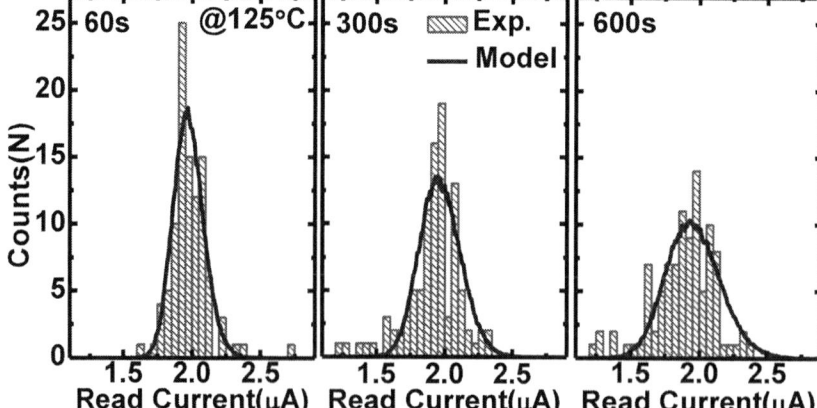

Fig. 6 The measured and calculated evolution of read current (I_{read}) distribution of the medium state with time (30s, 300s, 600s). The distribution becomes wide with time.

978-1-7281-1988-5/18 $31.00 © 2018 IEEE

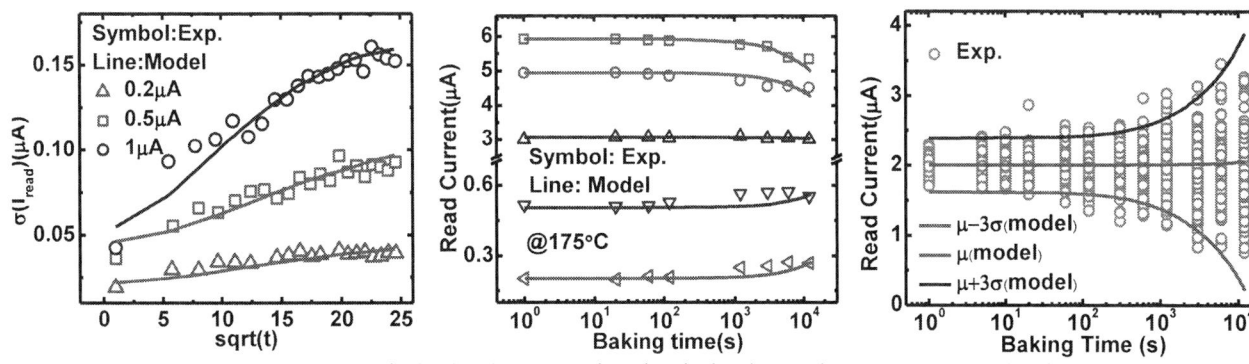

Fig. 7 Calculated and measured standard deviation of I_{read} vs baking time. $\sigma(I_{read})$ is linear with the square root of baking time.

Fig. 8 Measured and calculated retention behaviors for a long baking time @175°C.

Fig. 9 Measured and calculated evolution of I_{read} (2μA) distribution with time @175°C.

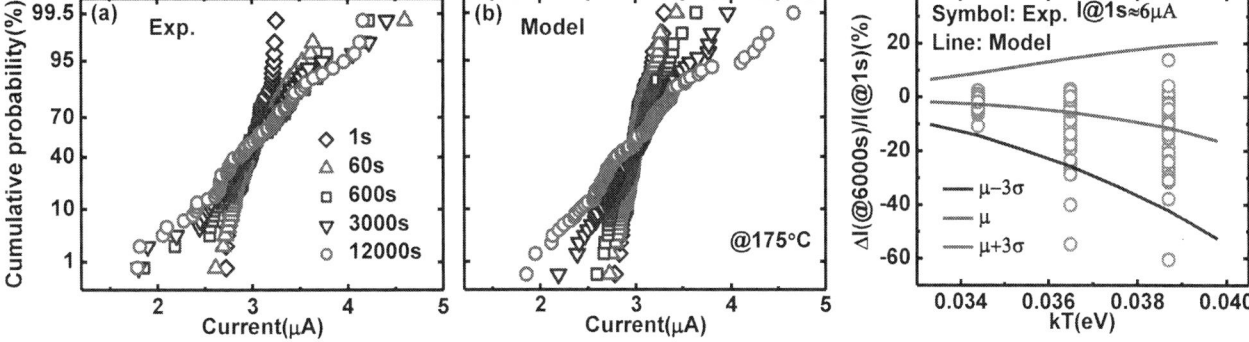

Fig. 10 (a) Measured and (b) calculated cumulative probability of I_{read}(3μA) at different baking time @175°C. The measured and calculated results show the same trend that the distribution of I_{read} is spread with time.

Fig. 11 Measured and calculated variation of μ(I_{read}) after baking for 12000s under different temperatures.

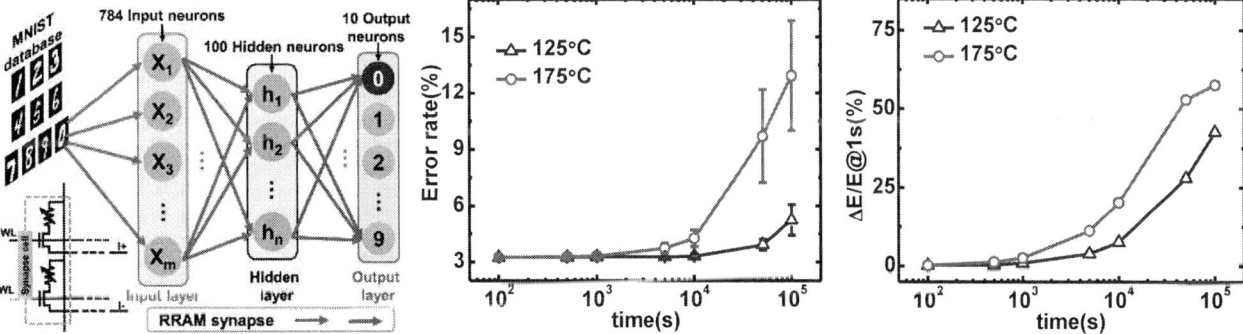

Fig. 12 Schematic of fully connected multi-layer neural network.

Fig. 13 Dependence of recognition error rate on the baking time @125°C and 175°C.

Fig. 14 Energy consumption on RRAM synapse during the inference process.

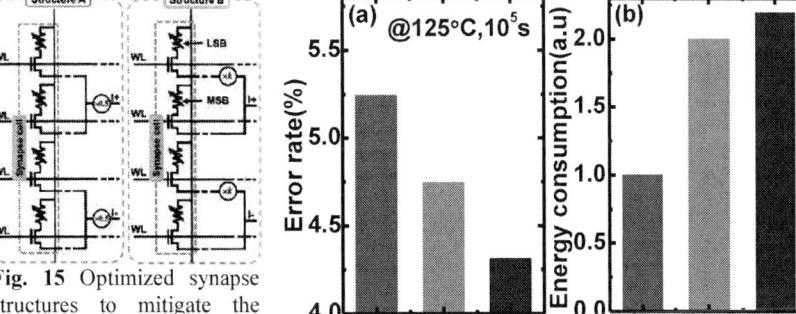

Fig. 15 Optimized synapse structures to mitigate the degradation of recognition rate induced by the state instability and retention degradation.

Fig. 16 Comparison of (a) error rate and (b) energy consumption on synapse of neural network constructed by different synapse cells as shown in **Fig. 15**.

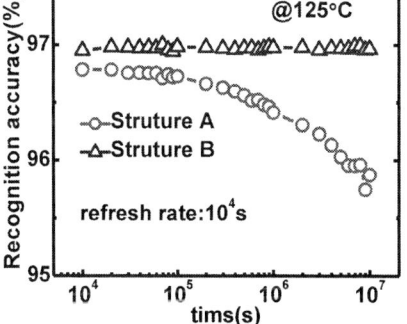

Fig. 17 Recognition accuracy as a function of time with refresh operation.

978-1-7281-1988-5/18 $31.00 © 2018 IEEE   934

Evidence of Magnetostrictive Effects on STT-MRAM Performance by Atomistic and Spin Modeling

K. Sankaran[1], J. Swerts[1], R. Carpenter[1], S. Couet[1], K. Garello[1], R. F. L. Evans[2], S. Rao[1], W. Kim[1], S. Kundu[1], D. Crotti[1], G. S. Kar[1], and G. Pourtois[1, 3]

[1]imec, Leuven, Belgium, email: sankaran@imec.be,
[2]University of York, York, UK, [3]University of Antwerp, Antwerp, Belgium

Abstract—For the first time, we demonstrate, using an atomistic description of a 30nm diameter spin-transfer-torque magnetic random access memories (STT-MRAM), that the difference in mechanical properties of its sub-nanometer layers induces a high compressive strain in the magnetic tunnel junction (MTJ) and leads to a detrimental magnetostrictive effect. Our model explains the issues met in engineering the electrical and magnetic performances in scaled STT-MRAM devices. The resulting high compressive strain built in the stack, particularly in the MgO tunnel barrier (t-MgO), and its associated non-uniform atomic displacements, impacts on the quality of the MTJ interface and leads to strain relieve mechanisms such as surface roughness and adhesion issues. We illustrate that the strain gradient induced by the different materials and their thicknesses in the stacks has a negative impact on the tunnel magneto-resistance (TMR), on the magnetic nucleation process and on the STT-MRAM performance.

I. INTRODUCTION

In the new generation of stand-alone and embedded memories, the STT-MRAM, based on an out-of-plane magnetized MTJ (pMTJ), is a strong contender to conventional non-volatile memories [1]. The performance of a pMTJ is evaluated based on its magnetic and electric coupling that occurs through the CoFe|MgO|CoFe interface, its associated TMR and coercivity of the magnetic CoFe layer (Fig. 1a). Unfortunately, in the typical bottom-pinned TaN|Ru|CoPt |CoFeTa|CoFe|t-MgO|CoFe|CoFeTa|CoFe|c-MgO|CoFe|Ta|Ru |TiN stack (Fig. 1b), there is a strong discrepancy between the performances obtained on thin-film blankets and patterned devices (Fig. 1c). Particularly, the annealing process required to crystallize amorphous MgO (400°C during 90min) before the patterning step leads to a strong reduction of the device TMR and coercivity. Maintaining the structural integrity of this multi-material/layer device at the sub-nanometer scale is challenging due to the differences in intrinsic mechanical properties (Fig. 2). Assuming a linear expansion for the materials, the expected mechanical strains induced by the thermal treatment in the different layers can be qualitatively assessed (Fig. 3). Among the different layers used in pMTJ devices, Ru undergoes the largest expansion with temperature, followed by MgO and CoFe (see Fig. 3 inset), which leads to a highly non-uniform strain gradient.The system hence releases the strain through interfacial processes such as surface roughening, delamination, material lift-off and diffusion [2]. These phenomena are enhanced during the patterning process after annealing and are

expected to be dependent on the thickness of the different layers.

This study aims at i) understanding the consequences of a non-uniform strain gradient on the pMTJ performances, ii) drawing guidelines for material/device improvements and iii) establishing how the resulting perturbation of material interfaces impacts on spin nucleation. For the first time, we report the presence of a non-uniform compressive strain profile in an etched STT-MRAM device of 30nm diameter using atomistic simulations and show that it impacts on the CoFe|MgO interfaces to result in edge damages, as evidenced by TEM (Fig. 4). This leads to a significant lowering of the TMR with respect to the blanket case. We propose strategies to reduce this effect and combine the atomistic structure of the pMTJ device with spin modeling to quantify its consequences on the M-H hysteresis loop.

II. METHODOLOGY

First-principles simulations are performed by coupling density functional theory (DFT) with Non Equilibrium Green functions (NEGF) to compute the TMR in perfect epitaxial CoFe|MgO|CoFe stack [3]. The atomistic structures of the relaxed MTJ stack are obtained by minimizing the atomic forces with a conjugated gradient algorithm [4]. The interatomic potentials of all materials and interfaces were described using the Buckingham-Coulomb and Lennard-Jones formalisms whose parametrizations were generated using a force-matching algorithm trained against DFT data [5] and carefully tested (Fig. 5). Atomistic spin dynamics simulations were performed by combining a spin Hamiltonian to describe the energetics and the Landau-Lifshitz-Gilbert formalism for the spin dynamics [6].

III. GATE STACK ATOMISTIC STUCTURE

A. Atomistic model and experimental links

Starting from the film morphology from an ideal epitaxial structure, the atomic structure of STT-MRAM devices has been built for a diameter of 30nm. The atomic forces of the stacks (Fig. 6a and c) have been relaxed with a conjugate gradient until the forces per atomic site reach 10^{-4} eV/ Å (Fig. 6b). The relaxation process leads to significant changes in the atomic morphology (Fig. 6d) with respect to the ideal epitaxial case (Fig. 6a and c). These are consistent with cross-section TEM (Fig. 4 a , b, and c) profiles which report the occurrence of a strong roughness at various interfaces[Fig. 4b I and Fig. 6d I] and the presence of edge damages in the t-MgO (Fig. 4b II and Fig. 6d II). We also observe a strong diffusion of the Ru layer

978-1-7281-1988-5/18 $31.00 © 2018 IEEE

in contact with the bottom TaN electrode (Fig. 4d III and Fig. 6d III) as evidenced by [7]. These deviations from idealities are expected to impact significantly on device performance.

B. Atomic volume deformation and relation with the TMR

The morphological changes induced by the relaxation of the epitaxial strain are captured by the atomic volume deformation (AVD) in the CoFe|MgO|CoFe layers which reflects the difference in atomic volume per site (computed using a Voronoi tessellation algorithm) between the relaxed and the epitaxial cases (Fig. 7a). The latter clearly suggests that a strong compressive strain occurs in the pMTJ, particularly in the core part of the t-MgO. Although the TMR is a macroscopic effect of the pMTJ, a qualitative link can be established with the evolution of the TMR in epitaxial stacks under different hydrostatic deformations (Fig. 8) and the AVD of t-MgO. Using the AVD, we establish a qualitative mapping of the top surface of t-MgO with its intensity for different components of the TMR. Fig. 10 (middle) reveals that the parallel (Rp) and anti-parallel (Rap) resistances are increasing, which results in a significantly reduced TMR with respect to the unstrained t-MgO (Fig. 10 top).

C. Impact of the top and bottom electrodes

A similar exercise was repeated by removing the top TiN and bottom TaN electrodes. The AVD map suggests that the strain in the t-MgO layer (Fig. 9) comes with an important i) MgO edge lift-off (~5nm), ii) surface roughness and iii) an increase of Rp and Rap resulting in a low TMR (Fig. 10 bottom) compared to the case where electrodes are present (Fig. 10 middle). This suggests that the mechanical compression from the electrodes prevents the t-MgO lift-off and that edge effects observed experimentally are not only related to etch issues.

IV. STACK ENGINEERING

A. Single MgO vs. dual MgO vs. tripple MgO stack

The deposition of a top thin MgO capping (c-MgO) (<1.0nm) on top of the free-layer has been proposed to improve the thermal robustness and to increase the free-layer coercivity [8]. Our simulations indicate that it also acts as a mechanical buffer which accommodates the compressive strain from the Ru layer and the top electrodes. The expansion of the pMTJ with an additional c-MgO is hereafter set as being our process of reference (POR) in Fig. 4c, 7 and 11. The utility of a thin c-MgO is further demonstrated by adding a third c-MgO cap on top of the POR. The resulting averaged compression in the t-MgO is reduced from about -9.1% (t-MgO) to -8.1% (t-MgO+1×c-MgO) and -5.3% (t-MgO+2×c-MgO) for the 30nm diameter device stack (Fig. 7b, 7c and 11). In the POR, the core of the MgO tunnel barrier is the object of a high compressive strain with respect to the edges, which is lessened by the introduction of a third c-MgO layer (Fig. 11).

B. Impact of the top Ru thickness

It is also possible to lower the compressive strain by optimizing the top Ru thickness (Fig. 12). As a proof of concept, we varied the Ru thickness (POR 5nm) from 2.5 to 10nm. The resulting averaged strain sampled over the t-MgO shows that a 2.5nm thick Ru leads to similar profile as the dual MgO [t-MgO+2×c-MgO] model (Fig. 11). Without any surprise, a thick Ru layer leads to a strong compressive strain (higher than in a single MgO). In that respect, reducing the thickness of the Ru layer is an excellent alternative to minimize the compression (Fig. 12) and to enhance the quality of the interface (Fig. 13).

C. Impact on the pMTJ interface

Figures 11, 12 and 13 summarize the overall interface quality of the pMTJ, which is assessed by quantifying the alignment of the CoFe|MgO interface by combining the Fe from CoFe and O from the MgO bond distortion and the amount of strain. Globally, changing the number of MgO layers in the stack has a limited impact on the interface quality, while it does strongly modulate the strain in the layer. On the other hand, thinning the thickness of Ru improves the interface quality by ~10% (Fig. 13) and reduces the strain (Fig. 12).

V. MAGNETIC RESPONSES OF THE STRAINED PMTJ

The magnetic response of the atomic structure of the CoFe free-layer (see Fig. 15 and 17 insets) and its M-H hysteresis loops were computed using atomistic spin dynamics simulations, which constitutes the ultimate limits of the special discretization (Fig. 14). At low T (1K), both unstrained and strained free-layers show a perpendicular magnetization, while the coercivity of the strained one is lowered due to an easy magnetic reversal (Fig. 15). This is confirmed by the weak torque for the strained free-layer (Fig. 16). At room T, the strained free-layer adopts an in-plane magnetization due to the high disorder of the atomic sites (Fig. 17 and 18) and to the change in magnetic reversal mechanism, induced by the strong thermal fluctuations and its associated spin modulations (see Fig. 17 inset). These findings indicate that the magnetic response of the free-layer is considerably altered due to the non-uniform strain gradient in the stack.

VI. CONCLUSIONS

Our atomistic model suggests that a strong non-uniform strain gradient is present in STT-MRAM devices. The resulting magnetostrictive effects is found to be detrimental for device performance. Moreover, our atomistic spin dynamics simulations confirm that the high disorder of the atomic sites in the CoFe free-layer leads to a strong perturbation of its magnetic response. The minimization of the strain gradient built in the blanket layer leads to edge effects upon patterning, such as a partial delamination and active area reduction of the MgO tunnel barrier. The introduction of additional thin MgO cap layers partially compensates the strain gradient by limiting edge effects and minimizing the compressive strain accumulated in the MgO tunnel barrier. Reducing the thickness of the Ru layer has a similar effect. It is therefore important to account for these effects in the design of real pMTJ stacks.

ACKNOWLEDGMENT

This work was carried out in the framework of the imec Core CMOS – MRAM Memory Program.

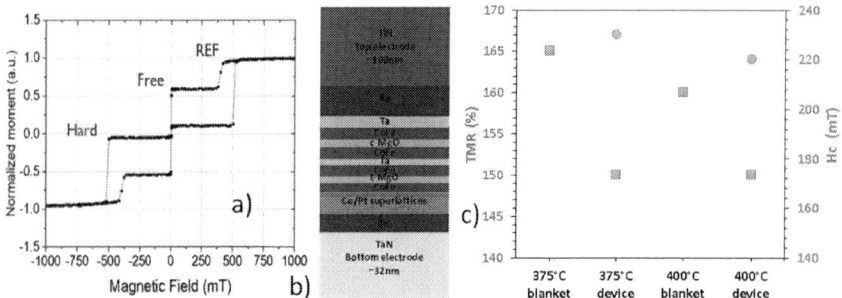

Fig. 1. a) Typical M-H loops of a STT-MRAM device. b) Schematics of the POR dual MgO with cap MgO (c-MgO) and tunnel MgO (t-MgO) device stacks. c) Experimental TMR (left) and free-layer coercivity (right) of blanket thin-film and 45nm electrical CD bottom-pinned devices annealed at 375 and 400°C.

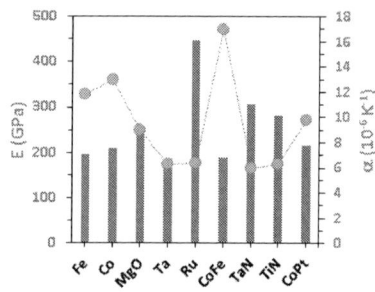

Fig. 2 Young modulus (histogram) and thermal expansion coefficients of the STT-RAM constituting bulk materials.

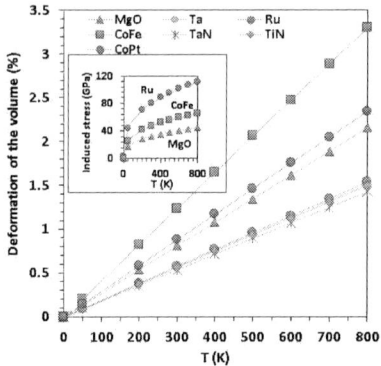

Fig. 3. Volume deformation expected upon thermal treatment based on bulk thermal expansion coefficients and Young modulus together with its induced stress for Ru, CoFe and MgO (inset).

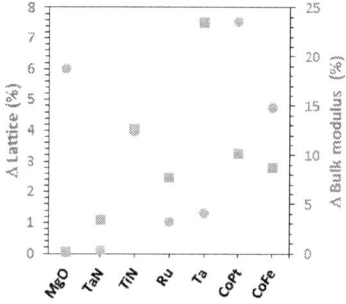

Fig. 5 Deviation of the computed lattice parameters and bulk Young modulus using the parameterized interatomic potentials with respect to DFT.

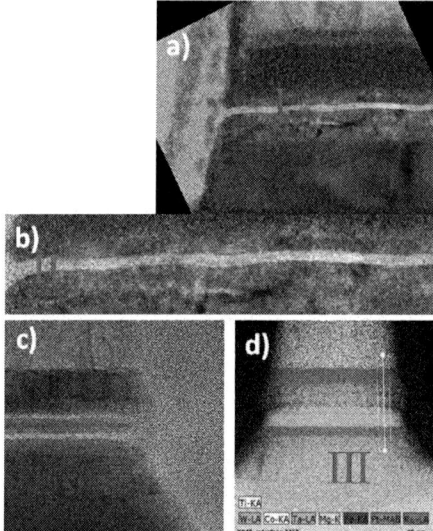

Fig. 4. a) TEM of a single MgO STT-MRAM device stack (a) and its zoom in pMTJ image (b). (c) TEM of a dual MgO STT-MRAM stack after etch (without BEOL thermal treatment) and its EDX profile (d). I corresponds to interface roughness, II to the MgO lift-off, and III to Ru-TaN in-diffusion.

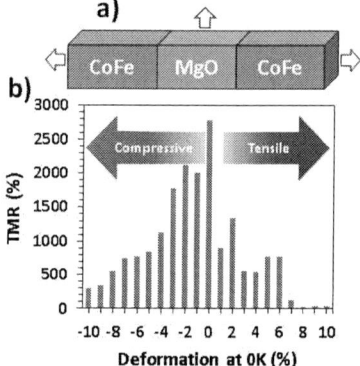

Fig. 8 a) Illustration of the hydrostatic deformations used to compute the TMR in epitaxial pMTJ stack. b) Resulting TMR obtained using DFT as a function of the strain applied in the pMTJ for an ideal CoFe|MgO|CoFe interface.

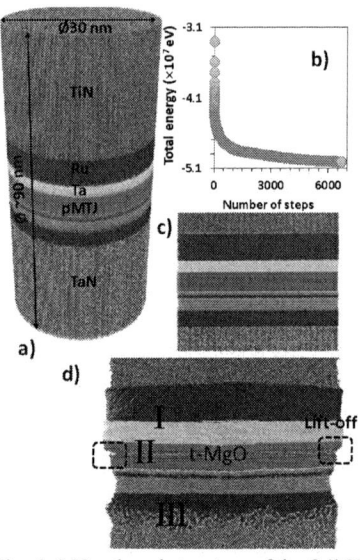

Fig. 6 a) Unrelaxed structure of the full STT-MRAM 30nm diameter device (a) and (b) the corresponding total energy minimization as a function of number of numerical steps. Transversal slice and zoomed view of stack before (c) and after atomic relaxation of single MgO (d) [t-MgO]. I corresponds to interface roughness, II to the MgO lift-off and III to Ru-TaN in-diffusion.

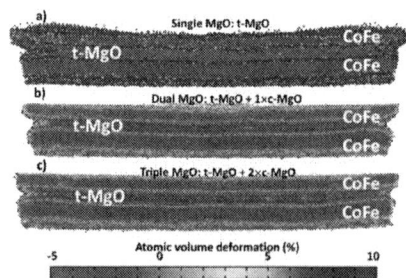

Fig. 7 Atomic volume deformation of pMTJ for a single MgO (a) [t-MgO], for a dual MgO (b) [t-MgO + 1×c-MgO] and for a triple MgO (c) [t-MgO + 2×c-MgO].

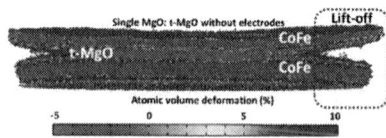

Fig. 9. Atomic volume deformation of the pMTJ of single MgO STT-MRAM 30nm-diameter device without top and bottom electrodes.

978-1-7281-1988-5/18 $31.00 © 2018 IEEE

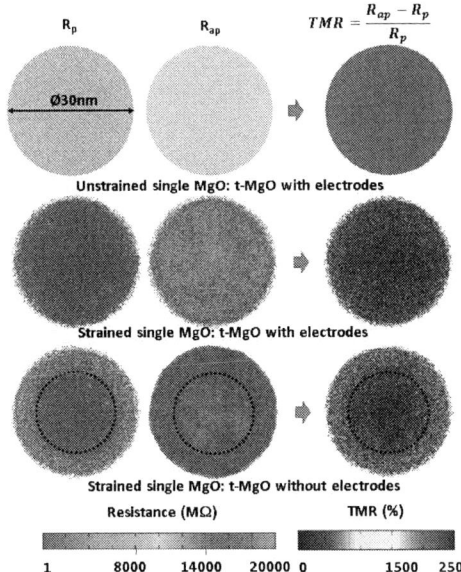

$$TMR = \frac{R_{ap} - R_p}{R_p}$$

Unstrained single MgO: t-MgO with electrodes

Strained single MgO: t-MgO with electrodes

Strained single MgO: t-MgO without electrodes

Resistance (MΩ)   TMR (%)

1   8000   14000   20000   0   1500   2500

Fig. 10. Parallel (Rp), anti-parallel (Rap) resistance and TMR mapping (top view) of the pMTJ of single MgO device with and without electrodes.

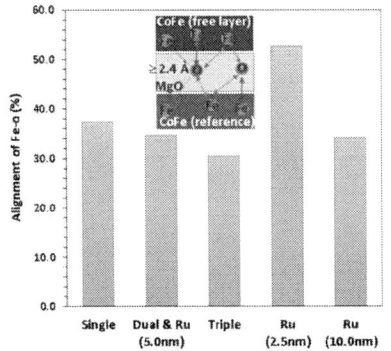

Fig. 13. Percentage of the available aligned Fe-O (see inset) at the interface of the pMTJ for different device stacks.

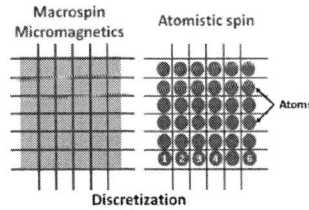

Fig. 14. Schematic description of the discretization for macrospin/ micromagnetics and atomistic spin simulations.

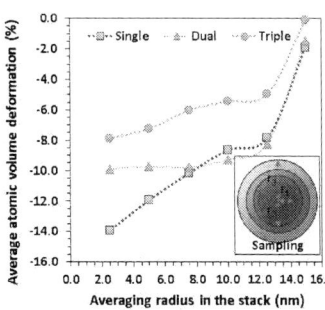

Fig. 11. Average atomic volume deformation of t-MgO in single, t-MgO+c-MgO and t-MgO+2 c-MgO) for 30nm diameter device as a function of the averaging radius r. The strain is sampled from the edge of the device to the center (inset).

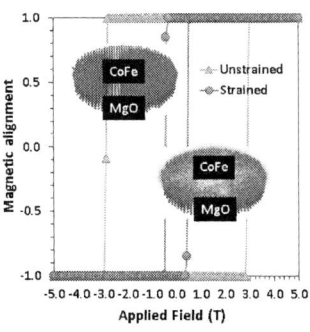

Fig. 15. M-H hysteresis loop at 1K of perfect epitaxial unstrained CoFe free-layer (top inset) compared with that of the strained one (bottom inset).

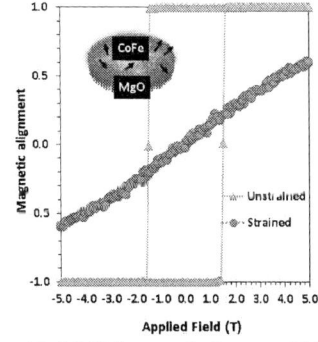

Fig. 17. M-H hysteresis loop at 300K of perfect epitaxial unstrained CoFe free-layer compared with that of the strained one (inset schematics with random spin).

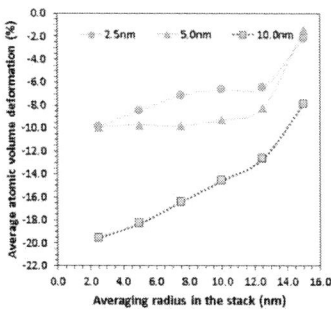

Fig. 12. Average atomic volume deformation of the MgO tunnel barrier in the dual MgO with Ru thickness of 2.5, 5.0 and 10.0nm for a STT-MRAM device of 30nm diameter as a function of the averaging radius in the stack (inset Fig. 11).

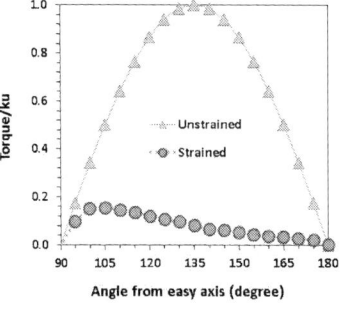

Fig. 16. Normalized torque at 1K of perfect epitaxial unstrained CoFe free-layer compared with that of the strained one.

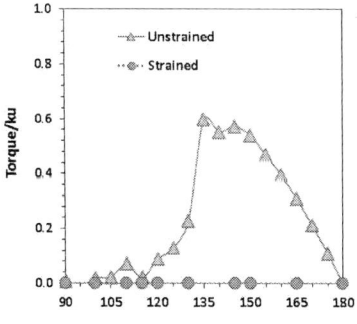

Fig. 18. Normalized torque at 300K of perfect epitaxial unstrained CoFe free-layer compared with that of the strained one.

REFERENCES

[1] S. Ikeda et al, "A perpendicular-anisotropy CoFeB-MgO magnetic tunnel junction" *Nat Mater.* 9, 721 (2010).

[2] J. Tersoff et al, "Competing relaxation mechanisms in strained layers" *Phys. Rev. Lett.* 72, 3570 (1988).

[3] M. Brandbyge et al, "Density-functional method for nonequilibrium electron transport" *Phys. Rev. B* 65, 165401 (2002); Atomistix Toolkit 2015.1, Synopsys QuantumWise A/S

[4] S. Plimpton, "Fast Parallel Algorithms for Short-Range Molecular Dynamics" *J. Comp. Phys.* 117, 1 (1995); LAMMPS package.

[5] Brommer et al, "Classical interaction potentials for diverse materials from ab initio data: a review of potfit" *Model. Simul. Mater. Sci. Eng.* 23, 074002 (2015); Potfit package.

[6] R. F. L. Evans et al, "Atomistic spin model simulations of magnetic nanomaterials" *J. Phys. Condens. Matter* 26, 103202 (2014; VAMPIRE package.

[7] L. Wang et al, "Electric field accelerating interface diffusion in Cu/Ru/TaN/Si stacks during annealing" *Electrochem. Solid. S. Lett.* 15, H188 (2012).

[8] S.Couet etal, "Impact of Ta and W-based spacers in double MgO STT-MRAM free layers on perpendicular anisotropy and damping" *Appl. Phys. Lett.* 111, 152406 (2017).

AUTHOR INDEX

Abe, T. 237
Abel, S. 540
Acosta-Alba, P. 153
Adelmann, Ch. 111
Adelmann, Christoph 831
Afzalian, A. 492
Agarwal, A. 201
Agarwal, Harshit 209
Agarwal, T. Kumar 512
Agarwal, V. 739
Agashiwala, Kunjesh 576
Ahn, Hyoshin 771
Ahn, Woojin 584
Ai, Kelvin 217
Aigner, R. 332
Akinwande, Deji 532
Alam, A. 683
Alam, Muhammad Ashraful 584
Alava, Thomas 281
Alessandri, C. 368
Alfieri, G. 440
Ali, M. 107
Alian, Alireza 304
Allain, P.E. 99
Allain, Pierre E. 281
Almeida, S. 75
Alzate, J. -G. 412
Amanapu, H. 819
Ambrosi, E. 927
Amin, P. 133
Amishiro, H. 185
Amisse, A. 141
Ando, T. 807, 911
Andre, F. 157
Andrieu, F. 153
Ang, Kah-Wee 580
Annema, A. J. 739
Annunzlata, R. 424
Anzai, Y. 284
Aoyagi, Y. 225
Appenzeller, J. 536
Arasawa, Ryo 312
Araujo, F. Abreu 616
Arimura, H. 496, 787
Arimura, Hiroaki 783
Arnal, V. 161
Arnaud, F. 424
Arnaud, L. 153, 157
Arnoux, M. 157
Arreghini, A. 43, 600
Arslan, U. 412
Arutchelvan, G. 512
Asai, Hidehiro 197
Asai, Y. 460
Asatsuma, T. 221

Asselberghs, I. 512
Atsumi, Tomoaki 312
Aubin, J. 153
Audoit, G. 500
Audran, S. 161
Aussenac, F. 153
Auth, Chris 636
Ayele, G. T. 277
Baba, Shotaro 79
Badami, Oves 759
Badaroglu, M. 408
Bae, B. J. 416
Bae, D.-I. 656
Bae, Geumjong 656
Bae, Jong-Ho 288
Baert, R. 420
Baeyens, Y. 552
Bai, P. 412
Bakir, Muhannad S. 672
Balan, V. 153, 157, 500
Banerjee, K. 43
Banerjee, Kaustav 564, 576, 799
Banniard, L. 99
Banniard, Louise 281
Bao, R. 807
Bao, Ruqiang 253, 648
Bao, T. Huynh 420
Barbato, A. 703
Barbato, M. 703
Barlage, D. 923
Barman, Soumendra 253
Baron, L. 161
Barr, R. 460
Barral, V. 424
Barraud, S. 141, 500
Barreto, J. 540
Barrois, Charlie 281
Basker, Veeraraghavan 652
Batude, P. 153
Baudin, F. 161
Baumgartner, Y. 907
Bavandpour, M. 476
Bayat, F. Merrikh 476
Bayha, B. 428
Bazizi, E.M. 428
Beaurepaire, S. 153
Beche, E. 161
Bedau, D. 867
Behin-Aein, B. 604
Bellando, F. 269
Bellini, M. 440
Belmonte, A. 51
Bender, H. 496
Bendersky, L.A. 536
Benedict, J. 107

AUTHOR INDEX

Beneyton, R. 424
Benoit, D. 424
Benschop, Jos 261
Bernard, M. 863
Bernier, N. 500
Berthelon, R. 424
Bertrand, B. 141
Besson, P. 153
Beyer, S. 428
Beyne, S. 111
Bhuiyan, M. A. 711
Bhuva, B. 795
Bhuwalka, K.K. 656
Bi, Z. 811
Bianda, E. 440
Bilgen, H. 157
Billiot, Gérard 281
Birkhahn, R. 460
Bishop, Douglas 292
Bocquet, M. 484
Boemmels, J. 149
Boeuf, F. 277
Bohr, M. 412
Bohuslavskyi, H. 141
Boivin, P. 424
Boland, J. 811
Boon, M.N. 628
Borchert, James W. 883
Borga, M. 703
Bortolotti, P. 616
Boter, J.M. 133
Botzem, T. 129
Boucard, F. 424
Bouchu, D. 157
Boulanger, Pascale 281
Bourdet, L. 141
Bourgeois, G. 863
Brems, S. 512
Brenac, Ariel 281
Breslin, C. M. 107
Bresson, N. 157
Breuil, L. 600
Brew, K. 819
Brew, Kevin 815
Bricalli, A. 927
BrightSky, M. 859
Brink, Markus 126
Brockman, J. 412
Bruce, R. 911
Bruce, R. L. 859
Bruley, J. 911
Brunelli, S. T. Šuran 324
Brunet, L. 153
Brus, S. 787, 843
Bu, H. 819

Buford, B. 412
Burghartz, Joachim N. 883
Burns, J.L. 632
Bury, E. 592
Bury, Erik 783
Cabout, T. 161
Cabral, C. 107
Cadien, K. 923
Cai, D. L. 620
Cai, Linlin 775, 779
Cai, Xiangbin 695
Cai, Zhimei 173
Caimi, D. 540, 899, 907
Caironi, Mario 883
Camlica, A. 755
Campbell, J.P. 536
Canales, F. 440
Canaperi, D. F. 811
Cao, Linjun 115
Cao, Wei 564, 576
Capogreco, E. 496
Capogreco, Elena 783
Cappelletti, P. 424
Carlson, E.P. 456
Carpenter, R. 935
Carr, A. 819
Carr, Adra 815
Carta, F. 859
Casella, A. 747
Cassé, M. 500
Castany, Olivier 281
Castellani, N. 472, 863
Caubet, V. 424
Cave, N. 819
Caymax, M. 512
Cazaux, Y. 229
Cha, Jungho 560
Cha, Moonhyun 771
Chae, Sujin 851
Chai, Yang 520, 524
Chakraborty, W. 364
Chalupa, Z. 500
Champenois, A. 161
Chan, B. T. 149
Chan, Kevin 292
Chan, Masun 520
Chan, Michael S 253
Chang, C.-Y. 504
Chang, Che-Chia 356
Chang, Chia-He 249
Chang, D. R. 640
Chang, F. 811
Chang, Huan-Lin 209
Chang, Jonathan 273
Chang, Ki Soo 288

AUTHOR INDEX

Chang, L. 632
Chang, Meng-Fan 340
Chang, Pengying 775, 779
Chang, Vincent S. 644
Chanrion, E. 141
Chao, R. 604
Chao, Robin 652
Chao, T.-S. 504
Charbon, E. 739
Charles, C. 157
Chasin, A. 508
Chasin, Adrian 783
Chatterjee, Korok 209
Chauhan, Y. S. 201
Che, Xiaoyu 839
Chen, An 855
Chen, B.-J. 751
Chen, Bing 173
Chen, Bo-Yuan 340
Chen, C.-L. 751
Chen, C.-Y. 751
Chen, Cheng 707
Chen, Chin-Hsuan 636
Chen, Chun-Chi 731
Chen, E. 492
Chen, Fan 348
Chen, H. P. 620
Chen, H.-W. 751
Chen, H.-Y. 735
Chen, Hsiu-Chih 340, 731
Chen, Huiming 847
Chen, Jiezhi 568
Chen, K.C. 31
Chen, K.-T. 735
Chen, Kevin J. 695
Chen, Kuan-Neng 249
Chen, Kun-Ming 340
Chen, L.-Y. 504
Chen, P.-G. 735
Chen, Peng 340
Chen, Pin-Chun 356
Chen, Pin-Guang 731
Chen, Po-An 532
Chen, S. 879
Chen, S.-L. 751
Chen, S.-Y. 735
Chen, Shaoyin 815
Chen, Shih-Wei 249
Chen, Siming 556
Chen, T.H. 396
Chen, T.K. 492
Chen, T.W. 31
Chen, Victoria 528
Chen, Wangyong 775, 779
Chen, Wei-Chen 39

Chen, Wei-Hao 340
Chen, Y. F. 620
Chen, Y.S. 624
Chen, Yen-Pu 584
Chen, Yi-Ju 731
Chen, Z. 508
Chen, Zhebo 253
Chen, Zhihong 516
Cheng, C. C. 31
Cheng, C. -W. 911
Cheng, C.W. 859
Cheng, H. Y. 859
Cheng, P. 791
Cheng, Ran 173
Cheng, S.-L. 751
Cheng, Yan 47
Cheramy, S. 157
Cherault, N. 161, 424
Cheung, K.P. 536
Chevalliez, S. 153
Chiang, Meng-Hsueh 532
Chiappe, D. 512
Chien, W. C. 859
Chih, Y.D. 624
Chih, Yue-Der 273
Chiu, H.P. 31
Chiu, Wen-Cheng 731
Chiu, Yu-Fan 731
Cho, Byung Jin 241
Cho, J. 265
Cho, Sunglae 851
Choi, Byoung Lyong 560
Choi, H. W. 640
Choi, Hyejung 851
Choi, Junho 532
Choi, K. 819
Choi, S. 811, 819
Choi, Samuel 815
Choi, Woosung 348, 767
Choi, Y.M. 656
Choi, Yongki 676
Chossat, J. 157
Chou, Y.-C. 735
Chou, Y.-T. 751
Chouteau, S. 424
Chow, Jerry M. 126
Chowdhury, S. 460
Chu, Jae Hwan 799
Chu, S. 811
Chu, S.-W. 751
Chua, Lye-Iling 823
Chudzik, M. 811
Chudzik, Michael P 253
Chueh, Yu-Lun 273
Chun, K.Y. 656

AUTHOR INDEX

Chun, Yunseok 851
Chung, Bryce 672
Chung, Hyein 767
Chung, Kisup 253
Chung, N. L. 604
Chung, S.-J. 656
Chung, Wonil 344
Chung, Won-Young 767
Ciofi, I. 111
Ciubotaru, Florin 831
Clarke, J.S. 133
Clement, H. 460
Clement, L. 424
Clevenger, L.A. 632
Clima, Sergiu 380
Cloarec, J-P. 277
Cogorno, M. 811
Cogorno, Matt 827
Cohen, G. M. 911
Collaert, N. 149, 496, 787
Collaert, Nadine 304
Collins, Philip G. 676
Colombeau, B. 811
Compagnoni, C. Monzio 35
Cong, H. 604
Connor, C. 412
Conti, R. A. 811
Conti, Richard A 253
Convertino, C. 899, 907
Copel, Matt 292
Cordero, E. Garcia 269
Costa, T. 91
Cott, D. 512, 787
Couet, S. 420, 592, 843, 935
Covi, E. 927
Crafton, Brian 300
Crippa, A. 141
Croes, K. 111
Cros, V. 616
Crotti, D. 420, 592, 935
Cunningham, D.W. 456
Cyrille, M. C. 863
Czornomaz, L. 899, 907, 911
Dangol, A. 508
Das, N. 412
Dasgupta, A. 201
Datta, S. 296, 364
Datta, Suman 55, 300
Davydov, A. V. 536
De Franceschi, S. 141
De Heyn, V. 149, 787
De Keersgieter, A. 496
De La Rosa, C. Lockhart 512
De Santi, C. 691, 703
De Wolf, I. 408

Debacker, P. 480
Decoutere, S. 703
Deen, Jamal 871
Defoort, Martial 281
Degraeve, R. 480, 592
Dehaene, W. 480
Dehollain, J.P. 133
Dekkers, H. 508
Del Alamo, Jesús A. 895
Delalleau, J. 161
Delaye, V. 500
Delhougne, R. 51, 600
Delmedico, S. 424
Deloffre, E. 157
Demarest, J. 107, 819
Demarest, James 815
D'Emic, C. 911
Demuynck, S. 149
Deshmukh, Sanchit 572
Deshpande, V. 149
Detzel, T. 703
Devolder, T. 843
Devolder, Thibaut 831
Devriendt, K. 149, 508
Di Piazza, L. 43, 380
Diaz, C.H. 492
Diaz, Carlos H. 624
Dietzel, B. 548
Disegni, F. 424
Divakaruni, Rama 648
Dmitry, Veinger 233
Do, N. 428
Doevenspeck, J. 480
Domengie, F. 424
Domengie, Florian 404
Dominauez-Medina, Sergio 281
Dona, Danian 47
Dong, Da Nian 464
Dong, P. 552
Dong, Yuan 544
Donnell, J.O 412
Doornbos, G. 492
Dourthe, L. 500
Doyle, B. 412
Drouin, D. 277
Droulers, G. 133
Du, Gang 775, 779
Duarte, Juan Pablo 209
Dube, A. 811
Duclaux, B. 161
Dumont, F. 332
Dünkel, S. 428
Durfee, C. 819
Durfee, Curtis 815
Duriez, B. 492

AUTHOR INDEX

Dutta, S. 364, 739
Dzurak, A.S. 129
Ecoffey, S. 277
Eenink, G. 133
El Kazzi, S. 512
Eleftheriou, E. 628
El-Falou, A. 755
Elloian, J. 91
Eltes, F. 540, 899
Endo, Kazuhiko 197
Endoh, T. 608
Eneman, G. 496
Enomoto, T. 237
Ernoult, M. 616
Ernst, T. 500
Ernst, Thomas 281
Escarabajal, Y. 161
Esseni, D. 213
Esseni, David 759
Euvrard, C. 157
Evans, R. F. L. 935
Everson, L. 352
Exbrayat, Y. 157
Ezhilarasu, G. 683
Fabris, E. 691
Fafin, A. 99
Fafin, Alexandre 281
Fan, J. 879
Fan, Zhiqiang 568
Fang, C.-C. 504
Fang, Y. 324
Fantini, A. 480
Farcy, A. 157
Fattinger, G. 332
Favennec, L. 424
Favero, I. 99
Favero, Ivan 281
Favia, P. 496
Federspiel, X. 153
Feng, L. 879
Feng, Philip X.-L. 87
Feng, Xue 668
Feng, Y. L. 488
Fenouillet-Beranger, C. 153, 472
Ferreira, P. 424
Fettweis, G. 11
Fischer, K. 412
Florent, K. 43, 380
Fompeyrine, J. 540
Fontaine, H. 153
Fontelaye, Caroline 281
Fostner, Shawn 281
Fournel, F. 153
Franco, J. 149, 787
Franco, Jacopo 783

Frank, M. M. 911
Fremont, H. 157
Friedrichs, Peter 436
Frougier, Julien 652
Fuji, Yoshihiko 79
Fujihara, Y. 225
Fujisaki, K. 284
Fukuda, Koichi 197
Fukuhara, K. 265
Fukui, M. 189
Fukushima, A. 616
Fukuzawa, Hideaki 612
Funck, C. 867
Furnemont, A. 420
Furnémont, A. 600
Furukawa, K. 189
Gabriel, Kristin N. 676
Gadigatla, Srinivasa Chaitanya 636
Gaidhane, A. 201
Gaillard, Frederic 404
Galatage, Rohit 652
Gallon, C. 424
Galpin, D. 161
Gan, K. W. 604
Gandolfo, A. 424
Gao, B. 488, 931
Gao, Bin 67, 468
Gao, Jianfeng 47
Gao, X. 691
Garello, K. 935
Garrione, J. 863
Garros, X. 153
Gaur, A. 512
Ge, Ruijing 532
Gely, M. 99
Gely, Marc 281
Gentile, A. A. 540
George, H.C. 133
George, Steven M. 895
Gertsch, Jonas 895
Ghani, T. 412
Ghatge, M. 95
Ghazavi, P. 428
Ghezzi, G. 161
Ghibaudo, Gerard 384, 404
Giannopoulos, I. 628
Gignac, L. 107, 859
Giorgio, Michele 883
Giraud, B. 472
Gluschenkov, O. 819
Gluschenkov, Oleg 815
Goda, A. 27
Goh, L. C. 604
Gokmen, Tayfun 292
Goldberg, C. 21

AUTHOR INDEX

Golonzka, O. 412
Gomez, Jorge 300
Gomiero, E. 424
Gong, Songbin 915
Gong, Tiancheng 47, 464
Gong, Xiao 544, 823
Gonzalez, M. 408
Gossmann, H.-J. 811
Goto, T. 660
Gouget, Gilles 384
Gouraud, P. 161
Goux, L. 51
Grant, Lindsay A. 217
Grasser, T. 787
Green, R. 448
Greene, Andrew 652
Grenier, J.C. 424
Grisafe, B. 296, 364
Gritters, J. 460
Grobis, M. 867
Groeseneken, G. 43, 787
Grollier, J. 616
Grosse, P. 99
Grossi, A. 472
Gu, S.-S. 735
Guerin, C. 153
Guérin, H. 269
Guiheux, Denis 404
Guillaumet, S. 157
Gul, O. Tolga 676
Guo, D. 807, 811, 819
Guo, Dechao 648, 652, 815
Guo, Hong 763
Guo, Shaofeng 388, 392
Guo, X. 879
Guo, Xuyun 520
Gupta, D. 819
Gupta, S. 364
Gupta, Sayak Dutta 803
Gupta, Sumeet Kumar 516
Ha, Kyoungho 560
Haeberlen, O. 703
Haensch, W. 59
Haensch, Wilfried 292
Haluska, Miroslav 83
Ham, B.H. 656
Hamori, H. 660
Hamzaoglu, F. 412
Han, G. 213
Han, Jin-Woo 432
Han, Kaizhen 823
Han, Qin 679
Han, S. H. 416
Hanna, A. 683
Haq, Jesmin 612

Hara, Michiko 79
Harada, S. 177, 181, 444
Haran, B. 811, 819
Haran, Bala S. 253, 648
Hart, A. 540
Hartmann, J.-M. 141, 500
Hashemi, P. 807, 911
Hashimoto, K. 265
Hatano, M. 265
Hatayama, T. 177
Hattori, Junichi 197
He, Jun 396
He, Renren 612
Heil, P. 412
Hellings, G. 149, 787
Henderson, R. K. 743
Henke, A. 428
Hennen, T. 867
Henrion, Y. 157
Henry, Todd 823
Hentges, P. 412
Hentz, S. 99
Hentz, Sébastien 281
Herment, G. 269
Hermouet, M. 99
Hermouet, Maxime 281
Herrmann, T. 428
Hertzberg, Jared 126
Heylen, N. 149
Heyns, M. 843
Heyns, Marc 831
Hiblot, G. 408
Hierold, Christofer 83
Higashi, Yoshihiro 79
Higashiki, T. 265
Hikavyy, A. 149, 496, 600
Hiramoto, T. 189
Hiramoto, Toshiro 372, 723
Hirtzlin, T. 484, 616
Ho, H. Y. 859
Ho, Paul S. 115
Hody, H. 51
Hoffmann, M. 727
Holland, M.C. 492
Holleitner, A.W. 245
Hong, H. S. 416
Hong, Hyeongsun 560
Hong, Jeongmin 847
Hong, R.-C. 735
Hong, Seongbin 288
Hong, T.-C. 504
Hong, Yoonki 288
Honjo, H. 608
Hopstaken, Marinus 253
Horiguchi, N. 496, 508, 787

AUTHOR INDEX

Horiguchi, Naoto 783
Horita, M. 687
Horng, J.J. 396
Hoshii, T. 189
Hosoda, T. 460
Hou, F.-J. 504
Hou, Tuo-Hung 356
Hsieh, H. D. 396
Hsieh, Tung-Ying 249
Hsiung, Alan Chih-Wei 217
Hsu, H.-S. 504
Hsueh, F.-K. 504
Hsueh, Fu-Kuo 340, 731
Hu, C.-K. 107
Hu, Chenming 209, 249, 731
Hu, Szu-Tung 115
Hu, Z. 691
Hu, Zongyang 193
Hua, Mengyuan 695
Huang, D.S. 396
Huang, G.-W 504
Huang, Guo-Wei 340
Huang, H.-F. 504
Huang, K.-P. 504
Huang, Kailiang 47
Huang, P. 376, 488, 931
Huang, Peng 464
Huang, Po-Tsang 249
Huang, Qianqian 707
Huang, Ru 388, 392, 707
Huang, Wen-Hsien 340
Huang, Y. 879
Huang, Y.-C. 504, 504
Huang, Y.-M. 504
Huang, Yi-Chiau 544
Hubert, Q. 161
Hudec, Boris 356
Hudson, F. 129
Hueting, R. J. E. 739
Hung, Steven 253
Hüselitz, R. 428
Hutin, L. 141
Hutin, Louis 281
Huyghebaert, C. 512
Hwang, Changyoun 851
Hwang, Cheol Seong 288
Hwang, Eung-Rim 851
Hwang, J. 604
Hwang, K.H. 416, 656
Hwang, S. H. 416
Hwang, S.M. 656
Hwang, Wan Sik 241
Ide, T. 165
Idekoba, T. 237
Ielmini, D. 927

Iida, S. 221
Ikeda, H. 237
Ikeda, S. 608
Ikegami, Tsutomu 197
Ilatikhameneh, Hesameddin 767
Im, Sung Gap 241
Imamura, T. 265
Imanishi, K. 460
Indiveri, G. 472
Inoue, F. 149
Inoue, H. 608
Inoue, M. 165
Inoue, S. 265
Ionescu, A.M. 269, 308
Ionescu, Adrian M. 304
Irrera, F. 747
Irwin, R. 683
Isobe, A. 284
Issakov, V. 328
Itoh, K.M. 129
Itoh, Kohei M. 137
Itoh, M. 265
Itou, K. 189
Iwai, H. 189
Iwai, Hiroshi 304
Iwamatsu, T. 185
Iwata-Harms, Jodi 612
Iyer, S. S. 683
Jadot, B. 141
Jagannathan, H. 807, 811
Jagannathan, Hemanth 253
Jahan, C. 424
Jahan, R. 412
Jakob, A.M. 129
Jamieson, D.N. 129
Jamieson, G. 149
Jan, Guenole 612
Jang, Dongkyu 288
Jang, E. 875
Jang, Inkook 771
Jang, Kyungmin 723
Jang, S. H. 604
Jannaud, A. 500
Jansen, S. 428
Jao, C.-Y. 504
Jehl, X. 141
Jena, D. 691
Jena, Debdeep 193
Jenni, Laura Vera 83
Jeon, H.Y. 656
Jeon, J. 75
Jeong, Chan Bae 288
Jeong, D. E. 416
Jeong, G. T. 416
Jeong, Yeon Joo 63

AUTHOR INDEX

Jeong, Yujeong 288
Jerry, M. 364
Ji, M. H. 620
Ji, Y. 416
Jiang, Junkai 576, 799
Jiang, Xiangwei 568, 919
Jiang, Zhengping 348, 767
Jiang, Zizhen 572
Jin, Chengji 723
Jin, Q. 811
Jin, Seonghoon 767
Jinno, Riena 193
Johnson, B.C. 129
Jonnalagadda, V.P. 628
Jönsson, Adam 903
Jourdan, G. 99
Jourdan, Guillaume 281
Jourdon, J. 157
Jousseaume, V. 153
Jouve, A. 157
Juge, André 384
Jung, E. S. 416
Jung, ES 1
Jung, Gyuweon 288
Jung, H. 640
Jung, M. 245
Jung, S-M 656
Jung, W. 265
Kachi, T. 687
Kachi, Tetsu 452
Kaczer, B. 787
Kaczer, Ben 783
Kaji, Shiori 79
Kaklin, F. 743
Kakushima, K. 189
Kamata, Y. 225
Kamineni, V. 819
Kanamitsu, S. 265
Kanechika, M. 687
Kang, H.K. 416, 656
Kang, Ho-Kyu 560, 771
Kang, J. F. 376, 488, 931
Kang, Jinfeng 779
Kang, M. 656
Kang, M.S. 656
Kang, S.- Y 608
Kao, K.-H. 504
Kao, Ming-Yen 209
Kar, G. 592
Kar, G. S. 935
Kar, G. Sankar 51, 420
Karg, S. 540
Karim, K. S. 755
Karnati, K. 332
Kato, H. 265

Kato, K. 791
Kato, Kiyoshi 312
Ke, M. 791
Kencke, D. 412
Kenis, K. 508
Kerdiles, S. 153
Keys, P. 133
Khan, Asif Islam 205, 300
Khare, M. 819
Khwa, W. S. 624
Kim, C. H. 352
Kim, D. 656
Kim, D.H. 656
Kim, D.-W. 656
Kim, Dae Sin 348, 767, 771
Kim, Donghoon 851
Kim, H. 352, 428
Kim, J. 352, 428
Kim, J.C. 656
Kim, Jae Hwan 241
Kim, Jin-Kook 851
Kim, Jongchol 767
Kim, Jun Shik 288
Kim, K. W. 552
Kim, M. 352, 508
Kim, Myoungsub 851
Kim, Myung Sun 827
Kim, Myungsoo 532
Kim, N. 508
Kim, Nam Sung 827
Kim, S. K. 640
Kim, S.S. 656
Kim, Sae-Jin 771
Kim, Seungkyu 767
Kim, Seyoung 292
Kim, Taehoon 851
Kim, W. 420, 592, 859, 935
Kim, W.D. 656
Kim, W.J. 656
Kim, Y. 265
Kim, Y.H. 656
Kim, Y.-J. 141
Kim, Yun Sang 241
Kimoto, T. 444, 687
King, Ya-Chin 273, 400
Kinoshita, M. 284
Kita, K. 185
Kittl, J. A. 296
Kizilyalli, I.C. 456
Klamkin, J. 324
Klauk, Hagen 883
Kleemeier, W. 819
Kleemeier, Walter 815
Klein, J.-O. 484
Knoll, L. 440

AUTHOR INDEX

Knorr, A.	819
Ko, Hyoungsoo	771
Kobayashi, K.	265
Kobayashi, Masaharu	372, 723
Kobayashi, S.	265
Kobayashi, Y.	181
Kocaay, D.	111
Koh, G. H.	416
Kohler, S.	424
Koike, H.	608
Komatsu, S.	284
Komiyama, Takaki	664
Komori, M.	265
Komukai, T.	265
Kong, Lisa	895
Kono, T.	265
Korndörfer, F.	548
Koseki, K.	444
Kosugi, R.	444
Kotani, J.	699
Kotlyar, R.	133
Kouemeni-Tchouake, F.	153
Kouwenhoven, Leo	145
Kranz, L.	440
Kreupl, F.	245
Krishnan, R.	604
Krishnan, Siddarth A	253
Krivokapic, Zoran	300
Krottenthaler, P.	428
Krylyuk, S.	536
Ku, S.H.	31
Kubo, T.	608
Kubota, H.	616
Kudo, S.	237
Kumagai, Y.	237
Kumar, A.	632
Kumar, Lalit	83
Kumar, Narendra	676
Kumar, Pushpendra	404
Kumar, Ranjith	636
Kumazawa, T.	177
Kundu, A.	396
Kundu, S.	420, 592, 935
Kunimune, Y.	165
Kunitake, Hitoshi	312
Kunitake, S.	237
Kuo, I. T.	859
Kuramata, Akito	193
Kuriyama, N.	225
Kuroda, R.	225, 660
Kushwaha, Pragya	209
Kuzum, Duygu	664
Kwok, Hoi Sing	871
Kwon, J.	604
Kwon, J.H.	596
Kwon, O. I.	416
Kwon, Ohseong	652
Kwon, T.Y.	656
Kwon, Uihui	767
Kyogoku, S.	181
La Rosa, F.	161
Lacord, J.	500
Lagrasta, S.	424
Lahav, Assaf	233
Lai, E. K.	859
Lai, Stefan	432
Lai, Tung-Yan	731
Lal, R.	460
Lalanne, F.	229
Lam, Vinh	612
Lamontagne, P.	157
Lampert, L.	133
Lanza, Mario	528
Lapras, V.	500
Larcher, L.	596
Larcher, Luca	588
Lardin, T.	153
Larrey, V.	153
Latessa, L.	747
Lau, Calvin J.	676
Laucht, A.	129
Laudato, M.	927
Lauwereins, R.	480
Lavizzari, S.	43
Lavoie, C.	819, 859, 911
Le Friec, Y.	424
Le Gallo, M.	628
Le, J.	735
Le, Son	612
Lee, C. H.	807
Lee, Chang Bum	560
Lee, Chun Ying	340
Lee, D. S.	416
Lee, F. M.	859
Lee, Feng-Min	39
Lee, H.-J.	316
Lee, H.-Y.	735
Lee, Hyunmin	851
Lee, J. H.	396, 416
Lee, J.-H.	656
Lee, Jack C.	532
Lee, Jaesung	87
Lee, Jong-Ho	288
Lee, K.	604
Lee, K. H.	416
Lee, K.H.	416, 596
Lee, Ko-Tao	292
Lee, Kyupil	560
Lee, M. H.	735, 859
Lee, M.J.	739

AUTHOR INDEX

Lee, Ming-Hsiu 39
Lee, Min-Hung 731
Lee, S. 352, 911
Lee, Shiuh-Wuu 173
Lee, T.J. 656
Lee, Tsung-Han 273
Lee, Y.-J. 504
Lee, Y.K. 416
Lee, Y.W. 396
Lee, Younghee 895
Lee, Yuan-Jen 612
Legrand, B. 99
Lelis, A. J. 448
Lemke, S. 428
Lemonnier, O. 99
Leo, K. 11
Leobandung, E. 911
Leonhardt, A. 512
Lepape, E. 161
Leroux, Charles 404
Lesniewska, A. 111
Letzkus, Florian 883
Levine, P. M. 755
Lhostis, S. 157
Li, Fuhai 67
Li, Gezi 67
Li, Haitong 572
Li, Huanglong 572
Li, J. 807
Li, J.-H. 504
Li, J.-Y. 504
Li, Junfeng 47, 679
Li, Juntao 652
Li, K.-S. 735
Li, Kai-Shin 340, 731
Li, Keshuang 556
Li, Ling 118
Li, Linsen 572
Li, Luping 253
Li, R. 133
Li, Ruofan 847
Li, W. 149, 691
Li, Weisheng 524
Li, Wenshen 193
Li, X. 620
Li, Xiaoqin 532
Li, Xin 847
Li, Xinyi 67, 468
Li, Xiuyan 715
Li, Y. 504
Li, Yudong 679
Li, Yun 775
Li, Z. 811
Lian, G. 107
Liang, C.-F. 751

Liang, Gengchiau 580
Liang, Y. 819
Liang, Zhongxin 707
Liao, C.-Y. 735
Liao, Mengya 556
Liao, T.-H. 504
Liao, Yu-Hung 209
Lie, F. 819
Lie, Fee-li 815
Likharev, K.K. 476
Lim, J.H. 596, 604
Lin, B. 412
Lin, C.K. 396
Lin, C.-T. 751
Lin, Chrong Jung 273, 400
Lin, D. 512
Lin, K.-L. 504
Lin, Ming-Huei 644
Lin, S. 811
Lin, W.L. 31
Lin, Y. 508
Lin, Y. F. 859
Lin, Y.-D. 735
Lin, Yen-Kai 209
Lin, Yongjing 253
Lin, Yudeng 67
Lin, Yu-Yu 39
Lin, Zhiqiang 217
Ling, T. 604
Linten, D. 592, 600, 787
Linten, Dimitri 783
Liou, Peng-Chun 273
Liu, B. 604
Liu, C. 376, 931
Liu, C. W. 735
Liu, F. 376
Liu, Fei 763
Liu, H.-D. 751
Liu, Huanlong 612
Liu, Huiyun 556
Liu, Jing 464
Liu, L. F. 488
Liu, Lixiang 887
Liu, Ming 47, 464
Liu, P. 811
Liu, Paul 612
Liu, Po-Tsun 356
Liu, Qi 47, 464
Liu, T.-J. K. 75
Liu, W.-H. 751
Liu, X. 428
Liu, X. H. 107
Liu, X. Y. 376, 488, 931
Liu, Xiaoyan 775, 779
Liu, Xin 664

AUTHOR INDEX

Liu, Y. 408
Liu, Yanghui 520
Liu, Yibo 871
Liu, Yue-Yang 919
Liu, Yuyi 468
Liu, Z. 711, 819
Liu, Zhaojun 871
Locatelli, N. 616
Lofaro, M. 911
Long, Shibing 47
Longo, J. 269
Loo, R. 496
Loubet, N. 807, 811
Loubet, Nicolas 652
Loup, V. 500
Loup, Virginie 404
Lovisi, N. 747
Low, R. 604
Lu, C.C. 31
Lu, Chih-Yuan 31, 39
Lu, Jen-Hsiang 644
Lu, Jiwu 67
Lu, M. 412
Lu, Ruochen 915
Lu, Ryan 396
Lu, T.C. 31
Lu, Wei D. 63
Lu, Wenjie 895
Lu, Yang 348
Lu, Yichen 664
Ludwig, J. 512
Ludwigs, Sabine 883
Lung, H. L. 859
Lung, Hsiang-Lan 39
Luo, Aileen 300
Luo, G.-L. 504
Luo, Q. 376
Luo, Qing 47, 464
Luo, S.-X. 504
Lv, H. B. 376
Lv, Hangbing 47, 464
Ly, D. R. B. 472
Lyu, Y.-F. 751
Ma, Haili 47
Ma, T. P. 711
Ma, W. C.-Y. 504
Ma, Xiaolei 568
Ma, Yinji 668
Ma, Zichao 520
Machida, S. 284
Machillot, J. 508
Madhavan, Atul 636
Madzik, M. 129
Maeda, Shigenobu 767
Maeda, T. 687

Magesan, Easwar 126
Magis, T. 863
Magyari-Köpe, Blanka 572
Mahajan, Bikram K. 584
Mahakik, K. N. A. 448
Maheshwari, Dinesh 432
Mahmoodi, M.R. 476
Mai, A. 548
Mai, C. 548
Mainuddin, M. 412
Maitrejean, S. 153
Maize, K. 536
Makiyama, K. 699
Malavena, G. 35
Malinge, P. 229
Manfrini, M. 843
Manipatruni, S. 843
Mannaert, G. 149, 508
Mantelli, M. 161
Mao, Duli 217
Marchioni, A. 747
Marinov, D. 512
Markman, B. 324
Martin, A. 157
Martin, Lane 300
Martinez, Eugenie 404
Martinie, S. 500
Marushchak, Denys 676
Marzaki, A. 161
Massa, L. 133
Masselon, Christophe 281
Masuda, T. 177
Masudy-Panah, Saeid 544
Masunishi, Kei 79
Masuoka, S. 656
Matasunaga, K. 265
Matsudal, T. 189
Matsumoto, Noriko 312
Matsuura, M. 165
Mattavelli, P. 424
Mattei, Paul 281
Maugain, F. 161
Maurand, R. 141
Max, B. 727
Mazen, F. 153
Mazur, M. 428
Mazzocchi, V. 141, 153
McCallum, J.C. 129
McCarthy, L. 460
McClellan, Connor 528
Mcdonald, M. 229
McHerron, D. C. 811
McHerron, Dalea 253
McKay, J. 460
McLaughlin, P. S. 107

AUTHOR INDEX

McMitchell, S.R.C. 43, 380
Meersschaut, J. 51
Mehrsa, Armaghan 664
Meier, N. 540
Meiling, Hans 261
Melikyan, A. 552
Memisevic, E. 308
Meneghesso, G. 691, 703
Meneghini, M. 691, 703
Menzel, S. 867
Mermoz, S. 157
Mertens, H. 508
Mertens, Hans 783
Meterelliyoz, M. 412
Metz, M. 133
Meunier, T. 141
Miao, X. 807
Miao, Xin 253, 652
Migita, Shinji 197, 719
Mihaila, A. 440
Mikolajick, T. 727
Mikolajick, Thomas 588
Minoura, Y. 699
Mishra, U. 460
Mistry, Kaizad 636
Mitard, J. 149, 496, 508
Mitra, A. 265
Mittal, Sushant 827
Mittmann, T. 727
Miura, N. 185, 225
Miura, S. 608
Miyamoto, Satoru 137
Miyashita, Toshihiko 827
Miyata, Noriyuki 169
Miyazoe, H. 911
Mizuno, H. 221
Mizuno, Ikuo 233
Mo, Fei 372
Mo, R.T. 911
Mochizuki, S. 807, 811, 819
Mocuta, A. 420
Mocuta, D. 149, 496, 508, 843
Mohammadi, R. 755
Mohiyaddin, F.A. 129
Molas, G. 472, 863
Moll, P. 428
Monfray, S. 277
Monnot, G. 229
Moon, Bum Ki 652
Moon, C. 656
Moore, M. 460
Morales, C. 153
Morand, Yves 404
Moreau, S. 157
Morello, A. 129

Mori, H. 237
Morimoto, T. 181
Mortemousque, P.-A. 141
Mothes, K. 428
Motokawa, T. 265
Motoyama, K. 107
Mourik, V. 129
Mukhopadhyay, S. 396
Müller, J. 428
Muneshwar, T. 923
Murakami, S. 660
Murata, M. 225
Murdzek, Jessica 895
Na, N. 751
Nabors, Marni 636
Naeemi, A. 103
Nagata, Tomohiko 79
Naik, Mehul 122
Naik, V.B. 596, 604
Nakajima, Shigeru 320
Nakamura, H. 284
Nakamura, N. 699
Nakamura, Y. 284
Nakasugi, T. 265
Nakayama, K. 444
Nakazawa, K. 237
Nara, Jun 169
Narayanan, V. 807, 911
Narayanan, Vijay 648
Narayanan, Vijaykrishnan 340
Narita, T. 687
Nasuno, T. 608
Natarajan, S. 811
Natarajan, Sanjay 253
Nauta, B. 739
Navarro, G. 863
Neeli, V. 316
Ney, D. 153
Nguyen, P. 412
Nguyen, Tu 432
Nguyen, V.D. 843
Ni, J. 107
Ni, K. 296, 364
Ni, Kai 55
Ni, Zhenyi 887
Niel, S. 161
Nikonov, D. 412
Nikonov, D.E. 843
Nili, H. 476
Niquet, Y.-M. 141
Nishi, Yoshiaki 233
Nishi, Yoshio 432
Nishida, T. 95
Nishizawa, S. 189
Niu, C. 819

AUTHOR INDEX

Niu, Chengyu 815
Niwa, M. 608
Nodin, J-F 472
Noé, P. 863
Noel, J-P 472
Noguchi, M. 185
Noguchi, Y. 608
Nohira, Hiroshi..169
Nolot, E. 863
Nomoto, K. 691
Nomoto, Kazuki 193
Nonglaton, Guillaume 281
Norwood, Christopher 432
Noudo, S. 237
Nowak, E. 472, 484, 819, 863
Numasawa, Y. 189
Nyns, L. 380
Obradovic, B. 296
O'brien, K. 412
Odaka, T. 284
Ogier, J.L. 424
Ogier, S. 879
Ogura, A. 189
Oguz, K. 412
Oh, H. 51
Oh, HR. 420
Oh, S. C. 416
Ohashi, H. 189
Ohba, N. 221
Ohki, T. 699
Ohno, K. 221, 237
Ohshima, Kazuaki 312
Oikawa, K. 656
Ok, I. 807
Oka, T. 221
Okamoto, Kazuaki 79
Okamoto, N. 699
Okishiro, K. 284
Okumura, H. 181, 444
Olsen, Tivoli J. 676
Olshausen, Bruno 71
Om'mani, H. 428
Omori, K. 165
Omura, I. 189
Ono, K. 284
Ono, Tomio 79
Ono, Y. 237
Oprins, H. 111
Or-Bach, Zvi 432
Osawa, N. 237
Ostrovski, Y. 107
O'Sullivan, B. J. 592
Ota, Hiroyuki 197, 719
Ott, J. A. 911
Ouellette, D. 412

Ozaki, S. 699
Pacelli, D. 424
Padovani, A. 596
Padovani, Andrea 588
Paiton, Dylan 71
Pak, Kwan Yong 241
Pal, Arnab 564, 576
Pal, Ashish 827
Pala, Marco G. 759
Palanchoke, Ujwol 281
Palayam, S. Vadakupudhu 600
Pan, Deng 676
Pan, Lei 839
Pan, Quanjun 839
Pandey, P. 368
Pandya, S. 879
Pang, Chin-Sheng 516
Parat, K. 27
Parikh, P. 460
Park, Byung-Gook 288
Park, C. 352
Park, C.-H. 656
Park, Hong Keun 241
Park, Hong-hyun 767
Park, Honglae 771
Park, J. 412
Park, J. H. 416
Park, K.C. 416
Park, K.J. 656
Park, S. 640
Park, S. O. 416
Park, S.H. 656, 656
Parmigiani, L. 161
Parto, Kamyar 564
Parvais, B. 149
Patel, J. 911
Patel, Sahil 612
Pathak, Kalpana 827
Paul, J. 428
Peczek, A. 548
Pedini, Jean-Michel 404
Pedreira, O. Varela 111
Pellegren, J. 412
Pena, V. 508
Peng, L. 149
Peng, Lian-Mao 763
Peng, Xiaochen 348
Perlas, A. 75
Peroulis, Dimitrios 336
Perreau, P. 153
Perumkunnil, M. 420
Pesic, M. 43
Pesic, Milan 588
Pey, K.L. 596
Pham, Anh-Tuan 767

AUTHOR INDEX

Phoa, K. 316
Phommahaxay, A. 512
Pi, Xiaodong 887
Pillarisetty, R. 133
Pin, J-B. 153
Pitera, J. W. 21
Pla, J.J. 129
Plantier, Christophe.............................281
Poiroux, Thierry..................................384
Ponthenier, F. 153
Pop, E. 572
Pop, Eric 528
Popovici, M. 43, 51, 380
Porret, C. 496
Portal, J.-M 484
Post, Ian 636
Posthuma, N. 703
Poth, J. 428
Potoms, G. 43, 51
Pourghaderi, M. Ali 767
Pourtois, G. 380, 512, 935
Pradeep, Krishna 384
Prakash, Somashekar Bangalore............636
Prasad, D. 103
Previtali, B. 500
Prezioso, M. 476
Prindle, C. 819
Pugliese, Kaitlin M. 676
Puls, C. 412
Qi, Weiyi 348
Qian, H. 488, 931
Qian, He 67, 468
Qiu, Chenguang 763
Quek, E. 604
Querlioz, D..................................484, 616
Quintero, P. 412
Radu, I. P. 512, 843
Radu, Iuliana P. 831
Raghavan, N. 596
Raghavan, Srinivasan 803
Ragnarsson, L......................................149
Ragnarsson, L.-Å. 496, 787
Rahimo, M. 440
Rahman, T. 412
Rajapakse, Arith J. 676
Rakshit, T. 296
Rambal, N..153, 500
Rami, S. 316
Ranica, R. 424
Rao, S. 420, 592, 935
Rassoul, N. 149
Rastogi, P. 201
Ravikumar, S. 316
Ray, A. 859
Raychowdhury, Arijit...........................300

Raymenants, E......................................843
Raynor, J. M. 743
Razavi, Seyed Armin 835
Razavieh, Ali 652
Realov, Simeon 636
Reboh, S. 153
Regnier, A. 161
Ren, Chi 664
Ren, Tian-Ling......................................572, 891
Reynard, J.P. 424
Richard, E. 424
Richard, O. 51
Richter, R. 428
Ristoiu, D. 424
Ritzenthaler, R. 149, 508
Ritzenthaler, Romain 783
Roberts, J. 133
Robison, Robert 652
Rodder, M. 296
Rode, J. 324
Rodwell, M.J.W. 324
Rolland, Emmanuel 281
Rollo, T. 213
Roman, A. 153
Roman, Cosmin 83
Romang, A. 412
Romano, G. 500
Romera, M. 616
Ronchi, N. 380
Rosca, T. 308
Rosenblatt, Sami 126
Rosseel, E. 149, 600
Rothemund, R. 332
Roussel, P. J. 592
Roux, N. 229
Roy, Deboleena 360
Roy, F. 229
Roy, Kaushik 360
Rozeau, O. 500
Rozen, John 292
Rupakula, M. 269
Rupp, J. A. J. 867
Rusch, M. 75
Russo, F. 747
Ryan, K. 819
Ryckaert, J. 149, 787
Rzepa, G. 787
Sabbagh, D. 133
Sachid, Angada 827
Saeidi, A. 308
Saeidi, Ali 304
Saga, Shiori 312
Saha, A. K. 364
Saha, D. 201
Saidi, B. 161

AUTHOR INDEX

Saito, M. 265
Saito, T. 608
Saito, W. 189
Saito, Y. 177
Sakamoto, K. 444
Sakano, Y. 221
Sakhare, S. 420
Salahuddin, Sayeef 209, 731
Samanni, G. 424
Samkharadze, N. 133
Samukawa, S. 504
Sankaran, K. 935
Sano, Ryousuke 169
Sanquer, M. 141
Sansa, Marc 281
Sant, S. 899
Santoro, G. 508
Santos, Eduardo Gil 281
Saraf, I. 819
Saraya, T. 189
Saraya, Takuya 372, 723
Sart, C. 157
Sasago, Y. 284
Sasaki, Kohei 193
Sassine, G. 472, 863
Sassoulas, P.O 424
Sato, H. 608
Sato, M. 237
Sato, N. 237
Satoh, K. 189
Savytskyy, R. 129
Sawai, Hiromi 312
Scappucci, G. 133
Scevola, D. 153, 157
Schaefer, M. 332
Scheer, Patrick 384
Schenk, A. 899
Scheuvens, L. 11
Schmid, H. 911
Schmidt, Alexander 771
Schmitt, V. 129
Schmitz, J. 739
Schneider, U. 11
Schram, T. 512
Schroeder, U. 727
Schwab, L. 99
Scibetta, C. 153
Scotti, L. 424
Seabaugh, A. C. 368
Sebaai, F. 43
Sebastian, A. 628
Seeds, Alwyn 556
Seet, C. S. 604
Sekhar, M. 412
Seki, Takako 312

Selarka, A. 412
Sell, B. 316
Sengupta, Abhronil 360
Seo, B. 656
Seo, B.Y. 416
Serrano-Guisan, Santiago 612
Seth, M. 412
Shakouri, A. 536
Shankar, Bhawani 803
Shao, Qiming 835, 839
Sharma, S. 811
Shen, Chang-Hong 249, 340
Shen, Dongna 612
Shen, L. 460
Shen, T.M. 492
Shen, W. S. 488
Shen, Y.-L. 504
Shepard, K. L. 91
Shi, C. 91
Shi, Jianping 532
Shi, Quan 636
Shi, Yi 524
Shi, Yuanyuan 528
Shibaguchi, T. 225
Shibata, H. 225
Shieh, J.-M. 504
Shieh, Jia-Min 249, 340, 731
Shigyo, N. 189
Shih, Jiaw-Ren 400
Shikha, Swati 803
Shim, Dongshik 560
Shimada, Y. 165
Shin, Changgyun 560
Shin, Dongjae 560
Shin, H. C. 416
Shin, H.J. 656
Shin, Jong Hoon 63
Shin, SangHoon 292
Shin, Yonghwack 560
Shobha, H. 107
Shono, K. 460
Shoute, G. 923
Shrestha, P.R. 536
Shrivastava, Mayank 803
Si, Mengwei 344
Siah, S. Y. 604
Siang, G.-Y. 735
Sikder, U. 75
Singh, K. 133
Singh, Sandeep 803
Slesazeck, Stefan 588
Slesazeck, S. 727
Smets, Q. 512
Smith, A. J. 412
Smith, A. K. 412

AUTHOR INDEX

Smith, J. A. 296, 364
Smith, P. 460
Sober, Samuel J. 672
Socquet-Clerc, C. 863
Sohn, Chang-Woo 652
Sohn, Joon 71
Solomon, Paul 292
Song, G. 656
Song, Jeongho 851
Song, Min 847
Song, S. 352
Song, Y. J. 416
Song, Z. T. 620
Soni, Ankit 803
Souhaite, A. 424
Souifi, A. 277
Souriau, L. 592, 843
Sousa, M. 628, 899, 907
Southwick, R. G. 807
Spence, C. 141
Spessot, A. 420
Spinella, Laura 115
Spinelli, A. S. 35
Spratt, W. 911
Srinivasa, Srivatsa 340
Srinivasan, Gopalakrishnan 360
Stahlbush, R. E. 448
Stark, P. 540
Steglich, P. 548
Stelzer, M. 245
Stoffels, S. 703
Stojanovic, V. 75
Stolfi, M. 811
Stolichnov, Igor 304
Strane, J. 819
Strukov, D.B. 476
Stucchi, M. 111
Su, C.-J. 504
Su, S.K. 492
Subirats, A. 43, 600
Suda, J. 687
Sugawa, S. 225, 660
Sugiyama, Y. 284
Suh, K. 416
Sulehria, Yasir 815
Sumita, Kyoko 169
Sun, S. 508
Sun, Shiyu 827
Sun, X. 911
Sun, Xiaowei 871
Sun, Xiaoyu 55, 468
Sun, Zixuan 392
Sundar, Vignesh 612
Sung, P.-J. 504
Sutar, S. 512

Suzuki, A. 221
Suzuki, K. 221
Suzuki, M. 660
Suzuki, S. 189
Svensson, Johannes 903
Swerts, J. 420, 592, 935
Tabone, Claude 281, 404
Tabrizian, R. 95
Tagawa, Yusaku 372
Tai, Lu 47, 464
Takagi, S. 791
Takahata, K. 265
Takakura, T. 189
Takami, M. 221
Takei, M. 181
Takenaka, M. 791
Takeuchi, K. 189
Takizawa, M. 221
Talatchian, P. 616
Talmelli, Giacomo 831
Tamura, R. 608
Tan, S. L. 604
Tanaka, T. 181
Tang, Jianshi 292
Tang, M. 735
Tang, Mingchu 556
Tang, W. 879
Tang, Wei 253
Tanigawa, T. 608
Tateshita, Y. 221, 237
Tellez, G.E. 632
Tenberg, S. 129
Teng, Zhongjian 612
Teugels, L. 149
Thakuria, Niharika 516
Thanigaivelan, Thirumal 815
Thean, Aaron Voon-Yew 580
Thiam, A. 843
Thiyagarajah, N. 604
Thomas, Luc 612
Thomas, N. 133
Thompson, M. G. 540
Tian, He 572, 891
Ting, J. W. 604
Tiwari, V. 428
Tkachev, Y. 428
Todorov, Teodor 292
Toh, E. H. 604
Tokei, Zs. 111
Tokue, H. 265
Tokumaru, Ryo 312
Tong, Ru-Ying 612
Toriumi, Akira 197, 257, 715, 719
Torres, J. 133
Tosi, G. 129

AUTHOR INDEX

Tournier, A.	229
Trastoy, J.	616
Trenteseaux, F.	161
Trentzsch, M.	428
Trotta, S.	328
Tsai, S.	819
Tsai, Wen-Jer	31
Tsai, Y.S.	396
Tsai, Yi-Pei	400
Tseng, J.C.	624
Tsuda, H.	265
Tsuda, Kazuki	312
Tsukuda, M.	189
Tsunegi, S.	616
Tsutsui, G.	819
Tsutsui, Gen	652
Tsutsui, K.	189
Tsutsui, Masafumi	233
Tu, Hailing	679
Tu, Thieu Quang	193
Tu, Yung-Ning	340
Ueda, H.	687
Uesugi, T.	687
Urdampilleta, M.	141
Urteaga, M	324
Usagawa, T.	284
Usai, Giulia	281
Ushifusa, N.	284
Van Beek, S.	592
Van Dal, M.J.H.	492
Van Den bosch, G.	600
Van Der Plas, G.	408
Van Elshocht, S.	51
Van Houdt, J.	43, 380
Vandemaele, Michiel	783
Vandersypen, L.M.K.	133
Vandooren, A.	149, 787
Vanherle, W.	149
Vanstreels, K.	408
Vardi, Alon	895
Vasen, T.	492
Vaysset, A.	843
Vecchio, E.	149
Vega, Reinaldo	652
Veldhorst, M.	133
Velenis, D.	408
Vellianitis, G.	492
Venezia, Vincent C.	217
Venitucci, B.	141
Verdy, A.	863
Verhulst, Anne S.	304
Verkest, D.	480
Vernhes, Emeline	281
Vernhet, A.	424
Verreck, D.	512

Vianello, E.	472, 484
Vici, A.	747
Villard, Patrick	281
Villaret, A.	424
Villringer, C.	548
Vincent, A.	476
Vinet, M.	141, 153, 500
Vizioz, C.	500
Vodenicarevic, D.	616
Voit, B.	11
Volk, C.	133
Vorenkamp, Pieter	432
Waldron, N.	149
Walke, A.	149
Walters, G.	95
Wan, D.	843
Wan, Weier	71
Wang, C.-J.	504
Wang, Chien-Ping	273
Wang, Ching-Hua	528, 572
Wang, H.	213
Wang, Hong	544
Wang, Huimin	707
Wang, Jian	763
Wang, Jing	348
Wang, Jingli	520
Wang, Joddy	388
Wang, Kang L.	835, 839
Wang, Kanwen	468
Wang, Keh-Chung	39
Wang, L.	620
Wang, Lin	580
Wang, Lingfei	580
Wang, M.	807
Wang, Miaomiao	253, 648, 652
Wang, Ning	695
Wang, Pannl	55
Wang, PoKang	612
Wang, Q.	620
Wang, Qingxue	388
Wang, Qiwen	63
Wang, Runsheng	388
Wang, Rusheng	392
Wang, Shu-Hui	644
Wang, T. Y.	624
Wang, Tahui	31
Wang, W.	819, 927
Wang, Wei	544, 823
Wang, Wenwu	679
Wang, Xinning	636
Wang, Xinran	520, 524
Wang, Xu	257
Wang, Xuefeng	891
Wang, Y.-H.	504
Wang, Y.-S.	504

AUTHOR INDEX

Wang, Yue 887
Wang, Yu-Jen 612
Wang, Z.-Y. 735
Wang, Zheng 300
Wang, Zhixuan 707
Waser, R. 867
Watanabe, H. 185
Watanabe, M. 189
Watanabe, T. 608
Watson, T.F. 133
Weber, O. 424
Webster, Eric A. G. 217
Wei, Feng 679
Wei, Jin 695
Wei, L. 412
Wei, Na 173
Wei, Qianhui 679
Wei, Yun-Jie 731
Weiss, Gregory A. 676
Welser, J. 21
Wemersson, Lars-Erik 903
Wernersson, L-E. 308
Widjaja, Yuniarto 432
Wiegand, C. 412
Wildhaber, F. 269
Willis, Jim 815
Wilson, C.J. 111
Wilson, James 432
Winstel, K. R. 811
Wirths, S. 440
Witters, L. 149, 496
Witters, Liesbeth 783
Wong, H.-S. Philip 71, 118, 528, 572
Wong, J. 604
Wong, Kin 839
Woo, S. T. 604
Wouters, D. J. 867
Wu, Bo-Wei 731
Wu, C. 111
Wu, C.-T. 504
Wu, Dehuang 388
Wu, Fan 891
Wu, H. 819
Wu, H. Q. 488, 931
Wu, Hao 835, 839
Wu, Heng 815
Wu, Huaqiang 67, 468
Wu, J. 324
Wu, J.Y. 624
Wu, Jiang 556
Wu, Jixuan 568
Wu, Meile 288
Wu, Meng-Chyi 249
Wu, Tai-Hsuan 636
Wu, W.-F. 504

Wu, Wan-Chi 249
Wu, Wei 67
Wu, Xiaohan 532
Wu, Xiu Long 464
Wu, Y. 460
Wu, Ying 823
Wu, Z. 149, 787
Wu, Z.Q. 492
Wu, Zhenhua 763
Wuetz, B. P. 133
Wurstbauer, U. 245
Xi, Yue 468
Xiang, Y. C. 488, 931
Xie, Tao 668
Xie, Xuejun 564
Xing, H. G. 691
Xing, Huili Grace 193
Xu, Nuo 348, 847
Xu, Ruijuan 300
Xu, Shengqiang 544
Xu, X. X. 376
Xu, Xiaoxin 47, 464
Xu, Yang 887
Xue, Chunling 679
Yagi, Y. 608
Yakimets, D. 420
Yakushiji, K. 616
Yamada, A. 699
Yamada, M. 221
Yamaguchi, K. 221
Yamaguchi, T. 165
Yamamoto, M. 660
Yamane, J. 237
Yamane, K. 596, 604
Yamasaki, Takahiro 169
Yamashita, K. 237, 608
Yamashita, T. 819
Yamashita, Tenko 652
Yamawaki, T. 284
Yamazaki, Shunpei 312
Yan, Jiang 679
Yan, Siwa 871
Yanagisawa, Yuichi 312
Yang, C. H. 859
Yang, Chih-Chao 249, 340
Yang, Deren 887
Yang, H. 596, 604
Yang, J. 819
Yang, M.-J. 751
Yang, M.S. 656
Yang, Mengxuan 707
Yang, Shyh-Horng 644
Yang, Song 695
Yang, Yansong 915
Yang, Yi 572, 612, 891

AUTHOR INDEX

Yang, Yixiong 253
Yao, Peng 67, 468
Yasin, F. 420, 592
Yasuda, T. 660
Yasuda-Masuoka, Y. 640
Yasuhira, M. 608
Yazdani, Armin 217
Ye, Fan 87
Ye, Peide D. 344
Ye, Z. A. 75
Yea, S. 460
Yeh, C. W. 859
Yeh, W.-K. 504
Yeh, Wen-Kuan 249, 340, 731
Yen, Anthony 261
Yeo, Yee-Chia 544
Yeoh, Andrew 636
Yeung, Chun Wing 652
Yi, Jaeyun 851
Yin, Huaxiang 47, 679
Yin, Jiahao 47, 464
Yokoyama, Toshifumi 233
Yonezawa, Y. 444
Yoo, J. 656
Yoon, Alexander 241
Yoon, J.S. 640
Yorita, C. 284
Yoshiba, I. 221
Yoshida, N. 508
Yoshiduka, T. 608
Yoshikawa, K. 284
Yoshita, R. 237
You, Long 847
You, Y. S. 604
Young, I. A. 843
Yu, Haoran 47
Yu, Jie 47, 464
Yu, L. 811, 819
Yu, Lan 815
Yu, S. M. 624
Yu, Shimeng 55, 67, 348, 468
Yu, Zhaoan 464
Yu, Zhihao 520, 524
Yuan, Peng 47
Yuasa, S. 616
Yuzawa, Akiko 79
Zagni, Nicolò 584
Zahedmanesh, H. 111
Zaka, A. 428
Zanoni, E. 691, 703
Zarcone, Ryan 71
Zeng, D. 604
Zeng, Yi 847
Zhan, Y. P. 620
Zhang, Chen 652

Zhang, F. 536
Zhang, Gang 524
Zhang, H. 536
Zhang, J.-R. 269
Zhang, Jiayang 392
Zhang, Jingyun 652
Zhang, Ke 871
Zhang, L. 604, 604
Zhang, Peng 839
Zhang, Qingtian 67, 468
Zhang, Qingzhu 47, 679
Zhang, Rui 173
Zhang, Shaoming 679
Zhang, Shuai 847
Zhang, T. 165
Zhang, W. 75
Zhang, X. 488
Zhang, Xiang 67
Zhang, Xiao 679
Zhang, Xing 707, 779
Zhang, Y. 316
Zhang, Yanfeng 532
Zhang, Ying 636
Zhang, Yingchao 668
Zhang, Z. 412
Zhang, Zexuan 193
Zhang, Zhaofu 695
Zhang, Zhaohao 679
Zhang, Zhe 388, 392
Zhang, Zhiyong 763
Zhang, Zuodong 392
Zhao, Hongbin 679
Zhao, J. 879
Zhao, M. 703
Zhao, Meiran 468
Zhao, Robert Chunhua 679
Zhao, Ruoyu 664
Zhao, Shuangyi 887
Zhao, Xiang 217
Zhao, Y. D. 376, 488, 931
Zhao, Yang 707
Zhao, Yi 173
Zheng, G. 133
Zheng, Ning 668
Zheng, Peng 636
Zheng, T. 149
Zheng, Xin 71, 528
Zheng, Zejie 173
Zheng, Zheyang 695
Zhong, Tom 612
Zhong, Yuan 707
Zhou, F. 428
Zhou, Huimei 648
Zhou, Z. 488
Zhu, Jian 612

AUTHOR INDEX

Zhu, Kunkun 707
Zhu, Xi 47, 464
Zhu, Y. 536
Zhu, Ye 520
Zhu, Ying 524
Zia, Muneeb..672
Zidan, Mohammed A......................................63
Zografos, O. 843
Zografos, Odysseas 831
Zota, C. 899
Zota, C. B. 907
Zou, Wei 823
Zschieschang, Ute...883
Zuk, P. 460
Zuliani, P. 424
Zwerver, A.-M..133

IEEE
445 Hoes Lane
Piscataway, NJ 08854-4141

ISBN 978-1-7281-1988-5

2018 IEEE International Electron Devices Meeting (IEDM 2018)

San Francisco, California, USA
1-5 December 2018

Pages 1-459

IEEE Catalog Number: CFP18IED-POD
ISBN: 978-1-7281-1988-5

2018 IEEE International Electron Devices Meeting (IEDM 2018)

San Francisco, California, USA
1-5 December 2018

Pages 1-459

IEEE Catalog Number: CFP18IED-POD
ISBN: 978-1-7281-1988-5

**Copyright © 2018 by the Institute of Electrical and Electronics Engineers, Inc.
All Rights Reserved**

Copyright and Reprint Permissions: Abstracting is permitted with credit to the source. Libraries are permitted to photocopy beyond the limit of U.S. copyright law for private use of patrons those articles in this volume that carry a code at the bottom of the first page, provided the per-copy fee indicated in the code is paid through Copyright Clearance Center, 222 Rosewood Drive, Danvers, MA 01923.

For other copying, reprint or republication permission, write to IEEE Copyrights Manager, IEEE Service Center, 445 Hoes Lane, Piscataway, NJ 08854. All rights reserved.

*** *This is a print representation of what appears in the IEEE Digital Library. Some format issues inherent in the e-media version may also appear in this print version.*

IEEE Catalog Number: CFP18IED-POD
ISBN (Print-On-Demand):   978-1-7281-1988-5
ISBN (Online):   978-1-7281-1987-8
ISSN: 0163-1918

Additional Copies of This Publication Are Available From:

Curran Associates, Inc
57 Morehouse Lane
Red Hook, NY 12571 USA
Phone: (845) 758-0400
Fax: (845) 758-2633
E-mail: curran@proceedings.com
Web: www.proceedings.com

TABLE OF CONTENTS

4TH INDUSTRIAL REVOLUTION AND BOUNDRY: CHALLENGES AND OPPORTUNITIES 1
ES Jung

VENTURING ELECTRONICS INTO UNKNOWN GROUNDS 11
G. Fettweis ; K. Leo ; B. Voit ; U. Schneider ; L. Scheuvens

FUTURE COMPUTING HARDWARE FOR AI 21
J. Welser ; J. W. Pitera ; C. Goldberg

SCALING TRENDS IN NAND FLASH 27
K. Parat ; A. Goda

ANALYSIS AND REALIZATION OF TLC OR EVEN QLC OPERATION WITH A HIGH
PERFORMANCE MULTI-TIMES VERIFY SCHEME IN 3D NAND FLASH MEMORY 31
C.C. Lu ; C. C. Cheng ; H.P. Chiu ; W.L. Lin ; T.W. Chen ; S.H. Ku ; Wen-Jer Tsai ; T.C. Lu ; K.C. Chen ; Tahui Wang ; Chih-Yuan Lu

IMPLEMENTING SPIKE-TIMING-DEPENDENT PLASTICITY AND UNSUPERVISED
LEARNING IN A MAINSTREAM NOR FLASH MEMORY ARRAY 35
G. Malavena ; A. S. Spinelli ; C. Monzio Compagnoni

A NOVEL VOLTAGE-ACCUMULATION VECTOR-MATRIX MULTIPLICATION
ARCHITECTURE USING RESISTOR-SHUNTED FLOATING GATE FLASH MEMORY DEVICE
FOR LOW-POWER AND HIGH-DENSITY NEURAL NETWORK APPLICATIONS 39
Yu-Yu Lin ; Feng-Min Lee ; Ming-Hsiu Lee ; Wei-Chen Chen ; Hsiang-Lan Lung ; Keh-Chung Wang ; Chih-Yuan Lu

VERTICAL FERROELECTRIC HFO$_2$ FET BASED ON 3-D NAND ARCHITECTURE:
TOWARDS DENSE LOW-POWER MEMORY 43
K. Florent ; M. Pesic ; A. Subirats ; K. Banerjee ; S. Lavizzari ; A. Arreghini ; L. Di Piazza ; G. Potoms ; F. Sebaai ; S. R. C. McMitchell ; M. Popovici ; G. Groeseneken ; J. Van Houdt

HYBRID 1T E-DRAM AND E-NVM REALIZED IN ONE 10 NM NODE FERRO FINFET DEVICE
WITH CHARGE TRAPPING AND DOMAIN SWITCHING EFFECTS 47
Qing Luo ; Tiancheng Gong ; Yan Cheng ; Qingzhu Zhang ; Haoran Yu ; Jie Yu ; Haili Ma ; Xiaoxin Xu ; Kailiang Huang ; Xi Zhu ; Danian Dona ; Jiahao Yin ; Peng Yuan ; Lu Tai ; Jianfeng Gao ; Junfeng Li ; Huaxiang Yin ; Shibing Long ; Qi Liu ; Hangbing Lv ; Ming Liu

HIGH-PERFORMANCE (EOT<0.4NM, JG~10^{-7} A/CM^2) ALD-DEPOSITED RU\SRTIO$_3$ STACK
FOR NEXT GENERATIONS DRAM PILLAR CAPACITOR 51
M. Popovici ; A. Belmonte ; H. Oh ; G. Potoms ; J. Meersschaut ; O. Richard ; H. Hody ; S. Van Elshocht ; R. Delhougne ; L. Goux ; G. Sankar Kar

EXPLOITING HYBRID PRECISION FOR TRAINING AND INFERENCE: A 2T-1FEFET BASED
ANALOG SYNAPTIC WEIGHT CELL 55
Xiaoyu Sun ; Panni Wang ; Kai Ni ; Suman Datta ; Shimeng Yu

ANALOG COMPUTING FOR DEEP LEARNING: ALGORITHMS, MATERIALS &
ARCHITECTURES 59
W. Haensch

HARDWARE ACCELERATION OF SIMULATED ANNEALING OF SPIN GLASS BY RRAM
CROSSBAR ARRAY 63
Jong Hoon Shin ; Yeon Joo Jeong ; Mohammed A. Zidan ; Qiwen Wang ; Wei D. Lu

DEMONSTRATION OF GENERATIVE ADVERSARIAL NETWORK BY INTRINSIC RANDOM
NOISES OF ANALOG RRAM DEVICES 67
Yudeng Lin ; Huaqiang Wu ; Bin Gao ; Peng Yao ; Wei Wu ; Qingtian Zhang ; Xiang Zhang ; Xinyi Li ; Fuhai Li ; Jiwu Lu ; Gezi Li ; Shimeng Yu ; He Qian

ERROR-RESILIENT ANALOG IMAGE STORAGE AND COMPRESSION WITH ANALOG-
VALUED RRAM ARRAYS: AN ADAPTIVE JOINT SOURCE-CHANNEL CODING APPROACH 71
Xin Zheng ; Ryan Zarcone ; Dylan Paiton ; Joon Sohn ; Weier Wan ; Bruno Olshausen ; H. -S. Philip Wong

DEMONSTRATION OF 50-MV DIGITAL INTEGRATED CIRCUITS WITH
MICROELECTROMECHANICAL RELAYS 75
Z. A. Ye ; S. Almeida ; M. Rusch ; A. Perlas ; W. Zhang ; U. Sikder ; J. Jeon ; V. Stojanovic ; T.-J. K. Liu

HIGHLY SENSITIVE SPINTRONIC STRAIN-GAUGE SENSOR BASED ON MAGNETIC
TUNNEL JUNCTION AND ITS APPLICATION TO MEMS MICROPHONE 79
Yoshihiko Fuji ; Yoshihiro Higashi ; Shiori Kaji ; Kei Masunishi ; Akiko Yuzawa ; Tomohiko Nagata ; Kazuaki Okamoto ; Shotaro Baba ; Tomio Ono ; Michiko Hara

INTERMIXING OF MOTIONAL CURRENTS IN SUSPENDED CNT-FET BASED RESONATORS 83
Lalit Kumar ; Laura Vera Jenni ; Miroslav Haluska ; Cosmin Roman ; Christofer Hierold

GLOWING GRAPHENE NANOELECTROMECHANICAL RESONATORS AT ULTRA-HIGH TEMPERATURE UP TO 2650K 87
Fan Ye ; Jaesung Lee ; Philip X.-L. Feng

MONOLITHIC INTEGRATION OF MICRON-SCALE PIEZOELECTRIC MATERIALS WITH CMOS FOR BIOMEDICAL APPLICATIONS 91
C. Shi ; T. Costa ; J. Elloian ; K. L. Shepard

A NANO-MECHANICAL RESONATOR WITH 10NM HAFNIUM-ZIRCONIUM OXIDE FERROELECTRIC TRANSDUCER 95
M. Ghatge ; G. Walters ; T. Nishida ; R. Tabrizian

COMPREHENSIVE OPTICAL LOSSES INVESTIGATION OF VLSI SILICON OPTOMECHANICAL RING RESONATOR SENSORS 99
L. Schwab ; P.E. Allain ; L. Banniard ; A. Fafin ; M. Gely ; O. Lemonnier ; P. Grosse ; M. Hermouet ; S. Hentz ; I. Favero ; B. Legrand ; G. Jourdan

INTERCONNECT DESIGN AND TECHNOLOGY OPTIMIZATION FOR CONVENTIONAL AND EMERGING NANOSCALE DEVICES: A PHYSICAL DESIGN PERSPECTIVE 103
D. Prasad ; A. Naeemi

MECHANISMS OF ELECTROMIGRATION DAMAGE IN CU INTERCONNECTS 107
C.-K. Hu ; L. Gignac ; G. Lian ; C. Cabral ; K. Motoyama ; H. Shobha ; J. Demarest ; Y. Ostrovski ; C. M. Breslin ; M. Ali ; J. Benedict ; P. S. McLaughlin ; J. Ni ; X. H. Liu

INTERCONNECT METALS BEYOND COPPER: RELIABILITY CHALLENGES AND OPPORTUNITIES 111
K. Croes ; Ch. Adelmann ; C.J. Wilson ; H. Zahedmanesh ; O. Varela Pedreira ; C. Wu ; A. Lesniewska ; H. Oprins ; S. Beyne ; I. Ciofi ; D. Kocaay ; M. Stucchi ; Zs. Tokei

MICROSTRUCTURE EVOLUTION AND EFFECT ON RESISTIVITY FOR CU NANOINTERCONNECTS AND BEYOND 115
Szu-Tung Hu ; Linjun Cao ; Laura Spinella ; Paul S. Ho

INTEGRATING GRAPHENE INTO FUTURE GENERATIONS OF INTERCONNECT WIRES 118
Ling Li ; H.-S. Philip Wong

INTERCONNECT TREND FOR SINGLE DIGIT NODES 122
Mehul Naik

DEVICE CHALLENGES FOR NEAR TERM SUPERCONDUCTING QUANTUM PROCESSORS: FREQUENCY COLLISIONS 126
Markus Brink ; Jerry M. Chow ; Jared Hertzberg ; Easwar Magesan ; Sami Rosenblatt

SCALABLE QUANTUM COMPUTING WITH ION-IMPLANTED DOPANT ATOMS IN SILICON 129
A. Morello ; G. Tosi ; F.A. Mohiyaddin ; V. Schmitt ; V. Mourik ; T. Botzem ; A. Laucht ; J.J. Pla ; S. Tenberg ; R. Savytskyy ; M. Madzik ; F. Hudson ; A.S. Dzurak ; K.M. Itoh ; A.M. Jakob ; B.C. Johnson ; J.C. McCallum ; D.N. Jamieson

QUBIT DEVICE INTEGRATION USING ADVANCED SEMICONDUCTOR MANUFACTURING PROCESS TECHNOLOGY 133
R. Pillarisetty ; N. Thomas ; H.C. George ; K. Singh ; J. Roberts ; L. Lampert ; P. Amin ; T.F. Watson ; G. Zheng ; J. Torres ; M. Metz ; R. Kotlyar ; P. Keys ; J.M. Boter ; J.P. Dehollain ; G. Droulers ; G. Eenink ; R. Li ; L. Massa ; D. Sabbagh ; N. Samkharadze ; C. Volk ; B. P. Wuetz ; A.-M. Zwerver ; M. Veldhorst ; G. Scappucci ; L.M.K. Vandersypen ; J.S. Clarke

SILICON ISOTOPE TECHNOLOGY FOR QUANTUM COMPUTING 137
Satoru Miyamoto ; Kohei M. Itoh

TOWARDS SCALABLE SILICON QUANTUM COMPUTING 141
M. Vinet ; L. Hutin ; B. Bertrand ; S. Barraud ; J.-M. Hartmann ; Y.-J. Kim ; V. Mazzocchi ; A. Amisse ; H. Bohuslavskyi ; L. Bourdet ; A. Crippa ; X. Jehl ; R. Maurand ; Y.-M. Niquet ; M. Sanquer ; B. Venitucci ; B. Jadot ; E. Chanrion ; P.-A. Mortemousque ; C. Spence ; M. Urdampilleta ; S. De Franceschi ; T. Meunier

MAJORANA QUBITS 145
Leo Kouwenhoven

FIRST DEMONSTRATION OF 3D STACKED FINFETS AT A 45NM FIN PITCH AND 110NM GATE PITCH TECHNOLOGY ON 300MM WAFERS 149
A. Vandooren ; J. Franco ; Z. Wu ; B. Parvais ; W. Li ; L. Witters ; A. Walke ; L. Peng ; V. Deshpande ; N. Rassoul ; G. Hellings ; G. Jamieson ; F. Inoue ; K. Devriendt ; L. Teugels ; N. Heylen ; E. Vecchio ; T. Zheng ; E. Rosseel ; W. Vanherle ; A. Hikavyy ; G. Mannaert ; B. T. Chan ; R. Ritzenthaler ; J. Mitard ; L. Ragnarsson ; N. Waldron ; V. De Heyn ; S. Demuynck ; J. Boemmels ; D. Mocuta ; J. Ryckaert ; N. Collaert

BREAKTHROUGHS IN 3D SEQUENTIAL TECHNOLOGY 153

L. Brunet ; C. Fenouillet-Beranger ; P. Batude ; S. Beaurepaire ; F. Ponthenier ; N. Rambal ; V. Mazzocchi ; J-B. Pin ; P. Acosta-Alba ; S. Kerdiles ; P. Besson ; H. Fontaine ; T. Lardin ; F. Fournel ; V. Larrey ; F. Mazen ; V. Balan ; C. Morales ; C. Guerin ; V. Jousseaume ; X. Federspiel ; D. Ney ; X. Garros ; A. Roman ; D. Scevola ; P. Perreau ; F. Kouemeni-Tchouake ; L. Arnaud ; C. Scibetta ; S. Chevalliez ; F. Aussenac ; J. Aubin ; S. Reboh ; F. Andrieu ; S. Maitrejean ; M. Vinet

HYBRID BONDING FOR 3D STACKED IMAGE SENSORS: IMPACT OF PITCH SHRINKAGE ON INTERCONNECT ROBUSTNESS 157

J. Jourdon ; S. Lhostis ; S. Moreau ; J. Chossat ; M. Arnoux ; C. Sart ; Y. Henrion ; P. Lamontagne ; L. Arnaud ; N. Bresson ; V. Balan ; C. Euvrard ; Y. Exbrayat ; D. Scevola ; E. Deloffre ; S. Mermoz ; A. Martin ; H. Bilgen ; F. Andre ; C. Charles ; D. Bouchu ; A. Farcy ; S. Guillaumet ; A. Jouve ; H. Fremont ; S. Cheramy

EMBEDDED SELECT IN TRENCH MEMORY (ESTM), BEST IN CLASS 40NM FLOATING GATE BASED CELL: A PROCESS INTEGRATION CHALLENGE 161

S. Niel ; F. La Rosa ; A. Regnier ; M. Mantelli ; F. Trenteseaux ; G. Ghezzi ; A. Marzaki ; Q. Hubert ; J. Delalleau ; T. Cabout ; F. Maugain ; E. Lepape ; L. Baron ; A. Champenois ; D. Galpin ; N. Cherault ; S. Audran ; L. Parmigiani ; P. Gouraud ; B. Duclaux ; Y. Escarabajal ; F. Baudin ; E. Beche ; B. Saidi ; V. Arnal

HIGHLY RELIABLE FERROELECTRIC $HF_{0.5}ZR_{0.5}O_2$ FILM WITH AL NANOCLUSTERS EMBEDDED BY NOVEL SUB-MONOLAYER DOPING TECHNIQUE 165

T. Yamaguchi ; T. Zhang ; K. Omori ; Y. Shimada ; Y. Kunimune ; T. Ide ; M. Inoue ; M. Matsuura

INTERFACE DIPOLE MODULATION IN HFO2/SIO2 MOS STACK STRUCTURES 169

Noriyuki Miyata ; Jun Nara ; Takahiro Yamasaki ; Kyoko Sumita ; Ryousuke Sano ; Hiroshi Nohira

GE-BASED NON-VOLATILE LOGIC-MEMORY HYBRID DEVICES FOR NAND MEMORY APPLICATION 173

Na Wei ; Bing Chen ; Zejie Zheng ; Zhimei Cai ; Rui Zhang ; Ran Cheng ; Shiuh-Wuu Lee ; Yi Zhao

0.63 MΩCM^2 / 1170 V 4H-SIC SUPER JUNCTION V-GROOVE TRENCH MOSFET 177

T. Masuda ; Y. Saito ; T. Kumazawa ; T. Hatayama ; S. Harada

FIRST DEMONSTRATION OF DYNAMIC CHARACTERISTICS FOR SIC SUPERJUNCTION MOSFET REALIZED USING MULTI-EPITAXIAL GROWTH METHOD 181

S. Harada ; Y. Kobayashi ; S. Kyogoku ; T. Morimoto ; T. Tanaka ; M. Takei ; H. Okumura

CHANNEL ENGINEERING OF 4H-SIC MOSFETS USING SULPHUR AS A DEEP LEVEL DONOR 185

M. Noguchi ; T. Iwamatsu ; H. Amishiro ; H. Watanabe ; K. Kita ; N. Miura

DEMONSTRATION OF 1200V SCALED IGBTS DRIVEN BY 5V GATE VOLTAGE WITH SUPERIORLY LOW SWITCHING LOSS 189

T. Saraya ; K. Itou ; T. Takakura ; M. Fukui ; S. Suzuki ; K. Takeuchi ; M. Tsukuda ; Y. Numasawa ; K. Satoh ; T. Matsudai ; W. Saito ; K. Kakushima ; T. Hoshii ; K. Furukawa ; M. Watanabe ; N. Shigyo ; K. Tsutsui ; H. Iwai ; A. Ogura ; S. Nishizawa ; I. Omura ; H. Ohashi ; T. Hiramoto

2.44 KV GA$_2$O$_3$ VERTICAL TRENCH SCHOTTKY BARRIER DIODES WITH VERY LOW REVERSE LEAKAGE CURRENT 193

Wenshen Li ; Zongyang Hu ; Kazuki Nomoto ; Riena Jinno ; Zexuan Zhang ; Thieu Quang Tu ; Kohei Sasaki ; Akito Kuramata ; Debdeep Jena ; Huili Grace Xing

MULTIDOMAIN DYNAMICS OF FERROELECTRIC POLARIZATION AND ITS COHERENCY-BREAKING IN NEGATIVE CAPACITANCE FIELD-EFFECT TRANSISTORS 197

Hiroyuki Ota ; Tsutomu Ikegami ; Koichi Fukuda ; Junichi Hattori ; Hidehiro Asai ; Kazuhiko Endo ; Shinji Migita ; Akira Toriumi

MODELING OF MULTI-DOMAIN SWITCHING IN FERROELECTRIC MATERIALS: APPLICATION TO NEGATIVE CAPACITANCE FETS 201

A. Dasgupta ; P. Rastogi ; D. Saha ; A. Gaidhane ; A. Agarwal ; Y. S. Chauhan

ON THE MICROSCOPIC ORIGIN OF NEGATIVE CAPACITANCE IN FERROELECTRIC MATERIALS: A TOY MODEL 205

Asif Islam Khan

EFFECT OF POLYCRYSTALLINITY AND PRESENCE OF DIELECTRIC PHASES ON NC-FINFET VARIABILITY 209

Yen-Kai Lin ; Ming-Yen Kao ; Harshit Agarwal ; Yu-Hung Liao ; Pragya Kushwaha ; Korok Chatterjee ; Juan Pablo Duarte ; Huan-Lin Chang ; Sayeef Salahuddin ; Chenming Hu

A SIMULATION BASED STUDY OF NC-FETS DESIGN: OFF-STATE VERSUS ON-STATE PERSPECTIVE 213

T. Rollo ; H. Wang ; G. Han ; D. Esseni

1.5μM DUAL CONVERSION GAIN, BACKSIDE ILLUMINATED IMAGE SENSOR USING STACKED PIXEL LEVEL CONNECTIONS WITH 13KE-FULL-WELL CAPACITANCE AND 0.8E-NOISE 217

Vincent C. Venezia ; Alan Chih-Wei Hsiung ; Kelvin Ai ; Xiang Zhao ; Zhiqiang Lin ; Duli Mao ; Armin Yazdani ; Eric A. G. Webster ; Lindsay A. Grant

A 0.68E-RMS RANDOM-NOISE 121DB DYNAMIC-RANGE SUB-PIXEL ARCHITECTURE CMOS IMAGE SENSOR WITH LED FLICKER MITIGATION 221

S. Iida ; Y. Sakano ; T. Asatsuma ; M. Takami ; I. Yoshiba ; N. Ohba ; H. Mizuno ; T. Oka ; K. Yamaguchi ; A. Suzuki ; K. Suzuki ; M. Yamada ; M. Takizawa ; Y. Tateshita ; K. Ohno

A 24.3ME- FULL WELL CAPACITY CMOS IMAGE SENSOR WITH LATERAL OVERFLOW INTEGRATION TRENCH CAPACITOR FOR HIGH PRECISION NEAR INFRARED ABSORPTION IMAGING 225

M. Murata ; R. Kuroda ; Y. Fujihara ; Y. Aoyagi ; H. Shibata ; T. Shibaguchi ; Y. Kamata ; N. Miura ; N. Kuriyama ; S. Sugawa

A HDR 98DB 3.2µM CHARGE DOMAIN GLOBAL SHUTTER CMOS IMAGE SENSOR 229

A. Tournier ; F. Roy ; Y. Cazaux ; F. Lalanne ; P. Malinge ; M. Mcdonald ; G. Monnot ; N. Roux

HIGH PERFORMANCE 2.5UM GLOBAL SHUTTER PIXEL WITH NEW DESIGNED LIGHT-PIPE STRUCTURE 233

Toshifumi Yokoyama ; Masafumi Tsutsui ; Yoshiaki Nishi ; Ikuo Mizuno ; Veinger Dmitry ; Assaf Lahav

BACK-ILLUMINATED 2.74 µM-PIXEL-PITCH GLOBAL SHUTTER CMOS IMAGE SENSOR WITH CHARGE-DOMAIN MEMORY ACHIEVING 10K E-SATURATION SIGNAL 237

Y. Kumagai ; R. Yoshita ; N. Osawa ; H. Ikeda ; K. Yamashita ; T. Abe ; S. Kudo ; J. Yamane ; T. Idekoba ; S. Noudo ; Y. Ono ; S. Kunitake ; M. Sato ; N. Sato ; T. Enomoto ; K. Nakazawa ; H. Mori ; Y. Tateshita ; K. Ohno

CONFORMAL, WAFER-SCALE AND CONTROLLED NANOSCALE DOPING OF SEMICONDUCTORS VIA THE ICVD PROCESS 241

Jae Hwan Kim ; Hong Keun Park ; Kwan Yong Pak ; Alexander Yoon ; Yun Sang Kim ; Sung Gap Im ; Wan Sik Hwang ; Byung Jin Cho

LOW TEMPERATURE SPUTTERED GRAPHENIC CARBON ENABLES HIGHLY RELIABLE CONTACTS TO SILICON 245

M. Stelzer ; M. Jung ; U. Wurstbauer ; A.W. Holleitner ; F. Kreupl

LOCATION-CONTROLLED-GRAIN TECHNIQUE FOR MONOLITHIC 3D BEOL FINFET CIRCUITS 249

Chih-Chao Yang ; Tung-Ying Hsieh ; Po-Tsang Huang ; Kuan-Neng Chen ; Wan-Chi Wu ; Shih-Wei Chen ; Chia-He Chang ; Chang-Hong Shen ; Jia-Min Shieh ; Chenming Hu ; Meng-Chyi Wu ; Wen-Kuan Yeh

NOVEL MATERIALS AND PROCESSES IN REPLACEMENT METAL GATE FOR ADVANCED CMOS TECHNOLOGY 253

Ruqiang Bao ; Steven Hung ; Miaomiao Wang ; Kisup Chung ; Soumendra Barman ; Siddarth A Krishnan ; Yixiong Yang ; Wei Tang ; Luping Li ; Yongjing Lin ; Michael S Chan ; Zhebo Chen ; Xin Miao ; Marinus Hopstaken ; Richard A Conti ; Hemanth Jagannathan ; Michael P Chudzik ; Dalea McHerron ; Bala S Haran ; Sanjay Natarajan

WHY GEO$_2$ GROWTH ON GE IS SUPPRESSED AND GEO$_2$/GE STACK IS MUCH IMPROVED IN HIGH PRESSURE O$_2$ OXIDATION? 257

Xu Wang ; Akira Toriumi

EUV LITHOGRAPHY AT THRESHOLD OF HIGH-VOLUME MANUFACTURING 261

Anthony Yen ; Hans Meiling ; Jos Benschop

HALF PITCH 14 NM DIRECT PATTERING WITH NANOIMPRINT LITHOGRAPHY 265

T. Nakasugi ; T. Kono ; K. Fukuhara ; M. Hatano ; H. Tokue ; M. Komori ; H. Tsuda ; T. Komukai ; K. Takahata ; H. Kato ; K. Kobayashi ; A. Mitra ; S. Kobayashi ; S. Inoue ; T. Higashiki ; T. Motokawa ; M. Saito ; S. Kanamitsu ; M. Itoh ; T. Imamura ; K. Matasunaga ; K. Hashimoto ; Y. Kim ; J. Cho ; W. Jung

ALL CMOS INTEGRATED 3D-EXTENDED METAL GATE ISFETS FOR PH AND MULTI-ION (NA$^+$, K$^+$, CA^{2+}) SENSING 269

J.-R. Zhang ; M. Rupakula ; F. Bellando ; E. Garcia Cordero ; J. Longo ; F. Wildhaber ; G. Herment ; H. Guérin ; A.M. Ionescu

HIGH RESOLUTION ION DETECTOR (HRID) BY 16NM FINFET CMOS TECHNOLOGY 273

Peng-Chun Liou ; Tsung-Han Lee ; Chien-Ping Wang ; Yu-Lun Chueh ; Yue-Der Chih ; Jonathan Chang ; Chrong Jung Lin ; Ya-Chin King

HIGHLY PERFORMANT INTEGRATED PH-SENSOR USING THE GATE PROTECTION DIODE IN THE BEOL OF INDUSTRIAL FDSOI 277

G. T. Ayele ; S. Monfray ; S. Ecoffey ; F. Boeuf ; J-P. Cloarec ; D. Drouin ; A. Souifi

VERY LARGE SCALE INTEGRATION OPTOMECHANICS: A CURE FOR LONELINESS OF NEMS RESONATORS? 281

Maxime Hermouet ; Marc Sansa ; Martial Defoort ; Louise Banniard ; Sergio Dominauez-Medina ; Shawn Fostner ; Ujwol Palanchoke ; Alexandre Fafin ; Marc Gely ; Louis Hutin ; Christophe Plantier ; Emmanuel Rolland ; Claude Tabone ; Giulia Usai ; Thomas Ernst ; Patrick Villard ; Gérard Billiot ; Paul Mattei ; Guillaume Nonglaton ; Caroline Fontelaye ; Charlie Barrois ; Olivier Castany ; Eduardo Gil Santos ; Pierre E. Allain ; Emeline Vernhes ; Pascale Boulanger ; Ariel Brenac ; Christophe Masselon ; Ivan Favero ; Thomas Alava ; Guillaume Jourdan ; Sébastien Hentz

SIC-FET-TYPE NOX SENSOR FOR HIGH-TEMPERATURE EXHAUST GAS...284

Y. Sasago ; H. Nakamura ; T. Odaka ; A. Isobe ; S. Komatsu ; Y. Nakamura ; T. Yamawaki ; C. Yorita ; N. Ushifusa ; K. Yoshikawa ; K. Ono ; Y. Anzai ; S. Machida ; M. Kinoshita ; K. Fujisaki ; T. Usagawa ; K. Okishiro ; Y. Sugiyama

A SI FET-TYPE GAS SENSOR WITH PULSE-DRIVEN LOCALIZED MICRO-HEATER FOR LOW POWER CONSUMPTION 288

Yoonki Hong ; Seongbin Hong ; Dongkyu Jang ; Yujeong Jeong ; Meile Wu ; Gyuweon Jung ; Jong-Ho Bae ; Jun Shik Kim ; Ki Soo Chang ; Chan Bae Jeong ; Cheol Seong Hwang ; Byung-Gook Park ; Jong-Ho Lee

ECRAM AS SCALABLE SYNAPTIC CELL FOR HIGH-SPEED, LOW-POWER NEUROMORPHIC COMPUTING 292

Jianshi Tang ; Douglas Bishop ; Seyoung Kim ; Matt Copel ; Tayfun Gokmen ; Teodor Todorov ; SangHoon Shin ; Ko-Tao Lee ; Paul Solomon ; Kevin Chan ; Wilfried Haensch ; John Rozen

SOC LOGIC COMPATIBLE MULTI-BIT FEMFET WEIGHT CELL FOR NEUROMORPHIC APPLICATIONS..296

K. Ni ; J. A. Smith ; B. Grisafe ; T. Rakshit ; B. Obradovic ; J. A. Kittl ; M. Rodder ; S. Datta

EXPERIMENTAL DEMONSTRATION OF FERROELECTRIC SPIKING NEURONS FOR UNSUPERVISED CLUSTERING 300

Zheng Wang ; Brian Crafton ; Jorge Gomez ; Ruijuan Xu ; Aileen Luo ; Zoran Krivokapic ; Lane Martin ; Suman Datta ; Arijit Raychowdhury ; Asif Islam Khan

NEAR HYSTERESIS-FREE NEGATIVE CAPACITANCE INGAAS TUNNEL FETS WITH ENHANCED DIGITAL AND ANALOG FIGURES OF MERIT BELOW VDD=400MV 304

Ali Saeidi ; Anne S. Verhulst ; Igor Stolichnov ; Alireza Alian ; Hiroshi Iwai ; Nadine Collaert ; Adrian M. Ionescu

AN EXPERIMENTAL STUDY OF HETEROSTRUCTURE TUNNEL FET NANOWIRE ARRAYS: DIGITAL AND ANALOG FIGURES OF MERIT FROM 300K TO 10K...308

T. Rosca ; A. Saeidi ; E. Memisevic ; L-E. Wernersson ; A.M. Ionescu

HIGH THERMAL TOLERANCE OF 25-NM C-AXIS ALIGNED CRYSTALLINE IN-GA-ZN OXIDE FET 312

Hitoshi Kunitake ; Kazuaki Ohshima ; Kazuki Tsuda ; Noriko Matsumoto ; Hiromi Sawai ; Yuichi Yanagisawa ; Shiori Saga ; Ryo Arasawa ; Takako Seki ; Ryo Tokumaru ; Tomoaki Atsumi ; Kiyoshi Kato ; Shunpei Yamazaki

INTEL 22NM FINFET (22FFL) PROCESS TECHNOLOGY FOR RF AND MM WAVE APPLICATIONS AND CIRCUIT DESIGN OPTIMIZATION FOR FINFET TECHNOLOGY 316

H.-J. Lee ; S. Rami ; S. Ravikumar ; V. Neeli ; K. Phoa ; B. Sell ; Y. Zhang

GAN HEMTS FOR 5G BASE STATION APPLICATIONS..320

Shigeru Nakajima

100-340GHZ SYSTEMS: TRANSISTORS AND APPLICATIONS 324

M.J.W. Rodwell ; Y. Fang ; J. Rode ; J. Wu ; B. Markman ; S. T. Šuran Brunelli ; J. Klamkin ; M Urteaga

CONSIDERATIONS ON DESIGN OF HIGHLY-INTEGRATED MILLIMETER-WAVE TRANSCEIVERS IN SIGE HBT 328

V. Issakov ; S. Trotta

BAW FILTERS FOR 5G BANDS...332

R. Aigner ; G. Fattinger ; M. Schaefer ; K. Karnati ; R. Rothemund ; F. Dumont

TUNABLE FILTER TECHNOLOGIES FOR 5G COMMUNICATIONS 336

Dimitrios Peroulis

ULTRA-LOW POWER 3D NC-FINFET-BASED MONOLITHIC 3D+ -IC WITH COMPUTING-IN-MEMORY FOR INTELLIGENT IOT DEVICES 340

Fu-Kuo Hsueh ; Wei-Hao Chen ; Kai-Shin Li ; Chang-Hong Shen ; Jia-Min Shieh ; Chun Ying Lee ; Bo-Yuan Chen ; Hsiu-Chih Chen ; Chih-Chao Yang ; Wen-Hsien Huang ; Kun-Ming Chen ; Guo-Wei Huang ; Peng Chen ; Yung-Ning Tu ; Srivatsa Srinivasa ; Vijaykrishnan Narayanan ; Meng-Fan Chang ; Wen-Kuan Yeh

FIRST DEMONSTRATION OF GE FERROELECTRIC NANOWIRE FET AS SYNAPTIC DEVICE FOR ONLINE LEARNING IN NEURAL NETWORK WITH HIGH NUMBER OF CONDUCTANCE STATE AND GMAX/GMIN 344

Wonil Chung ; Mengwei Si ; Peide D. Ye

STT-MRAM DESIGN TECHNOLOGY CO-OPTIMIZATION FOR HARDWARE NEURAL NETWORKS..348

Nuo Xu ; Yang Lu ; Weiyi Qi ; Zhengping Jiang ; Xiaochen Peng ; Fan Chen ; Jing Wang ; Woosung Choi ; Shimeng Yu ; Dae Sin Kim

A 68 PARALLEL ROW ACCESS NEUROMORPHIC CORE WITH 22K MULTI-LEVEL SYNAPSES BASED ON LOGIC-COMPATIBLE EMBEDDED FLASH MEMORY TECHNOLOGY..................352

M. Kim ; J. Kim ; C. Park ; L. Everson ; H. Kim ; S. Song ; S. Lee ; C. H. Kim

INTERCHANGEABLE HEBBIAN AND ANTI-HEBBIAN STDP APPLIED TO SUPERVISED LEARNING IN SPIKING NEURAL NETWORK 356

Che-Chia Chang ; Pin-Chun Chen ; Boris Hudec ; Po-Tsun Liu ; Tuo-Hung Hou

STOCHASTIC INFERENCE AND LEARNING ENABLED BY MAGNETIC TUNNEL JUNCTIONS 360

Abhronil Sengupta ; Gopalakrishnan Srinivasan ; Deboleena Roy ; Kaushik Roy

IN-MEMORY COMPUTING PRIMITIVE FOR SENSOR DATA FUSION IN 28 NM HKMG FEFET TECHNOLOGY 364

K. Ni ; B. Grisafe ; W. Chakraborty ; A. K. Saha ; S. Dutta ; M. Jerry ; J. A. Smith ; S. Gupta ; S. Datta

EXPERIMENTALLY VALIDATED, PREDICTIVE MONTE CARLO MODELING OF FERROELECTRIC DYNAMICS AND VARIABILITY 368

C. Alessandri ; P. Pandey ; A. C. Seabaugh

SCALABILITY STUDY ON FCRROCLCCTRIC-HFO2 TUNNEL JUNCTION MEMORY BASED ON NON-EQUILIBRIUM GREEN FUNCTION METHOD WITH SELF-CONSISTENT POTENTIAL 372

Fei Mo ; Yusaku Tagawa ; Takuya Saraya ; Toshiro Hiramoto ; Masaharu Kobayashi

ROLE OF OXYGEN VACANCIES IN ELECTRIC FIELD CYCLING BEHAVIORS OF FERROELECTRIC HAFNIUM OXIDE 376

C. Liu ; F. Liu ; Q. Luo ; P. Huang ; X. X. Xu ; H. B. Lv ; Y. D. Zhao ; X.Y. Liu ; J. F. Kang

FIRST-PRINCIPLES PERSPECTIVE ON POLING MECHANISMS AND FERROELECTRIC/ANTIFERROELECTRIC BEHAVIOR OF $HF_{1-X}ZR_XO_2$ FOR FEFET APPLICATIONS 380

Sergiu Clima ; S.R.C. McMitchell ; K. Florent ; L. Nyns ; M. Popovici ; N. Ronchi ; L. Di Piazza ; J. Van Houdt ; G. Pourtois

CHARACTERIZATION METHODOLOGY AND PHYSICAL COMPACT MODELING OF IN-WAFER GLOBAL AND LOCAL VARIABILITY 384

Krishna Pradeep ; Thierry Poiroux ; Patrick Scheer ; André Juge ; Gilles Gouget ; Gérard Ghibaudo

TOO NOISY AT THE BOTTOM? —RANDOM TELEGRAPH NOISE (RTN) IN ADVANCED LOGIC DEVICES AND CIRCUITS 388

Runsheng Wang ; Shaofeng Guo ; Zhe Zhang ; Qingxue Wang ; Dehuang Wu ; Joddy Wang ; Ru Huang

COMPREHENSIVE STUDY ON THE "ANOMALOUS" COMPLEX RTN IN ADVANCED MULTI-FIN BULK FINFET TECHNOLOGY 392

Jiayang Zhang ; Zhe Zhang ; Rusheng Wang ; Zixuan Sun ; Zuodong Zhang ; Shaofeng Guo ; Ru Huang

AN UNIQUE METHODOLOGY TO ESTIMATE THE THERMAL TIME CONSTANT AND DYNAMIC SELF HEATING IMPACT FOR ACCURATE RELIABILITY EVALUATION IN ADVANCED FINFET TECHNOLOGIES 396

S. Mukhopadhyay ; A. Kundu ; Y.W. Lee ; H. D. Hsieh ; D.S. Huang ; J.J. Horng ; T.H. Chen ; J.H. Lee ; Y.S. Tsai ; C.K. Lin ; Ryan Lu ; Jun He

7NM FINFET PLASMA CHARGE RECORDING DEVICE 400

Yi-Pei Tsai ; Jiaw-Ren Shih ; Ya-Chin King ; Chrong Jung Lin

DEVELOPMENT OF X-RAY PHOTOELECTRON SPECTROSCOPY UNDER BIAS AND ITS APPLICATION TO DETERMINE BAND-ENERGIES AND DIPOLES IN THE HKMG STACK 404

Pushpendra Kumar ; Charles Leroux ; Florian Domengie ; Eugenie Martinez ; Virginie Loup ; Denis Guiheux ; Yves Morand ; Jean-Michel Pedini ; Claude Tabone ; Frederic Gaillard ; Gerard Ghibaudo

IN-SITU INVESTIGATION OF THE IMPACT OF EXTERNALLY APPLIED VERTICAL STRESS ON III-V BIPOLAR TRANSISTOR 408

Y. Liu ; G. Hiblot ; M. Gonzalez ; K. Vanstreels ; D. Velenis ; M. Badaroglu ; G. Van der Plas ; I. De Wolf

MRAM AS EMBEDDED NON-VOLATILE MEMORY SOLUTION FOR 22FFL FINFET TECHNOLOGY 412

O. Golonzka ; J. -G. Alzate ; U. Arslan ; M. Bohr ; P. Bai ; J. Brockman ; B. Buford ; C. Connor ; N. Das ; B. Doyle ; T. Ghani ; F. Hamzaoglu ; P. Heil ; P. Hentges ; R. Jahan ; D. Kencke ; B. Lin ; M. Lu ; M. Mainuddin ; M. Meterelliyoz ; P. Nguyen ; D. Nikonov ; K. O'brien ; J.O Donnell ; K. Oguz ; D. Ouellette ; J. Park ; J. Pellegren ; C. Puls ; P. Quintero ; T. Rahman ; A. Romang ; M. Sekhar ; A. Selarka ; M. Seth ; A. J. Smith ; A. K. Smith ; L. Wei ; C. Wiegand ; Z. Zhang ; K. Fischer

DEMONSTRATION OF HIGHLY MANUFACTURABLE STT-MRAM EMBEDDED IN 28NM LOGIC 416

Y. J. Song ; J. H. Lee ; S. H. Han ; H. C. Shin ; K. H. Lee ; K. Suh ; D. E. Jeong ; G. H. Koh ; S. C. Oh ; J. H. Park ; S. O. Park ; B. J. Bae ; O. I. Kwon ; K. H. Hwang ; B.Y. Seo ; Y.K. Lee ; S. H. Hwang ; D. S. Lee ; Y. Ji ; K.C. Park ; G. T. Jeong ; H. S. Hong ; K. P. Lee ; H. K. Kang ; E. S. Jung

ENABLEMENT OF STT-MRAM AS LAST LEVEL CACHE FOR THE HIGH PERFORMANCE COMPUTING DOMAIN AT THE 5NM NODE 420

S. Sakhare ; M. Perumkunnil ; T. Huynh Bao ; S. Rao ; W. Kim ; D. Crotti ; F. Yasin ; S. Couet ; J. Swerts ; S. Kundu ; D. Yakimets ; R. Baert ; HR. Oh ; A. Spessot ; A. Mocuta ; G. Sankar Kar ; A. Furnemont

TRULY INNOVATIVE 28NM FDSOI TECHNOLOGY FOR AUTOMOTIVE MICRO-CONTROLLER APPLICATIONS EMBEDDING 16MB PHASE CHANGE MEMORY 424

F. Arnaud ; P. Zuliani ; J.P. Reynard ; A. Gandolfo ; F. Disegni ; P. Mattavelli ; E. Gomiero ; G. Samanni ; C. Jahan ; R. Berthelon ; O. Weber ; E. Richard ; V. Barral ; A. Villaret ; S. Kohler ; J.C. Grenier ; R. Ranica ; C. Gallon ; A. Souhaite ; D. Ristoiu ; L. Favennec ; V. Caubet ; S. Delmedico ; N. Cherault ; R. Beneyton ; S. Chouteau ; P.O. Sassoulas ; A. Vernhet ; Y. Le Friec ; F. Domengie ; L. Scotti ; D. Pacelli ; J.L. Ogier ; F. Boucard ; S. Lagrasta ; D. Benoit ; L. Clement ; P. Boivin ; P. Ferreira ; R. Annunziata ; P. Cappelletti

A COST-EFFICIENT 28NM SPLIT-GATE EFLASH MEMORY FEATURING A HKMG HYBRID BIT CELL AND HV DEVICE 428

R. Richter ; M. Trentzsch ; S. Dünkel ; J. Müller ; P. Moll ; B. Bayha ; K. Mothes ; A. Henke ; M. Mazur ; J. Paul ; P. Krottenthaler ; J. Poth ; S. Jansen ; R. Hüselitz ; H. Kim ; A. Zaka ; T. Herrmann ; E.M. Bazizi ; S. Beyer ; P. Ghazavi ; H. Om'mani ; S. Lemke ; Y. Tkachev ; F. Zhou ; J. Kim ; X. Liu ; V. Tiwari ; N. Do

A BI-STABLE 1- /2-TRANSISTOR SRAM IN 14 NM FINFET TECHNOLOGY FOR HIGH DENSITY / HIGH PERFORMANCE EMBEDDED APPLICATIONS 432

Yuniarto Widjaja ; James Wilson ; Tu Nguyen ; Jin-Woo Han ; Christopher Norwood ; Dinesh Maheshwari ; Stefan Lai ; Pieter Vorenkamp ; Zvi Or-Bach ; Yoshio Nishi

SIC DEVICES FOR MAINSTREAM ADOPTION 436

Peter Friedrichs

THE CURRENT STATUS AND FUTURE PROSPECTS OF SIC HIGH VOLTAGE TECHNOLOGY 440

A. Mihaila ; L. Knoll ; E. Bianda ; M. Bellini ; S. Wirths ; G. Alfieri ; L. Kranz ; F. Canales ; M. Rahimo

PROGRESS IN HIGH AND ULTRAHIGH VOLTAGE SILICON CARBIDE DEVICE TECHNOLOGY 444

Y. Yonezawa ; K. Nakayama ; R. Kosugi ; S. Harada ; K. Koseki ; K. Sakamoto ; T. Kimoto ; H. Okumura

EFFECTS OF BASAL PLANE DISLOCATIONS ON SIC POWER DEVICE RELIABILITY 448

R. E. Stahlbush ; K. N. A. Mahakik ; A. J. Lelis ; R. Green

GAN DEVICES FOR AUTOMOTIVE APPLICATION AND THEIR CHALLENGES IN ADOPTION 452

Tetsu Kachi

BARRIERS TO THE ADOPTION OF WIDE-BANDGAP SEMICONDUCTORS FOR POWER ELECTRONICS 456

I.C. Kizilyalli ; E.P. Carlson ; D.W. Cunningham

GAN POWER COMMERCIALIZATION WITH HIGHEST QUALITY-HIGHEST RELIABILITY 650V HEMTS-REQUIREMENTS, SUCCESSES AND CHALLENGES 460

P. Parikh ; Y. Wu ; L. Shen ; R. Barr ; S. Chowdhury ; J. Gritters ; S. Yea ; P. Smith ; L. McCarthy ; R. Birkhahn ; M. Moore ; J. McKay ; H. Clement ; U. Mishra ; R. Lal ; P. Zuk ; T. Hosoda ; K. Shono ; K. Imanishi ; Y. Asai

40× RETENTION IMPROVEMENT BY ELIMINATING RESISTANCE RELAXATION WITH HIGH TEMPERATURE FORMING IN 28 NM RRAM CHIP 464

Xiaoxin Xu ; Lu Tai ; Tiancheng Gong ; Jiahao Yin ; Peng Huang ; Jie Yu ; Da Nian Dong ; Qing Luo ; Jing Liu ; Zhaoan Yu ; Xi Zhu ; Xiu Long Wu ; Qi Liu ; Hangbing Lv ; Ming Liu

CHARACTERIZING ENDURANCE DEGRADATION OF INCREMENTAL SWITCHING IN ANALOG RRAM FOR NEUROMORPHIC SYSTEMS 468

Meiran Zhao ; Huaqiang Wu ; Bin Gao ; Xiaoyu Sun ; Yuyi Liu ; Peng Yao ; Yue Xi ; Xinyi Li ; Qingtian Zhang ; Kanwen Wang ; Shimeng Yu ; He Qian

IN-DEPTH CHARACTERIZATION OF RESISTIVE MEMORY-BASED TERNARY CONTENT ADDRESSABLE MEMORIES 472

D. R. B. Ly ; B. Giraud ; J-P Noel ; A. Grossi ; N. Castellani ; G. Sassine ; J-F Nodin ; G. Molas ; C. Fenouillet-Beranger ; G. Indiveri ; E. Nowak ; E. Vianello

MIXED-SIGNAL NEUROMORPHIC INFERENCE ACCELERATORS: RECENT RESULTS AND FUTURE PROSPECTS 476

M. Bavandpour ; M.R. Mahmoodi ; H. Nili ; F. Merrikh Bayat ; M. Prezioso ; A. Vincent ; D.B. Strukov ; K.K. Likharev

TEMPORAL SEQUENCE LEARNING WITH A HISTORY-SENSITIVE PROBABILISTIC LEARNING RULE INTRINSIC TO OXYGEN VACANCY-BASED RRAM 480

J. Doevenspeck ; R. Degraeve ; A. Fantini ; P. Debacker ; D. Verkest ; R. Lauwereins ; W. Dehaene

IN-MEMORY AND ERROR-IMMUNE DIFFERENTIAL RRAM IMPLEMENTATION OF BINARIZED DEEP NEURAL NETWORKS 484

M. Bocquet ; T. Hirtzlin ; J.-O. Klein ; E. Nowak ; E. Vianello ; J.-M. Portal ; D. Querlioz

A NEW HARDWARE IMPLEMENTATION APPROACH OF BNNS BASED ON NONLINEAR 2T2R SYNAPTIC CELL 488

Z. Zhou ; P. Huang ; Y. C. Xiang ; W. S. Shen ; Y. D. Zhao ; Y. L. Feng ; B. Gao ; H. Q. Wu ; H. Qian ; L. F. Liu ; X. Zhang ; X. Y. Liu ; J. F. Kang

GE CMOS GATE STACK AND CONTACT DEVELOPMENT FOR VERTICALLY STACKED LATERAL NANOWIRE FETS 492

M.J.H. van Dal ; G. Vellianitis ; G. Doornbos ; B. Duriez ; M.C. Holland ; T. Vasen ; A. Afzalian ; E. Chen ; S.K. Su ; T.K. Chen ; T.M. Shen ; Z.Q. Wu ; C.H. Diaz

ADVANTAGE OF NW STRUCTURE IN PRESERVATION OF SRB-INDUCED STRAIN AND INVESTIGATION OF OFF-STATE LEAKAGE IN STRAINED STACKED GE NW PFET 496

H. Arimura ; G. Eneman ; E. Capogreco ; L. Witters ; A. De Keersgieter ; P. Favia ; C. Porret ; A. Hikavyy ; R. Loo ; H. Bender ; L.-Å. Ragnarsson ; J. Mitard ; N. Collaert ; D. Mocuta ; N. Horiguchi

TUNABILITY OF PARASITIC CHANNEL IN GATE-ALL-AROUND STACKED NANOSHEETS 500

S. Barraud ; B. Previtali ; V. Lapras ; C. Vizioz ; J.-M. Hartmann ; S. Martinie ; J. Lacord ; M. Cassé ; L. Dourthe ; V. Loup ; G. Romano ; N. Rambal ; Z. Chalupa ; N. Bernier ; G. Audoit ; A. Jannaud ; V. Delaye ; V. Balan ; O. Rozeau ; T. Ernst ; M. Vinet

VOLTAGE TRANSFER CHARACTERISTIC MATCHING BY DIFFERENT NANOSHEET LAYER NUMBERS OF VERTICALLY STACKED JUNCTIONLESS CMOS INVERTER FOR SOP/3D-ICS APPLICATIONS 504

P.-J. Sung ; C.-Y. Chang ; L.-Y. Chen ; K.-H. Kao ; C.-J. Su ; T.-H. Liao ; C.-C. Fang ; C.-J. Wang ; T.-C. Hong ; C.-Y. Jao ; H.-S. Hsu ; S.-X. Luo ; Y.-S. Wang ; H.-F. Huang ; J.-H. Li ; Y.-C. Huang ; F.-K. Hsueh ; C.-T. Wu ; Y.-M. Huang ; F.-J. Hou ; G.-L. Luo ; Y.-C. Huang ; Y.-L. Shen ; W. C.-Y. Ma ; K.-P. Huang ; K.-L. Lin ; S. Samukawa ; Y. Li ; G.-W Huang ; Y.-J. Lee ; J.-Y. Li ; W.-F. Wu ; J.-M. Shieh ; T.-S. Chao ; W.-K. Yeh ; Y.-H. Wang

VERTICALLY STACKED GATE-ALL-AROUND SI NANOWIRE CMOS TRANSISTORS WITH REDUCED VERTICAL NANOWIRES SEPARATION, NEW WORK FUNCTION METAL GATE SOLUTIONS, AND DC/AC PERFORMANCE OPTIMIZATION 508

R. Ritzenthaler ; H. Mertens ; V. Pena ; G. Santoro ; A. Chasin ; K. Kenis ; K. Devriendt ; G. Mannaert ; H. Dekkers ; A. Dangol ; Y. Lin ; S. Sun ; Z. Chen ; M. Kim ; J. Machillot ; J. Mitard ; N. Yoshida ; N. Kim ; D. Mocuta ; N. Horiguchi

2D MATERIALS: ROADMAP TO CMOS INTEGRATION 512

C. Huyghebaert ; T. Schram ; Q. Smets ; T. Kumar Agarwal ; D. Verreck ; S. Brems ; A. Phommahaxay ; D. Chiappe ; S. El Kazzi ; C. Lockhart de la Rosa ; G. Arutchelvan ; D. Cott ; J. Ludwig ; A. Gaur ; S. Sutar ; A. Leonhardt ; D. Marinov ; D. Lin ; M. Caymax ; I. Asselberghs ; G. Pourtois ; I.P. Radu

FIRST DEMONSTRATION OF WSE2 BASED CMOS-SRAM 516

Chin-Sheng Pang ; Niharika Thakuria ; Sumeet Kumar Gupta ; Zhihong Chen

STEEP SLOPE P-TYPE 2D WSE$_2$ FIELD-EFFECT TRANSISTORS WITH VAN DER WAALS CONTACT AND NEGATIVE CAPACITANCE 520

Jingli Wang ; Xuyun Guo ; Zhihao Yu ; Zichao Ma ; Yanghui Liu ; Masun Chan ; Ye Zhu ; Xinran Wang ; Yang Chai

TOWARD HIGH-MOBILITY AND LOW-POWER 2D MOS$_2$ FIELD-EFFECT TRANSISTORS 524

Zhihao Yu ; Ying Zhu ; Weisheng Li ; Yi Shi ; Gang Zhang ; Yang Chai ; Xinran Wang

3D MONOLITHIC STACKED 1T1R CELLS USING MONOLAYER MOS$_2$ FET AND HBN RRAM FABRICATED AT LOW (150°C) TEMPERATURE 528

Ching-Hua Wang ; Connor McClellan ; Yuanyuan Shi ; Xin Zheng ; Victoria Chen ; Mario Lanza ; Eric Pop ; H.-S. Philip Wong

ATOMRISTORS: MEMORY EFFECT IN ATOMICALLY-THIN SHEETS AND RECORD RF SWITCHES 532

Ruijing Ge ; Xiaohan Wu ; Myungsoo Kim ; Po-An Chen ; Jianping Shi ; Junho Choi ; Xiaoqin Li ; Yanfeng Zhang ; Meng-Hsueh Chiang ; Jack C. Lee ; Deji Akinwande

AN ULTRA-FAST MULTI-LEVEL MOTE$_2$-BASED RRAM 536

F. Zhang ; H. Zhang ; P.R. Shrestha ; Y. Zhu ; K. Maize ; S. Krylyuk ; A. Shakouri ; J.P. Campbell ; K.P. Cheung ; L.A. Bendersky ; A. V. Davydov ; J. Appenzeller

FIRST CRYOGENIC ELECTRO-OPTIC SWITCH ON SILICON WITH HIGH BANDWIDTH AND LOW POWER TUNABILITY 540

F. Eltes ; J. Barreto ; D. Caimi ; S. Karg ; A. A. Gentile ; A. Hart ; P. Stark ; N. Meier ; M. G. Thompson ; J. Fompeyrine ; S. Abel

HIGH SPEED (F$_{3-DB}$ ABOVE 10 GHZ) PHOTO DETECTION AT TWO-MICRON-WAVELENGTH REALIZED BY GESN/GE MULTIPLE-QUANTUM-WELL PHOTODIODE ON A 300 MM SI SUBSTRATE 544

Shengqiang Xu ; Wei Wang ; Yi-Chiau Huang ; Yuan Dong ; Saeid Masudy-Panah ; Hong Wang ; Xiao Gong ; Yee-Chia Yeo

QUADRATIC ELECTRO-OPTICAL SILICON-ORGANIC HYBRID RF MODULATOR IN A PHOTONIC INTEGRATED CIRCUIT TECHNOLOGY 548

P. Steglich ; C. Mai ; A. Peczek ; F. Korndörfer ; C. Villringer ; B. Dietzel ; A. Mai

SILICON PHOTONICS: A SCALING TECHNOLOGY FOR COMMUNICATIONS AND INTERCONNECTS 552

P. Dong ; K. W. Kim ; A. Melikyan ; Y. Baeyens

INAS/GAAS QUANTUM DOT LASERS MONOLITHICALLY INTEGRATED ON GROUP IV PLATFORM 556

Keshuang Li ; Mingchu Tang ; Mengya Liao ; Jiang Wu ; Siming Chen ; Alwyn Seeds ; Huiyun Liu

HETEROGENEOUSLY INTEGRATED LIGTHT SOURCES ON BULK-SILICON PLATFORM 560

Dongjae Shin ; Jungho Cha ; Yonghwack Shin ; Kyoungho Ha ; Chang Bum Lee ; Changgyun Shin ; Dongshik Shim ; Byoung Lyong Choi ; Hyeongsun Hong ; Kyupil Lee ; Ho-Kyu Kang

INTERFACIAL THERMAL CONDUCTIVITY OF 2D LAYERED MATERIALS: AN ATOMISTIC APPROACH 564

Kamyar Parto ; Arnab Pal ; Xuejun Xie ; Wei Cao ; Kaustav Banerjee

COMPUTATIONAL DESIGN OF SILICON CONTACTS ON 2D TRANSITION-METAL DICHALCOGENIDES: THE ROLES OF CRYSTALLINE ORIENTATION, DOPING LEVEL, PASSIVATION AND INTERFACIAL LAYER 568

Xiaolei Ma ; Zhiqiang Fan ; Jixuan Wu ; Xiangwei Jiang ; Jiezhi Chen

FIRST PRINCIPLES STUDY OF MEMORY SELECTORS USING HETEROJUNCTIONS OF 2D LAYERED MATERIALS 572

Linsen Li ; Blanka Magyari-Köpe ; Ching-Hua Wang ; Sanchit Deshmukh ; Zizhen Jiang ; Haitong Li ; Yi Yang ; Huanglong Li ; He Tian ; E. Pop ; Tian-Ling Ren ; H.-S. Philip Wong

CAN KINETIC INDUCTANCE IN LOW-DIMENSIONAL MATERIALS ENABLE A NEW GENERATION OF RF-ELECTRONICS? 576

Kunjesh Agashiwala ; Arnab Pal ; Wei Cao ; Junkai Jiang ; Kaustav Banerjee

A SURFACE POTENTIAL- AND PHYSICS- BASED COMPACT MODEL FOR 2D POLYCRYSTALLINE-MOS$_2$ FET WITH RESISTIVE SWITCHING BEHAVIOR IN NEUROMORPHIC COMPUTING 580

Lingfei Wang ; Lin Wang ; Kah-Wee Ang ; Aaron Voon-Yew Thean ; Gengchiau Liang

DESIGN AND OPTIMIZATION OF ß-GA$_2$O$_3$ ON (H-BN LAYERED) SAPPHIRE FOR HIGH EFFICIENCY POWER TRANSISTORS: A DEVICE-CIRCUIT-PACKAGE PERSPECTIVE 584

Bikram K. Mahajan ; Yen-Pu Chen ; Woojin Ahn ; Nicolò Zagni ; Muhammad Ashraful Alam

DECONVOLUTING CHARGE TRAPPING AND NUCLEATION INTERPLAY IN FEFETS: KINETICS AND RELIABILITY 588

Milan Pesic ; Andrea Padovani ; Stefan Slcsazeck ; Thomas Mikolajick ; Luca Larcher

IMPACT OF SELF-HEATING ON RELIABILITY PREDICTIONS IN STT-MRAM 592

S. Van Beek ; B. J. O'Sullivan ; P. J. Roussel ; R. Degraeve ; E. Bury ; J. Swerts ; S. Couet ; L. Souriau ; S. Kundu ; S. Rao ; W. Kim ; F. Yasin ; D. Crotti ; D. Linten ; G. Kar

INVESTIGATING THE STATISTICAL-PHYSICAL NATURE OF MGO DIELECTRIC BREAKDOWN IN STT-MRAM AT DIFFERENT OPERATING CONDITIONS 596

J.H. Lim ; N. Raghavan ; A. Padovani ; J.H. Kwon ; K. Yamane ; H. Yang ; V.B. Naik ; L. Larcher ; K.H. Lee ; K.L. Pey

TRAP REDUCTION AND PERFORMANCES IMPROVEMENTS STUDY AFTER HIGH PRESSURE ANNEAL PROCESS ON SINGLE CRYSTAL CHANNEL 3D NAND DEVICES 600

A. Subirats ; A. Arreghini ; R. Delhougne ; E. Rosseel ; A. Hikavyy ; L. Breuil ; S. Vadakupudhu Palayam ; G. Van den bosch ; D. Linten ; A. Furnémont

22-NM FD-SOI EMBEDDED MRAM TECHNOLOGY FOR LOW-POWER AUTOMOTIVE-GRADE-L MCU APPLICATIONS 604

K. Lee ; R. Chao ; K. Yamane ; V. B. Naik ; H. Yang ; J. Kwon ; N. L. Chung ; S. H. Jang ; B. Behin-Aein ; J.H. Lim ; B. Liu ; E. H. Toh ; K. W. Gan ; D. Zeng ; N. Thiyagarajah ; L. C. Goh ; T. Ling ; J. W. Ting ; J. Hwang ; L. Zhang ; R. Low ; R. Krishnan ; L. Zhang ; S. L Tan ; Y. S. You ; C. S. Seet ; H. Cong ; J. Wong ; S. T. Woo ; E. Quek ; S. Y. Siah

14NS WRITE SPEED 128MB DENSITY EMBEDDED STT-MRAM WITH ENDURANCE>10^{10} AND 10YRS RETENTION@85^{0}C USING NOVEL LOW DAMAGE MTJ INTEGRATION PROCESS 608

H. Sato ; H. Honjo ; T. Watanabe ; M. Niwa ; H. Koike ; S. Miura ; T. Saito ; H. Inoue ; T. Nasuno ; T. Tanigawa ; Y. Noguchi ; T. Yoshiduka ; M. Yasuhira ; S. Ikeda ; S.- Y. Kang ; T. Kubo ; K. Yamashita ; Y. Yagi ; R. Tamura ; T. Endoh

STT-MRAM DEVICES WITH LOW DAMPING AND MOMENT OPTIMIZED FOR LLC APPLICATIONS AT OX NODES 612

Luc Thomas ; Guenole Jan ; Santiago Serrano-Guisan ; Huanlong Liu ; Jian Zhu ; Yuan-Jen Lee ; Son Le ; Jodi Iwata-Harms ; Ru-Ying Tong ; Sahil Patel ; Vignesh Sundar ; Dongna Shen ; Yi Yang ; Renren He ; Jesmin Haq ; Zhongjian Teng ; Vinh Lam ; Paul Liu ; Yu-Jen Wang ; Tom Zhong ; Hideaki Fukuzawa ; PoKang Wang

MICROWAVE NEURAL PROCESSING AND BROADCASTING WITH SPINTRONIC NANO-OSCILLATORS 616

P. Talatchian ; M. Romera ; S. Tsunegi ; F. Abreu Araujo ; V. Cros ; P. Bortolotti ; J. Trastoy ; K. Yakushiji ; A. Fukushima ; H. Kubota ; S. Yuasa ; M. Ernoult ; D. Vodenicarevic ; T. Hirtzlin ; N. Locatelli ; D. Querlioz ; J. Grollier

HIGH ENDURANCE PHASE CHANGE MEMORY CHIP IMPLEMENTED BASED ON CARBON-DOPED GE$_2$SB$_2$TE$_5$ IN 40 NM NODE FOR EMBEDDED APPLICATION 620

Z. T. Song ; D. L. Cai ; X. Li ; L. Wang ; Y. F. Chen ; H. P. Chen ; Q. Wang ; Y. P. Zhan ; M. H. Ji

A 40NM LOW-POWER LOGIC COMPATIBLE PHASE CHANGE MEMORY TECHNOLOGY 624

J.Y. Wu ; Y.S. Chen ; W. S. Khwa ; S. M. Yu ; T. Y. Wang ; J.C. Tseng ; Y.D. Chih ; Carlos H. Diaz

8-BIT PRECISION IN-MEMORY MULTIPLICATION WITH PROJECTED PHASE-CHANGE MEMORY 628

I. Giannopoulos ; A. Sebastian ; M. Le Gallo ; V.P. Jonnalagadda ; M. Sousa ; M.N. Boon ; E. Eleftheriou

SYSTEM PERFORMANCE: FROM ENTERPRISE TO AI 632

A. Kumar ; L. Chang ; G.E. Tellez ; L.A. Clevenger ; J.L. Burns

DESIGN-TECHNOLOGY CO-OPTIMIZATION OF STANDARD CELL LIBRARIES ON INTEL 10NM PROCESS 636

Xinning Wang ; Ranjith Kumar ; Somashekar Bangalore Prakash ; Peng Zheng ; Tai-Hsuan Wu ; Quan Shi ; Marni Nabors ; Srinivasa Chaitanya Gadigatla ; Simeon Realov ; Chin-Hsuan Chen ; Ying Zhang ; Kaizad Mistry ; Andrew Yeoh ; Ian Post ; Chris Auth ; Atul Madhavan

AN ACCURATE FINFET'S VMIN ESTIMATION METHOD FOR EXTREME LOW OPERATION VOLTAGE DESIGN 640

H. W. Choi ; S. K. Kim ; H. Jung ; D. R. Chang ; S. Park ; Y. Yasuda-Masuoka ; J.S. Yoon

TACKLING FUNDAMENTAL CHALLENGES OF CARRIER TRANSPORT AND DEVICE VARIABILITY IN ADVANCED SINFINFETS FOR 7NM NODE AND BEYOND 644

Ming-Huei Lin ; Vincent S. Chang ; Jen-Hsiang Lu ; Shu-Hui Wang ; Shyh-Horng Yang

EXTENDABLE AND MANUFACTURABLE VOLUME-LESS MULTI-VT SOLUTION FOR 7NM TECHNOLOGY NODE AND BEYOND 648

Ruqiang Bao ; Huimei Zhou ; Miaomiao Wang ; Dechao Guo ; Bala S Haran ; Vijay Narayanan ; Rama Divakaruni

CHANNEL GEOMETRY IMPACT AND NARROW SHEET EFFECT OF STACKED NANOSHEET 652

Chun Wing Yeung ; Jingyun Zhang ; Robin Chao ; Ohseong Kwon ; Reinaldo Vega ; Gen Tsutsui ; Xin Miao ; Chen Zhang ; Chang-Woo Sohn ; Bum Ki Moon ; Ali Razavieh ; Julien Frougier ; Andrew Greene ; Rohit Galatage ; Juntao Li ; Miaomiao Wang ; Nicolas Loubet ; Robert Robison ; Veeraraghavan Basker ; Tenko Yamashita ; Dechao Guo

3NM GAA TECHNOLOGY FEATURING MULTI-BRIDGE-CHANNEL FET FOR LOW POWER AND HIGH PERFORMANCE APPLICATIONS 656

Geumjong Bae ; D.-I. Bae ; M. Kang ; S.M. Hwang ; S.S. Kim ; B. Seo ; T.Y. Kwon ; T.J. Lee ; C. Moon ; Y.M. Choi ; K. Oikawa ; S. Masuoka ; K.Y. Chun ; S.H. Park ; H.J. Shin ; J.C. Kim ; K.K. Bhuwalka ; D.H. Kim ; W.J. Kim ; J. Yoo ; H.Y. Jeon ; M.S. Yang ; S.-J. Chung ; D. Kim ; B.H. Ham ; K.J. Park ; W.D. Kim ; S.H. Park ; G. Song ; Y.H. Kim ; M.S. Kang ; K.H. Hwang ; C.-H. Park ; J.-H. Lee ; D.-W. Kim ; S-M. Jung ; H.K. Kang

A CMOS PROXIMITY CAPACITANCE IMAGE SENSOR WITH 16μM PIXEL PITCH, 0.1AF DETECTION ACCURACY AND 60 FRAMES PER SECOND 660

M. Yamamoto ; R. Kuroda ; M. Suzuki ; T. Goto ; H. Hamori ; S. Murakami ; T. Yasuda ; S. Sugawa

3D EXPANDABLE MICROWIRE ELECTRODE ARRAYS MADE OF PROGRAMMABLE SHAPE MEMORY MATERIALS 664

Ruoyu Zhao ; Xin Liu ; Yichen Lu ; Chi Ren ; Armaghan Mehrsa ; Takaki Komiyama ; Duygu Kuzum

BIO-INSPIRED 3D NEURAL ELECTRODES FOR THE PERIPHERAL NERVES STIMULATION USING SHAPE MEMORY POLYMERS 668

Yingchao Zhang ; Ning Zheng ; Yinji Ma ; Tao Xie ; Xue Feng

FABRICATION AND CHARACTERIZATION OF 3D MULTI-ELECTRODE ARRAY ON FLEXIBLE SUBSTRATE FOR IN VIVO EMG RECORDING FROM EXPIRATORY MUSCLE OF SONGBIRD 672

Muneeb Zia ; Bryce Chung ; Samuel J. Sober ; Muhannad S. Bakir

BIOELECTRONICS AT THE SINGLE MOLECULE LEVEL 676

O. Tolga Gul ; Kaitlin M. Pugliese ; Yongki Choi ; Arith J. Rajapakse ; Calvin J. Lau ; Narendra Kumar ; Kristin N. Gabriel ; Denys Marushchak ; Tivoli J. Olsen ; Deng Pan ; Gregory A. Weiss ; Philip G. Collins

SI NANOWIRE BIOSENSORS USING A FINFET FABRICATION PROCESS FOR REAL TIME MONITORING CELLULAR ION ACTITIVIES 679

Qingzhu Zhang ; Hailing Tu ; Huaxiang Yin ; Feng Wei ; Hongbin Zhao ; Chunling Xue ; Qianhui Wei ; Zhaohao Zhang ; Xiao Zhang ; Shaoming Zhang ; Qin Han ; Yudong Li ; Robert Chunhua Zhao ; Jiang Yan ; Junfeng Li ; Wenwu Wang

A FLEXIBLE, HETEROGENEOUSLY INTEGRATED WIRELESS POWERED SYSTEM FOR BIO-IMPLANTABLE APPLICATIONS USING FAN-OUT WAFER-LEVEL PACKAGING 683

G. Ezhilarasu ; A. Hanna ; R. Irwin ; A. Alam ; S. S. Iyer

PARALLEL-PLANE BREAKDOWN FIELDS OF 2.8-3.5 MV/CM IN GAN-ON-GAN P-N JUNCTION DIODES WITH DOUBLE-SIDE-DEPLETED SHALLOW BEVEL TERMINATION 687

T. Maeda ; T. Narita ; H. Ueda ; M. Kanechika ; T. Uesugi ; T. Kachi ; T. Kimoto ; M. Horita ; J. Suda

DEMONSTRATION OF AVALANCHE CAPABILITY IN POLARIZATION-DOPED VERTICAL GAN PN DIODES: STUDY OF WALKOUT DUE TO RESIDUAL CARBON CONCENTRATION 691

C. De Santi ; E. Fabris ; K. Nomoto ; Z. Hu ; W. Li ; X. Gao ; D. Jena ; H. G. Xing ; G. Meneghesso ; M. Meneghini ; E. Zanoni

SUPPRESSED HOLE-INDUCED DEGRADATION IN E-MODE GAN MIS-FETS WITH CRYSTALLINE GAO$_X$N$_{1-X}$ CHANNEL...695

Mengyuan Hua ; Xiangbin Cai ; Song Yang ; Zhaofu Zhang ; Zheyang Zheng ; Jin Wei ; Ning Wang ; Kevin J. Chen

RECENT ADVANCEMENT OF GAN HEMT WITH INALGAN BARRIER LAYER AND FUTURE PROSPECTS OF A1N-BASED ELECTRON DEVICES 699

J. Kotani ; A. Yamada ; T. Ohki ; Y. Minoura ; S. Ozaki ; N. Okamoto ; K. Makiyama ; N. Nakamura

POWER GAN HEMT DEGRADATION: FROM TIME-DEPENDENT BREAKDOWN TO HOT-ELECTRON EFFECTS..703

M. Meneghini ; A. Barbato ; M. Borga ; C. De Santi ; M. Barbato ; S. Stoffels ; M. Zhao ; N. Posthuma ; S. Decoutere ; O. Haeberlen ; T. Detzel ; G. Meneghesso ; E. Zanoni

NEW INSIGHTS INTO THE PHYSICAL ORIGIN OF NEGATIVE CAPACITANCE AND HYSTERESIS IN NCFETS 707

Huimin Wang ; Mengxuan Yang ; Qianqian Huang ; Kunkun Zhu ; Yang Zhao ; Zhongxin Liang ; Cheng Chen ; Zhixuan Wang ; Yuan Zhong ; Xing Zhang ; Ru Huang

A CRITICAL EXAMINATION OF 'QUASI-STATIC NEGATIVE CAPACITANCE' (QSNC) THEORY 711

Z. Liu ; M. A. Bhuiyan ; T. P. Ma

DIRECT RELATIONSHIP BETWEEN SUB-60 MV/DEC SUBTHRESHOLD SWING AND INTERNAL POTENTIAL INSTABILITY IN MOSFET EXTERNALLY CONNECTED TO FERROELECTRIC CAPACITOR...715

Xiuyan Li ; Akira Toriumi

ASSESSMENT OF STEEP-SUBTHRESHOLD SWING BEHAVIORS IN FERROELECTRIC-GATE FIELD-EFFECT TRANSISTORS CAUSED BY POSITIVE FEEDBACK OF POLARIZATION REVERSAL 719

Shinji Migita ; Hiroyuki Ota ; Akira Toriumi

EXPERIMENTAL STUDY ON THE ROLE OF POLARIZATION SWITCHING IN SUBTHRESHOLD CHARACTERISTICS OF HFO$_2$-BASED FERROELECTRIC AND ANTI-FERROELECTRIC FET 723

Chengji Jin ; Kyungmin Jang ; Takuya Saraya ; Toshiro Hiramoto ; Masaharu Kobayashi

DEMONSTRATION OF HIGH-SPEED HYSTERESIS-FREE NEGATIVE CAPACITANCE IN FERROELECTRIC HF$_{0.5}$ZR$_{0.5}$O$_2$ 727

M. Hoffmann ; B. Max ; T. Mittmann ; U. Schroeder ; S. Slesazeck ; T. Mikolajick

NEGATIVE-CAPACITANCE FINFET INVERTER, RING OSCILLATOR, SRAM CELL, AND FT................... 731

Kai-Shin Li ; Yun-Jie Wei ; Yi-Ju Chen ; Wen-Cheng Chiu ; Hsiu-Chih Chen ; Min-Hung Lee ; Yu-Fan Chiu ; Fu-Kuo Hsueh ; Bo-Wei Wu ; Pin-Guang Chen ; Tung-Yan Lai ; Chun-Chi Chen ; Jia-Min Shieh ; Wen-Kuan Yeh ; Sayeef Salahuddin ; Chenming Hu

EXTREMELY STEEP SWITCH OF NEGATIVE-CAPACITANCE NANOSHEET GAA-FETS AND FINFETS 735

M. H. Lee ; K.-T. Chen ; C.-Y. Liao ; S.-S. Gu ; G.-Y. Siang ; Y.-C. Chou ; H.-Y. Chen ; J. Le ; R.-C. Hong ; Z.-Y. Wang ; S.-Y. Chen ; P.-G. Chen ; M. Tang ; Y.-D. Lin ; H.-Y. Lee ; K.-S. Li ; C. W. Liu

OPTOCOUPLING IN CMOS 739

V. Agarwal ; S. Dutta ; A. J. Annema ; R. J. E. Hueting ; J. Schmitz ; M.J. Lee ; E. Charbon ; B. Nauta

HIGH VOLTAGE GENERATION USING DEEP TRENCH ISOLATED PHOTODIODES IN A BACK SIDE ILLUMINATED PROCESS 743

F. Kaklin ; J. M. Raynor ; R. K. Henderson

THROUGH-SILICON-TRENCH IN BACK-SIDE-ILLUMINATED CMOS IMAGE SENSORS FOR THE IMPROVEMENT OF GATE OXIDE LONG TERM PERFORMANCE 747

A. Vici ; F. Russo ; N. Lovisi ; L. Latessa ; A. Marchioni ; A. Casella ; F. Irrera

HIGH-PERFORMANCE GERMANIUM-AN-SILICON LOCK-IN PIXELS FOR INDIRECT TIME-OF-FLIGHT APPLICATIONS...751

N. Na ; S.-L. Cheng ; H.-D. Liu ; M.-J. Yang ; C.-Y. Chen ; H.-W. Chen ; Y.-T. Chou ; C.-T. Lin ; W.-H. Liu ; C.-F. Liang ; C.-L. Chen ; S.-W. Chu ; B.-J. Chen ; Y.-F. Lyu ; S.-L. Chen

CMOS-INTEGRATED SINGLE-PHOTON-COUNTING X-RAY DETECTOR USING AN AMORPHOUS-SELENIUM PHOTOCONDUCTOR WITH 11×11-μM^2 PIXELS 755

A. Camlica ; A. El-Falou ; R. Mohammadi ; P. M. Levine ; K. S. Karim

TRANSPORT MODELS BASED ON NEGF AND EMPIRICAL PSEUDOPOTENTIALS: A COMPUTATIONALLY VIABLE METHOD FOR SELF-CONSISTENT SIMULATION OF NANOSCALE DEVICES..759

Marco G. Pala ; Oves Badami ; David Esseni

FIRST PRINCIPLES SIMULATION OF ENERGY EFFICIENT SWITCHING BY SOURCE DENSITY OF STATES ENGINEERING...763

Fei Liu ; Chenguang Qiu ; Zhiyong Zhang ; Lian-Mao Peng ; Jian Wang ; Zhenhua Wu ; Hong Guo

UNIVERSAL SWING FACTOR APPROACH FOR PERFORMANCE ANALYSIS OF LOGIC NODES...767

M. Ali Pourghaderi ; Anh-Tuan Pham ; Seungkyu Kim ; Hyein Chung ; Zhengping Jiang ; Hesameddin Ilatikhameneh ; Hong-hyun Park ; Seonghoon Jin ; Jongchol Kim ; Won-Young Chung ; Uihui Kwon ; Woosung Choi ; Dae Sin Kim ; Shigenobu Maeda

MULTI-DOMAIN PROCESS MODELING FOR ADVANCED LOGIC AND MEMORY DEVICES: FROM EQUIMPMENTS TO MATERIALS...771

Inkook Jang ; Hyoungsoo Ko ; Alexander Schmidt ; Sae-Jin Kim ; Moonhyun Cha ; Hyoshin Ahn ; Honglae Park ; Dae Sin Kim ; Ho-Kyu Kang

ENTIRE BIAS SPACE STATISTICAL RELIABILITY SIMULATION BY 3D-KMC METHOD AND ITS APPLICATION TO THE RELIABILITY ASSESSMENT OF NANOSHEET FETS BASED CIRCUITS..775

Wangyong Chen ; Yun Li ; Linlin Cai ; Pengying Chang ; Gang Du ; Xiaoyan Liu

A PHYSICS-BASED THERMAL MODEL OF NANOSHEET MOSFETS FOR DEVICE-CIRCUIT CO-DESIGN...779

Linlin Cai ; Wangyong Chen ; Pengying Chang ; Gang Du ; Xing Zhang ; Jinfeng Kang ; Xiaoyan Liu

UNDERSTANDING THE INTRINSIC RELIABILITY BEHAVIOR OF N -/P-SI AND P-GE NANOWIRE FETS UTILIZING DEGRADATION MAPS.......................................783

Adrian Chasin ; Erik Bury ; Jacopo Franco ; Ben Kaczer ; Michiel Vandemaele ; Hiroaki Arimura ; Elena Capogreco ; Liesbeth Witters ; Romain Ritzenthaler ; Hans Mertens ; Naoto Horiguchi ; Dimitri Linten

BTI RELIABILITY IMPROVEMENT STRATEGIES IN LOW THERMAL BUDGET GATE STACKS FOR 3D SEQUENTIAL INTEGRATION...787

J. Franco ; Z. Wu ; G. Rzepa ; A. Vandooren ; H. Arimura ; L. -Å Ragnarsson ; G. Hellings ; S. Brus ; D. Cott ; V. De Heyn ; G. Groeseneken ; N. Horiguchi ; J. Ryckaert ; N. Collaert ; D. Linten ; T. Grasser ; B. Kaczer

CHARACTERIZATION AND UNDERSTANDING OF SLOW TRAPS IN GEO$_X$-BASED N-GE MOS INTERFACES...791

M. Ke ; P. Cheng ; K. Kato ; M. Takenaka ; S. Takagi

SOFT ERROR TRENDS IN ADVANCED SILICON TECHNOLOGY NODES.......................795

B. Bhuva

CMOS-COMPATIBLE DOPED-MULTILAYER-GRAPHENE INTERCONNECTS FOR NEXT-GENERATION VLSI..799

Junkai Jiang ; Jae Hwan Chu ; Kaustav Banerjee

TIME DEPENDENT EARLY BREAKDOWN OF AIGAN/GAN EPI STACKS AND SHIFT IN SOA BOUNDARY OF HEMTS UNDER FAST CYCLIC TRANSIENT STRESS.....................803

Bhawani Shankar ; Ankit Soni ; Sayak Dutta Gupta ; Swati Shikha ; Sandeep Singh ; Srinivasan Raghavan ; Mayank Shrivastava

TOWARD HIGH PERFORMANCE SIGE CHANNEL CMOS: DESIGN OF HIGH ELECTRON MOBILITY IN SIGE NFINFETS OUTPERFORMING SI.......................................807

C. H. Lee ; R. G. Southwick ; S. Mochizuki ; J. Li ; X. Miao ; M. Wang ; R. Bao ; I. Ok ; T. Ando ; P. Hashemi ; D. Guo ; V. Narayanan ; N. Loubet ; H. Jagannathan

ADVANCED ARSENIC DOPED EPITAXIAL GROWTH FOR SOURCE DRAIN EXTENSION FORMATION IN SCALED FINFET DEVICES...811

S. Mochizuki ; B. Colombeau ; L. Yu ; A. Dube ; S. Choi ; M. Stolfi ; Z. Bi ; F. Chang ; R. A. Conti ; P. Liu ; K. R. Winstel ; H. Jagannathan ; H.-J. Gossmann ; N. Loubet ; D. F. Canaperi ; D. Guo ; S. Sharma ; S. Chu ; J. Boland ; Q. Jin ; Z. Li ; S. Lin ; M. Cogorno ; M. Chudzik ; S. Natarajan ; D. C. McHerron ; B. Haran

EXTERNAL RESISTANCE REDUCTION BY NANOSECOND LASER ANNEAL IN SI/SIGE CMOS TECHNOLOGY...815

Oleg Gluschenkov ; Heng Wu ; Kevin Brew ; Chengyu Niu ; Lan Yu ; Yasir Sulehria ; Samuel Choi ; Curtis Durfee ; James Demarest ; Adra Carr ; Shaoyin Chen ; Jim Willis ; Thirumal Thanigaivelan ; Fee-li Lie ; Walter Kleemeier ; Dechao Guo

PARASITIC RESISTANCE REDUCTION STRATEGIES FOR ADVANCED CMOS FINFETS BEYOND 7NM..819

H. Wu ; O. Gluschenkov ; G. Tsutsui ; C. Niu ; K. Brew ; C. Durfee ; C. Prindle ; V. Kamineni ; S. Mochizuki ; C. Lavoie ; E. Nowak ; Z. Liu ; J. Yang ; S. Choi ; J. Demarest ; L. Yu ; A. Carr ; W. Wang ; J. Strane ; S. Tsai ; Y. Liang ; H. Amanapu ; I. Saraf ; K. Ryan ; F. Lie ; W. Kleemeier ; K. Choi ; N. Cave ; T. Yamashita ; A. Knorr ; D. Gupta ; B. Haran ; D. Guo ; H. Bu ; M. Khare

SUB-10^{-9} Ω-CM^2 SPECIFIC CONTACT RESISTIVITY ON P-TYPE GE AND GESN: IN-SITU GA DOPING WITH GA ION IMPLANTATION AT 300°C, 25°C, AND -100°C 823

Ying Wu ; Lye-Hing Chua ; Wei Wang ; Kaizhen Han ; Wei Zou ; Todd Henry ; Xiao Gong

SELECTIVE FIN TRIMMING AFTER DUMMY GATE REMOVAL AS THE LOCAL FIN WIDTH SCALING APPROACH FOR N5 AND BEYOND 827

Toshihiko Miyashita ; Shiyu Sun ; Sushant Mittal ; Myung Sun Kim ; Ashish Pal ; Angada Sachid ; Kalpana Pathak ; Matt Cogorno ; Nam Sung Kim

FIRST EXPERIMENTAL DEMONSTRATION OF A SCALABLE LINEAR MAJORITY GATE BASED ON SPIN WAVES 831

Florin Ciubotaru ; Giacomo Talmelli ; Thibaut Devolder ; Odysseas Zografos ; Marc Heyns ; Christoph Adelmann ; Iuliana P. Radu

SPINTRONIC DEVICES FOR LOW ENERGY DISSIPATION 835

Kang L. Wang ; Hao Wu ; Seyed Armin Razavi ; Qiming Shao

ROOM TEMPERATURE HIGHLY EFFICIENT TOPOLOGICAL INSULATOR/MO/COFEB SPIN-ORBIT TORQUE MEMORY WITH PERPENDICULAR MAGNETIC ANISOTROPY 839

Qiming Shao ; Hao Wu ; Quanjun Pan ; Peng Zhang ; Lei Pan ; Kin Wong ; Xiaoyu Che ; Kang L. Wang

SCALED SPINTRONIC LOGIC DEVICE BASED ON DOMAIN WALL MOTION IN MAGNETICALLY INTERCONNECTED TUNNEL JUNCTIONS 843

E. Raymenants ; D. Wan ; S. Couet ; O. Zografos ; V.D. Nguyen ; A. Vaysset ; L. Souriau ; A. Thiam ; M. Manfrini ; S. Brus ; M. Heyns ; D. Mocuta ; D.E. Nikonov ; S. Manipatruni ; I. A. Young ; T. Devolder ; I. P. Radu

BINARY AND TERNARY TRUE RANDOM NUMBER GENERATORS BASED ON SPIN ORBIT TORQUE 847

Huiming Chen ; Shuai Zhang ; Nuo Xu ; Min Song ; Xin Li ; Ruofan Li ; Yi Zeng ; Jeongmin Hong ; Long You

HIGH-PERFORMANCE, COST-EFFECTIVE 2Z NM TWO-DECK CROSS-POINT MEMORY INTEGRATED BY SELF-ALIGN SCHEME FOR 128 GB SCM 851

Taehoon Kim ; Hyejung Choi ; Myoungsub Kim ; Jaeyun Yi ; Donghoon Kim ; Sunglae Cho ; Hyunmin Lee ; Changyoun Hwang ; Eung-Rim Hwang ; Jeongho Song ; Sujin Chae ; Yunseok Chun ; Jin-Kook Kim

A HIGHLY EFFICIENT AND SCALABLE MODEL FOR CROSSBAR ARRAYS WITH NONLINEAR SELECTORS 855

An Chen

ULTRA-HIGH ENDURANCE AND LOW IOFF SELECTOR BASED ON ASSEGE CHALCOGENIDES FOR WIDE MEMORY WINDOW 3D STACKABLE CROSSPOINT MEMORY 859

H. Y. Cheng ; W. C. Chien ; I. T. Kuo ; C. W. Yeh ; L. Gignac ; W. Kim ; E. K. Lai ; Y. F. Lin ; R. L. Bruce ; C. Lavoie ; C.W. Cheng ; A. Ray ; F. M. Lee ; F. Carta ; C. H. Yang ; M. H. Lee ; H. Y. Ho ; M. BrightSky ; H. L. Lung

OPTIMIZED READING WINDOW FOR CROSSBAR ARRAYS THANKS TO GE-SE-SB-N-BASED OTS SELECTORS 863

A. Verdy ; M. Bernard ; J. Garrione ; G. Bourgeois ; M. C. Cyrille ; E. Nolot ; N. Castellani ; P. Noé ; C. Socquet-Clerc ; T. Magis ; G. Sassine ; G. Molas ; G. Navarro ; E. Nowak

FORMING-FREE MOTT-OXIDE THRESHOLD SELECTOR NANODEVICE SHOWING S-TYPE NDR WITH HIGH ENDURANCE (> 10^{12} CYCLES), EXCELLENT V_{TH} STABILITY (5%), FAST (< 10 NS) SWITCHING, AND PROMISING SCALING PROPERTIES 867

T. Hennen ; D. Bedau ; J. A. J. Rupp ; C. Funck ; S. Menzel ; M. Grobis ; R. Waser ; D. J. Wouters

FULLY MULTI-FUNCTIONAL GAN-BASED MICRO-LEDS FOR 2500 PPI MICRO-DISPLAYS, TEMPERATURE SENSING, LIGHT ENERGY HARVESTING, AND LIGHT DETECTION 871

Zhaojun Liu ; Ke Zhang ; Yibo Liu ; Siwa Yan ; Hoi Sing Kwok ; Jamal Deen ; Xiaowei Sun

ENVIRONMENTALLY FRIENDLY QUANTUM DOTS FOR DISPLAY APPLICATIONS 875

E. Jang

SOLUTION PROCESSED HIGH PERFORMANCE SHORT CHANNEL ORGANIC THIN-FILM TRANSISTORS WITH EXCELLENT UNIFORMITY AND ULTRA-LOW CONTACT RESISTANCE FOR LOGIC AND DISPLAY 879

L. Feng ; Y. Huang ; J. Fan ; J. Zhao ; S. Pandya ; S. Chen ; W. Tang ; S. Ogier ; X. Guo

RECORD STATIC AND DYNAMIC PERFORMANCE OF FLEXIBLE ORGANIC THIN-FILM TRANSISTORS 883

James W. Borchert ; Ute Zschieschang ; Florian Letzkus ; Michele Giorgio ; Mario Caironi ; Joachim N. Burghartz ; Sabine Ludwigs ; Hagen Klauk

HYBRID STRUCTURE OF SILICON NANOCRYSTALS AND 2D WSE$_2$ FOR BROADBAND OPTOELECTRONIC SYNAPTIC DEVICES 887

Zhenyi Ni ; Yue Wang ; Lixiang Liu ; Shuangyi Zhao ; Yang Xu ; Xiaodong Pi ; Deren Yang

HIGH PERFORMANCE 2D PEROVSKITE/GRAPHENE OPTICAL SYNAPSES AS ARTIFICIAL EYES 891

He Tian ; Xuefeng Wang ; Fan Wu ; Yi Yang ; Tian-Ling Ren

FIRST TRANSISTOR DEMONSTRATION OF THERMAL ATOMIC LAYER ETCHING: INGAAS FINFETS WITH SUB-5 NM FIN-WIDTH FEATURING IN SITU ALE-ALD 895

Wenjie Lu ; Younghee Lee ; Jessica Murdzek ; Jonas Gertsch ; Alon Vardi ; Lisa Kong ; Steven M. George ; Jesús A. del Alamo

INGAAS-ON-INSULATOR FINFETS WITH REDUCED OFF-CURRENT AND RECORD PERFORMANCE 899

C. Convertino ; C. Zota ; S. Sant ; F. Eltes ; M. Sousa ; D. Caimi ; A. Schenk ; L. Czornomaz

BALANCED DRIVE CURRENTS IN 10–20 NM DIAMETER NANOWIRE ALL-III-V CMOS ON SI 903

Adam Jönsson ; Johannes Svensson ; Lars-Erik Wemersson

HIGH PERFORMANCE QUANTUM WELL INGAAS-ON-SI MOSFETS WITH SUB-20 NM GATE LENGTH FOR RF APPLICATIONS 907

C. B. Zota ; C. Convertino ; Y. Baumgartner ; M. Sousa ; D. Caimi ; L. Czornomaz

HIGH PERFORMANCE INGAAS GATE-ALL-AROUND NANOSHEET FET ON SI USING TEMPLATE ASSISTED SELECTIVE EPITAXY 911

S. Lee ; C. -W. Cheng ; X. Sun ; C. D'Emic ; H. Miyazoe ; M. M. Frank ; M. Lofaro ; J. Bruley ; P. Hashemi ; J. A. Ott ; T. Ando ; W. Spratt ; G. M. Cohen ; C. Lavoie ; R. Bruce ; J. Patel ; H. Schmid ; L. Czornomaz ; V. Narayanan ; R.T. Mo ; E. Leobandung

SCALING ACOUSTIC FILTERS TOWARDS 5G 915

Yansong Yang ; Ruochen Lu ; Songbin Gong

PHYSICS OF HOLE TRAPPING PROCESS IN HIGH-K GATE STACKS: A DIRECT SIMULATION FORMALISM FOR THE WHOLE INTERFACE SYSTEM COMBINING DENSITY-FUNCTIONAL THEORY AND MARCUS THEORY 919

Yue-Yang Liu ; Xiangwei Jiang

PARASITIC SURFACE REACTIONS IN HIGH-ASPECT RATIO VIA FILLING USING ALD: A STOCHASTIC KINETIC MODEL 923

T. Muneshwar ; G. Shoute ; D. Barlage ; K. Cadien

PHYSICS-BASED MODELING OF VOLATILE RESISTIVE SWITCHING MEMORY (RRAM) FOR CROSSPOINT SELECTOR AND NEUROMORPHIC COMPUTING 927

W. Wang ; A. Bricalli ; M. Laudato ; E. Ambrosi ; E. Covi ; D. Ielmini

ANALYTIC MODEL FOR STATISTICAL STATE INSTABILITY AND RETENTION BEHAVIORS OF FILAMENTARY ANALOG RRAM ARRAY AND ITS APPLICATIONS IN DESIGN OF NEURAL NETWORK 931

P. Huang ; Y. C. Xiang ; Y. D. Zhao ; C. Liu ; B. Gao ; H. Q. Wu ; H. Qian ; X. Y. Liu ; J. F. Kang

EVIDENCE OF MAGNETOSTRICTIVE EFFECTS ON STT-MRAM PERFORMANCE BY ATOMISTIC AND SPIN MODELING 935

K. Sankaran ; J. Swerts ; R. Carpenter ; S. Couet ; K. Garello ; R. F. L. Evans ; S. Rao ; W. Kim ; S. Kundu ; D. Crotti ; G. S. Kar ; G. Pourtois

Author Index

2018 International Electron Devices Meeting

WELCOME FROM THE GENERAL CHAIR

On behalf of the entire IEDM committee, I would like to welcome you to the 2018 IEEE International Electron Devices Meeting to be held December 1-5, 2018, in San Francisco, CA. This is the 64th annual IEDM and promises to continue its long tradition as the world's premier venue for presenting the latest breakthroughs in electron device technologies. See our new video at (https://youtu.be/BKwGWYyEkmQ).

IEDM 2018 will feature outstanding contributed and invited papers from industrial and academic leaders as well as students from around the world. An outline of the technical program and short summaries of all the papers are available on the IEDM website– http://www.ieee-iedm.org/. We will continue to distribute an abbreviated digest at the meeting, along with the full digest in electronic format. An IEDM smartphone and tablet app that supports Android and iOS platforms is also available. The full digest will also be available through IEEE Xplore after the conference.

The meeting's technical activities begin on Saturday afternoon, December 1 with our popular and highly successful tutorials. Now in their 8th year, they are targeted at students, engineers, or anyone who wants an introduction to, or review of, the basics of contemporary and emerging topics in semiconductor technology. Three tracks run in parallel, for a total of six tutorial topics; see the IEDM website for full information. On Sunday, two comprehensive short courses will be offered: "Scaling Survival Guide in the More than Moore Era", and "It's All About Memory, Not Logic!!!". These full-day courses are organized and presented by internationally recognized researchers and technologists from industry and academia active in these areas. The topics and instructors have been carefully chosen to have broad appeal to IEDM participants, and will include material suitable for both newcomers and experts.

The Plenary Session on Monday morning will feature presentation of awards from the IEEE Electron Devices Society (EDS), and three invited talks. ES Jung from Samsung Electronics will give us his perspectives in a talk entitled "4th Industrial Revolution and Foundry: Challenges and Opportunities," followed by Gerhard Fettweiss from TU Dresden who will describe efforts to go beyond current computing paradigms in his talk on "Venturing Electronics into Unknown Grounds." The third plenary talk, "Future Computing for AI" will be given by Jeff Welser from IBM Research.

In addition to the excellent contributed paper sessions, four special Focus Sessions will feature talks from leading experts in exciting new and rapidly advancing areas. The topics are Interconnects to Enable Continued Scaling, Quantum Computing Devices, Future Technologies Towards Wireless Communications: 5G and Beyond, and Challenges for Wide Bandgap Device Adoption in Power Electronics.

Tuesday night will feature an interactive panel session that promises to be both relevant and engaging. The panel: "The Next 25 years in Electronics, will be moderated by Sanjay Natarajan from Applied Materials. He will engage a panel of experts and industry veterans in a lively discussion on who will drive and lead innovation in the semiconductor industry in the coming decades.

There also will be an entertaining and informative IEDM Career Luncheon on Tuesday. In an informal setting, Professor Veena Misra, North Carolina State University, and John Chen of nVidia will share their perspectives on the current status and career paths in the semiconductor field and industry, in an interactive session with a focus on students and young professionals.

Also, for the third year, IEDM 2018 will host an Exhibits area during the conference where you can learn more about the latest products and publications. Stop by any time during the exhibition open hours to browse the booths and enjoy complimentary coffee. Also, we will host in this venue the 3rd edition of the MRAM poster session, a joint initiative with the IEEE Magnetics Society, as well as a second poster session to showcase SRC student research, which is a joint initiative with SRC.

On behalf of Mariko Takayanagi, Technical Program Chair, and Suman Datta, Technical Program Vice-Chair, as well as the entire IEDM committee, I want to express my sincere appreciation to all of the authors and speakers who contributed to the technical program. Your efforts are the engine that continues to make IEDM the premier conference in electron devices and related technologies. I also wish to thank each of the members of the IEDM executive and technical subcommittees whose dedication and efforts were critical in planning and organizing the 2018 conference.

IEDM is sponsored by the IEEE Electron Devices Society. If you are not already an IEEE and EDS member, please consider joining this great institution that has played such an important role globally for over 130 years. More detailed information regarding the IEEE is available at the conference and on their website – http://www.ieee.org.

It is again my great honor and pleasure to extend a warm welcome to everyone attending the 2018 IEEE International Electron Devices Meeting.

Ken Rim, General Chair
IEDM 2018

Ken Rim
General Chair

Mariko Takayanagi
Technical
Program Chair

Suman Datta
Technical Program
Vice Cha

AWARD PRESENTATIONS

PLENARY SESSION AWARDS

Monday, December 3

2017 Roger A. Haken Best Student Paper Award

To: Felix Eltes, IBM Research - Zurich

For the paper entitled: "A Novel 25 Gbps Electro-optic Pockels Modulator Integrated on an Advanced Si Photonic Platform"

2017 EDS Paul Rappaport Award

To: Liesbethy Johanna Witters, Hiroaki Arimura, F. Sebaai, A. Hikavyy, A. P. Milenin, Roger Loo, A. De Keersgieter, G. Eneman, Tom Schram, Kurt Wostyn, Katia Devriendt, A. Schulze, Ruben Lieten, Steven Bilodeau, Emanuel Cooper, Peter Storck, Eddie Chiu, Crista Vrancken, Paola Favia, Eric Vancoille, Jerome Mitard, Robert Langer, Ann Opdebeeck, F. Holsteyns, Niamh Waldron, Kathy Barla, Vincent De Heyn, Dan Mocuta, Nadine Collaert

For the paper entitled: "Strained Germanium Gate-All-Around pMOS Device Demonstration Using Selective Wire Release Etch Prior to Replacement Metal Gate Deposition"

2017 EDS George Smith Award

To: Ali Saeidi, Farzan Jazaeri, Francesco Bellando, Igor Stolichnov, Gia V. Luong, Qing-Tai Zhao, Siegfried Mantl, Christian C. Enz, Adrian M. Ionescu

For the paper entitled: "Negative Capacitance as Performance Booster for Tunnel FETs and MOSFETs: An Experimental Study"

2018 EDS Distinguished Service Award

To: Shuji Ikeda, Tei Solutions Co. Ltd., Tsukuba, Japan

"To recognize and honor outstanding service to the Electron Devices Society"

2018 EDS Education Award

To: Ashraf Alam, Purdue University, West Lafayette, IN, USA

"For educating, inspiring and mentoring students and electron device professionals around the world"

2018 EDS J.J. Ebers Award

To: Michael Shur, Rennselaer Polytechnic Institute, Troy, NY, USA

"For pioneering the concept of ballistic transport in nanoscale semiconductor devices"

2018 IEEE/EDS Fellows

This is a complete listing of the 2018 IEEE/EDS Fellows. Not all Fellows will be recognized at the 2017 IEDM.

Pamela Ann Abshire, Silver Spring, MD, USA
Timothy Boykin, University of Alabama Huntsville, AL, USA

Jeffrey Calame, Annapolis, MD, USA
Kun-yung Chang, Los Altos Hills, CA, USA
Kuan-neng Chen, National Chiao Tung Univeristy Hsinchu, Taiwan
Akira Fujiwara, NTT Basic Research Laboratories Atsugi, Japan
Michel Houssa, University of Leuven Leuven, Belgium
Jaroslav Hynecek, Allen, TX, USA
Thomas Kazior, Raytheon: Radio Frequency Components Andover, MA, USA
Michael Krames, Philips Lumileds Lighting Company San Jose, CA, USA
Isaac Lagnado, San Diego, CA, USA
Xiaobing Luo, Huazhong University of Science & Technology Wuhan, China
Chee Wee Liu, National Taiwan University Taipei, Taiwan
Ming Liu, Institute of Microelectronics of Chinese Academy of Sciences Beijing, China
Wei Lu, University of Michigan Ann Arbor, MI, USA
Zhenqiang Ma, University of Wisconsin-Madison Madison, WI, USA
Saibal Mukhopadhyay, Georgia Institue of Technology Atlanta, GA, USA
Hideo Ohno, Tohoku University Sendai, Japan
Hidetoshi Onodera, Kyoto University Kyoto, Japan
Philippe Paillet, University of Montpellier-CEA Paris, France
Joseph Pawlowski, Micron Technology, Inc. Boise, ID, USA
Seiji Samukawa, Tohoku University Sendai, Japan
Riichiro Shirota, National Chao-Tung University Hsinchu, Taiwan
Gregory Snider, University of Notre Dame Notre Dame, IN, USA
Shuji Tanaka, Tohoku University Sendai, Japan
Victor Veliadis, PowerAmerica Raleigh, NC, USA
Robert Weikle, University of Virginia-Charlottesville Charlottesville, VA, USA
Shien-yang Wu, Taiwan Semiconductor Manufacturing Company, Limited Hsinchu, Taiwan
Huikai Xie, University of Florida Gainesville, FL, USA
Jianbin Xu, Chinese University of Hong Kong Shatin NT, Hong Kong
Anthony Yen, ASML, San Jose, CA, USA

IEDM CAREER LUNCHEON

Tuesday, December 4

<u>2018 IEEE Cledo Brunetti Award</u>

To: Siegfried Selberherr

"For pioneering contributions to Technology Computer Aided Design."

<u>2018 IEEE Andrew S. Grove Award</u>

To: Gurtej Singh Sandhu

"For contributions to silicon CMOS process technology that enable DRAM and NAND memory chip scaling."

<u>2018 IEEE Leon K. Kirchmayer Graduate Teaching Award</u>

To: Mark S. Lundstrom

"For creating a global online community for graduate education in nanotechnology as well as teaching, inspiring, and mentoring graduate students."

SRC Student Showcase Poster Session
Tuesday, December 4th, 2:00 – 4:00 pm
Golden Gate Ballroom

New to IEDM this year, Semiconductor Research Corporation (SRC) is hosting a student research showcase. SRC is sponsored by industry members and government agencies with the mission to drive advances in materials, devices, processing, metrology, and modeling, among other areas. Because these topics are relevant to the IEDM community, SRC is organizing a poster session led by students highlighting fundamental semiconductor research in universities worldwide. This session will be a great opportunity to engage with student researchers and their budding ideas.

IEDM Panel
Tuesday, December 4
Continental 1-5
Moderator: Sanjay Natarajan, Applied Materials

Title: **The Next 25 Years in Electronics**

IEEE Magnetics Society MRAM Poster Session
Wednesday, December 5th, 2:00 – 5:00 pm
Plaza Room

For the 3rd consecutive year, the IEEE Magnetics Society is organizing a special MRAM poster session to foster closer interactions between the microelectronics and magnetism communities. The posters will cover topics including MRAM materials, phenomena, technology, testing, hybrid CMOS/MTJ technology and circuits, and spin-logic.

Standing from left to right: Masao Inoue, Asian Arrangements Co-Chair; Su Jin Ahn, Asian Arrangements Chair; Dina Triyoso, Courses Chair; Suman Datta, Technical Program Vice Chair; Ken Rim, General Chair; Mariko Takayanagi, Technical Program Chair; Zhiping Zhou, Optoelectronics, Displays, and Imagers Chair; Barbara De Salvo, Publications Co-Chair; Srabanti Chowduri, Power and Compound Semiconductor Devices Chair

Second row standing from left to right: Rihito Kuroda, Publicity Vice Chair; Meng-Fan Chang, Courses Co-Chair (Tutorials); Jan Hoentschel, European Arrangements Chair; Bernard Legrand, Sensors, MEMS, and BioMEMS Chair; Geert Eneman, Modeling and Simulation Chair; Tibor Grasser, Publications Chair; Clemens Ostermaier, European Arrangements Co-Chair; Curtis Tsai, Focus Session and Special Events Chair; Jungwoo Joh, Courses Co-Chair (Short Courses); Deji Akinwande, Nano Device Technology Chair

Third row standing from left to right: Anthony S. Oates, Characterization, Reliability, and Yield Chair; Martin Giles, Focus Session and Special Events Chair; Pierre Morin, Process and Manufacturing Technology Chair; Kang-ill Seo, Circuit and Device Interaction Chair

Not in attendance: Kristen Moselund, Publicity Chair; John Paul Strachan, Memory Technology Chair, Phyllis Mahoney, Senior Conference Manager and Polly Hocking, Conference Manager

EXECUTIVE COMMITTEE

Ken Rim
General Chair
Qualcomm

Mariko Takayanagi
Technical Program Chair
Toshiba

Suman Datta
Technical Program Vice Chair
University of Notre Dame

Tibor Grasser
Publications Chair
TU Wien

Barbara De Salvo
Publications Co-Chair
CEA-LETI

Dina Triyoso
Courses Chair
Avera Semiconductors

Jungwoo Joh
Courses Co-Chair (Short Courses)
Texas Instruments

Meng-Fan Chang
Courses Co-Chair (Tutorials)
Samsung

Martin Giles
Focus Session and Special Events Chair
Intel

Curtis Tsai
Focus Session and Special Events Chair
Intel

Kirsten Moselund
Publicity Chair
IBM Zurich

Rihito Kuroda
Publicity Vice Chair
Tohoku University

Su Jin Ahn
Asian Arrangements Chair
Samsung

Masao Inoue
Asian Arrangements Co-Chair
Renesas

Jan Hoentschel
European Arrangements Chair
GLOBALFOUNDRIES

Clemens Ostermaier
European Arrangements Co-Chair
Infineon

Phyllis Mahoney
Senior Conference Manager
Widerkehr and Associates

Polly Hocking
Conference Manager
Widerkehr and Associates

SUBCOMMITTEE ON CIRCUIT AND DEVICE INTERACTION

Kang-ill Seo
Circuit and Device Interaction Chair
Samsung

Yen-Ming (James) Chen
TSMC

Jie Deng
Qualcomm

Walid (Mac) Hafez
Intel Corporation

Tuo-Hung Hou
National Chiao Tung University

Chung-Hsun Lin
GLOBALFOUNDRIES

Anda Mocuta
imec

Shaloo Rakheja
New York University

Chang-Hong Shen
National Nano Devices Lab

Ronald Tetzlaff
TU Dresden

Runsheng Wang
Peking University

Yanfeng Wang
Nvidia

Shimeng Yu
Arizona State University

SUBCOMMITTEE ON CHARACTERIZATION, RELIABILITY, AND YIELD

Anthony S. Oates
Characterization, Reliability, and Yield Chair
TSMC

Hideki Aono
Renesas

Charles Cheung
NIST

Paul Hurley
Tyndall

MoonYoung Jeong
Samsung Electronics

Dmitri Linten
imec

Tanya Nigam
GLOBALFOUNDRIES

Kenji Okada
TowerJazz Panasonic Semiconductor

Hokyung Park
SK Hynix

Gilles Reimbold
CEA/Leti

Dhanoop Varghesse
Texas Instruments

Miaomiao Wang
IBM

Bonnie Weir
Broadcom

SUBCOMMITTEE ON OPTOELECTRONICS, DISPLAYS, AND IMAGERS

Zhiping Zhou
Optoelectronics, Displays, and Imagers Chair
Peking University

Peter De Dobbelaere
Luxtera

Chris Doerr
Acacia Communications

Lindsay Grant
Omnivision Tech

Changhee Lee
Seoul National University

Pawel Malinowski
imec

Becky Peterson
University of Michigan

Sara Pellegrini
STMicroelectronics

Toshikatsu Sakai
NHK

Vyshnavi Suntharalingam
MIT Lincoln Laboratory

Mingbin Yu
Shanghai Institute of Microsystem and IT

Li Zhou
BOE Technology

Lars Zimmermann
IHP

SUBCOMMITTEE ON MODELING AND SIMULATION

Geert Eneman
Modeling and Simulation Chair
imec

Stephen Cea
Intel

Yogesh Chauhan
IIT Kanpur

Raphael Clerc
Institute of Optics

Vihar Georgiev
University of Glasgow

Jin-Feng Kang
Peking University

Seongdong Kim
SK hynix

Marco Pala
CNRS, Université Paris-Saclay

Lee Smith
Synopsys

Chika Tanaka
Toshiba

William Vandenberghe
University of Texas, Dallas

Richard Williams
IBM

Nuo Xu
Samsung

SUBCOMMITTEE ON MEMORY TECHNOLOGY

John Paul Strachan
Memory Technology Chair
Hewlett Packard

Regina Dittman
FZ Juehlich

Pei-Ying (Penny) Du
Macronix

Daniele Ielmini
Politecnico di Milano

Seung Kang
Qualcomm

Masaharu Kobayashi
University of Tokyo

Jaeduk Lee
Samsung

Ming Liu
Chinese Academy of Science

Blanka Magyari-Köpe
Stanford

Gabriele Navarro
CEA-Leti

Chris Petti
SanDisk/Western Digital

Luc Thomas
Headway

Rainer Waser
RWTH Aachen

Takeshi Yamaguchi
Toshiba

SUBCOMMITTEE ON NANO DEVICE TECHNOLOGY

Deji Akinwande
Nano Device Technology Chair
University of Texas, Austin

Yang Chai
Hong Kong Poly

Kuan-Lun (Alan) Cheng
Taiwan Semiconductor Manufacturing
Company

Zhihong Chen
Purdue University

Wei-Chih Chien
Macronix

Hidenobu Fukutome
Samsung

Mario Lanza
Soochow University

Iuliana Radu
imec

Frank Schwiez
TU Ilmenau

Mayank Shrivastava
IISc

William Taylor
GLOBALFOUNDRIES

Lars-Erik Wernersson
Lund University

Tomohiro Yamashita
Renesas

Wenjuan Zhu
University of Illinois, Urbana-
Champaign

SUBCOMMITTEE ON POWER DEVICES / COMPOUND SEMICONDUCTOR AND HIGH SPEED DEVICES

Srabanti Chowduri
Power and Compound Semiconductor Devices Chair
University of California, Davis

Shinsuke Harada
AIST

Kei May Lau
Hong Kong University

Erik Lind
Lund University

Dong Seup Lee
Texas Instruments

Shankar Madathil
University of Sheffield

Kozo Makiyama
Fujitsu

Gaudenzio Meneghesso
Padua University

Erwan Morvan
CEA-LETI

Tetsuo Narita
Toyota Central R&D

Shyh-Chiang Shen
Georgia Tech

Jun Suda
Nagoya University

Alon Vardy
Massachusetts Institute of Technology

SUBCOMMITTEE ON PROCESS AND MANUFACTURING TECHNOLOGY

Pierre Morin
Process and Manufacturing Technology Chair
imec

Takashi Ando
IBM

Peter Baars
GLOBALFOUNDRIES

Hsin-Ping Chen
Taiwan Semiconductor Manufacturing Company

Serge Ecoffrey
University de Shebrooke

Sandy Liao
Intel

Chee-Wei Liu
National Taiwan University

Sylvain Maitrejean
CEA-LETI

Bich-Yen Nguyen
SOITEC

Qi Xie
ASM

Tadashi Yamaguchi
Renesas

Sam Yang
Qualcomm

Yi Zhao
Zhejiang University

SUBCOMMITTEE ON SENSORS, MEMS, AND BIOMEMS

Bernard Legrand
Sensors, MEMS, and BioMEMS
Chair
LAAS CNRS

Boyan Boyanov
Illumina

Edwin Carlen
Aichmis Tech

Melissa Cowan
Intel

Isodiana Crupi
Universite degli studi di Palermo

Jin-Woo Han
NASA

Duygu Kuzum
University of California, San Diego

Frances Perez-Murano
University Autonomous Barcelona

Niels Tas
University of Twente

Agnes Tixier-Mita
University of Tokyo

Xiaohong Wang
Tsignhua University

Jun-Bo Yoon
KAIST

4th Industrial Revolution and Foundry: Challenges and Opportunities

ES Jung

Foundry Business, Samsung Electronics, Yongin, Gyeonggi, Republic of Korea, 17113

Abstract—**Semiconductor has been the key enabler in the advancement of electronics for the past 50 years. With the coming of 4th industrial revolution, semiconductor will continue to play an even greater role as we invite a wide variety of new applications into our lives, including smart cars, smart factories, artificial intelligence, data centers, robots, etc. Such importance of semiconductor is attributed to its unique ability to copy and create everything human beings imagine. In this paper, the roles of foundry in the 4th industrial revolution, as the entity to turn ideas into reality along with electronic design automation (EDA), intellectual property (IP) vendor, and outsourced semiconductor assembly and test (OSAT) companies, as well as the need for global open innovation to overcome imminent challenges will be discussed.**

I. INTRODUCTION

Starting with the invention of transistors in 1947 followed by Sony's productization of the world's first transistor radio in 1957, semiconductors have influenced digital transitions for over 70 years, as shown in Fig. 1. [1] However, while particular electronic devices came and went, semiconductor – such as microprocessor, dynamic random access memory (DRAM), static random access memory (SRAM), NAND flash memory, application processor (AP), and high bandwidth memory (HBM) – has continually evolved, enabling the most technologically advanced, cutting-edge devices of its time.

The 1st and 2nd Industrial Revolution best harnessed mechanization with electricity, resulting in the successful start of mass production with use of engines and conveyor belts and laying the foundation for the modern society. The 3rd Industrial Revolution, which came about 50 years later, was driven by information technology involving computers and the internet. The coming era, the era of the 4th Industrial Revolution, will be characterized by a range of new technologies that are fusing the physical and cyber worlds. As such, usage of semiconductors is expected to be greater than ever before. Not only has the number of connected devices shown exponential growth, surpassing the world's population in 2008 and expected to reach 125 billion in 2030 (Fig. 2) [2], but also a range of new applications such as smart cars, smart factories, artificial intelligence, and data centers has arisen, necessitating the development of advanced semiconductor in an unprecedented scale. For example, smart cars, whose market opportunities are projected to be $140 billion in 2020, [3] require, at the least, smart sensors, ADAS, infotainment systems and battery management for their functionality [4]. A completely autonomous level 5 smart car will require more than 6,000 semiconductors to be fully functional [5].

This paper will examine the characteristics of semiconductor industry's ecosystem and focus on the importance of foundry in the realization of the 4th industrial revolution, key technological challenges of the era and ways to overcome them.

II. SEMICONDUCTOR INDUSTRY

The answer to how semiconductor has made possible the technological advances that electronics industry has seen lies on semiconductor's unique ability to copy and create every human beings imagine. Semiconductor has enabled human functions to be carried out by electronic systems. As human cognitive system remembers, makes decisions, and perceives, so does semiconductor, a few examples of which are DRAM for short-term memory, NAND flash for long-term memory, CPU/AP for reasoning, computation, and calculation, and CMOS image sensor (CIS) for perception.

Semiconductor evolution never stopped to match human capability and beyond. Over time, semiconductor has evolved to be faster in speed, higher in density, and lower in power consumption. In addition, new memory architectures and a completely new scheme of computation, such as neuromorphic computing, have been in development [6]. Ever improving cost, performance, and scale of integration has also allowed integrated circuits to continuously enable new products. With all the new applications of the 4th Industrial Revolution, fabless integrated circuits market is expected to grow to $142 billion in 2025. To give it perspective, the market was $5 billion in 1995 and $89 billion in 2017 (Fig. 3) [7]. Number of fabless companies has also increased from about 200 in 1995 to over 1000 in 2018 including some of the world's largest tech companies. In 2025, that number will grow exponentially as well with vast opportunities available in the 4th Industrial Revolution.

Semiconductor industry has evolved into one of the critical foundations for the global economy since the age of computers and the internet. Following many positive disruptions expected of the coming era, semiconductor industry will have a greater impact on the global economy. Not only has it reached a revenue level well in access of $420B per year, it is expected to have a continued long-term growth at a rate of 7.5% [8].

978-1-7281-1988-5/18 $31.00 © 2018 IEEE

III. FOUNDRY

Semiconductor industry's ecosystem is composed of 5 distinct key players: fabless design houses, foundry, EDA/IP, OSAT, and memory vendors. Among them, foundry plays a crucial role in the realization of 4th industrial revolution. Fabless design houses' key function is to birth brilliant, innovative ideas. With over 1000 fabless design houses in the world [9] with great ideas, and the number expected to increase rapidly, those ideas must be realized to be meaningful, and only foundries have the means to turn ideas into reality. When combined with memory chips, ideas become something that has definitive functions originally meant by the designers.

Traditionally, the notion of foundry was limited to wafer manufacturing, providing the process technology. Foundry was simply considered an outsourced manufacturing center that had the monetary resources for facilities and equipment.

However, modern day circuits have become far more complex and associate fabrication cost has drastically increased. For example, gate counts in the state-of-the-art application processor have increased fifteen-folds from 45nm technology node to the late 10nm technology node, thus making it difficult for a single company to do all the work necessary to bring a new product to the market (Fig. 4).

Therefore, some of these roles have been shifted to foundries. Especially, mid- to small-sized companies that might lack in resource started to rely more on foundries for added services, especially in three areas: design service and infrastructure, product engineering, and packaging/testing, as shown in Fig. 5.

Design complexities added at each technology node, along with layers of functions integrated into a single chip, transformed the semiconductor industry to call for more design solution partners (DSPs) and foundries with application-specific integrated circuit (ASIC) design capabilities. It has now become a challenge for a single company to focus all its resources into one project to design the circuitry and adequately meet the time-to-market requirement in their segments of interest. Hence, foundry has evolved to further provide requested on-site or proximate design support.

Secondly, product engineering services such as system-level optimization and failure analysis (FA) are being sought more and more. In-fab FA shortens TAT significantly and reduces cost while DFM and DFT support allows fabless companies to enjoy rapid production yield ramp-up and stringent reliability requirements.

Lastly, advanced packaging and testing, tailored for each individual application, can boost system-level performance and simplify supply chain. Fig. 6 shows Samsung Foundry's various advanced packaging types to best harness individual application's full system-level performance. Packaging for fingerprint sensor has shorter sensing distance for higher sensitivity while packaging for mobile is made for thinner overall thickness and higher performance. Packaging for servers and network addresses high bandwidth needs by adopting high bandwidth memory (HBM) and interposer solutions.

Foundry industry, with transformation of its character, is able to meet the needs of integrated circuit companies around the world. As the spectrum of applications grow, how many foundries will be needed to accommodate all the fabless design houses is a question to be answered.

IV. TECHNOLOGICAL CHALLENGES

Despite many technological advancements, 4th industrial revolution still inflicts great number of challenges with its never-seen-before connectivity and data generation. Especially three hardships are imminent in the field of innovation: complexity of sensor network; unprecedented amount of data generation; and latency of connectivity (Fig. 7).

Complexity of sensor network is a challenge in a world that always has to stay connected with ultimate accuracy. Smart homes require networks of various sensors such as thermostats and surveillance sensors to be always "on" and "connected" in order to function without human intervention. Likewise, autonomous vehicles need sensors, such as LiDAR, short range radar, and ultrasonic sensors, for seamless integration to ensure safety of its driver and passengers. Yet, having everything connected to everything else will monumentally increase the vulnerabilities of any given network. Network technologies must be able to provide hyper-immersive, hyper-seamless services. However, with little interaction in the flow of data from device to data center, rogue or misidentified devices on networks can cause outages, affect analytics and cause security risks.

Unprecedented amount of data generated and extracting meaningful information from the flux of data is also a challenge. One autonomous vehicle is expected to generate data in the order of terabytes per day. Assuming 5 million of these vehicles, data in excess of exabytes will need to be stored somewhere. Problems will arise with analytics and data storage as not all locations might be convenient in terms of access to secure data centers, capacity, or reliable network access. Furthermore, in order to correctly analyze in the amount of data in real-time, vast amount of computing resources will be needed.

The third challenge is the latency of connectivity. As machine to machine communication requires instantaneous connection, even a split second of delay can cause significant damages. For example, an autonomous vehicle driving at 100 kilometers per hour would have driven about 4 meters at best before it even makes a decision to stop on a 3G network whose latency is around 150 milliseconds. The latency becomes 10 millisecond, and therefore 2.8 centimeters at least on a 4G network, and almost real-time on a 5G network.

V. OPPORTUNITIES FOR INNOVATION

To address these challenges, many breakthroughs have recently been made. The issues aforementioned require faster speed and higher density of transistors, first and foremost. Fig.

978-1-7281-1988-5/18 $31.00 © 2018 IEEE

8 shows the advances made in speed, power consumption and density over the years, depicted as normalized to 28nm technology node down to 7nm.

These latest breakthroughs were made possible via EUV lithography, whose advantages are explained in Fig. 9. EUV gives higher fidelity to patterning, reducing line width and roughness variation, thereby improving reliability. The latest EUV lithography can pattern 13nm half pitch. By applying industry-wide double patterning technique (DPT)/ quadruple patterning technique (QPT) on EUV, 3.25nm patterning is possible, whereas the theoretical thermodynamic limit of transistor scaling with silicon is believed to be 1.5nm at room temperature. At the same time, EUV drastically reduces the number of process steps and cost by eliminating the need for DPT/QPT to nearly 1/3 compared to the LELELELE technique .

A great deal of emphasis has recently been placed on power consumption as well. It is expected that we will have about 44 Zetabytes of data in 2020, ten time that in 2013 [10]. More data means more energy consumption, so power-efficient electronics are crucial. One of the examples of new semiconductor devices that consumes less power compared to its counterpart is eMRAM. Its benefits are compared against eSRAM in Table I and Fig. 10. As memory density becomes higher, eMRAM's power benefits become more prominent, consuming only 0.5% of power compared to eSRAM at 1024Mb.

In achieving these technological advances, global open innovation has played a crucial role. Although still at a relatively nascent stage, open innovation has already managed to positively disrupt the semiconductor industry by encouraging innovation, reducing development costs, and accelerating time to market. None of the technological advances mentioned would have been possible had it not been for the joint collaboration across the entire semiconductor industry. Because of the increased difficulties of R&D, materials and equipment R&D is pivotal in extending the Moore's Law. Fig. 11 shows how the number of joint development projects and working groups with materials and equipment suppliers have changed over the past few years at Samsung Semiconductor R&D Center. It attests to the importance of industry-wide collaboration as device scaling becomes more difficult.

As a product of such industry-wide collaboration, many things were made possible such as EUV-compatible photoresist (PR) and a new-concept dryer. Fig. 12 shows the inorganic EUV PR for better resolution and sensitivity to the EUV light source. Metallic PR enables ultrafine critical dimensions, compared to the conventional carbon-based organic PR. Additionally, in order to prevent pattern collapse during cleaning, supercritical phase was employed, greatly reducing surface tension and leaving superfine dimension patterns intact (Fig. 13).

VI. CONCLUSIONS

Advancement of technology is changing our daily lives in various ways. A rapid surge of data generation and adaptation of artificial intelligence are some of the changes seen in this new 4th Industrial Revolution era. To realize these changes, however, imminent technological challenges, including complexity of network of sensors, unprecedented amount of data generation, and latency of connectivity, must be overcome.

Increase in complexities of semiconductor technology has altered the role of foundry from a conventional wafer manufacturing business to a total solution provider, pushing the sector's limits for broader engagement in the technological revolution. At the same time, many breakthroughs in technology have enabled transistors to be faster in speed, less power-consuming, and smaller in size, allowing a wide range of new applications.

Yet with noticeable challenges on the road, it is to note that global open innovation is essential. While foundry companies may be evolving within, open innovation will enable transformation of the entire ecosystem beyond current state of development to fully reify the 4th Industrial Revolution.

ACKNOWLEDGMENT

The author gratefully acknowledges Jongshik Yoon and SD Kwon for their technical contribution, and Youngchang Charlie Bae and Seungwon Augustine Lee for their industry insight.

REFERENCES

[1] Computer History Museum, "The Silicon Engine,"
[2] IHS Markit, "The Internet of Things: a movement, not a market,"
[3] Vontobel, "Smart Cars in a Connected World," Aug. 22, 2017.
[4] K. Heineke, P.. Kampshoff, et. al., "Self-driving car technology: When will the robots hit the road?," *McKinsey&Co., Automotive & Assembly*, May, 2017.
[5] Samsung, SK Hynix seek upper hand in car semiconductors," Korea Times, May, 2018.
[6] A. Calimera et. al., "The Human Brain Project and Neuromorphic Computing," *Functional Neurology, 28(3)*, pp 191-196, Jul. 2013.
[7] Len Jelinek, "Pure Play Foundry Market Tracker," IHS Markit Research, Jun. 29, 2018
[8] Gartner, "Semiconductor Forecast Database, Worldwide, 2Q18 Update," Jul., 2018.
[9] Rambus & GSA, "Charting a New Course for Semiconductors," Mar., 2016.
[10] D. Reinsel, J. Gantz, J. Rydning, "Data Age 2025: Thje Evolution of Data to Life-Critical," *IDC White Paper*, 2017.

978-1-7281-1988-5/18 $31.00 © 2018 IEEE

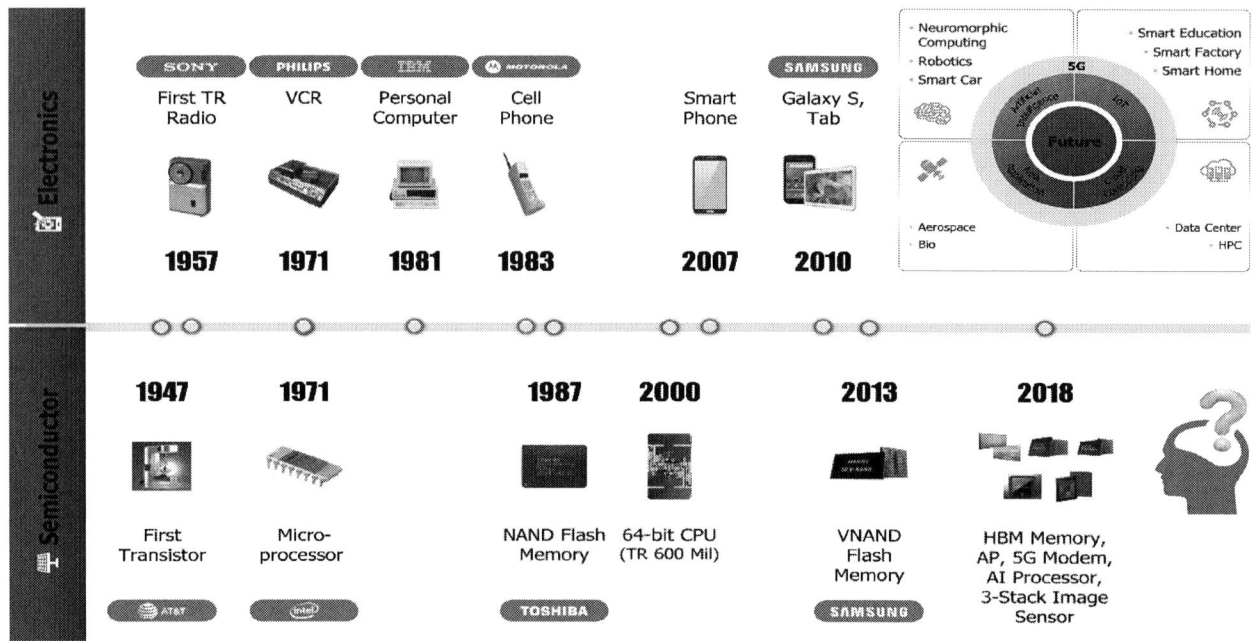

Fig. 1. History of Electronics and Semiconductor.

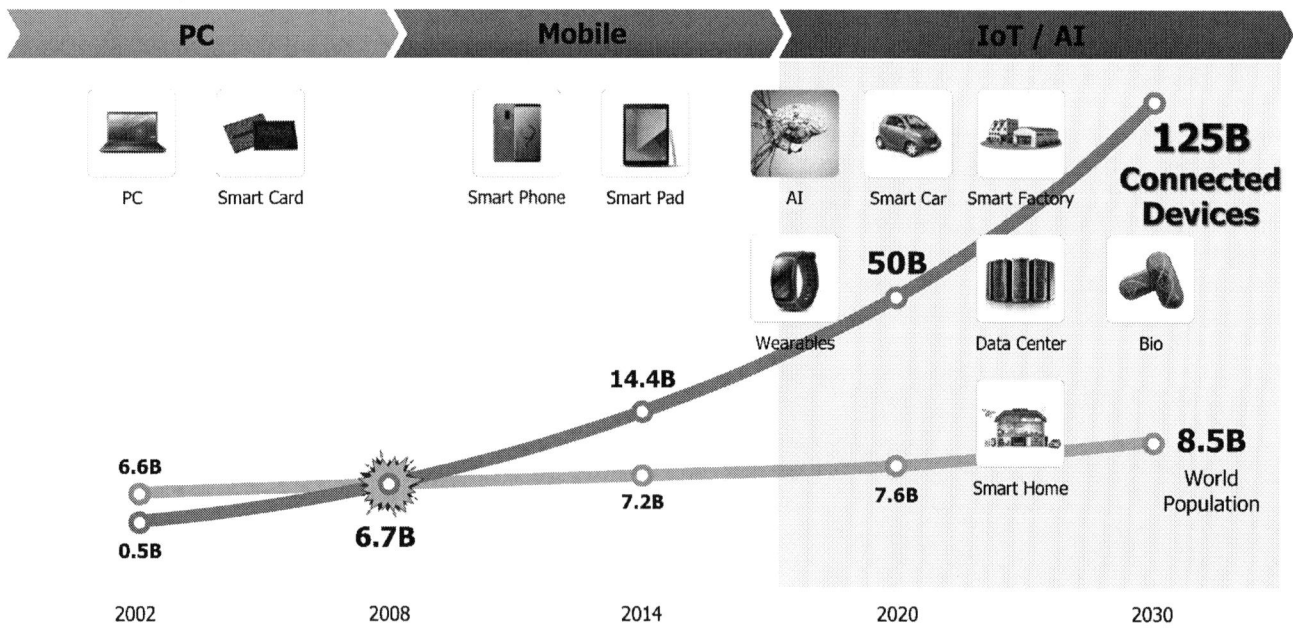

※ Reference : Cisco Global Cloud Index (2016), IHS(2017)
※ B(Billion) : 10⁹

Fig. 2. Explosive Growth of Connected Device in the 4th Industrial Revolution.

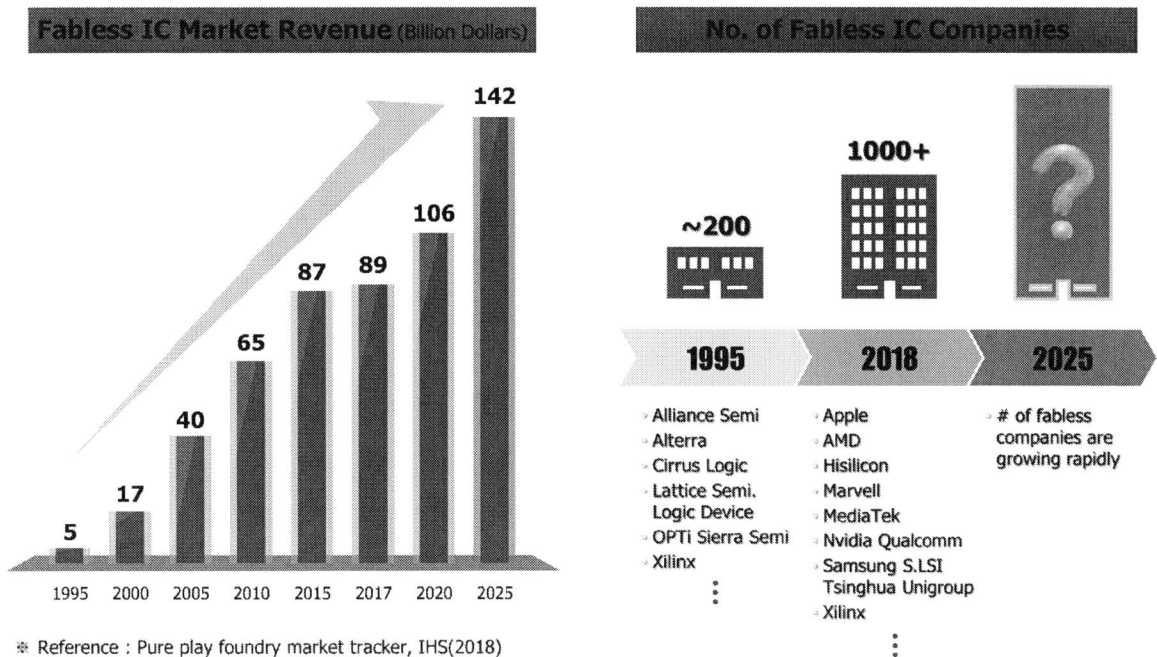

Fig. 3. Growth of Fabless IC Market.

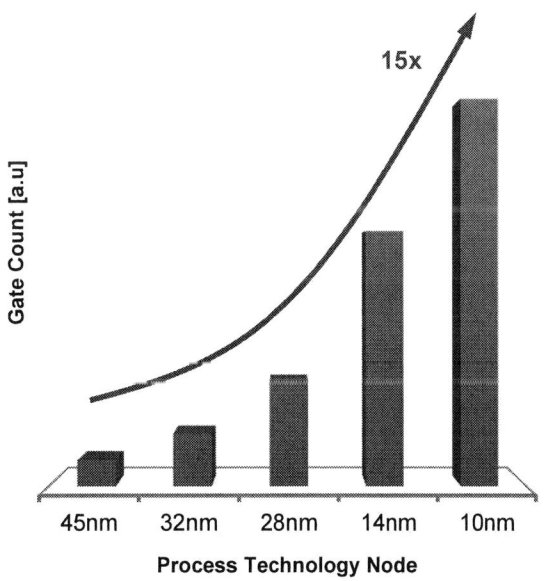

Fig. 4. Design Complexities by Technology Node

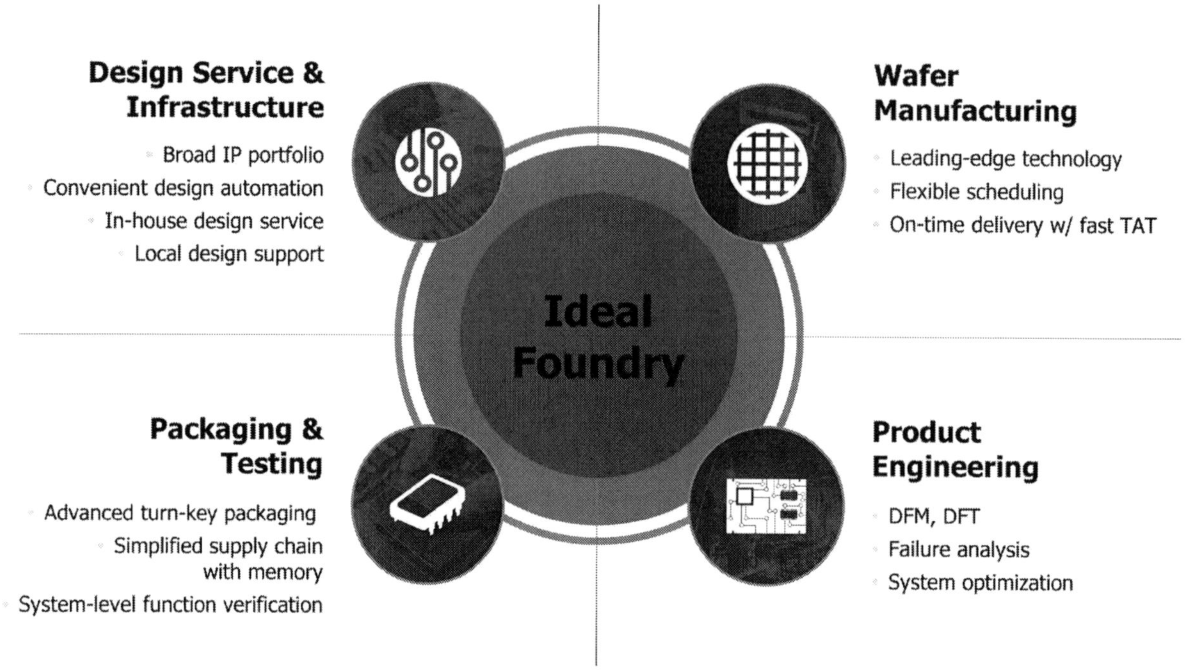

Fig. 5. Expanded Roles of Foundry.

Application	Fingerprint Sensor	Mobile	Server & Network
Structure	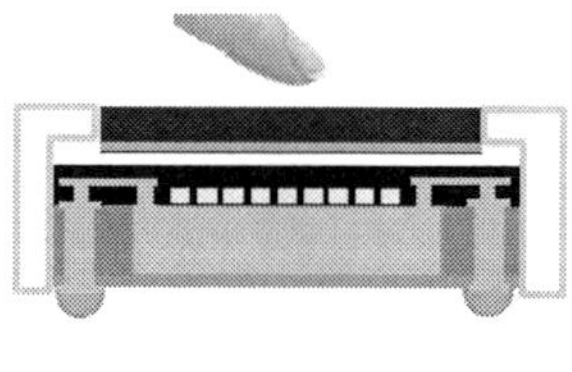		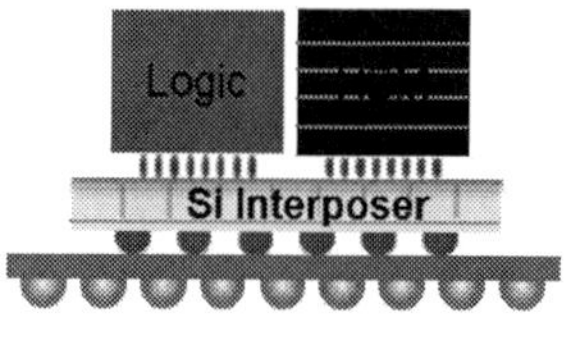
Features	**Fan-out type for Fingerprint Sensor** Shorter sensing distance supporting higher sensitivity	**Fan-out type Package on Package** For thinner chip and better performance	**I-Cube™** HBM and interposer solutions addressing higher bandwidth needs

Fig. 6. Advanced Packaging Optimized for Application Needs

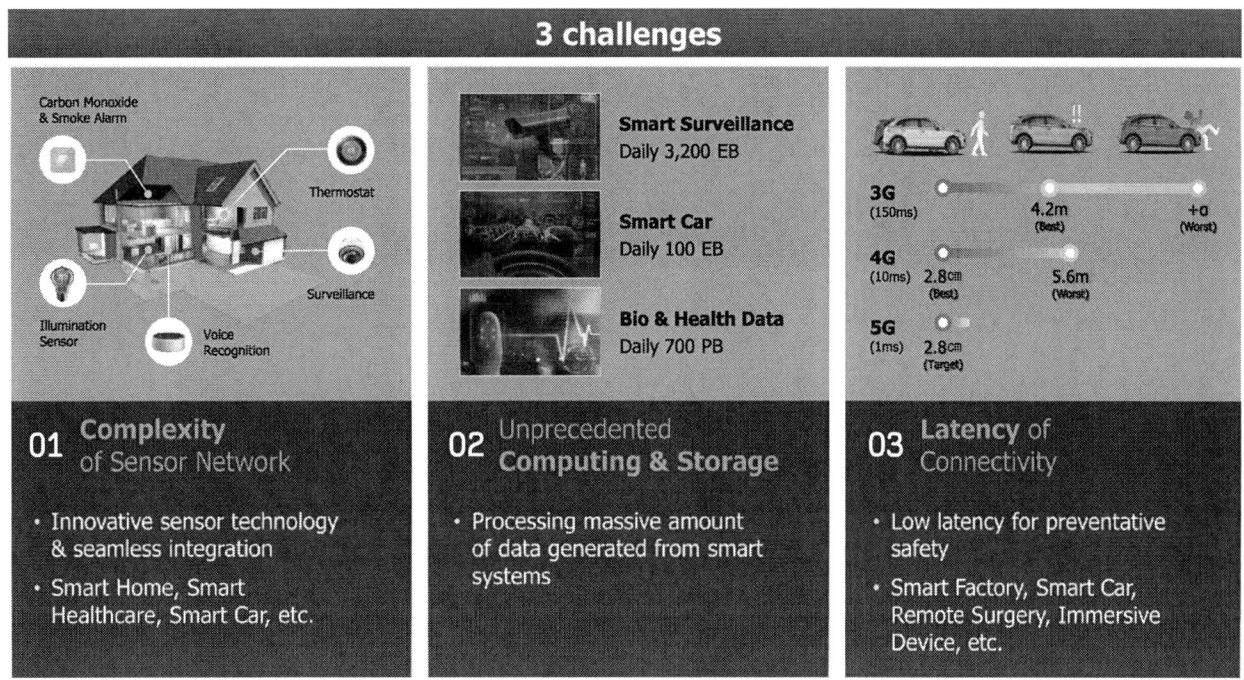

Fig. 7. 3 Key Challenges of 4th Industrial Revolution

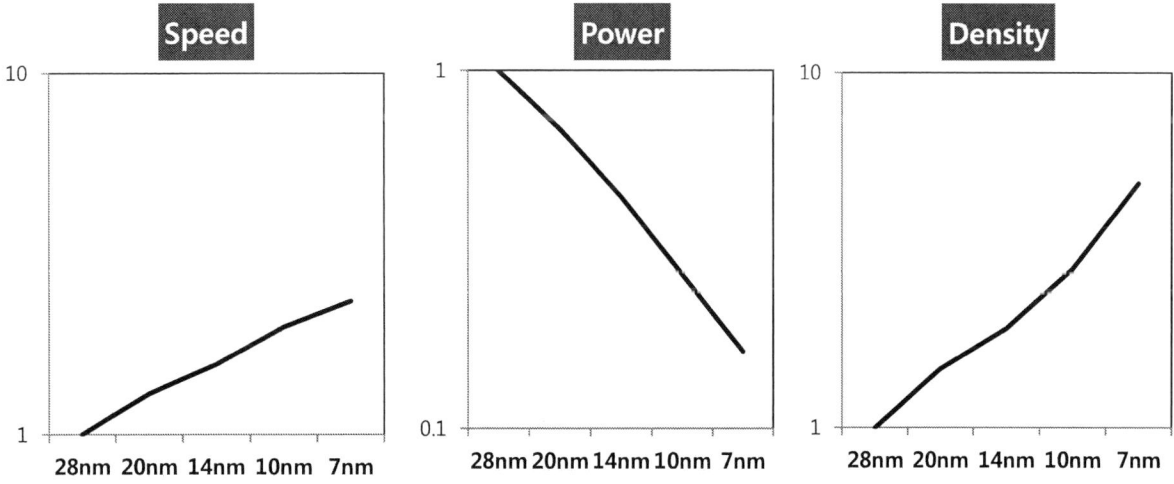

Fig. 8. Normalized Speed, Power, Density by Technology Node

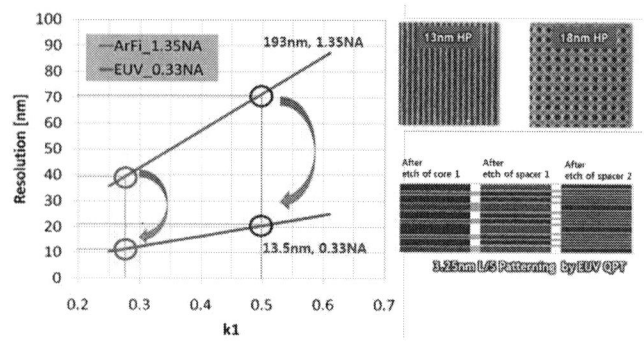

a) Resolution enhancement

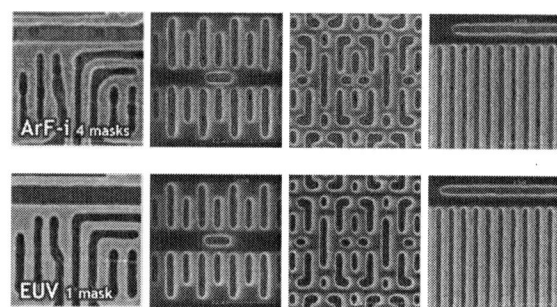

b) Better pattern fidelity

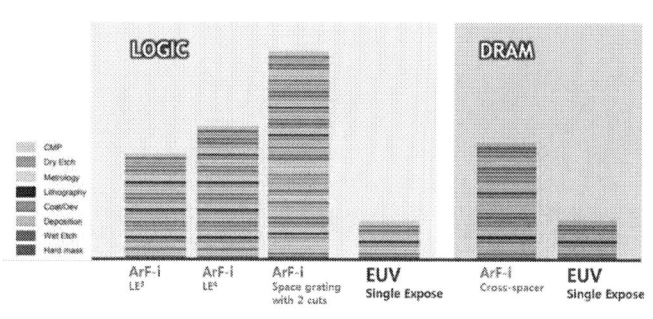

c) Reduction of patterning steps

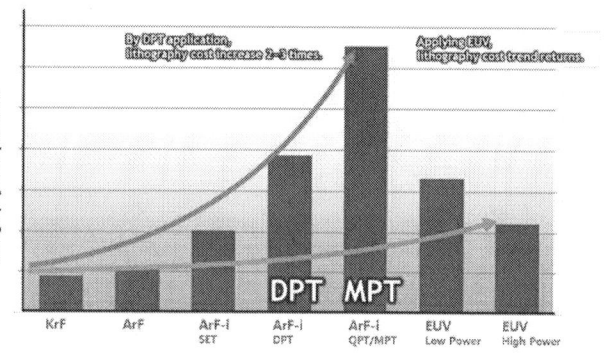

d) Cost effective process

Fig. 9. Advantages of EUV

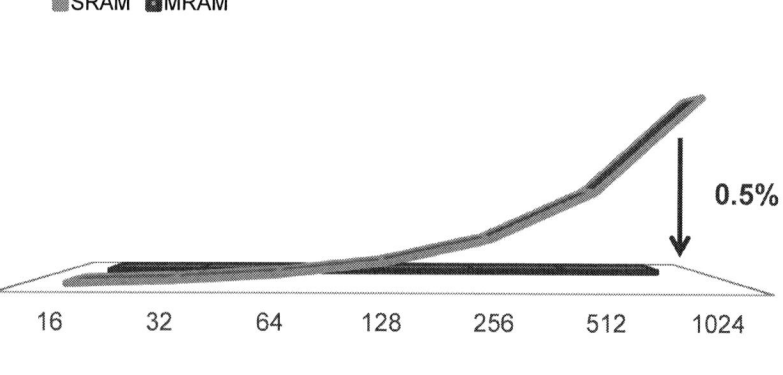

Fig. 10. Power Advantage of eMRAM compared to eSRAM

978-1-7281-1988-5/18 $31.00 © 2018 IEEE

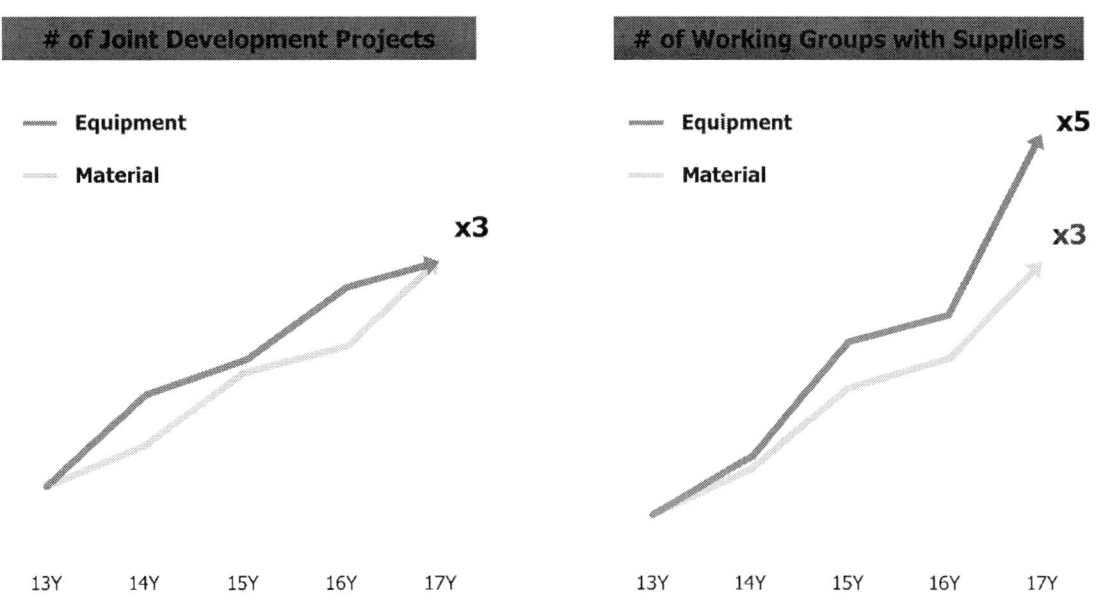

Fig. 11. Increase in Cross-Functional R&D

LOGIC

🧪 Inorganic EUV PR
High Resolution & Sensitivity

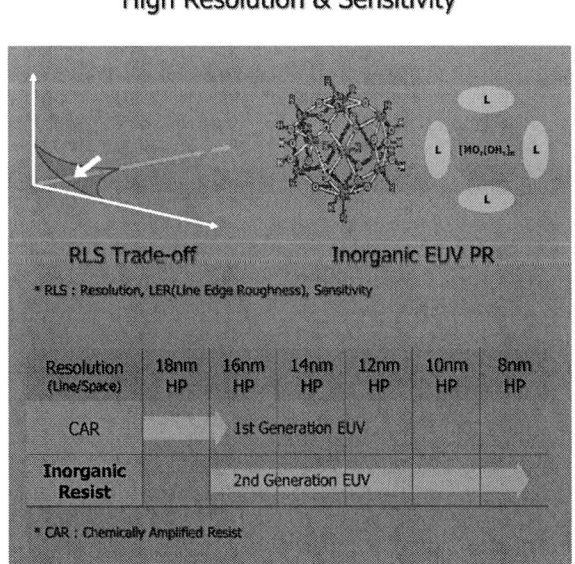

Fig. 12. Inorganic EUV PR

LOGIC / DRAM

⚙️ Super Critical Phase Dryer
Low Surface Tension

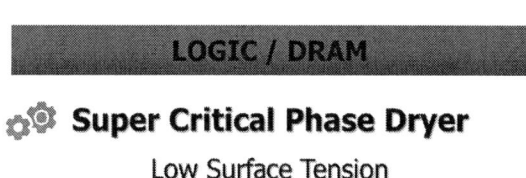

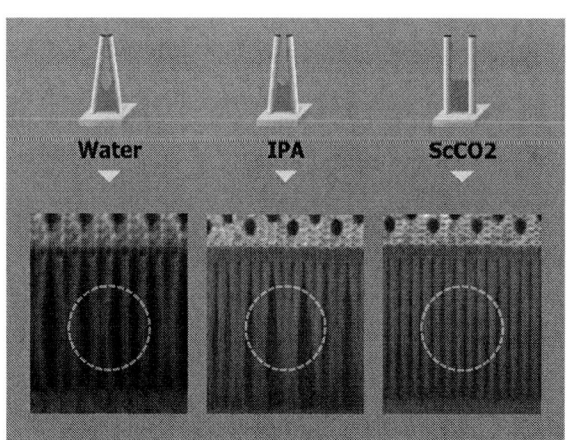

Fig. 13. Super Critical Phase Dryer

		eSRAM	eMRAM		
Area	**Cell** (um^2)	0.120	0.036		
Reliability	**Retention**	Volatile	85'C, 10yr	85'C, 1mo.	85'C, 1day
	Endurance	Unlimited	1.0.E+10	1.0.E+11	1.0.E+12
Performance	**Write (ns)**	< 10ns	25ns		
	Read (ns)	< 10ns	25ns		

Table I. Comparison of eMRAM and eSRAM

Venturing Electronics into Unknown Grounds

G. Fettweis[1], K. Leo[2], B. Voit[3], U. Schneider[1], L. Scheuvens[1]

[1]Center for Advancing Electronics Dresden (cfaed), Technische Universität Dresden (TUD), Germany,
email: {gerhard.fettweis, uta.schneider, lucas.scheuvens}@tu-dresden.de
[2]cfaed, TUD, Integrated Center for Applied Physics and Photonic Materials, Dresden, Germany, email: karl.leo@iapp.de
[3]cfaed, TUD, Leibniz Institute of Polymer Research Dresden, Dresden, Germany, email: voit@ipfdd.de

Abstract—**Electronics is a huge driver for economic success of today's societies. cfaed, the German Cluster of Excellence located in Dresden (Germany), aims at pushing the boundaries of electronics into unknown grounds. This includes not only current electronics but also scientist's and engineer's projection of how the electronics landscape will look like in the future. We pursue an approach that connects all layers from new materials to new system design (vertical) as well as across our Research Routes (horizontal) and ensure coherence through adequate measures. cfaed is centered at one location which, combined with our unique approach, places it above the highly funded Competitive Landscape.**

I. Introduction / Main Research Objective

Electronics is a crucial driver for innovation in modern societies. It is the basis for finding solutions to big societal problems. The innovation in electronics thrives on delivering ever-increasing capabilities that fuel and drive the success of many applications, e.g., wireless communications, machine learning, Industry 4.0, Internet of Things, and the Tactile Internet.

In traditional CMOS, Moore's law, i.e. doubling of the number of transistors per chip every 18 months, has been a good estimate for what to expect. However, as CMOS technology is reaching atomic boundaries, Moore's law is projected to end and new ways must be explored to enable continued innovation. In particular, innovating electronics requires addressing the following challenges: small physical size, speed, energy efficiency, new functionality, self-assembly/-organization, adaptivity, resilience and low cost.

Accurately estimating the performance of future electronics is so crucial that both industry and science communities identified the need to map the potential of its future, leading to the creation of the ITRS and, since 2016, the IRDS roadmap (International Technology Roadmap for Semiconductors/International Roadmap for Devices and Systems, https://irds.ieee.org). It not only outlines the potential of electronics innovation, but also points out the boundaries of what seems achievable based on current knowledge. With valuable insights gained within cfaed's current funding period (hereafter referred to as cfaed-1[1]), we are convinced that we can generate breakthroughs that

push some of the IRDS boundaries, enabling continued innovation in the coming funding period (cfaed-2).

Among many other smaller scientific and structural goals, cfaed's main research objective and vision is to **generate breakthroughs in electronics showing a path beyond current roadmaps**. This cannot be done by extending and optimizing existing ideas but only through exploiting radically new scientific discoveries.

Hence, an interesting research challenge arises. Can electronics be driven beyond the currently known boundaries, and if yes, what is an approach to do this? On the US side, DARPA has come to realize this untapped research ground, and rolled out the heavily funded ($216 Mio.) Electronics Resurgence Initiative (ERI) in 2017 [1]. Within the last 6 years, cfaed has started addressing this interesting research topic, and has come to the conclusion that we should be able to advance electronics "into unknown grounds".

This paper sketches cfaed's research approach and work program. In section II, the *research approach* is presented, i.e. we address the question: *How does cfaed plan to reach its main objective?* In section III, the Research Program of cfaed is presented. In section IV, cfaed's horizontal measures, which are put in place to ensure research coherence, are outlined. Finally, section V concludes the paper.

II. Research Approach

A. Vertical bridging of layers

cfaed has close to seven years of experience in carrying out inspiring collaborative research, where natural & material scientists jointly work with engineering scientists to generate impactful research results, targeted at showing a path for innovation. We learnt that a comprehensive approach, integrating research competency across layers from materials to systems (cf.

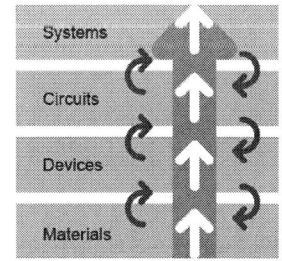

Fig. 1. Multi-layer research approach - bridging the gaps between layers is key to cfaed's success.

---

[1] cfaed-1 has received close to 6 years of funding as a Cluster of Excellence, one of three funding lines of the Excellence Initiative of the German Federal and State Governments, initiated to strengthen Germany's position as an outstanding place for research in the long term and further improve its international competitiveness.

978-1-7281-1988-5/18 $31.00 © 2018 IEEE

Fig. 1), requires tailored strategies and instruments, which will be an important focus for cfaed.

From Materials to Devices: Some of the recent significant advances in new materials have now reached a point that warrants exploring device fabrication and circuit design to build novel information processing systems for potential future applications. For this, one has to understand not only how to further explore new materials and their properties, but also how to engineer them for a targeted solution, i.e. a new device with absolutely innovative features.

From Devices to Circuits: The full potential of a new device and its benefits can only be unleashed when building circuits out of these devices. On the one hand, the understanding of the full potential of a circuit requires the analysis and measurement of its capabilities, and bridging the gap between devices and circuits. On the other hand, this guides the device design as well as materials research in finding better solutions.

From Circuits to Systems: Finally, the targeted novel electronic circuits need to be evaluated by exploring their potential for building new systems, which requires research reaching all the way to the area of computer engineering. During cfaed-1, we were reassured that crossing the layers is the road to success. Drawing from our experience and achievements, we evolved our structure and concept incorporating ideas from new investigators and emergent areas of research.

From our experience within cfaed-1, we discovered that it is beneficial to focus the research on its innovation potential by introducing vertical design targets with a circuit outcome, "V-Targets", as described below. They serve to show the measurable potential of our research results. Not all V-Targets are hardware demonstrators, as they might also be a simulation/emulation. They serve to generate focus and coherence between all involved team members. Details are given in the Route Descriptions in section III.

B. Research Structure

Each selected Route of cfaed is based on (i) research insights gained by our researchers during cfaed-1, from which we identified a great potential to push the frontiers ("wall") of IRDS beyond its boundaries, (ii) having identified an approach

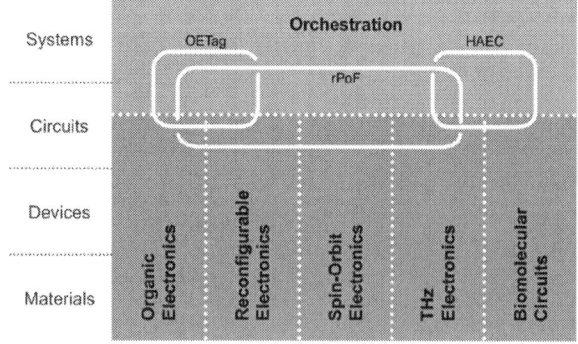

Fig. 2. cfaed's research structure consists of 5+1 Research Routes, linked by Horizontal Integration Targets (HITs)

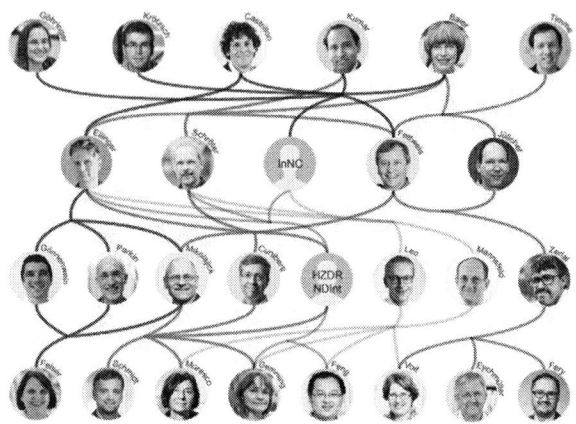

Fig. 3. Multi-layer research approach - bridging the gaps between layers is key to cfaed's success.

to move the "wall", (iii) the right team and favorable environment, and (iv) having identified a measurable benefit for society for advancing electronics.

To optimize our chances, we focus on those research directions that match the four above-mentioned criteria, not withholding others, e.g., quantum computing, which may also possess the potential for advancing electronics beyond IRDS.

This focus led us to establish six Research Routes within cfaed, depicted in Fig. 2. Five Routes envision breakthroughs in enabling new electronics by innovating from materials to circuits. This becomes the basis for the sixth Route to enable new systems.

C. Ensuring Coherence in Research

To foster interplay and coherence between the Routes, as well as to show additional potential when combining the results, we introduce *Horizontal Integration Targets* (HITs). The HITs are crucial for multiple Routes that join forces to achieve common systems goals. They span multiple Routes, as can be seen in Fig. 2. Details on the HITs are given in section IV.

This interplay is facilitated through organizing our research areas into a network of routes which provide the framework for our extensive teamwork in Dresden (see Fig. 3); the scientific achievements obtained in one Route will become applicable to help advance the others, and can be further catalyzed due to the involvement of many investigators active in multiple routes. Our joint research experience has also helped to build a strong team spirit that gives us the adaptability to improve the cluster's structure according to future findings.

In a nutshell, what makes cfaed's approach truly unique is that it firstly, aims at bridging all the involved abstraction layers, starting from materials, devices, circuits, reaching to (information processing) systems, within a new unifying framework and "vertical measures", explained in detail in Section III. cfaed, hence, links a wide range of sciences from material sciences to electrical engineering to computer science. And, secondly, our approach is carried out in multiple research directions by our Routes, which are linked with "horizontal

978-1-7281-1988-5/18 $31.00 © 2018 IEEE

measures". The combination of the two provides fertile ground for mutual exchange and inspiration.

D. Competitive Landscape

Below, we list important research activities that relate to cfaed as a whole.

USA Here we list just the main activities in the US in the area of cfaed: SRC has funded the STARnet program with DARPA and the NRI program with NSF and NIST until the end of 2017, which cover many aspects of cfaed, however distributed across the US. New SRC initiatives, JUMP co-funded with DARPA and nCORE co-funded with NSF/NIST, have started since 2018, with the same model of a distributed US academic research network. Some examples of large US research initiatives and facilities include: (1) DARPA-funded ERI (09/2017, $216 Mio.) tackling new disruptive materials, integration, circuit design and systems architectures, (2) E3S funded by NSF, tackling energy efficiency, centered in Berkeley and MIT; (3) MIT's NanoStructures Lab, focusing e.g., on nanostructuring, self-assembly, and nano-analysis; (4) E2CDA (energy-efficient computing from devices to architectures) funded by NSF and SRC and an outcome of the OSTP Grand Challenge to achieve energy efficiency by co-optimizing emerging devices and architectures; (5) Stanford SystemX funded by industry members, building energy efficient beyond CMOS circuits; (6) on a smaller scale, NEEDS funded by NSF and SRC, focusing on compact models (Purdue, Berkeley, MIT, Stanford); (7) CMU's NanoFab, also facilitating research around novel devices. While there is some overlap with cfaed's mission, many of the above US centers differ. Furthermore, as most are decentralized, strong structural differences apply. In the field of large-scale processing systems, the University of Berkeley has been very influential.

Japan Large programs are funded under CREST, however, geographically distributed. Among others, silicon nanowires and carbon nanotubes are considered. However, the projects lack a system component and are mainly limited to the materials level.

China The Universities in Beijing (e.g., Tsinghua University, Beijing University), Dalian, Chengdu, and other cities have a special focus on hiring talent in areas covered by cfaed. We shall expect research outcomes to become widely visible soon.

Europe Besides the large institutions IMEC, Leti, Tyndall, large research programs are the EU GRAPHENE Flagship, a large center at the University of Manchester. In Germany, two Centers of Excellence (in Erlangen and Munich) focus on materials and functionalization, though not covering devices and circuits. The very large French nano programs mainly focus on research to execute the IRDS. The Swiss Nano-Tera program is focused on health and environment applications, and is in the ramp-down phase. The Swiss BRIDGE program has a broad focus not overlapping with cfaed. The SiNANO program, with many high-profile members, focuses on nanoelectronic research doing both, "More Moore" and "Beyond Moore", however, scattered all over Europe.

In summary, our main differentiator is:

1. A comprehensive vertically integrated approach, spanning from materials to systems.

Furthermore, we are confident that cfaed uniquely demonstrates the combination of the above with the following six key ingredients:

2. One location, making it easy for researchers to meet, exchange, and collaborate;
3. An internationally recognized excellence in research, attracting the best researchers;
4. The local concentration of R&D expertise, attracting scientists and industry;
5. Experience in innovation and proven record of success, in particular, bridging from basic sciences to creating 60+ start-ups;
6. The "Dresden Spirit" of collaboration and joint research, including open sharing of labs;
7. A team with the ambition to push the boundaries of the IRDS roadmap.

III. Research Program

In this section, we give an outline for each of our six Research Routes, clearly stating the identified walls, our approach to overcome them, the benefit we expect from a solution, our recent success that makes us believe in our approach, the measures applied in the Route to reach our goals, and the answer to the question "Why Dresden?".

A. Organic Electronics Route "OE"

Wall Although with our results from cfaed-1 we are able to produce cutting-edge single devices, we cannot yet produce on-demand printable, low-cost circuits. There reason for this is a lack of processes for circuit-grade device fabrication (throughput and uniformity) and materials, more specifically, insufficient carrier mobility.

Approach In order to successfully and rapidly move towards our V-Target, our research is focused on several key areas; we will investigate new materials with unprecedented performance and novel vertical transistor device principles. These materials and device technologies will be employed in fully in-

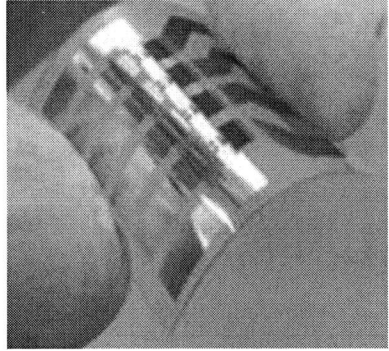

Fig. 2. photograph of a flexible PBT on a PET substrate

tegrated devices and complex organic circuits. One part of our materials research also specifically focuses on materials for printable transistor circuits. We plan to integrate lateral and vertical high-performance OFET devices into circuits in order to combine them with sensors such as our novel near-infrared organic sensors, and to ultimately demonstrate their integration in complex organic devices. We will also employ a new "system-level informed" approach to our research. Instead of developing device technologies without feedback regarding their "real world" (i.e. circuit level) utility, we will develop models and design circuits based on those models that allow us to evaluate our technologies in circuits of a complexity that, while we cannot yet fabricate them, will steer our device and material design. We believe that this approach will allow us to increase the practical impact of Organic Electronics research and to foster closer collaborations with the companies in the Organic Electronics Saxony network, or even kick-start new companies based on our technologies. We will benchmark our best single-device technologies in a far more realistic way than is currently possible.

Benefit Organic Electronics may become the basis for immediately, on-demand manufacturable organic circuits in the LSI range. This enables new types of electronic circuits, being lightweight, inexpensive, mechanically flexible, stretchable and energy efficient, that could be fabricated on demand through printing at small business places or even at home.

cfaed-1 success During cfaed-1, our scientists generated a number of world-leading results, especially in the area of single device design and electrical performance. Among these are a novel high current density vertical transistor with unrivalled current density (high kA/cm^2 regime) and transit frequency (>30MHz recently achieved) [2], and the first demonstration of an organic FET operating in inversion [3]. Furthermore, OFET devices and conducting polymer materials with record high mobility values and conductivities [4], and proof-of-concept devices with new materials such as 2D conjugated polymers were demonstrated [5-6]. Building on these significant advances at the material and device levels, OE will push towards demonstrator circuits containing up to hundreds of devices.

V-Targets OETag (cf. HITs): A unique, multifunctional, fully organic and printed demonstrator tag is explored combining sensor functionality, computing, wireless communication, positioning and power supply. We will design the organic optochemical sensors, which are used to enable the chemical analysis of materials. The sensed data can be processed by the printed processor unit comprising up to 10,000 transistors.

Why Dresden? Our team in Dresden has proven to be world class throughout recent years. We are the #1 location for organics in Europe and we have a proven track record in successful spin-offs.

B. Reconfigurable Electronics Route "RE"

Wall Today's electronics for computing rely on static devices and hardware. In arithmetic-logic-units (ALU) that build the basis for processors, for instance, information is routed to fixed functional blocks resulting in large inactive chip areas and considerable routing resources. In addition, a rigid separation between hardware (function) and software (command) results in a latency bottleneck as information is exchanged.

Approach Our approach of breaking those walls is to rethink the complete computation chain by enacting a flexible reconfiguration of hardware and enabling software defined hardware across all functional levels. We explore the unique window of opportunity that opens with reconfigurable Field Effect Transistors (RFETs). The functionality of the individual transistors is boosted from the current rigid and limited on/off switch function to enabling flexible reconfiguration between n- and p-type FETs, resulting in multiple deterministic functions as well as distinct logic operation and memory states.

The foundation of the RE Route is built on the successful work from cfaed-1 on Si nanowire RFET devices and circuits. The RFET merges two fundamental transistor types - electron- (n-channel) and hole- (p-channel) conduction - into one universal type of transistor allowing a flexible reconfiguration between either function [7] by a dedicated electric non-volatile program signal (cf. Fig. 5). Si and Ge nanowires are chosen as the vehicles to realize RFETs as they deliver the ultimate gating behavior [8] and show low integration barriers towards possible future implementation with the most aggressively scaled CMOS fabrication technology being currently developed for the sub-10 nm node at the industry level [9-10].

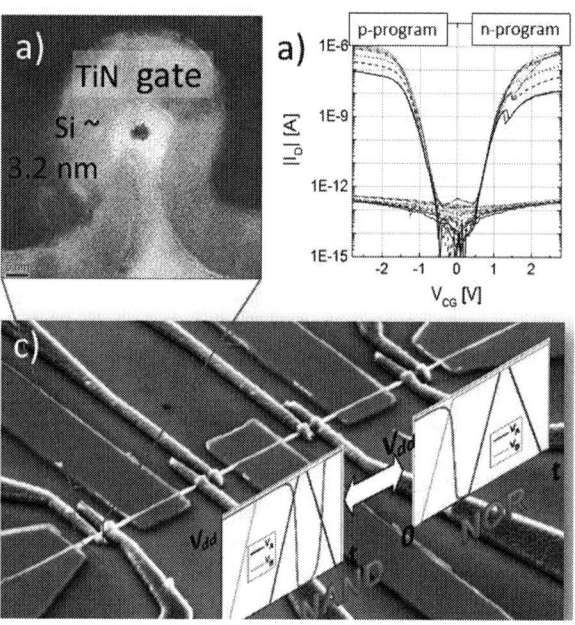

Fig. 5. Reconfigurable Si nanowire transistor. a) cross-section of device with omega gate and TiN electrode. b) Subthreshold transfer characteristics. c) 4T reconfigurable NAND-NOR cell with full output swing.

Benefit We expect benefits in substantial savings of power consumption, integration density and latency at the circuit and system levels as compared to CMOS benchmark systems. Inherent reconfigurability shall be exploited towards the design of hardware secure and resilient circuits as the functionality of these systems, i.e. the transistor's polarities, cannot be reversed engineered. With the fusion of memory and logic as well as non-volatile state memory enabled by ferroelectric HfO$_2$-based

978-1-7281-1988-5/18 $31.00 © 2018 IEEE

gate stacks [11], zero-boot-time systems enter the realm of possibility. Digital sensing, i.e. sensors that are able to detect quantum events and output these events digitally, renders ADCs useless and therefore save chip area and power consumption.

cfaed-1 success The RFET results achieved in cfaed-1 have significantly shaped the field of reconfiguration of logic operation states by proposing and maturing the Si nanowire RFET. The RFET concept was first demonstrated with bottom-up Si nanowires [7]. Important experimental and theoretical advancements in nano-material science involving solid state reactions and strain engineering of nanowires [8] were implemented to yield a scalable and integratable device concept for the adjustment of drain-current symmetry between n- and p-program [12]. The drastic disparity of 10x otherwise found was lifted. This breakthrough has made cfaed symmetric RFETs the only devices of its kind to facilitate practical complementary circuits with runtime reconfigurability. The symmetric bottom-up technology was recently successfully transferred to a top-down SOI platform with omega shaped gate to enable Si nanowire based RFETs for device prototyping and circuit demonstration with a higher yield, and to proof compatibility of our technology with modern manufacturing [13]. Device performance limitations were initially identified [14] especially for specific realizations, e.g., the planar one in accordance to literature [15]. In response to that, performance boosting and reduction of dynamic power consumption strategies were found and demonstrated with bottom-up Ge nanowires and high-k / metal gate stacks and advanced gate architectures [16], [14]. To increase device functionality and to enable in-memory computing, we devised the multi-independent gate RFETs, adding more logic inputs without comprising on-conductance [17], [13] and included non-volatile and multi-bit operable reconfiguration features [18-19]. Inherent device reconfigurability was exploited at the circuit level, defining a library of fine grain reconfigurable logic cells, e.g. such as NAND/NOR/MIN or XOR/XNOR runtime reconfigurable functions [17], [14]. These efficient logic representations have seeded logic circuit opportunities as in multi-bit adders [20] and multiplexers [13]. A logic and physical synthesis flow from the application behavior level to the layout was set up [21] enabling the evaluation of the impact at the system level in terms of chip area, performance and critical delay paths.

V-Targets Our Route's V-Targets are to demonstrate disruptive innovations by bringing new fully hardware-reconfigurable ALUs, zero-boot-time systems as well as digital sensor circuits with ultimate resolution that cannot be executed with CMOS.

Why Dresden? Dresden is cradle and world leader in RFETs and HfO_2 based ferroelectrics. We have proven intra-disciplinarity starting from material science to disruptive device and circuit innovations and yielding into modern logic and physical synthesis flows. The local facilities together with the RE team make this activity in Dresden world-leading.

C. Spin-Orbit-Torque Electronics Route "SPOT"

Wall A major roadblock to advancing CMOS based computing systems are (1) scaling of existing memory technologies beyond the 10 to 20nm technology nodes (depending on the type of memory), (2) limitations on the size of these memories due to the innate physical structure of these memories, and (3)

the need for a complex hierarchy of memory types that typically trade-off lower densities to achieve higher performance (speed and latency).

Approach Racetrack Memory (RTM) is a novel spintronic memory [22] that has a radically different operating principle than any charge-based memory – current or proposed – and which could allow for such a massive increase in high performance memory capacity whilst also offering non-volatility, and a reduction in the complexity of today's memory hierarchies. These unique properties can be realized by novel spin-orbitronic concepts, discovered only in the last three years. In particular, spin-orbitronics makes possible the motion of (anti-ferro-)magnetic domain walls in nanoscale structures at very high speeds exceeding 1km/s [23], thereby overcoming the limitations in propagation speed – and in magnetic domain size - that jeopardized the original RTM proposal [22][24]. Thus, Racetrack Memory now has very high potential as a unique memory concept.

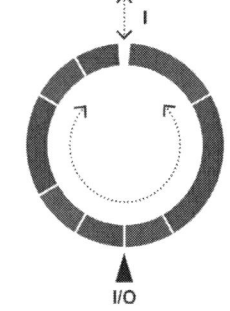

The basic principle of RTM is that data is encoded in boundaries – the domain walls – between regions of opposite magnetization (red or blue regions in Fig. 5) in a magnetic nanowire that forms the racetrack. The presence or absence of the domain wall corresponds to a "1" or a "0". Most importantly, the domain walls are mobile and the data – a series of "0"s and "1"s – are moved synchronously, to and from, around the racetracks by spin-orbit torques generated via current pulses fed into the racetracks. Using vertical racetracks that have a cross-sectional area of just a few nm^2 and that are tall enough to accommodate 100 to 500 domain walls, RTM has the potential for memory capacities far exceeding any known solid-state memory today. In simple terms, RTM thus is the equivalent of a high capacity hard disk drive, but on a chip and with no moving parts. Moreover, RTM is innately 3D, with its major active

Fig. 5. Racetrack memory concept: A series of magnetic domain walls (boundaries between magnetic regions (red and blue) of opposite magnetization) are moved by current pulses (I) to an I/O device for reading and writing the domain walls.

components, the racetracks, being vertically oriented, unlike, for example, vertical FLASH which is a stacked 2D technology that limits improvements in density.

Benefit Since RTM can uniquely trade off speed and latency, RTM has the potential to replace several of today's memories in one technology. This makes RTM attractive for future rPoF architectures (see HITs, section IV). Furthermore, RTM has the potential to increase the memory capacity close to the processor by 1 or 2 orders of magnitude and thereby significantly accelerate processing operation.

cfaed-1 success Over the past five years, major discoveries have led to a new major subfield of spintronics - spin-orbitronics - in which the phenomenon of spin-orbit coupling has led to novel and highly efficient means of creating pure spin currents.

978-1-7281-1988-5/18 $31.00 © 2018 IEEE

V-Target Racetrack Memory: Exploiting our recent spin-or-bit-torque breakthroughs and advancing beyond, we shall be able to monolithically integrate circular magnetic racetracks that will allow for domain wall bits that that can be driven in circles. Hence, one single read and write circuit shall be integrated to sequentially read/write a complete racetrack of, e.g., 128 bits. This shall allow for a breakthrough: monolithically embedding storage within circuits with a memory density of orders of magnitude beyond today's memories.

Why Dresden? Prof. Stuart Parkin, the inventor of RTM and 2014 Millenium Technology Prize winner, heads the team. We are leaders in spintronic materials and phenomena and in sprintronic circuit design.

D. Terahertz Electronics Route "THz"

Wall THz signal power generation at room temperature by either electronic or optical components presently suffers from poor device performance. Particularly, between 0.5–10THz the output power of both optics and electronics drops to very low levels. Similarly, signal detection in this THz gap (cf. Fig. 6) is plagued by high loss in the frequency conversion process.

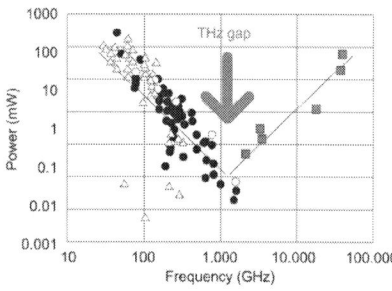

Fig. 6. The THz gap is generally defined by frequencies between 300GHz to 30THz (i.e. by wave lengths between 1mm and 10μm). (image: [25])

Approach Fig. 7 shows the cross section of an advanced npn HBT along with its relevant vertical layers and the carrier flow. According to the present roadmaps [26-27], the most critical issues limiting HBT performance are: increase of contact and series resistances with vertical and lateral scaling; vertical spreading of highly doped layers such as base and emitter; formation of parasitic energy barriers; leakage; thermal breakdown. Thus, this Route pursues two approaches in parallel: (i) circuit and system design using advanced HBT technology and (ii) exploring HBTs with new architectures and materials.

For both the transceiver and radar frontend, THz circuit and system concepts will be explored utilizing experimentally calibrated HBT models that enable circuit design at the process performance limit. For generating high power THz signals with high tuning range (i.e. bandwidth) we propose a novel quadrature oscillator topology where combining the phase shifted paths leads to frequency quadrupling. Reuse of inherent injection locking for frequency tuning will maximize oscillation frequency. Balancing peak oscillation frequency and tuning range with minimum phase noise will be aided by theory development. For THz transceiver-based communications systems, carrier recovery concepts will be investigated.

Sensitive detectors combined with high-power THz tuneable signal sources enable compact and extremely precise μm-scale resolution radar systems not being sensitive to dirt and dust. The requirements for realizing THz detectors are, among others, high responsivity, low noise, and lowest possible intrinsic and parasitic capacitances. Achieving these requirements favors 2D materials as well as 1D materials (preferably CNTs) from the aspect of their fundamental properties [28].

Possible 2D materials include, e.g., BP, GeS and other binary metal VI or IV-VI-compounds, which will be first screened for a high in-plane mean free path (at low field) by quantum chemical simulations. Promising candidate materials will be synthesized and experimentally characterized. Materials turning out to be suitable will then be integrated in actual detector structures with channel lengths

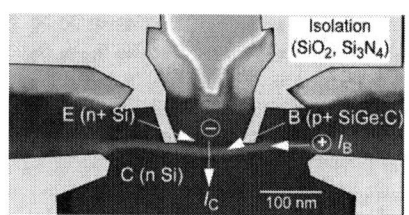

Fig. 7. TEM cross section of a 0.7THz SiGe:C npn HBT showing the relevant device regions and current flow directions.

below the carrier mean free path, which lets expect a strong interferometric enhancement of the sensitivity. Once a suitable channel material has been found, choosing proper contact materials and finding fabrication procedures for minimizing the contact resistivity will be addressed.

Benefit Using ground-breaking approaches (e.g. specially tailored material layers) towards transistor, detector, and circuit design, our goal is to generate at least 1 mW of output power at 1 THz and room temperature. This shall enable a long list of new applications, including, among others, non-destructive material inspection (safety), extremely advanced medical diagnostics, radar for autonomous driving, and extremely high data rate wireless communications of 1 Tb/s and beyond.

cfaed-1 success HBT transistor modeling related research has led to the SiGe HBT roadmap (ITRS/IRDS) [26], while driving the highly successful EU projects DOTFIVE and DOTSEVEN (as Technical Project Manager) [29] has resulted in the most advanced SiGe:C HBT technology worldwide at IHP (Innovations for High Performance Microelectronics, Frankfurt/Oder/ Germany). The heterojunction bipolar transistor model HICUM has been an industry-wide standard since 2003 and is worldwide available in all commercial circuit simulators. High-frequency CNTFET expertise was documented in [30].

First-principles simulation capability (Gemming) has been utilized in cfaed for explaining experimentally observed conductivity changes in nanostructures [31][32] and can be applied to tailoring materials towards specific in- and out-of-plane transport properties. Antenna structures have been developed for Graphene based detectors working at frequencies ranging from 0.4THz up to 400THz [33] and structures for probing in- and out-of-plane transport properties have been fabricated [34].

V-Targets An oscillator with ≥1 mW output power: We will start our circuit design with this basic component for signal generation. Integrated radar front-end/wireless communications transceiver: Exploiting the new device and circuit design methodology jointly with our superior transistor modeling capabil-

978-1-7281-1988-5/18 $31.00 © 2018 IEEE

ity, a THz emitter/detector "transceiver" and on-chip radar system for μm distance resolution shall be designed to operate at room temperature. For the first time, this shall enable commercially viable THz imaging, radar, as well as wireless communications.

Why Dresden? In our team in Dresden, we bring together long-term and complementary experience in devices, circuits, materials, fabrication, and systems.

E. Biomolecular Circuits Route "BIO"

Wall The rapid advances in electronics technology pose major walls with respect to the integration density of heterogeneous components, reliable operation despite unreliable components, and energy efficiency of both fabrication and operation. Biological systems - through billions of years of evolution - discovered unique solutions to these challenges on the level of molecules, cells and cellular networks. These solutions are fundamentally different from conventional electronic engineering approaches and often outmatch their performance in terms of size, energy efficiency, and resilience by orders of magnitude. The vision of this Route is to exploit these solutions for advancing electronics.

Approach Within the area of DNA-origami based optoelectronics, DNA-origami structures will direct the energy-efficient fab-free assembly of nano-objects into optoelectronic devices/circuits with massively reduced size and superior bandwidth. After assembly, DNA templates can be removed and devices/circuits can operate in a dry environment for on-chip computations. The produced optoelectronic devices/circuits are 100-1,000 times smaller than currently used elements (e.g. waveguides, modulators, antennas).

We will also tackle dynamic self-organization of biomolecules in wet environments exploiting two complementary strategies:

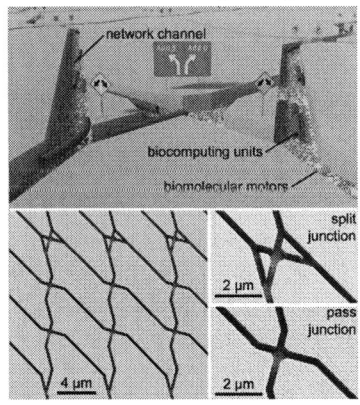

Fig. 8. Schematic of a programmable molecular-motor based computer

Our approach "Programmable molecular-motor based computer" (cf. Fig. 8) harnesses biomolecular transport to solve specific computational problems with substantially increased energy efficiency. Another approach, "Robust biomolecular signal processing" (cf. Fig. 9), exploits signal processing by collectively interacting biomolecules and bridging it with wet multimodal sensing to yield a robust biomolecular near-sensor processor that directly senses multichannel input in wet environments at few-molecule resolution. Of course, these systems shouldn't be standalone but integrated with conventional CMOS technologies providing guidelines for their rational design and translating generic bio-concepts to other routes. As functional DNA-origami structures can be op-

erative in biomolecular signal processing systems and in optoelectronic functional elements, they will serve as an interface between 'wet-state' and 'dry-state' electronics. Thus, the molecular sensitivity and collective processing capabilities of biochemical signaling will enable true bio-hybrid processing

Benefit If we succeed with our approach of DNA-origami based optoelectronics, we expect 100x denser high-bandwidth optoelectronics. Furthermore, our Route shall result in enabling

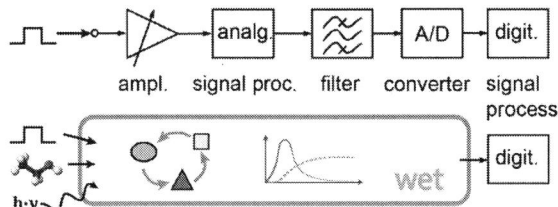

Fig. 9. The biomolecular near-sensor processor operates next to biomolecular sensors and simultaneously replaces sequential processing by standard engineered routines such as amplifier, analog signal processing, filter and A/D converter. It directly provides robust decision signals in response to multichannel inputs, even at few-molecule resolution

robust bio-molecular sensors and processors for new applications, e.g., in food and health, and, in particular for complex multi-channel sensing and low-energy processing under harsh energy constraints.

cfaed-1 success We were first to use DNA origami for templating nanoparticles into plasmonic waveguides [35-36] and motor proteins for low-energy computing [37]. We discovered fundamentals of molecular decision making of cells [38], now to be used for electronics.

V-Targets Plasmonic waveguides: Innovating on DNA templating, we shall enable 100x density improvements on optical chip-to-chip interconnects. Bio-hybrid computing: Innovating on our synthetic motor protein circuits, we plan to show 100x energy improvement for difficult computational tasks. Multi-channel biomolecular sensing: Incorporating our understanding of decision-making in a bio-sensor, we will build first bio-molecular hybrid sensing and signal processing circuits.

Why Dresden? Our team spans all layers from materials to systems level and includes scientists from all necessary research fields: biology, physics, materials, and engineering.

F. Orchestration Route "ORCH"

Our previous five Routes reach across layers from materials to circuits, exploiting fundamentally new concepts. Above the circuits layer, in ORCH, we envision a radically new system architecture with hardware reconfiguration everywhere, blurred boundaries between computing and memory, an extreme dense storage, acceleration via unconventional computing approaches, and unprecedented communication throughput. To make such a complex system usable, we jointly co-design novel hardware and software architectures. To this end, we follow a model-based approach that enables a systematic design space exploration. This is fascinating new ground, open for exploration with a perspective of adaptivity and efficiency, far beyond those of today's computing systems.

Wall Two walls need to be tackled. (1) Ever-widening HW-SW gap: The rapid increase in software (SW) complexity with numerous layers of abstraction and the pace of innovation in HW architectures make it difficult to exploit the full potential of computing systems. This is aggravated by the high disruption potential of novel cfaed technologies. Without jointly architecting HW and SW, the gap may become unsurmountable, rendering technological innovation of no use for applications. (2) Systems design wall: Classical architectures, including von-Neumann approaches, comprise separate computing, memory, interconnect and sensing components. This "systems design wall" hinders cfaed's materials-inspired innovations, which enable breaking classical component boundaries, to come to full fruition.

Approach In ORCH, we propose a hardware and software system-level research to unleash the full potential of new technologies developed by the materials-inspired Routes, in particular: (OE) We conceive a design framework aware of fabrication constraints such that organic electronics can be targeted efficiently. (RE) We exploit reconfigurable electronics via a new kind of reconfigurable fabric that allows switching among instruction set architectures to adapt to the workload at runtime. (SPOT) We develop ways to embed reconfigurable racetrack memories within as well as around the reconfigurable logic fabric. (THz) We contribute to integrating beyond 10 Tb/s I/O bandwidth enabled by THz transceivers. (BIO) We show a path to systems concepts exploiting dense optical interconnects and low-energy motor-driven filament accelerators with a new dimension of heterogeneity for unchallenged new applications.

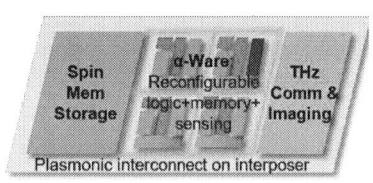

Fig. 10. α-Ware sketch

Our scientific approach is based on a system-level model-based HW/SW co-design methodology that imports early device models from the materials-inspired Routes and exports system-level insight to help steer development in physics, chemistry, material sciences and circuits. The main characteristics will be extracted from lower-level models of future components and translated into higher level operational models with quantitative annotations. The symbiotic effects of combining new technologies into a system will be investigated and optimized by dependability analysis, simulation and probabilistic model checking. This will allow us to perform HW/SW co-design and systematically rethink HW and SW architectures.

We envision a novel fabric with unprecedented reconfigurability and dissolved boundaries between memory, logic and sensing for unchallenged new applications, which we call **α-Ware** (cf. Fig. 10). With α-Ware, we leverage the technologies of the materials-inspired Routes: (1) new device reconfigurability (RE), (2) memory-in-logic (RE), (3) extremely fast local RTMs and dense storage-class RTMs (SPOT) as well as new register files, (4) wet computing for analog near-sensor computing and acceleration of computationally hard problems (BIO), and (5)

extremely fast and energy efficient interconnect (BIO) in collaboration with HAEC. α-Ware will break the systems design wall and open up new possibilities for HW/SW architectures. For example, fast logic reconfiguration will allow changing the data path close to memory, instead of moving data to the right data path, whereas RTM may flatten the memory hierarchy, making architectural innovations obsolete that hide memory latencies (e.g., certain types of caches or HW multi-threading). This, in turn, will lead to a new system SW, e.g., (1) simplified memory management by the Operating System (OS), (2) novel resource management for reserving areas of the α-Ware while ensuring isolation, and (3) novel languages and compilers that not only program but also define the architecture. These kinds of system SW considerations will use a systematic knowledge-based representation of system resources at design-, compile- and run-time. Only by jointly developing HW/SW architectures around cfaed technologies will we unleash the device breakthroughs at system level, disrupting the otherwise incremental innovation paradigm.

Benefit The expected benefits consist of measurable improvements for end-user applications, unleashing their full potential to the application cause. cfaed will not only deliver exciting devices and circuits but will also demonstrate their impact on real applications. Beyond the benefits of single technologies, we expect systems research in ORCH to lead to synergistic effects, e.g., combining embedded RTMs close to reconfigurable logic. Novel resource management at different levels such as intelligent data placement to RTMs will make SW considerably more efficient. Compared to classical SW systems for certain application domains (e.g., big data or wireless communications), we expect orders of magnitude improvement in selected figures of merit such as energy efficiency, area or performance.

cfaed-1 success We developed the concept of a wildly heterogeneous system design, incorporating software design concepts, operating systems, kernel scheduling for heterogeneous multi-processor systems as well as a resilient network-on-chip [39]. All was proven by designing and testing multiple "Tomahawk" chips fabricated in current CMOS [40].

V-Targets α-Ware: We will build multiple instantiations of our radically new processing platform ("α-Ware") and benchmark it using system-level simulation techniques. At the system level, this will allow us to quickly observe the impact of new software architectures, and of changes at the device and circuit level from Routes A-E (captured by their V-Targets). Early system-level analysis within cfaed will enable large leaps in technology viability and adoption, disrupting the otherwise incremental innovation paradigm.

Why Dresden? We bring together a strong interdisciplinary team with a proven track record, very strong in HW/SW design, theory and applications.

IV. HORIZONTAL INTEGRATION TARGETS "HITS"

To foster interplay and coherence between the Routes, as well as to show additional potential when combining the results, we introduce "Horizontal Integration Targets" (HITs, cf. Fig. 2). The HITs are crucial for multiple Routes that join forces to achieve common systems goals.

978-1-7281-1988-5/18 $31.00 © 2018 IEEE

G. Reconfigurable Processor-of-the-Future (rPoF)

Today, many different kinds of processors exist, e.g., DSPs, GPPs, MPUs, GPUs, just to name a few. Also, processor design frameworks exist that allow to add acceleration and custom instructions to make processors more efficient for a specific application area. We envision a whole new processor platform to be developed by exploiting the reconfigurability enabled by the RE Route, embedding Racetrack Memories of the SPOT Route, and Tb/s wideband I/O as enabled by the THz Route. By designing a "reconfigurable Arithmetic Logic Unit", for example, an rPoF can reconfigure every single clock tick from one instruction set to another, enabling unheard flexibility, and thereby fueling ORCH's research goals. Finally, the rPoF shall also be implementable in a non-reconfigurable way that incorporates the constraints of organic electronics from the OE Route (cf. OETag below).

H. Organic Electronics Tag (OETag)

Exploiting the rPoF HIT framework, we can generate specific hardwired processor versions targeted for organic electronics. By the exchange of results from OE/RE/ORCH Routes, we want to demonstrate printed individualized processors. Improving on the transit frequency achieved within cfaed-1, we shall be able to add the world's first printable active radio frequency interface. Adding sensing, an individualized tag "OETag" shall become feasible, allowing it to sense and tag objects as needed, and printed on-demand.

I. Highly Adaptive Energy Efficient Computing (HAEC)

Within cfaed-1, we did not have dedicated HITs but used CRC HAEC (separately funded DFG collaborative research center) to benchmark our results. Within cfaed-2, we will continue cooperation with HAEC, in particular to show the benefits of the BIO and THz Routes' outputs. With the explosion in the number of compute nodes, the bottleneck of future computing lies in the network architecture connecting the nodes. Addressing the bottleneck lies in replacing current rack backplanes. HAEC proposes to revolutionize computing electronics by realizing embedded optical waveguides for on-board networking, and wireless chip-to-chip links at 200 GHz carrier frequency connecting neighboring boards in a rack. This shall drive current backplane rates from Tb/s to Pb/s orders of magnitude. Driven by new interconnects using our plasmonic interconnects of the BIO and THz Route's output, we envision backplanes in the range of Eb/s. This extends the futuristic vision of HAEC by orders of magnitude. Incorporating ORCH's processing platform would enable innovation in edge cloud computing for future cellular communications for decades to come.

V. SUMMARY

In this paper, we outlined the approach of cfaed (Center for Advancing Electronics Dresden) for advancing electronics beyond the IRDS roadmap into unclaimed territory. We identified six Research Routes, in which we believe we can demonstrate outstanding potential through our excellence in Research and our favorable setting in Dresden. Our research measures brought in place for achieving our objectives are (1) vertical "bridging" of layers across materials, devices, circuits and systems through V-Targets as well as (2) horizontal coherence in research across our six Research Routes through Horizontal Integration Targets that combine the individual research outcomes to a system never seen before. Together, in a truly unique interdisciplinary manner, we will venture electronics into unknown grounds.

ACKNOWLEDGMENT

Firstly, we thank the German Research Council (Wissenschaftsrat) and the German Research Foundation (DfG) for giving us the opportunity to carry out our research in Dresden and funding us for the past 6 years. Secondly, we thank the Saxon Government for their continuous financial support, in particular their generous funding towards cfaed's new building and permanent basic funding. Thirdly, we thank TU Dresden for its large financial as well as organizational support. Also, we thank the German Federal Ministry of Education and Research for their funding of complementary research projects. Last but not least, we thank all investigators of cfaed for their outstanding work and cooperation over the last years, including their continuous ambition to overcome traditional boundaries of the sciences:

W. Aßmann, F. Baader, C. Baier, L. Baraban, J.W. Bartha, K. Bock, J. Castrillon-Mazo, G. Cuniberti, A. Deac, A. Deutsch, S. Diez, M. Dörpinghaus, F. Ellinger, L. Eng, A. Erbe, A. Eychmüller, C. Felser, X. Feng, A. Fery, G. Fettweis, C. Fetzer, W.-J. Fischer, F. Fitzek, B. Friedrich, J. Fröhlich, S. Gemming, G. Gerlach, T. Geßner (†), D. Göhringer, S.T.B. Gönnenwein, H. Härtig, T. Heine, M. Helm, S. Hermann, J. Howard, A.C. Hübler, K. Jamshidi, R. Jordan, E.A. Jorswieck, F. Jülicher, A. Kiriy, H. Kleemann, T. König, M. Krötzsch, A. Kumar, H. Lang, W. Lehner, K. Leo, S.C.B. Mannsfeld, C.G. Mayr, M. Mertig, T. Mikolajick, I. Minev, F. Moresco, W.E. Nagel, A. Nestler, K. Nielsch, S.S.P. Parkin, D. Plettemeier, S. Reineke, B. Rellinghaus, A. Richter, I. Sbalzarini, O.G. Schmidt, T.-L. Schmidt, M. Schröter, R. Schüffny, S. Schulz, R. Seidel, G. Seifert, S. Siegmund, J.-U. Sommer, M. Stamm, R. Stenzel, D. Tang, T. Strufe, M. Timme, A. Voigt, B. Voit, W. Weber, D. Zahn, C. Zechner, M. Zerial, M. Zimmerling, and E. Zschech

REFERENCES

[1] DARPA, 2017, [Online]. Available: https://www.darpa.mil/news-events/2017-09-13, [Accessed: 14-Sep-2018]

[2] M. P. Klinger, A. Fischer, F. Kaschura, J. Widmer, B. Kheramand-Boroujeni, F. Ellinger, and K. Leo, "Organic Power Electronics: Transistor Operation in the kA/cm2 Regime", Scientific Reports, vol. 7, p. 44 713, Mar. 2017.

[3] B. Lüssem, M.L. Tietze, H. Kleemann, C. Hoßbach, J.W. Bartha, A. Zakhidov, and K. Leo, "Doped Organic Transistors: Inversion and Depletion Regime", Nature Communications, vol. 4, Nov. 2013, Art. No. 2775

[4] R. Di Pietro, T. Erdmann, J.H. Carpenter, N. Wang, R.R. Shivhare, P. Formanek, C. Heintze, B. Voit, D. Neher, H. Ade, and A. Kiriy, "Synthesis of High-Crystallinity DPP Polymers with Balanced Electron and Hole Mobility", Chemistry of Materials, vol. 29, no. 23, pp. 10 220–10 232, Dec. 2017.

[5] H. Sahabudeen, H. Qi, B.A. Glatz, D. Tranca, R. Dong, Y. Hou, T. Zhang, C. Kuttner, T. Lehnert, G. Seifert, U. Kaiser, A. Fery, Z. Zheng and X. Feng, "Wafer-sized multifunctional polyimine-based two-dimensional conjugated polymers with high mechanical stiffness", Nature Communications, vol. 7, p. 13 461, Nov. 2016.

[6] X. Zhuang, W. Zhao, F. Zhang, Y. Cao, F. Liu, S. Bi and X. Feng, "A two-dimensional conjugated polymer framework with fully sp2-bonded carbon skeleton", Polym. Chem., vol. 7, no. 25, pp. 4176–4181, 2016.

[7] A. Heinzig, S. Slesazeck, Franz Kreupl, T. Mikolajick, and W. M. Weber, "Reconfigurable silicon nanowire transistors", Nano Letters, vol. 12, no. 1, pp. 119–124, 2012.

[8] W. M. Weber and T. Mikolajick, "Silicon and germanium nanowire electronics: Physics of conventional and unconventional transistors", Reports on Progress in Physics, vol. 80, no. 6, p. 066 502, 2017.

[9] H. Mertens, R. Ritzenthaler, A. Chasin, T. Schram, E. Kunnen, A. Hikavyy, L.-Å. Ragnarsson, H. Dekkers, T. Hopf, K. Wostyn, K. Devriendt, S.A. Chew, M.S. Kim, Y. Kikuchi, E. Rosseel, G. Mannaert, S. Kubicek, S. Demuynck, A. Dangol, N. Bosman, J. Geypen, P. Carolan, H. Bender, K. Barla, N. Horiguchi and D. Mocuta, "Vertically stacked gate-all-around Si nanowire CMOS transistors with dual work function metal gates", in 2016 IEEE International Electron Devices Meeting (IEDM), Dec. 2016, pp. 19.7.1–19.7.4.

[10] S. Barraud, V. Lapras, B. Previtali, M.P. Samson, J. Lacord, S. Martinie, M.-A. Jaud, S. Athanasiou, F. Triozon, O. Rozeau, J.M. Hartmann, C. Vizioz, C. Comboroure, F. Andrieu, J.C. Barbé, M. Vinet and T. Ernst, "Performance and design considerations for gate-all-around stacked-nanowires FETs", in 2017 IEEE International Electron Devices Meeting (IEDM), Dec. 2017, p. 29.2.

[11] J. Müller, T. S. Böscke, U. Schröder, S. Mueller, D. Bräuhaus, U. Böttger, L. Frey and T. Mikolajick, "Ferroelectricity in simple binary ZrO2 and HfO2", Nano Letters, vol. 12, no. 8, pp. 4318–4323, 2012.

[12] A. Heinzig, T. Mikolajick, J. Trommer, D. Grimm and W. M. Weber, "Dually active silicon nanowire transistors and circuits with equal electron and hole transport", Nano Letters, vol. 13, no. 9, pp. 4176–4181, 2013.

[13] M. Simon, J. Trommer, B. Linag, D. Fischer, T. Baldauf, M.B. Khan, A. Heinzig, M. Knaut, Y.M. Georgiev, A. Erbe, J.W. Bartha, T. Mikolajick, and W. M. Weber, "A wired-AND transistor: Polarity controllable FET with multiple inputs", IEEE "Conference Digest of the 76th Device Research Conference (DRC)", Santa Barbara CA, USA, 2018

[14] J. Trommer, A. Heinzig, T. Baldauf, S. Slesazeck, T. Mikolajick and W. M. Weber, "Functionality-enhanced logic gate design enabled by symmetrical reconfigurable silicon nanowire transistors", IEEE Transactions on Nanotechnology, vol. 14, no. 4, pp. 689–698, Jul. 2015.

[15] C. Navarro, S. Barraud, S. Martinie, J. Lacord, M.-A. Jaud and M. Vinet, "Reconfigurable field effect transistor for advanced CMOS: Advantages and limitations", Solid-State Electronics, vol. 128, pp. 155–162, 2017.

[16] J. Trommer, A. Heinzig, U. Mühle, M. Löffler, A. Winzer, P.M. Jordan, J. Beister, T. Baldauf, M. Geidel, B. Adolphi, E. Zschech, T. Mikolajick, and W. M. Weber, "Enabling energy efficiency and polarity control in germanium nanowire transistors by individually gated nanojunctions", ACS Nano, vol. 11, no. 2, pp. 1704–1711, 2017.

[17] J. Trommer, A. Heinzig, T. Baldauf, T. Mikolajick, W. M. Weber, M. Raitza and M. Völp, "Reconfigurable nanowire transistors with multiple independent gates for efficient and programmable combinational circuits", in 2016 Design, Automation Test in Europe Conference Exhibition (DATE), Mar. 2016, pp. 169–174.

[18] S. J. Park, D.-Y. Jeon, S. Piontek, M. Grube, J. Ocker, V. Sessi, A. Heinzig, J. Trommer, G.-T. Kim, T. Mikolajick and W. M. Weber, "Reconfigurable Si nanowire nonvolatile transistors", Advanced Electronic Materials, vol. 4, no. 1, 1 700 399–n/a, Dec. 2017.

[19] V. Sessi, H. Mulaosmanovic, R. Hentschel, S. Pregl, T. Mikolajick and W.M. Weber, "Junction tuning by ferroelectric switching in silicon nanowire Schottky field effect transistors". IEEE Nanotechnology Proc., 2018

[20] M. Raitza, J. Trommer, A. Kumar, M. Völp, D. Walter, T. Mikolajick and W. M. Weber, "Exploiting transistor-level reconfiguration to optimize combinational circuits", in Design, Automation Test in Europe Conference Exhibition (DATE), 2017, Mar. 2017, pp. 338–343.

[21] S. Rai, A. Rupani, D. Walter, M. Raitza, A. Heinzig, T. Baldauf, J. Trommer, C. Mayr, W. M. Weber and A. Kumar, "A physical synthesis flow for early technology evaluation of silicon nanowire based reconfigurable FETs", in Proceedings of the Conference on Design, Automation & Test in Europe, 2018.

[22] S. S. P. Parkin and S.-H. Yang, "Memory on the racetrack", Nature Nanotechnology, vol. 10, pp. 195–198, 2015.

[23] S. Meyer, Y.-T. Chen, S. Wimmer, M. Althammer, T. Wimmer, R. Schlitz, S. Geprägs, H. Huebl, D. Ködderitzsch, H. Ebert, G.E.W. Bauer, R. Gross and S. T. B. Goennenwein, "Observation of the spin Nernst effect", Nature Materials, vol. 16, p. 977, 2017.

[24] M. Hayashi, L. Thomas, R. Moriya, C. Rettner and S. S. P. Parkin, "Current-controlled magnetic domain-wall nanowire shift register", Science, vol. 320, pp. 209–211, 2008.

[25] U. T. Lab. (). Thz gap, [Online]. Available: https://www.uvate-rahertz.com/terahertz.html (visited on 02/18/2018).

[26] M. Schröter, T. Rosenbaum, P. Chevalier, B. Heinemann, S.P. Voinigescu, E. Preisler, J. Bock and A. Mukherjee, "SiGe HBT technology: Future trends and TCADBased roadmap", Proceedings of the IEEE, vol. 105, no. 6, pp. 1068–1086, Jun. 2017.

[27] M. Rodwell, "III-V HBT and (MOS) HEMT scaling", IEEE MTT IMS Symposium, May 2015.

[28] R. Kim, S. Datta, and M. S. Lundstrom, "Influence of dimensionality on thermoelectric device performance", Journal of Applied Physics, vol. 105, no. 3, p. 034 506, 2009.

[29] P. Chevalier, M. Schröter, C.R. Bolognesi, V. d'Alessandro, M. Alexandrova, J. Böck, R. Flückiger, S. Fregonese, B. Heinemann, C. Jungemann, R. Lövblom, C. Maneux, O. Ostinelli, A. Pawlak, N. Rinaldi, H. Rücker, G. Wedel and T. Zimmer, "Si/SiGe:C and InP/GaAsSb heterojunction bipolar transistors for THz applications", Proceedings of the IEEE, vol. 105, no. 6, pp. 1035–1050, Jun. 2017.

[30] M. Schroter, M. Claus, P. Sakalas, M. Haferlach and D. Wang, "Carbon nanotube FET technology for radio-frequency electronics: State-of-the-art overview", IEEE Journal of the Electron Devices Society, vol. 1, no. 1, pp. 9–20, Jan. 2013.

[31] Y. Karpov, T. Erdmann, I. Raquzin, M. Al-Hussein, M. Binner, U. Lappan, M. Stamm, KL Gerasimov, T. Beryozkina, V. Bakulev, DV Anokhin, DA Ivanov, F. Günther, S. Gemming, G. Seifert, B. Voit, R. Di Pietro and A. Kiriy, "High conductivity in molecularly p-Doped diketopyrrolopyrrole-based polymer: The impact of a high dopant strength and good structural order", Advanced Materials, vol. 28, no. 28, pp. 6003–6010, 2016.

[32] Y. Karpov, T. Erdmann, M. Stamm, U. Lappan, O. Guskova, M. Malanin, I. Raguzin, T. Beryozkina, V. Bakulev, F. Günther, S. Gemming, G, Seifert, M. Hambsch, S. Mannsfeld, B. Voit, and A. Kiriy, "Molecular doping of a high mobility diketopyrrolopyrrole– dithienylthieno [3,2-b]thiophene donor–acceptor copolymer with F6TCNNQ", vol. 50, Jan. 2017.

[33] M. Mittendorff, J. Kamann, J. Eroms, D. Weiss, C. Drexler, S. D. Ganichev, J. Kerbusch, A. Erbe, R.J. Suess, T.E. Murphy, S. Chatterjee, K. Kolata, J. Ohser, J.C. König-Otto, H. Schneider, M. Helm and S. Winnerl, "Universal ultrafast detector for short optical pulses based on graphene", Opt. Express, vol. 23, no. 22, pp. 28 728–28 735, Nov. 2015.

[34] H. Arora, T. Schönherr, and A. Erbe, "Electrical characterization of two-dimensional materials and their heterostructures", IOP Conference Series: Materials Science and Engineering, vol. 198, no. 1, p. 012 002, 2017.

[35] M. Mayer, A. M. Steiner, F. Röder, P. Formanek, T.A. König, and A. Fery, "Aqueous gold overgrowth of silver nanoparticles: Merging the plasmonic properties of silver with the functionality of gold", Angewandte Chemie International Edition, vol. 56, no. 50, pp. 15 866–15 870, 2017.

[36] F. N. Gür, F. W. Schwarz, J. Ye, S. Diez, and T. L. Schmidt, "Toward self-assembled plasmonic devices: High-yield arrangement of gold nanoparticles on DNA origami templates", ACS Nano, vol. 10, no. 5, pp. 5374–5382, 2016.

[37] D. V. Nicolau, M. Lard, T. Korten, F.C.M.J.M. van Delft, M. Persson, E. Bengtsson, A. Månsson, S. Diez, and H. Linke, "Parallel computation with molecular-motor-propelled agents in nanofabricated networks", Proceedings of the National Academy of Sciences, vol. 113, no. 10, pp. 2591–2596, 2016.

[38] I. Neri, É. Roldán, and F. Jülicher, "Statistics of infima and stopping times of entropy production and applications to active molecular processes", Physical Review X, vol. 7, no. 1, p. 011 019, 2017.

[39] J. Castrillon, M. Lieber, S. Kluppelholz, M. Voelp, N. Asmussen, U. Aßmann, F. Baader, C. Baier, G. Fettweis, J. Fröhlich, "A hardware/software stack for heterogeneous systems", IEEE Transactions on Multi-Scale Computing Systems, Nov. 2017.

[40] S. Haas, T. Seifert, …, C. Mayr, and G. Fettweis, "A heterogeneous SDR MPSoC in 28 nm CMOS for low-latency wireless applications", Austin, TX, USA: ACM, 2017, 47:1–47:6.

[41] cfaed, TU Dresden, proposal for a Cluster of Excellence "Center for Advancing Electronics Dresden, 2019-2025", 2018

978-1-7281-1988-5/18 $31.00 © 2018 IEEE

Future Computing Hardware for AI

J. Welser[1], J. W. Pitera[1], C. Goldberg[1]

[1]IBM Research – [Albany, Almaden, Yorktown Heights, Zurich], email: welser@us.ibm.com

Abstract—

Hardware has taken on a supporting role in the maturation and proliferation of narrow AI, but will take a leading role to enable the innovation and adoption of broad AI. The concurrent evolution of broad AI with purpose-built hardware will shift traditional balances between cloud and edge, structured and unstructured data, and training and inference. Heterogeneous system architectures are already being delivered where varied compute resources, including high-bandwidth CPUs, specialized AI accelerators, and high-performance networking are infused in each node to yield significant performance improvements. Looking to the future, we envision a roadmap of specialized technologies to accelerate AI, starting with heterogeneous digital von Neumann machines, exploring reduced-precision accelerator approaches, finding the limits of conventional device power-performance with analog AI devices, and finishing with quantum computing for AI.

I. INTRODUCTION

Narrow AI, marked by performance in a single domain with human or superhuman accuracy and speed for certain tasks, has been broadly adopted in applications from facial recognition to natural language translation. We are just at the beginning of Broad AI, which encompasses multi-task, multi-domain, multi-model, distributed and explainable AI. Transfer learning and reasoning are central to expanding AI to small datasets. Reducing the time and power requirements of AI computing is fundamental to the development and adoption of Broad AI solutions, and will enable exploration towards General AI [Table 1].

Narrow AI	Broad AI	General AI
Single task, single domain	Multi-task, multi-domain	Cross-domain learning and reasoning
Superhuman accuracy and speed for certain tasks	Multi-modal distributed AI	Broad autonomy
	Explainable	

Table 1: Sequence of evolution from Narrow AI to Broad AI to General AI (after Dario Gil, IBM)

In general, machine learning and deep learning methods can be divided into two distinct operation modes: training and inference. First, the training phase is an optimization problem in a multi-dimensional parameter space, to build a model that can be used to provide a wider generalization in the inference process. In deep learning, a model usually consists of a multilayer network with many free parameters (weights) whose values are set during the training process [1]. Second, the trained model needs to be deployed on real-world data in the inference mode. For many applications, this inference step needs a trained model that is fixed for consistency, reproducibility, liability, performance or regulatory reasons. Consequently, high-performance training and inference may drive different device and system requirements.

Creating a portfolio of compute hardware which can fluidly accommodate both high efficiency training and inference is an enabler to broadly deploy deep learning and address the diversity of data types, models, and domains in emerging broad AI techniques and applications.

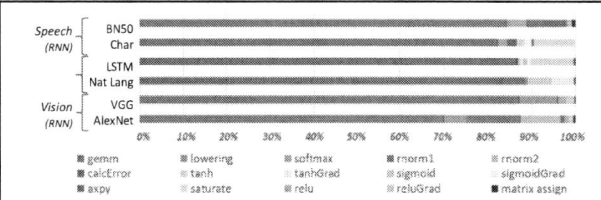

Figure 1: Deep learning comprised of a spectrum of operations. Although matrix multiplication is dominant, optimizing performance efficiency while maintaining accuracy requires the core architecture to efficiently support all of the auxiliary functions [2].

To start our discussion of future AI hardware, which bridges the evolution from narrow to broad AI, we perform an algorithmic autopsy, starting with deep neural networks, as shown in Figure 1. Clearly, matrix multiplications are at the core of deep learning networks. For fully connected networks, data travels through the network in the form of vector-matrix multiplications. For a better utilization of compute resources, several data points are usually batched together (mini-batch). In convolutional neural networks, the input convolution operation can be described as a matrix, and the first step is a matrix – matrix multiplication. Creating optimized systems for these workloads requires re-thinking how we innovate the end-to-end system, including devices, hardware, software, and programming – each element is potentially transformed.

In current AI systems, GPUs initially developed for gaming and 3D graphics were recognized to be a good fit to accelerate deep learning training. The simple mathematical structure of backpropagation can easily be parallelized and can

978-1-7281-1988-5/18 $31.00 © 2018 IEEE

therefore take advantage of the parallel architecture of GPUs in a natural way.

The next step is assembling GPU resources in the context of optimized heterogeneous systems that maximize performance across the CPU, GPU, memory system and network, such as the world-leading Summit supercomputer [3,4].

Moving forward, the first foundational progress in compute efficiency for AI model training, and even more for inference, can be made by exploiting the statistical and approximate nature of deep learning algorithms, leading to reduced-precision approaches [5,6].

Data movement in the system can introduce significant bottlenecks that slow down the whole training process. The second foundational approach to the future of AI hardware is rooted in addressing the performance efficiency loss from data movement at the system level and ultimately, creating new device technologies that eliminate the data transport entirely.

Finally, a fundamental re-formulation of the AI problem is central to harnessing the promise of quantum computing to unravel today's intractable AI problems.

II. HETEROGENEOUS SYSTEMS FOR AI – BEYOND HOMOGENEITY

The Summit supercomputer at Oak Ridge National Laboratory, shown in Figure 2, embodies multiple features of system-level purpose-built architecture for AI computation [3,4].

SUMMIT
- 200 Petaflops
- 9.216 IBM Power9 CPUs
- 27,648 NVIDIA Tesla GPUs
- 250 petabytes storage capacity
- 25 gigabytes per second between nodes

Figure 2: IBM built the Summit Supercomputer for Oak Ridge National Laboratories and it was ranked the #1 most powerful supercomputer in the world in June 2018. The Summit architecture is designed not only for raw performance, but specifically tailored for AI workloads.

Heterogeneous systems are characterized by a flexible assemblage of specialized components, in contrast to reliance on a single multi-purpose CPU. By recasting the CPU as the conductor of an AI orchestra, we get much more system performance by using accelerators (e.g. specialized processors like GPUs with their own memory nearby). The CPU now plays a critical role in feeding the GPUs with data, orchestrating the workload, running the serial parts of the code that do not run efficiently on the GPUs, and managing all of the other services it takes for the computer to work.

Summit employs multiple hardware and software approaches to address data transport, connectivity, and scalability. Summit's compute nodes each contain dual IBM POWER9 CPUs, six NVIDIA Volta GPUs, over half a terabyte

of coherent memory (high bandwidth memory + DDR4) addressable by all CPUs and GPUs, plus 1.6TB per node of non-volatile RAM that can be used as a burst buffer or as extended memory. Second generation NVLink, a proprietary interface, allows CPUs and GPUs to share data up to 4X faster than x86-based systems [7]. Dual-rail Mellanox EDR InfiniBand interconnects, used for both storage and inter-process communications traffic, deliver 200 Gb/s bandwidth between nodes.

Ultimately, the communications overhead associated with distributing machine learning model calculations across a scaled data center can be a source of diminishing returns in model build time. System level software tuning is also part of the solution to suppressing communications penalties by efficiently distributing learning with minimal noise across compute nodes. Custom algorithms and software were developed to automate and optimize the parallelization of deep learning training across hundreds of GPU accelerators, leveraging all available links in the system, as shown in Figure 3. Record low communication overhead and 95% scaling efficiency was achieved on the Caffe deep learning framework over 256 NVIDIA GPUs in 64 IBM POWER systems [8].

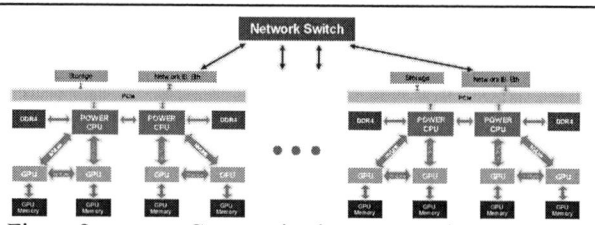

Figure 3: Communication mapping within heterogeneous systems is critical to fully utilize bandwidth for links within each node and across all nodes, as shown for the IBM Power AI Distributed Deep Learning architecture, in which learner communication efficiency is optimized.

III. BRAIN-INSPIRED ARCHITECTURE AND APPROXIMATE COMPUTING

The vast majority of computing machines evolved from the von Neumann architecture, which separates compute and memory. In contrast, biological brains tightly integrate low precision compute, memory, and communication, achieving extraordinary efficiency. Likewise, computation has historically relied on high precision 64- and 32-bit floating point arithmetic which is not typically required for AI workloads. New architectures have been demonstrated in conventional digital CMOS technology that derive performance efficiency based on a brain-inspired architecture and approximate computing.

Taking inspiration from the brain's structure to achieve efficiency, scalability and flexibility, the IBM TrueNorth architecture uses parallel digital neurons co-located with their associated synapses (providing the computation and memory elements of deep networks), adopts a low precision operating regime, and provides local and long-range communication

978-1-7281-1988-5/18 $31.00 © 2018 IEEE

substrates. The TrueNorth chip, built for low power inferencing, represents the largest digital neuromorphic chip to-date, integrating 4,096 neurosynaptic cores interconnected with an on-chip network for a total of over 1 million programmable spiking neurons and over 256 million configurable, low-precision synapses [9]. Chips can be tiled in two dimensions via an inter-chip communication fabric, enabling large-scale system composition, as shown in Figure 4 [10]. Low precision algorithm development, coupled with the TrueNorth architecture represents a prodigious step forward to delivering high efficiency neural network inference deployment in an end-to-end ecosystem [11].

Figure 4: A 64 million neuron, 16 billion synapse system with 64 TrueNorth chips.

Reduced precision can also benefit more conventional architectures. Computational building blocks with 16-bit precision engines are 4x smaller than comparable blocks with 32-bit precision; this gain in area efficiency becomes a boost in performance and power efficiency for both AI training and inferencing workloads. Simply stated, in approximate computing, we can trade numerical precision for computational efficiency, provided we also develop algorithmic improvements to retain model accuracy [12,13]. This approach also complements other approximate computing techniques—including recent work that described novel training compression approaches to cut communications overhead, leading to a 40-200x decrease of data communications volume over existing methods [14]. The choice of fixed-point or floating-point arithmetic is another lever that can be exploited.

Another IBM digital accelerator architecture based on approximate computing principles, demonstrated a multi-TeraOPS accelerator core building block for both training and inference across a broad range of AI hardware systems [2]. This digital AI core features a parallel architecture that ensures very high utilization, and efficient compute engines that carefully leverage reduced precision. Figure 5 shows a schematic of the scratch pad architecture, central to the customized dataflow. The matrix multiplication components are computed in the core architecture by using a customized dataflow organization of the Processing Elements shown in Figure 5, where reduced precision computations can be efficiently exploited. The remaining vector functions (all of the non-red bars in Figure 1) are executed in either the Processing

Elements or the Special Function Units, depending on the precision needs of the specific function. This AI core can be integrated into SoCs, CPUs, or microcontrollers and used for training, inference, or both. Chips using the core can be deployed in the data center or at the edge.

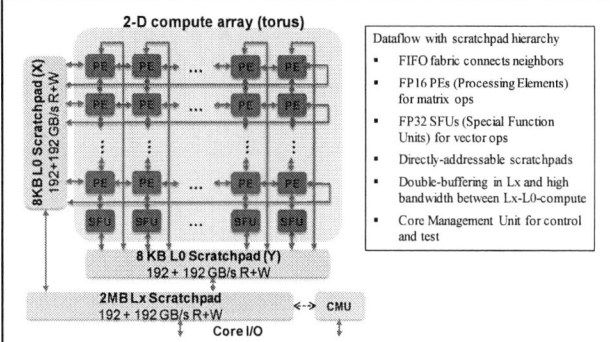

Figure 5: Example of digital accelerator architecture designed to exploit a new dataflow for approximate computing, using a scratchpad architecture [2].

IV. ANALOG AI CORES – BEYOND DIGITAL COMPUTING

Even with these improvements in operational efficiency, data movement still has a cost in energy and time. Mapping math operations to analog devices avoids the data transfer bottleneck between memory and processor. This bottleneck fundamentally stems from the fact that the processor can only access a finite amount of data at a time, which needs to be brought to and from memory. Analog accelerators avoid these bottlenecks by performing the computation in the memory itself.

The tolerance of reduced precision in AI models [12,13] is what opens the possibility to revisit analog computing, which is intrinsically noisy. If noise can be tolerated, it is possible to execute the matrix operations for deep learning in constant time on arrays of analog non-volatile memories

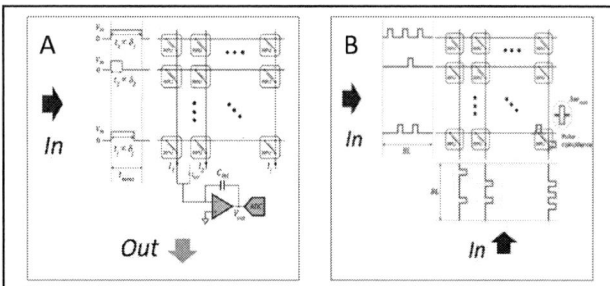

Figure 6: (A) Analog array read operation. The row input vector Vin is time encoded (DAC) and the column current integrated. The integrated charge is either converted in a digital signal (ADC) or a time encoded signal is transferred to the next layer. (B) Analog array update operation. Row (x) and column (d) signal are encoded in bit steams with length BL. Pulse coincidence at node changes the conductivity of the node element.

978-1-7281-1988-5/18 $31.00 © 2018 IEEE   23

(NVM). To take full advantage of this in-memory compute paradigm, current non-volatile memory materials are of limited use.

Analog computing for deep learning uses arrays of NVM to perform matrix operations with the weights imprinted in the memory nodes. Because the weights do not move between memory and the compute unit, matrix operations can be done in parallel at constant time. This enables mapping of the two-dimensional matrix into a physical array (Fig. 6) with the same number of rows and columns as the abstract mathematical object [15]. IBM teams are exploring both mixed digital-analog and pure analog approaches utilizing phase change materials (PCM) based NVM. In the mixed-precision, mixed analog-digital approach explored by one group, the NVM array stores the synaptic weights with a digital processing unit and an additional memory unit that stores the accumulated weight updates in high precision, as shown in Figure 7 [16,17,18]. Another IBM team also recently demonstrated a pure analog approach that achieved accuracy equivalent to GPUs, estimating potential computational efficiency gains of two orders of magnitude beyond today's GPUs for training fully connected networks on an integrated analog device [19].

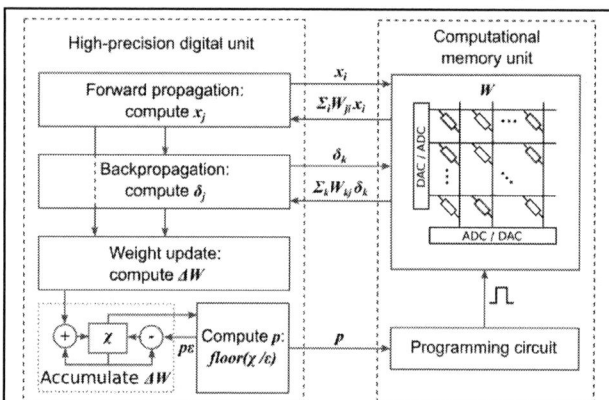

Figure 7: Mixed precision deep learning with computational memory approach. Synaptic weights are stored in computational memory. The matrix-vector multiplications associated with forward/backward propagation are performed in place with low precision. The desired weight updates are accumulated in high precision [16].

For all analog approaches, materials research is ground zero in the development of analog NVM elements that can meet the device specifications needed for high classification accuracy, thereby providing at least 1000 times better compute efficiency than currently possible during training [20]. As shown in Table 2, we estimate that analog NVM elements with about 1000 accessible conductance states are required. Furthermore, these devices need to respond symmetrically in up and down conductance changes and this needs to be achieved with appropriate yield, reduced device to device variability and minimum conductance drift.

These are challenging requirements, but research is ongoing to explore many materials systems, including chalcogenide based PCM and high-κ transition-metal (TM)

	Memory Cell	Analog NVM Cell
States	"0" & "1"	~1000
Ratio between Max and Min Conductance	>10,000	~10
Conductance response to pulses	Does not matter	Symmetrical

Table 2: Comparison of material requirements for a traditional memory cell application as compared to an Analog NVM cell for AI applications.

oxides [21,22], for analog computing applications. Resistance (conductance) drift is a concern for PCM devices, attributed to the structural relaxation of the amorphous (RESET) phase. We have demonstrated a novel memory cell structure using a metallic surfactant layer that provides a parallel conductive path to the amorphous region during the read operation, thereby suppressing resistance variation caused by the amorphous phase. The benefits of the surfactant are shown in Figure 8 [23]. This effectively creates a larger conductance ratio between adjacent conductance levels by minimizing the time-dependent drift. In parallel with mainstream NVMs, we are also developing and evaluating a new Electro-Chemical synaptic element (ECRAM) as a building block for AI computing, where we have demonstrated symmetric analog behavior at sub-μm scale and sub-μs speed in 3-terminal devices based on gated ionic exchange with a tunable oxide channel. [24].

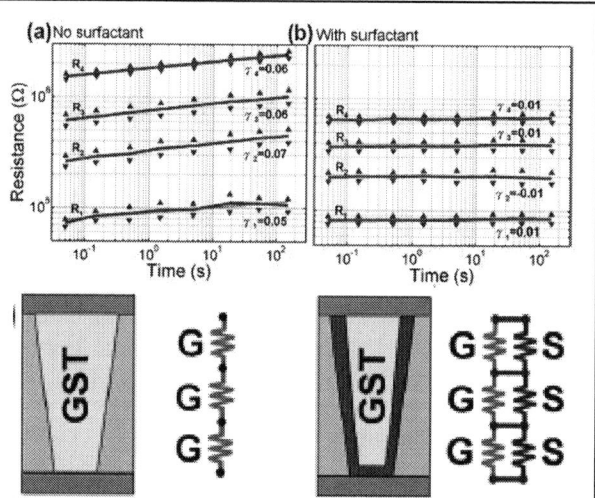

Figure 8: Surfactants have been employed in materials optimization for analog cross-point array switching elements. Resistance vs time plot comparison (a) without and (b) with the surfactant metallic liner. With the surfactant metallic liner, all resistance levels are stabilized with 6x reduction in the drift coefficient. Adapted from [23].

V. Quantum -- Beyond Classical Computing

Quantum computing applied to AI promises to solve problems we now simply cannot even attempt, even on leadership heterogeneous systems like the Summit supercomputer.

The power of quantum computing comes from qubits—quantum bits—which can express an exponential state space through superposition, and a computing process where problems are solved through entanglement and interference of quantum information to constructively amplify the correct answer. Our approach to quantum computing relies on an array of qubits constructed from and interconnected by superconducting circuits which support a universal gate set allowing all logical operations, much like a NAND gate can be used to build any conventional logic circuit [25]. Because quantum information is fragile, error mitigation or error correction codes are required to perform error-free or "fault-tolerant" quantum operations. Future machines will have a physical quantum processor executing low-level gates with RF controls and readout with information protected by a topological parity code (Fig. 9). This code forms "Logical" qubits out of many physical qubits. Algorithms executed on logical qubits will be error free—but, at present, the overhead of constructing such a machine is daunting and is the subject of active research.

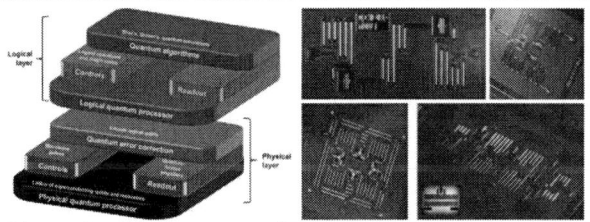

Figure 9: (A) A systems view of a quantum information processor. The physical layer provides the error correction and consists of a physical quantum processor that has both input and output lines that are controlled by the QEC processor. This processor is in turn controlled by the logical layer, where the encoded qubits are defined and the logical operations are performed for the desired quantum algorithm. (B) Images of four recent quantum devices fabricated at IBM, designed to explore different basic functions of the physical layer of a quantum computer.

Recent work has demonstrated a binary classification algorithm (Support Vector Machine, SVM) which exploits a quantum feature space and achieves classification rates up to 100%. Figure 10 shows a two label (red and blue) dataset classification as performed by a two-qubit quantum processor. The classification method is a SVM where quantum support is provided for the estimation of a non-linear kernel [26]. Previous work had demonstrated a more specialized algorithm to learn parity with noise (LPN) by consulting a quantum oracle [27]. With fault-tolerant quantum systems, a polynomial speedup for some AI problems is likely.

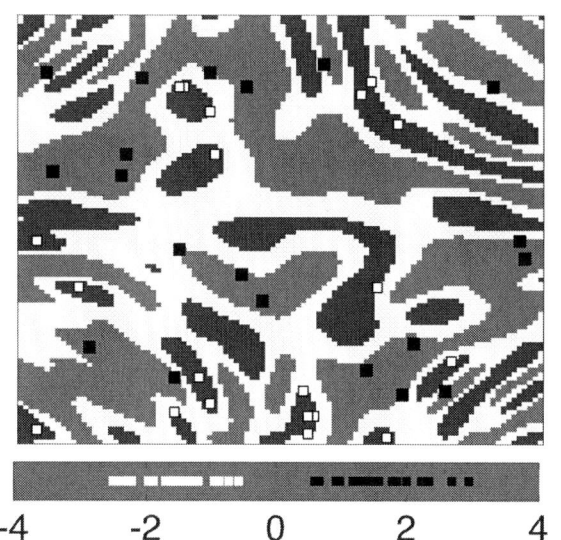

Figure 10: A two-label (red and blue) dataset classification as performed by a two-qubit quantum processor. The labeled data are separated by dataless white regions. Individual data points (squares) are successfully classified (bottom gray bar) by the quantum processor after a training phase [26].

VI. Summary

For decades, advances in computational technology have aimed to increase versatility while improving overall performance, so that hardware could be used for a wide variety of applications. The trade-off is that leading-edge hardware is not specifically optimized for any of its applications. Hardware and system accelerators may be the answer to this dilemma. These accelerators are tailored for a specific class of applications to give the best compute efficiency, in time or power. At IBM Research, we are pursuing algorithm, hardware, and system-level accelerators for AI that are rooted in conventional CMOS technology, while simultaneously taking a fresh look at how everything can be redesigned for better efficiency. This includes analog devices and quantum computing, opportunities that exploit strong relationships between the executable AI algorithms and the physics of the system components.

Acknowledgments

The paper surveys the work across the global team of IBM Research and IBM Cognitive Systems on AI hardware. The authors gratefully acknowledge the contributions of Vijay Narayanan, Dharmendra Modha, Mark Ritter, Antonio Corcoles-Gonzalez, Wilfried Haensch, Hillery Hunter, Jeffrey Burns, Evangelos Eleftheriou, Abu Sebastian, and Geoff Burr, as well as the leadership support of Mukesh Khare, Dario Gil, and Alessandro Curioni from IBM Research and Sumit Gupta and Bob Picciano from IBM Cognitive Systems.

REFERENCES

[1] W. Haensch, T. Gokmen, and R. Puri, "Next Generation of Deep Learning Hardware: Analog Computing" submitted to Proc. IEEE Non-silicon, Non-von Neumann Computing.

[2] B. Fleischer *et al.*, "A Scalable Multi-TeraOPS Deep Learning Processor Core for AI Training and Inference", 2018 Symposium on VLSI Circuits, C4-2, 2018.

[3] IBM press release, "Reaching the Summit: The World's Smartest Supercomputer", June 8, 2018. http://newsroom.ibm.com/Reaching-the-Summit-The-Worlds-Smartest-Supercomputer

[4] M. Rosenfield, "We've Reached the Summit", June 14, 2018, https://www.ibm.com/blogs/research/2018/06/summit/

[5] J. Nickolls and W. J. Dally, "The GPU Computing Era," IEEE Micro, vol. 30, no. 2, 2010.

[6] Y. LeCun *et al.*, "Gradient-based learning applied to document recognition," Proceedings of the IEEE, vol. 86, p. 2278–2324, 1998.

[7] IBM POWER9 NPU team, "Functionality and performance of NVLink with IBM POWER9 processors", IBM J. Res. & Dev., 25 June 2018, doi: 10.1147/JRD.2018.2846978

[8] M. Cho, U. Finkler, S. Kumar, D. Kung, V.Saxena, D. Sreedhar, "PowerAI DDL", arXiv:1708.02188, 7 Aug 2017.

[9] P.A. Merolla *et al.*, "A million spiking-neuron integrated circuit with a scalable communication network and interface", Science, 08 Aug 2014: Vol. 345, Issue 6197, pp. 668-673.

[10] IBM press release, "U.S. Air Force Research Lab Taps IBM to Build Brain-Inspired AI Supercomputing System", June 23, 2017, https://www-03.ibm.com/press/us/en/pressrelease/52657.wss

[11] S. Esser *et al.*, Convolutional networks for fast, energy-efficient neuromorphic computing, Proc. Natl. Acad. Sci. October 11, 2016. 113 (41) 11441-11446.

[12] S. Gupta, A. Agrawal, K. Gopalakrishnan and P. Narayanan, "Deep learning with limited numerical precision," in International Conference on Machine Learning, Lille, France, 2015

[13] M. Courbariaux, *et al.*, "Binarized Neural Networks: Training Neural Networks with Weights and Activations Constrained to +1 or −1," http://arXiv:1602.02830v3, 2016.

[14] C.-Y. Chen *et al.*, "AdaComp: Adaptive Residual Gradient Compression for Data-Parallel Distributed Training", AAAI, 2018.

[15] Steinbuch, K. "Die Lernmatrix", Kybernetik **1**, p. 36-45 (1961)

[16] S.R. Nandakumar *et al.*, "Mixed-precision architecture based on computational memory for training deep neural networks," IEEE International Symposium on Circuits and Systems, 2018.

[17] M. Le Gallo, *et al.*, "Mixed-precision in-memory computing", Nature Electronics **1**, p. 246-253 (2018)

[18] W.W. Koelmans *et al.*, "Projected phase-change memory devices", Nature Communications **6**, 8181 (2015).

[19] S. Ambrogio *et al.*, "Equivalent-accuracy accelerated neural-network training using analogue memory," *Nature* **558**, pp. 60–67 (2018).

[20] T.Gokmen and Y. Vlasov, "Acceleration of Deep Neural Network Training with Resistive Cross-Point Devices: Design Considerations," *Front. Neurosci.* **10**, 333 (2016).

[21] M. Salinga *et al.*, "Monatomic phase change memory," *Nature Materials* **17**, pp. 681–685 (2018).

[22] G. W. Burr *et al.*, "Neuromorphic computing using non-volatile memory," *Advances in Physics* X, 2(1), 89-124 (2017).

[23] S. Kim *et al.*, "A phase change memory cell with metallic surfactant layer as a resistive drift stabilizer", IEEE International Electron Device Meeting Proceedings, 2013.

[24] J. Tang *et al.*, "ECRAM as Scalable Synaptic Cell for High-Speed Low-Power Neuromorphic Computing" submitted to IEDM 2018.

[25] J. Gambetta *et al.*, "Building logical qubits in a superconducting quantum computing system," Nature Quantum Information, vol. 3, 2017.

[26] V. Havlicek *et al.*, "Supervised learning with quantum enhanced feature spaces," arXiv:1804.11326, submitted June 5, 2018.

[27] D. Rieste *et al.*, "Demonstration of quantum advantage in machine learning," Nature Quantum Information, **3**:16, (2017).

Scaling Trends in NAND Flash

K. Parat[1], and A. Goda[2]

[1]Intel Corporation, 2200 Mission College Blvd., Santa Clara, CA, USA, email: krishna.parat@intel.com
[2]Micron Technology, 8000 S. Federal Way, Boise, ID, USA

Abstract—As the 2D NAND Flash scaling plateaued due to physical and electrical scaling limitations, 3D NAND emerged as a strong successor to continue the scaling trend. 3D NAND has rapidly achieved maturity and is already in the 3rd and 4th generation of technology with the total number of layers reaching 96 active layers. The improved cell characteristics of 3D NAND have enabled 4bits/cell capability, allowing for further bit density scaling. This paper describes some of the recent innovations in the 3D NAND technology and the key challenges ahead for continued scaling.

I. INTRODUCTION

After delivering ~100x increase in die capacity (Gb) during the first decade of this century with the cell area scaling and the number of bits/cell increasing from 1bit/cell to 3bits/cell (Fig. 1), 2D NAND scaling plateaued around 15nm node due to physical as well as electrical scaling limitations [1-3]. This presented an opportunity for disruption and the 3D NAND emerged [4-13] as a strong successor for continued scaling (Fig. 2). 3D NAND came of age with the 32 layer technology by delivering >2X bit areal density (Gb/mm^2) of the 2D NAND [4,5]. While the 2D NAND chip capacity had saturated at 128Gb, 3D NAND chip capacity has now reached 1Tb [12].

Several 3D NAND architectures were initially considered to overcome the NAND scaling challenge [7-10]. The two primary approaches that have finally emerged are: Floating Gate (FG) 3D NAND with CMOS under array, and Charge Trap Flash (CTF) 3D NAND (Fig 3). The CTF cell was explored in the 2D NAND era as well, but it became viable only in the 3D configuration due to the assist from field enhancements of cylindrical channel [4,6]. The CTF cell has a simpler cell structure due to the continuous charge storage node, but the cell operation requires high work-function metal gate and replacement gate process. On the other hand FG NAND has had proven cell electrical capability from the 2D NAND, but required innovative ways for separating the floating gate charge storage node between the two cells on neighboring wordlines [5].

Some of the attributes are common to both the CTF 3D NAND technology and the FG 3D NAND technology. These include a surround gate cell structure with poly silicon channel and a staircase contact scheme for the wordlines. Both technologies require high aspect ratio memory hole etch as well as cell formation in deep high aspect ratio memory holes. While the 2D NAND scaling relied on lateral shrink of the cell geometries, the primary scaling path for the 3D NAND is vertical scaling by increasing the number of active layers in the technology.

This paper describes the innovations that have enabled Intel-Micron 2nd generation of 3D NAND Flash to achieve 64 layers with 512Gb capacity. This is followed by a brief description of 4bits/cell technology that is emerging as a viable scaling path for driving the bit areal density further and to achieve 1Tb capacity. The approach taken for the 3rd generation with 96 active layers is also described in brief. The paper concludes with describing some of the challenges to be overcome to continue the scaling trend.

II. 64 LAYER FG 3D NAND TECHNOLOGY

The very first generation of the FG 3D NAND with 32 layers delivered >2X scaling in the bit areal density compared to the 2D NAND predecessor [5]. Key attributes of this technology were – surround gate FG cell structure with 32 active layers and CMOS under the NAND Flash array. The 2nd generation delivered 64 active layers. Since the memory hole etch is one of the most difficult part of the technology, the 64 layer technology employed stacking of two 32 layers (Fig 4). While the memory hole etch and the associated cell formation were done separately for the two stacks, rest of the process that include wordline separation between blocks, wordline staircase formation, etc. were done in one step for the two stacks to minimize the overall process cost.

In the FG 3D NAND architecture with CMOS under the array, the NAND strings are connected to N$^+$ diffusions at both ends. During erase, the source as well as the bitlines are biased to the positive erase voltage and the source & drain select gates are biased to a slightly lower voltage than this to induce sufficient GIDL current to bias up the body of the NAND string to the desired erase voltage (Fig. 5). Biasing of the body by GIDL current from both ends achieves uniform erase voltage across the full NAND string [14].

In addition to increasing the total number of active layers from 32 to 64, another key innovation in the 2nd generation was further optimization of the CMOS circuitry. First generation had wordline & bitline decoders and sense-amp circuits sitting under the NAND array while rest of the periphery circuitry was kept outside the array. The second generation optimized the layout and moved majority of these circuitry to under the NAND array to improve the overall die efficiency. This enabled the 2nd generation to deliver the world's smallest 512Gb die size of 110.5mm^2 for a 64 layer NAND technology node (Fig 6).

III. 4BITS/CELL TECHNOLOGY

While increasing the number of active layers is one way for achieving scaling, availability of the process capability for high aspect ratio etches and the associated increase in the process

978-1-7281-1988-5/18 $31.00 © 2018 IEEE

cost due to the increased process times detract from the overall cost reduction. Increasing the number of bits/cell (bpc) at a given technology node on the other hand, can provide cost scaling without the added process complexity. This has been an approach used in 2D NAND for providing higher Gb/mm² scaling as the mainstream technology migrated from 1bpc to 2bpc and then to 3bpc. One advantage of the 3D NAND cell is the bigger physical cell size compared to a scaled 2D NAND cell. The bigger cell size results in larger cell capacitance leading to more number of stored electrons in the cell for a given change in the cell threshold voltage, which helps reduce number fluctuation effects [5]. The larger cell area results in better Vt distribution as well due to the reduced trap and dopant fluctuation effects. The surround gate structure of 3D NAND makes it less susceptible to interference from neighboring cells and also allows for large boosting for very good program disturb immunity. These advantages are summarized in Table I, where the superiority of the 3D NAND Cell over a scaled 2D NAND cell is shown. This makes the 3D NAND capable of extending to more bits per cell than what was possible with 2D NAND.

Going from 3bits/cell to 4bits/cell requires increasing the number of distinct Vt levels from 8 to 16. Fig. 7 shows the Program/Erase Vt as a function of the change in the applied gate to channel bias (for erase, the body is biased positive with respect to the gate). Greater than 10V Program/Erase Vt window with near unity Program/Erase slope is achieved, which is necessary and sufficient for accommodating the 16 Vt levels required for the 4bits/cell.

While tight placement of the cell Vt's can be achieved by using a smaller gate step during the programming operation, it is critical that the cell Vt does not drift from these placed values. A key advantage of the FG cell relative to the CTF cell is the good charge isolation and confinement of the injected charges in FG making it immune to the charge loss or charge migration and Vt drift issues inherent in the CTF cell [12].

Cell to cell interference from the programming of neighboring cells is kept to a minimum by using an 8-16 programming algorithm, where the cells on a wordline are first programmed to 8 levels based on the data of the 3 LSB bits. And the MSB bit is programmed after the neighboring wordlines have seen most of the Vt movements from LSB programming. Figure 8 describes the programming algorithm schematically and Fig. 9 shows the actual Vt distributions from the 4bit/cell product. As can be seen from this figure, excellent Vt distributions and low bit-error-rates are achieved.

Die photograph of the 1Tb 4bits/cell product built on the 64 active layer technology is shown in Fig. 10. The die size is 159.7mm² resulting in an areal density of 6.41Gb/mm² which is the highest Gb/mm² densities for a 64Layer 3D NAND die reported to date and is >6X of that achieved on 2D NAND.

IV. 96 LAYER FG 3D NAND TECHNOLOGY

The 64 layer process architecture was extended to 96 layers by increasing the number of active layers in each stack from 32 to 48. Figure 11 shows the scaling methodology adopted for the

successive generations of the FG 3D NAND utilizing stacking as well as increasing number of tiers. Figure 12 shows the SEM cross-section of the NAND array showing the two stacks of 48 active layers.

V. FUTURE SCALING CHALLENGES

Figure 13 shows the bit areal density (Gb/mm²) scaling trend. With transition to the 3D NAND, the scaling has continued at near historic rate. Continued scaling will have to come from increasing the number of layers with 4bits/cell providing about ~33% higher bit area density than the 3bits/cell products. Key process challenges will be – Memory hole etching, cell formation in the high aspect ratio memory holes. The net challenge here is somewhat alleviated by using two-step process which is being adopted for both FG 3D NAND as well as CTF 3D NAND [13].

Unlike the 2D case, where the cell physical scaling was decoupled from the NAND string length, in 3D case, increased number of layers results in a longer NAND string length as well, causing reduction in the string current, which can hurt sensing. This poses the device challenge of needing to improve the channel conductivity to maintain high string current.

Currently the wordline layer pitch in the 3D NAND is around 50-60nm. Since 2D NAND had scaled the pitch down to ~30nm one can expect wordline pitch scaling below 50nm for 3D case as well, which will help reduce the burden on the etch process. The key device challenge will be to maintain good cell electrical capability with wordline pitch scaling.

VI. CONCLUSION

3D NAND has emerged as a strong successor to 2D NAND providing continued scaling driven by increasing number of layers as well as leveraging the superior cell characteristics of the 3D NAND to increase the number of bits/cell from 3bits/cell to 4bits/cell. Extending this scaling trend will require improved process capability for memory hole etching and cell formation in high aspect ratio memory holes and improved channel conductivity.

ACKNOWLEDGMENT

Authors gratefully acknowledge the Intel-Micron NAND team that performed the work presented here.

REFERENCES

[1] M. Helm, et al, *ISSCC Tech Digest*, pp. 326-327, 2014.
[2] M. Sako, et al., *ISSCC Tech Digest*, pp. 128 - 129, 2015.
[3] P. Cappelletti, *IEDM Tech Digest*, pp. 241 - 244, 2015.
[4] J-W. Im, et. al, *ISSCC Tech Digest*, pp. 130 - 132, 2015.
[5] K. Parat, et. al, *IEDM Tech Digest*, pp. 48 - 51, 2015.
[6] R. Yamashita, et. al, *ISSCC Tech Digest*, pp. 130 - 132, 2017.
[7] H. Tanaka et al., *VLSI Tech Symp. Digest*, pp. 14-15, 2007.
[8] J. Jang, et al., *VLSI Tech Symp. Digest*, pp. 192-193, 2009.
[9] R. Katsumata, et al., *VLSI Tech Symp. Digest*, pp. 136-137, 2009.
[10] H-T. Liu, *VLSI Tech Symp. Digest*, pp. 131-132, 2010.
[11] J. Lee, et al., *IEDM Tech Digest*, pp. 284 - 287, 2016.
[12] S. Lee, et. al, *ISSCC Tech Digest*, pp. 340 - 342, 2018.
[13] S. Inaba, *IMW Tech Digest*, pp. 1 - 4, 2018.
[14] C. Caillat, et al., *IMW Tech Digest*, pp. 1 - 4, 2017.

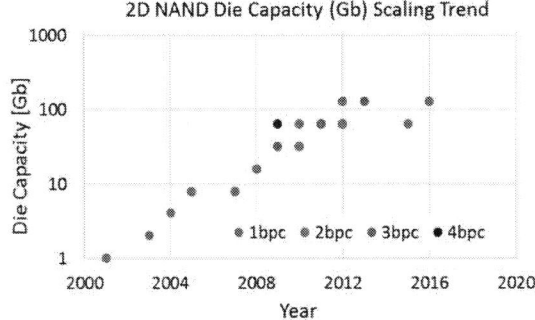

Fig. 1. 2D NAND die capacity (Gb) trend over the past decade showing plateauing at 128Gb.

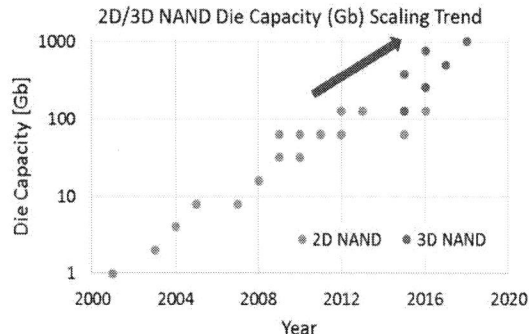

Fig. 2. Transition to 3D NAND has extended the die capacity to 1Terabit (Tb).

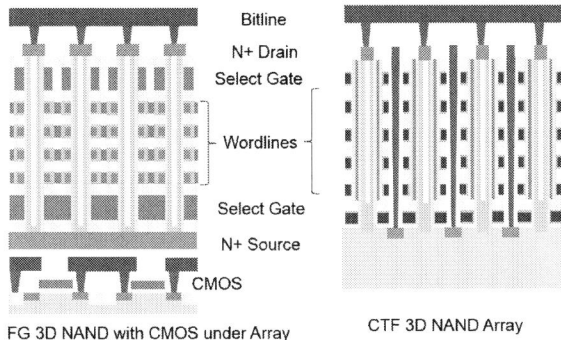

Fig. 3. The two primary 3D NAND technologies - FG Flash 3D NAND with CMOS under array & CTF 3D NAND.

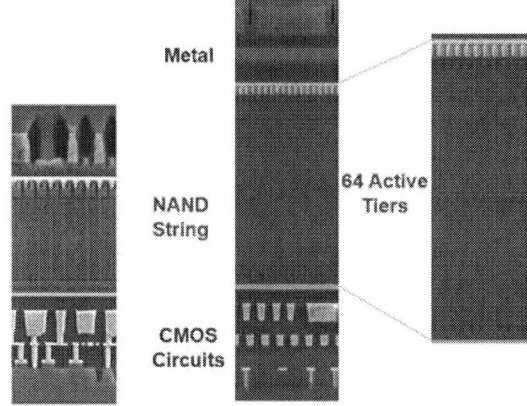

Fig. 4. SEM cross-sections of the 1st and 2nd Gen FG 3D NAND with 32 and 64 active Layers and CMOS under the array.

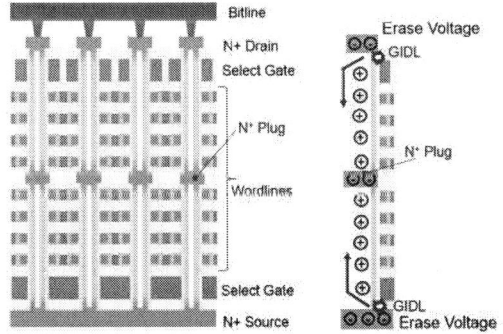

Fig. 5. GIDL current from the source and the bitline side is used for biasing the body to the desired erase voltage.

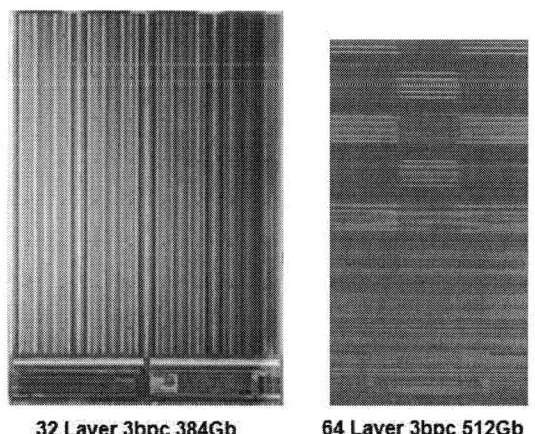

Fig. 6. Layout improvement between 1st and 2nd Gen to fit majority of circuits under the NAND array.

	Scaled 2D NAND	3D NAND
Number of Electrons/V of Vt change	1X	~6X
Cell Intrinsic Vt distribution	1X	~0.5X
Cell to Cell interference	1X	~0.2X
Program Vt fluctuation	1X	~0.4X
Program Disturb window	1X	~1.5X

Table I. Comparison of the key cell properties between 2D NAND and 3D NAND Flash Cell.

978-1-7281-1988-5/18 $31.00 © 2018 IEEE

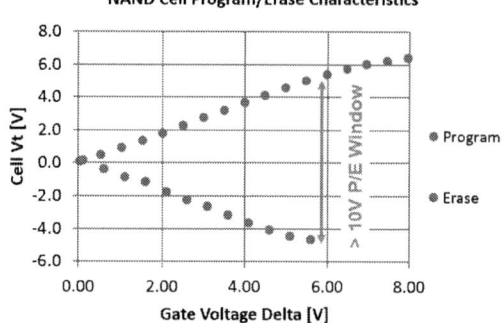

Fig. 7. NAND Cell Program/Erase Vt window.

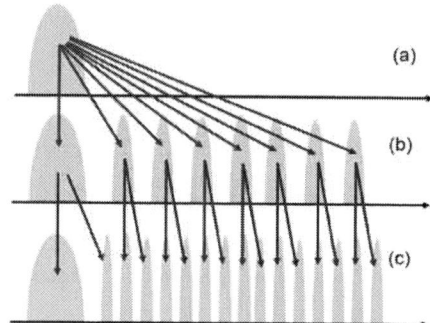

Fig. 8. 4 bits/cell programming algorithm.

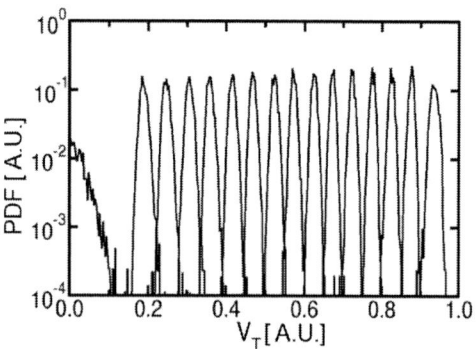

Fig. 9. Vt distribution of cells in the 4bits/cell die.

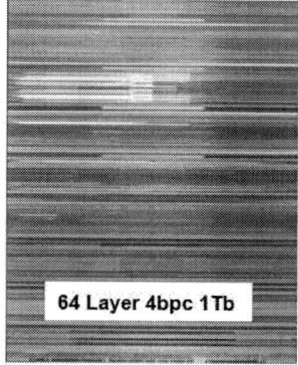

Fig. 10. Die photo of the 64 layer 4bits/cell 1Tb die.

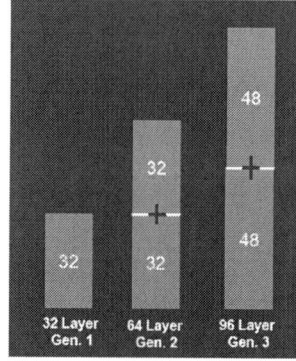

Fig. 11. Scaling methodology for the successive generations of the FG 3D NAND.

Fig. 12. SEM cross-section of the 3rd Gen FG 3D NAND with 96 active layers and CMOS under the array.

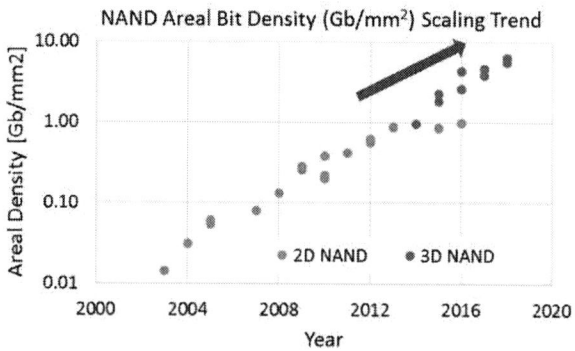

Fig. 13. NAND Areal bit density scaling. Transition to 3D NAND has enabled continuation of the historic scaling trend.

978-1-7281-1988-5/18 $31.00 © 2018 IEEE

Analysis and Realization of TLC or even QLC Operation with a High Performance Multi-times Verify Scheme in 3D NAND Flash memory

C.C. Lu[*], C.C. Cheng[*], H.P. Chiu[*], W.L. Lin[*], T.W. Chen[*], S.H. Ku[*], Wen-Jer Tsai[*], T.C. Lu[*], K.C. Chen[*], Tahui Wang[†,*], and Chih-Yuan Lu[*]

[*]Macronix International Co., Ltd. No. 16, Li-Hsin Road, Science Park, Hsin-Chu, Taiwan, R.O.C.
[†]Dept. of Electronics Engineering, National Chiao-Tung University, Hsin-Chu, Taiwan, R.O.C.
email: shku@mxic.com.tw

Abstract — Feasibility of multi-times verify (MTV) scheme on triple-level cell (TLC) and quad-level cell (QLC) operations of charge-trap storage 3D NAND memories is investigated comprehensively. Results reveal that random telegraph noise (RTN) and program noise are the major factors affecting lower (LB) and upper boundaries (HB) of Vt distribution, respectively. Enlargement of operation window and reduction of ECC usage with MTV scheme to mitigate RTN-induced LB tail are demonstrated on TLC and QLC operations. In addition, the impact of program noise on HB Vt under various process conditions and ISPP steps is studied experimentally and also explained by our Monte Carlo simulator. Finally, program performance and reserved margin with and without MTV scheme applied on TLC and QLC operation are demonstrated.

Introduction

NAND flash memory has attracted much interest as it is widely used as storage media of many electronic products, such as smart phones and solid-state drives (SSD) [1, 2]. To meet the continuous demand for high density and low bit cost, 3D NAND based on multi-level cell (MLC) or TLC operations has become the mainstream of flash memory technology development [3-7]. In the near future, QLC packs 33% capacity per NAND cell boost over TLC NAND, enabling enterprise solid state storage at a more affordable price point for a broader range of applications than ever before [8]. Previously, RTN is negligible in 3D NAND structure due to its larger cell dimension relative to 2D NAND [9, 10]. For now, it becomes more and more difficult to have the Vt distribution tighten for sufficient window under TLC or even QLC schemes. Hence, the impact of RTN should be re-considered. Recent research has demonstrated that MTV scheme is promising to alleviate Vt distribution broadening caused by RTN without scarifying the program throughput in MLC FG NAND technology [11]. Meanwhile, it also relaxes the difficulty during process development and relieves the requirement of ECC bit accordingly.

Continued from the preceding understanding, the implementation of MTV scheme on TLC and QLC operation is studied in this paper. Two different process conditions of 3D NAND flash memory are used to extract the λ of RTN [12] and the σdVt saturation level of program noise [13]. Here, λ corresponds to the slope of the exponential distributions and σdVt is the variation of program Vt shift. With applying these values in our Monte Carlo simulation, the ineffective movement (or two-slope behavior) of Vt HB with varying ISPP step on poor process split can be explained. By adopting MTV scheme, the "pass" area of Shmoo plots is extended no matter it is good or poor process condition. Proposed MTV scheme also guarantees both reserved window and program throughput under QLC operation.

Device structure and the Monte Carlo Simulator

A gate-all-around (GAA), vertical-channel device is used in this study. The composition of 3D NAND structure is illustrated in Fig. 1(a). In general, the interference between cells in 3D GAA structure is insignificant [14], and thus enhancing program throughput by a one-step full sequence method is adopted (Fig. 1(b)). To understand the impact of process conditions on operation window, two splits (Split-A and -B) are taken into comparison. Split-A, known for its better process controllability, exhibits smaller program noise and better read stability. Moreover, a Monte Carlo method [15] taking both RTN and program noise into consideration is developed. With the parameters extracted from the measured data, emulated Vt distribution looking alike real silicon one under TLC or QLC operation can be achieved.

Results and Discussion

(1) Window Loss Caused by RTN and Program Noise:

Fig. 2(a) and (b) show the TLC Vt distribution right after passing Program-verify (PV) and at the very first reading, respectively. The LB tail is contributed by those cells passing but still below PV level due to RTN. It becomes noticeable at the 1st reading after PV of all PGM states. In the nature of RTN, a single electron trap randomly switches between two discrete levels, corresponding to the occupied (high-Vt) and the empty (low-Vt) states. Those cells which just pass PV at high-Vt state are likely to be read at low-Vt state during the following operations, thus forming the RTN tail. Fig. 3 compares the Vt distributions of two splits. Theoretically, the distribution width can be approximated simply by ISPP slope and step, but in fact, RTN would ruin LB profile while HB is enlarged by program noise significantly. Split-A displays narrow Vt distributions for its modest RTN and program noise. A fine ISPP step is usually adopted for more compact Vt distribution due to less σdVt [13]. The Vt distributions of Split-B under various ISPP steps are measured and fitted by Monte Carlo simulation, as shown in Fig. 4(a). As expected, the retraction of Vt HB with LB unaltered leads to a compact Vt distribution. The resultant opened window between "E" and "F" states under various ISPP steps for Split-A and -B are compared in Fig. 4(b). It is easy to get sufficient window with Split-A, the good process condition. However, with Split-B in contrast, a modest window can hardly be achieved even by an ISPP step as tiny as 0.2V. Besides, the inconsistent slope of opened window versus ISPP step is observed in Split-B, which can be explained by the variation of program noise saturation levels [13]. The effect of saturation level on σdVt as a function of ISPP step is calculated and shown as Fig. 5(a). As the level increases, simulation (solid lines) converges to the measured data (symbols) as depicted in Fig. 5(b). The Vt HB evolves effectively as ISPP step less than

978-1-7281-1988-5/18 $31.00 © 2018 IEEE

0.4V, but then slows down drastically as exhibited in Fig. 5(c). The low saturation level, restricting the movement of Vt HB, can be explained by a stronger E-field reduction of tunnel oxide during program and is manifested by sub-Poissonian nature of the electron injection process [13].

(2) Multi-Times Verify Scheme:

The schematic diagram of MTV scheme is shown in Fig. 6(a). Cells are erased as usual and then programmed under ISPP scheme. Once any cell passes PV, the procedure would get into MTV stage. Let's assume that both Bit_I and _II pass PV at a certain ISPP shot but only Bit_II is detected as failure during MTV stage. Bit_II is set as pseudo-pass and will be re-programmed at the following ISPP shot while Bit_I is turned to solid-pass or inhibit phase. In this way, the overhead of extra program shot can be minimized, and so is the impact on program throughout. Fig. 7(a) shows that the cumulative tail bit count affected by MTV scheme continues to increase with progressive MTV settings, among which around 90% population can be detected with #MTV=3. Extra verifications can further eliminate the tail spreading but also degrade the program throughput seriously. Hence, #MTV=3 is chosen as the condition for comparison of program throughput and opened window under conventional and MTV schemes. As a result, enlargement of opened window with MTV scheme at a cost of slightly decreased program throughput can be observed as shown in Fig. 7(b).

(3) Analysis of TLC Operation with MTV and ECC

Fig. 8 shows the measured Vt distributions under TLC operation with and without MTV scheme for Split-A and -B, respectively. Narrow Vt packets are preferred to clearly differentiate from each other. MTV scheme is capable of shrinking LB tail and meanwhile keeping HB stable, which leads to narrower Vt packets and enlarged opened window.

Alternatively, an advanced ECC is usually employed to ensure reading data reliable enough. In Fig. 9, the amounts of opened windows between "E" & "F" states at a specific ECC strength with and without MTV scheme for both Split-A and –B are examined. Split-A can sustain read margin inherently as shown in Fig. 9(a), and the addition of MTV scheme further improves it by 25% at ECC strength of 0.1% fulfilled by Bose-Chaudhuri-Hocquenghem (BCH) ECC decoding [16]. As to Split-B demonstrated in Fig. 9(b), however, simple BCH scheme fails to provide reliable margin, and an advanced Low-density parity-check code (LDPC) must be adopted [17]. The cooperation of MTV scheme achieves 30% gain in read margin and hence relieves ECC burden. In Fig. 10, Shmoo plots represent the pass/fail situations with designated window under the criteria of (a) RBER=0.1% and (b) RBER=1%. Pass, Pass-with-MTV, and Fail regions are denoted by colors of green, light green, and red, respectively. As stated previously, RTN and program noise cause loss in Vt window simultaneously. The retrievable window by mitigating RTN noise can compensate the loss caused by program noise, which re-enables the feasibility of Split-B in RBER=1%.

(4) Implementation on QLC operation

In this section, the real-like Vt distributions under QLC operation with small ISPP step and intermediate ISPP step + MTV scheme are analyzed. In addition to more bits storing per cell, the number of Oxide/Nitride (ON) pairs also increases for high density demand [14]. To keep etching profile as vertical as possible, thickness of each pair needs to be thinned down, thus enhancing the interference in vertical direction. Consequently, a generally two-step programming is adopted to achieve a tight Vt distribution [18], as shown in Fig. 11. After erasing of all cells and then programming of low pages (page0/1), high pages (page2/3) are programmed thereafter with (a) small ISPP step or (b) intermediate ISPP step + MTV scheme, individually. Fig. 12 shows the window opened by these two program methods. Fig. 12(b) shows that MTV scheme can effectively suppress LB tail and reserve more operation window for QLC application. Regarding ECC usage, the opened windows versus allowable RBER with two kinds of methods are illustrated in Fig. 12(c) for comparison. The method with MTV scheme significantly improves the opened window in low RBER region, corresponding to utilization of simpler ECC code (like BCH). Accordingly, the iteration time and power consumption in ECC decoding for QLC operation can be reduced. In brief, the enlargement of Vt margin and the relaxation of ECC usage get benefit from MTV scheme. The resultant program throughput between the two program methods are also calculated as shown in Fig. 13. The equation of program throughput with the overhead of extra verification and additional program shot is also listed in the right of the figure to elucidate the 30% improvement from the proposed method. Intuitively, the implement of MTV scheme seems to lower the throughput due to extra verifications, however, other factors should be considered in further. Since QLC operation needs much tight Vt packet, ISPP step is chosen as small as possible. Such tiny step increases the program shot number and thus lowers the throughput. Once MTV scheme is implemented, an intermediate ISPP step can be utilized and the program shot number is no longer a negative factor in throughput calculation. The reduction of shot number gets much more benefit to throughput than the loss due to extra verifications, and as a result, around 30% performance boost can be achieved with MTV scheme. Finally, the influence of MTV scheme accompanied with intermediate ISPP step on QLC operation has been featured as Shmoo plot in Fig. 14, in which the extension of "pass" region with MTV scheme is studied at a fixed RTN. The higher RBER tolerance brings our Split-A back, which evidences that the demand for complicated ECC algorithm is not necessary.

Conclusions

The operational window loss caused by RTN and program noise under TLC and QLC operation in 3D NAND flash memories is studied experimentally. Process parameters from two kinds of splits are extracted for a Monte Carlo simulation, which can explain the two-slope window behavior induced by low program noise saturation level in poor process condition. To further enlarge the operational window, MTV scheme is proposed and implemented. Compared to the conventional program scheme with small ISPP step, an optimal reading time of #MTV=3 along with an intermediate ISPP step can achieve the narrow Vt packets without scarifying and even enhancing its program throughput. The Shmoo plots clearly demonstrate the impact of MTV scheme on process condition and ECC algorithms. Results show that the difficulty of process development and the burden of ECC decoding can greatly be alleviated by adopting MTV scheme.

Reference

[1] S. Aritome, IMW, p. 1, 2016.
[2] TrendForce, "NAND Flash Market Trend", Computex Taipei 2016.
[3] Y. Cai et al., ICCD, p. 123, 2013.
[4] K-T Park et al., VLSI, p. 188, 2007.
[5] K. T. Park et al., NVMTS, p. 1, 2014.
[6] S. Lee, et al., ISSCC, p. 138, 2016.
[7] C. Kim et al., ISSCC, p. 202, 2017.
[8] S. Lee, et al., ISSCC, 20.3, 2018.
[9] K. K. Hung et al., TED, p. 1323, 1990.
[10] S. Aritome et al., IRPS, p. 13.1, 2017.
[11] S. H. Ku et al., IMW, p. 50, 2018.
[12] Y-T Chung et al., TED, p. 1371, 2012.
[13] Neal Mielke et al., EDL, p. 769, 2009.
[14] H. Kim et al., IMW, p. 7, 2017.
[15] H. Li, TED, p. 3527, 2016.
[16] Technologic Systems, "SLC NAND: Secrets Exposed",
[17] E. F. Haratsch, "NAND Flash Media Management Algorithms," FMS, 2017.
[18] S. Aritome, NAND Flash Memory Technologies, Wiley-IEEE Press, 2016

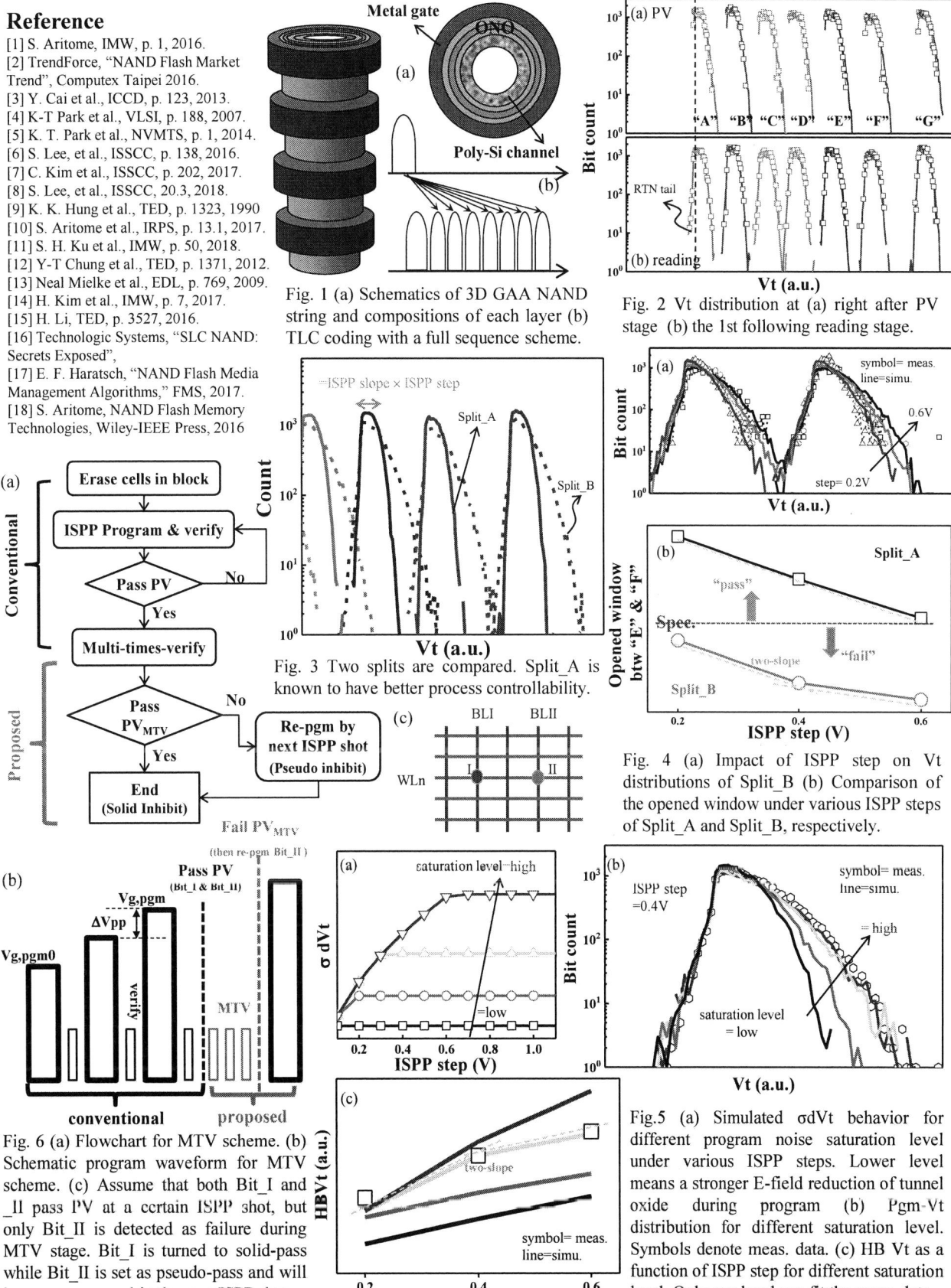

Fig. 1 (a) Schematics of 3D GAA NAND string and compositions of each layer (b) TLC coding with a full sequence scheme.

Fig. 2 Vt distribution at (a) right after PV stage (b) the 1st following reading stage.

Fig. 3 Two splits are compared. Split_A is known to have better process controllability.

Fig. 4 (a) Impact of ISPP step on Vt distributions of Split_B (b) Comparison of the opened window under various ISPP steps of Split_A and Split_B, respectively.

Fig. 6 (a) Flowchart for MTV scheme. (b) Schematic program waveform for MTV scheme. (c) Assume that both Bit_I and _II pass PV at a certain ISPP shot, but only Bit_II is detected as failure during MTV stage. Bit_I is turned to solid-pass while Bit_II is set as pseudo-pass and will be re-programmed in the next ISPP shot.

Fig.5 (a) Simulated σdVt behavior for different program noise saturation level under various ISPP steps. Lower level means a stronger E-field reduction of tunnel oxide during program (b) Pgm-Vt distribution for different saturation level. Symbols denote meas. data. (c) HB Vt as a function of ISPP step for different saturation level. Only one level can fit the meas. data.

978-1-7281-1988-5/18 $31.00 © 2018 IEEE   33

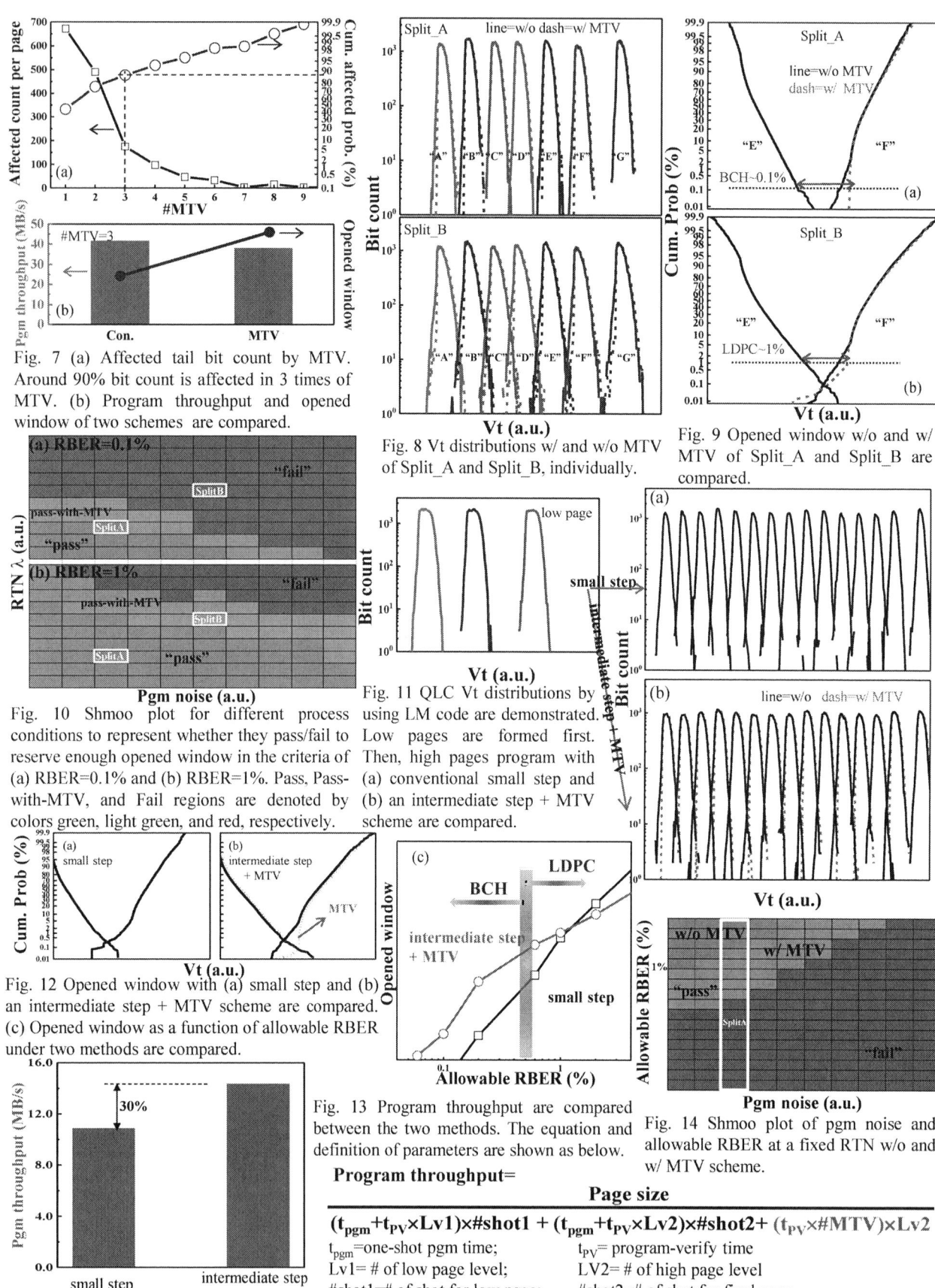

Fig. 7 (a) Affected tail bit count by MTV. Around 90% bit count is affected in 3 times of MTV. (b) Program throughput and opened window of two schemes are compared.

Fig. 8 Vt distributions w/ and w/o MTV of Split_A and Split_B, individually.

Fig. 9 Opened window w/o and w/ MTV of Split_A and Split_B are compared.

Fig. 10 Shmoo plot for different process conditions to represent whether they pass/fail to reserve enough opened window in the criteria of (a) RBER=0.1% and (b) RBER=1%. Pass, Pass-with-MTV, and Fail regions are denoted by colors green, light green, and red, respectively.

Fig. 11 QLC Vt distributions by using LM code are demonstrated. Low pages are formed first. Then, high pages program with (a) conventional small step and (b) an intermediate step + MTV scheme are compared.

Fig. 12 Opened window with (a) small step and (b) an intermediate step + MTV scheme are compared. (c) Opened window as a function of allowable RBER under two methods are compared.

Fig. 13 Program throughput are compared between the two methods. The equation and definition of parameters are shown as below.

Fig. 14 Shmoo plot of pgm noise and allowable RBER at a fixed RTN w/o and w/ MTV scheme.

Program throughput=

$$\frac{\textbf{Page size}}{(t_{pgm}+t_{PV}\times Lv1)\times\#shot1 + (t_{pgm}+t_{PV}\times Lv2)\times\#shot2+ (t_{PV}\times\#MTV)\times Lv2}$$

t_{pgm}=one-shot pgm time; t_{PV}= program-verify time
Lv1= # of low page level;   LV2= # of high page level
#shot1=# of shot for low page;   #shot2=# of shot for final page

Implementing Spike-Timing-Dependent Plasticity and Unsupervised Learning in a Mainstream NOR Flash Memory Array

G. Malavena, A. S. Spinelli, and C. Monzio Compagnoni

Politecnico di Milano, piazza L. da Vinci 32, 20133 Milano, Italy, e-mail: gerardo.malavena@polimi.it

Abstract—In this work, we present the first implementation of spike-timing-dependent plasticity (STDP) and unsupervised learning in a mainstream NOR Flash memory array based on floating-gate cells. A simple yet effective word-line and bit-line pulse scheme is proposed to make a common-ground double-polysilicon NOR array in 40 nm embedded technology work as an artificial synaptic array in a spiking neural network learning according to the STDP rule, with no change required either to the array or to the cell design. With this scheme, long-term potentiation and long-term depression of the synaptic weights are achieved, respectively, by hot-hole injection and channel hot-electron injection at the drain side of the cells. Unsupervised learning is experimentally demonstrated in the array, paving the way for the development of large-scale and high-density neuromorphic systems based on mainstream nonvolatile memory technologies.

I. INTRODUCTION

The idea of developing nonvolatile memory arrays working as artificial synaptic arrays in neuromorphic networks has been attracting considerable interest since its first proposal [1], [2]. Among the different cell structures fit for the purpose, those based on charge storage in floating-gate or charge-trap layers offer the benefits of virtually analog tuning of the synaptic weights, low-power consumption and excellent CMOS compatibility [3]–[7]. In spite of all these benefits, what may make a storage solution far more favorable than the others is the possibility to create the artical synaptic array directly from mature, reliable and highly-scaled mainstream nonvolatile memory technologies with just slight changes in the array design [4].

In this work, we demonstrate the operation of a mainstream common-ground double-polysilicon NOR Flash array in 40 nm embedded technology as an artificial synaptic array learning according to the spike-timing-dependent plasticity (STDP) rule [8]. With no change either in the cell or in the array design, long-term potentiation (LTP) and long-term depression (LTD) of the synaptic weights are achieved through a simple yet effective word-line (WL) and bit-line (BL) pulse scheme, triggering either hot-hole injection (HHI) or channel hot-electron injection (CHEI) at the drain side of the cells. Starting from this pulse scheme, unsupervised learning in the array is, finally, experimentally proved.

II. ARRAY STRUCTURE AND OPERATION

Fig. 1 shows the schematic structure of the common-ground NOR Flash array investigated in this work, featuring stacked-gate nonvolatile memory cells. Test elements of the array, developed in a mainstream 40 nm embedded technology by STMicroelectronics [9] and allowing flexible external biasing of the array lines, were experimentally tested for STDP and unsupervised learning implementation. A mandatory requirement for this implementation is the possibility to perform not only program but also erase operations with single-cell selectivity, overcoming the parallel cell erase typical of Flash arrays. Keeping the CHEI mechanism for cell programming, this was achieved by moving from Fowler-Nordheim (FN) tunneling erase to HHI erase (Tab. I). As schematically depicted in Fig. 2, in fact, both (a) CHEI and (b) HHI need simultaneous WL and BL biases (V_{WL} and V_{BL}, respectively), which guarantee the selectivity of the program and erase operations at the single-cell level in the NOR array. Besides, Fig. 3 shows that large threshold-voltage (V_T, extracted as the V_{WL} giving a constant BL current $I_{BL} = 10$ nA with $V_{BL} = 200$ mV) shifts can be achieved over comparable timescales during (a) CHEI program and (b) HHI erase with the same $V_{BL} = 4.5$ V and relatively low $|V_{WL}|$ ranging from 3 to 8 V. No significant change of cell V_T appears, instead, in Fig. 3(b) when looking at the results for an FN tunneling erase with $V_{WL} = -10$ V and grounded BL, source and p-well, confirming that the reduction of cell V_T during the tests with $V_{BL} = 4.5$ V and negative V_{WL} is due to HHI and not to FN tunneling over the channel or the drain area. In order to compare the efficiency of the CHEI and HHI mechanisms, we directly measured I_{BL} during the program and erase pulses (Fig. 4) and compared it with the injection current (I_{inj}) to the floating-gate extracted from the V_T transients of Fig. 3 [10], avoiding any indirect assessment through equivalent transistor analyses [10]. Results are shown in Figs. 5-6 and reveal that an injection efficiency I_{inj}/I_{BL} close to 10^{-6} can be extracted for both the mechanisms in the explored biasing conditions. This makes HHI an acceptable erase mechanism even from the power consumption standpoint.

III. STDP AND UNSUPERVISED LEARNING

A. STDP

When operating in the subthreshold regime, each cell in the NOR array can be considered as an artificial synapse with weight $w = \exp(-q\alpha_G \Delta V_T / mkT)$ [1], [2], where q is the elementary charge, α_G is the control-gate–to–floating-gate capacitive coupling ratio, m is the subthreshold slope ideality factor of the equivalent transistor, kT is the thermal energy and ΔV_T is the cell V_T shift from a reference condition. Starting from the CHEI and HHI results of the previous section, STDP of the synaptic weight w can be easily achieved with the pulse scheme depicted in Fig. 7. A presynaptic spike at time t_{pre} triggers a double-triangular pulse on the WL of the associated synapse, making V_{WL} linearly grow up to $V_{WL}^{max} = 4$ V, then suddenly drop to $V_{WL}^{min} = -7$ V and finally linearly return to zero. The total pulse duration was set to $t_{WL} = 2$ ms, equally split between the positive and the negative front of the waveform. A postsynaptic spike at time t_{post} triggers, instead, a rectangular pulse on the BL connected to the synapse, delayed by $t_{WL}/2 = 1$ ms and with duration $t_{BL} = 10$ μs. The pulse amplitude was set to $V_{BL} = 4.5$ V. Depending on the time delay $\Delta t = t_{post} - t_{pre}$ between the post- and the pre-synaptic spike, the scheme makes the BL pulse occur either during the negative (Fig. 7(a), $\Delta t > 0$) or during the positive (Fig. 7(b), $\Delta t < 0$) front of the WL waveform. As a result, HHI (reducing cell V_T) and CHEI (increasing cell V_T) take place, respectively, in the former and in the latter case, giving rise to LTP and LTD of w and reproducing the STDP learning rule. This is proved in Fig. 8, where the evolution of w when repeatedly applying the pulse scheme of Fig. 7 with Δt equal to 0^+ (max. LTP) or 0^- (max. LTD) is shown, taking $V_T = 4$ V as a reference for ΔV_T extraction. Results reveal that large changes of w can be achieved through the cumulative effect of the LTP and LTD pulses. Besides, Fig. 9 shows that the ratio between the final (w_f) and the initial (w_i) value of w when applying the STDP pulse scheme displays an exponential dependence on Δt, mimicking the behavior of biological synapses [8]. LTD and LTP, moreover, display a relevant dependence on w_i, with the former getting weaker for decreasing w_i (from part (a) to (c) of Fig. 9) and the latter showing the opposite trend. Finally, Figs. 10-11 show that synapses can withstand large changes of their w for at least 10^5 times with relatively low degradation and preserving their STDP learning capability.

B. Unsupervised learning

We implemented unsupervised learning in the NOR array by considering it as an artificial synaptic array which receives voltage pulses on its WLs as a result of the activity of N_i input neurons and produces an excitatory postsynaptic current (EPSC) on each of its BLs, increasing the membrane potential (V_m) of an output neuron (Fig. 12). Firing of the output neuron occurs when V_m overcomes a threshold value. Following [11], synchronous firing of the input neurons with time periodicity t_{WL} was assumed, switching the firing pattern between a signal pattern (SP) to be learned and a noise pattern (NP). To achieve LTP of the synapses excited by the SP and LTD of the other synapses, we modified the pulse scheme of Fig. 7 by: i) simplifying the WL waveform in a double-rectangular pulse of positive and negative amplitude equal to, respectively, V_{WL}^{max} and V_{WL}^{min}; and ii) introducing a second BL pulse delayed by $t_{WL}/2$ with respect to the first (Fig. 12). This allows to maximize CHEI and HHI in the presence of the BL pulse,

speeding up the learning process. Besides, timings allow to reproduce an unsupervised STDP rule thanks to the uniform firing of the output neuron during the positive interval of the WL waveforms. This results in the application of the first BL pulse during the negative interval of the WL waveforms of the same pattern, giving rise to the LTP of the synapses that fired before the output neuron. The second BL pulse, instead, occurs during the positive interval of the WL waveforms of the subsequent pattern, contributing to the LTD of the synapses that fired after the output neuron.

To prove the functionality of the proposed unsupervised learning scheme, Fig. 13 demonstrates, first of all, that no change in the w of a synapse occurs when (a) only BL pulses or (b) only WL pulses are applied, confirming that LTP and LTD take place just when a postsynaptic spike occurs in the presence of an excited synapse. LTP of the synapses excited by the SP and LTD of the other synapses are experimentally proved in Fig. 14, where the evolution of the w of 8 synapses is reported as a function of the learning epoch (number of SP and NP applied at the input). As done in [11], the definition of the firing patterns of the input neurons, the integration of the EPSC and triggering of the BL pulses were performed by an ad-hoc circuit board driven by a microcontroller. By keeping the number of input neurons firing during the NP low, firing of the output neuron occurs mainly in the presence of the SP. This results in the LTP of the synapses excited by the SP and in the LTD of the other synapses during the subsequent NP phases, giving rise to unsupervised learning in the array.

IV. CONCLUSIONS

In this work, we reported the first implementation of STDP and unsupervised learning in a mainstream NOR Flash memory array operated as an artificial synaptic array in a spiking neural network. LTP and LTD of the synaptic weights according to the STDP learning rule were achieved by a simple pulse scheme triggering HHI and CHEI at the drain side of the cells, without the need of changes either in the cell or in the array design. Results are an important step to the development of large-scale and high-density neuromorphic systems based on mainstream memory technologies.

V. ACKNOWLEDGMENTS

The authors would like to thank P. Cappelletti and F. Piazza from STMicroelectronics for support. This article received funding from the European Research Council (ERC) under the European Union's Horizon 2020 research and innovation programme (Grant Agreement No. 648635).

REFERENCES

[1] C. Diorio, et al., *IEEE-TED*, vol. 43, pp. 1972–1980, 1996.
[2] C. Diorio, et al., *IEEE-TED*, vol. 44, pp. 2281–2289, 1997.
[3] S. Ramakrishnan, et al., *IEEE-TBCS*, vol. 5, pp. 244–252, 2011.
[4] X. Guo, et al., in *IEDM Tech. Dig.*, pp. 151–154, 2017.
[5] H.-S. Choi, et al., *IEEE-TED*, vol. 65, pp. 101–107, 2018.
[6] H. Kim, et al., *IEEE-EDL*, vol. 39, pp. 630–633, 2018.
[7] C.-H. Kim, et al., *IEEE-TED*, vol. 65, pp. 1774–1780, 2018.
[8] G.-Q. Bi and M.-M. Poo, *J. Neurosci.*, vol. 18, pp. 10464–10472, 1998.
[9] C. Boccaccio, "Embedded 1T Flash NOR: still alive at 40 nm. And beyond?," in *LETI Memory Workshop*, 2013.
[10] B. Eitan and D. Frohman-Bentchkowsky, *IEEE-TED*, vol. 28, pp. 328–340, 1981.
[11] G. Pedretti, et al., in *IEDM Tech. Dig.*, pp. 653–656, 2017.

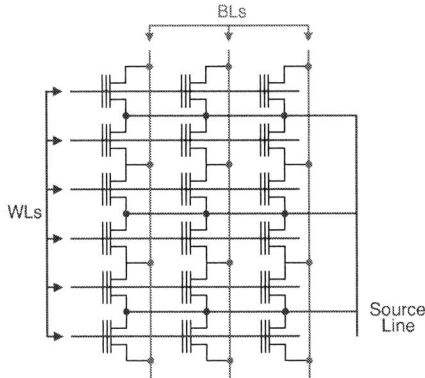

Fig. 1: Schematic for the connections of the stacked-gate cells in the common-ground NOR Flash array investigated in this work.

	Standard operation	This work
Program	CHEI (few bytes)	CHEI (single cell)
Erase	FN tunn. (block)	HHI (single cell)

Tab. I: Physical mechanisms used for cell programming and erasing in the standard operation of the NOR array and in this work.

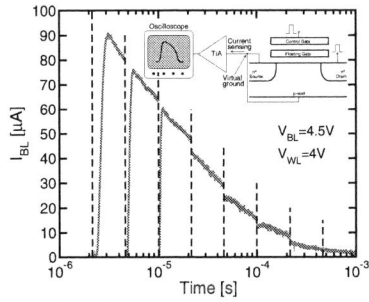

Fig. 4: I_{BL} during a CHEI program transient ($V_{BL} = 4.5$ V, $V_{WL} = 4$ V), as measured through the setup shown in the inset. The vertical dashed lines identify the stretches of time corresponding to the applied programming pulses. The slow rising front of I_{BL} at short times is due to the limited bandwidth of the transimpedance amplifier (TIA).

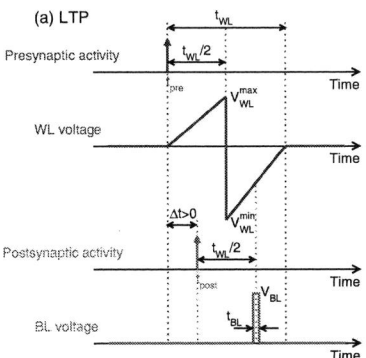

Fig. 7: Pulse scheme exploited to implement STDP in the investigated common-ground NOR Flash array, in the case of (a) presynaptic spike preceding the postsynaptic spike (LTP of w) and (b) presynaptic spike following the postsynaptic spike (LTD of w). By adopting HHI and CHEI for, respectively, LTP and LTD, the proposed pulse scheme is simpler than those previously assumed for other charge-storage artificial synapses [3], [5].

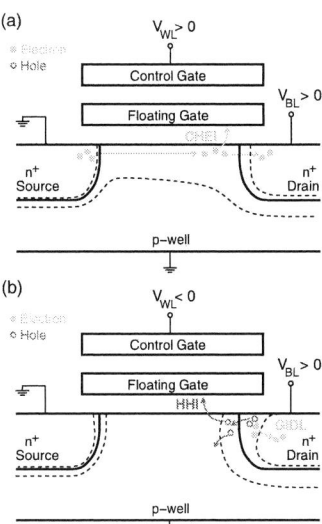

Fig. 2: Schematic for the stacked-gate cells in the investigated NOR array, highlighting the biasing conditions used for (a) CHEI program and (b) HHI erase (holes are generated by band-to-band tunneling at the drain and become hot by moving towards the p region). The source-line and the p-well were always grounded throughout our work.

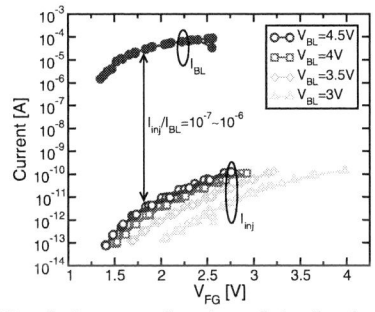

Fig. 5: I_{inj} as a function of the floating-gate potential V_{FG}, as extracted from the CHEI program V_T transients [10], for different V_{BL} and V_{WL}. Results for the average I_{BL} during the program pulses (see Fig. 4) are also reported for $V_{BL} = 4.5$ V.

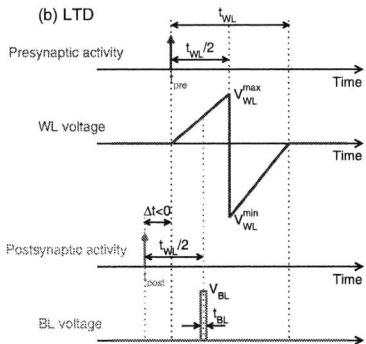

Fig. 6: I_{inj} as a function of the floating-gate potential V_{FG}, as extracted from the HHI erase V_T transients [10], for different V_{BL} and V_{WL}. Results for the average I_{BL} during the erase pulses (similar to Fig. 4) are also reported for $V_{BL} = 4.5$ V.

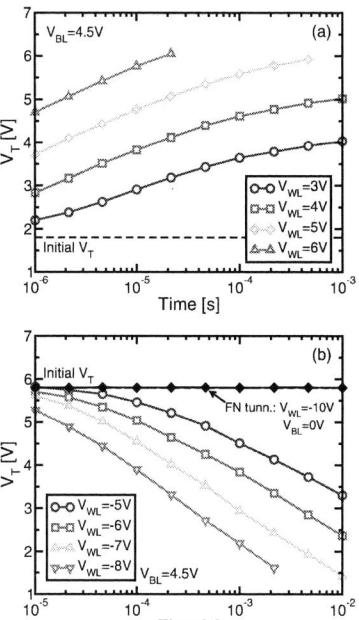

Fig. 3: Experimental cell V_T transients during (a) CHEI program and (b) HHI erase, for $V_{BL} = 4.5$ V and different V_{WL}. In (b) the results for an FN tunneling erase at $V_{WL} = -10$ V are also reported.

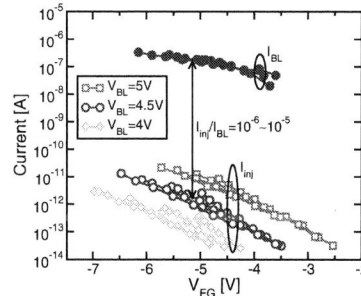

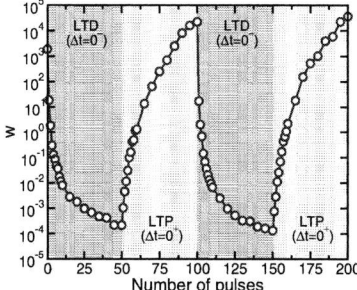

Fig. 8: Evolution of w when repeatedly applying the pulse scheme of Fig. 7 with Δt equal to 0^+ (max. LTP) or 0^- (max. LTD). $V_T = 4$ V was taken as reference for ΔV_T and $w = \exp(-q\alpha_G \Delta V_T / mkT)$ throughout this work.

978-1-7281-1988-5/18 $31.00 © 2018 IEEE   37

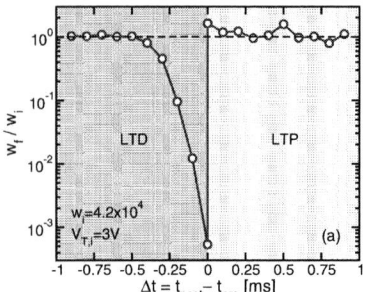 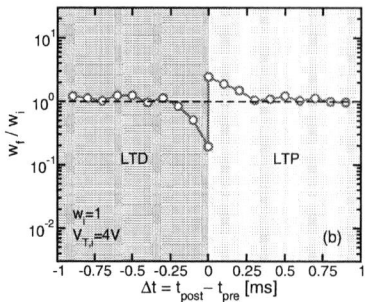 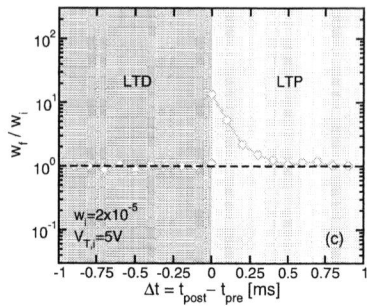

Fig. 9: Ratio between the final (w_f) and the initial (w_i) value of w when applying the STDP pulse scheme of Fig. 7, as a function of Δt (only one single pulse per each Δt value was applied). Results for different w_i (corresponding to a different initial threshold-voltage $V_{T,i}$ of the artificial synapse) are shown in parts (a), (b) and (c).

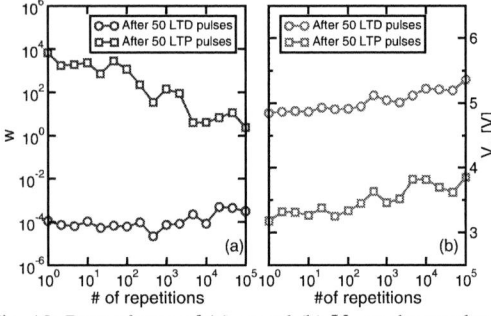

Fig. 10: Dependence of (a) w and (b) V_T on the number of repetitions of an STDP test made of 50 LTD pulses with $\Delta t = 0^-$ and 50 LTP pulses with $\Delta t = 0^+$.

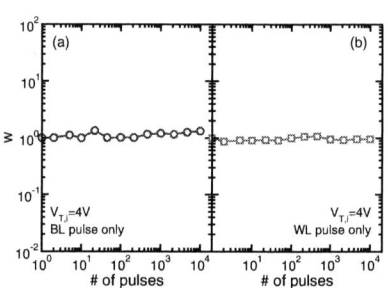

Fig. 11: w_f/w_i resulting from the STDP experiment (the cumulative effect of 50 pulses has been considered) on the synapse previously subjected to the test of Fig. 10.

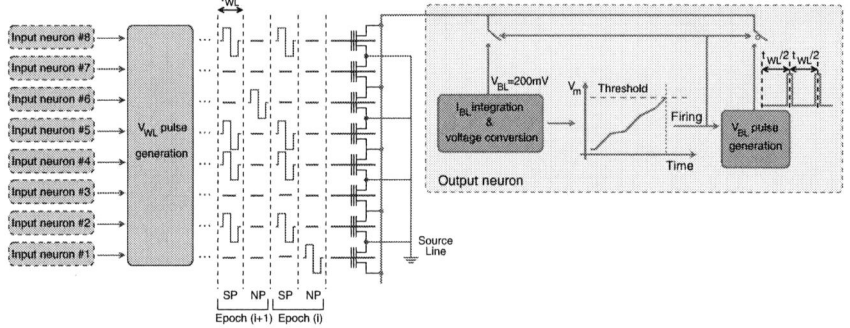

Fig. 12: Schematic for a spiking neural network making use of the investigated NOR array. Synchronous firing of the input neurons with period equal to t_{WL} is assumed, with firing pattern alternated between the SP and a NP. This firing triggers double-rectangular WL pulses, giving rise to an EPSC at the array BLs (for better clarity, only one BL is shown in the picture). The EPSC is integrated and converted into a membrane potential V_m by an output neuron, firing when V_m overcomes a selected threshold. This latter firing triggers two pulses on the BL, with a delay of $t_{WL}/2$, giving rise to LTP or LTD of the array synapses.

Fig. 13: Impact on w of (a) an increasing number of BL pulses (no WL pulses applied) and (b) an increasing number of WL pulses (no BL pulses applied).

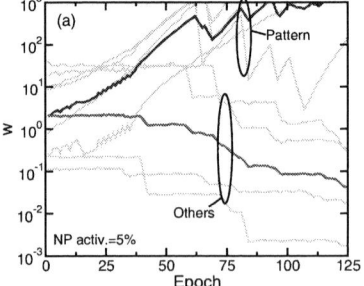 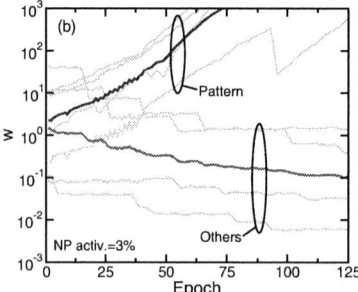 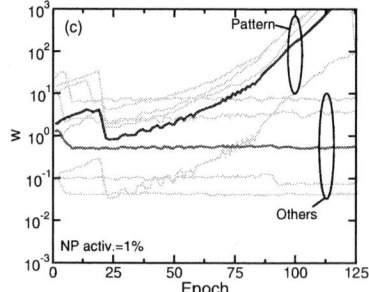

Fig. 14: Results of the unsupervised learning test, for different probability of firing of the input neurons during the NP phase, equal to (a) 5%, (b) 3% and (c) 1%. Shaded lines are the w of the 8 synapses involved in the test, blue lines are the average trend of the w of the synapses excited by the SP and red lines are the average trend of the w of the other synapses. At the beginning of the test, a random initialization of the w was adopted. Note that, in the explored range, the reduction of the probability of firing during the NP results in a slower learning.

978-1-7281-1988-5/18 $31.00 © 2018 IEEE   38

A Novel Voltage-Accumulation Vector-Matrix Multiplication Architecture Using Resistor-shunted Floating Gate Flash Memory Device for Low-power and High-density Neural Network Applications

Yu-Yu Lin, Feng-Min Lee, Ming-Hsiu Lee, Wei-Chen Chen, Hsiang-Lan Lung, Keh-Chung Wang, and Chih-Yuan Lu

Macronix Emerging Central Lab., Macronix International Co., Ltd.,
16 Li-Hsin Rd. Hsinchu Science Park, Hsinchu, Taiwan, ROC
TEL: +886-3-5786688 ext. 78158, Email: yylin01@mxic.com.tw

Abstract—We propose a novel processing-in-memory (PIM) architecture based on the voltage summation concept to accelerate the vector-matrix multiplication for neural network (NN) applications. The core device is formed by adding a buried shunt resistor to a floating gate Flash memory device. The NN string is constructed the same way as in NAND Flash by connecting the core devices in series. In perceptron operation the weighting factors are stored in the floating gate device and the sum-of-product is readily obtained by summing the voltage drop of the cells in each NN string. The energy consumption for 128 multiply-and-sum operations within a string can be as low as 0.2pJ. Finally, with the weight values stored in the non-volatile memory there is no need to move data around and this greatly improves the performance and energy efficiency for neural network applications.

I. INTRODUCTION

Moving data between the processing unit and memory components is the energy bottleneck in conventional von Neumann computer architecture. Even in neuro-network computation much energy is still needed to repeatedly access the weight values and to process the input information. Consequently, new computing approaches such as the processing-in-memory architecture [1, 2, 3] attracted much attention as a promising solution of providing better efficiency for the deep learning technique [4]. Most PIM proposals, however, use sum-of-current approach thus suffers from high leakage and high power consumption. In this work we explore a novel PIM using sum-of-voltage based on a Resistor-shunted Floating Gate (RS-FG) device to drastically improve the energy efficiency in a high-density array.

II. VOLTAGE-ACCUMULATION VECTOR-MATRIX MULTIPLICATION ARCHITECTURE

Figure 1 shows the architecture and RS-FG device of our novel voltage accumulating PIM proposal. Instead of just a conventional floating gate device, a resistor is connected to the source and drain in parallel to the normal surface channel, formed by ion implantation (Fig. 1(a)). A voltage accumulating string is constructed in serial connection the same way as in NAND Flash memory (Figs. 1(b), 1(c)). No contact is required within the voltage accumulating string and

the array can achieve high density like NAND Flash. The resistance of the parallel resistor (the buried channel) is designed to be much higher than that of the surface channel of floating gate device when it is turned on. Unlike in a NAND Flash, for which all devices in the string must be turned on to read the string current, the buried resistors allow the reading of the status without turning on all the floating gate devices in the string. This is a tremendous power advantage, since it eliminates the need to bring all word lines in the string to Vpass which consumes much energy.

Figure 2(a) shows the PIM sum-of-product concept. The output value of the "perceptron" is the sum of all "input times the weighting factor" for every element in a matrix. In the matrix shown in Fig. 2(b), the inputs are the voltages applied to the word lines while the weighting factors are the Vt's of array cells that are programmed into the RS floating gate transistors. For any single cell in this matrix (Fig. 1(a)) when the input voltage is higher than Vt, the transistor is turned on and current flows mostly through the transistor surface channel (transistor state, TS). When the input voltage is lower than Vt, the transistor is turned off and the current only flows through the buried resistor (resistor shunted state, RSS). For such a voltage-accumulating string, the string voltage detected by the sensing circuit is determined almost entirely by the number of transistors that are turned off. (Unlike conventional NAND, for which unselected transistors must be turned on to read the selected one, the parallel fixed value resistors allow many, even all, transistors to turn off yet can still read the state of the string.) The sum-of-product is then directly reflected in the number of transistors turned on (or off). By forcing a specific current through the bit line string the voltage drop can be measured (Fig. 2(b)), which is linearly correlated with the sum-of-product.

Figures 2(b), 3, and 4 show the schematic of device hierarchy, including the RS-FG cell, a voltage accumulating string, a block, a tile, and the sensing circuit. The input values are shared for one certain block. The sensing currents are applied on each bit line to detect the accumulated voltage which is used to do the sum-of-product calculation simultaneously. The number of inputs and outputs of the perceptron corresponds to the number of WL's on the voltage accumulation string and the number of bit lines in the array, respectively. One block is selected for each tile. The inputs and weight values are distributed into many tiles and the accumulating voltages are collected by the voltage manipulation function block which does the sum or average operation. The sensing circuit measures and transfers the voltage from the voltage manipulation block to the output digits with limited sensing levels and resolution, e.g. 3 bits or 4 bits. The input and output numbers of the perceptron and sum-of-product calculation can be easily extended to thousands in this RS-FG vector-matrix multiplication architecture and greatly improve the throughput performance. Handling the voltage sum or average before sensing scheme can reduce the error of the sum-of-product result from the limited sensing levels and resolution.

III. DEVICE DEMONSTRATION

A tester of voltage accumulating string based on 75 nm NAND Flash is used to demonstrate the proposed RS-FG vector-matrix multiplication architecture. The buried channel resistance as function of n-type doping is shown in Fig. 5. Lower buried channel resistance provides higher sensing current for neural network calculation with higher sensing speed and throughput, but leads to smaller resistance window and is more sensitive to variation. The program/erase/select operation scheme of the voltage accumulating string is similar to the NAND flash device, which is shown in table 1. Figure 6 is the measured Id-Vg curves from the voltage accumulating string tester with the buried channel parallel resistors. The device property in Fig. 6 is used to simulate a 9 unit cells voltage accumulating string and calculate a 3x3 matrix multiplication and summation, as shown in Fig. 7. The voltage drop summation is well correlated to the sum-of-product computation results. However, the slanting curves below 1.5uA at the lower left corner of Fig. 6 show the interference from the gate voltage to the buried channel parallel resistor and enlarges the variation of the sum-of-product calculation.

To reduce gate interference a guard layer is inserted between the surface channel and buried channel to shield the buried channel from the gate voltage, and we can obtain a more ideal Id-Vg curves as shown in Fig. 8. The buried channel current is smaller than 1nA and the incremental step pulse programming (ISPP) scheme can be applied to fine tune the Vt distribution, hence it obtains designated weight value which in turn increases the NN accuracy.

Neural network calculation throughput is proportional to the sensing current but the maximum applied sensing current is limited by the highest acceptable accumulated voltage on each string (~1V) to eliminate body effect on the cells. To raise the current of the RSS state an additional bias voltage is applied to the P-well (PWI) as in Fig. 9. (The PWI bias lowers the barrier of the P-type guard layer to N-type Source/drain.) Figure 10 shows the Id-Vg curves of different Vt for the PWI bias = 0.5V. Fig. 11 shows the cell resistance corresponding to the high or low Vt values (programmed weight) and various Vg (input). When Vg < Vt the transistor is turned off and the cell shows RSS. When Vg > Vt the transistor is turned on and the cell shows TS with low resistance. When Vg ~ Vt the transistor is not fully turned on and the cell can fall into some intermediate state, which can affect the accuracy of the NN.

The RSS is about 1M-ohm and about 23.5x the resistance of TS.

Fig. 12 shows the measured Id-Vg curves of 5 cells on one voltage accumulating string. Two program modes are used to demonstrate the sum-of-product calculation concept. (a) all 5 cells are at low Vt and (b) 3 cells are at low Vt and 2 cells at high Vt. All the 32 combinations of input values are applied on the 5 WL's. No pass voltage is needed on WL while doing the sum-of-product operation thus allows fast access response. The sum-of-product results range from 0 to 5 and 0 to 3 for case (a) and (b), respectively, as shown in figure 13. The applied sensing current is 30nA and is forced through the bit line string. Low sensing current results to low power consumption and highly parallel operation. The accumulated voltage can be measured and is linearly correlated with the sum-of-product result from 0 to 4. The mismatch of the point at the sum-of-product result value 5 is due to the non-optimized process condition.

Fig. 14 shows the results of simulating the accumulated voltage distributions of computed sum-of-products for 128 cells in a voltage accumulating string, using the device characteristics shown in Fig. 10. The distribution is divided to 8 groups and 16 groups based on the sum-of-product results for 3 bits and 4 bits sensing circuit resolution, respectively. Each group is separated with small overlap in Fig. 14(a, b) and the accumulated voltages and the computed sum-of-product are linear and well correlated in Fig. 14(c, d). These results show that the new approach is capable of dealing with a large number of inputs and weighting factors for larger scale NN calculation.

The energy efficiency can also be estimated here. For a string of 128 cells, the energy consumption could be 100nA(sensing current) * 1V(max. accumulation voltage) * 2us(read time) = 0.2pJ for computing the 128 sum-of-products. Such low energy consumption almost can be ignored as compared to the typical neutral network computation system [5]. That means we can provide a high energy efficiency array architecture, based on novel RS-FG device, for NN applications.

IV. SUMMARY

A novel vector-matrix multiplication architecture is proposed based on the voltage accumulating concept. The unit cell is implemented by a resistor connected in parallel to the floating gate device. This voltage accumulating string can be configured into a high density array. This device may be implemented in deep neural network systems and provide low power, high density, reliable, low cost, and high throughput PIM. It can substantially improve the performance and energy efficiency for NN applications.

REFERENCES

[1] D. Kuzum, et al., Nano Lett., 12 (5), pp 2179–86, 2012.
[2] P. Chi, et al., ISCA, pp. 27-39, 2016.
[3] F. M. Bayat, et al., ISCAS, pp. 1921-24, 2015.
[4] Y. LeCun, et al., Nature 521, 436–444, 2015.
[5] V. Sze, et al., Proc. IEEE, 105, 12, pp. 2295-2329, 2017.

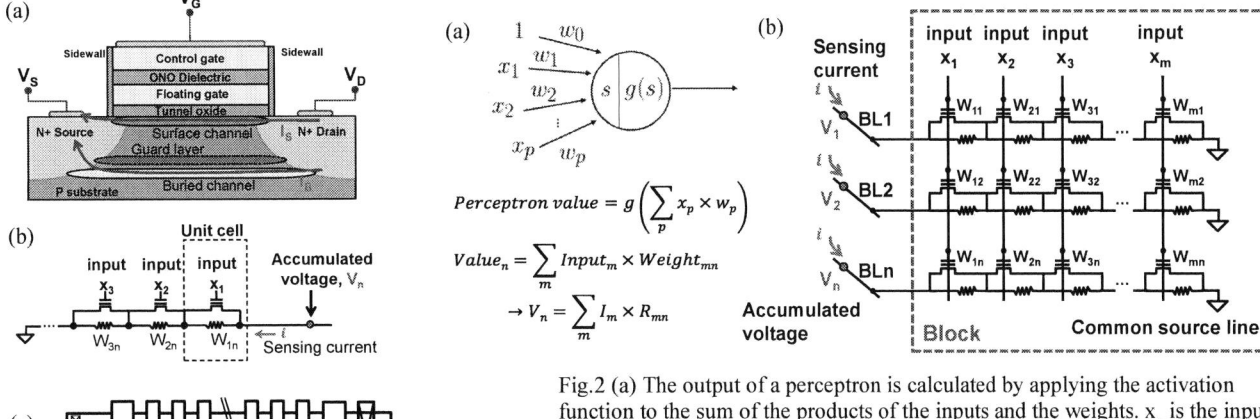

Fig.2 (a) The output of a perceptron is calculated by applying the activation function to the sum of the products of the inputs and the weights. x_p is the input value, w_p the weight value, and g(s) the activation function. (b) The schematic of one block of the voltage-accumulating neural network architecture. The cascaded voltage at each BL head represents the simultaneous sum-of-product operation from that BL string. (BL index : n, WL index : m)

Fig.1 (a) Schematic cross section of the floating gate device with a guard layer and a parallel buried channel resistor. I_S: surface channel current, I_B: buried channel current. When Vg < Vt, the surface channel is not formed and current (I_B) only goes through the resistor. When Vg > Vt, the surface channel dominates and current = I_S + I_B. (b) The voltage-accumulating matrix multiplier architecture is configured with series connected unit cells. (c) Schematic top view and cross section of the buried channel floating gate device. The layout is compatible with NAND Flash and the unit cell has $4F^2$ footprint. (SSL: string select line, GSL: ground select line)

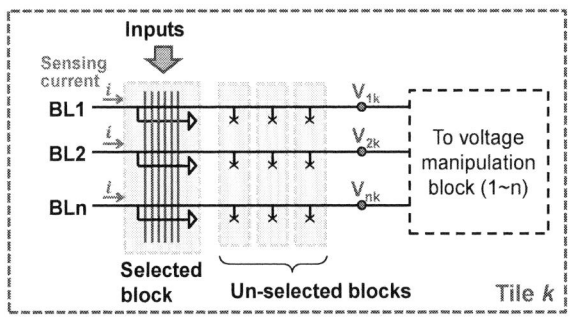

Fig.3 Schematic of many blocks in one tile. Multiple blocks may share the same BL and then select by SSL and GSL. The sensing current is applied to each bit line and the cumulated voltage is measured for every bit line in the selected block. (Tile index : k)

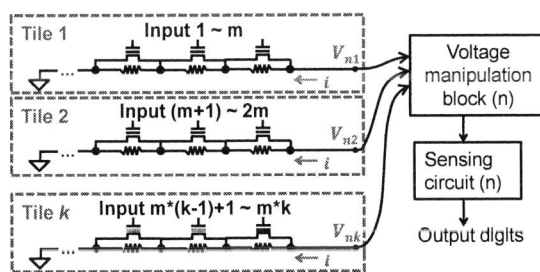

Fig.4 Schematic of the architecture from the inputs and tiles to the sensing circuits and output digits. The inputs and weight values are distributed into many tiles and the outputs are collected with the voltage manipulation function block. The sensing circuits measure and transfer the voltage sum or the average value to the output digits with definable resolution.

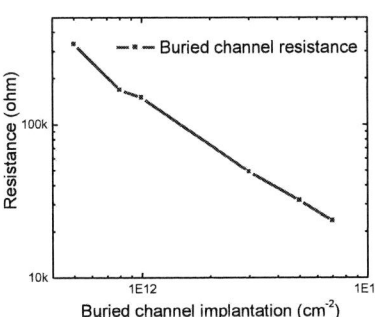

Fig.5 Experiment results on the buried channel resistance versus As implantation dosage. The buried channel resistance can be optimized for suitable resistance window and better power consumption.

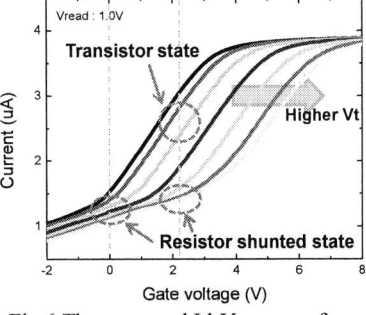

Fig.6 The measured Id-Vg curves for different Vt are measured on the 75nm floating gate device with a buried parallel resistor. The saturation of the cell current at high Vg is caused by the string configuration and could be ignored since the proposed NN sensing current is lower than it.

		GSL	SSL	WL	BL	Sub.
Program	Selected block	G	P	Sel. : Programming voltage (16~24V) Non-sel. : P	Sel. : G Non-sel. : Inhibit voltage	G
	Non-sel. block	G	G	G	None	
Erase	Selected block	F	F	0V	20V	20V
	Non-sel. block			F		
Neural network calculation	Selected block	P	P	Input voltage High : 3V Low : -1V	Sensing current at each bit line	PWI bias
	Non-sel. block	G	G	Prepare for input data	None	
	Stored weight values : (Vread = 0.1V) High Vt : 4.8V @ 10^{-7}A Low Vt : 0.4V @ 10^{-7}A					
	G: GND, F: Floating, P: Pass voltage					

Table 1 The operation conditions of the resistor-shunted floating gate device.

978-1-7281-1988-5/18 $31.00 © 2018 IEEE   41

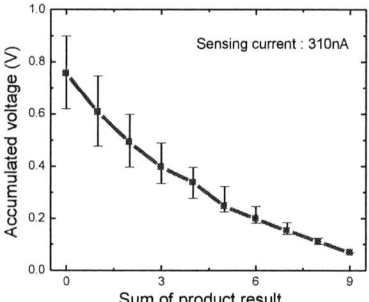

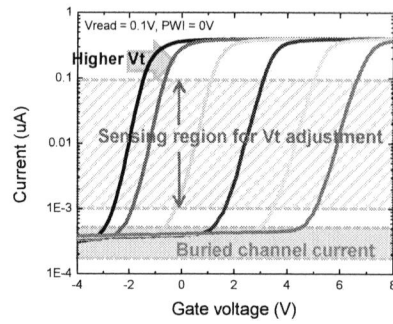

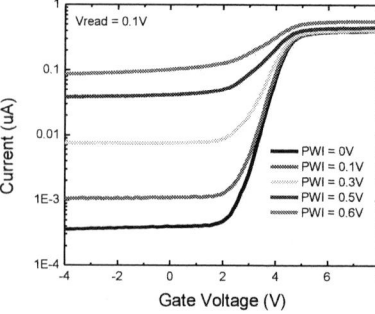

Fig.7 The voltage summation values versus the computed sum-of-product results. 0 to 9 unit cells are selected in one voltage-accumulating string.

Fig.8 The Id-Vg curves for different Vt are measured on 75nm floating gate device with a buried parallel resistor. A guard layer is inserted between the surface channel and buried channel. The buried channel current is below 1nA and there is sufficient window for ISPP Vt program-verify for weight adjustment.

Fig.9 The measured Id-Vg curves under different PWI bias. When the read BL voltage is 0.1V, the Vt is 3.4V at 10^{-8}A and PWI bias = 0V. The buried channel current increases with the increasing PWI bias.

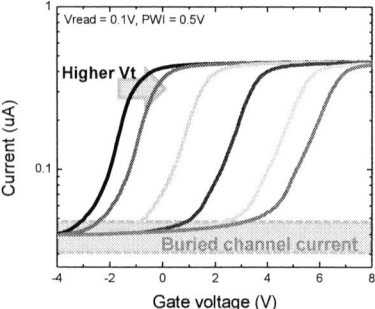

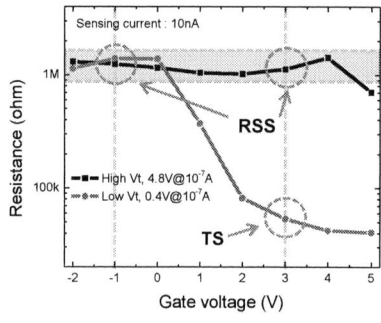

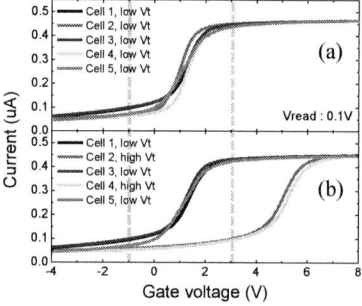

Fig.10 The measured Id-Vg curves for different Vt on the device with a guard layer and a buried channel resistor. The PWI bias is 0.5V and the buried channel current is about 40nA.

Fig.11 The equivalent resistance value of a unit cell that extracted from the data of Fig. 10. The resistance value of RSS has weak dependence on Vg and is about 23.5x of magnitude higher than the resistance of TS.

Fig.12 The Id-Vg curves of 5 cells on one voltage-accumulating string. Two programming scenarios are used to demonstrate the sum-of-product calculation concept. (a) All 5 cells are at low Vt, and (b) 3 cells are at low Vt and 2 cells are at high Vt. The range of sum-of-product result values are 0-5 and 0-3 for case (a) and (b), respectively.

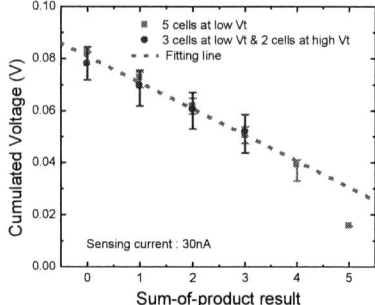

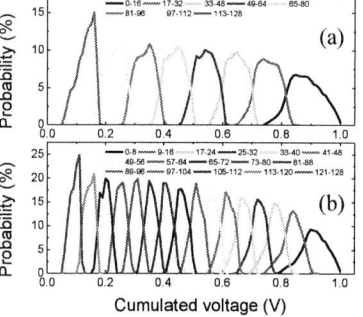

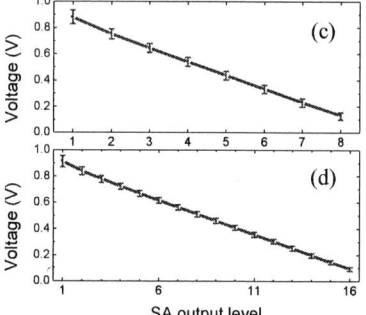

Fig.13 The accumulated voltage values for sum-of-product computation results for case (a) & (b) in Fig.12. The applied sensing current is 30nA. The accumulated voltage value is linearly dependent with the sum-of-product results from 0 to 4. The mismatch between sum-of-product result 0-4 and 5 is due to the junction resistance from non-optimized process condition.

Fig.14 The accumulated voltage distribution of a 128-cell string with the sum-of-product results ranging from 0 to 128. The distribution is divided into 8 groups and 16 groups based on the predefined 3-bit and 4-bit SA output resolution, respectively. The groups are separated with acceptable overlap, as shown in (a) and (b), and the accumulated voltage shows great output linearity with the group sequence ((c) and (d)).

978-1-7281-1988-5/18 $31.00 © 2018 IEEE   42

Vertical Ferroelectric HfO_2 FET based on 3-D NAND Architecture: Towards Dense Low-Power Memory

K. Florent[1], M. Pesic[2], A. Subirats, K. Banerjee, S. Lavizzari[3], A. Arreghini, L. Di Piazza, G. Potoms, F. Sebaai, S. R. C. McMitchell, M. Popovici, G. Groeseneken[1] and J. Van Houdt[1]

imec, Leuven, Belgium, email : karine.florent@imec.be

[1]also with ESAT- KU Leuven, Leuven, Belgium, [2]MDLSoft Inc, Santa Clara, CA, USA, [3]now with Prophesee, Paris, France

Abstract— A vertical ferroelectric HfO_2 field effect transistor based on 3-D macaroni NAND architecture is reported for the first time. Up to 2 V memory window was obtained after the application of 100 ns program/erase pulses. Flash-like endurance of 10^4 cycles is reported and first reliability assessments were performed.

I. INTRODUCTION

The memory market is currently in an era of tremendous growth: the big data explosion has yielded the need for more and more data storage. Although 3-D NAND memory is a high density and cost-effective technology, it still suffers from some drawbacks, *i.e.* speed, cell size and power consumption at system level due to required periphery (e.g. charge pumps). Addressing these issues while keeping the advantages of this technology would be very appealing.

Recent years have seen a growing interest in ferroelectric (FE) memory applications. The discovery in 2011 of FE hafnium oxide enabled the emergence of scaled FE devices [1]. Planar HfO_2-based FE capacitors and FeFET (Ferroelectric Field Effect Transistor) have received large interest either as a replacement for Dynamic Random-Access Memory (DRAM) or for embedded applications [2-3]. A vertical 3-D FE-capacitor (FeCap) has also been realized using silicon electrodes [4]. Implementation of doped HfO_2 - based 3-D FeCaps confirmed the presence of FE properties with 10-year retention [4]. The fabrication of these vertical 3-D FeCaps, using a 3-D NAND architecture, was the first step towards Vertical FeFET (V-FeFET) [5]. Such device could potentially have several benefits over the conventional 3-D NAND memory, such as lower power, periphery reduction, improved endurance and faster operations, while maintaining full CMOS compatibility and high-density.

In this study, a vertical macaroni-type 3-D FeFET with three transistors in series is reported. This is the first demonstration of such device.

II. DEVICE FABRICATION

Fig. 1a shows the process sequence for the test device, consisting in a vertical string of three transistors in series with a channel length (L_{ch}) of 50 nm. The fabrication steps until FE deposition are same as previous work [4]. Here, silicon was used as dopant in HfO_2 to form the FE material, deposited by atomic layer deposition (ALD). A thin a-Si layer was then deposited as gate stack protection from the subsequent etching of the bottom of the hole, required to allow the contact between

the source junction at the substrate and the channel. A HF clean followed by a TMAH etch are then performed to remove the remains of the protective layer. A 20 nm n-type amorphous silicon (4×10^{19} cm^{-3}) is then deposited as a channel. To crystallize the channel and the FE layer, a 30 min anneal is performed at 900 °C in an ambient mixture of N_2 and O_2. In the next step, the hole was filled with oxide and then recessed (hence macaroni structure). Highly doped n-type silicon was deposited on top of the structure to form the drain and then annealed. Finally, the staircase was formed, followed by metal contacts. A cross-section schematic of the complete test vehicle is shown in Fig. 1b. HR-TEM cross-section of a 100 nm diameter hole can be seen in Fig. 2. A deep recess in the substrate is visible due to the TMAH etch.

In parallel of the macaroni-type FeFET device fabrication, highly-doped full-channel 3-D devices were also made to confirm the presence of ferroelectricity in this configuration and extract FE properties, as previously described [4].

III. RESULTS AND DISCUSSIONS

A. Ferroelectric Capacitor:Material screening

9.5 nm- and 15 nm-thick Si:HfO_2 were deposited in the full-channel configuration and characterized using polarization-voltage (P-V) measurements. Cycling dependent P-V and corresponding current-voltage (I-V) characteristics are shown in Fig. 3. Both thicknesses under test exhibited FE behavior accompanied by a slight wake-up. Fig. 4 shows the endurance of these devices up to 10^5 cycles. An initial increase in remnant polarization was observed which was followed by a stable behavior. No fatigue was observed, as devices were no longer operational after 10^5 cycles. Both devices have comparable FE characteristics. Due to the presence of several etch steps during device fabrication, which could potentially damage the FE layer, a 15 nm-thick dielectric was preferred.

B. V-FeFET:Standard Electrical Characterization

Before investigating V-FeFET memory performance, standard transistor measurements were carried out. A pass voltage of 3 V was applied to both the top and the bottom transistors, while the source was grounded. Output characteristics (I_D-V_D) were recorded (Fig. 5) for a hole diameter of 70 nm. Typically for scaled MOSFET devices, an increase in current was observed at large drain and gate voltages. The I_D-V_G characteristics are shown in Fig. 6 for multiple devices with 70 nm hole diameter. The gate had negligible leakage current, as shown in the inset. An ON-

978-1-7281-1988-5/18 $31.00 © 2018 IEEE

current of ~1 μA was achieved, while maintaining an OFF-current of a few pA, giving an excellent ON/OFF ratio of ~6 decades. The statistical distributions of the threshold voltage (V_T), extracted using a constant current (CC) criterion (10 nA), the ON-current I_{ON}, taken at $V_T + 2$ V and the subthreshold swing STS are shown in Fig. 7 for various hole diameters. V_T and STS distributions were broad, which suggested the presence of a defect-rich HfO_2 / Poly-Si interface.

C. V-FeFET:Memory Characterization

As charge trapping at the interface can effect proper cell read-out, DC and pulsed I_D-V_G measurement methodologies were benchmarked. This first technique is relatively slow, as each measurement at a given voltage requires milliseconds. In the presence of traps, this can lead to a degradation of the characteristics and to a smaller FE memory window (MW). Pulsed I-V measurements, which are in the order of microseconds at every voltage, can reduce charge trapping. FE MW is shown here for a vertical macaroni-type FET with a hole diameter of 70 nm, a pass voltage of 3 V and ± 10 V program/erase (PRG/ERS) pulse with a width of 100 ns (Fig. 8). A MW of ~2 V was obtained with pulsed I-V. This was further utilized as the preferred method to characterize memory devices, as DC I-V resulted in an I_D-V_G degradation (charge trapping) as well as a smaller MW: MW pulsed ~2 V vs MW DC ~0.5 V.

To investigate the optimal operation conditions for the device, ISPP and ISPE on a 70 nm-diameter device were performed and are shown in Fig. 9 and 10, respectively. A ± 10 V with 100 ns pulse length was used to program and erase before the measurements. The ISPP showed a gradual increase in the MW with pulse amplitude, while ISPE yielded a more sudden step. This could be due to the presence of a small number of domains because of the small size of the device. The gradual programming could be the result of some trapping interferences and/or different domain growth kinetics. At larger pulse widths and high voltages, charge trapping starts to dominate FE.

The evolution of V_T with cycling is shown in Fig. 11. Closure of the MW was observed at 10^4 cycles. A slight wake up period of around 10 cycles was initially observed. This was followed by a decrease in the MW. While the ERS state decreased slightly with cycling, the PRG state was much more impacted by the endurance possibly due to trapping [6]. Fig. 12 shows the PRG and ERS density distributions after 10 cycles. The device-to-device variability may be caused by differences in defect distribution and charge injection as well as variable FE properties, that can be affected by a combination of factors such as film uniformity and mechanical stress. The ERS state was narrower than the PRG, confirming that one state was more stable than the other. Further integration improvements are required to enhance MW characteristics.

D. V-FeFET: Reliability Consideration

High fields over the interface can generate defects and can also cause injection of charges into defects, resulting in charge trapping which causes a reliability issue for FeFET devices. A way to study the trapping mechanisms in high-k material is to use single-pulse technique [7]. The shift between the rising and falling edges of the pulse ($\Delta V_{T21} = V_{T2}-V_{T1}$) is proportional to the number of trapped charges, here electrons (Fig. 13 inset). The trapping process was studied on a transistor in PRG state by changing the pulse width and amplitude. A positive ΔV_{T21} shift in I_D-V_G confirmed the occurrence of electron trapping (Fig. 13a) which increased with both pulse width and amplitude, as shown in Fig. 13b.

Retention measurements were also carried out. Short-term retention at 25 °C (related to FE relaxation) showed a stronger degradation of the PRG state compared to the ERS state (Fig. 14) nicely correlating to the broader V_T distributions of the PRG state compared to the ERS equivalent. The latter one stays relatively steady, with degradation occurring at higher voltages. The same remark held true for the PRG state but with a more intense degradation as a possible result of charge trapping and depolarization. Longer retention tests at 85 °C confirmed a larger degradation of the PRG pulse (Fig. 15). A clear separation of states was observed after 100h at 85 °C.

E. Path for improvements

The presence of high-k dielectric as the gate oxide results in charge trapping phenomenon, countering the ferroelectric behavior. This is a known issue for HfO_2 FeFET [6], which needs to be addressed through process integration optimization. A thick oxide, without special treatment or film improvement, was used, leading to large PRG/ERS voltages. This could be seen as too high for FE, however this corresponds to an electric field of ~ 6.5 MV/cm and is in agreement with the P-V loop shown in Fig. 3. A thinner gate oxide would reduce the operation voltage and additionally improve the reliability, i.e. reduce trapping and enhance endurance. Detailed reliability studies are required for better understanding of the device behavior, i.e. impact of pass voltages. The potential of 3-D FeFET and its advantages compared to the prior state-of-the-art are presented in Table 1. This technology offers low power high-density memory that decreases the speed gap between the central processing unit (CPU) and storage.

IV. CONCLUSIONS

A functional vertical macaroni-type 3-D FeFET with three gates was fabricated and characterized for the first time. Memory window up to 2 V was obtained with pulse width of 100 ns. Endurance of 10^4 cycles and reliability assessments are also reported. A decrease of the FE layer thickness could potentially decrease the operation voltage and increase the endurance and lifetime of these devices. This study paves the way for a high-density high-speed non-volatile memory that reduces the gap between CPU and storage.

REFERENCES

[1] T. S. Böscke *et al*, APL, vol. 99, no. 10, pp. 102903-1–102903.3, 2011
[2] S. Dunkel *et al.*, 2017 IEEE IEDM, pp. 19.7.1–19.7.4
[3] M. Pesic *et al.*, 2016 IEEE IEDM, pp. 11.6.1-11.6.4.
[4] K. Florent *et al.*, 2017 Symp. on VLSI Technology, pp. T158-T159.
[5] Patent US 2016/0181259 A1, 2016
[6] E. Yurchuk *et al.*, IEEE TED, vol. 63, no. 9, pp. 3501-3507, 2016.
[7] D. Heh *et al.*, 2006 IEEE IIRW, pp. 120-124
[8] H. Tanakaet al., 2007 Symp. on VLSI Technology, pp. 14-15

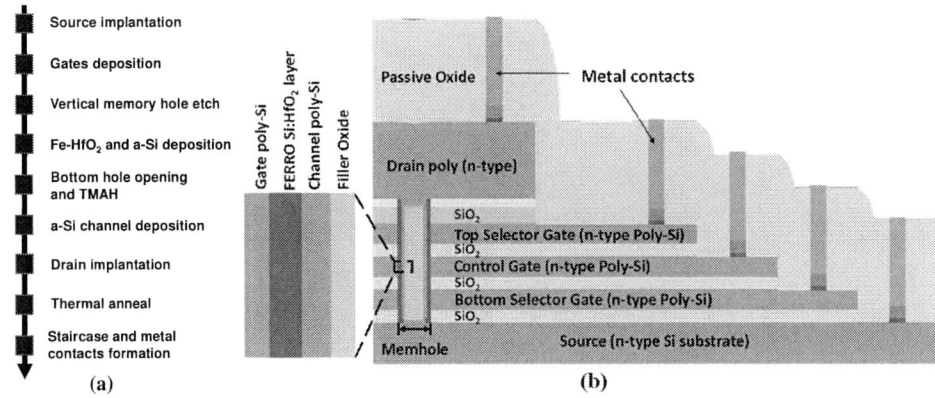

Figure 1. (a) Process Flow description. (b) Schematic cross-section of the macaroni-type 3-D FeFET with three cells in series.

Figure 2. TEM cross-section (Ø: 100nm)

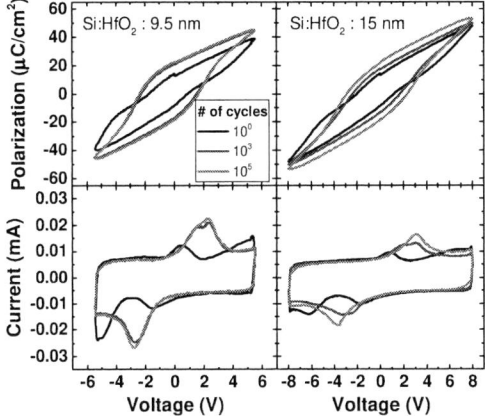

Figure 3. Polarization - voltage (P-V) and corresponding current - voltage (I-V) characteristics for 3-D capacitors with (left) 9.5 nm- and (right) 15 nm-thick Si:HfO$_2$.

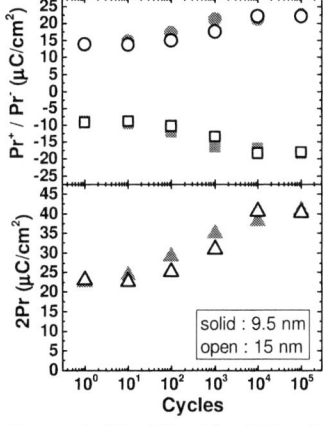

Figure 4. (Top) Positive P_r^+ and negative P_r^- remnant polarization for 3-D capacitors. (Bottom) $2P_r$ endurance showing wake-up behavior.

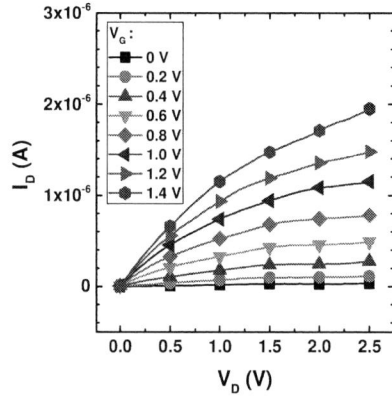

Figure 5. Output characteristics (I_D-V_D) of control gate with 3 V pass voltage, exhibiting typical behavior for scaled MOSFET devices. (L_{ch}= 50 nm)

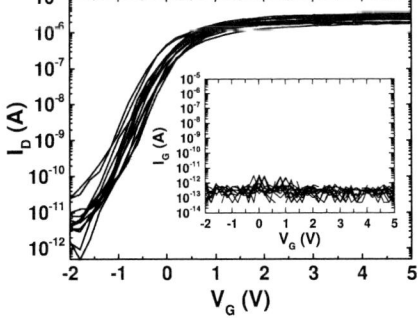

Figure 6. I_D-V_G of control gate transistors with $V_D = 1$ V and $V_{pass} = 3$ V for top and bottom selectors. (inset) I_G-V_G showing no leakage. ($L_{ch} = 50$ nm and Ø: 70 nm)

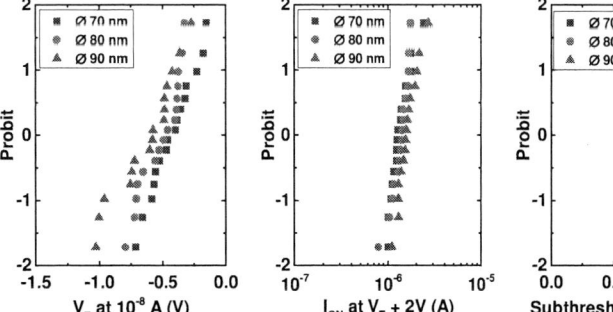

Figure 7. Statistical distribution of (a) V_T taken at CC criterion (10 nA), (b) I_{ON} at $V_T + 2V$, and (c) Subthreshold swing for various diameter holes. Broad distributions of V_T and STS could indicate the presence of traps at the HfO$_2$ / Poly-Si interface. (L_{ch}= 50 nm)

978-1-7281-1988-5/18 $31.00 © 2018 IEEE   45

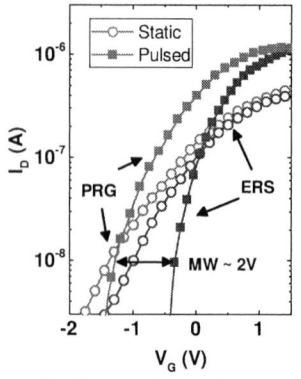

Figure 8. Comparison between DC I_D-V_G and Pulsed I_D-V_G after PRG and ERS (± 10 V – 100 ns). A MW of ~2 V was measured

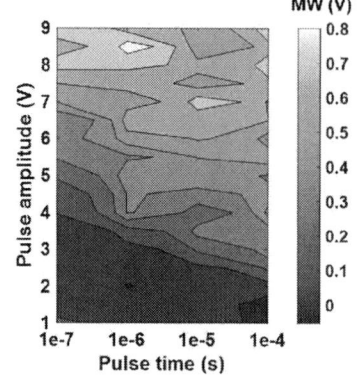

Figure 9. ISPP after -10 V/100 ns erase pulse. A gradual increase is observed with pulse amplitude. Blue: no MW, Yellow: Max MW

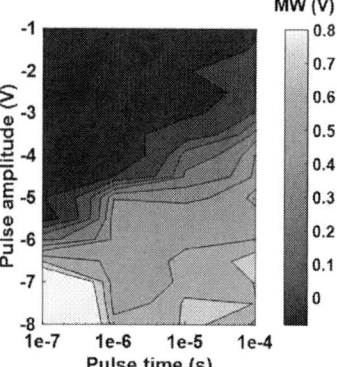

Figure 10. ISPE after +10 V/100 ns program pulse. The MW increase is more sudden. Blue: no MW, Yellow: Max MW

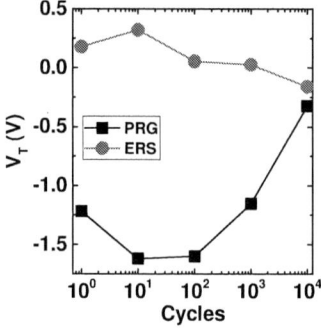

Figure 11. V_T evolution with cycling after PRG and ERS. A larger degradation of PRG is observed.

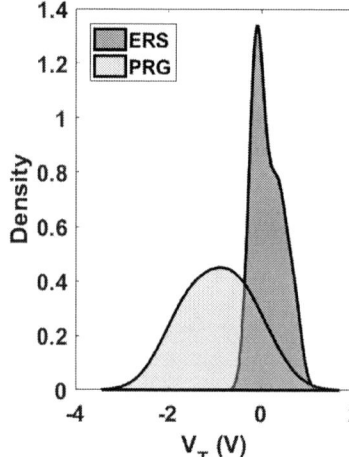

Figure 12. V_T after 10 cycles (± 10 V – 100 ns). ERS distribution is narrower than PRG distribution. (Ø: 70 nm)

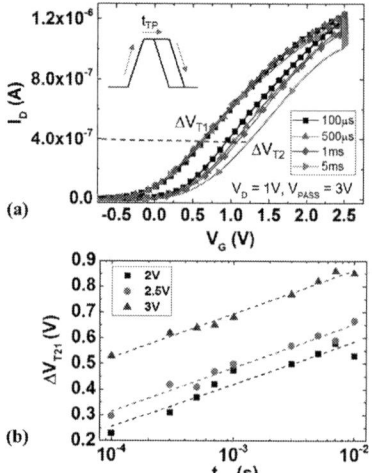

Figure 13. (a) Single pulse I_D-V_G (obtained from rising/falling edge) after PRG for incrementing pulse width t_{TP}, (b) ΔV_{T21} vs t_{TP} (ΔV_T extracted at $I_D = 4 \times 10^7$ A, pulse width: 100 μs to 10 ms, pulse amplitude: 2 V to 3 V)

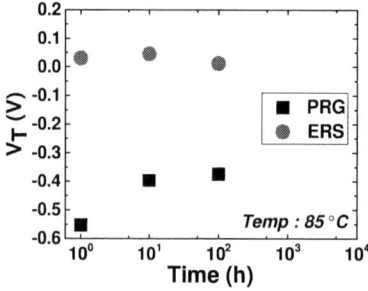

Figure 14. Short retention at 25 °C for (top) PRG and (bottom) ERS states, showing degradation of PRG state.

Figure 15. Retention at 85 °C showing some retention loss of PRG after 100 h.

	Planar FeFET[2]	Flash NAND[8]	This work
Energy	< 1fJ/bit	< 1nJ/bit	< 1fJ/bit
Endurance	10^5 cycles	10^4 cycles (MLC)	10^4 cycles
PRG/ERS Speed	10 ns / 10 ns	100 us / 10 ms	100 ns/100 ns
Retention at 85 °C	10 years *(expected)*	10 years	10 years *(expected)*
Cell area	0.025 μm²	0.026 μm² per N_{layer} *†	0.026 μm² per N_{layer} *
Multibit capability	NA	Yes	NA

*Stackable devices with pitch size of 160 nm
†Assumed similar to this work

Table 1. Comparison with prior art

Hybrid 1T e-DRAM and e-NVM Realized in One 10 nm node Ferro FinFET device with Charge Trapping and Domain Switching Effects

Qing Luo[1,2], Tiancheng Gong[1,2], Yan Cheng[3], Qingzhu Zhang[1,2], Haoran Yu[1], Jie Yu[1], Haili Ma[1], Xiaoxin Xu[1,2], Kailiang Huang[1,2], Xi Zhu[1], Danian Dong[1], Jiahao Yin[1,2], Peng Yuan[1], Lu Tai[1], Jianfeng Gao[1], Junfeng Li[1], Huaxiang Yin[1,2], Shibing Long[1,2], Qi Liu[1,2], HangbingLv[*1,2], Ming Liu[*1,2]

[1]Key Laboratory of Microelectronics Devices and Integrated Technology, Institute of Microelectronics, Chinese Academy of Sciences, Beijing, China; [2]University of the Chinese Academy of Sciences, Beijing, China. [3]Key Laboratory of Polar Materials and Devices, Ministry of Education, East China Normal University, Shanghai, China
[*]Email: lvhangbing@ime.ac.cn; liuming@ime.ac.cn

Abstract— For the first time, we experimentally demonstrated a 10nm node HfZrO based FE-FinFET device with both Charge Trapping and Domain Switching memory effect. Extreme high endurance ($>10^{12}$), high operation speed (<20ns), good data retention (10^4@85C), low operation voltage (<3V) were identified in charge trapping mode, which is quite promising for e-DRAM application. As the device working in domain switching mode, even more robust retention (>10 years) and read disturbance immunity were achieved, showing great potential for e-NVM application.

INTRODUCTION

Due to recent trends of increasing system core-count and memory bandwidth bottleneck, large size on-chip caches are pursued to effectively reduce the performance gap between processor and main memory. Suffering from the large cell size and increasing leakage of SRAM, frustrations with SRAM for majority real estate occupation and unsatisfying scaling performance were encountered by chip designer. To overcome the limitations of SRAM, alternate memory technologies, such as eDRAM and STT-RAM were explored as a substitution of SRAM(**Fig.1**).[1]However, eDRAM suffers from poor data retention and scaling limitation below 1z node. Other alternates with better scaling potential and non-volatility are needed. On the other hand, e-NVM in SOC chip is also suffering from the limitation of e-FLASH integration on the advanced technology node. BEOL based memories, such as STT-RAM and RRAM were proposed as the possible solutions at 40 nm and 22 nm node(**Fig.2**).[2, 3]However, for the e-NVM solution at 10 nm node or beyond, there still is no candidate identified.

In this work, we proposed a Fe-FinFET memory device with HfZrO (HZO) ferro layer as the gate dielectric for both 1T e-DRAM and e-NVM applications. The coexistence of charge trapping (CT) and domain switching (DS) phenomenon was observed. As the operation voltage <3V, charge trapping effect was dominant, and as the operation voltage >5V, domain switching of HZO layer occurred. Both of the CT and DS behavior exhibit high operation speed (<20ns). High endurance of more than $10^{12}$and good data retention of 10^4s were achieved in CT mode, which is quite promising for e-DRAM application. The DS mode exhibits even more robust retention (>10 years) and read disturbance immunity, showing good possibility for e-NVM applications.

EXPERIMENTAL

The process flow of this Fe-FinFET was shown in **Fig.3**. The Spacer-Image Transfer (SIT) method was implemented to form ultra-small fin structure. The STI oxide provides the basic body isolation between adjacent fins. After the junction isolated process, Punch through Stop Layer (PTSL) implantation in the PTSL flow, there is a low thermal anneal, followed by dummy gate formation and SD formation. The PTSL implantation was skipped in the self-aligned punch through stop pocket (PTSP) process flow. In following steps, the dummy gate, spacers, source/drain doping process, activation anneal and all-last high-k/metal gate process were implemented by regular bulk FinFETs integration process After the self-aligned PTSP IMP, the 1050C spike anneal is conducted to active SD and channel dopants at the same time. The $Hf_{0.5}Zr_{0.5}O_2$ layer was deposited by ALD. After the TiN/TaN/TiN/W metal gates deposition, rapid annealing process at 600℃ for 30s to crystallize HZO. The structure schematic, TEM image and the spectrum of EDX mapping over the cross-section of the device are shown in **Fig.4**. Fin structure with 9nm width/20 nm length and Si/IL/HZO (9.37nm)/TiN/TaN/TiN/W gate stack was confirmed.

RESULTS AND DISCUSSION

Coexistence of Charge Trapping and Domain Switching

To confirm the ferroelectric property of Si/IL/HZO/TiN structure, planar device with 5nm TiN top electron was characterized. Fig.5 shows the TEM image of planar Si/IL(0.8nm)/HZO(10.7nm)/TiN/W stack with 600°C/30s annealing. Interfacial layer (SiO$_2$) between the HZO layer and Si substrate can be observed.Fig.6 shows the XRD Spectra of HZO film before and after 30s annealing at 400, 500 and 600°C, respectively. The peak of the orthorhombic phase increased as the annealing temperature increases. Considering that various factors may contribute to the polarization which could be categorized as "remanent" (true or switchable) and "non-remanent" polarizations. The remanent hysteresis could be obtained by PUND test, as described in **Fig.7**. **Fig.8** exhibits the P-V characteristics of Si/IL/HZO/TiN structure after 30s annealing at 400, 500 and 600°C and the corresponding changes of2Pr as a function of TiN thickness. Higher annealing temperature will result in stronger true-remanent polarization. The result shows there is almost no influence of thin 5nm TiN TE on the ferroelectricity compared with the thicker TiN TE. On the contrary, a slight

978-1-7281-1988-5/18 $31.00 © 2018 IEEE

enhancement was observed. The coercive voltage of the Si/IL/HZO/TiN structure is larger than the control sample withTiN/HZO/TiN structure[3], which was resulted from the voltage drop on SiO_2 IL (**Fig.8**).The trap in the HZO layer could be analyzed by Random Telegraph Noise (RTN) which is generated by electron capture/emission in the trap site.**Fig. 9a and b** show the RTN signal of I_G measured under 2.5V with SR=10^5Hz. Five peaks RTN signal was observed (**Fig. 9c**).The PSD result shows that multi-traps existed in the HZO layer (**Fig.9d**).

As shown in **Fig. 10,** the charge trapping induces V_T shift, which is opposite to the V_T shift caused by the ferroelectric domain switching in the same polarity of the gate voltage. For example, holes trapped in the gate dielectric at positive gate voltage result in a positive shift of the V_T. On the other hand, positive ferroelectric polarization induced by the same positive gate voltage leads to a negative V_T shift. Both of charge trapping and ferroelectric switching was found in the same FinFET device with a Si/IL/HZO/TiN/TaN/TiN/W gate stack. As shown in **Fig. 11**, after a -5V voltage sweep, the polarization was poled up and holes were trapped in the HZO layer. The V_T was shift to -1.5V due to the domination of charge tapping effect. After 1 V voltage sweep, the V_T begins to move to the positive side. As the sweep voltage increased to around 3V, the charge trap memory window reached to the peak value. When higher voltage was applied on the device, ferroelectric switching behavior began to dominate the performance (negative shift of V_T). The voltage drop on SiO_2 IL results in the increase of ferroelectric switching voltage. The FE-FET memory window is larger than the CTM. These special characteristics of charge trapping and domain switching coexistence can be used for both 1T e-DRAM and e-NVM applications.

Charge trap behavior for 1T-DRAM application

The down-scaling in a conventional combination of 1T1C DRAM is becoming increasingly difficult at 1z node, in particular due to the high demand for non-scalable cell capacitance.1T-DRAM has been proposed to solve this issues. For the 1T-DRAM application, some critical properties need to be solved, such as high speed, high endurance and low power. **Fig. 12**represents consecutive DC cycles of the device for 100 cycles with a sweeping gate voltage of -3V to 2V and drain voltage of 100mV. A 0.6V ΔV_T window, low off state current ~10^{-11} and large Ion/Ioff ratio of >10^4 were achieved. The cumulative distribution of V_T of the '0'and '1' states (**Fig.13**) shows high cycle to cycle uniformity of this device. **Fig. 14a** and **14b** show the dependence of switching speed on pulse amplitudes for program and erase operations, respectively. Higher programming voltage corresponds to shorter switching time. Programming and erase speed as high as 20ns was achieved under 2.6V and -3.8V, respectively.**Fig.15** shows the endurance test result under successive voltage pulses with 3V/20ns for program and -4V/20ns for erase. After 10^{12} cycles, the ΔV~0.5V can still be well maintained. Stable retention was also confirmed at 85 ℃ for 10^4 s as shown in **Fig. 16.**Compared with other 1T-DRAM

options[4-6], The Fe-FinFET in CTM mode exhibit excellent switching speed (20ns), lower operation voltage (<3V), lower read current (<10 μA) and excellent endurance (**Fig.17**).

Domain switching effect of FE-FET for embedded NVM

As the conventional e-FLASH encountering severe limitations on integration into the high-k/metal gate (HKMG) process and FinFET process, the upcoming IoT era shows huge demands on the new types embedded non-volatile memories (eNVM) with low cost, low power and good compatibility with advanced logic process. The Fe-FinFET memory demonstrated here could well solve these problems.**Fig.18** shows Fe-FET behavior of the device by pulse operation with pulse height >6V. Positive gate voltage leads to a negative V_T shift, while negative gate voltage leads to a positive V_T shift. After 30 pulse cycles, a 1.2V memory window was confirmed and this device shows high uniformity by AC cycling (**Fig.19**). Under successive voltage pulses with 6V/20ns and -6V/20ns, 100k endurance was achieved (**Fig.20**). The device could retain over 10 years at 25C, which is sufficient to meet the requirement of eNVM. **Table 1** summarizes the specifications of different candidates of eNVM for low power SOC applications at 14 nm technology node and beyond.

CONCLUSION

In summary, the coexistence of charge trapping and domain switching phenomenon in one FE-FinFET device was demonstrated for the first time. Excellent switching speed (20ns), lower operation voltage (<3V), lower read current (<10 μ A) and high endurance (>10^{12}), the CT mode show great potential to be used as 1T e-DRAM. The domain switching behavior of this device exhibits high uniformity, good retention and low power consumption, which is quite suitable for eNVM application.

ACKNOWLEDGMENT

This work was supported in part by the MOST of China under Grants 2016YFA0203800, 2016YFA0201803, 2018YFB0407502, and in part by the National Natural Science Foundation of China under Grants 61522408, 61334007, 61521064, and in part by Beijing Municipal Science & Technology Commission Program (Z161100000216153) and by Huawei Data Center Technology Laboratory..†This authors contribute to this work equally.

REFERENCES

[1] J. S. Vetter et al., "Opportunities for nonvolatile memory systems in extreme-scale high performance computing," Computing in Science and Engineering, 2015.

[2] Y. J. Song, et al., "Highly functional and reliable 8 Mb STT-MRAM embedded in 28 nm logic," in Proc. IEEE Int. Electron Device Meeting, Dec. 2016, pp. 27.2.1–27.2.4.

[3] . H. B. Lv, et al., "BEOL based RRAM with one extra-mask for low cost, highly reliable embedded application in 28 nm node and beyond," in Proc. IEEE Int. Electron Device Meeting, Dec. 2017, pp. 36–39

[4] D.-I. Moon, et al., "A novel FinFET with high-speed and prolonged retention for dynamic memory," IEEE Electron Device Lett., vol. 35, no. 12, Dec. 2014,pp. 1236–1238.

[5] J. Müller, et al., "Ferroelectricity in HfO2 enables nonvolatile data storage in 28 nm HKMG," in Proc. IEEE VLSI Technol., Jun. 2012, pp. 25–26.

[6] Y.-C. Chiu, et al.,, "Low power 1T DRAM/NVM versatile memory featuring steep sub-60-mV/decade operation, fast 20-ns speed, and robust 85 °C-extrapolated 10^{16} endurance," in Proc. Symp. VLSI Technol., Kyoto, Japan, Jun. 2015, pp. T184–T185.

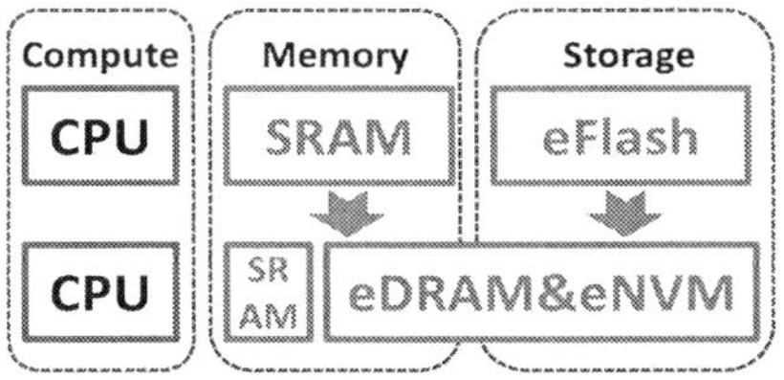

Compute	Memory	Storage
CPU	SRAM	eFlash
CPU	SR AM / eDRAM&eNVM	eDRAM&eNVM

Fig.1 Evolution of memory hierarchy in embedded system.

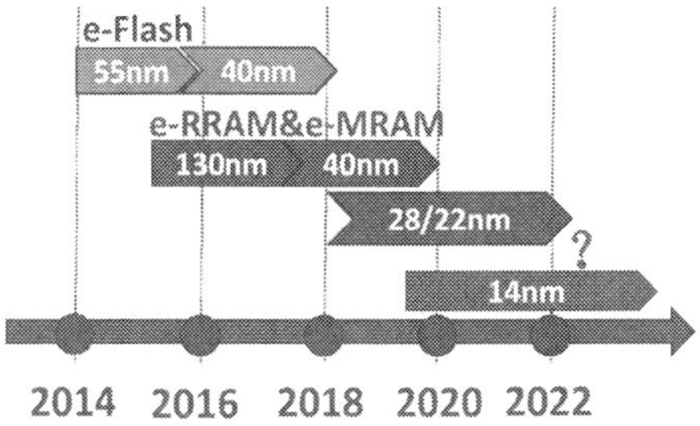

e-Flash: 55nm → 40nm

e-RRAM&e-MRAM: 130nm → 40nm

28/22nm ?

14nm

2014 2016 2018 2020 2022

Fig.2 New memory will become the main solution of embedded NVM in advanced technology nodes.

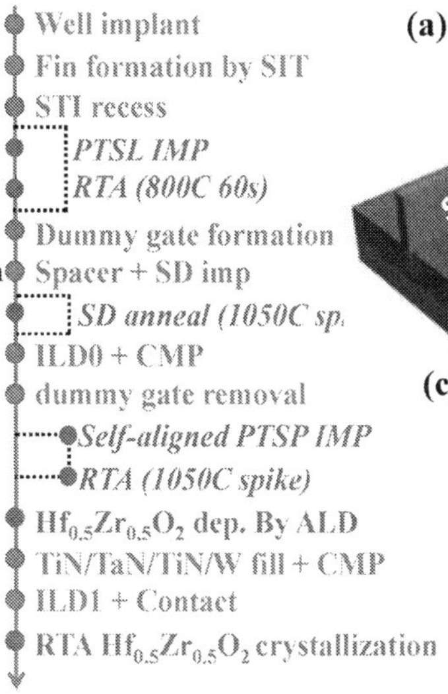

- Well implant
- Fin formation by SIT
- STI recess
- *PTSL IMP*
- *RTA (800C 60s)*
- Dummy gate formation
- Spacer + SD imp
- *SD anneal (1050C sp.*
- ILD0 + CMP
- dummy gate removal
- *Self-aligned PTSP IMP*
- *RTA (1050C spike)*
- $Hf_{0.5}Zr_{0.5}O_2$ dep. By ALD
- TiN/TaN/TiN/W fill + CMP
- ILD1 + Contact
- RTA $Hf_{0.5}Zr_{0.5}O_2$ crystallization

Fig.3 Process flow of FE-FinFET.

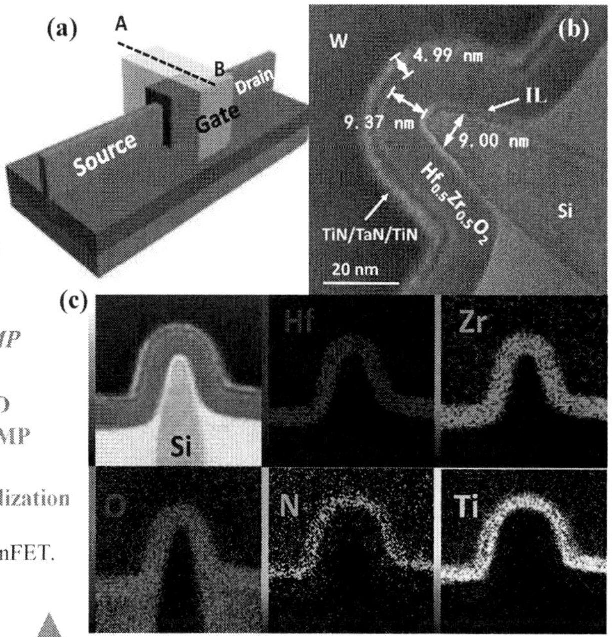

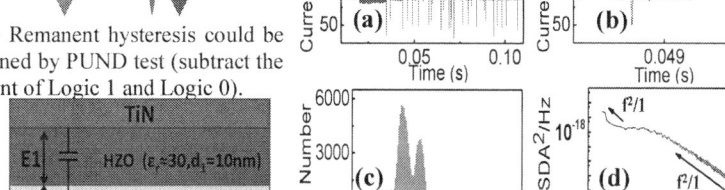

Fig. 4a)The structure schematic,(b)TEM image and c)the EDX mapping for Hf, Zr, O, N and Ti over the cross-section A-B of the device.

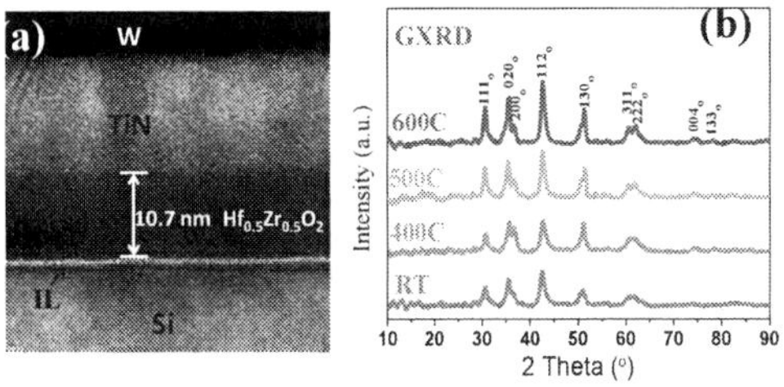

Fig.5 a) TEM of planar HZO capacitor with ~0.8 nm interfacial layer (IL) SiO2. b) GXRD of HZO.

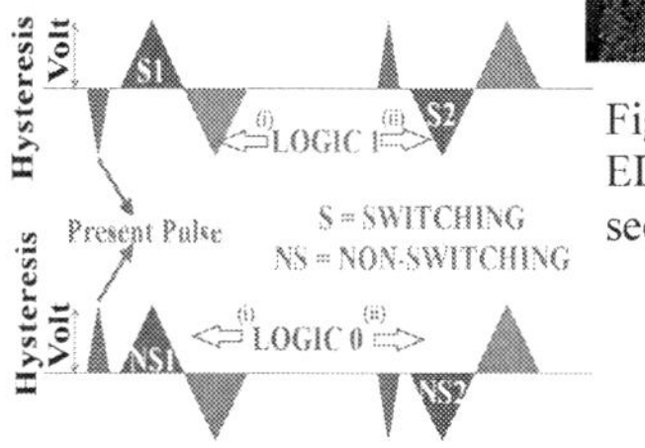

Fig.6 Remanent hysteresis could be obtained by PUND test (subtract the current of Logic 1 and Logic 0).

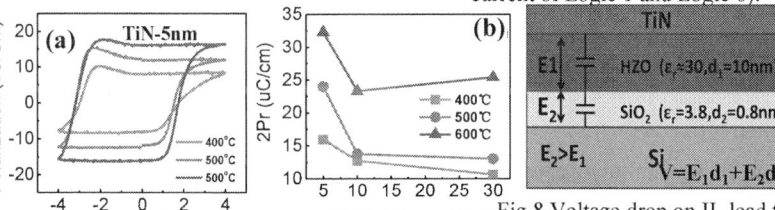

Fig.7 a)P-V characteristics of SI/IL/HZO/TiN with different annealing temperature.b) Summary of the changes in 2 Pr as a function of TiN thickness.

TiN

E_1 HZO ($\varepsilon_r \approx 30, d_1 = 10nm$)

E_2 SiO$_2$ ($\varepsilon_r = 3.8, d_2 = 0.8nm$)

$E_2 > E_1$ Si, $V = E_1 d_1 + E_2 d_2$

Fig.8 Voltage drop on IL lead to the large coercive voltage

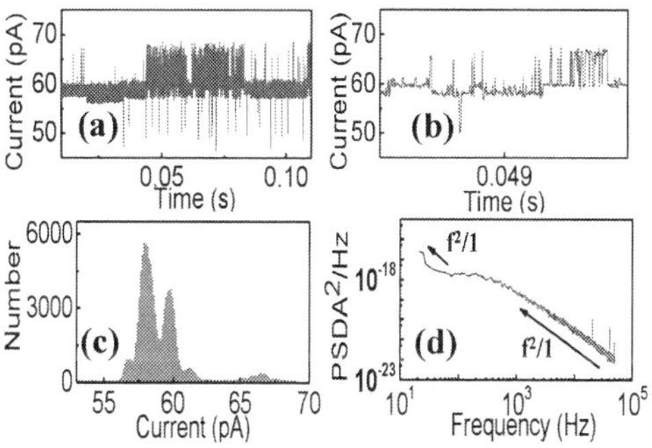

Fig.9 a-b) the RTN signal of IG, c) There are five peaks of the IG RTN signal. d) The PSD curve indicates the existence of multi-traps in HZO.

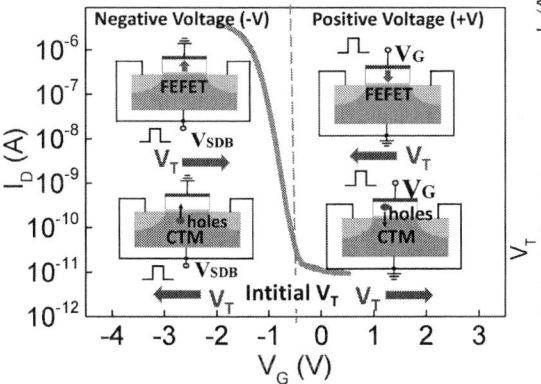

Fig.10 Impact of trapped electric charges and ferroelectric polarization on the threshold voltage of a field effect transistor.

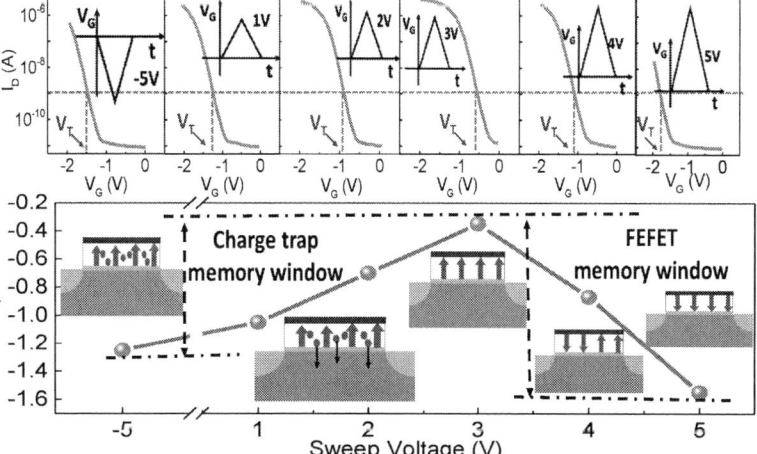

Fig.11 With different operation voltage (<3V or >5V), this device shows Charge trap memory behavior or Fe-FET behavior.

978-1-7281-1988-5/18 $31.00 © 2018 IEEE   49

For embedded 1T-DRAM application

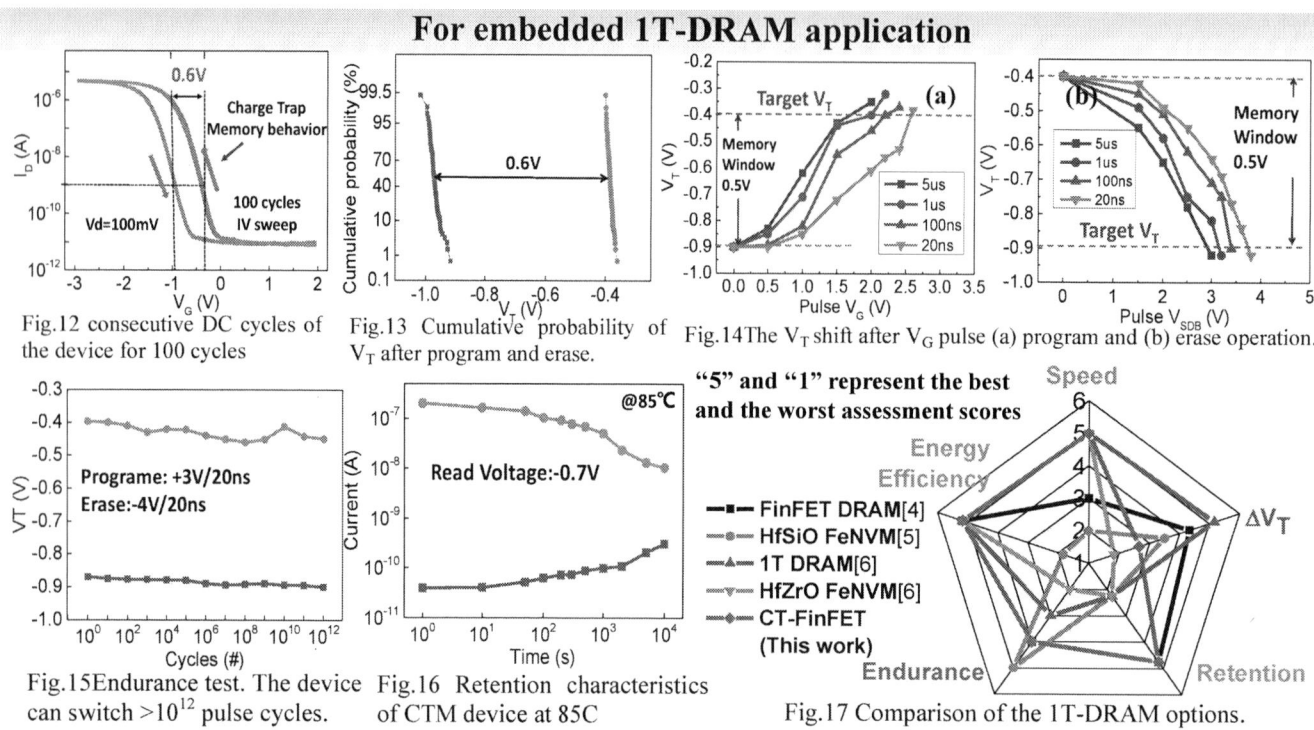

Fig.12 consecutive DC cycles of the device for 100 cycles

Fig.13 Cumulative probability of V_T after program and erase.

Fig.14 The V_T shift after V_G pulse (a) program and (b) erase operation.

Fig.15 Endurance test. The device can switch >10^{12} pulse cycles.

Fig.16 Retention characteristics of CTM device at 85C

Fig.17 Comparison of the 1T-DRAM options.

For embedded NVM application

Fig.18 Fe-FET behavior was performed by pulse operation with pulse height >6V. Positive gate voltage leads to a negative V_T shift, while negative gate voltage leads to a positive V_T shift.

Fig.19 Cumulative probability of V_T after program and erase with 30 pulse cycles. High uniformity by AC cycling was confirmed.

Fig. 20 Switching endurance test. The device can switch 100k pulse cycles.

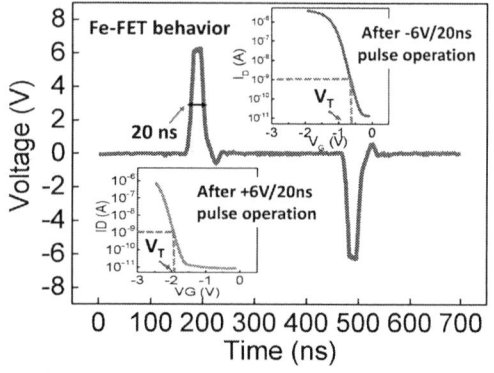

Fig.21 The device could retain over 10 years at 25C.which is sufficient to meet the requirement of eNVM.

Table 1. Comparison of embedded NVM options

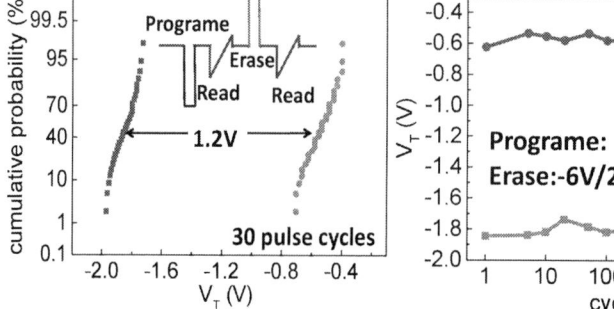

Parameter	Floating Gate e-Flash	Charge Trap e-Flash	e-MRAM	e-RRAM	FE-FinFET (This work)
Speed	20 ns	25ns	20ns	20ns	20ns
Tech. node	2x nm	2x nm	2x nm	40 nm	10 nm
Retention	10 years	10 years	10 years	10 years	10 years
Endurance	100k	100k	1M	100k	>100k
Write Voltage	12V	8V	2V	2V	6V
Write Power	Med	Low	High	Low	Low

978-1-7281-1988-5/18 $31.00 © 2018 IEEE

High-performance (EOT<0.4nm, Jg~10^{-7}A/cm^2) ALD-deposited Ru\SrTiO$_3$ stack for next generations DRAM pillar capacitor

M. Popovici[1], A. Belmonte[1], H. Oh[1], G. Potoms[1], J.Meersschaut[1], O. Richard[1], H. Hody[1],
S. Van Elshocht[1], R. Delhougne[1], L. Goux[1], and G. Sankar Kar[1]

[1]imec, Leuven, Belgium, email: Mihaela.Ioana.Popovici@imec.be

Abstract—We demonstrate the fabrication of strontium titanate (STO) based metal-insulator-metal (MIM) capacitors with very-high dielectric constant (k~118) and low leakage of 10^{-7} A/cm^2 at ±1V for a ~11nm thick dielectric using Ru as bottom electrode (BE) and top electrode (TE). The k enhancement is attributed to the formation of an ultrathin cubic SrRuO$_3$ phase at the Ru\STO bottom interface, acting as a template optimizing the STO crystal quality from the interface to the bulk. This interface quality is evidenced by the same k~118 extracted from STO thickness series and relating to the bulk-k value. *This achievement opens up an alternative integration roadmap for DRAM capacitors, moving from the current cup-shape to a denser pillar-shape design.*

I. INTRODUCTION

Current memory technologies, such as DRAM, SRAM, and NAND Flash, are encountering difficult challenges related to their continued scaling to and beyond the 16nm generation. DRAM technology is especially challenged as the leakage currents specs are the most severe. A key role in DRAM technology is played by the metal-insulator-metal (MIM) capacitors. As physical thickness scales down, the conventional TiN/ZAZ/TiN stack faces huge hurdles to fulfill the requirements of equivalent oxide thickness (*EOT*) and leakage current density *(Jg)*, which are *EOT*≤0.4nm and *Jg*≤10^{-7}A/cm^2 at +/-1V and dielectric physical thickness (t_{phys}~5nm) for the sub-20nm technology node [1]. On the other hand, these severe *EOT* and *Jg* requirements may be reached using the well-known SrTiO$_3$ (STO) perovskite, however at the cost of larger t_{phys}~10nm. For example we demonstrated *EOT*=0.40nm and *Jg*=10^{-7}A/cm^2 using Ru electrodes and 8.5nm STO showing *k*~85 [2]. However, as the physical thickness of STO is downscaled to 5nm, the drastic increase of leakage prohibits its use as DRAM dielectric, and so far no solution has been identified to circumvent this roadblock. *An alternative solution is to change the capacitor design from the double-sided cup-shape to a pillar shape (Fig.1).* The strong advantage of this design is that the STO physical thickness may be kept ~10nm for some generations. However the associated strong drawback is that the effective capacitor area is halved. *It implies that in this approach the k value of STO needs to be substantially boosted to compensate for this loss of capacitive signal.*

Hence, the purpose of this work is to optimize the STO microstructure and MIM quality using 3D-compatible Atomic-Layer-deposition (ALD) process to fabricate the MIM stack and *leverage material physics knobs to maximize the k value while keeping leakage under control.*

An epitaxial growth of STO would minimize the amount of defects at the interface and boost the k value of the dielectric. Moreover, the substrate on which the dielectric will be grown should be a fully metallic electrode layer, with a low lattice mismatch against the lattice of the dielectric. SrRuO$_3$ (SRO) is a conductive oxide that can act as electrode and is able to crystallize in a cubic structure with a lattice parameter of 3.910Å whereas the SrTiO$_3$ has a lattice parameter of 3.903Å. Due to this low lattice mismatch (~0.18%), SRO qualifies not only as electrode but also as epitaxial template.

Physical vapor deposition (PVD) of SRO/STO/SRO stack was shown to lead to EOT~0.4nm for t_{phys}~12nm [3]. However, the method of deposition should be a conformal one like ALD or chemical vapor deposition (CVD) for both SRO and STO. Only few attempts to grow SrRuO$_3$ by ALD were reported, due to the difficulty to grow Sr and Ru altogether. An ALD SrRuO$_3$ process developed by Han (2013) showed that the lowest resistivity of ~2,300μΩ·cm achieved was much larger in comparison to those of sputtered SrRuO$_3$ (~1,000μΩ·cm) and CVD SrRuO$_3$ (~470μΩ·cm). The extracted k value of STO grown on developed strontium ruthenate was only 44, indicating a poor interface with the dielectric or not fully crystallized SrTiO$_3$ [4].

In this work, we circumvent this SRO process limitation by optimizing the Ru\STO ALD process to favor the in-situ formation of SRO during post-deposition crystallization anneal.

II. EXPERIMENTAL

A. Process flow description

Planar (100x100)μm^2 MIM capacitors were fabricated via one mask process patterning of the metal top electrode (TE) on 300mm diameter wafers. The complete stack consisted of a sequence of layers deposited as such: 15nm ALD TiN + 5nm ALD Ru bottom electrode (BE), followed by 9-14nm ALD STO as dielectric and 5nm Ru + 35nm TiN as TE. These stacks will be called *SRT* stacks hereafter. For comparison we also fabricated stacks without the 5nm Ru BE layer, which will be called *ST* stacks hereafter.

The STO films were deposited in a two-step process: *firstly a Sr-rich STO seed layer was deposited on the Ru BE. Then we optimized a rapid thermal annealing (RTA) intended to induce the formation of an interfacial SRO phase.* This RTA step also allowed to create sufficiently small grain size and avoid micro-crack defects. Then, a second (Ti-rich) STO layer is deposited, and the resulting stack is submitted to a crystallization anneal performed at 600°C for 1min in N_2. During the second anneal, the inter-diffusion of the two STO layers and the crystallization of a STO layer with a final composition of ~48 at.% Sr occurred. After deposition of the Ru+TiN TE, the stacks were patterned and electrical characterization were performed. The EOT values were extracted at 0V and room temperature from the capacitance - voltage (C–V) curves measured at 10kHz, while the leakage is measured at ±1V using delay time of 0.1s.

B. Physical properties of STO

Physical properties were assessed after the post-deposition anneal at 600°C. The thickness of the STO/Ru/TiN layers was evaluated via X-ray reflectometry (XRR). We observed a gradual decrease in the density of Ru layer towards the top interface giving a first indication of interfacial reaction between Ru and STO films. The STO films have a uniform grain size distribution and exhibit the well-known cubic structure of the perovskite both for *SRT* and *ST* stacks. The only additional observation is the presence of a shoulder with low intensity at ~33.0° for the *SRT* stack (**Fig.2a**). From Rietveld refinement [6], the best model describing the *SRT* stack contains a small amount of tetragonal $Sr_3Ru_2O_7$ and cubic $SrRuO_3$, next to the expected phases of cubic $SrTiO_3$, hexagonal Ru and cubic TiN (**Fig.2b**). The fitting parameters (Rwp=3.95% and sigma of 0.87) showed a very good estimation of the phases present in the system. The calculated lattice parameter of $SrTiO_3$ is 3.888Å, in agreement with the composition, while the adjacent templating layer of cubic $SrRuO_3$ has a parameter of 3.838Å. We assume the $Sr_3Ru_2O_7$ phase accommodates the transition to the Ru BE phase. Although no distinct phase could be evidenced neither by TEM imaging of the SRT capacitor TiN/Ru/STO/Ru/TiN/Si (**Fig.3a**) nor for that of STO seed on Ru (**Fig.3b**), some degree of intermixing between STO and Ru BE was observed by energy dispersive X-ray analysis (EDS) (**Fig.3c**). The Ru profile has a less steep slope at the STO/Ru interface, suggesting the presence of a very thin SRO-based templating layer.

III. ELECTRICAL RESULTS & DESIGN CONSIDERATIONS

The *EOT vs Jg* extracted at +1V (**Fig.4a**), together with leakage-voltage (**Fig.4b**) and capacitance – voltage (**Fig.4c**) results obtained for *SRT* and *ST* capacitors all over the 300mm wafer are shown. STO composition is the same for both *SRT* and *ST* capacitors. Although the STO layer thickness is ~11nm and ~10nm for these *SRT* and *ST* stacks resp., substantially lower EOT is obtained for SRT capacitors. The difference in thickness is due to a slightly lower growth rate of STO on TiN as compared to Ru substrate. *In agreement, an effective dielectric constant value k_{eff}~118 is extracted for SRT while k_{eff}~80 is extracted for ST stacks.* **Fig.5** shows the excellent within-wafer (WIW) uniformity of *EOT*, and *Jg*, resulting in marginal k_{eff} variations in the range 116-119 attributed to slight physical thickness variation in the range of ± 0.2 nm. Note also that higher EOT-performance wafer locations (NE) also exhibit

lowest leakage areas. Concomitantly, the better developed SRO-induced XRD shoulder in these NE areas (**Fig.5b**) *indicates that the SRO templating benefits not only k_{eff}, but also Jg.* To confirm this excellent interfacial quality we measured the *EOT* of *SRT* capacitors having varied STO thickness and we extracted the bulk-k value (k_{bulk}) from the slope of EOT data vs. STO thickness (**Fig.6**). We obtained k_{bulk}~k_{eff}~118, *which means that the bulk dielectric properties of STO are not altered by the interface with the Ru BE and are fully sensed by C-V characterization.* Hence, further improvement of STO bulk dielectric performances will be entirely exploited as capacitor signal in DRAM devices. **Fig.7** shows the further leakage reduction of *SRT* capacitors of about two orders of magnitude obtained by trivalent-element doping of the 10 nm thick STO films, with *EOT* kept constant to 0.35 nm, which could allow to further reduce the thickness below 10nm. Doping solutions also exist to boost k_{bulk} up to ~200 without parasitic impact of low interface quality. Based on these results and projected progress the further downscaling of DRAM technology from D18 may consider a cell design move from the cup- to the pillar-type scheme. **Fig.8** shows extrapolations of capacitor aspect ratio (A/R) required for D16, D14 and D12 using both types of cell schemes, suggesting already realistic pillar A/R for D16 using data of this work. Next technologies D14 and D12 will require k increase to a range >200 together with dielectric t_{phys} scaling to <7nm (**Fig.9**). The former is at reach considering high-quality bulk properties of STO, while the latter is obtainable through *Jg* reduction by means of doping, as shown in **Fig.7**.

IV. CONCLUSIONS

In this work we demonstrate for the first time a 3D-compatible ALD process of Ru\SrTiO₃\Ru capacitor for DRAM technology, allowing the in-situ formation of a SrRuO₃ template phase at the Ru\SrTiO₃ interface during stack anneal. For 11nm STO this method results in boosted k_{eff}~118, which is unaltered from the bulk to the interface of $SrTiO_3$. Trivalent-element doping reduces further the leakage, which thus holds the promise of STO thickness scaling in the range <7nm and enabling the move to pillar cell scheme in next generation DRAM technologies from D16 to D12.

REFERENCES

[1] S. Kim, and M. Popovici, "Future of dynamic random-access memory as main memory," *MRS Bulletin*, vol. 43, pp. 334-339, May 2018.

[2] J. Swerts, M. Popovici, B. Kaczer, M. Aoulaiche, A. Redolfi, S. Clima, C. Caillat, W.C. Wang, V.V. Afanas'ev, N. Jourdan, C. Olk, H. Hody, S. Van Elshocht, and M. Jurczak, "Leakage Control in 0.4-nm EOT Ru/SrTiOx/Ru Metal-Insulator-Metal Capacitors: Process Implications", *IEEE Electron Device Lett.* 35, pp. 753–755, July 2014.

[3] S. Kupke, S. Knebel, U. Schroeder, S. Schmelzer, U. Böttger, and T. Mikolajick, " Reliability of SrRuO₃/SrTiO₃/SrRuO₃ Stacks for DRAM Applications," *IEEE Electron Device Lett.* 33, pp. 1699–1701, Dec. 2012.

[4] J. H. Han and C. S. Hwang, "ALD and pulsed CVD of Ru, RuO₂, and SrRuO₃," *ECS Transactions*, vol. 58 (1), pp. 171–182, 2013.

[5] M. Popovici, J. Swerts, A. Redolfi, B. Kaczer, M. Aoulaiche, I. Radu, S. Clima, J.-L. Everaert, S. Van Elshocht, and M. Jurczak, "Low leakage Ru-strontium titanate-Ru metal-insulator-metal capacitors for sub-20nm technology node in dynamic random access memory", *Appl. Phys. Lett.* 104 (082908), pp. 1–5, Feb. 2014.

[6] L. Lutterotti and P. Scardi, "Simultaneous Structure and Size-Strain Refinement by the Rietveld Method", *J. Appl. Cryst.* 23, pp. 246-252,1990.

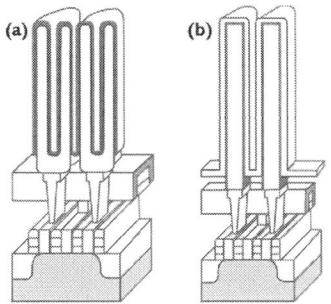

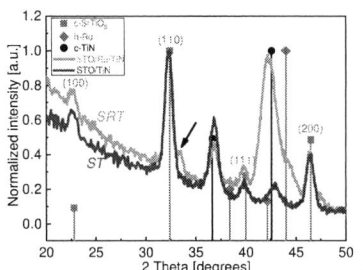

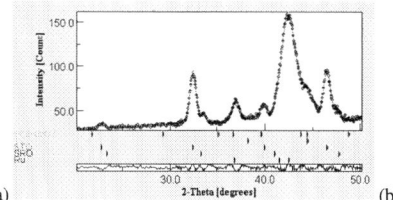

Fig.1: Schematics of double-sided cup-type (a) and pillar-type (b) DRAM capacitor designs.

Fig.2: X-ray diffraction diagrams obtained both for SRT and ST stacks by grazing incidence X-ray diffraction (GIXRD) at incident ω angle of 0.5°, exhibiting a shoulder of $SrRuO_3$ peak at ~33.0° for the SRT stacks (a), and SRT experimental data (circles) and fit (red line) with their difference (residual, line below) (b).

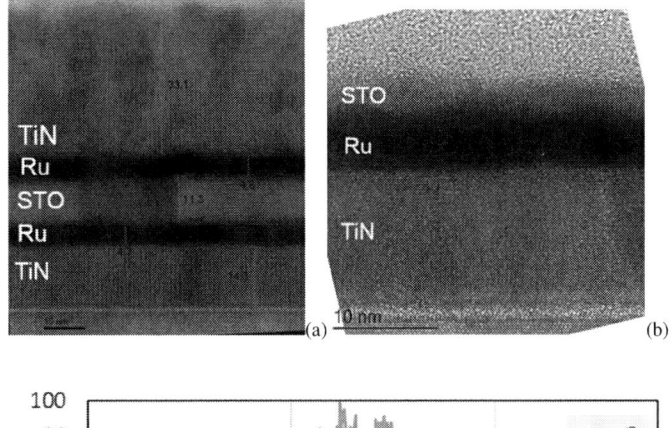

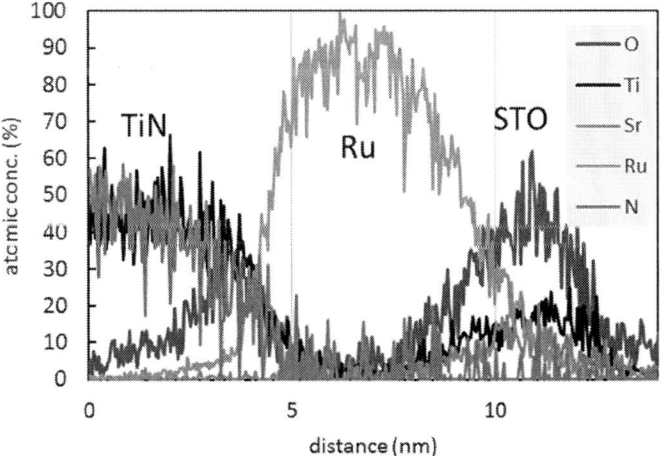

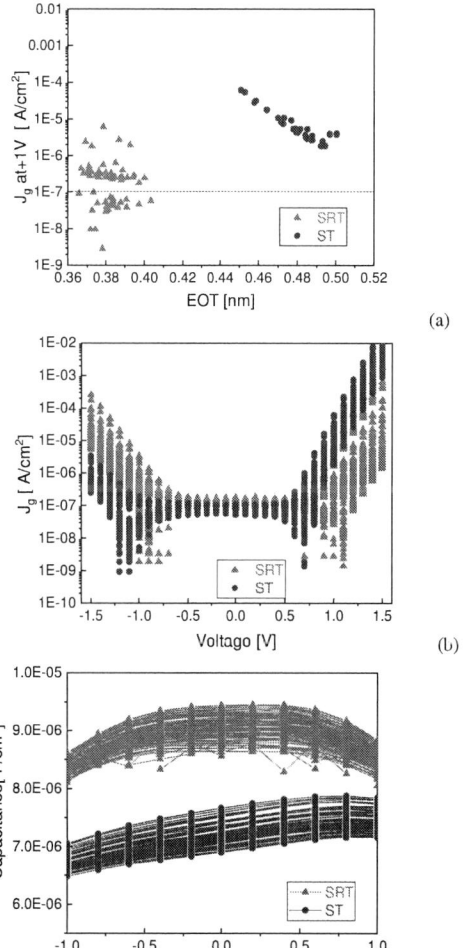

Fig.3: Cross-sectional TEM image of the TiN/Ru/STO/Ru/TiN capacitor(a), of the annealed STO seed/Ru/TiN (b) and Elemental depth profile obtained by EDS and indicating some intermixing at the Ru\STO interface (c).

Fig.4: Jg at +1V vs. EOT distribution across the 300 mm wafer area (a), Leakage-Voltage (b) and Capacitance-Voltage (c) dependence, collected for both SRT (triangles) and ST (squares) capacitors.

978-1-7281-1988-5/18 $31.00 © 2018 IEEE   53

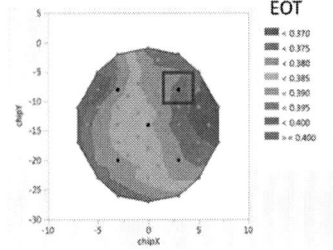

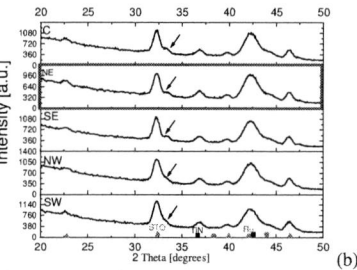

Position	EOT[nm]	Jg at+1V(A/cm²)	C(F/cm²)
NE	0.374	<1E-07	9.2E-06
SE	0.376	3.8E-07	9.2E-06
C	0.381	2.5E-07	9.1E-06
SW	0.386	<1E-07	8.9E-06
NW	0.388	2.5E-07	8.9E-06

Fig.5: EOT within wafer distribution, 5 dies locations highlighted and a summary table showing their EOT, Capacitance (C) and Jg at +1V (a); GIXRD recorded after post-deposition anneal for the 5 positions C(center), NE(north-east), SE(south-east), NW(north-east) and SW(south-west) (b).

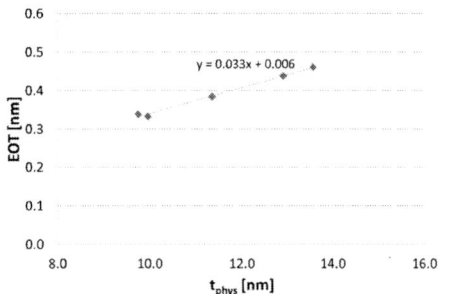

Fig.6: EOT obtained for various STO thicknesses in SRT stacks, and allowing to extract from the slope the Bulk-k value (k_{bulk}) of the STO dielectric, resulting in $k_{bulk} \sim k_{eff} \sim 118$.

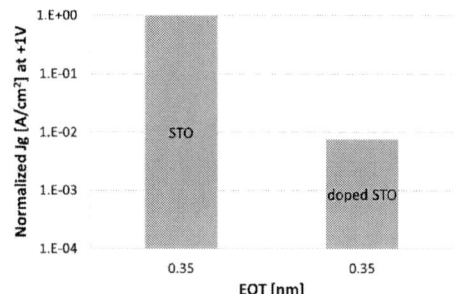

Fig.7: Leakage reduction at fixed EOT obtained by doping STO with a trivalent element, intended to minimize oxygen vacancy defects in the STO layer. For comparison, the reference STO Jg value at +1V was normalized to 1.

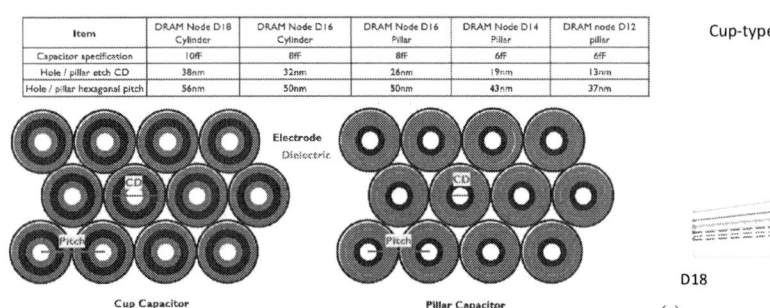

Fig.8: Physical assumption of CD and pitch of cup-type and pillar-type capacitor in DRAM scaling roadmap (a); calculated capacitor aspect ratio required to reach sufficient capacitance signal (from 8 to 15fF) for cup-type and pillar capacitor designs, and using either state-of-the-art ZAZ characteristics (k_{eff}=40 and t_{phys}=5.8m) or properties similar to those obtained in this work ($k \sim 120$, t_{phys}=11nm) (b).

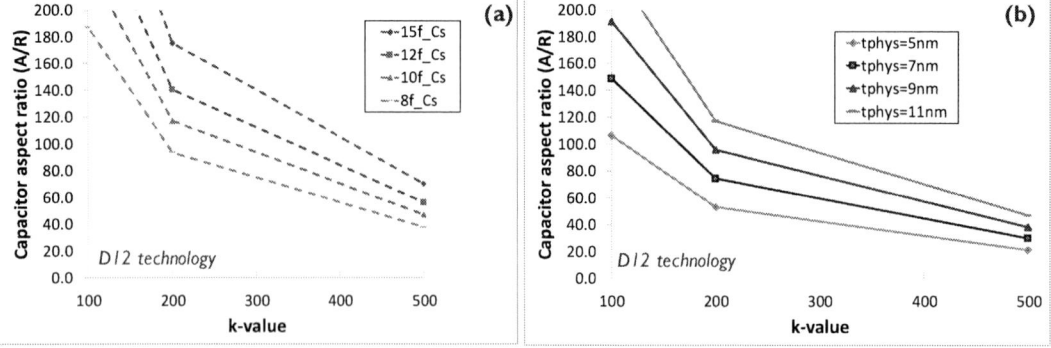

Fig.9: D12 technology: Pillar-type capacitor aspect ratio required as a function of the dielectric k value: (a) for fixed tphys=11nm and various capacitance signal levels (Cs), and (b) for fixed Cs=10fF and various t_{phys} values from 11nm to 5nm.

978-1-7281-1988-5/18 $31.00 © 2018 IEEE

Exploiting Hybrid Precision for Training and Inference: A 2T-1FeFET Based Analog Synaptic Weight Cell

Xiaoyu Sun[1], Panni Wang[1], Kai Ni[2], Suman Datta[2], and Shimeng Yu[3]

[1]Arizona State University, Tempe, AZ 85281, USA   [2]University of Notre Dame, Notre Dame, IN 46556, USA
[3]Georgia Institute of Technology, Atlanta, GA 30332, USA Email: shimeng.yu@ece.gatech.edu

Abstract— In-memory computing with analog non-volatile memories (NVMs) can accelerate both the in-situ training and inference of deep neural networks (DNNs) by parallelizing multiply-accumulate (MAC) operations in the analog domain. However, the in-situ training accuracy suffers from unacceptable degradation due to undesired weight-update asymmetry/nonlinearity and limited bit precision. In this work, we overcome this challenge by introducing a compact Ferroelectric FET (FeFET) based synaptic cell that exploits hybrid precision for in-situ training and inference. We propose a novel hybrid approach where we use modulated "volatile" gate voltage of FeFET to represent the least significant bits (LSBs) for symmetric/linear update during training only, and use "non-volatile" polarization states of FeFET to hold the information of most significant bits (MSBs) for inference. This design is demonstrated by the experimentally validated FeFET SPICE model and co-simulation with the TensorFlow framework. The results show that with the proposed 6-bit and 7-bit synapse design, the in-situ training accuracy can achieve ~97.3% on MNIST dataset and ~87% on CIFAR-10 dataset, respectively, approaching the ideal software based training.

I. INTRODUCTION

DNNs have made remarkable advances in cognitive tasks such as image and speech recognition. However, the energy-efficiency and speed of DNN training is highly limited by moving data back and forth between the memory and the processor in conventional von Neumann hardware. To overcome this challenge, in-memory computing, where computing is done at the location of the data storage, has been proposed to accelerate the computation. To store a large number of DNN weights on-chip, logic process compatible NVM devices are attractive where synaptic weights are encoded as their analog conductance values. There are array-level experimental demonstrations for training/inference with resistive random-access memory (RRAM) [1-2] and phase change memory (PCM) [3]. However, training with these NVMs suffers from unacceptable accuracy degradation due to various non-idealities including limited dynamic range, variation, and most importantly asymmetric/nonlinear weight update [4]. For example (Fig. 1), in filamentary RRAM [5], the excessive asymmetry/nonlinearity between positive and negative update leads to a poor accuracy ~41% for MNIST dataset. While interfacial RRAM [6] exhibits improved nonlinearity with higher accuracy ~73%, the programming pulse width is on the orders of ms due to the slow diffusion process of ions or vacancies. A recent discovery of partial polarization switching in ferroelectric-FET (FeFET) [7] provides highly symmetric weight update leading to an

accuracy ~90%, but "non-identical" pulses must be applied for conductance tuning, which increases the peripheral circuitry complexity. Despite recent progress, these hardware implementations are not competitive with the software training accuracy ~98% even for MNIST dataset.

Motivated by the observation that in a DNN algorithm a relatively higher precision (larger than 6-bit) is necessary during training to accumulate the incremental weight change, while a lower precision (less than 2-bit) is sufficient during inference to achieve a reasonably good accuracy [8], we introduce a synaptic weight cell design in this work that combines two CMOS transistors and one FeFET (2T1F) for training and inference with hybrid precision. During training, the "volatile" modulated gate voltage of FeFET is used to represent LSBs for symmetric and linear update. After training process is complete, the LSBs information is discarded, only MSBs are preserved by "non-volatile" polarization states of FeFET for inference. We demonstrate a 6-bit/7-bit synapse design (2-bit MSBs + 4-bit/5-bit LSBs) for MNIST/CIFAR-10 dataset and benchmark with a LeNet-5-like/VGG-like convolutional neural network (CNN). The SPICE simulation results with the experimental validated FeFET model and TSMC 65nm PDK is coupled with the TensorFlow framework, showing that the learning accuracy could achieve ~97.3%/~87%, approaching the ideal software training.

II. 2T1F SYNAPTIC WEIGHT CELL DESIGN

A. 2T1F Analog Synaptic Weight Cell

Fig. 2-3 show the schematic and fundamental principle of the proposed 2T1F synaptic weight cell. The FeFET gate capacitor serves as an analog memory for LSBs, which is charged/discharged by the corresponding pull-up pFET and pull-down nFET. Thus, the LSBs of the weight can be encoded to the channel conductance of the FeFET by modulating the gate voltage (V_G) while keeping the FeFET working in the triode region as shown in Fig. 3(a). During weight update, pulses are applied to the gate of pFET/nFET for positive/negative updates while keeping these two transistors working in saturation region to ensure the charging/discharging current is independent of V_G. With the pulse amplitudes that generate the balanced charging and discharging current, the positive/negative update of LSBs is expected to be symmetric. The MSBs can be encoded to different FeFET polarization (thus channel conductance) states without overlapping LSBs within each MSB state. For example, assuming 2-bit MSBs (i.e., 4 polarization states) as shown in Fig. 3(b), the V_G dynamic range $[V_A, V_B]$ which is constrained by the linear region overlap of multiple polarization states, determines the number of update steps (i.e., the bitwidth of LSBs). Fig. 3(c) illustrates different

978-1-7281-1988-5/18 $31.00 © 2018 IEEE

scenarios of updating LSBs and MSBs. If V_G increases beyond V_B, the consequential read-out current I_D will be larger than the reference current (ref. 2 in Fig. 3), requiring a programming towards S_2 state to transfer the weight information to MSBs, then the LSBs can be continuously updated within S_2 state and V_G prefers to be reset to the certain level that maintains the same I_D to prevent the information loss of LSBs. Similarly, if I_D decreases below ref. 2, the FeFET requires a programming towards S_1 state.

With the proposed synaptic weight cell, the modified DNN training flow is shown in Fig. 4. For each training batch, update LSBs by applying certain pulses to modulate V_G based on the value of ΔW calculated through stochastic gradient descent (SGD) based backpropagation algorithm [8]. Due to the limited V_G dynamic range and capacitor leakage, the information of LSBs needs to be occasionally transferred to MSBs to prevent the information loss. As a result, for every N batches, we need to transfer the weight, i.e., program the FeFET to the corresponding state according to the read-out current level. After the weight-transfer, V_G prefers to be reset to the certain level that maintains the same channel conductance to recover the residual information of LSBs. However, this step requires a high-precision ADC (equals the total bitwidth of weights) which is too power- and area-hungry in practice. Therefore, we only reset the V_G to $(V_A+V_B)/2$ to avoid high-precision ADCs at the expense of inducing possible residual errors. The impact of these errors on learning accuracy is investigated in Section III.

B. Implementation of 2-bit MSBs + 4-bit LSBs Synapse

First, we demonstrate the implementation of 6-bit synapse (2-bit MSBs + 4-bit LSBs) as an example. The FeFET utilizes multi-domain polarization switching dynamics in ferroelectric $Hf_{0.5}Zr_{0.5}O_2$ (HZO) gate dielectric to gradually tune the threshold voltage of the underlying channel by the application of programming pulses to the gate. Fig. 5 shows the measured I_D-V_G characteristics of our fabricated HZO FeFET with tunable V_{th}. We adopt the FeFET SPICE model from our prior work [9], where the model consists of a conventional MOSFET model by BSIM 4, and a ferroelectric switching by Preisach dynamic model. The SPICE model accurately captures the experimental P-V loop (Fig. 6). Fig. 7(a) shows the pulse scheme and the simulated corresponding remnant polarization charge that result in 4 states shown in Fig. 7(b), which serve as 2-bit MSBs. The dynamic range of V_G is set to be [1.44V, 1.76V], with a pulse width of 5ns that leads to ΔV_G of 20mV per update pulse, 4-bit LSBs can be achieved. The voltage bias schemes for update/read operations of the 2T1F weight cell are summarized in Fig. 8. The equivalent weight update curve of the 6-bit synapse is shown in Fig. 9. However, because the charging and discharging current cannot always be the same in practical circuits, ΔV_G per update pulse is not ideally the same at different V_G, resulting a slight nonlinearity as shown in Fig. 10, but the weight update is still symmetric as the maximum difference of ΔV_G between positive update and negative update is only ~5% of one LSB step. The nonlinearity is observed to be less than +1/+1 as defined in Fig. 11, which is much better than those asymmetric and nonlinear NVM devices [4] as compared in Fig. 12.

III. RESULTS AND DISCUSSION

We benchmark the performance of the proposed hybrid 6-bit 2T1F synapse by incorporating the aforementioned synaptic characteristics into TensorFlow simulation with a CNN, which is a variation of LeNet-5 (Fig. 13), for MNIST dataset. The learning accuracy of ~98.5% from ideal software training with 6-bit weights is utilized as the baseline. With the slight nonlinearity in 2T1F design, the accuracy can achieve ~98.3% (Fig. 14). Then we investigate the impact of residual errors caused by occasional weight-transfer on the accuracy. Fig. 15 shows the simulation results of V_G leakage with different starting V_G, the inset figure shows that it takes 1.64ms for V_G to leak by one LSB step (20mV) in the worst case (starting $V_G = 2V$). Assuming the training time per batch (forward + backward + update, batch size is 100) is ~7 μs, the maximum transfer interval becomes ~230 batches. Fig. 16 shows the training accuracy curve with transfer interval of 100, 200, and 300 batches. When the transfer interval is 100 batches, the accuracy can only achieve ~96%, when the transfer interval is 200 batches and 300 batches, the accuracy can reach ~97.3% and ~98.0% respectively, showing slight degradations compared to 98.3%. The reason is that if the absolute accumulated ΔW within one transfer interval is less than half of one MSB step (8 LSB steps), which fails to trigger the MSBs state change, the weight will be reset back after weight transfer as the V_G will be reset to $(V_A+V_B)/2$ as aforementioned. Fig. 17 shows the percentage of effective $|\Delta W|$ (>8 LSB steps) during first, second, and third weight-transfer operations as an example. A larger interval leads to a larger percentage of effective $|\Delta W|$. Given the fact that weights tend to be stabilized through training process, a dynamic transfer interval (increasing through training) is preferred to fully recover the accuracy. Fig. 18 shows the impact of FeFET polarization state variation on the learning accuracy. A small variation (<5%) does not hurt the accuracy as it may help on compensating the residual errors caused by non-ideal weight-transfer. The degradation becomes unacceptable when variation exceeds 10%. By directly reducing the LSBs tuning step to 10 mV, which results in a 7-bit synapse (2-bit MSBs + 5-bit LSBs), we estimated the learning accuracy on the more complex CIFAR-10 dataset with a VGG-like CNN. Fig 19 shows that the accuracy can achieve ~87% without noise, and ~88% with noise in one LSB step due to the random fluctuation of a 10mV step in practice.

Fig. 20 compares this work to recent works with "volatile" capacitor-based design. The work [10] using 1T1C is totally volatile thus could not support inference. While the work [11], which combines 2 PCM cells with a 3-transistor-1-capacitor structure, is suitable for both training and inference, it has relatively larger cell size and higher programming energy.

IV. CONCLUSION

We introduce a compact 2T1F synaptic weight cell design that combines the benefits of capacitor-based symmetric weight update for LSBs during training and NVM based long-term weight storage for MSBs during inference. A 6-bit/7-bit synapse is demonstrated for MNIST/CIFAR-10 dataset, which can achieve accuracy of ~97.3%/~88%, approaching that of the ideal software based training.

ACKNOWLEDGMENT: This work is supported by ASCENT, one of the six SRC/DARPA JUMP centers.

Analog Synapse Devices				
Type	Filamentary RRAM	Interfacial RRAM	Ferroelectric FET	This work
Weight update behavior	[5]	[6]	[7]	
Symmetry	Low	Medium	Medium	High
Programming	Identical Pulses	Identical Pulses	Non-identical Pulses	Identical Pulses
Speed	10-100ns	100µs-10ms	50ns-100ns	5ns
MNIST Accuracy	~41%	~73%	~90%	~97.3%

Fig. 1. Comparison of analog synapses for on-chip in-situ learning. The proposed 2T1F design exhibits the desired characteristics including highly symmetric/linear weight update, fast and identical update pulses, allowing fast training of neural networks with high accuracy.

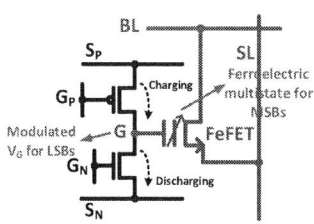

Fig. 2. Schematic of the proposed 2T1F weight cell design. The LSBs of weight are encoded to the conductance of the FeFET by modulating V_G while the MSBs are encoded to different FeFET polarization states.

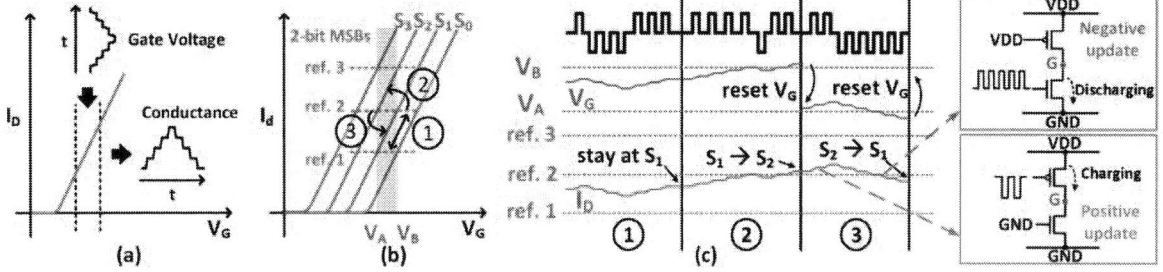

Fig. 3 (a) The LSBs of weight are linearly encoded to the conductance value of the FeFET by modulating the gate voltage while keeping the FeFET in the triode region. (b) The MSBs are encoded to different FeFET polarization states without overlapping of LSBs within each MSB state. (c) Illustration of updating LSBs within a FeFET polarization state and updating MSBs depending on the corresponding read-out current level. V_G is preferred to be reset to the certain level to maintain the same I_D after the update of MSBs to prevent the information loss of LSBs.

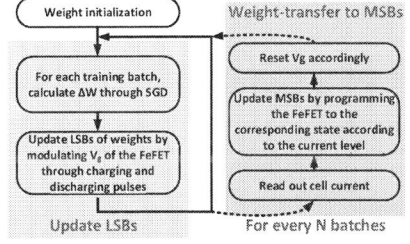

Fig. 4. The training flow chart. For each training batch, update LSBs by applying charging/discharging pulses to modulate V_G based on the value of ΔW. For every N batches, program the FeFET to the corresponding polarization state according to the read-out current level, namely weight-transfer.

Fig. 5. Measured I_D-V_G characteristics of our fabricated HZO FeFET [9] for program voltage from 2V to 4V, showing tunable threshold voltage.

Fig. 7. (a) The pulse scheme and the corresponding remnant polarization charge that generates 4 FeFET states. (b) Simulated I_D vs. V_G curve of different FeFET states. 4 polarization states serve as 2-bit MSBs. The dynamic range of V_G is set to be [1.44V, 1.76V], with a pulse width of 5ns that leads to ΔV_G of 20mV per update pulse, 4-bit LSBs can be achieved.

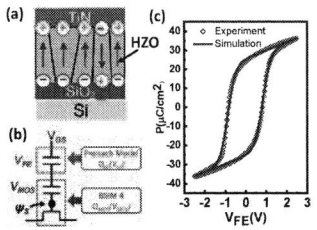

Fig. 6. (a) Schematic of partially switching of HZO ferroelectric domains. (b) The FeFET model [9] consists of the conventional MOSFET modeled by BSIM 4 and the ferroelectric modeled by the dynamic Preisach model. (c) The model accurately captures the experimental P-V loop.

Operation	Update LSBs	Update MSBs	Read
BL	GND	GND/4V	V_{IN}
SL	GND	GND/4V	GND
S_P	VDD	2V-4V/VDD	1.6V
G_P	Pulse/VDD	GND/VDD	GND
S_N	GND	GND/GND	GND
G_N	GND/Pulse	GND/VDD	GND
Substrate	GND	GND/4V	GND

Fig. 8. The voltage bias scheme for different operations for 2T1F weight cell. For the update of MSBs, V1/V2 means the voltage for program/erase operation respectively. For the update of LSBs, V1/V2 means the voltage for charging and discharging, respectively.

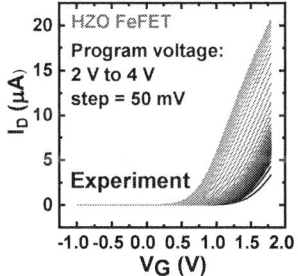

Fig. 9. The equivalent conductance update curve of the 6-bit synapse realized by 2T1F weight cell, showing much improved symmetry and linearity between positive update and negative update.

978-1-7281-1988-5/18 $31.00 © 2018 IEEE   57

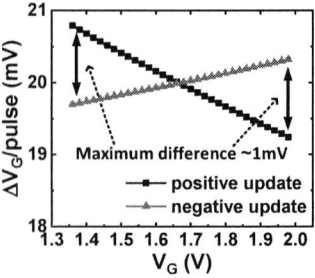

Fig. 10. The ΔV_G per pulse during positive update and negative update as a function of V_G. The maximum difference of ΔV_G between two directions is only ~1mV (5% of one LSB step), suggesting symmetry in weight update.

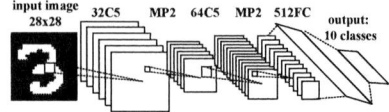

Fig. 13. Benchmark with a CNN on MNIST dataset. The adopted CNN is a variation of LeNet-5 with 32C5-MP2-64C5-MP2-512FC-10 configuration.

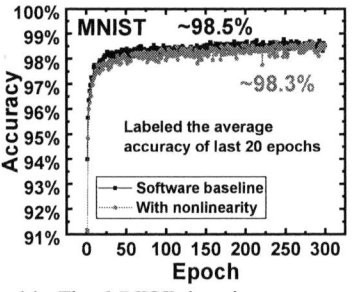

Fig. 14. The MNIST learning accuracy can achieve 98.3% with the slight nonlinearity of the proposed 2T1F design, showing 0.2% degradation compared to the ideal software training with 6-bit weights.

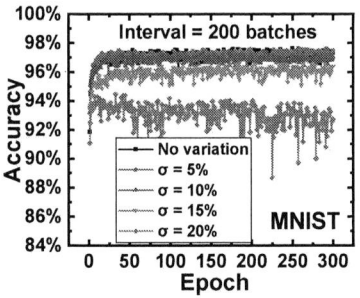

Fig. 18. The impact of FeFET polarization state variation on the MNIST learning accuracy. A small variation does not hurt the accuracy as it may help on compensating the residual errors caused by non-ideal weight-transfer. The degradation becomes unacceptable when variation exceeds 10%.

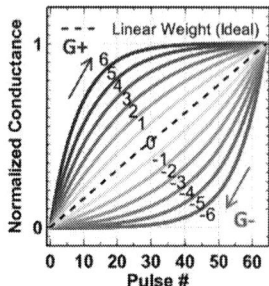

Fig. 11. Analog NVM device behavioral model [4] of the nonlinear/asymmetric weight update. The nonlinearity degree is labeled from +6 to -6.

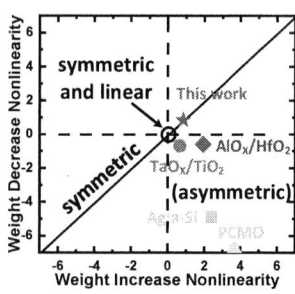

Fig. 12. Comparison of the asymmetry/linearity between this work and other NVM devices [4]. The nonlinearity of this work is fitted using the same model in Fig. 11 as used by other works.

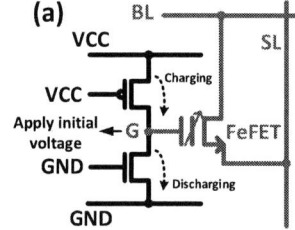

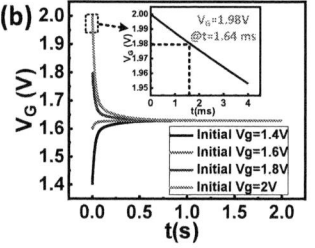

Fig. 15. (a) Circuit setup for leakage simulation. (b) Simulation results of V_G leakage with different starting V_G. The inset figure shows that it takes 1.64 ms for V_G to leak by one LSB step (20mV) in the worst case (starting $V_G = 2V$). Assuming the training time is ~7 μs/batch, the maximum transfer interval becomes ~230 batches, limited by the leakage.

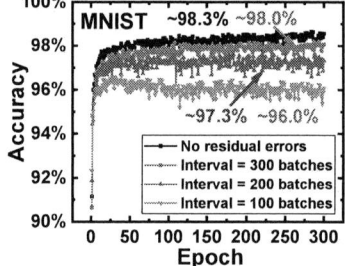

Fig. 16. The MNIST learning accuracy with weight-transfer interval of 100, 200, and 300 batches, achieving ~96.0%, ~97.3%, and ~98.0% respectively.

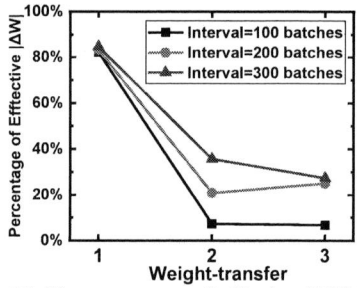

Fig. 17. The percentage of effective $|\Delta W|$ (>8 LSB steps) during first, second, and third weight-transfer with different number of interval batches. A larger interval leads to a larger percentage of effective $|\Delta W|$ to be accumulated, which benefits the training.

Fig. 19. The learning accuracy on CIFAR-10 dataset could achieve 87% w/o noise and 88% w/ noise using the proposed 7-bit synapse with a VGG-like CNN.

Work	[10]	[11]	This work
Weight Cell	3T1C	2PCM+ 3T1C	2T1F
Programming energy	Low	High	Medium
Area	Medium	High	Low
Training	✓	✓	✓
Inference	✗	✓	✓

Fig. 20. Comparison between this work and recent works with the capacitor-based design. This work supports both training and inference with relatively lower programming energy and area.

REFERENCES

[1] C.-C. Chang, et al., *IEDM*, 2017. [2] P. Yao, et al., *Nature Communications*, 2017. [3] G. W. Burr, et al., *IEDM*, 2014. [4] S. Yu, *Proc. IEEE*, 2018. [5] J. Woo, et al., *IEEE Electron Device Lett.*, 2016. [6] S. H. Jo, et al., *Nano Letters*, 2010. [7] M. Jerry, et al., *IEDM*, 2017. [8] S. Wu, et al., *ICLR*, 2018. [9] K. Ni, et al., *Symp. VLSI Tech.*, 2018. [10] Y. Li, et al., *Symp. VLSI Tech.*, 2018. [11] S. Ambrogio, et al., *Nature*, 2018.

978-1-7281-1988-5/18 $31.00 © 2018 IEEE

Analog Computing for Deep Learning: Algorithms, Materials & Architectures

W. Haensch

IBM Research, TJ Watson Research Center, Yorktown Heights NY email: whaensch@us.ibm.com

Abstract—Analog, or neuromorphic, computing for Deep Learning (DL) utilizes the fact that matrix manipulations that are inherent in the back-propagation algorithm, can be performed at constant time, in parallel, on arrays with non-volatile memory (NVM) elements in which the weights are encoded. We discuss the NVM material requirements that need to be met to achieve a classification accuracy on par with the conventional digital approaches, discuss advantages and drawbacks, and highlight opportunities that can take advantage using analog arrays.

I. INTRODUCTION

The success of deep learning rests on two facts: (1) GPUs are an excellent match for DL workloads due to the high degree of parallelization possible, (2) the availability of large data sets for training the models. Once the model is trained inference, or classification, can be run CPUs, FPGAs or other custom hardware [1]. At the core of DL are a few matrix manipulations, shown in Fig. 1, and in analog, or neuromorphic, computing for DL these operations are done on arrays of non-volatile memory (NVM) in parallel at constant time. Thus, these computations are performed locally in memory and therefore avoiding moving weights from memory to the compute unit and back. The array operations at constant time and the reduction in data movement created the recent interest for analog computing for DL [2], with the promise to provide further improvement in compute efficiency. Algorithmic considerations will create constraints for the possible material choices and mapping of the conventional neural network architectures needs special attention [3] [4] [5]. We show that for DL training these NVM elements need to have an incremental analog switching behavior and a symmetric response to potentiation and depression stimulation, Fig. 2.

II. DEEP NEURAL NETWORKS

Back-propagation is the algorithm of choice that is used to train deep neural networks. A deep neural network is a sequence of layers of which each element of a previous layer is connected to each element of the following layer. The strength of these connections are the weights of the network which are determined during the training process. This is a three-step process. Data is moved forward through the network (1). The output of the last layer is compared with the expected value and an error is calculated that is then propagated backward through the network (2) and weights are updated (3) accordingly at each layer. This process is repeated with all data in the data set, one epoch, and as many epochs that are required to reach convergence. The quality of convergence, or classification error, is than tested with part of the data set that

was not used for training. In mathematical terms, repeatedly performed are: (1) forward propagation – input vector x × matrix & activation, (2) backward propagation - error vector δ × transposed weight matrix & derivative of activation, (3) weight update – modification of individual matrix elements $w_{ij} \leftarrow w_{ij} + \eta x_i \delta_j$. If the application is inference, only operation (1) is needed. Convergence of the process is controlled by a set of network parameters of which the learning rate η is crucial. The learning rate determines the amount of weight change per update cycle. If the change is too large the process becomes unstable and if it is too small convergence is not reached. In the digital space the learning rate is a control parameter. For the analog application it is related to the material and stimuli properties. Since in the digital implementation changes are controlled in digital increments increase and decrease are completely symmetric and limited by digital precision. This is not the case in the analog case in which the switching behavior will depend on the material and can depend on the direction of state change and the state itself. Deep neural networks fall into three classes of architectures:

A. Fully connected Neural Netwroks (FCN)

All inputs of a layer are connected to all outputs, Fig. 3. The number of weights and operations is directly proportional to the dimensions of the layers.

B. Convolutional Networks (CNN)

In a convolutional networks spatial correlation of the input data is taking advantage of. A filter of size k x k containing k^2 weights is moved in increments across the input data to perform the convolution at each step. Each convolution is followed by an activation and the result is stored in a two-dimensional feature map. Usually the convolution process will be performed with N filters which produces a feature map dataset that is N layers deep. Before going into the next convolution layer further reduction of data is achieved by pooling or averaging. Convolution and pooling is shown in Fig. 4. CNN can achieve superb classification accuracy for image processing at much lower weight count than FCN.

C. Recurrent Networks (RNN)

Sequential or temporal data correlations are exploited in recurrent networks. Recurrent networks can be understood as temporal networks in which the input at each time step (or sequence) is comprised of the data at step j in conjunction with the network output from a previous step j-1, shown in Fig. 5. Back propagation is implemented after a sequence of J steps is performed. Gate functions in the network will emphasize or de-emphasize sequence elements in the data. Long-term-short-term memory (LSTM) is a frequently used RNN for text and speech.

III. MATERIAL CONSTRAINTS

Weights are encoded in the conductivity of NVM. Since conductivities are only positive and weights carry a sign a differential approach is needed. The encoding scheme will depend of the switching properties of the NVM. To find the algorithmic constraints we have developed a simulation tool that allows to study material impact for the three principle network types discussed above. The tool is built to capture matrix multiplication, weight update at constant time in analog space and the digital/analog interface in multilayer networks. We allow for spatial and temporal stochasticity in the weights, which capture node-to-node and cycle-to-cycle variations in each individual layer, Fig. 6. In addition, the digital-to-analog (DAC) conversion and the analog-to-digital (ADC) conversion at the input and output of each layer, respectively, is captured by the discretization level of the DAC and the ADC circuits. Read and write noise sensitivities that originate from the analog elements and the circuitry itself are captured as well. One key outcome of this analysis is that it is important that all material properties need to be considered simultaneously, Table 1. The key weight parameters we find are the number of incremental steps that connect the lowest to the highest state and the asymmetry in the response to potentiation (increasing weight) and depression (decreasing weight). Considering these two simultaneously we find for all three network configurations that a 10bit granularity (1000 steps) and a 2% asymmetry is necessary to obtain near floating point accuracy, Fig. 7. The impact of node-to-node and cycle-to-cycle variation is rather relaxed with 30%. These results were obtained using a model in which the incremental change was independent of the weight. More sophisticated models can be incorporated, for instance allowing a weight dependent incremental change, which widens the asymmetry spec somewhat. However, minimizing the asymmetry spec is the most important material parameter if analog computing is used for DL training. The class of NVM materials can be separated into uni-directional and bi-directional devices, Fig. 8. Uni-directional devices show incremental change either in the Set (potentiation) or Reset (depression) and bi-directional devices in both. The symmetry of the weight in uni-directional devices requires a matching linearity in the Set (Reset) branch for a differential pair of devices per node. Bi-directional devices do not require linearity but Set and Reset branch need to be mirror images around the incremental changes.

IV. ARCHITECTURAL SOLUTIONS

The natural function of an NVM array is to perform multiply-accumulate (MAC) of vector matrix multiplications in constant time, Fig. 9. This, however, is only one component in the sequence of operations needed. The activation, convolutions, pooling are operations that occur in training and inference. For training the weight update must be added. Not all operations can be performed with-in the analog array, therefore, imbedding the analog-array into a digital backbone needs to be considered. This will raise the question of how to design the analog/digital interfaces. For the array input and output these are digital-to-analog (DAC) and analog-to-digital (ADC) conversions, respectively, Fig. 9. While the conversion from analog to digital between tiles provides the flexibility to have access to a programable digital unit it comes at the cost of usually high power consuming ADCs. An alternative solution is to provide a time encoded signal between tiles [2]. The column output is integrated, and an analog circuit will provide a Relu equivalent activation and a time encoded signal that can directly feed into the next layer. We find that for the digital interface a DAC at 5bit and a ADC at 10bit resolution is required for training [4]. For inference the ADC resolution can possibly be reduced to 4bit. The ADC solution results in a one order of magnitude higher energy per conversion in the case of training, while for inference the energy per conversion is similar to the time encoded solution. To take advantage of the analog arrays the weight update needs to be performed in parallel across the entire array as well. The weight update operation in back-propagation, Fig. 10, is proportional to the product of the forward input vector (x_i) and the backward error vector (δ_j). Coincidence detection, Fig. 11, is used in which x_i. and δ_j is encoded in row and column pulse sequences at a given bit length BL and the node that sees a coincidence will change its state. Both deterministic and stochastic encoding schemes are proposed [3], [4]. Mitigating the short comings of the existing NVM materials hybrid nodes are suggested that separate trailing and leading digits in the weight [6], Figs. 12 and 13. This comes at the cost of area and power at improved accuracy but opens the choice of suitable NVM materials. The scalability of analog computing to larger more complex situations, is topic of current research, Fig 14.

V. SUMMARY

We have presented an overview of how arrays of NVM devices can be used for analog computing in DL. The complex interaction between the material requirements, array architecture and network configurations provide a wide design space. Symmetric response to potentiation and depression drives the exploration and optimization of NVM materials and array architectures.

References

[1] V. Sze et al., "Efficient Processing of Deep Neural Networks: A Tutorial and Survey," *http://arXiv:1703.09039v2 [cs.CV]*, 201

[2] G. Burr et al, "Neuromorphic computing using non-volatile memory," *Advances in Physics: X*, vol. 2, no. 1, pp. 89-124, 2017

[3] G. W. Burr et al., "Large-scale neural networks implemented with nonvolatile memory as the synaptic weight element: comparative performance analysis (accuracy, speed, and power)," in *IEEE International Electron Devices Meeting IEDM*, Washington, DC, 2015.

[4] T. Gokmen and Y. Vlasov, "Acceleration of Deep Neural Network Training with Resistive Cross-Point Devices: Design Considerations," *Frontiers in neuroscience*, vol. 10, p. 333, 2016.

[5] T. Gokmen et al, "Training LSTM Networks with Resistive Cross-Point Devices," *arXiv preprint arXiv:1806.00166*, 2018.

[6] S. Ambrogio,a. et, "Equivalent-accuracy Power-efficient Neuromorphic Hardware Acceleration of Neural Network Training using Analog Memory," *Nature*, vol. 558, pp. 60-67, 2018

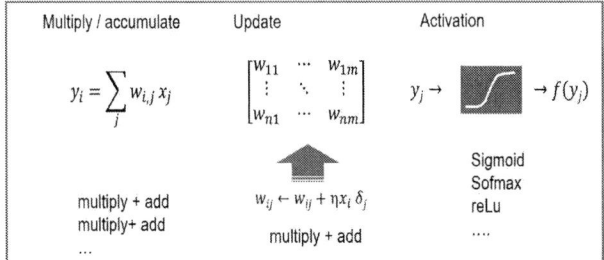

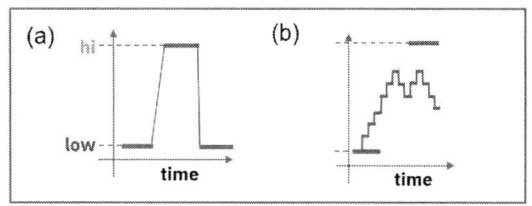

Fig. 1: Basic operations in deep learning: left: vector-matrix multiplication, with input vector x and output vector y. Matrix elements represent the weights w. middle: individual update of matrix (weight) elements, right: non-linear activation function

Fig. 2: Difference between memory and neuromorphic NVM element. (a) A memory element requires high hi/low ratio and fast switching between these states. (a) Neuromorphic elements require incremental symmetric switching for potentiation (up) and depression (down).

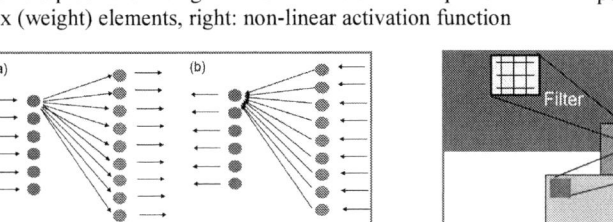

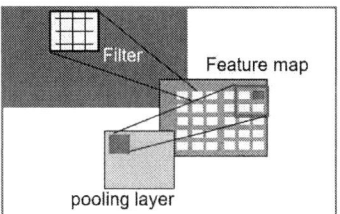

Fig. 3: Fully connected network layer. All input neurons (read) are connected with all out put neurons(green). Strength of the connections are encoded in the weights. (a) forward propagation (b) back ward propagation uses the transposed weight matrix

Fig. 4: Convolution and pooling layer. The input data is scanned by a filter of size kxk to create a feature map at reduced dimension. Further reduction of data is done by only retaining the maximum value in a pooling window.

Fig. 5: Recurrent network. The network feeds processed data back into itself. (a) The box represents a fully connected weight matrix. (a) Sequential data processing can be understood as unrolling of the network. Back propagation will be done after a finite set of J unrolling steps.

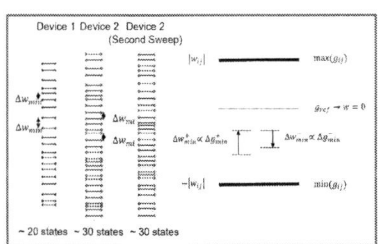

Fig. 6: Device parameters in included in model: range 2|w|, symmetry Δw^+ (up) & Δw^- (down), spatial variation (device 1 to device 2), temporal variation (cycle 1 device 2 to cycle 2 device 2). These variations are stochastic in nature. The reference g_{ref} allows for signed weights

Parameter	Variation	Individual	Combined
Δw_{min}	Global	0.01	0.001
$|w_{ij}|$	Global	0.3	0.6
Δw_{min}	Cycle	150%	30%
Δw_{min}	Device	110%	30%
$|w_{ij}|$	Device	80%	30%
Δw_{min}^-	Global	5%	small
Δw_{min}^+	Global	5%	small
$\frac{\Delta w_{min}^-}{\Delta w_{min}^+}$	Device	6%	2%
Noise	Cycle	10%	6%

Table 1: Device variation for 0.3% deviation from floating point accuracy for MNIST data set for a three-layer (784/256/128/10) FCC network for individual sensitivity and combined impact. A strong interaction of the number of available states (Δw) nad symmetry requirements ($\Delta w^+/\Delta w^-$) is shown. Cycle to cycle (temporal) and device to device (spatial) variability can be larger.

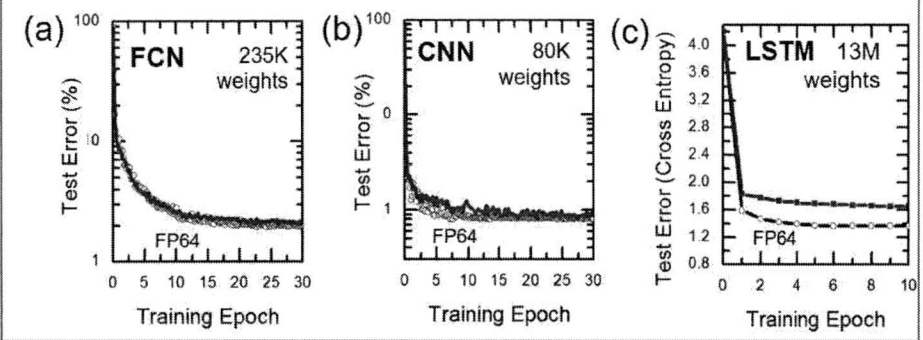

Fig. 7: Simulation results for different neural network topologies using the same material parameters. Base is combined parameter variation from Table 1. For CNN maximal accuracy is achieved by noise and bound management executed in the peripheral CMOS support circuitry. For LSTM stochastic rounding for the 5bit input DAC is used. Base line is a 64bit floating point software solution. SoftMax classifier is used for FCN and CNN and cross entropy for LSTM.

978-1-7281-1988-5/18 $31.00 © 2018 IEEE

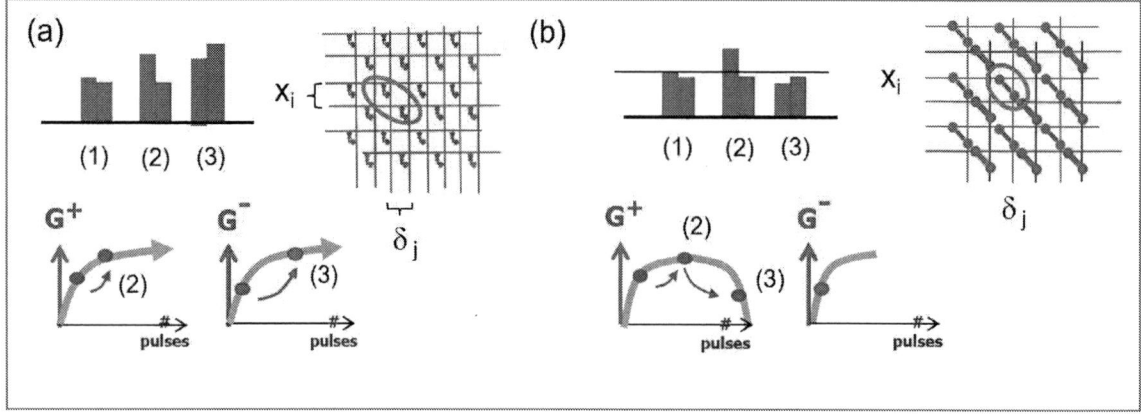

Fig. 8: Uni directional and bi-directional devices. (a) Device switches incrementally in one direction only. Successive updates (1)→(2)→(3) will increase conductivity in each element and eventual saturate. Symmetry is achieved by matching linear switching regime. (b) Device switches bi-directional. Reference is held constant. Active device switches up and down. Symmetry is achieved by symmetric incremental switching. Linearity is not required.

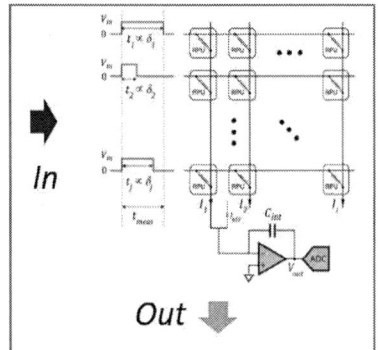

Fig. 9: Analog array read operation. The row input vector V_{in} is time encoded (DAC) and the column current integrated. The integrated charge is either converted in a digital signal (ADC) or time encoded signal is transferred to the next layer.

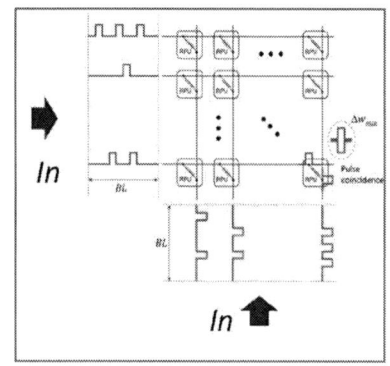

Fig. 10: Analog array update operation. Row (x) and column (δ) signal are encoded in bit steams with length BL. Pulse coincidence at node changes the conductivity of the node element.

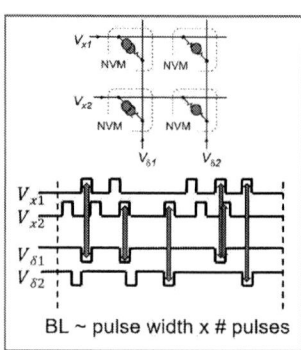

Fig. 11: Coincidence detection. A pulse encoded bit stream is applied to rows (V_x) and columns ($V_δ$). Pulse coincidence at node (blue & red) increase conductivity of node element.

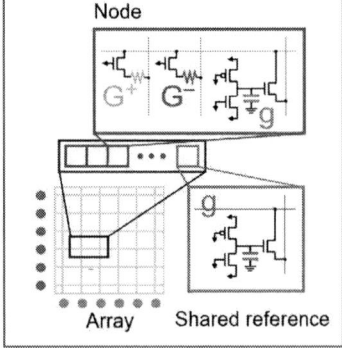

Fig.12: Hybrid cell. Separation of leading and trailing digits onto two elements. Trailing digits (fast changes) are encoded in the switch capacitor that switches symmetrically. Leading digits (slowly changing) are encoded in NVM element (PCM). Reference for switch capacitor can be shared across columns [6]

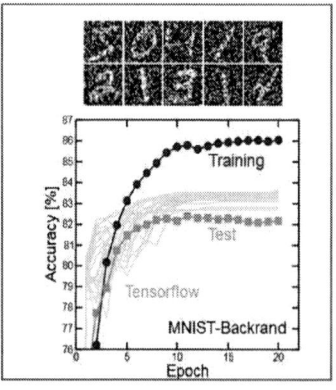

Fig. 13: Training with hybrid cell shows test accuracy on par with floating point solution [6]. Example used full MNIST Backrand dataset (12k data, 50k test) on a three layer 784/180/125/10 FCN network.

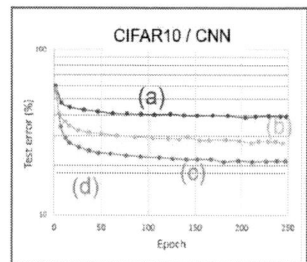

Fig. 14: Simulation of CIFAR 10 /CNN with 3 convolution layers and 1 fully connected layer (x 25 comp. complex than 3-layer MNIST/ FCC). (a) same parameters as used in Fig. 9 (b) improved symmetry (c) larger number of states. (d) Floating point reference.

978-1-7281-1988-5/18 $31.00 © 2018 IEEE

Hardware Acceleration of Simulated Annealing of Spin Glass by RRAM Crossbar Array

Jong Hoon Shin[1], YeonJoo Jeong[1], Mohammed A. Zidan[1], Qiwen Wang[1], and Wei D. Lu[1*]

[1]Department of Electrical Engineering and Computer Science, University of Michigan, Ann Arbor, Michigan 48109, USA

Email: wluee@eecs.umich.edu

Abstract—Simulated annealing (SA) was successfully implemented and accelerated by in-memory computing hardware/software package using RRAM crossbar arrays to solve a spin glass problem. Ta_2O_5-based RRAM array and stochastic Cu-based CBRAM devices were utilized for calculation of the Hamiltonian and decision of spin-flip events, respectively. A parallel spin-flip strategy was demonstrated to further accelerate the SA algorithm.

I. INTRODUCTION

The spin glass system is a representative combinational optimization problem (COP) which tries to find the globally optimal object in discrete space. Since COPs such as spin glass systems and the traveling salesmen problem are NP-hard, simulated annealing (SA), a metaheuristic algorithm that effectively search global optima, has been developed and widely used.[1] However, the convergence of SA may be slow because it involves compute-intensive operations within a massively connected interaction network and stochastic search rules that require random number generation (RNG) with an exponentially decaying probability distribution. Recently, there have been significant progress in RRAM-based acceleration of numerical computation such as partial differential equation and neural network using vector-matrix multiplication,[2,3] in-memory computing,[4] and stochastic computing using stochastic bit streams.[5,6] Inspired by the ability of RRAM devices for numerical computation, in this work, we utilized the ability for vector-matrix multiplication of Ta_2O_5 RRAM-based crossbar and stochastic switching observed in Cu-based CBRAM devices to accelerate an SA algorithm that solves a spin glass problem effectively.

II. SPIN GLASS PROBLEM AND SIMULATED ANNEALING

Finding the ground state of a two-dimensional (2D) spin glass, from randomly mixed states as shown in Fig.1a is a classical problem in COP. Although the interaction between two spins is simple such that the Hamiltonian is just a multiplication between neighboring spins weighted by the coupling strength, complex interactions between arbitrary spin pairs exist in the spin glass, as illustrated in Fig.1b and make the problem difficult to solve in polynomial time.[7] Fig.2a shows the flowchart of conventional SA that starts from initializing the spin configuration, followed by calculating the change of Hamiltonian ΔH_y due to flip of randomly selected y^{th} single spin, σ_y. The Hamiltonian of the spin glass is given as:

$$H = -J \sum_{<x,y>} \sigma_x \sigma_y = -\frac{1}{2} J \sum_{x,y} N_{xy} \sigma_x \sigma_y \qquad (1)$$

where J is the amplitude of the coupling strength, σ_x and σ_y are the xth and y^{th} spin in the spin glass. $<x,y>$ in Eq (1) indicates that the spin multiplication needs to be conducted only for neighboring spins. The introduction of N_{xy}, a coupling strength (CS) matrix, makes the expression more concise. Elements in N_{xy} are '1' if σ_x and σ_y are neighbors of each other, and '0' for non-neighboring spins. If a spin flip decreases energy, e.g. inversion of σ_y leading to negative ΔH_y, SA accepts the change because it stabilizes the spin system. If on the other hand ΔH_y is positive, the spin flip will happen with a probability proportional to the Boltzmann factor ($P = exp(-\Delta H_y/kT)$) where T is absolute temperature. After a fixed number of attempted spin flips, the temperature T is decreased following a cooling schedule, and the process is repeated at the new temperature. The stochastic hill climbing provided by the Boltzmann factor enables the spin glass to escape from local optima as depicted in Fig.2b, and the escape probability decrease to zero as time increases and temperature cools down.

III. SIMULATED ANNEALING ACCELERATED BY RRAM ARRAY AND STOCHASTIC CBRAM

During SA, calculations of the inner products in ΔH_y and the probability generated by the RNG function in the Boltzmann factor make the process compute-intensive. To reduce the computational cost and speed up SA, inner products between the spin vector $\vec{\sigma}$ and neighboring spins, as determined by the CS matrix, can be directly obtained in an RRAM array storing the CS matrix N_{xy}, as shown in Fig.3a-b. For example, when the yth spin attempts to be flipped, all x^{th} row ($\forall \sigma_x \in \vec{\sigma}$) in the RRAM array in Fig.3c are applied with a $V_x (=\sigma_x V_{read})$ pulse, and the output current I_y at the yth column is proportional to $\sum_{x,y} N_{xy} \sigma_x \sigma_y$, producing the desired value of ΔH_y. As a result, the inner-products can be readily obtained from read operations through the RRAM array.

Since only nearest neighbor interactions are non-zero, the CS matrix can be very large but sparse. The large CS matrix can be effectively mapped into smaller RRAM arrays where only the non-zero portions are stored, as illustrated in Fig.4,5. Here a 9×9 2D spin glass was chosen as an example. The 81×81 CS matrix of the spin glass represents all-to-all connection, and can be divided into three groups (top-edge row, mid rows, and bottom-edge row), representing the coupling strength of a spin in the top (middle, or bottom) row with its neighbors. The

978-1-7281-1988-5/18 $31.00 © 2018 IEEE

groups are 9 column wide (corresponding to the 9 spins in each row), and can be further divided into sub-groups of 3 spins (3 columns), for spins at the left-edge, middle columns, and right-edge, producing the patterns shown in Fig. 5. All the possible (non-zero) sub-matrix patterns can then be stored in a three-column RRAM array (11×3), as shown in Fig.5d. Experimentally, the 11×3 RRAM array was fabricated with a Pd/Ta/Ta$_2$O$_5$/Pd cell structure. The RRAM crossbar array is then wire-bonded and connected to a custom test board as shown in Fig.6.

Reliable switching characteristics and tight forming, set and reset voltage distribution can be obtained from all devices in the RRAM array (Fig. 7a,b). The cell-to-cell current variations shown in Fig.7c can be significantly improved to be lower than 1% using a write-verify method, as shown in Fig. 7d, enabling robust dot product operations to obtain ΔH_y .[8] The hill climbing probability was also obtained through hardware by using stochastic switching effects in a Cu-based CBRAM, as shown in Fig.8. The CBRAM device shows stochastic switching behavior at low programming voltage, with a switching probability $P(\Delta t) = 1 - \exp(\Delta t / \tau)$ for programming pulse width Δt , where τ is a time constant dependent on the voltage amplitude. A Cu/ALD Al$_2$O$_3$/Pd CBRAM structure is used in this experimental implementation, with $\tau = 24.9$ms for transition from HRS to LRS. After applying a single SET pulse, the probability of the device staying at HRS then follows the exponential decaying function $\exp(-\Delta t / \tau)$, which follows the Boltzmann factor required for SA, after converting ΔH_y to $\Delta t = \tau(\Delta H_y / kT(t))$.

IV. EXPERIMENTAL DEMONSTRATION OF RRAM-BASED SIMULATED ANNEALING

The flow chart of implementing SA to simulate a spin glass is shown in Fig.9. Starting from the initial spin configuration, a spin (i^{th} row and j^{th} column in the spin glass) is randomly selected for flip-trial. The spin vector is converted as input pulse vector based on its location and applied to the 11×3 Ta$_2$O$_5$ RRAM array. After the current measurement from the selected column I_y, the sign of I_y is compared with σ_y. The flip-event of σ_y is accepted if the signs match (corresponding to negative ΔH_y). If the signs of I_y and σ_y do not match, the flip-event is only accepted if a single SET pulse on a the CBRAM does not change its original HRS state, following discussions above. The data flow is illustrated in Fig.10.

A 15×15 2D ferromagnetic spin glass was tested to prove the concept of RRAM-based SA process. Fig.11 shows one test case with a fixed spin edge condition, where all the edge spins are fixed at the 'up'(+1) state and the rest of the spins are initialized to 'down'(-1) state at time = 0. Because the edge spins are always fixed, the only possible ground state of this problem is "all-up" configuration. The SA parameters such as *J, T(t)*, and N_T for the experiment are 1.0, $5/\sqrt[3]{t+1}$, and 100, respectively. As time flows, the initially down-spins get affected by the edge spin states due to ferromagnetic interaction that favors spins with same orientations. Note some of the down-spins surrounded by other down spins are also flipped to

up-spin (e.g. at time=5), although this event increases the total *E*. This is an example of hill climbing phenomenon which can speed up the optimization process by escaping from the local optima, as discussed in SA. The ground state is achieved at ~ time=200. Other cases with multiple ground states, i.e. initially random configurations without any fixed edges, were also tested using the RRAM-based SA, as shown in Fig.12. Due to the existence of two possible ground states with "all-up" and "all-down" spin configurations, the same initial condition can evolve to opposite results, as verified by the experiments. Note that the two solutions also show similar proportions of majority spin during the evolutions (e.g. at time=150), since the SA strategy leads to similar dynamic progress towards the respective ground state. Comparison between the experimental RRAM-based SA results and software results verifies the *E* and magnetization (*M*) of both cases show similar dynamics that converge to global optima near time = 200, further proving the successful experimental implementation of RRAM-based SA.

V. PARALLEL SPIN-FLIP STRATEGY USING MEMRITIVE SIMULATED ANNEALING

To further accelerate the RRAM-based SA, it is possible to flip multiple non-neighboring spins together simultaneously to take advantage of the parallel vector-matrix multiplication (vs. vector-vector inner product) offered by RRAM arrays, as illustrated in Fig 14. The flipped spins have to be non-neighboring to not affect the energy calculations compared with consecutive spin flips. The parallel spin-flip strategy was also implemented in the RRAM-based hardware. Comparisons of the experimental results obtained from the conventional single spin-flip and the parallel double spin-flip schemes are shown in Fig. 15, for the fixed edge test case. The *E* and *M* from double spin-flip scheme (red) show faster convergence than the single spin-flip scheme (blue). The single spin-flip scheme even fell into a local minimum near time=100 for a while before finally escaping, while the double spin-flip method already reached its ground state. Since the double spin flip should be equivalent to two consecutive spin flips (at the same temperature), the results are compared with another experiment where 2x iterations (i.e. 2N$_T$=200) are attempted at each time step using the single spin-flip scheme (black curves). This approach indeed produced results similar to those obtained from the double spin-flip experiments, and suggested possibility of further acceleration of SA with an N spin-flip scheme that can be calculated simultaneously in RRAM-based array.

ACKNOWLEDGMENT

This work was supported in part by NSF through grant CCF-1617315

REFERENCES

[1] Kirkpatrick, S., et. al., Science 220 (1983): 4598, 671-680
[2] Zidan, M. A., et. al., Nature Electronics 1.7 (2018): 411-420
[3] Chang, C., et al., 2017 IEDM, San Francisco, CA (2017):11.6.1-11.6.4
[4] W. Chen et al., 2017 IEEE IEDM, San Francisco, CA (2017): 28.2.1-28.2.4
[5] Gaba, S., et. al., Nanoscale 5.13 (2013): 5872-5878
[6] Knag, P., et. al., IEEE Trans. Nanotechnology 13.2 : 283–293
[7] F. Barahona, J. Phys. A: Math. Gen., 15 (1982) : 3241-3253
[8] Zidan, M. A., et. al., IEEE Trans. Multi-Scale Comput. Syst. (2017)

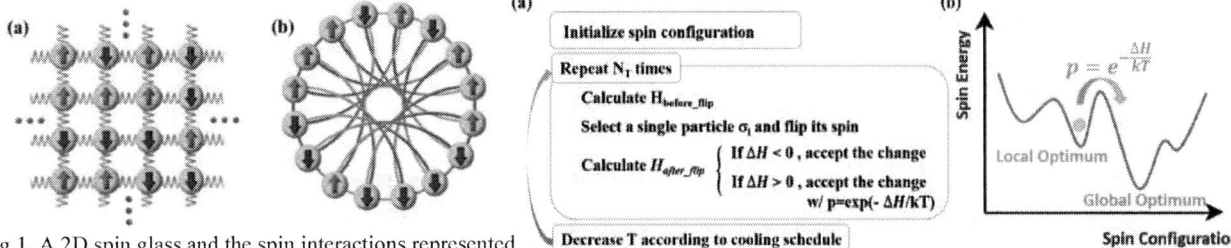

Fig.1. A 2D spin glass and the spin interactions represented by (a) connections to neighboring spins and (b) circular graph showing the complex couplings.

Fig. 2. (a) Flow chart of the SA algorithm. (b) Schematic showing finite spin flip probability even for positive ΔH can help the system escape from local optima.

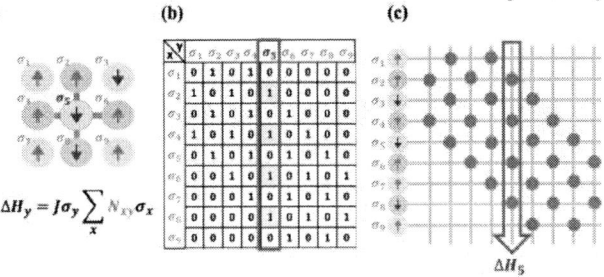

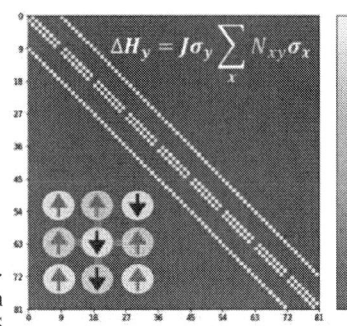

Fig. 3. (a) ΔH_y due to the change of σ_y surrounded by its neighbor spins. (b) CS matrix where the 5th column represents interaction between 5th spin and all the other spins (c) Schematic of inner product between the 5th CS column vector and spin vector $\vec{\sigma}$ conducted by RRAM array.

Fig. 4. 81×81 CS matrix of a 9×9 2D spin array. The large but sparse CS matrix can be sliced to fit into a smaller RRAM array.

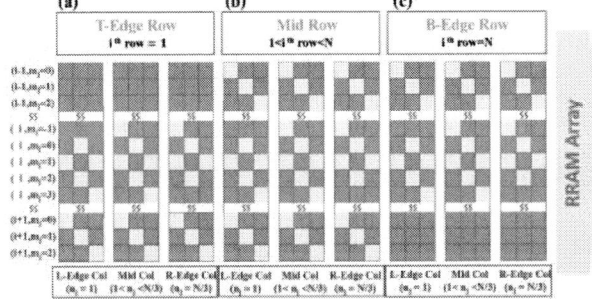

Fig. 5. 9 sub-patterns with three columns each from the 81×81 CS matrix, depending on the position of the spin in the 2D spin glass. (a) Top-Edge Row case, (b) Mid Row case, and (c) Bottom-Edge Row case. (d) All the non-zero and unique patterns in (a-c) can be stored in a single 11×3 RRAM array.

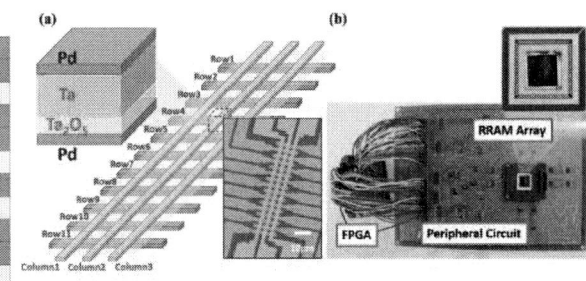

Fig. 6. (a) Schematic of the Ta2O5-based RRAM cell and array structure. SEM image of the RRAM crossbar array. (b) Test board comprised of FPGA, peripheral circuit, and the RRAM array chip for experimental implementation of simulated annealing.

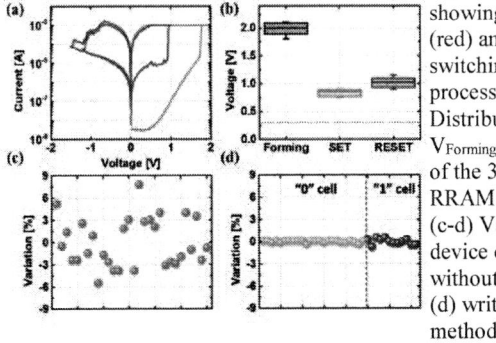

Fig. 7. (a) I-V curves showing the forming (red) and subsequent switching (blue) processes. (b) Distribution of $V_{Forming}$, V_{SET}, V_{Reset} of the 33 cells in the RRAM array. (c-d) Variation of device current without (c) and with (d) write-verify pulse method.

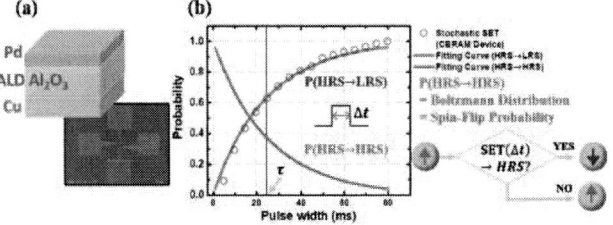

Fig. 8. (a) Structure and SEM image of Cu-based CBRAM devices. (b) Experimentally measured probability of HRS→LRS switching (blue). The Boltzmann factor (red) can be obtained by the probability of the device staying at HRS after applying a single SET pulse with pulse width Δt.

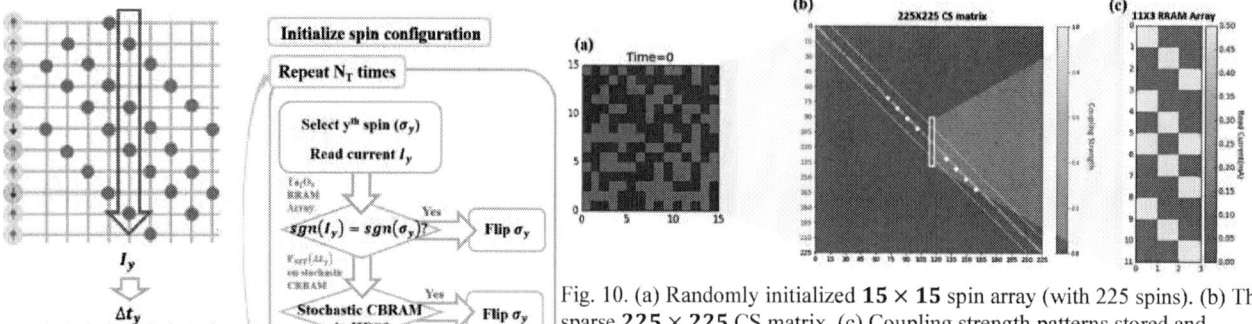

Fig. 9. Flowchart of implementing the SA algorithm using RRAM array for the 2D spin glass problem.

Fig. 10. (a) Randomly initialized 15×15 spin array (with 225 spins). (b) The sparse 225×225 CS matrix. (c) Coupling strength patterns stored and measured from the RRAM array used in the experimental setup.

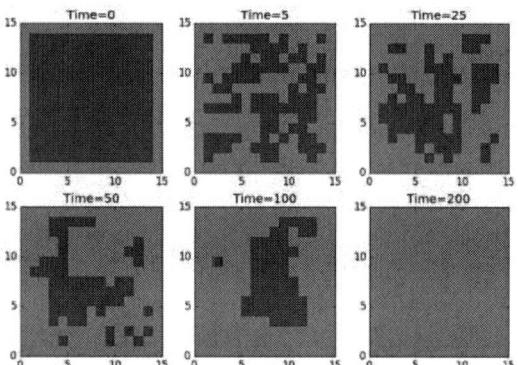

Fig. 11. Evolution of the spin configuration at different time steps for the fixed spin-edge case. Data obtained experimentally from the RRAM array-based hardware system.

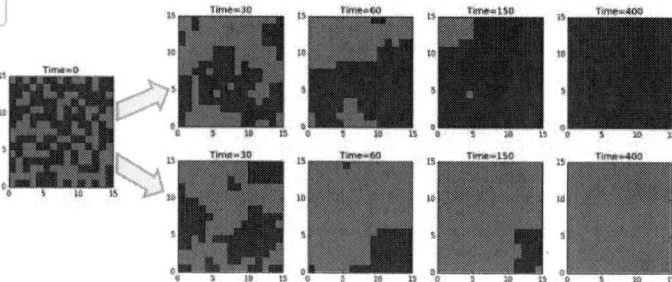

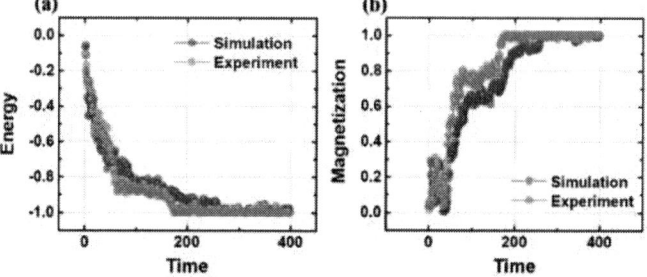

Fig. 12. Time-dependent evolution of the spin glass system solved by the RRAM hardware, for random initial states with no fixed spins. Two ground states with global energy minima, 'all-up' state and 'all-down' states, can be generated from the same initial state in different runs.

Fig. 14. Schematic illustration of multi-spin flip method that exploits parallel vector-matrix multiplications in RRAM crossbar array.

Fig. 13. (a) Average energy and (b) magnetization as a function of cooling schedule. Conventional software version of SA (red) and experimental SA results obtained from the RRAM array (blue) are compared.

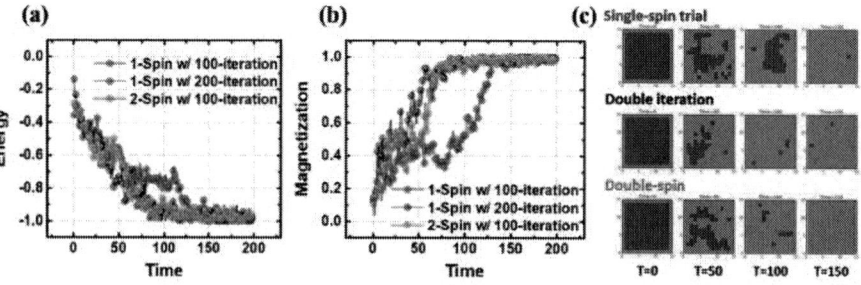

Fig. 15. Comparison of (a) energy, (b) magnetization, and (c) spin configuration snap shots, for results obtained using the single-spin method with 100 iterations per time step (blue), single-spin method with 200 iterations per time step (black), and double-spin method with 100 iterations per time step (red). All results are obtained from the RRAM hardware setup.

978-1-7281-1988-5/18 $31.00 © 2018 IEEE   66

Demonstration of Generative Adversarial Network by Intrinsic Random Noises of Analog RRAM Devices

Yudeng Lin[1,2], Huaqiang Wu[1]*, Bin Gao[1], Peng Yao[1], Wei Wu[1], Qingtian Zhang[1], Xiang Zhang[1], Xinyi Li[1], Fuhai Li[2], Jiwu Lu[2], Gezi Li[3], Shimeng Yu[4], and He Qian[1]

[1]Institute of Microelectronics, Tsinghua University, Beijing 100084, China; Email: wuhq@tsinghua.edu.cn
[2]College of Electrical and Information Engineering, Hunan University, Changsha 410082, China;
[3]Huawei Technologies CO., LTD.   [4]Georgia Institute of Technology, Atlanta, GA 30332, USA.

Abstract—For the first time, Generative Adversarial Network (GAN) is experimentally demonstrated on 1kb analog RRAM array. After online training, the network can generate different patterns of digital numbers. The intrinsic random noises of analog RRAM device are utilized as the input of the neural network to improve the diversity of the generated numbers. The impacts of read and write noises on the performance of GAN are analyzed. Optimized methodology is developed to mitigate the excessive noise effect on RRAM based GAN. This work proves that RRAM is suitable for the application of GAN. It also paves a new way to take advantage of the non-ideal effects of RRAM devices.

I. INTRODUCTION

Resistive random access memory (RRAM) is emerged as one of the most promising synaptic devices for neuromorphic computing applications. Various computing algorithms have been demonstrated on the RRAM array [1-3]. However, due to the non-ideal effects of RRAM, such as low weight tuning accuracy, nonlinearity, variability, etc., the computing accuracy of RRAM array is always lower than CPU/GPU, which compute with floating number [4-5]. On the other hand, Generative Adversarial Network (GAN), which has recently drawn tremendous attention in many artificial intelligence applications, might take advantage of the non-ideal effects of RRAM. As a class of generative neural networks which are different from conventional neural networks that classify patterns, GAN can solve tasks such as image generation, producing high resolution images from low resolution ones, natural language processing, etc [6]. However, GAN suffers from mode dropping and gradient vanishing issues. To fix these issues, adding continuous random noise externally to the inputs of the discriminator is very important and helpful [6].

Analog RRAM, which shows quite different resistive switching behavior compared to digital RRAM [1], is suffered from large random noises during write and read process. These noises are attributed to random ion hopping, and is different from the well-known random telegraph noise (RTN). So far few work has studied the characteristics of random noises in analog RRAM. Although the noises may decrease computing accuracy of conventional neural networks, they can be used as inputs of GAN to fix aforementioned GAN's issues. In this work, we demonstrate that the intrinsic random noises of RRAM can be utilized for the generative neural network. For the first time, we experimentally realized pattern generation function and online training of GAN based on fabricated 1kb analog RRAM array.

II. GENERATIVE ADVERSARIAL NETWORK

GAN system consists of two networks - a Generator (G) and a Discriminator (D). Typically, both G and D are fully connected neural networks. A typical training and inference process of the two networks are shown in Fig.1.

For the inference, G takes a noise series as input and generates fake samples. Then both fake and true samples feed into D, and D outputs a probability of a given sample coming from the real dataset or generated fake ones. In this operation, before D is fed, adding noise into input samples is the important key to fix aforementioned GAN's issues. For the online training, when training D, D is a binary classifier and learns how to discriminate real and generated data. D is expected to output high and low probability when true and fake samples to be fed, respectively. Here, D is trained and updated twice: one for true samples, and the other for fake samples. When training G, G is optimized to generate samples which resemble real data in order to cheat D into accepting its outputs as being real. Thus D is expected to output high probability when generated samples feed it. To achieve these goals, one network's weights are kept constant, while the other's weights are updated to minimize its cost function. Both networks aim to minimize their own cost, and the solution to the game is the Nash equilibrium, at which neither network can improve their cost unilaterally.

III. ANALOG RRAM ARRAY

To demonstrate a GAN, 1kb 1T1R arrays were fabricated. The fabrication process can be found in [1]. RRAM cell is integrated on top of transistor's drain contact. Bit line, word line, and source line connect to top electrode of RRAM, gate of transistor, and source of transistor, respectively. The structure of RRAM cell is TiN/TEL/HfO$_x$/TiN stack. HfO$_x$ is the switching layer, and thermal enhanced layer (TEL) contributes to the good analog switching behaviors. To implement a GAN, the analog RRAM array acts as synapses which connect the neurons at different layers of both D and G. The conductance of RRAM cell represents the weight in the network, which is tunable during the training process. Voltage pulses are applied on the bit lines, and the output currents from source lines are sent to computer for further processing, as shown in Fig.2.

The analog resistive switching behaviors under identical pulses during SET process and RESET process are measured. Fig.3 and Fig.4 show the analog switching of different cycles and different cells, respectively. The random conductance fluctuations are observed on each weight update curve. This fluctuation is defined as write noise in this work. The write

978-1-7281-1988-5/18 $31.00 © 2018 IEEE

noise is analyzed and captured as shown in Fig.5 and Fig.6. To simulate this characteristic, a device model is developed. The simulated continuous analog switching behaviors with different variation parameters of write noise during SET and RESET process are shown in Fig.7. The measured variation parameter of write noise on the RRAM device is between $10\sim15\times\sigma$.

Measured read current variations during repeated read process are shown in Fig.8. The variation is defined as read noise. For analog RRAM, this read noise is attributed to the random hopping of oxygen vacancies. Fig.9 shows the normal distribution fitting for the read noise. The standard deviation of fitting distribution is almost a constant in different resistance levels, as shown in Fig.10. This constant noise amplitude is important for the application of read noises on the computation.

IV. DEMONSTRATION OF PATTERN GENERATION

The performance of GAN is evaluated with two key parameters: peak signal noise ratio (PSNR) and diversity of pattern classes. Fig.11 illustrates how to calculate the PSNR and diversity of the generated pattern samples. A convolution neural network (CNN) is used to classify the generated patterns, as shown in Fig.12. The original MNIST images with 28×28 pixels are down-sampled as the real data pattern to train the 1kb RRAM-array based GAN as shown in Fig.13.

For the traditional training algorithm of GAN, D is updated twice in one training iteration. This method may magnify the negative effect of write noise, which harms the network convergence process. To mitigate this effect, we develop a modified training method for D, as shown in Fig.14a. This method sums the gradients of true and fake samples, and then modulates the device conductance depending on the sign of gradient's sum. The verification scheme is also introduced for the accurate tuning of conductance, as shown in Fig.14b. The mean values of D's outputs for true and fake samples can indicate whether the system reaches the Nash equilibrium or not. Fig.15 shows the simulated D's outputs for the two kinds of samples. The two output curves keep almost constant after 150 epochs, indicating the system has reached convergence.

Before experimental demonstration, we first investigate the impacts of RRAM noises on the GAN performance with simulation. The original MNIST dataset, which has a uniform percent of ten classes (Fig.16a), is used as the real data to train the network. When training without any noise, the generated patterns shows mode dropping phenomenon. Almost all the patterns fall into one class ("8") (Fig.16b). While training with different variation parameters of write noise or different standard deviations of read noise, the generated patterns shows diversity (Fig.16 c&d). Training GAN with low variation of write noise can be helpful to fix mode dropping issues. But with excessive write variations, G's performance becomes poor, as shown in Fig.17. High variation of write noise can also cause difficulty in the network to converge. As for read noise, it can be found that the standard deviation of read noise does not influence the diversity significantly (Fig.18). These results indicate that introducing small amplitude of noise is sufficient to improve diversity.

On the other hand, both write and read noise can introduce noisy pixels in the generated patterns. As the variation of write

noise increases, the number of noisy pixels increases (Fig.19). Whereas, the patterns still keep smooth and sharp as standard deviation of read noise increases (Fig.20). Write noise and read noise introduce noisy pixels in different modes (Fig.21). After training with write noise, the output current varies from the ideal output. This effect can introduce obvious noisy pixels in different generated patterns (Fig.21c). However, unlike write noise, the read noise introduces slight fluctuation on every pixel (Fig. 21b). Since GAN can be regarded as a large multi-layer network and the generated patterns are hidden layer's output, the patterns can be filtered to reduce the influence of noise. From this perspective, it inspires us that we can filter hidden layers' outputs to mitigate the noise effect between layers.

Based on the above optimization methodology, we finally implement GAN on the fabricated RRAM array successfully. The complete implementation platform is shown in Fig.21. Due to the limited array size, the two networks, D and G, are alternatively trained using three classes of the down-sampled patterns from MNIST dataset (digital number "0", "1" and "7") as the real data. The measured conductance evolution of some RRAM cells during the online training process is shown in Fig.22. Weight increase and weight decrease can be both observed on the same cell during training. The measured conductance maps of the RRAM array before and after training are shown in Fig.23b. The three classes of digital patterns are generated correctly from the RRAM array, as shown in Fig.24. This result indicates that RRAM is capable of realizing generative network by utilizing its intrinsic noise.

V. CONCLUSION

Key achievements: 1) For the first time, pattern generation was demonstrated experimentally on the RRAM array; 2) Unique behaviors of random noises in analog RRAM devices and their impacts on the generative network were investigated; 3) Optimization methods for the RRAM based generative network were proposed.

ACKNOWLEDGMENT

This work is supported in part by the MOST of China (2016YFA0201801), Beijing Innovation Center for Future Chip (ICFC), Beijing Municipal Science and Technology Project (D161100001716002, Z181100003218001), and NSFC (61674087, 61674089, 61674092, 61076115).

REFERENCES

[1] H. Wu *et al.*, "Device and circuit optimization of RRAM for Neuromorphic computing," *IEEE International Electron Devices Meeting (IEDM)*, pp. 274–277, 2017.

[2] Z. Wang *et al.*, "Fully memristive neural networks for pattern classification with unsupervised learning," *Nat. Electron.*, vol. 1, no. 2, pp. 137–145, 2018.

[3] P. M. Sheridan, F. Cai, C. Du, W. Ma, Z. Zhang, and W. D. Lu, "Sparse coding with memristor networks," *Nat. Nanotechnol.*, vol. 12, no. 8, pp. 784–789, 2017.

[4] W. Wu *et al.*, "A Methodology to Improve Linearity of Analog RRAM for Neuromorphic Computing," *Symp. VLSI Technology*, pp. 3–4, 2017.

[5] M. Zhao *et al.*, "Investigation of statistical retention of filamentary analog RRAM for neuromophic computing," *IEEE International Electron Devices Meeting (IEDM)*, 2017, pp. 39.4.1-39.4.4.

[6] A. Creswell *et al.*, "Generative Adversarial Networks: An Overview," arXiv: 1710.07035, 2017.

978-1-7281-1988-5/18 $31.00 © 2018 IEEE

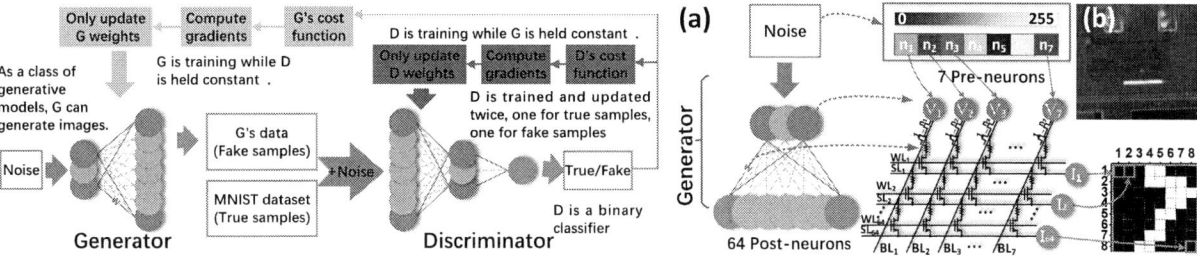

Fig. 1. Schematic of the working principle and architecture of GAN. GAN includes two networks. G takes a noise series as input and generates samples. D outputs a probability of a given sample coming from the real or generated data.

Fig. 2. (a) Mapping of GAN on the analog RRAM array. The RRAM devices act as synapses which connect the neurons at different layers of network. (b) The fabricated 1kb RRAM array.

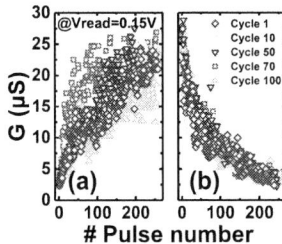

Fig. 3. Analog switching behavior under identical pulses during (a) SET process and (b) RESET process in different cycles.

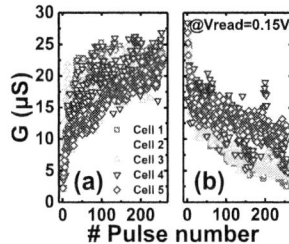

Fig. 4. Analog switching behavior under identical pulses during (a) SET process and (b) RESET process in different RRAM cells.

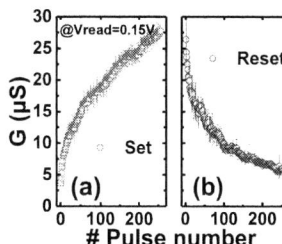

Fig. 5. Statistics of analog switching behaviors under identical pulses during (a) SET process and (b) RESET process.

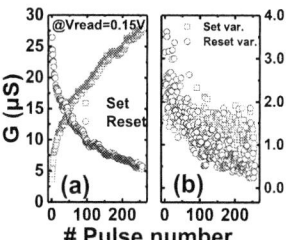

Fig. 6. Statistics of analog switching behaviors. (a) Nonlinear curve. (b) Variation of SET and RESET process.

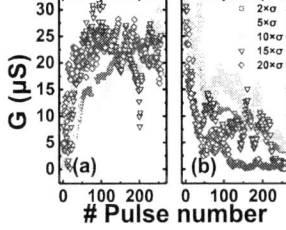

Fig. 7. Simulated analog switching behaviors with different variation parameters of write noise during (a) SET and (b) RESET process.

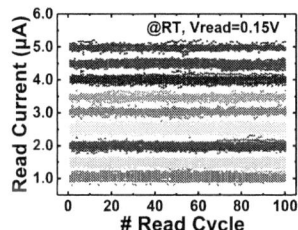

Fig. 8. Measurement of read current noise of 9 levels at room temperature. Every level is measured in multiple cells.

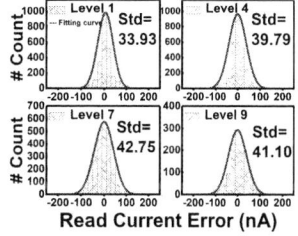

Fig. 9. Normal distribution fitting for read noise in 4 levels. Read current error is the error between every cell's read current and its average.

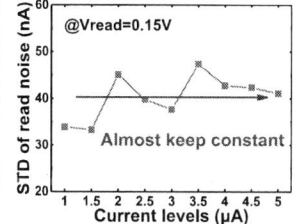

Fig. 10. Standard deviation of read noise in different levels. It keeps almost constant as current level changes.

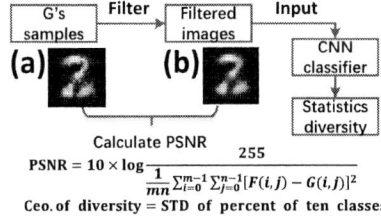

$$PSNR = 10 \times \log \frac{255}{\frac{1}{mn}\sum_{i=0}^{m-1}\sum_{j=0}^{n-1}[F(i,j)-G(i,j)]^2}$$

Ceo. of diversity = STD of percent of ten classes

Fig. 11. Schemtic of the method to calculate PSNR and diversity. (a) The generated pattern samples are filtered, then (b) the filtered patterns are classified by a CNN.

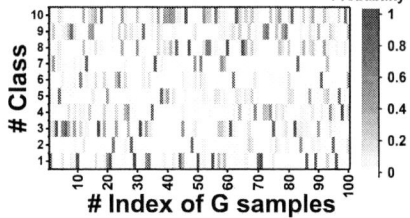

Fig. 12. Classification result of the CNN, which is used for classifying the generated patterns automatively. Recognition accuracy of this CNN for the standard MNIST dataset is 98%.

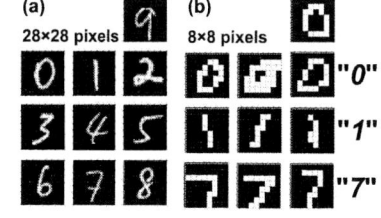

Fig. 13. (a) The original images with 28×28 pixels from the MNIST dataset. (b) The down-sampled patterns for training on the RRAM array.

978-1-7281-1988-5/18 $31.00 © 2018 IEEE

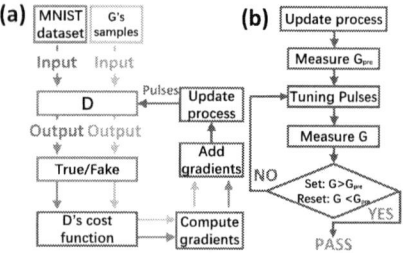

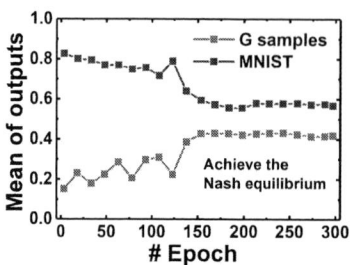

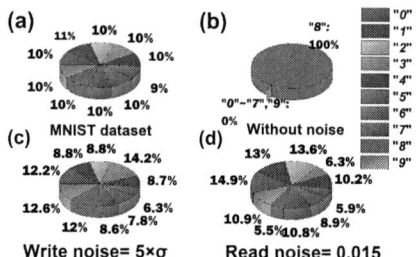

Fig. 14. (a) The modified training method for D. In one training iteration, the gradients are added for true and fake samples and execute update prosess for one times instead of updating twice. (b) Update prosess with verify scheme.

Fig. 15. The mean of the discriminator's outputs for true and fake samples during online taining process. The simulation of GAN is based on the device model with write and read noises.

Fig. 16. The percentage of every class in (a) the original MNIST dataset, (b) generated patterns when training without any noise, (c) generated patterns when training with write noise, and (d) generated data when training with read noise.

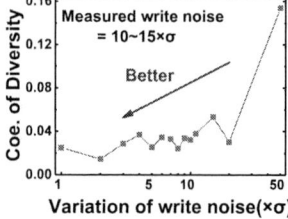

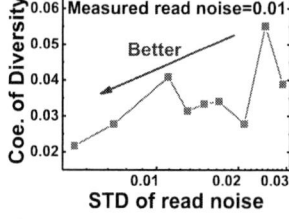

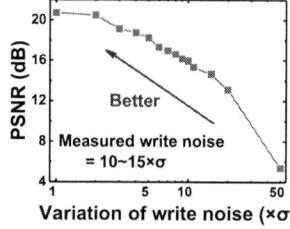

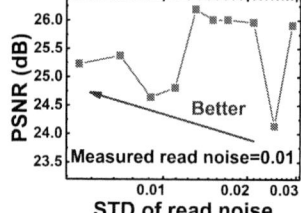

Fig. 17. Diversity of the generated patterns as a function of variation of write noise.

Fig. 18. Diversity of the generated patterns as a function of standard deviation of read noise.

Fig. 19. PSNR of the generated patterns as a function of variation of write noise.

Fig. 20. PSNR of the generated patterns as a function of standard deviation of read noise.

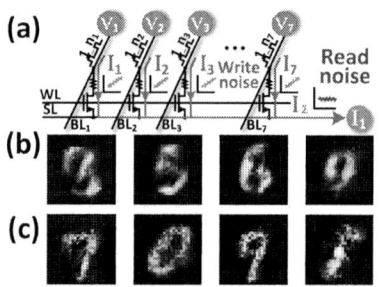

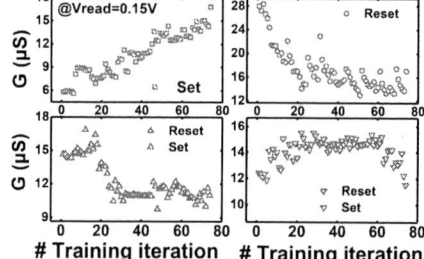

Fig. 21. (a) Write noise and read noise introduce noisy pixels in different modes. (b) Generated patterns with read noise.(c) Generated patterns with write noise.

Fig. 22. Measurement platform for implementing the RRAM array based GAN. A custom designed tester is used to measure the analog RRAM array.

Fig.23. Typical conductance evolution of 4 RRAM cells in the array during online training process of the GAN. The device conductance changes incrementally until network converges.

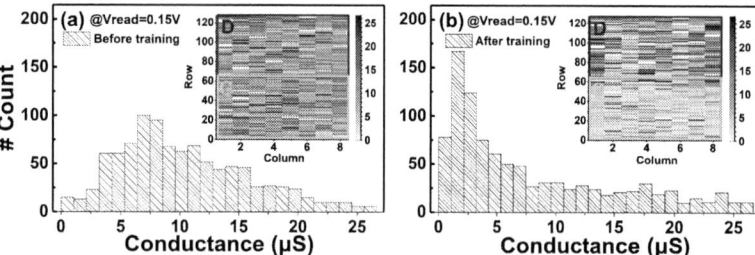

Fig. 24. Distribution of device conductance in the RRAM arrray (a) before training and (b) after training. Inset: conductance map of the 1kb array.

Fig. 25. Typical generated patterns from the GAN. Three classes of patterns are generated uniformly. Each class contains different patterns which can be classified by the CNN correctly.

978-1-7281-1988-5/18 $31.00 © 2018 IEEE

Error-Resilient Analog Image Storage and Compression with Analog-Valued RRAM Arrays: An Adaptive Joint Source-Channel Coding Approach

Xin Zheng[1*], Ryan Zarcone[2], Dylan Paiton[3], Joon Sohn[1], Weier Wan[1], Bruno Olshausen[3+] and H. -S. Philip Wong[1#]

[1]Department of Electrical Engineering, Stanford University, Stanford, CA, 94305, USA,

[2]Biophysics Graduate Group, [3]Vision Science Graduate Group, UC Berkeley, Berkeley, CA, 94720, USA.

E-mail: [*]xzheng3@stanford,edu, [+]baolshausen@berkeley.edu, [#]hspwong@stanford.edu

Abstract – We demonstrate by experiment an image storage and compression task by directly storing analog image data onto an analog-valued RRAM array. A joint source-channel coding algorithm is developed with a neural network to encode and retrieve natural images. The encoder and decoder adapt jointly to the statistics of the images and the statistics of the RRAM array in order to minimize distortion. This adaptive joint source-channel coding method is resilient to RRAM array non-idealities such as cycle-to-cycle and device-to-device variations, time-dependent variability, and non-functional storage cells, while achieving a reasonable reconstruction performance of ~ 20 dB using only 0.1 devices/pixel for the analog image.

I. INTRODUCTION

Much of today's data (e.g. video, audio, and images) are inherently analog. To store and compress these analog data onto digital memory, analog-digital data conversion (source coding) and digital compression (channel coding) are usually performed separately. Analog, non-volatile memory (NVM), such as RRAM and PCM, offer opportunities to directly store multi-dimensional analog data. The challenge is to perform reliable storage and retrieval of analog signals with non-ideal NVM devices. In this paper, we present a joint source-channel coding algorithm with a neural network to store and compress natural images onto an analog-valued RRAM array. Through jointly learning the data and memory statistics, our source-channel coding algorithm finds an optimal use of the memory's intrinsic data storage capacity [1]. Using this algorithm, we demonstrate by experiment that natural images can be reliably stored and retrieved with an analog-valued RRAM array, while having additional desirable properties of being resilient to array-level non-idealities such as cycle-to-cycle and device-to-device variations, time-dependent variability, and non-functional storage cells. Our work presents a way to use imperfect NVMs whereby cost and fabrication advantages are utilized while the non-idealities of the device technology are circumvented. This approach shows that it is fruitful to customize the way we use the device technology to suit the task at hand.

II. JOINT SOURCE-CHANNEL CODING ALGORITHM

To learn a mapping from natural images to the RRAM array, we constructed a multilayer autoencoder neural network (**Fig. 1**), analogous to a denoising autoencoder [2]. With a succession of linear/nonlinear operations, the encoder transforms an input image into a set of resistances to be written to the array. To retrieve the stored image, the decoder transforms the read-resistances into a reconstruction. Specifically, the encoder and decoder weights are parameterized as filter convolutions. For the non-linearities, or "activation functions", we used divisive normalization (Equation 3 in [3]), a population nonlinearity that implements a local form of gain control.

In order to train the network, we constructed a differentiable model of the RRAM channels, with two sources of channel noise: (1) a uniform noise induced by the write process and (2) a "sparse" noise induced by device failure. The latter was implemented during network training by randomly choosing a certain percentage of the devices and then sending them to the low resistance state (LRS) (i.e. setting R_T to 0). Thus, the noise model for each device was:

$$\log R_M (R_T) = \log R_T \cdot (1 - \eta) + \log \sigma \cdot \epsilon,, \epsilon \sim U(-0.5, 0.5) \quad (1)$$

where σ is the write process acceptance range, $U(-0.5, 0.5)$ is the uniform distribution function and $\eta \in \{0,1\}$ is the sparse noise parameter (by default $\eta = 0$, but $\eta = 1$ when the device fails). More importantly, this channel model is not specific to the type of memory device. The model can be adapted to any analog programmable memory device (even those with highly nonlinear input/output functions [4]).

For training, the network weights were learned by backpropagating the gradient of the objective function:

$$C = \langle \|X - \hat{X}\|^2 + \lambda \left[\max\left(0, \log \frac{R_T}{R_{T_{MAX}}}\right) + \max\left(0, \log \frac{R_{T_{MIN}}}{R_T}\right) \right] \rangle \quad (2)$$

The first term corresponds to the squared error between the original image, X, and its reconstruction, $\hat{X}$. The second term is a cost that penalizes the network for using values outside the range of acceptable R_T (λ being a scalar hyper-parameter that weights this term's importance). $\langle . \rangle$ indicates an average over an image-batch during training. The training set consisted of ~1.2×10^5 images from the 2016 ImageNet test set [5] and the Flickr Creative Commons set [6].

III. ANALOG VALUE STORAGE WITH RRAM ARRAY

To demonstrate by experiment this analog data storage method, we fabricated a CMOS-integrated TiN/HfOx/Pt 1K 1T1R array (**Fig. 2**). **Fig. 3** shows a cross-section schematic of the RRAM Stack: Pt (30 nm) / HfOx (5 nm) / TiN (50 nm). **Fig. 4** shows typical DC (**Fig. 4 (a)**) and AC (**Fig. 4 (b)**) cell characteristics in the 1T1R array with analog storage capability. We encoded 448 analog resistances, R_T (10 KΩ to 100 MΩ), from 64×64 pixels (8 bits per pixel) images onto

978-1-7281-1988-5/18 $31.00 © 2018 IEEE

448 cells, a ratio of 1 device per 10 pixels. We set the write acceptance range σ to 2. Ref. [7] showed that for a two-terminal RRAM device, any resistance level within its dynamic range can be achieved with high accuracy by applying a combination of incremental step RESET and SET pulse trains. Here we extend this write algorithm to 1T1R device programming. A double-direction incremental step pulse programming strategy (DD-ISPP) was used to precisely program cell resistances into the acceptance range, in spite of cycle-to-cycle and device-to-device variations. **Fig. 5 (a)** shows the DD-ISPP sequence. The device was initially SET to the LRS (around 10 kΩ). A full DD-ISPP writing pulse parameter set is listed in **Fig. 5 (b)**. The cell resistance was verified after each write pulse. The sequence continued until one out of three conditions was satisfied: (i) the programmed resistance fell into the acceptance range; (ii) the WL voltage exceeded the maximum value (5V for RESET, 1.2V for SET); (iii) the total number of write pulses reached the maximum number (100). If the final programmed resistance was still outside the acceptance range, we SET it back to the initial LRS (corresponding to the sparse noise in the RRAM model of Section II). **Fig. 5 (c) (d)** show an example writing with DD–ISPP, where RESET over-programming is fixed by the SET pulse train. Targeting at different R_T (100 KΩ, 300 KΩ, 1 MΩ, and 3 MΩ) with the acceptance range of 2, **Fig. 6** shows cycle-to-cycle (**Fig. 6 (a)**) and device-to-device (**Fig. 6 (b)**) resistance distributions written by DD-ISPP, both following the same uniform distribution centered at R_T with boundaries defined by the acceptance range (this being well-modeled by the uniform noise term in the RRAM model). **Fig. 7** shows the log-scale 448 resistance values encoded by the network (R_T, **Fig. 7 (a)**) and stored on the array (R_M, **Fig. 7 (b)**).

IV. IMAGE RECONSTRUCTION WITH DEVICE NON-IDEALITY

Device non-ideality brings different challenges to the system robustness of analog RRAM storage than digital (binary) storage. Specifically, there are two main kinds of device non-ideality that degrade analog array storage capability: error cells and resistance relaxation.

Error Cells – The full resistance dynamic range (max. and min. achievable resistance values) of different cells within an array may differ. For a specific cell, if the target resistance assumed by the algorithm is outside the full dynamic range of the cell by a distance larger than the acceptance range, error values are generated. Our experimental array had an error rate of 0.2% (1 out of 448). Different images could have different number of error bits based on the targeted resistance from the encoder. In order to achieve reliable reconstruction results, the algorithm needs to tolerate the worst-case scenario (maximum error rate). We trained the network with an estimated maximum error rate of 2%. Note that this is a very high error rate for a product technology and we used this as a way to illustrate the error resiliency of our methodology. To test the algorithm's error-resiliency, we explored the relation between reconstruction

mean squared error (MSE) and storage error rate. Various error rates (0.5%, 1%, 2%, 5%, and 10%) were dialed in by randomly selecting a set of cells and SETTING them to LRS (**Fig. 8**). **Fig. 9** shows reconstructions (top) for the original image and corresponding MSE (bottom) for the different error rates (all images are scaled to have pixel values of mean = 0, variance = 1). Interestingly, even though the network was trained with 2% error, the algorithm was able to tolerate error up to 5% before suffering from serious degradation.

Resistance Change After Programming – Our DD-ISPP method is capable of programming cell resistances efficiently within the acceptance range given by the target. However, when decoding from resistance values read from the array, time-dependent variability (TDV), which cannot be suppressed with write-verification methods [8][9], may cast the resistance values outside of acceptance ranges and degrade reconstructions. TDV must be properly considered when evaluating image reconstruction performance. After storing the resistances onto the array, resistance drift over time was monitored (**Fig. 10**). **Fig. 11 (a)** shows the final resistances programmed with DD-ISPP. Within 1 second after programming, we read the resistance again (**Fig. 11 (b)**) and observed some outliers due to read noise and short-term relaxation [8]. **Fig. 11 (c)** shows that more outliers appear 1 minute after programming. A broader distribution is observed when reading resistance 12 hours after programming (**Fig. 11 (d)**). **Fig. 12** shows the reconstructions (top) of the original image and their corresponding MSE (bottom) over time. **Fig 12 (a)** shows compression-only, where $R_T = R_M$. **Fig. 12 (b)-(e)** correspond to the reconstruction results from the resistance read values from the array shown in **Fig. 11 (a)-(d)**. Our scheme is resilient to outliers, where "verify-read", "direct read" and "1 minute after write" show similar MSE and resistance distribution broadening after 12-hours only causes a small increase in MSE.

It should also be noted that at this compression rate of 1 device/10 pixels and 1 bit/ channel, JPEG would produce a very poor reconstruction. As shown previously [4], if the goal is simply to perform compression with a set of additive-noise storage devices (neglecting the more realistic device failures) the neural-network algorithm is able to outperform JPEG.

V. CONCLUSION

Key achievements: (1) A joint source-channel coding algorithm was developed, adapting to RRAM array characteristics to store analog data directly onto the analog-valued RRAM array. (2) An image storage and compression task was demonstrated by experiments with an RRAM array. (3) Our method empirically showed error-resilience against device non-idealities such as error cells, cycle-to-cycle and device-to-device variations, and resistance relaxation.

ACKNOWLEDGMENT

Work supported in part by ASCENT, one of the six centers in JUMP, ENIGMA from the NSF/SRC, E2CDA, NSFGRFP, and the Stanford SystemX Alliance and Stanford NMTRI.

REFERENCES

[1] JH. Engel, *et al. IEDM* (2014)
[2] P. Vincent *et al. ICML* (2008)
[3] J. Balle *et al. ICLR* (2017)
[4] R. Zarcone, *et al. DCC* (2018)
[5] J. Deng *et al. CVPR* (2009)
[6] B. Thomee *et al, Communications of the ACM*, 59, 2, pp. 64-73 (2016)
[7] F. Alibart, *et al. Nanotechnology*, 23, 7, p.075201 (2012)
[8] A. Fantini, *et al. IEDM* (2015)
[9] S. Ambrogio, *et al. IEEE TED*, 62, 11, pp.3812-3819 (2015)

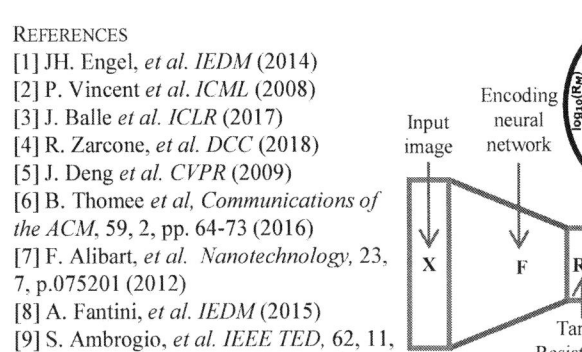

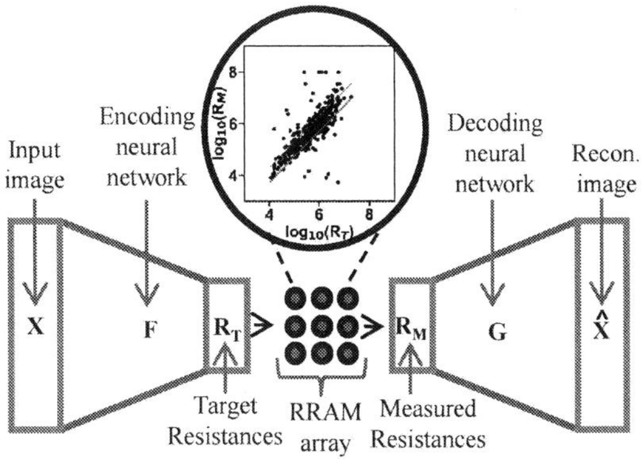

Fig. 1 Diagram of the autoencoder architecture. The input image, X, is transformed through the encoding neural network, F, into a set of target write resistances, R_T. These resistances are then written to the device array (purple circles). The devices are then read, yielding measured resistances R_M. R_M are passed through the decoding neural network, G, yielding the reconstructed image, $\hat{X}$.

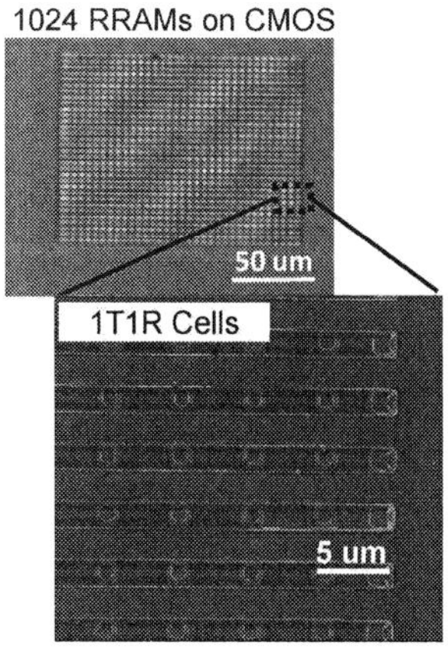

Fig. 2 SEM image of the 1K 1T1R analog-valued RRAM array used for storing image in this work

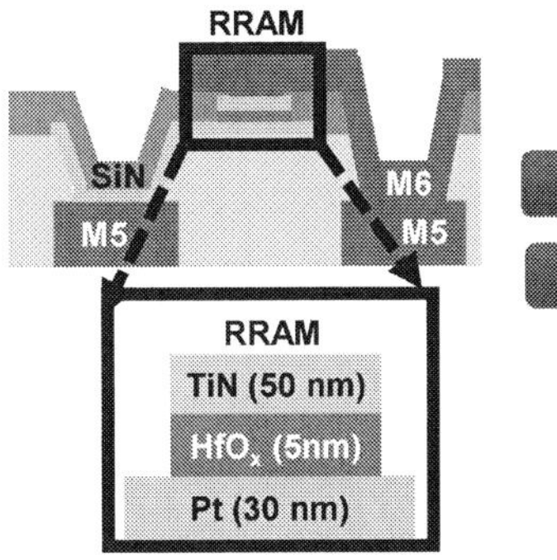

Fig. 3 The cross-section schematic of RRAM stack: Pt (30 nm) / HfOx (5 nm) / TiN (50 nm). SiN : passivation layer. Al : M7 layer

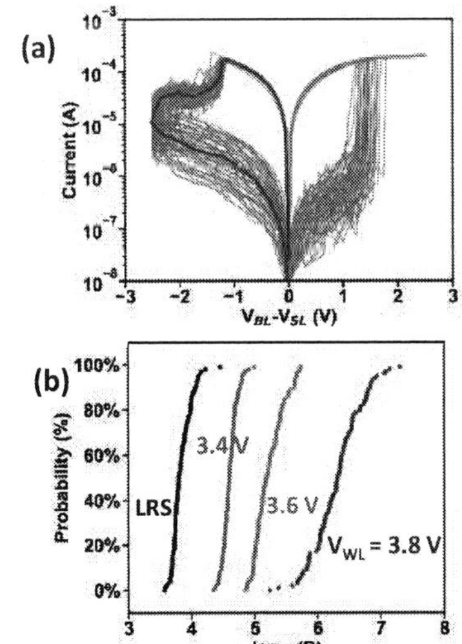

LL: Lower Limit of Acceptance Range
UL: Upper Limit of Acceptance Range
N: Maximum number of write pulses

DD-ISPP Programming Conditions		
	RESET	**SET**
V_{WL} **Start (V)**	2.5	0.5
V_{WL} **Stop (V)**	5	1.2
V_{WL} **Step (V)**	0.2	0.1
V_{BL} **(V)**	0	2.5
V_{SL} **(V)**	3.5	0
Pulse Width (ns)	100	50

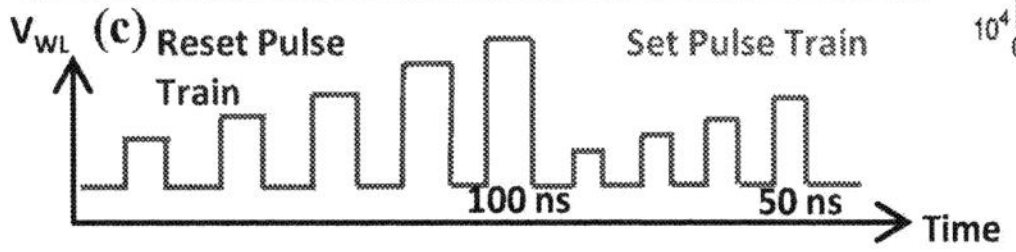

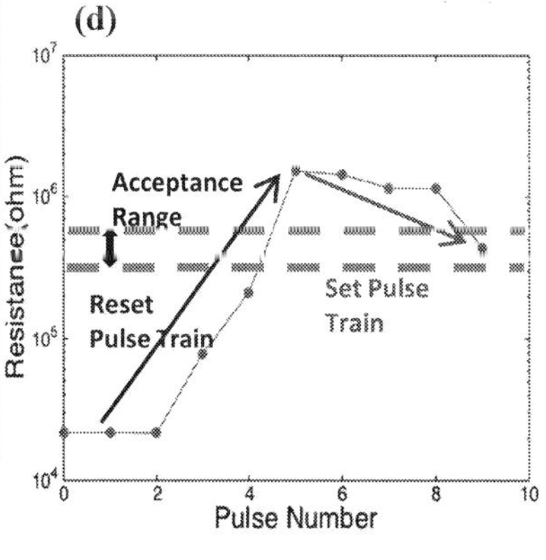

Fig. 4 Analog programmable capability of RRAM cell (a) Typical I-V curves of one cell in RRAM array. Median set (red) and reset (blue) of 50 set (V_{WL} = 1.2 V, V_{SL} = 0 V) /reset (V_{WL} = 4.5 V, V_{BL} = 0 V) cycles. (b) Multiple resistance levels achieved by pulse RESET with different V_{WL} (Pulse Width = 100 ns, V_{SL} = 3.5 V, V_{BL} = 0 V).

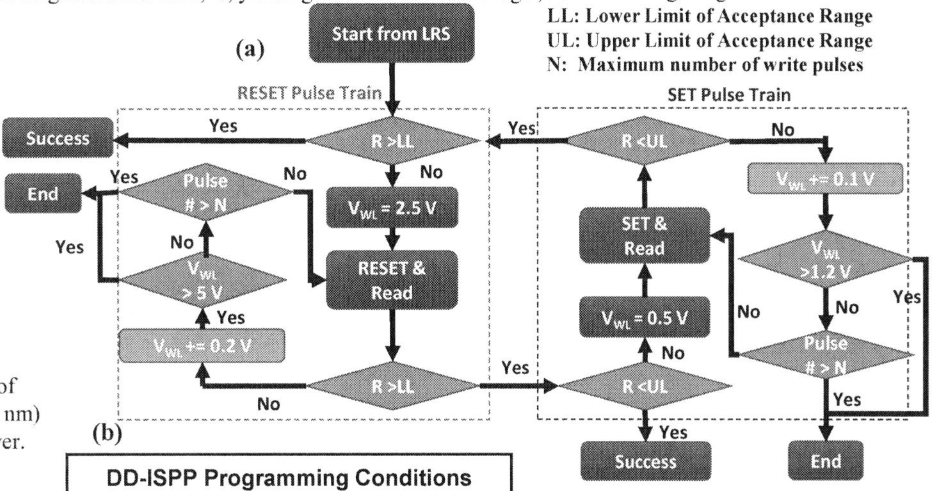

Fig. 5 DD - ISPP Scheme (a) Flow chart showing how to use DD-ISPP to fine tune the resistance into acceptance range using incremental step RESET pulse train and incremental step SET pulse train (b) Table: SET and RESET incremental pulse train parameters used in DD-ISPP. (c) -(d) Example writing with DD – ISPP showing the over-programming can be fixed by programming in the opposite direction, and starting from minimal voltage when changing the direction can minimize programming across the range and thus save energy. (c) write pulse train waveform (d) measured resistance as a function of pulse number.

978-1-7281-1988-5/18 $31.00 © 2018 IEEE

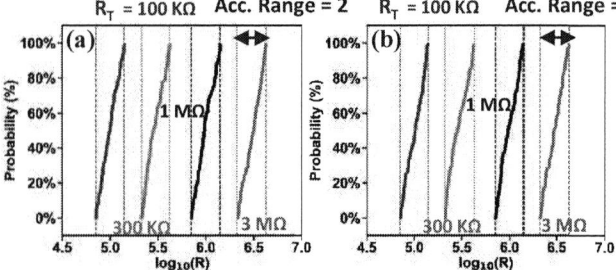

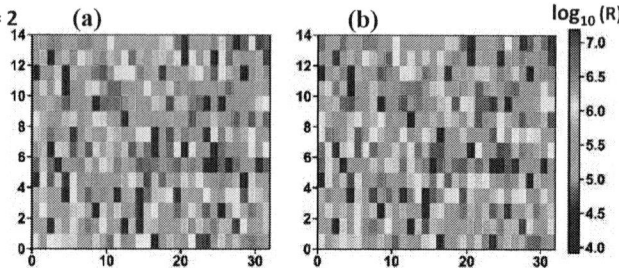

Fig. 6 Resistance generated from DD-ISPP scheme follows uniform distribution $\log_{10}(R_M) = \log_{10}(R_T) + \log_{10}(2) \cdot U(-0.5, 0.5)$, where $U(-0.5, 0.5)$ is the uniform distribution function. (a) 1 cell, 100 cycles for each R_T (b)100 cell, 1 cycle for each R_T

Fig. 7 448 Resistance values stored in (32×14) RRAM array (a) Target resistance log10 (R_T) (b) Programmed RRAM resistance log10 (R_M), with error rate of 0.2%

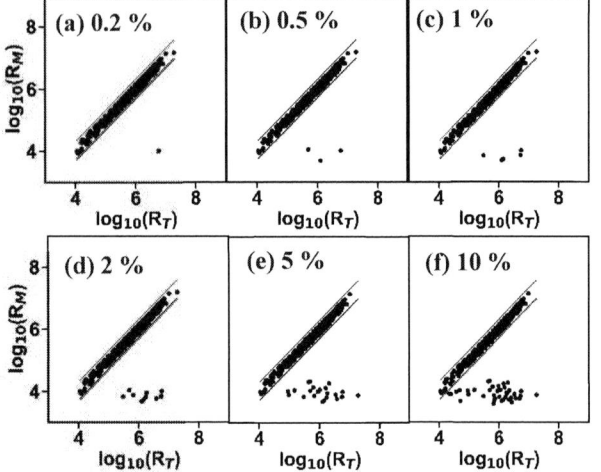

Fig. 8 (a) Writing error rate of 0.2% is achieved in experiment (b)-(f)Various writing error rates (0.5%, 1%, 2%, 5%, and 10%) are dialed into the measurement by randomly selecting a portion (error rate) of cells and SET them to LRS. (red line: upper limit of acceptance range, blue line: bottom limit of acceptance range)

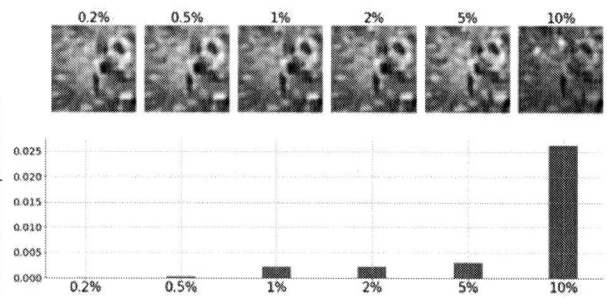

Fig. 9 Reconstructions (top) for the original image and their corresponding relative mean squared error (MSE) (bottom) to the MSE of 0.2% error rate (Fig. 12 (b)) for various error rates (0.2%, 0.5%, 1%, 2%, 5%, 10%) of devices. The network is trained with 2% noise, while it is able to tolerate an error up to 5% before suffering serious reconstruction degradation.

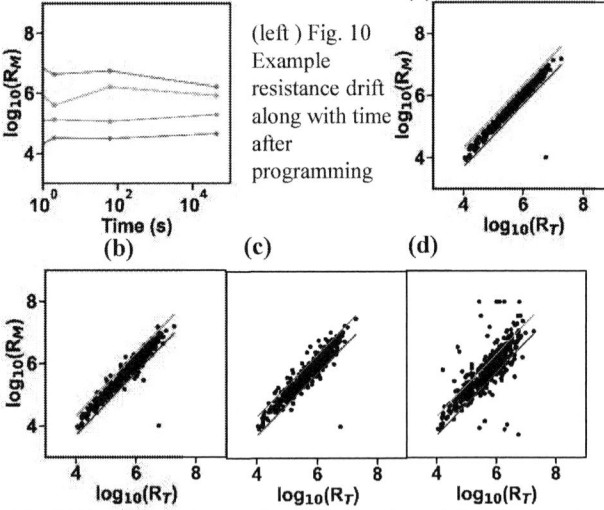

(left) Fig. 10 Example resistance drift along with time after programming

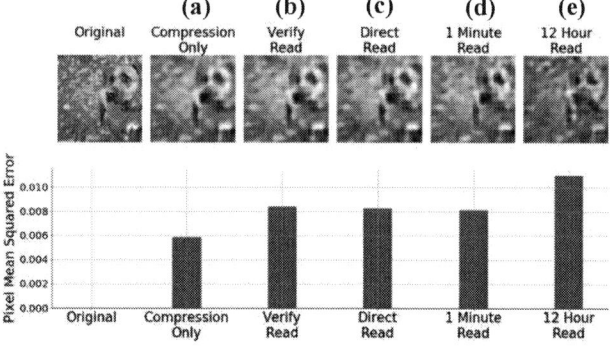

Fig. 12 Reconstructions (top) for the original image and their corresponding mean squared error (MSE) (bottom) for various conditions. (a) Compression only: where images are not passed through the devices and thus $R_T = R_M$, this corresponds to the case when the storage medium (the RRAM) is perfect, (b) Verify: where write-verified resistance values are used. (c) Direct read: where read noise and short-term resistance relaxation has been added. (d) 1 minute after write: where values are taken after the RRAM array has drifted for 1 minute. (e) 12 hour read: where the RRAM array has relaxed for 12 hours.

Fig. 11 Resistance change after programming (a) Resistance value after write-verify (b) Immediately read after write-verify with outliers contributed from read noise and short-term resistance relaxation. (c) More outliers appear when reading 1 minute after programming (d) Wider distribution is observed when reading resistance 12 hours after programming. (red line: upper limit of acceptance range, blue line: bottom limit of acceptance range)

978-1-7281-1988-5/18 $31.00 © 2018 IEEE

Demonstration of 50-mV Digital Integrated Circuits with Microelectromechanical Relays

Z. A. Ye[1], S. Almeida, M. Rusch, A. Perlas, W. Zhang, U. Sikder, J. Jeon, V. Stojanović, and T.-J. K. Liu

Electrical Engineering & Computer Sciences Dep't, University of California, Berkeley, CA, USA, [1]email: alice.ye@berkeley.edu

Abstract—50-mV operation of digital integrated circuits at room temperature is demonstrated for the first time using body-biased microelectromechanical relays. An improved relay design and self-assembled molecular coating provide for lower contact adhesion to reduce hysteretic switching behavior, so that the relays operate reliably with sub-50-mV gate voltage swing to provide for ultra-low active power consumption as well as zero static power consumption.

I. INTRODUCTION

Nanometer-scale mechanical relays are of interest for ultra-low-power digital computing applications because they in principle can achieve zero OFF-state leakage current (which provides for zero static power consumption) and abrupt switching characteristics so that they can be operated with very low gate voltage swing [1-2]. To minimize dynamic power consumption, a digital integrated circuit (IC) should be operated with a very low supply voltage (V_{DD}). V_{DD} reduction for a relay-based digital IC is limited by the switching hysteresis voltage (V_H) arising from contact adhesive force (F_{AD}). Piezoelectric relays have been demonstrated to operate with very low V_H, ~10 mV [3], but have a more complex structure (*i.e.*, are more costly to manufacture and difficult to miniaturize) than electrostatic relays. In this work, a body-biased electrostatic relay design for digital logic applications is improved to reduce F_{AD}. Furthermore, V_H is reduced with self-assembled molecular coating of the relays to enable stable switching operation with sub-50-mV gate voltage swing. Room-temperature operation of relay-based digital ICs with V_{DD} as low as 50 mV is demonstrated for the first time.

II. RELAY DESIGN AND FABRICATION

The schematic diagrams in **Fig. 1a** illustrate the body-biased microelectromechanical (MEM) relay designs used in this work. Each comprises a movable gate electrode suspended by four folded-flexure beams over a fixed body electrode, and two sets of source/drain electrodes (*i.e.*, two electrical switches). The relay design with four contact dimples (referred to herein as the 4C design) was first introduced in [4], and uses a conductive channel layer (attached underneath the gate electrode with an intermediary insulating layer) to form a bridge between a pair of source and drain electrodes (formed from the same layer as the body electrode) in the ON state. The new relay design introduced in this work has only two contact dimples (one for each electrical switch) for smaller total contact area and hence lower F_{AD} as well as lower ON-state resistance (R_{ON}) [5]; it is referred to herein as the 2C design. A plan-view scanning electron micrograph (SEM) of a fabricated 2C relay is shown in **Fig. 1b**.

As fabricated, air gaps exist between the conductive source and drain electrodes so that no current can flow between them, *i.e.*, $I_{DS} = 0$ (**Fig. 1c**). When a voltage (V_{GB}) is applied between the gate and body, the movable structure is actuated downward by the electrostatic force; if $|V_{GB}|$ is larger than a certain threshold pull-in voltage (V_{PI}), the conductive electrodes are brought into physical contact so that current can flow between the source and drain electrodes (**Fig. 1c**). When $|V_{GB}|$ is subsequently reduced toward 0 V, the spring restoring force of the suspension beams pulls the movable structure out of contact so that I_{DS} drops abruptly to zero at a certain release voltage (V_{RL}). V_H is defined as $V_{PI} - V_{RL}$. Depending on the body bias voltage (V_B), a relay can turn on with either increasing or decreasing gate voltage (V_G) because of the ambipolar nature of electrostatic force. If V_B is negative, then the relay turns on with increasing positive V_G, similarly as an n-channel MOSFET; in this case, it is referred to as a N-relay. If V_B is positive, then the relay turns on with increasingly negative V_G or decreasingly positive V_G, similarly as a p-channel MOSFET; in this case, it is referred to as a P-relay. Circuit symbols for N-relay and P-relay are shown in **Fig. 2**.

Fig. 3 illustrates key steps in the 2C relay fabrication process, after the patterning of the tungsten (W) layer (body and source electrodes), contact dimple regions, drain electrode via regions, drain electrode (W attached underneath the gate electrode with an intermediary insulating layer of Aluminum Oxide, Al_2O_3), and heavily doped p-type polycrystalline Silicon-Germanium ($Si_{0.4}Ge_{0.6}$) structural layer (gate electrode). (Patterning of the Al_2O_3 layer to define via/anchor regions prior to deposition of poly-$Si_{0.4}Ge_{0.6}$ is not shown since these regions lie beyond the *B-B'* cutline.) As fabricated, the relays in this work have an actuation-gap thickness (g_0) to dimple-gap thickness (g_d) ratio larger than three (*cf.*, **Fig. 3c**) so that they operate in non-pull-in mode to avoid unnecessarily large V_H [6]. To reduce V_H further, relays were coated with a hydrophobic anti-stiction self-assembled monolayer (SAM) of Perfluorooctyltriethyloxysilane (PFOTES) using a vapor-phase process [7] after the relay structures were released (by selectively removing sacrificial oxide layers) in vapor HF (*cf.*, **Fig. 3f**).

III. RESULTS AND DISCUSSION

Measured I_{DS}-V_G characteristics for 2C relays and for 4C relays operated as N-relay or P-relay are shown in **Fig. 4** and **Fig. 5**, respectively. Immeasurably-low OFF-state leakage current and >10^7 ON/OFF current ratio are observed, as expected. Note that V_H values (summarized in **Fig. 6**) are much lower for coated relays due to reduced surface adhesion energy between the contacting electrode surfaces. The reduction in V_H afforded

978-1-7281-1988-5/18 $31.00 © 2018 IEEE

by PFOTES coating comes with the tradeoff of increased subthreshold swing (SS), due to tunneling conduction through the PFOTES coating that is modulated by V_G. Average SS values for PFOTES-covered relays are compared for 2C and 4C relay designs in **Fig. 7**. (The full range of measured SS values across eight relays is also indicated.) SS is approximately twice as large for the 4C design as compared with the 2C design, since twice as much force is needed to compress the PFOTES in a 4C relay. (The PFOTES coating between the conducting electrodes can be mechanically modeled as a spring; springs in parallel combination—*cf.*, **Fig. 7** inset—have equivalent stiffness equal to the sum of the individual springs.)

Measured timing diagrams for a relay-based inverter circuit are shown in **Fig. 8**. (The glitches in the signals are due to noise from the power supply.) Various two-input logic functions can be implemented with only two relays (*vs.*, four transistors for CMOS implementations) as shown in **Fig. 9**. Note that pass-gate topology is employed in these circuits, as relays are functionally transmission gates. Since each relay comprises two electrical switches, it is straightforward to implement dual-polarity pass-gate logic, *i.e.*, generate complementary output signals, which eliminates the need for inversions (incurring additional mechanical delays) along the signal path. This topology minimizes the number of mechanical delays and also the number of relays per digital function [8]. (The signal propagation delay in a relay-based IC is dominated by mechanical switching delay, which is much larger than RC charging/discharging delay; therefore, an optimally designed relay circuit should minimize the number of mechanical delays, *i.e.*, all relays should switch simultaneously to achieve the fastest possible circuit operation.) The measured voltage waveforms in **Fig. 10** for 2C-relay-based ICs demonstrate their correct operation for NOT, AND, OR, and XOR functions for V_{DD} as low as 50 mV.

Fig. 11 shows the circuit diagram and measurement setup for a 2:1 multiplexer (MUX) implemented with two relays. The source electrode of each relay serves as an input signal line, the gates are interconnected together to form a select line, and the drain electrodes are interconnected together to form the output node. The measured voltage waveforms in **Fig. 12** confirm that this circuit functions properly for V_{DD} down to 50 mV.

In **Fig. 7**, it can be seen that there is significant variation in SS for the coated relays, which indicates that the coating process was non-uniform. For very low operating voltage, poor SS results in lower ON-state current (nA range for $V_{DD} = 50$ mV, *cf.*, **Fig. 4**) and hence high effective R_{ON}. In this work, the output voltage waveforms were measured using an oscilloscope probe with internal resistance of 10 MΩ (R_{OSC}), which is not much larger than R_{ON} for relays with poor SS operating at very low V_{DD}. As a result, the output voltage does not reach V_{DD} in some cases due to the resistive voltage divider effect modeled by the equivalent circuit shown in **Fig. 13**. When either the N-relay or

P-relay is turned on, the voltage divider consisting of R_{ON} connected in series with R_{OSC} limits the output voltage to be less than V_{DD}; when both relays are turned on, the output voltage swing increases closer to V_{DD} since the parallel combination of two resistors provides for lower resistance. It should be noted that, since relays have nearly infinite OFF-state resistance (in contrast to the R_{OSC} of the oscilloscope) due to zero OFF-state leakage, high effective R_{ON} should not prevent proper operation of relay-based ICs. Nevertheless, improvement in the uniformity of the SAM coating process is expected to provide for uniformly low SS (and thereby low R_{ON}) in the future.

It should be noted that body biasing is used for all of the relays in this work in order to minimize V_{DD}. This is necessary to minimize the switching energy of a relay across a range of values of mechanical switching delay [9]. **Fig. 14** shows how the mechanical turn-ON delay (τ_{ON}) of the 2C relay can be improved by increasing V_{DD}, albeit at a tradeoff of higher dynamic power consumption.

IV. Conclusion

A new body-biased relay design is introduced to reduce the number of contacts and thereby reduce the switching hysteresis voltage (V_H) to enable lower gate voltage operation. A self-assembled monolayer coating of PFOTES is effective for further reducing V_H, with a tradeoff of degraded subthreshold swing. Together, these improvements enable relay-based digital ICs to operate reliably with a supply voltage of 50 mV at room temperature. These results show the promise of nanoelectromechanical switches for ultra-low-power computing at the edge of the cloud.

Acknowledgments

The authors would like to thank B. Saha, B. Osoba, S. Fathipour, Prof. R. Muller, and Prof. E. Alon for useful discussions, and E. Acosta and A. Vidaña for help in sample preparation. This work was supported by the Center for Energy Efficient Electronics Science (NSF Award 0939514). Z. A. Ye gratefully acknowledges support from the NSERC of Canada. The relay devices and circuits were fabricated in the UC Berkeley Marvell Nanofabrication Laboratory.

References

[1] K. Akarvardar *et al.*, *IEDM* 2007, pp. 299-302.
[2] V. Pott *et al.*, *Proc. IEEE*, vol. 98, no. 12, pp. 2076-2094, 2010.
[3] U. Zaghloul and G. Piazza, *IEEE EDL*, vol. 35, no. 6, pp. 669-671, 2014.
[4] R. Nathanael *et al.*, *VLSI-TSA* 2012.
[5] Y.-H. Yoon *et al.*, *IEEE JMEMS*, vol. 27, no. 3, pp. 497-505, 2018.
[6] H. Kam *et al.*, *IEEE T-ED*, vol. 58, no. 1, pp. 236-250, 2011.
[7] B. Osoba *et al.*, *IEDM* 2016, pp. 655-658.
[8] F. Chen *et al.*, *ICCAD* 2008, pp. 750-757.
[9] C. Qian *et al.*, *IEDM* 2015, pp. 475-478.

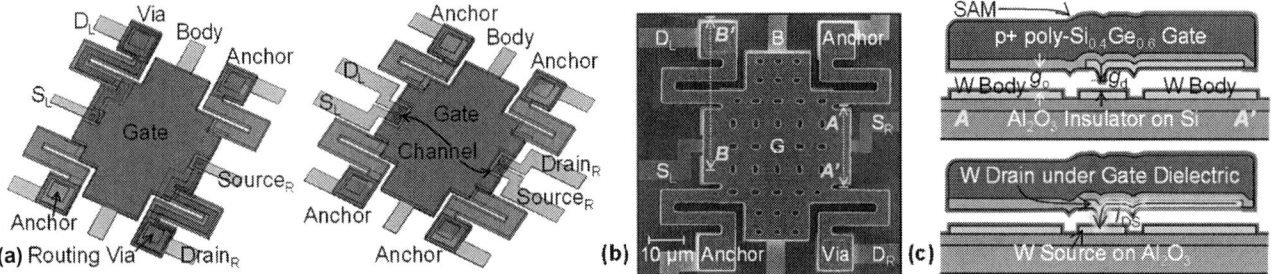

Fig. 1. (a) Isometric schematic views of the new two-contact (left) and old four-contact (right) relay designs. (b) Plan-view SEM of two-contact relay. (c) Cross-sectional views (along the *A-A'* cutline in Fig. 1b) in the off-state (top) and on-state (bottom). As fabricated g_o = 220 nm and g_d = 60 nm.

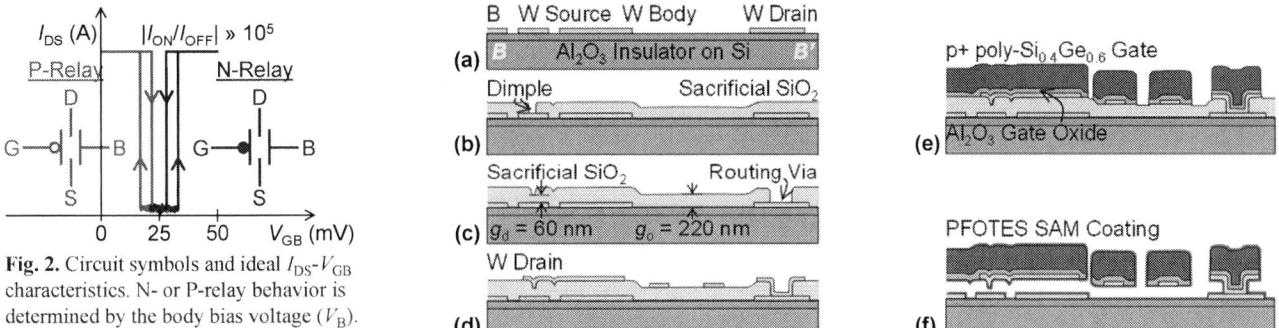

Fig. 2. Circuit symbols and ideal I_{DS}-V_{GB} characteristics. N- or P-relay behavior is determined by the body bias voltage (V_B).

Fig. 3. *B-B'* (*cf.*, **Fig. 1b**) cross-sections illustrating key steps of the 2C relay fabrication process: (a) 60 nm W source and body electrodes formed on 80 nm Al_2O_3 insulating layer. (b) 160 nm 1st sacrificial SiO_2 layer deposition followed by contact dimple definition. (c) 60 nm 2nd sacrificial SiO_2 layer deposition followed by routing via definition. (d) 60 nm W drain electrode formation. (e) 55 nm Al_2O_3 gate insulator deposition followed by anchor region definition (not shown) and 1.9 μm p+ poly-$Si_{0.4}Ge_{0.6}$ gate formation. (f) Release in HF vapor followed by SAM coating.

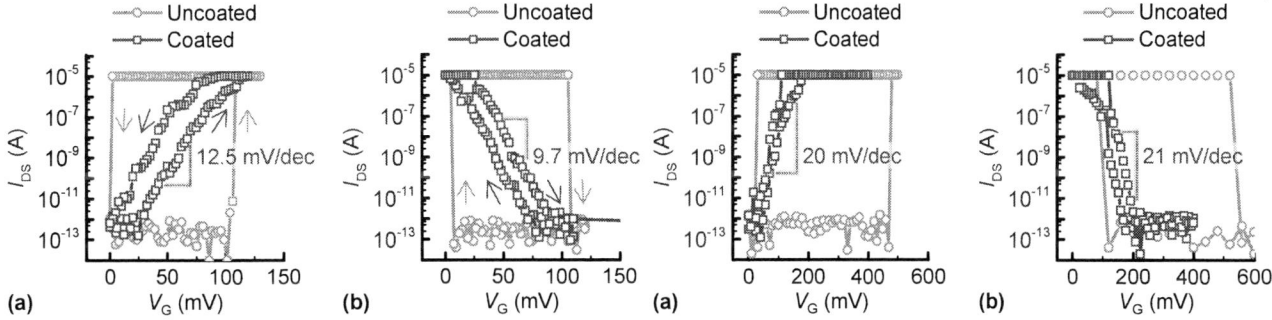

Fig. 4. I_{DS}-V_G curves for 2C (a) N-relay and (b) P-relay, with and without PFOTES coating, measured at 23 °C and 10 μTorr. V_{DS} = 200 mV. $|V_B|$ = ~15 V. The current compliance is set to limit I_{DS} to 10 μA.

Fig. 5. I_{DS} *vs.* V_G curves for 4C (a) N-relay (b) P-relay, with and without PFOTES coating, measured at 23 °C and 10 μTorr. V_{DS} = 200 mV. $|V_B|$ = ~15 V. The current compliance is set to 10 μA.

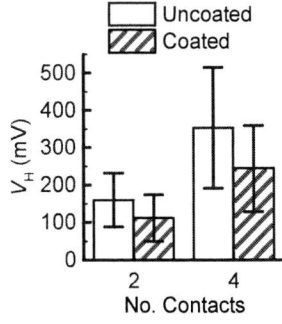

Fig. 6. Average V_H values for 2C and 4C relays, with and without PFOTES coating, measured at 23 °C and 10 μTorr. V_{DS} = 200 mV. $|V_B|$ = ~15 V.

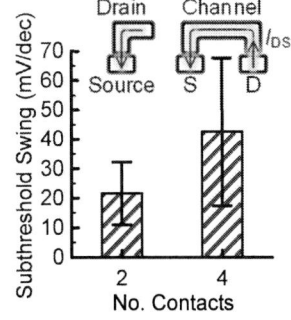

Fig. 7. Average SS values for 2C and 4C PFOTES-coated relays measured at 23 °C and 10 μTorr. V_{DS} = 200 mV. $|V_B|$ = ~15 V.

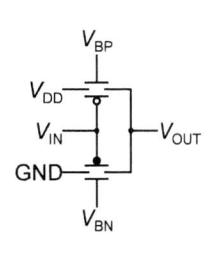

Fig. 8. (a) Schematic circuit diagram for an inverter. (b) Measured timing diagrams for V_{DD} 100 mV and 50 mV at 23 °C and 10 μTorr. $|V_B|$ = ~15 V.

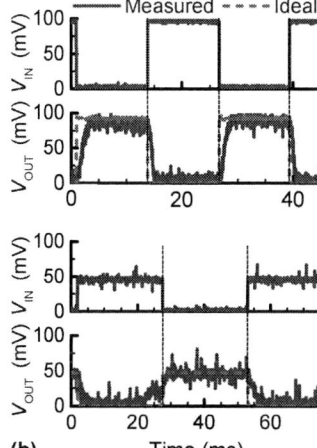

978-1-7281-1988-5/18 $31.00 © 2018 IEEE

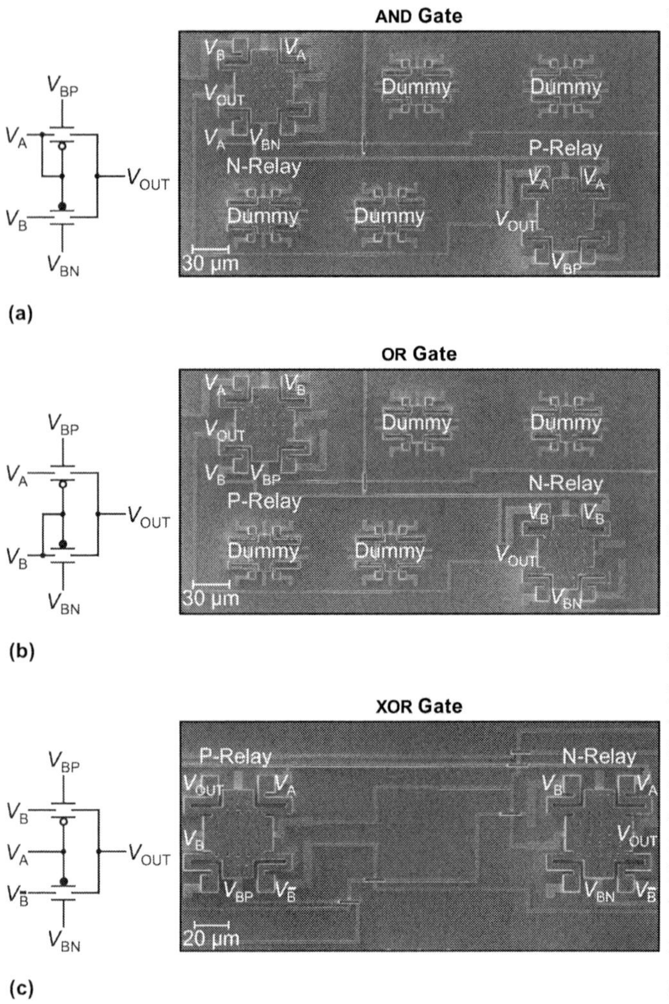

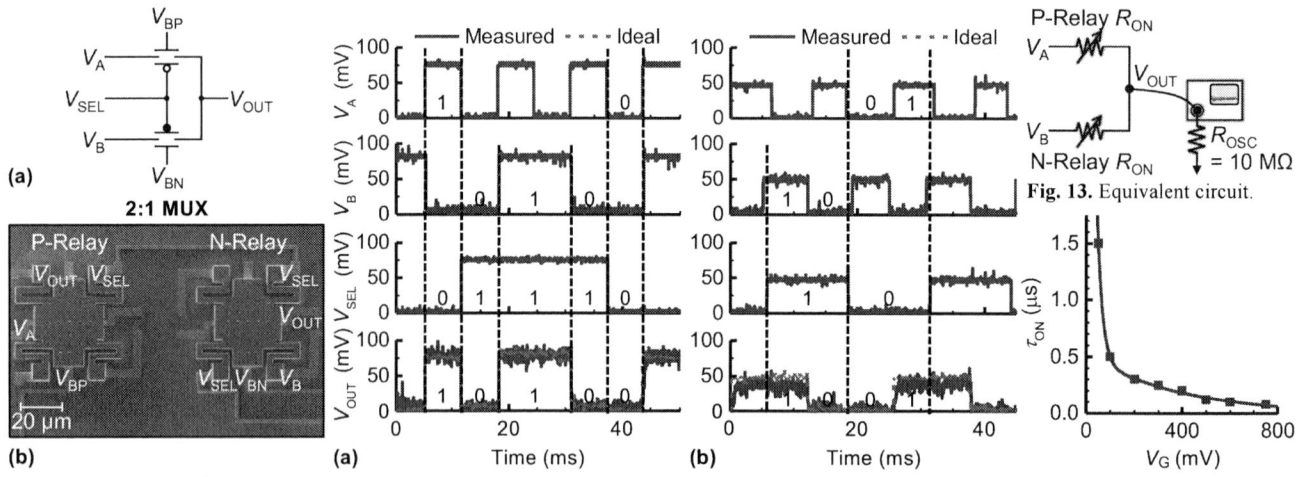

Fig. 9. Schematic circuit diagrams and plan-view SEM images of 2C-relay-based two-input logic gates: (a) AND. (b) OR. (c) XOR.

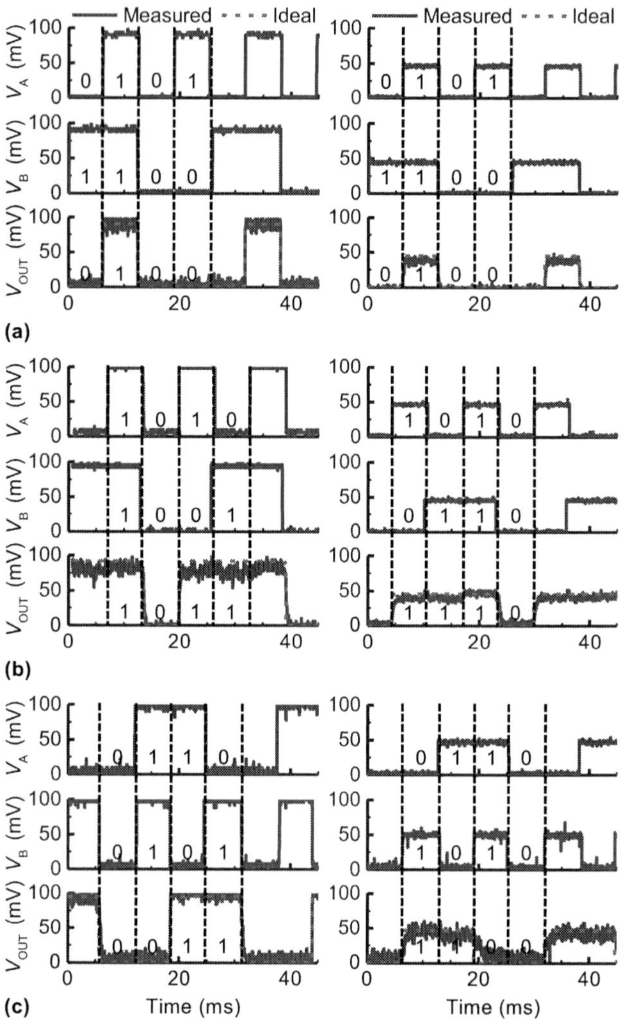

Fig. 10. Waveforms showing ultralow-voltage operation of relay logic gates at 23 ˚C and 10 µTorr. $|V_B|$ = ~15 V. (a) AND. (b) OR. (c) XOR.

Fig. 11. 2C-relay-based 2:1 MUX. (a) schematic circuit diagram. (b) plan-view SEM image.

Fig. 12. Measured voltage waveforms demonstrating operation of the 2:1 MUX for low V_{DD} (a) 80 mV and (b) 50 mV at 23 ˚C and 10 µTorr. $|V_B|$ = ~15 V.

Fig. 13. Equivalent circuit.

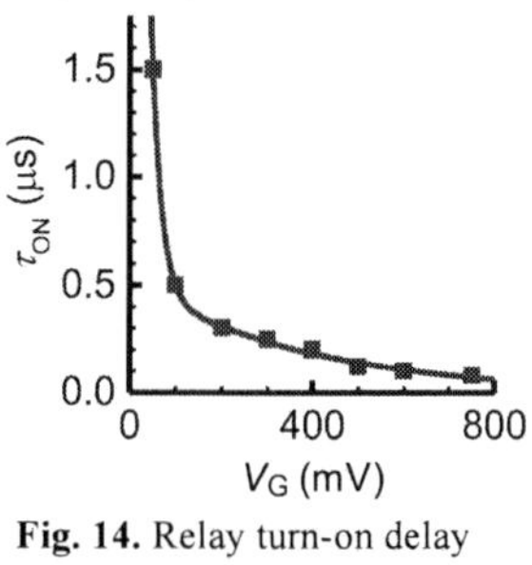

Fig. 14. Relay turn-on delay (τ_{ON}) vs. V_G at 23 ˚C and 10 µTorr. V_B = -14.3 V.

Highly sensitive spintronic strain-gauge sensor based on magnetic tunnel junction and its application to MEMS microphone

Yoshihiko Fuji, Yoshihiro Higashi, Shiori Kaji, Kei Masunishi, Akiko Yuzawa, Tomohiko Nagata, Kazuaki Okamoto, Shotaro Baba, Tomio Ono, and Michiko Hara

Corporate Research and Development Center, Toshiba Corp, Tokyo, Japan, email: yoshihiko.fuji@toshiba.co.jp

Abstract—Recently, new strain-gauge sensors based on magnetic tunnel junctions (MTJs) have attracted attention because they are promising for detecting extremely small strains. To further enhance the strain sensitivity of these spintronic strain-gauge sensors (Spin-SGSs) and obtain a linear response that is suitable for MEMS applications, we have previously developed an MTJ film structure and magnetic biasing for Spin-SGSs. This paper describes the strain-gauge properties of our Spin-SGSs with novel MTJs employing an amorphous magnetostrictive sensing layer, and its application to a MEMS microphone.

I. INTRODUCTION

Strain-gauge sensors utilizing the piezoresistive effect have been widely used in MEMS sensors. The sensitivity of strain-gauge sensors is defined by the gauge factor (GF) = $(\Delta R/R)/\Delta\varepsilon$, with poly-Si piezoresistive elements, which are the most commonly used in MEMS sensors, exhibiting GF as high as 40. Recently, new strain-gauge sensors based on magnetic tunnel junctions (MTJs) have attracted attention for exhibiting high GFs [1]. As shown in Fig. 1, MTJs have so far found applications in magnetic sensors for hard disk drive heads and magnetic random access memory (MRAM). It has also been reported that MTJs can be used as strain-gauge sensors by using a magnetostrictive material as a sensing layer [1]. In these spintronic strain-gauge sensors (Spin-SGSs), the magnetization configuration between the sensing layer (one magnetic layer) and the reference layer (the other magnetic layer) changes due to the inverse magnetostriction effect, which results in a change in the resistance of the MTJ due to the magnetoresistance (MR) effect. A high GF of 2150 under small tensile strain (0 to +0.02%) has been reported for a Spin-SGS employing an MTJ with MgO-BL, which offers high MR [1].

The combination of Spin-SGSs and MEMS technology to create spintronic MEMS (Spin-MEMS) sensors is promising for high-sensitivity detection of various mechanical quantities (Fig. 2). To further improve the GF potential of Spin-SGS, we have developed Spin-SGSs that employ an amorphous CoFeB and FeB alloy that exhibits high magnetostriction and soft magnetic properties as the SL [2], [3] and successfully confirmed an extremely high GF of 5072 [3]. We have also developed a "Spin-MEMS microphone" to demonstrate the feasibility of Spin-SGS in MEMS sensors [3], [4]. When strain-gauge sensors are applied to microphone devices that detect

small AC pressure, it is important to realize a linear response to tensile and compressive strain. In this paper, we review the MTJ film development toward realizing highly sensitive Spin-SGS and its application to MEMS microphones. We also discuss the magnetic properties and magnetic biasing design for obtaining both high GF and linear response to tensile and compressive strain.

II. SPINTRONIC STRAIN-GAUGE SENSOR

To enhance the GF of Spin-SGSs, it is essential to realize a strain-sensitive sensing layer in which the magnetization rotates easily under strain. We therefore developed a Spin-SGS employing an amorphous CoFeB alloy that exhibits high magnetostriction and soft magnetic property as the SL in MgO-MTJ [2]. To maintain an amorphous structure in the CoFeB sensing layer even after annealing, which is performed to fix the magnetization of the RL and improve the crystallinity of MgO-BL, we applied a MgO capping layer (MgO-cap) on the CoFeB-SL as a boron diffusion blocking layer, as shown in Fig. 3 (b). A MgO-MTJ with generic Ta capping layer (Ta-cap) is also fabricated for comparison (Fig. 3 (a)). From cross-sectional HRTEM analysis (Fig. 4), boron moves from the CoFeB sensing layer to the CoFeB/Ta-cap interface in the Ta-cap MTJ during annealing such that the CoFeB sensing layer with Ta-cap crystallizes. In the MgO-cap MTJ, the boron exists uniformly in the CoFeB sensing layer such that the CoFeB layer maintains an amorphous structure. Therefore, CoFeB-SL with a MgO-cap exhibits low coercivity (H_c) even after annealing (Fig. 5, Table 1). The strain-gauge sensor properties for these MTJs were evaluated by using a four-point bending apparatus (Fig. 6). The Spin-SGS with a MgO-cap exhibits a GF of 4016, whereas that with a Ta-cap exhibits a GF of 942. Therefore, we successfully confirmed that the GF of the Spin-SGS can be enhanced by adopting amorphous CoFeB-SL and that an MgO-cap is effective for maintaining the amorphous structure.

To obtain a linear response to tensile and compressive strain, it is necessary to set the initial SL magnetization obliquely to the strain axis (for example, at an angle of 45° relative to the strain axis) as shown in Fig. 7. Thus, as shown in Fig. 8, we evaluated the strain-gauge properties of the Spin-SGS with a MgO-cap under a bias magnetic field aligned at 45° relative to the applied strain axis and at 135° relative to the RL magnetization (135° BMF). Linear response to tensile and compressive strain was confirmed under a strong BMF of 12 kA/m. As the BMF decreases, the maximum GF increases because the SL can easily rotate with strain. However, the strain loop shifted to the tensile strain side, and high GF was not obtained for both tensile and compressive strain. In order to clarify the reason for the shift in the strain loop, we theoretically analyzed the SL rotation in response to strain by using the Stoner-Wohlfarth (SW) model shown in Fig. 9. As a result of calculations based on the magnetic properties of CoFeB-SL (Fig. 10), a similar strain loop shift under weak BMF was reproduced and it was found that the strain loop shift is caused by the induced magnetic anisotropy (IMA) of the CoFeB-SL (K_{ima}=1223 J/m^3), which forces the SL magnetization to an angle of 180°, and since this is not negligible compared with the 135° BMF.

Therefore we developed an SL that exhibits less IMA than the amorphous CoFeB-SL. As shown in Fig. 11, we confirmed that an amorphous FeB based SL exhibits small IMA (K_{ima} = 518 J/m^3) and large magnetostriction (λ = 26 ppm). As can be seen in Fig. 12, it was confirmed that strain loop shift did not occur in the Spin-SGS with amorphous FeB based SL and high strain sensitivity was obtained for both tensile and compressive strain. This kind of zero loop shift was also confirmed in the calculated strain loop as shown in the inset in Fig. 12. The experimentally confirmed GF in Fig. 12 was 5072 [3]. This GF is the highest to date for Spin-SGSs and around 100 times the value for conventional poly-Si piezoresistors (Table 2).

III. Spin-MEMS Microphone

We fabricated a Spin-MEMS microphone in which Spin-SGSs with an amorphous FeB-based SL are integrated onto a bulk micromachined diaphragm (Fig. 13). The diaphragm film was formed on the Si substrate and mainly consisted of a Si-N film prepared by chemical vapor deposition. Spin-SGSs were fabricated on the Si-N-based diaphragm film and then the substrate was etched using the Boshe process of deep reactive ion etching. The first resonance frequency of the diaphragm as evaluated by laser-Doppler vibrometer (LDV) was 72 kHz and a flat mechanical response up to 50 kHz was confirmed (Fig. 14) for the fabricated Spin-MEMS microphone (device A) [4].

Figure 15 shows the microphone response to a 94 dB input sound pressure level (SPL) at 1 kHz. Sinusoidal output voltages were clearly confirmed in response to the input sound [4]. Input SPL dependence is shown in Fig. 16. Output voltage increased linearly with input SPL. Input SPL at total harmonic distortion (THD) at 1 kHz = 1% is 116 dB SPL, which is similar to commercial capacitive MEMS microphones. Therefore, based on the linear response of the Spin-SGS employing amorphous FeB based SL and 135° BMF, the Spin-MEMS microphone is found to exhibit sufficient linear response to input sound.

The minimum detectable pressure at 1 kHz in a 1 Hz bin (MDP) and an A-weighted SNR of the fabricated Spin-MEMS microphone were evaluated and results are summarized in Table 3 and Fig. 17. We also fabricated a Spin-MEMS microphone with first resonance frequency of 37 kHz and flat mechanical response up to 20 kHz (device B). The results for device B are also described. In Fig. 17, the MDPs of the reported Si piezoresistive microphones, which are also strain-gauge-type microphones, are shown. It is known that strain-

gauge-type microphones with a simple single-plate diaphragm are advantageous for obtaining a wide bandwidth, and Si piezoresistive MEMS microphones with upper band limits in the ultrasonic range have been reported [5]-[7]. However, these Si piezoresistive MEMS microphones were less sensitive (MDP of over 50 dB SPL) than capacitive-type microphones. Thanks to the high GF of Spin-SGSs, Spin-MEMS microphones, which are a new strain-gauge-type microphone, offer the potential for realizing wide bandwidth and high sensitivity. In this study, we confirmed a small MDP of 15 dB SPL for a Spin-MEMS microphone with an upper band limit up to 50 kHz. Microphone properties like this are promising for acoustic health monitoring.

IV. Conclusion

In this paper, we presented recent advances in the development of Spin-SGSs and applications to MEMS microphones. We introduced a Spin-SGS employing an amorphous FeB-based SL that can detect tensile and compressive strain with high GF exceeding 5000, and also demonstrated a Spin-MEMS microphone in which highly sensitive Spin-SGS were integrated on a diaphragm, and exhibited a small MDP of 8–15 dB SPL.

Acknowledgment

Part of the MEMS fabrication and LDV measurement in this study was supported by Tohoku University Nanofabrication (MEMS) Platform and Tokyo University Nanofabrication Platform in Nanotechnology Platform Project sponsored by the Ministry of Education, Culture, Sports, Science and Technology (MEXT), Japan.

References

[1] A. Tavassolizadeh, K. Rott, T. Meier, E. Quandt, H. Hölscher, G. Reiss, and D. Meyners, "Tunnel Magnetoresistance Sensors with Magnetostrictive Electrodes : Strain sensors", Sensors 16, p. 1902 (2016).

[2] Y. Fuji, S. Kaji, M. Hara, Y. Higashi, A. Hori, K. Okamoto, T. Nagata, S. Baba, A. Yuzawa, K. Otsu, K. Masunishi, T. Ono, and H. Fukuzawa, "Highly sensitive spintronic strain-gauge sensor based on a MgO magnetic tunnel junction with an amorphous CoFeB sensing layer", Appl. Phys. Lett. 112, p. 062405 (2018).

[3] Y. Fuji, M. Hara, Y. Higashi, S. Kaji, K. Masunishi, T. Nagata, A. Yuzawa, K. Otsu, K. Okamoto, S. Baba, T. Ono, A. Hori and H. Fukuzawa, "An ultra-sensitive spintronic strain-gauge sensor with gauge factor of 5000 and demonstration of a Spin-MEMS Microphone", Digest on the 19th International Conference on Solid-State Sensors, Actuators and Microsystems (TRANSDUCERS 2017), pp. 63-66 (2017).

[4] Y. Higashi, Y. Fuji, S. Kaji, K. Masunishi, T. Nagata, A. Yuzawa, K. Otsu, K. Okamoto, S. Baba, T. Ono, and M. Hara, "SNR Enhancement of a Spin-MEMS microphone by optimum bias magnetic field and demonstration of operation sound monitoring of rototating equipment", Digest on the 31st IEEE International Conference on Micro Electro Mechanical Systems (MEMS 2018), pp. 1060-1063 (2018).

[5] C. Huang, A. Naguib, E. Soupos, and K. Najafi : "A silicon micromachined microphone for fluid mechanical research", J. Micromecha. Microeng., 12(6), pp. 767-774 (2002).

[6] D. P. Arnold, S. Gururaj, S. Bhardwaj, T. Nishida, and M. Sheplak, : "A MEMS microphone for aeroacoustic measurements", in Proceedings of International Mechanical Engineering Congress and Exposition, pp. 281-288 (2001).

[7] M. Sheplak, K. S. Breuer, and M. A. Schmidt : "A wafer-bonded, silicon nitride membrane microphone with dielectrically-isolated single crystal silicon pioreistors", in Proc. Solid-State Sens. Actuator workshop, pp. 23-26 (1998).

978-1-7281-1988-5/18 $31.00 © 2018 IEEE

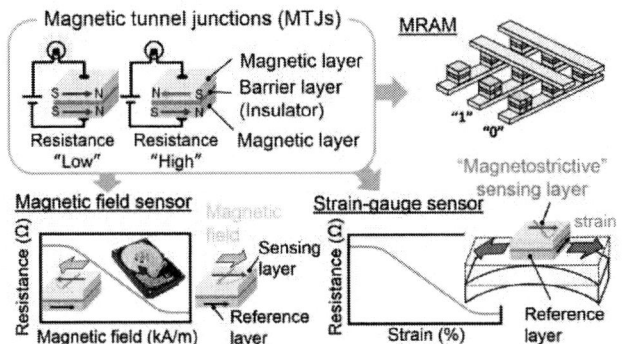

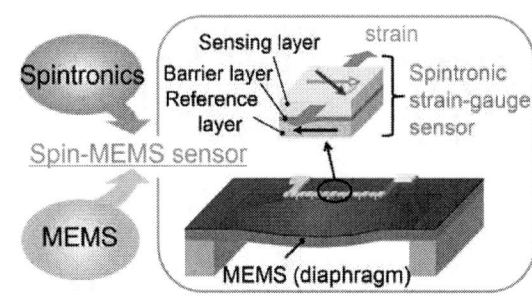

Fig. 1. Schematic diagram of magnetic tunnel junctions (MTJs) and their applications. Can be used as a strain sensor by using a magnetostrictive sensing layer (SL).

Fig. 2. Schematic diagrams of Spin-SGS and Spin-MEMS sensors. In the Spin-SGS, strain-induced rotation of the sensing layer (SL) magnetization changes the magnetization configuration between the SL and the reference layer (RL), resulting in a change in the resistance of the MTJ due to the TMR effect [3].

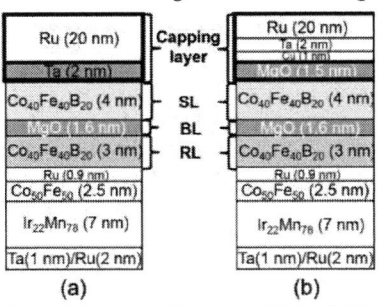

Fig. 3. Schematic diagrams of Spin-SGSs based on MgO-MTJs using a CoFeB sensing layer with (a) Ta-cap and (b) MgO-cap [2].

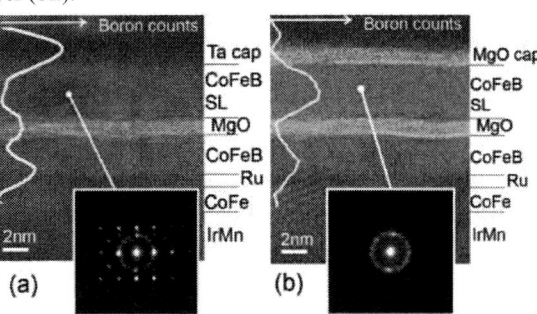

Fig. 4 Cross-sectional TEM images of MTJs with (a) Ta-cap and (b) MgO-cap after annealing at $T_a = 320$ °C for 1 h. Solid lines indicate EELS counts of boron distribution. Inset are diffraction patterns of the CoFeB sensing layers [2].

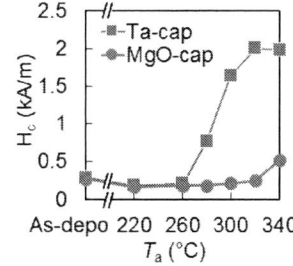

Fig. 5 Coercivity (H_c) of CoFeB sensing layer with Ta-cap and MgO-cap versus T_a [2].

	Ta-cap	MgO-cap
H_c (kA/m)	2.16	0.24
λ (ppm)	30	20
MR ratio (%)	186	164

Table 1. Coercivity (H_c), magnetostriction (λ) and MR ratio of Spin-SGSs with Ta–cap and MgO-cap after annealing at $T_a = 320$ °C for 1 h [2].

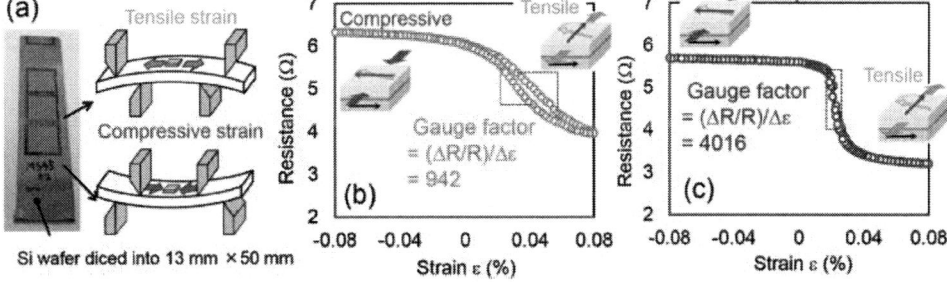

Fig. 6. (a) Measurement setup for gauge factor characterization. The change in resistance as a function of applied uniaxial strain (strain-loop) of Spin-SGS with (b) Ta-cap and (c) MgO-cap. In this measurement, a bias magnetic field (BMF) of 0.8 kA/m was applied anti-parallel to the RL magnetization (180° BMF) [2].

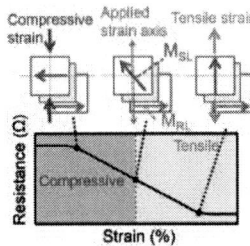

Fig. 7. Schematic diagrams of configuration of SL, RL magnetization (M_{SL}, M_{RL}) and applied strain axis for obtaining a linear response to tensile and compressive strain.

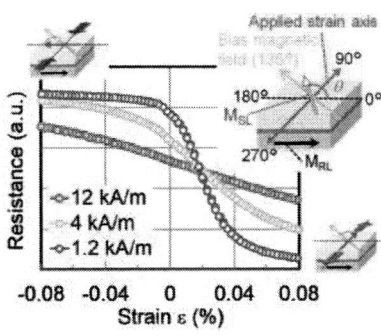

Fig. 8. Strain loop of the Spin-SGS with CoFeB-SL and MgO-cap under the 135° BMF.

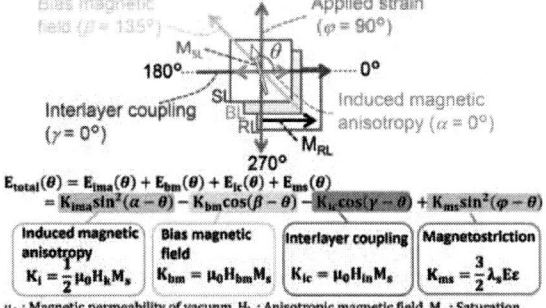

$$E_{total}(\theta) = E_{ima}(\theta) + E_{bm}(\theta) + E_{ic}(\theta) + E_{ms}(\theta)$$
$$= K_{ima}\sin^2(\alpha - \theta) - K_{bm}\cos(\beta - \theta) - K_{ic}\cos(\gamma - \theta) + K_{ms}\sin^2(\varphi - \theta)$$

Induced magnetic anisotropy	Bias magnetic field	Interlayer coupling	Magnetostriction
$K_i = \frac{1}{2}\mu_0 H_k M_s$	$K_{bm} = \mu_0 H_{bm} M_s$	$K_{ic} = \mu_0 H_{in} M_s$	$K_{ms} = \frac{3}{2}\lambda_s E\varepsilon$

μ_0 : Magnetic permeability of vacuum, H_k : Anisotropic magnetic field, M_s : Saturation magnetization, α : Induced magnetic anisotropy angle, H_{bm} : Bias magnetic field, β : Bias magnetic field angle, H_{in} : Interlayer coupling field, γ : Interlayer coupling angle, λ_s : Magnetostrictive constant, E : Young modulus, ε : Applied strain, φ : Applied strain angle

Fig. 9 Modified Stoner-Wohlfarth (SW) model for calculating the rotation of SL magnetization (MSL) in response to strain.

978-1-7281-1988-5/18 $31.00 © 2018 IEEE

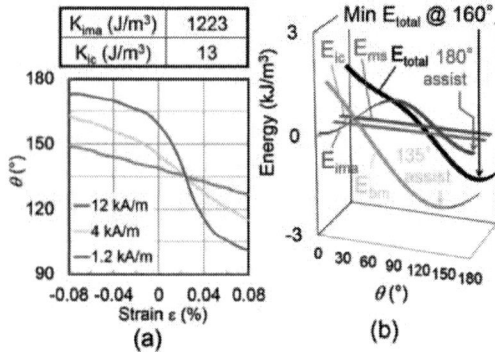

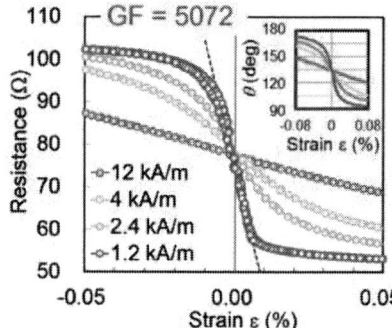

Fig. 10. (a) Calculated strain loop for the Spin-SGS with CoFeB-SL and MgO-cap under the 135° BMF. (b) Energy distribution of each term in SW-model at ε = 0% and H$_{bm}$ = 1.2 kA/m.

Fig. 11. (a) Schematic diagram of Spin-SGS with FeB based SL [3]. (b) Coercivity (H$_c$), magnetostriction (λ), induced magnetic anisotropy energy (K$_{ima}$), interlayer coupling energy (K$_{ic}$) and MR ratio of Spin-SGS with FeB based SL.

Fig. 12. Strain loop of Spin-SGS with FeB based SL under 135° BMF [3]. Inset is calculated strain loop.

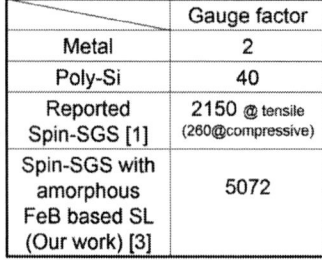

	Gauge factor
Metal	2
Poly-Si	40
Reported Spin-SGS [1]	2150 @ tensile (260@compressive)
Spin-SGS with amorphous FeB based SL (Our work) [3]	5072

Table 2. Comparison of the gauge factor of conventional strain-gauge and spintronic strain-gauge sensors (Spin-SGS) [3].

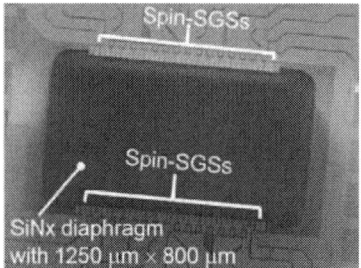

Fig. 13. SEM image of Spin-MEMS microphone. Spin-SGSs were fabricated on a rectangular diaphragm [4].

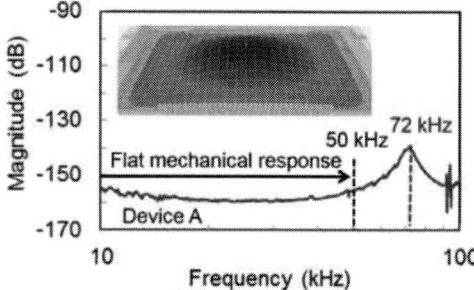

Fig. 14. Displacement as a function of frequency for Spin-MEMS microphone. Inset is typical vibrational behavior at first resonance frequency [4].

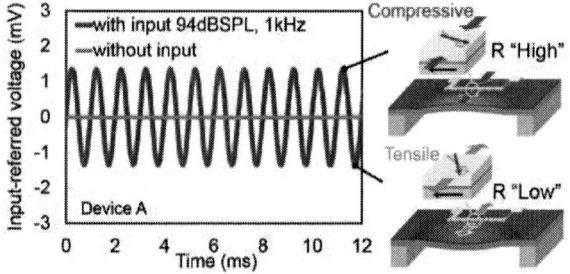

Fig. 15. Microphone response to 94 dB input sound pressure level at 1 kHz for Spin-MEMS microphone [4].

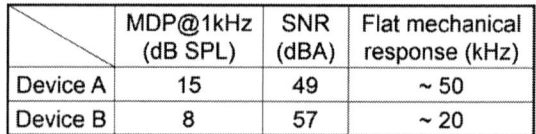

	MDP@1kHz (dB SPL)	SNR (dBA)	Flat mechanical response (kHz)
Device A	15	49	~ 50
Device B	8	57	~ 20

Table 3. Minimum detectable pressure (MDP), A-weighted SNR and flat mechanical response for two types of Spin-MEMS microphones [3], [4].

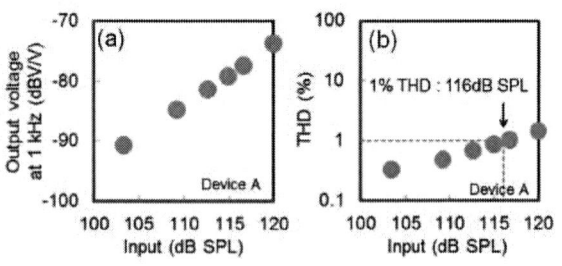

Fig. 16. (a) Output voltage at 1 kHz and (b) total harmonic distortion (THD) as a function of input sound pressure level for Spin-MEMS microphone.

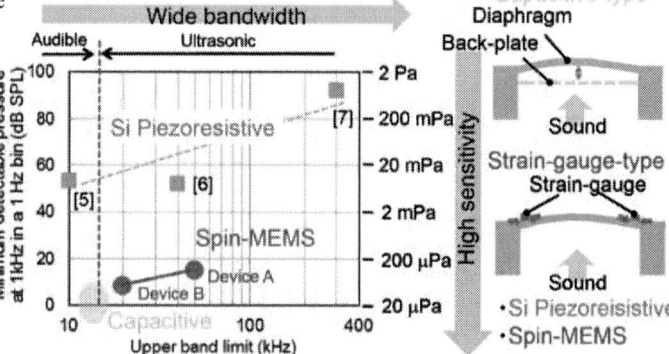

Fig. 17. Comparison of MDP at 1 kHz for capacitive MEMS, Si piezoresistive MEMS and Spin-MEMS microphone as a function of upper band limit [3], [4].

978-1-7281-1988-5/18 $31.00 © 2018 IEEE

Intermixing of motional currents in suspended CNT-FET based resonators

Lalit Kumar, Laura Vera Jenni, Miroslav Haluska, Cosmin Roman and Christofer Hierold
Micro and -Nanosystems, Department of Mechanical and Process Engineering, ETH Zurich, Switzerland
Email: lalit.kumar@micro.mavt.ethz.ch

Abstract—Here, we report the intermixing of piezoresistive and conduction modulation current in a carbon nanotube field effect transistor (CNT-FET) based resonator. We show that due to static displacement of the nanotube, as a result of electrostatic actuation, the motional current at the resonance frequency consist of both current components. For instance at a DC gate bias of 1.3 V, 3/4 of the motional current is conduction modulation current while the rest arises from piezoresistive effects. The intermixing effect due to asymmetry influences the fundamental harmonic response as well as the physical nature of the electrical signal being sensed; both of which are important for understanding frequency harmonics in nanoresonators and developing efficient readout schemes for nanoscale sensors.

I. INTRODUCTION

Various nanomaterials such as silicon nanowires (SiNWs), graphene, carbon nanotubes (CNTs) have been used as building blocks for resonant sensing applications. The ability of these materials in exhibiting multiple electrical characteristics (piezoresistivity or conductivity modulation) results in a mechanical resonance response with multiple harmonics. This also allows for multiple transduction readout schemes such as capacitive or piezoresistive readout among others. For implementation of a selective and efficient readout scheme, in particular for nanoscale sensors, where high bandwidth and SNR can be challenging, it is crucial to know the nature of the electrical signal being sensed. For nanoresonators, most transduction schemes for mechanical resonance characterization are reported to selectively sense either the piezoresistive current or the capacitance or conductance modulation current. Here we investigate the fundamental harmonics and the selectivity of the transduction schemes for the desired component and report a qualitative and quantitative analysis of mechanical motional currents in a nanoresonator based on a single walled carbon nanotube (SWCNT).

II. SIMULATION AND VERIFICATION

The two used transduction schemes for carbon nanotube FET based nanoresonators as shown in Fig. 1 are piezoresistive transduction based on strain modulation and conduction modulation transduction based on field effect induced charge modulation [1]. The transduction scheme uses an RF down mixing and lock-in detection [2] with CNT as a mixer as shown in Fig. 2. This scheme requires a low bandwidth readout and minimizes the effect of parasitic capacitances.

A. Piezoresistive motional current

Piezoresistivity in nanotube arises from the strain induced resistive changes. The beam elongation as a result of out-of-plane resonance induces periodic strain $\epsilon(t)$ given by [2]

$$\epsilon(t) = z(t)^2 \frac{1}{2L} \int_0^L \left(\frac{\partial \varphi(x)}{\partial x} \right)^2 dx \qquad (1)$$

where z is the nanotube displacement, L is the length of tube and $\varphi(x)$ is the beam mode shape. The time varying resistive change $\Delta R(t)$ due to strain modulation is then given by

$$\frac{\Delta R(t)}{R_0} = G_F \epsilon(t) \qquad (2)$$

where G_F is defined as the CNT's Gauge factor [3]. The resulting piezoresistive current can then be computed by [2]

$$i_p = V/[R_o + \Delta R(t)] \qquad (3)$$

$$i_p = \frac{V_{sd}}{R_0} \left(1 - z(t)^2 G_F \frac{1}{2L} \int_0^L \left(\frac{\partial \varphi(x)}{\partial x} \right)^2 dx \right) \qquad (4)$$

The piezoresistive current (II term) represents a quadratic transduction scheme w.r.t CNT's displacement z.

B. Conductance modulated motional current

Conductance modulation due to capacitive changes have been extensively used to characterize CNT nanoresonators. The charge fluctuations due to resulting nanotube motion can be read out as a mechanical current given by [4]

$$i_g = V_{sd} \frac{dG}{dV_g} \left(V_g^{ac} + z(t) \frac{V_g^{DC}}{C_g} \frac{dC_g}{dz} \right) \qquad (5)$$

where C_g is the CNT-gate capacitance and G is the CNT conductance. In contrast to piezoresistive current, the conductance modulation current (II term) exhibits a linear transduction w.r.t CNT's displacement z. ($z \ll g_0$; CNT-gate distance)

C. Intermixing of motional currents

For a nanotube resonating at frequency ω with displacement amplitude $z_o e^{i\omega t}$, the motional currents (considering the II terms only from Eq. 4-5) can be re-written as

$$i_g = z_o e^{i\omega t} V_{sd} \frac{dG}{dV_g} \frac{V_g^{DC}}{C_g} \frac{dC_g}{dz} = z_o e^{i\omega t} K_G \qquad (6)$$

$$i_p = z_o^2 e^{i2\omega t} \frac{V_{sd}}{R_0} G_F \frac{1}{2L} \int_0^L \left(\frac{\partial \varphi(x)}{\partial x} \right)^2 dx = z_o^2 e^{i2\omega t} K_P \quad (7)$$

where K_G and K_P are introduced as bias dependent proportionality parameters. From Eq. 6-7, the frequency component of motional currents, i_g and i_p, occur at ω and 2ω

978-1-7281-1988-5/18 $31.00 © 2018 IEEE

respectively and hence can be separated by two source techniques (Fig. 2). This model assumes a symmetrical displacement beam profile. However, in presence of a static bending of a beam, symmetry breaking can occur [5] and can affect the resonance behavior. In such a scenario, the nanotube displacement can be expressed as $z_o e^{i\omega t} = z_s + z_d e^{i\omega t}$ where z_s and z_d represents the static and time-varying dynamic displacement respectively. The modified expression for motional currents would then be given by

$$i_g = z_s K_G + z_d e^{i\omega t} K_G \tag{8}$$

$$i_p = z_s^2 K_P + 2 z_s z_d e^{i\omega t} K_P + z_d^2 e^{i2\omega t} K_P \tag{9}$$

In contrast to the symmetrical beam model, both currents are inseparable with motional current at ω consisting of terms proportional to conduction modulation and piezoresistivity, thereby hindering the selectivity of the widely employed ω transduction scheme. To study the impact of static displacement on intermixing of motional currents, we adapt the model to comprehensive CNT parameters experimentally obtained by Ning et. al. [6]; mentioned in Table I. We simulate the static and dynamic displacements and the resulting motional currents by considering electrostatic actuation on a harmonic beam resonator [4] through Simulink/MATAB.

Fig. 3 plots the conduction modulation current (Eq. 8) for various DC gate bias exhibiting spring hardening effect and increase in resonance current (Inset-Fig. 3) at resonance frequency ω. The higher order derivatives such as $d^2 C_g / d z^2$ lead to a negligible current at 2ω. In contrast, the piezoresistive current in Fig. 4 has components at both ω and 2ω as expected from Eq. 9. While the 2ω component of total motional current (Eq. 8+Eq. 9) is purely piezoresistive, ω component results from mixing of both currents as shown in Fig. 5. For the considered set of CNT model parameters, the intermixing effect increases the peak resonance signal as well as hinders the selectivity of the ω measurement technique.

III. EXPERIMENTS AND RESULTS

For experimental investigation of the intermixing effect, we use a small bandgap semiconducting (SGS) nanotube resonator device (Fig. 6) as they exhibit both gate dependent conductance modulation and high Gauge factor in contrast to purely semiconducting or metallic nanotubes [3]. We characterize the mechanical behavior of the nanotube by both ω and 2ω transduction at an RF down mixing frequency $\Delta\omega$=10 kHz exhibiting a resonance frequency at 72.6 MHz (Fig. 6f) at room temperature (300K) and low vacuum ($<10^{-3}$ mbar). The measured high resonance frequency in comparison to theoretical eigenmode suggest a slack-free CNT resonator. The Lorentzian fit to measured frequency spectral responses was used to extract the peak resonance current.

Fig. 7 shows purely piezoresistive current at various DC gate bias detected through 2ω scheme. To quantity the currents, the dependency $z_d^2 \propto \left(V_g^{DC}\right)^2$ was used for power fitting. Peak resonance currents obtained from ω scheme as shown in Fig. 8 were also fitted, with an additional dependency $z_s \propto \left(V_g^{DC}\right)^2$, to separate the motional currents. To account for the intermixing

effect, square and fourth power fit was used (the peak current was normalized to background current, I term of Eq. 5, to account for bias dependent conductance). The proportionality parameter K_P obtained with power fits from ω and 2ω measurements were extracted to be 0.268 and 0.263 respectively supporting the validity of fits and intermixing between motional currents. In addition, the term dG/dV_g extracted from K_G, 4.33×10^{-7} S/V and CNT-FET I_D-V_G transfer characteristic, 3.58×10^{-7} S/V shows a good agreement from the fit based on intermixing effect.

IV. DISCUSSION AND CONCLUSION

With the extracted parameters, we quantify the intermixing effect at resonance frequency ω for the measured nanotube device. As shown in Fig. 9, for low DC gate bias of 0.5 V, piezoresistive current constitutes only an estimated 4% of the total motional current. This increases to 26% while the remaining 74% arises from conduction modulation current at a bias of 1.4 V due to asymmetrical beam profile as a result of DC electrostatic force. Amplitude and Q-factor fluctuations, observed in nanoresonators, could lead to inaccuracies in the estimated current components. Piezoresistivity due to asymmetrical beam profile has also been observed in a 90 nm thick silicon nanowire resonator [2], however at a high gate bias of 15 V, suggesting the significance of intermixing in sub-nanometer resonators at low bias; especially in non-linear resonators operating at higher gate bias resulting in several harmonics. Differences in resonator characteristics like Gauge factor, chirality or conductance such as the one considered in Table I can also lead to an intermixing with 50-50% contribution, thereby effecting the physical nature of the electrical signal and the selectivity of ω transduction scheme, both of which are important for understanding the harmonic components of resonators and in the design of highly efficient and sensitive readout schemes for nanoscale sensors.

ACKNOWLEDGMENT

The authors acknowledge the support of cleanroom facility at Binnig and Rohrer Nanotechnology Center (BRNC) at IBM Rüschlikon and FIRST CLA at ETH Zurich. We would also like to thank the Swiss National Science Foundation (SNSF-Project No. 153292) for the financial support.

REFERENCES

[1] S. W. Lee, S. Truax, L. Yu, C. Roman and C. Hierold, "Carbon nanotube resonators with capacitive and piezoresistive current modulation readout," *Appl. Phys. Lett.*, 103, 033117, 2013.

[2] M. Sansa, M. F. Regulez, J. Llobet, A. S. Paulo and F. P. Murano, "High-sensitivity linear piezoresistive transduction for nanomechanical beam resonators," *Nat. Comm.*, 5, 4313, 2013.

[3] W. Obitayo and T. Liu, "A Review: Carbon nanotube based piezoresistive strain sensors," *J. Sensors*, 652438, 2012.

[4] V. Sazonova, Y. Yaish, H. Üstünel, D. Roundy, T. A. Arias and P. L. McEuen, "A tunable carbon nanotube electromechanical oscillator," *Nature*, 431, 2004, pp 284-287.

[5] A. Eichler, J. Moser, M. I. Dykman and A. Bachtold, "Symmetry breaking in a mechanical resonator made from a carbon nanotube," *Nat. Comm.*, 4, 2843, 2013.

[6] Z. Y. Ning, M. Q. Fu, T. W. Shi, Y. Guo, X. L. Wei, S. Gao and Q. Chen, "In situ multiproperty measurements of individual nanomaterials in SEM and correlation with their atomic structures," *Nanotechnology*, 25, 275703, 2014.

978-1-7281-1988-5/18 $31.00 © 2018 IEEE

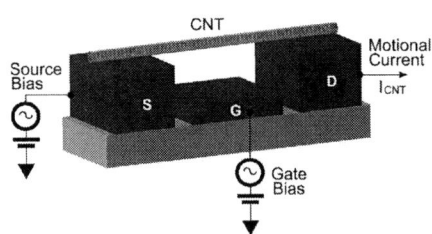

Fig. 1. Schematic of a resonator based on suspended carbon nanotube between source (S) and drain (D) electrode of a field effect transistor with bottom gate (G) architecture.

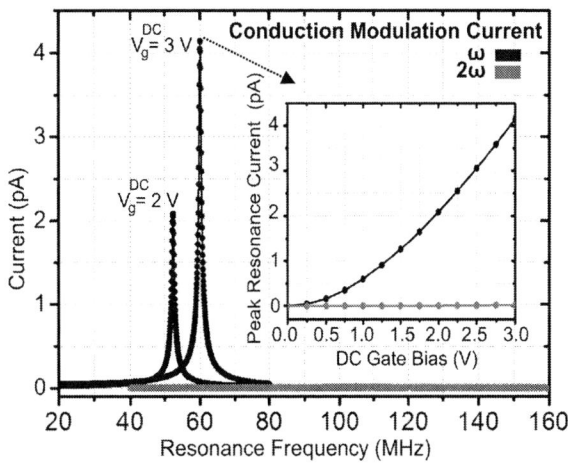

Fig. 2. (a) Experimental setup using RF mixing and lock-in technique for readout of motional current components at resonance frequency ω; (b) Modified setup to readout the current component at 2ω.

CNT parameter	Value	Unit
Radius (r)	1.1	nm
Length (L)	1.59	μm
Density (ρ)	1400	kg/m³
Mass (m)	2.54×10^{-21}	kg
Cross-sectional area (A$_c$)	2.29×10^{-18}	m²
Young's modulus (E)	154	GPa
Device resistance (R$_0$)	1.01	Mohm
CNT-gate distance (g$_0$)	300	nm
Zero-strain bandgap (E$_0$)	113.2649	meV
dE/dϵ	49.8	meV/%
dG/dV$_g$	1.43×10^{-7}	S/V
Spring constant	1.53×10^{-4}	N/m
Damping factor (b)	9.7×10^{-15}	kg/s
Source bias (V$_{sd}^{AC}$, V$_{sd}^{DC}$)	10	mV
Gate AC bias (V$_g^{AC}$)	10	mV

Table. I. CNT parameters experimentally extracted by Ning et.al.[6] through resonance frequency measurement, DC measurements, Raman spectroscopy, scanning electron microscopy (SEM), and transmission electron microscopy (TEM).

Fig. 3. Conduction modulation current; Eq. 8 simulated for various DC gate bias for CNT resonator at ω. The higher second order terms (d^2C/dz^2) results in a negligible 2ω component. Inset – Peak resonance current vs DC gate bias.

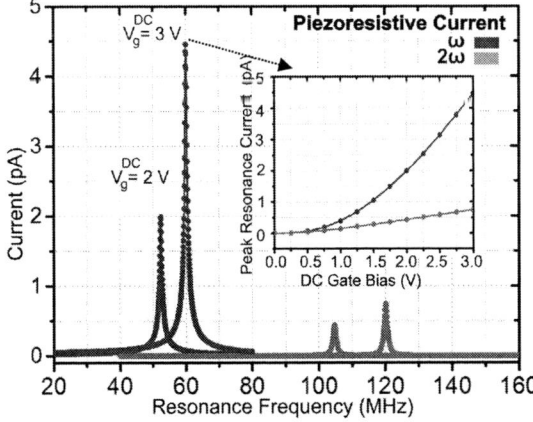

Fig. 4. Piezoresistive current; Eq. 9 simulated for various DC gate bias for CNT resonator at ω and 2ω due to $2z_s z_d$ and z_d^2 respectively. Inset – Peak resonance current vs DC gate bias.

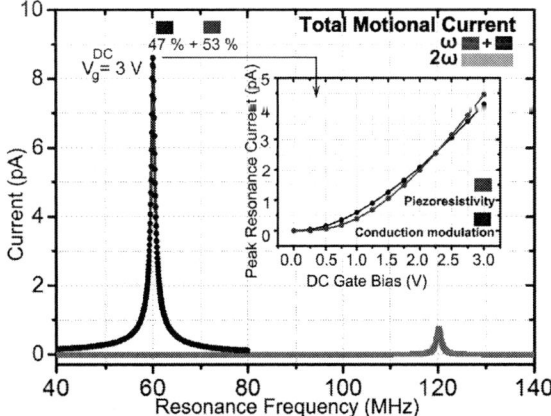

Fig. 5. Total motional current (Eq. 8+Eq. 9) at 3 V DC gate bias for CNT resonator at ω and 2ω. While 2ω is purely piezoresistive, the current at ω consist of 47% conduction modulation current and 53 % piezoresistive current (extracted from peak currents -Inset)

978-1-7281-1988-5/18 $31.00 © 2018 IEEE   85

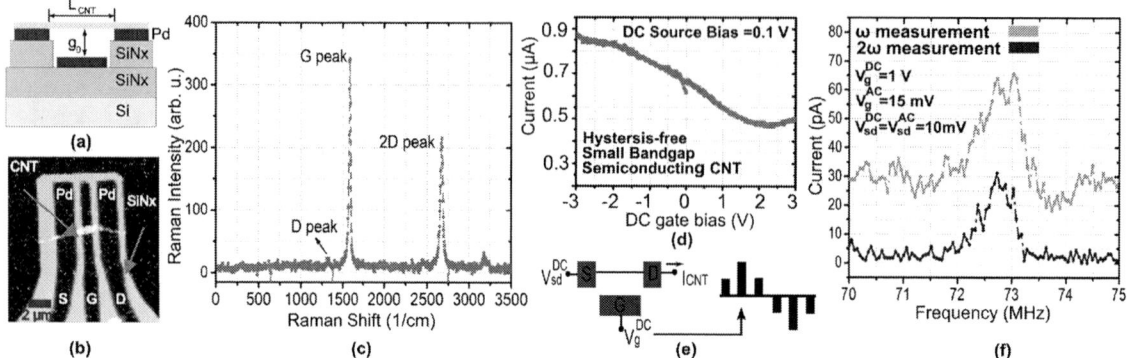

Fig. 6. (a) Device architecture of the CNT-FET based resonator with suspended channel length of 2 μm and CNT-gate distance of 275 nm. (b) Raman spectroscopy of the fabricated device with a nanotube obtained after dry-transfer process; (c) Raman spectrum measured for the suspended nanotube; (d) DC CNT-FET transfer characteristics exhibiting a small bandgap semiconducting (SGS) behavior; (e) Pulsed gate DC measurement for extracting hysteresis free performance; (f) Resonance frequency characterization through two source ω and 2ω detection technique. Both scheme detects the resonance frequency at 72.6 MHz with different motional current amplitudes.

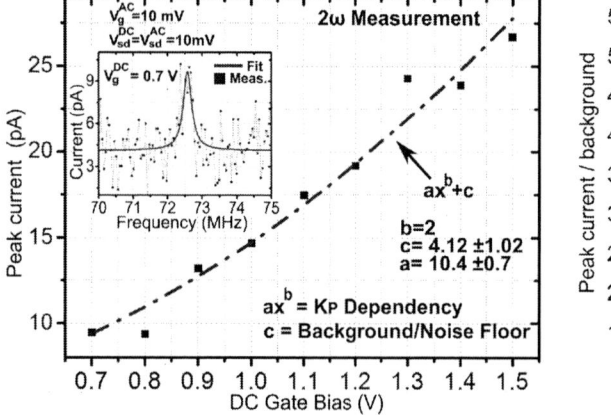

Fig. 7. Resonance peak current vs DC gate bias for 2ω measurement corresponding to piezoresistive current. Inset – Measured frequency response with Lorentzian fit to extract parameters.

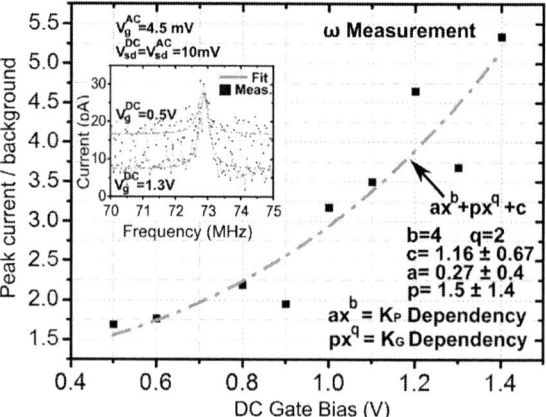

Fig. 8. Resonance peak current (normalized to background current-dG/dV dependency) vs DC gate bias for ω measurement corresponding to conduction modulation and piezoresistive current. Inset – Measured frequency response with Lorentzian fit.

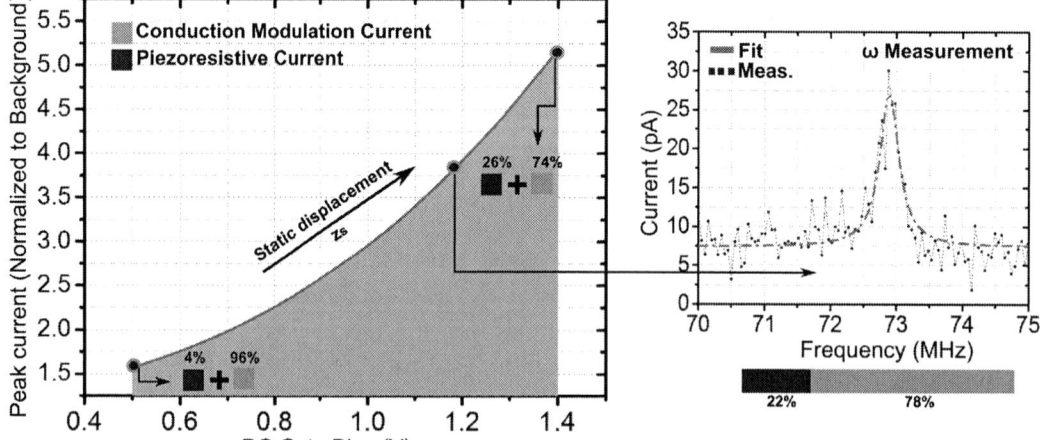

Fig. 9. Strength of intermixing of piezoresistive current and conduction modulation current as a function of DC gate bias. The increase in the nanotube's static displacement is estimated to increase the intermixing component $2z_s z_d K_P$ to one-fourth of the total motional current.

978-1-7281-1988-5/18 $31.00 © 2018 IEEE   86

Glowing Graphene Nanoelectromechanical Resonators at Ultra-High Temperature up to 2650K

Fan Ye, Jaesung Lee, and Philip X.-L. Feng

Electrical Engineering, Case School of Engineering, Case Western Reserve University, Cleveland, OH 44106, USA

Emails: fan.ye@case.edu, philip.feng@case.edu

Abstract—We report on the first experimental demonstration of electrothermally tuned few-layer graphene resonant nanoelectromechanical systems (NEMS) operating at high and very high frequency (HF/VHF) simultaneously with *strong visible light emission*. In tri-layer graphene resonators with carefully controlled Joule heating, we have demonstrated *ultra-wide frequency tuning* up to $\Delta f/f_0 \approx 1300\%$, which is the highest frequency tuning range known to date among reported 2D materials resonators. Simultaneously, device temperature variations imposed by Joule heating are monitored using Raman spectroscopy and emission spectrum; and we find that the device temperature increases from 300K up to 2650K, which is the highest operating temperature known to date for electromechanical resonators. When device temperature is above 1800K, the vibrating graphene NEMS starts glowing and emitting visible light with robust mechanical resonance. These results show that electromechanical resonance modes can be robustly sustained and read out at glowing temperatures with incandescent emission in graphene NEMS, suggesting new perspectives for integrating and configuring timing functions in light emitting graphene devices for harsh and extreme environment applications.

I. INTRODUCTION

Continuous and broad frequency tuning is an indispensable feature desired for state-of-the-art nano/micro-electro-mechanical systems (N/MEMS) and their applications, such as tunable oscillators, filters, and mixers. To realize wide frequency tuning, device material should be endowed with strong mechanical compliance thus its tension level could be modified efficiently. Besides large tunability, N/MEMS devices that can sustain and operate in harsh and extreme environments are also required in space, energy, and military applications. The extraordinary and unique thermal [1] and mechanical properties [2] make graphene an ideal candidate for highly tunable NEMS devices operating in HF and VHF regimes at ultra-high temperature. In previous studies on graphene NEMS, frequency tuning is mainly achieved by electrostatic effects, with the highest tuning range around ~200% [3]. Besides electrostatic method, another promising tuning mechanism is *electrothermal tuning* that alters the tension by modifying the temperature via Joule heating [3].

In this work, by utilizing electrothermal effects with graphene's special thermal and mechanical attributes, we demonstrate ultra-widely tunable graphene NEMS resonators. When the applied bias is higher than certain voltage, the graphene devices start to emit strong visible light. Using the

Raman spectroscopy and emission spectrum, we carefully calibrate the temperature and find the devices show robust resonances even at glowing temperatures (1800K to 2650K). As the applied voltage increases, the frequency increases from 7.4MHz to 102.2MHz, with a tuning range up to $\Delta f/f_0 \approx 1300\%$ (record high). Meantime, the device quality (Q) factors exhibit a five-fold enhancement as temperature increases. The remarkable, unprecedented frequency tuning range and operating temperatures show strong promises for novel graphene-based sensors and oscillators operating in harsh environments. The controllable and strong light emission also suggests perspectives for nanoscale flexible light emitters with low operation voltages.

II. CONCEPT, DESIGN AND METHODS

The graphene device in this work is illustrated in **Figure 1**. Using the all-dry transfer techniques, we fabricate both fully clamped circular and doubly clamped ribbon suspended graphene devices, with the microtrench depth around 2μm. By applying a DC voltage, the graphene crystal is heated up and the temperature of graphene increases due to Joule heating. In the meantime, another small AC voltage is superposed to the DC voltage, which drives the motion of the suspended graphene. As the temperature keeps increasing, the graphene device starts to emit visible light. During the electrothermal tuning, the variation of thermal expansion coefficients (TECs) comes into play, to affect the resonance characteristics of graphene device (**Figures 2 & 3**). In lower temperature regime, the TEC of the graphene is negative and thus the tension level increases as temperature ramps up, resulting in frequency boost. As the temperature continues increasing, at certain temperature, the TEC of graphene becomes positive and the tension level and frequency begin to reduce. As temperature keeps elevating, the graphene device gradually starts to emit visible light until sublimation. The measurement scheme is shown in **Figure 4**. The integrated Raman spectroscopy and laser interferometry system allow us to measure the resonances and precisely calibrate the temperature simultaneously.

III. ELECTROTHERMAL TUNING OF THE GRAPHENE NANOELECTROMECHANICAL RESOANTOR

Figure 5 shows a tri-layer graphene circular resonator contacted with gold electrodes. At room temperature, the device shows resonance frequency f_0 =7.4MHz and quality factor of Q =157. By increasing the applied voltage to 5.85V, the frequency increases to 102.2MHz, with a tuning range around $\Delta f/f_0 \approx 1300\%$. As the applied voltage keeps increasing, the frequency gradually reduces to 57.2MHz, due to the

978-1-7281-1988-5/18 $31.00 © 2018 IEEE

transition from negative TEC to positive TEC. When the applied voltage is higher than 6.0V, the graphene device starts to emit visible light. The light intensity increases as the voltage ramps up. **Figure 7** shows a microscopy image of the glowing graphene device at V_{DC}=6.35V. The emitted light is very strong and readily distinguishable even without illumination from other light sources. As the graphene is glowing, the device still vibrates robustly with resonance frequency around 57.2MHz and Q factor around 77. Besides the frequency shifts, the Q factor also shows an enhancement from 157 to 733 during Joule heating, which could be attributed to tension increase and self-annealing effect. In high voltage regime, due to the tension release and possible defects generation, Q factor gradually decreases (**Figure 9**).

IV. TEMPERATURE CALIBRATION DURING ELECTROTHERMAL HEATING

The temperature of the graphene device during electrothermal tuning is calibrated by both Raman spectroscopy and emission spectrum. In low voltage regime, as the temperature increases, the Raman peaks of the graphene exhibit redshift and broadening due to anharmonic effects. The Raman spectra of the graphene during Joule heating is shown in **Figure 8**. By fitting the Raman peaks, the temperature variation can be readily estimated using the following equation

$$\omega_T = \omega_{300K} + \chi\left(T - 300K\right),\tag{1}$$

where ω_T is frequency at certain temperature T, ω_{300K} is frequency at room temperature (300K), and χ is first-order temperature coefficient, in which χ = -0.016cm^{-1}/K for 1L graphene and χ = -0.015cm^{-1}/K for 2L and 3L graphene for G mode [1]. As V_{DC} increases to 4.7V, temperature of the graphene device increases to 1600K (**Figure 13**). It should be noted that the measured temperature from Raman is the weighted, spatially averaged temperature in the laser spot region, thus the average temperature of the graphene is slightly lower than the measured results, which could be corrected based on the temperature distribution equation [4].

As the voltage goes higher than 4.7V, the Raman background enhances significantly and peaks become broader, which makes it hard to estimate temperature accurately from Raman. Instead, in the high voltage regime, the temperature could be estimated by using emission spectrum based on the gray body emission theory. Combining the Raman and emission spectrum, the average temperature of the graphene during electrothermal tuning is estimated and shown in **Figure 14**. As the voltage increases to 6.4V, the average temperature of the device elevates to 2650K. It is worth noting that there exists some difference between temperature estimated from Raman and emission spectrum around V_{DC}=4.7V. This could be explained by the fact that the temperature measured from Raman is the lattice temperature, while that from the emission spectrum is the electron temperature. Normally, the electron temperature and lattice temperature are close but not exactly the same. Using the estimated average temperature from Raman and emission spectrum, the temperature distributions at various V_{DC} are estimated and shown in **Figure 15**. It can be seen that the temperature of transition from negative TEC to positive TEC of graphene is around 1600K.

V. POWER ANALYSIS AND BENCHMARKING OF FREQUENCY TUNING AND TEMPERATURE

Based on the temperature calibrated from Raman and emission spectra, we further estimate the power on the suspended graphene using the equation

$$T_{avg} = \frac{\int_0^{2\pi}\int_0^R \left[T_0^{1-\phi} + \frac{P(1-\phi)}{4\pi t \kappa T^\phi}\left(1 - \frac{r^2}{R^2}\right)\right]^{\frac{1}{1-\phi}} r dr d\theta}{\int_0^{2\pi}\int_0^R r dr d\theta},\tag{2}$$

where $\kappa(T)$ is the temperature-dependent thermal conductivity, R is the radius of graphene membrane, P is the applied power on suspended graphene, ϕ is a temperature-dependent power index, describing how the thermal conductivity is affected by temperature [4]. The estimated power versus voltage is shown in **Figure 18**. The light emission of the vibrating graphene NEMS occurs at $P\sim$8mW.

We further benchmark key performance metrics of graphene electrothermal tuning in this work by comparing with electrostatic gate tuning (**Figure 19**) [3,5,6,7,8,9,10]. We consider three important parameters: highest voltage applied, $V_{DC,max}$, across the suspended graphene, initial tension level γ_{300K}, and tuning range (defined as $\Delta f/f_0$). Our device exhibits tuning range almost eight times as high as the largest frequency tuning achieved using electrostatic coupling, demonstrating remarkable performance of electrothermal tuning. We also compare the highest operating temperature of the device with other N/MEMS devices [11,12,13,14,15,16], as shown in **Figure 20**. It can be seen that most devices are only demonstrated operating at less than 1000K, which is much lower than the record in this work (2650K). This demonstrates a significant improvement of robustness and durability of graphene NEMS in harsh and extreme environments.

VI. CONCLUSIONS

In conclusion, we have demonstrated, for the first time, a glowing graphene NEMS resonator with remarkable frequency tuning range up to $\Delta f/f_0\approx$1300% and operating temperature up to T = 2650K. The graphene NEMS emit strong visible light during simultaneous electrothermal tuning. This work paves a way to building durable graphene NEMS for harsh environment applications and flexible nanoscale light emitters.

REFERENCES

[1] A. Balandin, *et al.*, *Nano Letters* **8**, 902-907 (2008). [2] C. Lee, *et al.*, *Science* **321**, 385-388 (2008). [3] C. Chen, *et al.*, *Nature Nanotechnology* **4**, 861-867 (2009). [4] F. Ye, *et al.*, *Nano Letters* **18**, 1678-1685 (2018). [5] A. M. van der Zande, *et al.*, *Nano Letters* **10**, 4869-4873 (2010). [6] R. Barton, *et al.*, *Nano Letters* **12**, 4681-4686 (2012). [7] X. Song, *et al.*, *Nano Letters* **11**, 198-202 (2011). [8] P. Weber, *et al.*, *Nano Letters* **14**, 2854-2860 (2014). [9] T. Miao, *et al.*, *Nano Letters* **14**, 2982-2987 (2014). [10] A. Reserbat-Plantey, *et al.*, *Nature Communications* **7**, 10218 (2016). [11] B. Kim, *et al.*, *Journal of Microelectromechanical Systems* **17**, 755-766, (2008). [12] M. Suster, *et al.*, *Journal of Microelectromechanical Systems* **13**, 536-541, (2004). [13] D. Young, *et al.*, *IEEE Sensors Journal* **4**, 464-470 (2010). [14] T. Lee, *et al.*, *Science* **329**, 1316-1318 (2010). [15] R. Azevedo, *et al.*, *IEEE Sensors Journal* **7**, 568-576 (2007). [16] T. He, *et al.*, *Tech. Digest of IEDM 2013*, Paper No. 4.6, 108-111 (2013).

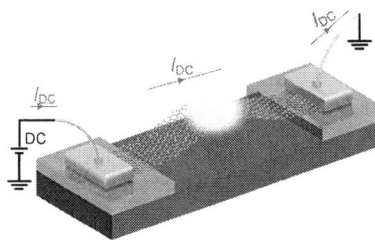

Fig. 1. Schematic illustration of doubly clamped glowing single-layer graphene NEMS resonator with Joule heating.

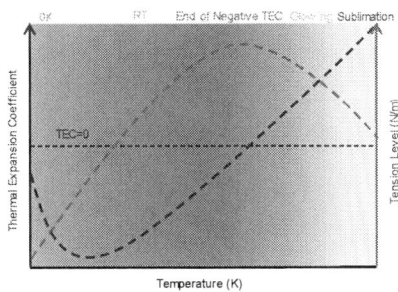

Fig. 2. Graphene thermal expansion coefficient and tension level of the graphene resonator with rising temperature.

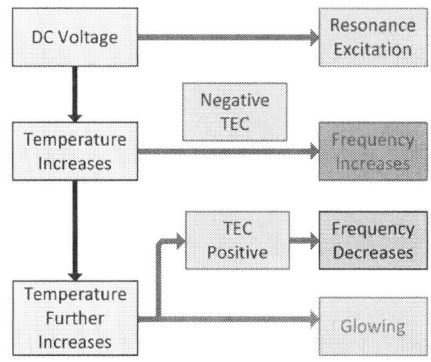

Fig. 3. Analysis of coupling effects of glowing graphene resonators under Joule heating.

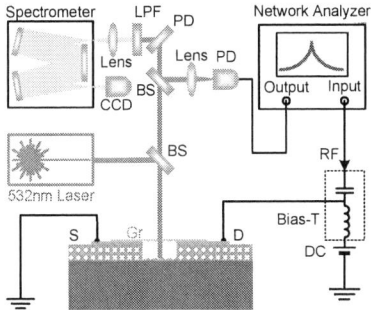

Fig. 4. Combined Raman spectroscopy-optical interferometry measurement system. LPF, PD, and BS represent long-pass filter, photodetector and beam splitter, respectively. All the measurements in this work are performed in moderate vacuum (~20 mTorr).

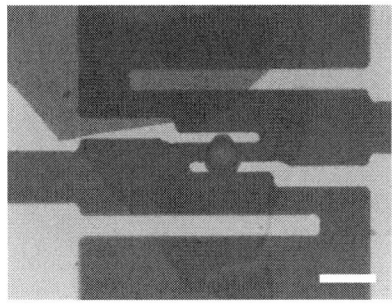

Fig. 5. Optical microscopy image of a tri-layer (3L) graphene resonator. Scale bar: 10μm.

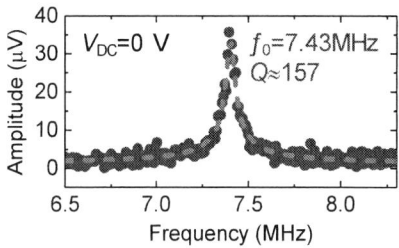

Fig. 6. Fundamental mode resonance of the graphene resonator in Fig. 5, measured at room temperature.

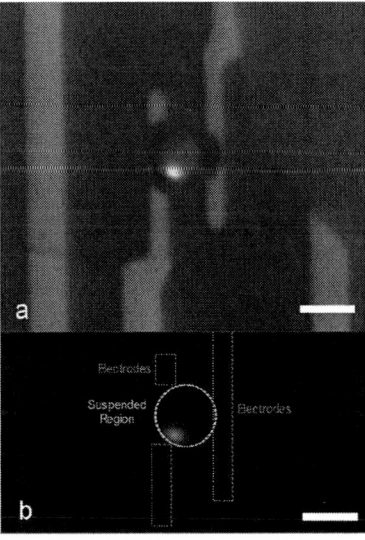

Fig. 7. Optical microscopy images of a glowing tri-layer (3L) graphene resonator under V_{DS}=6.35V, (a) with and (b) without white light. Scale bars: 5μm.

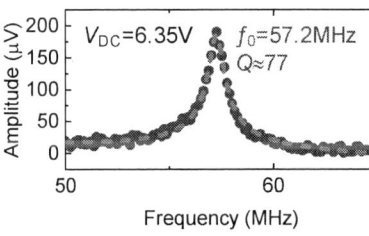

Fig. 8. Fundamental resonance measured from the tri-layer graphene resonator while it is glowing (as shown in Fig. 7).

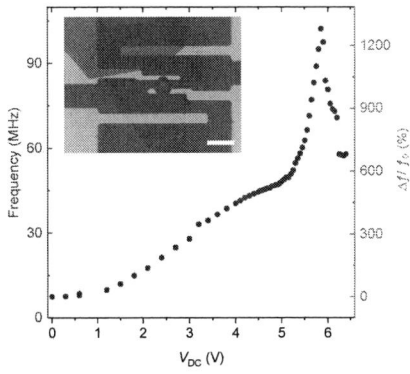

Fig. 9. Frequency tuning measured from the tri-layer graphene device under Joule heating.

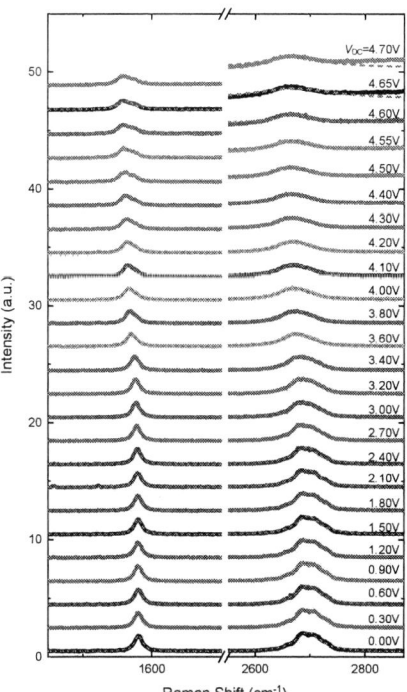

Fig. 10. Raman shift measured from the tri-layer graphene resonator as V_{DC} increases.

978-1-7281-1988-5/18 $31.00 © 2018 IEEE

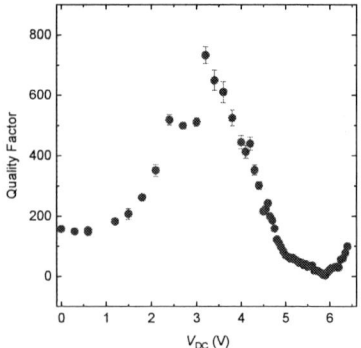

Fig. 11. Quality (Q) factor change of the 3L graphene resonator as V_{DC} increases.

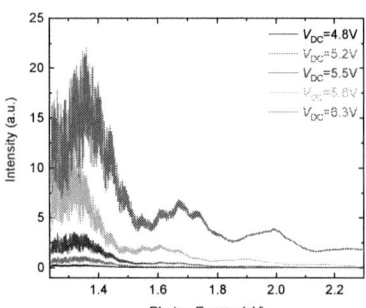

Fig. 12. Visible light emission spectra of the suspended mechanically exfoliated graphene in moderate vacuum.

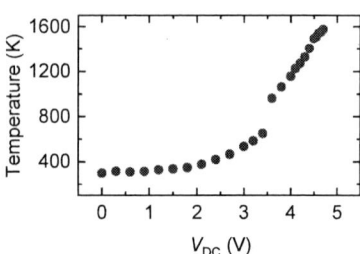

Fig. 13. Calibrated temperature from Raman measurement in Fig. 10.

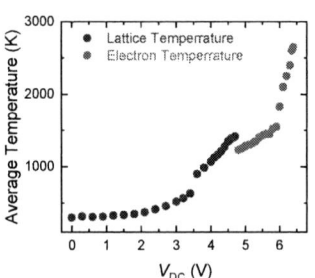

Fig. 14. Estimated average temperature from Raman (lattice temperature) and emission spectra (electron temperature).

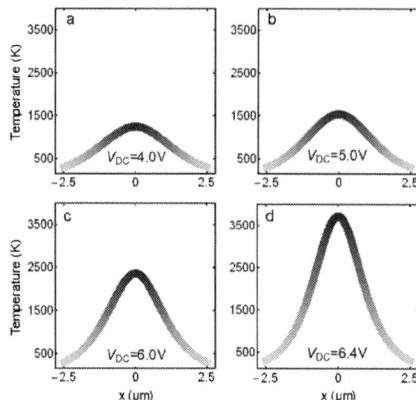

Fig. 15. Calculated temperature profiles with the applied DC voltage (V_{DC}) of (a) 4.0V, (b) 5.0V, (c) 6.0V and (d) 6.4V.

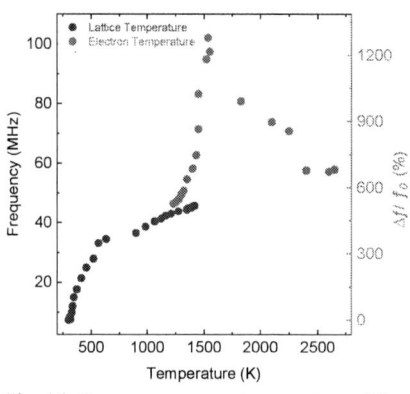

Fig. 16. Frequency versus temperature of the tri-layer graphene device under Joule heating.

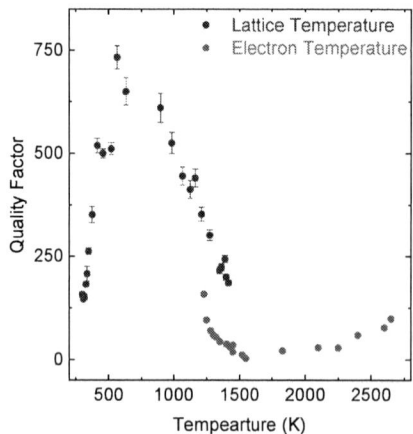

Fig. 17. Q factor versus temperature of the tri-layer (3L) graphene device under Joule heating.

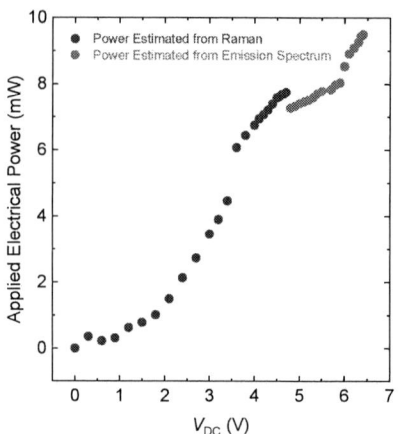

Fig. 18. Estimation of applied electrical power in the suspended graphene from measured Raman and emission spectra.

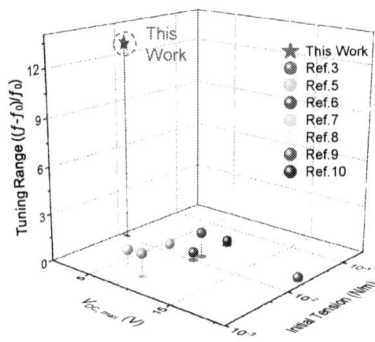

Fig. 19. Comparison of frequency tuning ranges between electrothermal tuning in this work and conventional electrostatic tuning.

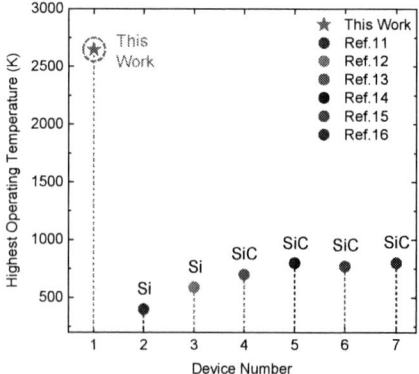

Fig. 20. Comparison of highest operating temperature between electrothermal tuning in this work and other N/MEMS devices.

978-1-7281-1988-5/18 $31.00 © 2018 IEEE

Monolithic Integration of Micron-scale Piezoelectric Materials with CMOS for Biomedical Applications

C. Shi, T. Costa, J. Elloian, and K. L. Shepard

Bioelectronics Systems Laboratory, Columbia University, New York, USA, email: cs3184@columbia.edu

Abstract— We present the development of micro-fabrication techniques achieving the monolithic integration of micron-scale piezoelectric ultrasonic transducers in both polyvinylidene difluoride (PVDF) and lead zirconate titanate (PZT) with complementary metal–oxide–semiconductor (CMOS) integrated circuits (ICs). PVDF-CMOS integration is driven by applications in energy harvesting and data telemetry for medical implants, while PZT-CMOS integration is applied to high-resolution two-dimensional (2D) ultrasound imaging. Both of these applications benefit from reduced parasitic capacitances and higher levels of integration possible with monolithic CMOS integration. Characterization results of micro-fabricated devices reveal the preservation of the piezoelectric properties of PVDF and PZT and transducer functionality with co-designed CMOS ICs.

I. Introduction

Piezoelectric materials, including PVDF and PZT, have unique properties that allow them to convert electrical energy into mechanical energy and, for that reason, have been utilized for decades as ultrasound transducers in medical imaging. These same materials can also be used to harvest energy from ultrasound waves to power implanted devices and transmit data by modulating information onto an ultrasound carrier. Nearly all of the existing implantable devices and ultrasound imaging systems suffer from large form factors because they primarily package functional piezoelectric transducers and the associated electronics discretely [1, 2]. Reducing transducer size to create piezoelectric ultrasonic transducers as small as ~100 μm can reduce displaced volume for implantable devices and enable superior spatial resolution for ultrasound imaging systems. To effectively interface with these micron-scale piezoelectric transducers, direct integration with CMOS ICs is critical both to achieve the requisite interconnect density and to reduce parasitic capacitances which can divide-down signal levels due to the low intrinsic capacitance of the micron-scale transducers. The far-back-end-of-the-line integration approach pursued here allows these devices to be viewed as "More than Moore" enhancements to CMOS. Here, we report approaches to monolithically integrate micron-scale PVDF and PZT piezoelectric transducers with co-designed CMOS ICs (in a standard 0.18 μm 1.8V/5V CMOS technology) for implantable applications and for imaging applications.

II. Fabrication Flow

PVDF is a flexible and biocompatible piezoelectric material, but it cannot tolerate temperatures greater than 80 °C due to a low Curie point. In comparison, PZT is a rigid piezoelectric ceramic material with superior electromechanical properties that can tolerate temperatures as high as 150 °C.

Process flows for integrating these materials must be tuned to these temperature requirements.

A. Piezoelectric transducer integration with CMOS

In the Van Dyke electrical model for a piezoelectric transducer (**Fig. 1a**), C_0 and R_0 represent the parallel plate capacitor and electrical losses of the transducer and the series R_1, L_1 and C_1 capture the mechanical behavior. The circuits interfacing with the PVDF transducers (**Fig. 1b**) for energy harvesting and with the PZT transducers (**Fig. 1c**) for ultrasound imaging greatly benefit from reduced interconnection parasitic capacitances (C_p) of less than 200 fF compared to those with discrete implementation (usually on the order of several pF), making impedance matching requirements for maximum power transfer between the PVDF transducers and the half-wave rectifier (**Fig. 1b**) easier and improving the transmit (TX) driver performance with the PZT transducers (**Fig. 1c**).

Fabrication in both cases starts with bulk PVDF and PZT films, which are mechanically and electrically connected to the CMOS ICs, followed by a subtractive fabrication process to obtain the desired micron-scale pillar structures. The co-designed CMOS ICs for PVDF integration (**Fig. 2**) and for PZT integration (**Fig. 3**) both provide signal and ground terminals on the top-level metal to be connected to the corresponding piezoelectric transducers. Both ICs are initially planarized by etching the polyimide passivation. PVDF and PZT integration then follow different fabrication flows specific to their particular properties.

B. Integration of PVDF with CMOS

The co-designed CMOS chip, contained in a 5 mm × 5 mm die (**Fig. 2a**) with 13 replicas (**Fig. 2b**), has a 150 μm × 95 μm signal pad and a 150 μm × 20 μm ground pad for transducer connections (**Fig. 2c**). A poled 33-μm PVDF film is used, delivering a mechanical resonance frequency of ~34 MHz but usable over a wide frequency range due to its inherently low quality factor. This PVDF film has its surface treated with O_2 plasma to enhance the interfacial adhesion [3]. A 1.6-μm-thick SU-8 2002 adhesive layer is then spun onto the die, bonded to a carrier substrate (**Fig. 4a**). While this introduces a series capacitance of 250 fF (negligible compared to a C_0 of 22.5 fF for the PVDF), the more preferable anisotropic conductive adhesives cannot be applied because they typically require curing temperatures above 100 °C. The substrate is then flipped over and gently placed on the pre-treated PVDF film, baked at 70 °C for 30 minutes, and flipped back to be UV cured (**Fig. 4b**). The now-exposed PVDF top surface is also treated with O_2 plasma and a ~10-μm layer of AZ-4620 photoresist (PR) is spun, followed by a baking step at 70 °C for 80 minutes. The same spinning-and-baking process is repeated two more times to create a ~30-μm-thick PR layer. Standard UV

978-1-7281-1988-5/18 $31.00 © 2018 IEEE

photolithography is applied to create PR patterns, used as etch masks, above the signal pads (**Fig. 2c**, **Fig. 4c**). RIE etching of the PVDF and the underlying SU-8 layer is used to define the micron-scale PVDF pillars (**Fig. 4d, Fig. 5a**) with the bottom terminals connected to the signal pads. Photolithography with a ~6-µm layer of AZ-4620 is performed to cover the entire die except the active chip area (**Fig. 4e**). DC sputtering is used to deposit ~1.2-µm of Cu onto the die. Finally, the Cu over the PR layer is removed with a lift-off process, leaving Cu connection between the top terminals of the PVDF pillars and the ground pads (**Fig. 2c, Fig. 4f, Fig. 5b**). The final PVDF transducers (**Fig. 6**) maintain their longitudinal charge sensitivity (d33) of ~19 pC/N, as measured with a PM300 d33 meter (Piezotest).

C. Integration of PZT with CMOS

The CMOS IC for PZT integration (**Fig. 3a**) features a 2D array of 26 × 26 transmit/receive (TX/RX) channels. Each channel includes a 75 µm × 75 µm signal pad for connection with the PZT transducers (**Fig. 3b**). The PZT material used here has a thickness of 267 µm, which produces a resonance around 8 MHz. The PZT fabrication steps are illustrated in **Fig. 7**, where the PZT transducer pillars are patterned using a mechanical dicing process [4, 5], which provides a much faster etch rate than RIE for the PZT thickness used in this work. The PZT bulk material is obtained as a 7.24 cm × 7.24 cm sheet with 50 nm of Ni on both sides of the film. The Ni is etched away by dipping the sheet into ferric chloride for 5 seconds. Next, a 10-nm chrome adhesion layer and a 50-nm gold layer are deposited by electron beam (e-beam) evaporation onto both sides of the PZT sheet (**Fig. 7a**) and patterned by lift-off of AZ-1512 PR (**Fig. 7b**). The bottom pattern matches the signal contacts on the CMOS chip (**Fig. 3b**). The PZT sheet is then diced into 4 mm × 4 mm individual dice using a dicing saw with a blade suitable for PZT ceramics (Z09-SD2000-Y1-90) with a kerf of 50 µm (14000 spindle, 3 mm/s feed speed). The PZT die is then diced from the bottom side to about 20% of its thickness by using the same blade and cutting settings (**Fig. 7c**). This defines the shape of the PZT elements from the bottom, which enables the dicing from the top side to go all the way through the PZT without risking damaging the CMOS chip. The PZT die is attached to the CMOS chip with an anisotropic conductive film (TFA220-8, H&S HighTech) as an adhesive and bonded in a FINEPLACER lambda die bonder (Finetech) with fine alignments, using a force of 150 N at 150 °C for 5 seconds (**Fig. 7c**). The PZT is then diced from the top to a depth of 90% of its thickness, which leaves the elements completely free-standing and fully etched (**Fig. 7d**). Microscope images of the 2D array of PZT pillars are shown in **Fig. 8a-c**, while a SEM image of the PZT pillars is shown in **Fig. 8d**. To provide mechanical stability to the array of PZT pillars, the kerfs are filled with the Epo-Tek 301-2 epoxy (Epoxy Technology), as shown in **Fig. 7e**. Finally, the same Cu deposition step, as detailed in Section II.B, is performed on top of the 2D array to implement the common ground connections among all the fabricated micron-scale PZT elements (**Fig. 7f**). In the fully fabricated chips (**Fig. 9**), d33 measurements before and after fabrication show only a 7% loss (398 pC/N for pristine PZT to 370 pC/N for post-fabricated PZT).

III. CHARACTERIZATION RESULTS

A. PVDF integrated on CMOS

The PVDF transducers were characterized with Raman spectroscopy and atomic force microscope (AFM). The background Raman spectrum due to the IC substrate was subtracted from the absorbance measurements to obtain the PVDF spectrum. The post-fabricated PVDF shows almost the same Raman spectrum as the pristine one with characteristic peaks at 511 cm^{-1}, 840 cm^{-1} and 1430 cm^{-1} (**Fig. 10**) [6], demonstrating the preservation of piezoelectric properties. Additionally, AFM was used to measure the surface morphology of the fabricated piezoelectric transducers under voltage excitation. **Fig. 11a** shows the roughness of the PVDF top surface to be close to 100 nm in the absence of voltage excitation. 10-V square wave excitation at 1 Hz and 10 Hz, however, resulted in observable mechanical oscillations in the AFM that scanned at 1 Hz (**Fig. 11b**, **Fig. 11c**). The interfacing on-chip half-wave rectifier (**Fig. 1b**) is functional with a 10 MHz, 1.8V peak-to-peak sinusoidal excitation (**Fig. 12**) after going through the PVDF fabrication processes.

B. PZT integrated on CMOS

To test the integration of PZT with CMOS, a fully-fabricated chip was wire-bonded to a custom printed circuit board in the test setup illustrated in **Fig. 13a**. A container was adhered to the board and then filled with deionized water. A HGL-0200 hydrophone connected to a AG-2010 amplifier (Onda) was placed at the center of the PZT array and connected to an oscilloscope to record the ultrasound waves generated by the transducers. The CMOS IC TX drivers (**Fig. 1c**) were configured to generate a 5-V, 8-MHz square wave and the hydrophone successfully recorded the incoming ultrasound waves for increasing distances from the CMOS chip, in steps of 2 mm (**Fig. 13b**). From the obtained data, the speed of sound in water was measured to be 1530 m/s, which is close to the reference value around 1540 m/s. The recorded pressure amplitude decreases with increasing propagation distance, as expected.

IV. CONCLUSION

This work reports on the fabrication approaches to monolithically integrate µm-sized PVDF and PZT transducers with CMOS. Different fabrications steps were employed according to the distinct characteristics of PVDF and PZT. Various characterization results show the successful integration for both materials, where the piezoelectric properties of PVDF and PZT as well as the functionality of the interfacing CMOS ICs are retained throughout both fabrication flows with the introduction of limited capacitive interconnect parasitics.

ACKNOWLEDGMENT

This work was funded by the DARPA ElectRx program.

REFERENCES

[1] C. Dagdeviren *et al.*, *Nat. Biomed. Eng.*, 1(10), pp. 807-817, 2017.
[2] T. Gang *et al.*, *J. Micromech Microeng* 22(6) p. 065017, 2012.
[3] J. S. Lee, *et al* *ACS Appl Mater* 1(12), pp. 2902-2908, 2009.
[4] C. Chen *et al.*, *IEEE Symp VLSI*, pp. 1-2. *2016*
[5] J.-S. Park et al., Sens Actuators A Phys, 108(1), pp. 206-211, 2003.
[6] P. Martins *et al*, *Prog. Polym. Sci.*, 39(4), pp. 683-706, 2014.

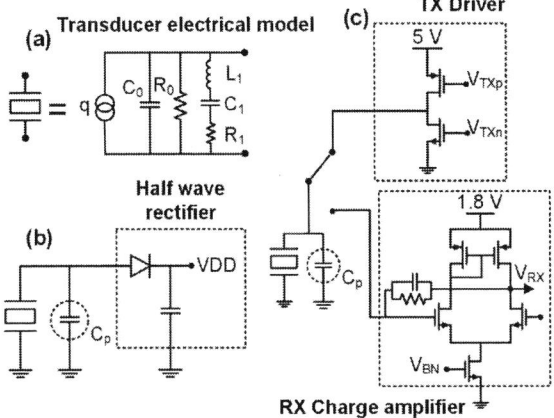

Fig. 1. Simplified schematics of the CMOS circuits that interface with the piezoelectric transducers. (a) electrical model and (b) half-wave rectifier for power harvesting from a PVDF transducer and (c) charge amplifier receiver and transmit driver for PZT-based ultrasound imaging.

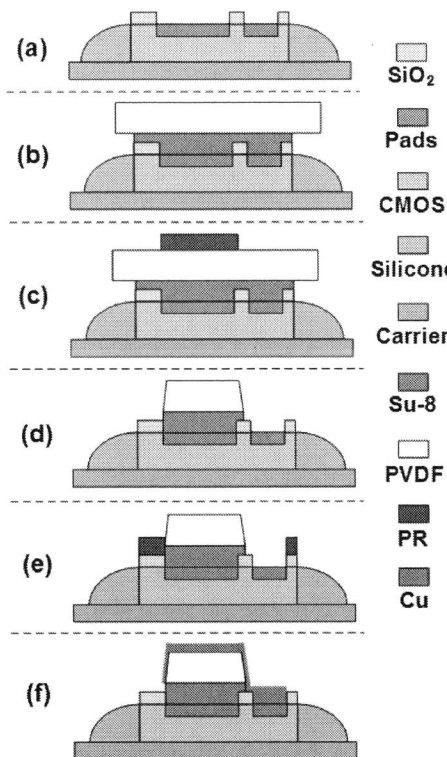

Fig. 4. Fabrication flow for integrating PVDF with the CMOS IC in Fig. 1. (a) Removal of polyimide and bonding of the IC die to a carrier with PMMA and silicone and (b) adhesion of pre-treated PVDF piece and the die with SU-8 and (c) patterning of PVDF areas on top of the chip signal pads and (d) RIE etching to create PVDF pillars and (e) patterning of chip areas and (f) metal deposition and lift-off for ground connection.

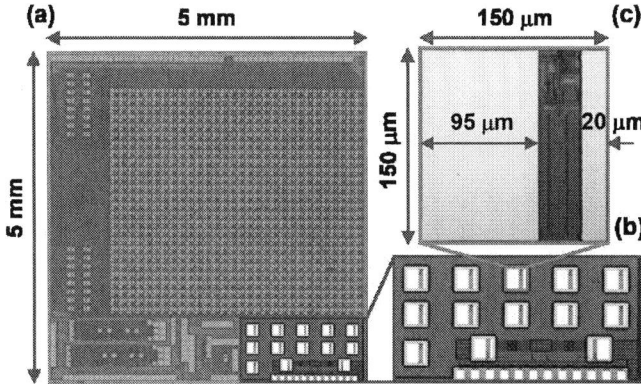

Fig. 2. CMOS IC microphotograph for PVDF integration. (a) Die image and (b) 13 chip replicas contained in one IC and (c) one individual chip.

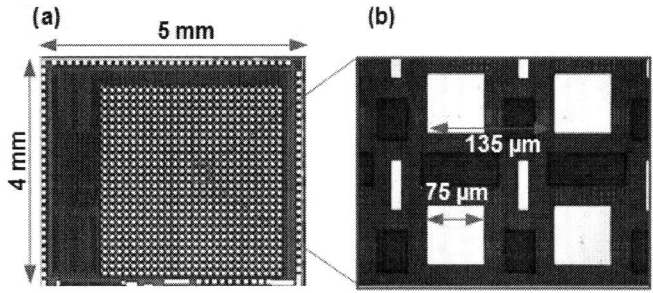

Fig. 3. CMOS IC microphotograph for PZT integration. (a) Die image and (b) transmitter/receiver circuits with the signal pads.

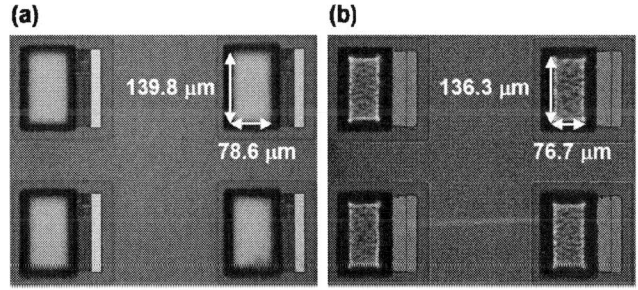

Fig. 5. Microscope photograph of (a) 4 PVDF pillar structures integrated with the chip input pads and (b) 4 fully-integrated PVDF structures with the chips.

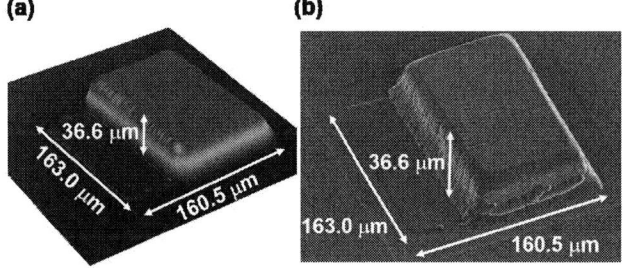

Fig. 6. (a) The profilometer map scan image and (b) the SEM image of one fully-fabricated micron-scale PVDF structure monolithically integrated on-chip.

978-1-7281-1988-5/18 $31.00 © 2018 IEEE   93

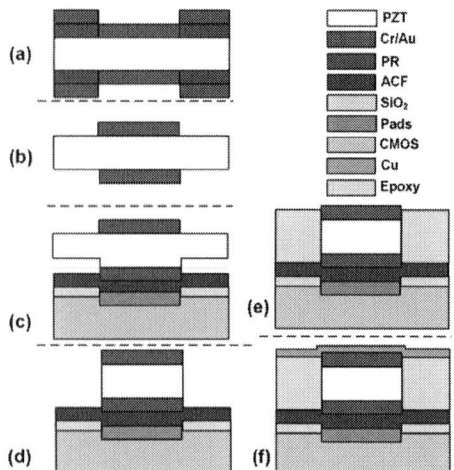

Fig. 7. Fabrication flow for integrating PZT with the CMOS IC in Fig. 2. (a) Patterning of the Cr/Au layer on both sides of the PZT sheet by spin-coating of the PR and the e-beam evaporation and (b) lift-off to define the Cr/Au contacts and (c) PZT diced from the bottom side and adhered to the CMOS IC with ACF and (d) PZT diced from the top side to define the pillar structures and (e) PZT kerfs filled with epoxy and (f) sputtering of Cu for PZT transducer ground connection.

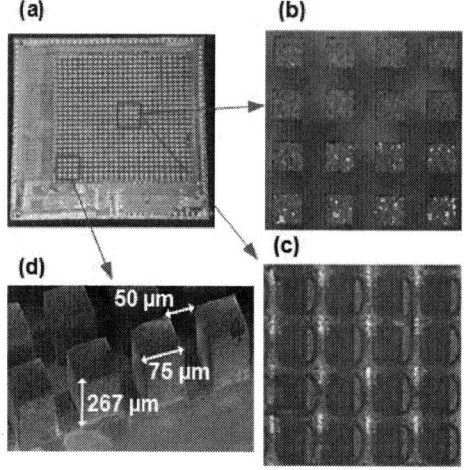

Fig. 8. Microscope photograph of PZT integrated with the CMOS IC chip after top-side dicing. (a) Full chip view and (b) top of the micro-fabricated PZT pillars and (c) surface of the CMOS IC and (d) SEM image of the PZT pillars.

Fig. 9. Microscope photograph of one fully-fabricated CMOS PZT chip.

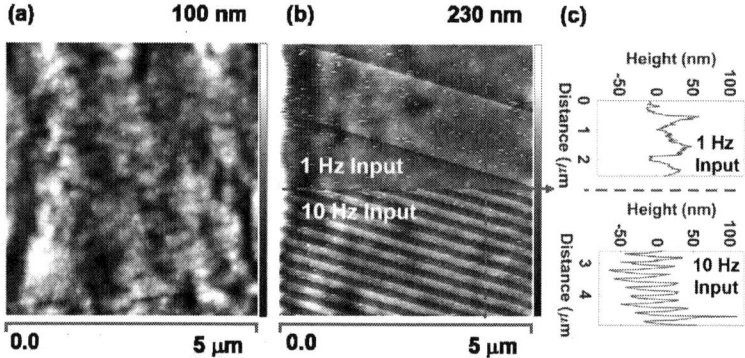

Fig. 11. AFM measurements on one fabricated PVDF structure (a) without voltage excitation and (b) with 1 Hz and 10 Hz 10 V peak-to-peak square wave excitations (c) a height plot along a vertical line on Fig. 11(b).

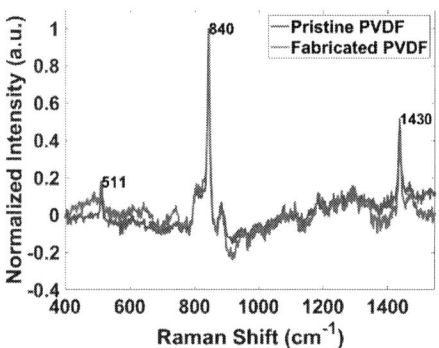

Fig. 10. Raman spectrum of one micro-fabricated PVDF structure in comparison with a pristine PVDF piece.

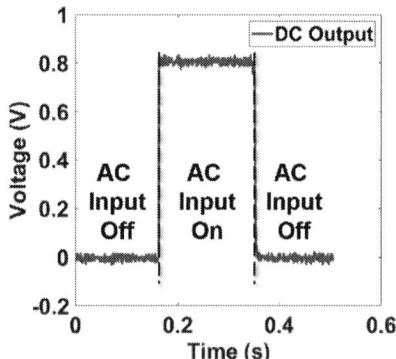

Fig. 12. Output voltage from a post-fabricated on-chip half-wave rectifier.

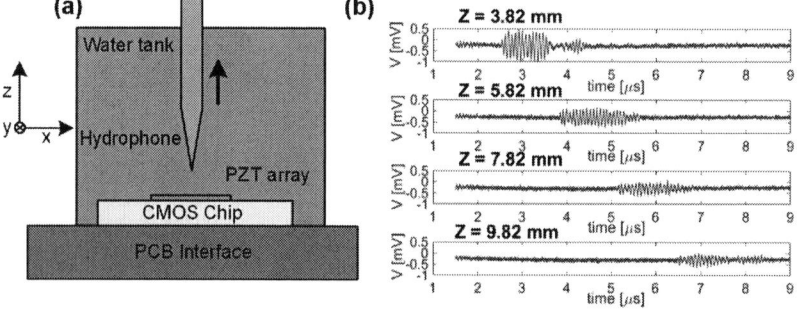

Fig. 13. (a) Measurement setup for characterizing the PZT-integrated CMOS chip and (b) ultrasound measurements with a hydrophone from different distances to the CMOS chip.

978-1-7281-1988-5/18 $31.00 © 2018 IEEE

A Nano-Mechanical Resonator with 10nm Hafnium-Zirconium Oxide Ferroelectric Transducer

M. Ghatge, G. Walters, T. Nishida and R. Tabrizian

Department of Electrical and Computer Engineering, University of Florida, Gainesville, FL, USA

email: ruyam@ufl.edu

Abstract—This paper reports, for the first time, on a 10nm hafnium-zirconium oxide ($Hf_{0.5}Zr_{0.5}O_2$) (HZO) piezoelectric transducer for nano-electromechanical systems (NEMS). The super-thin HZO films are engineered through atomic-level stacking, capping with titanium nitride (TiN) electrodes, and proper thermo-mechanical treatment, to realize ferroelectric transducers with large piezoelectric properties. The developed 10nm transducer is used for excitation of a silicon-based multi-morph nano-mechanical resonator, with an overall thickness of ~350nm, at ~4MHz. The developed resonator, along with 120nm aluminum-nitride (AlN) transduced counterparts, are also used as test-vehicles to characterize ferroelectric and piezoelectric properties. Benefiting from large piezoelectric coefficient ($e_{31,HZO} \approx 2.3 e_{31,AlN}$), fully conformal deposition, and CMOS-compatibility, ALD-deposited 10nm HZO transducer paves the way for realization of truly monolithic cm- and mm-wave RF front-ends for the emerging 5G wireless communication systems, and extreme / 3D integration of NEMS sensors and actuators.

I. INTRODUCTION

Ever since the advent of micro- and nano-electro-mechanical resonators, the need for large electromechanical coupling coefficient, extreme frequency scalability, and CMOS processing compatibility have been the governing drivers for the advancement of thin-film piezoelectric transducers. The development of high quality piezoelectric films realized high-performance bulk acoustic resonators and filters, and enabled RF front-end modules for wireless mobile systems.

However, the forthcoming 5G era, with ambitious target of the extension of wireless communication to mm-wave regime, has raised an unprecedented urgency for transformation of piezoelectric films and acoustic resonator architectures. To fulfill the demand for extreme frequency scaling to mm-wave regime, the quest for material and architectural transformation of the acoustic resonator technology continues. While the development of fin-based resonator architectures [1, 2] and the use of single crystal films and substrates help further the scaling limits beyond the current state, the ultimate bound of the frequency scaling is set by the technological limitations in piezoelectric film thickness minuaturization. With the frequency of bulk acoustic resonators inversely proportional to the thickness of the piezoelectric film, extreme frequency scaling to mm-wave regime requires radical thickness miniaturization to sub-100nm. Such a miniaturization is substantially inhibited by the nucleation, crystallization, and texture development processes in current piezoelectric film deposition techniques (e.g. magnetron-sputtering and MOCVD), and drastically degrade the electromechanical coupling and energy dissipation coefficients [3].

To address this technological gap, the paper demonstrates, for the first time, the use of piezoelectric properties of a 10nm atomically engineered ferroelectric hafnium-zirconium oxide (HZO) film for transduction of nano-mechanical resonators. Though HZO is widely studied for FeFET and FeRAM devices [4], it has not yet been applied in piezoelectric transducers. Three unique characteristics of ferroelectric HZO stand out in comparison with other ferroelectric materials that are profoundly transformative for advanced devices: CMOS compatibility, occurrence of ferroelectricity, and hence piezoelectricity, in sub-10nm films, and the capability to engineer the ferroelectric properties through atomic layering of mono-layers of dopants [5]. Furthermore, the extreme thickness scaling makes it a promising candidate for nano-acoustic resonators in the mm-wave regime. Finally, the conformal nature of ALD enables its 3D integration for the realization of fin-based resonators, and sidewall transducers for very-large-scale-integrated sensors and actuator arrays.

II. ATOMICALLY ENGINEERED HZO FILM

ALD deposited HZO films are typically amorphous due to the low thermal budget operation but can be engineered into crystallinity with rapid thermal annealing (RTA). The non-centrosymmetric orthorhombic crystal phase achieved after RTA exhibits a ferroelectric behavior. A capping layer, such as titanium nitride (TiN), helps suppress the monoclinic phase and promotes the orthorhombic phase during the RTA process [6, 7]. Doped HfO_2 have been explored extensively for ferroelectricity, however the choice of 1:1 binary HfO_2:ZrO_2 is driven by its low annealing temperature and high polarization [8]. The substrate or bottom electrode on which HZO is grown has been shown to have a pronounced effect on its ferroelectric response and thus piezoelectricity. The films grown on top of (002) c-axis oriented AlN and subsequently sputtered Mo are observed to show a higher polarization than films grown on Ge or TiN/Si as substrate or bottom electrode. The ferroelectric nature of HZO can be further exploited to tune the material properties with an applied DC voltage or to permanently reorient the spontaneous polarization. Unlike PZT, an order of magnitude higher coercive field strength of HZO increases its resilience to internal depolarization or signal fluctuations thus widening its material tuning capability for piezoelectricity [6]. The high-*k* dielectric nature of HZO along with its

978-1-7281-1988-5/18 $31.00 © 2018 IEEE

piezoelectricity can be used for dual electrostatic/piezoelectric hybrid-actuation for low-voltage operation.

III. MULTI-MORPH RESONATOR DESIGN

A multi-morph nano-mechanical resonator is used for characterization of HZO piezoelectric properties. The resonator is formed by stacking HZO/TiN, c-axis oriented AlN/Mo, and Si. Benefiting from two independent piezoelectric transduction ports (i.e. HZO and AlN), various drive/sense mechanisms are used. Two-port resonators with asymmetric transducer designs (port-1 AlN, port-2 HZO) are used (Fig. 1(a)) to evaluate the frequency response for HZO-actuate/AlN-sense driving mechanism. To aviod the potential interference of AlN, alternative architecture with HZO-only transducer is also studied (Fig. 1(b)). Fig. 2 details the resonator stack through the high-resolution cross-sectional TEM (HR-XTEM) image.

IV. FABRICATION PROCESS

Fig. 3 summarizes the fabrication process used for implementation of the multi-morph nano-mechanical resonator. AlN film (120nm) sandwiched between top and bottom 50nm molybdenum (Mo) layers are magnetron-sputtered on top of a 70nm device layer SOI substrate. Top Mo is patterned to serve as the ground (GND) for the HZO transduction port. An ALD 10nm-HZO / 10nm-TiN stack is then deposited using alternating series of pulses: tetrakis (dimethylamino) zirconium (IV), followed by a water (H_2O) to oxidize the layer and tetrakis (dimethylamido) hafnium (IV), again followed by H_2O for oxidation. By alternating equal cycles of HfO_2 and ZrO_2 a 1:1 binary of 10nm HZO is achieved. This is followed by tetrakis (dimethylamido) titanium (IV) and nitrogen plasma cycles to deposit the 10nm TiN. The TiN capping layer is selectively patterned for the HZO transduction regions using a Cl_2/H_2 based plasma process. HZO is then etched using BOE to get access to AlN in all regions except the HZO transducer area. Following HZO/TiN selective etching, RTA at 500^0C for 20 seconds is used to crystalize the HZO in its ferroelectric phase. The crystalline form of the HZO film is evident from the diffraction pattern shown in TEM image (Fig. 2(b)). 30nm platinum (Pt) $RF_{in/out}$ electrodes are sputtered and patterned using a liftoff process. Bottom Mo (serving as GND for the AlN transducer) is accessed by dry etch of the 120nm AlN outside the device area. The lateral geometry of the device is then defined with selective etching of the 120nm AlN / 50nm Mo / 70nm Si in a RIE/ICP process. Finally, the devices are released from the top side by etching the buried oxide layer through the trench and etch holes using HF acid.

V. CHARACTERIZATION RESULTS

Various electrical and optical characterization schemes are used to evaluate the performance of the resonator and the ferroelectric and piezoelectric properties of HZO (Fig. 4). Fig. 5 shows the two-port frequency response of the device in Fig. 1(a), highlighting the resonance peak with a Q_{max} of ~50 (in air) at ~4MHz. Beside two-port electrical characterization, the HZO-only transduced resonator (Fig. 1(b)) is optically characterized by monitoring the out-of-plane vibration amplitude using a digital holographic microscope (DHM) and the results are compared with AlN-only transduced counterpart (Fig. 6). Such a comparison enables characterization of piezoelectric coefficient of HZO, resulting in $e_{31,HZO} \approx 2.3 e_{31,AlN}$. Fig. 7 (a,b) show the short-span frequency response of the resonator in Fig. 1(a) for various excitation voltages applied to HZO- and AlN-transduction ports, respectively, highlighting the transition to nonlinear operation regime. Fig. 8 compares the improved P-V response of 10nm HZO on top of AlN/Mo with typical TiN/Si electrodes. Fig. 9 shows the P-V response of the HZO on top of the resonator after the device release using HF. The reduced polarization (P_r) compared to Fig. 8 is attributed to the effect of HF etch on HZO through the the trench and etch holes and the smaller ferroelectric capacitor area on resonator. The P-V response can be improved through exploiting backside dry-release and increasing the transduction area. Fig. 10 shows the permittivity response of HZO demonstrating tunable high-k properties.

VI. CONCLUSION AND DISCUSSION

This paper demonstrates the thinnest ever-reported CMOS-compatible piezoelectric transducer using 10nm ferroelectric HZO film. Atomically deposited ferroelectric HZO films are engineered to demonstrate large piezoelectric properties, and used for excitation of multi-morph nano-mechanical resonator. Various schemes, including isolated HZO- and AlN-transduction ports, along with different electrial and optical characterization are used to extract the ferroelectric and piezoelectric properties of HZO film. The demonstration of the 10nm atomically engineered HZO with a large piezoelectric response paves the way for extreme miniaturization of nano-mechanical resonators to mm-wave regime and for 5G applications. Besides, the low-temperature and truly conformal nature of ALD HZO process offers substantial advantages over conventional magnetron-sputtered / MOCVD films, including CMOS-compatibility and sidewall transducer integration. Finally, the capability to engineer the material properties by varying the dopant layering to enhance ferroelectricity and piezoelectricity or by applying DC bias to tune them, increases the potential of the proposed piezoelectric HZO many-fold.

ACKNOWLEDGMENT

This work was supported in part by NSF grants ECCS 1610387 and ECCS 1752206. The authors would like to thank Nanoscale Research Facility staff at the University of Florida and Nicholas Rudawski for help with TEM.

REFERENCES

[1] Ramezani, M., et. al. IEEE IEDM, *2017* (pp. 40-41).
[2] Bahr, B., et al. IEEE ISSCC, *2018* (pp. 348-350).
[3] Yarar, E., et al. AIP Advances 6.7 (2016): 075115.
[4] Müller, J., et al. *ECS J. Solid State Sci. & Techno.* 4.5 (2015): N30-N35.
[5] Lomenzo, P. et al., APL 105, 072906 (2014).
[6] Polakowski, P., Müller, J., APL 106, 232905 (2015).
[7] Ghatge, M., et. al. IEEE Transducers, *2017* (pp. 746-749).
[8] Hyuk Park, Min, et al. APL 104, 072901 (2014).

978-1-7281-1988-5/18 $31.00 © 2018 IEEE

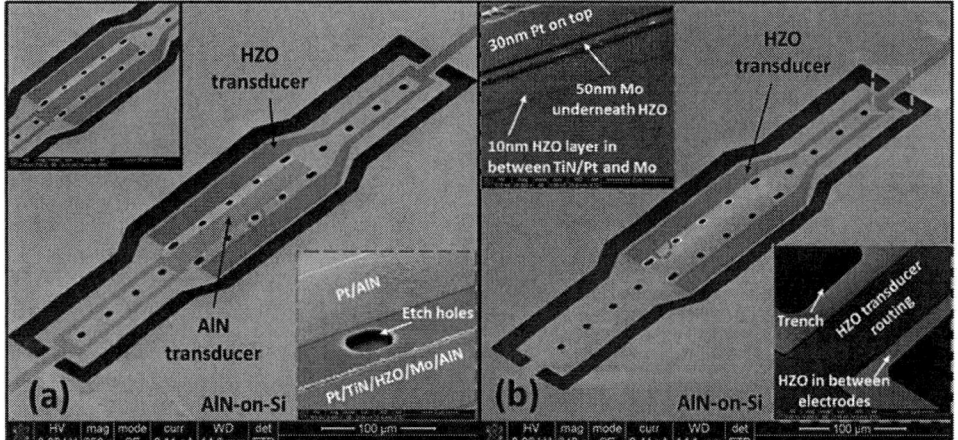

Fig. 1. (a) SEM image of two-port resonator. Port-1 (highlighted in pink) is an AlN transducer while port-2 (highlighted in green) is 10nm HZO transducer. (inset-top) shows the zoomed-in image of the transducer region. (inset-bottom) shows the close-up of the etch holes. (b) SEM image of 1-port HZO transduced resonator. (inset-top) highlights the HZO transducer stack. 10nm HZO is sandwiched between 30nm/10nm Pt/TiN on the top and 50nm Mo underneath.

The wider HZO region helps avoid plausible shorting between the electrodes thus enabling viable transduction. (inset-bottom) shows routing of HZO transducer following same strategy as (inset-top) to avoid shorting of electrodes.

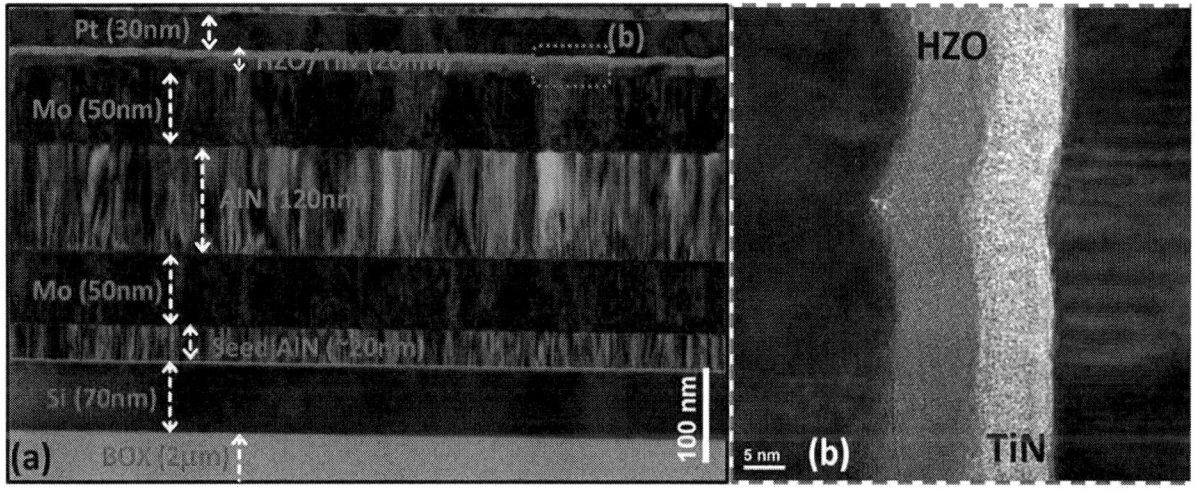

Fig. 2. (a) HR-XTEM image verifying the individual thicknesses of the materials in the stack. The c-axis oriented ~120nm AlN is evident on top of 50nm Mo. (b) Zoomed-in view of the 10nm/10nm HZO/TiN layers. The crystal diffraction patters are evident in HZO indicating the crystalline form of HZO. The conformal ALD deposition of HZO is unaffected by the topography of the bottom surface.

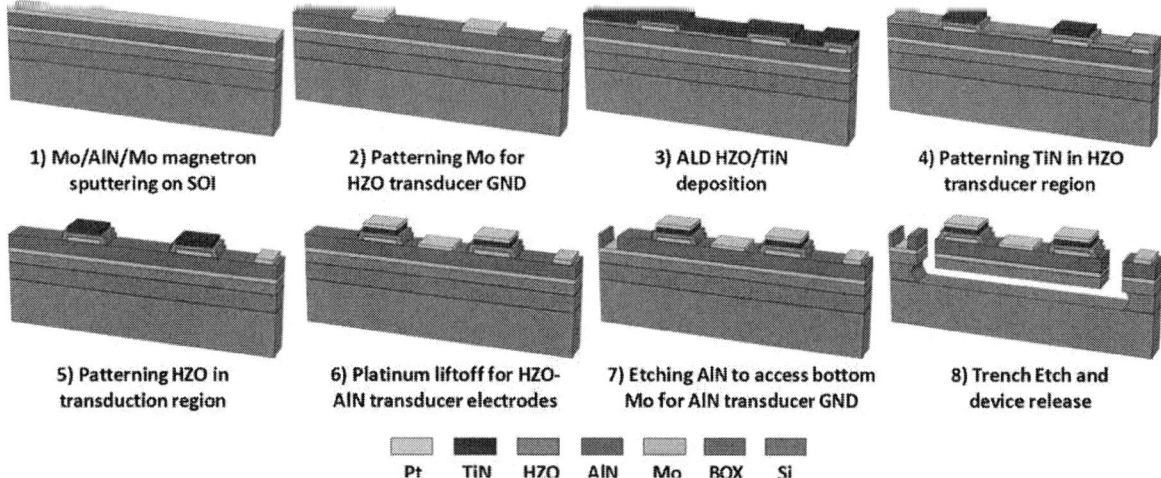

Fig. 3. The fabrication process flow for HZO/AlN dual-transduced 2-port resonator on 70nm Si. 120nm AlN is magnetron sputtered in between 50nm Mo layers. Top Mo acts as GND for HZO transducer while bottom Mo serves as GND for AlN transducer.

978-1-7281-1988-5/18 $31.00 © 2018 IEEE   97

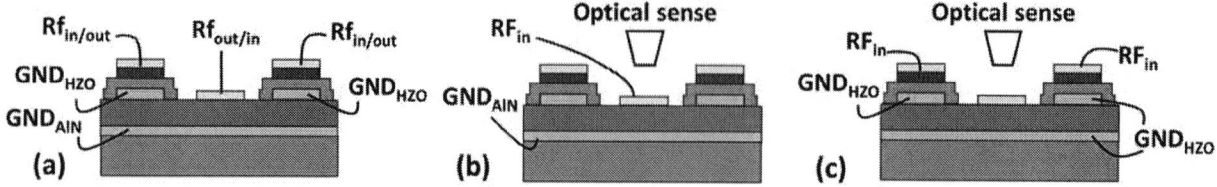

Fig. 4. (a) 2-port measurement set-up with AlN-drive and HZO-sense or vice versa. (b) AlN-drive and optical sense for the vibration using DHM. (c) HZO-drive and optical sense for vibrations using DHM.

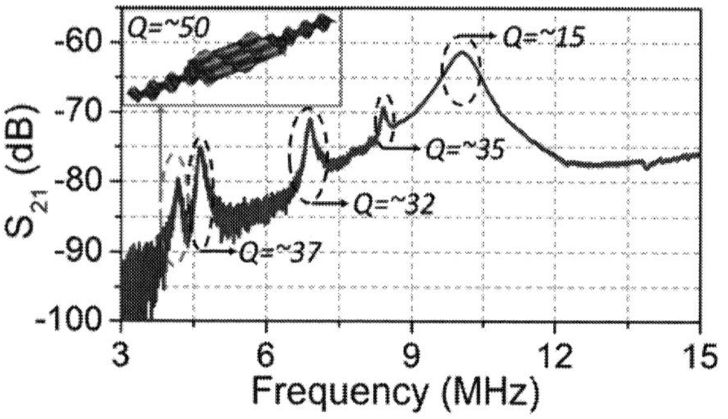

Fig. 5. Measured 2-port frequency response from 3MHz-15MHz operating for resonator design shown in Fig. 1(a). The maximum Q of ~50 (in air) at ~4.1MHz is observed. The resonator is actuated and sensed using the drive/sense mechanism shown in Fig. 4(a).

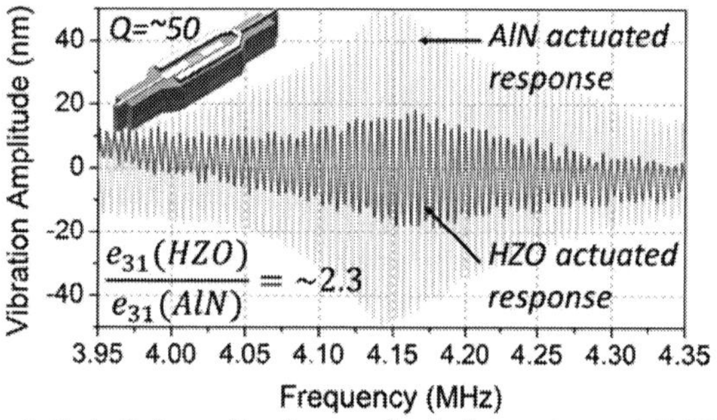

Fig. 6. Out-of-plane vibration amplitude for mode at ~4.1MHz actuated using driving mechanisms (Fig. 4(b) and Fig. 4(c)) with V_{drive}=2V. The AlN-actuated vibration amplitude (green) is ~50nm while HZO-actuated amplitude is ~20nm. The vibration amplitude comparison indicate $e_{31,HZO} \approx 2.3 e_{31,AlN}$.

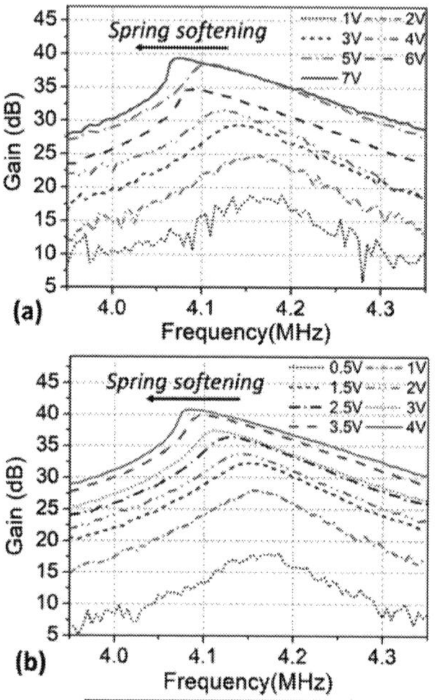

Fig. 7. (a) Short-span frequency response with power handling for ~4.1MHz mode with driving scheme shown in Fig. 4(c) (b) Short-span frequency response with power handling for the same mode with driving scheme shown in Fig. 4(b). AlN transducer is driven to non-linearity much sooner compared to HZO due to large vibration amplitude for the same bias voltage as shown in Fig. 6. The gain for the respective mode is derived using the vibration amplitudes captured using DHM at different input biases.

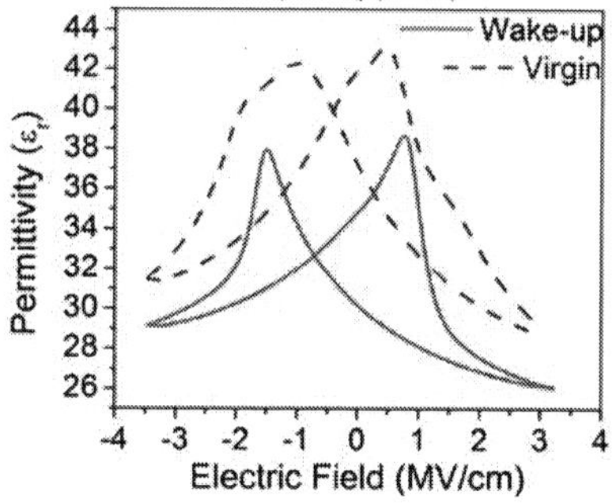

Fig. 10. Permittivity vs voltage for HZO reveals the signature dual CV peak for ferroelectric materials. The peak is more pronounced after wake-up. The tunable high-k dielectric nature of HZO is evident.

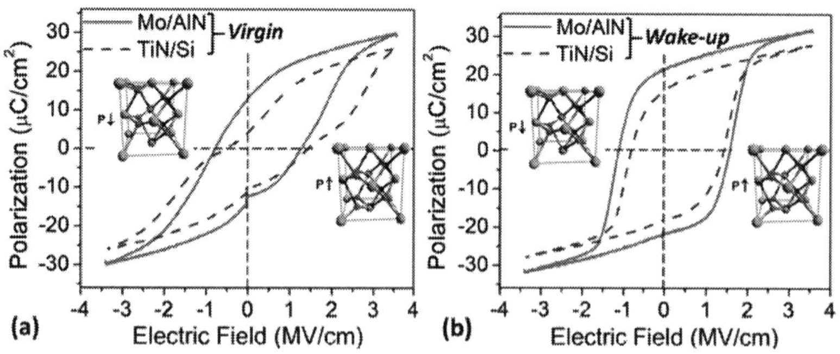

Fig. 8. Hysteresis curves for 10nm HZO on TiN/Si (dashed) and Mo/AlN (solid) in the virgin state (after RTA) (a) and after wake-up (b) achieved by 10k cycles of bipolar square wave at 1kHz. The remanent polarization (P_r) for HZO/Mo/AlN is higher than that of HZO/TiN/Si used typically signifying higher ferroelectricity.

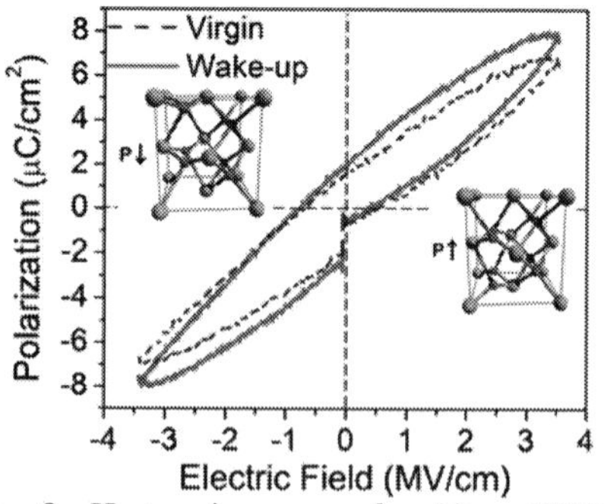

Fig. 9. Hysteresis curves for 10nm HZO transducer. The reduced P_r compared to Fig. 8(b) is attributed to smaller ferroelectric capacitor area and HF attacking HZO during release.

978-1-7281-1988-5/18 $31.00 © 2018 IEEE

Comprehensive optical losses investigation of VLSI Silicon optomechanical ring resonator sensors

L. Schwab[1], P.E. Allain[2], L. Banniard[3], A. Fafin[3], M. Gely[3], O. Lemonnier[3], P. Grosse[3], M. Hermouet[3], S. Hentz[3], I. Favero[2], B. Legrand[1] and G. Jourdan[3]

[1]LAAS-CNRS, Université de Toulouse, CNRS, Toulouse, France, email: lschwab@laas.fr
[2]MPQ, Univ. Paris Diderot, CNRS UMR 7162, Sorbonne Paris Cité, 75013 Paris, France
[3]Univ. Grenoble Alpes, CEA LETI, 38000 Grenoble, France

Abstract—Cavity optomechanics devices are leading edge candidates for a new generation of sensors both in the quantum and classical realms. Several single devices have been demonstrated in numerous labs, however large-scale integration capability necessary for industrial deployment is still an issue. In this paper, we present very-large-scale integrated (VLSI) optomechanical sensors fabricated from standard 200 mm Silicon-On-Insulator (SOI) wafers. Optical properties over a statistically significant sample size have been systematically investigated and show an excellent modeling to experiment agreement, a coupling parameter dispersion of 7% and a manufacturing yield larger than 98%. Controlled versatile sensors, such as these, could easily be embedded in any chip where mass or force sensing is needed.

I. INTRODUCTION

Optomechanical resonators have attracted recent interest for their record displacement sensitivity with different approaches down to a 10^{-18} m/$\sqrt{Hz}$ [1]. On-chip cavity optomechanics are key players towards quantum studies [2]. A typical example is the achievement of quantum mechanical ground state for mesoscopic objects [3, 4] and the prospect of performing quantum entanglement with such objects [5]. For non-quantum applications, this technology is mature enough to be applied to mass [6], gas [7] magnetic field [8] and acceleration [9] sensing and to be operated in highly dissipative environment such as liquids [10], opening new sensing prospects in microfluidics. In cavity optomechanics, displacement sensitivity is proportional to the product between the loaded optical quality factor Q_{loaded} and the transmission contrast C. In the past decade, canonical individually-chipped ring or disk resonators with high $Q_{loaded}C$, have been routinely investigated for academic interest [11-15]. Industrial applications however, will require reaching state-of-the-art optical performance along with Very Large Scale Integration processes and packaging.

In this study, we focused on integration of VLSI optomechanical rings evanescently coupled by photonic circuitry that could be inexpensively integrated sensing chips. Those structures are a first step towards the standardization of integrated optomechanical sensors. As a typical high-throughput measurement case study, we investigate the optical properties of such resonators. Our work on optomechanical properties of numerous variants of the ring resonator enables to identify dominant dissipation channels of the system as well as optimal coupling conditions. This knowledge of optimal optical properties allows us to build a ring resonator with excellent optical and optomechanical sensibility.

II. RESONATORS' DESIGN

The ring-shaped resonator was chosen for its high intrinsic optical Q-factor ($Q_i>10^5$) and high frequency radial breathing mechanical resonance ($f_m>100$ MHz). Also, the coupling factors between mechanical modes and the optical modes are larger in wheel-shaped structures compared to non-circular devices. In silicon material, this coupling is borne mainly by radiation pressure and photothermal forces. Our ring-shaped resonators are evanescently coupled to an optical waveguide with coupling losses parameterized by an extrinsic Q-factor Q_e. The ring is anchored by 3 suspended inner spokes, separated from each other by an angle of 120°, to a central pedestal.

A. Mechanics

The 1st order breathing mode of our structure reaches 130 MHz. The main dissipation channel of mechanical energy, parameterized by its mechanical Q-factor Q_m, is air damping. In vacuum, clamping losses through the pedestal and spokes become dominant. In the absence of viscous, sensor bandwidth (here f_m/Q_m ~100 kHz) can be tuned by altering the anchoring, which controls clamping losses.

B. Optics

To achieve ultimate detection, optomechanical sensors need very large cavity quality factors, and consequently very large intrinsic quality factor Q_i. To this end, all scattering sources must be kept as small as possible: anchors, surface rugosity, waveguide curvature. For our design, spokes are evidenced to hinder Q_i to reach its true potential. We have therefore carefully and massively investigated their influence by changing their widths from 200 nm down to the e-beam lithography limit of 50 nm. Our parametric study focused on 2 parameters: spokes width, ring-waveguide gap distances (Fig. 1).

The resonance operates in the optical C-band, which makes them compatible with off-chip optical fiber telecommunication devices. The optomechanical structures were mechanically and optically designed with finite-elements method (FEM) simulation (Fig. 3) to ensure that the coupling factor was large enough to probe the optical cavity and to control the optical quality factor Q_i.

978-1-7281-1988-5/18 $31.00 © 2018 IEEE

III. Device Fabrication Steps

Ring resonators have been fabricated in an industrial-grade clean room on 200 mm SOI wafers. Si device layer (SiTop) and buried oxide layer (BOx) are 220 nm and 1 μm thick respectively. The fabrication process, which is CMOS compatible, is based on three main steps depicted in Fig. 2. A deep UV lithography level followed by a partial silicon etching is first performed to pattern on chip optical grating couplers (Fig. 2a). A second electronic lithography level using high definition electron beam resist is then achieved by Variable Shape Beam (VSB). Optical waveguides and ring resonators are patterned by Induced Coupled Plasma (ICP) dry etching of the SiTop layer (Fig. 2b). Resist exposure during VSB as well as the ICP process have been optimized so as to minimize silicon wall roughness. Micro-rings are finally released using vapor HF etching (Fig. 2c) and are suspended above the silicon substrate by SiO_2 pedestals. It can be pointed out that VSB lithography makes it possible to fabricate more than 120,000 devices overnight on a single 200 mm wafer.

IV. Characterisation and Data Analysis

A high throughput prober (Fig. 5) has been used to acquire the optical response of 256 photonic devices, by aligning two optical fibers above them. The light is brought to the waveguide through a grating coupler to record each device transmission spectrum. To extract the $Q_{loaded} = (Q_i^{-1} + Q_e^{-1})^{-1}$ of each optical resonances without thermo-optic instability effect [16], 125 μW laser power was injected into the input waveguide from 1550 to 1600 nm with a step of 0.8 pm.

Using coupling mode theory transmission model, the finesse and contrast of each resonance are used to extract the intrinsic Q_i and extrinsic Q_e quality factors (Fig. 4). We note that Q_i is high enough to discriminate the degeneracy of the resonance due to co-existing clockwise and anti-clockwise modes [17] in the structure. Those two propagating modes hybridize into 2 stationary modes whose nodes are separated by a quarter of a wavelength. They are characterized by their own intrinsic Q-factor $Q_{i1/2}$ and can display a doublet in the spectral response. Owing to these high $Q_{loaded}C$, by analysis of the RF-spectrum of the optical transmission, we were able to measure the thermomechanical motion of our resonators (Fig. 6).

V. Results Discussion

We evidenced the spokes dissipative effect on the optical resonance. For a 4-fold spokes width reduction, the Q-factor is increased more than tenfold from Q_i=6000 to Q_i=80 000 (Fig. 7-8). We thus reach state of the art e-beam quality where intrinsic losses are nearly limited by surface scattering. FEM based simulations show that Q_i depends on the azimuthal mode order m. Due to the configuration of 3 anchoring spokes set 120° from each other, an unbalance appears for Q_i attached to the two optical modes of the doublet. We found that when the optical mode azimuthal number m is a multiple of 3, for the first mode of the doublet, the optical energy density is minimal at the intersection of the spoke and the ring. For the other mode of the doublet the energy density is maximum displaying ample scattering. In this configuration, Q_{i1} is maximum whereas the Q_{i2} of the quarter-wavelength translated mode of the doublet is low. We therefore reach a discrepancy as high as 2.3 between the Q_is of the doublet (Fig. 9). Those modes with m multiple of 3 are preferred to operate the mechanical resonators since their Q_i are the highest in this study (more than twice the average value).

Q_e and Q_i were compared, for identical devices, to evaluate the manufacturing dispersion. They are respectively in agreement with an estimated 7% and 16% difference (Fig. 10). This weak dispersion over the overall performances of these optomechanical sensors paves the way for tailored wavelength integrated devices. Mechanical experiments were also conducted in ambient conditions displaying a thermomechanical or Brownian-noise limited detection. Extremely low detection levels were calibrated down to 0.8 fm/$\sqrt{Hz}$. With a finite-element calculated k_{eff}=3.2 kN/m effective stiffness and m_{eff}=4.5 pg effective mass, minimal detectable force and mass can be evaluated around 8.6 fN/$\sqrt{Hz}$ and 450 ag/$\sqrt{Hz}$, respectively.

VI. Conclusion

We have comprehensively characterized industry compatible ring-shaped optomechanical sensors. This study along with earlier FEM simulations allows us to predict optimal coupling conditions ($Q_{loaded}C > 10^5$) and limited optical losses ($Q_i > 10^5$) for these optomechanical sensors, showing technological maturity. Our ring resonators provide very-high-frequency (>100 MHz) and low detection limit (0.8 fm/$\sqrt{Hz}$) in ambient air, almost more than two orders of magnitude below canonical electromechanical counterparts [18], paving the way for sensing devices with exquisite resolution.

Acknowledgments

This work was supported by the French National Research Agency (ANR) under the research project OLYMPIA, grant ANR-14-CE26-0019 and by the Délégation Générale de l'Armement (DGA).

References

[1] L. Ding, I. Favero et al., Appl. Phys. Lett., vol. 98, no. 11, 2011
[2] M. Aspelmeyer, K. Schwab et al., Phys. Today, vol. 65, no. 7, 2012
[3] J. Chan, O. Painter et al., Nature vol. 478, 2011
[4] R. W. Peterson, C. A. Regal et al. Phys. Rev. Lett. 116, 2016
[5] C. F. Ockeloen-Korppi, M. A. Sillanpaa, Nature 556, 2018
[6] F. Liu, M. Hossein-Zadeh et al., Opt. Express, vol. 21, no. 17, 2013
[7] A. Venkatasubramanian, W. K. Hiebert et al., Nano Lett., 2016
[8] S. Forstner, H. Rubinsztein-Dunlop et al., Phys. Rev. Lett. 108, 2012
[9] A. G. Krause, O. Painter et al., Nature Photonics vol. 6, 2012
[10] E. Gil-Santos, I. Favero et al, Nat. Nanotechnol., vol. 10, no. 9, 2015
[11] M. Borselli, O. Painter et al., Opt. Express, vol. 13, no. 5, 2005
[12] X. Sun, H. X. Tang et al., Appl. Phys. Lett., vol. 100, no. 17, 2012
[13] X. Sun, H. X. Tang et al., Optics Express vol. 19, Issue 22, 2011
[14] A. Gondarenko, M. Lipson et al, Optics Express vol. 17, Issue 14, 2009
[15] M Hamoumi, I Favero et al, PRL 120 (22), 2018
[16] P. E. Barclay, O. Painter, Optics Express vol. 13, Issue 3, 2005
[17] T. Kippenberg ; Nature Photonics vol. 4, 2010
[18] Mo Li, L. Roukes et al., Nature Nanotech. Vol. 2, 2007

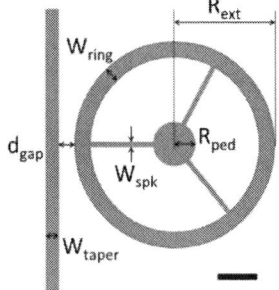 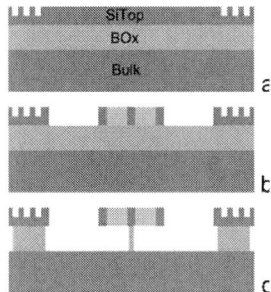

Figure 1: Scheme of an opto-mechanical resonator highlighting the design's important dimensions. Typical dimensions are R_{ext}=10 μm, R_{ped}=1.75 μm, W_{ring}=500 nm, W_{spk}=100 nm, W_{taper}=475 nm, d_{gap}=200 nm, <u>scale bar</u>: 4 μm.

Figure 2: Simplified fabrication steps. See section III in text for details.

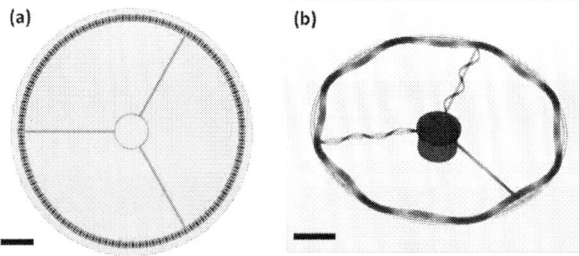

Figure 3: FEM simulated contour plots of (a) optical (λ~1550 nm) and (b) mechanical (f_m=134 MHz) mode shapes, <u>scale bars</u>: 3.5 μm.

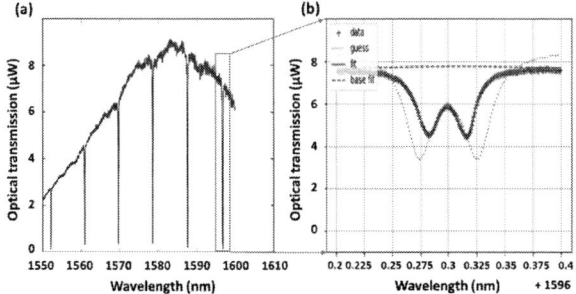

Figure 4: (a) Typical transmission spectrum of a ring resonator displaying whispering gallery modes with high-contrast. (b) Coupling-mode theory model allowing the precise fitting of the mode to extract optical parameters.

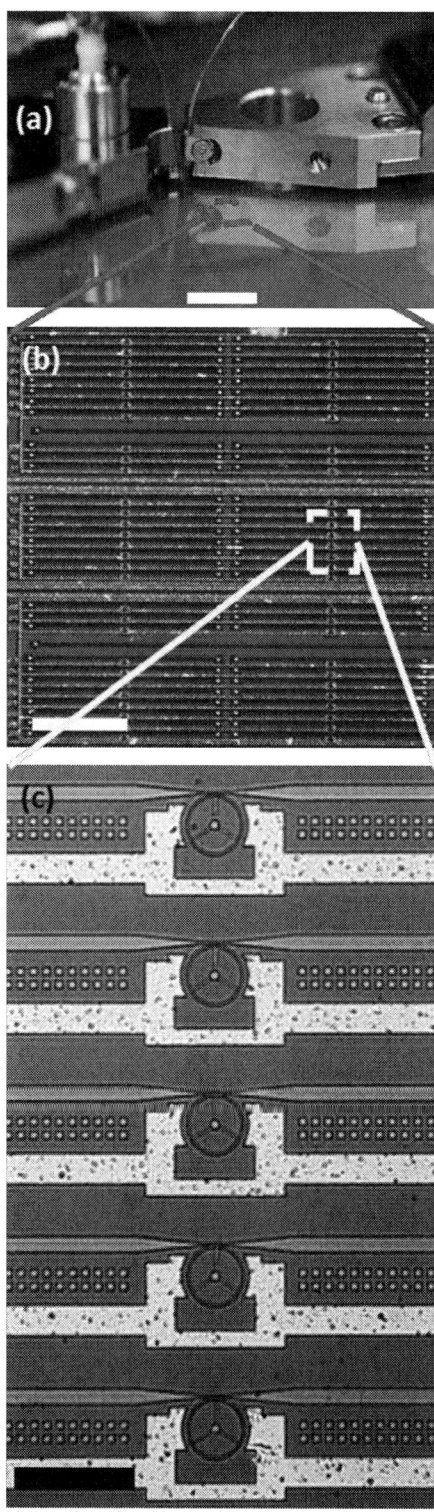

Figure 5: (a) Top–injection fiber prober, <u>scale bar</u>: 6 mm. (b) Battery of photonic circuits, <u>scale bar</u>: 430 μm. (c) Zoom on several optomechanical ring resonators, <u>scale bar</u>: 30 μm.

978-1-7281-1988-5/18 $31.00 © 2018 IEEE

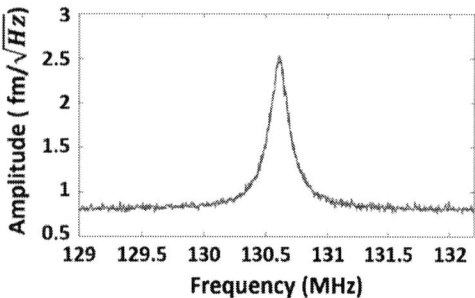

Figure 6: Selected thermomechanical vibrational spectrum of silicon ring resonator in air fitted with a Lorentzian function (in red) with Q_m=880. Calibration is made with on-resonant Brownian noise amplitude theoretical value: A_{th}= $2k_B T Q_m / \pi f_m k_{eff}$

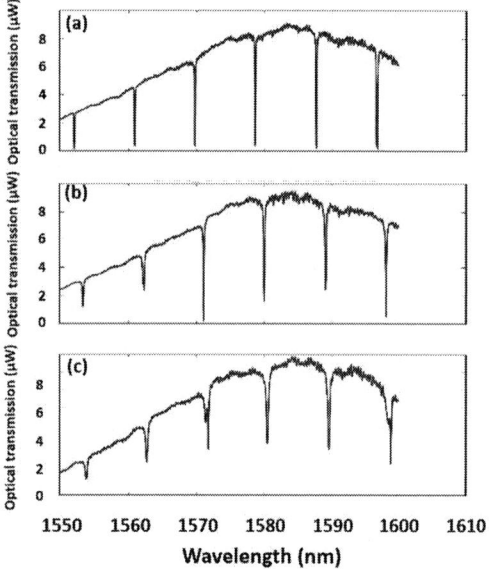

Figure 7: Transmission spectra of 3 resonators families with different spoke width demonstrating the scattering effect of the anchoring. The larger the spokes, the lower are the optical resonances intrinsic Q-factors. (a) w_{spk}=50 nm, $<Q_i>$=80 000 (b) w_{spk}=100 nm, $<Q_i>$=15 000 (c) w_{spk}=200 nm, $<Q_i>$=6000.

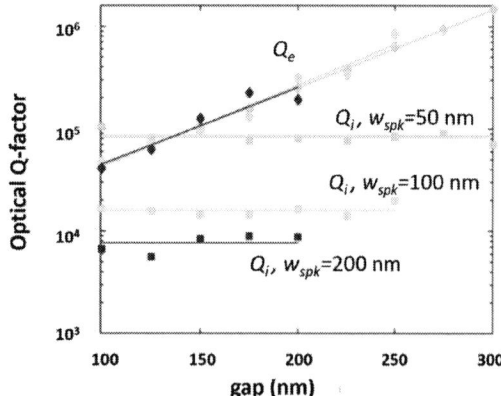

Figure 8: For w_{spk}=50 nm (green), w_{spk}=100 nm (cyan), w_{spk}=200 nm (blue), Q_i (■) and Q_e (♦) plotted versus gap distances between the waveguide and the ring. Trend of Q_is and Q_e are compatible with coupling mode theory for ring resonators. $1/Q_e$ decreases with an exponential decay constant of 58 nm. $<Q_i>$ decrease when w_{spk} increases.

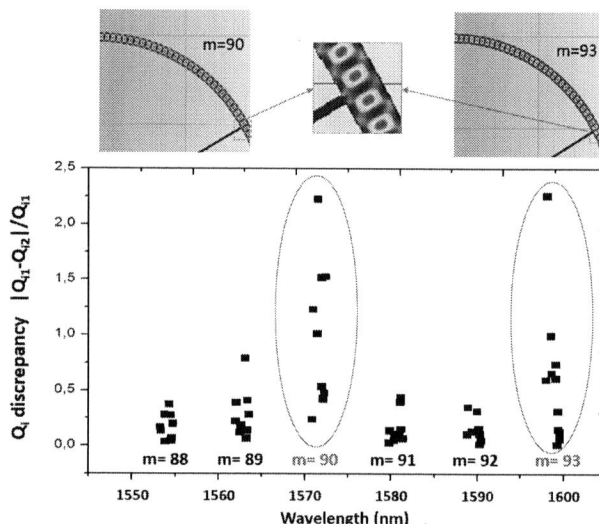

Figure 9: Discrepancy of intrinsic optical intrinsic Q-factors $Q_{i1/2}$ of the 2 modes of the doublet versus wavelength. For each m-number the scattered point correspond to several gap distances. For m=90 and m=93, one of the mode of the doublet has a high Q_i because its nodes sit where the spoke is anchored. Whereas, the other mode of the doublet has its antinode sitting at the anchoring point diminishing its Q_i.

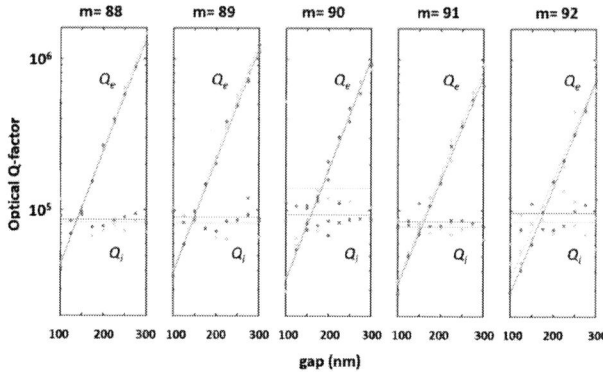

Figure 10: Global comparison of Q_i (■) and Q_e (♦) plotted versus gap distances between the waveguide and the ring, for two distinct dies (die 1: green, die 2: pink) showing good reproducibility of the optical losses for geometrically identical devices. The comparison is shown for different optical modes of different azimuthal order m. Solid lines represent averaged value for Q_i and linear fit for Q_e.

Interconnect Design and Technology Optimization for Conventional and Emerging Nanoscale Devices: A Physical Design Perspective

(*Invited*)

D. Prasad[1,2] and A. Naeemi[1]

[1]Department of Electrical and Computer Engineering, Georgia Institute of Technology, Atlanta, GA, USA.

Email:azad@gatech.edu

[2]Arm Research, Austin TX, USA.

Abstract—Interconnect parasitics severely limit the performance and power dissipation in modern circuits at the advanced process technology nodes. Hence, device-level advances must be complemented with appropriate interconnect technology and design innovations for effective enablement at the circuit and system levels. This paper highlights the impact of device technologies on the optimal interconnect design and circuit-level metrics. The FinFET and Tunnel-FETs are studied by building fully placed-and-routed physical designs. The impact of device and interconnect technology co-optimization on circuit performance, power, and variability is shown for a range of emerging devices.

I. INTRODUCTION

The evolution of technology, microarchitecture and design practices with technology scaling has introduced several challenges and opportunities for innovation in modern VLSI design. As the device and wire dimensions are scaled aggressively in the advanced process nodes, maintaining reliable device functionality and managing wire parasitics with efficient design implementation have caused the system-level performance to almost plateau. Copper interconnect resistivity has increased rapidly beyond the 14/16 nm technology node due to size effects, leading to the exploration of alternate materials for tight-pitch metal levels in the Back-End-Of-Line (BEOL) stack, like cobalt and ruthenium, which have less pronounced size effects [1]. On the device front, new exotic transistors have been proposed catering to diverse applications introducing new challenges in various facets of VLSI design to achieve the target Performance-Power-Area (PPA) [2]. With the increasing importance of interconnects in determining the technology scaling benefits, we argue the need for evaluating the device and interconnect interactions, and co-optimizing interconnect and device technologies to extend the Moore's scaling paradigm.

II. OPTIMAL INTERCONNECT DESIGN FOR HIGH-PERFORMANCE FINFET

The FinFET transistor that is designed for enhanced electrostatic control of the channel exhibits lower resistance and increased device capacitance due to its non-planar structure in comparison to planar technology. This has exacerbated the interconnect resistance problem, particularly at lower metal levels in the BEOL stack [3]. The increased

device capacitance amplifies the importance of wire resistance in circuit timing. Similarly, the reduced device resistance has alleviated the influence of wire capacitance on circuit timing [3]. In older technologies, prior to the onset of copper size effects, the interconnect design at the local metal levels was primarily focused on minimizing the wire capacitance while ensuring sufficient resistance to electromigration [4, 5]. The FinFET device Resistance-Capacitance (RC) combined with severe copper size effects have weighed the wire resistance to be far more important to circuit performance, while reducing the impact of wire capacitance at tight-pitch metal levels [3]. The shift in the interconnect design paradigm in the FinFET technology is demonstrated by studying fully placed-and-routed designs at the 7nm technology node (Tables I-III) [3]. Traditional interconnect geometry (Case I in Table III) is defined with equal interconnect width and spacings [3]. The definition of wire geometry is typically a tradeoff between interconnect parasitics, reliability, and manufacturability. However, to take advantage of the shift in the interconnect trends in the FinFET era, an alternate wire geometry is explored that alleviates wire resistance at the cost of larger wire capacitance (Fig. 2), primarily with the goal of improving performance [3]. Lowering the wire resistance increases the circuit performance by up to 1.5× (Fig. 3). In addition, the buffering required to meet the circuit timing is reduced with lower wire resistance; hence, lowering the overall power dissipation despite an increased wire capacitance [3].

Multi-patterning regimes using immersion lithography and EUV technology are employed to pattern narrow wires in the nano-scale regime which have rapidly increased the design rules making standard-cell design extremely difficult [1] Furthermore, variation in interconnect dimensions critically affects the system-level metrics, particularly due to the sensitivity of copper wire resistance to the wire dimensions [6]. In addition to the wire resistance problem that is worsened by the higher FinFET capacitance, it is also found that wire resistance variability is larger than wire capacitance variability [6]. Therefore, the proposed wire geometry with reduced wire resistance and higher wire capacitance (in Fig. 2-b) alleviates the impact of interconnect variability on circuit timing, in addition to improving the circuit timing at a target yield. This is studied by inducing lithography-dependent statistical interconnect variability in placed-and-routed designs (Fig. 3) [6]. A proposed novel hybrid quadruple-patterning lithography

978-1-7281-1988-5/18 $31.00 © 2018 IEEE

regime (LELE+SADP) can potentially lower wire resistance variation, improve the circuit performance at a target yield, and substantially improve the variation in timing (3σ) [6, 7].

III. INTERCONNECT DESIGN FOR LOW-POWER DEVICES

In contrast to the high-performance multi-gate transistors, the low-power transistors are characterized by low gate capacitance and the capability of operating at lower supply voltages. For example, the Tunnel-FET (TFET) devices have been widely investigated thanks to their potential steep subthreshold swing (SS). However, these devices exhibit substantially higher device resistances compared to thermionic FETs [8-10]. This increased device resistance amplifies the importance of wire capacitance in circuit delay, whereas the reduced device capacitance alleviates the influence of wire resistance on circuit delay [11]. This is validated by building physical designs based on vertical-TFET (vTFET) technology libraries (Fig. 5), and using an alternate wire geometry that alleviates wire capacitance at the cost of wire resistance (Fig. 6). It is demonstrated that the energy-delay product (EDP) is improved despite worsening the wire resistance with the proposed wire geometry at aggressively scaled dimensions (Fig. 7). Furthermore, Fig. 8 shows that, both, the performance and power improve with the proposed wire geometry (Case 2), beyond which, the performance and power begin to degrade because of the excessive wire resistance (Cases 3-5).

To evaluate the sensitivity of the TFET-based circuits to wire resistance, the copper wires are replaced with various tungsten wire configurations (Table IV). Tungsten is particularly chosen as it is more resistive than copper, but exhibits superior tolerance to electromigration at narrow wire dimensions [12]. The sheet resistance is computed in Fig. 9 for copper and tungsten interconnects with varied barrier liner thickness assumptions as tungsten-based wires do not demand barrier liners as thick as those needed for copper wires [3, 12]. It is illustrated in Fig. 10 that vTFET-based physical designs can tolerate higher resistance of the tungsten wires without taking a hit on circuit performance and power. This study showcases the opportunity to replace copper with higher resistance materials that exhibit superior reliability properties or those that are easier to fabricate, broadening the scope and advantage areas for ultra-low power devices. Furthermore, interconnect variation critically impacts FinFET-based circuits [6]. However, the TFET-based circuits are less sensitive to interconnect resistance variation, and can tolerate up to 2× variation in wire resistance (see Fig. 9 and Fig. 10).

IV. INTERCONNECT DESIGN TECHNIQUES FOR NOVEL DEVICE ARCHITECTURES

The potential (a priori) performances of a diverse set of emerging device proposals have been evaluated in [2, 17], and some of them are listed in Table V. These devices have unique implications on the interconnect design paradigm due to their diverse RC characteristics. The device RC values presented in Table V take parasitic RC into account, which play a pivotal role in the device performance. Based on a Fanout-Of-4 (FO4) circuit setup, the optimal interconnect geometry is obtained for each device technology to achieve maximum performance (Fig. 11). The spectrum of devices in Fig. 11, ranging from more resistive to more capacitance dominated require the wire geometry to be either optimized for wire capacitance or wire resistance, respectively, to achieve optimal performance. This trend is delineated in Fig. 11, showing the unique implications of device innovations on interconnect technology and design.

V. NEED FOR ACCURATE WIRE LENGTH MODELING

In recent years, accurate benchmarking of technology has gained substantial importance as a diverse set of potential technology options is being pursued. New devices are required to be evaluated in the context of large systems and demand accurate representation of interconnection networks. In prior sections, interconnects are modeled/optimized either based on fully placed-and-routed designs which are very time intensive and not always a plausible option, or they are based on simplistic FO4 circuits. However, in real systems, there are various circuit topologies which comprise unique device and wire configurations that simplistic representations fail to holistically capture. A popular method, based on Rent's rule (a relationship between number of gates and connections), has been employed in the past few decades to model the interconnect lengths in various design families [13]. This approach provides the capability to extrapolate the potential performance of a device in a large system. However, it has been shown that these methods are highly erroneous (~90% error) in projecting performance due to underestimation of the interconnect length in modern designs [13]. A modernized Rent's rule proposed in [13], aims to capture the critical facets of modern VLSI design which include standard-cell design, microarchitecture and the Electronic-Design-Automation tools which determine the design PPA [14]. The new Rent-based model facilitates accurate representation of wirelengths and is validated against a suite of state-of-the-art commercial designs across technology nodes (Fig. 12). These models substantially improve the existing benchmarking methodologies and facilitate futuristic technology pathfinding [14].

VI. CONCLUSIONS

In this paper, the implications of device characteristics on interconnect technology, design, and optimization is discussed, providing a holistic view of the impact of device innovation on interconnects, PPA, and reliability. Adapting interconnects to suit the device properties further enhances the device benefit at the circuit and system level. The FinFET-based designs are more sensitive to the interconnect resistance than the interconnect capacitance. However, low-power devices like TFET, reduce the impact of the wire resistance on the circuit and can tolerate higher variability in wire resistance, even at the nanoscale regime. Capturing the device and wire interactions, early co-optimization, and accurate wire length modeling is required for future technology pathfinding.

ACKNLEDGEMENTS

This work was supported by National Science Foundation and in part by the Semiconductor Research Corporation (SRC) and DARPA.

978-1-7281-1988-5/18 $31.00 © 2018 IEEE

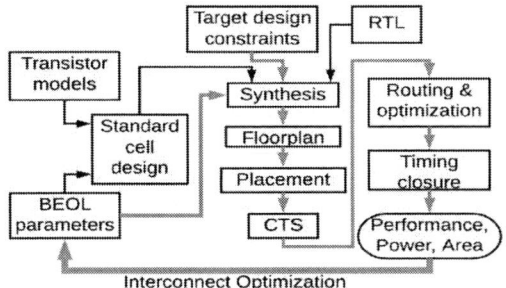

Fig. 1. VLSI-design flow utilized in this study to build GDSII-level layouts and optimize interconnects.

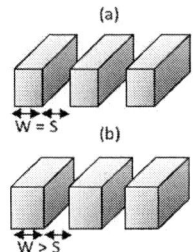

Fig. 2. (a) Wire geometry defined by the ITRS, and (b) optimized geometry proposed for high-performance FinFET designs at a constant wire pitch.

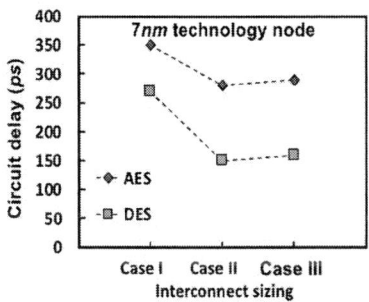

Fig. 3. Circuit delay obtained from fully placed-and-routed designs (AES and DES circuit blocks), for three cases of wire geometry.

Table I. BEOL stack assumptions at the 11- and 7-*nm* technology nodes.

11nm technology node			
Metal-levels	Metal width (W) in *nm*	Metal thickness (T) in *nm*	Metal pitch (P) in *nm*
M1-M3	17.5	35	35
M4-M6	35	70	70
7nm technology node			
M1-M3	10.8	21.8	21.8
M4-M6	21.8	43.6	43.6

Table II. FinFET Technology assumptions at the 11- and 7-nm nodes. Device models are obtained from [14].

Parameters	11nm node	7nm node
ITRS node	2018	2022
VDD	0.75	0.7
Transistor channel length	14	11

Table III. Interconnect geometries studied for FinFET-based circuits. The metal pitch (P), and wire thickness to width aspect ratio (AR), are maintained constant for the three cases (Case I-III).

Interconnect sizing at constant pitch	Wire width (W)	Wire spacing (S)
Case I	W	S=W
Case II	1.1xW	0.9xS
Case III	1.15xW	0.85xS

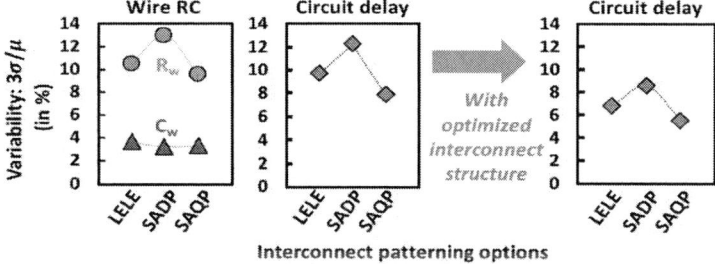

Fig. 4. Interconnect RC variability and its impact on circuit delay variability using the ITRS defined wire geometry, versus the optimized interconnect geometry (Case II on the right), in the 11*nm* technology node. The variability in circuit timing is studied on fully placed-and-routed designs for Litho-Etch-Litho-Etch (LELE) and Self-Aligned-Double/Quadruple-Patterning (SADP/QP).

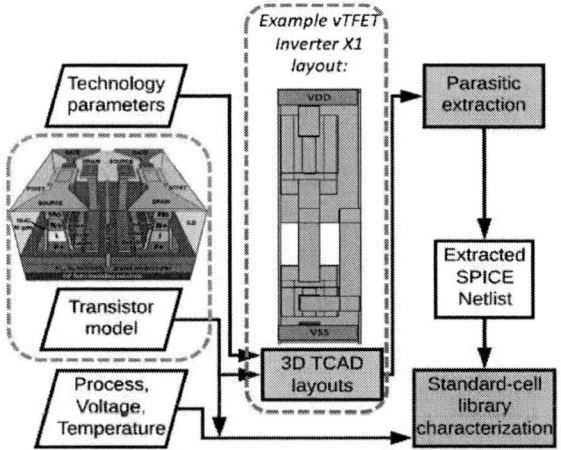

Fig. 5. Standard-cell library characterization methodology for vTFET-based layouts. The parasitic extraction is done using a 3D TCAD tool [15], and the library characterization is done using Silicon Smart Synopsys tool. The vTFET image is taken from [9].

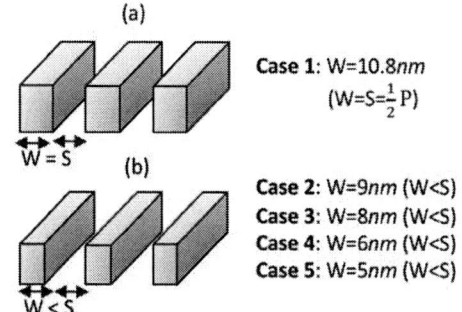

Fig. 6. (a) Wire geometry defined by ITRS, and (b) optimized geometry (Case 1), and (b) optimized wire geometry proposed for low-power vTFET-based designs, at constant P and AR (Cases 2-5).

978-1-7281-1988-5/18 $31.00 © 2018 IEEE   105

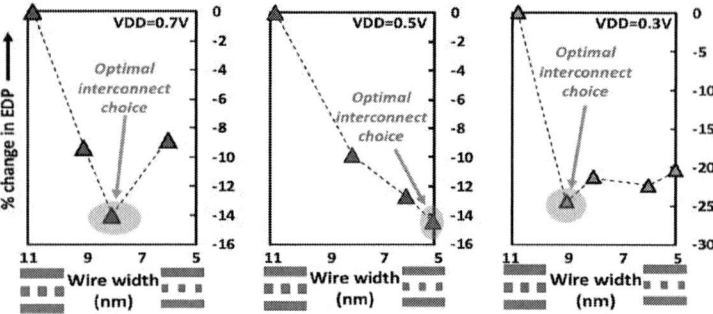
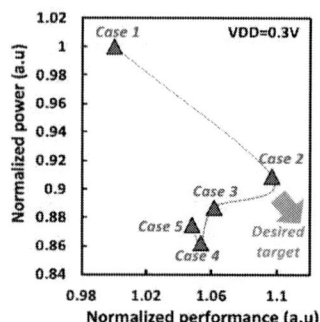

Fig. 7. EDP of a place-and-routed design (AES circuit block) based on vTFET libraries using the different interconnect geometries (Cases 1-5 in Fig. 6).

Fig. 8. Performance versus power plot for vTFET-based physical design (AES), for the wire geometries defined in Fig. 6 (Case 1-5).

Table IV. Interconnect assumptions to study the impact on vTFET-based placed and routed designs. Here, λ is the mean free path, and ρ is the bulk resistivity.

Wire composition	Wire material	Metal width (M1-M3)	Barrier thickness	(specularity p, reflectivity R)	λ	ρ
①	Copper	8nm	TB1: 1.8nm	(0, 0.43)	40nm	1.8μΩ-cm
②	Tungsten	8nm	TB1: 1.8nm	(0.3, 0.25)	28nm	8.7μΩ-cm
③	Tungsten	8nm	TB2: 1nm	(0.3, 0.25)	28nm	8.7μΩ-cm
④	Tungsten	8nm	TB3: 0.5nm	(0.3, 0.25)	28nm	8.7μΩ-cm

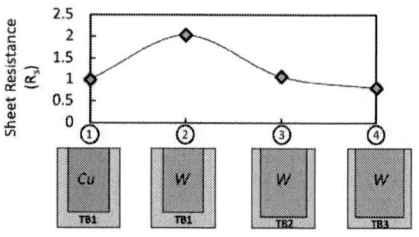

Fig. 9. Impact of interconnect setup on the sheet resistance of the interconnect (in a.u).

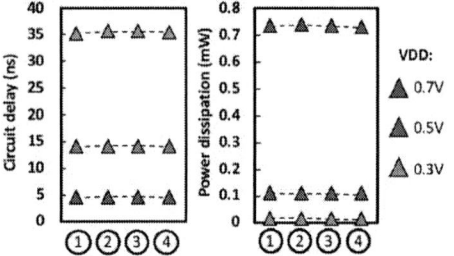

Fig. 10. Impact of interconnect material and barrier liner thickness on vTFET-based physical design timing and power.

Table V. List of the exotic devices considered in this paper. Device parameters are taken from [17] and parasitic capacitances are estimated by Raphael [16].

Device type	Naming convention	VDD (V)	Device resistance (R_o) in kΩ	Device capacitance (C_o) in aF
Planar high-performance CMOS	CMOS HP	0.73	9.4	61.4
Planar low-power CMOS	CMOS LV	0.3	97	40.3
Homo-junction III-V TFET	HomJTFET	0.2	139	66.6
Hetero-junction III-V TFET	HetJTFET	0.4	18.7	66.6
GaN TFET	GaNTFET	0.4	6.15	57.2
Transition Metal Dihalchogenide TFET	TMDTFET	0.2	46.5	76.5
Van der Walls FET (Black Phosphorous)	vdWFET-BP	0.3	148.4	39.3

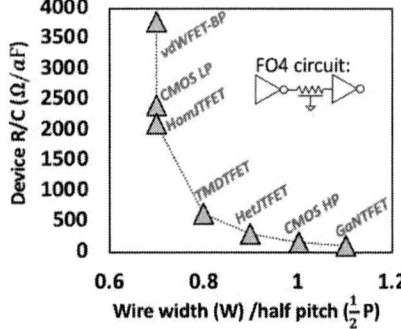

Fig. 11. Optimal interconnect geometry obtained by modifying the wire width and spacing, for a spectrum of exotic devices on the benchmarking landscape. The metal pitch and aspect ratio are maintained constant and the interconnects is assumed to be 5μm long.

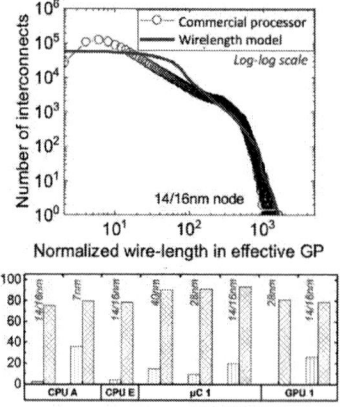

Fig. 12. Wirelength modeling: Histogram of wirelengths obtained from the model versus processor data (top), and accuracy of average wirelength projection (bottom) [14].

REFERENCES

[1] Z. Tokei, *IEDM Short Course*, 2017
[2] D. E. Nikonov et al., *JXCDC*, 1, pp 3-15, 2015.
[3] D. Prasad et al., *IEEE TED*, 62, pp 1-7, 2015
[4] D. Sekhar et al., *ICCAD*, 2007
[5] H. B. Bakoglu, Addison-Wesley, 1990.
[6] D. Prasad et al., *IEEE TED*, 64, pp 1236-1253, 2017.
[7] K. Lin et al., *IITC*, 2018.
[8] H. Lu, et al., Notredame TFET model, NanoHub.
[9] R. Pandey et al., *VLSI Technology*, 2015.
[10] H. Liu et al., *DRC*, 2012.
[11] D. Prasad et al., *IITC*, 2017.
[12] W. Steinhogl et al., *Microelectronic Engineering*, 82, pp 266-272, 2005.
[13] D. Prasad et al., *IEDM*, 2017.
[14] D. Prasad et al., *DAC*, 2018.
[15] S. Sinha et al., *DAC* 2012.
[16] Raphael, Synopsis 3D Field Solver.
[17] C. Pan, et al, *JXCDC*, 3, pp 101-110, 2017.

978-1-7281-1988-5/18 $31.00 © 2018 IEEE

Mechanisms of Electromigration Damage in Cu Interconnects

C.-K. Hu,[1] L. Gignac,[2] G. Lian,[3] C. Cabral,[2] K. Motoyama,[1] H. Shobha,[1] J. Demarest,[1] Y. Ostrovski,[1] C. M. Breslin,[2]
M. Ali,[3] J. Benedict,[3] P. S. McLaughlin,[1] J. Ni,[1] X. H. Liu,[1]

[1] IBM Research at Albany Nanotech, Albany, NY, USA, email: haohu@us.ibm.com
[2] IBM T. J. Watson Research Center, Yorktown Heights, NY, USA, [3] IBM Systems, Hopewell Junction, NY, USA

Abstract—Mechanisms of electromigration (EM) damage in Cu interconnects through various CMOS nodes are reviewed. Pure Cu and Cu alloy interconnects that were used down to 14 nm node can no longer satisfy the electrical current used for 10 nm node and beyond in high-performance ICs. Cu interconnects with a metal cap should be used. Cu interface diffusivity with EM activation energy of 1.6 eV was found to be the dominate EM factor in Cu lines with a Co liner and cap. The median lifetime of 7 or 10 nm node Cu with TaN/Co liner and Co cap is predicted to be over ten thousand years at 140°C with 1.5×10^7 A/cm^2. However, the resistivity size effect and the difficulty of scaling barrier/liner layer without defects can limit the Cu BEOL roadmap below the 7 nm node.

I. INTRODUCTION

On-chip Cu wiring in the back end of the line (BEOL) has been developed for several decades [1]. Scaling BEOL dimensions has degraded electromigration (EM) reliability with the same metallization [2] and increased Cu resistivity (ρ) [3]. EM is defined as atom diffusion under the influence of an electric potential gradient. The size effects in EM and ρ were caused primarily by increased contributions from EM-induced mass flow and electron scattering with interfaces and grain boundaries. The EM lifetime has further degraded by the decrease in the void volume required to cause EM failure. The Cu interconnect resistance was further increased by increasing the volume fraction of diffusion barrier/liner in wires which are required to a provide good reliable chip. In this review article, we will present the mechanisms of EM damage in Cu, the effect of barrier/liner on interconnect R, and the factors limiting the Cu interconnect roadmap.

II. EXPERIMENT

Damascene Cu BEOL were processed down to the 7 nm technology node with 36 nm line/space pitch. Single and dual damascene processes were used. The Cu damascene process consisted of physical vapor deposition (PVD) TaN/Ta and chemical vapor deposition (CVD) Co thin metal liner, followed by a PVD Cu seed layer, and by Cu electrochemical plating deposition (ECD). The Cu lines were capped with selective CVD Co on the Cu line surface. The microstructure was examined by transmission electron microscopy (TEM). The dimensions of the Ta, Cu, and Co layers were measured from TEM images and electron energy loss spectroscopy (EELS) maps. The Cu line cross sectional areas and resistivity

ρ were estimated by the temperature dependent coefficient of resistivity (TCR) technique. Two–level EM structures consisted of either M1 to V1/ M2 or W M0 to V0/M1, where V0 and V1 were the vias connecting M0 to M1 and M1 to M2, respectively. The current densities j were 10 to 300 mA/μm^2. The sample stress temperatures ranged from 233 to 435°C.

III. RESULTS AND DISCUSSION

A. Interconnect Microstructure and Resistance

EM mass flow is influenced by Cu microstructure. For Cu lines beyond the 65 nm node with height/width aspect ratios (A.R.) > 1, the Cu microstructure has changed from large bamboo grains to small grains mixed with the bamboo-like grain structures. For lines wider than 90 nm or A.R. $\leq$ 1, the extension of the overburden grains into damascene trenches will give rise to a near bamboo-like grain structure. Fig. 1(a) and (b) are TEM images along 50 nm wide lines with A.R. = 2 and 1, respectively. The polycrystalline-bamboo and bamboo-like grain structures were shown in (a) and (b), respectively. The line with A.R. ~ 1 in (b) had the advantage of forming a near-bamboo grained line which was also predicted by a Cu grain growth model [4]. Fig. 2 shows the Cu grain orientation map of a 24 nm wide line with A.R. = 2.5 with a polycrystalline-bamboo microstructure and a high fraction of twin boundaries.

Fig. 3 (a) shows a plot of R/L as a function of Cu or Co line cross sectional areas, *A*, at T = 21°C, where R is the line resistance and L is metal line length. R/L= ρ/A can be expressed as R/L = (ρ_o+B/A$^{0.5}$)/A. [5] The extracted values of ρ_o and B for Cu are 1.8 $\mu\Omega$-cm and $88 \times 10^2 \Omega$-nm^2, respectively, and for Co are 6.5 $\mu\Omega$-cm and $151 \times 10^2 \Omega$-nm^2, respectively. Fig. 3(b) shows the estimated R/L vs trench area for Cu and Co damascene lines; if 40% or 60% of the metal line were occupied by the barrier/liner in Cu and the Co line had only 30% or was linerless. If the Cu line trench was 40% occupied by barrier/liner, Co R/L would still be higher than Cu in 7 nm and 5 nm nodes. For the case of 5 nm node, Co interconnects with 30% of the trench occupied by liner would be close to Cu with 60% of the trench occupied by liner. Thus, the extendibility of Cu in 5 nm node for fine lines relies on a good 3 nm thick barrier/liner, that occupies 50-60% of the trench, and satisfies BEOL reliability requirements.

B. Electromigration

EM mass flow (drift velocity) for a long line with a diffusion blocking boundary at the cathode end can be

978-1-7281-1988-5/18 $31.00 © 2018 IEEE

expressed by $v_d = [D_{eff}/(k_B T)] Z^* e \rho j$, where e is the absolute value of the electronic charge, Z^* is the apparent effective charge number, ρ is the metallic resistivity, D_{eff} is the effective diffusivity of atoms diffusing through a metal line, T is the absolute temperature, and k_B is the Boltzmann constant. For a polycrystalline line, EM drift velocity can be $v_d = Z_{GB}^*(D_{GB}/k_B T)(\delta_{GB}/d)e\rho j$. For a bamboo-polycrystalline line with a metal cap and the length of two near bamboo grains < Blech short length [6], the dominated EM v_d is expressed as $[Z_I^*(D_I/k_B T)2\delta_I(1/h+1/w)+(1-n_I) Z_B^*(D_B/k_B T)]e\rho j$, [5]. If the same interface diffusivity were around the line, where h, w, and d are the metal height, width, and grain size, respectively; the subscripts B, I and GB refer to bulk, the Cu/metal liner and cap interfaces, and grain boundary, respectively; δ and D denote the effective width and diffusivity, respectively; and n_I is the fraction of atoms at interfaces. EM lifetime τ would be $L_{cr} = \int_0^\tau v_d dt$, where L_{cr} is the EM-induced critical void size to cause interconnect R increase and failure.

• EM in Cu-CNT composite or defected Cu interconnects

EM in 2 μm wide Cu/carbon nanotubes (CNT) composite single damascene two-level interconnects was investigated. For Cu/CNT wafer, the Cu and CNT were filled into the M2 trenches by using a co-electroplating technique. The M2 lines were passivated with 30 nm thick a-SiC$_x$N$_y$H$_z$ layer and annealed at 400°C. Fig. 4 (a) and (b) are top down SEM images of pure Cu and Cu/CNT composite lines after the passivation layer was removed by FIB, respectively. Near bamboo-like and polycrystalline Cu grains structures were revealed on pure Cu and Cu/CNT composite lines, respectively. As compared to pure Cu, t50 of Cu/CNT were degraded by factors of 2, 3 and 5 at T = 348, 300 and 250°C, respectively. The EM activation energy Ea in Cu/CNT composite lines was found to be 0.70 eV, corresponding to GB diffusion activation energy [7]. This result contradicted with a previous report finding decreased mass flow by 100x with Ea = 2 eV [8].

Fig. 5(a) is a TEM image of an unstressed 24 nm wide line with a Co cap. Many isolated circular Cu fill voids and CMP Cu surface gouges are shown. The EM lifetimes from these defected samples were found to be close to the samples without any visible voids. These defects apparently did not have a large impact on EM mass flow. The EM-induced V0 and M1 line voids were shown in Fig. 5 (b) and (c), respectively. Some circular voids seem to grow a large void size but did not migrate.

• EM in 36 and 48 nm pitch samples

Fig. 6 is a plot of cumulative lifetime distribution using a log-normal scale. The observed lifetimes follow a log-normal distribution having values of deviation, σ, around 0.4-0.5 for thick Co liner and multi-modal for the thin Co liner, respectively. Fig. 7 (a) shows a TEM image, and (b)-(c) show two EELS images taken at and near the cathode end of 23 nm wide M2 using TaN/3 nm thick Co liner with a selective Co cap tested at 340°C for 1046h. This data shows that some of the migrated Cu atoms were filled by Co atoms from Co liner and/or cap and formed Co precipitates. A Co liner thickness variation of only 1 nm significantly impacts the integrity of BEOL fabrication and consequently EM quality. These results

indicate that the present PVD TaN barrier layer, CVD Co liner, PVD Cu seed and ECD Cu for wafer fabrication can reach a limit for providing a high quality EM wafer.

• EM through various technology nodes

The measured t50 as a function of 1/T are plotted in Fig. 8 for various CMOS generation wafer types [5, 9]. The data of EM in 18 nm V0 to 54 nm wide A.R. < 1 line are also included. Ea of 0.75 eV was observed for polycrystalline lines with or without a CoWP cap. Pure Cu lines without metal cap had Ea of 1.0 eV and 0.9 eV for above and below 90 nm wide lines, respectively. For the case of the Cu(Mn) or Cu(Al) alloy lines without metal cap, Ea was found to be 1.2 and 1.0 eV for near bamboo and bamboo-polycrystalline lines, respectively [7]. Ea was measured to be about 2 eV for 0.2 μm wide near-bamboo lines with TaN/Ta and capped with CoWP. Ea of 2 eV was close to E_B in Cu bulk 2.1 eV [10]. For 10nm and 7nm nodes with TaN/Co barrier/liner and Co cap (stars), Ea of 1.5-1.7 eV was observed. Ea of 1.6 eV was also found in 18 nm diameter V0 to 54 nm wide M1. Similar Ea from 20 nm to 54 nm wide lines indicated Ea of 1.6 eV was determined by the Cu/Co interface diffusivity. For 20 to 24 nm wide Cu lines with TaN/Co liner and Co cap, t50 is predicted to be over ten thousand years at 140°C with j = 1.5×10^7 A/cm^2.

• Limiting Factors in Cu Interconnect Roadmap

EM in a single damascene Cu line with atomic layer deposition (ALD) MnO$_x$ or MnO$_x$/PVD Ta barrier/liner on M2 was studied. Fig. 9(a) and (b) are TEM images of EM tested lines with ALD MnO$_x$ and MnO$_x$/Ta liner, respectively. The t50 for TaN/Ta, MnO$_x$/Ta, and MnO$_x$ liner wafers degraded from 60h to 16h to 6 h, respectively. The Cu in V1 migration through MnO$_x$ to M2 shown in (a) indicated that MnO$_x$ was not a diffusion blocking boundary. Fig. 10 (a)-(c) are TEM line cross section image from 50 nm wide Cu line, EELS element map and TEM image of EM tested 20 nm wide line with Co cap, respectively. The interface sidewall voids in (a) and discontinuous barrier/liner in (b) were shown which were probably the main reason for poor EM. Increasing barrier/liner thickness by 1.5 nm would improve EM lifetime from 0.1 h to 100 h at T = 380°C. The barrier/liner thickness on sidewalls ≥ 3 nm at the present technology is probably too thick for the next node. Through Co self-forming barrier layer [11], metal-oxide or -nitride, a-TaWSiC [12], graphene [13], ALD barrier, Cu "seed reflow" [14], and improved ECD [15] etc. for thin barrier/liner have been reported.

IV. CONCLUSION

This paper reviews the impacts of on-chip interconnect scaling effects on microstructure, resistance, barrier/liner, and EM. EM mass transport in Cu bamboo-polycrystalline interconnects with Co cap is controlled by interface and grain boundary diffusivities. The net mass flow is dominated by Cu/Co interface diffusivity and EM-induced interconnect failure locations most likely occur in the weak Blech length links near the cathode end of the line. We also presentated that Cu/CNT composite line and Cu with ALD MnOx liner significantly degraded EM lifetime. Indentifying a thin barrier/liner ≤ 3nm combined with impoved Cu seed and ECD processes would allow the use of Cu for future nodes.

Fig. 1. Cross section TEM images along 50 nm wide lines with two A.R., a) A.R. = 2 and b) A.R. =1, where the grain structure in the line will depend on the line A.R values and the thermal budget. Sections of multiple small grains are observed through the thickness in (a) and a single Cu grain extending through the metal thickness in (b).

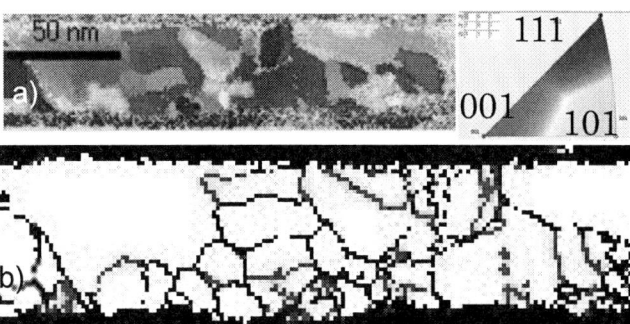

Fig. 2. a) Cu orientation map along 24 nm wide line with A.R.=2.5 using TEM precession-assisted diffraction recognition to obtain diffraction pattern information over a scanned area. The red curves in b) are the twin boundaries which are not the fast diffusion paths

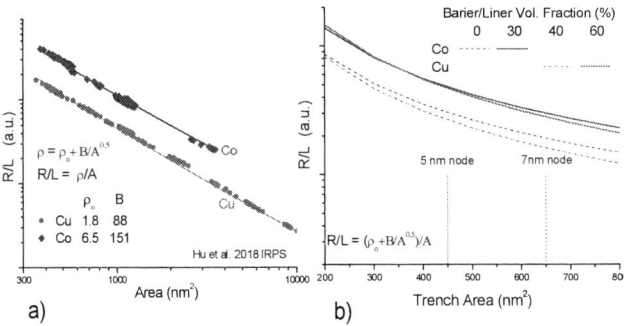

Fig. 3. (a) R/L vs. Co or Cu damascene line cross section area, A. (b) Estimated R/L vs. trench area for Cu and Co lines, the green and blue curves are the estimated curves of Cu and Co with various % of liner volumes occupied in the trenches, respectively. Cu and Co damascene lines were fabricated with a similar barrier/liner thickness that resulted in a similar main conductor line cross section area as shown in Fig. 3(a). The width and heights used in these calculations for (b) are 18 nm and 36 nm, and 13nm and 36 nm for 7 and 5 nm nodes, respectively.

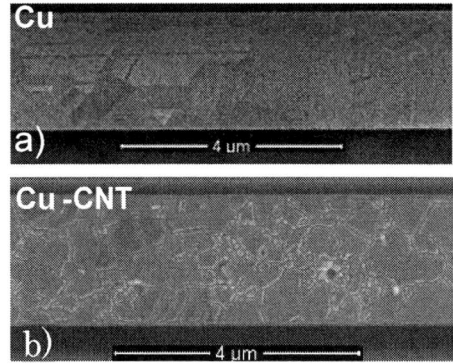

Fig. 4. Top-down SEM images of 2 μm wide line with a-SiC$_x$N$_y$H$_z$ cap removal by FIB: (a) pure Cu and (b) Cu/CNT composite line, respectively.

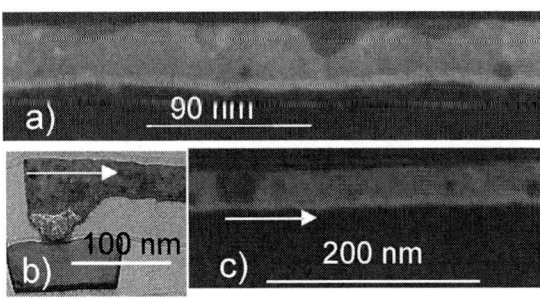

Fig. 5. TEM images along 24 nm wide Cu line with Co capped (10 nm node). a) is an unstressed sample. b) and c) are TEM images along the cathode end of V0/M1 line after EM stressed at T = 380°C with 100 mA/μm² for 128h. The gouged surface and isolated voids in (a) were believed from CMP and from Cu fill voids, respectively. The EM-induced via void and line voids located > 1 μm from the cathode end of the line were shown in (b) and (c), respectively. The arrows are the directions of electron flows.

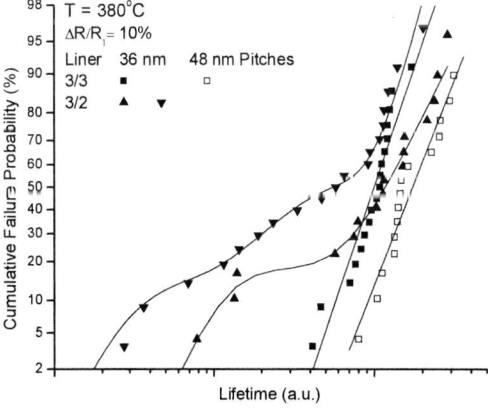

Fig 6. Cumulative lifetime probability vs EM lifetime from Cu lines with 36 nm (solid symbols) and 48 nm (open squares) pitches, plotted in a log-normal scale at T = 380°C. The barrier/liner thicknesses for two wafers were 3 nm TaN/3 or 2 nm Co for 36 nm pitch (7nm node) and 3 nm TaN/3 nm Co for 48 nm pitch (10nm node), respectively. The reported barrier/liner thicknesses were the nominal thicknesses on the field

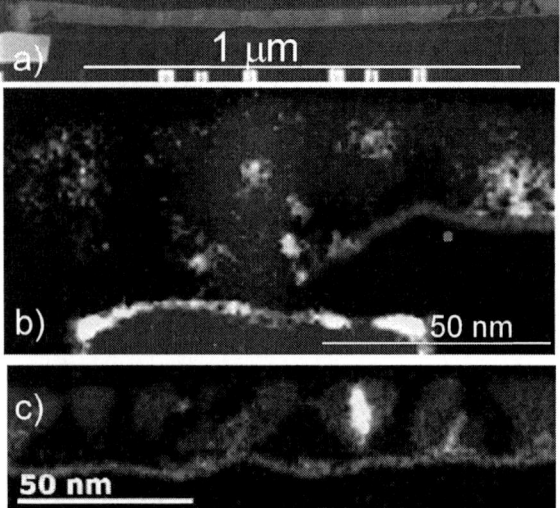

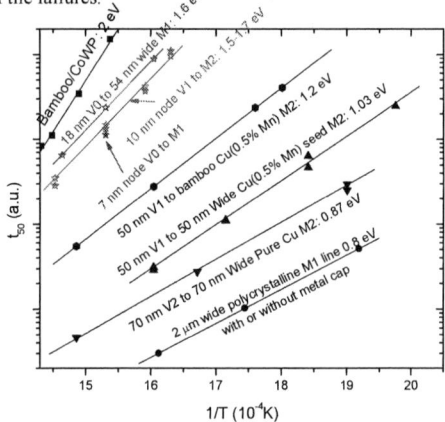

Fig 7. a)-c) show physical analyses of a 23 nm wide line after EM test at 340°C for 1046h; a) is a TEM image, b) and c) show EELS images of EM-induced voids at and located 1 μm away from the cathode end of M2, respectively. Cu, Co and Ta are in blue, yellow and magenta colors, respectively. These EM-induced voids locations indicate that the cathode end of the line/via contact does not have a critical effect on lifetime and that the weak Blech length links near line cathode end were most likely the cause of the failures.

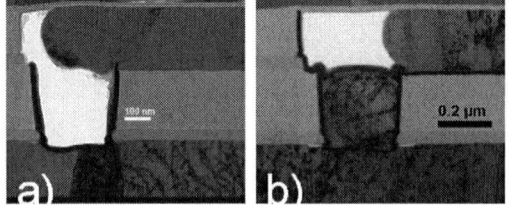

Fig. 8. Plot of t50 vs 1/T for various Cu microstructures with and without a metal cap. All data points are bamboo-polycrystalline lines, unless specified. Pure Cu and Cu alloys with barrier/liner TaN/Ta were generally used down to 45 nm node and 32 to 14 nm node, respectively. Data from 7 nm and 10 nm node wafers are in green and blue, respectively.

Fig. 9. TEM cross section images of EM tested two level Cu single damascene lines with M2 barrier/liner ALD MnOₓ for 6 h in (a) and ALD MnOₓ/Ta for 16 h in (b), respectively, at T = 300°C. M1 and V1 used the TaN/Ta liner and single damascene process.

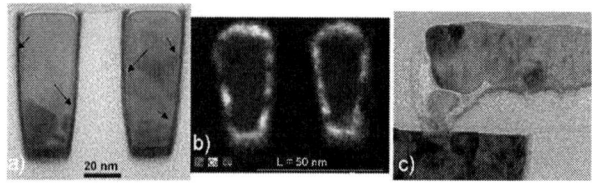

Fig. 10 are (a) TEM Cu line cross section images from 50 nm wide line (14 nm node) with 2-3 nm thick TaN/Ta on sidewall, (b) EELS map of Cu line with Co cap and (c) TEM image of 20 nm wide Cu line with Co cap tested at T = 380°C for 0.1h, respectively. The arrows show the interfacial sidewall voids run from the top to bottom of the line.

ACKNOWLEDGMENT

This work was performed by the Research Alliance Teams at various IBM Research and Development Facilities. We also like to express our gratitude to IBM Nanotech in Albany and Material Research Laboratory in Yorktown Heights, NY for wafer fabrication.

REFERENCES

[1] "Cu-based metallization and interconnects for ULSI applications", 1995 Thin Solid Films, vol. 262, Special issue

[2] C.-K. Hu, et al, "Effects of overlayers on electromigration reliability improvement for Cu/Low k Interconnects", 2004 IEEE IRPS p.222

[3] W. Zhang, et al.," Analysis of the size effect in electroplated fine copper wires and a realistic assessment to model copper resistivity", J. Appl. Phys. 101, 063703 (2007)

[4] J.-K. Jung et al., "Grain growth simulation of damascene interconnects: effect of Overburden Thickness", Japan. J. Appl. Phys. Vol. 43, No. 6A, p. 3346–3352, 2004

[5] C.-K. Hu, et al, "Future on-chip interconnect metallization and electromigration", 2018 IRPS 4.1

[6] I. A. Blech, "Electromigration in thin aluminum films on titanium nitride", *J. Appl. Phys.*, **47**, 1203 (1976)

[7] C.-K. Hu, et al., "Electromigration in Cu(Al) and Cu(Mn) damascene lines", J. Appl. Phys. Vol. 111, p. 093722, 2012

[8] C. Subramaniam, et al., "One-hundred-fold increase in current carrying capacity in a carbon nanotube-copper composite", Nature Communication. vol. 4, Article no 2202, Jul. 2013

[9] C.-K. Hu, et al., "Materials and scaling effects on on-chip interconnect reliability", 2013 MRS, vol 1559. doi:10.1557/opl.2013.872

[10] N. L. Peterson, J. Nucl. Mater. Vol. 69&70, p.3, 1970

[11] T. Nogami, et al.," Through-Co Self forming barrier for Cu/ULK BEOL", 2015 IEEE IEDM P.8.1.1

[12] R. Wongpiya, et al.,"Amorphous thin film TaWSiC as a diffusion barrier for copper interconnects, Appl. Phys. Lett. 103, 022104 (2013)

[13] L. Li, et al., "BEOL compatible graphene/Cu with improved EM lifetime for future interconnects", 2016 IEEE IEDM, pp.240-243

[14] K. Motoyama, et al., "PVD Cu Reflow Seed Process Optimization for Defect Reduction in Nanoscale Cu/Low-k Dual Damascene Interconnects", J. Electrochemical Soc. 160, (12) D3211-D3215 (2013)

[15] X. Sun et al., "Experimental study of PVD Cu/CVD Co bilayer dissolution for BEOL Cu interconnect applications", ECS Transactions, 80 (4) 297-309 (2017)

Interconnect metals beyond copper: reliability challenges and opportunities

K. Croes, Ch. Adelmann, C.J. Wilson, H. Zahedmanesh, O. Varela Pedreira, C. Wu, A. Leśniewska, H. Oprins, S. Beyne, I. Ciofi, D. Kocaay, M. Stucchi and Zs. Tőkei

imec, Kapeldreef 75, 3001 Leuven, Belgium, phone: +32 16 28 21 16, email: kristof.croes@imec.be

Abstract—Reliability challenges of candidate metal systems to replace traditional Cu wiring in future interconnects are discussed. From a reliability perspective, a key opportunity is electromigration improvement: due to their high melting point and slower self-diffusion kinetics, higher current carrying capabilities are possible. Also, the higher cohesive energy and better resistance to oxidation of some metals potentially allows for barrierless integration, although adhesion properties must be carefully optimized. Besides avoiding small grain pinning and enabling high aspect ratio trench fill, the main processing challenges are identified to be a) avoiding seam voids, b) adhesion, c) CMP and d) disruptive metal etch. Main reliability challenges are related to higher mechanical stresses and higher joule heating which could lead to delamination during further processing and packaging and to enhanced electromigration in nearby metal lines.

I. INTRODUCTION

Scaling of Cu interconnect dimensions is becoming increasingly difficult due to the fast rise in line/via resistances [1-2]. Strategies to extend Cu to smaller dimensions have focused on i) increasing the Cu cross-sectional area by thinning barrier/liner materials or using SFB's [3-4], ii) using thin liner materials to aid filling and compensate for thinner seeds [5], iii) using novel reflow or more conformal deposition to avoid voiding [6] and iv) breaking the traditional dual damascene fill using via "pre-fill" where the via is first selectively filled and then the process is completed with single damascene Cu [7-8]. Each of these process changes have an impact on interconnect reliability and performance gains must be carefully co-optimized with reliability in mind and often there is a trade-off.

Although the industry will try to extend Cu as long as possible [9], continued scaling of Cu interconnects using the aforementioned techniques cannot continue indefinitely. An option is to replace Cu with metals that show equivalent or improved via and/or line resistance as well as reliability at small dimensions [10-12].

In this paper, we focus on the processing/reliability challenges/opportunities of novel metal systems.

II. METAL SELECTION AND BENCHMARK

For the initial selection and benchmark of alternative metals, an approach has been developed that is based on a combination of *ab initio* calculations and the experimental assessment of the resistivity at small dimensions. First, the product of the mean free path and the bulk resistivity is calculated as a figure of merit using ab initio techniques using the method described in [13-14]. A small value of this product suggests low resistivity at small dimensions. In addition, the melting point is used as a proxy for electromigration performance. The result for all elemental metals is shown in fig. 1 and compared to literature results if available [13]. Second, the thin film resistivity is measured as a function of film thickness. The data in fig. 2 show that indeed a weaker thickness dependence of the resistivity is observed for Pt-group metals with smaller figure of merit values, leading to lower absolute resistivity below about 8-10nm. Third, narrow lines of the selected metals are fabricated using a metal-spacer process [15,16]. The resistivity vs. area is then deduced using the TCR method. Fig. 3 shows that the best performance has been obtained for Ir and Rh lines with Ru and Co showing similar performance.

III. PROCESSING CHALLENGES AND OPPORTUNITIES

Alternative metals enable fill techniques, such as CVD or ALD offer the prospect of barrierless integration. These techniques can fill narrow lines, but such conformal processes can leave seam voids that are revealed at CMP or remain buried. Although barrierless integration is appealing for resistance, the adhesion of the metal to the dielectric is critical. Another key challenge with the integration of these alternative metals is CMP, where corrosion or polish non-uniformity may limit yield. Continuing to scale damascene technology is still limited by aspect ratio fill and grain-pinning in small trenches where metal etch of blanket films can enable both grain size optimization and high aspect ratio lines. This metal etch has been tested in Cu [17] and recently in Ru [18] yielding lines of approximately 10nm width and 60nm height. Today, such a process is experimental and multilevel integration yet to be demonstrated however metal etch offers potentially very low resistance at scaled dimensions.

IV. RELIABILITY CHALLENGES AND OPPORTUNITES

A. Dielectric breakdown/metal drift

Cu metallization options require barrier and liner systems to prevent copper drifting into the dielectric and to enable Cu fill, respectively. While in wider lines and vias, barriers >2nm thickness could be used without much resistance loss, such thick barriers cannot be tolerated in small dimensions [2]. A reliability issue with scaled barriers is metal drift: ionized metal drifts through the dielectric due to electric fields and contaminate the dielectric leading to leakage and dielectric breakdown. This phenomenon requires careful characterization as the process of metal drift depends on the applied field and temperature during the test [19]. As an example, fig. 4 shows dielectric breakdown times vs. field for Co deposited on SiO_2. Regions linked to intrinsic breakdown, filament growth and filament formation can be observed depending on the chosen

978-1-7281-1988-5/18 $31.00 © 2018 IEEE

stress condition. As the conditions where these different mechanisms are dominant depend on the dielectric material and thickness, a revision of today's test methodologies is required to qualify technologies using alternative metals with thin barriers [19].

Metal drift also depends on the metal itself. Fig. 5 shows failure times vs. field of planar capacitors using varying top electrodes (Cu, Co or Ru) and different polarities for voltage stress. When applying a negative voltage to the top-electrode, the breakdown of the dielectric is intrinsic and leads to good reliability. When reversing the polarity, metal drift can occur. For the case with Cu and Co, a clear reduction in reliability is observed, where for Ru, no reliability reduction and thus no metal drift is observed [20]. Because of its high cohesive energy and its better resistance to oxidation, Ru allows for a barrierless integration although ensuring sufficient adhesion may require the implementation of thin glue layers or surface treatments.

B. Electromigration

Cu electromigration scaling limits are determined by a lower critical void volume and by higher contributions of grain boundaries in narrow lines. Fig. 6 shows intrinsic DC-scaling limits of Cu electromigration in long lines, typically used for power rails (methodology for predictions shown in [21]). Maximum current carrying capabilities decrease with line width with a ~75% decrease in J_{max} when scaling from 22nm to 10nm ½pitch.

Recently, extremely good electromigration reliability is demonstrated for Co and Ru for wider lines (>36nm pitch) [11-12]. Using a new test vehicle [15,16], we determined fuse currents of Co and Ru lines with area's <<100nm^2 (Fig. 7). Compared to Cu, fuse currents for Co and Ru are 2-8x higher even in these dimensions. Besides the proposed correlation between their higher melting points with its cohesive energy and therefore to the slower formation of vacancies and slower self-diffusion kinetics [10], the better electromigration performance of Ru could also be explained by p-type electromigration as voids have been observed at the anode end of the line (Fig. 8, [22]). Note that for wider lines with thicker barriers, a rather high E_a of 1.9eV for electromigration was reported [11], while, using our LF-noise measurements [23], our scaled samples without barrier and rough surface show an E_a of ~1.0eV [24] suggesting Ru-SiO$_2$ to become a dominant diffusion path.

C. Mechanical stress

During processing, high temperature steps are required. When cooling down, stresses are generated due to CTE-mismatches between different materials which can lead to delamination at weak interfaces. Our modelling suggests that the mechanical stresses induced by alternative metals are higher compared to Cu. Fig. 9 shows stresses induced in lines during processing for different alternative metal schemes (methodology shown in [25]). 10-20% higher stresses are expected for Ru and Co, respectively. Also, for via pre-fill with Co, a higher stress gradient in the via could lead to worse reliability in stress induced voiding tests (Fig. 10, [8]). Furthermore, metal intermixing must be avoided in order to limit stress build-up.

D. Heating

Given that, compared to Cu, Co and Ru can withstand higher currents and that the thermal conductivity of Co and Ru is 3x lower, the self-heating of the lines could cause a) higher temperatures in adjacent metal layers and thus a higher electromigration risk and b) stresses at the interface between the metal and its dielectric eventually leading to delamination. Initial quantification of this heating is done by sending high currents through single lines and by estimating the heating based on RT and RI trends (Fig. 11). Fig. 12 shows the heating slope for different materials as a function of area. Highest heating is observed for Ru, followed by Co and Cu with a >10x difference in slope. Also, Ru liners lead to higher joule heating (5x compared to Co). The effect of the surrounding dielectric also impacts the heating in the line, where a significantly higher heating is found for porous low-k or airgap compared to dense low-k and SiO$_2$ (Fig. 13).

E. Ru scheme with air gaps

Besides line/via-R and electromigration, dielectric breakdown is a potential showstopper for further scaling. Fig. 14 shows predicted lifetime contours for 10y dielectric breakdown lifetimes for 12nm ½pitch lines (methodology explained in [26]). Further scaling with existing schemes will become very difficult, where for non-spacer defined patterning (ρ=0), σ_{LER}<0.7nm and for spacer defined patterning (ρ=0.8), σ_{LER}<1nm is required for this pitch.

An alternative approach is to use a scheme with barrierless Ru interconnects and air gaps which can be enabled by direct etch [18]. Although the performance and reliability of this scheme needs to be assessed on Si hardware, potential reliability showstoppers have been studied through modelling, where field distributions and energy release rates have been assessed. Fig. 15 shows that fields in an air gap scheme are comparable to traditional damascene schemes with dielectric, now the highest fields are in the air gap. Although the breakdown strength of air in these narrow spaces might be lower than expected [27], the elimination of the metal CMP step is potentially leading to an improved reliability. Also, the simulated energy release rate G of an induced crack during packaging is 6x higher for a damascene scheme with dielectric compared to a direct etch scheme with air gap suggesting improvements in chip-package interaction for the airgap scheme (Fig. 16).

V. CONCLUSIONS

We assessed the processing and reliability challenges and opportunities of alternative metals. For processing, a) avoiding grain pinning/seam voids in high aspect ratio trenches, b) metal etch, c) CMP and d) providing adhesion in barrierless schemes have been identified as challenging steps. For reliability, electromigration, the possibility of barrierless Ru and an integration scheme with barrierless Ru with air gaps have been identified as opportunities. Identified challenges are a) the development of proper test methods to asses and understand of metal drift, b) the higher stresses and stress gradients generated by alternative metals and c) line heating which can have a negative impact on adjacent lines and which can induce normal and shear stresses at interfaces.

978-1-7281-1988-5/18 $31.00 © 2018 IEEE

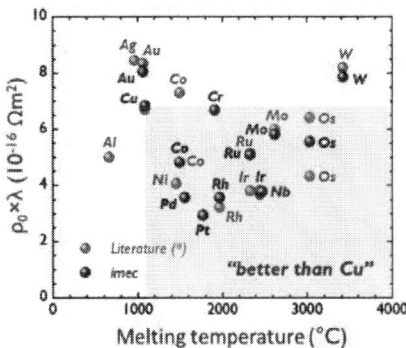

Figure 1. Mean free path times bulk resistivity versus melting temperature for different metals

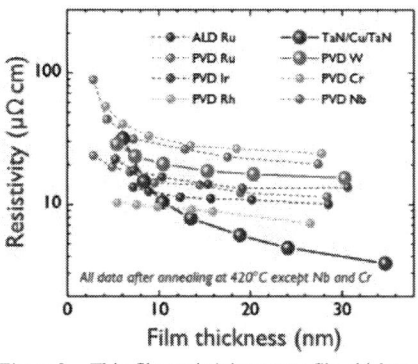

Figure 2. Thin film resistivity versus film thickness for different metals

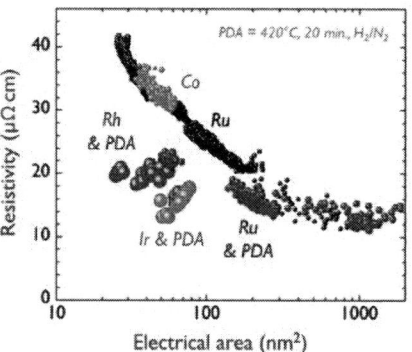

Figure 3. Resistivity versus electrical area for different metals

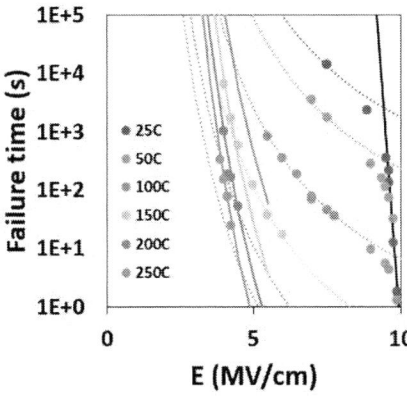

Figure 4. Dielectric breakdown time versus field for Co deposited directly on SiO$_2$ Full black line: Intrinsic breakdown. Dotted lines: Filament growth. Full coloured lines: Filament formation

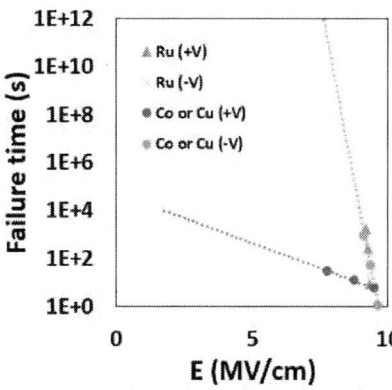

Figure 5. Dielectric breakdown time versus field for a) Cu or Co or b) Ru deposited directly on SiO$_2$ for different stress polarities: positive/negative bias to top electrode (+V/-V), where +V potentially leads to metal drift

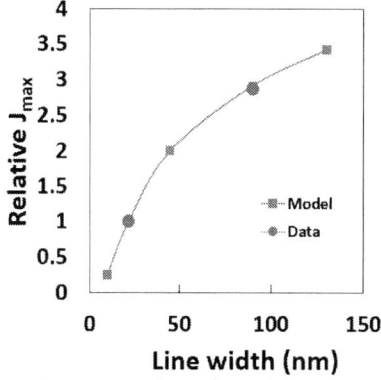

Figure 6. Predicted J$_{max}$ trends of long Cu lines as a function of line width with respect to 22nm ½pitch reference

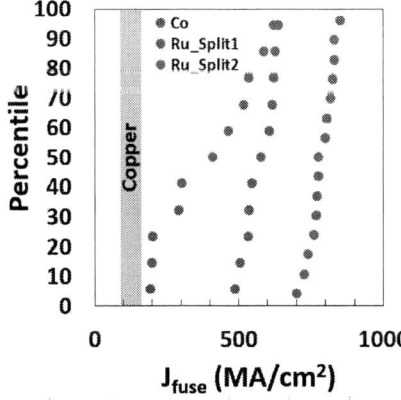

Figure 7. Fuse currents for Co and Ru lines in <<100nm^2 area lines. Cu benchmark is on wider lines.

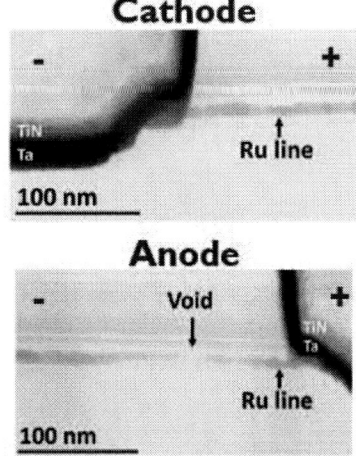

Figure 8. Ru void observed at anode end of the line in 50nm^2 lines without barrier

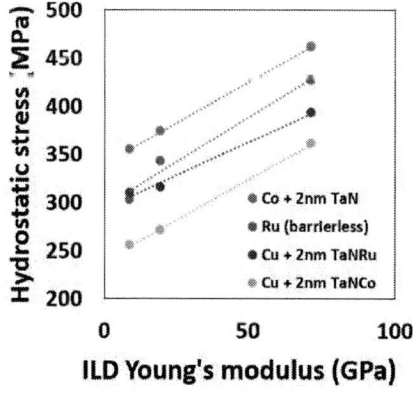

Figure 9. Stresses induced during processing for different alternative metal schemes benchmarked to Cu with a TaNCo or TaNRu barrier/liner system

978-1-7281-1988-5/18 $31.00 © 2018 IEEE

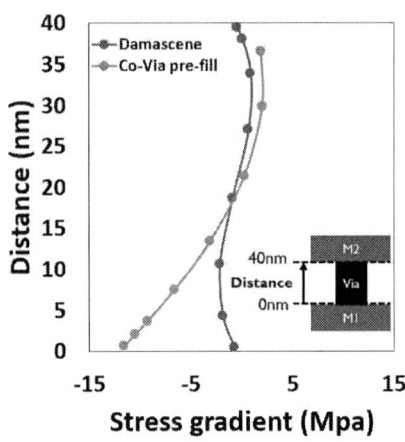

Figure 10. Stress gradients along the height of a Co via compared to standard TaNCo+Cu

ΔT calculated from TCR and R vs. I

$$\Delta T = \frac{\Delta R}{R_0} \cdot \frac{1}{TCR} = B \cdot I^2$$

$$B \sim f(\kappa, \rho, \text{metal/dielectric dimensions})$$

Joule heating slope(B) ↑ → heating ↑

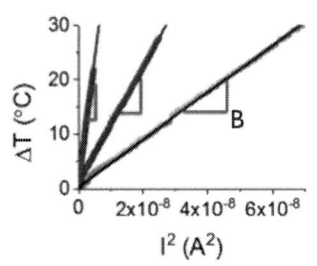

Figure 11. Methodology to quantify heating in lines

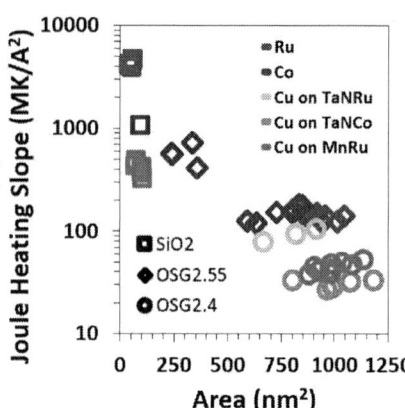

Figure 12. Joule heating slope for different materials as a function of cross-sectional area

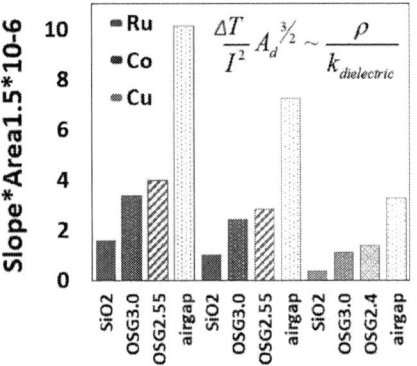

$$\frac{\Delta T}{I^2} A_d^{3/2} \sim \frac{\rho}{k_{dielectric}}$$

Figure 13. Effect of surrounding dielectric on heating in metal lines. SiO$_2$ and porous low-k is measured. Airgap and OSG3.0 is extrapolated.

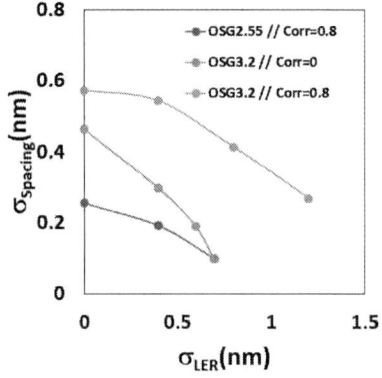

Figure 14. Predicted 10y dielectric lifetime contours for 12nm ½pitch lines. Correlated LWR is considered for spaced-defined patterning

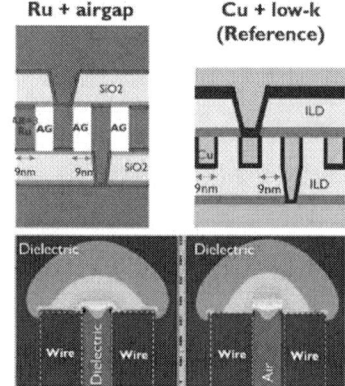

Figure 15. Field distributions for a damascene scheme with a dielectric compared to a direct etch scheme with airgaps

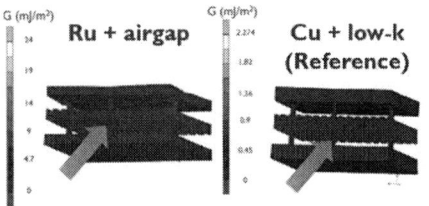

Figure 16. Simulated energy release rate G of an induced crack during packaging for a damascene scheme with a dielectric compared to a direct etch scheme with airgaps

REFERENCES

[1] I. Ciofi, et al, "Impact of wire geometry on interconnect RC and circuit delay", IEEE Transactions on Electron Devices, Vol. 63, Nr. 6, p. 2488, 2016

[2] I. Ciofi, et al., "Modeling of via resistance for advanced technology nodes", IEEE Transactions on Electron Devices, Vol. 64, Nr. 5, p. 2306, 2017

[3] C. Witt, et al, "Testing The Limits of TaN Barrier Scaling", IEEE IITC, p. 7.4, 2018

[4] N. Jourdan, et al, "CVD-Mn/CVD-Ru-based Barrier/Liner Solution for Advanced BEOL Cu/Low-k Interconnects", IEEE IITC, p. 37, 2016

[5] M.H. van der Veen, et al, "Conformal Cu electroless seed on Co and Ru liners enables Cu fill by plating for advanced interconnects", MAM, p. 247, 2016

[6] M.H. van der Veen, et al, "Extending the Cu metallization and alternatives", AMC, 2017

[7] M.H. van der Veen et al, "Cobalt bottom-up contact and via prefill enabling advanced logic and DRAM technologies", IEEE IITC, p. 25, 2015

[8] O. Varela Pedreira et al, "Electromigration and Thermal Storage study of Barrierless Co vias", IEEE IITC, p. 48, 2018

[9] D. Edelstein, "CMOS/Cu BEOL Technology in Manufacturing: 20 years and Counting", IEEE IITC, p. 39, 2018

[10] C. Adelmann, et al., "Alternative metals for advanced interconnects", IEEE IITC, p. 173, 2014

[11] C.-K. Hu, et al, "Future on-chip interconnect metallization and electromigration", IEEE IRPS, p. 4F.1, 2018

[12] F. Griggio, et al, "Reliability of dual-damascene local interconnects featuring Cobalt in 10nm logic technology", IEEE IRPS, p. 6E.1, 2018

[13] D. Gall, "Electron mean free path in elemental metals," Journal of Applied Physics, Vol. 119, No. 8, p. 085101, 2016

[14] S. Dutta et al., "Thickness dependence of the resistivity of platinum-group metal thin films," Journal of Applied Physics, Vol. 122, No. 2, p. 025107, 2017

[15] S. Dutta et al., "Highly Scaled Ruthenium Interconnects," IEEE Electron Device Letters, p. 949, 2017

[16] S. Dutta et al., "Sub-100 nm² Cobalt Interconnects," IEEE Electron Device Letters, Vol. 39, No. 5, p. 731, 2018

[17] L. Wen, "Direct etched Cu characterization for advanced interconnects", IEEE IITC, p. 173, 2015

[18] D. Wan, et al, "Subtractive Etch of Ruthenium for Sub-5nm Interconnect", IEEE IITC, p. 2.4, 2018

[19] C. Wu, et al, "New insights into metal drift induced failure in MOL and BEOL", IEEE IRPS, p. 3A.1, 2018

[20] O. Varela Pedreira, et al, "Reliability Study on Cobalt and Ruthenium as Alternative Metals for Advanced Interconnects", IEEE IRPS, p. 6B.2, 2017

[21] H. Zahedmanesh, et al, "Prediction of the electromigration limits of Cu nano-interconnects using a comprehensive physics-based model", to be submitted to IITC/IRPS 2019

[22] S. Beyne, et al, "The first observation of p-type electromigration failure in full Ru interconnects" IEEE IRPS, p. 6D.7, 2018

[23] S. Beyne, et al, "1/f noise measurements for faster evaluation of electromigration in advanced microelectronics interconnections", JAP, Vol. 199, Nr. 18, p. 184302, 2016

[24] S. Beyne, et al, "Validation and application of Low-frequency noise measurements for characterization of electromigration in highly scaled interconnects", Submitted to IEDM2018

[25] H. Zahedmanesh, et al, "Airgaps in advanced nano-interconnects; mechanics and impact on electromigration", JAP, Vol. 120, Nr. 9, p. 095103, 2016

[26] D. Kocaay, et al, "Impact of LER, L2L Overlay and Via Misalignment on BEOL TDDB Reliability down to 24nm Metal Pitch", to be published

[27] V. Babrauskas, "Arc breakdown in air over very small gap distances", Interflam, Vol. 2, p. 1489, 2013

Microstructure Evolution and Effect on Resistivity for Cu Nanointerconnects and Beyond

Szu-Tung Hu[1], Linjun Cao[2], Laura Spinella[3] and Paul S. Ho[1]

[1]Microelectronics Research Center and Texas Materials Institute, University of Texas at Austin, Austin, TX 78712

email: paulho@mail.utexas.edu

[2] GLOBALFOUNDRIES, Malta, NY 12020

[3]National Renewable Energy Laboratory, Lakewood, CO 80401

Abstract— In this paper, we investigate the microstructure evolution in Cu, Co and Ru nanointerconnects and the scaling effect on resistivity. The scaling effect on microstructure of Cu interconnects was analyzed to the 24 nm linewidth for the 14 nm node using a high-resolution TEM precession microdiffraction technique. The TEM study was supplemented by a Monte Carlo simulation to investigate grain growth in nanointerconnects based on local energy minimization. The scaling effect on electrical resistivity was analyzed for Cu, Ru and Co nanointerconnects, taking into account the contributions from surface and grain boundary scatterings. The results for Cu and Co are consistent with recent experiments.

Keywords—Cu nanointerconnects, scaling effects, microstructure, resistivity.

I. INTRODUCTION

The continued scaling of Cu low k technology is facing serious challenges imposed by basic limits from materials, processing and reliability. This has generated great interests recently to investigate the limit of Cu nanointerconnects and to develop alternatives, particularly Co and Ru nanointerconnects beyond the 10nm node. The performance and reliability are important for the development of nanointerconnects in addition to challenges from processing complexity and manufacturing cost. In this paper, we investigate the microstructure evolution in Cu, Co and Ru nanointerconnects and the scaling effect on resistivity, a key factor contributing to the RC delay and performance.

II. METHODS OF STUDY

The scaling effect on microstructure of Cu interconnects was analyzed to the 24 nm linewidth for the 14 nm node using a high-resolution TEM precession microdiffraction technique with capabilities to map the orientation and size distribution of individual grains with a resolution of 3-5 nm [1]. The TEM study was supplemented by a Monte Carlo simulation to study grain growth in nanointerconnects based on local energy minimization.

The scaling effect on electrical resistivity was analyzed for Cu, Ru and Co nanointerconnects, taking into account the contributions from surface and grain boundary scatterings.

III. RESULTS AND DISCUSSION

Results from the microstructure study have revealed a consistent trend of microstructure evolution with scaling in Cu nanointerconnects. In Figure 1, we show first the scaling effect on the grain size and orientation, viewing from the direction normal to the line width. At 120nm line width, grain growth starting from the overburden leads to relatively large grain size with near bamboo microstructure. At the 70nm line width, grain growth becomes constrained with less effects from the overburden while small grains emerge from the bottom as well as the sidewalls. This trend continues to the 22 nm node with 40 nm linewidth.

The results of the precession electron microscopy show a systematic scaling effect on the overall grain orientations as summarized in Figure 2. These results together with the change observed in grain size distributions provide a detailed description of the scaling effect on the Cu grain structure. At the 180 nm linewidth, the microstructure is dominated by the (111) grains grown from the trench bottom due to the lowest grain boundary energy. As linewidth reduces to 120 nm, the growth of the (111) grains begins to shift to the trench sidewalls, reflecting the increase in the surface-to-volume ratio and become interface energy controlled. With further reduction to the 45 nm linewidth, the growth of the (111) grains shifts again to along the trench length direction, a trend remaining to the 22 nm linewidth. At the same time, more small grain aggregates appeared, indicating further dominance of the interface energy in competition with the strain energy as the surface-to-volume ratio of the line increases.

Additional information was obtained from the precession microscopy about the twin boundaries which can be readily identified although only a small fraction was found to be the Σ3 coherent boundaries. In general, twin formation is driven by minimization of the strain energy due to Cu elastic anisotropy by converting (111) to (200) grains during annealing. The large elastic anisotropy of Cu (about 3x), however, makes the direct (111) to (200) transition energetically difficult; instead a series of twins with descending magnitude of the elastic energy are formed to facilitate the transition where a large fraction of the twins are not the coherent twins. Details of the energetics of twin formation and TEM observation have been reported [2]. The amount of twin boundaries continues to decrease with scaling, reaching to ~1% at 40nm linewidth as shown in Figure 3a. The overall grain size distribution is bi-modal, indicating an abnormal grain growth with increasing number of small

978-1-7281-1988-5/18 $31.00 © 2018 IEEE

grains emerging near the trench bottom as the linewidth decreases, as shown in Figure 3b.

The Monte Carlo simulation on grain structure was performed to analyze the scaling effect on the microstructure of Cu, Ru and Co nanointerconnects. The simulation is based on total energy minimization, taking into account the orientation-dependent grain boundary, strain, and interface energies in order to examine the effect of scaling and material properties on grain growth [3, 4]. For Cu interconnects, the simulation was extended to 7 nm linewidth and the results are consistent with the observed grain structure as a function of linewidth. Overall, there is a clear trend of degradation in the microstructure characteristics of Cu nanointerconnects due to scaling with less structural order and increasing number of small grains, both of which have significant implications on the electrical resistivity as analyzed below.

The scaling effect on electrical resistivity was analyzed for Cu, Ru and Co nanointerconnects, following the approach of a recent study [5], taking into account the contributions from surface and grain boundary scatterings:

$$\rho = \rho_0 \frac{3}{8} C (1 - p) \left(\frac{1}{h} + \frac{h}{A} \right) \lambda + \rho_0 \left[1 - \frac{3}{2}\alpha + 3\alpha^2 - 3\alpha^3 \ln(1 + \frac{1}{\alpha}) \right]^{-1} \quad (1)$$

Here the first term represents the contribution from surface scattering based on the Fuchs-Sondheimer model with ρ_0 : bulk resistivity, C: geometry based constant, p: specularity of electron collision with surface, h: line height, A: cross-section area of interconnect and λ: bulk electron mean free path. The second term is based on the Mayadas-Shatzkes model for grain boundary scattering with α=λR/[G(1-R)], G: Average Grain size and R: electron reflection coefficient of the grain boundaries. Calculation has been carried out for Cu, Ru and Co interconnects and the results are summarized in Figure 4.

The results obtained for Cu and Co nanointerconnects are in good agreement with the experimental results recently reported [5, 6]. The scaling effect on the resistivity was extended beyond 7 nm linewidth based on grain structures deduced from the Monte Carlo simulation. The calculated results of Ru show a similar trend but seem to be consistently below those measured and the reason is not clear. Overall, there is a clear trend of continued increase in the resistivity of nanointerconnect with scaling as expected, which can be traced to the degradation of the grain structure and increasing surface and grain boundary scattering.

Finally, microstructure evolution studies have carried out for Co and Ru by simulation and with some limited TEM observations. Results will be reported.

REFERENCES

[1] Linjun Cao, Lijuan Zhang, and Paul. S. Ho, "Scaling Effects on Microstructure and Electromigration Reliability for Cu and Cu(Mn) Interconnects," *IEEE IPRS* 2014.

[2] M. Hauschildt, MS Thesis, University of Texas at Austin (1999); MRS Symposium proceedings (2000).

[3] L.Spinella et.al. *IITC 2016*;L.Spinella Ph.D Thesis, University of Texas at Austin (2017)

[4] Szu-Tung Hu et.al. to be published.

[5] Pyzyna et.al, "Resistivity of copper interconnects at 28nm pitch and copper cross-sectional areal below 100nm^2," IITC 2017.

[6] C.K Hu et.al, "Electromigration and resistivity in on-chip Cu, Co and Ru damascene nanowires", IITC 2017

Technology	Line Width (nm)	Representative Grain Orientation Mapping
90nm	120	
45nm	70	
28nm	45	
22nm	40	

Fig. 1. Scaling effects on Cu grain size and orientation viewed from trench width direction.

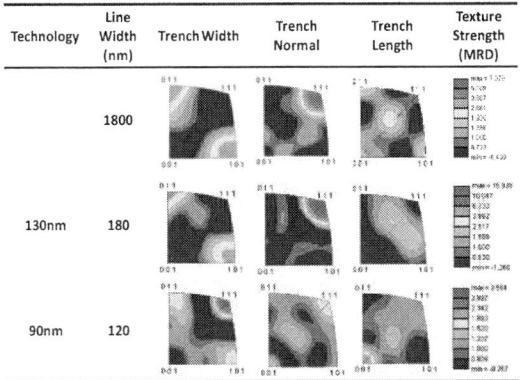

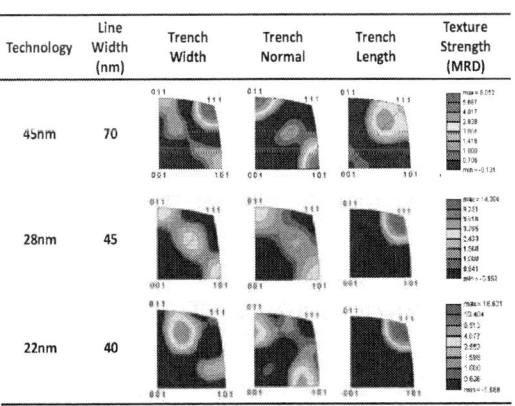

Fig. 2. Scaling effects on the grain orientation as observed from the trench width, trench normal and trench length directions. Here the results together with the change in the grain size provide a detailed description of the scaling effect on grain structure.

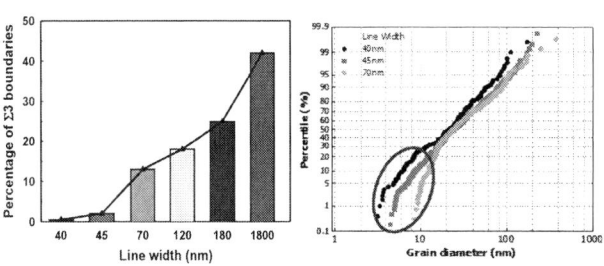

Fig. 3. Scaling effects on a. percentage of twin boundaries and b. grain size distribution.

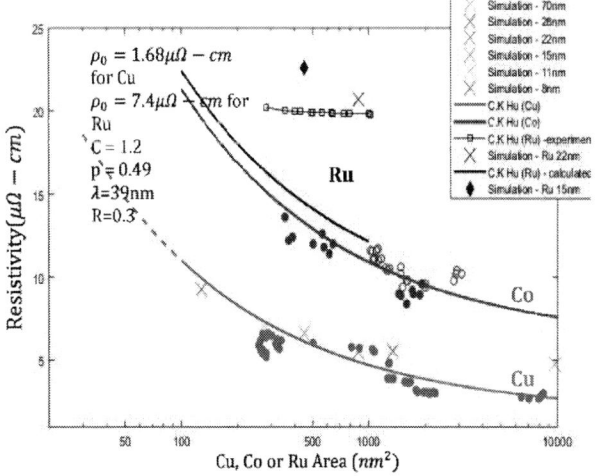

Fig. 4 Scaling effects on electrical resistivity calculated for Cu, Ru and Co interconnects and compared with results reported from recent studies [4, 5].

Integrating Graphene into Future Generations of Interconnect Wires

Ling Li and H.-S. Philip Wong

Department of Electrical Engineering and Stanford SystemX Alliance, Stanford University, Stanford, CA, USA

email: hspwong@stanford.edu

Abstract—The escalating RC delay and diminishing reliability of Cu interconnect present immense challenges for continued integrated circuits performance improvement. This paper reviews the use of single-layer graphene as the diffusion barrier and capping layer to extend scaling of Cu into future generations of interconnects. With graphene barrier/capping layer, processor core simulations predict an 8% speed boost or 12% energy saving, plus higher tolerance for process variations. Single-layer graphene (3.35 Å thick) provides 3.3× longer barrier lifetime than 2 nm TaN. Barrier reliability is expected to further improve with transfer-free and single-crystalline graphene. In-situ low-temperature grown graphene (<0.7 nm thick) improves Cu electromigration lifetime by 10× than Cu with 2 nm CoWP. For interconnect scaling beyond Cu, we discuss the potential benefits and challenges of employing multilayer graphene as a Cu replacement. Multi-layer graphene shows better resistivity scaling trend with $FeCl_3$ doping and higher immunity to electromigration. Replacing Cu with multi-layer graphene, processor cores achieve 9% higher speed or 16% less energy consumption. Spin-on-glass encapsulated multi-layer graphene shows two times longer electromigration lifetime than CoWP capped Cu.

I. INTRODUCTION

Future high performance and energy efficient systems call for high speed and robust interconnects with ever finer interconnect wire dimensions and pitch. [1-2] For sub-40 nm wire pitch, Co has been exploited to replace Cu as the local interconnect [3]. On the other hand, single-layer graphene (SLG) enables further Cu scaling as both the diffusion barrier and the capping layer (Fig. 1), to mitigate the resistivity and reliability problems simultaneously. Besides, without drastic change of current BEOL technologies, a possible path for direct integration of graphene with etched Cu processes has been identified [4]. Beyond the use of Cu, multilayer graphene (MLG) is a promising candidate for sub-10 nm width interconnects with a lower resistivity, longer electromigration (EM) lifetime and high frequency response. Nevertheless, there are several problems such as graphene synthesis, integration, and contact resistance, which must be solved before possible adoption by industry.

II. INTEGRATE GRAPHENE WITH CU INTERCONNECTS

Cu possesses the lowest bulk resistivity besides silver (Ag), and thus has been used as back-end of the line (BEOL) interconnects for over two decades. However, Cu is losing this advantage after generations of wire pitch scaling, because of the rising resistivity due to increasing surface and grain boundary scattering, non-scalability of the barrier/capping layer, and deteriorating reliability. Alternative metals such as Co and Ru

are being explored, with better reliability, higher stability with the dielectric, and smaller size effect [5]. SLG provides a means to preserve the Cu resistivity advantages for narrower line widths as shown in Fig. 2. The lower effective resistivity results from reduced surface scattering and more space for Cu with thinnest possible SLG (3.35 Å thick) barrier. Compared to Cu with traditional barrier material of 2 nm TaN, the most ideal case of barrierless Co or Ru shows lower resistivity beyond 20 nm wire width (though the exact crossover point is still uncertain without experimental measurements of the resistivity of Co/Ru at such small dimensions; besides, Co may still require a barrier and Co or Ru is still not compatible with the standard dual damascene process); whereas with graphene barrier, Cu/SLG outperforms ideal barrierless Co and Ru until around 2 nm wire width. Lee [2] reported that use of Cu/SLG wires lead to an 8% speed boost or 12% energy reduction for a processor core at 5 nm technology node, as compared to wires using Cu/TaN (Fig. 3a). These improvements are obtained mainly through resistance reduction using thinner barrier materials (Fig. 3b). Moreover, there is larger circuit design space for variations of the wire critical dimensions (CD) at the critical data paths with thinner diffusion barrier, as shown in Fig. 4. Having examined the potential benefits of graphene to Cu interconnects scaling and processor system performance, we will then discuss the effectiveness of graphene as the barrier and capping layer, direct integration scheme, and the remaining challenges.

A. Graphene as a Cu diffusion barrier

Fig. 5 shows the mean time to fail (MTTF) of graphene and TaN as the Cu diffusion barrier with various barrier thicknesses [6]. MTTF is extracted from time dependent dielectric breakdown (TDDB) tests. SLG exhibits impressive reliability and better scaling capacity than TaN: 3.3× longer lifetime than 2 nm TaN and comparable performance to 4 nm TaN. Furthermore, by transferring SLG two and three times, bilayer graphene (2-SLG) and tri-layer graphene (3-SLG) become more effective than even much thicker 4 nm TaN. However, multi-layer graphene (MLG) has shorter lifetime, even with larger thickness than SLG. This suggests defects play a critical role in barrier reliability since MLG has higher defect density than SLG due to different growth mechanism. Besides, defects in MLG are correlated across multiple layers [7].

To understand how to improve graphene barrier reliability, the roles of intrinsic and extrinsic defects are studied. Intrinsic defects are mostly graphene grain boundaries and extrinsic defects are mostly caused by the graphene transfer process. As shown in Raman mapping in Fig. 6, SLG with larger grain size (thus less grain boundaries) or SLG from a cleaner transfer method (modified RCA clean transfer [8]) shows lower defect

978-1-7281-1988-5/18 $31.00 © 2018 IEEE

density. With reduced grain boundaries, SLG with larger grains (10~15 μm) improves MTTF by 4× over SLG with smaller grains (2~3 μm). For SLG with less ion/polymer residues from a cleaner transfer process, 2× longer MTTF is obtained (Fig. 7).

These results indicate that further improvements for SLG barrier is possible through the use of single-crystalline SLG (without any grain boundary) [9] and a transfer-free approach that grows SLG directly on the Cu wire or on a dielectric. It should be noted that the SLG tested here is transferred from Cu foil with a growth temperature over 1000 °C, far above the BEOL thermal budget. In addition, development of transfer-free approaches is still in early stages, as discussed in the following section.

B. Graphene direct integration with etched Cu wire

Growth temperature and graphene quality are two key considerations for developing transfer-free direct integration of graphene with Cu. Though graphene growth on dielectric (Fig. 8a) is the most straightforward integration method with the damascene process, graphene quality from this approach [10] is far from satisfactory. Other two-dimensional materials such as hexagonal boron nitride (h-BN) and molybdenum disulfide (MoS_2) do grow directly on dielectrics. As diffusion barrier, their conductivity, thickness and effectiveness are not as good as graphene so far [11].

Another integration approach is to grow graphene directly on patterned Cu wire (Fig. 8b). This is compatible with an etched Cu processes [12], with one possible process flow shown in Fig. 9. The improved conduction across a thin layer of graphene possibly alleviates the via/contact resistance but this is still speculative. Since graphene only grows on Cu surface, the capping layer is self-aligned. However, the vias are not self-aligned with the wire above or below it (as a dual damascene process does). On patterned Cu nanowire, CVD growth of SLG has been demonstrated at 650 °C with decent quality [13]. At 400 °C with electron cyclotron resonance CVD, it is possible to grow multilayer graphene directly on Cu wire [14]. However, optimal graphene diffusion barrier/capping layer requires: 1. growth temperature below 400 °C; 2. high quality and single layer. Growth of monolayer graphene with high mobility at <420 °C has been demonstrated in [15] with plasma enhanced CVD. However, the monolayer graphene is only found on the bottom of the Cu foil. It is still challenging to obtain monolayer graphene on pattern Cu wire. The method introduced in [4] produces 1 – 2 layers of graphene on the surface of Cu wire as shown in transmission electron microscopy (TEM) image. A by-product of amorphous carbon (around 10 nm) is found on top of graphene.

C. Graphene as a capping layer

With the low temperature in-situ grown graphene from Fig. 8b as a capping layer, Cu shows 10× longer EM lifetime over Cu with a 2 nm CoWP cap, and comparable EM lifetime compared to 3 nm CoWP (Fig. 11) for global interconnects with 1-3 μm width. Similar trend is found when graphene capped Cu and CoWP capped Cu are encapsulated in SiN_x. [4] This implies graphene scales better than CoWP as a capping layer. Though capping layer scaling is not as critical as the diffusion barrier, the thin conductive graphene alleviates the parasitic capacitance compared to thicker SiN_x capping. For local/intermediate interconnects, 3× longer EM lifetime is seen

with graphene capped Cu with widths of 80 – 120 nm. [16] DFT calculations unveil the mechanism for the EM improvement: the pristine interface between graphene and Cu provides extra bonding to slow down the Cu migration [4].

Furthermore, graphene improves Cu wire (80 – 120 nm width) resistivity by 12 – 30% (Fig. 12) due to reduced surface scattering [13, 16]. Depending on the wire width, the breakdown current density (J_{BD}) (Fig. 13) increases by 3 – 30% after Cu is capped with graphene. Narrower wires show less J_{BD} improvement possibly because of larger surface roughness and poorer graphene uniformity.

Though improvement of the resistivity and reliability with SLG is only demonstrated on Cu wires so far, the surface scattering reduction and interface binding for EM lifetime enhancement is likely to be also applicable to other metals.

While the effectiveness and scaling advantages of graphene as both the diffusion barrier and capping layer for Cu interconnects are shown to be promising, it is equally important to have a viable process integration scheme. A possible integration scheme of graphene with etched Cu processes is shown in Fig. 9. However, there still remains several problems: 1. low-temperature graphene growth condition requires further optimization to improve graphene quality and eliminate the amorphous carbon by-product; 2. high quality graphene growth on dielectric, possibly the more attractive integration approach for the damascene process, is still lacking; 3. The potential damage on low-k dielectric from graphene growth is uncertain. 4. The variations of graphene quality on Cu electromigration performance is still unknown.

III. GRAPHENE AS A CU REPLACEMENT

Further down the technology node beyond Cu, multilayer graphene (MLG) is a promising candidate for future interconnects. Monolithic integration of graphene with conventional CMOS processes has been demonstrated almost a decade ago [18]. MLG interconnects operates up to 1.3 GHz in a ring oscillator. In the sub-10 nm regime, $FeCl_3$ intercalated MLG (8 nm width) has a resistivity of 3.2 μΩ·cm [19], which is lower than 11 μΩ·cm of Cu at the same width. Based on the ARM processor core simulation with 5 nm design rule by Lee *et al.* [2], this lower resistivity of MLG translates to 56% – 73% lower resistance at M2-M5 (Fig. 14). Processor cores with MLG interconnects achieve 9% higher speed or 16% less energy consumption (Fig. 15) over processor cores with Cu interconnects. Analysis of how the chip is wired by the place-and-route process also reveals that MLG interconnects lead to a smaller chip area, since the wires are longer and fewer vias are used without delay penalty due to the use of lower resistance wires (Fig. 16).

MLG also has better electromigration reliability than Cu [20]. As shown in Table 1, spin-on-glass (SOG) encapsulated MLG has 2 times longer lifetime than CoWP capped Cu. Further studies reveal that the MLG breakdown is originated from oxidation at the MLG grain boundaries and other defect sites, instead of electromigration. The extracted activation energy for EM for MLG is 1.97 eV, much higher than 1.2 eV for electroplated Cu [21]. Thus, MLG wire is more immune to electromigration process than Cu wire for normal use case.

978-1-7281-1988-5/18 $31.00 © 2018 IEEE   119

MLG overcomes the resistivity and reliability problems that scaling Cu wire faces. However, integrating MLG into the CMOS process flow needs radical changes from current BEOL technologies. Solutions to the large-scale high-temperature synthesis, transfer processes, high contact resistance, compatibility with low-k dielectrics are some of the more obvious barriers that must be overcome before possible adoption by industry. This may take quite some time and significant development effort.

IV. SUMMARY

SLG as barrier and capping is shown by various experimental demonstrations to be a favorable approach to extend Cu interconnects scaling, striking a balance between system performance and reliability. However, the window for adoption may be narrowing as competing options such as Co, Ru, and metal alloys are making progress. The mechanism for resistivity and reliability improvement from SLG may be applicable for other metals such as Co and Ru. MLG provides good scaling potential and reliable performance for future interconnect, but many barriers must be overcome. Challenges for graphene integration with the BEOL technology lie at graphene synthesis (aiming for large-scale, low-temperature,

and low defect), transfer-free process development, and lower contact resistance.

ACKNOWLEDGMENT

This work is supported in part by STARnet FAME, and member companies of the Stanford Initiative for Nanoscale Materials and Processes (INMP) industrial affiliate program. This work is made possible through collaborations with C. Lee, X. Chen at Stanford, Prof. M. Arnold (U. Wisconsin), A. Yoon and Z. Zhu from Lam Research Corporation, TSMC, and ARM. Critical review by J. Stathis (IBM), L. Clevenger (IBM), and Prof. Z. Chen (Purdue) are gratefully acknowledged.

REFERENCES

[1] ITRS http://www.itrs2.net/itrs-reports.html. [2] C. S. Lee, *et al. IEDM*, 2016. [3] C. Auth, *et al. IEDM*, 2017. [4] L. Li, *et al. IEDM*, 2016. [5] R. Brain. *IEDM*, 2016. [6] L. Li, *et al. ACS Nano*, 2015. [7] L. Li, *et al. VLSI-T*, 2015. [8] X. Li, *et al. ACS Nano*, 2011. [9] J.-H. Lee, *et al. Science*, 2014. [10] R. Mehta, *et al. Nanoscale*, 2017. [11] C.L. Lo, *et al. npj 2D Materials and Applications*, 2017. [12] L. Wen, *et al.* IITC/MAM, 2015. [13] R. Mehta, *et al. Nano Lett.*, 2015. [14] C.-H, Yeh, *et al. ACS Nano*, 2014. [15] D. A. Boyd, *et al.* Nat. Comm., 2015. [16] L. Li, *et al. EDL*, 2018 in preparation. [17] A. Pyzyna, *et al. VLSI-T*, 2015. [18] X. Chen, *et al. IEDM*, 2009. [19] D. Kondo, *et al. IITC*, 2014. [20] X. Chen, X, *et al. VLSI-T*, 2012 [21] D. C. Edelstein. *IEDM*, 2017. [22] C.–K. Hu, *et al, IITC, 2017*. [23] L. G. Wen, *et al, IITC, 2016*. [24] X. Zhang, *et al, IITC, 2016*

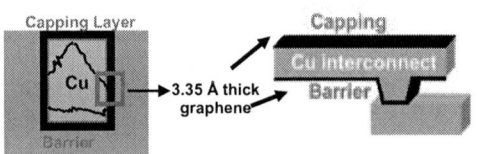

Fig. 1. Schematic cross-section and side view of the Cu interconnect structure with a single layer graphene (3.35 Å), as both diffusion barrier and capping layer.

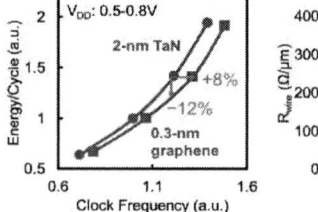 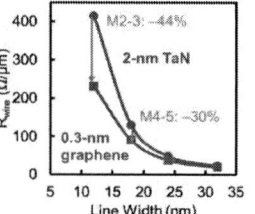

Fig. 3. a) Comparison of energy per cycle with barrier of 0.3 nm graphene and 2 nm TaN; b) wire resistances with graphene and TaN at different line widths. [2]

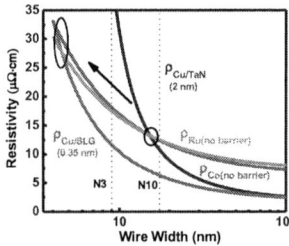

Fig. 2. Resistivity vs. wire width with various metal/barrier candidates, including Cu/SLG, Cu/2 nm TaN, barrierless Ru and barrierless Co, with parameters from [17, 22-24]

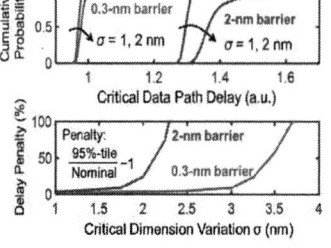

Fig. 4. a) Cumulative distribution function and b) penalty of critical path delay vs. critical dimension variation for 0.3 and 2 nm barrier. [2]

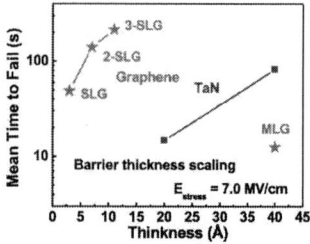

Fig. 5. Mean time to fail vs. various thicknesses of barriers, graphene and TaN. 2-SLG and 3-SLG are SLG transferred twice and thrice. [6]

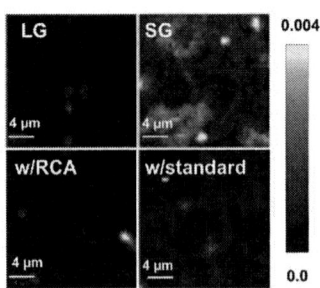

Fig. 6. Raman mapping of SLG D peak relative to silicon ref peak ratio: a) large-grain (10~15 μm); b) small-grain (2~3 μm), c) modified RCA clean transferred SLG, d) standard approach transferred SLG. [7]

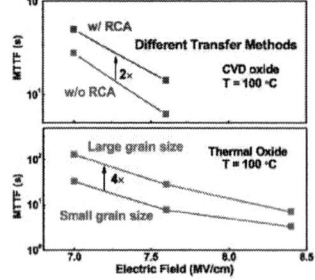

Fig. 7. Mean time to fail of Cu with graphene barrier a) transferred with RCA clean method vs. with standard method; b) of different grain sizes.

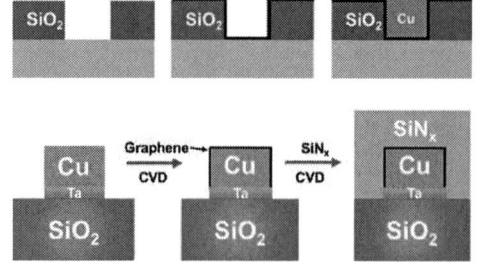

Fig. 8 a) schematic of graphene growth directly on SiO2 dielectric; b) the process flow of graphene growth on patterned Cu wire. [4]

978-1-7281-1988-5/18 $31.00 © 2018 IEEE   120

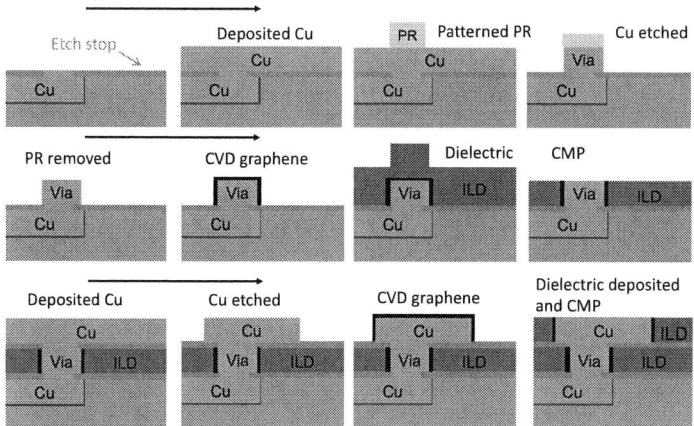

Fig. 9. Process flow of graphene integration with etched Cu interconnects.

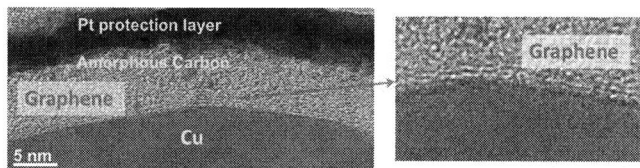

Fig. 10. Cross sectional TEM images of graphene capped Cu wire. 2-layer graphene is shown on Cu surface with another 10 nm amorphous carbon by-product. [16]

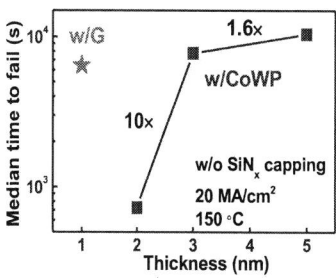

Fig. 11 Summarized Cu EM median time to fail shows scalability of graphene and CoWP. Cu wrapped with <1 nm graphene takes 10× longer to fail than 2 nm CoWP, comparable to 3 nm CoWP. [4]

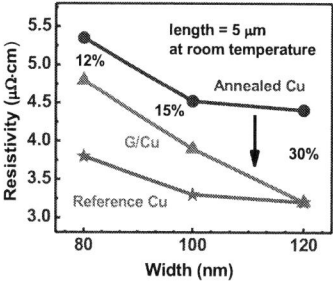

Fig. 12. Resistivity of Cu wire with various dimensions with graphene capping (red), annealed (blue), and reference Cu (pink) as reported in [17].

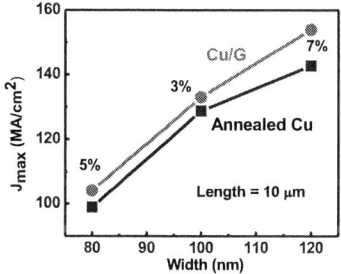

Fig. 13. Breakdown current density of graphene capped Cu (red) and anneal Cu (black) with various wire width. [16]

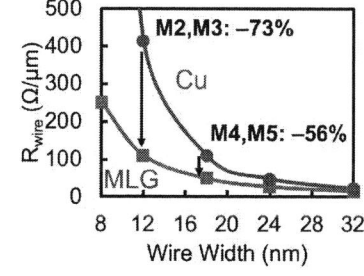

Fig. 14. Wire resistance with MLG and Cu at metal layers 2 to 5 (M2-M5). MLG reduces the wire resistance by 56 – 73%. [2]

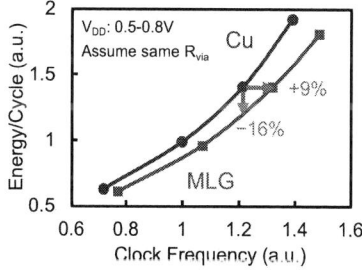

Fig. 15. Core energy vs. clock frequency with MLG and Cu interconnects. Total energy delay product (EDP) is lower by 11% with MLG interconnects. [2]

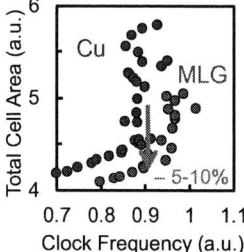

Fig. 16. Total cell area vs. clock frequency with MLG and Cu interconnects. Cell area reduces by 5 – 10% with MLG interconnects. [2]

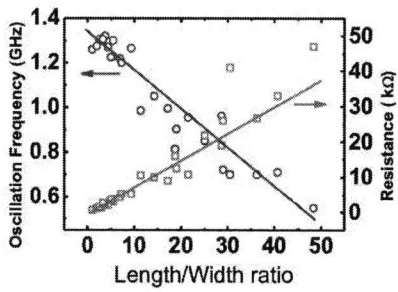

Fig. 17. Oscillation frequency for various interconnect lengths and widths. GHz range operation frequency is observed for interconnects up to ~80 μm long. [18]

	Length (μm)	Width (μm)	Capping	J_{stress} (MA/cm²)	Temp. (K)	MTTF (hr)
Copper	125	0.4	SiCN	20	595	5.2
	100	2	CoWP	20	650	8.5
Graphene	100	3	N/A	20	550	6.0
	100	3	SOG	20	550	18

Table 1. Electromigration lifetime comparison of Cu and graphene. [20]

978-1-7281-1988-5/18 $31.00 © 2018 IEEE

Interconnect Trend for Single Digit Nodes

Mehul Naik

Applied Materials Incorporated, Santa Clara, CA, USA
email: mehul_naik@amat.com

Abstract— Transistor performance continues to improve with density scaling and move to FinFET architectures. However, feature size reduction increases parasitic contact and interconnect resistance. This degrades the power-performance equation. To address the resistance bottleneck; new materials, new fill technologies and new integration schemes are in play. This paper reviews metallization trends for contact and interconnect as the industry prepares for 7nm node production and looks towards developing 5 and 3nm nodes.

I. INTRODUCTION

CMOS scaling continues to march to the drum beat of density scaling. To accommodate all the routing, contact and interconnect pitch has to be reduced node over node. With pitch reduction, resistance of vertical (contact and via) and horizontal interconnect (metal wires) increases resulting in excess IR drop (~power), and increased delay in the circuit. In a sense, all the drive current advantages gained through multiple generations of FinFET scaling are being negated by high contact/MOL and BEOL resistance. From a process perspective, as critical dimensions reduce, filling these vertical and horizontal interconnects becomes more and more difficult and reliability degrades as well. For the industry to continue on a pitch scaling pathway, these challenges and trade-offs need to be addressed. Industry is looking towards new conductors to solve some of these challenges In this paper, we will first review the reasons behind the change from tungsten to cobalt (Co) at contact and local interconnect, followed by a comprehensive look at M_x levels of interconnect where copper (Cu) continues to be of interest with conductors such as Co and ruthenium (Ru) vying to replace Cu at the lower M_x levels.

II. CONTACT AND LOCAL INTERCONNECT(LI)

With continual reduction in contact CD, R_{ext} limits transistor performance. Fig. 1 shows the various factors that contribute to R_{ext}. Over the last few years, there have been many papers published towards reducing the silicide contact resistance [1,2] with more than an order of magnitude reduction in n-MOS and p-MOS contact resistance. However, as contact CD scales below 20nm, a more holistic look is required to optimize the transistor performance and needs consideration of not only the silicide contact resistance, but, also the resistances associated with the contact metallization and local interconnect. Tungsten (W) has been the choice for contact metallization and local interconnect due to its ease of integration and excellent reliability. A key scaling bottleneck for tungsten is the need for a high resistivity chemical vapor deposited (CVD) TiN barrier layer that protects underlying titanium layer from the corrosive chemistry used to deposit

tungsten. As the TiN thickness does not scale with contact CD, the contact plug resistance increases and on-state current decreases node over node. For local interconnect, CD scaling will increase the resistance of the W lines due to the continuous reduction in volume of tungsten as Ti/TiN occupies a larger fraction of the trench, and due to the seam that is inherent with CVD W deposition. Interface resistance between the contact plug and the local interconnect will also increase resulting in an overall increase of the combined contact and LI structure.

Co is the leading candidate to replace both W contact and W local interconnect. The interest in Co is driven by the fact that while Co has similar bulk resistivity to W, CVD W resistivity is about 1.5-2x that of CVD Co due to impurities and grain boundary scattering in W. There are other advantages as well; CVD Co gives the opportunity to thin down the high resistivity TiN layer, and Co fill can be seamless, while W CVD fill has an inherent seam. This is clearly seen in Fig. 2 which shows 10-12nm CD Trenches filled with CVD W and CVD Co. These factors not only help reduce contact resistance through interface management, but, also help reduce LI resistance by increasing the volume of the conductor that has lower in-structure resistivity as well. Fig. 3 shows the improvement in line resistivity by replacing W with Co. These improvements are attributed to thinner TiN, seamless Co fill, larger grains with Co and better size effect with Co.

To understand contact/LI resistance scaling dependence on metallization schemes, the structure shown in Fig. 4(inset) was modeled. Here, contact and LI have the same CD's and the LI length is 5x CPP at the corresponding technology node. Results plotted in Fig. 4 show that the impact of switching from a W based contact/LI to a Co based contact/LI increases as one reduces the contact/LI CD. While, the increase in LI resistance is ameliorated somewhat by the reduction in length node over node, the contact resistance increases by more than factor of 20, and the overall impact is almost a 90% reduction in total resistance on replacing W contact/LI with Co contact/LI. How this MOL resistance improvement translates to drive current or delay improvements is heavily dependent on many factors including contact silicide resistance, transistor resistance and capacitance among others. Demuynck et al. [3] have modeled about a 4% improvement in drive current when reducing the TiN thickness from 2nm to 1nm and replacing W with Co for a 14nm CD contact at 42nm CPP lay-out with 3 fins at 52nm height.

It is anticipated that devices that use W contact and W LI will first replace W contact with Co, and then look into

978-1-7281-1988-5/18 $31.00 © 2018 IEEE

replacing W LI with Co LI to drive down MOL resistance even further to manage delay.

It should be noted that devices that use W for contact and Cu LI have already moved to Co contact and LI [4]. A 60% reduction in contact line resistance was reported on replacing W with Co. The primary reason to replace Cu LI with Co was lower via resistance due to the elimination of barrier required for Cu interconnects, and the higher electromigration reliability of Co.

III. BEOL INTERCONNECT

Copper has been the BEOL conductor for the last 20 years or so. A typical 22nm half-pitch Cu interconnect is shown in Fig. 5 along with a simplified metallization sequence. A state of the art Cu interconnect in production today has a barrier to prevent Cu diffusion into the dielectric, a CVD Co liner to promote metal fill, followed by a PVD Cu seed that acts as conducting layer for electroplated fill. Cu CMP is then followed by a selective CVD Co cap to improve electromigration. Due to the difficulty in scaling either the barrier or the liner, both Cu line and via resistance increase continually node over node.

A. Barrier and Fill Trends for Copper Interconnect

As the incumbent, there is a strong incentive for the industry to extend Cu as far as possible. Fig. 6 shows the approaches to reduce via resistance and Fig. 7 does the same for line resistance. For via resistance, barrier thickness reduction is the priority, while for line resistance, the priority is maximizing volume of Cu through barrier+liner thickness (=cladding) reduction. All of the above needs to be accomplished while maintaining perfect metal fill with good reliability.

Wu et al. [5] have proposed an ALD TaN based barrier to replace PVD TaN based barrier with a resulting 20% reduction in via resistance for a 20nm via (Fig. 8). They also demonstrated that this via resistance reduction does not come with a reliability or metal fill penalty. Fig. 9 shows that this new barrier approach has equivalent electromigration to the PVD based approach, while Fig. 10 validates the time dependent dielectric breakdown (TDDB) performance. A bonus is improvement of the metal fill process window down to 15nm CD as shown in Fig. 11 compared to the PVD based approach. This new ALD barrier approach provides a pathway to scale Cu down to 5nm node.

To further improve the fill process margin, Cu reflow based fill is under consideration with both Co and Ru based liners. While Ru based liner shows better Cu reflow characteristics, difficult reliability challenges [6] have prevented Ru going into production and focus remains on enabling Cu reflow with Co liner.

B. Metal Candidates to Replace Copper

While the industry remains focused on extending Cu, fundamental challenges remain. This has created interest in exploring conductors such as Co and Ru as possible replacements for Cu. While, both Co and Ru have higher bulk resistivity than Cu, their lower mean free path results in a lower $\rho_o\lambda$ product [7], and it is expected that materials with $\rho_o\lambda$ less than Cu will have less size induced resistivity increase, and will eventually have lower in-structure resistivity compared to Cu. Fig. 12 and 13 compare resistance scaling of Cu interconnect to Co and Ru respectively using data generated on single damascene trenches down to 10nm CD [8]. Fig. 12 shows that Co Interconnect has lower resistance than Cu interconnect at conductor area below 450nm^2. For a conductor aspect ratio of 2; this represents a cross-over at 15nm CD. Fig. 13 adds experimental data on Ru fill. It is seen that Cu and Co experimental data match with model predictions, while Ru experimental resistance is higher than model prediction. It is believed that this is due to incomplete recrystallization of Ru. From a materials property perspective, it is expected that Ru resistance will scale the same as Co. From a reliability perspective, both Co and Ru interconnects are better than Cu. Bekiaris et al. [9] have shown that Co-Low k interconnects pass TDDB requirements (Fig.14), and show Co fill feasibility down to 10nm CD (Fig. 15). Griggio et al. [10] have shown 50x electromigration improvement with a Cu-alloy based interconnect. For Ru, Wen et al. [11] have shown excellent electromigration performance as well. Which conductor wins will depend on ease of integration, manufacturing and cost. Since, Co has already been implemented at contact and LI; it has a head start to replace Cu in the BEOL.

IV. SUMMARY

Interconnect delay and power overhead is a bottleneck that can negate transistor advances. Changes in barriers, conductors and integration schemes are in play to improve interconnect performance. In this paper, we have reviewed conductor changes in contact and MOL from W to Co. Technologies such as ALD barrier and Cu reflow fill to extend Cu, and reviewed recent advances in Co and Ru interconnects as possible Cu replacement for BEOL.

ACKNOWLEDGMENT

The author gratefully acknowledgments Nam Sung Kim, Nikos Bekiaris, He Ren, Zhiyuan Wu, and members of the Metal Deposition Business Unit in Applied Materials.

REFERENCES

[1] C-N Ni et al; VLSI-TSA (2016)
[2] C-N Ni et al; IEEE symposium on VLSI Technology (2016)
[3] Demuynck et al; IEDM (2017)
[4] C. Auth et al, IEDM (2017)
[5] Z. Wu et al; IITC-AMC (2018)
[6] Z. Wu et al; IITC-AMC (2018)
[7] D. Gall et al; Journal of Applied Physics, 119, 085101 (2016)
[8] H. Ren et al; IITC-AMC (2018)
[9] N. Bekiaris et al; IITC (2017)
[10] Griggio et al; IRPS (2018)
[11] Wen et al; IITC-AMC (2016)

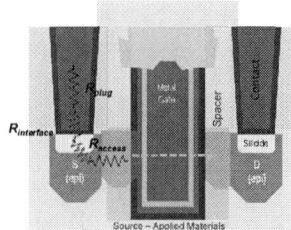

Fig.1: Simplified Schematic of factors impacting R_{ext}

Fig. 2: TEM x-section showing (a) seam and grain roughness in W, (b) Large grained seamless Co Fill

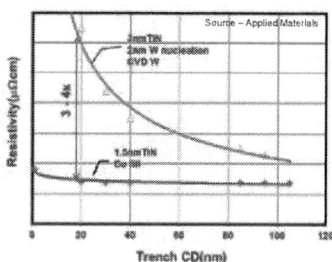

Fig. 3: Comparison of W and Co scaling shows 3-4x improvement in resistivity when Co replaced W

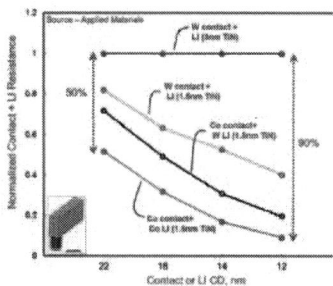

Fig.4: Comparison of various contact and LI fill scenarios shows 90% resistance reduction for W contact/LI to Co contact/LI

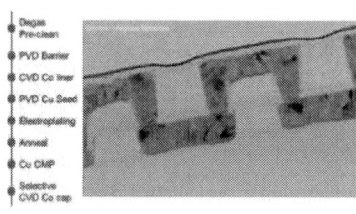

Fig. 5: 22nm half-pitch Cu two-level metal interconnect with simplified metallization flow

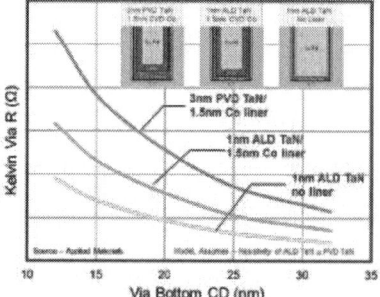

Fig. 6: Metallization scenarios to lower Cu via resistance highlight the importance of barrier thickness reduction

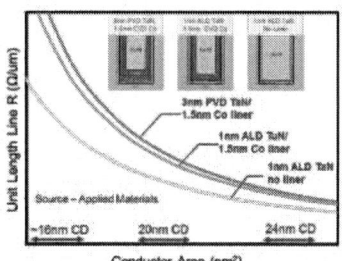

Fig. 7: Metallization scenarios to lower Cu line resistance highlight the importance of maximizing Cu volume

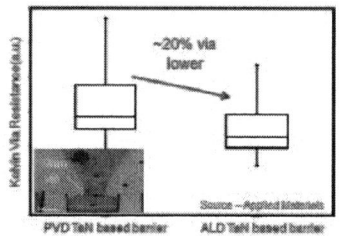

Fig. 8: ALD TaN based barrier shows 20% lower via resistance vs. PVD TaN based barrier on 22nm half-pitch test vehicle

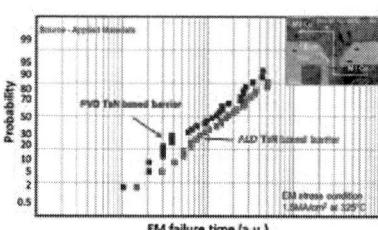

Fig.9: ALD TaN based barrier shows equivalent electromigration performance vs. PVD TaN based barrier on 22nm half-pitch test vehicle

978-1-7281-1988-5/18 $31.00 © 2018 IEEE

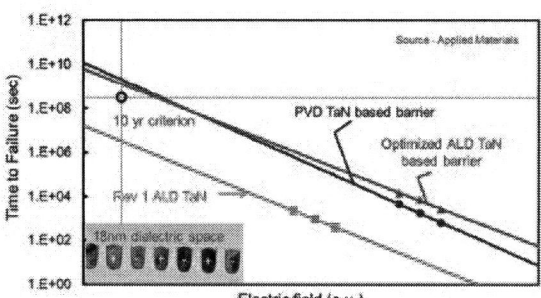

Fig. 10: Optimized ALD TaN based barrier passes TDDB requirement on 20nm CD test vehicle

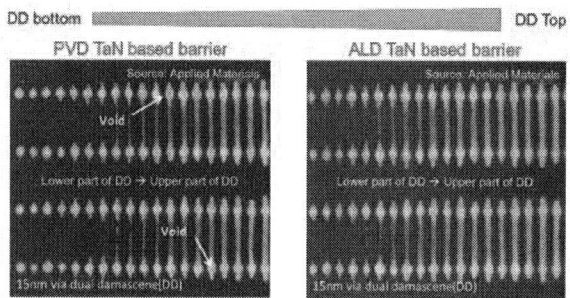

Fig. 11: Scanning electron microscope images of 15nm CD dual damascene (DD) structure showing better fill with ALD TaN based barrier

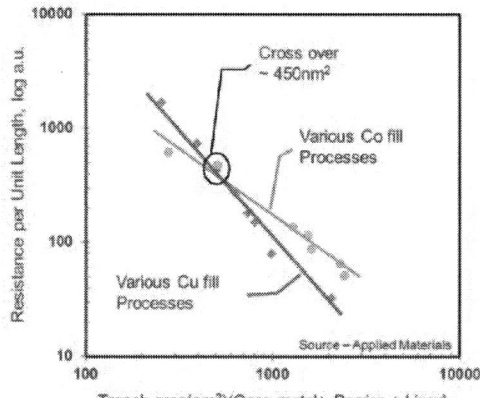

Fig. 12: A comparison of Cu vs. Co line resistance scaling shows that Co cross-over at ~450nm^2 area

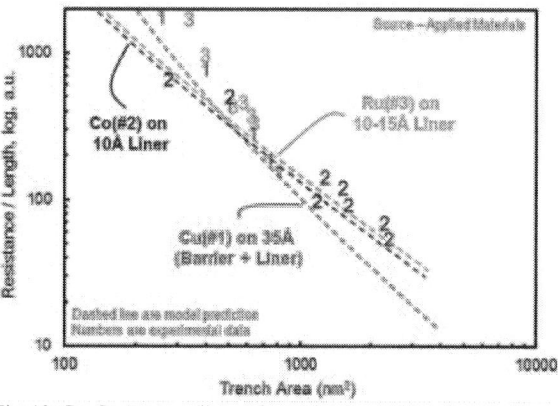

Fig. 13: Cu, Co, Ru overall experimental and model benchmark. Ru does not meet model prediction likely due to incomplete recrystallization and possible micro-voids

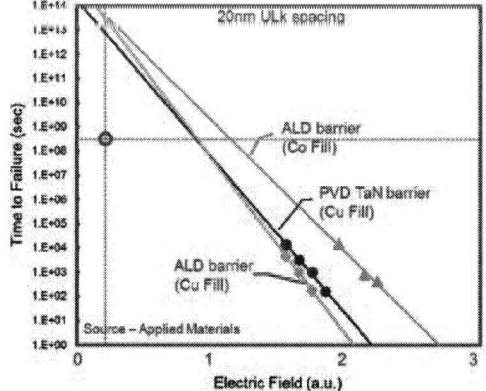

Fig. 14: TDDB performance comparison of Co-ULK and Cu-ULK shows No reliability degradation with Co

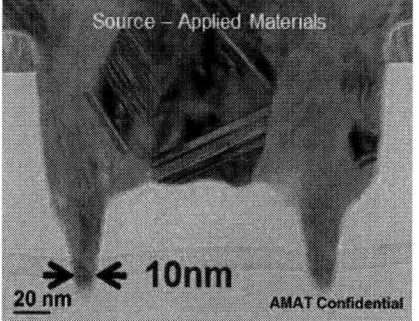

Fig. 15: Co interconnect fill demonstrated down to 10nm CD on a Co-ULK structure

978-1-7281-1988-5/18 $31.00 © 2018 IEEE   125

Device challenges for near term superconducting quantum processors: frequency collisions

Markus Brink[1], Jerry M. Chow[1], Jared Hertzberg[1], Easwar Magesan[1], Sami Rosenblatt[1],
[1]IBM T.J. Watson Research Center, Yorktown Heights, NY

Abstract— The outstanding progress in experimental quantum computing with superconducting Josephson-junction based qubits over the past few decades has pushed coherence times many orders of magnitude above that of the first measured. We are also in the midst of scaling towards complex architectures of multi-qubit processors where maintaining very low gate error rates at the limits supported by coherence times is extremely important. Here we will review some of the critical materials and device challenges for superconducting qubits from the perspective of improved coherence and improved error rates. In particular we will focus on the problem of frequency allocations in order to target multi-qubit lattices for fixed-frequency microwave-based gates.

I. INTRODUCTION

In quantum processors employing fixed-frequency superconducting Josephson-junction-based transmon qubits [1] and all-microwave cross-resonance two-qubit gates [2-3], 'frequency collisions' and 'frequency crowding' are a distinct challenge to attaining low gate error rates. The problem arises due to nearest-neighbor or next-nearest-neighbor qubits which are degenerate in one or another excitation energy. Josephson junctions for transmon qubits are typically patterned lithographically out of Al/AlOx/Al. Fabrication defines the critical current of the junction and in turn the frequency of the transmon qubit.

The cross-resonance gate is an all-microwave entangling gate between a control qubit and a target qubit. It involves a defined fixed coupling between two qubits, where a ZX interaction (generator of a controlled NOT, CNOT, gate) is activated by driving the control qubit at the transition frequency of the target qubit. The strength of the drive and the frequency detuning between the qubits affect the total CNOT gate time and effectiveness. In particular, the transmon is a weakly anharmonic qubit, meaning there are higher energy levels that are not too far away from the ground to first excited state energy. Such levels can also cause collisions that can be detrimental to the cross-resonance gate performance.

Therefore, as processors scale up in the number of qubits, allowed cross-resonance gates depend upon accuracy of fabricated Josephson junctions and where the frequencies of the qubits come out. The rest of this paper will describe the types of problems that can arise with respect to frequency collisions and we show a Monte Carlo modeling method to demonstrate the yield of devices with usable qubit frequencies.

II. TYPES OF GATE COLLISIONS

The transmon cross-resonance frequency collision conditions are not simply limited to qubits that participate together in a CR gate, but extends to non-nearest neighbors as well. Considering transmon qubits of frequency f and anharmonicity δ, we know of six degeneracy conditions that degrade gate fidelities and one that leads to unfavorably slow gate rates: $f_j = f_k$ (any two qubits j, k sharing a coupling), $f_j = f_k - \delta/2$ (gate pair of control j, target k qubits), $f_j = f_k - \delta$ (any two qubits j, k sharing a coupling), $f_j > f_k - \delta$ (gate pair of control j, target k will exhibit 'slow gate' behavior), $f_i = f_k$ (two target qubits i, k sharing a control j), $f_i = f_k - \delta$ (two target qubits i, k sharing a control j) and $2f_j + \delta = f_k + f_i$ (gate pair of control qubit j & target qubit k; spectator qubit i is coupled to control j). To ensure high gate fidelities in our devices, we must avoid any of these conditions. Assuming all qubits are transmons with frequencies on the order 5 GHz and anharmonicity $\delta \sim$ -340 MHz, we can assume an exclusion region of at least +/- 5 MHz around each condition. Exact bounds remain under study both experimentally and using effective-Hamiltonian modeling.

III. FREQUENCY ARRANGEMENT STRATEGIES AND MONTE CARLO MODELING

How can we be confident to meet these constraints in a lattice of 17 (e.g. for a distance 3 rotated surface code) or more qubits? A useful metric is the standard deviation σ_f in precisely setting the qubit frequency. One tactic to avoid frequency crowding would be to design all of the qubits in the lattice to be identical, and rely on the random scatter σ_f to avoid 'frequency collisions.' Another idea would be to arrange the lattice into a regular pattern of qubit frequencies. Figure 1 shows likely arrangements. For instance, a pattern of five frequencies should prevent any two adjacent qubits from sharing a frequency [4]. We can model the behavior in a Monte Carlo manner, as diagrammed in Figure 3: assign a mean frequency to each position in the pattern, populate the lattice with random frequencies from distributions σ_f around each mean, count the collisions, and repeat the process more than 10^3 times. The yield is the fraction of cases having no collisions. In Figures 4 and 5 we see that in order to achieve a useful yield (more than a few %) in a 16 or 17 qubit lattices we will have to use a 5-frequency pattern with σ_f well below 50 MHz. To produce a useful yield in a 49-qubit lattice will require a 5-frequency pattern with σ_f well below 20 MHz.

The connectivity of the lattice also has a measurable effect. For instance, figure 4 shows that in a 17-qubit device, for most values of σ_f, using a square-lattice layout will improve yield ~ 2x over a skew-symmetric lattice.

Statistical models can guide our designs for scaled-up qubit lattices if we know the fabrication precision σ_f. We expect the ground-state to first excited state qubit frequency difference to follow $f01 \sim (8\ E_J\ E_C)^{0.5}$ with Josephson energy E_J given by $E_J \sim I_c/2$ and Josephson junction critical current I_c related to junction resistance R_n by the Ambegaokar-Baratoff relation $I_c \sim \Delta/2eR_n$, while charging energy E_c derives from qubit capacitance C as $E_c = e^2/2C$. This implies that we can learn about the statistics of our qubit frequencies from measurements of junction resistances.

We will present correlations of room temperature resistance measurements with actual measured qubit frequencies and give a guide towards achievable device yields for larger qubit lattices.

ACKNOWLEDGMENT

The authors gratefully acknowledge support from Intelligence Advanced Research Projects Activity (IARPA) under W911NF-16-1-0114-FE.

REFERENCES

[1] Jens Koch et al., "Charge-insensitive qubit design derived from the Cooper pair box" *Physical Review A*, 76, 042319, 2007.

[2] Chad Rigetti and Michel Devoret, "Fully microwave-tunable universal gates in superconducting qubits with linear couplings and fixed transition frequencies" *Phyiscal Review B*, 81, 134507, 2010.

[3] Jerry M. Chow et al., "Simple All-Microwave Entangling Gate for Fixed-Frequency Superconducting Qubits" *Physical Review Letters*, 107, 080502, 2011 .

[4] Jay M. Gambetta, Jerry M. Chow, Matthias Steffen, Building logical qubits in a superconducting quantum computing system". *npj Quantum Information*, 3, 2, 2017.

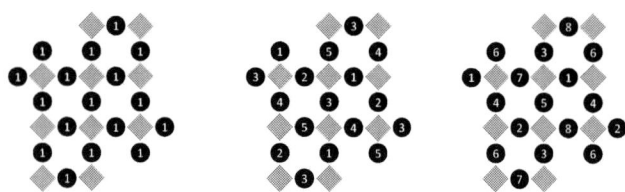

FIG. 1: Possible frequency patterns for 17Q skew-symmetric lattice. Qubits are indicated by solid circles, coupling buses by gray squares. Left: One frequency. Rely on random scatter to avoid collisions. Middle: Two frequencies. Right: Five frequencies. No two qubits on the same bus has the same frequency

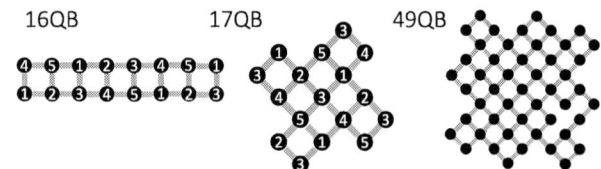

FIG. 2: Square lattices of three sizes. Five-frequency patterns illustrated in two cases.

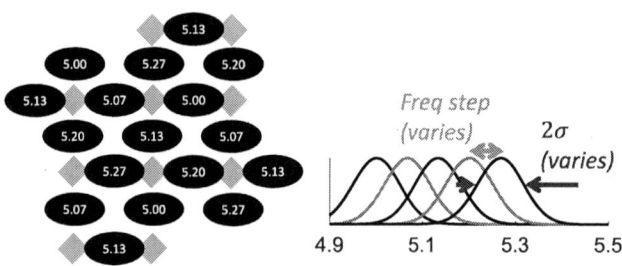

FIG. 3: Example of frequency collision statistical model. Five-frequency pattern. Lattice is populated from random distributions about five mean frequencies. We vary the frequency step between means (67 MHz in this example) as well as the width of the distributions. Left: Mean frequencies, Right: Distributions

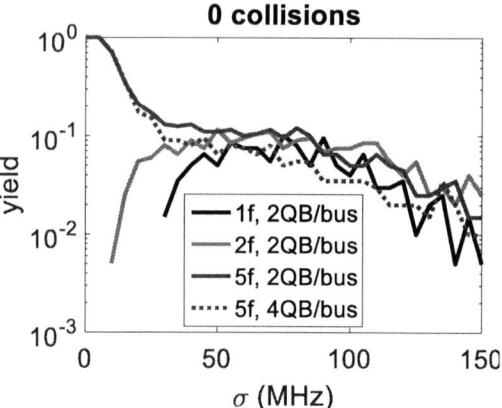

FIG. 4: Predicted yield of 17-qubit chips having no frequency crowding. Lattice connected either in skew-symmetric (4Q/bus) or square (2Q/bus) manner. Three possible patterns of qubit frequencies

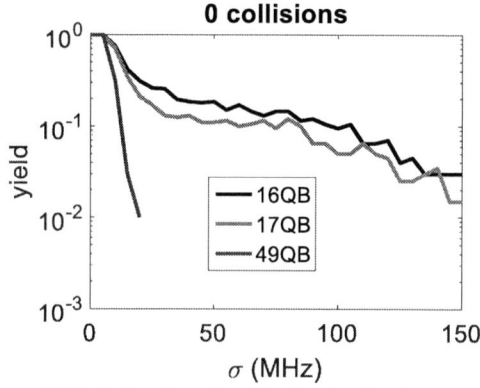

FIG. 5: Predicted yield of chips having no frequency crowding. Monte Carlo model predicts yield of square qubit lattices shown in FIG. 2, as a function of σ

Scalable quantum computing with ion-implanted dopant atoms in silicon

A. Morello[1], G. Tosi[1], F.A. Mohiyaddin[1], V. Schmitt[1], V. Mourik[1], T. Botzem[1], A. Laucht[1], J.J. Pla[1], S. Tenberg[1], R. Savytskyy[1], M. Madzik[1], F. Hudson[1], A.S. Dzurak[1], K.M. Itoh[2], A.M. Jakob[3], B.C. Johnson[3], J.C. McCallum[3] and D.N. Jamieson[3]

[1]School of Electrical Engineering & Telecommunications, UNSW Sydney, Australia, email: a.morello@unsw.edu.au
[2]Keio University, Japan [3]School of Physics, University of Melbourne, Australia

Abstract—We present a scalable strategy to manufacture quantum computer devices, by encoding quantum information in the combined electron-nuclear spin state of individual ion-implanted phosphorus dopant atoms in silicon. Our strategy allows a typical pitch between quantum bits of order 200 nm, and retains compatibility with the standard fabrication processes adopted in classical CMOS nanoelectronic devices. We theoretically predict fast and high-fidelity quantum logic operations, and present preliminary experimental progress towards the realization of a "flip-flop" qubit system.

I. INTRODUCTION

Quantum computers offer the prospect of performing certain complex calculations using radically different and more efficient algorithms compared to their classical counterparts. Intriguingly, the physical platform that underpins all of the existing classical computer hardware – silicon MOS – is likely to retain a prominent role even in quantum computing [1].

The first-ever qubit demonstrated in silicon was the electron spin bound to a single, ion-implanted phosphorus (P) donor, with basis states $|\uparrow\rangle$ and $|\downarrow\rangle$, integrated within a simple MOS device [2]. A second qubit can be defined within the same atom, using the quantum states $|\Uparrow\rangle$ and $|\Downarrow\rangle$ of the ^{31}P nuclear spin [3]. When hosted in isotopically enriched ^{28}Si, these spin qubits exhibit record-long coherence times – exceeding 30 seconds for the nuclear spin [4] – and quantum gate fidelities approaching 99.99%. Similar results have been obtained using lithographically-defined "artificial atoms" (quantum dots), wherein 2-qubit logic gates have been demonstrated [5] using the exchange interaction as the physical medium for the quantum logic. Both for donors and for quantum dots, coupling spins via exchange interactions requires fabricating structures (including interconnects and classical readout devices) on a scale of order 20 nm.

Here we discuss an alternative choice of encoding quantum information, which uses again individual P donors in MOS devices, but relaxes the fabrication tolerances and the qubit pitch by a factor 10.

II. THE "FLIP-FLOP" QUBIT - THEORY

A. Definition of the qubit

Let us consider a single P donor, with electron spin S and nuclear spin I, both of quantum number 1/2. The donor is implanted in silicon, ~15 nm below the interface with a SiO_2 dielectric, on top of which rest some metallic gates (Fig. 1). The electron and nucleus are mutually coupled by the contact hyperfine interaction A, which depends on the electron density at the nuclear site. We define a new qubit, called "flip-flop" qubit, as the two-level system with basis states $|\uparrow\Downarrow\rangle$ and $|\downarrow\Uparrow\rangle$ resulting from combining the electron and the nuclear spin [6] (Fig. 2). Arbitrary quantum superpositions of these states are achieved by modulating the hyperfine coupling A at a frequency corresponding to the energy splitting (Fig. 3). This modulation is achieved electrically, by applying an oscillating electric field on a metallic gate placed above the donor.

B. Electrical control

To control the qubit most efficiently, we propose to distort the donor electron wavefunction using a static, vertical electric field E_z, that displaces the electron halfway towards the Si/SiO_2 interface. At this bias point, A is very sensitive to small electric field variations, resulting in a very fast transitions between the qubit states. We estimate a 1-qubit logic gate time of order 20 ns, using a small ac electric field of amplitude 30 V/m. Importantly, this sensitivity to resonant ac electric fields is not accompanied by an excessive fragility of the quantum state, thanks to a second-order 'clock transition', i.e. the vanishing of the first and second derivative of the qubit energy splitting with respect to static electric field [6]. Using realistic models for charge noise in the device, we predict 1-qubit gate fidelities in excess of 99.9% (Fig. 4).

C. Electric dipole and qubit coupling

The key feature of our proposal is that, by pulling the donor-bound electron halfway towards the Si/SiO_2 interface, we create a large electric dipole (the negative charge of the electron is displaced from the positive charge of the P atom). This, in turn, allows coupling multiple flip-flop qubits using the electric dipole-dipole interaction (Fig. 5). This coupling can be of significant strength (of order 10 MHz) across distances in the range 100 – 500 nm. High-fidelity 2-qubit logic gates can be performed in 40 ns under these conditions, and the predicted gate fidelity exceeds 99% with realistic noise models (Fig. 6).

Importantly, the electric dipole can be turned completely off, by simply bringing the electron back to the donor. Therefore, one can envisage a dense two-dimensional array of P atoms where logic operations are interspersed with idle qubits, with zero cross-talk (Fig. 7).

978-1-7281-1988-5/18 $31.00 © 2018 IEEE

D. Coupling to microwave photons

The electric dipole associated with the flip-flop qubit allows coupling it to the vacuum electric fluctuations of microwave resonators, thereby implementing a circuit-quantum electrodynamics architecture with donor spins in silicon. Coupling strengths can exceed 1 MHz, and permit the entanglement of flip-flop qubits separated by centimeter distances.

III. EXPERIMENTAL PROGRESS

A. Device fabrication

We have fabricated several batches of devices designed for the electrical control and coupling of flip-flop qubits. The devices comprise an isotopically enriched ^{28}Si epilayer, thermally-grown SiO_2, diffused ohmic contacts, and aluminum gates fabricated by electron beam lithography, metal evaporation and lift-off. The P donors for hosting the qubits are introduced by ion implantation. The device design is optimized to allow maximum tunability of the key parameters that govern the physics of the flip-flop qubit, in particular the tunnel coupling between the donor and the quantum dot at the Si/SiO_2 interface (Fig. 8).

Superconducting microwave cavities have also been developed, wherein the flip-flop qubit tuning gates are integrated with the center conductor of a coplanar waveguide structure. In this way, energy quanta can be swapped between flip-flop qubit and cavity via the cavity electric field, whereas classical fields are used to fine-tune the parameters of the system, or switch the coupling on or off (Fig. 9).

B. Observation of flip-flop resonance

We have succeeded in observing the flip-flop resonance in a standard donor spin qubit device – not optimized for the electrical drive of flip-flop qubits. This device comprises a coplanar waveguide antenna, terminated by a short circuit in order to maximize the ac magnetic field at its tip, as required for conventional magnetic resonance experiments. Nonetheless, even the spurious electric field produced by such structure proved sufficient to observe the $|\downarrow\Uparrow\rangle \leftrightarrow |\uparrow\Downarrow\rangle$ resonance at the expected frequency (Fig. 10). This is a promising result for the coherent control of the flip-flop states in dedicated devices.

C. Deterministic single-ion implantation

Ion implantation is very mature process for the semiconductor industry [7]. For the purpose of creating regular arrays of P donor qubits in silicon, we are developing a method to deterministically implant single atoms, and control their final position to a precision of order 10 nm. This method is cited in the semiconductor roadmap for potential applications to ultra-scaled classical CMOS [8].

We have developed on-chip electrodes to register signals from electron-hole pairs induced by the implantation of a single ion [9]. The signals can be used for fast beam blanking and to reposition a nanostencil collimator, which acts as a movable implantation mask, and directs each implanted ion to the required location for the large scale architectures (Fig. 11).

IV. SCALE-UP AND INTEGRATION WITH CLASSICAL ELECTRONICS

The key advantages of the flip-flop qubit relate to the potential ease of integration with classical silicon MOS electronics, and the associated interconnect technology.

We envisage a planar, 2-dimensional array of deterministically-implanted P donors, with inter-donor distance of 200 nm. This is a factor 4 wider than the minimum pitch for interconnects in the 14 nm node [10] – a feature unmatched by any other silicon-based qubit technology. With this spacing, it becomes possible to intersperse between the qubits not only the interconnects, but also (semi-) classical qubit readout devices, such as single-electron transistors. These yield (essentially digital) baseband signals in response to the qubit state, thus avoiding the need for the microwave circuitry required in alternative readout methods.

Scaling up to useful size quantum computers also requires implementing quantum error correction protocols. The ability to form 2D arrays of qubits with nearest-neighbor coupling and associated read-out makes the flip-flop system naturally suited for the implementation of a surface code, which is among the most tolerant to errors in the physical qubit. However, our system could potentially go even further, by exploiting the ability to selectively couple pair of qubits well beyond the nearest neighbor. We will therefore explore the computational advantages of operation and correction schemes that exploit this unique feature of the flip-flop qubit system.

ACKNOWLEDGMENT

This research was funded by the Australian Research Council (CE11E0001027, CE170100012) and the US Army Research Office (W911NF-13-1-0024, W911NF-17-1-0200).

REFERENCES

[1] F. A. Zwanenburg et al., "Silicon Quantum Electronics," *Reviews of Modern Physics*, vol. 85, pp. 961-1019, July 2013.

[2] J. J. Pla et al., "A single-atom electron spin qubit in silicon," *Nature*, vol. 489, pp. 541-545, September 2012.

[3] J. J. Pla et al., "High-fidelity control and readout of a nuclear spin qubit in silicon," *Nature*, vol. 496, pp. 334-338, April 2013.

[4] J. T. Muhonen et al., "Storing quantum information for 30 seconds in a nanoelectronic device," *Nature Nanotech.*, vol. 9, pp. 986-991, December 2014..

[5] M. Veldhorst et al., "A two-qubit logic gate in silicon," *Nature*, vol. 526, pp. 410-414, October 2015.

[6] G. Tosi et al., "Silicon quantum processor with robust long-distance qubit couplings," *Nature Commun.*, vol. 8:450, September 2017.

[7] J.M. Poate, K. Saadatmand, "Ion beam technologies in the semiconductor world", *Rev. Sci. Instrum.*, vol. 73, pp. 868-872, February 2002.

[8] 2011 International Semiconductor Roadmap (ITRS - International Technology Roadmap for Semiconductors, Emerging Research Materials, Section 6: Single ion implantation, www.itrs2.net/2011-itrs.htm).

[9] J.A. van Donkelaar et al., "Single atom devices by ion implantation", *J. Phys.: Condens. Matt.*, vol. 27, pp. 154204, 2015.

[10] K. Fischer et al., "Low-k interconnect stack with multi-layer air-gap and tri-metal-insulator-metal capacitors for 14 nm high volume manufacturing", *2015 IEEE International Interconnect Technology Conference and 2015 IEEE Materials for Advanced Metallization Conference (IITC/MAM)*, Grenoble, May 2015, pp. 5-8.

978-1-7281-1988-5/18 $31.00 © 2018 IEEE

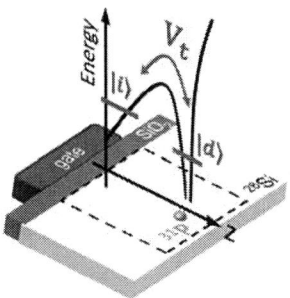

Fig. 1. Schematic depiction of the flip-flop qubit system. An ion-implanted ^{31}P donor atom is placed ~15 nm below a Si/SiO$_2$ interface. This creates two confined charge states: one at the donor, denoted with $|d\rangle$, one at the interface, $|i\rangle$. The two potential wells are coupled by a tunnel coupling V_t. A metallic gate controls the charge state of the donor.

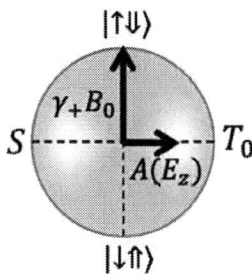

Fig. 2. Bloch sphere of the flip-flop qubit. The basis states are $|\uparrow\Downarrow\rangle$ and $|\downarrow\Uparrow\rangle$, while $|S\rangle$ and $|T_0\rangle$ are their singlet and triplet superpositions, respectively. In an external magnetic field B_0, the qubit energy splitting is $\hbar(\gamma_e + \gamma_n)B_0$, with γ_e and γ_n the electron and nuclear gyromagnetic ratios. The hyperfine coupling A is controlled by the vertical electric field E_z and appears as a transverse term in the flip-flop qubit subspace.

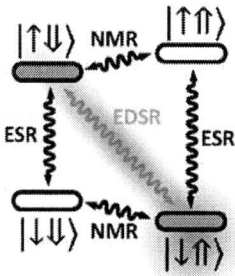

Fig. 3. Energy level diagram of a ^{31}P donor atom, showing the electron-nuclear basis states. Transitions that change only the nuclear state are obtained by Nuclear Magnetic Resonance (NMR), while those changing only the electron are implemented by Electron Spin Resonance (ESR). The flip-flop qubit transition between the $|\uparrow\Downarrow\rangle$ and $|\downarrow\Uparrow\rangle$ states is an Electric Dipole Spin Resonance (EDSR).

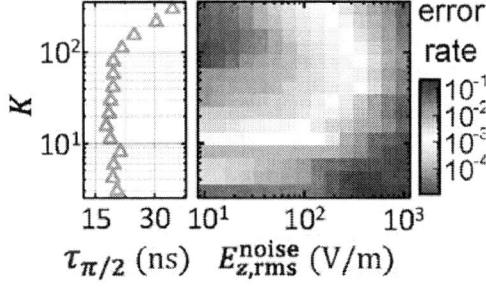

Fig. 4. Theoretical estimate of the error rate for a $\pi/2$ operation on the flip-flop qubit, as a function of the adiabaticity parameter K and the amplitude of the electric field noise, $E_{z,rms}^{noise}$. K describes the rate at which the control field is turned on and off. The horizontal axis of the left panel shows the total duration of the $\pi/2$ operation. With noise amplitudes of order 100 V/m, the control error is well below 10^{-3}, implying control fidelity > 99.9%.

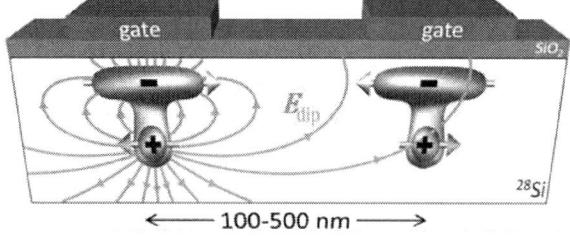

Fig. 5. Schematic depiction of a pair of flip-flop qubits coupled by electric dipole interaction. The electric dipole arises from pulling the electron wavefunction halfway towards the Si/SiO$_2$ interface. The resulting dipole field mediates a qubit-qubit interaction stronger than 1 MHz over distances up to 500 nm.

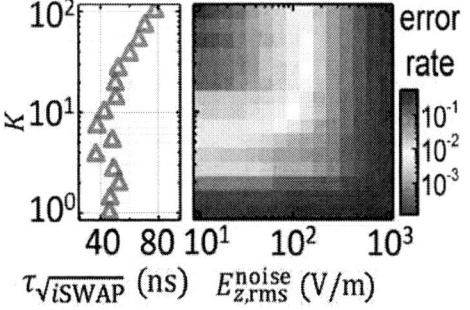

Fig. 6. Theoretical estimate of the error rate for an entangling 2-qubit $\sqrt{iSWAP}$ operation between flip-flop qubits. A typical electric field noise value of 100 V/m yields a 2-qubit logic gate fidelity in excess of 99%.

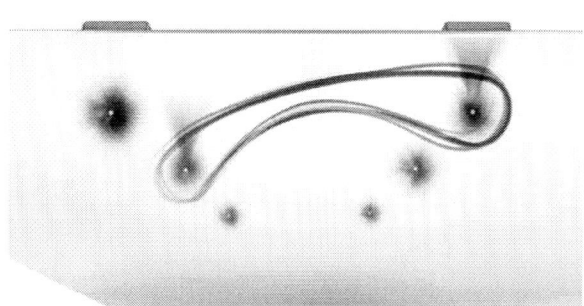

Fig. 7. Artist's impression of a two-dimensional array of flip-flop qubits, where pairs of qubits are selected for an entangling logic operation by inducing an electric dipole on each atom in the pair, whereas all other qubits are left idle and decoupled. Due to the long-range nature of the dipole interaction, quantum logic gates can be performed beyond the nearest-neighbor, without interfering with other qubits in the vicinity.

978-1-7281-1988-5/18 $31.00 © 2018 IEEE

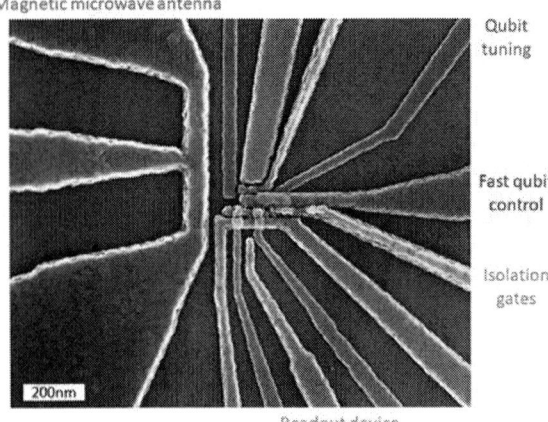

Fig. 8. False-color scanning electron micrograph of a prototype flip-flop qubit device. It comprises a magnetic microwave antenna and a single-electron transistor for spin readout, plus a set of gates to control the quantum state of the flip-flop qubit, and the tunnel coupling between the donor, the interface state, and the adjacent charge reservoirs.

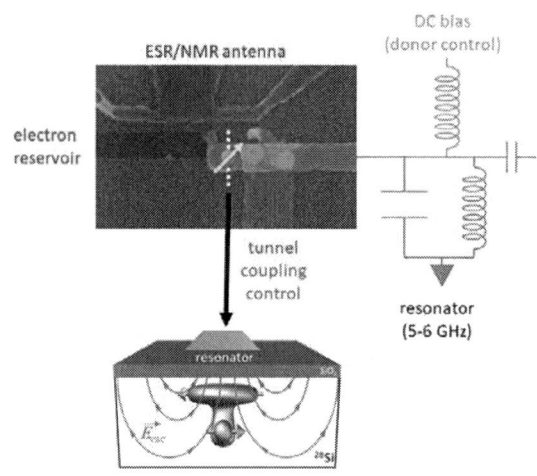

Fig. 9. False-color scanning electron micrograph and circuit schematic of a setup to couple a flip-flop qubit to a superconducting microwave resonator. The electric dipole induced on the ^{31}P atom couples to the vacuum electric field of the cavity, allowing long-distance entanglement with other qubits coupled to the same cavity field.

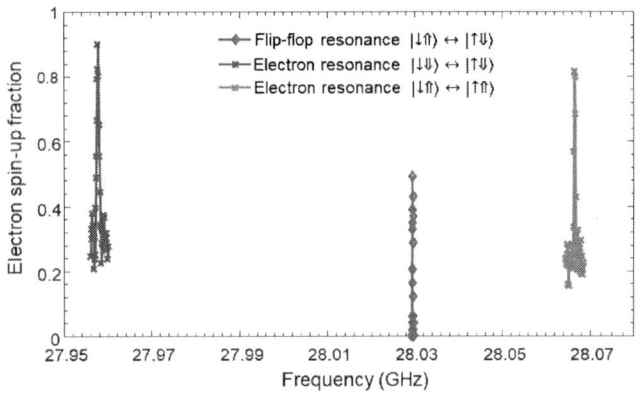

Fig. 10. Experimental observation of the flip-flop resonance in a non-optimized device. The resonances at the left and the right are the ordinary electron spin transitions.

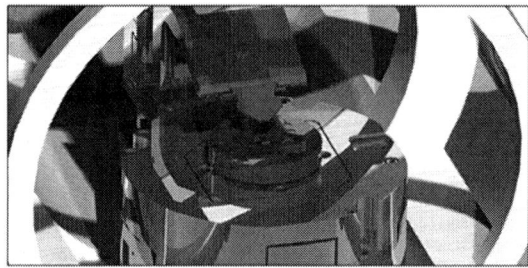

Fig. 11. Schematic depiction of the deterministic single-ion implantation system. The piezoactuated nanostencil is shown in red.

978-1-7281-1988-5/18 $31.00 © 2018 IEEE   132

Qubit Device Integration Using Advanced Semiconductor Manufacturing Process Technology

R. Pillarisetty, N. Thomas, H.C. George, K. Singh, J. Roberts, L. Lampert, P. Amin, T.F. Watson, G. Zheng, J. Torres, M. Metz, R. Kotlyar, P. Keys, J.M. Boter[*], J.P. Dehollain[*], G. Droulers[*], G. Eenink[*], R. Li[*], L. Massa[*], D. Sabbagh[*], N. Samkharadze[*], C. Volk[*], B. P. Wuetz[*], A.-M. Zwerver[*], M. Veldhorst[*], G. Scappucci[*], L.M.K. Vandersypen[*], J.S. Clarke

Intel Corporation, Technology and Manufacturing Group, Hillsboro, OR 97124, USA
[*]QuTech and Kavli Institute of Nanoscience, TU Delft, P.O. Box 5046, 2600 GA Delft, The Netherlands

Abstract

Quantum computing's value proposition of an exponential speedup in computing power for certain applications has propelled a vast array of research across the globe. While several different physical implementations of device level qubits are being investigated, semiconductor spin qubits have many similarities to scaled transistors. In this article, we discuss the device/integration of full 300mm based spin qubit devices. This includes the development of (i) a ^{28}Si epitaxial module ecosystem for growing isotopically pure substrates with among the best Hall mobility at these oxide thicknesses, (ii) a custom 300mm qubit testchip and integration/device line, and (iii) a novel dual nested gate integration process for creating quantum dots

Introduction

The phenomena of quantum superposition and it's potential to provide an exponential speedup in compute for certain applications, such as chemistry, machine learning, and cryptography, has propelled a vast array of industry, university, and government research across the globe. This research is expansive and spans the entire stack from the device/qubit level all the way to quantum algorithms (Figure 1). While several different physical implementations of device level qubits are being investigated [1-13], the semiconductor [1-9] and superconducting [10] based approaches are most compatible with integrated circuit manufacturing techniques. While the superconducting qubits have similarities with interconnect and BEOL integration [14], semiconductor spin qubits have many similarities to scaled transistors [15-18]. Figure 2 shows a schematic of a typical multi-gate quantum dot structure, with separate plunger and barrier gates to control the local electrostatic potential and confine electrons underneath the plunger gates. The accumulation gates, act as tips in a conventional transistor, and serve to space the implanted Source/Drain regions far from the active dots. Many of the challenges limiting spin qubit performance, such as CD variation, dielectric interfaces, nuclear spins, charge impurities, and epi defects, are process control related and require a high volume state-of-the-art manufacturing line to solve.

In this article, we discuss the device/integration of full 300mm based spin qubit devices. A ^{28}Si based materials ecosystem has been developed, which achieves high quality Si quantum well (QW) and Si MOS substrate growth with low temperature Hall mobility of 400,000 and 10,000 cm^2/V*s, respectively. To transition from transistor to qubit integration, a customized qubit testchip and process flow were created. Transistors fabricated based on qubit design rules exhibited ideal electrical characteristics. Additionally, a novel dual nested gate integration process for the formation of independent plunger and barrier gates was implemented to enable quantum dot devices.

^{28}Si Epitaxy and Substrate Growth

One of the grand challenges required for spin qubit devices to be scaled up to enable a viable quantum compute technology is improving the spin coherence lifetime [9]. Poor coherence times arising from nuclear spins interacting with qubit electron spins leads to loss of quantum information and errors when running quantum algorithms. In contrast, charge based transistor technology has no corresponding issue as drive current and mobility are not impacted by nuclear spin interactions. To address this problem Intel partnered with Urenco and Air Liquide to create an ecosystem to provide isotopically purified precursors for our 300mm epitaxial growth module at a scale that supports high volume manufacturing. Figure 3 provides high resolution cross sectional TEM images of our ^{28}Si based epitaxial growth process on 300mm substrates. Figure 3a shows a ^{28}Si MOS epilayer capped with thermal ^{28}SiO2. An electron diffraction pattern highlighting the film quality is provided in the inset. Figure 3b shows a biaxial tensile strained ^{28}Si quantum well (QW) structure with Si$_{70}$Ge$_{30}$ barriers, grown on a relaxed Si$_{70}$Ge$_{30}$ buffer. The majority of spin qubit research focuses on these silicon MOS [1-4] and QW [5-7] structures. To address film quality, a SIMS profile of a ^{28}Si epilayer grown on a natural Si 300mm substrate is shown in Fig. 4, which indicates a high isotopic purity > 99.9% in the ^{28}Si region. The corresponding electrical characterization of these device structures is given in Figure 5, which plots mobility vs charge density for the ^{28}Si MOS (a) and ^{28}Si QW structures shown in Fig 3a and 3b, respectively at temperature (T) = 1.7 K. The data are obtained from gated Hall bar devices with gate stacks

978-1-7281-1988-5/18 $31.00 © 2018 IEEE

having an equivalent oxide thickness (EOT) of 17 nm (a) and 22 nm (b). These data are among the highest mobility reported in the literature at these gate oxide thicknesses [19,20].

From Transistors to Qubits

While spin qubits are similar to transistors, there are significant complexities and challenges to creating a full 300mm qubit integration/device line. These are highlighted in Figure 6, which displays schematics of transistor, quantum dot, and qubit device structures along with the corresponding e-test metrics necessary to gauge device quality. Quantum dots require two sets of gates (plungers and barriers) to independently tune the electrostatic potential in the device and confine electrons underneath the plunger gates. Additionally, long accumulation gates are required to space the n+ S/D doping impurities away from the confined electrons. The e-test is also significantly more complex and involves investigating single electron charging signatures, charge stability scans, and charge noise all at low temperature < 2 K. Full qubits require additional process and design steps that involve the BEOL integration of interconnects that serve as RF transmission lines for electron-spin-resonance, which is needed to couple to and manipulate an individual electron spin in the device. These measurements are even more complex involving characterization of spin relaxation and coherence times, along with fidelity measurements of actual qubit operations, and are performed in dilution refrigerators at $T < 30$ mK.

In order to create a viable 300mm device/integration line, we have created a fully customized qubit testchip. The device designs and test structures were created based on customized design rules necessary to achieve the device and e-test requirements for quantum dots and qubits. An example of a wafer fully processed with this testchip is shown in Figure 7a. A single die showing a variety of different test structures, along with individually diced 55, 23, 15, and 7 gate arrays are shown in Figure 7b and 7c, respectively. A single wafer produces over 10000 distinct qubit test structures, which enables the potential for a significant scale up in volume statistical data to help accelerate qubit device research and development.

Transistors based on Qubit Design Rules

The initial steps in this qubit process flow involve the traditional transistor modules of shallow-trench-isolation (STI), gate dielectric formation, and low resistance Ohmic contacts. In that sense, a transistor flow is used to begin validating the testchip designs. The transistor test structures utilized were designed using the exact same design rules used for the qubit array test structures. Figure 8 shows a high resolution cross sectional TEM image of such a transistor structure with fin based STI and high-k metal gate electrodes. The design mimics a qubit array, where one of the fins would be used as a single electron transistor (SET) for read out of the qubit device. The electrical characterization of these transistors is shown in Figure 9, which plots an $I_D V_G$ trace of a typical device. Here devices fabricated from Si MOS substrates made with ^{28}Si and natural Si are overlaid showing both healthy and essentially identical results. Additionally, Figure 10, which plots the cross wafer distribution of V_T, subthreshold slope (SS), I_{GATE}, and I_{BODY} show clean distributions with no observable difference for ^{28}Si vs natural Si. These data validate that isotopically purified ^{28}Si and natural silicon process similarly through the various etches, cleans, polishes, implants, film depositions, and thermal Dt's used in standard front end integration.

Dual Nested Gate Quantum Dot Integration

A 3D model guiding quantum dot device design, which incorporates accurate effects of layout and stress is presented in Figure 11. A device structure consisting of plunger and barrier gates separated by dielectric spacers is provided in Figure 11a. The calculated conduction band (Fig. 11b) and electron density (Fig 11c) profiles at $T = 4$ K are provided predicting the ability to confine electrons under the plunger gates. A high resolution TEM image showing a dual nested gate integration process is shown in Figure 12a, highlighting the isolated plunger, barrier, and accumulation gates. A planar STEM image of a similar structure is presented in Figure 12b, which highlights the active nested gate device region, the gate fan out, and the interconnect bussing. One of the key advantages of a 300mm process line is the ability to generate a large number of devices, trend statistical volume data and screen the best devices for subsequent measurements and analysis. This is highlighted in Figure 12c, where we plot room temperature V_T matching data for the three plunger gates in the nested gate structure for many devices across an entire wafer. The gate uniformity in quantum dots at low temperature is well known to correlate to reliable qubit operation. As electrostatics is temperature independent, identifying good uniformity bits can help screen and identify promising devices for subsequent low temperature characterization. Figure 12d, shows one example of a quantum dot array with matched V_T and SS for all plunger gates.

Conclusion

Quantum computing is an emerging technology with the potential for an exponential compute speedup for certain applications. Semiconductor spin qubits, which have many similarities to transistors, are one promising potential qubit technology. Many of the challenges limiting spin qubit performance, are fundamentally process control related and require a high volume state-of-the-art manufacturing line to solve. To address this we have developed full 300mm semiconductor spin qubit devices. This includes (i) a ^{28}Si epitaxial module ecosystem for growing isotopically pure substrates with among the best Hall mobility at these oxide thicknesses, (ii) a custom 300mm qubit testchip and integration/device line (iii) a novel dual nested gate integration process for creating quantum dots. This 300mm line provides a large volume of devices to e-test, which is something new to the quantum computing research community. It allows for volume statistical data and screening of leading edge devices for subsequent measurement and analysis. Given the complexity and low temperature requirements of quantum dot and qubit e-test characterization, the field will require a significant ecosystem development to enable e-test to keep pace with the high volume output of a 300mm fab.

References

[1] M. Veldhorst et al., *Nature* **526**, 410-414 (2015)
[2] C.H. Yang et al., arXiv:1807.09500 (2018)
[3] W. Huang et al., arXiv:1805.05027 (2018)
[4] R Maurand et al., *Nature Communications* **7**, 13575 (2016)
[5] J. Yoneda et al., *Nature Nanotechnology* **13**, 102-106 (2018)
[6] T.F. Watson et al., *Nature* **555**, 633-637 (2018)
[7] D.M. Zajac et al., *Science* **359**, 439-442 (2018)
[8] J.T. Muhonen et al., *Nature Nanotechnology* **9**, 986-991 (2014)
[9] F.A. Zwanenburg et al., *Rev. Mod. Phys.* **85**, 961 (2013)
[10] M.H. Devoret et al., arXiv:cond-mat/0411174 (2004)
[11] S. Debnath et. al., *Nature* **536,** 63-66 (2016)
[12] N. Friis et al., *Phys. Rev. X* **8**, 021012 (2018)
[13] T.H. Taminiau et al., *Nature Nanotechnology* **9**, 171-176 (2014)
[14] K. Fischer *et al.*, *IITC Tech. Dig.*, pp. 5-8 (2015)
[15] S. Natarajan, et al, *IEDM Tech. Dig.*, pp. 3.7.1 – 3.7.3 (2014)
[16] C. Auth et al., *VLSI Tech. Dig.*, pp. 131-132 (2012)
[17] K. Mistry, et al., *IEDM Tech. Dig*, p. 247-250 (2007)
[18] T. Ghani, et al., *IEDM Tech. Dig.*, pp. 978-980 (2003)
[19] X. Mi, et al., *Appl. Phys. Lett.* **110**, 043502 (2017)
[20] J.S. Kim et al., *Appl. Phys. Lett.* **110**, 123505 (2017)

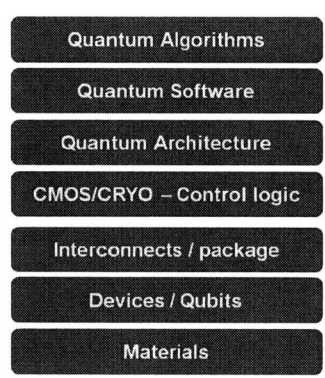

Fig 1: Stack of various elements being pursued as part of the Intel-QuTech collaboration spanning materials and devices to quantum algorithms.

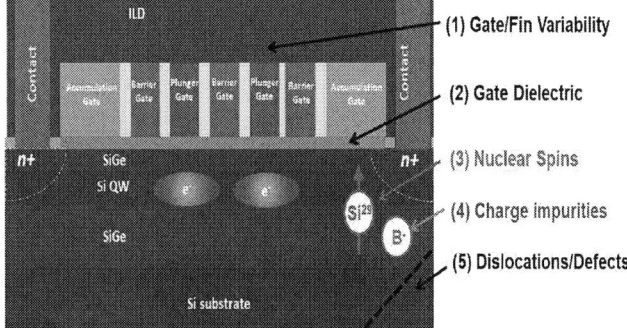

Fig 2: Schematic illustrating a multi-gate quantum dot structure. The plunger and barrier gates tune the electrostatic potential to confine electrons under the plunger gates. The accumulation gates space the n+ S/D implant from the active dots.

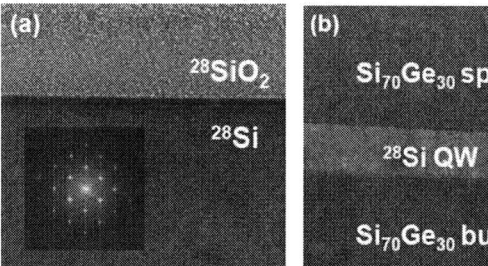

Fig 3: High resolution cross sectional TEM image of a ^{28}Si based epitaxial growth process on 300mm substrates. (a): ^{28}Si MOS epilayer capped with thermal ^{28}SiO$_2$. An electron diffraction pattern is provided in the inset (b) Tensile biaxial strained ^{28}Si QW structure grown on a relaxed Si$_{70}$Ge$_{30}$ buffer.

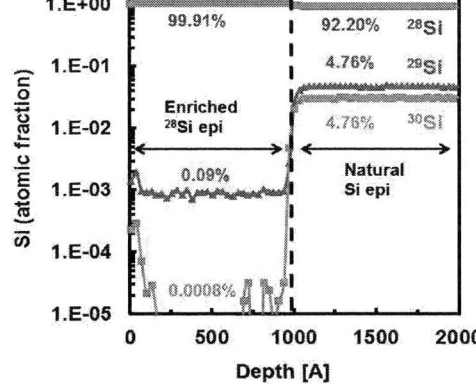

Fig 4: SIMS profile of ^{28}Si epilayer grown on a natural Si 300mm substrate, showing a high purity ^{28}Si concentration > 99.9%.

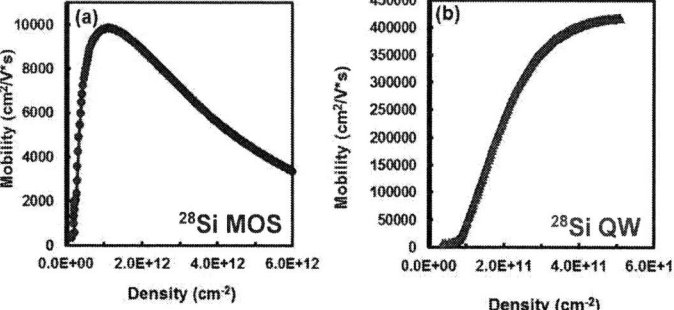

Fig 5: Mobility vs charge density for the ^{28}Si MOS (a) and ^{28}Si QW structure shown in Fig 5a and 5b, respectively at T = 1.7 K. The data are obtained from gated Hall bar devices with EOT of 17nm (a) and 22 nm (b). These data represent among the highest mobility observed at these gate oxide thicknesses.

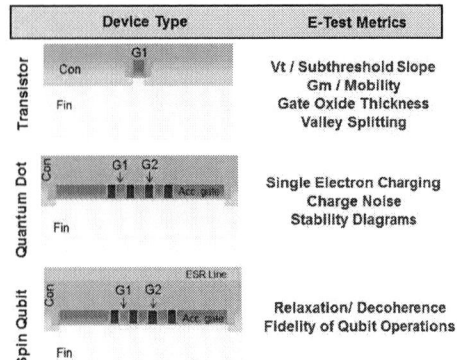

Fig 6: Schematics of transistor, quantum dot, and qubit device structures, and the corresponding e-test metrics needed for gauging device quality. Moving from transistors to qubits involves more complicated processing/masking, but also more complicated e-test techniques.

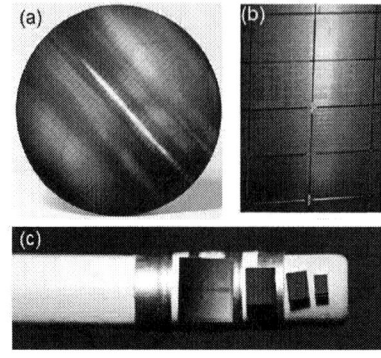

Fig 7: (a): Full 300mm wafer processed using Intel's customized qubit testchip. (b): single die, and (c): individually diced 55, 23, 15, and 7 gate arrays. Each wafer holds over 10,000 qubit test structures.

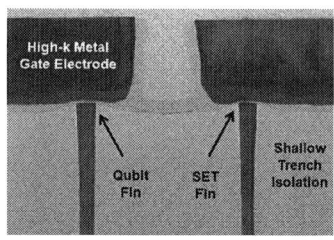

Fig 8: Cross sectional TEM image showing paired transistor structures with fin based shallow-trench-isolation and high-k metal gate electrodes. The design mimics a qubit array, where the adjacent device is used as a SET for read out of the qubit device.

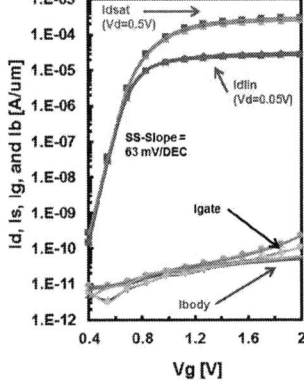

Fig 9: Transistor $I_D V_G$ trace for Si MOS transistors test structures designed with qubit design rules. ^{28}Si and natural Si devices are overlaid and matched.

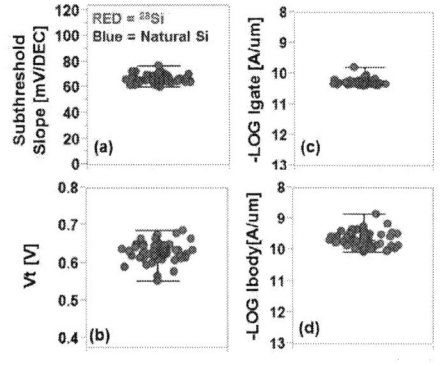

Fig 10: Cross wafer distributions comparing ^{28}Si and natural Si MOS based qubit design rule transistor structures for (a): V_T, (b): subthreshold slope, (c): gate leakage, and (d): body leakage

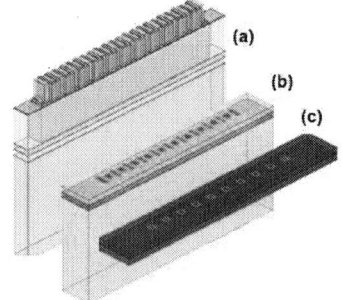

Fig 11: Si QW device structure consisting of plunger and barrier gates separated by dielectric spacers (a). The calculated conduction band (b) and electron density (c) profiles at T=4K are provided predicting the ability to confine electrons under the plunger gates.

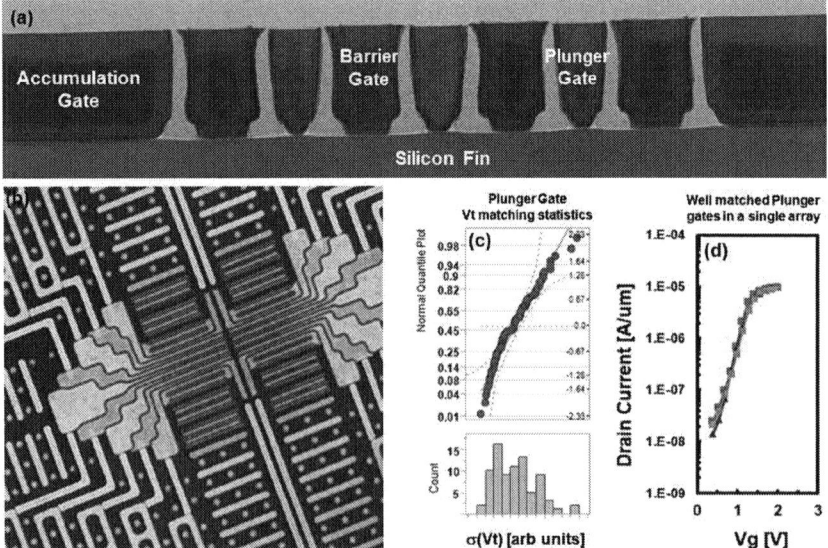

Fig 12: (a): High resolution TEM image showing dual nested gate integration with isolated plunger, barrier, and accumulation gates (b): A planar STEM image of the nested gate structure showing the fan out and interconnect bussing (c): Cross wafer distribution of V_T matching for three plunger gates within a single array and (d): quantum dot array with well-matched V_T and subthreshold slope for all three plunger gates.

Silicon Isotope Technology for Quantum Computing

Satoru Miyamoto and Kohei M. Itoh
School of Fundamental Science and Technology and Spintronics Research Center,
Keio University, Yokohama 223-8522, Japan

Abstract— We present isotopically engineered Si-28/SiGe heterostructures for development of silicon-based quantum computers using a standard silicon CMOS integration technology. Our Si-28 quantum-wells are well-strained and demonstrate high electron mobility and large valley-splitting. These properties provide promising platforms for realization of highly integrated spin qubits working together with silicon CMOS circuits.

I. INTRODUCTION

Isotope engineering of silicon semiconductors provides a solution to unwanted decoherence, i.e., loss of quantum information, in Si-based quantum computing [1]. At first, we adopted the state-of-the-art in isotope engineering to support the development of Si-MOS nanoelectronic devices, where two-qubit proof-of-concept experiments were demonstrated successfully using donors [2]. Furthermore, multiqubit architecture of quantum computing was proposed for the donor qubits embedded in the Si-28 epilayer on SiGe virtual substrates [3], since strain-induced valley repopulation and lattice distortion are accompanied with a local controllability of hyperfine coupling [4]. Following such a success, the isotope engineering was extended to support the development of standard CMOS lithography-based Si quantum computer development utilizing single-electron spin confined in Si quantum dots (QDs) as qubits [5]. An alternative approach is offered by strained-Si/SiGe quantum-well (QW) architectures, where QDs are placed away from gate-oxide interface traps that can work as the sources for temporally-fluctuating charges to decohere electron spin qubits. In fact, we achieved isotope enrichment of Si-QW layers and efficiently removed magnetic nuclear-spin noise, which enabled to demonstrate gate fidelity exceeding 99.9% on the basis of a fast electrical qubit control [6]. However, despite of the spatial separation from the interface traps, the spin coherence remains subject to charge noise whose origin is possibly embedded in the grown QW structures. Hence the removal of the noise sources is necessary to accomplish a large-scale quantum computing based on the Si-28/SiGe heterostructures. In this work, in addition to the conventional continuous growth from the virtual substrate to the top layer, we present promising, Si-28 epilayer and QW formed on atomically flat SiGe virtual substrates made by chemically-mechanically polishing (CMP).

II. SI-28 EPILAYERS FOR DONOR-BASED QUBITS

A series of sample fabrications began from growth of the SiGe on 6-inch Si(001) wafers by means of gas-source molecular beam epitaxy (GS-MBE). While the substrate temperature was kept at 570 °C, the graded SiGe buffer was grown by incrementally varying GeH_4 flow rate at fixed Si_2H_6 flow rate. Subsequently, the SiGe relaxed buffer layer was formed at the same Ge concentration x_{Ge} as the top of the graded buffer layer. For controlling the accumulated strain in Si-28 epilayers, the $Si_{1-x}Ge_x$ virtual substrates were grown at various Ge content x_{Ge} = 5 - 30%, and the CMP on the grown surface was once carried out to reset the surface roughness to $\sim$ 0.1 nm. Subsequently, the SiGe buffer was regrown with keeping x_{Ge}, which was followed by deposition of the 20-nm-thick Si-28 epilayer employing isotopically purified $^{28}SiH_4$ source at 620 °C. Figure 1 displays the cross-sectional transmission electron microscope (TEM) images of the grown Si-28/$Si_{0.90}Ge_{0.10}$ epilayer. The occurrence of dislocations is restricted in the graded SiGe layer so that they cannot reach up to the SiGe relaxed layer and Si-28 epilayer. In addition, the resulting Si-28 epilayer on the polished SiGe virtual substrates possesses the surface roughness suppressed at the level of $\sim$ 1 nm. In Figs. 2(a)-2(d), the (224) X-ray diffraction reciprocal space mappings (XRD-RSMs) are shown for the Si-28/SiGe epilayers having distinct x_{Ge}, which experience a various magnitude of strain induced by the SiGe virtual substrates. Moreover, the Raman scattering spectroscopy (RSS) with respect to the Si-Si modes was used to determine the in-plane strain in the Si-28 epilayers independently from the X-ray measurements [Fig. 3(a)]. As the SiGe-related peaks shift from the bulk-Si signal with increasing x_{Ge}, the RSS signals associated with the strained Si-28 layer appear on the shoulder of the SiGe peaks. The strained-Si peak shift provides the in-plane strain, which is given as a reciprocal lattice point on the XRD-RSMs [see Figs. 2(a)-2(d)]. Hence the Raman analysis represents good agreement with the XRD-RSM results, which indicates that the Si-28 epilayers are well-strained by relaxed SiGe virtual substrates.

III. SI-28 QUANTUM-WELLS FOR QUANTUM DOT QUBITS

A. Continuous growth on SiGe graded buffer

We first start from the continuous Si-28 QW growth with the thickness of $t_{Si-28} \sim$ 11 nm from the SiGe virtual substrate in the same growth sequence. Here, after the growth temperature was set back to 570 °C, the SiGe barrier layer was deposited with a thickness of $t_{cap} \sim$ 40 nm. Eventually, the growth surface was terminated with a thin Si layer of $\sim$ 3 nm to protect against oxidation of the SiGe layer. Figures 4(a) and 4(b) display the atomic force microscope (AFM) images of the sample surface grown at two different Ge contents, $x_{Ge} \sim$ 25% and $\sim$ 15%, with setting the grading rate at $R_{grad} \sim$ 10%/μm. Then, the cross-hatch patters can be observed reflecting inhomogeneous strain fields originating from the underlying SiGe buffer. The AFM analysis also reveals that the reduction in x_{Ge} suppresses considerably the root mean square (RMS) of surface roughness [see Figs. 4(a) and 4(b)].

978-1-7281-1988-5/18 $31.00 © 2018 IEEE

In addition, when the $Si_{0.85}Ge_{0.15}$ buffer layer is deposited at the higher growth rate $R_{SiGe} \sim 0.50$ nm/s, the surface roughness is lowered to a few nm as shown in Fig. 4(c). Figure 5 highlights the formation of the well-defined Si-28/$Si_{0.85}Ge_{0.15}$ layer on a thicker graded SiGe buffer layer ($R_{grad} \sim 5\%/\mu m$). In particular, the upper interface with the SiGe barrier layer seems to be fluctuated solely at the atomic level compared with the bottom interface. Figure 3(b) shows the Raman spectra of the Si-28 QW heterostructures grown at $x_{Ge} \sim 25\%$ and $\sim 15\%$. From the ^{28}Si-derived peak positions, the in-plane strain accumulated in the QW is estimated to be $\varepsilon_{//} \sim 0.53\%$ and $\sim 1.25\%$ for the respective samples. Again, these values are consistent with the XRD-RSM results [see Figs. 6(a) and 6(b)]. Moreover, the secondary ion mass spectroscopy (SIMS) was performed for the Si-28/$Si_{0.85}Ge_{0.15}$ QW [Fig. 7(a)], showing that the ^{29}Si isotopes having nuclear spins are effectively depleted in the Si-28 QW layers. Also, it is worth noting that Ge segregates into the QW from the bottom interface during the growth, displacing the Ge dip position towards a shallow depth. This fact can be consistently observed as a hidden bottom interface in Fig. 5(b). Furthermore, the low-temperature electron mobility in the undoped Si-28/$Si_{0.85}Ge_{0.15}$ QWs having different layer thickness is measured on the Hall-bar devices stacked by gate/Al_2O_3 structures. In Fig. 8(a), the high mobility of $\mu \sim 2 \times 10^5$ cm^2/V·s is demonstrated at the maximum density $n \sim 3 \times 10^{11}$ cm^{-2}. Here, a weak dependence on the SiGe capping-layer thickness t_{cap} implies a sufficient suppression of remote impurity scattering from the Al_2O_3 gate-insulator.

B. Regrowth on polished SiGe virtual substrates

Since the above continuous growth resulted in larger surface roughness at a higher x_{Ge} [see Fig. 4(a)], the polishing of SiGe virtual substrates by terminating the growth was adopted to form the Si-28/$Si_{0.70}Ge_{0.30}$ QW samples. The regrown surface was investigated after deposition of 200-nm-thick SiGe buffer layers [Fig. 9], suggesting that 1-nm-level roughness was achieved by tuning the growth rate and temperature. On such a flat virtual substrate, the Si-28/$Si_{0.70}Ge_{0.30}$ QW was grown at the fixed temperature of $T_{regrowth} = 620$ °C. Here the diffusion of background impurities from the regrown interface to the Si-28 QW region was negligible as seen from Fig. 7(b). The sample was processed into the gated Hall-bar devices for the low-temperature mobility measurement. As the result we obtained $\mu \sim 5 \times 10^4$ cm^2/V·s at the electron density $n \sim 6 \times 10^{11}$ cm^{-2} [Fig. 8(b)]. Also, because the regrowth onto the planarized SiGe virtual

substrates suppressed the upper QW interface fluctuation [see Fig. 10(a)], the valley-splitting measurements in an integer quantum-Hall regime found a large magnitude of several hundred μeV. When comparing with Fig. 10(b), the lower-temperature regrowth is likely to improve the upper interface even though the bottom interface is still affected by the Ge segregation. Additionally, Figures 11(a)-11(c) show the XRD-RSMs for the Si-28/$Si_{0.70}Ge_{0.30}$ QWs capped by the SiGe barrier layer at the lower temperature, along with the reference, uncapped Si-28 epilayer sample. Longitudinal expansion of the Si-28 signal from the SiGe-related peak also supports the formation of the high-quality upper QW interface. Such a controlled QW interface is essential for tuning the magnitude of valley splitting in the QD spin qubits. The development of Si-28/SiGe heterostructures with superior quality should lead to further suppression of magnetic/charge noise for the accelerated development of scalable quantum computers based on silicon.

ACKNOWLEDGMENT

We gratefully acknowledge Y. Hoshi, M. Kuroda, and N. Usami for fruitful collaborations, and S. Neyens, T. McJunkin, and M. A. Eriksson for the low-temperature transport measurements. This work has been supported by the KAKENHI (S) No. 26220602 and Spintronics Research Network of Japan.

REFERENCES

[1] K. M. Itoh and H. Watanabe, "Isotope engineering of silicon and diamond for quantum computing and sensing applications," MRS Communications vol. 4, p. 143 (2014).

[2] J. T. Muhonen et al., "Storing quantum information for 30 seconds in a nanoelectronics devices," Nat. Nanotechnol., vol. 9, pp. 986-991 (2014); A. Laucht et al., "Electrically controlling single-spin qubits in a continuous microwave field," Sci. Adv. vol. 1, p. e1500022 (2015).

[3] L. Dreher et al., "Electroelastic Hyperfine Tuning of Phosphorus Donors in Silicon," Phys. Rev. Lett. vol. 106, p. 037601 (2011).

[4] M. Usman et al., "Strain and electric field control of hyperfine interactions for donor spin qubits in silicon," Phys. Rev. B vol. 91, p. 245209 (2015).

[5] M. Veldhorst et al., "An addressable quantum dot qubit with fault-tolerant control-fidelity," Nat. Nanotechnol. vol. 9, pp. 981-985 (2014); M. Veldhorst et al., "A two-qubit logic gate in silicon," Nature vol. 526, pp. 410-414 (2015).

[6] J. Yoneda et al., "A quantum-dot spin qubit with coherence limited by charge noise and fidelity higher than 99.9%," Nat. Nanotechnol. vol. 13, pp. 102-106 (2018).

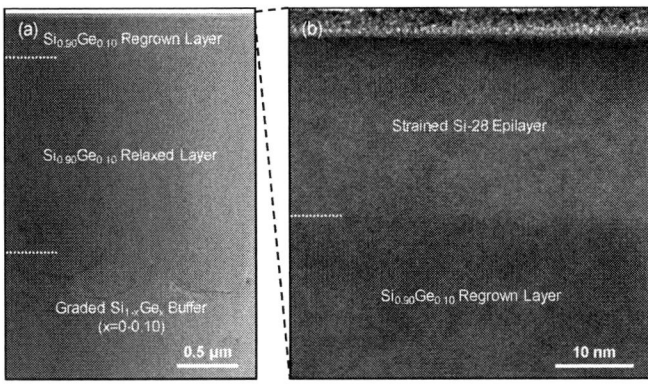

Fig. 1. (a) Overview and (b) enlarged cross-sectional TEM image of the strained Si-28 epilayer regrown on the polished $Si_{0.90}Ge_{0.10}$ virtual substrate.

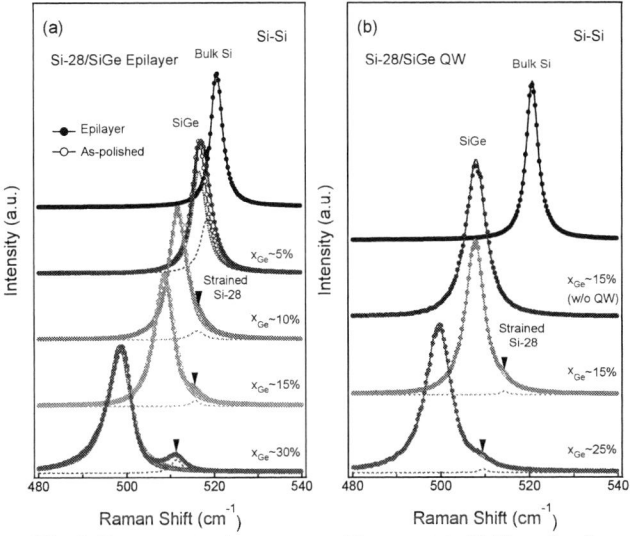

Fig. 3. Raman scattering spectra with respect to Si-Si modes for (a) the Si-28 epilayers regrown on the $Si_{1-x}Ge_x$ virtual substrates having x_{Ge} = 5 - 30 %, and (b) the Si-28/$Si_{1-x}Ge_x$ QW heterostructures grown at x_{Ge} ~ 15% and ~ 25%.

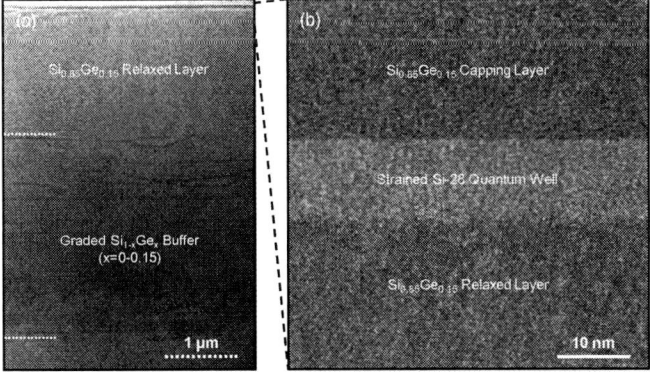

Fig. 5. (a) Overview and (b) enlarged cross-sectional TEM images of the grown Si-28/$Si_{0.85}Ge_{0.15}$ QW heterostructures.

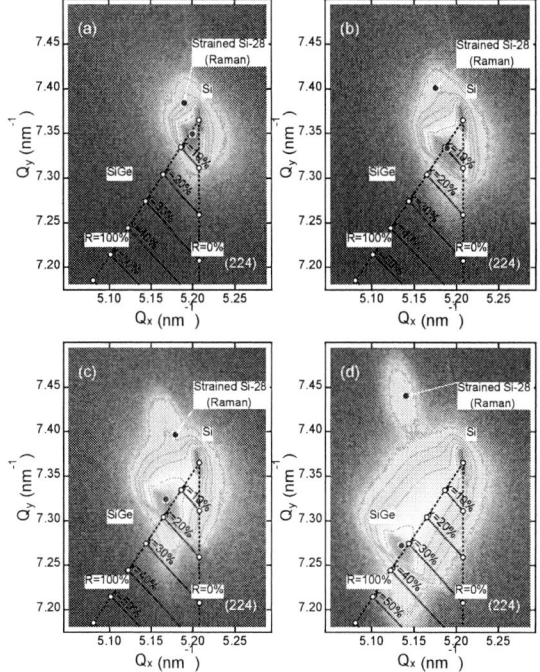

Fig. 2. (224) XRD-RSM of the Si-28/$Si_{1-x}Ge_x$ epilayers grown at various composition x_{Ge}: (a) ~ 5%, (b) ~ 10%, (c) ~ 15%, and (d) ~ 30%. The Si-28 epilayer thickness is around t_{Si-28} ~ 20 nm.

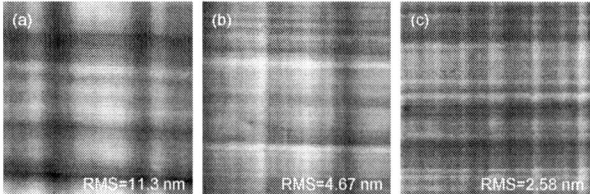

Fig. 4. AFM images of the Si-28/$Si_{1-x}Ge_x$ QW heterostructure surface at different growth conditions. The details of (a)-(c) are given in the main text, and all images are displayed in a scan range of 10 μm × 10 μm.

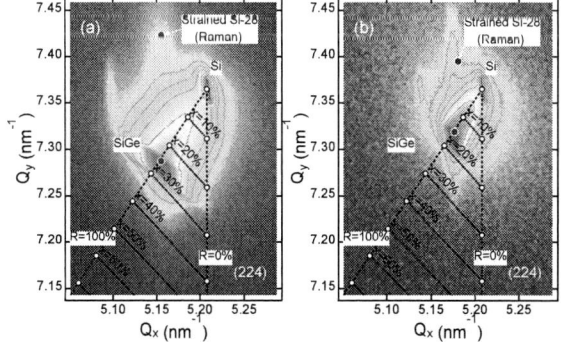

Fig. 6. XRD-RSM in the proximity of (224) reciprocal lattice points for the Si-28/$Si_{1-x}Ge_x$ QW heterostructures: (a) x_{Ge} ~ 25% and (b) x_{Ge} ~ 15%.

978-1-7281-1988-5/18 $31.00 © 2018 IEEE   139

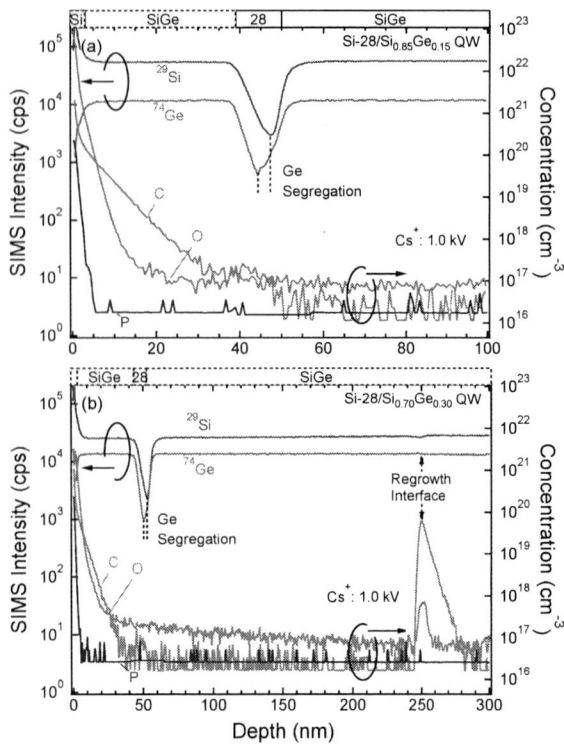

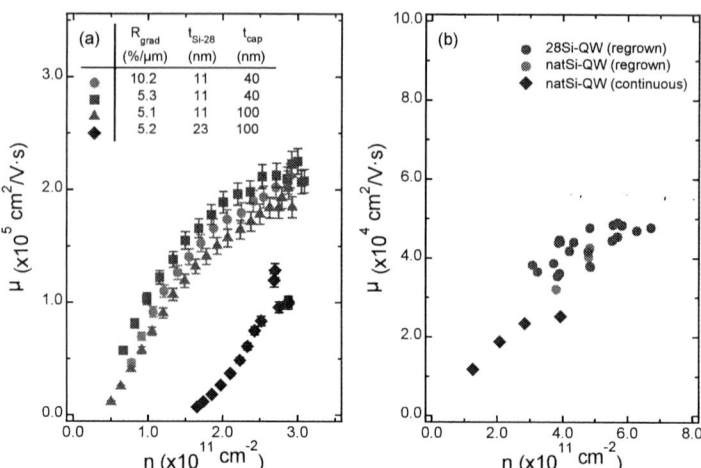

Fig. 8. Low-temperature electron mobility for Si-28/Si$_{1-x}$Ge$_x$ QW wafers: (a) continuously grown samples obtained at a fixed $x_{Ge} \sim 15\%$ and different R_{grad}, t_{Si-28}, and t_{cap}, and (b) a regrown sample at $x_{Ge} \sim 30\%$ compared with similar ^{nat}Si-QW heterostructures.

Fig. 7. SIMS profiles of Si-28/Si$_{1-x}$Ge$_x$ QW heterostructures: (a) a continuously grown sample at $x_{Ge} \sim 15\%$ and (b) a regrown sample at $x_{Ge} \sim 30\%$.

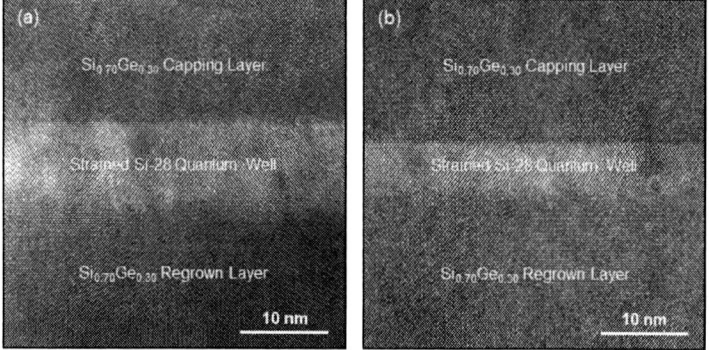

Fig.10. Cross-sectional TEM images of the Si-28/Si$_{0.70}$Ge$_{0.30}$ QW heterostructures regrown at fixed temperatures $T_{regrowth}$: (a) 620 °C and (b) 530 °C.

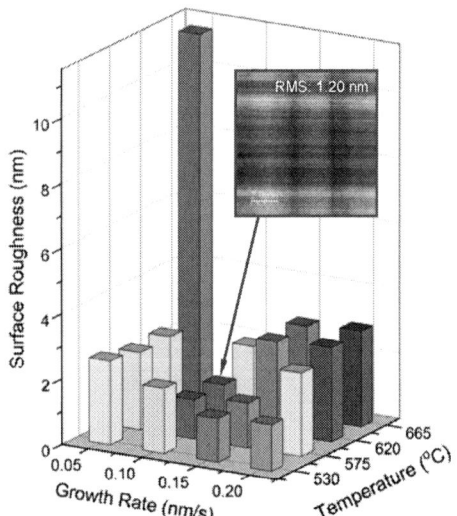

Fig. 9. Characterization of surface roughness on the 200-nm-thick Si$_{0.70}$Ge$_{0.30}$ buffer regrown with varying growth rate and temperature. The inset is the AFM image showing a reduced roughness on the cross-hatch-patterned surface grown at 620 °C.

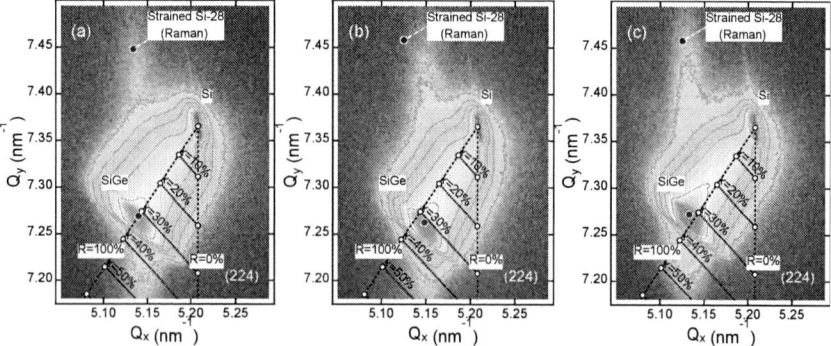

Fig.11. (224) XRD-RSM of the Si-28/ Si$_{0.70}$Ge$_{0.30}$ QW heterostructures obtained by controlling the temperatures for SiGe capping-layer growth ($t_{Si-28} = 11$ nm). (a) no capping layer, (b) $T_{cap} = 570$ °C, and (c) $T_{cap} = 500$ °C.

978-1-7281-1988-5/18 $31.00 © 2018 IEEE

Towards scalable silicon quantum computing

M. Vinet[1*], L. Hutin[1], B. Bertrand[1], S. Barraud[1], J.-M. Hartmann[1], Y.-J. Kim[1], V. Mazzocchi[1], A. Amisse[1,2], H. Bohuslavskyi[1,2], L. Bourdet[2], A. Crippa[2], X. Jehl[2], R. Maurand[2], Y.-M. Niquet[2], M. Sanquer[2], B. Venitucci[2], B. Jadot[3], E. Chanrion[3], P.-A. Mortemousque[3], C. Spence[3], M. Urdampilleta[3], S. De Franceschi[2] and T. Meunier[3]

Université Grenoble Alpes, France

[1] CEA, LETI, Minatec Campus, F-38054 Grenoble, [2] CEA, INAC, F-38054 Grenoble, [3] CNRS, Institut Néel, F-38042 Grenoble

*e-mail: maud.vinet@cea.fr / Phone: (+33) 438 78 99 00

Abstract— We report the efforts and challenges dedicated towards building a scalable quantum computer based on Si spin qubits. We review the advantages of relying on devices fabricated in a thin film technology as their properties can be in situ tuned by the back gate voltage, which prefigures tuning capabilities in scalable qubits architectures.

I. INTRODUCTION

It is now well acknowledged that quantum computing (QC) will extend high performance computing roadmap [1-2]. However, to be a serious contender to classical computers, digital QC will have to perform large number of quantum operations and thus manipulate large numbers of quantum bits. Typically, to solve problems beyond classical computer reach, quantum operations will be over a billion. To cure errors due to decoherence, quantum error correction techniques, which utilize the idea of redundant encoding, are needed to define logical qubits on which those operations are performed [3-5]. With state-of-the-art codes, error thresholds or fidelities (around 10^{-2} in Si spin qubits), it is believed that logical qubits will be made out of a few thousands or more of physical qubits [6] bringing the number of required physical qubits to perform relevant quantum calculations to at least a million.

Because of these large numbers of operations and physical qubits, Si-based QC appears as a promising approach due to the size of the qubits, the quality of the quantum gates and the VLSI ability to fabricate billions of closely identical objects. However, silicon spin qubits have a quite recent history. The first single qubit relying on a confined electron spin in silicon were only reported in 2012 [7]. Since then, their development in terms of gate realization has been comparatively fast. Moreover, thanks to the introduction of isotopically purified [28]Si, a large enhancement of spin coherence has been observed. Multiple research groups have realized single- and two-qubit quantum gates with already high and yet improving fidelities, see **fig.1** [8-11]. Recently, first qubit functionalities in a device fabricated in a 300mm CMOS platform have been demonstrated [12, 13]. All these results provide encouraging building blocks for scalable, fault-tolerant QC.

In this paper, we will discuss the figures of merit and review the challenges to be tackled to build a large-scale QC. We will show how FDSOI technology features can be leveraged for quantum chip optimization.

II. EFFORT TOWARDS RELIABLE AND REPRODUCIBLE SINGLE AND TWO QUBITS GATES

There is no consensus yet on the most relevant figures of merit and the required trade-offs optimizing quantum dot and qubit variability and fidelity mostly due to the lack of statistical data. Efforts are simultaneously made in several directions.

A. Material optimization

First from a pure material perspective, due to the absence of residual nuclear spins, isotopically purified [28]Si has already shown very promising improvement in spin coherence times [11]. We have grown [28]silicon epitaxial layers with an isotopic purity greater than 99.992 % on 300mm natural abundance silicon crystalline wafers. The quality of the mono-crystalline [28]Si epilayer respects the same drastic quality requirements as the natural epilayers used in standard CMOS technology. Synthesis and enrichment of SiF_4 took place in Russia at SC "PA Electrochemical Plant" (ECP). The synthesis of silane was carried out by the reaction of high-purity silicon tetrafluoride with calcium hydride. We profiled with secondary ion mass spectrometry (SIMS) the various Si isotopes in a 60 nm thick Si layer grown at 650°C, 20 Torr with [28]SiH_4 on a natural abundance Si(001) wafer **(fig. 2)** [15].

Then to reduce potential fluctuations in CMOS and ensure that quantum dots are actually gate-defined rather than by disorder, surface roughness and gate stack will have to be carefully engineered. Several calculations attempted to quantify the impact of physical dimensions on potential fluctuations have been performed [15, 16]. For qubits defined by nanowires in thin film devices, it means that both thickness and etched defined edges need to be as smooth as possible as presented in **Fig. 3**. **Fig. 4** quantifies surface roughness depending on several finishing preparations showing that Rms can be brought down to 0.1nm range [17]. Qubits will harness developments towards reducing D_{it} or gate stack granularity as well.

B. Hole and electrons CMOS compatible qubits

The first qubit implemented on a foundry-compatible Si CMOS platform was built using a SOI NanoWire MOSFET, it is in essence a compact two-gate pFET, **fig. 5** [12]. The inhomogeneous dephasing time T_2^* was limited to 60ns as the studied EDSR transition was involving an excited orbital state. In a recent experiment on a similar device we investigated the coherence of an EDSR transition between spin-orbit states within same orbital. It revealed an extended T_2^*=270ns close to the inhomogeneous dephasing time of electron in natural Si.

In addition, in SOI architecture, the back bias (V_{bb}) can be leveraged to in-situ tune the system properties. First we have evidenced that the inter-dot coupling, mediated by tunneling and Coulomb interaction, can be tuned over 6 orders of magnitude by means of V_{bb}. [18], **fig 6**. Then, the Rabi frequencies of qubits show complex dependence on V_{bb}, which results from the control of the shape and symmetry of the wave functions. The qubits may, therefore, be switched between bias points where they can be efficiently manipulated electrically and others where they are far less responsive but (as a consequence) decoupled from the electric noise and longer-lived.

978-1-7281-1988-5/18 $31.00 © 2018 IEEE

In the case of electrons, the spin-valley mixing has led to the experimental observation of electrically-induced spin resonance in Si without having to co-integrate micro-magnets [19]. As for holes, using V_{bb} to tune the vertical confinement (and thus valley splitting) enables continuously switching between a protected spin regime and an E-field addressable valley regime [20, 21]. Trade-offs for optimization still need to be found to take advantage of this in situ tuning capability in terms of number of operations per error or architecture.

C. Readout measurements

We have recently shown [22] an ultra-compact device fabricated in foundry-compatible Si MOS technology, with a built-in charge detector (SET) capacitively coupled to two Gate-defined QDs. Thanks to an energy-selective detection scheme, we have demonstrated single-shot readout of a two-electron spin-state in a "corner QD" by measuring the time trace of the SET current. As detailed in **Fig. 7**, the optimization of the readout speed/fidelity trade-off in this readout scheme is carried out by increasing the tunnel rates while keeping a large (Γ_{SET}-Γ_{QD}) window. Here again, V_{bb} was used to tune the cross-capacitance between the SET and the QD in order to improve the readout signal and shorten the minimum integration time.

In this geometry, we have further investigated the quality of the readout that can be estimated by measuring the amplitude of the detector signal in the two different spin configurations, as presented by the histogram in **fig. 8**. We extract the so-called readout fidelity which is as good as 99.9% for 1ms of integration time. Alternative methods such as gate-reflectometry are under investigation, it is considered a more compact and scalable readout method. In this technique, the charge sensing required to sense the qubit state is accomplished by measuring the dispersive response of an electromagnetic RF resonator connected to one of the qubit gates and excited at its resonance frequency. We have obtained the complete charge stability diagram of a coupled two-dot system [23].

D. Possible manipulation schemes

A first way of driving coherent rotation of a spin is through Electron Spin Resonance, or ESR. Experimentally, one can deposit in close proximity of the device a microstrip line that is used to flow a large AC current and generate an oscillating magnetic field resonant with the spin transition frequency. Coupling the spin to an RF magnetic field seems like the most straightforward method, although the excitation is hardly applied locally. This can be a drawback for maximizing the manipulation speed which depends on the coupling strength and is typically in the range of 1MHz for this scheme [8].

A second mechanism is the Electric Dipole Spin Resonance (EDSR). In this case, the spin rotation is induced by an oscillating electric field, which can be provided by a field-effect Gate placed directly above the QD. If the properties of the system are such that Spin-Orbit Coupling (SOC) is significant, the orbital motion caused by an RF E-Field alone can drive spin rotations [12, 13]. Otherwise a possible approach consists in embedding a micro-magnet as an auxiliary in the vicinity of the device, causing the particle traveling back and forth to perceive an oscillating B-field [9-11]. Although efficient for fast manipulation of a few qubits, this technique may become problematic for the design and integration at large-scale. Note that for both implementations of EDSR, a stronger coupling can

lead to ~100MHz spin rotations, but sensitivity to charge noise may limit the coherence time (**fig. 1**).

III. QUBITS IN AN ARRAY OF QUANTUM DOTS

A. Qudots definition and coupling to the nearest neighbours

The challenge towards large-scale integration comes down to the ability to individually control single spin qubits in arrays of millions of quantum dots, together with controlling the nearest-neighbor interaction. By resorting to a line/column addressing in a 2D architecture and a definition of the dots through the potential applied to their surrounding tunnel barriers (**fig. 9**), the number of gates can be scaled down proportionally to sqrt(N). In this layout scheme, individual quantum dots as well as the interaction between adjacent dots are defined by 4 gate voltages (**fig. 10**).

B. Cell for full quantum functionalities

In a CMOS-compatible technology, the only demonstrated qubit so far looks like two gates in series on a nanowire with one dot used as a qubit and the second as its readout or sensing dot [12, 22]. However, this charge detector presents some limitation in terms of scalability as the presence of two reservoirs in series to perform current measurements is required. To extend this principle and design a 2D array required for quantum error correction, we propose to locate the sensing dot in a layer below the qubit in order to build a compact and local unit-cell device containing a spin qubit with all the quantum functionalities (**fig. 11**). The top layer is used to encode the quantum information and the bottom layer is used to design local electrometers for readout, **fig. 12**. By using a spin-to-charge conversion and applying RF reflectometry techniques to the source of the detector, it provides a compact and scalable qubit & read-out design. The bottom layer is also connected to electron reservoirs to allow a scalable initialization to overcome the challenges of loading electrons from the sides of the array [24].

IV. QUANTUM CHIPS ARCHITECTURE

Because of Zeeman splitting, Si spin computing is limited to very low temperature operation. It impose very stringent conditions on power consumption for classical control electronics surrounding the quantum core, **fig. 14**. In view of their co-integration, we have demonstrated that unique back-biasing capability to FDSOI technology can be used at low temperature to compensate for Vt increase as shown in **fig. 15** and dynamically tune the performance/consumption trade-off [25].

Finally, we have demonstrated that package optimization through thermal coating or insertion of heat management modules can dramatically improve heat dissipation [26].

V. CONCLUSION

Electron and hole qubits have been fabricated in SOI nanowire-like integration. We have shown that the back gate voltage can be used to in situ adjust the system properties; i) to tune the tunneling coupling between dots over 6 orders of magnitude; ii) to optimize the readout speed/fidelity trade-off, we used it to demonstrate a read-out fidelity of 99.9% for 1ms of integration time; iii) to enable continuous switching between a protected regime and an E-field addressable regime. To upscale the number of qubits, we have proposed a vertical integration with sensing dots located below the qubits to enable a compact 2D array of qubits with full quantum functionalities.

978-1-7281-1988-5/18 $31.00 © 2018 IEEE

ref	techno	T_π	T_2^*	$Q = \dfrac{T_2^*}{T_\pi}$	1-qubit gate fidelity	2-qubit gate fidelity
Yoneda2018 (RIKEN)	^{28}Si/SiGe μ-magnet	25 – 100 ns	20 μs	200 – 800	> 99.9 %	
Watson2018 (QuTech)	Si/SiGe μ-magnet	250 ns	0.6 – 1 μs	4	98 – 99 %	~ 90 % ~ 800 ns CNOT
Zajac2018 (Princeton)	Si/SiGe μ-magnet	100 ns	1.2 – 1.4 μs	12	> 99 %	~ 90 % ~ 200 ns CNOT
Chan2018 (UNSW)	^{28}Si bulk ESR antenna	1 μs	30 μs	30	~ 99.9 %	
Huang2018 (UNSW)	^{28}Si bulk ESR antenna	1 μs	10 – 30μs	10 – 30		~ 90 % ~ 1 μs CROT
Maurand (CEA)	FDSOI – hole qubit - EDSR	5 ns	0.3 μs	60		

Fig. 1: Table summarized the Si spin qubits state of the art.

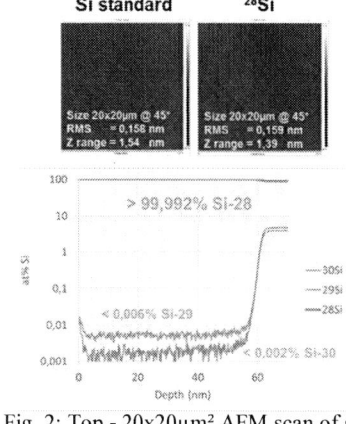

Fig. 2: Top - 20x20μm² AFM scan of standard Si surface compared to epitaxially grown ^{28}Si. Both surfaces exhibits same surface roughness. Bottom: SIMS characterization of ^{28}Si after growth showing about 99.992% of ^{28}Si.

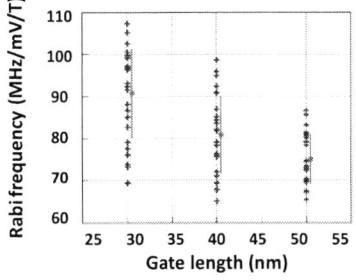

Fig. 3: Rabi frequency in rough hole qubit as a function of gate length. Each cross is a different realization of a Gaussian surface roughness profile with rms = 0.4 nm. The red dot and bar are mean and standard deviation.

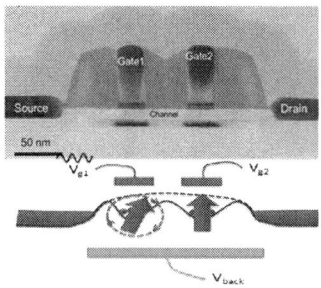

Fig. 4: SOI surface roughness illustrating that it can be decreased thanks to wafer fabrication process tuning.

Fig. 5: First qubit implemented on a foundry-compatible Si CMOS platform. It consists of a SOI NanoWire MOSFET with two-gate pFET in a 64nm pitch.

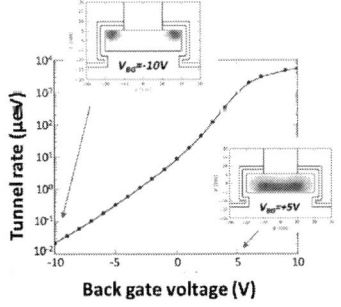

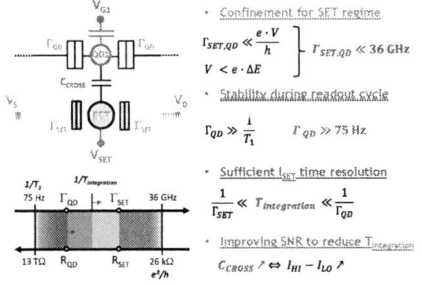

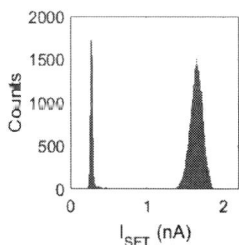

Fig. 6: A two-band k.p model accounting for valley-orbit coupling was used to calculate single-electron states. The tunnel coupling t was extracted versus V_{BG} from the anticrossing between the lowest single-electron states. This shows the tunability of inter-dot transport by the Back-Gate voltage.

Fig. 7: Readout by transport through a coupled SET charge detector. The charge events in QD1 are assumed to be spin-dependent. Design windows in terms of Γ tunnel rates, and corresponding tunnel resistance are represented (V=150μV, and T_1=13.5ms). Increasing the red area yields faster readout. Maximizing the SET-Dot C_{cross} leads to improving the readout signal and thus enables to reduce the integration time, i.e. squeeze the blue area. V_{bb} can be a handle for tuning C_{CROSS}.

Fig. 8: Histogram of the level of current of the single-shot traces for the singlet and triplet spin states after 1ms integration time. The detection bandwidth is fixed by room-T° electronics to 1 kHz. A spin read-out fidelity above 99.9% is achieved and the relaxation was measured close to 13.5 ms.

978-1-7281-1988-5/18 $31.00 © 2018 IEEE

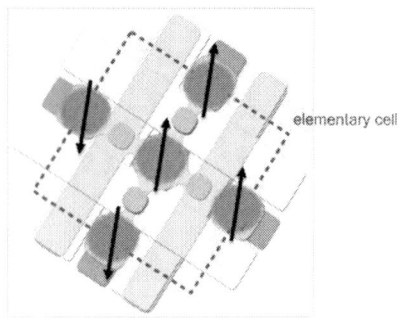

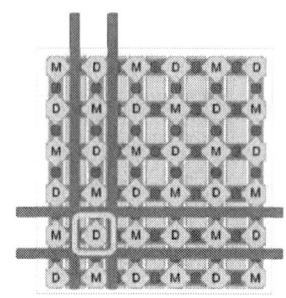

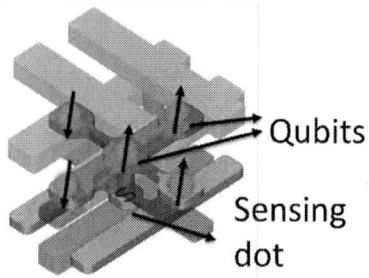

Fig. 9: Unit 2D cell where the cental dot is defined thanks to voltage applied through the grey gates on the constrictions with its nearest neighbors.

Fig. 10: 2D array with one dot selected thanks to gate bias applied to its 4 neighbouring tunnel junctions.

Fig11: Vertical unit cell with the sensing dot located below its qubits and coupled through a controllable and addressable tunnel junction.

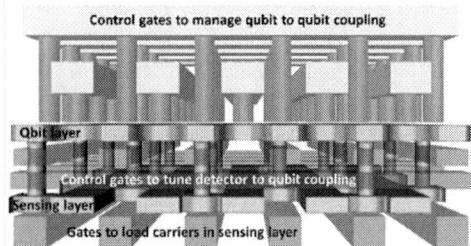

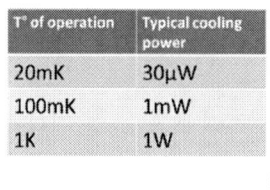

Fig. 12: The qubit contains the quantum hardware with spin qubits stored in an array of tunnel-coupled quantum dots. To control the quantum dots, a grid of long gates is designed and allows individual electrostatic and coherent control of the electron spin qubits. The lower layer is dedicated to the engineering of local reservoirs of electrons and local electrometers for the qubit layer. The two layers are coupled by a controllable tunnel barrier to allow electron transfer between them. This tunable tunnel coupling will be used to perform the read-out and the initialization of the quantum hardware.

T° of operation	Typical cooling power
20mK	30μW
100mK	1mW
1K	1W

Fig. 13: Cooling power table as a function of T(K).

Fig. 14: Close-up in the threshold voltage region of Id versus gate voltage curves for temperature ranging from 300 (green to blue shades) down to 4K. Left panel shows Vt increase around 0.5V both for N and PMOS that can be corrected thanks to Vbg as shown in right panel.

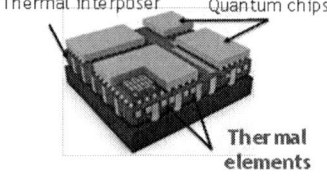

Fig. 15: Thermal elements can be used both as thermal insulator to force a gradient between the coldest part of the core quantum chip and some parts of classical electronics control. They can also be used to manage heat dissipation by damping heat burst into phase change materials.

ACKNOWLEDGMENT - The authors gratefully acknowledge financial support from the EU under Project MOS-QUITO (No. 688539) and the Marie Curie Fellowship within the Horizon 2020 program and from French Agence Nationale de la Recherche through the projects ANR-15-IDEX-02 and the ANR-16-ACHN-0029.

REFERENCES

[1] H. Riel, DRC (2018), [2] A. Steegen, IMEC Technology Forum, SemiconWest (2018), [3] EU quantum manifesto, [4] C. Jones, Phys. Rev. X 2, 031007 (2012), [5] S. J. Devitt *et al.*, Rep. Prog. Phys. 76, 076001, (2013), [6] M. Suchara, arXiv:1312.2316v1 (2013), [7] J. J. Pla *et al.*, Nature 489, 541 (2012), [8] M. Veldhorst *et al.*, Nature 526, 410 (2015), [9] D. M. Zajac *et al.*, Science, 359 (2018) [10] T. F. Watson *et al.*, Nature (2018), [11] J. Yoneda *et al.*, Nature Nanotechnology 13, 2 (2018), [12] R. Maurand *et al.,* Nature Communications 7, 13575 (2016), [13] L. Hutin, *et al.*, VLSI (2016), [14] V. Mazzocchi *et al.*, http://arxiv.org/abs/1807.04968 (2018), [15] S. Poli *et al.*, TED 2008, [16] A. Lherbier *et al.*, PRB 2008, [17] V. Deshpande, PhD manuscript thesis (2012), [18] S. De Franceschi *et al.*, IEDM (2016), [19] A. Corna, *et al.*, NPJ Quantum Inf. (2018), [20] L. Hutin, *et al.*, VLSI (2018), [21] L. Bourdet, *et al.*, PRB 97, 155433 (2018), [22] M. Urdampilleta *et al.*, VLSI (2017), [23] A. Crippa, *et al.*, Nano Lett. (2017), [24] H. Flentje *et al.*, Nature comm 8 (1) (2017), , [25] H. Bohuslavskyi, *et al.*, submitted to TED (2018), [26] P. Coudrain, *et al.*, VLSI (2018)

Majorana Qubits

Leo Kouwenhoven
Microsoft Quantum Labs Delft and QuTech, Delft University of Technology, 2600 GA Delft,
The Netherlands. Leo.Kouwenhoven@Microsoft.com

Abstract- We present an overview of Majorana qubits based on one-dimensional semiconducting nanowires partially covered with a conventional superconductor. Majorana zero modes emerge at the wire ends when this hybrid system transitions from a conventional superconducting phase to a topological phase, in general occurring on increasing a magnetic field. For sufficiently long wires different Majoranas are fully independent and Majorana-based qubit states become topologically protected, which make them insensitive to local sources of noise. We present qubit designs, materials and device development and ongoing experimental efforts.

I. INTRODUCTION

Majorana zero modes (MZM), or 'Majoranas', always come in pairs. To zeroth-order they can be viewed as "half-electrons", and thus a Majorana pair constitutes a "full electron", or a fermion. The ground state of a conventional superconductors consists of an even number of fermions, all paired up in Cooperpairs. A topological superconductor in addition to Cooperpairs hosts two Majoranas and thus can contain both an even and an odd number of fermions. These even and odd parities form a two-fold degenerate ground state (see introduction in Fig. 1), which can be used as the Majorana qubit states. Interestingly, measurements on one Majorana cannot provide any information on the parity of a pair of Majoranas. Also, local noise coupling to just one Majorana cannot extract information or influence the state of the qubit. This insensitivity due to absence of wave function overlap, or interactions in general, is provided by the large separation between the two "half-electrons". This is referred to as topological protection. The resulting enhanced qubit stability is the motivation for pursuing topological quantum computation [1].

II. THE SYSTEM

In conventional superconductors electrons with opposite spin pair up in Cooperpairs and collectively open a superconducting gap, Δ, in the energy spectrum. A magnetic field tends to polarize spins thereby breaking up Cooperpairs and lowering the value of Δ with gap-closure when the Zeeman energy $E_Z = g\mu_B B/2 = \Delta$. The same is true for induced superconductivity into a semiconductor which is electrically connected to a superconductor. The connection "proximitizes" the semiconductor with loosely-speaking having the Cooperpairs leaking into the semiconductor. The microscopic energy spectrum is known as the Andreev Bound State (ABS) spectrum. Figure 1 (lower left panel) illustrates that the gap in the ABS spectrum closes when increasing E_Z.

Interestingly, in the presence of strong spin-orbit interaction (SOI) spin polarization is negated by spatially rotating the spins. Since Cooperpairs in the parent superconductor are spatially extended, ABS states can still form but now with interesting spin structures. While the ABS are two-fold spin degenerate at $B = 0$, a finite B lifts the degeneracy causing level crossings of opposite spin states in the absence of SOI (lower left panel Fig. 1) and avoided level crossings in the presence of SOI (lower right panel Fig. 1). The Majorana mid-gap states have no spin, which is really peculiar since these states correspond to either even or odd fermion-parity and fermions always have non-zero spin. In addition to zero-spin, the defining property "particle-equals-antiparticle" implies that Majoranas also don't have electric charge. These are three easy to test Majorana properties, zero-energy, zero-spin and zero-charge.

978-1-7281-1988-5/18 $31.00 © 2018 IEEE

Figure 2 shows the basic characterization measurement to detect the presence of Majoranas [2]. Tunneling into a Majorana state should yield a resonance at zero energy in the conductance. If particle-equals-antiparticle holds, the conductance value is predicted to be quantized at $2e^2/h$, irrespective of precise values for magnetic field or electron density. Figure 2 indeed shows a resonance at zero voltage, V, that remains at zero while changing B from 0.7 to 0.9 T (implying the absence of spin-degrees of freedom). Shown in Ref. x is that also an electric field does not move the resonance away from zero energy (implying the absence of charge). Energy, spin and charge being zero, together with the quantized value of the resonance, is the current evidence for the existence of Majoranas [3].

II. MAJORANA QUBITS

Qubits can be formed with two Majorana pairs where each pair either contains zero or one fermion, i.e. encodes for either even or odd parity. In a Majorana-transmon qubit as illustrated in Figure 3 the two inner Majoranas are tunnel coupled. Suppose that at some moment the left pair $\gamma_1\gamma_2$ has even parity and the right pair $\gamma_3\gamma_4$ has odd parity. The tunnel coupling between γ_2 and γ_3 creates a superposition between the parity combination even/odd and odd/even. Probing such superposition can be done with a circuit-QED setup commonly used for superconducting transmon qubits. Our device layout including the nanowire structure is shown in Figure 3.

An alternative to a tunnel coupling geometry, is a measurement based approach to topological qubits. Figure 4 illustrates how measuring the parity of a particular Majorana pair can create a superposition of parity states in a different pair. The measurements can be done via simple conductance measurements or by using quantum dots as parity sensors. The latter can yield a scalable architecture for topological qubits. We will present state of the art experiments on both Majorana-transmons as well as measurement-based qubits.

Work done in collaboration with colleagues and collaborators at QuTech in Delft and the Microsoft Quantum Labs in Santa Barbara, Copenhagen and Delft.

REFERENCES

[1] Sarma, S. D.; Freedman, M.; Nayak, C. Majorana Zero Modes and Topological Quantum Computation. *Npj Quantum Inf.* **2015**, *1* (1).

[2] Zhang, H.; Liu, C.-X.; Gazibegovic, S.; Xu, D.; Logan, J. A.; Wang, G.; van Loo, N.; Bommer, J. D. S.; de Moor, M. W. A.; Car, D.; et al. Quantized Majorana Conductance. *Nature* **2018**, *556*, 74.

[3] Lutchyn, R. M.; Bakkers, E. P. A. M.; Kouwenhoven, L. P.; Krogstrup, P.; Marcus, C. M.; Oreg, Y. Majorana Zero Modes in Superconductor–semiconductor Heterostructures. *Nat. Rev. Mater.* **2018**, *3* (5), 52–68.

[4] Aguado, R.; Kouwenhoven, L. P.; review in preparation.

Figure 1. **Majorana Introduction** (a) Majorana nanowires consist out of a semiconducting wire with large spin-orbit interaction (e.g. InAs or InSb) partially covered by a conventional superconductor (e.g. Al). For appropriate electron densities and magnetic fields Majorana zero modes (MZM) appear at the two wire ends. The Majorana wave functions, γ_L and γ_R, decay and have vanishing overlap in the middle of long wires. Right side shows SEM photo and schematic of an elemental device, false colored to illustrate semiconducting

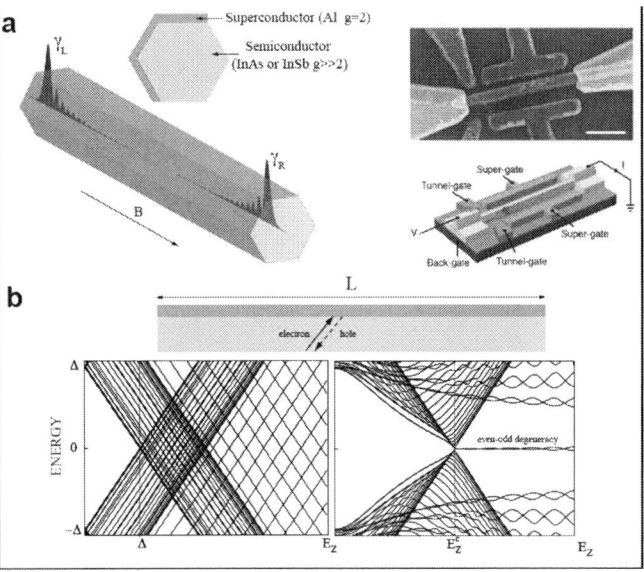

nanowire (gray), Al coverage (green), source and drain contacts (yellow), gates to induce a tunnel barrier (red) and gates to tune the electron density (purple). White bar indicates 1 micron. **(b)** Electrons and holes form particle-in-a-box states with one of the walls being the semiconductor-superconductor interface These states are known as Andreev Bound States (ABS) with the symmetry property that electron-like states at energy $+E$ have a hole-like partner at $-E$. When increasing the magnetic field, B, and thus the Zeeman energy for spin-splitting, $E_Z = g\mu_B B/2$, the superconducting gap, Δ, closes when $E_Z = \Delta$. Two energy spectra are shown. On the left (right) the case without (with) spin-orbit interaction. The spin-orbit interaction causes the gap to re-open, forming a topological superconducting phase with two Majorana mid-gap states very close to zero energy. Figure from Ref. 4.

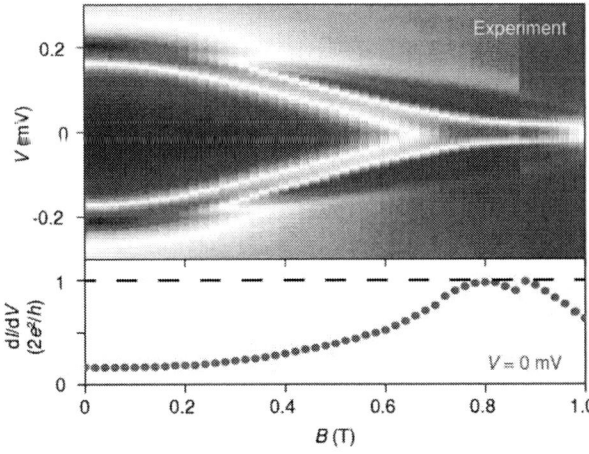

Figure from Ref. 2.

Figure 2. **Quantized Majorana Conductance**. Top panel: differential conductance, dI/dV, versus source-drain bias voltage, V, and magnetic field, B. The resonance with red color is electron-hole symmetric and decreases energy when increasing B until it reaches zero energy at 0.7 Tesla. Between 0.7 and 0.9 T the conductance resonance stays at zero energy and has a quantized value equal to $2e^2/h$. Bottom panel: line cut at $V = 0$. The quantization at $2e^2/h$ results from particle-equals-antiparticle, the defining symmetry for Majoranas.

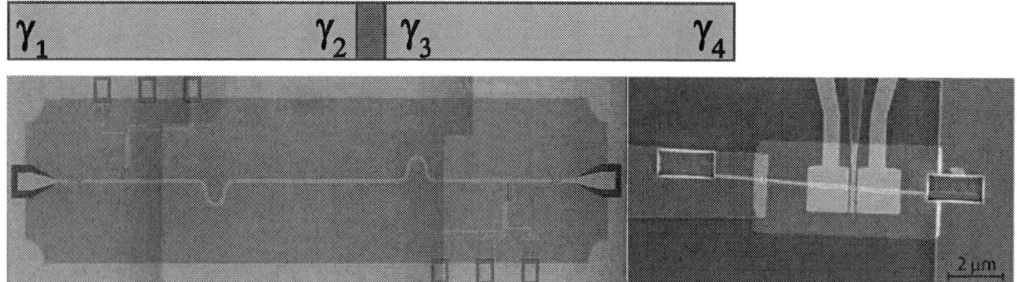

Figure 3. **Majorana-Transmon Qubits**. Top panel: schematic diagram of a qubit consisting out of 4 Majoranas in two topological sections (green) where γ_2 and γ_3 are coupled via a tunnel barrier (red). Left panel: Superconducting resonator optimized for measurements in a magnetic field. Qubit dc control-leads are visible in upper left and lower right. Right panel: zoom on Majorana wire containing two islands with charge occupation controlled by plunger gates (green). The Josephson coupling between islands is controlled by the central gate electrode (red).

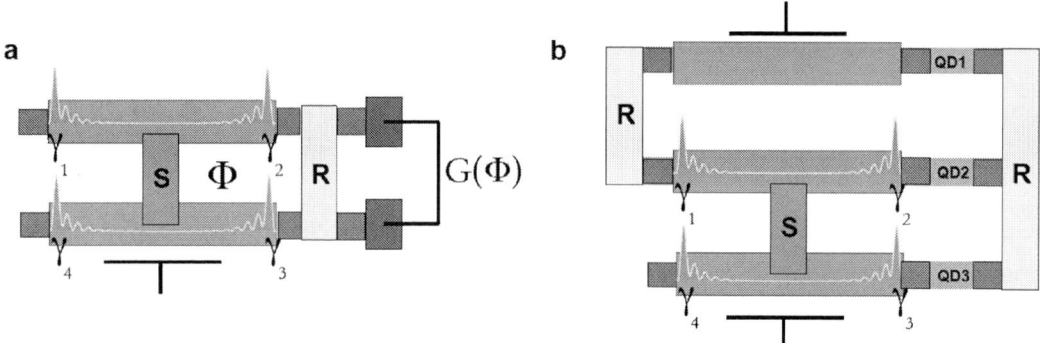

Figure 4. **Measurement-based Topological Qubits**. Nanowire networks for measurement-based qubit operations. Topological sections (green) each contain two Majoranas. Two pairs of two Majoranas form a qubit with topological sections connected by a conventional superconductor (orange). Different Majorana pairs form different qubit basis such as $\gamma_1\gamma_2$ for the x-basis, $\gamma_1\gamma_3$ for the y-basis, and $\gamma_2\gamma_3$ for the z-basis. (**a**) a simple conductance, G, measurement reads out the parity of the $\gamma_2\gamma_3$-pair and constitutes a measurement in the z-basis. (**b**) layout allowing for a full qubit characterization. QD stands for quantum dots serving as parity sensors. The combined state of QD1-QD2 constitutes a readout in the x-basis, QD1-QD3 for the y-basis, and QD2-QD3 for the z-basis. Figure from Ref. 4.

978-1-7281-1988-5/18 $31.00 © 2018 IEEE

First Demonstration of 3D stacked Finfets at a 45nm fin pitch and 110nm gate pitch technology on 300mm wafers.

A. Vandooren, J. Franco, Z. Wu[1], B. Parvais[2], W. Li, L. Witters, A. Walke, L. Peng, V. Deshpande, N. Rassoul, G. Hellings, G. Jamieson, F. Inoue, K. Devriendt, L. Teugels, N. Heylen, E. Vecchio, T. Zheng, E. Rosseel, W. Vanherle, A. Hikavyy, G. Mannaert, B. T. Chan, R. Ritzenthaler, J. Mitard, L. Ragnarsson, N. Waldron, V. De Heyn, S. Demuynck, J. Boemmels, D. Mocuta, J. Ryckaert and N. Collaert.

IMEC, Kapeldreef 75, 3001 Leuven, Belgium email: anne.vandooren@imec.be

[1] also with KULeuven, Leuven, Belgium, [2] also with VUB, Brussels, Belgium

Abstract—3D stacking using a sequential integration approach is demonstrated for finfet devices on 300mm wafers at a 45nm fin pitch and 110nm poly pitch technology. This demonstrates the compatibility of the 3D sequential approach for aggressive device density stacking at advanced nodes thanks to the tight alignment precision of the first processed top layer to the last processed bottom layer through the top silicon channel and bonding stack during 193nm immersion lithography. The top devices are junction-less devices fabricated at low temperature ($T \leq 525°C$) in a top Si layer transferred by wafer-to-wafer bonding with a bonding dielectric stack down to 170nm. The top devices offer similar performance as the high temperature bulk finfet technology for LSTP applications. The use of TiN/TiAl/TiN/HfO$_2$ gate stack provides the proper threshold voltage adjustment while the insertion of the LaSiO$_x$ dipole improves device performance and brings the BTI reliability within specification at low temperature.

I. INTRODUCTION

Conventional 2D feature size scaling is currently facing major limitations which led technologists to explore alternative paths to increase circuit functionality per area for next technology nodes, without requiring further reduction of the device dimensions. 3D sequential integration is a very attractive option which allows high device density per chip area when stacking devices on top of another owing to the high alignment accuracy dependent only on the lithography stepper performance. The design-technology co-optimization (DTCO) study of stacked devices for logic applications, such as 3D stacked devices [1] and complementary FET (CFET) [2] has shown a promising structural scaling up to 50% area reduction. Other benefits of 3D sequential integration are the reduced interconnect wire length, the simplified co-integration of heterogenous channel materials (such as Ge or III-V), channel orientation and device architecture, and hybrid technologies such as RF, low power, memory or optical IO on logic [1]. One of the main challenges of 3D sequential integration is the management of the process thermal budget. The top tier thermal budget needs to be reduced to avoid degrading what is below, namely the bottom devices, the interconnects and the bonding interface, while maintaining the top tier device performance

and reliability [3-6]. In this work, finfet devices are stacked using a 3D sequential integration flow with top nmos devices processed at a temperature as low as 525°C. The top devices are junction-less devices with channel doping set prior to top silicon layer transfer [6]. Finfet device stacking is achieved using a platform featuring a fin pitch of 45nm and a gate pitch of 110nm, demonstrating the suitability of the 3D sequential integration for large device density stacking.

II. DEVICE FABRICATION

A. Process flow

The process flow of the 3D stacked finfet devices is illustrated in **Fig.1**. The bottom devices are first processed using a 300mm CMOS Si bulk finfet flow featuring a 45*nm* fin pitch, a 110 *nm* gate pitch and a high-k last Replacement Metal Gate (RMG) [7]. The bottom tier processing ends with two layers of W local interconnects (Li1, Li2). The doped top Si layer is then transferred onto the bottom device layer by wafer to wafer bonding using oxide planarization followed by SiCN to SiCN bonding [8]. Scaling of the bonding stack down to a planarizing oxide thickness <*100nm* is achieved without impacting the integrity of the bonding interface and the silicon channel. Finfet devices are then processed in the top Si layer with a fin height of H$_{fin}$ ~40nm, a fin width of W_{fin} ~10nm and a minimum gate length of L_G ~24-30nm. Dense fin patterning is obtained using a *spacer defined double patterning* (SADP) process, resulting in a 45nm fin pitch. The first patterned top layer is aligned to the last processed local interconnect bottom layer using an immersion 193nm lithography stepper. The best alignment precision is achieved using a red, near-infrared or far-infrared light reaching a *mean + 3σ* value below 10nm. The next top layers are then aligned to the first patterned layer, as in a standard process flow. Various gate stacks are studied for the top tier finfet devices for threshold voltage (V$_{TH}$) adjustment and reliability purposes: TiN on HfO$_2$, TiN/TiAl/TiN on HfO$_2$ and TiN on HfO$_2$/LaSiO$_x$. 3D stacked devices as well as SOI devices were processed using the same low temperature process flow and are compared electrically. A TEM cross-section of the final device along the fin and across fins are shown in **Figs.2&3** and **Fig.4**, respectively.

978-1-7281-1988-5/18 $31.00 © 2018 IEEE

B. Bonding and scaling of the bonding dielectric

One main challenge of 3D sequential integration is the top layer transfer by bonding with good yield, maintaining Si film integrity and providing a scaled bonding dielectric for advanced nodes. **Fig.5a&b** shows the Patterned Wafer Geometry (PWG) of the carrier wafer after bottom tier processing and post-bonding scanning acoustic microscope (SAM) images for various amounts of planarization of the deposited oxide onto the carrier wafers. A minimum amount of oxide removal is needed to avoid pattern-related voids formation due to the remaining topography on the carrier wafers. The pattern-related voids appear first toward the edge of the wafer and gradually extend toward the center of the wafer when less oxide CMP is applied. Scaling of the bonding dielectric on the carrier wafer was also assessed. The bonding quality also degrades at the wafer edges when the final oxide thickness after CMP becomes too thin likely due to large oxide thickness variation from center to edge (**Fig.5c**).

III. ELECTRICAL RESULTS

A. DC electrical results

The I_{DS} vs V_{GS} characteristics of the top tier nmos devices (**Fig.6**) are well behaved for both long and short channel devices. The impact of the channel doping (N_{CH}) is low (**Fig.7a**) compared to single gate planar devices [6], as expected from TCAD simulations (**Fig.7b**), due to the stronger gate control in double gate devices. Short channel effects (SCE) are well controlled down to the shortest gate length ~24-30nm, with an average saturation subthreshold slope (SS) of ~72mV/dec (**Fig.8**) and Drain-induced barrier lowering (DIBL) of ~33mV/V (**Fig.9**) at a V_{DD}=1V. SCE are improved slightly by lowering the channel doping. Long channel V_{TH_LIN} adjustment from ~0.62V down to ~0.22V is obtained for a TiN/HfO_2 and $TiN/TiAl/TiN/HfO_2$ gate stack, respectively. When inserting a dipole layer of 0.2nm $LaSiO_x$ in the TiN/HfO_2 gate stack, the V_{TH} is lowered from ~0.62V down to ~0.38V (**Fig.10**). The extracted CET is ~1.45 and 1.25nm for the TiN/HfO_2 and $TiN/TiAl/TiN/HfO_2$ gate stack, respectively, while the 0.2nm $LaSiO_x$ dipole increases the CET by 0.1nm compared to the TiN/HfO_2 case. Increasing doping in the channel also results in a reduction of the mobility (**Fig.11**) due to an increase in carrier scattering. In long channel devices, the on-current at fixed off-state current reduces with increased channel doping due to degradation in mobility. Whereas, in short channel devices, the on-state current increases with channel doping due to a decrease in the resistance of the ungated extension region under the spacers (**Fig.12**). The extracted series resistance shows a gradual increase when lowering the channel doping due to an increase in the extension resistance (**Fig.13**). Comparison with SOI devices processed at low temperature shows a similar V_{TH}, SS and I_{ON} cumulative distribution (**Fig.14**), suggesting that the layer uniformity and variability is similar in both devices without impact from the top Si layer transfer.

B. Analog & mismatch

Mismatch $\sigma(\Delta V_{TH})$ was measured on matched device pairs (**Fig.15**) for different device geometries. The extracted A_{VT} coefficient is ~2mV.µm for N_{CH}~3E18at/cm³, which agrees with published values for doped channel bulk finfet technologies at same CET [9]. Analog figures of merit, such as voltage gain (A_v) and g_m/I_{DS} show overall good performance related to the good electrostatic control of the devices (**Fig.16**).

C. Reliability

PBTI measurements on the low temperature top tier nmos devices with the different gate stacks (**Fig.17**) show improvement for the junction-less devices over inversion-mode reference devices without reliability anneal thanks to the reduced operating electric field [10]. When combining junction-less devices with $LaSiO_x$ dipole [11], further improvement of the BTI reliability is obtained, meeting reliability target with considerable headroom and only ~0.1nm for 0.2nm $LaSiO_x$ (~0.2nm for 0.4nm $LaSiO_x$) EOT penalty.

D. Benchmarking

The low temperature (LT) top tier devices are compared to high temperature (HT) bulk finfet devices fabricated on the same platform with similar process conditions featuring ~13nm-wide spacer and Si:P1% selective epitaxially grown source-drains. The LT and HT devices show similar control of short channel effects. For LSTP applications, the LT devices have a similar or better on-state current than the HT devices, while for HP applications, the LT devices exhibit up to 25% on-state current reduction (**Fig.18**). This technology is therefore suitable for applications combining analog/LSTP top tier devices onto HP bottom tier devices. When inserting the $LaSiO_x$ dipole in the gate stack, the on-state current degradation can be reduced to less than 10% thanks to reduced interface traps and increased carrier mobility [11,12]. Further improvement of the HP device performance can be obtained by process optimization, including a reduced spacer width and an increase of the active doping level in the source/drain regions, by laser anneal for example [13].

IV. CONCLUSIONS

We have demonstrated 3D stacked finfets at 45nm fin pitch and 110nm gate pitch using a 3D sequential integration with tight alignment of the top layer to the bottom layer using 193nm immersion lithography. The top LT device process (T<525°C) offers functional devices with no performance degradation for LSTP applications over the HT reference devices. The use of $TiN/TiAl/TiN/HfO_2$ gate stack enables proper threshold voltage tuning while the insertion of the $LaSiO_x$ dipole improves device performance and brings reliability within specification in spite of a lower processing temperature.

REFERENCES

[1] A. Mallik *et al., Proc. IEDM Tech. Dig., 2017.* [2] J. Ryckaert et al. *VLSI Symp.,* 2018. [3] P. Batude *et al., Proc. IEDM Tech. Dig.,* 2011. [4] L. Pasini *et al. VLSI Symp.,* 2016. [5] L. Brunet *et al., , VLSI Symp.,*2016. [6] A. Vandooren *et al. VLSI Symp.,* 2018. [7] L. Ragnarsson et al *VLSI Symp.,* 2015. [8] S.-W. Kim et al., *3D System Intl. Conf.* 2015. [9] F. Arnaud et al. *IEDM Tech. Dig.,* 2009. [10] M. Toledano-Luque et al., *EDL* 35(12), 2014. [11] J. Franco et al.,*Proc. IRPS, 2017.* [12] J. Franco et al. submitted to IEDM 2018 [13] C. Fenouillet-Beranger et al., *IEEE S3S Conf.,* 2016.

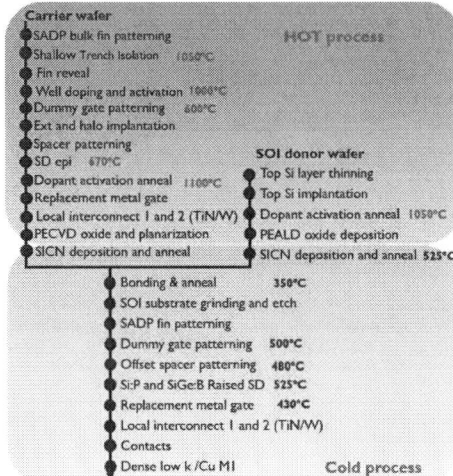

Fig. 1. Process flow of the 3D stacked finfets.

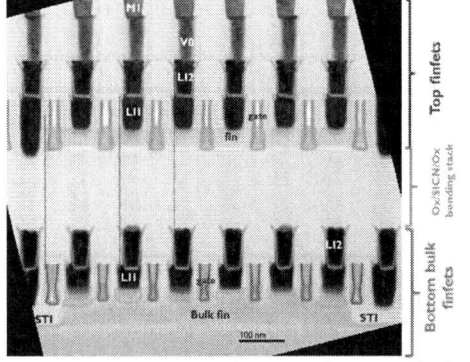

Fig. 2. Cross-sectional TEM image of the fabricated 3D stacked finfets along fins and across gates showing the tight alignment achieved by the top processed layers (Gate, Li1, Li2) toward the bottom layers.

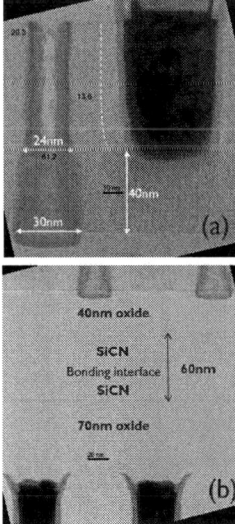

Fig. 3. Magnified Cross sectional TEM images of (a) the top finfet gate stack and (b) the bonding dielectric stack oxide/SiCN/SiCN/oxide.

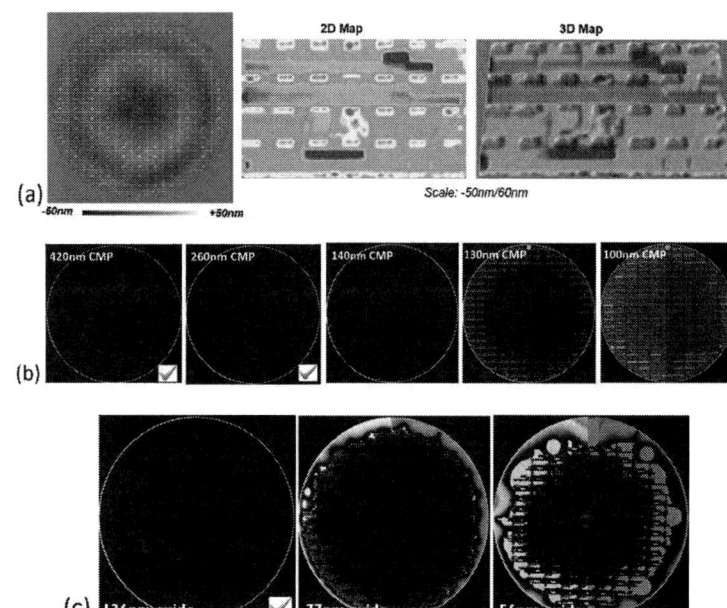

Fig. 5 (a) Patterned wafer geometry (PWG) of carrier wafer after processing. Map of the nano-topography peak to valley value on full wafer and 2D and 3D plots at die level (b) High resolution Scanning acoustic microscope (HR SAM) images after wafer to wafer bonding of a blanket donor wafer onto the carrier processed wafer. The amount of CMP to planarize the bonding oxide on the carrier wafer is varied from 420nm down to 100nm. Below 140nm oxide removal by CMP, pattern-related voids become visible because the planarization is not sufficient to fully remove the remaining topography of the carrier wafer (c) HR SAM images after oxide wafer to wafer bonding of a blanket donor wafer onto the carrier wafer. A same amount of planarization of 250nm is applied to all wafers, reaching different final oxide thickness. Bonding integrity is degraded at wafer edges when the final oxide thickness is reduced.

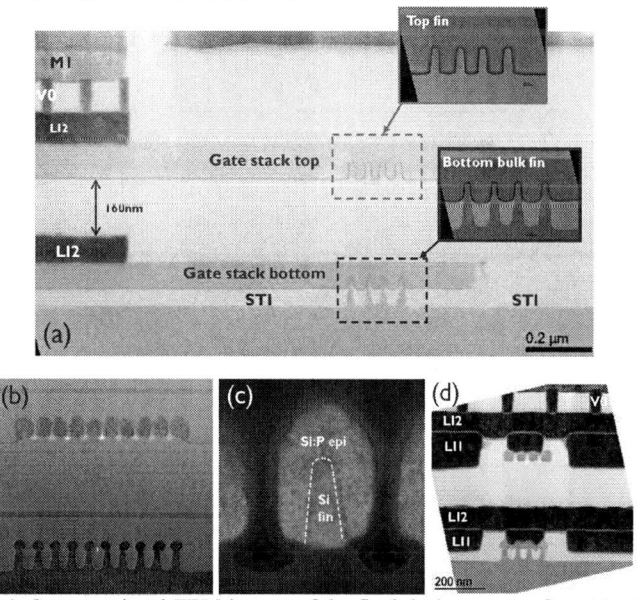

Fig. 4. Cross sectional TEM images of the final devices across fins: (a) with gates covering fins. Inset images show the fin profile of the bottom bulk finfets and the top finfets on bonding dielectric (b) without gates covering fins (c) magnified top fins with Si:P epi layer (d) with local interconnect layer contacting the fins.

978-1-7281-1988-5/18 $31.00 © 2018 IEEE   151

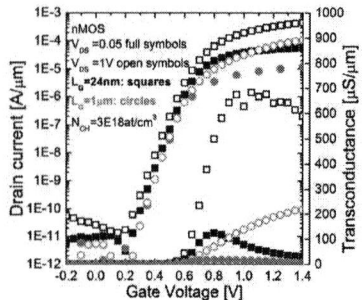

Fig. 6. I_{DS} and G_M *vs.* V_{GS} characteristics for long (L_G=1μm) and short channel (L_G~24nm) nmos top devices with TiN/HfO₂ gate stack.

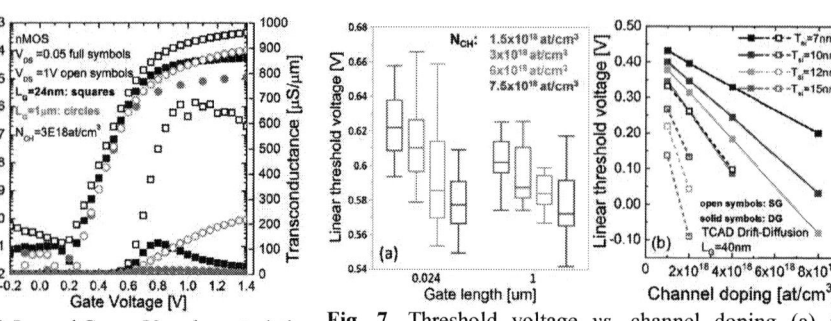

Fig. 7. Threshold voltage *vs.* channel doping (a) in experimental top tier nmos devices with TiN/HfO₂ gate (b) from TCAD simulations for different channel thickness in single gate (SG) and double gate (DG) devices.

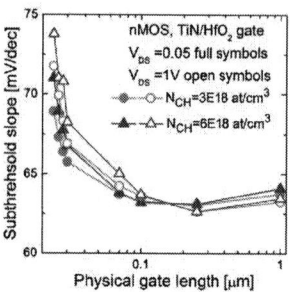

Fig. 8. Subthreshold Slope *vs.* gate length for different channel doping (N_{CH}). Slight SCE degradation is observed for higher N_{CH}.

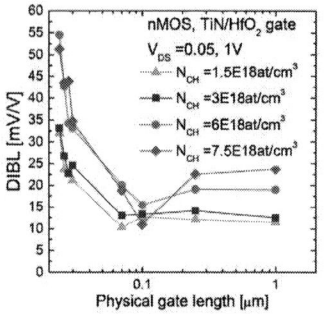

Fig. 9. DIBL vs. gate length for different channel doping (N_{CH}). Slight SCE degradation is observed for higher N_{CH}.

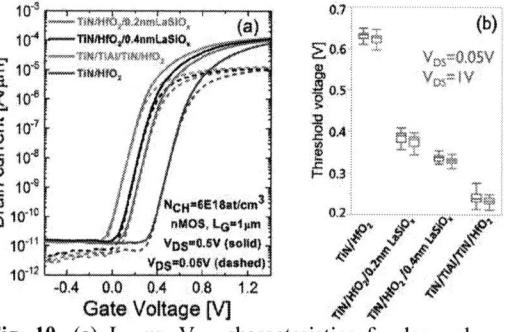

Fig. 10. (a) I_{DS} *vs.* V_{GS} characteristics for long channel (L_G=1μm) nmos top devices for different gate stacks (TiN/HfO₂, TiN/HFO₂/LaSiO$_x$ and TiN/TiAl/TiN/HfO₂) (b) V_T tuning by eWF change.

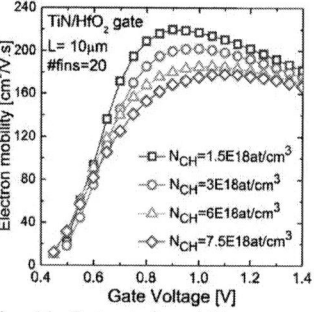

Fig. 11. Extracted mobility *vs.* gate voltage for different channel doping between 1.5 and 7.5E18at/cm³. Mobility degrades with doping level due to increase in carrier scattering.

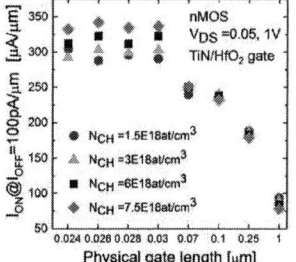

Fig. 12. I_{ON} at fixed I_{OFF} *vs.* gate length for different channel doping and TiN/HfO₂ gate stack.

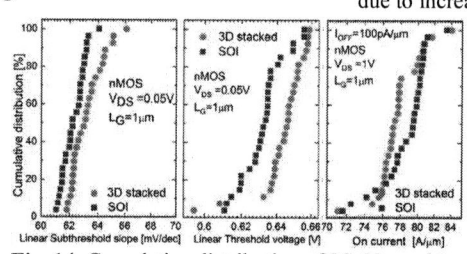
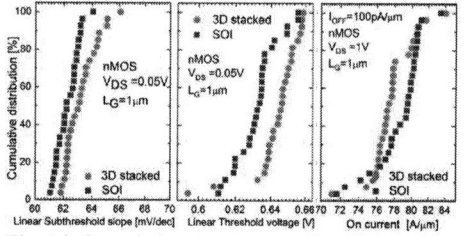

Fig. 13. Extracted external resistance for different channel doping, showing a slight resistance decrease with increasing channel doping.

Fig. 14. Cumulative distribution of SS, V_T, and I_{ON} comparing long channel low temperature nmos top 3D stacked finfets and SOI finfets at N_{CH}=6e¹⁸at/cm³ and TiN/HfO₂ gate.

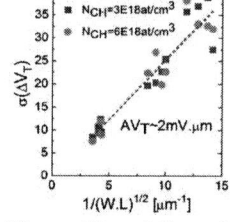

Fig. 15. Mismatch $\sigma(\Delta V_T)$ vs $(WL)^{-1.2}$ for different channel doping TiN/HfO₂ gate.

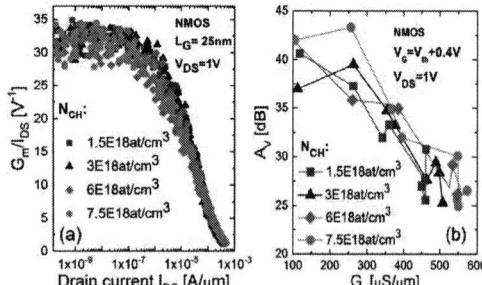

Fig. 16. (a) G_m/I_{DS} *vs.* I_{DS} and (b) analog gain A_v *vs.* G_m in junction-less top tier nmos devices for different channel doping.

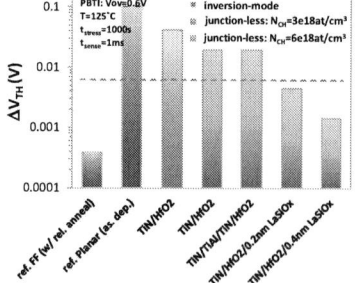

Fig. 17. PBTI V_{TH} shift at V_G=V_{th0}+0.6V for junction-less top tier low T devices with various gate stacks and reference inversion mode devices with and without reliability anneal.

Fig. 18. Benchmarking of the *best* low temperature (LT) top tier junction-less devices and high temperature (HT) bulk inversion-mode finfet devices for different gate stacks and V_{DD}=0.5V.

Breakthroughs in 3D Sequential technology

L. Brunet[1], C. Fenouillet-Beranger[1], P. Batude[1], S. Beaurepaire[1], F. Ponthenier[1,2], N. Rambal[1,2], V. Mazzocchi[1], J-B. Pin[3], P. Acosta-Alba[1], S. Kerdiles[1], P. Besson[2], H. Fontaine[1], T. Lardin[1], F. Fournel[1], V. Larrey[1], F. Mazen[1], V. Balan[1], C. Morales[1], C. Guerin[1], V. Jousseaume[1], X. Federspiel[2], D. Ney[2], X. Garros[1], A. Roman[1,2], D. Scevola[1,2], P. Perreau[1], F. Kouemeni-Tchouake[1], L. Arnaud[1], C. Scibetta[1], S. Chevalliez[1], F. Aussenac[1], J. Aubin[4], S. Reboh[1], F. Andrieu[1], S. Maitrejean[1], M. Vinet[1]

[1] CEA-LETI, Minatec Campus, 17 rue des Martyrs, 38054 Grenoble, France ; [2] STMicroelectronics, 850 rue Jean Monnet, F38926 Crolles ; [3] Applied Materials Inc., Sunnyvale, California 94085, USA ; [4] SCREEN LASSE, 92230 Gennevilliers, France

Abstract— The 3D sequential integration, of active devices requires to limit the thermal budget of top tier processing to low temperature (LT) (i.e. $T_{TOP}=500°C$) in order to ensure the stability of the bottom devices. Here we present breakthrough in six areas that were previously considered as potential showstoppers for 3D sequential integration from either a manufacturability, reliability, performance or cost point of view. Our experimental data demonstrate the ability to obtain 1) low-resistance poly-Si gate for the top FETs, 2) Full LT RSD epitaxy including surface preparation, 3) Stability of intermediate BEOL between tiers (iBEOL) with standard ULK/Cu technology, 4) Stable bonding above ULK, 5) Efficient contamination containment for wafers with Cu/ULK iBEOL enabling their re-introduction in FEOL for top FET processing 6) Smart Cut™ process above a CMOS wafer.

I. INTRODUCTION

3D-monolithic or 3D sequential CMOS technology (Fig. 1) is based on stacking active device layers on top of each other with very small 3D contact pitch (similar pitch as standard contact) [1,2,3,4,5]. This integration scheme offers a wide spectrum of applications [1] including for example i) increasing integration density beyond device scaling [5] ii) enabling neuromorphic integration where RRAM is placed between top and bottom tiers [3] and iii) enabling low-cost heterogeneous integration [4] for *e.g.* smart sensing arrays. However, such an integration process faces the challenge of fabricating high-performance devices in the top tier without degrading the electrical characteristics of the bottom tier. Therefore, **limiting the thermal budget to 500°C** is mandatory [6]. Owing to in-depth and exhaustive experimental studies, our previous work has highlighted two critical processing issues: i) Low-temperature gate stack integration [7] ii) Low-temperature selective epitaxy of silicon on source and drain. In parallel, in order to avoid routing congestion [8] and to benefit from the full 3D opportunities for a large domain of applications, there is a **need to implement local routing on the bottom tier**. However, using Cu metal lines and ULK dielectrics materials in the iBEOL is challenging for three reasons: i) copper metallization can result in contamination issues when wafers are reintroduced in a FEOL environment for top transistors processing, ii) the stability and reliability of iBEOL materials after a 500°C TB processing has not been demonstrated. iii) the use of standard ULK materials in the bottom tier for top wafer bonding has never been reported. Finally, using the Smart Cut™ process on bulk wafers instead of using SOI wafers would be a cost saver.

II. KEY LOW-TEMPERATURE PROCESS CHALLENGES

A- Gate resistance lowering: Low resistance, and good reliability are the main challenges for low-temperature gate stack integration.

In a gate-first scheme, neither the crystallization of the amorphous Si nor dopant activation can be efficiently obtained at 500°C. To solve this issue, we propose an original approach based on recrystallizing and activating dopants of *in-situ* doped amorphous-Si using UV Nanosecond (ns) Laser Anneal (UV-NLA). UV-NLA experiments have been performed using a SCREEN-LASSE excimer laser ($\lambda = 308$ nm) with a pulse duration of approximately 160 ns. The laser energy density is ranging from 0.4 to 1.1 J/cm² by incremental steps of 50 mJ/cm². UV-NLA is performed just after the deposition of 50nm of in-situ doped amorphous-Si (LPCVD, 475°C Si_2H_6/PH_3 gaseous mixture) on top of 5nm TiN and 2nm of HfO_2 as described in Fig. 2. After annealing, surface roughness and sheet resistance (Rs) of structures were characterized. Fig. 3 shows the evolution of haze intensity and sheet resistance as function of pulse laser energy density. The minimum sheet resistance value is obtained for an energy density of 1.05 J/cm² (as compared to the un-annealed reference) corresponding to the highest haze intensity. After UV-NLA the surface roughness of the poly-Si (Fig.4) is removed using Chemical Mechanical Polishing (CMP). Atomic Force Microscopy (AFM) scan (after CMP) coupled with cross-sectional TEM picture show the crystallization of amorphous Si into poly-Si (Figs 5 & 6). Low resistance and full recrystallization have been achieved at high temperature (>1200°C) by UV-NLA while not increasing the thermal budget of the bottom level [6].

B- Low-temperature silicon epitaxy: Epitaxy at low temperature is a key enabler for 3D sequential integration but it poses several challenges: control of contaminants on the starting surface, exponential decrease of growth rate with temperature, lower epitaxial layer quality or loss of selectivity, to give a few examples. Here we report for the first time a full 500°C process, including RSD growth and prebake treatments. Traditional surface preparation (HF-Last cleaning and 650°C prebake) was successfully replaced by surface preparation including a HF/HCl wet clean followed by an in-situ Siconi NH_3/NF_3 remote plasma process and a 500°C H_2 bake. Selective epitaxial growth is based on the cyclic use of a new Si precursor and Cl_2 etching, which was shown to provide an excellent etch selectivity of amorphous-Si versus single-crystal Si over a large range of Cl_2 partial pressures and temperatures (see Fig.7). The cross-sectional TEM image in Fig.8 shows that both selectivity and good crystallinity of the 500°C Si epi were obtained using either 650°C or 500°C H_2 bake temperatures.

III. iBEOL CONTAMINATION MANAGEMENT, STABILITY AND RELIABILITY

A- Re-introducing wafers in FEOL after iBEOL processing. When metal lines are present between the stacked devices, contamination management is one of the biggest challenge for manufacturing since, for top FET fabrication, the wafers will have to be re-introduced in a Front End (FE) contamination-free environment after having been processed in a BE contaminated environment for

978-1-7281-1988-5/18 $31.00 © 2018 IEEE

contact and metal lines fabrication. Typically iBEOL process introduces Ti, TiN, W, Ta, TaN and Cu on the wafers. The backside can be easily decontaminated [9]. On the front side, the materials can be encapsulated. However, at the bevel edge, simple encapsulation is not sufficient because this area is subjected to handling, potential scratch formation, etc. which all may release contaminants. Thus the most critical contamination source is the wafer bevel edge. Fig. 9 illustrates the contamination containment strategy developed including: 1- bevel & notch etch back on bulk Si, 2- decontamination and finally 3- encapsulation of these zones in order to cut any access to the contamination of BEOL metals. The first step was implemented in the bevel polishing tool Model EAC300bi-hv from EBARA. Fig.10 presents optical microscopy images of the notch and bevel after polishing. Using a tape with optimized grain size, a roughness below 1nm RMS can be obtained. The EDX spectrum of Fig.11, taken before bevel edge polishing of a CMOS wafer with four metal lines clearly shows presence of Ti and Ta. The bevel etch step will allow us to remove all deposited metals and expose the bare silicon wafer. Further edge decontamination by wet cleaning is now possible.

Decontamination is then carried out on the bevel edge via dual step wet cleaning process (Aqua Regia + nitric acid spiked with controlled HF solution). Results of bevel contamination before and after cleaning steps using VPD-CD-ICPMS technique are presented in Fig.12 [10]. Contamination has been efficiently decreased by a factor ten for most elements and can be further reduced using front side bevel cleanings. Finally, the encapsulation of the bevel can be realized by deposition of SiO_2 or SiN followed by planarization. The material needs to be thick enough to resist handling scratches and all cleanings necessary for the realization of a transistor level. This methodology ensures high-quality contamination control before wafer bonding.

B- Cu/ULK iBEOL thermal stability: For this study we have used as reference the standard 28nm design rules damascene BEOL integration of the state-of-the-art FDSOI technology [11]. The Inter Metal Dielectrics (IMD) materials are ULK material (20 nm SiCNH etch stop layer (k_{as-dep} = 5.6) and 100 nm porous SiOCH (k_{as-dep} = 2.7)). Four metal lines/vias levels have been integrated and encapsulated with SiCN layer in order to apply different annealing processes. The SiCN layer is afterward removed by dry etching and a CMP touch is realized in order to enable electrical tests on M1, V1 and M2. Figures 13 summarizes the process flow used to test the stability and reliability of iBEOL. Fig. 14 and 15 evidence for the first time the stability of the breakdown voltage for line to line and via to line, respectively; the fluctuations observed are within standard dispersion due to process variations and not split-related.

C- Bonding on Cu/ULK iBEOL & thermal stability: The morphology of Cu/ULK iBEOL must also be checked as porous ULK layers tend to shrink with increasing thermal budget in free surface condition (no capping) [12]. Indeed, when pore radius decreases, carbon-containing gases are released due to alkyl groups (CHx) bond breaks, which may lead to voids in the structure. For this study, a CMOS wafer with 4 metal lines, SiCN capping and a 120nm planarized oxide on top have been bonded to an oxidized

bulk silicon substrate, and annealing of 600°C 2h has been applied. Fig.16 shows no significant change of the acoustic signal before and after 600°C annealing of the whole 300mm wafer, which proves the good stability of the stack. The existing initial bonding defects were explained by non-optimized planarization as described in [13] putting this study in a worst case condition with thick encapsulation and initial bonding defects.

IV. SMART CUT™ FILM TRANSFER

Direct bonding of SOI wafer followed by etch-back above a CMOS level has already been demonstrated previously [14]. This approach benefits from the perfect crystallinity and thickness control of commercial SOI wafers. However, to reduce cost, to avoid grinding back the entire SOI wafer (usually a source of metallic contamination) and to limit post-grinding Si etching steps (which can results in bevel defects), Smart Cut™ can be directly performed on a bulk wafer [15] (illustrated in Fig. 17) to transfer a thin Si layer onto the CMOS wafer. Strategies for the post processing of the transferred film in order to limit the thermal budget imposed by the stability of silicide and iBEOL can be found in [16,17]. Here, for the first time, we demonstrate the feasibility of the Smart Cut™ directly over a 300 mm CMOS processed wafer. The transferred 10 nm thin silicon layer is shown in the scanning TEM cross section in Fig. 18. Its good uniformity is confirmed by spectroscopic ellipsometry mapping in Fig. 19 with a maximum variation of 0.7nm on 48 measurement points. A photography of the whole 300 mm wafer with top transferred Si film is presented in Fig. 20.

V. CONCLUSION

This work presents advances bringing 3D sequential integration closer to manufacturability. Here we present six achievements on process steps that were considered as potential showstoppers for this technology.

In order to obtain high performance top FETs, low gate access resistance has been achieved using UV nano-second laser recrystallization of in-situ doped amorphous silicon. Full 500°C selective Si epitaxy process is demonstrated owing to an advanced LT surface preparation with a combination of dry and wet etch preparation; the selective epitaxial growth is obtained with the cyclic use of a new Si precursor and Cl_2 etching.

In parallel, this work paves the way to manufacturability of 3D sequential integration including iBEOL with standard ULK and Cu metal lines. A bevel edge contamination containment strategy composed of 3 steps (bevel etch, decontamination, encapsulation) enable wafers re-introduction in FEOL environment after BEOL process. In addition, the stability of line to line breakdown voltage for interconnections submitted to 500°C anneals is demonstrated for the first time.

Finally, Smart Cut™ transfer of a crystalline silicon layer on a processed bottom level of FDSOI CMOS devices is demonstrated as an alternative to SOI bonding and etch back process scheme for top channel fabrication.

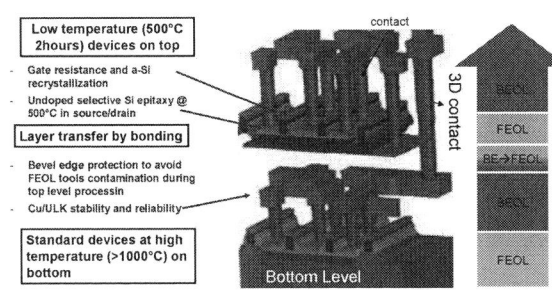

Fig.1 : 3D sequential Coolcube™ integration and remaining process issues

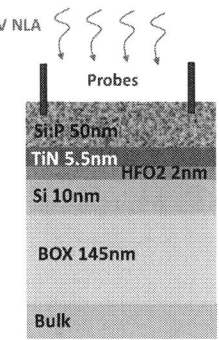

Fig.2 : Schematic of UV-NLA experiment on gate stack

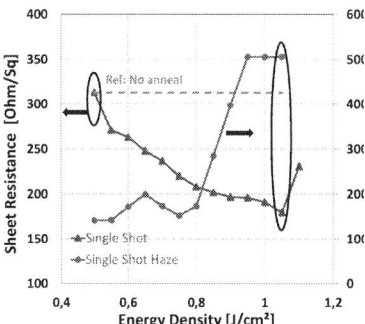

Fig.3 : Both Rs (left axis) and haze (right axis) evolution versus pulse laser energy density for a 50nm thick a-Si layer

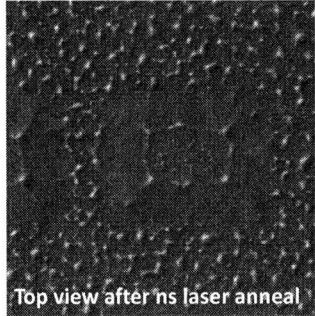

Fig.4: SEM image top view of an active area after 1 ns laser anneal pulse on a 50nm thick a-Si gate (laser energy density 1.05J/cm²).

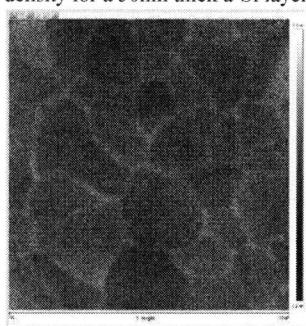

Fig.5: Cross sectional TEM micrographs of 50nm thick a-Si after ns laser anneal (top) and after CMP (bottom).

Fig.6: AFM 3D image of the surface of Fig.6 (bottom) after UV-NLA (1.05 J/cm²) and CMP. Rq=3.2A RMS.

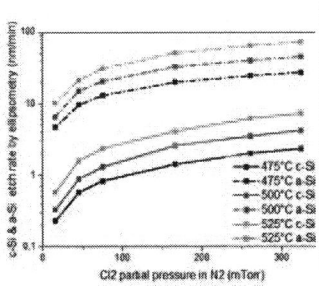

Fig.7: a-Si (a) and single-crystal (c) Si etch rates for various Cl₂ partial pressures at 475°C, 500°C and 525°C.

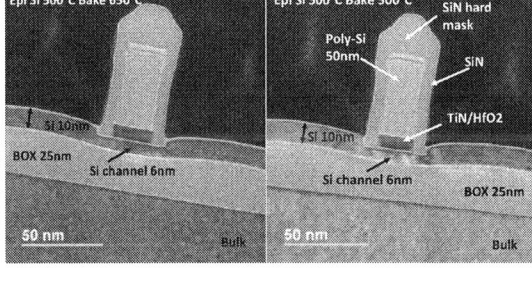

Fig.8: TEM cross sections of FDSOI devices with selective undoped Si epi in the source and drain @ 500°C. Bake 650°C (left) bake 500°C right.

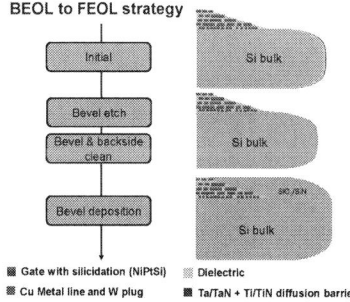

Fig.9: 3-step BEOL to FEOL strategy. 1- bevel etch-back on Si 2- backside and bevel decontaminaton 3- Encapsulation.

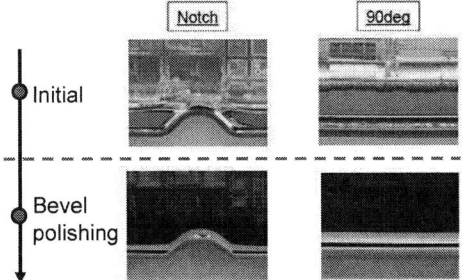

Fig.10: Optical microscopy images of the noch and bevel edges before and after bevel polishing using Model EAC300bi-hv from EBARA.

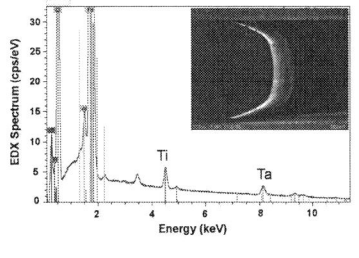

Fig.11: EDX spectrum on the bevel edge of a CMOS wafer with 4 metal lines evidenced Ti and Ta residues.

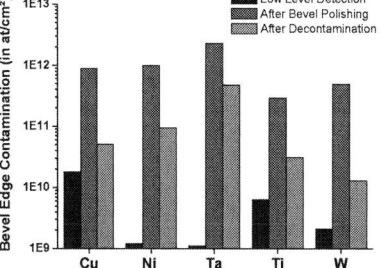

Fig.12. VPD-DC-ICMPMS results before and after backside decontamination steps.

978-1-7281-1988-5/18 $31.00 © 2018 IEEE   155

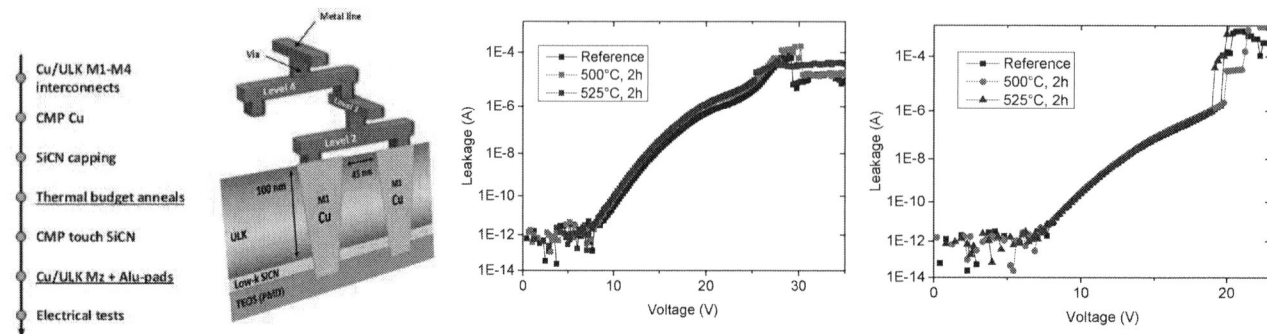

Fig. 13: Process flow used for testing line to line and via to line reliability (28 nm DR Cu/ULK BEOL).

Fig.14: Line to line breakdown voltage with and without anneals (500°C and 525°C). Unsignifcant degradation is observed with 500°C anneal.

Fig.15: Via to line breakdown voltage with and without anneals (500°C and 525°C). Unsignifcant degradation is observed with 500°C anneal.

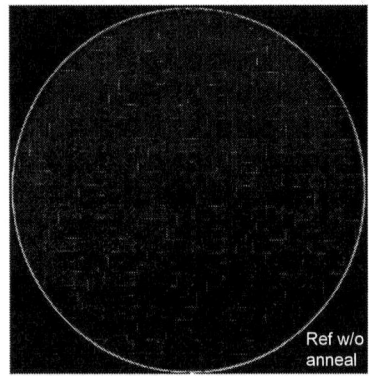

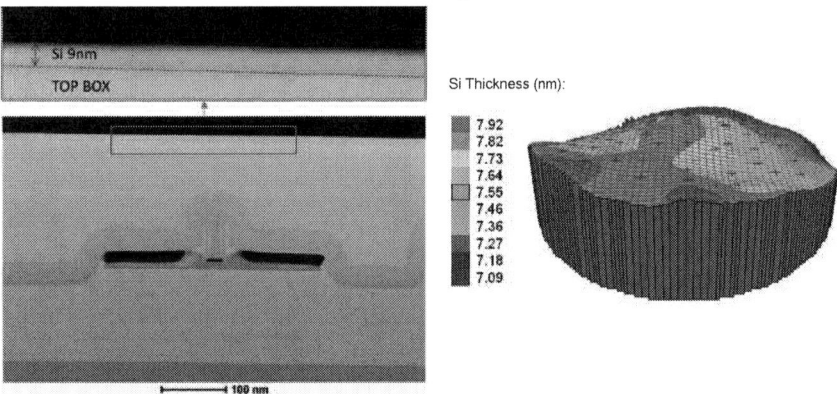

Fig.16: Bonding acoustic microscopy of CMOS wafers up to M4 before and after 600°C 2h annealing.

Fig.17: Layer transfer by Smart Cut™ process above CMOS wafer. Splitting done at 500°C.

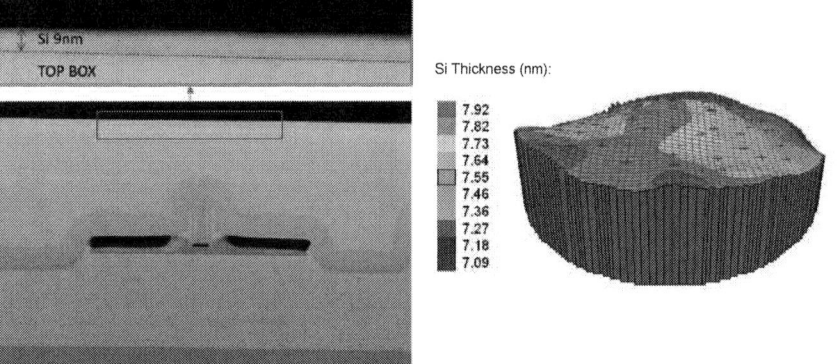

Fig.18: STEM cross-section after layer transfer by Smart Cut™ process and thinning above a 20nm gate length transistor.

Fig.19: 48-point ellipsometry map of reported Si: Min=7.15nm, Max=7.88nm, Range=0.73nm, Mean=7.55nm.

Fig.20: Photography of layer transfer above CMOS after layer transfer by wafer Smart Cut™ process.

AKNOWLEDGEMENTS

This work was partly funded by French Public Authorities through the NANO2017 & WAYTOGOFAST programs.

REFERENCES

[1] P. Batude et al, IEDM 2017 [2] A. Vandooren et al., VLSI 2018, [3] M. Shulaker et al., IEDM 2014 [4] T-T. Wu et al, IEDM 2015 [5] Y-H. son et al, VLSI 2007 [6] C. Fenouillet-Beranger et al., IEDM 2014 [7] C-M.V. Lu et al., VLSI 2017 [8] O. Billoint et al., DATE 2015 [9] P. Besson et al., UCPS 2014 [10] Devita et al., Solid State Phenomena pp 268-271, Vol. 219 2015 [11] N. Planes et al., VLSI 2012 [12] V. Larrey et al., WAFERBOND 2017 [13] M. Abdel Sater et al., IMAPS 2018 [14] L. Brunet et al., VLSI 2016 [15] M. Bruel, US patent 5374564 [16] I. Radu et al., ESSDERC 2013 [17] L. Di Cioccio et al., ECS Transaction 50 (7) 169-175 2012

978-1-7281-1988-5/18 $31.00 © 2018 IEEE

Hybrid bonding for 3D stacked image sensors: impact of pitch shrinkage on interconnect robustness

J. Jourdon[1,2,3], S. Lhostis[1], S. Moreau[3], J. Chossat[1], M. Arnoux[4], C. Sart[1], Y. Henrion[1], P. Lamontagne[1], L. Arnaud[3], N. Bresson[3], V. Balan[3], C. Euvrard[3], Y. Exbrayat[3], D. Scevola[1], E. Deloffre[1], S. Mermoz[1], A. Martin[4], H. Bilgen[1], F. Andre[1], C. Charles[1], D. Bouchu[3], A. Farcy[1], S. Guillaumet[1], A. Jouve[3], H. Fremont[2], and S. Cheramy[3]

[1]STMicroelectronics, Crolles, France, email: joris.jourdon@st.com, [2]University of Bordeaux, IMS Laboratory, Talence, France, [3]CEA-LETI, Grenoble, France, [4]STMicroelectronics, Grenoble, France.

Abstract— Hybrid bonding is a high-density technology for 3D integration but further interconnect scaling down could jeopardize electrical and reliability performance. A study of the influence of hybrid bonding pitch shrinkage on a 3D stacked backside illuminated CMOS image sensor was performed from a process, device performance and robustness perspectives, from 8.8 µm down to 1.44 µm bonding pitches. As a result no defect related to smaller bonding pads was evidenced neither by thermal cycling nor by electromigration, thus validating fine-pitch hybrid bonding robustness and introduction for next generation image sensors.

I. INTRODUCTION

The emerging of 3D stacked BackSide Illumination (BSI) makes hybrid bonding process a key solution for CMOS Image Sensors (CIS). A high level of maturity has been recently demonstrated for a pitch of 4 µm [1], justifying the prediction of a significant growth of this technology in the field of CIS for the next years [2]. A low interconnect pitch (<2 µm) would enable to increase interconnect density and to reduce pixel size for applications such as Single Photon Avalanche Diode (SPAD). Early improvements of hybrid bonding levels scaled the pitch down to 2 µm and below for a single Cu bonding pad level architecture [3]. In the case of a dual damascene approach which enables to isolate hybrid bonding pads from last Back End of Line (BEoL) levels [4], the shrinkage of hybrid bonding could be limited as new failure mechanisms may occur.

In this paper the impact of reducing hybrid bonding pitch is investigated by comparing bonding quality, electrical and optical performance and robustness of 3D stacked BSI CIS with 8.8 µm down to 1.44 µm hybrid bonding pitches.

II. TEST VEHICLE

The demonstrators are integrated on 300 mm wafers, stacking a BSI CIS on a digital CMOS technology node (Fig. 1). Wafers are connected thanks to Hybrid Bonding Metal pads (HBM) and Hybrid Bonding Vias (HBV) processed by a dual damascene architecture. The face-to-face hybrid bonding process is performed on an EVG GEMINI®FB system followed by a bonding annealing at 400 °C to strengthen the bonding interface and stabilize Cu microstructure.

The demonstrators consist in a BSI-CIS to detect the impact of HBM size on sensor performance and passive test structures designed for in-depth electrical and robustness studies of hybrid bonding levels. All these test vehicles are integrated with different hybrid bonding pitches (8.8, 7.2 and 1.44 µm). The other BEoL levels left unchanged enable to distinguish the physical HBM pitch from the electrical HBM pitch depending on how HBM are connected to BEoL (Fig. 1). FEM simulations of daisy chains identify HBV as the most resistive part of interconnects (Fig. 2a). Daisy chains with various geometries were designed as test structures to minimize or maximize the current density (Fig. 2b).

III. PROCESS VALIDATION

A. Topography mitigation at wafer level

Hybrid bonding relies on global flatness and local topography. Finite Element Method (FEM) simulations of Cu-Cu interface closure during annealing were carried out for HBM with a given dishing profile (Fig. 3). An overall decrease of the contact area is observed with HBM width reduction at a given dishing depth value in agreement with the literature [5-6]. Dishing compensation thanks to thermal expansion is thus a critical parameter. The dishing has to be uniform and light enough to enable the contact of HBM during the post-bonding annealing. Controlled planarization process is thus required as dishing variation is affected by Cu pad width [7].

A dedicated CMP process was developed in response to obtain identical dishing from 1.44 to 8.8 µm HBM pitch on the same wafer, as confirmed by Atomic Force Microscopy (AFM) measurements (Fig. 4). The process also compensates the topography induced by prior BEoL levels. Scanning Acoustic Microscopy (SAM) performed after bonding annealing reveals no unbonded zone within the 50 µm limit of resolution of the SAM. Backside process steps and final annealing at 400 °C were completed with no defectivity.

B. Cu-Cu interface closure

Cu-Cu bonding relies on thermal expansion of Cu and on atoms diffusion at bonding interface. This mechanism is enhanced by a high density of grain boundaries. Electron BackScatter Diffraction (EBSD) indicates that the mean grain size decreases from 1.1 to 0.3 µm, respectively for 4.4 and 0.5 µm HBM widths. TEM pictures (Fig. 5) evidence a good closing of the Cu-Cu interface whatever the pitch. In both cases, bonding voids, caused by Cu oxide demixion and vacancies migration [4], have similar sizes. A good bonding quality is also obtained for smaller Cu grains.

978-1-7281-1988-5/18 $31.00 © 2018 IEEE

Bonding quality is assessed thanks to resistance measurements of daisy chains with 100 or 30,000 links (respectively DC100 and DC30k). Each test structure reaches 100 % yield whatever HBM pitch (Fig. 6). Results are in agreement with FEM calculations indicating a negligible effect of interface resistance and of Cu microstructure on daisy chains resistance. Interconnects are simulated by FEM for different HBV diameters measured after the lithography step. According to extracted resistances, the dispersion evidenced at wafer edges is linked to HBV diameter variations (Fig. 7). The shift between experimental and theoretical values is attributed to process variations, such as HBV conical shape and diameter enlargement due to over-etching (Fig. 8).

C. Sensitivity to bonding overlay

Bonding misalignment reduces HBM contact area and brings interconnects laterally closer, potentially leading to resistance and capacitance increases. Full wafer overlay measurement gives a mean value of 200 nm ($\pm 3\sigma$). The Cu-Cu contact area decreases faster for small HBM but the resistance of 1.4 µm pitch interconnects is demonstrated to be insensitive to misalignment (Fig. 9). FEM parametrical study of resistance variations as a function of misalignment indicates an increase lower than 1 % for a J_{max}-1.44 µm interconnect (Fig. 2) at maximum overlay (200 nm) (Fig. 10). This variation is not measurable as long as it does not overcome resistance standard deviation. Considering the dual damascene integration, overlay is not expected to become critical for fine pitch until HBM contact area reaches HBV section. For a 200 nm misalignment this critical pitch is estimated at 0.8 µm.

The capacitance measured on 3D combs with a redundant HBV and redundant HBM is 12 ± 1 pF. Simulation of 3D combs shows that a 200 nm overlay increases the capacitance by 0.2 % (Fig. 11). This increase is twice higher for combs containing dummies but remains negligible. The 7.2 µm electrical HBM pitch is therefore too large for the capacitance to be significant.

IV. IMAGE SENSOR DEMONSTRATOR

The influence of hybrid bonding pitch on optical performance is evaluated using a 14 Mpixels BSI imager with 1.5 µm pixel pitch. The top die contains exclusively the array of pixels while the bottom die has all analog blocks: Analog-to-Digital Converter (ADC), column decoder and control. It also comprises a state of the art High Dynamic Range (HDR) image signal processing pipeline able to sustain up to 600 Mpixels/s and a large cluster of computer vision processing engines including a hexa-core CPU. Two versions of the 3D stacked sensor are studied comparing hybrid bonding pitches of 8.8 µm and 1.44 µm with respectively redundant HBV or redundant HBM. The integrity of the whole stack is validated for both pitches according to TEM cross sections (Fig. 12).

The standard optical performance such as sensibility and Photo Response Non-Uniformity (PNRU) are measured at wafer level (Fig. 13). Cumulative distributions have a low dispersion and are similar from one pitch to another. After packaging and mounting in a specific set-up, the optical properties of the 3D BSI-CIS for 8.8 or 1.44 µm HBM pitch

are tested based on the parameters described in Table 1 and are within the specifications. Identical values were obtained for both HBM widths. Both sensors can take pictures (Fig. 14), validating the use of fine pitch for 3D stacked CIS.

V. ROBUSTNESS OF BONDING INTERFACE

D. Thermal cycling

Since the bonding interface is made of Cu and SiO_2 with different Coefficients of Thermal Expansion (CTE), thermo-mechanical stresses can occur at ambient temperature or due to Joule heating. 2D FEM is carried out to localize the critical zones in an array of HBM. Fig. 15 shows that von Mises stress is higher in HBM corners and triple points formed by the bonding interface and HBM walls. At smaller pitch oxide spaces are not wide enough to release stress, so delamination is more inclined to occur at the Cu/SiO_2 bonding interface for the 1.44 µm pitch.

In order to evaluate this risk of delamination, thermal cycling tests at wafer level are conducted at -65/+150 °C and -55/+150 °C, 500 cycles, respectively for HBM pitches 7.2 and 1.44 µm. The resistance of the daisy chains is measured by 4-wire sensing before and after the test. All test structures remain fully functional after test and resistance variation is lower than 1 % (Fig. 16). Despite a higher stress pointed out by simulation for the smallest pitch, the bonding interface is robust.

E. Electromigration

Pitch reduction potentially associated with ever-increasing current densities [8] rekindles discussions about electromigration (EM) immunity of hybrid bonding level. For the 7.2 µm hybrid bonding pitch, EM-related failures localized in the BEoL confirm the previous studies on a demonstrator containing only daisy chains [9]. No specific hybrid bonding-related failure occurs with additional metal levels. For a BEoL made of four top metal lines and seven bottom metal lines, the bonding interface is immune (Fig. 17). The Black's parameters are typical ones of Cu interconnects [10]. EM test are conducted (350 °C, 30 mA, failure criterion: 10 % of electrical resistance increase) on DC100 (Fig. 18) with either J_{min}-7.2µm or J_{min}-1.44µm interconnects described on Fig. 2. The probability plots are similar and suggest the same failure mechanisms for both pitches.

CONCLUSION

The transition of hybrid bonding pitch from 8.8 to 1.44 µm is made possible thanks to the development of a new surface preparation process enabling flatness and uniform dishing. The physical characterizations and 100 % yield extracted on 30k daisy chains evidence the quality of this fine-pitch bonding. The electrical resistance was demonstrated to be insensitive to bonding overlay as long as the contact area remains larger than HBV section. The electrical and optical tests of a full BSI CIS demonstrate no fine-pitch hybrid bonding impact on performance. Aging tests point out no failure related to the HBM pitch shrinkage, thus confirming the dual damascene integration choice. These results evidence the robustness of dual damascene hybrid bonding down to 1.44 µm pitch.

978-1-7281-1988-5/18 $31.00 © 2018 IEEE

ACKNOWLEDGMENT

This work was funded thanks to the French national program "Programme d'Investissement d'Avenir IRT Nanoelec" ANR-10-AIRT-05 and to ENIAC Joint undertaking "Pilot Optical Line for Imaging and Sensing" (POLIS) project.

REFERENCES

[1] Y. Kagawa et al., "Novel stacked CMOS image sensor with advanced Cu2Cu hybrid bonding," IEDM, 2016, pp. 8.4.1-8.4.4.

[2] P. Cambou and J.-L. Jaffard, "Status of the CMOS Image Sensor Industry 2017," Yole Developpement, 2017.

[3] E. Beyne et al., "Scalable, sub 2µm pitch, Cu/SiCN to Cu/SiCN hybrid wafer-to-wafer bonding technology," IEDM, 2017, pp. 32.4.1-32.4.4.

[4] S. Lhostis et al., "Reliable 300 mm Wafer Level Hybrid Bonding for 3D Stacked CMOS Image Sensors," ECTC, 2016, pp. 869–876.

[5] L. Di Cioccio et al., "An overview of patterned metal/dielectric surface bonding: Mechanism, alignment and characterization," J. Electrochem. Soc., 2011, vol. 158, no. 6, pp. P81–P86.

[6] C. Sart et al., "Cu/SiO2 hybrid bonding: Finite element modeling and experimental characterization," ESTC, 2016, pp. 1-7.

[7] L. Arnaud et al., "Fine pitch 3D interconnections with hybrid bonding technology: From process robustness to reliability," IRPS, 2018, pp. 4D.4–1–4D.4–7.

[8] International Technology Roadmap for Semiconductors (ITRS), Interconnect, 2013.

[9] S. Moreau et al., "Mass Transport-Induced Failure of Hybrid Bonding-Based Integration for Advanced Image Sensor Applications," ECTC, 2016, pp. 1940–1945.

[10] J. Jourdon et al., "Effect of passivation annealing on the electromigration properties of hybrid bonding stack," IRPS, 2017, pp. MR-3.1-MR-3.6.

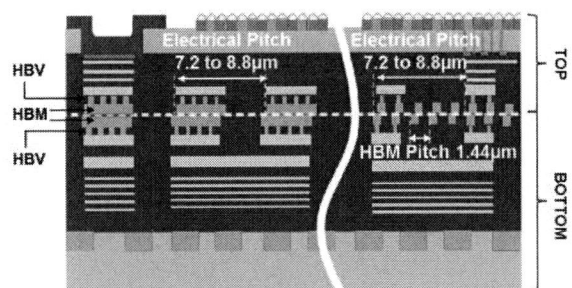

Fig. 1. Schematic cross section of the image sensor (top) stacked on advanced CMOS (bottom). The connection is achieved by interconnects with a hybrid bonding pitch of 8.8, 7.2 (left) or 1.44 µm (right). Electrical pitch is kept identical for a same set of electrical structures ranging from 7.2 to 8.8 µm.

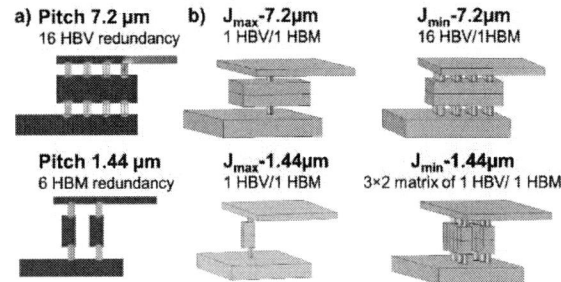

Fig. 2. (a) The current density extracted from FEM simulations is higher in HBV. (b) Schema of single interconnects. The current density can be maximized (J_{max}) or minimized (J_{min}) by varying width of HBM (7.2 or 1.44 µm) and redundancy (HBV for 7.2µm pitch or HBM for 1.44µm pitch).

Fig. 3. Thermo-mechanical finite element simulations of dished Cu-Cu interface closure during annealing at 400 °C for varying HBM width and dishing depth.

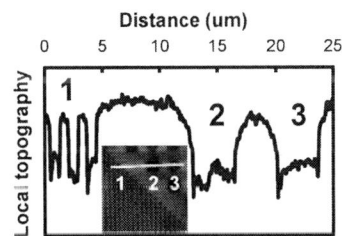

Fig. 4. AFM 2D scan on 0.72 and 3.6 µm-wide HBM with profile scan along white line showing identical dishing depth for both pitches.

Fig. 5. TEM image cross section of (a) 4.4 and (b) 0.72 µm-wide HBM. Bonding voids (black arrows) are small compared to the large well-bonded Cu/Cu interface.

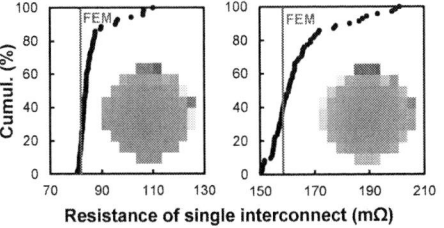

Fig. 6. Cumulative distributions of single interconnect resistances and mapping for DC30k (left) J_{min}-7.2µm and (right) J_{min}-1.44µm.

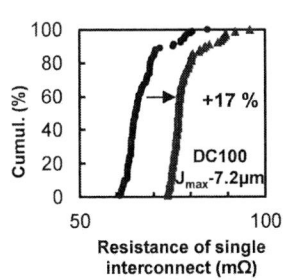

Fig. 7. Cumulative distributions of experimental (black circles) and FEM (blue triangles) resistance based on HBV diameters measured during HBV lithography step.

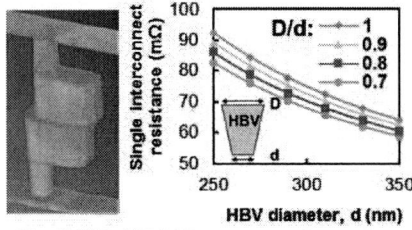

Fig. 8. FIB-SEM 3D reconstruction on J_{min}-1.44µm showing HBV conical shape (left). FEM simulation showing the resistance dependence to the HBV shape and diameter of HBV on J_{max}-7.2µm. (right).

978-1-7281-1988-5/18 $31.00 © 2018 IEEE   159

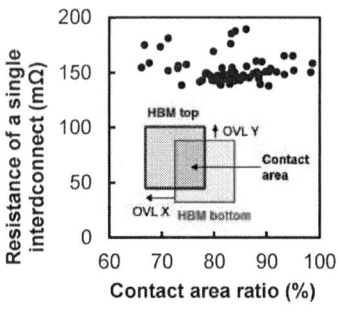

Fig. 9. Resistance of DC100 J_{min}-1.44μm as a function of contact area deduced from overlay (OVL) measurements.

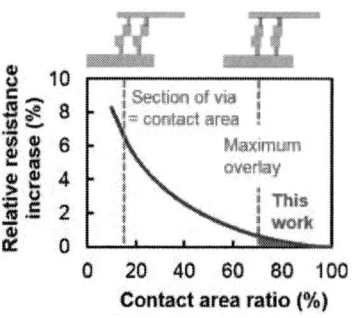

Fig. 10. FEM calculation of the resistance increase with contact area reduction (DC100 J_{min}-1.44μm).

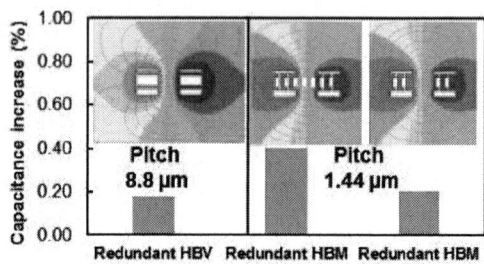

Fig. 11. FEM simulations of capacitance increase due to a 200 nm misalignment for various combs designs. Black lines represent the electric field between interconnect at a 1 V polarization.

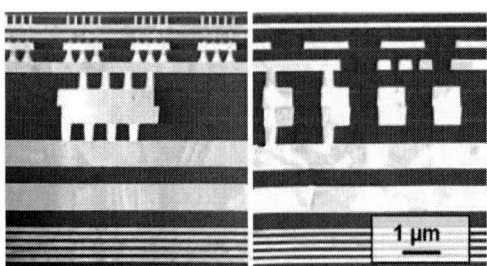

Fig. 12. TEM cross section of 3D stacked image sensor with 8.8 (left) and 1.44 μm-pitch (right). All BEoL levels are visible.

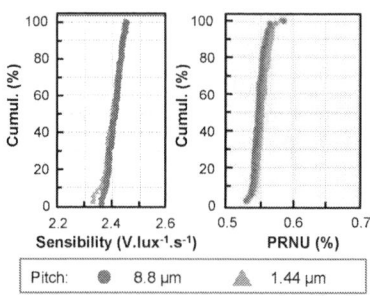

Fig. 13. Cumulative distributions of the sensibility and PRNU of 3D image sensor.

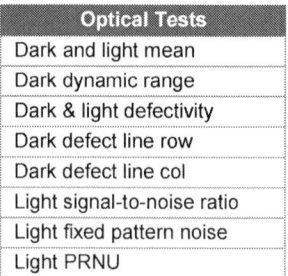

Optical Tests
Dark and light mean
Dark dynamic range
Dark & light defectivity
Dark defect line row
Dark defect line col
Light signal-to-noise ratio
Light fixed pattern noise
Light PRNU

Table 1. Electrical and optical parameters tested for 3D CIS with HBM pitches of 8.8 and 1.44 μm.

Fig. 14. Pictures taken with 3D stacked image sensor with hybrid bonding pitch (a) 8.8 and (b) 1.4 μm.

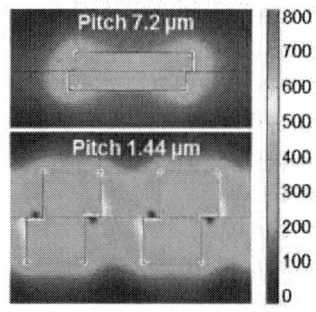

Fig. 15. 2D FEM simulations of thermo-mechanical stress due to CTE mismatch in an array of HBM.

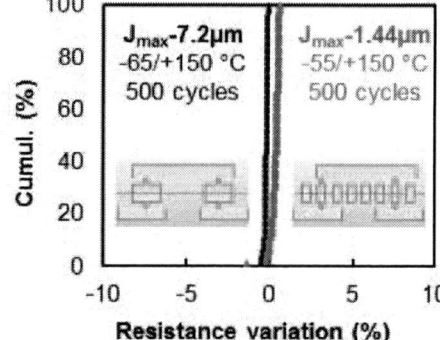

Fig. 16. Distribution of DC100 resistance variation after thermal cycling. J_{max}-1.72μm (black round) and J_{max}-1.44μm (blue triangle).

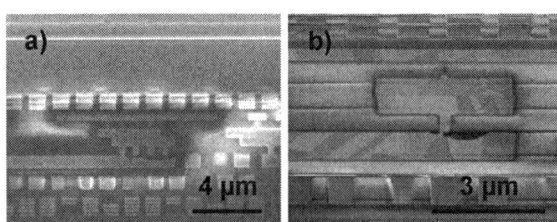

Fig. 17. Post-mortem cross-section of DC100 (a) J_{min}-7.2μm after an EM test at 350 °C/30 mA and (b) J_{max}-7.2μm after an EM test at 350 °C and 5 mA. Electrons flow from the top to the bottom of the test structure.

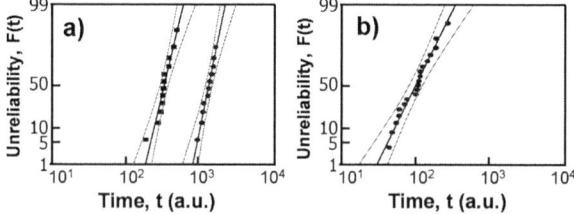

Fig. 18. Probability plot (lognormal distribution) for DC100 after EM tests at 350 °C (left) J_{max}-7.2μm with 20 mA (blue circle) or 30 mA (black square), and (right) J_{min}-1.44μm enduring an electromigration test at 350 °C and 30 mA. Confidence bounds: 90 %.

Embedded Select in Trench Memory (eSTM), best in class 40nm floating gate based cell: a process integration challenge

S. Niel[1], F. La Rosa[2], A. Regnier[2], M. Mantelli[2], F. Trenteseaux[1], G. Ghezzi[1], A. Marzaki[2], Q. Hubert[2],
J. Delalleau[2], T. Cabout[1], F. Maugain[2], E. Lepape[2], L. Baron[2], A. Champenois[2], D. Galpin[1], N. Cherault[1],
S. Audran[1], L. Parmigiani[1], P. Gouraud[1], B. Duclaux[1], Y. Escarabajal[1], F. Baudin[1], E. Beche[1], B. Saidi[1], V. Arnal[1]
STMicroelectronics, [1]Crolles- France, [2]Rousset France,

Abstract— This paper discusses an innovative architecture of charge storage NVM cell, which outpaces state-of-the-art in term of bit-cell area. This new concept of memory cell is used today in production for microcontrollers. After cell architecture and activation description, we will present process flow integration challenges, process optimizations and single cell characterizations.

I. INTRODUCTION

The embedded non volatile memory based microcontrollers (eNVM MCU) market is growing very fast thanks to the increasing demand on general purpose (GP), automotive (Auto), internet of things (IoT) and secure (SC) applications. These applications require a scalable eNVM cell with high cycling capability, good retention and low power consumption on a wide range of temperature. Many emerging NVM cells are under development for advanced nodes (MRAM, PCM, RRAM) but charge storage cells remain the most used since decades for their simpler process and proven reliability.

Table I compares performance and scalability of various charge storage/trapping eNVM cell. Today, despite its more complex process, split gate eNVM cell family is the preferred solution in 40nm node. The presence of select gate allows better performance in term of power consumption during write operations, but also NVM macrocell optimization (area gain and simpler erase operation management) [1-4]. In this paper, we present a new highly reliable and scalable 40nm 1.5T cell. This NVM cell, called Embedded Select in Trench Memory (eSTM cell), has been developed to optimize microcontroller's area, cost and performance versus 1T NOR cell. eSTM presents the advantages of split gate cell with a very good scalability and reliability performances.

II. CELL ARCHITECTURE DESCRIPTION

A. Cell architecture

e-STM cell is a Flash NOR cell formed by a conventional 1T cell coupled with a vertical select transistor (Fig.1). Being vertical, select transistor channel length can be adjusted to sustain very low leakage level, even at hot temperature, without impacting bitcell area which is the more aggressive eNVM cell in 40nm node ($0.049\mu m^2$) [5]. The cell is placed on Pwell isolated from bulk thanks to an n-iso buried implant

that plays also the role of source plate. Vertical select transistor is common for odd and even cell. Cell selection is performed by selecting odd or even Bit Line (BL) and the select transistor word line (WL). The layout of the e-STM cell is pictured in Fig.2. CG and vertical select transistor are not self-aligned. We have studied the cell-to-trench distance impact on programming efficiency and demonstrated that working window is larger than alignment tolerance.

B. Cell operations

eSTM cell is erased by Fowler-Nordheim (FN) tunneling and programmed using hot electron mechanism similar to source side injection (SSI). As in a conventional 1T NOR flash, FN erase is obtained by generating high electric field across tunnel oxide using high voltage (HV) on CG and bulk terminals.

eSTM programming is performed using fast and low power mechanism enabled by the 1.5T architecture [6]. Efficient programming is obtained by polarizing the vertical select trench transistor in low inversion regime to limit current consumption and generate hot carriers (impact ionization in the pinch-off region as shown on Fig.3). Electrons move toward the tunnel oxide interface, and those having sufficient energy are injected into the floating gate across tunnel oxide thanks to the vertical electric field. Injection current obtained by TCAD simulation shows a wide injection distribution along the tunnel oxide with a clear peak on the cell source side (Fig.4). This wide injection distribution is attributed to a spread of electrons in the region between tunnel oxide and select transistor.

Thanks to the very low select transistor off current, we can tolerate depleted cells (negative erase Vt distribution). This allows to perform read operation by keeping CG grounded and suppresses read disturb mechanism [7]. Table II shows bias applied for each operation.

III. PROCESS

A. Process flow

The eSTM specific trench definition steps are performed at the beginning of the flow. Active, Niso, HV and NVM well implants are defined similarly to a 1T NOR cell. Select Trench steps are then inserted in the flow prior to the HV & Tunnel oxide definition. The trench transistor process flow can be

978-1-7281-1988-5/18 $31.00 © 2018 IEEE

described as follow (Fig.5). First a nitride layer is deposited and will be used as polishing stop layer. Then, trench is patterned. It is perpendicular to the active layer and therefore crosses both silicon active and oxide Shallow Trench Isolation (STI) regions. (Fig.6). The trench depth control is a technological challenge of this integration. However, trench depth process window is comfortable enough for manufacturing, as presented in Fig.7, showing impact of trench depth (select length) on select transistor parameters. Minor saturation and leakage current impacts are observed with 5% of variation on select gate depth. Then gate oxide grows along sidewalls and bottom of the trench. Sidewalls roughness and steepness are critical to ensure maximum gate oxide lifetime performance. Trench filling with in situ doped polycrystalline silicon (poly0) deposition allows to generate the transistor gate. Chemical and Mechanical Polishing (CMP) is then performed to remove poly0 surplus. Last step is dry etch of poly0 in order to adjust height (recess) in the trench. Then rest of eSTM process flow is similar to the 1T NOR one's.

One of the challenge of integrating vertical select transistor in the memory process flow was to maintain manufacturing compatibility with the 1T NOR cell platform for IPs re-use and time to market speed up. Fig.8 shows a perfect parametric alignment of digital and HV MOS for both 1T NOR and eSTM technologies.

B. Process trial

Poly0 filling is a key step for vertical transistor integration. Poly voids may impact transistor reliability or induce word line connection issues (resistive or open contacts) (Fig.9). Several process optimizations were performed to improve poly0 filling and WL connections on trenches.

Trench aspect ratio is critical for poly0 filling. We have introduced a nitride pull back step after trench patterning to reduce nitride overhang, facilitate trench filling and to limit remaining voids (Fig.10). Combined with complementary process tuning, we successfully stabilized poly0 filling and removed all poly voids. Fig.11 compares contact resistance on trench between initial and optimized processes. Contact resistance mean values and spread are significantly improved with new processes.

IV. RESULTS

A. Select transistor characterization

Main critical point is to have a reproducible, and reliable vertical transistor MOS. Fig.12 shows single select drain current (Id) versus Gate voltage (Vg) curves for read (Vd~0.5V) and write (Vd~4V) operation conditions at 25°C and 125°C. Id(Vg) curves show standard MOS behavior with leakage current inside specification to sustain maximum bitline leakage criteria. Fig. 13 shows select transistor gate oxide lifetime extrapolation at ambient (25°C) and hot

temperature (125°C) demonstrating high margin versus specification.

B. eSTM Cell characterization

Fig.14 represents single eSTM Cell Id(Vg) curves measured at 25°C and 125°C showing the window between erased and programed states. Write threshold is positive whereas erased cell is totally depleted with negative threshold. Many characterizations have been performed on eSTM to evaluate the best way to activate the cell. For instance, Fig 15 shows programing kinetic versus drain (Vd) and control gate (Vcg) voltages. Best programming efficiency is obtained for select gate bias (V_{gSEL}) near threshold voltage as shown on Fig.16. In this condition, electron current during write operation is minimized and cell consumption is in the range of 1μA (Fig 17) which is in line with split gate consumption reported in literature [8].

Strong reliability and high cycling capability are required for embedded NVM in advanced product. Fig.18 represents the programming window evolution during cycling from 0 to 500kCycles. Threshold voltage degradation occurs only on Erase state and it is compensated at product level thanks to an adaptive algorithm on CG voltage during erase operation.

V. CONCLUSION

In this paper we disclosed the eSTM NVM cell developed to address microcontroller markets. Tens of millions units have been produced on 40nm node. This cell shows a very compact footprint with a bitcell area of $0.049\mu m^2$. It's claimed to be the smallest 40nm charge storage based NVM on the market, and should remain competitive compare to several 28nm conventional charge storage cells [9-11].

Scalability is supported by a simple cell architecture having only one contact per cell. On other hand, reduced drain-source voltage facilitates further reduction of storage MOS length. A shrunk version is currently under development for next microcontroller product generation.

ACKNOWLEDGMENT

Authors would like to thank all the manufacturing teams of STMicroelectronics for supporting this program.

REFERENCES

[1] Yong Kyu Lee et al., IMW proc., p145-148, 2016
[2] L. Q. Luo et al., IMW proc, p149-152, 2016
[3] L. Q. Luo et al., IMW proc., p123-126, 2017
[4] I. Kouznetsov et al, IMW proc., p187-190, 2018
[5] Pak et al., IMW proc., p121-124, 2018
[6] F. La Rosa et al., US Patent No 9224482 (2015, Dec)
[7] F. La Rosa, S. Niel, A. Regnier, US Patent No 9825186 (2017, Nov)
[8] Y. Tkachev et al., IMW proc., p48-51, 2017
[9] S. Tsuda et al., IEDM proc., p469-472, 2017
[10] Y. Taito et al., IEEE JSSC, VOL. 51, NO. 1, January 2016
[11] N.Do et al, IMW proc., p47-50, 2018

Cell Architecture	1T eFlash Planar	1.5T Split-gate Planar	2T UCP Flash Planar	1.5T eSTM Vertical
Device Schematic				
Technology node (Prod → R&D)	40nm	40 → 28nm (FG) 22nm (SONOS)	40nm	40 → 28nm
Memory layer	FG	FG or ONO	ONO	FG
cell Area (40nm)	0.059µm²	0.053µm²	0.079µm²	0.049µm²
P/E Mechanism	CHEI / FN	SSI / FN or SSI / HHI	FN / FN	SSI / FN
Prog. Time	~ µs	~ µs	~ ms	~ µs
Prog. Consumption	> 50µA	~ 1µA	~ pA	~ 1µA
Endurance	100K	100K-1M	100K	100K-1M
Scalability	-	+	--	+ +
Process complexity	Low	Medium (ONO) – High (FG)	Low	Medium
Applications	Auto / GP / SC	Auto / GP / SC	SC	Auto / GP / SC

Table I. 40nm cell architecture benchmark (CHEI stands for Channel Hot Electron Injection and HHI for Hot Hole Injection).

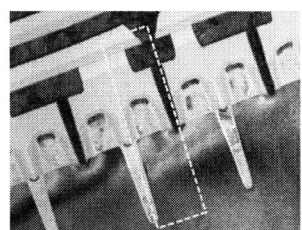

Fig.1. eSTM cell Transmission Electron Microscopy (TEM) cross section. Doted line highlight one bitcell.

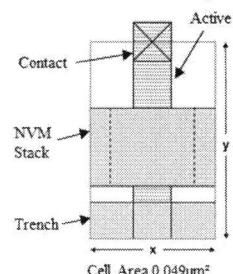

Fig.2. eSTM cell layout.

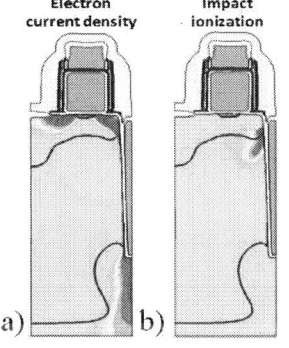

Fig. 3. TCAD simulations of (a) electron current density and (b) impact ionization during programming operation.

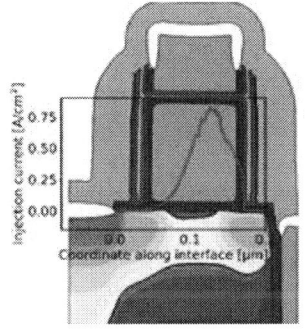

Fig. 4. TCAD simulation of injection current distribution along tunnel oxide interface during programming.

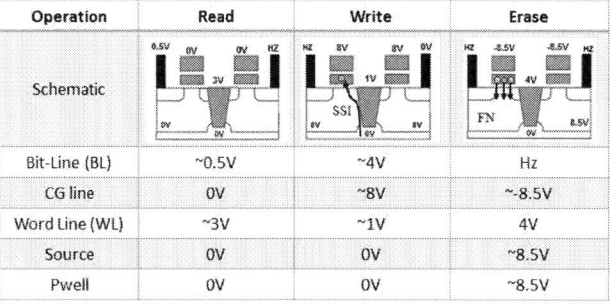

Operation	Read	Write	Erase
Schematic			
Bit-Line (BL)	~0.5V	~4V	Hz
CG line	0V	~8V	~-8.5V
Word Line (WL)	~3V	~1V	4V
Source	0V	0V	~8.5V
Pwell	0V	0V	~8.5V

Table II. eSTM cell activation conditions description.

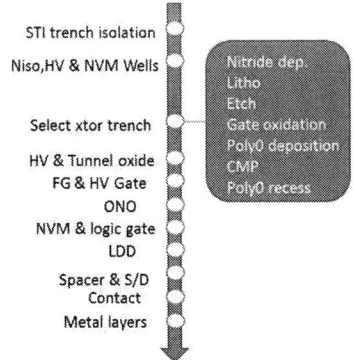

Fig. 5. Process flow description.

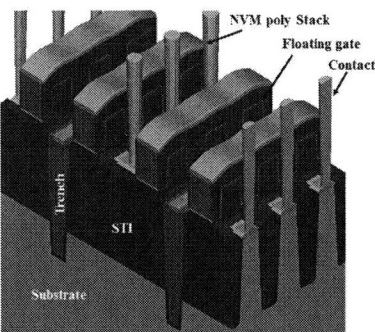

Fig.6. eSTM cell 3D view.

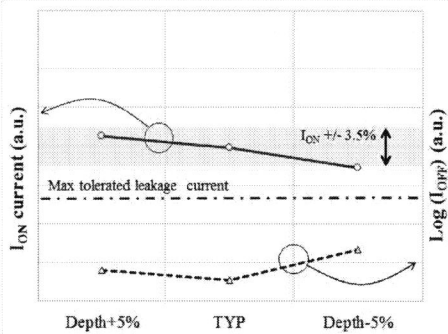

Fig. 7. vertical select transistor parameters (I_{ON} & I_{OFF}) versus trench depth.

978-1-7281-1988-5/18 $31.00 © 2018 IEEE

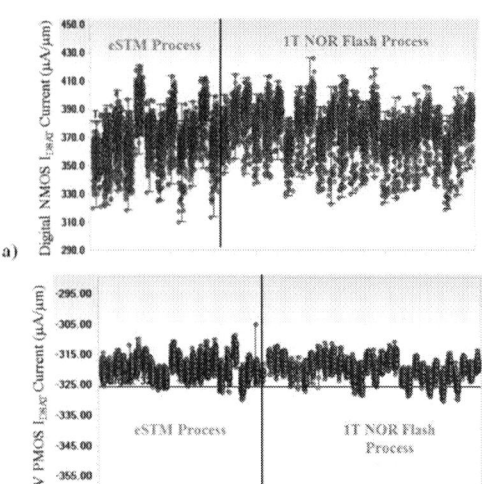

a)

b)

Fig. 8. Digital & HV MOS parametric (saturation current) trend on around 1000 wafers showing good alignement between 1T NOR and eSTM process flow.

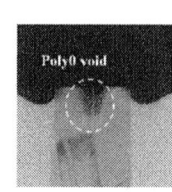

Fig.9. Poly0 void TEM cross-section

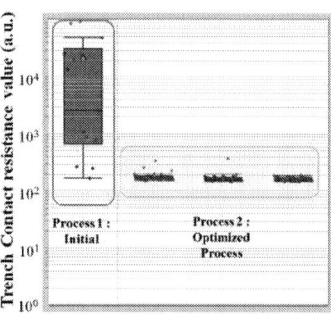

Fig.11. Voids impact on trench contact resistance with initial and optimized process.

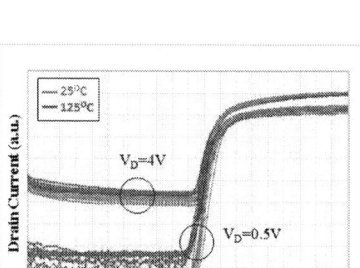

Fig.10. TEM cross-section (a) without and (b) with nitride pull-back process.

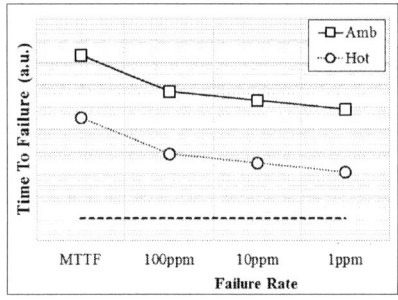

Fig. 12. Select transistor Id(Vg) curves for both read (Vd~0.5V) and write (Vd~4V) conditions at ambient (25°C) and hot temperature (125°C).

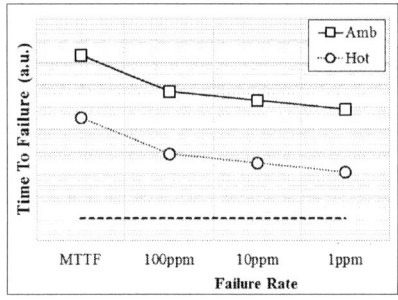

Fig. 13: Select transistor gate oxide lifetime extrapolation at ambient(25°C) and hot temperature (125°C) showing high margin versus specification.

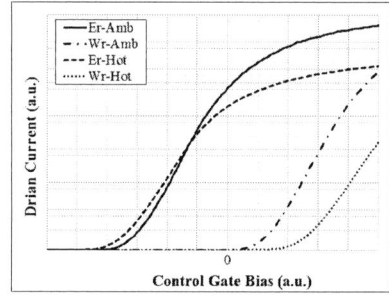

Fig.14. eSTM Cell Id(Vg) curves in both write and erase states at ambient (25°C) and hot temperature (125°C). Erase state threshold is negative.

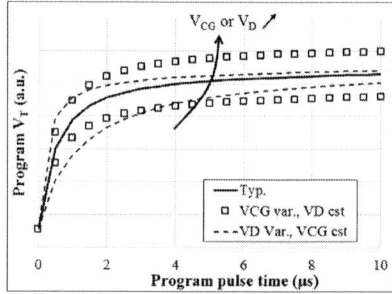

Fig.15. eSTM V_T kinetic during programming operation with typical conditions and V_{CG}, V_D bias variations.

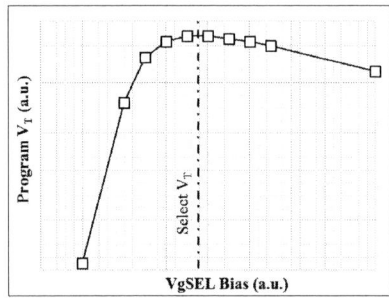

Fig.16. eSTM Write threshold voltage versus V_{gSEL} bias :optimum V_T Write obtained for V_{gSEL} bias around select threshold.

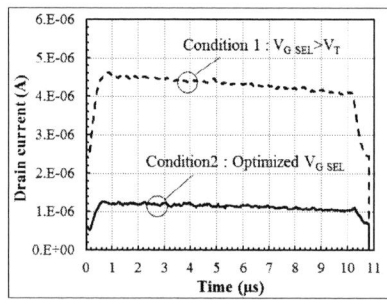

Fig. 17. Dynamic drain current measured during programing step: current consumption peak is in the range of 1µA.

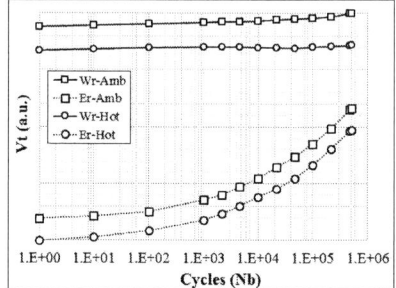

Fig.18. eSTM single cell erase and program threshold evolution during cycling at ambiant (25°C) and hot temperature (125°C).

978-1-7281-1988-5/18 $31.00 © 2018 IEEE   164

Highly Reliable Ferroelectric $Hf_{0.5}Zr_{0.5}O_2$ Film with Al Nanoclusters Embedded by Novel Sub-Monolayer Doping Technique

T. Yamaguchi[1], T. Zhang[1], K. Omori[1], Y. Shimada[2], Y. Kunimune[2], T. Ide[2], M. Inoue[1], and M. Matsuura[1]

[1]Device Technology Div., Renesas Electronics Corp., Hitachinaka, Ibaraki, Japan, email: tadashi.yamaguchi.pz@renesas.com
[2]Renesas Semiconductor Manufacturing Co., Ltd., Hitachinaka, Ibaraki, Japan

Abstract—Highly reliable ferroelectric (FE) $Hf_{0.5}Zr_{0.5}O_2$ (HZO) film with Al nanoclusters embedded by sub-monolayer doping technique is demonstrated for the first time. Al nanoclusters increase the remnant polarization (Pr) and reduce the voltage necessary for polarization switching. Furthermore, the program and erase endurance at the cycle of more than 250k and the Pr retention at 85°C for 10 years are achieved. Al nanoclusters are formed by the partial oxidation of sub-monolayer metallic Al embedded in HZO films. Al nanoclusters enhance the large grain growth of orthorhombic-phase HZO during FE-HZO crystallization annealing. The reduction of grain boundaries caused by the large grain growth with Al nanoclusters effectively reduces the leakage current in the HZO film. As a result, reliability of the FE HZO film is significantly improved.

I. INTRODUCTION

Recently, ferroelectric (FE) orthorhombic-phase HfO_2 film is one of the most attractive materials for future emerging memory because of high compatibility of CMOS integration together with high versatility as high-k gate dielectrics for CMOS devices. A lot of additives to enhance the ferroelectricity, such as Si, Y, Zr, Al, N, etc., have been examined [1-3]. In particular, Zr has advantages of large composition range showing ferroelectric behavior and the high remnant polarization (Pr) [3]. Furthermore, it has been reported that the Hf and Zr composition ratio of 50% shows high Pr [4]. The $Hf_{0.5}Zr_{0.5}O_2$ (HZO) film is one of promising candidates for the application to memory devices. On the other hand, although there were few papers on reliability of HZO films, the degradation of endurance and retention characteristics due to the depolarization was revealed [5, 6]. From the viewpoint of materials engineering, the orthorhombic phase formation and the grain size control in FE HZO films are key issues to improve the reliability. Indeed, the grain size engineering in thick HZO films by the laminated HZO structure using the insertion of the Al_2O_3 interlayer has been reported [7, 8]. However, the distinct fatigue was observed due to the formation of additional charge trap sites at the Al_2O_3/HZO interface [8]. Further novel and attractive methods to improve reliability of HZO films are strongly required for highly reliable FE memory.

In this paper, we demonstrate highly reliable HZO films with partially oxidized Al nanoclusters embedded by sub-monolayer metallic Al doping instead of inserting a layer. Precise control of Al doping ensures the stable formation of Al nanoclusters, and Al nanoclusters enhance the coalescence of FE HZO grains and enlarge the grain size. Reduction of grain boundaries due to the large grain growth and localization of Al nanoclusters at the grain boundary interrupt the leakage current pass. As a result, program and erase (P/E) endurance and retention characteristics can be significantly improved.

II. EXPERIMENTAL

Figure1 shows process flow of metal ferroelectric metal (MFM) capacitor. 10-nm-thick TiN film was deposited on B implanted Si wafers at the dose of 5×10^{15} cm^{-2} with 10 keV. 5-nm-thick HZO film (Hf : Zr = 1 : 1) was deposited by ALD system at 300°C. Hafnium chloride ($HfCl_4$) and zirconium chloride ($ZrCl_4$) were used as precursors of Hf and Zr, respectively. H_2O was used as a source of oxidizer. Sub-monolayer Al was doped at the dose ranging from 1×10^{13} to 1×10^{14} cm^{-2}. Figure 2 shows the dose rate of Al using the doping system designed for a low dose. The quite low dose rate of 2.4 $\times10^{12}$ cm^{-2}/sec can precisely control Al dose. Then, 5-nm-thick HZO film was deposited again. After 10-nm-thick TiN film was formed as a capping layer, rapid thermal annealing (RTA) at 600°C for 60s was carried out to form the FE crystal phase. After 100-nm-thick poly-crystalline silicon was deposited, P implantation at the dose of 5×10^{15} cm^{-2} with 10 keV and activation annealing were conducted. The high resolution TEM image of the HZO film after FE-HZO crystallization annealing is shown in Fig. 3. Al nanoclusters in the middle of the film are confirmed. P-V measurement characterized Pr and coercive voltage (Vc). C-V and I-V were also measured. P/E endurance and retention characteristics were evaluated as the reliability of HZO films. The condition of the endurance cycle test was fixed at ±2V with the frequency of 10kHz. For physical analyses, the distribution of grain size and crystal direction in HZO films with Al nanoclusters were characterized by scanning precession electron diffraction. The profile of Al and the binding energy of Al nanoclusters in HZO films were also measured by STEM-EDS system and XPS, respectively.

III. RESULTS AND DISCUSSION

A. Electrical properties of HZO films with Al nanoclusters

Figure 4 shows P-V hysteresis curves with and without Al nanoclusters. Pr with the Al nanocluster is larger than that without one. In addition, Vc with the Al nanocluster is lower than that without one. Pr and Vc as a function of Al dose are summarized in Fig. 5 and 6, respectively. Pr increases with increasing Al dose. At the Al dose of 5×10^{13} cm^{-2}, Pr shows maximum, then Pr drastically decreases with increasing Al dose. On the other hand, Vc monotonically decreases with increasing Al dose, which indicates lowering the voltage necessary for polarization switching. Figure 7 shows C-V characteristics of the HZO film with and without Al nanoclusters. Maximum capacitance (C_{max}) with Al nanoclusters is higher than that without one, and the voltage of C_{max} with Al nanoclusters is lower than that without one. C_{max} as a function of Al dose is shown in Fig. 8. C_{max} increases with increasing Al dose and the peak of C_{max} is shown at 5×10^{13} cm^{-2}. Resulting from the enhancement of Pr (or C_{max}) and the reduction of Vc, precise optimization of Al doping was found to be required. The Al dose of 5×10^{13} cm^{-2} was chosen in the following reliability study. Figure 9 shows P/E endurance of the HZO film with and without Al nanoclusters. Pr without Al nanoclusters monotonically decreases with the P/E cycle. Meanwhile, Pr with Al nanoclusters is stable up to 10k cycle. Pr window keeps about 12µC/cm^2 after 250k cycle. Furthermore, the retention characteristic of the HZO film with Al nanoclusters at the room temperature (RT) is superior to that without one, as shown in Fig. 10. Al nanoclusters can ensure the extrapolated Pr window of 15µC/cm^2 after 10 years. Moreover, retention characteristics after the P/E cycle of 10k and 200k indicate not only the window margin but also no change in the decay rate between 10k and 200k cycles, as shown in Fig. 11. The HZO film with Al nanoclusters also exhibits the retention characteristic at elevated temperature (85°C), as shown in Fig. 12. The extrapolated Pr window can be confirmed after 10 years, which should be contrasted to the reported value of 100s at 85°C for the single HZO film [5].

To understand the cause for the improvement of reliability in HZO films with Al nanoclusters, I-V characteristics are compared, as shown in Fig. 13. The leakage current in the HZO film with Al nanoclusters decreases with increasing Al dose. It has been well-known that fatigue effect during cycle test is due to defects in FE films and injected charges that pin the domain walls while retention loss results from the leakage followed by charge trapping [8]. The injection of charges into HZO films with Al nanoclusters is strictly suppressed and reliability is improved.

B. Physical properties of HZO films with Al nanoclusters

To clarify the physical properties of the HZO film with Al nanoclusters, crystal grain maps and mean grain size as a function of Al dose are shown in Fig. 14 and 15, respectively. The mean grain size increases with increasing Al dose. This result is consistent with the reduction of the leakage current since grain boundaries as leakage passes decrease due to the increase of grain size. Figure 16 shows crystal direction maps of orthorhombic phase grains in HZO films with and without Al nanoclusters. Here, the grains of tetragonal phase are indicated by black areas. The crystal direction of orthorhombic grains with Al nanoclusters tends to align with [001] direction, as shown in Fig. 17. This result suggests Pr increase and Vc decrease (see Figs. 5 and 6).

To understand the feature of Al nanoclusters after the FE-HZO crystallization annealing, STEM-EDS maps and the line profile are shown in Fig. 18. Most of Al are located at the middle of the HZO film despite 600°C annealing. Moreover, the detail analysis of Al nanoclusters was carried out with XPS. Figure 19 shows XPS spectra of Al 2p and 2s before and after FE-HZO crystallization annealing. Sub-monolayer Al was partially oxidized after top HZO deposition, and oxidation of Al might proceed during FE-HZO crystallization annealing. Metallic Al-Al bonds, however, remain. It is considered that nuclei of metallic Al are located at the center of Al nanoclusters. Although further study of Al nanoclusters is needed, creation of Al nanoclusters is a key to realize highly reliable FE films.

C. Role of Al nanoclusters in HZO films

To explain the cause for the improvement of reliability in the HZO film with embedded Al nanoclusters, feasible models are considered. Figure 20 shows schematic illustrations about the role of Al nanoclusters in HZO films. In the case of the HZO film without Al nanoclusters, during FE-HZO crystallization, poly-crystalline HZO grains grow on both TiN electrodes and contact each other, then grain boundaries are created. These grain boundaries increase leakage current, and charges are injected during cycle test. On the other hand, in the case of the HZO film with Al nanoclusters, Al are embedded in amorphous HZO films. During FE-HZO crystallization, clustering of Al proceeds. then, Al nanoclusters are surrounded by poly-crystalline FE HZO grains because of the grain growth. It is considered that micro-strain is generated at the interface between poly-crystalline HZO grains with Al nanoclusters due to the difference of the thermal expansion coefficient between Al nanocluster and HZO. The coalescence of FE HZO grains is enhanced to relax the strain energy. This coalescence causes the large HZO grain growth with a preferred [001] crystal direction. As a result, grain boundaries decrease. In addition, several of Al nanoclusters are localized at the grain boundary. These Al nanoclusters might interrupt leakage current.

IV. CONCLUSION

We demonstrated the highly reliable ferroelectric (FE) Hf$_{0.5}$Zr$_{0.5}$O$_2$ (HZO) film with Al nanoclusters embedded by sub-monolayer Al doping for the first time. Al nanoclusters in HZO films ensure the reliability of program and erase endurance and retention characteristics due to the reduction of the leakage current in the HZO film. Furthermore, detail physical analyses with scanning precession electron diffraction, STEM-EDS and XPS led to the feasible model about the role of Al nanoclusters in HZO films. The HZO film with embedded Al nanoclusters is a promising material for highly reliable FE memory.

REFERENCES

[1] T. S. Boscke, *et al.*, *Appl. Phys. Lett.* **99**, 112904 (2011).
[2] P. D. Lemenzo, *et al.*, *Appl. Phys. Lett.* **107**, 242903 (2015).
[3] J. Muller, *et al.*, *Appl. Phys. Lett.* **99**, 112901 (2011).
[4] J. Muller, *et al.*, *Nano Lett.* **12**, 4318 (2012).
[5] Y-C. Chiu, *et al.*, *2015 Symposium on VLSI Technology Digest of Technical Papers*, T184.
[6] K-Y. Chen, *et al.*, *2018 Symposium on VLSI Technology Digest of Technical Papers*, 119.
[7] H. J. Kim, *et al.*, *Appl. Phys. Lett.* **105**, 192903 (2014).
[8] S. Riedel and J. Muller, *AIP Advances*, **6**, 095123 (2016).

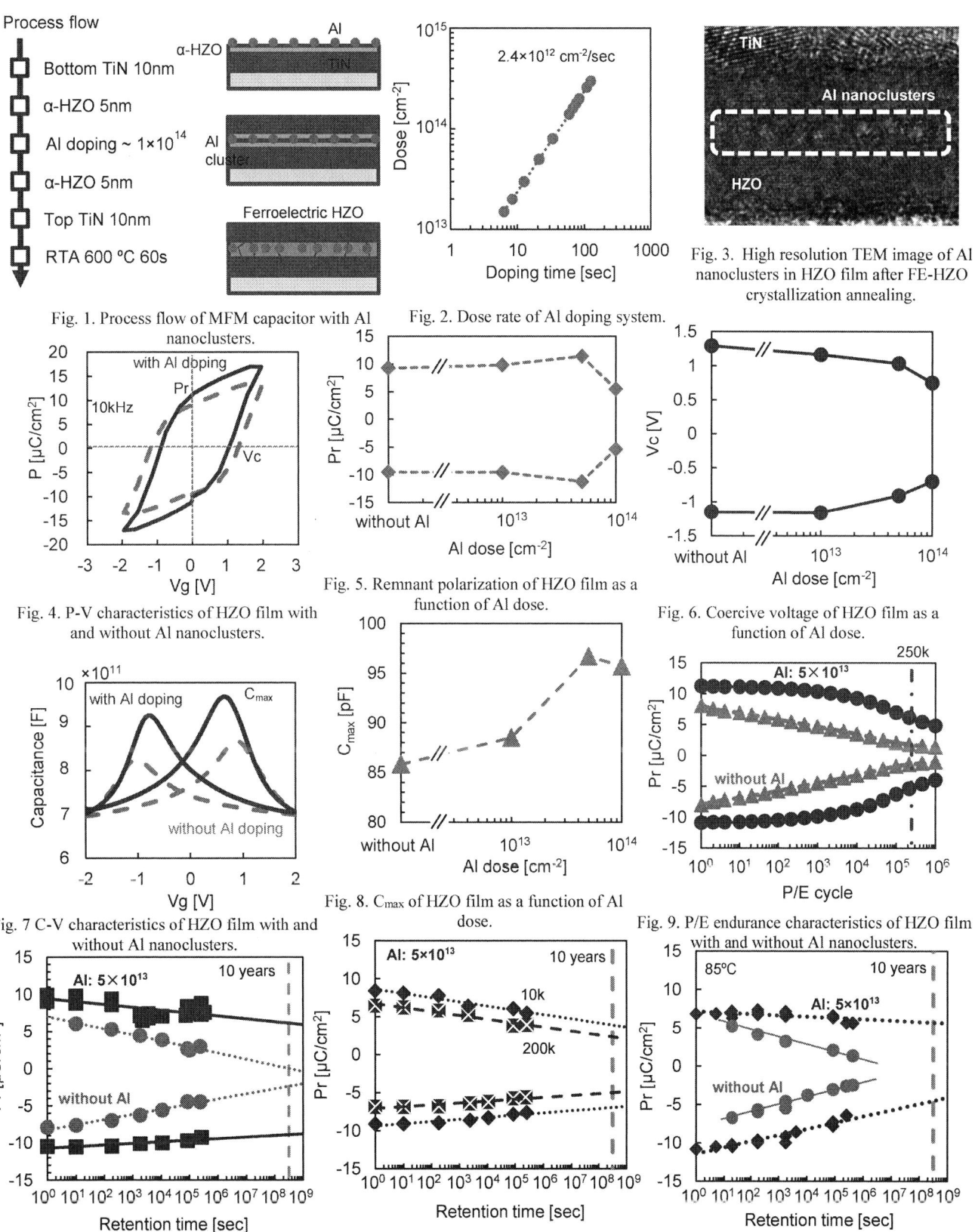

Fig. 1. Process flow of MFM capacitor with Al nanoclusters.

Fig. 2. Dose rate of Al doping system.

Fig. 3. High resolution TEM image of Al nanoclusters in HZO film after FE-HZO crystallization annealing.

Fig. 4. P-V characteristics of HZO film with and without Al nanoclusters.

Fig. 5. Remnant polarization of HZO film as a function of Al dose.

Fig. 6. Coercive voltage of HZO film as a function of Al dose.

Fig. 7 C-V characteristics of HZO film with and without Al nanoclusters.

Fig. 8. C_{max} of HZO film as a function of Al dose.

Fig. 9. P/E endurance characteristics of HZO film with and without Al nanoclusters.

Fig. 10. Retention characteristics of HZO film with and without Al nanoclusters at RT.

Fig. 11. Retention characteristics of HZO film with Al nanoclusters after P/E endurance of 10k and 200k cycles at RT.

Fig. 12. Retention characteristics of HZO film with Al nanoclusters at 85°C.

978-1-7281-1988-5/18 $31.00 © 2018 IEEE   167

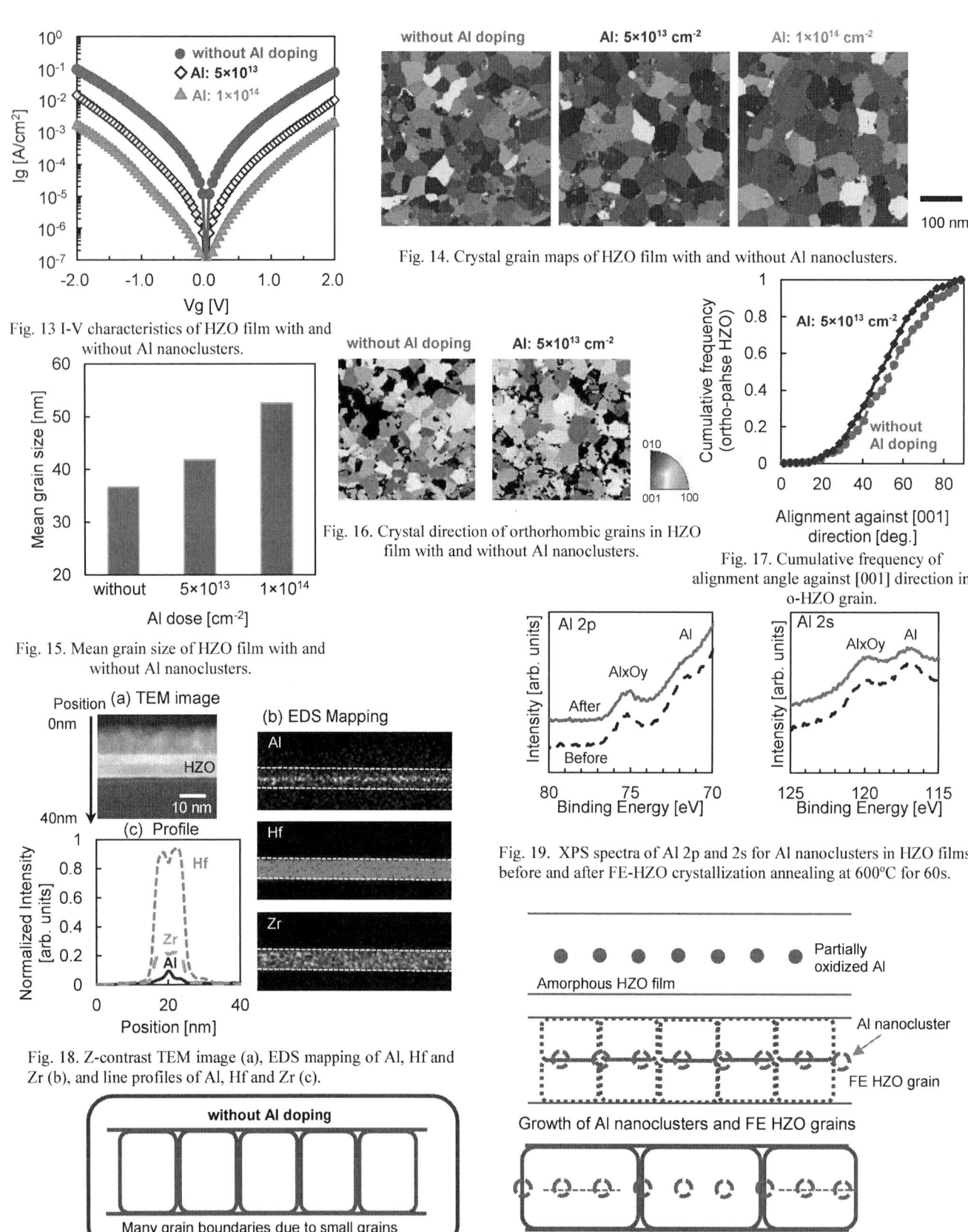

Fig. 13 I-V characteristics of HZO film with and without Al nanoclusters.

Fig. 14. Crystal grain maps of HZO film with and without Al nanoclusters.

Fig. 15. Mean grain size of HZO film with and without Al nanoclusters.

Fig. 16. Crystal direction of orthorhombic grains in HZO film with and without Al nanoclusters.

Fig. 17. Cumulative frequency of alignment angle against [001] direction in o-HZO grain.

Fig. 18. Z-contrast TEM image (a), EDS mapping of Al, Hf and Zr (b), and line profiles of Al, Hf and Zr (c).

Fig. 19. XPS spectra of Al 2p and 2s for Al nanoclusters in HZO films before and after FE-HZO crystallization annealing at 600°C for 60s.

Fig. 20. Schematic illustrations of feasible model about role of Al nanoclusters in HZO films

Interface Dipole Modulation in HfO$_2$/SiO$_2$ MOS Stack Structures

Noriyuki Miyata[1], Jun Nara[2], Takahiro Yamasaki[2], Kyoko Sumita[1], Ryousuke Sano[3], and Hiroshi Nohira[3]

[1]National Institute of Advanced Industrial Science and Technology (AIST), Tsukuba, Japan, email: nori.miyata@aist.go.jp
[2]National Institute for Materials Science (NIMS), Tsukuba, Ibaraki, Japan, [3]Tokyo City University, Setagaya-ku, Tokyo, Japan

Abstract We report an electric-field-induced interface dipole modulation (IDM) in HfO$_2$/1-ML TiO$_2$/SiO$_2$ MOS stack structures. Experimental evidence for IDM was exhibited, and rearrangement of interfacial Ti-O configuration by an electric field was theoretically demonstrated to cause the potential modulation. Multi-stack HfO$_2$/SiO$_2$ MOSFETs with multiple dipole modulation layers are promising in terms of a low temperature process, practical memory window, and stable potential switching.

I. INTRODUCTION

HfO$_2$-based nonvolatile memory devices attract attention mainly due to material compatibility with Si-CMOS technology [1-4]. In particular, ferroelectric field-effect transistors (FeFETs) with ferroelectric HfO$_2$ are promising as the flash-type memory operation of FeFET has been demonstrated [4]. Recently, interface dipole modulation (IDM) was observed in amorphous HfO$_2$/SiO$_2$ stack structures, and it was reported to show FeFET-like operation [5]. In this case, 1-monolayer (1-ML) TiO$_2$ inserted into HfO$_2$/SiO$_2$ interface is considered to act as potential modulator. However, IDM mechanism has not been investigated in detail. In this study, we carried out electrical measurement, photoelectron spectroscopy, and first-principle simulation to explore the IDM mechanism. We also report that stable switching is possible even for IDM structures prepared by using a low temperature process (<400°C).

II. EXPERIMENTALS

HfO$_2$/1-ML TiO$_2$/SiO$_2$ IDM structures were fabricated on Si substrates covered with thermally grown SiO$_2$ layers [5]. During the HfO$_2$, TiO$_2$, and SiO$_2$ deposition, the substrates were slightly heated by thermal radiation from the filament of an electron-beam evaporator, but the substrate temperature was kept below 100°C [6]. To prepare the MOS capacitors and MOSFETs, Ir electrodes were fabricated on the IDM structures. PMA was performed at 320–350°C in an Ar atmosphere. Figure 1 shows a TEM image of multi-stack IDM MOS structure with six 1-ML TiO$_2$ layers. Thin amorphous HfO$_2$ and SiO$_2$ layers are alternately stacked, suggesting that the effect of ferroelectric HfO$_2$ can be ignored in the following discussion. The IDM FETs were fabricated by using a gate last process in which n+ S/D region was formed on a p-Si substrate and then an IDM structures were fabricated in the same manner as described above.

Hard x-ray photoelectron spectroscopy (HAXPES) measurements were undertaken using synchrotron radiation (hv=7940 eV) at BL47XU at SPring-8 [7]. A HfO$_2$/1-ML TiO$_2$/SiO$_2$/n-Si MOS capacitor with 15-nm-thick Ir electrodes was prepared for the HAXPES measurement.

III. RESULTS AND DISCUSSION

A. MOS characteristics of HfO$_2$/SiO$_2$ IDM structures

Figure 2 shows a high-frequency *C-V* curve of IDM MOS capacitor including a TiO$_2$ modulation layer. In general, HfO$_2$-based MOS capacitors formed on n-Si substrates exhibit clockwise *C-V* hysteresis after applying high-electric field, as the electrons injected from the Si substrate were trapped inside the oxide stack structure. However, this *C-V* curve shows a weak counterclockwise hysteresis. This suggests that the charge distribution in the IDM stack was changed by gate bias similar to ferroelectric MOS capacitor. Note that the formation of ferroelectric HfO$_2$ can be excluded as the maximum

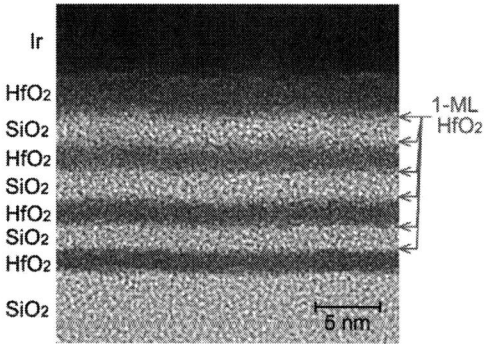

Fig. 1. TEM image of multi-stack HfO$_2$/SiO$_2$ structure with six 1-ML TiO$_2$ modulation layers.

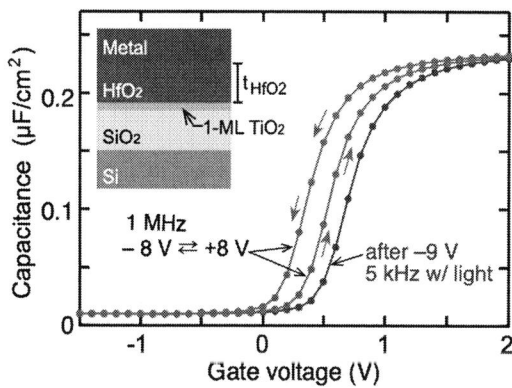

Fig. 2. *C-V* curves of HfO$_2$/1-ML TiO$_2$/SiO$_2$/Si MOS capacitor. Counter-clockwise hysteresis suggests dipole modulation.

978-1-7281-1988-5/18 $31.00 © 2018 IEEE

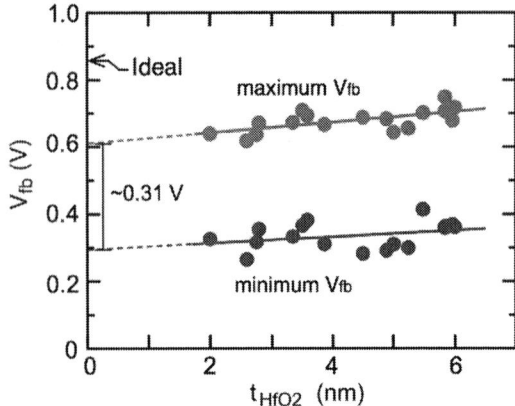

Fig. 3. HfO₂-thickness dependence of maximum and minimum V_{fb} for HfO₂/1-ML TiO₂/SiO₂/Si MOS capacitors. V_{fB} switching behavior with weak thickness dependence supports IDM mechanism.

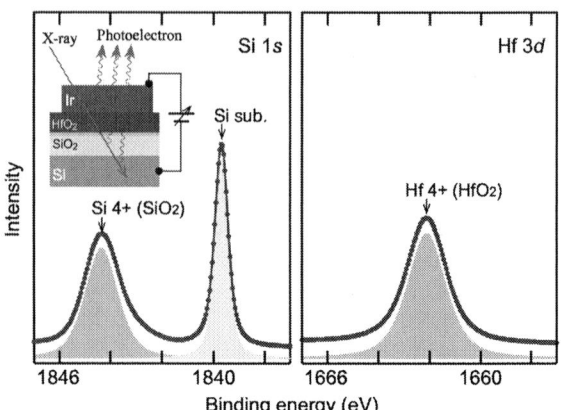

Fig. 4. Si 1s and Hf 3d photoelectron spectra observed from HfO₂/1-ML TiO₂/SiO₂ IDM MOS capacitor. Stoichiometric HfO₂ and SiO₂ can be recognized from MOS capacitor showing electrical switching.

temperature applied to this IDM structure is 350°C of PMA. In the high-frequency C-V measurement, the electric field in the oxide layers in the negative gate bias range does not increase sufficiently due to the formation of a depletion layer. To evaluate the C-V shifts by negative bias stress, an inversion condition was achieved at 5 kHz by weak light illumination. By this method, the maximum switching width of this IDM structure was estimated to be about 0.3 V.

To investigate the charge distribution in the IDM structure, the thickness dependence of the upper HfO₂ layer of HfO₂/1-ML TiO₂/SiO₂/Si MOS capacitors was investigated [Fig. 3]. As described above, the maximum and minimum V_{fb} shifts were estimated by using light-illuminated C-V measurement. Both V_{fb} behaviors exhibited similar small linear slopes. Here, we assume two types of charges at around the HfO₂/1ML-TiO₂/SiO₂ interfaces, a sheet charge [qS_I (cm⁻²)] and an interface dipole with a negative sheet charge on the HfO₂ side and positive on the SiO₂ side [Φ_D (V)]. HfO₂-thickness dependence of V_{fb} can be given by the following equation [6],

$$V_{FB} = \Phi_{MS} - \frac{qS_I t_{HfO2}}{\varepsilon_{HfO2}} - \Phi_D \quad (1),$$

where ε_{HfO2} is the dielectric constant of HfO₂ layer. From this equation, we can conclude that the small amount of negative charges (~10¹² cm⁻²) exist at around the HfO₂/SiO₂ interface. However, the charge density does not change by switching, so we concluded that the interface charges do not contribute to the switching. Conversely, the interface dipoles are changed between 0.61 V and 0.29 V. This is a strong evidence that the observed switching is due to a change in the interface dipole.

Switching ability of a single IDM structure is as small as 0.3 V, which is insufficient for nonvolatile memory applications. The interface dipole of the HfO₂/SiO₂ system has been reported to be 0.2-0.3 V [8, 9], which is the main reason for the small switching window. The switching characteristic in Fig. 3 suggest that a slightly larger dipole can be produced by inserting the TiO₂ layer. However, we cannot expect larger dipole switching as long as the same material system is used.

B. HAXPES analysis of IDM MOS capacitor

In order to investigate the chemical composition and the electric-field-induced phenomenon in HfO₂/1-ML TiO₂/SiO₂ MOS capacitor, photoelectron spectra existed by hard X-rays were measured. HfO₂ and SiO₂ layers under the Ir electrode were stoichiometric as shown in Fig. 4, and TiO₂ was detected as the main component of the interfacial Ti oxide (not shown). Therefore, the contribution of oxide defects to switch is considered to be small. This is consistent with the MOS characteristic shown in Fig. 3, which shows a small charge density in the oxide stacks.

When the gate bias applied, peak shifts of photoelectron spectra were observed. Figure 5 (a) summarizes gate bias dependence, where BE = Si 1s (4+) – Hf 4d (4+) and δBE = BE – BE (initial). Applying a gate bias of –5 V changes δBE, but it returns to 0 eV when V_g = 0 V. However, after applying a gate bias of +5 V, a small energy difference of about 0.06 eV remains. This is reasonably concluded that the interface dipole is changed by the gate bias as described in Fig. 5 (b). The negligible potential change after applying the negative gate bias likely means the weak electric field due to surface depletion. These

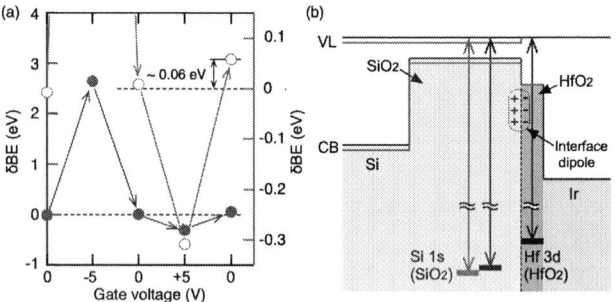

Fig. 5. Gate voltage dependent measurement of Si 1s (SiO₂) and Hf 3d (HfO₂) photoelectron peaks. (a) energy difference between Si 1s and Hf 3d peaks and (b) energy band diagram of HAXPES measurement for IDM MOS capacitor.

978-1-7281-1988-5/18 $31.00 © 2018 IEEE

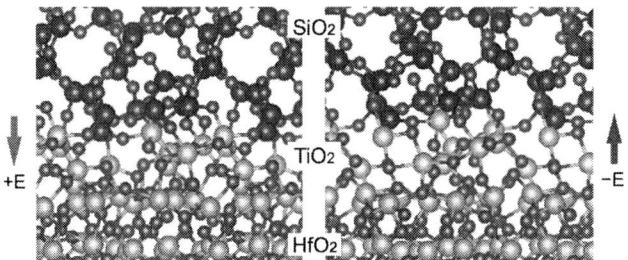

Fig. 6. HfO$_2$/TiO$_2$/SiO$_2$ IDM structures calculated under electric fields (±0.3 eV/Å)

results are in agreement with the MOS *C-V* characteristics, so we considered that the HAXPES result supports IDM in the HfO$_2$/1-ML TiO$_2$/SiO$_2$ structure.

C. First-principles simulation of IDM structure

In order to investigate electric-field-induced structural change of amorphous HfO$_2$/1-ML TiO$_2$/SiO$_2$ structure, a first-principles molecular dynamics (FPMD) simulation was performed [10]. Two values of electric field, *i.e.* ± 0.3 eV/Å, were employed. The temperature was set at 500K. Figure 6 shows the snap shots during FPMD calculation on the IDM structure. We found that the bonding around the Ti atoms largely vary depending on the electric field, while those around Hf and Si do not. This may be related to the stiffness of these materials. Figure 7 shows the potential profiles for the IDM structure under two values of electric field. We obtained the profiles in the following way. Some snap shots were selected from the FPMD calculations. Then, the electronic structures were calculated self-consistently for those snap shots with no electric field. The potential profiles are the average of those calculated for the selected snap shots. With this procedure, the potential drop around TiO$_2$ could be clear, as shown in the figure. In the figure, potentials around SiO$_2$ on the left side are aligned. Then, the potential difference between two electric fields in the HfO$_2$ region are clearly seen, as about 3 eV, which is quite larger than experiments. This may be because we do not optimize the structures for the calculation of the potential profile. Note that the potential profiles in SiO$_2$ and HfO$_2$ regions do not depend on the electric field so much, while those around TiO$_2$ largely vary. This is related to the stiffness of those materials denoted above. This fact means the electric field changes only the structure around TiO$_2$ which is the softest among the three. Other oxides with the similar stiffness

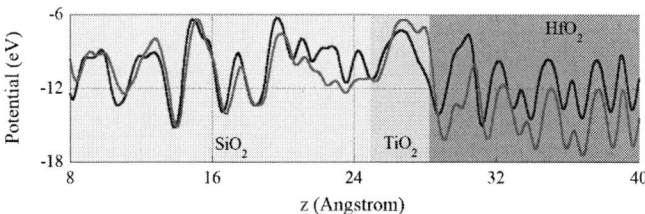

Fig. 7. Potential profile of HfO$_2$/TiO$_2$/SiO$_2$ structure. SiO$_2$ sits on the left side with its width of 17 Å and HfO$_2$ sits on the right side with its width of 12 Å. The red and blue curves are for E = 0.3 eV/Å and -0.3 eV/Å, respectively.

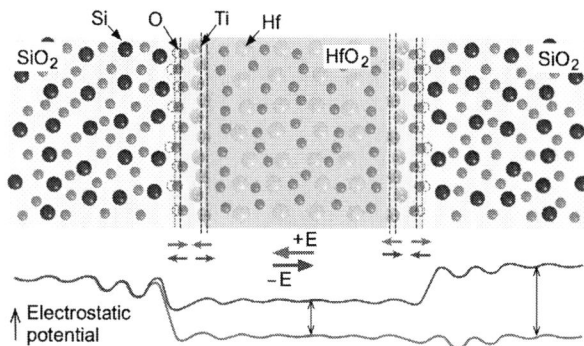

Fig. 8. IDM behavior in multi-stack HfO$_2$/SiO$_2$ IDM structures. IDM occurring at two facing interfaces are expected to enhance memory window.

to TiO$_2$ may exhibit the IDM operation. These atom/charge shifts are applicable for explaining the IDM operation.

D. IDM operation of multi-stack IDM structure

To enhance the switching window, we propose multi-stack IDM structure shown in Fig. 1. In this structure, IDM occurs simultaneously in two opposite interface structures of HfO$_2$/SiO$_2$ and SiO$_2$/HfO$_2$ interfaces. Thus, we need to consider the effect of each IDM behavior on the gate potential change. As shown in Fig. 8, Ti and O atoms at each interface move in the same directions under the same electric field. According to the above Ti-O rearrangement mechanism, it can be expected that as one interface dipole increases, the other interface dipole decreases. This means that IDM at two facing interfaces are superimposed and contribute to the enhanced memory window.

Figure 9 shows a *C-V* hysteresis curve of an IDM MOS capacitor including two TiO$_2$ modulation layers. Obviously, wider hysteresis can be obtained compared to a IDM structure with a TiO$_2$ layer shown in Fig. 2. Maximum switching width was estimated to be larger than 0.6 V. Therefore, we consider that the IDM mechanism shown in Fig. 9 is reasonable to explain the observed switching behavior.

A multi-stack IDM MOS capacitor with six TiO$_2$ modulation layers shown in Fig. 1 exhibits a larger *C-V*

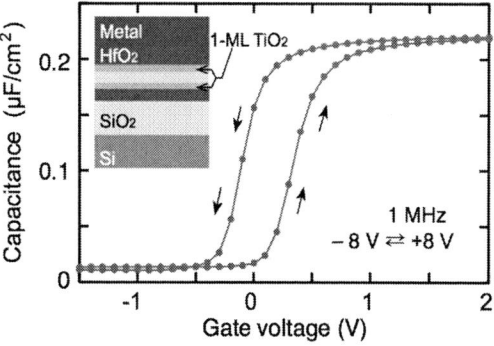

Fig. 9. *C-V* hysteresis curve of IDM MOS capacitor with two TiO$_2$ modulation layers. Wider hysteresis takes place compared with an IDM MOS capacitor with a TiO$_2$ layers shown in Fig. 2.

978-1-7281-1988-5/18 $31.00 © 2018 IEEE

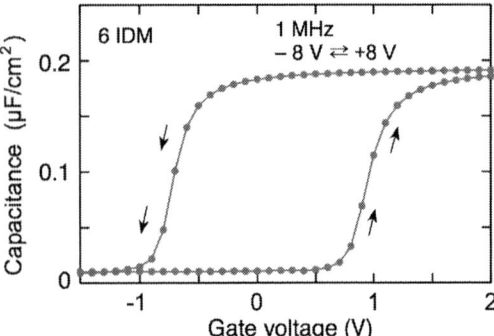

Fig. 10. *C-V* hysteresis curve of multi-stack HfO_2/SiO_2 IDM MOS capacitor including six TiO_2 modulation layers as shown in Fig. 1.

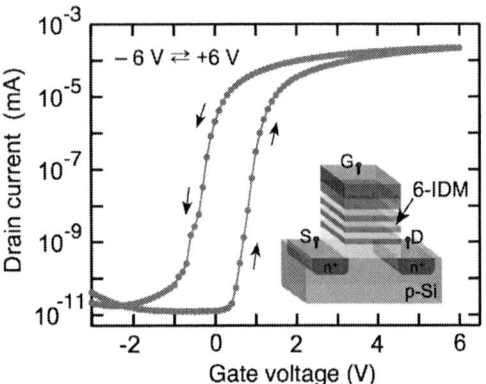

Fig. 11. I_d-V_g characteristics of IDM FET with six TiO_2 modulation layers.

hysteresis. Ideally, the maximum switching width of 6-IDM structure is expected to be about 1.8 V since one IDM structure has a 0.3 V modulation abilities. The observed *C-V* hysteresis is almost consistent with this prediction. In addition, this IDM structure was fabricated using PMA at 350°C. This low temperature process of IDM device fabrication is a clear advantage compared to ferroelectric HfO_2 technology.

Figure 11 shows drain current characteristics of an IDM FET including six TiO_2 modulation layers. Counterclockwise hysteresis takes place, which indicates that the IDM function can be integrated into the MOSFET. The change of V_{th} values, *i.e.*, the memory window, is estimated to be larger than 1 V under the sweeping condition of ±6 V. In addition, switching the drain current at about 5-digit ratio is also an advantage of IDM FET device [Fig. 12]. Therefore, we consider that multi-stack IDM structure is promising as a flash-type memory application.

In conclusion, the interfacial dipole modulation (IDM) in $HfO_2/1$-ML TiO_2/SiO_2 structures was demonstrated using electrical, spectroscopic, and theoretical means. The stable

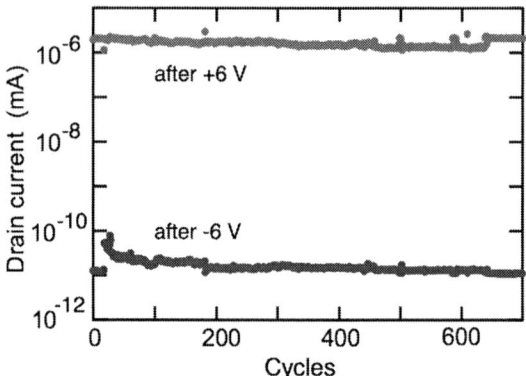

Fig. 12. Switching characteristics of drain current of IDM FET with six TiO_2 modulation layers.

memory operation of multi-stack IDM structure formed by a low temperature process have also been presented.

ACKNOWLEDGMENT

This work was supported by JSPS KAKENHI Grant Number 16H02335. Part of the FET fabrication was conducted at the AIST Nano-Processing Facility (AIST-NPF). The experiments were partly performed at SPring-8 with the approval of the program review committee (2016B0109). The calculations in this study were performed on Numerical Materials Simulator at NIMS.

REFERENCES

[1] M. Lanza, "A Review on Resistive Switching in High-*k* Dielectrics: A Nanoscale Point of View Using Conductive Atomic Force Microscope", *Materials* 7, 2155-2182 (2014).
[2] E. Yurchuk *et al.*, "Impact of scaling on the performance of HfO_2-based ferroelectric field effect transistors," *IEEE Trans. Electron Devices* 61, 3699 (2014).
[3] J. Müller, P. Polakowski, S. Mueller, and T. Mikolajick, "Ferroelectric Hafnium Oxide Based Materials and Devices: Assessment of Current Status and Future Prospects", *ECS Journal of Solid State Science and Technology* 4, N30-N35 (2015).
[4] M. Trentzsch, *et al.* "A 28nm HKMG super low power embedded NVM technology based on ferroelectric FETs", *IEEE International Electron Devices Meeting* (IEDM), San Francisco, CA, pp. 11.5.1-11.5.4. (2016).
[5] N. Miyata, "Electric-field-controlled interface dipole modulation for Si-based memory devices", *Sci. Rep.* 8, 8486 (2018).
[6] N. Miyata, "Study of Direct-Contact HfO_2/Si Interfaces", *Materials* 5, 512-527 (2012).
[7] E. Ikenaga, *et al.*, "Development of high lateral and wide angle resolved hard X-ray photoemission spectroscopy at BL47XU in SPring-8", *J. Electron Spectrosc. Relat. Phenom.* 190, 180 (2013).
[8] K. Kita, A. Toriumi, "Origin of electric dipoles formed at high-*k*/SiO_2 interface," *Appl. Phys. Lett.* 94, 132902 (2009).
[9] Y. Abe, N. Miyata, Y. Shiraki, T. Yasuda, "Dipole formation at direct-contact HfO_2/Si interface, " *Appl. Phys. Lett.* 90, 172906 (2007).
[10] See http://azuma.nims.go.jp for download

978-1-7281-1988-5/18 $31.00 © 2018 IEEE

Ge-based Non-Volatile Logic-Memory Hybrid Devices for NAND Memory Application

Na Wei[1], Bing Chen[1], Zejie Zheng[1], Zhimei Cai[1], Rui Zhang[1], Ran Cheng[1], Shiuh-Wuu Lee[1], Yi Zhao[1, 2, *]

[1]College of Information Science & Electronic Engineering, Zhejiang University, Hangzhou, China
[2]State Key Laboratory of Silicon Materials, Zhejiang University, Hangzhou, China
*E-mail: yizhao@zju.edu.cn

Abstract—In this work, novel Ge-on-Insulator (GeOI) MOSFETs with resistive-switchable gate stacks, named RFETs, are proposed and experimentally realized. The junctionless GeOI RFET and typical inversion-mode GeOI RFET are fabricated and both types of RFETs exhibit decent transistor behaviors and RRAM characteristics at the same time. Furthermore, by utilizing these two types of RFETs, a new GeOI RFET-based NAND memory is constructed and the memory functions of the arrays are experimentally demonstrated. This RFET-based NAND memory has a simple cell structure and very simplified I/O circuit in comparison with the conventional flash memory and non-volatile memory such as RRAM and MRAM. Therefore, RFETs should be promising for the applications of next-generation high density, low power memory and in-memory computing and neuromorphic computing.

I. INTRODUCTION

In recent years, the development of internet of things (IoT) has continuously driven semiconductor industry to deliver multifunctional and low power system-on-chip (SoC) chips and the exponentially increasing IoT applications have resulted in the explosive data growth issue[1]. As a consequence, the SoCs in IoT are facing great challenges in effectively addressing the storage and processing associated with "big data". Since traditional embedded-memory technologies such as NOR-Flash are limited by speed, scalability, and density [2-3], integrating an ultra-fast massive non-volatile memory (NVM) near or within the SoCs is a promising way to meat these challenges. Some emerging NVMs, such as RRAM and MRAM, with the fast operational speed, low power, high reliability and good scalability, have been proposed and experimentally demonstrated for high density storage and SoC applications[4-5]. However, different from the traditional flash memory, these NVMs are not transistor-based devices, which cannot implement the logic control and operations for memories by themselves. To perform these operations, additional circuits and devices, like selectors for RRAM, are required, which will increase not only the circuit design complexity but also the cost of fabrication process. Developing a non-volatile logic-memory hybrid device should be a feasible solution to this problem. Recently, it has been reported that the Metal/HfOx/GeOx/Ge MOS structure owns good RRAM characteristics[6]. On the other hand, Ge MOSFETs have been widely studied for replacing Si MOSFETs due to the high mobility of Ge channel[7-8]. Therefore, integrating Ge-based RRAM with Ge

MOSFET could be a possible technology for fabricating a non-volatile logic-memory hybrid device, as shown in Fig. 1.

In this work, we propose a novel Ge MOSFET combined with a resistive-switchable gate stack, named RFET, and experimentally fabricated the Junctionless RFET (JL-RFET) and typical inversion-mode RFET (IM-RFET) (Fig. 2). This study will also discuss the application of both types of RFETs for memory application where only the scheme of state erasing, programming and reading are different from each other. Employing TCAD simulation on RFETs (Figs. 3-4), it can be observed that the drain current of JL-RFET and the gate leakage current of IM-RFET are programmable, meaning that the RFET could work as a memory device. Both the JL-RFET and IM-RFET based NAND memory arrays are constructed and the memory array's function is also demonstrated in this study.

II. EXPERIMENTAL

Both JL-RFET and IM-RFET have been fabricated in this study, and the structures of these two types of RFETs are shown in Fig. 2. GeOI substrates were used to isolate the cells column by column in the memory array. The process flow for fabricating JL- and IM-RFETs are shown in Figs. 5 and 6. The initial Ge thickness of GeOI substrates is 20 nm and the Ge film was thinned by thermal oxidation process at 550°C for several cycles. The high-resolution TEM picture shows that the GeOI film was finally thinned down to ~6nm (Fig. 7(a)). After the active areas being defined by etching Ge into islands, HfOx/GeOx/Ge gate stack was formed *in-situ* by 2-step atomic layer deposition (ALD) using ozone post oxidation (OPO) [9]. The gate stack could be clearly observed in the EDS spectrum, as shown in Fig. 7(b). The gate metal was sputtered and patterned, followed by ion-implantation and Ni deposition respectively for JL- and IM-p-MOSFETs to form S/D. The dopant activation and NiGe metallization were performed at 400 °C 3min and 1min, respectively. Finally, the Ni contact pads were deposited on S/D and gate. A 2×2 array was also fabricated with the same process flow as mentioned above.

III. RESULTS AND DISCUSSION

A. RFETs based on GeOI Junctionless (JL) MOSFETs

The measured I_d-V_g and I_d-V_d characteristics of fresh *JL* HfOx/GeOx/GeOI p-MOSFET are shown in Figs. 8 and 9. Meanwhile, the Metal/HfOx/GeOx/GeOI gate stack also shows excellent resistive switching behavior, which is similar to that of a good RRAM device (Fig. 10). This is consistent with the previously reported results in the literature [10]. As a

978-1-7281-1988-5/18 $31.00 © 2018 IEEE

comparison, Metal/Al$_2$O$_3$/Ge gate stack cannot be reset after being set to the low resistance state (LRS) as shown in Fig. 11. Figure 12 shows the I_d-V_g and I_d-V_d characteristics of HfO$_x$/GeO$_x$/GeOI JL pMOSFETs when the gate stack is at high resistance state (HRS) and LRS. It could be observed that the drain current, I_d, could be effectively tuned by the set and reset operations. It can be further commented that in the memory application of JL-RFET, the state of the memory cell will be determined by the value of I_d. At the same time, the set and reset processes do not degrade the RFET's transistor characteristics (Figs. 14 and 15). In fact, the Metal/HfO$_x$/GeO$_x$/Ge gate stack could be set at multi-level resistances by applying different compliance currents during the set process (Fig. 16 and the inserted graph), which will be very useful for future memory applications. From the statistic analysis of I_d and I_g with thousands of cycles (Figs. 17 and 18), it can be found that there is an obvious window between LRS and HRS which can guarantee a low operation error rate in these devices. Meanwhile, as shown in Figs. 19-21, reasonable reliability behaviors, including I_d and I_g degradation of the gate stack and the endurance of devices, have also been confirmed.

B. RFETs based on Inversion Mode (IM) GeOI MOSFETs

As illustrated in Fig. 4, when operating in the inversion mode GeOI MOSFETs, the gate leakage, I_g, and the resulting I_{off} of the transistor could be tuned via the set and reset processes because the inversion mode transistor is usually at the OFF state when V_g is 0 V. Therefore, the IM-RFET based memory has the merit of low read power compared with one that is JL-RFET based since the state of memory in JL-RFET based memory is read out by the I_d when the transistor is at the ON state. Figs. 21 and 22 show the fresh I_d-V_g and I_d-V_d characteristics of IM Ge p-MOSFET. As can be seen from Fig. 23, the gate stack of JL Ge p-MOSFET could also be set into LRS and reset into HRS with good stability (Fig. 24). The set and reset processes do not degrade the transistor characteristics of RFET as shown in Fig. 25.

C. NAND Memory Array Constructed by GeOI RFETs

Finally, we constructed and experimentally fabricated a NAND memory array with the above two types of RFETs. Figure 26 shows the microscope photo of the fabricated 8×8 memory array and the bottom illustration in the figure is the circuit description of a 2×2 memory array, which is same for JL-RFET and IM-RFET. In a memory array, all gate terminals in the same raw of the array are connected to the word line (WL), and, simultaneously, the drain and source terminals in one column are defined as the bit lines (BL), named BLD and BLS, respectively. In this study, for both JL-RFET and IM-RFET, we define the low resistance state of the gate stack as "1", and the high resistance state of the gate stack as "0". The state of a memory cell is read out by measuring the current (I_d for JL-RFET and I_g for IM-RFET) and a threshold current is set to judge if the state is "1" or "0".

The memory cell operation schemes for both JL- and IM-RFETs are summarized in Fig. 27. In the JL-RFET based memory cell, first, all cells are erased and reset into "0" by applying a negative voltage on all WLs. Then, for example, in order to program cell3 into "1", a positive voltage is applied on

WL$_1$, and simultaneously other WLs, like WL$_2$ here, BLD$_2$, and BLS$_2$ are connected to 0V while all other BLD and BLS are connected to 1/2 V_{set}^{JL} to avoid the miss setting of unselected cells. In the case of reading the state of cell3, WL$_1$ is connected to V_{read}^{JL}, which is close to transistor's threshold voltage (V_{th}), and the current, I_{read}, through BLD$_2$ is read with a negative voltage of $-V_{read}^{JL}$, and all other terminals are connected to 0 V. Once the I_{read} is higher than V_{th} value, the cell state will be output as "1" and otherwise "0".

In the case of IM-RFET based memory cell, the erasing operation is performed with an erasing voltage set on all WLs while all other terminals could be floating. Similarly with that in the JL-RFET, in order to program cell3, a V_{set}^{IM} voltage is applied on WL$_1$, and a small negative voltage, V_{on}^{IM}, is applied to other WLs. Simultaneously, BLD$_2$ and BLS$_2$ are connected to 0V, while all other BLD and BLS are connected to 1/2 V_{set}^{IM} to avoid the miss setting of unselected cells. In case of reading the state of cell3, WL$_1$ is applied with V_{read}^{IM}, and WL$_2$ is applied with V_{on}, which is equal to V_{read}^{IM} to avoid the miss reading. At the same time, V_{on}^{IM} is also applied on other BLD and BLS to avoid the sneak path from the unselected cells. Once, the current, I_{read}, through WL$_1$ is higher than the OFF gate leakage, the cell state will be output as "1" and otherwise "0". Finally, we also experimentally demonstrated the work functionality of a 2×2 JL-RFET based NAND memory array, as shown in Fig. 28, to verify the operation scheme of the proposed NAND memory. It's demonstrated that correct program, erase, and read operations have been achieved in the array.

IV. CONCLUSION

A new type of transistor, named RFET in this study, has been proposed and experimentally realized on GeOI substrates. The RFET not only exhibits good RRAM characteristics, but also decent transistor behaviors. Furthermore, both junctionless GeOI RFET and inversion-mode GeOI RFET were fabricated. By utilizing these two types of RFETs, a new NAND memory array has been experimentally constructed and the resulting array shows good functionality. Since RFET is a logic-memory hybrid device, the constructed NAND memory have simple cell structure and superior scalability and enable greatly simplified memory's I/O.

ACKNOWLEDGMENT

This work was supported by National Natural Science Foundation of China (No. 61376097) and National Key Research and Development Program of China (2017YFA0207600), and Major Research Program of Zhejiang Lab, China.

REFERENCES

[1] J. Gubbi et al., *Future Gene. Comp. Systems*, pp. 1645-1660, 2013. [2] Marco A. A. Sanvido et al., *Proc. IEEE*, vol. 96, no. 11, 2008. [3] C. M. Compagnion et al., *Proc. IEEE*, vol. 105, no. 9, 2017. [4] H.-S. P. Wong et al., *Proc. IEEE* , vol. 100, no. 6, 2012. [5] H. Yoda et al., *IEEE IEDM.*, pp. 259-262, (2012). [6] B. Chen et al., *IEEE Elec. Dev. Lett.*, (2018). [7] Y. Zhao et al., *IEEE ICIDT*, 2018, PP. 153-156. [8] R. Zhang et al., *IEEE TED*, vol. 63, no. 7, pp. 2665-2670, 2016. [9] R. Zhang et al., *IEEE Elec. Dev. Lett.*, vol. 37, no. 7, pp. 831-834, 2016. [10] Y. Zhang et al., IEEE Symp. VLSI Techno. Tech. Dig., 2018.

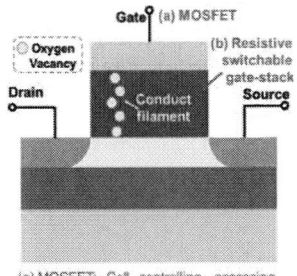

Fig. 1. The schematic and concept of RFET.

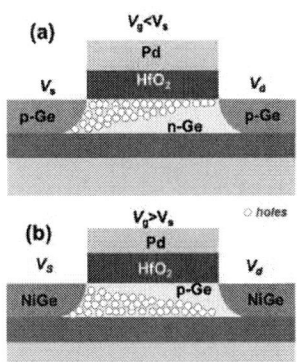

Fig. 2. Schematic of (a) Inversion-mode and (b) Junctionless RFET.

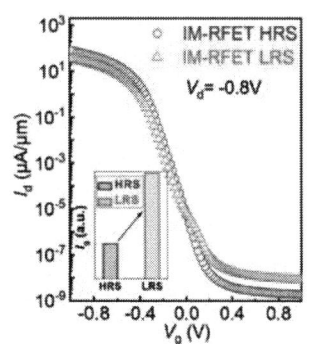

Fig.3. TCAD Simulated I_d-V_g plot for IM-RFET at HRS and LRS.

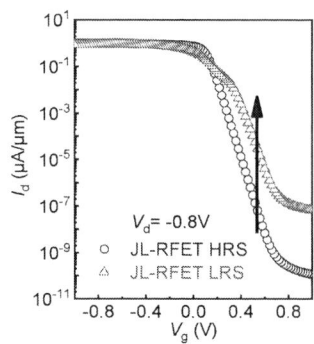

Fig.4. TCAD Simulated I_d-V_g plot for JL-RFET at HRS and LRS.

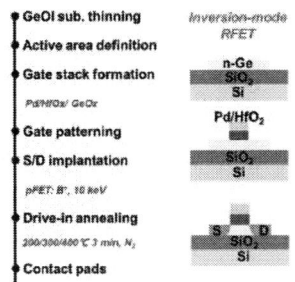

Fig. 5. Process flow of the GeOI IM- RFET.

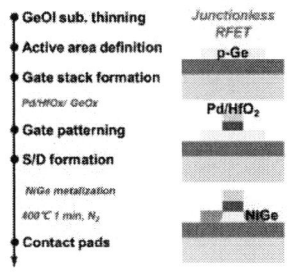

Fig. 6. Process flow of the GeOI JL-RFET.

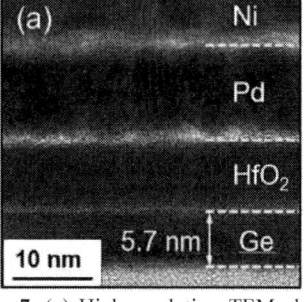

Fig. 7. (a) High-resolution TEM photo and (b) EDS spectra of the gate stack for the GeOI JL-RFET.

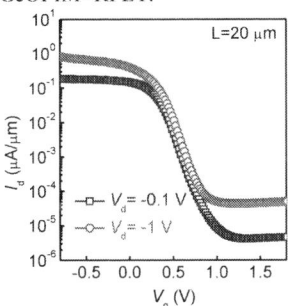

Fig. 8. Measured I_d-V_g characteristics of a fresh JL-RFET with $W/L = 8\ \mu m/20\ \mu m$.

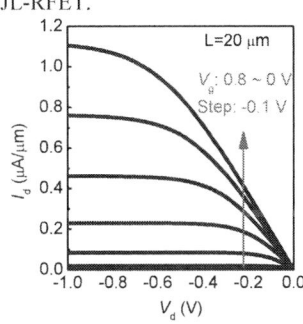

Fig. 9. Measured I_d-V_d cureve of the same fresh JL-RFET with $W/L = 8\ \mu m/20\ \mu m$.

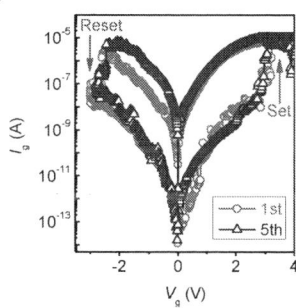

Fig. 10. I_g-V_g characteristics of the JL-RFET. The gate stack shows resistive switching behavior.

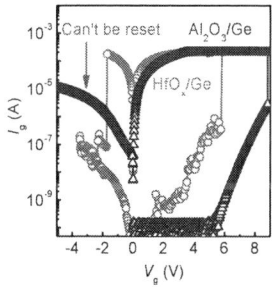

Fig. 11. I_g-V_g characteristics of the Metal/HfO$_x$/Ge and Metal/Al$_2$O$_3$/Ge gate stack. Only the former is switchable.

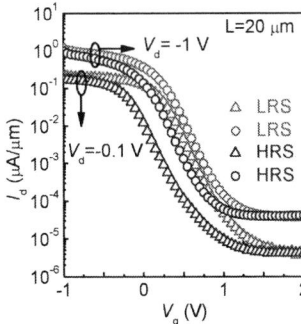

Fig. 12. I_d-V_g curves of the JL-RFET at HRS and LRS. Both show large current window when reading near the V_{th}.

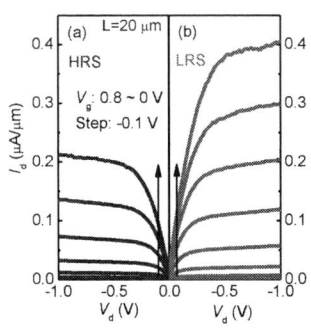

Fig. 13. I_d-V_d curves of the same JL-RFET at HRS and LRS, showing different output characteristics.

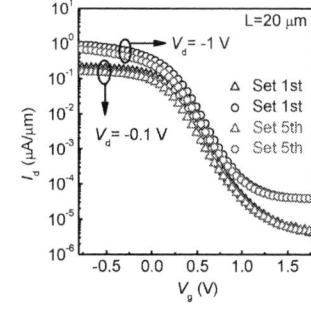

Fig. 14. I_d-V_g characteristics of the JL-RFET in LRS at the 1st and 5th switching cycles, showing good repeatability after several times of switching.

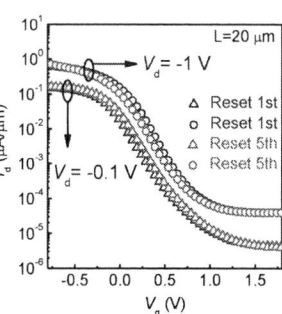

Fig. 15. I_d-V_g characteristics of the JL-RFET in HRS at the 1st and 5th switching cycles, showing no degradation on I_{on}, off-leakage, and subthreshold swing.

978-1-7281-1988-5/18 $31.00 © 2018 IEEE

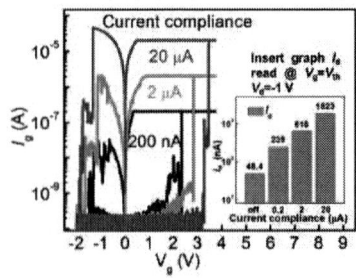

Fig. 16. *I-V* characteristics of the JL MOS structure under different current compliance. The insert graph shows two-bit data storage is achievable.

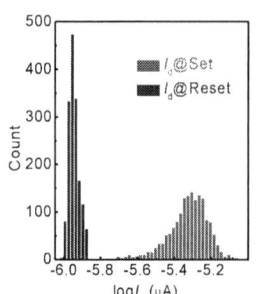

Fig. 17. Distribution of the readout I_d at $V_g=V_{th}$ at HRS and LRS.

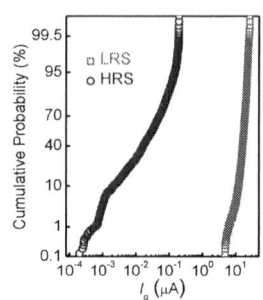

Fig. 18. Statistical distribution of I_g at HRS and LRS respectively

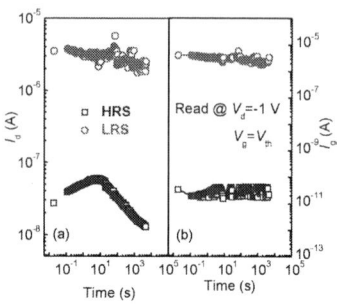

Fig. 19.. Read disturb of (a) I_d and (b) I_g under continuous V_g stress.

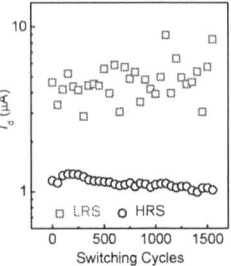

Fig. 20. Measured readout I_d at HRS and LRS for more than 1000 switching cycles.

Fig. 21. Measured I_g at HRS and LRS for more than 1000 switching cycles.

Fig. 22. Measured I_d-V_g characteristics of a fresh IM-RFET with W/L=50 μm/10 μm.

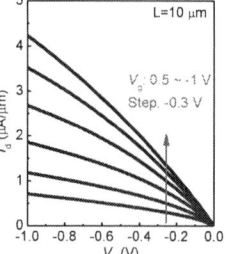

Fig. 23. Measured I_d-V_d curves of a fresh IM-RFET with W/L = 50 μm/10 μm.

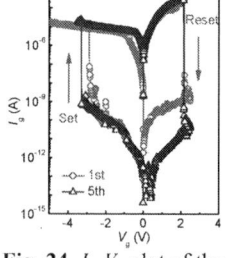

Fig. 24. I_g-V_g plot of the IM-RFET, showing a resistive switching behavior of the device.

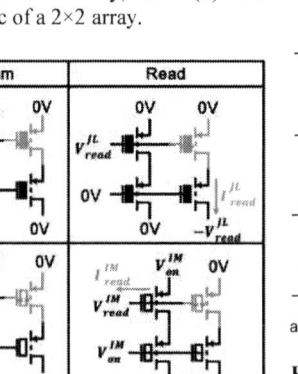

Fig. 25. I_d-V_d curves of the IM-RFET at LRS after the 1st and 5th switching cycles, showing good repeatability of the device.

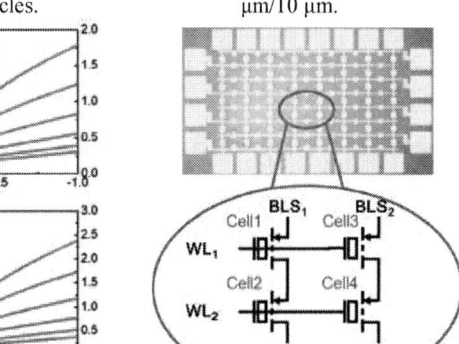

Fig. 26. (a) Microscopic photo of a fabricated 8×8 array, and (b) the schematic of a 2×2 array.

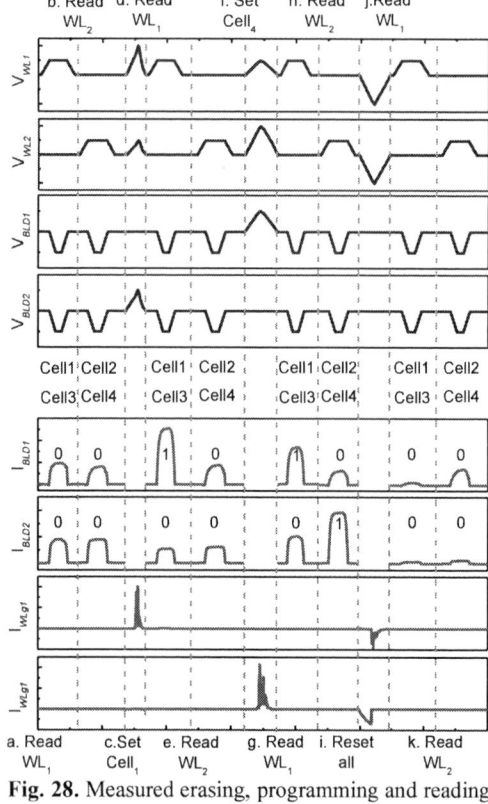

Fig. 28. Measured erasing, programming and reading process in the 2×2 NAND JL-RFET array, showing the change of (a) applied voltage, (b) response current on WLs, BLDs, and BLSs. Correct function of the array is demonstrated.

	Erase		Program		Read	
Junctionless(JL) p-MOSFET $V_{reset}^{JL} < 0$, $V_{set}^{JL} > 0$, $V_{read}^{JL} \approx V_{th}^{JL} > 0$.	0V	0V	$1/2V_{set}^{JL}$	0V	0V	0V
	V_{reset}^{JL}	V_{reset}^{JL}	V_{set}^{JL}		V_{read}^{JL}	
	V_{reset}^{JL}		0V	$1/2V_{set}^{JL}$	0V	I_{read}^{JL}
	0V	0V	0V	0V	0V	$-V_{read}^{JL}$
Inversion-Mode(IM) p-MOSFET $V_{reset}^{IM} > 0$, $V_{set}^{IM} < 0$, $V_{read}^{IM} = V_{on}^{IM} < 0$.	Floating	Floating	$1/2V_{set}^{IM}$	0V	I_{read}^{IM}	V_{on}^{IM} 0V
	V_{reset}^{IM}	V_{reset}^{IM}	V_{set}^{IM}		V_{read}^{IM}	
	V_{reset}^{IM}		V_{on}^{IM}	$1/2V_{set}^{IM}$	V_{on}^{IM}	0V
	Floating	Floating	0V	0V	V_{on}^{IM}	0V

Fig. 27. Schematic view of the erasing, programming and reading schemes of a 2×2 NAND JL RFET array.

978-1-7281-1988-5/18 $31.00 © 2018 IEEE 176

0.63 mΩcm^2 / 1170 V 4H-SiC Super Junction V-Groove Trench MOSFET

T. Masuda[1], Y. Saito[1], T. Kumazawa[1], T. Hatayama[1], and S. Harada[1]

[1]National Institute of Advanced Industrial Science and Technology, Ibaraki, Japan, email: takeyoshi-masuda@aist.go.jp

Abstract—4H-SiC super junction, 0.63 mΩcm^2 and 1170 V, V-groove trench MOSFETs (SJ-VMOSFET) were demonstrated. The specific on-resistance ($R_{on,\ sp}$) of the SJ-VMOSFET is the lowest ever among all the reported SiC-MOSFETs with the blocking voltage (B_v) over 600 V. Superior electrical properties were realized with the structural combination of the V-groove MOS channel and the charge balance at super junction area. The $R_{on,\ sp}$ analysis of SJ-VMOSFET was carried out after mounted on a TO-268 5pin package having a Kelvin source terminal. The V-groove {$0\bar{3}3$ 8} channels could keep a high inversion mobility even in increasing doping concentration over 1×10^{18} cm^{-3}. The excellent-static and –dynamic performances of SJ-VMOSFET were allowed to realize the ultra-low loss switching applications.

I. INTRODUCTION

Silicon carbide MOSFETs have been anticipated as a candidate for power devices owing to the superior material properties of wide band gap, high critical electric field and high electron saturation drift velocity [1]. The V-groove trench MOSFETs (VMOSFET) have reduced the channel resistance by utilized high quality SiO$_2$/SiC interface with a high channel mobility [2]. As a result, the major problem of the tradeoff relationship between the drift resistance and the breakdown voltage remain. A super junction structure brought a breakthrough to reduce the $R_{on,\ sp}$ of silicon (Si) MOSFETs [3]. Similarly, it can be effective for the SiC power MOSFETs. The SiC MOSFET with SJ structure was first demonstrated at the 11th European Conference Silicon Carbide Related Materials 2016 [4]. In this paper, we have improved drastically the trade-off relationship between on-resistance and the breakdown voltage by the narrowing SJ cell pitches and the higher doping concentration in SJ regions. In addition, we have reduced the parasitic resistance by a thin V-groove {$0\bar{3}3\bar{8}$ } channel.

II. FABRICATION OF 4H-SIC SJ-VMOSFET

Figure 1 shows a schematic cross-section view of SJ-VMOSFET which has utilized the calculated doping profiles by technology computer aided design (TCAD).

A buffer layer of 3.0μm thickness was located below the SJ structure not so as to prohibit a high electric field to the substrate. The SJ structure was formed by the 6 times n-type epitaxial growths and aluminum (Al) ion-implantations that the doping concentration was 1.0×10^{17} cm^{-3}. The SJ structure

was periodically arranged by 2.5 μm pitch. The p-pillar width was varied in the range from 1.00 to 1.45 μm considering that the dispersion of n-pillar doping concentration at the epitaxial growth. The upper p-pillar regions with a doping concentration of 1.0×10^{18} cm^{-3} were formed for the trench bottom oxide protection from the electric field. The upper p-pillar connects the p-pillars below the trench bottom with the p-pillars below source contact. Current spreading layers (CSLs) were formed by the phosphorus (P) ion-implantations in order to counteract the Al ion-implantation tails of the channels and connect the channels to the n-pillar with a low resistivity. The V-groove trench MOSFETs were formed in alignment with the SJ structure. The ultra-thin channel was a highly doped 2.0×10^{18} cm^{-3} in order to keep the high threshold voltage (V_{th}) and to suppress short channel effects. An activation anneal for the ion implantations was performed at 1800°C. A gate oxidation at oxide (O$_2$) ambient and a post oxidation anneal at nitric-oxide (NO) ambient was performed at 1350°C continually. The gate oxide thickness of approximately 50 nm was measured. Source electrodes were formed by a nickel (Ni) sputtering deposition and a post deposition annealing (PDA) at 1000°C. The substrate was grinded from 350 to 50 μm-thick. A drain electrode was formed by the Ni and silicon (Si) sputtering after grinding the substrate and laser-annealed.

Figure 2 shows a scanning electron microscopy (SEM) image of the fabricated SJ-VMOSFETs. The existence of the ultra-thin channel with high contrast can be seen. The channel thickness of approximately 0.17 μm was measured by the contrast of the SEM image.

III. DEVICE SIMULATION AND EXPERIMENTAL RESULTS

Figure 3 shows the electric field distribution simulated the unit cell of SJ-VMOSFET in cases with and without the connecting upper p-pillar regions. The maximum of electric field crowding around the bottom of p-pillar with connecting upper p-pillar and the B_V of 1200 V was calculated by TCAD.

Figure 4 shows the output characteristics of the fabricated SJ-VMOSFET with the active area of 0.0377 cm^2 in the chip size of 0.25 cm x 0.25 cm, which is mounted on a TO-268 5pin package having a Kelvin source terminal. An $R_{on,\ sp}$ of 0.63 mΩcm^2 at the gate voltage (V_G) of 25 V was measured. Figure 5 shows the detailed analysis of the $R_{on,\ sp}$ component for the SJ-VMOSFET. The extremely low $R_{on,\ sp}$ was considered to realize due to the ultra-thin channels, the multiple SJ drift layers and the other elements eliminating parasitic resistance (the low resistivity CSLs and the grinding substrate). Figure 6 shows the transfer characteristic of the SJ-VMOSFET, which has a high V_{th} of 4.7 V defined at $I_D = 50$ mA. A V_{th} lowering, and a gate leak were not seen even by using the ultra-thin channel of 0.17 μm thickness because of the high channel doping concentration of 2.0×10^{18} cm^{-3}. It means the SJ-VMOSFET is compatible with typical Si power devices. Figure 7 shows the blocking characteristic of the SJ-VMOSFET. The B_V of 1170 V was measured and soft breakdown. It indicates that the p-type and n-type pillar charges are balanced as expected.

As a result, the tradeoff relationship between $R_{on, sp}$ and B_V of the SJ structure has exceeded those of the conventional SiC MOSFETs with unipolar SiC drift layers in figure 8 [2], [5], [6], [7] and [8]. This SJ-VMOSFET static property was the best of the other reported SiC MOSFETs.

IV. CAPACITANCE AND SWITCHING CHARACTERISTICS

Figure 9 shows the capacitance properties of the SJ-VMOSFET and the VMOSFET without SJ structures which were fabricated by using the same reticle set. The values of input, output and reverse transfer capacitances (C_{iss}, C_{oss} and C_{rss}) at $V_D = 600$V were extracted in Table I. There was no difference for the C_{iss} in the two devices due to the same active area. On the other hand, the C_{oss} and C_{rss} dropped at the $V_D = 10$ V or 50 V which pinch off their drift layers and depend on their doping concentrations. The C_{oss} in the SJ-VMOSFET was 1.4 times higher than that of VMOSFET because of the pn-junction area increasing by using SJ pillars. The V_D dependence of C_{rss} in the SJ-VMOSFET has a distinctive feature: the C_{rss} spike at $V_D = 50$ V. It is supposed to relate to the SJ depletion layer expansion depend on the V_D.

Figure 10 shows inductive load circuit to evaluate MOSFET switching properties. The waveforms comparison with the SJ-VMOSFET and the VMOSFET without SJ structures in figure 11. For the turn-off characteristics, a fall time (t_f) of the SJ-VMOSFET was slightly larger than the conventional VMOSFET because the C_{oss} of the SJ-VMOSFET was 1.4 times larger. However, considering that the $R_{on, sp}$ of the SJ-VMOSFET is approximately a forth smaller than the other, the C_{oss} of the SJ-VMOSFET will be smaller and it becomes rather an advantage at the same current capacity. On the other hand, a rise time (t_r) of the SJ-VMOSFET was faster than that of the other because of the lower V_{th}. The calculated switching losses: a turn-off (E_{off}), a turn-on (E_{on}) and the sum of them (E_{total}) were summarized in the table. II. Subsequently, the SJ-VMOSFET switching properties indicate equal or better than those of the same size VMOSFET without SJ structures.

V. CONCLUSION

0.63 mΩcm^2, 1170 V SJ-VMOSFET was successfully confirmed as originally designed due to integrate the multiple SJ structure with the narrow pitch pillars and the ultra-thin channel with the high doping concentration at 2×10^{18} cm^{-3}. The static characteristics of the fabricated SJ-VMOSFET was equal to the expected values by TCAD simulation. Further the switching properties of them are equal in the condition of the same chip size. Under the same current capacity, the SJ-VMOSFET will be rather superior to other SiC- and Si-MOSFETs. Finally, SJ-VMOSFETs could achieve higher and lower voltage applications with a fine tuning of a high electric field area, which the SJ structure thickness increase. Especially in a lower voltage application, the process cost reduction due to a thinner SJ structure to maintain the excellent switching performance of V-groove channels is expected.

ACKNOWLEDGMENT

This work has been implemented under a joint research project of Tsukuba Power-Electronics Constellations (TPEC), Japan. The authors gratefully acknowledge the contributions of the AIST Nano-processing Facility and DISCO corp. for utilizing the facilities. Thanks to Mr. H. Notsu and Mr. H. Michikoshi for the support the chip mounting and the switching measurements. Thanks to Mr. Y. Mikamura Sumitomo Electric Industries, Ltd., for the discussion.

T. Masuda, Y. Saito and T. Hatayama are assigned from Sumitomo Electric Industries, Ltd., and T. Kumazawa is assigned from Toyota Motor corp.

REFERENCES

[1] T. Kimoto, "Material science and device physics in SiC technology for high-voltage power devices", *Jpn. J. Appl. Phys.*, vol. 54, 040103, 2015.

[2] K. Uchida, Y. Saito, T. Hiyoshi, T, Masuda, K. Wada, H, Tamaso, T, Hatayama, K. Hiratsuka, T. Tsuno, M, Furumai and Y. Mikamura, "The optimised design and characterization of 1200 V / 2.0 mΩcm^2 4H-SiC V-groove trench MOSFETs", *Int. Symp. Power Semicond. Devices & ICs*, pp. 85-88, 2015.

[3] T. Fujihira, "Theory of semiconductor superjunction devices", *Jpn J. Appl. Phys.*, vol. 36, pp. 6254-6262. 1997.

[4] T. Masuda, R. Kosugi and T. Hiyoshi, "0.97 mΩcm^2 / 820 V 4H-SiC super junction V-groove trench MOSFET", *Mater. Sci. Forum.*, vol. 897, pp. 483-488, 2017.

[5] Y. Nakano, R. Nakamura, H. Sakairi, S. Mitani and T. Nakamura, "690 V, 1.00 mΩcm2 4H-SiC double-trench MOSFETs", *Mater. Sci. Forum.*, vol. 717-710, pp. 1069-1072, 2012.

[6] T. Nakamura, Y. Nakano, R. Nakamura, S. Mitani, H. Sakairi, and Y. Yokotsuji, "High performance SiC trench devices with ultra-low Ron", *Tech. Digest. 2011 Int. Electron Device Meeting*, 2011.

[7] A. Ichimura, Y. Ebihara, S. Mitani, M.Noborio, Y. Takeuchi, S. Mizuno, T. Yamamoto and K.Tsuruta, "4H-SiC trench MOSFET with ultra-low on-resistance by using miniaturization technology.", *Mater. Sci. Forum.*, vol. 924, pp 707-710, 2018.

[8] Y. Ebihara, A. Ichimura, S. Mitani, M.Noborio, Y. Takeuchi, S. Mizuno, T. Yamamoto and K.Tsuruta, "Deep-P encapsulated 4H-SiC trench MOSFETs with ultra low RonQgd", *Int. Symp. Power Semicond. Devices & ICs*, pp. 44-47, 2018.

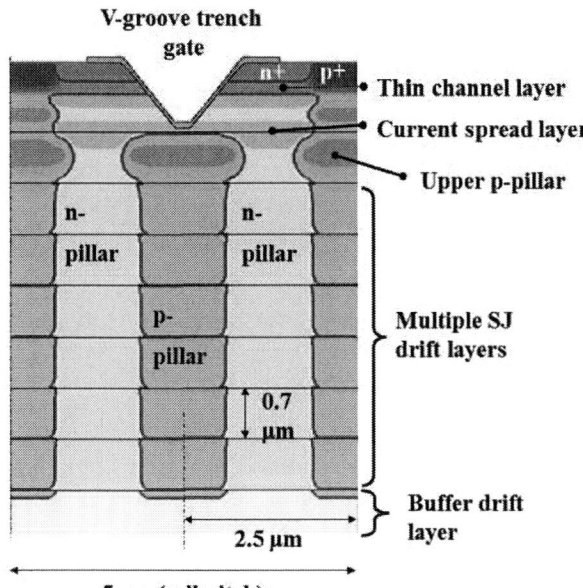

Fig. 1. A schematic cross section of the SJ-VMOSFET.

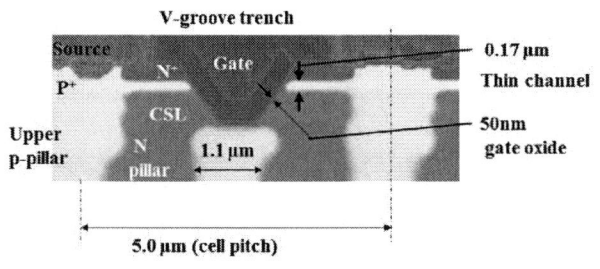

Fig.2. An SEM image of the fabricated SJ-VMOSFET with the ultra-thin channels.

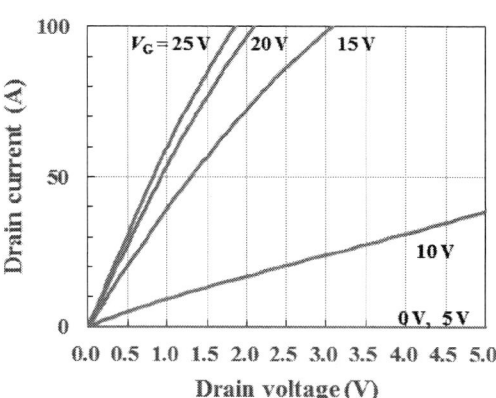

Fig. 4. Output characteristics of the SJ-VMOSFET with the active area of 0.0377cm².

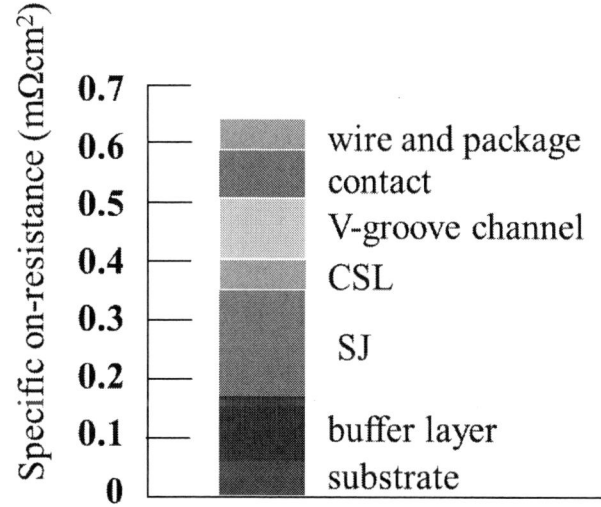

Fig. 5. A $R_{on, sp}$ analysis of the fabricated SJ-VMOSFET mounted in a TO-268 3pin package.

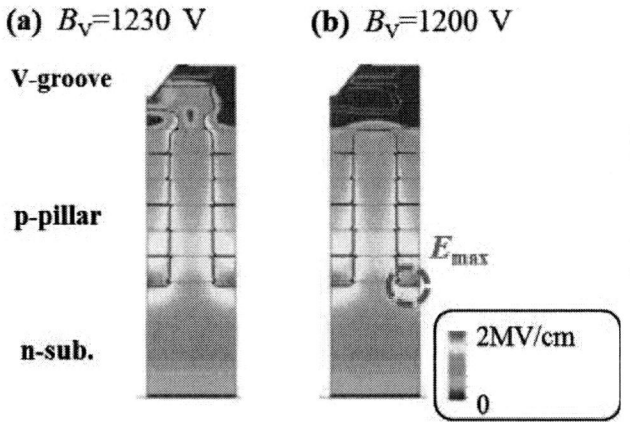

Fig. 3. Electric field simulations in the unit cell without (a) and with (b) connecting the upper p-pillar regions.

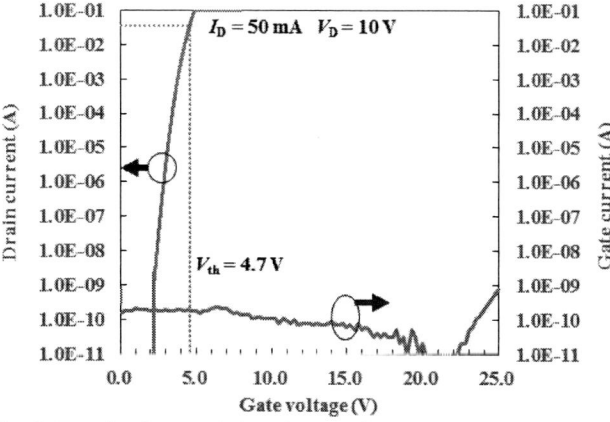

Fig. 6. Transfer characteristic and gate leakage current of the SJ-VMOSFET.

978-1-7281-1988-5/18 $31.00 © 2018 IEEE   179

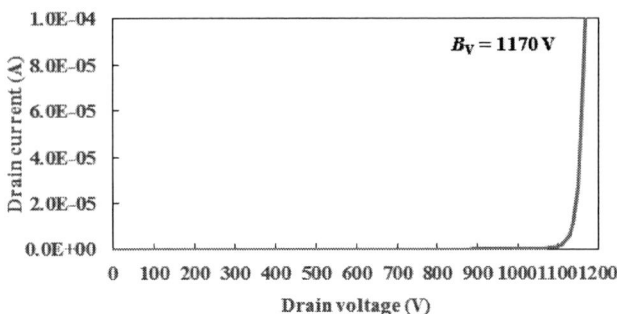

Fig. 7. A blocking characteristic of the SJ-VMOSFET.

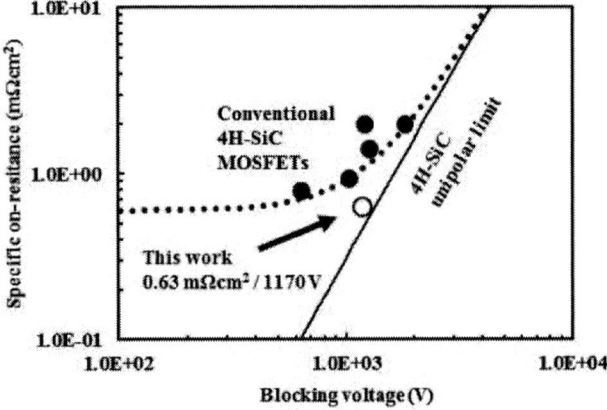

Fig. 8. Tradeoff relationship between $R_{on,\,sp}$ and B_V of 4H-SiC MOSFETs.

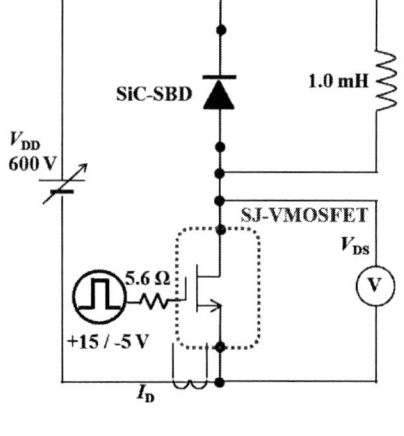

Fig. 10. Inductive load circuit for the SJ-VMOSFET switching.

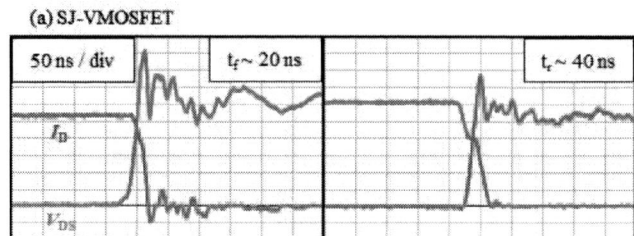

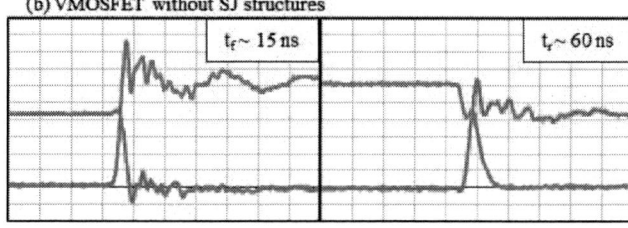

Fig. 11. Switching waveform comparison between the SJ-VMOSFET (a) and the VMOSFET without SJ structures (b).

Fig. 9. Capacitance properties comparison between the SJ-VMOSFET (solid-line) and the VMOSFET without SJ structures (dotted-line).

Capacitance properties at V_D=600V	This work SJ-VMOS	VMOSFET without SJ
C_{iss} (pF)	1.4	1.5
C_{oss} (pF)	70.0	52.0
C_{rss} (pF)	4.3	2.0

Table. I. Capacitances comparison at VD = 600 V between the SJ-VMOSFET and the VMOSFET without SJ structures.

Switching conditions and results	This work SJ-VMOS	VMOSFET without SJ
V_{DD} (V)	600	
I_{DS} (A)	25	20
V_{th} (V) at I_D=50mA	4.6	6.8
t_f (ns)	20	15
t_r (ns)	40	60
E_{off} (mJ)	0.28	0.15
E_{on} (mJ)	0.22	0.30
E_{total} (mJ)	0.50	0.45

Table. II. Switching conditions and results comparison between the SJ-VMOSFET and the VMOSFET without SJ structures.

First Demonstration of Dynamic Characteristics for SiC Superjunction MOSFET Realized using Multi-epitaxial Growth Method

S. Harada, Y. Kobayashi, S. Kyogoku, T. Morimoto, T. Tanaka, M. Takei, and H. Okumura

National Institute of Advanced Industrial Science and Technology (AIST), Tsukuba, Ibaraki, Japan, email: s-harada@aist.go.jp

Abstract—A 1.2 kV-class superjunction (SJ) UMOSFET was realized using a multi-epitaxial growth method. The dynamic characteristics were characterized, and the potential of a product level device was identified for the first time. The switching characteristics with Schottky barrier diode showed no degradation in spite of the large drain-source capacitance (C_{DS}). The reverse recovery characteristics of the body diode exhibited a soft recovery which may originate from the large C_{DS} and the short lifetime of minority carrier. A high short circuit capability comparable to a non-SJ device was demonstrated.

I. INTRODUCTION

Silicon carbide has a significant impact on reducing the specific on-resistance ($R_{on}A$) of a vertical power MOSFET owing to its excellent properties such as a high maximum electric field and high thermal conductivity. The reduction of $R_{on}A$ of a SiC MOSFET thus far has been achieved by reducing the dominant channel resistance. A trench gate MOSFET (UMOSFET) has succeeded in reducing the $R_{on}A$ owing to a small cell pitch and high channel mobility on the trench sidewall [1, 2]. However, such improvement is being saturated, and thus a reduction in the drift resistance is required for further improvement. For example, the drift resistance including the spread resistance with a breakdown voltage (V_B) of around 1.6 kV for a 1.2 kV-class MOSFET is approximately 1 mΩcm². A superjunction (SJ) is a novel structure that improves the trade-off between $R_{on}A$ and V_B beyond the unipolar limit, and a low $R_{on}A$ has been realized in a commercial silicon power MOSFET [3]. Thus, the application of an SJ structure is also expected for a SiC MOSFET. Although some studies have experimentally demonstrated the principle of the effects of an SJ on SiC [4, 5], detailed characterizations have not been conducted. This originates from the material properties, such as a high hardness and small diffusivity of the impurities, which makes the method established in a Si SJ-MOSFET difficult [6]. The influence of such difficulties and related instabilities during the process should be minimized to identify the potential of a SiC SJ-MOSFET. In this study, a multi-epitaxial growth method with a keV order energy implantation was developed for a 1.2 kV-class SiC SJ-MOSFET, and the dynamic characteristics were evaluated for the first time.

II. DEVICE DESIGN

Figure 1 shows cross-sectional schematics of the proposed 1.2 kV-class SJ-MOSFET. A UMOSFET with a cell pitch of 5 μm is employed as the MOSFET structure. This UMOSFET has a highly reliable structure, called IE-UMOSFET, which has implanted buried p-base regions with trench bottom p-regions functioning as a gate shielding structure [7]. To secure the current path, p-pillars with a width of 1.5 μm are connected to the buried p-base region in type-A (Fig. 1(a)), or p-pillars with a width of 0.75 μm are connected to both the buried p-base region and the trench bottom p-region in type-B (Fig. 1(b)). Thus, the p-pillar pitches are 5 μm in type-A and 2.5 μm in type-B. The SJ structure and a buffer layer as a semi-SJ structure have a thickness of 5.2 and 3.8 μm, respectively. Before fabricating the SJ-MOSFET, the process margin of the multi-epitaxial growth for the charge balance was analyzed. Figures 2 and 3 show the simulation results of $R_{on}A$ and V_B when the p/n charges are varied with the concentration of the n-epitaxial layer and the width of the p-pillar. Under the charge balance conditions indicated with the peak in Fig. 3, the reduction of $R_{on}A$ contributed by the SJ structure is 0.4 mΩcm² for type-A, and 0.5 mΩcm² for type-B, indicating a small difference based on the pillar pitch. In contrast, V_B is more sensitive to the charge imbalance than $R_{on}A$, and the sensitivity is drastically influenced by the p-pillar pitch. When the concentration of the n-epitaxial layer varies by ±10%, the margin of the p-pillar width for a breakdown voltage of over 1,400 V is 0.8 μm for type-A, but is only 0.1 μm for type-B. Therefore, to identify the potential of product-level devices, a type-A device was fabricated considering the process margin.

III. FABRICATION

A multi-epitaxial method for SiC is concerned with the misalignment caused by the off-axis of the SiC wafer, in which the alignment mark is deformed and shifts toward the [11-20] direction after the epitaxial growth. To suppress this effect, the method developed in this study applies a thin epitaxial film of 0.65 μm combined with keV order implantation. Further, a striped SJ structure toward the [11-20] direction was employed to ignore the misalignment. Figure 4 shows the fabrication flow of a 1.2 kV-class SJ-UMOSFET. First, an n-type buffer layer with a doping concentration of 1.8×10^{16} cm^{-3} was grown on a 4-inch, 4° off-axis 4H-SiC substrate, and the bottom of the p-pillar was formed through aluminum ion (Al$^+$) implantation. Next, multi-epitaxial growth was conducted through seven steps, resulting in a p-pillar with depth of 5.2 μm and an n-drift layer with concentration of 3×10^{16} cm^{-3}. The net acceptor concentration of the p-pillar after activation annealing was estimated to be 6×10^{16} cm^{-3}. Buried p-base regions and trench bottom p-regions were also formed through two steps of multi-epitaxial growth. The p-base for the channel region was formed through p-type epitaxial growth. The unit cell is a striped cell with {1-100} trench sidewalls. The total chip size is 3 mm x 3 mm. Figure 5 shows a cross-sectional SEM image of the fabricated device. Each p-region is well aligned in the p-pillar, and the p-pillar is also well aligned with the buried p-base region of the UMOSFET. The alignment accuracy for all ten multi-epitaxial steps are within 0.1 μm in the perpendicular to the stripe direction.

IV. STATIC CHARACTERISTICS

Figure 6 compares the typical output characteristics of the fabricated SJ-UMOSFET and the non-SJ-UMOSFET with similar values of V_B. The $R_{on}A$ was measured at a drain current (I_D) of 18 A and a gate voltage (V_{GS}) of 20 A, and V_{th} was defined as the V_G value at a drain voltage (V_{DS}) of 20 V and I_D of 18 mA. The $R_{on}A$ is 2.7 and 3.1 mΩcm^2 for the SJ and non-SJ devices with the V_{th} at around 4.0 V, respectively. The $R_{on}A$ reduction of 0.4 mΩcm^2 by the SJ structure is consistent with the design as is shown in the simulation results. Figure 7 shows off-state characteristics of these two devices. Almost the same avalanche V_B of above 1600 V is observed without an additional leakage current. Thus, the $R_{on}A$ of the 1.2 kV-class SiC MOSFET is successfully reduced in the SJ-UMOSFET without a sacrifice in the off-state characteristics. Figures 8 and 9 show the sensitivity of the static characteristics on the width of the p-pillars. Although the characteristics fluctuate owing to the doping non-uniformity of the drift layer in the wafer, both $R_{on}A$ and V_B qualitatively reproduce the tendencies by the simulation, indicating that the SJ structures are fabricated properly as designed.

V. DYNAMIC CHARACTERISTICS

Figure 10 shows the V_D dependences of the capacitances. In the non-SJ device, a drastic reduction is observed at a low V_D in the gate-drain capacitance (C_{GD}) and drain-source capacitance (C_{DS}), indicating a gate shielding effect through a pinch-off of the JFET region. On the other hand, in the SJ device, C_{DS} is large owing to the long pn-junction and has two shoulders indicating that the pinch-off progresses through two steps. As V_D increases, the pinch-off occurs in the JFET region first, followed by the SJ region. As a result, the large C_{DS} is sustained until a high V_D is reached and the drastic reduction is suppressed. The switching characteristics of the SJ-UMOSFET were tested under an inductive load at room temperature with conditions of V_{DS} = 600 V and V_{GS} = 20 V. Figures 11 and 12 show the turn-on and turn-off switching waveforms with a Schottky barrier diode (SBD) as a freewheeling diode (FWD). The external gate resistance is 22 Ω. No apparent differences between the non-SJ and SJ devices were observed in either the turn-on or turn-off waveforms. As shown in Fig. 13, the energy losses evaluated under various gate resistances were almost independent on the existence of the SJ structure, indicating that the large C_{DS} does not dominate the switching characteristics in the SiC SJ-MOSFET. Figures 14 and 15 show the turn-on and turn-off switching waveforms when the internal body diode is utilized as a FWD. The turn-off waveforms are similar, but the turn-on waveforms of the SJ device are delayed in V_D, which originates from the large recovery current of the body diode in the SJ device. As shown in Fig. 16, the energy losses for the turn-on are significantly increased in the SJ device. These behaviors are basically consistent with those of the Si SJ-

MOSFET. Figure 17 shows the reverse recovery waveforms of the body diode. As demonstrated in the turn-on energy loss, the reverse recovery charge is clearly larger in the SJ device than in the non-SJ device. Nevertheless, the recovery in the SJ device is rather soft. Such soft reverse recovery characteristic of the body diode must originate from the large C_{DS}, which is sustained until a high V_D is reached. This result also means that the reverse recovery characteristic of the body diode in a SiC SJ UMOSFET is not dominated by the minority carrier charges. This may be caused by the short lifetime of the minority carrier in case of the pn junction formed by implantation. Figure 18 shows short circuit waveforms measured at room temperature under V_{DS} = 600 V and V_{GS} = 20 V conditions. A long short circuit time of 7.2 μs, which is comparable to 7.6 μs of the non-SJ device, is obtained, indicating no degradation due to the SJ structure.

VI. CONCLUSION

In this study, dynamic characteristics such as the turn-on, turn-off, and short circuit capability of a SiC SJ-MOSFET were characterized for the first time. An evaluable 1.2 kV-class SJ-MOSFET was realized through the combination of a multi-epitaxial growth method with a trench MOSFET. The turn-on and turn-off with SBD exhibit no degradation by SJ structure, but the turn-on with body diode exhibit an increasing in switching loss. The soft recovery of the body diode and a short circuit time of 7.2 μs comparable to that for a non-SJ UMOSFET were achieved.

ACKNOWLEDGMENT

This work was implemented under a joint research project of Tsukuba Power Electrics Constellations (TPEC).

The co-authors Y. Kobayashi and M. Takei are assigned from Fuji Electric Co., Ltd., and S. Kyogoku is assigned from Toshiba Corporation, and T. Tanaka is assigned from Mitsubishi Electric Corporation.

REFERENCES

[1] T. Nakamura, Y. Nakano, M. Aketa, R. Nakamura, S. Mitani, H. sakairi, and Y. Yokotsuji, "High performance SiC trench devices with ultra-lon ron," *IEEE International Electron Devices Meeting (IEDM)*, 26.5.1, Dec. 2011.

[2] D. Peters, T. Aichinger, T. Basler, W. Bergner, D. Kueck, and R. Esteve, "1200V SiC Trench-MOSFET Optimized for High Reliability and High Performance," *Mater. Sci. Forum*, Vol. 897, pp. 489–492, 2017.

[3] T. Fujihira, "Theory of Semiconductor Superjunction Devices," *Jpn. J. Appl. Phys.*, Vol. 36, pp. 6254-6262, 1997.

[4] R. Kosugi, Y. Sakuma, K. Kojima, S. Itoh, A. Nagaka, T. Yatsuo, Y. Tanaka, and H. Okumura, "First experimental demonstration of SiC superjunction (SJ) structure by multi-epitaxial growth method," *International Symposium on Power Semiconductor Devices and ICs*, pp. 346-349, 2014.

[5] T. Masuda, R. Kosugi, and T. Hiyoshi, "0.97 mWcm2/820 V 4H-SiC Super Junction V-Groove Trench MOSFET," *Mater. Sci. Forum*, Vol. 897, pp. 483–488, 2017.

[6] S. Yamauchi, Y. Urakami, N. Suzuki, N. Tsuji, and H. Yamaguchi, "Fabrication of High Aspect Ratio Doping Region by Using Trench Filling of Epitaxial Si Growth," *International Symposium on Power Semiconductor Devices and ICs*, pp. 363-366, 2001.

[7] S. Harada, Y. Kobayashi, A. Kinoshita, N. Ohse, T. Kojima, M. Iwaya, H. Shiomi. H. Kitai, S. Kyogoku, K. Ariyoshi, Y. Onishi, and H. Kimura, "1200V SiC IE-UMOSFET with Low On-Resistance and High Threshold Voltage," *Mater. Sci. Forum*, Vol. 897, pp. 497–500, 2017.

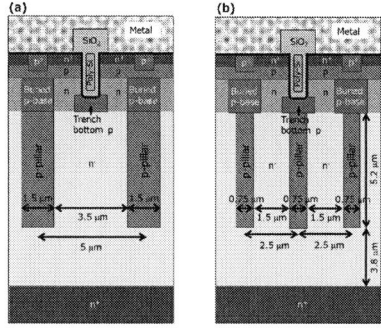

Fig. 1. Schematic cross-section of proposed SJ-UMOSFET with cell-pitch of 5 μm. p-pillar pitch is 5 μm in (a) type-A and 2.5 μm in (b) type-B.

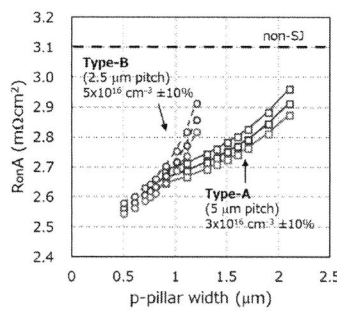

Fig. 2. Simulation results of specific on-resistance when the p/n charges are varied based on the concentration of the n-epitaxial layer and the width of the p-pillar.

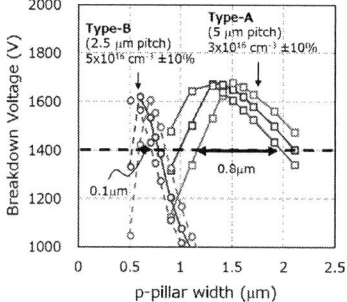

Fig. 3. Simulation results of specific breakdown voltage when the p/n charges are varied based on the concentration of the n-epitaxial layer and the width of the p-pillar.

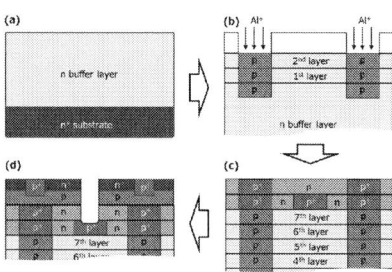

Fig. 4. Fabrication flow of SiC SJ-UMOSFET through multi-epitaxial growth method.

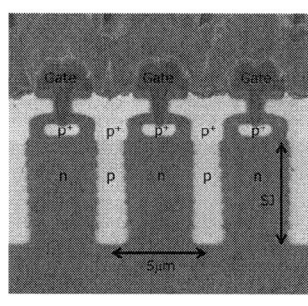

Fig. 5. SEM image of developed 1.2-kV class SiC SJ-UMOSFET.

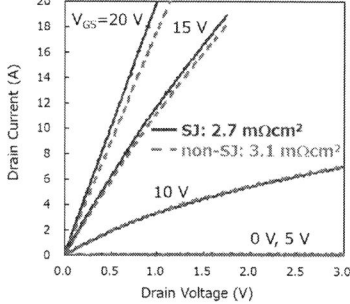

Fig. 6. Typical output characteristics of SJ- and non-SJ-UMOSFET. The RonA was measured at a drain current of 18 A and a gate voltage of 20 V.

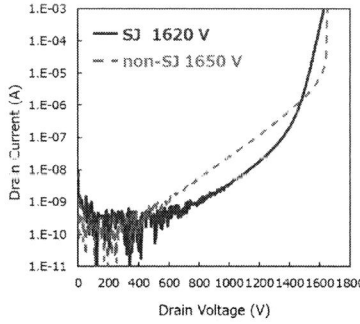

Fig. 7. Typical blocking characteristics of SJ- and non-SJ-UMOSFET.

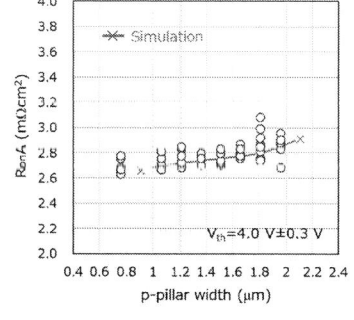

Fig. 8. p-pillar width dependence of $R_{on}A$ of SJ-UMOSFET.

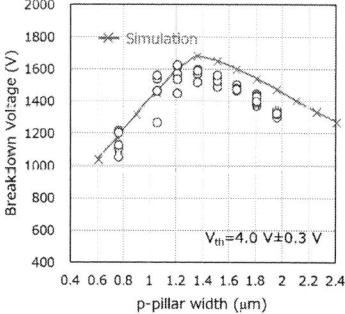

Fig. 9. p-pillar width dependence of breakdown voltage of SJ-UMOSFET.

978-1-7281-1988-5/18 $31.00 © 2018 IEEE   183

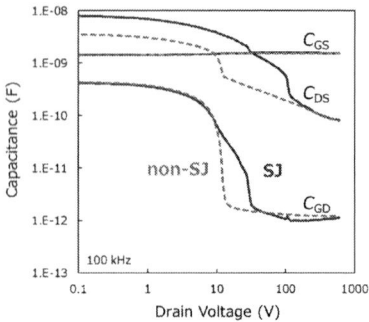

Fig. 10. Drain voltage dependence of capacitances of SJ- and non-SJ-UMOSFET.

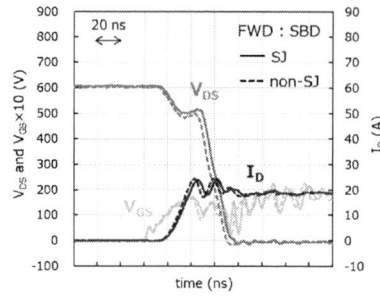

Fig. 11. Turn-on switching waveforms of SJ- and non-SJ-UMOSFET with SBD as freewheeling diode.

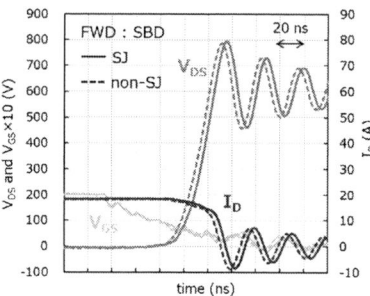

Fig. 12. Turn-off switching waveforms of SJ- and non-SJ-UMOSFET with SBD as freewheeling diode.

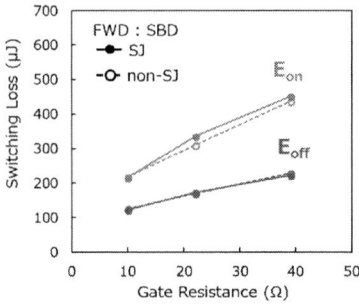

Fig. 13. Gate resistance dependences of switching energies for turn-on and turn-off of SJ- and non-SJ-UMOSFET with SBD as freewheeling diode.

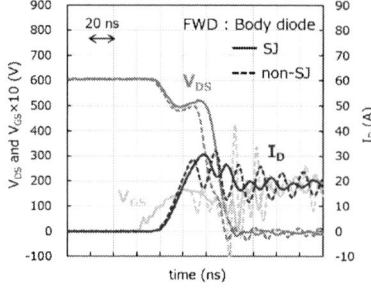

Fig. 14. Turn-on switching waveforms of SJ- and non-SJ-UMOSFET with internal body diode as freewheeling diode.

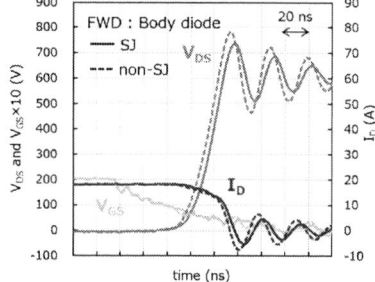

Fig. 15. Turn-off switching waveforms of SJ- and non-SJ-UMOSFET with internal body diode as freewheeling diode.

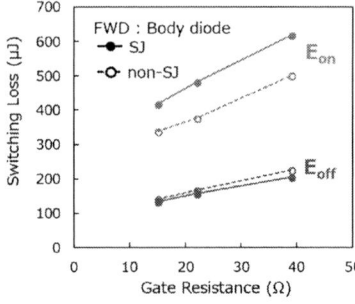

Fig. 16. Gate resistance dependences of switching energies for turn-on and turn-off of SJ- and non-SJ-UMOSFET with internal body diode as freewheeling diode.

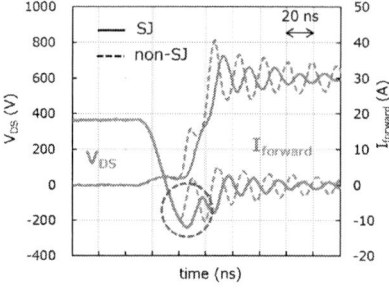

Fig. 17. Reverse recovery waveform of the body diode as freewheeling diode in SJ- and non-SJ-UMOSFET.

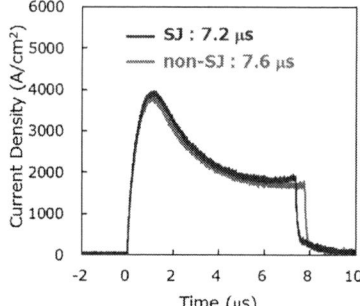

Fig. 18. Short circuit waveforms of SJ- and non-SJ-UMOSFET.

978-1-7281-1988-5/18 $31.00 © 2018 IEEE   184

Channel engineering of 4H-SiC MOSFETs using sulphur as a deep level donor

M. Noguchi[1], T. Iwamatsu[1], H. Amishiro[1], H. Watanabe[1], K. Kita[2] and N. Miura[1]

[1] Advanced Technology R & D Center, Mitsubishi Electric Corporation,
8-1-1 Tsukaguchi-Honmachi, Amagasaki City, Hyogo 661-8661, Japan
Phone: +81-6-6497-7096 E-mail: Noguchi.Munetaka@dh.MitsubishiElectric.co.jp
[2] Department of Materials Engineering, The University of Tokyo, 7-3-1 Hongo, Bunkyo-ku, Tokyo 113-8656, Japan

Abstract—We demonstrate Si-face 4H-SiC MOSFET using sulphur (S) as a deep level donor in channel region, for the first time. Contrary to general recognition that deep level donors are not suitable for device fabrication, S is found to be a promising deep level donor for the channel region of 4H-SiC MOSFETs. Compared with channels doped by shallow level donors, S-doped channel is found to provide lower channel resistance (R_{ch}) and higher threshold voltage (V_{th}). On the basis of simulations and experiments, this improvement is found to be ascribed to two inherit natures of S in 4H-SiC. One is the large ionization energy (E_{ion}), resulting in the increase of V_{th}. Another is that S act as a donor, improving inversion layer mobility in channel region. By applying this novel channel engineering to vertical 4H-SiC MOSFETs, 31 % reduction of specific on resistance (R_{on}) at high V_{th} of 4.0 V was achieved.

I. INTRODUCTION

Si-face 4H-SiC MOSFETs have been developed because of its material properties suitable for power devices such as high breakdown electric field and high thermal conductivity [1]. Although some gate oxidation processes for achieving low R_{ch} have been proposed [2-5], nitridation process [6, 7] is usually used in the industry to maintain sufficient stability of V_{th}. A key challenge for Si-face 4H-SiC MOSFETs with nitridation process is to improve the trade-off relationship between R_{ch} and V_{th}. R_{ch} increases severely in high V_{th} region, such as in V_{th} over 3 V (**Fig. 1**). Both low R_{ch} and high V_{th} is generally required for power electronic systems. This is because low R_{ch} contributes to reduce energy loss of them, providing better controllability owing to high V_{th}. Moreover, high V_{th} could have an opportunity to make power modules more compact by a single power supply source of the gate voltage (V_g). We consider that channel engineering in semiconductor side could give a solution for reducing R_{ch} in high V_{th} region. Counter-doping to channel region of SiC MOSFETs by shallow level donors such as N, P, As, and Sb is known to improve peak field effective mobility (μ_{FE}) [8-11]. However, the trade-off between R_{ch} and V_{th} keeps consistent when those impurity profiles are almost the same (**Fig. 2, Fig. 3**). To realize 4H-SiC MOSFETs with low R_{ch} and high V_{th}, an alternative method is strongly required.

In this study, we propose S doping in channel region to achieve low R_{ch} and high V_{th} for Si-face 4H-SiC MOSFETs with nitridation process for the gate oxide (**Fig. 4**). In general, deep level donors are not been used in SiC device technologies

due to large E_{ion} and low ionized dopant factor in bulk [1, 12-14] (**Fig. 5, Fig. 6**). However, it is not sure whether they can be adapted in channel region or not, which is different from bulk region. It is expected that they can capture electrons in weak inversion condition, which possibly results in the increase of V_{th}. In addition, doping donor in channel region improves inversion layer mobility [15]. We consider that these two effects of deep level donors could change the conventional trade-off between R_{ch} and V_{th} obtained by shallow level donors. Therefore, S was selected as an example. We demonstrate that low R_{ch} and high V_{th} can be achieved by Si-face 4H-SiC MOSFETs with S-doped channel, for the first time.

II. EXPERIMENTAL

To simply predict the effect of doping S in channel region, we simulated the electrical characteristics of lateral Si-face 4H-SiC MOSFETs when doping donors with large E_{ion} in channel region. E_{ion} was virtually changed between 0 to 500 meV. Then, we experimentally evaluated the trade-off between R_{ch} and V_{th} when S was doped in channel region. We fabricated lateral and vertical Si-face 4H-SiC MOSFETs with S-doped channel. S atoms were implanted in channel region and electrically activated at the same thermal annealing process for source, well, and p^+ contact region. Around 50 nm-thick gate oxide was formed by thermal oxidation followed by nitridation in diluted NO. Lateral 4H-SiC MOSFETs were fabricated on a standard p-type well region formed by Al-implantation. To fairly discuss the effect of S in channel region, the peak position of S was controlled to be almost the same as those of other donor (**Fig. 7**). Additionally, test element groups (TEGs) of vertical Si-face 4H-SiC MOSFETs with S-doped channel were fabricated on an n-type substrate. The optimization of unit cell pitch with JFET doping [16] were not applied to simply discuss the effect of S doping on channel characteristics.

III. LATERAL SiC MOSFETs WITH S-DOPED CHANNEL

A. Simulation of SiC MOSFETs with large E_{ion}

To simply understand the intrinsic effect of S doping in channel region of Si-face 4H-SiC MOSFETs, we simulated the surface carrier density (N_S) as a function of V_g, for various E_{ion} (**Fig. 8 (a)**). Here, interface traps are not considered. There are almost no change in N_S-V_g characteristics when E_{ion} is below 100 meV, corresponding to shallow level donors. On the other hand, subthreshold slope is found to become less steep when E_{ion} increases up to 300 meV. In addition, positive shifts of V_{th}

978-1-7281-1988-5/18 $31.00 © 2018 IEEE

are calculated when E_{ion} is larger than 300 meV. This boundary energy is close to E_{ion} of S, which is 260 meV. Distinct from conventional donors, S doping could enable positive V_{th} shift owing to less steep subthreshold slope and positive shift in the N_S-V_g curve. To further understand this phenomena, Fermi level (E_F) from conduction band (E_C) edge as a function of V_g was calculated with E_{ion} of 0 and 260 meV (**Fig. 8 (b)**). When E_F reaches the impurity level, elevation of E_F by the increase in V_g becomes gradual, resulting from electron trapping to impurity level. This change is attributed only to E_{ion}, which is one of the material properties. Thus, this result is common for 4H-SiC, and is the same for other faces like C-, m-, and a-faces.

B. Trade-off between R_{ch} and V_{th}

4H-SiC MOSFETs with intentionally doped S in channel region were demonstrated, for the first time. The relationship between R_{ch} and V_{th} of lateral Si-face 4H-SiC MOSFETs with S doping shows different behavior compared with the conventional trade-off (**Fig. 9**). Different from conventional donors, V_{th} does not monotonically decrease with the increase in S dose. With increase in S dose, V_{th} slightly decreases first, and then V_{th} starts to increase. Compared with the conventional trade-off, higher V_{th} at the same normalized R_{ch} is achieved. For example, S doping with the dose of 8×10^{12} cm^{-2} enables positive shift of V_{th} and increases drain current (I_d) (**Fig. 10**).

C. Characterization

To precisely understand the physical properties of S doping in the channel region of Si-face 4H-SiC MOSFET, we examined interface traps, inversion layer mobility, and the temperature dependence of R_{ch}. First, the effect of S doping on the SiO$_2$/SiC interface was evaluated by CV measurements. Interface traps near E_C edge, which is important for I_d-V_g characteristics, are not significantly affected by S doping even when S dose is as much as 8×10^{12} cm^{-2} (**Fig. 11**). Then, to investigate interface traps inside E_C and inversion layer mobility ($\mu_{inv, Hal}$), Hall effect measurements [15, 17] were carried out. S doping does not affect the energy distribution of interface traps inside E_C (**Fig. 12**). At room temperature (RT), the increase in S dose results in higher $\mu_{inv, Hall}$, which is similar to N doping in channel region (**Fig. 13**) [15]. This increase in $\mu_{inv, Hall}$ results from the relaxed electric field at the MOS interface because S atom acts as a donor in depletion layer. The role of S atoms in channel region is considered to depend on its charged states as follows. Whether S is neutralized or ionized changes by the energy position of E_F and its impurity level (**Fig. 14 (a)**). Neutralized S atoms increase V_{th}, trapping electrons to its impurity level. Note that this effect is quite small for shallow level donors. In addition, ionized S atoms increase $\mu_{inv, Hall}$ owing to the relaxed electric filed at the MOS interface. (**Fig. 14 (b)**). To explain the benefit of S doping in channel region, we compared μ_{FE} as a function of V_{th} with previously reported results [2-5] (**Fig. 15**). Although high μ_{FE} was reported by using gate oxide engineering at MOS interface, their V_{th} is limited below 2 V. On the other hand, compared with conventional channel engineering using popular nitridation process, S doping in channel region improves μ_{FE} at high V_{th} region over 4 V. The V_{th} shift under positive or negative bias stress of V_g tend to be

large for gate oxide engineering, but this is negligibly small for samples with S doping. Finally, temperature (T) dependence of normalized R_{ch} was examined between -25°C and 175°C for the samples with different S doses. Although R_{ch} of Si-face 4H-SiC MOSFETs suffers severe increase in R_{ch} at low temperature, S doping can make this temperature dependence weaker (**Fig. 16**).

IV. VERTICAL SiC MOSFETs WITH S-DOPED CHANNEL

To demonstrate the improvements of the trade-off between R_{on} and V_{th} by applying S doping in channel region, vertical Si-face 4H-SiC MOSFETs were fabricated. I_d-V_g characteristics with and without S doping are shown for samples with high and low V_{th} (**Fig. 17 (a), Fig. 17 (b)**). The samples with S doping do not show significant decrease of I_d in high V_{th} region, resulting from a parallel shift of + 2.9 V from the I_d-V_g curve of a low V_{th} sample. However, samples without S doping show severe decrease of I_d in high V_{th} region. This shows that S doping is beneficial in high V_{th} region. Relationship between R_{on} and V_{th} at RT is shown (**Fig. 18**). Note that total R_{on} includes not only R_{ch} but other resistances such as contact, substrate, drift layer, and JFET resistances. S doping provide a novel R_{on}-V_{th} relationship compared with conventional relationship given by only N doping. Improved trade-off by S doping can be found especially in the high V_{th} region over 3 V. R_{on} reduces from 8.3 mΩ·cm^2 to 5.7 mΩ·cm^2 at V_{th} of 4.0 V, corresponding to 31 % reduction of total R_{on}.

V. CONCLUSION

For the first time, we demonstrated Si-face 4H-SiC MOSFETs with nitridation process using S-doped channel. S in 4H-SiC acts as a donor and forms deeper impurity level than those of conventional donors. These two natures of S in 4H-SiC were found to realize a novel relationship between R_{ch} and V_{th}. It is considered that neutralized and ionized S atoms in channel region respectively increase V_{th} and inversion layer mobility. By applying this channel engineering to vertical 4H-SiC MOSFETs, trade-off between R_{on} and V_{th} was improved. The effect is more significant in higher V_{th} region over 3 V.

REFERENCES

[1] T. Kimoto *et al.* Jpn. J. Appl. Phys. **54** 040103 (2015).
[2] D. Okamoto *et al.*, IEEE Electron Device Lett. **31**, 7 (2010).
[3] D. Okamoto *et al.*, IEEE Electron Device Lett. **35**, 12 (2014).
[4] M. Cabello *et al.*, Spain Conference on Electron Devices (2017).
[5] D. J. Lichtenwalner *et al.*, Mater. Sci. Forum **897**, 163 (2017).
[6] G. Y. Chung *et al.*, IEEE Electron Device Lett. **22**, 4 (2001).
[7] K. Fujihira *et al.*, Solid-State Electron. **49**, 896 (2005).
[8] K. Ueno *et al.*, IEEE Electron Device Lett. **20**, 624 (1999).
[9] S. Harada *et al.*, IEEE Electron Device Lett. **20**, 272 (2001).
[10] A. I. Mikihaylov *et al.*, Mater. Sci. Forum **858**, 651 (2016).
[11] A. Modic *et al.*, IEEE Electron Device Lett. **35**, 894 (2014).
[12] S. A. Reshanov *et al.*, J. Appl. Phys. **99**, 123717 (2006).
[13] https://www.nist.gov/publications/arsenic-and-antimony-implantations-sic
[14] T. Kimoto and J. A. Cooper, *Fundamentals of Silicon Carbide Technology*, John Wiley & Sons, 2014.
[15] M. Noguchi *et al.*, Jpn. J. Appl. Phys. **57** 04FR13 (2018).
[16] K. Hamada *et al.* Jpn. J. Appl. Phys. **54** 04DP07 (2015).
[17] M. Noguchi et al., Tech. Dig. of IEEE Int. Electron Devices Meet., p. 219 (2017).

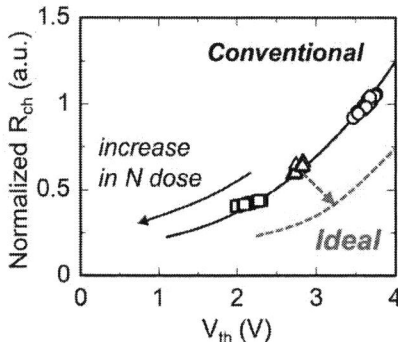

Fig. 1 Conventional and ideal relationship between normalized channel resistance (R_{ch}) and threshold voltage (V_{th}). Ideally, lower R_{ch} at higher V_{th} is required. Conventional trade-off is obtained by N doping into a standard p-type well. R_{ch} is normalized by that of p-type well region without N counter doping at V_g of 15 V. V_{th} was defined as V_g where I_d of 1×10^{-10} A/μm at V_d of 0.1 V.

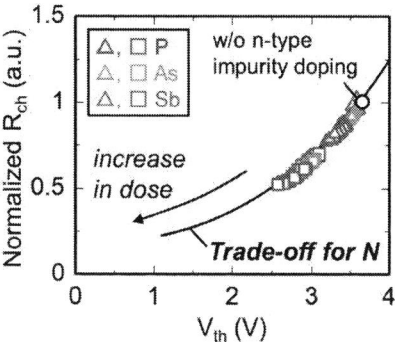

Fig. 2 Relationship between normalized R_{ch} and V_{th} when P, As, or Sb was doped. Compared with the trade-off when N was doped, they are plotted on the same trade-off irrespective of donors. Impurity profiles are in Fig. 3.

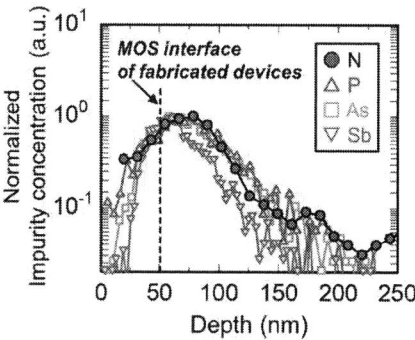

Fig. 3 Normalized impurity profiles of N, P, As, and Sb used for device fabrication. By controlling the ion implantation energy, the peak position was set to be similar to each other for the suppression of the difference in electric field at MOS interface.

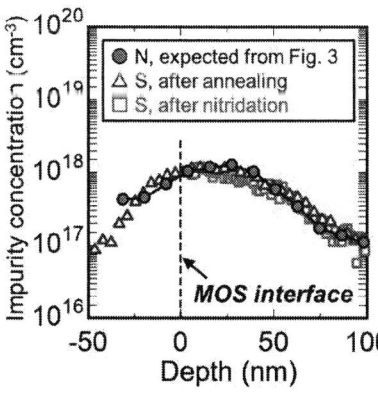

Fig. 4 Schematic cross section of vertical 4H-SiC MOSFETs. Different from the conventional approaches, S was doped in channel region in this work.

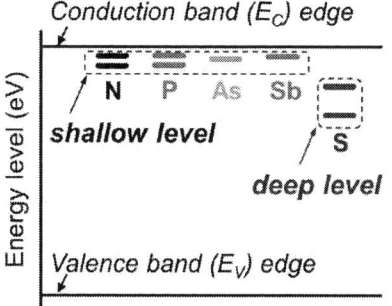

Fig. 5 Schematic donor levels of each dopant. N, P, As, and Sb [1, 13] have smaller ionization energy than S [12]. Donor level of S is deeper than that of N, P, As, and Sb.

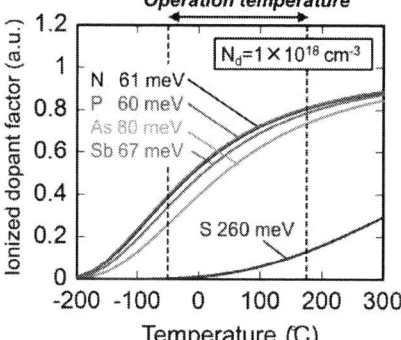

Fig. 6 Ionized dopant factor in bulk as a function of temperature for each donor. Compared with conventionally used shallow level donors, S shows much lower ionized dopant factor, meaning that S is not suitable for a donor in bulk region. Calculation was done as in Ref. 14.

Fig. 7 Impurity profiles of S after annealing and after nitridation analyzed by SIMS. Here, the dose of S was 8×10^{12} cm^{-2}. The peak position of S was almost the same as that of N, which is expected from Fig. 3 for the same dose. The impurity profile of S shows little change during oxidation followed by nitridation.

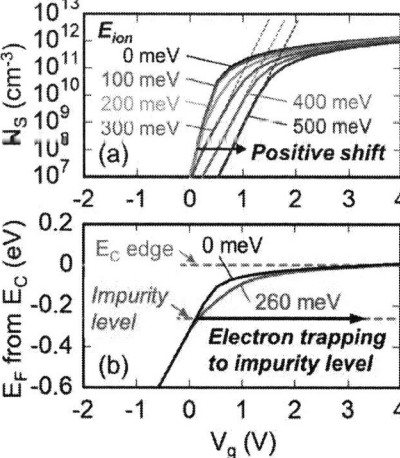

Fig. 8 (a) Simulated E_{ion} dependence of surface carrier density (N_S) as a function of V_g. Increase in E_{ion} results in positive shift of V_{th}, resulting in the increase of V_{th}. **(b)** Simulated Fermi level (E_F) from E_C as a function of V_g for E_{ion} of 0 and 260 meV. Increase in E_F becomes gradual when E_F reaches the impurity level ascribed to electron trapping to the impurity level.

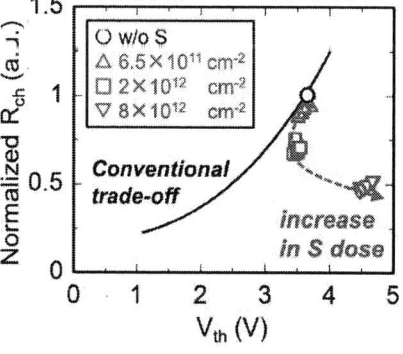

Fig. 9 Relationship between normalized R_{ch} and V_{th} when S was doped. It is found that the increase in S dose shows a different behavior from the conventional trade-off. At the same normalized R_{ch}, V_{th} of S-doped samples are higher than that of conventional trade-off. Here, the dose of S was 6.5×10^{11}, 2×10^{12}, and 8×10^{12} cm^{-2}.

978-1-7281-1988-5/18 $31.00 © 2018 IEEE

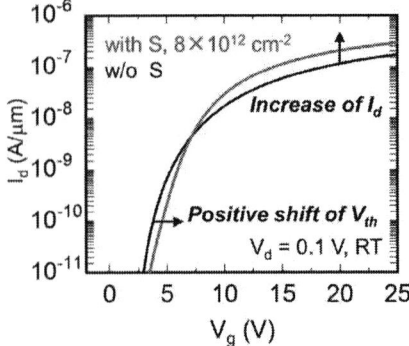

Fig. 10 I_d-V_g characteristics of planar-type 4H-SiC MOSFETs for samples with and without S doping in channel region at RT. Here, the dose of S was 8×10^{12} cm^{-2}. S doping in channel region enables positive shift of V_{th}. In addition, I_d increases in high V_g region.

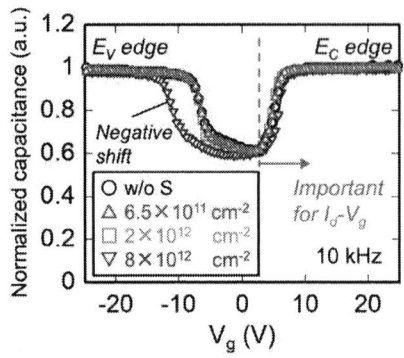

Fig. 11 S dose dependence of normalized capacitance of planar-type SiC MOSFETs as a function of V_g at 10 kHz. When S dose increases up to 8×10^{12} cm^{-2}, a CV curve near E_V edge shift negatively, suggesting the increase in interface traps near E_V. On the other hand, S doping does not significantly affect interface traps near E_C.

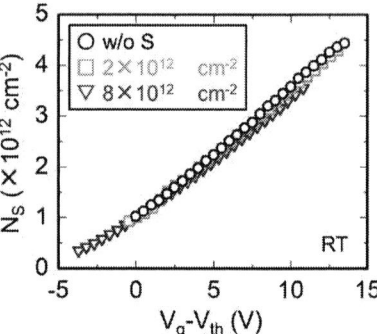

Fig. 12 S dose dependence of N_S as a function of $V_g - V_{th}$ for the samples with and without S doping. Here, V_{th} was determined at V_g where N_S is 1×10^{12} cm^{-2}. They are plotted on almost the same line, indicating that S doping does not affect interface trap distribution inside E_C.

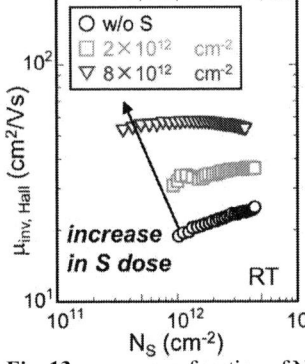

Fig. 13 $\mu_{inv, Hall}$ as a function of N_S for the samples with and without S doping. The increase in S dose results in higher $\mu_{inv, Hall}$. Relaxed electric field suppresses Coulomb scattering at the MOS interface. S dose was 2×10^{12} and 8×10^{12} cm^{-2}.

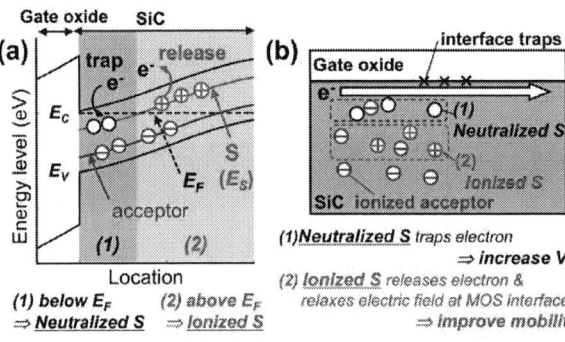

Fig. 14 (a) Schematic band diagram of S-doped channel. The charged states of S are determined by E_F and impurity level of S (E_S). When E_S is below E_F, S is neutralized. When E_S is above E_F, S is ionized. Different from S, shallow level donors show little neutralized states due to smaller E_{ion}. (b) Schematic cross section of S-doped channel of 4H-SiC MOSFETs. Two charged states of S enable V_{th} increase and $\mu_{inv, Hall}$ improvement.

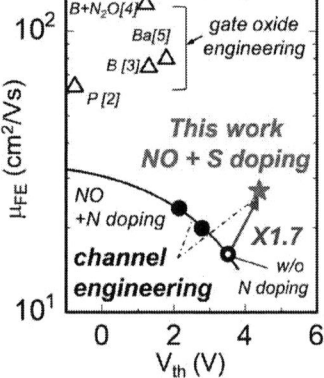

Fig. 15 Benchmark of μ_{FE} as a function of V_{th}. This work improves μ_{FE} at high V_{th} region by channel engineering and popular nitridation in NO. μ_{FE} is evaluated at the gate oxide electric field around 3 MV/cm.

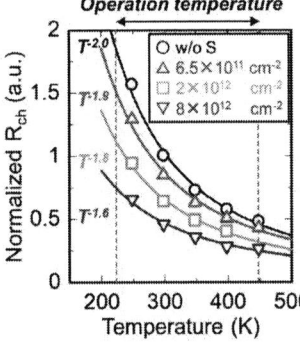

Fig. 16 S dose dependence of normalized R_{ch} as a function of temperature. Temperature power-law factor of R_{ch} increases -2 to -1.6 by S doping, meaning that the increase in R_{ch} at low temperature can be suppressed by S doping.

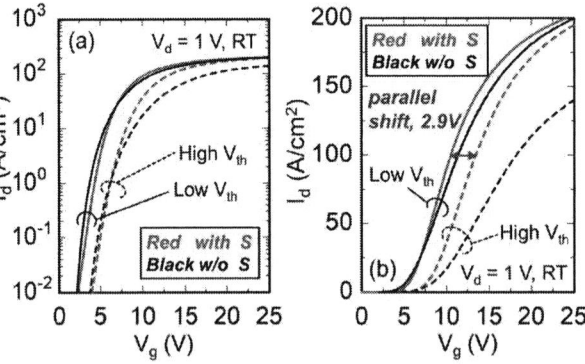

Fig. 17 I_d-V_g characteristics of vertical SiC MOSFETs for the samples with and without S doping in channel region. I_d is shown as in (a) log and (b) linear scale. Samples with S doping maintain I_d even at high V_{th}. Comparing samples with high and low V_{th}, the I_d-V_g curves of samples with S doping show parallel shift of 2.9 V. However, samples without S doping show severe decrease in I_d when V_{th} is increased.

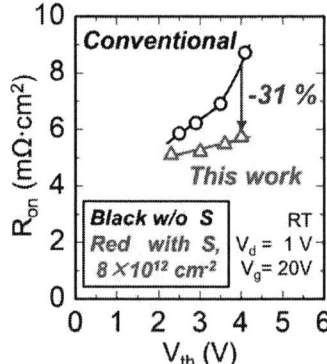

Fig. 18 Relationship between R_{on} and V_{th} for the samples with and without S doping. V_{th} was changed by additional N doping. Improved trade-off was achieved by doping S, especially at high V_{th} region over 3V. Total R_{on} reduces 31 % at V_{th} of 4.0 V at RT.

978-1-7281-1988-5/18 $31.00 © 2018 IEEE

Demonstration of 1200V Scaled IGBTs Driven by 5V Gate Voltage with Superiorly Low Switching Loss

T. Saraya[1], K. Itou[1], T. Takakura[1], M. Fukui[1], S. Suzuki[1], K. Takeuchi[1], M. Tsukuda[2], Y. Numasawa[3],
K. Satoh[4], T. Matsudai[5], W. Saito[5], K. Kakushima[6], T. Hoshii[6], K. Furukawa[6], M. Watanabe[6], N. Shigyo[6],
K. Tsutsui[6], H. Iwai[6], A. Ogura[3], S. Nishizawa[7], I. Omura[8], H. Ohashi[6], and T. Hiramoto[1]

[1]The University of Tokyo, Tokyo, Japan, email: saraya@nano.iis.u-tokyo.ac.jp
[2]Green Electronics Research Institute, Kitakyushu, Japan, [3]Meiji University, Kawasaki, Japan, [4]Mitsubishi Electric Corp.,
Fukuoka, Japan, [5]Toshiba Electronic Devices & Storage Corp., Tokyo, Japan, [6]Tokyo Inst. of Technology, Yokohama,
Japan, [7]Kyushu University, Fukuoka, Japan, [8]Kyushu Inst. of Technology, Kitakyushu, Japan.

Abstract—Functional trench-gated 1200V-10A class Si-IGBTs, designed based on a three dimensional (3D) scaling concept, were fabricated, and 5V gate voltage switching operation has been demonstrated for the first time. 33% reduction of turn-off loss and 100mV improvement of on-state voltage were achieved, while keeping 1.2kV forward blocking voltage.

I. INTRODUCTION

Today, Si-IGBTs (Insulated Gate Bipolar Transistors) are used as the main stream power switching device [1-3], and their improvement is still continued by various ways [4-7]. As an attempt to further enhance Si-IGBT performance, a 3D scaling concept was proposed (Fig.1 and Table.1) [8]. That is, similarly to the CMOS scaling, all the geometrical dimensions, both horizontal and vertical, as well as gate voltage, are scaled down proportionately, while keeping the cell pitch W constant. A major effect of this scaling is the reduction of on-state voltage drop (V_{cesat}), which is realized by the Injection Enhancement (IE) effect [8], and also lowered trench-gated region resistance. This leads to reduced chip size and cost. Another apparent effect is the reduction of gate voltage. Traditionally, the gate drive voltage has been 15V. However, if this voltage can be reduced to 5V, standard low cost CMOS logic devices with 5V I/O can be used for the gate drivers. This will reduce the driver power by around 1/10. In addition, high integration level of CMOS LSIs will open up new possibilities of smarter digital control [9], enabling more efficient and safer operations of IGBTs. Noise issue could be also solved by digital processing [9].

Recently, to confirm the validity of the 3D scaling concept, Si-IGBT test devices were fabricated [10]. Significant reduction of V_{cesat} by the scaling from k=1 to k=3 was confirmed. However, the devices were designed solely for low voltage dc characterization, and were fabricated on p+ substrates with limited size (I_{ce}~3mA). Hence, demonstration of 5V driven switching at practically high voltage with large current flow could not be accomplished. Therefore, in this work, fully functional 3D scaled 1200V-10A class, 5V driven Si-IGBT chips were fabricated, integrating controlled back side doping and guard ring (surface termination) structure. 600V switching operation with 33% less turn-off loss, 100mV better V_{cesat}, and 1.2kV forward blocking voltage (BV) are demonstrated. The results pave the way for the realization of scaled IGBTs.

II. DEVICE DESIGN AND FABRICATION

As briefly mentioned above, the effects of the IGBT scaling are two-fold. (i) Change of the dimensions as in Table.1 will decrease hole sinking efficiency and increase electron injection efficiency of the trench gated emitter (IE effect), and will selectively increase the on-state carrier concentration near the top surface. As a result, the n-base resistance can be lowered without increasing the bottom side carrier concentration. (ii) Shorter MOS inversion region and trench extrusion also contributes to reducing parasitic resistance. Through both (i) and (ii), the 3D scaling is expected to realize reduced V_{cesat}, while minimizing the turn-off loss (E_{off}) degradation caused by high bottom side carrier concentration, thus improving E_{off} and V_{cesat} trade-off.

A 3D scaled IGBT was designed starting from a 15V control device (k=1, Fig.1a). As for the emitter cells, following [10], all the key dimensions of the MOS-gated region were scaled as shown in Table.1 (k=1→3), while keeping the cell pitch W constant (16μm). To carry 10A current, the device area was set to 5mm², which consists of 324 emitter cells with 1mm×16μm area. To safely achieve stable high voltage switching with the large device area, p-float region was not fully floated, but resistively connected to ground (connected p-float design). To support a high off-state voltage, guard rings that surround the emitters are necessary. The layout and vertical structures of the rings were carefully designed using TCAD simulations, to achieve 1.2kV BV. For fast switching and low turn-off loss, controlling the p-collector concentration to a relatively low desired value is important. To achieve controlled back side doping, without affecting the front side doping profiles, laser annealing process was adopted.

Fig.2 summarizes the process flow for the fabricated IGBTs. To fabricate both k=1 and k=3 devices with different doping depth (p-base, p-float) and gate oxide thickness consistently, thermal budget and ion implantation conditions were carefully adjusted (Table.2). Table 3 shows measured minority carrier lifetime by quasi steady state photo conductance (QSSPC) method using bare wafers that received equivalent annealing to the real IGBTs. The bulk lifetime exceeds 10μs for both k=1 and 3, which is sufficient for the 120μm n-base length for 1.2kV blocking. Comparable lifetime for k=1 and 3 were obtained, since the depth of the guard ring was not scaled for k=3.

978-1-7281-1988-5/18 $31.00 © 2018 IEEE

Fig.3 shows simulated collector-to-emitter current density (J_{ce}) vs collector-to-emitter voltage (V_{ce}) characteristics for k=1 and 3 for the same p-collector doping. The p-float connection to ground is taken into account as shown in the inset. It can be confirmed that J_{ce} increases by the 3D scaling. Fig.4 shows the on-state carrier distributions for three different p-collector concentrations. As is well known, the back side carrier concentration can be easily controlled by the p-collector doping. In addition to this, 3D scaling provides additional means for controlling the carrier profile. The front side concentration can be selectively increased by the 3D scaling, owing to the enhanced IE effect. The higher J_{ce} for k=3 than k=1 in Fig.3 can be explained by this effect, combined with reduced MOS inversion layer length. The smaller difference than [10] between k=1 and 3 in Fig.3 is caused by the connected p-float design.

Three types of Si-IGBTs (k=1, k=3, and partially scaled k=3'; see Table.4) were fabricated on 3-inch wafers (Fig.5). P-collector concentration was also varied. Figure 6 shows cross sectional TEM images of the trench gate regions. While the trench depth and the distance between the two trenches (i.e. mesa width S) were reduced as designed for k=3, the trench width was kept constant (1μm) to avoid complication of the fabrication process. As a result, the trench extrusion D_T-D_P for k=3 was effectively reduced by the trench rounding at the bottom. N-buffer and p-collector doping profiles before and after the laser activation measured by secondary ion mass spectroscopy (SIMS) are shown in Fig.7. Boron concentration becomes almost flat since the surface region is molten. Ohmic contacts with sufficiently low contact resistance were obtained with low p+ concentration less than 1×10^{18}cm^{-3}.

III. DEVICE CHARACTERISTICS

Fig.8 shows collector current (I_{ce}) vs gate voltage (V_{ge}) characteristics of the fabricated k=1 and 3 IGBTs. Threshold voltage for k=3 was adjusted to be 1/3 of that for k=1 by tuning the p-base dose and diffusion. Subthreshold slope and transconductance for k=3 are improved than k=1 as expected. Fig.9 compares I_{ce} vs V_{ce} characteristics between k=1 and 3, keeping the backside doping the same. Despite the reduced gate voltage by 1/3, the V_{cesat} (defined as V_{ce} at J_{ce}=200A/cm^2) improvement from k=1 to 3 of 0.10V is obtained. Fig.10 shows forward blocking characteristics of both k=1 and 3 IGBTs. Both k=1 and 3 devices withstood 1.2kV; no significant change in I-V characteristics was observed.

The smaller V_{cesat} improvement than expected (0.2V, Fig.3) suggests that the IE effect for k=3 was weak. This could be attributed to the trench rounding for k=3 mentioned above (Fig.6). By extending a compact model proposed in [11], on-state carrier concentration n under flat carrier (i.e. equal top and bottom concentration) condition is approximated as

$$n \propto \frac{V_g}{St_{ox}} \left(\sqrt{1 + \frac{Wt_{ox}(D_T - D_P)}{aV_g}} - 1 \right), \quad (1)$$

where a is a constant not relevant to structural dimensions, and the other symbols are given in Table.1. Equation (1) suggests that the IE effect improvement by scaling S was excessively offset by the effectively reduced D_T-D_P. Figs.11 and 12 show

the characteristics of the k=3' device. Additional 0.1V improvement by simply extending the trench depth supports this assumption. Another possible reason would be excessive p-float connecting, since hole sinking by the p-float will weaken the IE effect.

Switching characteristics of the fabricated IGBTs were measured using the circuit in Fig.13. Fig.14 shows typical switching waveforms in the condition of 400V-140A/cm^2. Normal characteristics were obtained for both k=1 (V_{ge}=±15V) and k=3 (V_{ge}=±5V). No anomaly is found even if the gate is driven by 5V. Fig.15 compares switching waveforms by overlaying different back side doping cases, as well as k=1 and 3 for (a) V_{ce}=400V and (b) 600V. Higher back side doping results in longer switching time, owing to the increased charge accumulation, as was shown in Fig.4. In contrast, the difference between k=1 and k=3 devices is small. This suggests the amounts of accumulated charge in k=1 and 3 devices are similar, which is consistent with the earlier discussion.

Generally, as more charge is accumulated in the n-base, V_{ce} is improved, but turn-off loss (E_{off}) is degraded. Therefore, for fair comparison of IGBT performance, V_{ce} vs E_{off} tradeoff must be examined. Fig.16 shows tradeoff curves for both k=1 and k=3. It was found that k=1 and k=3 devices do not fall on the same tradeoff, but the scaled k=3 devices are noticeably improved. More than 30% reduction of E_{off} was achieved while keeping the same V_{cesat}. That is, clear improvement of performance by the 3D scaling with 5V gate drive was confirmed. The improvement can be primarily attributed to the fact that V_{cesat} for k=3 is improved by 0.1V than k=1 (Fig.9). Further improvement would be expected by optimizing the connected p-float design and trench depth control to enhance the IE effect.

IV. CONCLUSIONS

1.2kV-10A class scaled IGBTs for 5V gate drivers were fabricated, and their switching operations were demonstrated. It was found that the scaled k=3 devices show over 30% reduction of turn-off loss for the same on-state voltage. The results show that the 3D scaling concept is effective for further improving the performance of Si-IGBTs.

ACKNOWLEDGMENT

This work is based on results obtained from a project commissioned by the New Energy and Industrial Technology Development Organization (NEDO).

REFERENCES

[1] N. Iwamuro, and T. Laska, IEEE T-ED, 64, p.741, 2017.
[2] A. Kopta, IEEE T-ED, 64, p.753, 2017.
[3] A. Q. Huang, Proc. IEEE, 105, p. 2019, 2017.
[4] M. Kitagawa et al., IEDM Tech. Dig., p.679, 1993.
[5] T. Laska et al., Proc. ISPSD, pp.355-358, 2000.
[6] H. Feng et al., ISPSD, p. 203, 2016.
[7] K. Eikyu et al., ISPSD, p. 211, 2016.
[8] M. Tanaka et al., Solid-State Electron., 80, p.118, 2013.
[9] K. Miyazaki et al., IEEE Trans. Industry Applications, p.2350, 2017.
[10] K. Kakushima et al., IEDM, p.268, 2016.
[11] M. Tanaka et al., Microelectronics Reliability, 51, p.1933, 2011.

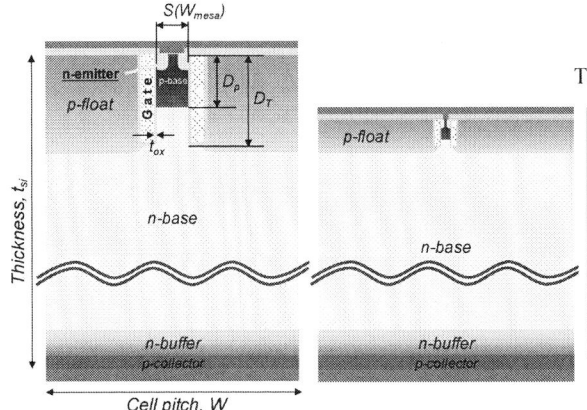

(a) k=1   (b) k=3

Fig. 1. Scaling principle of trench gate IGBTs.

Table 1. Structural parameters of scaled IGBTs.

	k=1	k=3
Cell pitch, W	1	1
Mesa width, S	1	1/3
Trench depth, D_T	1	1/3
p-base depth, D_P	1	1/3
Gate oxide thickness, t_{ox}	1	1/3
Gate voltage, V_g	1	1/3

- 1st oxidation
- Guard ring formation
- 2nd oxidation
- p-float
- Trench formation
- Gate oxidation
- p-base
- p+-contact
- n-emitter
- Passivation
- Contact formation
- Front side metallization
- Polyimide
- Wafer thinning
- Field stop
- p-collector
- Back side metallization

Fig. 2. Fabrication process flow used for fully integrated IGBTs.

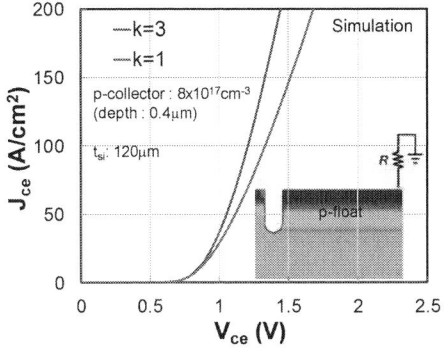

Fig. 3. Simulated on-state characteristics of k=1 and k=3 trench gate IGBTs.

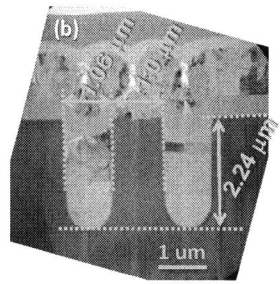

Fig. 6. Cross sectional TEM images at the trench gate of IGBT device. (a) k=1 IGBT with 6μm trench depth and 3μm mesa width. (b) k=3 IGBT with 2μm trench depth and 1μm mesa width.

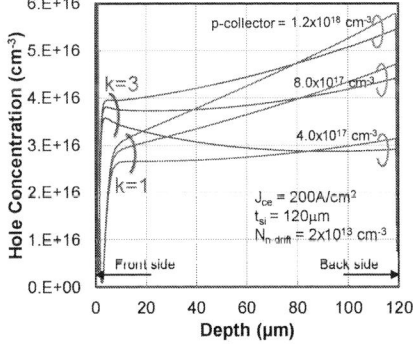

Fig. 4. Simulated carrier distribution on k=1 and k=3 IGBTs. Electron and hole are same amount in n-base region.

Table 2. Thermal budget for k=1 and k=3 full IGBT processes

Thermal process	k=1	k=3
1st oxidation	1100°C, 180 min, 1 μm	←
Guard ring	--	1100°C, 1300 min, 6 μm
2nd oxidation	1050°C, 48 min, 380 nm	←
p-float	1100°C, 1000 min, 6 μm	1100°C, 20 min, 2 μm
Sacrificial oxidation for trench	1000°C, 13 min, 100 nm	←
Gate oxidation	1000°C, 47 min, 100 nm	1050°C, 16 min, 33 nm
Poly oxidation	1000°C, 20 min, 50 nm	←
p base	1100°C, 240 min, 3 μm	1050°C, 50 min, 1 μm
Activation	950°C, 10 min	←
Field stop/ p-collector	1.8 J cm⁻² (Laser anneal)	←

Table 3. Lifetime measurement by QSSPC method for k=1 and k=3 test wafer with same thermal budget. Measurement has errors of ± 1μs.

	k=1	k=3
Lifetime	15 μs	24 μs

Table 4. Structural parameters of scaled IGBTs.

	A (k=1)	B (k=3')	C (k=3)
W (μm)	16	16	16
S (μm)	3	1	1
D_T (μm)	6	6	2
L_g (μm)	3	1	1
t_{ox} (nm)	100	33	33
V_g (V)	15	5	5

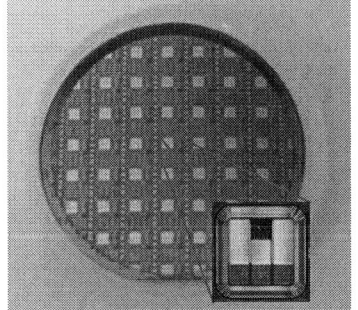

Fig. 5. Chip photograph of IGBT wafer fabricated in facilities of the Univ. of Tokyo.

978-1-7281-1988-5/18 $31.00 © 2018 IEEE

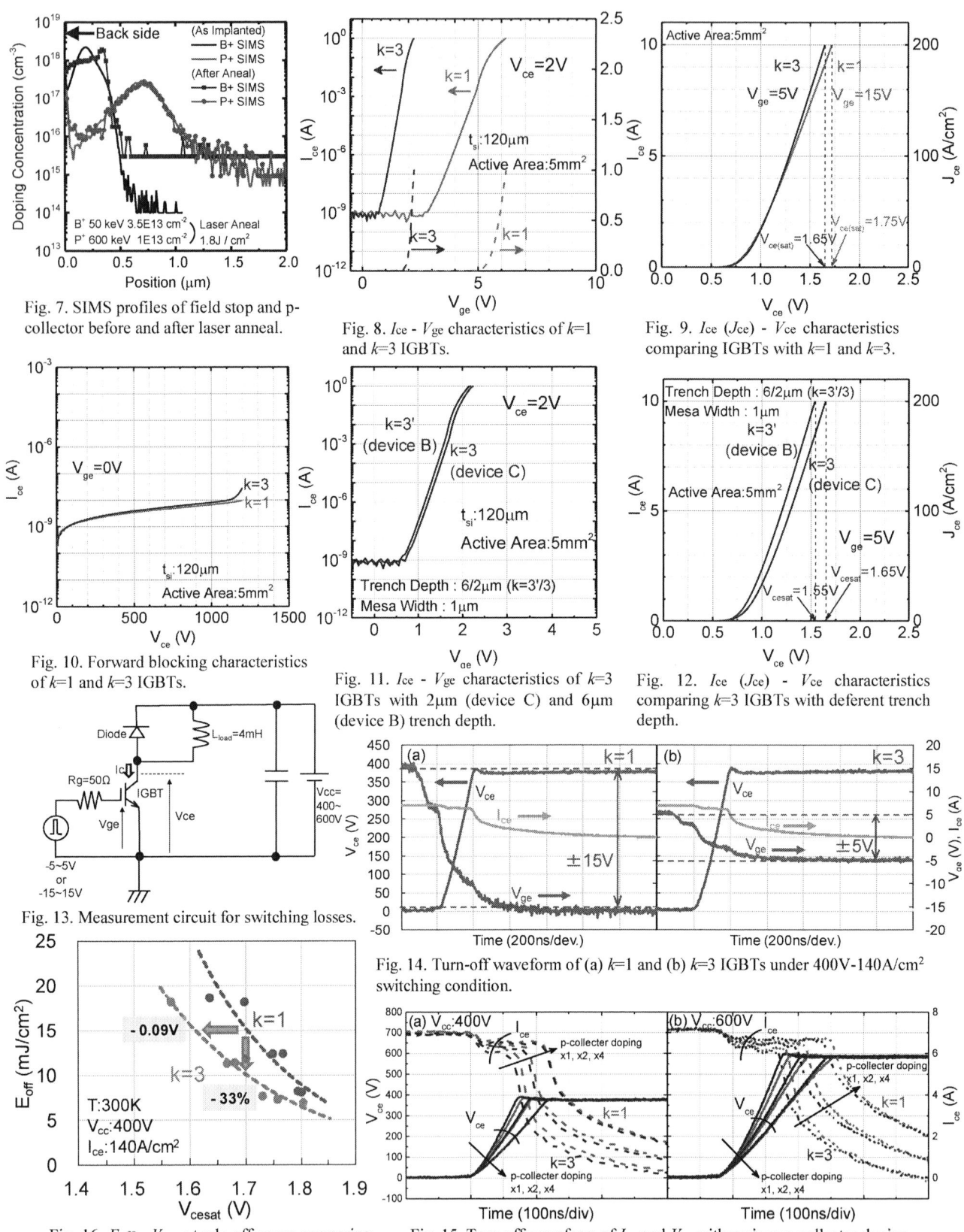

Fig. 7. SIMS profiles of field stop and p-collector before and after laser anneal.

Fig. 8. I_{ce} - V_{ge} characteristics of $k=1$ and $k=3$ IGBTs.

Fig. 9. I_{ce} (J_{ce}) - V_{ce} characteristics comparing IGBTs with $k=1$ and $k=3$.

Fig. 10. Forward blocking characteristics of $k=1$ and $k=3$ IGBTs.

Fig. 11. I_{ce} - V_{ge} characteristics of $k=3$ IGBTs with 2μm (device C) and 6μm (device B) trench depth.

Fig. 12. I_{ce} (J_{ce}) - V_{ce} characteristics comparing $k=3$ IGBTs with deferent trench depth.

Fig. 13. Measurement circuit for switching losses.

Fig. 14. Turn-off waveform of (a) $k=1$ and (b) $k=3$ IGBTs under 400V-140A/cm^2 switching condition.

Fig. 16. E_{off} - V_{cesat} trade-off curve comparing $k=1$ and $k=3$ IGBTs.

Fig. 15. Turn-off waveform of I_{ce} and V_{ce} with various p-collector doping.

978-1-7281-1988-5/18 $31.00 © 2018 IEEE   192

2.44 kV Ga$_2$O$_3$ vertical trench Schottky barrier diodes with very low reverse leakage current

Wenshen Li[1], Zongyang Hu[1], Kazuki Nomoto[1], Riena Jinno[1,2], Zexuan Zhang[1], Thieu Quang Tu[3], Kohei Sasaki[3], Akito Kuramata[3], Debdeep Jena[1,4,5] and Huili Grace Xing[1,4,5]

[1]School of Electrical and Computer Engineering, Cornell University, Ithaca, NY 14853, USA, email: wl552@cornell.edu
[2]Department of Electronic Science and Engineering, Kyoto University, Kyoto 615-8510, Japan
[3]Novel Crystal Technology, Inc., Sayama 350-1328, Japan
[4]Department of Material Science and Engineering, Cornell University, Ithaca, NY 14853, USA
[5]Kavli Institute at Cornell for Nanoscale Science, Ithaca, NY 14853, USA

Abstract—High-performance β-Ga$_2$O$_3$ vertical trench Schottky barrier diodes (SBDs) are demonstrated on bulk Ga$_2$O$_3$ substrates with a halide vapor phase epitaxial layer. A breakdown voltage (BV) of 2.44 kV, Baliga's figure-of-merit (BV^2/R$_{on}$) of 0.39 GW/cm^2 from DC measurements and 0.45 GW/cm^2 from pulsed measurements are achieved, all of which are the highest among β-Ga$_2$O$_3$-based power devices. A lowest reverse leakage current density below 1 μA/cm^2 until breakdown is observed on devices with a fin width of 1-2 μm, thanks to the reduced surface field (RESURF) effect provided by the trench SBD structure. The specific on-resistance is found to reduce with increasing area ratio of the fin-channels following a simple relationship. The reverse leakage current agrees well with simulated results considering the barrier tunneling and barrier height lowering effects. The breakdown of the devices is identified to happen at the trench bottom corner, where a maximum electric field over 5 MV/cm could be sustained. This work marks a significant step toward reaching the promise of a high figure-of-merit in β-Ga$_2$O$_3$.

I. INTRODUCTION

β-Ga$_2$O$_3$ has seen increasing research effort in recent years towards the realization of high-performance electronic devices. A Baliga's figure of merit (FOM) of β-Ga$_2$O$_3$ projected to well exceed that of GaN and 4H-SiC arises from a combination of an ultra-wide bandgap of ~4.5 eV, therefore a high expected critical electric field of up to 8 MV/cm [1] and a decent room temperature electron mobility of ~200 cm^2/V·s [2, 3]. These properties make β-Ga$_2$O$_3$ an excellent material candidate for next-generation power electronic devices, especially under harsh environment. The availability of melt-growth methods for single-crystal bulk substrate provides added advantage towards lower cost and a head start for fast development of epitaxial growth and device technologies. With the availability of high quality halide vapor phase epitaxial (HVPE) layers with a net doping concentration around ~2×10^{16} cm^{-3}, β-Ga$_2$O$_3$ power electronic devices has well-surpassed the unipolar material limit of Si, exemplified by the demonstration of kilovolt-class enhancement-mode transistors [4] and Schottky barrier diodes (SBDs) [5, 6]. However, the FOM of the present devices are still far from the projected material limit. In this

work, we demonstrate record high-performance vertical β-Ga$_2$O$_3$ SBDs through (i) effectively reduced leakage current with the reduced surface field (RESURF) effect offered by a trench MIS-type SBD structure [7-9] and (ii) smoothly-etched trench profile. A record high breakdown voltage (BV) of 2444 V, and a record high FOM (BV^2/R$_{on}$) are achieved without other dedicated field management techniques. The diodes show near ideal behavior in both forward and reverse operations well predicted by physics-based models.

II. DEVICE DESIGN AND FABRICATION

Fig. 1 shows the schematic cross-section of the Ga$_2$O$_3$ trench SBDs. The drift layer consists a 10-μm HVPE n$^-$-Ga$_2$O$_3$ grown on a (001) n-type Ga$_2$O$_3$ substrate. Majority of the fin widths (W$_{fin}$) are 1-4 μm and the trench depth is 1.55 μm. The fin area ratio is defined to be the fin width over the pitch size.

Fig. 2 illustrates the fabrication process of the trench SBDs. First, the trench was dry etched using a BCl$_3$/Ar gas mixture and Ti/Pt as the hard mask [10]. After the dry etch and mask removal, wet acid treatment was performed to remove the dry etch induced damage. A rounded trench corner profile is realized as a result of the dry and wet etching as shown in the scanning electron microscope (SEM) cross-section image (**Fig. 4**), which is desired to reduce field crowding. The back ohmic contact was formed by a Ti/Au deposition and a rapid thermal anneal for 1 min in N$_2$. Then, a 100-nm Al$_2$O$_3$ dielectric was deposited by atomic layer deposition (ALD) and opened by dry etching for the Schottky contact on top of the fins. Finally, Ni Schottky contact was deposited by e-beam evaporation, followed by Ti/Pt sputtering for the sidewall metal coverage. **Fig. 3** shows an optical graph of the top view of a fabricated device with a fin length of 150 μm and W$_{fin}$ of 2 μm. *The entire central anode area within the dashed lines is used for the calculation of current density throughout the discussion.*

III. RESULTS AND DISCUSSION

Capacitance-voltage measurements were performed on the regular SBDs co-fabricated on the same sample. **Fig. 5** plots the extracted net doping concentration profile from the measurements on both the SBD made on the original epitaxial surface as well as the SBD made on the etched planar surface formed by the trench-etch step. The doping concentration

978-1-7281-1988-5/18 $31.00 © 2018 IEEE

increases from ~1×10^{16} cm^{-3} near the epi-surface to ~2×10^{16} cm^{-3} beyond a depth of ~2.5 µm. The $1/C^2$ plot is shown in **Fig. 6**. A built-in potential (V_{bi}) of 1.25±0.1 V is extracted, corresponding to a Schottky barrier height ($q\phi_B$) of 1.4±0.1 eV.

Fig. 7 shows the forward I-V characteristics of the trench SBDs in comparison with the regular SBD measured with DC scans. The 2-4 µm trench SBDs shown in the figure all have a fin area ratio of 50%, while the 1-µm device has an area ratio of 33%. The trench SBDs and the regular SBD have a similar turn-on voltage of 1.25 V. From a fitting to the thermionic emission model, $q\phi_B$ is extracted to be 1.35 eV, which agrees with the extraction from C-V measurements. The differential R_{on} of the 2-4 µm fin devices are similar, being around 11.3 mΩ·cm^2 and higher than that of the regular SBD (7 mΩ·cm^2).

Fig. 8 shows the pulsed I-V measurements of the forward characteristics. In comparison with the DC measurements. The slightly lower turn-on voltage measured under the pulsed condition is believed to be related to the trapping effect that is more severe during the DC scan, while the higher current at >3 V is attributed to the mitigation of device self-heating under the pulsed measurement condition. Thus, we believe the pulsed I-Vs provide more accurate measurements of the R_{on} determined by the intrinsic conduction properties of the drift layer.

As shown in **Fig. 9**, the R_{on} values extracted using the pulsed measurements are compared among all types of SBDs. It is found that the SBD made on the etched surface with an 8.55-µm drift layer has the lowest R_{on} of 4 mΩ·cm^2, while the SBD with a 10-µm drift layer has an R_{on} of 7 mΩ·cm^2. The difference is due to the top 1.55 µm drift layer, which has a net doping concentration of ~1×10^{16} cm^{-3}. The trench SBDs have a R_{on} of 10 mΩ·cm^2, which matches well with the expected value considering the 50% area ratio leads to a doubled contribution from the 1.55 µm top layer to the specific R_{on}. A simple model for the specific R_{on} of the trench SBDs can be thus developed:

$$R_{on,sp} = 4 + \frac{3}{\text{area ratio}} \text{ mΩ} \cdot \text{cm}^2 \qquad (1)$$

Fig. 10 shows the statistics of the extracted R_{on} for the trench SBDs with different W_{fin} and area ratio. The R_{on} of the 2-8 µm devices follows the trend expressed by (1) very well. For the 1-µm devices, (1) has to be modified reflecting a reduced effective fin width for best fitting, which indicates that there may be some etching damage and/or trapping effect at the fin sidewall, which is most dominant for the 1-µm fin devices.

Fig. 11 plots the simulated electric field profile along vertical cut-lines at the fin center (dash line in Fig. 1). The field near the surface is effectively reduced by the trench structure. The RESURF effect is more prominent for smaller fin width.

Fig. 12 shows the representative reverse I-V characteristics of the trench SBDs in comparison with the regular SBDs. The reverse leakage current of the trench SBDs is much lower than the regular SBDs, and the BV is much higher, reaching a record value of 2.44 kV in the 1-µm devices. The reverse leakage current is lower than 1 mA/cm^2 before breakdown for the 1-3

µm devices, a typical value used to specify the reverse blocking voltage for commercial SBDs. Even lower leakage current beyond the detection limit is observed in some of the devices, as shown in **Fig. 13**. The leakage current of the lower leakage devices follows the simulated reverse I-V characteristics considering the barrier tunneling and barrier-height lowering effect. An electron effective mass of 0.3 m_0 [3] and the extracted barrier height of 1.4 eV is used in the simulation.

The statistics of the BV is shown in **Fig. 14**. The BV is found to increase with decreasing fin width. No correlation between the BV and the area ratio is observed. To identify the breakdown mechanism, the electric field profile is simulated. **Fig. 15** and **Fig. 16** shows the simulated electric field profile along a horizontal cutline across the trench bottom corner, at a fixed voltage of 2 kV and around the highest breakdown voltage, respectively. It is observed that the field crowding happens around the trench corner and the field peak increases with the fin width at a fixed voltage. As shown in **Fig.16**, the field peaks near a similar value of ~5.9 MV/cm at the highest BV for each fin width, indicating that the breakdown happens at the trench corner. Note that the 5.9 MV/cm value is simulated considering a trench corner angle of 90°. In reality the value should be lower due to the rounded trench corner profile.

IV. CONCLUSION

Exploiting pronounced RESURF effects in the trench SBD structure and the smoothly etched trench corner profile, the reverse leakage current is effectively suppressed in the Ga$_2$O$_3$ trench SBDs and high breakdown voltages are achieved. **Fig. 17** benchmarks the state-of-the-art β-Ga$_2$O$_3$ SBDs [5, 6, 8, 11-13]. Our 1-µm device achieved a record high BV and our best 2-µm device with a 50% area ratio achieved a record FOM of 0.39 GW/cm^2 (2096 V, 11.3 mΩ·cm^2) from DC scan and 0.45 GW/cm^2 (2096 V, 9.8 mΩ·cm^2) from pulsed measurements. The R_{on} can be further reduced by increasing the fin area ratio. The improved BV and FOM from the current state-of-the-art further unveils the excellent material properties of β-Ga$_2$O$_3$, hence an attractive platform for power electronic devices.

ACKNOWLEDGMENT

Supported in part by NSF DMREF 1534303 and AFOSR (FA9550-17-1-0048), carried out at CNF sponsored by the NSF NNCI program (ECCS-1542081), and CCMR Shared Facilities supported through the NSF MRSEC program (DMR-1719875).

REFERENCES

[1] M. Higashiwaki *et al.*, *Appl. Phys. Lett.*, vol. 100, no. 1, p. 013514, 2012.
[2] N. Ma *et al.*, *Appl. Phys. Lett.*, vol. 109, no. 21, p. 212101, 2016.
[3] Y. Zhang *et al.*, *Appl. Phys. Lett.*, vol. 112, no. 17, p. 173502, 2018.
[4] Z. Hu *et al.*, *IEEE EDL*, vol. 39, no. 6, pp. 869-872, 2018.
[5] K. Konishi *et al.*, *Appl. Phys. Lett.*, vol. 110, no. 10, p. 103506, 2017.
[6] J. Yang *et al.*, *Appl. Phys. Lett.*, vol. 110, no. 19, p. 192101, 2017.
[7] M. Mehrotra *et al.*, in *IEDM Tech. Dig.*, 1993, pp. 675-678.
[8] K. Sasaki *et al.*, *IEEE EDL*, vol. 38, no. 6, pp. 783-785, 2017.
[9] W. Li *et al.*, *Proc. 76th DRC*, pp. 289-290, June 2018.
[10] L. Zhang *et al.*, *Jap. J. Appl. Phys.*, vol. 56, no. 3, p. 030304, 2017.
[11] J. Yang *et al.*, *IEEE EDL*, vol. 38, no. 7, pp. 906-909, 2017.
[12] Z. Hu *et al.*, *IEEE JEDS*, vol. 6, pp. 815-820, 2018.
[13] J. Yang *et al.*, *Proc. 76th DRC*, pp. 291-292, June 2018.

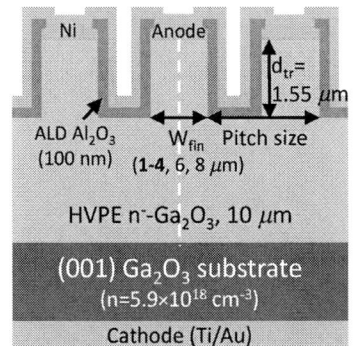

Fig. 1. Schematic cross-section of the Ga$_2$O$_3$ trench Schottky barrier diodes. Fin widths (W$_{fin}$) of 1-4, 6, 8 μm are designed with different fin area ratio (W$_{fin}$/pitch size). The trench depth (d$_{tr}$) is 1.55 μm.

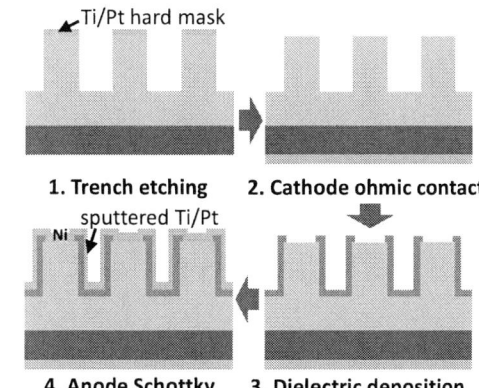

Fig. 2. Fabrication process flow of the trench SBDs.

Fig. 3. Optical top view image of a fabricated device with a W$_{fin}$ of 2 μm and a fin area ratio of 50%. The length of the fins is 150 μm. *Central anode area within the dashed lines is used for current density calculation.*

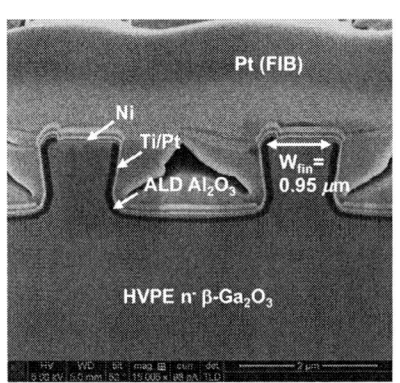

Fig. 4. Scanning electron microscopy (SEM) cross-section image of a device with a designed fin width of 1 μm. A slightly inward-slanted sidewall profile and rounded trench corners are observed, preferable for improved RESURF effect and reduced field crowding.

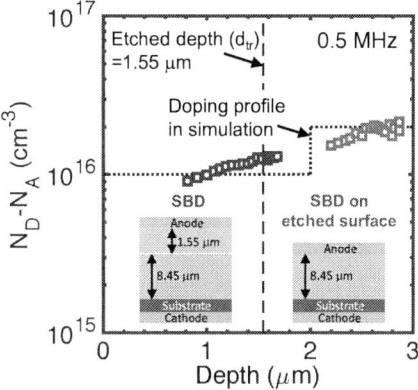

Fig. 5. Extracted net doping concentration (N$_D$-N$_A$) from C-V measurements. Two types of regular Schottky barrier diode are used for the measurements with the cross-sections shown in the insets. Dotted line shows the approximated doping profile used in the TCAD simulation.

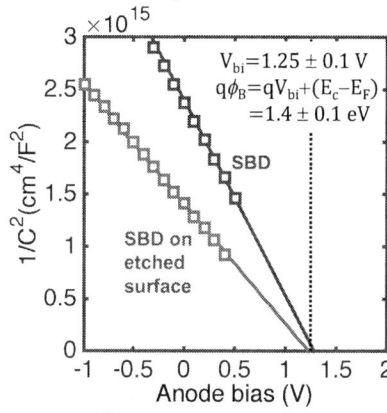

Fig. 6. 1/C^2 plot for the SBD and the SBD on the etched surface. The V$_{bi}$ is extracted to be ~1.25 V, corresponding to a barrier height (qϕ_B) of ~1.4 eV.

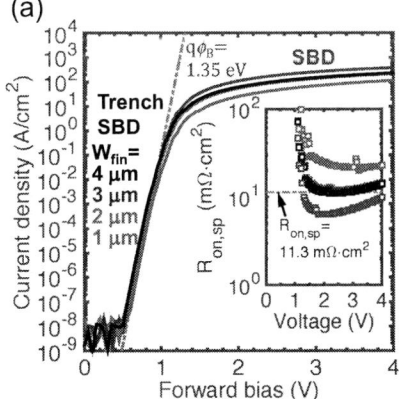

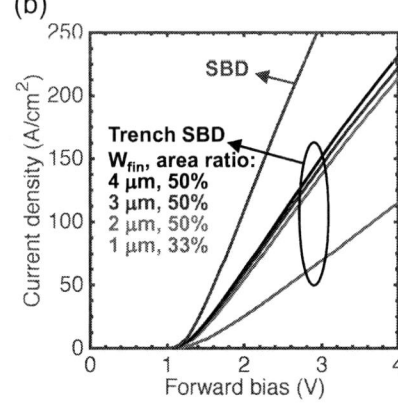

Fig. 7. Forward I-V characteristics (a) in log scale and (b) in linear scale of the trench SBDs in comparison with the regular SBD, measured by DC scans. *The current density of the trench SBDs is normalized by the central anode area (see Fig. 3).* A barrier height (qϕ_B) of 1.35 eV is extracted from the thermionic emission model using a reduced effective Richardson constant A** of 33.1 A/cm^2·K^2 [5], which agrees with the extracted barrier height by C-V.

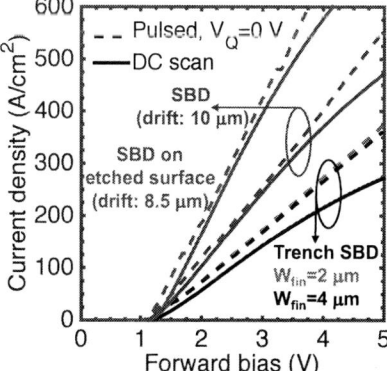

Fig. 8. Comparison of the forward I-V characteristics under pulsed condition versus DC. A pulse width of 8.4 μs and a duty cycle of 0.84% is used. The observed difference is likely due to a combination of device self-heating and trapping effect.

978-1-7281-1988-5/18 $31.00 © 2018 IEEE

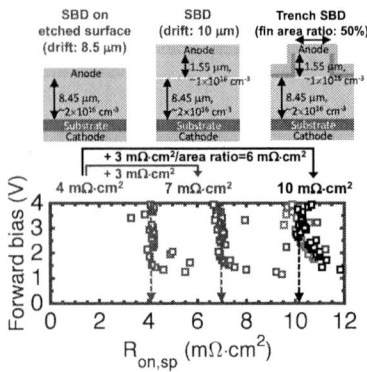

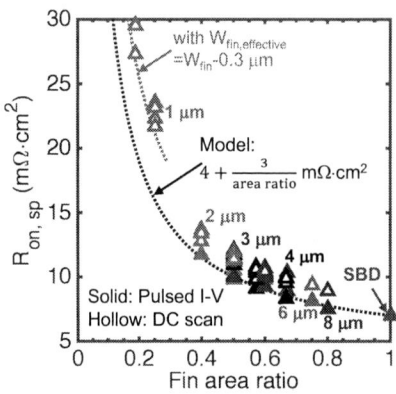

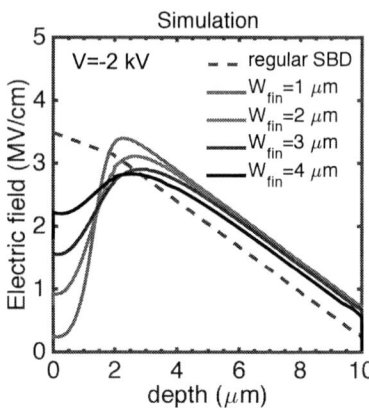

Fig. 9. Extracted specific differential on-resistance ($R_{on,sp}$) of the devices from pulsed I-V measurements. $R_{on,\,sp}$ of the trench SBDs is well-explained by considering the 50% fin area ratio.

Fig. 10. Statistics of the extracted specific R_{on} of the trench SBDs. The trend of R_{on} agrees well with a simple model considering the area ratio for the devices with 2-8 μm fin widths, while a modified model with a reduced effective W_{fin} is found to match the 1-μm device data.

Fig. 11. Simulated electric field profile along vertical cut-lines at the fin center (see the dash line in Fig. 1) at a reverse bias of 2 kV. The surface field is reduced effectively with the trench-MIS structure.

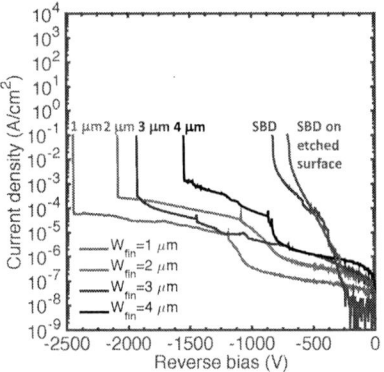

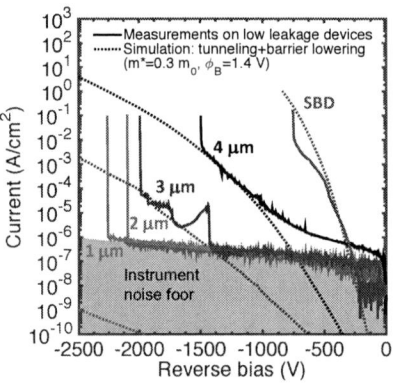

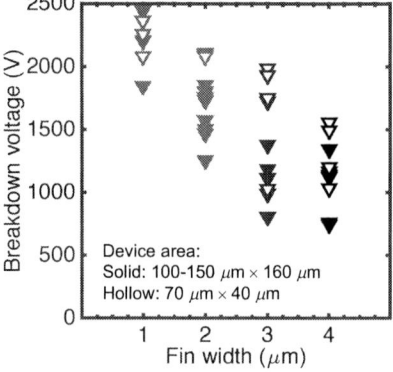

Fig. 12. Representative reverse I-V characteristics of the trench SBDs in comparison with the regular SBDs. Much lower leakage current below 1 mA/cm² is observed for the 1-3 μm devices in comparison with the regular SBDs, together with much higher breakdown voltages.

Fig. 13. Representative I-V characteristics of the trench SBDs with low leakage current. The leakage current profile agrees well with the simulation considering barrier tunneling and barrier height lowering. Effective mass of 0.3 m_0 and the extracted $q\phi_B$ of 1.4 eV is used in the TCAD Sentaurus simulation.

Fig. 14. Statistics of the measured breakdown voltage of the trench SBDs. The BV is found to increase with the reduction of the fin width.

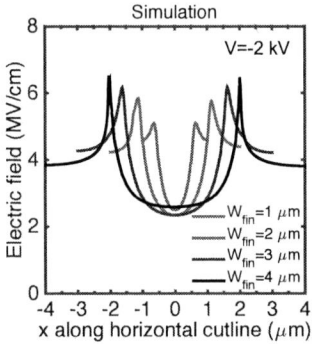

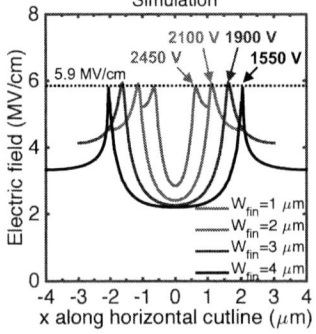

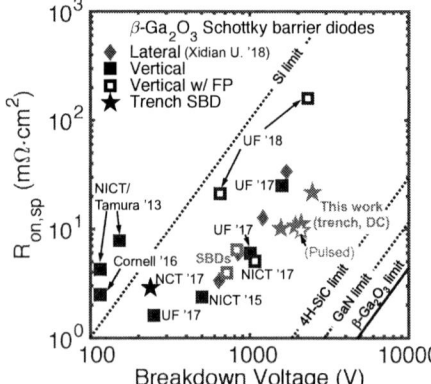

Fig. 15. Simulated electric field profile along a horizontal cutline across the trench bottom corner at 2 kV. The electric field at the trench corner is found to increase with increasing fin width.

Fig. 16. Simulated electric field profile along a horizontal cutline across the trench bottom *around the highest BV of each fin width*. The field peak is about the same for all fin widths, indicating the device breakdown is limited by the breakdown at the trench corner.

Fig. 17. Benchmark plot of β-Ga₂O₃ Schottky barrier diodes [5, 6, 8, 11-13]. Our 1-μm trench SBD achieves the highest BV, while the 2-μm trench SBD achieves the highest FOM, even without dedicated field management techniques.

978-1-7281-1988-5/18 $31.00 © 2018 IEEE

Multidomain Dynamics of Ferroelectric Polarization and its Coherency-Breaking in Negative Capacitance Field-Effect Transistors

Hiroyuki Ota[1], Tsutomu Ikegami[1], Koichi Fukuda[1], Junichi Hattori[1], Hidehiro Asai[1], Kazuhiko Endo[1]
Shinji Migita[1] and Akira Toriumi[2]

[1]National Institute of Advanced Industrial Science and Technology (AIST), Ibaraki, Japan, email: hi-ota@aist.go.jp
[2]Graduate School of Engineering, The University of Tokyo, Tokyo, Japan

Abstract—In this paper, for the first time, we clarified the multidomain dynamics of ferroelectric polarization in the Negative Capacitance Field-Effect Transistors (NCFETs) by an in-house Technology Computer-Aided Design (TCAD) module. It enables self-consistent simulations among the time-dependent Landau-Khalatnikov equation and the other equations that govern operation of FETs. Our simulation reveals that domain-wall thickness T_{dw}, which reflects a correlation strength between domains predominates coherency of the NC polarization. In a strong correlation (large T_{dw}) case, a coherent NC polarization is realized at least to a degree of T_{dw}. On the other hand, in a weak correlation (small T_{dw}) case, perturbation from peripherals easily leads to incoherence of polarization, where a uniform polarization is split into multiple spontaneous polarization domains. This coherency breaking found to give rise to deteriorated voltage amplification of NC. Design methodology to maintain the NC coherency is also proposed.

I. Introduction

The Internet-of-Thing (IoT) is expected to be a smart and highly efficient platform of every industry. To realize IoT network, ultralow energy consumption LSIs are a prerequisite. From this viewpoint, novel FETs with steep subthreshold swing (SS) less than 60 mV/decade - the physical minimum in conventional MOSFETs- have been intensively developed in the last decade. Among steep SS devices studied, the negative capacitance field-effect transistors (NCFETs) [1], which integrate the ferroelectric layer as the gate insulator, have come into the limelight. In the NCFETs, possibility of a steep SS by harnessing the NC states was predicted theoretically [1]. So far, although a number of excellent demonstrations of observing NC [2, 3] and of NCFETs [4, 5] have been reported, some papers casted doubt on interpretation of device characteristics [6-8]. To understand device physics involving elusive NC, TCAD simulation based on the accurate physics of the ferroelectricity is desired.

In general, multidomain dynamics of the ferroelectric polarization is described by the time-dependent Landau-Khalatnikov (LK) equation (1)[9].

$$\rho \frac{dP}{dt} = -\nabla_P G \tag{1}$$

$$G = \alpha P^2 + \beta P^4 + \gamma P^6 + \frac{1}{2}\delta(\nabla P)^2 - EP \tag{2}$$

where P, G and E are the polarization, the Gibbs free-energy, and the electric-field, respectively. α, β, γ, δ and ρ are the ferroelectric material parameters. Here, δ is a correlation strength. In the almost all TCAD simulations of NCFETs so far [10-12], the steady state, in which $d/dt = 0$ is assumed in (1), and the single domain ferroelectricity ($\delta = 0$) is widely assumed to simplify the modeling of NC. In reality, however, the ferroelectric polarization is inhomogeneous due to formation of multidomain (**Fig. 1**). Therefore, the TCAD simulation incorporating physics of the multidomain ferroelectricity is mandatory to predict performance of NCFETs. Moreover, as clearly stated in the previous paper [13], we found that the steady state LK equation cannot describe instability of NC, because it assumes that NC is always stable. Although a few papers have reported characteristics of NCFETs based on the physics of multidomain ferroelectricity [14], they assumed the steady state.

In this paper, we reveal the transient characteristics of NCFETs based on TCAD simulation, in which all equations describing NCFETs are solved in a coupled manner. An impact of coherency breaking on NCFET performance is described as well as inability of hysteresis-free operation in NCFETs due to instability of NC.

II. Modeling

Dynamical model of the ferroelectricity in our TCAD simulation is presented in **Fig. 2**. In describing the multidomain ferroelectricity, the correlation strength δ plays an important role, which involves in the domain wall thickness T_{dw} (**Figs. 2 and 3**) [15, 16]. Although T_{dw} is a static property at equilibrium, it also governs the dynamics of multidomain structure formation. In this model, ferroelectric polarization along z-axis, and dielectric ones along x- and y-axes are assumed. Numerical calculation was performed by in-house TCAD named "Impulse TCAD" [17], in which the time dependent LK equation, the Poisson equation, and the other device physics are self-consistently solved at each mesh point (**Fig. 4**).

III. Results and Discussionn

A. Coherency in the NC Polarization in MFIM Capacitors

Figure 5 shows the free-energy of a metal (M) / ferroelectric (F) /insulator (I)/M (MFIM) capacitor. Here, we assume the thickness and the remnant polarization P_r of the F layer to be 10 nm and 20 $\mu C/cm^2$, respectively. As seen in **Fig. 5(a)**, NC is stable if the thickness of I (IL) is 0.8 nm, but it becomes unstable in the case of IL=0.2 nm. The dynamic aspect of the stability of NC state is not such simple. To reveal this point, we simulated evolution of NC at 0 V under initial conditions of inhomogeneous or homogeneous polarization [**Fig. 5(b)**]. When δ is small and NC is unstable [**Fig.6(a)**], a coherent initial state evolves into a domain structure, as expected from **Fig. 5(a)**. In a NC stable case [**Fig.6(b)**], however, the coherent NC state is not achieved if we start from an incoherent initial state: recovery of the coherency is stopped in the halfway. On the contrary, when δ is large, a coherent NC state is quickly achieved irrespective of the stability of the NC state [**Fig. 6(c) and 6(d)**]. These tendencies become more clear-cut in the FET structure.

B. Coherency Breaking of NC Polarization in FETs

Next, we analyzed the effects of δ on transfer (I_D-V_{GS}) characteristics of double gate NCFETs. **Figure 7** schematically illustrates the device cross section of the double gate NCFETs with the internal gate oxide (IGO) and the internal metal gate (IMG). Relatively small P_r of 3 $\mu C/cm^2$ was chosen to achieve the capacitance matching for steep SS [12, 13]. The gate length is set to 10 nm. **Figures 8(a)** and **8(b)** compare I_D-V_{GS} characteristics of the NCFETs at 10 MHz of a V_{GS}-waveform, between $\delta = 10^{-7}$ and 10^{-9} m^3/F. In both cases, considerably large hysteresis can be seen with $\rho = 10$ m/S, which is attributable to the switching delay of the polarization against the external field. Indeed, "apparent" steep SS [6] was observed in the downward I_D-V_{GS} curves. The hysteresis for $\rho = 1$ m/S is drastically reduced. That is, practically, to operate the NCFETs at a higher rate, we have to design the material with smaller ρ. **Figures 8(c)** and **8(d)** shows snapshots of the polarization contour for $\delta = 10^{-7}$ and 10^{-9} m^3/F, respectively, calculated with $\rho = 1$ m/S. In the case of $\delta = 10^{-9}$ m^3/F, the ferroelectric layers are divided into a few domains. SS and hysteresis ($\rho = 1.0$ m/S) are summarized in **Fig. 9**. In materials with narrow domain walls, large hysteresis is inevitable due to multi-domain formation within the gate stack. This means that the gate length has to be comparable to or less than T_{dw} for keeping the coherency of NC polarization.

C. Designing the Area-Retio of Ferroelectric/Dielectrics in FETs

The constraints implied in **Fig. 9** have to be considered also in designing FET structures which integrate the larger P_r ferroelectrics like HfZrO$_2$. To resolve the incoherency problem, we propose the device architecture, where an area-ratio between the gate and ferroelectric layers is tunable. The device structure is schematically shown in **Fig. 10(a)**. For the ferroelectric parameters of HfZrO$_2$, LK parameters are fitted to the experimental data [**Fig. 10(b)**]. In this simulation, $\delta = 10^{-9}$ m^3/F ($T_{dw} = 2.5$ nm) was assumed. **Figures 11(a)** and **11(b)** show the simulated transfer characteristics for the case of ferroelectric length $L_{FERRO}= 3$ nm and the polarization at the gate bias of 0.5 V, respectively. Both the hysteresis of the

transfer characteristics and the formation of multidomain are well suppressed. The dependence of hysteresis and subthreshold swing on the length of the ferroelectric are summarized in **Figs. 12**. For short L_{FERRO} cases, negative capacitance condition becomes unstable which causes large hysteresis. For long L_{FERRO} cases, negative capacitance in the gate stack is never fulfilled, and SS is larger than 60 mV/decade. To realize the steep NC-FETs, it is required to find a solution in such a trade-off window.

IV. CONCLUSIONS

In this paper, the dynamics of NCFETs was predicted and elucidated based on the realistic and self-consistent model, in which the LK equation for the multidomain ferroelectric polarization and the device equations are solved simultaneously. Our numerical simulation clearly revealed that the domain wall thickness T_{dw}, which itself is a static property, is a critical parameter to determine the dynamical behavior of NCFETs. To avoid coherency breaking of NC polarization, we have shown that the gate size comparable to or less than T_{dw} is a key and that the elaborate design of the gate stack is effective to some extent, where area-ratio between the ferroelectric and gate is tunable. However, its design window is found to be narrow, indicating that NC FETs entail the intrinsic difficulties in hysteresis-free operation and steep SS in principle.

ACKNOWLEDGMENT

This work was supported by the Core Research for Evolutional Science and Technology (CREST, JPMJCR14F2) of Japan Science and Technology Agency (JST).

REFERENCES

[1] S. Salahuddin and S. Datta, Nano Lett. **8**, 405 (2008).
[2] A. I. Khan, K. Chatterjee, B. Wang, S. Drapcho, L. You, C. Serrao, S. R. Bakaul, R. Ramesh, and S. Salahuddin, Nat. Mater. **14**, 182 (2015).
[3] M. Kobayashi, N. Ueyama, K. Jang, and T. Hiramoto, IEDM Tech. Dig. 2016, p. 314.
[4] M. H. Lee, P.-G. Chen, C. Liu, K-Y. Chu, C.-C. Cheng, M.-J. Xie, S.-N. Liu, J.-W. Lee, S.-J, Huang, M.-H. Liao, M. Tang, K.-S. Li, and M.-C. Chen, IEDM Tech. Dig., 2015, p. 616.
[5] Z. Krivokapic, U. Rana, R. Galatage, A. Razavieh, A. Aziz, J.Liu, J.Shi, H.J. Kim, R. Sporer, C. Serrao, A.Busquet, P. Polakowski, J. Müller, W. Kleemeier, A. Jacob, D. Brown, A. Knorr, R. Carter, and S. Banna, IEDM Tech. Dig., 2017, p. 357.
[6] A. Cano and D. Jimenez, Appl. Phys. Lett. **97**, 133509 (2010).
[7] B. Obradovic, Symp. on VLSI Tech. Dig. 2018, p.51.
[8] J. A. Kittl, B. Obradovic, D. Reddy, T. Rakshit, R. M. Hatcher, and M. S. Rodder, Appl. Phys. Lett. **113**, 042904 (2018).
[9] L. D. Landau and I. M. Khalatnikov, Dokl. Akad. Nauk SSSR, vol. **96**, 469, (1954).
[10] A. I. Khan, C. W. Yeung, C. Hu, and S. Salahuddin, IEDM Tech. Dig., 2011, p. 255.
[11] C. Hu, S. Salahuddin, C.-I. Lin, and A. I. Khan, Proc. 73rd Annu. Device Research Conf. (DRC), 2015, 39.
[12] H. Ota, T. Ikegami, J. Hattori, K. Fukuda, S. Migita, and A. Toriumi, IEDM Tech. Dig. 2016, p. 318.
[13] H. Ota, K. Fukuda, T. Ikegami, J. Hattori, H. Asai, S. Migita and A. Toriumi, IEDM Tech. Dig., 2017, p. 361.
[14] A. K. Saha, P. Sharma, I. Dabo, S. Datta and S. K. Gupta, IEDM Tech. Dig., 2017, p. 326.
[15] J. Hattori, T. Ikegami, K. Fukuda, H. Ota, S. Migita, and H. Asai, SISPAD 2018 (submitted).
[16] P. Marton, I. Rychetsky, and J. Hlinka, Phys. Rev. B **81**, 144125 (2010).
[17] https://unit.aist.go.jp/neri/en/ImpulseTCAD/index.html.

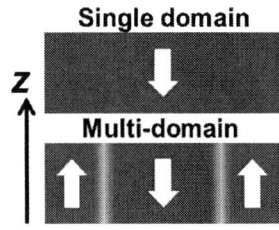

Single domain

Z

Multi-domain

Fig. 1 Schematic views of the polarization field in ferroelectrics: (top) single domain structure; (bottom) multidomain structure.

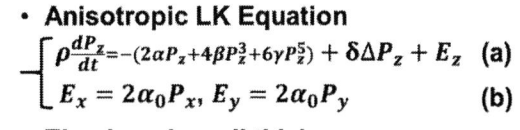

- **Anisotropic LK Equation**

$$\rho\frac{dP_z}{dt} = -(2\alpha P_z + 4\beta P_z^3 + 6\gamma P_z^5) + \delta\Delta P_z + E_z \quad \text{(a)}$$

$$E_x = 2\alpha_0 P_x, \quad E_y = 2\alpha_0 P_y \quad \text{(b)}$$

- **The domain wall thickness**

$$T_{dw} = \sqrt{\frac{2\delta}{\beta P_r^2 + 2\gamma P_r^4}} \quad \text{(c)}$$

Fig. 2 The multidomain ferroelectricity model in our simulation. Ferroelectricity along the z-axis and dielectric properties along the other axes are assumed. E_i and P_i (i=x,y,z) are the electric field and the polarization along the x, y, and z-axes. $\alpha, \beta, \gamma, \delta$ and ρ are ferroelectric material parameters. P_r is the remnant polarization. $2\alpha_0$ is the reciprocal permittivity along the x and y axes.

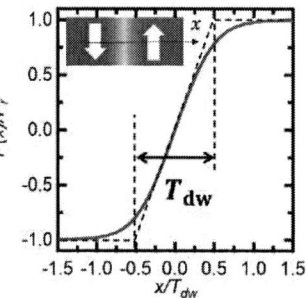

Fig. 3 Polarization field in ferroelectrics having two domains with opposite polarization vectors. A domain wall is formed at the interface of the two domains and its thickness is approximated as T_{dw} (see **Fig. 2**).

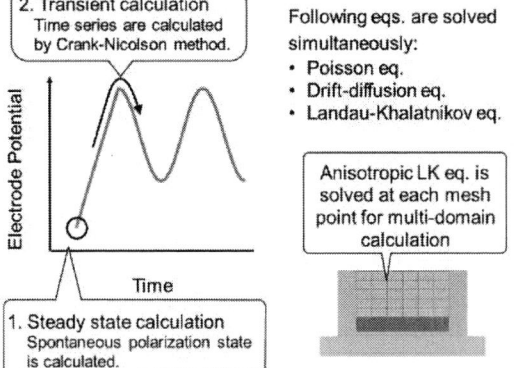

2. Transient calculation
Time series are calculated by Crank-Nicolson method.

1. Steady state calculation
Spontaneous polarization state is calculated.

Following eqs. are solved simultaneously:
- Poisson eq.
- Drift-diffusion eq.
- Landau-Khalatnikov eq.

Anisotropic LK eq. is solved at each mesh point for multi-domain calculation

Fig. 4 Diagram for the transient simulation of NCFETs. After calculating DC solution, time-evolution of NCFETs are simulated self-consistently.

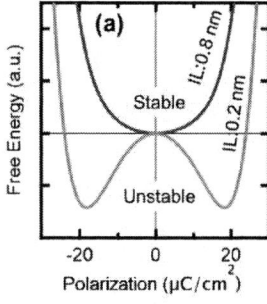

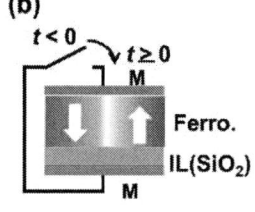

Fig. 5 (a) The simulated free energy for metal/ferroelectric (F) /SiO$_2$ (IL)/metal structure. P_r and the coercive field are assumed as 20 μC/cm^2, and 1 MV/cm. NC is unstable in the case of IL=0.2 nm while NC is stable in the case of IL=0.8 nm. The thickness of F layer is 20 nm. (b) A schematic setup of relaxation simulation in **Figs. 6**.

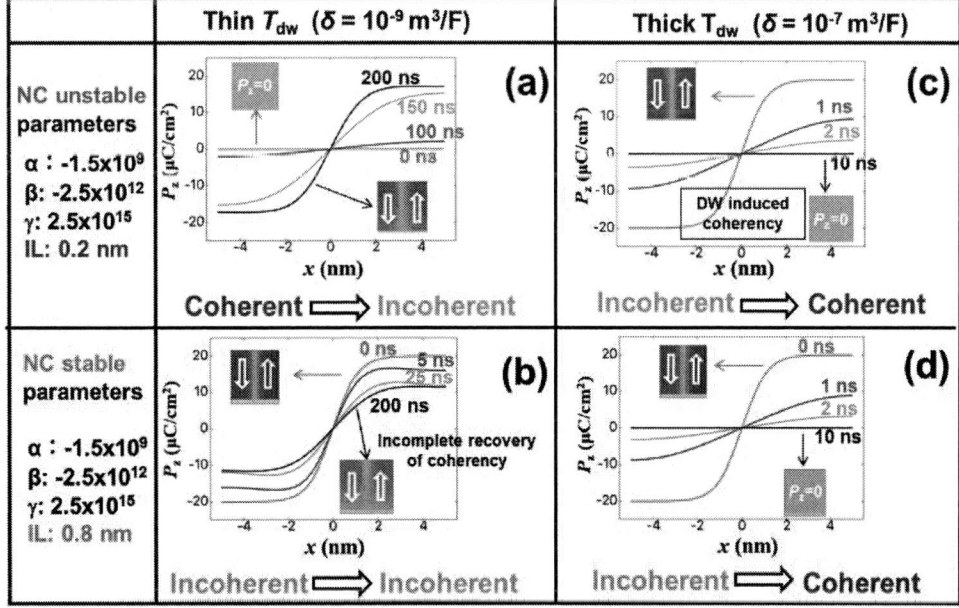

Fig. 6 Simulation of four types of distinctive polarization dynamics in metal/ferroelectric/SiO$_2$ (IL)/metal structure at 0 V. P_r and the coercive field are assumed as 20 μC/cm^2, and 1 MV/cm. NC is unstable in the case of IL = 0.2 nm while NC is stable in the case of IL = 0.8 nm. Simulation conditions are **(a)** NC is unstable and thin T_{dw}, **(b)** NC is stable and thin T_{dw}, **(c)** NC is unstable and thick T_{dw}, **(d)** NC is stable and thick T_{dw}.

978-1-7281-1988-5/18 $31.00 © 2018 IEEE

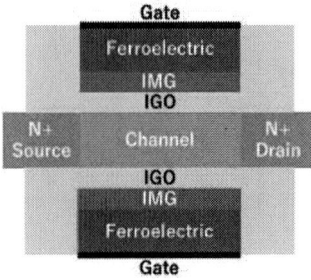

Fig. 7 Schematic of the double gate NCFET with the internal gate oxide (IGO) and the internal metal gate (IMG). The gate length is 9.8 nm. Thicknesses of fin, IGO, and ferroelectric layer are 3.1, 0.5 and 5.0 nm. P_r was assumed to be 3 $\mu C/cm^2$.

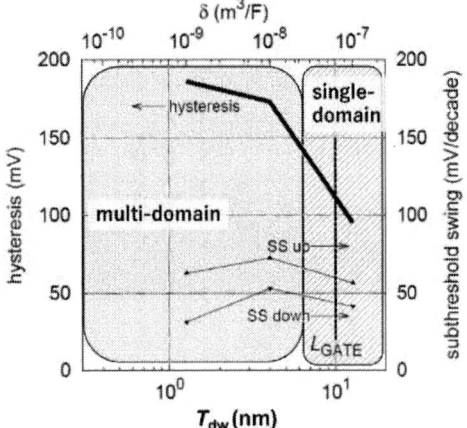

Fig. 9 Summary of the subthreshold swing (SS) and hysteresis (ρ = 1.0 m/S). In materials with narrow domain walls, large hysteresis is inevitable due to multidomain formation inside the gate stack. This means that the gate length has to be comparable to or less than T_{dw} for keeping the coherency of NC.

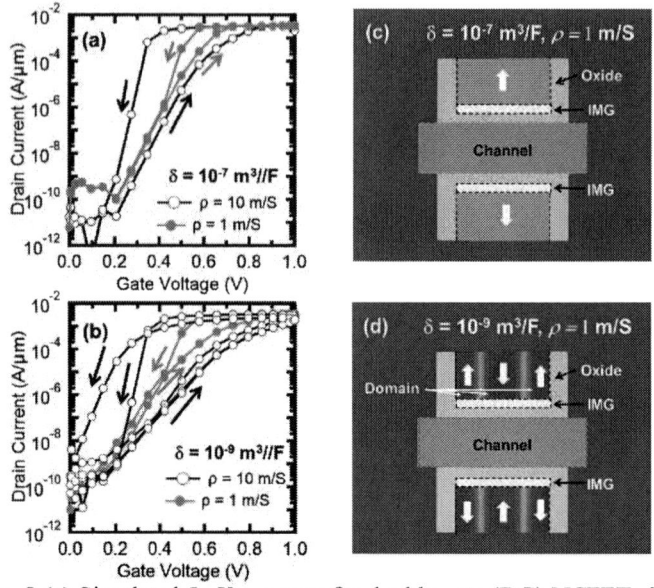

Fig. 8 (a) Simulated I_D-V_{GS} curves for double gate (DG)-NCFETs [$\delta = 10^{-7} m^3/F$]. ρ = 1.0 and 10.0 m/S. V_{DS} is 50 mV and V_{GS} frequency is 10 MHz. **(b)** Simulated I_D-V_{GS} curves for DG-NCFETs [$\delta = 10^{-9} m^3/F$]. **(c)** A contour plot for polarization for the DG-NCFET [$\delta = 10^{-7} m^3/F$, ρ = 1 m/S]. **(d)** A contour plot for polarization for the DG-NCFET [$\delta = 10^{-9} m^3/F$, ρ = 1 m/S].

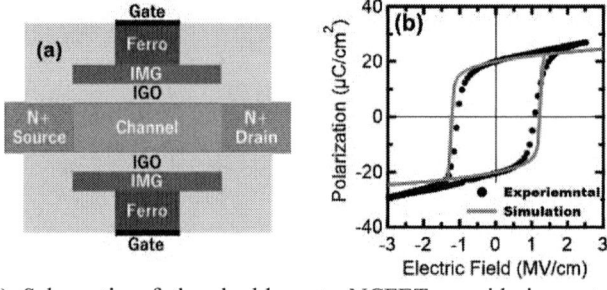

Fig. 10 (a) Schematic of the double gate NCFET considering gate to ferroelectrics area-ratio. While IMG covers the whole channel, the ferroelectric layers cover smaller area comparable to T_{dw}. The gate length is 9.8 nm. Thicknesses of fin, IGO, and ferroelectric layer are 3.1, 0.5 and 3.0 nm. **(b)** The P-E curve (red) for the device simulation reproduced by measured results of the actual $HfZrO_2$ MFM capacitors (black solid circles).

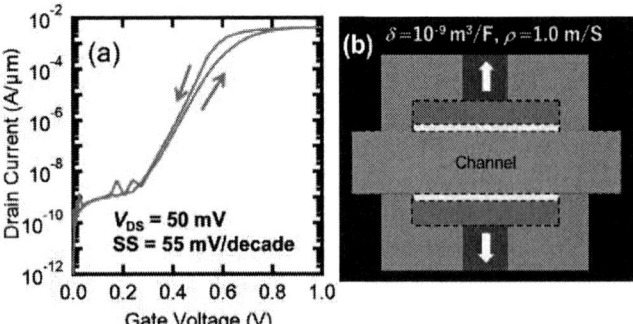

Fig. 11 (a) Transfer characteristics of the NCFET with area-ratio of 3 : 10, and **(b)** polarization in ferroelectrics for V_{GS} = 0.5 V. $\delta = 10^{-9} m^3/F$ (T_{dw} = 2.5 nm) was assumed. The small ferroelectrics comparable to T_{dw} prevent the multidomain formation.

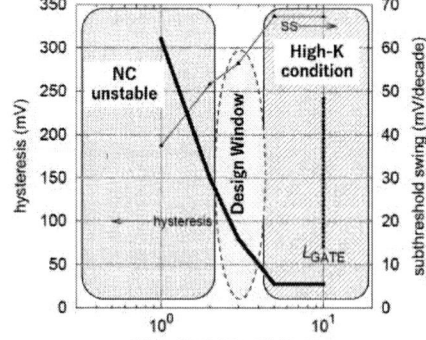

Fig. 12 Dependence of hysteresis and subthreshold swing on the length of ferroelectrics L_{FERRO}. Negative capacitance is unstable for short L_{FERRO}, and negative capacitance condition is not fulfilled for long L_{FERRO}. It is required to find a solution in a narrow window.

978-1-7281-1988-5/18 $31.00 © 2018 IEEE   200

Modeling of Multi-domain Switching in Ferroelectric Materials: Application to Negative Capacitance FETs

A. Dasgupta[1], P. Rastogi[2], D. Saha[2], A. Gaidhane[2], A. Agarwal[3] and Y. S. Chauhan[2]

[1]Dept. of Electrical Engineering and Computer Sciences, Univ. of California, Berkeley, USA. Email: avirup@berkeley.edu
[2]Dept. of Electrical Engineering, IIT Kanpur, India. Email: chauhan@iitk.ac.in
[3]Dept. of Physics, IIT Kanpur, India.

Abstract—We present a new multi-domain model for polarization switching in ferroelectric materials. The computationally efficient model captures the time evolution of multi-domain ferroelectrics with good accuracy along with the frequency dependent switching behavior. We have fabricated (PVDF) and measured P-E characteristics of PZT and PVDF capacitors and have validated the model with measurements. The model allows the visualization of time dependent domain switching allowing further physical insights. We have also proposed a method to extract the distribution of domain orientations experimentally.

I. Introduction

Ferroelectric materials with dielectric hysteresis [1] are used in numerous applications [2-4]. A prominent example is Negative Capacitance Field Effect Transistors (NCFETs) [4,5]. However, the performance gains obtained from NCFETs are limited by the domain behavior of the ferroelectric material [6]. Therefore a computationally efficient and accurate model for the effect of multiple domains on the ferroelectric device behavior is of utmost importance to analyze the impact of domains on the device performance and design better devices and circuits.

Accurate studies of ferroelectric devices generally have to use computationally expensive ab-initio simulations, molecular dynamics simulations [7-11], and/or phase field models [12-18] using a combination of multiple mono-domain blocks along with the time dependent Ginzburg-Landau equation. For circuit simulations, accuracy is usually sacrificed in favor of speed by using the mono-domain Landau–Khalatnikov (L K) model [5]. Fig. 1 shows that mono-domain theories predict a sharp polarization switching whereas multi-domain models show a smoother behavior. The impact of this on NCFET behavior is shown in Fig. 2. The mono-domain model predicts a much steeper switching. This results in faulty Q-V relation from mono-domain models and an overestimation of device capabilities, like switching time and maximum frequency of operation.

Motivated by these shortcomings, in this paper we present a computationally efficient multi-domain model for ferroelectric devices which correctly reproduces the experimentally observed domain dynamics in different materials. Our model correctly captures 1) the growth of domains around random defects and boundary effects, 2) the randomness of the polarization orientation of different domains, 3) the interaction between domains, and 4) the time dependent evolution of the polarization across different domains with applied electric field.

II. Domain structure

Defect sites act like seeds for domain growth [13]. Since the defect sites influence the dipoles through an electric field which is inversely proportional to the distance from the seed, we can argue that all dipoles within a domain are influenced more by its corresponding seed than by any other defect. Therefore, the domain structure should resemble a Voronoi tessellation [14] as shown in Fig. 3. The tessellation is created using a fixed number of defect seeds (white dots) randomly distributed on the plane. However, the seeds are arranged regularly along the perimeter to mimic a finite boundary since the edge can be thought of as a continuous defect. The more the number of these regular boundary seeds, the more accurately we can capture the boundary effect, although it will result in increased computational complexity. Therefore, the number of seeds at the boundary is a trade-off between required accuracy and computational efficiency.

The domains are filled with dipoles oriented along an angle θ ($0 \leq \theta \leq \pi$) from the z-direction (co-ordinate system shown in Fig. 4). Central Limit Theorem (CLT) suggests that the orientation angle θ will have a symmetric Gaussian distribution over $-\pi \leq \theta \leq \pi$ centered at $\theta = 0$ for an initial polarization along +z direction, as given in (1) . The color of the domains in Fig. 3 represents the projection of the normalized polarization of the domain on the z-axis (p_z) with red and blue denoting positive and negative p_z, respectively; while the intensity of the color depends linearly on $|p_z|$ (Black implies $|p_z| = 0$).

III. Quasi-static switching

Let us consider the polarization switching from +z to –z direction. We assume sufficiently large W and L and neglect any fringing fields in and x and y directions. Application of an electric field sweep to switch the net polarization (+z to –z) results in the dipoles with θ tending to π switching first and the orientation of the switching dipoles decreases from π to 0 with increasing field. Assuming all dipoles have uniform polarization, p, we can write the net polarization for a given field (E) where dipoles with orientation $\in (\theta, \pi]$ have switched, as (2). The term P_{diel} (3) captures the linear dielectric behavior of the material after switching, as shown in Fig. 1.

Consider the next dipole that will switch. The external field required to switch it should at least be equal to the internal field on it due to all the other dipoles, i.e. $E \geq P/2\epsilon_{fe}$. Also, due to the Gaussian nature of the orientation distribution, the switching happens for a narrow range of θ as shown from the plot of (2) in Fig. 5. These two criteria allow us to express θ as

978-1-7281-1988-5/18 $31.00 © 2018 IEEE

a function of the applied field (4), which when substituted in (2) gives us the net polarization as a function of the applied electric field.

IV. Domain dynamics and Simulation Methodology

The quasi-static model can be used to develop a simulation framework allowing us to visualize the time dependent switching of domains and predict the dynamic behavior.

Eq. (5) gives us a value of θ for an applied electric field which allows us to track the domains that have switched (initial orientation angle $>\theta$) in real time as the field is swept. Fig.6 shows the polarization distribution of the domains at different times as switching progresses (in order, from (a) to (c)). We can also obtain the net polarization at any instant by summing over all the domains, which gives us the P-E behavior as shown by the blue line in Fig. 1.

V. Frequency response

Each domain has a specific time constant based on the initial orientation of dipoles as well as the domain size. As a result, different domains respond at different frequencies. There is a maximum frequency (f_{max}) corresponding to the domain with the smallest time constant, above which the material does not respond. If we start measuring P-E curves at this maximum frequency, and subsequently go down in frequency; we can see that although only a few domains respond initially; with decreasing frequency of operation (f) more domains with time constants lower than f^{-1} start responding. This is evident from the increasing peak-to-peak polarization ($\propto$ number of dipoles responding) with decreasing f.

A. Number of dipoles responding at a given frequency

The derivation considers the domains arranged in a 2D z-x plane, with uniformity along the y-direction. The average domain size is given by $A_d = A/N_s$. Approximating the domains to be regular polygons with n sides as shown in Fig. 7, we get a relation between the perimeter (P_d) and the area (5). Since $N_d \propto A_d$ we get (6) if we assume a thin grain boundary (grey region in Fig. 7). If the net polarization of the domain is oriented along θ_0, CLT says that the dipole orientations will be a Gaussian centered at $\theta = \theta_0$, implying the relation (7) which in turn gives us σ (8).

Switching happens for a narrow window in θ for a Gaussian system as shown in Fig. 5, which allows us to write the net polarization as $P \propto erf\left(\frac{\theta_0}{\sigma}\right)$ for the considered domain. This equation predicts a behavior like the capacitive switching of an RC network, as shown in Fig. 8. Comparing this model with an RC network, along with (5), gives us $\theta_0(f)$ as in (9). Eq. (9) tells us that for a given f, domains with initial orientation $\geq \theta_0$ will respond. This allows us to get the exact frequency dependent behavior of the ferroelectric material.

B. Standard deviaion of polarization distribution (b)

The material parameter b defined in (1) can be extracted by fitting P-E curves for multiple frequencies with our model. However, an easier way to extract b is to measure the peak-to-peak polarization for multiple frequencies. We can use (9) to get θ_0 for all the frequencies used and plot the normalized peak-to-peak polarization against it, as shown in Fig. 9. This is effectively a plot of the normalized number of dipoles as a function of their orientation angle, and it can be seen to resemble a Gaussian distribution which validates our initial assumption. It is now a simple task to fit our Gaussian distribution (1) to this plot and extract b from the standard deviation.

VI. Results and Discussion

A. Experimental validation

We have fabricated (PVDF) and measured PZT and PVDF ferro-capacitors for frequency dependent P-E characteristics. The values of b for the two materials were extracted from the frequency dependent measurements as shown in Fig. 9. These were then used to validate our model as shown in Fig. 10.

B. Speed and accuracy

The significant improvement in accuracy of our model as compared to mono-domain L-K models is shown in Fig. 1. Our model is also able to capture experimental data accurately as shown in Fig. 10. It is interesting to note that we have been able to achieve an accuracy comparable to involved multi-domain simulations while retaining a simulation time comparable to mono-domain models, as shown in Fig. 11. We are able to get more than 80% reduction in simulation time while ensuring an accurate match with experimental data (Fig. 11 (b)).

VII. Conclusion

We have presented an experimentally validated multi-domain model for time dependent ferroelectric polarization switching including hysteresis. The accuracy of the model is comparable to more involved multi-domain TCAD/ab-initio/molecular dynamics simulations while the simulation times are of the order of mono-domain calculations. The model also allows us to visualize the domain switching, thus providing physical insights. We have also proposed a method to extract the distribution of polarization across the domains from frequency dependent measurements. Our computationally efficient model captures the time evolution of multi-domain ferroelectrics reasonable well, allowing for fast simulation of material and device characteristics.

References

[1] J. Valasek, *Phys. Rev., no. 17, pp. 475-481, 1921*
[2] G. H. Haertling, *J. Am. Ceram. Soc.*, no. 82, pp. 797-818, 1999.
[3] P. R. Potnis et.al., *Material*, no. 4, pp. 417-447, 2011.
[4] S. Salahuddin et.al., *Nano Lett.*, v. 8, no. 2, pp. 405-410, 2017.
[5] G. Pahwa, *IEEE TED*, v. 63, no. 12, pp. 4981-4985, 2016.
[6] A. Cano et.al., *Appl. Phys. Lett.*, 97, p. 133509, 2010.
[7] P. Paruch et.al., *Nature*, 534, pp. 331-332, 2016.
[8] S. Liu et.al., *Nature*, 534, pp. 360-363, 2016.
[9] B. Meyer et.al., *Phys. Rev. B.*, 65, pp. 13-24, 2002.
[10] I. I. Naumov et.al., *Nature*, 432, pp. 737-740, 2004.
[11] R. E. Cohen et.al., *Phys. Rev. B.*, 42, pp. 113-140, 2002.
[12] L. Q. Chen et.al., *Annu. Rev. Mater. Res.*, 32, pp. 113-140, 2002.
[13] D. Schrade et.al., *Comput. Method. Appl. M*, 196, p. 4365, 2007.
[14] E.-M. Anton et.al., *JAP*, 105, p. 024107, 2009.
[15] Y. L. Li, *Acta Mater.*, 50, p. 395, 2002.
[16] J. Wang et.al., *Acta Mater.*, 52, p. 749, 2004.
[17] W. Zhang et.al., *Acta Mater.*, 53, p. 185, 2005.
[18] S. Choudhury et.al., *Acta Mater.*, 53, p. 5313, 2005.
[19] "Sentaurus device user guide", 2018

$$F(\theta) = \frac{N_0}{b\sqrt{\pi}} \cdot exp\left(-\frac{\theta^2}{b^2}\right); \quad -\pi \leq \theta \leq \pi \tag{1}$$

$$P(\theta) = \underbrace{P_0}_{initial\ polarization} - \underbrace{2\int_\theta^\pi p\cos(\phi)F(\phi)d\phi}_{intial\ contribution\ of\ switched\ dipoles} + \underbrace{\left[-2p\int_\theta^\pi F(\phi)d\phi\right]}_{final\ contribution\ of\ switched\ dipoles} + P_{diel}$$

$$= -2J + \frac{N_0 p}{2}exp\left(-\frac{b^2}{4}\right)\left[erf\left(\frac{2\theta+ib^2}{2b}\right) + erf\left(\frac{2\theta-ib^2}{2b}\right)\right] + N_0 p \cdot erf\left(\frac{\theta}{b}\right) + P_{diel} \tag{2}$$

$$P_{diel} = H_f(2\epsilon_d(E-E_0),0,1) + H_n(2\epsilon_d(E-E1),0,1) \tag{3}$$

$$\theta = b \cdot erf^{-1}\left(H_f\left(\frac{2\epsilon_{fe}(E-E_{eff})}{abp\sqrt{\pi}}, erf\left(\frac{\pi}{b}\right), \Delta\right)\right) \tag{4}$$

$$P_d = 2n\sqrt{\frac{2A_d}{n\sin\left(\frac{2\pi}{n}\right)}}\sin\left(\frac{\pi}{n}\right) \tag{5} \qquad \frac{N_{P_d}}{N_d} = \frac{P_d t_d}{A_d} = \frac{P_d R}{\Gamma_{br} A_d} \ ; \ \Gamma_{br} = \frac{R}{t_d} \tag{6}$$

$$N_d = 2\int_{\theta_0}^{\pi+\theta_0}(N_d-N_{P_d})exp\left(-\frac{(\theta-\theta_0)^2}{\sigma^2}\right)d\theta = \sqrt{\pi}(N_d-N_{P_d})\sigma \cdot erf\left(\frac{\pi}{\sigma}\right) \tag{7}$$

$$\sigma = \pi\left[3 - \frac{3N_d}{2\pi(N_d-N_{P_d})}\right]^{-\frac{1}{2}} \tag{8} \qquad \theta_0(f) = \frac{\pi erf^{-1}\left(1 - exp\left(-\frac{\pi f}{f_{max}}\right)\right)}{\sqrt{3}\sqrt{1 - \frac{\Gamma_{br}\cos\left(\frac{\pi}{n}\right)}{2\pi\left(\Gamma_{br}\cos\left(\frac{\pi}{n}\right)-2\right)}}} \tag{9}$$

Fig. 0. : Equations referred to in the text.

Symbol	Meaning	Symbol	Meaning
N_0	Total number of dipoles	$H_{f/n}(x, x_M, c)$	Max/Min. func.: $\frac{x+x_M\pm\sqrt{(x-x_M)^2+4c}}{2}$
N_d	Number of dipoles in a domain	P_d	Perimeter of a domain
N_{P_d}	Number of dipoles in the domain wall	A_d	Area of a domain
p	Polarization of a single dipole	t_d	Thickness of domain wall
a	$N_0/(b\sqrt{\pi})$	t_{fe}	Thickness of ferroelectric
ϵ_{fe}	Ferroelectric permittivity	ϵ_d	Permittivity of material in dielectric state
E_{eff}	$-J/(2\epsilon_{fe})$	σ	Std. deviation for Gaussian distr. in domain

Table 1: Description of symbols used.

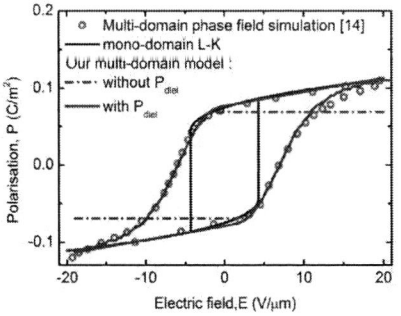

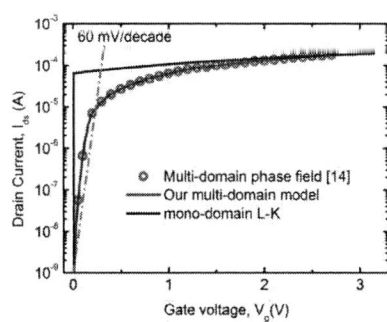

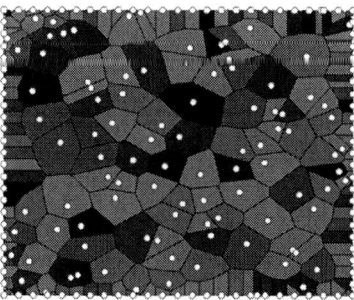

Fig. 1. Comparison between predictions from mono-domain Landau theory, multi-domain phase field simulations [14] and our multi-domain model for dynamic domain switching. Mono-domain Landau theory does not include the limitations and properties of multi-domain switching, resulting in a sharp switching. Our multi-domain model is comparable in accuracy to more involved multi-domain simulations. The impact of P_{diel} (3) is also shown. The fixed initial condition results in the pre-switching flat regions (blue dotted). P_{diel} captures the E dependence of the dielectric state.

Fig. 2. Comparison of the mono-domain model with our model for an NCFET (MFMIS: Metal-Ferroelectric-Metal-Insulator-Semiconductor) application, shows that the mono-domain model results in a much steeper switching, thus overestimating device capabilities. The characteristics obtained from our model match the predictions from multi-domain phase field simulations with high accuracy. The base MOSFET is simulated using the BSIM-BULK model (default parameter set.)

Fig. 3. Graphical representation of defect seeds (white dots) and resulting domain structure from Voronoi algorithm [14]. Regular arrangement of seeds at boundary captures the finite truncation of the material. The color of the domains represent the orientation angle, θ: red implies $|\theta| < \pi/2$, blue implies $|\theta| > \pi/2$ and black denotes $|\theta| = \pi/2$. The color intensity decreases linearly with decreasing $||\theta| - \pi/2|$. The complete set of values for θ follows a Gaussian distribution (1).

978-1-7281-1988-5/18 $31.00 © 2018 IEEE

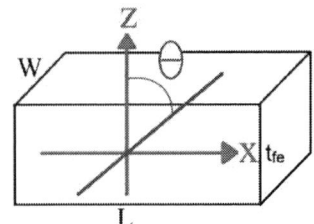

Fig. 4. Co-ordinate system used. W=width, L=length and t_{fe} = thickness. The orientation angle, θ, is measured with respect to the z-axis.

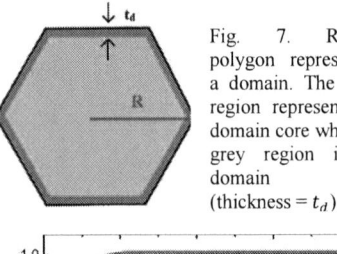

Fig. 7. Regular polygon representing a domain. The green region represents the domain core while the grey region is the domain wall (thickness = t_d).

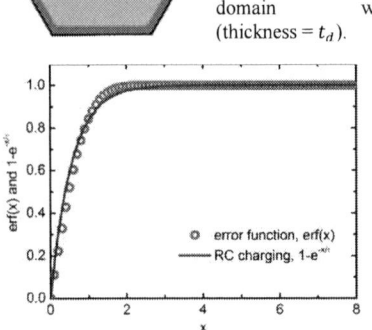

Fig. 8. Comparison between $erf(x)$ and $[1 - exp(-x/\tau)]$ shows P switching can be reasonably approximated to an RC network.

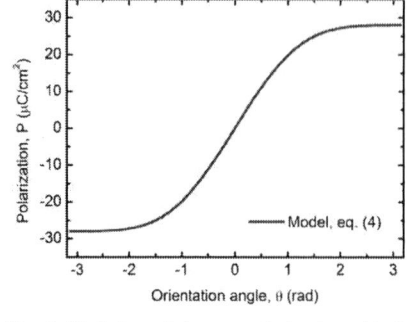

Fig. 5. Variation of the net polarization with θ, during switching, for the device in Fig. 5. Here θ is the limiting orientation angle such that all dipoles with orientation $\in (\theta, \pi]$ have switched. The switching happens over a narrow range of θ around $\theta = 0$, due to the Gaussian distribution.

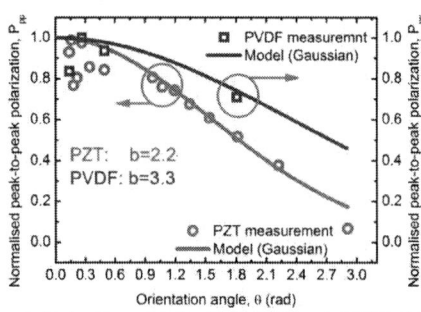

Fig. 9. Extraction of b for PZT (red) and PVDF (blue) devices. Peak-to-peak polarization is measured as a function of the frequency of operation. Eq. (9) is used to plot normalized P_{pp} with θ. Gaussian distribution confirms our initial assumption and allows extraction of b from the standard deviation. The ordered crystalline PZT has smaller b than the amorphous PVDF material, as expected.

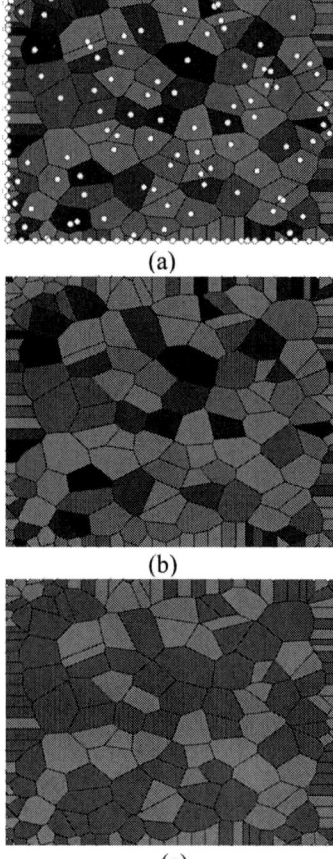

Fig. 6. The progression of domain switching is illustrated in (a), (b) and (c). We start with the orientation distribution shown in (a). The net polarization is along +z as denoted by the dominance of red (same color scheme as in Fig. 3). With increasing electric field (b) the domains start turning bright blue denoting their switching to the -z direction. Further increase in electric field increases the blue patch, as in (c), denoting an increased number of domains switching to -z direction. This continues till all domains are along -z. We can therefore study the growth of the blue patch during switching and also get the net polarization for any field by summing over the polarizations of all the domains, as shown in Fig. 5 (black line).

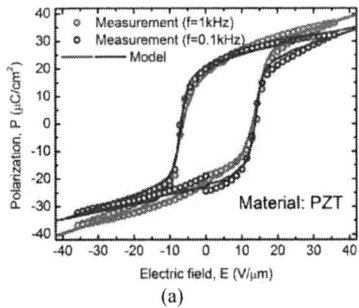 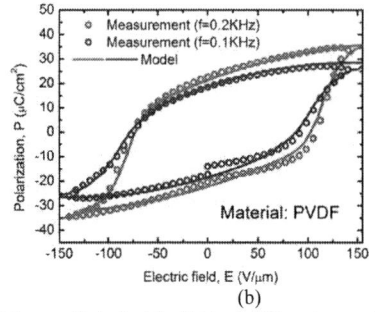

(a) (b)

Fig. 10. Polarization switching as a function of the applied electric field at different operating frequencies for the (a) PZT and the (b) PVDF devices from Fig. 9. The extracted value of b has been used for all the frequencies. The model fits the experimental data with good accuracy for all frequencies with a single parameter set.

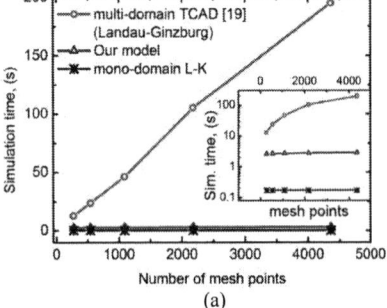 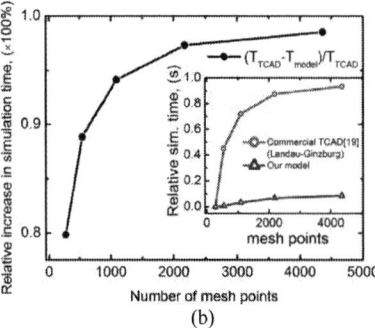

(a) (b)

Fig. 11. (a) Simulation time as function of the number of mesh points (number of dipoles). Commercial TCAD simulators are significantly slower than our multi-domain model. Mono-domain L-K models are fastest but compromise accuracy. (b) The relative increase in simulation time is shown. Our model is more than 80% faster. Also, the disparity between our model and TCAD increases with increasing number of mesh points which results in nearly 98.5% faster simulation for 4356 mesh points. For larger number of mesh points, the speed gain approaches 100%

978-1-7281-1988-5/18 $31.00 © 2018 IEEE   204

On the Microscopic Origin of Negative Capacitance in Ferroelectric Materials: A Toy Model

Asif Islam Khan

School of Electrical and Computer Engineering, Georgia Institute of Technology, GA 30329, USA
Email: asif.khan@ece.gatech.edu

Abstract—We present a simple, physical explanation of underlying microscopic mechanisms that lead to the emergence of the negative phenomena in ferroelectric materials. The material presented herein is inspired by the pedagogical treatment of ferroelectricity by Feynman and Kittel. In a toy model consisting of a linear one-dimensional chain of polarizable units (i.e., atoms or unit cells of a crystal structure), we show how simple electrostatic interactions can create a microscopic, positive feedback action that leads to negative capacitance phenomena. We point out that the unstable negative capacitance effect has its origin in the so called "polarization catastrophe" phenomenon which is essential to explain displacement type ferroelectrics. Furthermore, the fact that even in the negative capacitance state, the individual dipole always aligns along the direction of the local electrical field not opposite is made clear through the toy model. Finally, how the "S"-shaped polarization vs. applied electric field curve emerges out of the electrostatic interactions in an ordered set of polarizable units is shown.

I. INTRODUCTION

The negative capacitance effect in ferroelectric materials can be utilized to enable continued performance gains in the complementary metal-oxide-semiconductor (CMOS) platforms. When used in the gate dielectric stack of a metal-oxide-semiconductor field-effect transistor (MOSFET), a ferroelectric oxide owing to its negative capacitance properties can provide a passive voltage amplification of the gate voltage at the oxide-semiconductor interface. This effect can lower the sub-threshold slope below the fundamental Boltzmann limit of 60 mV/decade [1]. Such steep switching in negative capacitance field-effect transistors (NCFETs) can "restart" the aggressive scaling of the power supply voltage, thereby, allowing for significant reduction of power dissipation in the CMOS technology. To date, different aspects of ferroelectric negative capacitance phenomena have been demonstrated in different experimental set-ups such as ferroelectric capacitors, ferroelectric-dielectric heterostructures and superlattices and NCFETs.

The underlying theory of ferroelectric negative capacitance as proposed by Salahuddin *et al.* is based on the Landau's phenomenological theory of phase transitions. The free energy density of a ferroelectric material can be expanded in an even order polynomial of the polarization which through the Landau-

Khalatnikov equation, leads to the following relation between the applied electric field E and the polarization P.

$$E = \alpha_1 P + \alpha_{11} P^3 + \alpha_{111} P^5 + \cdots \qquad (1)$$

Here, $\alpha_1, \alpha_{11}, \alpha_{111}$ are anisotropy constants. Polarization is equivalent to the surface charge density. Furthermore, $\alpha_1 = a_0(T - T_c)$ where T is the temperature, T_c is the Curie temperature and a_0 is a positive constant. Below Curie temperature, $\alpha_1 < 0$, and the E vs. P curve has an "S"-shape where under certain range of E and P, the curve has a negative slope and hence a negative capacitance (*i.e.*, $C = \frac{1}{t_F}\left(\frac{dP}{dE}\right) < 0, t_F$ being the ferroelectric thickness). Between the coercive field ($+E_C$ and -E_C), the polarization P is a multiple valued function of E. The negative capacitance states are unstable in a free-standing ferroelectric material; hence in the absence of an applied electric field, the ferroelectric gets spontaneously polarized in either of the stable states ($+P_0$, -P_0) indicated by points A and B in fig. 1(a). For a detailed treatment of stability of the negative capacitance state, the readers are referred to Ref. [2].

First appeared in the classic 1937 paper [3,4], the Landau framework takes a phenomenological, mean field and symmetry-based approach to analyze phase transitions. Right after the discovery of ferroelectricity in $BaTiO_3$ in early 1940s, the Landau theory was first applied to ferroelectric oxides led

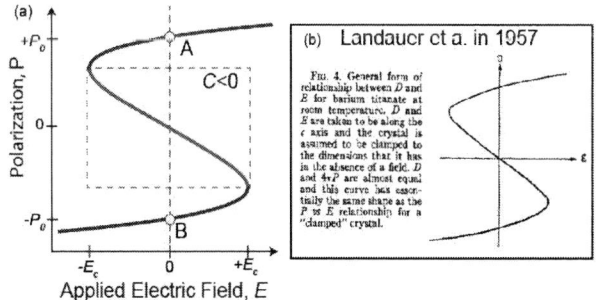

Fig. 1. (a) The "S"-shaped polarization P vs. applied electric field E curve of a ferroelectric material. (b) One of the earliest illustrations of the "S"-shaped P vs. E curve appeared in Landauer et al. in Ref. [7]. Similar illustration appeared in Ref. [8]. In these reports, the properties of the negative capacitance region were not explicitly discussed except for saying that this region is "thermodynamically unstable," and hence needs to be replaced by hysteretic jumps. Reproduced with permission from Ref. [7].

the works of Ginzburg and Devonshire [5, 6]. However, in none of these early works and the ones that followed, the negative capacitance region in ferroelectric was explicitly discussed except for asserting that this region is "thermodynamically unstable." Furthermore, while the Landau theory serves as a reliable, conceptual bridge between microscopic models and observed macroscopic phenomena, the theory itself leaves out the physical, microscopic details of the phenomenon it describes. These two facts in fact has created multitudes of the confusion whether the negative capacitance effect is a real, physical phenomenon or an unphysical, artificial construct for the convenience of the phenomenology (the Landau theory neither implicitly nor explicitly implies the latter). To date, all of the theoretical analysis of negative capacitance effects and modeling and simulation of NCFETs starts with Eq. 1 or its variants all based on the Landau framework. What is missing in the current discussion of negative capacitance is a physical picture that explains the microscopic origin of the negative capacitance in ferroelectric materials. As such, in this invited article, we take on this task to present a simple, atomic scale model that elucidates the underlying physical mechanisms responsible for this phenomenon. We point out that the unstable negative capacitance effect has its origin in the so called "polarization catastrophe" phenomenon which is generally used in the ferroelectric literature to explain the emergence of ferroelectricity of displacement type.

Another commonly held misconception is that, in the negative capacitance state even when it is stabilized by putting a positive capacitor in series, the ferroelectric dipoles align opposite to the electric field they feel thereby "violating" the fundamentals of thermodynamics. To resolve this issue, we make the point that it is important to distinguish between the applied electric field and the local electric field (the field that the dipoles feel). The local electric field is, in fact, the sum of the applied electric field and the effective dipole field (mean field) created by all the other dipoles. It is only when the polarizability of the ferroelectric dipoles attains a large enough value to create an atomic scale positive feedback mechanism such that the local field overcompensates the applied field and the unstable negative capacitance phenomena and the

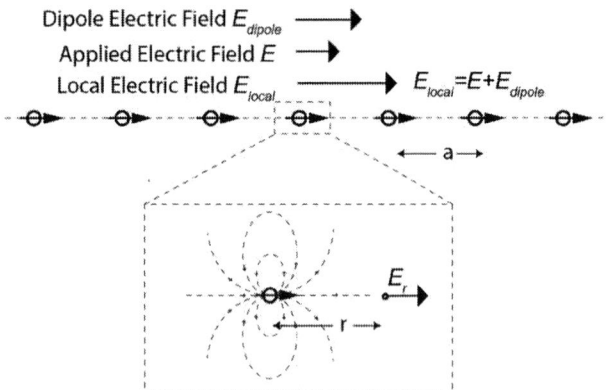

Fig.2. A linear, one dimensional chain of polarizable units with a spacing of a. The electric field created by a polarized dipole is also shown.

"polarization catastrophe" ensue. Even in such a case, the polarizability of the dipoles always remains positive and the dipoles always remain in the same direction of the local electric field—the dipole aligns opposite only to the applied electric field. Building upon the pedagogical treatment of ferroelectricity by Feynman, we show how an "S"-shaped dipole moment p vs. applied electric field E curve emerges from a toy model consisting of a linear one-dimensional chain of polarizable units.

II. A MICROSCOPIC TOY MODEL OF FERROELECTRICS AND NEGATIVE CAPACITANCE

Let us consider a linear, one dimensional chain of polarizable units with a spacing of a. on which an electric field E is applied along the chain axis (Fig. 2). The treatment presented herein is influenced by the pedagogical writing of Feynman in Ref. [9]. A more rigorous treatment along the same lines based on Clausius-Mossotti relation available in Ref. [10]. The polarizable unit here can represent an atom or a bond or a unit cell of a crystal. For the sake of simplicity, we first assume that these units have a linear polarizability α such that the dipole moment p and the local electric field E_{local} is relation by the following relation.

$$p = \alpha \epsilon_0 E_{local} \qquad (2)$$

Here, ϵ_0 is the vacuum permittivity. The field created by a dipole at a distance r from the along its axis is given by $E_r = \frac{1}{4\pi\epsilon_0} \frac{2p}{r^3}$. Hence, at a given dipole, the electric field due to the interaction with all the other dipoles in the chain is

$$E_{dipole} = 2 \times \frac{1}{4\pi\epsilon_0} \frac{2p}{a^3} \left(1 + \frac{1}{2^3} + \frac{1}{3^3} + \cdots \right) = \frac{p}{\epsilon_0} \frac{0.383}{a^3} = \zeta p \qquad (3)$$

This particular calculation of E_{dipole} is repeated from Ref. [9] Here, ζ is a structural factor. In the calculations that follow, we will not pay attention to the value of ζ, rather treat it as a variable that depends on the arrangement of the units. ζ will have a different value if the dipoles have a different arrangement (e.g. three dimensional cubic, tetragonal, or orthorhombic lattice). Note that the local electric field is the sum of the applied electric field E and the electric field created by all the other dipoles E_{dipole}, i.e.,

$$E_{local} = E_{dipole} + E \qquad (4)$$

It is interesting to note that, the dipole moment p depends on the local electric field E_{local} (Eq. 2), which in turn depends on p through Eq. 4 thereby creating a microscopic positive feedback. A block diagram representation is shown in Fig. 3. Combining Eq. (2), (3) and (4), we obtain the following expression.

$$p = \alpha \epsilon_0 (\zeta p + E) \qquad (5)$$

Rearranging this equation, we obtain,

$$p = \frac{\alpha \epsilon_0}{1 - \alpha \epsilon_0 \zeta} E \qquad (6)$$

978-1-7281-1988-5/18 $31.00 © 2018 IEEE

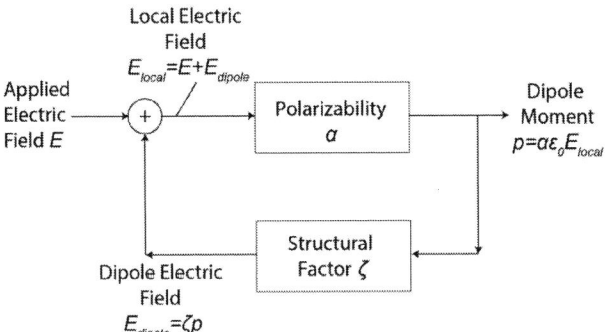

Fig.3. The positive feedback mechanism at the atomic scale.

and scaled dielectric constant, ϵ_r is given by the following relation.

$$\epsilon_r = \frac{1}{\epsilon_0}\frac{dp}{dE} = \frac{\alpha}{1-\alpha\epsilon_0\zeta} \quad (7)$$

The polarizability α is generally inversely proportionate to the temperature T (i.e., $\alpha \propto 1/T$). At Curie temperature T_c, $\alpha = 1/\epsilon_0\zeta$ for which the scaled dielectric constant ϵ_r shows a singularity. For $T>T_c$, $\alpha < 1/\epsilon_0\zeta$, and $p=0$ is stable solution of Eq. 5 for $E=0$. Physically, it means that in the absence of an applied electric field, the dipoles are not polarized and the dipole moment and the local electric field are both zero. It can be shown that Eq. 7 leads to the Curie-Weiss dependent of ϵ_r on temperature (i.e., $\epsilon_r \propto 1/T - T_c$).

Of particular interest to our analysis is the case when $\alpha > 1/\epsilon_0\zeta$ and the scaled dielectric constant ϵ_r becomes negative. As the ferroelectric material is cooled down from a temperature higher than T_C to below that, the resulting negative value of ϵ_r amplifies small thermal fluctuations in dipole moments through the positive feedback mechanism in the absence of an applied electric field E. As soon as a dipole electric field E_{dipole} however small emerges due to the fluctuations of p, it regeneratively increases p. In the ferroelectric literature, this situation is typically referred to as the "polarization catastrophe". This means that in the absence of an applied electric field, any thermal fluctuation of dipole moment sets up a local electric field that spontaneously polarizes the dipoles.

The point to note here is that it is the negative dielectric constant or the negative capacitance at $T<T_c$ that sets off the polarization catastrophe and leads to the emergence of the spontaneous polarization in ferroelectric material. By introducing time dependent, kinetic terms in Eq. (2) and (5), it can be shown that $p=0$ is no longer a stable solution of the system when $T<T_C$ _and_ $\alpha > 1/\epsilon_0\zeta$.

III. THE EMERGENCE OF THE SPONTANEOUS POLARIZATION AND THE "S"-SHAPED P-E CURVE

Now that we have seen at $T<T_c$, the negative capacitance sets off the positive feedback mechanism that tends to increase the dipole moment p in an unbounded fashion, we now address what stops this run-way process such that p settles down to

stable spontaneous polarization states indicated by points A and B in fig. 1 (in fig. 1, the polarization P is a scaled version of the dipole moment p). To explain the emergence of the stable spontaneously polarized state, we need to add the next level of details by considering the non-linearity in the polarizability in the dipole moment. An electric field stretches a dipole—however, a dipole is not infinitely stretchable. With the increase of the local electric field beyond a critical value, the dipole moment is not expected to increase any further. We assume that the saturation dipole moment is $p_{max} = \alpha\epsilon_0 E_{cr}$, E_{cr} being the critical saturation local electric field. For the sake of simplicity, we assume the following relation between p and E_{local}:

$$p = p_{max}\tanh\frac{E_{local}}{E_{cr}} \quad (8)$$

where $p_{max} = \alpha\epsilon_0 E_{cr}$. When $E_{local} \ll E_{cr}$, we get back the linear relation between p and E_{local} as in Eq. (2). Combining Eq. 2, 3, 4 and 8, we obtain

$$p = p_{max}\tanh\frac{\zeta p+E}{E_{cr}} \quad (9)$$

Using the identity: $\tanh^{-1} x = \frac{1}{2}\log\frac{1+x}{1-x}$ in Eq. (9), the following relation is obtained.

$$E = -\zeta p + \frac{E_{cr}}{2}\log\frac{1+p/p_{max}}{1-p/p_{max}} \quad (10)$$

Assuming $\alpha = 1/\varsigma\epsilon_0 T$, ς being a positive constant, and $\zeta = \varsigma T_C$, Eq. 10 can be simplified as follows.

$$\frac{E}{E_{cr}} = -\frac{T}{T_c}\frac{p}{p_{max}} + \frac{E_{cr}}{2}\log\frac{1+p/p_{max}}{1-p/p_{max}} \quad (11)$$

Fig. 4 plots p/p_{max} vs. E/E_{cr} curves using Eq. 11 for T_c/T=0.6, 1.01, 1.4 and 1.7. The "S"-shaped p-E curve clearly emerges for $T<T_c$. In fig. 4, the stable $p=0$ at $T>T_c$ is indicated point P. As the temperature reduces below T_c, the point P no longer remains stable due to the instability of negative capacitance. At this point, as soon as thermal fluctuations cause the dipole moment p to attain a small value, the dipole moment p traverses a path through the "S"-shaped curve in a transient path and settles at one of the stable spontaneously polarized states (for example, at points A or B for T/T_c=1.4 in fig. 4).

In the regime where $|E| < E_{cr}$ and $|p| < p_{max}$, we expand Eq. 10 to obtain the following relation.

$$E = -\zeta p + E_{cr}\left(\frac{p}{p_{max}} + \frac{1}{3}\left(\frac{p}{p_{max}}\right)^3 + \frac{1}{5}\left(\frac{p}{p_{max}}\right)^5 + \cdots\right)$$
$$= \left(\frac{E_{cr}}{p_{max}} - \zeta\right)p + \frac{E_{cr}}{3p_{max}^3}p^3 + \frac{E_{cr}}{5p_{max}^5}p^5 + \cdots \quad (12)$$

Thus, from our toy model, we arrive at an odd order expansion of the applied electric field E in terms of p as in Eq. 1. Comparing Eq. 12 with Eq. 1, $\alpha_1 = \frac{E_{cr}}{p_{max}} - \zeta = 1/\alpha\epsilon_0 - \zeta = \varsigma(T - T_c)$, $\alpha_{11} = E_{cr}/3p_{max}^3$, $\alpha_{111} = E_{cr}/5p_{max}^5$ and so on.

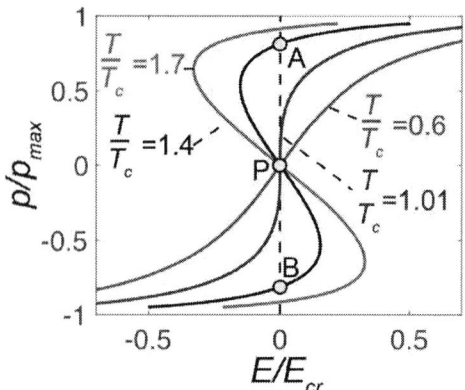

Fig.4. p/p_{max} vs. E/E_{cr} curves for T_c/T=0.6, 1.01, 1.4 and 1.7 plotted using Eq. 11: $\frac{E}{E_{cr}} = -\frac{T}{T_c}\frac{p}{p_{max}} + \frac{E_{cr}}{2}\log\frac{1 + p/p_{max}}{1 - p/p_{max}}$. Note that p_{max} is a function of T as well.

An important point to note in this analysis is that the polarizability of the polarizable units is always positive—i.e., the dipole always aligns along the direction of the local electric field. This is because p and E_{local} always have the same sign in Eq. 2 and 8. It is in fact the case, at T<Tc, the polarizability attains a large positive value which leads to the negative capacitance phenomena and the emergence of ferroelectricity.

Given the limited scope of the article, we intentionally excluded a discussion of the stabilization mechanism of the otherwise unstable negative capacitance states which is described in details elsewhere. In a stabilized negative capacitance state in a ferroelectric-dielectric series combination and an NCFET, the polarization and the external electric field in the ferroelectric are in opposite direction—however, the local electric field and the polarization are still in the same direction.

IV. CONCLUSIONS

We have presented a physical, microscopic picture of the emergence of negative capacitance in ferroelectric materials. We have shown that a large dielectric polarizability set off an atomic scale, positive feedback mechanism that aligns the dipoles opposite to the applied electric field but not to the local electric field. This situation is unstable and hence transient. The instability there caused leads to the "polarization catastrophe" and the dipoles become spontaneously polarized even in the absence of an applied electric field. By assuming a non-linear polarizability of the dipoles, we show how an "S"-shaped dipole moment p vs. applied electric field E curve emerges from a toy model consisting of a linear one-dimensional chain of polarizable units (i.e., atoms or dipoles or crystal unit cells).

It is important to note that polarization reversal generally occurs through domain nucleation and growth mechanisms especially in ferroelectric capacitors with lateral dimensions of tens of microns. In fact, that is why an "S"-shaped polarization vs. applied electric curve also had a hysteresis in recent

experimental work reported in Ref. [11, 12, 13]. The effects of domain mediated switching in such observation of negative capacitance was analyzed in Ref. [14]. In fact, the topic of intrinsic, homogeneous and single domain switching has appeared in the ferroelectric literature since the 1950 in the context of what is known as the "Landauer paradox." [7, 15, 16]. It is interesting that in nanoscale ferroelectric HfO$_2$ gated field-effect transistors where the dimensions are of a few tens of nanometers, the evidence of single domain switching has been claimed to have been observed in Ref. [17]. Further analysis of such nanoscale devices will elucidate more about the nature of negative capacitance and its stability of associate ferroelectrics.

ACKNOWLEDGMENT

This work was supported in part by the National Science Foundation under grant 1718671, in part by Semiconductor Research Corporation-Global Research Collaboration grant 2018-LM-2829 and in part by ASCENT, one of six centers in JUMP, a Semiconductor Research Corporation (SRC) program sponsored by DARPA.

REFERENCES

[1] S. Salahuddin et al., "Use of Negative Capacitance to Provide Voltage Amplification for Low Power Nanoscale Devices," Nano Lett., vol. 8, pp. 405-410, 2008.

[2] Khan et al. "Negative capacitance behavior in a leaky ferroelectric," IEEE Trans. Electron Dev., vol. 63, pp. 4416-4422, 2016.

[3] L. D. Landau, Phys. Z. Sowjun, vol. 11, pp. 545, 1937.; "On the theory of phase transitions. II," Zh. Eksp. Teor. Fiz. Vol. 7, pp. 627, 1937.

[4] D. ter Haar (Ed.): Collected Papers of L. D. Landau (Pergamon, Oxford 1965) contains Engl. transl. of Ref. [3].

[5] V. L. Ginzburg, Zh. Eksp. Teor. Fiz., vol. 15, pp. 739, 1945; V. L. Ginsburg, J. Phys. USSR, vol. 10, pp.107, 1946.

[6] A. F. Devonshire: Philos. Mag., vol. 40, pp. 1040, 1949.

[7] R. Landauer. "Electrostatic Considerations in BaTio$_3$ Domain Formation during Polarization Reversal," J. Appl. Phys, vol. 28, pp. 227-234, Feb. 1957.

[8] W. J. Merz, "Double Hysteresis Loop of BaTiO$_3$ at the Curie Point," Phys. Rev., vol. 91, pp. 513-517, Aug. 1953.

[9] R. P. Feynman, The Feynman Lectures of Physics, vol. II. New York: Basic Books, 2011, p. 11-10.

[10] C. Kittel, Introduction to Solid State Physics, New York: John Willey & Sons, 2005.

[11] A. I. Khan et al., "Negative capacitance in a ferroelectric capacitor," Nature Mater. vol. 14, pp. 182-186, Jan. 2015.

[12] M. Kobayashi et al., "Experimental Study on Polarization-Limited Operation Speed of Negative Capacitance FET with Ferroelectric HfO2," Proc. IEDM, 2016.

[13] M. Hoffmann et al., "Direct Observation of Negative Capacitance in Polycrystalline Ferroelectric HfO2", Adv. Func. Mater, Oct. 2016. https://doi.org/10.1002/adfm.201602869

[14] M. Hoffmann et al., "Ferroelectric negative capacitance domain dynamics," J. of Appl. Phys, vol. 123, pp. 184101-10-184101-10, May 2018.

[15] V. Janovec, "On the theory of the coercive field of single-domain crystals of BaTiO3". Cechoslovackij fiziceskij zurnal, vol. 8, pp. 3-15, 1958.

[15] J.F. Scott, "Switching of ferroelectrics without domains". Adv. Mater., vol. 22, pp. 5315-5317, 2010.

[16] Mulaosmanovic et al., "Switching Kinetics in Nanoscale Hafnium Oxide Based Ferroelectric Field-Effect Transistors", ACS Appl. Mater. Interfaces, vol. 9, 3792−3798, 2017.

Effect of Polycrystallinity and Presence of Dielectric Phases on NC-FinFET Variability

Yen-Kai Lin, Ming-Yen Kao, Harshit Agarwal, Yu-Hung Liao, Pragya Kushwaha, Korok Chatterjee,
Juan Pablo Duarte, Huan-Lin Chang, Sayeef Salahuddin, and Chenming Hu

Dept. of Electrical Eng. and Computer Sciences, University of California, Berkeley, CA, USA, email: yklin@berkeley.edu

Abstract—**A Monte Carlo TCAD simulation study of the impact of polycrystallinity and dielectric phases of the ferroelectric film on an 8/7 nm node NC-FinFET is presented. The study considers the random variation of ferroelectric remnant polarization (P_r) and the presence of dielectric phases. In order to keep the ferroelectric-film induced device variability less than those induced by other sources (RDF, GER, FER, and MGG), we found that the DE content must be less than 20%, which is theoretically possible, and the grain to grain P_r variations less than 27%. While uniform single-crystalline ferroelectric film would provide the least device variation, we found 4 nm grains to produce less device variability than 5.3 nm grains due to the larger number of grains in the channel area.**

I. INTRODUCTION AND METHODOLOGY

Negative capacitance field-effect transistor (NCFET) has the potential of achieving steeper subthreshold slope than simple MOSFET [1], [2]. The ferroelectric (FE) layer sandwiched by the metal gate and interfacial dielectric layer in a NCFET provides voltage amplification [3], which enable future supply voltage (V_{DD}) scaling. Doped HfO_2 [4] is the most promising ferroelectric material for this application due to its compatibility with modern CMOS process. However, the polycrystalline material may contain multiple phases including dielectric (DE) phases [4], [5]. Even the ferroelectric phase grains may have variance in ferroelectric characteristics such as the remnant polarization, P_r, due to local strain variance. In this work, the impact of the granular variability of the FE film on NC-FinFETs is studied.

The NC-FinFETs are simulated using Sentaurus TCAD tool [6], which simultaneously solves Poisson's equation with Landua-Khalatnikov (LK) equation [2]. The baseline FinFET is designed based on the low power (LP) 8/7 nm technology node of the International Roadmap for Devices and Systems (IRDS) [7]. The remnant polarization (P_r) of the doped HfO_2 is an assignable parameter [4]. To support fin spacing scaling, the FE thickness is kept as thin as 2 nm. We assume columnar grains and consider three variation sources: (1) P_r variation among the FE grains, (2) DE to FE grain ratio variation, and (3) grain size. Monte Carlo simulations are carried out to randomly assign the P_r (E_C variation is not considered for simplicity) and whether a grain is DE or FE within the confines of chosen μ and σ (means and sigmas). The grain size is kept constant for simplicity, but the effect of grain size on

FinFET sensitivity to the material variations is separately studied. Fig. 1 shows the simulated device structure and parameters. The channel area is divided into 45 to 92 tiles (grains) depending on the assumed grain size. Note that there is no internal gate in NC-FinFET in this work [8].

II. RESULTS AND DISCUSSION

A. Ferroelectric P_r Variations

Due to local material or strain variations, different FE grains may have varying properties. Fig. 2 shows I_{DS}–V_{GS} of the baseline FinFET and NC-FinFET with $\mu P_r = 20$ and $\sigma P_r = 5$ μC/cm^2 (grain size = 6.2 nm × 6.2 nm). The FE variation leads to spatially non-uniform FE amplification and thus non-uniform current flow (Fig. 3). The channel under the grain with smaller P_r (with better capacitance matching to the channel capacitance [9]) carries more current (Fig. 3). Fig. 4 exhibits the distributions of I_{ON}, I_{OFF}, V_{TH}, and SS. V_{TH} is defined as the V_{GS} where the current (I_{TH}) is equal to 100nA × W / L_G ($W = 2H_{FIN} + T_{FIN}$), and SS is the average subthreshold slope from I_{OFF} to $I_{TH}/10$. As expected, larger P_r variation results in larger device variation. Interestingly, increasing P_r variation also causes the mean values of I_{ON}, I_{OFF}, V_{TH}, and SS to shift in Fig. 4 although the mean P_r is kept constant. This can be understood from the fact that the voltage amplification is a nonlinear function of P_r [9] (Fig. 5). Since the grains with smaller P_r provides more voltage amplification advantage than the disadvantage due to the grains with larger P_r, P_r variations cause variations in I_{ON} and other device properties and shifts of their mean values. Based on Fig. 4, we suggest that P_r variation in the FE should be kept below 25% (e.g., $\mu P_r = 20$ and $\sigma P_r = 5$ μC/cm^2) to ensure that the device performance variations are negligible compared with other variation sources, e.g., random dopant fluctuation (RDF), in FinFETs [10], [11].

B. DE and FE Grain Ratio

Ignoring P_r variation for now, three DE (assumed permittivity is 16) tile probabilities are considered in Fig. 6: 33%, 50%, and 67%. These values are used because the literature suggested values as high as 70% [12] and as low as 0% [5]. The grain size is assumed to be 6.2 nm × 6.2 nm. Higher DE content broadens the device parameter distributions. Of course, one expects 100% DE to cause no variation as shown in Fig. 7. As the DE content increases in Fig. 7, the mean I_{ON} and the mean V_{TH} decrease, while the mean I_{OFF} and the mean SS increase. All four trends support the mental model that the presence of DE effectively increases

978-1-7281-1988-5/18 $31.00 © 2018 IEEE

the EOT of the NC-FinFET. Fig. 8 shows the current flow contour of the maximum I_{ON} case for 33% DE at $V_{GS} = V_{DS} = V_{DD}$. The low current density regions are over DE grains as expected, showing the effect of effective larger local EOT and the absence of the voltage amplification effect provided by FE. Note that the device variations due to the assumed DE content are much more harmful than that due to P_r variation (Fig. 4). Fortunately, a theoretical study [5] has shown that 0% DE (pure FE) is possible. The material should be carefully engineered to get rid of the DE grains. In general, the performances of NC-FinFET with DE grains can be understood as a baseline FinFET with thicker interfacial EOT (100% DE would mean the thickest EOT) and partial voltage amplification benefit of a 0% DE NC-FinFET. Empirically the relation between the DE content and device performance parameters can be described by the bowing equation

$$F(\mathrm{FE}_{1-x}\mathrm{DE}_x) = F(\mathrm{FE})\cdot(1-x) + F(\mathrm{DE})\cdot x - b\cdot x\cdot(1-x), \quad (1)$$

where F can be I_{ON}, I_{OFF}, V_{TH}, and SS, x is the content of DE ranging from 0 to 1, and b is the bowing parameter. Fig. 7 shows the empirical bowing parameters. Note that the device with 100% DE has a thicker EOT ($= 0.9$ nm $+ 2 \times 3.9 / 16$ nm ≈ 1.39 nm) than the baseline FinFET (EOT $= 0.9$ nm).

C. Combined Effects of DE and P_r Variation

The combined effect of DE phase and P_r variation is shown in Fig. 9 and Fig. 10. The DE variation is the dominant source of device parameter variations (Fig. 10) as mentioned earlier. In order to keep FE induced device variations below those due to other FinFET variation sources [e.g. gate edge roughness (GER), fin edge roughness (FER), and metal gate granularity (MGG)] [10], [11], [13], the DE content must be less than 20% and σP_r less than 27% of the mean. Fig. 11 further shows the coefficient of variation at different DE content. The normal Q-Q test on the I_{ON}, I_{OFF}, and V_{TH} distributions due to the DE content variation shows that they closely follow the Gaussian distribution (Fig. 12). Some outliers of I_{ON}, I_{OFF}, and V_{TH} are observed in 60% DE case due to stronger short channel effect.

D. Grain Size Effect on Device Variations

The effect of grain size (comparing 4 nm $\times$ 4 nm and 5.3 nm $\times$ 5.3 nm) are also investigated (Fig. 13 and Fig. 14). Grain size can be engineered through the stress and doping [5] and perhaps deposition method. Small grain reduces the device variations because the larger number of grains in each device reduces the number of outliers. Indeed, the smaller grain cases in Fig. 14 follows the Gaussian distribution well. The small grain can average out the variation from the adjacent grains. Similar results have been observed in metal gate granularity (MGG) effect [11]. In addition to I_{ON}, I_{OFF}, V_{TH}, and SS, the effective drive current (I_{EFF}) and output resistance (R_{out}) are also important from circuit perspective. Fig. 15 shows the quantile plot of I_{EFF} and R_{out}. FE induced I_{EFF} and R_{out} variations of NC-FinFET is less than the variations induced by other sourced in the baseline FinFET if the DE content is kept below 20% and P_r variation below 27 %.

III. CONCLUSION

Monte Carlo simulation of NC-FinFET variations induced by remnant polarization variation and the presence of dielectric phases in the ferroelectric film is presented. It is found that the DE variation is a more serious source of NC-FinFET variation than the P_r variaion. To keep the FE variation effects below those induced in FinFET by other sources (RDF, GER, FER, and MGG), the DE content must be less than 20% which is theoretically possible. The P_r variation must be less than 27%. Our simulation shows that 4 nm $\times$ 4 nm grains (92 grains in the channel) leads to less device variations than 5.3 nm $\times$ 5.3 nm grains (54 grains in the channel). While uniform single crystalline ferroelectric film would minimize the device variation, nano-crystalline ferroelectric film may be preferrable to poly-crystalline film.

ACKNOWLEDGMENT

The authors gratefully acknowledge the support of Berkeley Device Modeling Center (BDMC) and Berkeley Center for Negative Capacitance Transistors (BCNCT).

REFERENCES

[1] S. Sayeef and S. Datta, "Use of negative capacitance to provide voltage amplification for low power nanoscale devices," *Nano Lett.*, vol. 8, no. 2, pp. 405–410, 2008.

[2] J. P. Duarte *et al.*, "Compact models of negative-capacitance FinFETs: lumped and distributed charge models," in *IEDM Tech. Dig.*, Dec. 2016, pp. 754–757.

[3] K.-S. Li *et al.*, "Sub-60mV-swing negative-capacitance FinFET without hysteresis," in *IEDM Tech. Dig.*, Dec. 2015, pp. 620–623.

[4] M. H. Park *et al.*, "Ferroelectricity and antiferroelectricity of doped thin HfO2-based films," *Adv. Mater.*, vol. 27, no. 11, pp. 1811–1831, Mar. 2015.

[5] Y.-T. Tang *et al.*, "A comprehensive study of polymorphic phase distribution of ferroelectric-dielectrics and interfacial layer effects on negative capacitance FETs for sub-5nm node," in *VLSI Symp. Tech. Dig.*, Jun. 2018, pp. 45–46.

[6] *Sentaurus Device User Guide, Version N-2017.09*, Synopsys, Mountain View, CA, USA, Sep. 2017.

[7] (2016). More Moore White Paper–International Roadmap for Devices and Systems. [Online]. Available: http://irds.ieee.org/reports.

[8] M. Hoffmann *et al.*, "On the stabilization of ferroelectric negative capacitance in nanoscale devices," *Nanoscale*, vol. 10, pp. 10891–10899, May 2018.

[9] H. Agarwal *et al.*, "NCFET design considering maximum interface electric field," *IEEE Electron Device Lett.*, vol. 39, no. 8, pp. 1254–1257, Aug. 2018.

[10] X. Wang *et al.*, "Interplay between process-induced and statistical variability in 14-nm CMOS technology double-gate SOI FinFETs," *IEEE Trans. Electron Devices*, vol. 60, no. 8, pp. 2485–2492, Aug. 2013.

[11] X. Wang et al., "Statistical variability and reliability in nanoscale FinFETs," in *IEDM Tech. Dig.*, Dec. 2011, pp. 103–106.

[12] L. Xu *et al.*, "Kinetic pathway of the ferroelectric phase formation in doped HfO2 films," *J. Appl. Phys.*, vol. 122, no. 12, pp. 124104, Sep. 2017.

[13] S. Natarajan *et al.*, "A 14 nm logic technology featuring 2nd-generation FinFET transistors, air-gapped interconnects, self-aligned double patterning and a 0.0588µm^2 SRAM cell size," in *IEDM Tech. Dig.*, Dec. 2014, pp. 71–73.

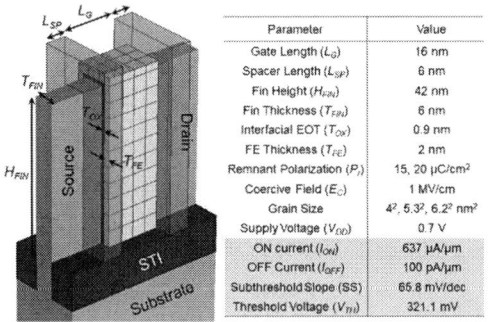

Fig. 1. Simulated 8/7 nm NC-FinFET structure and device parameters. The shadowed rows present the baseline FinFET performance.

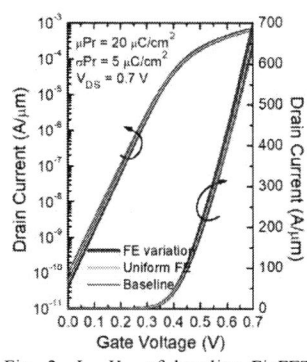

Fig. 2. I_{DS}–V_{GS} of baseline FinFET and NC-FinFET with P_r variation only.

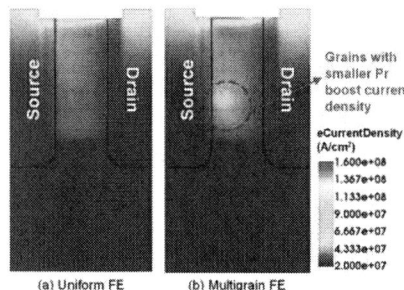

Fig. 3. Current density in NC-FinFETs with (a) uniform FE with 20 μC/cm² P_r and (b) FE variation with σP_r = 5 μC/cm² (maximum I_{ON} case shown).

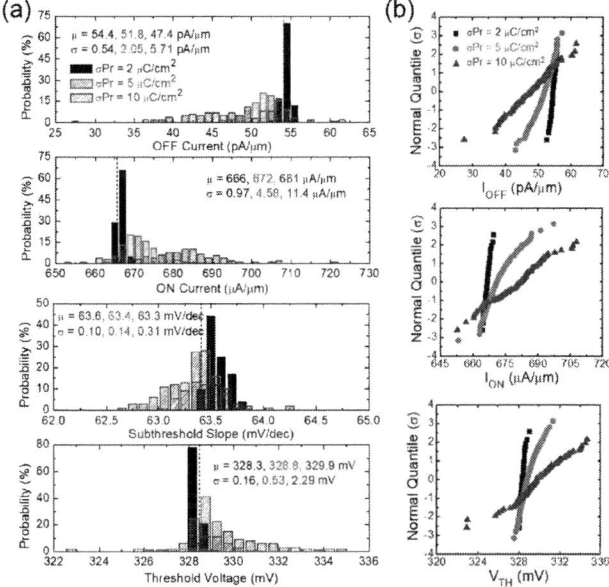

Fig. 4. (a) Distributions of I_{OFF}, I_{ON}, V_{TH}, and SS in NC-FinFET with μP_r = 20μC/cm², σP_r = 2, 5, and 10 μC/cm². The dashed lines represent NC-FinFET with uniform FE (P_r = 20μC/cm²). (b) Normal quantile plots.

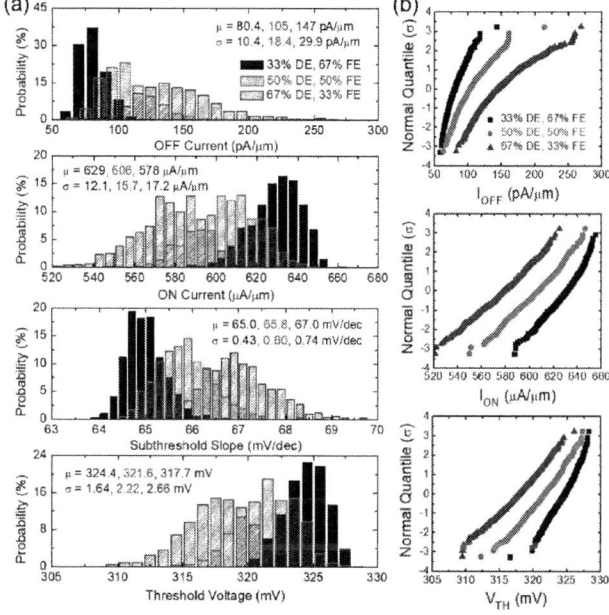

Fig. 6. (a) Distributions of I_{OFF}, I_{ON}, V_{TH}, and SS in NC-FinFET with DE variation (33%, 50%, and 67% DE). (b) Normal quantile plots.

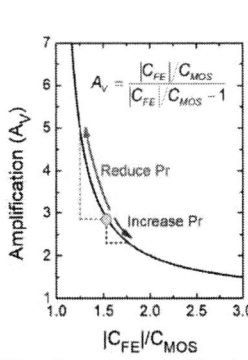

Fig. 5. Voltage amplification (A_V) as a function of $|C_{FE}|/C_{MOS}$ (P_r), showing nonlinearity of A_V when changing P_r.

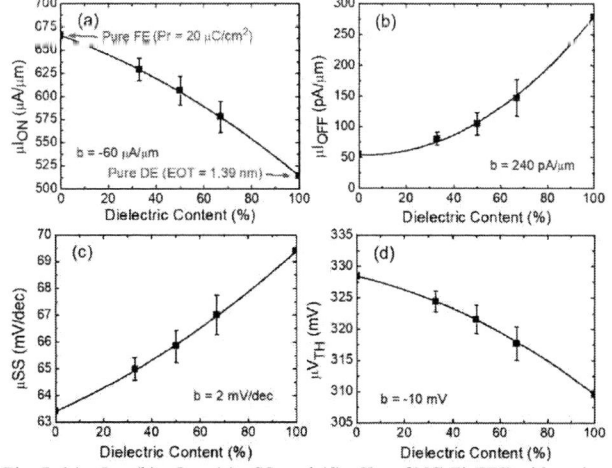

Fig. 7. (a) μI_{ON}, (b) μI_{OFF}, (c) μSS, and (d) μV_{TH} of NC-FinFET with various DE content. The bars show the ±σ. The solid lines represent the bowing equations with bowing parameter b. Note that 0% DE and 100% DE are NC-FinFET with uniform ferroelectric film and FinFET with thicker EOT (= 1.39 nm) than the baseline EOT (= 0.9 nm).

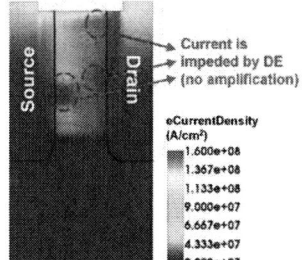

Fig. 8. Current density in NC-FinFET with 33% DE (maximum I_{ON} case shown). The dark blue regions are covered by DE.

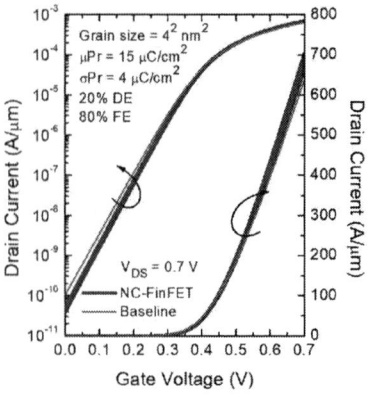

Fig. 9. I_{DS}–V_{GS} of baseline FinFET and NC-FinFET with 20% DE content and P_r variations.

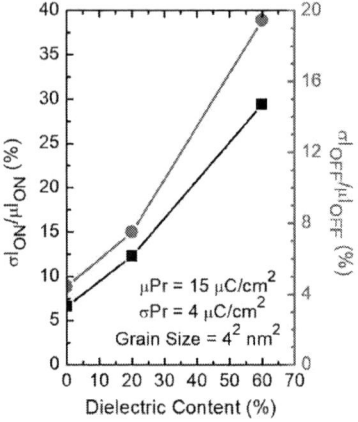

Fig. 11. Coefficient of variation of I_{ON} and I_{OFF} for different DE content

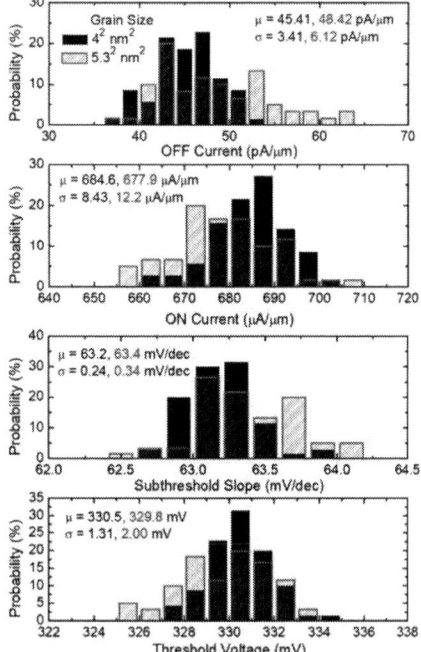

Fig. 13. Grain size effect on the distributions of device performances (20% DE, $\mu P_r = 15$ μC/cm² and $\sigma P_r = 4$ μC/cm²).

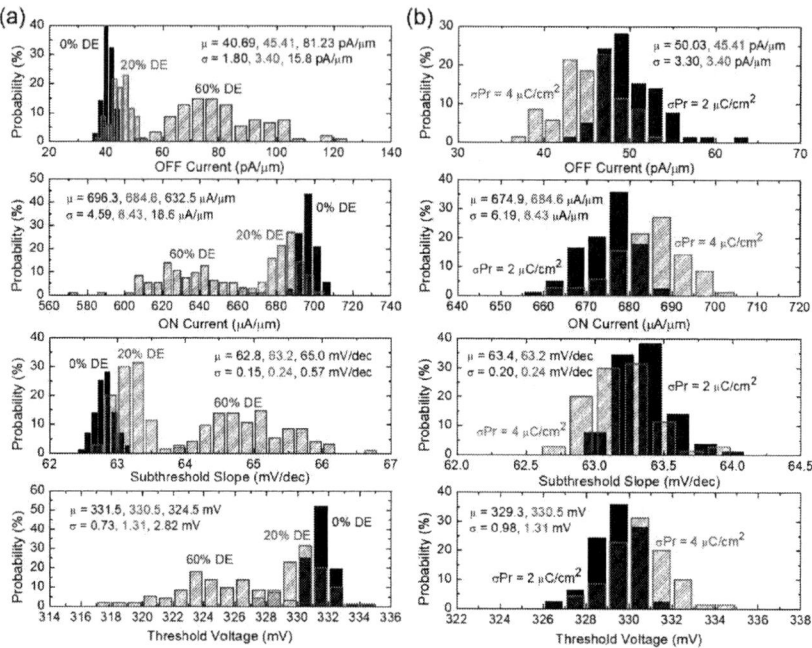

Fig. 10. Statistical distributions of NC-FinFET with (a) different DE content variations ($\mu P_r = 15$ μC/cm² and $\sigma P_r = 4$ μC/cm²) and (b) different P_r variations (20% DE and $\mu P_r = 15$ μC/cm²). The DE content variation has more impact on NC-FinFET.

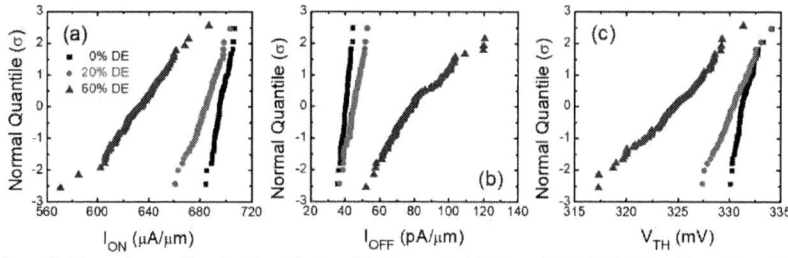

Fig. 12. Normal quantile plot for (a) I_{ON}, (b) I_{OFF}, and (c) V_{TH} of NC-FinFET with various DE content (grain size = 4² nm², $\mu P_r = 15$ μC/cm² and $\sigma P_r = 4$ μC/cm²).

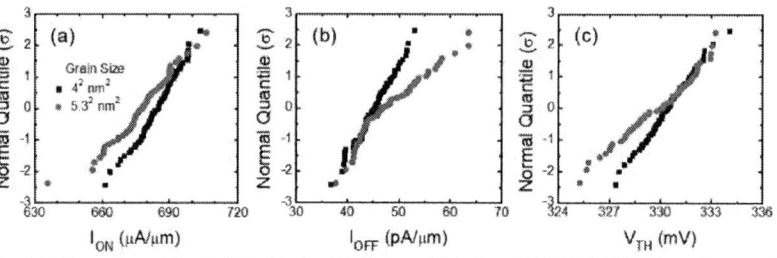

Fig. 14. Normal quantile plot for (a) I_{ON}, (b) I_{OFF}, and (c) V_{TH} of NC-FinFET with various grain sizes (20% DE, $\mu P_r = 15$ μC/cm² and $\sigma P_r = 4$ μC/cm²).

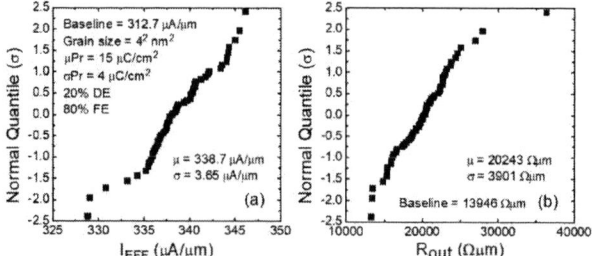

Fig. 15. Normal quantile plot for (a) I_{EFF} and (b) R_{out} of NC-FinFET considering both P_r and DE variations. The coefficient of variation of I_{EFF} (~10.77%) is less than the I_{ON} variation in Fig. 13 and is a better indicator of circuit speed variation.

A simulation based study of NC-FETs design: off-state versus on-state perspective

T. Rollo[1], H. Wang[1,2], G. Han[2], D. Esseni[1]

[1] DPIA, University of Udine, Via delle Scienze 206, 33100 Udine, Italy; email: david.esseni@uniud.it;
[2] School of Microelectronics, Xidian University, China, email: hangenquan@ieee.org;

I. Abstract

This paper presents new analytical and numerical models aiming at a better insight about the physics and design of ferroelectric NC-FETs. We argue that a design focused on the off-state and targeting steep slope with negligible hysteresis is unlikely to be successful. A design targeting an enhanced on-state capacitance is instead more feasible, and can improve both sub-threshold swing and on-current. Also, NC-FETs can reduce the temperature sensitivity compared to baseline FETs, but the sensitivity to dielectrics thickness is critical.

II. Introduction

Energy efficiency and aggressive V_{DD} scaling have become primary needs for CMOS technologies. These is made challenging by the contrasting need to keep a large ratio $[I_{on}/I_{off}]$, which is ultimately set by the average transconductance, g_m, of the FETs in the V_{DD} window. Steep slope transistors intend to improve $[I_{on}/I_{off}]$ thanks to a subthreshold swing, SS, below 60mV/dec at room temperature [1], [2], [3], [4].

A ferroelectric Negative-Capacitance FET (NC-FET) can be designed with a negative overall dielectric capacitance C_{di} (see **Fig.1(b)**) to have SS<60mV/dec [3], or with a positive C_{di} to enhance g_m and I_{DS} in the on-state [5].

We here report a comprehensive study about the design space, feasibility and challenges for the steep slope or g_m boosted operation of NC-FETs. Novel numerical and analytical models are developed to investigate the influence of traps, temperature and variations of the dielectric films thickness.

III. Numerical modelling of ferroelectric and NC-FETs

A. Model for the ferroelectric material and calibration

We describe the ferroelectric dynamics by using the multi-domain time-dependent Landau-Khalatnikov Equation (LKE) in **Eq.1** [5]. We calibrated the model against data for large area metal-ferroelectric-metal structures [6], where multi-domain effects are important [5]. **Fig.2** shows a good agreement with experiments at different temperatures (T), obtained thanks to a randomness of a_i, b_i, c_i corresponding to a $\pm 0.45 MV/cm$ dispersion of the coercive field [5]. From the fairly linear a versus T plot we extracted $\alpha_0 = 1.3 \cdot 10^6$ m/(F K) and $T_0 = 923$ K (see **Eq.2**), with T_0 being inside the reported range [7].

B. Model for a nanoscale NC-FET

For the double-gate UTB NC-FETs shown in **Fig.1(a)** we used a ballistic top-of-the-barrier (ToB) FET model [8], that grasps the essential MOSFET physics and compares well with experiments [9]. A single domain analysis is used for nanoscale NC-FETs, as the channel length is comparable to the estimated size of ferroelectric domains [10]. The semiconductor is described by a 1D, parabolic effective mass Schrödinger

solver **Eqs.3**, where (ν,n) indicate (valley,subband) and m_z, m_d correspond to a [100] silicon interface [11]. The n_S and n_D in **Eq.3** correspond to electrons with positive and negative velocity, which are taken in equilibrium with respectively the source, $E_{f,S}$, and drain Fermi level $E_{f,D} = (E_{fS} - qV_{DS})$ [8]. **Eqs.3** are solved self-consistently with the 1D Poisson **Eq.4**, where p, N_A are hole and dopant concentrations. The link between semiconductor and dielectrics is given by the gate voltage equation in **Eq.5** (see **Fig.3**), and by equating the overall semiconductor charge to the dielectrics charge $Q = C_{ox}V_{ox} = (P + \varepsilon_0 F_{fe})$. The C_S, C_D describe the influence of V_S, V_D on ϕ_s (see **Fig.1(b)**) [8]; I_{DS} is given by **Eq.6**.

C. Dynamic equations for interface traps

We solve traps continuity equation in **Eq.4**, where n_t is the electron density in the traps at energy E_t having a concentration N_t. We consider acceptor-type traps in the upper half of the energy gap, that exchange electrons with the conduction band with an emission, e_n, and capture rate c_n At any time step, Q_{it} entering **Eq.5** is given by $Q_{it} = -q \sum_{E_t} n_t(E_t)$.

All dynamic equations were solved using a Runge-Kutta algorithm with automatic time step adjustment. The I_{DS} versus V_g curves were obtained by simulating many cycles of a triangular V_g waveform so as to reach a periodic regime.

IV. NC-FET design for steep sub-threshold swing

In this paper we discuss two design options for NC-FETs identified by the sign of the capacitance C_{di} defined in **Fig.1**, with $C_{di}^{-1} = C_{ox}^{-1} + C_{fe}^{-1} = C_{ox}^{-1} - |C_{fe}|^{-1}$. For small Q values the static LKE in **Eq.2** allows us to write $C_{fe}^{-1} = (\partial V_{fe}/\partial Q) \sim \alpha_0(T - T_0)T_{fe}$, so that the condition $C_{di} \gtrless 0$ leads to the first inequality in **Eq.8**, where the steep slope design corresponds the lower case (i.e $C_{di} < 0$).

According to **Eq.9** an NC-FET with negative C_{di} can have an $SS = (K_B T)/q) \ln(10)(\partial \phi_s/\partial V_g)^{-1} < 60$mV/dec if $(C_{sct} + C_p + C_{it})$ is comparable to $|C_{di}|$, with C_{sct}, $C_p = (C_S + C_D)$ (see **Fig.1**) and C_{it} being the intrinsic semiconductor, parasitic and trap capacitance. In UTB devices C_{sct} is much smaller than $|C_{di}|$ in sub-threshold and, neglecting momentarily traps (i.e. $C_{it} \simeq 0$), the capacitance matching can be improved by increasing C_p. However C_p also degrades SS (see **Eq.9**) and DIBL of the baseline transistor, so that for well behaved FETs C_p is typically a small fraction of C_{ox}.

Eq.9 also suggests that, when C_{di} is negative, interface traps can improve SS in contrast to the detrimental effect in baseline FETs [15]. While the analysis of [15] assumed a quasi-static trap response, C_{it} in **Eq.9** is given only by those traps that can follow the V_g waveform. In this work we improve well beyond

[15], and we solve the dynamic SRH type model in **Eqs.7** self-consistently with the LKE and Schrödinger-Poisson problem (see Sec.III.C). **Fig.4** reports I_{DS} (per gate) versus V_g curves of a steep slope NC-FET and for different frequencies of the triangular V_g waveform. The ferroelectric resistivity was set to $\rho = 0.5\Omega m$ throughout the work, namely in the range necessary for a GHz operation of NC-FETs [12]. **Fig.4** confirms that a constant density D_{it} of acceptor traps in the upper half of the energy gap can induce a sub-60mV/dec SS value at low frequencies. However the beneficial effects of traps vanish already in the tens of MHz for the trap parameters from [13], because a progressively larger fraction of traps cannot contribute to C_{it}. This is clearly illustrated in **Fig.5** reporting the occupation versus time of some trap energy levels. As it can be seen deeper traps can capture an electron for positive V_g, but cannot emit for negative V_g, so that they behave similarly to negative fixed charges. The asymmetric capture and emission rates are inherent to the model in **Eq.3**, where e_n depends only on the trap depth while c_n becomes much larger than e_n when E_t is driven below the Fermi level $E_{f,S}$.

Fig.4 also shows that the sub-60mV/dec is accompanied by hysteresis, as actually observed in most experimental reports [17], [18], [19], because in inversion C_{sct} increases so as to make $(C_{sct}+C_p+C_{it})$ larger than $|C_{di}|$ [3], [5].

V. NC-FET design for on-state boosting

The upper case in the first inequality of **Eq.8** identifies the positive C_{di} design of NC-FETs discussed in this section. For this design the voltage gain $(\partial V_{ox}/\partial V_g)$ gives the g_m and I_{DS} gain in the on-state [5], and from the $(\partial V_{ox}/\partial V_g)$ expression in **Eq.10** one infers the second inequality in **Eq.8**. The inequalities in **Eq.8** define the design space illustrated in **Fig.6** for EOT=0.5 nm. As it can be seen the design space becomes narrower when G_{ox} increases, thus suggesting a sensitivity to the design parameters that is discussed below.

Fig.7(a) compares the $I_{DS}-V_g$ curves of NC-FETs with the corresponding baseline FET (having same EOT and C_p) at fixed $I_{off}=100nA/\mu m$. The large I_{DS} values are due to the fact that neither scattering nor series resistance are included. NC-FETs have improved g_m and I_{DS} in the on state (see $(\partial V_{ox}/\partial V_g)$ in **Eq.10**), and also an almost ideal SS (for $D_{it} \simeq 0$) thanks to the large $C_{di} > C_{ox}$ (see **Eq.9**). **Fig.7(b)** reports the I_{DS} gain compared to the baseline FET versus V_g: the I_{DS} improvements are consistent with recent experiments [14], and are fairly constant in the on state. For all simulations the maximum oxide field is smaller than 10 MV/cm [20].

A. Sensitivity to Temperature

Fig.8 reports I_{DS} versus V_g curves for three T values. In the off-state all FETs suffer an SS degradation with increasing T (not shown). In the on-state the I_{DS} curve for the baseline FET has an expected rigid shift, where V_T decreases with increasing T [16]. For the NC-FET, instead, by increasing T the I_{DS} is enlarged at small V_g but it is reduced at large V_g, so that a V_{GT0} exists where I_{DS} is insensitive to T. This interesting behavior is due to the fact that, when T is increased for a

fixed charge Q, the surface potential $\phi_s = -E_{C,s}/q$ decreases, but the negative V_{fe} increases. We developed an analytical model for $(\partial V_g/\partial T)$ at fixed Q by taking $C_S \simeq C_D \simeq 0$ and considering an equilibrium, quantum limit condition (i.e. entire Q carried by the lowest subband, see **Eq.11**). The $(\partial V_{fe}/\partial T)$ in **Eq.12** is readily given by **Eq.1**. To calculate $(\partial \phi_s/\partial T)$ we assumed that ε_0 moves rigidly with ϕ_s, and then derived $(\partial \varepsilon_0/\partial T)$ by setting $(\partial Q/\partial T)=0$ via **Eq.11**. We now go back to the gate voltage equation in **Eq.5** and notice that $(\partial V_{ox}/\partial T)=0$ at fixed Q, so that $(\partial V_g/\partial T)$ can be written as in **Eq.13** (for $(\partial \Phi_M/\partial T)\simeq 0$). By setting $(\partial V_g/\partial T)=0$ in **Eq.13** and substituting $(\partial \varepsilon_0/\partial T)$ from **Eq.12**, we obtain an equation that can be solved for the Q_{T0}. **Fig.9** compares the Q_{T0} versus α_0 and T_{fe} curves obtained from **Eq.13** with results from numerically calculated Q versus V_g curves. The analytical model tracks numerical results quite well and its accuracy improves at small T_{Si} that favours the quantum limit.

B. Sensitivity to EOT and T_{fe}

We now introduce the drive capacitance $C_G = (\partial Q/\partial V_g)$, that for $(C_{sct}+C_p) \gg |C_{fe}|$ can be approximated as in **Eq.14** [15], from which one readily derives the $(\partial C_G/\partial EOT)$ and $(\partial C_G/\partial T_{fe})$ expressions also reported in **Eq.14**. As it can be seen, designing for a large G_{ox} implies also a strong sensitivity to EOT, T_{fe}. This is clearly confirmed by the I_{DS} versus V_g curves of **Fig.10**, and then by the corresponding I_{ON} sensitivity to EOT and T_{fe} reported in **Figs.11** and **12**.

VI. Conclusions

We presented novel models and analyzed different design scenarios for NC-FETs. We conclude that a positive C_{di} design can improve both the off- and the on-state operation of NC-FETs compared to baseline FETs. Temperature dependence of the semiconductor and ferroelectric voltage is opposite, that may be used to obtain temperature insensitive $I_{DS}-V_g$ curves. The sensitivity to dielectrics thickness is instead critical.

REFERENCES

[1] A. C. Seabaugh *et al.*, *IEEE Proceedings*, pp. 2097-2107, 2010
[2] D. Esseni *et al.*, *Semicond. Science and Techn.*, vol. 32, p. 083005, 2017
[3] S. Salahuddin *et al.*, *Nano Letters*, vol.8, n.2, 2008
[4] H. Wang *et al.*, *IEEE EDL*, vol.39, n.3, 2018
[5] T. Rollo *et al.*, *IEEE EDL*, vol.39, n.4, pp.603-606, 2018
[6] D. Zhou *et al.*, *Acta Materialia*, vol.99, pp.240-246, 2015
[7] T. Shimizu *et al.*, *Scientific Reports*, vol.6, 2016
[8] A. Rahman *et al.*, *IEEE TED*, vol.50, n.9 , 2003
[9] S. Rakheja *et al.*, *IEEE TED*, vol.62, n.9, 2015
[10] A. Roelofs *et al.*, *Nanotechnology*, vol.14, n.2, pp.250-253, 2003
[11] D.Esseni *et al.*, Nanoscale MOS Transistors, *Cambr. Univ. Press*, 2011
[12] S.-C. Chang *et al.*, *IEEE Journal on Exploratory SSCDC*, 2018
[13] G. Brammertz *et al.*, *App. Physics Letters*, vol.91, n.13, 2007
[14] Z. Krivokapic *et al.*, *IEEE IEDM*, pp. 15.1.1-15.1.4, 2017
[15] T. Rollo *et al.*, *IEEE EDL*, vol.39, n.7, pp.1100-1103, 2018
[16] S.M. Sze *et al.*,Physics of Semic. Dev., *Wiley-Interscience*, 1981
[17] M.H. Lee *et al.*, *IEEE IEDM*, pp.12.1.1-12.1.4, 2016
[18] P. Sharma *et al.*, *VLSI Symposium*, pp.T154-T155, 2017
[19] E. Ko *et al.*, *IEEE EDL*, vol.38, n.4, pp.418-421, 2017
[20] H. Agarwal *et al.*, *IEEE EDL*, vol.39, n.8, pp.1254-1257, 2018

LKE Module	Multi domain LKE (large area structures): $\rho\, dQ_i/dt = -(a_i Q_i + b_i Q_i^3 + c_i Q_i^5) + V_{fe,i}/T_{fe} + k\sum_j (Q_j - Q_i)$	(1)
	Static LKE (single domain): $V_{fe} = T_{fe}(aQ + bQ^3 + cQ^5)$ Temperature dependence: $a(T) = a_0(T - T_0)$	(2)
Semiconductor	$\left[\dfrac{-\hbar^2}{2m_z}\dfrac{\partial^2}{\partial z^2} + U(z)\right]\xi = \varepsilon\,\xi,$ $n_{S(D)}(z) = \sum_{\nu,n}|\xi_{\nu,n}(z)|^2 \dfrac{\mu_\nu m_{d,\nu} K_B T}{\pi\hbar^2}\ln\left(1 + e^{\eta_{\nu,n}^{S(D)}}\right),$ $\eta_{\nu,n}^{S(D)} = \dfrac{E_{f,S(D)} - \varepsilon_{\nu,n}}{K_B T}$	(3)
	Poisson equation in Si: $\varepsilon_{Si}\dfrac{\partial^2\phi(z)}{\partial z^2} = q[n_S(z) + n_D(z) - p(z) + N_A]$	(4)
Transistor	$V_g = (\Phi_M - \chi_{sct}) + V_{fe} + V_{ox} + (\phi_s - \phi_{F,s}),$ $Q = -Q_{sct} - Q_{it} + C_S(\phi_s - V_S) + C_D(\phi_s - V_D)$	(5)
	$I_{DS} = \dfrac{q}{\sqrt{2}\hbar^2}\left[\dfrac{K_B T}{\pi}\right]^{3/2}\sum_{\nu,n} n_\nu \sqrt{m_y}\left[F_{1/2}(\eta_{\nu,n}^S) - F_{1/2}(\eta_{\nu,n}^D)\right]$ $F_{1/2}(\eta)$: Fermi integral order 1/2	(6)
Interface Traps	$\partial n_t/\partial t = c_n(N_t - n_t) - e_n n_t,$ $e_n = \sigma\, v_{th}\, N_C \exp[(E_t - E_C)/(K_B T)],$ $c_n = e_n \exp[(E_{f,S} - E_t)/(K_B T)]$	(7)
Design	$\boxed{C_{di} \gtrless 0}$: $a_0(T_0 - T)T_{fe} \lessgtr \dfrac{T_{ox}}{\varepsilon_{ox}} = \dfrac{EOT}{\varepsilon_{SiO_2}}$ $\boxed{\dfrac{\partial V_{ox}}{\partial \mathbf{V}_g} > G_{ox}}$: $a_0(T_0 - T)T_{fe} > \dfrac{EOT}{\varepsilon_{SiO_2}}\cdot\dfrac{G_{ox} - 1}{G_{ox}}$	(8)
equations	NC-FET: $\left(\dfrac{\partial\phi_s}{\partial V_g}\right)^{-1} = \left[1 \pm \dfrac{C_{sct} + C_p + C_{it}}{|C_{di}|}\right],$ $'\pm'$ for $C_{di} \gtrless 0$ Baseline FET: $\left(\dfrac{\partial\phi_s}{\partial V_g}\right)^{-1} = \left[1 + \dfrac{C_{sct} + C_p + C_{it}}{C_{ox}}\right]$	(9)
	NC-FET for $C_{di} > 0$ and $(C_{sct} + C_p) \gg |C_{fe}|$: $\dfrac{\partial V_{ox}}{\partial \mathbf{V}_g} \simeq \dfrac{|C_{fe}|}{|C_{fe}| - C_{ox}}$	(10)
Temperature	Equilibrium, quantum limit: $Q \simeq qD_0 K_B T \ln(1 + e^{\eta_0})$ with $D_0 = \dfrac{\mu_0 m_{d,0}}{\pi\hbar^2}$ $\eta_0 = \dfrac{E_f - \varepsilon_0}{K_B T}$	(11)
dependence	$\dfrac{\partial V_{fe}}{\partial T}\bigg|_Q = a_0 T_{fe} Q$ $\dfrac{\partial\phi_s}{\partial T}\bigg|_Q \simeq \dfrac{-1}{q}\dfrac{\partial\varepsilon_0}{\partial T}\bigg|_Q = \dfrac{-K_B}{q}\left\{\dfrac{Q/(qD_0 K_B T)}{1 - \exp[-Q/(qD_0 K_B T)]} - \ln\left[e^{Q/(qD_0 K_B T)} - 1\right]\right\}$	(12)
	$\dfrac{\partial V_g}{\partial T}\bigg|_Q \simeq -\dfrac{\partial\chi_{sct}}{\partial T} + a_0 T_{fe} Q - \dfrac{1}{q}\dfrac{\partial\varepsilon_0}{\partial T}\bigg|_Q$ $\implies$ solve for Q_{T0} such that $\dfrac{\partial V_g}{\partial T}\bigg|_Q = 0$	(13)
EOT, T_{fe}	$C_G \simeq C_{ox} G_{ox}$ with $G_{ox} = \dfrac{|C_{fe}|}{|C_{fe}| - C_{ox}}$ $\implies$ $\dfrac{\partial C_G}{\partial EOT} = \dfrac{-C_{ox}}{EOT}\left[G_{ox} + \dfrac{C_{ox}}{|C_{fe}|}G_{ox}^2\right]$ $\dfrac{\partial C_G}{\partial T_{fe}} = \dfrac{C_{ox}^2}{|C_{fe}|T_{fe}}G_{ox}^2$	(14)

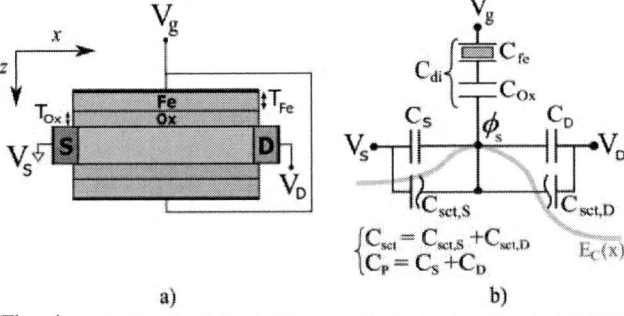

Fig. 1: (a) Sketch of the double-gate thin body, ferroelectric NC-FET used in the simulations of this work, where C_{di} is the series dielectric capacitance $C_{di}^{-1} = C_{ox}^{-1} + C_{fe}^{-1} = C_{ox}^{-1} - |C_{fe}|^{-1}$. (b) Physical picture behind the ToB ballistic FET model. The ToB potential, ϕ_S, is controlled by V_g and it is influenced by V_S, V_D through parasitic capacitances C_S, C_D (see **Eqs.5**). $C_{sct,S(D)} = \partial n_{S(D)}/(\partial\phi_s)$ (with n_S, n_D in **Eq.3**) are the intrinsic semiconductor capacitances, implicitely included via the self-consistent Schrödinger-Poisson solution.

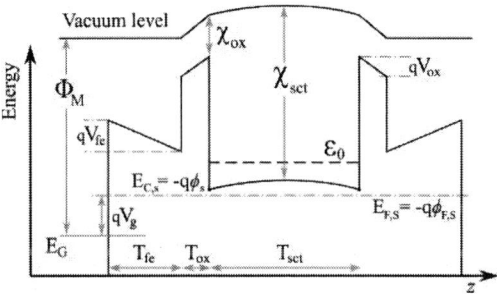

Fig. 3: Band diagram and vacuum level for an n-type NC-FET that supports **Eq.5**. Here χ_{sct}, Φ_M are the semiconductor electron affinity and the metal gate work-function. $E_{C,s} = -q\phi_s$, $E_{F,S} = -q\phi_{F,S}$ are the interface conduction band and source Fermi level.

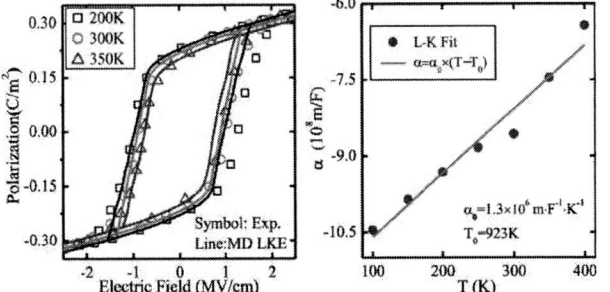

Fig. 2: Calibration of the multi domain LKE model in **Eq.1** (squares, circles and triangles are respectively for T=200, 300, 350K), by comparison to experimental polarization versus field curves at different temperatures from [6]. From the temperature dependence of the nominal a values we extracted a_0 and T_0 defined in **Eq.2**.

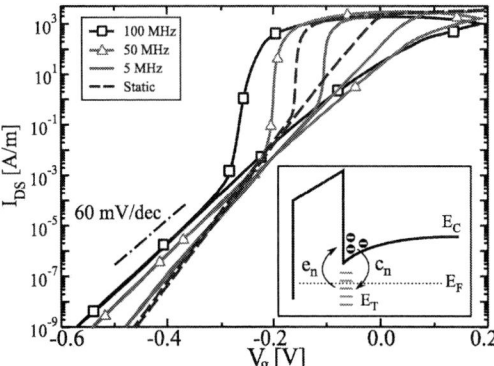

Fig. 4: I_{DS} (per gate) versus gate voltage for a steep slope NC-FET with T_{fe}=20 nm and a concentration D_{it}=10^{13}cm^{-2}/eV of acceptor type in the upper half of the energy gap and for different frequencies of the triangular V_g waveform. Inset shows a sketch of emission and capture processes. Parameters used in simulations are σ=10^{-15}cm^2, v_{th}=2.3·10^7 cm/s, N_C=3.2·10^{19} cm^{-3}, experimentally extracted for a Si-SiO$_2$ interface [13].

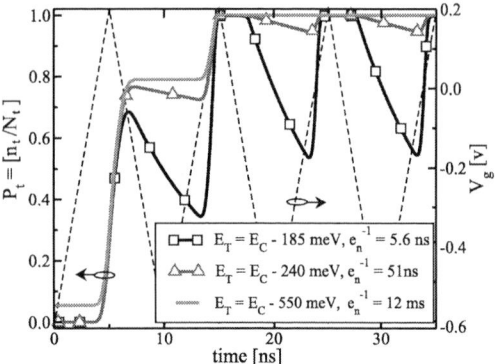

Fig. 5: Occupation probability $P_t = n_t/N_t$ for some traps corresponding to the 100MHz curve in **Fig.4**; the V_g waveform is also shown (right, y axis). Deeper traps have a larger emission time e_n^{-1} (see **Eq.7**) and so cannot follow the externally applied bias.

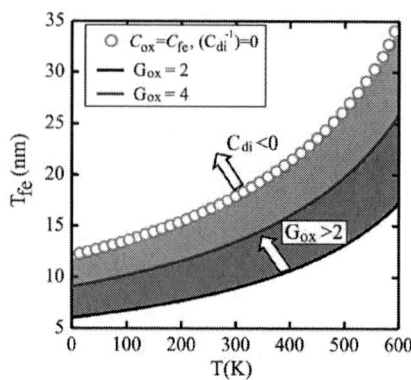

Fig. 6: Design space for EOT=0.5 nm in the T_{fe} versus temperature plane as defined by **Eq.8** and corresponding to $C_{di}>0$ and $\partial V_{ox}/\partial V_g > G_{ox}$. The design space is shown for fixed EOT=0.5nm and different G_{ox} values.

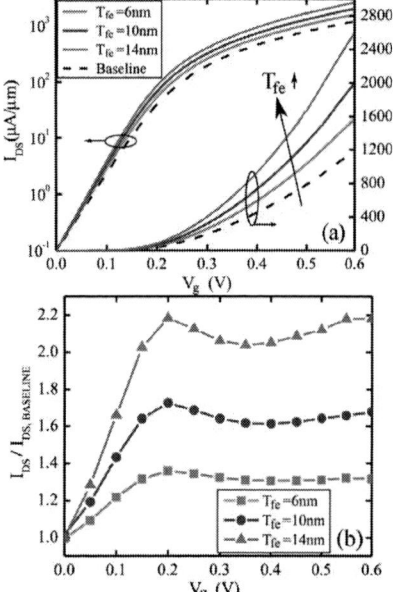

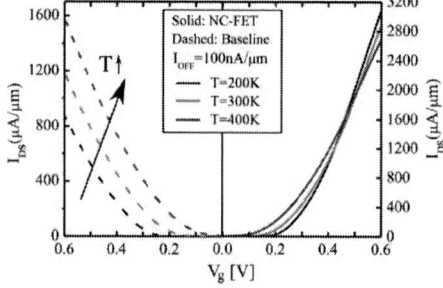

Fig. 8: I_{DS} versus V_g curves at $V_{ds}=0.6$V for different temperatures, for the baseline and a sample NC-FET with EOT= 0.5 nm, $T_{fe}=15$ nm and $\alpha_0=1.34 \cdot 10^6$ m/(F K).Si electron affinity χ_{Si} versus T from [16]. The curves at different T have a crossing point for the NC-FET.

Fig. 7: a) I_{DS} versus V_g curves at $V_{ds}=0.6$V and fixed $I_{off}=100$ $nA/\mu m$ for the baseline FET and NC-FETs with different T_{fe}; EOT=0.5nm and positive C_{di}. b) I_{DS} gain [$I_{DS}/I_{DS,baseline}$] versus V_g.

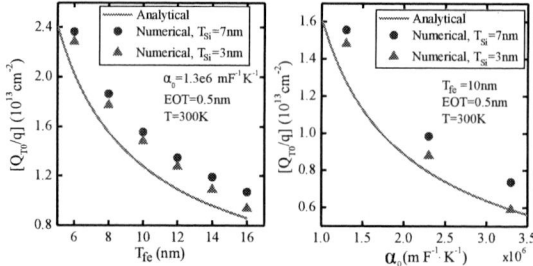

Fig. 9: Charge Q_{T0} corresponding to $\partial V_g/\partial T=0$ at fixed Q versus the material parameter T_{fe} (left) and versus α_0 (right).

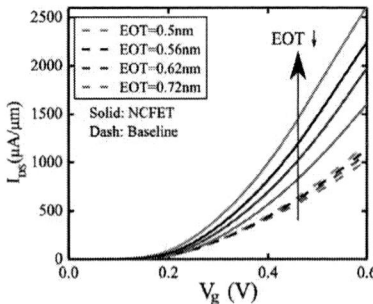

Fig. 10: I_{DS} versus V_g curves at $V_{DS}=0.6$V for either baseline FETs (dashed lines) or NC-FETs ($T_{fe}=15$nm, solid lines) and for different EOT values.

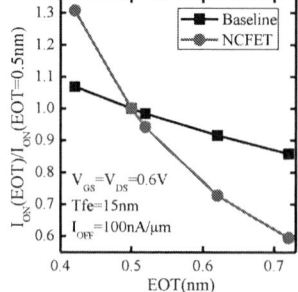

Fig. 11: I_{ON} (i.e. I_{DS} at $V_g=V_{DS}=0.6$ V) normalized to I_{ON} at EOT=0.5 nm as extracted from the results of **Fig.10**.

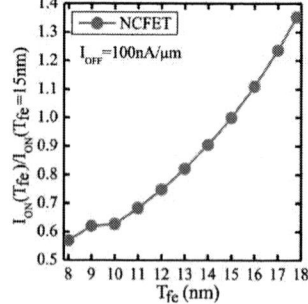

Fig. 12: I_{ON} (i.e. I_{DS} at $V_g=V_{DS}=0.6$ V) versus T_{fe} and normalized to the I_{ON} at $T_{fe}=15$ nm. All devices have EOT=0.5 nm.

978-1-7281-1988-5/18 $31.00 © 2018 IEEE   216

1.5µm Dual Conversion Gain, Backside Illuminated Image Sensor Using Stacked Pixel Level Connections with 13ke- Full-Well Capacitance and 0.8e- Noise

Vincent C. Venezia, Alan Chih-Wei Hsiung, Kelvin Ai, Xiang Zhao, Zhiqiang Lin, Duli Mao, Armin Yazdani, Eric A. G. Webster, and Lindsay A. Grant, email: vincentv@ovt.com
OmniVision Technologies, Santa Clara CA 95054

Abstract—A 1.5µm pixel size, 8 mega pixel density, dual conversion gain (DCG), back side illuminated CMOS image sensor (CIS) is described having a linear full-well capacity (FWC) of 13ke- and total noise of 0.8e- RMS at 8x gain. The sensor adopts a world smallest 1.5µm pitch, stacked pixel-level connection (SPLC) technology with greater than 8M connections, maximizing fill-factor of the photodiode and dimensions of the associated transistors to achieve a large FWC and low noise performance at the same time. In addition, by allocating transistors into two different layers, the DCG function can be realized with 1.5µm pixel size.

I. INTRODUCTION

In the past 10 years, images captured by mobile devices have made a profound impact on everyone. Internet of things (IoT) and artificial intelligence (AI) have become additional driving forces of technology and application development. Image sensors as edge components are the primary source of data collection for various applications including autonomous driving, surveillance, medical, as well as machine vision for deep learning [1]. Wafer stacking has become the key technology for next generation image sensors allowing independent optimization of pixels and readout/processing circuitry [2], especially as pixel scaling continues.

In many applications, a high-dynamic range (HDR) image sensor is required that simultaneously achieves excellent low and high light performance. Dual conversion gain (DCG) image sensors have been established as a method of achieving good HDR performance [3, 4]. In such a system, a high conversion gain is used under low illumination to reduce read noise, while a lower conversion gain is used under strong illumination, to accommodate a high FWC. However, DCG image sensors require a large capacitor to modulate the floating diffusion capacitance and an additional transistor to switch between low and high conversion gain. These additional components restrict pixel shrinkage and degrade the pixel performance due to scaled photo diode and transistor dimensions. However, the use of stacking technology mitigates such restriction. It creates flexibility to place pixels and associated transistors over two silicon layers and connect them with direct bonding. Both photo diode and other required devices have sufficient silicon real estate for optimized design and performance. In this work, a stacked back-side-illuminated (BSI), 1.5µm pixel, DCG CIS is described. It is

believed to be the smallest pixel pitch DCG CIS. Wafer stacking is achieved with a 1.5um pitch, pixel level connection.

II. PIXEL ARCHITECTURE

Two DCG pixel architectures are investigated in this work and are shown in Fig.1. For Design 1A, the photo-diode (PD), transfer gate (TX), floating diffusion (FD), source-follower (SF) and DCG switch transistor remain on the sensor layer, while the reset (RST), and row select (RS) are moved to the logic layer (Fig.1A). For Design 1B, the PD and TX are on the sensor layer, with all other transistors on the logic layer (Fig.1B). Both designs have an additional MOS capacitor on the logic layer to control the conversion gain via the DCG switch transistor. There are benefits to each design, for instance, Design 1A has a much lower FD capacitance for a high conversion gain, while Design 1B has a larger PD area, for increased FWC.

The sensor and logic layer wafers are processed separately, and are bonded together using stacking technology with a 1.5µm connection pitch, shown schematically in Fig. 2A. The sensor layer in a stacked BSI-CIS typically contains a PD array. However for this DCG device, the logic layer contains an array of supporting transistors and MOS capacitors, with controlling circuits in the periphery of the logic layer, schematically shown in Fig. 2B. The pixel level connections enable fast data transfer from the pixel array on the sensor layer to the array of supporting devices on the logic layer. The signal is read out row by row just as a conventional rolling shutter image sensor. The pixel and logic processes have been designed and optimized separately in order to achieve the best performance. OmniVision's 2nd Generation BSI stacking process, with composite metal grid and backside deep trench isolation [5], is used on this DCG device to achieve a high quantum efficiency and low optical cross talk for enhanced low light imaging.

III. RESULTS AND DISCUSSION

The conversion gain for the two DCG image sensor designs used in this work are shown in Fig. 3. In this figure the square of the photon shot noise vs. signal is plotted for two operation conditions, the DCG switch on and off. The slope of these curves is the conversion gain, and is labeled in the figures. When the DCG switch is on, low conversion gain mode, the conversion gain is determined by the large capacitance of the MOS capacitor, resulting in a low conversion gain. With the DCG switch off, high conversion gain mode, the conversion gain is determined by the much smaller FD junction capacitance and the metal routing parasitic capacitance, resulting in a high conversion gain. To achieve a large dynamic range, the conversion gain ratio between the high and low conversion gain modes should be as large as possible. The Design 1A has a conversion gain ratio of 10, while the Design 1B has a ratio 3.4, demonstrating the advantage of the Design 1A. The conversion gain for the low conversion gain mode is similar for both designs, since the MOS capacitor is the same for both devices. The much larger conversion gain of Design 1A, in high conversion gain mode, demonstrates the advantage of placing the SF on the sensor layer, much closer to the FD. Design 1A achieves a high conversion gain of 200μV/e which results in a low read noise of 0.8e- RMS. The noise histogram for this device in the high conversion gain mode is shown in Fig 4.

The FWC achieved in the low conversion gain mode for each design is shown in Fig. 5, where the square of the photon shot noise is plotted vs. increasing signal. The pixel linear FWC is achieved when the image saturates and is indicated in the figures. A larger FWC of 16.8ke is measured for the Design 1B, and a 13ke FWC is measured for the Design 1A. The larger FWC is due to the larger PD area used in the Design 1B. With the high FWC of the low conversion gain mode and the low noise of the high conversion gain mode, an 83.8dB dynamic range is achieved for the Design 1A.

In addition to increasing the conversion gain for low light conditions, the DCG image sensor in this work uses OmniVision's Gen2 BSI structure. The quantum efficiency is shown in Fig. 6. With the composite metal grid and deep backside oxide deep trench isolation, as described in [5], an 80% QE and low crosstalk is achieved. The above pixel performance parameters are listed in Table 1 for the Design 1A, DCG architecture.

Fig. 7 compares images taken with a Design 1A CIS using the high and low conversion gain modes. The low conversion gain mode image is dark, with a low signal to noise ratio (SNR) due to the high read out noise; indicated in the Fig.7B. However, the light emitting diode (LED) light sources area remains visible due to the high FWC in low conversion gain mode. Using the high conversion gain mode, a more detailed image is observed for the low light areas, with a 7db SNR improvement. In this case, the LED light source area is saturated, with no observable detail or color. Combining the images, shown in Fig.8, demonstrates the advantage of this DCG technology; a HDR images is observed with clear features in both the bright and dark regions of the image. This image is obtained from one exposure, two non-destructive readouts from both conversion gain modes. This approached has no motion artifacts, as compared to a conventional two exposure HDR approach.

IV. SUMMARY

Using stacked pixel level connections, a backside illuminated, 1.5μm pixel, DCG, CMOS image sensor was described. Two pixel designs were investigated. The first design placed only the PD and TX on the sensor layer, while the second design placed the PD, TX, SF, and DCG switch transistor on the sensor layer. In both cases a large MOS capacitor on the ASIC layer was used to lower the conversion gain. The first design has the advantage of a large PD area and therefore increased FWC. However the second design reduced the overall FD capacitance, making it possible to achieve a large conversion gain of 200 μV/e. and a low read noise of 0.8e- RMS This design achieved a conversion ratio of 10 between the high and low conversion gain modes and a dynamic range of 83.8db. The SPLC is the essential technology in realizing the DCG architecture for this 1.5μm pixel product.

REFERENCES

[1] Johannes Solhusvik, Jiangtao Kuang, Zhiqiang Lin, Sohei Manabe, Jeong-Ho Lyu, Howard Rhodes, "A comparison of high dynamic range CIS technologies for automotive applications," IISW 2013

[2] Shunichi Sukegawa, Taku Umebayashi, Tsutomu Nakajima, Hiroshi Kawanobe, Ken Koseki, Isao Hirota, Tsutomu Haruta, Masanori Kasai, Koji Fukumoto, Toshifumi Wakano, Keishi Inoue, Hiroshi Takahashi, Takashi Nagano, Yoshikazu Nitta, Teruo Hirayama, Noriyuki Fukushima, "A ¼ inch 8M pixel back-illuminated stacked CMOS image sensor" ISSCC session 27, 2013.

[3] Trygve Willassen, Johannes Solhusvik, Robert Johansson, Sohrab Yaghmai, Howard Rhodes, Sohei Manabe, Duli Mao , Zhiqiang Lin , Dajiang Yang, Orkun Cellek, Eric Webster, Siguang Ma, and Bowei Zhang, "A 1280x1080 4.2μm Split-diode Pixel HDR Sensor in 110 nm BSI CMOS Process", IISW 2015.

[4] Johannes Solhusvik , Sam Hu , Robert Johansson , Zhiqiang Lin , Siguang Ma , Keiji Mabuchi , Sohei Manabe , Duli Mao , Bill Phan, Howard Rhodes, Charles Shan, Eric Webster, and Trygve Willassen., "A 1392x976 2.8μm 120dB CIS with Per-Pixel Controlled Conversion Gain", IISW 2017

[5] Vincent C. Venezia, Alan Chih-Wei Hsiung, Wu-Zang Yang, Yuying Zhang, Cheng Zhao, Zhiqiang Lin and Lindsay A. Grant, "Second Generation Small Pixel Technology Using Hybrid Bond StackingGen2 IISW reference", Sensors 18(3), 667 February 2018

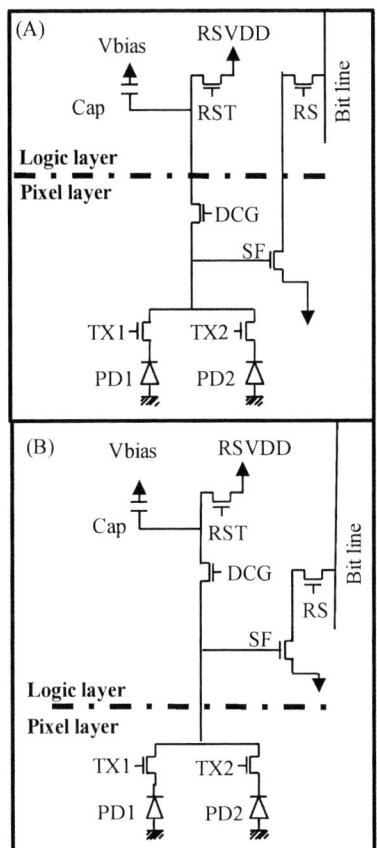

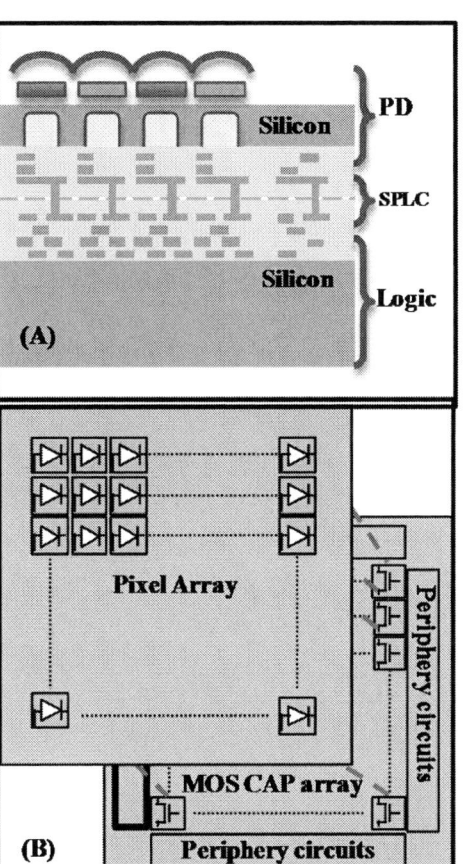

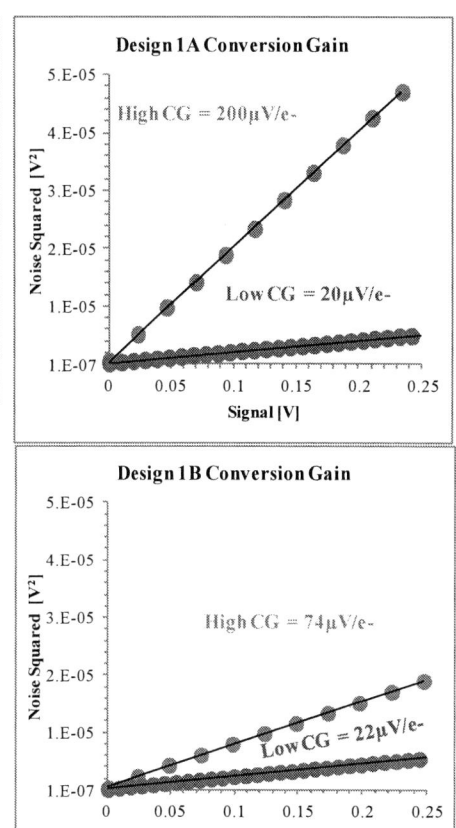

Fig. 1. Circuit diagram schematic showing Design 1A (A) and Design 1B (B) used in this work. Placement of devices on sensor or logic layers are indicated by dashed line.

Fig. 2. (A) Schematic showing the stacked pixel level connection between bonded sensor and logic layers. (B) Schematic showing the array of photo-diodes on the sensor layer (top) and the array of supporting transistors on the logic layer (bottom).

Figure3: Measured shot noise squared vs signal for the 1.5um, BSI, DCG CIS Design 1A (A) and Design 1B (B). The high and low conversion gain mode results are shown for each design; red and blue symbols, respectively. The conversion is the slope of each line and is indicated in the figures in μV/e-.

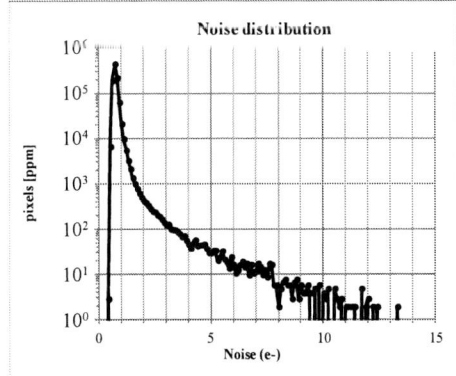

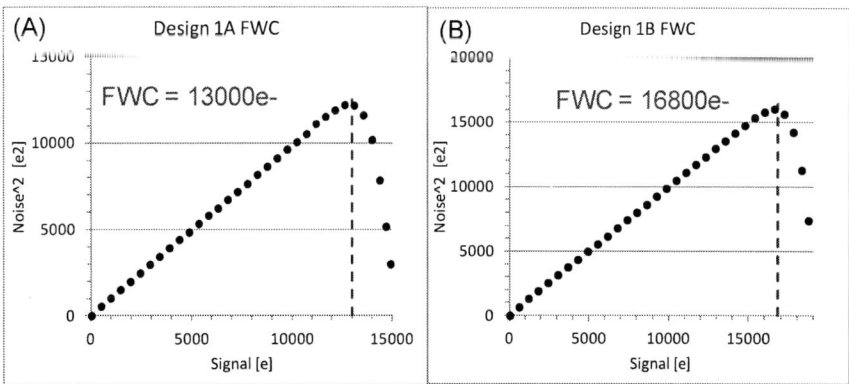

Fig. 4. Measured statistical read noise distribution at 8x gain for the 1.5μm, BSI DCG CIS of Design 1A using the high conversion gain mode.

Fig 5. Measurement of square of the shot noise vs signal for the Design 1A and Design 1B, 1.5μm BSI DCG CIS devices: The FWC in electrons and the point at which the linear FWC is achieved is indicated in the plots

978-1-7281-1988-5/18 $31.00 © 2018 IEEE   219

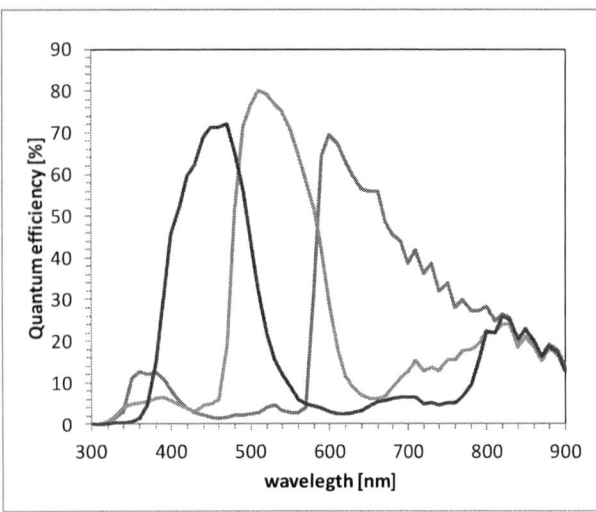

Fig. 6. Measured quantum efficiency spectra for the 1.5um, BSI DCG CIS of Design 1A.

Pixel design		Design 1A	
Pixel size		1.5um Pixel	
Measured parameters	**Units**	HiCG	LoCG
Conversion gain	μV/e-	200	20
Peak QE - R	%	70%	
Peak QE - Gb	%	80%	
Peak QE - Gr	%	80%	
Peak QE - B	%	72%	
Average Xtalk	%	13.5%	
Read Noise	e-	0.8	9.1
Linear FWC	e-	4526	13000
Inter-Scene Dynamic Range	dB	74.6	63.1
Intra-Scene Dynamic Range	dB	83.8	

Table 1. Measured performance parameters for the 1.5um, BSI DCG CIS of Design 1A.

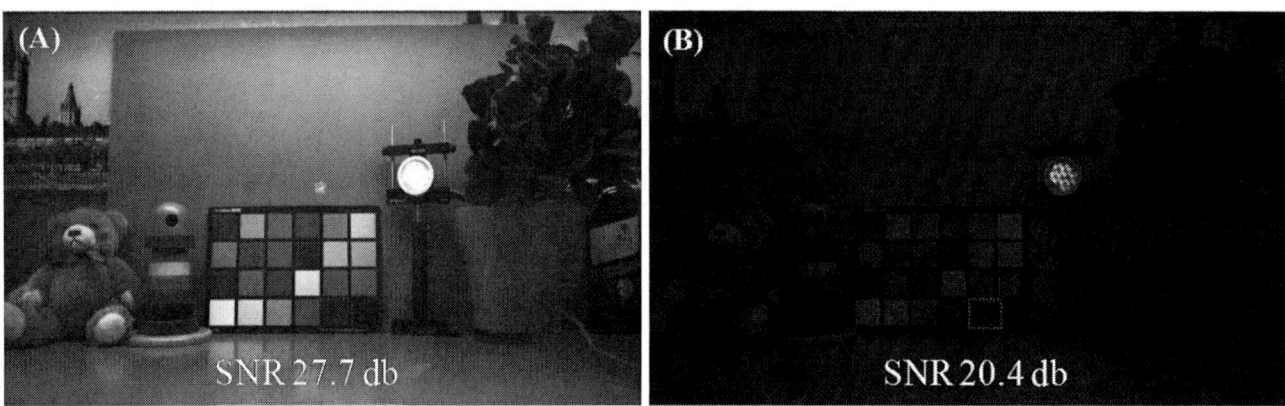

Fig. 7 Images from the 1.5um, BSI DCG CIS of Design 1A.in (A) high conversion gain mode and (B) low conversion gain mode, under the same frame rate of 30 fps. The SNR, indicated in each figure, is calculated from the region highlighted by the red box.

Fig. 8: HDR image combining high and low conversion gain images from Fig.7A and 7B.

A 0.68e-rms Random-Noise 121dB Dynamic-Range Sub-pixel architecture CMOS Image Sensor with LED Flicker Mitigation

S. Iida[1], Y. Sakano[1], T. Asatsuma[1], M. Takami[2], I. Yoshiba[1], N. Ohba[2], H. Mizuno[1], T. Oka[1],
K. Yamaguchi[1], A. Suzuki[1], K. Suzuki[1], M. Yamada[1], M. Takizawa[1], Y. Tateshita[1], and K. Ohno[1]

[1]Sony Semiconductor Solutions, Kanagawa, Japan. [2]Sony Semiconductor Manufacturing, Kumamoto, Japan.
E-mail: Satoko.Iida@sony.com

Abstract—This is a report of a CMOS image sensor with a sub-pixel architecture having a pixel pitch of 3 um. The aforementioned sensor achieves both ultra-low random noise of 0.68e-rms and high dynamic range of 121 dB in a single exposure, further realizing LED flicker mitigation.

I. INTRODUCTION

Recently, real-time sensing has been creating new businesses and social changes, specifically in the internet of things (IoT) and automotive fields. Clearly, image sensing is a critical function in these fields. The accurate perception of moving objects and obstacles as well as detection with high color reproducibility for all light conditions is a necessity.

For example, people, objects, or features must be recognized through sampling of high-sensitivity and low noise images, even when moving in darkness.

Additionally, using the multiple exposure method, a conventional high dynamic range (HDR) technique [1–2] can cause motion artifacts depending on the sampling time difference of dynamic subjects. This results in misrecognition.

Moreover, the signal lights of light-emitting diode (LED) actually blink; however, they must appear as though they are always on in the images. When using a method that extends the exposure time simply in order to capture such blinking signals, the signals become saturated within a short period of time, thus losing their luminance and color information. As a method of LED flicker mitigation (LFM), sampling multiple times in the time direction during the exposure period has been proposed. This method has a non-supplementing period and cannot completely mitigate the light flicker effect. [3]

We have developed a new image sensor to address these issues. The characteristic of this sensor is that it has been designed with a sub-pixel architecture that has a single large photodiode, a single small photodiode, and an in-pixel floating capacitor.

II. SENSOR ARCHITECTURE

A. Sensor Configuration

Fig.1 shows the block diagram of the image sensor. A pixel array with 1920×1200 pixels, read-out circuits (load MOS transistors, column ADC's, DAC), driver circuits (row driver, row decoder), image signal processor, and other circuits (PLL, regulator, MIPI I/F, CPU, etc.), are all mounted using a 90-nm process.

B. Pixel Circuit

Fig.2 shows the pixel schematic of the sub-pixel architecture. This circuit employs a single large photodiode (SP1), a single small photodiode (SP2), an in-pixel floating capacitor (FC), and seven transistors. The SP1 has a high sensitivity (Green) of 36000e-/lx · s and SP2 has 1/10 of SP1's sensitivity. SP1's linear full-well capacity (FWC) of 10000e- and SP2's linear FWC of 78500e- are attributed to the FC. The seven transistors are as follows: transfer gate of SP1 (TGL), transfer gate of SP2 (TGS), floating diffusion gate (FDG), floating capacitor gate (FCG), reset transistor (RST), select transistor (SEL), and the source follower amplifier (AMP). A floating diffusion (FD) is separated as FD1, FD2, and FD3 by FDG and FCG, which serve as a switch to connect FD1 with FD2, and FD2 with FD3, respectively. The two-electrodes of FC are connected to FD3 and the counter electrode of which supply voltage is FCVDD respectively.

As shown in the pixel top view of Fig. 3, one pixel has a large on-chip micro lens (OCL) and a small OCL. SP2's OCL is located in the gap section of SP1's OCL. This makes the sensitivity ratio of SP1 to SP2 equal to 10:1.

Fig.4 shows the pixel cross-sectional view corresponding to the dotted line in Fig.3. As shown in Fig.4, deep trench isolations are employed in the silicon substrate to prevent the leakage of electrical charges from SP1 to SP2.

C. Pixel Read-out Method

Fig.5 shows the pixel driving sequence. The signals that come from SP1 and SP2 are output serially. Additionally, the electrical charges accumulated in SP1 is converted to signal voltage in two modes, namely high conversion gain (HCG) and low conversion gain (LCG) by switching FDG. In this manner, three types of signals are read-out in a single exposure. First, an exposure of SP1 and SP2 begins by the reset of SP1, SP2, and FC. Then, LCG reset level 2 and HCG reset level 1 is sampled. Subsequently, HCG signal level 1 is sampled after switching TGL, LCG signal level 2 is sampled after switching TGL once again. By performing co-related double sampling (CDS) for each reset and signal level, two signals are read-out: SP1H and SP1L. Subsequently, the signal that comes from SP2 is read-out by performing delta reset sampling (DRS): SP2L, in which the signal level 3 is sampled first, followed by the reset level 3. Because the signal charges are accumulated in FD3, FD3 cannot be reset prior to sampling the signal level 3. The flaw of DRS is that kTC noise cannot be removed; however, it can be suppressed by securing the capacitance of FC sufficiently. The FCVDD in the accumulation period is lower than that in the read-out period to reduce the fixed pattern noise (FPN) of SP2L.

Fig.6 shows SP1's potential diagram that considers the driving sequence of SP1 in Fig.5. The cross section shows the path of the perforated line A-B in Fig.2. In the beginning of the exposure period, switching TGL, FDG, and RST resets the electrical charges of SP1 (Fig.6-a). During the exposure period, FD1 and FD2 are always reset (Fig.6-b), and after that, the LCG reset level 2 is sampled when RST is turned off (Fig.6-c). Subsequently, the HCG reset level 1 is sampled when FDG is turned off (Fig.6-d). Then, the HCG signal level 1 is sampled after TGL is switched and the electrical charges accumulated in SP1 are transferred to FD1 (Fig.6-e). After FD1 is connected to FD2 by turning FDG on, The LCG signal level 2 is sampled when TGL is switched once again and the remaining charges in SP1 are fully transferred to FD1 and FD2 (Fig.6-f); thus, SP1H and SP1L can be read-out from the electrical charges accumulated in SP1.

Fig.7 shows SP2's potential diagram that considers the driving sequence of SP2 in Fig.5. The cross section shows the path of the perforated line C-B in Fig.2. In the beginning of the exposure period, switching TGS, FCG, and RST resets the electrical charges of SP2 and FD3 (Fig.7-a). During the exposure period, the electrical charges that come from SP2 are accumulated in both SP2 and FD3 (Fig.7-b). After FD3 is connected to FD1 and FD2 by turning FCG on, the signal level of SP2 is sampled when TGS is switched and the electrical charges accumulated in SP2 are fully transferred to FD3 (Fig.7-c). Again, the RST is turned on and the electrical charges of FD1, FD2, and FD3 are reset and sampled as the reset level of SP2 when RST is turned off (Fig.7-d); thus, the SP2L can be read-out from the electrical charges accumulated in SP2 during the same exposure period of SP1.

III. SENSOR CHARACTERISTICS

A. Sensor Characteristics

Fig.8 shows the photo responses for SP1H, SP1L, and SP2L. The linear FWC of SP1 and SP2 are 10000e- and 78500e-, respectively. As mentioned earlier, the sensitivity ratio of SP1 to SP2 is 10:1. By multiplying SP2L and a gain of 10 , its linear FWC becomes equivalent to 785000e-, thus achieving a dynamic range of 121 dB.

Fig.9 shows the FCVDD dependency of SP2's linear FWC and the FPN of SP2L. The linear FWC is determined by the capacitance of the FC and the difference in electric potential between the reset level of SP2 and the threshold voltage of FCG. Hence, the FPN is caused by the variation of FD3 dark current. The cause of the FD3 dark current is an electric field between FD3 and its P-well, which is generated by FCVDD. Therefore, the supply voltage of FCVDD in the accumulation period is lower than that in the read-out period to reduce the FD3 dark current. The linear FWC and dark current have a trade-off relationship; nevertheless, a linear FWC of 78500e- can be secured.

Fig.10 shows the SNR curve of the synthesized signal. It consists of three types of signals. SP1H is used in low light

scenes, SP1L is for medium light scenes, and SP2L is for high light scenes. An SNR of 20 dB can be maintained even when connecting SP1L to SP2L at 60°C.

B. Synthesized Image

Fig.11 shows a synthesized image of a high dynamic range scene. Fig.11 (a) shows an image of SP1L with an exposure time of only 0.3 ms. There are crushed shadows inside the tunnel. Fig.11 (b) shows an image of SP1L with an exposure time of 11 ms. The outside of the tunnel has blown-out highlights. As shown in Fig.11 (c), by synthesizing SP1H, SP1L, and SP2L, the darkness of the tunnel and the outside landscape are accurately captured. Additionally, blinking LED signal lights and car headlights are captured without being switched off.

Fig.12 shows a synthesized image captured in a very low light scene of 0.1 lx. Fig.12 (a) shows the SP1H being applied and Fig.12 (b) shows an image in the absence of SP1H. When comparing the sections of the outer wall, it can be observed that the wall is covered in noise, and it is impossible to distinguish the wall's pattern and the white lines from it without SP1H. Furthermore, when comparing the sections with people, we can clearly observe their figures and contours if we apply SP1H.

Table 1 shows the sensor performance developed in this study and Table 2 shows a comparison of the characteristics reported previously [4]–[6]. The well-balanced characteristics have been achieved using a CMOS image sensor with a sub-pixel architecture.

IV. CONCLUSIONS

We have developed a new image using a sub-pixel architecture with a pixel pitch of 3 um that achieves both ultra-low noise of 0.68e-rms and a high dynamic range of 121 dB in a single exposure, further realizing LFM.

ACKNOWLEDGMENT

The authors would like to thank M. Torii, T. Machida, Y. Matsumura, and T. Toyofuku, for their support and advice regarding the sensor pixel design. The authors also appreciate the support from the members of Sony Semiconductor Solutions and Sony Semiconductor Manufacturing.

REFERENCES

[1] Trygve Willassen et al., "A 1280x1080 4.2μm Split-diode Pixel HDR Sensor in 110nm BSI CMOS Process" in IISW 2015.

[2] Sergey Velichko et al., "140 dB Dynamic Range Sub-electron Noise Floor Image Sensor" in IISW 2017, pp. 294-297,

[3] Chris Silsby et al., " A 1.2MP 1/3" CMOS Image Sensor with Light Flicker Mitigation" in IISW 2015,

[4] M. Takase et al., "An over 120 dB wide-dynamic-range 3.0 μm pixel image sensor with in-pixel capacitor of 41.7 fF/um2 and high reliability enabled by BEOL 3D capacitor process," in Symp. VLSI 2018, pp. 71-72.

[5] K. Nishimura et al., "An Over 120dB Simultaneous-Capture Wide-Dynamic-Range 1.6e- Ultra-Low-Reset-Noise Organic-Photoconductive-Film CMOS Image Sensor," in ISSCC 2016, pp. 110-112.

[6] Johannes Solhusvik et al., "A 1392x976 2.8μm 120dB CIS with Per-Pixel Controlled Conversion Gain," in IISW 2017, pp. 298-301

978-1-7281-1988-5/18 $31.00 © 2018 IEEE

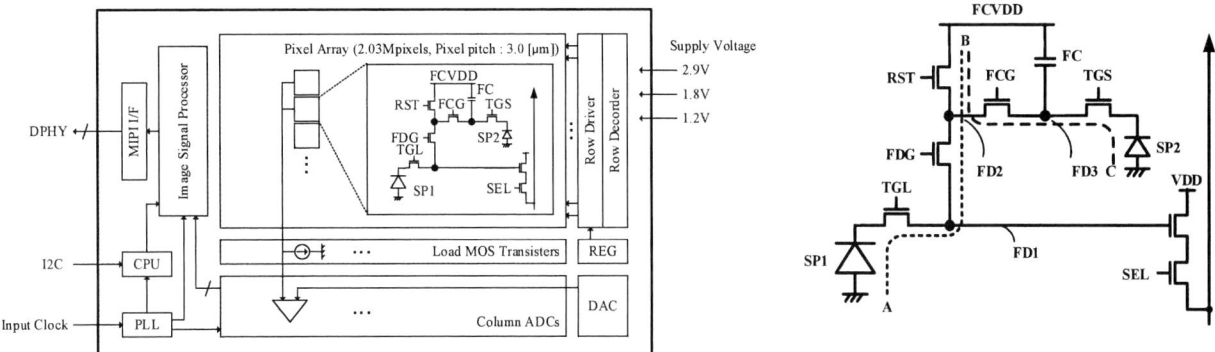

Fig. 1   Sensor block diagram

Fig. 2   Pixel schematic

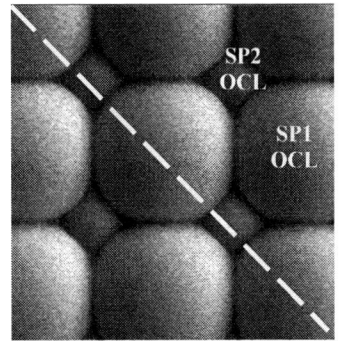

Fig. 3   Pixel top view

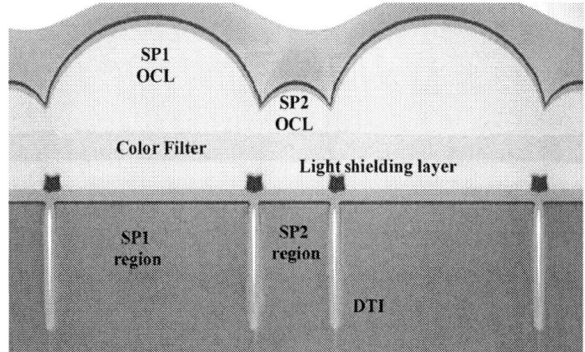

Fig. 4   Pixel cross-sectional view

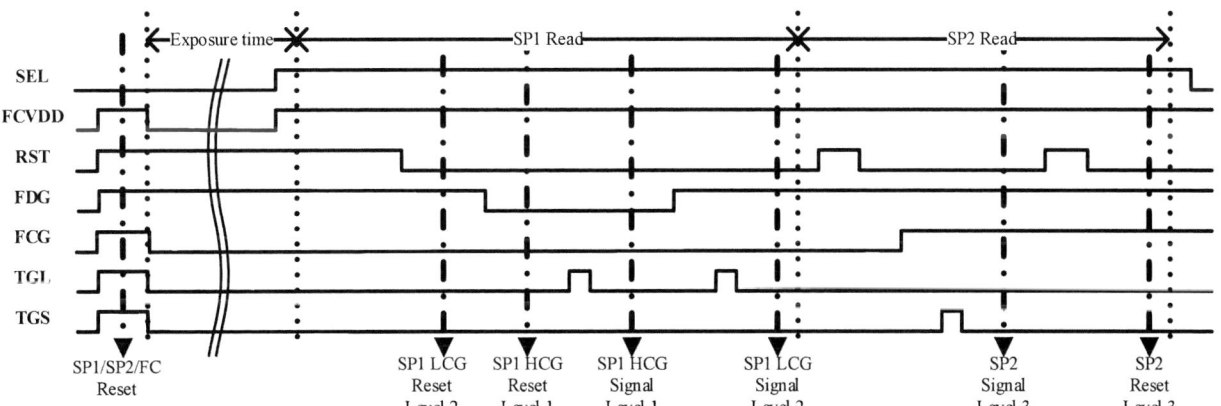

Fig. 5 Sensor driving sequence

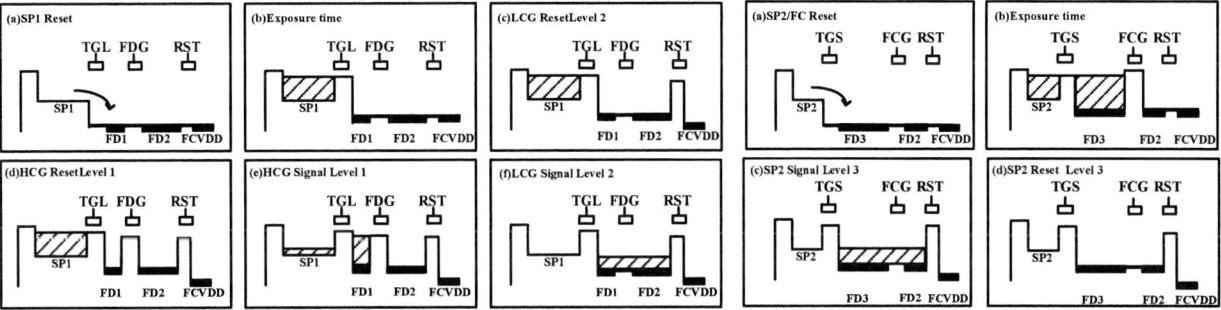

Fig. 6 SP1 potential diagram (A-B in Fig.2)   Fig. 7 SP2 potential diagram (C-B in Fig.2)

978-1-7281-1988-5/18 $31.00 © 2018 IEEE   223

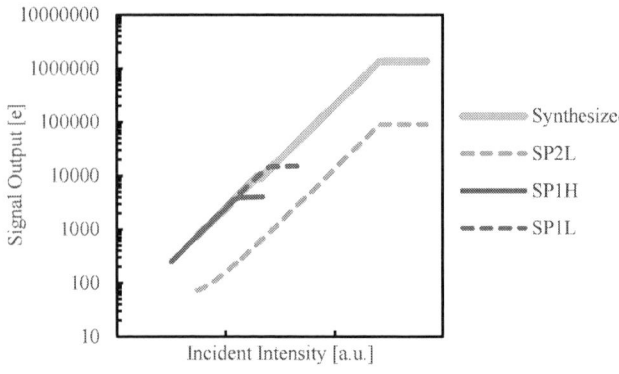

Fig. 8 Photo response of each signal

Fig. 9 FCVDD dependency of
full-well capacity and FPN

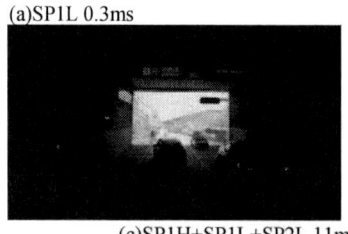

(a)SP1L 0.3ms   (b)SP1L 11ms

(c)SP1H+SP1L+SP2L 11ms

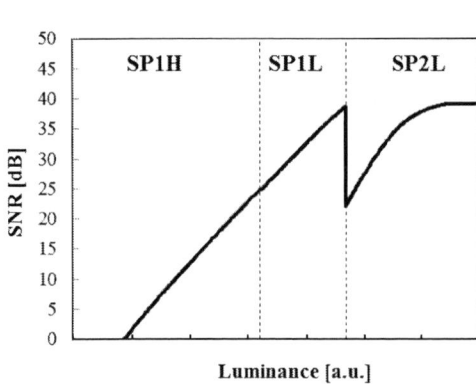

Fig. 10 SNR curve of synthesized signal

Fig. 11 Synthesized image (high dynamic range scene)

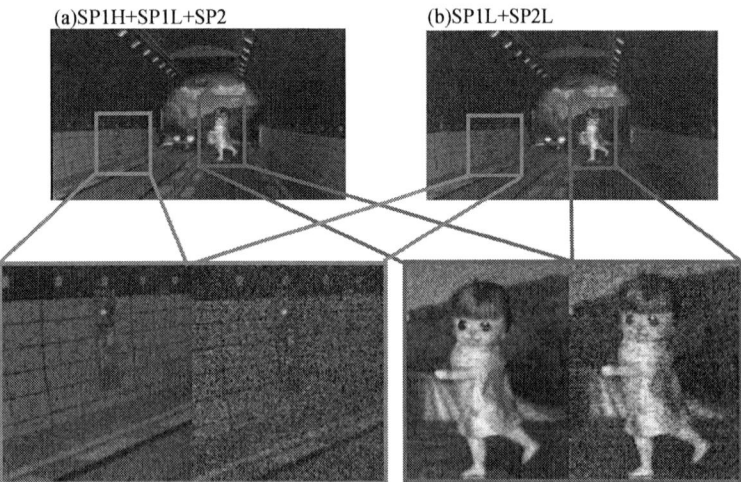

(a)SP1H+SP1L+SP2   (b)SP1L+SP2L

Fig. 12 Synthesized image (low light scene)

Table 1 Sensor performance

Parameter	Unit	Value
Power Supply	V	2.9/1.8/1.2
Process Technology	-	90nm 4Cu1AL
Pixel Array	pixels	1920x1200
Pixel Pitch	μm	3
SP1L Linear Full-well Capacity	e-	10000
SP2L Linear Full-well Capacity	e-	78500
Sensitivity(Green, 3200K with IR cut filter)	e/lx・sec	36000
Sensitivity Ratio (SP1:SP2)	-	10:1
Random Noise	e-rms	0.68
Dynamic Range	dB	121

Table 2 Comparison of sensor characteristics

	Unit	This work	VLSI 2018 [4]	ISSCC 2016 [5]	IISW 2017 [6]
Pixel pitch	um	3	3	6	2.8
Random noise@RT	e - rms	0.68	6.2	5.4	1
Full-well Capacity	e -	78500	489K	600k	50000
Dynamic Range(SingleExp.)	dB	121	121	123.8	94

978-1-7281-1988-5/18 $31.00 © 2018 IEEE

A 24.3Me⁻ Full Well Capacity CMOS Image Sensor with Lateral Overflow Integration Trench Capacitor for High Precision Near Infrared Absorption Imaging

M. Murata[1], R. Kuroda[1], Y. Fujihara[1], Y. Aoyagi[1], H. Shibata[2], T. Shibaguchi[2], Y. Kamata[2], N. Miura[2], N. Kuriyama[2] and S. Sugawa[1]

[1]Graduate School of Engineering, Tohoku Univ., Sendai, Miyagi, Japan, email: maasa.murata.t5@dc.tohoku.ac.jp

[2]LAPIS Semiconductor Miyagi Co., Ltd., Kurokawa, Miyagi, Japan

Abstract— This paper presents a 16μm pixel pitch CMOS image sensor exhibiting 24.3Me⁻ full well capacity with a record spatial efficiency of 95ke⁻/μm² and high quantum efficiency in near infrared waveband by the introduction of lateral overflow integration trench capacitor on a ~10^{12}cm⁻³ p-type Si substrate. A diffusion of 5mg/dl concentration glucose was clearly visualized by an over 71dB SNR absorption imaging at 1050nm.

I. INTRODUCTION

In the IoT era, sensing applications using CMOS image sensors (CIS) in healthcare, medical, food inspection, agriculture, disaster prevention fields and so on will play significant roles for the safe and sustainable society. For some sensing applications like absorption imaging and optical coherence tomography, the resolution of microscopic change of light level under high light illumination determines the sensing accuracy [1-3]. Thus a high signal-to-noise ratio (SNR) over 70dB as in analytical instruments is required for CIS in those usage. As the photon shot noise dominates the total noise under high light illumination, a high full well capacity (FWC) over 10Me⁻ is needed [1-2]. Moreover, in order to detect various phenomena with different spectral characteristics, a high quantum efficiency (QE) for ultraviolet (UV)-visible-near infrared (NIR) wavebands is important [4].

In general CIS, FWC is determined either by photodiode (PD) or floating diffusion (FD) capacitance both based on pn-junction; the spatial efficiency of FWC is limited low. In the lateral overflow integration capacitor (LOFIC) CISs with a single exposure wide dynamic range (WDR) capability [4-5], FWC is determined by capacitance of LOFIC, which is typically made of metal-oxide-silicon (MOS) or metal-insulator-metal (MIM) capacitors with a relatively high spatial efficiency. In organic photoconductive film (OPF) CISs [6-7], FWC is determined by FD capacitance. Recently, OPF CISs employing high-k MIM capacitors have been reported to be useful to achieve a high FWC [7-8]. Furthermore, a complementary carrier collection CIS utilizing the in-pixel capacitive trench isolation for WDR has been proposed recently [9]. For the NIR light sensitivity improvement, methods to increase the effective light path length in PD have been extensively researched [10-11].

We have previously developed a LOFIC CIS using a very low dopant concentration p-type Si substrate with 10Me⁻ FWC by 1pF LOFIC using MOS and MIM capacitors and a high NIR QE of PD due to the deep depletion width [1]. However, the fill factor (FF) was limited to the order of 10%. The purpose of this work is to develop a CIS with highly spatial efficient lateral overflow integration trench capacitor (LOFITreC) formed on the low

dopant concentration Si substrate to achieve an extremely high FWC for high precision NIR absorption imaging. The structure and performances of the developed CIS and its absorption imaging capability are described as follows.

II. DEVELOPED CMOS IMAGE SENSOR WITH LOFITreC

A. Circuit archtecture and operation

Fig. 1 shows the circuit block diagram of the developed CIS. The pixel consists of a pinned PD, a transfer gate (T), a FD, a source follower driver (SF), a select switch (X), an overflow switch (S), a LOFITreC, and a reset gate (R). In this work, the 1.6pF LOFITreC is integrated adjacent to the PD. Select switches choosing the reset voltage level (VR) either VR1 or VR2 by the pulse φVR are placed in each row. VR1 and VR2 are used for the PD reset level and the reference level in high FWC signal (S2), respectively [2]. Figs.2 and 3 show the operation timing and the potential diagrams of a normal LOFIC operation [4-5], and the dual VR mode for S2, respectively. In the normal LOFIC operation, overflow photoelectrons from PD and FD are accumulated in LOFITreC during the integration period (t_2). The high sensitivity voltage signal converted at small capacitance FD (S1) and a high FWC voltage signal converted at FD+LOFITreC (S2) are readout to achieve WDR under a single exposure. This operation is employed to capture images of low emission light as well as of high bright light in which coexistence of high sensitivity and WDR is required. In contrast, the dual VR mode is introduced to detect a fine difference of light level under high light illumination condition by a high gain readout for absorption imaging. In this mode, T and S are always turned ON to increase the saturation level. During the integration period (t_6), overflow photoelectrons from PD are accumulated in FD+LOFITreC. After reading out S2 (t_7), the PD, FD and LOFITreC are set to the reference level VR2 which is tuned close to the light signal level, and the reference signal is readout from the same pixel SF (t_8). By taking the difference of the light signal and the reference signal levels with a high gain, a slight change of absorption of light can be detected accurately.

B. LOFITreC integration

Fig. 4 shows the pixel cross sectional schematic of the 16μm pixel pitch CIS in this work. To enhance QE in NIR waveband by increasing the depletion width of PD, a very low dopant concentration Si substrate in the order of 10^{12}cm⁻³ based on a low oxygen concentration Czochralski growth method was employed. The developed CIS uses front side illumination so that the high NIR light illumination does not affect the charge storage floating nodes of the CIS. The surface p⁺ layer of the PD

978-1-7281-1988-5/18 $31.00 © 2018 IEEE

with steep dopant concentration profile was introduced in order to achieve high sensitivity and high robustness toward UV light band [12]. Consequently, a high QE in UV-visible-NIR waveband can be obtained. Recently a global shutter CIS with analog memories composed of $30fF/\mu m^2$ trench MOS capacitors formed aside the pixel array has been reported for ultra-high speed framerate by the authors' group [13]. In this work, the trench MOS capacitor was integrated inside each pixel as LOFITreC to achieve high FWC and high FF. The enlarged cross section diagram around the LOFITreC and PD and TEM images of LOFITreC are shown in Fig.4(b-d). To suppress the dark current at the LOFITreC, overflow photoelectrons are accumulated at the LOFITreC's gate electrode side and the silicon side electrode formed by the electron inversion layer is grounded. In order to suppress the leakage current between this inversion layer and the buried n-type layer of pinned PD, a deep p-well (DPW) was formed around the LOFITreC. DPW and the p-well under the transistors form a potential gradient so that the photoelectrons generated deep inside the PD by NIR light are drifted into the PD. The concentration of DPW was optimized to obtain the uniform capacitance in the signal range of LOFITreC (0.5~2.7V). Fig.5 shows the fabrication process flow of the developed CIS with LOFITreC by a 0.18µm 1-poly-Si 5-Metal layer CMOS process technology with pinned PD. Fig.6 shows the current-voltage characteristics of LOFITreC. A very low current was obtained for the signal voltage range due to the high integrity gate oxide film. Fig.7 shows the micrograph of the developed CIS chip with $128^H \times 128^V$ effective pixels. The number of pixels is easily extendable under the same design.

III. MEASURMENT RESULT

Fig.8 shows the measured photoelectric conversion characteristics of the developed CIS. In the normal LOFIC operation, a 130dB WDR with linear response was obtained by S1 and S2 signals under a single exposure. Fig.9 shows the number of signal photoelectrons and SNR of S2 signal in the dual VR mode as functions of light illuminance. A highly linear response with the FWC of 24.3Me⁻ and over 71 dB SNR were obtained. Fig.10 shows the measured PD QE of this work and our previous works as reference [1]. Here, the measured QE was normalized by the FF of CIS in this work (52.8%). A 200-1100nm wide spectral sensitivity with over 5 times improvement was obtained by increasing the FF, as a result, PD QE of 89.7, 78.2 and 26.7% were achieved at 860, 940 and 1050nm, respectively. Fig.11 shows the sample images of a flash light, a UV LED and a stuffed animal captured at 256fps with F# 4.0 lens. Figs.11(a) and 11(b) were captured by the S1 signal under background light and the S2 signal while the flash light and UV LED were turned on, respectively. The results show that the developed CIS exhibits a single exposure WDR performance with a UV-visible-NIR wide spectral response.

For the non-invasive blood glucose sensing application, an absorption imaging of glucose was experimented by the over 71dB SNR dual VR mode at 1050nm wavelength. Here the 1050nm corresponds to an absorption peak of glucose in spectral response range of Si. In order to accurately measure the blood glucose concentration from very low to high levels, the resolution of 5mg/dl is considered to be desirable [1]. Fig.12 shows the measurement setup. 1050nm LED was used as light source and the optical path length was set to 10mm. Fig.13 shows the absorbance as a function of glucose concentration measured by the developed CIS. Fig.14 shows the captured absorption images when 5mg/dl glucose aqueous solution was dropped into the cell with physiological saline solution captured at 30fps. A drop, diffusion and convection of glucose aqueous were clearly visualized by the developed CIS.

Table 1 shows the performance summary of the developed CIS. Fig.15 shows the relationship of FWC per unit area and DR with other linear response CIS. A record spatial efficiency of 95ke⁻/µm² with 130dB DR was successfully achieved.

IV. CONCLUSIONS

A 16µm pixel pitch CIS with LOFITreC fabricated on ~$10^{12}cm^{-3}$ p-type Si substrate was presented. The developed CIS achieved a linear response 24.3Me⁻ FWC with a record spatial efficiency of 95ke⁻/µm² and very high QE in NIR waveband. A diffusion of 5mg/dl concentration glucose was clearly visualized by an over 71dB SNR absorption imaging at 1050nm. The developed CIS is promising for sensing applications in medical and healthcare fields and more in the IoT era.

ACKNOWLEDGMENT

This work was partly supported by JSPS KAKENHI Grant Number 17H04921.

REFERENCES

[1] Y. Fujihara et al., "A Multi Spectral Imaging System with a 71dB SNR 190-1100nm CMOS Image Sensor and an Electrically Tunable Multi Bandpass Filter," *ITE Trans. MTA*, vol.6, pp.187-194, 2018.

[2] Y. Aoyagi et al., "Dual Pixel Reset Voltage CMOS Image Sensor for High SNR Ultraviolet Light Absorption Spectral Imaging," to be published in SSDM, Sep., 2018.

[3] G. Meynants et al., "700 frames/s 2 MPixel global shutter image sensor with 2 Me⁻ full well charge and 12 µm pixel pitch," *IISW*, pp.409-412, 2015.

[4] S. Nasuno et al., "A CMOS Image Sensor with 240 µV/e⁻ Conversion Gain, 200ke⁻ Full Well Capacity and 190-1000nm Spectral Response and High Robustness to UV," *ITE Trans. MTA*, vol.2, pp.116-122, 2016.

[5] S. Sugawa et al., "A 100dB Dynamic Range CMOS Image Sensor Using a Lateral Over Flow Integration Capacitor," *ISSCC*, pp.352-353, 2005.

[6] M. Mori et al., "Thin Organic Photoconductive Film Image Sensors with Extremely High Saturation of 8500 electrons/µm²," *Symp. VLSI Tech.*, pp.22-23, 2013.

[7] K. Nishimura et al., "An 8K4K-Resolution 60fps 450ke⁻-Saturation-Signal Organic-Photoconductive-Film Global-Shutter CMOS Image Sensor with In-Pixel Noise Canceller," *ISSCC*, pp.82-83, 2018.

[8] M. Takase et al., "An Over 120 dB Wide-Dynamic-range 3.0 µm Pixel Image Sensor with In-pixel Capacitor of 41.7 fF/µm² and High Reliability Enabled by BEOL 3D Capacitor Process," *Symp. VLSI Tech.*, pp.71-72, 2018.

[9] F. Lalanne et al., "A 750 K Photocharge Linear Full Well in a 3.2 µm HDR Pixel with Complementary Carrier Collection," *MDPI Sensors*, vol.18, 2018.

[10] J. Solhusvik et al., "A 1392x976 2.8µm 120dB CIS with Per-Pixel Controlled Conversion Gain," *IISW*, R35 (presentation material), 2017.

[11] I. Oshiyama et al., "Near-infrared Sensitivity Enhancement of a Back-illuminated Complementary Metal Oxide Semiconductor Image Sensor with a Pyramid Surface for Diffraction Structure," *IEDM*, pp.397-400, 2017.

[12] R. Kuroda et al., "A Highly Ultraviolet Light Sensitive and Highly Robust Image Sensor Technology Based on Flattened Si Surface," *ITE Trans. MTA*, vol.2, pp.123-130, 2014.

[13] M. Suzuki et al., "An Over 1Mfps Global Shutter CMOS Image Sensor with 480 Frame Storage Using Vertical Analog Memory Integration," *IEDM*, pp.212-215, 2016.

[14] Y. Sakano et al., "224-ke Saturation Signal Global Shutter CMOS Image Sensor with In-pixel Pinned Storage and Lateral Overflow Integration Capacitor," *Symp. VLSI Circ.*, pp.250-251, 2017.

[15] S. Sakai et al., "A 2.8 µm Pixel-Pitch 55 ke⁻ Full-Well Capacity Global-Shutter Complementary Metal Oxide Semiconductor Image Sensor Using Lateral Overflow Integration Capacitor," *Jpn. J. Appl. Phys.*, vol.52, pp.04CE01-1-5, 2013.

[16] M. Kobayashi et al., "A 1.8e⁻rms Temporal Noise Over 110dB Dynamic Range 3.4µm Pixel Pitch Global Shutter CMOS Image Sensor with Dual-Gain Amplifiers, SS-ADC and Multiple Accumulation Shutter," *ISSCC*, pp.74-75, 2017.

[17] B. Fowler et al., "Wide Dynamic Range Low Light Level CMOS Image Sensor," *IISW*, Paper 48, 2009.

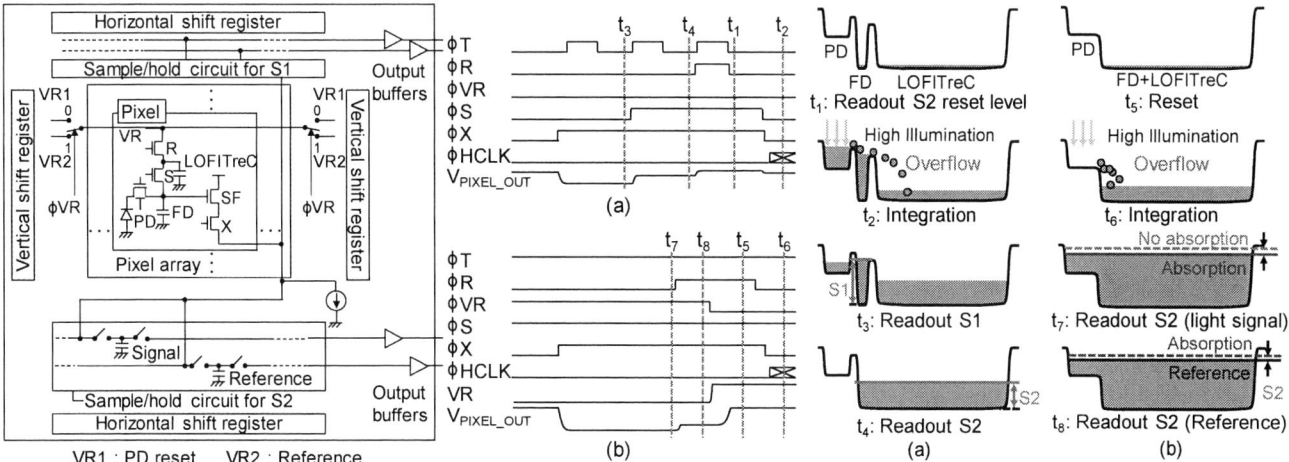

VR1 : PD reset   VR2 : Reference

Fig. 1. Circuit block diagram of the developed CMOS image sensor with LOFITreC.

Fig. 2. Operation timing diagrams of (a) the normal LOFIC operation and (b) the dual VR mode for S2.

Fig. 3. Potential diagrams of (a) normal LOFIC operation and (b) dual VR mode for S2. In (b) SF V_{th} variation is canceled and a high gain readout is applicable.

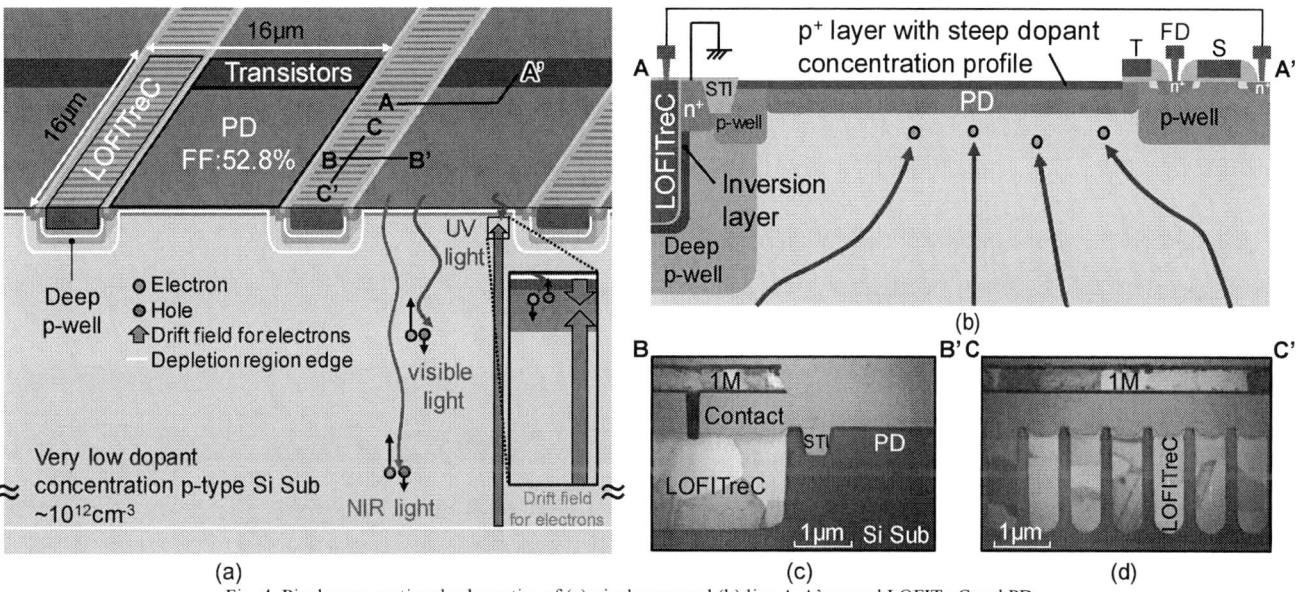

Fig. 4. Pixel cross sectional schematics of (a) pixel array and (b) line A-A' around LOFITreC and PD, and cross sectional TEM images of (c) line B-B' and (d) line C-C', respectively.

Very low dopant concentration Si substrate (~10^{12}cm^{-3}, low oxygen concentration Cz wafer)

○ STI formation
○ LOFITreC formation
  ● Hard mask deposition & patterning
  ● Si trench etching
  ● Hard mask removal
  ● Dielectric and electrode formation
  ● Deep p-well ion implantation
○ Transistor & photodiode formation
○ Metallization (5 Metal + Shield Metal)

Fig. 5. Process flow of the developed CMOS image sensor with LOFITreC.

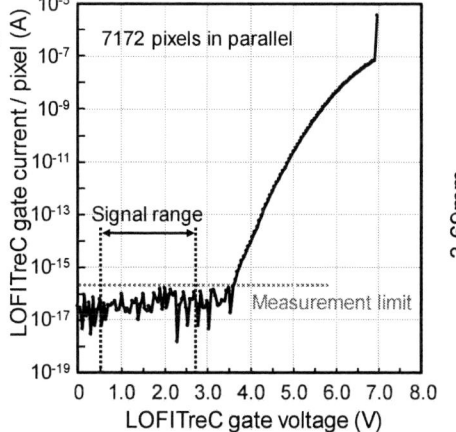

Fig. 6. Current-voltage characteristic of a LOFITreC test pattern.

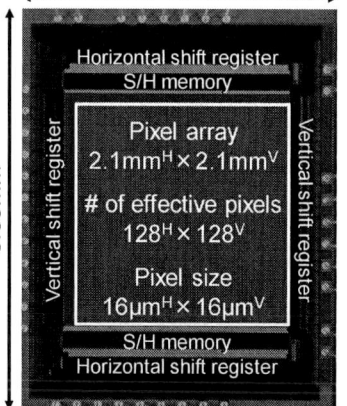

Fig. 7. Micrograph of the developed CMOS image sensor chip.

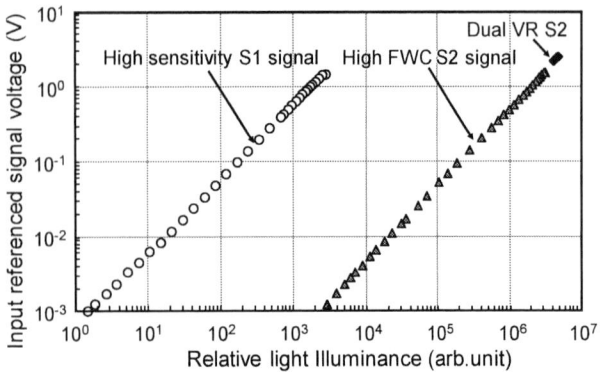

Fig. 8. Measured photoelectric conversion characteristics.

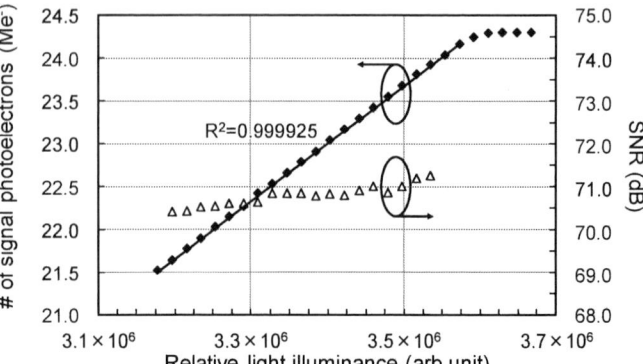

Fig. 9. Measured number of signal photoelectrons and SNR as function of light illuminance for S2 signal in the dual VR mode.

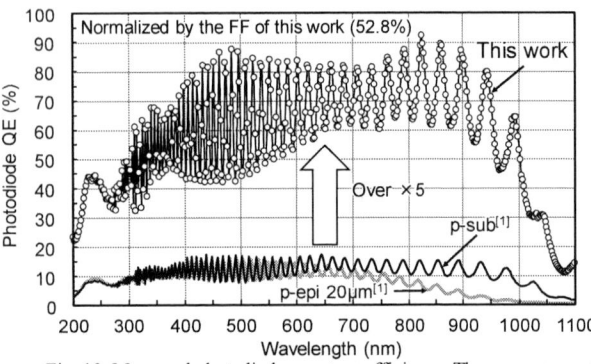

Fig. 10. Measured photodiode quantum efficiency. The measurement step and bandwidth were 2nm and 1nm, respectively.

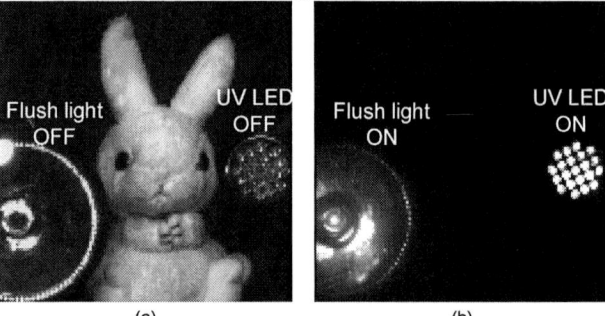

Fig. 11. Sample images by (a) S1 and (b) S2 signals captured at 256fps with F# 4.0 lens.

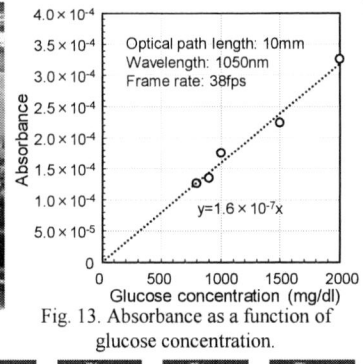

Fig. 12. Setup of glucose absorption imaging.

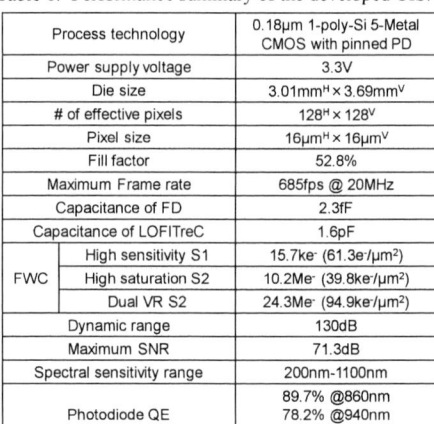

Fig. 13. Absorbance as a function of glucose concentration.

Table 1. Performance summary of the developed CIS.

Process technology		0.18µm 1-poly-Si 5-Metal CMOS with pinned PD
Power supply voltage		3.3V
Die size		3.01mm^H × 3.69mm^V
# of effective pixels		128^H × 128^V
Pixel size		16µm^H × 16µm^V
Fill factor		52.8%
Maximum Frame rate		685fps @ 20MHz
Capacitance of FD		2.3fF
Capacitance of LOFITreC		1.6pF
FWC	High sensitivity S1	15.7ke- (61.3e-/µm^2)
	High saturation S2	10.2Me- (39.8ke-/µm^2)
	Dual VR S2	24.3Me- (94.9ke-/µm^2)
Dynamic range		130dB
Maximum SNR		71.3dB
Spectral sensitivity range		200nm-1100nm
Photodiode QE		89.7% @860nm 78.2% @940nm 26.7% @1050nm

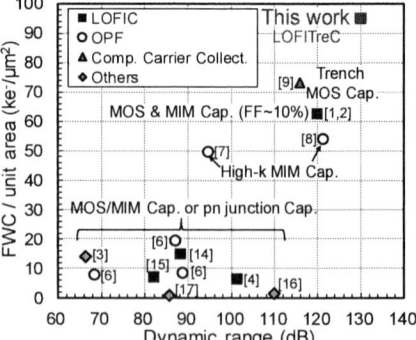

Fig. 14. Diffusion and convection of 5mg/dl glucose in physiological saline solution visualized by an over 71dB absorption imaging at 1050nm wavelength captured at 30fps. Based on the captured images, the concentration can be calculated accurately by removing inhibiting substances such as bubbles in this case.

Fig.15. FWC per unit are as a function of dynamic range with other linear response CIS. For the vertical axis, FWC is divided by pixel area.

978-1-7281-1988-5/18 $31.00 © 2018 IEEE

A HDR 98dB 3.2µm Charge Domain Global Shutter CMOS Image Sensor

A. Tournier[1], F. Roy[1], Y. Cazaux[2], F. Lalanne[1], P. Malinge[1], M. Mcdonald[1], G. Monnot[3], N. Roux[3]

[1]STMicroelectronics, Digital FMT, Technology for Optical Sensors, Crolles, France, email: arnaud.tournier@st.com
[2]CEA Leti, Département Optique et Photonique, Laboratoire Imagerie Visible, Grenoble, France
[3]STMicroelectronics, Analog, MEMS & Sensors Group, Imaging, Grenoble, France

Abstract— We developed a High Dynamic Range (HDR) Global Shutter (GS) pixel for automotive applications working in the charge domain with dual high-density storage node using Capacitive Deep Trench Isolation (CDTI). With a pixel size of 3.2µm, this is the smallest reported GS pixel achieving linear dynamic range of 98dB with a noise floor of 2.8e-. The pinned memory isolated by CDTI can store 2 x 8000e- with dark current lower than 5e-/s at 60°C. A shutter efficiency of 99.97% at 505nm and a Modulation Transfer Function (MTF) at 940nm better than 0.5 at Nyquist frequency is also reported.

I. INTRODUCTION

CMOS image sensors using pinned photodiode (PPD) are widely used in the imaging industry because of their good performances at low light level. A large majority of these sensors use a 4T pixel architecture, well suited for video recording, but at the same time suffering from motion artefacts inherent to rolling shutter operation. To overcome this issue, global shutter pixels were introduced [1]-[4]. Moreover, in order to manage complex lighting conditions and to ensure strong robustness against ambient, like sunlight or automotive headlights, high dynamic range capability is required. Both GS and HDR mode perfectly fit with applications like head pose detection, eyelids analysis or accurate gaze direction. In this paper, we present a global shutter pixel using pinned diodes for both collection and retention areas with charge transfer and storage assisted by a vertical gate to achieve low dark current and high full well capacity (FWC) performances. This architecture employs CDTI, which offers electrical and optical crosstalk reduction, low dark current and signal storage capability. The in-pixel memory node works in the charge domain with intrinsic key performance as good as pinned photodiode or pinned CCD [5]-[6].

II. DEEP TRENCH PIXEL ISOLATION

A. Deep Trench Isolation - DTI

As CMOS image sensor pixel size is reduced, electrical and optical crosstalk in the silicon layer becomes detrimental for color reconstruction. To overcome this issue, oxide filled deep trench isolation techniques were developed [7]-[8]. As shown in Fig. 1, the DTI is used as an in-silicon light guide to improve the pixel optical cross talk and at the same time to tackle pixel-to-pixel electron diffusion. This image quality to be maintained even with high Chief Ray Angle (CRA). A simplified DTI process flow is shown in Fig. 2a. DTI fabrication comprises photolithography and etch process, DTI sidewall passivation by implantations, oxide gapfill, anneal and finally planarization (CMP).

B. Capacitive Deep Trench Isolation - CDTI

Deep Trench Isolation is efficient for crosstalk control but suffers from pixel dark current figure due to the large Si-SiO2 surface introduced in the pixel. To contain this issue, specific surface treatment like boron implantation and forming gas anneal have been successfully introduced. To go further and eradicate this parasitic signal contribution, active MOS deep trench isolation was proposed [9]. By applying a suitable bias on the CDTI gate resulting in strong hole accumulation at the silicon interface, a dark current below 6e-/s at 60°C for a 1.4µm pixel pitch is measured. Compared to DTI pixel dark current, this figure has been improve by a factor of 5. A simplified CDTI process flow is shown in Fig. 2b. Vertical MOS capacitance fabrication consists of a photolithography process step for trench realization. Using a dedicated recipe for anisotropic etching step, an aspect ratio higher than 1:30 is feasible. The capacitance gate dielectric is obtained by oxide growth with subsequent oxide deposition. The trench is then filled with in-situ doped polysilicon deposition. The CDTI process is completed by CMP process before entering a classical CMOS imaging process.

III. GLOBAL SHUTTER DEVICES AND OPERATIONS

A. High Density Memory Device Description

The pixel charge memory is based on a fully depleted MOS capacitance with N doped buried channel for electron signal storage. A low-doped N-type layer is pinched by two CDTI gates. Similar to a pinned diode or planar CCD in fully depleted mode, a well potential is created and is now ready to store electrons. The Si-Poly gate electrode is biased to manage the electrostatic potential in the pinned memory and to control the charge transfer. This storage node device is formed by multiple high-energy N-type phosphorous implantations. TCAD simulation tools were used to adjust all doping profiles in both the photodiode and memory, but also the height of the barrier potential between them. Fig. 3 shows the electro-static equipotential lines for the memory device in empty and full states based on TCAD results. The pinned photodiode and pinned memory fully depleted electrostatic voltages were matched and targeted around 1.5V, a memory switching high voltage below 2.5V has been maintained.

B. Global Shutter Operations Description

The vertical MOS trench gates are designed and processed to manage the charge flow from the pinned photodiode to the storage transit zones and finally to the exit stage towards the readout circuitry [10]. The first part of the memory device, acting as the entrance for the charges, is designed like a buried channel MOS transistor used as charge transfer gate. The

---

978-1-7281-1988-5/18 $31.00 © 2018 IEEE   229

second part is the transit zone wherein the charge will be stored waiting for the timing pulse in the readout sequence. The third part is the memory exit, designed as a planar MOS transfer gate switching charge from memory to Floating Diffusion (FD). Fig. 4 describes the potential variation and charge transfer during GS pixel operation. With reference to the figure, the following sequence is followed for readout.

① After resetting all photodiodes of the pixel array thanks to the anti-blooming device, the global integration starts and the photo-generated charges are collected by the PPD until electron overflow occurs through the AB device, avoiding any memory signal corruption.

② The charges are globally transferred from the PPD to the memory by applying a high-level bias on the CDTI Gates (TG1 + MEM). This vertical MOS device is used for charge storage and also for photodiode to memory electrostatic barrier modulation.

③ The memory gate (TG1 + MEM) is lowered to store charge at lower potential.

④ Finally, the end memory transfer gate (TG2) is switched ON row by row, in rolling mode for pixel signal readout using the CDS readout noise cancellation technique.

IV. HDR 3.2μM GLOBAL SHUTTER PIXEL

A HDR 3.2μm GS pixel has been designed and implemented in a 2.3Mpix array. Wafer processing was performed using a Front-Side Illumination (FSI) dedicated imaging technology with CDTI and specific pixel process steps. The following section describes the architecture and details the measured Electro-Optical characteristics of this pixel.

A. Dual Storage Node Global Shutter Pixel Description

The pixel schematic is shown in Fig. 5: it comprises Pinned Photo-Diode (PPD), Source Follower (SF), Read (RD) and Reset (RST) transistors, coming from 4T pixel architecture technology and used in their conventional way. According to the electrostatic potential description provided in the previous section, CDTI gates are introduced for GS functionality: TG1 for global charge transfer and MEM for charge retention. The memory node (MEM) is ended by a classical transfer gate (TG2) operating in rolling readout operation sequence. An anti-blooming transistor (AB) is also implemented to control photodiode reset, integration start and photodiode overflow. In order to implement HDR functionality based on dual integration time, 2 GS paths are implemented (TG1/MEM/TG2). In this way, long and short exposure signals are acquired and stored in long and short GS paths, respectively. For the pixel described, Fig. 6 provides a TEM cross-section of one GS path and the readout transistors and Fig. 7 illustrates standard global video timing applied to extract the main electro-optical characteristics.

B. Characterization results

Table 1. summarizes the electro-optical characteristics of the fabricated pixel. The pixel linear full well capacity driven by the memory charge storage capability, reaches 7100e- per storage node, with good linearity as shown in Fig. 8. The memory device is designed to be fully depleted after charge transfer. In line with this operating condition, transfer from the PPD to the FD is a noiseless operation and GS low light

performance is greatly improved: The pixel presents no lag and, after CDS operation, a total noise floor of 2e- and 2.8e- has been demonstrated in single and dual integration mode (HDR), respectively. This results in a linear dynamic range of 98dB for an integration time ratio of 1:32 between long and short acquisition. The PRNU (Fig. 9) exhibits very good behavior with a value close to 0.4% maintained across the majority of the signal range. Maintaining the MOS CDTI interface in hole accumulation for long enough, the pixel side wall dark current contribution is very small, leading to a memory device dark current lower than 5e-/s at 60°C. The deep trench isolation is also used to control pixel-to-pixel cross talk, achieving high resolution in the whole wavelength spectrum including Near Infrared (NIR) and leading to an exceptional Modulation Transfer Function (MTF) performance shown in Fig. 10. Accordingly a MTF better than 0.5 at Nyquist frequency is reported at 940nm, close to maximum "ideal" theoretical value. Regarding the memory zone, the use of CDTI for charge transfer and storage is also useful for photodiode to memory isolation, the deep trench working as an efficient optical and electrical stop channel. Thanks to this, GS efficiency of 99.97% at 505nm is measured. Quantum efficiency has been measured in both integration times (short and long as shown in Fig. 11). Matching between both measurements over the full spectrum guarantees good HDR image reconstruction.

V. CONCLUSION

In this paper, we present a HDR GS automotive pixel. The pixel operates in the charge domain, using a dual storage node constructed by fully depleted pinned memories isolated by CDTI. The pixel performance presented is at the state of the art with exceptional MTF performance. Images in Fig. 12 are shown as illustration.

ACKNOWLEDGMENT

The authors thank the Digital Front End Manufacturing & Technology group of STMicroelectronics for wafer processing, physical characterization and measurement and CEA-Leti for their support.

REFERENCES

[1] N. Bock et al, "A wide vga cmos image sensor with global shutter and extended dynamic range," IEEE Workshop on CCDs and Adv. Image Sensors, 2005.

[2] I. Takayanagi et al, "A 600x600 pixel, 500 fps cmos image sensor with a 4.4 μm pinned photodiode 5- transistor global shutter pixel," IISW, 2007.

[3] S. Lauxtermann et al, "Comparison of global shutter pixels for cmos image sensors," IISW, 2007.

[4] S. Velichko et al, "Low noise high efficiency 3.75 μm and 2.8 μm global shutter cmos pixel arrays," IISW, 2013.

[5] J. Janesick, "Multi-pinned phase charge-coupled device," NASA Tech Brief, 1990

[6] N. Teranishi et al, "No image lag photodiode structure in the interline CCD image sensor," Electron Devices Meeting, 1982.

[7] B. J. Park et al, "Deep Trench Isolation for Crosstalk Suppression in Active Pixel Sensors with 1.7μm Pixel Pitch," JJAP, 2007.

[8] A. Tournier et al, "Pixel–to–pixel isolation by deep trench technology: Application to cmos image sensor," IISW, 2011

[9] N. Ahmed et al, "Mos capacitor deep trench isolation for cmos image sensors," IEDM, 2014.

[10] F. Roy and Y. Cazaux, "Image sensor" Patent US 9,236,407, Jan. 2, 2013. Available [Online] : https://patents.google.com/

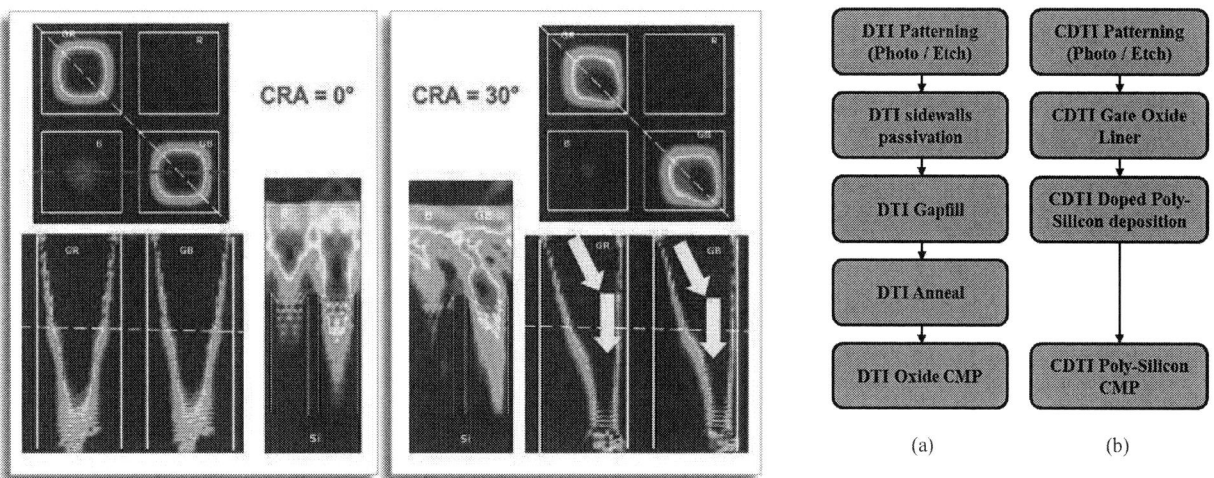

Fig. 1. On-axis (CRA=0°) and Off-axis (CRA=30°) optical simulation at 940nm for a pixel with oxide filled DTI

Fig. 2. Simplified DTI and CDTI process step sequence

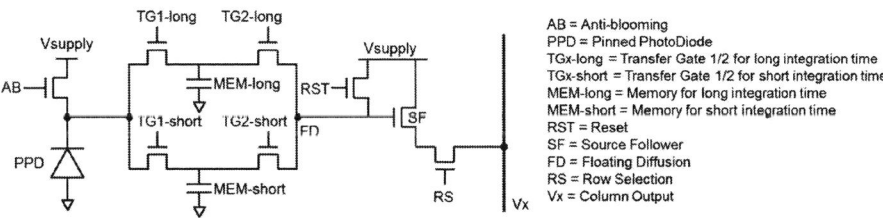

Fig. 3. 2D MEM equipotential lines, on the left: Empty MEM with fully depleted N-type zone, on the right: Full MEM with neutral N-type zone filed with e-.

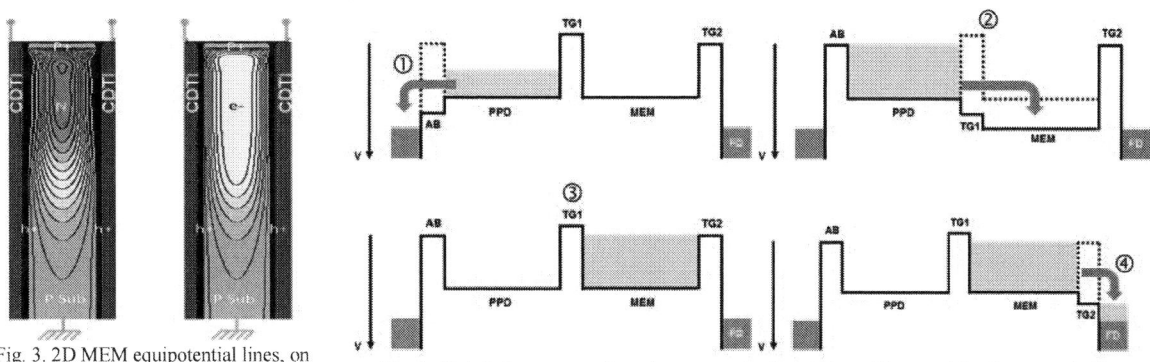

Fig. 4. Global shutter operation: charge transfer description with potential voltages evolution

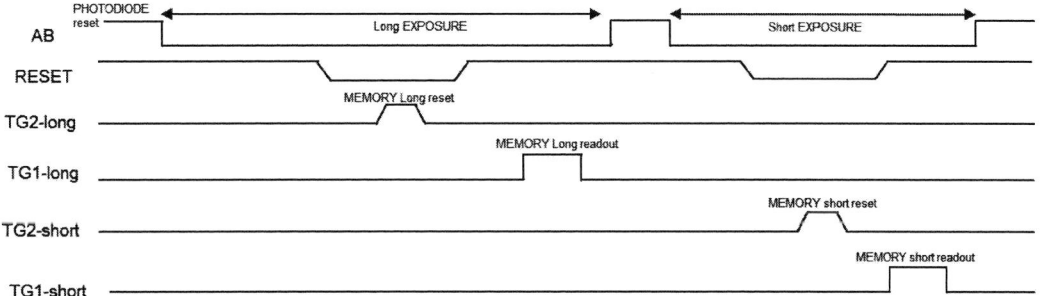

AB = Anti-blooming
PPD = Pinned PhotoDiode
TGx-long = Transfer Gate 1/2 for long integration time
TGx-short = Transfer Gate 1/2 for short integration time
MEM-long = Memory for long integration time
MEM-short = Memory for short integration time
RST = Reset
SF = Source Follower
FD = Floating Diffusion
RS = Row Selection
Vx = Column Output

Fig. 5. HDR GS pixel with dual memory schematic

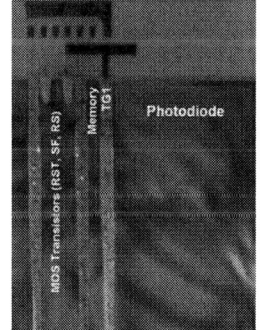

Fig. 6. Pixel TEM cross-section of one GS path

Fig. 7. Standard global video timing applied for integration and rolling readout

978-1-7281-1988-5/18 $31.00 © 2018 IEEE   231

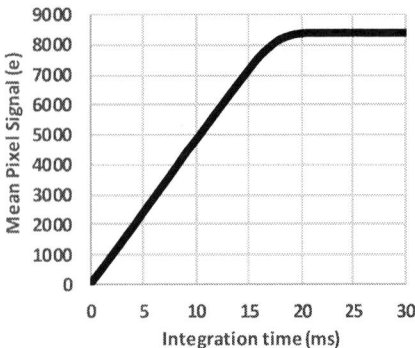

Fig. 8. Pixel signal transfer curve for one GS Path

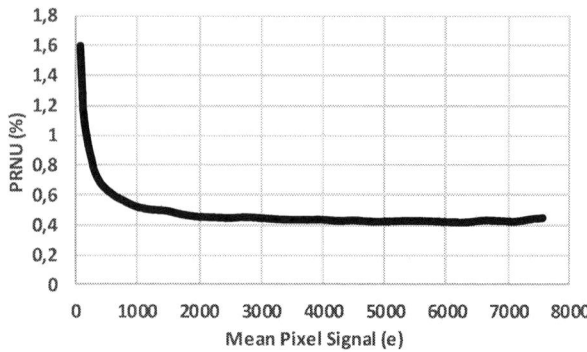

Fig. 9. Photo-Response Non-Uniformity (PRNU) versus Pixel Signal Level

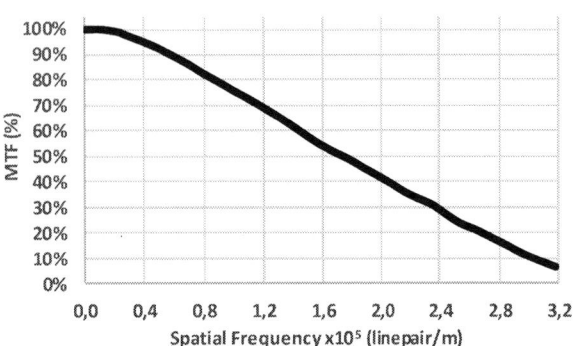

Fig. 10. Modulation Transfer Function (MTF) Curve at 940nm

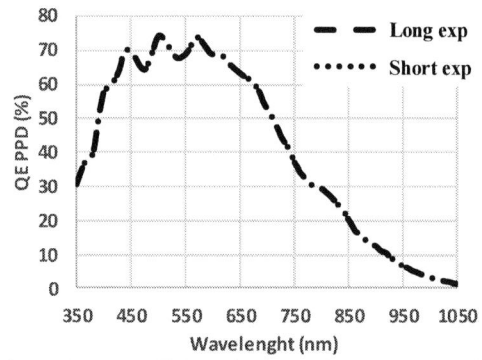

Fig. 11. Quantum efficiency for short and long integration time

Characteristics	Value	Test conditions
Pixel Pitch	3.2μm	HDR GS Pixel
Full Well Capacity	2 x 8300 e-	Tj=60C
Linear Full Well	2 x 7100 e-	Tj=60C
LAG	no LAG	Signal = 2000 e- and Tj=0C
Dark Current	22 e-/s 5 e-/s	Tj=60C, Photodiode Tj=60C, Memory
Total Noise Floor	2.0 e- 2.8 e-	Single Exposure, Tj=60C Dual Exposure, Tj=60C
Dynamic Range	68.2 dB 98.2 dB	Single Exposure Dual Exposure (L/S Ratio = 32)
PRNU	0.43 %	50% FWC, Tj=60C
Maximum SNR	39 dB	Tj=60C
Quantum Efficiency (QE)	72.9 % 19.9 % 8.0 %	λ = 505 nm, f/2 λ = 850 nm, f/2 λ = 940 nm, f/2
Shutter efficiency	99.97 % 99.89 % 99.84 %	λ = 505 nm, f/2 λ = 850 nm, f/2 λ = 940 nm, f/2
MTF Horizontal MTF Vertical	0.55 0.51	$F_{Nyquist}, \lambda$ = 940 nm, f/2

Table 1. Pixel performances summary

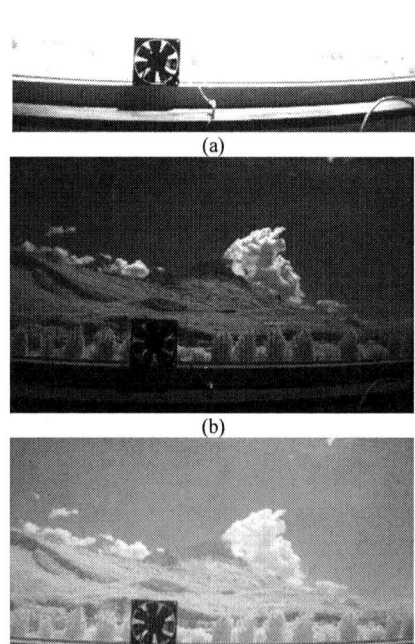

Fig. 12. Picture example using the described pixel (a) long exposure (b) short exposure and (c) HDR mode

978-1-7281-1988-5/18 $31.00 © 2018 IEEE

High Performance 2.5um Global Shutter Pixel with New Designed Light-Pipe Structure

[1*]Toshifumi Yokoyama, [1]Masafumi Tsutsui, [1]Yoshiaki Nishi, [1]Ikuo Mizuno, [2]Veinger Dmitry & [2]Assaf Lahav

[1]TowerJazz Panasonic Semiconductor Co,. Ltd. 800 Higashiyama, Uozu City, Toyama, Japan 937-8585.
[1*]E-mail address: yokoyama.toshifumi@tpsemico.com TEL: +81-50-3739-0551
[2] TowerJazz Migdal Haemeq 23105, Israel.

Abstract— We developed a 2.5um global shutter (GS) CMOS image sensor pixel using an advanced Light-Pipe (LP) structure designed with novel guidelines. To the best of our knowledge, it is the smallest reported GS pixel in the world. The developed pixel shows an excellent Quantum Efficiency (QE), Angular Responses (AR) and very low Parasitic Light Sensitivity (PLS). Also, even in oblique light condition of 10 degrees, the 1/PLS is maintained to about half value. These key characteristics allow development of ultra-high resolution sensors, industrial cameras with wide aperture lenses and low form factors optical modules for GS mobile applications.

I. INTRODUCTION

In recent years there is a strong market demand for high resolution, smaller pitch and high performance GS sensors. We already showed a low noise high QE 2.8um GS pixel using double lenses structure in 110nm technology [1]. In this paper, we report further reduction of pixel size to 2.5um, by using a 65nm process node and advanced LP structure.

Keeping and improving the optical characteristics is one of the main challenges in the development of smaller pixels. For GS pixel it is particularly important to keep the PLS small enough and the QE high enough even at high incident angles. High incident angles can occur either when low F# lenses are used, like in mobile applications, or when the pixel is used in large array formats for machine vision. For example when system lens with F#2.8 is required, the optical characteristics should be maintained up to ± 10 degrees.

Figure 1 shows a schematic cross section of the developed low noise charge domain pixel [2]. We used a fully buried Memory Node (MN) to store the charges till readout. PLS is generated when photo-electrons are generated in the MN. This parasitic signal is a major cause for image artifacts in highly dynamic scene. Commonly, the MN is shielded from incoming light with tungsten (W) plate, which is placed over and in close vicinity to MN [3]. It is also desired to keep MN small in area relative to the diode or pixel area. When the pixel size is shrunk, it is difficult to reduce the area of the MN, since area of MN controls the Full Well Capacity (FWC). Moreover, since W shield covers this area, the optical aperture

in a GS pixel is much smaller than a Rolling Shutter (RS) pixel in the same pitch. It is clear from above discussion that when the GS pixel size is shrunk, the optical aperture becomes small, as a result the AR and QE deteriorate. Another important effect in small GS pixel is the diffraction from and under the W shield, since it is another cause for PLS increase. It is necessary to investigate a structure that efficiently collects light into the optical aperture and minimizes the reflection from the W shield.

II. DEVICE STRUCTURE

In order to realize better AR of QE and PLS, a LP with a high refractive index material was used in both <1.4um RS devices [4] and for relatively large 3.45um GS pixel [5]. We adopted the LP structure as shown in Figure 1. The core material for the LP is Si_3N_4. Since the refractive index of Si_3N_4 is high (n=2), incident light is bent at the Si_3N_4 surface. In addition, incident light into the LP is totally reflected inside the LP, by utilizing the difference in refractive index between Si_3N_4 (n=2) and the oxide film (n=1.45), and is confined in the LP, which greatly improved AR performance.

For further immunity against oblique light it is necessary to reduce the optical height from the Si surface to the micro-lens, since the displacement of light circle from optical center per unit angle can be reduced. This task was achieved by developing a 65nm node with an ultra-thin backend process, and using only three level metals in the pixel.

Figure 2 shows the circuit schematic of the developed GS pixel. In order to maximize active area, there is no row select transistor, and Floating Diffusion (FD) is shared by two pixels in a diagonal direction. We also designed a narrow MN considering tradeoff between AR and FWC. Regarding the reduction of the dark current, we adopted the fully buried low dark current MN structure as already reported [6].

III. APPROACH

In this section, the novel design approach for LP, with target to maintain high QE and low PLS for light incident at 0 to 10 degrees is described. Design of LP is the one of the most important keys to realize high performance GS pixel.

978-1-7281-1988-5/18 $31.00 © 2018 IEEE

In order to improve the AR of QE, it is necessary to maximize the width of the upper surface of the LP (L_{upper}) to collect more light into the LP, as shown in Figure 1. The L_{upper} needs to be wide enough to capture incident light tilted by 10 degrees or more. On the other hand, in order to collect light in the optical aperture, it is necessary to make the width of the LP bottom (L_{bottom}) smaller than the W aperture ($L_{bottom} < L_w$). Therefore, the LP has a tapered shape, as shown in Figure 1.

In order to find the ideal LP shape, 1/PLS was calculated by optical simulation using the 3D-FDTD method, as shown in Figure 3. Comparison was made with the following three types of shapes, with varying taper angle α and light pipe depth. :

1）Type A: Long Light pipe (α was 16 degrees).

2）Type B: Large taper angle (α was 20 degrees).

3）Type C: Small taper angle (α was 10 degrees).

At first, incident angle of 0 degrees was simulated. The simulation was performed at a wavelength of 530 nm. Results for QE and 1/PLS are summarized in Figure 4. First, it is important to note that the simulated QE is almost equal for all three structures, but 1/PLS is greatly affected by LP topology. Type A had the worst PLS out of the three types. As shown in Figure 4, in a long light pipe with a taper angle, once light is fully reflected from the first interface, its angle is too small to be fully reflected in the second time and the light leaks out of the LP. Type C clearly gives much better results than type A. In Type C, the incident light is focused at a position distant from the MN.

Next, as shown in Figure 5, 1/PLS at the incident angle of 10 degrees was calculated. Type B with the largest tapering angle has clearly the worst 1/PLS. As the taper angle of interfaces increases, there is smaller range of the incident light that can be totally reflected. The Light is leaking out of the LP, which leads to degradation of 1/PLS.

From the simulation, the new design guidelines for the light pipe are as follows: First, the height of light pipe should be designed so that once totally reflected, light goes to Si surface through the bottom of light pipe. Single reflection in light pipe is preferred. Second, the tapered angle of the light pipe should be designed small considering oblique incident light.

Finally, we choose Type C as our LP structure. Our simulation showed QE at zero degree angle was 65%, QE degraded to 80% of its peak value only at 12.5 degrees, 1/PLS was 12500 at 0 degree (-81.9 dB) and 6000 at 10 degrees.

IV. EXPERIMENTAL RESULTS

Figure 6 shows x-section of fabricated 2.5um GS pixel. The LP has almost the same shape as designed. Figure 7 shows the QE curves of the color and mono samples. The QE for Green pixel (530nm) is close to 68%. The QE of blue (450nm) and red (600nm) pixels are 54% and 46%, respectively. The peak QE of the Mono sample was 78% at the wavelength of around 500 nm. 1/PLS of the Mono sample was 8100 (-77dB). Figure 8 shows the AR of QE. It is shown that the pixel can maintain 80% of its peak value at ±12.5 degrees. Figure 9 shows the AR of 1/PLS. 1/PLS is maintained at half value or more in the range of ± 10 degrees, which achieved our design targets. Figure 10 shows the F# dependence of 1/PLS. In F#2.8 1/PLS is 10400 (-80.3dB). 1/PLS is very stable with F # 2.8 or higher. Figure 11 shows snapshots from a 25Mpixel product [8] in different F#. Figure 11(a) is captured with F#4 lens and exposure time of 2.8msec. Figure 11(b) is captured with F#0.95 lens and exposure time of 0.24msec. The exposure time is changed to equalize the signal amount in the pixels and emphasize the dynamics of the scene. In Figure 11(b), there were no dynamic PLS artifacts even though F# is 0.95. Figure 12 shows the AR of QE in the diagonal direction. Although we used shared pixel structure, as shown in Figure 2, there is no difference between Gr and Gb pixels because the incident light is confined in the LP and almost no light hit metals or the poly-Si gates.

Table 1 is the summary of characteristics and comparison with previous reports [5, 7]. Despite the smaller pixel size compared with past reports, the best in class performances were achieved.

V. CONCLUSION

We developed the world's smallest 2.5um GS pixel which is suitable for large format machine vision sensors and for consumer applications due to its immunity to large incident light angles. In order to improve the AR a new LP structure was designed according to novel guidelines based on 65nm process. The achieved QE was 68% (color), 78% (mono). The pixel QE is maintained up to 80% from peak value at ±12.5 degrees. The 1/PLS of 12000 (81.6dB) was achieved at F#8 and degraded only to 10400 (-80.3dB) with F#2.8.

ACKNOWLEDGMENT

We gratefully acknowledge the contributions of Gpixel members.

REFERENCES

[1] T. Yokoyama, et al., A. Design of Double Micro Lens Structure for 2.8μm Global Shutter Pixel. In Proceedings of the 2017 International Image Sensor Workshop, Hiroshima, Japan, pp.398.

[2] S. Lauxtermann, et al., Comparison of Global Shutter Pixels for CMOS Image Sensors. In Proceedings of the 2007 International Image Sensor Workshop, Ogunquit, ME, USA, 7–10 June 2007.

[3] N. Teranishi, et al., Smear reduction in the interline CCD image sensor. IEEE Trans. Electron Devices 1987, 34, 1052–1056.

[4] H. Watanabe, et al., A 1.4μm Front-side Illuminated Image Sensor with Novel Light Guiding Structure Consisting of Stacked Lightpipes. In Proc of the IEDM2011, Washington, DC, USA, 2011.

[5] H. Sekine, et al., A High Optical Performance 3.4 um Pixel Pitch Global Shutter CMOS Image Sensor with Light Guide Structure. In Proc of the 2017 International Image Sensor Workshop, Hiroshima, Japan, pp. 394.

[6] M. Tsutsui, et al., Development of Low Noise Memory Node in a 2.8um Global Shutter Pixel with Dual Transfer. In Proceedings of the 2017 International Image Sensor Workshop, Hiroshima, Japan, 2017; pp.28.

[7] S. Velichko, et al., "CMOS Global Shutter Charge Storage Pixels With Improved Performance", IEEE Transactions on Electron Devices, vol. 63, no. 1, Jan. 2016, pp. 106-112.

[8] GMAX0505 product data sheet retrieved on 26/7/2018. http://www.gpixelinc.com/en/index.php?s=/b/112.html,

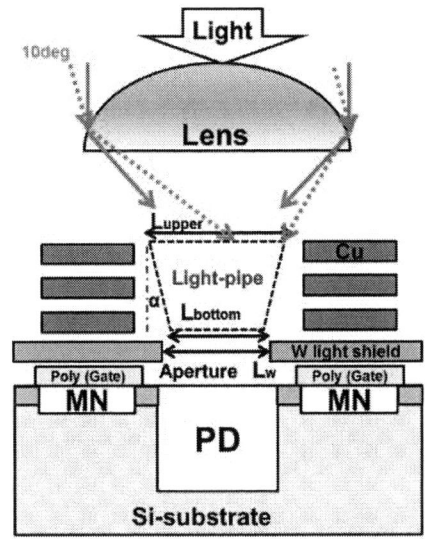

Figure 1: Cross-section of GS pixel.

Figure 2: Pixel circuit schematic of GS pixel.

	Type A Long pipe	Type B Large taper angle	Type C Small taper angle
Electric Field (cross section)			
1/PLS@0degree	7700	9800	12500
QE@0degree[%]	64.6	63.2	65.0

Figure 3: Comparison of three types of light pipes at incident angle 0 degree (3D-FDTD simulation).

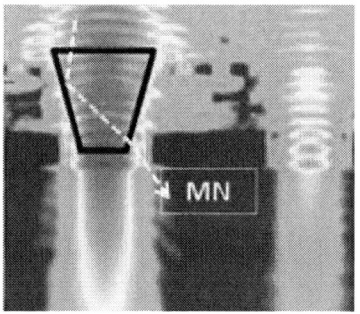

Figure 4: Enlaeged view of Type A@0degree

	Type A Long pipe	Type B Large taper angle	Type C Small taper angle
Electric Field (cross section)			
1/PLS@10degree	5600	4500	6000
QE@10degree[%]	55.5	54.8	55.3

Figure 5: Comparison of three types of light pipes at incident angle 10 degrees (3D-FDTD simulation).

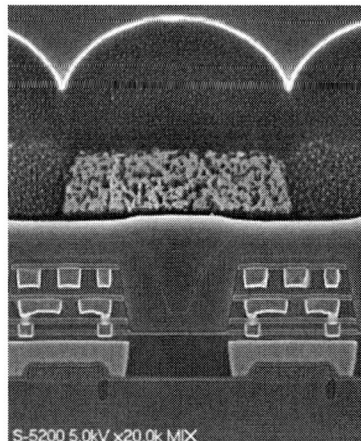

Figure 6: Cross-section of GS pixel.

978-1-7281-1988-5/18 $31.00 © 2018 IEEE

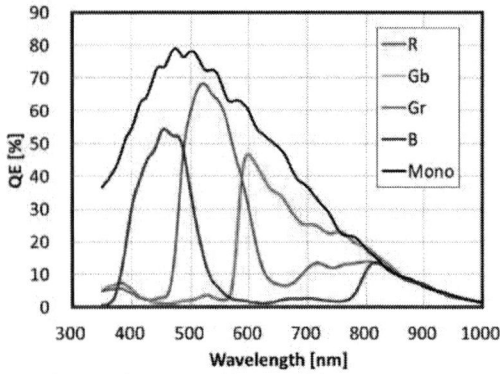

Figure 7: QE curves of 2.5 μm GS image sensor.

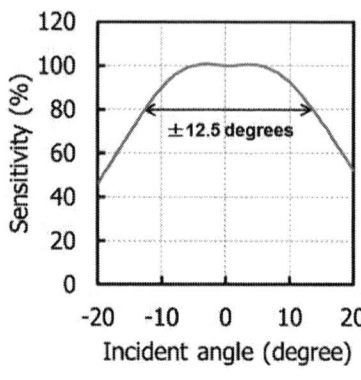

Figure 8: Angular response of QE.

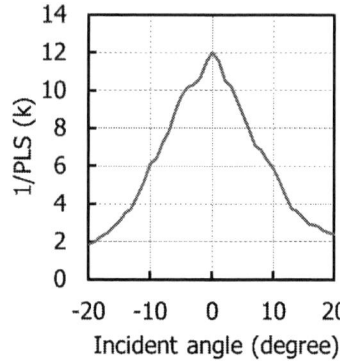

Figure 9: Angular response of 1/PLS.

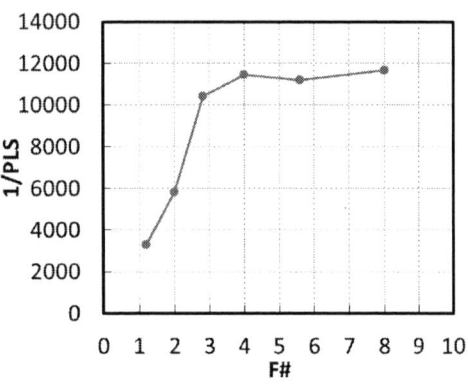

Figure 10: F#dependence of 1/PLS.

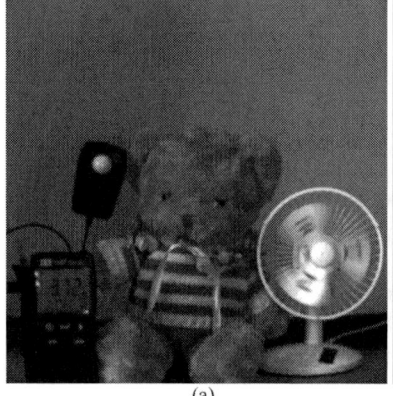

(a)   (b)

Figure 11: photographs in different F#.
(a) GS image with f#4 lens, exposure time was 2.8msec.
(b) GS image with f#0.95 lens, exposure time was 0.24msec.

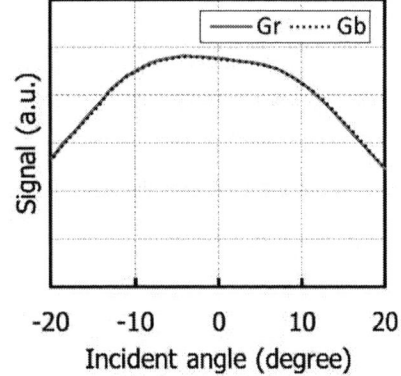

Figure 12: Angular response of QE (diagonal direction).

Table.1: Summary of characteristics and comparison with previous reports

	Unit	This work	[5]	[7]
Pixel Pitch	um	2.5	3.4	2.8
Peak QE	%	67(color) 78 (mono)	- -	- 70 (mono)
Linear Full Well Capacity	e-	6300	8100	6000
PLS	dB	-81.6(color)[a] -80.3(color)[b] -77(mono)[a]	-89(color) -	- -66.8(mono)
Angle with 50% 1/PLS	degree	10	-	-
Dark current	e-/s	17	-	60

[a]Collimated light. [b]F#2.8

Back-Illuminated 2.74 μm-Pixel-Pitch Global Shutter CMOS Image Sensor with Charge-Domain Memory Achieving 10k e- Saturation Signal

Y. Kumagai[1], R. Yoshita[1], N. Osawa[1], H. Ikeda[1], K.Yamashita[1], T. Abe[1], S. Kudo[2],
J. Yamane[3], T. Idekoba[3], S. Noudo[3], Y. Ono[3], S.Kunitake[3], M. Sato[3], N. Sato[3], T. Enomoto[3],
K. Nakazawa[1], H. Mori[1], Y. Tateshita[1], and K. Ohno[1]

[1]Research Division 1, Sony Semiconductor Solutions Corporation, Kanagawa, Japan, email: Yoshimichi.Kumagai@sony.com
[2] Imaging System Business Division, Sony Semiconductor Solutions Corporation, Kanagawa, Japan
[3] Sony Semiconductor Manufacturing Corporation, Kumamoto, Japan

Abstract— A 3208×2184 global shutter image sensor with back-illuminated architecture is implemented in a 90 nm/65 nm imaging process. The sensor, having 2.74 μm-pitch-pixels, achieves 10000 electrons full-well capacity and -80 dB parasitic light sensitivity. Furthermore, 13.8 e-/s dark current at 60°C and 1.85 erms random noise are obtained. In this paper, the structure of a pixel with memory along with saturation enhancement technology is described.

I. INTRODUCTION

CMOS image sensors (CIS) with a global shutter (GS) function are required in a variety of fields including broadcasting, cinema, surveillance, and factory automation [1]. In particular, application to factory automation requires the compatibility of high resolution while maintaining an electron full well capacity (FWC) of over 10000 and a high frame rate of more than 60 fps. Since GS-CIS with memory (MEM) structure has an advantage with respect to temporal noise over previously reported GS-CIS [2,3], front-illuminated GS-CIS with MEM has been developed [1, 4]. Although the FWC in front-illuminated GS-CIS can be enhanced by adopting the multiple read-out method [1], its application is still limited due to deterioration of the frame-rate and imperfect saturation when capturing moving images. A further increase in resolution and frame rate with front-illuminated GS-CIS, however, is virtually unattainable as an increased number of vertical signal lines and control lines for high speed reading in these CISs limit their photo-electrical conversion efficiencies due to decreased apertures. One way to resolve this technical issue is the utilization of back-illuminated sensor architectures in which signal lines are implemented below the active region, and are therefore free from the aforementioned optical restriction. In a back-illuminated GS-CIS, it is necessary to implement light-shielding structures over MEM to suppress the influence of parasitic light when signals are read in a column-wise sequence. However, an introduction of such a light-shielding structure reduces the effective area, thereby deteriorating the FWC of CIS. In this paper, technology for realizing compatible performance of FWC and parasitic light sensitivity (PLS) in recently developed back-illuminated GS-CIS with MEM is presented.

II. SENSOR ARCHITECTURE

A back-illuminated GS-CIS architecture with 2.74 μm-pitch-pixels is shown in Fig.1. A 2×2 pixel consists of a photo-diode (PD), over-flow gate (OFG), over-flow drain (OFD), and MEM with two transistors (TRY and TRX); each pixel shares a floating diffusion (FD) and a pixel circuit consisting of a reset transistor (RST), a select transistor (SEL), and a source-follower amplifier transistor (AMP). By switching TRY and TRX transistors, electrons accumulated in the PD are transferred to MEM which then are further transferred to FD via the TRX and transfer gate (TRG).

III. PARASITIC LIGHT SENSITIVITY SUPRESSION TECHNOLOGY

A key factor affecting the image quality of GS-CIS with MEM is the influence of parasitic light during the reading period. The degradation of PLS is mainly attributed to the stray light entering the MEM region, and is heavily dependent on light collection management. Thus, suppression of PLS when shrinking pixel size is a challenging issue regardless of front-illuminated or back-illuminated CIS architecture. In comparison to front-illuminated GS-CIS with MEM [1, 4], the light-shielding metal layer in conventional back-illuminated GS-CIS is located far above the MEM region as shown in Fig.2(a), in which the direct absorption of incident light by MEM and incoming electrons generated outside of the effective area are inevitable. Therefore, as shown in Fig. 2(b), MEM is encapsulated with a metal-embedded reverse deep trench isolation (RDTI) while providing a partial opening sufficient for transferring electrons from the PD to MEM. Additionally, to further suppress the effect of electrons generated by the stray light from partial openings, a potential well is designed in the region above MEM which allows trapping of generated electrons and subsequent discharging to OFD. With the PLS suppression technology mentioned above, a PLS of -80 dB is obtained, which is comparable to front-illuminated GS-CIS of similar pixel size (2.8 μm-pitch) [5].

978-1-7281-1988-5/18 $31.00 © 2018 IEEE

IV. SATURATION ENHANCEMENT TECHNOLOGY

A. Reverse-Transfer Prevention Gate

Fig.3(a) shows the potential diagram of the conventional single-gated (TRX) MEM in GS-CIS. Depending on the relative potential of MEM and the PD, electrons stored in MEM are unintentionally reverse-transferred to the PD upon switching the TRX off. This is described as follows. When the TRX is turned on, not only the depletion capacity of MEM but also the gate oxide capacitance contributes to charge accumulation, so that all charges can be stored in MEM. However, during the process of switching TRX off, the amount of charge that can be accumulated at the gate interface decreases concomitantly with decreasing TRX voltage. Therefore, the capacitance corresponding to the potential difference in MEM and the PD is only effective for storing charges, consequently leading to the reverse-transfer of electrons from MEM to the PD. To prevent nonlinearity of signals originating from this electron reverse-transfer, the gate in the MEM region is planarly divided in to two (TRX and TRY) as shown in Fig. 1. With this configuration, switching off a gate that is adjacent to the PD (TRY) first can effectively prevent reverse-transfer of electrons and high saturation can be realized [Fig. 3 (b)].

B. Vertical Gate

Vertical gates (VG) are implemented in the TRG and TRY for increasing the saturations of both PD and MEM as shown in a cross-sectional schematic of Fig. 4. Electrons accumulated in deep potentials of the PD and MEM are successfully transferred with the aid of VG [6], resulting in an increase in FWC up to 15% when compared with a back-illuminated GS-CIS without VG.

C. Sequential Charge Transfer Method

Fig. 5 shows a one-dimensional potential profile of the conventional MEM and the MEM in this work. In a conventional single-gated MEM, a monotonic potential gradient along MEM region should be designed to fully transfer electrons to the FD via the TRG. However, its saturation is limited due to decreasing potential along the MEM region. Utilization of the potential profile and transfer method in this work can effectively resolve the trade-off between saturation and full transfer of electrons. Sequential transfer of electrons in the double-gated MEM structure is depicted in Fig.6. Part of the accumulated charge in the entire MEM region is first transferred to the FD by switching on the TRG and TRX. The remaining charge in the TRX region is then transferred by switching off the TRX, followed by switching off the TRG for complete charge transfer. With this method, FWC is increased up to 20% without sacrificing transfer characteristics.

V. PIXEL PERFORMANCE

The effect of introducing the PLS suppression technology is presented in captured images in Fig.7 wherein the influence of parasitic light is drastically reduced. Furthermore, improved linearity by applying the saturation technology is also shown in Fig.8 (Note that only reverse-transfer prevention is applied in the comparison for this work). Other CIS characteristics are summarized in Table 1.

VI. SUMMARY AND CONCLUSION

Characteristics of the back-illuminated GS-CIS in this work in comparison with previously reported architecture [1]-[4] are summarized in Table 2. Our back-illuminated GS-CIS with 2.74 μm-pitch-pixels with charge-domain memory achieves 1331 e-/μm^2 in saturation while simultaneously suppressing PLS which is advantageous over conventional types of GS-CIS.

ACKNOWLEDGMENT

The authors gratefully acknowledge the support given to this work by the analog circuit design engineers of Sony Semiconductor Solutions Corporation, and I. Yamamura of Sony Corporation.

REFERENCES

[1] M. Kobayashi *et al.*, "A 1.8e− rms temporal noise over 110-dB-dynamic range 3.4 μm pixel pitch global-shutter CMOS image sensor with dual-gain amplifiers SS-ADC, light guide structure, and multiple-accumulation shutter" *IEEE Solid-State Circuits*, vol. 53, no. 1, pp. 219-228, Jan. 2018.

[2] M. Sakakibara *et al.*, "A back-illuminated global-shutter CMOS image sensor with pixel-parallel 14b subthreshold ADC," *2018 IEEE International Solid - State Circuits Conference - (ISSCC)*, San Francisco, CA, 2018, pp. 80-82.

[3] L. Stark, J. M. Raynor, F. Lalanne and R. K. Henderson, "Back-illuminated voltage-domain global shutter CMOS image sensor with 3.75μm pixels and dual in-pixel storage nodes," *2016 IEEE Symposium on VLSI Technology*, Honolulu, HI, 2016, pp. 1-2.

[4] Y. Oike *et al.*, "An 8.3M-pixel 480fps global-shutter CMOS image sensor with gain-adaptive column ADCs and 2-on-1 stacked device structure," *2016 IEEE Symposium on VLSI Circuits (VLSI-Circuits)*, Honolulu, HI, 2016, pp. 1-2.

[5] T. Yokoyama *et al.*, "Development of low noise memory node in a 2.8 um global shutter pixel with dual transfer," *Proc. of 2017 International Image Sensor Workshop*, Hiroshima, Japan, 2017, pp. 398-401

[6] T. Shinohara *et al.*, "Three-dimensional structures for high saturation signals and crosstalk suppression in 1.20 μm pixel back-illuminated CMOS image sensor," *2013 IEEE International Electron Devices Meeting*, Washington, DC, 2013, pp. 27.4.1-27.4.4.

978-1-7281-1988-5/18 $31.00 © 2018 IEEE

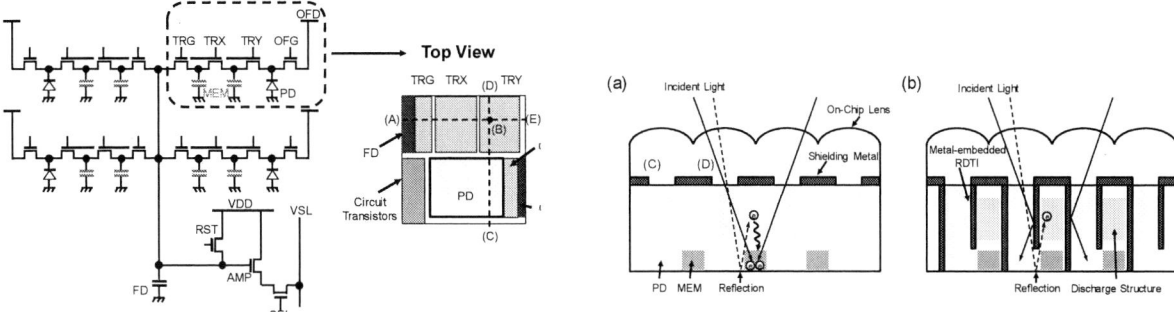

Fig.1 Schematic of 2×2 pixel with in-pixel memory

Fig.2 Cross-sectional schematics of back-illuminated GS-CIS with (a) conventional (b) light-shielding structure in this work

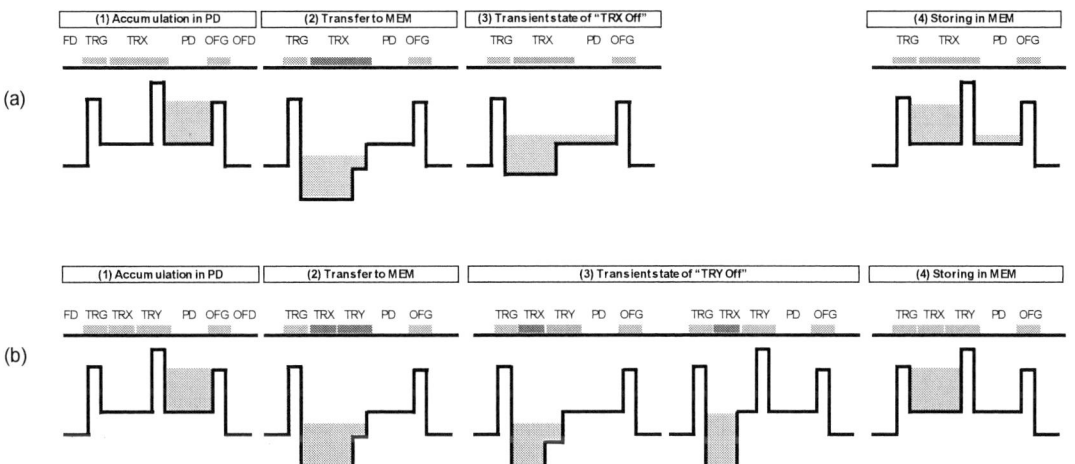

Fig.3 Potential diagram of charge storing process in MEM for (a) conventional back-illuminated GS-CIS (b) this work

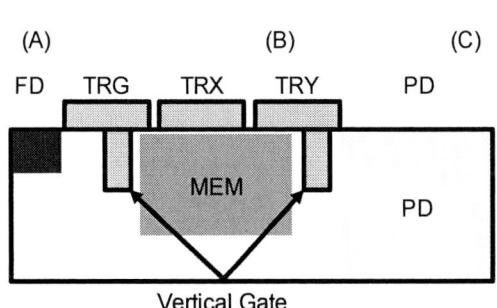

Fig.4 Cross-sectional schematics showing VG along (A) to (C) in Fig.1

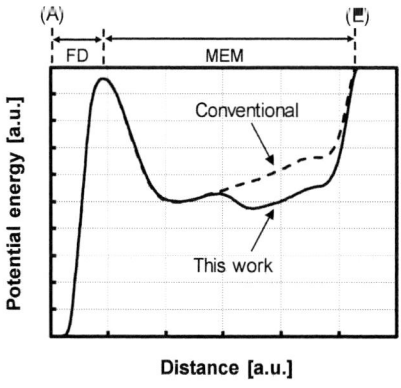

Fig.5 One-dimensional potential profile of (a) conventional back-illuminated GS-CIS (b) back-illuminated GS-CIS of this work

978-1-7281-1988-5/18 $31.00 © 2018 IEEE   239

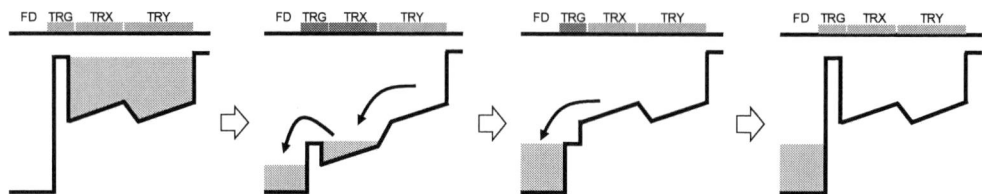

Fig.6 Potential diagram describing sequential charge transfer method

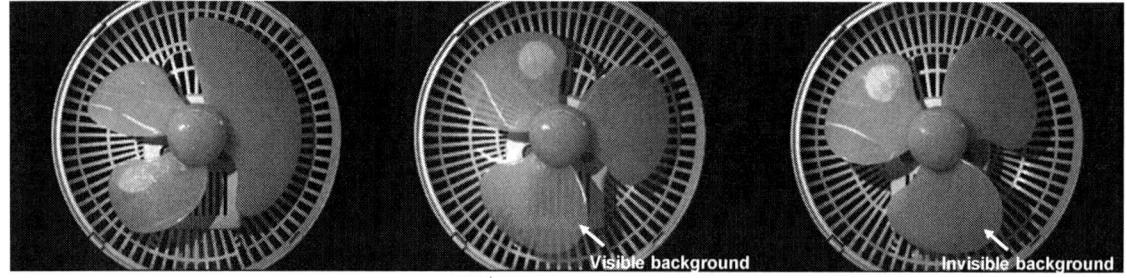

(a) Rolling-shutter   **(b) GS without light-shielding (PLS = -55dB)**   **(c) GS with light-shielding (PLS = -80dB)**

Fig.7 Captured images

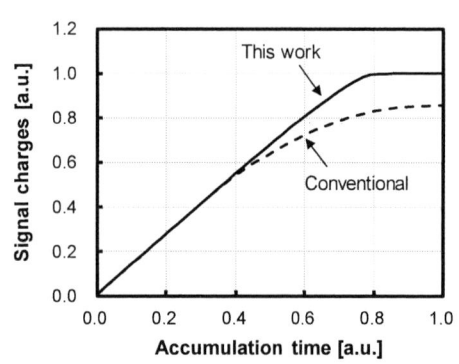

Fig.8 Linearity of signals

Parameter	Value
Process Tecnology	BI 90 nm(BE)/65 nm(FE) 1P4M
Pixel Array	3208 × 2184
Supply Voltage	3.3 V / 1.8V / 1.2V
Pixel Pitch	2.74 µm × 2.74 µm
Color Filter	Monochrome
Dark current @ 60℃,1s,1pix	13.8 e- (PD+MEM)
Temporal Noise	2.00 erms
Full Well Capacity	10000 e-
Sensitivity	23900 e-/lx s (3200K light with IR cut filter)
Parasitic Light Sensitivity	-80 dB
Dynamic Range	74 dB
Quantum Efficiency	85.5 %
Frame Rate	190 fps @ 12bit 260 fps @ 10bit

Table.1 CIS performance

	This work	[2] ISSCC 2018 M. Sakakibara et al	[3]VLSI 2016 L. Stark et al	[1]ISSCC 2018 M. Kobayashi et al	[4]VLSI 2016 Y. Oike et al
Structure	BI	BI	BI	FI	FI
	Charge domain	In-pixel ADC	Voltage domain	Charge domain	Charge domain
Cell Size [µm]	2.74	6.9	3.75	3.4	5.86
Num. of Eff. Pixels	3208 × 2184	1632 × 894	1024 × 800	2592 × 2054	3840 × 2160
Frame Rate [fps] (AD [bit])	190 (12)	660 (14)	50 (16)	120 (12)	480 (14)
Full Well Capacity [e-] (per pixel area [e-/µm²])	10000 (1331)	16600 (348)	8100 (576)	8100 (700)	30450 (890)
PLS [dB]	-80	-75	-80	-89	-99.6
Temporal Noise [erms]	1.85	5.15	8.5	1.8	4.6
Dynamic Range [dB]	74.6	70.2	59	73.1	76.3

Table.2 Comparison of CIS performance in a single read-out system

Conformal, Wafer-Scale and Controlled Nanoscale Doping of Semiconductors Via the iCVD Process

Jae Hwan Kim[1], Hong Keun Park[2], Kwan Yong Pak[2], Alexander Yoon[3], Yun Sang Kim[3], Sung Gap Im[2], Wan Sik Hwang[4] and Byung Jin Cho[1]

[1]Dept. of Electrical Engineering, KAIST, Daejeon 34141, Republic of Korea
Phone: +82-42-350-3485 Email: bjcho@kaist.edu
[2] Dept. of Chemical and Biomolecular Engineering, KAIST, Daejeon 34141, Republic of Korea
[3]Lam Research Corporation, Fremont, California 94538, USA
[4] Dept. of Materials Engineering, Korea Aerospace University, Gyeonggi-do 10540, Republic of Korea

Abstract— For the first time, a novel doping technique using an initiated CVD (iCVD) process was developed, facilitating the conformal, wafer-scale and controlled nanoscale doping of semiconductors at a high concentration. iCVD poly(boron allyloxide) (pBAO) and poly(triallyl phosphate) (pTAP) were used as a p-type and n-type dopant diffusion source, respectively. In detail, an optimized integration process was developed involving copolymer p(BAO-co-V3D3) passivation for pBAO and double-step deposition for pTAP. It was found that a dopant-containing polymer layer with a sub-10-nm thickness provided a high doping concentration at a shallow junction depth (10 nm) for both the p-type (10^{20} cm^{-3}) and the n-type (10^{21} cm^{-3}). Furthermore, the conformality and dopant distribution of the iCVD polymer layer were investigated using a high-aspect-ratio Si fin (5:1). The SOI nFET with iCVD doping at the source/drain regions exhibited better subthreshold swing and on-current values than a SOI nFET with conventional ion-implantation doping. Compared to other diffusion doping methods, the iCVD process could achieve lower sheet resistance.

I. INTRODUCTION

To continue the success of the semiconductor industry even beyond the 10-nm technology node, an ultra-thin body (UTB) in the form of a FinFET and/or a SOI would be indispensable to minimize the undesirable short-channel effect. However, this UTB approach faces more challenges with regard to process integration. Particularly, a conformal and high concentration doping technique within fin arrays has become a critical issue on next generation devices. Conventional ion-implantation causes non-conformal doping and structural damage to the UTB, resulting in the degradation of the device performance. Therefore, diffusion doping methods using solid sources such as phosphorus-doped silicate glass (PSG) and a self-assembled monolayer (SAM) have emerged and been discussed [1-5]. The surface-specific nature of SAM makes it difficult to realize both wafer-scale and mass production of the CMOS process. Moreover, solution-based approach suffer from insufficient doping concentrations and contamination. These alternatives require additional improvements to form an ultra-shallow junction with high dopant concentrations and conformal doping profile. In this work, a novel doping technique is proposed which uses the iCVD process to realize the conformal, wafer-scale and controlled nanoscale doping of semiconductors with high dopant concentration.

II. EXPERIMENTS

Dopant-containing polymer layers, poly(boron allyloxide) (pBAO) and poly(triallyl phosphate) (pTAP), with a sub-10-nm thickness were developed and deposited via the iCVD process as shown in Fig. 1(a). An optimized integration process was also developed with in-situ copolymer p(BAO-co-V3D3) passivation for pBAO and double-step deposition for pTAP, where pBAO and pTAP were used as a p-type and n-type dopant diffusion source, respectively (Fig. 1(b)). For the iCVD process, vaporized monomers and initiators were provided in the reactor and radicals were generated by the thermal dissociation of the initiator upon the heating filaments. Finally, free-radical polymerization was formed on the Si surface. After the deposition of the polymer film, a post-bake process was conducted to anneal the film at 230°C. Subsequently, rapid thermal annealing (RTA) was performed to drive in and activate the dopants. Fig. 2 shows the FTIR spectra of the pBAO/BAO and pTAP/TAP, indicating that the precursors of BAO and TAP were successfully polymerized into pBAO and pTAP, respectively, via the iCVD process. In addition, the p-type polymer film showed hygroscopic characteristics of the B-O-C bonds. It was noted that the bonds are the weakest point in the film and are easily decomposed by adsorbed molecules when exposed to ambient air.

III. RESULTS AND DISCUSSION

First, the feasibility of high-concentration doping via the iCVD process was demonstrated using a 40-nm-thick polymer film. It was found that surface concentrations of 1×10^{20} cm^{-3} for the p-type and 1×10^{21} cm^{-3} for the n-type were obtained (Fig. 3). In addition, the post-bake effect was investigated, as shown in Fig. 3. The results showed that the post-bake process was critical to increase the doping concentration for B and P atoms in the Si substrate. To explore the post-bake effect further, the EELS and back-side SIMS profiles were investigated (Figs. 4 and 5). Fig. 4 showed that an 8.3-nm-thick layer remained from the 40-nm-thick polymer layer after the drive-in annealing process. It was interesting to note that thin Si oxide layers grew on the Si substrate, which most likely affected the enhancement of dopant concentration. Fig. 5 shows the B and P SIMS concentration profiles. It was found that the dopant concentrations of B and P were highest at the Si oxide layer which grew on the Si substrate regardless

978-1-7281-1988-5/18 $31.00 © 2018 IEEE

of the post-bake condition. The post-bake effect was significant for the P concentration compared to that for the B concentration, as shown in Fig. 5.

Meanwhile, the technology node was scaled down the sub-10-nm regime, and the thickness of the diffusion source layer was restricted to less than 10 nm. Fig. 6 shows the B concentration with different thicknesses of the diffusion source. It showed that the B concentration was sharply reduced as the film became thinner. For this reason, an additional process scheme is required to achieve a high doping concentration using a diffusion source layer for which the thickness is less than 10 nm. Fig. 7 shows the B peak intensity changes with the exposure time in ambient air with and without an in-situ passivation layer. As indicated in Fig. 2(a), the p-type polymer film (pBAO) was easily decomposed by the adsorbed molecules when exposed to the ambient air, and it caused the degradation of the film. It revealed that the XPS peak intensity of B in the 10-nm-thick p-type polymer film decreased rapidly at an exposure time of 120 sec (Fig. 7(a)). This rapid decrease of the B dopant required an in-situ passivation layer to minimize the loss of B. In contrast to the sample without a passivation layer, a significant loss of B was not observed in the sample with an in-situ passivation layer, as shown in Fig. 7(b). Fig. 7(c) shows the high B SIMS concentration profiles which were observed with 5-nm pBAO followed by the use of in-situ 3-nm of copolymer. As in Fig. 6, the P concentration decreased as the film became thinner due to the degradation of the film uniformity (Fig. 8(a)), indicating that a high doping concentration cannot be obtained with a sub-10-nm diffusion source thickness. Therefore, an additional process scheme is required and is proposed here, as shown in Fig. 8(b). The double-step deposition approach showed a higher doping concentration than the single-step deposition process. Based on preliminary results, the RTA condition was optimized for the p-type and n-type dopants. The corresponding SIMS profiles are shown in Fig. 9.

Fig. 10 shows the conformal deposition of the 8.3-nm-thick n-type polymer on high-aspect-ratio fins (5:1) with a 40-nm pitch, where the double-step deposition method was adopted. After an oxide capping process, drive-in annealing and an etch-back process of the unreacted layer, the dopant distribution within the Si fin was characterized via dopant EDS mapping and spreading resistance profiling (SRP). Fig. 11 displays the B signal map within the fin structure and the corresponding SRP profile in the lateral/vertical direction. The results showed that B atoms at a concentration of $1 \times 10^{20} cm^{-3}$ were distributed uniformly throughout the fin structure. As on similar occasions of B doping, a uniform distribution of P atoms at a concentration up to $1 \times 10^{21} cm^{-3}$ was also observed, as shown in Fig. 12. These results suggest that doping technology using iCVD could provide high doping level with high conformality from a sub-10 nm diffusion source layer on a 3D-structured wafer.

Finally, this novel doping technology was implemented in SOI nFETs of 33 nm and 8 nm UTB. Fig. 13 shows the process flow to fabricate the SOI nFET with iCVD n-doping at the source/drain regions. For a comparison, identical SOI nFETs with conventional ion-implantation at the source/drain regions were also fabricated. Fig. 14 shows the transfer characteristics of the SOI nFET with the iCVD doping method (a) and the conventional ion-implantation method (b). It indicates that device performance such as the on-state current with iCVD doping were indistinguishable regardless of whether the 33 nm or 8 nm UTB was used. Unlike the iCVD doping method, the on-state current of the 8-nm SOI nFET with conventional ion-implantation was severely degraded compared to that of the 33-nm SOI nFET. The device degradation found in the 8-nm SOI nFET was attributed to the structural damage caused by ion bombardment during ion-implantation process. This suggests that the 8-nm SOI was not fully recrystallized, whereas the 33-nm SOI was fully recrystallized after the RTA process (Fig. 14 (b) insets). Fig. 14 reveals that the iCVD doping process is beneficial to UTB devices as it causes no structural damage.

Fig. 15 summarizes the benchmarks of state-of-the-art doping techniques for both the n-type and p-type depending on sheet resistance and doping concentration. The proposed iCVD doping method showed the lowest sheet resistance in both the n-type and the p-type. This lowest sheet resistance via the iCVD method was attributed to the highest n-type doping concentration, which is an intrinsic advantage of the iCVD method compared to the PSG and SAM approaches. Like n-type doping, the iCVD method showed the lowest sheet resistance compared to the SAM method for the p-type. While the p-type doping concentration of the SAM was comparable to of the iCVD method, the latter showed lower sheet resistance than the former.

IV. CONCLUSION

The authors have developed a novel doping technique using the iCVD process, which allows the conformal, wafer-scale and controlled nanoscale doping of semiconductors with high concentrations. To achieve a high doping concentration from a sub-10-nm layer, an optimized integration process was developed in which copolymer p(BAO-co-V3D3) passivation for pBAO and double-step deposition for pTAP were used. This doping technique was successfully implemented with a SOI nFET. The proposed iCVD doping method showed the low enough sheet resistance for both the n-type and the p-type.

ACKNOWLEDGMENT

This work was supported by a research grant from the Lam Research Corporation.

REFERENCES

[1] Yao-Jen Lee, et al., *IEDM*, 2014, pp. 32.7.1-32.7.4
[2] Yao-Jen Lee, et al., *IEDM*, 2015, pp. 6.2.1-6.2.4
[3] Liang Ye et al., *ACS Appl. Mater. Interfaces*, 2015, 7, pp. 3231-3236
[4] Thi. Alphazan, et al., *Chem. Mater.*, 2016, 28(11), pp. 3634-3640
[5] Y. Sasaki, et al., *IEDM*, 2015, pp. 21.8.1-21.8.4
[6] Hanul Moon, et al., *Nature materials*, 2015, 14(6), pp. 628-635

978-1-7281-1988-5/18 $31.00 © 2018 IEEE

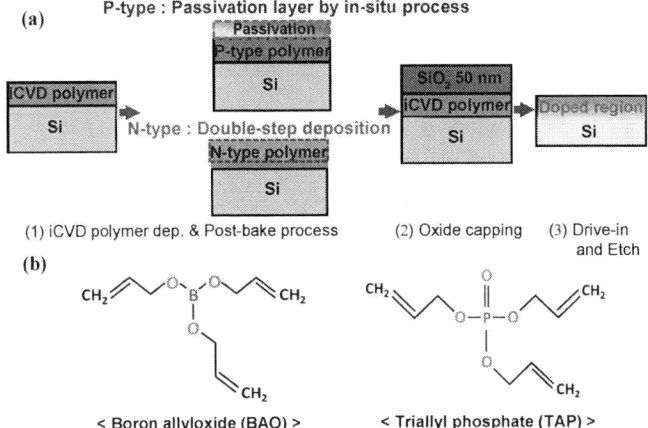

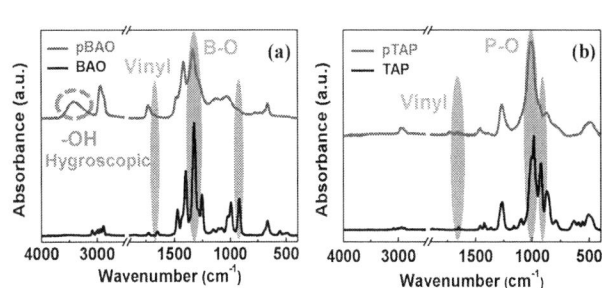

Fig. 1. (a) Illustration of diffusion doping process using iCVD polymer, and (b) Chemical structures of BAO monomer for p-type and TAP monomer for n-type doping.

Fig. 2. FT-IR spectra of (a) pBAO and (b) pTAP films. Disappearance of the vinyl groups (-CH$_2$) indicates the complete polymerization of BAO and TAP.

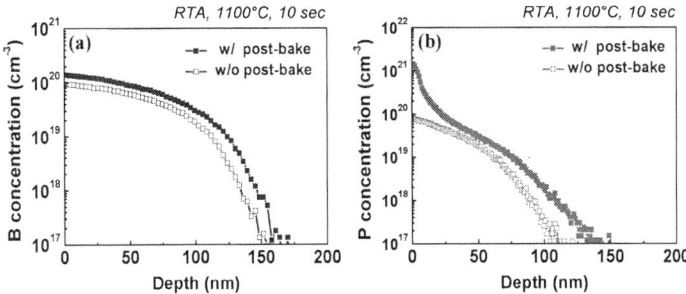

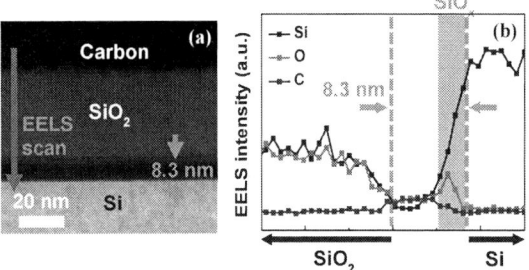

Fig. 3. SIMS profiles of (a) B and (b) P in Si substrates. The doping levels of B and P, 1×10^{20} cm^{-3} and 1×10^{21} cm^{-3} respectively, are obtained from 40-nm-thick polymer film with a post-bake process.

Fig. 4. Cross-sectional scanning TEM image of (a) the remaining layers of the stack after RTA and (b) EELS spectra. SiO$_x$ is observed on the Si surface. An 8.3-nm-thick layer remains after drive-in annealing using 40 nm polymer film.

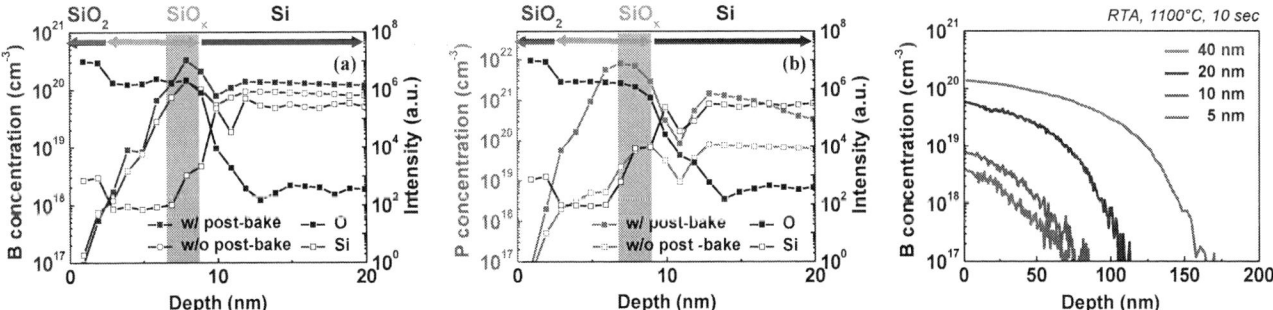

Fig. 5. Back-side SIMS profiles of (a) B and (b) P in layers to determine the dopant distribution from the Si substrate into the SiO$_2$ capping layer. A higher dopant concentration is observed when post-annealed polymer is used.

Fig. 6. SIMS profiles of B with various p-type polymer film thicknesses.

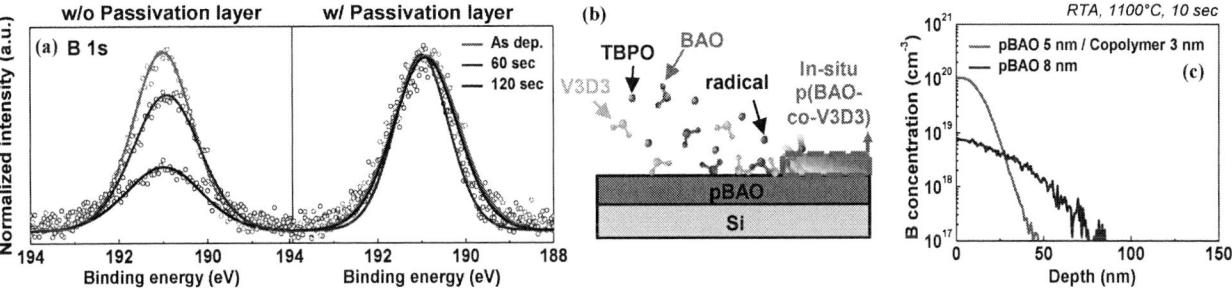

Fig. 7. (a) B peak signal intensity of the 10-nm-thick p-type polymer with the exposure time in an air ambient environment. After the p-type polymer film is passivated, the ratio of B was maintained. (b) Schematic illustration of the passivation layer by copolymerization of BAO with 1,3,5,-trivinyl-1,3,5-trimethylcyclotrisiloxane (V3D3) [6] via the in-situ iCVD process. (c) SIMS profiles of B showing that the passivation layer enhances the doping level.

978-1-7281-1988-5/18 $31.00 © 2018 IEEE   243

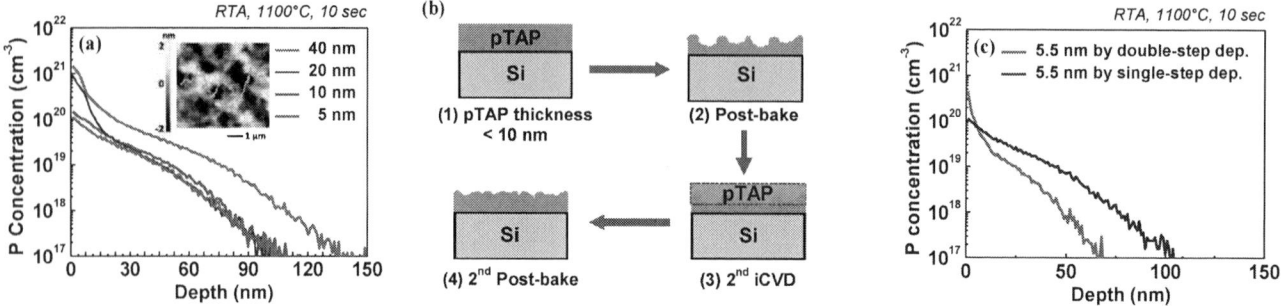

Fig. 8. (a) SIMS profiles of P with various film thicknesses. The inset is AFM image of n-type polymer film surface after the post-bake process. (b) Schematic illustrations of the double-step deposition process. The iCVD deposition and post-bake process are repeated in sequence. Compared to the initial deposition, the final polymer layer is densified while maintaining a high P concentration and thickness of less than 10 nm. (c) A higher P concentration is shown when using polymer film with a final thickness of 5.5 nm and double-step deposition (1st dep.: 5 nm + 2nd dep.: 3 nm) compared to the same thickness with single-step deposition.

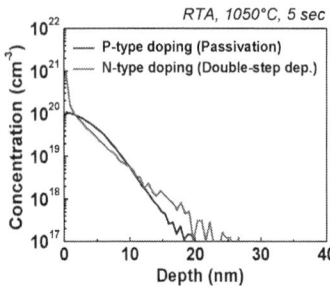

Fig. 9. SIMS profiles of both B and P subjected to the optimized RTA conditions (1050°C, 5 sec). A shallow junction depth of 10 nm is achieved while maintaining a high doping level.

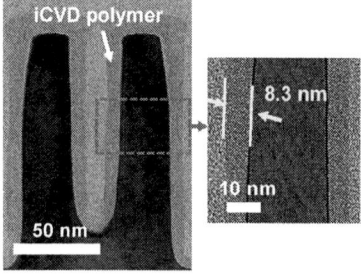

Fig. 10. Cross-sectional bright-field TEM images of a conformally deposited 8.3-nm-thick n-type polymer layer created by the double-step deposition method on a Si fin (Pitch: 40 nm, Width: 22 nm, Height: 100 nm)

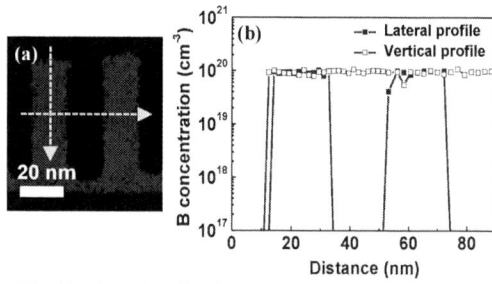

Fig. 11. (a) EDS B signals. (b) SRP profile of B. Uniform B distribution with high activation level is observed inside the overall fin.

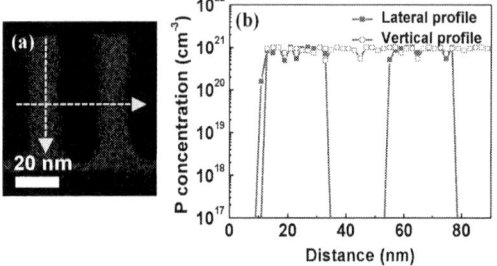

Fig. 12. (a) EDS P signals. (b) SRP profile of P. A uniform P distribution with high activation level is observed inside of the fin overall. Conformal doping profile on the fin with a 40-nm pitch is confirmed for both p-type and n-type doping.

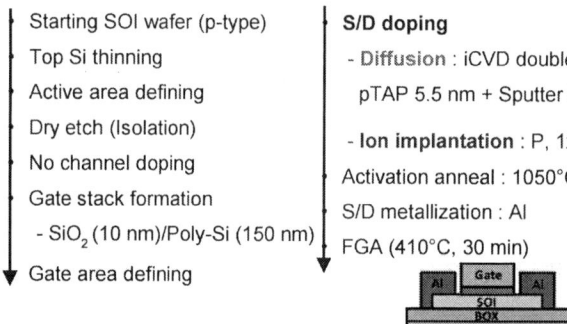

Starting SOI wafer (p-type)

Top Si thinning

Active area defining

Dry etch (Isolation)

No channel doping

Gate stack formation

  - SiO$_2$ (10 nm)/Poly-Si (150 nm)

Gate area defining

S/D doping

- Diffusion : iCVD double step.

  pTAP 5.5 nm + Sputter SiO$_2$ 50 nm

- Ion implantation : P, 1x10^{16} cm^{-2}

Activation anneal : 1050°C, 5 sec

S/D metallization : Al

FGA (410°C, 30 min)

Fig. 13. Process flow for the fabrication of the SOI NFET with diffusion or implant doping.

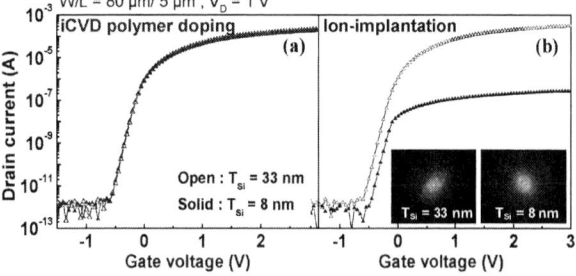

Fig. 14. I$_D$-V$_G$ of the n-channel SOI nFET (T$_{Si}$ = 33 nm & 8 nm) by diffusion and implant doping. The insets in (b) are images of the TEM diffraction pattern at the implanted S/D regions in the 33 nm (left) and 8 nm (right) SOI nFETs. The 8-nm UTB SOI nFETs doped by iCVD doping exhibits a higher on-current level due to better crystallinity in S/D regions.

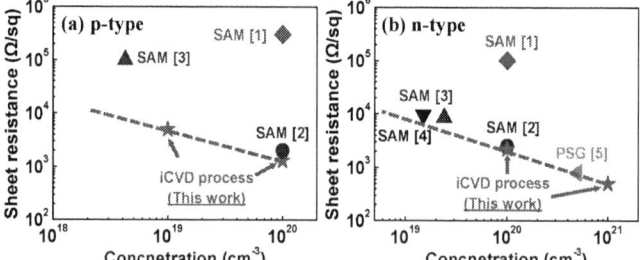

Fig. 15. Comparison of the sheet resistance levels and doping concentrations of solid sources in previous studies and in this work. The iCVD process provides the lowest sheet resistance with the highest B and P concentrations among the diffusion processes from doped layers.

978-1-7281-1988-5/18 $31.00 © 2018 IEEE   244

Low Temperature Sputtered Graphenic Carbon Enables Highly Reliable Contacts to Silicon

M. Stelzer[1], M. Jung[1], U. Wurstbauer[2], A.W. Holleitner[2], and F. Kreupl[1]

[1]Department of Electrical and Computer Engineering, Technical University of Munich, Arcisstr. 21, 80333 Munich, Germany, email: max.stelzer@tum.de
[2]Walter Schottky Institute and Physics Department, Technical University of Munich, Am Coulombwall 4a, 85748 Garching, Germany

Abstract—Titanium silicide (TiSi) contacts are commonly used metal-silicon contacts [1]-[3] but are known to diffuse into the active region under high current stress. Recently we demonstrated [4],[5] that graphenic carbon (GC) deposited by CVD has the same low Schottky barrier on silicon as TiSi, but a much improved reliability against high current stress. The drawback of the CVD-GC is the required deposition temperature of ~ 900 °C. In this paper we demonstrate now that **the deposition of graphenic carbon is possible at 100 - 400 °C by a modified sputter process.** We show that the **sputtered carbon-silicon (SC-Si) contact is over 10^9 times more stable** against high current stress pulses than the conventionally used TiSi-Si junction, while it has the same or even a lower Schottky barrier. **Doping SC by nitrogen (CN) leads to an even lower resistivity** and improved stability. The finding that there is **a low temperature approach** for using the superb carbon properties has important consequences for the reliability of contacts to silicon and opens up the use of GC in a plethora of other applications.

I. INTRODUCTION

Common S/D contacts in state-of-the-art FinFETs rely on the formation of titanium silicide (TiSi) [1]-[3]. TiSi provides a very low barrier height to silicon that's why it is also highly appropriate for Schottky diodes in zero-barrier mixer and detector applications [6]. The low Schottky barrier height (SBH) makes TiSi very promising for further contact resistance reductions [2],[7] but the increased current density in small devices leads to an increased Joule heating. Ti-based contacts have a low thermal stability [7],[8], and high temperatures, especially during electrostatic discharges (ESD), can force a diffusion and migration of Ti or TiSi in silicon [4],[9]. This can degrade the electrical properties or even cause a failure of the device due to a junction burnout.

We reported earlier [4],[5] that a CVD-deposited graphenic carbon (GC) [10],[11] contact to silicon (CVD-C-Si) has similar electrical properties to TiSi-Si but a much higher temperature stability and a more than 100 million times improved stability against ESD current pulses. **Unfortunately the deposition temperatures of > 850 °C are not compatible** with many applications, where temperature budget restrictions for middle and back end of line (~ 400 °C) of advanced nodes are in place.

In this work we demonstrate the electric characteristics and reliability of low temperature sputtered carbon (SC) and nitrogen-doped carbon (CN) contacts to silicon and compare them to TiSi-Si contacts. The characterization and reliability tests were conducted on a Schottky diode structure based on the Infineon BAT15 diode vehicle (**Fig. 1(a)**). All doping and dimensions are the same as the commercial available TiSi-Si-BAT15 diode to allow a direct comparison of SC, CN and TiSi contacts (like in [4],[5]). The BAT15 has a thin n⁻-epitaxial layer (150 nm), which is very sensitive in its electrical reverse bias characteristics when contaminations like metal ions enter this epitaxial region. Pulses with high current density (**Fig. 1(b)**) can force the diffusion of Ti/TiSi into the epitaxial layer, which then can lead to a short-circuit in the diode (**Fig. 1(c)**).

II. TiSi-Si AND SPUTTERED CARBON RELIABILITY

The commercially available TiSi-Si-BAT15 diode was pulsed with different current densities and a degradation can be detected as an increased leakage current at a reverse voltage $V_r = 1$ V. Even a single 3.5 MA/cm² event can degrade the original behavior of the diode, indicated by the over 100x increase of the reverse current (**Fig. 2(a)**). This signals the inter-diffusion of Ti/TiSi within the epitaxial layer. The device is short-circuited after 3 pulses as a metal or silicide filament might be created. **Fig. 2(b)** clearly highlights that **2-4 pulses are sufficient to destroy the TiSi-Si junction at 3.5 MA/cm²** [4],[5]. In addition, it illustrates that the maximum number of transient stress pulses depends on the current density level used. For 1.35 MA/cm², which is slightly above the permissible pulse load of 1.21 MA/cm² specified by the manufacturer, the diode is able to withstand up to 1 M events, but the failure probability is spread over a wide range [4],[5].

III. SPUTTERED CARBON – SILICON CONTACTS

The electrical properties of the carbon-silicon contact were evaluated by sputtering a uniform film of SC or CN on the BAT15 structure. A stack of Ti (50 nm), Cu (1.2 µm) and Au (40 nm) was deposited through a shadow mask on top of the carbon, which also acts as hard mask for a H_2 plasma etching of carbon (**Fig. 1(a)**).

The dc characteristics in **Fig. 3** illustrate that the behavior of the SC-Si diode sputtered at 20 °C presents non-ideal diode characteristics. However, a substrate temperature of 100 °C is sufficient to change the dc characteristics to that of a diode having a low ideality factor. It is even close to the electrical characteristics of a TiSi-Si and a CVD-C-Si diode as it also shows a low SBH (**Fig. 4**). The CN-Si deposited on a 400 °C substrate exhibits a much lower SBH, which is necessary for the creation of low ohmic contacts but has a higher ideality factor. The breakdown voltage and the reverse leakage depend on the SBH of the contact and the used temperature budget (**Fig. 5**). The CN-Si diode even performs better than a CVD-C-Si diode as it has a lower SBH but the breakdown

978-1-7281-1988-5/18 $31.00 © 2018 IEEE

voltage is the same. It has lower leakage current at $V_r > 1$ V because the CN was deposited only at 400 °C in contrast to the 1000 °C used in the CVD process. No dopant diffusion occurs at low temperatures, which would otherwise change the reverse breakdown voltage.

The diodes were stressed with current pulses having a current density of 3.5 MA/cm^2 and duration of 100 ns to allow for a direct comparison with the results obtained from TiSi-Si diodes. The **CN-Si-diode outperforms the stability of the TiSi-Si contact by more than a factor of one billion** and that of the CVD-C-Si diode by a factor of 3 (**Fig. 6**). These values are lower bounds as the diodes didn't fail, but the tests were stopped due to accrued testing time of several months. After billions of stress pulses, the diodes only showed an increased SBH, which might be caused by the formation of silicon carbide at the interface. As a consequence, the reverse leakage is decreased (**Fig. 7**), but even after 10^9 pulses, the breakdown behavior did not deteriorate, which reveals that there was no diffusion of the interface metal or of dopants from the highly doped substrate into the thin epi-layer. This excellent diffusion barrier property of carbon may also be one reason to use carbon as an interconnect material to the phase change material in the 3D XPoint Memory [12].

For longer pulse durations, CN-Si is not as reliable as CVD-C-Si (**Fig. 8**). The CN-Si diodes have a higher resistance and, as the amount of current has been kept constant, more Joule heating is generated in longer pulses. The evaluation of the top metallization after pulsing (**Fig. 9**) indicates that longer pulse durations lead to a higher temperature in the device and greater damage on the surface. On the other hand, the high number of stress events for a pulse width of 100 ns already led to an oxidization of the metallic surface.

IV. SPUTTERED CARBON PROCESS AND PROPERTIES

The sputtered carbon was deposited from a 4'' graphite target (purity 99.999 %) with an RF magnetron plasma source on a non-biased but heated substrate, which was 70 mm away from the C-target. The plasma was operated in a **very low-pressure (2-16 µbar) regime** where it interacts with many parts in the chamber, not only with the target. In order to achieve the superb reliability results, it is very important that all parts near the sputtering source and especially the sample holder and shutter are well coated with carbon. The thin epi-layer of the diode is so sensitive to any metallic trace contaminants that already a few milliseconds long exposure of a badly coated shutter to the plasma during the shutter open/close operation is enough to induce early failures (**Fig. 10**).

We compared SC films made with pure argon (Ar) gas to nitrogen doped CN films, sputtered in a gas flow mixture of $Ar/N_2 = 56$. At lower Ar/N_2 ratios, the film has an increased resistivity. The films exhibit a very good adhesion to Si, SiO_2 and SiN as if covalent bonds are formed. The Raman spectra in **Fig. 11** highlight that SC sputtered on a substrate with a temperature of 20 °C is more of amorphous structure, while

that deposited at 400 °C has strong D and G peaks indicating graphenic sp^2-bonds. An annealing at 550 °C for 1 h is necessary to similarly graphitize the SC deposited at 20 °C. The CN films display D and G peaks at all temperatures showing a higher content of sp^2-bonds [13]. The C≡N peak in the Raman spectrum indicates that nitrogen is incorporated into the film. **Fig. 12** underlines that the SC sputtered at 20 °C shows a very low surface roughness while at 400 °C the surface is rougher. The CN features a much smaller grain size associated with a smoother surface.

In **Fig. 13**, we evaluated the film resistivity ρ as a function of the used RF power P, working pressure p and substrate temperature T. From **Fig. 13(a)**, one can conclude that a further increase of power could lead to an even lower resistivity. The CN film exhibits a 2-times lower resistivity at the same sputter conditions as nitrogen acts as an electron-donor and therefore approaches the value of CVD-C (1.1 mΩ·cm), which was deposited at 1000 °C. When exposed to atmosphere, the resistivity of SC drifts to higher values whereas CN was stable at all times. As conclusion from **Fig. 13**, we argue that high temperature, high power density and low pressures lead the way to an even improved carbon film. However, the equipment needs to be able to support these conditions. Furthermore, the sputtered carbon films show a negative temperature coefficient (NTC) like the CVD-C (**Fig. 14**). The higher the resistivity, the higher is the NTC.

ACKNOWLEDGMENT

The authors gratefully thank Infineon Technologies AG for providing the used silicon vehicle. Supported by DFG through the TUM International Graduate School of Science and Engineering (IGSSE).

REFERENCES

[1] C. Auth *et al.*, "A 10nm high performance and low-power CMOS technology featuring 3rd generation FinFET transistors, Self-Aligned Quad Patterning, contact over active gate and cobalt local interconnects," *IEDM*, 2017.

[2] S.-A. Chew *et al.*, "Ultralow resistive wrap around contact to scaled FinFET devices by using ALD-Ti contact metal," *IITC*, 2017.

[3] C. Lavoie *et al.*, "Contacts in Advanced CMOS: History and Emerging Challenges," *ECS Transactions*, 2017.

[4] M. Stelzer and F. Kreupl, "Graphenic carbon-silicon contacts for reliability improvement of metal-silicon junctions," *IEDM*, 2016.

[5] M. Stelzer *et al.*, "Graphenic Carbon: A Novel Material to Improve the Reliability of Metal-Silicon Contacts," *IEEE J-EDS*, 2017.

[6] A. Mai and A. Fox, "Reliability aspects of TiSi-Schottky barrier diodes in a SiGe BiCMOS technology," *ESSDERC*, 2015.

[7] J. Zhang *et al.*, "Thermal Stability of TiN/Ti/p$^+$-Si$_{0.3}$Ge$_{0.7}$ Contact with Ultralow Contact Resistivity," *IEEE Electron Device Letters*, 2018.

[8] H. Yu *et al.*, "Thermal Stability Concern of Metal-Insulator-Semiconductor Contact: A Case Study of Ti/TiO$_2$/n-Si Contact," *IEEE TED*, 2016.

[9] K. Banerjee *et al.*, "Characterization of contact and via failure under short duration high pulsed current stress," *IRPS*, 1997.

[10] F. Kreupl, "Carbon-based materials as key-enabler for 'more than Moore'," *MRS Proceedings*, 2011.

[11] S. Huebner *et al.*, "High performance X-ray transmission windows based on graphenic carbon," *IEEE TNS*, 2015.

[12] J. Choe, "Comparing XPoint memory architecture with NAND and DRAM products," *TechInsights*, 2017. [Online; Accessed July 30, 2018]. https://electroiq.com/2017/10/comparing-xpoint-memory-architecture-with-nand-and-dram-products/.

[13] A.C. Ferrari *et al.*, "Interpretation of infrared and Raman spectra of amorphous carbon nitrides," *Physical Review*, 2003.

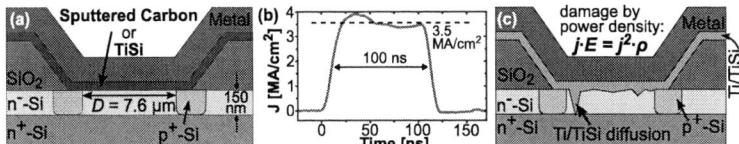

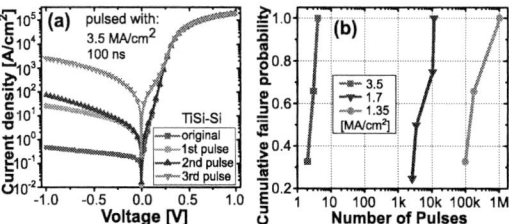

Fig. 1. (a) A schematic cross-section of the used Schottky diode with sputtered carbon or TiSi as interface material to silicon. The active region (indicated with D) is 45.36 μm^2. (b) illustrates a measured waveform of a 100 ns current pulse with a rise/fall time of < 15 ns and an average current density j of 3.5 MA/cm^2. (c) Schematic cross-section of a TiSi-Si Schottky diode where Ti or TiSi is driven into the n$^-$-Si epitaxial layer and it could even reach the highly doped substrate by the application of a high current pulse [5]. The power per volume (power density) in a conductor during a current density pulse j, which leads to Joule heating, is given by $j \cdot E = j^2 \cdot \rho$, where E is the local electric field and ρ the electrical resistivity.

Fig. 2. The J-V curves in (a) show the degradation of a TiSi-Si diode pulsed with j = 3.5 MA/cm^2 and a pulse width of 100 ns. Every single pulse led to a deterioration of the reverse behavior of the diode and after the third pulse the reverse current has increased severely [5]. (b) illustrates the failure probability versus number of current pulses at different current densities for commercial BAT15 diodes with TiSi-Si interface [4]. At least three devices were stressed for a given pulse current density, a pulse width of 100 ns and a duty cycle < 0.0001. The failure of the diode was defined when the diode degrades to a reverse current > 220 A/cm^2 @ V_r = 1V.

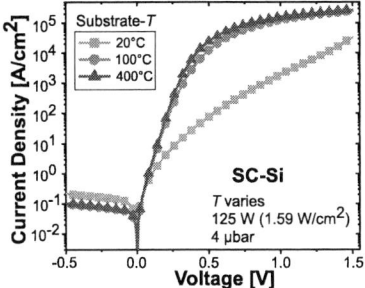

Fig. 3. J-V curves of SC-Si diodes sputtered at different substrate temperatures whereas other sputter parameters were kept constant. The diode sputtered at 20 °C has a lower SBH (0.45 eV) but the forward behavior is far from ideal. A substrate temperature of 100 °C is sufficient to change the conduction properties of the carbon leading to dc characteristics close to a device sputtered at 400 °C.

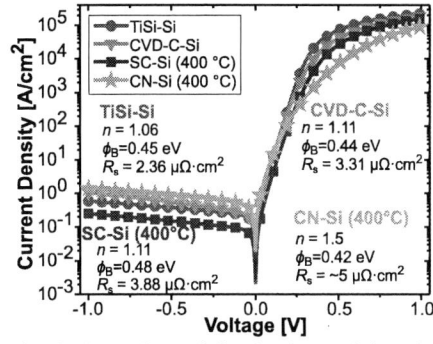

Fig. 4. Comparison of the dc characteristics of a TiSi-Si, a CVD-C-Si, a SC-Si and a CN-Si diode. The thickness of the CVD-C is 28 nm and of the SC and CN is 35 nm. Ideality factor n, SBH ϕ_B and area-normalized series resistance R_s are displayed. The SBHs are all close to TiSi-Si. The CN-Si has even a lower SBH but also an increased ideality factor.

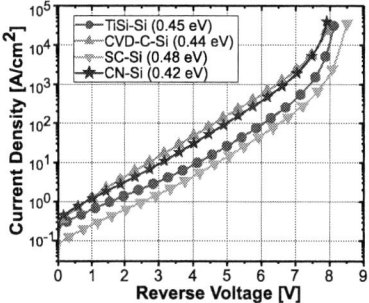

Fig. 5. Reverse current characteristics of TiSi-Si and different carbon-Si diodes. The curves are shown up to the voltage where breakdown occurs. The higher the SBH, the higher is the observable reverse breakdown voltage. The CN-Si has a lower SBH but the breakdown voltage is almost the same as for the CVD-C-Si. In addition it shows a lower leakage current at a voltage > 1 V. The CN was deposited at 400 °C only where no dopant diffusion occurs. In contrast to this, the 1000 °C of the CVD process reduces the epi-layer thickness a bit, which gives rise to an earlier breakdown.

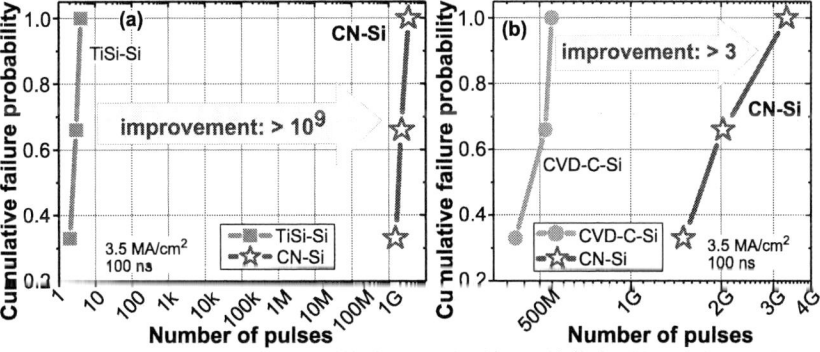

Fig. 6. Pulse endurance comparison of TiSi-Si (BAT15) with CN-Si diodes (a), and CVD-C-Si with CN-Si diodes (b) for a current density of 3.5 MA/cm^2 and a pulse width of 100 ns. The **CN-Si diodes** can withstand at least **over 1 billion pulses more than the TiSi-Si diodes** and 3 times more than the CVD-C-Si. The CN-Si values are only lower bounds as the pulsing of the CN-Si contact (non-filled stars) has been aborted not due to failure but due to the excessive testing time. The duty cycle was < 0.00015.

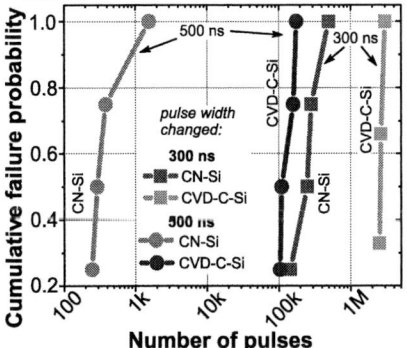

Fig. 8. Pulse length dependent stress pulse endurance of CN-Si and CVD-C-Si diodes for a current density of 3.5 MA/cm^2 and a pulse width of 300 ns and 500 ns. The duty cycle was < 0.0001. Compared to the behavior at 100 ns pulses, the CN-Si diodes fail earlier than CVD-C-Si diodes for longer pulse durations. The reason might be the higher resistance of the carbon diode due to carbon thickness and carbon type. As the current through the diode is kept constant, more resistance leads to more dissipated power, which leads to an earlier destruction.

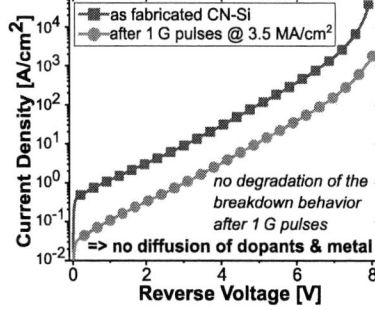

Fig. 7. Reverse dc characteristics of a CN-Si diode as fabricated and after 1 G pulses (3.5 MA/cm^2, 100 ns). The SBH was increased due to the interface reactions due to the high interface temperature induced by the pulse. The reverse blocking capability was improved as the SBH has increased. The onset of the reverse breakdown is not shifted to lower voltages. This would point to a diffusion of dopants as it would decrease the total thickness of the epi-layer, or it would point to the creation of a filamentary current by diffused metals.

978-1-7281-1988-5/18 $31.00 © 2018 IEEE   247

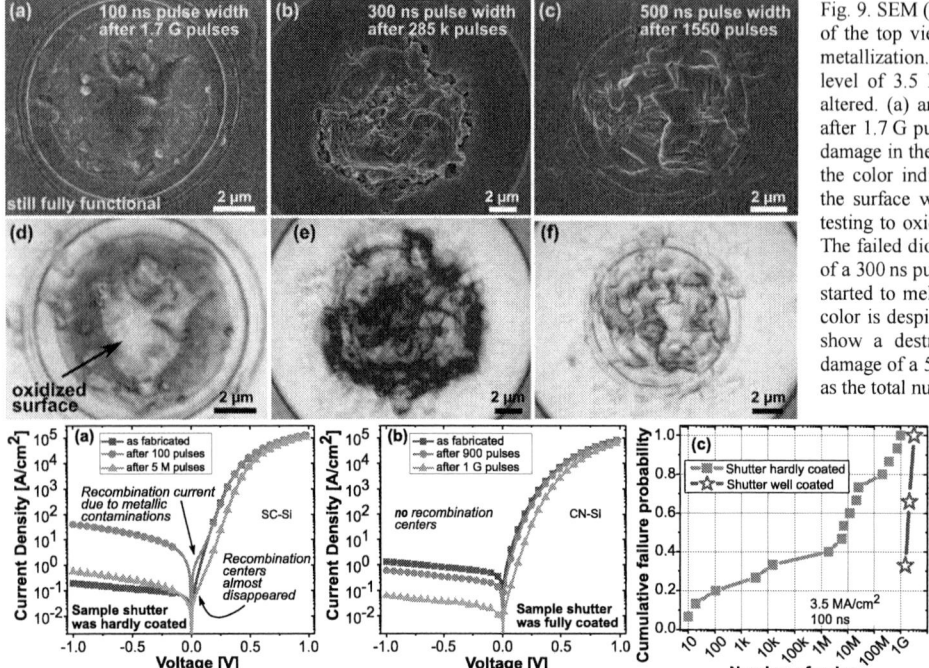

Fig. 9. SEM (a-c) and true color microscope images (d-f) of the top view of CN-Si diodes with Ti/Cu/Au as top metallization. All the devices where stressed at a current level of 3.5 MA/cm^2 while only the pulse length was altered. (a) and (d) show a still fully functional device after 1.7 G pulses of a width of 100 ns where almost no damage in the top metallization is visible. The change in the color indicates most likely that the temperature on the surface was high enough during the two weeks of testing to oxidize the Cu even through the Au capping. The failed diode in (b) and (e) demonstrates the damage of a 300 ns pulse after 285 k pulses where the metal even started to melt as it cracks and piles up. The change in color is despite the strong damage quite low. (c) and (f) show a destroyed sample where the current-induced damage of a 500 ns pulse just slightly melted the surface as the total number of stress events was lower.

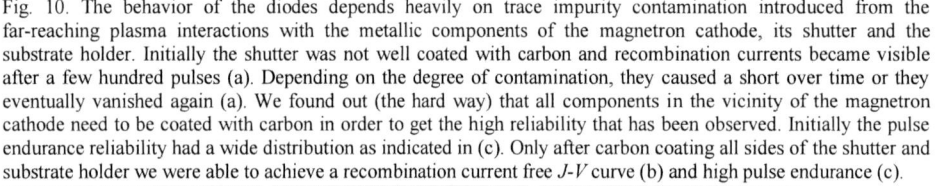

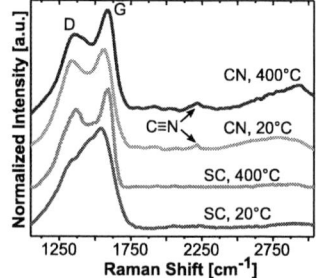

Fig. 10. The behavior of the diodes depends heavily on trace impurity contamination introduced from the far-reaching plasma interactions with the metallic components of the magnetron cathode, its shutter and the substrate holder. Initially the shutter was not well coated with carbon and recombination currents became visible after a few hundred pulses (a). Depending on the degree of contamination, they caused a short over time or they eventually vanished again (a). We found out (the hard way) that all components in the vicinity of the magnetron cathode need to be coated with carbon in order to get the high reliability that has been observed. Initially the pulse endurance reliability had a wide distribution as indicated in (c). Only after carbon coating all sides of the shutter and substrate holder we were able to achieve a recombination current free J-V curve (b) and high pulse endurance (c).

Fig. 11. Raman spectra (offset for clarity) of SC and CN films sputtered at a substrate temperature of 20 °C and 400 °C, respectively. The used excitation wavelength is 532 nm. The pronounced G-peak is always higher than the D peak, which indicates the prevalence of graphenic sp^2-bonds.

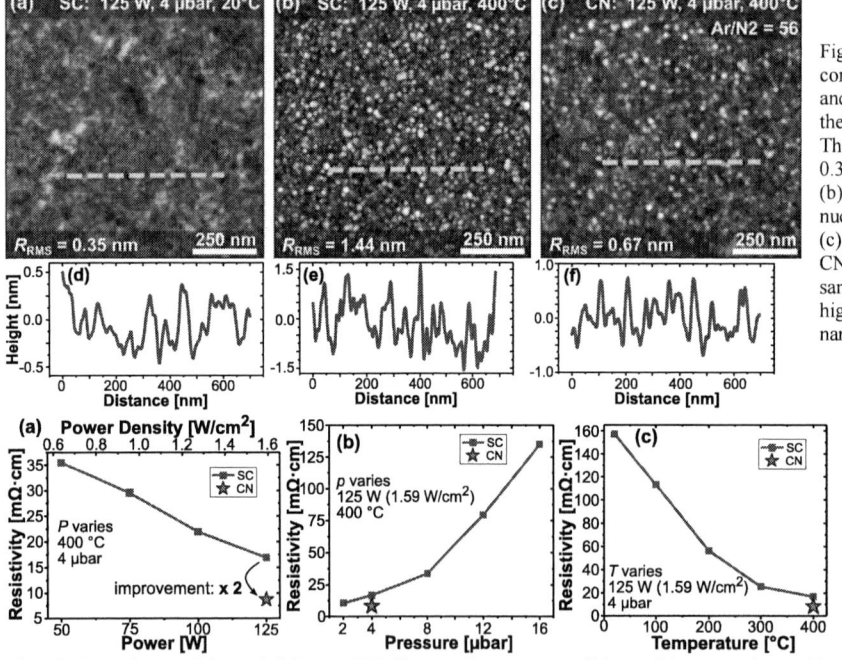

Fig. 12. Comparison of the surface roughness and the corresponding height profile along the dashed line of SC and CN films measured with an AFM. (a) and (d) show the surface and profile of a SC film sputtered at 20 °C. The root-mean-square surface roughness (R_{RMS}) is only 0.35 nm. The profiles of a SC film sputtered at 400 °C in (b) and (e) show a rougher surface and the surface nucleation of carbon leads to a polycrystalline texture. (c) and (f) illustrate the surface and height profile of a CN film deposited at 400 °C. It is smoother than the SC sample deposited at 400 °C as CN seems to have a higher nucleation density and growths more in a nanocrystalline fashion.

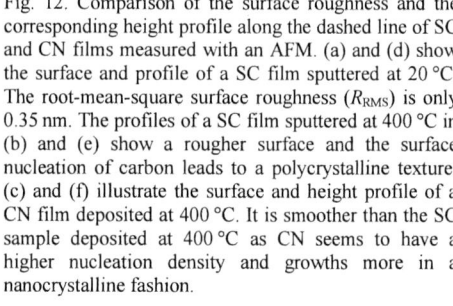

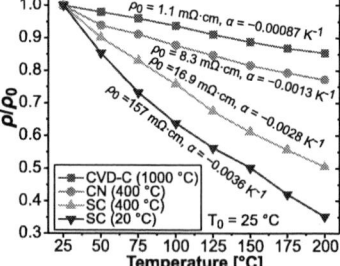

Fig. 13. Dependence of the resistivity ρ of SC films on the sputter condition and in comparison with a CN film. The values of power P, pressure p and the deposition temperature T are 125 W (1.59W/cm^2), 4 µbar and 400 °C, respectively, if they were not varied. The ratio of the sputter gas flow for the CN film is Ar/N$_2$ = 56, as a lower ratio leads to higher resistivity. (a) shows ρ of SC for different sputtering power P and compares it to the CN film. (b) and (c) show ρ as a function of pressure p and of the substrate temperature T. Further improvement might be achievable with better equipment enabling even lower pressures and higher power.

Fig. 14. Temperature-dependent relative electrical resistivity ρ/ρ_0 of CVD-C deposited at 1000°C, CN deposited at 400 °C and SC deposited at 400 °C and 20 °C, respectively. ρ_0 is the initial resistivity at $T_0 = 25$°C. All of them feature a NTC α, which was fitted for simplicity with a linear behavior over the measured T range.

978-1-7281-1988-5/18 $31.00 © 2018 IEEE

Location-controlled-grain Technique for Monolithic 3D BEOL FinFET Circuits

Chih-Chao Yang[1*], Tung-Ying Hsieh[1,4], Po-Tsang Huang[2], Kuan-Neng Chen[2], Wan-Chi Wu[2], Shih-Wei Chen[1], Chia-He Chang[1], Chang-Hong Shen[1], Jia-Min Shieh[1], Chenming Hu[2,3], Meng-Chyi Wu[4], and Wen-Kuan Yeh[1]

[1] National Nano Device Laboratories, No.26, Prosperity Road 1, Hsinchu, Taiwan;
[2] National Chiao Tung University, Hsinchu, Hsinchu, Taiwan;
[3] Department of Electrical Engineering and Computer Science, University of California, Berkeley, CA, USA;
[4] Department of Electrical Engineering, National Tsing Hua University, Hsinchu, Taiwan;
[*] Tel:+886-3-5726100-7565, Fax:+886-3-5722715, E-mail: samyang@narlabs.org.tw;

Abstract

A location-controlled-grain technique is presented for fabricating BEOL monolithic 3D FinFET ICs over SiO_2. The grain-boundary free Si FinFETs thus fabricated exhibit steep sub-threshold swing (<70mV/dec.), high driving currents (n-type: 363 $\mu A/\mu m$ and p-type: 385 $\mu A/\mu m$), and high I_{on}/I_{off} (>10^6). According to simulation, the thickness of the interlayer dielectric plays an important role and shall be thicker than 250nm so that the sequential pulse laser crystallization process does not heat the bottom devices and interconnects to more than 400 °C.

I. INTRODUCTION

Monolithic 3D integration can provide efficient connectivity of circuits and decrease power consumption, enhance system performance and reduce chip size. Managing the thermal impact of multi-tiered processes, such as 3D stackable semiconductor formation, source/drain dopant activation, and silicidation, [1][2] is essential to fabricating high performance monolithic 3D integrated circuits. Research groups, e.g. CEA-Leti, IBM, AIST, Stanford and UMC [3]-[7], have published their monolithic 3D integration technologies using low thermal budget processes.

A BEOL location-controlled technique using pulse laser anneal process for fabricating monolithic 3D FinFET circuits within Si grains is proposed (**Fig. 1**). Spatially separating devices and grain boundaries provides a promising solution for developing practical monolithic 3D ICs

II. DEVICE FABRICATION

Placing the poly-Si grain boundaries at permitted locations according to the circuit layout can enhance circuit performance and yield and minimize variability. **Fig. 2** illustrates monolithic 3DIC fabrication using the location-controlled grain technique. To fabricate high quality controlled-grain epi-like Si film, the most important steps are partial "via hole" etching into the interlayer-dielectric (ILD) to form selective cooling and crystallization seeding points, conformal a-Si deposition (250-nm-thick), green-nanosecond laser crystallization (GNS-LC), and chemical-mechanical planarization (CMP) (**Fig. 2(a)-(d)**). FinFETs were fabricated inside the Si grains using low thermal budget plasma doping and far-infrared laser anneal

(FIR-LA, **Fig. 2(e)-(g)**). Metal interconnects would connect the BEOL stacked integrated circuit and the bottom FEOL integrated circuits (**Fig. 2(h)**) through via holes.

III. RESULTS AND DISCUSSION

Fig. 3(a)-(c) illustrates the shiny controlled-grain epi-like Si film prepared by GNS-LC followed by CMP to reduce the Si thickness (from 250 to 40nm), and surface roughness (from 14.6 to 1.17nm) After the CMP, the uniformity of epi-like Si thickness, which roughly defines the fin height of FinFETs, was 5.4% within 200-mm Si wafer. 3D FinFETs (fin with~16nm) were fabricated inside the Si area without grain boundaries (**Fig. 4**). The device exhibits high I_{on}/I_{off} (>10^6) and high driving currents (n-type: 363 $\mu A/\mu m$ and p-type: 385 $\mu A/\mu m$), about four-fifths of the monocrystalline Si device announced by CEA-Leti [8][9]. Several important topics for monolithic 3D IC are discussed below:

(a) Controlled-grain epi-like Si film

The location of the controlled-grain Si island is determined by the pattern of partial "via holes" etched into the interlayer-dielectric with the hole depth $H_{via} < T_{ILD}$. The grain size is determined by the distance between "vias" due to lateral grain growth as illustrated in **Fig. 5 (a) and (b)**. Compared to the random grain growth poly-Si channel without the partial "via hole" in **Fig. 5 (d)**, we obtained a regularly patterned array of single-grain Si matrix in **Fig. 5 (c)** with pre-determined location of grain boundaries marked by the yellow line. This predictability allows the transistors and circuits to stay away from the grain boundaries by design.

(b) Low thermal impact of pulse laser crystallization

The 3D stackable Si quality directly determines the device and circuit performance. The melt-regrowth technique using pule laser and "partial via etch" determines the location and size of the regrown Si grain. However, more heat may penetrate the ILD layer and degrades the underlying FEOL devices/circuits and may cause interconnect reliability issue during the green-nanosecond laser process under the partially etched "via hole". The temperature of the underlying devices and interconnect is still below 400ºC, as illustrated in **Fig. 6(a)**. In **Fig. 6(b)**, we utilized computational fluid dynamics tools to simulate the temperate distribution during pulse laser anneal process. According to our simulation results, the device layer

978-1-7281-1988-5/18 $31.00 © 2018 IEEE

temperature only reach ~209°C (T_{max} @200ns) while the recrystallized poly-Si film is cooled down from 1500°C to 276 °C with 300nm ILD thickness (**Fig. 6(c)**). To keep the temperature of the FEOL device layer below 400°C, the ILD thickness shall be thicker than 250nm (**Fig. 6(d)**).

(c) Plasma doping and FIR-LA dopant activation

A plasma doping process provides shallow and conformal doping profiles of the fin structure and pulse FIR-LA sufficiently activates dopants without high thermal budget. It reduces the contact and series resistances and suppresses short channel effects. **Fig. 7** shows the measured sheet resistances and dopant profiles before and after FIR-LA dopant activation. The epi-like Si film by plasma doing (B, 1E16, 1keV) and FIR-LA (140W) shows lower sheet resistance (1.3 kohm/sq.) and almost diffusionless dopant profile compared to the result by RTA process (950 °C, 5s).

(d) BEOL SRAM with location-controlled-grain technique

Without the location-controlled-grain technique, the intra-die local variation of BEOL FinFETs is large since both I_{on} and I_{off} are affected by the presence of grain boundaries. I_{off}, the leakage current of transistors containing a grain boundary would vary from 10^{-8}A/μm to 10^{-5}A/μm. For an SRAM array, read/write-assist circuits, timing tracking circuits, and sensing amplifiers (**Fig. 8**) are sensitive to local variation. Based on the proposed controlled-grain process, the variation of BEOL FinFETs can be reduced without placing transistors at the grain boundary. The difference between the controlled-grain and random-grain FinFETs is the former's layout coordination between circuits and grain-boundary patterns. **Fig. 9** presents the floorplans of SRAM in the random-grain film (top row figures) and the grain-controlled film, (bottom row figures). To guarantee no FinFETs over the grain boundaries, an SRAM array is divided into mini-arrays of m x n SRAM cells based on cell size and the achievable grain size. The SRAM bitcell size [10-13] and required grain width and length are listed in Table I to eliminate SRAM cell overlapping the grain boundaries.

For understanding the boundary effect on SRAM, a 256x32 (8kb) SRAM array with timing tracking read replica circuits, keepers, and sensing amplifiers was simulated at 1V and 1GHz. Without adopting the location-controlled-grain technique, leakage current of the tracking bitcells and keepers is increasing to the order of 10^{-5}A/μm to evaluate the timing mismatch of the read replica circuits by the increasing local variation of BEOL FinFETs. **Fig. 10, right column,** presents the simulated waveform of read errors without the controlled-grain process. The data "0" (Dout) cannot be readout as to the timing mismatch of the sensing pulse (SEN). The enable signal, SEN, of sensing amplifiers is utilized as a sampling window to capture the voltage difference between bitline (BL) and BLB, and generated by the read replica circuits in the tracking array. If the tracking time of SEN is too early or too late, the read errors or read disturbs occurs, respectively.

IV. CONCLUSION

In this work, we propose a location-controlled-grain technique and pulse laser anneal processes to manufacture monolithic BEOL FinFETs and SRAMs in Si islands free of grain boundaries over ILD. The simple manufacturing method provides a promising path to low power high density monolithic stacked 3D IC products.

V. ACKNOWLEDGEMENTS

This work was financially supported by the National Applied Research Laboratories (NARLabs) of the Republic of China and the "Center for the Intelligent Semiconductor Nano-system Technology Research" from The Featured Areas Research Center Program within the framework of the Higher Education Sprout Project by the Ministry of Education (MOE) in Taiwan. Also supported in part by the Ministry of Science and Technology, Taiwan, under Grant MOST-107-3017-F-009-002-.

REFERENCES

[1] R. Ishihara et al., "Electrical property of coincidence site lattice grain boundary in location-controlled Si island by excimer-laser crystallization", Thin Solid Films, vol. 487 p. 97, 2005.

[2] C. Y. Liao et al., "Location-controlled single-crystal-like silicon thin-film transistors by excimer laser crystallization on Recessed-Channel Silicon Strip With Under-Layered Nitride", IEEE ElDL, vol. p. 1135, 2016.

[3] P. Batude et al., "3D VLSI-CoolCube process: an alternative path to scaling, ", Symposia on VLSI Technology and Circuits, p. 5-2, 2015.

[4] V. Deshpande et al., "Advanced 3D monolithic hybrid CMOS with sub-50 nm gate inverters featuring replacement metal gate (RMG)-InGaAs nFETs on SiGe-OI fin pFETs", IEDM Tech. Dig., 2015, p. 8.8.

[5] K. Usuda et al., "High-performance tri-gate poly-Ge junction-less p- and n-MOSFETs fabricated by flash lamp annealing process", IEDM Tech. Dig., 2014, p. 422.

[6] M. M. Shulaker et al., "Monolithic 3D integration of logic and memory: Carbon nanotube FETs, resistive RAM, and silicon FETs", IEDM Tech. Dig., 2014, p. 638.

[7] S. H. Wu et al., "Extremely low power C-axis aligned crystalline In-Ga-Zn-O 60 nm transistor integrated with industry 65 nm Si MOSFET for IoT normally-off CPU application", Symposia on VLSI Technology and Circuits, p. 58, 2016.

[8] C. C. Yang et al., "Enabling low power BEOL compatible 3D⁺IC for IoTs using local and selective far-infrared ray", IEDM Tech. Dig., 2015, p. 8.7.

[9] C. C. Yang et al., "Footprint-efficient and power-saving monolithic IoT 3D⁺ IC constructed by BEOL-compatible sub-10nm high aspect ratio (AR>7) single-grained Si FinFETs with record high Ion of 0.38 mA/μm and S.S. of 65 mV/dec. and Ion/Ioff ratio of 8", IEDM Tech. Dig., 2016, p. 9.1.

[10] E. Karl et al., "A 4.6GHz 162Mb SRAM design in 22nm trigate CMOS technology with integrated active VMIN-enhancing assist circuitry," IEEE ISSCC, 2012, p. 230.

[11] Y.-H. Chen et al., "A 16 nm 128 Mb SRAM in high-k metal-gate FinFET technology with write-assist circuitry for low-Vmin applications," IEEE ISSCC, 2015, p.170.

[12] Zheng Guo et al., "A 23.6Mb/mm2 SRAM in 10nm FinFET technology with pulsed PMOS TVC and stepped-WL for low-voltage applications," IEEE ISSCC, 2018, p.196.

[13] Jonathan Chang et al., "A 7nm 256Mb SRAM in high-k metal-gate FinFET technology with write-assist circuitry for low-VMIN applications, IEEE ISSCC, 2017, p. 206.

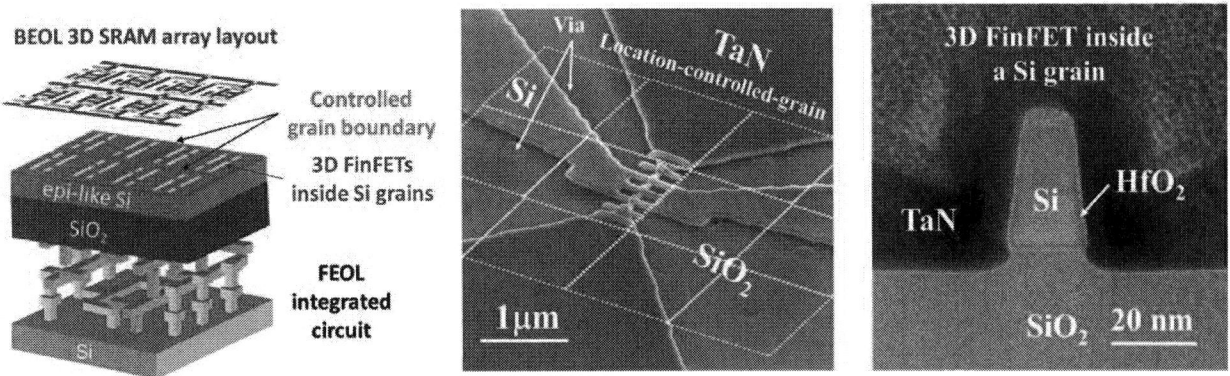

Fig 1. Illustration of BEOL 3D FinFET IC using location-controlled-grain technique to keep grain boundaries away from circuits, such as a memory-centric computing platform.

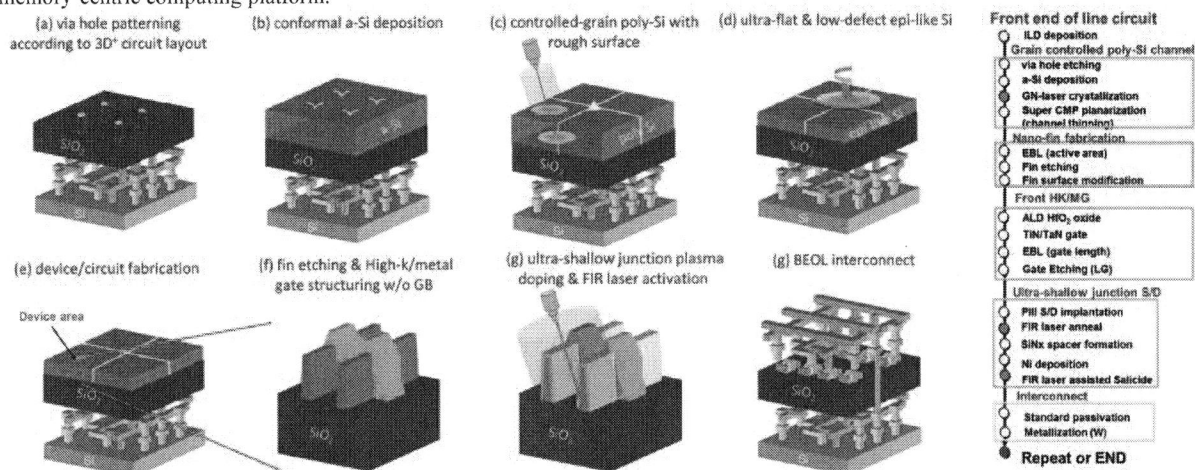

Fig 2. Process flow of BEOL controlled-grain monolithic 3DIC on FEOL/BEOL integrated circuit.

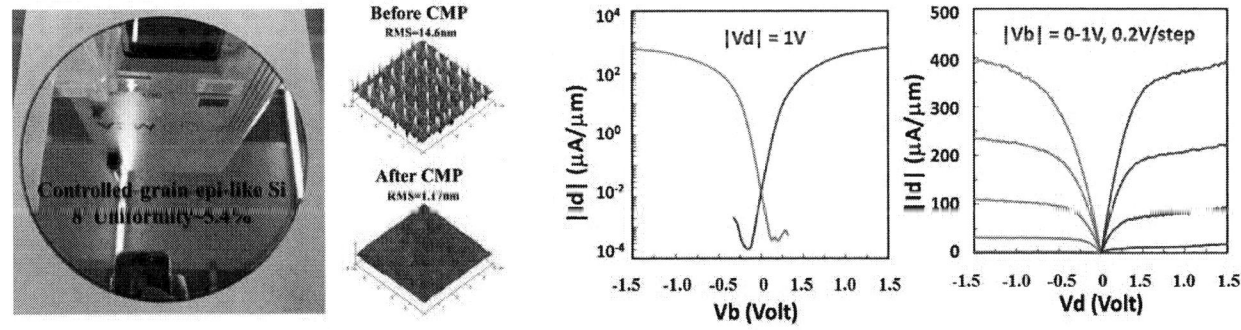

Fig 3. (a) 8 inch epi-like Si channel; AFM after (b) GNS-LC and (c) CMP process.

Fig 4. (q) Id-Vg curves (b) Id-Vd of the location-cintrolled-grain 3D FinFETs.

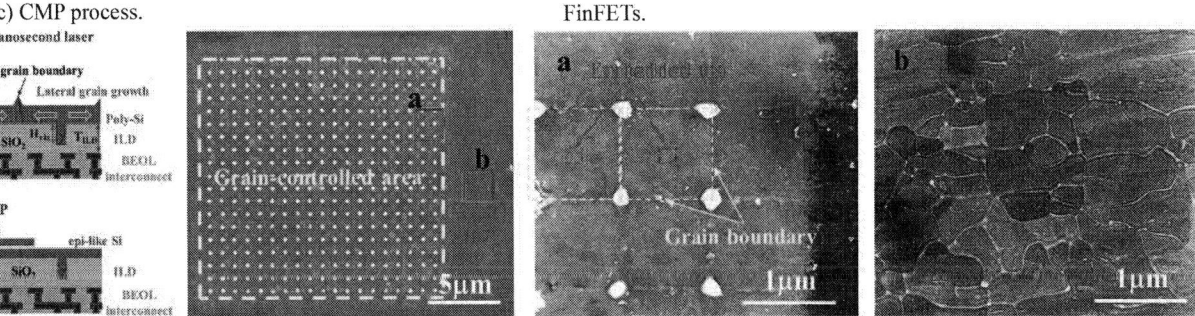

Fig 5. (a) Schematics of grain-controlled epi-like poly-Si by GNS-LC technique; (b) poly-Si film before CMP, the light dots are Si spikes over the pre-patterned partially etched "via hole"; (c, d) SEM top views of poly-Si films with and without pre-patterned holes.The yellow lines in (c) shows the grain boundaries at predictable locations.

978-1-7281-1988-5/18 $31.00 © 2018 IEEE   251

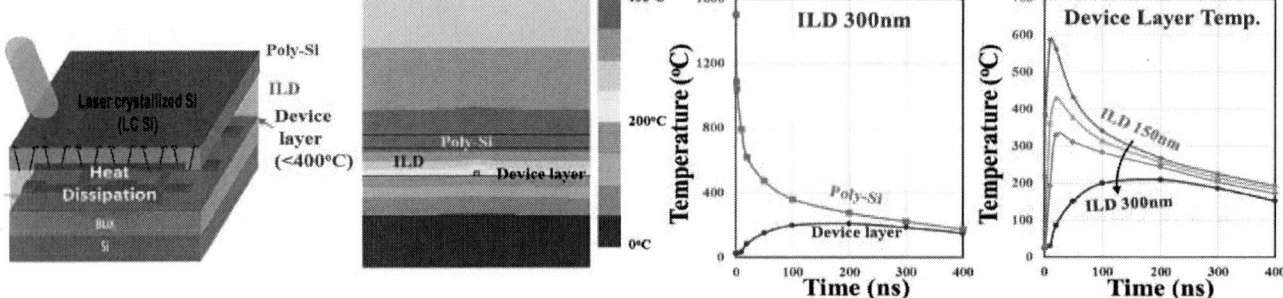

Fig 6. (a) Illustration of heat dissipation during GNS-LC; (b) temperature distribution after GNS-LC; (c) Temperature versus time in recrystallized Si and underlying device layer and (d) underlying device layer with various interlayer-dielectric layer thicknesses.

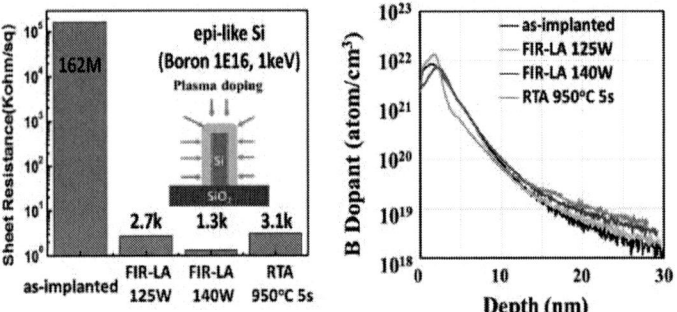

Fig 7. (a) Conformal fin doping by plasma doping; (b) sheet resistances and (c) dopant profiles before and after FIR-laser and rapid thermal anneals.

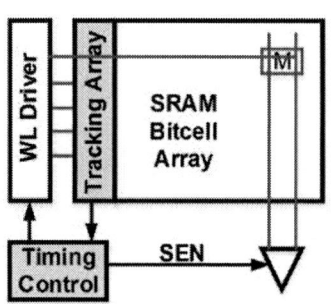

Fig 8. Sensitive circuit blocks in SRAM

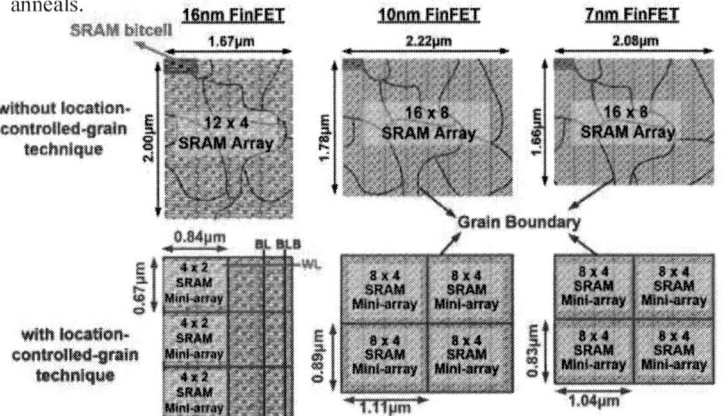

Fig 9. SRAM mini-arrays in poly-Si film, Top: random grain boundaries intersect many SRAM cells. Bottom: the SRAM layout can completely avoid the grain boundaries in the location-controlled-grain film.

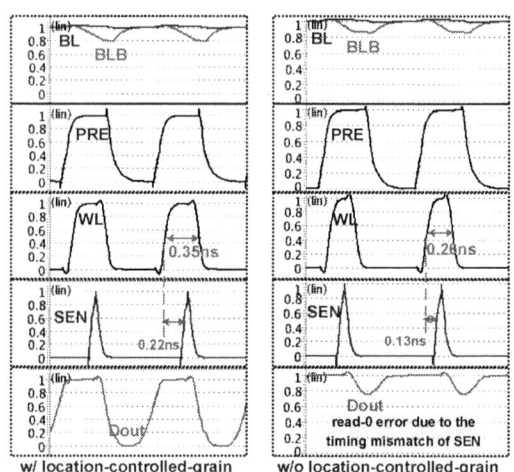

Fig 10. Examples of read success (read-0) with location controlled process and failure without it.

	ISSCC'12 [10]	ISSCC'15 [11]	ISSCC'18 [12]	ISSCC'18 [12]	ISSCC'17 [13]
Technology	22nm HKMG FinFET	16nm HKMG FinFET	14nm HKMG FinFET	10nm HKMG FinFET	7nm HKMG FinFET
Bit cell size (µm²)	HDC:0.092 LVC:0.108	HDC:0.070	HDC:0.050 LVC:0.059	HDC:0.031 LVC:0.037	HDC:0.027
# of cells in a grain	4 x 2	4 x 2	8 x 2	8 x 4	8 x 4
Grain length (µm)	0.77	0.67	1.13	0.89	0.83
Grain width (µm)	0.96	0.84	0.71	1.11	1.04

HDC: High-Density Cell   LVC: Low-Voltage Cell

Table I. SRAM bitcell size and required grain width and length to eliminate SRAM cell overlapping the grain boundaries.

Group	NDL			CEA-Leit	AIST	UMC
Reference	this work	2016IEDM [9]	2015IEDM [8]	2015 VLSI [3]	2014IEDM [5]	2016VLSI [7]
Thermal budget (°C)	<400	<400	<400	<650	400	<500
3D stackable device	controlled-grain Si FinFET	single-grained Si FinFET	epi-like Si NWFET	SOI-Si UTB	poly-Si/Ge FinFET	IGZO OSFET
W_{Fin}	16	7.2	20	60	16	60
H_{Fin} or T_{ch} (nm)	42	53.4	16	7	47	-
L_g (nm)	30	30	30	10	117	60
SS (mV/dec.)	<70	65	<90	-	64	-
Ion/W_{eff} (n-type/p-type)	363/385	386/352	310/220	620/530	119/330	<20
Ion/Ioff	>10^6	>10^7	>5x10^5	>10^7	>10^7	>10^{21}

Table II. Benchmark of monolithic 3D stackable device.

978-1-7281-1988-5/18 $31.00 © 2018 IEEE

Novel Materials and Processes in Replacement Metal Gate for Advanced CMOS Technology

Ruqiang Bao, Steven Hung[1], Miaomiao Wang, Kisup Chung, Soumendra Barman[1], Siddarth A Krishnan[1], Yixiong Yang[1], Wei Tang[1], Luping Li[1], Yongjing Lin[1], Michael S Chan[1], Zhebo Chen[1], Xin Miao, Marinus Hopstaken, Richard A Conti, Hemanth Jagannathan, Michael P Chudzik[1], Dale McHerron, Bala S Haran, Sanjay Natarajan[1]

IBM Semiconductor Technology Research, Albany, NY, USA; [1]Applied Materials, Sunnyvale, CA, USA, email: rbao@us.ibm.com

Abstract—This paper addresses novel approaches at material and integration fronts for gate applications. Material wise, new n work function metal (WFM) material is explored to address the need for reducing gate resistance and maintaining proper Vt at 20Å or less WFM thickness. Integration wise, next generation dipole is tested with various process sequences to address the need in lowering overall thermal budget at the gate level for advanced architectures, such as scaled FinFET and Nanosheets.

I. INTRODUCTION

As the device architecture evolves toward ultra-scaled FinFET and Nanosheets, extreme constraints are placed on the gate stack processes. Overall, the constraints can be divided into two parts. First, the reduction in gate level dimension requires thinner metal gate layers for given performance. Second, the ever-increasing complexity in junction and channel engineering requires the reduction in overall gate level thermal budget for preserving junction and channel qualities. Given the daunting task, this work explores novel approaches that address the following: (1) validating new dipole materials and integration sequences that can offer lower overall gate level thermal performance, (2) exploring nWFM with 20Å or less thickness that offers better bandedge performance over baseline material system, (3) exploring the possibility of co-optimization the gate level resistance and channel length dependence of nWFM material.

II. EXPERIMENTAL

For dipole study, Si nFET, Si pFET, and SiGe pFET devices were fabricated using a standard high-κ replacement metal gate (RMG) FinFET process flow on 300mm wafers. Three process sequences were proposed to test the response of dipole performance with respect to the level of thermal cycle in gate stack (Fig. 1). Process flow "A" represents the standard RMG process without dipole formation, where reliability anneal is usually conducted at high temperature (>850 °C) and is the critical step for RMG thermal budget. Process flow "B" represents a departure from process flow "A", where the low temperature (LT) drive-in anneal is performed after the standard reliability anneal. The purpose is to decouple the reliability module and the dipole formation module so each can be optimized separately. Process flow "C" represents another alternative, where the LT dipole drive-in is merged with the reliability anneal at ~700 °C, offering lower overall thermal budget at the gate level, benefiting source and drain junction

formation and other device parameters. Effect of encapsulation on Mg-based dipole layer is examined through the capping layer above the Mg-based dipole layer prior to thermal drive-in for two capping materials (cap1 and cap2). Results both with and without encapsulation are included. Gate stack without Mg-based dipole is included as reference for comparison purpose. Device level parameters are extracted. However, the key parameter of interest is Vt adjustment range for various process flows mentioned above.

For nWFM study, standard RMG Si FinFETs were fabricated. Integration with self-aligned contact (SAC) has also been performed as part of the activity to ensure compatibility with complete gate process flow. Electrical characterization on short channel FinFETs with both CPP = 64nm and CPP = 48nm was carried out with M1 readout.

III. RESULTS AND DISCUSSION

A. *ALD Mg-based dipole*

ALD Mg-based dipole process was developed to provide the capability for LT dipole formation suiting the need for advanced device architecture since it was found that PVD Mg is one of the dipole materials which can define band edge nFET work function [1,2] at lower temperature [3]. The ALD growth of Mg-based dipole is physically calibrated with XRF characterization (Fig. 2), demonstrating ALD-like liner growth behavior. The thickness range of interest for ALD Mg-based dipole is between 3 to 9Å. The nFET Vt with process flow "B" is summarized in Fig. 3. First of all, the experiment demonstrated that a lower thermal budget of 700 °C is capable of achieving dipole formation. The overall range of Vt adjustment is around 120mV with adjustment sensitivity of ~11mV/Å. It is good to see that there is no capping material impact. Data also suggests that the Mg-based dipole stack with encapsulation provides ~20 to 30mV lower Vt than the stack without encapsulation.

A physical model is proposed to explain the mechanism. As shown in Fig. 4, where the case of Mg-based dipole without encapsulation (a) progresses through downward diffusion during the post dipole thermal anneal (b), where majority of diffused Mg atoms are merged into the HfO_2 layer and establish the dipole. In contrast, the case of Mg-based dipole with encapsulation (c) progresses through bidirectional diffusion during the post dipole thermal anneal (d), which reduces the final concentration of Mg in the HfO_2 layer and thus lowers the dipole strength, thereby lowering the Vt adjustment range.

978-1-7281-1988-5/18 $31.00 © 2018 IEEE

The interface traps density (N_{it}) measured by charge pumping from nFET with process flow "B" is analyzed and shown in Fig. 5. The results parallel with the physical model proposed earlier, that the dipole with encapsulation reduces the level of Mg diffusion into the HfO_2 layer, thus resulting in lower N_{it} value than the case without encapsulation. The difference in N_{it} is ~ 20% for 3Å Mg-based dipole, while the difference disappears as Mg-based dipole thickness increases to 9Å. (Based on N_{it} alone, one can project that 3 to 6Å thickness range with inclusion of encapsulation can be a good dipole stack candidate). The mobility of nFET with process flow "B" is summarized in Fig. 6, where minor difference between Mg-based dipole device and no Mg-based dipole device is detected, suggesting Mg-based dipole has minimal impact on interface quality. However, the case of thicker Mg-based dipole without encapsulation resulted in 6% mobility degradation at the inversion charge density (N_{inv}) of 1×10^{13} cm^{-2}, likely due to the earlier finding of N_{it} impact. Similar comparison also shows minor difference in CV and IV profiles (Fig. 7) at over drive (OD), suggesting the use of Mg-based dipole with necessary thermal has minimal impact on overall gate quality. Compatibility check with BEOL thermal cycle (400 °C 2hr) has been included and summarized in Fig. 8, where Id/Vg profiles of w/ and w/o BEOL thermal are plotted side by side for comparison. The level deployment of Vt with respect to Mg-based dipole thickness can be seen on both cases, furthermore, the range of Vt adjustment is shown to be independent of BEOL thermal, suggesting Mg-based dipole formation is stable through BEOL processes.

Compatibility check with CMOS integration has been included and validated through both Si pFET and SiGe pFET with process flow "B" since SiGe is new channel material for pFET to improve device performance and reliability [4]. The resultant Vt adjustments are plotted in Fig. 9 and 10. While Si nFET resulted in bandedge displacement with Mg-based dipole, both Si and SiGe pFET resulted in midgap displacement. This is expected since the polarity of electrostatic dipole is independent of metal stack. However, interestingly the range of Vt adjustment from Si pFET (~80mV) and SiGe pFETs (~100mV) are about 10 to 20% lower than that from Si nFET. The asymmetry in Vt adjustment between n and p FET can be an issue while engineering CMOS multi-Vt.

Finally, the option of lowering overall gate thermal cycle was explored with process flow "C" (merging the reliability module with Mg-based dipole drive-in at a lower temperature ~ 700 °C). The resultant Si nFET Vt adjustment is plotted in Fig. 11. Vt adjustment is demonstrated, however the range of adjustment (~100mV) is less than the one through process "B". The concept while demonstrated to be feasible requires further optimization to expand the range of Vt adjustment, likely through post Mg-based dipole encapsulation or thermal budget.

B. *ALD Ta-based nWFM*

Validation on FinFET electrical performance of ALD Ta-based nWFM was carried out with RMG process flow "A" without the inclusion of dipole module (Fig. 1). In terms of material property, the current industry nWFM baseline ALD

TiAl bears a resistivity value of ≥ 2000 μΩ-cm for 100Å film. In contrast, as shown in Fig. 12, an ALD Ta-based nWFM demonstrates significantly lower resistivity in comparison to ALD TiAl over a wide range of thicknesses down to 20Å level. This can provide gate level RC benefit, especially for architectures with extremely small gate where WFM thickness is limited to sub 20Å level (Stacked Nanosheets as an example). Existing work on gate level simulation (Fig. 13) is used to highlight the importance of resistivity contribution from nWFM layer [5], where it shows the overall gate level resistance can be significantly reduced if the nWFM resistivity is properly suppressed, and the benefit is more for smaller gate length (L_g).

In term of process integration, compatibility with SAC has been tested and is shown in Fig. 14. Properly tuned RIE process was used to optimize the recess depth of ALD Ta-based nWFM, and the recess profile has been checked for both long and short channel devices.

Fig. 15 summarizes the Vt as a function of nWFM thickness over 15 to 25Å range for both material systems. The data clearly demonstrates Ta-based nWFM provides lower Vt (lower effective work function) in comparison to the baseline TiAl over the entire thickness range, suggesting that the film provides better capability to establish lower Vt for gate level architectures that require sub-20Å WFM thickness. Furthermore, the Vt dependence of Ta-based nWFM on L_g and Fin-number is summarized in Fig. 16. The novel film retains its Vt value over extreme wide range of Lg and fin-number, suggesting ALD Ta-based nWFM is insensitive to device size and Fin population.

IV. CONCLUSIONS

We developed and validated ALD Mg-based dipole process to address low temperature multi-Vt application and for fine Vt tuning or Vt centering. In addition, we successfully demonstrated band edge nWFM (Ta-based material) that offers advantage in low resistivity, channel length independency and compatibility with SAC process for highly scaled CMOS.

ACKNOWLEDGMENT

This work was performed by the joint research and development project teams at IBM and Applied Materials research and development facilities.

REFERENCES

[1] V. Narayanan et al., *VLSI Technical Digest,* 2006, pp 178.
[2] V. K. Paruchuri et al., *VLSI-TSA Technical Digest*, 2007, pp. 1.
[3] R. Ritzenthaler, et al, *IEDM Technical Digest* 2014, 773.
[4] D. Guo et al., *VLSI Technical Digest*, 2016, p14.
[5] R. Bao et al., *IEDM Technical Digest*, 2015, p.884.

978-1-7281-1988-5/18 $31.00 © 2018 IEEE

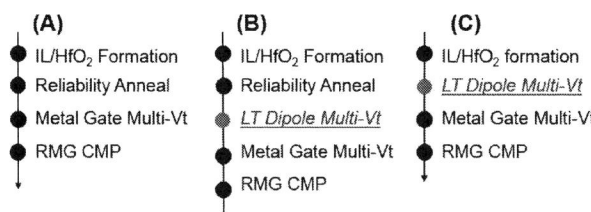

Fig. 1. RMG process flow investigated by this work (A) Standard RMG process flow for reference, (B) RMG flow with low temperature dipole (LT ~700°C), (C) RMG flow with reliability anneal combined with low temperature dipole (LT ~700°C).

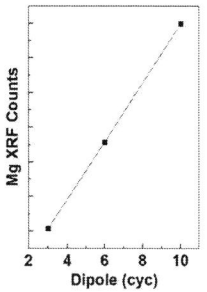

Fig. 2. Growth rate curve of ALD Mg-based dipole.

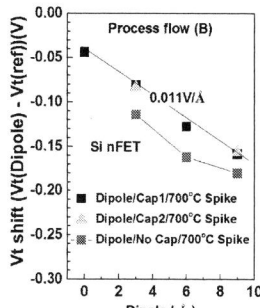

Fig. 3. Vt shift by Mg-based dipole for Si nFET with process flow (B).

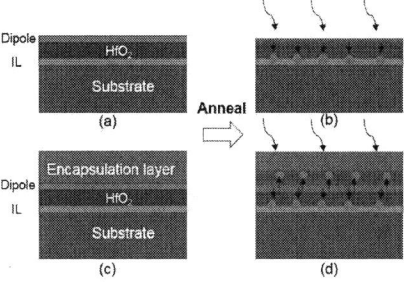

Fig. 4. Proposed diffusion mechanism for Mg-based dipole. (a & b) Without encapsulation layer, Mg-based dipole only diffuses into HfO2, (c & d) With encapsulation layer, Mg-based dipole diffuse bidirectionally.

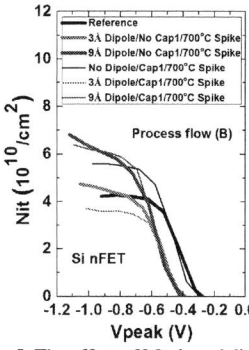

Fig. 5. The effect of Mg-based dipole on interface traps density (N_{it}).

Fig. 6. The effect of Mg-based dipole on electron mobility.

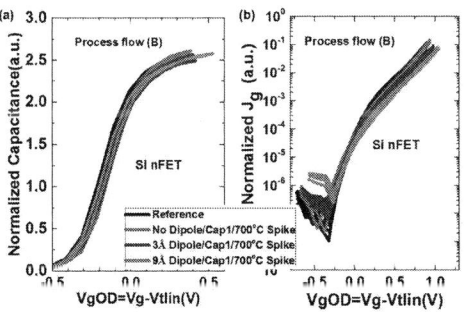

Fig. 7. The effect of Mg-based dipole on gate capacitance(a) and gate leakage (b) at gate over drive (OD).

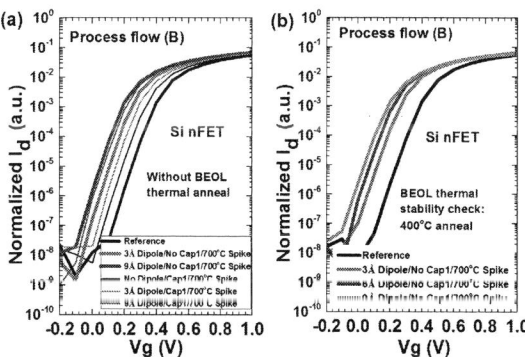

Fig. 8. The effect of BEOL thermal budget on Vt stability of Mg-based dipole.

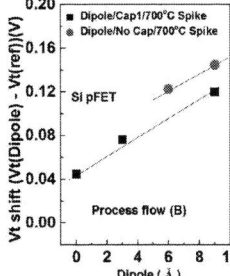

Fig. 9. Vt shift by Mg-based dipole for Si pFET with process flow (B).

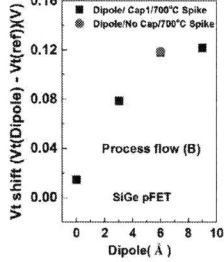

Fig. 10. Vt shift by Mg-based dipole for SiGe pFET with process flow (B).

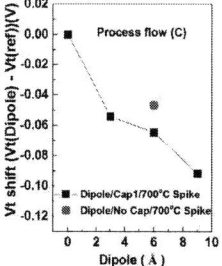

Fig. 11. Vt shift by Mg-based dipole for Si nFET with process flow (C).

978-1-7281-1988-5/18 $31.00 © 2018 IEEE
255
IEDM18-255

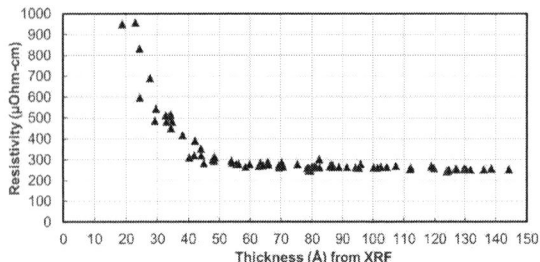

Fig. 12. Resistivity of ALD Ta-based nWFM, significantly lower than ALD TiAl (typically ~2000uΩ-cm for 10nm film).

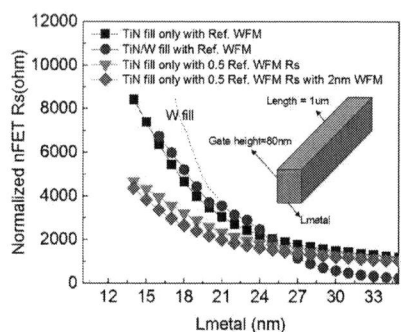

Fig. 13. The impact of metal fill scheme, nWFM film resistivity and nWFM film thickness on nFET gate resistance.

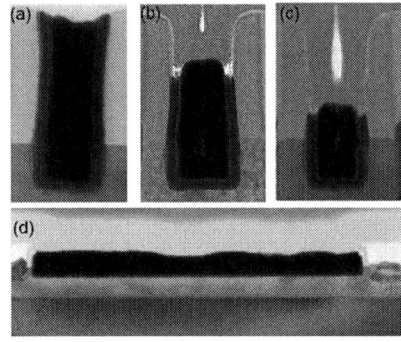
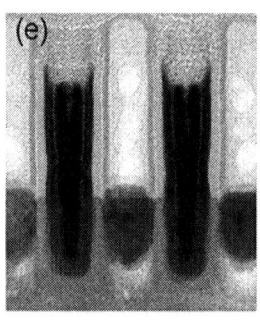
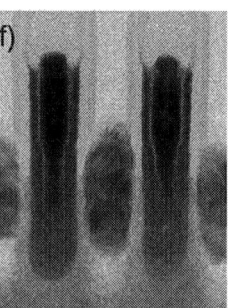

Fig. 14. Gate metal recess with ALD Ta-based nWFM. (a) w/o recess, (b) short-time recess, (c) extended recess, (d) long channel recess profile, and gate metal recess for SAC enablement with CPP=64nm (e) and with CPP=48nm (f).

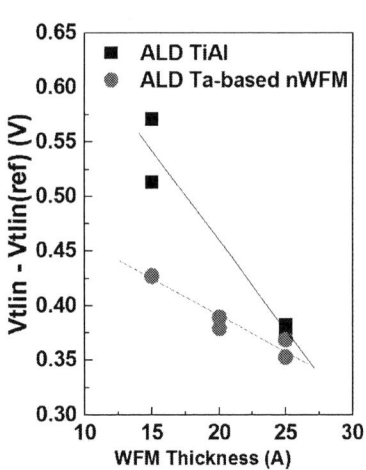

Fig. 15. Comparison on thickness dependence of Vt comparison between ALD TiAl and ALD Ta-based nWFM.

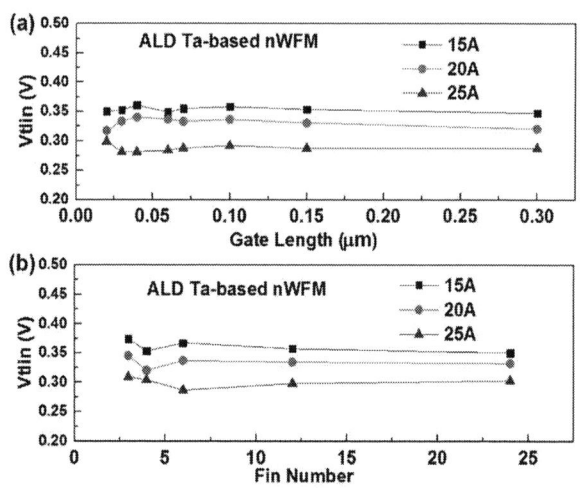

Fig. 16. Vt as a function of (a) gate length and (b) Fin number. Both plots derived from device with SAC enablement at CPP=64nm.

Why GeO$_2$ growth on Ge is suppressed and GeO$_2$/Ge stack is much improved in high pressure O$_2$ oxidation?

Xu Wang and Akira Toriumi

Department of Materials Engineering, The University of Tokyo
7-3-1 Hongo, Tokyo 113-8656, Japan
Phone & Fax: +81-3-5841-7120, e-mail: toriumi@material.t.u-tokyo.ac.jp

Abstract— This paper reports for the first time a new kinetic model of thermal oxidation of Ge that considers both O-vacancy and atomic O diffusion as a function of O$_2$ pressure. The model is based on newly obtained results that Ge oxidation is described by kinetics completely different from the Deal-Grove model and that it exhibits anomalous O$_2$ pressure dependence. Furthermore, new experimental results have been obtained in the oxidation of SiO$_2$/GeO$_2$/Ge, GeO$_2$/SiO$_2$/Si and GeO$_2$/SiO$_2$/Ge stacks. They also strongly support new kinetic model of Ge oxidation. This is critically important for high quality Ge gate stacks, as the Deal-Grove model have played a significant role in Si technology.

I. INTRODUCTION

Ge CMOS has been intensively studied from various interface engineering viewpoints [1]. The oxidation process is one of the most important processes in semiconductor technology. Although Si has been well described by the Deal-Grove (D-G) model [2], it is not self-evident that Ge is also the case. Nevertheless, people generally believe Ge oxidation should be described by the same mechanism because Ge is quite similar to Si in any aspects. There were reports in which O$_2$ molecule diffusion assumed in the D-G model would not be the case for Ge by using high-resolution RBS [3] or the nuclear reaction analysis [4]. Recently, a theoretical calculation also suggested non-D-G type oxidation of Ge [5], and experimental work using ^{18}O with SIMS was also reported [6].

The objective of this paper is to understand and to model Ge oxidation kinetics as a function of temperature and O$_2$ pressure (p-O$_2$). The results indicate that the dominant oxidation mechanism changes dependent on p-O$_2$ and temperature. This is critically important, because great GeO$_2$/Ge gate stacks is achievable by high p-O$_2$ (HPO) oxidation [7].

II. EXPERIMENTS

A. Anomalous oxidation of Ge under high p-O$_2$

Fig. 1 shows GeO$_2$ thickness vs. p-O$_2$ relationship in a wide range of oxidation temperature, in which the oxidation time was fixed at 30 min. A bell-shaped reverse p-O$_2$ dependence was reproduced at all temperatures, though we reported the results phenomenologically [7]. The p-O$_2$ corresponding to the fastest oxidation (the peak position) at a given temperature shifts to a higher p-O$_2$ at a higher

temperature (from 0.1 atm at 480°C to 20 atm at 580°C). This fact is really anomalous, because the oxidation is depressed under a huge amount of O$_2$ ambient. The result directly shows that Ge oxidation is quite different from Si one, and not simply caused by the D-G type kinetics schematically described in **Fig. 2**.

B. ^{18}O isotope marker experiment in high p-O$_2$

In the present study, the isotope tracing analysis on Ge was carried out in (HPO) oxidation. Two kinds of samples were prepared. For the sample-A, Ge wafer was first oxidized in ^{16}O$_2$ at 520°C at 1 atm to form 85-nm-thick GeO$_2$. Then, it was re-oxidized in ^{18}O$_2$ at 520°C at 1 atm for 50 min. The total oxide thickness was 95 nm. It is referred as (APO+APO). The same two-oxidation steps at 520°C was performed to prepare the sample-B by increasing p-O$_2$ to 40 atm (at room temperature). 89-nm-thick GeO$_2$ was initially formed, and then 93-nm-thick GeO$_2$ was totally formed in the second HPO oxidation. (HPO+HPO). Finally, 30-nm-thick Ge^{16}O$_2$ was deposited on the top of the stacks to minimize the surface effect in SIMS measurement. The depth profile of ^{18}O and ^{16}O in both stacks were analyzed by the SIMS.

^{18}O profiles in sample-A and sample-B grown in APO and HPO, respectively, are so different from each other. The sample-A in **Fig. 3(a)** shows almost the same results as those oxidized at 550°C [6]. On the other hand, the sample-B in **Fig. 3(b)** shows significantly different profile, in which ^{18}O is accumulated at the interface with a long tail towards the oxide. The result indicates that some diffusion species may directly transport to the interface, and that ^{18}O may diffuse in GeO$_2$ through the exchange or interstitial process with ^{16}O in the existing GeO$_2$ film. Therefore, atomic O is another important diffusion species in Ge oxidation under HPO.

C. Thermal oxidation of SiO$_2$/GeO$_2$/Ge, GeO$_2$/SiO$_2$/Si and GeO$_2$/SiO$_2$/Ge stacks

Next, three kinds of stacks, GeO$_2$/SiO$_2$/Ge, SiO$_2$/GeO$_2$/Si, and GeO$_2$/SiO$_2$/Ge were prepared to compare the oxidation of Ge different from that of Si. The film thickness was estimated by the grazing incidence X-ray reflectivity (GIXR) measurement. **Fig. 4** shows **(a)** SiO$_2$/GeO$_2$/Ge structure and **(b)** GeO$_2$ thickness increase as a function of time in O$_2$ at 550°C for various SiO$_2$ thicknesses. GeO$_2$ thickness increases dependent on initially exiting SiO$_2$ thickness which limits O$_2$ diffusion in SiO$_2$. **Fig. 5** shows **(a)** GeO$_2$/SiO$_2$/Si schematic structure and **(b)** each oxide thickness increase as a function of time in O$_2$ at 700°C. It shows no oxidation of Ge, which

978-1-7281-1988-5/18 $31.00 © 2018 IEEE

implies that the interface reaction at GeO$_2$/Ge is needed for Ge oxidation, because appreciable oxidation is observed in SiO$_2$/Si at a same temperature. These results strongly suggest GeO$_2$/Ge reaction is needed in the Ge oxidation in APO.

In case of GeO$_2$/SiO$_2$/Ge stacks, Ge was slightly oxidized. Interestingly, the oxidation exhibits a kind of the incubation characteristics, dependent on the capped GeO$_2$ thickness, as shown in **Fig. 6**. This is not discussed in this paper, because it is just related to the initial oxidation kinetics on Ge.

D. Effects of surface orientation on oxidation of Ge

Fig. 7 show the surface orientation dependences of **(a)** Ge oxidation and **(b)** GeO desorption from GeO$_2$/Ge, respectively. Both results indicate that Ge (100) is the easier reaction surface than Ge (111). Considering that the oxygen vacancy (Vo) generated at GeO$_2$/Ge interface is the key in GeO desorption [9], it is strongly supported that Vo is associated with the Ge oxidation as well, and degrades the GeO$_2$/Ge interface.

III. DISCUSSION AND MODELING

A. Intuitive discussion of Ge oxidation

We assume there are typically three kinds of oxidation species. (1) Vo (oxygen vacancy), (2) O (atomic O through exchange and/or interstitial), (3) O$_2$. First of all, in the experiments using ^{18}O$_2$ isotope, we can rule out O$_2$ diffusion model for the dominant oxidation mechanism. In fact, O$_2$ diffusion in GeO$_2$ is quite limited because of relatively high diffusion energy in GeO$_2$ [5]. It is also reasonable when a small ring structure in GeO$_2$ is considered [10].

There is evidence that Vo diffusion can form GeO$_2$ at GeO$_2$/Ge interface in UHV [9]. In fact, Vo is efficiently formed in a low p-O$_2$ region, as expected thermodynamically (**Fig. 7(c)**) [11]. It is intuitively understandable that with an increase in p-O$_2$, Vo-induced oxidation may be reduced, while atomic O oxidation may be more dominant.

In the reaction point of view, a simplest model for Ge oxidation associated with Vo generation at the interface will be described as follows.

$$O_2+2GeO_2+2Vo \rightarrow 2GeO_2 \text{ (surface)}. \qquad (1)$$
$$Ge+GeO_2 \rightarrow 2GeO \rightarrow 2GeO_2 +2Vo \text{ (interface)}. \qquad (2)$$

This model indicates that ambient O$_2$ reacts with GeO$_2$ with the help of Vo, while GeO$_2$ might interact with Ge at the interface, forming Vo in GeO$_2$. GeO$_2$ thermodynamically becomes more stable and GeO formation is significantly reduced under HPO condition, which can quantitatively explain the experimental results in Fig. 3.

This view is, however, too simple to describe the total oxidation process of Ge, because ^{18}O profile in HPO seems to be rather flat in GeO$_2$, and be accumulated at GeO$_2$/Ge interface, as shown in Figs. 3 and 4. Therefore, we would propose a model as follows. O$_2$ diffuses into the film, followed by O$_2$ decomposition to atomic O with the help of defects (including Vo) in the film. The atomic O diffuses in the film through the exchange and/or interstitial process, and Vo generated at the interface may recombine with O as follows in place of the reaction (1).

$$Vo+O \rightarrow x \text{ (inside GeO}_2). \qquad (3)$$

Atomic O density is limited by O$_2$ cracking process to O with the help of defects in the film. Meanwhile, defects including Vo should be much lower in HPO-grown GeO$_2$ than in APO-one. This is also supported by the fact that HPO-grown GeO$_2$ has a slightly but systematically higher density than APO-one [1]. Furthermore, as shown in **Fig. 8**, HPO-grown GeO$_2$ suppresses further oxidation by APO, and vice versa. This fact also strongly supports our view.

Above view on Ge oxidation can explain most of our experimental results shown in II. Concerning the Ge diffusion, we know oxygen is much faster diffusion species than Ge in GeO$_2$ [9]. Schematic view for above model is shown in **Fig. 9**.

B. Temperature dependence of Ge oxidation in APO and HPO

By confining our attention to the steady state, simple two limiting cases are discussed in this abstract. (i) Vo diffusion-limited case, and (ii) O diffusion-limited one. By and large they correspond to APO and HPO oxidations, respectively. If this view is right, it is interesting to see the temperature dependence. **Fig. 10** shows the temperature dependence of GeO$_2$ formation in APO and HPO cases. $\mathit{\Delta}_{APO}$ in APO is ~2.6 eV, while $\mathit{\Delta}_{HPO}$ is ~3.2 eV, in which $\mathit{\Delta}$ denotes the activation energy in the oxidation. This activation energy difference clearly indicates that the dominant oxidation mechanism in Ge is dependent on O$_2$ pressure.

The p-O$_2$ dependent oxidation mechanism definitely has a great impact of high-pressure O$_2$ on GeO$_2$/Ge stack qualities. One is to suppress the Vo formation <u>thermodynamically</u>, and the other is to <u>kinetically</u> passivate the remaining defects both in the bulk film and at the interface.

IV. CONCLUSION

Non Deal-Grove type characteristics as well as anomalous p-O$_2$ dependence in Ge oxidation are well explained for the first time by the new kinetic model that both Vo and atomic O diffusion (through exchange and interstitial) are involved. Although an analytical formula for the Ge oxidation as parameters of temperature and p-O$_2$ has not been shown, the present view can provide a powerful and critical guideline for achieving high quality gate stacks on Ge.

Acknowledgment

This work is partly supported by JSPS-Kakenhi. The authors are grateful to T. Nishimura for his kind suggestions and discussions.

References

[1] A. Toriumi and T. Nishimura, JJAP. **57**, 010101 (2018).
[2] B. E. Deal and A. S. Grove, JAP **36**, 3770 (1965).
[3] A. Toriumi et al., ECS-Trans. **28**(2), 171 (2010).
[4] S. R. M. da Silva et al., APL **100**, 191907 (2012).
[5] H. Li, and J. Robertson, APL **110**, 222902 (2017).
[6] X. Wang, et al., APL **111**, 052101 (2017).
[7] C. H. Lee et al., TED-**58**, 1295 (2011).
[8] C. H. Lee et al., APEX **2**, 071404 (2009).
[9] S. K. Wang et al., JAP **108**, 054104 (2010).
[10] R. L. Mozzlt et al., J. Appl. Cryst. **2**, 164 (1969).
[11] K. Nagashio et al., MRS Proc. **1155**, C06-02 (2009).

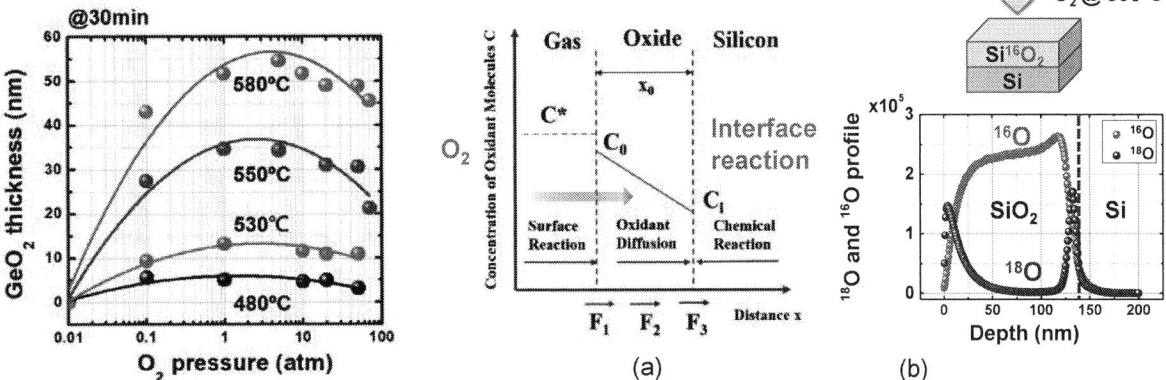

Fig. 1 GeO$_2$ thickness grown at several temperatures for 30 min as a function of O$_2$ pressure. Bell shaped behavior is observed. The O$_2$ pressure was measured at room temperature. This is anomalous from the viewpoint of the Deal-Grove type of oxidation.

Fig. 2 **(a)** Schematic view of the Deal-Grove model established in Si oxidation. **(b)** ^{18}O isotope tracer experiment in Si with SIMS, in which it is clearly reproduced that ^{18}O atoms are accumulated at the interface, while only a slight amount of ^{18}O in the film. The same experiments were carried out for Ge oxidation, which are shown in the following.

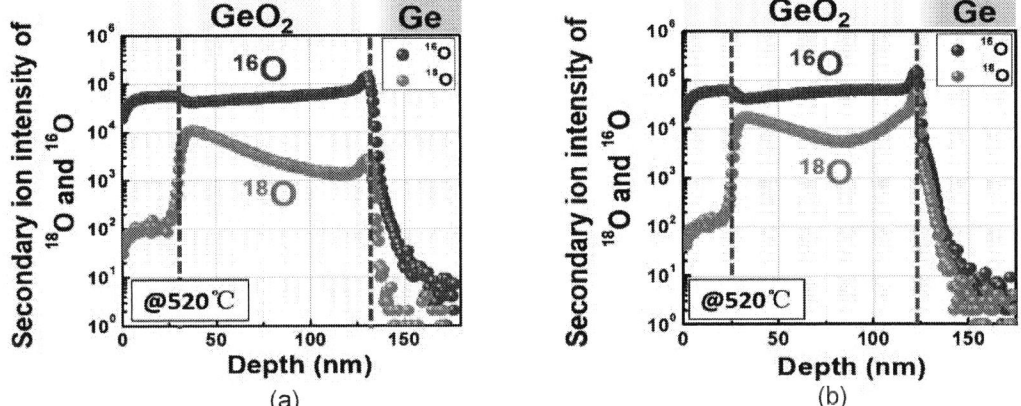

Fig. 3 ^{18}O isotope marker experiments in Ge oxidation with SIMS, **(a)** in APO and **(b)** in HPO cases, are compared to each other. There is striking contrast between two cases in ^{18}O profile. In APO, ^{18}O atoms are rather flat inside the GeO$_2$ film, while in HPO, they are accumulated at the interface with a long tail to the bulk. It shows that Ge oxidation cannot be described simply by the Deal-Grove model. Top GeO$_2$ layers (20~30 nm) were deposited Ge^{16}O$_2$ for both cases to minimize the surface ambiguity effect in SIMS (see the text).

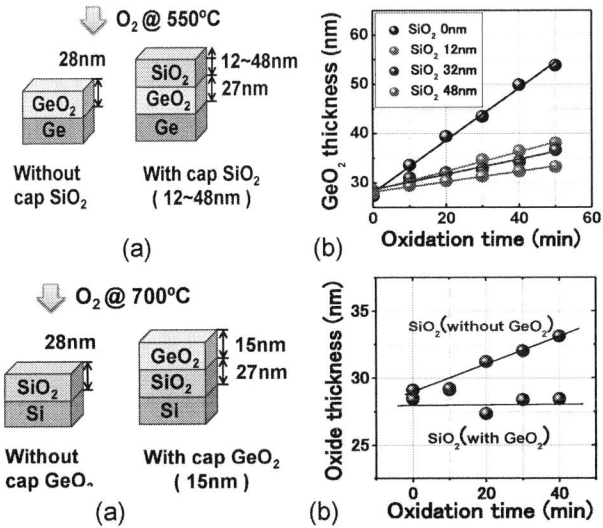

Fig. 4 Thermal O$_2$ Oxidation in SiO$_2$/GeO$_2$/Ge stack to investigate the diffusion species in GeO$_2$. **(a)** Schematic stack structure, **(b)** GeO$_2$ thickness increase as a function of the time at 550°C. The top SiO$_2$ was deposited by rf-sputtering. GeO$_2$ thickness increase depends on the cap-SiO$_2$ thickness. It is understandable from the fact that SiO$_2$ works as the "O$_2$ filter" in GeO$_2$/Ge oxidation.

Fig. 5 Thermal O$_2$ Oxidation in GeO$_2$/SiO$_2$/Si stack to investigate the diffusion species in GeO$_2$. **(a)** Schematic stack structure, **(b)** GeO$_2$ thickness increase as a function of the oxidation time w/cap (black) and w/o cap (blue). With cap-GeO$_2$, SiO$_2$ thickness does not further increase. It means that O$_2$ molecule can hardly diffuse in GeO$_2$.

978-1-7281-1988-5/18 $31.00 © 2018 IEEE

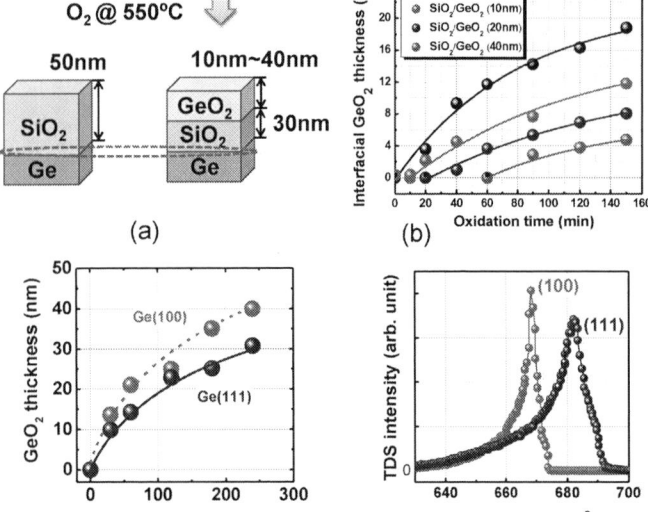

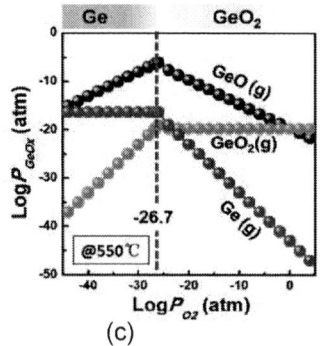

Fig. 6 Thermal O_2 Oxidation in $GeO_2/SiO_2/Ge$ stack to investigate the diffusion species in GeO_2. **(a)** Schematic stack structure, **(b)** GeO_2 thickness increase as a function of the oxidation time. With cap-GeO_2, Ge is slightly oxidized with an incubation time dependent on cap-GeO_2 thickness. SiO_2 thickness does not further increase. It means that O_2 molecule can hardly penetrate in GeO_2.

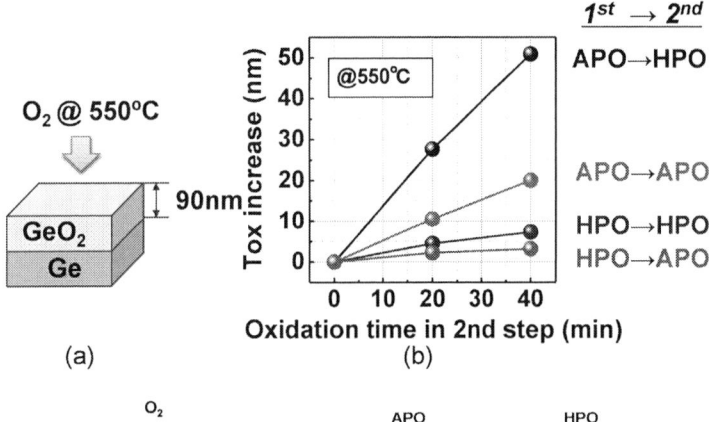

Fig. 7 (a) Surface orientation dependence of thermal oxidation in Ge (100) and Ge (111), **(b)** surface orientation dependence of GeO desorption from GeO_2/Ge stack in UHV between on Ge (100) and Ge (111). It shows the tight relationship between GeO desorption and oxidation [1]. **(c)** The partial GeOx pressure providing the Gibbs free energy as a function of O_2 pressure. By increasing O_2 pressure, GeO partial pressure is sharply decreased [11].

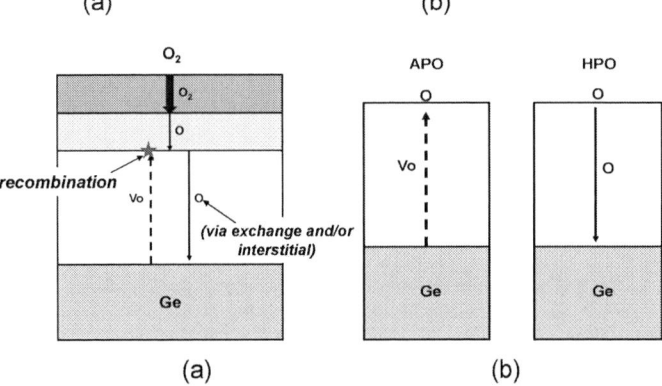

Fig. 8 Thermal O_2 Oxidation in GeO_2/Ge stack to investigate effects of initially existing GeO_2 quality on further oxidation. **(a)** Schematic GeO_2/Ge stack structure with 90-nm-thick GeO_2, **(b)** Four kinds of combinations using APO and HPO are shown. This result clearly shows that the oxidation rate is considerably dependent on how initially existing GeO_2 was grown. HPO-grown GeO_2 has a smaller defect density.

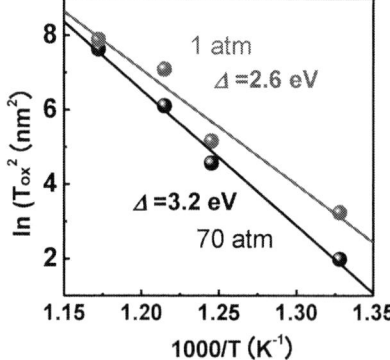

Fig. 9 (a) Schematic description of thermal oxidation of Ge. Now the steady state is considered. **(b)** Two extreme cases are shown, in which Vo diffusion-limited (APO-like) and atomic O diffusion via exchange and/or interstitial (HPO-like) are shown. This process explains the deceleration of oxidation of Ge in HPO condition. At higher temperatures, Vo oxidation still powerful, so the peak position in Fig. 1 shifts to higher p-O_2.

Fig. 10 Temperature dependence of Ge oxidation for two kinds of O_2 pressures. It is clearly seen that HPO-grown GeO_2 exhibits higher activation energy than APO one. This activation energy difference strongly indicates the dominant mechanism should change, dependent on O_2 pressure.

978-1-7281-1988-5/18 $31.00 © 2018 IEEE

EUV Lithography at Threshold of High-Volume Manufacturing*

Anthony Yen, *Fellow, IEEE,* Hans Meiling, and Jos Benschop

Abstract— **A throughput of >140 wph at a dose of 20 mJ/cm² has been achieved on NXE:3400B EUV exposure systems, using a source power of 250W. Power degradation rate has been concurrently driven down so that high system throughput can be maintained. Improvement in mask-area cleanliness has resulted in over 2,000 exposures per fall-on particle, and solid progress on the pellicle has enabled it to provide an EUV transmission of 83% at high source power. ASML continues to improve the performance of EUV scanners with higher throughput and tighter overlay specifications to further enhance their productivity and capability. Further improvements in resist and mask absorber materials are required to extend EUV single patterning to low k_1. ASML has also started to develop a next-generation $NA = 0.55$ EUV exposure tool to enable continued scaling in semiconductor manufacturing.**

I. INTRODUCTION

The fundamental equation governing the resolution of a projection imaging system such as a lithographic scanner, in terms of the minimum half-pitch (HP_{min}), is given by

$$HP_{min} = k_1 \lambda / NA$$

where λ is the wavelength of the light source and NA is the numerical aperture of the imaging optics. The value of k_1 varies depending on the quality of the imaging optics, the way the mask is illuminated, and the mask pattern itself. The theoretical minimum value of k_1 for forming an image is 0.25. For 193-nm wavelength water-immersion lithography, the highest practical NA is 1.35 and the lowest workable k_1 (with imaging restricted to defining lines and spaces) is about 0.27. Both have been realized in state-of-the-art immersion scanners produced by ASML. Putting these numbers into the above equation results in a minimum printable pitch of 77 nm. Double patterning brings the minimum pitch down to 38 nm. Another mask featuring cuts or blocks geometries is usually required to complete the patterning process of such a lines/space layer. Four masks were used to complete a contact-hole layer using immersion lithography[1]. In theory, more masks could be added to further reduce the minimum pitch using 193-nm immersion lithography. In practice, patterning processes beyond four masks per layer becomes unworkable in manufacturing due to the likelihood of severely degraded CD uniformity, overlay, and defectivity. Hence 38 nm is about the minimum pitch in logic integrated circuits that immersion lithography can practically support, by placing many restrictions on design rules. In contrast, at $NA = 0.33$, an EUV scanner can easily pattern 38-nm-pitch features at a relatively high $k_1 = 0.46$. In fact, single-mask patterning of the same features using EUV lithography results in a considerably larger process window compared to multiple patterning using 193-nm immersion lithography[1], due to higher image contrast at the high k_1 value, a larger depth of focus, and the use of a single mask. Higher values of the k_1-factor in EUV lithography also give layout designers more flexibility by offering design rules with fewer restrictions. For EUV lithography, a 0.27 k_1 will result in a minimum pitch of 22 nm. We are quite far from this number; more work is required before we can practice low k_1 imaging in EUV, as we shall address later in this paper. However, the lithography roadmap of logic integrated circuits going forward will be EUV-based.

II. EUV LITHOGRAPHY ENTERS MANUFACTURING

Why have end users waited to implement EUV lithography in manufacturing until now? Thirty-two years have passed since the very first results of EUV imaging were made public by Kinoshita and co-workers of NTT Japan[2]. It turns out that the technological barriers that had to be overcome to put EUV in high-volume manufacturing (HVM) have been very high; technology infrastructure had to mature[1]. Besides the superior imaging capability, EUV scanners must have good throughput numbers so that the increase in production cost from generation to generation can be contained. Two decisive technology milestones were reached in the first half of 2018 that enabled EUV to enter HVM in 2019. One of them is the 250W output power of the EUV source, as shown in Fig. 1. Fig. 2 shows that this source power can provide the exposure tool with a throughput exceeding 140 wafers per hour (wph) at the resist dose of 20 mJ/cm². Implementation of this level of power on existing EUV tools in the field has started. Another one is the significant reduction, at 250W, of the rate of degradation of the collector mirror which directs the EUV light into the scanner proper, as shown in Fig. 3. In addition, we have made progress on protecting the EUV mask from fall-on particles while it is chucked inside the scanner. Here we work on two approaches: minimizing the fall-on rate of such particles and eliminating the effect of these particles by covering the front-side of the mask with a pellicle. On the former, improvements in the cleanliness of the environment of the mask have resulted in a rate of under 5 fall-on particles per 10,000 exposures, as shown in Fig. 4. Our goal is to achieve less than 1 particle per 10,000 exposures. On the latter, solid progress on the mask-protecting pellicle has been made, resulting in a 83% EUV transmission at high source power. Our goal is to get close to 90% transmission. Such a transmission will give us a 125 wph throughput with the pellicle attached to the mask (see Fig. 2). Although much more work is required on EUV photoresists (see below), performances of currently available materials are adequate for first-generation applications[3].

---

¹Anthony Yen is with ASML US, LP, San Jose, CA 95131 USA (tel: 669-265-3518; e-mail: tony.yen@asml.com).

Hans Meiling is with ASML NV, 5504 DR, Veldhoven, The Netherlands (e-mail: hans.meiling@asml.com).

Jos Benschop is with ASML NV, 5504 DR, Veldhoven, The Netherlands (e-mail: jos.benschop@asml.com).

978-1-7281-1988-5/18 $31.00 © 2018 IEEE

III. CONTINUOUS IMPROVEMENTS AND HIGH-NA SYSTEMS

ASML continues to improve the performance of EUV scanners towards higher throughput and tighter overlay specifications to further enhance capability and productivity. Fig. 5 shows our latest EUV product roadmap. For integrated-circuit manufacturing with EUV lithography beyond the first generation, progress in several areas are still necessary to enable single patterning at lower k_1 values. One area concerns EUV resist materials. On much-discussed stochastic behavior of EUV resists, photon stochastics (counting statistics and where photon absorptions occur) is only part of the cause; various components in chemically amplified resists also contribute to the behavior of the developed resist. For example, simulation studies show that the spatial distributions of photon absorption and chemical components determine whether a hole is closed or open even when the number of absorbed photons is the same[4]. An example of this is shown in Fig. 6. Our results agree with the presence-of-chemical-effect-in-resist assertion made by the author of Ref. [3]. EUV resists with better resolution can be obtained if materials of higher absorbance and simpler resist chemistry are adopted, to combat both photon shot noise and chemical fluctuations[4].

Another area is the absorber material of the EUV mask. For low-k_1 imaging, dipole-like illuminators are used to raise image contrast. However, due to mask-side non-telecentricity of the imaging optics and the 3D nature of the reflective EUV mask, the two component images resulting from the monopoles, though individually maybe of high contrast, do not coincide completely in the wafer plane, leading to reduced contrast and possible edge-placement errors (EPE) in the final image. These adverse effects can be lessened by adopting more optimized absorber materials. For details see Ref. [5]. One possible candidate is Ni. At the EUV wavelength of 13.5 nm, it has a similar index of refraction to but more than twice the optical absorbance of the currently used material TaBN. Progress has been made on reactive-ion etching of Ni to form absorber patterns[6]. Finally, the overall optimization to minimize EPE requires optimizing the illuminator shape and intensity, along with optimizing mask pattern corrections and assisting feature placements, while taking the phase effect of the mask into account[7].

In addition, ASML and our partner Carl Zeiss are developing a next-generation EUV exposure system with $NA = 0.55$ and tightened specifications to allow continued scaling in semiconductor manufacturing well into the next decade[8]. The goal here is to enable EUV single-patterning at a minimum pitch below 20 nm. Fig. 7 shows this can be realized with the high-NA imaging.

IV. SUMMARY

EUV lithography is entering HVM of logic integrated circuits. This is out of necessity for the semiconductor industry to continue on the Moore's Law curve, coupled with the readiness of the technology. Decisive technology milestones were reached in 2018. Source power of 250W has now been realized in the field, providing a tool throughput capability of >140 wph at a dose of 20 mJ/cm^2. Rate of source power degradation has been driven down so that high system throughput can be maintained. Improvement in chamber cleanliness has resulted in under 5 fall-on particles per 10,000 exposures, and solid progress on the pellicle has enabled it to provide an EUV transmission of 83% at high source power. Meanwhile, ASML continues to improve the performance of EUV scanners, with higher throughput and tighter overlay specifications to further enhance productivity and capability. Further improvements in resist and mask absorber materials are required to extend EUV single patterning to low k_1. Finally, ASML has started to develop a $NA = 0.55$ EUV exposure system to enable continued scaling in semiconductor manufacturing well into the next decade.

ACKNOWLEDGMENTS

We thank our ASML colleagues who are engaged in EUV related work for their contributions to the material and results presented in this paper. These results could not have been obtained without their dedication and innovative thinking.

REFERENCES

[1] A. Yen, , "EUV lithography: from the very beginning to the eve of manufacturing," Proc. of SPIE Vol. 9776, 977632 (2016).

[2] H. Kinoshita, "30 years have passed from the first experiment," 2015 International Symposium on EUV Lithography, Maastricht, The Netherlands, October 6, 2015.

[3] B. Turkot, "Current status, challenges, and outlook of EUV lithography for high volume manufacturing," 2018 International Workshop on EUV Lithography, Berkeley, CA, June 12, 2018.

[4] S. Hansen, ASML, unpublished.

[5] M. Burkhardt, "Investigation of alternate mask absorbers in EUV lithography," Proc. of SPIE Vol. 10143, 1014312, 2017.

[6] T. Kim, J. K.-C. Chen, and J. P. Chang, "Thermodynamic assessment and experimental verification of reactive ion etching of magnetic metal elements," J. Vac. Sci. Tech. A 32, 041305 (2014); E. Chen, N. D. Altieri, and J. P. Chang, "Plasma-enhanced chemical etching of nickel," unpublished.

[7] S. Hsu, "The challenges of resolution enhancement techniques (RET) for advanced technology development," 2018 SPIE Advanced Lithography symposium, San Jose, CA, February 26, 2018.

[8] J. v. Schoot et al., "The future of EUV lithography: continuing Moore's law into the next decade," Proc. SPIE 10583, 105830R, 2018.

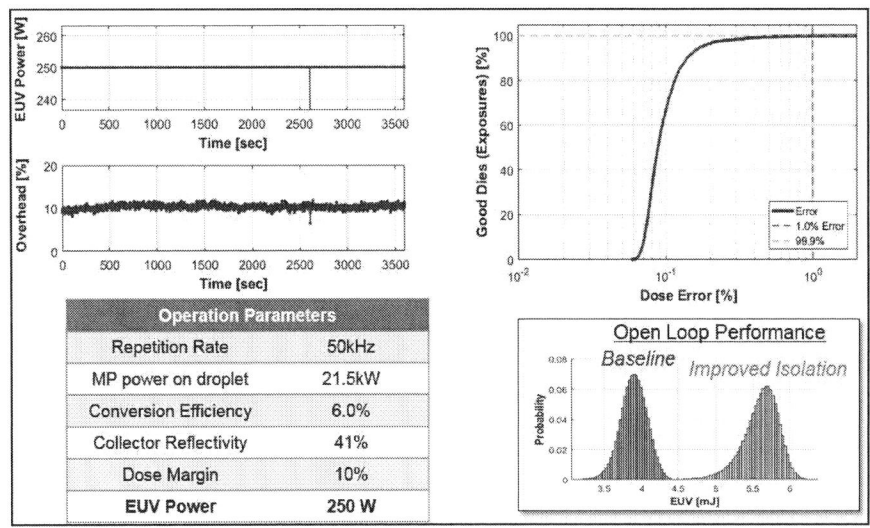

Fig. 1 Realization of 250W of stable EUV power in ASML's laser-produced plasma source.

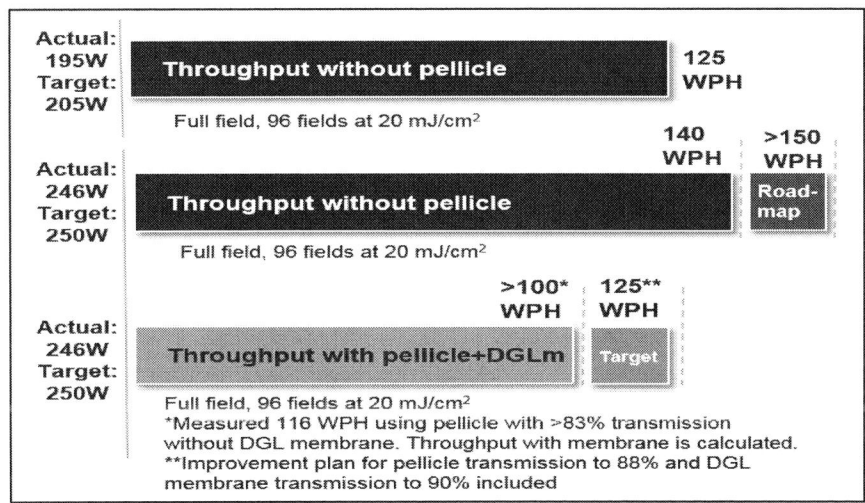

Fig. 2 A throughput of 140 wph was realized with 246W of source power. With a pellicle and a dynamic gas lock membrane (DGLm) in place, a throughput of >100 wph can be obtained; our target is 125 wph.

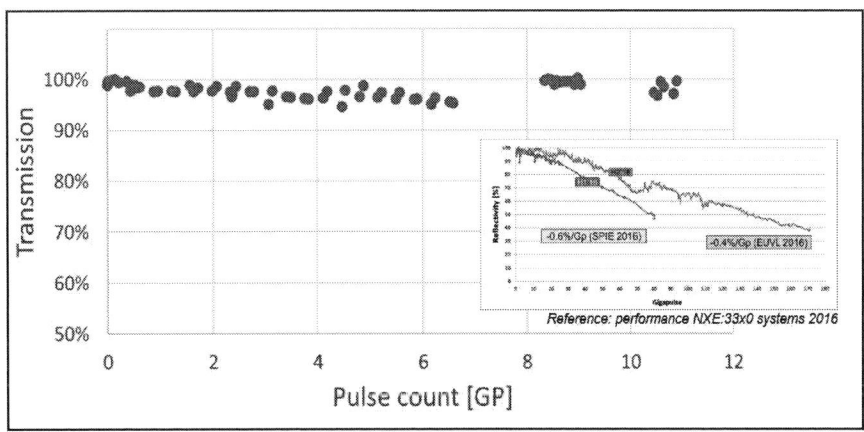

Fig. 3 Significant progress in reducing the rate of degradation of the collector mirror was achieved in the first half of 2018. The 250W source configuration shows 0.25% of reflectivity loss per giga-pulse of EUV light. Insert shows our previous results.

978-1-7281-1988-5/18 $31.00 © 2018 IEEE

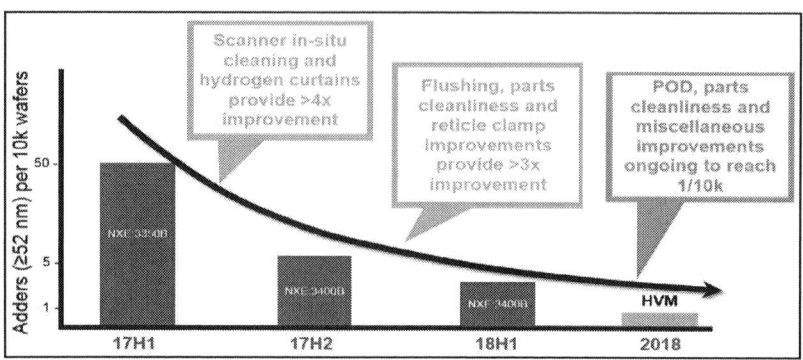

Fig. 4   Progress in minimizing fall-on particles on the chucked mask. The number has been reduced to under 5 per 10,000 wafer exposures.

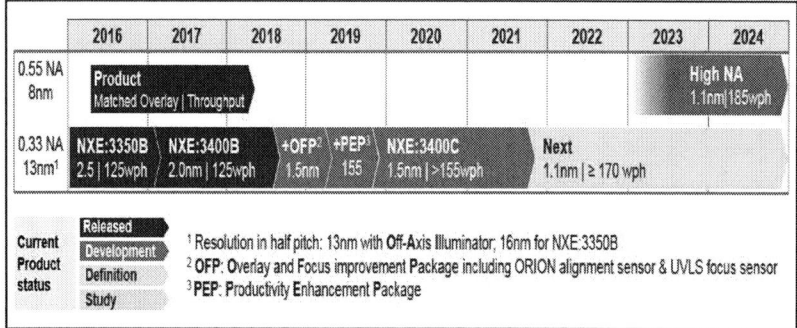

Fig. 5   ASML's roadmap of EUV exposure tools, including high-NA scanners.

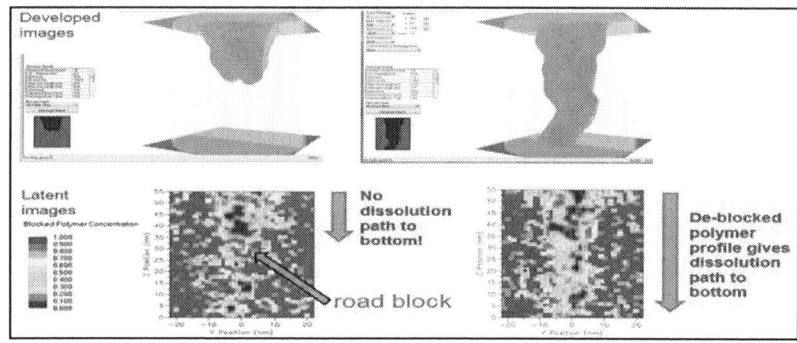

Fig. 6   Simulation study shows that a contact hole in developed EUV resist can be either closed or open, with the same number of absorbed photons. 48-nm-pitch holes are underexposed to result in a 15-nm mean CD. Photon absorption and subsequent chemical interactions are treated stochastically[4].

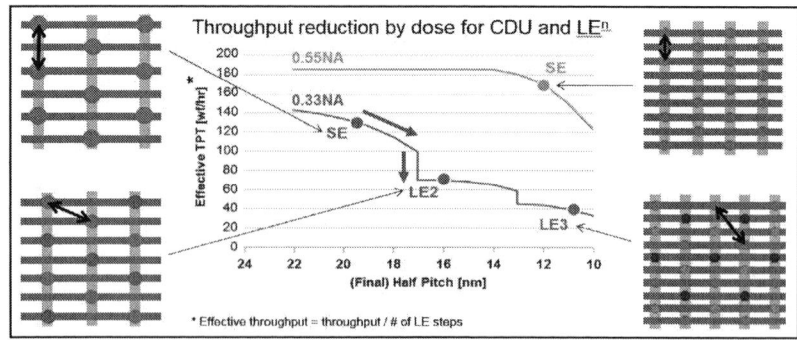

Fig. 7   Single patterning of contact holes at $NA = 0.55$ vs. multiple patterning of the same pattern at $NA = 0.33$. Notice the clear advantage in the effective tool throughput for the high-NA case.

Half pitch 14 nm direct pattering with Nanoimprint lithography

T. Nakasugi[1], T. Kono[1], K. Fukuhara[1], M. Hatano[1], H. Tokue[1], M. Komori[1], H. Tsuda[1],T. Komukai[1], K. Takahata[1], H. Kato[1],
K. Kobayashi[1], A. Mitra[1], S. Kobayashi[1], S. Inoue[1], T. Higashiki[1], T. Motokawa[2], M. Saito[2], S. Kanamitsu[2], M. Itoh[2],
T. Imamura[3], K. Matsunaga[3], K. Hashimoto[3], Y. Kim[4], J. Cho[4], and W. Jung[4]

[1]Institute of Memory Technology Research & Development, Toshiba Memory Corp, Yokohama, [2]Kawasaki, and
[3]Yokkaichi, Japan, email: tetsuro.nakasugi@toshiba.co.jp
[4]SK hynix Inc, Icheon, Korea

Abstract—we developed a nanoimprint lithography (NIL) technology including NIL system, template and resist process for half pitch (hp) 14 nm direct pattering. The latest NIL system NZ2C shows the mix and match overlay (MMO) of 3.4 nm (3σ) and the template life around 125 lots. Throughput of 80 wafers per hour (wph) was demonstrated using throughput enhancement solutions, such as gas permeable spin-on-carbon (GP-SOC) and multi field dispense (MFD). The hp 14 nm template was fabricated by a self-aligned double patterning (SADP) on a template. Using this template, we fabricated hp 14 nm dense Si lines with a depth of 50 nm on a 300 mm wafer.

I. INTRODUCTION

The challenges of nanoimprint lithography (NIL) are overlay, defects, throughput, template life and template patterning [1-2]. The overlay and defects must satisfy the requirements of the products applied. The throughput needs to provide adequate cost of ownership (CoO). Since NIL is a contact process, its template damage by the particles on a wafer is inescapable and a longer template life is required for mass production. The defect-free template having 1x-nm feature size is also needed. In general, high resolution template patterning can be performed by electron beam (EB) lithography. But, a risk of a defect exists due to insufficient resolution of EB resist.

These challenges of NIL are sufficient to divert attention of most of the lithographers away from NIL. However, we are coming to the place to overcome these challenges by the progress of NIL system, process technology and template manufacturing technology. In this paper, we report on the latest lithography performance of NIL including half pitch (hp) 14nm pattering with single mask exposure.

II. LITHOGRAPHY PERFORMANCE

A. NIL system

FPA1200 NZ2C (Canon Corp.) shown as Fig.1 is the latest NIL tool [3-6]. NZ2C has 4 imprint heads and each head has ink-jet nozzle for dispensing resist drops in a required area. Thereby, the resist thickness becomes constant regardless of pattern density. By repeating sequence of a series of resist dispense and imprinting, a desired pattern is formed all over a wafer (we call this mode single field dispense (SFD)).

B. Overlay

Unlike conventional photo lithography, NZ2C has adapted die-by-die alignment system. Canon has reported that the single machine overlay (SMO) and the mix and match overlay (MMO) is 2.5 nm (3σ) and 3.4 nm (3σ), respectively [3]. A high order distortion correction (HODC) system is one of the unique features of NZ2C. The HODC system consists of two different approaches; one is mag actuator which applies force using an array of piezo actuators, and another is heat input to correct distortion on a field by field basis.

Fig. 2 shows the champion data of the MMO result. The overlay accuracy of 3.4 nm (mean + 3σ) is obtained. It has been confirmed that HODC system is effective

C. Defects

Most of the defects are non-fill defects due to bubbles between the resist drops. In order to remove the bubbles, Spin-On-Carbon (SOC) film and resist drop optimization are effective. In the case of the silicon wafer coated with SOC (200 nm thick), the typical adder defect density (DD) of NIL is less than 0.5 pcs/cm^2 for hp 28 nm dense pattern.

D. Throughput

Spread time is the time required to completely fill the template pattern with the resist. In order to achieve the throughput of 80 wafers per hour (wph), it is necessary to reduce the spread time to 1.1 sec/shot. On a flat wafer, a filling time of 1.1 sec/shot is achievable. But, on an actual product wafer with steps, the spread time becomes longer at the deep trenches as the spreading speed of resist depends on the resist thickness. Therefore, we need additional throughput enhancement solutions for an actual product wafer.

Multi field dispense (MFD) is one of the solution. Fig. 3 shows an imprint sequence of MFD. First, resist is dropped at two or more fields, and subsequently is imprinted. In the case of 6 shots MFD with the shot map of Fig. 3 (b), the number of stage movement for resist dispensing can be reduced from 78 times to 20 times. By applying 6 shots MFD, the throughput of 80 wph can be attained with a maximum spread time of 1.25 seconds.

Another solution is quick removal of bubble. The effective means are gas permeable spin-on-carbon (GP-SOC). Fig. 4 shows the situation where the bubble disappears in longer spread time. In the case of GP-SOC, it is possible to remove bubble in 1.2 seconds, and it becomes possible to obtain the throughput of 80 wph.

E. Template life

Template life is the key of CoO. In order to avoid template damage, some unique technologies such as surface treatment of parts, electrostatic cleaning plate and air curtains keeping

978-1-7281-1988-5/18 $31.00 © 2018 IEEE

particles away from the template and wafer have been applied to NZ2C. Fig. 5 shows the historical trend of template life. The actual template life by imprint durability test is around 125 lots, on the condition that the defect density is less than 1 pcs/cm^2.

F. Template technology

The template for imprinting on wafer is called a replica template, and is fabricated by NIL process from a master template. The outer shape is the same as a photomask and it can conform to most of manufacturing tools adapting a photomask. The typical specifications of commercially available 2x nm template are shown in Table 1 [7].

The resolution of NIL is determined by that of a master template. It has been reported that full field hp 14 nm dense line master template is successfully fabricated using a multi-beam EB writer [7]. On the other hand, a self-aligned double patterning (SADP) on a template is a promising method. Fig.6 shows the hp 14 nm dense line master template fabricated by SADP; (a) template fabrication flow by SADP, (b) top-down scanning electron microscope (SEM) image and (c) cross-section transmission electron microscope (TEM) image. The space width is around 14 nm, and the depth is around 30 nm. The aspect ratio is 2.14.

G. HP 14nm patterning on wafer by single mask exposure

Fig.7 (a) shows the hp 14 nm dense line on a 300 mm wafer by single mask exposure. The mean value of critical dimension (CD) is 13.7 nm, and the variation is 0.4 nm (3σ). The line width roughness (LWR) is 2.7 nm. The cross-sectional image shows good shape as shown in Fig.7 (b).

Fig.8 shows the hp 14 nm etched silicon patterns. The depth of silicon is about 50 nm. It has been found that hp 14 nm pattern can be etched with the resist pattern of aspect ratio 2.14. It's also clear that NIL is an effective solution for 14 nm direct pattering.

III. COMPATIBIITY TO CONVENTIONAL LITHOGRAPHY AND ITS APPLICAIONS

A. Pillar and contact-hole Application

Fig.9 (a) and (b) show the template with a hole of 25 nm size and the resist pillar patterns formed by it. NIL can be applied to contact-hole process using the pillar patterns and an image reversal technology. Fig.9 (c) shows the flow of the image reversal process. The pillar patterns formed by NIL, is covered with an image reversal material containing silicon by spin coating. Then the SOC and silicon dioxide is etched by using an image reversal material as a mask. We have obtained the intra-shot CD variation less than 2 nm. The dominant process step for the CD variation is the replica manufacturing. With improving replica process, the CD variation of the intra-shot will be certainly less than 1.5 nm (3σ).

B. Other Applications

Fig. 10 (a) shows 24 nm dense hole-patterns on spin-coating resist by NIL [8]. The 2x nm complicated features like

14nm node SRAM metal can also be formed without optical proximity correction as shown in Fig. 10 (b)

C. Enhancement of cost of ownership

Even though the throughput of NIL (80 wph) is still lower than that of optical lithography, the CoO of NIL is estimated to be about 40% of that of multi-patterning using optical lithography [3-6]. However, in order to make the CoO of lithography process step equivalent to optical lithography, the throughput enhancement techniques are indispensable for NIL.

A spin-coating NIL is one of the strong candidates. Fig. 11 is (a) schematic diagram of ink-jet NIL, (b) schematic flow of spin-coating NIL and (c) contact-hole patterns by spin-coating NIL. As opposed to ink-jet NIL which has bubbles between drops, there is no bubble at spin-coating NIL. Therefore, even at a spread time of 0.5 seconds, there is no non-filling defects. In this case, the throughput of spin-coating NIL is estimated to be over 110 wph by skipping ink-jet process [8].

IV. CONCLUSION

We developed NIL technology for hp 14 nm direct pattering. The latest NIL system NZ2C shows the overlay accuracy around 3.4 nm (3σ). By GP-SOC and MFD, the throughput of 80 wph became clear. We have fabricated the hp 14 nm template by SADP on a mask-blanks-template. Using this template, we conducted hp 14 nm dense lines patterning by single mask exposure and etched silicon wafer with the depth of 50 nm. We demonstrated 2x nm dense hole-patterns and 2x nm complicated features like 14nm node SRAM metal, and showed the possibility of the improvement in throughput by spin-coating NIL.

ACKNOWLEDGMENT

The authors gratefully acknowledge the contributions of Canon Corp, Dai Nippon Printing Co Ltd and Fuji film Corp.

REFERENCES

[1] I. Yoneda, et al., "A study of filling process for UV nanoimprint lithography using a fluid simulation," *Proc. of SPIE*, vol. 7271 72712A, 2009.

[2] T. Higashiki, et al., "Nanoimprint lithography and future patterning for semiconductor devices," *Journal of Micro/Nanolithography, MEMS, and MOEMS* 10(04), 043008, 2011.

[3] M. Hiura, et al, "Overlay improvements using a novel high order distortion correction system for NIL high volume manufacturing," *Proc. of SPIE*, 2018, in press.

[4] S.V. Sreenivasan, "Nanoimprint lithography steppers for volume fabrication of leading-edge semiconductor integrated circuits", *Microsystems & Nanoengineering*, 17075, 3 (2017)

[5] Y. Takabayashi, et al, "Nanoimprint system development for high-volume semiconductor manufacturing and the status of overlay performance," *Proc. of SPIE*, vol. 10144, 1014405 (2017)

[6] T. Takashima, et al, "Nanoimprint System Development and Status for High Volume Semiconductor," *Proc. of SPIE*, vol. 9777, 977706 (2015)

[7] K. Ichimura, et al, "Fabrication of full-field 1z nm template using multi-beam mask writer," *Proc. of SPIE*, 10584-24, 2018, in press.

[8] W. Jung, et al, "The opportunity and challenge of spin coat based nanoimprint lithography," *Proc. of SPIE*, vol. 10144 1014412, 2017

Fig.1. Canon's NIL tool (FPA1200 NZ2C) corresponding to a 300 mm wafer.

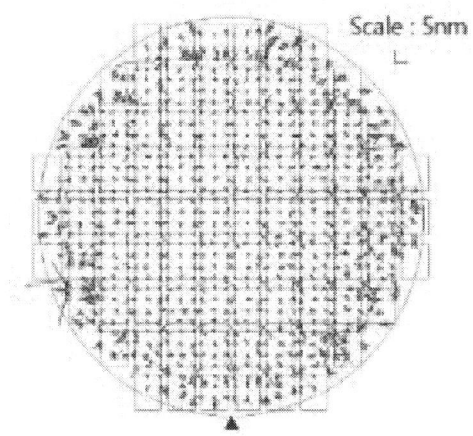

Fig.2. MMO results of NZ2C for device-like contact-hole layer. The overlay accuracy (mean + 3σ) in x and y are3.4 nm and 3.4 nm, respectively (12 points/shots, 84 shots/wafer, and 23 wafers).

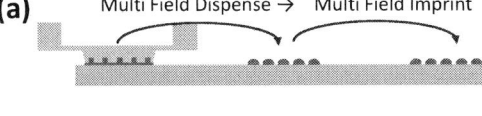

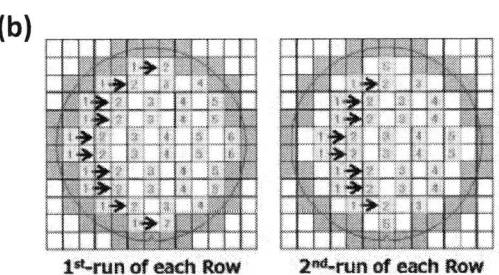

Fig.3. (a) Imprint sequence of MFD and (b) stage moving strategy. Each row is processed by 2 imprint sequences (multi-field dispense and multi-field imprint). The pink regions are the 1st run of each row and the white regions are the 2nd run of each row.

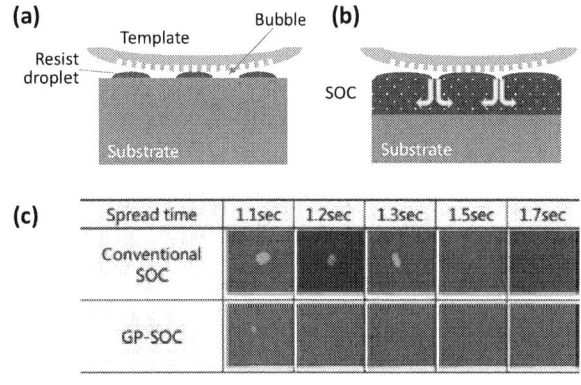

Fig.4. (a) A bubble is trapped among the resist droplets, (b) SOC film has a function to erase bubbles by gas diffusion, and (c) relationship between a bubble and spread time. A bubble disappears as spread time becomes longer. GP-SOC can provide shorter spread time.

Fig.5. Historical trend of template life. The actual template life is around 125 lots at the middle of 2018.

Item	Specification
Size	6″x6″ square and 0.25 in thick
Image placement error	$\leqq$ 1.83 nm (3δ)
CD error	$\leqq$ 0.95 nm (3δ)
Defect density	$\leqq$ 0.8 pcs/cm^2

Table 1. Specifications of replica template.

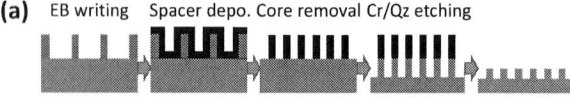

(a) EB writing Spacer depo. Core removal Cr/Qz etching

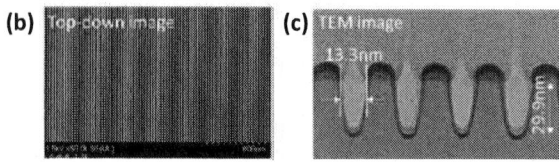

(b) Top-down image **(c)** TEM image

Fig.6. HP 14 nm master template fabricated by SADP. (a) Process flow, (b) top-down SEM image, and (c) cross-section TEM image.

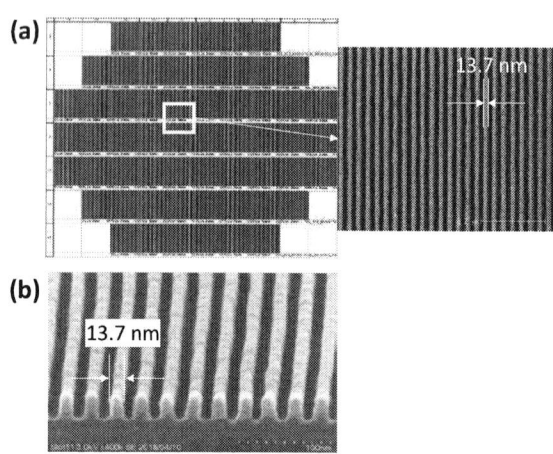

(a) 13.7 nm

(b) 13.7 nm

Fig.7. HP 14 nm resist patterns on a 300 mm wafer; (a) top-down SEM image, and (b) cross-section SEM image. The mean value of CD is 13.7 nm and the variation 0.4 nm (1,152 points/wafer =18 points/shot × 64 shots/wafer). The LWR is 2.7 nm.

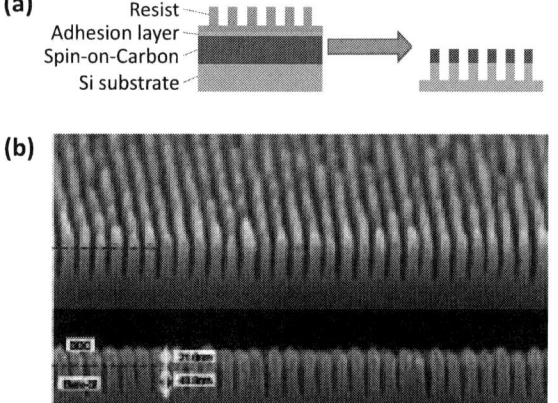

(a) Resist
Adhesion layer
Spin-on-Carbon
Si substrate

(b)

Fig.8. HP 14 nm etched silicon patterns; (a) schematic diagram of a sample, and (b) cross-section SEM image of etched silicon wafer.

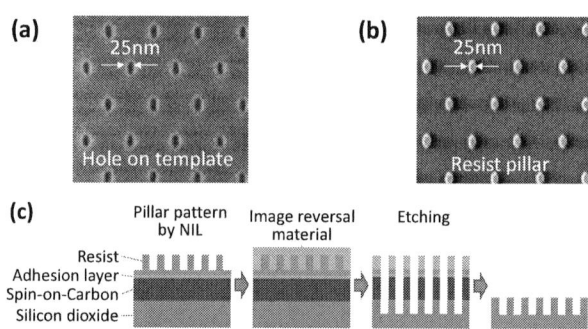

(a) 25nm Hole on template

(b) 25nm Resist pillar

(c) Pillar pattern by NIL Image reversal material Etching
Resist
Adhesion layer
Spin-on-Carbon
Silicon dioxide

Fig.9. (a) Top-down SEM image of hole template, (b) top-down SEM image of resist pillars, and (c) Process flow of image reversal process.

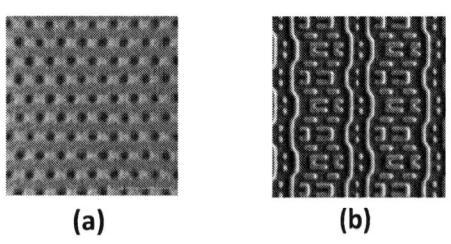

(a) **(b)**

Fig.10. (a) 24nm dense hole-patterns, and (b) 2x nm complicated features like 14 nm node SRAM metal.

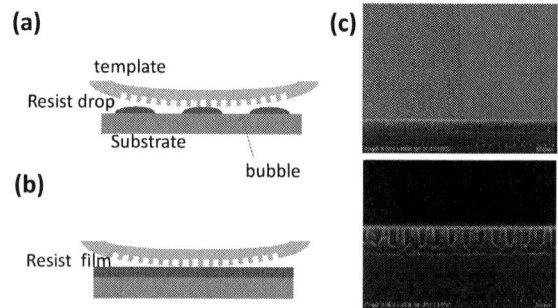

(a) template Resist drop Substrate bubble

(b) Resist film

(c)

Fig.11. (a) Schematic flow of ink-jet NIL, (b) schematic flow of spin-coating NIL and (c) contact-hole patterns by spin-coating NIL.

978-1-7281-1988-5/18 $31.00 © 2018 IEEE

All CMOS Integrated 3D-Extended Metal Gate ISFETs for pH and Multi-Ion (Na^+, K^+, Ca^{2+}) sensing

J.-R. Zhang[1], M. Rupakula[1], F. Bellando[1], E. Garcia Cordero[1], J. Longo[2], F. Wildhaber[2], G. Herment[1], H. Guérin[2], A.M. Ionescu[1]

[1] Nanolab, Ecole Polytechnique Fédérale de Lausanne, Switzerland, email: junrui.zhang@epfl.ch

[2] Xsensio S.A., EPFL Innovation Park, 1024 Ecublens, Switzerand

Abstract—This paper reports for the first time, smart 3D-Extended-Metal-Gate Ion-Sensitive-Field-Effect-Transistors (3D-EMG-ISFETs), with unique figures of merit: (i) extremely-low-power (down to a record value of 2 pW per sensor under excellent linearity), (ii) all CMOS integrated, (iii) high performance pH and multi-ion (Na^+, K^+, Ca^{2+}) sensing, and, (iv) uniquely low cross sensitivity experimentally proven. Detailed electrical DC and dynamic characterizations show excellent sensitivities (56.8 mV/pH, -58mV/dec for Na^+, -49.5 mV/dec for K^+, and -21.9 mV/dec for Ca^{2+}) and high selectivity of each ion sensor against 4 different ions that usually coexist in biofluids, all achieved on same CMOS die. Furthermore, unprecedented results show that the threshold voltage (V_{th}) variability of such CMOS ISFET is reduced by 78 times. We report a V_{th} drift rate in liquid conditions of 0.67 mV/h, decreased by one order of magnitude compared to other state of the art CMOS ISFETs. Overall, the reported experimental achievements, supported by SPICE calibrated behavioral model simulations results shown in this paper, are expected to greatly enhance the predictability of high performance multi-analyte ISFETs, which is a big step towards ISFET sensor system mass production.

I. INTRODUCTION

Recent advances in electrochemistry and bioelectronics have led to novel definitions for next generation wearables [1, 2], non-invasive and real-time monitoring of a large variety of biomarkers in human bodyfluids for personalized and preventive healthcare. Monitoring biomarkers such as Na^+, K^+, H^+, NH_4^+, Ca^{2+}, Cl^- is of high interest for a large variety of applications, ranging from complex body states such as hydration, to heart and metabolic diseases [3-6].

In future wearable and Internet of Things (IoT), energy efficiency [7] is setting more and more demanding requirements for the underlying sensing devices. Among all the commonly used sensing technologies, Ion-Sensitive-Field-Effect-Transistors (ISFETs), offer a good opportunity for both highly compact integration and ultra-low power consumption [4, 5]. However, the fabrication of a high performance ISFET has the drawbacks of low yield, low robustness and difficulty of integration, especially when multi-analyte sensing is required. Having this in mind, commercial CMOS process has been exploited in its Back-End-of-the-Line (BEOL) for more than 2 decades [8, 9, 10], in order to realize cost-efficient, high performance ISFETs (CMOS ISFETs). However, the performance of CMOS ISFETs is limited by low sensitivity, large spread in threshold voltage (V_{th}), large drift rate, and high power consumption.

In this work, we report for the first time, a fully CMOS integrated, only BEOL functionalized, smart 3D-Extended-Metal-Gate ISFET (3D-EMG-ISFET), which is capable of: (i) Highly sensitive and selective sensing of multiple ions (Na^+, K^+, Ca^{2+}); (ii) pico-watt operation, without degrading the sensitivity; (iii) greatly enhanced predictability of ISFETs behavior before fabrication.

II. FABRICATION

A. From MOSFET to 3D-EMG-ISFET

The 3D-EMG-ISFETs are fabricated by post-processing MOSFET devices designed in a commercial 0.18 μm CMOS chip. The gate of the MOSFET (10 μm × 20 μm) is vertically extended in 3D to the top metal layer through stacks of vias and metal layers with SiO_2 as inter-metal dielectric (IMD), as shown in the cross section in Fig. 1a. A Reactive Ion Etching (RIE) post-process (2 steps, Fig. 1b) is utilized to open the passivation layers (Si_3N_4 and SiO_2) sitting above (not shown here) the top metal [8]. The exposed top metal (Aluminum) is oxidized to form a thin Al_2O_3 which is used as a sensing layer. These post-process steps can easily be replaced by using a PAD mask in the layout design phase before sending to foundry fabrication, thus making the fabrication of 3D-EMG-ISFET an unmodified commercial CMOS process.

In order to know exactly the sensing dielectric composition, two sets of Transmission Electron Microscopy (TEM) images are taken, Figs. 2b – 2g and Figs. 3b – 3d, showing the composition of material layers over the top metal gate, for devices before (Fig. 2a) and after (Fig. 3a) the RIE steps, respectively. A comparison between these two sets of TEM images clearly shows that there is a layer of 7 nm Al_2O_3 formed after exposing the Al top metal with the RIE steps.

An observation of Figs. 2a and 3a show exactly the same FEOL devices, with the same stacks of metal and via layers, interspersed with IMDs, whereas the only difference is in the last passivation layers at the BEOL.

B. Controlled sensing surface

After etching the passivation layers above the top metal layer, an Atomic layer deposition (ALD) step is added in order to compare the difference in the 3D-EMG-ISFET's behavior with/without a controlled oxide surface. Using thermal ALD process at 200°C, 8nm HfO_2 is deposited to cover the extended metal gate.

C. Wet etching of the native oxide

In order to reduce the effect of trapped charge in the sensing dielectric, the native Al_2O_3 is etched by a dedicated process step. Phosphoric acid (85% wt. in H_2O) diluted in DI

978-1-7281-1988-5/18 $31.00 © 2018 IEEE

with a ratio 1:1 is deposited onto the chip surface at room temperature, while it is biased at 0V with a commercial Ag/AgCl reference electrode (16-702, Microelectrodes Inc.). During this step, the Ag/AgCl reference electrode is immersed in 3 M KCl buffer (protecting the reference electrode). A trapped-charge free (TC-free) Al_2O_3 is grown afterwards.

D. Sensor functionalization for ion sensing

As mentioned earlier, the 3D-EMG-ISFETs use the Al_2O_3 or HfO_2 dielectric layer for pH sensing, as shown in Fig. 4a. In order to make it selectively sensitive to different ions, the 3D-EMG-ISFET is smartly functionalized with an Ion Selective Membrane (ISM) for each ion species. The ISM is an effective functionalization method for chemical sensing, made through the embedding of a specific ion receptor (called ionophore) in a polyvinyl chloride (PVC) based membrane. It is drop-casted on top of the 3D-EMG-ISFET's sensing dielectric. The ionophore from the membrane selectively interacts with its target ion while shielding other ions' interaction with the surface of the sensor. In this work, there are three types of membranes deposited for Na^+, K^+ and Ca^{2+} sensing, resulting in 3D-EMG-(Na^+, K^+, Ca^{2+}) sensitive FET, respectively (Figs. 4b – 4d).

The Sodium-selective membrane was prepared by mixing Na ionophore X (1% w/w), Sodium tetrakis[3,5-bis(trifluoromethyl)phenyl] borate (Na-TFPB) (0.55% w/w), polyvinyl chloride (PVC) (33% w/w), and bis(2-ethylhexyl) sebacate (DOS) (65.45% w/w). 100 mg of the membrane cocktail was dissolved in 660 μl of tetrahydrofuran (THF). The Potassium-selective membrane cocktail was composed of valinomycin (2% w/w), Sodium tetraphenylborate (NaTPB) (0.5% w/w), PVC (32.7% w/w), and DOS (64.7% w/w). 100 mg of the membrane cocktail was dissolved in 350 μl of cyclohexanone. The Calcium-selective membrane was composed of calcium ionophore IV (1.2% w/w), 2-nitrophenyl octyl ether (66% w/w) and PVC (32.5% w/w), Na-TFPB (0.3% w/w). 100 mg of the membrane cocktail was dissolved in 660 μl of THF. Ion-selective membranes were then prepared by drop-casting 10 μl of the membrane cocktails onto the respective groups of ISFETs.

III. EXPERIMENTS AND RESULTS

A. Experimental Setups

The CMOS chip has been bonded to a readout and NFC communication ready PCB, with a specific setup for measurements of a drop of ~ 100μL liquid under test (LUT).

B. Pico-Watt 3D-EMG-ISFET for pH Sensing

A SPICE behavioral model [9, 10], sketched in Fig. 4e, is used to simulate the pH sensing characteristics of the 3D-EMG-ISFET. The drain current (I_D) to Reference voltage transfer characteristics (V_{Ref}) simulation results, together with the measurement results, are shown in Fig. 6. On-Off current ratio (I_{on}/I_{off}) ~ 10^6. Subthreshold slope in liquid is 85.3 mV/dec. The pH sensitivity is extracted from Fig. 6, in terms of shift in V_{ref} at various constant I_D in weak inversion, Fig. 7. The sensitivity is close to the Nernstian limit (57.2 mV/pH), with no degradation down to 20 pA of operation current.

C. Trapped Charge Elimination

In a statistical study of 4 devices in different chips (Fig. 8), the CMOS ISFETs which utilize the Si_3N_4 passivation layer for sensing (Fig. 2a), exhibit a V_{th} spread of 4.39 V. The 3D-EMG-ISFET reduced this spread to 184 mV, due to removal of the passivation layers, together with charges trapped in them during the foundry process. The wet etching step introduced in Section II.C further squeezed this span to 56 mV.

D. Dynamics and Drifts

The sensor's dynamic response is in Fig. 9, showing excellent repeatability and fast response (< 5 s). The stability of the 3D-EMG-ISFET with native oxide and controlled oxide layers are compared in long term stability measurements of 20 hours (Figs. 10 & 11). The 3D-EMG-ISFETs are immersed in pH 4 buffer solution with constant bias voltages, while I_D is monitored from the beginning. A drift rate of 0.67 mV/h is extracted for the ISFET with native oxide after 14th hour in pH 4 solution, whereas 2 mV/h is extracted for the ISFET with controlled oxide for the same period of time.

E. Pico-Watt 3D-EMG-ISFET for Na^+, K^+ and Ca^{2+} sensing

$I_D V_G$ characteristics for 3D-EMG-(Na^+, K^+, Ca^{2+}) sensitive FET are shown in Figs. 12, 13, 14, respectively. Sensitivities of each sensor to the corresponding ion species are extracted at constant I_D = 10 nA, and shown in Figs. 15, 16, 17, respectively. Excellent sensitivities (-58mV/dec for Na^+, -49.5 mV/dec for K^+, and -21.9 mV/dec for Ca^{2+}) are achieved. The same method is applied to extract cross sensitivities among the 3 ions as well as with pH and NH_4^+. Excellent cross sensitivities without any need for post-sensing computation are shown in Table 1. Dynamic measurements for the 3D-EMG-(Na^+ and K^+) sensitive FETs are shown in Figs. 18 and 19, respectively. Fast dynamic response (sub-5s), low operation power and high selectivity are fully demonstrated.

IV. CONCLUSIONS

This paper presented, the first to our knowledge, pico-Watt, all-CMOS-integrated, highly sensitive and selective 3D-EMG-ISFETs for highly selective multi-ion sensing platform that can be extended to many other types of analytes in the future. The reported results are expected to greatly accelerate the sensor system design based on standard CMOS technologies, with a significant enhancement in device performance, yield and robustness.

REFERENCES

[1] H. Nyein, et al, ACS Nano 10, pp. 7216-7224, (2016).
[2] S. Nakata, et al, ACS Sens. 2, pp. 443-448, (2017).
[3] H. Li, et al, DRC, pp. 1-2, (2017).
[4] F. Bellando, et al, IEDM, pp. 18.1.1-18.1.4, (2017).
[5] M. Wipf, et al, ACS Nano 7, 7, pp. 5978-5983, (2013).
[6] J. Heikenfeld, Electroanalysis, 28, pp. 1-9, (2016).
[7] A. M. Ionescu, IEDM, pp. 1.2.1-1.2.8, (2017).
[8] P. Bergveld, Sens. Act. B, 88, p. 1, (2003).
[9] P.A. Hammond, et al. Sens. Act. B, 111 p.254, (2005).
[10] N. Moser, et al., IEEE Sens. J., 16, pp. 6496-6514, (2016).
[11] J. Bausells et al., Sens. Act. B, 57, pp. 56, (1999).
[12] E. Shahrabi et al., PRIME conf, p. 1, (2016).
[13] P. Georgiou et al., Sens. Act. B, 143, p. 211, (2009).
[14] S. Martinoia, et al., Sens. Act. B, 62, pp. 182-189, (2000).
[15] S. Jamasb, IEEE Sen. J., 4, p795, (2004)

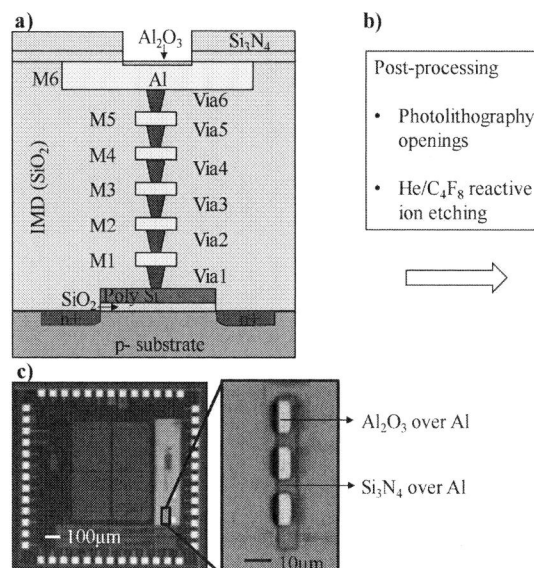

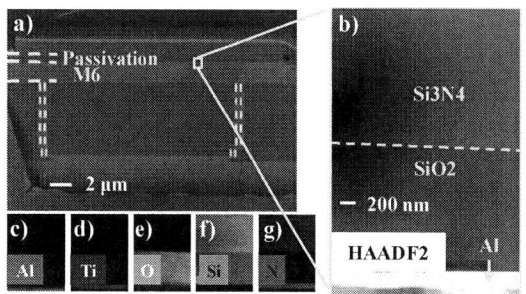

Fig. 2. a) FIB Cross section of CMOS ISFET before RIE, b) HAADF STEM indicating passivation Si_3N_4/SiO_2 layers above the top metal. c) – g) EDX analysis confirm constituent elements of Al, Ti, O, Si, N in the cross section.

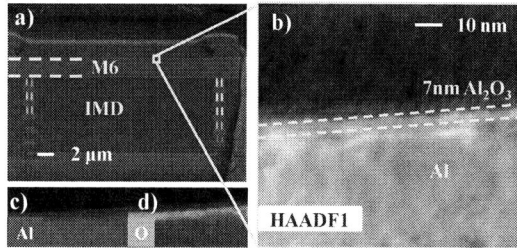

Fig. 3. a) FIB cross section of 3D-EMG-ISFET with etched openings. b) HAADF STEM image indicating 7nm native oxide on exposed top metal used for sensing. c) – d) EDX analysis confirms elemental composition of Al_2O_3.

Fig. 1. a) Schematic cross section of 3D-EMG-ISFET b) main process steps listed c) Optical image overview of chip, zoom in of openings exposing top metal layer (Al) and its natively formed Al_2O_3 layer used for sensing.

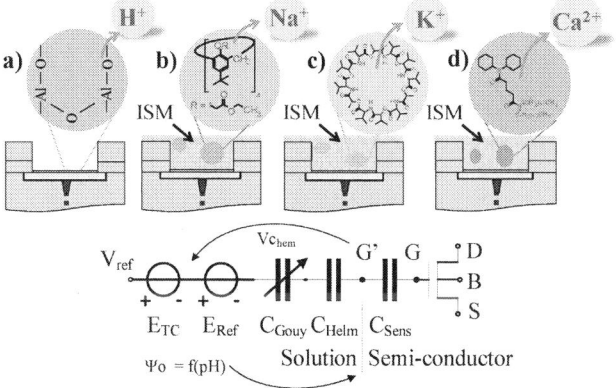

Fig. 4. a) 3D-EMG-ISFET with Al_2O_3 as the sensing dielectric. b) 3D-EMG-Na^+ sensitive FET. c) 3D-EMG-K^+ sensitive FET. d) 3D-EMG-Ca^{2+} sensitive FET. All ion sensitive FETs use specific ion selective membrane (ISM). e) Schematic of ISFET model applied to simulate the pH sensing characteristics. E_{TC}: voltage drop due to trapped charge; E_{ref}: voltage drop from reference electrode to the liquid-ISFET interface.

Fig. 5 a) A photo with the measurement setup, PCB with test connectors, NFC communication capability and readout circuits. b) Zoom in of the epoxy well created to contain a drop of ~ 100 μL LUT. c) Sketch of the measurement setup. R.E: Reference electrode.

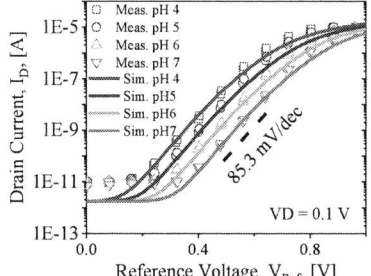

Fig.6. $I_D V_G$ Characteristics of the 3D-EMG-ISFET in various pH buffers. SPICE behavioral model [14] simulations agree well with measurements. A subthreshold slope of 85.3 mV/dec is shown in pH measurement.

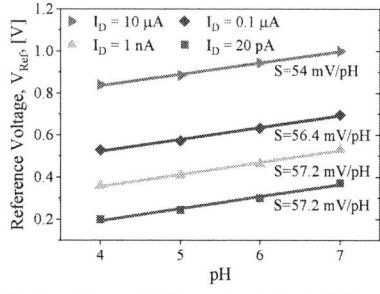

Fig.7. pH sensitivities extracted at different drain current levels with data from Fig. 6. From this figure, we state that the 3D-EMG-ISFET can work at 2 pW power consumtion without degradation in sensitivity.

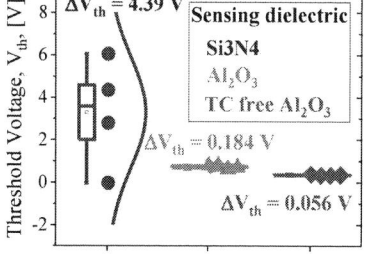

Fig.8. V_{th} spread in pH 7 buffer solution. Before RIE: the Si_3N_4 passivation layer is used as sensing dielectric. After RIE: the Al_2O_3 layer is used for sensing. After wet etch: trapped charge free (TC free) Al_2O_3 is used for sensing.

978-1-7281-1988-5/18 $31.00 © 2018 IEEE   271

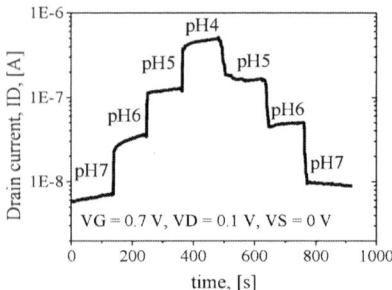

Fig.9. Dynamic measurement of pH in various buffer solutions, showing 3D-EMG-ISFET's response time < 5 s.

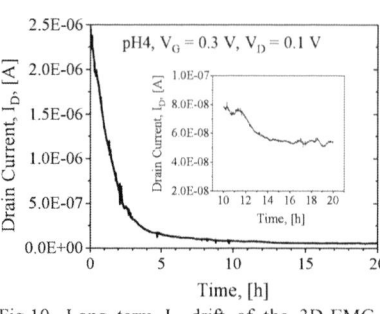

Fig.10. Long term I_D drift of the 3D-EMG-ISFET with Al_2O_3 as sensing layer. Inset: zoom in of the drift plot from 10h to 20h.

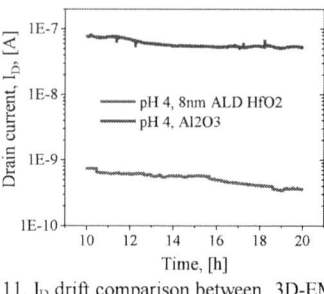

Fig.11. I_D drift comparison between 3D-EMG-ISFETs with Al_2O_3 and ALD layer of HfO_2. Converted drift rate: 0.67 mV/h – Al_2O_3, 2 mV/h – HfO_2.

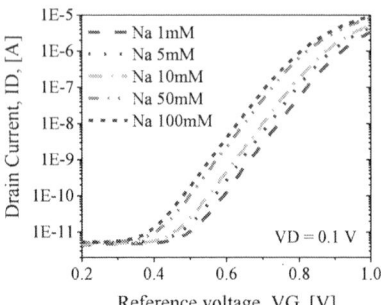

Fig.12. I_DV_G characteristics of the 3D-EMG-Na^+ sensitive FET, in NaCl solutions of various concentrations.

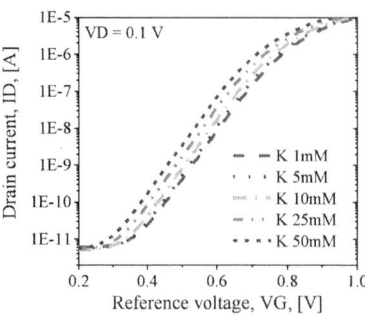

Fig.13. I_DV_G characteristics of the 3D-EMG-K^+ sensitive FET, in KCl solutions of various concentrations.

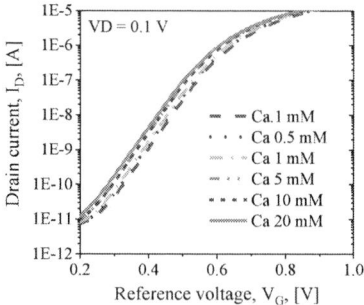

Fig.14. I_DV_G characteristics of the 3D-EMG-Ca^{2+} sensitive FET, in $CaCl_2$ solutions of various concentrations.

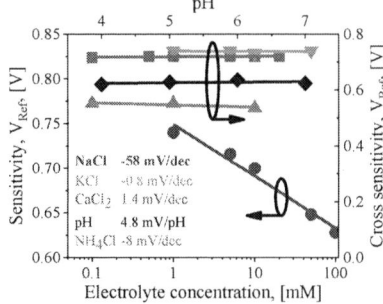

Fig.15. Sensitivity and Cross sensitivity plot for the 3D-EMG-Na^+ sensitive FET, extracted from Fig. 13 at constant I_D = 10 nA.

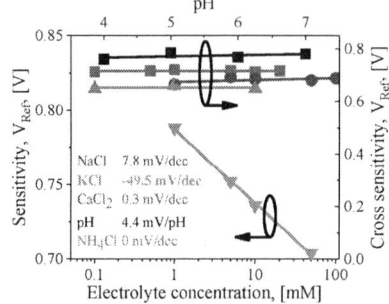

Fig.16. Sensitivity and Cross sensitivity plot for the 3D-EMG-K^+ sensitive FET, extracted from Fig. 14 at constant I_D = 10 nA.

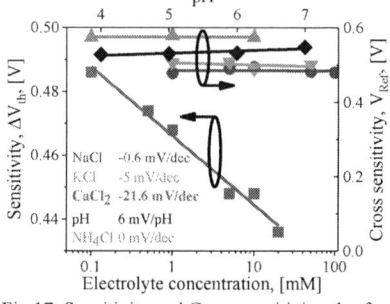

Fig.17. Sensitivity and Cross sensitivity plot for the 3D-EMG-Ca^{2+} sensitive FET, extracted from Fig. 15 at constant I_D = 10 nA.

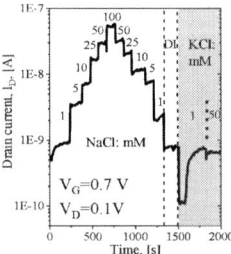

Fig.18. Dynamic measurement of the 3D-EMG-Na^+ sensing FET, in various salt buffer solutions. Showing fast response < 5s, repeatibility and high selectivity.

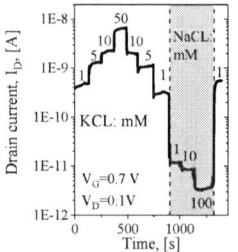

Fig.19. Dynamic measurement of 3D-EMG-K^+ sensing FET, in various salt buffer solutions. Showing fast response, repeatibility and high selectivity.

FOM SoA	Ion	S¹	CS²				min. Power³	RT⁴	DR⁵	CMOS Int⁶
			H	Na	K	Ca				
This work	H	-56		NA	NA	NA	2 pW	<5	0.7	Yes
	Na	-58	-4.8		-0.8	1.4				
	K	-49	-4.4	7.8		0.3				
	Ca	-22	-6	-0.6	-5					
[2]	H	-51		NA	NA	NA	900 nW *	<5	NA	No
[4]	Na	-38	-46 **		NA	NA	1 nW *	<5	NA	No
[3]	K	-64	NA	-8 *		0 *	320 μW *	NA	NA	No
[1]	Ca	-32	NA	NA	NA		NA	<5	0.7	No

Table 1. State-of-the-Art (SoA) in electrochemical ion sensors

1. S: Sensitivity (mV/dec)
2. CS: Cross sensitivity (mV/dec)
3. min. Power: Minimum power consumption per sensor
6: CMOS Int: All integrated in CMOS
* by calculation from published figures

4. RT: Response time (s)
5. DR: Drift rate (mV/h)
** Nullified CS with differential meas.

978-1-7281-1988-5/18 $31.00 © 2018 IEEE

High Resolution Ion Detector (HRID) by 16nm FinFET CMOS Technology

Peng-Chun Liou[1], Tsung-Han Lee[2], Chien-Ping Wang[1], Yu-Lun Chueh[2], Yue-Der Chih[3], Jonathan Chang[3],
Chrong Jung Lin[1], Ya-Chin King[1]

[1]Institute of Electronics Engineering, National Tsing Hua University, Hsinchu, Taiwan
[2]Institute of Materials Science and Engineering, National Tsing Hua University, Hsinchu, Taiwan
[3]Design Technology Division, Taiwan Semiconductor Manufacturing Company, Hsinchu, Taiwan
Phone/Fax: +886-3-5162219/+886-3-5721804, E-mail: ycking@ee.nthu.edu.tw

Abstract— A novel approach for the ion-sensing of electrolyte solution, using specially designed CMOS FinFET process compatible floating-gate (FG) device is proposed. With the self-balancing readout scheme, the floating gate based pH sensor shows a maximum pH readout sensitivity of 115 mV/pH. Through a laterally coupling structure to the metal floating gate of a FinFET, its channel potential can be controlled both by the read gate as well as the sensing gate. In additional, this novel scheme also enables high linearity pH sensing in the target sensing range, readily adjusted by coupling ratios and biasing levels.

I. INTRODUCTION

Field Effect Transistor (FET)-based ion sensors are used in various applications as chemical sensors [1]. With widely accepted working principle of FETs, these sensors are designed to feature high sensitivity, compactness and low costs. FET-based ion sensors typically operate in the sub-threshold regime, where the threshold voltage V_T levels critically affects the corresponding sensing results. V_T depends on the material properties of gate, dielectric layer, and channel of a transistor. The most common FET-based sensors are ion-sensitive FETs (ISFETs) which was introduced in the 1970s by Bergveld [2]. The first ISFETs is without a gate, where the channel regions with the dielectric layer on top was directly exposed to electrolyte solution while sensing. An improved ISFET add a sensing gate to detect to the change of interface charges between the sensing pad and electrolyte solution, as shown in *Figure 1(a)*. To control the channel potential to reach a desirable sensing region, an external reference electrode is used to control its operating point [3]. ISFETs are commercially available as pH meters and are also widely used for genome sequencing and biomolecule sensing in array configurations [4]. ISFETs offer several advantages such as small dimensions, fast response time, compatibility with integrated circuits, and the direct signal amplification [5].

The first ISFET using the FG structure, ion-sensitive floating gate FETs (ISFGFETs) was proposed in [6]. The gate of the ISFET is replaced by a floating node which is capacitively coupled to two inputs: a read gate (RG) and a sense gate (SG) connected to sensing PAD, both of which can be used to modulate the current flow through the channel of transistor, as illustrated in *Figure 1(b)*. Comparing to the conventional ISFETs, the channel potential can be regulated through the read gate rather than the reference electrode. With the separated potential of RG and SG, the pH value can be sensed by modulating the FG potential. In this paper, a laterally coupled ISFGFETs is first time implemented in FinFET technologies, featuring high sensitivity and high linearity through a new self-balancing readout scheme.

II. DEVICE STRUCTURE AND OPERATION PRINCIPLES

The proposed CMOS compatible ISFGFET is an n-channel MOSFET with two input gates that are capacitively coupled to a common FG on the top of the shallow trench isolation (STI) region. Two contact slots on both sides are placed closely to the floating metal gate serves as the read gate (RG), while the other coupling electrode is designed as the sense gate (SG) connected to the sensing pad on top of the passivation layer. *Figure 2(a)* shows the 3D illustration and the 2D layout of the FG ISFET. The cross-sections TEM of the contact coupling structure along the AA' line and the device along the BB' and lines are shown in *Figure 2 (b)* and *(c)*, respectively. *Figure 3(a)* shows the illustration of sensing mechanism. *Figure 3(b)* shows the TEM picture of the sensing device and the corresponding connecting metal layers to the sensing pad.

It is expected that potential difference on SG can be detected by the channel current response when one sweeps the RG voltage (V_{RG}). The measured I-V curve of the FG FinFET with different SG voltage (V_{SG}) are compared in *Figure 4(a)*. As revealed by the measured data, the potential on the sense gate indirectly affect its transfer characteristics. On the other hand, *Figure 4(b)* shows drain current response with sweeping V_{SG} under a fixed V_{RG}. Different coupling ratio α_{SG} and α_{RG} can be readily obtained by controlled the coupling capacitance, C_{SG} and C_{RG}, through changing the contact length of SG and RG, L_1 and L_2, as illustrated in the inset of *Figure 5*. The relation between coupling ratios and contact length of both gates and the measured coupling ratios are summarized in *Figure 5*.

To accommodate the new FG FinFET-based ion sensor, a new self-balancing readout scheme and the corresponding circuit is proposed, as illustrated in *Figure 6*. The circuit layout of the self-balancing readout circuit with FG ISFET is shown in *Figure 7*. Through a feedback network, a constant channel current can be kept at reference levels by a simple 2-stage operation amplifier. As the V_{SG} increases in response to the ion level, V_{RG} must reduce to keep a contact channel current by a self-balancing mechanism. In a close-loop operation, the output voltage V_{OUT} decreases linearly to ensure a constant floating gate voltage. Consequently, the difference of V_{SG} can be reflected by V_{OUT}, proportionally. In addition, the gain and

---

978-1-7281-1988-5/18 $31.00 © 2018 IEEE   273

sensitivity of this sensor can be adjusted by designing the coupling ratios of α_{RG} and α_{SG}, as suggested in *Figure 5*.

The gain and sensing range of the proposed scheme has been investigated by simulation, comprehensively. As shown in *Figure 8*, V_{OUT} decreases linearly with the raising V_{SG} and the circuit gain can be extracted from the slopes of respective curves. One of the unique features of this ion sensor is its adjustable sensing range and sensitivity (gain). As found in the following analysis, different sensing range and sensitivities can be achieved by the designing ratios of α_{SG} and α_{RG}. Increasing α_{SG} to α_{RG} ratio promotes the coupling strength from SG, while reducing that from RG. Thus, the circuit exhibits narrower sensing range with higher voltage gain. In contrast, decreasing α_{SG} and increasing α_{RG} leads to weaker coupling from SG, stronger coupling from RG. Consequently, a wider sensing range with lower sensitivities is observed on the output response. *Figure 9* summarizes the sensing range corresponding to different L_1/L_2 with respect to raising V_{REF}. To emulate the effective on potential change on SG, the response on output node is measured with voltage applied on V_{SG} directly. *Figure 10* shows the measured output voltage characteristics, which I_{REF} is set to *100nA* on the circuit with different α_{SG}/α_{RG} designs. High linearity on the output response can be demonstrated in the measured data in *Figure 10*, where the gain can be extracted by the slopes of the curves. In *Figure 11*, the circuit gain can be effectively increased by raising α_{SG} and reducing α_{RG}, precisely control with the corresponding contact slots length in the coupling structure.

III. MEASUREMENT RESULTS AND DISCUSSION

A testing structure is prepared for sensing the pH level of a solution. The sensing structure consists of the sensing layer and reference electrode (RE) as shown in *Figure 12*. The test liquid drop covers both electrodes, see the picture in *Figure 13*. The cross-sectional view of Au electrode obtained and that of the sensing pad with coating layer obtained by focus ion beam imaging are arranged in *Figure 14 (a)* and *Figure 14(b)*, respectively. A coating layer of 50nm aluminum oxide (Al_2O_3) on Au electrode by electron beam evaporation is deposited on the sensing pad, to absorb H+ ions (for pH sensing). In addition, a closely placed the Au reference electrode in the test structure to provide a reference potential during ion sensing operations [7]. Both the sensing pad and Au reference electrode are placed on the printed circuit board connected to the underline FG ISFET circuit. In pH sensing measurements, *20µL* of pH 3-9 electrolyte solutions were applied on the testing structure. The Au reference electrode on SP is connected to common ground electrode of the circuit.

First, the channel current of a FG ISFET with response to the pH levels of the testing electrolyte are measured. *Figure 15* compares the I_D-V_{RG} curve of the FG ISFET when different solutions of pH values from 3 to 9 are tested. The curves shift left with the decreasing pH value. This is caused by the change in sensing gate potential as the number of H+ ions increase in test solution. Therefore, the smaller V_{RG} is required to achieve the same sensing current. Under constant current condition,

$I_{REF} = 100nA$ set by V_F on the reference transistor, the V_{OUT} responding to the solutions of each pH value can be obtained. *Figure 16* compared the sense current change per pH at fix read gate voltage of 2V. Measured current responses reveals that devices with lower α_{RG} can enhance its sensitivities.

The pH sensitivity of FG ISFET circuit in electrolyte pH solutions are compared in *Figure 17*. At different V_{REF}, the output sensitivities are 115, 103.3, and 71.7 mV/pH. Choosing an appropriate V_{REF}, desirable sensing range can be obtained to meet different application needs. As a result of the floating gate structure, the sensitivity of this work can reach 115 mV/pH without additional amplifying stage. The output characteristics of readout circuit by conventional ISFET and FG ISFET are compared in *Figure 19*. As revealed, the output voltage response is much wider on the circuit with FG ISFET. By adapting a FG ISFET in this ion sensor, multiple ion sensing with sensing pad coated with corresponding material for different ions can be achieved, as illustrated in *Figure 20(a)*. This enable multiple sensing of different ion to be incorporated in one single device, as shown in *Figure 20(b)*. *Table 1* list the performance comparison of the conventional ISFETs with that has been demonstrated in this work. Implemented in advanced FinFET technologies, the proposed ion sensor has shown enhanced sensitivity, flexibility and high compatibility to advance logic circuits.

IV. CONCLUSIONS

A FG ISFET based sensor is proposed and demonstrated in advanced FinFET technologies. High sensitivity and linearity can be obtained by adjusting coupling ratios from the sensing node. Voltage readout can be obtained through a self-balancing scheme for maintaining constant read current in a FG device. Furthermore, sensing range and sensitivity of the circuit can be readily designed by layout and bias design to the targeted range adaptively.

ACKNOWLEDGEMENTS

The authors would like to thank the support from the Ministry of Science and Technology (MOST), Taiwan, and Taiwan Semiconductor Manufacturing Company (TSMC).

REFERENCES

[1] P. Bergveld, Biosensors 2.1 (1986): 15-33.
[2] P. Bergveld, IEEE Transactions on Biomedical Engineering 1 (1970): 70-71.
[3] Janata, Jiri, Chemical Reviews 90.5 (1990): 691-703.
[4] T. Sakurai and Y. Husimi, Anal. Chem., vol. 64, no. 17, pp. 1996–1997, 1992.
[5] Chen, Yu, et al. Applied Physics Letters 91.24 (2007): 243511.
[6] Shen, NY-M., et al. IEEE Transactions on Electron Devices 50.10 (2003): 2171-2178.
[7] Vonau, W., et al. Journal of Solid State Electro-chemistry 10.9 (2006): 746-752.
[8] Kim, Sungho, et al. Nano/Micro Engineered and Molecular Systems,2011 IEEE International Conference.
[9] Lee, Jieun, et al. IEEE Electron Device Letters 33.12 (2012): 1768-1770.
[10] Zeng, Ruixue, et al. Sensors and Actuators B: Chemical 254 (2018): 102-10

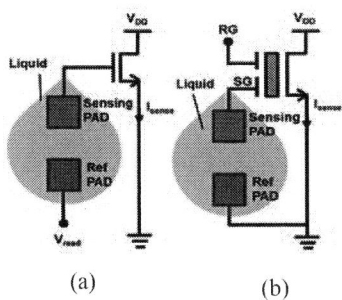

(a)   (b)

Figure 1 Illustration and comparison of (a) a conventional FET-based ion sensor and (b) a FG FinFET ion sensor with indirect coupling to a metal FG from the sensing pad.

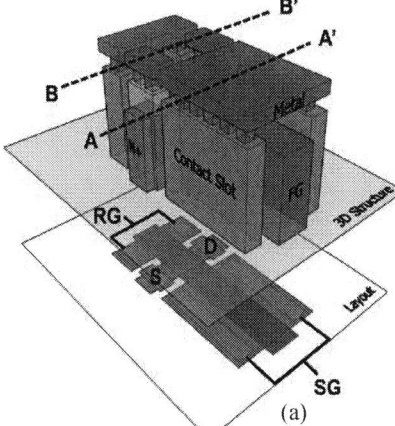

(a)

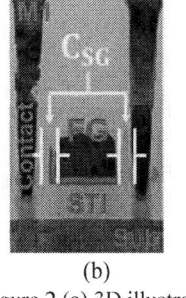

(b)

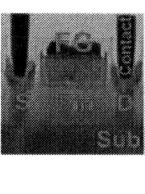

(c)

Figure 2 (a) 3D illustration and FG-ISFET layout of the laterally coupled floating gate structure and (b) the cross-sectional TEM of the sensing gate (SG) along the AA' cross-line, (c) the readout device along the BB' cross-sectional line.

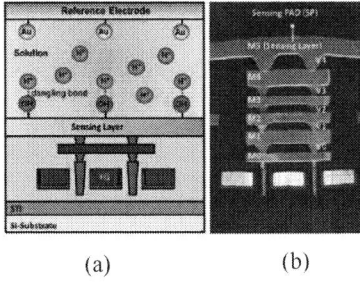

(a)   (b)

Figure 3 (a) Illustration of Sensing mechanism. (b) TEM picture of the sensing device and the corresponding connection to the sensing pad through multiple metal/via layers.

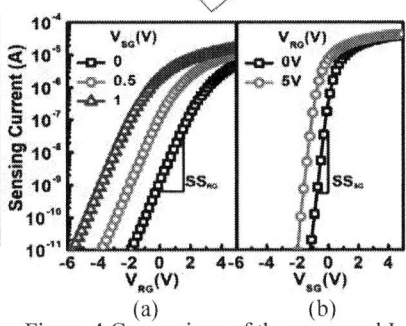

(a)   (b)

Figure 4 Comparison of the measured I_D-V_G characteristics of the floating-gate device with (a) different V_{SG} and (b) different V_{RG}, where V_{DS} is set to 0.1V, where the device are designed with α_{RG}=9% and α_{SG}=25%, respectively.

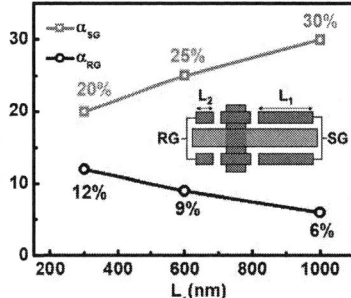

Figure 5 Measured coupling ratios from each input node can be precisely controlled and optimized by adjusting the length of the sense gate, when the read gate length, L_2 is fixed at 150nm.

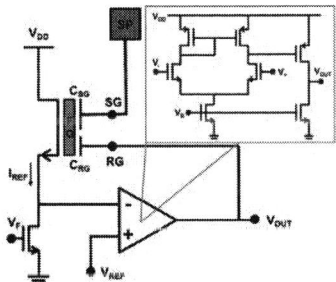

Figure 6 Schematic of proposed readout circuit with a proposed FG ISFET in a feedback loop for high linearity output through the self-balancing technique.

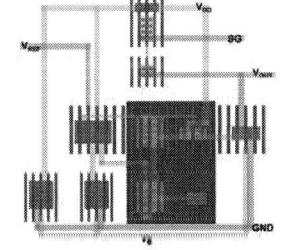

Figure 7 Layout of the proposed readout circuit by a nano-scaled CMOS process with FG ion sensor integrated

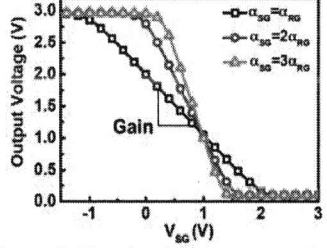

Figure 8 The simulated output voltage V_{OUT} with the different V_{SG}. It is found that the slope of the transfer function can be controlled by coupling ratios.

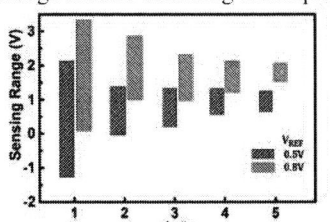

Figure 9 Sensing range corresponds to FG ISFETs with different L_2/L_1 ratio are compared. V_{REF} can be used to shift the desirable sensing range.

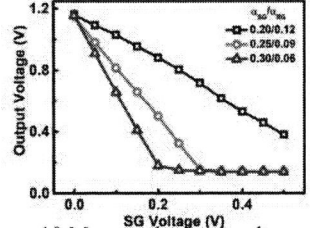

Figure 10 Measured output voltage with respect to different sense gate voltage with three coupling ratio conditions, under a fix reference current of *100nA*.

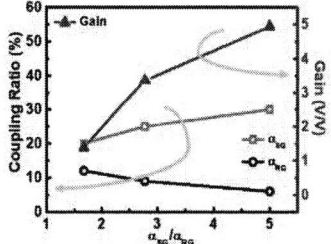

Figure 11 Measured output voltage gain with different coupling ratio for the read gate and sense gate changes.

978-1-7281-1988-5/18 $31.00 © 2018 IEEE   275

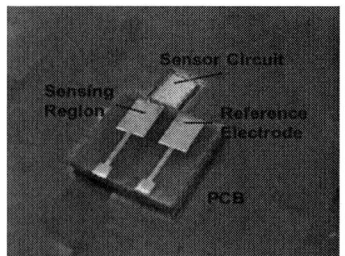

Figure 12 A picture of the test sample prepared with sensor circuits connected to sensing pad for electrolyte pH level testing.

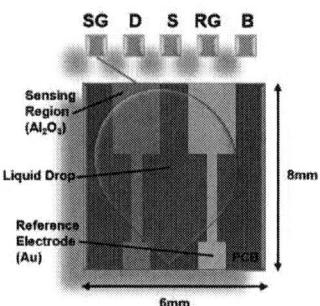

Figure 13 A sensing pad consists of Al_2O_3 coating for ion sensing and a closely place Au reference electrodes under the electrolyte drop of $20\mu L$.

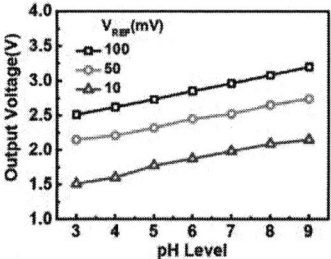

Figure 17 Comparison of output voltage in response to pH value change under different V_{REF} levels, as simulated on the readout circuit with $\alpha_{SG} = 25\%$ and $\alpha_{RG} = 9\%$, respectively.

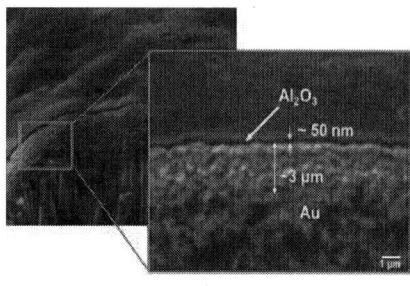

(a)

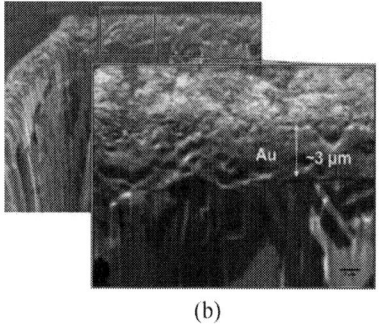

(b)

Figure 14 Cross-sectional view of (a) Al_2O_3 sensing layer deposited on the sensing pad by electron beam evaporation of 50nm thick and (b) an image of the Au reference electrode of $3\mu m$ thick obtained by Focused Ion Beam system.

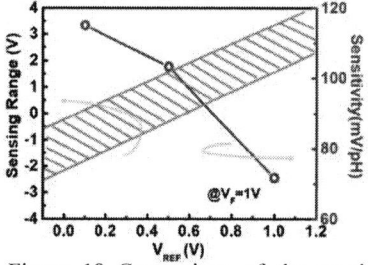

Figure 18 Comparison of the sensing range and the corresponding sensitivities as V_{REF} shifts. Both positive potential and negative potential shifts be detected for different ion detection applications.

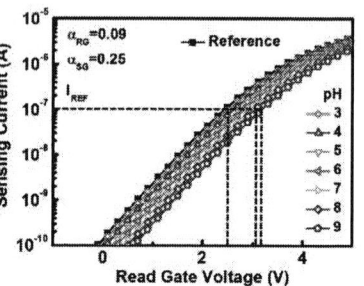

Figure 15 The measured transfer characteristics of an FG ISFET with electrolyte solutions of pH value ranging from 3 to 9 applied on the Al_2O_3 coated Au sensing pad.

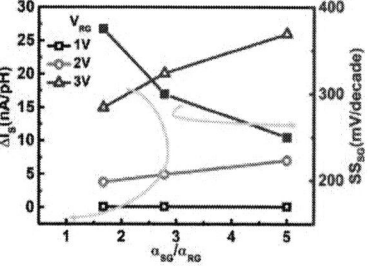

Figure 16 Responses based by current sensing mode, where lowering α_{RG} can effectively enhances the read current sensitivities toward pH level changes.

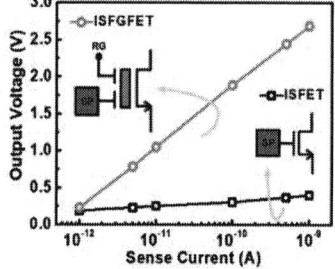

Figure 19 Comparison of the output voltage responses. Data reveals that a much wider response range on FG ISFET can effectively obtained.

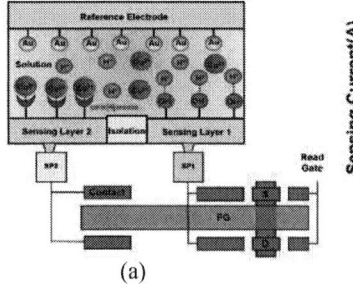

(a)

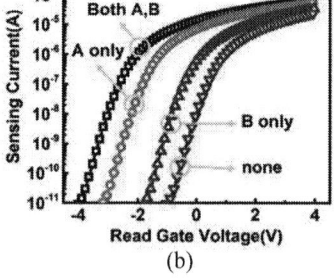

(b)

Figure 20 (a) A illustration of multiple ion sensing capability through adding metal pads with different coatings to the coupling gates to FG ISFET. (b) Measurement results showing the effect of multiple inputs, A, B,.... etc. can all affect the read current response of FG ISFET.

	Ref[8]	Ref[9]	Ref[10]	This work
Type	ISFET	ISFET	EGFET	FG FinFET
Technology	Si-nanowire	Si-nanowire	glass substrate	16nm FinFET
Sensitivity (mV/pH)	30	58.3	57.9	115
PH range	2~10	5~9	4~9	3~9
Multi-ion	No	No	No	Yes

Table 1 Performance comparison of other reported ISFETs with results obtained from this work. Data shows that sensitivity toward pH level can be greatly improved with the new readout and sensing scheme proposed.

978-1-7281-1988-5/18 $31.00 © 2018 IEEE

Highly Performant Integrated pH-Sensor Using the Gate Protection Diode in the BEOL of Industrial FDSOI

G. T. Ayele[1,2,3], S. Monfray[1], S. Ecoffey[3], F. Boeuf[1], J-P. Cloarec[2], D. Drouin[3], and A. Souifi[2]

[1]STMicroelectronics, Crolles, France, email: getenet-tesega.ayele@insa-lyon.fr
[2]INL-Université de Lyon, Lyon, France, [3]LN2, 3IT-Université de Sherbrooke, Sherbrooke, Canada

Abstract—This is the first demonstration of a CMOS pH-sensor using the gate protection diode of standard FDSOI transistors in the BEOL. The extremely steep switching of the drain current induced by an exploitation of the DIBL effect is used for fabrication of extremely sensitive pH-sensors. The back gate voltage at which the abrupt switching of drain current occurs depends on the potential at the gate protection diode. Integrating the pH sensing film on this diode BEOL metal, the shift depends on the pH value of the liquid which creates a proportional potential. The abrupt switching (as small as 9 mV/decade) of the drain current can give a theoretical maximum sensitivity of 6.6 decade of drain current change per unit pH. In this paper, we report an experimental sensitivity of 1.25 decade/pH which is superior to state-of-the-art CMOS pH sensors which have a maximum sensitivity of 0.9 decade/pH.

I. INTRODUCTION

Sensors constitute the backbone of the rapidly growing Internet of Things (IoT): the next technological revolution, forming the interface between the physical world and the computer network. Highly sensitive and small sized CMOS sensors are ideal candidates to meet the fundamental requirements of the IoT such as ultralow power operation, compactness and autonomous sensor networks.

pH value of blood, gastric juice, urine, soil, pharmaceutical inventory, food and beverage indicate health condition of a person and quality of an industrial product. Therefore, pH measurement is highly important in a wide range of industries including health, agriculture, pharmaceutical, water quality and environmental monitoring, food processing, and other bio-chemical applications. In the DNA sequencing industry, the MOSFET based pH sensors are mentioned as one of the three next generation sequencing (NGS) platforms [1].

The widely used glass electrode suffers from several limitations such as bulky size, toxicity, fragility, and CMOS incompatibility. A solid state solution, commonly referred to as ion-sensitive field-effect transistor (ISFET), has been under research and development for nearly five decades. However, ISFETs are still at the stage of infancy towards commercialization due to sensitivity and stability issues [2]. The thermally limited subthreshold slope (59.6 mV/pH) of conventional ISFETs falls short of providing ultrasensitive detection for fixed bias readout circuits.

In this paper, we report for the first time a CMOS pH sensor in the back-end-of-line (BEOL) of industrial FDSOI devices where the sensing is made through the diode BEOL metal. The abrupt switching (as small as 9 mV/decade) of the drain current can provide an extremely high sensitivity of 6.6 decade/pH which is more than 7 times higher than the state-of-the-art CMOS pH sensors [2-4].

II. WORKING PRINCIPLE

This sensor is developed in the BEOL of industrial FDSOI devices where the pH sensing is made through the protection diode of the transistor. While the pH detection is carried out at the P-terminal of the protection diode, the opposite terminal of the diode is tied to the front gate. The schematic diagram of the sensor and the equivalent electrical connections are given on Fig. 1 and Fig. 2 respectively.

The diode-front gate connection provides an NPN junction with the back gate BG (in case of N-well FDSOI devices). When the back gate voltage (V_{BG}) $\ll$ 0, the NPN junction becomes conductive and the back gate bias becomes applied at the front gate of the FDSOI NMOS transistor. Such negative biases both at the front gate and at the back gate turn off the FDSOI (no drain current).

At small negative voltages and for positive biases at the back gate, the NPN junction gets turned off. In this condition, the front gate of the FDSOI becomes floating and if the drain voltage (V_D) increases, the FDSOI transistor becomes turned on very abruptly due to the drain induced barrier lowering (DIBL) effect in very short devices. DIBL is a short-channel effect referring to a number of phenomena encompassing a simple change of threshold voltage at higher drain voltages, change of the subthreshold slope, and entire failure of the gate to turn the device off at extremely short gate lengths [5-6]. The profile of the potential at the different regions of the FDSOI and the NPN structure is shown on Fig. 3.

By changing slightly the potential at P area of the NPN junction, the entire I_D-V_{BG} curve shifts. As the I_D-V_{BG} curve is extremely steep, any small change of the potential at the diode P-area leads to large variation in the drain current providing an extremely sensitive detection. This effect will be used to sense small pH variation on an electrode connected to this P-area.

At the electrolyte-dielectric interface of the pH sensing film, a pH dependent surface potential builds up [3, 7], resulting in a

978-1-7281-1988-5/18 $31.00 © 2018 IEEE

proportional potential at the BEOL metal of the diode (P zone of the NPN junction). Consequently, integrating the pH sensing film on the BEOL metal of the diode, the back gate voltage at which the drain current switches depends on the pH value of the electrolyte. By changing the pH value, the potential at P area of the NPN junction changes and the entire I_D-V_{BG} curve shifts. Fig. 4 shows the site binding model of the surface charging mechanism.

III. FABRICATION AND CHARACTERIZATION

Fabrication of the sensor is carried out in the BEOL of FDSOI devices fabricated by STMicroelectronics. Fig. 5 shows SEM cross sectional image of these transistors. 36 sensors were developed per die where the electrical contacts were extended away from the liquid area, and different sizes of sensing areas are designed. The outline of the sensors on the die is shown on Fig. 7. Top-down and cross sectional SEM images of the sensor are provided on Fig. 8 and Fig. 9 respectively.

For extension of the electrical contacts, passivation of the die surface is done first. The photosensitive polyimide, HD-4104 is coated, patterned and cured for passivation. Ti/Al metal is deposited by sputtering, patterned by photolithography and etched for the metal extension of source, drain, and back gate electrical contacts. A BEOL process with Cu metallization and associated barriers can replace the Ti/Al stack used here for the demonstration without affecting the sensor performance. The aluminum oxide pH-sensing film is deposited by PVD, and finally encapsulation is made. The process flow for the fabrication of the sensor in the BEOL of the above described transistors is presented on Fig. 6.

The characterization is carried out using a Keithely semiconductor characterization instrument. Fig. 10 shows a picture of the sensor with the electrical connection made in the probe station. Before starting the pH-sensing characterization, the samples were cleaned with acetone-isopropanol-deionized water and dried with nitrogen. The pH solutions are then dispensed on the sensing area, and I_D-V_{BG} sweeps are taken at different pH solutions. The dispensing and removal of the pH solutions are made with micropipettes, where the volume of the measured pH solution is 10 µl.

IV. RESULT AND DISCUSSION

We characterized the switching behavior of the FDSOI transistors as a function of device geometry and at increasing values of drain voltages. Since DIBL due to the floating gate is a short channel effect, the behavior of abrupt switching induced by the connected diode is much more significant in shorter gate lengths. This measurement result is presented on Fig. 11. In addition, the DIBL effect increases with increasing drain voltage when the gate is floating as shown on Fig. 12.

For the sensing application, we measured the shift of the drain current switching at different biases of the diode. When the diode bias is changed from -0.1 V to 0 and 0.1 V, the back gate voltage at which the drain current switches from OFF to ON state changes from -0.52 V to -0.42 V and -0.32 V respectively (Fig. 13). This feature is exploited for the sensing as the external bias applied to the diode can be substituted by a pH dependent potential that develops at the surface of the pH sensing film.

For ultrahigh resolution pH monitoring, and for detection of trace amount of charged molecules, this shift in back gate voltage becomes very small so that the change in drain current stays on the extremely steep "subthreshold" region. The switching of drain current at 9 mV/decade can give a maximum sensitivity of 6.6 decade/pH corresponding to a Nernst responsive sensing film. This is more than 7 times higher sensitivity than that of state-of-the-art CMOS pH sensors.

Fig. 14 illustrates the pH sensing result using (W x L= 170 nm x 100 nm) devices. The lower pH solutions result in more positive potentials shifting the I-V curve to the right. In addition to the theoretically expected shift, very good linearity is obtained in wide pH range (pH 1.68 to pH 12.46) as shown on Fig. 15.

Increasing the drain bias from 0.1 V to 0.6 V, steeper drain current switching is obtained which is also in confirmation with theory. The pH sensing result at drain voltage of 0.6 V is given on Fig. 16 demonstrating higher sensitivity for fixed bias readout circuits.

We also characterized pH sensors developed on very short gate length FDSOI transistors. On Fig. 17 and Fig. 18, we present pH sensing result of sensors based on 20 nm gate length devices. This demonstrates very high sensitivity (1.25 decade/pH) which validates our proposed approach and is superior to state-of-the-art performance.

The selectivity of the sensor depends on the sensing film. We chose aluminum oxide for the pH sensing film as it shows very good selectivity [8].

V. CONCLUSION

We fabricated and tested a novel CMOS pH sensor exploiting the DIBL effect on a floating gate controlled by an NPN junction connected to the sensing area made in the BEOL metal of the FDSOI technology. Exploiting the extremely abrupt current switching, an ultrahigh sensitive pH sensor is demonstrated. Corresponding to a Nernstian responsive pH sensing film, sensitivities as high as 6.6 decade/pH is achievable which is more than 7-times better than state-of-the-art solid state pH sensors. As a first demonstration, we report an experimental sensitivity of 1.25 decade/pH.

ACKNOWLEDGMENT

Région Auvergne Rhône-Alpes is acknowledged for a Coopera project funding.

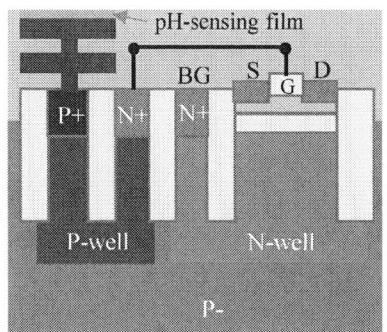

Fig. 1. Schematics of the pH sensor using the protection diode in BEOL.

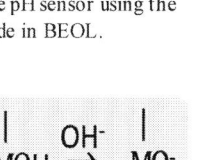

Fig. 4. Site binding model of surface charging for the pH sensing (M is for metal).

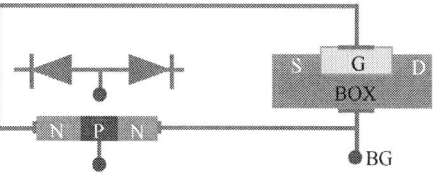

Fig. 2. Equivalent electrical schematics of the FDSOI and the protection diode. pH sensing area in the BEOL is connected to P-Area of the diode.

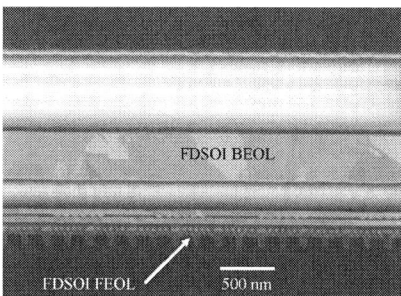

Fig. 5. SEM cross sectional image of the FDSOI devices.

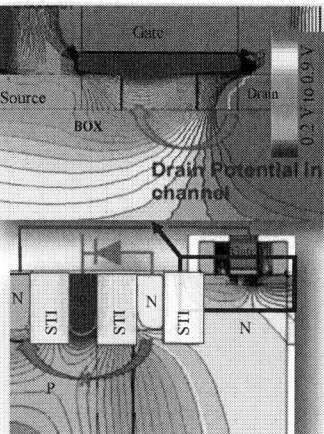

Fig. 3. Voltage profile in the device illustrating the short channel effect.

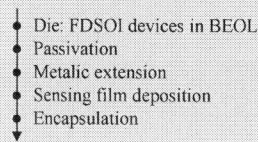

Fig. 6. Process flow for developing the pH sensor in the BEOL.

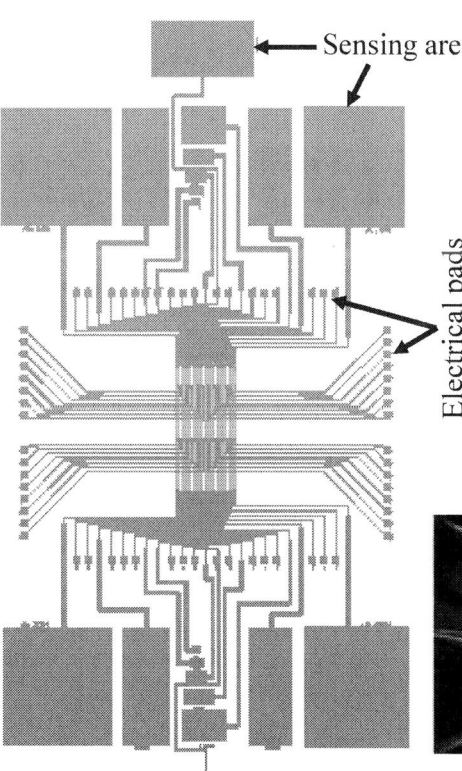

Fig. 7. Top-down layout of the sensors showing sensing areas and electrical connection pads of 36 pH sensors on a die.

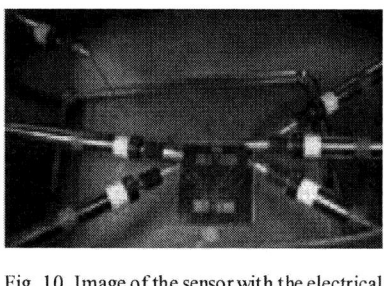

Fig. 8. SEM top-down image of the sensor indicating sensing areas.

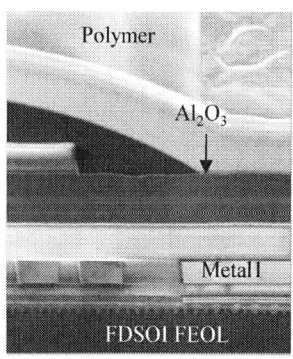

Fig. 9. SEM cross sectional image of the sensor.

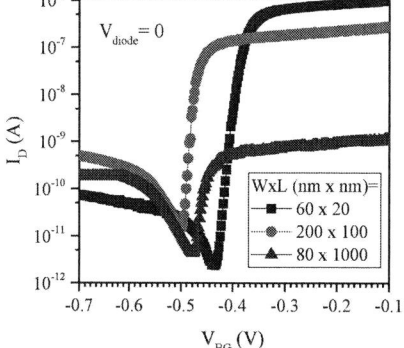

Fig. 10. Image of the sensor with the electrical connection made in a probe station.

Fig. 11. Measured drain current for different gate length.

978-1-7281-1988-5/18 $31.00 © 2018 IEEE

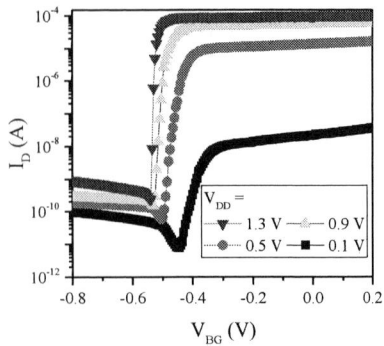

Fig. 12. Measured drain current at different drain voltages.

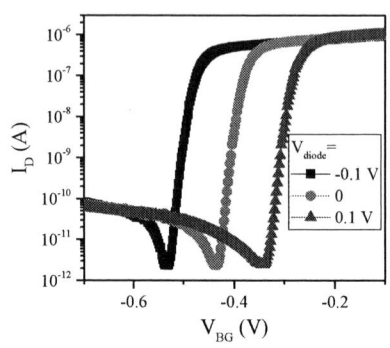

Fig. 13. Measured drain current at different diode biases.

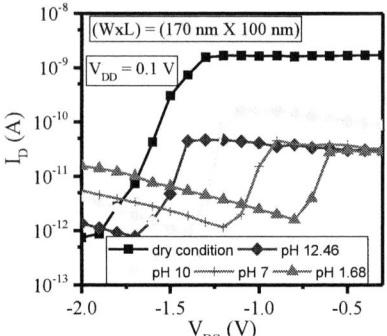

Fig. 14. pH response of sensor with (170 nm x 100 nm) geometry and non-elevated drain voltage.

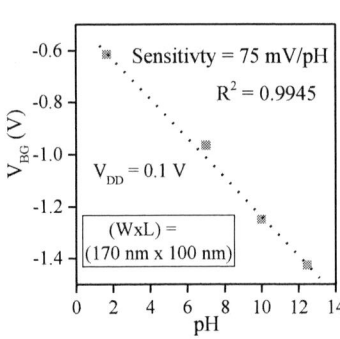

Fig. 15. Sensitivity of (170 nm x 100 nm) sensors at fixed current (I_D= 12 pA).

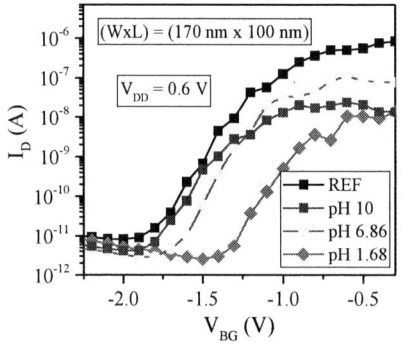

Fig. 16. pH response of sensors with (170 nm x 100 nm) geometry and higher drain voltage.

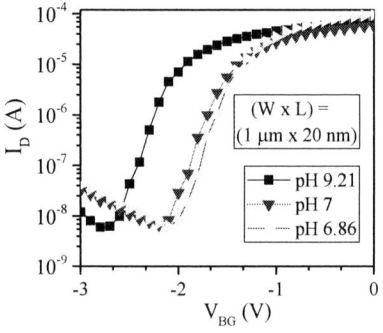

Fig. 17. pH response of sensors with very short gate length (1 μm x 20 nm).

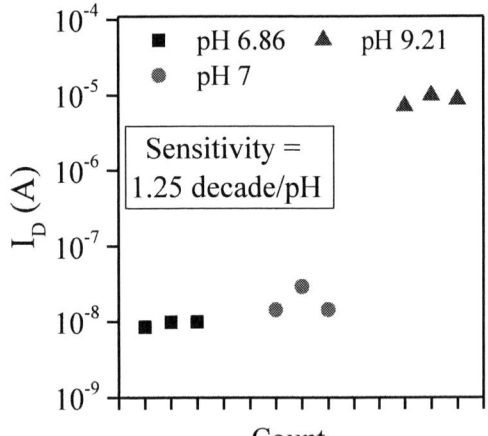

Fig. 18. Sensitivity of very short gate length (1 μm x 20 nm) sensors at fixed bias (V_{BG}= -2 V).

REFERENCES

[1] Quail, Michael A., et al. "A tale of three next generation sequencing platforms: comparison of Ion Torrent, Pacific Biosciences and Illumina MiSeq sequencers." BMC genomics 13.1 (2012): 341.

[2] G. T. Ayele et al., "Ultrahigh-Sensitive and CMOS Compatible ISFET Developed in BEOL of Industrial UTBB FDSOI," in 2018 Symposia on VLSI Technology and Circuits, 2018.

[3] G. T. Ayele et al., "Development of ultrasensitive extended-gate Ion-sensitive-field-effect-transistor based on industrial UTBB FDSOI transistor," in 2017 47th European Solid-State Device Research Conference (ESSDERC), 2017, pp. 264–267.

[4] Huang, Y-J., et al., "High performance dual-gate ISFET with non-ideal effect reduction schemes in a SOI-CMOS bioelectrical SoC." Electron Devices Meeting (IEDM), 2015 IEEE International. IEEE, 2015.

[5] S. G. Chamberlain and S. Ramanan, "Drain-induced barrier-lowering analysis in VSLI MOSFET devices using two-dimensional numerical simulations," IEEE Trans. Electron Devices, vol. 33, no. 11, pp. 1745–1753, Nov. 1986.

[6] N. Arora, Mosfet Modeling for VLSI Simulation: Theory and Practice. World Scientific, 2007.

[7] D. E. Yates, S. Levine, and T. W. Healy, "Site-binding model of the electrical double layer at the oxide/water interface," Journal of the Chemical Society, Faraday Transactions 1: Physical Chemistry in Condensed Phases 70 (1974): 1807-1818.

[8] H. Abe, M. Esashi, and T. Matsuo, "ISFET's using inorganic gate thin films," IEEE Trans. Electron Devices, vol. 26, no. 12, pp. 1939–1944, Dec. 1979.

978-1-7281-1988-5/18 $31.00 © 2018 IEEE

Very Large Scale Integration Optomechanics: a cure for loneliness of NEMS resonators ?

Maxime Hermouet[1], Marc Sansa[1], Martial Defoort[1], Louise Banniard[1], Sergio Dominguez-Medina[2,3], Shawn Fostner[1], Ujwol Palanchoke[1], Alexandre Fafin[1], Marc Gely[1], Louis Hutin[1], Christophe Plantier[1], Emmanuel Rolland[1], Claude Tabone[1], Giulia Usai[1], Thomas Ernst[1], Patrick Villard[1], Gérard Billiot[1], Paul Mattei[1], Guillaume Nonglaton[1], Caroline Fontelaye[1], Charlie Barrois[1], Olivier Castany[1], Eduardo Gil Santos[4], Pierre E. Allain[4], Emeline Vernhes[5], Pascale Boulanger[5], Ariel Brenac[6], Christophe Masselon[2,3], Ivan Favero[4], Thomas Alava[1], Guillaume Jourdan[1] and Sébastien Hentz[1]*

[1]Univ. Grenoble Alpes, CEA LETI, 38000 Grenoble, France, email: sebastien.hentz@cea.fr
[2]CEA, BIG, Biologie à Grande Echelle, F-38054 Grenoble, France
[3]Inserm, Unité 1038, F-38054 Grenoble, France
[4]Université Paris Diderot, CNRS, 75013 Paris, France
[5] Institute for Integrative Biology of the Cell (I2BC), CEA, CNRS, Univ Paris-Sud, Université Paris-Saclay, 91198, Gif sur Yvette cedex, France
[6] Univ. Grenoble Alpes, CEA, CNRS, Grenoble INP, INAC-Spintec, 38000 Grenoble

Abstract—The first Very Large Scale Integration process with variable shape beam lithography for optomechanical devices is presented. State of the art performance was obtained with silicon microdisk resonators showing 1 million optical quality factors and 10^{-17} m.Hz$^{(-1/2)}$ displacement resolution. Single-particle mass spectrometry could be performed with these optomechanical resonators in vacuum. The devices retained high performance when directly immersed in liquid media, allowing for biosensing experiments. These results open the door to large, dense arrays of optomechanical sensors.

I. INTRODUCTION

After two decades of pioneering work, Nano Electro Mechanical Systems (NEMS) are only starting to fulfil (some) of their huge promises, in particular for sensing. A few start-up companies have been created in the last few years, but NEMS are still far from the industrial success of their micro-counterparts. Among others, one reason is the increasing difficulty to interface the "real-world" quantities to sense with the extremely small size of nanomechanical resonators. This is an acute issue for chemical, biological and single-particle mass sensing [1]–[5]. Commercial applications may require the use of large arrays comprising from 10's to 10000's resonators. In the last decade, the NEMS group at LETI have been developing nanoresonators for a number of applications with Very Large Scale Integration processes. Much effort has been dedicated to the development of electrical transduction at the nanoscale to reach state of the art performance with single-crystal silicon resonators adapted to VLSI processes, in terms of signal to background ratio, signal to noise ratio, frequency stability and mass resolution [6], [7]. The real strength of VLSI though, is the possibility to process a large number of devices operating in sync with great reproducibility and control.

II. ARRAYS OF NANORESONATORS

We first demonstrated arrays of NEMS comprising a few 1000's resonators collectively addressed, with thermomechanical actuation and metallic piezoresistive transduction. Collective addressing is well suited for chemical analysis where target species cover the whole array area homogenously. With such devices, we achieved ppb concentration detection of a chemical warfare simulant [2].

The case of single-particle mass spectrometry (MS) is more complex, as particles may land at any position on every resonator at any time. This requires the real-time monitoring of at least two resonance modes simultaneously, for each resonator within the array. We proposed arrays of silicon resonators with semiconductor piezoresistive transduction comprising a few 10's of resonators. The inputs/outputs of all resonators are connected to each other, while each resonator has a distinct length, hence a distinct resonance frequency. Each of them can then be frequency-addressed sequentially in time. We demonstrated mass spectrometry of metallic nanoparticles with more than an order improvement in analysis time [8]. We also recently performed the first measurements of viruses with NEMS-MS, and the highest molecular mass ever measured [9]. Using such arrays, these experiments could be performed in a few hours instead of a few weeks. Improving further detection efficiency or analysis time requires a much larger number of resonators. Co-integrating NEMS with a CMOS circuitry is an elegant solution, and we have investigated such schemes for more than a decade [10]. This led to the recent demonstration of a dense array of 1024 NEMS co-integrated with 130nm CMOS circuit and fabricated with 3D sequential monolithic co-integration (based on the Coolcube™ process [11]).

III. VLSI OPTOMECHANICS

Reading out arrays of resonators with electrical transduction requires solving highly complex issues. Silicon photonics technologies on the other hand, have been developed specifically for signal transmission with staggering data rate and are quickly emerging. Moreover, on-chip optomechanics has been a privileged route towards fundamental studies in the last 15 years and it has recently become mature enough to reach a more applied realm, like

978-1-7281-1988-5/18 $31.00 © 2018 IEEE

chemical sensing [12]. The possibility to use standard optical multiplexing schemes [13] opens the way towards fast readout of very large and dense arrays. Optomechanical systems' performance depends both on the coupling between the optical and the mechanical resonators and on the optical cavity, which needs to be finely optimized. We showed that a two optical modes model allows the unambiguous identification of the losses and the optical coupling parameters, with good agreement between experimental data and theoretical expectations [14]. On top of this model, we have developed a VLSI-compatible process to optimize the performance of optical cavities like silicon optomechanical microdisks. The devices were the first optomechanical resonators fabricated on 200mm SOI wafers with variable shape beam lithography, allowing for both high fabrication throughput and high patterning resolution: each wafer contained more than 120,000 optomechanical devices. Most measured loaded quality factors were in the high 100,000s, and many resonances showed quality factors above 1 million [15]. These figures are among the best measured in the literature with silicon disks of a few μm radius [16]. We were able to resolve the thermomechanical motion of our resonators at a few 100 MHz frequencies at ambient pressure without the need for an optical amplifier [15], demonstrating motional sensitivity down to 10^{-17} m.Hz$^{(-1/2)}$. Demanding applications could thus be addressed: using an optomechanical transduction, the motion of a plate-like mechanical resonator coupled to a distinct optical cavity could be transduced with high sensitivity. We have demonstrated single particle mass spectrometry in real time with such devices, with a large gain in capture cross-section. The second application is biosensing in liquid. A sensitive resonator for mass sensing directly immersed in liquid has eluded research for many years. Following a recent experiment [17], we first showed that our silicon VLSI microdisks retained high optical performance in liquid, as well as exquisite motional sensitivity [18]. Next, we performed the first biosensing demonstration with optomechanical resonators directly immersed in liquid. These demonstrations, along with our first multiplexed optomechanical resonators, show that VLSI optomechanics is a viable route for sensing.

ACKNOWLEDGMENT

The authors acknowledge support from the LETI Carnot Institute NEMS-MS and Oppladiag projects, from the DGA Astrid NEMS-MS project, and from the European Union through the ERC Enlightened project (616251) and the Marie-Curie Eurotalents incoming fellowships.

REFERENCES

[1] S. Guillon, S. Salomon, F. Seichepine, D. Dezest, F. Mathieu, A. Bouchier, L. Mazenq, C. Thibault, C. Vieu, T. Leïchlé, and L. Nicu, "Biological functionalization of massively parallel arrays of nanocantilevers using microcontact printing," *Sensors Actuators B Chem.*, vol. 161, no. 1, pp. 1135–1138, 2012.

[2] I. Bargatin, E. B. Myers, J. S. Aldridge, C. Marcoux, P. Brianceau, L. Duraffourg, E. Colinet, S. Hentz, P. Andreucci, and M. L. Roukes, "Large-scale integration of nanoelectromechanical systems for gas sensing applications," *Nano Lett.*, vol. 12, no. 3, pp. 1269–74, 2012.

[3] M. S. Hanay, S. Kelber, A. K. Naik, D. Chi, S. Hentz, E. C. Bullard, E. Colinet, L. Duraffourg, and M. L. Roukes, "Single-

protein nanomechanical mass spectrometry in real time.," *Nat. Nanotechnol.*, vol. 7, no. 9, pp. 602–8, 2012.

[4] E. Sage, A. Brenac, T. Alava, R. Morel, C. Dupré, M. S. Hanay, M. L. Roukes, L. Duraffourg, C. Masselon, and S. Hentz, "Neutral particle mass spectrometry with nanomechanical systems," *Nat. Commun.*, vol. 6, p. 6482, 2015.

[5] O. Malvar, J. J. Ruz, P. M. Kosaka, C. M. Dominguez, E. Gil-Santos, M. Calleja, and J. Tamayo, "Mass and stiffness spectrometry of nanoparticles and whole intact bacteria by multimode nanomechanical resonators," *Nat. Commun.*, vol. 7, p. 13452, 2016.

[6] E. Mile, G. Jourdan, I. Bargatin, S. Labarthe, C. Marcoux, P. Andreucci, S. Hentz, C. Kharrat, E. Colinet, and L. Duraffourg, "In-plane nanoelectromechanical resonators based on silicon nanowire piezoresistive detection.," *Nanotechnology*, vol. 21, no. 16, p. 165504, 2010.

[7] M. Sansa, E. Sage, E. C. Bullard, M. Gély, T. Alava, E. Colinet, A. K. Naik, L. G. Villanueva, L. Duraffourg, M. L. Roukes, G. Jourdan, and S. Hentz, "Frequency fluctuations in silicon nanoresonators," *Nat. Nanotechnol.*, vol. 11, no. June, pp. 552–559, 2016.

[8] E. Sage, M. Sansa, S. Fostner, M. Defoort, M. Gély, A. K. Naik, E. Colinet, C. Masselon, A. Brenac, and S. Hentz, "Single-particle Mass Spectrometry with arrays of frequency-addressed nanomechanical resonators," *Nat. Commun.*, p. to appear, 2018.

[9] S. Dominguez-Medina, S. Fostner, M. Defoort, M. Sansa, A.-K. Stark, M. A. Halim, E. Vernhes, M. Gely, G. Jourdan, T. Alava, P. Boulanger, C. Masselon, and S. Hentz, "Charge-independent mass spectrometry of single virus capsids above 100MDa with nanomechanical resonators," *arXiv:1804.02340 [physics.app-ph]*, pp. 1–15, 2018.

[10] J. Arcamone, J. Philippe, G. Arndt, C. Dupré, M. Savoye, S. Hentz, L. Duraffourg, and E. Ollier, "Nanosystems monolithically integrated with CMOS : emerging applications and technologies," in *IEEE International Electron Devices Meeting (IEDM)*, 2014, pp. 550–553.

[11] P. Batude, T. Ernst, J. Arcamone, G. Arndt, P. Coudrain, and P. Gaillardon, "3-D Sequential Integration : A Key Enabling Technology for Heterogeneous Co-Integration of New Function With CMOS," *IEEE J. Emerg. Sel. Top. CIRCUITS Syst.*, vol. 2, no. 4, pp. 714–722, 2012.

[12] A. Venkatasubramanian, V. T. K. Sauer, S. K. Roy, M. Xia, D. S. Wishart, and W. K. Hiebert, "Nano-Optomechanical Systems for Gas Chromatography," *Nano Lett.*, 2016.

[13] V. T. K. Sauer, Z. Diao, M. R. Freeman, and W. K. Hiebert, "Wavelength-division multiplexing of nano-optomechanical doubly clamped beam systems," *Opt. Lett.*, vol. 40, no. 9, pp. 1948–1951, 2015.

[14] L. Banniard, M. Hermouet, M. Sansa, A. Fafin, M. Gely, P. E. Allain, I. Favero, S. Hentz, and G. Jourdan, "Intermodal coupling in an optomechanical system," in *MME 28th Micromechanics and Microsystems Europe workshop*, 2017.

[15] M. Hermouet, L. Banniard, M. Sansa, A. Fafin, M. Gely, S. Pauliac, P. Brianceau, J. Dallery, P. E. Allain, E. G. Santos, I. Favero, T. Alava, G. Jourdan, and S. Hentz, "1 Million-Q Optomechanical Microdisk Resonators with Very Large Scale Integration," in *Eurosensors*, 2017.

[16] M. Borselli, T. Johnson, and O. Painter, "Beyond the Rayleigh scattering limit in high-Q silicon microdisks: theory and experiment.," *Opt. Express*, vol. 13, no. 5, pp. 1515–30, Mar. 2005.

[17] E. Gil-Santos, C. Baker, D. T. Nguyen, W. Hease, C. Gomez, A. Lemaître, S. Ducci, G. Leo, and I. Favero, "High-frequency nano-optomechanical disk resonators in liquids," *Nat. Nanotechnol.*, vol. 10, no. 9, pp. 810–816, 2015.

[18] M. Hermouet, M. Sansa, L. Banniard, A. Fafin, M. Gely, P. E. Allain, E. Gil-Santos, I. Favero, T. Alava, G. Jourdan, and S. Hentz, "Ultra sensitive optomechanical microdisk resonators operating in liquid with very large scale integration process," in *Micro Electro Mechanical Systems (MEMS), IEEE 31th International Conference on. IEEE*, 2018.

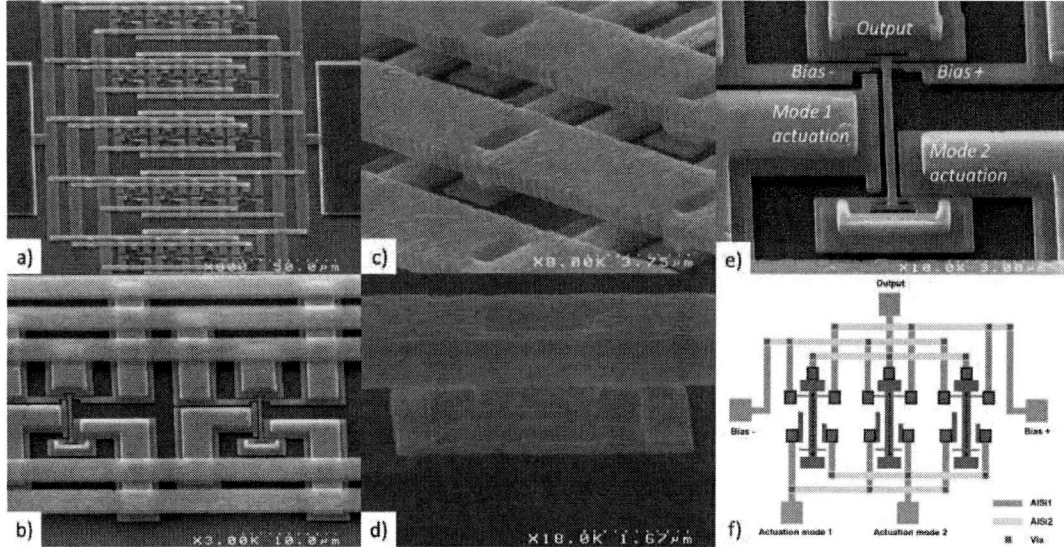

Fig.1. SEM images of NEMS array used for nanomechanical mass spectrometry. a) General view of the array b) zoom on two resonators. c) and d) zoom-in on interconnects and via. e) Typical doubly clamped in-plane resonator. In-plane motion transduction is performed using piezoresistive nanogauges in a bridge configuration to allow background cancellation. Electrodes are specifically patterned for efficient mode 1 and mode 2 actuation. f) Schematic of the interconnect layout. Each resonator has a unique beam length, hence a unique resonance frequency.

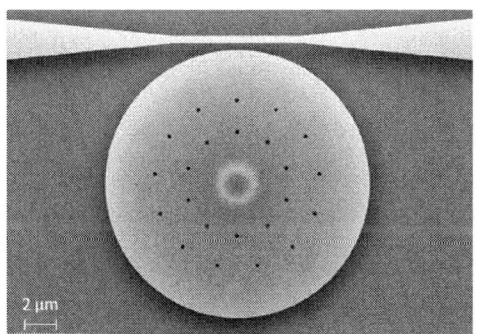

Fig.2. Scanning electron microscope image of a Si optomechanical disk and its waveguide. The black circles are release holes.

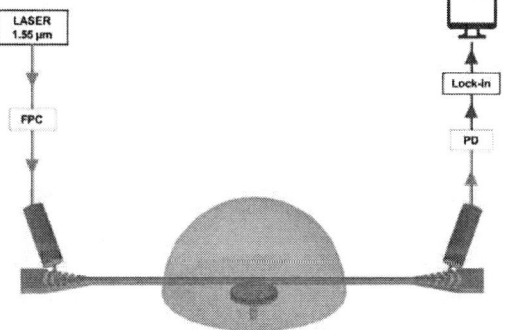

Fig.3. Optomechanical measurement set-up. FPC stands for fiber polarization controller and PD for photodetector.

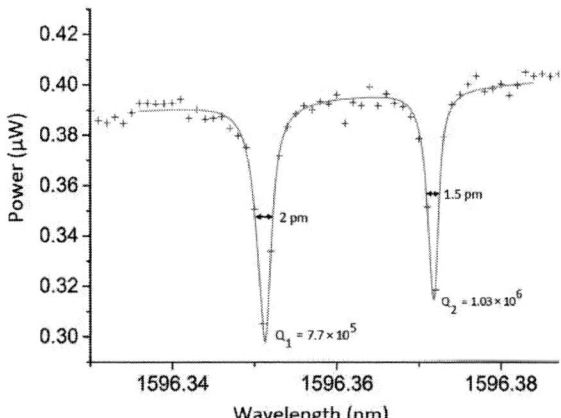

Fig. 4. Narrowband transmission spectrum of a disk with optical quality factors up to 1 million.

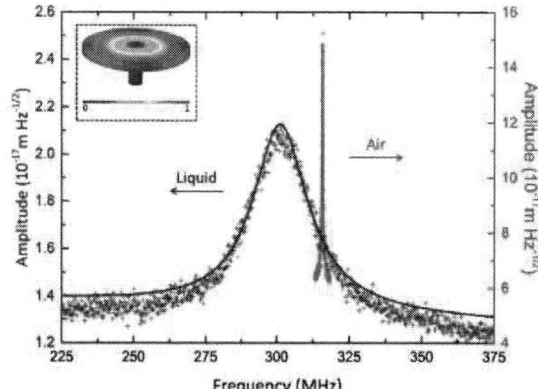

Fig. 5. Brownian noise spectra of the microdisk vibrating in air (red) and deionized water (blue).

SiC-FET-type NOx Sensor
for High-Temperature Exhaust Gas

Y. Sasago[1], H. Nakamura[1], T. Odaka[1], A. Isobe[1], S. Komatsu[1], Y. Nakamura[1], T. Yamawaki[1],
C. Yorita[1], N. Ushifusa[1], K. Yoshikawa[1], K. Ono[1], Y. Anzai[1], S. Machida[1], M. Kinoshita[1],
K. Fujisaki[1], T. Usagawa[1], K. Okishiro[2], and Y. Sugiyama[3]

[1]R&D Group, Hitachi, Ltd., 1-280 Higashi-Koigakubo, Kokubunji, Tokyo 185–8601, JAPAN
[2]Global Research & Innovative Technology Center, Hitachi Metals, Ltd., 5200 Mikajiri, Kumagaya, Saitama, 360-8577 JAPAN
[3]Cable Materials Company, Hitachi Metals, Ltd., 880 Isagozawa-cho, Hitachi, Ibaraki 319–1418, JAPAN
E-mail: yoshitaka.sasago.tx@hitachi.com

Abstract— We have developed a SiC-FET-type gas sensor that enables highly sensitive NO detection in high-temperature exhaust gas. The gate of the FET is a gas detection layer consisting of yttria-stabilized zirconia, nickel oxide, and platinum, which are deposited on the SiC substrate. The threshold voltage of the FET depends on the NO concentration. Experimental results demonstrate that the FET-type sensor can detect NO concentration less than 1 ppm, thus meeting the specifications required to satisfy the strict regulations for exhaust gas in the next generation.

I. INTRODUCTION

Recently, global awareness of environmental conservation has been expanding, and automobile technologies accordingly have to be improved. Electrification is one of the most promising ways to reduce the impact on the environment, but several issues need to be resolved before all automobiles can be completely electrified.

It is well known that the diesel engine has good fuel economy and a small carbon footprint per mileage. However, lean burn at high temperatures produces toxic nitric oxide (NOx), which pollutes the atmosphere and has a harmful effect on human health. Diesel-engine cars process NOx by catalytic reaction in a purification device, transforming NOx in exhaust gas into nontoxic components. The NOx sensor is a key part to control the purification device. As the regulations become more stringent, more accurate NOx sensing, which can detect NOx less than several ppm, will be required.

In diesel-engine cars, limiting-current-type ceramic sensors [1] have been utilized for NOx sensing. Although this type of sensor can detect NOx in a wide range of concentrations, it is difficult to detect in concentrations less than several ppm because the sensing current becomes very small. We have therefore developed a novel SiC-FET-type NOx sensor that has the potential to enable smaller size and lower cost compared to the conventional sensors [1] by taking advantage of the wafer processing of semiconductor products. The biggest issues for the new NOx sensor are ensuring high sensitivity and high heat resistance. We have resolved these issues by merging the chemical sensor with semiconductor technologies.

II. DEVICE STRUCTURE AND OPERATION PRINCIPLE

In a diesel-engine car, two NOx sensors are attached at the downstream side of the engine: one at the front part of the purification device and the other at the back end (Fig. 1).

The SiC-FET-type NOx sensor chip includes a sensor FET and a reference FET, as shown in Fig. 2(a). The chip size is 2×2 mm^2. Both FETs have a gate layer stack of yttria-stabilized zirconia (YSZ), nickel oxide (NiO), and platinum (Pt) on a gate oxide layer (Fig. 2(b)). The composition of YSZ in our FET-type sensor is $(ZrO_2)_{0.97}(Y_2O_3)_{0.03}$. The sensing of NOx is possible at high temperatures above around 500°C. In order to increase and stabilize the sensor temperature, the heater should be equipped nearby the sensor. The heater can be included in a module along with the sensor.

Figure 3 shows the process flow of the SiC-FET-type NOx sensor. First, a p-type well and n- and p-type diffusion layers are formed on an n-type SiC epitaxial layer. Next, the SiO$_2$ layer is deposited. After that, contact holes to the diffusion layers are formed (Fig. 3(a)). Photo resist patterns are formed for the contact metal (W/TiN) layer (Fig. 3(b)). After depositing the contact metal layer (Fig. 3(c)), resist patterns are removed by lift-off process (Fig. 3(d)). Similarly, gate layers (Fig. 3(e)(f)) are formed by lift-off processes. Lift-off can prevent etching damage to the gate oxide. After depositing inter-layer dielectric film, contact holes, platinum lines, and a silicon nitride (SiN) passivation layer are formed. The surface of the platinum lines is revealed only on the square-shaped electrode pads near the edges of the chip. These pads are necessary for the external power supply. Finally, the gate layer of the sensor FET is exposed by removing the dielectric film (Fig. 3(f)). At that time, the dielectric film on the reference FET remains.

Figure 4 shows the operation principles of the SiC-FET-type sensor. NOx is the generic name of nitric oxides and includes NO, NO$_2$, N$_2$O, and so on. Similar to a limiting-current-type ceramic sensor [1], our sensor targets the sensing of NO. The threshold voltage (Vth) of the sensor FET shifts to the positive direction by increasing NO concentration. We presume oxygen ions (O^{2-}) are produced from NO molecules with the aid of catalytic reactions of NiO layers. The oxygen ions accumulated in the YSZ layer have negative charges and thus shift the Vth of the FET. NO concentration can be calculated from the Vth shift. Although the SiC-FET-type gas sensor has been studied [2], the gate structure is different from our device.

The Vth of a MOS capacitor also shifts depending on the NO concentration, which is similar to that of FETs. The FET- and capacitor-type gas sensors are classified as work function-type sensors [3]. In this study, we utilized both FETs and capacitors to investigate the gas sensing characteristics.

978-1-7281-1988-5/18 $31.00 © 2018 IEEE

III. HEAT RESISTANCE

In diesel-engine cars, the temperature of the exhaust gas may increase while the diesel particulate filters are refreshing. Since the sensor may be exposed to at most 900°C atmosphere for several minutes, we examined the heat resistance characteristics.

Figure 5 shows the heat resistance of SiC-FET with TiN contacts and a poly-Si gate. Figure 5(a) shows the current-voltage characteristics of the FET for as-fabricated and after 950°C for 1 hour. Figure 5(b) shows the channel length dependence of the Vth. We found that the characteristics of the FET are not affected by the annealing. The contact resistivity of the Ti-contacts increases from 10^{-2} to 10^5 Ωcm^2 after 900°C annealing for 10 minutes. In contrast, that of the TiN-contacts is not degraded after 950°C for 1 hour.

Although TiN-contact improves the heat resistance, the as-fabricated contact resistance is larger than that of the Ti-contacts at around room temperature. The temperature dependence of the TiN-SiC contact resistance is shown in Fig. 6. The resistance decreases as the temperature increases and is reduced sufficiently above around 500°C. Because the contact resistance for the n-type diffusion layer is smaller than that for the p-type one, we utilize n-type FET to reduce the source and drain contact resistance.

We investigated the heat resistance of the gate stacks by SEM micrographs, as shown in Fig. 7. The gate stacking samples were fabricated by depositing layers on a SiO_2 layer. The platinum layer on the YSZ layer is easily aggregated and peeled off above 700°C. The insertion of the NiO layer between the YSZ and Pt layers enhances the heat resistance up to 900°C. The NiO layer is essential not only for improving the heat resistance but also for NO sensing itself, as discussed in the next section.

IV. GAS SENSING CHARACTERISTICS

The current-voltage characteristics of the sensor FET are shown in Fig. 8. By virtue of the SiC-substrate, on- and off-switching at 500°C is possible and the Vth is well defined. The time dependence of the Vth was measured for the sensor FET. As shown in Fig. 9, a SiC-FET with NiO layer shows Vth depending on NO concentration down to 1 ppm.

Figure 10 summarizes the NO and oxygen responses of SiC-FET sensors with and without a NiO layer. The sensor without the NiO layer hardly responds to NO. This indicates that the catalytic reaction for NO stems mainly from the NiO layer rather than the Pt layer. The SiC-FET sensors respond to oxygen both with and without the NiO layer. The oxygen response of the Si-FET-type sensor with a gate stack of Pt and YSZ was reported previously [4]. This result indicates that the catalytic reaction for oxygen stems mainly from the Pt layer rather than the NiO layer. As shown in Fig. 10(c), the selectivity of the sensors for NO and oxygen depends on NiO thickness, and the optimum thickness appears to be around 10 nm.

We have clarified the sensor response to other gases typically contained in diesel exhaust gas. The Vth increases for SO_2, O_2, and NO but decreases for NH_3 and CO (Fig. 11(a)(b)). The solid lines in Fig. 11 are the fitted results by Langmuir's theory [5, 6]. Since we utilize the device for an NO sensor, NO sensing is

disturbed by the response to other gases. This disturbance needs to be suppressed (the method to do so will be described later).

To enhance NO sensitivity, we improved the gate stacks. Figure 12 shows the NO response of the sensor FET with YSZ and composite material of NiO and Pt. The ratio of NiO and Pt is 35: 65 for Fig. 12(a). The Vth shift is increased for 1-ppm NO compared to the result of Fig. 9. In addition, the Vth responds to 0.1-ppm NO. Similarly, as shown in Fig. 12(b), the Vth shift is also increased with composite material of NiO: Pt = 25: 75. We can attribute the high sensitivity with the composite-material gate to the enhanced three-phase boundary of the NiO, YSZ, and gas phases, as described in Ref. [7].

To reduce the disturbance from other gases contained in diesel exhaust gas, preprocessing the exhaust gas is known to be effective. We have fabricated an on-chip ion pumping device that has a similar function to the one in Ref. [1]. Figure 13 shows the concept of suppressing disturbance from other gases by the ion pumping device. The inflammable components in exhaust gas (HC, CO, H_2, etc.) are oxidized with the aid of the catalytic effect of the platinum layer of the bottom surface of the ion pumping device, producing H_2O and CO_2. H_2O and CO_2 do not affect the FET. The remaining oxygen gas is removed by the oxygen ion pumping, and thus the sensor FET can detect NO without the disturbance.

We have fabricated the ion pumping device on a Si substrate for a feasibility study. The top and cross-sectional views of the ion pumping device on a Si substrate are shown in Figs. 14(a) and (b), respectively. By applying 1.0 and 0 V on the top and bottom electrodes, respectively at 500°C, the ion pumping current by oxygen ions is observed, as shown in Fig. 14(c).

V. SUMMARY

We have developed a SiC-FET-type NOx sensor that has high sensitivity and enables the detection of NO less than 1 ppm. The feasibility of an on-chip ion pumping device was confirmed, and we found that disturbance from other gases in exhaust can be removed by the ion pumping device. Our developed sensor will be beneficial for next-generation diesel-engine cars corresponding to stricter regulations.

ACKNOWLEDGMENTS

The authors gratefully acknowledge Dr. T. Sekiguchi and Dr. H. Kurata of Hitachi Ltd. for their continuous encouragement. The authors also thank all staff at the clean room of Hitachi's R&D group for the device fabrication. The authors thank Dr. R. Goto, Dr. H. Sano, Dr. T. Shimada, Mr. K. Kato, Mr. N. Ashizuka, and Mr. Y. Nozawa of Hitachi Metals, Ltd. for insightful discussion from the early stage of the research. The authors also thank Mr. Y. Nagai of TOMOE SHOKAI Co., LTD. for valuable assistance with the gas response experiments.

REFERENCES

[1] N. Kato, K. Nakagaki, and N. Ina, SAE Technical Paper 960334 (1996).
[2] R. Loloee et al., Sensors and Actuators B 129 (2008) pp. 200–210.
[3] T. Hübert et al., Sensors and Actuators B 157, pp. 329–352 (2011).
[4] Y. Miyahara, K. Tsukada, and H. Miyagi, J. Appl. Phys. 63, 2431 (1988).
[5] T. Usagawa and Y. Kikuchi, J. Appl. Phys. 108, 074909 (2010).
[6] Y. Sasago et al., Symp. on VLSI Technol. 2017, T106-T107.
[7] B. Wang et al., ACS Appl. Mater. Interfaces, 8, pp. 16752–16760 (2016).

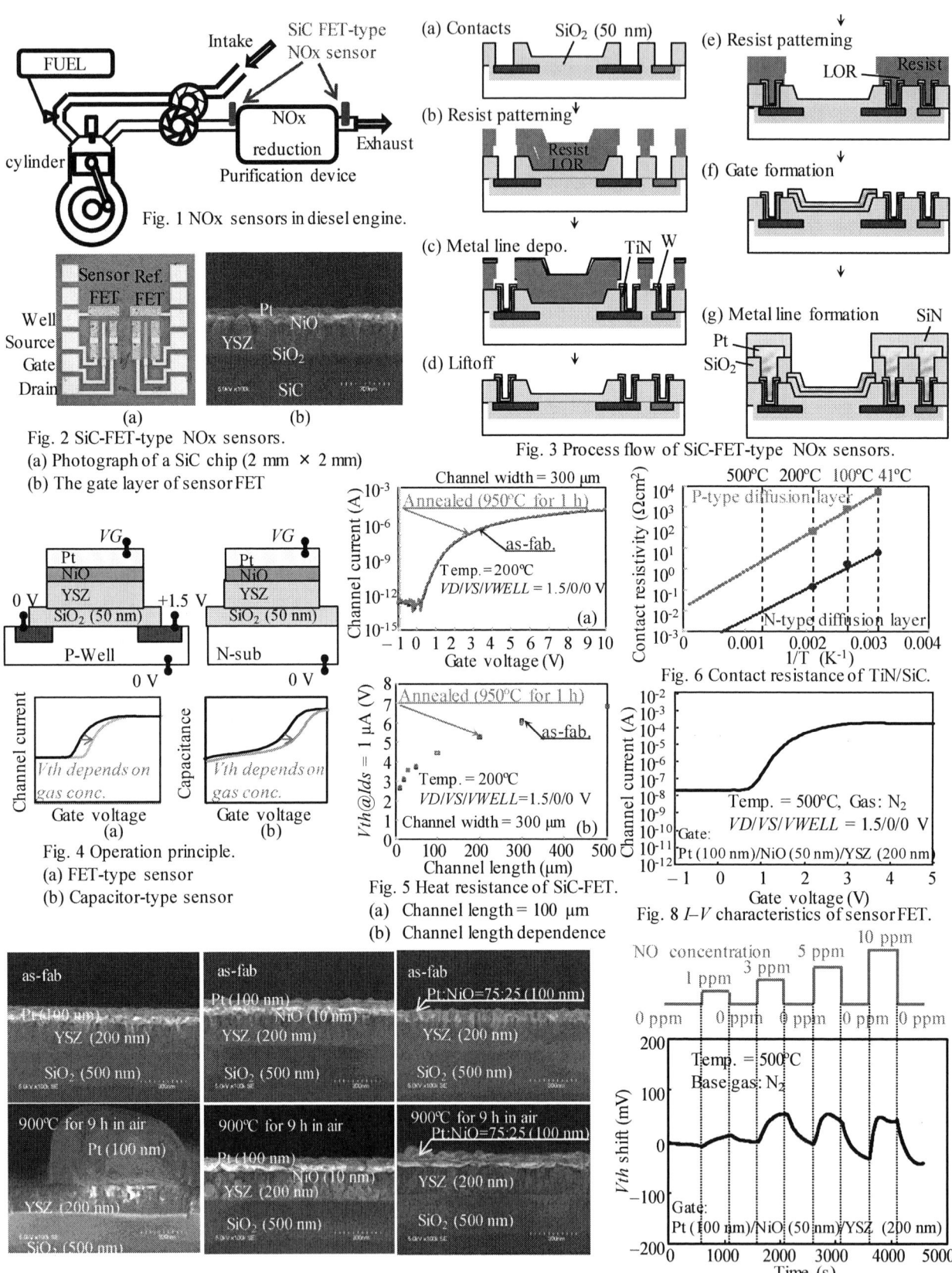

Fig. 1 NOx sensors in diesel engine.

Fig. 2 SiC-FET-type NOx sensors.
(a) Photograph of a SiC chip (2 mm × 2 mm)
(b) The gate layer of sensor FET

Fig. 3 Process flow of SiC-FET-type NOx sensors.

Fig. 4 Operation principle.
(a) FET-type sensor
(b) Capacitor-type sensor

Fig. 5 Heat resistance of SiC-FET.
(a) Channel length = 100 μm
(b) Channel length dependence

Fig. 6 Contact resistance of TiN/SiC.

Fig. 8 I–V characteristics of sensor FET.

Fig. 7 Heat resistance of sensor gate including platinum.

Fig. 9 NO response of SiC-FET-type sensor.

978-1-7281-1988-5/18 $31.00 © 2018 IEEE   286

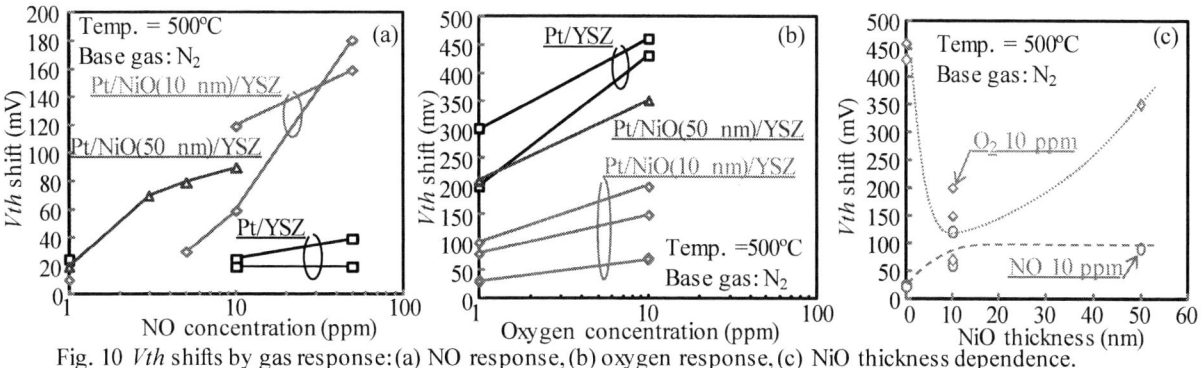

Fig. 10 *Vth* shifts by gas response: (a) NO response, (b) oxygen response, (c) NiO thickness dependence.

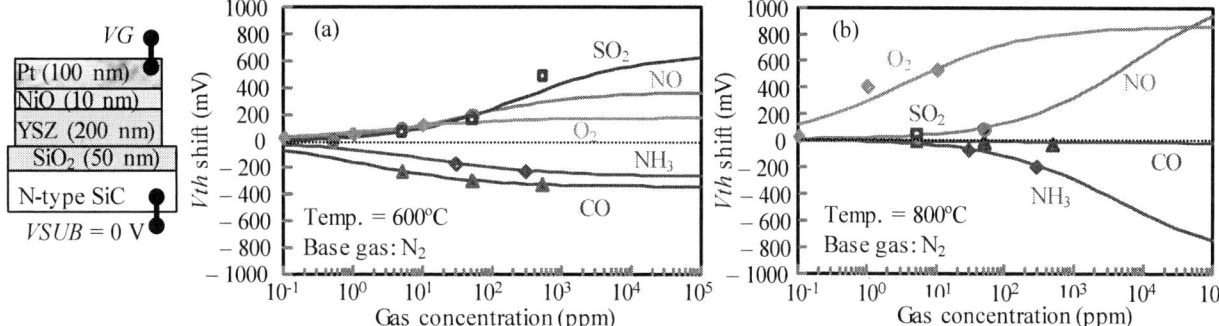

Fig. 11 *Vth* shifts of capacitor-type sensor by NO, O_2, SO_2, CO, and NH_3: (a) 600°C, (b) 800°C.

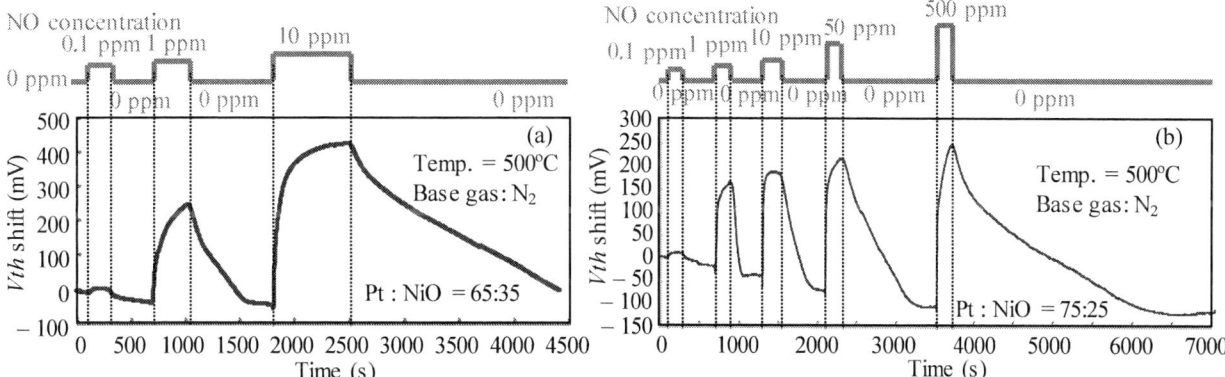

Fig. 12 NO response of capacitor-type sensors with composite-material gates: (a) Pt : NiO = 65:35, (b) Pt : NiO = 75:25.

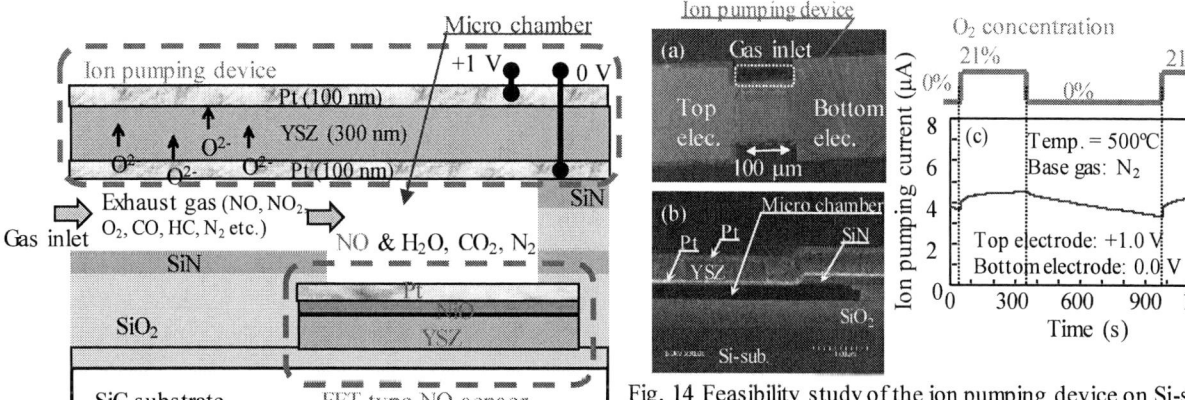

Fig. 13 Concept of suppressing influence from other gases.

Fig. 14 Feasibility study of the ion pumping device on Si-sub.
(a) Plane view of ion pumping device
(b) Cross-sectional image of ion pumping device
(c) Ion pumping current

978-1-7281-1988-5/18 $31.00 © 2018 IEEE   287

A Si FET-type Gas Sensor with Pulse-driven Localized Micro-heater for Low Power Consumption

Yoonki Hong, Seongbin Hong, Dongkyu Jang, Yujeong Jeong, Meile Wu, Gyuweon Jung, Jong-Ho Bae, Jun Shik Kim[1], Ki Soo Chang[2], Chan Bae Jeong[2], Cheol Seong Hwang[1], Byung-Gook Park, and Jong-Ho Lee
School of ECE, [1]MSE and ISRC, Seoul National University, Seoul 08826, Korea,
[2]Division of Scientific Instrumentation, Korea Basic Science Institute, Daejeon 34133, Korea,
Phone: +82-2-880-1727; Fax: +82-2-882-4658; E-mail: jhl@snu.ac.kr

Abstract— A poly-Si localized micro-heater for Si FET-type gas sensor is proposed. The gas sensor has an air gap under the heater to prevent the heat dissipation. It is verified that the heater temperature can be read by reading the heater resistance immediately after the heating pulse is turned on or off. A heater temperature of 112°C is achieved at a heat pulse bias of 2 V, which consumes ~0.92 mW. The heating and cooling times of the heater are ~200 μs and ~100 μs, respectively. NO_2 and H_2S sensing are successfully performed by using a pulse-driven micro-heater in the proposed gas sensor.

I. INTRODUCTION

Humans have been constantly exposed to a variety of gases in everyday life and there are many kinds of harmful gases among them. Many studies have been carried out to develop a reliable gas sensor to prevent damage from harmful gases. Recently, studies on gas-sensing at room temperature are increasing [1], but report low reliability. Most of the gas sensors still show better sensing performance at high temperatures [2]. To raise the operating temperature, hot plate or micro-heater has been used in general. In most studies, resistor-type gas sensors have been widely used and need a large sensing area for reliability. The large sensing area requires a large size of heater (~mm²) to raise the temperature during sensing, resulting in higher power consumption. In this work, we propose a Si FET-type gas sensor having a localized poly-Si micro-heater to reduce heater size and power consumption. The temperature of the micro-heater driven by a series of pulse-type heating bias (V_H) is identified by the proposed method and the infrared thermal microscopy. We also investigate gas sensing properties by applying heating pulses to fabricated a gas sensor with the micro-heater.

II. DEVICE FABRICATION

A Si FET-type gas sensor having a localized poly-Si micro-heater is fabricated using 6 masks. The fabricated gas sensor features an air gap under the micro-heater. Air has lower thermal conductivity (0.026 W/(m·K)) than Si and SiO_2 (148 and 1.4 W/(m·K)) [3]. Therefore, the air gap prevents the heat generated by the micro-heater from escaping through the Si substrate. The gas sensor is based on *p*MOSFET because it has less flicker noise than *n*MOSFET [4]. Fig. 1 (a)-(e) show schematic bird's eye views of key fabrication process steps of the gas sensor. The active region is defined on a (100) *n*-type Si wafer and a field oxide is grown. Then, the buried channel implantation is performed for reducing flicker noise. After a 10 nm thick gate oxide is grown, a 350 nm thick n^+ poly-Si is

deposited and patterned to form a floating-gate (FG) and the micro-heater (a). After the S/D implantation, a $SiO_2/Si_3N_4/SiO_2$ (O/N/O, 10 nm/20 nm/10 nm) passivation layer is formed (b). Since the sensing material and the heater are spaced apart by the thickness of the passivation layer, the heat of the heater is effectively transferred to the sensing material. Then, the metal contact holes are defined and the Cr/Au electrodes are deposited (c). To form an air gap under the micro-heater, O/N/O passivation layer, field oxide, and Si substrate are sequentially dry-etched (d). A 15 nm thick *n*-type semiconducting ZnO film as a sensing material is then formed to cover the part of CG and FG by using atomic layer deposition (e). Fig. 1 (f) depicts a schematic cross-sectional view of the gas sensor cut along a white dotted line A-A' in Fig. 1 (e). The air gap is formed at a depth of ~10 μm under and around the sensing part, and the FET is formed on the bulk substrate. Fig. 2 shows a top SEM image of the gas sensor. The width/length of the channel is 2 μm/2 μm. There are bridges between the etching holes (white dotted lines), which sustain the sensing part structurally. The heater size is 15×5 μm².

III. RESULT AND DISCUSSION

Fig. 3 (a)-(c) show I_D-V_{CG}, I_D-V_{DS} curves, and micro-heater current (I_H) versus V_H of the gas sensor at 25°C measured by using DC and pulsed *I-V* (PIV) methods. Fig. 4 shows the change in temperature of the micro-heater (T_H) as the temperature of hot plate (T_{plate}) rises from 30°C to 100°C. The T_H is measured by using infrared thermal microscopy. The difference between T_{plate} and T_H is ~2°C and all T_H data obtained in this work are calibrated. Fig. 5 (a) shows heater resistance (R_H) versus T_{plate} as a parameter of the bias for reading the temperature (V_{rT}). In Fig. 5 (a), the R_H increases with increasing T_{plate}, which is characteristic of heavily doped poly-Si micro-heater. The slope of R_H-T_{plate} curve becomes steeper as the V_{rT} applied to the heater increases from 0.5 to 2.0 V and the curve becomes more nonlinear with increasing V_{rT} at > 150°C. The reason is that the increase of V_{rT} leads to the increase of Joule heating, so that T_H rises. This is also inferred by the result, as shown in Fig. 5 (b), that temperature coefficient of resistance (TCR), denoted by α, increases with the increase of V_{rT}. The αs are extracted from the equations in the inset of Fig. 5 (a). In this work, the $|V_{rT}|$ of 0.5 V or less is applied to read the T_H, because Joule heating is negligible. Using the calibrated R_H versus T_H, the T_H can be extracted from the measured R_H after the V_H pulse (> 1 V) is applied to the heater at a fixed T_{plate}. Fig. 6 (a) depicts transient R_H and T_H behaviors as a function of V_H after a 500 μs-long V_H pulse is applied at $t = 0$ s. The R_H and the T_H are saturated when the heat generated by the V_H above 1 V is equal to the heat dissipated. Here, we define heating time (t_h) as a time duration for which T_H rises by 90% of the temperature difference between 30°C and the saturated temperature. Fig. 6 (b) shows the t_h with respect to V_H. A pulse width of V_H longer than 200 μs is required to reach the maximum T_H at $V_H = 2$ V. Fig. 7 (a) depicts transient R_H and T_H behaviors as a function of V_H after a 500 μs-long V_H pulse is turned off at $t = 0$ s. Note a V_{rT} pulse of 0.5 V with a period of 30 μs is applied to read the R_H at $t = 5$ μs. Cooling time (t_c) is defined as a time duration for which T_H falls by 90% of the temperature difference between the temperature at $t = 0$ s and 30°C in Fig. 7 (a). Fig. 7 (b) shows the t_c with respect to V_H. A 100 μs is sufficient to cool down the heated heater with an applied 2 V of V_H to 30°C. Fig. 8 (a) and (b) display maximum value of T_H ($T_{H,max}$) versus frequency as parameters of V_H (@ 50% of duty cycle) and duty cycle (@ $V_H = 2$ V), respectively. The T_H data are obtained by using infrared thermal microscopy. In Fig. 8 (a), the $T_{H,max}$s at frequency below 1 kHz are kept at the same value due to the reason that the pulse width of V_H is long enough to reach the maximum temperature, as mentioned in Fig. 6 (a). In Fig. 8 (b), the $T_{H,max}$ increases as the duty cycle of V_H pulse increases, since the pulse width becomes longer at a certain frequency. Fig. 9 shows the DC I_D-V_{CG} characteristics of the gas sensor at different T_{plate}s (symbols) and the same T_{HS} as T_{plate}s by applying the V_{HS} to the heater (lines). The V_{DS} is fixed at -0.1 V. As the T_{plate} rises, the T_H also increases, as shown in Fig. 4, and the drain off-current increases due to drain-to-substrate junction leakage. As the V_H increases at a fixed T_{plate} of 25°C, only the T_H of the micro-heater rises. Therefore, there is no increase in drain off-current by virtue of the FET temperature being kept at 25°C. To demonstrate the result in Fig. 9, the temperature distribution data of the gas sensor is obtained by using thermoreflectance microscopy in Fig. 10 (a) and (b). Here, the V_H of the pulse applied to the heater is 2 V. The distance between the micro-heater and the FET is ~ 20 μm. In Fig. 10 (b), no temperature increase is observed in the vicinity of the FET. It is verified that heating by applying V_H up to 2 V to the heater has no influence on the FET operation of the gas sensor. Fig. 11 illustrates the experimental setup for gas-sensing measurement. Three pulse signals are used and applied to the CG, the drain, and the micro-heater. Fig. 12 indicates an operation scheme adopted for gas-sensing measurement. In the heating period, the V_H are applied to the heater with a pulse width of t_H while the pre-bias (V_{pre}) is applied to the CG [5]. In the read period, the drain bias (V_{rDS}) is synchronized with the V_{rCG} for a pulse width of t_{read}. All the gas-sensing measurements in this work are done at fixed V_{rCG} and V_{rDS} of -0.1 V and a frequency of 1 kHz in linear region of the FET-type gas sensor. Figs. 13 and 14 show the transient I_D behaviors of the gas sensor as parameters of V_H and NO_2 concentration, respectively. In Fig. 13, the response and the recovery characteristics are improved as the V_H increases. The increase of V_H leads to the rise in T_H. NO_2 molecules then acquire more energy, resulting in higher response and faster recovery. In Fig. 14, as the NO_2 concentration increases, the $|I_D|$ increases. The gas response versus NO_2 concentration as a parameter of V_H is shown in Fig. 15. The gas response is obtained by the absolute value of the difference between the I_{DS} at $t = 70$ s and $t = 10$ s. At a 500 ppb NO_2 and a V_H of 2 V, the gas response is 43.7 nA. Fig. 16 shows the transient I_D behaviors of the gas sensor as a parameter of t_H. The I_D behavior with a t_H of 500 μs is similar to that with a t_H of 700 μs. As shown in Fig. 8(b), a $T_{H,max}$ at a duty cycle of 50% is nearly the same as that of 70% at a f up to 1 kHz. Thus, using a 500 μs-long V_H pulse is reasonable for low power operation of the gas sensor with the micro-heater. The power consumption is ~0.92 mW at 112°C ($V_H = 2$V). The transient responses to H_2S gas are illustrated as parameters of V_H, gas concentration, and t_H in Figs. 17, 18, and 20, respectively. Fig. 19 shows the gas response versus H_2S concentration as a parameter of V_H. At a 10 ppm H_2S and a V_H of 2 V, the gas response is 38.3 nA. In Figs. 17-20, all the results can be explained in the same way as in Figs. 13-16, except for the direction of transient I_D behaviors, since H_2S is well known as a reducing gas, unlike NO_2, known as an oxidizing gas. Since H_2S recovery requires reversible oxygen adsorption, the recovery seems to be delayed in Figs. 17, 18, 20, and 22. The transient NO_2 and H_2S responses are illustrated as a parameter of V_{pre} as shown in Figs. 21 and 22, respectively. Note the responses are negligible during finite t_{read} when $V_H = 0$ V as represented by open symbols, even though V_{pre} is applied, since the gas reaction takes a long time at 25°C. In Fig. 21, the response increases with negatively increasing V_{pre}, whereas the recovery time is shortened with positively increasing V_{pre} [5]. Fig. 22 shows the V_{pre} effect on H_2S response. The $|I_D|$ behavior with the polarity of V_{pre} is opposite to that of NO_2. Fig. 23 indicates the gas response in 500 ppb NO_2 ambience versus total heating time ($n \times t_H$) where n is the number of applied pulses ($t_H = 500$ μs, $f = 1$ kHz). A pulse scheme used in this measurement is explained in the inset of Fig. 23. We think a ΔI_D of ~2 nA is large enough to be detected even when there is noise. Since the ΔI_D of ~2 nA needs a total heating time of ~10 ms, the energy consumption is estimated to be ~18.4 μJ.

IV. CONCLUSIONS

A Si FET-type gas sensor having a localized poly-Si micro-heater has been proposed and fabricated. A method to identify the temperature (T_H) of the heater was verified. The T_H of 112°C was obtained at a heater pulse bias of 2 V and the heating/cooling times of the heater were 200 μs/100 μs. Low power consumption was achieved, which is attributed to the reduced distance between the heater and sensing material, small heater size, and pulse operation. The proposed sensor showed excellent performance in NO_2 and H_2S gas detection.

ACKNOWLEDGMENT

This work was supported by the National Research Foundation of Korea (NRF-2016R1A2B3009361) and the Brain Korea 21 Plus Project in 2018.

REFERENCES

[1] M. Kaur, *et al.*, *Sensors and Actuators B*, 242, p. 389, 2017
[2] K. Suematsu, *et al.*, *Anal. Chem.*, 87, p. 8407, 2015
[3] M. Asheghi, *et al.*, *J. Heat Transfer*, 120, p. 30, 1998
[4] J.H. Lee, *et al.*, *J. Semicond. Techno. Sci.*, 6, p. 32, 2006
[5] J. Shin, *et al.*, *IEEE IEDM*, p. 472, 2016

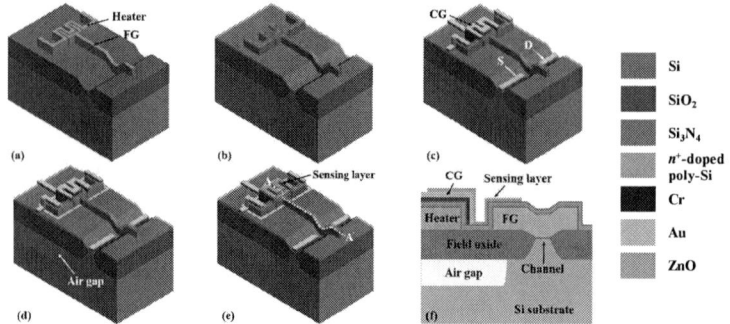

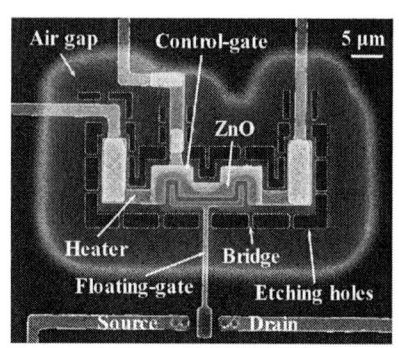

Fig. 1. (a)-(e) Schematic bird's eye views of key fabrication process steps and (f) cross-sectional view of the fabricated gas sensor having a localized poly-Si micro-heater cut along a white dotted line (A-A'). The micro-heater and the FG are simultaneously formed and covered by the O/N/O passivation layer. A ZnO thin film partly covers the CG and the FG.

Fig. 2. Top SEM image of the fabricated gas sensor. Brown and green shaded areas represent an air gap formed under the micro-heater and a ZnO sensing layer, respectively.

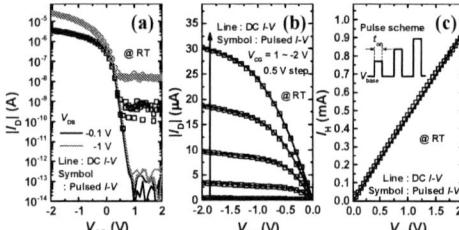

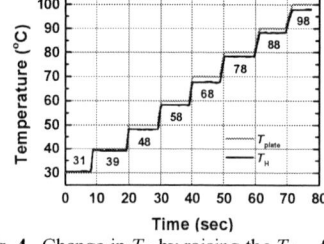

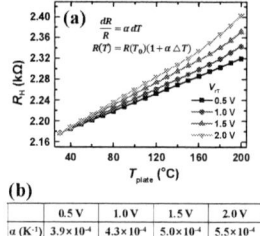

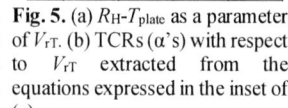

Fig. 3. (a) I_D-V_{CG}, (b) I_D-V_{DS} and (c) I_H-V_H curves of the gas sensor. The curves measured by DC and PIV (t_{on} = 30 μs, f = 1 kHz and V_{base} = 0 V) methods are represented by lines and symbols, respectively. The PIV pulse scheme is depicted in the inset of (c).

Fig. 4. Change in T_H by raising the T_{plate} from 30 to 100°C (10°C step). No bias voltage is applied to the micro-heater. The T_H is measured by using infrared thermal microscopy.

Fig. 5. (a) R_H-T_{plate} as a parameter of V_{rT}. (b) TCRs (α's) with respect to V_{rT} extracted from the equations expressed in the inset of (a).

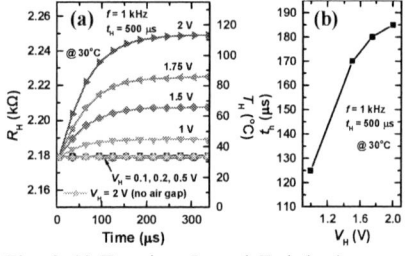

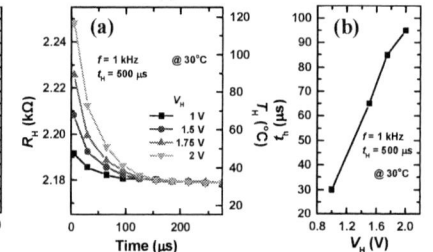

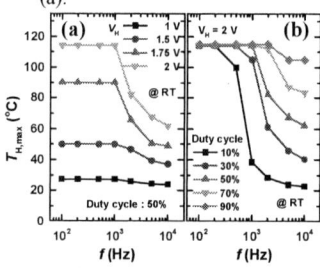

Fig. 6. (a) Transient R_H and T_H behaviors as a function of V_H. A 500 μs-long V_H pulse is applied at t = 0 s. The T_Hs are extracted from the R_H-T_{plate} curves in Fig. 5 (a). (b) t_h with respect to V_H.

Fig. 7. (a) Transient R_H and T_H behaviors as a function of V_H. A 500 μs-long V_H pulse is applied and turned off at t = 0 s. The T_Hs are extracted from the R_H-T_{plate} curves in Fig. 5 (a). (b) t_c with respect to V_H.

Fig. 8. $T_{H,max}$ versus frequency as a parameter of (a) V_H (50% of duty cycle) and (b) duty cycle (V_H = 2 V). The temperatures are obtained by using infrared thermal microscopy.

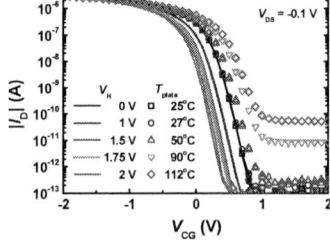

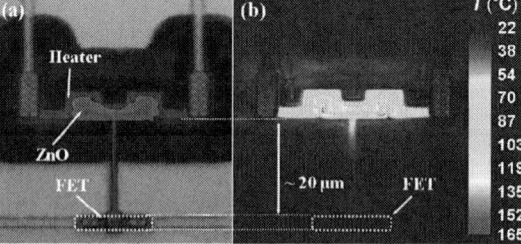

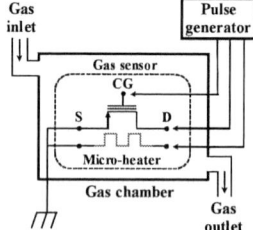

Fig. 9. I_D-V_{CG} curves as a parameter of V_H (line) and T_{plate} (symbol) measured by DC method. The T_Hs at V_Hs of 0, 1, 1.5, 1.75 and 2 V are 25, 27, 50, 90 and 112°C, respectively.

Fig. 10. (a) Optical microscopic image and (b) temperature distribution image of the gas sensor. Temperature distribution image is obtained by using thermoreflectance microscopy. The distance between the micro-heater and the channel is ~20 μm. A bias voltage of 2 V is applied to the micro-heater.

Fig. 11. Experimental setup for gas-sensing measurement. The generated pulse signals are applied to the CG, the drain and the micro-heater.

978-1-7281-1988-5/18 $31.00 © 2018 IEEE

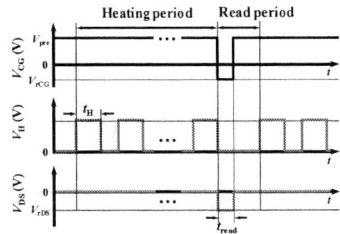

Fig. 12. Operation scheme of the fabricated gas sensor. The V_{pre} and the V_{rCG} are alternately applied to the CG. The V_{rDS} is synchronized with the V_{rCG} with a pulse width of t_{read}. The V_H pulses are applied with a pulse width of t_H.

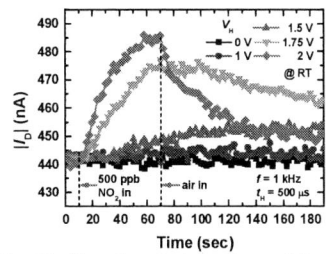

Fig. 13. Transient I_D behaviors of the gas sensor as a parameter of V_H. 500 ppb of NO$_2$ and dry air are injected at $t = 10$ s and $t = 70$ s, respectively. The pulse width of the V_H is 500 μs ($V_{rCG} = V_{rDS} = -0.1$ V, $V_{pre} = 0$ V).

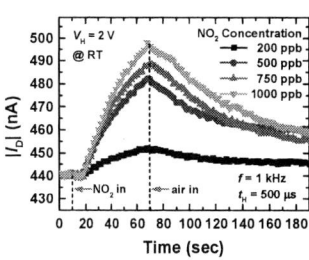

Fig. 14. Transient I_D behaviors of the gas sensor with different NO$_2$ concentrations. NO$_2$ and dry air are injected at 10 s and 70 s, respectively. The amplitude of the V_H pulse is 2 V ($t_H = 500$ μs, $V_{rCG} = V_{rDS} = -0.1$ V, $V_{pre} = 0$ V).

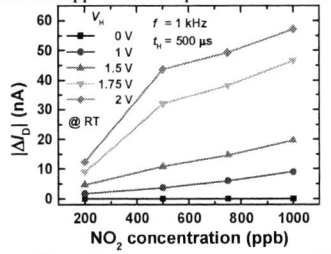

Fig. 15. Gas response versus NO$_2$ concentration as a parameter of V_H (0, 1, 1.5, 1.75 and 2 V). The width of the V_H pulse is 500 μs.

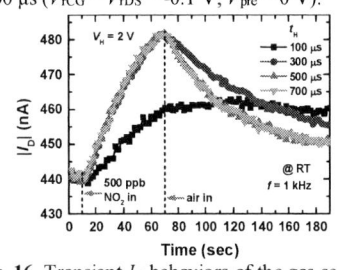

Fig. 16. Transient I_D behaviors of the gas sensor as a parameter of t_H. 500 ppb of NO$_2$ and dry air are injected at $t = 10$ s and $t = 70$ s, respectively. The amplitude of the V_H pulse is 2 V ($V_{rCG} = V_{rDS} = -0.1$ V, $V_{pre} = 0$ V).

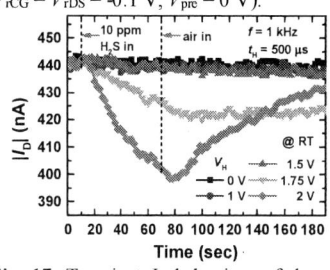

Fig. 17. Transient I_D behaviors of the gas sensor as a parameter of V_H. 10 ppm of H$_2$S and dry air are injected at $t = 10$ s and $t = 70$ s, respectively. The width of the V_H pulse is 500 μs ($V_{rCG} = V_{rDS} = -0.1$ V, $V_{pre} = 0$ V).

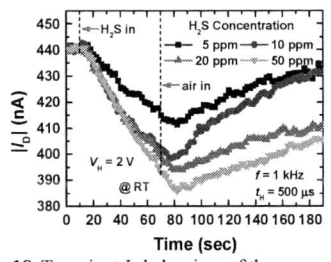

Fig. 18. Transient I_D behaviors of the gas sensor with different H$_2$S concentrations. H$_2$S and dry air are injected at 10 s and 70 s, respectively. The amplitude of the V_H pulse is 2 V ($t_H = 500$ μs, $V_{rCG} = V_{rDS} = -0.1$ V, $V_{pre} = 0$ V).

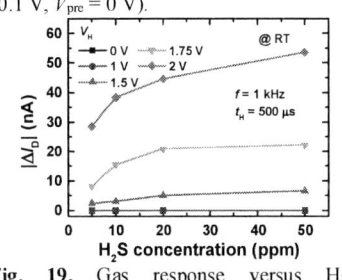

Fig. 19. Gas response versus H$_2$S concentration as a parameter of V_H (0, 1, 1.5, 1.75 and 2 V). The width of the V_H pulse is 500 μs.

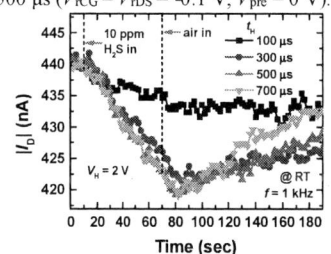

Fig. 20. Transient I_D behaviors of the gas sensor as a parameter of t_H. 10 ppm of H$_2$S and dry air are injected at $t = 10$ s and $t = 70$ s, respectively. The amplitude of the V_H pulse is 2 V ($V_{rCG} = V_{rDS} = -0.1$ V, $V_{pre} = 0$ V).

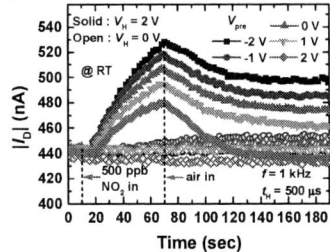

Fig. 21. Transient I_D behaviors of the gas sensor as a parameter of V_{pre} (-2 ~ 2 V). The NO$_2$ concentration is 500 ppb. Solid and open symbols represent NO$_2$ responses at a V_H of 2 and 0 V, respectively ($t_H = 500$ μs, $V_{rCG} = V_{rDS} = -0.1$ V).

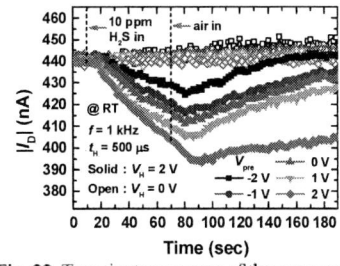

Fig. 22. Transient responses of the gas sensor as a parameter of V_{pre} (-2 ~ 2 V). The H$_2$S concentration is 10 ppm. Solid and open symbols represent H$_2$S responses at a V_H of 2 and 0 V, respectively ($t_H = 500$ μs, $V_{rCG} = V_{rDS} = -0.1$ V).

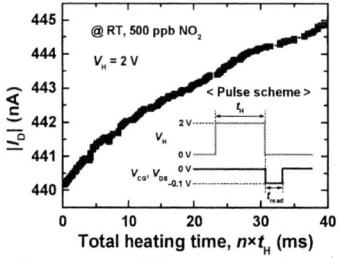

Fig. 23. Response of the gas sensor in 500 ppb of NO$_2$ ambience as a function of $n \times t_H$. Here, t_H is 500 μs. The gas sensor starts operating in the stable NO$_2$ ambience. The amplitude of the V_H pulse is 2 V ($V_{rCG} = V_{rDS} = -0.1$ V, $V_{pre} = 0$ V).

978-1-7281-1988-5/18 $31.00 © 2018 IEEE

ECRAM as Scalable Synaptic Cell for High-Speed, Low-Power Neuromorphic Computing

Jianshi Tang*, Douglas Bishop, Seyoung Kim, Matt Copel, Tayfun Gokmen, Teodor Todorov,
SangHoon Shin, Ko-Tao Lee, Paul Solomon, Kevin Chan, Wilfried Haensch, John Rozen
IBM Thomas J. Watson Research Center, Yorktown Heights, New York, USA, *Email: jtang@us.ibm.com

Abstract—We demonstrate a nonvolatile Electro-Chemical Random-Access Memory (ECRAM) based on lithium (Li) ion intercalation in tungsten oxide (WO_3) for high-speed, low-power neuromorphic computing. Symmetric and linear update on the channel conductance is achieved using gate current pulses, where up to 1000 discrete states with large dynamic range and good retention are demonstrated. MNIST simulation based on the experimental data shows an accuracy of 96%. For the first time, high-speed programming with pulse width down to 5 ns and device operation at scales down to 300×300 nm^2 are shown, confirming the technological relevance of ECRAM for neuromorphic array implementation. It is also verified that the conductance change scales linearly with pulse width, amplitude and charge, projecting an ultralow switching energy ~1 fJ for 100×100 nm^2 devices.

I. INTRODUCTION

The success of deep learning is related to the availability of large data sets and the use of GPUs to implement the training with the back-propagation algorithm. While different solutions are pursued to increase computing efficiency, the von Neumann bottleneck may eventually prevent further progress. Neuromorphic computing has emerged as a new computing paradigm to enable massively parallel analog computing for deep learning. For example, a new architecture of resistive processing unit (RPU) could provide 30,000× acceleration compared to state-of-the-art CPU/GPU in training deep neural networks [1]. Experimentally, various nonvolatile memories (NVMs), such as resistive random-access memory (ReRAM) and phase-change memory (PCM), have been evaluated as synaptic elements to build prototype neural networks [2-3]. While such NVMs have recently shown encouraging results for inference, their success in training neural network is hampered by their non-ideal switching characteristics, such as asymmetric weight update, stochasticity, and limited endurance.

To circumvent those intrinsic flaws, nonvolatile electrochemical switches have been proposed as artificial synapses for neuromorphic computing [4-5]. As a trade-off for cell complexity using three-terminal device, their read and write operations are decoupled, allowing for better endurance and low-energy switching while maintaining nonvolatility. More importantly, the electrochemically driven intercalation or redox reaction can be precisely and reversibly controlled by the amount of charge through the gate, so they can provide symmetric switching with plentiful discrete states and reduced stochasticity. Indeed, symmetric second-scale switching has been shown on millimeter-size redox transistors with $Li_{1-x}CoO_2$

channel [4]. Meanwhile, organic electrochemical transistors have shown millisecond switching and low switching energy of 10 pJ [5]. However, a clear path to nanosecond switching and device scaling in solid-state electrochemical transistors with sufficient dynamic range remained to be demonstrated to make them technologically relevant. In this work, a nonvolatile ECRAM with up to 1000 discrete conductance levels and large dynamic range is fabricated. Sub-micron devices and high-speed write with pulse down to 5 ns are demonstrated for the first time. An ultralow switching energy ~1 fJ is projected for scaled devices. ECRAM emerges as a promising candidate for high-speed, low-power neuromorphic computing.

II. SWITCHING CHARACTERISTICS OF ECRAM

Fig. 1 illustrates the device structure of ECRAM, where Li ions are electrochemically driven by the gate to (de)intercalate into WO_3 to change its conductance for synaptic weight update. Lithium phosphorous oxynitride (LiPON) is used as a solid-state electrolyte. The amount of Li ions intercalated in WO_3 is precisely controlled by the gate current and this process is reversible, enabling symmetric update. In operation, series of positive (negative) current pulses are fed into the gate for potentiation (depression). As shown in **Fig. 2**, a typical ECRAM is sequentially programmed with 50 up then 50 down pulses (amplitude $I_G = \pm 100$ pA and width $t_w = 5$ s), featuring good symmetry and a large conductance dynamic range ~ 40. The zoom-in view shows discrete conductance states and good retention during read ($t_r = 5$ s) when the gate is floating. It should be noted that the conductance G and its change per pulse ΔG can be tuned by pulse width/amplitude (see **Figs. 10-12** later), device geometry, and material engineering. Such a wide tunability provides advantages over filamentary-type NVMs and previously reported electrochemical switches.

Cycling within a smaller dynamic range achieves a better linearity and symmetry in switching (**Fig. 3**). Here on average 55 up/down pulses ($I_G = \pm 100$ pA and $t_w = 1$ s) are used to cycle G between 1 nS and 3 nS, showing good reproducibility. In **Fig. 4a**, the switching nonlinearity analysis yields nonlinearity of $\nu = 0.347$ and 0.268 for potentiation and depression, respectively, representing near-ideal symmetry and linearity compared to literature (see **Table 1**) [6-7]. The cycle-to-cycle variations from **Fig. 3** (assuming device-to-device variation of $\sigma = 30\%$) is also used in neural network training simulations to yield a more practical classification accuracy. **Fig. 4b** shows the extracted ΔG^+ and ΔG^- per up/down pulse as a function of G over 100 cycles. The asymmetry $AS = |\Delta G^+/\Delta G^-|$ is also plotted in the inset, varying between 0.6 and 1.6. Building on Ref. [1], a modified weight model that captures ΔG dependency on G along with noise mitigation yields a better trade-off between asymmetry requirement, noise, and number of states. As shown in **Fig. 4c**, the simulated classification accuracy based on our device data (black line) for MNIST dataset is about 96%, close to the ideal numerical floating-point accuracy at 98% (black circle). Increasing the number of states to $N = 110$ (blue line) or reducing variations to $\sigma = 0\%$ (green line) does not affect the result, but the accuracy starts to degrade when N is reduced to 32 (red line). This implies that the accuracy is predominantly determined by the switching asymmetry, which needs to be reduced by about 2× to match the floating-point value as shown in **Fig. 5**. Such reduction in asymmetry is possible by shrinking the dynamic range of G, which can be done at constant N by reducing ΔG with smaller pulses; however, this may increase the noise in the vector-matrix multiplications. Here we use Gaussian noise of $\sigma = 0.06$ as baseline noise [1], and **Fig. 6** shows that MNIST simulations can tolerate up to 7× more noise if appropriate noise management is performed [8]. It should be noted that having access to 1000 levels or more in ECRAM is projected to be significant in larger neural networks for complex datasets beyond MNIST benchmarking.

III. HIGH-SPEED PULSE MEASUREMENTS

To demonstrate high-speed programming of ECRAM, we implement a circuit in which the drain terminals of discrete PFET and NFET are connected at the gate of ECRAM as shown in **Fig. 7**. Here the PFET (NFET) serves as a current source to supply fast current pulses to potentiate (depress) ECRAM by turning on the FETs exclusively. This circuit can be used as a unit cell in an ECRAM-based cross-point array [7]. Using this setup, we demonstrate the first successful sub-µs programming of electrochemical switches. **Fig. 8** shows reproducible cycling through nonvolatile discrete levels with 1 µs pulses with $I_G = \pm 100$ µA. **Fig. 9** shows reproducible cycling with 5 ns pulses with $I_G = \pm 1$ mA. Here the programmed conductance levels are read 1.5 s (pulse period) after each pulse. Write-induced transients that can affect the update frequency and implications of current vs voltage operation are discussed elsewhere [9]; material optimization and improved device design are needed to further accelerate read and update in scaled ECRAM. We note that the switching symmetry is slightly degraded compared to **Fig. 3**, due to the non-ideal characteristics of current source FETs: finite output resistance and non-zero drain leakages.

To further understand the programmability of our ECRAM, we perform systematic cycling tests with various pulse widths (t_w) and amplitudes (I_G). **Fig. 10** displays ΔG as a function of t_w down to 5 ns while keeping the identical $I_G = \pm 1$ mA. In **Fig. 11**, we show ΔG as a function of I_G down to ± 20 µA with fixed $t_w = 100$ ns. Both trends clearly reveal a linear scaling relation between ΔG and pulse charge $Q (= I_G \times t_w)$, which reaffirms the charge-driven nature of ECRAM programming mechanism as shown in **Fig. 12**. Here the data from **Fig. 3** with $t_w \sim 1$ s is also plotted, showing consistency over a wide range of pulse widths. The inset in **Fig. 12** shows that the linear scaling holds down to 2 pC, corresponding to a low switching energy of ~2 pJ per

update (assuming an average gate voltage of 1 V). In addition, our ECRAM has shown excellent endurance and no degradation in symmetry across 1000 levels after being cycled with 10^5 pulses, as shown in **Fig. 13**.

IV. SCALING AND PROJECTION

To shed light on device scaling, **Fig. 14** demonstrates the switching on a ECRAM device with 1 µm channel length, where much smaller current pulses of $I_G = \pm 1$ pA with $t_w = 5$ s are used for weight update. The scaled device still shows discrete conductance states with good retention and an even larger dynamic range $> 10^3$. **Fig. 15** further shows switching on a 300×300 nm^2 ECRAM with 100 discrete conductance levels. This demonstrates, for the first time, multi-level operation of ECRAM at scales relevant for large-array implementation. **Fig. 16** shows that the average ΔG scales roughly linearly with the normalized charge by area Q/A. Note that the shortest device channel length on this graph is 500 nm; the smaller devices exhibit an offset ΔG scaling due to geometry effects and distinct patterning. The switching energy E, normalized by ΔG, is plotted as a function of the device size A in **Fig. 17**, which also exhibits linear scaling. It is then extrapolated to yield $E/\Delta G \sim 10^{-4}$ J/S for an ultra-scaled device of 100×100 nm^2 (assuming no geometry effects). Using this number, we can estimate the required current pulse amplitude I_G for a given pulse width t_w to achieve the target ΔG, as shown in **Fig. 18**. For example, given $t_w = 10$ ns, $I_G = 10$ µA is needed to yield $\Delta G = 1$ nS, or just 100 nA for $\Delta G = 0.01$ nS, which corresponds to an ultralow switching energy of about 1 fJ, matching the ultimate energy efficiency of the human brain (~1−10 fJ per synaptic event).

V. CONCLUSION

In conclusion, we have fabricated a nonvolatile WO$_3$-based ECRAM that relies on electrochemically driven Li-ion intercalation for neuromorphic computing. Compared to conventional NVMs, ECRAM has shown many unique merits in switching, including superior symmetry and linearity, discrete conductance states with less stochasticity, large dynamic range, and excellent endurance. It is verified that the weight update scales with pulse charge and device size, projecting an ultralow switching energy down to 1 fJ. For the first time, programming with sub-µs pulses and sub-micron devices have both been demonstrated. As summarized in **Table 1**, our work, representing near-ideal symmetry and linearity compared to the literature surveyed in Refs. [6-7], paves the path for using ECRAM in future neuromorphic computing.

ACKNOWLEDGMENT

The authors gratefully acknowledge T.-C. Chen, and Z. Lemnios for executive support; W. Green, V. Narayanan, G. Burr, T. Ando, J. Hannon and J. Tersoff for helpful discussions; and J. Bucchignano and J. Yurkas for technical assistance.

REFERENCES

[1] Gokmen & Vlasov, *Front. Neurosci.*, **10**, 333 (2016). [2] Burr *et al.*, *IEEE Trans. Electron Devices*, **62**, 3498, (2015). [3] P. Yao *et al.*, *Nat. Commun.*, **8**, 15199, (2017). [4] Fuller *et al.*, *Adv. Mater.*, **29**, 1604310, (2017). [5] van de Burgt *et al.*, *Nat. Mater.*, **16**, 414, (2017). [6] P. Y. Chen *et al.*, *ICCAD*, 194 (2016). [7] Li *et al.*, *VLSI Tech. Dig.*, 25 (2018). [8] Gokmen *et al.*, *Front. Neurosci.* **11**, 538, (2017). [9] Bishop *et al.*, *SSDM*, accepted (2018).

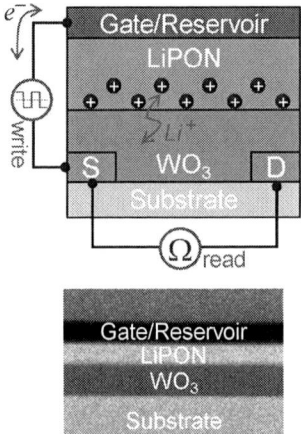

Fig. 1. ECRAM device schematic (top) and cross-sectional TEM image (bottom)

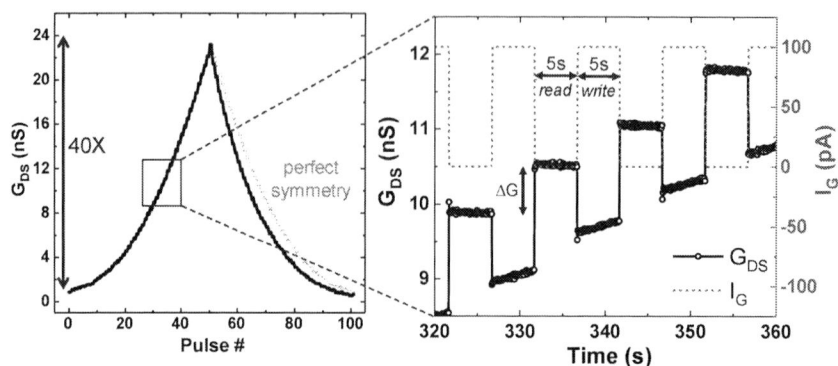

Fig. 2. Plot of the source-drain conductance G_{DS} during gate current pulses (50 up then 50 down pulses with amplitudes $I_G = \pm100$ pA and width $t_w = 5$ s), showing good symmetry and a large on/off ratio ~ 40. This device has a channel size of $L \times W = 10 \times 60$ μm^2. The green dotted line illustrates the reflection of up trace as a guide to the eye. The zoom-in view (right) shows discrete conductance states with good retention during read. A constant source-drain bias of $V_{DS} = 0.1$ V is applied to monitor the channel conductance during read and write.

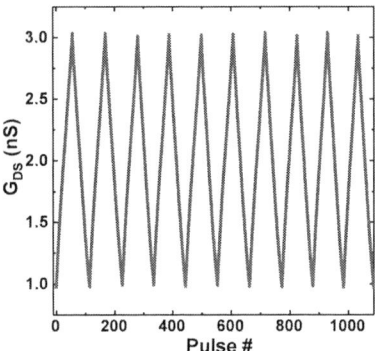

Fig. 3. Reproducible cycling demonstrates both good symmetry and linearity within a conductance range of 1–3 nS. Each cycle has 55 up/down pulses on average with $I_G = \pm100$ pA and $t_w = 1$ s. This device has a channel size of 80×100 μm^2.

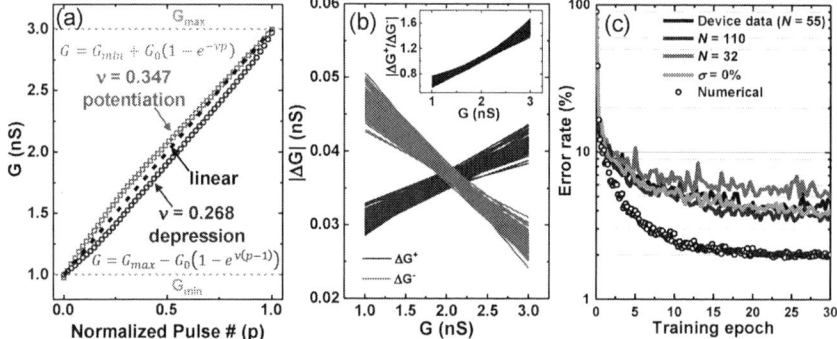

Fig. 4. (a) Nonlinearity analysis on the switching characteristics yields small nonlinearity factors of $\nu = 0.347$ and 0.268 for potentiation and depression, respectively. (b) Plots of ΔG per up/down pulse and asymmetry $|\Delta G^+/\Delta G^-|$ (inset) as a function of G. (c) Taking the cycle-to-cycle variation into consideration, MNIST simulation (assuming another device-to-device variation of $\sigma = 30\%$) shows an accuracy of about 96%, approaching the ideal numerical accuracy. The accuracy is not affected when increasing the number of states ($N = 110$) and reducing variations ($\sigma = 0\%$), but it starts to degrade when N is reduced to 32.

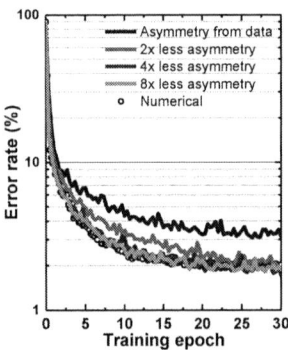

Fig. 5. MNIST simulation accuracy shows clear dependence on the switching asymmetry, which needs to be reduced by just 2× to match the ideal numerical accuracy.

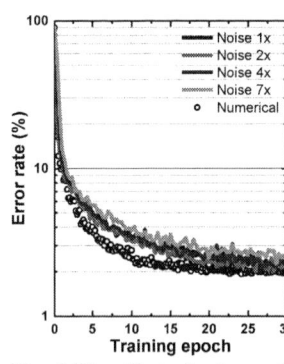

Fig. 6. The effect of noise on the MNIST simulation accuracy, which is shown to be able tolerate up to 7× of the Gaussian noise specified by RPU requirement.

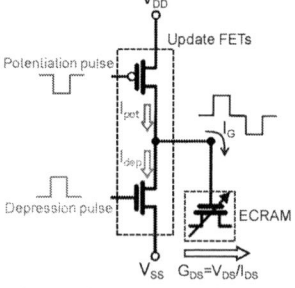

Fig. 7. ECRAM unit cell design for high-speed programming, in which discrete PFET and NFET serve as current source to ECRAM for positive (potentiation) and negative (depression) weight update, respectively.

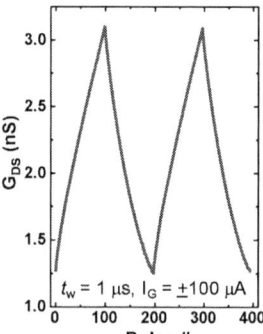

Fig. 8. Reproducible cycling through nonvolatile discrete levels with 1 μs pulses with amplitude of ±100 μA. The pulse period is 1.5 s.

978-1-7281-1988-5/18 $31.00 © 2018 IEEE

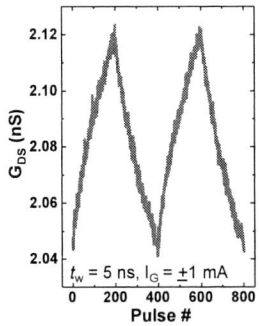

Fig. 9. Reproducible cycling with 5 ns pulses with amplitude of ±1 mA. The pulse period is 1.5 s. The slightly degraded switching symmetry is due to the non-ideal characteristics of current source FETs.

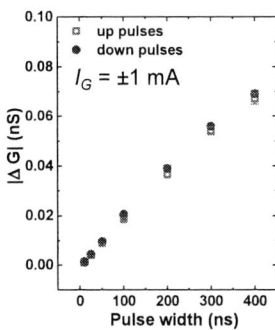

Fig. 10. The average change in conductance ΔG is shown to scale linearly with pulse width from 400 ns down to 5 ns while keeping the same $I_G = \pm 1$ mA.

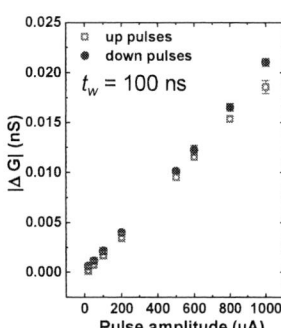

Fig. 11. The average change in conductance ΔG also scales linearly with pulse amplitude from 1 mA down to 20 μA while keeping the identical $t_w = 100$ ns.

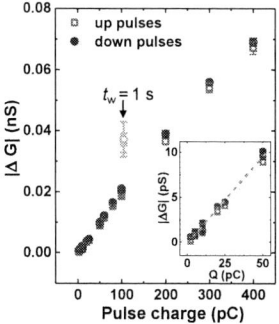

Fig. 12. ΔG scales with pulse charge $Q (= I_G \times t_w)$. The two green data points ($I_G = \pm 100$ pA, $t_w = 1$ s) are from Fig. 3. The "zoom-in" inset shows a linear scaling down to 2 pC per pulse.

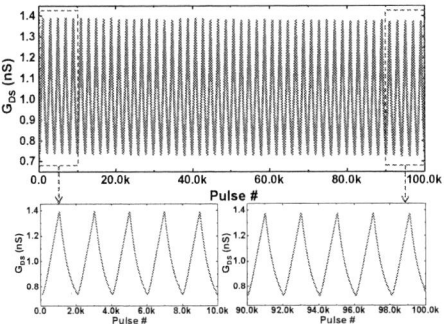

Fig. 13. Endurance test on an ECRAM with 10^5 pulses ($I_G = \pm 100$ μA, $t_w = 100$ ns), showing no degradation in symmetry. There are 10^3 up/down pulses in each cycle while reading the device conductance after every 10^2 pulses.

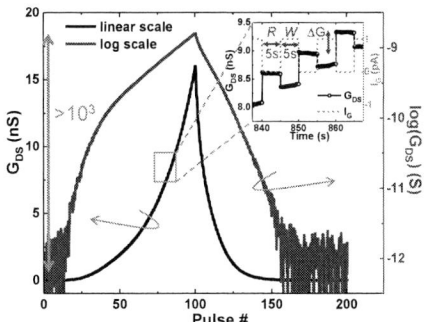

Fig. 14. Demonstration of switching on a scaled ECRAM (1×10 μm^2) using ±1 pA current pulses (pulse width $t_w = 5$ s) for positive/negative weight update. The scaled device still shows good retention and an even larger on/off ratio > 10^3.

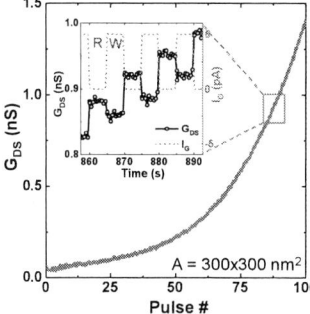

Fig. 15. Programming of a 300×300 nm^2 ECRAM with 100 up pulses ($I_G = 5$ pA, $t_w = 5$ s), showing multi-level operations at scale. Here the ΔG vs A scales differently due to geometry effects.

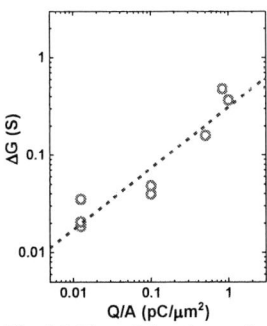

Fig. 16. Plot of the change in conductance ΔG versus pulse charge per device area Q/A, showing a roughly linear scaling.

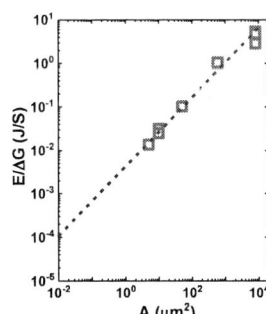

Fig. 17. The switching energy per conductance change $E/\Delta G$ scales linearly with device area A (assuming an average gate voltage ~ 1V).

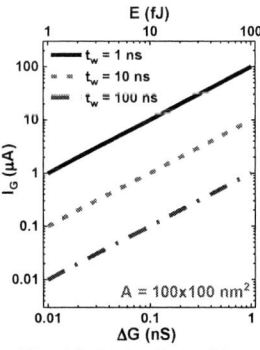

Fig. 18. Projection of the required current pulse amplitudes I_G for given pulse widths t_w in a 100×100 nm^2 device, where the switching energy can be as low as 1 fJ to make a weight update of ΔG = 0.01 nS.

on/off ratio	$40-10^3$
# of states	1000 (tunable)
G range	0−24 nS (tunable)
$\|\Delta G^+/\Delta G^-\|$	0.6−1.6
Nonlinearity ν	0.347 / 0.268
Write pulse width	down to 5 ns
Smallest channel size	0.09 μm^2

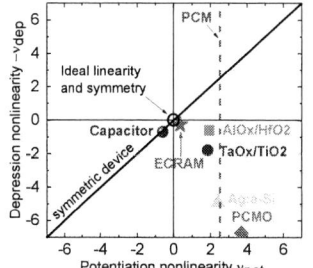

Table 1. Summary of ECRAM key metrics for neuromorphic computing and comparison with other technologies surveyed in Ref. [6-7]. The sign of ν_{dep} is corrected for consistency with literature.

SoC Logic Compatible Multi-Bit FeMFET Weight Cell for Neuromorphic Applications

K. Ni[1], J. A. Smith[1], B. Grisafe[1], T. Rakshit[2], B. Obradovic[2], J. A. Kittl[2], M. Rodder[2] and S. Datta[1]

[1]University of Notre Dame, Notre Dame, IN, USA; [2]Samsung Advanced Logic Lab, Austin, TX, USA; email: kni@nd.edu

Abstract—We demonstrate an SoC logic compatible ferroelectric-metal field effect transistor (FeMFET) digital 2-bit weight cell by monolithic BEOL integration of a ferroelectric (FE) capacitor with the gate of a conventional Si HK/MG MOSFET. Through optimization of the area ratio between the FE capacitor and the MOSFET, we show: 1) program/erase write voltages can be scaled down to logic compatible level, ± 1.8 V, simplifying write circuitry; 2) write speed of 100ns; 3) write endurance $>10^{10}$ cycles without degradation due to elimination of charge trapping in FE; 4) 2 bits/cell achieving software levels of accuracy for inference on MNIST training database; 5) state retention approaching 10^4 s for a depolarization field of 0.3 MV/cm; 6) Multi-port (independent read and write) operations.

INTRODUCTION

Deep neural networks (DNNs) are ubiquitously applied in tasks such as image recognition, natural language processing and self-driving cars (Fig.1(a)). Various neuromorphic accelerators based on the cross-bar architecture with nonvolatile weight cells have been investigated (Fig.1(b)). Neuromorphic accelerator performance is improved via elimination of data movement between the processing unit and off-chip memory by storing the synaptic values locally (on-chip) in the weight cells and performing the multiply-accumulate (MAC) operation in the analog domain. The performance metrics of the on-chip weight cells vary depending on whether an in-situ training is supported or not. To support in-situ training, a weight with high bit precision is necessary to handle the incremental weight update during training, and complex peripheral circuitry is required to implement back-propagation. However, for inference only applications, low precision weight cell can achieve near-analog levels of accuracy. Therefore, by relegating the training to the cloud domain, we focus on a logic compatible embedded weight cell capable of excellent inference performance targeting edge devices [1].

The dynamics of multi-domain ferroelectric polarization switching in FeFET can be utilized for an analog weight cell [2]. However, all reported FeFETs suffer from high write voltages (± 4.0 V) and limited endurance of $\sim 10^5$ cycles (Fig.2) [3]. This is because of the non-uniform field distribution across the gate stack, with the majority of the voltage drop across the interlayer (IL) and the semiconductor channel [4]. This increases the write voltage and degrades the reliability due to enhanced charge trapping in the FE [4]. This inefficiency can be overcome by integrating the FE capacitor in the back-end, electrically coupled to the gate of a conventional MOSFET (Fig.1(d)), rather than direct integration within the MOS gate stack. By independently optimizing the A_{FE}/A_{MOS} (AR: area ratio between MFM and MOSFET) of the metal-ferroelectric-metal (MFM)

capacitor and the MOSFET, the voltage drop across the FE can be maximized, which in turn reduces the write voltage, eliminates charge trapping and improves endurance(Fig.2). In this work, we experimentally demonstrate a 2bit FeMFET weight cell that has a logic compatible write voltage of 1.8V and endurance cycle $>10^{10}$, and achieves the same inference accuracy as floating point weight.

II. DEVICE FABRICATION PROCESS

The key fabrication steps are shown in Fig.3(a). Boron doped (3×10^{17} cm^{-3}) p-type silicon wafers are used to achieve enhancement mode MOSFET operation. After alignment mark etch, source/drain ion implantation and junction activation follow. SiO$_2$ chemical oxide passivation and thermal ALD of 10nm thick HfO$_2$ with 50nm thick tungsten (W) metal electrode completes the gate stack. CVD SiO$_2$ is deposited for MFM capacitor isolation and a W via process is used to connect to the MOSFET gate. W bottom electrode, 10nm thick PEALD Hf$_{0.5}$Zr$_{0.5}$O$_2$ deposited at 300°C and W top electrode forms the MFM stack before 600°C FE crystallization anneal. CVD SiO$_2$ is used for MFM isolation. The FeMFET device fabrication is completed by via etch for connection to W top electrode and source/drain and Ti/Al contact metallization and anneal to form low-resistivity Ti-Silicide contacts.

Top and tilted view SEM images (Fig.3(b)) show the top MFM electrode and the internal gate, which are used to probe the C-V, I-V characteristics of the MFM and MOSFET individually. A zoomed-in view shows the W via that connects the MFM cap to the underlying MOSFET gate. Cross-section TEM images (Fig.3(c)) of FeMFET clearly show the connections between the MFM and MOSFET gate, highlighting the monolithic integration of the MFM with the MOSFET. The HRTEM images reveal poly-crystalline nature of the FE HZO in the MFM.

III. RESULTS AND DISCUSSION

A. FeMFET Operation: Program, Erase and Read

The fundamental device operating principle is first demonstrated by electrically coupling a discrete MFM cap to the gate of a MOSFET (Fig.4(a)). The Q_{FE}-V_{FE} characteristics of MFM cap show the transition from saturated to non-saturated hysteresis loops as the V_{FE} sweep range decreases (Fig.4(b)). When connected with a MOSFET, the DC I_D-V_G characteristics exhibit large hysteresis window (~ 2.0V) for a V_G sweep voltage of -1.8V to 1.8V. The lowering of V_G sweep range reduces the current ratio between the two FE polarization states, which corresponds to minor loop operation of the ferroelectric.

To quantitatively understand the program/erase operation of the FeMFET, a model is developed by introducing the AR design parameter to the Preisach FeFET model [5]. Fig.5(a) shows the bias conditions to operate the FeMFET weight cell.

978-1-7281-1988-5/18 $31.00 © 2018 IEEE

The select transistor eliminates the disturb to half-select cells and pass the write voltage to the FeMFET gate. Fig.5(b) shows the waveform for the cell write and read operations. After erase, the internal gate voltage, V_{MOS}, remains high, whereas it becomes low after program. V_{MOS} after program/erase corresponds to the two intersection points between the MOSFET loadline and the FE subloop during program/erase cycle (Fig.6(a)). Their separation determines the current ratio between the two states and the available cell read window. By decreasing the AR, the MFM capacitance is reduced and the voltage across it is increased (Fig.6(b)). This mitigates the undesirable voltage division issue encountered in FeFET and reduces the write voltage from 4.0V to 1.8V. However, too small of an AR decreases the total polarization charge (Fig.6(c)), which in turn decreases the read window, leading to an optimal AR range maximizing memory window (Fig.6(d)). Furthermore, unlike other nonvolatile memories such as STT-MRAM or RRAM, FeMFET can be operated as a multi-port device due to the separation of read and write path.

B. Integrated FeMFET Device

The integrated FeMFET device characteristics are shown in Fig.7. The MFM capacitor C_{FE}-V_{FE} curve exhibits double peaks, indicating ferroelectric polarization switching. The MOSFET characteristics are shown in Fig.7(b). The C_{MOS}-V_{MOS} measurements indicate that the capacitance ratio C_{FE}/C_{MOS} is 1/8, which increases the voltage drop across the MFM cap, as shown in Fig.6. The FeMFET I_D-V_G characteristics are similar to the discrete MFM and MOSFET measurement. It exhibits ~1.5V hysteresis window for V_G range of -1.8V to 1.8V. Fig.8 shows measured program/erase/read operation waveform of FeMFET with logic compatible write voltages.

The thick gate oxide and small voltage drop (due to small AR) in the MOSFET, ensures that there is no charge trapping in the FE or the internal floating node, thereby markedly enhancing the FeMFET endurance[4]. Fig.9(a) shows that the FeMFET endurance exceeding 10^{10} cycles without degradation. Fig.9(b) shows the read current, I_D, corresponding to the programmed/erased state as a function of write pulse width for write voltages of 2V, 1.8V, and 1.5V. FeMFET achieves fast write operation (of 100ns for a program/erase voltage of 1.8V, while maintaining read current ratio of >10). The retention state of FeMFET is shown in Fig.10. During retention after erase, the internal node voltage V_{MOS} is approximately the MOSFET V_{TH} to have enough sensed current. This would result in voltage drop of $-V_{TH}$ across the MFM cap, which acts as depolarization field, decreasing the FE polarization(Fig.10(a)). Fig.10(b) shows the measured polarization retention under externally applied depolarization field. It suggests that, in order to reduce the polarization loss, it is necessary to design the underlying MOSFET V_{TH} close to 0V.

C. FeMFET Multi-Bit Weight Cell

In this section, we demonstrate multi-bit programming of the weight cell using partial polarization switching in the MFM cap[2]. Fig.11(a) shows a well-tempered distribution of the cycle-to-cycle variation for the four distinct I_D levels (2 bits/cell) for FeMFET. By tuning the erase voltage amplitude, four I_D levels are achieved. The I_D-V_{DS} measurement after write indicates that the programmed conductance values follow approximately a linear behavior (Fig.11(b)), critical for neuromorphic inference application. Higher V_{DS} and shorter channel length can further enhance the FeMFET conductance values and accelerate the read speed.

The deviation of the FeMFET channel conductance values or weights from ideal linearity and its impact on inference accuracy is illustrated in Fig.12(a). A nonlinear fit is performed on the measured weights and used for hardware aware quantization. Fig.12(b) shows the inference error incurred for the MNIST digit recognition task compared with a floating point weight, resulting from the weight quantization effect (1-bit, 2-bits, and 4-bits). It is clear that 2bits/cell is suffice to achieve analog-level accuracy, albeit with increase in number of the hidden layer neurons.

Fig.13(a) shows the FeMFET array architecture for inference. The pseudo-crossbar architecture performs the MAC operation in the analog domain. Two FeMFETs are used to represent positive and negative weights, respectively, which can be achieved by current subtractor (e.g. current mirror). The inference accuracy of the FeMFET array matches that of the floating point weight. These results highlight the promising application of the multi-bit FeMFET weight cell for neuromorphic inference. Fig. 14 benchmarks the FeMFET with other types of embedded nonvolatile memory candidates [6][8],. FeMFET has matched or superior performance in almost all the aspects of write voltage, write energy, latency, and cycle endurance, etc. Thus, FeMFET is a promising weight cell candidate for low power, high performance neuromorphic inference applications for edge devices.

IV. CONCLUSIONS

In summary, we demonstrate an SOC-logic-compatible multi-bit FeMFET weight cell via monolithic BEOL integration of a MFM capacitor with the gate of front-end MOSFET. We show that, by optimizing the area ratio between MFM and MOSFET, we can reduce the program/erase voltage to logic-compatible level of <1.8V, improve write latency to <100ns, improve the endurance beyond 10^{10} cycles. Further, 2bits/cell FeMFET weight cell is demonstrated with high linearity and has no loss in inference accuracy. These characteristics render FeMFET a promising candidate for implementing low power, high performance inference task on edge devices.

ACKNOWLEDGEMENT

This work supported in part by ASCENT, one of six centers in JUMP, Semiconductor Research Corporation (SRC) program sponsored by DARPA

REFERENCES

[1] B. Obradovic, et al., "A multi-bit neuromorphic weight cell using ferroelectric FETs, suitable for SoC integration," JEDS 2018.

[2] M. Jerry, et al., "Ferroelectric FET analog synapse for acceleration of deep neural network training," IEDM 2017.

[3] S. Dunkel et al., "A FeFET based super-low-power ultra-fast embedded NVM technology for 22nm FDSOI and beyond," IEDM 2017

[4] K. Ni et al., "Critical role of interlayer in $Hf_{0.5}Zr_{0.5}O_2$ ferroelectric FET nonvolatile memory performance," TED 2018.

[5] K. Ni et al., "A circuit compatible accurate compact model for ferroelectric-FETs" VLSI 2018

[6] D. Takashima et al., "A 100MHz ladder FeRAM design with capacitance-coupled-bitline cell," JSSC 2011

[7] C. Park et al., "Systematic optimization of 1Gbit perpendicular magnetic tunnel junction arrays for 28nm embedded STT-MRAM and beyond," IEDM 2016

[8] S. R. Lee et al., "Multi-level switching of triple layered TaOx RRAM with excellent reliability for storage class memory," VLSI 2012

978-1-7281-1988-5/18 $31.00 © 2018 IEEE

Motivation: FeMFET as Weight Cell for Neural Network Inference

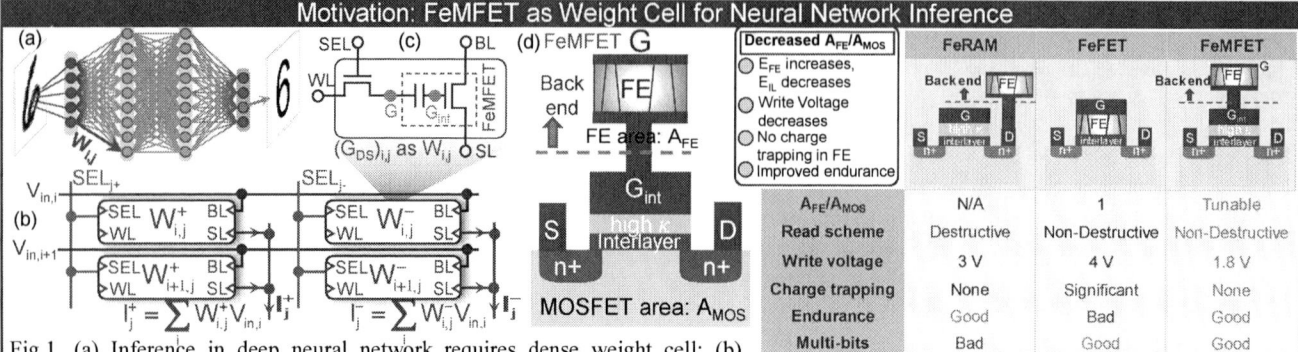

Fig.1. (a) Inference in deep neural network requires dense weight cell; (b) FeMFET pseudo-crossbar array for inference accelerator; (c) FeMFET weight cell circuit diagram, including a select transistor and FeMFET weight cell; (d) FeMFET structure, including a MOSFET and back-end MFM capacitor.

Fig.2. Advantages of FeMFET compared with other types of ferroelectric memories, including write voltage scaling and endurance improvement.

	FeRAM	FeFET	FeMFET
A_{FE}/A_{MOS}	N/A	1	Tunable
Read scheme	Destructive	Non-Destructive	Non-Destructive
Write voltage	3 V	4 V	1.8 V
Charge trapping	None	Significant	None
Endurance	Good	Bad	Good
Multi-bits	Bad	Good	Good

Integrated FeMFET Fabrication Process and Physical Structure

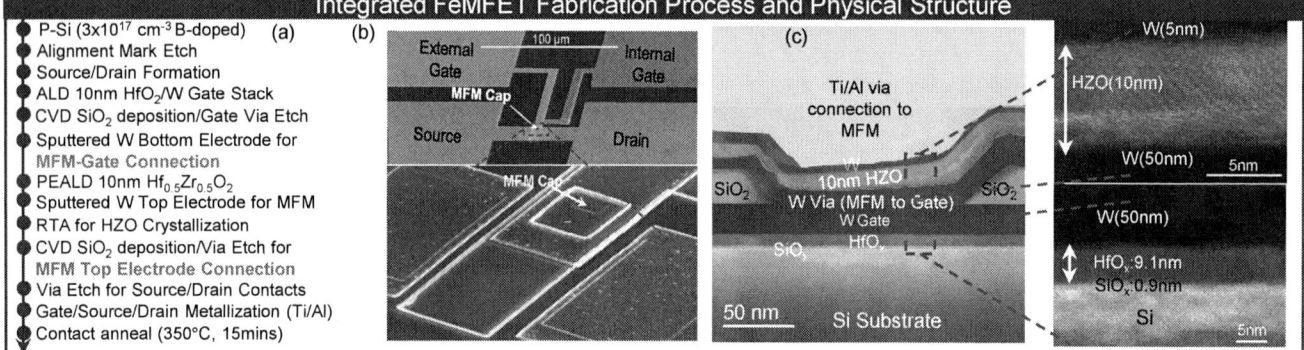

- P-Si (3×10^{17} cm^{-3} B-doped)
- Alignment Mark Etch
- Source/Drain Formation
- ALD 10nm HfO$_2$/W Gate Stack
- CVD SiO$_2$ deposition/Gate Via Etch
- Sputtered W Bottom Electrode for MFM-Gate Connection
- PEALD 10nm Hf$_{0.5}$Zr$_{0.5}$O$_2$
- Sputtered W Top Electrode for MFM
- RTA for HZO Crystallization
- CVD SiO$_2$ deposition/Via Etch for MFM Top Electrode Connection
- Via Etch for Source/Drain Contacts
- Gate/Source/Drain Metallization (Ti/Al)
- Contact anneal (350°C, 15mins)

Fig.3. (a) Key processing steps in the fabrication of FeMFET; (b) SEM images of one FeMFET device (the zoomed in image shows the MFM capacitor); (c) TEM images of the whole stack (W/HZO/W MFM, W via, and W/HfO$_2$/SiO$_2$/Si MOSFET). High-resolution TEM images show poly-crystalline HZO in MFM capacitor.

Proof of Concept: Discrete MFM Cap + MOSFET

FeMFET Program/Erase Operation

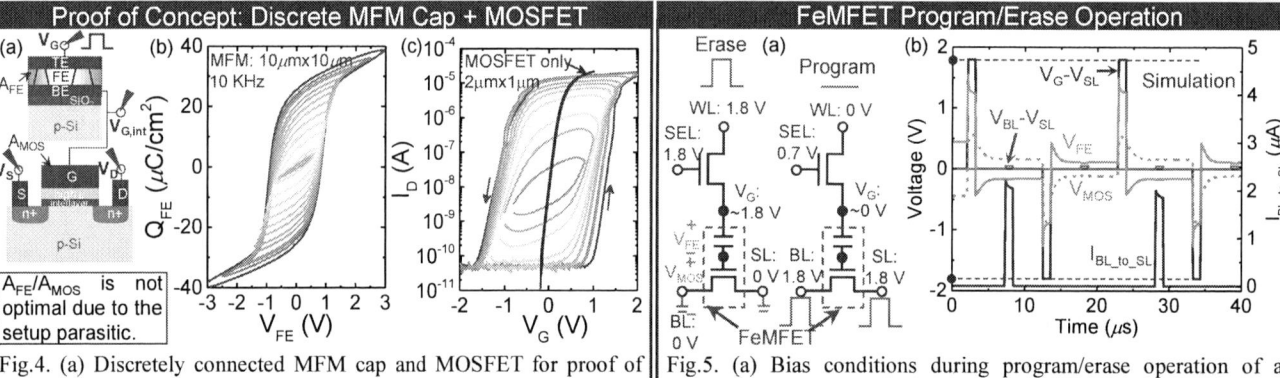

Fig.4. (a) Discretely connected MFM cap and MOSFET for proof of concept; (b) MFM Q_{FE}-V_{FE} loops for different V_{FE} ranges; (c) I_D-V_G characteristics showing hysteresis window ~2V for V_G (-1.8V to 1.8V).

Fig.5. (a) Bias conditions during program/erase operation of a FeMFET weight cell; (b) Simulated waveform of write and sensing operation on the FeMFET cell.

FeMFET Operation Principle and Modeling

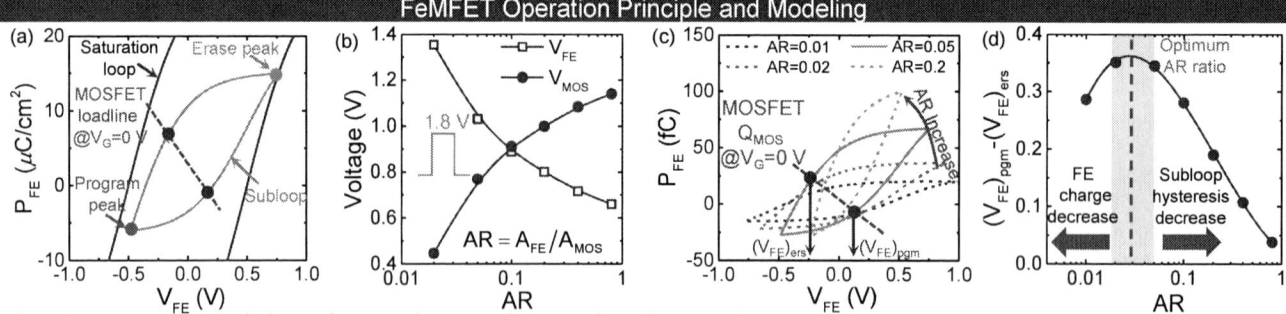

Fig.6. (a) P_{FE}-V_{FE} hysteresis loops for several program/erase cycles. The saturation loop is shown in black line. The MOSFET loadline at V_G=0 V is shown in red; (b) V_{FE} increases (V_{MOS} decreases) with the decrease of A_{FE}/A_{MOS} ratio; (c) P_{FE}-V_{FE} subloop modulation as a function of the area ratio (intersection points between the P_{FE}-V_{FE} loop and Q_{MOS}-V_{FE} loadline, red dashed line, sets the memory window); (d) Memory window vs. area ratio characteristic shows a peak (small ratio reduces the total charge and large ratio reduces FE voltage)

978-1-7281-1988-5/18 $31.00 © 2018 IEEE

FeMFET Device Characteristics

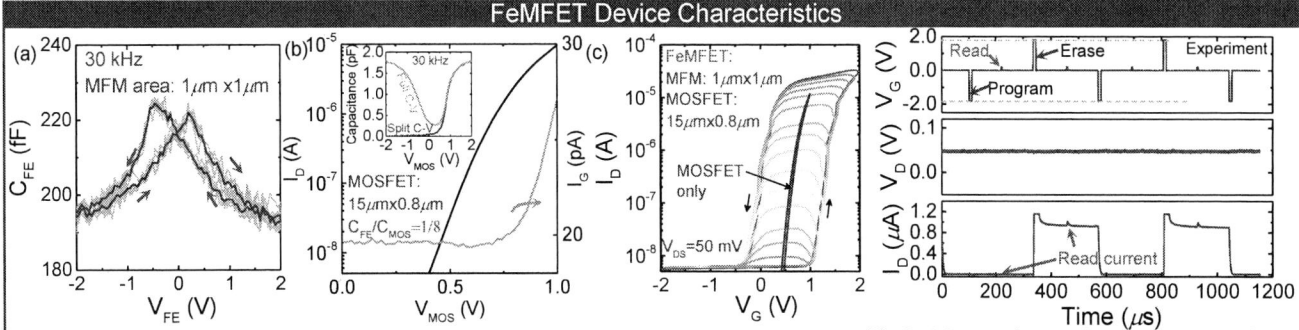

Fig.7. (a) Measured MFM cap C_{FE}-V_{FE} characteristics shows good ferroelectric behavior; (b) MOSFET I-V and C-V characteristics. C_{FE}/C_{MOS}=1/8 implying efficient voltage dividing; (c) I_D-V_G characteristics of FeMFET for different V_G sweep ranges, showing ~1.5V hysteresis window. The reduction of I_D with V_G range corresponds to FE operating on minor loops.

Fig.8. Measured waveform of program/erase operation and sensing of the memory state on a FeMFET device. ±1.8 V write voltage is applied. Successful memory operation is shown.

Endurance and Speed

Retention

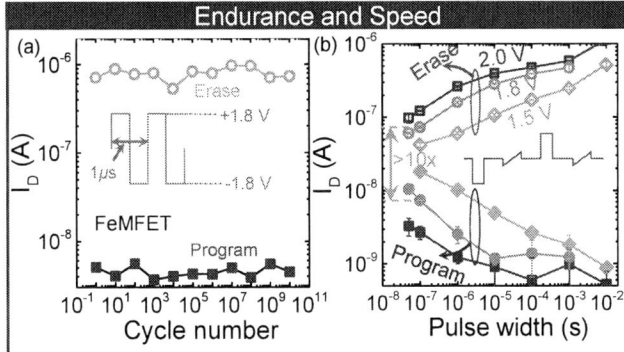

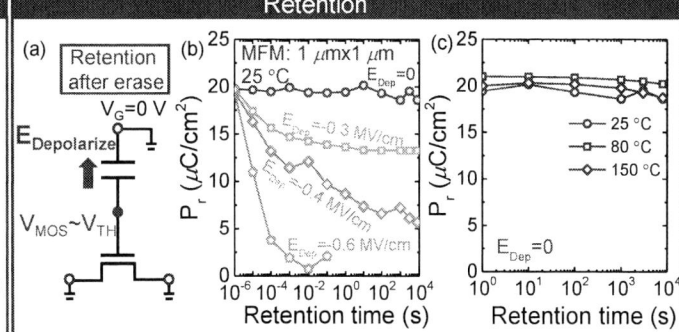

Fig.9.(a) Endurance of FeMFET device. No degradation is observed after 10^{10} cycles due to small write voltage (1.8 V) and minimized charge trapping; (b) I_D, as a function of write pulse width. I_D ratio >10 between program/erase state under ±1.8V, 100ns is achieved.

Fig.10.(a) After erase, finite depolarization field is present in the MFM cap, which could cause retention loss; Polarization retention of MFM cap under (b)externally applied depolarization field at 25 °C and (c) different temperature. Properly designed V_{TH} close to 0 V is necessary.

FeMFET Weight Cell and System

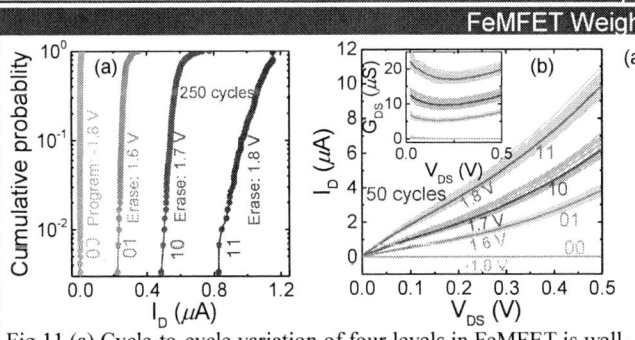

Fig.11.(a) Cycle-to-cycle variation of four levels in FeMFET is well-behaved; (b) Measured I_D-V_D characteristics show small non-linearity. Large I_D can be obtained with large V_{DS} for fast read speed.

Fig.12.(a) Comparison between the measured FeMFET weights and ideal linear weight; (b) Errors induced by quantization of the ideal linear weight (2 bits) is negligible compared with floating point weight by increasing the number of hidden neurons.

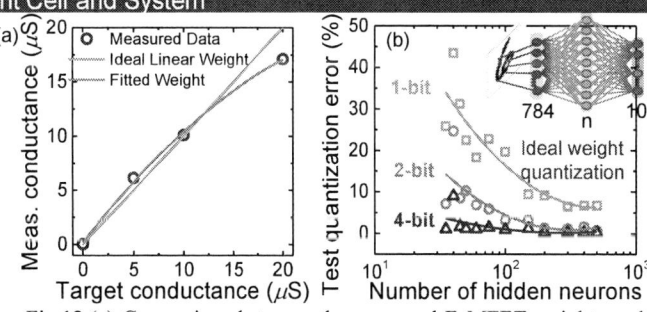

Fig.13.(a) FeMFET array architecture for inference; (b) Inference error due to FeMFET weight quantization compared with floating point weight.

Fig.14. Benchmarking FeMFET with other nonvolatile memories. FeMFET achieves excellent performance in all aspects.

	FeRAM [6]	FeFET [3]	FeMFET (this work)	STT-MRAM [7]	RRAM [8]
Structure	1T-1C (drain)	1T	1T-1C (gate)	1T-1MTJ	1T-1R
Read scheme	Destructive	Non-destructive	Non-destructive	Non-destructive	Non-destructive
Write voltage	3.3 V	4.0 V	1.8 V	1.5 V	4 V
Write energy	~0.1 pJ	~0.1 pJ	~0.1 pJ	~5 pJ	~ 10 pJ
Multi-bit	Bad	Good	Good	Bad	Good
Read speed	Medium	Good	Good	Good	Good
Charge trapping	None	Significant	None	None	None
Endurance	~10^{14}	~10^5	>10^{10}	~10^{15}	>10^6
Variation	Good	Good	Good	Good	Bad
Multi-port	No	Yes	Yes	No	No

978-1-7281-1988-5/18 $31.00 © 2018 IEEE

Experimental Demonstration of Ferroelectric Spiking Neurons for Unsupervised Clustering

Zheng Wang[1,Γ], Brian Crafton[1], Jorge Gomez[2], Ruijuan Xu[3], Aileen Luo[3], Zoran Krivokapic[4], Lane Martin[3,5], Suman Datta[2], Arijit Raychowdhury[1], Asif Islam Khan[1,Ω]

[1]School of ECE, Georgia Institute of Technology, GA 30329, USA, Email: {Γzw@, Ωasif.khan@ece.}gatech.edu
[2] Dept. of ECE, University of Notre Dame, Notre Dame, IN 46556, USA, [3]Dept. of MSE, University of California, Berkeley, CA 94720, USA [4]2321 De Varona Pl., Santa Clara, CA 95050, USA, [5]Material Science Division, Lawrence Berkeley National Laboratory, Berkeley, CA 94720, USA.

Abstract—We report the first experimental demonstration of ferroelectric field-effect transistor (FEFET) based spiking neurons. A unique feature of the ferroelectric (FE) neuron demonstrated herein is the availability of both excitatory and inhibitory input connections in the compact 1T-1FEFET structure, which is also reported for the first time for any neuron implementations. Such dual neuron functionality is a key requirement for bio-mimetic neural networks and represents a breakthrough for implementation of the third generation spiking neural networks (SNNs)—also reported herein for unsupervised learning and clustering on real world data for the first time. The key to our demonstration is the careful design of two important device level features: (1) abrupt hysteretic transitions of the FEFET with no stable states therein, and (2) the dynamic tunability of the FEFET hysteresis by bias conditions which allows for the inhibition functionality. Experimentally calibrated, multi-domain Preisach based FEFET models were used to accurately simulate the FE neurons and project their performance at scaled nodes. We also implement an SNN for unsupervised clustering and benchmark the network performance across analog CMOS and emerging technologies and observe (1) unification of excitatory and inhibitory neural connections, (2) STDP based learning, (3) lowest reported power (3.6nW) during classification, and (4) a classification accuracy of 93%.

I. INTRODUCTION

In spite of staggering successes of deep neural networks (DNNs), the second generation of neural networks (NNs), we have come to the realization that true advances in cognitive systems will require autonomous agents to learn from the environment without the need for labelled data. Unsupervised learning provides such a paradigm. In particular, unsupervised learning and clustering in spiking neural networks (SNNs)– which represent the third generation of NNs–emulate neural properties via their coupled dynamics. Recent advances in neurosciences as well as estimation theory have revealed the advantages of data-encoding through spike-timing: compactness, sparsity, and the ability to learn via local updates only (STDP), all of which have led to efficient hardware implementations of *at-scale* SNNs [1,2].

In parallel, emerging nanodevices such as resistive RAMs, memristors, spin and metal-insulator transition devices offer significant benefits in terms of power, performance and area as physical hardware platforms for implementing NNs—thanks to their unique properties that are not intrinsic to the CMOS technology. A template leaky-integrate-and-fire (LIF) neuron implemented with any of these technologies provides promising opportunities, but they all suffer from a fundamental shortcoming. All these neurons are excitatory, which means that inputs coming to these neurons result in a spike generation. However, it is well known in neurobiology and also in biomimetic neuromorphic architectures that excitatory neurons need to be paired with inhibitory connections to enable homeostasis, high accuracy in unsupervised learning and increased sparsity in spiking–all of which are essential to implement functionally correct and efficient compute models.

In this paper, we introduce ferroelectric field-effect transistor (FEFET) as the underlying device technology for implementing SNNs, and demonstrate, for the first time, ferroelectric spiking neurons—the functional unit of SNNs— with built-in excitatory (exc.) and inhibitory (inh.) input connections, which (1) inherently demonstrate bio-mimetic dynamics, and (2) leads to compact and efficient implementation of neurons and hence the synaptic weights.

II. EXPERIMENTAL DEMONSTRATION OF FERROELECTRIC NEURON

The core structure of the FE neuron consists of a ferroelectric FET (FEFET) and a MOSFET (the discharge FET) (fig. 1(a)). An important feature of our demonstration is that, FEFET being a three terminal neuromorphic device with an intrinsic transistor gain, allows for handling both excitatory and inhibitory input connections in this simple, area efficient two transistor neuron structure. The excitatory and the inhibitory inputs (V_x and V_i, respectively) are connected to the gate terminals of the discharge FET (V_{GM}) and FEFET (V_{GF}), respectively, through respective leaky integrators (fig. 2(a)). The output of neuron V_N is at the voltage across the capacitor C which is digitized by an inverter at V_O. The FEFET was consists of a L_G=80 nm n-FinFET with 14 fins with its gate terminal connected to an epitaxial 100 nm thick $Pb(Zr_{0.2}Ti_{0.8})O_3$ (PZT) ferroelectric. The key to our demonstration of spike generation and inhibition in FE neuron is twofold: (1) the existence of abrupt transitions in the hysteresis edges of the current-voltage characteristics of our FEFET much below the thermal limit with

978-1-7281-1988-5/18 $31.00 © 2018 IEEE

no stable states therein, and (2) the dynamic tunability of the location and the width of the FEFET hysteresis by bias conditions which allows for inhibition—both of which are achieved by careful design of the experiments (Fig. 2(a)).

To explain the operating principle, we refer to the load-line analysis shown in fig. 3(b). Fig. 3(b) plots the measured d.c. drain current of the FEFET I_{DF} as a function of its source voltage (V_N) while keeping the gate and drain voltages (V_{GF} and V_D) fixed. Also plotted in fig. 3(b) are the output characteristics of the discharge FET (I_{DM}-V_N curves at different V_{GM}).

<u>Resting state:</u> When at rest (no exc. V_x and inh. V_i input spikes, V_{GM} and V_{GF} are constant), the operating point of the system is the intersection of the I_{DF}-V_N (FEFET) and I_{DM}-V_N (discharge FET) curves which is point F in fig. 3(b).

<u>Neuron firing:</u> Upon the arrival of an exc. spike train at V_x, V_{GM} reaches the threshold (V_{GM}=1.15 V for V_{GF}=1 V as in fig. 3(b)) for firing when the corresponding I_{DM}-V_N curve intersects the abrupt transition regions of the I_{DF}-V_N curve of the FEFET. Note that there are no stable states in these transition regions, and hence, the system oscillates through the loop ABCD in fig. 3(b) [3]. Fig. 4(a) shows the measured waveform of the neuron output voltage V_N in response to an exc. input spike train V_x. of period T= 28 ms and no inh. input spikes (thereby, V_{GF}=1 V). Output spikes have a period of 27.9 ms. During this input spike train, V_{GM} varies between 1.225 V and 1.425 V (corresponding load-lines drawn in fig. 3(b)). In the zoomed-in version of the output spikes in fig. 4(a), note that the capacitor C is discharged during the transition A→B through the discharge FET, and during C→D, the FEFET charges the capacitor C. Fig. 4(b) shows the evolution of the V_N waveform as the exc. input spike period T is changed from 26 ms to 300 ms. The decrease of the period decreases the output firing rate and at T=300 ms, no firing is observed.

<u>Neuron Inhibition:</u> Fig. 5(a) shows the measured neuron output voltage waveform V_N and the digital output voltage V_O in response to an exc. input V_x spike train of period T=32 ms starting at t=0 and an inh. input V_i spike train of duration 0.18 s starting at t=1.025 s. When both exc. and inh. inputs spikes are present, the output spikes are inhibited and the average voltage level of V_N moves to a higher value. The digitized outputs are also shown in fig. 4(a). The key to neuron inhibition is the fact that the hysteresis in the I_{DF}-V_N curve (FEFET) becomes narrower and shifts to the right when V_{GF} increases (fig. 3(b)). This effect arises due to that the location and the width of the hysteresis in I_D-V_{GS} curve of the FEFET depends on the value of V_D (fig. 2(a)). Moreover, the hysteretic transition is actually not abrupt enough when V_{GF}=1.6V to travel around the hysteresis. The inh. input V_i spikes raise V_{GF} from 1 V to 1.6 V; the corresponding the I_{DF}-V_N curve with V_{GF}=1.6 V is shown in fig. 3(b). The I_{DM}-V_N curves corresponding to V_{GM} swing (1.225 V to 1.425 V) intersects to the I_{DF}-V_N curve at V_{GF}=1.6 V at point S and P where hysteretic transition is not steep enough. As such, in this case, the neuron does not fire, and V_N moves back and forth between S and P. Fig. 5(b) shows that a relatively large, single exc. input V_x pulse can generate a series of chirped output spike train which is inhibited when an inh. input V_i pulse arrives (fig. 5(c)).

III. SPICE SIMULATION & PERFORMANCE PROJECTION

A SPICE model for the FEFET was developed using multi-domain Preisach model [4] (fig. 6(a)) and calibrated with experimental results by considering the device geometry, measured FE hysteresis loop, and parasitic capacitances (fig. 6(b)). An accurate and quantitative agreement between the simulated and experimentally measured FE neuron waveforms and the rate coding is observed in fig. 6(d) and 6(e), respectively. The performance of a scaled FE neuron was projected at the 45 nm node using the PTM 45nm model (PSD shown in fig. 6(f)). The scaled 45 nm node FE neuron dissipates 0.36 nJ/cycle which is an improvement of 390x compared to the experimental one (0.13 μJ per cycle). The SPICE FE Neuron model was also used to simulate a specific topology of neuromorphic networks (Fig. 6(g) and 6(f)) which is used to implement and benchmark the spiking neural network (SNN) described in section IV.

IV. SNN IMPLEMENTATION AND BENCHMARKING

The experimentally calibrated SPICE models have been used to evaluate network performance for unsupervised learning. We emulate the dynamics of a fully connected network (Fig. 7a) with 784 input neurons, 400 excitatory and 400 inhibitory neurons. We apply images from the MNIST data-set and use STDP to learn synaptic weights over. Fig. 7(b) illustrates how the network performs unsupervised clustering over the data-set over training examples and clusters become stronger as training progresses. This is also reflected in Fig. 7(c) and (d) where the square of the change of weights decreases and the classification and clustering accuracy increases. We benchmark the network performance across analog CMOS and emerging technologies and observe (1) unification of excitatory and inhibitory neural connections, (2) STDP based learning, (3) lowest reported power (3.6nW) during classification, and (4) a classification accuracy of 93%.

Acknowledgement: This work was supported by NSF (grant# 1718671, DM-1708615, DMR-1608938), by ASCENT and C-BRIC, two of six centers in JUMP, a Semiconductor Research Corporation (SRC) program sponsored by DARPA and by ARO (grant# W911NF-14-1-0104).

REFERENCES

[1] Merolla et al., "A million spiking-neuron integrated circuit with a scalable communication network and interface," Science, 345, 668-673, 2014.

[2] M. Davies et al., "Loihi: A Neuromorphic Manycore Processor with On-Chip Learning," in IEEE Micro, 38, 82-99, 2018.

[3] Z. Wang, et al., "Ferroelectric oscillators and their coupled networks," IEEE Trans. Electron Dev., 38, 1614-1617, 2017.

[4] K. Ni, et al., "A Circuit Compatible Accurate Compact Model for Ferroelectric-FETs," Proc. Symp. VLSI Tech., 2018.

[5] G. Indiveri, "A low-power adaptive integrate-and-fire neuron circuit," Proc. ISCAS 2003, 4, IV-IV.

[6] A. Amravati et al. "A 55nm time-domain mixed-signal neuromorphic accelerator with stochastic synapses and embedded reinforcement learning for autonomous micro-robots." Proc. ISSCC 2018.

[7] M. Jerry, et al. "Ultra-low power probabilistic IMT neurons for stochastic sampling machines." Proc. 2017 Symp. onVLSI Tech., pp. T186-T187

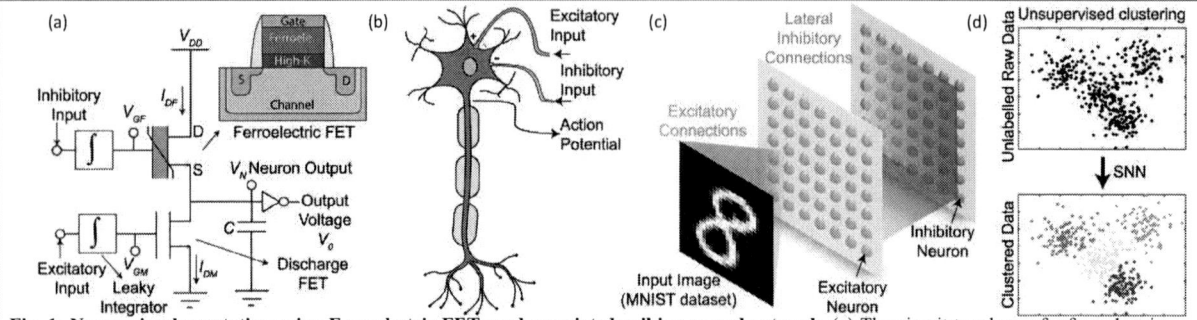

Fig. 1: Neuron implementation using Ferroelectric FETs and associated spiking neural network. (a) The circuit topology of a ferroelectric spiking neuron with excitatory and inhibitory inputs. (b) A biological neuron. (c) A schematic representation of spiking neural network (SNN) with an excitatory and an inhibitory neuron layer. (d) The concept of unsupervised clustering on unlabeled, raw data using an SNN .

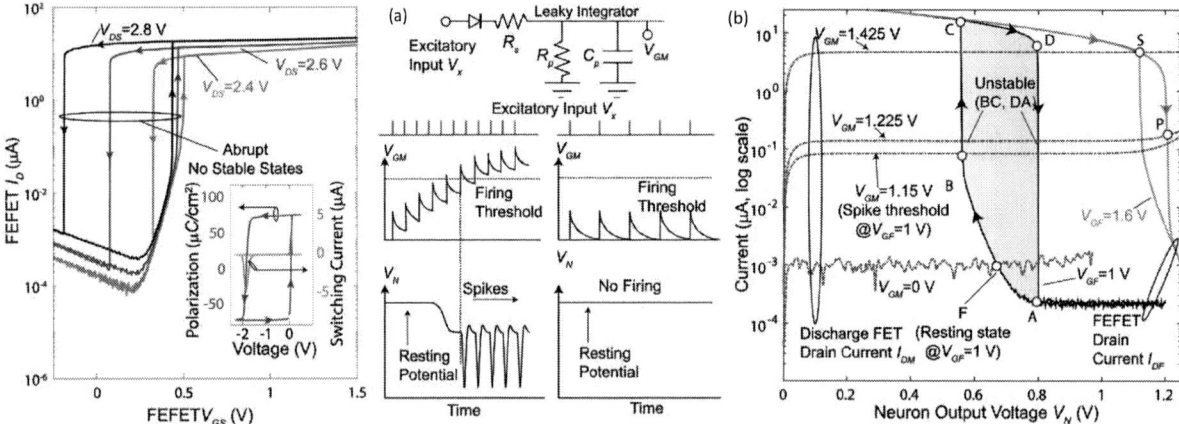

Fig. 2: Device characterization. Measured d.c. I_D-V_{GS} characteristics of the FEFET. Polarization-voltage and switching current-voltage characteristics of epitaxial PZT ferroelectric in the FEFET structure at 150 Hz (inset). The transitions for V_{DS}=2.6 V and 2.8V are abrupt. However, for V_{DS}=2.4V, only the downward transition is abrupt.

Fig. 3: Operating principle and Load line analysis of the experimental ferroelectric neuron (a) Operating principle of the ferroelectric neuron and circuit diagram of the leaky integrator. (b) Load-line analysis. Measured d.c. I_D-V_N characteristics of FEFET at V_{GF}=1 V and 1.6 V and V_D=V_{DD}=3.3 V and output characteristic (I_M-V_S) of the discharge FET. At V_{GF}=1.6 V, the hysteresis in I_D-V_S characteristics shifts to the right compared to that at V_{GF}=1 V. The neuron generates spikes when the discharge FET load lines intersect the FEFET I_D-V_S curves in the unstable transition regions (i.e., BC and DA @V_{GF}=1 V). For V_{GF}=1.6V, the hysteretic transition is not abrupt enough to travel around the hysteresis (*i.e.*, move back and forth between S and P).

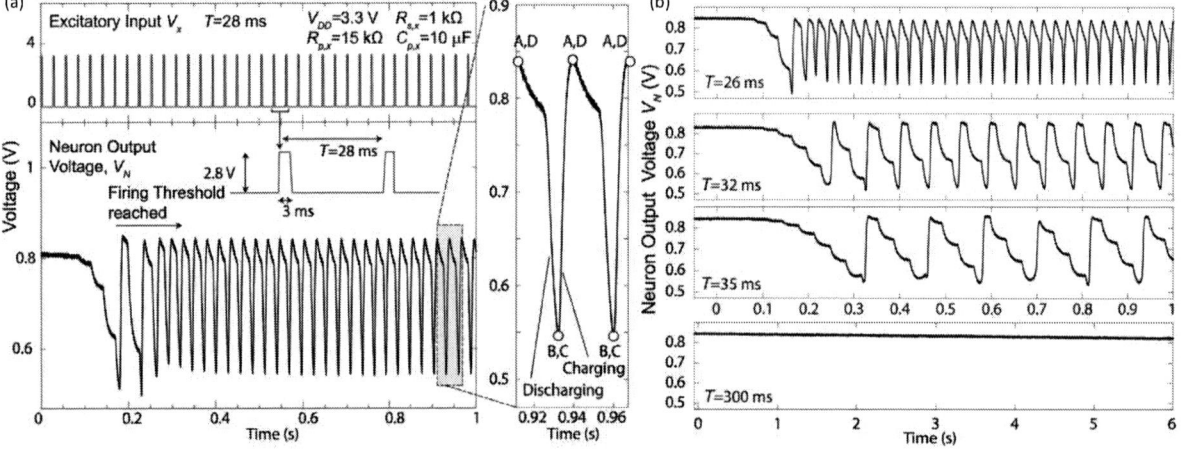

Fig. 4: Experimental demonstration of a ferroelectric neuron. (a) Measured waveforms of the excitatory input V_x and the neuron output voltage V_N in response to excitatory input V_x spike train with period T=28 ms and no inhibitory input spikes at V_i. The output spikes have a reverse polarity compared to the usual polarity of biological neurons. The output spike period is 27.9 ms. Also shown in a zoomed in version of two spikes. During a spike, the state of the FEFET approximately traverses the path ABCD as shown in the load line analysis shown in fig. 3(b). (b) The neuron output voltage V_N in response to excitatory input spike trains with periods T=26, 32, 35, and 300 ms and no inhibitory input spikes. The output spike period is 24.8, 63.8, 114.1 ms for T=26, 32, 35 ms, respectively. The neuron do not output any spikes for T>100 ms.

978-1-7281-1988-5/18 $31.00 © 2018 IEEE

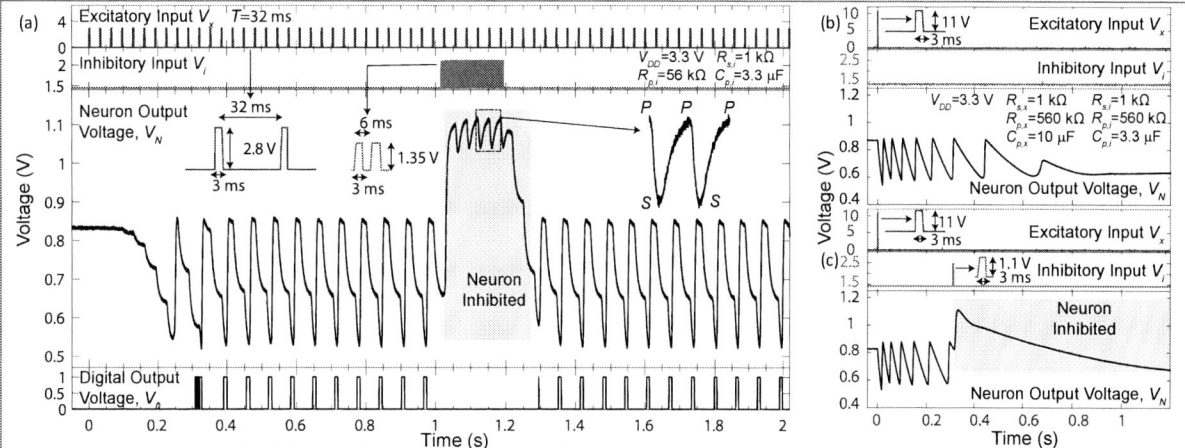

Fig. 5: Demonstration of Inhibition in Ferroelectric Neuron. (a) Measured waveforms of the excitatory input V_x, inhibitory input V_i, and the neuron output voltage V_N in response to excitatory input spike train with period T=32 ms and inhibitory input spike train of 0.18 s duration starting at t=1.025 s. When both inhibitory and excitatory input is present, the neuron do not fire and do not generate any spikes. The crests and troughs of the inhibited neuron output correspond to point S and P in the load-line analysis shown in Fig. 2(b). (b,c) A large spike in the excitatory input V_x generates a chirped output spike train (fig. c) which is be inhibited when an inhibitory pulse at V_i is arrives during this spiking mode (fig. c).

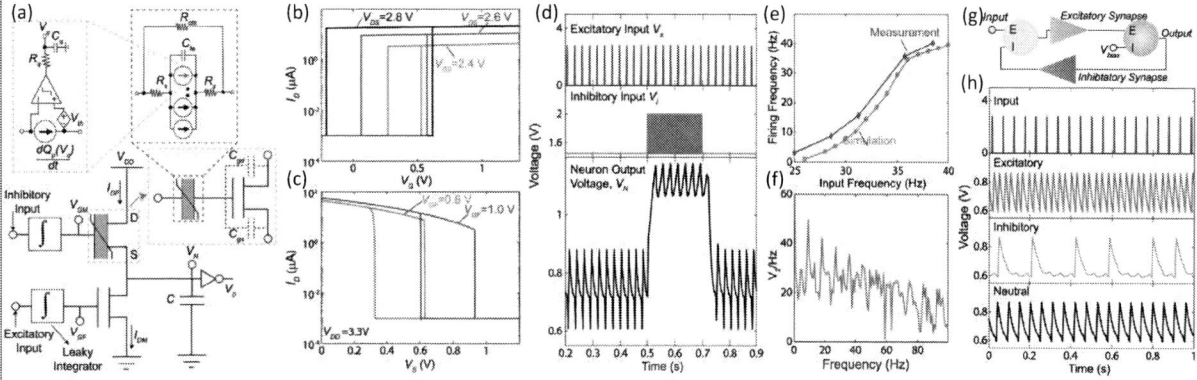

Fig. 6: SPICE Simulation of Ferroelectric Neuron. (a) SPICE model of the FE neuron. (b, c, d) Simulated the I_D-V_G (b) and I_D-V_S (c) characteristics for the FEFET and the neuron waveforms (c) under similar experimental conditions presented in Fig. 5(a) showing reasonable agreement between experiment and simulation. (e) Input Frequency versus firing frequency of the neuron. (f) Simulated power spectrum density of a FE neuron projected at 45 nm node. (g,h) The neuromorphic topology of interconnected neurons (g) used in SNN simulation in Fig. 7 and its SPICE simulated behavior (h) when either of the excitatory (red) and inhibitory (green) connections are active or both of them are neutral (black).

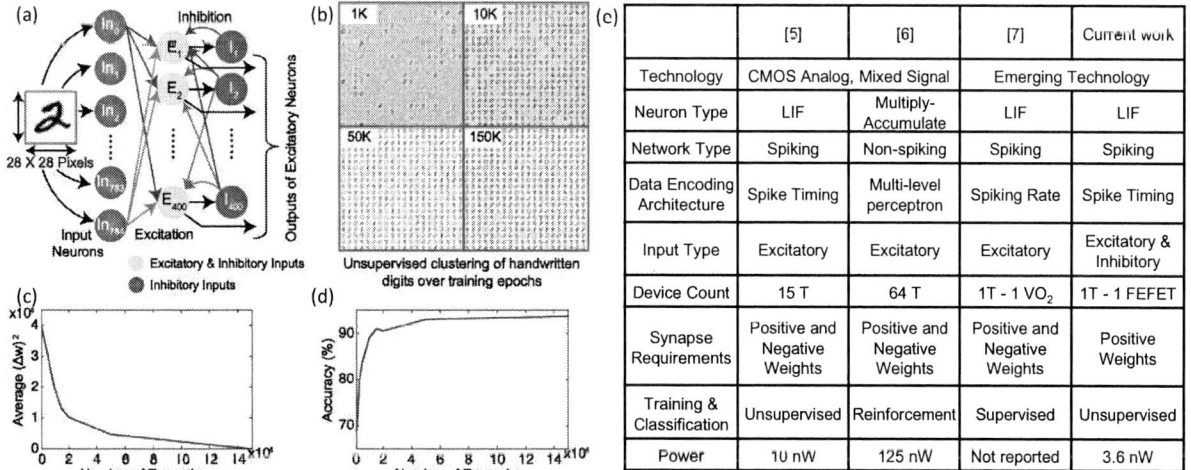

	[5]	[6]	[7]	Current work
Technology	CMOS Analog, Mixed Signal		Emerging Technology	
Neuron Type	LIF	Multiply-Accumulate	LIF	LIF
Network Type	Spiking	Non-spiking	Spiking	Spiking
Data Encoding Architecture	Spike Timing	Multi-level perceptron	Spiking Rate	Spike Timing
Input Type	Excitatory	Excitatory	Excitatory	Excitatory & Inhibitory
Device Count	15 T	64 T	1T - 1 VO$_2$	1T - 1 FEFET
Synapse Requirements	Positive and Negative Weights	Positive and Negative Weights	Positive and Negative Weights	Positive Weights
Training & Classification	Unsupervised	Reinforcement	Supervised	Unsupervised
Power	10 nW	125 nW	Not reported	3.6 nW

Fig. 7: Ferroelectric Spiking Neural Network. (a) Ferroelectric spiking neural network architecture. (b) Illustration of MINST over training epochs of 1k, 10k, 50k, and 150k. (c,d) Average $(\Delta w)^2$ (c) and %accuracy (d) versus number of examples. (e) Benchmark table.

978-1-7281-1988-5/18 $31.00 © 2018 IEEE   303

Near Hysteresis-Free Negative Capacitance InGaAs Tunnel FETs with Enhanced Digital and Analog Figures of Merit below V_{DD}=400mV

Ali Saeidi[1], Anne S. Verhulst[2], Igor Stolichnov[1], Alireza Alian[2], Hiroshi Iwai[3], Nadine Collaert[2], and Adrian M. Ionescu[1]

[1]Nanoelectronic Devices Laboratory, EPFL, Lausanne, Switzerland, [2]imec, Leuven, Belgium, [3]Tokyo Institute of Technology, Tokyo, Japan
Email: ali.saeidi@epfl.ch, anne.verhulst@imec.be, adrian.ionescu@epfl.ch

Abstract— We report the universal boosting impact of a true negative capacitance (NC) effect on digital and analog performances of Tunnel FETs (TFETs), mirrored for the first time in near hysteresis-free experiments and exploiting the S-shaped polarization characteristics. Well behaved InGaAs TFETs with a minimum swing of 55 mV/dec at room temperature are combined with high-quality single crystalline PZT capacitors, placed in series with the gate. When fully satisfying the exact NC matching conditions by a single crystalline ferroelectric that can perform a mono-domain state, *a hysteresis-free (sub-10 mV over 4 decades of current) NC-TFET* with a sub-thermionic swing and an SS_{min} of 40 mV/dec is demonstrated. In other devices, improvement in the subthreshold swing, down to 30 mV/dec, and analog current efficiency factor, up to 180 V^{-1}, are achieved in NC-TFETs with a hysteresis as small as 30 mV. Importantly, the I_{60} FoM of the TFET is improved up to 2 orders of magnitude. The supply voltage is thereby reduced by 50%, down to 300 mV, providing the same drive current. Our results show that NC can open a new direction as a *universal performance booster* in the FET design by significantly improving the low I_{60} and low overdrive of TFETs.

I. INTRODUCTION

Energy-efficient logic devices are required for the enablement of the Internet of Things (IoT) platform. However, the Boltzmann electron energy distribution imposes a fundamental limit to lowering the power dissipation of conventional MOS devices: a minimum increase of the gate voltage, i.e. 60 mV at room temperature, is required to have a 10-fold increase in the current. Among alternative structures, Tunnel Field-Effect Transistors (TFETs) [1] and Negative Capacitance (NC) MOSFETs [2] have attracted a great deal of attention for achieving a sub-60 mV/dec subthreshold swing (SS). TFETs go sub-60 mV/dec by using quantum mechanical band-to-band tunneling (BTBT), rather than thermionic injection, to inject charge carriers into the channel. Meanwhile, the NC of ferroelectric materials is proposed as an important way to bypass the noted fundamental limit. Ferroelectrics are traditionally modeled with a double well energy function (Fig. 1). In equilibrium, the ferroelectric resides in one of the wells and gives rise to a positive capacitance ($C_{FE}^{-1} = d^2 U_{FE}/dQ_{FE}^2$). Nevertheless, it shows an effective NC while switching from one stable polarization state to the other one [3]. It has been proposed that the concept of NC when integrated into the gate stack of TFETs, would be highly beneficial for energy band bending due to the internal voltage amplification, enhancing the BTBT probability. Previous works have mainly focused on the theoretical investigation of NC-TFETs and there are few experimental demonstrations, showing a limited performance [4].

A major drawback of most reported NC devices is that they show large hysteresis, as little effort was dedicated to the analytical design to correctly fulfill the conditions of negative capacitance, which in theory should provide hysteresis-free characteristics. Many of the reported hysteretic and claimed NC devices are, in fact, exploiting a ferroelectric polarization switching and do not meet the initial NC theory of Salahuddin [2], later refined by our group [5]. The important value of the NC effect, with a proper design and matched conditions, is that it can act *as a universal swing and on-current booster for any Field-Effect Device.*

In this report, we have combined the tunneling of carriers as the operation principle with a NC gate and experimentally demonstrated NC-TFETs. InGaAs ring TFETs are investigated as the baseline transistors with a sub-thermionic swing down to 55 mV/dec at RT. Single crystalline PZT capacitors with the ability to form a mono-domain state are employed as the NC booster. Firstly, a NC-TFET with a hysteresis as small as 30 mV shows a significant performance boosting in NC operation conditions. An improved SS down to 30 mV/dec together with an enhanced current efficiency factor, g_m/I_d, up to 180 V^{-1} highlights the important point that the NC effect simultaneously enhances both digital and analog performances of TFETs. In a fully matched design of capacitances, a hysteresis-free NC-TFET with a sub-60 mV/dec swing down to 40 mV/dec, an enhanced g_m/I_d factor with a maximum value of 120 V^{-1}, and an improved I_{60} by nearly 2 orders of magnitude, is achieved.

II. EXPERIMENTAL

The experimental configuration of the NC-TFET of this work is depicted in Fig. 2a where a ferroelectric capacitor is externally connected to the gate of a TFET. To ensure a sufficient enhancement together with a non-hysteretic characteristic, the negative value of the ferroelectric NC (C_{FE}) should be close to the gate intrinsic capacitance of the baseline TFET (C_{MOS}) while the total capacitance of the structure should remain positive in the whole range of operation [4, 6].

For this experiment, 46 nm of high-quality epitaxial $Pb(Zr,Ti)O_3$ (PZT) was grown by pulsed laser deposition on a (110) $DyScO_3$ (DSO) substrate. A 20 nm $SrRuO_3$ (SRO) layer was grown between the substrate and PZT film to serve as the bottom electrode. A 50 nm Pt layer is sputtered and patterned using shadow masking technique. Fig. 3a depicts the schematic of the PZT capacitor (top) along with the crystal structure of PZT in its two stable polarization states (bottom). Reflection high-energy electron diffraction (RHEED) (Fig. 3b) and XRD (Fig. 3c) analysis of the deposited layer confirm the crystallinity as well as the coherency of the epitaxial interface, without any grain boundaries or other extended defects. The polarization hysteresis loop measured at 100 Hz shows a remanent polarization of 80 $\mu C/cm^2$ and coercive voltages of ± 1.2 V (Fig. 3d). The surface of the PZT layer is conformal and the polarization in all c-domains is oriented from bottom to top interface as confirmed by piezoelectric force microscopy (Fig. 3e).

InGaAs homojunction TFETs with 53% In content and a sub-60 mV/dec swing are used as the baseline devices [7]. The fabrication process starts with the Metal Organic Chemical Vapor Deposition (MOCVD) growth of the III-V stack on an InP substrate. The stack consists of a 10 nm InP seed layer, a 90 nm thick doped $In_{0.53}Ga_{0.47}As$ layer as the channel material, 3 nm of InP etch stop layer, and a 50 nm $n+$ $In_{0.53}Ga_{0.47}As$ drain layer. A SiO_2 layer is then deposited as the hard mask and the drain is defined through wet etching (Fig. 4a). The gate stack is deposited next with ALD and consists of a bilayer

978-1-7281-1988-5/18 $31.00 © 2018 IEEE

Al$_2$O$_3$ (1 nm) and HfO$_2$ (2 nm) (Fig. 4b). The EOT for the gate stack is about 0.8 nm. A 100 nm TiN layer is then deposited as the gate metal and is etched to define the source, which is doped via Zn diffusion. The room temperature transfer characteristic and the gate current on the same plot are depicted in Fig. 4c. The TFET achieves a sub-thermionic swing, down to 55 mV/dec, for all drain voltages.

III. RESULTS AND DISCUSSIONS

A. Sub-30 mV Hysteretic Negative Capacitance TFET

In this section, we report a hysteretic NC-TFET where the gate stack of the reference TFET (L$_g$ = 6 µm and W$_g$ = 4×94 µm) is loaded with a PZT capacitor with an estimated area of 15×15 µm^2. The V$_g$ is swept from -1 V to 2 V and back to the starting point while the drain voltage is set to 200 mV. In order to decouple the effect of the threshold voltage (V$_{TH}$) variation, curves are plotted with respect to the effective gate voltage: V$_{gs\_eff}$ = V$_{gs}$ − V$_{TH}$. Fig. 5a illustrates the I$_d$-V$_g$ plot of the NC-TFET compared to the baseline device. The NC of the ferroelectric is partially stabilized, as is schematically explained in Fig. 1, leading to a small hysteresis (ΔV$_{TH}$) of 30 mV, while V$_{TH}$ is extracted at I$_d$ = 10^{-3} µA/µm (the positive going branch is considered for the NC-TFET). The small hysteresis of 30 mV suggests that this condition is not fulfilled only in a limited region [4]. The subthreshold swing is significantly lowered, down to 30 mV/dec, and a sub-60 mV/dec swing over about 3 decades of current is obtained (Fig. 5b).

The I$_{60}$ FoM, which is the maximum current below which the TFET still has a sub-60 mV/dec swing, is improved by more than one decade. Fig. 5c demonstrates that the transconductance (g$_m$) also has a steeper transition and its maximum value is improved by a factor of 10. The extracted g$_m$/I$_d$ factor, an analog FoM, is remarkably improved and reaches a maximum value of 180 V^{-1} (Fig. 5d). To quantitatively determine the voltage amplification of the NC, the internal node is measured and the dV$_{int}$/dV$_g$ curve, defined as the internal gain of NC, is plotted in Fig. 5e. A voltage amplification up to 2 with an average of 1.5 is observed over 40% of the operation range. This internal amplification allows the surface potential to change faster than the gate voltage, leading to an enhanced tunneling probability and reduced subthreshold swing. Due to the voltage amplification, the NC-TFET can demonstrate the same output current at a supply voltage of 0.3 V, which is 50% lower than the one of the reference TFET. The existence of NC effect is confirmed by extracting the P-V characteristic of the PZT capacitor during the NC-TFET operation, which displays a clear S-shape (Fig. 5f). Fig. 5g and 5h summarize the impact of the source-to-drain electric field on the electric performance of the same NC-TFET. The drain voltage is varied from 200 mV to 400 mV while the source contact is grounded. Besides the common impact of V$_d$ on the increase of on-current, it does not severely affect the electrical properties of the NC-TFET. Generally, the variation of V$_d$ changes the operation point of the transistor, the intersection of the TFET charge line and the negative slope of the polarization, which is expected to affect the NC-TFET electrical properties [8]. However, opposite to polycrystalline ferroelectrics, the employed single crystalline PZT capacitor provides a uniform NC over a wide range of applied gate voltages. Therefore, the variation of the operation point does not change the value of the ferroelectrics NC which leads to similar subthreshold swing and voltage gain for different drain voltages.

B. Hyteresis-Free (sub-10 mV) Negative Capacitance TFET

In a different structure, the gate stack of the InGaAs planar TFET is loaded with a PZT capacitor, having the same thickness with a smaller area of 10×10 µm^2, close to the ideal NC matching conditions, supposing a negligible effect of interface trap capacitance [9]. Thereby, a hysteresis-free transfer characteristic of the NC-TFET, around 5 mV at I$_d$ = 10^{-2} µA/µm, is achieved (Fig. 6a). A sub-10 mV hysteresis over the whole range of operation is evidenced (Fig. 6b). A single crystalline ferroelectric that can exhibit a mono-domain state is essential to fulfill the matching conditions. The gate voltage was swept from -1 V to 2 V and back to -1 V at the drain voltage of 400 mV. The gate leakage is negligible compared to I$_d$ and is not impacting the reported effects.

Fig. 6c compares the I$_d$-V$_g$ curves of the NC-TFET and its baseline transistor, showing a remarkable improvement in the steepness of the off-to-on transition, which allows to reduce the supply voltage by 50%. Fig. 6d depicts the SS, showing a sub-60 mV/dec swing down to 40 mV/dec. The I$_{60}$ parameter is improved by 2 orders of magnitude and reaches a value of 10 nA/µm. The transconductance of the NC-TFET exhibits an improvement in the overdrive region, up to 6 times of its original value (Fig. 6e). The internal node measurement confirms the non-hysteretic operation of PZT as the step-up voltage transformer (Fig. 6f). The internal gain, dV$_{int}$/dV$_g$, demonstrates an effective gain higher than 1 over 50% of the NC-TFET operation range (Fig. 6g). The polarization characteristic of the PZT capacitor is extracted (Fig. 6h), showing an effective NC over a wide range of operation.

Overall, NC effect can be employed as an effective universal performance booster of FETs, significantly improving the SS and overdrive. By properly satisfying the matching condition, a hysteresis-free NC-TFET, suitable for logic applications, can be achieved. A well-designed negative capacitor integrated to the gate stack of a TFET significantly increases both I$_{60}$ and the drive current, which are the main challenges involved in the fabrication of TFETs (see Table 1). Extrapolating the improvements to state-of-the-art TFETs [10], the NC effect brings a MOSFET competitive I$_{60}$ = 10 µA/µm within reach. The novelty and universality of this approach relate to the fact that the gate stack is not anymore a passive part of a field-effect transistor and contributes to the signal amplification. Therefore, a NC booster can be applied in parallel with other conventional performance boosters of TFETs.

IV. CONCLUSION

Near hysteresis-free NC-TFETs with a sub-thermionic swing over about 4 decades of current and a minimum of 30-40 mV/dec are experimentally demonstrated by fulfilling the NC matching conditions. The supply voltage can be reduced by 50%, down to 0.3-0.4 V, as a result of the overdrive improvement by NC. The NC effect also enhances the analog FoM of TFETs, in addition to digital performances, showing a new path to advance the performance engineering of TFETs for improved efficiency.

ACKNOWLEDGMENT

The authors thank the ERC Grant MilliTech (695459) for the financial support. This work was also supported by imec's Industrial Affiliation Program. PZT films used in this study were grown by Dr. Ludwig Feigl.

REFERENCES

[1] A. Ionescu, H. Riel. Nature (2011). [2] S. Salahuddin, S. Datta. Nano letters (2008). [3] Asif I. Khan et al. Nature materials (2014). [4] A. Saeidi et al. Nanotechnology (2018). [5] A. Rusu et al. IEDM (2010). [6] G. Pahwa et al. IEEE TED (2018). [7] A. Alian et al. APL (2016). [8] A. Rusu et al. Nanotechnology (2016). [9] W. Chung et al. IEDM (2017). [10] E. Memisevic et al. IEDM (2016).

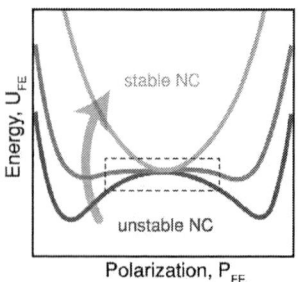

Fig. 1. Energy density of a ferroelectric. NC is unstable by itself, but it can be partially or fully stabilized if it is placed in series with a positive capacitor of proper value.

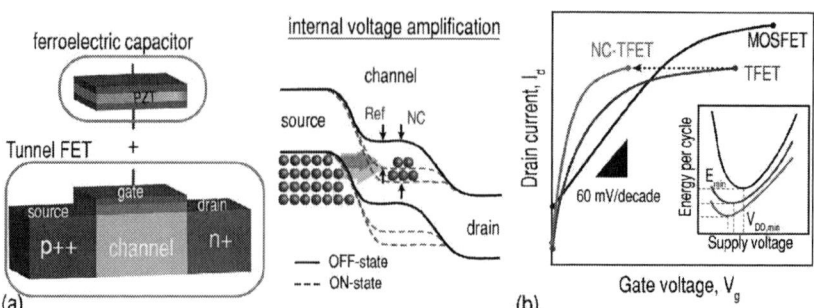

Fig. 2. (a) Schematic of an NC-TFET (left) where a ferroelectric capacitor is connected to the gate of a conventional TFET. This series connected NC booster amplifies the gate voltage and increases the tunneling probability (right). (b) I_d-V_g plot and energy efficiency of a MOSFET, a TFET, and a NC-TFET.

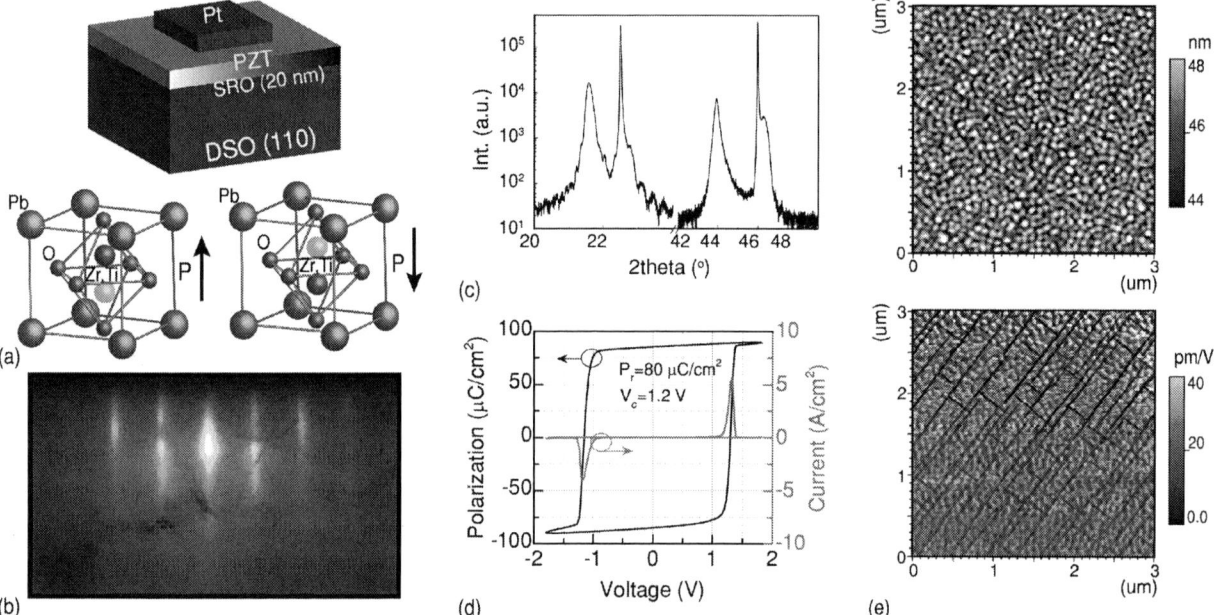

Fig. 3. (a) Schematic of a PZT capacitor (top) and the two stable polarization states of PZT, which occur due to ionic movement (bottom). 46nm of PZT was grown on a (110) DyScO₃ (DSO) substrate. RHEED (b) and XRD (c) analysis of the surface of the deposited PZT confirm the crystallinity of the layer as well as the coherency of the epitaxial interface. (d) The P-V and I-V curves of the PZT capacitor show a sharp and coherent switching. (e) The AFM measurement demonstrates a smooth surface (top) and the piezo-force microscopy confirms that the polarization in all c-domains of PZT are oriented from bottom to top interface (bottom).

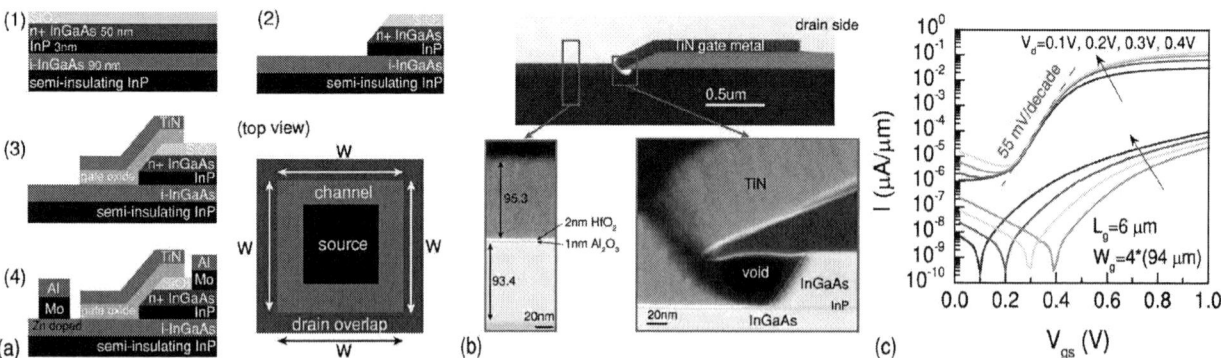

Fig. 4. (a) Process flow and the cross section schematic of the baseline TFET: 1) MOCVD growth and SiO₂ deposition. 2) Drain isolation etch. 3) gate stack (1 nm Al₂O₃ + 2 nm HfO₂) ALD deposition and patterning. 4) Zn diffusion from the gas phase @ 500 °C for 60 s and source/drain contacts with lift-off process. Finally, devices are annealed in forming gas at 400 °C for 15 minutes. (b) TEM image of the cross-section of the TFET, showing a clean surface along the channel and source and a void at the drain to avoid ambipolar current. (c) I_d-V_g and I_g-V_g plots of a TFET with a gate length of 6 μm for different drain voltages, showing a sub-thermionic swing down to 55 mV/decade at 300 °K.

978-1-7281-1988-5/18 $31.00 © 2018 IEEE   306

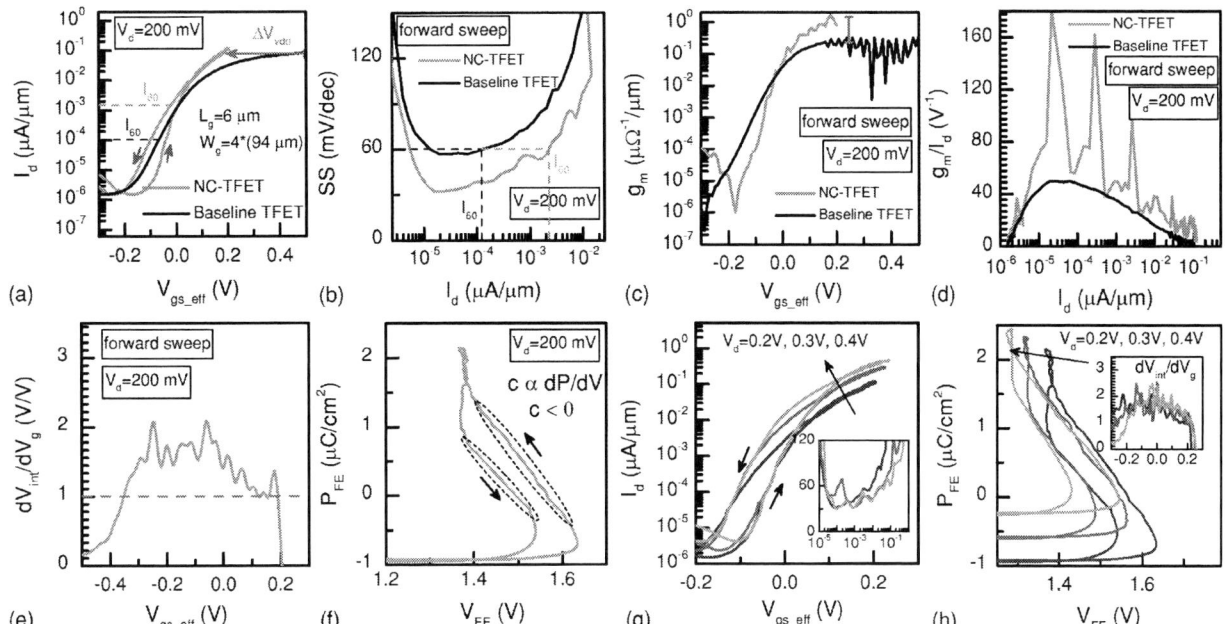

Fig. 5. **Small-hysteresis NC-TFET.** (a) I_d-V_g curve of the NC-TFET compared the base TFET (V_d=200 mV). (b) The SS is significantly reduced, down to 30 mV/dec. (c) g_m is enhanced, up to 10 times of its original value. (d) g_m/I_d is also improved, having a peak of 180 V^{-1}. (e) A voltage gain, dV_{int}/dV_g >1, over about 50% of the device operation range is observed. The extracted S-shaped P-V of the PZT capacitor confirms the existence of an effective NC (f). The variation of V_d on the NC effect shows no major impact on the I_d-V_g (g) and P-V (h) curves of the NC-TFET, confirming a sub-60 mV/dec swing over almost 4 decades of current.

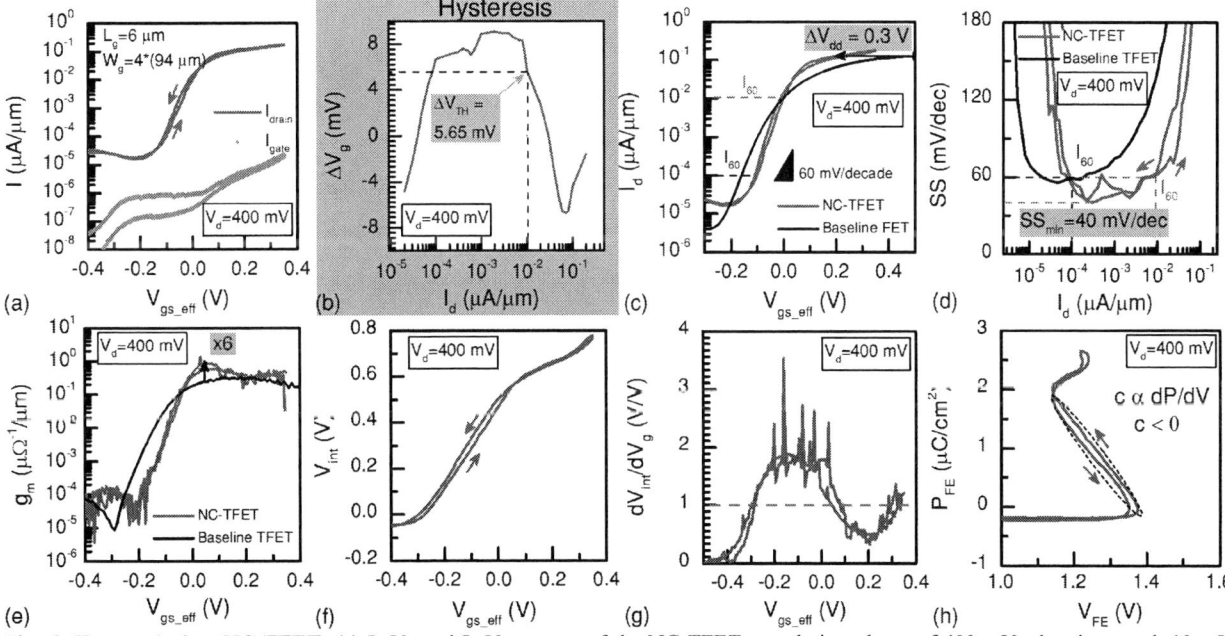

Fig. 6. **Hysteresis-free NC-TFET.** (a) I_d-V_g and I_g-V_g curves of the NC-TFET at a drain voltage of 400 mV, showing a sub-10 mV hysteresis over the whole range of operation (b). Transfer characteristic of the NC-TFET is compared with the base device (c), confirming a significant improvement in the swing (d). Transconductance is also improved (e). The voltage of the internal node is measured (f) and a clear voltage gain over 50% of the operation range is observed (g). The extracted P-V plot confirms the existence of an effective NC with a negligible hysteresis (h).

Table 1. Boosting of digital and analog performances of the hysteresis-free NC-TFET compared the baseline TFET at V_d = 400mV.

	SS_{min}	sub-60mV/dec range	Hysteresis, ΔV_g (10^{-5}-10^{-1} μA/μm)	V_{dd} (I_{on}=0.1 μA/μm)	I_{60}	$(g_m)_{max}$	$(g_m/I_d)_{max}$
TFET	55 mV/dec	10^{-5}-10^{-4} μA/μm	-	0.6 V	0.1 nA/μm	0.18 μΩ^{-1}/μm	50 V^{-1}
NC-TFET	40 mV/dec	8×10^{-5}-10^{-2} μA/μm	±10 mV	0.3 V (1/2)x	10 nA/μm 100x	1.0 μΩ^{-1}/μm 6x	120 V^{-1} > 2x

978-1-7281-1988-5/18 $31.00 © 2018 IEEE

An Experimental Study of Heterostructure Tunnel FET Nanowire Arrays: Digital and Analog Figures of Merit from 300K to 10K

T. Rosca[1], A. Saeidi[1], E. Memisevic[2], L-E. Wernersson[2] and A.M.Ionescu[1]

[1]Nanolab, EPF Lausanne, 1015 Lausanne, Switzerland, email: teodor.rosca@epfl.ch, adrian.ionescu@epfl.ch
[2]Department of Electrical and Information Technology, Lund University, Sweden

Abstract— In this work, we experimentally report the figures of merit of state-of-the-art heterostructure Tunnel Field-Effect-Transistor (TFET) arrays from room (300K) down to cryogenic temperature (10K) at supply voltages below 400mV. We demonstrate here, for the first time, that InAs/InGaAsSb/GaSb Nanowire (NW) TFETs are robust enough to maintain excellent figures of merit over a large temperature range even in devices with a large number arrayed nanowires (here, from 4 to 184 nanowires per device), accounting for technological variability. The investigated Tunnel FETs have temperature-independent min and average subthreshold swings of 45mV/dec/67mV/dec in large NW arrays, versus ~36/45mV/dec in smaller arrays, once the trap-assisted tunneling is removed (from 150K down to 10K). In all NW arrays we observe improvement of the on-current and of maximum transconductance, g_{max}, at cryogenic temperatures, with very little dependence of temperature, from 150K to 10K. The paper reports that in the range 150K to 10K only band-to-band-tunneling dominates the analog figures of merit of Tunnel FETs; we measured transconductance efficiencincies higher than $60V^{-1}$ for small arrays (breaking the limit of CMOS at RT) and close to $42V^{-1}$ for large arrays, for supply volrages smaller than 100mV, offering the possibility to design future energy efficient readouts and analog-to-digital converters. In contrast with cryogenic MOSFETs, Tunnel FETs show almost no hysteresis (<24mV), steep transfer characteristics, are free of kinks in output characteristics, with a unique stability of the swing drift with T, and negligible threshold voltage drift in all arrays configurations.

I. INTRODUCTION

The study of low temperature electronics (LTE) has been motivated by many reasons [1]. It is known that low temperature has a strong beneficial influence on electronic transport and some of the main CMOS parameters such as current, conductance, drift velocity, and noise margins improve. Traditionally, cryogenic electronics has relied on CMOS technology, as the performance of MOSFETs improves at sub-ambient temperatures and down to liquid Nitrogen temperatures (80K). On the other hand, LTE is domain that enables certain explorations and applications that are not yet possible for room temperature electronics (RTE).. That is, certain electronics only run at low temperatures. For instance, superconductors and the Josephson Effect need deep cryogenic temperatures. More recently, there is an increased interest in using semiconducting technology, such as CMOS for Quantum Computing operation below 10K [2,3]. However, challenges become different in the sub-20K range, where CMOS experiences significant performance degradation due to effects such as dopant deactivation and carrier freeze-out, kinks in their output characteristics, hysteresis [4], etc. In addition, key device parameters and subthreshold swing are highly temperature dependent, which adds certain limitations and complexity to the resulting CMOS circuitry.

Recently, steep slope Tunnel FETs – a class of devices based on quantum mechanical band-to-band-tunneling (BTBT), which has very limited dependence on temperature – has shown an interesting potential for medium and deep cryogenic operation. Tunnel FETs are promising solutions for energy efficient digital design and have unique stability in temperature, but their full digital and analog figures of merit at low temperature have not been systematically studied until now. One of their main limitations at room temperature results form a combination of Band-To-Band-Tunneling (BTBT) and Trap-Assisted-Tunneling (TAT), the second being an undesired effect that heavily degrades the steep subthreshold slope normally attainable by BTBT alone. Such effects have been studied mostly in individual Tunnel FET devices with limited current capability and the TAT in arrays of nanowire Tunnel FETs, necessary to produce a useful high level of current, is scarcely reported. Additionally, some authors [5] have demonstrated that at low temperatures, the effect of TAT in Tunnel FETs is minimized and band-to-band-tunneling could prevail as the main carrier transport mechanism. However, until now there was no reported robust tunnel FET technology systematically evaluated down to 10K. This work reports for the first time such detailed characterization on heterostructure tunnel FETs in arrayed configurations, including both digital and analog figures of merit across a wide range of temperatures, with emphasis on values below 150K.

II. DEVICES

The devices investigated in this study are Vertical III-V Nanowire TFETs engineered to achieve sub-thermionic swings below 60mV/dec at room temperature and exhibiting excellent electrostatic control, achieved through advanced scaling and Gate-All-Around (GAA) geometry (Fig. 1). Two device architectures are systematically under test: arrays of 4, and 184 nanowires. In the current study, we will refer to these arrays as the "Small Arrays" and the "Large Arrays", for the 4 and 184 nanowire arrays, respectively. Each individual nanowire used in such arrays has three segments: $InAs/In_xGa_{1-x}As_ySb_{1-y}/GaSb$ with lengths of 160/80/300nm (large array) and 190/90/330nm (small array), and diameters of 25/27/53nm (large array) and 20/22/53nm (small array), respectively. Starting at the InAs/InGaAsSb heterojunction (Fig. 3a) and moving towards the top of each nanowire, the composition of the InGaAsSb segment changes with x= 0.7-0.32 and y= 0.84-0.72. Top half of the InAs segment i.e. channel region (90/80 nm) is the only undoped region of the nanowire. The rest of the nanowire is either n- or p-doped [6]. In case of the large arrays, the nanowires are grown in a zig-zag pattern with a pitch of 300nm, while the small arrays exhibit a pitch of 1.5um between the individual nanowires. Final devices have a 15nm thick silicon dioxide film as bottom spacer. A bilayer of Al_2O_3/HfO_2 is used as high-k dielectric with an estimated effective oxide thickness of 1.4nm. Physical gate length of the final devices is 240nm and

280nm for the 184 NW, and 4NW arrays, respectively, which is formed using a 30nm thick sputtered tungsten film and reactive ion etching. Top-spacer is formed using photoresist and the top-contacts are formed using sputtered Ni/Au-films.

III. EXPERIMENTS AND RESULTS

A. Transfer characteristics and Subthreshold Slope

Figs. 2 and Fig. 4 report measured sets of transfer characteristics and their extracted subthreshold slope values for the small and large arrays, at RT (300K) and 50K. It is clearly noticed a reduction in off-current of both devices, accompanied by a reduction in subthreshold slope, without impacting on-current performance. While the SHR leakage decreases with temperature, the reduction of the TAT results in significantly steeper slope associated with BTBT. It is worth noting that many measurements performed on individual NW Tunnel FETs (Fig. 3) show a significant variability at room temperature due to multiple factors such as dimensional, trap density, and defect variability. In NW Arrays of Tunnel FETs the 'golden device' (such as a deep sub-thermionic swing) effect is averaged out and the real control of a technology can be evaluated. However, the important aspect of our low temperature study is that it removes the variability effect of TAT even in large arrays, which enables a real evaluation of the technology potential. In support of this statement, Fig. 5 and Fig. 7 depict full sets of transfer characteristics measured from 300K to 10K. It is confirmed that below 150K, once the detrimental effect of Trap-Assisted-Tunneling is removed, the characteristics show *outstandingly stable subthreshold swings*, with minimum values saturating near 45mV/dec (small arrays) and 60mV/dec (large arrays), as seen in Fig. 6 and Fig. 7. The stability of the subthreshold slope (SS) with temperature can be explained by the low temperature coefficient of the BTBT current, only via the quasi-linear temperature dependence of the semiconductor bandgap [5]. We have extracted minimum and average values of SS between 10K and 150K and reported the results in Fig. 9; although there is a variation of the minimum SS, from 36 mV/dec to around 44 mV/dec, the average SS values over two decades of current show a very limited variation.

Moreover, as expected, our experiment shows that in the BTBT regime there is little variation of the threshold voltage in both devices. Fig. 10 outlines measured values of the threshold voltage for both devices, showing stable values at cryogenic temperatures, and much lower values at room temperature.

Finally, it is worth noting that we have performed I_D-V_G double-sweep measurements, for hysteresis evaluation at low temperature (Fig. 11). The extracted median value is very low (~25mV) and the hysteresis is decreasing as the temperature is increasing.

B. Output characteristics and Ion and Ioff dependence on Temperature

Both small and large arrays of Tunnel FETs show well-saturated output characteristic and no noticeable cryogenic kink effect. Output characteristics are shown in Fig.12. Fig. 13a depicts the Arrhenius plot of the small and large arrays for various drain voltages between 100 mV to 300 mV. The on-current is measured at a gate bias that assures the linear operation of transistors. The on-current is exponentially increasing by raising the temperature from 150K to 300K, consistently indicating activation energies of 0.26 eV and 0.21 eV for the trap-assisted tunneling of the small and large arrays, respectively. In the range of 10K to 150K, the on-current shows no dependency on the temperature. Fig. 13b

confirms that the off-current of both small and large array transistors, reduces with lower temperatures.

C. Analog Performance

Fig. 14 and Fig. 15 depict the extracted device transconductance (g_m) values, showing that maximum values of g_m are reached below 150K, and saturating near 10K. Another key analog figure of merit, g_m/I_D, is reported for the first time here down to cryogenic temperatures (Fig. 16 and Fig. 17), demonstrating the capability of tunnel FETs to reach values higher than $60V^{-1}$, which is breaking the CMOS limit of $40V^{-1}$ for g_m/I_D at 300K. The maximum values of transconductance and transconductance efficiency are summarized in Fig. 18, showing that tunnel FETs have a remarkable stability of the analog figures of merit at low temperature below 400mV, which can enable low power interface electronic IC design for quantum computing applications. Finally, a full benchmarking of main digital and analog figures of merit of Tunnel FETs at 300K, 150K and 10K is summarized in Table 1. Main analog performance parameters, transconductance and transconductance efficiency, are given, in addition to digital figures of merit, all showing a remarkable stability with temperature when TAT is eliminated, for both small and large arrays of NW tunnel FETs.

IV. CONCLUSIONS

In this work, we explored for the first time the experimental characteristics of heterostructure Tunnel FET nanowire arrays, operating from room temperature (300K) down to 10K for operation voltages below 400mV. We have demonstrated that all major digital and analog figures of merit for TFETs – subthreshold swing, threshold voltage, on-current, maximum g_m, g_m/I_D - exhibit great stability and little variation with respect to temperature, once the detrimental effect of Trap-Assisted-Tunneling is removed, as a consequence of low-temperature operation. The variability of individual TFETs does not influence the overall stability below 150K of large arrays (180NWs), which may suggest that TAT is the main source of device characteristics variability and not other parameters related to BTBT. The excellent device characteristics are supported by extremely low hysteresis found in our devices. Even though the traditional CMOS platform can show better performance at medium-low temperatures and higher voltage, but deep cryogenic operation impacts its performance. Thus, the advantage of the reported Tunnel FET arises from its remarkable consistency across a wide range of temperatures at very low voltage operation, making it an ideal candidate for the fundamental building blocks of low temperature electronics.

REFERENCES

[1] W.F. Clark et al, "Low temperature CMOS – a brief preview", Electronic Components and Technology Conference, 1991. Proceedings., 41st. IEEE, 1991
[2] Charbon, E., et al. "Cryo-CMOS for quantum computing." IEDM 2016
[3] Patra, Bishnu, et al. "Cryo-CMOS circuits and systems for quantum computing applications." IEEE-JSSC 53.1 (2018): 309-321
[4] Ghibaudo, G., and F. Balestra. "Low temperature characterization of silicon CMOS devices." IEEE, 1995
[5] Sajjad, Redwan N., et al. "Trap assisted tunneling and its effect on subthreshold swing of tunnel FETs." IEEE Trans. Electron Devices 63.11 (2016): 4380-4387
[6] Memisevic, Elvedin, et al, *Nano Lett.* 17.7 (2017): 4373-4380.

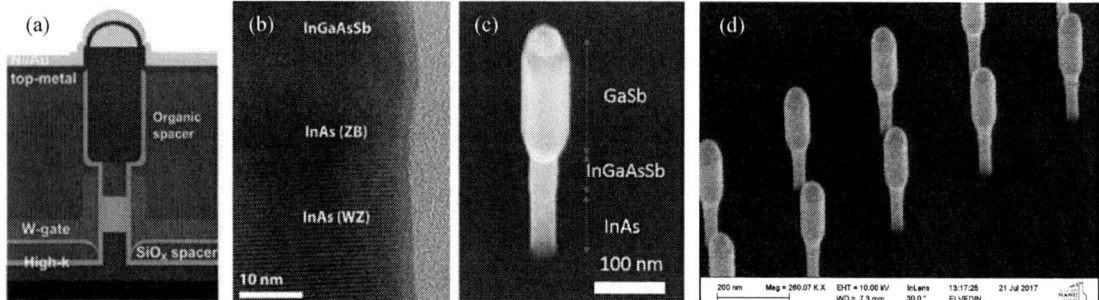

Fig. 1 – Fabrication details. (a) – Cross-section view of InAs/In$_x$Ga$_{1-x}$As$_y$Sb$_{1-y}$/GaSb Nanowire Tunnel FET; (b) – TEM micrograph of a single nanowire; (c) – SEM micrograph of a single nanowire; (d) – SEM micrograph of a section of the 184-nanowire array

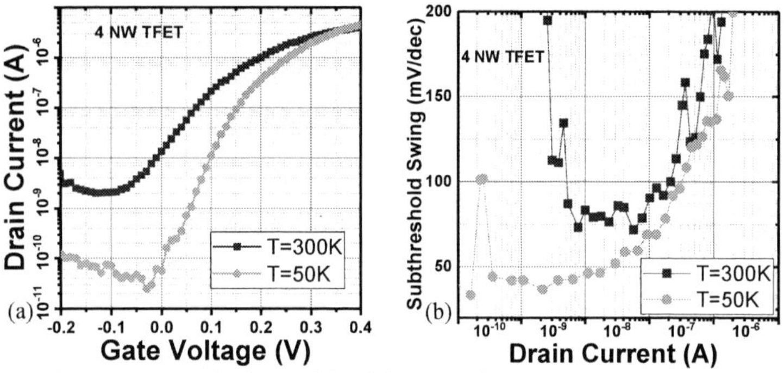

Fig. 2 – (a) – Transfer characteristics of the 4-nanowire TFET at 300K and 50K ; (b) – associated Subthreshold Swing values

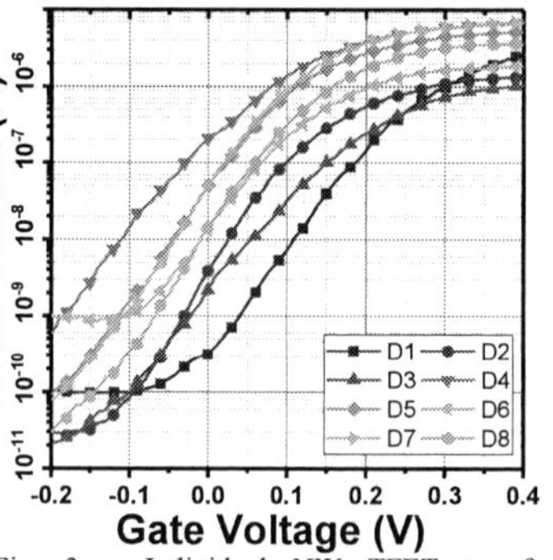

Fig. 3 – Individual NW TFET transfer characteristics, showing significant variability

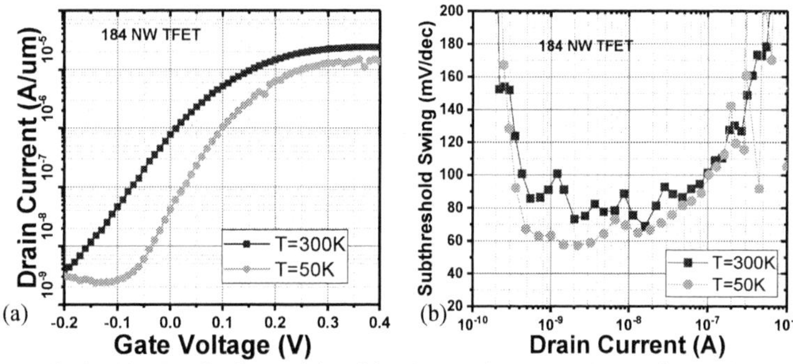

Fig. 4 – (a) – Transfer characteristics of the 184-nanowire TFET at 300K and 50K ; (b) – associated Subthreshold Swing values

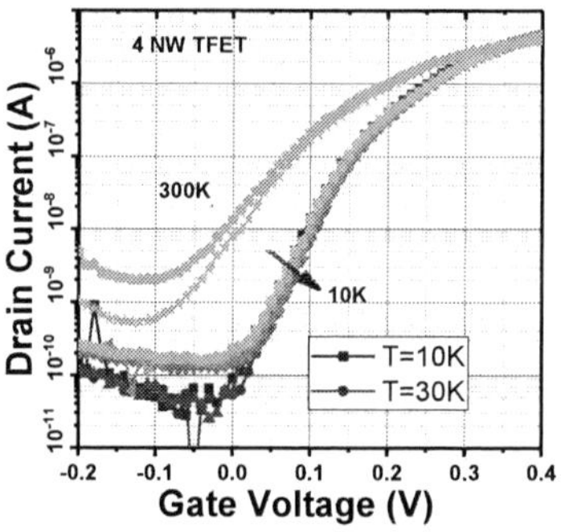

Fig. 5 – Full set of transfer characteristics of the small array

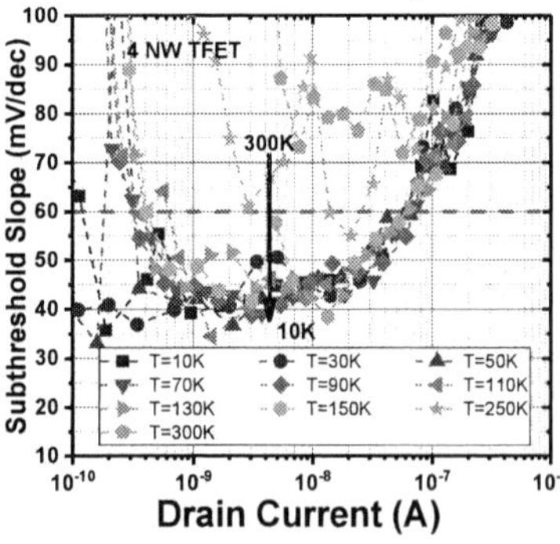

Fig. 6 – Subthreshold slope values obtained with the small array

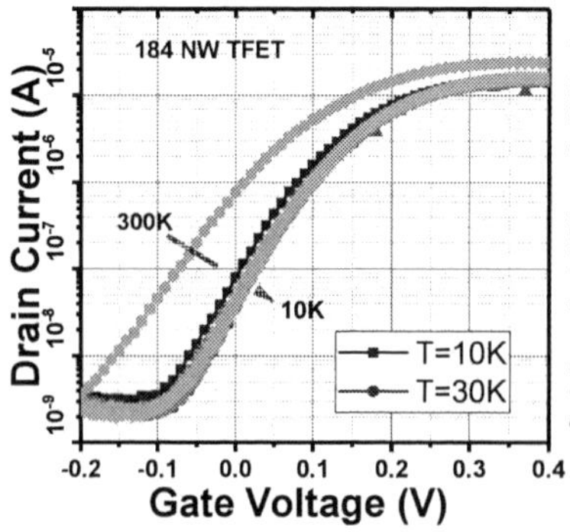

Fig. 7 – Full set of transfer characteristics of the large array

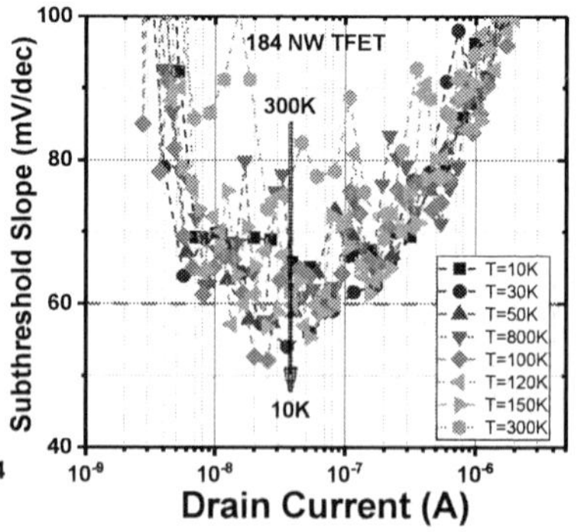

Fig. 8 – Subthreshold slope values obtained with the large array

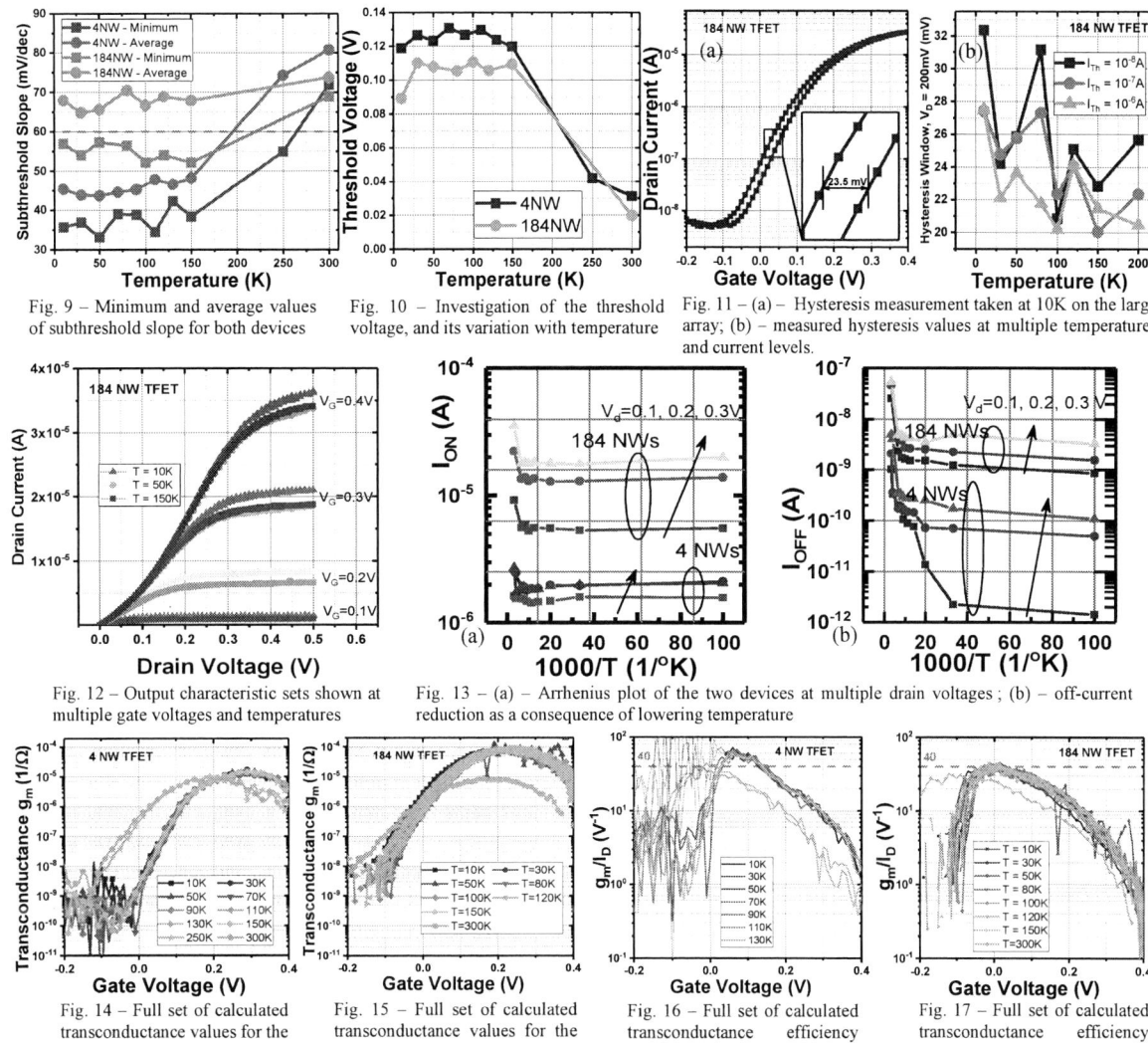

Fig. 9 – Minimum and average values of subthreshold slope for both devices

Fig. 10 – Investigation of the threshold voltage, and its variation with temperature

Fig. 11 – (a) – Hysteresis measurement taken at 10K on the large array; (b) – measured hysteresis values at multiple temperatures and current levels.

Fig. 12 – Output characteristic sets shown at multiple gate voltages and temperatures

Fig. 13 – (a) – Arrhenius plot of the two devices at multiple drain voltages ; (b) – off-current reduction as a consequence of lowering temperature

Fig. 14 – Full set of calculated transconductance values for the small array.

Fig. 15 – Full set of calculated transconductance values for the large array.

Fig. 16 – Full set of calculated transconductance efficiency values for the small array.

Fig. 17 – Full set of calculated transconductance efficiency values for the large array

Fig. 18 – Maximum values of transconductance and transconductance efficiency for both arrays at different temperatures

Table 1 – Summary: Extracted TFET figures of merit at key temperatures: 300K, 150K and 10K

Temperature	300K		150K		10K	
	4 NW	184 NW	4 NW	184 NW	4 NW	184 NW
I_{off} (nA), V_D=200mV	2.07	3.68	0.23	2.43	0.09	3.55
I_{on} (μA), V_D=200mV	3.98	23.89	4.51	15.93	4.68	15.79
I_{On}/I_{off}	$1.92 \cdot 10^3$	$6.49 \cdot 10^3$	$1.96 \cdot 10^4$	$6.56 \cdot 10^3$	$5.2 \cdot 10^4$	$4.45 \cdot 10^3$
Threshold Voltage (V)	0.031	0.020	0.120	0.109	0.119	0.089
Minimum SS (mV/dec)	71.94	68.92	38.38	52.14	35.69	56.92
Average SS (mV/dec)	80.73	73.76	48.16	67.90	45.36	67.94
Maximum g_m (μS)	10.28	107.40	14.70	88.63	14.76	81.15
Maximum g_m/I_D (V^{-1})	38.22	35.43	63.23	44.71	68.01	41.46

978-1-7281-1988-5/18 $31.00 © 2018 IEEE   311

High thermal tolerance of 25-nm *c*-axis aligned crystalline In-Ga-Zn oxide FET

Hitoshi Kunitake, Kazuaki Ohshima, Kazuki Tsuda, Noriko Matsumoto, Hiromi Sawai, Yuichi Yanagisawa,
Shiori Saga, Ryo Arasawa, Takako Seki, Ryo Tokumaru, Tomoaki Atsumi,
Kiyoshi Kato and Shunpei Yamazaki

Semiconductor Energy Laboratory Co., Ltd., 398 Hase, Atsugi, Kanagawa, 243-0036, Japan
Phone: +81-46-248-1131, Fax: +81-46-270-3751, E-mail: hk1372@sel.co.jp

Abstract—We developed FETs having gate lengths of 25 and 60 nm that are suited for high-temperature operation, using c-axis aligned crystalline In-Ga-Zn oxide (CAAC-IGZO) as its channel material. The FETs with a gate length of 60 nm achieved off-state leakage currents of 10^{-20} A at 150°C. Furthermore, cutoff frequency the FETs with a gate length of 25 nm was 33 GHz at room temperature and changing the temperature from room temperature to 150°C changed the cutoff frequency by only -13% against -36% in Si FET. The CAAC-IGZO FET enables integrated circuits that consume little power even under high-temperature environments.

I. Introduction

Research that targets the commercialization of artificial intelligence (AI) is rapidly accelerating. Low-power devices are necessary to achieve this target. CMOS scaling was accompanied by an increase in its off-state leakage current to a point at which standby power is more substantial than active power[1]. CMOS carries another issue that the off-state leakage current increases as the FET becomes hotter.

An FET having a channel layer of *c*-axis aligned crystalline In-Ga-Zn oxide (CAAC-IGZO)[2] has an extremely low off-state leakage current of 1.35×10^{-22} A/μm[3]. In addition, its mobility does not degrade in high temperatures[4]. This means that even at high temperatures, the CAAC-IGZO FET can still exhibit both its low off-state leakage and small gate delays.

For this work, we have prototyped CAAC-IGZO FETs with gate lengths of 25 nm and 60 nm. This work demonstrates that even under high temperatures (150°C), the off-state leakage current of the CAAC-IGZO FET is extremely low, and its cutoff frequency is high.

II. Structure and Performance of *c*-axis-aligned In-Ga-Zn oxide FET

This FET is formed on an insulating film, and uses CAAC-IGZO as the channel layer. The FET is fabricated with a trench-gate-self-aligned (TGSA) structure[5]. Figs. 1 and 2 show a process flow of the FET and a bird's-eye view of the FET, respectively. A TGSA FET forms a gate between a source and a drain (S/D) in a self-aligned manner. The gate does not overlap with S/D, and hence the TGSA FET has a lower parasitic capacitance than an FET that has a gate-to-S/D overlap. Figs. 3 and 4 show an L-direction and a W-direction

cross section of a CAAC-IGZO FET with a gate length of 25 nm, respectively. The physical gate length and width of the FET are 25 nm and 21 nm, respectively. For this work, the gate insulator (GI) is formed with an equivalent oxide thickness (EOT) of approximately 6 nm, and the power supply voltage is set at 2.5 V. The FET's I_d-V_g curves are shown on Fig. 5. When the drain voltage is 1.2 V, the FET exhibits a subthreshold slope (*S.S.*) of 79 mV/dec., saturation field-effect mobility of 10.2 cm^2/Vs, and an I_{on} of 2.8 μA/FET when gate voltage is 2.5 V.

The lower measurement limit of measurement equipment is usually around 10^{-13} A. However, the off-state leakage current of the CAAC-IGZO FET is extremely low, which is lower than the measurement limit and thus cannot be measured from a single FET. Figs. 6 and 7 illustrate the off-state leakage measurement methodologies. One of the defining features of the CAAC-IGZO FET is its off-state leakage current on the order of 10^{-24} A. Multiple FETs are connected in parallel to multiply the leakage current, but even this method is not realistic for the CAAC-IGZO FET as 10^{11} FETs need to be connected in parallel to reach the measurement limit. Such a small current is measured using a special circuit as shown in Fig. 8. This measurement does not measure the drain current; instead, it measures the voltage change at a floating node over time to estimate the off-state leakage current. Changing the connection of the device under test (DUT), we can also measure the leakage current from the gate as well as from the drain. The FET of this work has a thinner GI than those in past IGZO-related works[2-6]. Thinner GIs could result in higher gate leakage. In order to confirm the thinner GI film does not cause higher gate leakage current, we measured the gate leakage current of the CAAC-IGZO FET prototype (Fig. 9). The gate leakage current was 3.3×10^{-20} A at 150°C, which is sufficiently low.

In addition, this CAAC-IGZO FET prototype has a back gate (BG) underneath. This configuration allows adjustment of the FET's V_{th} with voltage applied on the back gate (V_{bg}). This prototype has a back gate insulator (BGI) with an EOT of 31 nm. I_d-V_g measurement results confirmed that changing V_{bg} changed the V_{th} (Fig. 10). The V_{th} change represented by $\partial V_{th}/\partial V_{bg}$ is -0.15 V/V. There is little change in maximum value of transconductance g_m in accordance with V_{bg} changes. Applying V_{bg} does not change the maximum g_m value, and the V_g that enables the maximum g_m value shifts in a similar manner to V_{th}. Providing BG enables dynamic V_{th} adjustments of the CAAC-IGZO FET for different circuits without

978-1-7281-1988-5/18 $31.00 © 2018 IEEE

changing the IGZO process. There is one disadvantage to introducing BG to the CAAC-IGZO FET, however. The presence of BG adds a parasitic capacitance between BG and S/D, increasing the gate delay. We compared f_T of CAAC-IGZO FETs with and without BG to verify the above. f_T is derived from the following expression (1):

$$f_T = g_m / 2\pi(C_{tg} + C_{bg}) = g_m / 2\pi C_g \qquad (1)$$

In (1), C_{tg} is a gate capacitance, C_{bg} is a back gate capacitance C_g is a sum of C_{tg} and C_{bg}. we can say that if the f_T values and g_m values are comparable with and without BG, C_g is also comparable. CAAC-IGZO FETs with gate lengths of 25 nm, both with and without BG, were evaluated for their f_T (Fig. 12). FETs with BG were measured under various V_{bg} bias conditions. At room temperature, samples with BG under V_{bg} = 0 V exhibited 30 GHz, and samples without BG exhibited 27 GHz. Furthermore, the f_T was 27.7 GHz under $V_{bg} = -6$ V, which is comparable to the f_T obtained when $V_{bg} = 0$ V. These results demonstrate that adding BG enables V_{th} shift with no substantial change to the f_T. As stated above, introduction of BG to the CAAC-IGZO FET enables dynamic V_{th} adjustments with V_{bg}. This dynamic adjustment cancels V_{th} shifts in reaction to temperature changes.

III. HIGH-TEMPERATURE PERFORMANCE OF C-AXIS-ALIGNED CRYSTALLINE IN-GA-ZN OXIDE

There are many works on analog RF applications using the IGZO FET[7, 8]. CAAC-IGZO FETs can be fabricated in BEOL process, and thus can be stacked on Si FETs[9, 10]. Fig. 13 shows cross sections showing a stack of Si and CAAC-IGZO layers. This stack structure allows the coexistence of low-leakage CAAC-IGZO circuits and high-speed CMOS circuits on one die.

Fig. 14 compares the temperature dependence of I_d–V_g performance in a bulk-Si FET and a CAAC-IGZO FET, both having a gate length of 60 nm. The Si FET's off-state leakage current increases as the temperature increases, whereas the CAAC-IGZO FET's off-state leakage current is consistently below the lower measurement limit. To understand the temperature-dependent change of the CAAC-IGZO FET's S/D off-state leakage current, we measured the above-described test circuit for S/D off-state leakage current estimation (Fig. 8) and compared the results to that of a single 60 nm Si FET. Fig. 15 shows the measurement results and conditions. At 150°C, the Si FET's off-state leakage current is approximately 2.2 × 10^{-6} A, whereas the CAAC-IGZO FET's off-state leakage current is 3.9×10^{-20} A. The CAAC-IGZO FET can maintain a low off-state leakage current even at high temperatures. Applying V_{bg} can lower the off-state leakage current even further.

Next, the temperature dependence of Hall mobility in a single CAAC-IGZO film is shown (Fig. 16). The Hall mobility of the film hardly changes in reaction to temperature changes. This is possibly because Coulomb scattering is a more dominant factor in determining the mobility of CAAC-IGZO than phonon scattering[11]. We then compared the temperature dependence of g_m in a CAAC-IGZO FET and a Si

FET (Fig. 17). High temperatures reduce g_m in the Si FET, conversely, g_m in the CAAC-IGZO FET does not change in reaction to temperature changes. Lastly, we measured and compared the temperature dependence of f_T in a CAAC-IGZO FET and a Si FET. For this measurement, we measured bulk-Si FET with a gate length of 60 nm and a CAAC-IGZO FET with a gate length of 25 nm. The measured result is shown in Fig. 18. The CAAC-IGZO FET's f_T, in a similar tendency to that of its g_m, hardly changed in reaction to temperature changes. The CAAC-IGZO FET with a gate length of 25 nm achieved an f_T of 33 GHz at 150°C and the drain voltage is 2.5 V. The CAAC IGZO FET's f_T changes by -13% in reaction to temperature changes.

IV. CONCLUSION

We have prototyped CAAC-IGZO FETs with gate lengths of 25 nm and 60 nm. Comparison of its high-temperature performance with that of Si FETs showed that CAAC-IGZO FETs' performance has little temperature dependence (Table I). Introducing BG to the CAAC-IGZO FET increases V_{th} controllability, which enables canceling of temperature-dependent V_{th} changes. The CAAC-IGZO FET has lower g_m than the Si FET. However, its reduction of f_T due to temperature changes is only -13%, which is lower than that in the Si FET. In addition, its off-state leakage current is extremely low even at 150°C, achieving drain leakage current of 10^{-20} A.

The CAAC-IGZO FET enables integrated circuits and memory devices that consume little power under various temperature conditions, which is optimal for hardware implementations of AI.

REFERENCES

[1] A. Ionescu *et al.*, "Energy efficient computing and sensing in the Zettabyte era: from silicon to the cloud," *IEDM*, pp. (1.2.1)-(1.2.8), 2017.

[2] S. Yamazaki *et al.*, "Properties of crystalline In–Ga–Zn-oxide semiconductor and its transistor characteristics," *JJAP*, vol.53, pp. (04ED181)-(04ED1810), 2014.

[3] K. Kato *et al.*, "Evaluation of Off-State Current Characteristics of Transistor Using Oxide Semiconductor Material, Indium Gallium Zinc Oxide," *JJAP*, vol.51, pp. (021201-1)-(021201-7), 2012.

[4] H. Kunitake *et al.*, "High-temperature Electrical Characteristics of 60nm CAAC-IGZO FET: Comparison with Si FET" *SSDM*, 2018 (to be published).

[5] D. Matsubayashi *et al.*, "20-nm-node trench-gate-self-aligned crystalline In-Ga-Zn-Oxide FET with high frequency and low off-state current," *IEDM*, pp. (6.5.1)-(6.5.4), 2015.

[6] R. Honda *et al.*, "Mechanism of Threshold Voltage Control by Back Gate in CAAC-IGZO FETs," *SSDM*, 2018 (to be published).

[7] K. Ishida *et al.*, "3-5 V, 3-3.8 MHz OOK Modulator with a-IGZO TFTs for Flexible Wireless Transmitter," *IEEE COMCAS*, pp. 1-4, 2017.

[8] M. Hung *et al.*, "Ultra Low Voltage I-V RFID Tag Implement in a-IGZO TFT Technology on Plastic," *IEEE RFID*, pp. 193-197, 2017.

[9] T. Onuki *et al.*, "Embedded Memory and ARM Cortex-M0 Core Using 60-nm C-Axis Aligned Crystalline Indium–Gallium–Zinc Oxide FET Integrated With 65-nm Si CMOS," *IEEE JSSC*, Vol. 52, No. 4, pp. 925-932, 2017.

[10] S. Maeda *et al.*, "A 20ns-Write, 45ns-Read, and 1014-Cycle Endurance Memory Module Composed Only of 60nm Crystalline Oxide Semiconductor Transistors," *IEEE ISSCC*, pp. 484-486, 2018.

[11] Y. Kang *et al.*, "Cation disorder as the major electron scattering source in crystalline InGaZnO," *Appl. Phys. Lett,*. Vol.102, p. 152104-1, 2013.

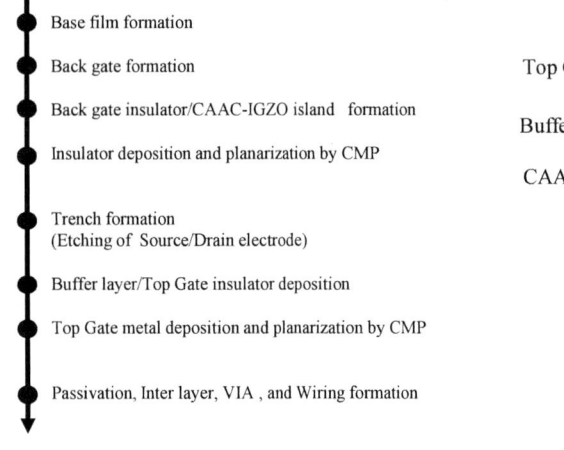

Base film formation

Back gate formation

Back gate insulator/CAAC-IGZO island formation

Insulator deposition and planarization by CMP

Trench formation
(Etching of Source/Drain electrode)

Buffer layer/Top Gate insulator deposition

Top Gate metal deposition and planarization by CMP

Passivation, Inter layer, VIA , and Wiring formation

Fig.1. The CAAC-IGZO FET

process flow (TGSA structure).

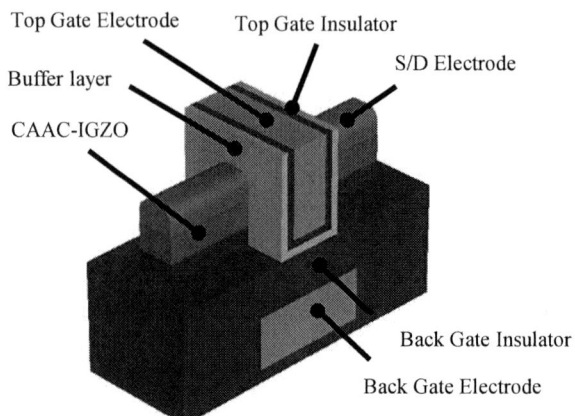

Fig.2. Bird's-eye view of the CAAC-IGZO FET.

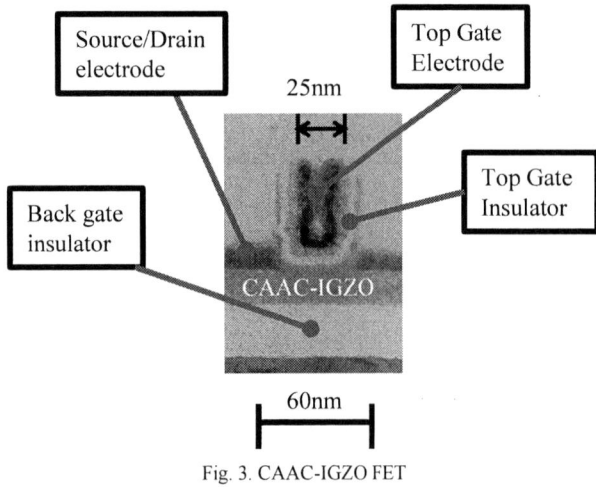

Fig. 3. CAAC-IGZO FET

STEM image in L-direction

(gate length = 25 nm).

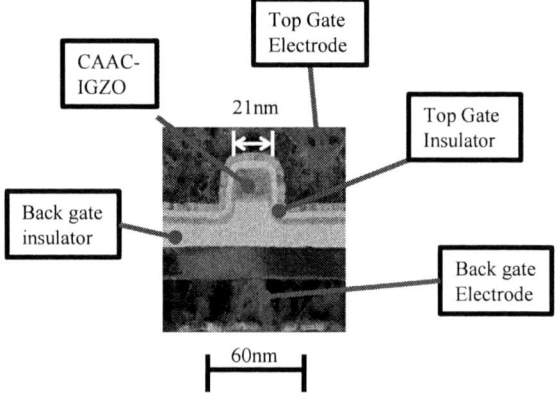

Fig. 4. CAAC-IGZO FET

STEM image in W-direction

(gate width = 21 nm).

Fig. 6. Lower measurement limit of leakage current in various test circuits.

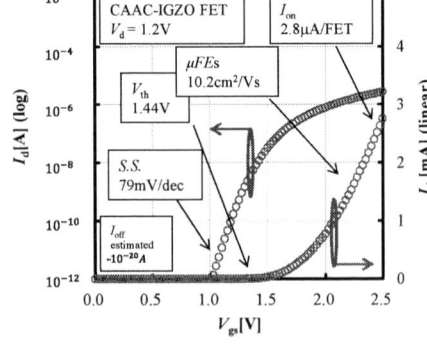

Fig. 5. Initial I_d-V_g performance
(CAAC IGZO FET L = 25 nm).

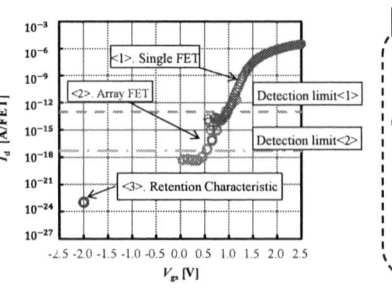

Fig. 7. Lower measurement limit of

leakage current

in various test circuits (I_d-V_g).

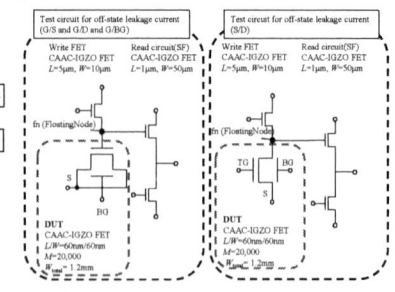

Fig. 8. Test circuit for off-state leakage

current measurement.

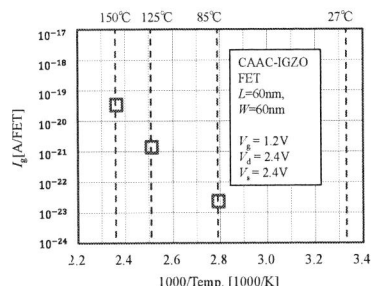

Fig. 9. Temperature vs. gate leakage current.

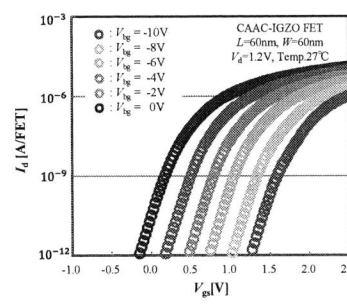

Fig. 10. V_{th} shift with V_{bg} application.

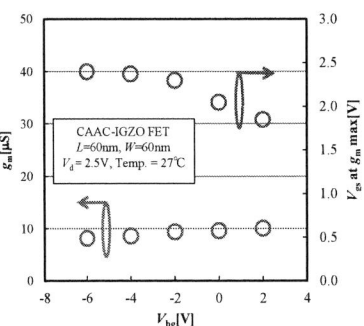

Fig. 11. V_{bg} dependence of g_m.

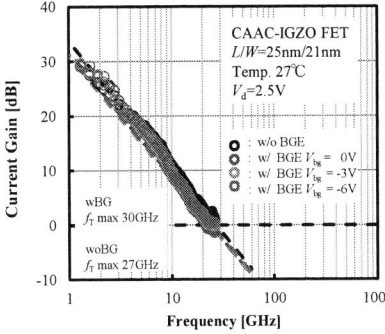

Fig. 12. CAAC-IGZO FET

f_T comparisons with and without BG.

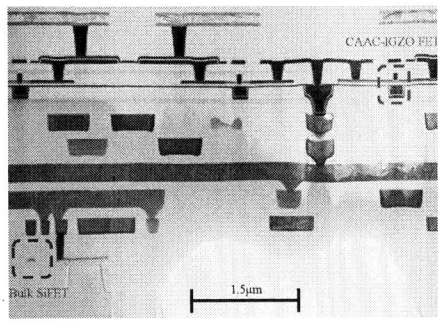

Fig. 13. Cross section of bulk-Si FET

and CAAC-IGZO FET stack.

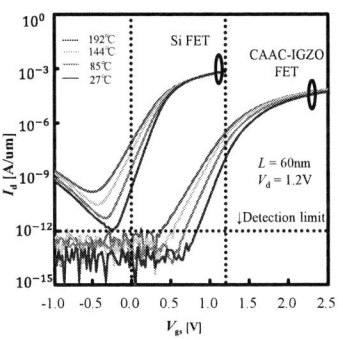

Fig. 14. Si FET vs. CAAC-IGZO FET (I_d-V_g).

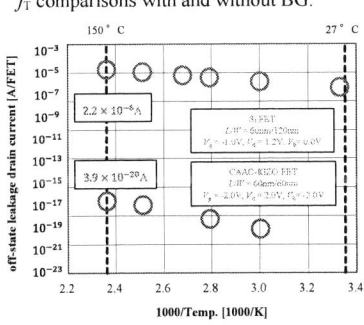

Fig. 15. Si FET vs. CAAC-IGZO FET

(off-state leakage current).

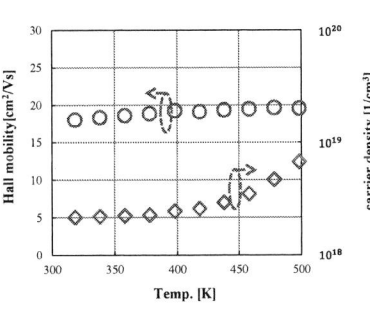

Fig. 16. Hall-effect measurement on the

CAAC-IGZO film.

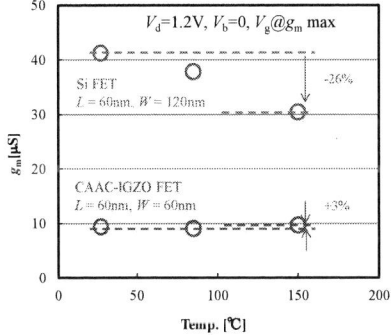

Fig. 17. Si FET vs. CAAC-IGZO FET

(temperature dependence of g_m).

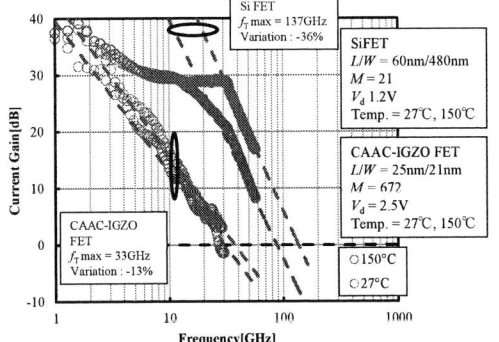

Fig. 18. Si FET vs. CAAC-IGZO FET

(Temperature dependence of f_T).

Table I. Summary table.

		This work	CMOS
Channel		CAAC-IGZO	Bulk Si
V_{dd}		2.5 V	1.2 V
L_g		25 nm	60 nm
EOT		6.0 nm	2.6 nm
I_d (off) (V_d = 1.2 V)	27°C	1.0×10^{-24} A	2.8×10^{-12} A
	150°C	3.9×10^{-20} A	2.2×10^{-6} A
f_T	Temperature change	-13%	-36%
	Maximum value	33 GHz	137 GHz

978-1-7281-1988-5/18 $31.00 © 2018 IEEE   315

Intel 22nm FinFET (22FFL) Process Technology for RF and mmWave Applications and Circuit Design Optimization for FinFET Technology

H.-J. Lee, S. Rami, S. Ravikumar, V. Neeli, K. Phoa, B. Sell, and Y. Zhang

Logic Technology Development, Intel Corporation, Hillsboro, Oregon, USA, email: hyung-jin.lee@intel.com

Abstract— Intel 22FFL is a unique FinFET process technology optimized for RF and mmWave applications supporting superior RF performance to planar technologies with both f_t and f_{max} of NMOS above 300 GHz and 450 GHz respectively. Flicker noise improvement over planar technologies and excellent gain-power efficiency enabling low-power wireless applications are demonstrated.

I. INTRODUCTION

Recently, an advanced fabrication technology, known as FinFET technology [1], is needed to overcome the physical limitations of planar devices and continue the 50-year of Moore's law scaling.

In this paper, a performance summary of 22FFL process [2] and its suitability for RF/mmWave designs are demonstrated. Figure of Merits (FoMs) will be presented and contrasted with respect to planar CMOS technology. Furthermore, design methodologies to exploit advantages of FinFET technology will be briefly illustrated.

II. OVERVIEW OF FINFET TECHNOLOGY

This section will briefly introduce how FinFET technologies extend the longevity of silicon technology scaling and the difference of the FinFET devices due to the structural change.

A. Drain-Induced Barrier Lowering

As modern fabrication technology drives the device scaling to an extreme level, the drain starts interacting with the source directly through region beneath the channel regardless gate potential, which is called Drain Induced Barrier Lowering (DIBL). DIBL in planar technology starts dominating drain current beyond around 30nm process node as drain-source interaction exceeds the gate-channel control, resulting in higher leakage current. Planar technology damps the DIBL issue with high channel doping and halo channel doping, but has to sacrifice channel mobility and the noise performance.

FinFET technologies, however, virtually isolate the drain from the source by physically carving out the parasitic channel material outside of the gate influence forming a three-dimensional channel structure called 'fin' as shown in Fig. 1. Such a fully depleted fin separates the drain and source and prevents the drain field from encroaching into the source, therefore dramatically reduces DIBL effect. This enables

scaling down the channel length beyond 30nm gate length without the listed degradation. At any given leakage current, transistor with stronger drive current, lower threshold voltage and higher output resistance and therefore higher intrinsic analog gain can be achieved as suggested in Fig. 2.

B. Nonlinear Gate Resistance

Unlike the planar devices, gate resistance in FinFET shows a non-linear relation with channel width as shown in Fig. 3. In addition to the horizontal components, because of the 3D nature of the 'fin', the gate material wraps around the fin generating a vertical component of the gate resistance. As channel width increases from the narrowest, the total gate resistance drops as vertical resistance, which dictates the total resistance, are connected in a parallel configuration. As the channel width continues to increase, the horizontal resistance becomes stronger, and the trend of the gate resistance turns into linear scaling by channel width as depicted in Fig. 4 [3].

As RF and mmWave circuit design are sensitive to gate resistance for quality factor (Q), stabilization and power delivery, device sizing for optimum gate resistance and performance balance will be discussed in section IV.

III. FINFET VS. PLANAR FOR RF/MMWAVE

This section will compare key RF performance Figure-of-Merits (FoMs) between FinFET and planar technologies to highlight the benefit of FinFET technology for RF and mmWave applications in addition to the scaling benefit.

A. Parasitic and RF Performance

One caveat of FinFET technologies is the increased lateral parasitic capacitance by the three-dimensional gate structure. The gate material between fins negatively impact on total gate capacitance without contributing to transconductance (g_m). Hence, as the poly pitch gets tighter, the parasitic capacitance increases, and thus the unity gain frequency (f_t) is lowered. Fig. 5 shows the recent f_t trend in silicon technology by process node.

FinFET technologies, however, have a potential to reach a higher maximum oscillation frequency (f_{max}) thanks to the vertical gate resistance components, and higher output resistance (R_{out}). Fig. 6 compares peak f_t and peak f_{max} between 22FFL FinFET technology and 32nm planar process technology.

978-1-7281-1988-5/18 $31.00 © 2018 IEEE

B. Flicker Noise Performance

As hinted in Section II.A., halo implant (Pocket implant) is the widely used remedy to compensate the DIBL effect at the cost of flicker degradation due to V_T non-uniformity and extra traps at the interface [4]. The FinFET structure inherently suppresses DIBL effect, hence negates the need of halo implant and flicker noise improvement thereafter. Flicker noise silicon data collected from 22FFL thin oxide device are shown in Fig. 7 [2].

C. Gain-Power Efficiency

The less DIBL and fully depleted operation in FinFET technologies improves device gain per power dissipation efficiency, as FinFET devices can drive stronger drain current with less short-channel effect. The DIBL improvement drastically enhances the device current usability. Fig. 8 illustrates the gain-power efficiency FoM (GPFoM) of FinFET and Planar for comparison purpose. GPFoM is defined as $GPFoM = U \cdot g_m / I_d \, [dB/V]$.

Note that Mason's gain (U or Unilateral gain) is used for the FoM to accommodate the performance metrics at mmWave frequency range. The current density reaching the peak FoM is the optimum bias condition for the maximum gain-power efficiency. As shown in Fig. 8, 22FFL FinFET devices offer about 600dB/V improvement over planar technologies in the gain-power efficiency at 30GHz.

22FFL process offers five flavors of RF transistor, which are low-leakage nominal and low V_T (LL_nom, LL_lvt), high-density nominal and low V_T (HD_nom, HD_lvt), and high-performance (HP). The RF performances of and HP RF transistor in peak f_t and peak f_{max} configurations are provided in Fig. 9 and Fig. 10. Both f_t and f_{max} measurement include up to Metal 2 routing; the second lowest metal layer for realistic usage condition. As shown in Fig. 9 and 10, NMOS RF transistors reach above 300GHz of f_t and 450GHz of f_{max} respectively, while PMOS achieves slight below 300GHz of f_t and f_{max}. The results demonstrate a superior f_{max} and a competitive f_t performance to leading-edge RF/mmWave silicon based technologies such as Fully-Deleted Silicon-on-Insulator (FDSOI) [5].

IV. DESIGN METHODOLOGY WITH FINFET

This section introduces examples of how to utilize the distinguishable device properties of FinFET technologies for RF performance optimization, and also how to avoid reliability issue associating with FinFET structure.

A. Performance Optimization with Fin Self Heat Awareness

Fin-Self-Heat (FiSH in short) should be considered for circuit design. As FinFET strives for narrow fins for excellent electrostatics, this results in limited thermal conductance to dissipate heat generated in the channel.

Recent work has demonstrated an LNA design in 22FFL at 71 ~ 76GHz frequency range with FiSH limit consideration [6]. The authors swept MAG and NF_{min} for I_{DS} and V_{DS} accordingly under the maximum power limit for *FiSH*, and

searched for the optimum supply voltage (V_{DS}) and the current (I_{DS}) for a single stage as shown in Fig. 11. The work achieves the target performance at the lower V_{DS} of 0.5V per single stage, and total two stage stacking with current sharing was suggested for higher gain performance with less power dissipation.

B. Device Sizing Consideration

The non-linear gate resistance in FinFET devices explained in section II. B. winds up various input impedance condition by the number of fin choice, despite the total equivalent device size; the product of the number of gate fingers and the number of fins. Therefore, it is strongly advised to consider the different number of fins for the device sizing in order to achieve higher correlation between optimum noise matching and the power matching by input impedance for low-noise amplifier designs.

Fig. 12 demonstrates three cases of fins configuration maintaining total device size by modulating the number of gate fingers. As one can notice, the maximum available gain G_{max} and the minimum noise figure NF_{min} are adjusted by the number of the fin, and the delta between G_{max} and NF_{min} is maximized by 4 or 6-fin device at J_d of 0.3mA/um.

V. CONCLUSION

FinFET technologies offer significant performance boosts for not only logic but also RF/mmWave over planar technologies. The three-dimensional fabrication technologies allow keeping Moore's law alive beyond the physical limit of the two-dimensional device fabrication. 22FFL is specially engineered to support both RF and mmWave applications with the best-in-class f_t and f_{max} reaching over 300GHz and 450GHz respectively, which provide the best opportunity to enable low-power mmWave applications in silicon technology.

ACKNOWLEDGMENT

The authors gratefully acknowledge the contributions of *Wireless Integration & Circuit Technology team* and *22FFL Program* in Intel to RF device enhancements.

REFERENCES

[1] C.-H. Jan, and et al., "A 22nm SoC Platform Technology Featuring 3-D Tri-Gate and High-k/Metal Gate Optimized for Ultra Low Power, High Performance and High Density SoC Applications," *2012 International Electron Devices Meeting*, pp. 44-47

[2] B. Sell, and et al., "22FFL: A high performance and ultra low power FinFET technology for mobile and RF applications," *2017 International Electron Devices Meeting*, pp. 685-688

[3] A.J. Scholten, and et al., "FinFET compact modelling for analogue and RF applications," *2010 International Electron Devices Meeting*, pp. 190-193

[4] J.W. Wu, and et al., "Pocket Implantation Effect on Drain Current Flicker Noise in Analog nMOSFET Devices," *Proc. 2004 IEEE Transactions on Electron Devices*, vol. 51, no. 8, pp. 1262-1266

[5] S.N. Ong, and et al., "A 22nm FDSOI Technology Optimized for RF/mmWave Applications," *Proc. 2018 IEEE Radio Frequency Integrated Circuits Symposium*

[6] W. Shin, and et al., "A Compact 75 GHz LNA with 20 dB Gain and 4 dB Noise Figure in 22nm FinFET CMOS Technology," *Proc. 2018 IEEE Radio Frequency Integrated Circuits Symposium*

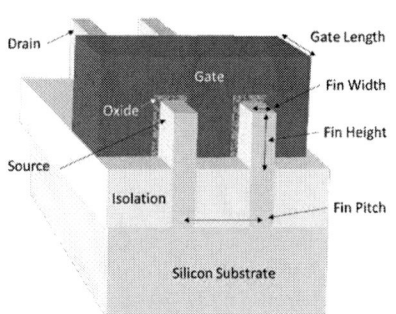

Figure 1: 3D view of FinFET device structure

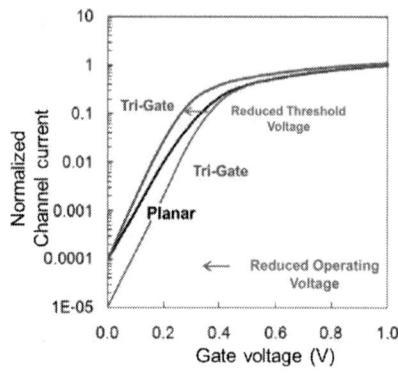

Figure 2:I-V curves of FinFET and Planar demonstrating DIBL improvement by FinFET technologies

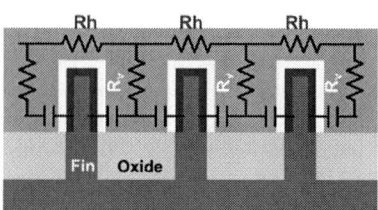

Figure 3: FinFET gate resistance structure: Horizontal resistance (R_h) and Vertical resistance (R_v) surrounding fin structures

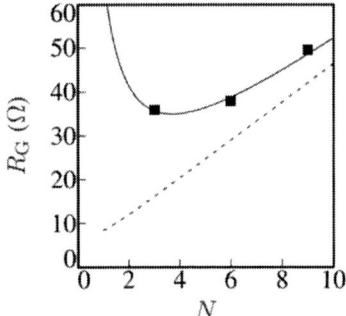

Figure 4: Gate resistance trend by channel width of FinFET devices [3]

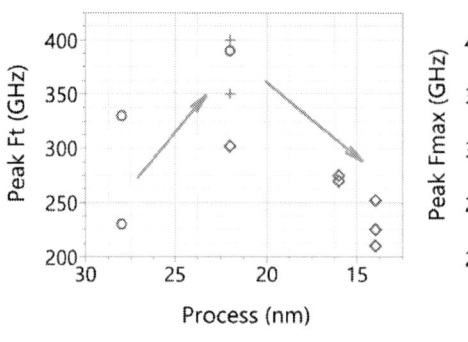

Figure 5: f_t and f_{max} trends by process node: both ft and fmax reach the peak performance around 20 ~ 25nm due to the excessive parasitic capacitance by high density interconnect

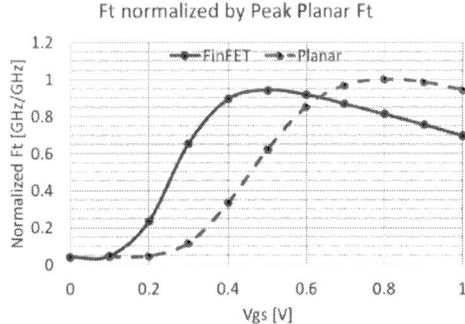

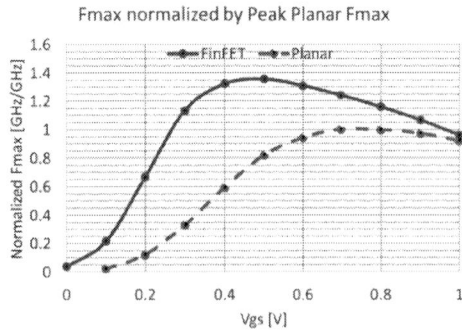

Figure 6: Peak f_t and f_{max}: Planar reaches 20% higher ft than FinFET, but FinFET reaches 40% higher fmax than Planar

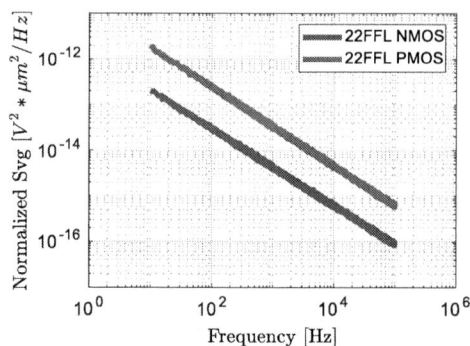

Figure 7: 1/f noise of 22FFL [2]

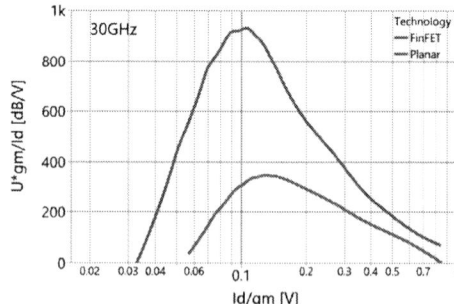

Figure 8: Gain-power efficiency FoM of 22FFL FinFET and Planar device at 30GHz

978-1-7281-1988-5/18 $31.00 © 2018 IEEE   318

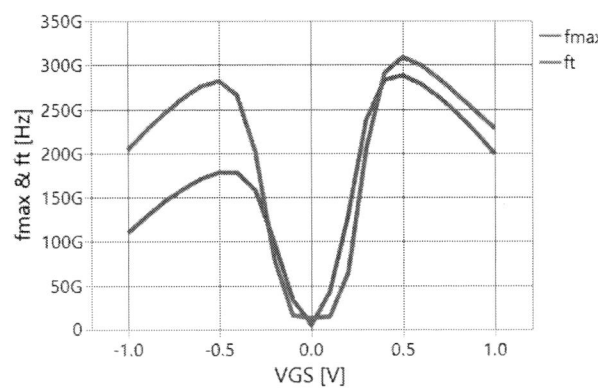

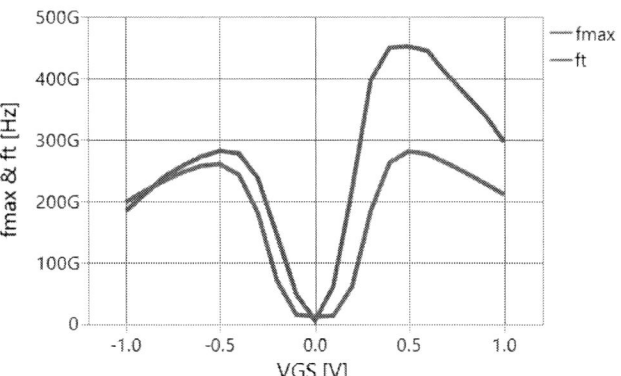

Figure 9:Optimized for peak ft of 22FFL RF transistors

Figure 10: Optimized for peak fmax of 22FFL RF transistors

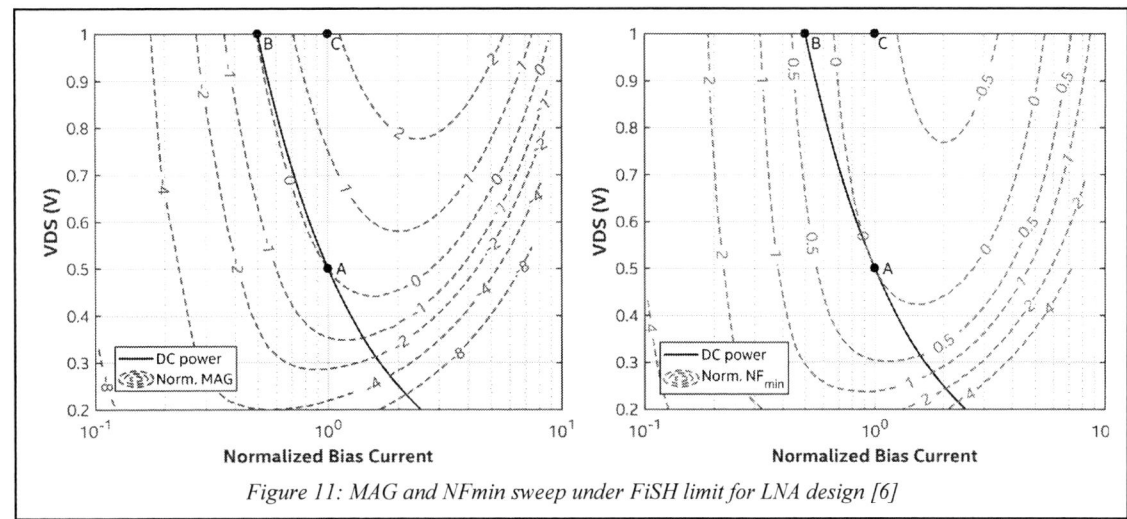

Figure 11: MAG and NFmin sweep under FiSH limit for LNA design [6]

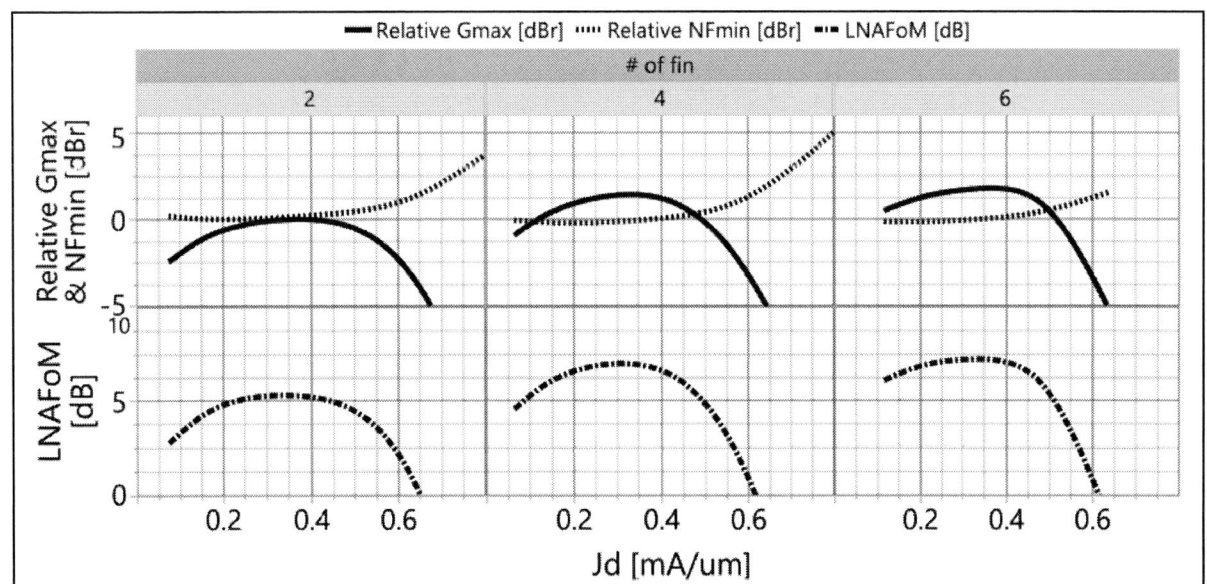

Figure 12: Device sizing for optimum LNA FoM: 4 or 6-fin devices reach highest equivalent LNAFoM at ~ 0.3mA/um current density. LNAFoM is defined by Gmax[dB] – NFmin[dB]. Gmax and NFmin are normalized to Gmax and NFmin of 2-fin device

GaN HEMTs for 5G Base Station Applications

Shigeru Nakajima

Semiconductor Innovation Business Unit, Sumitomo Electric Industries, Ltd., Yokohama, Japan, email: snakajm@sei.co.jp

Abstract—Many challenges have been overcome in developing highly reliable, cost effective and excellent performance GaN HEMTs. We have focused on GaN HEMT on SiC, and have been shipping commercial GaN HEMTs for the base station market since 2005. The state of the art GaN HEMT has penetrated into the 4G/LTE base station. The efficiency advantage, based on its material properties will also attract 5G power amplifier designers. This paper explains our development history, and overviews the GaN HEMT power amplifiers in the 5G era.

I. INTRODUCTION

The features of 5G network are high density, high speed, and low latency, so that this technology is expected to develop IOT (Internet of Things) applications. The base station amplifier with broadband and high efficiency, is one of the most critical issues to realize higher bit rate communication. GaN HEMT power amplifiers have been studied for almost 20 years [1,2], and we released the first commercial GaN HEMT in 2005. The combination of Doherty configuration and GaN HEMT devices, which feature high efficiency and sufficient reliability has been a promising solution for the 4G base station PAs.

This paper describes how GaN HEMTs are attractive devices for 5G base station applications.

II. CONCEPT OF 5G BASE STATION SYSTEM

Fig. 1 shows the basic concept for 5G base station network. It is planned that there are two signals that pass from base station to 5G handset. One is C (Control) -plane which signal sends to handsets from macro cell type base station without beam forming. The other is U (User) -plane which sends signals from small cell type base station with beam forming by using massive-MIMO technologies.

Table 1 shows the allocation of frequencies for 5G applications. It is planned that sub-6 GHz bands will be used for both C-plane and also U-plane in the early stages. Above 20 GHz bands will be developed in the future.

III. COMPARISON OF MATERIAL PROPERTIES

The material properties of GaN and competing materials are shown in Table 2. GaN, featuring high breakdown field and high electron mobility, allows high voltage operation of around 50V with rather low on-resistance. Thus, GaN HEMT realizes higher efficiency. The superior power density of GaN HEMT, originating from high operation voltage with excellent current density, contributes to wider frequency response. In addition, the thermal conductivity of SiC is more than three times higher than that of Si. Therefore, GaN on SiC is a most promising combination for high frequency and high power device applications.

IV. QUALITY OF GAN HEMTs

The typical cross section of GaN HEMT structure is shown in Fig. 2. In the dawning of the first commercial GaN HEMT device of more than 10 years ago, the material of GaN was quite new and most of the base station amplifier designers felt uncertain about its commercial use. We have focused on the establishment of sufficient reliability for infrastructure applications. The robustness of GaN HEMTs is based on the quality control of SiC substrate, epitaxial layer, electrode fabrications, and so on [3]. The understanding and the ascertainment of the trade-off between performance and reliability is essential for the commercial devices. Various tests and challenges have been met in order to successfully satisfy both at high levels of performance.

Fig. 3 shows the typical reliability results for our L-band GaN HEMTs. The MTTF of our devices shows 10^7 hours at Tj=200℃ [4]. We have shipped more than 20 million devices so far, with a market failure rate below 1 FIT.

V. COST OF GAN HEMTs

Cost is also important factor for new devices to spread to the markets. Especially in the base station amplifier markets, Si LD-MOS has been used for long time. Fig. 4 shows our approach to cost reduction of GaN HEMTs. We have realized cost reduction by introducing larger diameter substrate, low cost package and so on. Non-hermetic package was one of the important technologies to reduce cost. Thus, we have developed moisture resistant chip by optimizing the electrode configuration and passivation film. Fig.5 shows the results of the highly accelerated temperature and humidity stress test (HAST) with reversed bias [5]. The developed moisture resistant chip has realized sufficient reliability to satisfy JEDEC standard criteria.

GaN HEMTs have a superior efficiency characteristic especially at higher frequency range, therefore GaN HEMTs intrinsically greatly improve total operation cost considering smaller device size and lower power consumption.

VI. PERFORMANCE OF GAN HEMTs FOR 5G BASE STATION APPLICATIONS

Improving the amplifier efficiency will be the primary performance requirement in 5G base station amplifiers. The utilization of switching mode operation, for example inverse class F operation, and Doherty configuration will be the promising solutions.

A. Inverse Class F Operation

It is well known that switching mode operation is useful to improve power efficiency. Switching mode operation requires higher breakdown voltage. As the operation voltage of our GaN HEMTs is 50V with the breakdown voltage of more than

200V, we have introduced the inverse class F operation. Fig.6 shows the power performance of GaN HEMTs with inverse class F operation. We have obtained saturated power of 50 dBm, drain efficiency of 70 % at 2.14 GHz.

B. Doherty Amplifier Performance

In the LTE and 5G applications, amplifiers need to operate in deep back-off condition for required modulation. Doherty amplifiers are widely used to improve the efficiency at large (6-8dB from saturation power) back-off condition, and we also have eagerly studied Doherty amplifiers for a long time [6-8].

Fig. 7 shows the Doherty amplifier performance for C-plane GaN HEMTs at 2.6 GHz. Psat of 56 dBm and drain efficiency of 52.2% at 7dB back off from P1dB are obtained [8]. Fig. 8 also shows the Doherty amplifier for sub-6GHz band U-plane (M-MIMO) [9]. For a two-way compact Doherty structure using two internally matched 30-W GaN HEMT transistor dies in a single plastic package with asymmetric loading for the carrier and peaking amplifiers, the peak output power of greater than 47 dBm at 2-dB gain compression point, peak drain efficiency of greater than 60%, and drain efficiency exceeding 45% at 8-dB power backoff with a linear power gain about 12 dB were obtained across the frequency bandwidth of 4.35-4.85 GHz. Fig. 9 shows the Doherty MMIC amplifier performance for >20GHz band U-plane (M-MIMO) GaN HEMTs [10]. The device shows PAE of more than 13 % at Pout of 26 dBm, fractional bandwidth of within 7 % and linear gain of 16 dB.

Fig. 10 shows prospect of mm-wave GaN HEMT for 5G applications. Higher transmission data rate requires higher frequency operation. GaN HEMTs have realized over 100GHz of fT and fmax with 0.1 μ m class gate, which will satisfy over 10Gbps operation and further expand 5G business.

These results demonstrate that GaN HEMTs are quite promising for 5G Base Station Applications.

VII. SUMMARY

GaN HEMT on SiC substrate has already been a proven technology with the MTTF of the order of 10^7 hours at Tj=200 ℃. Various innovations to improve the cost competitiveness have been realized, including the moisture resistant chip structure. GaN HEMT has already been widely employed for 4G/LTE base station amplifiers because of efficiency superiority. Further efficiency improvement has been explored for both the macro cell and massive MIMO amplifiers, and several current achievements are introduced. GaN HEMTs are attractive devices for 5G base station applications.

ACKNOWLEDGMENT

The author gratefully acknowledges the contributions of K. Inoue, H. Hirata, M. Kobayashi, and Y. Hasegawa to support the completion of this paper.

REFERENCES

[1] T. Kikkawa, M. Nagahara, N. Okamoto, Y. Tateno, Y. Yamaguchi, N. Hara, K. Joshin and P.M. Asbeck, "Surface-charge controlled AlGaN/GaN-power HFET without current collapse and gm dispersion," International Electron Devices Meeting. Technical Digest, 2001, pp. 25.4.1-25.4.4..

[2] T. Kikkawa, T. Maniwa, H. Hayashi, M. Kanamura, S. Yokokawa, M. Nishi, N. Adachi, M. Yokoyama, Y. Tateno and K. Joshin, "An over 200-W output power GaN HEMT push-pull amplifier with high reliability," 2004 IEEE MTT-S International Microwave Symposium Digest, 2004, pp. 1347-1350

[3] F. Yamaki, K.Ishii, M.Nishi, H.Haematsu, Y-Tateno and H-Kawata, "Leakage Current Screening for AlGaN/GaN HEMT Mass-Production," CS MANTECH Conference, 2007, pp.95-98.

[4] K. Osawa, H. Yoshikoshi, A. Nitta, T. Tanaka, E. Mitani and T. Satoh, "Over 74% efficiency, L-band 200W GaN-HEMT for space applications," 2016 46th European Microwave Conference (EuMC), London, 2016, pp. 397-400.

[5] F.Yamaki, and S.Sano, "Mass-Production of High Reliability GaN HEMT for Wireless Communication", CS MANTECH Conference, May 2018

[6] H. Deguchi, N. Ui, K. Ebihara, K. Inoue, N. Yoshimura and H. Takahashi, "A 33W GaN HEMT Doherty amplifier with 55% drain efficiency for 2.6GHz base stations," 2009 IEEE MTT-S International Microwave Symposium Digest, Boston, MA, 2009, pp. 1273-1276.

[7] Andrei Grebennikov, "A High-Efficiency 100-W Four-Stage Doherty GaN HEMT Power Amplifier Module for WCDMA Systems", IEEE-IMS2011 Digest, June 2011.

[8] H. Deguchi, K. Ebihara and Y. Hasegawa, "Consistent growth and successive development of GaN-HEMT for wireless communication," *2014 Asia-Pacific Microwave Conference*, Sendai, Japan, 2014, pp. 185-187.

[9] Naoki Watanabe, James Wong, Andrei Grebennikov, Gaku Nishio, " A High-Efficiency 4.35-4.85 GHz Doherty Amplifier for Base Station Applications," *2018 Asia-Pacific Microwave Conference*, Kyoto, Japan, 2018, *in submitted.*

[10] T. Kawai, "GaN for 5G Base Stations", 2016 CS International Conference, March 2016

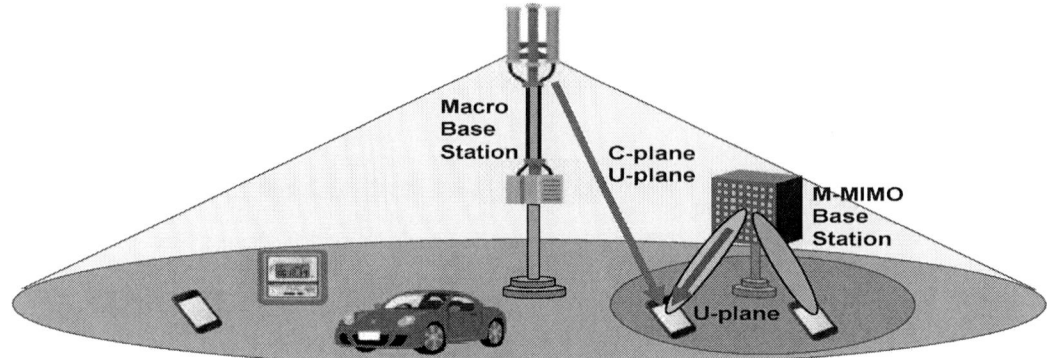

Fig. 1. 5G radio access network.

Country or Region	Sub6 GHz(MHz)	mm wave (GHz)
US	663 - 698 / 617 - 652, 2,496-2,690, 3,550-3,700, 3,700-4,200, 5,925-7,125	24.25-24.45, 24.75-25.25, 27.5-28.35, 37.0-40.0,47.2-48.2
EU	703 – 748 / 758-803, 3,400-3,800	24.5-27.5, 31.8-33.4, 40.5-43.5
China	3,300-3,400, 3,400-3,600, 4,800-5,000	24.75-27.5, 37.0-42.5
Korea	3,400-3,700	26.5-29.5
Japan	3,600-4,200, 4,400-4,900	27.0-29.5

Table. 1. 5G frequency band candidates.

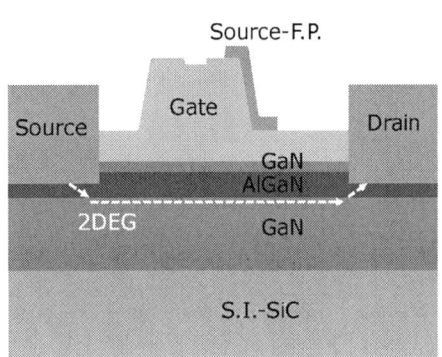

Fig. 2. Cross section of GaN HEMT.

Material	Band Gap Energy (eV)	Critical Breakdown Field (MV/cm)	Thermal Conductance (W/cm/K)	Mobility (cm²/V/s)	Saturated Velocity (x 10⁷cm/s)
Si	1.1	0.3	1.5	1300	1.0
GaAs	1.4	0.4	0.5	6000	1.3
SiC	3.2	3.0	4.9	600	2.0
GaN	3.4	3.0	1.5	1500	2.7

Table. 2. Comparison of material properties.

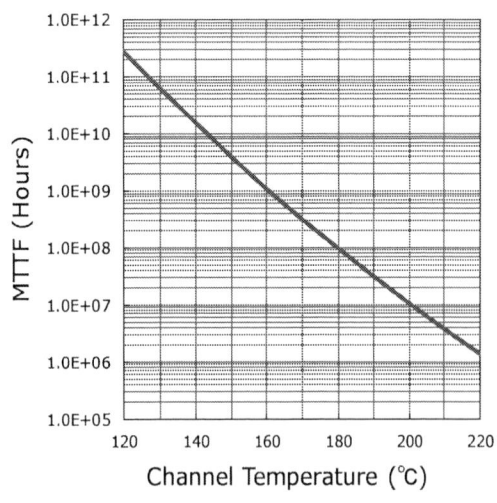

Fig. 3. MTTF of L-band GaN HEMT.

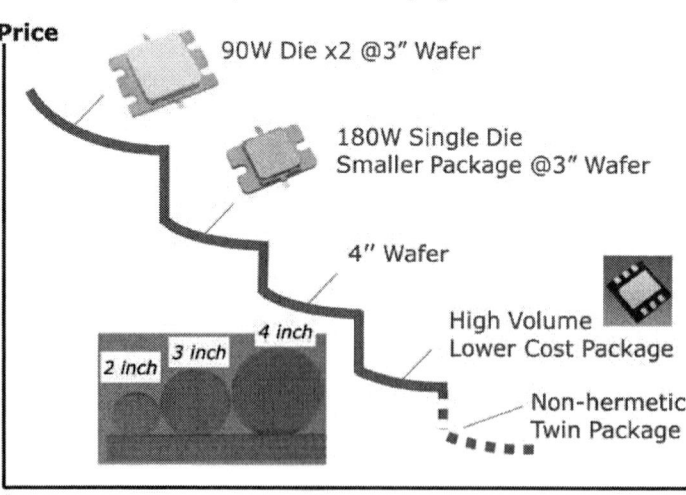

Fig. 4 Cost down method of GaN HEMT.

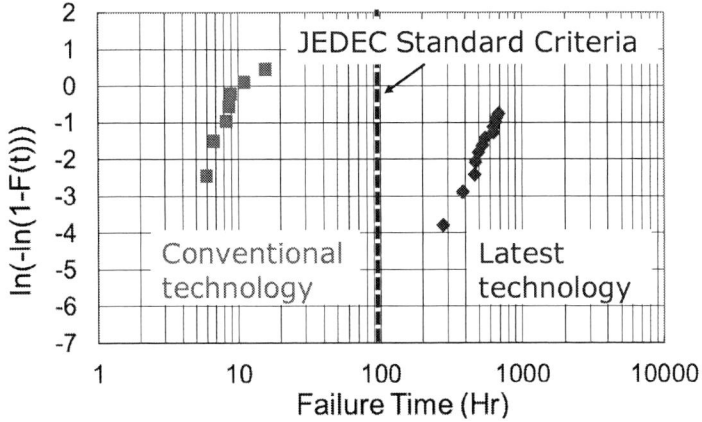

Fig. 5. HAST results.

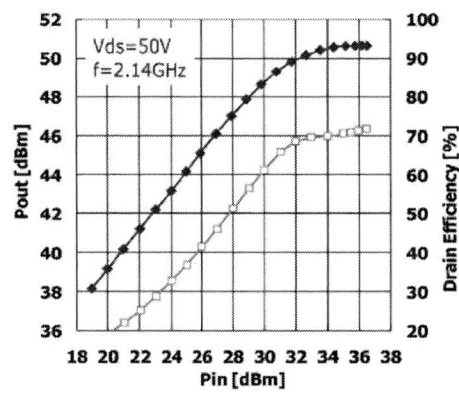

Fig. 6. Pin-Pout profile of inverse class-F GaN HEMT.

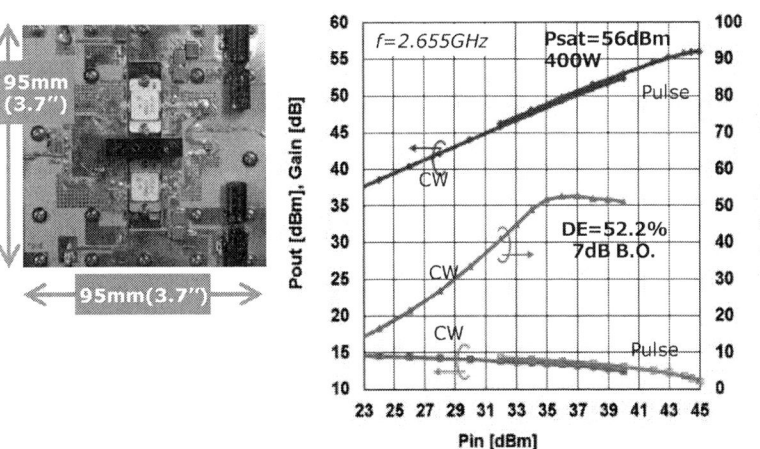

Fig. 7. 400W Doherty amplifier for C-plane.

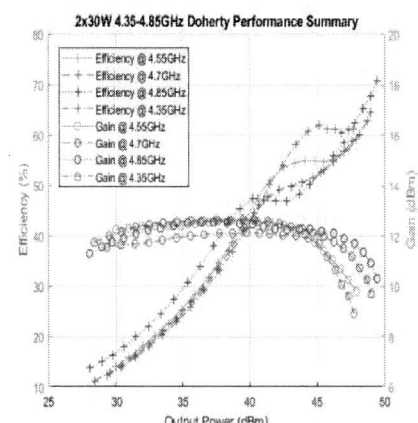

Fig. 8. 4.35-4.85 GHz Doherty amplifier for U-plane (M-MIMO).

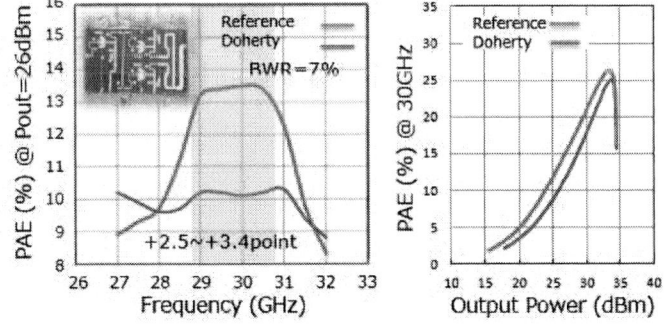

Fig. 9. 30 GHz Doherty MMIC.

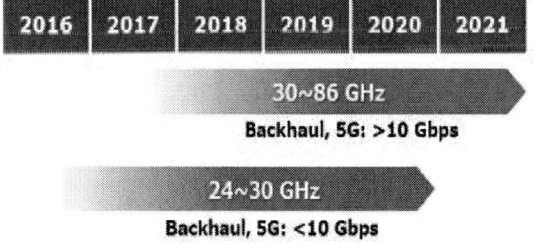

Fig. 10. Prospect of mm-wave GaN HEMT.

978-1-7281-1988-5/18 $31.00 © 2018 IEEE   323

100-340GHz Systems: Transistors and Applications

M.J.W. Rodwell[1], Y. Fang[1], J. Rode[1,2], J. Wu[1], B. Markman[1], S. T. Šuran Brunelli[1], J. Klamkin[1], M Urteaga[2]

[1]University of California, Santa Barbara, USA, email: rodwell@ece.ucsb.edu

[3]Teledyne Scientific Company, Thousand Oaks, CA,USA, [3]now with Intel, Portland, OR, USA

Abstract— We examine potential 100-340 GHz wireless applications in communications and imaging, and examine the prospects of developing the mm-wave transistors needed to support these applications.

I. INTRODUCTION

Wireless networks face exploding demand; the available spectrum is nearly exhausted. Industry is responding by introducing 5G systems at 28, 38, 57-71(WiGig), and 71-86GHz. Research now considers next-generation systems between 100-340 GHz. These will access a much larger spectrum. The short wavelengths provide massive spatial multiplexing in hub and backhaul communications. The short wavelengths will permit high-resolution imaging, from small apertures, to assist driving or flying in foul weather. Such applications will drive THz transistor development.

II. 100-340 GHz SYSTEMS

Consider three applications: a wireless communications base station or hub serving 100's of mobile users, a spatially multiplexed point-point backhaul link, connecting such hubs to the internet, and an imaging system, with TV-like resolution, enabling driving in very heavy fog or rain.

Fig. 1 shows a 140GHz spatially multiplexed hub, supporting 512 users. Each of 4 hub faces has 256 antennas, each $0.6\lambda \times 2.4\lambda$, in a 0.35m length phased array. Assume a handset with an 8×8 array, at $\lambda/2$ spacing (9mm×9mm), 20 dB margins for packaging, manufacturing, aging, and partial beam obstruction, and uncoded QPSK modulation at 10^{-3} error rate. A 45 mW transmitter output power per element and 3 dB handset noise figure can then support 128 users, at up to 100m range, and 10Gb/s downlink rate per user, even in 50mm/hr. rain. At 75GHz, the same parameters would provide 160m range, but the arrays would have 4:1 larger areas.

Fig. 2 shows a spatially multiplexed backhaul link. N transmitters, carrying independent data, form an array of length L. The receiver, at distance R, has a similar array but uses MIMO [1] beamforming. If the array angular resolution λ/L is smaller than the element apparent angular separation L/NR, then the signals can recovered with high SNR. Link capacity is increased N:1. Short wavelengths are of great advantage, as a short array can then carry many channels; at 500m range, an 8-element array must be 1.6m long at 340GHz, 2.6m at 140GHz, and 3.5m at 75GHz. At 340GHz, if each array element is an 8×8 subarray of 7λ by 7λ elements (for small beam angle adjustment) then, with 20 dB total margins and QPSK at 10^{-3} error rate, transmitting 640Gb/s over 500m range in 50mm/hr. rain requires only 80

mW/element output power, and 4dB receiver noise figure. Using two polarizations, the capacity is 1.2Tb/s in the same length array. At 140GHz, only 2mW/element is required, but the array is longer.

The third example (Fig. 3) is 340GHz imaging radar, for driving in e.g. heavy fog, providing a TV-like 64×512 pixel image refreshed at 60Hz. A linear 1×64 array steers the beam vertically, while frequency scanning and a frequency-selective Fresnel lens steers it horizontally. Given a 35cm×35cm aperture, 10% pulse duty factor, and 10dB SNR from a 1 ft^2 target at 300m range in heavy fog, the necessary peak output power is 50mW/element given 6.5dB receiver noise. The high 340GHz carrier permits a sharp 0.14 degree resolution from an array that can fit behind a car's radiator grille.

III. THZ TRANSISTORS

These and similar systems will drive THz transistor development. CMOS VLSI can provide the baseband and mm-wave signal processing, with a few *application specific mm-wave transistors* (SiGe, InP, GaN) providing the low-noise amplifiers (LNAs), efficient power amplifiers (PAs), and, for >200GHz system, signal conversion to the final carrier frequency. GaN [2] provides very high power and efficiency at 94GHz, while SiGe HBTs have reached 720GHz f_{max} [3]. Here we consider the prospect for further improvement in InP THz transistors, serving applications to 340GHz and above.

InP HBTs serve in mm-wave frequency converters and PAs [4]; 130nm node InP HBTs attain 1.1THz f_{max}. Though 650GHz InP HBT transceivers have been demonstrated [4], even at 220GHz, PA efficiency is impaired by limited gain. Higher-bandwidth HBTs are desirable for efficient 220 and 340GHz PAs, and for 650GHz systems.. Bandwidth is increased by scaling, decreasing semiconductor thicknesses, reducing junction widths, increasing current densities, and decreasing contact resistivities; [5] gives scaling laws and roadmaps. Yet, further improving f_{max} is difficult.

One challenge is parasitics distributed along the emitter stripe (Fig. 4); as with FETs, base metal resistance [6] decreases f_{max}. If the emitter length is reduced to reduce metal resistance, then capacitances from the base pad and the emitter ends become more significant, and f_{max} is again reduced. The base metal must be made thicker, and the base pad and inactive regions at the emitter ends made smaller

A second challenge is obtaining adequately low base contact resistance. Though base contacts of resistivity sufficiently low for the 32nm (3THz f_{max}) node have been demonstrated (Fig. 5) [7] in TLM test structures, it difficult

978-1-7281-1988-5/18 $31.00 © 2018 IEEE

realize such contacts in a processed transistor, where the contact is deposited on a surface exposed to several prior process steps. Refractory base contacts penetrate negligibly into even an 18nm thick base, and have shown excellent resistivity in TLMs [7]. Unfortunately, their resistivities in processed HBTs has been high, possibly due to their failure to penetrate residual surface oxides. Addressing this Rode [6] (Fig. 6, Fig. 7) used a 1nm Pt surface contact metal layer below a refractory Ru barrier. The Pt reacts with the base semiconductor and penetrates 2.7nm into it. This contact cannot be used for the 64 and 32nm nodes: the resistivity is too high. Further, the base doping is graded, being extremely high at the surface for low contact resistivity, but moderate towards the collector for reduced Auger recombination. As the base is made thinner with scaling, less base contact penetration can be tolerated if the metal is to contact heavily-doped semiconductor.

The first step in developing 90nm node HBTs was developing the scaled emitter junction and its tall contact via (Fig. 8). This uses a sputter-deposited, REI-etched $Ti_{17\%Wt}W_{83\%}$ alloy. Device images, and DC data, are shown in Fig. 9 and Fig. 10. RF data was poor due to a process failure during isolation. Given the difficulties with base contact, scaled emitter and base widths will not alone provide further increased f_{max}. We are therefore exploring regrowth processes.

Extrinsic base regrowth (Fig. 11) can indirectly provide the low base contact resistivity required for increased f_{max}. After forming the emitter-base junction, an extrinsic base is regrown upon the intrinsic base. Here, the motivation for regrowth differs from earlier work [8]; the extrinsic base can be very heavily doped ($\sim 2 \cdot 10^{20}$ cm^{-3}) (fig xx) for low contact resistivity, yet the intrinsic base can be more lightly doped ($\sim 5 \cdot 10^{19}$ cm^{-3}) for low Auger recombination hence acceptable β. The extrinsic base can be made moderately thick (~30nm), permitting a thicker Pd layer to penetrate more deeply through surface oxides for low contact resistivity, yet the intrinsic base can be made thin (10-15nm) for low base transit time. β will also increase because of reduced Auger recombination in the intrinsic base and reduced electron diffusion to the base contacts. High β improves noise figure in LNAs and improves DC precision in ADCs and DACs.

Regrowth can also bury dielectrics into the base-collector junction, as is common in SiGe HBTs [3] (Fig. 12). This permits wide base contacts for low base resistance, yet a narrow collector junction for low collector capacitance. Such structures can be formed by template assisted selective epitaxy (TASE) [9]. A dielectric template is first formed on the N+ subcollector. Upper portions of the N+ subcollector, plus the N- drift collector, are then grown within the template. The template top is removed, and the HBT base and emitter grown on the drift collector. The emitter-base junction is formed normally. Fig. 13 shows wide InP lateral overgrowth on a (111) InP wafer. This structure can be combined with extrinsic base regrowth to further enhance bandwidth.

Among transistors, InP FETs have the lowest noise, and serve in LNAs. For lower noise or higher frequencies, the cutoff frequencies must be increased. For this, the gate length must be reduced, but the $g_m R_{ds}$ product must remain constant, and, given the fixed parasitic source-gate and gate-drain capacitances, the transconductance per unit gate width must be increased. These factors require a thinner channel and a thinner gate-channel insulator. In present InP HEMTs, this insulator, ~6nm InAlAs, cannot be made much thinner without unacceptably increasing the gate leakage. This gate dielectric scaling limit impairs further bandwidth improvements.

High-K ZrO_2 gate dielectrics can now be deposited with low interface trap density onto InAs. Using these, even 2.5nm ZrO_2 provides low gate leakage and 4.8:1 greater capacitance density than a 6nm InAlAs layer [10,11]. This will permit further scaling of mm-wave InAs MOS-HEMTs. Unlike in III-V MOSFETs, where the source and drain lie close to the gate, in a THz HEMT, for low parasitic capacitances, high-mobility modulation-doped spacers separate the gate from the source and drain. Fig. 14 shows a target device, formed using multiple MOCVD regrowths. Fig. 15 and Fig. 16 show DC and RF data of a preliminary device fabrication effort [11].

ACKNOWLEDGMENT

This work was supported in part by the Semiconductor Research Corporation and DARPA under the JUMP program

REFERENCES

[1] C. Sheldon, M. Seo, E. Torkildson, M. Rodwell and U. Madhow, "Four-channel spatial multiplexing over a millimeter-wave line-of-sight link," 2009 IEEE International Microwave Symposium, Boston, MA,.

[2] S. Wienecke et al., "N-Polar GaN Cap MISHEMT With Record Power Density Exceeding 6.5 W/mm at 94 GHz," IEEE Electron Device Letters, vol. 38, no. 3, pp. 359-362, March 2017.

[3] B. Heinemann et al., "SiGe HBT with ft/fmax of 505 GHz/720 GHz," 2016 IEDM, San Francisco, CA, 2016, pp. 3.1.1-3.1.4.

[4] M. Urteaga, Z. Griffith, M. Seo, J. Hacker and M. J. W. Rodwell, "InP HBT Technologies for THz Integrated Circuits," IEEE Proceedings, vol. 105, no. 6, pp. 1051-1067, June 2017.

[5] M. J. W. Rodwell, M. Le and B. Brar, "InP Bipolar ICs: Scaling Roadmaps, Frequency Limits, Manufacturable Technologies," IEEE Proceedings, vol. 96, no. 2, pp. 271-286, Feb. 2008.

[6] J. C. Rode et al., "Indium Phosphide Heterobipolar Transistor Technology Beyond 1-THz Bandwidth," IEEE Trans Electron Devices, vol. 62, no. 9, pp. 2779-2785, Sept. 2015.

[7] A. Baraskar, A. C. Gossard , M. J. W. Rodwell, " Lower Limits To Metal-Semiconductor Contact Resistance: Theoretical Models and Experimental Data", Journal of Applied Physics, 114, 154516 (2013)

[8] M. Ida et al. , "Enhancement of fmax in InP/InGaAs HBTs by selective MOCVD growth of heavily-doped extrinsic base regions," IEEE Trans Electron Devices, vol. 43, no. 11, pp. 1812-1818, Nov. 1996.

[9] H. Schmid et al., " Template-assisted selective epitaxy of III–V nanoscale devices for co-planar heterogeneous integration with Si" Appl. Phys. Lett. 106, 233101 (2015);

[10] A. Leuther, et al. , "80 nm InGaAs MOSFET W-band low noise amplifier," 2017 IEEE International Microwave Symposium (IMS), Honolulu, HI.

[11] J. Wu, et al. "Lg=30 nm InAs Channel MOSFETs Exhibiting f_{max} =410 GHz and f_τ=357 GHz," IEEE EDL, vol. 39, no. 4, pp. 472-475, April 2018.

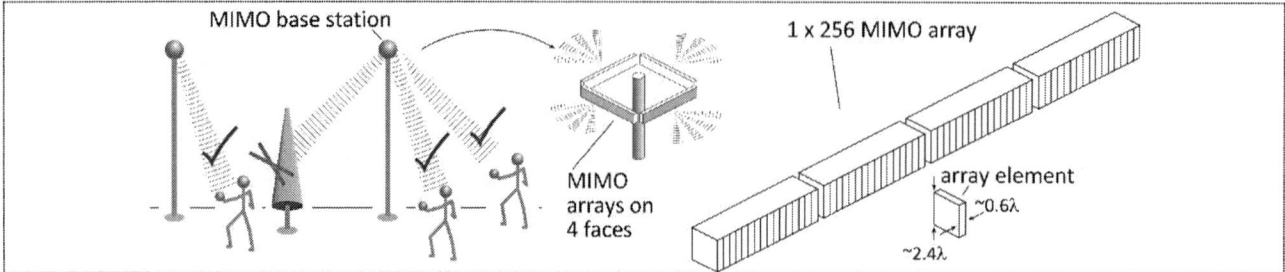

Fig. 1: Spatially multiplexed network hub. The hub has 4 faces, each a 256-element MIMO array, providing up to 128 independent signal beams. Link SNR analysis suggests that, with a 140GHz carrier at 100m range, 10Gb/s transmission per beam is feasible.

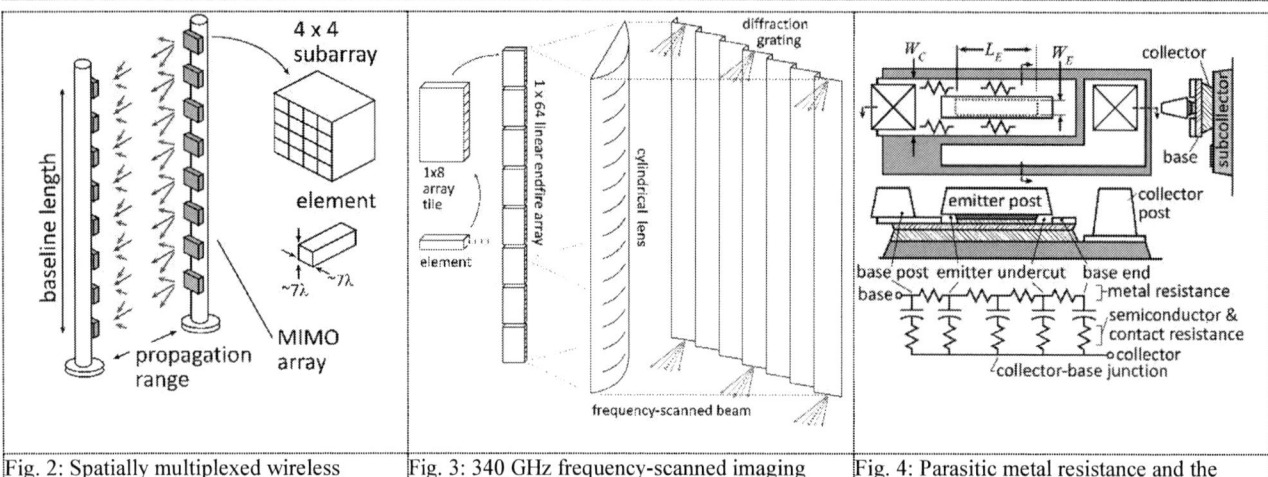

Fig. 2: Spatially multiplexed wireless backhaul link, using linear transmitter and receiver array, with each element being a 4×4 subarray.

Fig. 3: 340 GHz frequency-scanned imaging radar for driving in foul weather. The figure shows a separate imaging lens and diffraction grating for lateral beamsteering; these can be combined into a Fresnel lens.

Fig. 4: Parasitic metal resistance and the junction capacitances of inactive device regions are both distributed along the length of the emitter stripe. These significantly reduce the f_{max} of THz HBTs.

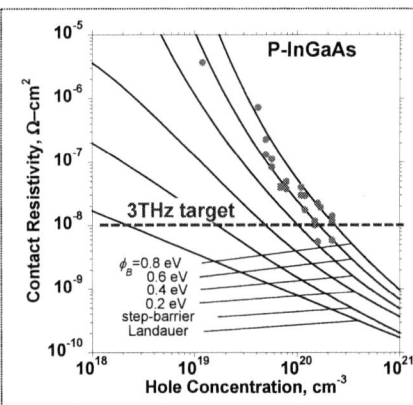

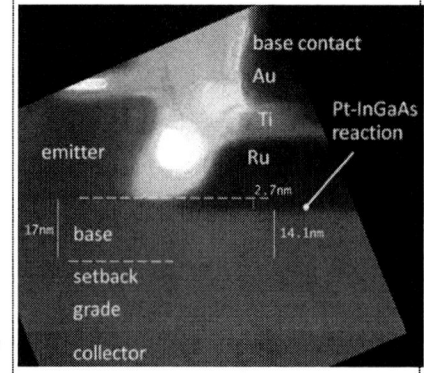

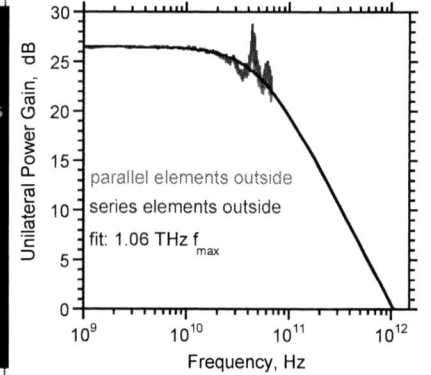

Fig. 5: Measured (dots) and computed (lines) contact resistance to P-InGaAs as a function of doping and barrier height. 10^{-8} Ω-cm^2 is sought for the 32nm (3THz f_{max}) node, and has been demonstrated in TLM test structures but not in HBTs.

Fig. 6: FIB/SEM cross section of the emitter-base junction of an HBT nominally at the 130nm node. The HBT base contacts are 1nm Pd below refractory Ru, and have penetrated 2.7nm into the base.

Fig. 7: Measured unilateral power gain of the HBT of figure 6, with the open- and short-circuit pads de-embedded in either order. A least-squares fit to the data de-embedded with the parallel elements outside yields slightly more than 1THz f_{max}.

978-1-7281-1988-5/18 $31.00 © 2018 IEEE   326

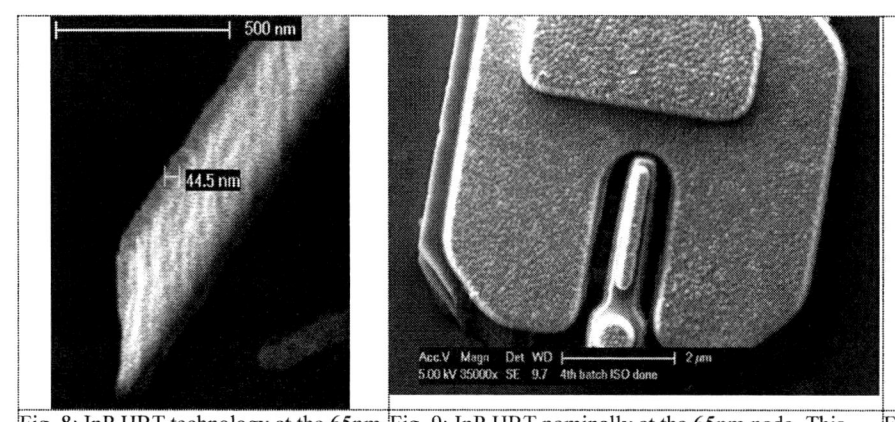
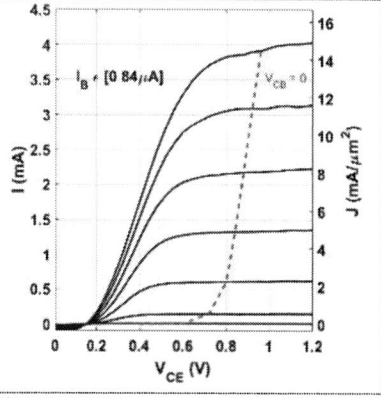

Fig. 8: InP HBT technology at the 65nm node: a ~400nm height, 44nm width dry-etched TiW emitter contact/via.

Fig. 9: InP HBT nominally at the 65nm node. This particular device has a 90nm emitter junction width.

Fig. 10: DC common-emitter characteristics of an InP HBT with a 90nm emitter. The DC current gain is 45.

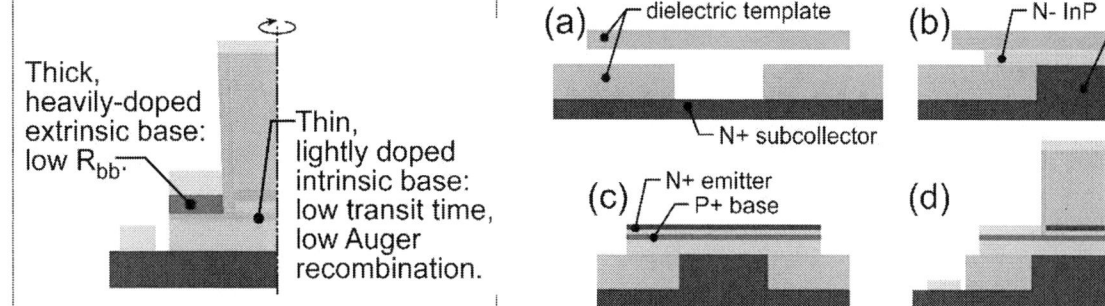

Fig. 11: Regrowth can provide a extrinsic base thicker and much more heavily-doped than the intrinsic base, enabling reduced contact resistance.

Fig. 12: Process flow for fabrication of THz HBTs with dielectrics buried within the base-collector junction: (a) dielectric template, (b) regrowth, (c) emitter and base growth, and (d) completed device.

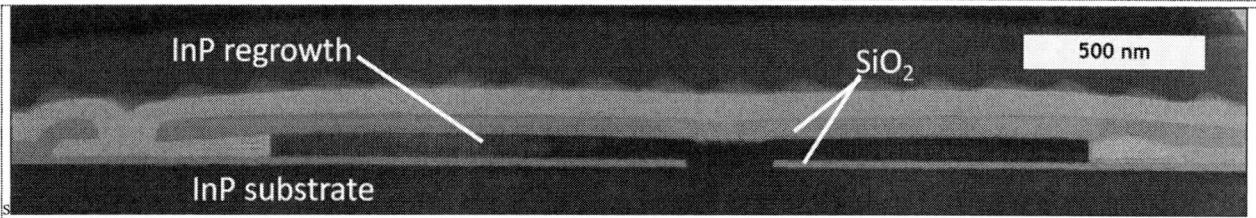

Fig. 13: FIB/TEM image of a ~100nm thick InP film grown by TASE on a (110) InP substrate.

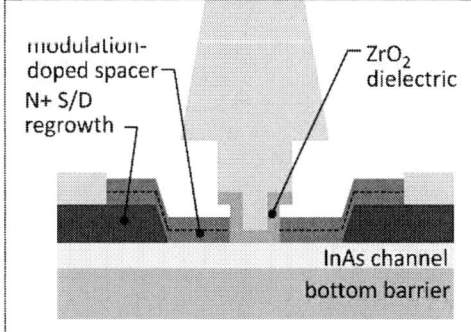
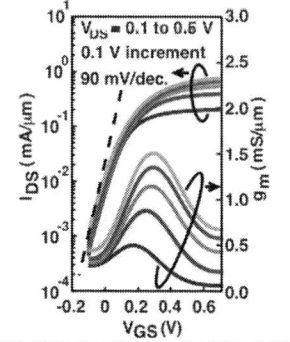
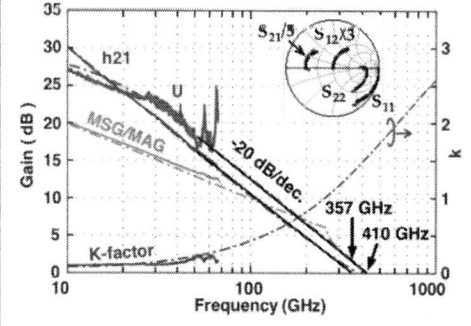

Fig. 14: THz InP MOS-HEMT with a ZrO2 gate dielectric and high-mobility modulation-doped source-gate and gate-drain spacers.

Fig. 15: DC characteristics of a InAs-channel MOSFET at 30nm gate length.

Fig. 16: RF characteristics of a InAs-channel MOSFET at 30nm gate length.

978-1-7281-1988-5/18 $31.00 © 2018 IEEE

Considerations on Design of Highly-Integrated Millimeter-Wave Transceivers in SiGe HBT

(Invited Paper)

V. Issakov and S. Trotta

Infineon Technologies AG, Am Campeon 1-12, D-85579 Neubiberg, Germany, email: Vadim.Issakov@infineon.com

Abstract— This paper addresses considerations on design of highly-integrated transceivers at mm-wave frequencies. Several aspects are discussed such as SiGe HBT scaling and co-design optimization. A highly-integrated chip operating at V-band for backhaul communication is shown as an example.

I. INTRODUCTION

The advances in silicon-based technologies and packaging platforms enable high-level integration of system on chip (SoC) and system in package (SiP) solutions. These solutions find growing interest due to the increasing demand for a cost reduction. Additionally, for a product to be competitive on the market, one needs to provide more functionality at the same or even lower price. Hence, the amount of external components shall be reduced by integrating more analog, digital, power management and RF functional blocks on the same chip, on a smallest chip area. This trend is demonstrated in Fig. 1 on example of the evolution of 77 GHz automotive radar chips. Additionally, use of embedded passives and antennas in package may simplify the applicability of the product in the target application, driving the trend towards SiP.

The demonstration of Silicon-Germanium (SiGe) HBT for millimeter-wave applications such as 77 GHz automotive radar [1] or V-band (57 – 64 GHz) backhaul [2] has given rise to the sales volume of systems for radar and communication applications. Advances of SiGe HBT have enabled replacing the expensive III-V components. However, in the next step more integration of complex digital and mixed-signal blocks, such as a microcontroller, memory, serial peripheral interface (SPI), FPGA, ADC and a digital PLL on the same SoC is required. Furthermore, digital controllability is needed for higher flexibility of the transceiver chips and realization of enhanced functionality. This requires use of MOS transistors.

Advanced CMOS is particularly attractive for high-level integration. Further, nano-scale CMOS nodes demonstrate competitive performance in terms of transit frequencies (f_T) and maximum oscillation frequencies (f_{max}) compared to SiGe HBT processes. However, it is still a challenge for CMOS technologies to achieve the noise performance, linearity, output power and temperature robustness, necessary for automotive radar products. Particularly, MOS transistors suffer from very high flicker noise corner compared to HBT. Therefore, BiCMOS technologies are very attractive for mm-wave applications as they combine the excellent RF properties of the HBT transistor with the advanced digital integration capabilities of a CMOS node. However, BiCMOS usually offers only moderate CMOS scaling due to economic reasons.

This work focusses on various implementation aspects of highly-integrated mm-wave transceivers for wireless small-cell backhaul applications. As an example we discuss V-Band chipset realized in Infineon's SiGe HBT technology. The V-band range is attractive for small-cell backhaul due to high frequency reuse. However, this frequency range has two major drawbacks: interference with other wireless standards and the question whether the licensing will stay cheap in the future. In this paper we address circuit design considerations, as well as co-design challenges during design of the transceiver [3]. Finally, measurement results are presented and discussed. The transceiver supports spectrally-efficient modulation schemes up to QAM128 and enables transmit rates up to 3.5 Gb/s.

II. TECHNOLOGY

The first family of the single channel direct conversion backhaul transceivers BGT60/70/80 was realized in Infineon's SiGe:C process B7HF200, which is a bipolar-only technology with a 350 nm lithographic feature size similar to the one described in [4]. It is based on a double-polysilicon self-aligned (DPSA) HBT using selective epitaxial growth (SEG) concept with shallow and deep trench isolation. This technology achieves f_T of 200 GHz and f_{max} of 250 GHz.

The next implementation of the V-band transceiver BGT60P for backhaul uses the Infineon's B11HFC BiCMOS technology with a lithography node of 130 nm, similar to the one described in [5]. This technology has a strongly improved npn transistor achieving f_T of 250 GHz and f_{max} of 385 GHz, as shown in Fig. 2. Variation of f_{max} versus lithography size is compared in Fig. 3 for these two HBT technologies versus advanced nano-scale CMOS nodes. The CMOS processes offer excellent f_{max}, however HBT remains superior with the scaling of the HBT lithographic feature, which is still relaxed compared to CMOS. Further, CMOS down-scaling continues towards FinFET transistors, which results in f_{max} drop for gate length reduction. In B11HFC the scaled HBT transistor is integrated into a 130 nm CMOS platform. This technology offers three thick top metal layers for high-quality mm-wave passives and four thin metal layers for digital routing.

Typically cut-off frequencies f_T, f_{max} should be at least five times larger than the operating frequency. E.g. for 77 GHz applications this would mean $f_{max} > 400$ GHz. Additional improvements of the HBT transistors can be achieved using such approaches as Non-Selective Epitaxial Growth (N-SEG) of the base and Elevated Extrinsic Base (EEB) [6]. For example, f_{max} of 720 GHz is demonstrated in [6]. Optimization of the technology is required to achieve larger design margins,

978-1-7281-1988-5/18 $31.00 © 2018 IEEE

higher gain, higher output power, better linearity, lower noise and lower power consumption. Finally, availability of accurate scalable, physics-based, compact models for active and passive devices is crucial for realization of mm-wave circuits.

III. CO-DESIGN CONSIDERATIONS

As the level of complexity of the highly-integrated RF and mm-wave systems rises, package and PCB play an ever increasing role on the overall system performance. Package poses several challenges on realization of highly-integrated mm-wave systems. First, it needs to support high routing density in the redistribution layer (RDL). Second, even small variations due to process tolerances can have a significant impact on the end-product performance. Finally, it needs to support implementation of passives. Hence, in the recent years several packaging technologies have been developed to address the SiP needs of highly integrated mm-wave transceivers. One possible package platform is Infineon's embedded wafer-level ball-grid array (eWLB), which offers fan-out area adaptable to the needs and thus more space for complex interconnect routing. Furthermore, it has outstanding electrical capabilities suitable for radio-frequency (RF) and mm-wave applications [7]. The cross-section of eWLB technology is shown in Fig. 4. Mold compound is used to carry the fan-out area and to protect the chip back-side. Additionally, fan-out area can be used for integration of passive components, such as interdigitated capacitors, transmission lines, inductors and antennas. A 3D view of the eWLB package with an integrated chip is shown in Fig. 5.

The parasitic wave effects such as impedance mismatch, signal reflections, crosstalk, and radiation become very critical at mm-wave frequencies. There are unwanted interactions between different parts of the system on chip, package or board conducted via electromagnetic fields. Hence, design of highly-integrated packaged transceivers requires careful chip-package-board co-design, co-simulation and co-optimization using accurate full-wave electromagnetic (EM) modelling [3]. To facilitate modelling and system design, it is recommended to integrate the physical layout data of board, package and chip in the same design environment. This approach is known as *concurrent editing* [3]. By this means one can identify the critical nets and predict possible interferences paths. This method is successfully applied for development of Infineon's mm-wave transceiver products. An example of the chip-package layout co-design is shown in Fig. 6. Unfortunately, EM modelling of the chip-package-PCB interactions poses a challenge, since there is no obvious way to define an interface at which separate parts can be "stiched". Chip-package-board interactions and layout cross-section are shown Fig. 7.

IV. CIRCUIT DESIGN CONSIDERATIONS

Fig. 8 shows the block diagram of the V-band transceiver BGT60P used as an example of a highly-integrated chip in this work. Similarly as the realization in [2], the transmit chain includes a single sideband (SSB) mixer, an I/Q modulator, a variable gain amplifier (VGA) controlled by a 6 bits DAC, a driver, and a cascode power amplifier. A peak detector monitors the signal level at the PA output. The receive chain includes consists of a differential two-stages cascade LNA and current-driven double balanced I/Q mixers. The local

oscillator (LO) signal generation uses a push-push Colpitts VCO based on the topology proposed in [8] and shown in Fig. 9. The VCO is operated at low current density and achieves a state-of-the-art phase noise better than -80 dBc/Hz at 100 kHz offset. This allows achieving high order modulation schemes. A detailed schematic of the mixer cell used in I/Q modulator is shown in Fig. 10. The mixer consists only of a switching-quad. The IF signals are applied at the common emitter nodes. Each differential pair is biased by an independent current source. In this way it is possible to add or subtract differential current provided by the current DAC to balance the modulator and suppress the LO feedthrough [2]. The chip is programmable via a serial-peripheral interface (SPI) that can be clocked up to 50 MHz. This allows fast switching in TDD mode. Additionally, phase shifter are added on the LO signal.

V. MEASUREMENT RESULTS

Fig. 11 shows micrograph of the chip packaged in the eWLB process. The package size is 6×6 mm^2. The receiver noise figure (NF) for the chip version BGT60 realized in B7HF200 is compared to the new version BGT60P realized in B11HFC is shown in Fig. 12. We see a major NF improvement of 1.5 dB, which can be attributed to a lower base resistance and larger frequency margins. Comparison of the transmitter output power for the two realizations is shown in Fig. 13. The B11HFC realization achieves 2 dB higher output power and higher flatness over the entire frequency band.

To evaluate the maximal achievable modulation scheme and data rate, a link was setup using two BGT60P chips (one as RX, one as TX) connected using a waveguide attenuator to model the propagation loss. The transmitted sequence is provided by Keysight's Arbitrary Waveform Generator M8190A, while on the receiving side the demodulated signals are analyzed using Keysight's MSOS404A Mixed-Signal Oscillocope. Figs. 14 and 15 show system measurements of BGT60P at 60 GHz using an external VCO. Using modulation scheme of QAM64 and at data rate of 3 Gbps an error-vector magnitude (EVM) of 29.8 dB was achieved. At QAM128 and data rate of 3.5 Gbps an EVM of 30 dB was shown.

REFERENCES

[1] H. Knapp *et al.*, "Three channel 77 GHz automotive radar transmitter in plastic package," in RFIC Symp., pp. 119 - 122, June 2012.

[2] S. Trotta *et al.*, "A V and E-band packaged direct-conversion transceiver chipset for mobile backhaul application in SiGe technology," in European Radar Conference (EuRAD), Oct. 2014.

[3] V. Issakov *et al.*, "Co-simulation and co-design of chip-package-board interfaces in highly-integrated RF systems", in BCTM, Nov. 2016.

[4] J. Böck *et al.*, "SiGe Bipolar Technology for Automotive Radar Applications,", in IEEE Proc. on BCTM, pp. 84-87, Oct. 2004.

[5] J. Böck *et al.*, "SiGe HBT and BiCMOS process integration optimization within the DOTSEVEN project" in BCTM, pp. 121 - 124, Oct. 2015.

[6] B. Heinemann *et al.*, "SiGe HBT with ft/fmax of 505 GHz/720 GHz," in IEDM, pp. 3.1.1 – 3.1.4, 2016.

[7] M. Wojnowski *et al.*, "High Frequency Characterization of Thin-Film Redistribution Layers for Embedded Wafer Level BGA," EPTC, 2007.

[8] Chakraborty *et al.*, "A Low Phase Noise Monolithically Integrated 60-GHz Push-Push VCO for 122 GHz Applications in a SiGe Bipolar Technology" in IEEE BCTM 2013.

[9] K. Aufinger *et al.*, "Advances in SiGe HBT Technology for mm-Wave Applications in the DOTSEVEN Project", Workshop, EuMIC 2016.

[10] M. Wojnowski, V. Issakov, "Packaging Trends", Workshop, EuMC2017.

978-1-7281-1988-5/18 $31.00 © 2018 IEEE   329

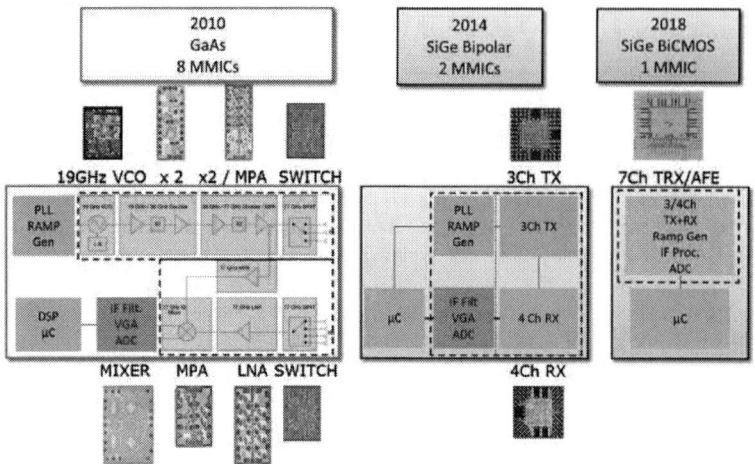

Fig. 1. Evolution of integration level of 77 GHz automotive radar solutions [9].

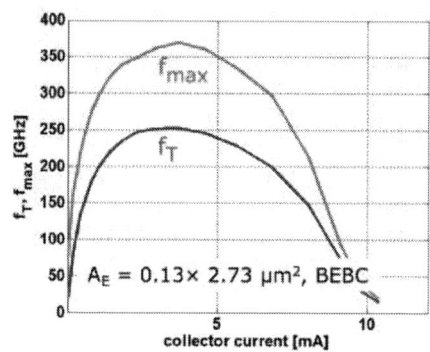

Fig. 2. f_T, f_{max} vs. collector current characteristics [9].

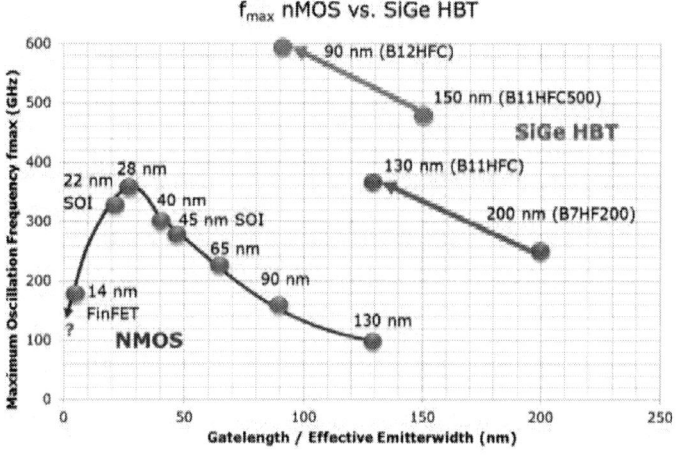

Fig. 3. f_{max} comparison of CMOS and SiGe HBT technologies.

Fig. 4. eWLB cross-section [9].

Fig. 5 eWLB 3D model.

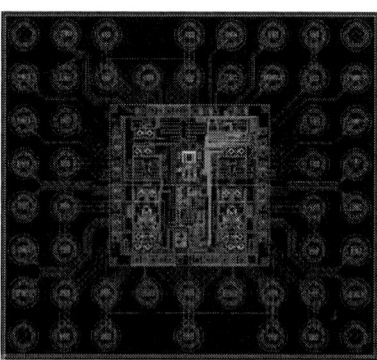

Fig. 6. Co-design example [10].

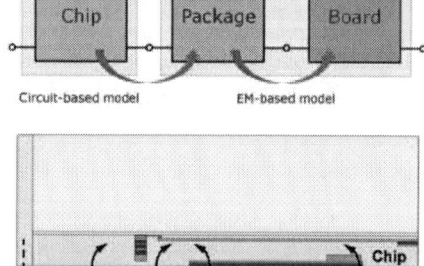

Fig. 7 Chip-package interactions [10].

978-1-7281-1988-5/18 $31.00 © 2018 IEEE   330

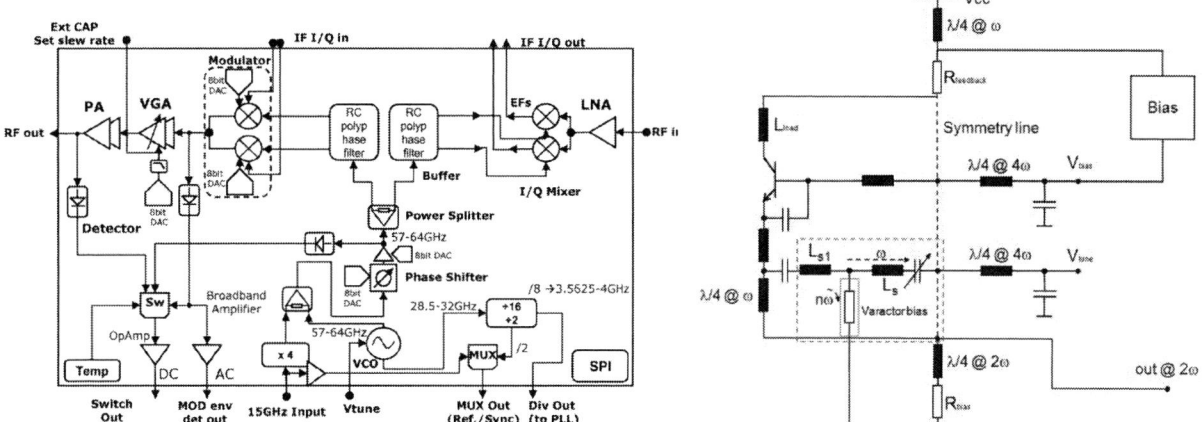

Fig. 8 Block diagram of the V-band transceiver BGT60P.

Fig. 9 Simplified VCO schematic.

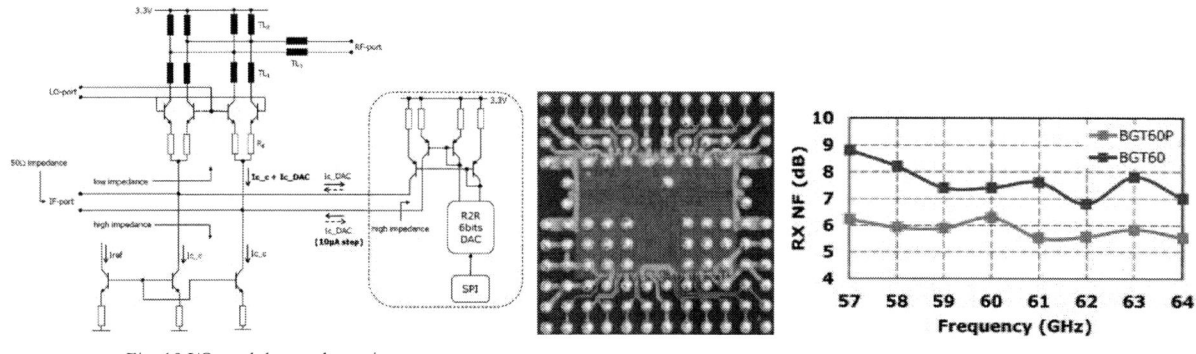

Fig. 10 I/Q modulator schematic.

Fig. 11. Chip micrograph.

Fig. 12. Receiver Noise Figure.

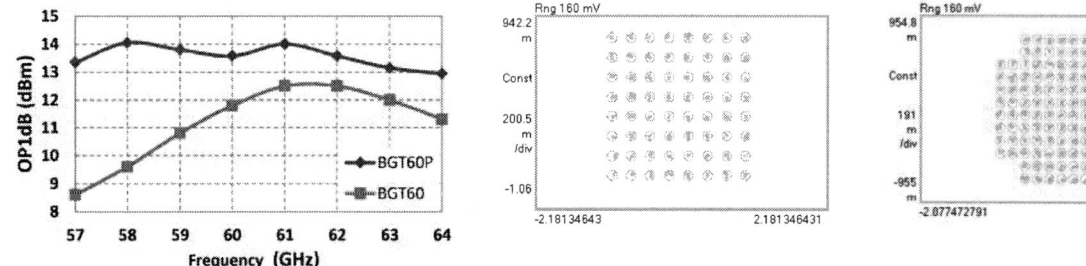

Fig. 13. Transmitter output power.

Fig. 14. System measurement QAM64.

Fig. 15. System measurement QAM128.

978-1-7281-1988-5/18 $31.00 © 2018 IEEE

BAW Filters for 5G Bands

R. Aigner, G. Fattinger, M. Schaefer, K. Karnati, R. Rothemund, F. Dumont

email: robert.aigner@qorvo.com

Qorvo Inc, Apopka, FL, USA

Abstract— The number of frequency bands requiring BAW filters is expected to grow significantly with the launch of 5G mobile applications. BAW is in particular well suited to address the new radio bands below 6 GHz also referred to as nr-1. RF integration is the only path forward to achieve full connectivity in LTE + 5G wireless due to the complexity of the antenna systems and the resulting coexistence challenges. The article gives an overview and describes recent trends regarding BAW for high frequency bands, wide bandwidth filters, miniaturization and thermal management. Evolution of RF content and the challenges of developing complex RF modules are discussed.

I. INTRODUCTION

5th Generation (5G) wireless and eLAA (Enhanced Licensed Assisted Access) will require RF filters currently unachievable with classical SAW or BAW technology. While many of the designated 5G n-bands (n for 'New Radio') will be re-purposed 3G and 4G LTE bands, there is a number of new bands falling into high frequency ranges from 3.3 GHz to 5.9 GHz. Some of those bands have very large relative bandwidth (for example n79 passband from 4.4 to 5GHz). The expectation is that acoustic filters will be available to support initial 5G deployment which is currently expected to start in year 2019. The 5G bands in the mmWave range above 24 GHz are attractive to serve fixed connections, however for mobile wireless there are many practical hurdles which will limit the range and utility for the user. Nearly all service providers foresee a rapid deployment of 5G in the sub 6 GHz bands, while the use of mmWave spectrum is mostly driven by chipset companies. For smartphone designers the introduction of 5G is another assault on battery life and board space. Pressure to integrate and shrink will be even greater than it is at present. Operating at the higher frequencies means power amplifiers are less efficient while antennas and lines have higher loss. RF-switches also have higher loss and there will be more switching needed to accommodate additional bands and antennas. In order to avoid severe degradation of sensitivity there is an expectation that filters will exhibit similar level of loss than for low frequency bands. Acoustic filters will (again) be crucial for the success of an entire wireless ecosystem. The specific challenges and how BAW-SMR aims to address them will be described in the subsequent sections.

Development work has progressed at a fast pace for the past two years. Prototypes for 3.5 GHz, 4.5 GHz, 5.2 GHz and 5.6 GHz filters have passed rigorous testing. Results are presented in later sections of the paper.

II. HIGH FREQUENCY BAW

A. Performance challenges

Acoustic losses (viscous, scattering, thermo-elastic...) relevant to BAW devises increase at least linear with frequency, often with the square of frequency or even worse. Just from this fact it is difficult to maintain the same Q values in BAW resonators going from 2.5 GHz to 5 GHz. Even very large resonators will not approach the same asymptotic Q values as seen at lower frequencies. The situation gets a lot worse when comparing resonators suitable in size for 50 ohm port impedances. The piezo-thickness will have to scale with $1/f$ and the desired capacitance per branch also decrease with $1/f$. Consequently the resonator area will scale in size with $1/f^2$. While this benefits die-shrinks it turns out to be a significant problem regarding performance. An approx. square resonator at ¼ of the area still has ½ of the circumference and is therefore more prone to lateral acoustic leakage. It also has more pronounced edge effects and spurious modes. When a BAW resonator in a branch of the filters gets too small to perform, a filter designer will typically replace it with a cascaded pair of resonators (C → 2C in series with 2C). This method has the drawback of doubling the series resistance in a branch. The series resistance of a resonator is mainly driven by sheet resistance of the electrodes. If all layers in a BAW stack are linearly scaled in thickness (as needed to achieve a higher frequency device) the sheet resistance will become a major contributor to filter loss. Fig. 1 shows the scaling behavior of key parameters [1]. The cascading of resonators in filter branches makes a bad situation even worse. An alternative is to increase the ratio of electrode thickness to piezo-thickness in order to reduce sheet resistance. Unfortunately this will reduce the effective coupling of the resonator and the size of a 50 ohm resonator will shrink even further. The coupling can be recovered by enhancing the intrinsic coupling of the piezo-layer as discussed in a later section.

B. Reliability

Smaller resonator size and lower Q values lead to higher density of dissipated power in a device. Power handling at high frequencies is more difficult to achieve. Fortunately the thickness of the reflector layers in a BAW-SMR also scales $1/f$ and therefore improves the heat extraction through the reflector.

While most layers in a BAW stack can be scaled without problems, there are some layers which have to maintain a minimum absolute thickness to function properly. One example is the passivation layer which protects the top surface of the device from corrosion and moisture attack.

978-1-7281-1988-5/18 $31.00 © 2018 IEEE

The minimum thickness of the passivation layer is defined by reliability tests. At 5 GHz it is highly desirable to use a very thin passivation layer which must be of superior quality and pin-hole free.

C. Manufacturability

Finally, the challenges manufacturing a 5 GHz BAW devices are exacerbated by the fact that deposition processes, clean-up processes and subtractive trimming processes introduce an error in absolute thickness. As Fig. 1 shows the sensitivity to changes in thickness [MHz/nm] scales with $1/f^2$. 5 GHz devices will require tighter tolerances on all process steps and more advanced trimming.

III. ENHANCED COUPLING

Wide bandwidth filters such as n79 (4.4 to 5 GHz) and n78 (3.3 to 3.8 GHz) benefit significantly from higher effective coupling. It is possible to address wideband filters with regular AlN piezo [2], but compromises have to be made for rejections in certain out-of-band regions. Those compromises are no longer possible in Carrier Aggregation (CA) multiplexers. Even for narrower bands which have traditionally used regular AlN [3] increased coupling is now necessary in order to meet the additional rejection requirements while maintaining passband return loss.

Sc-doped AlN piezolayers have come to the rescue of BAW and FBAR. The adoption of CA in many band combinations would probably have happened a lot slower without Sc-doped AlN entering our field in year 2015. Manufacturing of ScAlN is more challenging than regular AlN and requires better control of seeding conditions and layer stress (along with many other parameters). While the enhancement in coupling is highly appreciated it comes with a degradation of Q values as shown in fig. 2 [4]. For Sc content >6% the degradation is significant and only bands with moderate nearby rejection requirements will show a net benefit. Most of the 5G bands below 6 GHz fall into this category.

IV. COMPLEXITY CHALLENGE

The rapidly increasing complexity of RF front-end architectures requires an unprecedented level of accuracy in simulation and modeling. For success in the marketplace it is necessary - but no longer sufficient - to have access to the best-in-class technologies for filters, switches, power amplifiers and packaging. The time allotted to develop a complex integrated RF module from the definition of functionality to a final qualified sample is very short and leaves no room for trial and error. The most significant step up in complexity came from the deployment of CA. In order to create the required multiplexers, the acoustic resonators need to have significant headroom in coupling coefficient. Excess bandwidth can then be traded in for isolation in several out-of-band regions where the other filters have their respective passbands. Maintaining a competitive system-sensitivity and power consumption is mandatory. As multiplexers show higher insertion loss compared to duplexers or single filters it is imperative to optimize Q

values [5] in resonators and reduce losses in matching elements.

V. DEVICE SIZE: µBAW

Present generation RF modules are extremely densely packed. Size of filter dies and auxiliary elements associated with a filter (inductances, matching elements, RF lines, …) determines what overall module-size can be achieved. The total die area of a year 2016 Wafer-Level-Packaged BAW filter (example shown in fig. 3) is less than 50% occupied by active resonators. The periphery (outside of the red dashed line) accounts for about 40% and contains the Cu-Sn-pillars as I/Os and probe-pads for on-wafer-probing. The trick of uBAW is to completely eliminate peripheral area by placing the I/Os on top of the WLP-roof, overlapping with active area.

The sketches in fig. 4 and 5 outline the change in construction from WLP to µBAW. The vias needed to connect the I/O pads to the BAW device have to be as small as possible in order to maintain most of the size advantage. Fig. 5 shows the I/Os overlaying the active resonator area on the BAW die. The distance between active resonators can be further reduced from the previous version shown in fig. 4 and some of the inner walls (on which the roof rests) can be entirely omitted. All in all the µBAW approach reduces die sizes by an average of 35% relative to a year 2016 BAW WLP.

VI. RESULTS AND SMMARY

Measurement results shown in fig. 6 to 9 proof BAW-SMR technology is uniquely positioned to achieve low loss and high selectivity for 5G bands. The example shown is for a bandpass filter centered at 5.25 GHz. The passband insertion loss this filter maintains is below 2 dB over full temperature range while rejection in adjacent bands is as high as 45 dB. The filter is very well matched to 50 ohm with return losses better than 15 dB. The challenges making this filter are representative for the other bands between 3 and 6 GHz. Product development for all bands needed in the initial deployment of sub-6GHz 5G systems is in full swing. Volume production is expected to start in year 2019.

REFERENCES

[1] S. Mahon," The 5G Effect on RF Filter Technologies", CS Mantech conference, May 23, 2017, Indian Wells, CA

[2] S. Kreuzer, A.Volatier, G.Fattinger, F. Dumont," Full band 41 filter with high Wi-Fi rejection – design and manufacturing challenges", Proceedings of 2015 IEEE Ultrasonics Symposium, Taipei, Oct. 2015.

[3] M. Li, M. El-Hakiki, D. Kalim, A. Link, B. Schumann, R. Aigner, "A Fully Matched LTE-A Carrier Aggregation Quadplexer Based on BAW and SAW Technologies", Proceedings of IEEE IUS 2014, Chicago, Sep. 3 – 6, 2014

[4] J. Sadhu,"Doped AlN Thin Films for Enhanced BAW Filters in Mobile", 2018 PiezoMEMS workshop, Jan 15-16th, 2018, Orlando, FL

[5] Press release BAW5: http://www.qorvo.com/newsroom/news/2017/qorvo-solves-advanced-carrier-aggregation-challenges-with-new-multiplexers

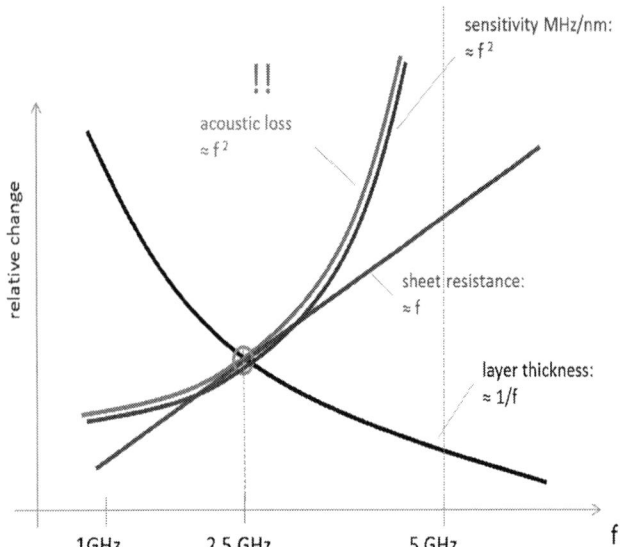

Fig. 1: Frequency scaling of key parameters for BAW

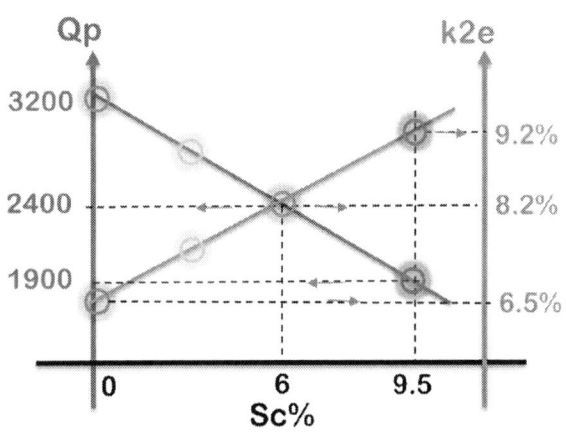

Fig. 2: Measured Qp for BAW-SMR at various Sc-contents.

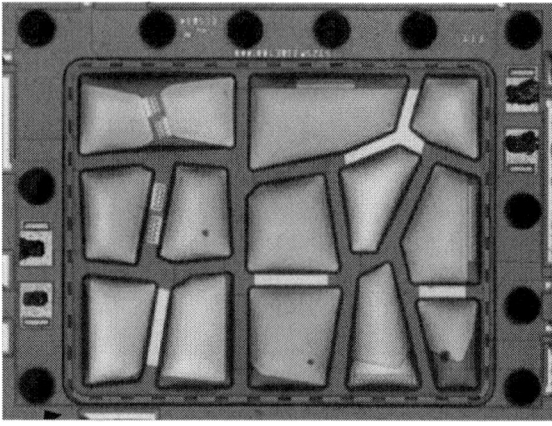

Fig. 3: Example of BAW filter in conventional WLP package (Generation 5 BAW, year 2016). The polymer used for the roof above the resonators is transparent. Red dashed line indicates boundary between active area and periphery.

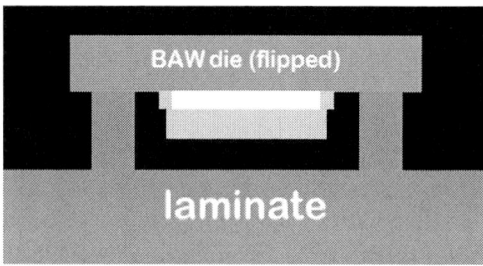

Fig. 4: simplified cross-section of year 2016 WLP BAW. Black: mold compound. Orange: Cu-interconnect. Yellow: Polymer-structure.

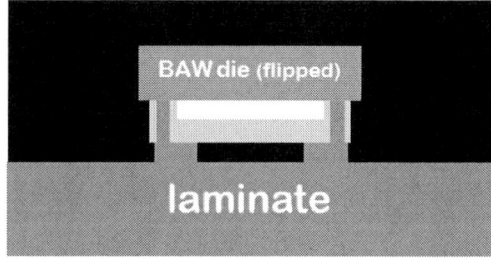

Fig. 5: cross-section of µBAW device in package. Color definition: see fig. 4.

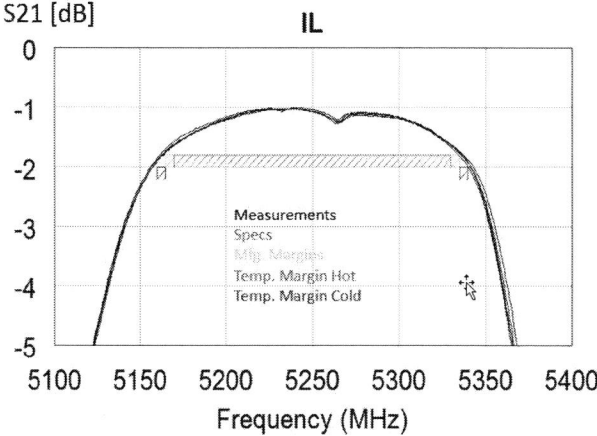

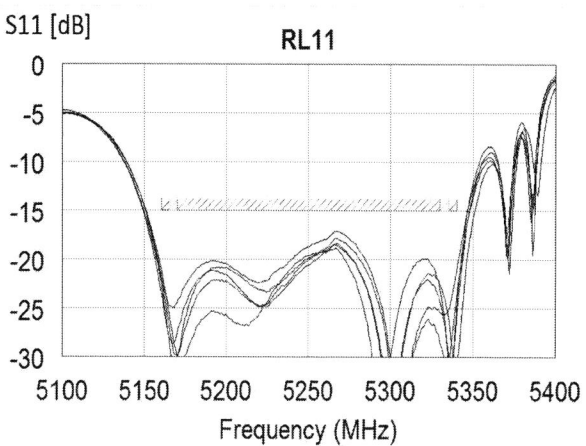

Fig. 6:. Measured passband insertion loss of 5.25 GHz BAW filter. Graph shows 5 traces from individual samples, fully packaged and on evaluation board. Specifications are shown as lines bordering hashed areas, see legend.

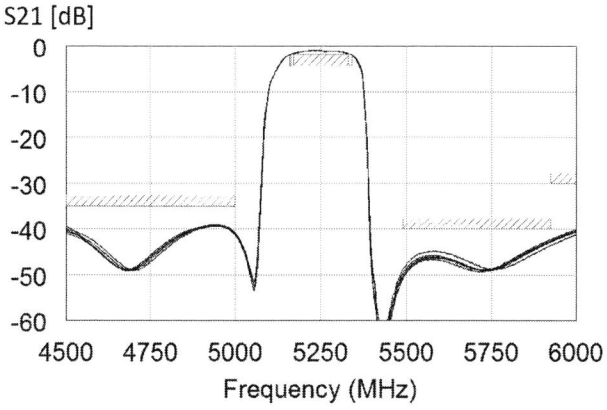

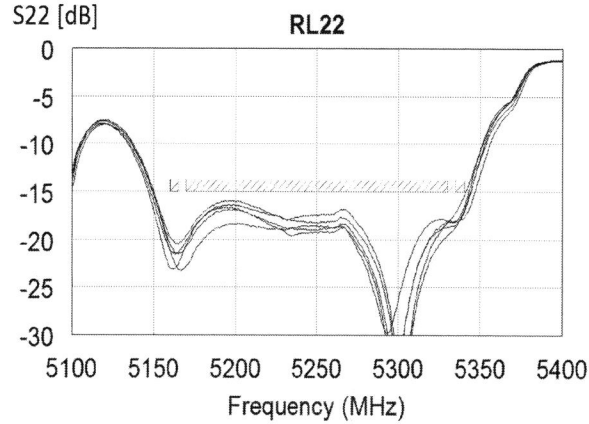

Fig. 8a and 8b:. Input and output port matching of of 5.25 GHz BAW filter.

Fig. 7:. Passband insertion loss and nearby rejection of 5.25 GHz BAW filter.

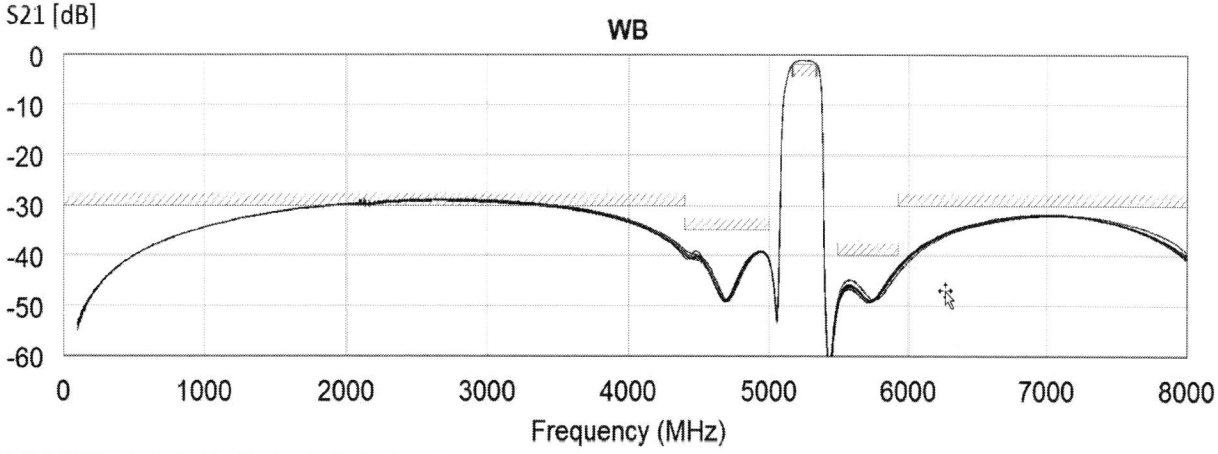

Fig. 9:. Wideband rejection of of 5.25 GHz BAW filter.

978-1-7281-1988-5/18 $31.00 © 2018 IEEE

Tunable Filter Technologies for 5G Communications

Dimitrios Peroulis

School of Electrical and Computer Engineering, Birck Nanotechnology Center, Purdue University, West Lafayette, IN47907, USA
email: dperouli@purdue.edu

Abstract—This paper presents an overview of available technologies for manufacturing three-dimensional front-end tunable filters for 5G systems. Specifically, we discuss three main technologies: a) RF MEMS, b) Printed Circuit Board (PCB), and c) injection molding. The advantages and drawbacks of each technology are discussed along with relevant proof-of-concept demonstrations. Future directions and improvements are also presented.

I. INTRODUCTION

The coming generation of 5G communication systems presents unique challenges for front-end filter design and manufacturing. Front-end filters will likely need to cover a much wider range of bands than ever before. Bands from below 1 GHz to over 70 GHz are currently being considered. In the United States, for example, licensed bands have been created at 600 MHz, 3.1-3.55 GHz, 3.7-4.2 GHz, 27.5-28.35 GHz, and 37-40 GHz. Operators are also exploring the 64-71 GHz band. When we also consider backward compatibility, it is not hard to realize the challenges presented for hardware designers and manufacturers.

To address these requirements, there are three main approaches: a) eliminate all front-end filters, b) utilize static filters banks, and c) employ tunable high-quality filters. It is not clear yet if the first approach will be feasible and it will largely depend on future integrated circuit technology and system-level design approaches. The second approach is largely in use today, but it will become increasingly more complex as discussed above. It is also not clear what static filter technologies can adequately cover the mm-wave and sub-mm-wave 5G bands. In this paper, we focus on the third approach and present promising technologies for 5G base stations.

II. TUNABLE RESONATOR CONCEPT

Fig. 1 shows the main tunable filter resonator concept [1,2]. The tunable resonator is a substrate-integrated cavity that is heavily loaded at its center with a metallic post. This loading scheme results in an evanescent-mode (EVA) cavity with two critical advantages: a) significant volume miniaturization (>90%), and b) the ability to tune its resonant frequency since the vast majority of the electric field energy is concentrated over the post. Consequently, this results in a quasi-static gap-tunable capacitor (Fig. 1).

We can tune the resonator's frequency by accurately controlling the critical gap between the top of the post and the ceiling of the cavity (Fig. 2). This critical gap can range from few micrometers to several hundreds of micrometers

depending on other requirements (e.g. tuning technology and power handling) [3-5]. Typical tuning range is typically one octave (Figs. 3 and 4), although up to two octaves have been demonstrated [6]. Despite this relatively wide frequency tuning, this resonator's quality factor is in the range of 500-1,500 depending on the required power handling and fabrication quality. Thus, unlike most other technologies that yield either wide frequency tuning or high quality factor (but not both), EVA tunable resonators present a nearly-ideal frequency tuning-quality factor trade off. Table I presents the main design considerations. Furthermore, they can be matched to nearly all impedance ranges, from few Ohms to a few kilo Ohms including 50 Ohms.

III. RF MEMS TECHNOLOGY

Silicon micromachined cavities integrated with RF MEMS tuners have resulted in high-quality EVA tunable filters up to 110 GHz [6-12]. Relatively deep silicon-micromachined cavities up to 1 mm are needed for quality factors in the order of 500-1,000. This requirement is reduced to 0.5 mm for W-band resonators. The RF MEMS tuners have similar diameter requirements while they are typically 1-2 micrometer thick. They are often corrugated membranes to satisfy the tuning range requirements. Octave-tunable filters with less than 3 dB loss and notch filters with over 60 dB isolation have been experimentally shown with this technology (Figs. 5,6).

The first main advantage of this approach is that it is a precise technology with sub-micron fabrication resolution and few-micrometer fabrication uncertainty. As a result, filters well above 100 GHz can be fabricated with this technology. The second main advantage is that the resulting filters can be integrated on the same substrate with other integrated circuits (ICs) to result in low-cost systems.

On the other hand, the biggest drawback of this technology is the relatively high up-front investment required in building the needed RF MEMS technology. While this may be justified for high-volume products, it could end up being a significant barrier for low and low/medium-volume applications.

IV. PCB TECHNOLOGY

Early development efforts at Purdue University in building EVA tunable filters were based on low-cost PCB technology [13]. Various successful bandpass, bandstop, as well as significantly more complicated filters have been demonstrated in this technology (Figs. 7-10) [14-16]. In this technology, the EVA resonators are readily defined with via holes, while the loading posts are defined with blind vias. Tuners are implemented with commercial piezoelectric actuators [13-16]. Excellent quality factors (>700) and tunable ranges in the order of 2:1 have been demonstrated in PCB technology. Furthermore, a few different techniques for closed-loop feedback and control have been shown with excellent results [3,17,18].

The main advantages of this approach are a) its low fabrication cost and b) mature manufacturing basis. This makes it suitable for a wide range of applications. On the other hand, the main drawback is the relatively high fabrication tolerances for conventional PCB fabrication lines. These limitations restricted early efforts to below 6 GHz. It is critical to remember though that tunable filters are forgiving since their tuning mechanisms can compensate for fabrication imperfections. Recently, for instance, Purdue researchers demonstrated PCB-based tunable filters up to 40 GHz with excellent performance [19]. Consequently, this technology is very promising even for high-frequency 5G bands.

V. INJECTION MOLDING

Several well-established polymer printing technologies can be considered as low-cost prototyping options. Selective laser sintering and Stereolithography are two such technologies. Conventional fabrication tolerances for these methods are in the order of 100 micrometers, while high-precision methods can reduce them by an order of magnitude. On the other hand, it is hard to consider these techniques in manufacturing due to their low throughput.

Injection molding (IM) and microinjection molding (µIM) [20] are great alternatives for large-scale production of high-precision parts. Although the initial production cost of the mold may be high ($10,000s), the cost per part can be very low ($0.1), making these methods particularly cost effective. Tiny gears, printer heads and connectors, all with fabrication tolerances of a few micrometers, are examples of µIM parts. A wide variety of high-quality plastics can be employed including, for instance, Acrylonitrile Butadiene Styrene (ABS), Polyether Ether Ketone (PEEK), Polyethylene Terephthalate (PET), and Thermoplastic Elastomer (TPE).

We have recently started exploring the IM technology for medium and large-volume manufacturing of the EVA filters. As a preliminary study, we demonstrated a 2.2-4.4 GHz bandpass filter machined on a ABS polymer substrate (Fig. 11). After machining, the plastic part was cleaned and prepared for metallization. The part's surface was coated with a 5-µm thick gold layer (and a thin titanium adhesion layer). The metallization cost can be significantly reduced by employing a copper electroplating process instead of a sputtering process. The remaining filter assembly is similar to the PCB technology. The filter showed a measured 0.77-0.26 dB loss across the entire band and an extracted unloaded quality factor of 500-700 (Fig. 12) [18]. Despite using a CNC process and not IM in [18], we implemented the filter with a common IM substrate (ABS). Furthermore, all materials and surface finish parameters are very similar to an IM process, thus demonstrating the future potential of this technology.

REFERENCES

[1] D. Peroulis, E. Naglich, M. Sinani and M. Hickle, "Tuned to Resonance: Transfer-Function-Adaptive Filters in Evanescent-Mode Cavity-Resonator Technology," *IEEE Microwave Magazine*, vol. 15, no. 5, pp. 55-69, May-June 2014.

[2] P. Blondy and D. Peroulis, "Handling RF Power: The Latest Advances in RF-MEMS Tunable Filters," *IEEE Microwave Magazine*, vol. 14, no. 1, pp. 24-38, Jan.-Feb. 2013.

[3] M. Abdelfattah and D. Peroulis, "A Novel Independently-Tunable Dual-Mode SIW Resonator with a Reconfigurable Bandpass Filter Application," in *IEEE MTT-S Int. Microwave Symp. Dig.*, June 2018, pp. 1-3.

[4] X. Liu, L P.B. Katehi, W.J. Chappell and D. Peroulis, "Power Handling of Electrostatic MEMS Evanescent-mode (EVA) Tunable Bandpass Filters," *IEEE Trans. Microwave Theory Tech.*, vol. 60, no. 2, pp. 270–283, Feb. 2012.

[5] K. Chen, H. Sigmarsson and D. Peroulis, "Power Handling of High-Q Evanescent-mode Tunable Filter with Integrated Piezoelectric Actuators," in *IEEE MTT-S Int. Microwave Symp. Dig.*, June 2012, pp. 1–3.

[6] M.S. Arif and D. Peroulis, "A 6 to 24 GHz Continuously Tunable, Microfabricated, High-Q Cavity Resonator with Electrostatic MEMS Actuation," in *IEEE MTT-S Int. Microwave Symp. Dig.*, June 2012, pp. 1-3.

[7] X. Liu, L.P.B. Katehi, W.J. Chappell and D. Peroulis, "High-Q Tunable Microwave Cavity Resonators and Filters Using SOI-based RF MEMS Tuners," *IEEE/ASME J. Microelectromech. Syst.*, vol. 19, no. 4, pp. 774-784, Aug. 2010.

[8] X. Liu, L.P.B. Katehi, W.J. Chappell and D. Peroulis, "A 3.4-6.2 GHz Continuously Tunable Electrostatic MEMS Resonator with Quality Factor of 460–530," in *IEEE MTT-S Int. Microwave Symp. Dig.*, June 2009, pp. 1149-1152.

[9] M.S. Arif and D. Peroulis, "All-silicon Technology for High-Q Evanescent Mode Cavity Tunable Resonators and Filters," *IEEE/ASME J. Microelectromech. Syst.*, vol. 23, no. 3, pp. 727-739, June 2014.

[10] Z. Yang, D. Psychogiou and D. Peroulis, "Design and Optimization of Tunable Silicon-Integrated Evanescent-Mode Bandpass Filters," *IEEE Trans. Microwave Theory Tech.*, vol. 66, no. 4, pp. 1790-1803, Apr. 2018.

[11] Z. Yang and D. Peroulis, "A 23-35 GHz MEMS Tunable All-silicon Cavity Filter with Stability Characterization up to 140 Million Cycles," in *IEEE MTT-S Int. Microwave Symp. Dig.*, June 2014, pp. 1-4.

[12] M.D. Hickle, M.D. Sinanis and D. Peroulis, "Tunable High-isolation W-band Bandstop Filters," in *IEEE MTT-S Int. Microwave Symp. Dig.*, June 2015, pp. 1-4.

[13] H. Joshi, H.H. Sigmarsson, S. Moon, D. Peroulis and W.J. Chappell, "High Q Fully Reconfigurable Tunable Bandpass Filters," *IEEE Trans. Microwave Theory Tech.*, vol. 57, no. 12, pp. 3525-3533, Dec. 2009.

[14] M.D. Hickle and D. Peroulis, "Tunable Constant-Bandwidth Substrate-Integrated Bandstop Filters," in IEEE *Trans. Microwave Theory Tech.*, vol. 66, no. 1, pp. 157-169, Jan. 2018.

[15] J. Lee, E.J. Naglich, H.H. Sigmarsson, D. Peroulis and W. J. Chappell, "Tunable Inter-Resonator Coupling Structure with Positive and Negative Values and Its Application to the Field-Programmable Filter Array (FPFA)," *IEEE Trans. Microwave Theory Tech.*, vol. 59, no. 12, pp. 3389-3400, Dec. 2011.

[16] E.J. Naglich, J. Lee, D. Peroulis and W.J. Chappell, "Switchless Tunable Bandstop-to-all-pass Reconfigurable Filter," *IEEE Trans. Microwave Theory Tech.*, vol. 60, no. 5, pp. 1258–1265, May 2012.

[17] M.A. Khater and D. Peroulis, "Real-Time Feedback Control System for Tuning Evanescent-Mode Cavity Filters," *IEEE Trans. Microwave Theory Tech.*, vol. 64, no. 9, pp. 2804-2813, Sept. 2016.

[18] M.D. Sinanis, M. Abdelfattah, M. Cakmak and D. Peroulis, "A 2.2-4.2 GHz Low-loss Tunable Bandpass Filter Based on Low Cost Manufacturing of ABS Polymer," in *Proc. IEEE 19th Wireless and Microwave Technology Conference (WAMICON)*, Apr. 2018, pp. 1-4.

[19] P. Adhikari, W. Yang, Y-C. Wu and D. Peroulis, "A PCB-technology-based 22-42 GHz Quasi-Absorptive Bandstop Filter," submitted to the *IEEE Microwave Wireless Comp. Lett.*, 2018.

[20] B. Bhushan, "Microinjection Molding," *Encyclopedia of Nanotechnology*, DOI: https://doi.org/10.1007/978-94-017-9780-1_101021, 2016 edition.

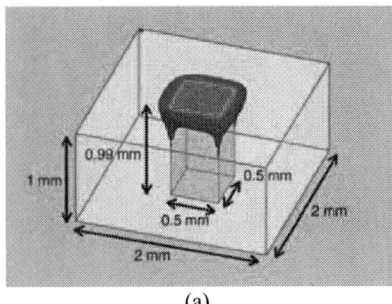
(a)

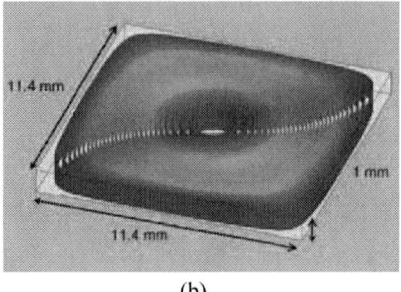
(b)

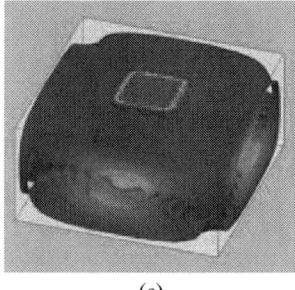
(c)

Fig. 1. Electric field distribution in (a) a loaded cavity resonator (18.6 GHz) and (b) an unloaded cavity resonator at the same frequency (18.6 GHz). Notice the different dimensions. (c) First spurious mode of loaded cavity (152 GHz) [1].

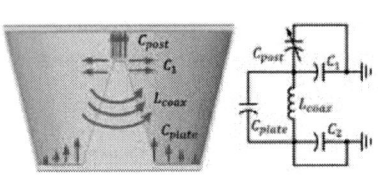

Fig. 2. Equivalent circuit of an all-silicon EVA resonator [10].

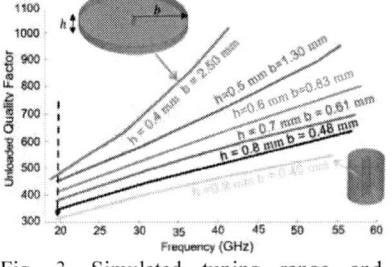

Fig. 3. Simulated tuning range and unloaded quality factor. The critical cap changes from 1 to 20 μm [10].

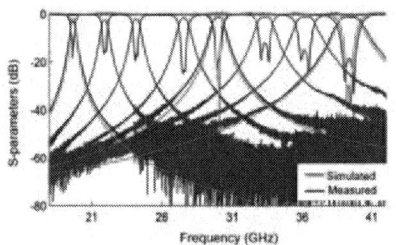

Fig. 4. Measured and simulated results of a typical 20-40 GHz MEMS EVA filter [10].

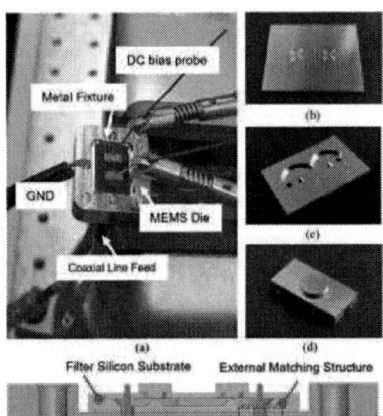

Fig. 5. (a) Fabricated all-silicon MEMS bandpass filter. (b) Front view of the MEMS tuner die. (c) Front view of the gold-coated silicon micromachined EVA mode cavities. (d) Side view of the MEMS biasing electrode. (e) Cross-section view of the fully-packaged device [10].

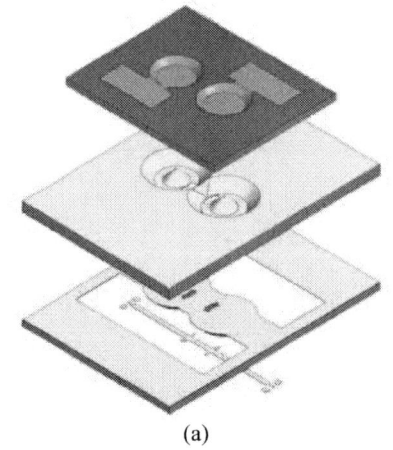
(a)

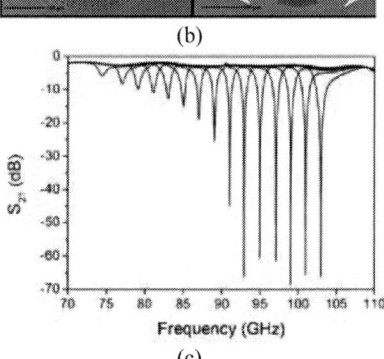
(b)

(c)

Fig. 6. (a) Schematic of W-band tunable RF MEMS EVA mode absorptive bandstop filter. (b) Fabricated RF MEMS tuners and silicon micromachined cavities. The black bar length is 0.5 mm. (c) Measured notch response from 75-103 GHz. The response below 90 GHz becomes progressively more reflective and less absorptive. Consequently, the isolation is progressively reduced [12].

Design Constraints	Root-of-Cause	Performance Impacts
Maximun tuner deflection	Tuner specifications	Tuning range, stability and tuning voltage
Minimun RF gap	Substrate and assembly tolerance	Tuning range
Cavity side wall slope	Substrate machining method	Resonator and coupling structure modeling
Minimun post top radius	Fabrication tolerance	Tuning range, quality factor

Table I: General design constraints of tunable EVA filters [10].

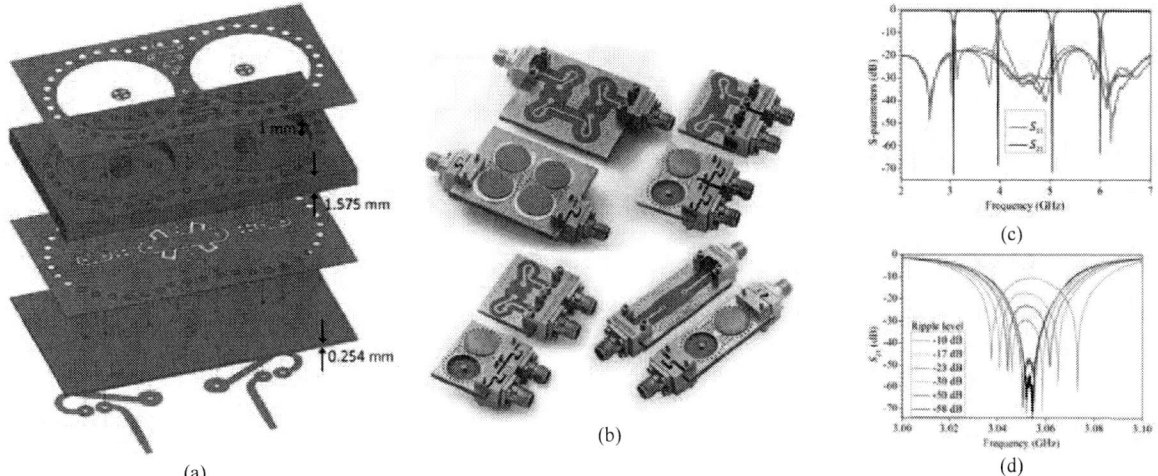

(a)   (b)   (d)

Fig. 7. (a) Two-pole constant bandwidth filter. (b) Fabricated two- and four-pole filters in low-cost PCB technology. (c) Measured performance of four-pole constant-absolute-bandwidth filter. (d) Measured performance of four-pole filter when tuned for increased notch bandwidth [14].

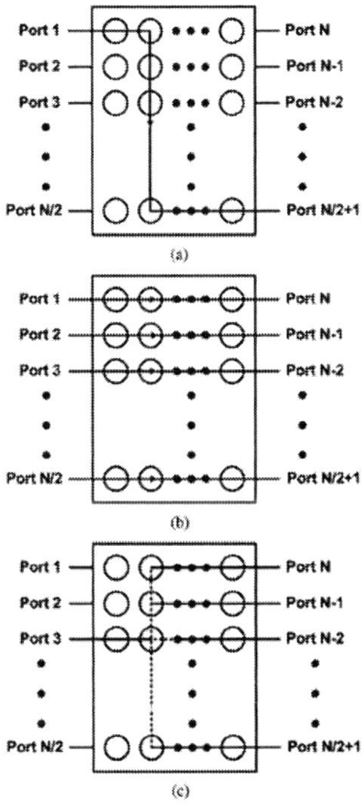

Fig. 8. Filter-programmable filter array (FPFA) and its operational modes [15].

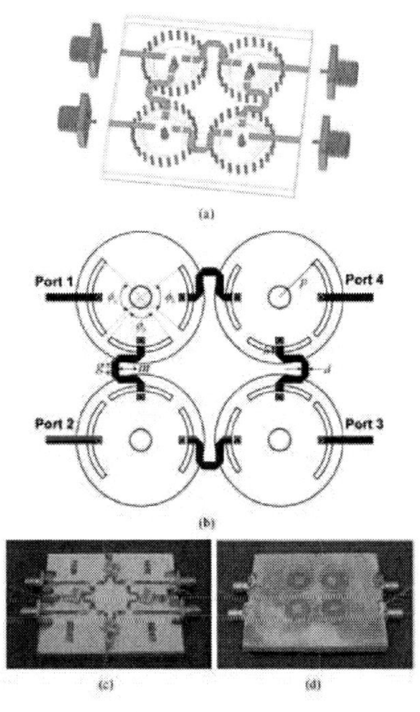

Fig. 9. PCB-fabricated 2x2 FPFA based on EVA mode resonators [15].

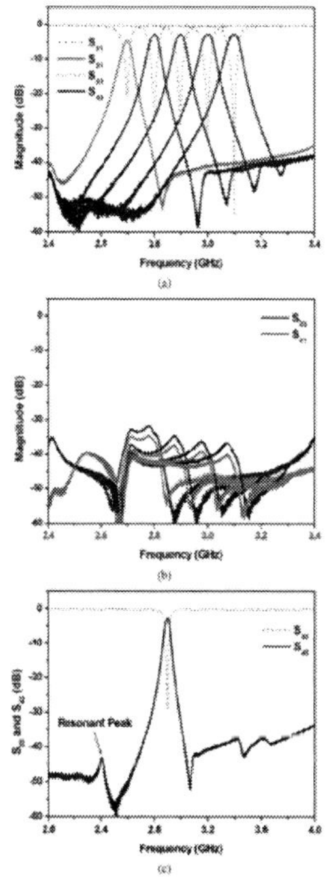

Fig. 10. Measured frequency responses of the 2x2 FPFA [15].

Fig. 11. ABS-manufactured EVA mode filter with LCP flexible tuners controlled by commercial piezoelectric actuators [18].

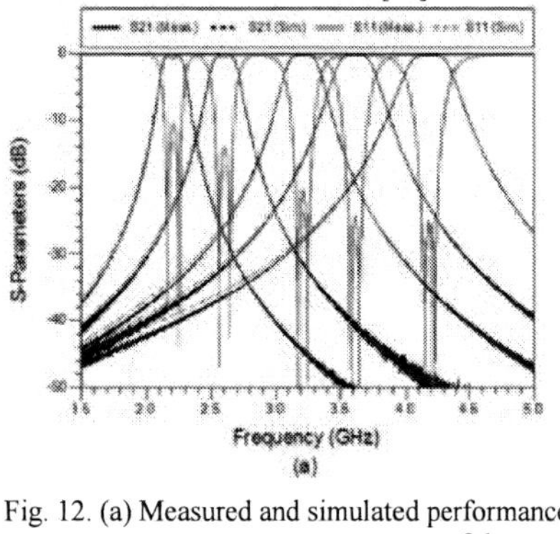

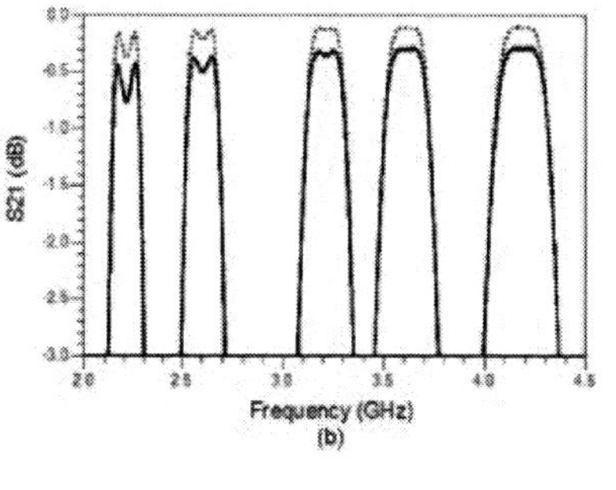

Fig. 12. (a) Measured and simulated performance of the filter shown in Fig. 11. (b) Enlarged view of the passband [18].

978-1-7281-1988-5/18 $31.00 © 2018 IEEE   339

Ultra-Low Power 3D NC-FinFET-based Monolithic 3D[+]-IC with Computing-in-Memory for Intelligent IoT Devices

Fu-Kuo Hsueh[1], Wei-Hao Chen[2], Kai-Shin Li[1], Chang-Hong Shen[1*], Jia-Min Shieh[1*], Chun Ying Lee[2], Bo-Yuan Chen[1], Hsiu-Chih Chen[1], Chih-Chao Yang[1], Wen-Hsien Huang[1], Kun-Ming Chen[1], Guo-Wei Huang[1], Peng Chen,[2], Yung-Ning Tu[2], Srivatsa Srinivasa[3], Vijaykrishnan Narayanan[3], Meng-Fan Chang[2*], and Wen-Kuan Yeh[1]

[1]National Nano Device Laboratories, No.26, Prosperity Road 1, Hsinchu, Taiwan;
[2]National Tsing Hua University, Hsinchu, Taiwan;
[3]Pennsylvania State University, University Park, PA, USA;
*Tel:+886-3-5726100, E-mail: chshen@narlabs.org.tw ; jmshieh@narlabs.org.tw ; mfchang@ee.nthu.edu.tw

Abstract

For the first time, ultra-low power ferroelectric FinFET-based monolithic 3D[+]-IC technology was demonstrated for near memory computing (NMC) circuit. Key enablers are ICP-SiO$_2$ interfacial layer, doped hafnia ferroelectric gate dielectric layer (HfZrO$_2$), and far-infrared laser activation. The proposed stackable 3D NC-FinFETs thus fabricated exhibit record-low sub-threshold swing (NC-nFinFET: 45mV/dec and NC-pFinFET: 50mV/dec) and high I_{on}/I_{off} (>10^6) that enable ultra-low power operation (V_{DD}=100mV) of CMOS inverter and SRAM. Moreover, above mentioned features of NC-FinFETs and the differential output of SRAM readout enable 50+% area reduction in the near-memory computing circuitry.

I. Introduction:

Low-cost energy-efficient computing is essential for intelligent IoT device. Fig.1 presents a TSV-free monolithic three dimensional integrated-circuits (3D[+]-IC) technology [1] combined with low power NC-FinFET [2] promises low manufacturing cost and low operation voltage. Fig.2 presents the conventional computing systems between CPU and memory, which is known as von Neumann structure. Limited by the narrow bandwidth between CPU and memory, von Neumann structure faces bottlenecks in performance and energy consumption. Therefore, Computing-in-Memory structure such as In-Memory-Computing (IMC) [1],[3-4] and Near-Memory-Computing (NMC) is being developed to deal with this challenge.

Recent monolithic 3D[+]-ICs, combining with intelligent memory, have already shown the significant improvement in area reduction, high bandwidth interconnection and power consumption of signal transport in the inter-connect, thanks to 3D vertical connectivity [5]. However, stackable device performance, such as sub-threshold swing and threshold voltage, still limits voltage scaling as well as power consumption in monolithic 3D circuit. In this work,

the first single grain Si stackable 3D NC-FETs was demonstrated using 600°C-process doped hafnia ferroelectric gate dielectric layer with ICP-deposited SiO$_2$ interfacial layer. High thermal stability W metallization process is necessary to prevent metal inter-connection from 600°C-process thermal damage [6-7]. 3D NC-FinFETs exhibit steep sub-threshold swing and high I_{on}/I_{off} that enable ultra-low voltage operation (V_{DD}=100mV) of CMOS inverter, low-leakage SRAM, and compact-area near-memory computing circuits.

II. Device Fabrication:

Figs. 3 illustrate the TSV-free ultra-low power ferroelectric NC-FinFET based monolithic 3D[+]-IC technology. First, stackable epi-like Si film prepared using low-thermal budget laser-processed/CMP-thinned process. Thin ICP-deposited SiO$_2$ serve as interfacial layer for ALD-deposited HfZrO$_x$ over the Si fin. The HfZrO$_x$ film was crystallized by rapid thermal annealing for 600 $^{\circ}$C/30sec. Low thermal budget CO$_2$ far-infrared laser activation was used to activate the highly doped and diffusionless source/drain region and prevent deleterious effects of high temperature of the orthorhombic phase in the HfZrO$_2$ (HZO) film. Finally, high thermal stability W metallization process was used for circuit interconnects.

III. Results and Discussion:
(A) Stackable 3D NC-FinFETs characteristics

Doped hafnia ferroelectric layers were successfully integrated into stackable sub-30nm single-grained Si FinFETs shown in the high-resolution TEM images (Fig. 4 (a) and (b)). The capacitance matching between C_{FE} and C_{MOS} leads to SS improvement and influenced hysteresis behavior [2]. A thick interfacial oxide (2.2 nm) provides not only surface defect reduction and V_{th} adjustment but also controls the capacitance matching between C_{FE} and $C_{MOS.}$ Ferroelectric HZO film was obtained by 600°C/30sec annealing process, confirmed by GI-XRD spectrum with strong orthorhombic phase (111) signal (Fig. 5). Furthermore, XPS spectra shows Hf(Hf4f) and Zr(Zr3d)

978-1-7281-1988-5/18 $31.00 © 2018 IEEE

atomic ratio with 1:1 that provides the desired ferroelectric property in HZO thin film (Fig. 6). P-V characteristics of HZO measured on the metal/HZO/metal capacitor shows the hysteresis loop and $12\mu C/cm^2$ of remanence polarization in the HZO capacitor (Fig. 7).

The first stackable ferroelectric Si n/p FinFETs as 3D NC-FinFETs were demonstrated with high on/off ratio ($>10^6$) and ultra-low subthreshold swing (NC-nFinFET: 45mV/dec and NC-pFinFET: 50mV/dec) due to the steep I_d-V_g characteristics (Fig. 8 and 9)that is a characteristic of the negative capacitance FinFETs. The on-currents of the ferroelectric Si n/p FinFETs were 172 and 142 μA/μm at $|V_d|$=1V and $|V_g|$=1.1V as shown in Fig. 10. Fig.11 presents the voltage transfer characteristics (VTC) of the fabricated stackable NC-FinFET and the voltage gain (= -dVout/dVin) of stackable ferroelectric CMOS inverter owing to excellent well-behaved transitions operated under V_{DD} as low as 0.1 V manifests the feasibility in enabling low power and high stability $3D^+$ IoE devices. Fig. 12 shows 3D NC-FinFET-based 6T SRAM operation at V_{DDmin} of 0.2V.

(B) PROPOSED NMC CIRCUITS

Fig.13 (a) presents the NC-FinFET-based Full-adder for near-memory computing, which use differential-input from SRAM cell array and PMOS-only NC-FinFET pass-gate logic. This full-adder circuit for SRAM NMC comprises only 10 transistors, Including an differential input XNOR gate (XNOR1: P1, N1 and N3), a single-ended XNOR-gate (XNOR2: P2, N2, N4 and inv1), and two PMOS pass-gates (P3, P4). In the proposed circuit, A and Cin are external inputs. Unlike typical full adder with single-ended input, the differential input DO and DOB are from the differential nodes of typical sense amplifier of SRAM. The ultra-low threshold-voltage (UL-V_{TH}) NC-FinFET enable small voltage drop for passing signal through a NC-FinFET pass-gate. Fig.13 (b) presents the truth table of this NMC full-adder.

For example, when A=1, DO=1, DOB=0 and Cin=0, N1 and N3 are both turned on and make D=VDD-V_{UL_VTH}. This turns off P4 and turns on N4. At the meantime, P3 is on and pass data A to the carry-output COUT (Cout=1). Finally, the output Sum result becomes zero.

Table I. presents the comparison table of various full adder. Our proposed full-adder for SRAM NMC using differential-input and PMOS low-Vth NC-FinFET pass-gate can achieve 1.6X and 2.8X area reduction compared to previous CMOS XNOR-gate based full adder and conventional CMOS full adder [8-9].

Fig.14 presents the captured waveform of our NMC full adder. This NMC full adder employ different input (DO, DOB) to compute In one memory access cycle. Fig.15 presents the die photo of the proposed SRAM based NMC circuit. Table II is show the comparison of circuit performance with monolithic $3D^+$ process.

IV. Conclusion:

In this work, we propose an ultra-low power ferroelectric FinFET-based monolithic $3D^+$IC technology with integration of logic, SRAM, and near memory computing (NMC) circuit. The 3D NC-FinFETs exhibit record-low sub-threshold swing and high I_{on}/I_{off}.

Acknowledgements

The authors would like to thank the support of Ministry of Science and Technology (MoST) and National Applied Research Laboratories (NARLabs) in Taiwan.

References:

[1] F.-K. Hsueh et al., "TSV-free FinFET-based Monolithic 3D$^+$-IC with computing-in-memory SRAM cell for intelligent IoT devices", International Electron Devices Meeting (IEDM) Tech. Dig., p.12.6.1, 2017.

[2] K.-S. Li et al., "Sub-60mV-Swing Negative-Capacitance FinFET without Hysteresis", International Electron Devices Meeting (IEDM) Tech. Dig., p.22.6.1, 2015.

[3] V. Khwa et al, "A 65nm 4Kb Algorithm-Dependent Computing-in-Memory SRAM Unit-Macro with 2.3ns and 55.8 TOPS/W Fully Parallel Product-Sum Operation for Binary DNN Edge Processors," IEEE International Solid-State Circuits Conference (ISSCC) Tech. Dig., pp. 496-497, Feb. 2018.

[4] W.-H. Chen et al., "A 65nm 1Mb Nonvolatile Computing-in-Memory ReRAM Macro with sub-16ns Multiply-and-Accumulate for Binary DNN AI Edge Processors," IEEE International Solid-State Circuits Conference (ISSCC) Tech. Dig., pp. 494-495, Feb. 2018

[5] Srivatsa Srinivasa et al, "Compact 3-D-SRAM Memory With Concurrent Row and Column Data Access Capability Using Sequential Monolithic 3-D Integration," IEEE Transactions on Very Large Scale Integration (VLSI) Systems, vol. 26, No. 4, April 2018.

[6] P. Batude et al., "3D VLSI-CoolCube Process: An Alternative Path to Scaling", Symp. VLSI Technology (VLSIT), p. 5-2, 2015.

[7] O. Turkyilmaz et al., "3D FPGA using high-density interconnect Monolithic Integration", Proceedings of the conference on Design, Automation & Test in Europe (DATE) Article No. 338 (2014).

[8] Neil H. E. Weste and David Harris, CMOS VLSI Design: A Circuits and Systems Perspective, 4th Ed., Addison Wesley, pp. 430-433, 2010.

[9] Partha Bhattacharyya et al, "Performance Analysis of a Low-Power High-Speed Hybrid 1-bit Full Adder Circuit,"IEEE Transactions On Very Large Scale Integration (VLSI) Systems, vol. 23, No. 10, October 2015.

[10] L. Pasini et al., "High Performance CMOS FDSOI Devices activated at Low Temperature", Symp. VLSI Technology (VLSIT), p. 13-2, 2016.

[11] F.-K. Hsueh et al., "First Fully Functionalized Monolithic 3D$^+$ IoT Chip with 0.5 V Light-electricity Power Management, 6.8 GHz Wireless-communication VCO, and 4-layer Vertical ReRAM", International Electron Devices Meeting (IEDM) Tech. Dig., p.2.3.1, 2016.

978-1-7281-1988-5/18 $31.00 © 2018 IEEE

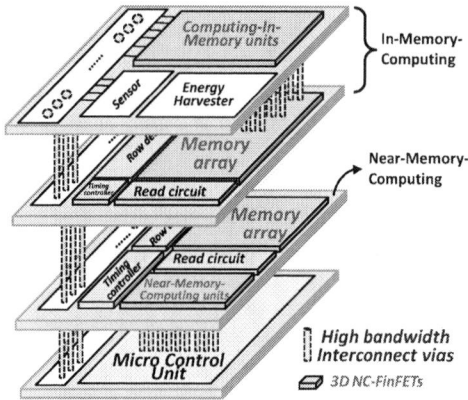

Fig. 1. Schematic illustration of proposed 3D NC-FinFETs monolithic 3D⁺-IC, including stackable in-memory-computing and near memory computing circuit with high-bandwidth vertical interconnect for IoT device.

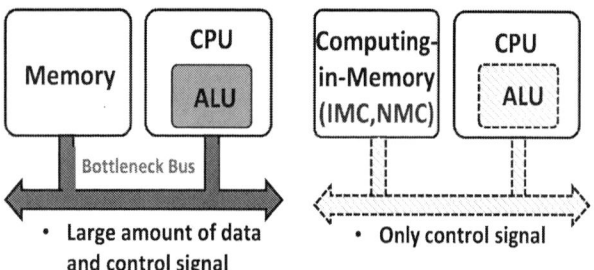

Fig. 2. (a) von Neumann structure and (b) Computing-In-Memory enabled structure.

3D NC-FinFET Process

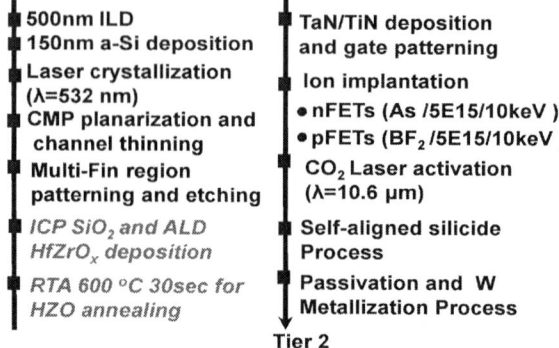

- 500nm ILD
- 150nm a-Si deposition
- Laser crystallization (λ=532 nm)
- CMP planarization and channel thinning
- Multi-Fin region patterning and etching
- *ICP SiO₂ and ALD HfZrOₓ deposition*
- *RTA 600 °C 30sec for HZO annealing*

- TaN/TiN deposition and gate patterning
- Ion implantation
  - nFETs (As /5E15/10keV)
  - pFETs (BF₂ /5E15/10keV)
- CO₂ Laser activation (λ=10.6 μm)
- Self-aligned silicide Process
- Passivation and W Metallization Process

Tier 2

Fig. 3. Process flow of 3D NC-FinFETs monolithic 3D⁺-IC technology.

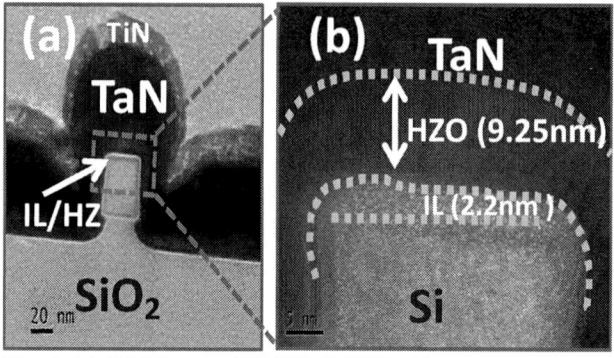

Fig. 4. (a) TEM image of 3D NC-FinFET with W_{Fin} = 26nm and H_{Fin} = 34nm (b) 9.25nm HZO / 2.2nm IL gate stack for 3D NC-FinFET.

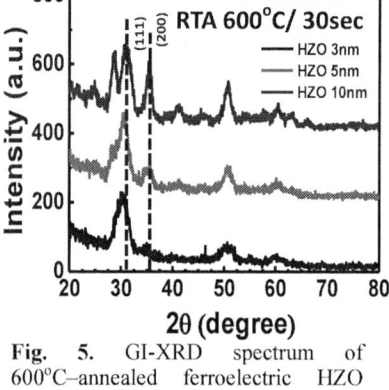

Fig. 5. GI-XRD spectrum of 600°C–annealed ferroelectric HZO film.

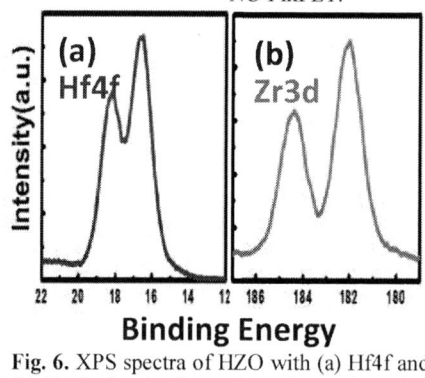

Fig. 6. XPS spectra of HZO with (a) Hf4f and (b) Zr3d signal. The estimated atomic ration of HZO film is 1:1.

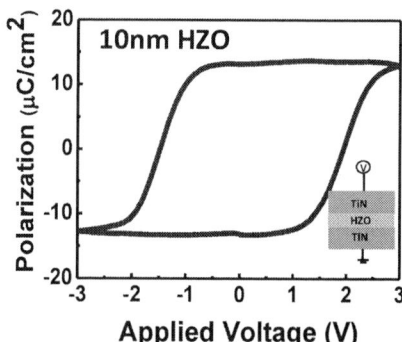

Fig. 7. The PV curves for 600°C–annealed metal/10nm HZO/metal films.

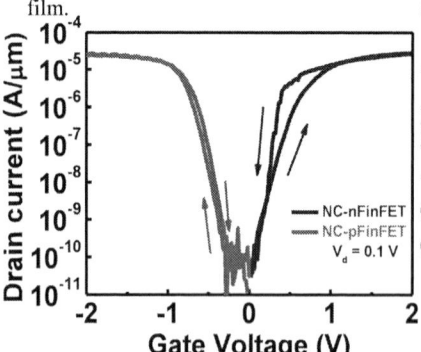

Fig. 8. Id-Vg characterization of stackable n/p 3D NC-FinFETs.

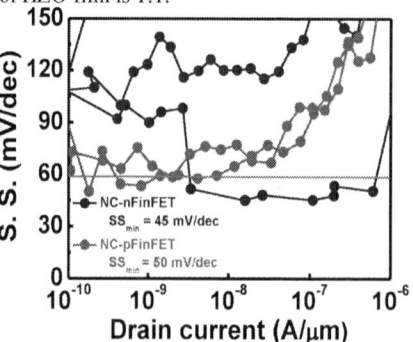

Fig. 9. S.S. versus drain current of the 3D NC-FinFETs.

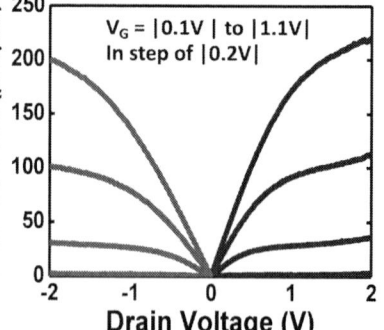

Fig. 10. Id-Vd curves of stackable 3D NC-FinFETs.

978-1-7281-1988-5/18 $31.00 © 2018 IEEE   342

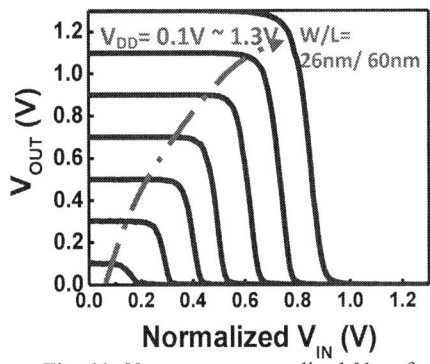

Fig. 11. V_{OUT} versus normalized V_{IN} of the 3D NC-FinFETs CMOS inverter.

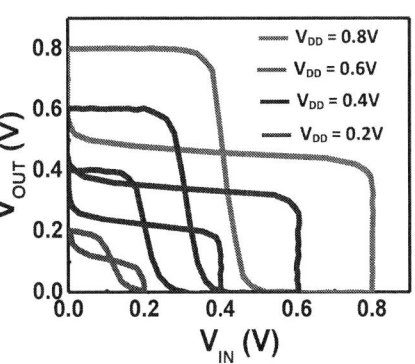

Fig. 12. Butterfly curve of stackable 6T-SARM operated at V_{DD}=0.8V, 0.6V, 0.4Vand 0.2V.

Fig. 15. Die photo of SRAM based NMC circuit.

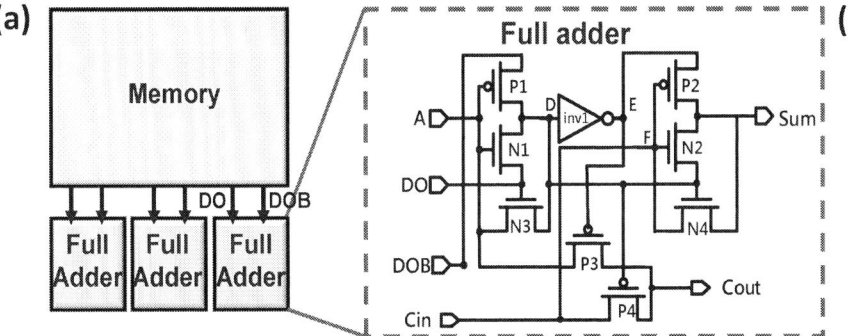

Fig. 13. Proposed Near-Memory Computing circuit: Full adder (a) schematic and (b) truth table.

Reference	[8]	[9]	This work
Operation	$S = a \oplus b \oplus c$ $C_{out} = ab + \text{bc+ac}$	$S = a \oplus b \oplus c$ $Cout = a \bullet b + c(a \oplus b)$	$S = a \oplus b \oplus c$ $Cout = a \bullet b + c(a \oplus b)$
Structure	CMOS	XNOR	XNOR
Input signal	Single-End	Single-End	Differential
Device	MOSFET	MOSFET	NC-FinFETs (Ultra low Vth)
Pass-gate type	CMOS	CMOS	Single NCFET
Area overhead	2.8x	1.6x	1x

Table I. Comparison table of various full adder.

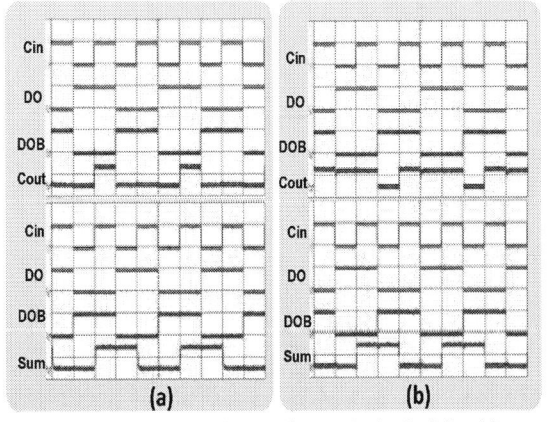

Fig. 14. The captured waveform of NMC full adder (a) external data A is 0 and (b) external data A is 1.

	2016 VLSI [10]	2016 IEDM (our work) [11]	2017 IEDM (our work) [1]	This work
Channel	SOI	Epi-like Si	Epi-like Si	Epi-like Si
Anneal	Solid Phase Epitaxy (SPE)	CO_2 laser anneal (CO$_2$-FIR-LA)	CO_2 Laser activation & CO_2 Laser Silicide	CO_2 Laser activation (CO$_2$-FIR-LA)
Gate Oxide	ALD-HfO$_2$	ICP-SiO$_2$/ALD-HfO$_2$	ICP-SiO$_2$/ALD-HfO$_2$	ICP-SiO$_2$/ALD-HfZrO$_2$
Thermal Budget	500-600 °C	400 °C	400 °C	600 °C
Device Architecture	FDSOI MOSFET	Multi-channel Epi-like Si UTB-MOSFETs	Multi-fin Epi-like Si FinFETs	NC-FinFETs
I_{on}/I_{off} ratio	>10^8	>10^6	>10^6	>10^5
S.S. (mV/dec)	N/A	P : 120 mV/dec N: 110 mV/dec	P : 80 mV/dec N: 75 mV/dec	P : 50 mV/dec N: 45 mV/dec
Circuit	N/A	• Wireless-communication : VCOs 6.8GHz • Power Management Units: 0.5V operation voltage	•3D 9T CIM SRAM (NAND/OR/XOR/XNOR)	NMC: Full adder (0.25V operation voltage)
Memory	N/A	4-layer vertical 3bits/cell ReRAM	SRAM ($V_{DD\,min}$ = 0.4V)	SRAM ($V_{DD\,min}$ = 0.2 V)

Table II. Summary of monolithic 3D processes and variation circuits.

978-1-7281-1988-5/18 $31.00 © 2018 IEEE   343

First Demonstration of Ge Ferroelectric Nanowire FET as Synaptic Device for Online Learning in Neural Network with High Number of Conductance State and G_{max}/G_{min}

Wonil Chung, Mengwei Si, and Peide D. Ye*

School of Electrical and Computer Engineering, Purdue University, West Lafayette, IN 47907, USA
*Tel: 1-765-494-7611, Fax: 1-765-496-6443, email: yep@purdue.edu

Abstract— In this paper, optimum weight update scheme for improved linearity and asymmetry of channel conductance potentiation and depression in a Germanium ferroelectric (FE) nanowire FET (NWFET) was experimentally demonstrated and simulated for the first time. It was found that -5 V, 320 pulses and +5 V, 256 pulses both with 50 ns pulse width were the optimum pulsing conditions for potentiation and depression process, respectively. With the optimized scheme, non-linearity for potentiation and depression were extracted to be $\alpha_p = 1.22$ and $\alpha_d = -1.75$, respectively resulting in asymmetry ($|\alpha_p - \alpha_d|$) of 2.97 based on models embedded in MLP simulator and NeuroSim [1]. G_{max}/G_{min} ratio (few hundreds) and number of conductance states (> 256) are both very large. 9 alternating consecutive conductance updates (potentiation followed by depression) were executed to observe variability in conductance profiles. Multilayer perceptron neural network was simulated over 1 million MNIST images with extracted experimental parameters which yielded in online learning accuracy of ~ 88 %.

I. INTRODUCTION

Due to introduction of processors with higher performance and faster parallel computing capabilities, brain-inspired synaptic device networks caught much attention for various real-life applications. Specifically, e-NVM (emerging non-volatile memory) such as resistive [2], [3], phase change [4] or ferroelectric [5]–[7] devices are studied towards non-von Neumann architectures. Specifically, ferroelectric (FE) devices, mostly with $Hf_{0.5}Zr_{0.5}O_2$ (HZO) due to its high compatibility with CMOS platform, are being studied actively related with negative capacitance (NC) devices [8], [9]. Studies on FE switching speed follow this trend [10], [11]. Partial polarization of a FE device makes it possible to serve as a synaptic device when optimized properly [5]. Motivation for using e-NVM for online learning in deep neural network (DNN) is related to their ability to retain the weight data and to locally process the input via multiplication of weights in the form of conductance. Subsequent addition of weighted currents are read with peripheral circuitry [1] and the processed data is used as the input of the following layer after non-linearization. Sequential processing of data from one layer to the following layer can be noted as forward propagation as shown in Fig. 1. During online training process, back propagation is used to update the less-accurate synaptic devices' weights. Iterative cycles of forward and back propagation increase the accuracy as it undergoes training epochs. For efficient and adaptive operation of the network, linear and symmetrical conductance profile is highly preferred. Fig. 2 depicts various possible non-ideal effects related to e-NVM's programming process including nonlinearity, asymmetry, stochasticity and large variability in conductance values [4]. Although it was reported that the e-NVM-based neural networks are relatively robust to conductance's stochasticity and variability [4], nonlinear and asymmetric

profiles significantly limit the accuracy. To overcome such detrimental influence, different pulsing schemes were proposed as depicted in Fig. 3 [5], [7]. However, if pulses are not *identical* throughout the programming process, an additional step of accessing the weight value is needed every time an update takes place to find the appropriate pulse at that specific level compromising the efficiency. In this paper, we demonstrate Germanium FE NWFET as a synaptic device with high number of conductance states and G_{max}/G_{min}. *Identical* conductance update pulsing schemes were optimized for improved linearity and asymmetry. Ge FE NWFETs can then be configured into pseudo-crossbar array as shown in Fig. 4 for practical implementation towards the DNN.

II. EXPERIMENTS

Germanium-on-insulator (GeOI) wafer with 100 nm of Ge layer on top of SiO_2 (400 nm) and Si handling wafer was used for the fabrication of the Ge FE NWFETs. Typical wafer cleaning and mesa isolation was processed, followed by ion implantation. Fin structure was first defined with SF_6-based dry etching and nanowire was released by partial etching of the underlying SiO_2. 3-step ALD deposition was conducted starting with 1 nm of Al_2O_3 and subsequent post-oxidation to form thin GeO_x (~1 nm) under the Al_2O_3 for better interface quality. 10 nm of HZO and additional 1 nm of Al_2O_3 capping layer were deposited. Then 500 °C post deposition annealing (PDA) was performed to enhance the ferroelectricity within the HZO stack. Source and drain were formed using recessed S/D technique [12]. Ohmic annealing and gate, source and drain metallization was finished in the last step. Fig. 5 presents more details related to the fabrication steps. Fig. 5 (a) and (b) visualize the 3D structure of the device and false-colored SEM images are shown in Fig. 6 (a)-(c). Multiple parallel nanowires form a single device and a typical transfer curve of such device can be seen in Fig. 7. Large negative hysteresis (-4 V) is observed showing ferroelectric switching. Fatigue measurement on our ALD HZO was done with 10^9 PUND (Positive Up, Negative Down) pulses to verify our HZO's reliability (Fig. 8). Details on our HZO film and devices' operation are elaborated in our previous reports [8], [13], [14].

III. RESULTS AND DISCUSSION

As seen in Fig. 4, FeFETs can be configured into a pseudo-crossbar design. When a row is selected, and subsequent potentiation and depression pulses are fed into the gate of a FeFET, already-saved conductance value can be newly programmed through partial polarization switching (Fig. 9). Input data fed (Fig. 4, purple line) into the FeFET's channel results in weighted current and can be read externally. To optimize the programming pulses, measurement set-ups were prepared as configured in Fig. 10. Fig. 11 is the real-time monitored potentiation process after applying a single -8.75 V pulse (75 ns) to the gate of FeFET. Current was measured using an oscilloscope through current amplifier (Fig. 10 (a)).

978-1-7281-1988-5/18 $31.00 © 2018 IEEE

Then, pulse voltage optimization was done with fixed pulse widths. As seen in Fig. 12, with a fixed pulse width (1 µs), increasing the potentiation voltage from -1 V to -4.5 V reduces the number of pulses needed to reach the maximum conductance (G_{max}, ~ 80 µS). Fixing the pulse level at -4.5 V, pulse widths were swept from 100 ns to 1 µs and similar trend in conductance profile could be observed (Fig. 13) where longer pulses reduce the number of pulses to maximize the conductance. Same procedure was executed for depression pulses (Fig. 14). It could be clearly seen from Fig. 15 that if the pulse conditions (either voltage or pulse width) are not properly optimized, it gives highly non-linear and asymmetric conductance profiles. Non-linearity for potentiation and depression were extracted to be α_p = 5.72 and α_d = -9, respectively yielding asymmetry ($|\alpha_p - \alpha_d|$) of 14.72. These values were extracted through curve-fitting model embedded in MLP (Multilayer perceptron) simulator (+NeuroSim) [1] for fair benchmarking among reported HZO-based FeFET synapse devices.

Freezing of a device (stuck-on conductance) were occasionally observed when excessively strong pulses (significantly larger number of pulses, long pulse widths, or larger pulse voltages) were delivered similar to the reported PCRAM-based study [4] limiting subsequent conductance programming. The report concludes that the contribution of these non-ideal circumstances doesn't compromise the accuracy too much. Restoration was possible via initialization (#1 in Fig. 9) but using intermediate conductance values in the first place rather than exploiting the whole maximum conductance range would be helpful in preventing the devices from becoming nonresponsive in the cost of reduced G_{max} and number of states.

Considering various factors, pulse width was reduced to 50 ns as seen in Fig. 16 for better $\alpha_{p,d}$ and lower asymmetry. Pulse period was 500 µs and conductance sampling was done 50 µs after each pulse. Conductance values were sampled at single point (V_G = 0 V) instead of sweeping V_G to minimize the unwanted effect of measurement V_G on the programmed polarization state. It was found that when our Ge FE pNWFET (L= 105 nm, W= 32 nm, H= 26 nm) was subjected to 50 ns, V_G = ∓5 V (- for potentiation, + for depression), it resulted in significantly improved linearity (Fig. 17) of α_p= 1.22 and α_d= -1.75 (Asymmetry = $|\alpha_p - \alpha_d|$ = 2.97). Respective pulse numbers were chosen to acquire symmetric operation between potentiation (320 pulses) and depression (256 pulses) resulting in effective control of the conductance without introducing non-responsive devices. Smaller pulse number for depression implies that the polarization switching is more sensitive to depression (+5 V) than potentiation (-5 V). To investigate the variability of the optimized pulse scheme, 9 consecutive cycles of continuously alternating potentiation (-5 V, 50 ns, 320 cycles) and depression (+5 V, 50 ns, 256 cycles) were executed. Fig. 18 (a) is the accumulated conductance profiles and Fig. 18 (b) shows the overlapped profiles. It could be observed that this pulse scheme was effective repetitively yielding reliable conductance profiles without serious non-ideal curves. Since number of conductance states were high in our case (320 for potentiation and 256 for depression), multiple pulses could be tied in the form of a pulse train to reduce the number of states for applications that require lower

number of states. If 10 pulses (-5 V, 50 ns) are delivered as a pulse train, it will result in 320/10 = 32 states (5 bits) with 10 times larger ΔG step. The G_{max}/G_{min} ratio is also considered as an important parameter in training simulation. Low ratio causes degradation in training accuracy which makes higher ratio more preferable [1], [3]. Our devices show excellent ratio in the range of hundreds because of low G_{min} < 1 µS and high G_{max} ~ 200 µS (Fig. 18).

Fig. 19 shows the improvement in non-linearity due to transition from non-optimized (Fig. 15) to optimized (Fig. 18) scheme. Both α_p and α_d approach the desired targeted values of +1 and -1 (ideally both 0) [1]. Fig. 20 compares the asymmetry and training accuracy of online learning through multilayer perceptron (MLP) neural network architecture (400 input, 100 hidden, 10 output neurons) [1] before and after conductance profile optimization. 1 million hand written digit images (MNIST, Modified National Institute of Standards and Technology, cropped 20 × 20 pixels) were trained for 125 epochs of training. With dramatic improvement in linearity (and thus asymmetry), high accuracy of ~ 88 % could be achieved. It could be further improved if various parameters in the simulator such as learning rates between layers, number of synapses, number of hidden layers are optimized more precisely. Fig. 21 summarizes device performance metrics of reported HZO-based FeFET synaptic devices. All three studies use HZO as main FE dielectric layer and results using identical pulses are shown for comparison.

IV. CONCLUSION

In this paper, we have reported the first experimental and simulation demonstration of Ge FE NWFET as a synaptic device for online learning in a neural network with optimized pulse schemes. Separate *identical* pulsing conditions for potentiation (-5V, 50 ns, 320 cycle) and depression (+5V, 50 ns, 256 cycle) were found respectively which gave significantly improved linearity and symmetry in conductance profiles. These conditions were effective in preventing devices from becoming non-responsive since the combination of pulse voltage, pulse width and number of pulses was optimized to prevent the freezing. As a result, improved linearity (α_p/α_d= 5.72/-9 → 1.22/-1.75) and asymmetry (14.72 → 2.97) in conductance profiles could be observed. Learning accuracy after training 1 million MNIST images over 125 epochs gave ~ 88 %. It can be concluded that precise optimization of pulsing conditions can affect the conductance update profiles significantly.

ACKNOWLEDGMENT

The authors appreciate Pragya R. Shrestha, Jason P. Campbell and Kin P. Cheung at National Institute of Standards and Technology (NIST) for demonstration of real-time monitoring of conductance during potentiation pulses. The work is supported by SRC and Lam Research.

REFERENCES

[1] P. Y. Chen et al., *IEDM*, 2017. [2] J. Woo et al., *EDL*, vol. 37, no. 8, pp. 994–997, 2016. [3] S. Yu et al., *IEDM* ,2015 [4] G. W. Burr et al., *IEDM*, 2014. [5] M. Jerry et al., *IEDM*, 2017. [6] M. Seo et al., *EDL*, vol. 39, no. 9, pp. 1445-1448, 2018. [7] S. Oh et al., *EDL*, vol. 38, no. 6, pp. 732–735, 2017. [8] W. Chung et al., *IEDM*, 2017. [9] M. Si et al., *Nat. Nanotechnol.*, vol. 13, no. 1, pp. 24–28, Jan. 2018. [10] Z. Krivokapic et al., *IEDM* ,2017. [11] J. Muller et al., *EDL*, vol. 33, no. 2, pp. 185–187, 2012. [12] H. Wu et al., *IEDM*, 2014. [13] W. Chung et al., *VLSI*, 2018. [14] M. Si et al., *IEDM*, 2017.

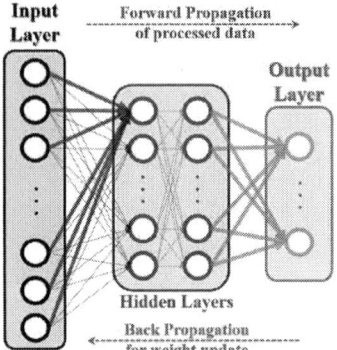

Fig. 1. Basic operation diagram of online training scheme with back propagation for weight update.

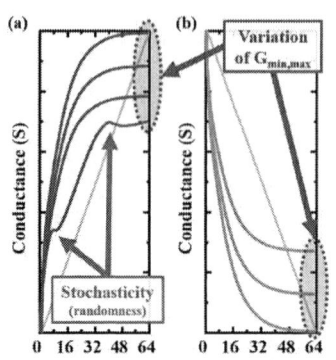

Fig. 2. Non-ideal effects such as variation in $G_{max,min}$, stochasticity, non-linear and asymmetric G profile are shown.

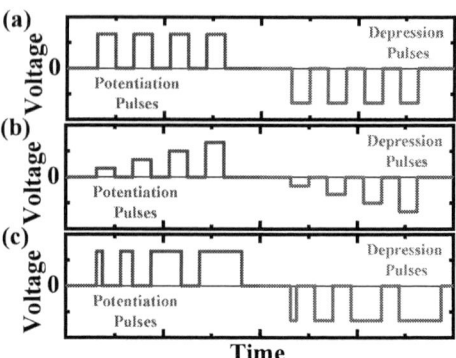

Fig. 3. Various possible pulses for potentiation and depression. **(a)** Identical pulses, **(b)** different pulse levels and **(c)** pulse widths.

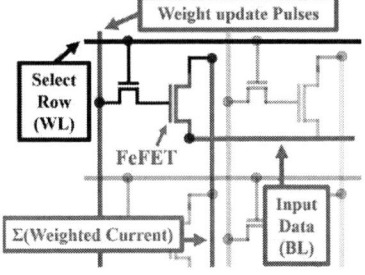

Fig. 4. Pseudo-crossbar scheme showing weight update (blue) and data processing (purple → green) in a row selected by WL.

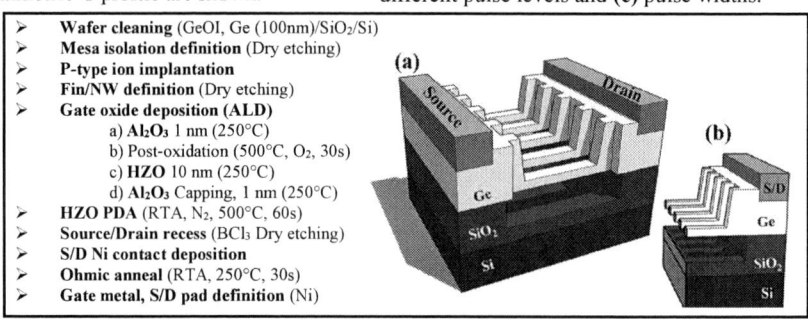

- ➤ **Wafer cleaning** (GeOI, Ge (100nm)/SiO₂/Si)
- ➤ **Mesa isolation definition** (Dry etching)
- ➤ **P-type ion implantation**
- ➤ **Fin/NW definition** (Dry etching)
- ➤ **Gate oxide deposition (ALD)**
  - a) **Al₂O₃** 1 nm (250°C)
  - b) Post-oxidation (500°C, O₂, 30s)
  - c) **HZO** 10 nm (250°C)
  - d) **Al₂O₃** Capping, 1 nm (250°C)
- ➤ **HZO PDA** (RTA, N₂, 500°C, 60s)
- ➤ **Source/Drain recess** (BCl₃ Dry etching)
- ➤ **S/D Ni contact deposition**
- ➤ **Ohmic anneal** (RTA, 250°C, 30s)
- ➤ **Gate metal, S/D pad definition** (Ni)

Fig. 5. Key process steps for fabrication of a Ge FE NWFET neuromorphic device. **(a)** 3D Structure of the device and **(b)** its cross-sectional view of nanowires are illustrated.

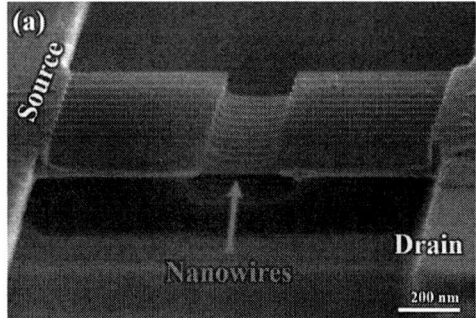

Fig. 6. SEM images of fabricated Germanium nanowire structures **(a)** viewed from the side before ALD ferroelectric oxide deposition and **(b)** top after completion of all fabrication processes. Nanowires connect source and drain regions and air-gap exists below the nanowire. **(c)** Zoom-in image of (b) shows multiple nanowires in parallel.

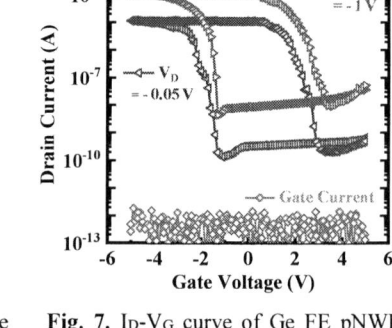

Fig. 7. I_D-V_G curve of Ge FE pNWFET (L= 105 nm, W= 50 nm, H= 26 nm) shows a clear negative ferroelectric hysteresis of approximately -5 V and negligible I_G.

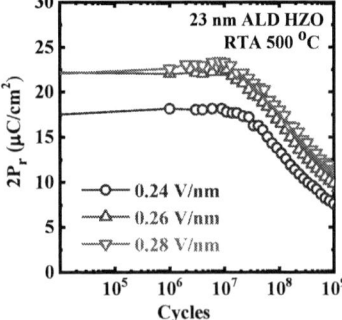

Fig. 8. Fatigue measurement on ALD Ferroelectric HZO capacitor was done with PUND pulses.

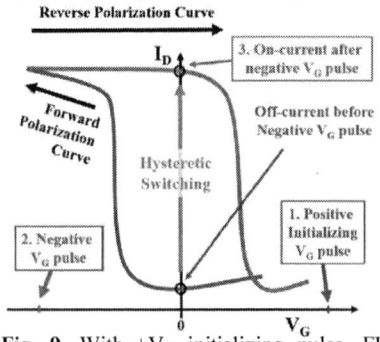

Fig. 9. With +V_G initializing pulse, FE pFET follows the forward polarization curve but a large -V_G switches the polarization and raises the I_D.

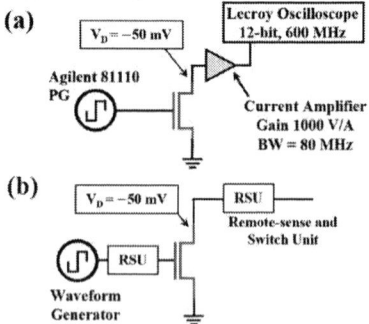

Fig. 10. (a) Set-up used for real-time conductance update during potentiation. **(b)** Set-up for conductance sampling between potentiation and depression pulses.

978-1-7281-1988-5/18 $31.00 © 2018 IEEE

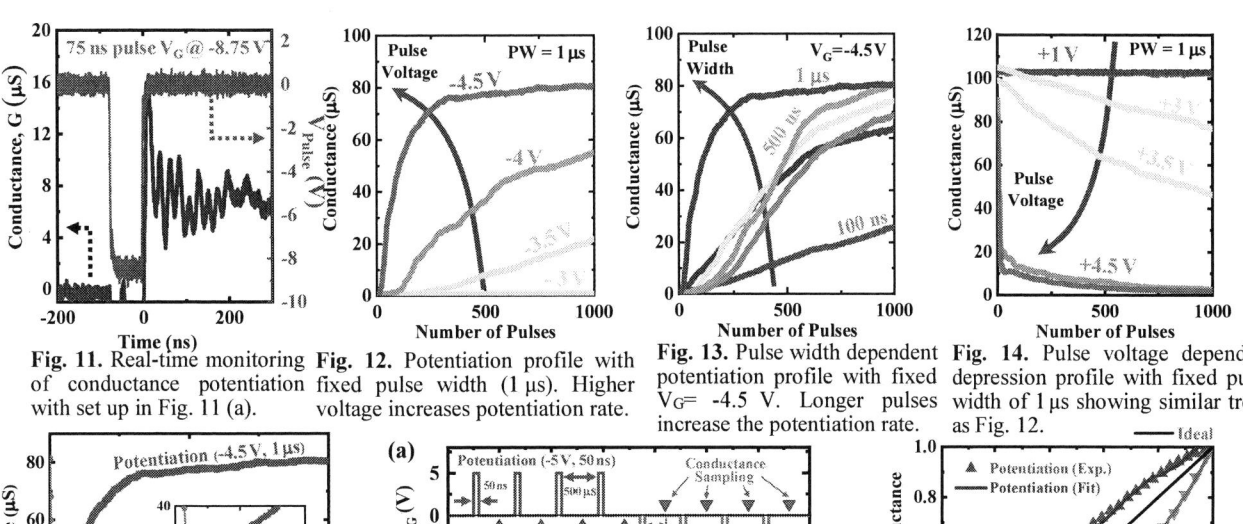

Fig. 11. Real-time monitoring of conductance potentiation with set up in Fig. 11 (a).

Fig. 12. Potentiation profile with fixed pulse width (1 µs). Higher voltage increases potentiation rate.

Fig. 13. Pulse width dependent potentiation profile with fixed V_G= -4.5 V. Longer pulses increase the potentiation rate.

Fig. 14. Pulse voltage dependent depression profile with fixed pulse width of 1 µs showing similar trend as Fig. 12.

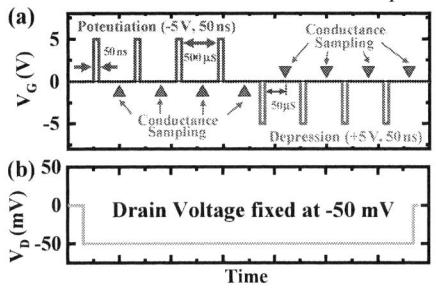

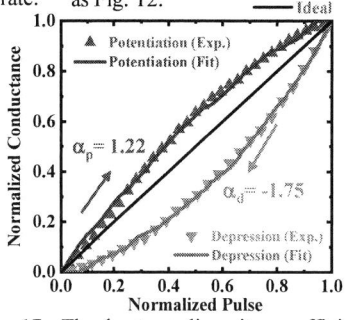

Fig. 15. Without optimized pulse conditions (±4.5 V, 1 µs), it results in highly non-linear and asymmetric profile

Fig. 16. (a) Optimized potentiation (-5 V, 50 ns) and depression (+5 V, 50 ns) pulses for the fabricated Ge FE pNWFET (L = 105 nm, W = 32 nm, H = 26 nm) **(b)** V_D is fixed at -50 mV.

Fig. 17. The best nonlinearity coefficient from pulse scheme in Fig. 16 using reported model [1]. Only 10 % of pulses are displayed for better visualization

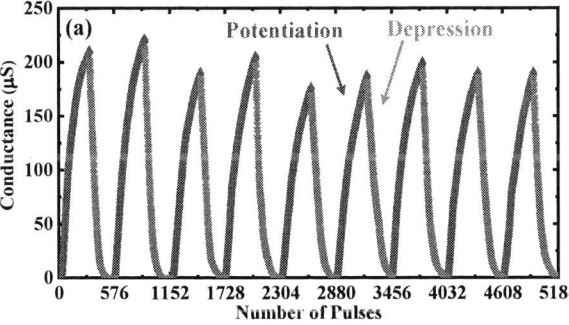

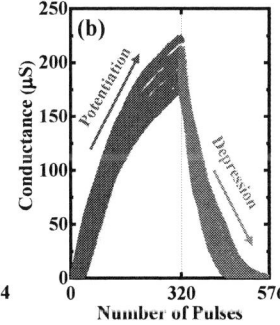

Fig. 18. (a) 9 cycles of consecutive alternating potentiation (-5 V, 50 ns, 320 pulses) and depression (+5 V, 50 ns, 256 pulses) give highly repetitive conductance profiles. **(b)** Overlapped curves from (a) show some conductance variation over multiple programming cycles.

Fig. 19. Non-linearity comparison extracted from both optimized and not optimized pulses.

Fig. 20. Improvement in training accuracy after optimization. Total of 1 million cropped 20×20-pixel MNIST images were trained using MLP+NeuroSim V2.0 [1].

	This Work	[5]		[6]
Device	Ge FE Nanowire pFET	Si FE Planar nFET		Si FE Junctionless nFinFET
Gate Stack (Thickness, nm)	GeOx (~1) + HZO (10) + Al₂O₃ (2)	HZO (10) + SiO₂ (0.8)		HZO (8.5) + SiO₂ (1.5)
Device Dimension	L = 105 nm, W = 32 nm	L = 600 nm, W = 20,000 nm		L = 120 nm, W = 50 nm
# States (Pot./Dep.)	320 / 256	20	32	> 32
Pot. Pulse (Type)	(Identical) 50 ns, 5 V	(Identical) 75 ns, 3.7 V	(Varying) 75 ns, 2.85 ~ 4.45 V	(Identical) 100 µs, 3.7 V
Dep. Pulse (Type)	(Identical) 50 ns, -5 V	(Identical) 75 ns, -3.2 V	(Varying) 75 ns, -2.1 ~ -3.8 V	(Identical) 100 µs, -3.2 V
Non-linearity (α_p/α_d)	1.22 / -1.75	5.54 / -8.08	1.75 / 1.46	1.58 / -7.57
Asymmetry ($|\alpha_p-\alpha_d|$)	2.97	13.62	0.29	9.15
G_{max}/G_{min}	Few hundreds	~ 8	45	4.98
Accuracy (# of trained images)	~ 88 % (1 Million)	N/A	~ 90 % (1 Million)	~ 80 % (3 Million)

Fig. 21. Benchmark of various reported FeFET-based synapse devices for online learning. Lower non-linearity, asymmetry coefficients and higher on/off ratio are preferred. Cycle to cycle variation during our potentiation and depression process is < 1 %. Accuracy can be further increased with better optimized simulation conditions including various learning rates and circuit parameters.

STT-MRAM Design Technology Co-optimization for Hardware Neural Networks

Nuo Xu[1,*], Yang Lu[1], Weiyi Qi[1], Zhengping Jiang[1], Xiaochen Peng[2], Fan Chen[1], Jing Wang[1], Woosung Choi[1], Shimeng Yu[2], Dae Sin Kim[3]

[1]Device Lab, Samsung Semiconductor Inc., San Jose, CA, USA, [*]E-mail: nuo.xu@samsung.com;
[2]School of Electrical and Computer Engineering, Georgia Institute of Technology, Atlanta, GA, USA;
[3]Semiconductor R&D Center, Samsung Electronics, Hwasung-si, Gyeonggi-do, Korea.

Abstract—the potential of embedded STT-MRAM technology for designing large-scale multiply-and-accumulation (MAC) array circuits are evaluated by comprehensive and holistic design-technology co-optimizations. After careful calibrations with experimental data, post-layout circuit simulations together with GPU-enabled massively parallel *Monte Carlo* evaluations are conducted to guarantee the designs at rare failure rates. With all critical device and design non-idealities included, architectural emulations are performed to examine the hardware neural network (HNN)'s accuracies and estimate system-level power, performance and area specs. Results indicate the amount of process variation, parasites and error levels to control in order to achieve a feasible solution for STT-MRAM based HNNs.

I. Introduction

The advent of artificial intelligence has brought semiconductor industry great opportunities for emerging architecture and technology innovations, among which computing-in memory (CIM) concepts arose extensive interests due to the inherent flexibility to carry on matrix-vector multiply-and-accumulation (MAC) operations in an architecture similar to the embedded memory system. Spin-transfer-torque magnetic random access memory (STT-MRAM) is regarded as one of the most promising candidates for embedded non-volatile memories (eNVM) due to its advantageous low operating voltage, fast switching speed and extremely high endurance [1-5]. So far, remaining technology challenges rise in two folds. First, the probabilistic switching nature of a ferromagnetic layer causes the measurable Write Error Rate (WER) and Read Disturbance Rate (RDR) under a certain operating condition (pulse width and bias). Second, the relatively low tunnel magnetoresistance ratio (TMR) imposes extra constraints for Read (sensing) margin design to maintain low Sense Error Rate (SER) between parallel (P) and anti-parallel (AP) states. Furthermore, there exist complicated trade-offs amongst these failure rates when designing large-scale systems, as illustrated by Fig.1. Due to the bi-stability nature of a magnetic tunnel junction (MTJ), multi-level conductance capability is infeasible to demonstrate in a single MTJ, making this technology not favored for "analog-type" CIM. Nevertheless, compared with existing embedded memory counterparts (*i.e.* SRAM and eDRAM), STT-MRAM technology can still provide significant advantages for the energy-efficient systems with extra efforts spent on the design-technology co-optimizations (DTCO) and for a domain-specific architecture, *i.e.* hardware neural networks (HNN).

This work performs a comprehensive examination of aforementioned challenges via (1) MAC core circuit and layout designs based on mature process development kits (PDK) [6] and pragmatic STT-MRAM technology [1]; (2) advanced algorithms to calculate and optimize the very rare failure rates; and (3) HNN architecture emulations [8, 9] to capture non-idealities at the system level. The overall evaluation flow is proposed as Fig.2, with key methodologies and findings to be discussed in later sessions.

II. STT-MRAM Technology Evaluation

The physics-based multi-domain (electrical and magnetic) compact model [10] is used in this work to calibrate with the state-of-the-art STT-MRAM technology [1], which achieves good TMR ratio (~200%) and AP/P-state resistance margin under process variations [~20, defined as $(R_{AP}-R_P)/\sigma R_P$], as shown in Fig.3. It has been justified that the edge damages occurred at the MTJ's free layer contribute to the major MTJ resistance variability source [11] and are included in this work, as shown in Fig.4(a). The resistance-area (RA) product increases with reducing MTJ diameter, exemplifying this effect [Fig.4(b)]. Geometry variations on MTJs cause RA fluctuations, which, associated with the reduced TMR at higher Read voltage (V_{Read}), degrades resistance margin for scaled MTJs, as shown in Fig.5. To accurately calculate the very rare WER and RDR caused by the intrinsic thermal noise and process variations, GPU-based computing grids are used to perform massively parallel *Monte Carlo* (MC) simulations, at a rate resolution of 1E-9 [11]. Fig.6 shows experimental *vs.* calculated WER for an STT-MRAM with different Write pulse duration and voltage values. Note that the increased RA for smaller area devices also causes the degradation (slope and critical voltage) on WER *vs.* voltage curves. By further including variations in magnetic material properties (interfacial energy K_s and polarization P), Write voltages at WER=1E-6 (V_{1E-6}) and their fluctuations are predicted and compared with measurement results, as shown in Fig.7.

III. STT-MRAM based MAC Design

To examine the impacts of STT-MRAM device characteristics and process variability on circuit designs, the MAC array is designed based on FreePDKs [6, 7] developed for logic technology at 15nm and 45nm nodes, with some customized design rules to better fit for MAC bit cells. Fig.8 (a) illustrates the integration schemes for the embedded STT-MRAM, in which the MTJ is inserted at a via layer higher than the Word-line (WL) metal layer, followed by the Bit-line (BL) metal layer deposition. As compared to standard 1-transistor-1-MTJ (1T1M) cell structure [Fig.8(b)] for embedded memories, the MAC cell structure can be designed by rotating the BLs by 90° [8], as shown in Fig.8 (c). For the MAC operation, the input (voltage) vector arrives at a BL each cycle, and multiple Source Lines (SL) can be sensed to digital bits in parallel. Access transistor's width is designed to accommodate the minimum spacing to include the contacts and metal lines while keeping reasonably small conducting resistance. The critical geometries as well as designed cell areas are shown in Fig.9. After the layout generation, LVS and

978-1-7281-1988-5/18 $31.00 © 2018 IEEE

parasitic RC extraction (PEX) are performed, followed by the post-layout circuit simulations to examine the designs. As the MAC array size grows, the parasitic RC components result in considerable degradations to each cell's performance. Fig.10 shows the IR drop effect when designing long SL (or BL). The worst case scenario is studied with all accessed cells switched to low resistance (R_P) states. Post-layout simulations suggest that the SL length should be no longer than 256 cell units to keep the voltage drop ratio below 5%, to avoid significant WER and Read margin degradation. Fig.11 shows the gate propagation delay effect when designing the metal bypass in order to assist the underlying metal gate local interconnects to deliver WL signals, indicating a bypass period of 128 cell units is optimal to leverage between delay and the increased area penalty, with the total WL length is set to 256 cell units for an MAC block. Therefore, preferred MAC design has an array size of 128×256 (32Kb) to guarantee a decent performance for advanced STT-MRAM technologies.

Fig.12 shows the WL decoder circuit's performance based on designs in [8]. Fig.13(a) describes the current sense amplifier (CSA) design for Read operations of an STT-MRAM array with R_{ref} as $1/2 \cdot (R_P + R_{AP})$. The bit array and reference cells can be shared with the same SL columns [14], as shown in Fig.13(b). Simulated CSA waveforms for Read operations are plotted in Fig.14, with its inset showing the averaged sensing time (t_{Sense}) versus SL length for 45nm and 15nm technologies, indicating the overall performance improvement at descending technology nodes. The MC circuit simulations (10^3 samples) are used to optimize the SER, due to the strong non-*Gaussian* distributions of t_{Sense}. As shown in Fig.15 inset, t_{Sense} increases drastically in order to lower the SER, incurring issues in two folds: first, it generates significant latency overhead during MAC inference operations; second, the elongated t_{Sense} exacerbates the RDR of an STT-MRAM. Thus, an under design on t_{Sense} is applied herein to guarantee an SER ~ 1E-3, whose effectiveness remains to be examined during the HNN emulation stage, based on practical problems.

With extracted IR-voltage drop and delay values, V_{Read} and t_{Write}/V_{Write} can be further optimized for STT-MRAM based MAC arrays. Another critical concern results from the voltage window to achieve both low WER and RDR, as fulfilled by GPU-based rare event simulations in Fig.16. V_{Read} is set below 0.3V for a reasonable RDR < 1E-6 under 10ns t_{Sense}. Similarly, 100ns/0.6V pulse scheme is selected as the Write condition to accommodate WER < 1E-6 under process variability. The operating conditions, technology parameters together with various failures (*i.e.* SER, WER and RDR) will be incorporated into architecture-level HNN emulations later.

IV. STT-MRAM based HNN Emulation

Fig.17 overviews the high-level HNN architectures as well as the MAC/neuron periphery components. The emulator developed in [8] is used for evaluating a MAC unit including both core array and neuron peripheries, which has been calibrated with circuit simulations [10] and is capable to predict the power-performance-area (PPA) specs for advanced technology nodes. MC simulations are needed to study the impacts from either stochastic failures or process variations during HNN's training/inference operations. Fig.18 identifies two realistic problems to be studied in this work, *i.e.* hand-written digits recognition based on a Multi-layer Perceptron (MLP) evaluated using the MNIST database; and the image classification using Convolutional Neural Networks (CNN) tested by the CIFAR-10 database [16].

For the MLP study, the network contains only the stacked fully-connected (FC) layers having a topology of [400, 100, 10]. Depending on the system-level requirement, fewer (<16) physical bits can be used for each arithmetic weight as the "low-precision" computing. The resulted architecture is referred as the Quantized NN (Q=1, 2,…). Fig.19 shows the Q-MLP inference accuracy as a function of MTJ conductance variation ($\sigma R/R$), with pre-trained weights from 1 (binary) to 16 bits. For each case, more than 10^2 MC simulations were conducted to showcase the distribution and average value of accuracy for each condition. Results indicate that down to a 4-bit weight precision, the inference accuracy is not affected as long as $\sigma R/R \leq 0.1$, which has been fulfilled by the SER design in this work as well as achieved by state-of-the-art STT-MRAM process [1]. With increasing $\sigma R/R$, the accuracy values are fluctuated, which is likely due to the quickly degraded SER (Fig.15). Fig.20 shows the on-line training accuracy versus training epochs, with low (4- and 8-bit) precision weights and a $\sigma R/R = 0.1$. Interestingly, with higher than 8-bit precision schemes, the WER degradation from 1E-6 to 1E-3 causes only slight decrease of the training accuracy, which is attributed to the "self-adaptive" feature of an NN during back-propagation stage.

For the Q-CNN evaluations, the networks proposed in [9] are studied due to their superior energy-efficiency demonstrated at a hardware level. Fig.21 shows Q-CNN inference accuracy as a function of $\sigma R/R$, with 256 kernels per convolutional layer. Similarly, acceptable accuracy loss is expected in the region of $\sigma R/R \leq 0.1$. Fig.22 benchmarks various Q-CNN designs in terms of inference accuracy and required memory size, suggesting a >8bit weight precision and >256 kernels per layer design to achieve robustness to process variations and random failures.

Fig.23 shows the estimated inference-stage latency and energy (including both MAC core and neuron periphery) versus their total layout area, based on 45nm and 15nm technology nodes. The circuit area is dominated by the MAC core which is further proportional to the number of weight precision bits. For specific applications, the latency and energy reduction can be realized simultaneously by using low-precision weights. On the other hand, it suggests an effective way to mitigate the lower cell efficiency of STT-MRAM as compared to "analog-type" CIM.

V. Conclusion

STT-MRAM-based MAC array DTCO has been performed for advanced technology nodes. With further architecture-level emulations considering various parasitic effects, failure types and process variations, guidelines are provided to better utilize STT-MRAM technology for HNNs.

References: [1] Y. J. Song, *et al.*, IEDM, pp.663-666, 2016. [2] Y.K. Lee, *et al.*, VLSI Symp, pp.181-182, 2018. [3] G. Jan, *et al.*, VLSI Symp, pp.65-66, 2018. [4] C. Park, *et al.*, IEDM, pp.664-667, 2015. [5] K. Lee, *et al.*, VLSI Symp. Tech., pp.183-184, 2018. [6] FreePDK 45 & 15, NCSU. [7] PTM, ASU. [8] P.-Y. Chen, *et al.*, IEDM, pp.135-138, 2017. [9] B. Moons, *et al.*, arXiv:1711.00215. [10] N. Xu, *et al.*, IEDM, pp.735-738, 2015. [11] N. Xu, *et al.*, VLSI Symp, pp.187-188, 2018. [12] J.J. Nowak, *et al.*, *Magnetic Letters*, 3102604, 2016. [13] H.-C. Yu, *et al.*, ICSIC, 2012. [14] Y.-C. Shih, *et al.*, VLSI Symp, pp.79-80, 2018. [15] H. Noguchi, *et al.*, ISSCC, 2016. [16] MNIST and CIFAR-10 data set can be found at: http://yann.lecun.com/exdb/mnist/; https://www.cs.toronto.edu/~kriz/cifar.html.

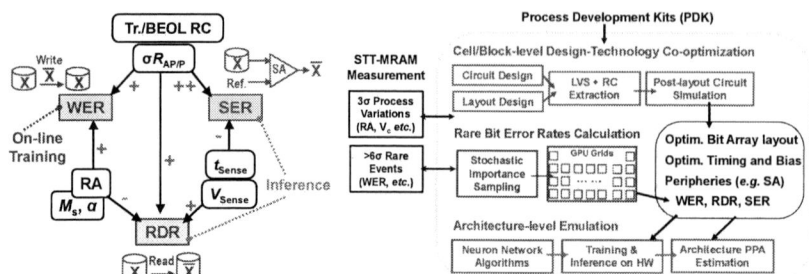

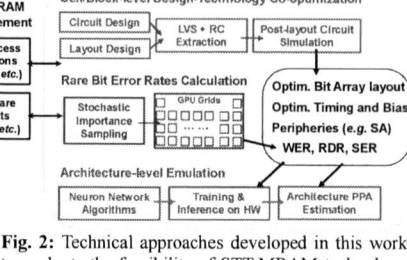

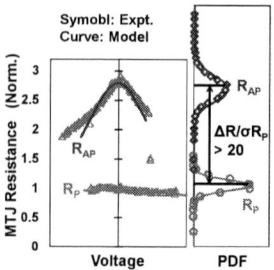

Fig. 1: STT-MRAM design trade-offs accommodating Write Error Rate (WER), Read Disturbance Rate (RDR) and Sense Error Rate (SER).

Fig. 2: Technical approaches developed in this work to evaluate the feasibility of STT-MRAM technology for hardware neural network (HNN) applications.

Fig. 3: Parallel (P) and anti-parallel (AP) state MTJ resistance *vs.* voltage and their resistance margin $(R_{AP}\text{-}R_P)/\sigma R_P$ under process variations.

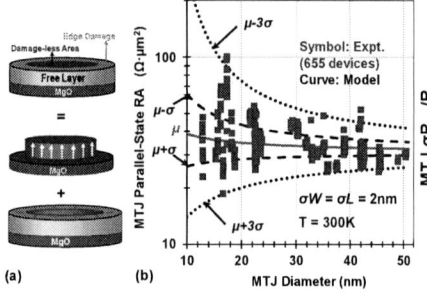

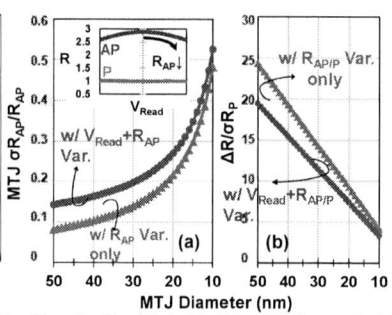

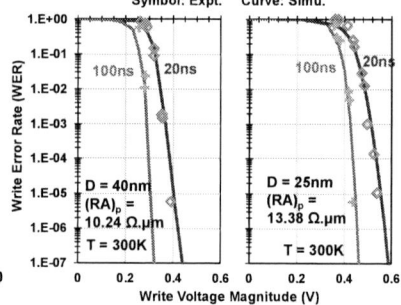

Fig. 4: (a) Edge damage effect during STT-MRAM device fabrication and **(b)** modeling *vs.* experimental [12] results for an MTJ's RA product and its variability as a function of its diameter.

Fig. 5: Simulated (a) R_{AP} variation and (b) resistance margin *vs.* MTJ's diameter, after calibrations with Fig.3, 4; the inset shows under V_{Read} variations on MTJ, reduced R_{AP} at higher voltage causes additional degradations.

Fig. 6: Simulation *vs.* experimental [12] results for STT-MRAM WER as a function of Write voltage and pulse duration for MTJs with diameters of 40nm and 25nm.

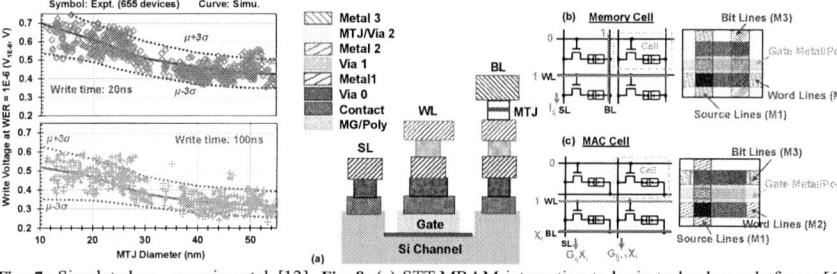

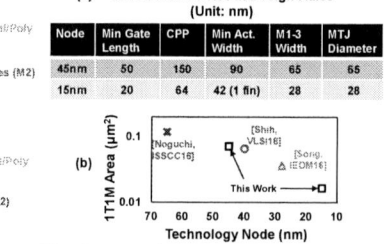

Fig. 7: Simulated *vs.* experimental [12] results of STT-MRAM Write voltage to achieve WER=1E-6 (V_{1E-6}) and its fluctuations under process variability, as a function of the MTJ diameter.

Fig. 8: (a) STT-MRAM integration to logic technology platforms [6], with major layers identified; and array circuit and layout designs for **(b)** memory and **(c)** MAC 1T1M cell structures. Note that WLs are contacted with Gate regions via WL bypass structures (shown later).

Fig. 9: (a) FEOL and BEOL critical features based on FreePDK15&45 for STT-MRAM MAC core designs; **(b)** designed 1T1M cell area with comparisons from literature [1, 14, 15].

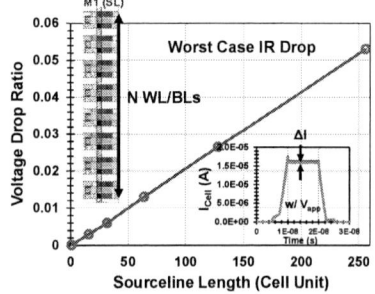

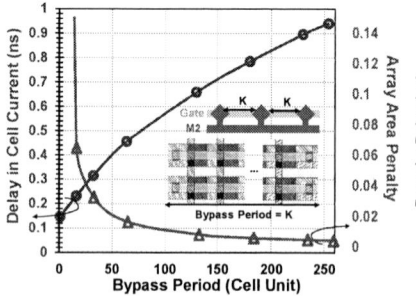

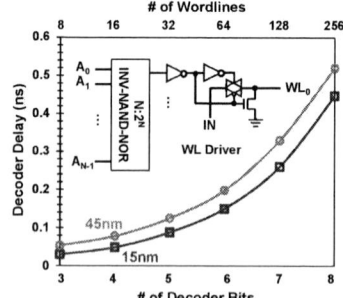

Fig. 10: Post-layout circuit simulations for voltage drop ratio *vs.* Source-line (SL) length, with the inset showing the current variations due to parasitic IR drop effect.

Fig. 11: Post-layout circuit simulations showing enhanced delay for cell current response *vs.* Word-line (WL) bypass period, in which the total WL length is set to be 256 cell units.

Fig. 12: Simulated WL decoder's delay *vs.* number of decoder bits N (corresponding to 2^N WL size); the inset illustrates the decoding matrix and WL driver circuit design.

978-1-7281-1988-5/18 $31.00 © 2018 IEEE

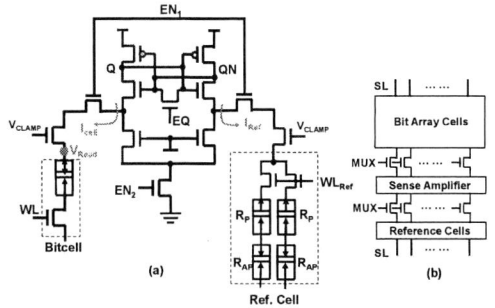

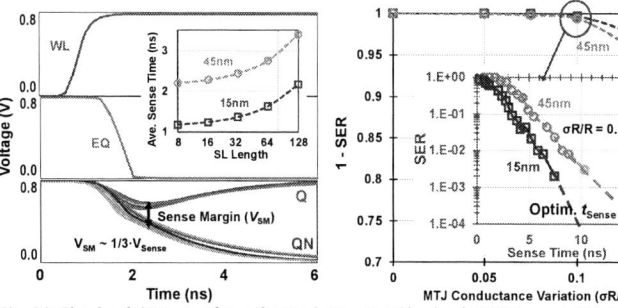

Fig. 13: (a) Current sense amplifier (CSA) design for STT-MRAM array Read/sense operations, together with the self-reference cell's resistance R_{ref} fixed at $1/2 \cdot (R_P + R_{AP})$; **(b)** CSA and bit array/reference cell configurations to save for extra SL columns.

Fig. 14: Simulated CSA waveforms for Read operations under MTJ conductance variations; the inset shows the averaged sense time *vs.* SL length (in cell units).

Fig. 15: Simulated 1-SER *vs.* MTJ $\Delta R/R$ for t_{Sense} designs, with the inset showing the SER *vs.* t_{Sense} at an MTJ's $\Delta R/R$=0.1. Results are collected from 10^3 MC circuit simulations.

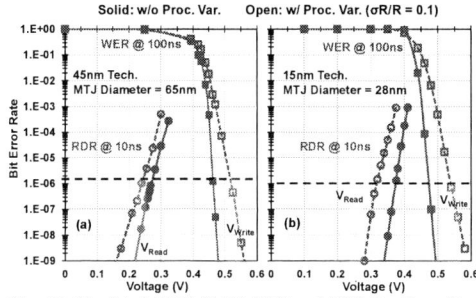

Fig. 16: Simulated STT-MRAM WER and RDR for V_{Read}, V_{Write} and t_{Write} designs, sufficing to a 1E-6 bit error rate (BER) for both Read/sense and Write operations. $\Delta R/R = 0.1$ is used to evaluate the impacts of process variation.

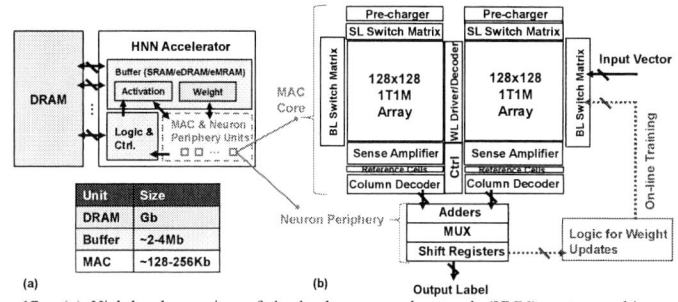

Fig. 17: (a) High-level overview of the hardware neural network (HNN) system architecture, including different kinds of memories and processing units; (b) circuit components for an MAC core and its associated neuron peripheries, as well as the dataflow for inference and on-line training.

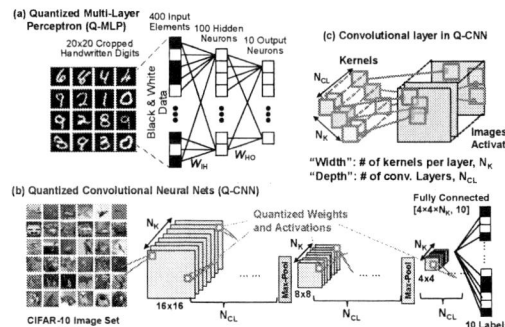

Fig. 18: (a) *MNIST* dataset and the Quantized-Multi-Layer Perceptron (Q-MLP); **(b)** *CIFAR-10* dataset and the Quantized-Convolutional-Neural-Network (Q-CNN) with **(c)** showing the designs for each convolution layer.

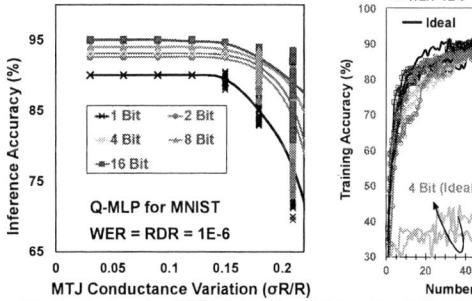

Fig. 19: Evaluation for STT-MRAM based Q-MLP in terms of inference accuracy *vs.* MTJ accuracy *vs.* number of epochs for 4- and 8-bit conductance variation under different weight weight precision digits, with different WER levels. precision digits.

Fig. 20: STT-MRAM based Q-MLP training

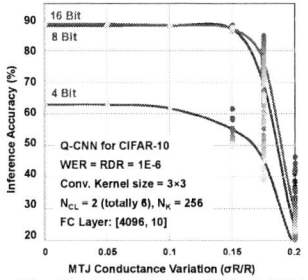

Fig. 21: Evaluation of STT-MRAM based Q-CNN inference accuracy *vs.* MTJ conduction variation. It is found that high accuracy levels can only be achieved at above-8-bit weight precision digits.

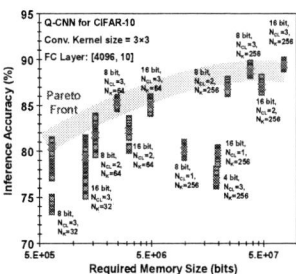

Fig. 22: Q-CNN inference accuracy *vs.* required memory size for various designs, in which a *Pareto* front exists (×: average, □: distribution of 100 MC simulations).

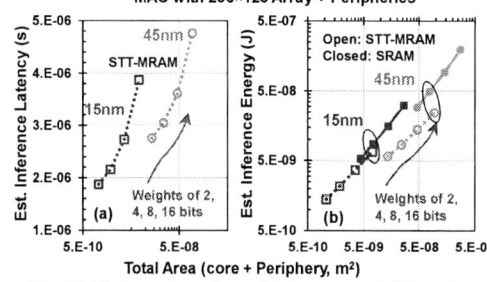

Fig. 23: Estimated per-frame **(a)** latency and **(b)** total energy (including MAC core and neuron peripheral circuits) for the inference process of STT-MRAM based MAC. Estimated values of 6T-SRAM based architectures are also included as a reference.

978-1-7281-1988-5/18 $31.00 © 2018 IEEE   351

A 68 Parallel Row Access Neuromorphic Core with 22K Multi-Level Synapses Based on Logic-Compatible Embedded Flash Memory Technology

M. Kim[1], J. Kim[1], G. Park[1], L. Everson[1], H. Kim[1], S. Song[1,2], S. Lee[2], and C. H. Kim[1]
[1]Dept. of ECE, University of Minnesota, Minneapolis, MN, USA email: chriskim@umn.edu
[2]Anaflash Inc, San Jose, CA, USA

Abstract – A neuromorphic core utilizing logic-compatible embedded flash technology for storing multi-level synaptic weights is demonstrated in a 65nm standard CMOS process. A carefully-designed program-verify sequence along with a bitline voltage regulation scheme allows the individual cell currents to be programmed precisely. This makes it possible to enable a large number of rows in parallel without impacting the current summation accuracy. Furthermore, eflash based synapses are non-volatile and hence consumes zero standby power and supports instant on/off operation. Our design stores excitatory and inhibitory weights in adjacent bitlines whose voltage levels are regulated for accurate current programming and measurement. Output spikes are generated by comparing the excitatory and inhibitory bitline currents. Our logic-compatible eflash-based spiking neuromorphic core achieves a 91.8% handwritten digit recognition accuracy which is close to the accuracy of the software model with the same number of weight levels. The maximum throughput of the core is 1.28G pixels/s and the average power consumption of a single neuron circuit is 15.9µW.

I. INTRODUCTION

Deep neural networks contain multiple computation layers each performing a massive number of multiply-and-accumulate operations (i.e. $\sum x_i w_i$) between the input data and trained weights. The computation is typically performed by a digital processor while the data and weight are transferred back and forth between the DRAM and the on-chip buffer memory. To overcome the memory bottleneck, there is growing interest in so-called "compute-in-memory" architectures where the weights are stored in a dense memory while the multiply-and-accumulate function is performed in the analog domain. Here, the input data is typically loaded on to the memory wordlines, activating multiple cell currents at the same time. The individual cell currents are summed up and compared to a pre-defined threshold by the local "neuron" circuit. Ideally, memory cells used for compute-in-memory architectures should be non-volatile as this obviate the need for reloading the weights after a power down period. It is also highly desirable if the memory cell can support multi-level storage as this can enhance the accuracy of inference tasks.

Both volatile and non-volatile memories have been considered for synaptic weight storage including SRAM,

magnetic tunnel junctions (MTJs) and resistive RAM (RRAM), flash, and phase-change memory (PCRAM) [1]-[6]. Each type of memory has its advantages and disadvantages. For instance, SRAM based synapses can be readily implemented in a standard CMOS process, but suffers from process variation which cannot be corrected after the chip has been fabricated. MTJs, RRAM, and PCRAM are non-volatile and dense. However, an MTJ can only store a 1 bit weight and the difference between the high resistance and low resistance states is only about 2X, rendering analog computing impractical. RRAM provides a wider resistance range, but the technology remains immature. Furthermore, robust multi-level programming has proven to be challenging for RRAM and PCRAM due to low controllability of the filament formation and heat diffusion [7]. Flash memory technology can easily store multiple levels by adjusting the number of electrons stored on a floating gate through row-by-row program-verify operation. However, conventional flash memory requires a specialized dual-poly or split-gate process which doesn't scale well below 40nm.

In this work, we demonstrate a logic-compatible eflash based spiking neuromorphic core in a 65nm standard CMOS process featuring multi-level non-volatile weight storage, and single cycle current integration and spike generation. The weights were tuned precisely using a carefully-designed program-verify sequence, allowing 68 individual cell currents to be summed up simultaneously, which to our knowledge, is the highest number ever reported.

II. EFLASH-BASED NEUROMORPHIC CORE DESIGN

Fig. 1 shows a comparison between dual-poly and single-poly eflash cells. Dual-poly eflash cell stores charge on a floating gate fabricated between the control gate and channel. Single-poly eflash cell is implemented using back-to-back connected transistors and hence does not require any modification to the process. The detailed schematic of the 5T eflash cell used in this work is shown in Fig. 2 where two asymmetrically sized PMOS devices are used for high voltage program and erase operation while the NMOS read device is accessed through two additional NMOS switches [8][9]. Different cell currents can be programmed as shown in Fig. 2 (right).

Synaptic weights stored in two adjacent bitlines as illustrated in Fig. 3. If the weight is positive then the cell current of the left bitline is increased accordingly while the

978-1-7281-1988-5/18 $31.00 © 2018 IEEE

cell current on the right bitline is programmed to <0.1µA, and vice versa. Four flash cells are reserved on each bitline for the spiking threshold. The input data is simultaneously loaded onto the wordline which activates multiple memory cell currents at the same time. The sum of the individual cell currents flows through each bitline. The bitline pair generates two currents: excitatory and inhibitory currents. The neuron circuit generates a spike depending on which bitline current is higher. Note that the weight multiplication, accumulation, and spike generation are all performed in a single cycle, which speeds up the overall computation.

The core architecture is shown in Fig. 4 (left). It contains high voltage switches (HVSs) for driving wordlines, neuron sensing circuits for current comparison and spike generation, and scan chains for data input/output. The HVS circuit must withstand a voltage as high as 10V during program and erase modes. We employed the multi-story latch based HVS circuit [8][9] because it is inherently immune to overstress issues and implementable using standard IO devices. The circuit diagram and layout of the unit 5T eflash cell are shown in Fig. 4 (right).

The weights are written to the array by first erasing the entire array and then adjusting the threshold voltage of each individual eflash cell. This is done by simultaneously programming each row through a selective program-verify operation. The cell currents were set to either 0, 5, or 10 µA while keeping the gate bias (VRD=0.8V) and drain bias (VBL=0.6V) fixed. This translates into five distinct weight levels (i.e. -10, -5, 0, 5, or 10 µA) using the bitline pair configuration described earlier in Fig. 3. Retention characteristics shown later in Fig. 15 confirm that a 5 µA margin between the different levels is sufficient to overcome charge loss issues. The wordline and bitline bias condition for erase, program, and program inhibition modes are denoted in Fig. 5. To obtain a precise cell current corresponding to the weight value, the bitline voltages were regulated to 0.6V during both verify and inference modes. Our neuron circuit shown in Fig. 6 employs a feedback loop to maintain a fixed 0.6V bitline voltage regardless of the amount of current flowing through the bitline. The bitline current is indirectly measured by reading out the feedback voltage driving the PMOS load. This makes it possible to compare the two bitline currents using a simple voltage sense amplifier circuit. The same neuron circuit was used for current-verify operation as shown in Fig. 6 (right). Here, the cell currents of the left and right bitlines were verified separately by activating one bitline at a time. The overall operation sequence of our neuromorphic core is shown in Fig. 7.

III. EXPERIMENTAL RESULTS

We first measured the program and program inhibition characteristics of the 5T eflash cell. The average current of 100 cells was measured for different program voltages, program pulse widths, and program pulse counts. Fig. 8 confirms excellent program and program inhibition results. Based on the test results, we designed the program sequence

in Fig. 9 for configuring the cell currents. To minimize cell disturbance, we first programmed the weight 0 cells (i.e. <0.1µA) while inhibiting the program of weight 1 and 2 cells. To ensure that the cell currents of all weight 0 cells are below 0.1µA, we use a high voltage (8.8V), long pulse width (20µs), and large pulse count (8). Next, the rest of the cells were programmed to an intermediate current level of about 15µA using a single 7.4V and 40µs pulse. Then, using smaller and shorter pulses (i.e. 7.1V, 5µs), we adjusted the weight 2 cells to 10µA, and finally the weight 1 cells to 5µA. Note that the unselected wordlines are not driven to a high voltage preventing the cells on those wordlines from being disturbed. Fig. 10 shows the cell currents for trained weights of the MNIST handwritten digit recognition algorithm. The variation for weight 0 cells is less than 0.1µA. Weight 1 and 2 cells also have a variation of only 0.8µA. The total number of program pulses applied to each wordline ranges from 25 to 32 as shown in Fig. 11. The average power consumption of a single neuron during inference mode is 15.9µW (Fig. 12). Fig. 13 provides further insight on how the cell current changes with more program pulses for weight 1 and weight 2 cells. It can be seen that that intrinsic cell current variation of 7µA is reduced to 0.8µA after the proposed program-verify sequence. This offers a significant advantage over SRAM or MRAM based implementations which do not have any post-silicon tuning capabilities.

Fig. 14 shows the overall work flow for demonstrating the handwritten digit recognition application on our test chip. During training phase, weights were trained based on 60,000 handwritten digit images from the MNIST dataset [10] and downloaded to the test chip. During inference phase, the neuromorphic core generates a spike signal based on the 16x16 pixel data and the programmed weights. 10,000 MNIST test images were processed to calculate the prediction accuracy. The accuracy measured from the test chip was 91.8% (Fig. 14) which is close to the software accuracy of 93.8% for the same number of distinct weight levels (i.e. 5). The small discrepancy can be attributed to noise effects and sense amplifier offset. Retention characteristics were measured after baking the chip for 16 hours at 150°C. Measured results in Fig. 15 confirm that the margin between the different current levels is not compromised, suggesting that storing more than 5 levels is also possible. Comparison with previous SRAM, RRAM, and NOR flash based neuromorphic core designs underscores the promising features of our logic-compatible eflash-based design (Fig. 16). The die photo and chip feature summary are given in Fig. 17.

REFERENCES

[1] D. Kuzum, R. Jeyasingh, H.S. Wong, IEDM, pp.693-696, Dec. 2011. [2] W. Chen, W. Lin, L. Lai, et al., IEDM, pp. 657-660, Dec. 2017. [3] X. Guo, F. Merrikh Bayat, M. Bavandpour, et al., pp. 6.5.1-6.5.4, IEDM, Dec. 2017. [4] W. Khwa, J. Chen, J. Li, et al., ISSCC, pp. 496-497, Feb. 2018. [5] S. Gonugondia, M. Kang, N. Shanbhag, ISSCC, pp. 490-491, Feb. 2018. [6] W. Chen, K. Li, W. Lin, et al., pp. 494-495, ISSCC 2018. [7] D. Nminibapiel, D. Veksler, P. Shrestha, et al., pp. 736-739, IEEE EDL, June 2017. [8] S. Song, K. Chun, C. Kim, pp. 1302-1314, JSSCC, May 2013. [9] S. Song, K. Chun, C. Kim, pp. 1-4, CICC, 2013. [10] MNIST dataset, http://yann.lecun.com/exdb/mnist/index.html

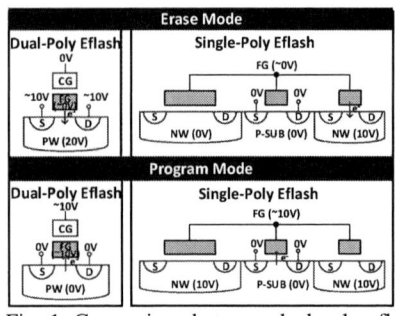

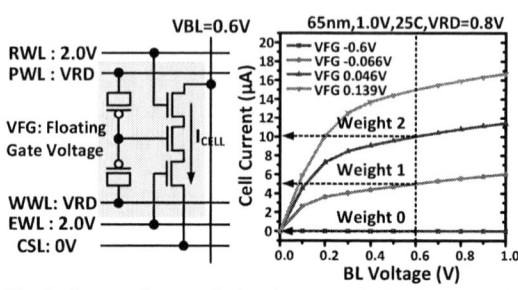

Fig. 1. Comparison between dual-poly eflash (left) and single-poly eflash (right).

Fig. 2. Output characteristic of proposed 5T eflash cell for different floating gate (FG) node voltages. Multi-level weights can be stored precisely through program-verify.

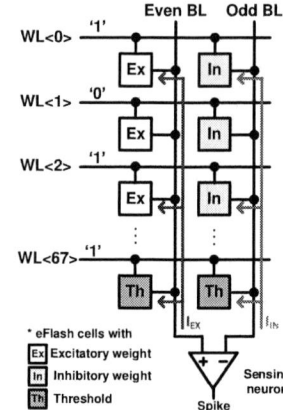

Fig. 3. Excitatory and inhibitory weight values are stored in two adjacent bitlines. Currents are summed up and compared for spike generation.

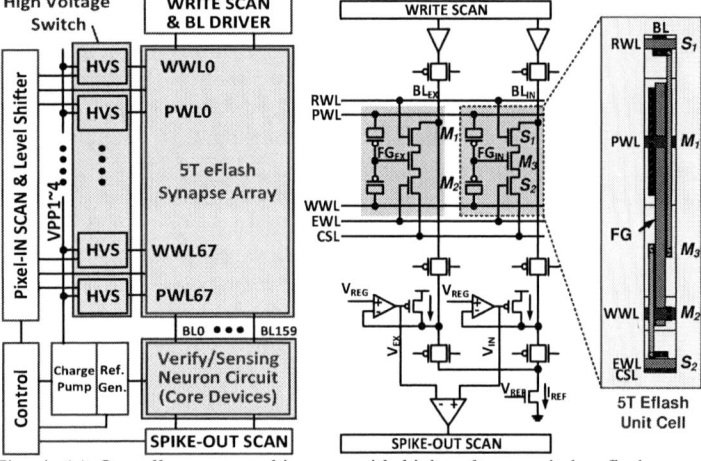

Fig. 4. (a) Overall neuromorphic core with high voltage switch, eflash array, and neuron sensing circuit. (b) Single column pair and 5T unit cell layout.

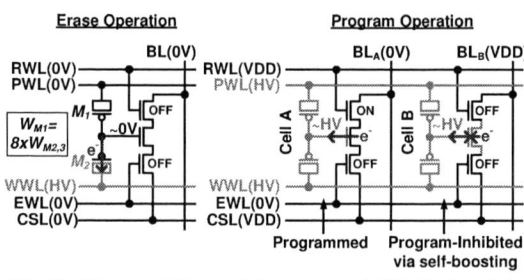

Fig. 5. Bias conditions of the proposed 5T eflash cell for erase and program operations [8].

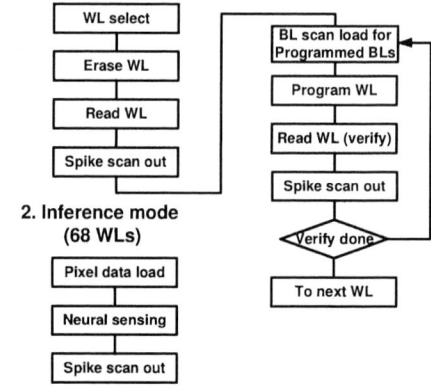

Fig. 6. Neuron circuit with regulated bitline voltage; (left) inference mode for spike generation and (right) weight programming mode with program-verify operation.

Fig. 7. Overall neuromorphic core operation sequence: Weight programming mode (upper) and inference mode (lower).

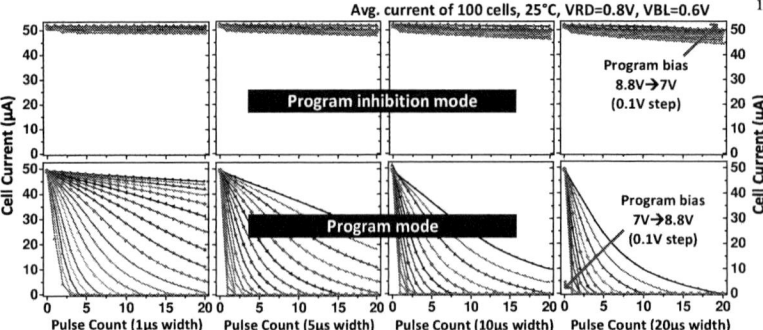

Fig. 8. Cell current versus number of program pulses for program inhibited cells (upper row) and programmed cells (low row). Results are shown for different pulse widths (i.e. 1µs, 5µs, 10µs, 20µs) and 0.1V program bias increments from 7.0V to 8.8V. Test chip data shows reliable programming and minimal program disturbance.

978-1-7281-1988-5/18 $31.00 © 2018 IEEE   354

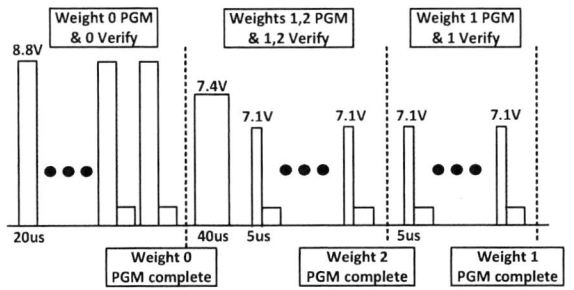

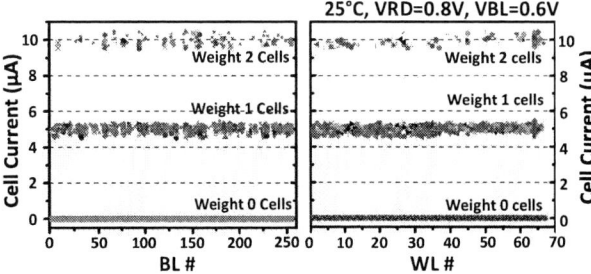

Fig. 9. Pulse sequence for programming weights 0, 1, and 2 into the eflash neuromorphic core. Multi-level weights can be programmed precisely owing to the carefully-design program-verify sequence.

Fig. 10. Individual cell currents in bitline and wordline direction for MNIST trained weights. The exceptionally tight current distribution suggests that storing 3 or more levels of weights is possible.

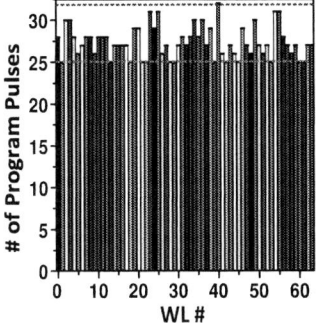

Fig. 11. The number of program pulses applied to each wordline for MNIST handwritten digit recognition.

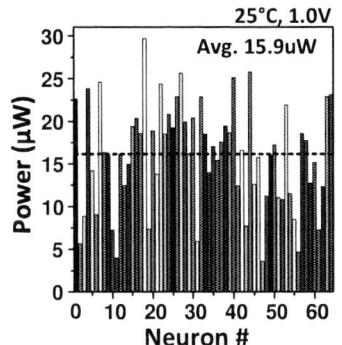

Fig. 12. Power consumption of each neuron during inference mode at VDD=1.0V.

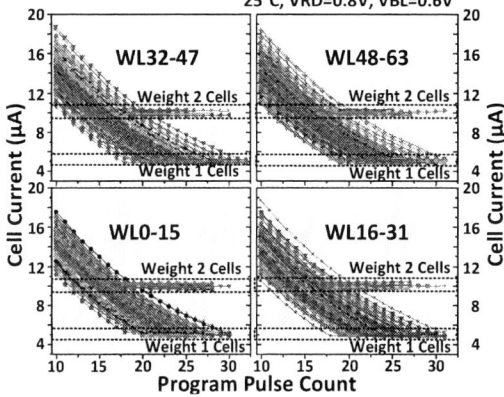

Fig. 13. Cell current versus program pulse count when applying the program-verify sequence.

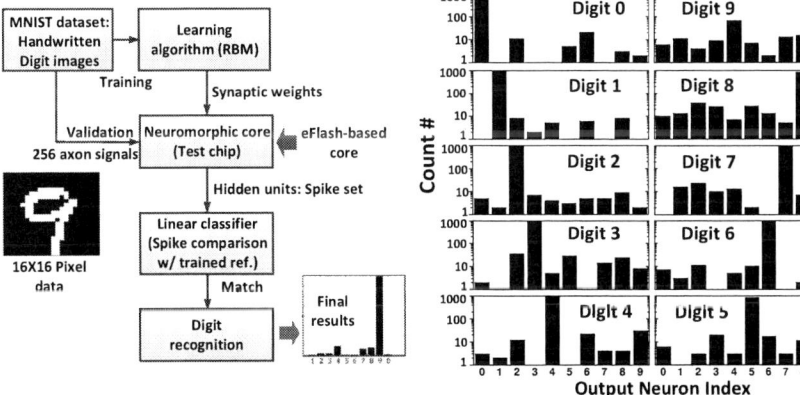

Fig. 14. (Left) Demonstration flow of hand-written digit recognition algorithm using the proposed neuromorphic core. (Right) Histograms of neuron output for 10,000 MNIST test images measured from the eflash-based neuromorphic core chip.

Fig. 15. Retention characteristics of weight 0, 1, and 2 cell currents confirm that the margin between the different states remains constant. Baking temperature was 150°C.

	This work	ISSCC'18 [7]	ISSCC'18 [4]	ISSCC'18 [5]	IEDM'17 [2]	IEDM'17 [3]
Application	Handwritten digit recognition	Handwritten digit recognition	Handwritten digit recognition	Machine learning classifier	Computing in memory	Handwritten digit recognition
Technology	65nm	65nm	65nm	65nm	150nm	180nm
Voltage	1.0V	1.0V	1.0V	1.0V	1.8V	2.7V
Non volatile?	YES (Eflash)	YES (ReRAM)	NO (SRAM)	NO (SRAM)	YES (ReRAM)	YES (Eflash)
Logic Compatible?	YES	NO	YES	YES	NO	NO
Program verify?	YES	NO	NO	NO	NO	YES
Weight Resolution	2.3 Bits (5 levels)	3 Bits	1 Bit	1 Bit	2 Bits	N/A
# of Currents Summed Up	68 Cells	14 Cells	30 Cells	4 Cells	2 Cells	N/A

Fig. 16. Comparison with prior art.

Technology	65nm Logic
Core Size	1100 X 600 µm²
VDD (Core, IO)	1.0V / 2.5V
# of Neurons	320
# of Synapses	22K (=68x320)
Throughput	1.28G pixels/s per core (tREAD : 50ns)
Power	15.9µW (per neuron)

Fig. 17. Die microphotograph and test chip feature summary.

978-1-7281-1988-5/18 $31.00 © 2018 IEEE   355

Interchangeable Hebbian and Anti-Hebbian STDP Applied to Supervised Learning in Spiking Neural Network

Che-Chia Chang, Pin-Chun Chen, Boris Hudec, Po-Tsun Liu[#], and Tuo-Hung Hou[*]

Department of Electronics Engineering and Institute of Electronics, National Chiao Tung University, Hsinchu, Taiwan
[#] Department of Photonics, National Chiao Tung University, Hsinchu, Taiwan
[*]Tel: +886-3-5712121 ext 54261; E-mail: thhou@mail.nctu.edu.tw

Abstract—This work provides a complete framework, including device, architecture, and algorithm, for implementing bio-inspired supervised spiking neural networks (SNNs) on hardware. An analog synapse with atypical dual bipolar resistive-switching (D-BRS) modes demonstrates interchangeable Hebbian spiking-timing-dependent plasticity (STDP) and anti-Hebbian STDP, and it is capable of implementing supervised ReSuMe SNNs in crossbar arrays. By using an "exchange" update scheme, accurate supervised learning (~96% for MNIST) is achieved in a compact network.

I. Introduction

In biological neural networks, each synapse is characterized by a synaptic weight, which determines the connecting strength between pre- and postsynaptic neurons (Fig. 1). Both the direction and magnitude of the nonvolatile synaptic weight change are sensitive to the precise relative timing of pre- and postsynaptic spikes [1, 2]. In Hebbian STDP, presynaptic spikes arriving prior to postsynaptic spikes result in increment of synaptic weight (potentiation) while those arriving later than postsynaptic spikes result in decrement of synaptic weight (depression). Anti-Hebbian STDP is the opposite of the Hebbian STDP [1, 2]. Only a few millisecond differences in the relative timing between pre- and postsynaptic spikes could drastically alter the tendency of forming connections in neural networks [1, 2]. Recently these new insights of synaptic plasticity in neuroscience propel active development of artificial neural networks (ANNs) for computing. In particular, SNN, which processes information by using temporal configuration of spikes, is regarded as the third-generation ANN (Fig. 2), not only because their design principles take full advantage of the biological spiking neuron model but also because they are computationally more powerful than their predecessors, such as perceptron neuron model [3]. Unsupervised winner-take-all (WTA) algorithms are widely used in STDP-based, bio-inspired SNN [4, 5]. However, supervised learning from instructions is considered a fundamental feature of human brain necessary to acquire new knowledge and skills. Remote supervised method (ReSuMe) is one of the most successful supervised SNN algorithms [6]. In ReSuMe SNN, both remote "teacher neuron" and "student neuron" fire with input neuron, but the weights connecting to them follow STDP and anti-STDP learning rules, respectively, producing desirable output pattern in response to the target pattern after training (Fig. 3).

Synaptic crossbar implementation of ANNs is known to accelerate neural network computation by exploiting local weight storage [7, 8]. In this work, we propose a compact hardware implementation of ReSuMe SNN by using synaptic crossbars, where the same type of synapses is used for both STDP and anti-STDP weights (W_{STDP} and $W_{anti-STDP}$) (Fig. 4). We report the first analog synapse with such dual STDP weights, and show that the STDP and anti-STDP modes are interchangeable. Finally, ReSuMe SNN is cascaded with an unsupervised WTA convolutional layer, and it shows promising training accuracy by taking real device properties into account. The proposed analog synapse paves the way for developing efficient supervised learning that expands the capability of future SNN.

II. Synapse with Interchangeable STDP and Anti-STDP

Fig. 5a shows the TEM image and process flow of the Ta/HfO$_x$/Al-doped TiO$_2$/TiN synapse fabricated in this work. The Al doping was introduced in-situ using TMA during plasma-enhanced ALD (PEALD) TiO$_2$ growth to improve the TiO$_2$ breakdown. HfO$_x$ film was deposited using in-situ NH$_3$ remote plasma treatment during PEALD HfO$_2$ growth. EDX depth profiling (Fig. 5b) shows negligible nitrogen content in HfO$_x$. XPS valence-band spectra shows an apparent defect band at approximately 3.1 eV above the valence band of HfO$_x$, which is absent in the untreated HfO$_2$ (Fig. 6). The generation of defects, most likely oxygen vacancies, could be attributed to the reduction effect of hydrogen radicals during plasma treatment. Fig. 7 illustrates the energy band alignment of the device. By using the bandgap value extracted from spectroscopic ellipsometry, we estimate the band offset of defect band to Ta and Al-doped TiO$_2$ is approximately 0.5 eV.

Fig. 8a shows the DC switching curves when operating the device as RRAM. Without any forming, D-BRS modes are present simultaneously, which is atypical in normal RRAM devices. In "– mode" (blue curve), the device is set to low resistance state (LRS) by applying negative voltage sweep on the Ta top electrode, and reset to high resistance state (HRS) by applying positive voltage sweep with a lower magnitude. In "+ mode" (red curve), the device is set and reset using voltage polarities opposite to those in "– mode". The device can be repeatedly cycled in either mode with tight resistance distributions (Fig. 8b). Most interestingly, the device can be operated in these two modes alternately with no need of any pre-settlement (Fig. 9). Retention of LRS in both modes lasts for at least few hours at room temperature before merging to HRS, while retention of HRS is much longer. The extracted activation energy value of LRS retention is around 0.5 eV similarly in both modes, and the value agrees with the band offset of HfO$_x$ defect band (Fig. 10). To further support that the D-BRS is originated from the defect band in HfO$_x$, we substituted Al-doped TiO$_2$ with ZnO or used simply a single HfO$_x$ layer. In both cases, D-BRS can be still observed but

978-1-7281-1988-5/18 $31.00 © 2018 IEEE

with worse stability because of early device breakdown (not shown). Furthermore, D-BRS does not appear in the Ta/HfO$_2$/Al-doped TiO$_2$/TiN device, where HfO$_2$ is not treated by NH$_3$ plasma (not shown). These results suggest that the HfO$_x$ layer acts as a resistive-switching layer while the Al-doped TiO$_2$ layer acts as a series resistance for preventing device breakdown. Both HRS and LRS currents scale with the device area (Fig. 11), indicating a non-filamentary switching mechanism. Fig. 12 illustrates an electron trapping-detrapping model explaining the non-filamentary D-BRS. The amount of electron trapping in preexisting shallow defects affects the HfO$_x$ conductance [9]. The region with electron-filled defects possesses lower resistance (R$_{on}$) while the region with empty defects possesses higher resistance (R$_{off}$). Depending on the direction of electron injection, the R$_{on}$ region of HfO$_x$ near the top electrode or near the Al-doped TiO$_2$ expands during set of "− mode" or "+ mode", respectively. Detrapping is energetically more favorable so that reset by detrapping electrons occurs when applying an opposite but lower voltage, and LRS retention is worse than HRS. By controlling the amount of trapped electrons, the device shows analog synaptic properties of long-term potentiation/ depression (LTP/LTD) under pulse stimuli in "− mode", "+ mode", and mixing of two (Fig. 13). Anti-STDP and STDP can be achieved in "− mode" and "+ mode", respectively, by using an identical presynaptic spike (Fig. 14).

III. ReSuMe SCNN Using Dual STDP Synapse

The co-existence of W$_{STDP}$ and W$_{anti-STDP}$ enables compact hardware implementation of the ReSuMe SNN layer shown in Fig. 4. To investigate its performance, we simulate a spiking convolutional neural network (SCNN) with one convolutional, one pooling and one fully-connected layer (Fig. 15). The convolutional feature-extraction layer is trained in an unsupervised manner using WTA, while the fully-connected classifier layer is trained using ReSuMe. The experimental plasticity characteristics in Fig. 14, including bounded conductance range, are taken into consideration. In original ReSuMe, the conductance change is unidirectional, meaning W$_{STDP}$ (W$_{anti-STDP}$) always increases (decreases) monotonically. This creates a problem in hardware with finite conductance range. One potential solution to this problem is to read out the respective weight values frequently and write back the net weight (W$_{STDP}$ + W$_{anti-STDP}$) to prevent weight saturation [10]. However, this inevitably increases the complexity of implementation and also degrades the performance. By exploiting the unique interchangeable STDP and anti-STDP, we propose a new "exchange" scheme where W$_{STDP}$ and W$_{anti-STDP}$ are exchanged after a fixed number of training iterations. This is done simply by swapping the postsynaptic spikes (as shown in Fig. 14) provided from peripheral circuits, and the net weight remains the same after exchange with no additional read-out and write-back steps. Because the same weight is used for both STDP and anti-STDP, the conductance change becomes bidirectional to overcome the limit of bounded conductance. Fig. 16 shows the learning accuracy of the single ReSuMe classifier layer on MNIST data. The "exchange" scheme substantially reduces the number of weights reaching the bounds during training and achieves the accuracy close to the ideal case with unbounded weights. The exchange is needed extremely infrequently, for example only after a complete training epoch of 60000 MNIST images (Fig. 17), representing negligible design overhead. The additional unsupervised feature-extraction layer is beneficial to improve learning accuracy (Fig. 18), and the convolutional scheme that extracts more detailed local features is superior to a fully-connected extraction layer (i.e. using a 28×28 kernel size). We achieved 95.7% MNIST classification accuracy by using 600 10×10 kernels. Although some SNN studies reported even higher MNIST accuracy by using non-STDP, backpropagation learning rules [11], their approaches are more suitable for software rather than hardware implementation. Table 1 shows the comparison of SNN classification performance for MNIST. Our ReSuMe SCNN approach demonstrates very high accuracy by using a relatively compact hardware.

IV. Conclusion

Inspired by the biological nervous system, we propose an implementation framework for supervised SNN based on a new dual-STDP synapse. This device could be a critical building block for future versatile and efficient SNNs on hardware.

ACKNOWLEDGMENT

This work was supported by Ministry of Science and Technology of Taiwan, ROC, under grant MOST 106-2119-M-009-008, and in part by Research of Excellence program (MOST-107- 2633-E-009 -003). C.-C. Chang and P.-C. Chen contributed equally to this work.

REFERENCES

[1] P. D. Roberts and C. C. Bell, "Spike timing dependent synaptic plasticity in biological systems," *Biol. Cybern.*, vol. 87, pp. 392–403, 2002.

[2] C. Zamarreño-Ramos, *et al.*, "On spike-timing-dependent-plasticity, memristive devices, and building a self-learning visual cortex," *Front. Neurosci.*, vol. 5, 26, 2011.

[3] W. Maass, "Networks of spiking neurons: The third generation of neural network models," *Neural Networks*, vol. 10, pp. 1659–1671, 1997.

[4] P. U. Diehl and M. Cook, "Unsupervised learning of digit recognition using spike-timing-dependent plasticity." *Front. Comput. Neurosci.*, vol. 9, 99, 2015.

[5] G. Pedretti, *et al.*, "Modeling-based design of brain-inspired spiking neural networks with RRAM learning synapses," *in IEDM Tech. Dig.*, 2017, pp. 653–656.

[6] F. Ponulak and A. Kasinski, "Supervised learning in spiking neural networks with ReSuMe: Sequence learning, classification, and spike shifting," *Neural Comput.*, vol. 22, pp. 467–510, 2010.

[7] S. Yu, *et al.*, "Scaling-up resistive synaptic arrays for neuro-inspired architecture: challenges and prospect," *in IEDM Tech. Dig.*, 2015, pp. 451–454.

[8] C.-C. Chang, *et al.*, "Challenges and opportunities toward online training acceleration using RRAM-based hardware neural network," *in IEDM Tech. Dig.*, 2017, pp. 278–281.

[9] J. H, Yoon, *et al.*, "Pt/Ta$_2$O$_5$/HfO$_{2-x}$/Ti resistive switching memory competing with multilevel NAND flash," *Adv. Mater.*, vol. 27, pp. 3811–3816, 2015.

[10] G. W. Burr *et al.*, "Experimental demonstration and tolerancing of a large-scale neural network (165,000 synapses), using phase change memory as the synaptic weight element," *in IEDM Tech. Dig.*, 2014, pp. 697–700.

[11] P. O'Connor and M. Welling, "Deep spiking networks," *arXiv:1602.08323*, pp. 1–16, 2016.

Supervised SNN Concept

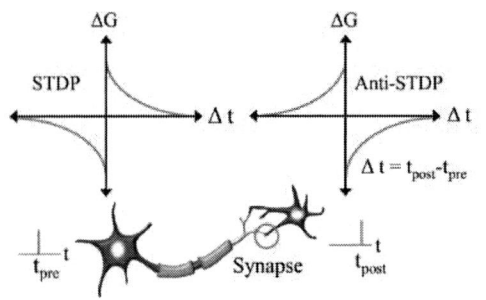

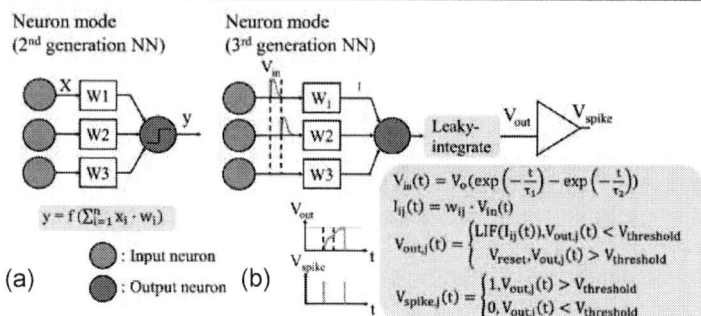

Fig. 1 STDP and anti-STDP observed in the biological neural system. Synaptic plasticity is modulated based on the relative timing of pre- and postsynaptic spike

Fig. 2 (a) Non-spiking neuron model cannot accumulate temporal information. (b) Spiking neuron model is more powerful with temporal summation capability.

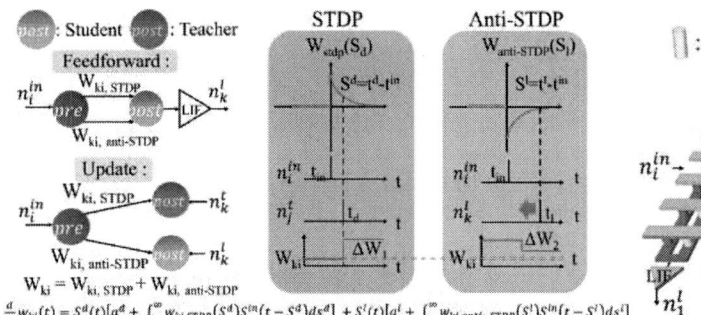

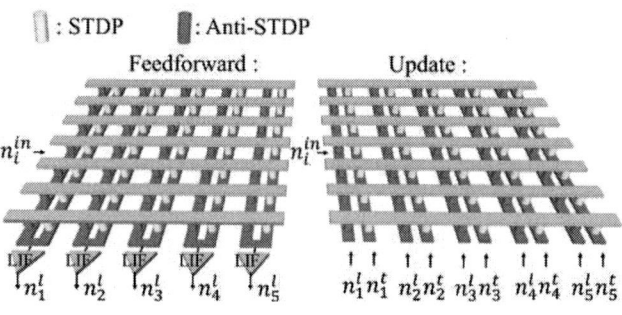

Fig. 3 Student neuron collects the information from both STDP and anti-STDP weights during feedforward, but only fire back to reduce anti-STDP weight. Teacher neuron increases STDP weight according to target spike. The net weight change is zero when Student neuron fires together with Teacher neuron.

Fig. 4 Hardware implementation of ReSuMe SNN by using synaptic crossbars. Each net weight is composed of one STDP and one anti-STDP weight. During update, the fire-back spike and target spike are fed into the columns of anti-STDP and STDP weights, respectively.

Dual Bipolar RS Modes

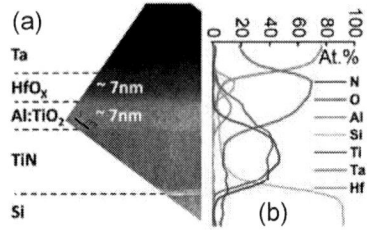

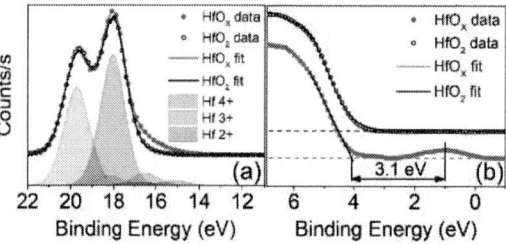

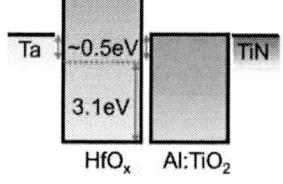

Fig. 5 (a) TEM image of the Ta/HfOₓ/Al-doped TiO₂/TiN synapse. HfOₓ was deposited by PEALD using NH₃ plasma treatment. (b) EDX depth profiling shows negligible nitrogen in HfOₓ layer.

Fig. 6 XPS spectra comparison of the HfOₓ film (w/ NH₃ plasma) with the control HfO₂ film (w/o NH₃). (a) Hf 4f spectrum of HfO₂ only contains Hf 4+ doublet (characteristic for HfO₂), while additional sub-oxide doublets (Hf 3+,2+) are present in HfOₓ spectrum. b) HfOₓ spectrum also shows a defect band located 3.1 eV above VBM, which we assign to V_O.

Fig. 7 Corresponding band alignment of each layer regarding the bandgap of HfOₓ (5.6eV) and Al:TiO₂ (3.6eV) from spectroscopic ellipsometry (SE) data. The defect band locates at ~0.5eV below E_c of both Ta and Al:TiO₂ calculated from the information of SE and XPS valence-band spectra.

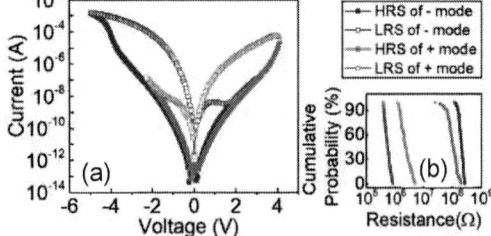

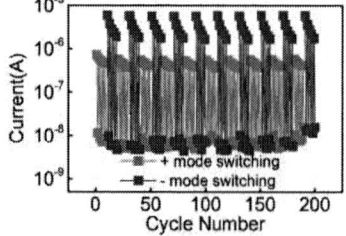

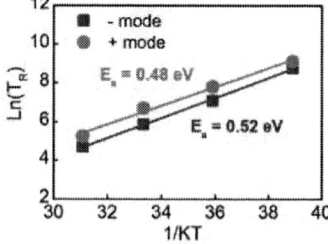

Fig. 8 (a) DC switching curve for "+ mode (red curve)" and "– mode (blue curve)". (b) Cumulative distribution of HRS and LRS resistance extracted from 100 switching cycles in both modes.

Fig. 9 Transition between two modes is demonstrated for ten repeated super-cycles. Each has five "+ mode" switching cycles, followed by immediate five "– mode" cycles.

Fig. 10 Extracted activation energy value of LRS retention is around 0.5 eV in both modes, similar to the band offset of HfOₓ defect band.

978-1-7281-1988-5/18 $31.00 © 2018 IEEE   358

Switching Mechanism

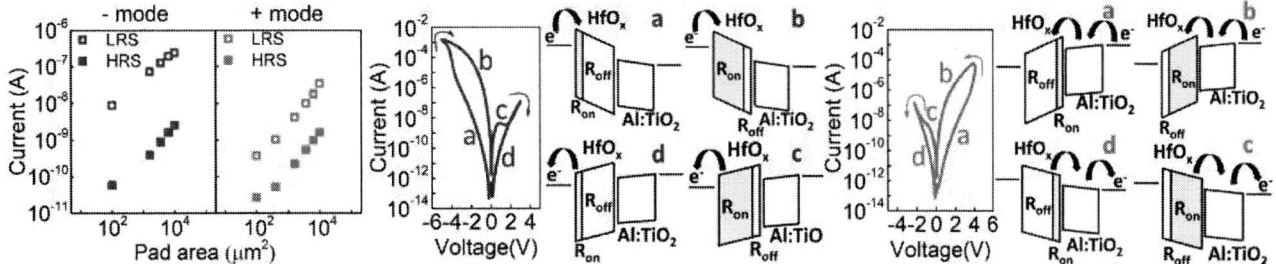

Fig. 11 Strong area dependence of cell current is shown in both modes, indicating a non-filamentary switching mechanism.

Fig. 12 Electron trapping-detrapping model: R_{on} region expands during electron injection and trapping, and the device is set to LRS (from a to b) in both modes. The device is reset to HRS (from c to d) by giving opposite bias to detrap electrons, and R_{off} region expands.

Interchangeable STDP / Anti-STDP

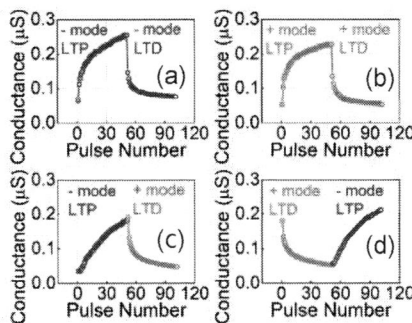

Fig. 13 Analog synaptic plasticity of LTP and LTD obtained by applying consecutive potentiation and depression pulses of (a) "– mode", (b) "+ mode", or (c & d) mixing of these two modes.

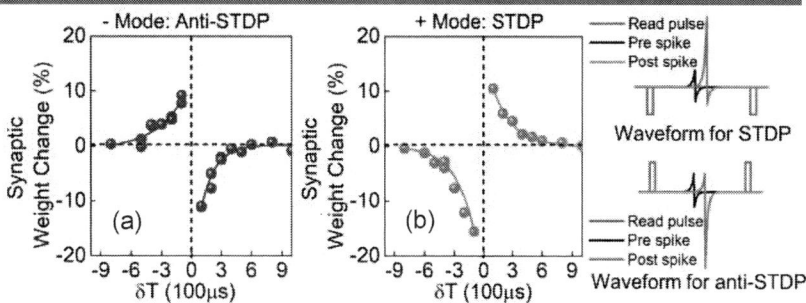

Fig. 14 Anti-STDP and STDP are achieved using (a) "– mode" and (b) "+ mode", respectively. The solid lines show the fitting curves of STDP and anti-STDP used for the later ReSuMe SNN simulation. The diagrams at right show the corresponding identical presynaptic spike and tailored postsynaptic spike waveforms.

ReSuMe SCNN Simulation

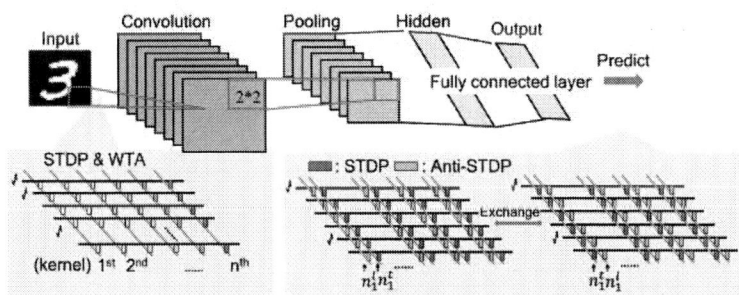

Fig. 15 Hybrid unsupervised and supervised convolutional spiking neural network. The convolutional layer is trained using STDP and unsupervised WTA. Supervised classifier is trained using ReSuMe and an "exchange" update scheme to prevent weight saturation.

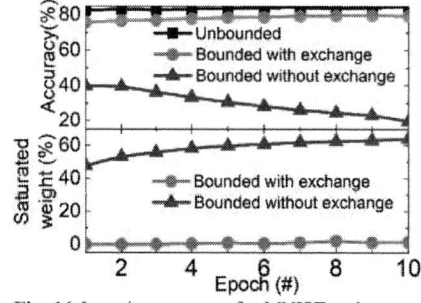

Fig. 16 Learning accuracy for MNIST and saturated weight can be improved using the "exchange" scheme.

Fig. 17 Effect of exchange rate. STDP and anti-STDP weights need to be swapped only after an entire epoch of 60000 MNIST images.

Fig. 18 Learning accuracy is improved by using additional convolutional feature-extraction layer and more kernels.

Table 1. Comparison of some recent SNN studies for MNIST classification. Our work uses complete STDP-based learning and a compact network to achieve comparable accuracy.

	Network	Update Scheme	# of Weight (k)	Accuracy
Diehl *et al.*,2015 [4]	784*6400	STDP unsupervised	5017.6	95 %
Pedretti, G., *et al.*,2017[5]	784*50000*10	STDP unsupervised + Non-STDP Supervised	39700	92 %
O'Connor *et al.*,2016 [11]	784*300*300*10	Non-STDP Supervised	328.2	96.4 %
This work	100*600+ 9600*10	STDP unsupervised + STDP supervised	156	95.72 %

978-1-7281-1988-5/18 $31.00 © 2018 IEEE

Stochastic Inference and Learning Enabled by Magnetic Tunnel Junctions

Abhronil Sengupta, Gopalakrishnan Srinivasan, Deboleena Roy and Kaushik Roy

All authors contributed equally to this work

School of ECE, Purdue University, West Lafayette, IN 47907, USA, Email: asengup@purdue.edu

Abstract—Neuromorphic computational paradigms that exploit the stochastic switching behavior of devices in the presence of thermal noise is bringing about a wave of change in the way we perceive brain-inspired computing. In this article, we present proposals of spintronics enabled neuromorphic computing systems that perform probabilistic inference and online learning. Such stochastic neuromimetic hardware has the potential of enabling a new generation of state-compressed, low-power computing platforms, which can be significantly more efficient and scalable than their deterministic counterparts.

I. Introduction

Autonomous Intelligence platforms are increasingly using "Artificial Neural Networks" as their underlying computing framework. However, deploying such networks in resource constrained edge devices in the current Internet of Things (IoT) era require significant rethinking of conventional von-Neumann model based computing scheme. To that effect, non-von Neumann architectures enabled by several post-CMOS technologies are being actively explored to overcome the bottlenecks of current CMOS-based implementations.

In addition to the hardware architecture and organization, the nature of computing - deterministic versus stochastic also plays an important role. Brain-inspired computing frameworks typically rely on computational units - the neurons and the synapses, that are deterministic in nature. However, there is increasing evidence that the brain performs probabilistic computation through its noisy neurons, synapses and dendrites. While such stochastic computing models are prevalent in the computational neuroscience field, they have received limited attention from the neuromorphic hardware community. This is probably due to the fact that the underlying CMOS hardware used to mimic neuronal and synaptic functionalities have been deterministic and would require costly random number generators to implement stochastic operations. In contrast, post-CMOS technologies like spintronic devices are characterized by inherent randomness during the switching process. Stochasticity in such single-bit hardware units can be harnessed to implement compressed synaptic memory storage and simplified neuronal hardware [1].

Work has started in earnest to demonstrate state-compressed stochastic spin-based neuromorphic hardware. Previous work has shown supervised learning frameworks to realize a network with stochastic spiking neurons on handwriting recognition problems [2] along with networks composed of binary synapses programmed stochastically in an online unsupervised

learning framework [3]. Both functionalities were enabled by the underlying stochastic magnetization dynamics of spin devices in the presence of thermal noise. Such stochastic spin-based neuromorphic hardware can be potentially an order of magnitude energy-efficient in contrast to corresponding deterministic CMOS implementations [2], [3]. In this article, we discuss probabilistic neuronal and synaptic computational paradigms that can inherently exploit behavior of the MTJ as a "stochastic bit" and can potentially pave the way for Stochastic "Binary Neural Networks". We demonstrate competitive accuracies of a deep stochastic "Binary Neural Network" on complex vision datasets for inference and demonstrate $8\times$ state compression in synaptic storage compared to iso-architecture networks with full precision for unsupervised online training.

II. Spintronic Device as a Stochastic Neuron and Synapse

Device Structure: Fig. 1(a) shows a typical Magnetic Tunnel Junction (MTJ) structure where a "free layer" (FL: ferromagnet whose magnetization can be switched) is separated from a "pinned layer" (PL: ferromagnet whose magnetization is fixed) by a tunneling oxide barrier (TB). We consider spin-Hall effect induced MTJ switching with in-plane magnetic anisotropy (due to the possibilities of achieving energy-efficient switching and decoupled "write" and "read" current paths), where current flowing through an underlying heavy-metal (HM) layer probabilistically switches the FL. The magnetization dynamics are described by the Landau-Lifshitz-Gilbert (LLG) equation in the presence of thermal noise, which induces randomness during the switching process. The switching probability is a function of the magnitude and duration of the input current flowing through the HM (see Fig. 1(b)). A typical set of device parameters have been outlined in Table I. Experimental measurements are shown in Fig. 3.

Stochastic Neuron: Fig. 2(a) depicts the neuron MTJ device interfaced with appropriate circuit peripherals to realize the stochastic spike generation functionality. The device is operated in a synchronous fashion in successive "write" and "read" phases as explained in Fig. 2(a). The variation of the output spiking rate of the MTJ neuron (i.e. the number of spikes produced over a large enough time window) with the input pulse current magnitude follows a non-linear variation (as depicted in Fig. 1(b)). Such a neuron can be abstracted in functionality as a stochastic firing rate model for a spiking

978-1-7281-1988-5/18 $31.00 © 2018 IEEE

neuron and has been used previously for unsupervised [4] and supervised [2] learning.

There has been increased research interest lately on super-paramagnetic switching with barrier heights scaled down to limits below $5k_BT$, where k_B is Boltzmann constant and T is absolute temperature [5]. As the magnet barrier height is scaled down, it undergoes faster telegraphic switching which can be also harnessed inherently (without the synchronous mode of MTJ operation discussed in the previous sub-section) to realize "stochastic bit" functionality in response to input bias current. The "read" circuit peripherals in this scenario entail a much higher constrained design space exploration since the "write" and "read" paths are activated simultaneously (the device behaves in a volatile manner at such low barrier heights). The "read" current has to be maintained at much lower levels in order not to bias the highly sensitive superpara-magnetic switching behavior. While fabrication of such highly scaled devices might be challenging, the highly sensitive stochasticity can also result in performance degradation for recognition problems in the presence of non-idealities [6].

Stochastic Synapse: Spike Timing Dependent Plasticity (STDP) [7] is commonly used to achieve unsupervised online learning in SNNs. STDP-based learning rules modulate the synaptic weight based on the correlation between the spike times of the input (pre) and output (post) neurons. Note that synapses are typically represented by multi-bit precision that can be implemented by domain-wall motion based devices in spintronics technology [8]. However, with aggressive scaling in device dimensions, multi-bit resolution may not be achiev-able. Hence, researchers in [3], [9], [10] have proposed binary synapses, where spike-based plasticity changes are embedded in the synaptic switching probability. In our previous work in [3], we presented a biologically plausible learning rule, where the synaptic switching probability varies exponentially with the timing correlation between a pair of pre- and post-spikes [11]. The device operation and associated circuit peripheral operation are described in Fig. 2(b).

III. ALL-SPIN BINARY NEURAL NETWORKS

Probabilistic Inference: We propose an "All-Spin" binary neural network implementation where both synapses and neu-rons are single-bit elements with the neurons being stochastic in nature. Taking advantage of recent efforts at training binary networks with backpropagation [12], we use the methodology outlined in Ref. [2] to train a network with binary weights and sigmoid neurons and subsequently convert it to stochastic sigmoid neurons (that can be potentially enabled by the MTJ devices) for inference. We achieve a testing accuracy of 75.53% on the CIFAR-10 dataset and 46.3% on the CIFAR-100 dataset [13] for a 5-layer deep convolutional network. The conversion error in the process is $< 1\%$. With such a network configuration (see Fig. 4 for basic functional unit), we can implement the core hardware fabric in a state-compressed "All-Spin" crossbar array fashion, as shown in Fig. 5, where a single-bit MTJ synaptic crossbar array is driven by stochastic MTJ spiking neurons. The accuracy results are depicted in

Fig. 6. Note that this is a functional simulation of a Stochastic SNN and does not include approximations and non-idealities arising from the underlying hardware implementation.

Online Learning: Fig. 7 represents a spin-based synaptic crossbar array for stochastic STDP learning. In order to scale up the prospects for stochastic online learning enabled by MTJ synaptic crossbar arrays, we explore the following hardware-compatible modifications to the basic version of the probabilistic STDP learning rule [9], [10] - (i) We propose a variant of STDP-based probabilistic learning rule, referred to as hybrid (HB)-STDP, incorporating Hebbian and anti-Hebbian learning mechanisms as shown in Fig. 8. (ii) We also explore learning in a ternary weight scenario where the synapses can be potentially represented by two binary MTJ-based crossbar arrays. To this effect, we propose quantized SNN, which uses two-bit synaptic weights, and the associated quantized STDP (Q-STDP) learning rule (shown in Fig. 9). Every synapse in the quantized SNN is capable of encoding four different logic states, namely, '11', '10', '01', and '00'. Our experiments on the MNIST dataset [14] indicate that quantized SNN achieves higher classification accuracy than its binary counterpart for smaller network sizes as shown in Fig. 10. Fig. 10 also indicates that binary SNN containing 6400 neurons offers comparable accuracy to that provided by 32-bit full precision fully-connected SNN containing 1600 neurons, leading to $8\times$ synaptic memory compression. Note that peripheral circuitry will be required to implement such learning rules in addition to the ones depicted in Fig. 7 for the original stochastic STDP implementation.

In order to further improve the accuracy, we explore bi-nary convolutional network architectures. We train the binary weight kernels interconnecting successive convolutional lay-ers in a greedy layer-wise unsupervised manner using HB-STDP. The fully-connected layers are initialized with real-valued synaptic weights and trained using gradient descent error backpropagation algorithm proposed in [15]. The trained weights are then binarized using BinaryConnect [16], a par-ticular variant of network binarization algorithm. We achieve a classification accuracy of 95.57% on the MNIST dataset with a 4 layer network architecture. Table II compares the performance of the proposed models with existing works. Note that the online learning results are based on deterministic neurons in order to have a fair comparison with prior work. However, stochastic learning in synaptic crossbar arrays can be potentially driven by stochastic spin neurons [17].

IV. CONCLUSION

Spintronic devices, characterized by enhanced reliability, endurance and lower operating voltage levels in comparison to other post-CMOS technologies, can provide a promising pathway to neuromimetic hardware. Stochastic SNNs charac-terized by online probabilistic learning in binary MTJ synaptic crossbar arrays driven by MTJ neurons performing stochastic inference can not only result in state-compressed, energy efficient hardware but can also open up new avenues to inherently implement more brain-like neuromorphic hardware.

978-1-7281-1988-5/18 $31.00 © 2018 IEEE

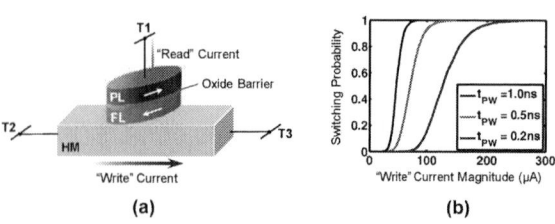

(a)   **(b)**

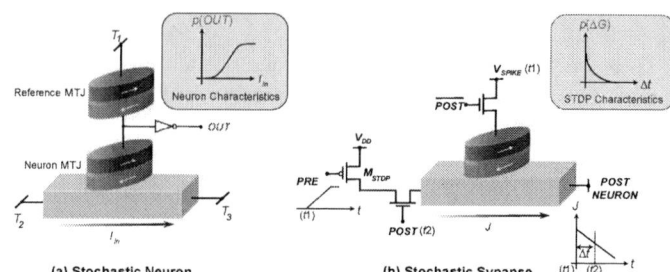

(a) Stochastic Neuron   **(b) Stochastic Synapse**

FIG. 1. (a) The core device structure being used as a stochastic neuron/synapse is a Magnetic Tunnel Junction where a "free layer" (FL) is separated from a "pinned layer" (PL) by an oxide barrier. FL magnetization gets manipulated by "write" current flowing through an underlying heavy-metal (HM) layer while PL magnetization is fixed in a particular direction. (b) Typical stochastic switching characteristics are shown for a CoFe-βW multilayer. The device modeling framework and parameters are shown below. The probability of switching has been plotted as a function of the input "write" current magnitude for different pulse width durations, t_{PW}.

FIG. 2. (a) The Neuron MTJ is interfaced with a Reference MTJ to realize a resistive divider circuit. In the initial "write" phase, charge current through the HM layer stochastically switches the magnet lying on top. In the subsequent "read" phase (after a relaxation period), the magnet state is read and the output is transmitted to the fan-out neurons. (b) The "write" and "read" paths of the stochastic synapse is decoupled by the POST control signal which is activated only when post-neuron fires. Pre-neuron spike signals, V_{SPIKE}, are transmitted from pre-neuron through the synaptic device to the post-neuron. Simultaneously, the gate voltage of the programming transistor, M_{STDP}, starts increasing and the programming path is activated only at post-neuron firing instant ($t2$). Hence, the programming current is a function of the delay, Δt, between the pre- and post-neuronal spike times.

LLG Equation for solving FL magnetization dynamics in the presence of spin-orbit torque & thermal noise

$$\frac{d\widehat{\mathbf{m}}}{dt} = -\gamma(\widehat{\mathbf{m}} \times \mathbf{H}_{eff}) + \alpha(\widehat{\mathbf{m}} \times \frac{d\widehat{\mathbf{m}}}{dt}) + \frac{1}{qN_s}(\widehat{\mathbf{m}} \times \mathbf{I}_s \times \widehat{\mathbf{m}})$$

$\widehat{\mathbf{m}}$: the unit vector of magnetization at each grid point,
γ: the gyromagnetic ratio for electron,
α: Gilbert's damping ratio.
$\mathbf{H}_{eff}$: the effective magnetic field including the shape anisotropy field for elliptic disks and the thermal field $\mathbf{H}_{thermal}$,
$I_s = \theta_{SH} \frac{W_{MTJ}}{t_{H,M}} I_Q$: Input spin current generated due to charge current I_Q (θ_{SH} is spin-Hall angle, dimensions W_{MTJ} and t_{HM} are MTJ width and HM thickness respectively),
$N_s = \frac{M_s V}{\mu_B}$: number of spins in free layer of volume V and,
$\mathbf{H}_{thermal} = \sqrt{\frac{\alpha}{1+\alpha^2} \frac{2K_B T}{\gamma \mu_0 M_s V \delta_t}} G_{0,1}$, where $G_{0,1}$ is a Gaussian distribution with zero mean and unit standard deviation, k_B is Boltzmann constant, T is the temperature and δ_t is the simulation time step.

TABLE I. DEVICE SIMULATION PARAMETERS[1]

Parameters	Value
Free layer area	$\frac{\pi}{4} \times 100 \times 40 nm^2$
Free layer thickness	$1.2nm$
Heavy-metal thickness, t_{HM}	$2nm$
Saturation Magnetization, M_S	$1000\ KA/m$ [18]
Spin-Hall Angle, θ_{SH}	0.3 [18]
Gilbert Damping Factor, α	0.0122 [18]
Energy Barrier, E_B	$20\ k_B T$
MgO Thickness, t_{MgO}	$1.2nm$
Resistivity of HM, ρ_{HM}	$200\ \mu\Omega.cm$
CMOS technology	45nm SOI CMOS
Supply voltage, V_{DD}	1V
Pulse width, t_{PW}	$0.2, 0.5, 1ns$
Temperature, T	$300K$

[1] The parameters are obtained from measurements in Ref. [18].

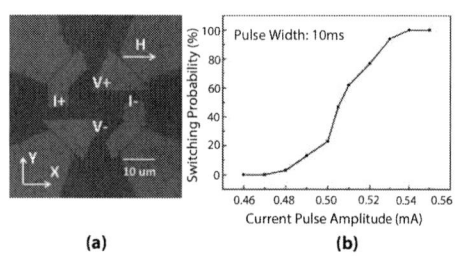

(a)   **(b)**

FIG. 3. (a) Hall-bar structure consisting of Ta (10nm) / CoFeB (1.3nm) / MgO (1.5nm) / Ta (5nm) (from bottom to top) material stack [19]. Input current flows between terminals $I+$ and $I-$ while the magnetization state is detected by change in the anomalous Hall-effect resistance measured between terminals $V+$ and $V-$. (b) Experimental measurements of the switching probability of the Hall-bar with variation in amplitude of the current pulse flowing through the HM layer. The pulse width is fixed at $10ms$ [19].

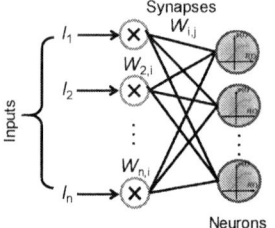

FIG. 4. The basic building block of an artificial neural network is shown where a set of neurons in a particular layer receive spikes from the previous layer through weighted synaptic junctions that encode the importance values of different inputs. We are considering probabilistic spiking neurons and synapses enabled by stochastic weight updates in this work. The core computing kernel for inference is a dot-product between inputs and synaptic weights followed by neuronal processing.

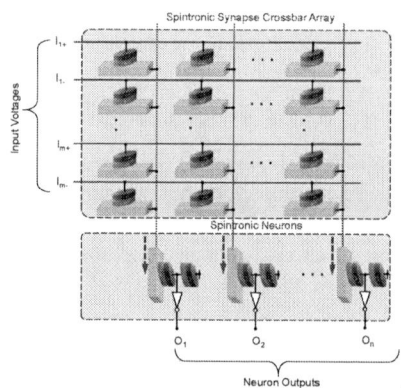

FIG. 5. The core computing kernel can be enabled by the "All-Spin" neuromorphic crossbar array architecture shown above. Input spikes are applied as voltages along the rows that modulate the current flowing through each synaptic device at the cross-point. The currents get added up along the column to realize the dot-product operation. The crossbar arrays are driven by magneto-metallic spin neurons that behave as stochastic neuronal elements.

978-1-7281-1988-5/18 $31.00 © 2018 IEEE

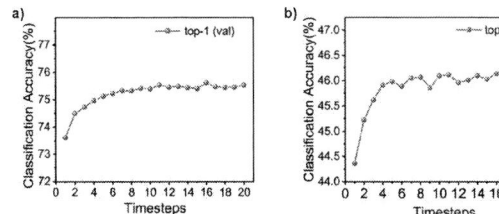

FIG 6. Stochastic SNN testing accuracy is depicted over time-steps for (a) CIFAR-10 and (b) CIFAR-100 datasets. The network topologies are 32x32x3-64C5-2S-128C5-2S-512C3-2S-512FC-10FC for CIFAR-10 dataset and 32x32x3-64C5-2S-128C5-2S-512C3-2S-1024FC-100FC for CIFAR-100 dataset. C, S and FC represent convolution, sub-sampling and fully connected layers respectively.

TABLE II. CLASSIFICATION ACCURACY OF DIFFERENT STDP TRAINED SNN MODELS

SNN Models	Synaptic Precision	Training Methodology	Accuracy
Fully-connected SNN [20]	32 bits	Exponential STDP	95%
Fully-connected SNN [21]	7 bits	Rectangular STDP	78%
Fully-connected SNN [3]	1 bit	Probabilistic STDP	70%
Fully-connected SNN (this work)	1 bit	HB-STDP	92.14%
Fully-connected SNN (this work)	2 bits	Q-STDP	92.36%
ConvNet (this work[2])	1 bit	HB-STDP + Backpropagation	95.57%

[2] We are using a 4 layer deep network architecture with a topology (16C5-36C5-2S-500FC-10FC configuration).

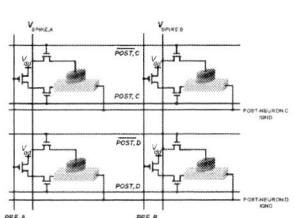

FIG 7. The "All-Spin" crossbar array with synaptic devices at each cross-point, shown in Fig. 5, do not account for on-chip learning. In order to implement probabilistic STDP, the synaptic device can be interfaced with peripheral transistors as shown above and arranged in array structures. The figure shows pre-neurons A and B connected to post-neurons C and D.

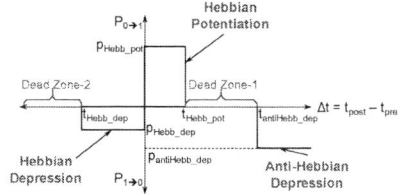

FIG 8. Illustration of the HB-STDP learning rule, where the spike timing information is encoded in the switching probability of the binary synapses.

$$P_{0 \to 1} = \begin{cases} p_{Hebb\_pot}, & \text{if } 0 < \Delta t \leq t_{Hebb\_pot} \\ 0, & \text{for all other } \Delta t \end{cases} \quad (1)$$

$$P_{1 \to 0} = \begin{cases} p_{antiHebb\_dep}, & \text{if } \Delta t \geq t_{antiHebb\_dep} \\ p_{Hebb\_dep}, & \text{if } t_{Hebb\_dep} \leq \Delta t \leq 0 \\ 0, & \text{for all other } \Delta t \end{cases} \quad (2)$$

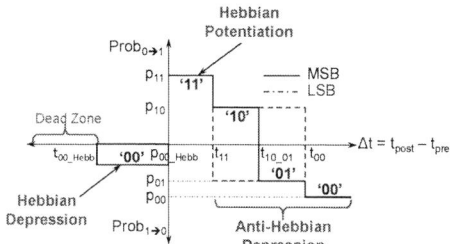

FIG 9. Illustration of the quantized-STDP (Q-STDP) learning rule for training an SNN composed of two-bit synaptic weights (w).

$$w = \begin{cases} \text{`11'} \text{ with } p_{11}, & \text{if } 0 < \Delta t \leq t_{11} \\ \text{`10'} \text{ with } p_{10}, & \text{if } t_{11} < \Delta t \leq t_{10\_01} \\ \text{`01'} \text{ with } p_{01}, & \text{if } t_{10\_01} < \Delta t \leq t_{00} \\ \text{`00'} \text{ with } p_{00}, & \text{if } \Delta t > t_{00} \\ \text{`00'} \text{ with } p_{00\_Hebb}, & \text{if } t_{00\_Hebb} \leq \Delta t \leq 0 \end{cases} \quad (3)$$

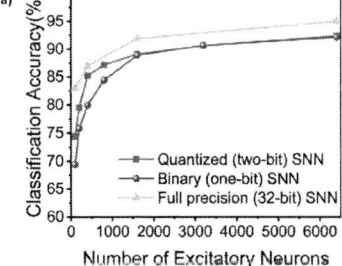

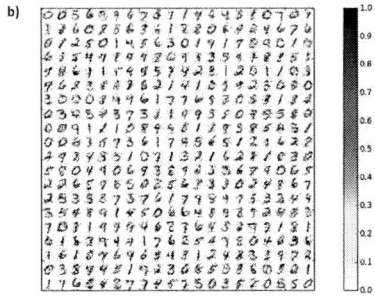

FIG 10. (a) Classification accuracy of quantized and binary fully-connected SNN against the number of excitatory neurons. The accuracy of 32-bit full precision fully-connected SNN is also plotted for comparison [20]. (b) Input representations encoded by the 784 binary synapses (re-arranged in 28×28 format) connecting the input to each of the 400 excitatory neurons (organized in 20×20 grid).

ACKNOWLEDGMENTS

The work was supported in part by, Center for Brain-inspired Computing Enabling Autonomous Intelligence (C-BRIC), a DARPA sponsored JUMP center, SRC, NSF, Intel Corporation and by the DoD Vannevar Bush Fellowship.

REFERENCES

[1] K. Roy *et al.*, *Journal of Applied Physics*, vol. 123, no. 21, p. 210901, 2018.
[2] A. Sengupta *et al.*, *IEEE Transactions on Electron Devices*, vol. 63, no. 7, pp. 2963–2970, 2016.
[3] G. Srinivasan *et al.*, *Scientific Reports*, vol. 6, p. 29545, 2016.
[4] A. Sengupta *et al.*, *Scientific Reports*, vol. 6, p. 30039, 2016.
[5] K. Y. Camsari *et al.*, *Physical Review X*, vol. 7, no. 3, p. 031014, 2017.
[6] C. M. Liyanagedera *et al.*, *Physical Review Applied*, vol. 8, no. 6, p. 064017, 2017.
[7] G.-q. Bi *et al.*, *Annual Review of Neuroscience*, vol. 24, no. 1, pp. 139–166, 2001.
[8] A. Sengupta *et al.*, *Physical Review Applied*, vol. 6, no. 6, p. 064003, 2016.
[9] M. Suri *et al.*, *IEEE Transactions on Electron Devices*, vol. 60, no. 7, pp. 2402–2409, 2013.
[10] A. F. Vincent *et al.*, *IEEE Transactions on Biomedical Circuits and Systems*, vol. 9, no. 2, pp. 166–174, 2015.
[11] D. Hebb, 1949.
[12] M. Rastegari *et al.*, in *European Conference on Computer Vision*. Springer, 2016, pp. 525–542.
[13] A. Krizhevsky *et al.*, Citeseer, Tech. Rep., 2009.
[14] Y. LeCun *et al.*, *Proceedings of the IEEE*, vol. 86, no. 11, pp. 2278–2324, 1998.
[15] C. Lee *et al.*, *IEEE Transactions on Cognitive and Developmental Systems*, 2018.
[16] M. Courbariaux *et al.*, in *Advances in Neural Information Processing Systems*, 2015, pp. 3123–3131.
[17] G. Srinivasan *et al.*, in *2017 Design, Automation & Test in Europe Conference & Exhibition (DATE)*. IEEE, 2017, pp. 530–535.
[18] C.-F. Pai *et al.*, *Applied Physics Letters*, vol. 101, no. 12, p. 122404, 2012.
[19] Y. Shim *et al.*, *Scientific Reports*, vol. 7, no. 1, p. 14101, 2017.
[20] P. U. Diehl *et al.*, *Frontiers in Computational Neuroscience*, vol. 9, p. 99, 2015.
[21] D. Querlioz *et al.*, *Proceedings of the IEEE*, vol. 103, no. 8, pp. 1398–1416, 2015.

In-Memory Computing Primitive for Sensor Data Fusion in 28 nm HKMG FeFET Technology

K. Ni[1], B. Grisafe[1], W. Chakraborty[1], A. K. Saha[2], S. Dutta[1], M. Jerry[1], J. A. Smith[1], S. Gupta[2], and S. Datta[1]

[1]University of Notre Dame, Notre Dame, IN, USA;

[2]Purdue University, West Lafayette, IN, USA; email: kni@nd.edu

Abstract—In this work, we exploit the spatio-temporal switching dynamics of ferroelectric polarization to realize an energy-efficient, and massively-parallel in-memory computational primitive for at-node sensor data fusion and analytics based on an industrial 28nm HKMG FeFET technology [1]. We demonstrate:(i) the spatio-temporal dynamics of polarization switching in HfO$_2$-based ferroelectrics under the stimuli of sub-coercive voltage pulses using experiments and phase-field modeling; (ii) an inherent rectifying conductance accumulation characteristic in FeFET with a large dynamic range of $G_{max}/G_{min} > 100$ in the case of 3.0V, 50ns gate pulses; (iii) transition to more abrupt accumulation characteristics due to single/few domain polarization switching in scaled FeFET (34nm L$_G$); and (iv) successful detection of physiological anomalies from real-world multi-modal sensor data streams.

I. INTRODUCTION

In the era of Internet of Things (IoT), the ubiquity of sensors and the continuous flow of data streams require sensor data fusion and processing in real time to discover patterns for predictive analytics. Statistical correlation detection is one approach to extract the global information distributed among the sensor signals. A key application of statistical correlation is the detection of physiological anomalies for distributed health monitoring in the form of wearable devices (Fig.1). Anomalies can be detected based on the correlation among different physiological signals (e.g., blood pressure, heart rate, blood oxygen saturation, etc.). The correlation between the sensors change in accordance with a person's health condition (highly correlated when anomaly occurs). In order to process large streams of data in real time, an energy-efficient hardware accelerator suitable for edge device is necessary.

In the conventional Von-Neumann approach, sensor data is loaded from the memory buffer into the computation unit which computes the distance norm between the data stream (correlation as one type of distance norm). Clustering algorithms (such as K-means clustering) are then applied to classify the data based on the calculated distance. This approach involves substantial data movement between the memory and the arithmetic logic unit, limiting the throughput and energy efficiency (Fig.2(a)), often called the von Neumann bottleneck. On the other hand, the in-memory computing approach exploits the physical dynamics of emerging memories to perform the computation within the memory with co-located computation and storage (Fig.2(b)) [2]. In the case of correlation detection with FeFET, each signal is associated with an exclusive memory cell. The signal data is fed to an encoder, which translates the correlation among signals to the rate of the pulse

applied to the FeFET cells. As a result, only cells associated with highly correlated signals receive frequent pulse update, whose ferroelectric polarization accumulates and eventually switches the total polarization along with the FeFET conductance. Thus, a conductance read can differentiate between the correlated and uncorrelated signals, thereby detecting the anomaly. In this work, we apply statistical correlation detection to detect physiological anomaly from physiological signal streams based on an industrial 28nm HKMG FeFET technology, using the spatial-temporal polarization switching dynamics of ferroelectrics (FE).

II. RESULTS AND DISCUSSION

A. Spatio-Temporal Dynamics of Ferroelectric Switching in MFM Capacitor

The channel conductance accumulation in FeFET results from the polarization accumulation properties of the ferroelectric when exposed to identical consecutive pulses with amplitudes below the coercive voltage threshold. It involves two physical processes, domain nucleation and domain growth (Fig.3). When sub-coercive field pulses are applied to the FE, a critical number of pulses are required for the reverse domain nuclei to be stabilized. The polarization remains unchanged until the domain reaches the critical nuclei size, beyond which the domain growth happens. When integrated into the gate stack of the FeFET, the polarization accumulation in FE manifests as a shift in the device threshold voltage (V_{TH}) and conductance accumulation characteristics.

We first measure the polarization accumulation in an MFM capacitor. Fig.4 shows the measured Q_{FE}-V_{FE} hysteresis loops, which exhibit a saturation loop and non-saturated minor loops. Fig.5(a-c) show the polarization accumulation with the number of consecutive pulses for varying pulse amplitude, pulse width, and inter-pulse delay. A modified pulsed measurement (PUND) is used to measure the switched polarization by the accumulation pulses. The results show that the reduction of either pulse amplitude or pulse width increases the required pulse number for domain nucleation [3]. Moreover, the reduction in the amount of switched polarization with longer inter-pulse delay suggests polarization relaxation between the pulses. This is because the interaction from the unswitched domains switches back the flipped domain, as discussed later.

To understand the voltage and time dependent domain nucleation and growth, we solve the time-dependent Landau-Ginzburg-Devonshire (LGD) equation in a phase-field framework (Fig.6) [4]. We consider the domain-domain interactions across the cross-section (x-y plane) via the Laplacian of polarization (P), which can be interpreted as an interaction field (E_{INTR}). In addition, we account for the random fluctuations in P switching due to thermal processes using an

978-1-7281-1988-5/18 $31.00 © 2018 IEEE

effective thermal field $E_{ACCU,}$ which increases with pulse number. The total effective field (E_{EFF}) in the ferroelectric is a sum of the applied external field (E_{APP}), and the internal E_{INTR} and E_{ACCU}. The latter two fields are self-consistently updated in our simulations at each time step as polarization switching causes field redistribution (Fig.6). The interactions between these different terms leads to domain nucleation and growth.

The phase field model is calibrated with the measured Q_{FE}-V_{FE} hysteresis loop showing an acceptable match (Fig.7). Simulated polarization accumulation in a 200nmx200nm MFM capacitor (10000 domains) is shown in Fig.8. The simulation reproduces the measured accumulation characteristics, such as the delayed onset of polarization increase with the reduction in pulse amplitude and pulse width, and the increase in inter-pulse delay. Fig.9 shows the simulated spatial-temporal evolution of polarization distribution in the FE for different pulse numbers. It exhibits two stages for polarization accumulation. In the initial stage, when the reverse domain volume is smaller than the critical size (10nmx10nm in simulation), increasing number of pulses yields larger thermal fluctuations due to dissipative processes associated with polarization switching. This raises E_{ACCU} and E_{EFF}, increasing the probability of reverse domain nucleation and stabilization. The second stage involves domain growth from the nucleation site, which propagates to the whole grain until a grain boundary is encountered. This process is mainly driven by E_{APP} and E_{INTR}. The polarization relaxation is attributed to E_{INTR} from the surrounding unswitched domains as polarizations from different domains tend to align with each other. For scaled MFM cap (20nmx20nm, 100 domains), the polarization accumulation exhibits abrupt switching due to small number of domains in switching process (Fig.10).

B. *Conductance Accumulation Characteristics of FeFET*

Conductance accumulation and threshold behavior is characterized on an 28 nm HKMG FeFET platform [1]. The device cross-sectional TEM images show a poly-Si/TiN/Si:HfO$_2$/SiON/p-Si gate stack (Fig.11(a)). Conductance response to input gate pulses are characterized for FeFETs with W/L=450/450nm (Fig.11(b-f)) and W/L=72/34nm (Fig.11(g-k)). Figs.11(b-c, g-h) show the erased state I_D-V_G charactcristics with progressively increasing erase voltage amplitudes. Polarization switching decreases the device V_{TH}, increasing the channel conductance. Compared with the large device, which shows continuous V_{TH} shift, the scaled FeFET shows an abrupt V_{TH} decrease. This is consistent with the phase field modeling and related with single/few domain switching in scaled FeFETs [3]. The number of domains in scaled device is small enough such that single/few domain switching is evident, whereas it is averaged out by the large number of domains for large device.

The conductance accumulation characteristics for different pulse amplitudes, pulse widths and pulse delays are shown in Fig.11(d-f, i-k). Similar to the erase amplitude dependence, the increase of the number of consecutive pulses cause a continuous shift in conductance for the large FeFET, while the scaled device displays an abrupt conductance increase. The conductance accumulation characteristics reflect the domain nuclei stabilization and growth processes shown in Fig.9. For scaled device, the threshold number of pulses required for domain nucleation is exponentially dependent on the applied pulse amplitude and pulse width (Fig.11(i,j)), consistent with the nucleation physics [5]. Unlike MFM cap, the inter-pulse delay has a negligible effect on the accumulation characteristic of FeFET. This is likely due to the depolarization field in FeFET, which causes fast polarization relaxation at timescale less than 10μs. Therefore, the increase in delay from 10μs to 1ms does not have any effect. The pulse width can be decreased to as low as 50 ns for a pulse amplitude of 3.0V, indicating high speed operation of FeFET.

C. *Statistical Correlation Detection*

The polarization accumulation properties of the FeFET are applied to detect anomalies in physiological signals (for scaled geometry, multiple FeFETs are combined as a single device to have continuous accumulation characteristics). The gate pulse encoder translates the correlation among signals to the pulse rate(Fig.12). A stochastic translator (STR) is used to translate real value signals into random binary bit streams so that the multiplication operation can be greatly simplified [6]. At each time instance, the collected momentum is calculated (sum of all the signals) and a MUX is used to achieve multiplication operation necessary for correlation calculation.

The physiological signals of patient 221 from the MIMIC database show three anomalies (Fig.13(a)) [7]. By translating the real signals to gate pulses to FeFET, a conductance read can differentiate the anomalies from normal signals reliably as the anomalies are highly correlated among different signals. A majority voting on the number of FeFET conductance over the threshold identifies the anomaly faithfully. Fig.13(b) shows that detection based on 3 signals cause information loss while in the case of 7 signals the detection is successful as more global information is extracted. An array level implementation with the FeFET cells is shown in Fig. 14. Compared with a CMOS ASIC based counter [8] or PCM cell [2], FeFET exhibits high speed, large G_{max}/G_{min} ratio and low write energy (Fig.15).

CONCLUSIONS

In summary, we demonstrate an in-memory computational primitive for statistical correlation detection based on an industrial 28nm HKMG FeFET technology. The spatio-temporal polarization switching dynamics, involving the domain nucleation and growth, are responsible for polarization accumulation, which causes continuous conductance accumulation in large FeFET. Extremely scaled FeFET exhibits abrupt switching but can be compensated by multiple devices grouping. Fast operation with 3.0V, 50 ns is also demonstrated. The conductance accumulation characteristic is successfully applied to detect real-world physiological signal anomalies. These results make FeFET based correlation detection an ultra-dense, highly energy-efficient, and massively parallel system for real-time signal processing for sensor analytics.

ACKNOWLEDGEMENT

We would like to thank M. Trentzsch, S. Dunkel, S. Beyer, and W. Taylor at Globalfoundries Dresden, Germany for providing 28nm HKMG FeFET test devices. This work was supported in part by the Semiconductor Research Corporation (SRC) and DARPA.

REFERENCES

[1] M. Trentzsch et al., IEDM 2016 [2] A. Sebastian, Nature Comm. 2017 [3] H. Mulaosmanovic et al., Appl. Mater. Interfaces 2017 [4] Y. H. Shin, et al., Nature 2007 [5] H. Mulaosmanovic et al., Appl. Mater. Interfaces 2018 [6] B. R. Gaines et al., 1969 [7] G. B. Moody et al., Computers in Cardiology 1996 [8] S. Mathew, ISSCC, 2004

In-Memory Computing for Sensor Analytics

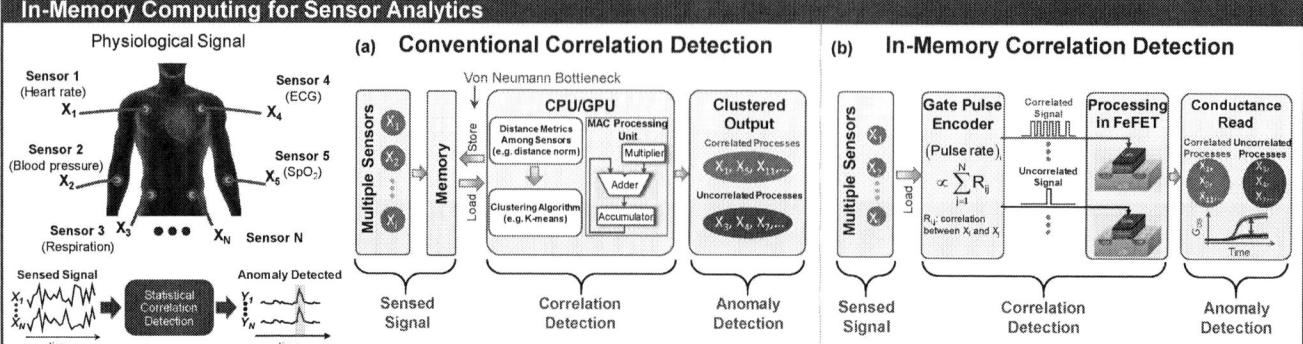

Fig. 1: Physiological anomaly detection based on the change of correlation among signals.

Fig. 2: (a) Conventional approach for correlation detection involves data transfer between the memory and computing unit, limiting the throughput and energy efficiency. (b) In-memory computing approach utilizing conductance accumulation of FeFET maps correlation among signals to FeFET gate pulse rate.

Physics of Polarization Switching Dynamics (MFM): Experiment

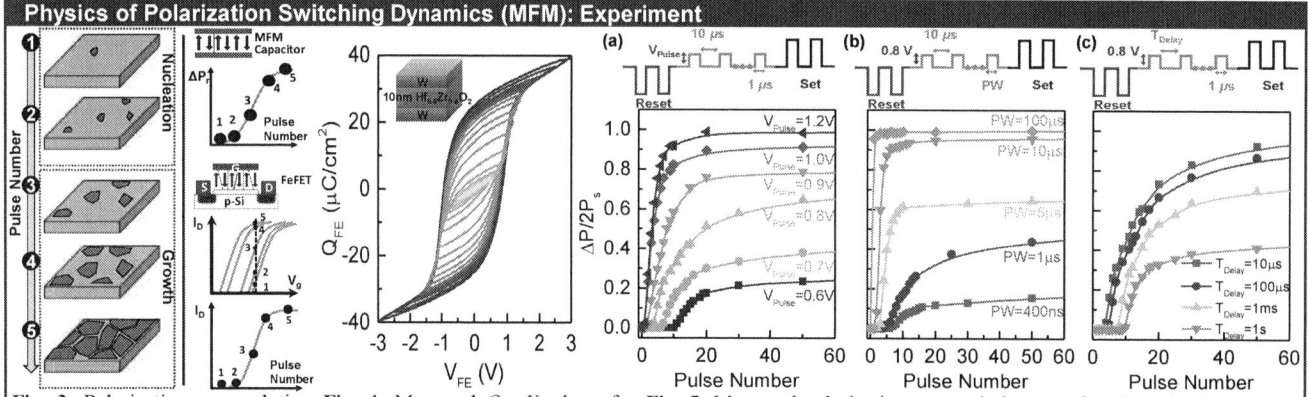

Fig. 3: Polarization accumulation due to domain nuclei stabilization and growth causes conductance accumulation in FeFET.

Fig. 4: Measured Q_{FE}-V_{FE} loop for W/HZO/W MFM capacitor showing non-saturated minor loops due to partial polarization switching

Fig. 5: Measured polarization accumulation as a function of pulse number on W/HZO/W MFM capacitor for different (a) pulse amplitude, (b) pulse width and (c) inter-pulse delay. Decrease of amplitude and pulse width delays the onset of polarization accumulation.

Phase Field Modeling of Polarization Switching Dynamics (MFM)

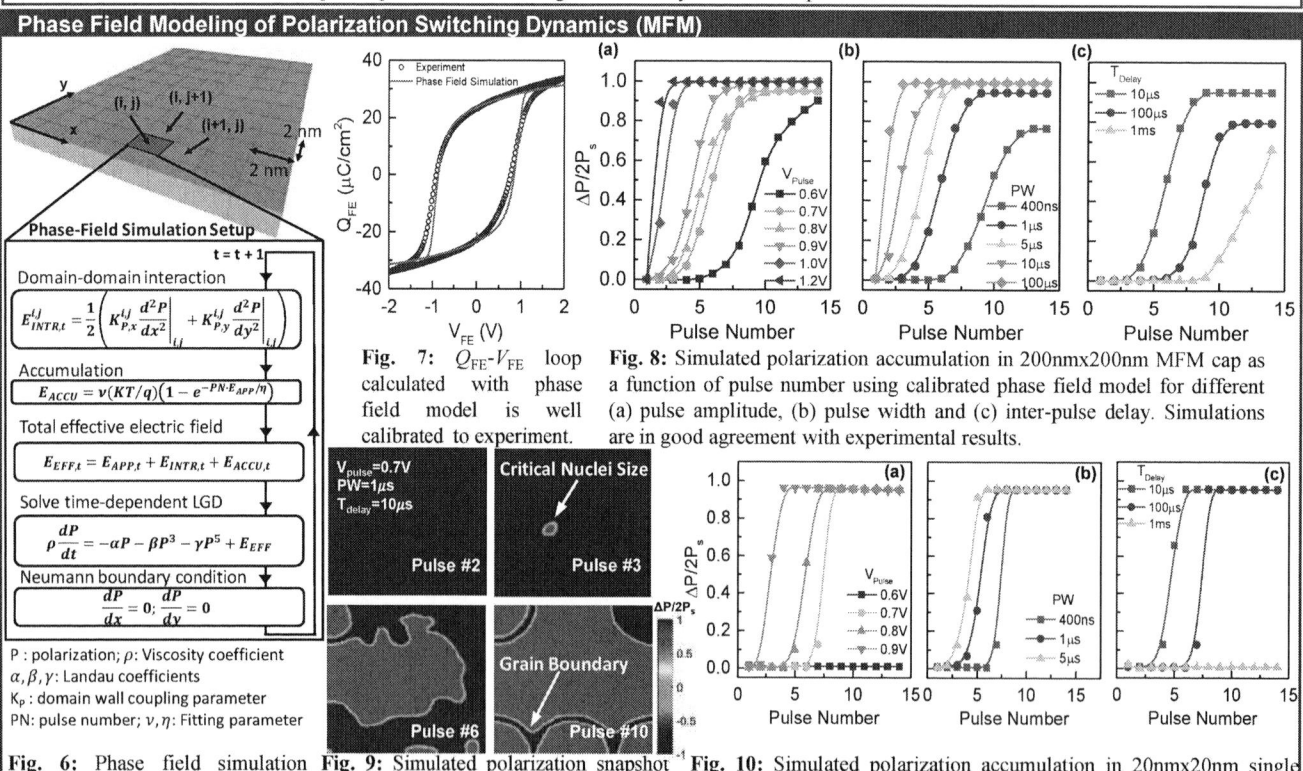

Fig. 6: Phase field simulation framework.

Fig. 7: Q_{FE}-V_{FE} loop calculated with phase field model is well calibrated to experiment.

Fig. 8: Simulated polarization accumulation in 200nmx200nm MFM cap as a function of pulse number using calibrated phase field model for different (a) pulse amplitude, (b) pulse width and (c) inter-pulse delay. Simulations are in good agreement with experimental results.

Fig. 9: Simulated polarization snapshot at different pulse numbers showing the domain nuclei stabilization and growth.

Fig. 10: Simulated polarization accumulation in 20nmx20nm single grain MFM cap as a function of pulse number for different (a) pulse amplitude, (b) pulse width and (c) inter-pulse delay.

978-1-7281-1988-5/18 $31.00 © 2018 IEEE

FeFET Conductance Accumulation Characteristics

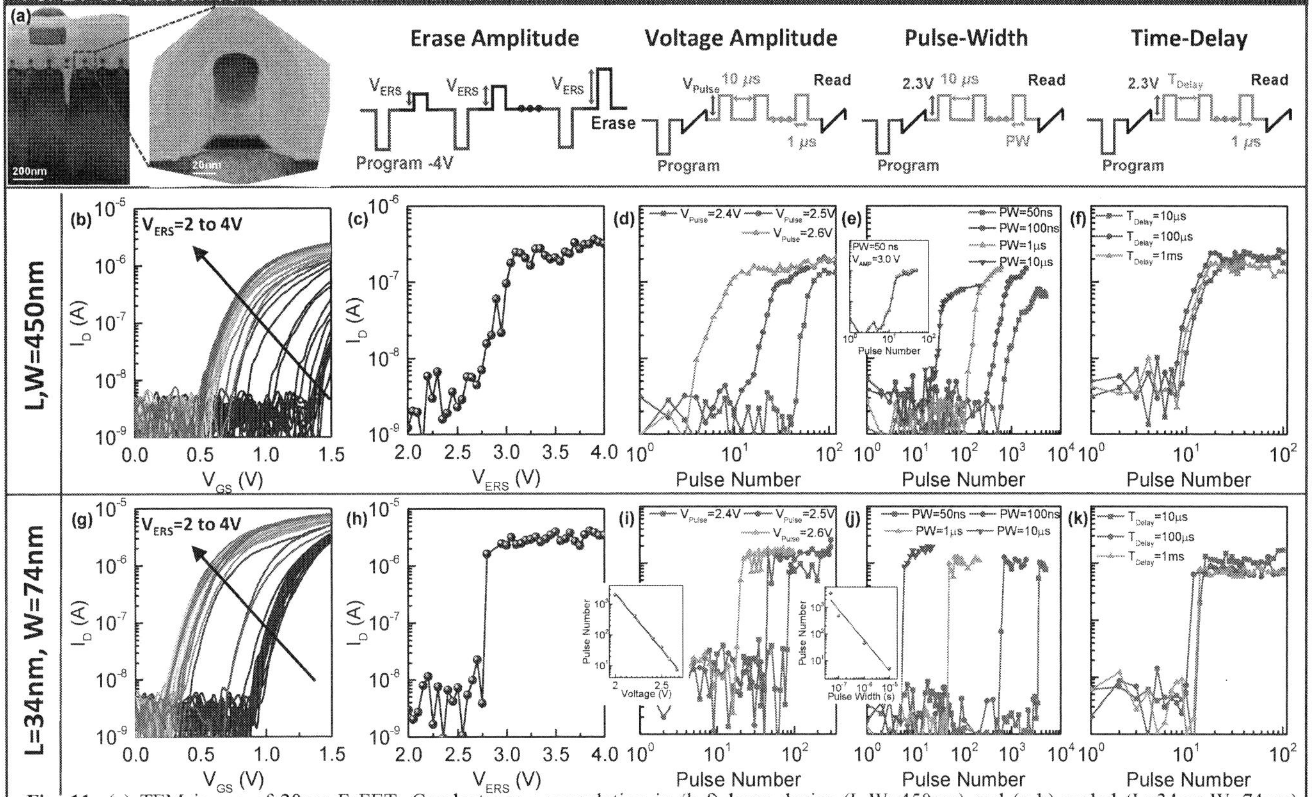

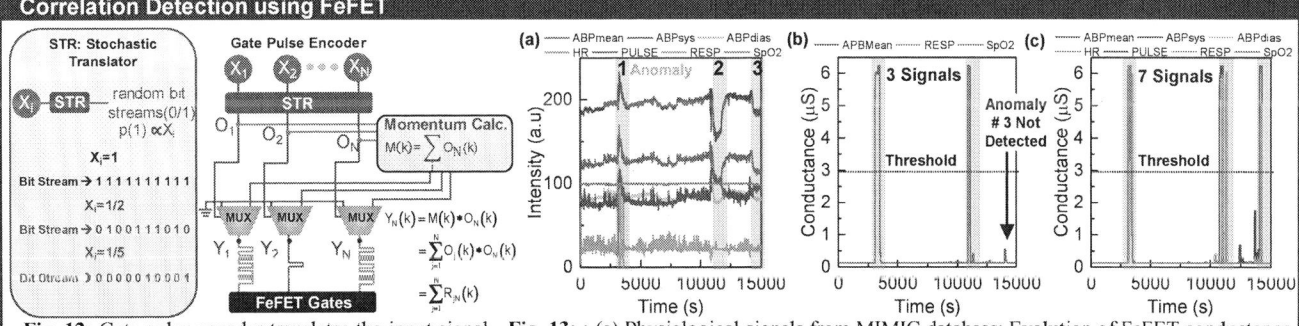

Fig. 11: (a) TEM image of 28nm FeFET; Conductance accumulation in (b-f) large device (L,W=450nm) and (g-k) scaled (L=34nm,W=74nm) device. Continuous V_{TH} shift and abrupt V_{TH} shift are observed for large and small device, respectively. The abrupt change is a signature of single-domain switching. 50ns operation is demonstrated.

Correlation Detection using FeFET

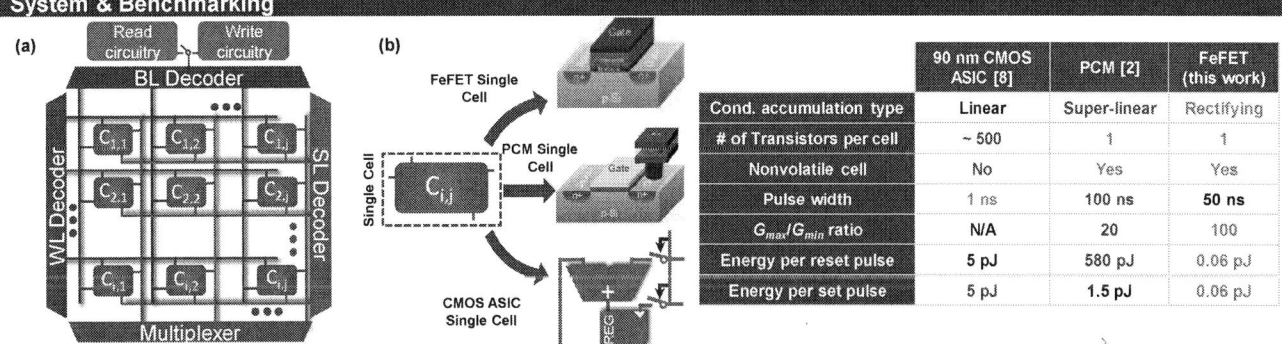

Fig. 12: Gate pulse encoder translates the input signal to random bit streams so that the multiply operation can be simplified to MUX gate.

Fig. 13: : (a) Physiological signals from MIMIC database; Evolution of FeFET conductance corresponding to each signal for (b) 3 signals input and (c) 7 signals input. Detection based on 3 signals lose information and miss one anomaly, which is avoided in 7 signals detection.

System & Benchmarking

Cond. accumulation type	90 nm CMOS ASIC [8]	PCM [2]	FeFET (this work)
Cond. accumulation type	Linear	Super-linear	Rectifying
# of Transistors per cell	~ 500	1	1
Nonvolatile cell	No	Yes	Yes
Pulse width	1 ns	100 ns	**50 ns**
G_{max}/G_{min} ratio	N/A	20	100
Energy per reset pulse	5 pJ	580 pJ	0.06 pJ
Energy per set pulse	5 pJ	1.5 pJ	0.06 pJ

Fig. 14: (a) Array level implementation of the correlation detection. The accumulation pulse update is performed row-wise. (b) The individual cell is either FeFET, phase-change memory (PCM) or CMOS counter.

Fig. 15: Unit cell for statistical correlation detection benchmarking. 28 nm FeFET exhibits good speed, G_{max}/G_{min} ratio and write energy efficiency compared with CMOS ASIC or PCM.

978-1-7281-1988-5/18 $31.00 © 2018 IEEE

Experimentally Validated, Predictive Monte Carlo Modeling of Ferroelectric Dynamics and Variability

C. Alessandri[*], P. Pandey[*], and A. C. Seabaugh

Department of Electrical Engineering, University of Notre Dame, Notre Dame, IN 46556, USA

Emails: calessan@nd.edu, ppandey@nd.edu. [*]Equal Contribution

Abstract—A physics-based, circuit-compatible Monte Carlo simulation framework, capable of predicting the dynamic response of a ferroelectric (FE) under any arbitrary input waveform, is developed by extending the nucleation-limited switching model. Measured polarization reversal data from fabricated FE W/Hf$_{0.5}$Zr$_{0.5}$O$_2$ (HZO)/W capacitors is used to extract the statistical distribution of FE grains, which show negligible variation with film thickness. After parameter extraction, the model is able to predict the dynamics of HZO and bilayer HZO/HfO$_2$ (FE-DE) thin films without further calibration. Unlike prior models, the proposed model is able to predict device-to-device variability, and quantify the resultant reduction in the memory window for highly scaled devices, revealing a significant reduction for FE capacitors having < 20 grains (~40×40 nm^2). The memory window is further reduced in FE-DE stacks for the same programming voltage and pulse duration due to the dielectric depolarizing field.

I. INTRODUCTION

Interest in ferroelectric (FE) devices has increased significantly after the discovery of ferroelectricity in the CMOS-compatible HfO$_2$ material system [1], having a variety of applications including memory, steep slope transistors, and neuromorphic computing [2]. Describing the switching behavior of thin-film polycrystalline FEs is complicated by the fact that they are composed of a multitude of grains having different switching thresholds, the distribution of which is highly dependent on the growth conditions. Prior models based on the static Preisach model [3,4] approximate the multi-domain polarization-voltage (P-V) hysteresis loops by a hyperbolic tangent function, while the dynamic component is included by using equivalent circuits having either fixed or bias dependent time constants. Due to these approximations, such models do not keep track of the distributions of switching thresholds, and require interpolation and scaling of parameters to replicate the history dependence of partially polarized FEs. While such models generate fits that match experimental data, they are unable to accurately predict the FE response under arbitrary input waveforms and the effects of grain variations in small-area devices.

Here, we implement a Monte Carlo based simulation framework for describing the dynamic, history dependent switching behavior of a multi-domain FE. After a parameter extraction procedure, the model is able to accurately predict the dynamical behavior of FE HZO under various applied waveforms, both with and without the presence of an additional dielectric (DE) HfO$_2$ layer.

II. MODEL DESCRIPTION

The model presented in this approach is based on the nucleation limited switching (NLS) polarization reversal model [5,6], which considers the FE as being composed of an ensemble of grains experiencing variations in the local fields when a uniform external field is applied, due to grain boundaries, surface roughness, defects, trapped charges, etc. This is mathematically equivalent to the grains having different activation fields, which is computationally convenient. The cumulative probability of switching for any grain is governed by a stretched exponential law with a field dependent time constant (Fig. 4). The net polarization of the entire ensemble at any time is obtained by calculating the expectation value of the cumulative probability over the distribution function of activation fields. However, the NLS model is limited as it is a polarization reversal model, and can only describe the field dependent switching dynamics of an FE from one fully polarized state to another, under the application of a constant field.

In this Monte Carlo based simulation framework, a set of FE grains is instantiated with fixed activation fields drawn from a distribution obtained using measured polarization reversal characteristics. These grains can have one of two possible orientations, and the probability of transition from one orientation to the other is governed by a Weibull process [7], thereby extending the NLS model. The applied voltage waveform is then divided into discrete time intervals, and the conditional probability of transition for each unswitched grain at any time is calculated based on its accumulated time constants (Fig. 4). These time constants are in turn dependent on the activation field of each individual grain, and the applied field (Fig. 4). Therefore, the model is able to fully capture the hysteretic time and field dependent behavior of an FE.

III. PARAMETER EXTRACTION AND TRENDS

FE W/HZO/W capacitors were fabricated with HZO thickness being 8.3, 10.6, and 15 nm. The growth rate and thicknesses were characterized using both *in situ* ellipsometry and transmission electron microscopy (TEM), while a Hf:Zr ratio of 1:1 was verified with energy dispersive X-ray linescans. To validate the model for device and circuit simulations, ferroelectric-dielectric (FE-DE) W/HZO/HfO$_2$/W bilayer capacitors were also fabricated on the 10.6 nm HZO film with 6 nm and 8 nm HfO$_2$ (Fig. 1).

Polarization reversal measurements were carried out by applying pulses with varying amplitudes, and widths ranging from 300 ns to 10 ms (Fig. 5(a)), to the FE capacitors. The switching parameters were then extracted by fitting the NLS model (Fig. 5(c-d)) to the polarization reversal measurements, and these parameters did not significantly vary if the distribution of activation fields was also left as a fitting parameter. The accuracy of the fits, and the weak dependence of the extracted parameters on the method of fitting underscores the physicality of the model.

As previously shown [8], the remnant polarization (P_R) decreases with increasing film thickness without significant change in the extracted activation field distributions. The extracted minimum switching time τ_0 (i.e. the time constant when an infinite field is applied) is of the order of ~100 ns, imposing a hard limit on the switching speed of these FEs. Furthermore, decreasing the FE thickness results in a higher P_R and a higher field for the same applied voltage, but τ_0 increases (Fig. 5(d)), leading to a speed trade-off.

IV. MODEL VALIDATION

The response of the FE to an applied external waveform was simulated using the extracted parameters, and the model closely replicates the measured behavior (Fig. 6(a-b)). Differences between the measured and simulated characteristics occur in part due to the assumption of a constant FE capacitance, whereas the measured capacitance exhibits the well-known butterfly shape (Fig. 3). The capability of the proposed model to accurately predict the behavior of the FE as it enters and exits the minor loops, as well as the drifting of the minor loops with field cycling, further highlights its advantage over the dynamic Preisach models [3,4], which scale the major loops to generate the minor loops.

Furthermore, due to the domain switching being a Weibull process with $n > 1$, the FE switching is faster upon application of a single long pulse, as compared to a pulse train having the same amplitude and the same cumulative duration (Fig. 6(c)). The model is able to accurately predict this difference since the switching probability of a grain is dependent on both its present and previous time constants, which Preisach models are unable to fully capture.

In the fabricated FE-DE bilayers, TEM showed that HfO_2 crystallized on the grain structure of the HZO film (Fig. 2), indicating a strain transfer from the latter to the former, leading to a thicker region of ferroelectricity with a lower percentage of FE grains having a thin DE layer in series. With the model parameters adjusted to account for these physical differences (increased FE thickness and reduced P_R), the model again shows close agreement with measurements (Fig. 6(d-e)). The fact that the sample with a thinner HfO_2 layer experienced a lower reduction in P_R and required a smaller increase of the FE thickness for accurate fits validates these changes. The capability of the model to predict these unexpected results further highlights its utility.

V. MODEL PREDICTIONS

A. Low Power Memory and Negative Capacitance FETs

FE-DE stacks form integral components of many proposed FE devices, in both memory and logic [2]. The widely used Landau-Khalatnikov (L-K) based models predict a decrease in the switching voltage upon placing a DE in series with an FE, albeit with a reduced P_R (Fig. 7(a)), suggesting the possibility of low power memory devices and hysteresis free Negative Capacitance FETs. However, measurements performed on fabricated FE-DE bilayers do not show this trend, and the measured behavior is explained well using our model (Fig. 7(a)). The depolarizing field of the DE aids switching only when the magnitude of FE polarization is decreasing (i.e. from $\pm P_R$ to 0), but opposes the switching when its magnitude is increasing (i.e. from 0 to $\pm P_R$). Thus, irrespective of the pulse duration, the FE-DE starts switching earlier than the FE, but takes a longer time to fully switch, and the maximum switched charge is also reduced (Fig. 7(b)). These results also support the hypothesis that the enhancement in hysteresis-free FE FETs is due to the higher dielectric constant of HZO relative to HfO_2, rather than FE switching.

B. Variability

The Monte Carlo modeling approach allows the investigation of the effects of device-to-device variability due to the grains having a distribution of activation fields. This variability in FE switching characteristics increases with a reduction both in lateral dimensions and programming voltage (Fig. 8(a)), so much so that a device consisting of 20 grains (~40 nm x 40 nm) exhibits a 50% reduction in memory window when programmed with 1.5 V pulses, and no memory window if programmed by 1.25 V pulses of 10 μs duration. A series DE reduces this variability in absolute terms, but there is no increase in the memory window due to the reduced maximum polarization, which is quite significant even for a very thin DE (Fig. 8(b)).

VI. CONCLUSIONS

An experimentally validated, circuit-compatible, physics based Monte Carlo model of FE behavior is used to make predictions regarding the variability, scaling, and memory application space of FE HZO and FE-DE bilayers by accurately simulating their response to arbitrary waveforms. The model highlights the need to develop processes yielding smaller, faster switching FE grains having activation field distributions with small standard deviations, and provides the tools to quantify the necessary improvements.

ACKNOWLEDGMENT

We gratefully acknowledge the contributions of T. Orlova and S. Rouvimov in the Notre Dame Integrated Imaging Facility for the TEM sample preparation and analysis.

REFERENCES

[1] T. S. Böscke et al., IEDM, 547, 2011. [2] A. Aziz et al., DATE 1289, 2018. [3] K. Ni et al., VLSI Tech. Symp., pp. 131-132, 2018. [4] B. Obradovic et al., VLSI Tech. Symp., pp. 51-52, 2018. [5] A. K. Tagantsev et al., Phys. Rev. B, vol. 66, 214109, 2002. [6] S. Zhukov et al., J. Appl. Phys., vol. 108, 014106, 2010. [7] W. Lee et al., Adv. Funct. Mater., 1801162, 2018. [8] M. H. Park et al., Appl. Phys. Lett., vol. 102, 242905, 2013.

FE HZO Samples	FE-DE HZO/HfO₂ Samples
DC Sputtering 200 nm W Bottom Electrode	
ALD HZO, 300 C, TEMAH, TEMAZ, O₂ Plasma	
DC Sputtering 40 nm top W	
RTA in N₂ at 500 C for 30 s	
Selective wet etch of top W layer	
	ALD HfO₂, 300 C, TEMAH, H₂O
Photolithography and liftoff 200 nm sputtered W top electrode	

Figure 1: Process flow for ferroelectric HZO (left) and bilayer HZO/HfO₂ (right) capacitors.

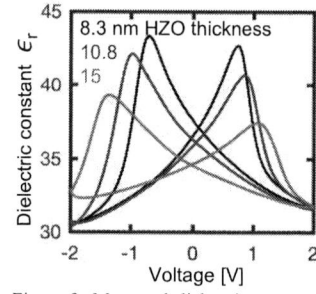

Figure 2: TEM of bilayer capacitor showing crystalline grain across HZO/HfO₂ interface.

Figure 3: Measured dielectric constant from capacitance-voltage data (0.2 V/s sweep rate and 25 mV, 100 kHz AC)

Nucleation limited switching (NLS) model [5, 6]

Time constant τ is a function of applied field E and activation field E_a

$$\tau(E, E_a) = \tau_0 \exp\left\{\left(\frac{E_a}{E}\right)^\alpha\right\}$$

Cumulative probability of domain switching for constant applied field E

$$P_{sw}(t, E, E_a) = 1 - \exp\left\{\left(\frac{t}{\tau(E, E_a)}\right)^n\right\}$$

For a distribution of activation fields $f(E_a)$, the mean polarization is

$$P(t, E) = P_R - 2P_R \int_0^\infty P_{sw}(t, E, E_a) f(E_a) dE_a$$

P_R: remanent polarization. Fitting parameters: n, α, τ_0

Monte Carlo simulation framework

1. Extract distribution of activation field $f(E_a)$ and parameters P_R, n, α, τ_0 from polarization reversal data.
2. Instantiate a set of N grains with activation field E_a drawn from probability distribution $f(E_a)$.
3. For each time interval and grain $i = 1..N$, compute switching probability according to Weibull process

$$p_{sw}^{(i)}(t_s < t + \Delta t | t_s > t) = 1 - \exp\left\{-h_i(t + \Delta t)^n + h_i(t)^n\right\}$$

with accumulated time constant: $h_i(t) = \int_{t_0}^t \frac{dt}{\tau(E(t), E_{ai})}$

4. Update accumulated time constants $h(t) \to h(t) + \Delta t/\tau$. If grain switches, update polarization state $(p_i = \pm 1)$ and reset $h(t)$.

$$Q_{FE}(t) = \frac{P_R}{N} \sum_{i=1}^N p_i(t) + \epsilon_{FE} E(t)$$

Figure 4: The NLS model (left) describes the FE as an ensemble of regions (grains) switching independently with a field dependent time constant. The FE statistics are captured by a distribution of activation field. Upon parameter extraction from the NLS model, the FE polarization dynamics can be predicted for arbitrary input waveforms using a Monte Carlo approach (right). A constant permittivity ϵ_{FE} was assumed to compute the total charge per unit area.

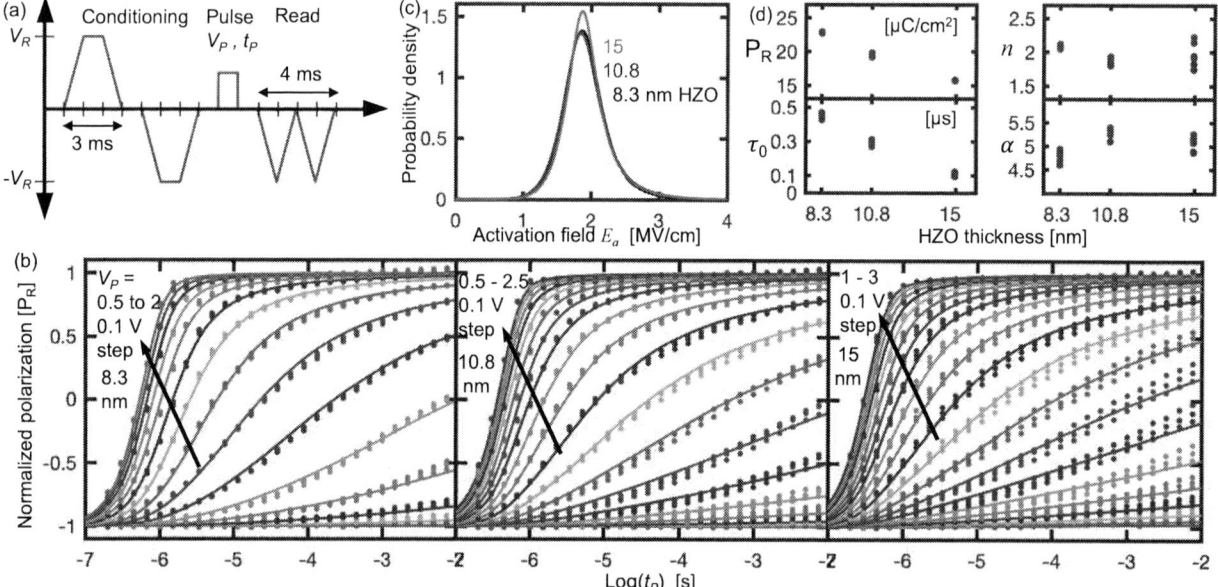

Figure 5: (a) Measurement protocol for parameter extraction. Pulse width (t_p) was stepped from 200 ns to 10 ms in increments of 1.5×, then amplitude (V_p) was stepped in increments of 100 mV. Reset and read amplitude (V_R) of 2.5, 3 and 3.5 V were used for 8.3, 10.8 and 15 nm capacitors, respectively. The procedure was repeated 3 times for each sample. (b) Partial polarization data for 3 runs (dots) show close agreement with fitted NLS model (solid line) over 5 decades. (c) Extracted distributions of activation field reflect minor variations in the statistical properties with film thickness. (d) Extracted parameters: P_R and τ_0 decrease by 0.6× and 0.25× respectively for thickness from 8.3 to 15 nm, whereas n and α show variations below 10% for different samples and thickness.

978-1-7281-1988-5/18 $31.00 © 2018 IEEE

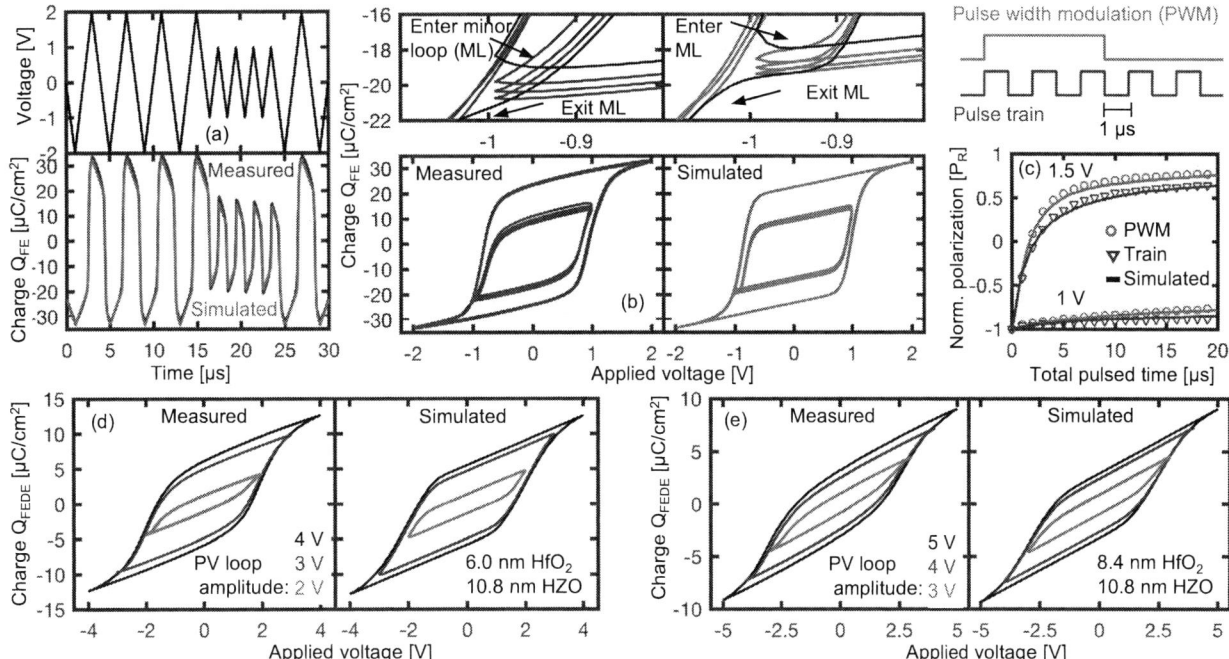

Figure 6: Experimental validation of Monte Carlo simulation framework. (a) Measured and simulated polarization vs. time for an 8.3 nm HZO capacitor with a triangular input waveform of varying amplitude (top). (b) Measured and simulated minor loops obtained from (a), with detail of the transition between minor loops and the saturated loop (top). (c) Experimental and simulated polarization obtained by pulse width modulation and a train of pulses with equivalent "on" time. Close agreement of measured and simulated PV loops for FE-DE structures with (d) 6/10.8 nm HfO₂/HZO and (e) 8.4/10.8 nm HfO₂/HZO.

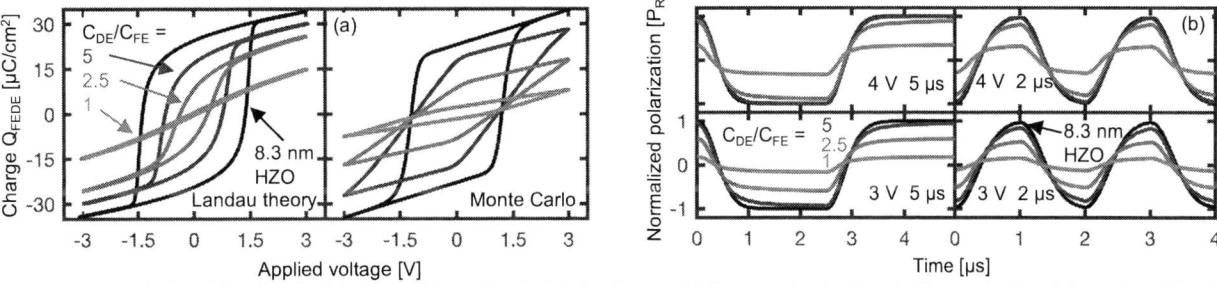

Figure 7: (a) Simulation of ferroelectric-dielectric P-V loops: Landau-Khalatnikov model shows a reduction in switching threshold with increasing dielectric thickness (decreasing capacitance). Monte Carlo simulation exhibits the behavior observed in experiments. (b) Simulated polarization vs. time of a FE capacitor and FE-DE structures with different dielectric capacitance, programmed with square waveforms of amplitudes 3 and 4 V with 2 μs and 5 μs period.

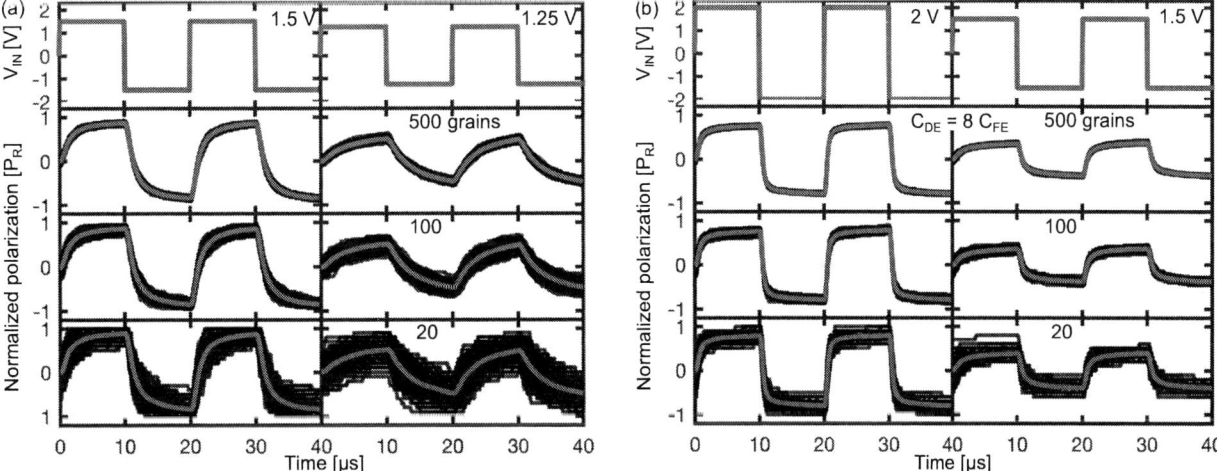

Figure 8: Simulated device-to-device variations of 200 devices (black) with 500, 100 and 20 grains for (a) 8 nm-thick FE and (b) FE-DE capacitor with $C_{DE} = 8\ C_{FE}$. With 20 grains, the memory window of the FE is reduced by 50% with respect to the mean value (red) for a 1.5 V programming voltage, and is completely lost with 1.25 V. With the same number of grains, the FEDE requires a programming voltage above 1.5 V to a obtain a memory window.

978-1-7281-1988-5/18 $31.00 © 2018 IEEE   371

Scalability Study on Ferroelectric-HfO$_2$ Tunnel Junction Memory Based on Non-equilibrium Green Function Method with Self-consistent Potential

Fei Mo, Yusaku Tagawa, Takuya Saraya, Toshiro Hiramoto, and Masaharu Kobayashi

Institute of Industrial Science, The University of Tokyo, Tokyo, Japan, email: mofei@nano.iis.u-tokyo.ac.jp

Abstract—We have investigated scalability and design guideline of HfO$_2$-based Ferroelectric Tunnel Junction (FTJ) memory by employing numerical simulation which is based on Non-Equilibrium Green Function (NEGF) method and self-consistent potential, and calibrated by our experimental FTJ data, for the first time. Metal-Ferroelectric-Insulator-semiconductor (MFIS) FTJ shows a higher TER than Metal-Ferroelectric-Insulator-Metal (MFIM) FTJ with almost the same read current because of the large asymmetry of dielectric screening property in top and bottom electrodes. High read current can be obtained by thinner layers while high TER and low depolarizing field are maintained by adjusting bottom semiconductor electrode property. Based on these results, a guideline for designing MFIS structure FTJ to achieve high read current and high TER has been proposed. We have shown a potential for scaling the FTJ down to sub-20 nm diameter.

I. INTRODUCTION

Non-volatile memories (NVM) have been key components for internet of things (IoT) devices under the constraint of power consumption. Due to their low battery capacity, IoT devices require low power NVMs to store and process data at low cost. Among various NVMs, ferroelectric tunnel junction (FTJ) memory has become one of the promising candidates for its non-destructive readout, field-driven low voltage operation and high on/off current ratio. FTJ is a two-terminal device, in which, a thin ferroelectric (FE) film is sandwiched by two electrodes as shown in Fig. 1. Tunneling barrier height can be modulated by polarization switching driven by external voltage and tunneling electroresistance (TER) can be low/high at on/off state. Experimental results of FTJ with various electrodes and FE film materials have been reported [1, 2]. Thanks to the recent discovery of CMOS-compatible FE-HfO$_2$, HfO$_2$-based FTJ has been attracting attentions [3, 4]. In particular, Metal-Ferroelectric-Insulator-Semiconductor (MFIS) FTJ shows high TER ratio ($\equiv (I_{ON}-I_{OFF})/I_{OFF}$) because of large asymmetry of dielectric screening property in top and bottom electrodes as reported [5].

Tunneling current of FTJ in the form of Metal-Ferroelectric-Metal (MFM) and Metal-Ferroelectric-Insulator-Metal (MFIM) can be analytically calculated by Wenzel-Kramer-Brillouin (WKB) approximation [6], or numerically calculated by non-equilibrium Green's function (NEGF) method [7, 8]. However, for MFIS structure FTJ, simulation method has not been established yet. It is not clear if FTJ has competitive scalability to other emerging NVMs. Thus, it is crucial to build

simulation framework for designing practical MFIS structure FTJ with regard to read current and TER ratio considering the FE property, interfacial layer (IL), and substrate doping concentration (N_d), as well as layer thickness.

In this paper, we develop a simulation framework to calculate the current-voltage characteristic of MFIS structure FTJ with FE-HfO$_2$ by using NEGF method and self-consistent potential including FE property and semiconductor band bending. A design guideline for HfO$_2$-based FTJ is proposed with regard to read current and TER ratio.

II. SIMULATION APPROACH

The device structure of MFIS FTJ to be simulated in this work is shown in Fig. 2. From left side to right side, there are metal electrode, HfZrO$_2$, interfacial layer and N$^+$-Si. Parameters are defined in this figure. V_a is an external bias voltage, a is the grid length for calculation. The proposed simulation framework consists of two key modules: (1) self-consistent potential calculation sub-module (Fig. 3) and (2) current density calculation main-module (Fig. 4).

For potential calculation, we consider voltage across each layer (metal, FE layer, interfacial layer, and N$^+$ Si), band offset, and build-in potential. The voltage across each layer and potential profile are determined by self-consistently solving analytical equation of FE polarization charge (P) versus voltage (V) relationship, Poisson equation, Thomas-fermi screening potential, and analytical equation of semiconductor charge and surface potential (ψ_s) [9].

For current calculation, NEGF method [10] is used at each bias voltage. f is the Fermi-Dirac function. T is the transmission coefficient. G is the Green's function, Γ and Σ are the broadening function and self-energy of metal and semiconductor. H is the Hamiltonian matrix. The Hamiltonian matrix and self-energy are made using the potential profile calculated by the sub-module at each bias voltage as a function of energy. The total current density is the sum of current density at each energy level in a certain energy range. Local Density of State (LDOS) can be also calculated at each bias voltage.

III. RESULTS AND DISCUSSIONS

A. Calibration of the Simulator by Experimental Current-voltage Curve of MFIS Structure FTJ

In order to calibrate the simulator, we used a measured *I-V* curve of MFIS structure FTJ. This FTJ was fabricated on N$^+$ Si substrate by using the process we developed [5]: 50% Zr-doped

HfO$_2$ (HZO) was deposited by ALD and it was annealed with TiN capping for crystallization. Then TiN was stripped by wet etching. Al was deposited and patterned as a top electrode. The FE and interfacial layer thickness were obtained from TEM measurement and remanent polarization (P_r) was determined by Positive-Up-Negative-Down (PUND) measurement. The default device parameters are summarized in Fig. 6. Fig. 7 shows that the experimental data of current density (J) is well fitted by the simulation results for both ON and OFF states with high TER ratio. Fig. 8 illustrates LDOS and transmission coefficient. Higher transmission was obtained with lower tunneling barrier height for ON-state. OFF-state is vice versa.

B. Comparison between the FTJ Structure

MFIS and MFIM structure FTJs are compared by simulation. Simulated ON-state current density (J_{ON}) and OFF-state current density (J_{OFF}) $-V$ curves are shown in Fig. 9. MFIS structure FTJ has a higher TER ratio than MFIM structure. As shown in Fig. 10, in the practical range of P_r, the J_{ON} of MFIM and MFIS structure FTJs are nearly the same, while J_{OFF} of the MFIS structure FTJ are smaller than MFIM structure FTJ due to the large asymmetry of dielectric screening property in top and bottom electrodes in MFIS structure. As layers are thinned down, P_r will become smaller. Even in small P_r region, the TER ratio of MFIS structure FTJ remains higher than MFIM structure FTJ as shown in Fig. 11.

C. ON-State Current and TER ratio of MFIS FTJ

In order to achieve high ON-state read current, FE and interfacial layer needs to be thinned down. However, as FE layer gets thin, coercive voltage (memory window, V_c) becomes small. Then read voltage (V_{read}) should be chosen so that V_{read} does not disturb the stored memory state. Fig. 12 shows voltage across FE layer (V_{FE}) as a function of V_{read}. As V_{read} increases, V_{FE} increases and voltage margin to V_c becomes small. To realize sufficiently long retention time and avoid read/write operation disturbance and error, we choose the absolute value of V_{read} to be less than 0.2 V.

Before thinning down FE and interfacial layers, first we consider the impact of P_r and N_d. Fig. 13 shows P_r dependence of J_{ON}, J_{OFF} and TER ratio with default device parameters. As P_r decreases, J_{ON} remains nearly constant, while J_{OFF} is more influenced and increases. TER ratio is monotonically reduced because tunneling barrier height modulation becomes small with very small P_r. Fig.14 shows N_d dependence of J_{ON}, J_{OFF} and TER ratio with default device parameters. As N_d decreases, J_{ON} remains nearly constant, while J_{OFF} is more influenced and decreases. TER ratio is monotonically improved because tunneling barrier height modulation becomes large due to wider depletion layer and weak screening in N$^+$-Si compared to top metal electrode with small N_d. As results of Fig. 13 and 14, it is possible to compensate TER ratio reduction with small P_r by reducing N_d which is an additional knob for MFIS structure FTJ and to be used later.

D. Scaling Potential

The challenge for scaling FTJ is to solve the trade-off among (1) ON state read current for faster access speed, (2) TER ratio for high read sensitivity and low error rate, (3) depolarizing field (E_{dep}) for reliability such as retention characteristics, and so on. We examined FE and interfacial layer thickness (t_{FE} and t_{ox}) dependence of (1) – (3) in Fig. 15-17, respectively. Here we introduce empirical relationship between P_r and t_{FE}: P_r is proportional to t_{FE}, $P_r = k \cdot t_{FE}$, so that the trend of P_r reduction with a decrease in t_{FE} is taken into account. Obviously, as t_{FE} and t_{ox} are thinned down, J_{ON} increases (Fig. 15). However, because of the smaller P_r, TER ratio becomes small (Fig. 16). TER ratio should be practically larger than 1 (on/off ratio is larger than 200%) to distinguish on and off state. TER ratio in thin t_{FE} and t_{ox} region could be very marginal. E_{dep} becomes small and does not exceed coercive field (E_c) (Fig. 17), which is preferable. Note that, with fixed P_r, E_{dep} is supposed to be larger as t_{FE} gets thinner [4]. In this work, however, with the empirical model of $P_r = k \cdot t_{FE}$, E_{dep} becomes smaller as t_{FE} gets thinner. t_{FE} dependence of E_{dep} at fixed t_{ox} is confirmed by plotting voltage across each layer at zero bias as a function of t_{FE} in Fig. 18. E_{dep} ($=V_{FE}/t_{FE}$) decreases and remains less than E_c.

Based on the understanding of thickness dependence, lastly, we examine scalability of FTJ cell size for target ON-state read current (I_{read}) and TER ratio. t_{ox} should be as thin as possible but finite. It is set to be 0.2nm so that interface state density of FE layer and Si is suppressed. Fig. 19 and 20 show TER ratio and required t_{FE} for target I_{read} as a function of FTJ cell size. Fig. 19 and Fig. 20 target 10nA and 100nA, respectively. I_{read}=10nA and TER ratio >1 (on/off ratio>200%) can be obtained down to 20nm size with default N_d=3e19cm^{-3}. I_{read}=100nA and TER ratio >1 is challenging but can be obtained down to 20nm with technologically possible FE thickness ~1.5nm [11] by decreasing N_d as it works in Fig. 14.

IV. CONCLUSIONS

We have developed simulation framework for simulating MFIS FTJ using NEGF method and self-consistent potential. The design guideline and competitive scalability are shown in this paper. Taking advantage of large asymmetry of dielectric screening property of MFIS structure, HfO$_2$-based FTJ can be scalable down to 20nm cell size with high I_{read} and TER ratio by thinning FE and interfacial layer and adjusting N_d. FE-HfO$_2$ based MFIS structure FTJ can be a low-cost NVM solution.

ACKNOWLEDGMENT

This work was supported by JST PRESTO and Tokyo Electron Ltd.

REFERENCES

[1] Vincent Garcia et al, *Nature Communication*, **5**, 4289 (2014).
[2] Zheng Wen et al, *Nature Mat.*, **12**, pp.617–621, (2013).
[3] Luca Larcher et al, IEDM 2017 pp.282-285, (2017).
[4] Shosuke Fujii et al., VLSI Symp. 2016, pp.148 (2016).
[5] Masaharu Kobayashi et al, IEEE SNW 2018, pp.29-30, (2018).
[6] D. Pantel and M. Alexe, *Physical Review B*, **82**, 134105, (2010).
[7] Zhipeng Dong et al, *J. of Applied Physics*, **123**, 094501 (2018).
[8] Sou-Chi Chang et al, *Physical Review Applied*, **7**, 024005 (2017).
[9] Yuan Taur and Tak H. Ning, *"Fundamentals of Modern VLSI Devices"*, Cambridge (2009).
[10] Supriyo Datta, *"Quantum Transport: Atom to Transistor"*, Cambridge (2006).
[11] M. H. Lee et al, IEDM 2016, pp.306-309, (2016).

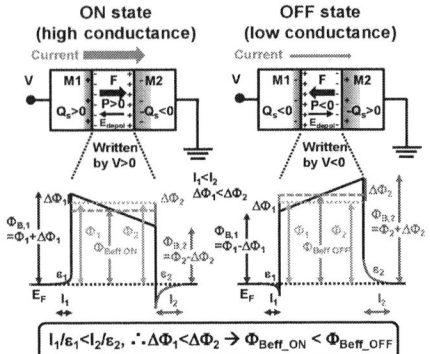

Fig.1 Operation principle of FTJ. Tunneling barrier height is modulated by polarization switching.

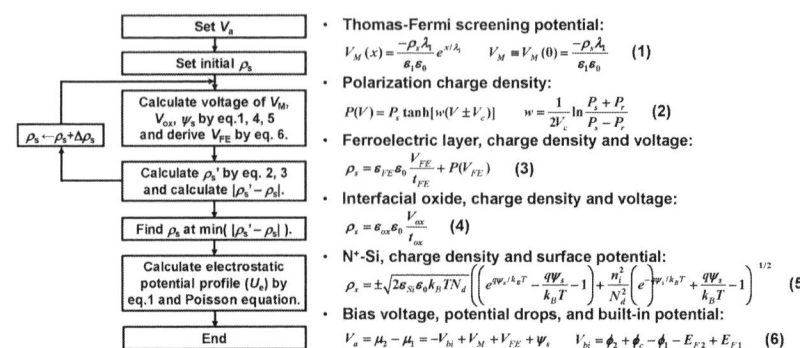

Fig.3 Self-consistent potential calculation sub-module of MFIS structure FTJ at given bias voltage. Output is total potential profile which consists of electrostatic potential and band offset.

- Thomas-Fermi screening potential:

$$V_M(x) = \frac{-\rho_s \lambda_1}{\varepsilon_1 \varepsilon_0} e^{x/\lambda_1} \qquad V_M = V_M(0) = \frac{-\rho_s \lambda_1}{\varepsilon_1 \varepsilon_0} \quad (1)$$

- Polarization charge density:

$$P(V) = P_s \tanh[w(V \pm V_c)] \qquad w = \frac{1}{2V_c} \ln \frac{P_s + P_r}{P_s - P_r} \quad (2)$$

- Ferroelectric layer, charge density and voltage:

$$\rho_s = \varepsilon_{FE} \varepsilon_0 \frac{V_{FE}}{t_{FE}} + P(V_{FE}) \quad (3)$$

- Interfacial oxide, charge density and voltage:

$$\rho_s = \varepsilon_{ox} \varepsilon_0 \frac{V_{ox}}{t_{ox}} \quad (4)$$

- N+-Si, charge density and surface potential:

$$\rho_s = \pm \sqrt{2\varepsilon_{Si}\varepsilon_0 k_B T N_d} \left(\left(e^{q\psi_s/k_B T} - \frac{q\psi_s}{k_B T} - 1 \right) + \frac{n_i^2}{N_d^2} \left(e^{-q\psi_s/k_B T} - \frac{q\psi_s}{k_B T} - 1 \right) \right)^{1/2} \quad (5)$$

- Bias voltage, potential drops, and built-in potential:

$$V_a = \mu_2 - \mu_1 = -V_{bi} + V_M + V_{FE} + \psi_s, \quad V_{bi} = \phi_2 + \phi_c - \phi_1 - E_{F2} + E_{F1} \quad (6)$$

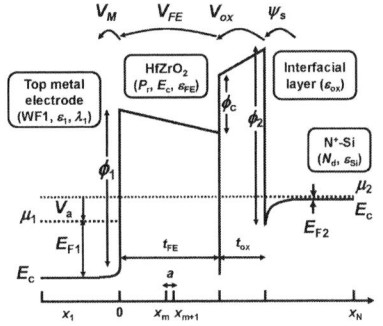

Fig.2 Band diagram of MFIS structure FTJ to be simulated, where device parameters are defined.

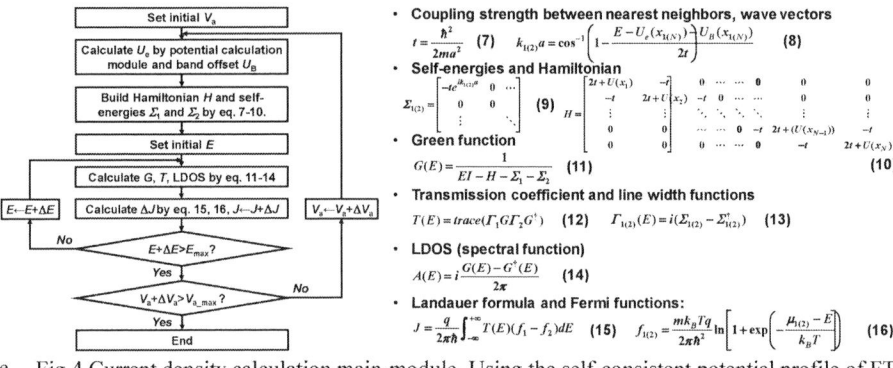

Fig.4 Current density calculation main-module. Using the self-consistent potential profile of FTJ calculated by the module shown in Fig. 3, tunneling current can be calculated by NEGF method.

- Coupling strength between nearest neighbors, wave vectors

$$t = \frac{\hbar^2}{2ma^2} \quad (7) \qquad k_{1(2)}a = \cos^{-1}\left(1 - \frac{E - U_e(x_{1(N)}) - U_B(x_{1(N)})}{2t} \right) \quad (8)$$

- Self-energies and Hamiltonian

$$\Sigma_{1(2)} = \begin{bmatrix} -te^{ik_{1(2)}a} & 0 & \cdots & 0 \\ 0 & 0 & & \vdots \\ \vdots & & \ddots & \\ 0 & 0 & \cdots & 0 \end{bmatrix} \quad (9) \qquad H = \begin{bmatrix} 2t + U(x_1) & -t & 0 & \cdots & 0 & 0 \\ -t & 2t + U(x_2) & -t & 0 & \cdots & 0 \\ 0 & -t & \ddots & & & \vdots \\ \vdots & & & & 2t + U(x_{N-1}) & -t \\ 0 & 0 & \cdots & 0 & -t & 2t + U(x_N) \end{bmatrix} \quad (10)$$

- Green function

$$G(E) = \frac{1}{EI - H - \Sigma_1 - \Sigma_2} \quad (11)$$

- Transmission coefficient and line width functions

$$T(E) = trace[\Gamma_1 G \Gamma_2 G^\dagger] \quad (12) \qquad \Gamma_{1(2)}(E) = i(\Sigma_{1(2)} - \Sigma_{1(2)}^\dagger) \quad (13)$$

- LDOS (spectral function)

$$A(E) = i \frac{G(E) - G^\dagger(E)}{2\pi} \quad (14)$$

- Landauer formula and Fermi functions:

$$J = \frac{q}{2\pi\hbar} \int_{-\infty}^{+\infty} T(E)(f_1 - f_2) dE \quad (15) \qquad f_{1(2)} = \frac{mk_B Tq}{2\pi\hbar} \ln\left[1 + \exp\left(-\frac{\mu_{1(2)} - E}{k_B T} \right) \right] \quad (16)$$

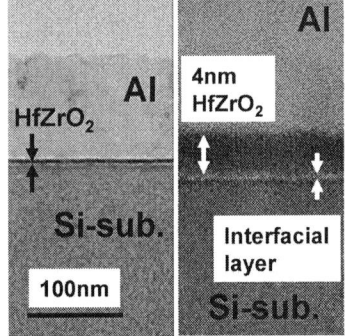

Fig.5 Cross sectional TEM image of FTJ, from which t_{FE} and t_{ox} were estimated.

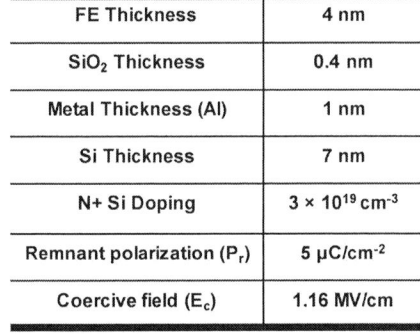

FE Thickness	4 nm
SiO₂ Thickness	0.4 nm
Metal Thickness (Al)	1 nm
Si Thickness	7 nm
N+ Si Doping	3×10^{19} cm⁻³
Remnant polarization (P_r)	5 µC/cm⁻²
Coercive field (E_c)	1.16 MV/cm

Fig.6 Default device parameters used for simulating HfO₂-based MFIS structure FTJ.

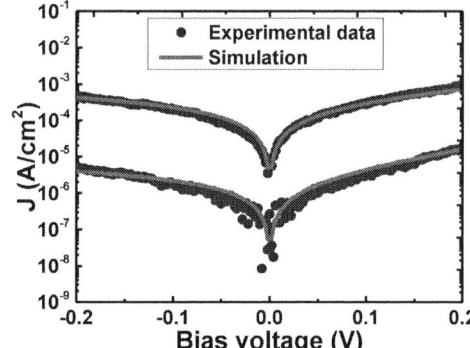

Fig.7 Measured and simulated J-V characteristics of MFIS structure FTJ with default device parameters

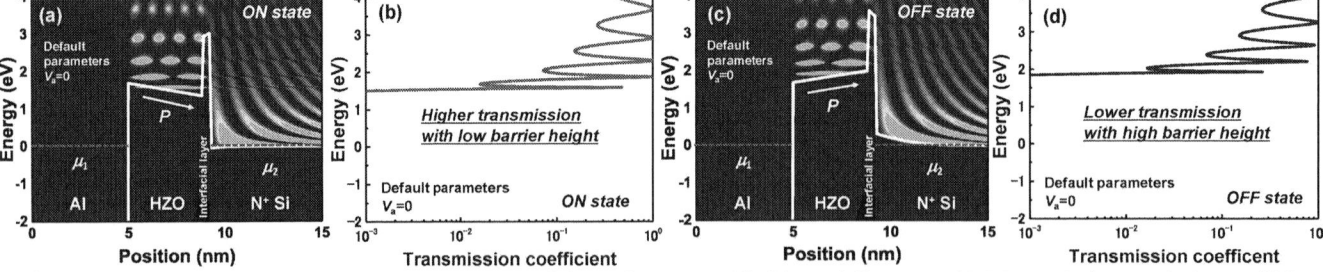

Fig.8 (a) Band diagram and LDOS, and (b) transmission coefficient, of ON state FTJ. (c) Band diagram and LDOS, and (d) transmission coefficient, of OFF state FTJ. Higher transmission with lower tunneling barrier height was confirmed for ON state, while lower transmission was for OFF state.

978-1-7281-1988-5/18 $31.00 © 2018 IEEE

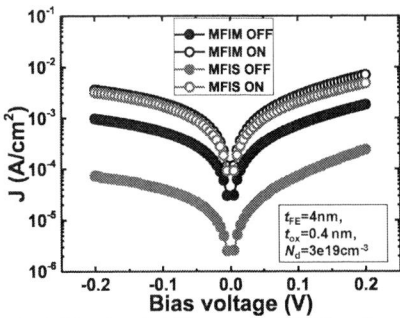

Fig.9 Simulated *J-V* curve for MFIS and MFIM FTJs with default device parameters.

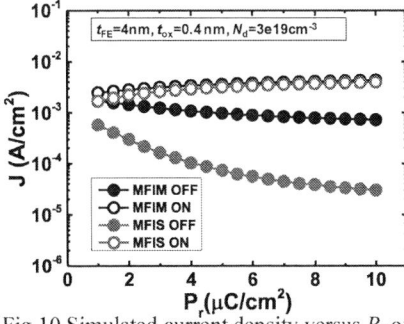

Fig.10 Simulated current density versus P_r of ON and OFF state of MFIM and MFIS FTJ.

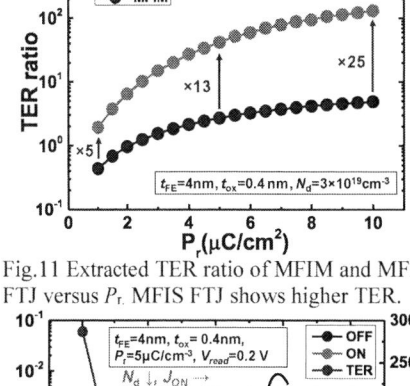

Fig.11 Extracted TER ratio of MFIM and MFIS FTJ versus P_r. MFIS FTJ shows higher TER.

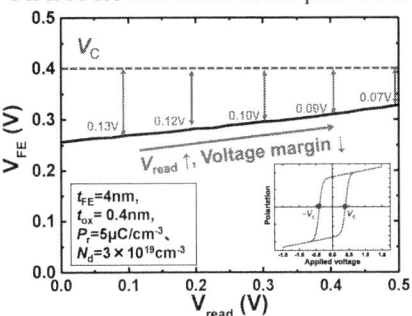

Fig.12 Simulated V_{FE} versus V_{read} and *P-V* curve (inset). Voltage margin from V_c should be sufficiently large to keep data.

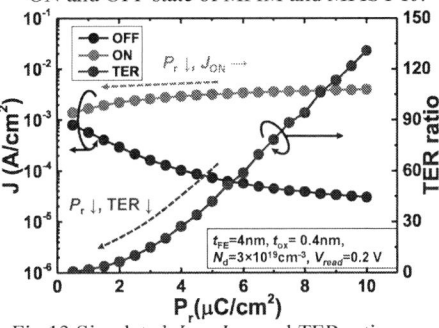

Fig.13 Simulated J_{ON}, J_{OFF} and TER ratio versus P_r for MFIS FTJ at V_{read}=0.2V. TER ratio is smaller for smaller P_r, while J_{ON} is maintained.

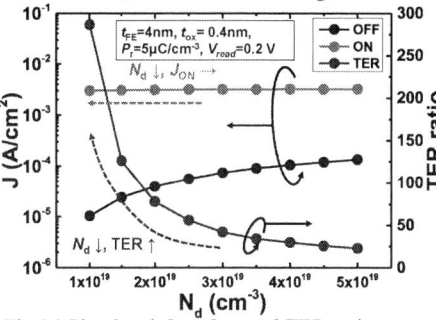

Fig.14 Simulated J_{ON}, J_{OFF} and TER ratio versus N_d for MFIS FTJ at V_{read}=0.2V. TER ratio is larger for lower N_d, while J_{ON} is maintained.

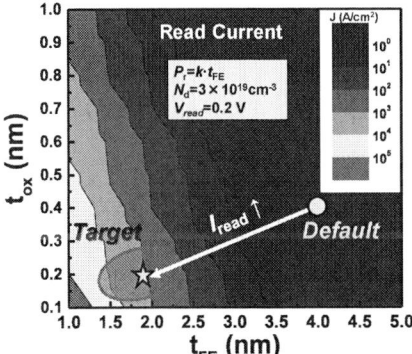

Fig.15 Simulated contour plot of J_{ON} versus t_{ox} and t_{FE}. J_{ON} increases with thinner t_{ox} and t_{FE}. A large read current means a shorter read time.

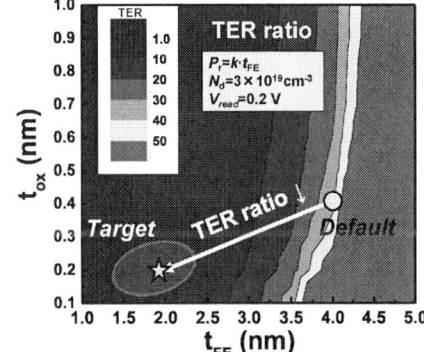

Fig.16 Simulated contour plot of TER ratio versus t_{ox} and t_{FE}. TER ratio decreases with thinner t_{ox} and t_{FE}. TER ratio should be higher than 1(>200%).

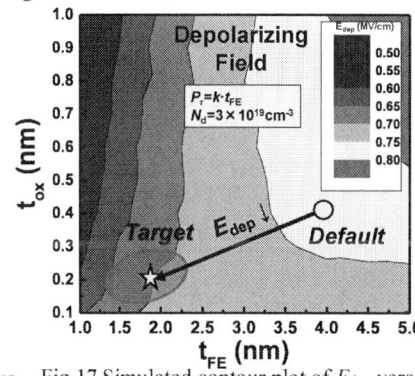

Fig.17 Simulated contour plot of E_{dep} versus t_{ox} and t_{FE}. E_{dep} decrease with thinner t_{ox} and t_{FE}. E_{dep} should be smaller than E_c~1 MV/cm.

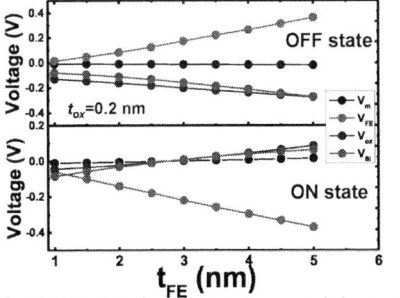

Fig.18 Simulated voltages across each layer at ON state and OFF state versus t_{FE} at retention state (zero bias). V_{FE} does not exceed V_c.

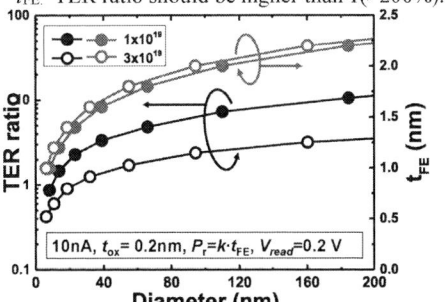

Fig.19 TER ratio and required t_{FE} versus FTJ size to target I_{read} = 10nA at ON state with different N_d. For 20 nm size, TER > 1 with N_d = 3x19 cm^{-3}.

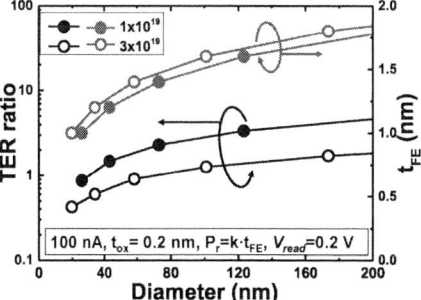

Fig.20 TER ratio and required t_{FE} versus FTJ size to target I_{read} = 100nA at ON state with different N_d. For 20 nm size, TER > 1 with N_d = 1x19 cm^{-3}.

Role of Oxygen Vacancies in Electric Field Cycling Behaviors of Ferroelectric Hafnium Oxide

C. Liu[1], F. Liu[1,3], Q. Luo[2], P. Huang[1*], X. X. Xu[2], H. B. Lv[2], Y. D. Zhao[1], X.Y. Liu[1] and J. F. Kang[1,*]

[1]Institute of Microelectronics, Peking University, Beijing, China, [2] Key Laboratory of Microelectronics Devices and Integrated Technology, Institute of Microelectronics, Chinese Academy of Sciences, Beijing, China, [3]Department of physics, University of Hong Kong, Hong Kong, email: phwang@pku.edu.cn, Kangjf@pku.edu.cn

Abstract—Based on the density functional theory (DFT) calculations, a new mechanism about the oxygen vacancies(Vo) in the HfO$_2$-based ferroelectric devices is presented. In this mechanism, the Vo in m-phase HfO$_2$ not only serve as the electron traps but also emerge ferroelectricity besides the known o-phase HfO$_2$. And the increased remanent polarization during the "wake-up" process is mainly attributed to this part of Vo-m-phase HfO$_2$ ferroelectric cells. Based on the new mechanism, a Kinetic Monte Carlo (KMC) simulator is developed to quantify the typical electric field cycling behaviors observed in the HfO$_2$-based ferroelectric devices, including the wake-up, fatigue, split-up, and breakdown effects. This new understanding establishes relationship between the Vo and the cycling behaviors, and further shows the connection between the dopant and the wake-up characteristics of HfO$_2$-based ferroelectric device.

I. INTRODUCTION

Since the discovery of the ferroelectricity in Si-doped HfO$_2$ films in 2011[1], the HfO$_2$-based ferroelectric devices have attracted fast-growing research attention in the following years. Compared with the conventional ferroelectric material, HfO$_2$-based ferroelectric devices exhibit lots of distinct merits, including the extremely thin film (<10nm), good compatibility to the CMOS technology and a relatively low dielectric constant, and has a broad application future [2-4]. It's believed that the origin of ferroelectricity in HfO$_2$-based films is the formation of o-phase HfO$_2$[5]. Various cycling behaviors have been observed and several mechanisms around the evolution of o-phase HfO$_2$ have been proposed to explain the phenomena [6,7]. It is acknowledged that oxygen vacancies(Vo) is tightly related to the various field cycling phenomena, such as "wake-up" and fatigue [8,9]. Therefore, deep understanding about Vo in ferroelectric HfO$_2$ is necessary. In this work, based on the DFT calculation, we bring up a new mechanism that the m-phase HfO$_2$ with Vo also offers ferroelectricity besides the o-phase HfO$_2$. The DFT calculations indicate that the Vo at different locations in the supercells have different spontaneous polarization. The increased remanent polarization (Pr) during the "wake-up" is attributed to the generation of Vo in m-phase HfO$_2$. Considering the evolution of the Vo, the depolarization field, the screening effect induced by oxygen ions and the electron trapping by the Vo states, a KMC simulator for the electric field cycling behaviors are developed and roles of the Vo in the cycling behaviors can be quantitatively explained. Furthermore, we present the relationship of the various dopants and the wake-up behaviors, and exhibit a trade-off between the wake-up efficiency and the device endurance.

II. DFT CALCULATION AND MECHANISMS

Fig.1 shows four typical electric cycling behaviors observed in the HfO$_2$-based ferroelectric devices with various kinds of dopants. "Wake-up": the increase of the Pr by the electric field applied to the pristine devices. "Fatigue": the gradual decrease of the Pr after amount of switching cycles. "Split-up"/"merging": the split/merging of the current peak(s) in the transient response after certain switching cycles. "Breakdown": the permanent and abrupt increase of the leakage current.

To investigate the impact of the Vo on the behaviors above, we first made DFT calculations of the spontaneous polarization of the m-HfO$_2$ phase with Vo. It has been experimentally proved that the ferroelectric is an intrinsic property of HfO$_2$[10], so we construct the atomic structure of Vo phase based on m-phase HfO$_2$ without any dopant. Considering the reasonable Vo concentration, we choose a 2x2x2 supercell as a unit structure for computing and pick Vo at 2 inequivalent positions respectively, as shown in Fig.2. The construction and calculation of the centrosymmetric reference phase required by the Berry-phase method is shown in the Fig.3[11]. By subtracting the spontaneous polarization of the computing unit and the reference phase, the polarization of the Vo nearby the symmetry center (P2&P3 in Fig.2) is 5.623μC/cm^2, while the result of faraway Vo (P1&P4 in Fig.2) is 29.237μC/cm^2, which is smaller than the theorical o-phase result (~50μC/cm^2) [5] but comparable with the experimental results of the increased Pr in wake-up process [6]. Fig.4 shows the switching barrier of the opposite $\vec{P}$ states. The calculation results indicate that the separated Vo-m-phase HfO$_2$ cells can act as the ferroelectric domains in the HfO$_2$ films, besides the known o-phase HfO$_2$. We further calculate the Hf$_{0.5}$Zr$_{0.5}$O$_2$ case, the results show about 40% higher polarization than the pure HfO$_2$. Based on the calculations, physical mechanisms are proposed to describe the impacts of Vo on the electric field cycling behaviors, and Fig.5a is the overall schematic diagram. The generation of Vo in the separated m-phase HfO$_2$ cells is the origin of the increase of ferroelectric domains in wake-up process; the absorption of the O^{2-} ions by the electrode forms a dead layer besides the electrodes, which will not only decrease the ferroelectric domains, but also induce a screening field partially offsetting the external field, as shown in Fig.5b. Fig.5c is the simplified unit cell used in our KMC simulation and possible distributions of Vo in a single supercell. As shown in Fig.5c, accompanied with further increase of the Vo, Vo-phase with rotational symmetry can be transformed to center symmetry again, meaning the disappearance of the Vo-phase ferroelectric cells. This process and the screening field are both part of the origin

978-1-7281-1988-5/18 $31.00 © 2018 IEEE

of "fatigue". The trapping of electrons from electrodes by the Vo may induce the "split-up" phenomenon because of the built-in field in ferroelectric cells. As the continuous generation of Vo, the breakdown arises from the formation of the Vo filaments connecting the electrodes, alike the resistive switching mechanism in resistive random access memory (RRAM) [12]. A numerical KMC simulator consisting of the processes above has been established. Equations describing the processes are summarized in Table I.

III. THE FIELD CYCLING BEHAVIORS

The stochastic simulation of the Vo evolution in the ferroelectric device and the corresponding Pr evolution is shown in Fig.6(a)&(b). The Vo continuously generates during the process till the formation of Vo filaments. To verify our mechanisms, we made some experimental measurements to compare with the simulation results. The measured devices are 10nm thick $Hf_{0.5}Zr_{0.5}O_2$ film with both TiN electrodes. Combining our KMC simulation and the KAI theory [13], the P-V hysteresis curve after wake-up process can be simulated and is also shown in Fig.7 along with the measured results.

To avoid the inconformity of the pristine states of different devices, we focus on the variation of the switchable remanent polarization, $\Delta Pr_{sw}(\Delta P_{r+}-\Delta P_{r-})$. We found that the DC bias without switching can also induce the wake-up and Fig.8(a)&(b) are the results of the wake-up process under different DC voltages and temperatures. Higher voltages and temperatures both enhance the wake-up efficiency in speed and the highest Pr value. This is mainly because both high voltage and temperature can increase the Vo generation probability, according to equation (1) in Table I. Fig.9(a)&(b) are results of AC wake-up and fatigue with different pulse frequencies and amplitudes. Our simulations agree well with the experiments under different conditions. Fig.10(a) shows result of wake-up with DC bias then AC pulses. The difference of the dead layers and Vo distribution of DC and AC wake-up is schematically shown in Fig.10(b). DC wake-up under normal temperature is a partially wake-up process and can be further woken up by AC pulses. That should be attributed to the accumulation of the oxygen ions at the same electrode, which will induce the screening field and limit the generation of Vo. In the case of AC wake-up, dead layers are formed beside both electrodes and the two induced screening fields will compensate. Fig.11 shows the calculated screening field, which is proportional to the difference of surface charge density of the dead layers, and as the scaling down of the device or the growth of the dead layer, it cannot be neglected any longer. The measured and simulated results about the fatigue and the following breakdown are shown in Fig.12. The devices often suffer an abrupt breakdown without Pr gradually decreasing to a low level, unlike the traditional ferroelectric materials. Fig.13(a) shows the measured and simulated leakage current evolution during the wake-up, fatigue and breakdown. The current rises whether the Pr increases or decreases, meaning the increase of defects in the HfO_2 matrix, such as Vo. Fig.13(b) is the simulated Vo count and ΔPr_{sw} during the wake-up and fatigue. Comparing the increased leakage current with the increased Vo, this is also an evidence for our Vo generation

mechanism. Fig.14 shows the measurements of the split-up and merging effect. Fig.15(a) shows the two dominate electron transportations: trap-assisted tunneling (TAT) and FN tunneling. Fig.15(b) shows the theorical probabilities of electron trapping in TAT and tunneling in FN. At low voltage, the trapping probability is larger than tunneling probability, and at high voltage this relationship reverses. So pulses with low voltages will charge the Vo-phase ferroelectric cells and change the coercive field because of the induced internal field, and high voltage pulses transport the electrons through FN tunneling.

At last, using the KMC simulator, we investigate the relationship between the formation energy of Vo (E_{form}) and the cycling behaviors, as shown in Fig.16(a). The results indicate that devices with lower E_{form} can be woken up with fewer cycles. To verify this relationship, E_{form} with various dopants is calculated by DFT and some experimental works about the wake-up of various dopants are summarized in Fig.16(b). The wake-up speed increases as the decrease of E_{form}, which is roughly in line with the prediction based on our simulation. According to Fig.16(a), there's a trade-off between the wake-up and the endurance: the higher speed of the wake-up also means a faster formation of the filaments, i.e. the breakdown. This is valuable not only in theory but also with reference value in the optimization of device design and operations.

IV. CONCLUSION

Based on DFT calculations, a new physical mechanism is proposed in which the m-phase HfO_2 with Vo can offer ferroelectricity. Roles of Vo in cycling behaviors are explained: 1) the wake-up is attributed to the generation of Vo-m-phase HfO_2; 2) the fatigue is partly attributed to the screening field induced by the O^{2-} ions and larger amount of generated Vo; 3) the breakdown originates from the formation of Vo filaments; 4) the split-up originates from the electron trapping by Vo. A KMC simulator for the cycling behaviors has been developed and is verified by experiments. The simulations establish relationship between the dopants and the cycling behaviors and may offer guideline for the device optimization.

ACKNOWLEDGEMENT

This work was supported in part by the NSFC (61421005,61334007) and the MOST of China (2016YFA0203800).

REFERENCE

[1] T. S. Böscke, et al, Appl. Phys. Lett. 2011, 99, 102903.
[2] M. H. Park, et al, Adv. Mater. 2015, 27, 1811–1831
[3] J. Müller et al., VLSI Tech. Dig., pp. 25-26, 2012.
[4] H. Mulaosmanovic, et al, Symp. on VLSI Tech. Dig., 2017, T176-177
[5] Tran Doan Huan, et al, Physical Review B 90, 064111 (2014)
[6] Min Hyuk Park, et al, ACS Appl. Mater. Interfaces 2016, 8, 15466–15475
[7] M.Pesic, et al, Adv. Funct. Mater. 2016,26,4601-4612
[8] T. Schenk, et al, ACS Appl. Mater. Interfaces 2015, 7, 20224–20233
[9] S. Starschich, et al, Appl. Phys. Lett. 108, 032903 (2016)
[10] Ashish Pal, et al, Appl. Phys. Lett. 110, 022903 (2017)
[11] J. B. Neaton, et al, Physical Review B 71, 014113 (2005)
[12] P. Huang, et al, IEDM. Tech.Dig., 2012, pp.605-608
[13] Y. Ishibashi, et al, J. Phys. Soc. Jpn.,31,2 (1971)
[14] P. D. Lomenzo, et al, JOURNAL OF APPL. PHYS. 117, 134105 (2015)
[15] P. Polakowski, Appl. Phys. Lett. 106, 232905 (2015)
[16] K. Florent, IEEE Trans. Electron Devices, 64,10, 2017
[17] S. Shibayama, Symp. on VLSI Tech. Dig., 2016
[18] A. G. Chernikova, Appl. Phys. Lett. 108, 242905 (2016)

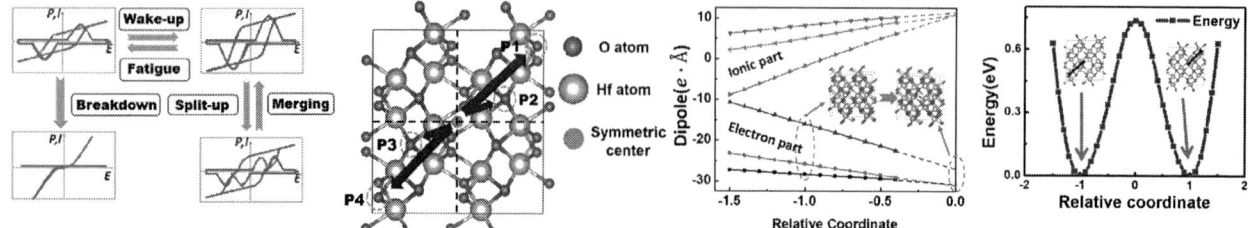

Fig.1 Schematic diagram of the electric field cycling behaviors and their relationship. The polarization cannot be measured when the leakage current is too large(breakdown).

Fig.2 The structure of HfO₂ supercell in DFT. We choose 2 inequivalent Vo positions: faraway (P1&P4) and nearby (P2&P3) the symmetric center.

Fig.3 Calculation of polarization of the reference phase. We gradually move the asymmetrical O atom to the center to get a centrosymmetric reference phase and in this way its polarization can be derived.

Fig.4 Switching barrier of the two Vo-m-phase HfO₂ dipoles with opposite directions (P1 and P4).

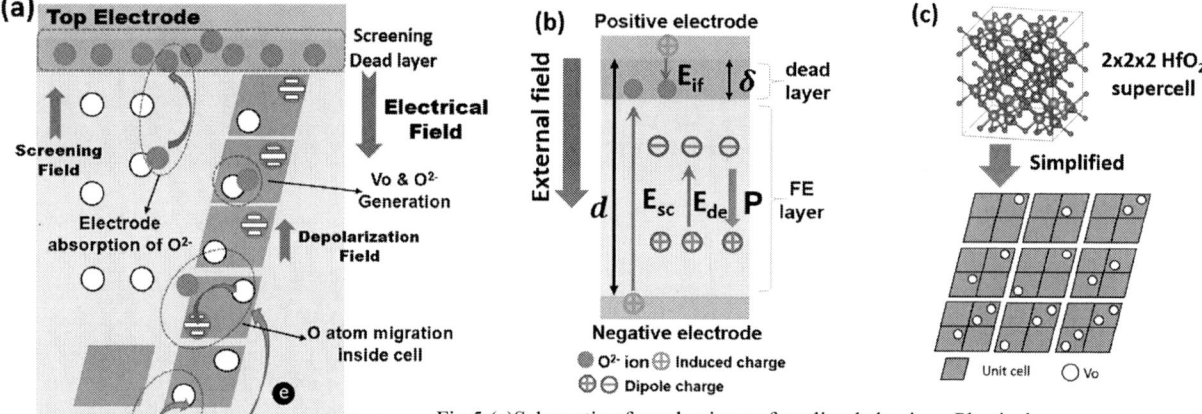

Fig.5 (a)Schematic of mechanisms of cycling behaviors. Physical processes: generation of Vo in separated m-phase HfO₂ cells; Vo migration; absorption/release of O²⁻ ions by electrodes (dead layer) and trapping of electrons by Vo. Screening field and depolarization field are taken into consideration. (b)Schematic of the screening field. P: polarization, E_{de} & E_{sc}: depolarization & screening field, d & δ: thickness of film & dead layer. (c)The simplified unit cell in KMC simulation and possible Vo distributions. We suppose 4 possible Vo positions in a single cell. As the Vo generation, rotational symmetry can be transformed to center symmetry again, meaning the disappearance of Vo-phase ferroelectric cells.

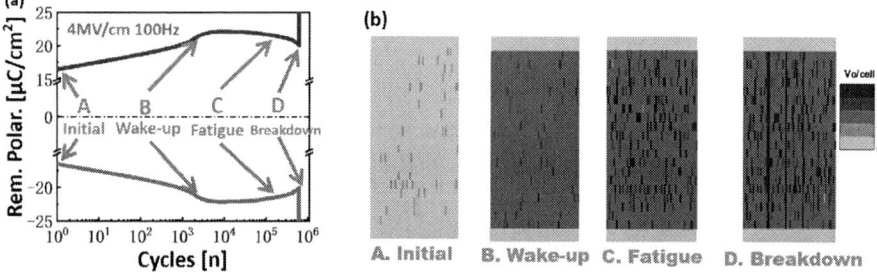

Fig.6 (a) Simulated evolution of remanent polarization during the electric cycles. (b) Simulated Vo distribution at different device states corresponding to the points in Fig.6(a). This schematic does not include the o-phase HfO₂ in the film.

Fig.7 Measured and simulated hysteresis curves after wake-up. In this work the measurement pulses are all 3.5V 1kHz triangle pulses.

Table I. Equation of the evolution of Vo & O atoms

Description	Equations	
Generation probability of Vo	$P_g(E,T,dt,P)=f\cdot dt\cdot\exp(-\dfrac{E_a-\alpha_a ZeE-\alpha_b PE}{k_B T})$	(1)
Absorption/release probability of O²⁻ by the electrode from the dielectric layer	$P_{ox}(V,T,dt)=f\cdot dt\cdot\exp(-\dfrac{E_{cr}-\gamma ZeV}{k_B T})$	(2)
Probability of O²⁻ hopping in the ferroelectric cell	$P_h(E,T,dt)=f\cdot dt\cdot\exp(-\dfrac{E_h-\alpha_h ZeE}{k_B T})$	(3)
Probability of recombination between O²⁻ and depleted Vo	$P_r(T,dt)=f\cdot dt\cdot\exp(-\dfrac{\Delta E_r}{k_B T})$	(4)
Electric field in the ferroelectric cell	$E=E_{ex}-E_{de}-E_{sc}$	(5)
Screen field (E_{sc}) induced by O²⁻ absorption of the electrodes	$E_{sc}=\dfrac{\Delta\sigma+P}{\varepsilon_0(\varepsilon_{fe}\dfrac{d-\delta}{\delta}+\varepsilon_{de})}$	(6)
Variation of switching remanent polarization (ΔPr_sw)	$\Delta P=n_1\overline{P_1}+n_2\overline{P_2}$	(7)

Parameters description: f: vibration frequency of O atom; P: the polarization of Vo phase; E_a: active energy of Vo. According to DFT results, Vo at different position has different E_a; E_i & E_o: energy barrier between the electrode and oxide; E_h: hopping barrier of O²⁻; ΔE_r: relaxation energy of recombination process; E_{ex} & V: external electric field & voltage; Z: charge number; e: unit charge; α_a&α_h: enhancement factor of electric field in lower of E_a&E_h; γ: coefficient of external voltage in the O²⁻ release; α_b: coefficient of work done by the electric field in the dipole formation; T: local temperature; E_{de}: depolarization field; $\Delta\sigma$: surface charge density difference between the top and bottom electrodes; d&δ: thickness of the film & dead layer; ε_{de}&ε_{fe}: relative dielectric constant of the dead layer and the ferroelectric film; n_1&n_2: proportion of cells with 1&2 Vo generated; $\overline{P_1}$ & $\overline{P_2}$: average switchable remanent polarization of cells with 1&2 Vo.

978-1-7281-1988-5/18 $31.00 © 2018 IEEE   378

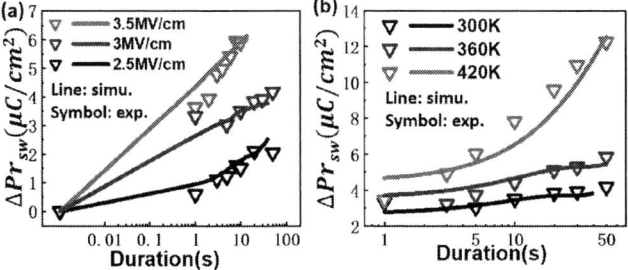

Fig.8 The measured and simulated evolution of switchable remanent polarization (ΔPr_{sw}) in DC wake-up processes: (a) various DC voltages (@300K); (b) Under various temperatures (@3V).

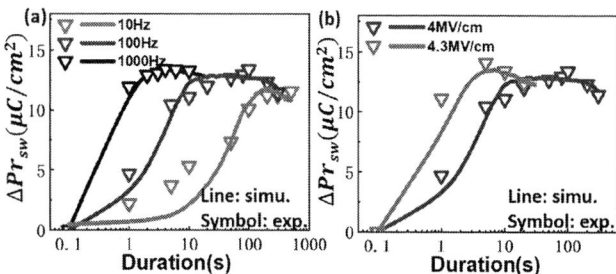

Fig.9 The measured and simulated ΔPr_{sw} in AC wake-up and fatigue. (a) Various frequencies (@4MV/cm); (b) various AC voltages (@100Hz). The pulse width in this work is 0.05ms.

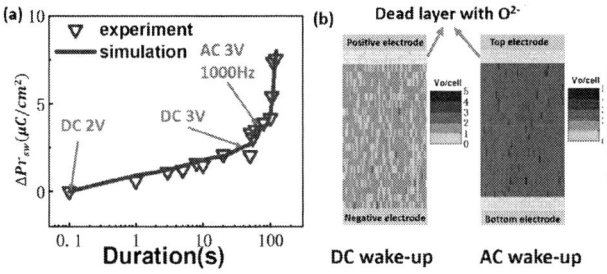

Fig.10 (a) The measured and simulated evolution of ΔPr_{sw} using DC then AC method for the wake-up. Obviously AC pulses have better effects. (b) The schematic of simulated Vo distribution of the device after DC and AC wake-up respectively. Blue dots in the dead layer refer to absorbed O^{2-} ions. The main difference is the dead layers besides the electrodes.

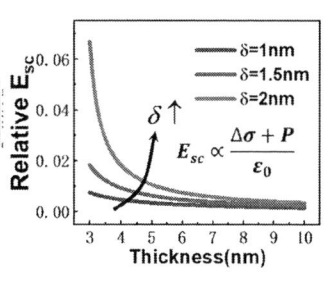

Fig.11 The relationship of the screening field and the thickness of film & dead layer.

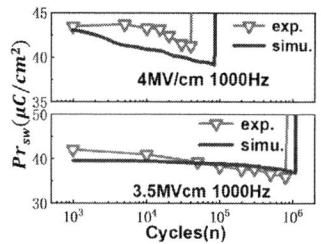

Fig.12 The measured and simulated fatigue and breakdown curves. Formation of Vo filaments is the origin of the breakdown.

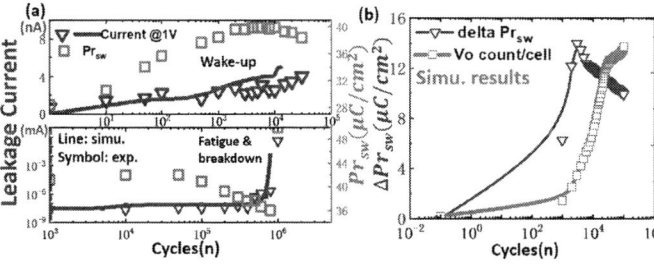

Fig.13 (a) Experimental and simulated results of the leakage currents and experimental Pr during the wake-up, fatigue and breakdown. The wake-up pulses are 4MV/cm 10Hz and the fatigue is 4MV/cm 1000Hz. (b) Simulated Pr evolution and Vo count per cell during the whole process.

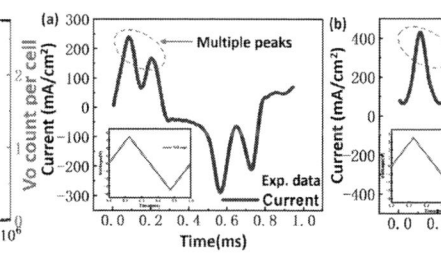

Fig.14 The current responses of the split-up (a) and merging effect (b). The split-up is observed after 10^5 cycles of 2MV/cm pulses and the peaks merges after 1000 cycles of 4MV/cm pulses. Inset are the measurement signals.

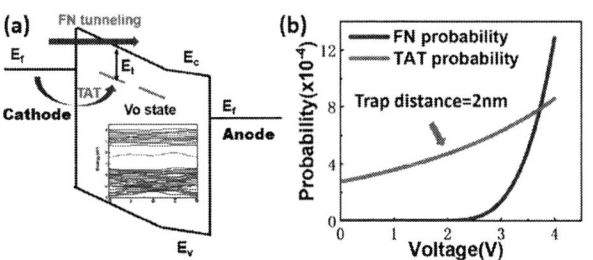

Fig.15 (a) Schematic of two electron transport mechanisms. Inset is the calculated bandstructure of HfO_2 with Vo. The trap state of Vo lies about 1.3eV beneath the conduction band in the gap. (b) The probabilities of electron trapping by the Vo state and the FN tunneling.

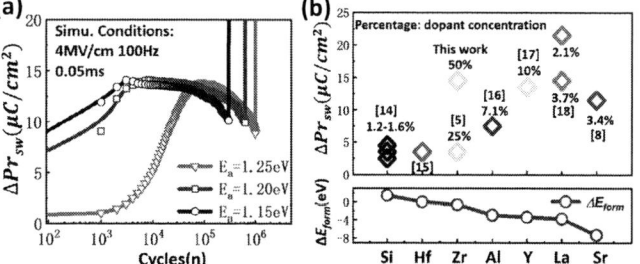

Fig.16 (a) Simulated impact of the E_{form} (E_a in equation (1)) of Vo on the cycling behaviors. (b) DFT calculated influence of various dopants on Vo formation energy change in relation to the pure m-phase (down). And the summarized experiments of wake-up of various dopant (up). We choose the data points of dopants with similar concentration and ΔPr_{sw} with 3MV/cm pulses at the 10^4th cycle for comparing the wake-up speed. The wake-up speed increases as the decrease of E_{form}.

978-1-7281-1988-5/18 $31.00 © 2018 IEEE   379

First-Principles Perspective on Poling Mechanisms and Ferroelectric/Antiferroelectric Behavior of $Hf_{1-x}Zr_xO_2$ for FEFET Applications

Sergiu Clima[1]*, S.R.C. McMitchell[1], K. Florent[1,2], L. Nyns[1], M. Popovici[1], N. Ronchi[1], L. Di Piazza[1], J. Van Houdt[1,2], G. Pourtois[1,3]

[1] imec, Kapeldreef 75, B-3001 Leuven, Belgium; [2] University of Leuven, 3001 Leuven, Belgium;
[3]PLASMANT, University of Antwerp, 2610 Antwerpen, Belgium *e-mail: sergiu.clima@imec.be

Abstract— We investigate at the atomic level the most probable *phase transformations* under strain, that are *responsible for the ferroelectric/ antiferroelectric behavior in $Hf_{1-x}Zr_xO_2$* materials. Four different crystalline phase transformations exhibit a polar/non-polar transition: monoclinic-to-orthorhombic requires a gliding strain tensor, orthorhombic-to-orthorhombic transformation does not need strain to polarize the material, whereas tetragonal-to-cubic cell compression and tetragonal-to-orthorhombic cell elongation destabilizes the non-polar tetragonal phase, facilitating the transition towards a polar atomic configuration, therefore changing the polarization-electric field loop from antiferroelectric to ferroelectric. *Oxygen vacancies can reduce drastically the polarization reversal barriers.*

I. INTRODUCTION

The Ferroelectric Field Effect Transistor (FEFET) is one of intensely developed emerging memory concept [1]. The built-in spontaneous polarization of the FE insulating layer (Fig. 1 left) shifts the threshold voltage (V_t) in a positive or negative direction (hysteretic window Fig. 1 right), as a function of the polarization direction, which defines the memory window. Some FE films show during initial cycling (wakeup/ poling stage) a pinched hysteresis window (characteristic of antiferroelectric AFE materials) that opens-up to a single Polarization-Electric field (P-E) loop (characteristic of FE switching) with increased remanent polarization (Fig. 2). The origin of the poling in FE-HfZrOₓ is not clear. Since the typical film thicknesses are of the size of a single grain and correspond to a column of ~ 10 switching HfO_2 unit cells between electrodes [2, 3], it is reasonable to investigate the process of single phase transformations from the atomic model point of view, before diving into complex domain dynamics.

The purpose of this study was to investigate the atomistic mechanisms that could be responsible for the wakeup/poling behavior in FE $Hf_{1-x}Zr_xO_2$. The polarization reversal process in FE/AFE materials is usually represented as a double/ triple-well potential energy curve (blue curves Fig. 3a-d and most of other figures presented here). Since the first derivative (dU/dP) is proportional to the electric field, the observed P-E loops will resemble the red loops in Fig. 3 (red area - the unstable region) [4]. The experimental evolution of the P-E loops is a strong indication that wakeup/poling should come from a AFE-FE phase transformation.

For the first time, we show how the evolution of the polarization reversal kinetic barriers with the applied strain, phase transformations kinetic barriers (m P2₁/c → o Pca2₁, o Pbca → o Pca2₁) and/or the (de)stabilization of the tetragonal atomic arrangement (P4₂nmc) with strain can explain the poling (AFE→FE) behavior. We discover *four different phase transformations* that *can reproduce the AFE pinched P-E loop*. Also, we show that defects such as V_O *can significantly lower the polarization barriers* in these materials. The results show the importance of the defects, film morphology/strain control on the nm scale for designing high-performance FEFET devices.

II. TEST STRUCTURES AND METHODOLOGY

Thin films *down to 4nm thickness* have been deposited and characterized in 100x100μm metal-insulator-metal (MIM) configuration between 10nm TiN electrodes for short-loop assessment of ferroelectric properties.

First-Principles simulations were carried out using Density Functional Theory in the QuantumEspresso package [5]. We combined ultrasoft pseudo-potentials with planewaves and PBE functional to compute the Nudged-Elastic Band kinetic barriers on 13 images with statically applied strain (mechanical from electrode interfaces or from piezoelectric response of the material) on the unit cell.

III. HFₓZR₁₋ₓO₂ FILM MORPHOLOGY

$Hf_xZr_{1-x}O_2$ film morphology is rather rich in crystallographic phases. Lowest-energy structures for HfO_2 are listed in Fig. 4a, labeled by their symmetry/energy ordering [6]. *o-I/o-III/t-I* phases do not show a low-energy transformation to a non-centrosymmetric system and most importantly, their diffraction pattern is distinctive and missing in the experimental measurements. Therefore, we will refer to the *o-II* (Pbca)/o-*IV* (FE Pca2₁) as *o* phase (their XRD patterns are difficult to distinguish), *t-II* (AFE P4₂nmc) as *t* and cubic/monoclinic as *c/m*. Strained systems can have several different atomic positions (*c/t/*FE) for the same cell shape (*c/t/o*). Thus, when discussing phase transformations, we will refer to cell shape changes under strain, but within each case, a FE↑ → FE↓ polarization reversal proceeds along A-path (Pbcm) or B-path (Fm3m or P4₂nmc centrosymmetric atomic arrangement, Fig. 4b)

By means of a XRD intensity fitting procedure (Fig. 5), we can estimate the approximate crystalline phase ratio in our films (Fig. 6). The amount of *c* phase is insignificant, only a hint of *o* phase, more *m* and predominantly *t* phase is found in the film (Fig. 6).

IV. POLARIZATION REVERSAL PATHS

Polarization pathways have been under scrutiny in recent years [7, 8]. Maeda et.al. identified six ferroelectric switching pathways in *o-IV* (Pca2₁) FE HfO_2 [8], two of them have high energy barriers, the other four pass through three types of centrosymmetric structures: A-path (Pbcm) and B-path (tetragonal P4₂nmc and cubic Fm3m shown in Fig. 4b) during the switching of polarization. For the relaxed o-HfO_2 structure, the B-path (P4₂nmc mechanism) is lowest in energy (Fig. 7), but is strongly influenced by the strain in the system: under in-plane compressive strain stabilizes the centrosymmetric

978-1-7281-1988-5/18 $31.00 © 2018 IEEE

tetragonal structure (Fig. 13b). The A-path mechanism is the least influenced by the strain (not shown).

V. V_O DEFECTS

Defects in materials break the local symmetry, hence can dramatically change the energy landscape for the polarization reversal. Here we investigated the effect of ~ 2% oxygen vacancy on the kinetic barriers for polarization reversal following A/B-path (Fig. 8). The B-path (only P4$_2$nmc mechanisms Fig. 8a-c) is energetically favored when V_O is positively charged, with a 50-70% lower barrier. The A-path also shows reduced kinetic barriers with V_O and charge injection (Fig. 8d-f), but the barriers are higher if compared to B-path.

VI. PHASE TRANSFORMATION UNDER STRAIN

Hf/Zr-O bonds in HfZrO$_x$ are very sensitive to the strain applied to the system, be that external (substrates/ interfaces) or internal (Hf:Zr ratio or other dopants). High-resolution TEM show several domains of different phases ($c/o/t/m$) in the same crystalline grain [2]. Here we computed the kinetic barriers for some of these phase transformations and polarization reversal barriers (where applicable) in the absence of external fields. All the described phase transformations are possible mechanisms for wakeup/poling.

A. Monoclinic-orthorhombic (m P2$_1$/c → o Pca2$_1$)

As experimentally observed, the m-o transformation can proceed with a gliding mechanism, given the right strain tensor (Fig. 9). This process requires less than 230meV/f.u., which is of the same order of magnitude with FE polarization reversal barrier (Fig. 10). The m-o transformation creates polarized phase, the potential energy curve resembles the triple-well AFE curve (nonpolar m between FE↑ and FE↓).

B. Orthorhombic-Orthorhombic (o-II Pbca → o-IV Pca2$_1$)

The atomic configuration in the o-I structure was found to be incompatible with polarization of any type (FE or AFE). o-II, on the other hand, consists of two o-IV FE cells with the polarization anti-aligned (AFE phase). FE alignment of the two sub-cells must cross the same energy barrier as the normal FE switching, however, in this case the A-path mechanism is expected to predominate over B-path (Fig. 11a). The triple-well potential energy curve should experimentally exhibit P-E loops, resembling the FE ones (Fig. 3a-b) – after the first FE polarization switching under electric field, the AFE minimum will be in the unstable region, and hence is inaccessible. Spontaneous FE-AFE relaxation is prohibited by large kinetic barriers. Since it is difficult with XRD to discriminate between the two orthorhombic phases, it is reasonable to assume that both are present. The initially o-II (AFE) o-phase will transform into o-IV (FE) during the poling /strain relaxation stage, therefore increasing the total polarization of the film. Interestingly, an increase in size of the in-plane cell lattice of the model results in the stabilization of the FE alignment (Fig. 11b). We can regard the energy difference of the two phases as the lateral dipole coupling energy, which is very weak and disappears under small in-plane tensile strain.

C. Tetragonal-Cubic

A compressive strain along the C axis in the t phase destabilizes the centrosymmetric arrangement, leading to FE atomic arrangement (Fig. 12) [10]. In other words, a tensile strain on the tetragonal long axis leads to the stabilization of a non-polar system, but upon the strain relaxation, the FE atomic arrangement becomes more stable. The potential energy curve, corresponding to the experimentally fitted strain of the t phase shows two metastable local minima (Fig. 12 brown arrows/ configurations) that are non-polar. Hence a strain relaxation process in the tetragonal P4$_2$nmc system or polarization of the nonpolar distorted t/c configuration can also account for the poling effect.

D. Tetragonal-Orthorhombic

Applying a uniaxial elongation of the tetragonal in-plane cell parameter, we can observe a strong destabilization of the P4$_2$nmc centrosymmetric atomic arrangement (Fig. 13a blue arrow) during polarization reversal along the B-path. It indicates that t phase under tensile in-plane strain would polarize (t-o transition). The Fm3m mechanism of the same strain shows reduction of the NEB barrier (Fig. 13b), whereas the A-path mechanism is not impacted (not shown). Since the deposited film show ~75% tetragonal (AFE) phase (Fig. 6), we conclude that this is the predominant mechanism for P-E loop wakeup.

The change in Hf/Zr ratio has a linear impact on the cell size (Fig. 13c), therefore an increase in Zr content in HfO$_2$ leads to an internal strain: in our fixed-cell polarization reversal NEB simulations, the Zr content affects the barrier in the same way as the compressive strain does (Fig. 13d,a).

VII. CONCLUSIONS

The non-polar m phase under a gliding strain can transform into the FE o-Pca2$_1$ phase with a kinetic barrier, comparable to the A-path polarization reversal barrier. We also show that in the o-II (Pbca) and t (P4$_2$nmc) phases, the initial AFE atomic configuration can be transformed under certain strain into o-Pca2$_1$ FE and destabilizes the triple-well towards a double-well potential, which can explain the observed poling effect. Experimental film composition suggests the $tetragonal$ → $orthorhombic$ phase transformation is the dominant mechanism.

Additionally, oxygen vacancies have the capability to break the symmetry and switch successively the O atoms instead of simultaneously, in some cases reducing the polarization reversal barriers by >50%. It highlights the important role of the O vacancies as domain nucleation centers, especially when positively charged.

In conclusion, strain/defect control is key to a stable FEFET operation. The described mechanisms show the possibility of an AFE pinched loop even before considering the domain-wall pinning/defect/interface/relaxation effects.

ACKNOWLEDGMENTS

This work was carried out in the framework of the imec Ferroelectric Program.

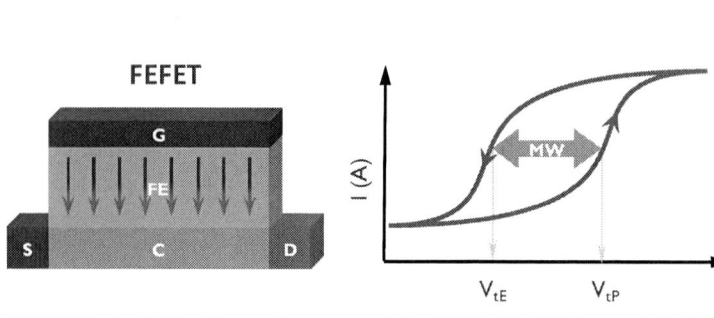

Fig. 1 FEFET device (left) and hysteretic memory window (MW) of the I-V characteristic (right).

Fig. 2 Experimentally measured I-E (top) and extracted P-E loops (bottom) on a 5.4nm $HfZrO_2$ film after 1-10^5 cycles.

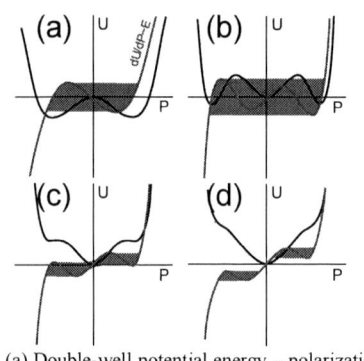

Fig. 3 (a) Double-well potential energy – polarization (U-P) curves (black), dU/dP (blue) and unstable regions (red area, P-E loops), (b) triple-well comparable energy, (c-d) more and more stable central minimum (AFE)

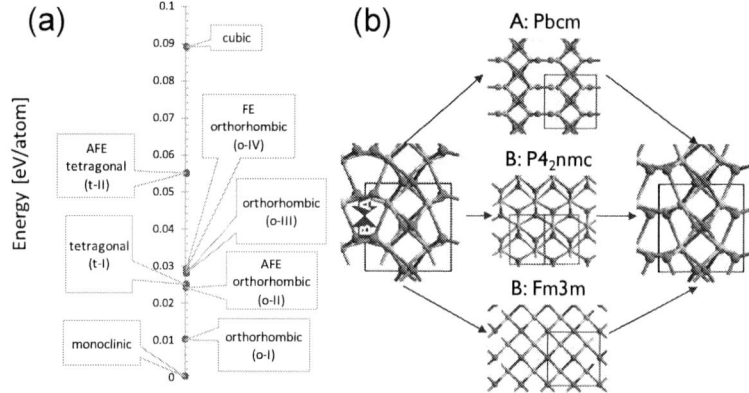

Fig. 4 (a) Energy ordering of crystalline phases in HfO_2. (b)Polarization reversal path in FE phase of $HfZrO_x$ can proceed through three centrosymmetric atomic arrangements: A-path: Pbcm and B-path: $P4_2nmc$, Fm3m.

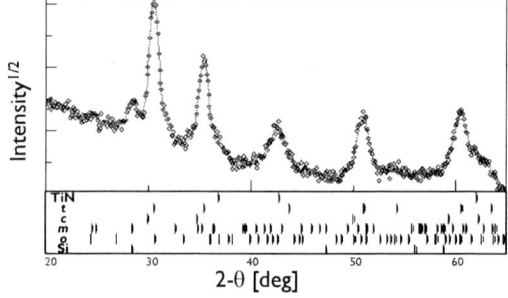

Fig. 5 XRD pattern fitting to a mixture of *t, c, m* and *o* phases.

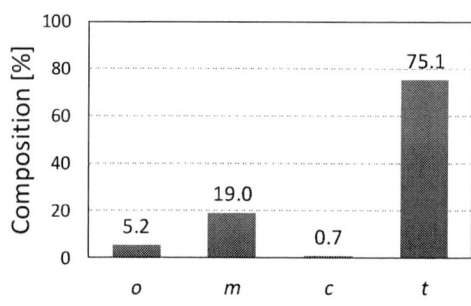

Fig. 6 Crystalline phase composition of $HfZrO_2$ film.

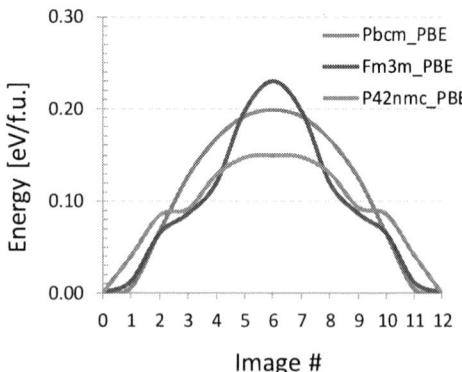

Fig. 7 NEB kinetic barriers in *o*-HfO_2 relaxed FE phase for A-path (Pbcm) and B-path (Fm3m and $P4_2nmc$) lowest-energy mechanisms (polarization in C direction)

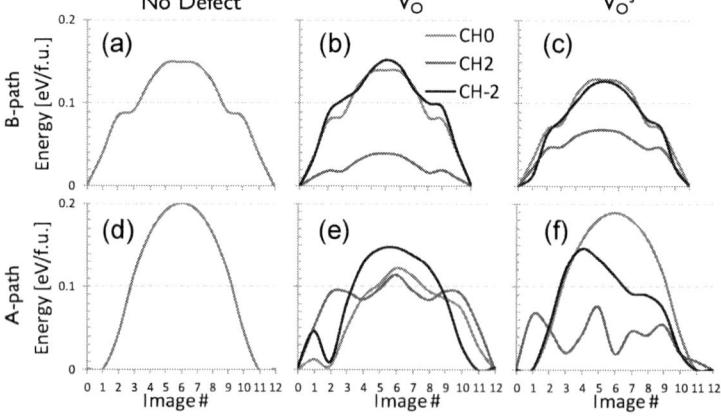

Fig. 8 NEB kinetic barriers for polarization reversal with 0/+/-2 charged (CH0/CH2/CH-2) systems in non-defective system (a,d), non-switching V_{O4} (b,e) and switching V_{O3} (c,f).

978-1-7281-1988-5/18 $31.00 © 2018 IEEE

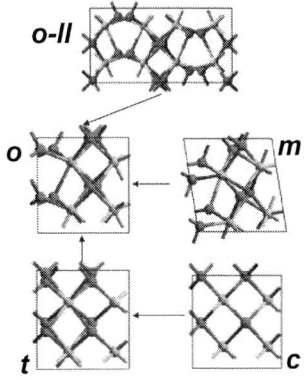

Fig. 9 Schematics for the strain / phase transformations between monoclinic (m), orthorhombic (o), tetragonal (t) and cubic (c) phases in Hf$_{1-x}$Zr$_x$O$_x$.

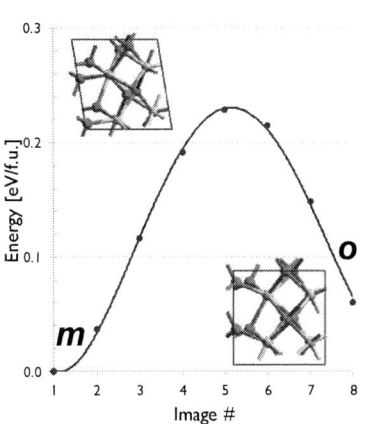

Fig. 10 NEB kinetic barrier for the monoclinic-orthorhombic phase transition.

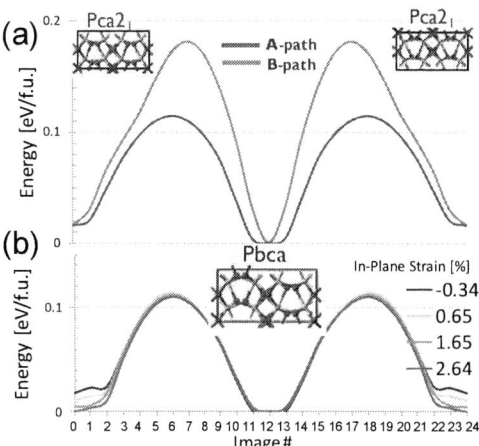

Fig. 11 NEB kinetic barriers for (a) Pbca (AFE) – Pca2$_1$ (FE) phase transformation and (b) in-plane tensile strain stabilization of FE phase. A-path mechanism is most probable.

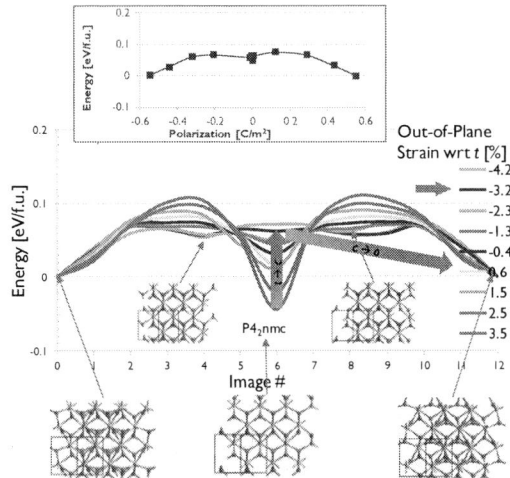

Fig. 12 NEB kinetic barriers in t phase for the FE↑ → FE↓ polarization reversal process under compressive out-of-plane strain on C-axis (positive for tensile) with B-path mechanism. Blue arrow: tetragonal destabilized to unstable cubic phase that polarize towards o phase. Top Inset: five central images have no polarization.

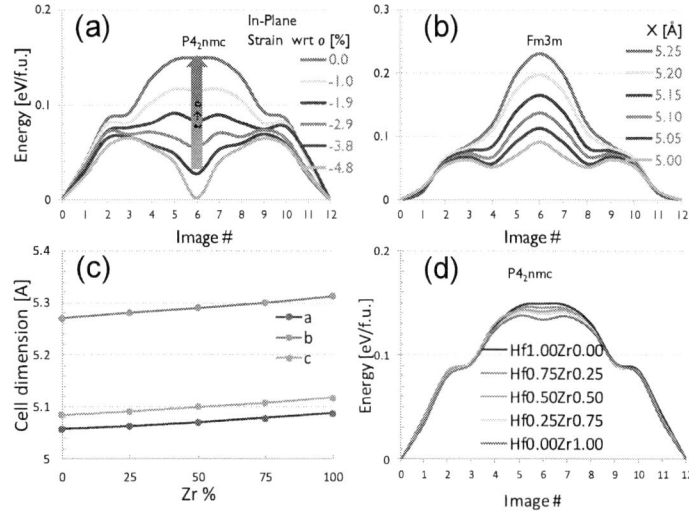

Fig. 13 B-path NEB kinetic barriers in o phase for the FE↑ → FE↓ polarization reversal process under B-axis strain (C-axis relaxed) for (a) P4$_2$nmc and (b) Fm3m mechanisms. (c) Increasing Zr % leads to a linear increase of equilibrium cell parameters and (d) slightly stabilizes tetragonal atomic arrangement (fixed cell).

REFERENCES

[1] S. Dunkel *et al.*, "A FeFET based super-low-power ultra-fast embedded NVM technology for 22nm FDSOI and beyond," (in English), *2017 Ieee International Electron Devices Meeting (Iedm)*, Proceedings Paper p. 4, 2017.

[2] E. D. Grimley, T. Schenk, T. Mikolajick, U. Schroeder, and J. M. LeBeau, "Atomic Structure of Domain and Interphase Boundaries in Ferroelectric HfO2," (in English), *Advanced Materials Interfaces*, Article vol. 5, no. 5, p. 9, Mar 2018, Art. no. 1701258.

[3] H. Mulaosmanovic *et al.*, "Switching Kinetics in Nanoscale Hafnium Oxide Based Ferroelectric Field-Effect Transistors," *Acs Applied Materials & Interfaces,* vol. 9, no. 4, pp. 3792-3798, Feb 2017.

[4] X. Y. Lu, H. Li, and W. W. Cao, "Landau expansion parameters for BaTiO3," (in English), *Journal of Applied Physics,* Article vol. 114, no. 22, p. 6, Dec 2013, Art. no. 224106.

[5] P. Giannozzi *et al.*, "QUANTUM ESPRESSO: a modular and open-source software project for quantum simulations of materials," (in English), *Journal of Physics-Condensed Matter,* Review vol. 21, no. 39, p. 395502, Sep 2009.

[6] A. Jain *et al.*, "Commentary: The Materials Project: A materials genome approach to accelerating materials innovation," *Apl Materials,* vol. 1, no. 1, Jul 2013, Art. no. 011002.

[7] T. D. Huan, V. Sharma, G. A. Rossetti, and R. Ramprasad, "Pathways towards ferroelectricity in hafnia," (in English), *Physical Review B,* Article vol. 90, no. 6, p. 5, Aug 2014, Art. no. 064111.

[8] T. Maeda, B. Magyari-Kope, Y. Nishi, and Ieee, "Identifying Ferroelectric Switching Pathways in HfO2: First Principles Calculations Under Electric Fields," (in English), *2017 Ieee 9th International Memory Workshop (Imw)*, Proceedings Paper pp. 107-110, 2017.

[9] E. D. Grimley *et al.*, "Structural Changes Underlying Field-Cycling Phenomena in Ferroelectric HfO2 Thin Films," (in English), *Advanced Electronic Materials,* Article vol. 2, no. 9, p. 7, Sep 2016, Art. no. 1600173.

[10] S. E. Reyes-Lillo, K. F. Garrity, and K. M. Rabe, "Antiferroelectricity in thin-film ZrO2 from first principles," (in English), *Physical Review B,* Article vol. 90, no. 14, p. 5, Oct 2014, Art. no. 140103.

Characterization Methodology and Physical Compact Modeling of in-Wafer Global and Local Variability

Krishna Pradeep[1,2], Thierry Poiroux[3], Patrick Scheer[1], André Juge[1], Gilles Gouget[1], and Gérard Ghibaudo[2]

[1]STMicroelectronics, 850 rue Jean Monnet, 38926 Crolles Cedex, France, [2]IMEP-LAHC, MINATEC Campus, 3 Parvis Louis Néel, 38016 Grenoble Cedex 1, France, [3]CEA-LETI, MINATEC Campus, 38054 Grenoble Cedex 9, France

Email : krishna.pradeep@st.com

Abstract— A unified, industrially compatible methodology to characterize and model in-wafer variability at different spatial scales, with addressable array test structures is proposed. Using a physics-based compact model, a single statistical model for both local and global variability is developed for the first time. The proposed method and model are validated using 28 nm FD-SOI devices and the dependence of dominant sources of variability on bias and device geometry is evaluated.

I. INTRODUCTION

Managing variability is a major concern in advanced process nodes [1]. From circuit design perspective, the in-wafer variability can be divided into inter-die global variability (GV) and intra-die local variability (LV) [2]. While a lot of discussion on LV, often modeled using inverse proportionality to √WL [3], can be traced in literature, analysis and modeling of GV is less reported. This work proposes a novel, systematic methodology to electrically characterize and model all in-wafer variabilities, both LV and GV, using addressable array test structures. The wafer to wafer GV is ignored in this work as in-wafer GV is reported to be more significant [4]. The following sections present the three stages of the proposed methodology: data acquisition, data analysis and physics-based compact modeling.

II. DATA ACQUISITION

All measurements are done on 28 nm FD-SOI n-MOSFETs in a single 300 mm wafer. The variability in thicknesses is electrically extracted from split C-V data on the same wafer, using robust extraction methods [5]. A low leakage addressable array test structure, which permits access to each transistor in an array of transistor pairs using an address decoder, while limiting the leakage current from unselected devices, is used to measure I-V data. In-wafer total variability (TV), which is the quadratic sum of GV and LV, is studied by measuring the same device (DUT) in the array on all 73 dies, whereas LV is analyzed by measuring all 256 DUT pairs in the same die. Multiple geometry devices are measured, at different biases, in both cases to analyze the impact of scaling. This approach enables statistically relevant data acquisition to study variability at different spatial scales in a single wafer.

III. DATA ANALYSIS

The drain current mismatch [6], $\Delta I_d / I_d \equiv \ln(I_{d2}/I_{d1})$ is used to analyze the variability at all spatial scales. In case of TV and GV, this metric is computed with respect to the median die of the wafer. The same metric is also computed for Y-function

$= I_d / \sqrt{g_m}$ (less sensitive to S/D series resistance (R_{sd})) [7] and g_m / I_d (less sensitive to low field mobility (μ_0), L, W), to help segregate different sources of variability. The bias dependent variance of these metrics is used to quantify variability.

1) In-wafer total variability (TV)

The computed metrics for two geometries (W/L = 1 μm/1 μm & 0.21 μm/0.1 μm) on all dies are plotted in Figs 1a-c. While GV is expected to scale weakly with device area, LV is known to scale strongly [3]. Here the spread of the metric on I_d, Y and g_m / I_d are observed to be slightly higher for the small device compared to the big one, corroborating the fact that the measured quantity is indeed a combination of GV and LV. This is better evidenced in Figs 2a,d,g demonstrating a clear gradient among the five different device sizings. Interestingly, very similar variance is observed in linear and saturation regimes (Figs 2b,e,h) and for different body bias, V_b (except body effect) (Figs 2c,f,i).

2) Intra-die local variability (LV)

Compared to TV, the spread of LV mismatch metrics is larger in the tiny device (W/L = 0.08 μm/0.03 μm) and smaller in the big device (W/L = 1 μm/1 μm) (Figs 1d-f). As reported in [3, 6, 8], a clear area dependence is observed in Figs 3a,d,g for variance of the metrics, irrespective of the applied V_g. The V_d and V_b dependences of the variances (Figs 3b,e,h and Figs 3c,f,i respectively) are similar to those observed for TV. It is interesting to note that the shape of V_g dependent variance of the metrics are similar for both TV and GV, suggesting same sources of variation in both spatial scales.

3) Die-to-die global variability (GV)

Removing intra-die LV from measured in-wafer TV gives the die-to-die GV. The addressable array test structure can be used to estimate this GV. The LV contribution is considerably reduced by averaging over 4 DUTs from the same die, and the remaining variance is mainly due to GV. Here a good compromise is achieved between measurement complexity (time to measure all DUTs in all dies) and accuracy of the estimated GV. Fig. 4 compares LV, GV & TV variances of the metrics at sub threshold and moderate inversion. The estimated GV variance is observed to be almost constant for the bigger devices, while it increases slightly for the tiny device (where the estimation is not as accurate due to the magnitude of LV). A Pelgrom-like [3] trend can be observed for LV. On the other hand TV, as expected, shows a trend dominated by a constant for large area and Pelgrom-like for small area devices.

978-1-7281-1988-5/18 $31.00 © 2018 IEEE

IV. PHYSICS-BASED COMPACT MODELING

4) Split C-V extracted effective mobility

The bias dependent effective mobility is extracted from split C-V [9] on all dies in the wafer at different temperatures. The metric $\Delta\mu_{eff}/\mu_{eff}$ and its variance can be used to study the TV of the effective mobility. A weak temperature dependence is observed in measured mobility TV (Fig. 5b, symbols).

Considering that mobility is affected mainly by Coulomb (CS), phonon (PS) and surface roughness (SR) scattering, a simple model based on Matthiessen's rule can be derived as (1), where E is the transverse electric field (in V/cm), T is the temperature (in K) and N is the oxide charge density (in cm^{-2}) regulating CS. In Fig. 5a, the proposed model is shown to accurately reproduce the measured effective mobility on the median die. The statistical model (2) [10], where X is any property whose variability is being analyzed, along with (1), can be used to model the measured mobility TV. By considering statistical variations on N ($\sigma_N \approx 10^{-10}$ cm^{-2}), it is observed that (2) can reproduce most of the mobility variance. Some variation in PS-limited mobility ($< 1\%$) is required to model the remaining variance in moderate inversion (Fig. 5b).

This observed dependence of mobility on oxide charge density is further studied using PBTI stress measurements on the same wafer. A simple model given by (3) [11], where ΔQ_{ox} is the change in oxide charge density due to PBTI stress is used to study this effect. A V_g stress of 2.5 V is applied to the W/L = 5 μm/4 μm device for 2000 s. Both I_d-V_g and C-V measurements are performed at regular intervals and effective mobility is calculated as explained earlier (Fig. 6b). The V_T shift caused by the stress is verified to be same in both I_d-V_g and C-V data (Fig. 6a). The value of the reciprocal CS parameter, $1/\alpha$, extracted using (3) at each V_g bias, is plotted in Fig. 7 (symbols). It is observed that, interestingly, the effect of oxide charge density is not restricted to weak inversion but propagates up to the onset of strong inversion. This effect is not accounted for in (1), requiring the small variation of PS-limited mobility in (2) to reproduce measurements. This extra variation can be avoided in a model adapted to reproduce the observed CS effects up to the onset of strong inversion, as explained in next section.

5) Leti-UTSOI compact model (CM)

The Leti-UTSOI CM [12] is an industry accepted physics-based model for FD-SOI devices. A recent work [4] reported in-wafer V_T variation reproduction using this CM. The CM is calibrated on C-V and I-V curves from different geometries at different biases (Fig. 8). In Leti-UTSOI, CS is controlled by the parameter CSO, and hence ($\partial\mu^{-1}/\partial CSO$) behaves like α in (3), which is verified in the calibrated model (Fig. 7). Also, the CM is confirmed to reproduce general trends of electrical FOMs with thickness variations (Fig. 9). The reduced spread in CM implies the presence of other variability sources on Silicon.

6) Complete statistical model

Equation (2) along with sensitivities from Leti-UTSOI CM can be used to model all in-wafer variabilities. This approach was reported recently to model inter-die electrostatic variability in terms of thickness variations [10]. Equation (2) is modified as (4) to include correlations between sources of variability, if any. Starting with split C-V extracted variances of thicknesses, the TV, GV and LV of I_d, Y-function and g_m/I_d are modeled with (2) by including additional sources of variability (consistent with previously reported works, Table II), namely effective gate work function (Φ_m), oxide charge density (through CS, as seen in the previous section), R_{sd}, sub-threshold slope (SS) and lateral dimensions (L & W). A negative correlation (r ≈ -0.5) is observed between Φ_m & SS and also between CS & R_{sd}. The same set of variability sources is found to be sufficient to reproduce both TV and GV, while the vertical thicknesses can be ignored (in geometries studied, up to W/L = 1 μm/1 μm) for LV (Fig. 10). On the other hand, DIBL variation needs to be accounted for in LV of devices with L = 30 nm. The accuracy of the model is verified on different geometries and different V_b, V_d biases. The geometry dependence of σ of the variability sources is shown in Figs 11 & 12. While Φ_m, CS, DIBL and SS follow a Pelgrom-like trend, R_{sd} (similar trends reported in [13]) and L scale with W, and W scales with L.

Percentage contributions (5) help identify dominant sources of variability at each operating region (Fig. 13 (GV) & 14 (LV)). As expected, irrespective of geometry, bias or spatial scale of variability, g_m/I_d is affected mainly by SS below V_T, Φ_m around V_T and CS & R_{sd} in moderate/strong inversion. DIBL plays a noticeable role in LV of I_d at saturation, below V_T. For I_d on all geometries, Φ_m plays a dominant role in variability below and around V_T (Silicon channel thickness (T_{Si}) becoming significant in weak inversion for GV), while CS & R_{sd} is prevailing in moderate/strong inversion. An increasing influence of SS in g_m/I_d variance is observed in smaller devices below V_T. Interestingly the role of L & W is $< 20\%$ in all cases.

V. CONCLUSION

A novel, unified methodology to electrically characterize and model all in-wafer variabilities (TV, GV & LV) is introduced and applied to 28 nm FD-SOI MOSFETs. Simple mobility models and PBTI studies are used to identify the physical sources of variability and calibrate Leti-UTSOI CM. Finally, a complete statistical model using Leti-UTSOI is demonstrated, which can reproduce in–wafer variability, irrespective of its spatial scale (global and local variability), device geometry and operating region, with a single, reduced set of variability sources. Employing this model, the dominant sources of variability in each case are identified.

ACKNOWLEDGMENT

The authors would like to thank the technology team for processing samples.

REFERENCES

[1] K. J. Kuhn, ASMC 2010 [2] A. Juge et al., SISPAD 2016 [3] M. Pelgrom et al., JSSC 1989 [4] K. Pradeep et al., SSE 2018 [5] K. Pradeep et al., ICMTS 2017. [6] E. G. Ioannidis et al., EDL 2015 [7] G. Ghibaudo, Electron. Lett 1988 [8] P. Drennan et al., IEDM 1999 [9] K. Romanjek et al., EDL 2004 [10] K. Pradeep et al., EUROSOI-ULIS 2017 [11] K. Bennamane et al., Electron. Lett 2013 [12] T. Poiroux et al., IEDM 2013 [13] E. G. Ioannidis et al., SSE 2016 [14] A. Paul et al., IEDM 2013 [15] S. Makovejev et al., JLPEA 2014 [16] T. A. Karatsori et al., TED 2017

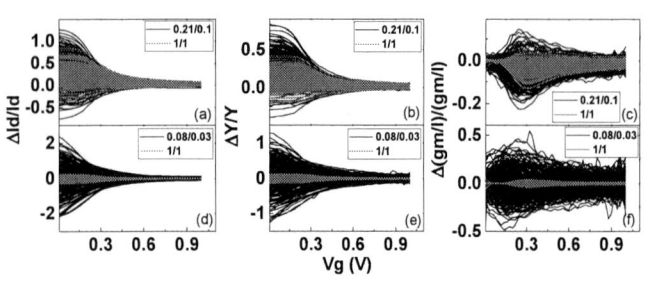

Fig. 1. $\Delta I_d/I_d$, $\Delta Y/Y$ and $\Delta(g_m/I_d)/(g_m/I_d)$ plots for 2 different geometries (W/L ratios in legend), in case of die-to-die TV (a,b,c) and intra-die LV (d, e, f) showing its bias dependent spread.

$$\begin{Bmatrix}\dfrac{1}{\mu_{eff}} = \dfrac{1}{\mu_{CS}} + \dfrac{1}{\mu_{PS}} + \dfrac{1}{\mu_{SR}} \\[2mm] \mu_{CS} = \mu_{0CS} * \left(\dfrac{1e10}{N}\right)\left(1 + \dfrac{E}{10^4}\right)\left(\dfrac{T}{300}\right)^{1.5} \\[2mm] \mu_{PS} = \mu_{0PS}\left(\dfrac{T}{300}\right)^{-1.3} \qquad \mu_{SR} = \mu_{0SR}\left(\dfrac{E}{10^6}\right)^{-2}\end{Bmatrix} \quad (1)$$

$$\sigma^2\left(\frac{\Delta X}{X}\right) = \sum_j \left(\frac{\partial lnX}{\partial P_j}\right)^2 \sigma_{P_j}^2 \quad (2) \qquad \frac{1}{\mu_{eff}} = \frac{1}{\mu_{0eff}} + \alpha \times \Delta Q_{ox} \quad (3)$$

$$\sigma^2\left(\frac{\Delta X}{X}\right) = \sum_j \left(\frac{\partial lnX}{\partial P_j}\right)^2 \sigma_{P_j}^2 + 2\sum_{i,j} \rho_{i,j} \frac{\partial lnX}{\partial P_j} \frac{\partial lnX}{\partial P_i} \sigma_{P_i}\sigma_{P_j} \quad (4)$$

$$\%P_i = \frac{(\partial lnX/\partial P_i)^2 \sigma^2(P_i)}{\sum_j (\partial lnX/\partial P_j)^2 \sigma^2(P_j)} \times 100 \quad (5)$$

Table I. Equations (1) Simple mobility model. (2) Variance statistical model (3) Oxide charge effects on mobility. (4) Modified statistical model with correlations (5) Percentage contribution of each variability source.

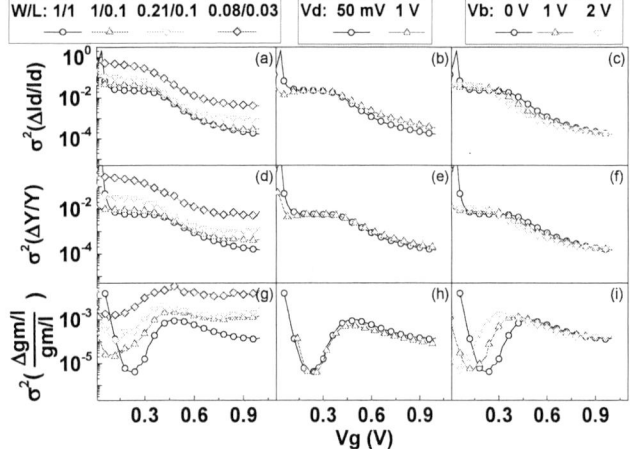

Fig. 2. Variance plots of I_d, Y & g_m/I_d for TV. (a, d, g) Scaling with device sizing at $V_b = 0$ V & $V_d = 50$ mV, showing increase for smaller devices. (b, e, h) V_d dependence at $V_b = 0$ V for W/L = 1 μm/1 μm device, showing similar trends. (c, f, i) V_b dependence at $V_d = 50$ mV for W/L = 1 μm/1 μm device, showing only body effect.

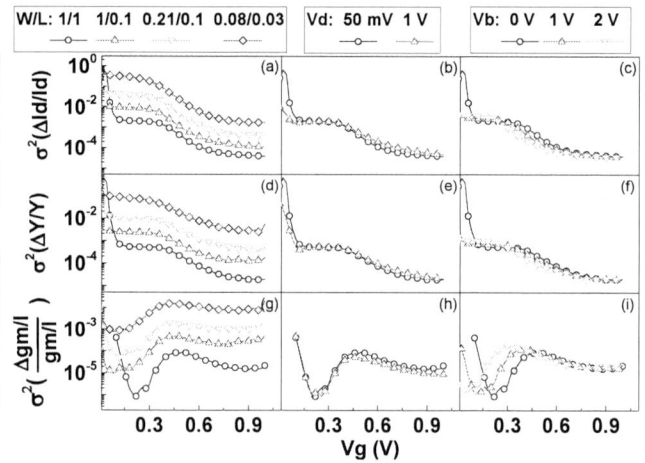

Fig. 3. Variance plots of I_d, Y & g_m/I_d for LV. (a, d, g) Scaling with device sizing at $V_b = 0$ V & $V_d = 50$ mV, showing increase for smaller devices. (b, e, h) V_d dependence at $V_b = 0$ V for W/L = 1 μm/1 μm device, showing similar trends. (c, f, i) V_b dependence at $V_d = 50$ mV for W/L = 1 μm/1 μm device, showing only body effect.

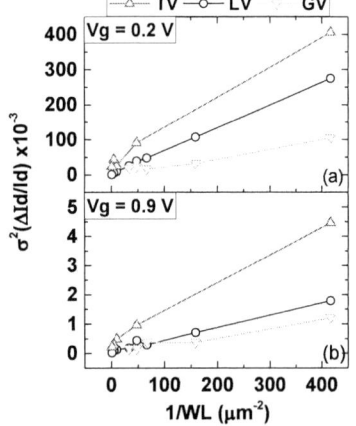

Fig. 4. Pelgrom plot of $\Delta I_d/I_d$ @ (a) $V_g = 0.2$ V and (b) 0.9 V for LV, GV and TV @ $V_b = 0$ V and $V_d = 50$ mV, showing its area scaling. While GV variance is almost a constant, a Pelgrom-like trend is observed for LV.

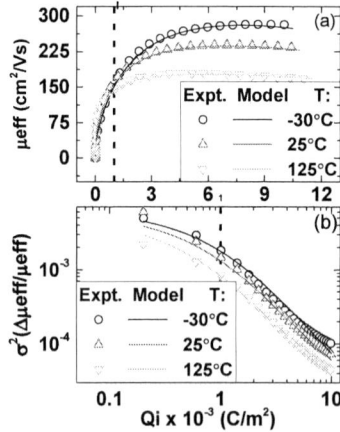

Fig. 5. Fitting of the (a) Split C-V extracted mobility (collapsing to 0 in deep sub threshold, as in [9]) and (b) the in-wafer variance of mobility on W/L = 1 μm/1 μm devices with simple mobility model (1). The value of Q_i at V_T ($\approx$ 1e-3 C/m²) is marked with dashed line. CS dominance is expected in low fields (around V_T). $V_b = 0$ V & $V_d = 50$ mV.

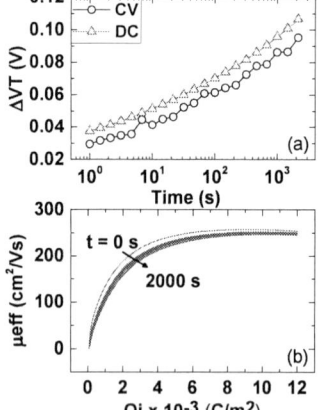

Fig. 6. Results from PBTI measurement on W/L = 5 μm/4 μm device, showing the time dependent shift in (a) V_T extracted on C-V and I-V data (b) and split C-V extracted mobility. Reciprocal CS parameter (1/α) is calculated from this shift of effective mobility using (3) at each Q_i.

978-1-7281-1988-5/18 $31.00 © 2018 IEEE

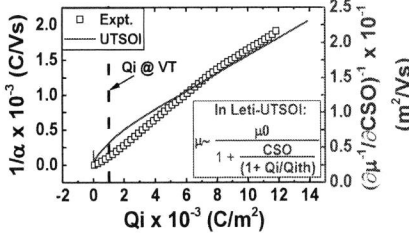

Fig. 7. Equation (3) extracted $1/\alpha$ from PBTI measurements compared to $(\partial\mu^{-1}/\partial CSO)^{-1}$ calculated from the calibrated Leti-UTSOI (CM mobility dependence in inset). The measured dependence of oxide charge density is reproduced in the calibrated CM. The different scales & units in both axes are due to parameter normalizations in Leti-UTSOI.

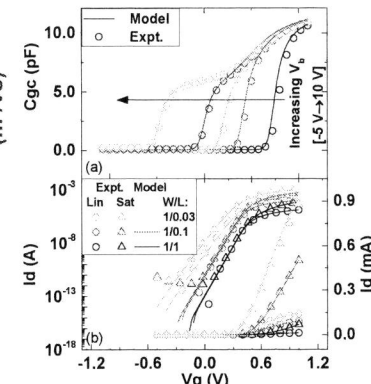

Fig. 8. (a) CV and (b) DC calibration of Leti-UTSOI, performed at different biases and different device sizings.

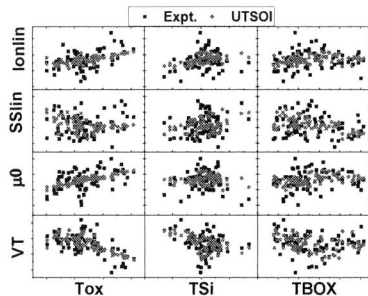

Fig. 9. As-measured (black squares) versus Leti-UTSOI (red dots) extracted wafer level correlation of DC FOMs with thicknesses, for a device geometry of $W/L = 1\,\mu m / 1\,\mu m$. A fair reproduction of general trends is achieved. Wider spread in measurement implies other sources of variability on Silicon.

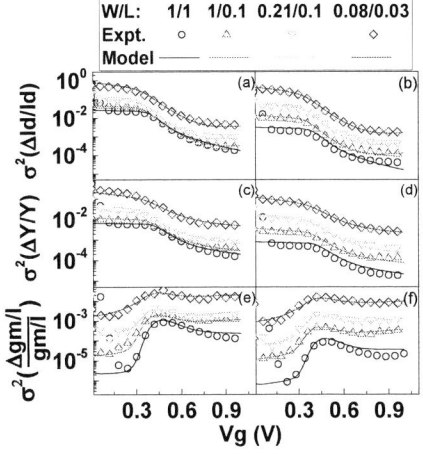

Fig. 10. Fit of proposed statistical model (4) on measurements. (a, c, e) TV and (b, d, f) LV. @ $V_b = 0$ V and $V_d = 50$ mV. Parameters used: Φ_m, CS, SS, R_{sd}, T_{Si}, T_{ox}, L, W & DIBL.

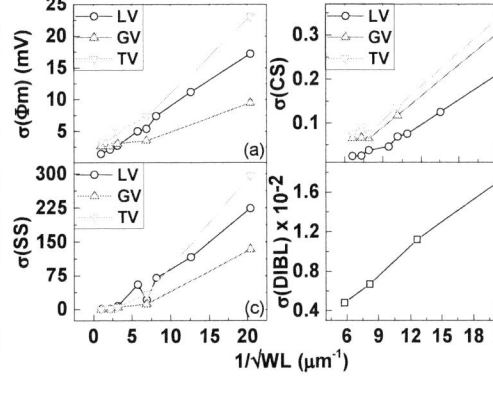

Fig. 11. Pelgrom plots for variability sources obtained from the statistical model (4) using Leti-UTSOI CM.

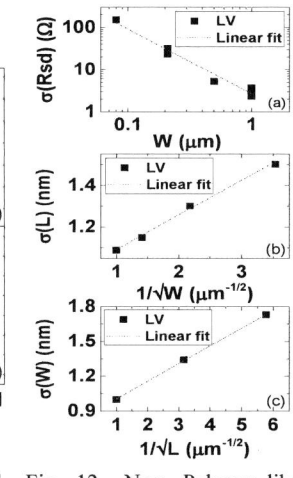

Fig 12. Non Pelgrom-like behaviour found in (a) R_{sd}, (b) L & (c) W for LV. Linear trend lines are also plotted.

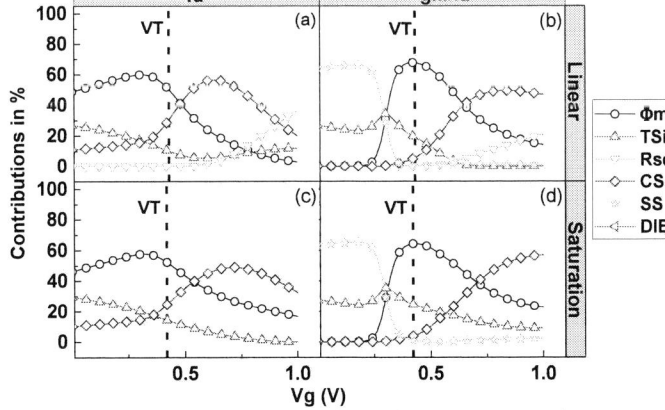

Fig. 13. Contributions (in % (5)) of identified major sources (> 20% in any region of operation) to GV at $V_b = 0$ V in $W/L = 0.21\,\mu m / 0.1\,\mu m$ device. (a, b) @ $V_d = 50$ mV. (c, d) @ $V_d = 1$ V.

Fig. 14. Contributions (in % (5)) of identified major sources (> 20% in any region of operation) to LV at $V_b = 0$ V in $W/L = 0.08\,\mu m / 0.03\,\mu m$ device. (a, b) @ $V_d = 50$ mV. (c, d) @ $V_d = 1$ V.

Ref:	Spatial scale	Analysis made on:	Identified variability sources
[8]	Local	Full I_d-V_g curves.	W, T_{ox}, V_{fb}, μ_0, L, V_{tl}, R_{sd}, N_{sub}
[14]	Local	Transistor FOMs	R_{sd}, β (=$\mu_0 C_{ox} W/L$), V_T
[15]	Global	Subthreshold I_d-V_g curves.	V_T, SS
[16]	Local	Full I_d-V_g & I_d-V_d curves.	R_{sd}, V_T, β, SS
This work	Global & Local	Full I_d-V_g, Y-V_g & g_m/I_d-V_g curves.	Φ_m, T_{ox}, T_{Si}, W, L, CS, R_{sd}, SS, DIBL. Dominant sources identified for each operating region (Figs 13 & 14)

Table II. Comparison of proposed methodology and identified sources of variability to those in literature. A unified methodology to study and model LV and GV using the same statistical model (with a unique, physical set of variability sources) is pioneered.

978-1-7281-1988-5/18 $31.00 © 2018 IEEE   387

Too Noisy at the Bottom? —Random Telegraph Noise (RTN) in Advanced Logic Devices and Circuits

Runsheng Wang[1]*, Shaofeng Guo[1], Zhe Zhang[1], Qingxue Wang[2], Dehuang Wu[2], Joddy Wang[2], Ru Huang[1]

[1]Institute of Microelectronics, Peking University, Beijing 100871, China
[2]Synopsys, Inc., Mountain View, CA 94043, USA
*Email: r.wang@pku.edu.cn

Abstract—In this paper, the recent advances of our studies on RTN are presented from device, circuit, and EDA perspectives. RTN characteristics in FinFETs are investigated and compared with planar devices. The AC RTN effect is discussed for understanding RTN impacts in practical circuit applications. Then, a new and efficient circuit simulation platform for RTN is presented for the first time, which has been implemented in HSPICE using OMI/TMI. In addition, some open questions related to RTN are discussed with outlooks.

I. INTRODUCTION

Random telegraph noise (RTN) in MOSFETs is not new actually, which has been observed in the 1980's [1]. However, recently it has attracted much more attention (e.g., [2-11]), due to its increasing amplitude as device size scaling [2], which can hurt circuit reliability due to the dynamic variability induced by RTN [12]. Therefore, it is becoming "a designer's nightmare" rather than "a device physicist's dream" [13], especially for advanced logic technologies with smaller headroom (V_{DD}–V_{TH}) and less design margin.

In this paper, the recent advances of our studies on RTN will be presented from device, circuit, and EDA perspectives. The properties of RTN in FinFETs are discussed with comparisons to planar devices. Then, AC RTN effect is highlighted for understanding RTN impacts on digital circuits. Furthermore, a new efficient circuit simulation platform for RTN is presented *for the first time*, which has been implemented in HSPICE using OMI/TMI (CMC open model interface/TSMC model interface [14]). Finally, some open questions related to RTN are discussed with outlooks.

II. RESULTS AT DEVICE LEVEL

A. Basic RTN properties: a short review

As shown in Fig. 1, the RTN in a MOSFET originates from the stochastic trapping/de-trapping of the channel carrier into/from the switching oxide traps (or border traps) in the gate dielectrics, which is a Markov process between two states (trap empty or occupied) [15,16]. The capture and emission time follow the exponential (*Exp.*) distribution, thus the RTN time constants can be extracted (e.g., Fig. 2). The RTN amplitude can be expressed with either $\Delta I_D/I_D$ or its equivalent ΔV_{TH}.

Since trap position/energy have variations (and can further interplay with device/process variations), RTN parameters have wide distributions. (1) Time constants of different RTN is found to be almost uniformly distributed in log scale [17,18], which could from nanoseconds to months. The bias dependence of time constants also varies [4], and on average τ_c reduces (τ_e increases) with increasing V_G. (2) Amplitudes of different RTN also have varied bias dependence [Fig. 3(a)], and on average it reduces with increasing V_G. Under each bias, it approximately follows a lognormal (*Logn.*) or *Exp.* distribution [19]. For one

device with multiple traps, people usually assume the traps are independent from each other, and thus treat multi-level RTN as the linear superposition of individual RTNs. Fig. 3(b) shows the measured amplitude distribution of independent RTN in FinFETs, which indicates that on average RTN in PMOS has larger amplitude than in NMOS.

B. FinFET vs. planar devices

If with the same gate stack process, the major difference of RTN properties, especially its amplitude, between FinFET and planar devices, comes from the unique 3-D geometry of Fin structure. Actually, FinFET benefits from it, because the effective width ($2H_{Fin}+W_{Fin}$) of FinFET is much larger than its layout width (~Fin pitch). Thus, RTN amplitude is smaller in FinFET than in planar device with the same layout area, as shown in Fig. 4.

Regarding the amplitude distribution, FinFETs agree better with *Logn.*, while planar FETs agree better with *Exp.* in low voltages (Fig. 5). The best way to understand RTN distribution, is using the concept of channel percolation path (PP) [20] induced by device/process variations. As shown in Fig. 6, RTNs caused by the traps on and off PP, both follow a single *Logn.* distribution, which indicates a two-stage *Logn.* function [19]. While a two-stage *Logn.* can be apparently fitted by a single *Exp.* function, which is exactly the case in planar devices. For FinFET, it actually has "intrinsic" PP even for the uniform device due to its structural nature, as shown in Fig. 7(a). Since in low V_G the majority of Fin region is quasi-uniform, while in high V_G the PP effect is clearer, the experimental results showed that the RTN amplitude distribution function changes with increasing V_G in FinFETs [Fig. 7(b)]. It is worth noting that the exact form of distribution function is important for the accurate prediction of RTN impacts on circuits towards high-σ tails.

III. UNDERSTANDING RTN EFFECT IN DIGITAL CIRCUITS: FROM DC RTN TO AC RTN

The RTN amplitude (together with the trap number) is a direct measure of the degradation of circuit static performance and static noise margin. The RTN time constants, together with the trap occupancy probability [$p=\tau_e/(\tau_e+\tau_c)$], determine the transient failure probability of the circuits. Thus for transient performance, understanding the latter is very important.

From practical circuit application perspectives, RTN strongly depends on device working conditions. It is found that RTN characteristics under AC large signal excitation (AC RTN) deviates from conventionally studied RTN under constant bias (DC RTN) [21,22]. As shown in Fig. 8, RTN time constants (and thus the trap occupancy probability) are different under DC and AC workloads with different frequencies. While RTN amplitude is frequency independent (Fig. 9).

Therefore, instead of trap occupancy probability under DC bias conditions (p_{DC}), AC trap occupancy probability (p_{AC}) is

978-1-7281-1988-5/18 $31.00 © 2018 IEEE

the key to investigate and model the dynamic trapping/de-trapping behavior of RTN [23]. The impacts of RTN on digital circuit performances, e.g., jitters of ring oscillators (RO) and transient read failures of SRAM have also been studied in details [23-25].

IV. NEW EFFICIENT CIRCUIT SIMULATION PLATFORM FOR RTN WITH OMI/TMI AND HSPICE

Accurate transient circuit simulation including RTN effect is helpful for designers. The traditional method proposed in [23,26] has an efficiency problem. As mentioned above, time constants of RTN in a design can vary in a wide range. Since the transient simulation time step (Δt) should be much smaller than the RTN time constants [Fig. 10(a)], and the total simulation time should be sufficiently large on the other hand, the traditional method can only cover a limited range of time constants in the simulation [Fig. 10(b)]. Thus it is very time consuming and impractical to conduct real-time circuit simulation that take all RTNs into consideration in the traditional framework. In this work, a two-stage RTN simulation platform, together with a newly-proposed trap occupancy probability prediction (OPP) model, is proposed and implemented in HSPICE using OMI/TMI, which enables the efficient and accurate simulation of RTN impacts on circuit functionality and timing/power characteristics.

The new RTN simulation flow is illustrated in Fig. 11, which is based on the RTN model (including time constants and amplitude models, integrated with BSIM-CMG) deployed in OMI/TMI and utilizes the interoperation of OMI/TMI with HSPICE. The two-stage simulation compose of two regular transient simulation (T_{sim}) for given circuit type. In the first simulation, bias dependent RTN time constants is evaluated based on the given testing vectors and the OPP model is generated. In this stage, the RTN effect does not impact circuit characteristics and the simulation result is the same as without RTN. In the second simulation, the trap initial state at the expected operation time (T_{exp}) is first determined based on the OPP model from stage 1, then Monte Carlo based RTN simulation [Fig. 10(a)] is conducted to capture the trap states switching event impact to circuit characteristics. The operation time, T_{exp}, is in general a much longer time span in order to take various RTN time constants into consideration.

Other techniques are also proposed and implemented into HSPICE to further improve simulation performance and RTN event coverage. This includes adaptive adjustment of simulation time step based on RTN time constant during transient simulation, RTN time constant scaling and dynamic initial state management, aiming at improving RTN events coverage within typical transient circuit simulation time for the given circuit type.

Fig. 12 shows the proposed OPP model, which can predict the trap occupancy probability under arbitrary waveform. It has been verified by both simulation (Figs. 13-14) and experimental (Fig. 15) results. The results in Figs. 16-18 show the new simulation method is able to capture the RTN events accurately at any user-specified operation time, without running the simulation from time-zero for a long time. This OPP-based analysis largely improves the simulation efficiency, which is important for designers to adopt RTN analysis in design and verification flow. Fig. 19 shows that the results of

the new method are well consistent with the brute force real-time simulation results using traditional method.

V. OPEN QUESTIONS AND OUTLOOK
A. "Anomalous" complex RTN

Some experimental data of RTN exhibit "anomalous" complex patterns, which cannot be explained by simple RTN theory. We have found two categories of complex RTNs, which are induced by the metastable trap-states and the trap coupling effect, respectively [27].

The RTN with metastable states are observed based on the "reversal RTN" data (Figs. 20-21) [28,29], which provides direct evidence of the complete 4-state trap model (Fig. 22). A detailed statistical study on this "anomalous" reversal RTN has also been performed in advanced FinFET technology [30].

The trap coupling in multi-level RTN changes both the amplitude and time constants of one trap when another trap is occupied (e.g., Fig. 23) [31]. The trap coupling mechanisms, including Coulomb repulsion and channel percolation effects have been revealed [31,32]. Strong coupling on time constants (with largely reduced trap occupancy probability) can cause RTN state loss (i.e., RTN level missing) in statistics (Fig. 24) [32]. Statistical study also shows that FinFETs have larger trap coupling strength than planar devices (Fig. 25), due to the double-side coupling mechanism (Fig. 26) [33]. A non-monotonic bias dependence of coupling strength is observed, and the strongest coupling is found around the $V_G@G_{m,max}$ (Fig. 27). In addition, the trap coupling effect can change the form of amplitude distribution functions [19,33].

Once have a proper model for the complex RTN, it can be embedded straightforwardly in the circuit simulation platform proposed in Section IV, as second-order effects.

B. BTI reliability modeling from RTN perspectives

It is believed that RTN is closely linked to the bias temperature instability (BTI) [15-18]. Therefore, on one hand, one can utilize RTN to characterize single trap for studying BTI physics, especially the metastable states [34,28,35]. On the other hand, one can utilize RTN-like model for simulating BTI aging and statistical BTI in circuits [26]. Since the RTN simulation flow proposed in Section IV is compatible with aging simulation flow, it is easy to build a unified trap-based circuit simulation platform.

C. Flicker noise modeling from RTN perspectives

It is believed that RTN is the main contributor to flicker noise in the frequency domain, especially in low-V_{DD} region, which is important for analog/RF/mixed-signal circuits. However, traditional flicker noise modeling needs updates. One thing is that, as discussed above, since RTN is workload dependent (Fig. 28), flicker noise will in turn has different features, as shown in Fig. 29, which have not been included in the standard model, such as in BSIM. Another thing is the statistical flicker noise modeling and circuit simulation, which could also be improved according to RTN understandings. All these work are being developed.

ACKNOWLEDGMENT

This work was partly supported by NSFC (61522402 and 61421005). The authors thank Y.C. Liang and K.-W. Su (TSMC) for the collaborative discussions. R. Wang thanks former students P. Ren, J. Zou, Z. Zhang, M. Luo, J. Ji, X. Jiang, D. Mao, P. Hao, M. Li, C. Liu etc. for the input, and thanks T. Grasser, Y. Cao, K.P. Cheung, Z. Ji, X. Wang, B. Cheng, A. Asenov for the helpful discussions.

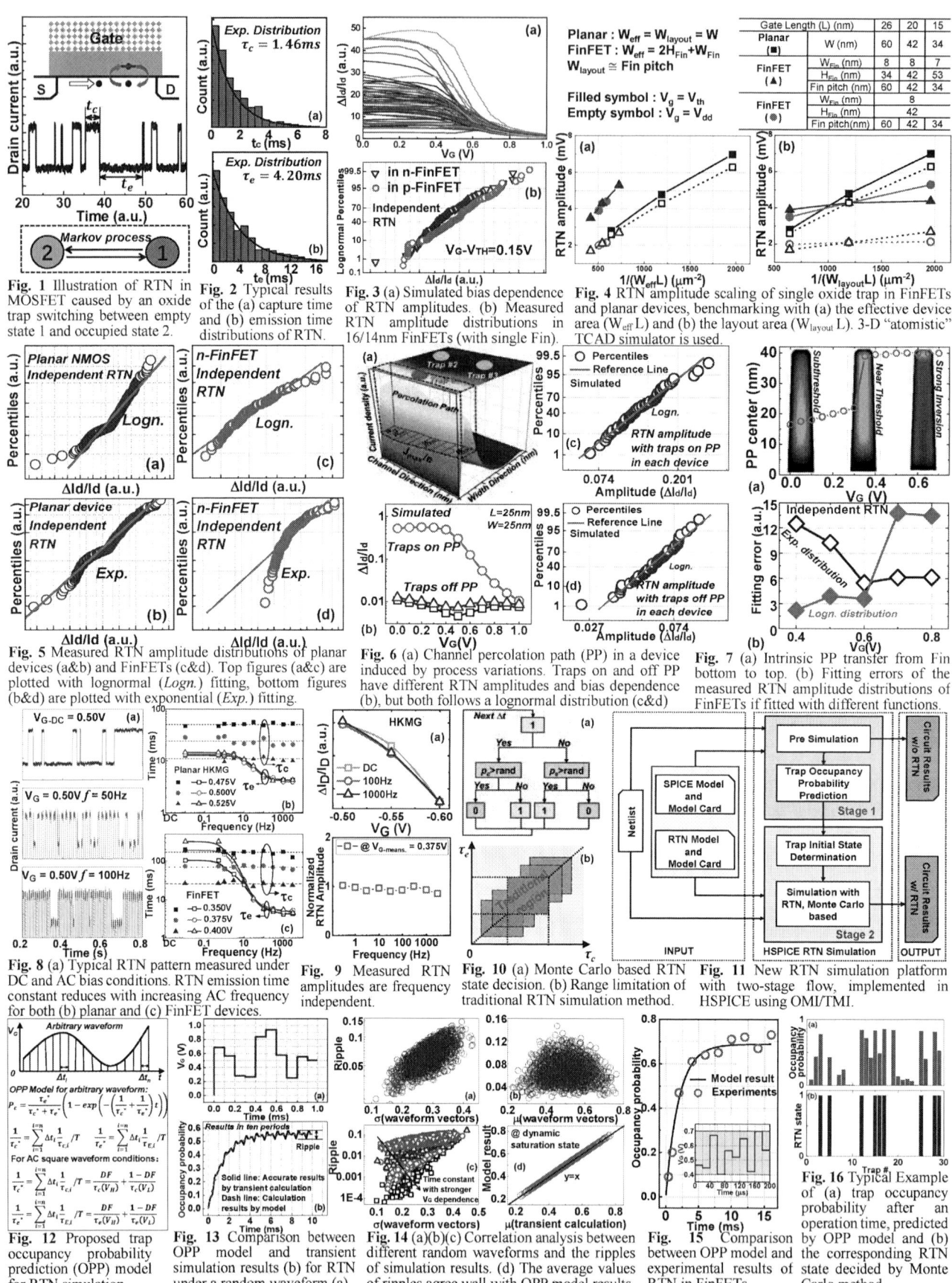

Fig. 1 Illustration of RTN in MOSFET caused by an oxide trap switching between empty state 1 and occupied state 2.

Fig. 2 Typical results of the (a) capture time and (b) emission time distributions of RTN.

Fig. 3 (a) Simulated bias dependence of RTN amplitudes. (b) Measured RTN amplitude distributions in 16/14nm FinFETs (with single Fin).

Fig. 4 RTN amplitude scaling of single oxide trap in FinFETs and planar devices, benchmarking with (a) the effective device area ($W_{eff}L$) and (b) the layout area ($W_{layout}L$). 3-D "atomistic" TCAD simulator is used.

Gate Length (L) (nm)		26	20	15
Planar (■)	W (nm)	60	42	34
FinFET (▲)	W_{Fin} (nm)	8	8	7
	H_{Fin} (nm)	34	42	53
	Fin pitch (nm)	60	42	34
FinFET (●)	W_{Fin} (nm)		8	
	H_{Fin} (nm)		42	
	Fin pitch(nm)	60	42	34

Fig. 5 Measured RTN amplitude distributions of planar devices (a&b) and FinFETs (c&d). Top figures (a&c) are plotted with lognormal (*Logn.*) fitting, bottom figures (b&d) are plotted with exponential (*Exp.*) fitting.

Fig. 6 (a) Channel percolation path (PP) in a device induced by process variations. Traps on and off PP have different RTN amplitudes and bias dependence (b), but both follows a lognormal distribution (c&d)

Fig. 7 (a) Intrinsic PP transfer from Fin bottom to top. (b) Fitting errors of the measured RTN amplitude distributions of FinFETs if fitted with different functions.

Fig. 8 (a) Typical RTN pattern measured under DC and AC bias conditions. RTN emission time constant reduces with increasing AC frequency for both (b) planar and (c) FinFET devices.

Fig. 9 Measured RTN amplitudes are frequency independent.

Fig. 10 (a) Monte Carlo based RTN state decision. (b) Range limitation of traditional RTN simulation method.

Fig. 11 New RTN simulation platform with two-stage flow, implemented in HSPICE using OMI/TMI.

Fig. 12 Proposed trap occupancy probability prediction (OPP) model for RTN simulation.

Fig. 13 Comparison between OPP model and transient simulation results (b) for RTN under a random waveform (a).

Fig. 14 (a)(b)(c) Correlation analysis between different random waveforms and the ripples of simulation results. (d) The average values of ripples agree well with OPP model results.

Fig. 15 Comparison between OPP model and experimental results of RTN in FinFETs.

Fig. 16 Typical Example of (a) trap occupancy probability after an operation time, predicted by OPP model and (b) the corresponding RTN state decided by Monte Carlo method.

978-1-7281-1988-5/18 $31.00 © 2018 IEEE

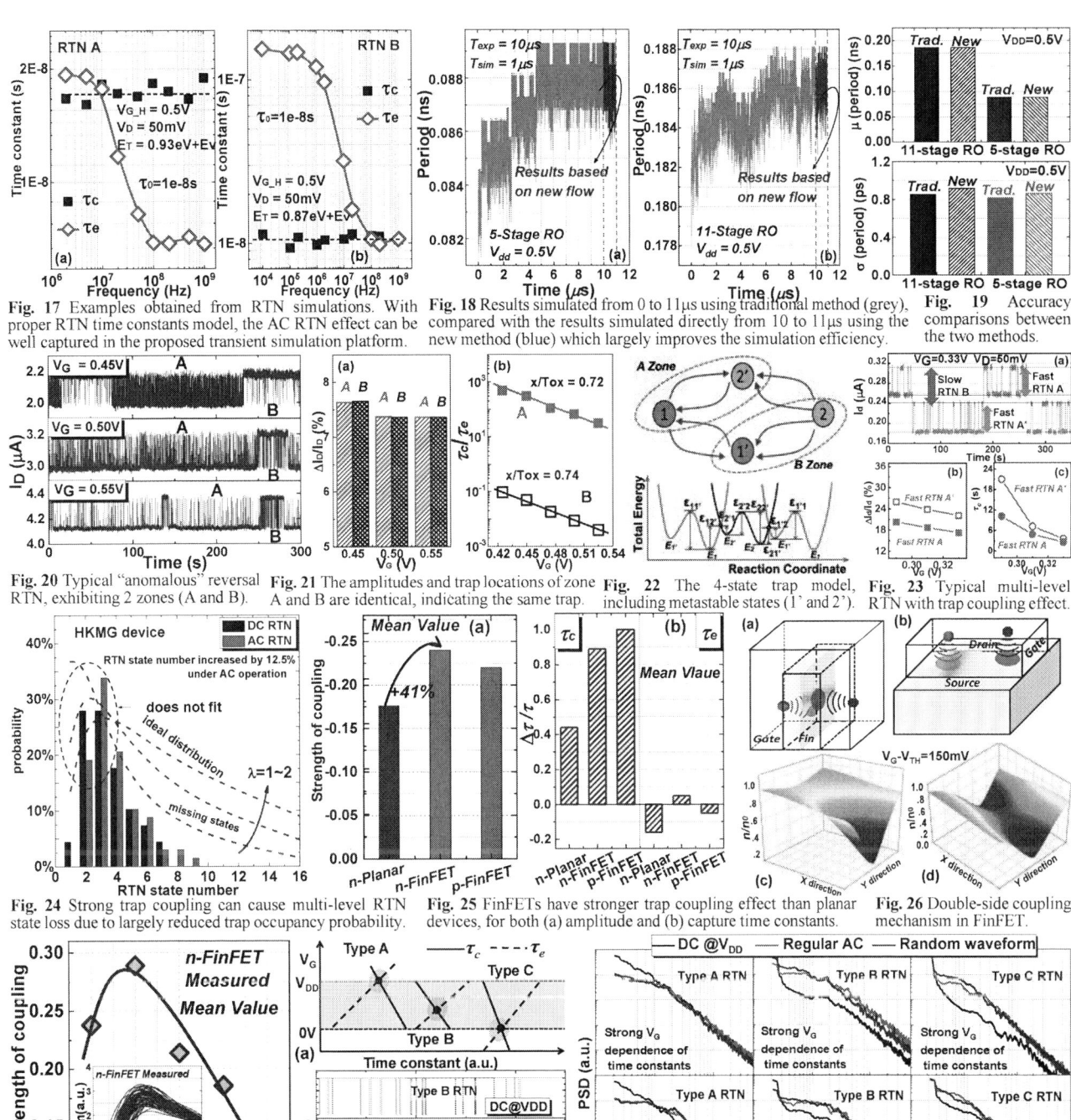

Fig. 17 Examples obtained from RTN simulations. With proper RTN time constants model, the AC RTN effect can be well captured in the proposed transient simulation platform.

Fig. 18 Results simulated from 0 to 11μs using traditional method (grey), compared with the results simulated directly from 10 to 11μs using the new method (blue) which largely improves the simulation efficiency.

Fig. 19 Accuracy comparisons between the two methods.

Fig. 20 Typical "anomalous" reversal RTN, exhibiting 2 zones (A and B).

Fig. 21 The amplitudes and trap locations of zone A and B are identical, indicating the same trap.

Fig. 22 The 4-state trap model, including metastable states (1' and 2').

Fig. 23 Typical multi-level RTN with trap coupling effect.

Fig. 24 Strong trap coupling can cause multi-level RTN state loss due to largely reduced trap occupancy probability.

Fig. 25 FinFETs have stronger trap coupling effect than planar devices, for both (a) amplitude and (b) capture time constants.

Fig. 26 Double-side coupling mechanism in FinFET.

Fig. 27 Bias dependence of trap coupling strength measured in FinFETs.

Fig. 28 (a) Consider 3 types of RTN with different time constant features. (b) RTN pattern is workload dependent.

Fig. 29 The corresponding PSD of different types of RTN under different types of waveforms.

References: [1] M.J. Kirton, et al., Adv. Phys., vol. 38, p. 367, 1989. [2] N. Tega, et al., VLSI, 2009, p. 50. [3] K. Takeuchi, et al., VLSI, 2009, p. 54. [4] H. Miki, et al., IEDM, 2012, p. 450. [5] C. Liu, et al., IRPS, 2014, p. XT.17.1. [6] S. Guo, et al., IEDM, 2014, p. 319. [7] S. Dongaonkar, et al., VLSI, 2016, p. 176. [8] J. Chen, et al., VLSI, 2016, p. 46. [9] H. Qiu, et al., VLSI, 2017, p. 50. [10] M. Simicic, et al., VLSI, 2017, p. 132. [11] Q. Tang, et al., JSSC, vol. 52, p. 1655, 2017. [12] R. Huang, et al., IEDM, 2017, p. 298. [13] E. Simoen, et al., ECS Trans., vol. 39, p. 3, 2011 [14] http://www.si2.org/cmc [15] T. Grasser, MR, vol. 52, p. 39, 2012. [16] E. Simoen and C. Claeys, *Random Telegraph Signals in Semiconductor Devices*, IOP Publishing, 2016. [17] T.H. Both, et al., MR, vol. 80, p. 278, 2018. [18] T. Grasser, et al., IRPS, 2014, p. 4A.5.1. [19] Z. Zhang, et al., IRPS, 2017, p. 3C-3.1. [20] Z. Zhang, IEDM, 2016, p. 172. [21] J. Zou, et al, VLSI, 2012, p. 139. [22] J. Zou, et al., VLSI, 2013, p. 186. [23] M. Luo, et al., T-ED, vol. 62, p. 1725, 2015. [24] D. Mao, et al., VLSI-TSA, 2016, p. 1. [25] S. Guo, et al., ISCAS, 2018, p. 1. [26] R. Wang, et al., IEDM, 2013, p. 834. [27] R. Wang, et al., IPFA, 2018, 8B.1. [28] J. Ji, et al., IEDM, 2014, p. 542. [29] S. Guo, et al., Sci. Rep., 7(1), 6239, 2017. [30] J. Zhang, et al., IEDM, 2018, submitted. [31] P. Ren, et al., IEDM, 2013, p. 778. [32] J. Zou, et al., IEDM, 2014, p. 832. [33] S. Guo, et al., IRPS, 2018, p. TX-6.1. [34] T. Grasser, et al., IEDM, 2014, p. 530. [35] D. Mao, et al., INEC, 2016, p. 1.

978-1-7281-1988-5/18 $31.00 © 2018 IEEE

Comprehensive Study on the "Anomalous" Complex RTN in Advanced Multi-Fin Bulk FinFET Technology

Jiayang Zhang[1], Zhe Zhang[1], Rusheng Wang[1,2*], Zixuan Sun[3], Zuodong Zhang[1], Shaofeng Guo[1], Ru Huang[1,2]

[1]Institute of Microelectronics, Peking University, Beijing 100871, China (*Email: r.wang@pku.edu.cn)
[2]National Key Laboratory of Science and Technology on Micro/Nano Fabrication, Beijing 100871, China
[3]College of Physics and Information Engineering, Fuzhou University, Fujian 350002, China

Abstract—In this paper, random telegraph noise (RTN) in advanced multi-Fin bulk FinFETs are comprehensively studied for the first time. Based on the statistical experiments, the complete categories of simple and complex RTNs are identified and analyzed in details. Especially, the anomalous "reversal RTN" induced by 2 metastable states in single oxide trap, are found not rare, but appears at a certain percentage, which provides a unique opportunity for statistically studying the metastable states directly from RTN measurements. In addition, anomalous layout dependence of RTN amplitudes are observed, with respects to Fin number. The results are helpful for deep understanding of reliability physics and robust circuit design against RTN.

I. INTRODUCTION

Recently, random telegraph noise (RTN) has attracted much attention, not only because its increasing impact on circuits, but also because it acts as an atomistic probe for characterizing single oxide trap [1] which is useful for understanding reliability physics. So far, most studies relied on "normal" simple RTN measurements. However, in reality some observed RTNs exhibit "anomalous" complex characteristics (e.g., [2,3]), which cannot be explained by conventional simple RTN theory, indicating more complex mechanisms during the charging/discharging behaviors of traps and thus unexpected impacts on circuits. Therefore, complex RTN should be studied in more detail.

In this paper, RTNs in advanced multi-Fin bulk FinFETs are comprehensively studied for the first time. The complete categories of simple and complex RTNs are identified and statistically studied in details. Especially, we have recently observed an anomalous RTN pattern (named "reversal RTN") in [3], induced by the 2 metastable states in single oxide trap, which is essential for modeling BTI reliability [4,5]. In this work, we found it is not a sole case or rare phenomena, but actually appears at a certain percentage in experiments, which provides a unique opportunity for statistically studying the metastable states directly from RTN measurements for the first time. In addition, anomalous layout dependence of RTN amplitudes are observed, with respects to Fin number. The results are helpful for deep understanding of trap physics and robust circuit design against RTN.

II. DEVICES AND CHARACTERIZATION

The devices measured in this work are foundry-level advanced bulk n-type FinFETs with multiple Fins (#Fin=2, 3, 4, 8, 12). RTNs were measured under 4 different bias conditions: (1) $V_G=V_{DD}$, $V_D=V_{DD}$; (2) $V_G=V_{DD}$, $V_D=0.5V_{DD}$; (3) $V_G=0.5V_{DD}$, $V_D=V_{DD}$; (4) $V_G=V_{TH}$, $V_D=50mV$. RTN amplitudes ($\Delta I_D/I_D$) and time constants (τ_c, τ_e) are extracted by

advanced hidden Markov model (HMM) method [6], which is capable for all kinds of complex raw data extraction and tolerant of background noise, as shown in Fig. 1 as examples. Multi-level RTNs are decomposed into individual 2-level RTNs to count the single-trap induced RTN amplitudes. The statistical results of extracted RTN amplitudes are shown in Fig. 2. As shown in Fig. 2(f), RTN amplitudes are uncorrelated with fresh device V_{TH} (V_{TH0}).

III. CATEGOTIES OF COMPLEX RTN AND RTN PATTERN PROPORTION ANALYSIS

According to our statistical experimental results, all the RTNs in FinFETs can be categorized into several patterns, with respects to the presents of trap coupling and/or metastable state effects, as shown in Table 1.

Fig. 3(a) shows that, most RTNs observed are with 4 levels, while 3-level RTNs (with strong trap coupling) have a non-negligible percentage. Further compare the proportion between coupling (CE) and no-coupling (NCE) RTNs in Fig. 3(b), showing that CE-RTN exceeds 40%. The results indicate that coupling between 2 traps is a frequent mechanism, even in multi-Fin structures. Because for multi-Fin devices, the device performance is actually determined by the critical Fin with minimum local threshold voltage.

The number of "reversal RTN" in FinFETs is counted in Fig. 4. If counted among devices, reversal RTN appears in about 20% devices on average, and the percent can even reach 50% in FinFETs with 12Fins when biased at $V_G=V_{TH}$. If counted among all the NCE-RTNs, about 10% on average are observed as reversal RTN. Therefore, the many observations of reversal RTN provide direct evidences of the 4-state trap model [4] as shown in Table 1.

Fig. 5 compares the distribution of NCE-RTN and CE-RTN in 12Fin-FinFETs, showing that the amplitudes of NCE-RTN distribute relatively on the right side of CE-RTN. This because that the charged trap can change the distribution of channel percolation path (PP) [7] in the Fin and thus change the RTN induced by another trap. The coupling mechanism in multi-Fin devices is similar to that in single-Fin devices [x], because multi-Fin device performance is determined by the critical Fin as mentioned above. For further quantitative analysis of the coupling effect on RTN amplitudes, it can be described by $\ln(\lambda_{CE}/\lambda_{NCE})$, where λ_{CE} and λ_{NCE} are the amplitude of the trap with and without coupling, respectively. The bias and Fin-number dependences of $\ln(\lambda_{CE}/\lambda_{NCE})$ are plotted in Fig. 6. Both positive and negative impacts are observed, but at $V_G=V_{TH}$ negative impacts on amplitudes are dominate, which is consistent with the results in Fig. 5.

978-1-7281-1988-5/18 $31.00 © 2018 IEEE

IV. LAYOUT DEPENDENCE OF RTN AMPLITUDE

To fairly compare the RTN amplitudes in FinFETs of different layouts with various Fin numbers, NCE-RTN are taken from the whole data set, as shown in Fig. 7. The NCE-RTN amplitudes are found following a two-stage lognormal distribution. The two stages originate from the interplays with process variation induced channel PP [6], where the high-stage is corresponding to the trap located above the PP and the low-stage representing the trap away from the PP.

Figs. 8-9 show the Fin-number dependence of NCE-RTN amplitudes at 1 and 2σ. With increasing Fin-number, the amplitudes of NCE-RTN reduces when $V_G=V_{DD}$, while it surprisingly increases when $V_G=V_{TH}$. Furthermore, if plotted with the intrinsic RTN amplitude per Fin, i.e., normalizing to the current per Fin as #Fin $\cdot \Delta I_D/I_D$, both bias conditions show an increasing tendency with increasing #Fin, as shown in Figs. 10-11.

This anomalous layout dependence of RTN amplitude comes from the interplays with process variations. As shown in Fig. 12, the device fresh V_{TH} variation σV_{TH0} deviates from Pelgrom law in devices with multiple Fins, which indicates a larger variation with increasing #Fin. This observation is consistent with previous reports of multi-Fin variations [8,9], which could due to the fact that the weight of Fin-edge roughness (FER) is increasing in multi-Fin structure [10]. As a result, the normalized process variations per Fin, i.e., normalizing to device area as $\sqrt{\text{#Fin}} \cdot \sigma V_{TH0}$, increases with #Fin rather than a constant predicted by Pelgrom law, as shown in Fig. 13. Therefore, the enhanced process variations in FinFETs with larger Fin-numbers cause the enlarged RTN amplitudes in the devices.

V. STATISTICAL ANALYSIS OF REVERSAL RTN WITH MEATSTBALE STATES

A. RTN Amplitude and Trap Occupancy Probability

The amplitude distributions of reversal RTN and normal RTN in 12-Fin devices are compared in Fig. 14. It shows that the reversal RTNs closely distribute to the high-stage distribution of normal RTN. Thus, under the $V_G=V_{TH}$ bias condition, the mean amplitude of reversal RTN is a little bit larger than NCE RTN (Fig. 15). On average, reversal RTNs have a little bit smaller amplitudes with normal RTN under other biases, as shown in Fig. 16. The results indicate that the reversal RTNs cannot be ignored due to the similar amplitudes with normal RTNs.

As shown in Table 1, reversal RTN exhibits two zones (A and B). The time constants in each zone are extracted, and thus the distributions of the trap occupancy probability [$p = \tau_e/(\tau_c + \tau_e)$] can be analyzed. Fig. 18 compares the trap occupancy probabilities of normal RTN, A-Zone, and B-Zone. The occupancy probability of NCE follows a U-shape distribution, because the RTN time constants are almost uniformly distributed in log-scale. The occupancy probability in A-Zone (P_A) mainly concentrates in the right edge of the distribution, but the occupancy probability in B-Zone (P_B) widely distributes from 0 to about 0.9. However, it is found that the P_A and P_B are highly correlated positively, as shown in Fig. 17.

B. Switching Characteristics bewteen A-Zone and B-Zone

The switching events between A-Zone and B-Zone in reversal RTN, can be equivalently taking as a very slow RTN.

The "capture time" τ_B and "emission time" τ_A are extracted. It is found that the time constants of the equivalent slow RTN follow a lognormal distribution as shown in Fig. 19. The mean time of the switching events between A-zone and B-zone is in the order of seconds, which is very slow events for circuits. The "occupancy probability" has a strong V_D dependence and weak V_G dependence, as shown in Fig. 20, which indicates that the traps that causing reversal RTNs may be located near the drain region.

C. Microsopic Trap State Transition Energy

To further study the transition between the 4 states (1, 1', 2, 2'), the corresponding reaction coordinate of reversal RTN is used [11]. The transition energy barriers can be calculated according to the extracted time constants. Fig. 21 shows the energy of each states extracted from the reversal RTNs in 12Fin devices biased at $V_G=V_{TH}$. Since the transition energy has a wide distribution, the correlation analysis is further done performed as shown in Fig. 22. Since 1' and 2' are metastable states, it is not surprising to see the strong correlations between $\varepsilon_{2'2}$ and $\varepsilon_{22'}$, $\varepsilon_{1'1}$ and $\varepsilon_{11'}$. However, interestingly, there are also strong correlations between $\varepsilon_{2'1}$ and $\varepsilon_{1'2}$, $\varepsilon_{1'2}$ and $\varepsilon_{12'}$. The results provide experimental support to the theoretical study on the microscopic origin of oxide traps [5].

D. Impact on SRAM Stability

With the above understanding of reversal RTN, its impact on SRAM can be precisely predicted. As shown in Fig. 23, three kinds of RTNs are considered. The transient simulation results show that A-Zone and B-Zone have different failure probabilities (FPs) under the same V_{CS}, and the FP change sharply when the reversal RTN switching between A and B zone [Fig.24 (a)&(b)]. FPs in B-Zone and A-Zone depend on their trap occupancy probability. Only when P_B trends to 0, the reversal RTN can be equivalent as a very slow normal RTN [Fig. 24 (b)&(c)]. Otherwise, the complete pattern of the reversal RTN should be taken into account [Fig. 24 (a)&(c)]. Since the time constants of A and B zones are in the order of seconds, which is much longer than the periods of SRAM, it is important to accurately characterize and model the reversal RTNs.

VI. SUMMARY

In this paper, a complete category of complex RTN was defined and studied in details based on the experiment results. Anomalous layout dependence of RTN amplitudes are found in devices with multiple fins due to the process variation. The general characteristic of reversal RTN is also investigated statistically. The results are helpful for understanding the mechanisms in complex RTN and CMOS circuit design.

ACKNOWLEDGMENT

This work was supported by NSFC (61522402 and 61421005). The author would like to thank Xiaobo Jiang and Zhuoqing Yu for the helpful discussions.

REFERENCES

[1]. E. Simoen, et al., Random Telegraph Signals in Semiconductor Devices, IOP Publishing. [2]. T. Grasser, et al., IRPS 2010. [3]. J. Ji, et al., IEDM 2013. [4]. T. Grasser, et al., MR 2012. [5]. W. Goes, et al., MR 2018. [6]. Z. Zhang, et al., IRPS 2017. [7]. Z. Zhang, et al., IEDM 2016. [8]. A. Pual, et al., IEDM 2013. [9]. H. Rhee, et al., VLSI 2018. [10]. X. Jiang, et al., IEDM 2015. [11]. S. Guo, et al., SR 2017.

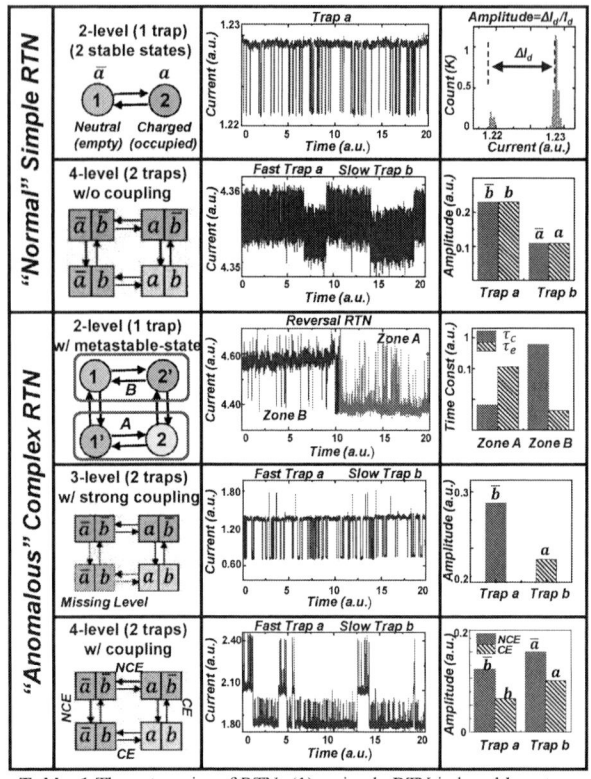

Table. 1 The categories of RTN: (1) a simple RTN induced by a trap; (2) RTN induced by two traps without coupling; (3) Reversal RTN; (4) RTN with a missing level; (5) RTN induced by two trap with coupling.

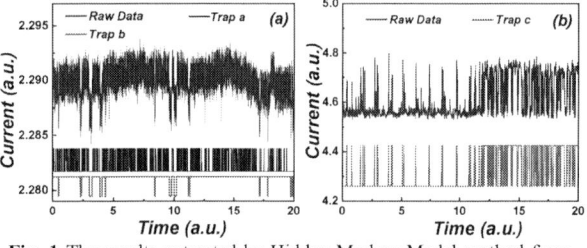

Fig. 1 The results extracted by Hidden Markov Model method from "noisy" measured data agree with the original data location indicate.

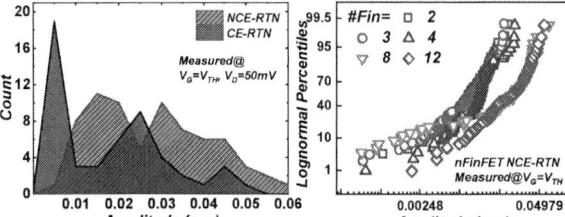

Fig. 5 The distributions of the amplitude of NCE-RTN and CE-RTN. The distribution of NCE-RTN is on the right side of CE-RTN

Fig. 7 The distributions of the amplitude of NCE-RTN in different kind of devices when $V_G=V_{TH}$, which show a clean two-stage distribution.

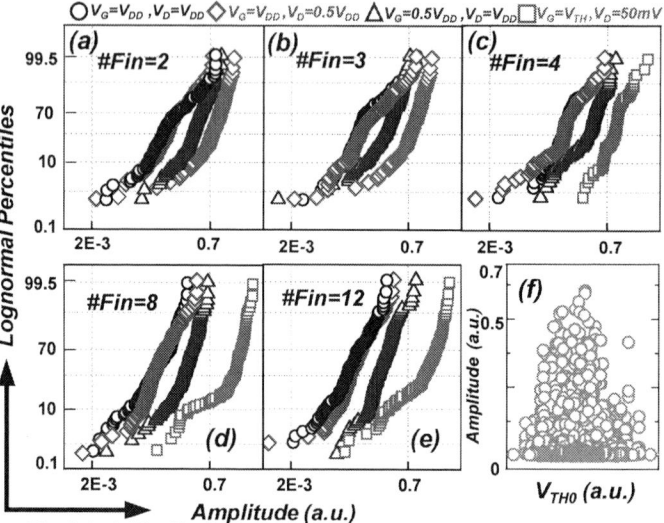

Fig. 2 (a-e) The distributions of the extracted RTN amplitude in n-FinFET with #Fin=2,3,4,8,12 under four bias condition; (f) the correlation between RTN amplitude and V_{TH0}.

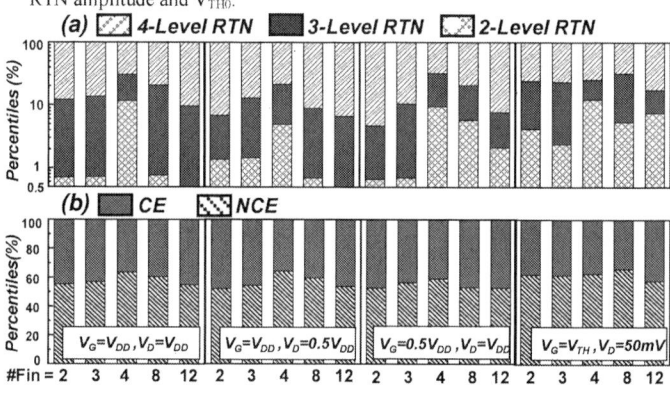

Fig. 3 Statistics of each kind of RTN: (a) percentiles of the RTN with different current levels; (b) percentiles of CE-RTN and NCE-RTN.

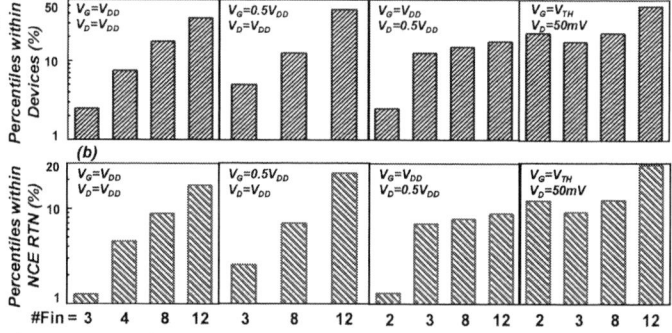

Fig. 4 Statistics of Reversal RTN: (a) percentiles of devices observed reversal RTN, reversal RTN appears in about 20% devices; (b) percentiles of reversal RTN in NCE RTN, it is found that about 10%~20% observed NCE-RTN was reversal RTN

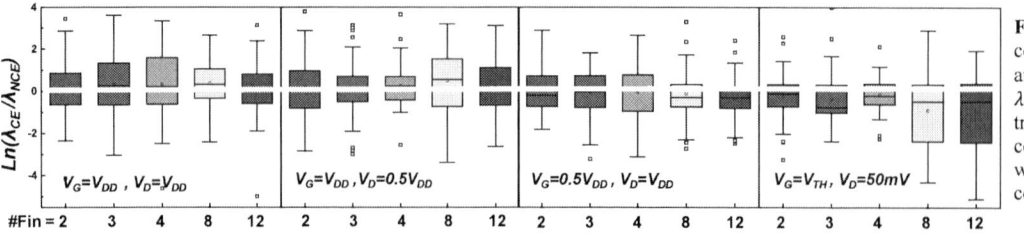

Fig. 6 The influence of coupling effects on the amplitude of RTN. λ_{CE} or λ_{NCE} is the amplitude of one trap with or without coupling, only four-level with coupling RTN is counted.

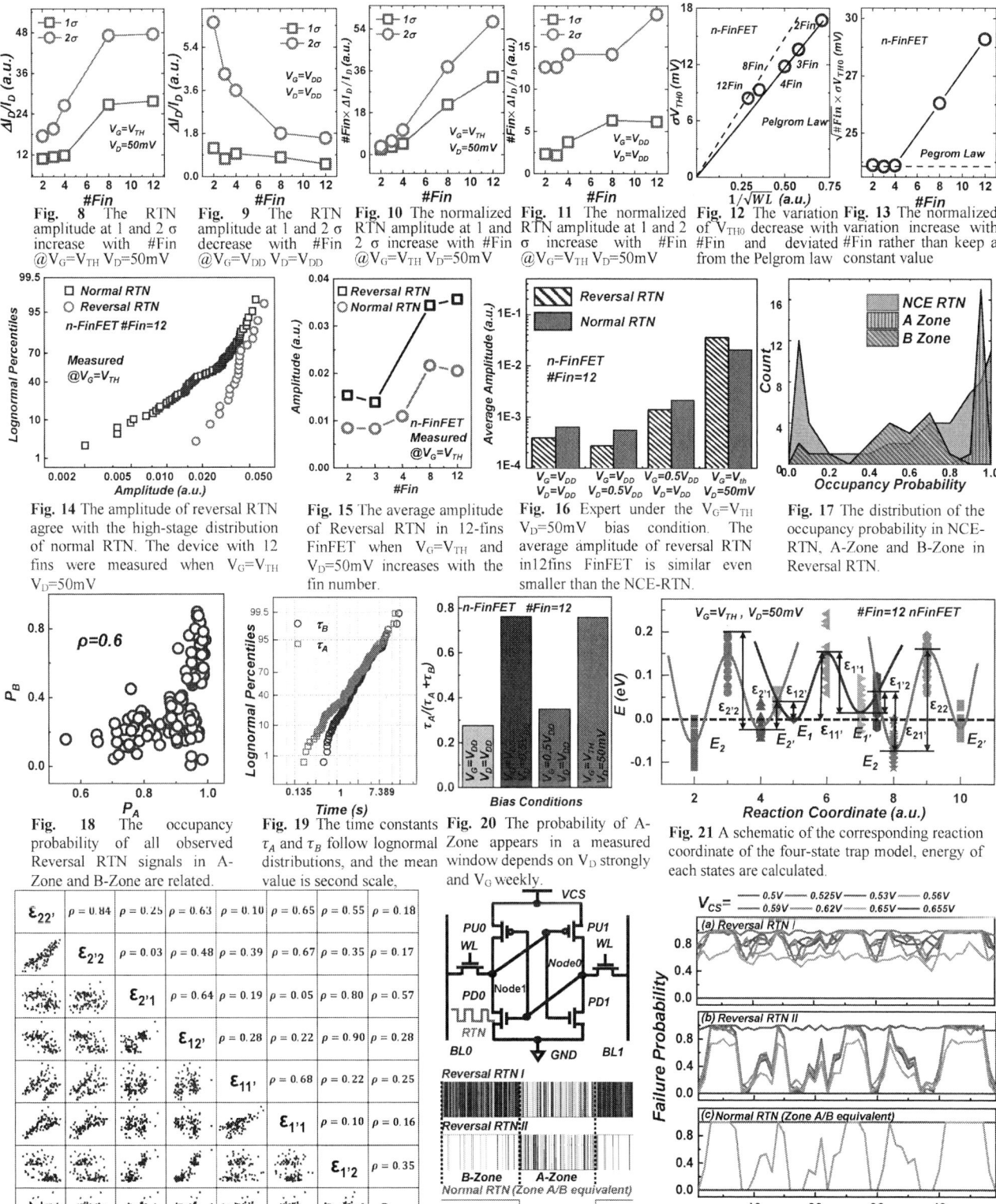

Fig. 8 The RTN amplitude at 1 and 2 σ increase with #Fin @V_G=V_{TH} V_D=50mV

Fig. 9 The RTN amplitude at 1 and 2 σ decrease with #Fin @V_G=V_{DD} V_D=V_{DD}

Fig. 10 The normalized RTN amplitude at 1 and 2 σ increase with #Fin @V_G=V_{TH} V_D=50mV

Fig. 11 The normalized RTN amplitude at 1 and 2 σ increase with #Fin @V_G=V_{TH} V_D=50mV

Fig. 12 The variation of V_{TH0} decrease with #Fin and deviated from the Pelgrom law

Fig. 13 The normalized V_{TH0} decrease with variation increase with #Fin rather than keep a constant value

Fig. 14 The amplitude of reversal RTN agree with the high-stage distribution of Normal RTN. The device with 12 fins were measured when V_G=V_{TH} V_D=50mV

Fig. 15 The average amplitude of Reversal RTN in 12-fins FinFET when V_G=V_{TH} and V_D=50mV increases with the fin number.

Fig. 16 Expert under the V_G=V_{TH} V_D=50mV bias condition. The average amplitude of reversal RTN in12fins FinFET is similar even smaller than the NCE-RTN.

Fig. 17 The distribution of the occupancy probability in NCE-RTN, A-Zone and B-Zone in Reversal RTN.

Fig. 18 The occupancy probability of all observed Reversal RTN signals in A-Zone and B-Zone are related.

Fig. 19 The time constants τ_A and τ_B follow lognormal distributions, and the mean value is second scale,

Fig. 20 The probability of A-Zone appears in a measured window depends on V_D strongly and V_G weakly.

Fig. 21 A schematic of the corresponding reaction coordinate of the four-state trap model, energy of each states are calculated.

Fig. 22 The correlation of the transition energy barriers extracted from the time constants. The lower triangular part shows the scatter plot of each barrier. The correlation coefficients were shown in the upper triangular part. It is found that there are strong correlations between $\varepsilon_{2'2}$ and $\varepsilon_{22'}$, $\varepsilon_{1'1}$ and $\varepsilon_{11'}$, $\varepsilon_{2'1}$ and $\varepsilon_{1'2}$, $\varepsilon_{1'2}$ and $\varepsilon_{12'}$.

Fig. 23 Three types of RTN is added on the PD0, Reversal RTN I&II have same τ_A and τ_B, but different time constants in A-Zone and B-Zone, Normal RTN is an equivalent of A/B Zone.

Fig. 24 The simulation results of the three types RTN, the failure probability is averaged for previous 30-iteration results. The Reversal RTN's failure probability range between an upper level (determined by A-Zone) and a lower level (determined by B-Zone).

An Unique Methodology to Estimate The Thermal Time Constant and Dynamic Self Heating Impact for Accurate Reliability Evaluation in Advanced FinFET Technologies

S. Mukhopadhyay, A. Kundu, Y.W. Lee, H. D. Hsieh, D.S.Huang, J.J.Horng, T.H.Chen, J.H. Lee, Y.S. Tsai, C.K.Lin, Ryan Lu, and Jun He

Taiwan Semiconductor Manufacturing Company, Hsinchu, Taiwan,

email: smukhopa@tsmc.com

Abstract—The increasing impact of self-heating effect (SHE) in complex FinFET structure is a serious reliability concern. Although the evaluation of SHE has become extremely arduous; this work proposes an *in-situ layout based experimental solution* to find out the precise thermal time constant (T_{TH}) due to SHE on advanced FinFET devices, even with the application of very pragmatic 'circuit-like' gate and drain input waveforms. Using this precise T_{TH}, the accurate dynamic thermal profile is found out from SPICE simulations. Finally, the *true* degradations due to different reliability mechanisms are evaluated including SHE impact and successfully compared with measured FinFET silicon data.

I. INTRODUCTION

The scaling of advanced technology nodes is closely associated with the invention of non-planar architecture, denser layout schemes and even with new material system in place. Such physical engineering increases the complexity of transistor reliability evaluation, and the self-heating effect (SHE) is one of such intrinsic penalty, that especially FinFET architectures suffer a lot [1-4]. Various literatures speak about the impact of SHE in device and circuit reliability [1, 4] and transistor geometry impact on it [2-3]. However, the time domain behavior of self-heating during the pragmatic device/circuit input waveform is extremely difficult to measure and no comprehensive report is presented so far due to complexity of temperature extraction [1]. This work proposes a unique methodology to extract the "time-dependent dynamic temperature change" due to self-heating with the application of practical input voltage waveform on TSMC advanced FinFET devices. Such precise SHE estimation leads to the evaluation of correct reliability impact in terms of bias temperature instability (BTI), hot carrier injection (HCI) or time dependent dielectric breakdown (TDDB) in device and circuit operation condition.

Typically, the device temperature change (ΔT) due to SHE is estimated using the evaluation of thermal resistance R_{TH} [3]. One such example is depicted in Fig. 1 for advanced FinFET transistors. The gate resistance is electrically measured for different chuck temperatures and under different power consumption scenarios (left fig). Using this data, the device temperature rise and the thermal resistance R_{TH} is calculated (right fig). Once the R_{TH} is calibrated for a process, the ΔT due to SHE can be calculated for that process and for any DC or

AC stress/use condition. Note that, such SHE ΔT estimation is only 'average' in nature as shown by the dashed line in Fig. 2(a). Hence, this calculation may suffice the DC SHE estimation but for AC stress (especially at higher frequency) such average approximation may be questionable. In this paper, for the first time, a unique in-situ layout test-key design is shown to experimentally measure the self-heating generated thermal time constant and the accurate dynamic temperature change is evaluated with complex input signal. Before discussing the test-key structure in detail, the thermal time constant explained in Fig. 2. A typical gate and drain input waveform is shown in Fig. 2(a) and as compared to average ΔT behavior (dashed line), the SHE driven temperature change can be severe within a very short time and needs to be captured before being dissipated. Fig 2(b) shows a first order RC phenomenon with the ΔT increasing (T_{heat}) and decreasing (T_{diss}) and the thermal time constant can be extracted from this behavior. The thermal time constant (T_{TH}) extraction test-key structure is explained next in Fig. 3. A frequency generator unit is designed as the input block and this is controlled by the Enable, VDD and VSS (GND) signals. This stage is then fed into the pulse duty cycle (PDC) control unit with VDC and P/N MOS control unit signals. Finally, the PDC block output is fed to the output buffer, at the end of which the gate sensor is connected to measure for the self-heating generated ΔT. The gate sensor consists of 4-pt kelvin sensors for source and drain and for each PMOS & NMOS. Note that, the thermal time constant is difficult to measure due to quick heat dissipation and hence this test-key is designed to operate in ~GHz range to find the SHE generated T_{TH}. Also, the T_{TH} evaluated here is for TSMC's advanced FinFET devices probing for the worst case SHE scenario. Fig. 4(a) shows the change in temperature measured from kelvin sensors with the change in duty cycle and this data is measured for ~3.1GHz. The average ΔT is also measured separately and compared with the dynamic change. This is very interesting to see that the ΔT, measured dynamically from the test-key structure, increases with PDC and eventually saturates; whereas the average change in ΔT remains almost constant for the entire PDC range. The increase in SHE impact with increasing PDC (i.e. with increasing ON time), especially for higher frequency range is a very important experimental observation and indicates that the average ΔT can be the incorrect SHE indicator for many such pragmatic waveform inputs. The PDC is further translated to the time domain and in Fig. 4(b), it is shown that

978-1-7281-1988-5/18 $31.00 © 2018 IEEE

the thermal time constant, for this given process, is ~0.2ns. Such experimental determination of thermal time constant (T_{TH}) further helps to evaluate the true SHE impact for any complex input waveform and for any given technology. The T_{TH} is next used in a SPICE simulator to evaluate the change in device temperature with varying stress conditions. The SPICE simulation algorithm shown in table-I. This calculation includes the power consumption and R_{TH} information to evaluate the dynamic thermal profile. Fig. 5 shows two such examples where the dynamic thermal profile is extracted from the SPICE calculation, one for high and one for low pulse duty cycle (PDC). The thermal profile closely follows the high frequency gate voltage waveform, unlike the average thermal calculation (c.f. dashed line in Fig 2(a)). In this case, drain voltage is used as V_{CC}. This is imperative that only such accurate estimation of thermal profile enables us to calculate the impact of other reliability mechanisms like BTI, HCI etc. considering the self-heating effect.

II. RESULTS AND DISCUSSION

Fig. 6 summarizes the entire procedure of evaluation of different reliability items with accurate (new method) and with average (old method) SHE impact. The average ΔT can be found out as shown in Fig. 1 and then using the standard BTI/HCI calculator, the reliability margins can be evaluated for a complex waveform. On the other hand, as proposed in this paper, the accurate thermal time constant (T_{TH}) is first measured using the unique test-key structure for a given technology and then the SPICE extracted dynamic thermal profile is fed to the reliability calculator. Finally, the comparison of reliability degradation is performed using both of these procedures and along with the advanced FinFET degradation data. Note that, the following measured data and model prediction is done keeping in mind of the pragmatic circuit waveforms with higher frequency and different pulse duty cycles and also different V_{GATE} and V_{DRAIN} applied simultaneously (c.f. Fig. 2(a)) to bring out both the BTI and HCI related degradation mechanisms. V_{GATE} was carefully chosen to avoid any gate dielectric breakdown (TDDB) condition. In Fig. 7, the measured data and model predictions are shown for three frequencies and for each frequency, two different duty cycles are shown. Fig. 7(a)-(c) are plotted for PDC-10% and Fig. 7(d)-(f) are plotted for PDC-50%. For each PDC, measured time evolution for I_{DSAT} degradation are plotted for three different frequencies and the model prediction using dynamic and average thermal profile is shown for each cases. In each model prediction, the BTI and HCI components are shown for dynamic and average thermal profiles. It needs to be explicitly mentioned that, depending on the arbitrary input waveform (with time varying V_{GATE} and V_{DRAIN}), the dynamic thermal profile and hence the BTI and HCI are estimated and thus a direct comparison of these two mechanisms is beyond the scope of this work. For both PDCs and all frequencies, the model prediction using the dynamic thermal profile shows better match as compared to the degradation modeled using the average thermal profile. Such unique model prediction of reliability degradation for advanced FinFET technology using experimentally measured thermal time constant and dynamic thermal profile was never shown before and justifies the significant impact of self-

heating with such realistic circuit like input. Fig. 8 shows the difference in peak-ΔT measured from the dynamic thermal profile and the average thermal profile used. In each of the cases, the dynamic profile shows more peak-ΔT rise indicating the SHE impact. Fig. 8(a) and (b) show the peak-ΔT change for different frequencies and for two PDCs. Because of lesser pulse on-time, the peak difference is more prominent in Fig. 8(a) (PDC-10%) as compared to Fig. 8(b) (PDC-50%). This is more clearly shown in Fig. 8(c) for a fixed frequency. Although not shown here explicitly, in Fig. 8(d), the bias dependent peak-ΔT rise shows more SHE impact for dynamic estimation as compared to the average one.

It is also important to understand the impact on lifetime (LT) calculation for a given technology due to SHE impact. Fig. 9 (a) depicts a general case, where the dynamic to average lifetime (LT) ratio due to BTI and TDDB stress is compared with variation of thermal time constant. For worst conductive channels (like SiGe) or complex transistor architecture, where the SHE impact can be detrimental, the thermal time constant (T_{TH}) can change. In that case, the accurate dynamic LT becomes smaller with more SHE, decreasing the overall dynamic to average LT ratio. The advanced FinFET device degradation data is shown under BTI and TDDB stress and both of them show decrease in dynamic LT with decreasing thermal time constant. A final summary is presented in Fig. 9(b) to show how the transistor LT can be impacted with SHE along with BTI or HCI degradation w.r.t. PDC. The LT impact ratio is plotted calculated from the average and dynamic thermal profiles. As compared to the average thermal LT, dynamically calculated LT would show upto ~2-4X LT benefit in lower duty cycles indicating lower SHE impact (lesser ON-time, more OFF-time to dissipate). On the other hand, higher PDC would suffer serious SHE impact (more ON-time, less OFF-time to dissipate) and the dynamic thermal profile calculation would result in lower LT as compared to averaged one, resulting in ~10-15% drop in LT impact ratio.

III. CONCLUSION

A very unique in situ layout based test-key design is proposed to measure the self-heating effect generated thermal time constant (T_{TH}, in sub nanosecond range) of a transistor under dynamic stress condition. When the circuit like 'realistic' V_{GATE} and V_{DRAIN} is applied on a device, the precisely extracted T_{TH} is used to evaluate the accurate dynamic thermal profile using holistic SPICE simulation considering the precise SHE impact. Such dynamic thermal profile is further used to estimate the BTI/HCI related degradations using the standard AC/DC calculator. It is conclusively shown that, the dynamic thermal profile calculation predicts silicon data more accurately as compared to the average SHE estimation for complex input waveforms. For a given advanced FinFET technology, the T_{TH} is reported here is ~0.2ns. However, for a poor conducting channel or complex FinFET architecture, the SHE would be more and the T_{TH} and the dynamic LT can even be worse. Thus this work not only gives a new experimental technique to measure the accurate device level thermal time constant under stress, but also equivocally establishes the importance of estimation of accurate dynamic thermal profile calculation with increasing self-heating impact.

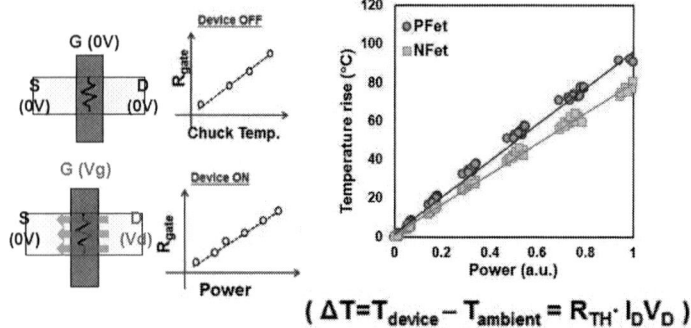

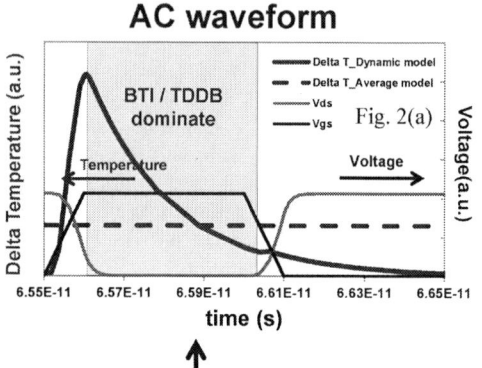

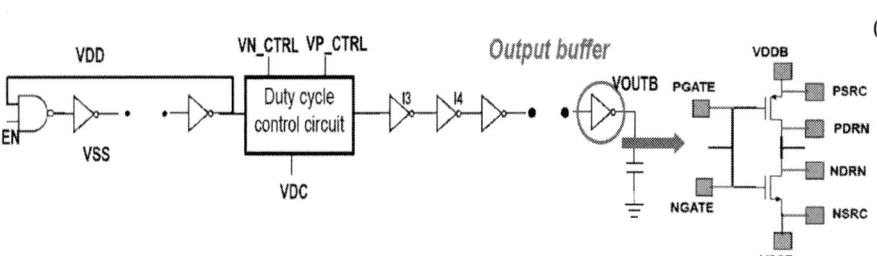

$$(\Delta T = T_{device} - T_{ambient} = R_{TH} \cdot I_D V_D)$$

Fig. 1: Classical technique to extract the Thermal Resistance (R_{TH}) for FinFET devices [3]. R_{gate} is measured with chuck temperature change in OFF state and then with the transistor power ($I_d.V_d$) change in ON state. Combining the two, the change in SHE ΔT is measured and also R_{TH} is estimated form the slope.

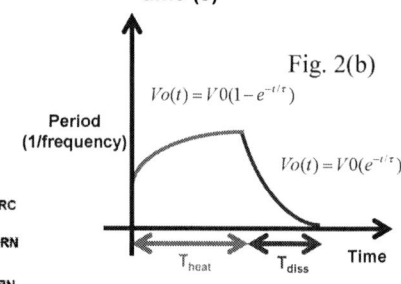

Fig. 2(b)

$$Vo(t) = V0(1 - e^{-t/\tau})$$

$$Vo(t) = V0(e^{-t/\tau})$$

Fig. 3: A frequency generator unit is designed as the input block and this is controlled by the Enable, VDD and VSS (GND) signals. This stage is then fed into the pulse duty cycle (PDC) control unit with VDC and P/N control unit signals. Finally, the PDC block output is fed to the output buffer, at the end of which the gate sensor is connected to measure for the self-heating generated ΔT. The gate sensor consists of 4-pt kelvin sensors for source and drain and for each PMOS & NMOS. For this work, the operating frequency was ~GHz.

Fig. 2: (a) A typical gate and drain input waveform is shown with dynamic (solid line) and average (dashed line) thermal ΔT behavior, the dynamic SHE driven ΔT can be severe within very short time. Fig 2(b), shows a first order RC phenomena with the ΔT increasing (T_{heat}) and decreasing (T_{diss}) and the thermal time constant can be found out from this.

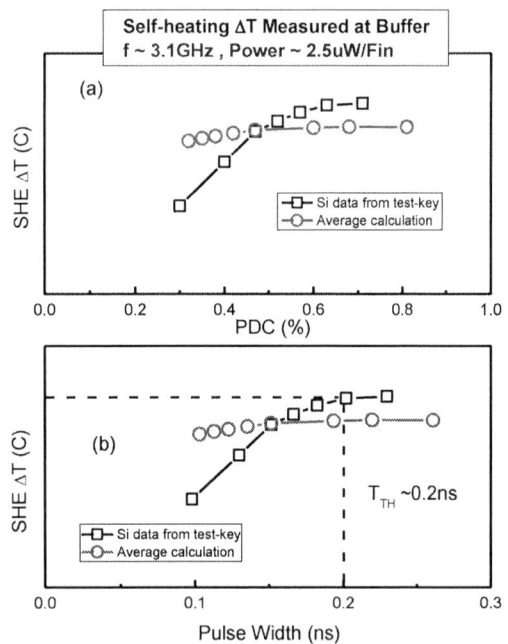

Fig. 4: (a) shows the change in temperature measured from kelvin sensors with the change in duty cycle and this data is measured for ~3.1GHz. The average ΔT is also compared with the dynamic change. (b) The PDC is further translated to the time domain and it can be shown that the T_{TH}, for this given process, is ~0.2ns

Table-I
SPICE simulation For Thermal Profile

$$P \sim C_{th} \frac{d(\Delta T)}{dt}$$

$$P \cdot R_{th} = T_{TH} \frac{d(\Delta T)}{dt} + K_1$$

$$d\Delta T_{i+1} = K_2 + (d\Delta T_i - K_2) \cdot f(\Delta t_i, T_{TH})$$

T_{TH} = Thermal time constant
$f(\Delta t_i, T_{TH})$ = function of time and thermal time constant
K_1, K_2 = constants

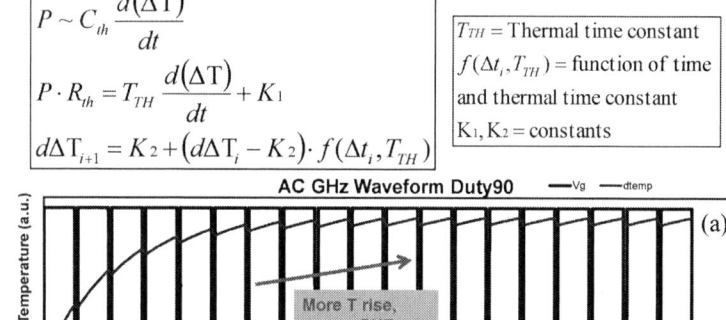

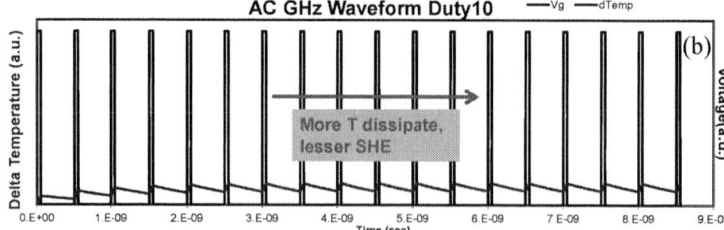

Fig. 5: Precise extraction of Dynamic thermal profile from the SPICE simulation with (a) high and (b) low PDC with ~GHz of frequency. SPICE simulation follows the algorithm in Table-I with the extracted correct thermal time constant (T_{TH}) evaluation.

978-1-7281-1988-5/18 $31.00 © 2018 IEEE

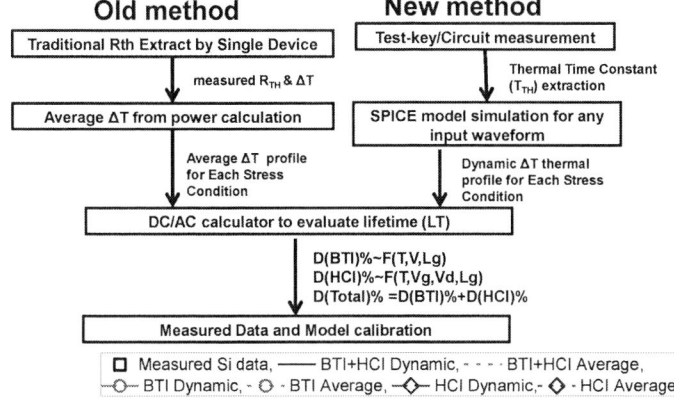

Old method

Traditional Rth Extract by Single Device

↓ measured R_{th} & ΔT

Average ΔT from power calculation

↓ Average ΔT profile for Each Stress Condition

New method

Test-key/Circuit measurement

↓ Thermal Time Constant (T_{TH}) extraction

SPICE model simulation for any input waveform

↓ Dynamic ΔT thermal profile for Each Stress Condition

DC/AC calculator to evaluate lifetime (LT)

↓ D(BTI)%~F(T,V,Lg)
D(HCI)%~F(T,Vg,Vd,Lg)
D(Total)% =D(BTI)%+D(HCI)%

Measured Data and Model calibration

Fig. 6: It summarizes the entire procedure of evaluation of different reliability items with accurate (New Method) and with average (Old Method) SHE impact. The dynamic thermal profile (ΔT vs time) is extracted from SPICE and fed to BTI/HCI calculator. SHE impacted model calculations (in either methods) and measured silicon results for advanced FinFET devices are shown below for the direct comparison.

References: [1] C. Prasad, IRPS, 2017, [2] S.H.Shin, IEDM, 2016, [3] S.E.Liu, IRPS, 2014, [4] Prof. Alam, IRPS, Tutorial, 2017.

Fig. 7: The measured data and model predictions are shown for three frequencies and for each frequency, two different duty cycles are shown. (a)-(c) are plotted for PDC-10% and (d)-(f) are plotted for PDC-50%. For each PDC, measured time evolution for I_{DSAT} degradation are plotted for three different frequencies and the model prediction using dynamic and average thermal profile is shown for each cases. In each model prediction BTI and HCI components are shown for dynamic and average thermal profiles.

Note: contributions from BTI and HCI components can't be directly compared due to complexity of the input waveform and extracted thermal profile. However, direct comparison can be done between dynamic and average thermal profile impact on BTI and HCI.

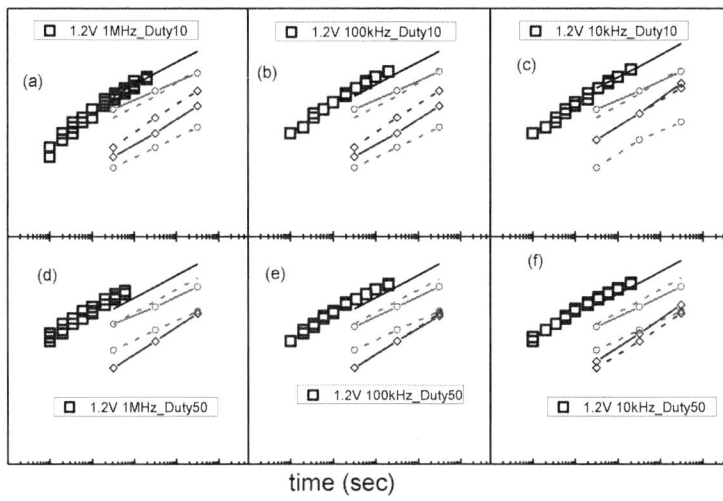

□ Measured Si data, —— BTI+HCI Dynamic, - - - - BTI+HCI Average, -○- BTI Dynamic, -○- BTI Average, -◇- HCI Dynamic, -◇- HCI Average

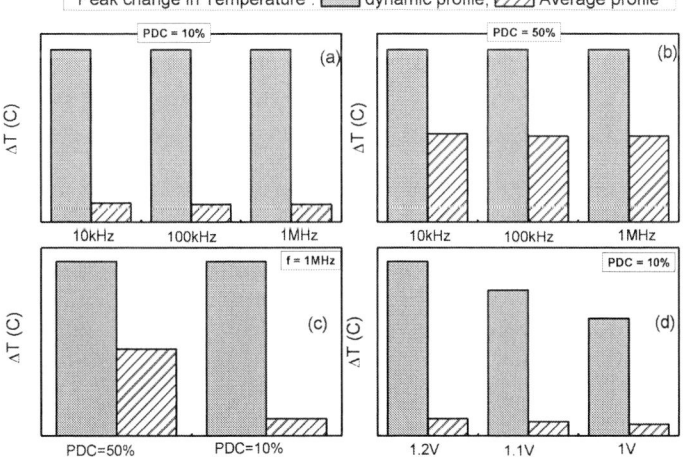

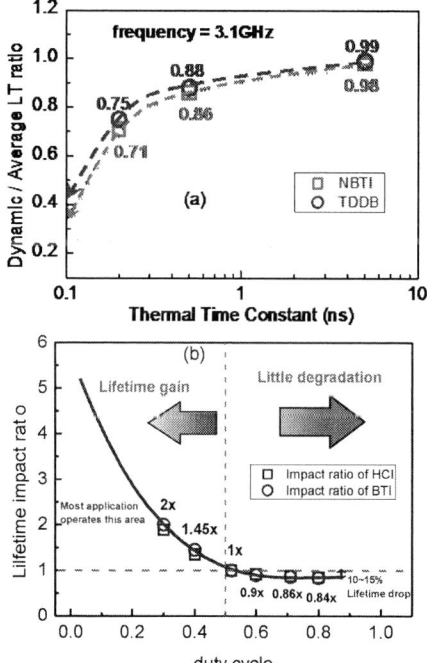

Fig. 8: shows the difference in peak-ΔT extracted from the dynamic thermal profile and the average thermal profile. Fig. 8(a) and (b) show the peak-ΔT change for different frequencies and for two PDCs. (c) plots the data for a given frequency and two different PDCs. Although not shown here explicitly, in Fig. 8(d) peak-ΔT rise shows bias dependent SHE impact for dynamic estimation as compared to the average one.

Fig. 9: (a) depicts a case, where the dynamic to average lifetime (LT) ratio due to BTI and TDDB stress is compared with variation of T_{TH} and both the degradations show decrease in dynamic LT with decreasing thermal time constant. 9(b) shows that the transistor LT gains margin (as compared to average) for dynamic profile in low PDCs, whereas LT decreases with more PDCs indicating sever SHE impact.

978-1-7281-1988-5/18 $31.00 © 2018 IEEE   399

7nm FinFET Plasma Charge Recording Device

Yi-Pei Tsai[1], Jiaw-Ren Shih[2], Ya-Chin King[1] and Chrong Jung Lin[1]

[1]Microelectronics Laboratory, Institute of Electronics Engineering, National Tsing Hua University (NTHU), Hsinchu, Taiwan

[2]Reliability Analysis Division, Taiwan Semiconductor Manufacturing Company (TSMC), Hsinchu, Taiwan

Phone: +886-3-5162182, Email: cjling@ee.nthu.edu.tw

Abstract— A new wafer-level coupling plasma charge recorder fabricated with 7nm FinFET CMOS logic process is presented in this paper. This plasma ion charge recording device provides the historic and quantitative plasma ion charges of damascene metallization steps in advanced 7nm FinFET COMS logic processes. The high-resolution plasma ion recorder is formed by an accurate FinFET coupling structure to store the plasma ion level and distribution of the whole wafer. By a simple wafer-level WAT measurement, the promising plasma charge recording device can efficiently collect the accumulated ion charges, ion polarization, and tiny plasma fluctuation of each metallization process step in 7nm FinFET CMOS logic technologies, which definitely provides a superior device and method in developing a reliable and non-latent plasma damage process for 7nm FinFET technology and beyond.

I. INTRODUCTION

In recent years, continuous dimension downsizing for semiconductor devices is the driving force for advancement of technology [1]. Plasma process is widely used in producing semiconductor devices such as etching, depositions, cleaning [2]-[3]. Therefore, during fabricating the vulnerable advanced FinFET devices, plasma damage is one of the major reliability and damage issues which severely affect the device and chip performance [4]. The conventional method for judging plasma charging level is to connect a series of large metallic antenna to a test gate made by MOSFET. Plasma Induced Damage (PID) can cause some dislocation and pinholes in the stressed gate dielectric layer by charging the antenna. The damage level is only correlated with the potential of the charged gate during plasma processing period, which is commonly detected by the destructive measurement such as Stressed Induced Leakage Current (SILC) and Time Dependent Dielectric Breakdown (TDDB) test [5]. However, in terms of the vulnerable devices in 7nm FinFET and beyond, to analyze the plasma charging procedure and polarization is becoming more important in advanced CMOS logic technology development. In this paper, a new 7nm coupling plasma charge recorder is proposed to accurately record and analyze the magnitude and polarization of plasma charges for each plasma step in metallization. The high-resolution recording device is realized by a high efficient coupling structure composed of FinFET slot contact and a floated gate for accurately recording the plasma potential of antennas during etch or deposition process. Moreover, since the charge storage gate is floated and non-volatile, the recorded data can be well analyzed by a quick Wafer Acceptance Test (WAT) measurement after Fab-out [6]. In addition, some latent damage in the inter-metal dielectric (IMD) by the damascene plasma processes [7] can also be detected and recorded by the advanced plasma charge recorder. The conventional PID on the 7nm frontend devices, the correlation of the latent IMD damage and the recorded charges is also firstly reported in this paper.

II. CELL STRUCTURE AND OPERATION PRINCIPLES

A. Device Structure of Plasma Charge Recorder

The new plasma recording device is fully compatible with the 7nm FinFET CMOS logic process, which can be easily built-in the testline as a WAT monitor pattern. This real-time coupling plasma charge recorder consists of an n-type floated metal gate structure and an n-type 7nm FinFET as shown in the TEM photo of Fig.1. As RF chamber activated, the antennal potential induced by ions directly transfer to the coupling gate, then boosts up the potential of the floated metal gate by the capacitance between slot contact and floating gate as depicted in the cross-section "A" of Fig.1. The boosted floating gate will induce some electrons or holes to jump into the floating gate by FN tunneling mechanism. The magnitude and polarization of the collected charges of the floating gate is determined by the plasma ion type and plasma density at this processing moment. To apply the new FinFET plasma recorders, the distribution of positive and negative ion charges on 12-inch 7nm wafer can be established for micro plasma behavior analysis. Fig.2 illustrates the 3D structures of the plasma charge recorder compared with a conventional pattern of plasma damage. Fig.3 shows the size comparison of the new charge recorders in different technology nodes, by taking the advantage of FinFET coupling structures, the area of the plasma recorders can be significantly shrunk on the FinFET nodes of 16nm, 7nm, and beyond.

B. Operation Principles

To acquire the best resolution and accuracy level of plasma damage, the coupling plasma charge recorder is designed to have a series of slot contact lengths for different coupling ratio in 7nm and 16nm FinFET technologies. Fig.4 demonstrates the basic $I_D V_G$ characteristics with different lengths of slot contact, which are used to extract the coupling ratios by increasing the subthreshold swing of coupling structure and summarized in Fig.5. The result clearly displays that higher coupling ratio is obtained by longer slot contact lengths which can be applied to record a tiny variation of plasma ion at the wafer surface. It further reveals that, by the evolution of FinFET technologies, 7nm FinFET can adopt a much smaller recorder with a same coupling ratio in recording plasma charges. Fig.6 shows the FinFET threshold shift (ΔV_T) influenced by the stored plasma charges. As some plasma ion charges are stored, the charge recorder has a parallel threshold shift (ΔV_T) and different plasma ion polarizations on antenna can result in different charge types stored into the floating metal gate, which will cause a specific direction of threshold voltage (ΔV_T) shift in device readout measurements of WAT.

978-1-7281-1988-5/18 $31.00 © 2018 IEEE

III. RESULTS AND DISCUSSION

A. Conventional Plasma Damage Test Pattern

The TDDB of the conventional plasma test patterns in 7nm technology has been measured and shown in Fig.7 and Fig.8, as prediction, higher antenna ratios (ARs) induces much severer gate dielectric damage and shorter TDDB (T_{BD}). In terms of the distribution of plasma induced damage (PID) in the FinFET wafer, to combined the gate dielectric damage TDDB (T_{BD}) from conventional patterns and the threshold shift (ΔV_T) from the charge recorders, the result is summarized and shown in Fig.9, there is a direct and strong correlation between the two methodologies in plasma induced damage (PID) analysis, the new superior recorder not only precisely reflect TDDB (T_{BD}) of gate dielectric damage and further collect more plasma valuable information, such as accumulated charges, tiny variation of ion distribution on wafer surface, and the polarization of plasma.

B. Plasma Charge Record on Different Metal Layers

In order to analyze the capability of plasma charge recorders in different metal layers, Fig.10 depicts the antenna structures of metal layers and its connection path of the coupled floating gate to the antenna layers. Fig.11 shows the recorder threshold shift (ΔV_T) of M1 to M5 across a 12-inch wafer. The data shows M5 performs a larger positive threshold shift (ΔV_T), as a result of high positive ion plasma inducing more negative charges into floating gate at wafer central region. Furthermore, the 12-inch wafer contour is characterized and depicted in Fig.12 to analyze the distribution of charging levels and polarization of plasma ions for M1 to M5 layers in the whole wafer, where M1 to M4 behave a similar plasma distribution with positive charges at the center and negative is around, but M5 shows an opposite distribution in this measurement. The 3D wafer contours of M2 and M5 are extracted and depicted in Fig.13 to totally compare the threshold voltage fluctuations for different metal layers on the wafer. Summarily, the coupling plasma charge recorder can accurately record tiny variation of ion plasma on wafer surface, which greatly help in analyzing the plasma ion characteristics: plasma distribution, ion polarization, and accumulated charges.

C. Plasma Charge Accumulative Model (pCAM)

Fig.14 summarizes the cumulative threshold voltage shift (ΔV_T) of the antenna charge recorders with different Cu metal thickness, which varies plasma time in processing of different metal thickness of damascene process. With varying the plasma process, threshold voltage shift (ΔV_T) widely ranges from -4V to 2V for different metal thickness. To consider the work function difference in metal gate and FIN substrate, the positive and negative ΔV_T are separated to analyze and shown in Fig.15, a thicker metal results in a larger ΔV_T due to longer plasma process time, the damage time is then converted to specific charge injection and stored in the recorders. Furthermore, a simulation of plasma charge accumulative model (pCAM) by FN tunneling mechanism has been developed and shown a very matching result with the mean and medium values of ΔV_T as compared in Fig.16. The equivalent voltage on antenna predicted by the pCAM simulation is extracted and plotted in Fig.17 and Fig.18, where the positive voltage on antenna attracts some negative charges into floating gate then causes a positive threshold voltage shift ($\Delta V_T>0$), and vice versa. The plasma charge accumulative model (pCAM) can precisely simulate the conversion of the magnitude of plasma ion to the induced voltage distribution on the wafers, which can help us to efficiently optimize the plasma conditions of 7nm FinFET and advanced nodes.

D. Detection of Dielectric Damage in BEOL

Some latent damage of inter-metal dielectric (IMD) in the damascene metallization process [7] can be accurately detected and recorded by the advanced plasma charge recorder. Fig.19 shows the TEM photo of finger-type metal structure with narrow pitches of 7nm FinFET wafer. Fig.20 compares the wafer mapping of the recorded ΔV_T level by plasma ion charge and the Stressed Induced Leakage Current (SILC) result of the of finger-type metal structures. In summary, larger ΔV_T of the recorder is correlated with higher SILC due to the latent damage being from plasma process. The accumulative plot of SILC is further characterized and shown in Fig.21. A continuous ΔV_T distribution of the recording devices as Fig.21(a), which is different from SILC result with a turning point as shown in Fig.21(b). It implies the SILC of IMD could have a critical electrical field and make the stress-induced leakage increase suddenly. Fig.22 shows the Pearsons value of 0.6 between SILC from the stressed IMD and ΔV_T taken from the plasma recorders. Summarily, the new charge recorder can provide a good monitor to the backend dielectric damage to avoid the latent reliability problem from plasma processes in FinFET.

IV. CONCLUTION

In our research, a novel coupling plasma charge recorder fabricated by a pure 7nm FinFET CMOS logic process has been successfully demonstrated and developed. Instead of the one-shot data and time-consuming TDDB test pattern and method, the new plasma recording device can reliably provide a superior wafer-level monitor of plasma ion distribution, ion polarization, and accumulative damage from plasma processes of advanced FinFET CMOS logic technologies. In summary, the precise coupling plasma charge recorder is surely capable of replacing the conventional measurement method for more and more complicate plasma process steps in the near future of FinFET CMOS technology of 7nm and beyond.

ACKNOWLEDGMENT

The authors gratefully acknowledge the contributions of Taiwan Semiconductor Manufacturing Company (TSMC) and Ministry of Science and Technology (MOST).

REFERENCES

[1] B. Narasimham, et al., in IEEE IRPS, Burlingame, CA, 2018, pp. 4C.1-1-4C.1-4.
[2] H. J. Roh, et al., in IEEE TSM, vol. 31, no. 2, pp. 232-241, May 2018.
[3] H. Zongjie, et al., ICEPT, Chengdu, 2014, pp. 360-362.
[4] G. Hiblot and G. Van der Plas, in IEEE EDL, vol. 39, no. 7, pp. 927-930, July 2018.
[5] L. Cheng, et al., RAMS, FL, 2015, pp. 1-5.
[6] C. H. Wu, et al., in IEEE IEDM, Washington, D.C., 7-9 Dec. 2015.
[7] T. Kouno, et al., IEEE IEDM Technical Digest., Washington, DC, 2005, pp. 187-190.

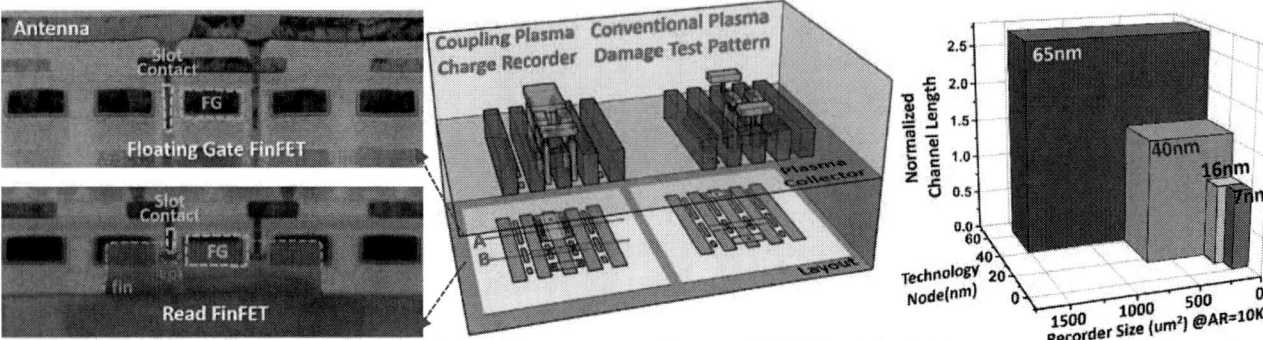

Fig. 1. TEM photo of the coupling plasma charge recorder composed of a floated metal gate and an n-type FinFET.

Fig. 2. 3D structures of the new FinFET plasma charge recorder and the conventional plasma induced damage (PID) pattern.

Fig. 3. The area comparison of plasma charge recorders in different technology nodes.

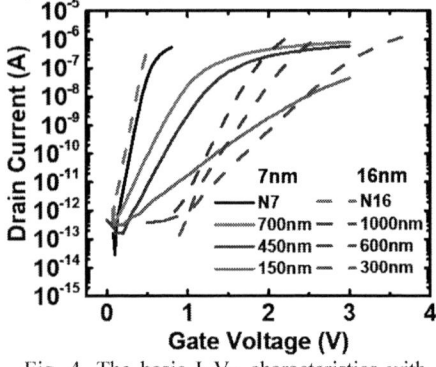

Fig. 4. The basic I_DV_G characteristics with different lengths of slot contact and standard swing of normal FinFET.

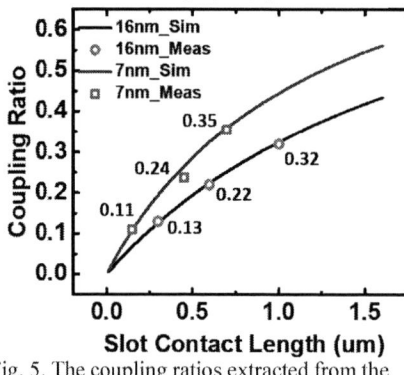

Fig. 5. The coupling ratios extracted from the increasing subthreshold swing of the charge coupling FinFET structures.

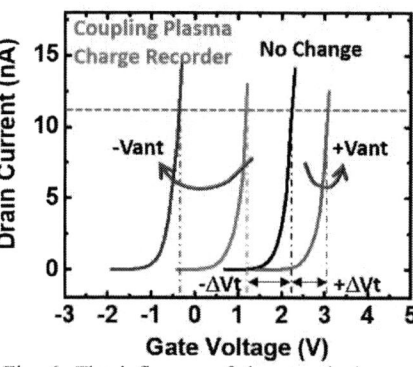

Fig. 6. The influence of the stored plasma charges on the read-out current in this I_DV_G plot.

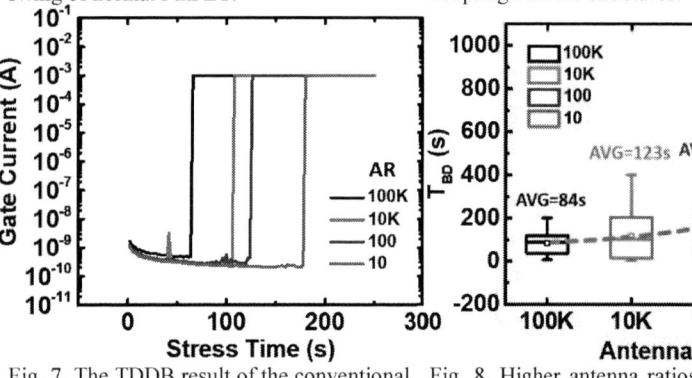

Fig. 7. The TDDB result of the conventional plasma test patterns in 7nm technology under 16.3MV/cm gate dielectric stress.

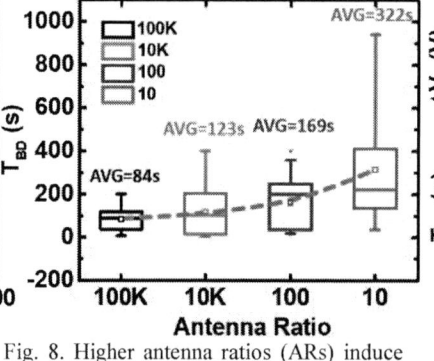

Fig. 8. Higher antenna ratios (ARs) induce much severer gate dielectric damage and shorter TDDB (T_{BD}).

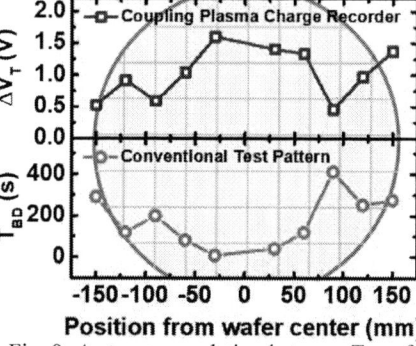

Fig. 9. A strong correlation between T_{BD} of the conventional patterns and the threshold shift (ΔV_T) of the new charge recorders.

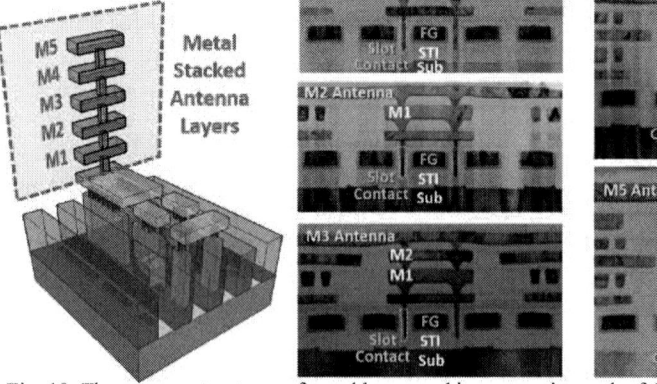

Fig. 10. The antenna structures of metal layers and its connection path of the recording floating gate to the plasma collecting layers.

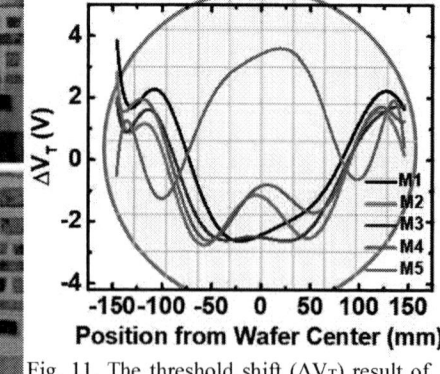

Fig. 11. The threshold shift (ΔV_T) result of recorders from metal 1 (M1) to metal 5 (M5) across the whole wafer.

978-1-7281-1988-5/18 $31.00 © 2018 IEEE

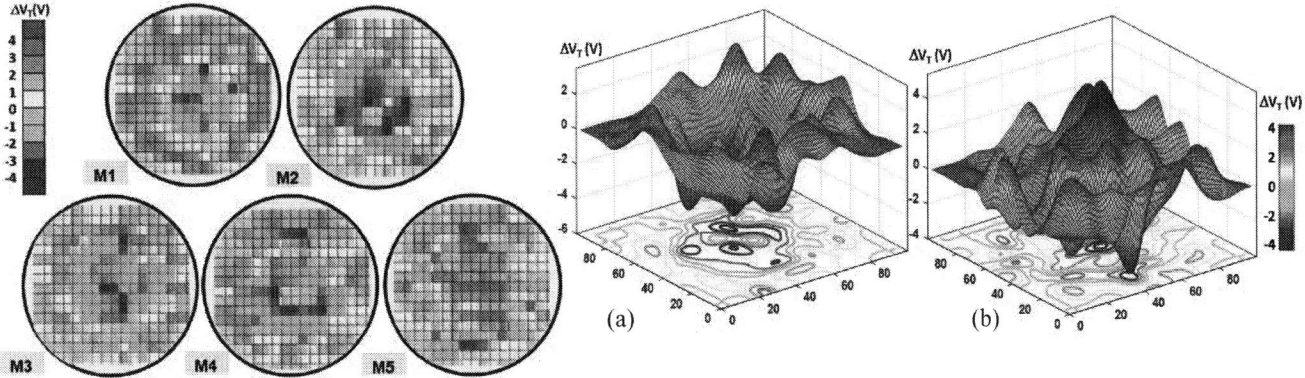

Fig. 12. A 12-inch wafer contour is inhibited, showing plasma charges and polarization distributions of M1 to M5.

Fig. 13. 3D contours of (a) M2 and (b) M5 are summarized to show the comparison of the threshold voltage shift for different metal layers.

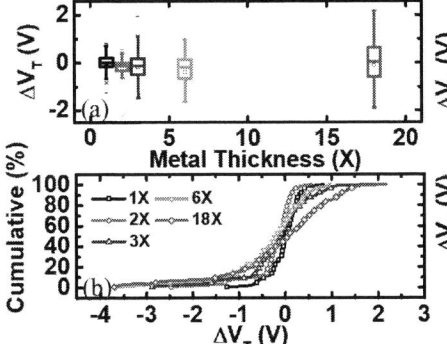

Fig. 14. (a) ΔV_T and (b) cumulative threshold voltage shift (ΔV_T) of the recorders with different metal thickness.

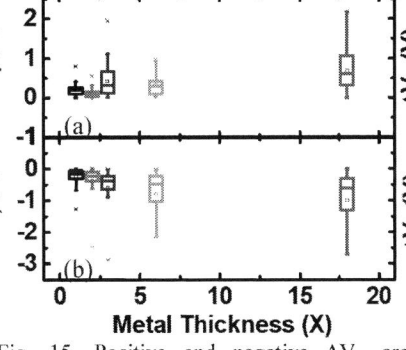

Fig. 15. Positive and negative ΔV_T are separated to consider work function difference of metal gate and substrate tunneling.

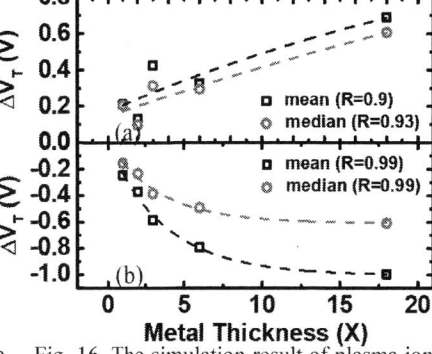

Fig. 16. The simulation result of plasma ion charge accumulative model (pCAM) by FN tunneling mechanism.

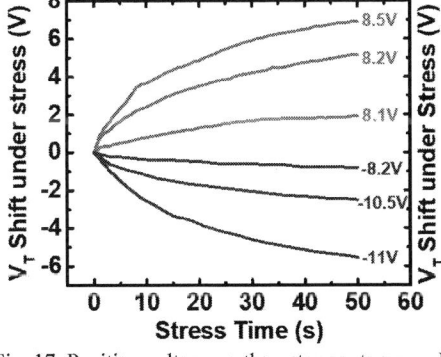

Fig. 17. Positive voltage on the antenna stores negative charges into floating gate and results in positive ΔV_T, and vice versa.

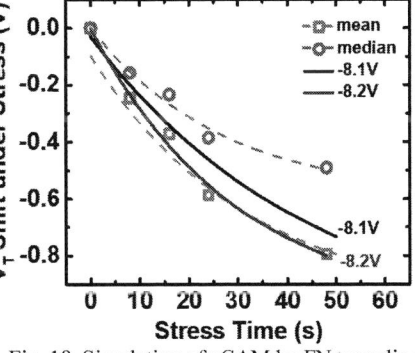

Fig. 18. Simulation of pCAM by FN tunneling mechanism is matching with the mean and medium values of ΔV_T.

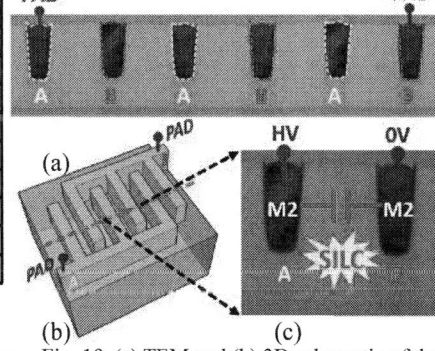

Fig. 19. (a) TEM and (b) 3D schematic of the finger-type metal structure with narrow pitches (c) Illustration of SILC on dielectric.

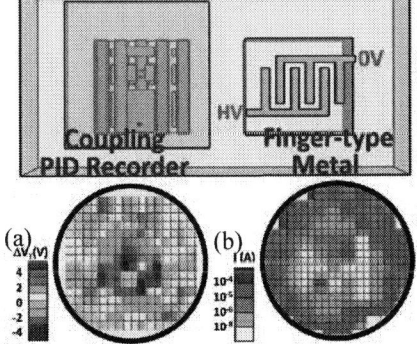

Fig. 20. Wafer mapping of the recorded ΔV_T level of plasma charge and the SILC result of finger-type metal structures.

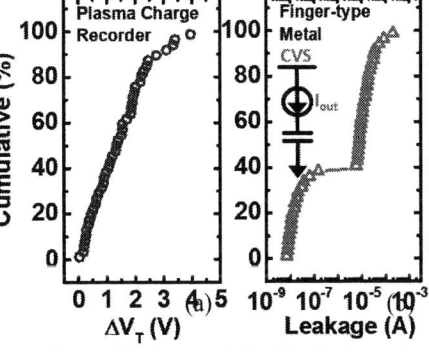

Fig. 21. (a) Continuous ΔV_T distribution of the recording devices, however, (b) shows a turning point in SILC result.

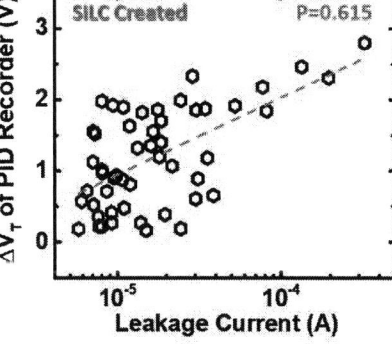

Fig. 22. Pearsons value of 0.6 between SILC from the stressed IMD and ΔV_T taken from the plasma recorders.

978-1-7281-1988-5/18 $31.00 © 2018 IEEE   403

Development of X-ray Photoelectron Spectroscopy under bias and its application to determine band-energies and dipoles in the HKMG stack

Pushpendra Kumar[1,2,3], Charles Leroux[2], Florian Domengie[1], Eugenie Martinez[2], Virginie Loup[2], Denis Guiheux[1], Yves Morand[2], Jean-Michel Pedini[2], Claude Tabone[2], Frederic Gaillard[2], Gerard Ghibaudo[3]

[1]STMicroelectronics, 850 rue Jean Monnet, 38926, Crolles Cedex, France, email : pushpendra.kumar@st.com
[2]Univ. Grenoble Alpes, CEA, LETI, 38000 Grenoble, France, email : pushpendra.kumar@cea.fr; charles.leroux@cea.fr
[3]IMEP-LAHC, Minatec/INPG, BP 257, 38016 Grenoble, France

Abstract— In this paper, we present for the first time specific methodology and test structures authorizing an accurate analysis of XPS under bias measurements. Such analysis which identifies effective biasing across the device, allows to determine the absolute energy levels of the different layers in the HKMG stack at any bias. This enables an accurate band diagram identification and it is applied to analyze the physical mechanisms at work in the threshold voltage (V_T) engineering of HKMG stacks. We demonstrate that V_T shift induced by La and Al additives or metal gate thickness variations originates by the modifications of the dipole at SiO_2/high-k interface.

I. INTRODUCTION

Advanced CMOS device development requires a comprehensive knowledge of band-energies of the different layers in the gate stack, particularly in High-κ Metal Gate (HKMG) stacks consisting of many ultra-thin layers and their interfaces. Moreover, process changes, such as introduction of La or Al additives [1, 2], interface states (Dit) and metal gate thickness variation [3, 4, 5] can modulate the relative band energy levels in the gate stack leading to a shift of metal gate effective workfunction (WF_{eff}) and thus the threshold voltage (V_T). Electrical measurements are very useful to obtain many key parameters of the gate stack but are unable to give information on the band energies inside the gate stack. In contrast, X-ray Photoelectron Spectroscopy (XPS) is able to measure the core levels (CL) and valence bands (VB) binding energies (BE) of different materials within its analysis depth of around 10 nm. However, there are many problems with XPS in determining the absolute energy levels of the gate stack layers, such as charging effects and correct determination of effective biasing across the device. Some studies have been conducted to determine gate stack Dit or dipoles [6, 7, 8], but unsatisfactory or contradictory results have been obtained that could be due to the improper control of device biasing.

In this work, for the first time, we present a set of test structures and electrical modelling which combines XPS and electrical measurements and enables us to accurately determine the exact biasing of the gate stack allowing its band diagram determination. This technique is used to analyze the physical mechanisms at work in the V_T engineering of HKMG stacks.

II. EXPERIMENTAL PROCEDURE

A. Process flow

HKMG stack deposition: The gate dielectrics consists of 1 nm SiO_2 as IL and 2 nm HfON or HfSiON as High-k. Sacrificial La/TiN or Al/TiN gate stack with different La or Al additive thicknesses were deposited over the High-k, followed by annealing to activate additive diffusion into the High-k/SiO_2 stack. The sacrificial gate stack is then removed by wet etching and a Poly-Si/TiN electrode is deposited, followed by S/D annealing as schematized in Fig. 1. *Test structures*: Fig. 2 depicts the test structures that enable to bias the device with simultaneous XPS measurements. 500 nm STI SiO_2 is deposited onto Si substrate and cavities are patterned, defining the active area or device. Then, HKMG stack is deposited, followed by gate patterning. Lastly, Poly Si is removed in order to analyze the gate stack by XPS. The device is biased on the HKMG/STI SiO_2 and XPS is performed inside the cavity by using a Versaprobe II spectrometer from ULVAC-PHI equipped with a 4-contacts sample holder.

B. Device biasing and issues

Fig. 3 demonstrate the biasing of the device, by applying a gate voltage Vg and grounding the backside Si substrate. However, we must account for the difference between Vg and effective bias appearing on the device gate stack V_{DUT} (between gate bias V_{TiN} and substrate bias V_b). This difference DV (Vg-V_{DUT}) occurs due to a combination of series resistance of the metal gate (Rs1) and of the bulk Si substrate (Rs2). DV must be accounted to know the exact device biasing. An ideal condition for band energies analysis is the flat band condition requiring a device biasing at Vfb. Moreover, XPS under bias requires relatively large devices (mm range) and compared to nominal size devices (μm range), they suffer from large substrate serial resistance effect that increases DV and consequently decreases the leakage current density J(V) (Fig. 4). Moreover, capacitance characteristics C(V) (Fig. 5) are also impacted [9]. Thus, exact biasing of the device (control of V_{DUT}) and analysis of biasing inside the gate stack from C(V) become great issues.

Smaller devices have negligible bias drops due to smaller leakage current. V_{DUT} on bigger devices can be identified by comparing its J(V) characteristics with that of 10*10 μm² device (Fig. 4) and so DV=Vg-V_{DUT}. To calculate DV1 separately, bias drop along the active area boundary is measured and then subtracted from Vg. Then, DV2 is calculated as DV2=DV-DV1. This way V_{DUT}, DV, DV1 and DV2 can be calculated for each value of leakage current for bigger devices, as shown in Fig. 6. Accurate determination of DV2 is necessary as it will shift band energies at the flatband condition (Fig. 7). Thus, DV2 contribution must be subtracted from raw BE values to calculate the correct band energies. Vfb was extracted by fitting experimental C(V) measurements of a $10*10$ μm^2 device with Poisson Schrödinger quantum simulations [10]. Then during XPS measurements of bigger devices, a constant leakage current value corresponding to $V_{DUT}=Vfb$ was forced in order to be at the flatband condition.

III. RESULTS

A. XPS under bias technique validation

Before we can apply XPS under bias to measure HKMG energy bands and dipoles, it must be shown that the shifts of the XPS BE related to the different gate stack layers, induced by device biasing, follows what is expected from electrical modelling. Ti-2p, Hf-4f and Si-2p XPS spectra were fitted to account of the different compounds and doublets and a specific BE was related to each gate stack layer: TiN (TiN 2p), high-k (HfO_2-4f), SiO_2 (SiO_2-2p), Si surface (Si^0-2p) (Fig. 8 and 9). A $200*200$ μm^2 device was biased from 0 V to strong accumulation condition and XPS spectra for Ti, Hf and Si were measured each time. Fig. 10 shows the shift of Ti signal with bias and Fig. 11 reports the same spectra shifted by D(TiN, Vg). The shifted spectra coincides, confirming that the same signal fitting applies at every bias. The same is true for Hf (not reported here). Fig. 12 shows similar analysis for Si with shift D (Si^0, Vg). We notice that SiO_2 peaks do not coincide, confirming that Si spectra corresponds to different layers in the gate stack. Fig. 13 summarizes the BE shifts with applied bias, related to each gate stack layer. For large bias, the shift of TiN BE is lower than the expected shift of the gate (dotted line), revealing parasitic bias drops in TiN gate. The difference between TiN and Si^0 corresponds to bias drop inside the dielectrics, and the signal related to SiO_2 is intermediate as expected. Fig. 14 compares TiN BE shifts, at different biases, with its corresponding TiN gate bias $V_{TiN} = V_{DUT} + DV2$, an accurate correlation is seen. Fig 15. compares Si^0 BE shift with the Si surface bias V_{Si} shift ΔV_{Si} (Fig. 3; $V_{Si} = \Psi_{Si(10*10)} + DV2$). $\Psi_{Si (10*10)}$ is the Si surface potential obtained from a CV modelling of the $10*10$ μm^2 device with or without Dit (Fig .15&16). Inset of Fig. 16 gives the Dit charge (Qit) obtained from the difference between ΔV_{Si} without Dit and Si^0 BE shift.

B. Dipole location by XPS under bias

After having validated an XPS under bias methodology, we apply it now to localize various dipoles in the gate stack. **La & Al dipoles**: Various doses of La and Al additives were introduced into the gate stack by sacrificial gate process (Fig. 1) and 4 different samples have been compared. They lead respectively to a relative shift in Vfb ($\Delta Vfb = \Delta WF_{eff}/q$) of 0.1, 0, -0.135 and -0.225 V (Vfb for reference sample = -0.553 V). BE for TiN-2p, HfO_2-4f, SiO_2-2p and Si^0-2p are extracted and in Fig. 17 we report their shift (from reference device) at Vfb as

a function of ΔVfb. No clear pattern is observed, especially that at Vfb Si^0 shift is expected to be equal to 0. This is due to the parasitic bias drop in Si substrate DV2 which is different among the 4 samples. Fig.18 shows the corrected BE ($BE_{corrected} = BE_{raw}$ - DV2) (see Fig. 7) at Vfb. Si^0 and SiO_2 BE remains then quite constant, meaning that no dipole modulation takes place at this interface. On the other hand, TiN and HfO_2 BE shift by the same amount and follows the line $\Delta BE = q\Delta Vfb$. We prove here for the first time that Al and La additives induce no dipole modification at high-k/metal gate interface, but only at the SiO_2/high-k interface that is equal to $q \cdot \Delta Vfb$ as shown by the band diagram in Fig. 19. XPS under bias measurements are easier to perform at 0 bias: no leakage current and no need of specific biasing connection inside the XPS tool. So here we define a strategy to extrapolate the expected BE values at Vfb from BE values measured at 0 bias. Raw BE data at 0 bias is reported on Fig. 20. To extrapolate to Vfb, as done in Fig. 21, The BE at 0 bias are shifted by δ, defined for each layer. For TiN we consider that BE must be shifted by Vfb (δBE=Vfb). Fig. 13 suggests that the same shift must be applied to HfO_2 BE levels. For Si^0, δ is obtained from Ψ_{Si} at 0 bias (δBE= -$\Psi_{Si(Vg=0)}$). For SiO_2 we must also add half of the bias drop in the oxide (Vox), since BE measurement is an average over the SiO_2 layer (δBE= -$\Psi_{Si(Vg=0)}$-Vox/2). The differences x between Fig. 18 and Fig. 21 are very small ($\bar{x}$=-1meV, σ=25 meV), demonstrating a very close correlation between the two methods. **TiN thickness modulation**: For HKMG, Vfb usually increases with TiN thickness. However the main mechanism behind this Vfb modulation remains ambiguous. To solve this ambiguity, we performed XPS under bias measurements on devices with a range of TiN thickness (3.5, 5, 6.5 and 10 nm) at 0 bias. Extrapolated results at Vfb are presented in Fig. 22. As with La/Al additives, Si & SiO_2 BE remains constant and TiN & HfO_2 shift by the same amount and follows the line $\Delta BE = q \cdot \Delta Vfb$. This proves here for the first time that it is the modulation of dipole at the SiO_2/High-k interface which is responsible for Vfb shift with TiN thickness.

IV. CONCLUSION

New test structures are presented that enable to perform XPS under bias measurements. Moreover, an electrical model and appropriate methodology are introduced for the first time in order to determine correct biasing for the flatband condition and at other biases. XPS under bias technique is validated by comparing XPS binding energy shifts with what is expected from our electrical model. Then, this technique has been applied to analyze the origin of effective work function shift related to La & Al or to TiN thickness modulation. It has been shown for the first time, by relative band energy measurements, that the dipole modulation at the SiO_2/High-k interface is responsible for these observed WF_{eff} change. Moreover, it has been shown that band energies at flatband condition can be obtained from measurements at only 0 bias condition. This technique can be used to characterize devices beyond CMOS.

ACKNOWLEDGMENT

This work has been performed in the context of the STMicroelectronics-CEA LETI collaboration.

REFERENCES

[1] C.Suarez et al., IEEE EDL, vol. 38, no. 3, pp. 379-382, 2017
[2] C. Leroux et al., Solid-State Electron. Vol. 88, 21-26, 2013
[3] R. Singanamalla et al., IEEE EDL, VOL. 27, no. 5, 2006
[4] K. Choi et al., VLSI-TSA-Tech, pp. 103-104, 2005
[5] M. Kadoshima et al., IEEE EDL, vol. 30, pp. 466–468, 2009
[6] Y. Yamashita et al., JAP 115, 043721, 2014
[7] Yugo Chikata et al., JJAP. Vol. 52, 021101, 2013
[8] H. Sezen et al., Thin Solid Films, vol. 534, 2013
[9] K.J. Yang et al.,IEEE TED, Vol.46, pp.1500-1502, 1999
[10] C. Leroux et al., Micro. Eng., vol. 84, pp. 2408–2411, 2007

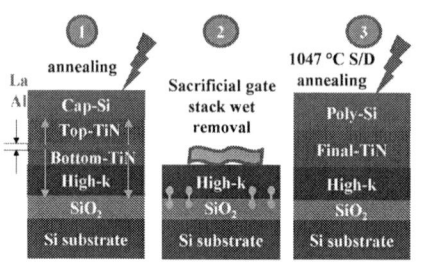

Fig. 1. HKMG stack deposition by sacrificial gate-first approach [1]

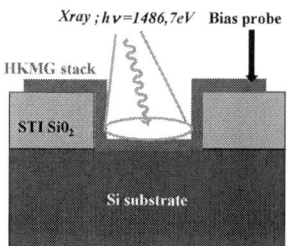

Fig. 2. Test structures used for XPS under bias measurements

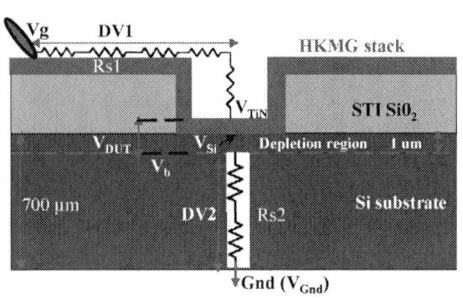

Fig. 3. Description of serial resistances and bias drops present in the test structure used for XPS under bias. $DV1 = Vg - V_{Ti}$; $DV2 = Vb - V_{Gnd}$; $V_{DUT} = V_{Ti} - V_b$

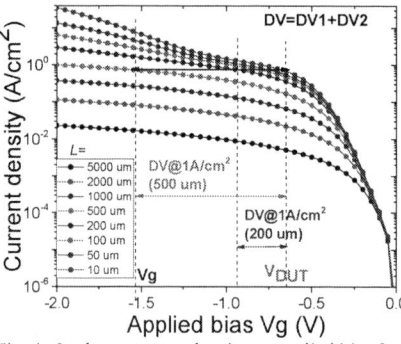

Fig. 4. Leakage current density vs applied bias for different square capacitances of edge size L, a same current density corresponds to a same V_{DUT}

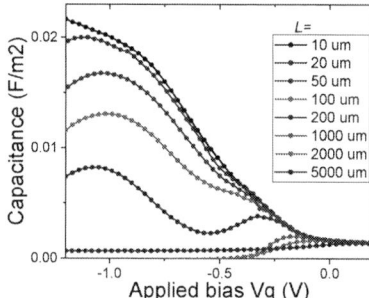

Fig. 5. C-V measurements showing capacitance degradation with device size (L is the edge size) due to serial resistance

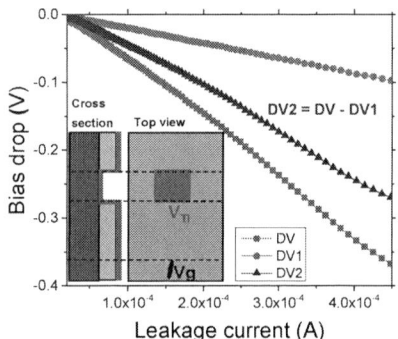

Fig. 6. Bias drop evaluations in the gate (DV1) and the Si substrate (DV2); inset showing device top and cross section view

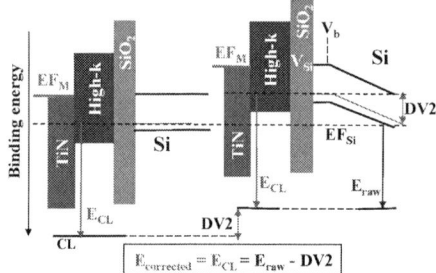

Fig. 7. Effect of substrate bias drop DV2 on energy bands at flatband condition during XPS under bias measurements. Backside Si substrate is grounded to the XPS spectrometer, BE measured are referenced from Si fermi level. Corrected CL BE can be calculated from the raw BE measurements by subtracting DV2 (< 0 for a p substrate sample)

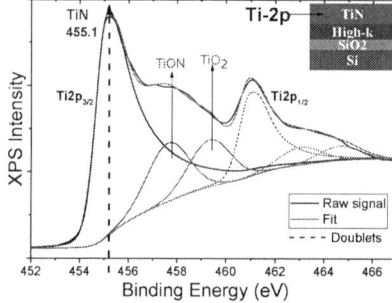

Fig. 8. XPS spectra of Ti-2p showing fitting with different components, TiN-2p BE is used as reference for the metal gate

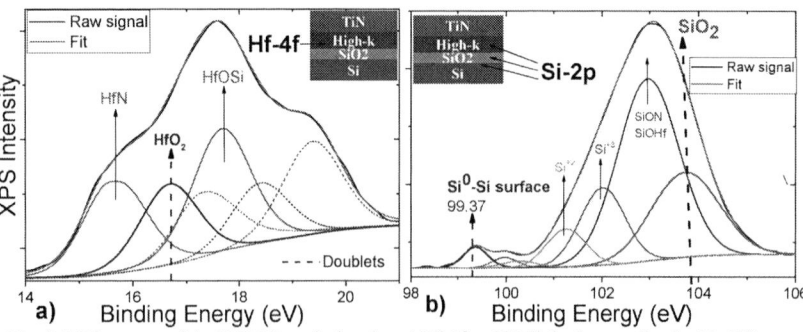

Fig. 9. XPS spectra of the HKMG stack showing a) Hf-4f and b) Si-2p Spectra fitted with different components, specific BE for HfO_2-4f, Si0-2p, SiO_2-2p are used as reference for high-k, Si substrate and SiO_2 IL respectively

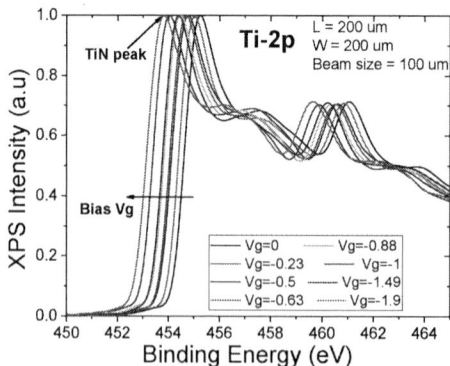

Fig. 10. Shift of Ti 2p XPS spectra with bias

978-1-7281-1988-5/18 $31.00 © 2018 IEEE

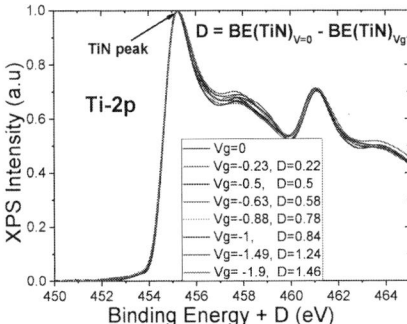

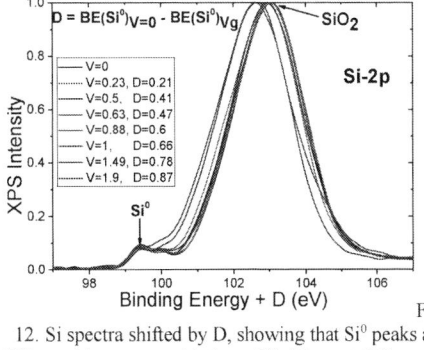

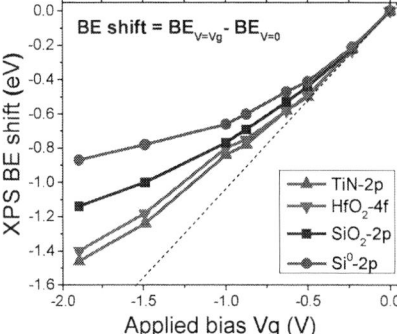

Fig. 11. Ti spectra shifted by D, showing that TiN peaks at different bias merge with that at V=0 and thus the validity of the fitting technique at every bias

Fig. 12. Si spectra shifted by D, showing that Si^0 peaks at different bias merge with that at V=0 but not SiO_2 peaks

Fig. 13. Variation of BE of the gate stack layers, extracted from their respective signal fitting, with applied bias Vg

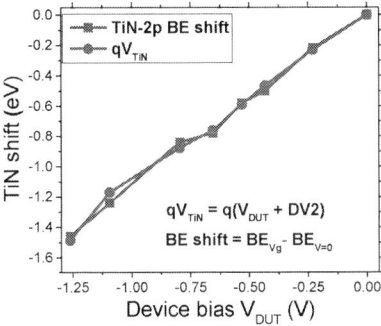

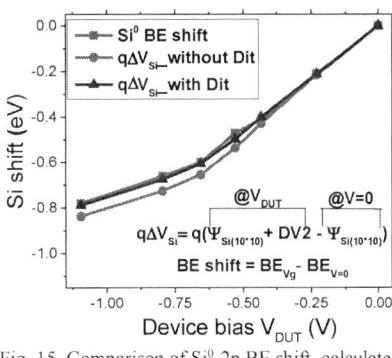

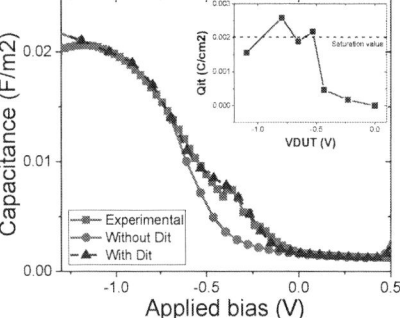

Fig. 14. Comparison of TiN-2p BE shift, extracted from Ti-2p XPS signal fitting, with TiN bias V_{TiN} (calculated from current modelling)

Fig. 15. Comparison of Si^0-2p BE shift, calculated from Si-2p XPS signal, with Si surface bias (V_{Si}) shift (calculated from CV modelling of a 10*10 device)

Fig. 16. Fitting of C-V curves by Poisson-Schrodinger model with and without Dit effect; inset showing the Dit charge (Qit) obtained from the difference between ΔV_{Si} without Dit and Si BE shift

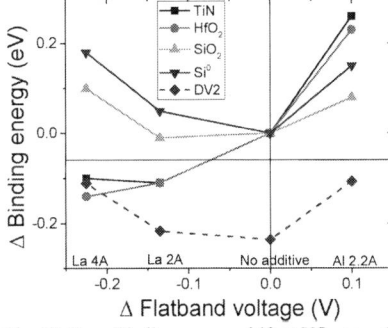

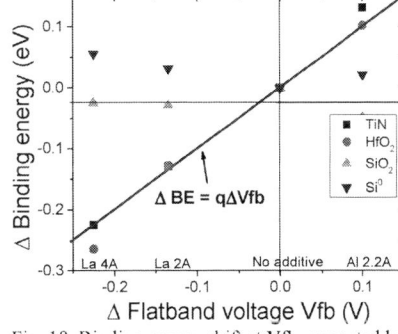

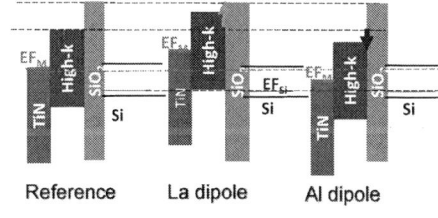

Fig. 17. Raw Binding energy shift at Vfb, as well as DV2, versus flatband voltage shift

Fig. 18. Binding energy shift at Vfb, corrected by substrate bias DV2, versus flatband voltage shift for samples with La and Al additives

Fig. 19. HKMG stack band diagram at Vfb, corrected by substrate bias DV2, showing the relative energy bands shifts. Relative energies change at the SiO_2/High-k interface due to La/Al dipole modulation that is responsible for WF_{eff} change

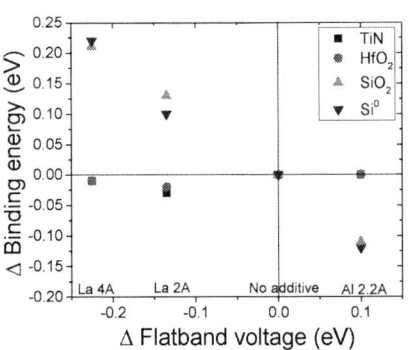

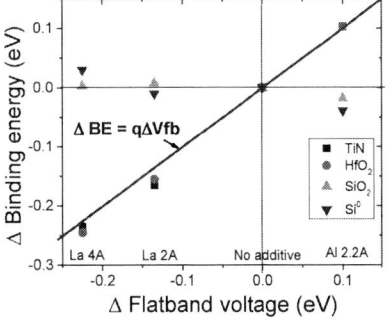

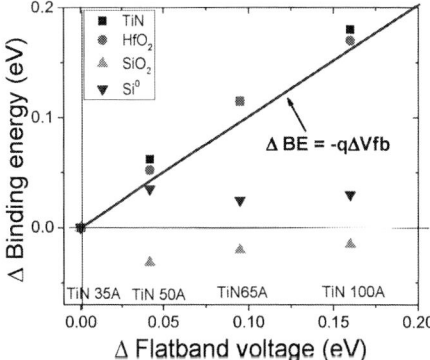

Fig. 20. Binding energy shift at 0 bias versus flatband voltage shift

Fig. 21. Extrapolated Binding energy shift at Vfb, from data at 0 bias, versus flatband voltage shift for samples with La and Al additives

Fig. 22. Extrapolated Binding energy shift at Vfb, from data at 0 bias, versus flatband voltage shift for samples with different TiN thickness

In-situ Investigation of the Impact of Externally Applied Vertical Stress on III-V Bipolar Transistor

Y. Liu[1,2], G. Hiblot[1], M. Gonzalez[1], K. Vanstreels[1], D. Velenis[1], M. Badaroglu[3], G. Van der Plas[1], I. De Wolf[1,2]

[1]IMEC, Leuven, Belgium, email: yefan.liu@imec.be
[2]Dept. Materials Science, KU Leuven, Leuven, Belgium
[3]Qualcomm Inc, Leuven, Belgium

Abstract—This work presents a new methodology to investigate in-situ the impact of vertical stress on the electrical characteristics of semiconductor devices. It is applied for the first time on III-V Heterojunction Bipolar Transistors (HBT). It combines a nanoindenter, which is used to apply controlled vertical forces on the sample surface, with in-situ electrical measurements using micro probes. The HBT devices are shown to be significantly affected by vertical stress: both the current and the capacitance show a reduction with increasing compressive vertical stress. The observations are confirmed by TCAD simulations This method can be employed to extract the sensitivity of advanced devices to vertical (out-of-plane stress) which is a growing concern in packaging and 3D integration.

I. INTRODUCTION

3D IC integration technology is a potential way for further improvement of CMOS performance beyond Moore's Law [1]. However, it is well-known that 3D processing (wafer thinning, through silicon vias, micro bumps, stacking, ...) induces mechanical stress that may affect the device characteristics due to piezo-resistive effects and deformation of the band structure [2]. While a significant amount of work has been dedicated to investigate the impact of in-plane stress induced by TSVs on devices [3], little attention has been paid to vertical stress. However, this stress component cannot be neglected in advanced 3D processes featuring extreme thinning and multi-tiers stacking. Four-point bending (4-pb) has been widely used to extract the transistors sensitivity to in-plane stress [4], but there is a lack of a corresponding method and experimental data for vertical (out-of-plane) stress. In this paper, we report for the first time an in-situ probing nanoindentation set-up and successfully use it to investigate the vertical stress impact on a radio-frequency (RF) HBT devices.

II. EXPERIMENT AND DEVICE

A. In-situ Nanoindentation Probing Set-up

The set-up is based on a commercial nanoindenter (Hysitron TriboIndenter 950) (Fig. 1), which allows precise control of force and displacement exerted on the sample's surface. A flat tip of 100 μm diameter is used to apply vertical force without indenting the sample. In-situ electrical measurements of the device under test are enabled by remotely controlled movable probes (miBot^M, Imina Technologys SA), visible as red robots in Fig. 1. These provide electrical access to the device, even in the very confined space below the nanoindenter tip and the

probe pads. The device characteristics are measured by Keithley 2602A SMUs and HP4284 LCR.

During indentation, we use a fixed force mode in which the force is increased in 10 continuous steps and we measure the device characteristics at each step. Typical force and displacement curves of the indentation are shown in Fig. 2(a). The rounded profile of the displacement at each step is attributed to material relaxation, requiring a larger indentation depth to keep the same force and thus stress on the device. Finite element simulations (FEM) are used to convert the applied force into vertical stress induced by the nanoindeter tip at the device (Fig. 2(b)). Below 1000mN, the compressive vertical stress increases linearly with the applied force (~0.14MPa/mN).

B. III-V Heterogeneous Bipolar Transistor

This nanoindentation probing set-up is applied on a n-p-n HBT whose collector, base, and emitter materials are InGaAs, InGaAs and InGaP, respectively. The subcollector is made of a heavily doped InGaAs layer on a GaAs substrate. An intrinsic layer of InGaAs is inserted between the base and the emitter. The three terminals of the HBT are led out of the BEOL with 40μm thick Cu pillars. The nanoindenter tip lands on the top surface of these Cu pillars (Fig. 3). FEM shows that the stress is uniform and vertical under the tip. It is efficiently transmitted to the device by the Cu pillar (Fig. 4).

III. RESULTS AND DISCUSSIONS

A. Stress impact on characteristics of HBT

Measurement results without indentation are given in Fig. 5. The bias of the collector-base junction is fixed at 1.2V. The intrinsic exponential dependence of the current on base-emitter voltage (V_{BE}) is observed between 0.8V to 1.2V (exponential regime in green). When V_{BE} goes above 1.2V, the parasitic resistance has a major impact on the measurements (ohmic regime in orange). Fig. 6 shows the emitter currents change in both exponential regime (V_{BE}=1.2V) and ohmic regime (V_{BE}=1.4V) with applied vertical stress. The emitter current clearly decreases under compressive vertical stress. Fig. 7 shows the base current change (same V_{BE} bias as in Fig. 6) where the stress impact is similar as the emitter current. In Fig. 6 and Fig. 7, all currents show a residual shift after the measurement, which is due to a permanent deformation of the copper pillar induced by deep indentation and the resulting mechanical stress left in the sample. By fitting the currents and stress, a clear linear dependence of stress is observed for both

978-1-7281-1988-5/18 $31.00 © 2018 IEEE

emitter current and base current. Fig. 8 shows the current density change induced by vertical stress in two devices with different dimensions: 6x30μm (device 1) and 24x30μm (device 2). The variations are similar in the "exponential regime", but differ in the "ohmic regime". This discrepancy is attributed to parasitic resistances, which explains why the largest transistor (device 2) is more sensitive to this effect.

As shown in Fig. 9, emitter and collector currents decrease by ~10% under ~70MPa while the base current decreases by only half of this percentage. Although the stress sensitivities are different, all currents variations are linear with stress. In Fig. 10, the beta variation under stress is represented, its decrease is also linear. Compared to current variations, the beta value decreases by less than 2.5% under 70MPa, which means that the HBT current gain is less sensitive than the currents to external stress. Finally, Fig. 11 shows the change of the I-V slope of HBT devices. Although they follow a linear trend, the magnitude of the change is negligible: The slope of the emitter current increases by 0.2% under 70MPa while the slope of the base current increases by 0.3%.

These effects of vertical stress on the base and emitter currents can be explained by an increase of the bandgap under compressive stress, as represented schematically in Fig. 12. The rapid increase of the conduction band outweighs the small variations of the valence bands due to the large value of the Gamma valley deformation potential $\Xi_{d,\Gamma}$ (Table. 2). This results in bandgap widening, which reduces the carrier concentration and hence the diffusion current (eq. 1). In comparison, the impact of stress on the electrons mobility is very small because they are concentrated in the single non-degenerate Gamma valley of III-V material. Thus, in contrast to Si CMOS, the stress sensitivity of III-V HBT comes mainly from bandgap variations rather than piezo resistive effects.

The base-emitter capacitance C_{BE} is measured with an LCR at 1 MHz at as function of bias conditions. It can be fitted by a depletion capacitance model (eq. 2) at low bias and diffusion capacitance (eq. 3) at high bias. In Fig. 14, the impact of vertical stress on this capacitance under different base-emitter bias is plotted. For low values of V_{BE}, C_{BE} shows less than 0.5% variation because it is dominated by the depletion capacitance, which is only affected by the "geometrical" deformation of the depletion width induced by vertical stress. At higher values of V_{BE}, C_{BE} becomes dominated by the diffusion capacitance which is proportional to the current. The reduction of the current due to stress therefore induces a similar reduction of the diffusion capacitance, which is indeed observed in Fig 14.

B. Simulation and comparison

These findings are confirmed with TCAD simulations, which include both energy shifts (deformation potential model) and mobility change (piezo resistive model). The key parameters used in TCAD model are given in Table. 2. The bandgap shift of the base and emitter materials is plotted as a function of stress in Fig. 15. The bandgap change is larger in the base material InGaAs (21.6meV under 150MPa) than in InGaP (12.9meV under 150MPa), which explains why the

emitter current decreases more than the base current. The simulated current variation versus vertical stress is plotted in Fig. 16, confirming the degradation observed experimentally. Discrepancies with the experimental values are attributed to inaccuracies in the estimated value of the vertical stress obtained from simulations. Notwithstanding these differences, the ratio of the relative changes of the emitter current and the base current relative change is 2.5, very close to the experimental value of 2.1. Simulated stress-induced variations of I-V slopes are shown in Fig. 17 and are very close to the experiment values. Fig. 18 shows the comparison of the stress impact by considering individually the contributions of mobility (piezo resistive model) and bandgap (deformation potential). It confirms that bandgap widening indeed contributes more to the change of the device current than the change in mobility. Fig. 19 shows the base-emitter capacitance variation with stress, which is also confirming the reduction observed experimentally.

IV. SUMMARY

For the first time, we have investigated the effect of vertical stress on a n-p-n III-V HBT by using a novel in-situ nanoindentation probing set-up. We have observed a reduction of the emitter and base current under compressive vertical stress (~10% under -70MPa vertical stress). This change is mainly due to the widening of the bandgap in the emitter and base layers. The beta of the HBT decreases by a lower amount ~2.5%, and the change of the I-V slope is negligible. The base-emitter capacitance is not affected by stress at low current level due to the insensitivity of the depletion capacitance. In contrast, at a higher bias, C_{BE} changes with stress due to the contribution from the diffusion capacitance which varies linearly with the current. It decreases by ~10% for ~-130MPa vertical stress at V_{BE}= 1.2V. Simulations combining the piezo resistivity model and the deformation potential theory confirm these trends.

ACKNOWLEDGMENT

The authors would like to thank Qualcomm for providing the sample used in this study.

REFERENCES

[1] E. Beyne, "The 3-D interconnect technology landscape", *IEEE Design and Test*, vol. 33, pp. 8, Mar. 2016.
[2] C. Herring and E. Vogt, "Transport and deformation-potential theory for many-valley semiconductors with anisotropic scattering", *Physics Review*, vol. 101, pp. 944-961, Feb. 1956.
[3] S. Suthram et al., "Strain additivity in III-V channels for CMOSFETs beyond 22nm technology node", *IEEE VLSI*, pp. 182-183, Jun. 2008.
[4] Y. Sun, S. E. Thompson and T. Nishida, "Physics of strain effects in semiconductors and metal-oxide-semiconductor field-effect transistor", *Journal of Applied Physics*, 101, 104503, Oct. 2007.
[5] P. A. Khomyakov et al., "Compositional bowing of band energies and their deformation potentials in strained InGaAs ternary alloys: a first-principles study", *Applied Physics Letters*, Jun. 2015.
[6] I. Vurgaftman et al., "Band parameters for III-V compound semiconductors and their alloys", *Journal of Applied Physics*, 89, 5815, Jun. 2007.
[7] B. W. Hakki et al., "Band Structure of InGaP from Pressure Experiments", *Journal of Applied Physics*, 41, 5291, Dec. 1970.

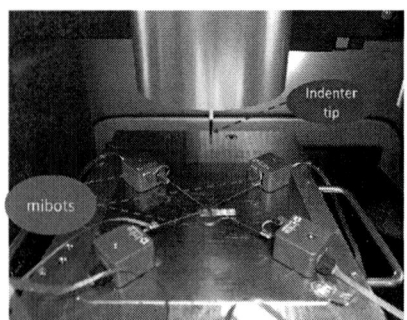

Fig. 1. In-situ nanoindentation set-up with mibots probing

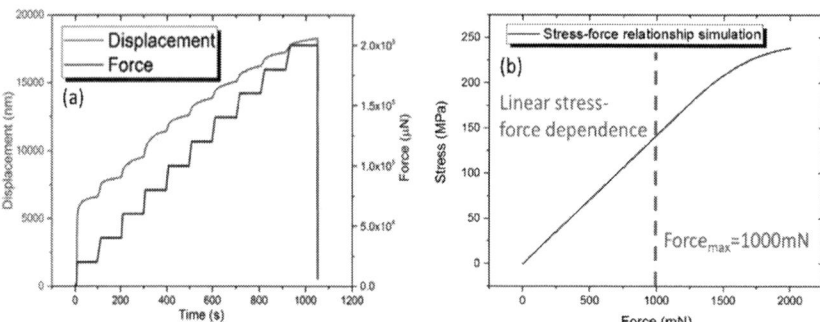

Fig. 2. (a) Force/displacement curves of a step by step fixed force mode, (b) mechanical simulation results of stress-force relationship which shows a linear stress-force dependence in experiment

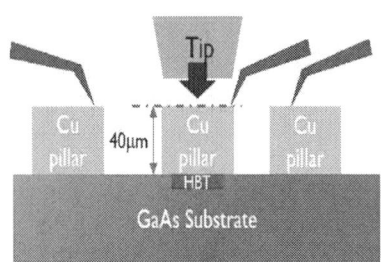

Fig. 3. Schematic of lead out copper pillar above the HBT and indentation position

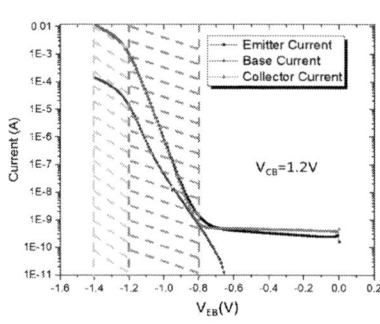

Fig. 5. Gummel plot of HBT (6μmx30μm), exponential regime (green) and ohmic regime (orange)

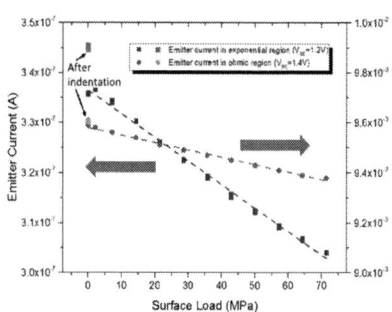

Fig. 6. Emitter currents vs. vertical stress in both exponential regime and ohmic regime

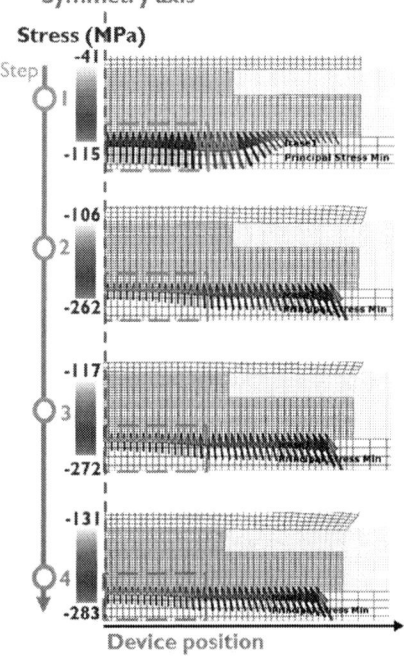

Fig. 4. FEM of half device showing the principle stress level and direction (arrows) Step (1) 300mN, (2) 750mN, (3) 850mN, (4) 950mN. The HBT location is indicated by green squares.

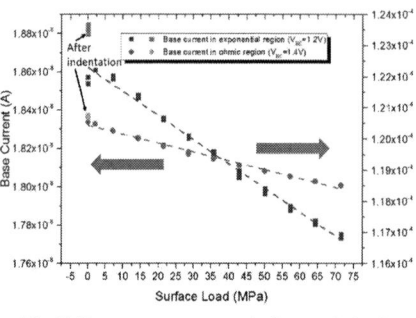

Fig. 7. Base currents vs. vertical stress in both exponential regime and ohmic regime

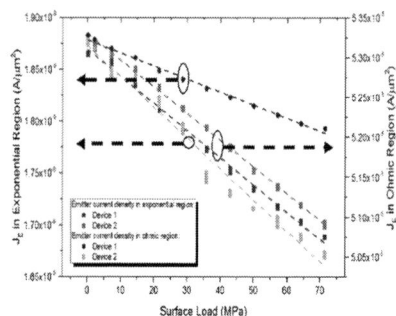

Fig. 8. Stress impact on emitter currents density in devices with different sizes. (device 1; 6μmx30μm; device 2: 24μmx30μm)

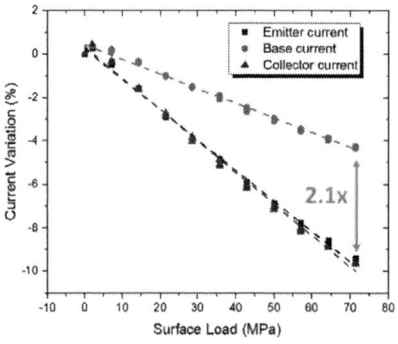

Fig. 9. Current sensitivity to vertical stress in exponential regime (V_{BE}=1.2V)

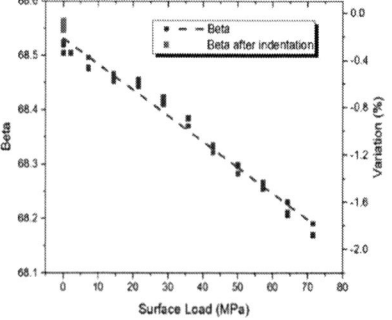

Fig. 10. Beta change vs. vertical stress in exponential regime

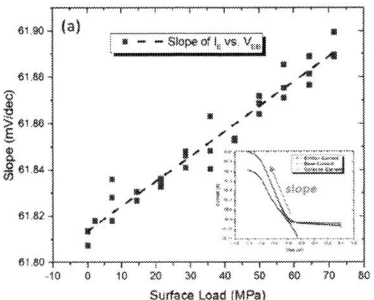
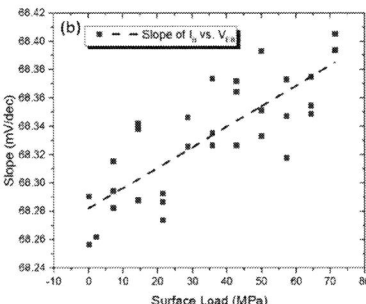

Fig. 11. I-V slope changes vs. vertical stress in exponential regime (a) emitter current, (b) base current

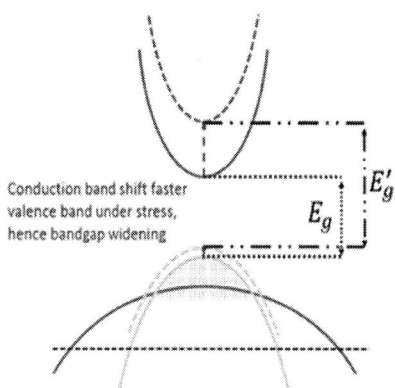

Fig. 12. Schematic of stress induced band structure alternation

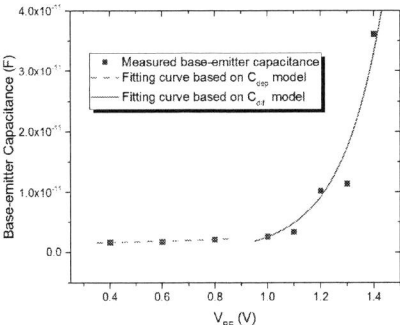

Fig. 13. C_{BE} vs. V_{BE}, fitting by the depletion capacitance and the diffusion capacitance models

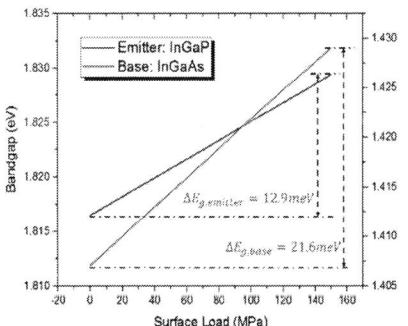

Fig. 14. C_{BE} vs. vertical stress under different V_{BE}

Eq. 1	$I_C, I_E \propto n_{iB}^2/N_B \propto e^{-E_{g,b}/kT}$ $I_B \propto n_{iE}^2/N_E \propto e^{-E_{g,e}/kT}$
Eq. 2	$C_{dep,BE} = A\varepsilon_s/W_{dep,BE}$
Eq. 3	$g_m = dI_C/dV_{BE} \approx qI_C/kT$ $C_{dif,BE} = dQ_F/dV_{BE}$ $= \tau_F dI_C/dV_{BE} = \tau_F g_m \propto I_C$

Table. 1. Equations

Parameters	InGaAs	InGaP
$\Xi_{d,\Gamma}$	-8.5	-9.5
$\Xi_{u,L}$	15.1	8.6
$\Xi_{d,L}$	-7.22	-
$\Xi_{u,X}$	8.22	0.74
$\Xi_{d,X}$	-0.19	-
a_v	1.16	1.66

Table 2. Key parameters used in TCAD model for stress impact simulation [5-7]

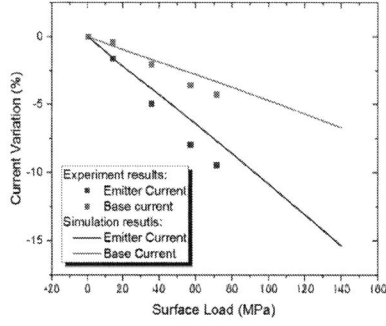

Fig. 15. Bandgap changes of base and emitter materials under vertical stress

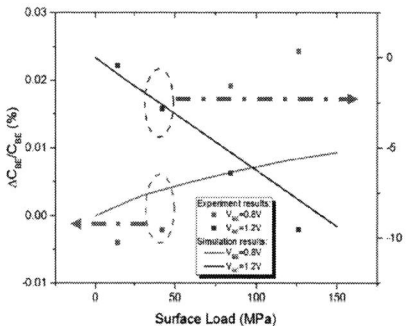

Fig. 16. Simulation of I_E and I_B vs. vertical stress

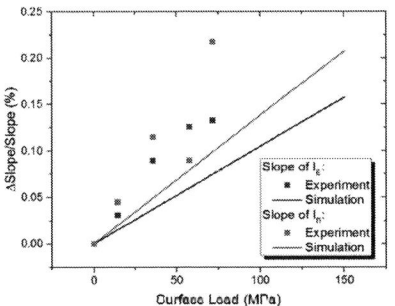

Fig. 17. Simulation results of I-V slope vs. vertical stress

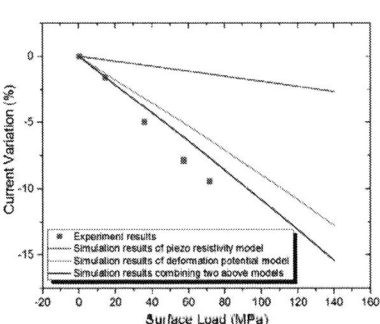

Fig. 18. Comparison of simulation results based on piezo resistivity and deformation potential models

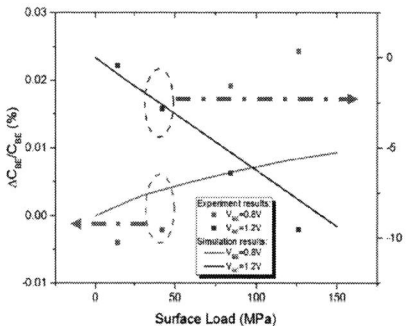

Fig. 19. Simulation results of C_{BE} vs. V_{BE}

978-1-7281-1988-5/18 $31.00 © 2018 IEEE   411

MRAM as Embedded Non-Volatile Memory Solution for 22FFL FinFET Technology

O. Golonzka, J. -G. Alzate, U. Arslan, M. Bohr, P. Bai, J. Brockman, B. Buford, C. Connor, N. Das, B. Doyle,
T. Ghani, F. Hamzaoglu, P. Heil, P. Hentges, R. Jahan, D. Kencke, B. Lin, M. Lu, M. Mainuddin, M. Meterelliyoz,
P. Nguyen, D. Nikonov, K. O'brien, J. ODonnell, K. Oguz, D. Ouellette, J. Park, J. Pellegren, C. Puls, P. Quintero,
T. Rahman, A. Romang, M. Sekhar, A. Selarka, M. Seth, A. J. Smith, A. K. Smith, L. Wei, C. Wiegand, Z. Zhang
and K. Fischer

Intel Corporation, Santa Clara, CA, USA, email: oleg.golonzka@intel.com

Abstract— This paper presents key features of MRAM-based non-volatile memory embedded into Intel 22FFL technology. 22FFL is a high performance, ultra low power FinFET technology for mobile and RF applications with extensive high voltage and analog support, and a high level of design flexibility at low cost[1]. Embedded NVM technology presented here achieves 200°C 10-year retention capability combined with $>10^6$ cycle endurance and high die yield. Technology data retention, endurance and yield capabilities are demonstrated on 7.2Mbit arrays. We describe device-level MTJ characteristics, key integration features, cell characteristics, array operation specifics, as well as key yield milestones.

I. INTRODUCTION

Embedded Non-Volatile Memory (e-NVM) technology has generated interest as a potential solution for several key market segments. Internet of Things applications, field-programmable arrays and chipsets with on-chip boot data are among potential customers for e-NVM. Existing solutions are provided by external flash and embedded flash memories. These technologies suffer from latency delay in the case of external flash and high manufacturing costs for embedded flash. Several new technologies are emerging as competitive replacements for flash memory. Among them, Magnetoresistive Random-Access Memory (MRAM) offers low manufacturing cost as well as exceedingly competitive data retention and switching endurance capabilities. In addition to embedded nonvolatile applications, Magnetic Tunnel Junction (MTJ)-based memory has the potential to be competitive as higher level SRAM or e-DRAM replacement, as well as a basic building block for future logic devices[2]. What makes MTJ-based technology unique is the large range of switching energy vs. retention vs. endurance tunability, allowing for large application flexibility. These features of MRAM have recently attracted investments from major semiconductor companies[3,4,5].

II. DEVICE CHARACTERISTICS

This work uses dual-MgO MTJs with a CoFeB-based free layer (Fig. 1). Typical Resistance-vs-Voltage (R-V) characteristics are shown on Fig. 2. Tunneling Magnetoresistance Ratio (TMR) is >180% with Resistance-

Area product (RA) of 9 Ωum². Target device size is between 60nm and 80nm. Coercivity of the device can be easily tuned to achieve the desired retention or switching characteristics. Fig. 3 shows write-error-rate curves for a typical device. Write-error-rate curves were collected with 20ns, 80ns, 500ns and 1µs pulses and show relatively flat Vc values for pulses down to 80ns and transition to increasingly precessional switching at 20ns pulses.

III. TEST VEHICLE

Fig. 4 shows a cross-sectional TEM of MTJ array embedded between Metal 2 and Metal 4 of the 22FFL logic process. Fig. 5 shows the layout of the 1T-1R MRAM cell used in this work. The MTJ device is centered on a Metal 2 pad placed on a 216nm x 225nm pitch grid (cell area of 0.0486µm²). We employ single-ended sensing and 128b-Triple error correction. We use an adaptive Write-Verify-Write (WVW) scheme with a sequence of write pulses of increasing pulse lengths and amplitudes. The WVW scheme can be tuned to trade off write energy vs retention vs endurance depending on the intended application. Read sensing is done with a short (<10ns) low amplitude pulse and data readout is accomplished by detecting the differential current signal between the MTJ and a thin film precision resistor tuned to provide optimal read margin.

IV. ENDURANCE AND RETENTION

Fig. 6 shows wafer-level bit error rates observed on full 7.2Mbit arrays subjected to 10^6 switching cycles with two switching protocols optimized for high retention or low switching energy, correspondingly. The data demonstrate cycling induced fallout <1E-6 rate for the high retention switching protocol and <5E-7 for the low energy switching protocol. Fig. 7 shows the projected wafer-median array-level 10-year retention capability for stacks with three different free layer thicknesses. The data demonstrate 155°C, 175°C and 200°C retention capability depending on the free layer thickness choice. Retention projections were made using raw bit failure rates observed at multiple bake temperatures and times, followed by extrapolation of those data to extract the temperature at which the fail rate would fall below 1E-5 after 10 years.

978-1-7281-1988-5/18 $31.00 © 2018 IEEE

V. YIELD

A. Shorting across the MgO barrier

One of the key issues in developing a high yielding MRAM process is overcoming shorting across the extremely thin MgO barrier layer. Several mechanisms contribute to this defect mode: shorting around the perimeter of the MTJ device due to remaining conducting metal residue, direct shorting between the free and reference layers due to non-uniformity of the MgO barrier layer, and shorting due to MTJ top contact wrap around. Fig. 8 shows the 12-month time trend of the "Shorting across the MgO barrier" defect indicator. With meticulous optimization of the MTJ stack, MTJ etch and integration we were able to achieve <1E-6 wafer-level bit error rate due to this defect mode, which is well below the ECC budget.

B. Thermal stability of the MTJ stack

Integrating MTJ-based memory into CMOS process brings an additional challenge of preserving magnetic properties of the stack through the thermal cycle of the subsequent back-end-of-line (BEOL) processing. Typical cumulative thermal cycle of modern BEOL will add close to one hour of exposure at temperatures >400°C. An implementable stack needs to show little-to-no degradation in TMR, coercivity, and switching voltage distributions when subjected to such exposure. Fig. 9 shows blanket-level TMR data of the stack developed for this process as a function of anneal temperature, demonstrating best-in-class capability to withstand 440°C one hour exposure.

C. Synthetic antiferromagnet locking

Device parametrics play a key role in achieving high yields. Fig. 10 shows a comparison of full M-H sweeps for two stacks: a typical stack (Stack A) and an advanced stack (Stack B). Stack A exhibits 3 stable states at zero external field: parallel and antiparallel states of the free and reference layers under antiparallel configuration of the Synthetic Antiferromagnet (SAF) as well as the state originating from the undesired parallel configuration of the bottom and top parts of the SAF. A small percentage of Stack A devices are manufactured with the SAF locked in the parallel state, leading to a strong stray field on the free layer, ultimately resulting in the free layer magnetization being locked parallel to the reference layer magnetization. This defect can be overcome by subjecting the devices to the appropriate external magnetic field at the end of manufacturing process, see Fig. 11. The downside of Stack A is that its functionality will be compromised if the device is maliciously subjected to a very large "positive" magnetic field. This vulnerability is completely eliminated in case of Stack B due to substantially stronger coupling in the SAF.

D. Existence of states with intermediate resistances

Another key consideration affecting MRAM yield is the switching time of MTJ devices. Fig. 12 shows two examples of time-resolved traces for AP to P switching under applied low-voltage 1μs pulses, showing typical switching behavior (left) and rare behavior in which the device exhibits a stalled state having an intermediate resistance value, in this case

persisting for several hundred nanoseconds (right). We will further refer to these states as "intermediate" states. A bit can remain in the intermediate state well after the end of the write pulse and ultimately complete switching into the P state or return back to the AP state. With an adaptive WVW switching scheme, returning back to the original AP state after falsely passing a verify read directly contributes to the write error rate. We attribute the physical origin of the intermediate states to the formation of a metastable, multi-domain magnetic configuration of the free layer. As expected, experimentally observed occurrence rate of the intermediate states drops dramatically with reducing the size of the device (Fig. 13). For a given device size, among other effects, formation of the intermediate states is driven by the inherent non-uniformity of the stray dipole field resulting from the incomplete cancellation of fields produced by the top and the bottom halves of the SAF structure. Fig. 14 shows simulations of the stray dipole field experienced by the free layer of an 80nm device of Stack A and Stack B discussed in the previous paragraph. Simulations were normalized to produce zero average stray field integrated across the full device and show that Stack A exhibits substantially larger non-uniformity in the radial distribution of the stray field with device edges favoring P configuration. This non-uniformity ultimately results in early nucleation of AP to P switching on some devices and formation of a metastable domain structure with P configuration around the device edges and AP configuration in the device center. Fig. 13 shows experimental results comparing the intermediate states metric for Stack A vs Stack B at the same device CD and matched average stray field. As expected, due to the improved within-bit stray field uniformity Stack B shows a significant reduction in the intermediate states.

VI. SUMMARY

An industry leading combination of high-retention, high-endurance MRAM-based e-NVM and high performance, ultra low power FinFET CMOS logic technology has been developed. e-NVM technology uses a 216nm x 225nm 1T-1R memory cell and demonstrates 200°C 10-year retention capability and >10^6 write endurance. Technology retention, endurance and high yield capabilities were shown on 7.2Mbit arrays and full 300mm wafers.

REFERENCES

[1] B. Sell et al., "22FFL: A High Performance and Ultra Low Power FinFET Technology for Mobile and RF Applications", IEDM, 2017.

[2] D. E. Nikonov et al., "Proposal of a Spin Torque Majority Gate Logic", IEEE Electron. Device Lett., v. 32, n. 8, pp. 1128-30, 2011.

[3] M-C. Shih, et al., "Reliability study of perpendicular STT-MRAM as emerging embedded memory qualified for reflow soldering at 260 C", VLSI Symposium, 2016.

[4] Y. J. Song, et al., "Highly Functional and Reliable 8Mb STT-MRAM Embedded in 28nm Logic", IEDM, 2016.

[5] D. Shum, et al., "CMOS-embedded STT-MRAM Arrays in 2x nm Nodes for GP-MCU applications", VLSI Symposium, 2017.

978-1-7281-1988-5/18 $31.00 © 2018 IEEE

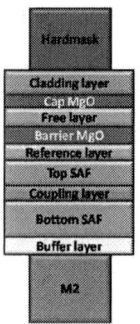

Fig. 1. Simplified stack schematic.

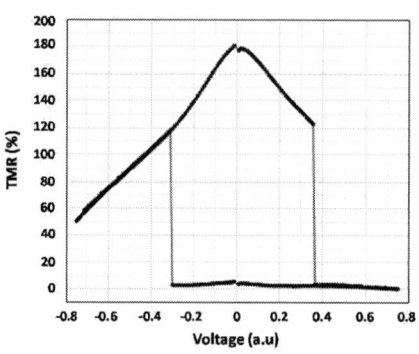

Fig. 2. Typical R-V curve. TMR is 180%.

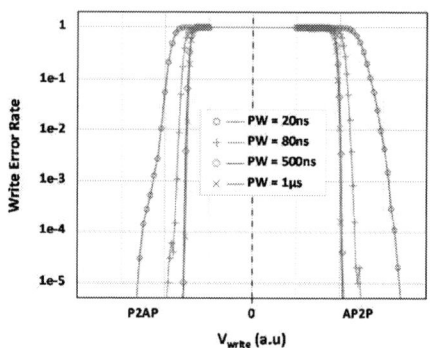

Fig. 3. Write error rates collected with 20ns, 80ns, 500ns and 1μs pulses. Significant increase in Vc between 80ns and 20ns data demonstrates transition into recessional switching regime.

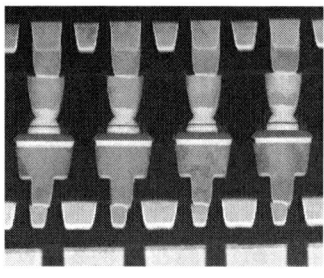

Fig. 4. Cross-sectional TEM of an MTJ array embedded between Metal 2 and Metal 4 of the 22FFL logic process.

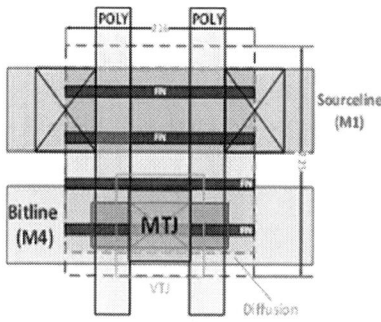

Fig. 5. Layout of the 1T-1R MRAM cell used in this work.

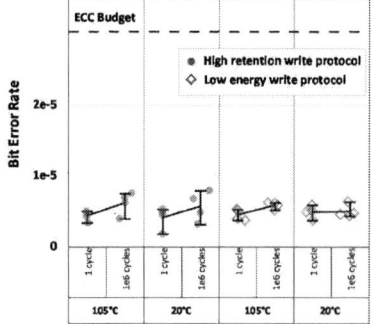

Fig. 6. Median wafer-level bit error rates before and after 10^6 write cycles at 105°C and 20°C with high retention and low energy switching protocols.

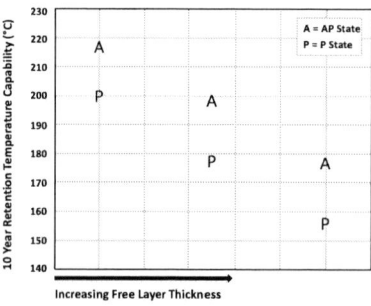

Fig. 7. Projected array level retention capability for stacks with different free layer thicknesses.

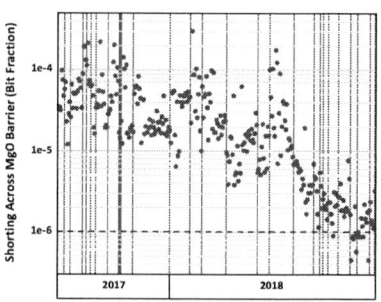

Fig. 8. 12-month time trend for the "Shorting across MgO Barrier" defect.

978-1-7281-1988-5/18 $31.00 © 2018 IEEE

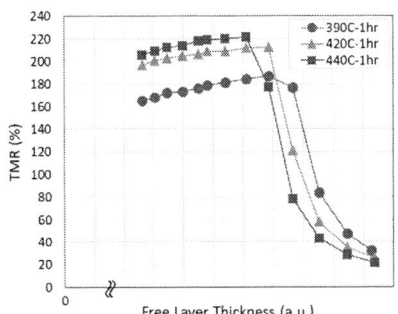

Fig. 9. TMR vs free layer thickness for a blanket film MTJ stack, with RA= 9 Ωμm² annealed at different temperatures.

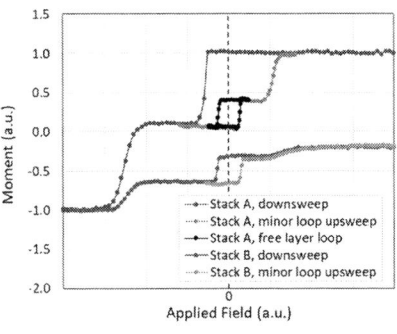

Fig. 10. M-H curves of patterned MTJ arrays. Stack A exhibits three states under no external magnetic field, while Stack B eliminates the undesired parallel SAF configuration.

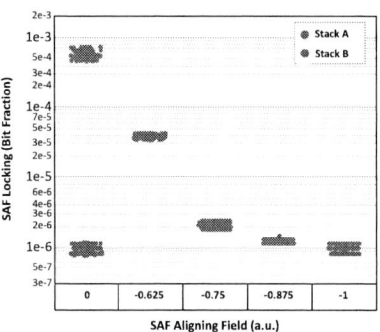

Fig. 11. Fraction of bits showing SAF locking vs the magnitude of SAF-aligning magnetic field. Stack B shows immunity to SAF locking behavior.

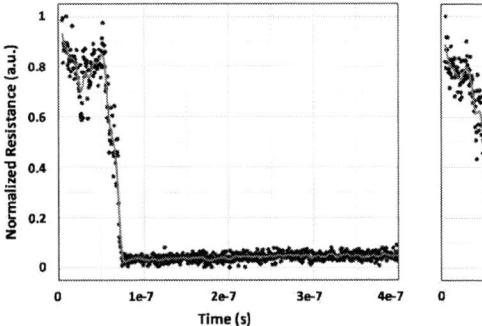
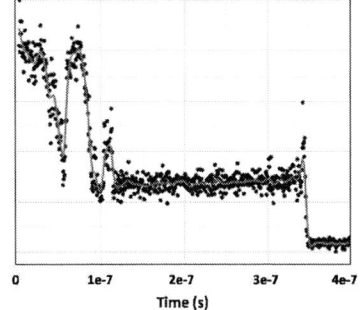

Fig. 12. Examples of time resolved resistance traces collected during AP to P device switching with low-voltage 1μs pulses. Typical device switching (left) vs device switching which exhibits a 250ns-long metastable state having an intermediate resistance value (right).

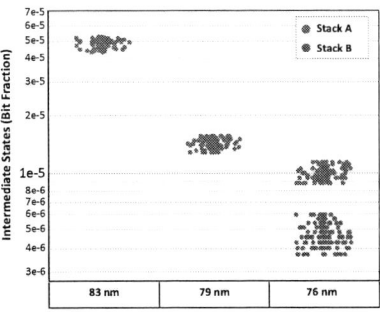

Fig. 13. Fraction of bits showing intermediate states vs device size. For a given device size Stack B shows lower intermediate state counts due to improved uniformity of the uncompensated stray field profile arising from the SAF.

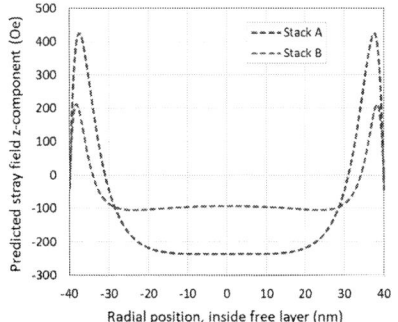

Fig. 14. Simulations of the stray field profile experienced by the free layer of an 80nm device with Stack A and Stack B.

978-1-7281-1988-5/18 $31.00 © 2018 IEEE

Demonstration of Highly Manufacturable STT-MRAM Embedded in 28nm Logic

Y. J. Song[1,a], J. H. Lee[1], S. H. Han[1], H. C. Shin[1], K. H. Lee[1], K. Suh[1], D. E. Jeong[1], G. H. Koh[1], S. C. Oh[1], J. H. Park[1], S. O. Park[1], B. J. Bae[1], O. I. Kwon[1], K. H. Hwang[1], B.Y. Seo[2], Y.K. Lee[2], S. H. Hwang[2], D. S. Lee[2], Y. Ji[2], K.C. Park[2], G. T. Jeong[2], H. S. Hong[1], K. P. Lee[1], H. K. Kang[1], and E. S. Jung[2]

[1]R&D Center, Samsung Electronics Co., Hwasung, Korea, [a]email: yoonjong.song@samsung.com, Tel.:(82)31-325-9319
[2]Foundry Business, Samsung Electronics Co., Kiheung, Korea

Abstract— We successfully demonstrated the manufacturability of 8Mb STT-MRAM embedded in 28nm FDSOI logic platform by achieving stable functionality and robust package level reliability. Read margin were greatly improved by increasing TMR value and also reducing distribution of cell resistance using advanced MTJ stack and patterning technology. Write margin was also increased by improving the efficiency using novel integration process. Its product reliability was confirmed in package level with passing HTOL 1000 hours tests, 10^6 endurance test, and retention test. For a wider application, we also demonstrated the feasibility of high density 128Mb STT-MRAM. Based on these results, we clearly verified the product manufacturability of embedded STT-MRAM.

I. INTRODUCTION

In recent decade, STT-MRAM (spin transfer torque magnetic RAM) has attracted great attention due to its ideal memory properties such as non-volatility, fast write speed, high endurance, and strong retention. Especially, it is more suitable for embedded non-volatile memory application than stand-alone application because of its low cost and great compatibility with other peripheral logic transistors. [1, 2] Recently, we successfully demonstrated highly functional and reliable 8Mb STT-MRAM embedded in 28nm FDSOI as well as LPP logic as illustrated in Figure 1. [3, 4] Since it is embedded between backend Cu layers, it was easily transported from 28nm bulk logic platform to FDSOI platform as shown in Figure 2. It can be merged into given logic platform without major change in peripheral transistors. Along with the friendly compatibility of process integration, it showed very attractive non-volatile memory properties such as high endurance of 10^6 cycles, strong retention property of 10 years, and a wide functionality at operating temperature ranging from -40°C to 125°C for industrial application. Regardless of its successful demonstration, in order to guarantee manufacturability of STT-MRAM product, it is strongly desired to improve device electrical and magnetic properties further for better read and write margin. In this paper, we propose advanced integration scheme and process to achieve highly manufacturable embedded MRAM whose product reliability is also evaluated in package level. In addition, we also demonstrate the feasibility of high density 128Mb MRAM embedded in 28nm FDSOI and unified memory showing high endurance of 10^{10} cycles with strong retention property of 85°C 10 years.

II. PRODUCT TECHNOLOGY

In order to develop robust STT-MRAM product technology, it is necessary to increase read and write margin. The approaches to be taken for a wide read margin are to increase TMR value and reduce the resistance distribution of cell array. The read margin was effectively increased by improving TMR value for median shift in the resistance distribution. Figure 3 shows the TMR improvement for different MTJ processes. It was found that TMR value was 220% after full integration process. Figure 4 illustrates the distribution of 8Mb main cell array. It was found that peak to peak separation was more than 25 sigma, which is a wide read margin even for high temperature operation. Since the tail bits play dominant role in determining the read sensing margin, it is also very crucial to improve the resistance distribution of main cells. Figure 5 shows the coefficient value (sigma/median) as a function of MTJ CD for evaluating the uniformity of main cell array. Typically, the CV was increased in the decrease of MTJ CD so that it is critical to reduce the distribution of cell resistance at small CD for achieving high and robust yield performance. As shown in Figure 5, by optimizing MTJ patterning process, the CV was improved at small MTJ CD.

In view point of write and retention property, there is a trade-off between these two device properties. Figure 6 shows the write pass rate and retention pass rate as a function of switching current. The window was observed in the trade-off. In order to achieve the manufacturability, it is necessary to increase the window by improving efficiency property defined as retention property divided by switching current. Figure 7 shows the correlation with retention and switching current. Using several optimization processes, the efficiency was increased further, which means that switching current can be decreased with maintaining the same retention property. From

978-1-7281-1988-5/18 $31.00 © 2018 IEEE

the correlation, it was found that the higher efficiency, the larger write operation margin. Figure 8 shows the fail bit count (FBC) for 8Mb device, operating from -40°C and 105°C with 10 years retention for the different process technologies including MTJ stack optimization, novel MTJ patterning process, and advanced integration process. It was observed that FBC was greatly improved for individual process scheme. Figure 9 shows the pass rate of 8Mb IP as a function of operating temperature after satisfying 10 years retention. It was found that high pass rate was maintained regardless of changing the operation temperature from -40°C to 105°C. We also confirmed its product reliability in package level including HTOL, HTS, endurance, and retention for 1000 hrs. Figure 10 displays the resistance distribution of 8Mb STT-MRAM package after 10^6 cycles. It was observed that there is no degradation after the endurance test. Critical reliability tests such as HTOL and HTS were also performed for 1000 hours in massive package volume. It was found that all the package samples passed the reliability tests. Based on these package level results, we successfully verified the manufacturability of our 8Mb STT-MRAM embedded in 28nm FDS platform.

III. NEAR FUTURE TECHNOLOGY

Near future technology will focus on fully utilizing its inherent attractive properties such as fast endurable non-volatility like unified memory and developing high density embedded STT-MRAM. We demonstrated the feasibility of unified memory. Figure 11 shows the correlation with endurance and retention property. It was found that we can achieve 10^{10} endurance cycles with 85°C 10 years retention property. Based on the correlation, by improving the efficiency property, we can shift the line, which means we can increase the retention property further without sacrificing

the endurance property. We also evaluated the function and reliability of high density 128Mb STT-MRAM. Figure 12 illustrates the yield trend as a function of memory density. It was found that we can achieve appreciable yield of 128Mb STT-MRAM whose functionality was confirmed at operation temperature from -40C to 85°C.

IV. SUMMARY

We successfully demonstrated highly manufacturable 8Mb STT-MRAM embedded in 28nm FDSOI logic platform. Read margin was improved by increasing TMR value and decreasing the cell distribution. A wide write margin was obtained by improving the efficiency using advanced integration process. In addition, its package level reliability tests were confirmed. For near future technology, we demonstrated the feasibility of unified memory and high density 128Mb STT-MRAM. It provides great feasibility of eMRAM commercialization for various applications.

REFERENCES

[1] S. H. Kang, " Embedded STT-MRAM for Energy-efficient and Cost-effective Mobile Systems," *VLSI*, pp. 1-2, 2014.

[2] Yu Lu et. al., "Fully Functional Perpendicular STT-MRAM Macro Embedded in 40nm Logic for Energy-efficient IOT Applications," *IEDM*, 26.1.1-1.4, 2015.

[3] Y. J. Song et. al., " Highly Functional and Reliable 8Mb STT-MRAM Embedded in 28nm Logic," *IEDM*, pp. 27.2.1-27.2.4, 2016.

[4] Y. K. Lee, et. al. " Embedded STT-MRAM in 28-nm FDSOI Logic Process for Industrial MCU/IoT Application," *VLSI*, pp. T17.1-17.2, 2018.

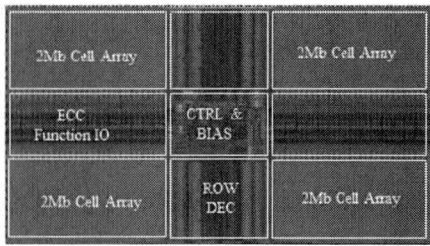

Fig. 1 Micrograph of 8Mb STT-MRAM embedded in 28nm FDSOI. [4]

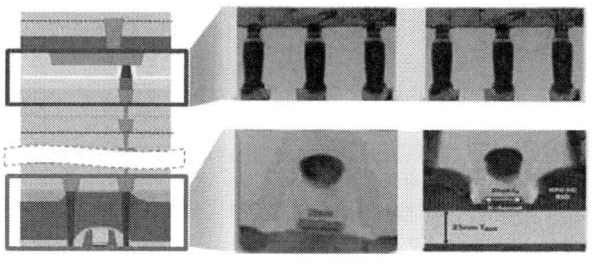

28nm LPP   **28nm FDSOI**

Fig. 2 Schematic view of integrating STT-MRAM into various logic platforms. [4]

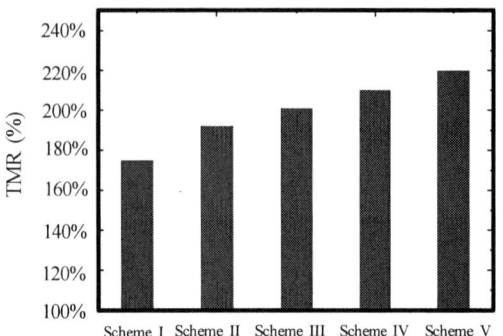

Fig. 3 TMR improvement trend as a function of various MTJ process schemes.

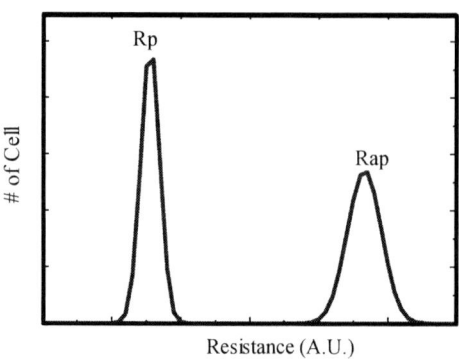

Fig. 4 Resistance distribution of 8Mb cell array.

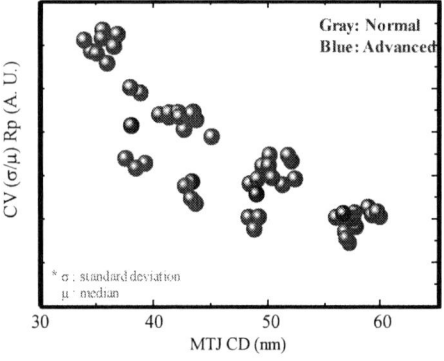

Fig. 5 Trend of resistance distribution of MTJ cells as a function of MTJ CD. CV R_p was dramatically improved by optimal integration scheme at small CD region.

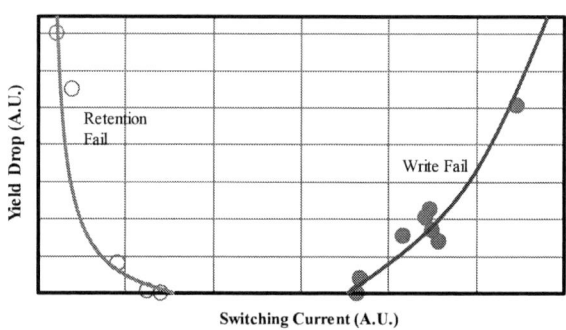

Fig. 6 Write and retention pass rate as a function of switching current. There is a write margin in the trade-off graph.

978-1-7281-1988-5/18 $31.00 © 2018 IEEE   418

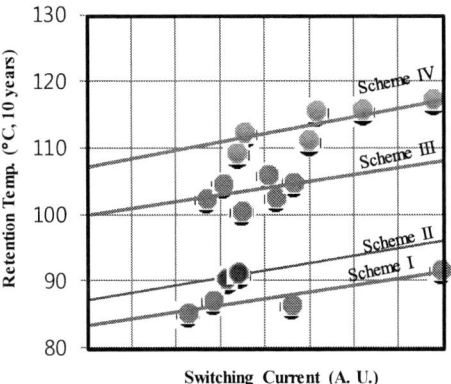

Fig. 7 Correlation with retention property and switching current for various integration schemes.

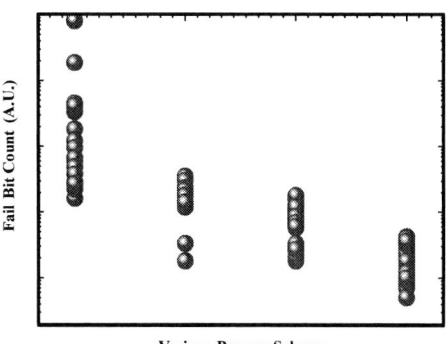

Fig. 8 Fail bit count (FBC) improvement trend for different process integration schemes.

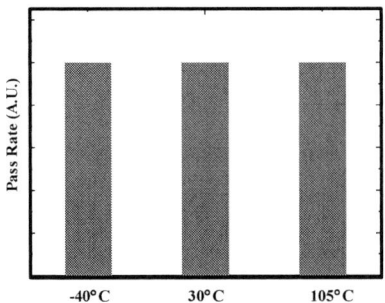

Fig. 9 Pass rate as a function of operation temperature after passing retention test. High pass rate was obtained from -40°C to 105°C.

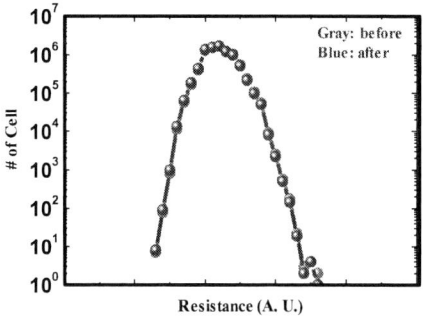

Fig. 10 Resistance distribution of 8Mb PKG before and after 10^6 endurance cycles. No degradation was observed after the endurance test.

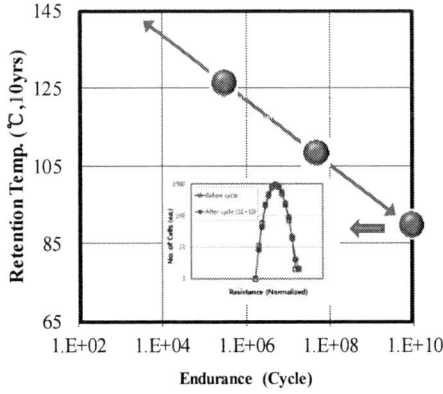

Fig. 11 Feasibility of unified memory with 85°C 10 years and 10^{10} endurance property.

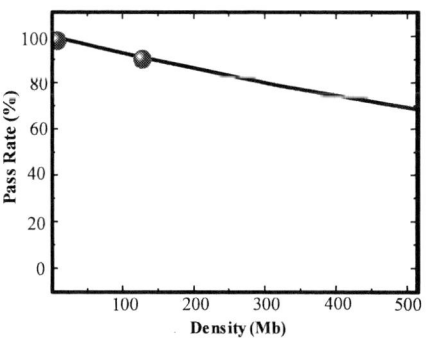

Fig. 12 Expectation as a function of MRAM density and Si results of 128Mb high density eMRAM.

978-1-7281-1988-5/18 $31.00 © 2018 IEEE

Enablement of STT-MRAM as last level cache for the high performance computing domain at the 5nm node

S. Sakhare*, M. Perumkunnil, T. Huynh Bao, S. Rao, W. Kim, D. Crotti, F. Yasin, S. Couet, J. Swerts, S. Kundu, D. Yakimets, R. Baert, HR. Oh, A. Spessot, A. Mocuta, G. Sankar Kar & A. Furnemont
*imec, Leuven, Belgium, email: sushil.sakhare@imec.be

Abstract—The increased complexity of CMOS transistor processing has led to limited scaling of high density SRAM cell at advanced technology nodes. STT-MRAM appears to be a promising candidate for replacing last level caches (LLC). This paper addresses design technology co-optimization (DTCO) of STT-MRAM technology and analyzes its viability as a LLC (compared to SRAM) for the high performance computing (HPC) domain (while maintaining a constraint of occupying merely 43.3% of SRAM macro area at identical capacities). This is the first study that breaks down a power, performance and area (PPA) comparison between SRAM and STT-MRAM based LLCs at the 5nm node. The STT-MRAM design and analysis is based on a silicon verified compact model and can be realized using 193i single patterning at the 5nm node. Our STT-MRAM design manages to achieve a nominal access latency <2.5ns and <7.1ns for read and write operations respectively. We also observe a clear and significant trend of increasing energy gains with respect to SRAM for increasing LLC sizes with the crossover points for STT-MRAM read and write operations at 0.4MB and 5MB respectively.

Keywords— STT-MRAM, SRAM, HPC, 5nm node, Embedded last level cache, Design Technology Co-optimization

I. INTRODUCTION

The traditional path for improving energy efficiency and functionality of digital systems through silicon CMOS scaling has become increasingly difficult due to a multitude of reasons (physical, technological, and economical). Beyond the 10nm node, we need to address challenges of SRAM scaling while coupling it with benefits in terms of PPAC [1]. STT-MRAM has been perceived as a memory technology alternative that provides an option to enable large ultra-high density last level caches for systems with at reduced area and energy (Fig 15). In this paper, we have analyzed the feasibility of introducing STT-MRAM at 5nm node for the HPC domain.

II. DESIGN ENABLEMENT

There are two key design choices to realize the MRAM bit-cell while maintaining a uniform MTJ pitch (Fig. 1) in X and Y direction. The 1.5*Contacted Gate Pitch (CPP) option and 2*CPP option which is 40% larger than the former. We highlight these two flavors of the STT-MRAM bit-cell in Table I and compare it with relevant SRAM flavors in Fig. 4 (for any technology node).

A. Pitch estimation for the 5nm technology node

Determining the pitch of scaled nodes (below 7nm) for each foundry is cumbersome since of decisions are driven by extensive DTCO. As shown in Fig. 2, we extrapolate the device sizing constraints of different foundries to estimate possible SRAM area targets for the 5nm CMOS node. If we normalize the reported area

of SRAM bit-cells from the different foundries against their respective metal pitch (MP) and CPP, we obtain a trend line for SRAM scaling as shown in Fig. 3. The extension of the normalized area of SRAM to the 5nm node results into targeted CPP & MP product. Going from the14nm node to the 5nm node, MP scales more aggressively relative to CPP[2][5]. The resulting estimates for CPP and MP numbers for each foundry are captured in Table II (for target 5nm CMOS technology). For our STT-MRAM DTCO and circuit simulations, a device with CPP of 45nm is assumed with IMEC predictive models (that are tuned for 5nm CMOS performance and energy targets) [7].

B. DTCO

In order to realize a uniform circular pillar of STT-MRAM, it is preferable to use uniform pitches. Since CPP is always greater than MP, CPP will govern both the MRAM bit-cell area and MTJ pitch. At the 5nm node, most of the technology components of Back-End Of Line (BEOL) and Front-End Of Line (FEOL) are realized using EUV (Extreme Ultraviolet) and multi-patterned via 193i. The HD MRAM bit-cell (1.5*CPP) with 66-69nm MTJ pitch can be realized only via 2xSADP (Self Aligned Double Patterning) or EUV whereas the HP MRAM (2*CPP) bit-cell with 90-92nm MTJ pitch can be realized using 193i single patterning (lower technology cost) (Fig. 5). This is the most adept pitch for Ion Beam Etching (IBE), which is essential for reliable MTJ processing [8]. Although, the 1.5*CPP option seems promising, it will soon land-up in minimum area for BEOL resulting in limited scaling possibilities. This option also requires an additional super-low threshold voltage (VT) transistor with discontinued fin structure that will in-turn result in increased variability (Fig. 1). The 2*CPP option, on the other hand, is more feasible due to uniform fins and gate resulting in reduced complexity and variability for a MRAM bit-cell. Hence, HP MRAM (2*CPP) bit-cell is the preferred solution for the LLCs in the high performance computing domain at the 5nm node.

The typical Resistance-Voltage (RV) curve for STT-MRAM is captured in Fig. 6. For any given technology node, the positive tail of write voltage and negative tail of break-down voltage serve as the boundary constraints that limit the maximum voltage across stack (V_{DD} of technology). In order to a maintain reasonable read current (for target LLC access frequencies; Table III), the negative tail of the write switching voltage should be higher than that of the maximum read voltage. Considering the above listed constraints, it is advisable to read MRAM cell resistance in the positive biased region. This is because the selector transistor of the bit-cell offers minimum resistance under positive bias (hence minimum voltage drop), and thus ensures maximum voltage across the MTJ. In the negative bias region, the NFET is in source degeneration mode and offers maximum on-resistance resulting in a lower operating

978-1-7281-1988-5/18 $31.00 © 2018 IEEE

current. Thus, the factor that puts a cap on STT-MRAM specifications is the parallel to anti-parallel (p2ap) switching operation (negative bias over MTJ), when the transistor is in degeneration mode.

To maximize the current flowing through the MTJ stack in the source degeneration mode, we require an LVT device (Fig. 7). In the STT-MRAM bit-cell, the MTJ and NFET (a 2 fin & 2 finger selector device for 2*CPP) are connected in series (as shown in Fig. 6). The LVT NFET in source-degeneration mode offers a resistance of 12.8kΩ and for a V_{DD} of 0.65V there is almost a ~0.5V drop across the device (and hence a resulting drive current of 39µA at the Slow-Slow corner). This leaves merely 0.15V of potential across the MTJ. Keeping these specifications in mind, from the RV curve, we deduce that a Parallel (P) state resistance of 3.84kΩ would be required to switch the MTJ (p2ap).

To speed-up the MTJ switching, the current passing through the MTJ needs to be increased. There are 3 possible ways to achieve this. The first option is to increase the number of fins of the device. However, this leads to an increase in the bit-cell area, which will disturb the aspect ratio. Increasing the supply voltage of the memory macro is another option, but this is limited by the breakdown of MgO barrier of the MTJ (during positive bias). The last and the most favorable option is that of boosting the gate voltage of the device. This will increase an effective current through the MTJ (for p2ap as well as ap2p switching), making certain that no more that V_{DD} is applied between SL & BL. This ensures the reliability of the MgO barrier.

By boosting the WL voltage by 100mV, the current through the source degenerated transistor can be increased by 40%, resulting in fast switching of the MTJ (Fig. 8). At 90nm pitch, to achieve an optimum aspect ratio between pillar height and spacing we choose a diameter of pillar of 34nm (and assume 10% diameter variability) [8]. This results in a resistance area (RA) product that ranges from 3.1 to 4.7 $\Omega.\mu m^2$ for the target current densities of 3.8 to 5.4MA/cm^2 (Fig. 10).

C. STT-MRAM realization on Silicon

For our design, a high performance p-MTJ stack was fabricated on 300mm Si wafers according to the process detailed in [3] (Fig. 9). The pMTJ stacks comprise of dual MgO interfaces interspersed by magnetic CoFeB electrodes, to improve the interfacial perpendicular magnetic anisotropy. The free layer and reference layer consist of CoFeB-based multilayer stacks. The RL is coupled to a Co/Pt-based hard layer (HL) to form a synthetic anti-ferromagnet (iSAF), which allows us to keep our offset field near zero. The optimized pMTJ stack, annealed at 400 C for 3 hours in a 1T magnetic field. The TMR is around 150% along with low resistance-area (RA) product (4.04 $\Omega.\mu m^2$) from the device measurement. The STEM analysis of the pMTJ array depicts 34nm pillars at 90nm pitch between the Mx+3 (where Mx=Metal) and Mx+4 layers with sharp interfaces in the Co/Pt multi-layers structure and well-defined crystalline phase across the film in the MgO barrier layers (Fig. 10). The measured MTJ characteristics are captured in Table IV & Fig. 11.

III. CIRCUIT DESIGN AND ARCHITECTURE

A typical SRAM bit-cell finishes at Mx+2 level but due to increasing resistance of BEOL, additional metal routing at Mx+3 level is required to reduce the WL-resistance [4]. The WL length for a MRAM bit-cell is 40% of SRAM cell height resulting in lower capacitance as well as resistance (Fig. 16). As, stated earlier, the

MRAM stack is present between Mx+3 & Mx+4, bit-line (BL) is routed in Mx+4, SL in Mx+2 and WL in Mx+3.

We have realized a simple butterfly architecture for the STT-MRAM macro with the necessary circuits for read, write and control (Fig. 14). The macro has a maximum bank size of 1Mb with shard local IOs for the top and bottom array. The structure is repeated using repeater logic between two banks. A maximum loading of 512 bit-cells per BL and WL are considered in an array. The local IO to global IO data-in and data-out are routed in Mx+6 with necessary buffers. The data-in, data-out, clock and address pins are located at the bottom of the macro, where internal tracking of bit-cell parasitics and independent tracking of read and write operation is realized to accommodate asymmetry of read & write access-time of MRAM technology [9].

For the typical-typical (TT) corner of the device & MTJ, the switching events are captured in Fig. 12. The p2ap switching is faster here since the bit-cell resistance is minimum when the P state resistance is combined with reduced device resistance (due to boosted WL voltage) and maximum current flows. In Fig. 13, a nominal differential of 100mV is observed for a TMR=150% at 1.5-2.5ns for a applied maximum voltage of 650mV.

IV. SRAM VS MRAM COMPARISON

Figures 16 and 17 highlight the comparison between SRAM and MRAM LLC designs for the HPC domain at the 5nm node with respect to access latency and energy consumption. For a fair comparison, the SRAM architecture is similar to ours (with HP SRAM cell). Our MRAM design manages to achieve a nominal access latency < 2.5ns and <7.5ns for read and write operations respectively. From Table III we know that these access latencies are sufficient to meet the performance requirements of LLCs for commercial HPC systems. A closer look at the energy profile for both SRAM and STT-MRAM designs reveal 2 major crossover points that can impact system energy consumption: when STT-MRAM read and write energy becomes more energy efficient as compared to SRAM @0.4MB and @5MB. This can be attributed to the exponential increase in the contribution of SRAM standby power with increasing capacity. Thus, at LLC capacities (<12MB) in present HPC systems, STT-MRAM is clearly more beneficial regardless of its inherent read-write asymmetry and irrespective of access profile of the applications in the target domain space.

V. CONCLUSION

We have analyzed the feasibility of the STT-MRAM as a LLC for the HPC domain at the 5nm node. A thorough peer to peer comparison between STT-MRAM and SRAM designs optimized for the target domain reveals that a drop in STT-MRAM replacement can still offer significant energy gains (for both read and write accesses) while meeting target frequencies (>100MHz) at 43.3% the area of the SRAM macro.

REFERENCES

[1] S. S. Sakhare et. al, *IEEE* **TED**, vol. 62, no. 6, pp. 1716-1724, 2015
[2] T. Huynh-Bao et al., in IEEE **TED**, vol. 63, no. 2, pp. 643-651, Feb. 2016.
[3] S. Van Beek et al., AIP **ADV**., 2018, 8, 055909.
[4] J. Chang et al., IEEE ISSCC, 2017, pp. 206-207.
[5] Z. Guo et al., IEEE **ISSCC**, 2018, pp. 196-198.
[6] S. Y. Wu et al., IEEE **IEDM**, 2013, pp. 9.1.1-9.1.4.
[7] D. Yakimets et al., IEEE **IEDM**, 2017, pp. 20.4.1-20.4.4.
[8] V. Ip et al., IEEE **TM**, vol. 53, no. 2, pp. 1-4, Feb. 2017.
[9] Q. Dong et al., IEEE - **ISSCC**, 2018, pp. 480-482.

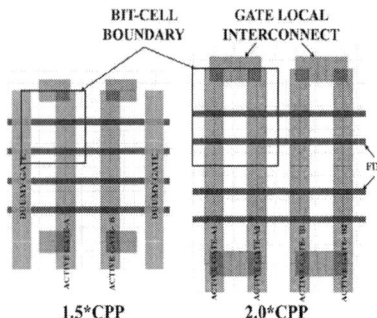

Fig. 1: The 1.5*CPP bit-cell has a dummy gate under which the FIN is cut whereas 2.0*CPP cell has uncut gate and uncut FINs.(CPP = Gate Pitch, MP=Metal Pitch).

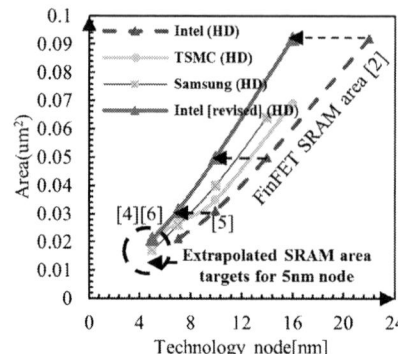

Fig. 2: Extrapolation of reported SRAM area to estimate targets for 5nm node. Intel [revised] here is obtained by pitch matching with other foundries.

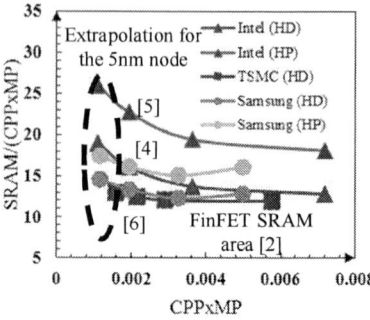

Fig. 3: Extrapolating normalized SRAM /(CPPxMP) for each foundry to estimate target for the 5nm node.

Table I: HD and HP cell for SRAM & MRAM.

Bit-cell	HD cell	HP cell
SRAM (PU:PG:PD)	1:1:1	1:1:2/1:2:2
MRAM	1.5*CPP	2.0*CPP

Table II: Estimates of pitches for 5nm node

5nm node	CPP[nm]	MP [nm]
INTEL	46	24
TSMC	45	34
SAMSUNG	44	26

Table III: LLC Characteristics for HPC domain

Parameters	AMD HPC	Intel HPC
Frequency	~100MHz	~100MHz
Size	20-38 MB	12-40MB
TECH - NODE	12nm	14nm
Associativity	16 way	16 way

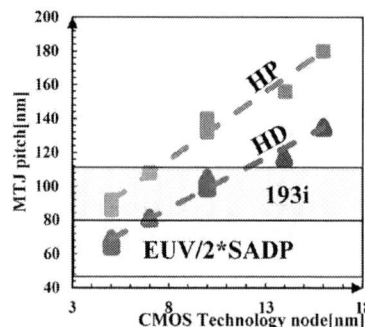

Fig. 4: Area comparison between HD and HP-MRAM bit-cell with reported HD and HP SRAM cells.

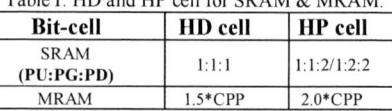

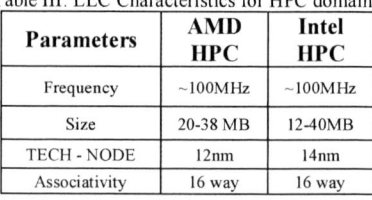

Fig. 5: For the 5nm node, a 2.0*CPP cell can be realized using 193i single patterning whereas a 1.5*CPP will require single patterning EUV or 2x SADP.

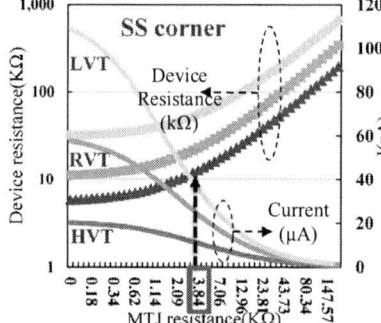

Fig. 6: Target specs for the pMTJ: where degenerated NFET governs switching from p2ap state.

Fig. 7: Minimum resistance is offered by the LVT device at an MTJ parallel state resistance of 3.84KOhm (ensuring optimum switching current).

Table IV: Characteristics of the device fabricated at imec.

Parameters	Units	IMEC BP stack
Diameter	nm	35
RA	Ωum^2	4.04
TMR	%	145
H_C	Oe	1400
J_{SW}	MA/cm^2	7.70
T_{SW}	ns	5
Δ_{avg}	a.u.	61
STE		0.76
V_{BD}	V	0.8 (DC)
Thermal Robustness		400˚C 180 mins

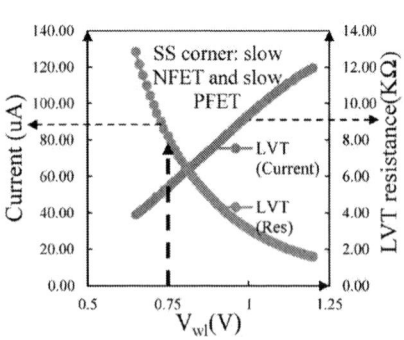

Fig. 8: Boosting the WL voltage to 0.75V, boosts the maximum drive current of the LVT to 55uA from 39uA, thus ensuring MTJ write operation.

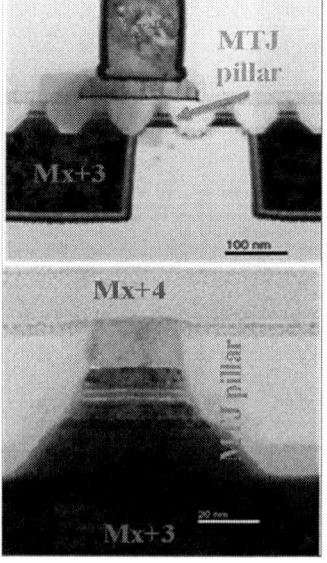

Fig. 9: MTJ arrays (34nm diameter with 90nm pitch, RA=4.04Ω.µm^2 & J_{sw}= 7.7MA/cm^2 at 5ns switching) realized on 300mm silicon wafer. A close-up reveals sharp interfaces in the Co/Pt multi-layers structure and well-defined crystalline phase across the film in the MgO barrier layers.

978-1-7281-1988-5/18 $31.00 © 2018 IEEE

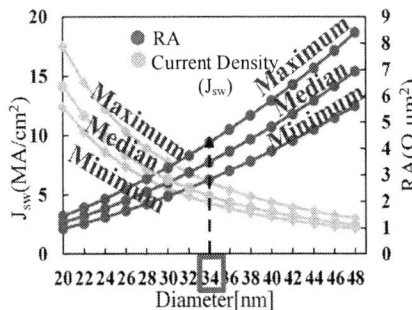

Fig. 10: At diameter of 34nm and 92-90nm pitch, the target RA≈3-5Ω.µm² & target J_{sw}=4 to 6MA/cm².

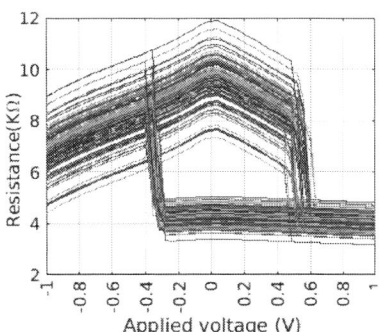

Fig. 11: The Resistance -Voltage distribution of the fabricated devices with a composite variability of Jsw, MTJ size, RA, TMR and variation of temperature.

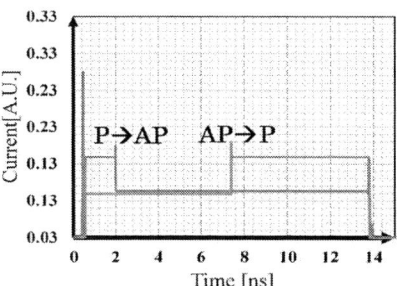

Fig. 12: Simulated write operation where AP to P switching is governed by TMR of the measured device. WL boosting improves p2ap switching but not ap2p switching.

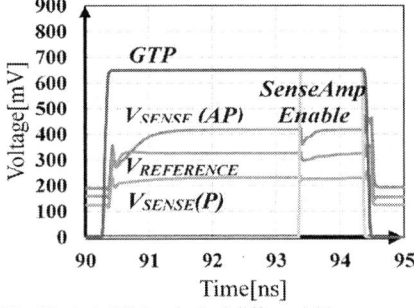

Fig. 13: At 1.5-2.5ns desired differential is generated to sense the state of the cell, resulting in possible read access-time of 2-5ns.

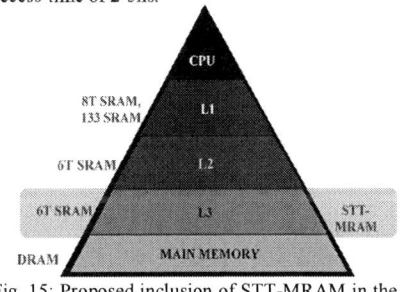

Fig. 15: Proposed inclusion of STT-MRAM in the memory hierarchy (at the L3) replacing SRAM.

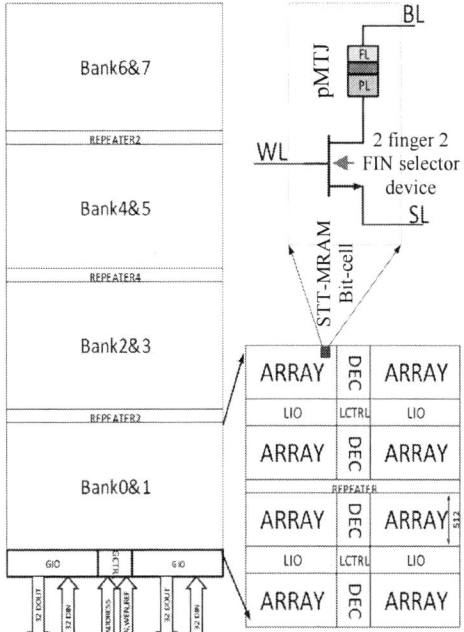

Fig. 14: Architecture of MRAM macro designed at 5nm node (targeting LLC) with layout for 2 banks and GPIO. Each bank contains 4 sets of MRAM arrays containing 512x512 bit-cells. The same architecture is used for SRAM for a fair comparison.

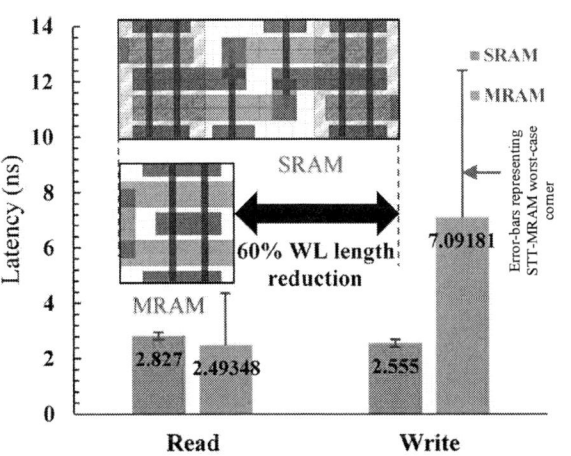

Fig 16: Access Latency comparison between SRAM and STT-MRAM for a single bank. The error bars indicate the worst case latencies for both SRAM and MRAM. The relatively higher SRAM read latency is due to the larger parasitic overhead on the WL.

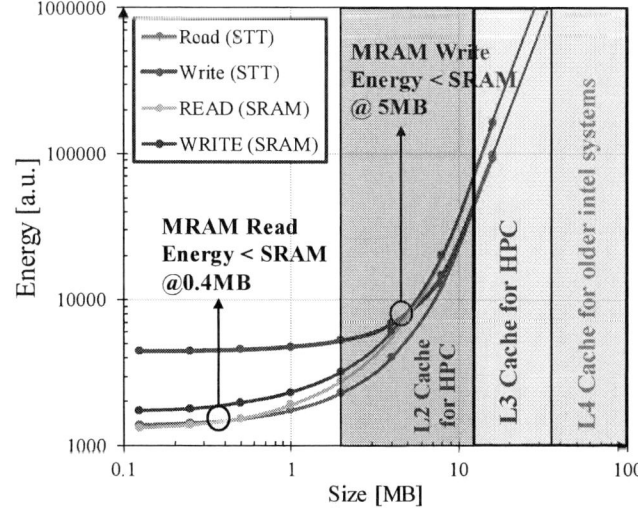

Fig. 17: Energy comparison between SRAM and STT-MRAM for varying sizes. STT-MRAM becomes more energy efficient compared to SRAM @0.4MB for read and @ 5MB for write.

978-1-7281-1988-5/18 $31.00 © 2018 IEEE   423

Truly Innovative 28nm FDSOI Technology for Automotive Micro-Controller Applications embedding 16MB Phase Change Memory

F.ARNAUD[1], P.ZULIANI[2], J.P.REYNARD[1], A. GANDOLFO[2], F.DISEGNI[2], P.MATTAVELLI[2], E.GOMIERO[2], G.SAMANNI[2], C.JAHAN[3],
R.BERTHELON[1], O.WEBER[3], E.RICHARD[1], V.BARRAL[3], A.VILLARET[1], S.KOHLER[1], J.C.GRENIER[1], R.RANICA[1], C.GALLON[1],
A.SOUHAITE[3], D.RISTOIU[1], L.FAVENNEC[1], V.CAUBET[1], S.DELMEDICO[1], N.CHERAULT[1], R.BENEYTON[1], S.CHOUTEAU[1],
P.O.SASSOULAS[1], A.VERNHET[1], Y.LE FRIEC[1], F.DOMENGIE[1], L.SCOTTI[2], D.PACELLI[2], J.L.OGIER[1], F.BOUCARD[1], S.LAGRASTA[1],
D.BENOIT[1], L.CLEMENT[1], P.BOIVIN[4], P.FERREIRA[1], R.ANNUNZIATA[2], P.CAPPELLETTI[2]

[1]STMICROELECTRONICS, [3]CEA-LETI, 850 rue Jean Monnet 38926 Crolles, France
[2]STMICROELECTRONICS via Camillo Olivetti 2, Agrate Brianza, Italy
[4]STMICROELECTRONICS, zone industrielle, 190 avenue Coq, 13106 Rousset, France

Abstract—for the first time we propose a 28nm FDSOI e-NVM solution for automotive micro-controller applications using a Phase Change Memory (PCM) based on chalcogenide ternary material. A complete array organization is described exploiting body biasing capability of <u>F</u>ully <u>D</u>epleted <u>S</u>ilicon <u>O</u>n <u>I</u>nsulator (FDSOI) transistors. Leveraging triple gate oxide integration with high-k metal gate (HKMG) stack, a true 5V transistor with high analog performance has been demonstrated. Reliable PCM 0,036um^2 analytical cell with 2 decades programming window after 1 Million of cycles has been demonstrated. Finally, current distributions based on a fully integrated 16MB macro-cell is presented achieving Bit Error Rate (BER) < 10^{-8} after multiple bakes at 150°C and 10k cycling of code storage memory.

I. INTRODUCTION

The automotive industry is undergoing a deep transformation, driven by disruptive technological innovation in the fields of Electrification. Despite the diverse areas of innovation, they rely on Microcontrollers to perform safe computation and secure data transmission. As the applications evolve, the demand on high computation, lower power consumption and larger memory footprint is driving the automotive MCU technology development. Embedded <u>N</u>on <u>V</u>olatile <u>M</u>emory (e-NVM) is one of the most challenging topics: MCUs are requested to store more data to host more complex firmware. OTA (<u>O</u>ver <u>The</u> <u>A</u>ir) capability requires to host multiple firmware images for recovery purposes. Real time requires fast access to code memory and more complex security software requires peculiar NVM array partitioning. In case of advanced CMOS platform, facing huge process complexity growth while shrinking, a storage element integrated in the BEOL of the process provides a tailored solution. Among all the resistive memories proposed nowadays [1] as innovative solutions able to replace Floating Gate cells, PCM is the only one demonstrated to be compliant simultaneously with automotive requirements (several years of Data Retention at 150°C) and able to guarantee code integrity after soldering reflow thermal profile (260°C peak temperature) on multimegabit array [2]. The aim of this paper is to propose a unique solution for a 28nm technology platform leveraging back biasing opportunity of FDSOI [3] co-integrated with true 5V transistors for high performance analogue designs and PCM

addressing the next generation of automotive embedded non-volatile memory products.

II. TECHNOLOGY DESCRIPTION

This paper presents for the first time the co-integration of Phase Change Memory (PCM) as non-volatile memory solution with advanced CMOS technology based on FDSOI substrate. Co-integration scheme is described by figure 1. PCM is inserted between contact and metal1 levels ensuring an excellent e-NVM cell size (area per cell=0,036um^2). TEM cross-sections of both PCM and SRAM arrays are exhibited in figure 2. Active devices are built on <100>/(100) Si substrate crystal orientation construct on 25nm buried oxide. A chalcogenide ternary alloy $Ge_xSb_yTe_z$, is used as the storage element with a specific heater based architecture as already published [2].

III. CMOS DEVICES SUITE

This technology proposes a unique triple gate oxide solution with HKMG, gate first scheme. High Voltage (5V), IO and logic transistor gates TEM images are presented in figure 3. Entire devices suite is summarized in the table of figure 4. Thin and medium oxide transistors are located in FDSOI side, whereas 5V one, dedicated to analog design blocks, is built on bulk area. Leveraging FDSOI capability, a fully mixable VT solution has been developed covering a wide performance /leakage range, as shown in figure 5. The entire logic transistors suite enables forward body bias technique thanks to the flip-well architecture [3]. Extremely low leakage logic transistors option reaching 30pA/stage on ring oscillator at room temperature is proposed sustaining the always-ON design blocks requirement in term of quiescent current. PCM process brick has been successfully integrated with high density SRAM array without degrading defect density (D0) trend, as shown in figure 6. Same performance of native 28FDSOI has been reached with PCM co-integration scheme. The output and transfer characteristics of 5V transistors are plotted in figure 7 evidencing ultra-low leakage capability (~1pA/um) and high drivability (550-300uA/um for NMOS and PMOS respectively). Excellent analog performance, aligned with 40nm previous node without PCM, in term of matching and low frequency noise has been demonstrated (see figure 8). Finally, as shown in figure 9, 5V transistor reliability criteria compatible with automotive application (10years at 150°C) have been successfully achieved for gate oxide break-down.

IV. PCM ANALYTICAL CELL

FDSOI based NMOS transistor is proposed to drive the current inside the non-volatile memory cell. Leveraging thin buried oxide substrate [3], a wide Reverse Body Bias (RBB) has been applied on the selector dropping by more 3 order of magnitude the leakage of the unselected transistors along the bit-line without any cell area degradation, as depicted in figure 10. Thus, RBB technique demonstrated much superior capability to optimize leakage/area compromise versus regular gate length modulation. Process structure of the PCM is presented in figure 11. A specific PCM cell architecture, so called "Wall" already published [5] is adopted, where a strip of chalcogenide material lays on top of a thin heater element. This solution is very effective allowing PCM integration in the 28nm technology platform and being fully respectful of the FEOL peculiarities, enabling an entire reuse of 28FDSOI CMOS model cards and IPs. In the chalcogenide ternary alloy $Ge_xSb_yTe_z$, the fraction of each element has been properly tuned to match the applications requirements, especially in terms of data retention and soldering reflow compliance [4], with definite improvement in thermal stability with respect to conventional $Ge_2Sb_2Te_5$. TEM images of $0,036um^2$ PCM cell are presented in figure 11, showing the heater and the GST scheme in both X and Y directions, corresponding to BL and WL views, respectively. In order to modify the phase change material from amorphous to crystalline and the reverse, a specific heater system has been developed, as shown in the TEM image of figure 11. Both SET (low resistive state) and RESET (high resistive state) current distributions measured on the $0,036um^2$ analytical cell are presented in the figure 12. Wide programming window of 2 decades has been reached with an electronic threshold switch close to 1V. Electrical measurements have been completed for high resistive state at different temperature and time allowing the extraction of the data retention of the memory element (figure 13) across temperature. An activation energy of 2,46eV has been measured being compatible with data retention criteria for automotive (10y at 150°C). Finally, endurance tests have been performed on PCM cell maintaining an excellent programming window close to 2 decades even after 1 Million of cycles, as shown in figure 14. Those measurements confirmed the low fatigue of the heater integration proposed in 28FDSOI-PCM solution, full-filling the request from automotive applications.

V. 16MB PCM ARRAY

PCM array organization is presented in figure 15. Classical bit-cell addressing scheme with Word-Line (WL) driving the gate of the selected FDSOI NMOS transistor and Bit-Line (BL) connected to the top electrode of the PCM storage element is proposed. Internally regulated programing voltages are supplied to the array to generate the currents needed for setting and resetting the memory cells. Programming voltages on selected BL are carefully regulated in order to take into account IR drop due to the storage element in series with the selector itself. For programming, selected WL are pulsed for short time

to higher voltages than the regular logic device gate bias in order to boost selected current capability with no gate oxide reliability issues. Much lower voltages are applied (below 1V) to the memory array for reading operation with a significant benefit for active power reduction. Optimized algorithms for both programming and reading phases have been implemented on 16MB test chip designed to qualify the PCM-FDSOI solution and using only low voltage devices. Layout of the test chip is presented in figure 16. A full electrical characterization has been completed on the 16MB array at time zero, before any stress, in order to collect the current distributions of both low and high resistive states (SET and RESET respectively). As shown in figure 17, a cumulative plot of current demonstrates a wide reading window (about 10uA) at normalized cells count of 10^{-7}, measured on a sub-array of 2MB. On top of the fresh current distributions, reliability of the PCM has been verified carefully by checking both the fatigue of the device and the capability to retain the information after bake. Figure 18 describes the endurance capability. Current distributions have been compared to the time-0 behavior after multiple cycling combining SET and RESET current pulses. As shown in figure 18, very stable reading window has been maintained after 10k cycles, the automotive specification for memory code storage. At this stage of stress exacerbating the fatigue of the device, no occurrences of defectiveness have been observed. Data retention efficiency has been monitored after multiple bakes with different time at 150°C, to be compatible with automotive criteria. As exhibited in the figure 19, no critical material recrystallization has been found for RESET cells.

VI. CONCLUSIONS

Leveraging RBB technique of 28FDSOI, triple gate oxide HK/MG architecture and the wide SET/RESET window of PCM, a full co-integration of one unique technology has been successfully demonstrated. Extremely low leakage selector and high analog performance 5V transistor have been presented. PCM with very low Bit Error Rate (down to 10^{-8}) has been measured on 16MB PCM array with retention and endurance behavior compatible with automotive needs.

ACKNOWLEDGMENT

The authors would like to warmly thank all the stakeholders of this project. Design, test, process and characterization teams from ST sites, Agrate, Rousset & Crolles, and especially our colleagues from CEA/LETI in Grenoble research facility.

REFERENCES

[1] N.Ciocchini et al,"Modeling Resistance Instabilities of Set and Reset States in Phase Change Memory With Ge-Rich GeSbTe", IEEE Trans.El.Dev., 61, 2136 (2014).

[2] P.Zuliani *et al.*, "Overcoming temperature limitations in Phase Change Memories with optimized $Ge_xSb_yTe_z$", *IEEE TED*, 2013,pp.4020-4026.

[3] N.Planes *et al*, "28nm FDSOI Technology Platform for High-Speed Low-Voltage Digital Applications" on Symposium on VLSI Technology Digest Technical Papers,2012.

[4] V.Sousa et al, "Operation fundamentals in 12Mb Phase Change Memory based on innovative Ge-rich GST materials featuring high reliability performance",Symp.VLSI Tech. 2015

[5] R.Bez, "Chalcogenide PCM: a Memory Technology for Next Decade", in *IEDM Tech.Dig.*, 2009, pp.89-9

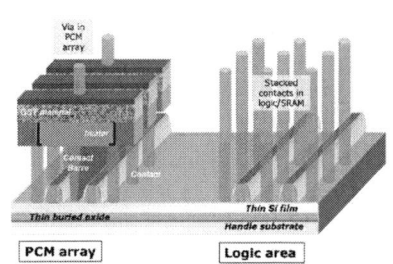

Fig1: description of NVM/Logic co-integration scheme on FDSOI substrate.

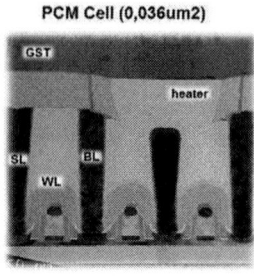

Fig2: TEM image of HD SRAM bit-cell co-integrated with advanced low area phase change based non volatile memory.

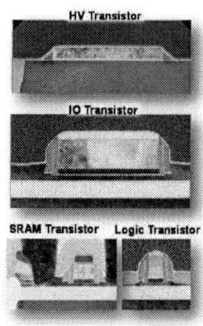

Fig3: Gate morphology of all transistors suite.

Devices	Logic devices	SRAM devices	I/O devices
VDDnom (Volt)	1	1	1,5 & 1,8
Lmin (um)	0,028	0,036	0,1 & 0,15
Tinv (nm)	1,6	1,6	3,4
VT options	HVT & LVT	LL & HS	RVT LVT
Substrate	FDSOI	FDSOI	FDSOI

Devices	HV/Analog devices	ESD devices	
VDDnom (Volt)	5	1	1,8
Lmin (um)	0,6	0,048	0.15
Tinv (nm)	14	1,6	3,4
VT options	HVT	RVT	RVT
Substrate	BULK	FDSOI	BULK

Fig4: transistors table description based on triple gate oxide HKMG FDSOI flow.

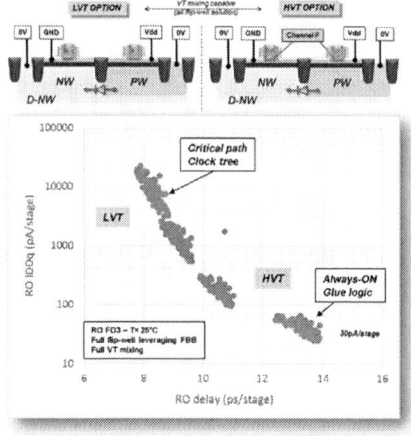

Fig5: extended range of mixable VT options using flip-well solution reaching ultra-low leakage capability for always-ON domains.

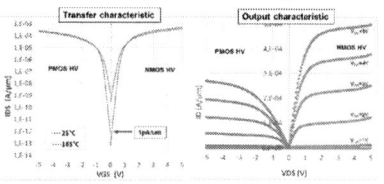

Fig6: HD SRAM D0 trend with embedded PCM inside 28FDSOI reference process flow

Fig7: output (right) and transfer (left) charcateristics for 5V transistor showing low leakage current (pA).

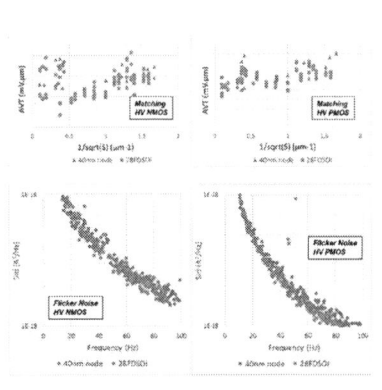

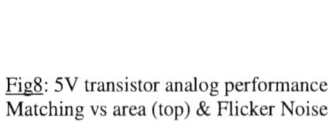

Fig8: 5V transistor analog performance. Matching vs area (top) & Flicker Noise.

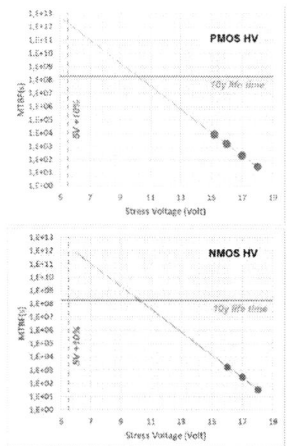

Fig9: 5V transistor gate oxide reliability. 10y criteria achieved for N and PMOS.

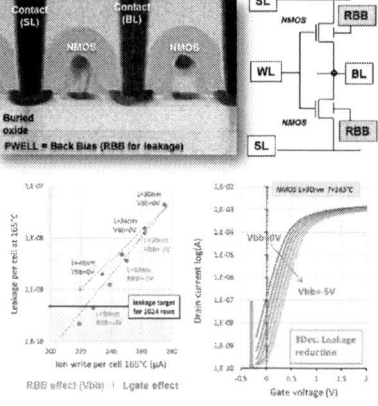

Fig10: PCM cell size reduction without leakage penalty thanks to RBB capability on FDSOI NMOS selector.

978-1-7281-1988-5/18 $31.00 © 2018 IEEE

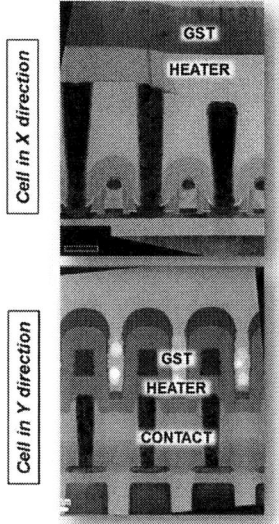

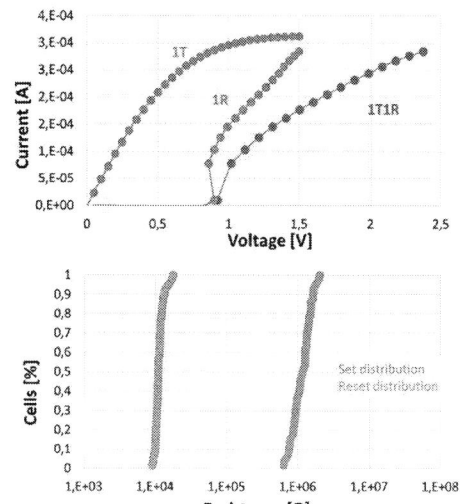

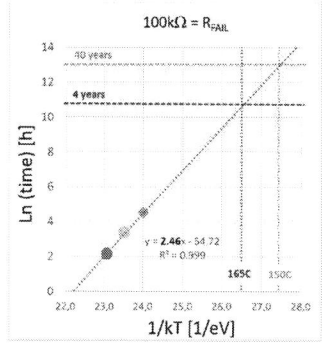

Fig11: 0,036um2 PCM cell images along BL (X) and WL (Y) direction.

Fig12: 1T1R analytical cell distribution showing 2 decade window. I-V plot with electronic threshold switching (top).

Fig13: Arrhenius plot & extrapolation at 150°C RESET data retention (2,46eV).

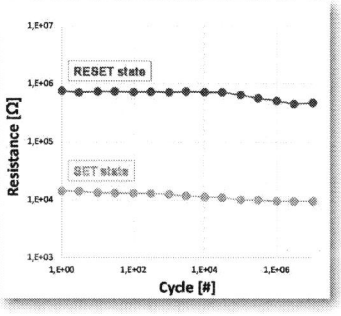

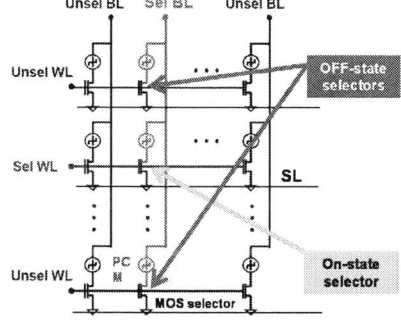

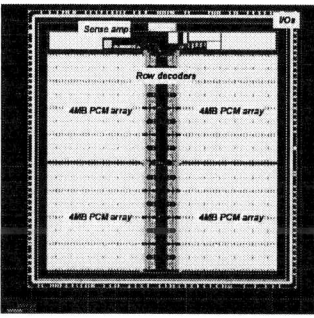

Fig14: 1T1R cell endurance. 2 decades programming window after 1Mcycle.

Fig15: PCM array organization. Gate of NMOS selector as WL & drain as BL. RBB applied.

Fig16: Full adressable 16MB PCM array allowing programming & reading.

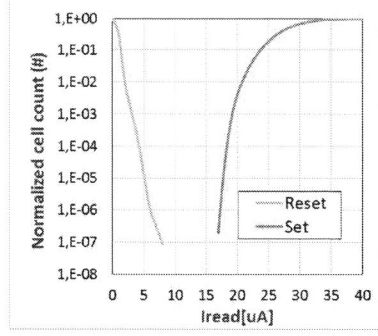

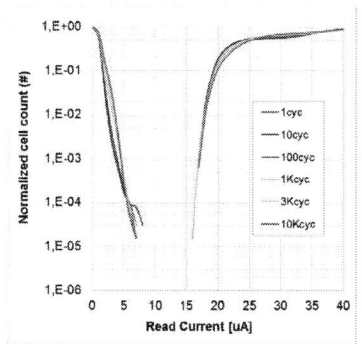

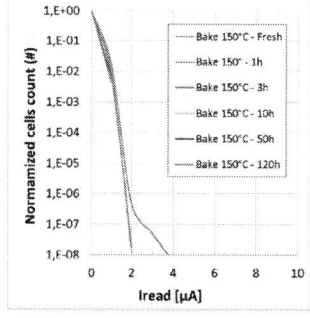

Fig17: SET and RESET current distributions extracted on 2MB. Wide reading window at 10uA

Fig18: SET and RESET distribution after endurance test. No reading window closure observed after 10kcycles.

Fig19: RESET data retention behavior on 16MB PCM array after 150°C bake. No recrystallization observed.

978-1-7281-1988-5/18 $31.00 © 2018 IEEE   427

A cost-efficient 28nm split-gate eFLASH memory featuring a HKMG hybrid bit cell and HV device

R. Richter[1], M. Trentzsch[1], S. Dünkel[1], J. Müller[1], P. Moll[1], B. Bayha[1], K. Mothes[1], A. Henke[1], M. Mazur[1], J. Paul[1],
P. Krottenthaler[1], J. Poth[1], S. Jansen[1], R. Hüselitz[1], H. Kim[2], A. Zaka[1], T. Herrmann[1], E.M. Bazizi[1], S. Beyer[1],
P. Ghazavi[3], H. Om'mani[3], S. Lemke[3], Y. Tkachev[3], F. Zhou[3], J. Kim[3], X. Liu[3], V. Tiwari[3], and N. Do[3]

[1]GLOBALFOUNDRIES Fab1 LLC & Co. KG, Dresden, Germany, email: ralf.richter@globalfoundries.com
[2]GLOBALFOUNDRIES, Malta, NY, USA
[3]Silicon Storage Technology, Inc., San Jose, CA, USA

Abstract— We demonstrate for the first time the integration of the proven SuperFlash® bit cell into 28 nm High-K Metal Gate (HKMG) technology, incorporating logic HKMG into the flash cell. Flash cell and high-voltage (HV) devices are implemented into a cost-optimized process flow saving seven masks compared to other 28nm eFLASH technologies. Comparable program/erase (P/E) endurance of up to one million cycles at 125°C is shown and program disturb characteristics meets array operation requirements. The Wordline transistor exhibits no degradation in sub-threshold slope of the post 100k P/E cycling, demonstrating robust reliability despite the introduction of HKMG into the flash cell. Additionally, the HKMG based HV devices demonstrate performance similar to platforms without HKMG material.

I. Introduction

More and more applications require embedded non-volatile memories accompanying the logic circuits on the same chip. A cost-efficient eNVM solution is especially important for IoT and automotive products. A fully CMOS-compatible integration of the matured third-generation embedded SuperFlash® memory cell (ESF3) concept with HV transistors [1] into a 28 nm HKMG super-low-power technology enables embedded memory products with ultra-low power consumption [2]. This paper discusses the process integration, electrical characteristics and intrinsic reliability of the hybrid ESF3 cell incorporating HKMG in the Wordline (WL). Additionally, electrical and intrinsic reliability results of HV devices with high-k (HK) oxide are demonstrated, which completes this highly competitive eNVM platform.

II. Device Fabrication

The hybrid ESF3-flash cell was embedded into GLOBAL-FOUNDRIES' high-volume gate-first 28 nm HKMG super low power platform 28SLP [3], (Table I). A high-voltage device is needed in order to manage voltages > 3.3 V which are required for flash cell operation. HV devices were developed using a triple gate oxide process with HKMG in the top-layer and implant mask layers to form the n-, p- and zero-Vth HV flavors. A TEM cross-section of the HV device is shown in Fig. 1. HV transistors for 5 V and up to 12 V operation are available.

A schematic flow of the embedded hybrid flash integration including HV devices is shown in Fig. 2. Compatibility to the

logic platform was achieved by flash cell formation before the logic process. Floating Gate (FG), Coupling Gate (CG) and Erase Gate (EG) are formed using flash-specific masks and a Poly-CMP process step before the well implant process module. Logic, HV devices and flash cell WL share the same HKMG patterning process. HV device and flash cell LDD implants are followed by the conventional 28SLP platform source/drain and silicide formation. Contact height for flash cell is adjusted to accommodate the cell stack height. All these optimizations enable a cost efficient eFLASH integration with a reduced mask count (2x ArF immersion and 5x DUV). Fig. 3 shows the ESF3 cell realized in 28SLP, highlighting the HKMG in the bottom part of the select gate (WL) in contrast to [2] and [4]. A top view SEM image of the 128 bitlines x 40 wordlines mini-array containing electrically addressable 8x8 cells in NOR architecture used for electrical testing is shown in Fig. 4. Key integration challenges including the HKMG process and topography interaction with the hybrid flash cell were predicted by TCAD process emulation (Fig. 5) and solved by integration, process or design adjustments. TCAD process and device simulation was used to optimize the program/erase and read scheme of the ESF3 cell (Fig. 6) as well as the HV device.

III. Results and Discussion

A. 28SLP Platform Matching

The embedded ESF3 integration scheme (Fig. 2) is fully CMOS-compatible and non-invasive. Critical process steps such as high thermal-budget treatments are done prior to the logic device formation. Other process steps could be shared between the flash cell and logic, such as the HKMG formation. The electrical device parameters of the 28SLP platform are not shifted by ESF3 integration, as shown exemplarily for the logic device performance in Fig. 7.

B. ESF3 Cell Electrical Data

The cell uses highly-efficient source-side hot electron injection for programming [4], and floating gate corner-enhanced Fowler-Nordheim electron tunneling for erase [5]. The typical cell operating conditions are shown in Table II. The EG oxide of the bit cell does not include any HK material. Therefore, erase performance is comparable to previous ESF3 technologies at the 28 nm node [4]. At typical erase conditions (e.g. 11.5 V, 10 ms) the cell is deeply erased with a CG Vt ~ −7 V and cell read current ~30 μA as illustrated in

978-1-7281-1988-5/18 $31.00 © 2018 IEEE

Fig. 8. Additionally, the cell matches the fast programming characteristics of conventional ESF3 bit cells at the 28 nm node [4]. For a typical programming current, Iprog=1 µA, it takes less than a microsecond for the cell transition from low to high Vt state with the read current in the 100 nA range (Fig. 9). In order to be able to see the entire cell programming kinetics, we intentionally slowed the programming down by using a very low programming current Iprog, namely 10 nA [6].

The time-to-disturb (T2D) dependencies vs. bit-line voltage are shown in Fig. 10. Applying ~1.8 V to the unselected BL effectively suppresses WL subthreshold current (represented by the steep region in the T2D-Vbl characteristics) even in the worst case (selected row, high temperature). The experimental T2D values extracted here are in the order of minutes, which far exceeds the actual program disturb time in NOR flash arrays, which ranges from several tens µs for Column disturb, to several and tens ms for Row and Diagonal disturb correspondingly, see Fig. 11.

We also evaluated the amount of charge stored in the FG, as depicted in Fig. 12, by comparing the single-electron-induced modulations of cell current with the effect of CG voltage [7]. We found that 1 V of Vt shift is equivalent to ~140 electrons on the FG. The "0"-"1" cell operating window is represented by a ~10 V Vt shift, i.e. by ~1400 electrons, which is high enough to guarantee there is no noticeable low-electron-count effects [8].

Single cell program/erase (P/E) for up to one million cycles at 125°C is shown in Fig. 13. Even at 1 million cycles there is still three orders of magnitude difference in cell current between program and erase state. This provides the large read current window required for high-end automotive with high speed and high reliability specifications. The incorporation of HKMG in the flash cell tends to raise concerns on reliability, due to the well-known high trap density in the HK material. This cell demonstrated for the first time the integration of HKMG in the WL transistor, and electrical data shows no degradation in the subthreshold slope or the Vt of the WL transistor after 100k cycling, as shown in Fig. 14. This indicates the absence of degradation or charge trapping associated with the HKMG. This result represents a key finding for the reliability of this new cell concept, which for the first time realizes a WL HK gate oxide. Additionally, CG Vt distribution of the bit cell post cycling matches conventional ESF3 cell data. In this context Fig. 15 shows the CG Vt distribution pre and post 10k cycling measured on the 8x8 mini array. After cycling a ~3.0 V tail-to-tail memory window with stable programmed state CG Vt can still be observed.

Data retention characteristics of this new cell matching conventional ESF3-technologies are illustrated in Fig. 16. A 72 hours data retention bake at 250°C for virgin and 100k pre-cycled bit cells was performed. A CG Vt shift of about 1.0 V for pre-cycled and of about 0.5 V for virgin devices was observed. The observed bake-induced Vt change is due to charge detrapping from the FG oxide for the cycled cells, and due to thermionic emission [9] for both cycled and fresh cells. The Vt shift distribution behaves in a predictable manner without any abnormally fast-moving bits.

C. High-Voltage Transistor

The HV devices of this technology contain an additional interfacial layer and HKMG layer above the flash HV thick oxide. These HK-HV devices demonstrate similar performance to prior platforms without HK material. Fig. 17 shows the HK-HV device gate leakage for both NMOS and PMOS transistors which are comparable to prior technologies without HKMG material (non-HK-HV). Fig. 18 further compares the high-voltage NMOS (HV N) BV_{DSS} and I_{OFF} at 10 V V_{DS} versus gate length (L) for HK-HV and non-HK-HV devices. At minimum gate length (L_min), 0.8 (au), BV_{DSS} near 14 V and $I_{OFF} < 1$ pA/µm are observed for both concepts. The reference non-HK-HV device has an electrical oxide thickness within a few angstroms of the HK-HV device and similar drain implantations. The intrinsic reliability of the HV N and PMOS (HV P) was evaluated for memory write operations near 12 V. Design practices are used to reduce the node-node potentials and disallow HV V_{DS} at $V_{GS} \geq V_t$ and HV V_{GS} at $V_{DS} > 5$ V. HV N and HV P reliability for allowed HV conditions is represented in Fig. 19 as BTI and I_{OFF}, non-conducting state (NCS) lifetimes. BTI stress, V_G=HV, 10 ms ac, V_D=V_S=V_B=0 V, lifetime to 100 mV V_t change for HK-HV transistors is >10 times required [10]. HK-HV versus non-HK-HV N, the voltage factors are within 5 %, -2.31 vs. -2.42. NCS stress with V_D=HV 10 ms ac, V_G=V_S=V_B=0 V compares HK-HV with the non-HK process, "a", and a non-HK with 2 nm thicker oxide, "b" [11]. The HK-HV life exceeds the requirement with 8.9 times margin, which is longer and within a magnitude of the non-HK devices.

IV. SUMMARY

A HKMG hybrid integration of a competitive ESF3 flash cell and HV devices into a 28 nm low power HKMG technology was demonstrated. Third-generation embedded SuperFlash® memory cell electrical results were reported. Endurance, program/erase kinetics and data retention capability are comparable to non-HKMG ESF3 cells while the total mask count of this hybrid integration was reduced by up to seven. No additional cycling-induced degradation mechanisms, associated with the introduction of HK gate dielectric into the cell, have been detected. HV devices with electrical and reliability performance matched to prior platforms without HKMG were additionally demonstrated.

ACKNOWLEDGMENT

This work has been supported by funding from the Free State of Saxony. We thank Fraunhofer IPMS for support within PHOENIX and especially acknowledge B. Pätzold and S. Bott for process development.

REFERENCES

[1] L. Q. Luo et al., *IMW*, 2017, pp. 123–126.
[2] Y. K. Lee et al., *VLSI Technology*, 2017, pp.202–203.
[3] M. Trentzsch et al., *IEEE IEDM*, 2016, pp. 11.5.1–11.5.4.
[4] N. Do et al., *IMW*, 2018, pp. 47–49.
[5] Y. Tkachev et al., *IEEE Trans. El. Dev.*, 2012, Vol. 53, pp.5–11.
[6] Y. Tkachev et al., *IMW*, 2017, pp. 48–51.
[7] Y. Tkachev, *ICMTS*, 2016, pp. 110–115.
[8] G. Molas et al., *IEEE Trans. El. Dev.*, 2006, V.53, pp. 2610–2619
[9] S. Aritome et al., *Proc. IEEE*,1993, V.81, pp.776–788.
[10] S. Pae et al., *IEEE IRPS*, 2008, pp. 352–357.
[11] C. Hu et al., *IEEE J. Solid-State Circuits*, 1985, V.20, pp. 295–305.

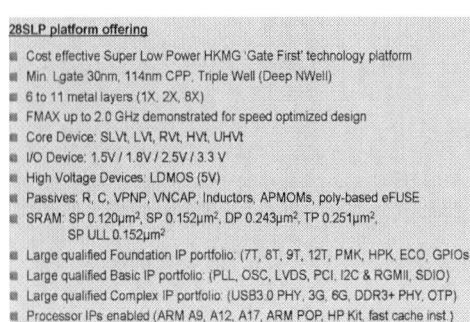

28SLP platform offering

- Cost effective Super Low Power HKMG 'Gate First' technology platform
- Min. Lgate 30nm, 114nm CPP, Triple Well (Deep NWell)
- 6 to 11 metal layers (1X, 2X, 8X)
- FMAX up to 2.0 GHz demonstrated for speed optimized design
- Core Device: SLVt, LVt, RVt, HVt, UHVt
- I/O Device: 1.5V / 1.8V / 2.5V / 3.3 V
- High Voltage Devices: LDMOS (5V)
- Passives: R, C, VPNP, VNCAP, Inductors, APMOMs, poly-based eFUSE
- SRAM: SP 0.120μm², SP 0.152μm², DP 0.243μm², TP 0.251μm², SP ULL 0.152μm²
- Large qualified Foundation IP portfolio: (7T, 8T, 9T, 12T, PMK, HPK, ECO, GPIOs)
- Large qualified Basic IP portfolio: (PLL, OSC, LVDS, PCI, I2C & RGMII, SDIO)
- Large qualified Complex IP portfolio: (USB3.0 PHY, 3G, 6G, DDR3+ PHY, OTP)
- Processor IPs enabled (ARM A9, A12, A17, ARM POP, HP Kit, fast cache inst.)

Table I. ESF3 eFLASH extends the 28nm super low power platform offering with an eNVM option.

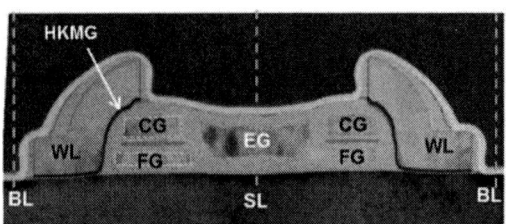

Fig. 3. TEM cross-section of the HKMG hybrid ESF3 flash cell (two bits, 0.047 μm² per bit) before contact formation with Coupling Gate (CG), Floating Gate (FG), Erase Gate (EG), Wordline (WL) and Bitline (BL). Two bit cells share EG and Source Line (SL) each. The WL uses HKMG as indicated by the arrow whereas CG and FG do not include HKMG.

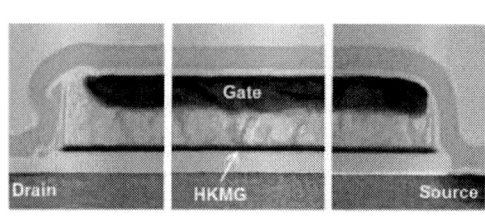

Fig. 1. Cross sectional TEM of the high-voltage (HV) transistor (n-type) featuring HKMG in the gate stack. Drain with silicide pullback.

Fig. 4. Tilted top-view SEM of the ESF3 8x8 mini array.

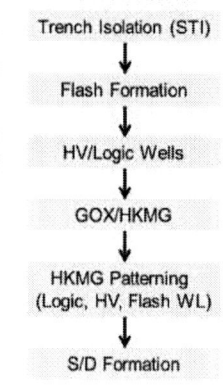

Process flow

Trench Isolation (STI)
↓
Flash Formation
↓
HV/Logic Wells
↓
GOX/HKMG
↓
HKMG Patterning (Logic, HV, Flash WL)
↓
S/D Formation

Fig. 2. Schematic process flow of the embedded ESF3 cell and HV devices into 28nm SLP platform. Saving seven masks compared to other 28nm eFLASH technologies.

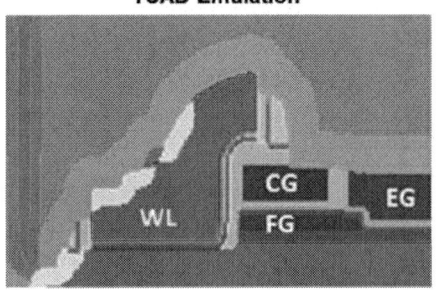

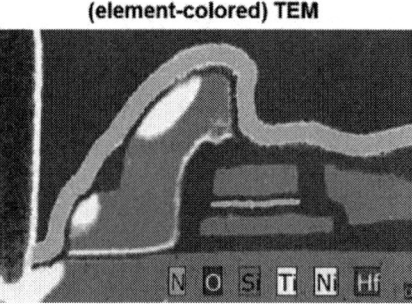

Fig. 5. Comparison of the final ESF3 structure (right) to the initial simulation (left). For the process emulation Sentaurus (Synopsys) software was used. Elemental compositions are colored in the TEM cross-section for better clarity.

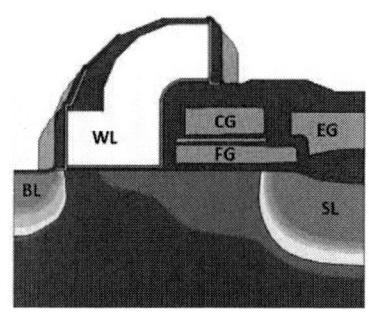

Fig. 6. Doping simulation in order to optimize erase/program/read conditions using Sentaurus (Synopsys).

Ioff-Ion Performance Curve

Fig. 7. I_{on}-I_{off} performance curves of PMOS logic devices and TEM cross-sections of PMOS and NMOS logic devices The embedded 28nm SLP-ESF3 flow is matched to the reference 28nm SLP platform.

	Program	Erase	Read
Bit-Line (BL)	1uA	0V	0.8V
Word-Line (WL)	0.8V	0V	1.8V
Coupling Gate (CG)	10.5V	0V	1.8V
Erase Gate (EG)	4.5V	11.5V	0V
Source-Line (SL)	4.5V	0V	0V

Table II. Typical 28nm ESF3 cell operating conditions.

978-1-7281-1988-5/18 $31.00 © 2018 IEEE

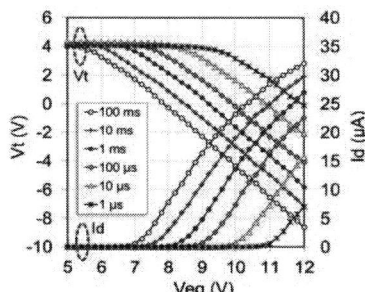

Fig. 8. The family of cumulative erase characteristics measured at different erase time.

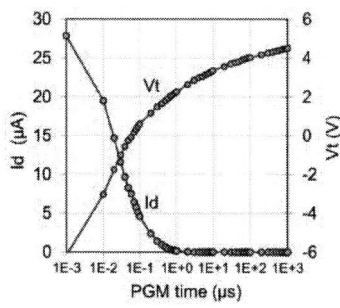

Fig. 9. The cell programming kinetics: Programming current is Iprog=10 nA. The X-axis is normalized to Iprog=1 µA.

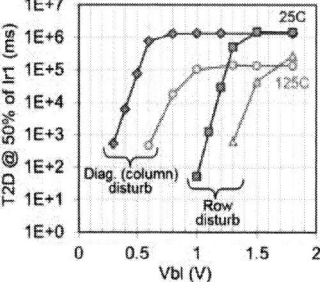

Fig. 10. (T2D) measured at 50% of Ir1 vs. BL voltage. Program disturb conditions: Row: WL=1 V; Diagonal (Column): WL=0 V. SL=4.5 V, CG=10.5 V for both types of disturb.

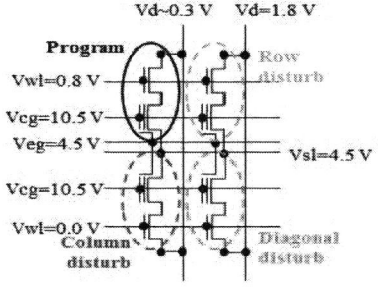

Fig. 11. Definition of different program disturb modes shown in this 2x2 array schematic, with appropriate bias conditions.

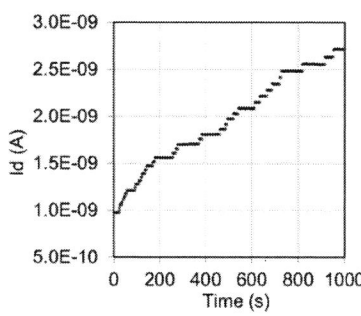

Fig. 12. The kinetics of cell current under EG stress 4.8V. The steps in the cell current are caused by FG potential modulations associated with the tunneling of individual electrons from FG to EG.

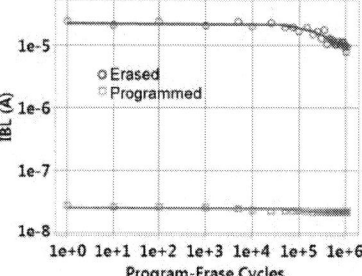

Fig. 13. Single cell program-erase cycling characteristics at 125°C. There is still operational window between program and erase state even at 1 M cycles.

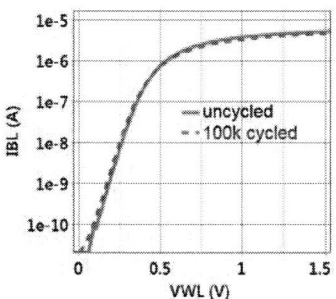

Fig. 14. IBL-VWL, BL=0.1 V, before and after 100k cycling for a single cell.

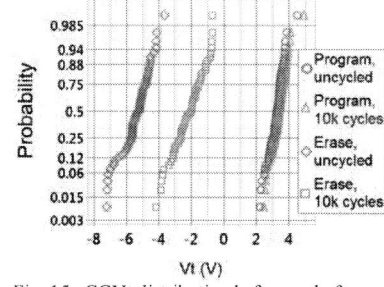

Fig. 15. CGVt distribution before and after program-erase cycling, 64 cell mini-array.

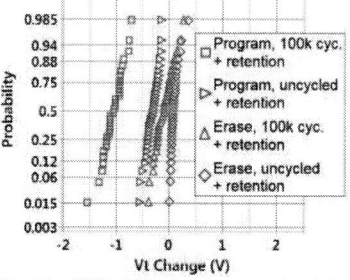

Fig. 16. CGVt shift after data retention bake (72 h at 250°C), pre and post 100k program-erase cycles.

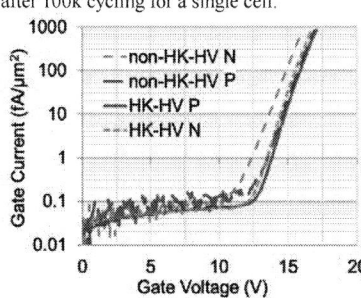

Fig. 17. Gate leakage vs. Gate voltage for large area HV oxides measured at 150°C.

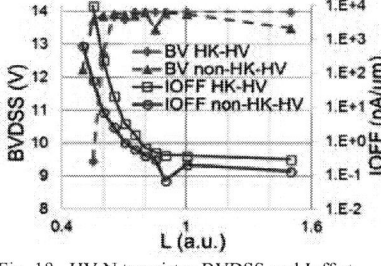

Fig. 18. HV N transistor BVDSS and Ioff at VD = 10 V, VG = VS = VSUB = 0 V, for devices with and without HKMG.

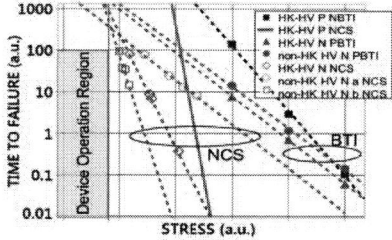

Fig. 19. HV transistor AC NBTI, PBTI, and NCS 150°C time to failure at 100 mV Vt shift. X-axis is stress voltage for BTI, VGS = HV, and $\log_{10}$(Ioff) for NCS, VDS = HV.

978-1-7281-1988-5/18 $31.00 © 2018 IEEE   431

A Bi-stable 1- /2-Transistor SRAM in 14 nm FinFET Technology for High Density / High Performance Embedded Applications

Yuniarto Widjaja[1], James Wilson[1], Tu Nguyen[1], Jin-Woo Han[1], Christopher Norwood[1], Dinesh Maheshwari[1], Stefan Lai[1], Pieter Vorenkamp[1], Zvi Or-Bach[1], Yoshio Nishi[2]

[1] Zeno Semiconductor Inc., Cupertino, CA 95014 USA
[2] Department of Electrical Engineering, Stanford University, Stanford, CA 94305 USA
E-mail: ywidjaja@zenosemi.com

Abstract

1-transistor and 2-transistor (1T/2T) SRAM are fabricated using 14 nm baseline foundry process without any process modifications. A bi-stable self-latch mechanism is established in a single transistor where its p-type body becomes electrically floating by reverse biased, buried depletion regions from adjacent n-wells. The bit cell operation and the disturb immunity are verified. A unit cell size of 0.039 μm^2 is achieved, offering >2x area reduction over 6T-SRAM and providing comparable power and performance.

Introduction

Embedded SRAM has been a key component in SoC. However, cell size scaling of conventional 6T based SRAM is increasingly challenging due to not only process complexity but also stability and leakage current requirements. In order to address these scaling issues, we previously demonstrated 1T-SRAM in 28 nm node, resulting in ~3.5x and ~5x cell size reduction using standard logic and SRAM design rule, respectively [1]. At 14 nm FinFET process, the 1T-SRAM cell size of 0.022 um^2 is achieved using standard logic design rule, offering up to 65% area reduction compared to high-density 6T-SRAM. The 1T-SRAM was previously demonstrated using a buried n-well layer to form a floating p-type body. In this work, we demonstrated a second generation 1T-SRAM in 14 nm baseline foundry process FinFET technology without the buried n-well implant process, and thus is fully compatible with the baseline FinFET process. Using standard logic design rule, the 1T-SRAM cell size is 0.039 um^2 (without buried n-well layer). **Fig. 1(a)** compares the cell size for various technology nodes and **Fig. 1(b)** compares the layout of 6T versus 1T cell, illustrating that the Gen-1 and Gen-2 1T-SRAM is approximately 2x and 3x smaller than 6T-SRAM, respectively, using standard logic design rule [1]-[3]. The schematic cross-sections of Gen-1 and Gen-2 1T-SRAM are shown in **Fig. 2(a)**. Within 1T footprint, the self-latch mechanism is accomplished by harnessing open base vertical

n-p-n bipolar device [4]. As illustrated in **Fig. 2(b)**, due to the device scaling, buried depletion region formed underneath of STI due to reverse biased junction from the adjacent standard n-wells (CI) is found to be sufficient to form an open base. Simultaneously, the n-wells function as charge injector (collector) of vertical n-p-n device for the self-latching. The SRAM operation is experimentally demonstrated, and T-CAD simulation is performed to provide further scalability. The cell operation and immunity against various disturb conditions are verified.

Results and Discussion

Fig. 3 shows the layout of cell array and SEM images of the fabricated cell array. I_d-V_g and I_d-V_d characteristics for two different V_{CI}=0V and V_{CI}=V_{mem} where V_{mem} is a positive voltage to enable memory effect are shown in **Fig. 4**. The memory effect is explained by two combined mechanism. First, the open base is formed electrically by buried depletion region. At V_{CI}=0V, ordinary p- substrate and n+ source/drain diode curve is observed in **Fig. 5(a)**. However, the forward current at V_{sub} = 1V starts to be turned off as V_{CI} increases as shown in **Fig. 5(b)**, which indicates that the buried depletion regions near STI are merging and isolating the n^+ S/D and p-substrate. The formation of open base (fin) by buried depletion boundary due to two adjacent n-wells is verified by T-CAD simulation as shown in **Fig. 6**. The second mechanism is that the bias voltage to n-well (V_{CI}) is sufficient to cause impact ionization near the buried depletion region, and the n-well functions as charge injector of vertical n(S/D)-p(open base)-n(n-well) bipolar device for self-latching.

The memory operation conditions are summarized in **Fig.7**. I_d-V_g curves for state '0' and '1' at read condition and associated energy band diagram are shown in **Fig. 8** and **Fig. 9**, respectively. For the state '0', the open base becomes neutral, showing ordinary I_d-V_g curve. For the state '1', majority carriers are accumulated in the open base through impact ionization at the n-well/p-open base junction, which promotes vertical bipolar action [5]. As this positive feedback

978-1-7281-1988-5/18 $31.00 © 2018 IEEE

continues, the device becomes a static memory. When the state '1' is sustained, the positive charge lowers the energy barrier of the lateral MOS device, thus resulting in a higher current flow. Non-destructive read is verified by constant read over 1 hour (**Fig. 10**). The smaller cell size as well as the low operating voltage (especially for read operation) results in a lower dynamic power compared to 6T-SRAM.

The write mechanism is illustrated in **Fig. 11** and the voltage operating region for write '1' and '0' is shown in **Fig. 12**. When $\beta \times (M - 1) \sim 1$ condition is met (β is the emitter current gain and $M - 1$ is the multiplication factor of the vertical bipolar device), write '1' is accomplished when V_g and V_d are sufficiently high such that capacitive coupling to the open base region activates the vertical bipolar action [1]. When the majority carrier is removed by forward junction current injection, write '0' is accomplished. The pulsed measurement result of repeated write-read and hold-read are shown in **Fig. 13**, which demonstrates the static memory operation. A write pulse width of 20 ns was used in the measurement due to our pulse generator limitation. Sub-ns write '1' and write '0' are verified using T-CAD simulation as shown in **Fig. 14**. **Fig. 15** shows simulation results of two stable body (fin) potentials. The net body current is determined by the competition between forward junction current and the impact ionization current. Due to the nature of open base vertical bipolar devices, stable states are obtained at the base potential showing zero net base current [6]. As a result, **Fig. 15** illustrates that the open-base vertical n-p-n bipolar device can have two equilibrium potentials. BL and WL disturb in half-selected cell in an array are characterized as shown in **Fig. 16**. No disturb was observed up to 0.4 V higher voltage than write conditions.

Although the sensing window of ~8 µA/fin at minimum V_{mem} is reasonable for SRAM, some applications may require extremely low latency. For high performance SRAM application, the read current and the read current window can be amplified by raising V_{mem}, which is similar to V_{dd} dependence of 6T-SRAM. Because overdrive of V_{mem} amplifies the base current, the state '1' current can be improved to 20 µA/fin due to additional lateral bipolar current as measured in **Fig. 17**. However, the lateral bipolar current increases not only the read current but also the unselected BL current because the lateral bipolar current is not readily shut off by gate field. As a result, the bipolar leakage remains as background current as shown in **Fig. 18**, which in turn limits the array size and consumes power. This bipolar leakage from unselected WL may add to the BL current as illustrated in **Fig.**

19. Therefore, in order to remove the leakage to BL, a select transistor is connected in series of memory cell transistor. By employing select transistor, the BL leakage is inhibited as measured in **Fig. 20**. Because the select transistor can be accommodated within the active region for the adjacent n-well, there is no area penalty for 2T compared to the 1T.

As the clock operating frequency increases and the transistor density increases, the chip is easily exposed to high temperature. The bi-stability is confirmed up to 125 °C as measured in **Fig. 21**. The read current slightly increases for both state '0' and state '1'. The increase of read current with the temperature indicates that the read voltage for $V_g = 0.4$ V is less than the zero-temperature-coefficient bias point so the diffusion current dominates. As the temperature increases, the leakage current increases as expected. **Fig. 22** shows the BL leakage current (unselected BL current during read operation) and **Fig. 23** shows the n-well leakage current (standby current) at various temperature.

In order to demonstrate scalability down to 7 nm, T-CAD simulation was conducted as shown in **Fig. 24**. Finally, **Table 1** compares the 1T-SRAM and other competing alternative memories including floating body DRAM and thyristor cell. Among all emerging embedded memory, the present 1T-SRAM is the only technology that is manufacturable without any specialty processes.

Conclusions

A Gen-2 1T and 2T-SRAM is implemented on a baseline 14 nm FinFET technology <u>without</u> any process modification. An isolated p-well is electrically formed by buried depletion region induced from neighboring n-wells. The intrinsic vertical n-p-n bipolar device enables self-latch mechanism at 1T footprint. The write and read characteristics are demonstrated and the disturb immunities are verified. The functionality at high temperature is confirmed up to 125 °C. T-CAD is used to investigate the optimization and scalability to 7 nm FinFET. The proposed 1T/2T-SRAM demonstrate feasibility as a 6T-SRAM alternative for SoC. The 14 nm 1T/2T-SRAM bit cell size of 0.039 µm² (standard design rules) is >2x smaller than the 6T-SRAM with sub-ns access time and low dynamic power.

References

[1] J.-W. Han, *IEDM Dig. Tech.* **2015**, pp. 26.7.1. [2] T. Song, *ISSCC Dig. Tech.* **2014**, pp. 13.2. [3] T. Song, *ISSCC Dig. Tech.* **2017**, pp. 12.2. [4] J.-W. Han, *VLSI Tech.* **2010**, pp. 171. [5] K. Sakui, *IEEE TED* **1989**, pp. 1215. [6] T. Ohsawa, *IEEE TED* **2009**, pp. 2302. [7] R. Ranica, *VLSI Tech.* **2005**, pp. 38. [8] T. Sugizaki, *IEDM Dig. Tech. 2006*, pp. 933.

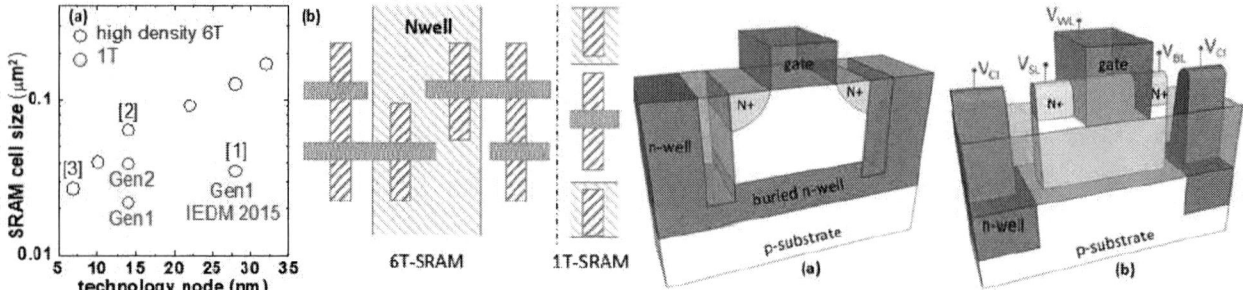

Fig. 1 (a) Unit cell size comparison and (b) unit cell layout (yellow: NWELL, red: POLY, green: DIFF) between 6T-SRAM and 1T-SRAM. 14 nm 1T-SRAM is 20% smaller than 7 nm 6T-SRAM.

Fig. 2 Schematic illustration of (a) the first generation and (b) the novel second generation 1T-SRAM. Whereas additional process step to form buried n-well is required in the first generation, the foundry baseline process can be used in the second generation 1T-SRAM.

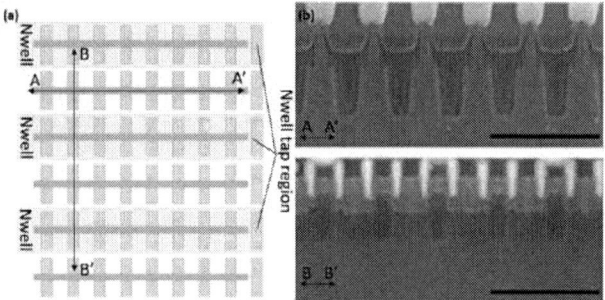

Fig. 3 (a) array layout of the second generation 1T-SRAM and (b) cross-sectional view of scanning electron microscopy images of fabricated array cut along WL (AA') and BL (BB') directions (scale bar is 300 nm).

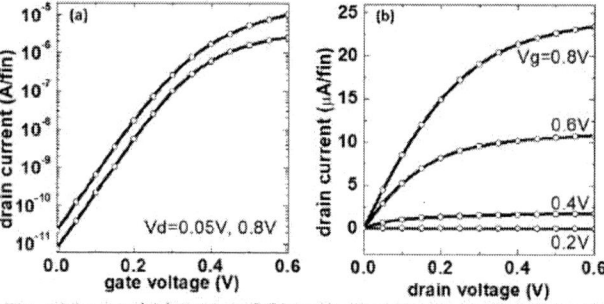

Fig. 4 Measured (a) I_d-V_g and (b) I_d-V_d characteristics of the fabricated 1T-SRAM cell for two charge injector n-well voltages; V_{CI} = 0V (line) and V_{CI} = V_{mem} (symbol). When the body is neutral, no difference in current-voltage characteristics are seen.

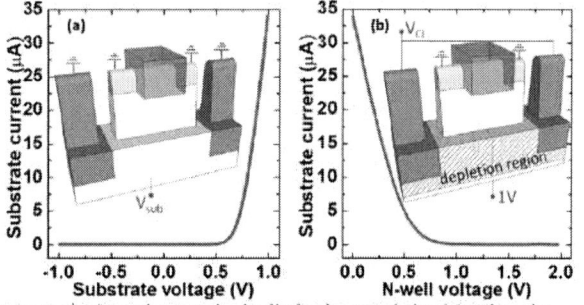

Fig. 5 n^+ S/D and p-type body diode characteristics (a) When the adjacent n-well is grounded, ordinary pn junction behavior is seen. (b) At forward junction condition, pn diode is turned off as the n-well voltage increases due to the depletion region forming isolated p-body.

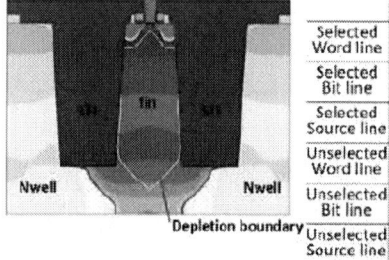

Fig. 6 T-CAD simulation demonstrating formation of isolated body through reversed biased depletion boundaries.

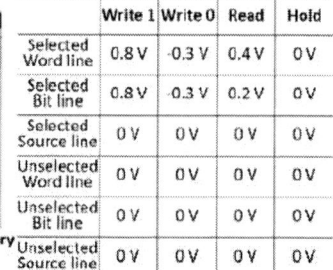

	Write 1	Write 0	Read	Hold
Selected Word line	0.8 V	-0.3 V	0.4 V	0 V
Selected Bit line	0.8 V	-0.3 V	0.2 V	0 V
Selected Source line	0 V	0 V	0 V	0 V
Unselected Word line	0 V	0 V	0 V	0 V
Unselected Bit line	0 V	0 V	0 V	0 V
Unselected Source line	0 V	0 V	0 V	0 V

Fig. 7 Memory write, read and hold operation conditions for 1T-SRAM. All other unselected terminals are grounded.

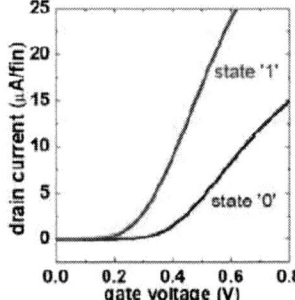

Fig. 8 Measured I_d-V_g (V_d=0.2V) characteristics for state '0' and state '1', showing threshold voltage shift.

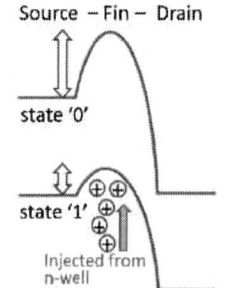

Fig. 9 Energy band diagram for states '0' and '1'. At state '1', positive charges are injected from adjacent n-well.

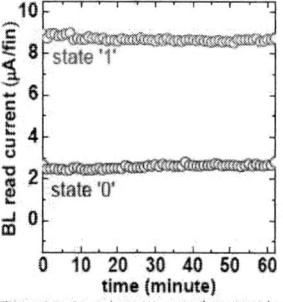

Fig. 10 Continuous read operation for 1 hours showing the static as well as non-destructive read.

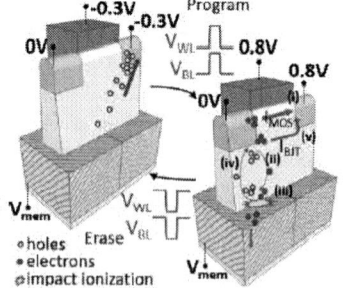

Fig. 11 Schematic illustration of impact ionization write '1' and forward junction current write '0'.

978-1-7281-1988-5/18 $31.00 © 2018 IEEE

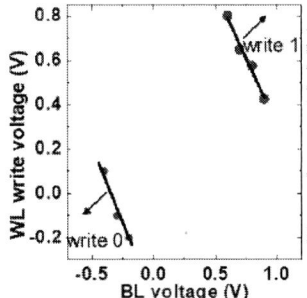

Fig. 12 Measured write '1' and write '0' voltage map. The voltage conditions indicated by the arrow results in write.

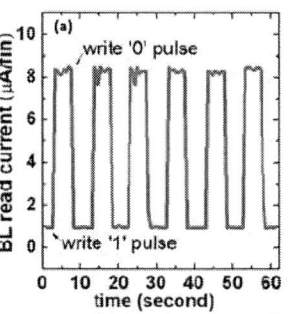

Fig. 13 Pulsed measurement of repeated (a) write-read and (b) hold-read characteristics. The static memory characteristics are demonstrated as long as V_{CI} is biased.

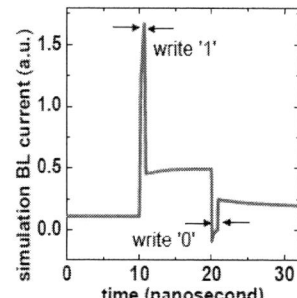

Fig. 14 T-CAD transient simulation results of writing with pulse width less than 500 ps.

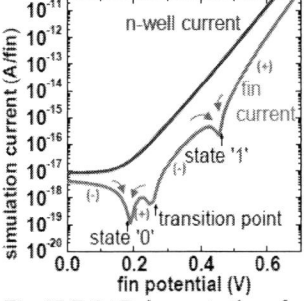

Fig. 15 T-CAD demonstration of bistable states and pA n-well leakage current while all terminals are grounded.

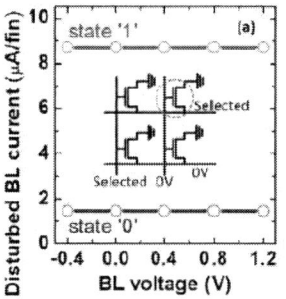

Fig. 16 (a) BL and (b) WL disturb cell current measured from half-selected cells as illustrated array in inset. No half-selected disturb is observed for both BL and WL.

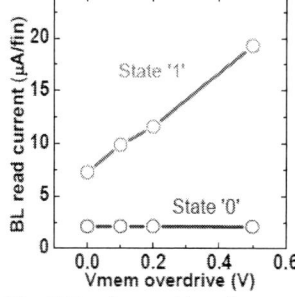

Fig. 17 Read current boosting by overdriving V_{mem}. The sensing window is increased with state '1' current.

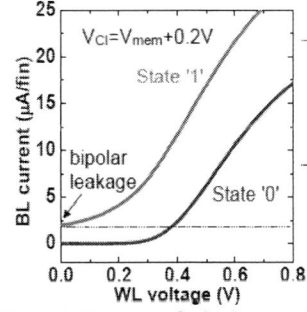

Fig. 18 I_d-V_g curves for both states with 0.2V overdrive V_{mem}. The bipolar leakage is seen at state '1'.

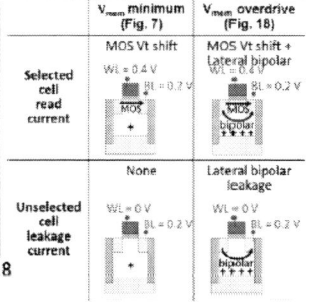

Fig. 19 Schematic illustration of mechanism in read current boosting and bipolar leakage due to V_{mem} overdrive.

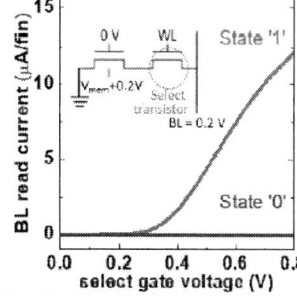

Fig. 20 2T cell scheme (1 cell + 1 select tr.) to suppress the bipolar leakage for high performance application.

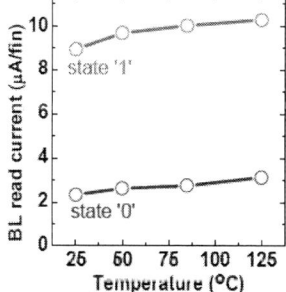

Fig. 21 High temperature characteristics of read current for both states.

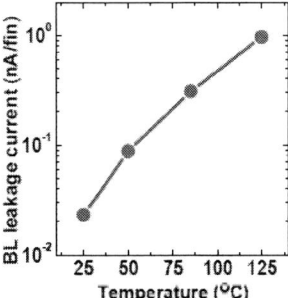

Fig. 22 High temperature characteristics of BL leakage current (V_g=0V & V_d=0.2V)

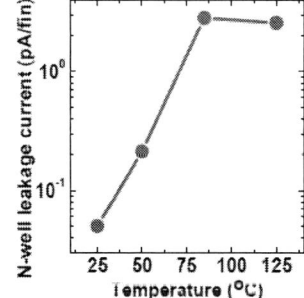

Fig. 23 High temperature characteristics of n-well leakage current.

Fig. 24 T-CAD simulation demonstrating memory hysteresis based on 7 nm FinFET technology.

	[This work] 1T SRAM	1st Gen. [1] 1T SRAM	6T SRAM
Type	Static	Static	Static
Process	Standard	Specialty	Standard
Size	12F2	8F2	~120F2
	1T DRAM [7]	Biristor [4]	TRAM [8]
Type	Dynamic	Dynamic	Static
Process	Specialty	Specialty	Specialty
Size	8F2	4F2	16F2

Table 1 Comparison of 1T-SRAM and other alternative memories.

SiC Devices for Mainstream Adoption

Peter Friedrichs

Infineon Technologies AG, Erlangen, Germany, email: peter.friedrichs@infineon.com

Abstract—SiC power devices enter more and more applications today. This process is supported by a couple of factors. Maturity levels and cost/performance of diodes and transistors are interesting enough for many users to consider the new components. Furthermore, devices are more and more fine-tuned to target application requirements which make it easier to enter the most promising target applications. The contribution will sketch on various examples how this philosophy can be rolled out. A new generation of power didoes will be discussed as well as corresponding more powerful package technologies. Finally, a similar assessment for SiC power MOSFETs will be presented.

I. INTRODUCTION

Silicon carbide based power devices are meanwhile a key element for surmounting existing barriers towards highest power density and/or efficiency. Most of the leading power semiconductor manufacturers in the power semiconductor supplier landscape is engaged in this. The sales worldwide are still dominated by ultrafast SiC diodes, offering negligible switching loses and high maturity. The main drawback of those diodes is the conduction loss, mainly its increase with temperature. Diodes offering more attractive conduction losses are a powerful technology in combination with novel IGBT's in mixed modules which could serve as both, a transition technology towards full SiC or even the best cost-performance solution for a given application.

An outstanding potential to empower semiconductor based energy conversion circuits like inverters and converters was shown by using SiC based transistors. For a while 1200 V rated devices are commercially released as discrete parts and in numerous power module based solution. Mainly the cost-performance level possible today, enabled e.g. by the roll out of the 150 mm SiC wafer technology pushed right now the implementation in various applications. However, even more than in the diodes besides cost and performance additional considerations are seen to be decisive for the actual success in real products. Factors like reliability, compatibility with today's control circuits as well as stable manufacturing are equally important aspects in a row with performance which have to be considered resp. fulfilled.

In the contribution it will be discussed which measures can be taken to further improve SiC diodes with respect to on-performance and how application specific trade off can be sued beneficially to achieve added value. Mainly for highest power densities the wish from the users is to use SMD packages for those components. While those offer very good

dynamic performance due to low stray inductance they are limited traditionally in its thermal performance. A new concept, addressing this dilemma by offering top side cooling with a TO like power handling capability was developed and will be discussed in the contribution. Related to SiC MOSFET devices it will be sketched how an application landscape can look like and which conclusions can be drawn for device and product design under various trade off aspects.

II. SCHOTTKY BARRIER DIODE ADOPTION TO MEET APPLICATION REQUIREMENTS

The major driver for the use of SiC diodes in power electronic circuits is the elimination of any storage charge effect usually coming along with high voltage pin based silicon diodes. Figure 1 demonstrates this by comparing the turn off current waveforms of a SiC SBD and silicon based high voltage (1200 V) pin diodes.

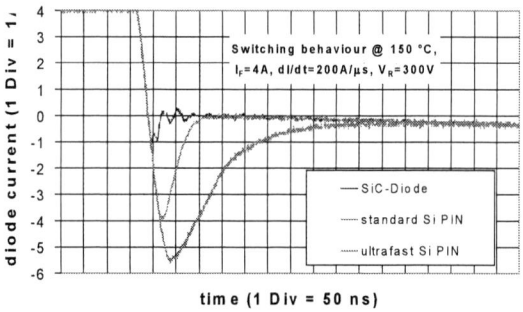

Fig. 1. Turn off current waveform of a SiC SBD compared with silicon based bipolar pin-diodes

This behavior can significantly reduce switching losses, however, due to the unipolar character of the forward current flow such diodes exhibit a major drawback it its conduction loss behavior since with temperature the forward current drop increases (Figure 2) and thus, at operating temperature static losses are much higher than in silicon based components. This fact was for many use cases a significant drawback, especially if there was an equal distribution of switching and conduction losses in place. Thus, most of the research dedicated to next generations of SiC diodes was addressing the forward conduction behavior either by improving the temperature dependence or by working on the differential resistance [1].

978-1-7281-1988-5/18 $31.00 © 2018 IEEE

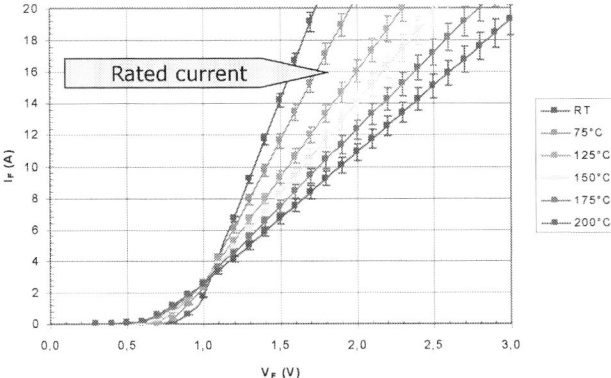

Fig. 2. Typical forward I-V characteristics of SiC 1200 V SBD's as a function of temperature

More recently Infineon introduced a new generation (G6) of didoes going a fundamentally new way. In this new generation of diodes the Schottky barrier height was reduced in order to enable lower forward voltage drop. Figure 3 shows it on an example of 650V diodes (old generation G5 vs new G6).

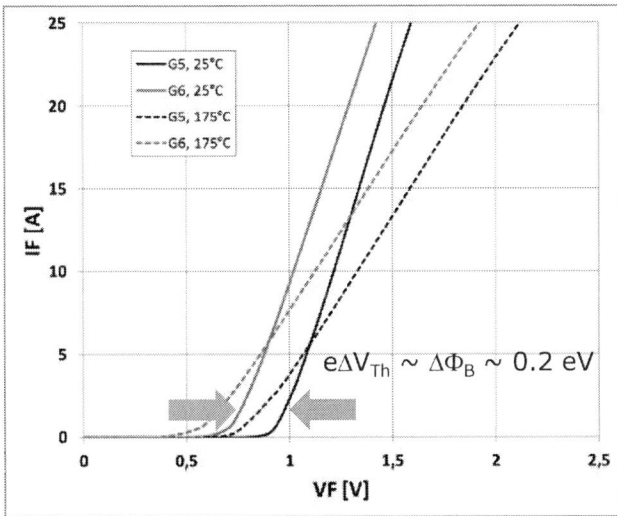

Fig. 3. Improved forward behavior of the new G6 based diode technology (red) vs. previous G5 based parts, at room temperature (solid) and Tjmax=175°C (dashed)

It can be clearly seen that now about 200 mV lower voltage drops can be realized, In the major use cases like PFC stages the conduction losses can be reduced by 22 % up to 38 %, depending on load conditions and case temperature [2]. This step now enables a next level of penetration for fast SiC based diodes in applications which could not be addressed before due to the inferior on performance compared to available fast silicon based diodes.

III. PACKAGE INNOVATION FOR SiC COMPONENTS

SiC devices usually are being used for fast switching applications. Thus, low stray inductance is one of the key requirements for the package selection. In addition, for discrete solutions, in many cases SMD style housings are preferred due to the easy automated assembly. However, the thermal performance of today's SMD packages is often inferior to the one delivered by through hole parts like TO220 or TO247. SiC components need a very good thermal performance in the package due to the high power density generated at the chips based on the fact that the die size is significantly smaller than for silicon components and thus, despite lower absolute losses the loss power density increases. For this reason up to now only quite small power handling capabilities could be realized by using standard SMD packages while for higher currents only TO220 or TO247 could be used. Infineon recently launched a new style of SMD high voltage packages which enable direct cooling from the top side – the DDPAK (Figure 4) [3].

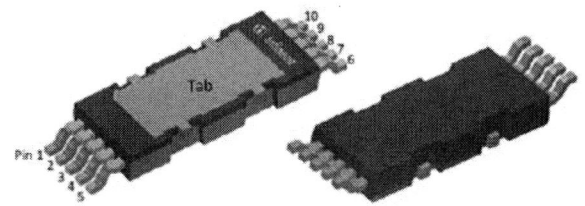

Package dimensions: 21.1x6.6x2.35mm (LxBxH)

Fig. 4. Outline of the new top side cooled SMD package DDPAK – an ideal solution for SiC diodes and transistors

With this new solution another hurdle in the mainstream adoption of SiC components could be addressed. Now the thermal performance of through-hole packages can be achieved also in a SMD style configuration. Thus also at higher power levels very compact and easy to assemble design are now possible. The large number of pins opens up the opportunity to implement transistors as well, including the valuable feature of a Kelvin contact.

IV. SiC TRANSISTORS

In contrast to the diodes it took much longer to commercialize SiC transistors. It was partially also related to the ongoing conceptual discussion over many years caused by insufficient performance offered by the most straightforward concept a MOSFET. Thus, JFETs and BJT's appeared and disappeared on the market until finally in 2012 the first commercial MOSFETs became available. SiC MOSFETs with blocking voltages of 1200 V are devices offering significant added value in application fields such as solar converters,

978-1-7281-1988-5/18 $31.00 © 2018 IEEE

UPS, fast EV chargers as well as industrial drives or train propulsion. These applications benefit from both, a reduction of dynamic losses as well as static losses which can be translated into the opportunity to use higher switching frequencies. This might be beneficial I order to shrink passive components. Additionally the cooling effort might be smaller due to better efficiency, and in the end weight and cost can be lower at system level. However, a major obstacle in th development of powerful SiC MOSFET is define the right trade-off between performance on the one hand side and other application relevant aspects on the other side as indicated in Figure 5.

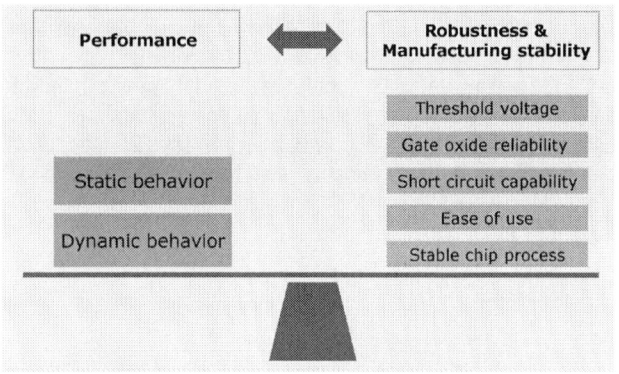

Fig.5 : Sketch of the typical trade of to be solved in case of SiC MOSFET designs which have the chance to find mainstream adoption – performance must be balanced against other application relevant requirements

As an example, In order to enable a safe normally off behavior at V_{GS}=0 V under all operating conditions and even under critical modes like short circuit the threshold voltage $V_{th(GS)}$ needs to be sufficiently high, e.g. >4 V. However, due to the trap structure at SiC-SiO$_2$ interfaces [4] and the related week transconductance curve a higher threshold voltage leads to higher on-state resistance R_{on} for a given V_{GS_ON}. It could be compensated by increasing V_{GS_ON}, however, this would result in very high absolute values being incompatible with typical driver circuit outputs and it would increase the electric field across the oxide in on state what can degrade the oxide reliability significantly with respect to lifetime and FIT rates. A similar approach is valid when it comes to the implementation of short circuit robustness as e.g. required for applications like motor drives. By a proper design it can be implemented, however, R_{on} will suffer and thus – the right balance needs to be found in order to fulfill the requirements in the individual target application but not wasting too much of performance since this influences the cost performance ratio and thus, the implementation speed in applications. Infineon decided to solve this dilemma by focusing on certain target applications and their individual mission profiles when defining the design point within a given technology curve for a SiC MOSFET. In order to do so of course a technology resp. concept needs to be selected which offers a technology positioning at various design points and at high performance levels in general. For several of such rationales a trench based

concept was preferred which e.g. with respect to performance vs. robustness allows a positioning as indicated in Figure 6.

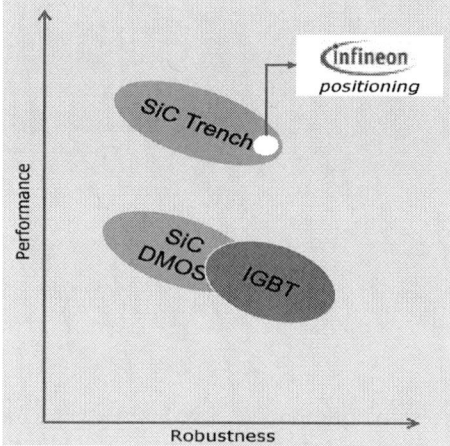

Fig.6 : Trade-off between performance and robustness for selected power switch technologies and Infineon's positioning

This is predominantly enabled by the higher channel conductivity in Trench based SiC MOSFETs compared to planar MOSFETs. Due to this difference it needs lower fields across the sensitive gate oxide to achieve attractive on performance and thus, effective screening for extrinsic failures becomes possible and low FIT rates can be assured [5].

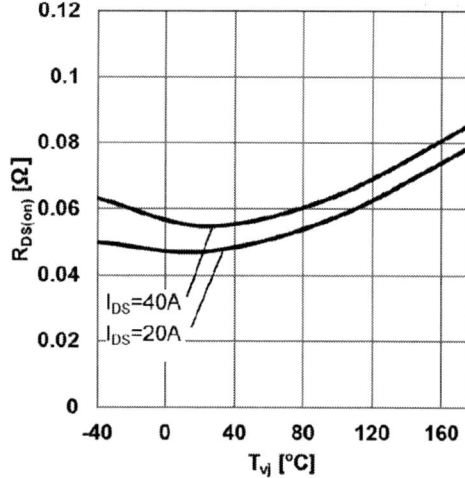

Fig.7 : Temperature dependence of a SiC MOSFET with a low channel defect density

The low defect density in the channel of Trench MOSFETs is also reflected in the temperature behavior of the on resistance. The physically expected dependence of a well designed MOSFET is a clear increase of the R_{on} with temperature since in case of unipolar devices the drift zone as the major contributor to R_{on} has a positive temperature coefficient above room temperature due to the reduction of carrier mobility with temperature (see Figure 7). However, occasionally SiC MOSFETs appear with very low R_{on} increase with temperature. This is possible only since a compensating element for the drift zone behavior is in place. A channel with a high defect density can act like this due to its clearly negative temperature coefficient when it comes to the impact on R_{on}. From an application point of view it looks initially like a favorable behavior, but since in such a case the resistance at lower temperatures is too high due to the large channel resistance. Thus, under partial load such a device will create more losses than a part with a positive temperature coefficient which has a lower resistance at temperatures below T_{j_max}. Furthermore, the small dependence on temperature easily turns into a negative slope if the applied gate source voltage is below the stated datasheet V_{GS_max} resp. V_{GS_use}. Thermal runaway and difficult paralleling will be the consequence. Thus, an on the first glance nice temperature independence for the on resistance in fact brings only disadvantages in the final application.

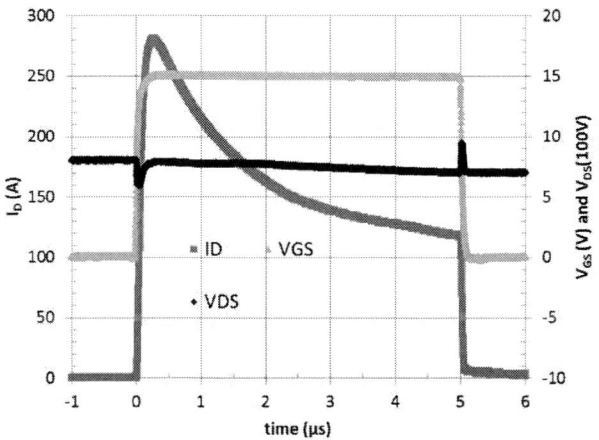

Fig. 8 : Typical short circuit waveforms for a SiC MOSFET

Another parameter relevant for certain applications is the short circuit robustness. Competing silicon based IGBT's do offer today up to 10 µs short circuit withstand times. MOSFETs, in particular the current SiC based high voltage parts have either no specified short circuit time or only a very short one what is sufficient for the initial target applications like solar or charger. Since e.g. motor drive applications require a certain robustness in short circuit mode this situation

might limit the applicability of the new technology for a wider scope of use cases. The reason for the still limited withstand time is again due to a strong focus on on-performance. A typical behavior is shown in Figure 8. As one can see a significant and steep increase of the current takes place which later will be limited due to internal heat generation caused by the tremendous power losses generated during this event. Reports indicate thermal limitations of the metallization as one possible root cause for failures under short circuit [6]. To counteract one has to limit the peak current what could be done by certain design measures, however, most of them leads to an increased R_{on}. Thus, as for other parameters as well a careful trade off decision is mandatory to optimize the device for a certain target application.

V. CONCLUSION

For the design of power devices intended to be used in a wide range of applications it is essential to have the option for adjusting certain parameters according to individual application needs. Thus, a base technology is needed offering those options without losing major performance aspects. For both, diodes and MOSFETs based on SiC performance is important, but beside on resistance a deeper look into further application relevant functions has to be done. Only an application specific well balanced technology will be able to penetrate mainstream applications successfully.

ACKNOWLEDGMENT

The author gratefully acknowledges the work of the whole SiC team at Infineon which contributed to the discussed results.

REFERENCES

[1] M. Draghici et al., "A New 1200V SiC MPS Diode with Improved Performance and Ruggedness", Materials Science Forum, Vols. 821-823, pp. 608-611, 2015
[2] R. Elpelt et al., "SiC MPS Devices: One Step Closer to the Ideal Diode", Materials Science Forum, Vol. 924, pp. 609-612, 2018
[3] https://www.infineon.com/dgdl/Infineon-ApplicationNote_Package_DoubleDPAK_DDPAK-AN-v01_00-EN.pdf?fileId=5546d462636cc8fb01638da8649d29c6
[4] S. T. Pantelides et al., "Si/SiO2 and SiC/SiO2 Interfaces for MOSFETs – Challenges and Advances", Materials Science Forum, Vols. 527-529, pp. 935-948, 2006
[5] D. Peters et al., " The New CoolSiC™ Trench MOSFET Technology for Low Gate Oxide Stress and High Performance", Proceedings of the PCIM 2017
[6] R. Green et al., "Short-Circuit Robustness of SiC Trench MOSFETs", Materials Science Forum, Vol. 924, pp. 715-718, 2018

978-1-7281-1988-5/18 $31.00 © 2018 IEEE

The current status and future prospects of SiC high voltage technology

A. Mihaila[1], L. Knoll[1], E. Bianda[1], M. Bellini[1], S. Wirths[1], G. Alfieri[1], L. Kranz[1], F. Canales[1], and M. Rahimo[2]

[1]ABB Switzerland Ltd, Corporate Research Centre, 5405, Baden-Dättwil, Switzerland, email: andrei.mihaila@ch.abb.com
[2]ABB Switzerland Ltd., Semiconductors CH-5600, Lenzburg, Switzerland

Abstract—This paper reviews the recent progress of SiC MOSFETs rated above 3.3kV. The static and dynamic performance of 3.3 and 6.5kV-rated MOSFETs will be evaluated and benchmarked against similarly rated state-of-the-art Si IGBTs. A numerical comparison between high voltage (15kV) SiC MOSFETs and IGBTs will also be provided. The paper will also attempt to comment on the future challenges facing high voltage (HV) devices in SiC technology.

I. INTRODUCTION

In recent years, Silicon Carbide (SiC) MOSFETs have provided a competitive alternative to the well-established Silicon (Si) IGBT technology in various low voltage applications. For the higher voltage ranges, in applications such as HVDC and FACTS, high voltage Drives, Traction Converters or high power Renewable Energy Conversion and Storage (see fig. 1), higher voltage MOSFETs (3.3kV to 6.5kV) have the potential to deliver similar improvements to those in the lower power ranges [1]. However, while HV SiC MOSFETs offer distinct benefits in terms of static and dynamic losses, their fault handling capability and long-term reliability are still somewhat short of the typical industry-required specifications. Moreover, the reliable operation of the body diode remains a question of concern.

This paper reviews the current status of 3.3 and 6.5kV-rated SiC MOSFETs. Both static and dynamic experimental results are presented. A numerical analysis of static losses for 15kV-rated SiC devices is also provided. The future prospects and challenges of high voltage devices in SiC technology are also briefly discussed.

II. HIGH VOLTAGE SiC MOSFETs

SiC MOSFETs have demonstrated that, in addition to drastically reduced switching losses, they can also offer low conduction losses for voltage ratings up to 10kV (fig. 2) [2]. In the lower voltage ranges (≤1.7kV), SiC MOSFETs specific on-resistance (R_{on-sp}) values are still quite far from the "ideal" unipolar limit, as here the inversion channel resistance is the dominant component of the static losses. For the ratings above 3.3kV, experimental R_{on-sp} values sit much closer to the SiC unipolar limit, indicating that in these voltage regimes the static losses are dominated by the doping and thickness of the drift layer.

A. Static behaviour

The quality and stability of the SiC MOS interface have been one of the major factors limiting the overall performance of SiC MOSFETs. A commonly used process to improve the MOS interface behavior involves the use of N_xO annealing, which reduces the interface defect density (D_{it}) [3]. Fig. 3 shows the forward/backward sweep and frequency-dependent CV characteristics of an optimized SiC MOS capacitor. Compared to thermally grown SiO_2 films and N_2O annealed stacks, our developed deposited oxide process exhibits lower D_{it} values below $10^{12}cm^{-2}eV^{-1}$ (Fig. 4).

The output characteristics of a 3.3kV MOSFET with two different pitches, 14 and 26um, respectively, are shown in fig. 5, for V_{GS}=15V and T_j=125°C [4]. For benchmarking reasons, the output IV of Si IGBTs and BIGTs are plotted on the same figure [5]. As can be seen, at the I_{NOM} of 60A/cm², both MOSFET pitches provide lower static losses compared to the bipolar Si devices. For this design, at 125°C, both pitches deliver similar R_{DS-on} values, as the JFET effect is stronger for p14. The MOSFET static behavior in the third quadrant is displayed in fig. 6. Reference curves for both Si PiN and BIGT diode are also shown. The body diode has a turn on threshold of -2.5V for negative V_{GS} values, but shows linear conduction for V_{GS}=15V. The high turn-on voltage of the MOSFET body diode makes it difficult to compete against the Si PiN in terms of static losses. The on-state drop seems to be independent of the pitch size, suggesting that the losses are controlled by the lifetime values in the epi layer.

Figure 7 shows the static characteristics of 6.5kV MOSFETs [6]. Several cell pitches are shown (p26, p14 and p12) and compared to a state of the art 6.5kV Si IGBT at T_j=125°C and V_{GS}=15V. The horizontal line indicates the rated current of 37.5A/cm² (I_{NOM}) of the IGBT. The SiC MOSFETs provide comparable (p26) or even higher (p14, p12) current densities compared to the Si IGBT. At the I_{NOM} values of the IGBT, the smaller pitch MOSFETs offer a substantial reduction in on state voltage (≈1V). The body diode characteristics, for different V_{GS} values, are shown in fig. 8. For reference, a Si PiN diode is also plotted. At the rated diode current of 75Acm⁻², the silicon diode provides the lowest voltage drop of 3.5V followed by p12 (3.8V), p14 (4.0V) and p26 (4.3V). A smaller cell pitch reduces the body diode on state by a 0.5V margin.

B. Dynamic behaviour

Surge current capability is an essential requirement in the specifications of the freewheeling diode. The surge event behavior of 4x 3.3kV MOSFETs connected in parallel is shown in fig. 9 (for V_{GS}=0V and V_{GS}=15V). A comparison with SiC JBS diodes (4x in parallel) and a Si diode is made on the same graph. As a pass criterion, each curve was measured twice and was not allowed to differ from the previous one. Compared to

978-1-7281-1988-5/18 $31.00 © 2018 IEEE

the Si diode (last pass not shown), the MOSFET body diode shows a positive temperature coefficient and has larger conduction losses. However, despite the inherent high built-in voltage and probably low plasma levels, the MOSFETs successfully demonstrate the required $10 \times I_{NOM}$.

The short-circuit waveforms, under nominal conditions of $V_D=1800V$ and $V_{GS}=15V$, for 3.3kV MOSFETs with small and large pitches are pictured in fig. 10. The strong trade-off between static losses (high density of MOS cells) and short circuit capability is immediately evident, as designs with lower conduction losses (p14) cannot meet the required 10us industry standard. Optimizing the MOS cell density and adjusting the V_{th} values could help improving the aforementioned trade-off.

The use of the MOSFET body diode also offers benefits in terms of increased power density in a power module, as more chips would fit inside the module. The turn-off characteristics of a 3.3kV MOSFET body diode is analyzed in fig. 11. Compared to a SiC JBS diode, the body diode has a similarly fast response and only shows a moderate increase in the reverse peak current. Nevertheless, the reliability of HV SiC MOSFETs body diodes remains a concern and special buffers layers designs are needed to improve the V_F stability [7, 8].

The possibility of chip paralleling is a key aspect for achieving the high current levels required in high voltage applications. The turn-off characteristics 4x 6.5kV MOSFETs connected in parallel are indicated in fig. 12 at 125°C (V_D =3600V, I=40A; the inset shows the MOSFETs bonded onto a test substrate). Different dV/dt levels have been tested by changing the turn-off gate resistor. Although larger oscillations are observed for the faster switching case, the MOSFETs successfully turn-off the required I_{NOM} level.

Figure 13 depicts the short circuit waveforms of 4x 6.5kV MOSFETs connected in parallel, for a pitches p12 design, at Tj=125°C for $V_D=3600V$, $V_{GS}=15V$. The MOSFETs show a short circuit capability of about 5µs, with the saturation current reaching levels in excess of 600A. Assuming equal current sharing between the four chips, each MOSFET takes more than 150A, which is a remarkable performance for a relatively small $5 \times 5 mm^2$ device.

The Reverse Bias Safe Operating Area (RBSOA) capability of the 6.5kV MOSFETs mentioned above is plotted in fig. 14. The MOSFETs successfully turn-off 80A, with a DC-link voltage of 4.4kV, with no failures observed during the testing. This level of performance is similar to the one provided by the latest state of the art 6.5kV Si IGBTs [9].

III. FUTURE PROSPECTS OF SiC HV TECHNOLOGY

SiC unique material properties enable the possibility of designing very high voltage devices (>10kV) with acceptable losses [2, 10]. The simulated output IV of 15kV-rated MOSFETs and IGBTs are shown in fig. 15. A comparison to the series connection of 5x 3.3kV and 3x 6.5kV MOSFETs is made. The parallel connection of 2x 15kV MOSFETs is also indicated on the same graph.

A simplified summary of the devices shown in fig. 15 is presented in Table 1. While it is obviously difficult to identify

a clear winner, the anticipated benefits of HV SiC devices in terms of reduced total losses as well as simplified topologies are evident [11]. However, challenges associated to the cost/Amp ratio, termination design, stable blocking performance, reliability and over-current capability still remain. Moreover, further advances in passivation materials and packaging technology would also be needed before HV SiC technology could be considered as a mature candidate in MW range applications.

IV. CONCLUSIONS

The performance of HV SiC MOSFETs has been reviewed with an emphasis on their dynamic and fault-condition operations. Using numerical results, an outlook into ultra HV SiC devices has also been presented. The benefits of ≥3.3kV SiC MOSFETs in terms of lower static and dynamic losses are substantial with a potentially strong impact on system performance. The road to market entry is, however, still paved with many challenges in terms of reliability, performance/cost ratio and, finally yet importantly, the continuous parallel developments made in Si technology.

REFERENCES

[1] M. Rahimo, "Performance Evaluation and Expected Challenges of Silicon Carbide Power MOSFETs and Diodes for High Voltage Applications", *Material Science Forum*, vol. 897, pp. 649-654, 2017.

[2] J. Palmour et al., "Silicon Carbide Power MOSFETs: Breakthrough Performance from 900 V up to 15 kV", *Proceedings of the 26th International Symposium on Power Semiconductor Devices & IC's*, pp. 79-82, June 2014.

[3] T. Kimoto et al., "Progress and Future Challenges of SiC Power Devices and Process Technology", *IEEE International Electron Devices Meeting (IEDM)*, pp. 227-230, December 2017.

[4] L. Knoll et al., "Robust 3.3kV Silicon Carbide MOSFETs with Surge and Short Circuit Capability", *Proceedings of the 29th International Symposium on Power Semiconductor Devices & IC's*, pp. 243-246, May 2017.

[5] M. Rahimo et al., "The Bi-mode Insulated Gate Transistor (BIGT) A Potential Technology for Higher Power Applications", *Proceedings of the 21th International Symposium on Power Semiconductor Devices & IC's*, pp. 283-286, June 2009.

[6] L. Knoll et al., "Dynamic Switching and Short Circuit Capability of 6.5kV Silicon Carbide MOSFETs", *Proceedings of the 30th International Symposium on Power Semiconductor Devices & IC's*, pp. 451-454, May 2018.

[7] Y. Fursin et al., "Reliability aspects of 1200V and 3300V silicon carbide MOSFETs", *IEEE 5th Workshop on Wide Bandgap Power Devices and Applications (WiPDA)*, pp. 373-377, May 2017.

[8] Y. Ebiike et al., "Reliability investigation with accelerated body diode current stress for 3.3 kV 4H-SiC MOSFETs with various buffer epilayer thickness", *Proceedings of the 30th International Symposium on Power Semiconductor Devices & IC's*, pp. 447-450, May 2018.

[9] C. Papadopoulos et al., "The third generation 6.5kV HiPak2 module rated 1000A and 150°C", *PCIM Europe 2018; International Exhibition and Conference for Power Electronics, Intelligent Motion, Renewable Energy and Energy Management*, pp. 273-280, June 2018.

[10] Y. Yonezawa et al., "Low Vf and Highly Reliable 16 kV Ultrahigh Voltage SiC Flip-Type n-channel implantation and epitaxial IGBT", *IEEE International Electron Devices Meeting (IEDM)*, pp. 164-167, December 2013.

[11] K. Vechalapu et al., "Comparative Evaluation of 15-kV SiC MOSFET and 15-kV SiC IGBT for Medium-Voltage Converter under the Same dv/dt Conditions," *IEEE Journal of Emerging and Selected Topics in Power Electronics*, Vol.: 5, Issue: 1, March 2017.

978-1-7281-1988-5/18 $31.00 © 2018 IEEE

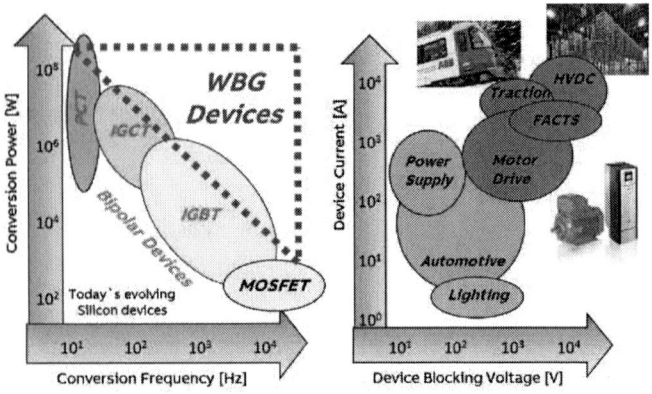

Fig. 1. SiC devices are targeting various applications in the high-power semiconductor market, which are currently served by a portfolio of silicon products.

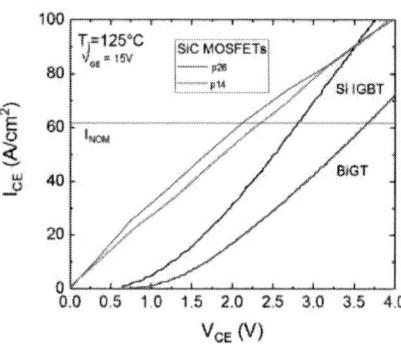

Fig. 2 Specific on-resistance of various Si and SiC devices at 25°C; the respective unipolar limits are also shown.

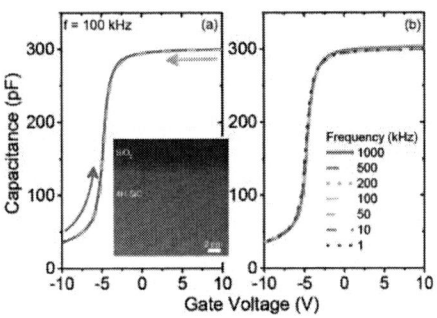

Fig. 3 Room-temperature C-V characteristics, The hysteresis and frequency dependent V_{FB}-shift are both negligible.

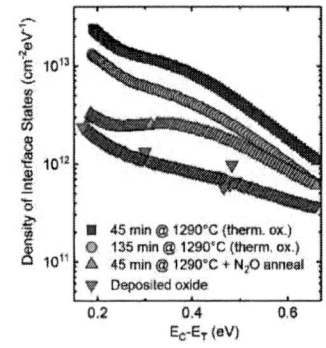

Fig. 4 D_{IT} distribution on MOS capacitors fabricated using different oxidation techniques.

Fig. 5 Output IV of a 3.3kV MOSFET at 125C; reference Si IGBT and BIGT are also shown.

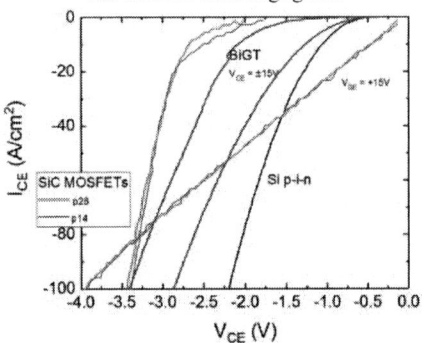

Fig. 6 Body diode characteritics of a 3.3kV MOSFET at 125C; reference Si PiN diode and BIGT diodes are also shown.

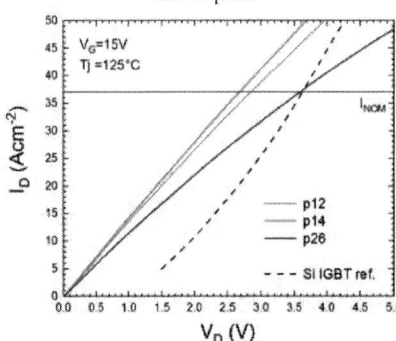

Fig. 7 Output IV of a 6.5kV MOSFET at 125C; a reference Si IGBT is also shown.

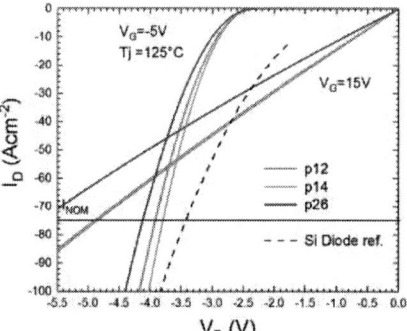

Fig. 8 Body diode characteristics of a 6.5kV MOSFET at 125C; a reference Si PiN diode is also shown.

978-1-7281-1988-5/18 $31.00 © 2018 IEEE   442

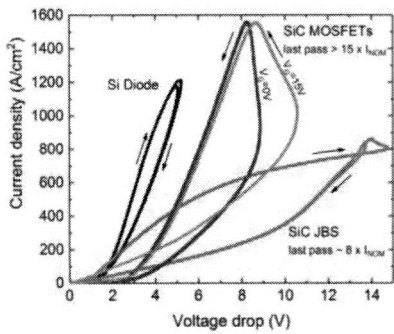

Fig. 9 Surge current density versus source - drain voltage drop of the body diode of 3.3kV SiC MOSFETs (4x in parallel), SiC JBS diodes (4x in parallel) and a Si diode.

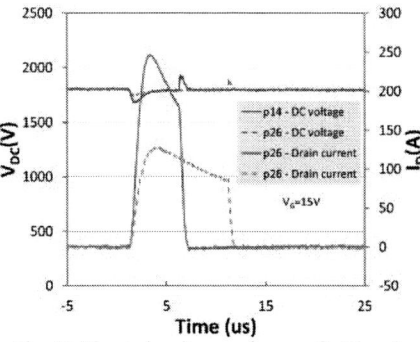

Fig. 10 Short circuit waveforms of p14 and p26 MOSFETs demonstrating 5µs and 10µs capability at 125°C, V_D=1800V and V_G=15V.

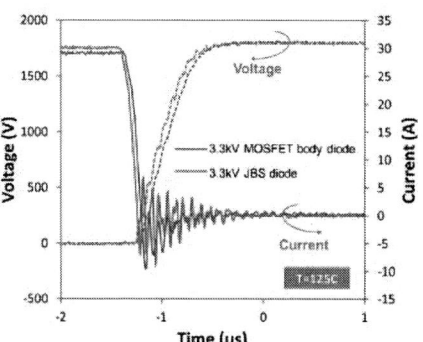

Fig. 11 Comparison between the turn-off curves of a 3.3kV MOSFET body diode and a JBS diode at 125°C; R_G= 33Ω; V_{DC}=1800V, I_D=30A.

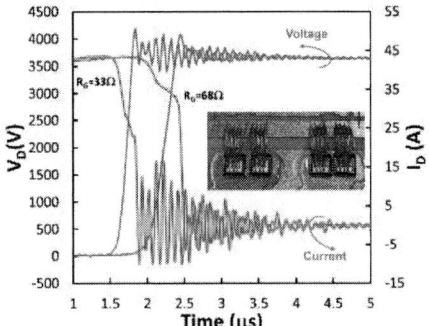

Fig. 12 Turn-off of 4x 6.5kV MOSFETs in parallel at 125°C, V_D =3600V, I_{NOM}=10A(per chip), R_G has been varied from 33Ω up to 68Ω.

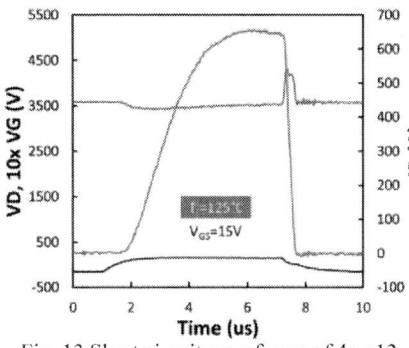

Fig. 13 Short circuit waveforms of 4x p12 MOSFETs in parallel at 125°C, V_D=3600V, V_{GS}=15V.

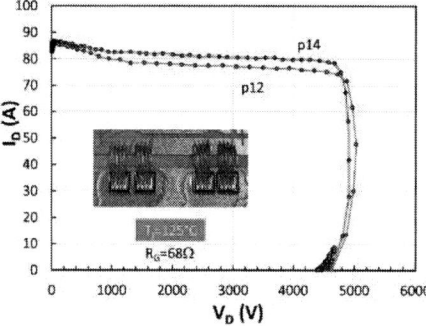

Fig. 14 Turn-off RBSOA for 4x 6.5kV MOSFETs in parallel with R_G=68Ω at 125°C; I_D=80A, V_D=4400V.

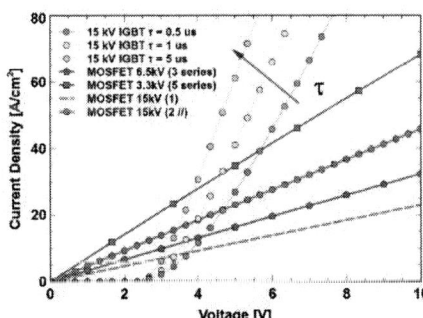

Fig. 15 Simulated output IV of 15kV-rated MOSFETs and IGBTs at 150°C; for comparison, series connections of 5x 3.3kV and 3x 6.5kV MOSFET are shown; the IV of 2x 15kV MOSFETs in parallel are also displayed.

Device type	Current costs	Material issues	Main challenges	Expected benefits
15kV SiC MOSFET	High	Quality of thick epi layer (≥100µm)	Static losses Large area devices-> yield issues Reliable body diode operation Packaging High dV/dt	Low switching losses (higher frequency operation) Body diode integrated Opportunity for simpler topologies
15kV SiC IGBT	Very high	Quality of thick epi layer (≥100µm) Lack of p+ substrates Low lifetime	Fabrication processes No body diode Negative T coefficient Packaging High dV/dt	High current ratings Enabler for ≥20kV applications Opportunity for simpler topologies
Series connection of lower voltage MOSFETs (e.g. 3.3-6.5kV)	Moderate	Quality of moderately thick epi layer (30 -60µm)	Voltage sharing Control Added complexity Moderate dV/dt	Lower cost Competitive static losses Improved redundancy Body diode integrated
Parallel connection of 15kV MOSFETs	High	Quality of thick epi layer (≥100µm)	Current sharing Reliable body diode operation Packaging High dV/dt	Lower static losses Low switching losses Body diode integrated Opportunity for simples topologies

Table 1. Summary of expected challenges and projected benefits of high voltage (15kV-rated) SiC devices

978-1-7281-1988-5/18 $31.00 © 2018 IEEE

Progress in High and Ultrahigh Voltage Silicon Carbide Device Technology

Y. Yonezawa[1], K. Nakayama[1], R. Kosugi[1], S. Harada[1], K. Koseki[1], K. Sakamoto[1], T. Kimoto[2], H. Okumura[1]

[1]Advanced Power Electronics Research Center, National Institute of Advanced Industrial Science and Technology (AIST),
Tsukuba, Ibaraki, Japan, email: yoshiyuki-yonezawa@aist.go.jp
[2]Department of Electronic Science and Engineering, Kyoto University, Kyoto, Japan

Abstract— The current developments in silicon carbide (SiC) device technology in various voltage ranges are introduced. These developments correspond to, in particular, next-generation high to ultrahigh-voltage devices, SiC super-junction metal oxide semiconductor field effect transistors, SiC insulated gate bipolar transistors, and the fundamental bipolar degradation suppression technology. We expect that these next generation devices will trigger a paradigm shift in power electronics.

I. INTRODUCTION

Global warming is an urgent matter, and measures must be adopted for early realization of a low-carbon society. With the large-scale introduction of renewable energy and energy storage, the sophistication of energy management technologies is required in a smart linkage inside and outside the smart grid. Consequently, the role of power electronics becomes more significant in the energy value chain, with fusion of energy and information. Power electronics and power devices are the two sides of evolution in energy distribution and consumption. Thus far, power electronics have been supported by the evolution of Si power devices, especially insulated gate bipolar transistors (IGBTs). However, the performance improvement of Si-IGBTs has reached the physical limit and expectations for wide-bandgap semiconductor devices are increasing. This paper highlights recent progress in SiC device technologies.

II. SI-IGBTS AND SIC MOSFETS

The bandgap of SiC is three times larger than that of Si, which results in a 10 times higher breakdown electrical field. Thus, SiC devices can achieve 10 times higher breakdown voltages (BVs) than Si devices with the same device structure, as shown in Fig. 1. Replacing bipolar Si-IGBTs with unipolar 600 V–3.3 kV SiC-MOSFETs will lead to a significant reduction in size and cost of power electronics (PE) components with lowering conduction and switching loss [1,2].

Another advantage of using SiC MOSFETs for Si IGBTs in the respective circuit is the possibility of omitting the external freewheeling diode by using the internal body diode. However, since it is composed of a pn junction, measures against the forward degradation induced by bipolar operation is required [3]. In particular, when surge current enters in the body diode, it is possible that the basal plane dislocation (BPD) in the substrate expands and form single Shockley stacking faults (Fig. 2a). As a countermeasure, we proposed the use of a recombination enhancement (RE) layer, which suppresses the number of holes reaching the substrate (Fig. 2b), and can successfully prevent forward degradation up to 600 A/cm^2 [4]. Another approach for preventing the forward degradation is the use of the built-in Schottky barrier diode (SBD). An example of this idea is the SWITCHMOS (SBD-wall integrated trench MOSFET), shown in Fig. 3. Built-in SBD suppresses the turn on of the body diode up to 2800 A/cm^2 [5].

III. NEXT GENERATION HIGH VOLTAGE UNIPOLAR DEVICE: SIC SUPER JUNCTION MOSFET DEVICE TECHNOLOGY

For high voltage unipolar devices, 6.5 kV SiC-MOSFETs have been reported [6]. A 13 kV MOSFET was also developed, expecting a dramatic size reduction for ultrahigh voltage power supplies with small current [7]. With regard to next-generation high voltage unipolar devices, we are developing a super junction MOSFET (SJ-MOS), which can allow dramatic reduction of the drift layer resistance. We are aiming at the 6.5 kV class SJ-MOSFET, which is expected to be applied to the traction systems of next generation high-speed trains (Fig. 4). We have developed the fabrication method of SJ structures by a multi epitaxial (ME) growth method with high energy implantations [8]. I-V characteristics of the SJ resistivity test TEG and BV curve of pn-junction TEGs are shown in Fig. 5 a) and b); these were constructed using ME growth steps of 3 and 4 times [8]. The resistivities of the drift layer were calculated as 0.50 and 0.63 mΩcm^2, which are 36 and 29 % of SiC limit at the same BV-class, respectively. To obtain a higher voltage SJ structure we must consider the use of the trench filling regrowth method. Fig. 6 a) shows the cross-sectional SEM image of 10 μm p-type regrowth toward n-drift layer trenches. The acceptor profile of a p-type regrowth region was estimated by a scanning non-linear dielectric microscope (SNDM) as shown in Figs. 6 b), c). A large distribution of concentration, which occurred owing to the difference in incorporation ratios in different axes, was observed. By optimizing the growth condition and controlling the trench direction toward step-flow direction, we successfully filled deep-trenches with no voids and high growth rates [9]. Fig. 7 shows a cross-sectional SEM image of 6.5 kV partial SJ structure wafer. The total resistance of the drift layer was estimated to be 13.3 mΩcm^2, which corresponds to less than half of the unipolar limit of SiC [9].

IV. SIC BIPOLAR DEVICE: PIN DIODE AND IGBTS

As mentioned before, if we apply the IGBT structure to SiC, MOS-controlled switching devices of more than 10 kV can be

978-1-7281-1988-5/18 $31.00 © 2018 IEEE

realized, which is difficult to attain for Si devices [12-16]. The specific-on-resistance target of SiC-IGBT, which is less than half of the series of Si-IGBTs, is also shown in Fig. 4. Fig. 8 shows the cross-sectional image of a 20-kV class ultrahigh voltage PiN diode with space-modulated JTE [10]. Fig. 9 shows the effect of space-modulated JTE on the BV compared with conventional JTE simulated with TCAD [11]. A BV of 29.6 kV was achieved with the fabricated PiN diode (Fig. 10).

Fig. 11 shows the schematic process flow of flip-type SiC implantation and epitaxial IGBTs (IE-IGBTs) [14]. Because a high-quality p-type substrate is not available presently, thick epitaxial p^{++} layer was used as a substrate. The pulsed on-state I–V characteristics of the fabricated 8×8 mm IE-IGBT with BV of 16.5 kV are shown in Fig. 12. An on-state current (I_{CE}) of 20 A and 60 A was obtained at the low V_{on} of 4.8 V and 7.2 V, respectively [15]. Field-Effect mobility was 75cm^2/Vs at the peak and 60 cm^2/Vs at $V_G = 20$V with carbon face. Turn off switching waveforms of the IE-IGBT with the 300 Ω gate resistance in the temperature range from RT to 250°C are shown in Fig. 13 a). Smooth turn-off waveforms were successfully obtained at $V_{CE} = 6.5$ kV and $I_{CE} = 60$ A. The turn-off current fall time increases as the temperature is increased (Fig. 13b). Fig 14 shows the turn-off characteristics using four 16 kV, 5.3 mm $\times$ 5.3 mm SiC IE-IGBTs arranged in parallel in one module. A very high-voltage switching of 10 kV, 100A (1 MVA) was achieved at temperatures from 20 to 200 °C [3].

Currently, we are working on developing 20 kV class IGBTs. Fig. 15 shows the on-state characteristics of the 20 kV IE-IGBT. The differential specific on resistance at room temperature was 67 mΩcm^2 and it increased to 107 mΩcm^2 at 200 °C. The on-resistance should be further improved, and one obvious way is to utilize the lifetime-enhancement process for the voltage-blocking region. To obtain sufficient conductivity modulation, a Shockley–Read–Hall lifetime of approximately 30 µs is required [10], considering a 13–33 kV bipolar device with a drift layer thickness between 150–300 µm. Moreover, improving the tradeoff between conduction losses and switching losses is a challenge in IGBTs. We must work on not only the lifetime enhancement but also device design optimizations, such as injection control of the collector side and injection enhancement, which is similar to the history of Si, in order to obtain satisfactory device performance. Another approach to reduce SW loss is as follows. By separating the current path for the gate driving circuit from the main circuit (Fig. 16), stable operation could be possible [17]. As a result, low gate resistance of 10 and 1 Ω could be utilized, resulting in extremely fast switching of over 300 kV/µs with low switching loss (Figs. 17 and 18).

V. APPLICATION OF HIGH AND ULTRAHIGH VOLTAGE SiC DEVICES

In Japan, society 5.0 has been proposed in the 5th Science and Technology Basic Plan, "Optimization of energy value chain based on IoT". For this purpose, we must actively control the electrical power flow by energy management system (Fig. 19). In an intelligent electricity network, several PE components, such as a solid state transformer (SST), intelligent power switch used in smart grids [18], static var compensator (SVC), and loop balance controller (LBC) must be made more compact,

low cost, and intelligent. To realize this, we believe that high voltage SJ-MOS, ultrahigh voltage SiC-PiN, and IGBT will be key devices. Utilizing these SiC devices promotes the downsizing of the PE components and reduction in the total cost including operation cost. Reduction in the stages of modular multilevel converters (MMC) in HVDC is also expected. We hope that SiC power devices will contribute further to the super smart society that adopts efficient energy usage, including energy storage, rather than the expansion of energy creation.

ACKNOWLEDGMENT

Part of this work was supported by Council for Science, Technology and Innovation (CSTI), Cross-ministerial Strategic Innovation Promotion Program (SIP), "Next-generation power electronics" (funding agency: NEDO).

REFERENCES

[1] T. Kimoto and J.A. Cooper, "Fundamentals of silicon carbide technology: Growth, Characterization, Devices and Applications" John Wiley & Sons, Singapore, 2014.

[2] D. Peters, et al., "Performance and Ruggedness of 1200V SiC–Trench–MOSFET", *Proceedings of ISPSD*, pp. 239–242, 2017.

[3] T. Kimoto and Y. Yonezawa, "Current status and perspectives of ultrahigh-voltage SiC power devices", *Materials Science in Semiconductor Processing*, 78, pp. 43–56, 2018.

[4] T. Tawara, et al., "Short minority carrier lifetimes in highly nitrogen-doped 4H-SiC epilayers for suppression of the stacking fault formation in PiN diodes", *J. Appl. Phys.* vol. 120, pp. 115101–1, 2016.

[5] Y. Kobayashi, et al., "Body PiN diode inactivation with low on-resistance achieved by a 1.2 kV-class 4H-SiC SWITCH-MOS," *IEEE International Electron Devices Meeting (IEDM)*, pp. 9-1, 2017.

[6] K. Kawahara, et al., "6.5 kV Schottky-barrier-diode-embedded SiC-MOSFET for compact full-unipolar module", *Proceedings of the 29th ISPSD*, pp. 41–44, 2017.

[7] H. Kitai, et al., "Low on-resistance and fast switching of 13-kV SiC MOSFETs with optimized junction field-effect transistor region", *Proceedings of the 29th ISPSD*, 2017, pp. 343–346, 2017.

[8] R. Kosugi, et al., "Current status of SiC super-junctions (SJ) device technology", The 4th Advanced Power Semiconductors, vol. 4, no. 1, pp. 27-28, 2017 (in Japanese) and references therein.

[9] R. Kosugi, et al., "Strong impact of slight trench direction misalignment from on deep trench filling epitaxy for SiC super-junction devices", Jpn. J. Appl. Phys., 56.4S (2017): 04CR05.

[10] N. Kaji, et al, "Ultrahigh-voltage SiC pin diodes with improved forward characteristics", *IEEE Trans. Electron Devices*. vol. 62, no. 2, pp. 374–381, 2015

[11] K. Nakayama, et al., "27.5 kV 4H-SiC PiN diode with space-modulated JTE and carrier injection control.", *Proceedings of the 30th ISPSD*, pp. 395–398, 2018.

[12] X. Wang and J. A. Cooper, "High-Voltage n-Channel IGBTs on Free-Standing 4H-SiC Epi layers", *IEEE Trans. Electron Devices*, vol. 57, no. 2, pp. 511–515, 2010.

[13] S. H. Ryu, et al., "Development of 15 kV 4H-SiC IGBTs", *Mater. Sci. Forum*. Vol. 717, pp. 1135-1138, 2012.

[14] Y. Yonezawa, et al., "Low V$_f$ and highly reliable 16 kV ultrahigh voltage SiC flip-type n-channel implantation and epitaxial IGBT", *IEEE International Electron Devices Meeting (IEDM)*, pp. 6.6.1–6.6.4, 2013.

[15] Y. Yonezawa, et al., "Device Performance and Switching Characteristics of 16 kV Ultrahigh-Voltage SiC Flip-Type n-channel IE-IGBTs", *Mater. Sci. Forum*, vol. 821-823, pp. 842–846, 2015.

[16] E. van Brunt et al., "27 kV, 20 A 4H-SiC IGBTs", *Mater. Sci. Forum*, vol. 821, pp. 847–850, 2015.

[17] K. Koseki, et al. "Dynamic Behavior of a Medium-Voltage N-Channel SiC-IGBT With Ultrafast Switching Performance of 300 kV/µs.", IEEE Transactions on Industry Applications, vol. 54.4, 2018.

[18] K. Mainali, et al., "A Transformer less Intelligent Power Substation: A three-phase SST enabled by a 15-kV SiC IGBT", *IEEE Power Electron. Mag*, vol. 2, no. 3, pp. 31–43, 2015.

978-1-7281-1988-5/18 $31.00 © 2018 IEEE

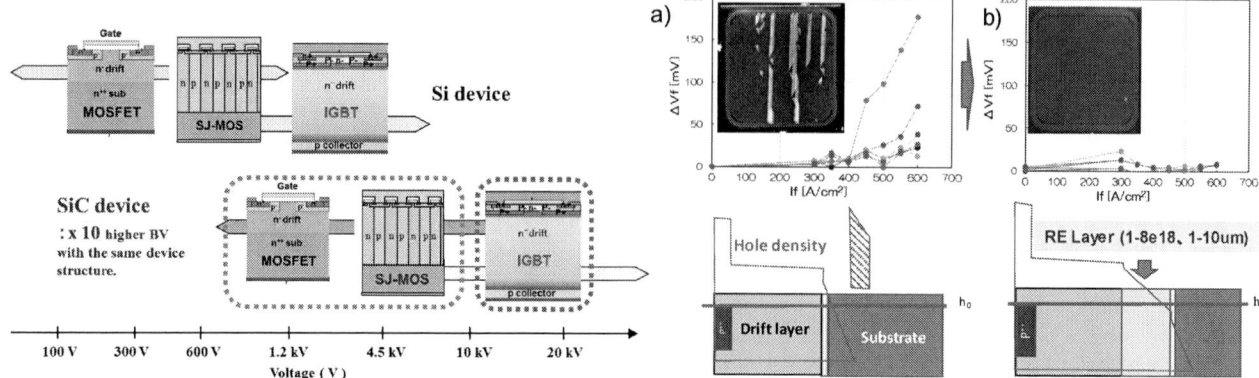

Fig. 1. Comparison of advanced Si and SiC power devices.

Fig. 2. a) BPD in the substrate causes stacking fault expansion resulting in forward degradation. b) Suppression of the number of holes to reach to the substrate by recombination enhancement (RE) layer

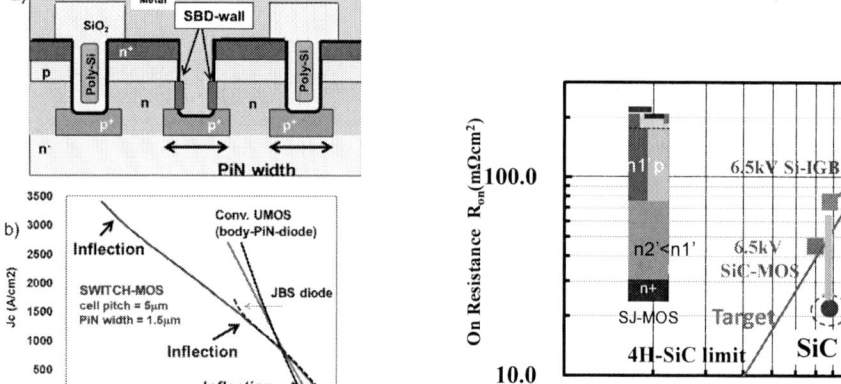

Fig. 3. a) Schematic cross-sectional illustration of the SWITCH-MOS and b) Forward characteristics of the body diode in the SWITCH-MOS.

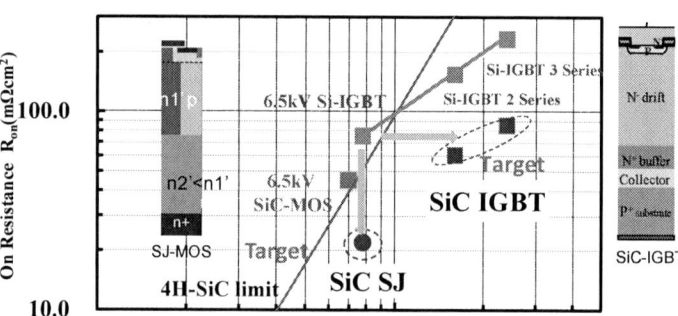

Fig. 4. Target of the next generation SiC SJ-MOS and IGBTs.

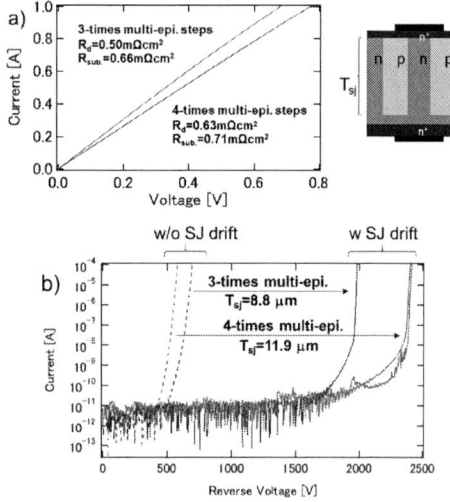

Fig. 5. a) IV characteristics of resistivity-TEGs on SJ drift-layer by multi-epi. b) BV curve of SJ-PN-TEG compared with normal drift layer.

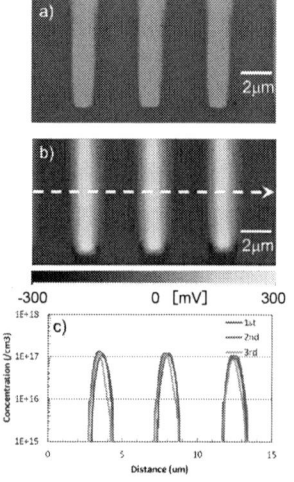

Fig. 6 a) Cross-sectional SEM image of 10 um p-type regrowth toward n-drift layer trenches. b) SNDM image of p-type regrowth. c) Estimated concentration profile of p-type regrowth.

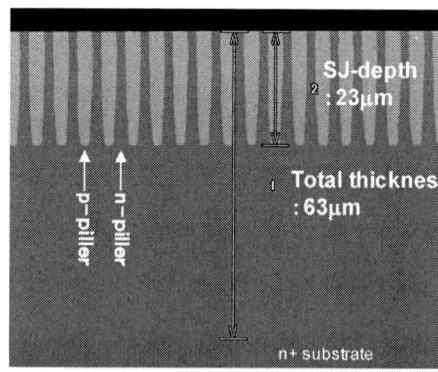

Fig. 7 Cross-sectional SEM image of 6.5 kV partial SJ structure wafer with 23 μm SJ drift structure and 40 μm n-drift layer

978-1-7281-1988-5/18 $31.00 © 2018 IEEE

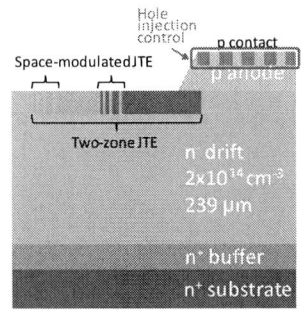

Fig. 8. Cross sectional image of fabricated 20 kV class PiN diode with space modulated JTE.

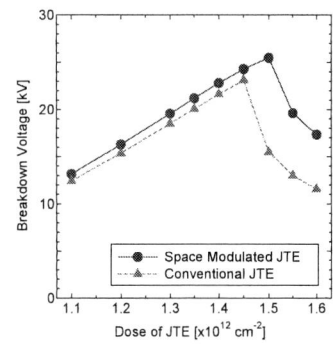

Fig. 9. Effect of space modulated JTE simulated with TCAD.

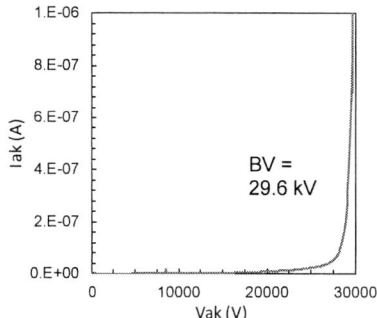

Fig. 10. Blocking voltage characteristic of the fabricated 5.3 mm x 3.5 mm ultra-high voltage PiN diode.

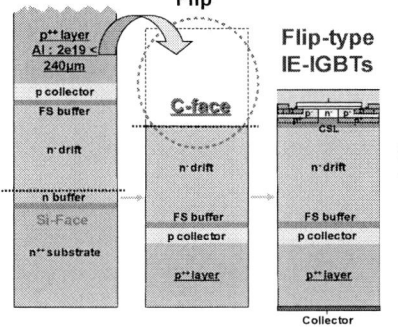

Fig. 11. Process flow of Flip-type IE-IGBT.

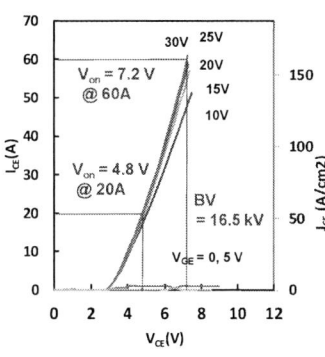

Fig. 12. On state characteristics of the 8 mm × 8 mm 16 kV IE-IGBT.

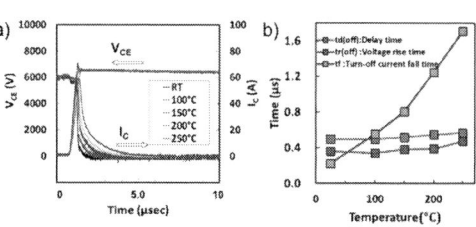

Fig. 13. a) Turn-off switching waveforms and b) Turn-off current fall time of the 8 mm × 8 mm IE-IGBTs.

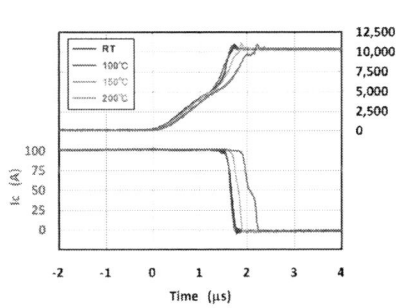

Fig. 14. Demonstration of 1 MVA turn-off switching wave form of four IGBTs parallel in one module.

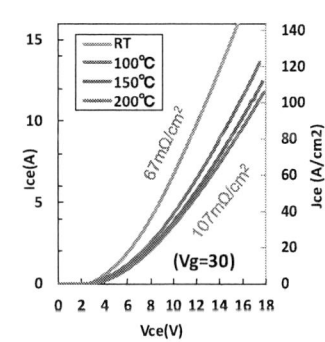

Fig. 15. On-state characteristics of the 20-kV class IE-IGBTs.

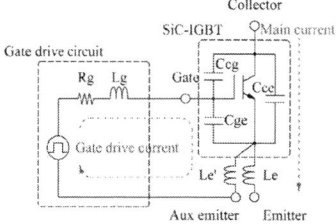

Fig. 16. Equivalent circuit of the power module and the gate drive circuit with the gate-loop separation.

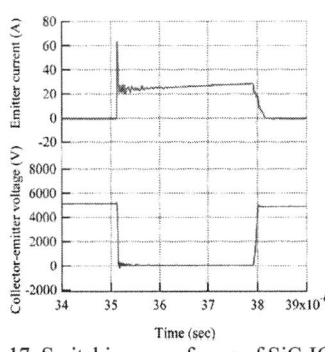

Fig. 17. Switching waveforms of SiC-IGBT in a high-power operation. Gate resistances for turn-on and turn-off were 10 Ω and 1.1 Ω.

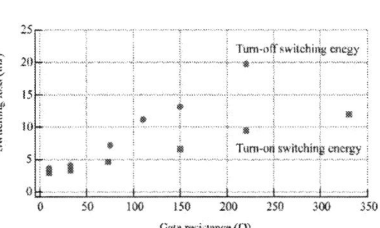

Fig. 18. Measured switching loss for various gate resistances.

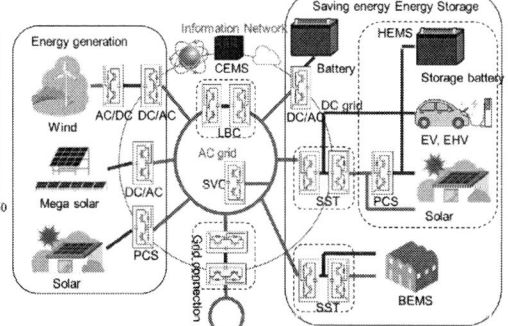

Fig. 19. Intelligent electricity network with power electronics components based on high and ultrahigh SiC power device.

978-1-7281-1988-5/18 $31.00 © 2018 IEEE

Effects of Basal Plane Dislocations on SiC Power Device Reliability

R. E. Stahlbush[1], K. N. A. Mahakik[1], A. J. Lelis[2] and R. Green[2]

[1]U.S. Naval Research Laboratory, 4555 Overlook Ave., SW, Washington, DC, USA, email: stahlbush@nrl.navy.mil

[2]U.S. Army Research Laboratory, 2800 Powder Mill Rd., Adelphi, MD, USA

Abstract—As silicon carbide power devices enter the commercial power electronics market there is a strong interest in all aspects of their reliability. This work discusses the degradation of MOSFETs due to basal plane dislocations (BPDs). During the forward bias of the MOSFET body diode, electron-hole recombination causes BPDs to fault and the resulting stacking faults in the drift layer degrade the MOSFET. As the stacking faults grow, the on-state conductivity of the MOSFET drift layer decreases, the off-state leakage of the drift layer increases, and the forward voltage of the body diode increases. Commercial 1200 V MOSFETs were stressed with a body current of 5 A or 10 A. The first generation of commercial MOSFETs showed significant degradation within minutes of stress time, whereas more recent MOSFETs did not show degradation for over 5 hours of stress time.

I. INTRODUCTION

The potential advantages of using silicon carbide to fabricate power electronics switches were recognized several decades ago. SiC devices that are now commercially available include MOSFETs and JBS diodes, and due to their superior capabilities, they have established a small and growing fraction of the power electronics market. Compared to silicon, SiC has a 3X larger bandgap, which results a 10X larger electric breakdown field and a lower intrinsic carrier density. The higher breakdown field decreases the required thickness of the device drift region and makes higher switching frequencies possible. The thermal conductivity of SiC is 3X higher than silicon's value so it more effectively conducts heat away from the devices active area. The development of SiC power devices also have the advantage that a majority of the processing steps developed for Si can be used with SiC. The oxidation of SiC forms a SiC/SiO_2 interface and similar implantation processes can be used to dope SiC either n-type of p-type. These characteristics make it possible to use similar equipment and techniques to fabricate familiar power devices such as MOSFETs, IGBTs, JBS diodes and PiN diodes with performance characteristics that are superior to their Si-based counterparts.

II. BPDS AND RELIABILITY

The effects of BPDs are more complicated than other materials extended defects in SiC. Other extended defects such as the various types of in-grown stacking faults can introduce a leakage path through the drift layer, but the

This work was supported by the Office of Naval Research.

leakage does not change due to device operation [1]. The adverse effects of BPDs were first reported by researchers working with ABB to develop 4.5 kV SiC PiN diodes [2]. ABB found that the forward voltage drop of SiC PiNs increased during forward voltage operation and identified the faulting of BPDs in the epitaxial drift layer as the source of the degradation, and they concluded that the stacking faults decreased the carrier lifetime. During the on state, the BPDs divide into two partial dislocations and form a stacking fault between them. These stacking faults are quantum wells trapping electrons and serve as a recombination site for electrons and holes that degrade the local lifetime, which reduces the carrier flow [3]. Later work showed that the stacking faults from BPDs are also potential barriers that impede current flow in majority carrier devices such as MOSFETs [4], [5]. The stacking faults originating from the BPDs also affect the leakage of SiC devices including MOSFETs by increasing the leakage in the off-state [5], [6].

The essence of the distinction between the effects of BPDs and other extended defects is that before faulting, the BPDs have a negligible effect on device performance. The degradation to both bipolar and unipolar devices is due to the stacking faults that develop from the BPDs. Degradation from other extended defects is primarily a yield challenge. Their effects are much more quickly and easily observed by screening of the dies on a wafer. In contrast, BPDs pose more of a reliability challenge. The development of stacking faults in a MOSFET or other device depends on the details of the switching cycle being used so the BPD degradation requires reliability testing.

With the maturing of SiC bulk (substrate) and epitaxial growth techniques, the concentrations of BPDs originating in the substrate and entering the epitaxial drift layer has been decreased several orders of magnitude from the original observations by ABB. BPD density in substrates has dropped two orders of magnitude over the last decade and is now typically 300 BPD/cm2 [7]. Another improvement was the change from 8° to 4° of the offcut angle of wafers to increase the number of substrates that could be cut from a boule [8].

On 8° substrates about 10% of the BPDs in the substrate propagate into the epitaxial layer and continue through the whole thickness of epitaxial layer. In contrast, on 4° substrates, a similar fraction enter the epitaxial layer but they convert into benign threading edge dislocations (TEDs) within the first few microns of epitaxial growth. However, a BPD causes degradation, even if only a short part of it is in the epitaxial drift layer, because it forms a stacking fault that expands to the top of the drift layer. Epitaxial growths for power devices normally start with a buffer doped nearly as high as the substrate to keep the BPD out of the drift layer and to keep e-h recombination away from the BPD. By optimizing the epitaxial growth, it is possible to reduce the BPD concentration in the drift layer to 0.1 BPD/cm^2 [9]. . The geometry of a BPD entering the epitaxial layer is shown in Figs. 1 and 2. Figure 1 is an ultraviolet photo luminescence (UVPL) plan-view image of several BPDs. Figure 2 is a 3D schematic showing the path of BPDs within the epitaxial layer. The two schematics compare a BPD that fully traverses the epitaxial layer with a BPD that converts into a TED during the epitaxial growth.

Contrary to earlier assumptions, it has been shown that stacking faults entering the drift layer can originate at BPDs in the substrate or epitaxial buffer layer. Using UV illumination corresponding to ~1000 A/cm^2, BPDs from a nitrogen-doped buffer layer, $3x10^{18}$/cm^3, were observed to slowly fault until they reached the drift layer, doped $1x10^{15}$/cm^3, and then rapidly expanded to the surface of the drift layer [10]. The buffer layer had a minority carrier lifetime of 100 ns. Recent work has also shown that a BPD that converts into TED at the substrate/epi interface can create stacking faults in the epitaxial layer [11], [12]. One difference is that many of the stacking faults originating from BPDs in at the substrate/epi interface expand to cover a larger area of a device so each BPD leads to more device degradation.

The control of BPDs is complicated by the fact that BPDs can be created and introduced into the device drift layer in multiple ways. Inclusions resulting from downfalls during epitaxial growth are a combination of misoriented 4H and 3C polytypes [13]. They introduce a local stress field that produces a cluster of BPDs surrounding the inclusion. An example is shown in Fig. 3.

Another important source of BPDs that has recently been observed is BPDs introduced during wafer processing [14]. These BPDs are formed by the combination of Al implantation and the activation anneal used to form p+ contacts in MOSFETs or other devices. The BPDs originate from the implanted region and glide through the epitaxial layer during the annealing as shown in Fig. 4. Their formation is enhanced at corners formed by RIE etching through the implanted area, but oxidation removes surface damage at the corners and suppresses BPD creation [15]. Figures 4 and 5 illustrate the formation of BPDs during the implantation/anneal processes. They show BPDs introduced into areas originally free of BPDs. If the Al implantation is sufficiently high, such as for p+ contacts, BPDs can be introduced. The BPDs are thought to be created around small defect areas in the implanted layer and expand by gliding through the epitaxial layer during the annealing.

Examples of BPD degradation on early commercial 1200 V MOSFETs (gen 1) and the improvement in more recent 1200 V MOSFETs (gen 2) are shown in Figs.6-11. Three characteristics of the MOSFETs are shown: body diode I_{DS}-V_{DS} curves, MOSFET on-state I_{DS}-V_{DS} curves, and MOSFET off-state leakage. The MOSFETs were stressed by turning on the body diode as shown in the figures to cause BPDs to fault and degrade the MOSFETs. The case temperature during stressing was kept below 100°C. Gen 1 MOSFETs have significant degradation, while no degradation is observed in gen 2 MOSFETs.

III. CONCLUSION

While SiC power devices have significant advantages over their Si based counterparts, the issue of extended defects, especially BPDs, must be carefully managed to produce reliable SiC devices. Examination of the characteristics of BPDs has resulted in improved understanding of their development and their effects on device performance. The insights gained through the study of the properties and behavior of BPDs have enabled decreasing the concentrations of BPDs, preventing the faulting of BPDs within the device drift layer and thereby limiting degradation to SiC device reliability.

REFERENCES

[1] R. E. Stahlbush, and N. A. Mahadik, "Defects affecting SiC device reliability," *Proc IEEE 2018 International Reliability Physics Symposium,* p. 2B.4-1, 2018.
[2] H. Lendenmann, F. Dahlquist, J. P. Bergman, H. Bleichner, and C. Halin, "High-power SiC diodes: characteristics, reliability and relation to material defects," *Mater. Sci. Forum,* vol. 389-393, pp. 1259, 2002..
[3] R. E. Stahlbush, M. Fatemi, J. B. Fedison, S. D. Arthur, L. B. Rowland, and S. Wang, "Stacking-fault formation and propagation in 4H-SiC PiN diodes," *J. Elec. Mater. ,* vol. 31, pp. 370–375, May 2002.
[4] M. Skowronski and S. Ha, "Degradation of hexagonal silicon-carbide-based bipolar devices," *J. Appl. Phys.,* vol. 99, p. 011 101, . 2006.
[5] A. Agarwal, "A case for high temperature, high voltage SiC bipolar devices," *Mater. Sci Forum,* vols. 556-557, pp. 687–692. , July 2007.
[6] R. E. Stahlbush, Q. J. Zhang and A. Agarwal, "Effects of stacking faults from half loop arrays on electrical behavior of 10 kV 4H-SiC PiN diodes," *Mater. Sci. Forum,* vols. 717-720, p.387, 2012.
[7] A. Bhalla, "Recent developments accelerating CiC adoption," *Mater. Sci. Forum,* vol. 924, pp.793-798, 2018.
[8] W. Chen, M. A. Capano, "Growth and characterization of 4H-SiC epilayers on substrates with different off-cut angles", *J. Appl. Phys.,* vol. 98, no. 114907, 2005.
[9] T. Kimoto, A. Iijima, H. Tsuchida, T. Miyazawa, et al., "Understanding and reduction of degradation phenomena in SiC power devices," *Proc. 2017 IEEE Reliability Physics Symposium,* p. 2A-1.1, 2017
[10] N. A. Mahadik, R. E. Stahlbush,M. G. Ancona and E. A. Imhoff, "Observation of stacking faults from basal plane dislocations in highly doped 4H-SiC epilayers," *Appl. Phys. Lett.,*vol.100, .042102, 2012.
[11] K. Konishi, S. Yamamoto, S. Nakata, Y. Nakamura, Y. Nakanishi, et al., "Stacking fault expansion from basal plane dislocations converted into threading edge," *J. Appl. Phys.,* vol. 114, 014504, July 2013.
[12] A. Tanaka, H. Matsuhata, N. Kawabata, D. Mori, K. Inoue, et al., "Growth of Shockley type stacking faults upon forward degradation in 4H-SiC p-i-n diodes," *J. Appl. Phys.* vol. 119, 095711, March 2016.
[13] N. A. Mahadik, R. E. Stahlbush, S. B. Qadri, O. J. Glembocki, D. A. Alexson, "Structure and Morphology of Inclusions in 4° Offcut 4H-SiC Epitaxial Layers,"*J. Elec. Mater.,* vol. 40, pp. 413–418, April 2011
[14] R. E. Stahlbush, N. A. Mahadik,. J. Zhang, A. A. Burk, B. A. Hull, J. Young, "Basal plane dislocations created in 4H-SiC epitaxy by implantation and activation anneal," *Mater. Sci. Forum,* vols. 821-823, pp 387-390, June 2015.
[15] Y. Bu, H. Yoshimoto, N. Watanabe, and A. Shima, "Fabrication of 4H-SiC PiN diodes without bipolar degradation by improved device processes," *J. Appl. Phys.,* vol. 122, 244504, Dec. 2017.

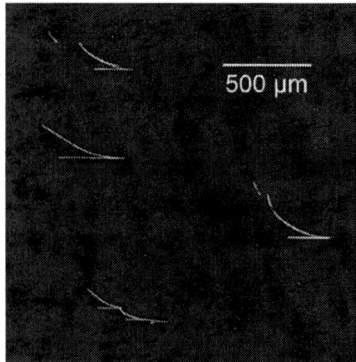

Fig. 1. UVPL image of five individual BPDs originating from the substrate. Shorter BPDs convert to TEDs during epi growth.

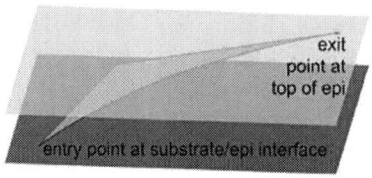

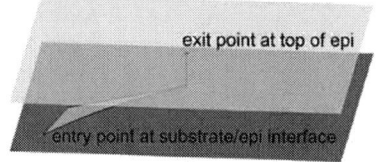

Fig. 2. Schematics of BPDs originating from the substrate: top BPD spans total epi layer, bottom BPD converting to a TED during epi growth.

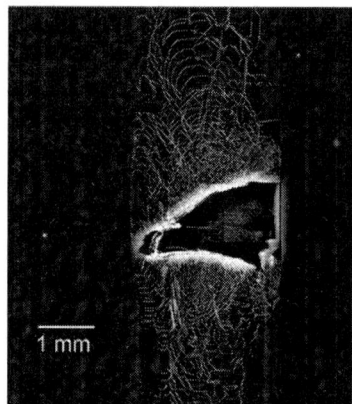

Fig. 3. UVPL image of an inclusion formed during epi growth and the cluster of BPDs the inclusion induces around it due to local stress.

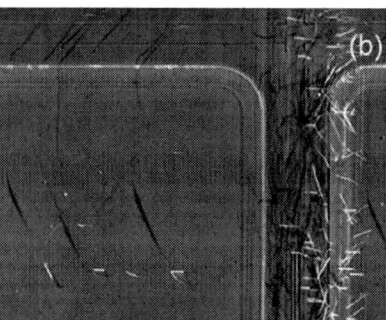

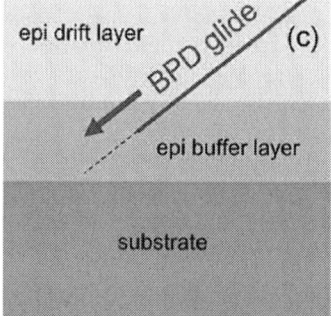

Fig. 4. BPDs introduced by the combination of Al implantation and activation anneal: (a) and (b) are UVPL images of the same area of a wafer before and after processing. The bright and dark lines in (b) are BPDs introduced by the processing. None of these BPDs are present in (a). Frame (c) illustrates the observation that the BPDs originate in or near the implanted layer and glide towards the substrate during the anneal.

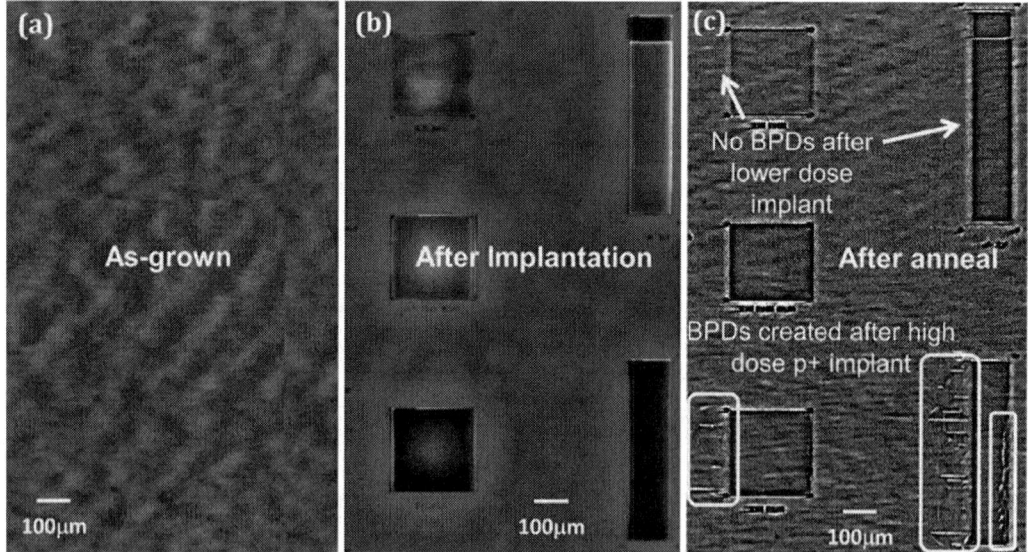

Fig. 5. UVPL from the same wafer area showing the introduction of BPDs during processing. There are no BPDs in (a) as-grown and (b) after-implantation. In (c) BPDs appear after annealing (see boxes). Note that BPDs are only introduced at the higher Al p$^+$ dose and not during the lower p dose.

978-1-7281-1988-5/18 $31.00 © 2018 IEEE   450

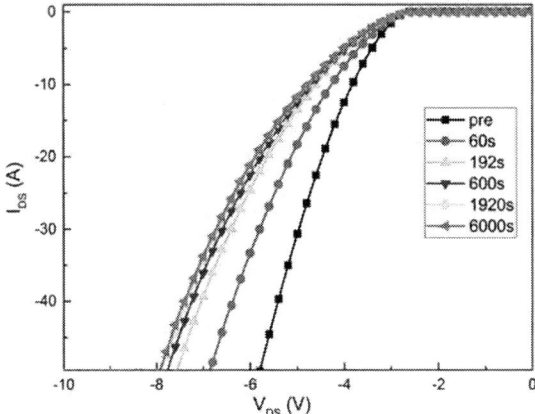

Fig. 6. Increases to the MOSFET (gen 1) body diode I-V curve due to BPD faulting caused by current stressing of 5 A for the accumulated times shown. $V_{gs} = -5V$ to eliminate current through the channel.

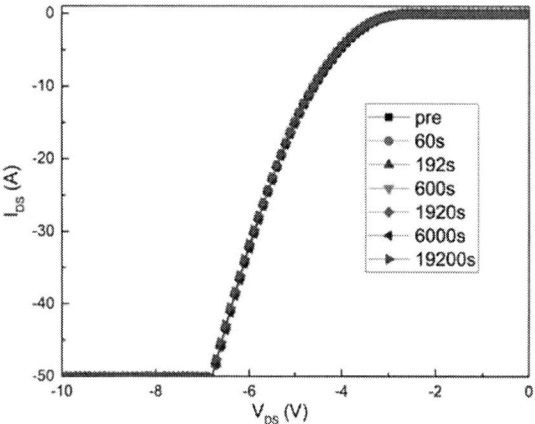

Fig. 7. No changes in the MOSFET (gen 2) body diode I-V curve during stressing of 10 A for the accumulated times shown. $V_{gs} = -5V$ to eliminate current through the channel.

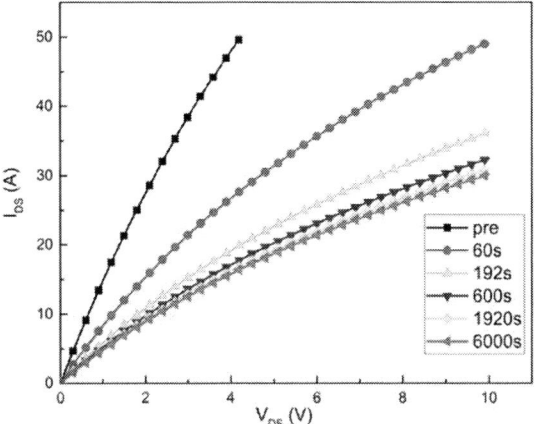

Fig. 8. Decreases to the MOSFET (gen 1) I-V conduction curve at $V_{gs} = 20$ V due to BPD faulting caused by current stressing of 5 A for the accumulated times shown.

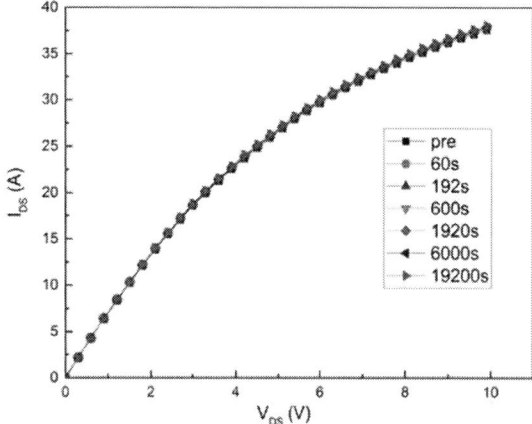

Fig. 9. No changes in the MOSFET (gen 2) I-V conduction curve at $V_{gs} = 20$ V during stressing of 10 A for the accumulated times shown.

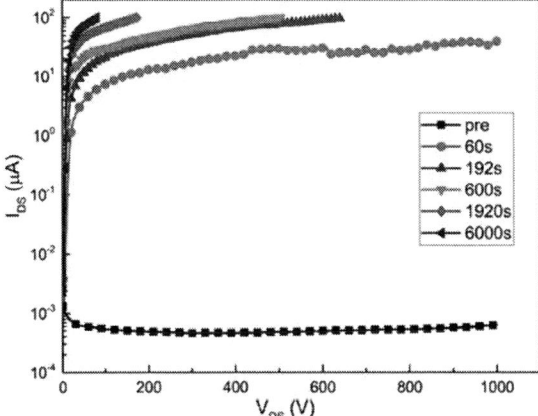

Fig. 10. Strongly increasing leakage in gen 1 MOSFET due to BPD faulting caused by current stressing of 5 A for the accumulated times shown.

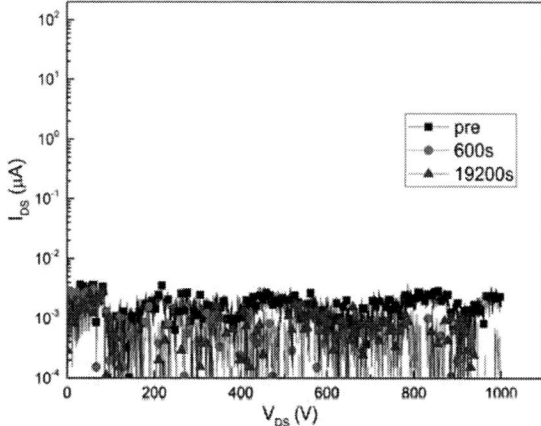

Fig. 11. No leakage increase in gen 2 MOSFET induced by current stressing of 5 A for the accumulated times shown.

978-1-7281-1988-5/18 $31.00 © 2018 IEEE   451

GaN devices for automotive application and their challenges in adoption

Tetsu Kachi

Institute of Materials and Systems for Sustainability, Nagoya University,
Nagoya, 464-8601, Japan, email: kachi@imass.nagoya-u.ac.jp

Abstract— Currently, electrification of automobiles is an urgent task, and high-performance power devices are indispensable items for their electrification. Wide bandgap semiconductors are powerful candidates for power devices used in electric vehicles (EV) and fuel cell vehicles (FCV) in the near future, and recent advances in GaN power devices are prominent in particular. Lateral GaN power devices on Si substrates began to be commercialized, and they are moving to system development. Research and development of vertical GaN power devices are also accelerating. Such high-performance devices are expected to greatly contribute to the electrification of automobiles, and interest in GaN power devices is increasing.

I. INTRIDUCTION

In recent years, automobiles have entered a period of paradigm shift. The key word is electrification and autonomous driving. For such a paradigm shift, innovative technology is strongly demanded, and for electrification, batteries and power electronics are key technologies. Electric vehicles (EVs) are required to have small electrical losses and to be small and lightweight due to long mileage. The main problem of the current power device (Si - IGBT) is on - resistance, which is approaching the theoretical limit. Therefore, high-performance GaN power devices are very expected for automotive applications as next generation power devices. In this paper, we will explain the recent progress of GaN power devices and automotive applications expected for GaN power devices.

II. POWER ELECTRONICS IN ELECTRIC VEHICLE

In the EV system, as schematically shown in Fig. 1, many power supply modules are used. Electrification includes EV and FCV as well as hybrid cars (HV), but the power electronics used are basically the same in all systems. These power modules can be divided into high power modules and medium and low power modules as shown in Fig.1. These modules require high efficiency and compact sizes. For power devices, these requirements can be described as low on-resistance and high-speed operation. GaN power devices have both performances, in particular, lateral GaN devices have high speed operation and vertical GaN devices have low on-resistance. Automobiles have many applications that utilize these high-performances.

III. LATERAL GAN POWER DEVICES

A. Automotive Application of Lateral Devices

There are many electric subsystems in the EV system. The subsystem of maximum output in the EV is an air conditioner, its power is about 5 kW. The air conditioner's compressor is driven by an inverter with a battery voltage of 200-300 V. Other one is a charging system. Fig. 2 shows a block diagram of the household charging system currently in use. The AC voltage is converted to a DC voltage using an AC-DC converter and the DC voltage is converted to high frequency and transferred to the rectifier via the isolation transformer for charging. The system power is about 3 kW. In addition, a DC-DC buck converter shown in Fig. 3 which steps down the battery voltage to 14 V, is also used, of which power is 1-2kW. In these medium power applications, Si-MOSFETs with 400-600V ratings are currently used as power devices. Since these subsystems are mounted in the car, it is required to reduce the size and weight as well as possible. As capacitance C and reactance L used in the subsystem circuit are proportional to $1/f^2$ and $1/f$, in which f is frequency, respectively, high frequency will contribute to reduce the sizes, which means that high speed operation of the power device is essential for future subsystem applications.

B. Recent Progress of Lateral GaN Power devices

Lateral GaN power devices have already been on sale. This makes it possible to manufacture more efficient converters than that using conventional Si-MOSFETs. However, it has been well known that there were two major problems with lateral GaN power devices. One is normally-off operation and the other is an increase in dynamic on resistance. Various gate structures for normally-off operation have been proposed, but now p-GaN gate structure and cascode connection structure are mainstream. Although high performance gate insulator for MOSFET has been reported in recent years [1,2], it seems difficult to reduce interface state density on surfaces with high dislocation density like GaN-on-Si. Fig.4 shows an example of C-V curves of MOS capacitors in which insulators (AlSiO) were formed on GaN-on-Sapphire and GaN-on-GaN under the same conditions. Frequency dispersion was observed only for the GaN-on-Sapphire sample. This result supports the difficulty of MOSFET on GaN-on-Si. The two typical p-GaN gate structures on sale are shown in Fig. 5, which are different contact types: Schottky contact and ohmic contact. A comparison of these specifications is shown in Table 1. It is understood that it is faster than the Si super junction

978-1-7281-1988-5/18 $31.00 © 2018 IEEE

MOSFET. Moreover, merit of p-GaN gate is very small temperature dependence of the threshold voltage. However, since the threshold voltage is around 1 V for both devices, attention must be paid to the gate drive, and then own gate drive circuit of each device was also being developed. Detailed technical contents are explained in each company's website [3-6]. On the other hand, the increase in dynamic on resistance has been overcome against 600V rating by improving crystal quality and device structures. The main characteristics of the lateral GaN device is high operating speed. Therefore, subsystems in the automobile are the best use of lateral GaN devices. The challenge is reliability and cost, but now trials of these new devices are beginning.

IV. VERTICAL GAN POWER DEVICES

A. Automotive Applications of Vertical GaN Power Devices

The high output power module consists of a bidirectional DC-DC converter and an inverter as shown in Fig. 1, of which circuit is shown in Fig. 6. The maximum output of the main motor is 60 to 150 kW depending on the model. In addition, since the module must provide guaranteed operation under any driving conditions, these power devices are required a large current capacity, such as 200 A or more. Currently, Si-IGBTs and pin diodes with sufficient current capacity and reliability are used in the high-power module. The breakdown rating of the device is 1.5kV due to the maximum motor source voltage 650 V. The efficiency of the inverter is very high (> 98%) at maximum output conditions. However, the average power used for driving is lower than half of the maximum output power. The efficiency of the Si-IGBT at low-power is lower than at high-power conditions due to junction voltage which is a feature of bipolar transistor. Therefore, unipolar devices such as MOSFETs with sufficient low on-resistance are needed to improve the overall efficiency of the inverter. In the inverter operation with pulse width modulation (PWM) control, increasing the carrier frequency makes the shape of the sine wave of the output voltage smoother and it improves motor efficiency. The high-speed performance of unipolar power devices is also desirable in DC-DC converters. In the present DC-DC converter, the switching frequency is around 10 kHz, and then a large capacitor and a large inductor are required as shown in Fig.7. Higher frequency operation allows the use of smaller capacitors and smaller inductors.

The main problem with high power modules is that the amount of heat generation increases and a water-cooling system is required. If a sufficiently low on-resistance of power devices is realized, for example, the cooling system is simplified to air-cooling system, which is an ultimate goal.

B. Recent Progress of Vertical GaN Power Devices

Vertical structure has the advantages of small chip size, easy wiring, high breakdown-voltage. These characteristics are highly suitable for high-power applications. A SiC-MOSFET is another candidate for the post Si-IGBT, and is ahead of the development. However, the on-resistance of SiC MOSFETs is limited long time by the low channel mobility of 20~30 cm^2/Vs. Though higher channel mobilities were observed in SiC MOS channels, they could

not be applicable to realistic devices because of low reliability or low threshold voltage. Low on-resistance is the greatest feature of wide bread gap semiconductors, but SiC can't make full use of its physical properties at the present. On the other hand, in GaN, high channel mobilities have been reported recently. Panasonic Group developed high performance vertical GaN device with 1.7kV breakdown voltage and 1mΩ·cm^2 on resistance [7]. The device had an AlGaN / GaN channel with a p-GaN gate as shown in Fig. 8, of which channel mobility was >500cm^2/Vs. This performance is beyond that of SiC-MOSFET for the first time. Furthermore, a high channel mobility exceeding 100 cm^2/Vs has been reported by the two groups. Fuji Electric Group showed high mobility of 120 cm^2/Vs in the inverted channel of the MOSFET [8]. UC Davis group reported 185 cm^2/Vs with a GaN channel regrown on the trench sidewall shown in Fig.9 in the trench gate MOSFET [9]. Fig.10 shows the specific on-resistance vs breakdown voltage of the drift layer for Si, SiC, GaN devices, which also shows on-resistances of channels for various channel mobilities calculated assuming 5μm cell pitch. If the channel mobility exceeds to 100 cm^2/Vs, vertical GaN devices may have lower on-resistance than 1mΩ·cm^2 over the wide breakdown voltage range from 600V to 2kV. Recent data show the high potential of low on-resistance of vertical GaN power devices due to high channel mobility. If such a GaN device with low on-resistance is realized, it not only improves the efficiency of the conventional EV system but also gives us the possibility of air-cooling system. Improvement of channel mobility and construction of a stable fabrication process are the current main issues in vertical GaN devices.

V. SUMMARY

To be adopted a new power device requires performance which enables a new innovative system. For example, the air-cooling system enables a four wheels drive in-wheel motor system as shown in Fig.11. GaN power devices have the potential to fulfill such demands and have great expectations to change the future of automobile system.

ACKNOWLEDGMENT

This work was supported by MEXT GaN R&D Project.

REFERENCES

[1] T. Yonehara, Y. Kajiwara, D. Kato, K. Uesugi, T. Shimizu, Y. Nishida, H. Ono, A. Shindome, A. Mukai, A. Yoshioka and M. Kuraguchi, 33.3, Proceedings in IEDM2017, San Francisco, 2017.

[2] M. Hua, Z. Zhang, J. Wei, J. Lei, G. Tang, K. Fu,Y. Cai, B. Zhang and K. J. Chen, 10.4, Proceedings in IEDM2016, San Francisco, 2016.

[3] https://gansystems.com/

[4] https://industrial.panasonic.com/ww/products/semiconductors/powerics/ganpower

[5] https://epc-co.com/epc/

[6] https://www.transphormusa.com/

[7] D. Shibata, R. Kajitani, M. Ogawa, T. Tanaka, S. Tamura, T. Hatsuda, M. Ishida, and T. Ueda, Proceedings in IEDM2016, San Francisco, 2016.

[8] S. Takashima, K. Ueno, H. Matsuyama, T. Inamoto, M. Edo, T. Takahashi, M. Shimizu, and K. Nakagawa, Appl. Phys. Express **10**, 121004, 2017.

[9] D. Ji, C. Gupta, S. H. Chan, A. Agarwal, W. Li, S. Keller, U. K. Mishra, and S. Chowdhury, 9.4, Proceedings in IEDM2017, San Francisco, 2017.

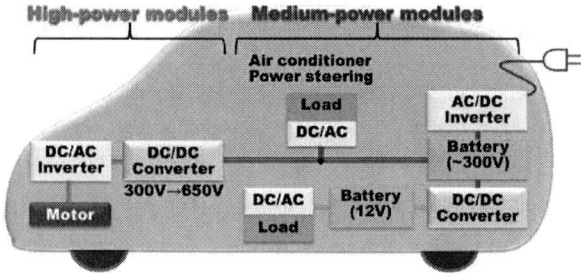

Fig. 1. Schematic diagram of power modules used in EVs.

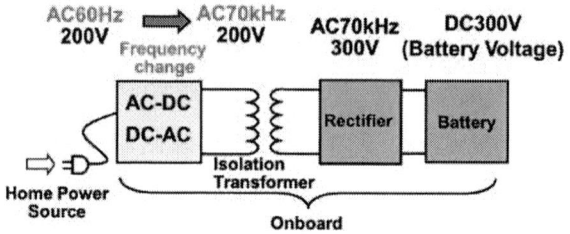

Fig. 2. Block diagram of the household charging system. Controlled maximum power is about 3kW.

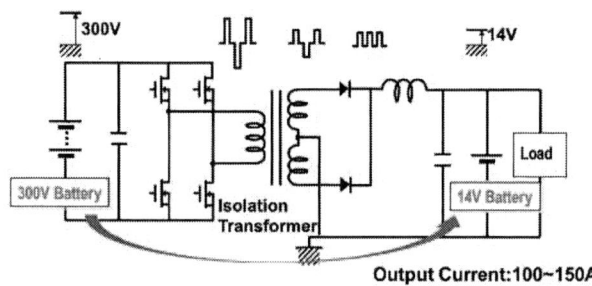

Fig. 3. DC-DC buck converter circuit which step down the battery voltage to 14V. The controlled power is 1-2kW.

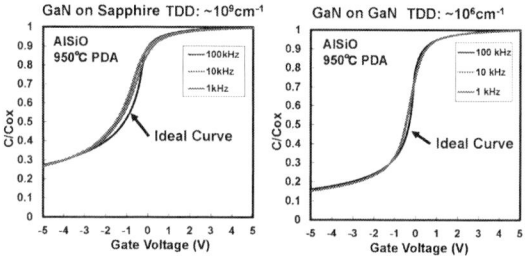

Fig. 4. C-V characteristics of AlSiO / GaN-on-Sapphire and AlSiO-/GaN-on-GaN MOS capacitors. Frequency dispersion was observed in GaN-on-Sapphire sample.

	Shottky contact (GaN System)	Ohmic contact (Panasonic)	Si SJ-MOS
VDSS(V)	650	600	650
Ron (mΩ)	50	56	62
Qg (nC)	5.8	5.0	64
Qrr (nC)	0	0	10000
Coss(pF)@400V	140	106	1100

Table 1. Comparison of electrical characteristics for p-GaN gate devices and Si super junction MOSFET.

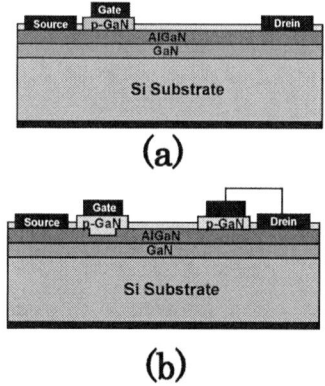

Fig. 5. Cross sectional device structures of p-GaN gate normally-off devices which are different contact types, (a) Schottky contact [3], (b) ohmic contact [4].

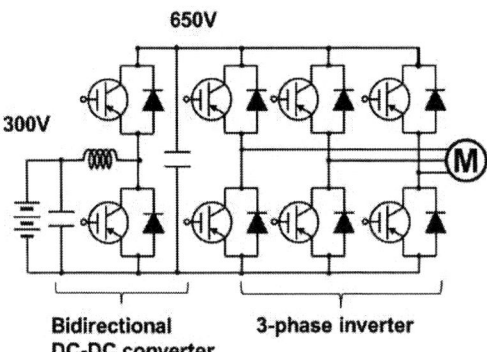

Fig. 6. Circuit of high-power module which consist of a bidirectional DC-DC converter and a 3-phase inverter.

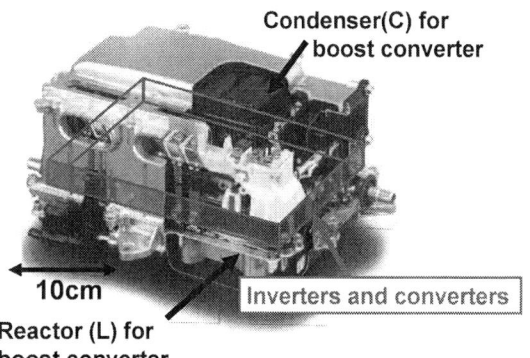

Fig. 7. Power control unit (high-power module) of HV (Prius). DC-DC converter requires large size capacitor and reactor.

μ_c : >500 cm²/Vs
V_T: 2.5V
1.7kV, 1mΩ•cm²

Fig. 8. Gate structure of AlGaN/GaN channel with p-GaN gate which has high channel mobility [7].

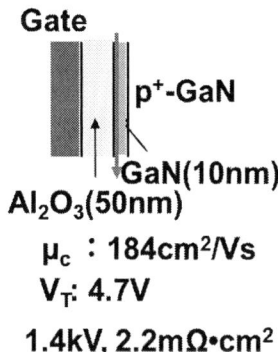

μ_c : 184cm²/Vs
V_T: 4.7V
1.4kV, 2.2mΩ•cm²

Fig. 9. Gate structure of regrown-GaN and Al₂O₃ gate insulator produced in MOCVD reactor simultaneously [9].

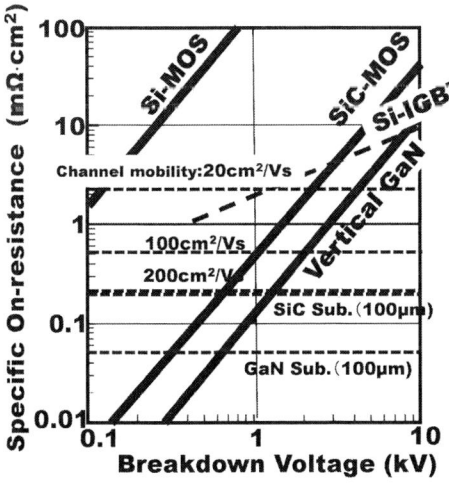

Fig. 10. On-resistance limits for Si, SiC and GaN power devices. Theoretical limit of channel resistances assuming 5μm cell pitch for various channel mobilities are also shown.

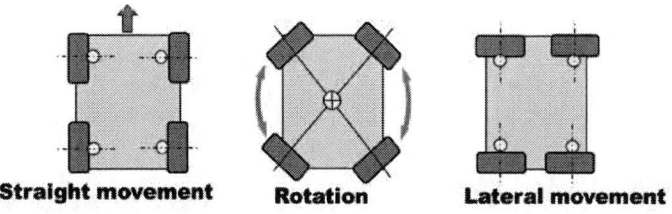

Fig. 11. Unique motion of 4 wheels in-wheel motor EV. If the air cooling system of the inverter can be achieved, such a new concept system will be realized.

978-1-7281-1988-5/18 $31.00 © 2018 IEEE   455

Barriers to the Adoption of Wide-Bandgap Semiconductors for Power Electronics

I.C. Kizilyalli[1], E.P. Carlson[2], and D.W. Cunningham[1]

[1]Advanced Research Projects Agency-Energy, U.S. Department of Energy, Washington, DC, email: Isik.Kizilyalli@hq.doe.gov
[2]Booz Allen Hamilton, Washington, DC

Abstract—Wide-bandgap power semiconductor devices offer enormous energy efficiency gains in a wide range of potential applications. As silicon-based semiconductors are fast approaching their performance limits for high power requirements, wide-bandgap semiconductors such as gallium nitride and silicon carbide with their superior electrical properties are likely candidates to replace silicon in the near future. Along with higher blocking voltages wide-bandgap semiconductors offer breakthrough relative circuit performance enabling low losses, high switching frequencies, and high temperature operation. However, even with the considerable materials advantages, a number of challenges are preventing widespread adoption of power electronics using WBG semiconductors.

I. INTRODUCTION

Electricity generation currently accounts for ~38% of primary energy consumption in the U.S. [1] and over the next 25 years is projected to increase more than 50% worldwide [2]. As a result, electricity continues to be the fastest growing form of end-use energy. Power electronics play a significant and growing role in the delivery of electricity as they are utilized to control and convert electrical power to provide optimal conditions for transmission, distribution, and load-side consumption. Estimates suggest that the fraction of electricity processed through some form of power electronics could be as high as 80% by 2030 (including generation and consumption), approximately a twofold increase over the current percentage [3]. Therefore, advances in power electronics have the potential for enormous energy efficiency improvements.

A key element of any power electronic system is the semiconductor switching device which determines the frequencies and power levels at which the electronic system may operate. A significant portion of the losses in power electronic converters is dissipated in the power semiconductor devices. Silicon (Si) has been the semiconductor material of choice for power devices for quite some time due to cost, ease of processing, and the vast amount of information available about its material properties. Si devices are, however, reaching their operational limits in blocking voltage capability, temperature of operation, and switching frequency due to the intrinsic material properties of Si. Wide-bandgap (WBG) power semiconductors, with their superior electrical properties, are an attractive emerging alternative to Si in many applications and can enable power converters with higher efficiency and higher power conversion densities.

II. BENEFITS OF WIDE-BANDGAP SEMICONDUCTORS

Achieving high power conversion efficiency requires low-loss power semiconductor switches. Today's incumbent power switches, typically metal oxide field effect transistors (MOSFET), insulated gate bipolar transistors (IGBT) and thyristors, are Si based and are quickly approaching their limits due to the fundamental material properties of Si and have several important limitations:

High Losses: The relatively low bandgap (1.1 eV) and critical electric field (0.3 MV/cm) of Si require high voltage devices to have substantial thickness. The large thickness translates to devices with higher specific on-resistance and results in higher conduction losses.

Low Switching Frequency: Si power MOSFETs require large die areas to keep conduction losses low. As the die size increases the gate capacitance and charge increases resulting in increases in switching losses for high-frequency applications. Si IGBTs have smaller die sizes compared to MOSFETs as they utilize minority carriers for conductivity modulation, but the long lifetime of minority carriers in Si reduces the usable switching frequency range of IGBTs to <10kHz.

Poor High-Temperature Performance: The relatively low Si bandgap contributes to higher intrinsic carrier concentrations at elevated junction temperatures which produces high leakage currents in p-n junctions. Additionally, the temperature variation of the bipolar gain in IGBTs amplifies the leakage and limits the maximum junction temperature to ~150°C.

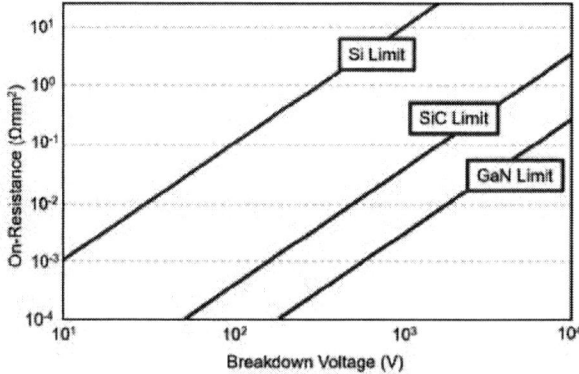

Figure 1. Semiconductor On-Resistance ($\Omega.mm^2$) versus breakdown voltage for silicon, silicon carbide and gallium nitride

New opportunities for higher efficiency power electronics have emerged with the development of wide-bandgap power semiconductor devices, driven by the fundamental differences

978-1-7281-1988-5/18 $31.00 © 2018 IEEE

in material properties between Si and the WBG semiconductors. Figure 1 illustrates an advantage of WBG semiconductors and shows the unipolar limit relationship as a function of on-resistance and breakdown voltage for Si, Silicon Carbide (SiC), and Gallium Nitride (GaN). The lower right region of the plot represents higher performance devices and thus is more desired.

Higher critical electric fields in WBG materials (>2 MV/cm) enable thinner, more highly doped voltage-blocking layers, which can reduce on-resistance by an order of magnitude in majority carrier architectures relative to equivalent Si devices [4]. High breakdown electric field and low conduction losses mean that WBG materials can achieve the same blocking voltage and on-resistance with a smaller form factor. This reduced capacitance allows higher frequency operation compared with a Si device. The low intrinsic carrier concentration of WBG materials enables reduced leakage currents and robust high-temperature performance. WBG semiconductors permit devices to operate at much higher temperatures, voltages, and frequencies therefore providing a pathway to more efficient, lighter, smaller, and higher temperature capable power electronics than those made from conventional semiconductor materials.

III. BARRIERS TO ADOPTION

However, even with the considerable materials advantages, a number of challenges are preventing widespread adoption of power electronics using WBG semiconductors

- *Cost*: The high cost and small size of GaN and SiC substrates lead to higher costs for WBG power devices compared to similarly rated silicon power devices.

- *Device Design and Fabrication*: Novel device designs are required that effectively exploit the properties of WBG materials to achieve the voltage and current ratings required in certain applications. Alternative packaging materials or designs are also needed to enable higher temperature and higher frequency operation.

- *Proven Reliability*: Demonstration of WBG device reliability in the field is needed before large scale industrial adoption of WBG based power electronics

- *Systems Integration*: WBG devices are not always suitable drop-in replacements for Si-based devices. The larger, more complex systems must be redesigned to integrate the WBG devices in ways that deliver unique capabilities.

High cost, challenging fabrication of practical devices, demonstrated reliability, and system integration remain important barriers to the widespread adoption of WBG devices. SiC and GaN substrates are expensive and limited in size, 150 mm and 100 mm, respectively. This is illustrated in Figure 2 which shows WBG substrates have a per area cost an order of magnitude higher compared to a 200 mm Si substrate. This results in higher cost for SiC and GaN power devices compared to similarly rated silicon power devices.

In order to take full advantage of the superior properties of the WBG semiconductors devices have to be developed that address key materials, device fabrication, and device architecture

issues that impact the cost and reliability of the devices. One such challenge in GaN is the requirement of selective area doping to fabricate normally-off vertical device architectures. The current lack of a viable selective area doping processes in GaN limits the possible devices architectures and ultimately limits the performance of the devices. Additionally advanced device packaging needs to be developed to enable the higher temperature capability of the WBG semiconductors.

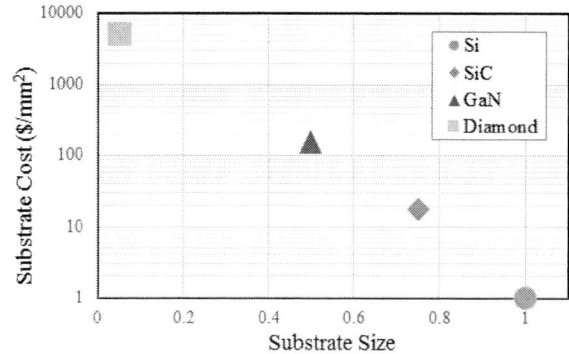

Figure 2. Wafer cost per area for varous WBG semiconductors versus wafer size. The plot has been normalized to a 200 mm silicon wafer.

WBG semiconductors and devices are far less mature than their Si counterparts with limited availability; and have yet to be proven reliable in the field which further limits their adoption. Transistors have limited availability and manufacturer data sheets often do not specify important application-level reliability parameters such as the dV/dt rating, avalanche rating, and the safe-operating area, especially at elevated temperatures. The only commercial WBG power devices with more than 10 years of market performance are SiC Schottky diodes. As such, they are the only devices with proof of their reliability on the scale required for high-end applications.

WBG devices cannot simply be swapped for Si devices in a circuit due to the dramatically different properties of the devices. Circuit designers will have to redesign systems to account for the WBG devices. Gate-driving circuits for SiC and GaN are more complex than those required for Si devices due to the high switching speed and common-mode noise can be more of an issue for WBG circuit designs. This leads to a large design inertia or reluctance to change in the circuit designers. Systems with significantly superior price to performance need to be demonstrated to overcome this inertia.

In order to "unlock" the enormous potential of WBG power electronics and promote widespread adoption, devices need to be fabricated cost effectively to improve the device price to performance ratio and confidence in the reliability of WBG power devices in actual power electronics is needed. Failure mechanisms of WBG devices need to be understood and a fundamental understanding needs to be developed that links the basic material properties of commercially available WBG semiconductors to the reliability characteristics of power devices in power electronics. Systems need to be designed and demonstrated that take full advantage of the superior performance of the WBG devices to provide evidence that more efficient, lighter, smaller, and higher temperature capable power electronics than those

978-1-7281-1988-5/18 $31.00 © 2018 IEEE

made from conventional semiconductor materials can be cost effectively fabricated.

IV. OVERCOMING THE BARRIERS

The U.S. Department of Energy's Advanced Research Project Agency - Energy (ARPA-E) has invested in WBG semiconductors to enable a new generation of power semiconductor devices that far exceed the performance of silicon-based devices. From materials and devices to modules and circuits to application-ready systems integration, ARPA-E projects have demonstrated the potential of WBG semiconductors to lower the cost of high-efficiency power electronics to enable broad adoption in energy applications [5].

To address the high cost barrier of WBG devices, ARPA-E launched the SWITCHES (Strategies for Wide Bandgap, Inexpensive Transistors for Controlling High-Efficiency Systems) program [6] in 2014. The SWITCHES program was aimed at the key materials and device fabrication, & architecture issues that drive costs for SiC and GaN devices. The goal was to enable the development of high voltage (>1200 V), high current (100 A) single die WBG power semiconductor devices that would have the potential to reach functional cost parity with Si power transistors while also offering breakthrough relative circuit performance. The SWITCHES technologies would reduce the barriers to ubiquitous deployment of low-loss WBG power semiconductor devices in many cost sensitive applications.

Monolith Semiconductor, under the SWITCHES program, made great strides in developing low-cost, large-area, SiC devices. Before the SWITCHES program, planar and vertical SiC MOSFET power devices were typically fabricated in relatively low volume, dedicated facilities that utilized unique process steps. Monolith partnered with an automotive qualified Si manufacturing foundry, to develop SiC power MOSFET devices on commercially available 150mm SiC wafers. Working in an active Si foundry provided significant cost benefits and by utilizing existing Si manufacturing infrastructure, the overhead costs to manufacture SiC devices were reduced [7]. Monolith successfully demonstrated fabrication of SiC MOSFETs with breakdown voltage (V_{BR}) of 1700V and a specific on-resistance as low as 3.1 mΩ-cm^2. Monolith anticipates reaching ¢10/A for their devices in the near future.

One of the main drivers for the cost of WBG devices are the high cost of the substrates. In the case of GaN, wafers are expensive, limited in size & availability, and have a quality that depends on the fabrication method. Bulk GaN produced by the hydride vapor phase epitaxy (HVPE) method can provide substrates with diameters of up to 100 mm, but crystallographic quality and orientation is variable across the substrate. The ammonothermal method produces higher quality substrates but diameters are limited to 2-inch. Several projects under the SWITCHES program investigated reducing the cost and increasing the size of GaN substrates produced by the ammonothermal method.

Potential pathways toward low cost GaN devices was investigated under the SWITCHES program. Two projects (Avogy,

Inc. and Cornell University) were able to demonstrate near theoretical, high-power vertical GaN p-n diodes exhibiting breakdown voltages >4 kV and figures-of-merit (V_{BR}^2/R_{ON}) greater than 3 GW/cm^2 [8]. These vertical GaN devices were avalanche capable [9] indicating the ruggedness of such devices in breakdown, a critical requirement for power switching and rectifying applications. The projects demonstrated 80% process yield for the large area p-n junctions indicating the pathway towards ¢10/A for GaN devices is promising. With the current cost of 100 mm GaN wafers and a die size of 12-16 mm^2 for a 100A, 1200V device, vertical GaN devices should be capable of reaching the cost range of ¢5/A to ¢7/A.

Before the SWITCHES program, the majority of GaN power device development had been directed toward lateral architectures. There were simply no vertical GaN devices available. The lateral devices suffered from well-known issues such as current-collapse, dynamic on-resistance, inability to support avalanche breakdown [10], and usable breakdown voltages no greater than 650V. Vertical devices on the other hand have the possibility to realize the material-limited potential of GaN including true avalanche-limited breakdown. Tackling the device design and fabrication barrier the SWITCHES program created a new field of vertical GaN device designs to exploit the properties of the WBG semiconductor.

Under the SWITCHES program the University of California, Santa Barbara demonstrated a modified vertical trench MOSFET, named OG-FET, which takes advantage of a regrown un-intentionally doped GaN interlayer followed by an in-situ dielectric deposited in the trench for enhanced electron mobility. For a single unit cell OG-FET, V_{BR} as high as 700 V corresponding to a breakdown electric field of 1.4 MV/cm was reported with $R_{on,sp}$ of 0.98 mΩ-cm^2 [11]. Columbia University demonstrated a vertical fin power field-effect-transistor structure (VFET) on bulk GaN substrates. The VFET consists of fin-shaped channels etched into an 8-μm-thick n- doped GaN drift layer surrounded by metal gate pads which pinch-off the channel. Fabricated VFETs demonstrated threshold voltage of 1V, 10^{11} on/off current ratio, and a blocking voltage of 800V [12]. Vertical trench MOSFETs were demonstrated by HRL using an AlN/SiN dielectric stack employed as the gate "oxide" yielding a device with a threshold voltage of 4.8 V, blocking voltage of 600 V at gate bias of 0 V, and on-resistance of 1.7 Ω at gate bias of 10 V [13]. Avogy demonstrated 2.5A vertical transistors using buried p-layers and a hexagonal layout with breakdown voltages exceeding 1.5-kV and specific on resistance of 2.2 mΩ-cm2 [14]. Reliability of vertical GaN devices is an area of interest with some preliminary initial investigations [15]. However, similar to other WBG semiconductors more work is needed to understand the degradation mechanisms in the material and in-field reliability demonstrations are still lacking.

A barrier in GaN device fabrication experienced by many SWITCHES project teams was the lack of a selective area p-type doping process in GaN. To confront this barrier the PNDIODES (Power Nitride Doping Innovation Offers Devices Enabling SWITCHES) program [6] was launched in 2017. The most obvious selective area p-type doping approaches, ion implantation and diffusion, have not produced p-type regions or

978-1-7281-1988-5/18 $31.00 © 2018 IEEE

satisfactory p-n junctions in GaN. Seven projects were selected for funding as part of the PNDIODES program. Three of the projects will focus on ion implantation of p-type dopants along with innovative annealing processes to remove the implantation damage and activate the dopants. The innovative annealing processes, including laser spike and Gyrotron annealing, are needed to overcome the thermodynamic limits of GaN. Three of the projects will focus on a non-traditional selective area doping process using patterned etch and regrowth to form selective p-type regions. These projects will use low damage etching methods, interface impurity control, and optimization of regrowth on the different crystal directions to produce defect free regrowth interfaces. The remaining project will focus on the development of neutron transmutation doping to fabricate a uniformly doped n-type GaN wafer by exposing the wafers to neutron radiation to create a stable network of Ge dopants.

The CIRCUITS (Creating Innovative and Reliable Circuits Using Inventive Topologies and Semiconductors) program [6] was launched in 2017 to surmount the systems integration barrier. Previous efforts by ARPA-E and others have primarily focused on WBG material and device development without consideration of the circuit topology. The circuit design is also critical to the large-scale implementation of more efficient WBG devices as a result of their ability to operate at higher voltage, higher frequency, and higher temperature. New circuit topologies and designs are needed that optimize the properties of the WBG semiconductor devices in the circuit while minimizing the size and costs of auxiliary circuit components such as cooling systems. The CIRCUITS program seeks to accelerate the development of a whole new class of efficient, lightweight, and reliable power converters based on WBG semiconductors. With an explicit focus on novel circuit topologies, advanced control and drive electronics, and innovative packaging, CIRCUITS aims to catalyze disruptive improvements for power electronics afforded by WBG semiconductors.

Twenty-one projects were selected for funding as part of the CIRCUITS program. The CIRCUITS project teams will develop efficient, lightweight, and reliable power converters for various applications including motor drives, automotive, power supplies, data centers, aerospace, distributed energy, and the grid. The circuit topologies employed by the CIRCUITS teams will be optimized for WBG semiconductors to maximize overall electrical system performance and offer significant direct and indirect energy savings. The CIRCUITS projects will establish the building blocks for WBG enabled power converters with higher efficiency, enhanced reliability, and superior total cost of ownership. In addition, a reduced form factor will drive adoption of higher performance and more efficient power converters relative to today's state-of-the-art systems.

V. Summary

Wide-bandgap power semiconductor devices offer breakthrough relative circuit performance enabling low losses, high switching frequencies, and high temperature operation in a wide range of potential applications. However, even with the considerable materials advantages, a number of challenges are preventing widespread adoption of power electronics using WBG semiconductors. Significant work still remains to overcome these barriers and realize the full potential of WBG materials in improving energy efficiency. As mentioned above, fundamental research into material properties and processing to bring down the cost of the devices, and continued development down the power electronics value chain into circuits and systems, are vital steps in ensuring that America can maintain its technological lead in these promising materials, and reap the energy benefits through wide-ranging applications.

Acknowledgment

The authors gratefully acknowledge the contributions of Dr. Timothy Heidel to the SWITCHES program.

References

[1] U.S. Energy Information Administration, *Monthly Energy Review*, May, 2018. Available at: https://www.eia.gov/totalenergy/data/monthly/

[2] U.S. Energy Information Administration, *International Energy Outlook 2017*, September, 2017. Available at: https://www.eia.gov/outlooks/ieo/

[3] L.M. Tolbert, et al. *Power Electronics for Distributed Energy Systems and Transmission and Distribution Applications: Assessing the Technical Needs for Utility Applications*, Eng. Sci. Technol. Div., Oak Ridge Nat. Lab., pg. 21-22, Oak Ridge, TN (2005)

[4] A. Heffner, "Recent Advances in High-Voltage, High-Frequency Silicon-Carbide Power Devices", *Industry Applications Conference*, 41st IAS Annual Meeting, 2006

[5] I.C. Kizilyalli, Y.A. Xu, E. Carlson, J. Manser, D.W. Cunningham. "Current and Future Directions in Power Electronic Devices and Circuits based on Wide Band-Gap Semiconductors", *5th Workshop on Wide Bandgap Power Devices and Applications (WiPDA)*. Albuquerque, NM USA. October 2017

[6] ARPA-E program listing information available at: https://arpa-e.energy.gov/?q=program-listing

[7] S. Banerjee, K. Matocha, K. Chatty, J. Nowak, B. Powell, D. Guttierrez, et al. "Manufacturable and rugged 1.2 KV SiC MOSFETs fabricated in high-volume 150mm CMOS fab.", *2016 28th International Symposium on Power Semiconductor Devices and ICs (ISPSD)*, Prague, Czech Republic, June 2016

[8] K. Nomoto, B.Song, Z. Hu, M. Zhu, M. Qi, N. Kaneda, , et al. "1.7-kV and 0.55- mΩ·cm2 GaN p-n Diodes on Bulk GaN Substrates With Avalanche Capability", *IEEE Electron Device Letters*, vol. 37, pp 161-164, 2016

[9] O.Aktas, and I. Kizilyalli, "Avalanche capability of vertical GaN pn junctions on bulk GaN substrates", *IEEE Electron Device Letters*, vol. 36, pp. 890-892, 2015

[10] E. Zanoni, M. Meneghini, A. Chini, D. Marcon, and G. Meneghesso, "AlGaN/GaN-based HEMTs failure physics and reliability: Mechanisms affecting gate edge and Schottky junction", *IEEE Trans. Electron Devices*, vol. 60, pp. 3119–3131, Nov. 2013

[11] D.Ji, C.Gupta, A. Agarwal, S. Chan, C. Lund, S. Chowdhury, et al. "Large-Area In-Situ Oxide, GaN Interlayer-Based Vertical Trench MOSFET (OG-FET)" *IEEE Electron Device Letters*, vol: 39, pp. 711-714, 2018

[12] M. Sun, Y. Zhang, X. Gao, and T. Palacios, "High-Performance GaN Vertical Fin Power Transistors on Bulk GaN Substrates" *IEEE Electron Device Letters*, vol: 38, pp. 509-512, 2017

[13] R. Li, Y. Cao, M. Chen, and R. Chu, "600 V/1.7 Ω Normally-Off GaN Vertical Trench Metal–Oxide–Semiconductor Field-Effect Transistor", *IEEE Electron Device Letters*, vol. 37, pp. 1466–91469, 2016

[14] H. Nie, Q. Diduck, B. Alvarez, A. P. Edwards, B. M. Kayes, M. Zhang, et al., "1.5-kV and 2.2-m_cm2 vertical GaN transistors on bulk-GaN substrates," *IEEE Electron Device Letters*, vol. 35, pp. 939–941, 2014

[15] I.C. Kizilyalli, P. Bui-Quanga, D. Disney, H. Bhatia, and O. Aktas, "Reliability studies of vertical GaN devices based on bulk GaN substrates", *Microelectronics Reliability*, vol. 55, pp. 1654-1661, 2015

978-1-7281-1988-5/18 $31.00 © 2018 IEEE

AUTHOR INDEX

Abe, T. 237
Abel, S. 540
Acosta-Alba, P. 153
Adelmann, Ch. 111
Adelmann, Christoph 831
Afzalian, A. 492
Agarwal, A. 201
Agarwal, Harshit 209
Agarwal, T. Kumar 512
Agarwal, V. 739
Agashiwala, Kunjesh 576
Ahn, Hyoshin 771
Ahn, Woojin 584
Ai, Kelvin 217
Aigner, R. 332
Akinwande, Deji 532
Alam, A. 683
Alam, Muhammad Ashraful 584
Alava, Thomas 281
Alessandri, C. 368
Alfieri, G. 440
Ali, M. 107
Alian, Alireza 304
Allain, P.E. 99
Allain, Pierre E. 281
Almeida, S. 75
Alzate, J. -G. 412
Amanapu, H. 819
Ambrosi, E. 927
Amin, P. 133
Amishiro, H. 185
Amisse, A. 141
Ando, T. 807, 911
Andre, F. 157
Andrieu, F. 153
Ang, Kah-Wee 580
Annema, A. J. 739
Annunziata, R. 424
Anzai, Y. 284
Aoyagi, Y. 225
Appenzeller, J. 536
Arasawa, Ryo 312
Araujo, F. Abreu 616
Arimura, H. 496, 787
Arimura, Hiroaki 783
Arnal, V. 161
Arnaud, F. 424
Arnaud, L. 153, 157
Arnoux, M. 157
Arreghini, A. 43, 600
Arslan, U. 412
Arutchelvan, G. 512
Asai, Hidehiro 197
Asai, Y. 460
Asatsuma, T. 221

Asselberghs, I. 512
Atsumi, Tomoaki 312
Aubin, J. 153
Audoit, G. 500
Audran, S. 161
Aussenac, F. 153
Auth, Chris 636
Ayele, G. T. 277
Baba, Shotaro 79
Badami, Oves 759
Badaroglu, M. 408
Bae, B. J. 416
Bae, D.-I. 656
Bae, Geumjong 656
Bae, Jong-Ho 288
Baert, R. 420
Baeyens, Y. 552
Bai, P. 412
Bakir, Muhannad S. 672
Balan, V. 153, 157, 500
Banerjee, K. 43
Banerjee, Kaustav 564, 576, 799
Banniard, L. 99
Banniard, Louise 281
Bao, R. 807
Bao, Ruqiang 253, 648
Bao, T. Huynh 420
Barbato, A. 703
Barbato, M. 703
Barlage, D. 923
Barman, Soumendra 253
Baron, L. 161
Barr, R. 460
Barral, V. 424
Barraud, S. 141, 500
Barreto, J. 540
Barrois, Charlie 281
Basker, Veeraraghavan 652
Batude, P. 153
Baudin, F. 161
Baumgartner, Y. 907
Bavandpour, M. 476
Bayat, F. Merrikh 476
Bayha, B. 428
Bazizi, E.M. 428
Beaurepaire, S. 153
Beche, E. 161
Bedau, D. 867
Behin-Aein, B. 604
Bellando, F. 269
Bellini, M. 440
Belmonte, A. 51
Bender, H. 496
Bendersky, L.A. 536
Benedict, J. 107

AUTHOR INDEX

Beneyton, R. 424
Benoit, D. 424
Benschop, Jos 261
Bernard, M. 863
Bernier, N. 500
Berthelon, R. 424
Bertrand, B. 141
Besson, P. 153
Beyer, S. 428
Beyne, S. 111
Bhuiyan, M. A. 711
Bhuva, B. 795
Bhuwalka, K.K. 656
Bi, Z. 811
Bianda, E. 440
Bilgen, H. 157
Billiot, Gérard 281
Birkhahn, R. 460
Bishop, Douglas 292
Bocquet, M. 484
Boemmels, J. 149
Boeuf, F. 277
Bohr, M. 412
Bohuslavskyi, H. 141
Boivin, P. 424
Boland, J. 811
Boon, M.N. 628
Borchert, James W 883
Borga, M. 703
Bortolotti, P. 616
Boter, J.M. 133
Botzem, T. 129
Boucard, F. 424
Bouchu, D. 157
Boulanger, Pascale 281
Bourdet, L. 141
Bourgeois, G. 863
Brems, S. 512
Brenac, Ariel 281
Breslin, C. M. 107
Bresson, N. 157
Breuil, L. 600
Brew, K. 819
Brew, Kevin 815
Bricalli, A. 927
BrightSky, M. 859
Brink, Markus 126
Brockman, J. 412
Bruce, R. 911
Bruce, R. L. 859
Bruley, J. 911
Brunelli, S. T. Šuran 324
Brunet, L. 153
Brus, S. 787, 843
Bu, H. 819

Buford, B. 412
Burghartz, Joachim N. 883
Burns, J.L. 632
Bury, E. 592
Bury, Erik 783
Cabout, T. 161
Cabral, C. 107
Cadien, K. 923
Cai, D. L. 620
Cai, Linlin 775, 779
Cai, Xiangbin 695
Cai, Zhimei 173
Caimi, D. 540, 899, 907
Caironi, Mario 883
Camlica, A. 755
Campbell, J.P. 536
Canales, F. 440
Canaperi, D. F. 811
Cao, Linjun 115
Cao, Wei 564, 576
Capogreco, E. 496
Capogreco, Elena 783
Cappelletti, P. 424
Carlson, E.P. 456
Carpenter, R. 935
Carr, A. 819
Carr, Adra 815
Carta, F. 859
Casella, A. 747
Cassé, M. 500
Castany, Olivier 281
Castellani, N. 472, 863
Caubet, V. 424
Cave, N. 819
Caymax, M. 512
Cazaux, Y. 229
Cha, Jungho 560
Cha, Moonhyun 771
Chae, Sujin 851
Chai, Yang 520, 524
Chakraborty, W. 364
Chalupa, Z. 500
Champenois, A. 161
Chan, B. T. 149
Chan, Kevin 292
Chan, Masun 520
Chan, Michael S 253
Chang, C.-Y. 504
Chang, Che-Chia 356
Chang, Chia-He 249
Chang, D. R. 640
Chang, F. 811
Chang, Huan-Lin 209
Chang, Jonathan 273
Chang, Ki Soo 288

AUTHOR INDEX

Chang, L. 632
Chang, Meng-Fan 340
Chang, Pengying 775, 779
Chang, Vincent S. 644
Chanrion, E. 141
Chao, R. 604
Chao, Robin 652
Chao, T.-S. 504
Charbon, E. 739
Charles, C. 157
Chasin, A. 508
Chasin, Adrian 783
Chatterjee, Korok 209
Chauhan, Y. S. 201
Che, Xiaoyu 839
Chen, An 855
Chen, B.-J. 751
Chen, Bing 173
Chen, Bo-Yuan 340
Chen, C.-L. 751
Chen, C.-Y. 751
Chen, Cheng 707
Chen, Chin-Hsuan 636
Chen, Chun-Chi 731
Chen, E. 492
Chen, Fan 348
Chen, H. P. 620
Chen, H.-W. 751
Chen, H.-Y. 735
Chen, Hsiu-Chih 340, 731
Chen, Huiming 847
Chen, Jiezhi 568
Chen, K.C. 31
Chen, K.-T. 735
Chen, Kevin J. 695
Chen, Kuan-Neng 249
Chen, Kun-Ming 340
Chen, L.-Y. 504
Chen, P.-G. 735
Chen, Peng 340
Chen, Pin-Chun 356
Chen, Pin-Guang 731
Chen, Po-An 532
Chen, S. 879
Chen, S.-L. 751
Chen, S.-Y. 735
Chen, Shaoyin 815
Chen, Shih-Wei 249
Chen, Siming 556
Chen, T.H. 396
Chen, T.K. 492
Chen, T.W. 31
Chen, Victoria 528
Chen, Wangyong 775, 779
Chen, Wei-Chen 39

Chen, Wei-Hao 340
Chen, Y. F. 620
Chen, Y.S. 624
Chen, Yen-Pu 584
Chen, Yi-Ju 731
Chen, Z. 508
Chen, Zhebo 253
Chen, Zhihong 516
Cheng, C. C. 31
Cheng, C. -W. 911
Cheng, C.W. 859
Cheng, H. Y 859
Cheng, P. 791
Cheng, Ran 173
Cheng, S.-L. 751
Cheng, Yan 47
Cheramy, S. 157
Cherault, N. 161, 424
Cheung, K.P. 536
Chevalliez, S. 153
Chiang, Meng-Hsueh 532
Chiappe, D. 512
Chien, W. C. 859
Chih, Y.D. 624
Chih, Yue-Der 273
Chiu, H.P. 31
Chiu, Wen-Cheng 731
Chiu, Yu-Fan 731
Cho, Byung Jin 241
Cho, J. 265
Cho, Sunglae 851
Choi, Byoung Lyong 560
Choi, H. W. 640
Choi, Hyejung 851
Choi, Junho 532
Choi, K. 819
Choi, S. 811, 819
Choi, Samuel 815
Choi, Woosung 348, 767
Choi, Y.M. 656
Choi, Yongki 676
Chossat, J. 157
Chou, Y.-C. 735
Chou, Y.-T. 751
Chouteau, S. 424
Chow, Jerry M. 126
Chowdhury, S. 460
Chu, Jae Hwan 799
Chu, S. 811
Chu, S.-W. 751
Chua, Lye-Hing 823
Chudzik, M. 811
Chudzik, Michael P 253
Chueh, Yu-Lun 273
Chun, K.Y. 656

AUTHOR INDEX

Chun, Yunseok 851
Chung, Bryce 672
Chung, Hyein 767
Chung, Kisup 253
Chung, N. L. 604
Chung, S.-J. 656
Chung, Wonil 344
Chung, Won-Young 767
Ciofi, I. 111
Ciubotaru, Florin 831
Clarke, J.S. 133
Clement, H. 460
Clement, L. 424
Clevenger, L.A 632
Clima, Sergiu 380
Cloarec, J-P. 277
Cogorno, M. 811
Cogorno, Matt 827
Cohen, G. M. 911
Collaert, N. 149, 496, 787
Collaert, Nadine 304
Collins, Philip G. 676
Colombeau, B. 811
Compagnoni, C. Monzio 35
Cong, H. 604
Connor, C. 412
Conti, R. A. 811
Conti, Richard A 253
Convertino, C. 899, 907
Copel, Matt 292
Cordero, E. Garcia 269
Costa, T. 91
Cott, D. 512, 787
Couet, S. 420, 592, 843, 935
Covi, E. 927
Crafton, Brian 300
Crippa, A. 141
Croes, K. 111
Cros, V. 616
Crotti, D. 420, 592, 935
Cunningham, D.W. 456
Cyrille, M. C. 863
Czornomaz, L. 899, 907, 911
Dangol, A. 508
Das, N. 412
Dasgupta, A. 201
Datta, S. 296, 364
Datta, Suman 55, 300
Davydov, A. V. 536
De Franceschi, S. 141
De Heyn, V. 149, 787
De Keersgieter, A. 496
De La Rosa, C. Lockhart 512
De Santi, C. 691, 703
De Wolf, I. 408

Debacker, P. 480
Decoutere, S. 703
Deen, Jamal 871
Defoort, Martial 281
Degraeve, R. 480, 592
Dehaene, W. 480
Dehollain, J.P. 133
Dekkers, H. 508
Del Alamo, Jesús A 895
Delalleau, J. 161
Delaye, V. 500
Delhougne, R. 51, 600
Delmedico, S. 424
Deloffre, E. 157
Demarest, J. 107, 819
Demarest, James 815
D'Emic, C. 911
Demuynck, S. 149
Deshmukh, Sanchit 572
Deshpande, V. 149
Detzel, T. 703
Devolder, T. 843
Devolder, Thibaut 831
Devriendt, K. 149, 508
Di Piazza, L. 43, 380
Diaz, C.H. 492
Diaz, Carlos H 624
Dietzel, B. 548
Disegni, F. 424
Divakaruni, Rama 648
Dmitry, Veinger 233
Do, N. 428
Doevenspeck, J. 480
Domengie, F. 424
Domengie, Florian 404
Dominauez-Medina, Sergio 281
Dona, Danian 47
Dong, Da Nian 464
Dong, P. 552
Dong, Yuan 544
Donnell, J.O 412
Doornbos, G. 492
Dourthe, L. 500
Doyle, B. 412
Drouin, D. 277
Droulers, G. 133
Du, Gang 775, 779
Duarte, Juan Pablo 209
Dube, A. 811
Duclaux, B. 161
Dumont, F. 332
Dünkel, S. 428
Durfee, C. 819
Durfee, Curtis 815
Duriez, B. 492

AUTHOR INDEX

Dutta, S. 364, 739
Dzurak, A.S. 129
Ecoffey, S. 277
Eenink, G. 133
El Kazzi, S. 512
Eleftheriou, E. 628
El-Falou, A. 755
Elloian, J. 91
Eltes, F. 540, 899
Endo, Kazuhiko 197
Endoh, T. 608
Eneman, G. 496
Enomoto, T. 237
Ernoult, M. 616
Ernst, T. 500
Ernst, Thomas 281
Escarabajal, Y. 161
Esseni, D. 213
Esseni, David 759
Euvrard, C. 157
Evans, R. F. L. 935
Everson, L. 352
Exbrayat, Y. 157
Ezhilarasu, G. 683
Fabris, E. 691
Fafin, A. 99
Fafin, Alexandre 281
Fan, J. 879
Fan, Zhiqiang 568
Fang, C.-C. 504
Fang, Y. 324
Fantini, A. 480
Farcy, A. 157
Fattinger, G. 332
Favennec, L. 424
Favero, I. 99
Favero, Ivan 281
Favia, P. 496
Federspiel, X. 153
Feng, L. 879
Feng, Philip X.-L. 87
Feng, Xue 668
Feng, Y. L. 488
Fenouillet-Beranger, C. 153, 472
Ferreira, P. 424
Fettweis, G. 11
Fischer, K. 412
Florent, K. 43, 380
Fompeyrine, J. 540
Fontaine, H. 153
Fontelaye, Caroline 281
Fostner, Shawn 281
Fournel, F. 153
Franco, J. 149, 787
Franco, Jacopo 783

Frank, M. M. 911
Fremont, H. 157
Friedrichs, Peter 436
Frougier, Julien 652
Fuji, Yoshihiko 79
Fujihara, Y. 225
Fujisaki, K. 284
Fukuda, Koichi 197
Fukuhara, K. 265
Fukui, M. 189
Fukushima, A. 616
Fukuzawa, Hideaki 612
Funck, C. 867
Furnemont, A. 420
Furnémont, A. 600
Furukawa, K. 189
Gabriel, Kristin N. 676
Gadigatla, Srinivasa Chaitanya 636
Gaidhane, A. 201
Gaillard, Frederic 404
Galatage, Rohit 652
Gallon, C. 424
Galpin, D. 161
Gan, K. W. 604
Gandolfo, A. 424
Gao, B. 488, 931
Gao, Bin 67, 468
Gao, Jianfeng 47
Gao, X. 691
Garello, K. 935
Garrione, J. 863
Garros, X. 153
Gaur, A. 512
Ge, Ruijing 532
Gely, M. 99
Gely, Marc 281
Gentile, A. A. 540
George, H.C. 133
George, Steven M. 895
Gertsch, Jonas 895
Ghani, T. 412
Ghatge, M. 95
Ghazavi, P. 428
Ghezzi, G. 161
Ghibaudo, Gerard 384, 404
Giannopoulos, I. 628
Gignac, L. 107, 859
Giorgio, Michele 883
Giraud, B. 472
Gluschenkov, O. 819
Gluschenkov, Oleg 815
Goda, A. 27
Goh, L. C. 604
Gokmen, Tayfun 292
Goldberg, C. 21

AUTHOR INDEX

Golonzka, O. 412
Gomez, Jorge 300
Gomiero, E. 424
Gong, Songbin 915
Gong, Tiancheng 47, 464
Gong, Xiao 544, 823
Gonzalez, M. 408
Gossmann, H.-J. 811
Goto, T. 660
Gouget, Gilles 384
Gouraud, P. 161
Goux, L. 51
Grant, Lindsay A. 217
Grasser, T. 787
Green, R. 448
Greene, Andrew 652
Grenier, J.C. 424
Grisafe, B. 296, 364
Gritters, J. 460
Grobis, M. 867
Groeseneken, G. 43, 787
Grollier, J. 616
Grosse, P. 99
Grossi, A. 472
Gu, S.-S. 735
Guerin, C. 153
Guérin, H. 269
Guiheux, Denis 404
Guillaumet, S. 157
Gul, O. Tolga 676
Guo, D. 807, 811, 819
Guo, Dechao 648, 652, 815
Guo, Hong 763
Guo, Shaofeng 388, 392
Guo, X. 879
Guo, Xuyun 520
Gupta, D. 819
Gupta, S. 364
Gupta, Sayak Dutta 803
Gupta, Sumeet Kumar 516
Ha, Kyoungho 560
Haeberlen, O. 703
Haensch, W. 59
Haensch, Wilfried 292
Haluska, Miroslav 83
Ham, B.H. 656
Hamori, H. 660
Hamzaoglu, F. 412
Han, G. 213
Han, Jin-Woo 432
Han, Kaizhen 823
Han, Qin 679
Han, S. H. 416
Hanna, A. 683
Haq, Jesmin 612

Hara, Michiko 79
Harada, S. 177, 181, 444
Haran, B. 811, 819
Haran, Bala S. 253, 648
Hart, A. 540
Hartmann, J.-M. 141, 500
Hashemi, P. 807, 911
Hashimoto, K. 265
Hatano, M. 265
Hatayama, T. 177
Hattori, Junichi 197
He, Jun 396
He, Renren 612
Heil, P. 412
Hellings, G. 149, 787
Henderson, R. K. 743
Henke, A. 428
Hennen, T. 867
Henrion, Y. 157
Henry, Todd 823
Hentges, P. 412
Hentz, S. 99
Hentz, Sébastien 281
Herment, G. 269
Hermouet, M. 99
Hermouet, Maxime 281
Herrmann, T. 428
Hertzberg, Jared 126
Heylen, N. 149
Heyns, M. 843
Heyns, Marc 831
Hiblot, G. 408
Hierold, Christofer 83
Higashi, Yoshihiro 79
Higashiki, T. 265
Hikavyy, A. 149, 496, 600
Hiramoto, T. 189
Hiramoto, Toshiro 372, 723
Hirtzlin, T. 484, 616
Ho, H. Y. 859
Ho, Paul S. 115
Hody, H. 51
Hoffmann, M. 727
Holland, M.C. 492
Holleitner, A.W. 245
Hong, H. S. 416
Hong, Hyeongsun 560
Hong, Jeongmin 847
Hong, R.-C. 735
Hong, Seongbin 288
Hong, T.-C. 504
Hong, Yoonki 288
Honjo, H. 608
Hopstaken, Marinus 253
Horiguchi, N. 496, 508, 787

AUTHOR INDEX

Horiguchi, Naoto 783
Horita, M. 687
Horng, J.J. 396
Hoshii, T. 189
Hosoda, T. 460
Hou, F.-J. 504
Hou, Tuo-Hung 356
Hsieh, H. D. 396
Hsieh, Tung-Ying 249
Hsiung, Alan Chih-Wei 217
Hsu, H.-S. 504
Hsueh, F.-K. 504
Hsueh, Fu-Kuo 340, 731
Hu, C.-K. 107
Hu, Chenming 209, 249, 731
Hu, Szu-Tung 115
Hu, Z. 691
Hu, Zongyang 193
Hua, Mengyuan 695
Huang, D.S. 396
Huang, G.-W 504
Huang, Guo-Wei 340
Huang, H.-F. 504
Huang, K.-P. 504
Huang, Kailiang 47
Huang, P. 376, 488, 931
Huang, Peng 464
Huang, Po-Tsang 249
Huang, Qianqian 707
Huang, Ru 388, 392, 707
Huang, Wen-Hsien 340
Huang, Y. 879
Huang, Y.-C 504, 504
Huang, Y.-M. 504
Huang, Yi-Chiau 544
Hubert, Q. 161
Hudec, Boris 356
Hudson, F. 129
Hueting, R. J. E. 739
Hung, Steven 253
Hüselitz, R. 428
Hutin, L. 141
Hutin, Louis 281
Huyghebaert, C. 512
Hwang, Changyoun 851
Hwang, Cheol Seong 288
Hwang, Eung-Rim 851
Hwang, J. 604
Hwang, K.H. 416, 656
Hwang, S. H. 416
Hwang, S.M. 656
Hwang, Wan Sik 241
Ide, T. 165
Idekoba, T. 237
Ielmini, D. 927

Iida, S. 221
Ikeda, H. 237
Ikeda, S. 608
Ikegami, Tsutomu 197
Ilatikhameneh, Hesameddin 767
Im, Sung Gap 241
Imamura, T. 265
Imanishi, K. 460
Indiveri, G. 472
Inoue, F. 149
Inoue, H. 608
Inoue, M. 165
Inoue, S. 265
Ionescu, A.M. 269, 308
Ionescu, Adrian M. 304
Irrera, F. 747
Irwin, R. 683
Isobe, A. 284
Issakov, V. 328
Itoh, K.M. 129
Itoh, Kohei M. 137
Itoh, M. 265
Itou, K. 189
Iwai, H. 189
Iwai, Hiroshi 304
Iwamatsu, T. 185
Iwata-Harms, Jodi 612
Iyer, S. S. 683
Jadot, B. 141
Jagannathan, H. 807, 811
Jagannathan, Hemanth 253
Jahan, C. 424
Jahan, R. 412
Jakob, A.M. 129
Jamieson, D.N. 129
Jamieson, G. 149
Jan, Guenole 612
Jang, Dongkyu 288
Jang, E. 875
Jang, Inkook 771
Jang, Kyungmin 723
Jang, S. H. 604
Jannaud, A. 500
Jansen, S. 428
Jao, C.-Y 504
Jehl, X. 141
Jena, D. 691
Jena, Debdeep 193
Jenni, Laura Vera 83
Jeon, H.Y 656
Jeon, J. 75
Jeong, Chan Bae 288
Jeong, D. E. 416
Jeong, G. T. 416
Jeong, Yeon Joo 63

AUTHOR INDEX

Jeong, Yujeong 288
Jerry, M. 364
Ji, M. H. 620
Ji, Y. 416
Jiang, Junkai 576, 799
Jiang, Xiangwei 568, 919
Jiang, Zhengping 348, 767
Jiang, Zizhen 572
Jin, Chengji 723
Jin, Q. 811
Jin, Seonghoon 767
Jinno, Riena 193
Johnson, B.C. 129
Jonnalagadda, V.P. 628
Jönsson, Adam 903
Jourdan, G. 99
Jourdan, Guillaume 281
Jourdon, J. 157
Jousseaume, V. 153
Jouve, A. 157
Juge, André 384
Jung, E. S. 416
Jung, ES 1
Jung, Gyuweon 288
Jung, H. 640
Jung, M. 245
Jung, S-M 656
Jung, W. 265
Kachi, T. 687
Kachi, Tetsu 452
Kaczer, B. 787
Kaczer, Ben 783
Kaji, Shiori 79
Kaklin, F. 743
Kakushima, K. 189
Kamata, Y. 225
Kamineni, V. 819
Kanamitsu, S. 265
Kanechika, M. 687
Kang, H.K. 416, 656
Kang, Ho-Kyu 560, 771
Kang, J. F. 376, 488, 931
Kang, Jinfeng 779
Kang, M. 656
Kang, M.S. 656
Kang, S.- Y 608
Kao, K.-H. 504
Kao, Ming-Yen 209
Kar, G. 592
Kar, G. S. 935
Kar, G. Sankar 51, 420
Karg, S. 540
Karim, K. S. 755
Karnati, K. 332
Kato, H. 265

Kato, K. 791
Kato, Kiyoshi 312
Ke, M. 791
Kencke, D. 412
Kenis, K. 508
Kerdiles, S. 153
Keys, P. 133
Khan, Asif Islam 205, 300
Khare, M. 819
Khwa, W. S. 624
Kim, C. H. 352
Kim, D. 656
Kim, D.H. 656
Kim, D.-W. 656
Kim, Dae Sin 348, 767, 771
Kim, Donghoon 851
Kim, H. 352, 428
Kim, J. 352, 428
Kim, J.C. 656
Kim, Jae Hwan 241
Kim, Jin-Kook 851
Kim, Jongchol 767
Kim, Jun Shik 288
Kim, K. W. 552
Kim, M. 352, 508
Kim, Myoungsub 851
Kim, Myung Sun 827
Kim, Myungsoo 532
Kim, N. 508
Kim, Nam Sung 827
Kim, S. K. 640
Kim, S.S. 656
Kim, Sae-Jin 771
Kim, Seungkyu 767
Kim, Seyoung 292
Kim, Taehoon 851
Kim, W. 420, 592, 859, 935
Kim, W.D. 656
Kim, W.J. 656
Kim, Y. 265
Kim, Y.H. 656
Kim, Y.-J. 141
Kim, Yun Sang 241
Kimoto, T. 444, 687
King, Ya-Chin 273, 400
Kinoshita, M. 284
Kita, K. 185
Kittl, J. A. 296
Kizilyalli, I.C. 456
Klamkin, J. 324
Klauk, Hagen 883
Kleemeier, W. 819
Kleemeier, Walter 815
Klein, J.-O. 484
Knoll, L. 440

AUTHOR INDEX

Knorr, A. 819
Ko, Hyoungsoo 771
Kobayashi, K. 265
Kobayashi, Masaharu 372, 723
Kobayashi, S. 265
Kobayashi, Y. 181
Kocaay, D. 111
Koh, G. H. 416
Kohler, S. 424
Koike, H. 608
Komatsu, S. 284
Komiyama, Takaki 664
Komori, M. 265
Komukai, T. 265
Kong, Lisa 895
Kono, T. 265
Korndörfer, F. 548
Koseki, K. 444
Kosugi, R. 444
Kotani, J. 699
Kotlyar, R. 133
Kouemeni-Tchouake, F. 153
Kouwenhoven, Leo 145
Kranz, L. 440
Kreupl, F. 245
Krishnan, R. 604
Krishnan, Siddarth A 253
Krivokapic, Zoran 300
Krottenthaler, P. 428
Krylyuk, S. 536
Ku, S.H. 31
Kubo, T. 608
Kubota, H. 616
Kudo, S. 237
Kumagai, Y. 237
Kumar, A. 632
Kumar, Lalit 83
Kumar, Narendra 676
Kumar, Pushpendra 404
Kumar, Ranjith 636
Kumazawa, T. 177
Kundu, A. 396
Kundu, S. 420, 592, 935
Kunimune, Y. 165
Kunitake, Hitoshi 312
Kunitake, S. 237
Kuo, I. T. 859
Kuramata, Akito 193
Kuriyama, N. 225
Kuroda, R. 225, 660
Kushwaha, Pragya 209
Kuzum, Duygu 664
Kwok, Hoi Sing 871
Kwon, J. 604
Kwon, J.H. 596

Kwon, O. I. 416
Kwon, Ohseong 652
Kwon, T.Y. 656
Kwon, Uihui 767
Kyogoku, S. 181
La Rosa, F. 161
Lacord, J. 500
Lagrasta, S. 424
Lahav, Assaf 233
Lai, E. K. 859
Lai, Stefan 432
Lai, Tung-Yan 731
Lal, R. 460
Lalanne, F. 229
Lam, Vinh 612
Lamontagne, P. 157
Lampert, L. 133
Lanza, Mario 528
Lapras, V. 500
Larcher, L. 596
Larcher, Luca 588
Lardin, T. 153
Larrey, V. 153
Latessa, L. 747
Lau, Calvin J. 676
Laucht, A. 129
Laudato, M. 927
Lauwereins, R. 480
Lavizzari, S. 43
Lavoie, C. 819, 859, 911
Le Friec, Y. 424
Le Gallo, M. 628
Le, J. 735
Le, Son 612
Lee, C. H. 807
Lee, Chang Bum 560
Lee, Chun Ying 340
Lee, D. S. 416
Lee, F. M. 859
Lee, Feng-Min 39
Lee, H.-J. 316
Lee, H.-Y. 735
Lee, Hyunmin 851
Lee, J. H. 396, 416
Lee, J.-H. 656
Lee, Jack C. 532
Lee, Jaesung 87
Lee, Jong-Ho 288
Lee, K. 604
Lee, K. H. 416
Lee, K.H. 416, 596
Lee, Ko-Tao 292
Lee, Kyupil 560
Lee, M. H. 735, 859
Lee, M.J 739

AUTHOR INDEX

Lee, Ming-Hsiu 39
Lee, Min-Hung 731
Lee, S. 352, 911
Lee, Shiuh-Wuu 173
Lee, T.J. 656
Lee, Tsung-Han 273
Lee, Y.-J. 504
Lee, Y.K. 416
Lee, Y.W. 396
Lee, Younghee 895
Lee, Yuan-Jen 612
Legrand, B. 99
Lelis, A. J. 448
Lemke, S. 428
Lemonnier, O. 99
Leo, K. 11
Leobandung, E. 911
Leonhardt, A. 512
Lepape, E. 161
Leroux, Charles 404
Lesniewska, A. 111
Letzkus, Florian 883
Levine, P. M. 755
Lhostis, S. 157
Li, Fuhai 67
Li, Gezi 67
Li, Haitong 572
Li, Huanglong 572
Li, J. 807
Li, J.-H. 504
Li, J.-Y. 504
Li, Junfeng 47, 679
Li, Juntao 652
Li, K.-S. 735
Li, Kai-Shin 340, 731
Li, Keshuang 556
Li, Ling 118
Li, Linsen 572
Li, Luping 253
Li, R. 133
Li, Ruofan 847
Li, W. 149, 691
Li, Weisheng 524
Li, Wenshen 193
Li, X. 620
Li, Xiaoqin 532
Li, Xin 847
Li, Xinyi 67, 468
Li, Xiuyan 715
Li, Y. 504
Li, Yudong 679
Li, Yun 775
Li, Z. 811
Lian, G. 107
Liang, C.-F. 751

Liang, Gengchiau 580
Liang, Y. 819
Liang, Zhongxin 707
Liao, C.-Y. 735
Liao, Mengya 556
Liao, T.-H. 504
Liao, Yu-Hung 209
Lie, F. 819
Lie, Fee-li 815
Likharev, K.K. 476
Lim, J.H. 596, 604
Lin, B. 412
Lin, C.K. 396
Lin, C.-T. 751
Lin, Chrong Jung 273, 400
Lin, D. 512
Lin, K.-L. 504
Lin, Ming-Huei 644
Lin, S. 811
Lin, W.L. 31
Lin, Y. 508
Lin, Y. F. 859
Lin, Y.-D. 735
Lin, Yen-Kai 209
Lin, Yongjing 253
Lin, Yudeng 67
Lin, Yu-Yu 39
Lin, Zhiqiang 217
Ling, T. 604
Linten, D. 592, 600, 787
Linten, Dimitri 783
Liou, Peng-Chun 273
Liu, B. 604
Liu, C. 376, 931
Liu, C. W. 735
Liu, F. 376
Liu, Fei 763
Liu, H.-D. 751
Liu, Huanlong 612
Liu, Huiyun 556
Liu, Jing 464
Liu, L. F. 488
Liu, Lixiang 887
Liu, Ming 47, 464
Liu, P. 811
Liu, Paul 612
Liu, Po-Tsun 356
Liu, Qi 47, 464
Liu, T.-J. K. 75
Liu, W.-H. 751
Liu, X. 428
Liu, X. H. 107
Liu, X. Y. 376, 488, 931
Liu, Xiaoyan 775, 779
Liu, Xin 664

AUTHOR INDEX

Liu, Y. 408
Liu, Yanghui 520
Liu, Yibo 871
Liu, Yue-Yang 919
Liu, Yuyi 468
Liu, Z. 711, 819
Liu, Zhaojun 871
Locatelli, N. 616
Lofaro, M. 911
Long, Shibing 47
Longo, J. 269
Loo, R. 496
Loubet, N. 807, 811
Loubet, Nicolas 652
Loup, V. 500
Loup, Virginie 404
Lovisi, N. 747
Low, R. 604
Lu, C.C. 31
Lu, Chih-Yuan 31, 39
Lu, Jen-Hsiang 644
Lu, Jiwu 67
Lu, M. 412
Lu, Ruochen 915
Lu, Ryan 396
Lu, T.C. 31
Lu, Wei D. 63
Lu, Wenjie 895
Lu, Yang 348
Lu, Yichen 664
Ludwig, J. 512
Ludwigs, Sabine 883
Lung, H. L. 859
Lung, Hsiang-Lan 39
Luo, Aileen 300
Luo, G.-L. 504
Luo, Q. 376
Luo, Qing 47, 464
Luo, S.-X. 504
Lv, H. B. 376
Lv, Hangbing 47, 464
Ly, D. R. B. 472
Lyu, Y.-F. 751
Ma, Haili 47
Ma, T. P. 711
Ma, W. C.-Y. 504
Ma, Xiaolei 568
Ma, Yinji 668
Ma, Zichao 520
Machida, S. 284
Machillot, J. 508
Madhavan, Atul 636
Madzik, M. 129
Maeda, Shigenobu 767
Maeda, T. 687

Magesan, Easwar 126
Magis, T. 863
Magyari-Köpe, Blanka 572
Mahajan, Bikram K. 584
Mahakik, K. N. A. 448
Maheshwari, Dinesh 432
Mahmoodi, M.R. 476
Mai, A. 548
Mai, C. 548
Mainuddin, M. 412
Maitrejean, S. 153
Maize, K. 536
Makiyama, K. 699
Malavena, G. 35
Malinge, P. 229
Manfrini, M. 843
Manipatruni, S. 843
Mannaert, G. 149, 508
Mantelli, M. 161
Mao, Duli 217
Marchioni, A. 747
Marinov, D. 512
Markman, B. 324
Martin, A. 157
Martin, Lane 300
Martinez, Eugenie 404
Martinie, S. 500
Marushchak, Denys 676
Marzaki, A. 161
Massa, L. 133
Masselon, Christophe 281
Masuda, T. 177
Masudy-Panah, Saeid 544
Masunishi, Kei 79
Masuoka, S. 656
Matasunaga, K. 265
Matsudai, T. 189
Matsumoto, Noriko 312
Matsuura, M. 165
Mattavelli, P. 424
Mattei, Paul 281
Maugain, F. 161
Maurand, R. 141
Max, B. 727
Mazen, F. 153
Mazur, M. 428
Mazzocchi, V. 141, 153
McCallum, J.C. 129
McCarthy, L. 460
McClellan, Connor 528
Mcdonald, M. 229
McHerron, D. C. 811
McHerron, Dalea 253
McKay, J. 460
McLaughlin, P. S. 107

AUTHOR INDEX

McMitchell, S.R.C. 43, 380
Meersschaut, J. 51
Mehrsa, Armaghan 664
Meier, N. 540
Meiling, Hans 261
Melikyan, A. 552
Memisevic, E. 308
Meneghesso, G. 691, 703
Meneghini, M. 691, 703
Menzel, S. 867
Mermoz, S. 157
Mertens, H. 508
Mertens, Hans 783
Meterelliyoz, M. 412
Metz, M. 133
Meunier, T. 141
Miao, X. 807
Miao, Xin 253, 652
Migita, Shinji 197, 719
Mihaila, A. 440
Mikolajick, T. 727
Mikolajick, Thomas 588
Minoura, Y. 699
Mishra, U. 460
Mistry, Kaizad 636
Mitard, J. 149, 496, 508
Mitra, A. 265
Mittal, Sushant 827
Mittmann, T. 727
Miura, N. 185, 225
Miura, S. 608
Miyamoto, Satoru 137
Miyashita, Toshihiko 827
Miyata, Noriyuki 169
Miyazoe, H. 911
Mizuno, H. 221
Mizuno, Ikuo 233
Mo, Fei 372
Mo, R.T. 911
Mochizuki, S. 807, 811, 819
Mocuta, A. 420
Mocuta, D. 149, 496, 508, 843
Mohammadi, R. 755
Mohiyaddin, F.A. 129
Molas, G. 472, 863
Moll, P. 428
Monfray, S. 277
Monnot, G. 229
Moon, Bum Ki 652
Moon, C. 656
Moore, M. 460
Morales, C. 153
Morand, Yves 404
Moreau, S. 157
Morello, A. 129

Mori, H. 237
Morimoto, T. 181
Mortemousque, P.-A. 141
Mothes, K. 428
Motokawa, T. 265
Motoyama, K. 107
Mourik, V. 129
Mukhopadhyay, S. 396
Müller, J. 428
Muneshwar, T. 923
Murakami, S. 660
Murata, M. 225
Murdzek, Jessica 895
Na, N. 751
Nabors, Marni 636
Naeemi, A. 103
Nagata, Tomohiko 79
Naik, Mehul 122
Naik, V.B. 596, 604
Nakajima, Shigeru 320
Nakamura, H. 284
Nakamura, N. 699
Nakamura, Y. 284
Nakasugi, T. 265
Nakayama, K. 444
Nakazawa, K. 237
Nara, Jun 169
Narayanan, V. 807, 911
Narayanan, Vijay 648
Narayanan, Vijaykrishnan 340
Narita, T. 687
Nasuno, T. 608
Natarajan, S. 811
Natarajan, Sanjay 253
Nauta, B. 739
Navarro, G. 863
Neeli, V. 316
Ney, D. 153
Nguyen, P. 412
Nguyen, Tu 432
Nguyen, V.D. 843
Ni, J. 107
Ni, K. 296, 364
Ni, Kai 55
Ni, Zhenyi 887
Niel, S. 161
Nikonov, D. 412
Nikonov, D.E. 843
Nili, H. 476
Niquet, Y.-M. 141
Nishi, Yoshiaki 233
Nishi, Yoshio 432
Nishida, T. 95
Nishizawa, S. 189
Niu, C. 819

AUTHOR INDEX

Niu, Chengyu 815
Niwa, M. 608
Nodin, J-F 472
Noé, P. 863
Noel, J-P 472
Noguchi, M. 185
Noguchi, Y. 608
Nohira, Hiroshi 169
Nolot, E. 863
Nomoto, K. 691
Nomoto, Kazuki 193
Nonglaton, Guillaume 281
Norwood, Christopher 432
Noudo, S. 237
Nowak, E. 472, 484, 819, 863
Numasawa, Y. 189
Nyns, L. 380
Obradovic, B. 296
O'brien, K. 412
Odaka, T. 284
Ogier, J.L. 424
Ogier, S. 879
Ogura, A. 189
Oguz, K. 412
Oh, H. 51
Oh, HR. 420
Oh, S. C. 416
Ohashi, H. 189
Ohba, N. 221
Ohki, T. 699
Ohno, K. 221, 237
Ohshima, Kazuaki 312
Oikawa, K. 656
Ok, I. 807
Oka, T. 221
Okamoto, Kazuaki 79
Okamoto, N. 699
Okishiro, K. 284
Okumura, H. 181, 444
Olsen, Tivoli J. 676
Olshausen, Bruno 71
Om'mani, H. 428
Omori, K. 165
Omura, I. 189
Ono, K. 284
Ono, Tomio 79
Ono, Y. 237
Oprins, H. 111
Or-Bach, Zvi 432
Osawa, N. 237
Ostrovski, Y. 107
O'Sullivan, B. J. 592
Ota, Hiroyuki 197, 719
Ott, J. A. 911
Ouellette, D. 412

Ozaki, S. 699
Pacelli, D. 424
Padovani, A. 596
Padovani, Andrea 588
Paiton, Dylan 71
Pak, Kwan Yong 241
Pal, Arnab 564, 576
Pal, Ashish 827
Pala, Marco G. 759
Palanchoke, Ujwol 281
Palayam, S. Vadakupudhu 600
Pan, Deng 676
Pan, Lei 839
Pan, Quanjun 839
Pandey, P. 368
Pandya, S. 879
Pang, Chin-Sheng 516
Parat, K. 27
Parikh, P. 460
Park, Byung-Gook 288
Park, C. 352
Park, C.-H. 656
Park, Hong Keun 241
Park, Hong-hyun 767
Park, Honglae 771
Park, J. 412
Park, J. H. 416
Park, K.C. 416
Park, K.J. 656
Park, S. 640
Park, S. O. 416
Park, S.H. 656, 656
Parmigiani, L. 161
Parto, Kamyar 564
Parvais, B. 149
Patel, J. 911
Patel, Sahil 612
Pathak, Kalpana 827
Paul, J. 428
Peczek, A. 548
Pedini, Jean-Michel 404
Pedreira, O. Varela 111
Pellegren, J. 412
Pena, V. 508
Peng, L. 149
Peng, Lian-Mao 763
Peng, Xiaochen 348
Perlas, A. 75
Peroulis, Dimitrios 336
Perreau, P. 153
Perumkunnil, M. 420
Pesic, M. 43
Pesic, Milan 588
Pey, K.L. 596
Pham, Anh-Tuan 767

AUTHOR INDEX

Phoa, K. 316
Phommahaxay, A. 512
Pi, Xiaodong 887
Pillarisetty, R. 133
Pin, J-B. 153
Pitera, J. W. 21
Pla, J.J. 129
Plantier, Christophe...............................281
Poiroux, Thierry.....................................384
Ponthenier, F. 153
Pop, E. 572
Pop, Eric 528
Popovici, M. 43, 51, 380
Porret, C. 496
Portal, J.-M 484
Post, Ian 636
Posthuma, N. 703
Poth, J. 428
Potoms, G. 43, 51
Pourghaderi, M. Ali 767
Pourtois, G. 380, 512, 935
Pradeep, Krishna 384
Prakash, Somashekar Bangalore............636
Prasad, D. 103
Previtali, B. 500
Prezioso, M. 476
Prindle, C. 819
Pugliese, Kaitlin M.................................676
Puls, C. 412
Qi, Weiyi 348
Qian, H. 488, 931
Qian, He 67, 468
Qiu, Chenguang 763
Quek, E. 604
Querlioz, D.....................................484, 616
Quintero, P. 412
Radu, I. P. 512, 843
Radu, Iuliana P. 831
Raghavan, N. 596
Raghavan, Srinivasan 803
Ragnarsson, L...149
Ragnarsson, L.-Å. 496, 787
Rahimo, M. 440
Rahman, T. 412
Rajapakse, Arith J. 676
Rakshit, T. 296
Rambal, N..153, 500
Rami, S...316
Ranica, R. 424
Rao, S. 420, 592, 935
Rassoul, N. 149
Rastogi, P. 201
Ravikumar, S. 316
Ray, A. 859
Raychowdhury, Arijit..............................300

Raymenants, E. 843
Raynor, J. M. 743
Razavi, Seyed Armin...............................835
Razavieh, Ali 652
Realov, Simeon 636
Reboh, S. 153
Regnier, A. 161
Ren, Chi..664
Ren, Tian-Ling. 572, 891
Reynard, J.P. 424
Richard, E. 424
Richard, O. 51
Richter, R. 428
Ristoiu, D. 424
Ritzenthaler, R..............................149, 508
Ritzenthaler, Romain 783
Roberts, J. 133
Robison, Robert......................................652
Rodder, M. 296
Rode, J. 324
Rodwell, M.J.W. 324
Rolland, Emmanuel 281
Rollo, T. 213
Roman, A. 153
Roman, Cosmin 83
Romang, A. 412
Romano, G. 500
Romera, M. 616
Ronchi, N..380
Rosca, T. 308
Rosenblatt, Sami 126
Rosseel, E. 149, 600
Rothemund, R. 332
Roussel, P. J. 592
Roux, N. 229
Roy, Deboleena 360
Roy, F. 229
Roy, Kaushik 360
Rozeau, O. 500
Rozen, John 292
Rupakula, M. 269
Rupp, J. A. J. 867
Rusch, M. 75
Russo, F. 747
Ryan, K. 819
Ryckaert, J. 149, 787
Rzepa, G. 787
Sabbagh, D. 133
Sachid, Angada 827
Saeidi, A. 308
Saeidi, Ali 304
Saga, Shiori 312
Saha, A. K. 364
Saha, D. 201
Saidi, B. 161

AUTHOR INDEX

Saito, M. 265
Saito, T. 608
Saito, W. 189
Saito, Y. 177
Sakamoto, K. 444
Sakano, Y. 221
Sakhare, S. 420
Salahuddin, Sayeef..............209, 731
Samanni, G. 424
Samkharadze, N. 133
Samukawa, S. 504
Sankaran, K. 935
Sano, Ryousuke 169
Sanquer, M. 141
Sansa, Marc 281
Sant, S. 899
Santoro, G. 508
Santos, Eduardo Gil 281
Saraf, I. 819
Saraya, T. 189
Saraya, Takuya 372, 723
Sart, C. 157
Sasago, Y. 284
Sasaki, Kohei 193
Sassine, G. 472, 863
Sassoulas, P.O. 424
Sato, H. 608
Sato, M. 237
Sato, N. 237
Satoh, K. 189
Savytskyy, R. 129
Sawai, Hiromi 312
Scappucci, G. 133
Scevola, D. 153, 157
Schaefer, M. 332
Scheer, Patrick 384
Schenk, A. 899
Scheuvens, L. 11
Schmid, H. 911
Schmidt, Alexander 771
Schmitt, V. 129
Schmitz, J. 739
Schneider, U. 11
Schram, T. 512
Schroeder, U. 727
Schwab, L. 99
Scibetta, C. 153
Scotti, L. 424
Seabaugh, A. C. 368
Sebaai, F. 43
Sebastian, A. 628
Seeds, Alwyn 556
Seet, C. S. 604
Sekhar, M. 412
Seki, Takako 312

Selarka, A. 412
Sell, B. 316
Sengupta, Abhronil 360
Seo, B. 656
Seo, B.Y. 416
Serrano-Guisan, Santiago 612
Seth, M. 412
Shakouri, A. 536
Shankar, Bhawani 803
Shao, Qiming 835, 839
Sharma, S. 811
Shen, Chang-Hong 249, 340
Shen, Dongna 612
Shen, L. 460
Shen, T.M. 492
Shen, W. S. 488
Shen, Y.-L. 504
Shepard, K. L. 91
Shi, C. 91
Shi, Jianping 532
Shi, Quan 636
Shi, Yi 524
Shi, Yuanyuan 528
Shibaguchi, T. 225
Shibata, H. 225
Shieh, J.-M. 504
Shieh, Jia-Min 249, 340, 731
Shigyo, N. 189
Shih, Jiaw-Ren 400
Shikha, Swati 803
Shim, Dongshik 560
Shimada, Y. 165
Shin, Changgyun 560
Shin, Dongjae 560
Shin, H. C. 416
Shin, H.J. 656
Shin, Jong Hoon 63
Shin, SangHoon 292
Shin, Yonghwack 560
Shobha, H. 107
Shono, K. 460
Shoute, G. 923
Shrestha, P.R. 536
Shrivastava, Mayank 803
Si, Mengwei 344
Siah, S. Y. 604
Siang, G.-Y. 735
Sikder, U. 75
Singh, K. 133
Singh, Sandeep 803
Slesazeck, Stefan 588
Slesazeck, S. 727
Smets, Q. 512
Smith, A. J. 412
Smith, A. K. 412

AUTHOR INDEX

Smith, J. A. 296, 364
Smith, P. 460
Sober, Samuel J. 672
Socquet-Clerc, C. 863
Sohn, Chang-Woo 652
Sohn, Joon 71
Solomon, Paul 292
Song, G. 656
Song, Jeongho 851
Song, Min 847
Song, S. 352
Song, Y. J. 416
Song, Z. T. 620
Soni, Ankit 803
Souhaite, A. 424
Souifi, A. 277
Souriau, L. 592, 843
Sousa, M. 628, 899, 907
Southwick, R. G. 807
Spence, C. 141
Spessot, A. 420
Spinella, Laura 115
Spinelli, A. S. 35
Spratt, W. 911
Srinivasa, Srivatsa 340
Srinivasan, Gopalakrishnan 360
Stahlbush, R. E. 448
Stark, P. 540
Steglich, P. 548
Stelzer, M. 245
Stoffels, S. 703
Stojanovic, V. 75
Stolfi, M. 811
Stolichnov, Igor 304
Strane, J. 819
Strukov, D.B. 476
Stucchi, M. 111
Su, C.-J. 504
Su, S.K. 492
Subirats, A. 43, 600
Suda, J. 687
Sugawa, S. 225, 660
Sugiyama, Y. 284
Suh, K. 416
Sulehria, Yasir 815
Sumita, Kyoko 169
Sun, S. 508
Sun, Shiyu 827
Sun, X. 911
Sun, Xiaowei 871
Sun, Xiaoyu 55, 468
Sun, Zixuan 392
Sundar, Vignesh 612
Sung, P.-J. 504
Sutar, S. 512

Suzuki, A. 221
Suzuki, K. 221
Suzuki, M. 660
Suzuki, S. 189
Svensson, Johannes 903
Swerts, J. 420, 592, 935
Tabone, Claude 281, 404
Tabrizian, R. 95
Tagawa, Yusaku 372
Tai, Lu 47, 464
Takagi, S. 791
Takahata, K. 265
Takakura, T. 189
Takami, M. 221
Takei, M. 181
Takenaka, M. 791
Takeuchi, K. 189
Takizawa, M. 221
Talatchian, P. 616
Talmelli, Giacomo 831
Tamura, R. 608
Tan, S. L. 604
Tanaka, T. 181
Tang, Jianshi 292
Tang, M. 735
Tang, Mingchu 556
Tang, W. 879
Tang, Wei 253
Tanigawa, T. 608
Tateshita, Y. 221, 237
Tellez, G.E. 632
Tenberg, S. 129
Teng, Zhongjian 612
Teugels, L. 149
Thakuria, Niharika 516
Thanigaivelan, Thirumal 815
Thean, Aaron Voon-Yew 580
Thiam, A. 843
Thiyagarajah, N. 604
Thomas, Luc 612
Thomas, N. 133
Thompson, M. G. 540
Tian, He 572, 891
Ting, J. W. 604
Tiwari, V. 428
Tkachev, Y. 428
Todorov, Teodor 292
Toh, E. H. 604
Tokei, Zs. 111
Tokue, H. 265
Tokumaru, Ryo 312
Tong, Ru-Ying 612
Toriumi, Akira 197, 257, 715, 719
Torres, J. 133
Tosi, G. 129

AUTHOR INDEX

Tournier, A. 229
Trastoy, J. 616
Trenteseaux, F. 161
Trentzsch, M. 428
Trotta, S. 328
Tsai, S. 819
Tsai, Wen-Jer 31
Tsai, Y.S. 396
Tsai, Yi-Pei 400
Tseng, J.C. 624
Tsuda, H. 265
Tsuda, Kazuki 312
Tsukuda, M. 189
Tsunegi, S. 616
Tsutsui, G. 819
Tsutsui, Gen 652
Tsutsui, K. 189
Tsutsui, Masafumi 233
Tu, Hailing 679
Tu, Thieu Quang 193
Tu, Yung-Ning 340
Ueda, H. 687
Uesugi, T. 687
Urdampilleta, M. 141
Urteaga, M 324
Usagawa, T. 284
Usai, Giulia 281
Ushifusa, N. 284
Van Beek, S. 592
Van Dal, M.J.H. 492
Van Den bosch, G. 600
Van Der Plas, G. 408
Van Elshocht, S. 51
Van Houdt, J. 43, 380
Vandemaele, Michiel............................783
Vandersypen, L.M.K. 133
Vandooren, A. 149, 787
Vanherle, W 149
Vanstreels, K. 408
Vardi, Alon 895
Vasen, T. 492
Vaysset, A. 843
Vecchio, E. 149
Vega, Reinaldo 652
Veldhorst, M. 133
Velenis, D. 408
Vellianitis, G. 492
Venezia, Vincent C.............................217
Venitucci, B. 141
Verdy, A. 863
Verhulst, Anne S. 304
Verkest, D. 480
Vernhes, Emeline 281
Vernhet, A. 424
Verreck, D. 512

Vianello, E. 472, 484
Vici, A. 747
Villard, Patrick 281
Villaret, A. 424
Villringer, C. 548
Vincent, A. 476
Vinet, M.............................141, 153, 500
Vizioz, C. 500
Vodenicarevic, D................................616
Voit, B. 11
Volk, C. 133
Vorenkamp, Pieter 432
Waldron, N. 149
Walke, A..149
Walters, G. 95
Wan, D. 843
Wan, Weier 71
Wang, C.-J. 504
Wang, Chien-Ping 273
Wang, Ching-Hua 528, 572
Wang, H. 213
Wang, Hong 544
Wang, Huimin 707
Wang, Jian 763
Wang, Jing 348
Wang, Jingli 520
Wang, Joddy 388
Wang, Kang L. 835, 839
Wang, Kanwen 468
Wang, Keh-Chung 39
Wang, L. 620
Wang, Lin 580
Wang, Lingfei 580
Wang, M. 807
Wang, Miaomiao 253, 648, 652
Wang, Ning 695
Wang, Panni 55
Wang, PoKang 612
Wang, Q. 620
Wang, Qingxue 388
Wang, Qiwen 63
Wang, Runsheng 388
Wang, Rusheng 392
Wang, Shu-Hui 644
Wang, T. Y. 624
Wang, Tahui 31
Wang, W. 819, 927
Wang, Wei 544, 823
Wang, Wenwu 679
Wang, Xinning 636
Wang, Xinran 520, 524
Wang, Xu 257
Wang, Xuefeng 891
Wang, Y.-H. 504
Wang, Y.-S. 504

AUTHOR INDEX

Wang, Yue 887
Wang, Yu-Jen 612
Wang, Z.-Y. 735
Wang, Zheng 300
Wang, Zhixuan 707
Waser, R. 867
Watanabe, H. 185
Watanabe, M. 189
Watanabe, T. 608
Watson, T.F. 133
Weber, O. 424
Webster, Eric A. G. 217
Wei, Feng 679
Wei, Jin 695
Wei, L 412
Wei, Na 173
Wei, Qianhui 679
Wei, Yun-Jie 731
Weiss, Gregory A. 676
Welser, J. 21
Wemersson, Lars-Erik 903
Wernersson, L-E. 308
Widjaja, Yuniarto 432
Wiegand, C. 412
Wildhaber, F. 269
Willis, Jim 815
Wilson, C.J. 111
Wilson, James 432
Winstel, K. R. 811
Wirths, S. 440
Witters, L. 149, 496
Witters, Liesbeth 783
Wong, H.-S. Philip 71, 118, 528, 572
Wong, J. 604
Wong, Kin 839
Woo, S. T. 604
Wouters, D. J. 867
Wu, Bo-Wei 731
Wu, C. 111
Wu, C.-T. 504
Wu, Dehuang 388
Wu, Fan 891
Wu, H. 819
Wu, H. Q. 488, 931
Wu, Hao 835, 839
Wu, Heng 815
Wu, Huaqiang 67, 468
Wu, J. 324
Wu, J.Y. 624
Wu, Jiang 556
Wu, Jixuan 568
Wu, Meile 288
Wu, Meng-Chyi 249
Wu, Tai-Hsuan 636
Wu, W.-F. 504

Wu, Wan-Chi 249
Wu, Wei 67
Wu, Xiaohan 532
Wu, Xiu Long 464
Wu, Y. 460
Wu, Ying 823
Wu, Z. 149, 787
Wu, Z.Q 492
Wu, Zhenhua 763
Wuetz, B. P. 133
Wurstbauer, U. 245
Xi, Yue 468
Xiang, Y. C. 488, 931
Xie, Tao 668
Xie, Xuejun 564
Xing, H. G. 691
Xing, Huili Grace 193
Xu, Nuo 348, 847
Xu, Ruijuan 300
Xu, Shengqiang 544
Xu, X. X. 376
Xu, Xiaoxin 47, 464
Xu, Yang 887
Xue, Chunling 679
Yagi, Y. 608
Yakimets, D. 420
Yakushiji, K. 616
Yamada, A. 699
Yamada, M. 221
Yamaguchi, K. 221
Yamaguchi, T. 165
Yamamoto, M. 660
Yamane, J. 237
Yamane, K. 596, 604
Yamasaki, Takahiro 169
Yamashita, K. 237, 608
Yamashita, T. 819
Yamashita, Tenko 652
Yamawaki, T. 284
Yamazaki, Shunpei 312
Yan, Jiang 679
Yan, Siwa 871
Yanagisawa, Yuichi 312
Yang, C. H. 859
Yang, Chih-Chao 249, 340
Yang, Deren 887
Yang, H. 596, 604
Yang, J. 819
Yang, M.-J. 751
Yang, M.S. 656
Yang, Mengxuan 707
Yang, Shyh-Horng 644
Yang, Song 695
Yang, Yansong 915
Yang, Yi 572, 612, 891

AUTHOR INDEX

Yang, Yixiong 253
Yao, Peng 67, 468
Yasin, F. 420, 592
Yasuda, T. 660
Yasuda-Masuoka, Y. 640
Yasuhira, M. 608
Yazdani, Armin 217
Ye, Fan 87
Ye, Peide D. 344
Ye, Z. A. 75
Yea, S. 460
Yeh, C. W. 859
Yeh, W.-K. 504
Yeh, Wen-Kuan 249, 340, 731
Yen, Anthony 261
Yeo, Yee-Chia 544
Yeoh, Andrew 636
Yeung, Chun Wing 652
Yi, Jaeyun 851
Yin, Huaxiang 47, 679
Yin, Jiahao 47, 464
Yokoyama, Toshifumi 233
Yonezawa, Y. 444
Yoo, J. 656
Yoon, Alexander 241
Yoon, J.S. 640
Yorita, C. 284
Yoshiba, I. 221
Yoshida, N. 508
Yoshiduka, T. 608
Yoshikawa, K. 284
Yoshita, R. 237
You, Long 847
You, Y. S. 604
Young, I. A 843
Yu, Haoran 47
Yu, Jie 47, 464
Yu, L. 811, 819
Yu, Lan 815
Yu, S. M. 624
Yu, Shimeng 55, 67, 348, 468
Yu, Zhaoan 464
Yu, Zhihao 520, 524
Yuan, Peng 47
Yuasa, S. 616
Yuzawa, Akiko 79
Zagni, Nicolò 584
Zahedmanesh, H. 111
Zaka, A. 428
Zanoni, E. 691, 703
Zarcone, Ryan 71
Zeng, D. 604
Zeng, Yi 847
Zhan, Y. P. 620
Zhang, Chen 652

Zhang, F. 536
Zhang, Gang 524
Zhang, H. 536
Zhang, J.-R. 269
Zhang, Jiayang 392
Zhang, Jingyun 652
Zhang, Ke 871
Zhang, L. 604, 604
Zhang, Peng 839
Zhang, Qingtian 67, 468
Zhang, Qingzhu 47, 679
Zhang, Rui 173
Zhang, Shaoming 679
Zhang, Shuai 847
Zhang, T. 165
Zhang, W. 75
Zhang, X. 488
Zhang, Xiang 67
Zhang, Xiao 679
Zhang, Xing 707, 779
Zhang, Y. 316
Zhang, Yanfeng 532
Zhang, Ying 636
Zhang, Yingchao 668
Zhang, Z. 412
Zhang, Zexuan 193
Zhang, Zhaofu 695
Zhang, Zhaohao 679
Zhang, Zhe 388, 392
Zhang, Zhiyong 763
Zhang, Zuodong 392
Zhao, Hongbin 679
Zhao, J. 879
Zhao, M. 703
Zhao, Meiran 468
Zhao, Robert Chunhua 679
Zhao, Ruoyu 664
Zhao, Shuangyi 887
Zhao, Xiang 217
Zhao, Y. D. 376, 488, 931
Zhao, Yang 707
Zhao, Yi 173
Zheng, G. 133
Zheng, Ning 668
Zheng, Peng 636
Zheng, T. 149
Zheng, Xin 71, 528
Zheng, Zejie 173
Zheng, Zheyang 695
Zhong, Tom 612
Zhong, Yuan 707
Zhou, F. 428
Zhou, Huimei 648
Zhou, Z. 488
Zhu, Jian 612

AUTHOR INDEX

Zhu, Kunkun 707
Zhu, Xi 47, 464
Zhu, Y. 536
Zhu, Ye 520
Zhu, Ying 524
Zia, Muneeb 672
Zidan, Mohammed A. 63
Zografos, O. 843
Zografos, Odysseas 831
Zota, C. 899
Zota, C. B. 907
Zou, Wei 823
Zschieschang, Ute 883
Zuk, P. 460
Zuliani, P. 424
Zwerver, A.-M. 133

IEEE
445 Hoes Lane
Piscataway, NJ 08854-4141

ISBN 978-1-7281-1988-5

2025 IEEE 26th International Conference of Young Professionals in Electron Devices and Materials (EDM 2025)

Altai, Russia
27 June - 1 July 2025

Pages 1-619

IEEE Catalog Number: CFP25500-POD
ISBN: 978-1-6654-7738-3